A TEXTBOOK OF THEORY OF MACHINES

A TEXTBOOK OF THEORY OF MACHINES

(In S.I. Units)

[For Degree U.P.S.C. (Engg. Services), A.M.I.E. (India)]

By

Dr. J.S. BRAR
B.E., M.E., Ph.D. (I.I.T., Delhi)
Formerly
Head Department of Mechanical Engineering
Delhi College of Engineering, Delhi

Dr. R.K. BANSAL
B.Sc. Engg. (Mech.), M.Tech., Hons. (I.I.T., Delhi)
Ph.D., M.I.E. (India)
Formerly
Professor in Mechanical Engineering
Department of Mechanical Engineering
Delhi College of Engineering, Delhi

LAXMI PUBLICATIONS (P) LTD
(An ISO 9001:2015 Company)
BENGALURU • CHENNAI • GUWAHATI • HYDERABAD • JALANDHAR
KOCHI • KOLKATA • LUCKNOW • MUMBAI • RANCHI
NEW DELHI

A TEXTBOOK OF THEORY OF MACHINES

Printed and bound in India

Typeset at : Goswami Associates, Delhi.

First Edition : Jan. 2000, *Reprint* : Jan. 2001, Oct. 2002, *Second Edition* : 2003, *Third Edition* : 2004

Fourth Edition : 2006, 2007, 2008, 2009, 2010, *Fifth Edition* : 2011, 2012, 2013, 2014, 2015

Revised Edition : 2016, *Reprint* : 2017, *Sixth Edition* : 2023

ISBN : 978-81-318-0805-4

Branches

Bengaluru	080-26 75 69 30
Chennai	044-24 34 47 26
Guwahati	0361-254 36 69
Hyderabad	040-27 55 53 83
Jalandhar	0181-222 12 72
Kochi	0484-405 13 03
Kolkata	033-40 04 77 79
Lucknow	0522-430 36 13
Ranchi	0651-224 24 64

PUBLISHED IN INDIA BY

Laxmi Publications (P) Ltd.

(An ISO 9001:2015 Company)

113, GOLDEN HOUSE, GURUDWARA ROAD,

DARYAGANJ, NEW DELHI - 110002, INDIA

Telephone : 91-11-4353 2500, 4353 2501

www.laxmipublications.com

info@laxmipublications.com

C—00412/022/08

Printed at: Ajit Printing Press, Delhi.

Contents

Dedicated
to the loving memory of
my daughter, Babli.

PREFACE TO THE SIXTH EDITION

The popularity of the fifth edition of the textbook of **Theory of Machines** amongst the students and the teachers of the various Universities in the country, has prompted the bringing out of the sixth edition of the book so soon. The sixth edition has been thoroughly revised and brought up-to-date.

In this edition, the three chapters entitled Computer Aided Analysis, Synthesis of Mechanism and Gyroscopic Couple and its effects have been added. The first two chapters have been written by Dr. J.S. Brar (Professor and Head of Deptt. of Mechanical Engg., Delhi College of Engg., Delhi–110042) and the third chapter by Dr. J.S. Brar and Dr. R.K. Bansal in a simple and easy-to-follow language so that even an average student can grasp the subject by self-study. The computer programmes for the chapter "Computer Aided Analysis" was initially in FORTRAN. Now, these programmes have also been written in C++ by Puneet Bansal in this edition. Many good examples have been solved using C++ programme. The input and output data for all these examples have been included in this edition. With these additions, it is hoped that the book will be quite useful for the students of various Engineering Institutions.

The objective type multiple-choice questions are often asked in various competitive examinations. Hence, a large number of objective type questions with answers have been added in the end of the book.

We express our sincere thanks to our colleagues, friends, students and the teachers of different Indian Universities for their valuable suggestions and recommending the book to their students.

Suggestions for the improvement of this book are most welcome and would be incorporated in the next edition with a view to make the book more useful.

—**R.K. BANSAL**
—**J.S. BRAR**

PREFACE TO THE FIRST EDITION

I am glad to present the book entitled, 'A Textbook of Theory of Machines' to the engineering students and to the students preparing for Section B of A.M.I.E. Examination of Institution of Engineers (India). The course-contents have been planned in such a way that the general requirements of all engineering students are fulfilled.

During my long experience of teaching to the engineering students for the past 30 years, I have observed that the students face difficulty in understanding clearly the basic principles, fundamental concepts and theory without adequate solved problems along with the text. To meet this very basic requirement of the students, a large number of the questions taken from the examinations of the various Universities of India and from other professional and competitive examinations (such as Institution of Engineers) have been solved along with the text in S.I. units.

The book is written in a simple and easy-to-follow language, so that even an average student can grasp the subject by self-study. At the end of each chapter highlights, theoretical questions and many unsolved numerical problems with answer are given for the students to solve them.

I am thankful to my colleagues, friends and students who encouraged me to write this book. I am grateful to Institution of Engineers (India), various Universities of India and those authorities whose work have been consulted and gave me a great help in preparing the book.

I express my appreciation and gratefulness to my publisher, Shri R.K. Gupta (a Mechanical Engineer) for his most co-operative, painstaking attitude and untiring efforts for bringing out the book in a short period.

Mrs. Nirmal Bansal deserves special credit as she not only provided an ideal atmosphere at home for book writing but also gave inspiration and valuable suggestions.

Though every care has been taken in checking the manuscripts and proof reading, yet claiming perfection is difficult. I shall be very grateful to the readers and users of this book for pointing any mistakes that might have crept in. Suggestions for improvement are most welcome and would be incorporated in the next edition with a view to make the book more useful.

—R.K. BANSAL

1

Introduction

1.1. DEFINITION

Theory of Machine is that branch of science which deals with the study of relative motion between the various parts of a machine, and forces which act on them. Theory of machine may be divided into *kinematics* and *dynamics*.

Kinematics is that branch of theory of machine which deals with the study of relative motion between the various parts of the machines. Here the various forces involved in the motion, are not considered. Thus kinematics is the study to know the displacement, velocity and acceleration of a part of the machine.

Dynamics is that branch of theory of machine which deals with the study of various forces involved in the various parts of the machine. The forces may be either static or dynamic.

Dynamics is further divided into *kinetics* and *statics*. Kinetics is that branch of theory of machine which deals with various forces when the body is moving whereas statics is that branch of theory of machine which deals with various forces when the body is stationary as shown in Fig. 1.1.

Theory of Machine

Kinematics

(*i*) This deals with the relative motion between the various parts of the machine.
(*ii*) Various forces involved in the motion are not considered.

Dynamics

(*i*) This deals with the study of various forces involved in the various parts of the machine.
(*ii*) Forces may be static or dynamic.

Kinetics
(This deals with the various forces when body is moving.)

Statics
(This deals with the various forces when body is stationary.)

Fig. 1.1

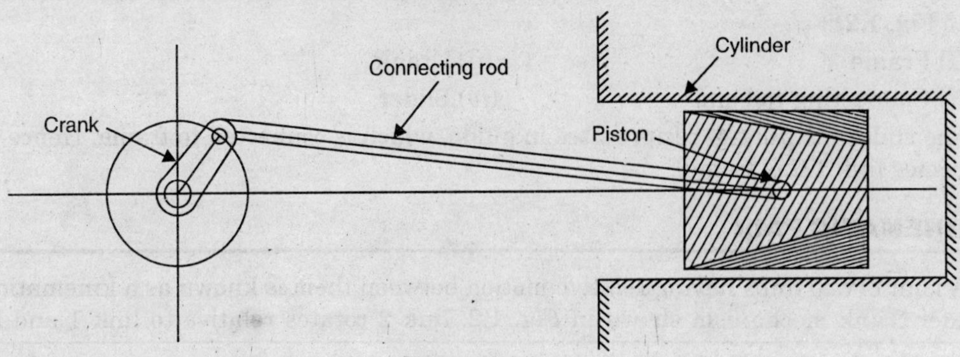

Fig. 1.1 (*a*)

1.2. MECHANISM AND MACHINE

Let us now define mechanism and machine which are commonly used in theory of machine. Fig. 1.1 (*a*) shows the parts of a reciprocating engine.

1.2.1. Mechanism. A combination of rigid* or restraining** bodies which are so shaped and connected that they move upon each other with definite relative motion, is known as a mechanism.

Fig. 1.2 shows a mechanism which is known as slider-crank mechanism. It is a combination of rigid or restraining bodies namely crank, connecting rod and slider. They are so shaped and connected that they move upon each other with definite relative motion.

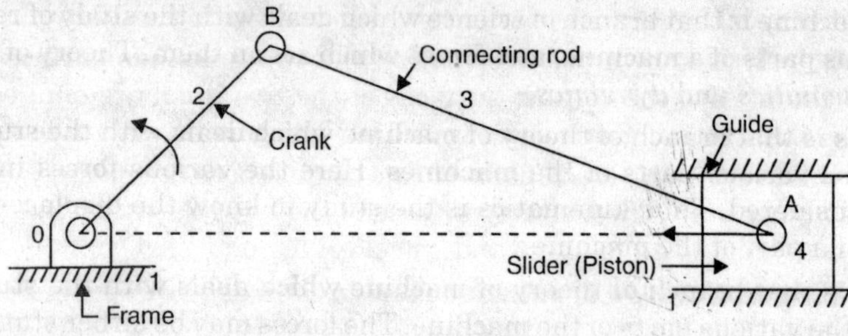

Fig. 1.2

The slider-crank mechanism shown in Fig. 1.2 converts the reciprocating motion of the slider into a rotary motion of the crank or vice-versa.

1.2.2. Machine. A machine is a mechanism or a combination of mechanisms which not only imparts definite motions to the parts but also transmits and modifies the available mechanical energy into some kind of useful energy. This useful energy may be in the shape of some kind of desired work.

The slider-crank mechanism shown in Fig. 1.2 will become a machine when it is used in automobile engine by adding valve mechanism etc. In that case it will convert the available energy (force on the piston) into the desired energy (*i.e.* torque on the crank shaft). This torque will move the vehicle.

1.3. LINK

A link is defined as a member or a combination of members, connecting other members and having motion *relative* to them. A slider-crank mechanism consists of following four links (Refer to Fig. 1.2) :

(*i*) Frame (*ii*) Crank
(*iii*) Connecting rod and (*iv*) Slider.

The slider (*i.e.* link 4) reciprocates in guide, which is connected to frame. Hence guide also becomes link 1 (*i.e.* frame).

1.4. KINEMATIC PAIR

A joint of *two links* having relative motion between them is known as a kinematic pair. In a slider-crank mechanism shown in Fig. 1.2, link 2 rotates relative to link 1 and hence

*Rigid body means a body with no deformation when the required force is transmitted.

**A restraining body means a body which is capable of transmitting the required forces with negligible deformation.

link 1 and 2 is a kinematic pair. Similarly link 2 is having motion relative to link 3 and hence links 2 and 3 is also a kinematic pair. Link 3 is having motion relative to link 4. Also link 4 is having motion relative to link 1. Hence links 3, 4 and 4, 1 constitute kinematic pairs.

1.4.1. Classifications of Kinematic Pairs. Kinematic pairs can be classified :

(*i*) According to nature of contact between the links

(*ii*) According to type of relative motion between the links

(*iii*) According to nature of mechanical constraint between the links.

(*i*) ***According to nature of contact***, the kinematic pairs are classified as

(*a*) Lower pair and

(*b*) Higher pair.

A kinematic pair is known as *lower pair* if the two links has surface contact or area contact between them. Also the contact surfaces of the two links are similar. Examples for lower pairs are (*i*) shaft rotating in a bearing and (*ii*) Nut turning on a screw.

If the two links (or a pair) has a point or line contact between them, then the kinematic pair is known as *higher pair*. The contact surfaces of the two links are not similar. Examples for higher pairs are (*i*) cam and follower and wheel rolling on a surface.

(*ii*) ***According to the type of relative motion between the two links***, the kinematic pairs are classified as

(*a*) Sliding pair (*b*) Turning pair

(*c*) Rolling pair (*d*) Screw pair (Helical pair)

(*e*) Spherical pair.

(*a*) *Sliding pair*. A kinematic pair is known as sliding pair if the two links have a sliding motion relative to each other. In Fig. 1.2, the links 4 and 1 are having sliding motion relative to each other and hence they form a sliding pair. Another example is a rectangular rod in a rectangular hole as shown in Fig. 1.3 (*a*).

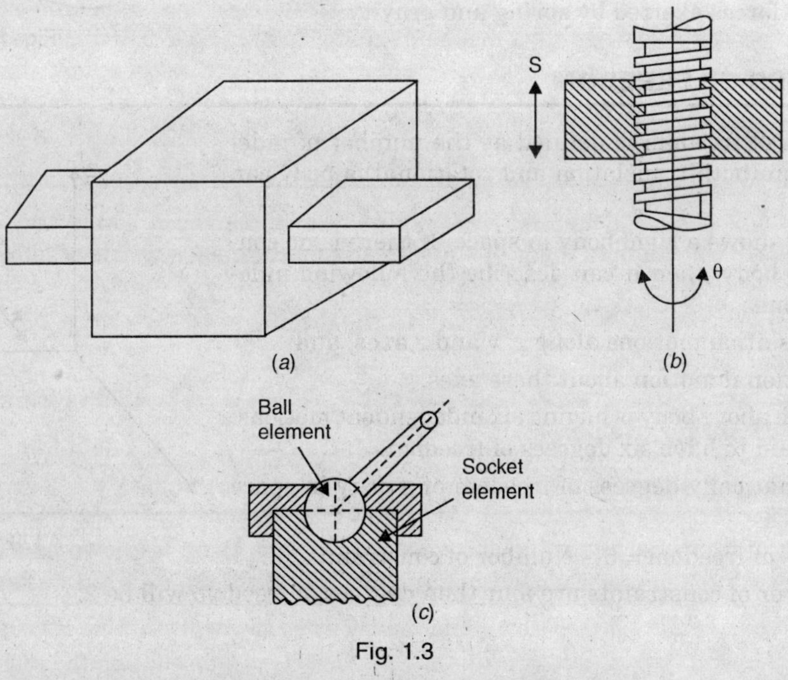

Fig. 1.3

(*b*) *Turning pair*. A kinematic pair is known as turning pair if one link has a turning or revolving motion relative to the other. In Fig. 1.2, link 2 is having revolving motion relative to link 1 hence links 2 and 1 constitutes a turning pair. Also the link 3 is having turning motion relative to link 4 and hence links 3 and 4 forms a turning pair. Similarly links 2 and 3 also forms turning pair.

(*c*) *Rolling pair*. A kinematic pair is known as rolling pair if one link has a rolling motion relative to the other. A rolling wheel on a flat surface forms a rolling pair. In a ball bearing, the ball and bearing forms one rolling pair whereas the ball and shaft forms second rolling pair.

(*d*) *Screw pair*. A kinematic pair is known as screw pair if the two links have a turning as well as sliding motion between them. The lead screw and the nut of a lathe is a screw pair. Fig. 1.3 (*b*) shows another screw pair. Also bolt with a nut is another example of screw pair. If bolt is kept fixed, nut will have sliding as well as rotational motion.

(*e*) *Spherical pair*. A kinematic pair is known as spherical pair if one link in the form of a sphere turns inside a fixed link. The ball and socket joint shown in Fig. 1.3 (*c*) is a spherical pair. The pen stand is another example of spherical pair.

(*iii*) ***According to nature of mechanical constraint between two links***, the kinematic pairs are classified as

(*a*) Closed pair, and

(*b*) Unclosed pair.

(*a*) *Closed pair*. In case of closed pair, the two elements of the pair are held together mechanically whereas in case of unclosed pair, the elements of the pair are in contact due to force of gravity or due to some spring action. The elements are not held together mechanically for example pair of cam and follower as shown in Fig. 1.4 is an example of unclosed pair as it is kept in contact by the forces exerted by spring and gravity.

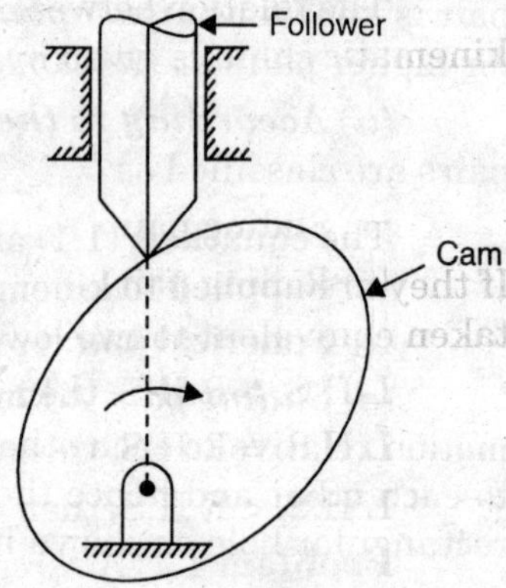

Fig. 1.4

1.5. DEGREES OF FREEDOM

Degrees of freedom is defined as the number of independent motion (both translation and rotational) a body can have.

Fig. 1.5 shows a rigid body in space. If there is no constraint on the body, then it can describe the following independent motions :

1. Translation motions along x, y and z axes, and
2. Rotational motion about these axes.

Thus the above body is having six independent motions. This body is said to have six degrees of freedom.

Mathematically degrees of freedom of a body in space is given by

Degrees of freedom = 6 – Number of constraints.

If number of constraints are four then degrees of freedom will be 2.

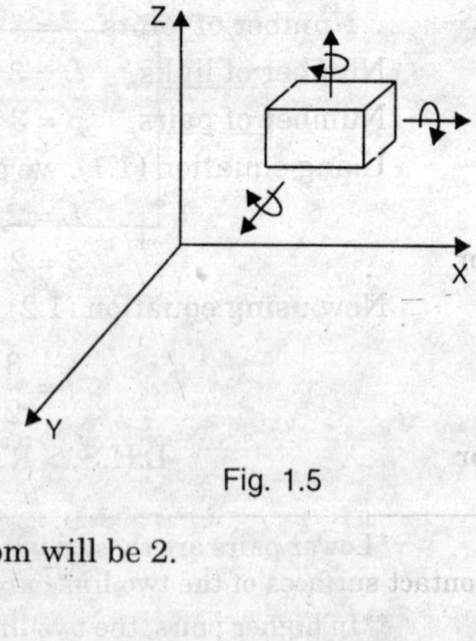

Fig. 1.5

1.6. KINEMATIC CHAIN

A kinematic chain is defined as the combination of kinematic pairs, joined in such a way that each link forms a part of two pairs and the motion of each relative to other is definite. Or when the kinematic pairs are coupled in such a way that the last link is joined to the first link to transmit definite motion, it is called a kinematic chain. In slider-crank mechanism shown in Fig. 1.2 link 1 is connected to link 2 and also to link 4. Now links 1 and 2 is a kinematic pair also links 1 and 4 is a kinematic pair. Hence link 1 forms a part of two pairs. Similarly link 2 forms a part of two pairs (*i.e.* kinematic pair of links 2 and 1 and kinematic pair of links 2 and 3). Similarly links 3 and 4 each forms a part of two pairs. Hence in slider-crank mechanism each link forms a part of two pairs and motion of each relative to other is definite. Hence the total combination of these links is a kinematic chain.

The relation between the number of pairs (p) and number of links (L) in a four link kinematic chain is given by

$$L = 2p - 4 \qquad \text{...(1.1)}$$

The relation between the number of links (L) and number of joints (j) forming a four link kinematic chain is given by

$$j = \frac{3}{2} L - 2 \qquad \text{...(1.2)}$$

The equations (1.1) and (1.2) are applied only to kinematic chains having lower* pairs. If they are applied to kinematic chains having higher** pairs, then each higher pair must be taken equivalent to two lower pairs and an additional link. In equations (1.1) and (1.2), if

L.H.S. > R.H.S. then chain is locked

L.H.S. = R.H.S. then chain is constrained

L.H.S. < R.H.S. then chain is unconstrained.

Problem 1.1. *A three links chain with three joints is shown in Fig. 1.6. Prove that the chain is locked.*

Sol. Given :

Three links 1, 2 and 3.

Three joints A, B and C.

∴ Number of joints, $j = 3$

Number of links, $L = 3$

Number of pairs, $p = 3$

Using equation (1.1), we get

$$L = 2p - 4$$

or $$3 = 2 \times 3 - 4 = 2 \text{ or L.H.S.} > \text{R.H.S.}$$

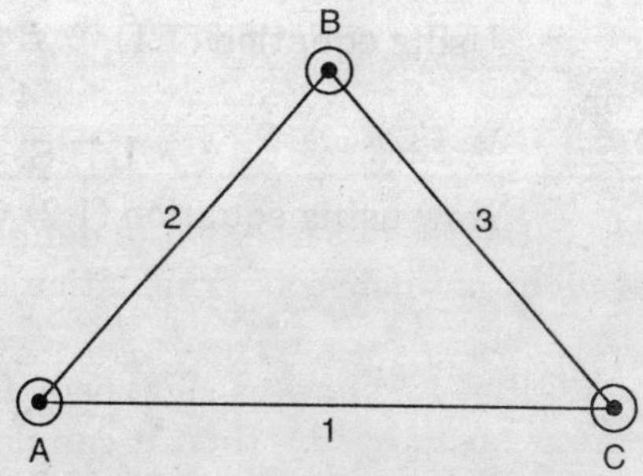

Fig. 1.6

Now using equation (1.2), we have

$$j = \frac{3}{2} \times L - 2 \text{ or } 3 = \frac{3}{2} \times 3 - 2 = 2.5$$

or L.H.S. > R.H.S.

---

*Lower pairs are those pairs in which two links have surface contact or area contact and also the contact surfaces of the two links are similar.

**In higher pairs, the two links have a point contact or line contact. Also contact surfaces of two links are not similar.

By using equations (1.1) and (1.2) for the three links with three joints shown in Fig. 1.6, we get left hand side (L.H.S.) is greater than the right hand side (R.H.S.). Hence it is not a kinematic chain. Hence no relative motion is possible and the chain is known as **locked chain**. This chain forms a rigid frame or structure which is used in bridges and trusses.

Problem 1.2. *What is a structure and what is the difference between a machine and a structure.*

Sol. The assemblage of a number of resistant bodies having no relative motion between them, is known as a structure. This assemblage carries loads and has straining action but there is no relative motion between the members (*i.e.* resistant bodies) of the assemblage.

The important differences between a machine and a structure is given in Table 1 as

Table 1

Machine	*Structure*
1. The parts of a machine move relative to one another.	1. The members of a structure do not move relative to one another.
2. A machine transforms the available energy into some useful work.	2. A structure does not transform any energy into useful work.
3. The links of a machine may transmit both power and motion.	3. The members of a structure transmit forces only.

Problem 1.3. *A four links chain with four joints is shown in Fig. 1.7. Prove that it is a constrained kinematic chain.*

Sol. Given :

Number of links, $L = 4$

Number of joints, $j = 4$

Number of pairs, $p = 4$

Using equation (1.1), $L = 2p - 4$

or $4 = 2 \times 4 - 4 = 4$

$\therefore$ L.H.S. = R.H.S.

Now using equation (1.2),

$$j = \frac{3}{2} L - 2$$

or $$4 = \frac{3}{2} \times 4 - 2 = 4$$

$\therefore$ L.H.S. = R.H.S.

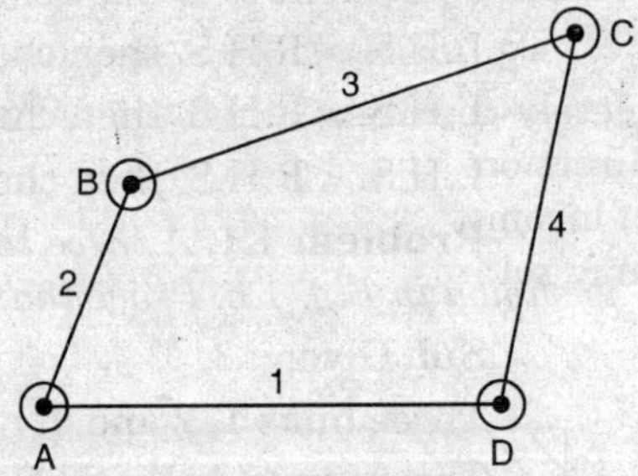

Fig. 1.7

By using equations (1.1) and (1.2), we get L.H.S. = R.H.S., hence the arrangement of four links shown in Fig. 1.7 is a constrained kinematic chain.

The four link (or bar) chain shown in Fig. 1.7 is a constrained kinematic chain, may also be proved as shown in Fig. 1.8. If a definite displacement (say θ) is given to the link AB, keeping the link *AD* fixed, then the displacements of links *BC* and *CD* are also prefectly definite. Hence in this case the relative motion is completely constrained. This kinematic

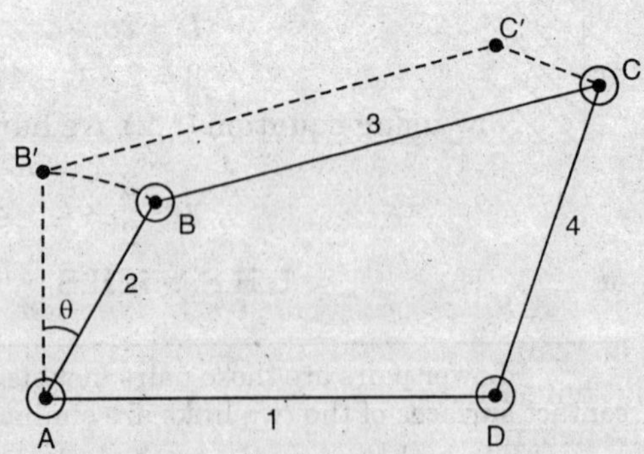

Fig. 1.8

chain is having one degree of freedom as a single link AB is sufficient to define the positions of other links.

Problem 1.4. *What are the types of constrained motions ?*

Sol. The constrained motions are of three types, namely :

1. Completely constrained motion,
2. Incompletely constrained motion, and
3. Successfully constrained motion.

1. *Completely constrained motion*. The motion between a pair in a definite direction irrespective of the direction of force applied, is known as a completely constrained motion. For example, the motion of a square bar in a square hole as shown in Fig. 1.9 (*a*) and the motion of a shaft with collars at each end in a circular hole as shown in Fig. 1.9 (*b*), are in a definite direction. Hence these are the examples of completely constrained motion.

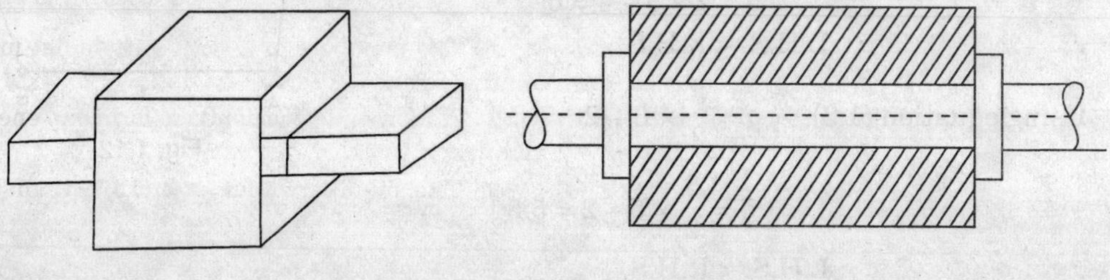
(*a*) (*b*)

Fig. 1.9

2. *Incompletely constrained motion*. The motion between a pair is known as an incompletely constrained motion if the motion between the pair can take place in more than one direction. The motion of a circular shaft in a circular hole as shown in Fig. 1.10, is an example of incompletely constrained motion. The shaft is having motions in two different directions. It may rotate or slide in the hole.

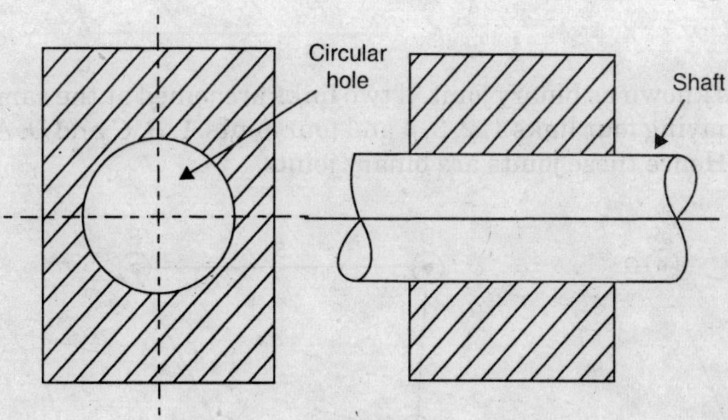

Fig. 1.10

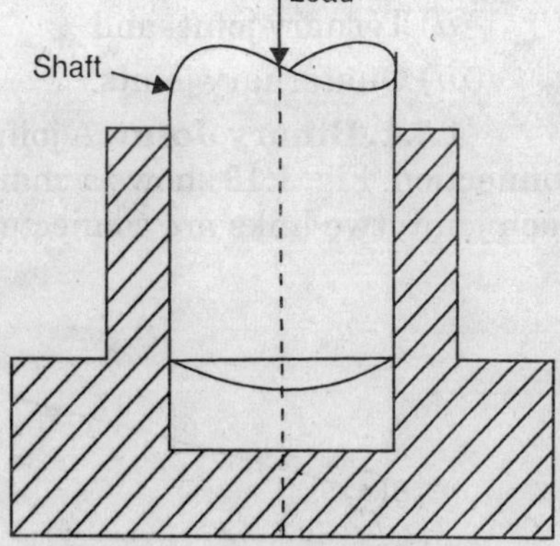

Fig. 1.11

3. *Successfully constrained motion*. The motion between a pair is known as a successfully constrained motion if the motion between the pair is not completely constrained by itself, but by some other means. Fig. 1.11 shows a foot-step bearing with a shaft. The shaft may rotate in the bearing and also may move upwards. This motion is incompletely constrained. But if a load is placed on the shaft, then shaft cannot have upward movement. Then the motion of the pair

becomes completely constrained. This completely constrained motion is obtained by some other means *i.e.* by placing load on the shaft. This type of motion is known as successfully constrained motion. The other examples of successfully constrained motion are the motion of an I.C. engine valve (which are kept on their seat by spring) and the piston reciprocating inside an engine cylinder.

Problem 1.5. *A five links chain with five joints is shown in Fig. 1.12. Prove that it is a unconstrained chain.*

Sol. Given :

Number of links, $L = 5$

Number of joints, $j = 5$

Number of pairs, $p = 5$

Using equation (1.1), $L = 2p - 4$

or $5 = 2 \times 5 - 4 = 6$

$\therefore$ L.H.S. < R.H.S.

Using equation (1.2), $j = \frac{3}{2} \times L - 2$

or $5 = \frac{3}{2} \times 5 - 2 = 5.5$

$\therefore$ L.H.S. < R.H.S.

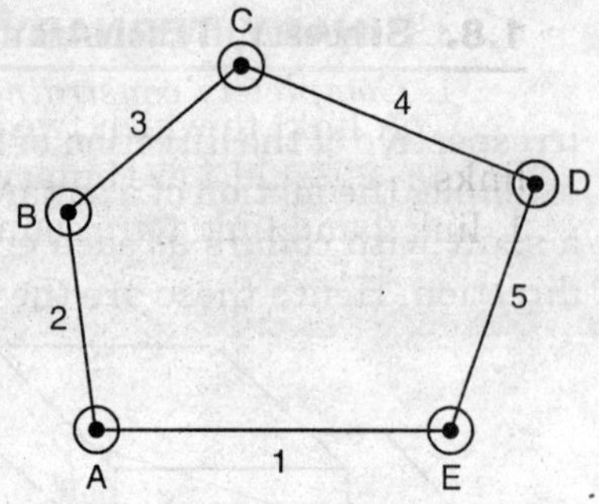

Fig. 1.12

By using equations (1.1) and (1.2), we get L.H.S. < R.H.S., hence the arrangement of five links shown in Fig. 1.12 is not a kinematic chain but a unconstrained chain. The relative motion between the links is not completely constrained.

1.7. BINARY, TERNARY, QUATERNARY JOINTS

In a chain, the following types of joints are usually found :

(*i*) Binary joint,

(*ii*) Ternary joint, and

(*iii*) Quaternary joints.

1.7.1. Binary Joint. A joint is known as binary joint, if two links are joined at the same connection. Fig. 1.13 shows a chain having four links 1, 2, 3, 4 and four joints *A*, *B*, *C* and *D*. At each joint, two links are connected. Hence these joints are binary joints.

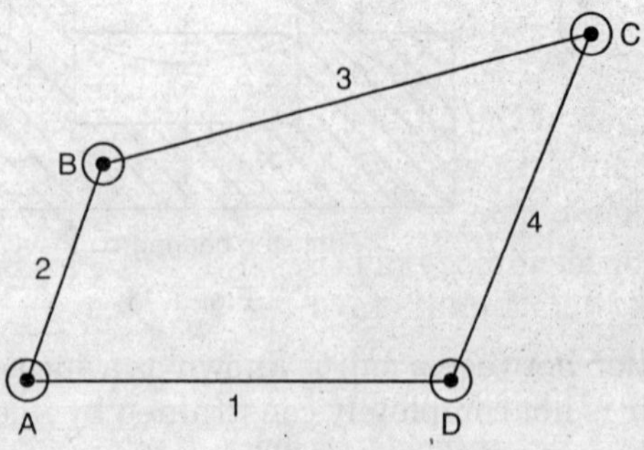

Fig. 1.13

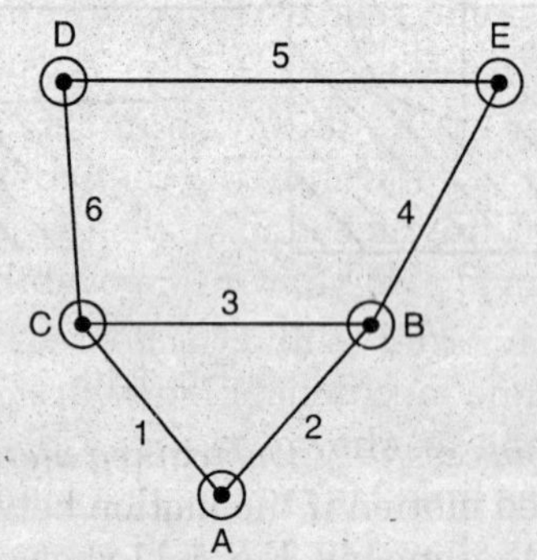

Fig. 1.14

1.7.2. Ternary Joint. A joint is known as ternary joint if three links are joined at the same connection. Fig. 1.14 shows a chain having six links and five joints. At joints C and B, three links are connected and hence these joints are ternary joints. But joints A, D and E are binary joints.

1.7.3. Quaternary joints. A joint is known as quaternary joints if four links are connected at the same connection.

1.8. BINARY, TERNARY AND QUATERNARY LINKS

A link, to which two links are connected, is known *a binary link*. In Fig. 1.15 (*a*), the links 1, 2, 3 and 4 are binary links, as to each link two links are connected. For example to link 1, link 2 and link 4 are connected. Similarly to link 2, link 3 and link 1 is connected.

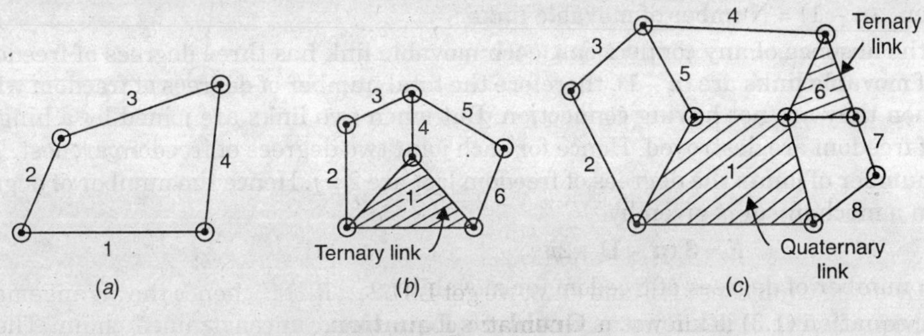

Fig. 1.15

A link, to which three links are connected, is known *a ternary link*. In Fig. 1.15 (*b*), the link 1 is a ternary link, as to link 1, three links namely 2, 4 and 6 are connected. Link 6 in Fig. 1.15 (*c*) is also a ternary link, as three links (*i.e.* link 4, link 7 and link 1) are connected to link 6.

A link, to which four links are connected, is known *a quaternary link*. In Fig. 1.15 (*c*), the link 1 is a quaternary link, as to link 1, four links namely link 2, link 5, link 6 and link 8 are connected.

1.9. DEGREES OF FREEDOM FOR PLANE MECHANISM

The plane mechanism means the mechanism in a plane (*i.e.* *x-y* plane). The three-dimensional relative motions between the links are not considered.

Degrees of freedom of plane mechanism means the number of inputs (*i.e.* number of independent co-ordinates) needed to determine the configuration or position of the all the links of the mechanism with respect to fixed link. Degrees of freedom is also known as movability of the mechanism.

In order to find the degrees of freedom (or movability) for any plane mechanism let us first consider two links 1 and 2 in *x-y* plane as shown in Fig. 1.16. Link 1 is fixed along *x*-axis whereas link 2 is moving in a plane. At any time the position of link 2 is shown in Fig. 1.16.

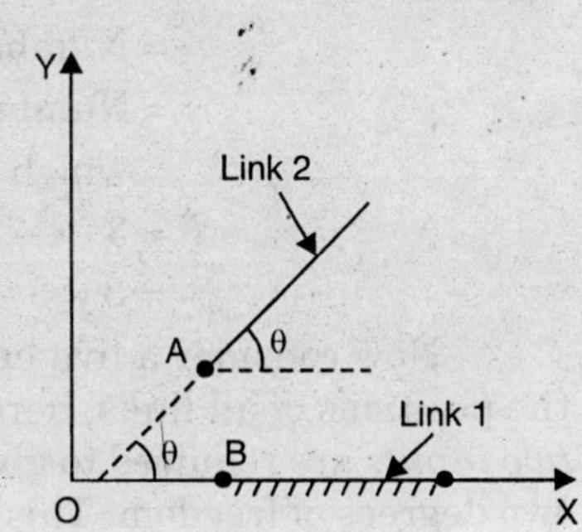

Fig. 1.16

The location of point A of link 2 can be completely specified by three variables *i.e.* co-ordinates x and y of the point A

and inclination θ of the link 2 with x-axis. Thus three independent variables *i.e.* two translations (x, y co-ordinates) and one rotation (θ) are necessary to completely locate link 2 in x-y plane. Hence the link 2 has three degrees of freedom.

But if point A is joined by a pin to point B on a fixed link 1, then the two-independent variables (x and y for point A) are fixed. The position of link 2, now is determined by a single variable θ only. Hence now the link 2 will have only one degree of freedom. This shows that two degrees of freedom (translational) are lost at every pin (or hinge) joint of any link with fixed link.

Let us now express the number of degrees of freedom of a mechanism in terms of number of links and number of joints.

Let n = Number of links in a mechanism out of which one link is fixed.

j = Number of simple joints (*i.e.* joints which connects two links)

Then $(n - 1)$ = Number of movable links.

In the absence of any connections, each movable link has three degrees of freedom. As number of movable links are $(n - 1)$, therefore the total number of degrees of freedom will be $3(n - 1)$ when they are not having connection. But when two links are joined by a hinge, two degrees of freedom are destroyed. Hence for each joint two degrees of freedom are lost. Therefore for j number of joints the degrees of freedom lost are $2 \times j$. Hence the number of degrees of freedom in a mechanism is given by,

$$F = 3(n - 1) - 2j \qquad \text{...(1.3)}$$

where F = number of degrees of freedom (or movability).

The equation (1.3) is known as **Grubler's Equation**.

(*i*) If $F = 0$, it means there is no movability of mechanism and hence mechanism is known as a *structure* and there is no relative motion between the links.

(*ii*) If $F = 1$, the mechanism is having only one degree of freedom. Hence only one co-ordinate is required to specify the positions of all links of the mechanism. One input gives a unique output.

(*iii*) If $F = 2$, the mechanism is having two degrees of freedom, which means two inputs are required to give a unique output [or two co-ordinates are required to specify the positions of all links of a mechanism].

Fig. 1.17 shows a four bar chain, having link 1 as fixed. To specify the positions of all links, only one co-ordinate *i.e.* θ is required. Hence the four bar chain is having only one degree of freedom. This can also be verified by applying equation (1.3) in which

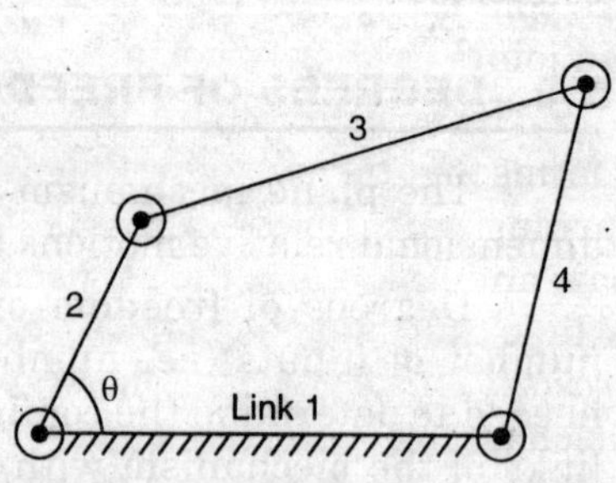

Fig. 1.17

n = Number of links = 4

j = Number of simple joints (*i.e.* joints which connects two links) = 4

$$\therefore \quad F = 3(n - 1) - 2j = 3(4 - 1) - 2 \times 4$$
$$= 3 \times 3 - 8 = 1$$

Now consider a five bar chain as shown in Fig. 1.18 in which link 1 is fixed. To specify the positions of all links, here two co-ordinates such as θ_1 and θ_2 are required. This means the *two inputs* are required to give a unique output. Hence the mechanism shown in Fig. 1.18, has two degrees of freedom. This can also be verified by applying equation (1.3) in which $n = 5$ and $j = 5$.

Hence $F = 3(n - 1) - 2j = 3(5 - 1) - 2 \times 5$

$= 12 - 10 = 2$

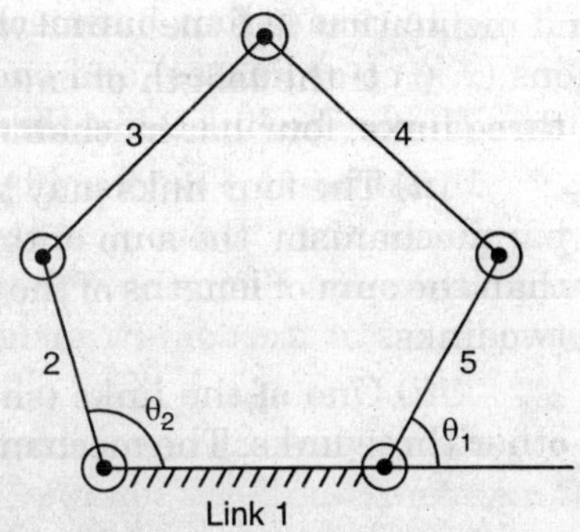

Fig. 1.18

A sliding pair is having only one degree of freedom. This is shown in Fig. 1.2 where link 4 (slider) can only move in horizontal direction. It cannot move in vertical direction and also cannot rotate in the plane. Hence the **slider joint** also substracts two degrees of freedom, one translation and one rotation. Hence j in equation (1.3) also includes slider joints.

Thus j = Sum of number of simple pin joints plus the number of slider joints.

1.10. INVERSION OF MECHANISM

Mechanism is a kinematic chain in which one link is fixed. By fixing the links of a kinematic chain one at a time, we get as many different mechanism as the number of links in the chain. This method of obtaining different mechanism by fixing different links of the same kinematic chain, is known as inversion of the mechanism. In the process of inversion, the relative motions of the links of the mechanism produced remain unchanged.

1.11. DIFFERENT TYPES OF KINEMATIC CHAINS AND THEIR INVERSIONS

The simplest kinematic chain is a chain consisting of four kinematic pairs, each pair being a sliding pair or a turning pair. The important types of kinematic chains with four kinematic pairs, are :

(*i*) Four bar chain,

(*ii*) Single slider crank chain, and

(*iii*) Double slider crank chain.

1.11.1. Four Bar Chains. This is the simplest kinematic chain. It consists of four rigid links which are connected in the form of a quadrilateral by four pin joints as shown in Fig. 1.19.

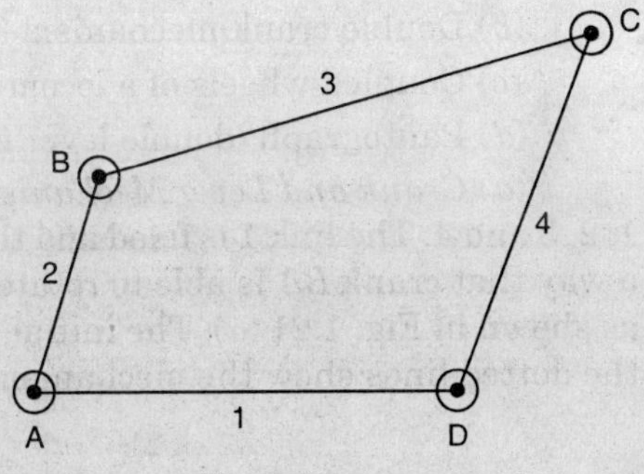

Fig. 1.19

It consists of four turning pairs. Link 1 and link 2 forms first turning pair, link 2 and link 3 form second turning pair, link 3 and link 4 forms third turning pair and link 4 and link 1 forms fourth turning pair. A link that makes complete revolution is known as *crank*. The fixed link is known as *frame* of the mechanism. The link opposite to the fixed link is known as *coupler* or *connecting rod*. The fourth link is known as *lever* or *rocker* (if it oscillates) or an another crank (if it rotates).

If different links of the four-bar mechanism are fixed, four different mechanism (known as inversions) as shown in Fig. 1.20 will be obtained.

In case of four-bar mechanism, the following points must be remembered :

(*i*) If the length of one of the links is greater than the sum of the lengths of the other three links, four-bar mechanism is not possible.

(*ii*) The four links may be of different lengths. But according to Grashof's law for a four-bar mechanism, the sum of the lengths of the shortest and longest link should not be greater than the sum of lengths of the remaining two links for continuous relative motion between the two links.

(*iii*) One of the links (shortest link) should make a complete revolution relative to the other three links. The mechanism in which no link makes a complete revolution is not useful.

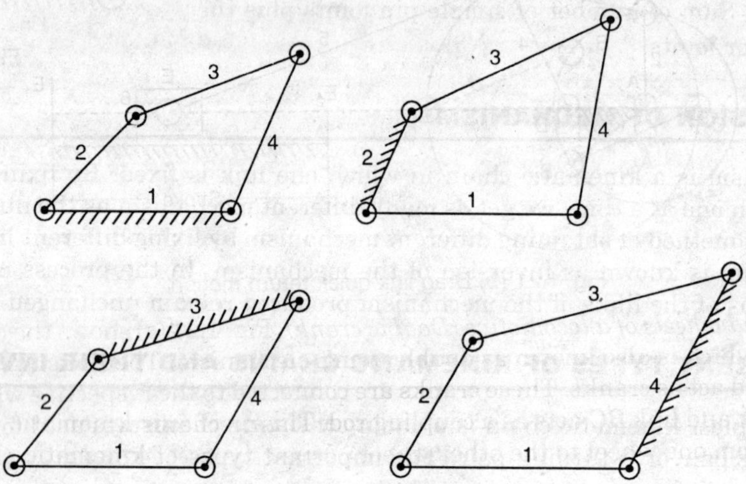

Fig. 1.20

Applications of Four Bar Chain. The followings are the important applications of four bar chain :

(*a*) Crank and lever mechanism (oscillatory motion)

(*b*) Double crank mechanism (complete rotation of the crank and the follower)

(*c*) Coupled wheels of a locomotive (double crank)

(*d*) Pantograph (double lever mechanism)

(*a*) *Crank and Lever Mechanism (oscillatory motion)*. The four links of the bar chain are 1, 2, 3 and 4. The link 1 is fixed and the lengths of the links 2, 3 and 4 are proportionate in such a way that crank *BA* is able to rotate completely. The follower *CD* only oscillate from C_1 to C_2 as shown in Fig. 1.21 (*a*). The initial position of the mechanism is shown by full lines whereas the dotted lines show the mechanism for two extreme different positions.

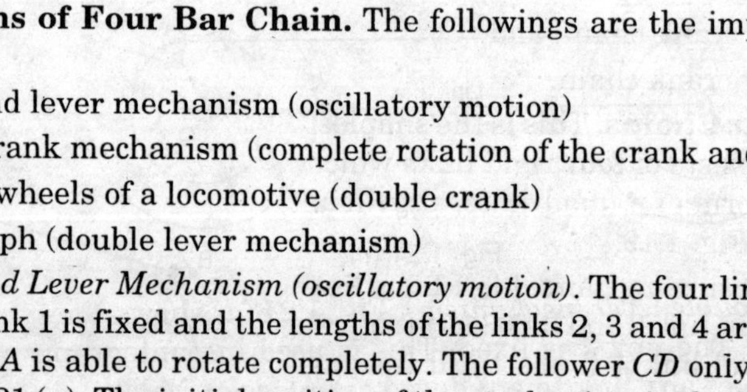

Fig. 1.21 (*a*) Crank and lever mechanism.

(*b*) *Double Crank Mechanism (complete rotation of the crank and the follower)*. The shortest link 1 is always stationary link. The sum of the shortest and the longest links should be less than the sum of the other two links.

For a drag link quick return motion, link 5 and link 6 are also connected as shown in Fig. 1.21 (*b*). The initial position of the mechanism is shown by full lines *i.e.* *ABCD* alongwith *CE*. The two extreme positions (*i.e.* AB_1C_1D alongwith C_1E_1 and position AB_2C_2D alongwith C_2E_2) are shown by dotted lines.

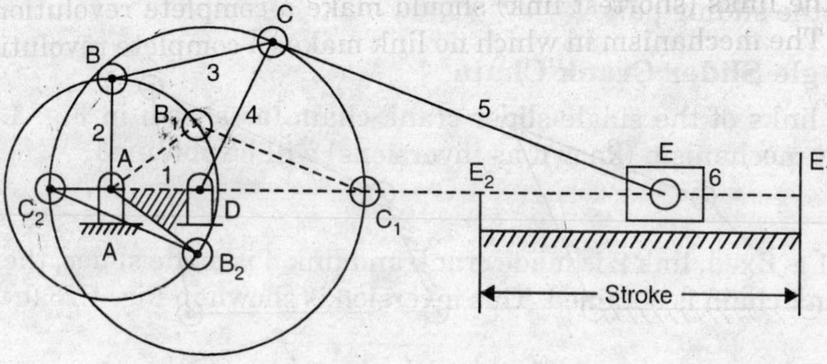

Fig. 1.21 (*b*) Drag link quick return motion.

(*c*) *Coupled wheels of a locomotive (Double crank)*. Fig. 1.21 (*c*) shows the coupled wheels of a locomotive which is also known as double crank mechanism. The links *AB* and *CD* are of equal lengths and act as cranks. These cranks are connected to the respective wheels. The link *AD* is a fixed link and link *BC* acts as a coupling rod. This mechanism is used for transmitting rotary motion from one wheel to the other wheel.

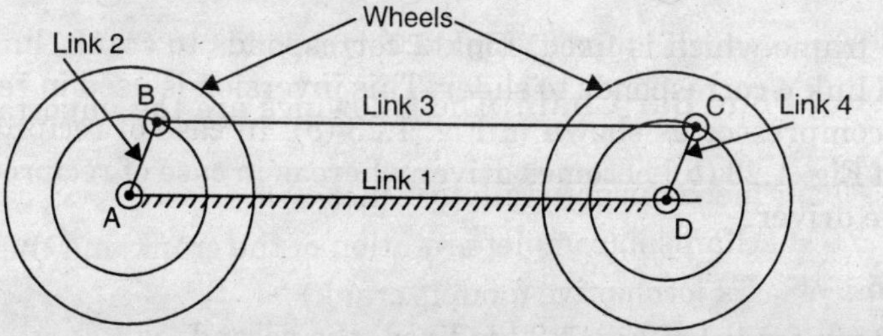

Fig. 1.21 (*c*)

(*d*) *Pantograph (Double lever mechanism)*. Fig. 1.21 (*d*) shows the four bar chain in the form of a parallelogram. The link 1 is fixed. This is used in pantograph which is a copying device. The exact motion of any link to a reduced or enlarged scale can be obtained with the help of this device.

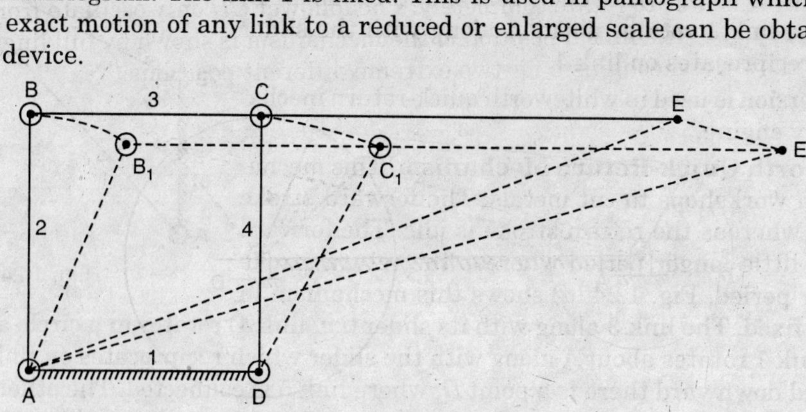

Fig. 1.21 (*d*) Pantograph.

1.11.2. Single-Slider Crank Chain. It is the modification of basic four bar chain. It consists of four kinematic pairs out of which one is sliding pair and three are turning pairs as shown in Fig. 1.22. Link 1 and link 2 form one turning pair, link 2 and link 3 form the second turning pair, link 3 and link 4 form the third turning pair. Link 4 and link 1 form the sliding pair.

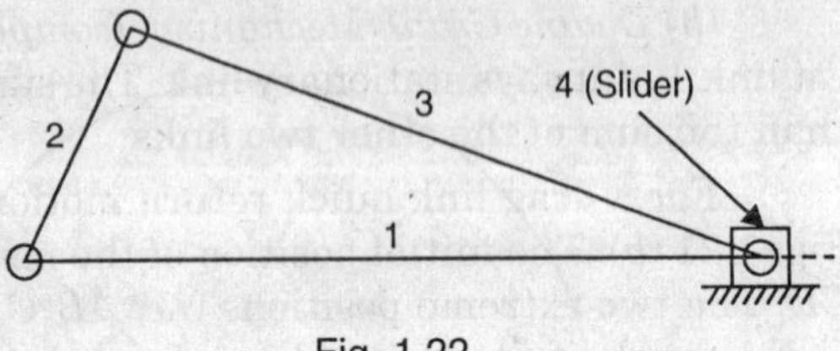

Fig. 1.22

Inversion of Single Slider Crank Chain

If different links of the single-slider crank chain (as shown in Fig. 1.22) are fixed in turn four different mechanism (known as inversions) will be obtained.

First Inversion

When link 1 is fixed, link 2 is made crank and link 4 is made slider, then first inversion of single slider crank chain is obtained. This inversion is shown in Fig. 1.23 (*a*). Here the link 1

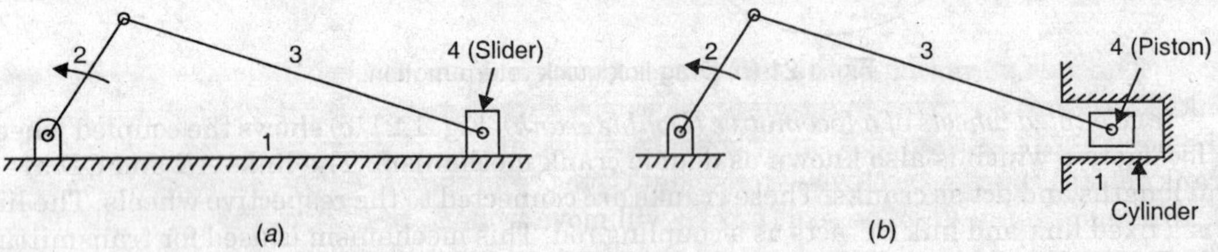

Fig. 1.23

corresponds to the frame which is fixed. Link 2 corresponds to crank, link 3 corresponds to connecting rod and link 4 corresponds to slider. This inversion is used in reciprocating engine and reciprocating compressor as shown in Fig. 1.23 (*b*). In case of reciprocating engine, the link 4 *i.e.* piston [of Fig. 1.23 (*b*)] becomes driver whereas in case of reciprocating compressor, link 2 (crank) is the driver.

Second Inversion

When link 2 (or crank) of Fig. 1.22 is fixed, the second inversion of single slider crank chain is obtained as shown in Fig. 1.24. Link 3 along with the slider at its end *C*, becomes a crank. Hence link 3 along with slider (link 4) rotates about *B*. By doing so, the link 1 rotates about *A* along with the slider (link 4) which reciprocates on link 1.

This inversion is used in whiteworth quick-return mechanism and rotary engine.

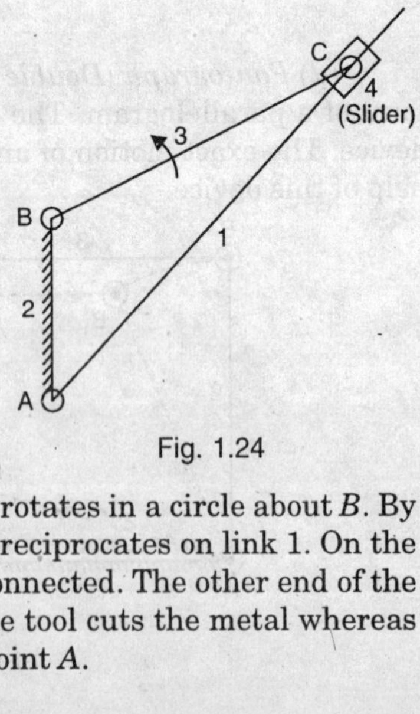

Fig. 1.24

Whiteworth Quick-Return Mechanism. This mechanism is used in workshops to cut metals. The forward stroke cuts the metal whereas the return stroke is idle. The forward stroke takes a little longer period whereas the return stroke takes a shorter period. Fig. 1.24 (*a*) shows this mechanism in which link 2 is fixed. The link 3 along with its slider (*i.e.* link 4) rotates in a circle about *B*. By doing so, the link 1 rotates about *A* along with the slider which reciprocates on link 1. On the link 1, produced downward there is a point *D*, where link 5 is connected. The other end of the link 5 is connected to the tool (link 6). The forward stroke of the tool cuts the metal whereas the return stroke is idle. The point *D* rotates in a circle about point *A*.

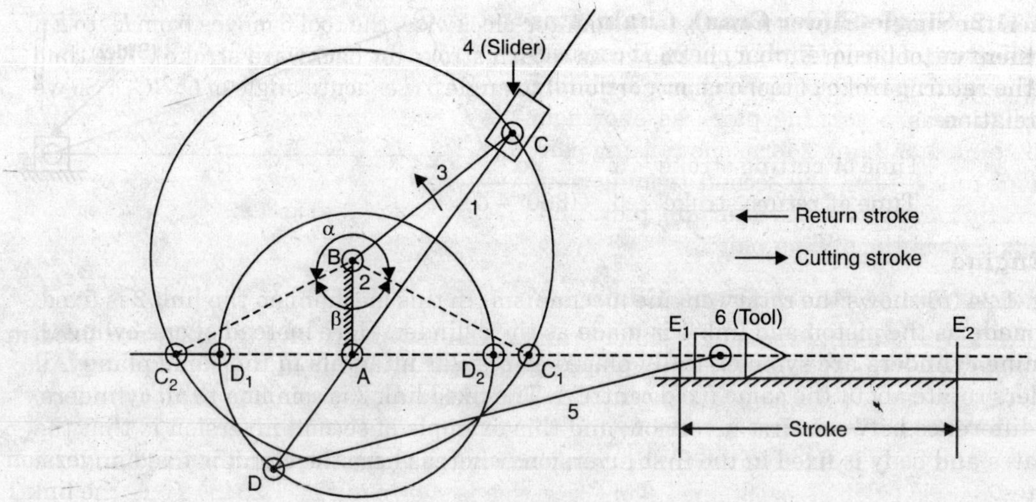

Fig. 1.24 (*a*)

The axis of motion of the tool passes through A and is perpendicular to AB, the fixed link. The crank 3 rotates in a counter-clockwise direction. Let, initially the slider 4 is at C_1. The point D will be then at D_1 and the tool 6 will be at E_1 (*i.e.* the extreme left position). As the crank rotates counter-clockwise from point C_1 to C and then to C_2, the point D will move from D_1 to D and then to D_2. The tool (link 6) will move from E_1 to E_2 (*i.e.* extreme right position). The movement of tool 6 from E_1 to E_2 is known as forward stroke. The time taken for the forward stroke of tool 6 is proportional to the angle α (*i.e.* obtuse angle C_1BC_2).

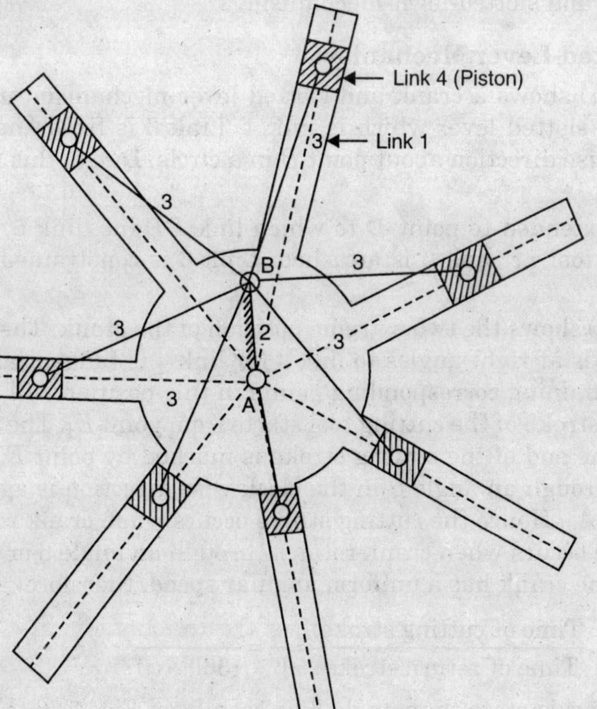

Fig. 1.24 (*b*)

When the slider 4 moves from C_2 to C_1 counter-clockwise, the tool 6 moves from E_2 to E_1. The movement of tool from E_2 to E_1 is known as return stroke (or backward stroke). The time taken for the return stroke of tool 6 is proportional to angle β (*i.e.* acute angle of C_2BC_1). So we have the relation as

$$\frac{\text{Time of cutting stroke}}{\text{Time of return stroke}} = \frac{\alpha}{\beta} = \frac{\alpha}{(360° - \alpha)}.$$

Rotary Engine

Fig. 1.24 (*b*) shows the rotary engine mechanism. In this mechanism the link 2 is fixed. Link 4 is made as the piston and link 1 is made as the cylinder. Here instead of one cylinder, seven or nine cylinders are symmetrically placed at regular intervals in the same plane. All the cylinders rotate about the same fixed centre A. The fixed link 2 is common to all cylinders. The only difference between first inversion, and this example of second inversion is that the crank rotates and body is fixed in the first inversion, whereas here the crank is fixed and the body rotates.

Third Inversion

When link 3 of Fig. 1.22 is fixed, the third inversion of single slider crank is obtained as shown in Fig. 1.25.

Link 2 acts as a crank and is rotating about point B. Link 4 oscillates.

This inversion is used in oscillating cylinder engine and crank and slotted-lever mechanism.

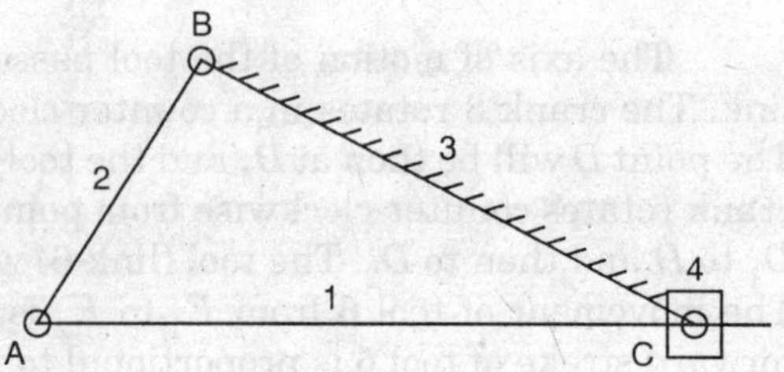

Fig. 1.25

Crank and Slotted-Lever Mechanism

Fig. 1.26 (*a*) shows a crank and slotted lever mechanism, in which link 1 is a slider which slides in a slotted lever which is link 4. Link 3 is fixed and link 2 is a crank which rotates in clock-wise direction about point B in a circle. Due to this the link 4 oscillates about the point C.

Link 4 is extended to point D to which link DE (*i.e.* link 5) is connected. At the end of link 5, cutting tool *i.e.* link 6 is attached. Link 6 is constrained to slide as shown in the Fig. 1.26 (*a*).

Fig. 1.26 (*b*) shows the two extreme position of the crank. The first position is when the crank (*i.e.* link 2) is at right angles to link 4 (or link 4 is tangential to crank radius circle at point A_1). The remaining corresponding points in this position will be D_1 and E_1 as shown in Fig. 1.26 (*b*). The stroke of the cutting tool starts from point E_1. The crank is rotating in clock-wise direction. The end of the cutting stroke is marked by point E_2, when again crank after having rotated through an angle α in the clock-wise direction is again at right angles to the link 4 at position A_2. Hence the cutting stroke occurs when crank rotates through an angle α and return stroke occurs when crank rotates through an angle β or (360 – α) in the clockwise direction. Since the crank has a uniform angular speed, therefore

$$\frac{\text{Time of cutting stroke}}{\text{Time of return stroke}} = \frac{\alpha}{\beta} = \frac{\alpha}{(360 - \alpha)}.$$

Since angle α is more than angle β as seen from Fig. 1.26 (*b*), the cutting time will be more than return time.

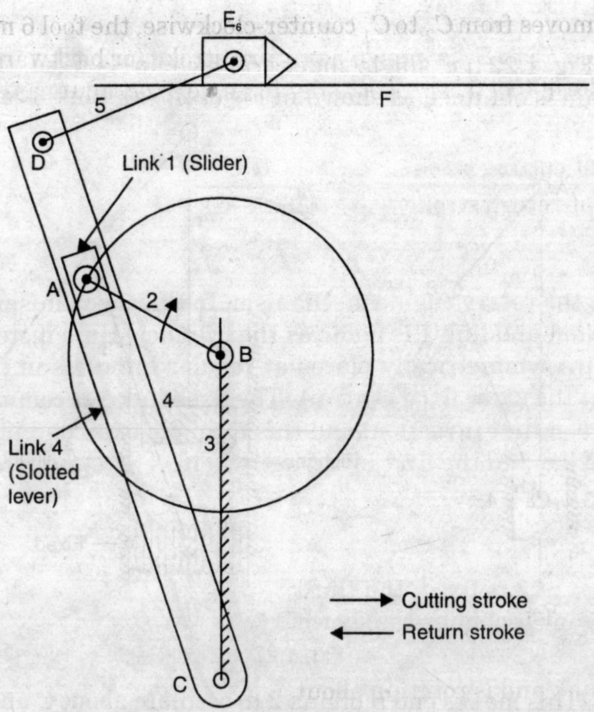

Fig. 1.26 (*a*)

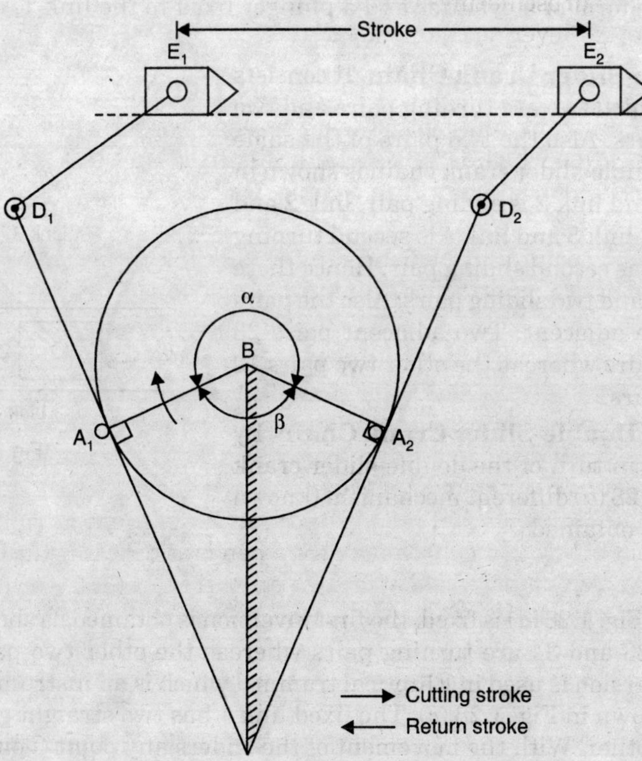

Fig. 1.26 (*b*)

Fourth Inversion

When link 4 of Fig. 1.22 (*i.e.* single slider crank chain) is fixed, the fourth inversion of single slider crank chain is obtained as shown in Fig. 1.27 (*a*). Link 3 can oscillate about the fixed point *C* on link 4. This makes end *B* of link 2 to oscillate about *C* and end *A* to reciprocate along the axis of fixed link 4. This inversion is used in hand-pump as shown in Fig. 1.27 (*b*). Link 4 is made in the form of a cylinder and a plunger fixed to the link 1, reciprocates in the cylinder.

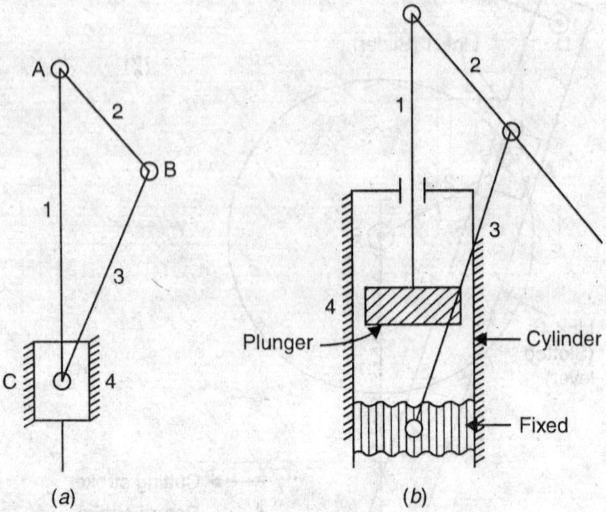

Fig. 1.27

1.11.3. Double-Slider Crank Chain. It consists of four pairs out of which two are turning pairs and two others are sliding pairs. Also the two pairs of the same kind are adjacent. Double-slider crank chain is shown in Fig. 1.28 (*a*). Link 1 and link 2 is sliding pair, link 2 and link 3 is turning pair, link 3 and link 4 is second turning pair, link 4 and link 1 is second sliding pair. Hence there are two turning pairs and two sliding pairs. Also the pairs of the same kind are adjacent. Two adjacent pairs 23 and 34 are turning pairs whereas the other two pairs 12 and 14 are sliding pairs.

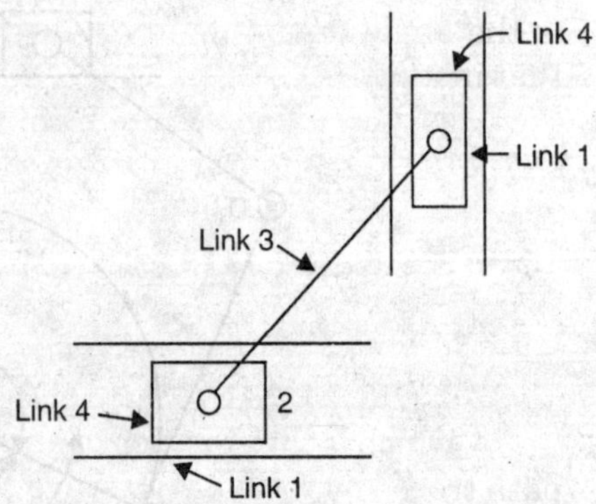

Fig. 1.28 (*a*)

Inversions of Double-Slider Crank Chain. By fixing different links in turn of the double-slider crank chain shown in Fig. 1.28 (*a*) different mechanism (known as inversions) will be obtained.

First Inversion

When link 1 of Fig. 1.28 (*a*) is fixed, the first inversion is obtained as shown in Fig. 1.28 (*b*). Two adjacent pairs 23 and 34 are turning pairs whereas the other two pairs 12 and 14 are sliding pair. This inversion is used in Elliptical trammel which is an instrument used for drawing ellipses and is shown in Fig. 1.28 (*c*). The fixed link 1 has two straight grooves cut in it, at right angles to each other. With the movement of the sliders any point *C* on the link 3 (except the mid-point of AB) will trace an ellipse on a fixed plate. The mid-point of AB will trace a circle.

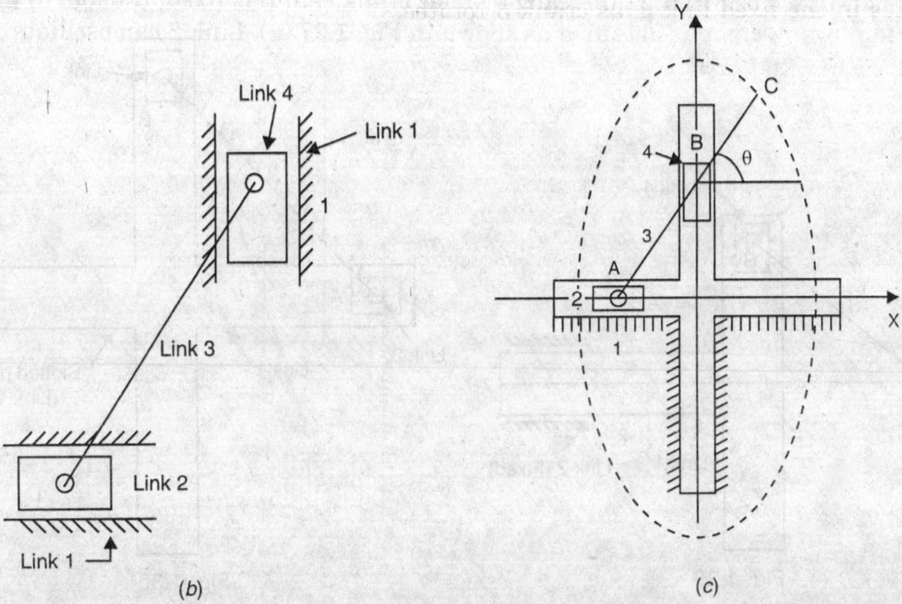

Fig. 1.28

This is proved as given below :

Let the co-ordinates of the point C at any instant are x and y as shown in Fig. 1.28 (d). Let θ is the angle made by link AC with horizontal at that instant.

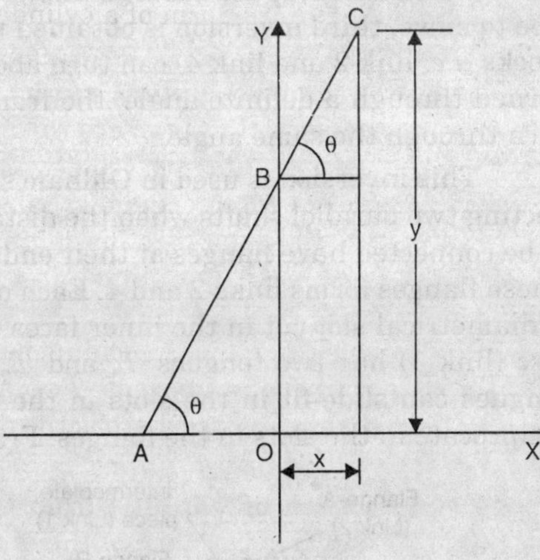

Fig. 1.28 (d)

Then $x = BC \cos \theta$ and $y = AC \sin \theta$

$$\therefore \quad \frac{x}{BC} = \cos \theta \text{ and } \frac{y}{AC} = \sin \theta$$

Squaring and adding, we get

$$\frac{x^2}{(BC)^2} + \frac{y^2}{(AC)^2} = \cos^2 \theta + \sin^2 \theta = 1$$

which is the equation of an ellipse. Hence the path traced by the point C is an ellipse whose semi-major axis is AC and semi-minor axis is BC.

If C is the mid-point of AB, then $AC = BC$.

$$\therefore \quad \frac{x^2}{(AC)^2} + \frac{y^2}{(AC)^2} = 1$$

or

$$x^2 + y^2 = (AC)^2$$

which is the equation of a circle whose radius is AC or BC.

Second Inversion

When link 2 or link 4 of the double-slider crank chain is fixed, the second inversion is obtained as shown in Fig. 1.29. Here link 2 is fixed and end B of link 3 rotates about A and link 1 reciprocates in the horizontal direction. This inversion is used in Scotch Yoke, which is used to convert the rotary motion into a sliding motion. The horizontal portion of link 1 reciprocates in the fixed link 2, as crank 3 rotates.

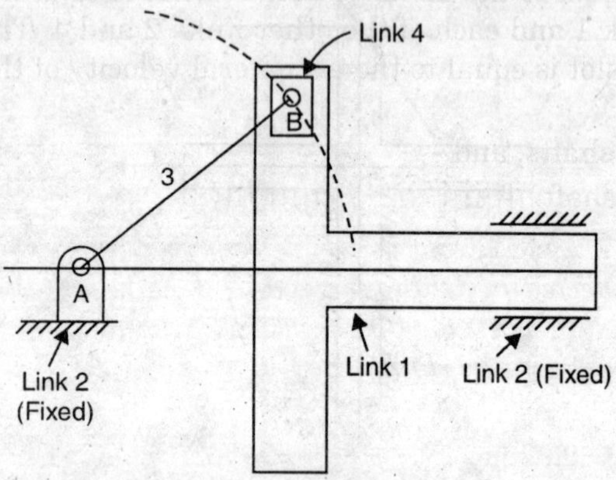

Fig. 1.29

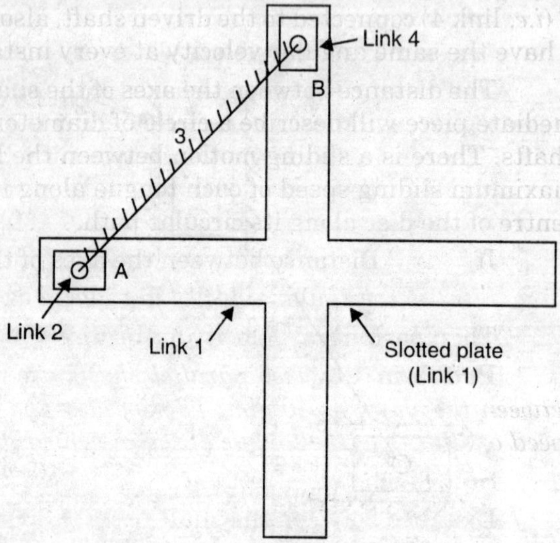

Fig. 1.30

Third Inversion

When link 3, of the double slider crank chain shown in Fig. 1.28 (*c*), is fixed and link 1 is free to move, third inversion is obtained which is shown in Fig. 1.30. In this case each of slide blocks (*i.e.* link 2 and link 4) can turn about the pins *A* and *B*. If one slide block (say link 2) is turned through a definite angle, the frame (*i.e.* link 1) and the other block (*i.e.* link 4) must turn through the same angle.

This inversion is used in Oldham's Coupling (shown in Fig. 1.31) which is used for connecting two parallel shafts when the distance between the two shafts is small. The two shafts to be connected have flanges at their ends which are rigidly fastened by forging to the shafts. These flanges forms links 2 and 4. Each of these links forms a turning pair with link 3. There is diametrical slot cut in the inner faces of these flanges. An intermediate piece is a circular disc (link 1) has two tongues T_1 and T_2 on each face at right angles to each other. These tongues can slide-fit in the slots in the two flanges (link 2 and 4). The link 1 can slide or reciprocate in the slots in the flanges. Frame and bearings form the link 3 which is fixed.

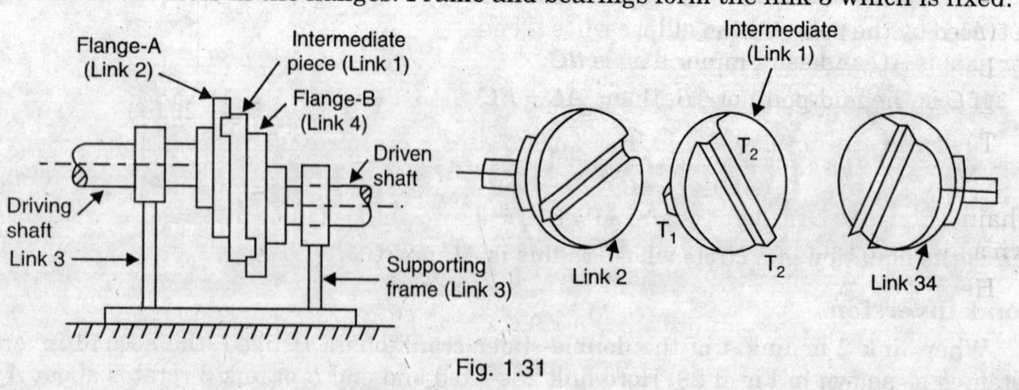

Fig. 1.31

When the driving shaft is rotated, the flange A (*i.e.* link 2) connected rigidly to the driving shaft also rotates by the same angle, the intermediate piece also rotates by the same angle through which flange A has rotated. Due to the rotation of intermediate shaft, the flange B (*i.e.* link 4) connected to the driven shaft, also rotates by the same angle. Hence links 2, 4 and 1 have the same angular velocity at every instant.

The distance between the axes of the shaft is constant and hence the centre of the intermediate piece will describe a circle of diameter equal to the distance between the axes of the shafts. There is a sliding motion between the link 1 and each of the other links 2 and 4. The maximum sliding speed of each tongue along its slot is equal to the peripheral velocity of the centre of the disc along its circular path.

If x = Distance between the axes of the shafts, and

ω = Angular velocity in rad/s of each shaft.

Then peripheral velocity of the disc = $\omega \times x$.

Problem 1.6. *Two parallel shafts are connected by Oldham's coupling. The distance between the shafts is 30 mm. The speed of the driving shaft is 500 r.p.m. Find the maximum speed of sliding of the tongue of the intermediate piece in the slot in the flange.*

Sol. Given :

Distance between the shaft, $x = 30 \text{ mm} = 0.03 \text{ m}$

Speed of driving shaft, $N = 500$ r.p.m.

$\therefore$ Angular speed of the shaft, $\omega = \dfrac{2\pi N}{60} = \dfrac{2\pi \times 500}{60} = 52.36$ rad/s

Maximum speed of sliding $= \omega \times x = 52.36 \times 0.03$ m/s = **1.5708 m/s. Ans.**

Problem 1.7. *Some four-bar linkages are shown in Fig. 1.32 where the number indicate the respective link lengths in cm. Identify the nature of each mechanism, i.e. whether (i) double crank, (ii) crank rocker or (iii) double rocker. Give reasons in brief.* (AMIE, Winter 1993)

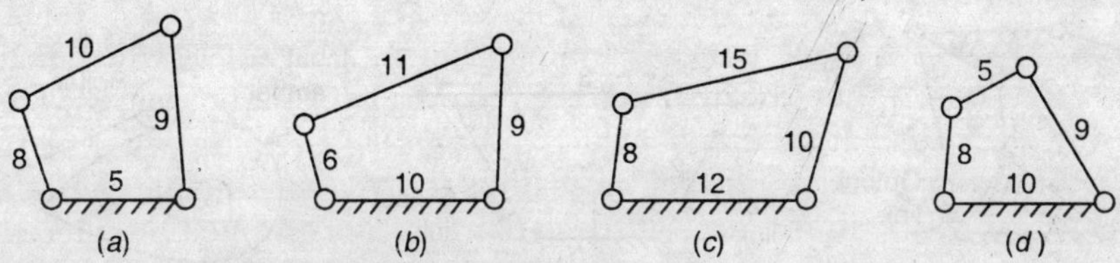

Fig. 1.32

Sol. Given :

(A) Length of longest link, $l = 10$

Length of shortest link, $s = 5$

Length of the other links, $p = 8$ and $q = 9$

To identify the nature of each mechanism, Grashoff's law is used. According to this law if

$$l + s \le p + q$$

the chain belongs to Grashoff's chain. All other chains, which do not satisfy the above law, are known as "non-Grashoff's" type and all inversions of such chains give rocker-rock mechanism.

Here $l + s = 10 + 5 = 15$

and $p + q = 8 + 9 = 17$

$$l + s < p + q$$

∴ Chain at (A) satisfies Grashoff's law. Since in this case, the shortest link is fixed, the mechanism belongs to the category of double crank mechanism.

(B) For chain shown in Fig. 1.32 (B), we have

$$l + s = 11 + 6 = 17$$

and $$p + q = 10 + 9 = 19$$

$$\therefore \quad l + s < p + q$$

Hence chain at (B) satisfies Grashoff's law. Since in this case neither shortest link nor the link opposite to shortest is fixed, the mechanism is of the category of "Crank-Rocker" type.

(C) For chain shown in Fig. 1.32 (C), we have

$$l + s = 15 + 8 = 23$$

$$p + q = 12 + 10 = 22$$

$$\therefore \quad l + s > p + q$$

∴ Chain at (C) does not satisfy Grashoff's law. Hence mechanism at (C) is Rocker-Rocker type.

(D) For chain shown in Fig. 1.32 (D), we have

$$l + s = 10 + 5 = 15$$

$$p + q = 8 + 9 = 17$$

$$\therefore \quad l + s < p + q$$

Hence chain at (D) satisfies Grashoff's law. In this case, neither the shortest link nor the link opposite to shortest link is fixed, hence mechanism belongs to the category of crank-rocker mechanism.

Problem 1.8. *Examine the mechanism shown in Fig. 1.33 and indicate the cases where unique relation between the motions of the input and output link exists. Give reasons for your answers.* (AMIE, Winter 1993)

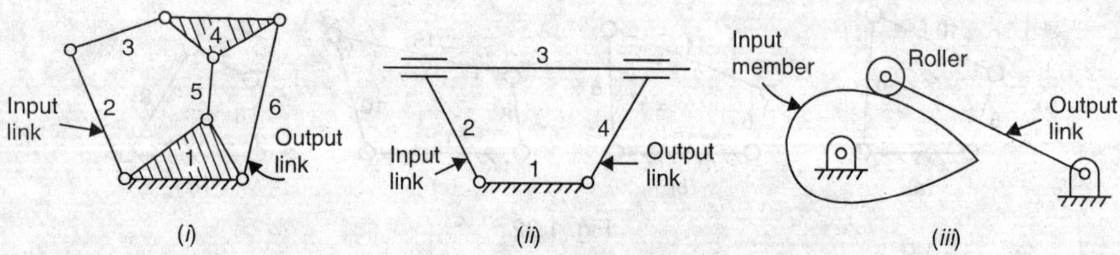

Fig. 1.33

Sol. (*i*) The Grubler's criterion for planer mechanism is given by

$$F = 3(n - 1) - 2j - h \qquad \ldots(i)$$

where F = degree of freedom

n = no. of links

j = no. of simple hinges or joints or lower pairs of degree of freedom = 1

h = no. of higher pair.

For the mechanism shown in Fig. 1.33 (*i*), we have

$n = 6$

$j = 7$ (pairs are links 1—2, 1—5, 1—6, 2—3, 4—3, 4—5, 4—6)

$h = 0$

Substituting these values in equation (*i*), we get

$$F = 3(6 - 1) - 2 \times 7 = 15 - 14 = 1$$

As in this case, degree of freedom is one, hence unique solution is possible.

(*ii*) In the mechanism shown in Fig. 1.33 (*ii*), there are four links (*i.e.* links 1, 2, 3 and 4) and four lower pairs. But link 3 is capable of sliding without the help of remaining links. Hence the mechanism in Fig. 1.33 (*ii*) is a locked system. Also link 3 has redundant degree of freedom.

(*iii*) The mechanism shown in Fig. 1.33 (*iii*), has a roller (carried at the end of output link) which can be rotated without causing any motion in the rest of the mechanism. Thus roller is a link with redundant degree of freedom. The roller can thus be considered welded to the output link. Hence in this case there are three links 1, 2, 3 together with two twining pairs and one higher pair. Thus

$$F = 3(3 - 1) - 2(2) - 1 \qquad (\because n = 3, l = 2 \text{ and } h = 1)$$

$$= 6 - 4 - 1 = 1$$

Hence in this case also there is a unique relation between input and output motion.

Problem 1.9. *Fig. 1.34 shows a plane mechanism with link lengths given in some unit. If slider A is the driver, will link CG revolve or oscillate ? Justify your answer.*

(AMIE, Summer 1993)

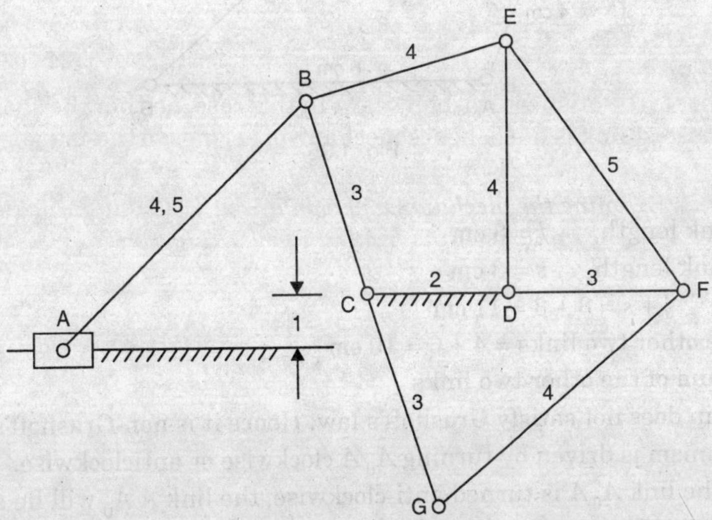

Fig. 1.34

Sol. The mechanism involves three sub-chains :

(*i*) A four-bar chain *CBED*,

(*ii*) A four-bar chain *CDFG* and

(*iii*) A slider-crank chain *ABC*.

(*i*) In four-bar chain *CBED*, longest link length, $l = 4$ units and shortest link length $s = 2$ units. The remaining two links are having lengths as 3 and 4 units.

$$\therefore \qquad l + s = 4 + 2 = 6$$

Sum of lengths of other two links = 3 + 4 = 7

$\therefore$ $l + s <$ sum of lengths of other two links.

Hence the above chain satisfies Grashoff's law and hence it is a Grashoff 's chain. In this case as the shortest link *CD* is fixed, it represents a double crank mechanism.

(*ii*) In four-bar chain *CDFG*, longest link length, $l = 4$ and shortest link length, $s = 2$ units. The remaining two links are having lengths as 3 and 3 units.

$\therefore$ $l + s$ = sum of other two link lengths.

Hence this chain also satisfies Grashoff's law and hence it is a Grashoff's chain. Here also the shortest link is fixed and hence it represents a double crank mechanism.

(*iii*) The third sub-chain is the slider-crank chain *ABC*. Since length $AB = 4.5$ units is greater than the length BC + offset 1 unit, the crank *CB* can revolve fully.

Since *CDEB* and *CDFG* are double-crank mechanism, link *CG* is capable of having full revolution.

Problem 1.10. *The mechanism shown in Fig. 1.35 is driven by turning A_0A. Find out geometrically the maximum and minimum transmission angles.* (AMIE, Summer 1993)

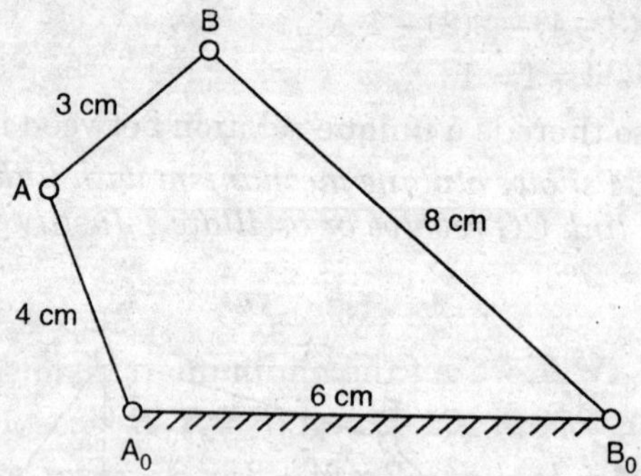

Fig. 1.35

Sol. Here

Longest link length, $l = 8$ cm

Shortest link length, $s = 3$ cm

$\therefore$ $l + s = 8 + 3 = 11$ cm

Sum of the other two links = 4 + 6 = 10 cm

$\therefore$ $l + s >$ sum of the other two links

Hence chain does not satisfy Grashoff's law. Hence it is non-Grashoff's type.

The mechanism is driven by turning A_0A clockwise or anticlockwise.

(*i*) When the link A_0A is turned anti-clockwise, the link AA_0 will lie along B_0A_0. This will give the maximum transmission angle (*i.e.* α_{max}) as shown in Fig. 1.36.

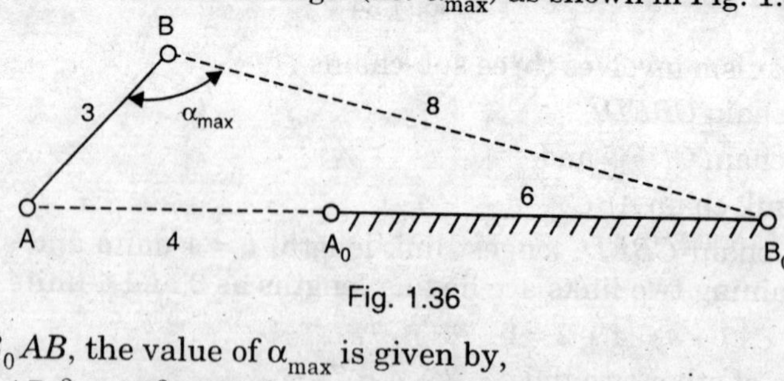

Fig. 1.36

In triangle B_0AB, the value of α_{max} is given by,

$$AB_0^2 = AB^2 + BB_0^2 - 2 \times AB \times BB_0 \cos \alpha_{max}$$

or $$(6 + 4)^2 = 3^2 + 8^2 - 2 \times 3 \times 8 \times \cos \alpha_{max}$$

or $$100 = 9 + 64 - 48 \cos \alpha_{max}$$

$$= 73 - 48 \cos \alpha_{max}$$

or $\cos \alpha_{max} = \dfrac{73 - 100}{48} = \dfrac{-27}{48}$

$\therefore \quad \alpha_{max} = \mathbf{124.23°}$. **Ans.**

(*ii*) When link A_0A in Fig. 1.35 is turned clockwise, the extreme possible position of link A_0A will be when A_0A and AB become collinear as shown in Fig. 1.37.

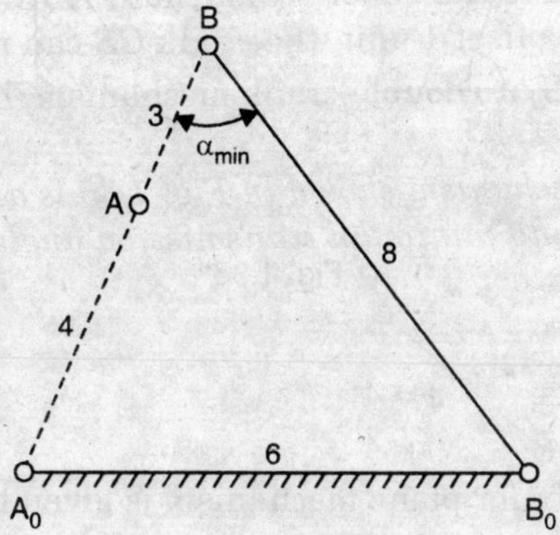

Fig. 1.37

Now from triangle B_0A_0B, we get the minimum transmission angle (α_{min}) as

$$A_0B_0^2 = A_0B^2 + B_0B^2 - 2 \times A_0B \times B_0B \times \cos \alpha_{min}$$

or $$6^2 = (4 + 3)^2 + 8^2 - 2 \times (4 + 3) \times 8 \times \cos \alpha_{min}$$

or $$36 = 49 + 64 - 112 \cos \alpha_{min}$$

$$36 = 113 - 112 \cos \alpha_{min}$$

$$\therefore \quad \cos \alpha_{min} = \frac{113 - 36}{112} = \frac{-77}{112} = \frac{11}{16}$$

$\therefore \quad \alpha_{min} = \mathbf{46.58°}$. **Ans.**

Problem 1.11. *Find the 'degrees of freedom' in each of the following cases :*

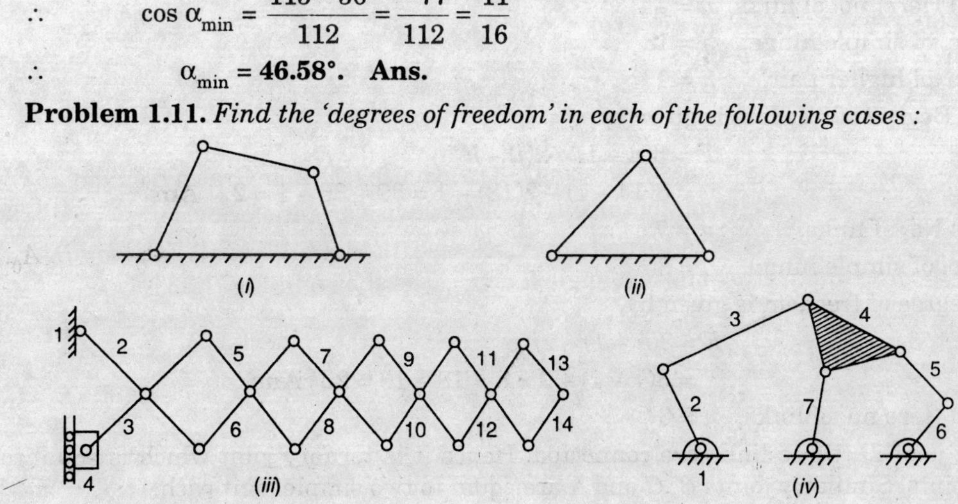

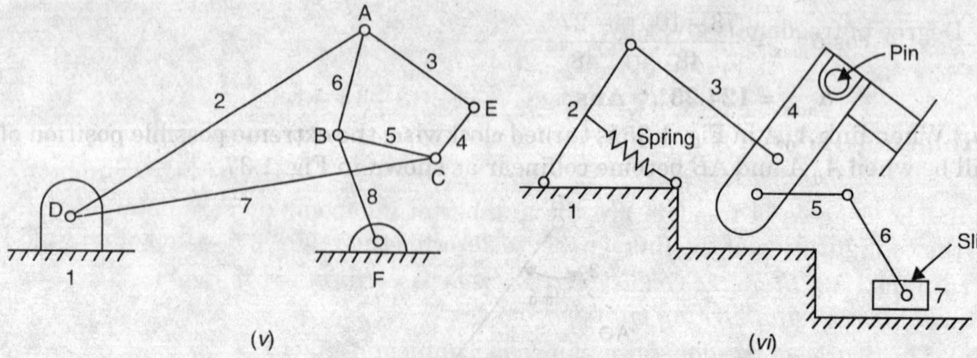

Fig. 1.38

(AMIE, Summer 1994)

Sol. (*i*) Here, no. of links, $n = 4$

No. of simple hinges, $j = 4$

Degree of freedom (d.o.f.) for plane mechanism is given by

$$F = 3(n - 1) - 2j$$

$$= 3(4 - 1) - 2 \times 4 = 9 - 8 = \mathbf{1}.$$ **Ans.**

(*ii*) Here, no. of links, $n = 3$

No. of simple hinge, $j = 3$

$\therefore$ Degree of freedom is given by,

$$F = 3(n - 1) - 2j$$

$$= 3(3 - 1) - 2 \times 3 = 6 - 6 = \mathbf{0}.$$ **Ans.**

(*iii*) Here, no. of links, $n = 14$

No. of simple hinge, $j = 18$

No. of higher pair, $h = 1$

$\therefore$ Degree of freedom is given by,

$$F = 3(n - 1) - 2(j) - h$$

$$= 3(14 - 1) - 2(18) - 1 = 39 - 36 - 1 = \mathbf{2}.$$ **Ans.**

(*iv*) No. of links, $n = 7$

No. of simple hinge, $j = 8$

Degree of freedom is given by,

$$F = 3(n - 1) - 2j$$

$$= 3(7 - 1) - 2 \times 8 = 18 - 16 = \mathbf{2}.$$ **Ans.**

(*v*) Here no. of links, $n = 8$

At joint D, three links are connected. Hence it is ternary joint which is equal to two simple joints. Similarly joints B, C and A are equal to two simple joint each.

$\therefore$ No. of simple joints, $j = 2$ (at D) + 2 (at B) + 2 (at A) + 2 (at C) + 1 (at E) + 1 (at F) = 10

$\therefore$ Degree of freedom is given by,

$$F = 3(n - 1) - 2j$$

$$= 3(8 - 1) - 2 \times 10 = 21 - 20 = \mathbf{1}.$$ **Ans.**

(*vi*) No. of links, $n = 7$

No. of simple joints, $j = 7$

No. of higher pair, $h = ?$

∴ Degree of freedom is given by,

$$F = 3(n-1) - 2(j) - h$$
$$= 3(7-1) - 2 \times 7 - 1 = 18 - 14 - 1 = \mathbf{3.} \quad \textbf{Ans.}$$

Problem 1.12. *What is meant by the 'degrees of freedom' of a linkage.*

(AMIE, Summer 1994)

Sol. The degrees of freedom are the number of independent parameters required to describe the configuration of the linkage (*i.e.* to describe the relative position of all links).

Problem 1.13. *State the Gribler's criterion for ascertaining the 'degree of freedom' of a planer mechanism having turning pairs only.* (AMIE, Summer 1994)

Sol. The degree of freedom for a planer mechanism is given by

$$F = 3(n-1) - 2j$$

where n = the number of links in a mechanism

j = number of simple hinges.

For a mechanism, one link is fixed. Hence the number of movable links remain as $(n-1)$. Each movable link has three degrees of freedom in a plane mechanism. Hence $(n-1)$ movable link will have $3 \times (n-1)$ degrees of freedom. When two links are connected by a simple hinge, two degrees of freedom are destroyed. Hence for j number of simple hinges, $2 \times j$ of freedom are lost. Hence degrees of freedom remaining will be $3(n-1) - 2j$.

$$\therefore \quad F = 3(n-1) - 2j$$

For constrained motion, $F = 1$ and this leads to Gribler's criterion as (substituting $F = 1$ in the above equation)

$$1 = 3(n-1) - 2j = 3n - 3 - 2j$$

or $$3n - 2j - 4 = 0.$$

Problem 1.14. *Extend Gribler's criterion for planer mechanism to obtain the degree of freedom of a space mechanism as*

$$F = 6(n-1) - 5g - 4C - 3s$$

where g = *total number of sliding pair,*

C = *total number of cylindrical pair,*

s = *total number of spherical pairs and*

n = *total number of links in the mechanism.*

Sol. For space mechanism, every link has six degrees of freedom. Also in a mechanism, one link is fixed. Hence number of movable links are $(n-1)$. The $(n-1)$ movable links will have $6(n-1)$ degrees of freedom. But some of the joints will have only one degree of freedom. Hence for these joints 5 degrees of freedom will be lost for each joint. Similarly some joints will have only two degrees of freedom which means four degrees of freedom will be lost for each such joints. Hence in a general way, we can say the degrees of freedom for a space mechanism is given by,

$$F = 6(n-1) - 5j_1 - 4j_2 - 3j_3 - 2j_4 - j_5 \quad \text{...(1)}$$

where n = Total number of links

j_1 = Number of joints having (or permitting) 1 d.o.f.

j_2 = Number of joints having 2 d.o.f.

j_3 = Number of joints having 3 d.o.f.

j_4 = Number of joints having 4 d.o.f.

j_5 = Number of joints having 5 d.o.f.

We know that : (*i*) sliding pairs (or joints) are having one d.o.f.

(*ii*) cylindrical pairs (or joints) are having 2 d.o.f.

(*iii*) spherical joints are having 3 d.o.f.

The most usual joints in space mechanism are the revolute, screw and prismatic joints (d.o.f. = 1), cylindrical joints (d.o.f. = 2) and spherical joints (d.o.f. = 3). Hence the terms j_4, j_5 can be eliminated in equation (1) above.

$$\therefore \quad F = 6(n-1) - 5j_1 - 4j_2 - 3j_3 = 6(n-1) - 5g - 4c - 3s$$

where g = total number of sliding pairs (d.o.f. = 1)

C = total number of cylindrical pair (d.o.f. = 2)

s = total number of spherical pair (d.o.f. = 3).

Problem 1.15. *The link lengths of a quadratic cycle chain, taken in order are 10 cm, 40 cm, 30 cm and 25 cm as shown in Fig. 1.39. Find all the inversions of the given chain and classify them.*

(AMIE, Winter 1994)

Sol. Given :

Longest link length, $l = 40$ cm

Shortest link length, $s = 10$ cm

Other link lengths, $p = 30$ cm, $q = 25$ cm.

Now $l + s = 40 + 10 = 50$ cm

and $p + q = 30 + 25 = 55$ cm

$\therefore$ $(l + s) < (p + q)$

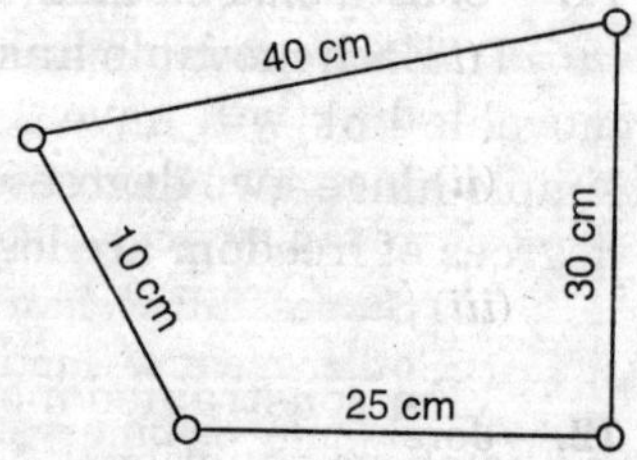

Fig. 1.39

The given chain satisfies Grashoff's criterion. Hence only three distinct inversions are possible.

1st Inversion. By fixing shortest link of length $s = 10$ cm, we get a double crank mechanism in which links of lengths 25 cm and 40 cm will be capable of having complete revolutions.

2nd Inversion. By fixing the link opposite to shortest link (*i.e.* the link of length 30 cm), we get a double rocker mechanism in which links of lengths 25 cm and 40 cm, can only have the oscillatory motion *i.e.* they cannot have complete revolutions.

3rd Inversion. By fixing any of the two remaining links (say of lengths 25 cm and 40 cm) we always get a crank rocker mechanism in which only one link (*i.e.* the shortest link) can have complete revolutions while the link of length 30 cm can only have oscillatory motion.

HIGHLIGHTS

1. The branch of science which deals with the study of relative motion between the various parts of a machine, and forces which act on them, is known as theory of machine.
2. Mechanism is a combination of rigid bodies which are so shaped and connected that they move upon each other with definite relative motion.
3. A machine is a mechanism which not only imparts definite motions to the parts but also transmits and modifies the available mechanical energy into some kind of useful energy.
4. A link is a member which is connected to other members and has motion relative to the other member.
5. A joint of two links having relative motion between them is known as a kinematic pair.
6. Kinematic pairs may be :

   (*i*) Lower pair or higher pair

   (*ii*) Closed pair or unclosed pair

   (*iii*) Sliding pair, turning pair, rolling pair, screw pair and spherical pair.

7. Degree of freedom is defined as the number of independent motion (both translation and rotational) a body can have.
8. Kinematic chain is the combination of kinematic pairs which are joined in such a way that the last link is connected to the first link to transmit definite motion.
9. The relation between the number of pairs (p) and number of links (n) in a kinematic chain is given by,

$$n = 2p - 4$$

whereas the relation between number of links (n) and number of joints (j) is given by,

$$j = \frac{3}{2} \times n - 2.$$

10. In the above two equations, if
L.H.S. > R.H.S. then chain is locked
L.H.S. = R.H.S. then chain is constrained
L.H.S. < R.H.S. then chain is unconstrained.
11. Constrained motions are of three types :
(*i*) Completely constrained motion—motion of a square bar in a square hole (motion in a definite direction).
(*ii*) Incompletely constrained motion—motion is in more than one direction—motion of a circular rod in a circular hole.
(*iii*) Successfully constrained motion—motion is not completely constrained by itself, but by some other means—motion of a shaft in a foot-step bearing with a load on the shaft.
12. Joints may be binary, ternary or quaternary.
Binary joints—Two links are joined at the same connection.
Ternary joints—Three links are joined at the same connection.
Quaternary joints—Four links are joined at the same connection.
13. Links may be binary, ternary or quaternary.
Binary links—A link to which two links are connected.
Ternary links—A link to which three links are connected.
Quaternary links—A link to which four links are connected.
14. Plane mechanism means the mechanism in a plane (*i.e.* *x-y* plane). Degrees of freedom of a plane mechanism means the number of independent co-ordinates needed to determine the positions of all links of the mechanism with respect to fixed link.
The number of degrees of freedom in a mechanism is given by Grubler's equation as

$$F = 3(n - 1) - 2j$$

where n = Number of links in a mechanism
j = Number of simple joints (*i.e.* joints which connects two links)
F = Number of degrees of freedom.
15. Two degrees of freedom (translational) are lost at every pin joint with fixed link. Hence pin joint is having only one degree of freedom.
16. The important types of kinematic chains with four kinematic pairs are :
(*i*) Four bar chains, (*ii*) Single slider crank chain, and
(*iii*) Double slider crank chain.
17. If different links of four-bar mechanism are fixed, four different mechanism (known as inversions) will be obtained. Four bar mechanism consists of four turning pairs.
18. Single slider crank mechanism consists of one sliding pair and three turning pairs. The first inversions is used in reciprocating engine and reciprocating compressor. The second inversion is used in Whiteworth quick-return mechanism and rotary engine. The third inversion is used in oscillating cylinder engine and fourth one is used in hand pump.

19. Double-slider crank mechanism consists of two turning pairs and two sliding pairs. Also the two pairs of the same kind are adjacent. The four different inversions are :
First inversion—used in elliptical trammel.
Second inversion—used in Scotch Yoke which is used to convert the rotary motion into a sliding motion.
Third inversion—used in Oldham's coupling.

EXERCISE 1

(A) Theoretical Questions

1. Define and explain the terms : mechanism, machine, link and kinematic pair.

2. Write a short note on the classification of kinematic pairs.

3. Differentiate between :
(*i*) Lower pair and higher pair, (*ii*) Turning pair and sliding pair,
(*iii*) Screw pair and spherical pair, (*iv*) Closed pair and unclosed pair.

4. What do you understand by degrees of freedom ? For a plane mechanism, derive an expression for Grubler's equation.

5. Define and explain the term kinematic chain. For a kinematic chain, what is the relation between number of pairs and number of links. Also write the equation, showing the relation between number of links and number of joints.

6. What do you mean by constrained motion ? What are the different types of constrained motions ? Explain each type with examples and neat sketches.

7. Define and explain the terms : binary joints, ternary joints, quaternary joints, binary links, ternary links and quaternary links.

8. Differentiate between :
(*i*) Completely constrained motion and incompletely constrained motion
(*ii*) Binary joints and binary links
(*iii*) Mechanism and machine.

9. Prove that the pin joint is having only one degree of freedom.

10. What are the important kinematic chains with four kinematic pair ?

11. What do you mean by inversion of a mechanism ? Explain with sketches all the inversions of single slider crank mechanism. Where these inversions are used.

12. Explain in details the double-slider crank mechanism. Name all the inversions of double-slider crank mechanism.

13. With a neat sketch explain the inversion of double-slider crank mechanism which is used in elliptical trammel.

14. Explain with the help of neat sketch a quick return mechanism using four bar chain.

15. Explain the following :
(*i*) Scotch Yoke mechanism, (*ii*) Elliptical Trammel, and
(*iii*) Oldham's coupling.

(B) Objective Questions

1. A kinematic pair is a joint of
(*a*) two links which are fixed
(*b*) two links having same velocity
(*c*) two links having relative motion between them
(*d*) none of the above.

2. Piston and cylinder of a reciprocating steam engine forms a
(*a*) turning pair (*b*) rolling pair
(*c*) sliding pair (*d*) none of the above.

3. A ball and a socket joint forms a
(*a*) turning pair (*b*) rolling pair
(*c*) sliding pair (*d*) spherical pair.

4. Shaft with collars at both ends fitted into a circular hole forms a
(*a*) turning pair (*b*) rolling pair
(*c*) sliding pair (*d*) spherical pair.

5. Ball-bearing and roller bearing forms a
(*a*) turning pair (*b*) rolling pair
(*c*) sliding pair (*d*) spherical pair.

6. A bolt and nut form a
(*a*) turning pair (*b*) rolling pair
(*c*) screw pair (*d*) spherical pair.

7. When the two elements of a pair have a surface contact when in motion and the surface of an element slides over the surface of the other element, the pair formed is called a
(*a*) higher pair (*b*) lower pair
(*c*) forced-closed pair (*d*) none of the above.

8. The function of an element is to
(*a*) transmit motion (*b*) to serve as a support
(*c*) to guide other elements (*d*) all of the above
(*e*) none of the above.

9. Oldham's coupling and elliptic trammels are the inversions of
(*a*) double slider crank chain (*b*) single slider crank chain
(*c*) four bar chain (*d*) none of the above.

10. Whiteworth quick return mechanism is an inversion of
(*a*) double slider crank chain (*b*) single slider crank chain
(*c*) four bar chain (*d*) none of the above.

11. In a kinematic chain with four lower pairs, if all the four lower pairs are turning pairs, the mechanism is classified into
(*a*) four bar chain (*b*) crossed slider crank chain
(*c*) slider crank chain (*d*) double slider crank chain.

12. In a kinematic chain with four lower pairs, if one is sliding pair and three turning pairs, the mechanism is classified into
(*a*) crossed slider crank chain (*b*) four bar chain
(*c*) slider crank chain (*d*) double slider crank chain.

13. If in the above question, there are two sliding pairs and two turning pairs and two similar pairs are adjacent then mechanism is classified into
(*a*) crossed slider crank chain (*b*) four bar chain
(*c*) slider crank chain (*d*) double slider crank chain.

14. Choose the wrong statement :
(*a*) A chain consisting of three links and three joints is known as locked chain.
(*b*) A chain consisting of four links with four joints is known as kinematic chain.
(*c*) Quaternary joint is equivalent to three binary joints.
(*d*) Rectangular bar in a rectangular hole is the example of partially constrained motion.

15. If 'n' links are connected at the same joint, the joint is equivalent to

(*a*) $(n - 1)$ binary joints (*b*) $(n - 2)$ binary joints

(*c*) $(n - 3)$ binary joints (*d*) $(2n - 1)$ binary joints.

16. If a kinematic chain has 'n' links then number of mechanism obtained are

(*a*) $(n - 1)$ (*b*) $(n - 2)$

(*c*) $(n + 1)$ (*d*) n.

Answers

1. (*c*)	**2.** (*c*)	**3.** (*d*)	**4.** (*a*)	**5.** (*b*)	**6.** (*c*)
7. (*b*)	**8.** (*d*)	**9.** (*a*)	**10.** (*b*)	**11.** (*a*)	**12.** (*c*)
13. (*d*)	**14.** (*d*)	**15.** (*a*)	**16.** (*d*).		

2

Motion of Inertia

2.1. INTRODUCTION

The parts of a machine may have a linear motion or a circular motion or a motion along a curved path whose radius is changing. For example, in case of reciprocating engine, the parts such as piston, piston rod and crosshead move in a straight lines and the crank and crank-pin move along circular arcs with centre on the axis of the crank-shaft. The connecting rod moves along an oval path, the radius of curvature of which changes from instant to instant. This chapter deals with these types of motions and also with the forces which act on these parts.

2.2. DISPLACEMENT

As displacement involves direction as well as magnitude, it is a vector quantity, and therefore can be represented by a straight line. The magnitude is represented by the length of the line and the direction is represented by the arrowhead, placed on the line.

Every particle of every link of a machine must move along a definite constrained path which may be a straight path or curvilinear. There are mostly three types of paths namely straight path, circular path with a fixed radius and curved path in which radius is different from instant to instant. Whatever the path followed, it is possible to determine successive displacements of the particles for equal interval of time.

2.3. VELOCITY

The rate of change of displacement with respect to time is known as the velocity of a particle. Velocity is also a vector quantity.

If the direction of displacement is known then the direction of velocity can be easily obtained. If the displacement is constant in direction, the velocity will be in the same direction. But if the displacement is not constant in direction, then the velocity at a given instant will be in the same direction as the displacement at that instant. As displacement at that instant is in tangential direction hence the direction of velocity be tangential to the path of the particle.

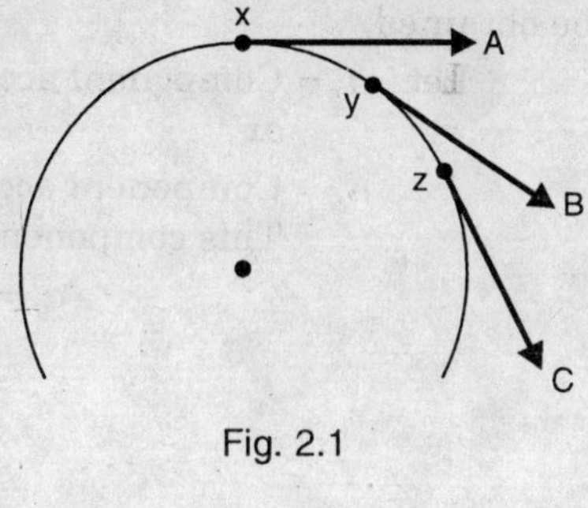

Fig. 2.1

If the displacement takes place along a circular path *XYZ* as shown in Fig. 2.1, the velocities of the particle at the instant it occupies positions *X*, *Y* and *Z* will be in the directions given by the tangents *XA*, *YB* and *ZC* respectively.

2.4. ACCELERATION

The rate of change of velocity with respect to time is known as acceleration. Acceleration is also a vector quantity. The direction of acceleration is not necessarily the same as that of the velocity and the displacement vectors. This will be clear in Art. 2.5, in which the general case of the acceleration of a particle moving in a circular path is considered. Even when a particle is moving in a straight path, the directions of velocity and acceleration may be in opposite direction (retardation).

2.5. ACCELERATION OF A PARTICLE MOVING ALONG A CIRCULAR PATH

Consider a particle moving along a circular path as shown in Fig. 2.2. Let x and y are the two positions of the particle along the circular path.

Let v = Velocity of particle at position x,

$v + \delta v$ = Velocity of particle at position y,

r = Radius of circular path, and

δt = Time required by the particle to move from x to y.

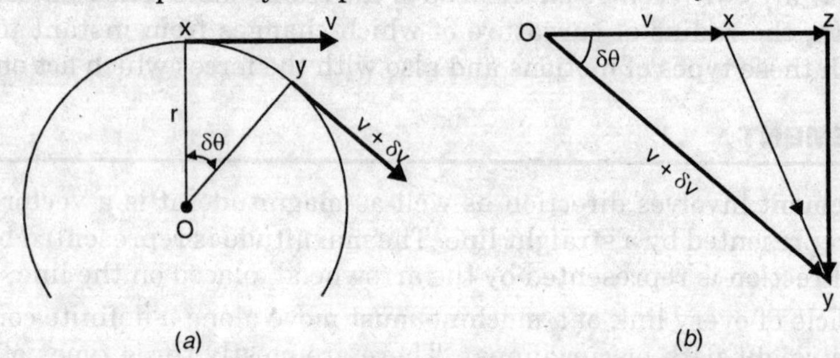

Fig. 2.2

To find the change in velocity, as the particle moves from x to y, a vector triangle oxy is drawn as shown in Fig. 2.2 (b) in which ox is equal to v in magnitude and direction (*i.e.* velocity at x) and oy is equal to $v + \delta v$ in magnitude and direction (*i.e.* velocity at y). In Fig. 2.2 (b) the vector xy represents the change of velocity in time δt. The vector xy is resolved in two components xz and zy. The component xz is parallel to ox whereas the component zy is $\perp$ to ox.

Now $\quad xz = oz - ox = oy \cos \delta\theta - v = (v + \delta v) \cos \delta\theta - v$

and $\quad zy = oy \sin \delta\theta = (v + \delta v) \sin \delta\theta.$

Now the components acceleration of the particle in tangential and normal direction can be obtained.

Let a_t = Component acceleration in tangential direction *i.e.*, in the direction parallel to ox

a_c = Component acceleration in normal direction *i.e.*, in the direction normal to ox. This component is also known as centripetal component of acceleration.

$\therefore$ a_t = Rate of change of velocity in tangential direction

$$= \frac{\text{Change of velocity in tangential direction}}{\text{Time}}$$

$$= \frac{xz}{\delta t} = \frac{(v + \delta v)\cos \delta\theta - v}{\delta t}$$

In the limit as δt approaches zero, $\cos \delta\theta \to 1$ and the above equation becomes as

$$a_t = \frac{dv}{dt} \quad \text{...(2.1)}$$

Similarly the component acceleration in normal direction (*i.e.*, centripetal acceleration) is obtained as

$$a_c = \text{Rate of change of velocity in normal direction}$$

$$= \frac{\text{Change of velocity in normal direction}}{\text{Time}}$$

$$= \frac{zy}{\delta t} = \frac{(v + \delta v)\sin \delta\theta}{\delta t}$$

In the limit as δt approaches zero, $\sin \delta\theta \to \delta\theta$ and the above equation becomes as

$$a_c = \frac{(v + dv)\, d\theta}{dt}$$

$$= v\,\frac{d\theta}{dt} + \frac{dv \times d\theta}{dt}$$

$$= v\,\frac{d\theta}{dt}$$

(The product of two small terms *i.e.*, dv and $d\theta$ are negligible)

$$= v \times \omega \qquad \left(\because \frac{d\theta}{dt} = \omega\right)$$

$$= \frac{v^2}{r} \text{ or } \omega^2 r \quad (\because \quad v = w \times r) \quad \text{...(2.2)}$$

From equation (2.1), it is clear that tangential component of acceleration is equal to the rate of change of the magnitude of the velocity of the particle whereas from equation (2.2), it is clear that the normal acceleration (or radial acceleration or centripetal acceleration) depends upon the velocity and radius of the circular path.

Note. (*i*) For the displacement along a straight path, the radius of the straight path is infinitely great and hence the centripetal component $\left(a_c = \frac{v^2}{r}\right)$ will be zero. There will be only tangential component of acceleration $\left(i.e.,\ a_t = \frac{dv}{dt}\right)$. The direction of tangential component will be same as the direction of displacement and of the velocity.

(*ii*) For the displacement along a circular path with constant speed, the tangential component of acceleration $\left(i.e.,\ a_t = \frac{dv}{dt}\right)$ will be zero. There will be only centripetal component of acceleration $\left(i.e.,\ a_c = \frac{v^2}{r}\right)$. The direction of the centripetal acceleration will be normal to the velocity and the displacement. As velocity and displacement are normal to the radius of the circular path, hence direction of centripetal acceleration will be along the radius towards the centre.

2.6. ANGULAR DISPLACEMENT

Fig. 2.3 (A) shows a line *OA* rotating about the centre *O*. Let the line *OA* is inclined at θ radians with fixed line *OX*. After a small interval of time δt, this line has moved anticlock-wise to the position *OB*. Then the angle $\delta\theta$ is the angular displacement of the line. The angular displacement is a vector quantity as it has magnitude (*i.e.*, $\delta\theta$ here) and direction (*i.e.*, anticlockwise here).

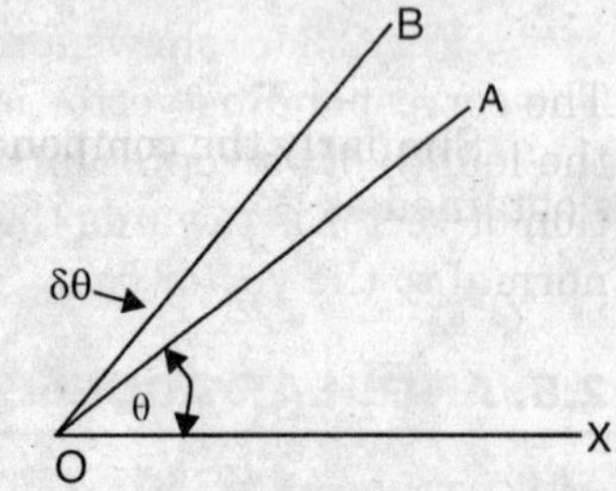

Fig. 2.3 (A)

2.7. ANGULAR VELOCITY

The rate of change of angular displacement with respect to time is known as angular velocity. It is also a vector quantity as it has magnitude and direction (*i.e.*, clockwise or anti-clockwise). The angular velocity is represented by a Greek letter ω (omega).

Mathematically it is given by, $\omega = \dfrac{d\theta}{dt}$.

Velocity is a vector quantity. Also the angular velocity (ω) of a rotating body is a vector quantity. But a vector of a rotating body is represented by magnitude, direction and sense of rotation (*i.e.*, clockwise or anticlockwise). Hence angular velocity is represented by magnitude, direction and sense of rotation of rotor. The magnitude is represented by length of the vector, direction by drawing vector parallel to the axis of the rotor and sense of rotation by the right handed screw. This rule states that if the screw is rotated in clockwise direction, it goes away from the viewer. Hence for clockwise rotation, the direction of vector is away from the viewer whereas for anti-clockwise rotation, the direction of vector is towards the viewer.

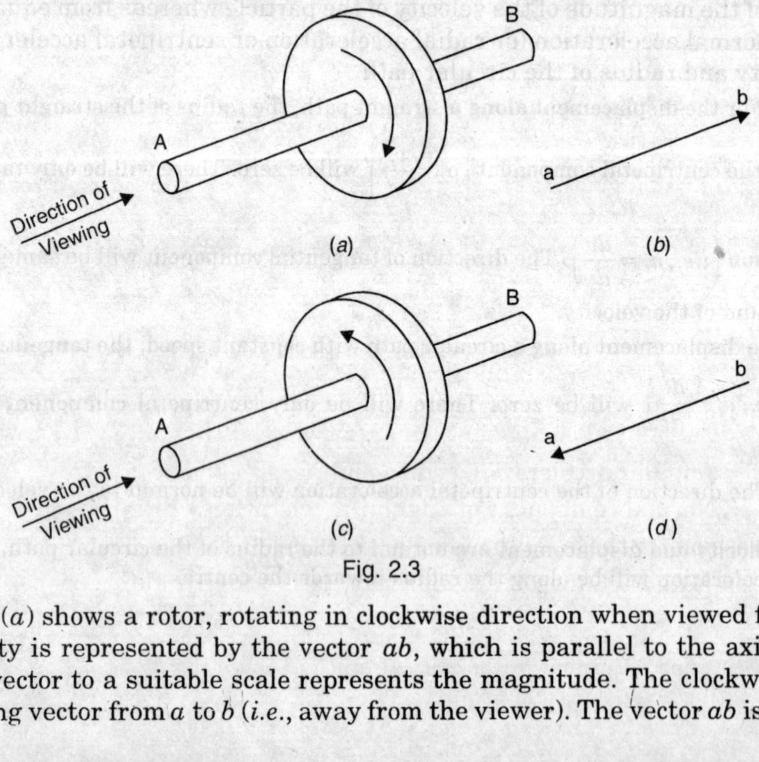

Fig. 2.3

Fig. 2.3 (*a*) shows a rotor, rotating in clockwise direction when viewed from end *A*. Its angular velocity is represented by the vector *ab*, which is parallel to the axis of rotor. The length of the vector to a suitable scale represents the magnitude. The clockwise direction is shown by taking vector from *a* to *b* (*i.e.*, away from the viewer). The vector *ab* is parallel to the axis of the rotor. The plane of rotation is normal to the axis (or the plane* of rotation is normal to the vector ab).

Fig. 2.3 (c) shows a rotor, rotating in anti-clockwise direction when viewed from end A. The corresponding vector is shown by ba where ba is parallel to the axis of rotation and also the length of ba represents the magnitude. As the rotation is anti-clockwise hence the direction of vector is towards the viewer *i.e.*, from b to a. The plane of rotation is normal to axis (or normal to the vector ba).

2.8. ANGULAR ACCELERATION

The rate of change of angular velocity with respect to time is known as angular acceleration. It is also a vector quantity. But the direction of angular acceleration is not necessarily the same as that of the angular displacement and the angular velocity.

Fig. 2.4 (a) shows a disc spinning or rotating with angular velocity ω about its axis of spin at a given instant. The plane of rotation of the disc is normal to the axis of spin. After a short interval of time δt, the disc is spinning with angular velocity $\omega + \delta\omega$ and the axis of spin has changed direction by the amount $\delta\theta$. Now the plane of rotation of the disc is normal to the new axis of spin.

Fig. 2.4 (b) shows the vector triangle oab in which vector oa represents angular velocity ω in magnitude and direction and vector ob represents the angular velocity $\omega + \delta\omega$. The vector oa is parallel to the axis of spin (or normal to the plane of rotation) whereas the vector ob is parallel to the new axis of spin (or normal to the new plane of rotation). Vector ab represents the change of angular velocity in time δt in magnitude and direction. The vector ab can be resolved in two components. Vector ac along oa is the component in tangential direction of the change of angular velocity whereas vector cb normal to oa is the component in the normal direction of the change of angular velocity. The vectors are perpendicular to the corresponding planes of rotations. Or the planes of rotation are perpendicular to the corresponding vectors.

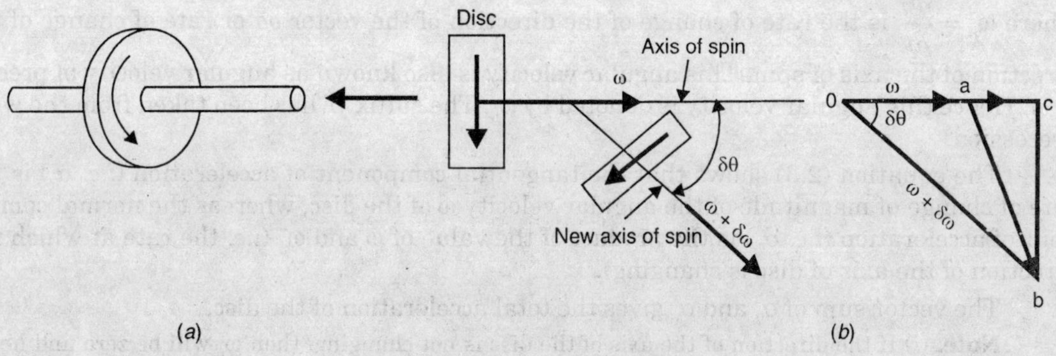

Fig. 2.4

Let α_t = Tangential component of angular acceleration along oa.

= Angular acceleration component parallel to oa

α_c = Normal component of angular acceleration which is normal to oa.

The expression for these components are obtained as :

α_t = Rate of change of angular velocity in tangential direction (or along oa)

*Plane of rotation is normal to the axis. If a body is rotating along x-axis, then the plane of rotation of the rotating body is normal to x-axis.

$$= \frac{\text{Change of angular velocity in tangential direction (or along } oa)}{\text{Time}}$$

$$= \frac{ac}{\delta t} = \frac{oc - oa}{\delta t} = \frac{ob \cos \delta\theta - oa}{\delta t} = \frac{(\omega + \delta w) \cos \delta\theta - \omega}{\delta t}$$

In the limit as δt approaches zero, $\cos \delta\theta \to 1$.

$$\therefore \qquad \alpha_t = \frac{d\omega}{dt} \qquad \text{...(2.3)}$$

The direction of vectors ac and oa are same. Hence the direction of the above acceleration and angular velocity ω are same. Also the above acceleration and angular velocity ω are acting in the same plane. This plane is normal to vector oa or oc.

Now α_c = Rate of change of angular velocity in normal direction (or normal to oa)

$$= \frac{\text{Change of angular velocity in normal direction}}{\text{Time}}$$

$$= \frac{cb}{\delta t} = \frac{ob \sin \delta\theta}{\delta t} = \frac{(\omega + \delta\omega) \sin \delta\theta}{\delta t}$$

In the limit, as δt approaches zero, $\sin \delta\theta \to \delta\theta$

$$\therefore \qquad \alpha_c = \frac{(\omega + d\omega)\, d\theta}{dt}$$

$$= \omega \times \frac{d\theta}{dt} + \frac{d\omega \times d\theta}{dt}$$

$$= \omega \times \frac{d\theta}{dt}$$ (As the product of two small terms is negligible *i.e.*, $d\omega \times d\theta$ is negligible)

$$= \omega \times \omega_p \qquad \text{...(2.4)}$$

where $\omega_p = \dfrac{d\theta}{dt}$ is the rate of change of the direction of the vector oa or rate of change of the direction of the axis of spin. This angular velocity is also known as angular velocity of precession. Hence this angular velocity is denoted by ω_p. The suffix 'p' has been taken from the word precession.

The equation (2.3) shows that the tangential component of acceleration (*i.e.* α_t) is the rate of change of magnitude of the angular velocity ω of the disc, whereas the normal component of acceleration (*i.e.* α_c) is the product of the value of ω and ω_p (*i.e.* the rate at which the direction of the axis of disc is changing).

The vector sum of α_t and α_c gives the total acceleration of the disc.

Note. (*i*) If the direction of the axis of the disc is not changing, then ω_p will be zero and hence α_c (which is equal to $\omega \times \omega_p$) will also be zero. The total acceleration of the disc will be equal to α_t or $\dfrac{d\omega}{dt}$ = 0 and hence.

(*ii*) If the direction of the axis of the disc is changing, but the angular velocity of the disc is constant, then $\dfrac{d\omega}{dt} = 0$ and hence α_t will be zero. The total acceleration of the disc will be equal to α_c or $\omega \times \omega_p$. The direction of this acceleration will be at right angles to the direction of angular velocity ω.

(*iii*) The angular velocity ω_p is the rate of change of the direction of the axis of the disc. This angular velocity is also known as the angular velocity of precession. The angular acceleration α_c is known as gyroscopic acceleration of the disc. Hence gyroscopic acceleration acts normal to the direction of angular velocity ω. Also the plane of gyroscopic acceleration is normal to vector *cb*, whereas the plane of angular velocity ω is normal to vector *oa*. Hence the plane of angular velocity ω and plane of gyroscopic acceleration are at right angles to each other.

2.9. FORCE, MASS AND WEIGHT

Force is defined as an agent which produces or tends to produce, destroy or tends to destroy motion. According to Newton's second law of motion, the magnitude of the applied force on a body is directly proportional to the rate of change of momentum which it produces.

Let F = Applied force

m = Mass of the body

v = Velocity of the body

Then $m \times v$ = Momentum of the body.

According to Newton's second law of motion,

Applied force ∝ Rate of change of momentum

or
$$F \propto \left(\frac{d}{dt}\right)(mv)$$

$$\propto m \times \frac{dv}{dt} \qquad (\because \text{ Mass is constant})$$

$$\propto m \times a \qquad \ldots(i)$$

where a is the acceleration or rate of change of velocity and hence $f = \frac{dv}{dt}$.

Now the above equation (i) can be written as

$$F = k \times m \times a \qquad \ldots(ii)$$

where k is constant of proportionality. The unit of force is so chosen that it produces a unit acceleration to a body of unit mass. In S.I. system the force is expressed in Newton (N). One Newton is a force acting on a mass of one kg and produces an acceleration of one meter per second square. Hence in equation (ii)

$$F = 1 \text{ N}, m = 1 \text{ kg and } a = 1 \text{ m/s}^2.$$

Substituting these values in equation (ii), we get

$$1 = k \times 1 \times 1 \qquad \therefore \quad k = 1$$

Substituting the value of $k = 1$ in equation (ii), we get

$$F = m \times a \qquad \ldots(2.5)$$

Weight. Weight is defined as the force with which a body is attracted towards the earth. The acceleration due to gravity is given by 'g'. Hence a body of mass 'm' will be attracted towards earth by a force ($m \times g$). Hence the product $m \times g$ becomes the weight of the body. If W is the weight of the body, then we have

$$W = m \times g. \qquad \ldots(2.6)$$

2.10. CENTRIPETAL AND CENTRIFUGAL FORCE

Consider a body moving along a circular path of radius 'r' with a constant velocity v. The magnitude of the velocity is constant, but its direction is changing, as the body is moving along the circular path. Velocity being a vector quantity, hence due to change of direction, some acceleration must act on the body so that body can move along a circular path with constant velocity. This acceleration is known as radial acceleration or centripetal acceleration (Refer to

Article 2.5). The direction of this acceleration is at right angles to the direction of the velocity and is acting radially inwards. Due to this acceleration, a force is acting radially inwards. This radially inward force is known as centripetal force. The magnitude of centripetal force (F_c) is given by,

Centripetal Force (F_c) = mass × centripetal acceleration

or $F_c = m \times a_c$ (where a_c = centripetal acceleration)

$$= m \times \frac{v^2}{r} \qquad \left[\because \text{ From equation (2.2), } a_c = \frac{v^2}{r} \right] \quad ...(2.7)$$

$$= m \times \omega^2 r \qquad (\because \; V = \omega r)$$

$$= (m \times v) \times \omega$$

$$= \text{Momentum} \times \text{Angular Velocity} \quad ...(2.8)$$

Fig. 2.5 (*a*) shows a body of mass m moving along a circular path of radius r with uniform velocity v. Let an inextensible string connects the body to the centre O of the circular path. In order that the body should move with uniform velocity along the circular path, then a radially inward centripetal force must act on the body. This centripetal force will be provided by the string. Hence string is exerting radially inward centripetal force (F_c) on the body as shown in Fig. 2.5 (*a*). In this figure, the centripetal acceleration is represented by a_c.

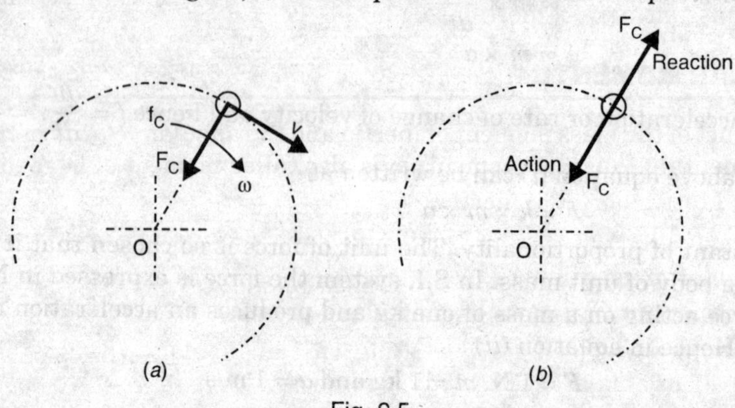

Fig. 2.5

According to Newton's third law of motion, action and reaction are equal and opposite. The force exerted radially inward by string on the body is action. The reaction will be the radially outward force exerted by the body on the string. The magnitude of the reaction will be same as that of action. The radially outward force exerted by body on the string is known as *centrifugal force*.

$$\therefore \quad \text{Centrifugal force, } F_c = m \times \frac{v^2}{r} \quad ...(2.9)$$

$$= m \,.\, \omega^2 .\, r \quad ...(2.10)$$

2.11. MASS MOMENT OF INERTIA

A body is composed of small elements. If the mass of every element of the body is multiplied by the square of its distance from a axis, the sum of the product of the mass of the element and square of the distance is known as the mass moment of inertia of the body about that axis. It is denoted by I.

Fig. 2.6 shows a body of small elements.

Let m_1 = Mass of first element of the body,

k_1 = ⊥ Distance of the first element from the axis,

m_2, m_3, m_4 etc. = Mass of other elements,

k_2, k_3, k_4 etc. = Perpendicular distances of respective elements from the axis.

Now the mass moment of inertia of the whole body is given by,

$$I = m_1k_1^2 + m_2k_2^2 + m_3k_3^2 + ...$$

Let m is the total mass of the body. The total mass m of body is assumed to be concentrated at a point, known as centre of mass or centre of gravity. Let k is the radius of such length that mk^2 is the mass moment of inertia, then

$$I = mk^2 = m_1k_1^2 + m_2k_2^2 + m_3k_3^2 + ...$$

The distance k is known as radius of gyration of the body. Hence radius of gyration is defined as the distance from the axis where the total mass of the body is assumed to be concentrated to give the same value of I.

$$\therefore \quad I = mk^2 \quad ...(2.11)$$

In the above equation, unit of mass is kg, unit of k is m and unit of mass moment of inertia is kg m².

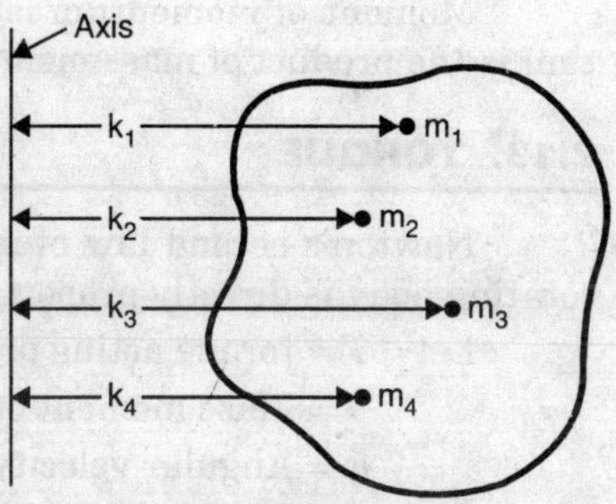

Fig. 2.6

2.12. ANGULAR MOMENTUM

The product of the mass moment of inertia and the angular velocity of the body is known as angular momentum. Angular momentum is also called moment of momentum.

Let I = Mass moment of inertia,

ω = Angular velocity,

Then angular momentum = $I \times \omega$

or Angular momentum = Moment of momentum = $I \times \omega$ ...(2.12)

Fig. 2.7 shows a body of mass m rotating with an angular velocity ω rad/s about an axis O. The body is composed of small elements. Let one of small element is having a mass dm. This mass dm is at a distance r from the axis of rotation *i.e.* from O. (The axis of rotation is ⊥ to the plane of paper and passes through O). Let v is the tangential velocity of small element of mass dm.

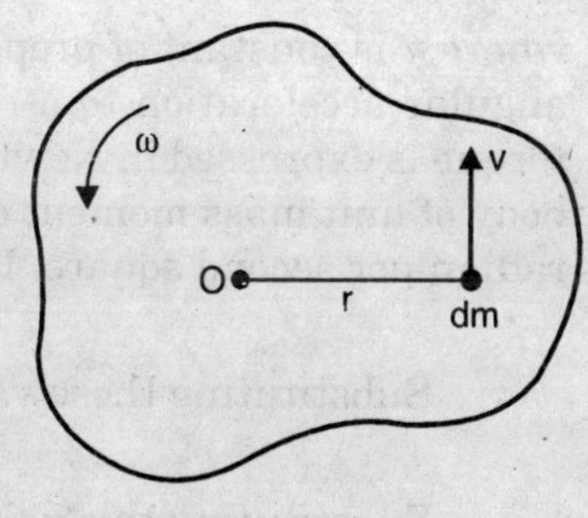

Fig. 2.7

∴ Momentum of the element

= Mass of element × Velocity

$= dm \times v$

$= dm \times \omega \times r \qquad (\because \; v = \omega \times r)$

Moment of the momentum of this element about O

= Moment × distance from O

$= (dm \times \omega \times r) \times r = dm \times \omega \times r^2$

$= (dm \times r^2) \times \omega$

$= I_m \times \omega \quad (\because \; dm \times r^2 = \text{Mass moment of inertia of the element} = I_m)$

$\therefore$ Moment of momentum (or angular momentum) of the whole body about O

$$= \int I_m \times \omega = \omega \int I_m$$

$$= \omega \times I$$

$$\left(\because \int I_m = \text{Mass moment of inertia of the whole body about } O = I\right)$$

Moment of momentum is also known as angular momentum. Hence angular momentum is the product of mass moment of inertia and angular velocity.

2.13. TORQUE

Newton's second law of motion for rotating bodies states that the torque acting on a rotating body is directly proportional to the rate of change of angular momentum.

Let T = torque acting on a rotating body

I = Mass moment of inertia,

ω = Angular velocity of the rotating body.

Then $\omega \times I$ = Angular momentum of the body.

According to Newton's second law of motion for rotating body,

Torque acting $\propto$ Rate of change of angular momentum

or $$T \propto \frac{d}{dt}(\omega \times I)$$

$$\propto I \times \frac{d\omega}{dt} \qquad (\because I \text{ is generally constant})$$

$$\propto I \times \alpha \qquad ...(i)$$

where α is the angular acceleration or rate of change of angular velocity. The above equation (*i*), can also be written as

$$T = k \times I \times \alpha \qquad ...(ii)$$

where k is constant of proportionality. The unit of torque is so chosen that it produces a unit angular acceleration to a rotating body of unit mass moment of inertia. In S.I. system, the torque is expressed in Newton meter (Nm). One Newton meter is a torque acting on a rotating body of unit mass moment of inertia (unit kg m^2) and produces an angular acceleration of one radian per second square. Hence in equation (*ii*),

$$T = 1 \text{ Nm} ; I = 1 \text{ kg m}^2 \text{ and } \alpha = 1 \text{ rad/s}^2.$$

Substituting these values in equation (*ii*), we get

$$1 = k \times 1 \times 1 \qquad \therefore k = 1$$

Substituting this value of k in equation (*ii*), we get

$$T = I \times \alpha. \qquad ...(2.13)$$

2.14. WORK

The product of force (acting on a body) and displacement of the body in the direction of force is known as *work done* on the body.

Let F = Force acting on the body and

x = Displacement of the body in the direction of the force.

Then work done = $F \times x$.

If the magnitude of the force changes continuously during the displacement, the work done during any small displacement is given by $F \times dx$. And total work done is given by,

$$\text{Work done} = \int F \times dx \quad \text{...(2.14)}$$

where dx = small displacement.

In S.I. system, the unit of work is Newton metre (Nm). Hence one Newton metre (Nm) is the work done by a force of one Newton acting on a body whose displacement is one metre in the direction of force.

Work done by Torque

The product of torque (acting on a rotating body) and angular displacement in the direction of the torque is known as work done on the rotating body.

Let T = Torque acting on the body, and

θ = Angular displacement.

$$\text{Then work done} = T \times \theta \quad \text{...(2.14A)}$$

If the magnitude of the couple changes continuously during the angular displacement, then work done during any small displacement is equal to $T \times d\theta$. And total work done is given by,

$$\text{Work done} = \int T \times d\theta$$

where $d\theta$ = small angular displacement.

Hence one Newton metre is also the work done by a torque of one Newton metre acting on a rotating body, whose angular displacement is one radian.

2.15. POWER

The rate of doing work or work done per unit time is known as power.

$$\therefore \quad \text{Power} = \frac{\text{Work done}}{\text{Time}} \quad \text{...}(i)$$

In S.I. system, work done is expressed in Nm whereas time is expressed in second(s). Hence unit of power is Nm/s. This unit is known as Watt (W). Generally a bigger unit of power known as kilo watt (kW) is used which is equal to 1000 W.

$\therefore$ One Watt (W) = One Nm/s or W = Nm/s

and kW = 1000 W = 1000 Nm/s

In equation (i), work done may be by the force F or by the torque T. If the work done is due to force F, then

$$\text{Power} = \frac{F \times x}{t} = F \times \frac{x}{t}$$

$$= F \times v \text{ watt} \left(\because \frac{x}{t} = v \right) \quad \text{...(2.15)}$$

where v is the velocity of the body in m/s.

Similarly, if the work done is due to torque T, then

$$\text{Power} = \frac{T \times \theta}{t} = T \times \frac{\theta}{t}$$

$$= T \times \omega \quad \left(\because \frac{\theta}{t} = \omega \right) \quad \text{...(2.16)}$$

$$= T \times \frac{2\pi N}{60} \qquad \left(\because \; \omega = \frac{2\pi N}{60}\right)$$

$$= \frac{2\pi NT}{60} \qquad ...(2.17)$$

where ω = Angular velocity

N = speed in r.p.m.

2.16. ENERGY

The capacity of doing work is known as energy. The energy exists in many different forms such as mechanical, electrical, chemical, light and heat etc. The mechanical forms of energy are :

1. Potential Energy,
2. Strain Energy, and
3. Kinetic Energy.

2.16.1. Potential Energy. The energy possessed by a body for doing work, by virtue of its position, is known as potential energy. If a body of weight W is raised through a vertical height 'h' above some datum level, then the body is said to possess potential energy of amount $W \times h$, since it is able to do an amount of work $W \times h$ in falling to the datum level.

$\therefore$ Potential Energy, P.E. $= W \times h = m.g.h.$ $\qquad (\because \; W = m \times g)$

2.16.2. Strain Energy. The energy stored in a body due to strains produced in the body is known as strain energy. The body is capable of doing work, when the body is unstrained and comes to original position. For example, if a spring of stiffness s newton per unit extension or compression is extended or compressed through a distance x by a load W, it is said to possess strain energy of amount $\frac{1}{2}sx^2$, since the spring is capable of doing an amount of work $\frac{1}{2}sx^2$, when unstrained and it comes to original position.

$\therefore$ Strain energy $= \text{Work done} = \frac{1}{2} W \times x$

$= \frac{1}{2} \times s \times x \times x \qquad (\because \; W = s \times x)$

$= \frac{1}{2} . s . x^2$

In the above expression, actually the mean force exerted by the spring is $(0 + W)/2$ or $\frac{1}{2} W$ or $\frac{1}{2}$ sx. The distance through which mean force acts is x. Hence work done = Mean force × distance $= \frac{1}{2} sx . x = \frac{1}{2} sx^2$.

Similarly, the strain energy possessed by a torsional spring of stiffness 'q' Nm per unit of angular displacement when twisted through an angle of θ radians, will be equal to $\frac{1}{2} q \times \theta^2$. This spring will be able to do this amount of work $\left(i.e. \frac{1}{2} q\theta^2\right)$ when returned to the unstrained position.

$\therefore$ Strain energy $= \text{Work done} = \frac{1}{2} q . \theta^2$.

2.16.3. Kinetic Energy. The energy possessed by a body for doing work by virtue of its mass and velocity, is known as kinetic energy.

Let m = Mass of the body

v = Velocity of the body.

Then kinetic energy = $\frac{1}{2}mv^2$.

The body is capable of doing an amount of work equal to $\frac{1}{2}mv^2$ when brought to rest. This is proved as given below :

Let F = Force acting on a body,

δx = Small displacement in time δt,

δt = Small interval of time,

δv = Change of velocity in time δt,

v = Initial velocity of body and final velocity is zero.

Now, Force = Mass × Acceleration

or $$F = m \times \frac{\delta v}{\delta t}$$

Also Work done = Force × Displacement

or $$\delta W = F \times \delta x$$

$$= \left(m \times \frac{\delta v}{\delta t}\right) \times \delta x \qquad \left(\because F = m \times \frac{\delta v}{\delta t}\right)$$

$$= m \times \delta v \times \frac{\delta x}{\delta t}$$

$$= m \times \delta v \times v \qquad \left(\because \frac{\delta x}{\delta t} = v\right)$$

$$= m \times v \times \delta v$$

In the limit, $$dW = m \times v \times dv$$

Integrating, $$W = \int_0^v m \times v \times dv = m\int_0^v v \times dv = \frac{mv^2}{2} = \frac{1}{2}mv^2.$$

Similarly, the kinetic energy possessed by a rotating body of mass moment of inertia I about a given axis, when rotating with an angular velocity ω about that axis, will be $\frac{1}{2} \times I \times \omega^2$. When this rotating body is brought to rest, it is capable of doing an amount of work equal to $\frac{1}{2} \times I \times \omega^2$.

$$\therefore \quad \text{Kinetic energy due to rotation} = \frac{1}{2} \times I \times \omega^2 \qquad \text{...(2.18)}$$

If a body having both linear and angular motions such as wheels of a moving car, then the total kinetic energy of the body is equal to the sum of kinetic energies of translation and rotation.

$$\therefore \quad \text{Total kinetic energy} = \frac{1}{2}mv^2 + \frac{1}{2}I\omega^2 \qquad \text{...(2.19)}$$

2.17. LAW OF CONSERVATION OF ENERGY

It states that the energy can neither be created nor destroyed though it can be transformed from one form to another form. The second statement of this law is :

"The total energy possessed by a body remains constant provided no energy is added to or taken from it."

Fig. 2.8 shows a body resting on the top of a tower.

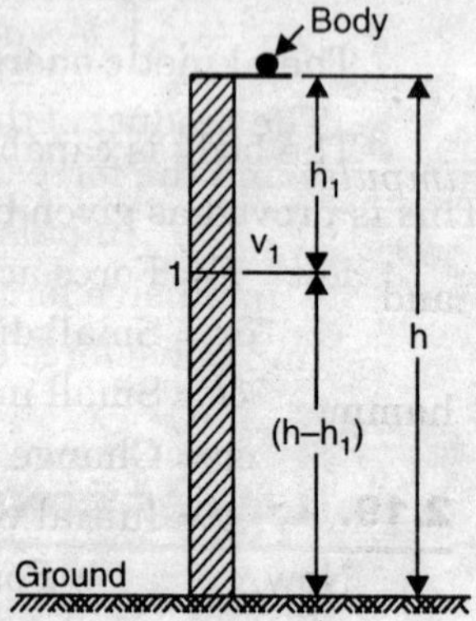

Fig. 2.8

Let W = Weight of body

m = Mass of body = $\frac{W}{g}$

h = Height of tower

Potential energy of the body with respect to ground

$= W \times h$ Nm ...(*i*)

K.E. of the body

$= \frac{1}{2}mv^2 = 0 \qquad (\because \ v = 0)$

$\therefore$ Total energy at the top of tower

= Potential energy + Kinetic energy

$= Wh + 0 = Wh$

Suppose the body falls down by a height h_1 to the position 1 as shown in Fig. 2.8. Let v_1 is the velocity at position 1.

Using equation $v^2 - u^2 = 2gh$, we get

$$v_1^2 - 0^2 = 2gh_1 \qquad (\because \ u = 0, h = h_1)$$

$$v_1^2 = 2gh_1$$

$\therefore$ K.E. of the body at position 1

$$= \frac{1}{2}mv_1^2 = \frac{1}{2}m \times 2gh_1 \qquad (\because \ v_1^2 = 2gh_1)$$

$$= mgh_1 = Wh_1 \qquad (\because \ W = mg)$$

Potential energy at position 1

= $W \times$ Height of body at 1 w.r.t. ground

$= W \times (h - h_1)$

$\therefore$ Total energy at position 1

= Potential energy + Kinetic energy

$= W(h - h_1) + Wh_1$

$= Wh - Wh_1 + Wh_1 = Wh$

And it is the same energy, the body was possessing at the top of the tower.

Similarly, it can be proved that total energy of the body on the ground will be equal to 'Wh'.

2.18. IMPULSE AND IMPULSIVE FORCE

In Art. 2.9, it is mentioned that the force applied on a body is directly proportional to the rate of change of momentum. If F is the applied force on a body of mass 'm' which is moving with a velocity v then we have

$$F = \frac{d}{dt}(mv)$$

or $$F \times dt = d(mv)$$

Integrating both sides, we get

$$\int F \times dt = \int d(mv) = \int m\, dv$$

The product of the force and time interval (when time interval is very short) is known as *impulse* and the force is known as impulsive force. Hence

$$\text{Impulse} = F \times dt \quad \text{...... if } dt \text{ is very short}$$

and $\quad$ impulsive force = F

Impulsive force occurs in collisions, in explosions, in the striking of a nail or a pile by a hammer.

2.19. LAW OF CONSERVATION OF MOMENTUM

It states that the total momentum of moving masses in a system in any one direction remains constant unless acted upon by an external force in that direction. This law is applied to problems of impact *i.e.* collision of two bodies. If two bodies of masses m_1 and m_2 with linear velocities v_1 and v_2 are moving in the same straight lines, and they collide and begin to move together with a common velocity v, then

Momentum before impact = Momentum after impact

or
$$m_1 v_1 + m_2 v_2 = (m_1 + m_2)v \quad \text{...(2.20)}$$

Similarly for rotating bodies, we have

$$I_1 \omega_1 + I_2 \omega_2 = (I_1 + I_2)\omega \quad \text{...(2.20A)}$$

where I_1 = Mass moment of inertia of first rotating body,

I_2 = Mass moment of inertia of second rotating body,

ω_1 = Angular velocity of first body,

ω_2 = Angular velocity of second body, and

ω = Common angular velocity.

2.20. SIMPLE HARMONIC MOTION

A body is said to have simple harmonic motion, if it moves or vibrates about a mean position such that its acceleration is always proportional to its distance from the mean position and is directed towards the mean position or equilibrium position. The simple harmonic motion is written as S.H.M.

Consider a body moving round the circumference of a circle in the anticlockwise direction as shown in Fig. 2.9.

Let ω = Constant angular velocity of the body,

O = Centre of the circle,

AC = Fixed diameter of the circle,

P = Position of the body,

r = Radius of the circle,

M = Projection of P on diameter AC.

v_P = Tangential velocity of point P.

When the body moves round the circumference of the circle from A to B, the projection of the body on the diameter AC (*i.e.* point M) moves from A to O, and when the body moves

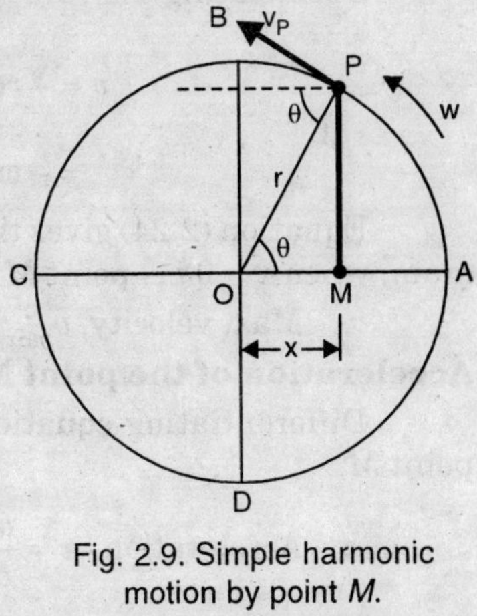

Fig. 2.9. Simple harmonic motion by point *M*.

round the circumference from B to C, the projection of the body on the diameter AC moves from O to C. Similarly when the body moves along the circumference from C to D, the point M moves from C to O on diameter AC and finally when the body moves from D to A, the point M moves from O to A. The motion of M along the diameter AC is called *simple harmonic motion*.

Let $\angle AOP = \theta$

t = Time taken by the body when moving from A to P.

Let the distance of the point M from the centre of the circle is equal to OM or x.

or $$OM = x = OP \cos \theta$$

or $$x = r \cos \theta \qquad (\because \; OP = r)$$

But θ is angular displacement and is equal to the product of angular velocity (ω) and time (t) *i.e.* $\theta = \omega t$.

$$\therefore \quad x = r \cos \omega t \qquad (\because \; \theta = \omega t) \qquad ...(2.21)$$

Equation (2.21) gives the displacement of point M from the centre of the circle.

Differentiating equation (2.21) with respect to t, we have

$$\frac{dx}{dt} = \frac{d}{dt}[r \cos \omega t]$$

$$= - r\omega \times \sin \omega t$$

But $$\frac{dx}{dt} = \text{Velocity}$$

$$\therefore \quad v = - r\omega \times \sin \omega t \qquad ...(2.22)$$

$$= - r\omega \times \sin \theta \qquad (\because \; \theta = \omega t) \qquad ...(2.23)$$

Equation (2.22) gives the velocity of point M in terms of angular velocity and time whereas the equation (2.23) gives the velocity in terms of angular displacement (θ). The – ve sign in equations (2.22) and (2.23) shows that with the increase of x (*i.e.* the distance from the centre of the circle), the velocity decreases.

In right angled triangle OPM,

$$\sin \theta = \frac{MP}{OP} = \frac{\sqrt{PO^2 - OM^2}}{OP} \qquad (\because \; MP = \sqrt{OP^2 - OM^2}\,)$$

$$= \frac{\sqrt{r^2 - x^2}}{r} \qquad (\because \; OP = r, \; OM = x)$$

Substituting the value of $\sin \theta$, in equation (2.23), we get

$$v = - r\omega \, \frac{\sqrt{r^2 - x^2}}{r}$$

$$= - \omega \sqrt{r^2 - x^2} \qquad ...(2.24)$$

Equation (2.24) gives the velocity in terms of x and radius of circle. The velocity is maximum, when $x = 0$ *i.e.* point M is at O.

$$\therefore \quad \text{Max. velocity, } v_{max} = - \omega \times r \qquad ...(2.25)$$

Acceleration of the point M

Differentiating equation (2.22) with respect to time 't', we get the acceleration of the point M.

$$\therefore \quad \text{Acceleration, } a = \frac{dv}{dt} = \frac{d}{dt}(- r\omega \sin \omega t)$$

$$= -r\omega^2 \cos \omega t \text{ or } -r\omega^2 \cos \theta \qquad (\because \ \theta = \omega t)$$

$$= -\omega^2 \times r \cos \omega t \qquad ...(2.26)$$

$$= -\omega^2 \times x \quad [\because \ x = r \cos \omega t \text{ from equation (2.21)}] \qquad ...(2.27)$$

In equation (2.27), as ω is constant hence acceleration is directly proportional to the distance of M from the centre. The – ve sign shows that acceleration is directed towards the centre of the circle. Hence the point M moves simple harmonic motion. The value of acceleration for different values of θ are obtained from equation (2.26) as :

(*i*) When $\theta = 0$, $a = -r\omega^2 \cos 0° = -r\omega^2$ $(\because \ \cos 0° = 1)$

(*ii*) When $\theta = 90°$, $a = r\omega^2 \cos 90° = -r\omega^2 \times 0 = 0$ $(\because \ \cos 90° = 0)$

(*iii*) When $\theta = 180°$, $a = -r\omega^2 \cos 180° = -r\omega^2 (-1) = r\omega^2$.

$\therefore$ Acceleration is maximum when $\theta = 0$ or $180°$,

$$\therefore \quad a_{max} = \pm \omega^2 r. \qquad ...(2.28)$$

Differential Equation of Simple Harmonic Motion

Displacement of M from mean position is given by,

$$x = r \cos \omega t$$

Then velocity,

$$v = \frac{dx}{dt} = -\omega r \sin \omega t$$

and acceleration,

$$a = \frac{dv}{dt} = \frac{d}{dt}\left(\frac{dx}{dt}\right) = \frac{d^2x}{dt^2} = -\omega^2 r \cos \omega t$$

$$= -\omega^2 x \qquad (\because \ x = r \cos \omega t)$$

$$\therefore \quad \frac{d^2x}{dt^2} = -\omega^2 x$$

or

$$\frac{d^2x}{dt^2} + \omega^2 x = 0 \qquad ...(2.29)$$

The above equation is known as differential equation of simple harmonic motion. The solution of the above equation is given by

$$x = A \cos \omega t + B \sin \omega t$$

where A and B are constants of integration. The values of A and B are determined from given boundary conditions.

If the equation of motion of a body is of the form $\frac{d^2x}{dt^2} + \omega^2 x = 0$, where ω is constant, then it may be assumed that the motion of the body is a simple harmonic motion.

The displacement, velocity and acceleration can also be written as :

Displacement, $x = r \cos \omega t = r \cos \theta$ $(\because \ \omega t = \theta)$

Velocity, $v = -\omega r \sin \omega t = -\omega r \sin \theta$

$= \omega r \cos (90 + \theta)$

Acceleration, $a = -\omega^2 r \cos \omega t = -\omega^2 r \cos \theta = \omega^2 r \cos (180 + \theta)$

The above three equations shows that the velocity leads the displacement by 90° and acceleration leads the displacement by 180°.

Alternate Method of Finding Velocity and Acceleration for S.H.M.

Refer to Fig. 2.9 in which case point P is moving round the circumference of a circle in the anticlockwise direction with a uniform angular velocity. The projection M of the point P on a diameter AC has simple harmonic motion (S.H.M.).

Let v_P = the peripheral speed of point P,

r = radius of the circle along which P moves, and

ω = Angular velocity.

When PO makes an angle θ with OA, the velocity and acceleration of M are respectively equal to the components of the velocity and acceleration of P parallel to AC.

$$\therefore \quad v_M = \text{Velocity of } M = \text{Component of } v_P \text{ parallel to } AC$$

$$= v_P \sin\theta$$

We know that the acceleration of P is the centripetal acceleration whose magnitude is $\omega^2 r$. The acceleration of M is the component of the acceleration of P parallel to AC and is directed towards the centre O.

$$\therefore \quad a_M = \text{Acceleration of } M = \text{Component of acceleration of } P \text{ parallel to } AC$$

$$= \omega^2 r\cos\theta$$

$$= \omega^2 \times x \qquad (\because \ x = OP\cos\theta = r\cos\theta)$$

From the above equation, we have

$$\frac{f_M}{x} = \omega^2 \qquad \ldots(i)$$

$$\text{or} \quad \frac{\text{Acceleration of } M}{\text{Displacement of } M} = \omega^2 \qquad (\because \ x = \text{Displacement of } M)$$

$$= (2\pi n)^2 \qquad \left(\text{where } n^* = \text{Frequency} = \frac{\omega}{2\pi}\right)$$

$$\therefore \quad n = \frac{1}{2\pi}\sqrt{\frac{\text{Acceleration of } M}{\text{Displacement of } M}} \qquad \ldots(2.30)$$

Note. (*i*) The product of angular velocity (ω) and time (t) gives angular displacement (θ). Hence $\theta = \omega t$. The value of θ so obtained is in radians.

(*ii*) The angular displacement (θ) from radians is converted into degrees by multiplying by $\frac{180}{\pi}$.

2.20.1. Definitions of Some terms used with S.H.M. The terms which are generally used with S.H.M. are defined as :

1. Amplitude. The maximum displacement of the point M (which is having simple harmonic motion) from its mean position is called the amplitude. In Fig. 2.9, O is the mean position of the point M and maximum displacement of M from the mean position is OA or OC.

Hence amplitude = OA = Radius of the circle.

---

*Frequency n is defined as the number of cycles per second. Mathematically it is equal to $\frac{1}{T}$

where T = period of S.H.M. = Time taken by P for one revolution = $\frac{2\pi}{\omega}$

$$\therefore \quad n = \frac{1}{T} = \frac{1}{2\pi/\omega} = \frac{\omega}{2\pi} = . \text{ Hence } 2\pi n = \omega.$$

2. Oscillation. The motion of M from A to C and back from C to A is called an oscillation.

3. Period of S.H.M. The period of simple harmonic motion is the time taken by the point M for one complete oscillation *i.e.* time taken by M in moving from A to C and then from C to A. The time taken by the point M for one complete oscillation is the same as time taken by the body (P) moving along the circumference for one complete revolution.

Let T = Period of S.H.M.

ω = Constant angular velocity of the body P.

Now by definition,

T = Time taken by the body (P) moving along the circumference for one complete revolution

$$= \frac{\text{Angular displacement for one revolution}}{\text{Constant angular velocity}}$$

$$= \frac{2\pi}{\omega} \quad \text{...(2.31)}$$

4. Frequency. Frequency is defined as the number of cycles per second and mathematically it is given as

$$= \frac{1}{T} = \frac{1}{2\pi/\omega} = \frac{\omega}{2\pi} \quad \text{...(2.32)}$$

Problem 2.1. *The piston of an engine moves with simple harmonic motion. The crank rotates at 100 r.p.m. and the stroke is 180 cm. Find the velocity and acceleration of the piston, when it is at a distance of 60 cm from the centre.* (AMIE, Summer 1974)

Sol. Given :

Speed of crank, $N = 100$ r.p.m.

$\therefore$ Angular velocity, $\omega = \frac{2\pi N}{60} = \frac{2\pi \times 100}{60} = 10.47$ rad/s.

Stroke of piston $= 180$ cm $= 1.8$ m

$\therefore$ Amplitude, $r = \frac{\text{Stroke}}{2} = \frac{1.8}{2} = 0.9$ m.

Displacement of piston from centre, $x = 60$ cm $= 0.6$ m.

Using equation (2.21) for displacement,

$$x = r \cos \omega t$$

$$\therefore \quad 0.6 = 0.9 \cos \omega t$$

or $$\cos \omega t = \frac{0.6}{0.9} = \frac{2}{3} = 0.667.$$

$$\therefore \quad \omega t = \cos^{-1} 0.667 = 48.164°.$$

Velocity of the piston is given by equation (2.22), as

$$v = -\omega r \sin \omega t$$

$$= -10.47 \times 0.9 \sin (48.164°) \qquad (\because \omega t = 48.164°)$$

$$= -10.47 \times .9 \times 0.745 = \mathbf{-7.02\ m/sec.} \quad \textbf{Ans.}$$

Acceleration of the piston is given by equation (2.26), as

$$a = -\omega^2 r \cos \omega t$$

$$= -(10.47)^2 \times 0.9 \times 0.667 \qquad (\because \cos \omega t = .667)$$

$$= \mathbf{-65.8\ m/sec^2.} \quad \textbf{Ans.}$$

Problem 2.2. *In a mechanism, a cross-head moves in a straight guide with simple harmonic motion. At distances of 125 mm and 200 mm from its mean position, it has velocities of 6 m/s and 3 m/s respectively. Find the amplitude, maximum velocity and period of vibration. If the cross-head weights 2N, calculate the maximum force on it in the direction of motion.*

(AMIE, Winter 1979)

Sol. When distance from mean position = 125 mm or .125 m, velocity v_1 = 6 m/s

When distance from mean position = 200 mm or 0.200 m, velocity v_2 = 3 m/s

Let r = Amplitude of cross-head

T = Period of vibration

v_{max} = Maximum velocity.

Distance from mean position, x_1 = .125 m, when v_1 = 6 m/s

Distance from mean position, x_2 = .200 m, when v_2 = 3 m/s

Using equation (2.24), $v = -\omega\sqrt{r^2 - x^2}$

For 1st velocity, $v_1 = -\omega\sqrt{r^2 - {x_1}^2}$

$$6 = -\omega\sqrt{r^2 - .125^2}$$

or $$= -\omega\sqrt{r^2 - .015625} \quad ...(i)$$

For 2nd velocity, $v_2 = -\omega\sqrt{r^2 - {x_2}^2}$

or $$3 = -\omega\sqrt{r^2 - .200^2}$$

$$= -\omega\sqrt{r^2 - .04} \quad ...(ii)$$

Dividing equation (*i*) by (*ii*),

$$\frac{6}{3} = \frac{-\omega\sqrt{r^2 - .015625}}{-\omega\sqrt{r^2 - .04}}$$

or $$2 = \frac{\sqrt{r^2 - .015625}}{\sqrt{r^2 - .04}}$$

Squaring both sides, $$4 = \frac{r^2 - .015625}{r^2 - 0.04}$$

or $$4(r^2 - .04) = r^2 - .015625$$

or $$4r^2 - 4 \times .04 = r^2 - .015625$$

or $$3r^2 = 4 \times .04 - .015625 = .1444.$$

(*i*) $\therefore$ $$r = \sqrt{\frac{0.1444}{3}} = .2194 \text{ m} = \mathbf{21.94\ cm.}\quad \textbf{Ans.}$$

Substituting the value of r = .2194 m in equation (*ii*),

$$3 = -\omega\sqrt{.2194^2 - .04} = -\omega \times 0.9$$

$$\omega = -\frac{3}{.09} = -33.33 \text{ rad/s} = 33.33 \text{ rad/s Numerically.}$$

(*ii*) Period of vibration is given by equation (2.37),

$$T = \frac{2\pi}{\omega} = \frac{2\pi}{33.33} = \mathbf{0.1885 \simeq 0.19\ sec.} \quad \textbf{Ans.}$$

(*iii*) Maximum velocity (v_{max}).

Maximum velocity is given by equation (2.25), as

$$v_{max} = \omega \times r$$

$$= -\ 33.33 \times .2194 \text{ m/s} \quad (\because \quad \omega = 33.33 \text{ radians, } r = .2194 \text{ m})$$

$$= \mathbf{-\ 7.31\ m/s.} \quad \textbf{Ans.}$$

(*iv*) Maximum force on cross-head in the direction of motion

Weight of cross-head, $W = 2.0$ N

or Mass of cross-head, $m = \frac{W}{g} = \frac{2.0}{9.81}$.

Now force, $F = \text{Mass} \times \text{Acceleration} = m \times a$.

Force will be maximum, when acceleration (f) is maximum.

$\therefore$ Maximum force, $F_{max} = m \times a_{max}$ ...(*iii*)

Maximum acceleration is given by equation (2.28), as

$$a_{max} = \pm\, \omega^2 r = \pm\, (33.33)^2 \times .2194 \text{ m/s}^2$$

$$= \pm\, 243.73 \text{ m/s}^2.$$

Substituting this value in equation (*iii*),

$$F_{max} = m \times 243.73 = \frac{2.0}{9.81} \times 243.73 = \mathbf{49.7\ N.} \quad \textbf{Ans.}$$

2.20.2. Oscillation of the Bodies having Simple Harmonic Motion. The oscillation of the following bodies, which are having simple harmonic motion will be considered :

1. Oscillation of a vertical elastic string or spring.
2. Oscillation of a simple pendulum.

1. Oscillation of a Vertical Elastic String or Spring. Consider a vertical elastic string or a vertical spring with upper end fixed as shown in Fig. 2.10 (*a*).

If a weight W is attached at the lower end A of the spring, the spring will be extended (stretched) in the downward direction and finally the load will be in the equilibrium position. Let 1–1 be the equilibrium position of the spring. If the spring is further stretched upto 2—2 in the downward direction and then released, the weight will start moving up and down with a simple harmonic motion. This is shown below :

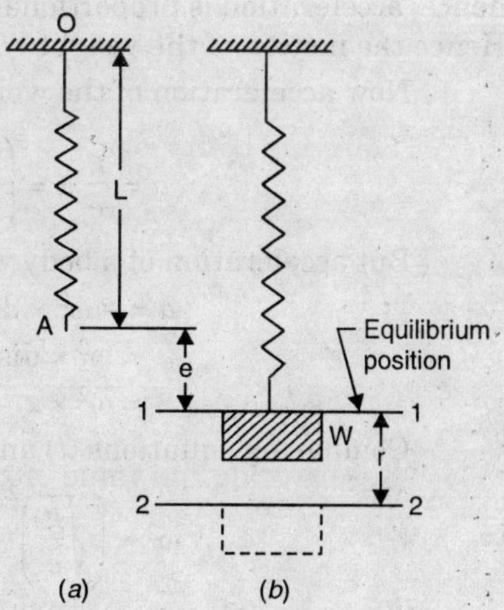

Fig. 2.10. Oscillation of spring.

Let. L = Length of spring

e = Extension of spring when weight is attached to the spring and is also called static extension

m = Mass attached to the spring

W = Weight attached = $m \times g$

r = Extension of spring from 1—1 or from equilibrium position

a = Acceleration of the weight W

T = Time period

Consider the weight W at a distance of x in the downward direction from 1—1.

Total extension of spring when weight W is at a distance x in the downward direction $= e + x$.

But the extension 'e' is produced by the force W. Also the extension is proportional to the tension in the spring and hence extension $(e + x)$ will be produced by the force or by tension in the spring

$$= \frac{W}{e}(e + x)$$

The tension $\frac{W}{e}(e + x)$ in the spring will try to bring back the weight W to equilibrium position *i.e.* to the position 1—1.

$\therefore$ Net force on the weight W in the upward direction

$$= \text{Tension in the spring} - \text{Weight } (W)$$

$$= \frac{W}{e}(e + x) - W$$

$$= \frac{m \times g(e + x)}{e} - mg \qquad (\because \; W = \text{Mass} \times \text{Acc. due to gravity} = m \times g)$$

$$= \frac{mg \times e}{e} + \frac{mgx}{e}$$

$$= m \times g + \frac{mgx}{e} - mg$$

$$\therefore \qquad = \frac{mgx}{e}.$$

$$\therefore \quad \text{Acceleration of the weight } W = \frac{\text{Net force}}{\text{Mass}}$$

$$= \frac{mgx}{e \times m} = \frac{gx}{e}.$$

But for a given weight W, e (the static extension) is constant. Also 'g' is constant and hence, acceleration is proportional to x (or to the displacement from the equilibrium position). Hence the motion of the weight W is S.H.M. about the equilibrium position.

Now acceleration of the weight W,

$$a = \frac{g}{e}x = \left(\sqrt{\frac{g}{e}}\right)^2 x \qquad ...(i)$$

But acceleration of a body with simple harmonic motion is given by equation (2.23), as

$$a = -\omega^2 \times \text{distance from the centre}$$

$$= \omega^2 \times \text{distance from the centre (numerically)}.$$

$$= \omega^2 \times x. \qquad ...(ii)$$

Comparing equations (i) and (ii), we get

$$\omega^2 = \left(\sqrt{\frac{g}{e}}\right)^2$$

or $$\omega = \sqrt{\frac{g}{e}} \qquad ...(2.33)$$

Time period (T) of oscillation is given by equation (2.31), as

$$T = \frac{2\pi}{\omega}$$

$$= \frac{2\pi}{\sqrt{g/e}} = 2\pi\sqrt{\frac{e}{g}}$$

$$= 2\pi\sqrt{\frac{\text{Static extension}}{\text{Acceleration due to gravity}}} \quad \text{...(2.34)}$$

Note. The weight W, should be pulled down from the equilibrium position by a displacement less than the static extension. This means that 'r' should be less than 'e'. If r is greater than 'e', the motion is not wholly simple harmonic.

Definitions

1. Stiffness of a Spring. It is defined as the weight or force required to stretch the spring by unit deflection.

If W = Weight attached to the spring

e = Extension of spring due to the weight W

Then stiffness of spring $= \frac{W}{e}$. It is denoted by C or S.

$$\therefore \quad C \text{ or } S = \frac{\text{Weight attached}}{\text{Static extension}}.$$

Problem 2.3. *Determine the stiffness in N/cm of a vertical spring to which a weight of 50 N is attached and is set vibrating vertically. The weight makes 4 oscillations per second.*

Sol. Given :

Weight attached, W = 50 N

No. of oscillation per second = 4

$\therefore$ Period of oscillation, $T = \frac{1}{4} = 0.25$ sec.

Using equation (2.34), $T = 2\pi\sqrt{\frac{\text{Static extension}}{\text{Acceleration due to gravity}}}$

$$\therefore \quad 0.25 = 2\pi\sqrt{\frac{\text{Static Extension}}{9.81}}$$

$$\therefore \text{ Static extension} = \left(\frac{0.25}{2\pi}\right)^2 \times 9.81 = 0.0155 \text{ m} = 1.55 \text{ cm}$$

$$\therefore \text{ Stiffness of spring} = \frac{\text{Weight attached}}{\text{Static extension}}$$

$$= \frac{50}{1.55} = \textbf{32.26 N/cm.} \quad \textbf{Ans.}$$

Problem 2.4. *Find the weight, which is to be attached to a spring of stiffness 1.5 N/cm so that the weight is oscillating with a periodic time of 1.5 sec.*

Sol. Given :

Stiffness, $C = \frac{1.5 \text{ N}}{\text{cm}} = \frac{1.5 \times 100 \text{ N}}{\text{m}} = \frac{150 \text{ N}}{\text{m}}$

Periodic time, T = 1.5 sec.

Let W = Weight to be attached to the spring.

Using equation (2.34), $T = 2\pi\sqrt{\dfrac{\text{Static extension}}{9.81}}$

$\therefore$ Static extension $= \left(\dfrac{T}{2\pi}\right)^2 \times 9.81 = \left(\dfrac{1.5}{2\pi}\right)^2 \times 9.81 = 0.559$ m.

But stiffness, $C = \dfrac{\text{Weight attached}}{\text{Static extension}}$

$$C = \frac{W}{0.559}$$

$\therefore$ $150 = \dfrac{W}{0.559}$

$\therefore$ $W = 150 \times 0.559 = $ **83.85 N. Ans.**

2.20.3. Oscillation of a Simple Pendulum. Fig. 2.11 shows a simple pendulum, which consists of a light, in elastic, flexible string with a heavy bob suspended at the lower end and the upper end is fixed at O.

The bob oscillates freely in a vertical plane.

Let l = Length of string = OA or OB

m = Mass of bob

W = Weight of bob = $m \times g$.

The pendulum is in equilibrium, when bob is at A. But when bob is displaced to B such that $\angle AOB = \theta$ = a very small angle and released, it starts oscillating between B and C with A as mean position.

Consider the bob at B. The forces acting on the bob are :

(*i*) Weight of bob, $W = mg$ acting vertically downward,

(*ii*) Tension T in the string.

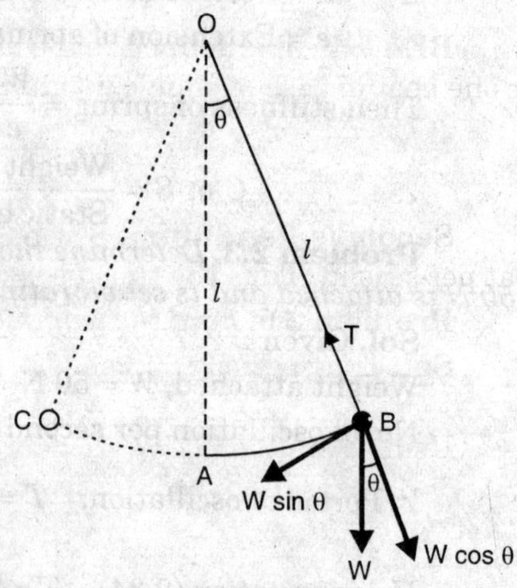

Fig. 2.11. Oscillation of a simple pendulum.

The weight W can be resolved into two components *i.e.*, $W \cos\theta$ along the string OB and $W \sin\theta$ at right angles to OB. For the equilibrium of the bob at B, $T = W\cos\theta$.

But the component $W\sin\theta$ acting towards A will be unbalanced and will give rise to an acceleration (f) in the direction BA.

$\therefore$ Acceleration, $a = \dfrac{\text{Force}}{\text{Mass}} = \dfrac{W\sin\theta}{m}$

$= \dfrac{mg\sin\theta}{m} = g\sin\theta$ $\qquad (\because \; W = mg)$

If θ is very small, $\sin\theta = \theta$

$\therefore$ $a = g\theta$

$= g\,\dfrac{\text{Length of arc } AB}{\text{Radius}}$ $\qquad \left(\because \; \theta = \dfrac{\text{Arc } AB}{l}\right)$

$= g \times \dfrac{\text{Arc } AB}{l} = g \times \dfrac{\text{Distance of } B \text{ from } A}{l}$ ...(*i*)

In equation (*i*), g and l are constant and hence acceleration is proportional to AB (or to the distance from A). Also acceleration is directed towards A. Hence the bob moves with simple harmonic motion about the mean position A.

But the acceleration of a body with S.H.M. is given by equation (2.27), as

$$a = -\omega^2 \times \text{Distance from centre}$$

$$= \omega^2 \times \text{Distance from the centre numerically} \qquad \ldots(ii)$$

Comparing (*i*) and (*ii*), we get

$$\omega^2 = \frac{g}{l}$$

$$\therefore \qquad \omega = \sqrt{\frac{g}{l}}$$

Time period (T) of oscillation is given by equation (2.31), as

$$T = \frac{2\pi}{\omega} = \frac{2\pi}{\sqrt{g/l}} = 2\pi\sqrt{\frac{l}{g}}. \qquad \ldots(2.35)$$

BEAT. Beat is defined as the half of the oscillation. It is also called *swing*. Thus the time for one beat in case of simple pendulum

$$= \frac{T}{2} = \pi\sqrt{\frac{l}{g}}.$$

Second's Pendulum. Seconds pendulum is defined as that pendulum which has one beat per second. Thus the time period for second's pendulum will be equal to two seconds.

Problem 2.5. *Find the length of a second's pendulum.*

Sol. Time period of seconds pendulum, $T = 2$ sec.

Using equation (2.35), $T = 2\pi\sqrt{\dfrac{l}{g}}$ or $2 = 2\pi\sqrt{\dfrac{l}{980}}$ $\qquad (\because \; g = 980 \text{ cm/sec}^2)$

$$\therefore \qquad l = \left(\frac{2}{2\pi}\right)^2 \times 980 = \mathbf{99.29\ cm.} \quad \textbf{Ans.}$$

If g is taken = 981 cm/sec^2, then

$$l = \left(\frac{2}{2\pi}\right)^2 \times 981 = \mathbf{99.39\ cm.} \quad \textbf{Ans.}$$

2.21. THE COMPOUND PENDULUM

Compound pendulum is a rigid body which is suspended vertically so as to oscillate with small amplitude under the action of gravity. Fig. 2.12 shows a rigid body suspended at point O. The body is given a small angular displacement θ and it starts oscillating under the action of gravity. This body becomes as a compound pendulum.

Let W = Weight of body = $m \times g$,

m = Mass of the body,

k = Radius of gyration about an axis through centre of gravity G and perpendicular to the plane of motion,

a = Distance of the point of suspension O from the centre of gravity G.

Now mass moment of inertia about an axis through G,

$$I_G = mk^2$$

Mass moment of inertia about the axis of suspension is obtained by parallel axis theorem as

$$I = I_G + m \times a^2$$
$$= mk^2 + ma^2 \qquad (\because \quad I_G = mk^2)$$
$$= m(k^2 + a^2)$$

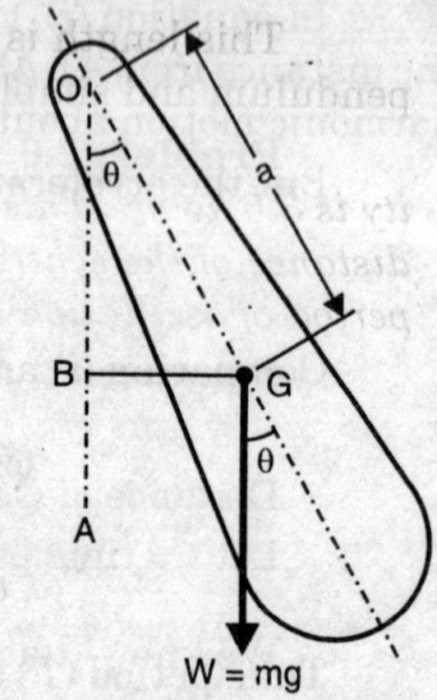

Fig. 2.12

After the body is given a small angular displacement θ, the restoring couple to bring the pendulum into equilibrium position OA, is given by,

$$T = \text{Weight} \times \perp \text{ distance between the line of action of weight and axis } OA$$
$$= W \times BG$$
$$= mg \times a \sin\theta \qquad (\because \quad W = mg \text{ and } BG = a \sin\theta)$$
$$= mg \times a \times \theta \qquad (\because \quad \theta \text{ is very small. Hence } \sin\theta \approx \theta)$$

Also we know that torque is equal to $I \times \alpha$ where α is angular acceleration

or $$T = I \times \alpha$$

$$\therefore \quad \alpha = \frac{T}{I} = \frac{mg \times a \times \theta}{m(k^2 + a^2)} \qquad [\because \quad I = m(k^2 + a^2)]$$
$$= \frac{g \times a \times \theta}{(k^2 + a^2)}$$
$$= \text{constant} \times \theta$$

[$\because$ g, a and k are constant and hence $g \times a/(k^2 + a^2)$ is constant]

In the above equation, θ is the angular displacement and α is angular acceleration. Hence angular acceleration is directly proportional to angular displacement. Hence the compound pendulum performs simple harmonic motion. We know that the time period for simple harmonic motion is given by

$$T = 2\pi \sqrt{\frac{\text{Angular Displacement}}{\text{Angular Acceleration}}}$$
$$= 2\pi \sqrt{\frac{\theta}{\alpha}}$$
$$= 2\pi \sqrt{\frac{(k^2 + a^2)}{g \times a}} \qquad \left(\because \text{ From equation } \frac{\theta}{\alpha} = \frac{(k^2 + a^2)}{g \times a}\right)$$

Now the frequency of oscillation is given by,

$$n = \frac{1}{t} = \frac{1}{2\pi} \sqrt{\frac{g \times a}{(k^2 + a)^2}} \qquad ..(2.36)$$

But the frequency of a simple pendulum $= \frac{1}{2\pi} \sqrt{\frac{g}{L}}$

The frequency of a simple pendulum will be same as of a compound pendulum if the length of simple pendulum is made as

$$L = \frac{k^2 + a^2}{a} \qquad ... (2.36 \text{ A})$$

This length is obtained by equating the two frequencies (*i.e.* frequency of a compound pendulum and simple pendulum) and finding the value of L.

Problem 2.6. *The moment of inertia of a flywheel of mass 90 kg through centre of gravity is 3.6 kg m². This flywheel is suspended in a vertical plane as a compound pendulum. The distance of the centre of gravity from the point of suspension is 120 mm. Determine : (i) time period of oscillation and (ii) frequency of oscillation.*

Sol. Given :

$$m = 90 \text{ kg}, I_G = 3.6 \text{ kg m}^2.$$

Distance of C.G. from point of suspension, $a = 120 \text{ mm} = 0.12 \text{ m}$

Let t = time period of oscillation and n = frequency

We know that $I_G = mk^2$ $\quad\therefore\quad k^2 = \frac{I_G}{m} = \frac{3.6}{90} = 0.04 \text{ m}^2$

Time period is given by,

$$t = 2\pi \sqrt{\frac{(k^2 + a^2)}{g \times a}}$$

$$= 2\pi \sqrt{\frac{0.04 + 0.12^2}{9.81 \times 0.12}} \qquad (\because \quad k^2 = 0.04)$$

$$= 2\pi \sqrt{\frac{0.04 + 0.0144}{9.81 \times 0.12}} \text{ seconds} = \textbf{1.35 seconds.} \quad \textbf{Ans.}$$

Frequency is given by $n = \frac{1}{t} = \frac{1}{1.35} = 0.74$ Hertz = **0.74 Hz. Ans.**

or 0.74 oscillations per second.

Problem 2.7. *Find the moment of inertia of a connecting rod of an engine if the weight of the connecting rod is 600 N and its length between the centres is 1 m. The connecting rod oscillates in a vertical plane about the centre of the small end. The distance of its C.G. from the point of oscillation is 700 mm. From the expt., it is observed that this rod completes 25 oscillation in 48 seconds. Find :*

(i) the moment of inertia of the rod about an axis through the C.G. and

(ii) length of the equivalent simple pendulum.

Sol. Given :

$W = 600$ N $\quad\therefore\quad m = \frac{W}{g} = \frac{600}{9.81} = 61.162$ kg.

Length* between the centres = 1 m

Distance of C.G. from the point of suspension, $a = 700 \text{ mm} = 0.7 \text{ m}$.

No. of oscillations = 25 in time 48 seconds

$\therefore$ No. of oscillations per second $= \frac{25}{48} = 0.52$

$\therefore$ Frequency, $n = 0.52$

For a simple pendulum, the frequency is given by,

$$n = \frac{1}{2\pi} \sqrt{\frac{g}{L}}$$

*This data is extra in this problem. Hence it was not used.

$\therefore$ Length of the simple pendulum which has the same frequency as the given connecting rod is obtained by substituting the value of $n = 1.92$ in the above equation.

$$\therefore \quad 0.52 = \frac{1}{2\pi}\sqrt{\frac{9.81}{L}}$$

Squaring both sides, we get

$$(0.52)^2 = \frac{1}{4\pi^2} \times \frac{9.81}{L}$$

or
$$L = \frac{9.81}{4\pi^2 \times (0.52)^2} = \mathbf{0.9189\ m.} \quad \textbf{Ans.}$$

To find I_G, first calculate the value of k^2. Hence using equation (2.36 A), we get

$$L = \frac{k^2 + a^2}{a} \quad \text{(As frequencies of the two pendulum are same)}$$

$$\therefore \quad 0.9189 = \frac{k^2 + (0.7)^2}{0.7}$$

or
$$0.9189 \times 0.7 = k^2 + 0.49$$

$$\therefore \quad k^2 = 0.9189 \times 0.7 - 0.49 = 0.6432 - 0.49 = 0.15323$$

Now
$$I_G = mk^2$$
$$= 61.162 \times 0.15323 = \mathbf{9.37\ kg\ m^2}. \quad \textbf{Ans.}$$

Problem 2.8. *Find : (i) the moment of inertia of a connecting rod about an axis through the C.G. and (ii) the distance of its C.G. from the small end centre if this connecting rod of mass 60 kg and of length 1 m between centres, is suspended vertically. The rod makes 100 oscillations in 189 seconds when suspended through the centre of the small end, and 162 seconds when suspended through big-end centre.*

Sol. Given :

$m = 60$ kg ; length between centres = 1 m ;

Let a_1 = distance of C.G. of the rod from small-end centre

a_2 = distance of C.G. from big-end centre

$a_1 + a_2$ = distance between the two centres = 1 m

n_1 = frequency* of oscillation when suspended from small-end centre

$$= \frac{\text{No. of oscillation}}{\text{Time}} = \frac{100}{189} = 0.529$$

n_2 = Frequency of oscillation when suspended from big-end centre

$$= \frac{\text{No. of oscillation}}{\text{Time}} = \frac{100}{162} = 0.617$$

L_1 = Length of equivalent simple pendulum when suspended from small-end centre

L_2 = Length of equivalent simple pendulum when suspended from big-end centre.

For a simple pendulum, the frequency is given by

$$n = \frac{1}{2\pi}\sqrt{\frac{g}{L}}$$

$\therefore$ Length of the simple pendulum (or length of equivalent simple pendulum), which has the same frequency as the given rod when suspended from the small-end centre is given by

*Frequency is number of oscillations per second.

$$n_1 = \frac{1}{2\pi}\sqrt{\frac{g}{L_1}}$$

or $$(n_1) \times 2\pi = \sqrt{\frac{g}{L_1}}$$

or $$L_1 = \frac{g}{n_1^2 \times (2\pi)^2}$$

$$= \frac{9.81}{0.529^2 \times 4\pi^2} \qquad (\because \ n_1 = 0.529)$$

$$= 0.888 \text{ m}$$

Similarly the length of the simple pendulum, which has the same frequency as the given rod when suspended from the big-end centre is given by

$$n_2 = \frac{1}{2\pi}\sqrt{\frac{g}{L_2}}$$

or $$L_2 = \frac{g}{n_2^2 \times 4\pi^2}$$

$$= \frac{9.81}{0.617^2 \times 4 \times \pi^2} \qquad (\because \ n_2 = 0.617)$$

$$= 0.652 \text{ m}$$

Now using equation (2.36 A) when suspended from small-end centre, we get

$$L_1 = \frac{k^2 + a_1^2}{a_1}$$

$\therefore$ $$a_1 \times L_1 = k^2 + a_1^2$$

or $$k^2 = a_1 \times L_1 - a_1^2 \qquad ...(i)$$

Similarly using equation (2.36 A) when suspended from big-end centre, we get

$$k^2 = a_2 \times L_2 - a_2^2 \qquad ...(ii)$$

Equating the two equations (*i*) and (*ii*), we get

$$a_1 \times L_1 - a_1^2 = a_2 \times L_2 - a_2^2 \qquad ...(iii)$$

But $a_1 + a_2$ = distance between two centres = 1 m (given)

$\therefore$ $$a_2 = 1 - a_1$$

Substituting this value of a_2 in equation (*iii*), we get

$$a_1 \times L_1 - a_1^2 = (1 - a_1) \times L_2 - (1 - a_1)^2$$

Substituting the values of L_1 and L_2, we get

$$a_1 \times 0.888 - a_1^2 = (1 - a_1) \times 0.652 - (1 + a_1^2 - 2a_1)$$

or $$0.888a_1 - a_1^2 = 0.652 - a_1 \times 0.652 - 1 - a_1^2 + 2a_1$$

or $$0.888a_1 - a_1^2 - 0.652 + 0.652a_1 + 1 + a_1^2 - 2a_1 = 0$$

or $$a_1(0.888 + 0.652 - 2) = -1 + 0.652$$

or $$a_1(-0.46) = -0.348$$

$\therefore$ $$a_1 = \frac{0.348}{0.46} = \mathbf{0.756 \text{ m.}}$$ **Ans.**

Substituting this value in equation (*i*), we get

$$k^2 = 0.756 \times 0.888 - 0.756^2 \qquad (\because \; a_1 = 0.756, L_1 = 0.888)$$
$$= 0.6713 - 0.5715 = 0.0998$$

∴ Mass moment of inertia of the rod $= mk^2$

$$= 60 \times 0.0998 = \mathbf{5.988\ kg\ m^2. Ans.}$$

Distance of C.G. of the rod from the small-end centre, a_1 = **0.756 m. Ans.**

2.22. TORSIONAL PENDULUM (OR TRIFILAR SUSPENSION)

It is used to find the moment of inertia of a disc or flywheel experimentally. The disc or flywheel whose moment of inertia is to be determined, is suspended with its axis vertical from a ceiling by three long flexible parallel wires of equal length L as shown in Fig. 2.13. The wires are attached at A, B and C which are the corners of an equilateral triangle. The points A, B and C are each at a distance 'a' from the axis of the disc (*i.e.*, from O). Hence $AO = BO = CO =$ 'a'. When the disc is twisted about its axis through a small angle θ and then released, it will oscillate with simple harmonic motion.

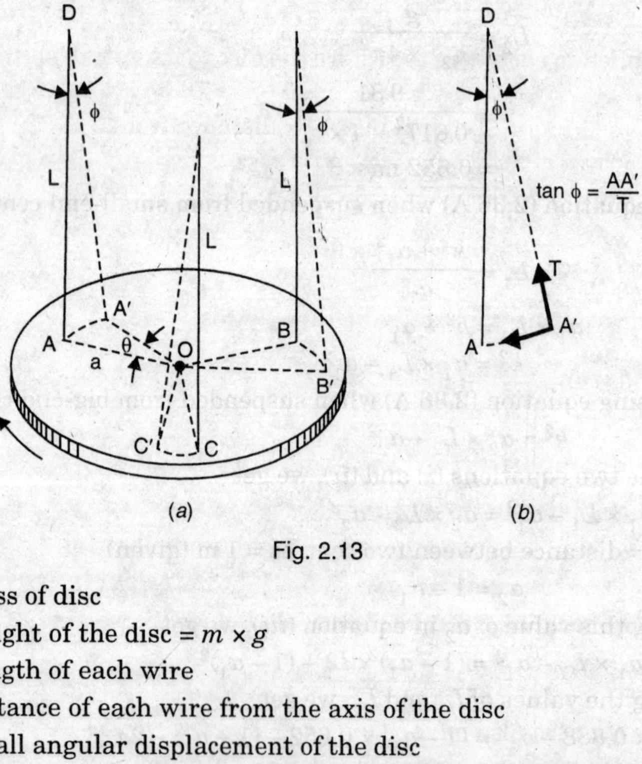

Fig. 2.13

Let m = Mass of disc

W = Weight of the disc = $m \times g$

L = Length of each wire

a = Distance of each wire from the axis of the disc

θ = Small angular displacement of the disc

ϕ = Corresponding angular displacement of each wire from the vertical

α = Angular acceleration

k = Radius of gyration.

For small angular displacement, $AA' = a\theta = L \times \phi$

$$\therefore \qquad \phi = \frac{a \times \theta}{L}$$

Since the three wires are attached symmetrically with respect to axis, the tension in each wire will be one-third to the weight of the disc.

$\therefore$ Tension in each wire, $T = \dfrac{W}{3} = \dfrac{m \times g}{3}$

In Fig. 2.13 (*b*), $\tan\phi = \dfrac{AA'}{A'D}$

$$= \frac{\text{Tension in } AA'}{\text{Tension in } A'D'}$$

$$= \frac{\text{Tension in } AA'}{T}$$

$\therefore$ Tension in $AA' = T \times \tan\phi$

$$= \text{Tension in each wire} \times \tan\phi$$

$$= \frac{m \times g}{3} \times \tan\phi \qquad \left(\because\ T = \frac{m \times g}{3}\right)$$

$$= \frac{m \times g}{3} \times \phi \qquad (\because\ \phi \text{ is very small } \therefore\ \phi \simeq \tan\phi)$$

$$= \frac{m \times g}{3} \times \frac{a \times \theta}{L} \qquad (\because\ \phi = a \times \theta/L)$$

$\therefore$ Torque applied to each wire to restore the disc to its initial equilibrium position *i.e.*, restoring torque

$$= (\text{Tension in } AA') \times \text{distance from } O$$

$$= \left[\frac{m \times g}{3} \times \frac{a \times \theta}{L}\right] \times a$$

$$= \frac{m \times g \times a^2 \times \theta}{3L}$$

$\therefore$ Total restoring torque applied to the three wires

$$= 3 \times \frac{m \times g \times a^2 \times \theta}{3L}$$

$$= \frac{mga^2\theta}{L} \qquad \text{...}(i)$$

But total torque applied to three wires

$$= I \times \alpha \text{ where } \alpha = \text{Angular acceleration}$$

$$= (mk^2) \times \alpha \qquad (\because\ I = mk^2) \qquad \text{...}(ii)$$

Equating the two torques given by equations (*i*) and (*ii*), we get

$$(mk^2) \times \alpha = \frac{m \times g \times a^2 \times \theta}{L}$$

From equation (*iii*), we also have

$$\frac{\theta}{\alpha} = \frac{L \times k^2}{g \times a^2}$$

But $\dfrac{\theta}{\alpha} = \dfrac{\text{Angular displacement}}{\text{Angular acceleration}}$

$$\therefore \quad \alpha = \frac{mga^2\,\theta}{mk^2\,L} = \left(\frac{g}{L} \times \frac{a^2}{k^2}\right) \times \theta \qquad \text{...}(iii)$$

where θ = angular displacement.

Hence the angular acceleration towards the equilibrium position is directly proportional to the angular displacement from equilibrium position. So the oscillation is simple harmonic.

$$\therefore \quad \frac{\text{Angular displacement}}{\text{Angular acceleration}} = \frac{\theta}{\alpha} = \frac{Lk^2}{g \times a^2}$$

We know that for simple harmonic motion, periodic time is given by,

$$t = 2\pi \sqrt{\frac{\text{Angular displacement}}{\text{Angular acceleration}}}$$

$$= 2\pi \sqrt{\frac{Lk^2}{g \times a^2}} = \frac{2\pi \times k}{a} \sqrt{\frac{L}{g}}$$

and frequency, $$n = \frac{1}{t} = \frac{a}{2\pi k} \sqrt{\frac{g}{L}} \quad \text{...(2.37)}$$

In equation (2.37), the values of L and a are known. From experiments, for fifty or hundred oscillations, the time is noted. Then the number of oscillations per second (*i.e.* frequency, n) is obtained. By substituting the values of n, a and L is equation (2.37), the radius of gyration k can be determined. Hence moment of inertia which is equal to mk^2 can be obtained.

2.23. MOMENT OF INERTIA FOR CONNECTING ROD

For finding the moment of inertia for connecting rod experimentally, first find the value of radius of gyration, k. This is obtained by using the same equation *i.e.* equation (2.37). But for connecting rod, two wires of equal lengths would be used. The wires would be attached to the connecting rod at equal distances from the mass centre (*i.e.* C.G. of the connecting rod), so that the tension is the same in each wire. Then $I = mk^2$.

Problem 2.9. *Determine the moment of inertia of a connecting rod of mass 60 kg about an axis through the C.G. The rod is suspended by two parallel vertical wires of lengths 2.5 m each. The wires are attached to the rod at points 400 mm from the C.G. The rod makes 50 oscillations in 165 seconds.*

Sol. Given :

$m = 60$ kg, $L = 2.5$ m ; Distance of each wire from C.G. of the rod, $a = 400$ mm $= 0.4$ m

No. of oscillations = 50 in time = 165 seconds

$\therefore$ No. of oscillation per second $= \frac{50}{165} = 0.303$

$\therefore$ Frequency, n = No. of oscillation/second = 0.303

Using equation (2.37), we get $n = \frac{a}{2\pi k} \sqrt{\frac{g}{L}}$

or $$0.303 = \frac{0.4}{2\pi \times k} \sqrt{\frac{9.81}{2.5}}$$

or $$k = \frac{0.4}{2\pi \times 0.303} \sqrt{\frac{9.81}{2.5}} = 0.416 \text{ m}$$

$\therefore$ Moment of inertia of the rod $= mk^2$

$= 60 \times 0.416^2$ kg m^2 = **10.383 kg m^2. Ans.**

HIGHLIGHTS

1. The rate of change of displacement with respect to time is known as velocity. The velocity will be in the same direction in which displacement takes place.
2. The rate of change of velocity with respect to time is known as acceleration. The acceleration is not necessarily in the same direction in which velocity or displacement take place.
3. When a body is moving along a circular path then its tangential (f_t) and radial accelerations (f_c) are given by $f_t = \frac{dv}{st}$ and $f_c = \frac{v^2}{r}$ or $\omega^2 r$.
4. Force is defined as an agent which produces or tends to produce the motion. It is given by

   $F = m \times a$ where a = acceleration.
5. The centripetal force (F_c) is given by, $F_c = m \times \frac{v^2}{r}$.
6. Mass moment of inertia (I) is given by, $I = mk^2$, where k = radius of gyration.
7. Angular momentum (or moment of momentum) is the product of mass moment of inertia and angular velocity (*i.e.* $I \times \omega$).
8. Torque (T) acting on a body is given by, $T = I \times \alpha$

   where I = Mass moment of inertia, and

   α = Angular acceleration.
9. The work done on a moving body is the product of the force acting on the body and its displacement in the direction of force.

   $\therefore$ Work done by a force $= F \times x$

   But work done on a rotating body is the product of the torque acting on the body and its angular displacement in the direction of torque.

   $\therefore$ Work done by a torque $= T \times \theta$.
10. The rate of doing work or work done per unit time is known as power.

$$\therefore \quad \text{Power} = \frac{\text{Work done}}{\text{Time}}$$

$$= \frac{F \times x}{t} \quad \text{or} \quad = \frac{T \times \theta}{t}$$

$$= F \times \frac{x}{t} \quad \text{or} \quad = T \times \frac{\theta}{t}$$

$$= F \times v \quad \text{or} \quad = T \times \omega.$$

11. The capacity of doing work is known as energy.

   (*i*) Potential Energy $= m \times g \times h$

   (*ii*) Strain Energy $= \frac{1}{2} \cdot s \cdot x^2$

   where s = Stiffness of spring = Force/Deflection

   x = Displacement of spring

   $= \frac{1}{2} \cdot q \cdot \theta^2$

   where q = Stiffness of torsional spring

   = Torque/angular deflection

   θ = Angular displacement

   (*iii*) K.E. $= \frac{1}{2} mv^2$.

12. The product of the force and time interval (when time interval is very short) is known as *impulse* and force is known as impulsive force.

13. The ratio of the relative velocities of colliding bodies after impact to their relative velocities before impact, is known as co-efficient of restitution which is denoted by 'e'.

14. A body is said to have simple harmonic motion (S.H.M.) if it moves or vibrates about a mean position such that its acceleration is always proportional to its distance from the mean position and is directed towards the mean position.

15. Period of S.H.M. is the time required for one complete oscillation $T = \dfrac{2\pi}{\omega}$.

16. Frequency is the no. of oscillations (or number of cycles) per second.

Frequency, $n = \dfrac{1}{T} = \dfrac{\omega}{2\pi}$.

17. For a simple pendulum, $T = 2\pi\sqrt{\dfrac{L}{g}}$ and $n = \dfrac{1}{2\pi}\sqrt{\dfrac{g}{L}}$.

18. For a compound pendulum, $T = 2\pi\sqrt{\dfrac{(k^2 + a^2)}{g \times a}}$ and $n = \dfrac{1}{2\pi}\sqrt{\dfrac{g \times a}{(k^2 + a^2)}}$

where k = radius of gyration about an axis passing through C.G.

a = distance of point of suspension from C.G.

19. Equivalent length of a simple pendulum for the same frequency as of a compound pendulum is given by $L = \dfrac{k^2 + a^2}{a}$

where L = Equivalent length of simple pendulum.

20. The moment of inertia of a disc (or flywheel) and a connecting rod is obtained by using equation,

$$n = \frac{a}{2\pi k}\sqrt{\frac{g}{L}}$$

where n = Frequency or no. of oscillations per second

L = Length of three vertical parallel wires for disc

= Length of two vertical parallel wires for connecting rod

a = Distance of each wire from the axis of oscillation

k = Radius of gyration about the axis passing through C.G.

The value of 'k' is obtained experimentally. Then $I = mk^2$.

EXERCISE 2

(A) Theoretical Questions

1. Define the terms : velocity, acceleration, tangential acceleration and centripetal acceleration.
2. Derive an expression for tangential acceleration and centripetal acceleration when a body is moving along a circular path.
3. Explain the following statements :

(*i*) When a particle moves along a straight path, the centripetal acceleration is zero.

(*ii*) When a particle moves along a circular path with a uniform velocity, tangential acceleration is zero.

4. Explain the following terms : Force ; weight, centripetal and centrifugal force, mass moment of inertia.
5. Prove that the angular momentum is the product of mass moment of inertia and angular velocity.
6. Define torque and prove that torque is equal to the product of mass moment of inertia and angular acceleration.
7. Find an expression for the work done by a force and by a torque.
8. Define and explain the terms : energy, strain energy, kinetic energy and potential energy.
9. What do you mean by impulse and impulsive force ?
10. Explain the law of conservation of momentum.
11. Define the terms : impact of two bodies and co-efficient of restitution.
12. Prove that the two perfect elastic bodies of equal masses exchange velocities in the case of direct central impact.
13. Find an expression for the loss of kinetic energy during the impact of two bodies.
14. Define and explain simple harmonic motion.
15. For a simple harmonic motion, prove that the velocity vector leads the displacement vector by 90° and acceleration vector leads the displacement vector by 180°.
16. Define the following terms which are used in case of S.H.M. : amplitude ; oscillation ; period of S.H.M. and frequency.
17. For a simple pendulum, prove that the time period is given by $T = 2\pi \sqrt{\frac{L}{g}}$

    whereas frequency (n) is given by $n = \frac{1}{2\pi} \sqrt{\frac{g}{L}}$.
18. Define a compound pendulum. Prove that the frequency of oscillation of a compound pendulum is given by $n = \frac{1}{2\pi} \sqrt{\frac{g \times a}{(k^2 + a^2)}}$

    where a = distance of C.G. from the point of suspension

    k = radius of gyration about an axis through C.G.
19. Find an expression for the equivalent length of a simple pendulum which has the same frequency as compound pendulum.
20. How will you find experimentally the moment of inertia of a (*i*) disc and (*ii*) connecting rod ?
21. What do you mean by torsional pendulum ? Prove that the frequency of oscillation of torsional pendulum is given by

$$n = \frac{a}{2\pi k} \sqrt{\frac{L}{g}}$$

where a = distance of each wire from the axis of oscillation

L = vertical length of each wire

k = radius of gyration.

(B) Numerical Problems

1. A tangential force of 900 N is acting on a shaft of diameter 20 mm. Find the work done by the force for one revolution of the shaft. [**Ans.** 56.54 Nm]

2. If in the above question, the shaft is rotating at 100 r.p.m., find the power of the shaft. [**Ans.** 94.25 W]

3. A flywheel, of mass 2500 kg and having a radius of gyration 1.414 m, looses its speed from 200 r.p.m. to 140 r.p.m. in 60 seconds. Calculate : (*i*) the retarding torque acting on it, (*ii*) change in the kinetic energy during the above period, and (*iii*) change in its angular momentum during the period. [**Ans.** (*i*) 523 Nm ; (*ii*) 558920 Nm ; (*iii*) 31400 Nm/s]

4. A cylinder rolls without slip along a horizontal level surface with a velocity of 30 cm/s. The mass of the cylinder is 1 kg and its radius is 10 cm. Find the total kinetic energy of the cylinder. Radius of gyration of cylinder is given as 8 cm. [**Ans.** 0.0738 Nm]

3

Velocity of Points in Mechanism

(A) INSTANTANEOUS CENTRE METHOD

3.1. Introduction

The velocity of various points in a mechanism, is determined by (*i*) the instantaneous centre method and (*ii*) relative velocity method, provided we know the velocity at one point. The velocity analysis is necessary for determining the acceleration of the points in the mechanism. The choice of the method depends upon the nature of the mechanism and accuracy required. In this chapter both the above methods are discussed.

3.2. Instantaneous Centre Method

This method is convenient and easy to apply in simple mechanism. Before discussing this method, let us consider the following types of motion that a body can have :

(*i*) Motion of translation,

(*ii*) Motion of rotation, and

(*iii*) Combination of motion of translation and motion of rotation.

3.2.1. Motion of Translation. If a body moves in such a way that all its particle move in parallel planes and travel the same distance, then the body is said to have motion of translation. When a rigid body is in translation, all the points of the body have the same velocity and same acceleration at any particular instant. The motion of the rigid link AB, from its initial position AB to A_1B_1, shown in Fig. 3.1 is an example of motion of translation.

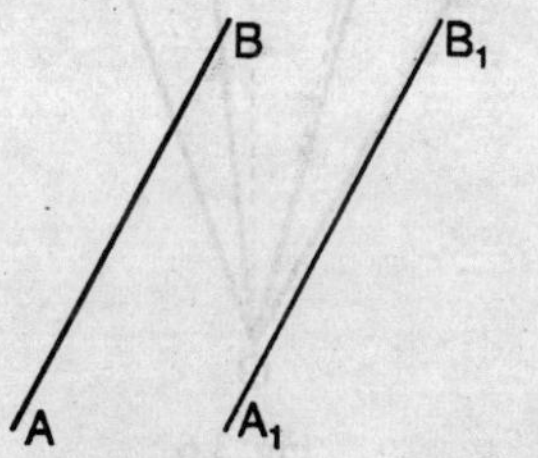

Fig. 3.1

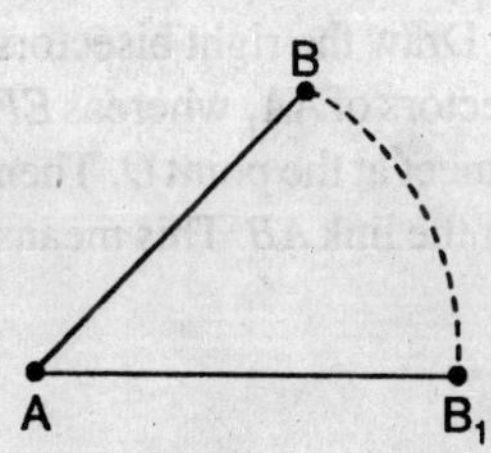

Fig. 3.2

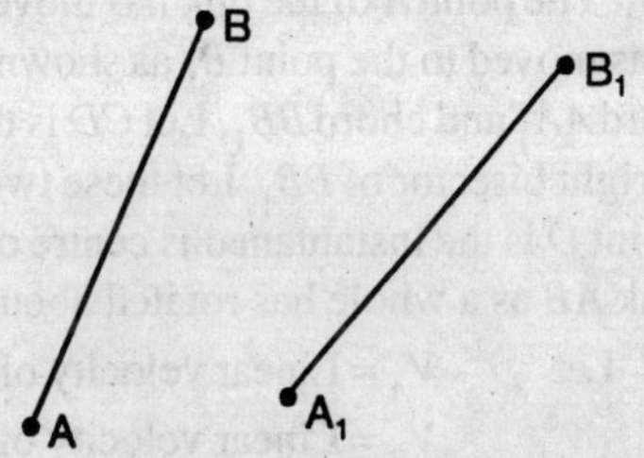

Fig. 3.3

3.2.2. Motion of Rotation. If a body rotates about a fixed point in such a way that all its particle move in circular path, the body is said to have the motion of rotation. The fixed point about which the body rotates is called the point of rotation and the axis, passing through the fixed point, is known as axis of rotation. The particles, lying on the axis of rotation, have zero velocity and zero acceleration. The motion of the link AB, from its initial position AB to AB_1, shown in Fig. 3.2 is an example of motion of rotation.

3.2.3. Combined Motion of Translation and Rotation. Consider a link AB, which moves from its initial position AB to A_1B_1 in a short interval of time as shown in Fig. 3.3. The link has neither entirely motion of translation nor entirely rotation, but a combination of the two. The motion of the link from the position AB to the position A_1B_1 may be regarded as to consists of :

1st Case

(*i*) A motion of entirely translation from the position AB to the position A_1B^* so that A_1B^* is parallel to AB as shown in Fig. 3.4, and

(*ii*) A motion of entirely rotation about A_1 from the position A_1B^* to the position A_1B_1.

2nd Case

(*i*) A motion of entirely rotation about A from the position AB to the position AB^* as shown in Fig. 3.5, and

(*ii*) A motion of entirely translation from the position AB^* to the position A_1B_1.

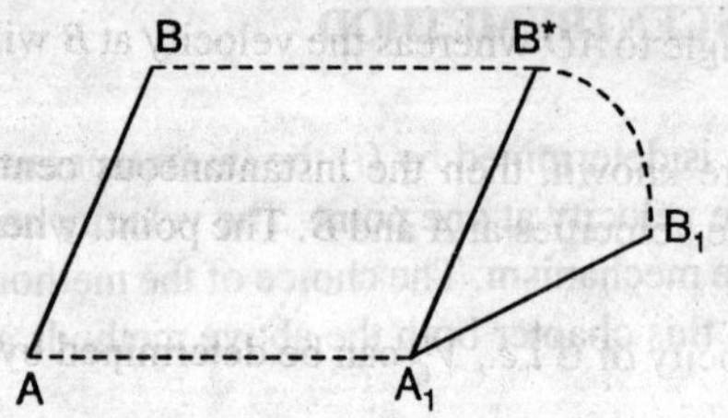

Fig. 3.4

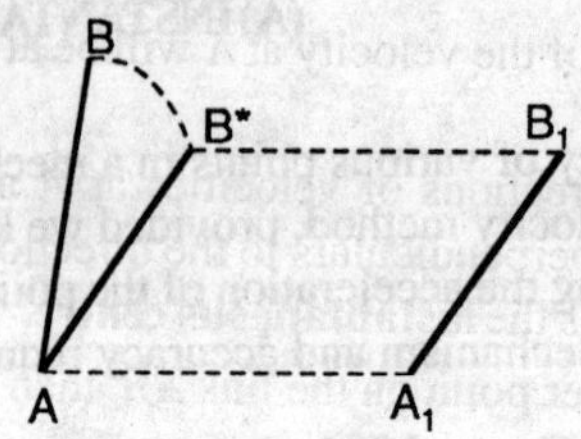

Fig. 3.5

3rd Case

The combined motion of translation and rotation of the link from its initial position AB to the position A_1B_1, may be assumed to be a motion of entirely rotation about a certain point. This point is known as **instantaneous centre of rotation.**

Location of the Instantaneous Centre. The position of the instantaneous centre of rotation is determined as given below :

Let the link AB in a short interval of time changes its position from AB to A_1B_1. The point A of the link has moved to point A_1 where the point B of the link has moved to the point B_1 as shown in Fig. 3.6. Draw the right bisectors of chord AA_1 and chord BB_1. Let CD is the right bisectors of AA_1 whereas EF is the right bisector of BB_1. Let these two bisectors meet at the point O. Then the point O is the instantaneous centre of rotation of the link AB. This means the link AB as a whole has rotated about O.

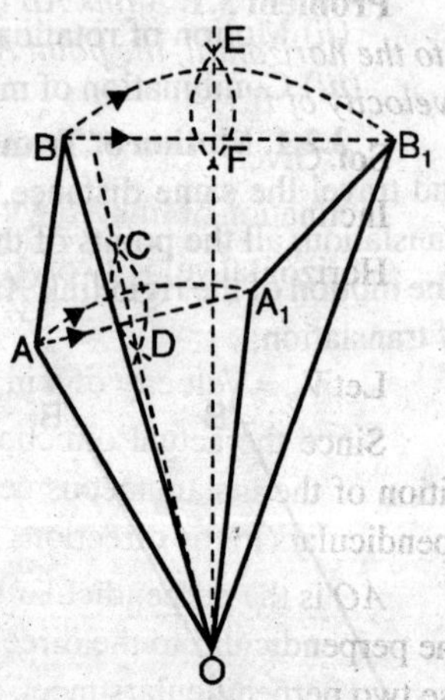

Fig. 3.6

Let V_A = Linear velocity of point A

V_B = Linear velocity of point B

ω = Angular velocity of link AB about O.

This means the angular velocity of point A and point B about O will also be ω.

We know, Linear velocity = Angular velocity $\times r$.

$\therefore$ Linear velocity of point A is given by

V_A = Angular velocity of point A about O

$\times$ Distance of A from centre of rotation (*i.e.*, from O)

$$= \omega \times AO$$

$$\omega = \frac{V_A}{AO} \quad \text{...(i)}$$

Similarly, the linear velocity of point B is given by

V_B = Angular velocity of point B about O × Distance of B from centre of rotation (*i.e.*, from O)

$= \omega \times BO$

$$\therefore \quad \omega = \frac{V_B}{BO} \quad \text{...(ii)}$$

Equating equations (*i*) and (*ii*), we get

$$\frac{V_A}{AO} = \frac{V_B}{BO} = \omega \quad \text{or} \quad \frac{V_A}{V_B} = \frac{AO}{BO} \quad \text{...(3.1)}$$

The direction of the velocity at A will be at right angle to AO whereas the velocity at B will be at right angle to BO.

Thus if the directions of velocities at A and B are known, then the instantaneous centre of AB is obtained by drawing perpendiculars to the directions of the velocities at A and B. The point, where these two perpendiculars, meet is the **instantaneous centre.**

If C is any other point on the link AB, then the velocity of C *i.e.*, V_C can be determined by

$$\frac{V_A}{AO} = \frac{V_B}{BO} = \frac{V_C}{CO} = \omega \quad \text{...(3.2)}$$

where CO = Distance of C from instantaneous centre O.

Problem 3.1. *A link AB is moving in a vertical plane. At a certain instant, when the link is inclined at 30° to the horizontal, the point A is moving horizontally at 4 m/s, while B is moving vertically upwards. Find the velocity of B.*

Sol. Given :

Inclination of link AB with horizontal = 30°

Horizontal velocity of A,

$V_A = 4$ m/s

Let V_B = Velocity of B in the vertical upward direction.

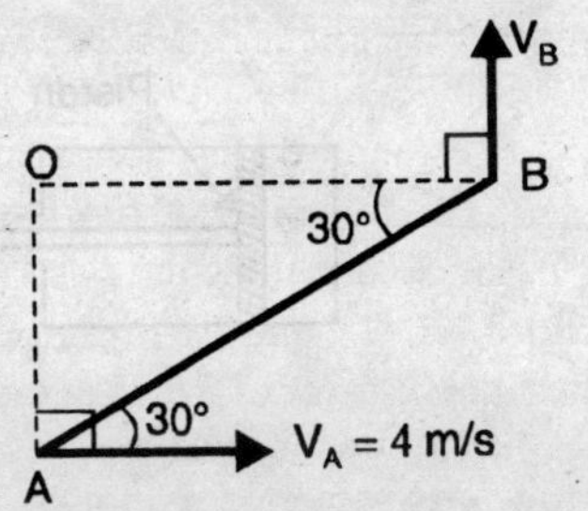

Fig. 3.7

Since the actual directions of motion of A and B are known, the position of the instantaneous centre can be easily determined by drawing perpendiculars to the directions of motions at A and B as shown in Fig. 3.7.

AO is the perpendicular to the direction of motion of A whereas BO is the perpendicular to the direction of motion of B. The point O, at which these two perpendiculars meet, is the instantaneous centre of the link AB.

Now using equation (3.1), we get

$$\frac{V_A}{V_B} = \frac{AO}{BO}$$

$$= \tan 30° \quad \left(\because \text{ In } \Delta AOB, \tan 30° = \frac{AO}{BO}\right)$$

or $$\frac{4}{V_B} = \frac{1}{\sqrt{3}}$$

$\therefore \quad V_B = 4 \times \sqrt{3} = 4 \times 1.732 =$ **6.928 m/s. Ans.**

Important Points

From the above problem, it is clear that

(*i*) If V_A is known in magnitude and direction and V_B in direction only, then velocity of point *B* may be determined in magnitude.

(*ii*) The velocity of any other point *C* lying on the link *AB* can be determined in magnitude and direction.

(*iii*) The magnitude of velocities of the points on a rigid link is inversely proportional to the distances of the points from the instantaneous centre and is ⊥ to the line joining the points to the instantaneous centre.

3.3. Analysis of Reciprocating Engine Mechanism By Instantaneous Centre Method

Fig. 3.8 shows a reciprocating engine mechanism in which *AB* is the connecting rod and *BC* is the crank. The crank *BC* is rotating at a uniform angular velocity in the clockwise direction about point *C*. The point *A* connected to the piston rod and connecting rod is having to and fro motion in the horizontal plane. The connecting rod is having a combined motion of translation and rotation.

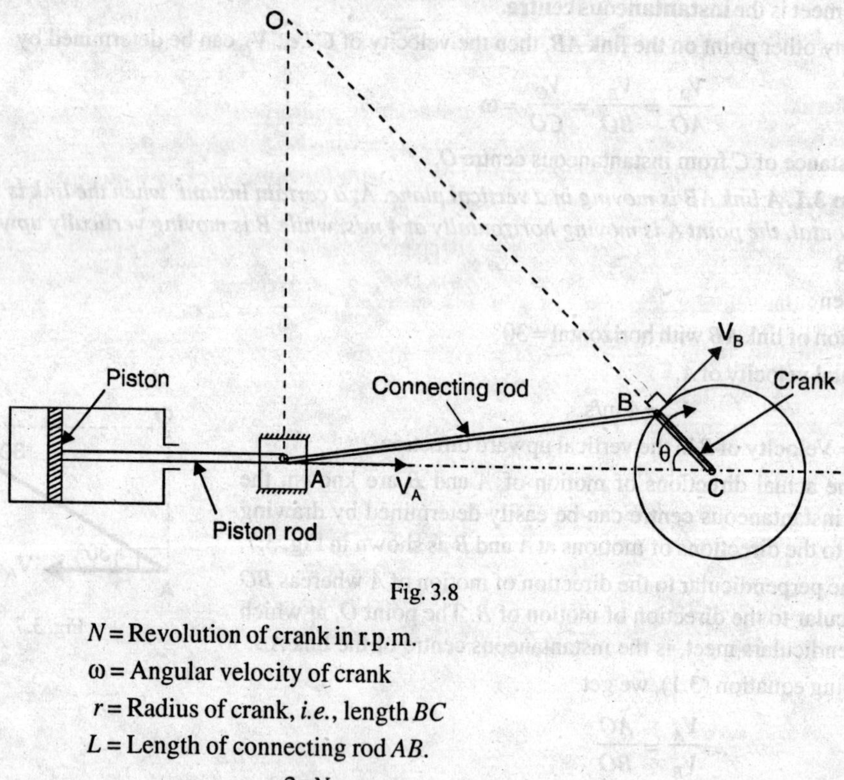

Fig. 3.8

Let N = Revolution of crank in r.p.m.

ω = Angular velocity of crank

r = Radius of crank, *i.e.*, length BC

L = Length of connecting rod AB.

Then $$\omega = \frac{2\pi N}{60}.$$

Linear velocity of point $B = V_B = \omega \times BC = \omega \times r$ ...(*i*)

The direction of the linear velocity at *B* is along the tangent at *B* to the crank circle. Hence V_B will be acting at right angles to *BC* as shown in Fig. 3.8. The velocity at *A* (*i.e.*, V_A) is in the horizontal direction acting along *AC* for the given position of the crank. Since the directions of V_A and V_B are known, the position of the instantaneous centre can be determined by drawing perpendiculars to the directions of the velocities at *A* and *B*. The two perpendiculars meet at *O*, which is the instantaneous centre for the connecting rod *AB*. Thus for the given position, the connecting rod can be considered as having a motion of entirely rotation about the instantaneous centre *i.e.*, about O.

Let ω_{AB} = Angular velocity of the connecting rod AB about O

$$\therefore \quad V_A = \omega_{AB} \times AO \quad ...(ii)$$

and

$$V_B = \omega_{AB} \times BO \quad ...(iii)$$

Equating equations (*i*) and (*iii*), we get

$$\omega \times r = \omega_{AB} \times BO$$

$$\therefore \quad \omega_{AB} = \frac{\omega \times r}{BO}$$

As the values of ω and r are known, hence the value of ω_{AB} will be obtained if length BO is measured to the scale.

Substituting the value of ω_{AB} in equation (*ii*), we get the velocity of the point A as

$$V_A = \frac{\omega \times r}{BO} \times AO \quad ...(3.3)$$

Hence by measuring to the scale the lengths AO and BO, the velocity at A can be determined from equation (3.3).

Problem 3.2. *The crank of a reciprocating engine is rotating at 210 r.p.m. The lengths of the crank and connecting rod are 20 cm and 100 cm respectively. Find the velocity of the point A (i.e., velocity of piston), when crank has turned through an angle of 45° with the horizontal as shown in Fig. 3.9.*

Sol. Given :

Rotation of crank, $N = 210$ r.p.m.

$\therefore$ Angular velocity of crank, $\omega = \frac{2\pi N}{60} = 2 \times \frac{22}{7} \times \frac{210}{60} = 22$ rad/s

Crank length or radius, $r = 20\text{ cm} = 0.20\text{ m}$

Length of connecting rod, $L = 100\text{ cm} = 1.0\text{ m}$

Angle turned by crank, $\theta = 45°$.

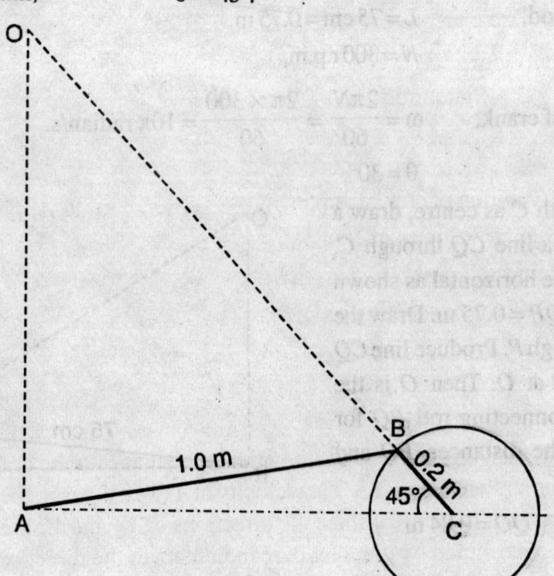

Fig. 3.9

The velocity of the point A is determined graphically as shown in Fig. 3.9. The procedure of finding velocity graphically is

1. With C as centre, draw a circle of radius 0.2 m.
2. Draw a line BC through C, making an angle of 45° with the horizontal and cutting the circle at B.
3. Draw a horizontal line AC through C.
4. With B as centre, cut off AB equal 1.0 m.
5. Produce the line CB. And draw a normal to the line AC passing through A. The intersection of the normal through A and produced line BC, gives the instantaneous centre O for the rod AB.

Measuring lengths AO and BO from the diagram, we get

$$AO = 1.15 \text{ m and } BO = 1.41 \text{ m.}$$

Now using equation (3.3), we get

$$V_A = \frac{\omega \times r \times AO}{BO} = \frac{22 \times 0.2 \times 1.15}{1.41} = \mathbf{3.58 \text{ m/s.}} \quad \textbf{Ans.}$$

Problem 3.3. *A steam engine has a crank radius of length 15 cm and connecting rod of length 75 cm as shown in Fig. 3.10. The crank CQ rotates in a clockwise direction with a constant speed of 300 r.p.m. Calculate the velocity of the piston P at the instant when the angle θ = 30°.* (AMIE, Summer 1980)

Sol. Given :

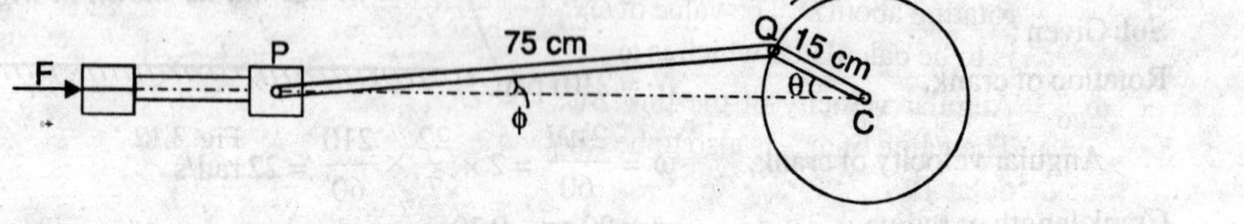

Fig. 3.10

Crank radius, $r = 15 \text{ cm} = 0.15 \text{ m}$

Length of connecting rod, $L = 75 \text{ cm} = 0.75 \text{ m.}$

Speed of crank, $N = 300 \text{ r.p.m.}$

$\therefore$ Angular velocity of crank, $\omega = \dfrac{2\pi N}{60} = \dfrac{2\pi \times 300}{60} = 10\pi \text{ radian/s.}$

Angle, $\theta = 30°.$

Graphical Method. With C as centre, draw a circle of radius 0.15 m. Draw a line CQ through C, making an angle of 30° with the horizontal as shown in Fig. 3.11. Take the distance $QP = 0.75$ m. Draw the normal to line PC passing through P. Produce line CQ to meet the normal through P at O. Then O is the instantaneous centre for the connecting rod PQ for the given position. Measure the distances PO and QO. They are :

$$PO = 0.495 \text{ m} \quad \text{and} \quad QO = 0.84 \text{ m}$$

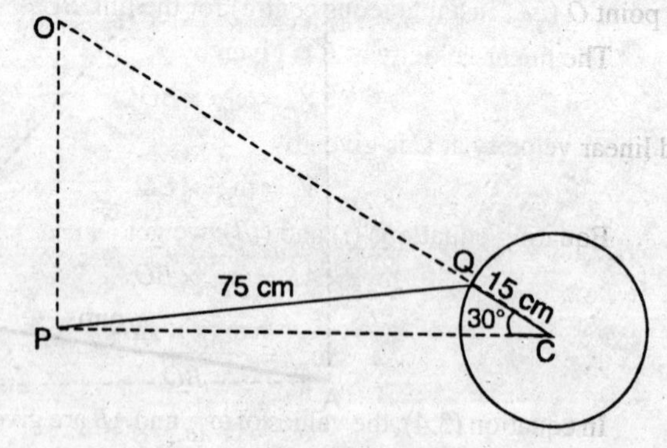

Fig. 3.11

Then the velocity of piston P is given by equation (3.3).

$$\therefore \quad \text{Velocity of piston} = \frac{\omega \times r \times PO}{QO} \qquad (\because AO = PO \text{ and } BO = QO \text{ in equation 3.3})$$

$$= \frac{10\pi \times 0.15 \times 0.495}{0.84} = \mathbf{2.766\ m/s.} \quad \textbf{Ans.}$$

3.4. Analysis of Four Bar Mechanism By Instantaneous Centre Method

The Fig. 3.12 shows a four bar mechanism, which consists of a fixed link AD, two movable links AB and CD rotating about points A and D respectively, and a connecting link BC (which is also known as coupler BC). Let the link AB is rotating at a uniform angular velocity, and it is required to find the corresponding motions of the two links BC and CD.

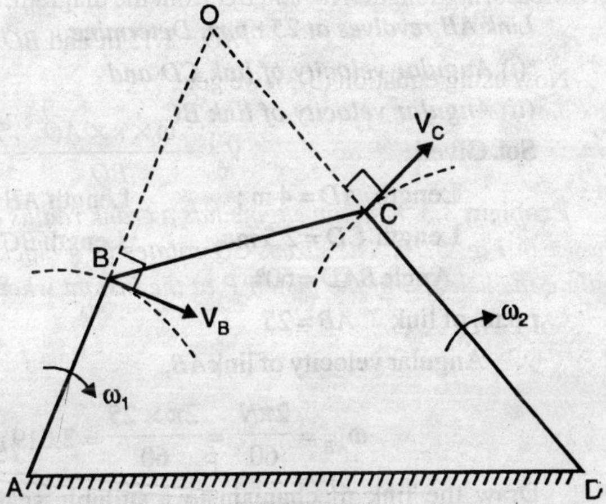

Fig. 3.12

Let ω_{AB} = Angular velocity of link AB, rotating about A in the clock-wise direction. The value of ω_{AB} is given as ω_1.

ω_{CD} = Angular velocity of link CD, rotating about D. The value of ω_{CD} is to be calculated which is ω_2.

ω_{BC} = Angular velocity of the link BC. The value of ω_{BC} is also to be calculated.

V_B = Linear velocity of point B in the direction perpendicular to AB.

$= \omega_{AB} \times AB$ ...(*i*)

V_C = Linear velocity of point C in the direction perpendicular to CD.

$= \omega_{CD} \times CD$ ...(*ii*)

The link AB and link CD are having motion of rotation, whereas the link BC is having motion of translation as well as rotation. The instantaneous centre of link BC for the given position is determined by drawing normals to the directions of velocity V_B and V_C. The normal to the direction V_B is the line AB whereas the normal to the direction V_C is line CD. Hence produce line AB and CD. The intersection of these lines gives the point O (*i.e.*, instantaneous centre) for the link BC.

The linear velocity at B is given by

$$V_B = \omega_{BC} \times BO \qquad \text{...(iii)}$$

and linear velocity at C is given by

$$V_C = \omega_{BC} \times CO \qquad \text{...(iv)}$$

Equating equations (*i*) and (*iii*), we get

$$\omega_{AB} \times AB = \omega_{BC} \times BO$$

$$\therefore \quad \omega_{BC} = \frac{\omega_{AB} \times AB}{BO} \qquad \text{...(3.4)}$$

In equation (3.4), the values of ω_{AB} and AB are given. The value of BO is obtained from Fig. 3.12. Hence ω_{BC} (angular velocity of link BC) can be determined. Again equating equations (*ii*) and (*iv*), we get

$$\omega_{CD} \times CD = \omega_{BC} \times CO$$

$$\therefore \quad \omega_{CD} = \frac{\omega_{BC} \times CO}{CD} \quad ...(3.5)$$

In the above equation value of CD is given, value of CO is obtained from Fig. 3.12 and the value of ω_{BC} is known from equation (3.4). Hence the value of ω_{CD} (angular velocity of link CD) can be determined.

Problem 3.4. *Fig. 3.13 shows a pin-joined four bar linkage having the following dimensions :*

Fixed link $AD = 4$ m ; Driving link $AB = 1.5$ m

Driven link $CD = 2.5$ m ; Connecting link $BC = 3$ m

Angle $BAD = 60°$.

Link AB revolves at 25 r.p.m. Determine :

(i) Angular velocity of link CD and

(ii) Angular velocity of link BC.

Sol. Given :

Length $AD = 4$ m ; Length $AB = 1.5$ m

Length $CD = 2.5$ m ; Length $BC = 3$ m

Angle $BAD = 60°$

r.p.m. of link $AB = 25$

$\therefore$ Angular velocity of link AB,

$$\omega_{AB} = \frac{2\pi N}{60} = \frac{2\pi \times 25}{60} = 2.619 \text{ rad/s.}$$

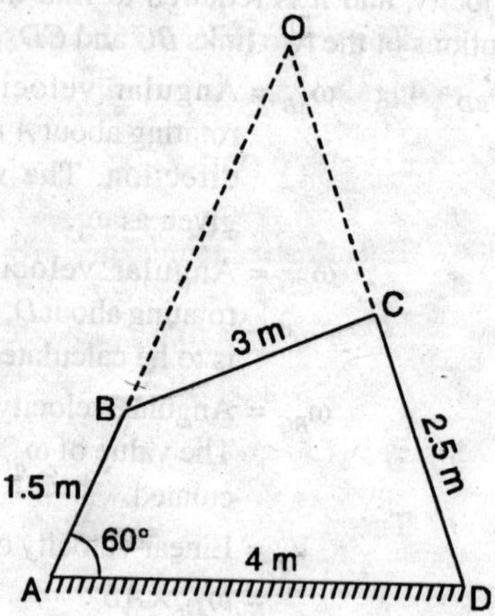

Fig. 3.13

Draw the link mechanism to a suitable scale as shown in Fig. 3.13. The instantaneous centre O of link BC is obtained by producing line AB and DC. Now measure the distance CO and BO from Fig. 3.13.

By measurement, $CO = 3$ m and $BO = 4.65$ m.

First calculate the angular velocity of link BC.

Let ω_{BC} = angular velocity of link BC.

Using equation (3.4), $\omega_{BC} = \frac{\omega_{AB} \times AB}{BO} = \frac{2.619 \times 1.5}{4.65}$ rad/s $\quad (\because AB = 1.5 \text{ m and } BO = 4.65)$

$= \mathbf{0.845}$ **rad/s. Ans.**

The angular velocity of link CD (*i.e.*, ω_{CD}) is obtained by using equation (3.5),

$$\therefore \quad \omega_{CD} = \frac{\omega_{BC} \times CO}{CD} = \frac{0.845 \times 3.0}{2.5} = \mathbf{1.014 \text{ rad/s.}} \quad \textbf{Ans.}$$

Problem 3.5. *In a four bar mechanism ABCD, points A and C are fixed points 30 cm apart and AB, CD are bars 60 cm and 70 cm long respectively, which are connected by a rod BD which is 50 cm long. If AB rotates with uniform speed of 60 r.p.m., determine : (i) velocity of D when AB is perpendicular to AC and also when it makes 10° on either side of the perpendicular, and (ii) the instantaneous centre of the bar BD and its angular velocities in the three positions.*

Sol. Given :

Length of fixed link AC = 30 cm

Length of link AB = 60 cm

Length of link CD = 70 cm

Length of link BD = 50 cm

Speed of link AB = 60 r.p.m.

∴ Angular velocity of link AB,

$$\omega_{AB} = \frac{2\pi N}{60} = \frac{2\pi \times 60}{60} = 2\pi \text{ rad/s.}$$

The link mechanism is drawn to a suitable scale as shown in Fig. 3.14. The instantaneous centre O for the rod BD is obtained by producing lines BA and DC.

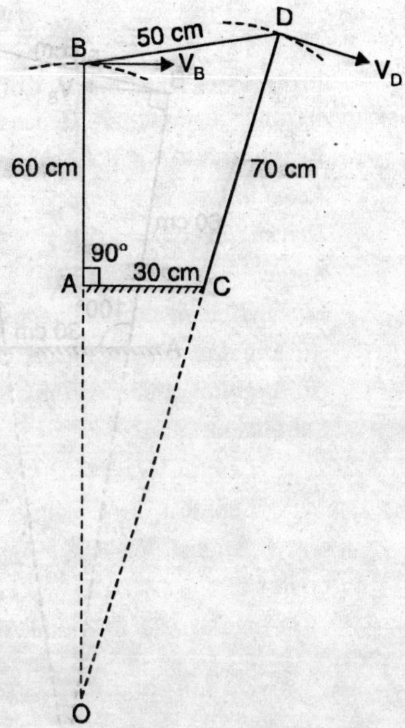

Fig. 3.14

By measurement,

$$OC = 107 \text{ cm and } AO = 103 \text{ cm}$$

∴ $$OD = OC + CD = 107 + 70 = 177 \text{ cm}$$

$$OB = OA + AB = 103 + 60 = 163 \text{ cm}$$

Now $$V_B = \omega_{AB} \times AB = 2\pi \times 60 \text{ cm/s}$$

$$= 376.8 \text{ cm/s} = 3.768 \text{ m/s.}$$

The velocity V_B is also equal to $\omega_{BD} \times OB$ where ω_{BD} = angular velocity of point B or of rod BD about O.

∴ $$V_B = \omega_{BD} \times OB = \omega_{BD} \times \frac{163}{100} \text{ m/s}$$

Equating the two values of V_B,

$$3.768 = \omega_{BD} \times \frac{163}{100}$$

∴ $$\omega_{BD} = \frac{3.768 \times 100}{163} = \mathbf{2.3116 \text{ rad/s.}} \quad \textbf{Ans.}$$

The velocity V_D at point D is given by

$$V_D = \omega_{BD} \times OD$$ Taking O as centre of rotation

$$= 2.3116 \times \frac{177}{100} \quad \left(\because \quad OD = 177 \text{ cm} = \frac{177}{100} \text{ m}\right)$$

$$= \mathbf{4.09 \text{ m/s.}} \quad \textbf{Ans.}$$

Second Case. When the angle between AB and AC = 100°. Draw the link mechanism to a suitable scale as shown in Fig. 3.15. The instantaneous centre O of the rod BD is obtained by producing lines BC and DC.

By measurement, $AO = 98$ cm and $OC = 97$ cm

∴ $$OB = OA + AB = 98 + 60 = 158 \text{ cm} = 1.58 \text{ m}$$

and $$OD = OC + CD = 97 + 70 = 167 \text{ cm} = 1.67 \text{ m}$$

Now $$V_B = \omega_{AB} \times AB$$ With A as centre of rotation

$$= \omega_{BD} \times OB$$ With O as centre of rotation

∴ $$\omega_{BD} \times OB = \omega_{AB} \times AB$$

or $$\omega_{BD} = \frac{\omega_{AB} \times AB}{OB} = \frac{2\pi \times 60}{158} = \mathbf{2.386 \text{ rad/s.}} \quad \textbf{Ans.}$$

For point D, $$V_D = \omega_{BD} \times OD$$ with O as centre of rotation

$$= 2.386 \times 1.67 = \mathbf{3.984 \text{ m/s.}} \quad \textbf{Ans.}$$

Third Case. When the angle between AB and BC = 80°. Draw the link mechanism to a suitable scale as shown in Fig. 3.16. This instantaneous centre O of the rod BD is obtained by producing lines BA and DC.

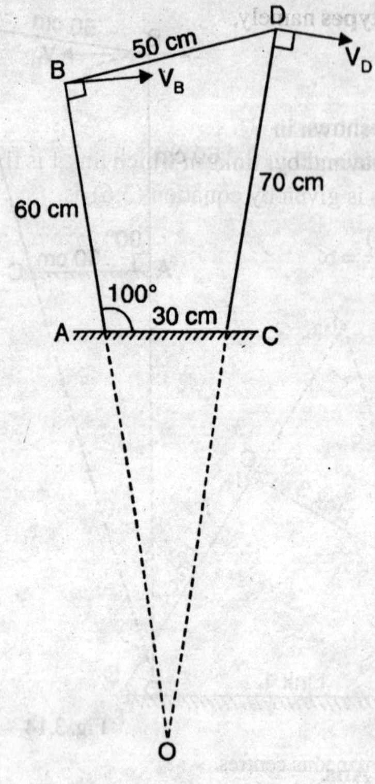

Fig. 3.15

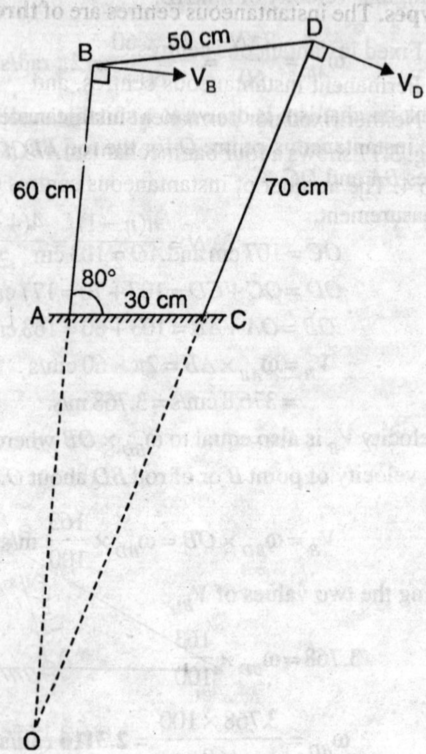

Fig. 3.16

By measurement, $AO = 100$ cm and $OC = 109$ cm

$\therefore \quad OB = OA + AB = 100 + 60 = 160$ cm or 1.6 m

and $\quad OD = OC + CD = 109 + 70 = 179$ cm or 1.79 m

Now $\quad V_B = \omega_{AB} \times AB$ with A as centre of rotation

$= \omega_{BD} \times OB$ with O as centre of rotation

$\therefore \quad \omega_{BD} \times OB = \omega_{AB} \times AB$

or $$\omega_{BD} = \frac{\omega_{AB} \times AB}{OB} = \frac{2\pi \times 60}{160} = \mathbf{2.356\ rad/s.}$$ **Ans.**

And $\quad V_D = \omega_{BD} \times OD$ with O as centre of rotation

$= 2.356 \times 1.79 = \mathbf{4.217\ m/s.}$ **Ans.**

3.5. Number and Types of Instantaneous Centres in a Mechanism

In a mechanism the number of instantaneous centres is the number of possible combinations of two links. Or in other words it is the number of combinations of n links taken two at a time which is mathematically equal to ${}^nC_2$.

But $$ {}^nC_2 = \frac{n!}{2!\,(n-2)!} = \frac{n(n-1)}{2}$$

If N = No. of instantaneous centres, then number of instantaneous centre is given by

$$N = \frac{n(n-1)}{2} \quad \text{...(3.6)}$$

where n = Number of links.

Types. The instantaneous centres are of **three types** namely,

1. Fixed instantaneous centres,
2. Permanent instantaneous centres, and
3. Neither fixed nor permanent instantaneous centres.

Fig. 3.17 shows a four bar mechanism *ABCD*, having four links in which link 1 is fixed. Here value of *n* is equal to 4. The number of instantaneous centres (*N*) is given by equation (3.6) as,

$$N = \frac{n(n-1)}{2} = \frac{4(4-1)}{2} = 6.$$

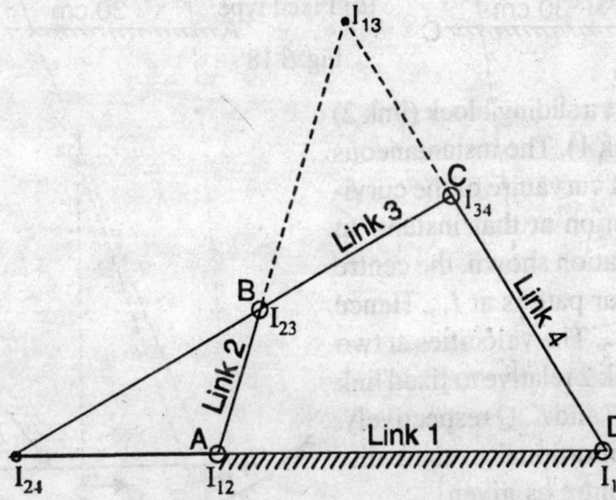

Fig. 3.17. Instantaneous centres.

Let us locate these six instantaneous centres. The instantaneous centre of two links such as link 1 and link 2 is represented by I_{12} and so on.

For the above four bar mechanism *ABCD*, shown in Fig. 3.17,

1. Instantaneous centres I_{12} and I_{14} are fixed instantaneous centres as for all the configurations of the mechanism, they remain at the same place.

2. Instantaneous centres I_{23} and I_{34} are permanent instantaneous centres as they move when the mechanism moves, but the joints are of permanent nature.

3. Instantaneous centres I_{13} and I_{24} are neither fixed nor permanent instantaneous centres as with the change of configuration of the mechanism, they also vary.

3.6. Method for Locating an Instantaneous Centre

1. The instantaneous centre for the two links, which are connected by a pin, will always be at the centre of the pin. This instantaneous centre will be of the permanent type as shown in Fig. 3.18 (*a*). But if one of the links is fixed, then the instantaneous centre will be of the fixed type as shown in Fig. 3.18 (*b*).

2. The instantaneous centre for the two links, which are having a pure rolling contact (*i.e.*, link 2 rolls without slipping upon the fixed link 1 which may be straight or curved) will always be on their point of contact as shown in Fig. 3.18 (*c*). The velocity of any point *P* on the link 2 relative to fixed link 1 will be perpendicular to $I_{12}P$ and is proportional to $I_{12}P$.

$$\therefore \qquad V_P \propto I_{12}P$$

Similarly the velocity at point *Q* will be

$$V_Q \propto I_{12}Q$$

$$\therefore \qquad \frac{V_P}{V_Q} = \frac{I_{12}P}{I_{12}Q}$$

3. The instantaneous centre for the two links, which are having sliding contact (*i.e.*, link 2 slides on the fixed link 1) is located as given below :

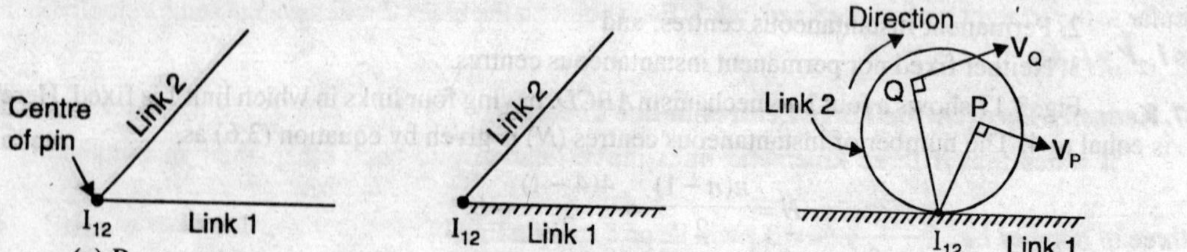

(*a*) Permanent type (*b*) Fixed type (*c*)

Fig. 3.18

(*a*) Fig. 3.18 (*d*) shows a sliding block (link 2) on a fixed curved surface (link 1). The instantaneous centre will be on the centre of curvature of the curvilinear path of the configuration at that instant. In Fig. 3.18 (*d*) for the configuration shown, the centre of curvature of the curvilinear path is at I_{12}. Hence instantaneous centre lies at I_{12}. The velocities at two points P and Q lying on the link 2 relative to fixed link 1 will be perpendicular to $I_{12}P$ and $I_{12}Q$ respectively. Their magnitude will be proportional to their distances from instantaneous centre as given

$$\frac{V_P}{V_Q} = \frac{I_{12}P}{I_{12}Q}.$$

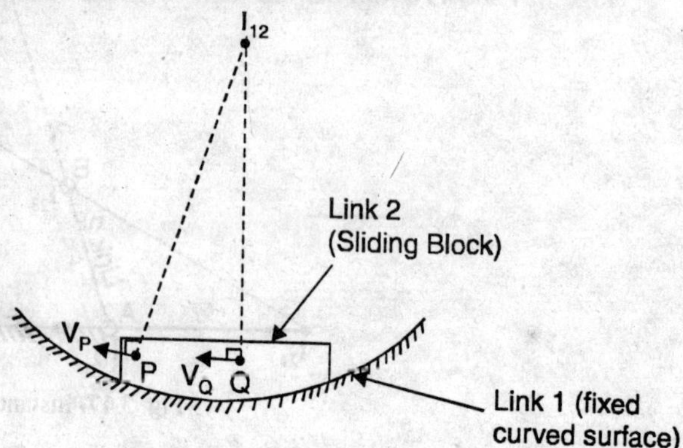

Fig. 3.18 (*d*)

Contrary of the above is also true *i.e.*, if the block 2 is fixed the link 1 will be moving relative to link 2 about the same instantaneous centre.

(*b*) Fig. 3.19 shows the sliding block link 2, moving on a fixed link 1 which has a constant radius of curvature. The instantaneous centre will lie at the centre of curvature *i.e.*, at the centre of the circle for all configurations of the links.

The velocities at two points P and Q lying on the link 2 relative to fixed link 1 will be perpendicular to $I_{12}P$ and $I_{12}Q$ respectively. The velocities are given by

$$\frac{V_P}{V_Q} = \frac{I_{12}P}{I_{12}Q}.$$

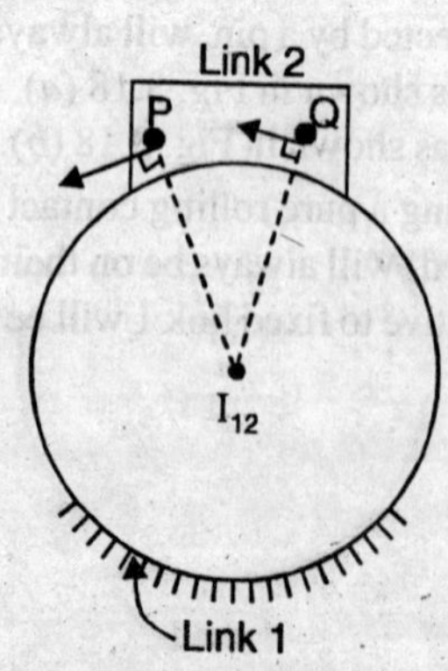

Fig. 3.19

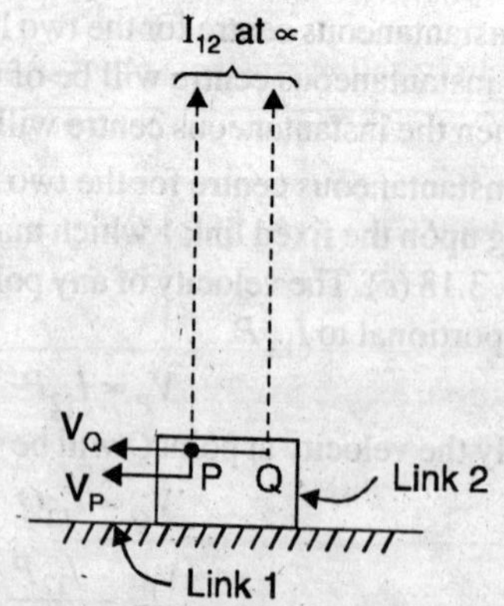

Fig. 3.20

(*c*) Fig. 3.20 shows the sliding block (link 2) moving on a fixed straight surface (link 1). We know that a straight line is a circle of radius equal to infinity. Hence the instantaneous centre will be at infinity perpendicular to the straight path of the fixed link 1. Each point on the block 2 will have the same velocity in this case (as $I_{12}P = I_{12}Q$).

3.7. Kennedy Theorem (or Three Centres-in-line Theorem)

It states that if three kinematic links move relatively to each other, their instantaneous centres $\left(\text{three in number } i.e., \frac{3 \times (3-1)}{2} = 3\right)$ will lie on a straight line.

3.8. Procedure of Locating Instantaneous Centres

Fig. 3.21 shows a four bar mechanism *ABCD*. The procedure of locating instantaneous centre is given as :

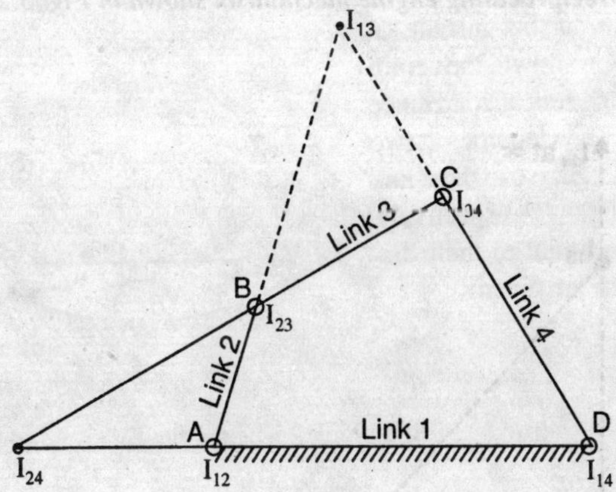

Fig. 3.21

1. Determine the number of instantaneous centres (*N*) by using the relation

$$N = \frac{n(n-1)}{2}$$

where n = Number of links = 4 (here)

$$\therefore \qquad N = \frac{4(4-1)}{2} = 6.$$

2. Prepare a list of all the instantaneous centres in a tabular form (known as book-keeping table).

Links	1	2	3	4
Instantaneous Centres	12 13 14	23 24	34	—

3. By inspection, locate fixed and permanent instantaneous centres. For the four bar mechanism shown in Fig. 3.21, these are I_{12}, I_{14}, I_{23} and I_{34}. The instantaneous centres I_{12} and I_{14} are fixed whereas I_{23} and I_{34} are permanent.

4. Locate the remaining neither fixed nor permanent instantaneous centres by Kennedy's theorem. Consider three links 2, 3 and 4. The instantaneous centre for links 2 and 3 is I_{23} and for links 3 and 4 is I_{34}. The

instantaneous centre for links 2 and 4 (*i.e.*, I_{24} according to Kennedy's theorem should be in a line of I_{23} and I_{34}. Hence draw a straight line passing through I_{23} and I_{34} and produce if necessary. Now consider the three links 2, 1 and 4. The instantaneous centres for links 1 and 2 is I_{12} whereas for links 1 and 4 is I_{14}. The instantaneous centre for links 2 and 4 (*i.e.*, I_{24}) according to Kennedy's theorem should be in line of I_{12} and I_{14}. Hence draw a straight line passing through I_{12} and I_{14} and produce. This line intersects the line already drawn through I_{23} and I_{34} at I_{24}. Thus the instantaneous centre I_{24} is located.

5. Similarly the instantaneous centre I_{13} is located by considering three links 1, 4 and 3. The instantaneous centre for links 1 and 4 is at I_{14} whereas the instantaneous centre for links 3 and 4 is at I_{34}. Draw a straight line passing through I_{14} and I_{34} and produce this line. According to Kennedy's theorem, the instantaneous centre for links 1 and 3 must lie on this straight line. Now consider three links 1, 2, and 3. The instantaneous centre for links 1 and 2 is at I_{12} whereas for links 2 and 3 is at I_{23}. The instantaneous centre for links 1 and 3 must lie on a straight line passing through I_{12} and I_{23}. Draw the straight line passing through I_{12} and I_{23}. Produce this line which intersects the straight line drawn through I_{14} and I_{34} at I_{13}. Thus the instantaneous centre I_{13} is located.

Problem 3.6. ***For the reciprocating engine mechanism shown in Fig. 3.22, locate all the instantaneous centres.***

Sol. Refer to Fig. 3.22.

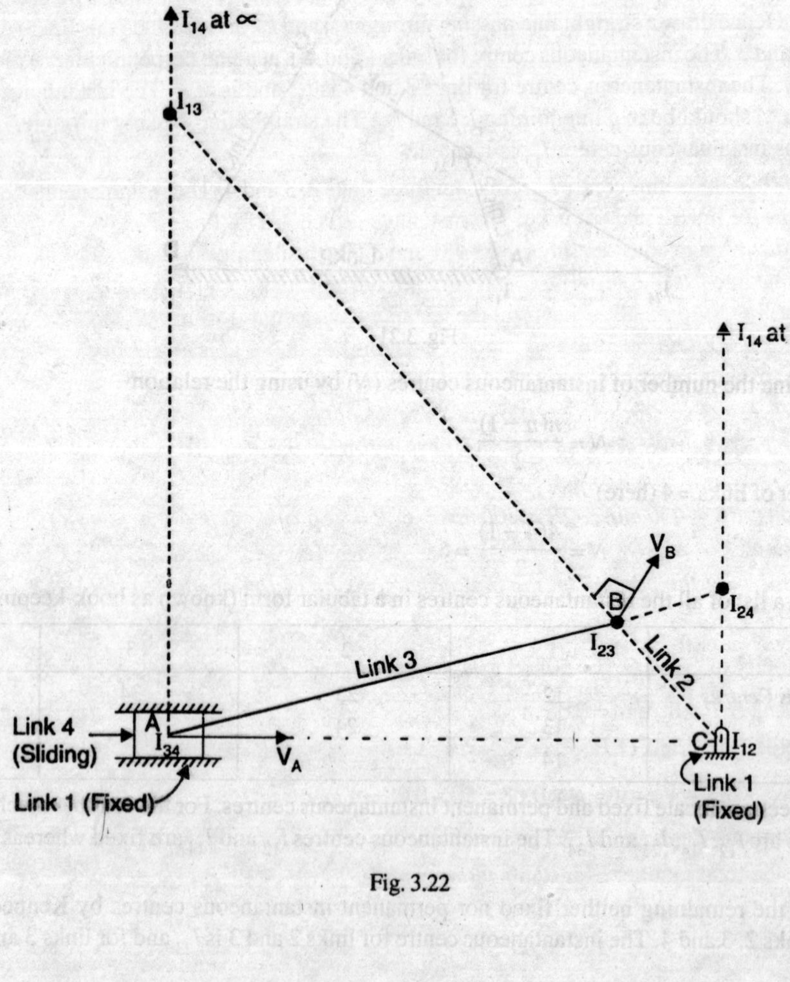

Fig. 3.22

No. of links, $n = 4$

∴ Number of instantaneous centres,

$$N = \frac{n(n-1)}{2} = \frac{4(4-1)}{2} = 6$$

List of all instantaneous centres in a tabular form (*i.e.*, book-keeping table) is given as :

Links	1	2	3	4
Instantaneous centres	12 13 14	23 24	34	—

By inspection, locate fixed and permanent instantaneous centres. These are I_{12}, I_{23} and I_{34} as shown in Fig. 3.22. The link 4 slides on a straight surface (link 1). Hence the instantaneous centre I_{14} will be at infinity (Refer Article 3.6 point 3 part *c*) perpendicular to the direction of V_A.

Using Kennedy's theorem, now locate the other two remaining neither fixed nor permanent instantaneous centres. Consider three links 1, 2 and 3. The instantaneous centre for links 1 and 2 is I_{12} whereas for links 2 and 3 is I_{23}. The instantaneous centre for links 1 and 3 (*i.e.*, I_{13}) according to Kennedy's theorem should be in a line of I_{12} and I_{23}. Hence draw a straight line passing through I_{12} and I_{23} and produce this line. Again consider the three links 1, 4 and 3. The instantaneous centre for links 1 and 4 is at a line perpendicular to the direction of V_A and is at infinity. The instantaneous centre for links 3 and 4 is I_{34} and is at A. The instantaneous centre for links 1 and 3 (*i.e.*, I_{13}) should be in a line joining I_{14} and I_{34}. The straight line passing through I_{12} and I_{23} cuts this line at I_{13}. Thus instantaneous centre I_{13} is located.

To locate the instantaneous centre I_{24}, consider three links 2, 3 and 4. The instantaneous centre for links 2 and 3 is I_{23} whereas for links 3 and 4 it is I_{34}. The instantaneous centre for links 2 and 4 (*i.e.*, I_{24}) should be in a line joining I_{34} and I_{23}. Again consider three links 4, 1 and 3. The instantaneous centre for links 4 and 1 *i.e.*, I_{14} will be at infinity on a line passing through C and parallel to line A and I_{14}. The instantaneous centre for links 1 and 2 is I_{12}. According to Kennedy's theorem, the instantaneous centre for links 2 and 4 (*i.e.*, I_{24}) should be on a line of I_{12} and I_{14} at ∞. The line joining I_{34} and I_{23} cuts this line at I_{24}. This locates the instantaneous centre I_{24}. Thus all the six instantaneous centres are located.

Problem 3.7. *Fig. 3.23 shows a mechanism for a wrapping machine in which pins O_1, O_2 and O_3 are fixed. The crank O_1P is a driving crank rotating at a uniform speed of 50 rad/s. The dimensions of the mechanism are :*

O_1P = 80 mm ; PR = 650 mm ; QR = 200 mm ; O_3R = 180 mm ; O_2T = 350 mm ; O_2S = 175 mm and QS = 125 mm.

Find the velocity of the point T on the bell-crank liver by instantaneous centre method.

Sol. Given :

Angular velocity of crank O_1P = 50 rad/s *i.e.*, $\omega_{O_1P} = 50$ rad/s

Length of crank O_1P = 80 mm = 0.08 m

∴ Linear velocity of crank O_1P,

$$V_{O_1P} = V_P = \omega_{O_1P} \times 0.08 = 50 \times 0.08 = 4 \text{ m/s.}$$

Location of Instantaneous Centres

1. First find the total number of instantaneous centres.

No. of links, $n = 6$

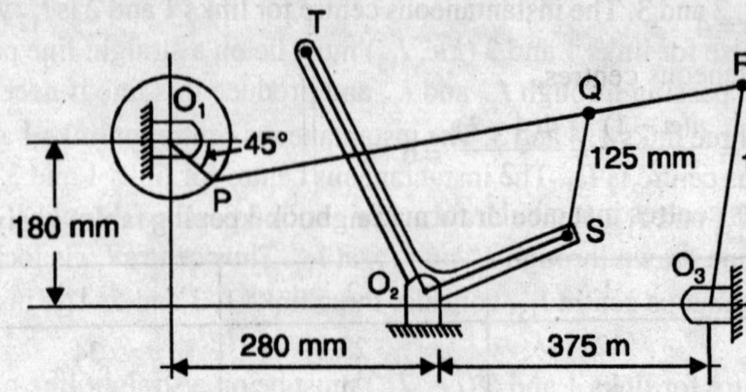

Fig. 3.23

∴ **No. of instantaneous centres,**

$$N = \frac{n(n-1)}{2} = \frac{6(6-1)}{2} = 15.$$

2. List all the 15 instantaneous centres in a tabular form (*i.e.*, book-keeping table) as given below :

Links	1	2	3	4	5	6
Instantaneous Centres	I_{12}	I_{23}	I_{34}	I_{45}	I_{56}	—
	I_{13}	I_{24}	I_{35}	I_{46}		
	I_{14}	I_{25}	I_{36}			
	I_{15}	I_{26}				
	I_{16}					

3. By inspection, locate the fixed and permanent instantaneous centres. These are I_{12}, I_{23}, I_{34}, I_{35}, I_{56}, I_{14} and I_{16} as shown in Fig. 3.24.

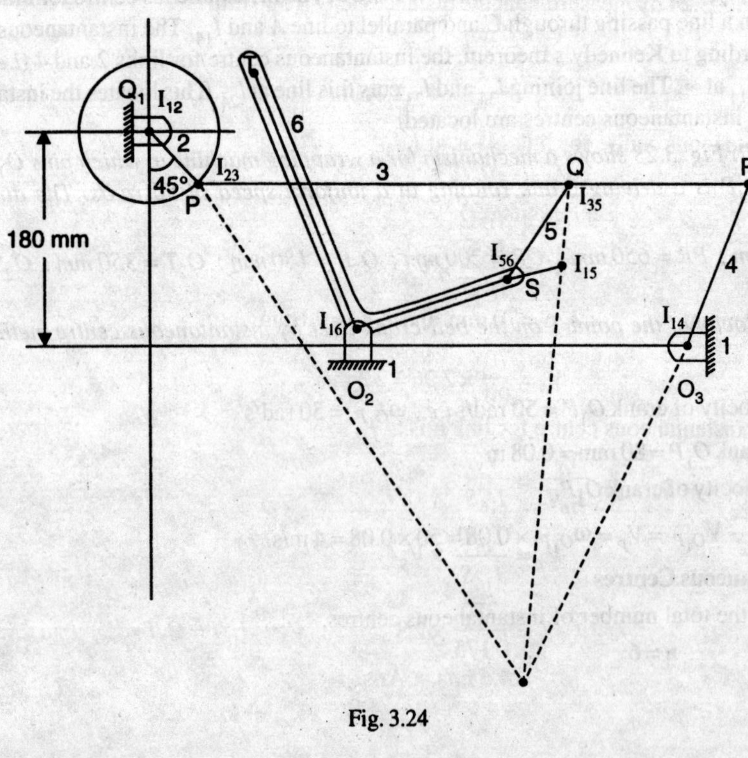

Fig. 3.24

4. Now using Kennedy's theorem, locate the remaining neither fixed nor permanent instantaneous centres. Consider three links 1, 2 and 3. The instantaneous centre for links 1 and 2 is I_{12} whereas for links 2 and 3 is I_{23}. The instantaneous centre for links 1 and 3 (*i.e.*, I_{13}) must lie on a straight line passing through I_{12} and I_{23}. Hence draw a straight line passing through I_{12} and I_{23} and produce this line if necessary.

5. Again consider the three links 1, 4 and 3. The instantaneous centre for links 1 and 4 is I_{14}, whereas for links 3 and 4, the instantaneous centre is I_{34}. The instantaneous centre for links 1 and 3 (*i.e.*, I_{13}) must lie on a straight line passing through I_{34} and I_{14}. Hence draw a straight line passing through I_{34} and I_{14} and produce this line, which intersect the line drawn through I_{12} and I_{23} at I_{13}. Thus centre I_{13} is located.

6. To locate the instantaneous centre I_{15}, consider three links 1, 3 and 5. The instantaneous centres I_{13} and I_{35} are already located.

The instantaneous centre for links 1 and 5 (*i.e.*, I_{15}) must be on a straight line passing through I_{13} and I_{35}. Now consider three links 1, 5 and 6. The instantaneous centre for links 1 and 6 is I_{16} whereas for links 5 and 6 is I_{56}. According to Kennedy's theorem, the instantaneous centre for links 1 and 5 (*i.e.*, I_{15}) should be on a straight line passing through I_{16} and I_{56}. Join the centres I_{16} and I_{56} by a straight line and produce this line. This line intersects the line already drawn through I_{13} and I_{35} at I_{15}. Thus centre I_{15} is located. For the given problem, there is no need of finding other instantaneous centres.

Velocity at Point T

Let V_T = velocity of point T

V_Q = Velocity of point Q

V_S = Velocity of point S.

The instantaneous centre for link PQ is at I_{13}.

$$\therefore \quad \frac{V_Q}{I_{13}Q} = \frac{V_P}{I_{13}P}$$

$$\therefore \quad V_Q = \frac{I_{13}Q}{I_{13}P} \times V_P$$

But from Fig. 3.24 by measurement the distances $I_{13}Q = 500$ mm and $I_{13}P = 675$ mm

Also $V_P = 4$ m/s

$$\therefore \quad V_Q = \frac{500}{675} \times 4 = 2.96 \text{ m/s}$$

The instantaneous centre for link QS is at I_{15}

$$\therefore \quad \frac{V_S}{I_{15}S} = \frac{V_Q}{I_{15}Q}$$

$$\therefore \quad V_S = \frac{I_{15}S}{I_{15}Q} \times V_Q$$

By measurement, $I_{15}S = 70$ mm and $I_{15}Q = 90$ mm

$$\therefore \quad V_S = \frac{70}{90} \times 2.96 = 2.3 \text{ m/s} \qquad (\because \; V_Q = 2.96 \text{ m/s})$$

Again the instantaneous centre for link 6 is at I_{16}

$$\therefore \quad \frac{V_T}{I_{16}T} = \frac{V_S}{I_{16}S}$$

$$\therefore \quad V_T = \frac{I_{16}T}{I_{16}S} \times V_S$$

$$= \frac{.350}{0.175} \times 2.3 \qquad (\because \; I_{16}T = O_2T = 0.350 \text{ m}, I_{16}S = O_2S = 0.175 \text{ m})$$

$$= \mathbf{4.6 \text{ m/s}}. \quad \textbf{Ans.}$$

(B) RELATIVE VELOCITY METHOD

3.9. Relative Velocity Method

The relative velocity method is very useful and rapid method for determining linear and angular velocities in the mechanism. This method has the additional advantage over instantaneous centre method that it can be used for making acceleration analysis. Before discussing this method, let us consider the following important cases :

(*i*) Relative velocity of two bodies moving in straight lines,

(*ii*) Relative velocity of any point on a link with respect to another point.

3.9.1. Relative Velocity of Two Bodies Moving in Straight Lines. Fig. 3.25 (*a*) shows the two bodies *A* and *B* which are moving along parallel lines in the same direction.

Let v_A = Absolute velocity of *A*

v_B = Absolute velocity of *B*

Let $v_A > v_B$. To find the relative velocity of A with respect to *B*, make the boody *B* stationary. This is possible if a velocity equal to v_B is applied on the body *B* in the opposite direction. Then we shall have to apply a velocity equal to v_B on the body *A* also in the opposite direction. As $v_A > v_B$, hence the resultant velocity will be in the direction of v_A and equal to $(v_A - v_B)$. This resultant velocity is the relative velocity of *A* with respect to *B*.

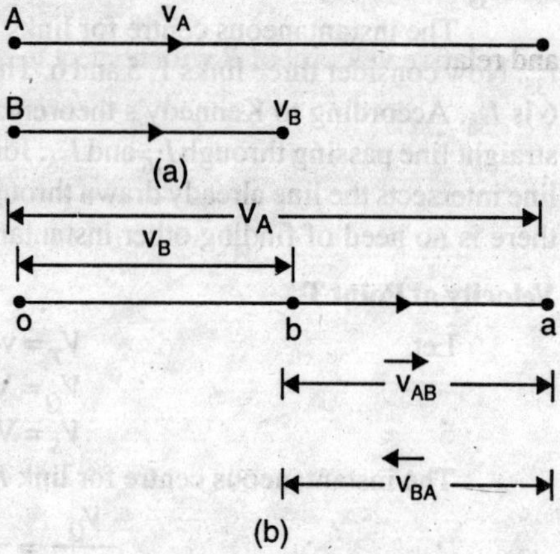

Fig. 3.25

The relative velocity of *A* with respect to *B* is also given by,

$$v_{AB} = \text{Vector difference of } v_A \text{ and } v_B$$

$$= \vec{v}_A - \vec{v}_B \quad ...(i)$$

Graphically, the relative velocity of *A* with respect to *B* can also be determined as :

(*i*) Take any point o. From o, drow *oa* parallel to velocity v_A and take $oa = v_A$ to some suitable scale.

(*ii*) From o, also draw *ob* parallel to velocity v_B. Take $ob = v_B$. Join *b* to *a*.

Then *ba* represents the relative velocity of *A* with respect to body *B* in magnitude and direction. Hence relative velocity of *A* with respect to *B* may also be written in the vector form as

$$\vec{ba} = \vec{oa} - \vec{ob}.$$

From Fig. 3.25 (*b*), the relative velocity of *A* with respect to *B* may be written in the vector form as

$$\vec{ba} = \vec{oa} - \vec{ob}$$

Please note that the velocity of *A* with respect to *B* (*i.e.,* v_{AB}) in vector form is written as $\vec{ba}$ and not $\vec{ab}$. And mathematically it is written as v_{AB}. Hence to find v_{AB} in magnitude and direction, start from point *b* and reach towards point *a*.

Similarly the relative velocity of *B* with respect to *A* is given by

$$v_{BA} = \text{Vector difference of } v_B \text{ and } v_A$$

$$= \vec{v}_B - \vec{v}_A \quad ...(ii)$$

The relative velocity of *B* with respect to *A* in vector form is written as

$$\vec{ab} = \vec{ob} - \vec{oa}$$

From equations (*i*) and (*ii*), it is clear that the relative velocity of point *A* with respect to *B* (v_{AB}) and the relative velocity of point *B* with respect to *A* (v_{BA}) are equal in magnitude but opposite in direction *i.e.,*

$$v_{AB} = -v_{BA} \quad \text{or} \quad \vec{ba} = -\vec{ab}$$

Relative Velocity of Two Bodies Moving along Inclined lines

Fig. 3.26 (*a*) shows the two bodies *A* and *B* moving along inclined lines with absolute velocities v_A and v_B respectively. The vector *oa* in Fig. 3.26 (*b*) represents the velocity v_A in magnitude and direction, whereas the vector *ob* represents the velocity v_B in magnitude and direction. Join points *b* and *a*. Then relative velocity of *A* with respect to *B* is written as v_{AB} and is given by,

$$v_{AB} = \text{Vector } ba$$

and relative velocity of *B* with respect to *A* is given by,

$$v_{BA} = \text{Vector } ab.$$

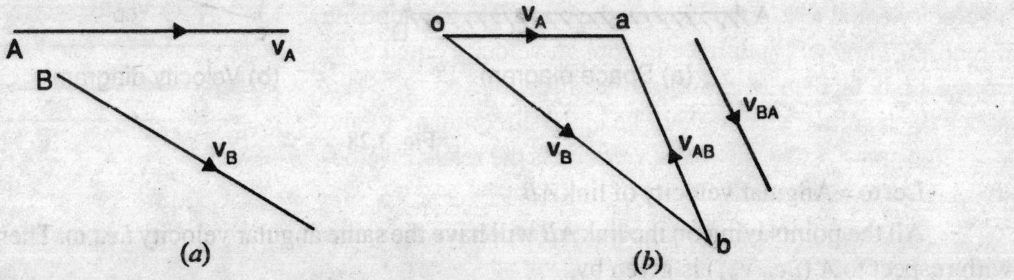

Fig. 3.26

3.9.2. Relative Velocity of a Rigid Link. Fig. 3.27 shows a rigid link *AB*. In case of a rigid link, the velocity of any point on the link with respect to another point on the same link is always perpendicular to the line joining these points on the space diagram. Hence the velocity of *B* with respect to *A* will be perpendicular to *AB* as shown in Fig. 3.27 (*a*).

Fig. 3.27 (*b*) shows the velocity diagram for the rigid link *AB*. The relative velocity of *B* with respect to *A* (*i.e.,* v_{BA}) is represented by the vector *ab* which is perpendicular to the line *AB*.

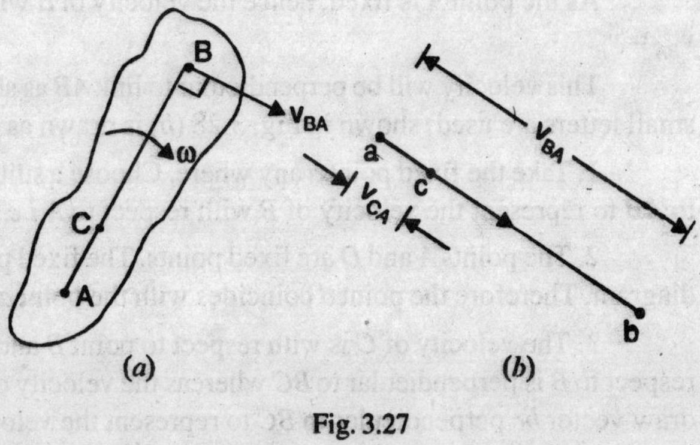

Fig. 3.27

$$\therefore \quad v_{BA} = \vec{ab} = \omega \times AB \qquad \text{...(i)}$$

where ω = Angular velocity of link *AB* about *A*.

Now the velocity of *C* (which is any point on link *AB*) with respect to *A* (*i.e.,* v_{CA}) is perpendicular to *AC* and is represented by vector *ac*.

$$\therefore \quad v_{CA} = \vec{ac} = \omega \times AC \qquad \text{...(ii)}$$

Now from equations (*i*) and (*ii*), we get

$$\frac{v_{CA}}{v_{BA}} = \frac{\vec{ac}}{\vec{ab}} = \frac{\omega \,.\, AC}{\omega \,.\, AB} = \frac{AC}{AB}.$$

From the above equation it is clear that the point C divides the link AB in the same ratio as the corresponding point of c in the velocity diagram divides the vector ab (*i.e.*, point c divides the vector ab).

3.10. Velocities in Four Bar Chain

A four bar chain $ABCD$ is shown in Fig. 3.28. The link AD is fixed and link AB is rotating anti-clockwise with a known angular velocity.

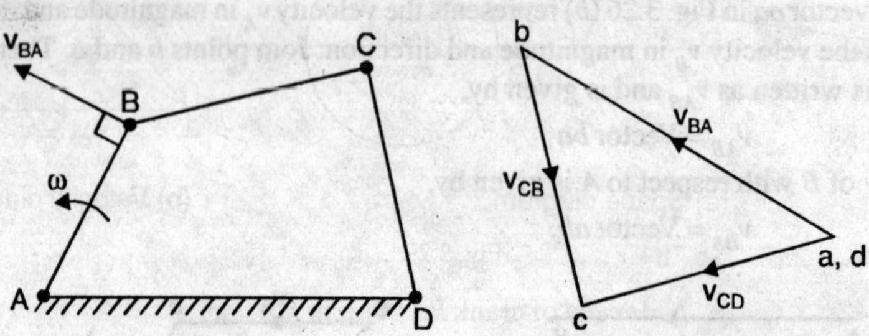

(a) Space diagram (b) Velocity diagram

Fig. 3.28

Let ω = Angular velocity of link AB.

All the points lying on the link AB will have the same angular velocity *i.e.*, ω. Then linear velocity of B with respect to A (*i.e.*, v_{BA}) is given by,

$$v_{BA} = \text{Angular velocity of link } AB \times \text{Length } AB = \omega \times AB.$$

As the point A is fixed, hence the velocity of B with respect to A will be equal to velocity of B also *i.e.*, $v_{BA} = v_B$.

This velocity will be perpendicular to link AB as shown in Fig. 3.28 (*a*). The velocity diagram (in which small letters are used) shown in Fig. 3.28 (*b*) is drawn as discussed below :

1. Take the fixed point a any where. Choose a suitable velocity scale and draw vector ab perpendicular to AB to represent the velocity of B with respect to A *i.e.*, v_{BA}.

2. The points A and D are fixed points. The fixed points of the mechanism are coincident in the velocity diagram. Therefore the point d coincides with the point a in the velocity diagram.

3. The velocity of C is with respect to point B and also with respect to point D. The velocity of C with respect to B is perpendicular to BC whereas the velocity of C with respect to D is perpendicular to CD. From b draw vector bc perpendicular to BC to represent the velocity of C with respect to B *i.e.*, v_{CB}. Also from a or d draw vector dc perpendicular to DC to represent the velocity of C with respect to D *i.e.*, v_{CD}. The vectors bc and dc (or ac) intersect at c. Then abc represents the velocity diagram for the given four bar chain.

Now, we can determine the angular velocities of links BC and CD if required as :

$$\text{Angular velocity of } BC = \frac{\text{Vector } bc}{\text{Length } BC}$$

$$\text{and angular velocity of } CD = \frac{\text{Vector } cd}{\text{Length } CD}.$$

Note. (*i*) For velocity diagram small letters are used.

(*ii*) For spece diagram capital letters are used.

3.11. Velocities in Slider Crank Mechanism

Fig. 3.29 shows a slider crank mechanism, in which the crank *AB* is rotating clockwise with a given speed. The slider *C* is moving along the line *CA*. The connecting rod is represented by *CB*.

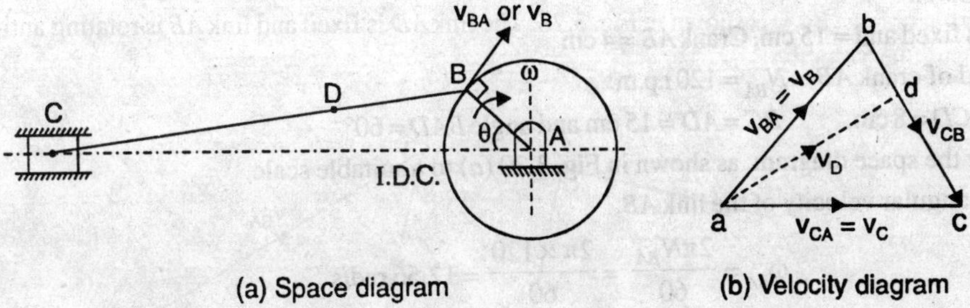

Fig. 3.29

Let N = speed of crank BA in r.p.m.

ω = angular velocity of crank BA

$$= \frac{2\pi N}{60}.$$

Then linear velocity of *B* with respect to *A* (or only velocity of *B* as point *A* is fixed) is given by,

$$v_{BA} \quad \text{or} \quad v_B = \omega \times AB$$

This velocity will be perpendicular to *AB* as shown in Fig. 3.29 (*a*). The velocity diagram shown in Fig. 3.29 (*b*) is drawn as discussed below:

1. Take the fixed point *a* anywhere. Choose a suitable velocity scale and draw vector *ab* perpendicular to *AB* to represent the velocity of *B* with respect to *A i.e.*, v_{BA} such that vector $ab = v_{BA}$.

2. The velocity of *C* is with respect to *B* and also with respect to *A*. The velocity of *C* with respect to *B* is perpendicular to *BC* whereas the velocity of *C* with respect to *A* is in horizontal direction along *CA*. Hence from *b* draw vector *bc* perpendicular to *BC* to represent the velocity of *C* with respect to *B i.e.*, v_{CB}. But relative to *A*, *C* can move horizontally, hence *c* must lie somewhere on a horizontal line through *a*. Hence from point *a* draw vector *ac* horizontally. The vectors *bc* and *ac* intersect at *c*. Then *abc* represents the velocity diagram for the given slider crank mechanism.

3. The absolute velocity of any point *D* on the connecting rod *BC* can be obtained by dividing vector *bc* such that

$$\frac{bd}{bc} = \frac{BD}{BC}$$

$$\therefore \quad bd = \frac{BD}{BC} \times bc$$

As *BD*, *BC* and vector *bc* are known, hence *bd* can be calculated. Then location of point *d* on vector *bc* is obtained. Now join point *a* to *d*. Then vector *ad* represents in magnitude and direction the absolute velocity of the point *D*.

If *D* is the mid-point of connecting rod *BC*, then corresponding point *d* will also be the mid-point of vector *bc*.

Notes. (*i*) Small letters are used in velocity diagram.

(*ii*) All the fixed points in a mechanism, are located at a common point in the velocity diagram.

(*iii*) Vector *ab* represents the velocity of *B* with respect to *A* and not velocity of *A* with respect to *B*.

(*iv*) The absolute velocities of the points are measured from *a*, the fixed point of the velocity diagram.

(*v*) The velocity diagram should be started from such a link whose velocity in magnitude and direction are known.

Problem 3.8. *In a four bar chain ABCD, AD is fixed and is 15 cm long. The crank AB is 4 cm long and rotates at 120 r.p.m. clockwise, while the link CD (= 8 cm) oscillates about D. BC and AD are of equal length. Find the angular velocity of link CD when angle BAD = 60°.* (AMIE, Summer 1984)

Sol. Given :

AD is fixed and = 15 cm, Crank AB = 4 cm

Speed of crank AB, $N_{BA} = 120$ r.p.m.

Link $CD = 8$ cm, $BC = AD = 15$ cm and angle $BAD = 60°$

Draw the space diagram, as shown in Fig. 3.30 (*a*) to a suitable scale.

Now angular velocity of the link AB,

$$\omega_{BA} = \frac{2\pi N_{BA}}{60} = \frac{2\pi \times 120}{60} = 12.56 \text{ rad/s}$$

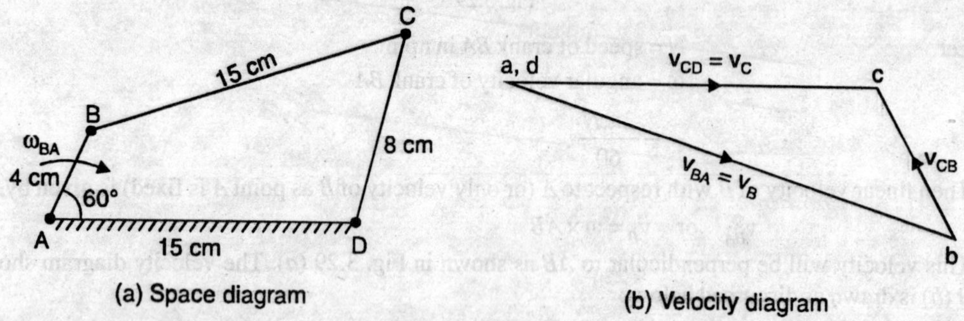

Fig. 3.30

∴ Velocity of B with respect to A (*i.e.*, v_{BA}) or simply velocity of B as A is fixed (*i.e.*, v_B) is given by

$$v_{BA} \text{ (or } v_B) = \omega_{BA} \times \text{Length } AB = 12.56 \times 4 = 50.24 \text{ cm/s}$$

The direction of this velocity is perpendicular to AB.

Now draw the velocity diagram as shown in Fig. 3.30 (*b*) as per method given below :

1. Take the fixed point a anywhere. Choose a suitable scale and draw vector ab perpendicular to AB to represent the velocity of B with respect to A (or simply the velocity of B) *i.e.*, v_{BA} (or v_B) such that vector ab $= v_{BA}$ or $v_B = 50.24$ cm/s.

2. Since points A and D are fixed hence the corresponding points a and d in velocity diagram will coincide. Now the point C moves with respect to point B and also with respect to D. The velocity of C with respect to B is perpendicular to BC whereas the velocity of C with respect to D is perpendicular to CD. Hence from point b, draw vector bc perpendicular to BC to represent the velocity of C with respect to B (*i.e.*, v_{CB}) and from point d (or a) draw vector dc (or ac) perpendicular to DC to represent the velocity of C with respect to D (*i.e.*, v_{CD}) or simply velocity of C *i.e.*, v_C (as point D is fixed). The vectors bc and dc (or ac) intersect at c. Then abc is velocity triangle.

On measurement from velocity triangle, velocity of C with respect to D is given by,

$$v_{CD} = \text{vector } dc \text{ (or } ac) = 38 \text{ cm/s.}$$

∴ Angular velocity of link CD is given by,

$$\omega_{CD} = \frac{v_{CD}}{\text{Length } CD} = \frac{38}{8} = \mathbf{4.75 \text{ rad/s (clockwise).}} \quad \textbf{Ans.}$$

Problem 3.9. *The crank of a slider crank mechanism is 15 cm and the connecting rod is 60 cm long. The crank makes 300 r.p.m. in the clockwise direction. When it has turned 45° from the inner dead centre position, determine :*

(i) Velocity of slider C,

(ii) Angular velocity of connecting rod and

(iii) Linear velocity of the mid-point of the connecting rod.

Sol. Given :

Crank length, AB = 15 cm

Connecting rod length, BC = 60 cm

Crank speed, N = 300 r.p.m.

Angle turned by crank from I.D.C. = 45°

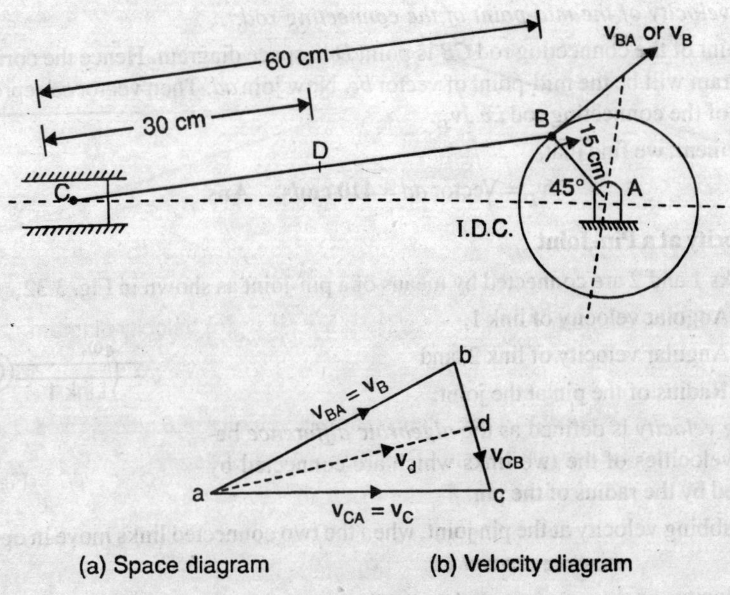

Fig. 3.31

Now angular velocity of crank is given by,

$$\omega = \frac{2\pi \times N}{60} = \frac{2\pi \times 300}{60} = 31.42 \text{ rad/s.}$$

∴ Linear velocity of B with respect to A (or velocity of B) is given by,

$$v_{BA} = v_B = \omega \times AB = 31.42 \times 15 = 471.3 \text{ cm/s}$$

The direction of v_{BA} will be at right angles to AB.

First draw the space diagram with the given dimensions to a suitable scale as shown in Fig. 3.31 (*a*). Now draw the velocity diagram as shown in Fig. 3.31 (*b*) as per method given below :

1. Take the fixed point *a* anywhere. Choose a suitable velocity scale and draw vector *ab* perpendicular to AB to represent the velocity of B with respect to A or simply the velocity of B *i.e.*, v_{BA} or v_B such that vector $ab = v_{BA} = v_B = 471.3$ cm/s.

2. Now the point C is moving with respect to B and also with respect to A. The velocity of C with respect to B is perpendicular to BC whereas the velocity of C with respect to A is along CA. Hence from point *b*, draw vector *bc* perpendicular to BC to represent the velocity of C with respect to B *i.e.*, v_{CB} and from point *a* draw

vector *ac* horizontally (*i.e.*, parallel to the path of motion of *C* which is along *CA*) to represent the velocity *C* with respect to *A* or simply the velocity of *C i.e.*, v_{CA} or v_C. The vectors *bc* and *ac* intersect at *c*. Then *abc* represents the velocity diagram for the given slider crank mechanism.

By measurement, we find velocity of *C* with respect to *B*,

$$v_{CB} = \text{Vector } bc = 340 \text{ cm/s}$$

and velocity of *C* with respect to *A* or simply velocity of *C*,

$$v_{CA} \text{ or } v_C = \text{Vector } ac = 400 \text{ cm/s.}$$

(*i*) ∴ Velocity of slider $C = v_C =$ **400 cm/s. Ans.**

(*ii*) Angular velocity of connecting rod *CB* is given by,

$$\omega_{CB} = \frac{v_{CB}}{\text{Length } BC} = \frac{340}{60} = \mathbf{5.67 \text{ rad/s.}} \quad \textbf{Ans.}$$

(*iii*) *Linear velocity of the mid-point of the connecting rod.*

The mid-point of the connecting rod *CB* is point *D* in space diagram. Hence the corresponding point *d* in the velocity diagram will be the mid-point of vector *bc*. Now join *ad*. Then vector *ad* represents the velocity of the mid-point *D* of the connecting rod *i.e.*, v_D.

By measurement, we find that,

$$v_D = \text{Vector } ad = \mathbf{410 \text{ cm/s.}} \quad \textbf{Ans.}$$

3.12. Rubbing Velocity at a Pin-Joint

The two links 1 and 2 are connected by means of a pin-joint as shown in Fig. 3.32.

Let ω_1 = Angular velocity of link 1,

ω_2 = Angular velocity of link 2, and

r = Radius of the pin at the joint.

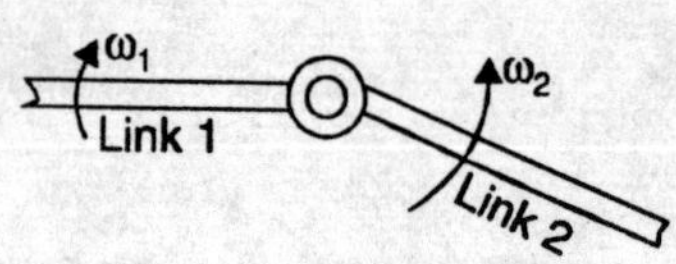

Fig. 3.32

The rubbing velocity is defined as the *algebraic difference* between the angular velocities of the two links which are connected by pin-joints, multiplied by the radius of the pin.

Hence the rubbing velocity at the pin joint, when the two connected links move in opposite direction is given by,

Rubbing velocity $= r(\omega_1 + \omega_2)$ ...(3.7)

But if the two connected links move in the same direction, then

Rubbing velocity $= r(\omega_1 - \omega_2)$ ...(3.8)

If a pin connects one sliding member and the other turning member (For example gudgeon pin of a connecting rod) the angular velocity of the sliding member is zero and hence the velocity of rubbing will be given by,

Rubbing velocity $= r \cdot \omega$ ...(3.9)

where r = Radius of the pin and

ω = Angular velocity of the turning member.

Problem 3.10. *The oscillating link OAB of a mechanism shown in Fig. 3.33, is pivoted at O and is moving at 90 r.p.m. anti-clockwise. If OA = 15 cm, AB = 7.5 cm and AC = 25 cm, then calculate :*

(*i*) *the velocity of the block C,*

(*ii*) *angular velocity of the link AC and*

(*iii*) *the rubbing velocities of the pins at O, A and C assuming that these pins are of equal diameter of 2 cm. The oscillating link OAB makes an angle of 15 with the vertical as shown in Fig. 3.33(a).*

Sol. Given :

Lengths $OA = 15$ cm, $AB = 7.5$ cm and $AC = 25$ cm

Speed of AO, $N_{AO} = 90$ r.p.m. anti-clockwise

$$\therefore \quad \omega_{AO} = \frac{2\pi \times 90}{60} = 3\pi \text{ rad/s.}$$

Hence the velocity of A with respect to O *i.e.*, v_{AO} or simply the velocity of A *i.e.*, v_A is given by

$$v_{AO} \text{ or } v_A = \omega_{AO} \times \text{Length } AO = 3\pi \times 15 = 45 \times \pi \text{ cm/s}$$

$$= \frac{45 \times \pi}{100} \text{ m/s} = 1.47 \text{ m/s}$$

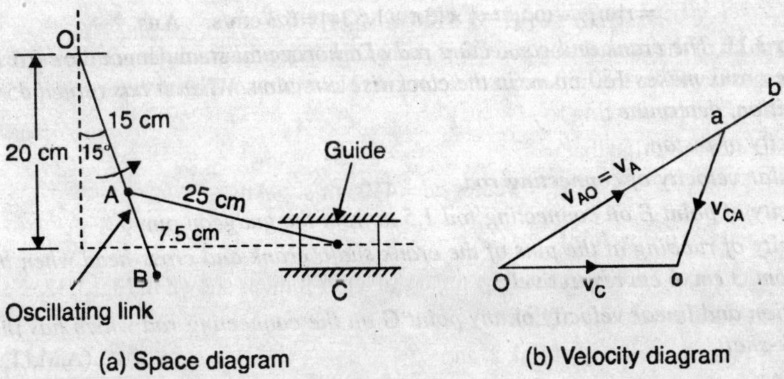

Fig. 3.33

The direction of velocity v_A is perpendicular to AO.

First draw the space diagram with the given dimensions to a suitable scale as shown in Fig. 3.33 (*a*). Now the velocity diagram can be drawn as shown in Fig. 3.33 (*b*) as per method given below :

1. Take the fixed point o anywhere. Choose a suitable velocity scale and draw vector oa perpendicular to OA to represent the velocity of A with respect to O or simply the velocity of A *i.e.*, v_{AO} or v_A such that vector $oa = v_{AO} = v_A = 1.47$ m/s.

2. Now the point C is moving with respect to point A and also moving horizontally in guides. The velocity of C with respect to A is perpendicular to AC whereas the velocity of C in guide is in horizontal direction. Hence from a, draw vector ac perpendicular to AC to represent the velocity of C with respect to A *i.e.*, v_{CA} and also from o draw vector oc horizontally (*i.e.*, along the path of motion of C in guides). The vectors ac and oc intersect at c. Then oac is the velocity diagram for the given mechanism.

From measurement, we get

(*i*) Velocity of block C, $\quad v_C$ = Vector oc = **1.32 m/s. Ans.**

(*ii*) Velocity of C with respect to A, $\quad v_{CA}$ = Vector ac = 0.38 m/s

$$\therefore \text{ Angular velocity of link } AC, \quad \omega_{CA} = \frac{v_{CA}}{\text{Length } AC} = \frac{0.38}{0.25} \qquad (\because AC = 25 \text{ cm} = 0.25 \text{ m})$$

= **1.524 rad/s (clockwise). Ans.**

(iii) Rubbing velocities of the pins at O, A and C

Diameters of pins at O, A and C = 2 cm

$$\therefore \quad \text{Radius of pins at } O, A \text{ and } C = \frac{2}{2} = 1.0 \text{ cm}$$

The pins at O and C are having only one link at their joint. Hence rubbing velocity of pins at O and at C will be obtained by multiplying the radius of the pin with their angular velocity.

Hence rubbing velocity of pin at O

$$= \text{Radius of pin} \times \text{Angular velocity of } OA = 1 \times 3\pi = 3\pi \text{ cm/s.} \quad \textbf{Ans.}$$

Similarly rubbing velocity of pin at C

$$= \text{Radius of pin} \times \text{Angular velocity of } AC = 1 \times 1.524 = \mathbf{1.524 \text{ cm/s.}} \quad \textbf{Ans.}$$

In order to calculate the rubbing velocity of pin at A, we must know the angular velocities of links OA and AC in magnitude and directions. The magnitudes of these angular velocities are known. The directions are obtained as explained here. With respect to pin A, the link OA is moving clockwise* whereas the link AC is also moving clockwise. Hence using equation (3.8), we get

The rubbing velocity of pin at A

$$= r[\omega_{AO} - \omega_{CA}] = 1 \times [3\pi - 1.524] = \mathbf{5.3 \text{ cm/s.}} \quad \textbf{Ans.}$$

Problem 3.11. *The crank and connecting rod of a horizontal steam engine are 0.5 m and 2 m long respectively. The crank makes 180 r.p.m. in the clockwise direction. When it has turned 45° from the inner dead centre position, determine :*

(i) Velocity of piston,

(ii) Angular velocity of connecting rod,

(iii) Velocity of point E on connecting rod 1.5 m from the gudgeon pin,

(iv) Velocity of rubbing at the pins of the crank shaft, crank and cross-head when the diameters of their pins are 5 cm, 3 cm, 6 cm respectively,

(v) Position and linear velocity of any point G on the connecting rod which has the least velocity relative to crank-shaft. (A.M.I.E., Summer 1979)

Sol. Given :

Crank length, $OB = 0.5$ m

Connecting rod length, $BP = 2$ m

Crank speed = 180 r.p.m.

Angle turned by crank from inner dead centre = 45°

Fig. 3.34 (*a*) shows the space diagram to some suitable scale.

Now $N_{BO} = 180$ r.p.m. $\quad \therefore \quad \omega_{BO} = \dfrac{2\pi \times 180}{60} = 18.85$ rad/s.

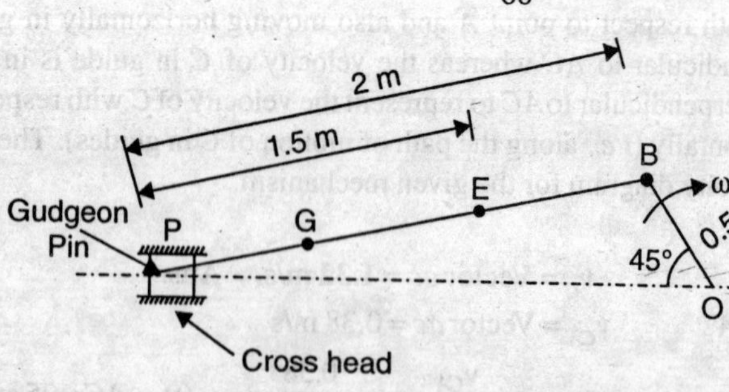

(a) Space diagram

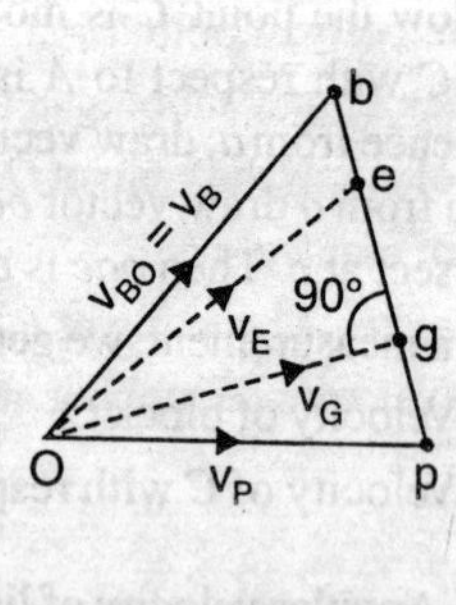

(b) Velocity diagram

Fig. 3.34

---

*The link AO with respect to O is moving anticlock-wise. Hence the link OA with respect to A will be moving in the opposite direction *i.e.*, clockwise.

Hence the velocity of B with respect to O is given by,

$$v_{BO} = \omega_{BO} \times \text{Length } OB = 18.85 \times 0.5 = 9.425 \text{ m/s}$$

The direction of v_{BO} is perpendicular to BO.

Now draw the velocity diagram as shown in Fig. 3.34 (*b*) as per method given below :

1. Take any point *o*. Choose a suitable velocity scale and draw vector *ob* perpendicular to *OB* to represent the velocity of *B* with respect to *O* (*i.e.*, v_{BO}) such that vector $ob = v_{BO} = 9.425$ m/s.

2. Now the point *P* is moving with respect to *B* and also with respect to *O*. The velocity of *P* with respect to point *B* is perpendicular to *PB* whereas the velocity of *P* with respect to *O* is along *PO*. Hence from point *b*, draw vector *bp* perpendicular to *BP* to represent the velocity of *P* with respect to *B* and from point *o* draw vector *op* parallel to the path of motion of *P* (which is along *PO*) to represent the velocity of *P* with respect to *O* or simply the velocity of *P i.e.*, v_P. The vectors *bp* and *op* intersect at *p*.

3. The point *E*, which is at a distance of 1.5 m from *P*, will have a corresponding point *e* in the velocity diagram. The location of *e* in the velocity diagram is obtained by dividing the vector *bp* at *e* in the same ratio, as *E* divides *PB* in space diagram. This means

$$\frac{ep}{bp} = \frac{EP}{BP}$$

$$\therefore \quad ep = \frac{EP}{BP} \times bp = \frac{1.5}{2.0} \times bp = 0.75 bp$$

Hence location of *e* in vector *bp* can be determined. Now join *o* to *e*.

Then vector *oe* represent the velocity of *E i.e.*, v_E.

By measurements from velocity diagram, we have :

(*i*) Velocity of piston (or point *P*),

$$v_P = \text{Vector } op = \mathbf{8.15 \text{ m/s.}} \quad \textbf{Ans.}$$

(*ii*) Angular velocity of connecting rod *BP*,

$$\omega_{BP} = \frac{v_{BP}}{BP}$$

But $\quad v_{BP} = \text{Vector } pb = 6.8 \text{ m/s and } BP = 2 \text{ m}$

$$\therefore \quad \omega_{BP} = \frac{6.8}{2} = \mathbf{3.4 \text{ rad/s (clockwise).}} \quad \textbf{Ans.}$$

(*iii*) Velocity of point *E*, $\quad v_E = \text{Vector } oe = \mathbf{8.5 \text{ m/s.}}$ **Ans.**

(*iv*) Velocity of rubbing at the pins :

Dia. of crank shaft pin at *O*, $\quad do = 5$ cm

Dia. of crank pin at *B*, $\quad db = 3$ cm

Dia. of cross head pin at *P*, $\quad dc = 6$ cm

Now velocity of rubbing at the pin of crank-shaft

$$= \frac{do}{2} \times \omega_{BO} = \frac{5}{2} \times 18.85 = \mathbf{47.125 \text{ cm/s.}} \quad \textbf{Ans.}$$

Velocity of rubbing at the pin of the cross-head

$$= \frac{dc}{2} \times \omega_{BP} = \frac{6}{2} \times 3.4 = \mathbf{10.2 \text{ cm/s. Ans.}}$$

In order to calculate the rubbing velocity of pin at *B*, we must know the angular velocities of links *OB* and *BP* in magnitude and direction. With respect to *B*, the link *OB* is moving anti-clockwise whereas the link *BP* is moving clockwise. Hence using equation (3.7), we get rubbing velocity of pin at *B*.

Velocity of rubbing at the pin of crank (*i.e.*, at *B*)

$$= \frac{do}{2}(\omega_{BO} + \omega_{BP}) = \frac{3}{2}(18.85 + 3.4) = \mathbf{33.375\ cm/s.} \quad \textbf{Ans.}$$

(*v*) *Position and linear velocity of any point G on the connecting rod which has the least velocity relative to crank shaft.*

The position of point *G* on the connecting rod which has the least velocity relative to crank shaft, is determined by drawing perpendicular from point *o* to vector *bp*. Since the length of *og* will be the least, hence point *g* represents the required position of *G* on the connecting rod. The position of *G* is obtained as

$$\frac{bg}{bp} = \frac{BG}{BP}$$

$$\therefore \quad BG = \frac{bg}{bp} \times BP = \frac{5}{6.8} \times 2 = \mathbf{1.47\ m.} \quad \textbf{Ans.}$$

($\because$ $bg = 5$ and $bp = 6.8$ from velocity diagram)

Now linear velocity of G, v_G = Vector og = **8 m/s. Ans.**

Problem 3.12. *In Fig. 3.35 shown below, the angular velocity of the crank OA is 600 r.p.m. Determine the linear velocity of slider 6 and angular velocity of the link 5.* (A.M.I.E. Winter 1986)

Sol. Given :

$$N_{OA} = 600\text{ r.p.m.}, \theta = 75°,$$
$$OA = 28\text{ mm}, AB = 44\text{ mm}, BC = 49\text{ mm}$$
$$BD = 46\text{ mm, and } OC = 65\text{ mm.}$$

Now $$\omega_{OA} = \frac{2\pi N_{OA}}{60} = \frac{2\pi \times 600}{60} = 20\pi\text{ rad/s}$$

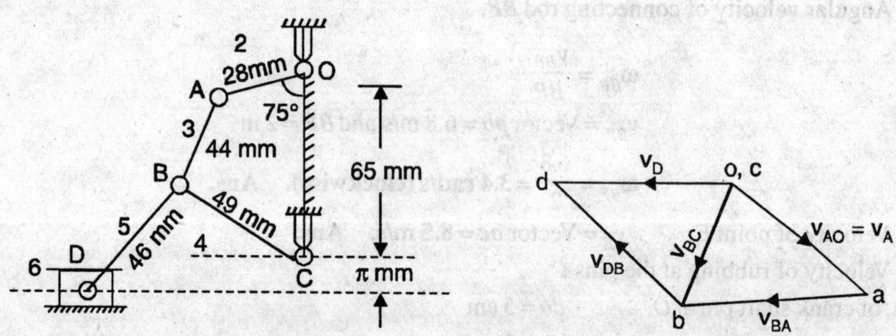

(a) Space diagram (b) Velocity diagram

Fig. 3.35

$\therefore$ Velocity of *A* with respect to *O*,

$$v_{AO} = w_{OA} \times OA = 20\pi \times 28\text{ mm/s} \qquad (\because \quad OA = 28\text{ mm})$$

$$= \frac{20\pi \times 28}{1000}\text{ m/s} = 1.758\text{ m/s}$$

The points *O* and *C* are fixed in the given mechanism. Hence these points will be located at a common point in the velocity diagram.

As point *O* is fixed, hence the velocity of *A* with respect to *O* is also equal to velocity of *A*.

$$\therefore \quad v_{AO} = v_A = 1.758\text{ m/s.}$$

First draw the space diagram to a suitable scale as shown in Fig. 3.35 (*a*). Now draw the velocity diagram as shown in Fig. 3.35 (*b*) as per method given below :

1. Take any point *o*. [Here fixed point should be taken]. Choose a suitable velocity scale and draw vector *oa* perpendicular to *OA* to represent the velocity of *A* with respect to *O* (*i.e.*, v_{AO}). Cut *oa* = 1.758 m/s.

2. The velocity of *B* with respect to *A* is perpendicular to *AB* whereas the velocity of *B* with respect to *C* is perpendicular to *BC*. From point *a*, draw vector *ab* perpendicular to *AB* to represent the velocity of *B* with respect to *A* (*i.e.*, v_{BA}). But the magnitude of v_{BA} is not known.

3. Since *OC* is fixed, hence points *O* and *C* in velocity diagram will coincide. Now from point *c*, draw vector *cb* (or *ob*) perpendicular to *BC* so as to intersect vector *ab* in *b*. Now vector *ob* represent the velocity of *B* with respect to *C* (*i.e.*, $v_{BC} = v_B$ as point *C* is fixed).

4. The velocity of point *D* with respect to point *B* is perpendicular to *BD*. Also the point *D* is moving in a horizontal direction. From point *b*, draw vector *bd* perpendicular to *BD* to represent the velocity of *D* with respect to *B* (*i.e.*, v_{DB}). The magnitude of v_{DB} is not known.

5. The velocity of the slider 6 or of point *D* is in the horizontal direction. Hence draw vector *od* parallel to the path of motion of the slider 6 (*i.e.*, parallel to horizontal line) to represent the velocity of *D* (*i.e.*, v_D). The vectors *bd* and *od* intersect at *d*. This completes the velocity diagram.

By measurement from velocity diagram, we get,

$$v_D = \text{Vector } od = 6.6 \text{ m/s and}$$

$$v_{DB} = \text{Vector } bd = 7 \text{m/s}.$$

(i) Linear velocity of slider 6

The linear velocity of slider 6 is equal to v_D.

∴ Linear velocity of slider 6 = v_D = **6.6 m/s. Ans.**

(ii) Angular velocity of link 5 or link BD

$$v_{DB} = \text{Vector } bd = 7 \text{ m/s}$$

But $$v_{DB} = \omega_{DB} \times BD$$

∴ $$\omega_{DB} = \frac{v_{DB}}{BD} = \frac{7}{\left(\frac{46}{1000}\right)} \qquad \left(\because \; BD = 46 \text{ mm} = \frac{46}{1000} \text{ m}\right)$$

$$= \frac{7 \times 1000}{46} = 152.17 \text{ rad/s.}$$

Problem 3.13. *In the mechanism shown in Fig. 3.36, the crank AB rotates about A at uniform speed of 120 r.p.m. in the clockwise direction. The lever DC oscillates about the fixed point D, which is connected to AB by the coupler BC. The block F moves in the horizontal guides, being driven by the link EF. Determine :*

(i) Velocity of the block F ;

(ii) Angular velocity of lever DC ;

(iii) Rubbing velocity at the pin C which is 5 cm in diameter.

The lengths of various links are : AB = DE = 15 cm ; BC = CD = 45 cm and EF = 37.5 cm.

(A.M.I.E., Winter 1982)

Sol. Given :

Lengths of links *AB* = *DE* = 15 cm ; *BC* = *CD* = 45 cm and *EF* = 37.5 cm

Speed of crank *AB* = 120 r.p.m.

∴ Angular velocity of crank *AB* is given by,

$$\omega_{BA} = \frac{2\pi N}{60} = \frac{2\pi \times 120}{60} = 12.56 \text{ rad/s.}$$

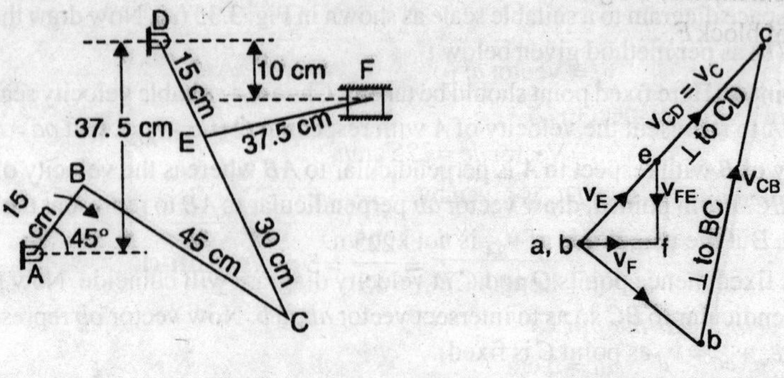

(*a*) Space diagram (*b*) Velocity diagram

Fig. 3.36

Now the velocity of *B* with respect to *A* (or simply velocity of *B* as point *A* is fixed) is given by,

$$v_{BA} \text{ or } v_B = \omega_{BA} \times \text{Length} AB = 12.56 \times 15 = 188.4 \text{ cm/s.}$$

The direction of v_{BA} or v_B will be at right angles to *AB*.

In the given mechanism, the points *A* and *D* are fixed. Hence these points will be located at a common point in the velocity diagram. The Block *F* is moving in a horizontal direction.

First draw the space diagram of the given mechanism to any suitable scale as shown in Fig. 2.36 (*a*). Now draw the velocity diagram as shown in Fig. 3.36 (*b*) as per method given below :

1. The points *A* and *D* in the mechanism are fixed, hence the corresponding points *a* and *d* in the velocity diagram will coincide. This serves as the starting point in the construction of the velocity diagram.

2. Take any point *a* or *d*. Choose a suitable velocity scale and from point *a*, draw vector *ab* perpendicular to *AB* to represent the velocity of *B* with respect to *A* such that vector $ab = v_{BA}$ or $v_B = 188.4$ cm/s.

3. The point *C* moves relative to *B* and also relative to *D*. The velocity of *C* with respect to point *B* is perpendicular to *BC* whereas the velocity of *C* with respect to *D* is perpendicular to *DC*. Hence from point *b* draw vector *bc* perpendicular to *BC* to represent the velocity of *C* with respect to *B i.e.*, v_{CB} and from point *d* (or *a*), draw vector *dc* (or *ac*) perpendicular to *DC* to represent the velocity of *C* with respect to *D* or simply velocity of *C* (*i.e.*, v_{CD} or v_C). These two vectors *i.e.*, *bc* and *dc* intersect at *c*.

4. The point *E* lies on *DC* and hence the corresponding point *e* in velocity diagram will lie on *dc*. To find the location of the point *e* on vector *dc*, divide the vector *dc* in the same ratio as *E* divides *DC* in space diagram. Hence

$$\frac{de}{dc} = \frac{DE}{DC}$$

$$\therefore \quad de = \frac{DE}{DC} \times dc = \frac{15}{45} \times dc = \frac{dc}{3}$$

$$= \frac{226}{3} = 75.33 \text{ cm/s} \qquad (\because \text{ By measurement } dc = 226 \text{ cm/s})$$

Hence take vector *de* = 75.33 cm/s. Now the point *e* is known.

5. The location of point *f* in velocity diagram is obtained as given below :

Since the block *F* moves in the horizontal guides and also the point *F* is moving relative to point *E*.

The velocity of *F* with respect to *E* is perpendicular to *EF*. Also the velocity of *F* is in horizontal direction. From point *e*, draw vector *ef* perpendicular to *EF* to represent the velocity of *F* with respect to *E i.e.*, v_{FE} and from point *d* (or *a*) draw vector *df* horizontally to represent the velocity of *F i.e.*, v_F. The vectors *ef* and *df* intersect at *f*.

Then by measurements, we get

(*i*) Velocity of block *F*, v_F = Vector *df* (or *af*) = **70 cm/s. Ans.**

(*ii*) Velocity of *C* with respect to *D*,

v_{CD} = Vector *dc* = 225 cm/s

∴ Angular velocity of lever *DC* is given by,

$$\omega_{CD} = \frac{v_{CD}}{\text{Length } CD} = \frac{225}{45} = \textbf{5 rad/s (Anti-clockwise).} \quad \textbf{Ans.}$$

(*iii*) *Rubbing velocity at the pin C*

Dia. at pin *C*, $dc = 5$ cm

∴ Radius of pin at *C* $= \frac{5}{2} = 2.5$ cm

In order to calculate the rubbing velocity of pin at *C*, we must know the angular velocities of links *BC* and *CD* in magnitude and directions. The magnitude and direction of angular velocity of link *CD* is known (*i.e.*, 5 rad/s anticlockwise).

Now the angular velocity of link *BC* is given by

$$\omega_{CB} = \frac{v_{CB}}{\text{Length } BC} = \frac{225}{45} = 5 \text{ rad/s (Anti-clockwise)}$$

(By measurement, v_{CB} = Vector *bc* = 225 and length *BC* = 45 cm)

Now using equation (3.7), we get

Rubbing velocity of pin at *C* $= r(\omega_{CB} - \omega_{CD}) = 2.5(5 - 5) = \mathbf{0}$. **Ans.**

Problem 3.14. *In a steam engine mechanism, shown in Fig. 3.37, the crank AB rotates at 200 r.p.m. The dimensions of the various links are : AB = 12 cm, BC = 48 cm, CD = 18 cm, DE = 36 cm and EF = 12 cm and FP = 36 cm. Find the velocities of C, D, E, F and P.*

Sol. Given : $AB = 12$ cm, $BC = 48$ cm, $CD = 18$ cm, $DE = 36$ cm, $EF = 12$ cm and $FP = 36$ cm.

Speed of crank *AB* = 200 r.p.m.

∴ Angular velocity of crank *AB* is given by,

$$\omega_{BA} = \frac{2\pi \times 200}{60} = 20.94 \text{ rad/s}$$

Now the velocity of point *B* with respect to *A* is given by,

$$v_{BA} = \omega_{BA} \times \text{Length } AB = 20.94 \times 12 = 251.3 \text{ cm/s}$$

The direction of this velocity will be perpendicular to *AB*. As point *A* is fixed, hence this is also the velocity of point *B*.

∴ $v_{BA} = v_B = 251.3$ cm/s

In the given mechanism, the point *A* is fixed. The points *C* and *P* move horizontally whereas the point *E* moves vertically.

First draw the space diagram of the given mechanism to some suitable scale as shown in Fig. 3.37 (*a*). Now draw the velocity diagram as shown in Fig. 3.37 (*b*) as per method given below :

1. Take any point *a* (fixed point) Choose a suitable velocity scale and draw vector *ab* perpendicular to *AB* to represent the velocity of *B* with respect to *A* or simply velocity of *B* (*i.e.*, v_{BA} or v_B), such that vector $ab = v_{BA} = v_B = 251.3$ cm/s.

2. Now point C is moving with respect to point B and also with respect to A. The velocity of C with respect to B is perpendicular to BC whereas the velocity of C with respect to A is in horizontal direction along CA. Hence from point b, draw vector bc perpendicular to BC to represent the velocity of C with respect to B and from point 'a', draw vector ac parallel to the path of motion of C (which is horizontal or along CA) to represent the velocity of C with respect to A (or simply velocity of C as point A is fixed) *i.e.*, v_{CA} or v_C. The vectors bc and ac intersect at c.

3. To find the location of D in the velocity diagram, divide vector bc at d in the same ratio as D divides BC in space diagram shown in Fig. 3.37 (a). This means that

$$\frac{bd}{bc} = \frac{BD}{BC}.$$

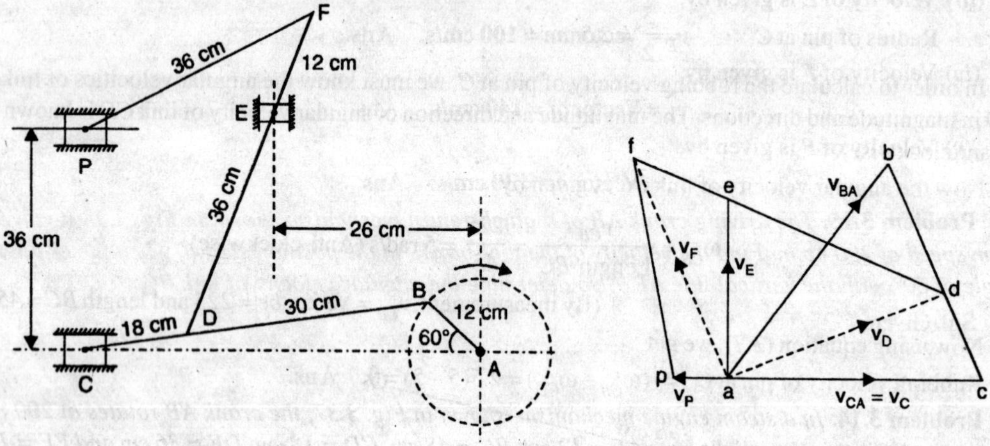

(a) Space diagram (b) Velocity diagram

Fig. 3.37

$$\therefore \qquad bd = \frac{BD}{BC} \times bc = \frac{30}{48} \times bc = 0.625 bc$$

Hence position of d is known. Now join point a to d. Then the vector ad represents the velocity of D *i.e.*, v_D.

4. The slider at E moves with respect to D and also in the vertical direction. The velocity of E with respect to D is perpendicular to DE also the velocity of E is in vertical direction. The location of E in velocity diagram is obtained by drawing from point d vector de perpendicular to DE to represent the velocity of E with respect to D and from point a drawing vector ae parallel to the path of motion of E (which is along the vertical direction) to represent the velocity of E *i.e.*, v_E. The vectors de and ae intersect at e.

5. To find the location of F in the velocity diagram, divide vector de produced in the same ratio as F divides DE produced in space diagram. This means

$$\frac{ef}{de} = \frac{EF}{DE}$$

$$\therefore \qquad ef = \frac{EF}{DE} \times de = \frac{12}{36} \times de = 0.33 de$$

Hence position of f in velocity diagram is known. Join point a to f. Then vector af represents the velocity of F *i.e.*, v_F.

6. The slider at P moves with respect to F and also in the horizontal direction. The velocity of P with respect to F is perpendicular to PF. Also the velocity of P is in horizontal direction. Hence from point f, draw vector fp perpendicular to PF to represent the velocity of P with respect to F *i.e.*, v_{PF}. Also from point a, draw vector ap parallel to the path of motion of P (which is along the horizontal direction) to represent the velocity of P *i.e.*, v_P. The vectors fp and ap intersect at p.

Now by measurements from velocity diagram, we get :

(*i*) Velocity of C is given by,

$$v_C = \text{Vector } ac = \mathbf{242.5\ cm/s.} \quad \textbf{Ans.}$$

(*ii*) Velocity of D is given by,

$$v_D = \text{Vector } ad = \mathbf{235\ cm/s.} \quad \textbf{Ans.}$$

(*iii*) Velocity of E is given by,

$$v_E = \text{Vector } ae = \mathbf{100\ cm/s.} \quad \textbf{Ans.}$$

(*iv*) Velocity of F is given by,

$$v_F = \text{Vector } af = \mathbf{140\ cm/s.} \quad \textbf{Ans.}$$

(*v*) Velocity of P is given by,

$$v_P = \text{Vector } ap = \mathbf{30\ cm/s.} \quad \textbf{Ans.}$$

Problem 3.15. ***The driving crank AB of a quick-return mechanism shown in Fig. 3.38 revolves at a niform speed of 200 r.p.m. Find the velocity of the tool-box R, in the position shown, when the crank makes an angle of 60° with the vertical line AP. Also determine the angular velocity of link PQ.***

Sol. Given :

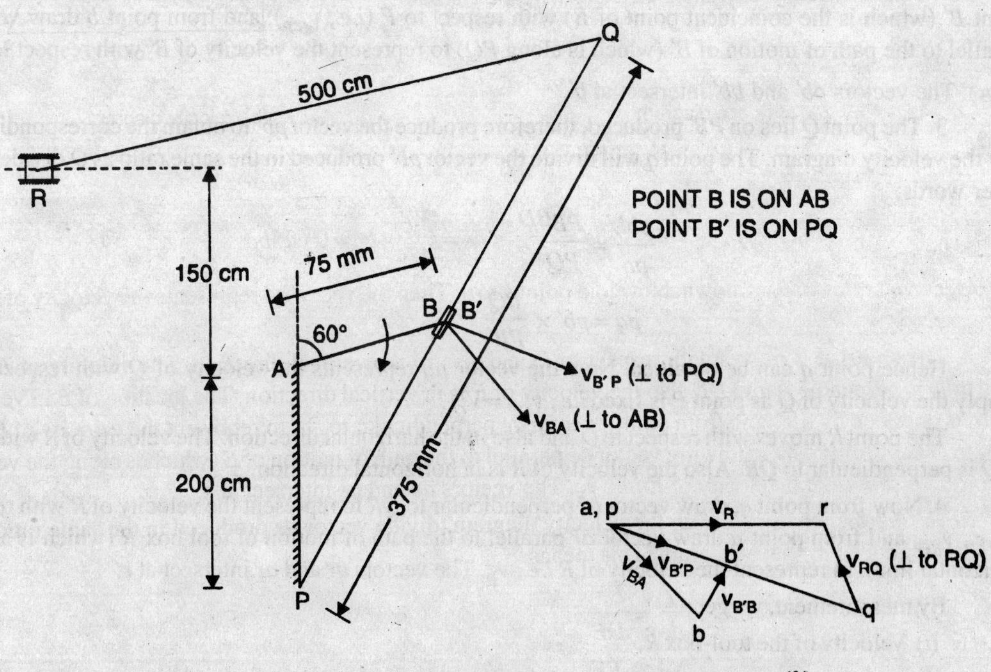

(*a*) (*b*)

(*a*) Space diagram (*b*) Velocity diagram

Fig. 3.38

Speed of crank $AB = 200$ r.p.m.

$\therefore$ Angular velocity of crank, $\omega_{BA} = \dfrac{2\pi \times 200}{60} = 20.95$ rad/s

$\therefore$ Linear velocity of point B (which is on crank AB) with respect to A is given by,

$$v_{BA} = \omega_{BA} \times \text{Length } AB = 20.95 \times 75 \text{ mm/s} = \frac{20.95 \times 75}{1000} \text{ m/s} = 1.57 \text{ m/s.}$$

The direction of v_{BA} will be perpendicular to the crank AB.

In the given mechanism, the points A and P are fixed. Hence these points will be located at a common point in the velocity diagram. The point R is moving horizontally. The point B is on crank AB and point B' is on PQ.

First draw the space diagram with the given dimensions to some suitable scale as shown in Fig. 3.38 (*a*). The point B lies on crank AB and its coincident point B' is on PQ. Now draw the velocity diagram as shown in Fig. 3.38 (*b*) as per method given below :

1. As points A and P are fixed points, hence the corresponding points a and p will coincide in the velocity diagram. Take any point a or p anywhere. Now choose a suitable velocity scale and from point a draw vector ab perpendicular to AB to represent the velocity of B with respect to A *i.e.*, v_{BA}. Hence take vector $ab = v_{BA} = 1.57$ m/s.

2. The point B' is on PQ. The point B' moves with respect to P and also with respect to B. The velocity of B' with respect to P is perpendicular to PB'. The velocity of B' with respect to B is along PQ. From point p which is at the same place as point a, draw vector pb' perpendicular to PB' (or PQ) to represent the velocity of point B' (which is the coincident point of B) with respect to P (*i.e.*, $v_{B'P}$), and from point b draw vector bb' parallel to the path of motion of B' (which is along PQ) to represent the velocity of B' with respect to B (*i.e.*, $v_{B'B}$). The vectors pb' and bb' intersect at b'.

3. The point Q lies on PB' produced, therefore produce the vector pb' to obtain the corresponding point q in the velocity diagram. The point q will divide the vector pb' produced in the same ratio as Q divides PB'. In other words,

$$\frac{pb'}{pq} = \frac{PB'}{PQ}$$

$$\therefore \quad pq = pb' \times \frac{PQ}{PB'}$$

Hence point q can be obtained. Now the vector pq represents the velocity of Q with respect to P or simply the velocity of Q as point P is fixed *i.e.*, $v_{QP} = V_Q$.

The point R moves with respect to Q and also in the harizontal direction. The velocity of R with respect to Q is perpendicular to QR. Also the velocity of R is in horizontal direction.

4. Now from point q, draw vector qr perpendicular to QR to represent the velocity of R with respect to Q *i.e.*, v_{RQ}, and from point a draw vector ar parallel to the path of motion of tool box R (which is along the horizontal line), to represent the velocity of R *i.e.*, v_R. The vectors qr and ar intersect at r.

By measurement, we get :

(*i*) Velocity of the tool-box R,

$$v_R = \text{Vector } ar = 1.6 \text{ m/s.} \quad \textbf{Ans.}$$

(*ii*) Velocity of Q with respect to P,

$$v_{QP} = \text{Vector } pq = 1.7 \text{ m/s}$$

∴ Angular velocity of link PQ,

$$\omega_{QP} = \frac{v_{QP}}{\text{Length } PQ} = \frac{1.7}{0.375} \qquad (\because \quad PQ = 375 \text{ mm} = 0.375 \text{ m})$$

$$= \mathbf{4.533 \text{ rad/s.}} \quad \textbf{Ans.}$$

3.13. Forces Acting in a Mechanism

The velocity diagram for a mechanism may be used to find the forces acting at a point when the force at another point is known.

Consider a rigid link AB as shown in Fig. 3.39.

Let F_A = External force acting at end A,

F_B = External force acting at end B,

v_A = Velocity of end A, and

v_B = Velocity of end B.

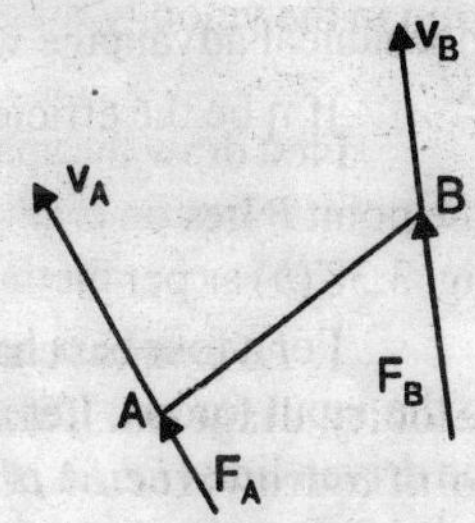

Fig. 3.39

Also assume that at the end A, the force F_A and velocity v_A are in the same direction. Similarly at the end B the force F_B and v_B are acting in the same direction.

Let us consider that A is the driving end and B is the driven end, in other words we assume that energy input is at A and energy output is at B. If we neglect the effect of friction and change of kinetic energy of the link (which means the efficiency of transmission is 100%) then by the principle of conservation of energy, we get

Energy input to the link during a particular time interval

= Energy output from the link during the same time interval

or Input work per unit time = Output work per unit time

∴ Work supplied to end A per sec = Work transmitted to end B per sec

or
$$F_A \times v_A = F_B \times v_B \qquad ...(3.10)$$

∴
$$F_B = \frac{F_A \times v_B}{v_B} \qquad ...(3.11)$$

If we consider the effect of friction, then efficiency of transmission will be given by,

$$\eta = \frac{\text{Output}}{\text{Input}} = \frac{F_B \times v_B}{F_A \times v_A} \qquad ...(3.12)$$

∴
$$F_B = \eta \times F_A \times \frac{v_A}{v_B} \qquad ...(3.13)$$

The force F_A and velocity v_A at end A should be in the same direction. If force F_A is not in the direction of v_A, then the component of force in the direction of velocity should be used in the above equations. Same is true for force F_B and velocity v_B.

3.14. Mechanical Advantage

The ratio of load to effort is known as mechanical advantage. The mechanical advantage of a mechanism can be obtained from the velocity analysis. Let us consider a rigid link AB as shown in Fig. 3.40, in which A is the driving end and B is the driven end. Then force F_A will be effort and the force F_B will be load or the resistance to overcome.

$$\therefore \quad \text{Mechanical advantage} = \frac{\text{Load}}{\text{Effort}} = \frac{F_B}{F_A} \qquad ...(i)$$

Now from the principle of conservation of energy (neglecting effect of friction), we get

$$F_A \times v_A = F_B \times v_B \quad \text{or} \quad \frac{F_B}{F_A} = \frac{v_A}{v_B}$$

But from equation (*i*), we know that $\frac{F_B}{F_A}$ = Mechanical advantage

$$\therefore \quad \text{Mechanical advantage} = \frac{F_B}{F_A} = \frac{v_A}{v_B} \qquad ...(3.14)$$

The mechanical advantage given by equation (3.14) is known as ideal mechanical advantage.

If we consider the effect of friction, then with a given effort, less resistance will be overcome and hence mechanical advantage will be less. Thus we see that effect of friction is to reduce the mechanical advantage.

If η be the efficiency of the mechanism, then actual mechanical advantage (M.A.) will be given by,

$$\text{M.A.} = \eta \times \frac{v_A}{v_B} \qquad ...(3.15)$$

For a four bar chain shown in Fig. 3.40, the mechanical advantage is defined as the ratio of output torque to the input torque. Let us assume that *AD* is the driving link and *BC* is the driven link (*i.e., A* is the driving joint and *B* is the driven joint of the link *AB*).

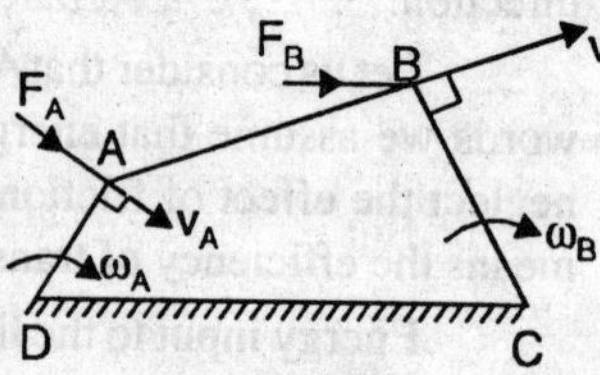

Fig. 3.40

Let T_A = Driving torque = $F_A \times AD$

T_B = Resisting torque = $F_B \times BC$

ω_A = Angular velocity of driving link *AD*

ω_B = Angular velocity of driven link *BC*.

Then by the principle of conservation of energy (neglecting effect of friction), we get

Input energy per unit = Output energy per unit time

$$T_A \times \omega_A = T_A \times \omega_B \qquad \text{or} \qquad \frac{T_B}{T_A} = \frac{\omega_A}{\omega_B}$$

$$\therefore \qquad \text{M.A.} = \frac{\text{Output Torque}}{\text{Input Torque}} = \frac{T_B}{T_A} = \frac{\omega_A}{\omega_B} \qquad ...(3.16)$$

The M.A. given by equation (3.16) is known as ideal mechanical advantage.

If the effect of friction is considered, then

$$\text{M.A.} = \eta \times \frac{\omega_A}{\omega_B} \qquad ...(3.17)$$

Equation (3.17) gives the actual mechanical advantage.

Problem 3.16. (S.I. Units) *Fig. 3.41 shows a four bar chain in which link AD is the driver and link BC is the driven. The driver rotates uniformly at a speed of 120 r.p.m. and the constant driving torque is 50 Nm. At a particular phase of the mechanism, the angular velocity of the driven link is 2 rad/s. Calculate the ideal mechanical advantage. What will be the resisting torque at this instant if the efficiency of the four bar chain is 70%.*

Sol. Given :

Speed of driver link = 120 r.p.m.

∴ Angular velocity of the driver,

$$\omega_A = \frac{2\pi \times 120}{60} = 12.56 \text{ rad/s}$$

Angular velocity of the driven link,

$$\omega_B = 2 \text{ rad/s}$$

Driving torque, $T_A = 50$ Nm

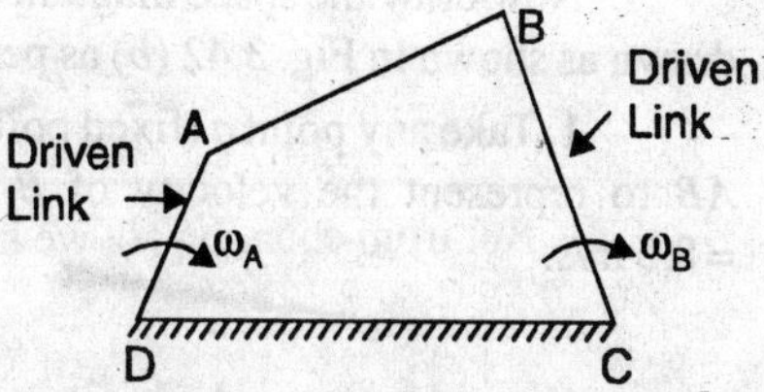

Fig. 3.41

Now ideal mechanical advantage (M.A.) in terms of angular speed is given by equation (3.16). Hence using equation (3.16), we get

$$\text{M.A. (ideal)} = \frac{\omega_A}{\omega_B} = \frac{12.56}{2} = \mathbf{6.28}. \quad \textbf{Ans.}$$

(ii) Resisting torque when efficiency (η) is 70%

Actual mechanical advantage is given by equation (3.17).

$$\therefore \quad \text{Actual M.A.} = \eta \times \frac{\omega_A}{\omega_B} = 0.7 \times 6.28 \qquad \left(\because \frac{\omega_A}{\omega_B} = 6.28\right)$$

$$= 4.396 \simeq 4.4$$

$$\text{But} \quad \text{M.A.} = \frac{\text{Output Torque}}{\text{Input Torque}} = \frac{\text{Resisting Torque}}{\text{Driving Torque}}$$

$$\text{or} \quad 4.4 = \frac{\text{Resisting Torque}}{50}$$

$$\therefore \quad \text{Resisting torque} = 4.4 \times 50 = \mathbf{220 \text{ N m}}. \quad \textbf{Ans.}$$

Problem 3.17. *A mechanism in which the crank AB is rotating uniformly at 180 r.p.m. in the clockwise direction is shown in Fig. 3.42. The blocks at D and E are working in frictionless guides. The dimensions cf the mechanism are : AB = 450 mm ; BD = 1500 mm ; BC = 900 mm and CE = 900 mm.*

(i) Draw the velocity diagram and determine the velocities of blocks D and E in their guides,

(ii) Find the turning moment at A if a force of 1000 N acts on D in the direction of arrow X and a force of 1500 N acts on E in the direction of arrow Y. **(London University)**

Sol. Given :

Speed of crank AB = 180 r.p.m.

$$\therefore \quad \text{Angular velocity of crank } AB = \frac{2\pi N}{60} = \frac{2\pi \times 180}{60} = 18.85 \text{ rad/s}$$

∴ Linear velocity of B with respect to A or velocity of B,

$$v_{BA} \text{ or } v_B = \text{Angular velocity of } AB \times \text{Length } AB$$

$$= 18.85 \times 0.45 \qquad (\because AB = 450 \text{ mm} = 0.45 \text{ m})$$

$$= 8.5 \text{ m/s}$$

This velocity will be at right angles to AB.

Dimensions of mechanism are : $AB = 450$ mm $= 0.45$ m, $BD = 1500$ mm $= 1.5$ m and $BC = CE = 900$ mm $= 0.9$ m.

In the given mechanism, the point A is fixed. The point E is moving vertically and the point D is moving at an angle of 45° to the horizontal.

Now draw the space diagram to a suitable scale as shown in Fig. 3.42 (*a*). Then the velocity diagram is drawn as shown in Fig. 3.42 (*b*) as per method given below :

1. Take any point *a* (fixed point). Choose a suitable velocity scale and draw vector *ab* perpendicular to *AB* to represent the velocity of *B* with respect to *A* or simply velocity of *B* such that vector $ab = v_{BA} = v_B = 8.5$ m/s.

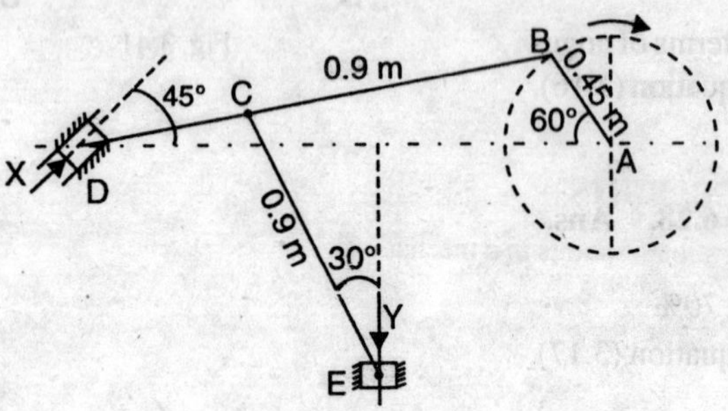

(*a*) Space diagram

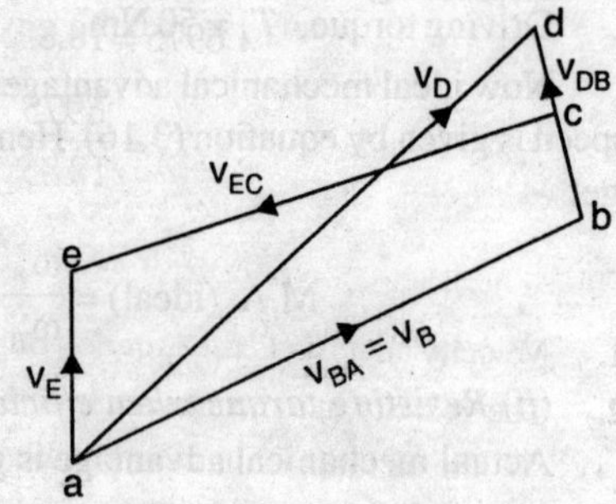

(*b*) Velocity diagram

Fig. 3.42

2. Now point *D* is moving with respect to *B* and also in the guides at an angle of 45° with the line *DA*. The velocity of *D* with respect to *B* is perpendicular to *BD*. Also the point *D* is moving at an angle of 45° with horizontal. Hence from point *b*, draw vector *bd* perpendicular to *BD* to represent the velocity of *D* with respect to *B i.e.*, v_{DB} and from point *a* draw vector *ad* parallel to the motion of *D* (*i.e.*, at an angle of 45° with line *DA*) to represent the velocity of *D i.e.*, v_D. The vectors *bd* and *ad* intersect at *d*.

3. To find the location of *C* in the velocity diagram, divide vector *bd* at *c* in the same ratio as *C* divides *BD*, which means

$$\frac{bc}{bd} = \frac{BC}{BD} \quad \text{or} \quad bc = \frac{BC}{BD} \times bd = \frac{0.9}{1.5} \times bd = 0.6\, bd$$

∴ The location of *c* is known.

4. The point *E* is moving with respect to *C* and also in the vertical direction in guides. The velocity of *E* with respect to *C* is perpendicular to *CE*. Also the point *E* is moving in vertical direction in guide. Hence from point *c*, draw vector *ce* perpendicular to *CE* to represent the velocity of *E* with respect to *C i.e.*, v_{EC} and from point *a* draw vector *ae* parallel to the path of *E* (which is vertical) to represent the velocity of *E i.e.*, v_E. The vectors *ce* and *ae* intersect at *e*.

(*i*) Now from velocity diagram, we get

Velocity of block *D*, v_D = Vector *ad* = **9.45 m/s to the right. Ans.**

Velocity of block *E*, v_E = **1.65 m/s upwards. Ans.**

(*ii*) ***Turning Moment at A*** (T_A)

Given : Force at *D*, $F_D = 1000$ N ; Force at *E*, $F_E = 1500$ N

Now power input $= F_D \times v_D - F_E \times v_E$

(The external forces are acting at *D* and *E*. Hence power input is at *D* and *E*. Also F_D and v_D are acting in the same direction whereas F_E and v_E are acting in opposite direction. Hence negative sign is taken with F_E.)

∴ Power input $= 1000 \times 9.45 - 1500 \times 1.65$

$= 9450 - 2475 = 6975$ Nm/s

Let T_A = Turning moment at crank shaft or at A

ω_A = Angular velocity of crank AB

= 18.85 rad/s

∴ Power output $= T_A \times 18.85$ Nm/s

Neglecting losses, the power input should be equal to power output.

$$\therefore \quad 6975 = 18.85 \times T_A$$

$$\therefore \quad T_A = \frac{6975}{18.85} = 370 \text{ Nm.} \quad \textbf{Ans.}$$

HIGHLIGHTS

1. Velocity analysis is used for determining acceleration of points in a mechanism.
2. The velocity in a mechanism is determined by :
   (*i*) the instantaneous centre-method and
   (*ii*) relative velocity method.
3. If a body is having a combined motion of translation and rotation, then body is assumed to be rotating about a certain point which is known as instantaneous centre of rotation.
4. The instantaneous centre of a link AB, having motion of translation and rotation, is determined if the directions of velocities at A and B are known. The perpendiculars to the directions of velocities at A and B are drawn. The point of intersection of two perpendiculars gives the instantaneous centre for the link AB.
5. If V_A and V_B are the velocities at points A and B of a link AB, and AO and BO are the distances of the points A and B from the instantaneous centre, then we have
$$\frac{V_A}{V_B} = \frac{AO}{BO}.$$
6. For a reciprocating engine mechanism, the angular velocity of the connecting rod AB is given by,
$$\omega_{AB} = \frac{\omega \times r}{BO}$$
   where ω = Angular velocity of crank
   r = Radius of crank
   BO = Distance of crank pin from instantaneous centre.
7. The velocity of B with respect to A is represented by v_{BA}. But in vector form it is written as $\vec{ab}$ and not $\vec{ba}$.
$$\therefore \quad v_{BA} = \vec{ab}.$$
8. The velocity of any point on a rigid link with respect to any other point on the same link, is perpendicular to the link.
9. If c is any point on a link AB, then corresponding point c in the velocity diagram will divide the vector ab in the same ratio as the point C divides link AB *i.e.*, $\dfrac{ac}{ab} = \dfrac{AC}{AB}$.
10. The fixed points in a mechanism are represented at one point in the velocity diagram.
11. Angular velocity of a link $AB = \dfrac{\text{Vector } ab}{\text{Length } AB}$ *i.e.*, $\omega = \dfrac{V}{r}$.
12. Small letters are used in velocity diagram whereas capital letters are used for mechanism (or space diagram).
13. Absolute velocities of the points are measured from a fixed point in the velocity diagram.
14. The velocity diagram should be started from such a link whose velocity in magnitude and direction are known.

15. Rubbing velocity at a pin-joint is the algebraic difference between the angular velocities of the two links which are connected by pin-joint, multiplied by the radius of the pin. It is given by,

(*i*) $r \cdot (\omega_1 + \omega_2)$—if the links move in opposite direction

(*ii*) $r \cdot (\omega_1 - \omega_2)$—if the links move in the same direction.

16. For a sliding and turning pair, rubbing velocity $= r \times \omega$

where r = Radius of pin and

ω = Angular velocity of turning pair.

17. The ratio of load to effort is known as mechanical advantage. If for a rigid link AB, driving end is A whereas driven end is B then M.A. is given by,

$$\text{M.A.} = \frac{F_B}{F_A} = \frac{v_A}{v_B} \quad \text{if effect of friction is neglected}$$

$$= \eta \times \frac{v_A}{v_B} \quad \text{if effect of friction is considered}$$

where F_B = Force at end B and is known load or resistance

F_A = Force at A and is known as effort

η = Efficiency of transmission.

4

Acceleration in Mechanism

4.1. Introduction

If the velocities of various points in a mechanism are known, then the acceleration of these points can be easily determined. After knowing acceleration the external force, which is the product of mass and acceleration, can be obtained. From the forces, the stresses (which are equal to force divided by area) at the various points of the mechanism will be known. These stresses are in addition to the stresses caused by the working loads. With increasing speeds, higher and higher accelerations are being called for. Hence the forces and stresses due to higher accelerations are sometimes more than the stresses caused by the working loads. As acceleration is proportional to the square of the speed, *i.e.*, acceleration = $\omega^2 \times r$, hence if the speed of the machines becomes two times, the centripetal force will become four times. Thus the acceleration diagrams are, therefore, fundamental to stress analysis of mechanism.

4.2. Acceleration of a Body Moving along a Circular Path

Acceleration is defined as the rate of change of velocity. Acceleration and velocity are vector quantities.

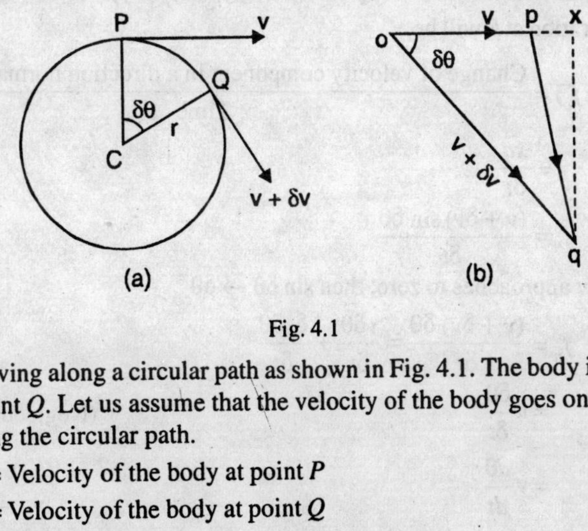

Fig. 4.1

Let a body is moving along a circular path as shown in Fig. 4.1. The body is initially at point P and in time δt moves to the point Q. Let us assume that the velocity of the body goes on changing in magnitude as the body is moving along the circular path.

Let v = Velocity of the body at point P

$v + \delta v$ = Velocity of the body at point Q

r = Radius of the circle

δt = Time taken by the body in moving from P to Q

$\delta\theta$ = Angle covered by the body in moving from P to Q.

The change of velocity, as the body moves from P to Q can be determined by drawing the vector triangle opq, as shown in Fig. 4.1 (b), in which op represents the velocity v and oq represents the velocity $v + \delta v$. In this triangle pq represents the change of velocity in time δt. The vector pq is resolved in two components px and xq parallel to op and perpendicular to op respectively.

Now,
$$px = ox - op = oq\cos\delta\theta - v$$
$$= (v+\delta v)\cos\delta\theta - v \qquad (\because \; oq = v + \delta v)$$

and
$$xq = oq\sin\delta\theta.$$
$$= (v+\delta v)\sin\delta\theta.$$

The change of velocity (*i.e.*, vector *pq*) has two components (*i.e.*, *px* and *xq*) which are mutually perpendicular. Hence the rate of change of velocity will also have two components which will be mutually perpendicular. But rate of change of velocity is acceleration. Hence a body moving in a circular path has two components of acceleration which are perpendicular to each other. These two components are tangential component and normal component of acceleration.

(*i*) *Tangential component of the acceleration* (f_t). The component of acceleration in the tangential direction is known as tangential acceleration and is denoted by 'f_t'.

∴ Tangential acceleration at *P* will be

$$f_t = \frac{\text{Change of velocity component in tangential direction}}{\text{Time}} = \frac{px}{\delta t} = \frac{(v+\delta v)\cos\delta\theta - v}{\delta t}$$

In the limit, when δt approaches to zero, then $\cos\delta\theta \to 1$.

$$\therefore \quad f_t = \frac{(v+\delta v)-v}{\delta t} = \frac{\delta v}{\delta t} = \frac{dv}{dt} = \frac{d(\omega\times r)}{dt} = \frac{r\,d\omega}{dt} = r\times\alpha \qquad \text{...(4.1)}$$

(*ii*) *Normal component of the acceleration* (f_n). The component of the acceleration in a direction normal to the tangent at that instant is known as normal component of the acceleration. This component is directed towards the centre of the circular path (*i.e.*, in the direction from *P* to *C*). It is also called the *radial acceleration* or *centripetal acceleration*. It is represented by f_n or f_c or f_r.

∴ Normal acceleration at *P* will be,

$$f_n\,(\text{or } f_r \text{ or } f_c) = \frac{\text{Change of velocity component in a direction normal to tangent}}{\text{Time}}$$
$$= \frac{xq}{\delta t}$$
$$= \frac{(v+\delta v)\sin\delta\theta}{\delta t}$$

In the limit, when δt approaches to zero, then $\sin\delta\theta \to \delta\theta$

$$\therefore \quad f_n = \frac{(v+\delta v)\,\delta\theta}{\delta t} = \frac{v\delta\theta}{\delta t} + \frac{\delta v\delta\theta}{\delta t}$$
$$= v\frac{\delta\theta}{\delta t} \qquad \text{(Neglecting product of } \delta v \text{ and } \delta\theta)$$
$$= v\frac{d\theta}{dt} \qquad \left(\text{But } \frac{d\theta}{dt} = \omega\right)$$
$$= v\times\omega = (\omega\times r)\times\omega \qquad (\because \; v = \omega\times r)$$
$$= \omega^2\times r \quad \text{or} \quad \frac{v^2}{r} \qquad \text{...(4.2)}$$

Total acceleration of the body will be the vector sum of the tangential component and normal component as shown in Fig. 4.2.

∵ Total acceleration or resultant acceleration is given by,

$$f = \sqrt{f_t^2 + f_c^2}.$$

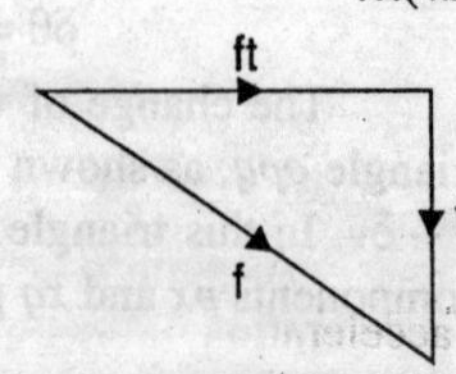

Fig. 4.2

Important Points:

1. If the body is moving with **uniform velocity**, then $\frac{dv}{dt}$ will be zero. Hence from equation (4.1), f_t *i.e.*, tangential acceleration will become zero. Hence the body will have only normal acceleration or radial acceleration or centripetal acceleration. This value is given by,

$$f_c = f_n = f_r = \frac{v^2}{r} = \omega^2 \times r$$

Also the total acceleration will be equal to radial acceleration or centripetal acceleration.

2. If the body is moving on a straight path, the radius r will be infinity and $\frac{v^2}{r}$ will be equal to zero. Hence there will be no normal (or radial) acceleration. Only tangential acceleration will exist. The value of tangential acceleration is given by,

$$f_t = \frac{dv}{dt} = \alpha \times r.$$

Also the total acceleration will be equal to tangential acceleration.

Hence for a body moving on a straight path, total acceleration is equal to $\alpha \times r$ whereas for a body moving along a circular path with uniform velocity, the total acceleration will be equal to radial acceleration *i.e.*, equal to $\omega^2 \times r$.

4.3. Acceleration Diagram for a Link

The acceleration of any point in a mechanism can be determined by two important methods, which are:

(*i*) Analytical method and

(*ii*) Graphical or acceleration diagram method.

If an expression for displacement in terms of time (*t*) is known, then analytical method can be applied. But for most of mechanism, the expression for displacement cannot be determined easily. Hence graphical or acceleration diagram method is generally used.

4.3.1. Acceleration Diagram Method

Fig. 4.3 (*a*) shows a rigid link *AB*. Let the point *B* is moving with respect to point *A*. To find the acceleration of *B* relative to *A*, find acceleration of *B* assuming point *A* to be fixed. And if *A* is assumed fixed, then only possible motion of *B* will be rotation about *A* as centre. For a particular instant,

Let ω = Angular velocity of link *AB*

α = Angular acceleration of link *AB*.

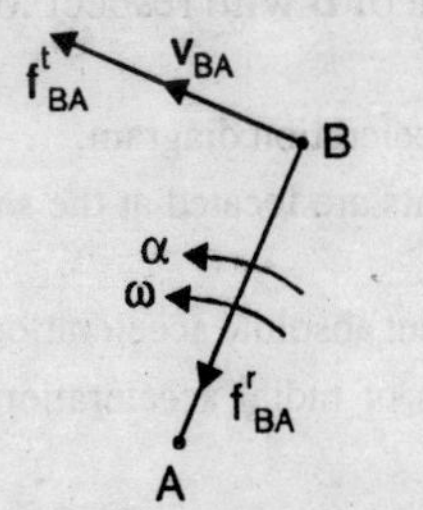

(*a*) Space diagram

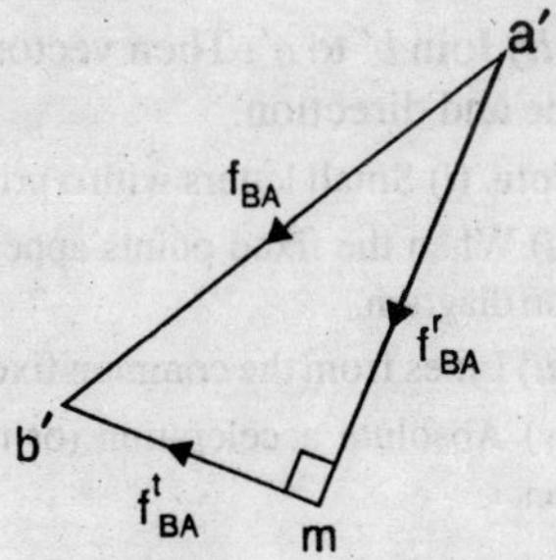

(*b*) Acceleration diagram

Fig. 4.3

As point *B* is rotating about *A*, the velocity of *B* is changing in magnitude* and direction. Hence the acceleration of the point *B* will have two components:

---

*The change in magnitude of velocity is due to angular acceleration. If point *B* is rotating with uniform angular velocity, then angular acceleration will be zero.

1. Radial component (or centripetal component) which is due to angular velocity. It will be acting along BA (or parallel to BA) and will be directed from B towards A. The magnitude of this component will be $\frac{v^2}{r}$ or $\omega^2 \times r$. Hence radial component of B with respect to A,

$$f^r_{BA} = \frac{V^2_{BA}}{BA} \quad \text{or} \quad \omega^2 \times BA \qquad (\because \; r = BA) \qquad ...(4.3)$$

2. Tangential component which is due to angular acceleration (α). This acts parallel to the velocity or it is perpendicular to AB. The magnitude of this component is $\alpha \times BA$. Hence tangential component of B with respect to A, is given by

$$f^t_{BA} = \alpha \times BA \qquad ...(4.4)$$

The total acceleration of B with respect to A is the vector sum of the above two components. This is represented by f_{BA}. Hence total acceleration of B with respect to A, is given by

$$f_{BA} = \text{vector sum of } f^r_{BA} + f^t_{BA} \qquad ...(4.5)$$

$$= \text{vector sum of } \frac{v^2_{BA}}{BA} + \alpha \, . \, BA \qquad ...(4.6)$$

Note. If point B is rotating with uniform angular velocity with respect to A, then angular acceleration (α) will be zero. And hence the tangential component (which is equal to $\alpha \, . \, BA$) will be zero. Then total acceleration of B with respect to A will be given by, $f_{BA} = \frac{v^2_{BA}}{BA}$.

Acceleration Diagram for the link AB

The acceleration diagram for the link AB is shown in Fig. 4.3 (*b*). This is drawn as given below :

(*i*) Take any point a' (Here fixed point should be taken). The radial component of acceleration is acting along BA whereas tangential component is acting perpendicular to BA in the direction of velocity. From a', draw vector $a'm$ parallel to BA to represent the radial component of acceleration of B with respect to A *i.e.*, f^r_{BA} such that vector

$$a'm = f^r_{BA} = \frac{v^2_{BA}}{BA}.$$

(*ii*) From point m, draw vector mb' perpendicular to AB (or at right angles to vector mb') to represent the tangential component of acceleration of B with respect to A *i.e.*, f^t_{BA} such that vector $mb' = f^t_{BA}$.

The tangential component of acceleration acts in the direction of velocity.

(*iii*) Join b' to a'. Then vector $a'b'$ represents the total acceleration of B with respect to A (*i.e.*, f_{BA}) in magnitude and direction.

Note. (*i*) Small letters with a prime such as a', b' etc. will be used for acceleration diagram.

(*ii*) When the fixed points appear in more than one place, all these points are located at the same place in the acceleration diagram.

(*iii*) Lines from the common fixed points in acceleration diagram represent absolute accelerations.

(*iv*) Absolute acceleration (or total acceleration) of a point is the sum of radial acceleration and tangential acceleration.

4.4. Acceleration Diagram for a Slider Crank Mechanism

Fig. 4.4 (*a*) shows a slider crank mechanism in which the crank AB is rotating clockwise with a *uniform* angular velocity ω about fixed point A. Let the crank makes an angle θ with the inner deed centre (I.D.C.). The connecting rod is represented by CB.

Now the velocity of B with respect to A (or only the velocity of B as point A is fixed) is given by,

$$v_{BA} \quad \text{or} \quad v_B = \omega \times AB$$

This velocity will be perpendicular to AB as shown in Fig. 4.4 (a). Now the velocity diagram can be drawn as shown in Fig. 4.4 (b). The procedure of drawing velocity diagram is given in Art. 3.11.

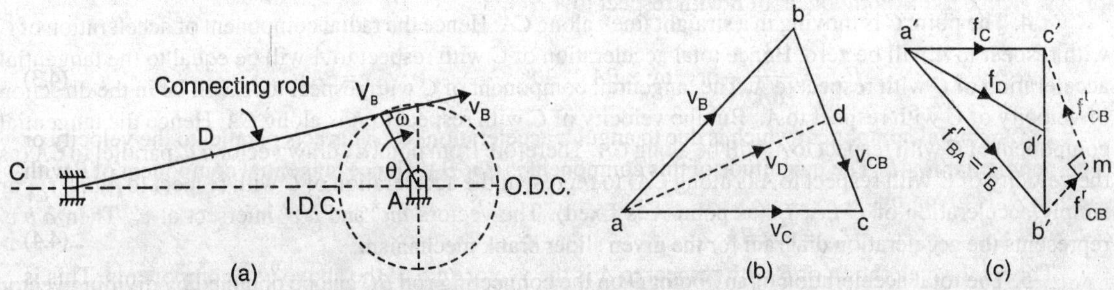

Fig. 4.4. (a) Space diagram (b) Velocity diagram (c) Acceleration diagram.

For drawing the acceleration diagram, first calculate the radial components of acceleration of different links.

Now radial acceleration of B with respect to A (or simply radial acceleration of B as point A is fixed) is given by,

$$f^r_{BA} \quad \text{or} \quad f^r_B = \frac{v^2_{BA} \text{ or } v^2_B}{AB} = \frac{v^2_B}{AB}$$

As the point B is rotating with uniform angular velocity with respect to A, hence angular acceleration of B is zero. Therefore the tangential acceleration of B (which is equal to angular acceleration × BA) with respect to A, will be zero. Hence total acceleration of B with respect to A [which is given by equation (4.3)] will be,

$$f_{BA} \quad \text{or} \quad f_B = f^r_{BA} \qquad (\text{as } f^t_{BA} = \text{Tangential acceleration is zero})$$

$$= \frac{v^2_{BA}}{AB} \quad \text{or} \quad \frac{v^2_B}{AB}$$

$$\therefore \qquad f_B = \frac{v^2_B}{AB}$$

This acceleration is acting along BA. Hence the magnitude and direction of total acceleration of B with respect to A (or simply total acceleration of B) is known.

The radial component of acceleration of C with respect to B is given by, $f^r_{CB} = \dfrac{v^2_{CB}}{BC}$.

The magnitude of v_{CB} ($i.e.$, velocity of C with respect to B) will be obtained from velocity diagram by vector bc as shown in Fig. 4.4 (b).

Now the acceleration diagram can be drawn as shown in Fig. 4.4 (c) as per method given below :

1. Choose a suitable scale for drawing the acceleration diagram. Take any point a' corresponding to fixed point A. Total acceleration of point B with respect to A is known. It is equal to $\dfrac{v^2_B}{AB}$ and is acting along BA. From a' draw vector $a'b'$ parallel to BA to represent the radial component of acceleration of B with respect to A (which is in this case is equal to f_B $i.e.$, total acceleration of B with respect to A) such that vector $a'b' = f_B$.

2. The point C is having radial component of acceleration with respect to B and also tangential component of acceleration with respect to B. Radial component is acting along CB whereas tangentical component is acting perpendicular to BC. Hence from b', draw vector $b'm$ parallel to CB to represent the radial component of acceleration of C with respect to B ($i.e.$, f^r_{CB}). As the magnitude of f^r_{CB} is known, therefore vector $b'm = f^r_{CB} = \dfrac{v^2_{CB}}{BC}$.

3. From point m, draw vector mc' perpendicular to BC to represent the tangential component of the acceleration of C with respect to B (*i.e.*, f^t_{CB}). The magnitude of f^t_{CB} is not yet known. The line mc' will also be at right angles to $b'm$.

4. The point C is moving in a straight line* along CA. Hence the radial component of acceleration of C with respect to A will be zero. Hence total acceleration of C with respect to A will be equal to the tangential acceleration of C with respect to A. The tangential component of C with respect to A, will be in the direction of velocity of C with respect to A. But the velocity of C with respect to A is along CA. Hence the tangential component of C with respect to A will be along CA. Therefore from point a', draw vector $a'c'$ parallel to CA (as the velocity of C with respect to A is along CA) to represent the acceleration of C with respect to A *i.e.*, f_{CA} or simply acceleration of C *i.e.*, f_C (as point A is fixed). The vectors mc' and $a'c'$ intersect at c'. Then $a'b'c'$ represents the acceleration diagram for the given slider crank mechanism.

5. The total acceleration of any point D on the connecting rod BC can be obtained by dividing vector $b'c'$ at d' in the same ratio as D divides BC. This means that

$$\frac{b'd'}{b'c'} = \frac{BD}{BC}.$$

$$\therefore \quad b'd' = \frac{BD}{BC} \times b'c'.$$

6. The angular acceleration of connecting rod CB is given by,

$$f^t_{CB} = \alpha_{CB} \times CB$$

$$\therefore \quad \alpha_{CB} = \frac{f^t_{CB}}{CB} \quad \text{where } \alpha_{CB} = \text{Angular acceleration of } CB.$$

Problem 4.1. *The crank of a slider crank mechanism is 15 cm and the connecting rod is 60 cm long. The crank makes 300 r.p.m. in the clockwise direction. When it has turned 45° from the inner dead centre position, determine :*

(*i*) *Acceleration of the mid-point of the connecting rod and*

(*ii*) *Angular acceleration of the connecting rod.*

Sol. Given :

Crank length, $AB = 15$ cm

Connecting rod length, $BC = 60$ cm

Crank speed, $N = 300$ r.p.m.

Angle turned by crank from I.D.C. $= 45°$

Angular velocity of crank, $\omega = \frac{2\pi N}{60} = \frac{2\pi \times 300}{60} = 31.42$ rad/s

$\therefore$ Velocity of B with respect to A (or velocity of B only as point A is fixed) is given by,

$$v_{BA} = v_B = \omega \times AB = 31.42 \times 15 = 471.3 \text{ cm/s}$$

The direction of v_B will be perpendicular to AB.

First draw the space diagram to a suitable scale as shown in Fig. 4.5 (*a*). Now draw the velocity diagram as shown in Fig. 4.5 (*b*) as per method given below :

1. Take any point a (fixed point). Choose a suitable velocity scale and draw vector ab perpendicular to AB to represent the velocity of B with respect to A or simply the velocity of B such that vector $ab = v_{BA}$ or v_B $= 471.3$ cm/s.

2. As the point C is moving with respect to B and also with respect to A. The velocity of C with respect to B is perpendicular to BC whereas the velocity of C with respect to A is along CA (*i.e.*, horizontally). Hence

*When a point moves along a straight line, then its radial acceleration will be zero.

from point b, draw vector bc perpendicular to BC to represent the velocity of C with respect to B and from point a draw vector ac horizontally to represent the velocity of C with respect to A. The vectors bc and ac intersect at point c. Then abc is the velocity diagram.

3. As the point D is the middle point of CB, hence corresponding point d will be the middle point of vector bc. Join ad. Then vector ad represents the velocity of point D.

From velocity diagram, we have :

(*i*) Velocity of D, v_D = vector ad = 410 cm/s

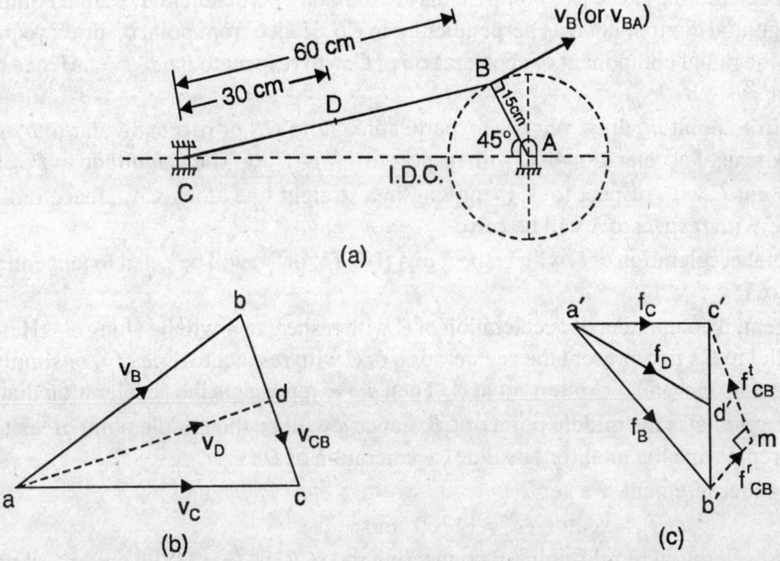

Fig. 4.5. (*a*) Space diagram (*b*) velocity diagram (*c*) Acceleration diagram.

(*ii*) Velocity of C, v_C = vector ac = 400 cm/s

(*iii*) Velocity of C with respect to B, v_{CB} = vector bc = 340 cm/s.

Acceleration diagram

Before drawing acceleration diagram, we must calculate the radial acceleration $\left(i.e., \frac{v^2}{r}\right)$ of each link.

∴ Radial acceleration of B with respect to A (or simply radial acceleration of B as point A is fixed) is given by,

$$f^r_{BA} \text{ or } f^r_B = \frac{v^2_B}{AB} = \frac{471.3^2}{15} = 14808 \text{ cm/s}^2$$

Radial acceleration of C with respect to B is given by,

$$f^r_{CB} = \frac{v^2_{CB}}{CB} = \frac{340^2}{60} \qquad (\because \; v_{CB} = 340 \text{ and } CB = 60 \text{ cm})$$

$$= 1926.7 \text{ cm/s}^2$$

As the point B is rotating with uniform angular velocity with respect to A, hence angular acceleration of B with respect to A is zero. Therefore the tangential acceleration of B with respect to A will be zero. Hence total acceleration of B with respect to A will be equal to radial acceleration of B with respect to A and will be acting along BA.

$$\therefore \quad f_{BA} \text{ or } f_B = f^r_{BA} \text{ or } f^r_B = 14808 \text{ cm/s}.$$

Now the acceleration diagram can be drawn as shown in Fig. 4.5 (*c*), as per method given below :

1. Choose a suitable scale for drawing acceleration diagram. Take any point a' corresponding to fixed point A. Total acceleration of B with respect to A is known. It is equal to radial acceleration of B and is acting along BA. From a' draw vector $a'b'$ parallel to BA to represent the radial acceleration of B with respect to A (which is in this case is equal to f_{BA} or f_B *i.e.*, total acceleration of B). Hence cut vector $a'b' = f_B$ or $f^r{}_{BA} = 14808$ cm/s.

2. The point C with respect to B is having radial component of acceleration and also tangential component of acceleration. These two components are mutually perpendicular. Radial component acts along CB whereas tangential component acts perpendicular to CB. Hence from point b', draw vector $b'm$ parallel to CB to represent the radial component of acceleration of C with respect to B *i.e.*, $f^r{}_{CB}$. Hence take vector $b'm = f^r{}_{CB} = 1926.7$ cm/s^2.

3. Now from point m, draw vector mc' perpendicular to CB or perpendicular to $b'm$ to represent the tangential component of acceleration of C with respect to B (*i.e.*, $f^t{}_{CB}$). The magnitude to $f^t{}_{CB}$ is not yet known.

4. The point C with respect to A, is moving in a straight line along CA. Hence radial component of acceleration of C with respect to A will be zero.

Hence total acceleration of C with respect to A (*i.e.*, f_{CA} or f_C) will be equal to tangential acceleration of C with respect to A.

The tangential component of acceleration of C with respect to A will be along CA. Hence from a', draw vector $a'c'$ parallel to CA to represent the acceleration of C with respect to A *i.e.*, f_{CA} or simply acceleration of C *i.e.*, f_C. The vectors mc' and $a'c'$ intersect at c'. Then $a'b'c'$ represents the acceleration diagram.

5. As the point D is the middle point of CB, hence d' will be the middle point of vector $b'c'$. Join $a'd'$. The vector $a'd'$ represents the total (or absolute) acceleration of D *i.e.*, f_D.

Hence by measurement, we get

$$f_D = \text{vector } a'd' = 11700 \text{ cm/s}^2$$

(*i*) ∴ Acceleration of mid-point of connecting rod (CB) is $f_D = \mathbf{11700\ cm/s^2}$. **Ans.**

(*ii*) *Angular acceleration of connecting rod (CB).* We know that

$$f^t{}_{CB} = \alpha_{CB} \times CB$$

where $f^t{}_{CB}$ = Tangential acceleration of C with respect to B (or tangential acceleration of connecting rod CB).

= Vector mc' = 10300 cm/s^2 (by measurement)

α_{CB} = Angular acceleration of connecting rod

$$= \frac{f^t{}_{CB}}{\text{Length } CB}$$

$$= \frac{10300}{60} = \mathbf{171.67\ rad/s^2}. \quad \textbf{Ans.}$$

Problem 4.2. *A reciprocating engine mechanism is shown in Fig. 4.6 (a). The crank CB = 10 cm and connecting rod BA = 30 cm with the centre of gravity G, 10 cm from B. In the position shown, the crank has a velocity of 75 rad/s and an angular acceleration of 1200 rad/s². Find : (a) the velocity and acceleration of G and (b) the angular velocity and angular acceleration of AB.* (AMIE, Summer 1981)

Sol. Given :

Crank length, $CB = 10$ cm, connecting rod, $BA = 30$ cm

Distance, $BG = 10$ cm

Angular velocity of crank, $\omega_{BC} = 75$ rad/s

Angular acceleration of crank, $\alpha_{BC} = 1200$ rad/s^2.

∴ Linear velocity of crank, $v_{BC} = \omega_{BC} \times \text{Length } BC$

$$= 75 \times 10 = 750 \text{ cm/s} = \frac{750}{100} = 7.5 \text{ m/s}$$

This velocity will be perpendicular to BC.

Now tangential acceleration of crank = Angular acceleration of crank × Crank length

$$\therefore \quad f^t_{BC} = \alpha_{BC} \times \text{Length } BC$$
$$= 1200 \times 10 = 12000 \text{ cm/s}^2$$
$$= \frac{1200}{100} = 120 \text{ m/s}^2$$

Please note that in this problem, the crank is having tangential acceleration whereas in problem 4.1 the tangential acceleration of crank was zero.

Now draw the space diagram as shown in Fig. 4.6 (*a*) to some suitable scale. Now draw the velocity diagram as shown in Fig. 4.6 (*b*) as per method given below :

1. Choose a suitable velocity scale. Take any point c corresponding to fixed point C. The velocity of B with respect to C is perpendicular to CB. Draw vector cb perpendicular to BC to represent the velocity of B with respect to C (or simply the velocity of B) *i.e.*, v_{BC} or v_B such that vector $cb = v_{BC} = v_B = 7.5$ m/s.

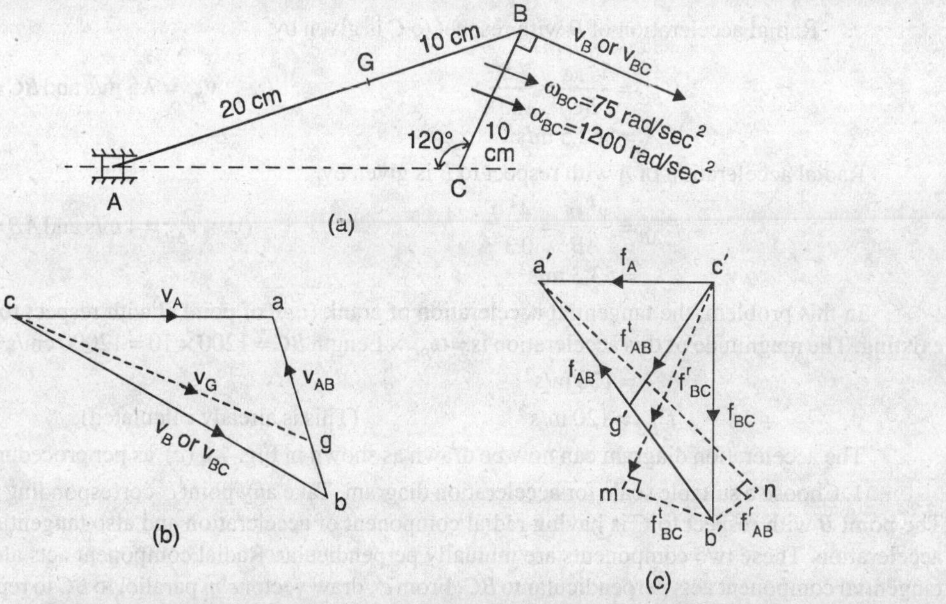

Fig. 4.6. (*a*) Space diagram (*b*) Velocity diagram (*c*) Acceleration diagram.

2. The point A is moving with respect to B and also with respect to C. The velocity of A with respect to B is perpendicular to AB whereas the velocity of A with respect to C is along AC. Hence from the point b, draw vector ba perpendicular to AB to represent the velocity of A with respect to B *i.e.*, v_{AB} and also from point c, draw a vector ca along the path of motion of A (which is horizontal) to represent the velocity of A with respect to C. The vectors ba and ca intersect at a.

3. The point G lies in AB and hence corresponding point g will lie in vector ab of velocity diagram. To find the location of g, divide vector ab at g in the same ratio as G divides AB. This means

$$\frac{bg}{ab} = \frac{BG}{AB}$$

$$\therefore \quad bg = \frac{BG}{AB} \times ab = \frac{10}{30} \times ab = \frac{ab}{3}$$

The vector cg represents the velocity of G *i.e.*, v_G.

By measurement, vector $cg = 6.8$ m/s.

$\therefore$ Velocity of G, v_G = vector cg = **6.8 m/s. Ans.**

The angular velocity of connecting rod is given by,

$$\omega_{AB} = \frac{v_{AB}}{\text{Length } AB}$$

But v_{AB} (*i.e.*, the velocity of A with respect to B) from velocity diagram is given by vector ba. By measurement vector $ba = 4$ m/s.

$\therefore$ v_{AB} = vector $ba = 4$ m/s.

$\therefore$ $\omega_{AB} = \frac{4}{0.3}$ ($\because$ $AB = 30$ cm $= 0.3$ m)

= 13.3 rad/s. Ans.

Acceleration of G and angular acceleration of AB

These will be obtained from acceleration diagram. Before drawing acceleration diagram, we must calculate the radial acceleration $\left(i.e., \frac{v^2}{r}\right)$ of each link.

$\therefore$ Radial acceleration of B with respect to C is given by,

$$f^r_{BC} = \frac{v^2_{BC}}{BC} = \frac{7.5^2}{0.1} \qquad (\because\ v_{BC} = 7.5 \text{ m/s and } BC = 10 \text{ cm} = 0.1 \text{ m})$$

$= 562.5$ m/s^2

Radial acceleration of A with respect to B is given by,

$$f^r_{AB} = \frac{v^2_{AB}}{AB} = \frac{4^2}{0.3} \qquad (\because\ v_{AB} = 4 \text{ m/s and } AB = 30 \text{ cm} = 0.3 \text{ m})$$

$= 53.3$ m/s^2

In this problem, the tangential acceleration of crank (*i.e.*, of point B with respect to C, f^t_{BC}) is also existing. The magnitude of this acceleration is $= \alpha_{BC} \times$ Length $BC = 1200 \times 10 = 12000$ cm/s^2

$= 120$ m/s^2

$\therefore$ $f^t_{BC} = 120$ m/s^2 (This is already calculated)

The acceleration diagram can now be drawn as shown in Fig. 4.6 (c) as per procedure given below :

1. Choose a suitable scale for acceleration diagram. Take any point c' corresponding to fixed point C. The point B with respect to C is having radial component of acceleration and also tangential component of acceleration. These two components are mutually perpendicular. Radial component acts along BC whereas tangential component acts perpendicular to BC. From c', draw vector $c'm$ parallel to BC to represent the radial component of acceleration of point B with respect to C *i.e.*, f^r_{BC} such that vector $c'm = f^r_{BC} = 562.5$ m/s^2.

2. From m, draw vector mb' perpendicular to $c'm$ (or perpendicular to BC) to represent the tangential component of acceleration of B with respect to C *i.e.*, f^t_{BC} such that vector $mb' = f^t_{BC} = 120$ m/s^2. [mb' is in the direction of velocity of B]

3. Join $c'b'$. Then vector $c'b'$ represents the total acceleration of B with respect to C *i.e.*, f_{BC}.

4. The point A with respect to B is having radial component of acceleration and also tangential component of acceleration. These two components are mutually perpendicular. Radial component acts along AB whereas tangential component acts perpendicular to AB. From point b', draw vector $b'n$ parallel to AB to represent radial component of acceleration of A with respect to B *i.e.*, f^r_{BA} such that vector $b'n = f^r_{AB} = 53.3$ m/s^2. And from point n, draw vector na' perpendicular to $b'n$ (or perpendicular to AB) to represent tangential component of acceleration of A with respect to B *i.e.*, f^t_{AB}. But the magnitude of f^t_{AB} is not known.

5. The point A with respect to C is moving in a straight line along AC. Hence total acceleration of A will be along AC. Now from point c', draw vector $c'a'$ parallel to AC to represent the acceleration of A with respect to C *i.e.*, f_{AC} or simply the acceleration of A *i.e.*, f_A. The vectors na' and $c'a'$ intersect at a'. Join $b'a'$. The vector $b'a'$ represent the total acceleration of A with respect to B *i.e.*, f_{AB}.

6. The point G lies in AB and hence corresponding point g' in acceleration diagram will lie in vector $a'b'$. To find the location of g', divide vector $a'b'$ at g' in the same ratio as G divides AB. This means

$$\frac{b'g'}{a'b'} = \frac{BG}{AB}$$

$$\therefore \quad b'g' = \frac{BG}{AB} \times a'b' = \frac{10}{30} \times a'b' = \frac{a'b'}{3}$$

Join c' to g'. Then vector $c'g'$ represents the total acceleration of G *i.e.*, f_G.

By measurement, vector $c'g' = 415 \text{ m/s}^2$.

$\therefore$ Acceleration of G, f_G = vector $c'g'$ = **415 m/s². Ans.**

The angular acceleration of connecting rod AB is given by,

$$f^t_{AB} = \alpha_{AB} \times \text{Length } AB$$

where f^t_{AB} = Tangential acceleration of A with respect to B

= Vector $na' = 550 \text{ m/s}^2$ (By measurement)

$$\therefore \quad \alpha_{AB} = \frac{f^t_{AB}}{\text{Length } AB} = \frac{550}{0.3} \qquad (\because \; AB = 30 \text{ cm} = 0.3 \text{ m})$$

= **1833.34 rad/s². Ans.**

Problem 4.3. *A four bar kinematic chain is represented by a quadrilateral ABCD in which AD is fixed and is 400 mm long. The crank AB 75 mm long rotates in a clockwise direction at 120 r.p.m. and drives the link CD 125 mm long by means of the connecting link BC 350 mm long. Determine the angle through which CD oscillates and find the angular velocities of the links BC and CD in one of the positions when BC is perpendicular to AB.* (AMIE, Summer 1987)

Sol. Given :

AD is fixed. Length AD = 400 mm = 40 cm, AB = 75 mm = 7.5 cm,

N = 120 r.p.m., CD = 125 mm = 12.5 cm, BC = 350 mm = 35 cm

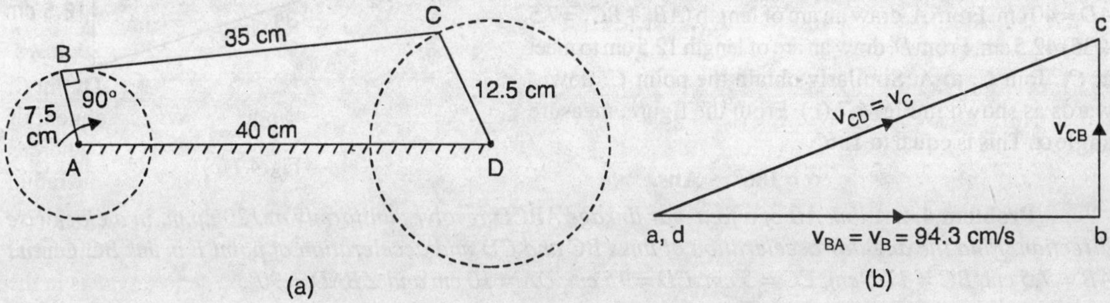

Fig. 4.7. (*a*) Space diagram (*b*) Velocity diagram.

Now $\quad N_{BA} = 120$ r.p.m.

$$\therefore \quad \omega_{BA} = \frac{2\pi \times 120}{60} = 12.57 \text{ rad/s}$$

Now velocity of B with respect to A,

$$v_{BA} = \omega_{BA} \times \text{Length } BA$$
$$= 12.57 \times 7.5 = 94.3 \text{ cm/s}$$

As point A is fixed, hence velocity of B (or velocity of B with respect to A) will be 94.3 cm/s.

$$\therefore \quad v_{BA} = v_B = 94.3 \text{ cm/s.}$$

First draw the space diagram to a suitable scale as shown in Fig. 4.7 (*a*). Now draw the velocity diagram as shown in Fig. 4.7 (*b*) as per method given below :

1. Since *A* and *D* are fixed points, therefore these points (*a, d*) lie at one place in velocity diagram. Choose a suitable velocity scale and from point *a* draw vector *ab* perpendicular to *AB* to represent the velocity of *B* with respect to *A* (*i.e.,* v_{BA}). Take *ab* = 94.3 cm/s.

2. The point *C* moves with respect to point *B* and also with respect to point *D*. The velocity of *C* with respect to *B* is perpendicular to *BC* whereas the velocity of *C* with respect to *D* is perpendicular to *DC*. Hence from point *b*, draw vector *bc* perpendicular to *BC*. Since link *AD* is fixed hence point *a* and *d* will coincide in the velocity diagram. From point *d* (or point *a*), draw vector *dc* (or *ac*) perpendicular to *DC* so as to intersect vector *bc* in *c*. Then *abc* is the velocity diagram in which :

Vector *bc* = Velocity of *C* with respect to *B* (*i.e.,* v_{CB})

Vector *dc* = Velocity of *C* with respect to *D* (*i.e.,* v_{CD}). As point *D* is fixed, this is also the velocity of point *C* (*i.e.,* v_C).

By measurement, from velocity diagram, we get

$$v_{CB} = \text{vector } bc = 36.75 \text{ cm/s}$$

and $$v_{CD} \text{ or } v_C = \text{Vector } dc = 100.8 \text{ cm/s}$$

(*i*) *Angular velocities of the links BC and CD*

Let ω_{CB} = Angular velocity of link *CB*

ω_{CD} = Angular velocity of link *CD*

Then $$\omega_{CB} = \frac{v_{CB}}{CB} = \frac{36.75}{35} = \mathbf{1.05 \text{ rad/s.}} \quad \textbf{Ans.}$$

and $$\omega_{CD} = \frac{v_{CD}}{CD} = \frac{100.8}{12.5} = \mathbf{8.06 \text{ rad/s.}} \quad \textbf{Ans.}$$

(*ii*) *Angle through which CD oscillates*

Let α = angle through which *CD* oscillates. Take *AD* = 40 cm. From *A*, draw an arc of length ($AB_1 + BC_1 = 7.5 + 35$) 42.5 cm. From *D*, draw an arc of length 12.5 cm to meet at C_1. Join C_1 to *A*. Similarly obtain the point C_2 downwards as shown in Fig. 4.7 (*c*). From the figure, measure angle α. This is equal to 186°.

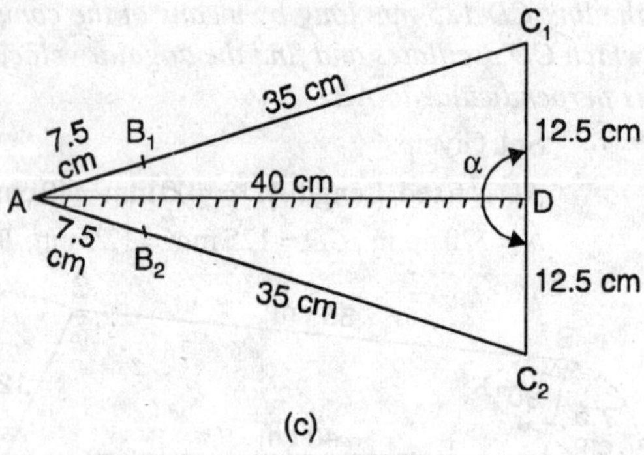

Fig. 4.7 (*c*)

∴ $$\alpha = \mathbf{186°.} \quad \textbf{Ans.}$$

Problem 4.4. *A link AB of a four-bar linkage ABCD revolves uniformly at 120 r.p.m. in a clockwise direction. Find the angular acceleration of links BC and CD and acceleration of point E in link BC. Given : AB = 7.5 cm, BC = 17.5 cm, EC = 5 cm, CD = 15 cm, DA = 10 cm and ∠BAD = 90°.*

(AMIE, Winter 1979)

Sol. Given :

Speed of link *AB*, N_{BA} = 120 r.p.m.

∴ Angular velocity of link *AB*,

$$\omega_{BA} = \frac{2\pi \times N_{BA}}{60} = \frac{2\pi \times 120}{60} = 12.57 \text{ rad/s}$$

Length *AB* = 7.5 cm

∴ Velocity of *B* with respect to *A* (or velocity of *B* as point *A* is fixed),

$$v_{BA} = v_B = \omega_{BA} \times \text{Length } BA = 12.57 \times 7.5 = 94.3 \text{ cm/s}$$

First draw the space diagram of the four-bar linkage *ABCD* with the given dimensions as shown in Fig. 4.8 (*a*) to same suitable scale. Now draw the velocity diagram as shown in Fig. 4.8 (*b*) as per method given below :

1. Since *A* and *D* are fixed points, therefore, these points lie at one place in velocity diagram. Choose a suitable velocity scale and from point *a* draw vector *ab* perpendicular to *AB* and cut off this length equal to 94.3 cm/s. Then this vector *ab* represents the velocity of *B* with respect to *A* (*i.e.*, v_{BA}). As the point *A* is fixed, hence the vector *ab* also represents the velocity of *B* (*i.e.*, v_B).

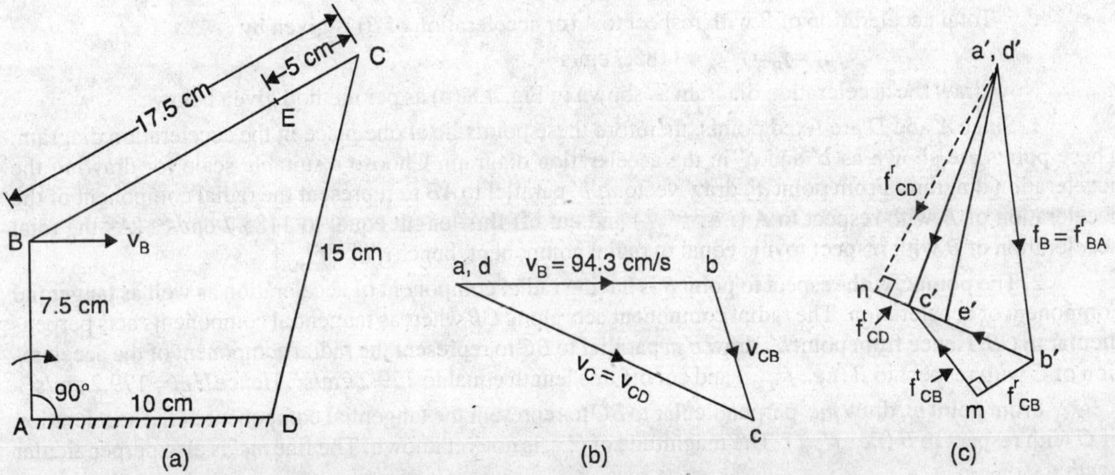

Fig. 4.8. (*a*) Space diagram (*b*) Velocity diagram (*c*) Acceleration diagram.

2. The point *C* moves with respect to point *B* and also with respect to point *D*. The velocity of *C* with respect to *B* is perpendicular to *BC* whereas the velocity of *C* with respect to *D* is perpendicular to *DC*. Hence from point *b*, draw vector *bc* perpendicular to *BC*. Since *AD* is fixed hence point *a* and *d* in the velocity diagram will coincide. From point *d*, draw vector *dc* perpendicular to *DC* so as to intersect vector *bc* in *c*. Then *abc* is the velocity diagram for the given mechanism. In this diagram :

Vector *bc* = Velocity of point *C* with respect to point *B* or v_{CB}

Vector *dc* = Velocity of point *C* with respect to point *D* (*i.e.*, v_{CD}) or velocity of *C* as point *D* is fixed (*i.e.*, v_C).

Vector *ab* = Velocity of point *B* with respect to point *A* (*i.e.*, v_{BA}) or velocity of point *B* as point *A* is fixed *i.e.*, v_{BA} or v_B.

From the velocity diagram, on measurement we find that

$$v_{CB} = \text{Vector } bc = 56 \text{ cm/s},$$
$$v_C = v_{CD} = \text{Vector } dc = 124 \text{ cm/s}.$$

Acceleration Diagram

Before drawing the acceleration diagram, we must calculate the radial components of the acceleration $\left(i.e. \dfrac{v^2}{r}\right)$ of each link. Here *v* is the velocity and *r* is the length of the link.

∴ The radial component of the acceleration of *B* with respect to *A* is given by,

$$f^r_{BA} = \frac{v^2_{BA}}{BA} = \frac{94.3^2}{7.5} = 1185.7 \text{ cm/s}^2$$

Similarly the radial component of the acceleration of *C* with respect to *B*,

$$f^r_{CB} = \frac{v^2_{CB}}{CB} = \frac{56^2}{17.5} = 179.2 \text{ cm/s}^2$$

and radial component of the acceleration of C with respect to D,

$$f^r_{CD} = \frac{v^2_{CD}}{CD} = \frac{124^2}{15} = 102.5 \text{ cm/s}^2$$

As the link AB revolves uniformly, the angular acceleration of AB is zero, therefore there will be no tangential component of acceleration of B with respect to A.

∴ Total acceleration of B with respect to A (or acceleration of B) is given by

$$f_{BA} = f_B = f^r_{BA} = 1185.7 \text{ cm/s}^2$$

Now draw the acceleration diagram as shown in Fig. 4.8 (*c*) as per method given below :

1. Since A and D are fixed points, therefore these points lie at one place in the acceleration diagram. These points are shown as a' and d' in the acceleration diagram. Choose a suitable scale for drawing the acceleration diagram. From point a', draw vector $a'b'$ parallel to AB to represent the radial component of the acceleration of B with respect to A (*i.e.*, f^r_{BA}) and cut off this length equal to 1185.7 cm/s². [As the total acceleration of B with respect to A is equal to radial component, hence $a'b' = f^r_{AB}$]

2. The point C with respect to point B is having radial component of acceleration as well as tangential component of acceleration. The radial component acts along CB whereas tangential components acts perpendicular to CB. Hence from point b', draw $b'm$ parallel to BC to represent the radial component of the acceleration of C with respect to B (*i.e.*, f^r_{CB}) and cut off this length equal to 179.2 cm/s². Hence $b'm$ = 179.2 cm/s².

From point m, draw mc' perpendicular to BC to represent the tangential component of the acceleration of C with respect to B (*i.e.*, f^t_{CB}). The magnitude of f^t_{CB} is not yet known. The line mc' is also perpendicular to mb'.

3. The point C with respect to point D is having radial component of acceleration as well as tangential component of acceleration. The radial component acts along CD whereas tangential component acts perpendicular to CD. Hence from point d', which coincides with point a', draw $d'n$ parallel to CD to represent the radial component of acceleration of C with respect to D (*i.e.*, f^r_{CD}) and cut off this length = 1025 cm/s². Hence $d'n$ = 1025 cm/s².

4. From point n, draw nc' perpendicular to CD (or perpendicular to nd') to represent the tangential component of acceleration of C with respect to D (*i.e.*, f^t_{CD}) intersecting mc' at c'. Then $a'b'c'$ represents the acceleration diagram.

By measurement, we find that

$$f^t_{CB} = \text{Vector } mc' = 250 \text{ cm/s}^2 \text{ and}$$
$$f^t_{CD} = \text{Vector } nc' = 185 \text{ cm/s}^2.$$

(i) Angular accelerations of links BC and CD

$$\text{Angular acceleration of } BC = \frac{\text{Tangential acceleration of } BC}{\text{Length of link } BC}$$

$$= \frac{f^t_{CB}}{CB} = \frac{250}{17.5} = \mathbf{14.3 \text{ rad/s}^2}. \quad \textbf{Ans.}$$

$$\text{Angular acceleration of } CD = \frac{\text{Tangential acceleration of } CD}{\text{Length of } CD}$$

$$= \frac{f^t_{CD}}{CD} = \frac{185}{15} = \mathbf{12.3 \text{ rad/s}^2}. \quad \textbf{Ans.}$$

(ii) Acceleration of point E

In order to find the acceleration of the point E, first locate the corresponding point e' in the acceleration diagram such that

$$\frac{b'e'}{b'c'} = \frac{BE}{BC}.$$

Now join $a'e'$ as shown. Then $a'e'$ is the acceleration of the point E.

By measurement, we find that $a'e' = 1075$ cm/s^2.

$\therefore$ Acceleration of E = **1075 cm/s^2. Ans.**

Problem 4.5. *In the steam engine mechanism, shown in Fig. 4.9, the crank AB rotates at 200 r.p.m. Find the velocities of C, D, E, F and P. Also find the acceleration of the slider at C. The dimensions of the various links are : AB = 12 cm, BC = 48 cm, CD = 18 cm, DE = 36 cm and EF = 12 cm and FP = 36 cm.*

(AMIE, Winter 1976)

Sol. Given :

$$N_{BA} = 200 \text{ r.p.m.}$$

$$\therefore \quad \omega_{BA} = \frac{2\pi \times 200}{60} = 20.94 \text{ rad/s}$$

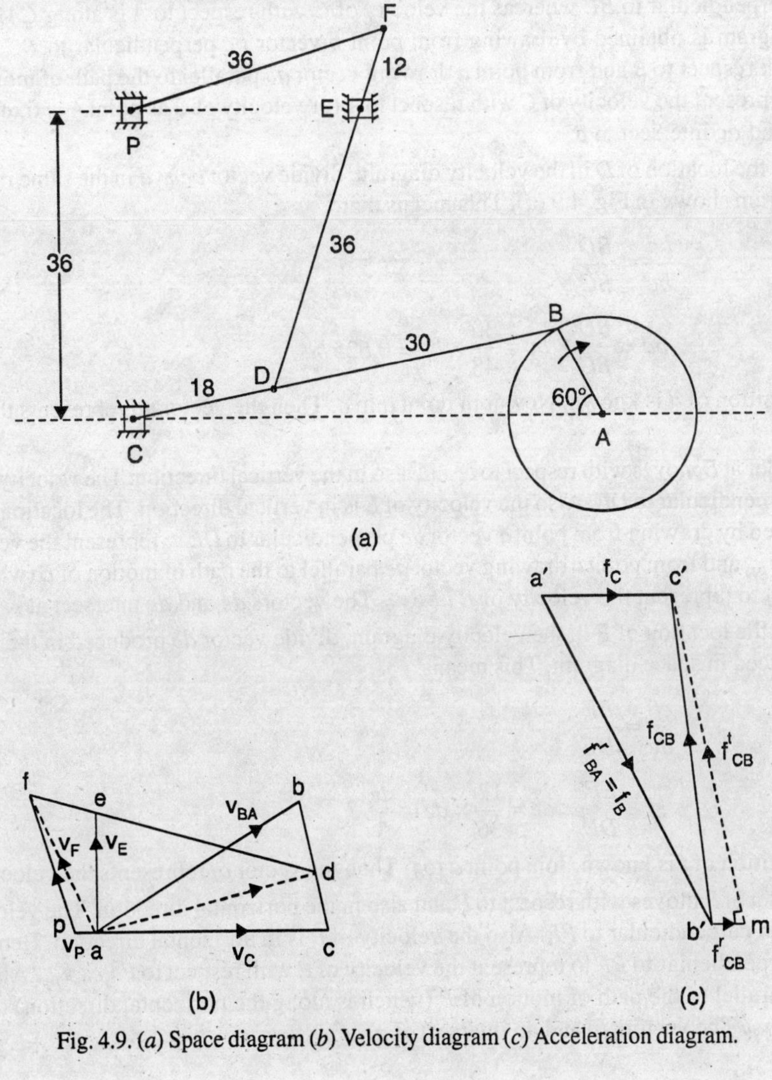

Fig. 4.9. (*a*) Space diagram (*b*) Velocity diagram (*c*) Acceleration diagram.

Now the velocity of point B with respect to A is given by,

$$v_{BA} = \omega_{BA} \times \text{Length } AB$$
$$= 20.94 \times 12 \text{ cm/s} \qquad (\because \ AB = 12 \text{ cm})$$
$$= 251.3 \text{ cm/s}$$

The direction of this velocity will be perpendicular to AB. As point A is fixed, hence this is also the velocity of point B.

$$\therefore \qquad v_{BA} = v_B = 251.3 \text{ cm/s.}$$

First draw the space diagram of the given mechanism to some suitable scale as shown in Fig. 4.9 (a). Now draw the velocity diagram as shown in Fig. 4.9 (b) as per method given below :

1. The point A is fixed hence the corresponding point will be a in velocity diagram. Take point a any where choose a suitable velocity scale and from point a draw vector ab perpendicular to AB to represent the velocity of B with respect to A or velocity of B (*i.e.*, v_{BA} or v_B), such that vector $ab = v_{BA} = v_B = 251.3$ cm/s.

2. Now the point C is moving with respect to B and also with respect to A. The velocity of C with respect to B is perpendicular to BC whereas the velocity of C with respect to A is along CA. The location of C in velocity diagram is obtained by drawing from point b vector bc perpendicular to BC to represent the velocity of C with respect to B and from point a drawing vector ac parallel to the path of motion of C (which is along CA) to represent the velocity of C with respect to A (or velocity of C as point A is fixed) *i.e.*, v_{CA} or v_C. The vectors bc and ac intersect at c.

3. To find the location of D in the velocity diagram, divide vector bc at d in the same ratio as D divides BC in space diagram shown in Fig. 4.9 (a). This means that

$$\frac{bd}{bc} = \frac{BD}{BC}$$

$$\therefore \qquad bd = \frac{BD}{BC} \times bc = \frac{30}{48} \times bc = 0.625\, bc$$

Hence position of d is known. Now join point a to d. Then the vector ad represents the velocity of D *i.e.*, v_D.

4. The slider at E moves with respect to D and also in the vertical direction. The velocity of slider E with respect to D is perpendicular to DE. Also the velocity of E is in vertical direction. The location of E in velocity diagram is obtained by drawing from point d vector de perpendicular to DE to represent the velocity of E with respect to D *i.e.*, v_{ED} and from point a drawing vector ae parallel to the path of motion of E (which is along the vertical direction) to represent the velocity of E *i.e.*, v_E. The vectors de and ae intersect at e.

5. To find the location of F in the velocity diagram, divide vector de produced in the same ratio as F divides DE produced in space diagram. This means

$$\frac{ef}{de} = \frac{EF}{DE}$$

or
$$ef = \frac{EF}{DE} \times de = \frac{12}{36} \times de = \frac{de}{3}$$

Hence position of f is known. Join point a to f. Then the vector af represents the velocity of F *i.e.*, v_F.

6. The slider at P moves with respect to F and also in the horizontal direction. The velocity of slider P with respect to F is perpendicular to PF. Also the velocity of P is in horizontal direction. Hence from point f draw vector fp perpendicular to PF to represent the velocity of P with respect to F *i.e.*, v_{PF}. Also from point a draw vector ap parallel to the path of motion of P (which is along the horizontal direction) to represent the velocity of P *i.e.*, v_P. The vectors fp and ap intersect at p.

Now by measurements from velocity diagram, we get :

(*i*) Velocity of C, v_C = Vector ac = **242.5 cm/s. Ans.**

(*ii*) Velocity of D, v_D = Vector ad = **235 cm/s. Ans.**

(*iii*) Velocity of E, v_E = Vector ae = **100 cm/s. Ans.**

(*iv*) Velocity of F, v_F = Vector af = **140 cm/s. Ans.**

(*v*) Velocity of P, v_P = Vector ap = **30 cm/s. Ans.**

Acceleration of the slider at C

This is obtained by drawing acceleration diagram. Before drawing acceleration diagram, let us calculate the radial components of acceleration $\left(i.e., \frac{v^2}{r}\right)$ of points B and C.

The radial component of acceleration of B with respect to A,

$$f^r_{BA} = \frac{v^2_{BA}}{AB} = \frac{251.3^2}{12} \qquad (\because \quad v_{BA} = 251.3 \text{ cm/s})$$

$$= 5240 \text{ cm/s}^2$$

The radial component of acceleration of C with respect to B,

$$f^r_{CB} = \frac{v^2_{CB}}{BC}$$

But from velocity diagram v_{CB} = Vector bc = 130 cm/s.

$$\therefore \qquad f^r_{CB} = \frac{130^2}{48} = 352 \text{ cm/s}^2$$

Now draw the acceleration diagram as shown in Fig. 4.9 (*c*) as per method given below :

1. The point A is fixed hence the corresponding point will be a' in acceleration diagram. Start the acceleration diagram from point a'. Choose a suitable acceleration scale and from point a draw vector $a'b'$ parallel to AB to represent the radial component of acceleration of B with respect to A (*i.e.*, f^r_{BA}) such that Vector $a'b' = f^r_{BA} = 5240$ cm/s^2.

As the total acceleration of B with respect to A is equal to radial component, hence $a'b' = f^r_{AB}$.

2. The point C with respect to point B is having radial component of acceleration as well as tangential component. The radial component acts along CB whereas the tangential component acts perpendicular to CB. Now from point b', draw vector $b'm$ parallel to BC to represent the radial component of acceleration of C with respect to B (*i.e.*, f^r_{CB}) such that, vector $b'm = f^r_{CB} = 352$ cm/s^2.

From point m, draw vector mc' perpendicular to the vector $b'm$ to represent the tangential component of acceleration of C with respect to B (*i.e.*, f^t_{CB}). But the magnitude of f^r_{CB} is not known.

3. The point C with respect to A, is moving in a straight along CA. The radial component of C with respect to A will be zero. The total acceleration will be along CA. Now from point a', draw vector $a'c'$ parallel to the path of motion of slider C (which along AC) to represent the acceleration of C (*i.e.*, f_C). The vectors mc' and $a'c'$ intersect at c'. The vector $a'c'$ represents the acceleration of C.

Hence by measurement from acceleration diagram, we find the acceleration of the slider at C,

f_C = vector $a'c'$ = **1900 cm/s². Ans.**

4.5. Coriolis Acceleration Component

When a slider is sliding along a rotating link, Coriolis component of acceleration comes into existence.

Consider a link OP rotating clockwise. There is a slider (link 3), sliding along the link OP. Hence the link 3 is moving along a straight line. At any instant there is a point A on the link OP and the corresponding point on the slider at that instant is B. After a small interval of time 'δt' the point on the link OP is A_1 and the point on slider is B_1 as shown in Fig. 4.10.

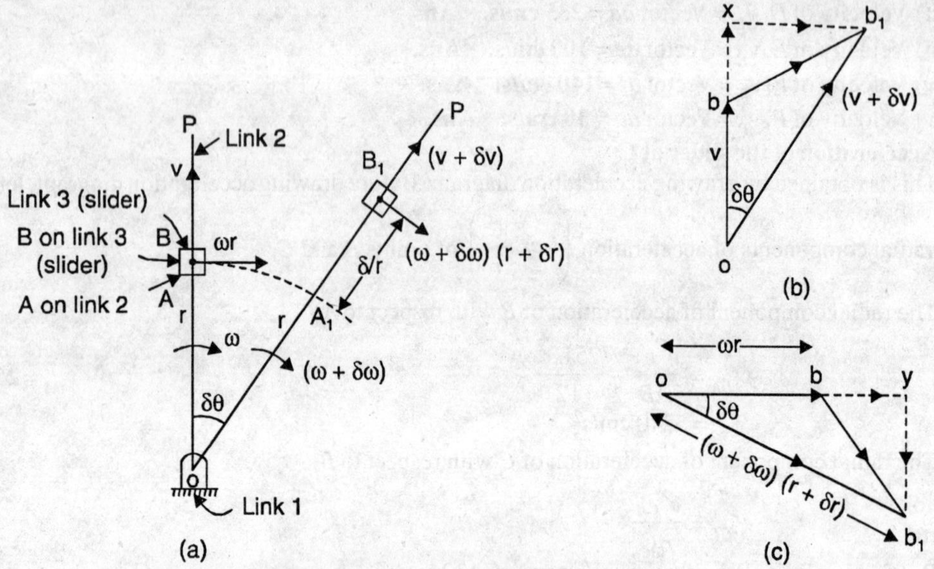

Fig. 4.10. Coriolis acceleration component.

Link 1 is fixed, Link 2 is rotating, Link 3 is sliding. Point A is on rotating link and point B is on slider.

Let ω = Angular velocity of the link OP. This is also the angular velocity of the slider at any time 't'

v = Linear velocity of the slider B along the link OP at time 't'

r = Radial distance of point B from O at time t.

Then tangential velocity of the slider B *i.e.*, velocity perpendicular to OP at time $t = \omega \times r$ ...(*i*)

In a short interval of time δt, let $\delta\theta$ be the angular displacement of the link OP and δr, the radial displacement of the slider in the outward direction. Let us assume that the link OP is rotating with variable angular velocity and the slider B is also moving with variable linear velocity along the link OP.

After the short interval of time δt, let

$(\omega + \delta\omega)$ = Angular velocity of the link OP. This is also the angular velocity of the slider after time δt

$v + \delta v$ = Linear velocity of the slider on the link OP

$r + \delta r$ = Radial distance of the slider.

Then tangential velocity of the slider after time δt

$$= (\omega + \delta\omega)\,(r + \delta r) \qquad \text{...(ii)}$$

Let us find the acceleration of the slider B with respect to O and with respect to coincident point A on the link OP.

The slider is rotating side by side it is sliding also. Hence slider is having rotational motion and also linear motion along the link OP. To find the acceleration of the slider B along OP and perpendicular to OP we must know the total change of velocity along OP and perpendicular to OP. Total change of velocity in any direction consists of the change of velocity due to sliding and due to rotational.

Change of velocity along OP and perpendicular to OP due to linear velocity (i.e., sliding velocity). Fig. 4.10 (*b*) shows the velocity diagram when the *linear velocities* (*i.e.*, sliding) are considered [*i.e.*, velocities v and $(v + \delta v)$]. The vector bb_1 represents the change in linear velocity in time δt. The vector bx represents the change in velocity along OP (or along radial direction) whereas the vector xb_1 represents the change in velocity perpendicular to OP (or in tangential direction).

Hence

$$\begin{aligned} bx &= ox - ob \\ &= ob_1 \cos \delta\theta - ob \\ &= (v + \delta v) \cos \delta\theta - v && (\because \quad ob_1 = v + \delta v) \\ &= (v + \delta v) \times 1 - v && (\because \quad \delta\theta \text{ is very small, hence } \cos \delta\theta = 1) \\ &= \delta v \uparrow && \text{...(iii)} \end{aligned}$$

and

$$\begin{aligned} xb_1 &= ob_1 \sin \delta\theta \\ &= (v + \delta v) \sin \delta\theta \\ &= (v + \delta v) \times \delta\theta && (\because \quad \delta\theta \text{ is very small, hence } \sin \delta\theta = \delta\theta) \\ &= v \times \delta\theta + \delta v \times \delta\theta \\ &= v \times \delta\theta \ (\delta v \times \delta\theta \text{ is negligible, being the product of two small terms}) && \text{...(iv)} \\ &= v \times \delta\theta \rightarrow \end{aligned}$$

Change of velocities along OP and perpendicular to OP due to tangential velocities which are due to rotation.

Fig. 4.10 (c) shows the velocity diagram of the point B (*i.e.*, slider) when its *tangential velocities* (due to rotation) are considered [*i.e.*, velocities $\omega\, r$ and $(\omega + \delta\omega)\ (r + \delta r)$]. The vector bb_1 represents the change in tangential velocity of the slider in time δt. The vector yb_1 represents the change in velocity in radial direction whereas the vector by represents change in velocity perpendicular to OP (or in tangential direction).

Hence $yb_1 = ob_1 \sin \delta\theta$ where

$$\begin{aligned} ob_1 &= (\omega + \delta\omega)(r + \delta r) \\ &= \omega r + \omega \delta r + r \delta\omega + \delta\omega \times \delta r \\ &= \omega r + \omega \delta r + r \delta\omega \ (\delta\omega \times \delta r \text{ is negligible}) \end{aligned}$$

$$\begin{aligned} &= (\omega r + \omega \delta r + r \delta\omega) \sin \delta\theta \\ &= (\omega r + \omega \delta r + r \delta\omega) \times \delta\theta && (\because \quad \sin \delta\theta = \delta\theta \text{ as } \delta\theta \text{ is very small}) \\ &= \omega r \delta\theta + \omega \delta r \delta\theta + r \delta\omega \delta\theta \\ &= \omega r \delta\theta && [\because \ \delta r \,.\, \delta\theta \text{ and } \delta\omega \,.\, \delta\theta \text{ are negligible}] \\ &= \omega r \times \delta\theta \downarrow \quad [\downarrow \text{ shows that change of velocity is in dounward direction}] && \text{...(v)} \end{aligned}$$

and

$$\begin{aligned} by &= oy - ob \\ &= ob_1 \cos \delta\theta - ob \\ &= (\omega r + \omega \delta r + r \delta\omega) \cos \delta\theta - \omega r && (\because \quad ob_1 = \omega r + \omega \delta r + r \delta\omega) \\ &= (\omega r + \omega \delta r + r \delta\omega) \times 1 - \omega r && (\because \quad \cos \delta\theta = 1 \text{ as } \delta\theta \text{ is very small}) \\ &= (\omega \delta r + r \delta\omega) && \text{...(vi)} \end{aligned}$$

$\therefore \qquad by = (\omega \delta r + r \delta\omega) \rightarrow$

[$\rightarrow$ shows that the change of velocity is perpendicular to OP in x-direction] ...(*vii*)

Now let us find the total change in velocity in radial direction (*i.e.*, along OP) and tangential direction *i.e.*, perpendicular to OP of the point B (or of the slider) with respect to O in time δt.

Total change in velocity of point B in radial direction (*i.e.*, along OP)

$= bx - yb_1$ (bx is upward but yb_1 is downward)

$= (\delta v - \omega r \times \delta\theta)$ $\quad (\because$ From (*iii*), $bx = \delta v$ and from (*v*), $yb_1 = \omega r \delta\theta)$

This total change of velocity of slider B with respect to O is acting radially outwards from O to A.

$\therefore$ *Component of acceleration of point B in radial direction (or parallel to OP)*

= Rate of total change of velocity in radial direction

$$= \frac{\text{Total change of velocity in radial direction}}{\delta t} = \frac{\delta v - \omega r \,\delta\theta}{\delta t}$$

$$= \left(\frac{\delta v}{\delta t} - \omega r \frac{\delta \theta}{\delta t} \right) \uparrow$$

In the limit, as $\delta t \to 0$

Component of acceleration of point B in radial direction (parallel to OP) with respect to O becomes as

$$f^r_{BO} = \left(\frac{dv}{dt} - \omega r \frac{d\theta}{dt} \right) \uparrow$$

$$= f - \omega r \times \omega \qquad \left(\because \frac{dv}{dt} = \text{Linear acceleration} = f \text{ and } \frac{d\theta}{dt} = \omega \right)$$

$$= (f - \omega^2 r) \uparrow$$

(This component of acceleration is acting radially outwards from O to A) ...(4.7)

where f = Linear acceleration of the slider and

$\omega^2 r$ = Centripetal acceleration.

The equation (4.7) shows that the acceleration of the slider (*i.e.*, point B) with respect to O perallel to OP (or in radial direction) consists of two accelerations *i.e.*, linear acceleration of the slider and centripetal acceleration.

Similarly the acceleration of point B perpendicular to OP (or in tangential direction) with respect to O

f^r_{BO} = Rate of change of total velocity of B perpendicular to OP

$$= \frac{\text{Total change of velocity of } B \text{ perpendicular to } OP}{\delta t}$$

$$= \frac{xb_1 + by}{\delta t} \rightarrow$$

$$= \frac{v\delta\theta + (\omega\delta r + r\delta\omega)}{\delta t} \qquad (\because \text{ From } (iv), xb_1 = v\delta\theta \text{ and from } (vii), by = \omega\delta r + r\delta\omega)$$

$$= v\frac{\delta\theta}{\delta t} + \omega\frac{\delta r}{\delta t} + r\frac{\delta\omega}{\delta t} \rightarrow$$

In the limit, as $\delta t \to 0$, acceleration of point B in tangential direction *i.e.*, perpendicular to OP with respect to O becomes

$$f^r_{BO} = v\frac{d\theta}{dt} + \omega\frac{dr}{dt} + r\frac{d\omega}{dt}$$

$$= v \times \omega + \omega \times v + r \times \alpha \qquad \left(\because \frac{d\theta}{dt} = \omega, \frac{dr}{dt} = v, \frac{d\omega}{dt} = \alpha \right)$$

$$= (2v\omega + r\alpha) \rightarrow$$

(This component of acceleration is acting perpendicular to OP and towards right) ...(4.8)

Let us now find the radial component and tangential component of acceleration of the point A which is on the link OP.

The radial component of acceleration of the point A with respect to O,

$$f^r_{AO} = \omega^2 r \downarrow$$

This is acting radially in a direction from C to O.

And the tangential component of acceleration of the point A with respect to O,

$$f^t_{AO} = \alpha \times r \rightarrow$$

This is acting perpendicular to AO and towards right.

Radial and Tangential components of Acceleration of slider B with respect to point A on the link OP.

Let us now find the radial and tangential components of acceleration of slider B with respect to its coincident point A on the link OP.

These components are given as

$$f^r_{BA} = f^r_{BO} - f^r_{AO}$$
$$= (f - \omega^2 r) - (-\omega^2 r) \qquad [\because f^r_{BO} = (f - \omega^2 r) \uparrow \text{ and } f^r_{AO} = \omega^2 r \downarrow]$$
$$= f \uparrow$$

(This is acting radially upwards)

and

$$f^t_{BA} = f^t_{BO} - f^t_{AO}$$
$$= (2v\omega + r\alpha) - \alpha \times r \qquad [\because f^t_{BO} = (2v\,\omega + r\alpha) \rightarrow \text{ and } f^t_{AO} = r\alpha \rightarrow]$$
$$= 2v\omega \rightarrow$$

[This is acting perpendicular to OP and towards right]

The above tangential component of acceleration of slider B with respect to the coincident point A on the link is known as *coriolis component of acceleration.*

$\therefore$ $2v\omega$ = Coriolis acceleration component.

The equation (4.8) shows that the acceleration of the slider (*i.e.*, point B) perpendicular to OP (or in tangential direction) consists of two accelerations *i.e.*, tangential acceleration and coriolis acceleration component.

$\therefore$ Coriolis acceleration component

$$= 2 \times v \times \omega$$
$$= 2 \times \text{Linear velocity of slider} \times \text{Angular velocity of rotating link.}$$

The Coriolis component of acceleration is always perpendicular to the link.

Equation (4.8) gives the acceleration of point B perpendicular to OP. The component '$2v\omega$' is known as Coriolis acceleration component. It is denoted by f^c_{BA}. It is positive if both ω and v are either positive or negative. *The direction of the Coriolis acceleration component is obtained by rotating the* **linear velocity vector** *v through 90° in the direction of rotation of the link.* The direction of ω and v are taken positive if link is rotating clockwise and slider is moving radially outwards. But if the link is rotating anticlockwise, direction of ω is taken – ve. And if slider is moving radially inwards, the direction of v is taken – ve.

The direction of Coriolis component of acceleration for all four possible cases, is shown in Fig. 4.10 (*d*). The direction of Coriolis component of acceleration is the direction of the **velocity vector** rotated by 90° in the direction of the angular velocity of the rotation of the link. In Fig. 4.10 (*d*) (*i*), velocity of slider is radially outward and rotation of link is clockwise. Hence Coriolis component (*i.e.*, $2 \times v \times \omega$) will be at right angle to velocity vector v in the clockwise direction.

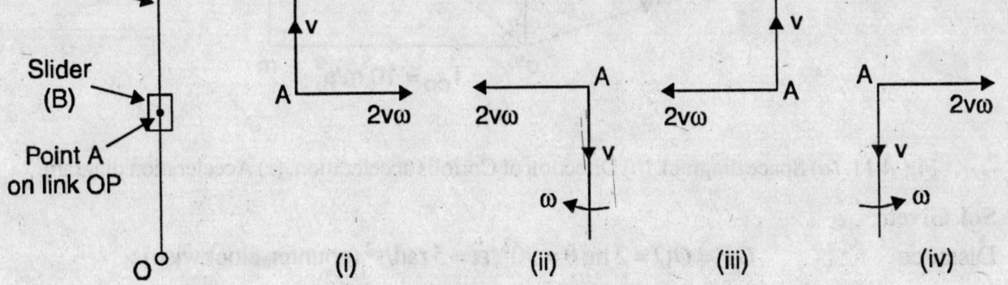

Fig. 4.10 (*d*)

In Fig. 4.10 (*d*) (*ii*), the velocity of slider is radially inward and rotation of the link is clockwise. Hence the Coriolis component will be at right angles to velocity vector *v* in the clockwise direction.

In Fig. 4.10 (*d*) (*iii*), the velocity of slider is radially outward and rotation of the link is anticlockwise. Hence the coriolis component will be at right angles to velocity vector in anticlockwise direction.

In Fig. 4.10 (*d*) (*iv*), the velocity of slider is radially inward and rotation of link is anticlockwise. Hence coriolis component will be at right angles to velocity vector in anticlockwise direction.

From all the four cases it is clear that the direction of coriolis component of acceleration is obtained by rotating the velocity vector through 90° in the direction of the rotation of the link.

Problem 4.6. *A link OR is rotating about point O. At a particular instant when $\theta = 90°$ with x-axis, the angular velocity of the link is 2 rad/s counter-clockwise and angular acceleration is 5 rad/s² counter-clockwise as shown in Fig. 4.11. A slider P is sliding along the rod radially outward and at that instant, the distance of the slider from point O is 2 m (i.e., distance OP = 2 m). The corresponding point on the rod is Q. The velocity and acceleration of the point P at that instant w.r.t. to point Q are 3 m/s and 4 m/s² both radially outwards. Find the acceleration of point P relative to point O. Also find (i) the acceleration of Q relative to O and (ii) acceleration of P relative to Q.*

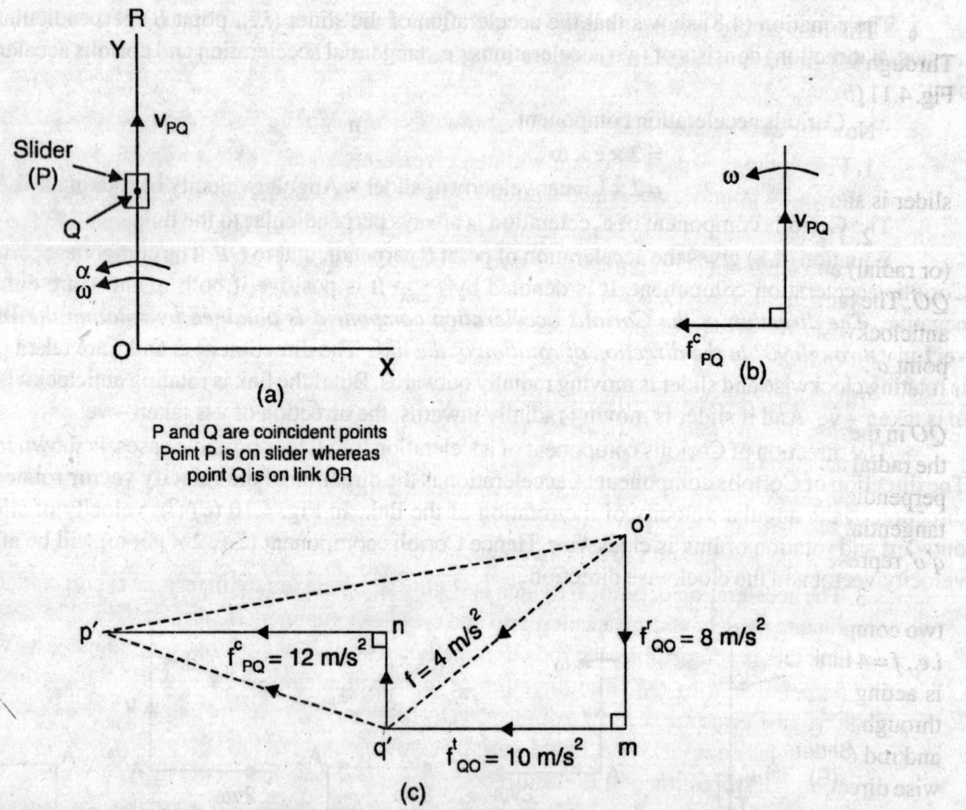

Fig. 4.11. (*a*) Space diagram. (*b*) Direction of Coriolis acceleration. (*c*) Acceleration diagram.

Sol. Given :

Distance $OP = OQ = 2$ m, $\theta = 90°$, $\alpha = 5$ rad/s² (counter-clockwise) ;

$\omega = 2$ rad/s (counter-clockwise) ; $v_{PQ} = 3$ m/s radially outward ;

Acceleration of P w.r.t. $Q = f = 4$ m/s² radially outward.

The point Q is fixed to link OR whereas the point P is sliding along OR.

The **point Q** which is on link has two components of acceleration namely :

(*i*) Centripetal (or radial) component,

$$f^r_{QO} = \omega^2 \times r \qquad (\because \quad r = OP = OQ = 2\text{ m})$$
$$= 2^2 \times 2 = 8\text{ m/s}^2.$$

The centripetal acceleration acts along the link towards centre of rotation. Hence it is acting along QO.

(*ii*) Tangential component, $f^t_{QO} = \alpha \times r = 5 \times 2 = 10\text{ m/s}^2$.

This is acting perpendicular to QO and in the direction of α.

The point P which is sliding on the link, has also two components of acceleration namely :

(*i*) The component parallel to $OP, f = 4\text{ m/s}^2$.

This is also known as linear acceleration of slider.

This component is acting parallel to OP and in the radially outward direction (given).

(*ii*) Coriolis component of acceleration which is perpendicular to OP. The magnitude of this component is given by

$$f^c_{PQ} = 2 \times \omega \times v_{PQ} = 2 \times 2 \times 3 = 12\text{ m/s}^2$$

This is acting perpendicular to OP. This direction is obtained by rotating velocity vector v_{PQ}. Through 90° in the same sense as the sense of rotation of link OP (*i.e.*, in anticlockwise direction) as shown in Fig. 4.11 (*b*).

Now the acceleration diagram can be drawn as shown in Fig. 4.11 (*c*) as per procedure given below :

1. First draw the space diagram with the given dimensions as shown in Fig. 4.11 (*a*). The position of slider is shown by point P, whereas the point Q shows the position of coincident point on the rod.

2. The acceleration of the **coincident point Q** with respect to O has two components namely centripetal (or radial) and tangential components. The radial component *i.e.*, $f^r_{QO} = 8\text{ m/s}^2$ and acting in the direction of QO. The tangential component *i.e.*, $f^t_{QO} = 10\text{ m/s}^2$ and is acting perpendicular to QO in the direction of α *i.e.*, anticlockwise. These two components are mutually perpendicular. Start the acceleration diagram from the fixed point o'.

Take any point o' in acceleration diagram as shown in Fig. 4.11 (*c*). From o', draw vector $o'm$ parallel to QO in the downward direction. Take this vector $o'm = 8\text{ m/s}^2$ to some suitable scale. Then this vector represents the radial acceleration of Q with respect to O *i.e.*, f^r_{QO}. From point m, draw vector mq' perpendicular to $o'm$ (or perpendicular to OQ in space diagram). This vector should be in the direction of α. This vector represents the tangential acceleration of Q with respect to O *i.e.*, f^t_{QO}. Take this vector $mq' = 10\text{ m/s}^2$. Join q' to o'. The vector $q'o'$ represent the acceleration of Q with respect to O.

3. The acceleration of **point P** (which is sliding along the rod) with respect to coincident point Q has also two components namely sliding acceleration and coriolis component of acceleration. The sliding acceleration *i.e.*, $f = 4\text{ m/s}^2$ and is acting along the rod OP radially outward. The coriolis component *i.e.*, $f^c_{PQ} = 12\text{ m/s}^2$ and is acting perpendicular to OP. The direction of this component is obtained by rotating *velocity vector* v_{PQ} through 90° in the same sense as that of the rod OQ rotates. The velocity vector v_{PQ} is acting radially outward and rod OQ rotates anticlockwise, hence coriolis component should act perpendicular to v_{PQ} in the anticlockwise direction. The direction of this coriolis component is shown in Fig. 4.11 (*b*).

From point q' in the acceleration diagram, draw vector $q'n$ parallel to OP in the upward direction from q'. Take $q'n = f = 4\text{ m/s}^2$. This vector $q'n$ represents the sliding acceleration. Now from point n, draw vector np' perpendicular to $q'n$ (or perpendicular to OP in space diagram) to represent the coriolis acceleration *i.e.*, f^c_{PQ}. Take $np' = 12\text{ m/s}^2$. The direction of np' should be same as shown in Fig. 4.11 (*b*). Join point p' to o'. Then $o'p'$ represents the absolute acceleration of P with respect to O.

By measurement :

Vector $o'p'$ = Acceleration of P with respect to O = **22.4 m/s²**. **Ans.**

(*i*) Acceleration of Q relative to O *i.e.*, f_{QO} = vector oq' = **12.8 m/s²**. **Ans.**

(*ii*) Acceleration of P relative to Q *i.e.*, $f_{PQ} = q'p'$ = **12.65 m/s²**. **Ans.**

Alternative Method

The acceleration of Q relative to O and acceleration of P relative to Q can also be determined **analytically** as given below :

(*i*) Acceleration of Q relative to O is the vector sum of f^r_{QO} and f^t_{QO}

$$\therefore \quad f_{QO} = \sqrt{(f^r{}_{QO})^2 + (f^t{}_{QO})^2}$$

$$= \sqrt{8^2 + 10^2} = \sqrt{64 + 100} = \sqrt{164} = \mathbf{12.8\ m/s^2}. \quad \textbf{Ans.}$$

(*ii*) Acceleration of P relative to Q is the vector sum of sliding acceleration of P (relative to Q) and coriolis component of acceleration of P with respect to Q.

$$\therefore \quad f_{PQ} = \sqrt{f^2 + (f^c{}_{PQ})^2}$$

$$= \sqrt{4^2 + 12^2} = \sqrt{16 + 144} = \sqrt{160} = \mathbf{12.65\ m/s^2}. \quad \textbf{Ans.}$$

Note. (*i*) If the link OR rotates with a uniform angular velocity, then the angular acceleration of the link is zero. Then point Q will have only one component of acceleration *i.e.*, f^r_{QO} only. The position of q' in acceleration diagram will be obtained from this component only.

(*ii*) As in the above case, the magnitude and direction of sliding acceleration and coriolis component of acceleration are known, hence the position of sliding point (*i.e.*, of point P) can be obtained easily in acceleration diagram p'.

(*iii*) In some machines, the position of the sliding point in acceleration diagram can be obtained directly. The coriolis component of acceleration is known in direction and magnitude. But the magnitude of sliding component is unknown. Then the coriolis component should be taken in proper direction with the given magnitude. As the position of sliding point is known in acceleration diagram, hence the coriolis component *must finish* at that point with proper direction. This means that coriolis component must not start from that point.

Problem 4.7. *In Fig. 4.12 (a) shown below, the washer is sliding outward on the rod with a velocity of 1.2 m/sec when its distance from point O is 0.6 m. Its velocity along the rod is increasing at the rate of 0.9 m/s². The angular velocity of the rod is 5 rad/s counter clockwise and its angular acceleration is 10 rad/s² clockwise. Determine the absolute acceleration of a point on the washer.* (AMIE, Winter 1986)

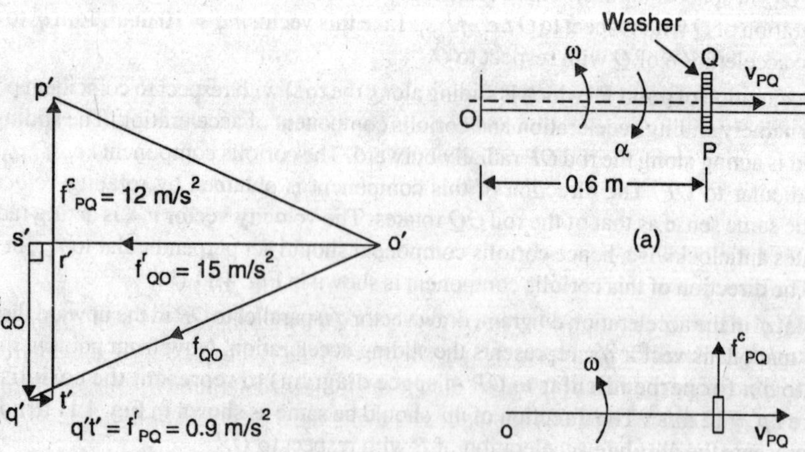

Fig. 4.12. (*a*) Space diagram (*b*) Direction of Coriolis acceleration (*c*) Accelerating diagram.

Sol. Given :

Let P shows the position of washer and Q is the coincident point on the rod.

Distance OP, $r = 0.6$ m

Velocity of washer P with respect to Q, $v_{PQ} = 1.2$ m/s

Angular velocity of rod, $\omega = 5$ rad/s (counter clockwise)

Angular acceleration of rod, $\alpha = 10$ rad/s^2 (clockwise)

Acceleration of washer along the rod, $f = 0.9$ m/s^2.

or Acceleration of slider, $f = 0.9$ m/s^2

Hence it is the acceleration of P with respect to Q in radial direction *i.e.*, f^r_{PQ}.

$\therefore$ $f^r_{PQ} = 0.9$ m/s^2

Let us first calculate the acceleration components.

Radial component of acceleration of Q with respect to O,

$$f^r_{QO} = \omega^2 \times r = 5^2 \times 0.6 = 15 \text{ m/s}^2$$

Tangential component of acceleration of Q with respect to O,

$$f^t_{QO} = \alpha \times r = 10 \times 0.6 = 6 \text{ m/s}^2$$

Coriolis component of acceleration of P with respect to Q,

$$f^r_{PQ} = 2 \times \omega \times v_{PQ} = 2 \times 5 \times 1.2 = 12 \text{ m/s}^2$$

The direction of this component is obtained by rotating velocity vector v_{PQ} through 90° in the same sense as that of rod OQ rotates. As velocity vector v_{PQ} is acting radially outward and rod OQ rotates anticlockwise, hence v_{PQ} should be rotated through 90° anticlockwise to get the direction of f^c_{PQ} as shown in Fig. 4.12 (*b*).

Now draw the acceleration diagram as shown in Fig. 4.12 (*c*) as per procedure given below :

1. First draw the space diagram with the given dimensions as shown in Fig. 4.12 (*a*). Position of washer is shown by point P, whereas the point Q shows the position of coincident point on the rod.

2. The acceleration of the coincident point Q with respect to O has two components namely :

(*i*) Radial component of Q with respect to O *i.e.*, $f^r_{QO} = 15$ m/s^2. This component is in the direction of QO.

(*ii*) Tangential component of Q with respect to O *i.e.*, $f^t_{QO} = 6$ m/s^2. This component is perpendicular to QO. As the angular acceleration of the rod is clockwise, hence this component will be acting downward with respect to space diagram.

These two components are mutually perpendicular.

Take any point o' in acceleration diagram as shown in Fig. 4.12 (*c*). From o', draw vector $o's'$ parallel to QO and towards the left from o'. Take this vector equal to 15 m/s^2. Then this vector represents the radial acceleration (*i.e.*, f^r_{QO}) of Q with respect to O. From point s', draw vector $s'q'$ perpendicular to vector $o's'$ [or perpendicular to OQ in space diagram]. This vector should be in downward direction. It represents the tangential acceleration of Q with respect to O *i.e.*, f^t_{QO}. Take vector $s'q' = 6$ m/s^2. Join q' and o'. Then $o'q'$ represents the acceleration of Q with respect to O.

3. The acceleration of washer P with respect to coincident point Q, has also two components, namely :

(*i*) Radial acceleration *i.e.*, $f^r_{PQ} = 0.9$ m/s^2. This is also known as acceleration of slider *i.e.*, acceleration of washer. This component will be parallel to OQ (or OP). It is given in the question that velocity of the washer along the rod, is increasing at the rate of 0.9 m/s^2. Hence the radial acceleration of washer (or of point P) with respect to Q is 9 m/s^2. The direction of this component is along the rod and acting towards right from point O. Hence in acceleration diagram, draw vector $q't'$ parallel to OP and towards right from q'. Take $q't' = f^r_{PQ} = 0.9$ m/s^2.

(*ii*) Coriolis component of acceleration *i.e.*, $f^c_{PQ} = 12$ m/s^2. This component is perpendicular to vector $q't'$ (or perpendicular to OP). The direction of this component is obtained by rotating v_{PQ} through 90° in the same sense as that of rod OQ rotates. As the rod OQ rotates anticlockwise, hence v_{PQ} should also be rotated through 90° anticlockwise to get the direction of f^c_{PQ}. Direction of this Coriolis component is shown in Fig. 4.12 (*b*).

From point t' in acceleration diagram, draw vector $t'p'$ perpendicular to $q't'$ (or perpendicular to OP) to represent the coriolis acceleration *i.e.*, f^c_{PQ}. The direction of $t'p'$ should be same as shown in Fig. 4.12 (*b*) *i.e.*, this direction is upwards. Take $t'p' = 12$ m/s^2. Join point p' to o'. Then vector $o'p'$ represents absolute acceleration of washer P with respect to O.

(*iii*) By measurement :

Vector $o'p'$ = Acceleration of washer P with respect to O

= 15.3 m/s². Ans.

Analytical Method

In acceleration diagram, shown in Fig. 4.12 (*c*), the length $o'p'$ can be obtained analytically, as given below :

The triangle $o'p'r'$ is a right angle.

$$\therefore \quad o'p' = \sqrt{(o'r')^2 + (p'r')^2}$$

$$= \sqrt{(o's' - s'r')^2 + (p't' - r't')^2} = \sqrt{(f^r{}_{QO} - f^c{}_{PQ})^2 + (f^c{}_{PQ} - f^t{}_{QO})^2}$$

$$= \sqrt{(15 - 0.9)^2 + (12 - 6)^2} = \sqrt{14.1^2 + 6^2} = \sqrt{198.81 + 36}$$

= 15.32 m/s². Ans.

Problem 4.8. *The driving crank AB of the quick-return mechanism shown in Fig. 4.13 revolves at a uniform speed of 200 r.p.m. Find the velocity and acceleration of the tool-box R, in the position shown, when the crank makes an angle of 60° with the vertical line of centres PA. What is the acceleration of sliding of the block at B along the slotted lever PQ.* (AMIE, Winter 1980)

Sol. Given :

$$N_{BA} = 200 \text{ r.p.m.} \qquad \therefore \qquad \omega_{BA} = \frac{2\pi \times 200}{60} = 20.95 \text{ rad/s}$$

$$\therefore \qquad v_{BA} = \omega_{BA} \times \text{Length } BA$$

$$= 20.95 \times 75 \text{ mm/s} = \frac{20.95 \times 75}{1000} \text{ m/s}$$

$$= 1.57 \text{ m/s}$$

The direction of v_{BA} will be perpendicular to the crank AB.

First draw the space diagram with the given dimensions to some suitable scale as shown in Fig. 4.13 (*a*). Point B lies on crank AB and coincident point B' is on PQ. Point B is a slider also. Now draw the velocity diagram as shown in Fig. 4.13 (*b*) as per method given follows :

1. As A and P are fixed points, hence the corresponding points a and p will coincide in the velocity diagram. Take point a and p at a common place anywhere. Now choose a suitable velocity scale and from point a draw vector ab perpendicular to AB to represent the velocity of B with respect to A *i.e.*, v_{BA}. Hence take vector $ab = v_{BA} = 1.57$ m/s.

2. The point B' is on PQ. The point B' moves with respect to P and also with respect to B. The velocity of B' with respect to P is perpendicular to PB' (or PQ). And the velocity of B' with respect to B is along PQ. From point p which is at the same place as point a, draw vector pb_1 perpendicular to PB' to represent the

velocity of point B' (which is the coincident point of B) with respect to P (*i.e.*, $v_{B'P}$), and from point b draw vector bb' parallel to the path of motion of B' (which is along PQ) to represent the velocity of slider B with respect to the coincident point B' (*i.e.*, $v_{BB'}$). The vectors pb_1 and bb_1 intersect at b_1.

3. The point Q lies on PB' produced, therefore produce the vector pb_1 to obtain the corresponding point q in the velocity diagram. The point q will divide the vector pb_1 produced in the same ratio as Q divides PB'. In other words,

$$\frac{pb_1}{pq} = \frac{PB'}{PQ}$$

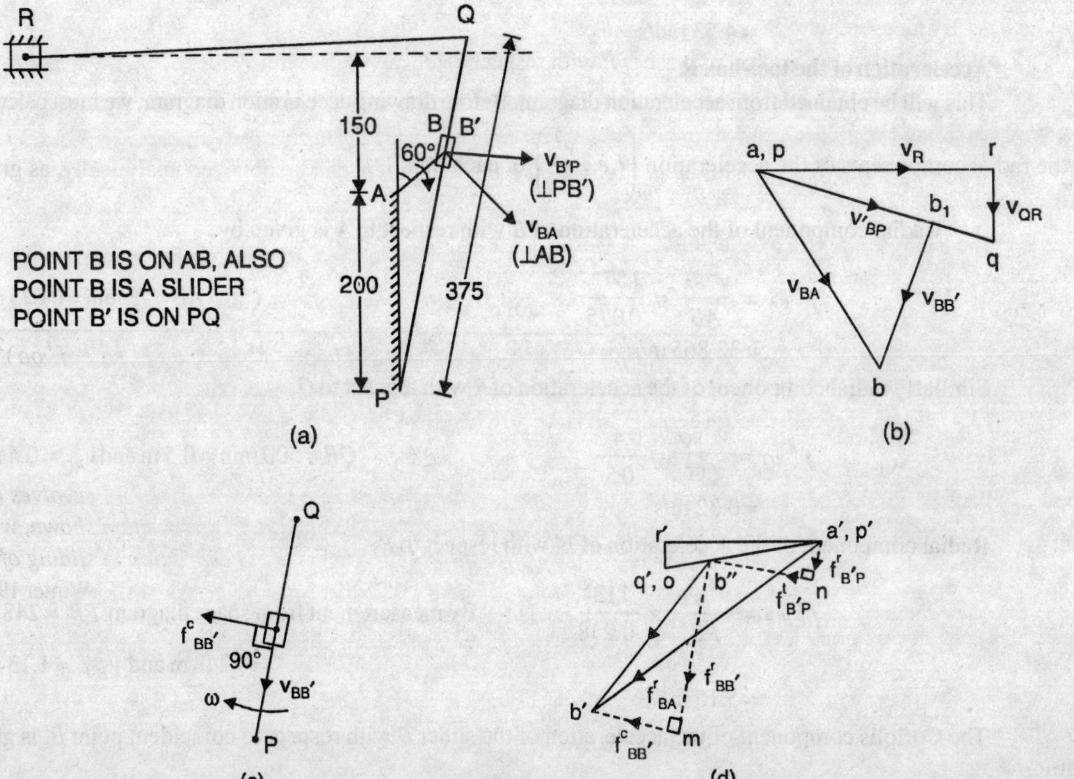

Fig. 4.13. (*a*) Space diagram (*b*) Velocity diagram (*c*) Direction of Coriolis component (*d*) Acceleration diagram.

$$\therefore \qquad pq = pb_1 \times \frac{PQ}{PB'}$$

The vector pq represents the velocity of Q with respect to P (*i.e.*, v_{QP}).

The point R moves with respect to Q and also in the horizontal direction. The velocity of R with respect to Q is perpendicular to RQ. Also the velocity of R is in horizontal direction.

4. Now from point q, draw vector qr perpendicular to QR to represent the velocity of R with respect to Q (*i.e.*, v_{RQ}), and from point a draw vector ar parallel to the path of motion of the tool-box R (which is along the horizontal line), to represent the velocity of R *i.e.*, v_R. The vectors qr and ar intersect at r.

By measurement, we get

$$v_R = \text{Vector } ar = 1.6 \text{ m/s}$$

$$v_{BB'} = \text{Vector } bb_1 = 1.06 \text{ m/s}$$

$v_{RQ} = \text{Vector } r = 0.4$ m/s

$v_{B'P} = \text{Vector } pb_1 = 1.13$ m/s

$v_{QP} = \text{Vector } pq = 1.7$ m/s

Now the velocity of the tool-box R is given by,

$v_R = \mathbf{1.6}$ **m/s. Ans.**

Angular velocity of link PQ,

$$\omega_{PQ} = \frac{v_{PQ}}{PQ} = \frac{1.7}{0.375} \qquad (\because \ PQ = 375 \text{ mm} = 0.375 \text{ m})$$

$$= 4.53 \text{ rad/s}$$

Acceleration of the tool-box R

This will be obtained from acceleration diagram. Before drawing acceleration diagram, we must calculate the radial components of the acceleration $\left(i.e., \frac{v^2}{r}\right)$ of each link.

∴ Radial component of the acceleration of B with respect to A is given by,

$$f^r_{BA} = \frac{v^2_{BA}}{AB} = \frac{1.57^2}{0.075} \qquad (\because \ AB = 75 \text{ mm} = 0.075 \text{ mm})$$

$$= 32.865 \text{ m/s}^2$$

Similarly radial component of the acceleration of R with respect to Q,

$$f^r_{RQ} = \frac{v^2_{RQ}}{QR} = \frac{0.4^2}{0.5} \qquad (\because \ QR = 500 \text{ mm} = 0.5 \text{ m and } v_{RQ} = 0.4 \text{ m/s})$$

$$= 0.32 \text{ m/s}^2$$

Radial component of the acceleration of B' with respect to P,

$$f^r_{B'P} = \frac{v^2_{B'P}}{PB'} = \frac{1.13^2}{0.248} \qquad [\because \ \text{By measurement from space diagram } PB' = 248 \text{ mm} = 0.248 \text{ m and } v_{B'P} = 1.13 \text{ m/s}]$$

$$= 5.15 \text{ m/s}^2$$

The Coriolis component of the acceleration of the slider B with respect to coincident point B' is given by,

$$f^c_{BB'} = 2\omega \times v \qquad [\text{where } \omega = \omega_{QP} = 4.53 \text{ rad/s and } v = v_{BB'} = 1.08 \text{ m/s calculated}]$$

$$= 2 \times \omega_{QP} \times v_{BB'} = 2 \times 4.53 \times 1.06 = 9.6 \text{ m/s}^2$$

As crank AB is revolving at a uniform speed, the tangential acceleration *i.e.*, f^t_{BA} will be zero. Hence total acceleration of B w.r.t. to A will be equal to f^r_{BA}.

The acceleration diagram can now be drawn as shown in Fig. 4.13 (*d*) as per procedure given below :

1. As A and P are fixed points, hence the corresponding points a' and p' will coincide. Take points a' and p' at a common place anywhere. Choose a suitable scale for acceleration diagram. From point a', draw vector $a'b'$ parallel to BA to represent the radial component of acceleration of B with respect to A *i.e.*, $f^r_{BA} = 32.865$ m/s². Hence vector $a'b' = f^r_{BA} = 32.865$ m/s². The point b' gives the position of sliding point B in acceleration diagram. Hence in this case the position of sliding point in acceleration diagram is known. The position of coincident point (B') in acceleration diagram is unknown. This position is shown by point b'' in acceleration diagram. Let us locate the point b'' in acceleration diagram.

2. The acceleration of the slider B with respect to coincident point B' has the two components namely :

(*i*) Coriolis component of the acceleration of B with respect to B' *i.e.*, $f^c_{BB'}$. This component will be perpendicular to PB'.

(*ii*) Sliding acceleration of B with respect to B' *i.e.*, f or $f^r_{BB'}$. This component will be parallel to PB.

These two components are mutually perpendicular. The direction and magnitude of coriolis component ($f^c_{BB'}$) are known. The magnitude of sliding acceleration (f or $f^r_{BB'}$) is not known. But sliding acceleration is perpendicular to coriolis component. The direction of $f^c_{BB'}$ is obtained by rotating velocity vector $v_{BB'}$ (represented by vector b_1b in the velocity diagram) through 90° in the same sense as that of the link PQ which rotates in the clockwise direction. The direction of $f^c_{BB'}$ is shown in Fig. 4.13 (*c*).

The Coriolis component acts perpendicular to PB' towards left and must finish at b'. Hence from point b', draw vector $b'm$ perpendicular to $B'P$ to represent $f^c_{BB'} = 9.6$ m/s^2. Now from point m, draw vector mb'' perpendicular to vector $b'm$ (or parallel to BP) to represent $f^r_{BB'}$. The magnitude of $f^r_{BB'}$ is not known yet.

3. The acceleration of the coincident point B' with respect to P has the two components namely :

(*i*) Radial component of the acceleration of B' with respect to P *i.e.*, $f^r_{B'P}$ and

(*ii*) Tangential component of the acceleration of B' with respect to P *i.e.*, $f^t_{B'P}$.

These two components are mutually perpendicular. The radial component is known in magnitude and direction. The magnitude of tangential component is not known. But the tangential component is perpendicular to radial component.

Hence from point p' draw vector $p'n$ parallel to $B'P$ to represent $f^r_{B'P} = 5.15$ m/s^2, and from point n draw vector nb'' perpendicular to vector $p'n$ to represent $f^t_{B'P}$. The vector mb'' and nb'' intersect at b''. The vector $p'b''$ (or $a'b''$) represents the total acceleration of B' with respect to P *i.e.*, $f_{B'P}$ and the vector $b''b'$ represents the total acceleration of B with respect to B' *i.e.*, $f_{BB'}$.

4. The point Q lies on PB' produced, therefore produce the vector $p'b''$ to obtain the corresponding point q' in the acceleration diagram. The point q' will divide the vector $p'b''$ produced in the same ratio as Q divides PB'. In other words,

$$\frac{p'b''}{p'q'} = \frac{PB'}{PQ}$$

$$\therefore \quad p'q' = p'b'' \times \frac{PQ}{PB'}$$

The vector $p'q'$ represents the acceleration of Q with respect to P *i.e.*, f_{QP}.

5. The acceleration of the tool-box R with respect to Q has the two components, namely :

(*i*) Radial component of acceleration of R with respect to Q *i.e.*, $f^r_{RQ} = 0.32$ m/s^2.

(*ii*) Tangential component of acceleration of R with respect to Q *i.e.*, f^r_{RQ}.

These two components are mutually perpendicular. Hence from point q', draw vector $q'o$ parallel to QR to represent $f^r_{RQ} = 0.32$ m/s^2. As the magnitude of this component is very small, hence the points q' and o coincides as shown in Fig. 4.13 (*d*). Now from point o (same as q'), draw vector or' (or $q'r'$) perpendicular to QR to represent f^t_{RQ}. The magnitude of f^t_{RQ} is not known yet.

6. From a', draw vector $a'r'$ parallel to the path of motion of tool-box R (*i.e.*, along the horizontal) which intersects the vector or' (or $q'r'$) at r'. The vector $a'r'$ represents the acceleration of the tool-box R *i.e.*, f_R.

(*i*) By measurement, we find that

$$f_R = \text{Vector } a'r' = \mathbf{22\ m/s^2}. \quad \textbf{Ans.}$$

(*ii*) The acceleration of sliding of the block *B* along the slotted lever

$$PQ = f^r{}_{BB'} = \text{Vector } b''m = \mathbf{18\ m/s^2}. \quad \textbf{Ans.}$$

Problem 4.9. *A mechanism of a crank and slotted quick return motion is shown in Fig. 4.14. If the crank rotates counter-clockwise at 120 r.p.m., determine (for the configuration shown) the velocity and acceleration of the ram. Also determine the angular acceleration of the slotted lever. Crank OA = 15 cm, Slotted arm O_1C = 70 cm and link CD = 20 cm.* (AMIE, Summer 1986)

Sol. Given :

Speed of crank AO = 120 r.p.m. (counter-clockwise)

$\therefore \quad N_{AO} = 120$ r.p.m., crank OA = 15 cm, slotted arm O_1C = 70 cm,

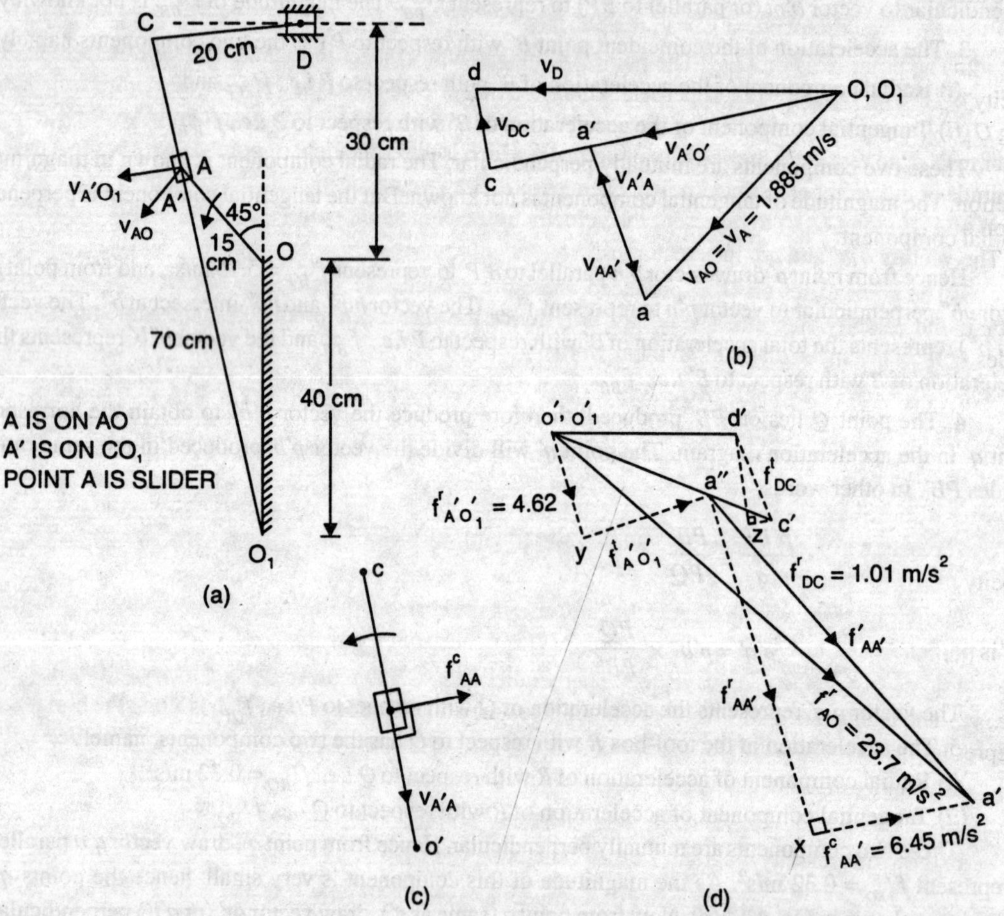

Fig. 4.14. (*a*) Space diagram (*b*) Velocity diagram (*c*) Direction of Coriolis component (*d*) Acceleration diagram.

Link CD = 20 cm, length OO_1 = 40 cm

Now $\qquad \omega_{AO} = \dfrac{2\pi N_{AO}}{60} = \dfrac{2\pi \times 120}{60} = 4\pi$ rad/s

Velocity of *A* with respect to *O* is v_{AO}.

$$\therefore \quad v_{AO} = \omega_{AO} \times AO = 4\pi \times 15 \text{ cm/s} = \frac{4\pi \times 15}{100} \text{ m/s} = 1.885 \text{ m/s}$$

Hence velocity of A with respect to O (*i.e.*, v_{AO}) or velocity of A (*i.e.*, v_A) as point O is fixed is equal to 1.885 m/s.

The direction of v_{AO} will be perpendicular to the crank OA. The point A lies on the crank whereas the coincident point A' lies on the slotted arm O_1C. The point A is slider.

First draw the space diagram with the given dimensions to some suitable scale as shown in Fig. 4.14 (*a*). Now draw the **Velocity diagram** as shown in Fig. 4.14 (*b*) as per method given below :

1. As O and O_1 are fixed points, therefore the corresponding points o and o_1 will coincide at one point in the velocity diagram. Take the points o and o_1 at one place anywhere. Now choose a suitable velocity scale. From point o draw vector oa perpendicular to AO to represent the velocity of A with respect to O *i.e.*, v_{AO} (or v_A). Hence take

$$\text{Vector } oa = v_{AO} = v_A = 1.885 \text{ m/s.}$$

2. The point A' is on O_1C. The point AA' moves with respect to O, and also with respect to A. The velocity of A' with respect to O_1 is perpendicular to O_1A' (or O_1C) And the velocity of A' with respect to A is along O_1C. From point o_1 which is at the same place as point o in velocity diagram, draw vector o_1a' perpendicular to O_1A_1 to represent the velocity of point A' (which is the coincident of point A) with respect to O_1. The magnitude of the vector o_1a' is not known. From point a, draw vector aa' parallel to the path of motion of A' (the motion of A' is along O_1C) to represent the velocity of the coincident point A' with respect to the point A *i.e.*, $v_{A'A}$. The vectors o_1a' and aa' intersect at a'.

3. The point C lies on O_1A_1 produced, therefore produce the vector o_1a' to obtain the corresponding point c in the velocity diagram. The point c will divide the vector o_1a' produced in the same ratio as point C divides O_1A' in the space diagram. In other words,

$$\frac{a'c}{o_1c} = \frac{A'C}{O_1C}$$

$$\therefore \quad a'c = \frac{A'C}{O_1C} \times o_1c$$

The vector o_1c represents the velocity of C with respect to O_1. Also the vector $a'a$ represents the velocity of A with respect to A'.

The point D moves with respect to C and also in the horizontal direction. The velocity of D with respect to C is perpendicular to CD. Also the velocity of D is in horizontal direction.

4. Now from point c, draw vector cd perpendicular to CD to represent the velocity of D with respect to C *i.e.*, v_{DC} and from point o draw vector od parallel to the path of motion of D (which is along the horizontal line) to represent the velocity of D *i.e.*, v_D. The vectors cd and od intersect at d.

5. Velocity of D is obtained by measuring the vector od.

$$\therefore \quad v_D = \text{vector } od = \mathbf{2.15 \text{ m/s.}} \quad \textbf{Ans.}$$

6. From the velocity diagram, we also find that

$$v_{AA'} = \text{velocity of } A \text{ respect to } A = \text{vector } a'a = 1.05 \text{ m/s}$$

$$v_{DC} = \text{velocity of } D \text{ with respect to } C = \text{vector } cd = 0.45 \text{ m/s}$$

$$v_{A'O_1} = \text{velocity of } A' \text{ with respect to } O_1 = \text{vector } o_1a' = 1.55 \text{ m/s}$$

$$v_{CO_1} = \text{velocity of } C \text{ with respect to } O_1 = \text{vector } o_1c = 2.15 \text{ m/s}$$

Now the angular velocity of the link O_1C can be obtained as :

$$\omega_{CO_1} = \frac{v_{CO_1}}{CO_1} = \frac{2.15}{0.7} = 3.07 \text{ rad/s (Anticlockwise)}$$

Acceleration Diagram

Before drawing the acceleration diagram, let us find the radial acceleration $\left(i.e., \frac{v^2}{r}\right)$ of each link.

The radial acceleration of A with respect to O,

$$f^r{}_{AO} = \frac{v^2{}_{AO}}{OA} = \frac{1.885^2}{0.15} = 23.7 \text{ m/s}^2$$

Radial acceleration of coincident point A' with respect to O_1,

$$f^r{}_{A'O_1} = \frac{v^2{}_{A'O_1}}{A'O_1} = \frac{1.55^2}{0.52} = 4.62 \text{ m/s}^2 \qquad (v_{A'O_1} = 1.55 \text{ m/s calculated above})$$

($\because$ From space diagram by measurement, $A'O_1 = 0.52$ m)

Radial acceleration of point D with respect to C,

$$f^r{}_{DC} = \frac{V^2{}_{DC}}{CD} = \frac{0.45^2}{0.20} = 1.01 \text{ m/s}^2$$

The Coriolis acceleration component of the slider A with respect to the coincident point A' is given by,

$$f^c{}_{AA'} = 2 \times \omega \times v$$

$$= 2 \times \omega_{CO_1} \times v_{AA'} \qquad (\because \omega_{CO_1} = \omega_{A_1O_1} = 3.07 \text{ rad/s and } v_{AA'} = 1.05 \text{ m/s})$$

$$= 2 \times 3.07 \times 1.05 = 6.45 \text{ m/s}^2$$

The acceleration diagram can now be drawn as shown in Fig. 4.14 (*d*) as per procedure given below :

1. As O and O_1 are fixed points hence the corresponding points o' and o_1' will coincide at one point in acceleration diagram. Take O′ and O_1', at one place anywhere. Now choose a suitable scale for acceleration diagram. From point O', draw vector $o'a'$ parallel to AO to represent the radial component of acceleration of A with respect to O *i.e.*, $f^r{}_{AO} = 23.7$ m/s². Hence vector $o'a' = 23.7$ m/s². The point a' gives the position of sliding point A in acceleration diagram. Hence in this case the position of sliding point in acceleration diagram is known. The position of coincident point (A') in acceleration diagram is unknown. This position is shown by point a'' in acceleration diagram. Let us locate the point a'' in acceleration diagram.

2. The acceleration of the slider A with respect to coincident point A' has two components, namely :

(*i*) Coriolis component of acceleration *i.e.*, $f^c{}_{AA'} = 6.45$ m/s². This component will be perpendicular to $A'O_1$.

(*ii*) Sliding acceleration of A with respect to A' *i.e.*, for $f^r{}_{AA'}$. This component will be parallel to $A'O_1$.

These two components are mutually perpendicular. The direction and magnitude of Coriolis component ($f^c{}_{AA'}$) are known. The magnitude of sliding acceleration (for $f^r{}_{AA'}$) is not known. But sliding acceleration is perpendicular to Coriolis component. The direction of $f^c{}_{AA'}$ is obtained by rotating $v_{AA'}$ through 90° in the same sense as that of link CO_1 rotates. As link CO_1 rotates anticlockwise hence $v_{AA'}$ should also be rotated through 90° anticlockwise to get the direction of $f^c{}_{AA'}$. The direction of $f^c{}_{AA'}$ is shown in Fig. 4.14 (*c*).

The Coriolis component acts perpendicular to $A'O_1$ towards right and must finish at a'. Hence from point a', draw vector $a'x$ perpendicular to $A'O_1$ to represent $f^c{}_{AA'} = 6.45$ m/s².

Take $a'x = 6.45$ m/s^2. Now from point x, draw vector xa'' perpendicular to vector $a'x$ (or parallel to $A'O_1$) to represent the sliding acceleration, $f^r{}_{AA'_1}$. The magnitude of this vector xa'' is not known.

3. The acceleration of the coincident point A' with respect to O_1 has also the two components, namely :

(*i*) Radial acceleration of A' with respect to O_1 *i.e.*, $f^r{}_{A'O'_1} = 4.62$ m/s^2. This component is in the direction of $A'O_1$.

(*ii*) Tangential acceleration of A' with respect to O_1 *i.e.*, $f^t{}_{A'O'_1}$. This component is perpendicular to $A'O_1$.

These two components are mutually perpendicular. The radial component is known in magnitude and direction. The magnitude of tangential component is not known. But the tangential component is perpendicular to radial component. Hence from point o_1' in acceleration diagram, draw vector $o_1'y = f^r{}_{A'O'_1} = 4.62$ m/s^2 in the direction of $A'O_1$. From point y, draw vector ya'' perpendicular to the vector $o_1'y$. This represents $f^t{}_{A'O'_1}$. The vectors ya'' and xa'' intersect at a''. Join vector $o_1'a''$. This vector represents the acceleration of A' with respect to O_1 *i.e.*, $f_{A'O'_1}$.

4. The point C lies on O_1A' produced, therefore produce the vector $o_1'a''$ to obtain the corresponding point c' in the acceleration diagram. The point c' will divide the vector $o_1'a''$ produced in the same ratio as C divides O_1A' produced in space diagram. In other words

$$\frac{o_1'a''}{o_1'c'} = \frac{O_1A'}{O_1C}$$

$$\therefore \quad o_1'c' = \frac{O_1C}{O_1A'} \times o_1'a''.$$

5. The acceleration of ram D with respect to C has the two components, namely :

(*i*) Radial acceleration of D with respect to C *i.e.*, $f^r{}_{DC} = 1.01$ m/s^2. This component will be in the direction of DC.

(*ii*) Tangential acceleration of D with respect to C *i.e.*, $f^t{}_{DC}$. This component will be perpendicular to DC.

These two components are mutually perpendicular. Hence from c', draw vector $c'z$ to represent $f^r{}_{DC}$ parallel to DC. Cut vector $c'z = 1.01$ m/s^2. From point z, draw vector zd' perpendicular to vector $c'z$ (or perpendicular to DC) to represent $f^t{}_{DC}$. The magnitude of $f^t{}_{DC}$ is not known.

6. From point o' (or o_1'), draw vector $o'd'$ in the direction of motion of the ram D (*i.e.*, in the horizontal direction). The vectors zd' and $o'd'$ intersect at d'. The acceleration of the ram D *i.e.*, f_D is represented by vector $o'd'$.

(*i*) From acceleration diagram, acceleration of ram D = Vector $o'd'$ = **8.4 m/s²**. **Ans.**

(*ii*) To find the angular acceleration of the slotted lever, we must find the tangential acceleration of A' with respect to O_1 *i.e.*, $f^t{}_{A'O_1}$. This is given by vector ya'' in acceleration diagram. Hence by measurement,

$$f^t{}_{A'O_1} = \text{Vector } ya'' = 6.4 \text{ m/s}^2.$$

∴ Angular acceleration of the slotted lever

$$= \frac{f^t{}_{A'O_1}}{O_1A'} = \frac{6.4}{0.52} = \textbf{12.3 rad/s}^2\ \textbf{(Anticlockwise).} \quad \textbf{Ans.}$$

($\because$ O_1A' by measurement in space diagram = 0.52 m)

Problem 4.10. *A four-bar is shown in Fig. 4.15. The angular velocity and angular acceleration of link AB is given as 50 rad/s and 2200 rad/s² respectively. (i) Draw the velocity and acceleration diagrams. (ii) Find the acceleration of F and E. (iii) Find the angular acceleration of the link 3 (i.e., link BCE).*

Sol. Given :

$$\omega_{AB} = 50 \text{ rad/s}, \qquad \alpha_{AB} = 2200 \text{ rad/s}^2$$

The four links are 1 (link *AD* fixed), 2 (link *AB*), 3(link *BCE*) and 4 (link *CD*). Length $AB = 60$ mm $= 0.06$ m ; $BC = 59$ mm $= 0.059$ m ; $CD = 20$ mm $= 0.02$ m, $AD = 100$ mm $= 0.1$ m, $\angle BAD = 60°$.

Velocity of *B* w.r.t. to *A*, $V_{BA} = V_B = \omega_{AB} \times AB = 50 \times 0.06 = 3$ m/s.

The points *A* and *D* are fixed.

Velocity Diagram

First draw the space diagram with the given dimensions as shown in Fig. 4.15 (*a*). The velocity diagram is drawn as explained below : (let us first draw the velocity diagram for *ABCD*)

(*i*) Points *A* and *D* are fixed. Hence in velocity diagram these points will coincide as *a* and *d* as shown in Fig. 4.15 (*b*). Take these points at one place anywhere choose a suitable scale for velocity diagram.

(*ii*) The velocity of B with respect to *A* is perpendicular to *AB*. From point *a*, draw velocity vector *ab* perpendicular to *AB*. Take *ab* = 3 m/s to a suitable scale.

The location of point *C* in velocity diagram is obtained by considering the velocity of *C* with respect to *B* and also with respect to *D*. The velocity of *C* with respect to *B* is perpendicular to *BC* whereas with respect to *D* it is perpendicular to *CD*.

(*iii*) From point *b*, draw vector *bc* perpendicular to *BC*. And from point *d* (or *a*) draw *dc* perpendicular to *DC* to intersect the vector *bc* at point *c*. Thus *abc* represents the velocity diagram for *ABCD*.

Velocity diagram for link 3 (BCE)

To locate the point *e* in velocity diagram for link *BCE*, the point *e* should be in the same sense as point *E* is in *BCE* (*i.e.*, in anticlockwise direction).

(*iv*) From point *b*, draw vector *be* perpendicular to *BE*. Also from point *C*, draw vector *ce* perpendicular to *CE* to intersect the vector *be* at *e*. Then *bce* is the velocity vector for link *BCE*. [**OR** Draw Δbce similar to ΔBCE on vector *bc* in the same sense to locate point *e*].

(*v*) The point *f′* in vector *bc* is obtained from

$$\frac{BF}{BC} = \frac{bf}{bc}$$

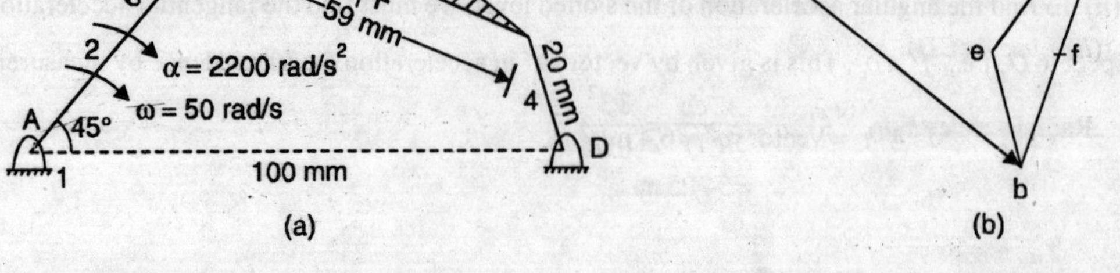

(a) (b)

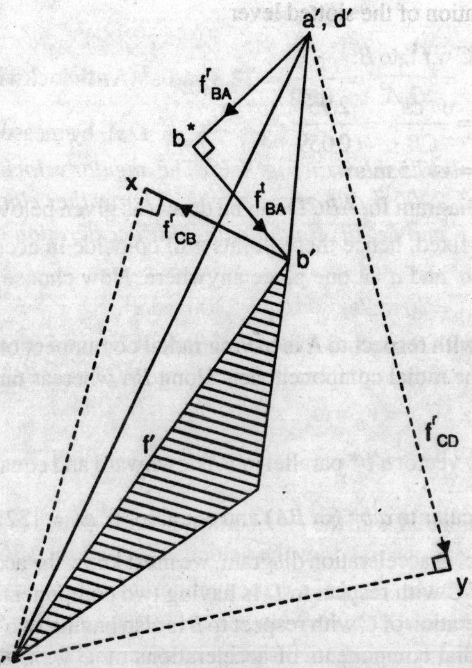

(c)

Fig. 4.15. (*a*) Space diagram (*b*) Velocity diagram (*c*) Acceleration diagram.

$$\therefore \quad bf = \frac{BF}{BC} \times bc$$

As lengths of *BF*, *BC* and *bc* are known, hence point '*f*' can be obtained.

Acceleration Diagram

Let us first draw acceleration diagram for *ABCD*. First calculate the tangential and radial accelerations of the different links as given below :

(*i*) *Link 2 (AB)*

Tangential acceleration of *B* w.r.t. to *A*,

$$f^t_{BA} = \alpha_{BA} \times BA$$

$$= 2200 \times 0.06 \qquad (\because \ \alpha_{BA} = \alpha_{AB} = 2200 \text{ rad/s}^2)$$

$$= 132 \text{ m/s}^2$$

Radial acceleration of *B* w.r.t. to *A*,

$$f^r_{BA} = \frac{v^2_{BA}}{BA} = \frac{3^2}{0.06} \qquad (\because \ v_{BA} = 3 \text{ m/s given})$$

$$= 150 \text{ m/s}^2$$

(*ii*) *Link 4 (CD)*

$$\text{Radial acceleration,} \quad f^r_{CD} = \frac{v^2_{CD}}{CD} = \frac{3.3^2}{0.02} \qquad (v_{CD} = \text{vector, } cd = 3.3 \text{ m/s})$$

$$= 544.5 \text{ m/s}^2$$

(iii) Link BC

Radial acceleration of C w.r.t. to B,

$$f^r_{CB} = \frac{v^2_{CB}}{CB} = \frac{2.95^2}{0.059} \qquad (\because \quad v_{CB} = \text{vector}, cb = 2.95 \text{ m/s})$$
$$= 147.5 \text{ m/s}^2$$

Now the acceleration diagram for $ABCD$ can be drawn as given below :

(i) Points A and D are fixed, hence these points will coincide in acceleration diagram as a' and d' as shown in Fig. 4.15 (*c*). Take a' and d' at one place anywhere. Now choose a suitable scale for acceleration diagram.

The acceleration of B with respect to A is having radial component of acceleration and also tangential component of acceleration. The radial component acts along BA whereas tangential component acts perpendicular to BA.

(ii) From point a', draw vector $a'b^*$ parallel to BA downward and equal to $f^r_{BA} = 150$ m/s^2. From point b^*, draw vector b^*b' perpendicular to $a'b^*$ (or BA) and equal to $f^t_{BA} = 132$ m/s^2. Join b' to a'.

(iii) To locate the point c' in acceleration diagram, we must know the acceleration of C w.r.t. to D and also w.r.t. to B. The acceleration of C with respect to D is having two components of accelerations *i.e.*, radial and tangential. Similarly the acceleration of C with respect to B is also having two components of acceleration (*i.e.*, radial and tangential). The radial components of accelerations of C w.r.t. D and also w.r.t. B are known in magnitude and direction.

(iv) From d', draw vector $d'y$ parallel to CD downward to represent the radial component of acceleration of C w.r.t. D. Take vector $d'y = f^r_{CD} = 544.5$ m/s^2. From y, draw vector yc' perpendicular to $d'y$. The magnitude is unknown. This vector represents the tangential component of acceleration of C w.r.t. D.

(v) From point b', draw vector $b'x$ parallel to BC to represent the radial acceleration of C w.r.t. B. Hence, take $b'x = f^r_{CB} = 147.5$ m/s^2. From point x, draw vector xc' perpendicular to $b'x$ (or BC) to represent the tangential acceleration of C w.r.t. B. But the magnitude is not known. The vector xc' cuts the vector yc' at c'. Join c' to d'. Then $a'b'c'd'$ represents the acceleration diagram for $ABCD$.

Acceleration diagram for link 3 (BCE)

To locate the position of e' in acceleration diagram, draw $\Delta\, b'c'e'$ similar to ΔBCE on vector $b'c'$ such that both triangles are read in the same sense *i.e.*, anti-clockwise.

(vi) The point f' in vector $b'c'$ is obtained from

$$\frac{BF}{BC} = \frac{b'f'}{b'c'}$$

$$\therefore \qquad b'f' = \frac{BF}{BC} \times b'c'$$

Acceleration of F is given by,

$$f_F = a'f' = \mathbf{410 \text{ m/s}^2}. \quad \textbf{Ans.}$$

Acceleration of E is given by,

$$f_E = a'e' = \mathbf{410 \text{ m/s}^2}. \quad \textbf{Ans.}$$

Angular acceleration of link 3 (link BCE) is obtained by knowing the tangential acceleration of link 3. As BC is a part of link 3, the tangential acceleration of C w.r.t. B is known from acceleration diagram, which is equal to vector $xc' = 585$ m/s^2.

∴ Angular acceleration of link 3 is given by

$$\alpha_3 = \frac{f^t{}_{CB}}{CB} = \frac{\text{Vector } xc'}{CB}$$

$$= \frac{585}{0.059} = \mathbf{9915\ rad/s^2.}\ \ \textbf{Ans.}$$

Problem 4.11. *The crank AB of the mechanism shown in Fig. 4.16 is rotating with a speed of 30 r.p.m. (i) Draw velocity and acceleration diagrams. (ii) Find the linear acceleration of slider F. (iii) Find the angular acceleration of the links BC, CED and FD.*

Sol. Given :

Link lengths, $AB = 200$ mm $= 0.2$ m ; $BC = 700$ mm $= 0.7$ m, $DF = 670$ mm $= 0.67$ m, link CED ($CE = 160$ mm ; $ED = 160$ mm and $CD = 100$ mm), vertical distance $AE = 700$ mm $= 0.7$ m and crank angle with vertical $= 45°$.

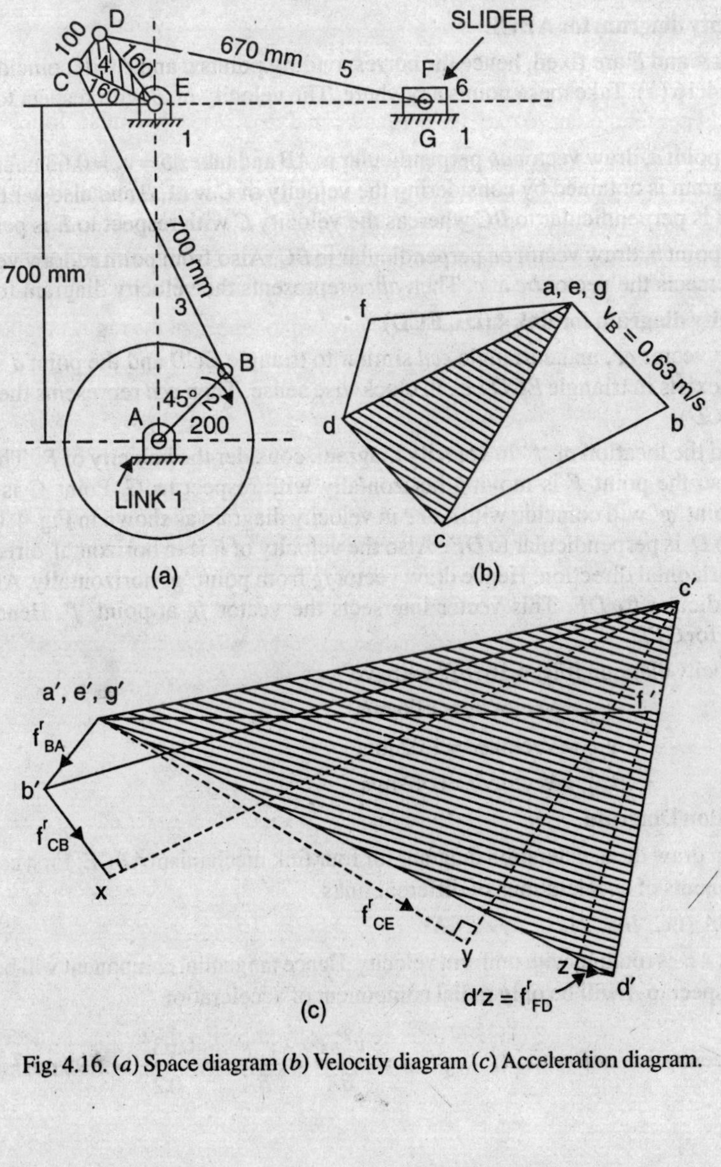

Fig. 4.16. (*a*) Space diagram (*b*) Velocity diagram (*c*) Acceleration diagram.

Speed of crank = 30 r.p.m.

Angular velocity of crank AB, $\omega_{AB} = \frac{2\pi N}{60} = \frac{2\pi \times 30}{60} = \pi$ rad/s

$\therefore$ Velocity of B w.r.t. to A, $v_{BA} = \omega_{AB} \times AB = \pi \times 0.2 = 0.63$ m/s

As point A is fixed, hence this is also the velocity of B.

$\therefore$ $v_B = v_{BA} = 0.63$ m/s.

In the given mechanism, the points A and E are fixed. The slider F is moving horizontally. Also CED is only one link. This is link No. 4.

First draw the space diagram as shown in Fig. 4.16 (*a*).

In the given mechanism, there are six links in which link 1 is fixed. Link 2 is crank AB, link 3 is BC, link 4 is CED, link 5 is DF and link 6 is slider.

Let us first consider the four-link mechanism $ABCE$.

(A) **Velocity diagram for ABCE**

(*i*) Points A and E are fixed, hence the corresponding points a and e will coincide in velocity diagram as shown in Fig. 4.16 (*b*). Take these points anywhere. The velocity of B with respect to A is perpendicular to AB.

(*ii*) From point a, draw vector ab perpendicular to AB and take $ab = v_B = 0.63$ m/s. The location of point c in velocity diagram is obtained by considering the velocity of C w.r.t. B and also w.r.t. E. The velocity of C with respect to B is perpendicular to BC whereas the velocity C with respect to E is perpendicular to EC.

(*iii*) From point b, draw vector bc perpendicular to BC. Also from point e, draw vector ec perpendicular to EC which intersects the vector bc at c. Then $abce$ represents the velocity diagram for $ABCE$.

(B) **Velocity diagram for link 4 (i.e., ECD)**

(*iv*) On the vector ec, make triangle ecd similar to triangle ECD and the point d should be in the same sense as point D exists in triangle ECD *i.e.,* in clockwise sense. Then ecd represents the velocity diagram for link 4 (*i.e.,* link ECD).

(*v*) To find the location of 'f' in velocity diagram, consider the velocity of F. The point E moves with respect to D. Also the point F is moving horizontally with respect to G. Point G is fixed and hence the corresponding point 'g' will coincide with a or e in velocity diagram as shown in Fig. 4.16 (*b*). The velocity of F with respect to D is perpendicular to DF. Also the velocity of F is in horizontal direction. The velocity of F w.r.t. G is in horizontal direction. Hence draw vector fg from point 'g' horizontally. Also from point d draw vector df perpendicular to DF. This vector intersects the vector fg at point 'f'. Hence gfd represents the velocity diagram for GFD.

From velocity diagram [Fig. 4.16 (*b*)], we have

$$v_{CB} = \text{Vector } bc = 1.06 \text{ m/s}$$

$$v_{CE} = \text{Vector } ec = 1.02 \text{ m/s}$$

$$v_{FD} = \text{Vector } df = 0.42 \text{ m/s.}$$

Acceleration Diagram

Let us first draw the acceleration diagram for four link mechanism $ABCE$. First calculate the tangential and radial components of accelerations of different links.

(*i*) *Link BA (i.e., link 2)*

The crank AB is rotating with uniform velocity. Hence tangential component will be zero. Total acceleration of B with respect to A will be only radial component of acceleration

Radial acceleration of B w.r.t. A, $f^r_{BA} = \frac{v^2_{BA}}{BA} = \frac{v^2_B}{BA} = \frac{0.63^2}{0.2} = 1.98 \text{ m/s}^2$

(ii) Link CB (i.e., link 3)

Radial acceleration of C w.r.t. B, $f^r_{CB} = \frac{v^2_{CB}}{CB}$

$$= \frac{1.06^2}{0.70} \quad (\because \; v_{CB} = \text{vector } bc = 1.06 \text{ m/s})$$

$$= 1.6 \text{ m/s}^2$$

(iii) Link CE

Radial acceleration of C w.r.t E, $f^r_{CE} = \frac{v^2_{CE}}{CE}$

$$= \frac{1.02^2}{0.16} \quad (\because \; v_{CE} = \text{vector } ec = 1.02 \text{ m/s})$$

$$= 6.5 \text{ m/s}^2$$

Now acceleration diagram for $ABCE$ can be drawn as explained below :

(*i*) Points A and E are fixed and hence these points will coincide in acceleration diagram as a' and e' as shown in Fig. 4.16 (*c*). Take these points at one place anywhere. Now choose a suitable scale for acceleration diagram.

(*ii*) Total acceleration of B with respect to A is only radial component and is acting along BA. From point a', draw vector $a'b'$ parallel to BA downwards and equal to $f^r_{BA} = 1.98 \text{ m/s}^2$. As there is no angular acceleration given for crank AB, hence f^r_{BA} (or vector $a'b'$) represents the total acceleration of B w.r.t. A.

(*iii*) To locate the point c' in acceleration diagram, we must know the acceleration of C w.r.t. B and also w.r.t. E. The acceleration of C with respect of B is having radial component of acceleration and also tangential component of acceleration. Radial component acts along CB whereas tangential component acts perpendicular to CB. Similarly the acceleration of C with respect E is having radial component of acceleration and also tangential component of acceleration. Radial component acts along CE whereas tangential components acts perpendicular to CE. The radial components of acceleration of C w.r.t. B and also w.r.t. E are known in magnitude and direction.

(*iv*) From e', draw vector $e'y$ parallel to CE downward to represent the radial component of acceleration of C w.r.t. E. Take vector $e'y = f^r_{CE} = 6.5 \text{ m/s}^2$. From y, draw vector yc' perpendicular to $e'y$ (or perpendicular to CE). The magnitude is unknown. This vector represents the tangential component of acceleration of C w.r.t. E.

(*v*) From point b', draw vector $b'x$ parallel to CB downward to represent the radial component of acceleration of C w.r.t. B. Hence take $b'x = f^r_{CB} = 1.6 \text{ m/s}^2$. From point x, draw vector xc' perpendicular to $b'x$ (or BC) to represent the tangential component of acceleration of C w.r.t. B. But magnitude is not known. The vector xc' and vector yc' meet at c'. Join c' to e'. Then $a'b'c'e'$ represents the acceleration diagram for $ABCE$.

Acceleration Diagram for link 4 (link CED)

To locate the position of d' in acceleration diagram, draw triangle $e'c'd'$ similar to triangle ECD on vector $e'c'$ such that both triangles are read in the same sense *i.e.,* clockwise.

(*vi*) To locate the position of f' in acceleration diagram, first determine the value of radial component of acceleration of F w.r.t. D *i.e.,* f^r_{FD}. This is given by

$$f^r_{FD} = \frac{v^2_{FD}}{FD}$$

$$= \frac{0.42^2}{0.67} \qquad (\because \quad v_{FD} = \text{Vector } df \text{ in velocity diagram} = 0.42 \text{ m/s})$$
$$= 0.263 \text{ m/s}^2$$

(*vii*) The point *F* with respect to *D* is having radial component of acceleration and also tangential component of acceleration. Radial component acts along *FD* whereas tangential component acts perpendicular to *FD*. From point *d'*, draw vector *d'z* parallel to *DF* to represent the radial component of *F* w.r.t. *D*. Hence take $d'z = f^r_{FD} = 0.263 \text{ m/s}^2$. From point *z*, draw vector *zf'* perpendicular to *d'z* (or perpendicular to *DF*) to represent the tangential component of acceleration of *F* w.r.t. *D*. But magnitudes is not known. The acceleration of *F* w.r.t. *G* is in horizontal direction. The point *G* is fixed and hence corresponding point *g'* in acceleration diagram will coincide with *a'* or *e'* as shown in Fig. 4.16 (*c*). From *g'*, draw vector *g'f'* horizontally, which cuts the already drawn vector *zf'* at *f'*.

(*i*) Linear acceleration of slider *F* = vector $g'f'$ = **9.1 m/s²**. **Ans.**

(*ii*) Angular acceleration of link $BC = \frac{f^t_{BC}}{BC} = \frac{\text{vector } xc'}{BC}$

or $$\alpha_{BC} = \frac{10.3}{0.7} \qquad (\because \quad \text{From acceleration diagram } xc' = 10.3)$$
$$= \mathbf{14.714 \text{ rad/s}^2}. \quad \textbf{Ans.}$$

(*iii*) Angular acceleration of link *CED*,

$$\alpha_{CED} = \frac{f^t_{CE}}{CE} = \frac{\text{vector } yc'}{CE}$$
$$= \frac{5.8}{0.16} \qquad (\because \quad \text{From acceleration diagram } yc' = 5.8)$$
$$= \mathbf{36.25 \text{ rad/s}^2}. \quad \textbf{Ans.}$$

(*iv*) Angular acceleration of link *FD*,

$$\alpha_{FD} = \frac{f^t_{FD}}{FD} = \frac{\text{vector } zf'}{FD}$$
$$= \frac{3.6}{0.67} \qquad (\because \quad \text{From acceleration diagram } zf' = 3.6)$$
$$= \mathbf{5.373 \text{ rad/s}^2}. \quad \textbf{Ans.}$$

HIGHLIGHTS

1. For the stress analysis of a mechanism, the accelerations at different points of the mechanism, must be known.
2. A body moving in a circular path, has two components of acceleration, which are perpendicular to each other. These components are tangential component and normal component (or radial component). They are given by :

   f_t = Tangential component = $\alpha \times r$

   f_r = Radial component = $\omega^2 r$ or $\frac{v^2}{r}$.
3. If a body is moving in a circular path, with uniform velocity, the body will have only radial acceleration.
4. If a body is moving on a straight path, it will have only tangential acceleration.
5. Acceleration of any point in a mechanism is determined by :

   (*i*) Analytical method and (*ii*) acceleration diagram method.
6. For analytical method, displacement should be known in terms of time '*t*'.

7. Acceleration diagram method is mostly used for determining acceleration.
8. The acceleration of a point B with respect to A (when the points A and B lie on the same rigid link) is given as

$$f_{BA} = f^r_{BA} + f^t_{BA}$$

$$= \frac{v^2_{BA}}{BA} + \alpha \times BA \quad \text{where } v_{BA} = \text{velocity of } B \text{ w.r.t. to } A.$$

9. For acceleration diagrams :
   (*i*) Small letters with prime, such as a', b' etc. are used.
   (*ii*) All fixed points are located at the same place.
   (*iii*) Lines from fixed point, represents absolute acceleration.
10. Coriolis component of acceleration only acts when a slider is sliding along a rotating link. Mathematically it is given by :

Coriolis acceleration = $2 \times v \times \omega$,

where v = Linear velocity of slider along the link, and

ω = Angular velocity of the rotating link.

11. The Coriolis component of acceleration is + ve if both ω and v are either + ve or – ve. The direction of this component is obtained by rotating the linear velocity v through 90° in the direction of rotation of the link. The direction of ω and v are taken + ve if link is rotating clockwise and slider is moving radially outward.

EXERCISE 4

(A) Theorctical Questions

1. What is the importance of finding accelerations of various points in a mechanism ? Name the two different methods of finding acceleration in a mechanism.
2. Find an expression for tangential acceleration and radial acceleration when a body is moving in a circular path. What will happen to these accelerations if : (*i*) body is rotating with uniform velocity and (*ii*) body is moving on a straight path.
3. A rigid link AB is rotating anticlockwise about point A with angular velocity 'ω' and angular acceleration 'α'. Describe the method of drawing the acceleration diagram and find the total acceleration of B w.r.t. to A.
4. Describe the method of drawing the acceleration diagram for a slider-crank mechanism.
5. What do you mean by Coriolis component of acceleration ? When it will exist ? Prove that this component of acceleration is equal to $2 \times v \times \omega$,

   where v = Linear velocity of the slider along the link and

   ω = Angular velocity of the rotating link.
6. How will you find the direction of Coriolis component of acceleration ?
7. If a mechanism has n-links, then the number of instantaneous centres would be equal to

   (*a*) n (*b*) $\frac{n(n-1)}{2}$

   (*c*) $\frac{n}{2}$ (*d*) $(n-1)$. **[Ans. (*b*)]**
8. If a point moves along a straight line which is rotating, then the tangential component of acceleration is

   (*a*) $\frac{V^2}{r}$ (*b*) $\frac{dv}{dt} - \omega^2 r$

   (*c*) $\frac{dv}{dt}$ (*d*) $2v\omega + r \times \alpha$

   where v = Velocity of point along the straight line

   ω = Angular velocity of the line

α = Angular acceleration of the line

r = Radius of the point

V = Tangential velocity of line = $\omega \times r$. **[Ans. (*d*)]**

9. Coriolis component is considered if

(*a*) the point considered moves on a path that rotates

(*b*) the point considered moves along a path which is stationary

(*c*) the point considered moves along a circular path

(*d*) the point considered moves in any curvillinear path. **[Ans. (*a*)]**

10. The direction of Coriolis acceleration

(*a*) is always directed to the centre of curvature of the path

(*b*) is sometimes directed to the centre of curvature

(*c*) depends upon the direction of the velocity of the point on the path

(*d*) none of the above. **[Ans. (*c*)]**

11. A slider sliding at 10 cm/s on a link, which is rotating at 60 r.p.m., is subjected to Coriolis acceleration of magnitude

(*a*) $40\pi^2$ cm/s^2 (*b*) $0.4\,\pi$ cm/s^2

(*c*) $40\,\pi$ cm/s^2 (*d*) $4\,\pi$ cm/s^2. **[Ans. (*c*)]**

12. In which mechanism, the Coriolis component of acceleration will be considered

(*a*) four-bar chain mechanism (*b*) quick return motion mechanism

(*c*) slider-crank mechanism (*d*) none of the above. **[Ans. (*b*)]**

(B) Numerical Problems

1. The crank of a slider crank mechanism is 150 mm and connecting rod is 750 mm. The crank rotates at a constant speed of 300 r.p.m. clockwise. Calculate the velocity and acceleration of the slider when crank has turned 30° from the inner dead centre position. **[Ans.** (*i*) 2.76 m/s (*ii*) 143.3 m/s^2**]**

2. The crank of a slider crank mechanism is 480 mm and connecting rod is 1600 mm long. The crank rotates at a uniform angular velocity of 20 rad/s clockwise. When the crank has turned 60° from the inner dead centre position, determine :

(*i*) the acceleration of the connecting rod and

(*ii*) the acceleration of the slider. **[Ans.** (*i*) 104 rad/s^2, (*ii*) 72 m/s^2**]**

3. The crank of an engine 200 mm long and connecting rod length to crank radius is 4. The crank has turned through 45° from inner dead centre position. The instantaneous speed of rotation of the crank is 240 r.p.m. clockwise and it is increasing at the rate of 100 rad/s^2. Determine :

(*i*) acceleration of the mid-point of connecting rod,

(*ii*) angular acceleration of connecting rod, and

(*iii*) acceleration of slider. **[Ans.** (*i*) 110.5 m/s^2 ; (*ii*) 88 rad/s^2 ; (*iii*) 108.2 m/s^2**]**

4. A four-bar chain is represented by a quadrilateral *ABCD* in which *AD* is fixed and is 200 cm long. The crank *AB* 6.25 cm long rotates in a clockwise direction at 95.5 r.p.m. and drives the link *CD* 11.25 cm long by means of the connecting link *BC* 17.5 cm long. Draw the velocity and acceleration diagrams when angle *BAD* = 60° and *B* and *C* lie on the same side of *AD*. Determine the angular velocity and angular accelerations of links *BC* and *CD*.

[Ans. 1.9 rad/s ; 3.78 rad/s ; 23.43 rad/s^2 ; 147.1 rad/s^2**]**

5. A four-bar chain is represented by a quadrilateral *ABCD* in which *AD* is fixed and is 0.6 m long. The crank *AB* = 0.3 m long rotates in a clockwise direction at 10 rad/s and an angular acceleration of 30 rad/s^2, both clockwise. The crank drives the link *CD* (= 0.36 m) by means of the connecting link *BC* (= 0.36 m). Determine : (*i*) angular velocities and angular accelerations of *CD* and *BC* and (*ii*) the velocity and acceleration of point *C*.

[Ans. (*i*) ω_{CD} = 6.1 rad/s ; ω_{BC} = 5.7 rad/s ; α_{CD} = 73.8 rad/s^2 ; α_{BC} = 37.8 rad/s^2 ; (*ii*) v_{CD} = 2.2 m/s ; f_{CD} = 29.6 m/s^2**]**

6. A four-bar chain of the same type (but of different dimensions) as shown in Fig. 4.14 is having link *AD* as fixed. The angular velocity and angular acceleration of link *AB* are 100 rad/s and 4400 rad/s² respectively. The link lengths are : link *AD* (*i.e.*, link 1) = 125 mm ; link *AB* (*i.e.*, link 2) = 75 mm, link *BCE* (*i.e.*, link 3 with *BC* = 80 mm, *BE* = 30 mm, *CE* = 70 mm) ; link *CD* (*i.e.*, link 4) = 37 mm and $\angle BAD = 53°$.
   (*i*) Draw the velocity and acceleration diagrams.
   (*ii*) Find the acceleration of *F* and *E* when *BF* = 30 mm on *BC* as shown in Fig. 4.14.
   (*iii*) Find the angular acceleration of link 3 (*i.e.*, link *BCE*).

[Ans. (*ii*) $f_F = 1340$ m/s² ; $f_E = 1340$ m/s² ; (*iii*) $\alpha_3 = 18875$ rad/s²]

7. A six link mechanism having link 1 as fixed is of the same type (but of different dimensions) as shown in Fig. 4.15. The crank *AB* of the mechanism is rotating with a speed of 60 r.p.m. The dimensions of the links are : *AB* (link 2) = 150 mm, *BC* (link 3) = 600 mm, *CED* (link 4 with *CE* = *ED* = 145 mm, *CD* = 125 mm), *DF* (link 5) = 500 mm, distance *AE* = 625 mm and link 6 is a slider. Angle of crank with vertical = 45°.
   (*i*) Draw velocity and acceleration diagrams.
   (*ii*) Find the linear acceleration of slider *F*.
   (*iii*) Find the angular acceleration of links *BC*, *CED* and *FD*.

[Ans. (*ii*) $f_F = 7$ m/s² (*iii*) $\alpha_{BC} = 23$ rad/s² ; $\alpha_{CED} = 13.8$ rad/s² ; $\alpha_{FD} = 14.4$ rad/s²]

5

Mechanism with Lower Pairs

5.1. Introduction

When the two elements of a pair have a surface contact and a relative motion takes place, the surface of one element slides over the surface of the other, the pair is known as lower pair. This chapter deals with such mechanism having lower pairs. Lower pairs generally comprise turning pairs and sliding pairs. The mechanism having lower pairs are pantographs, an exact straight-line mechanism, an approximate straight-line mechanism, steering gear mechanism and universal or Hooke's joint.

5.2. Pantograph

A pantograph is a mechanism consisting of lower pairs, used to produce paths on an enlarged or a reduced scale exactly similar to the one traced out by a point on the linkage.

Fig. 5.1 shows the diagram of a pantograph in which four links *AB*, *BC*, *CD* and *DA* are arranged in such a way that a parallelogram *ABCD* is formed in which *AB* = *CD* and *AD* = *BC*. The link *CD* is extended to *O* and link *CB* is extended to *E* such that *OAE* becomes a straight line.

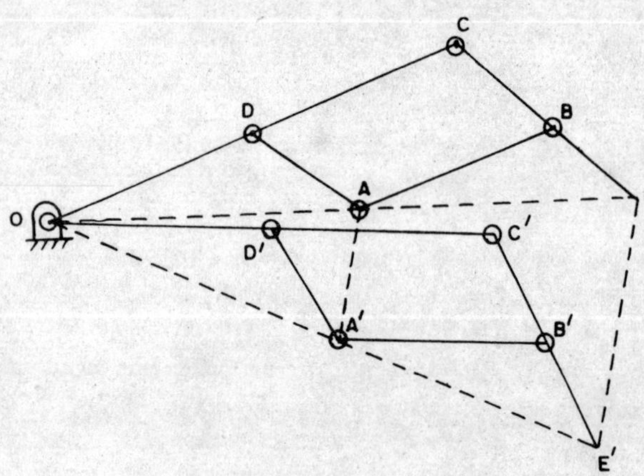

Fig. 5.1

For all relative positions of the links, the triangles *ODA* and *OCE* are similar and points *O*, *A* and *E* are in one straight line. It may be proved that the point *E* traces out the same path as described by point *A*.

ABCD is the initial assumed position as shown in Fig. 5.1. Also *OAE* is the initial straight line.

In triangles *OAD* and *OEC*

$$\angle AOD = \angle EOC \qquad \text{(Common angle)}$$

$$\angle ODA = \angle OCE \qquad (\because\ DA \parallel CE)$$

Therefore the triangles are similar and

$$\frac{OD}{OC} = \frac{OA}{OE} = \frac{DA}{CE} \quad ...(i)$$

Let the point O be fixed and the links are moved to another position so that A moves to A', B to B', E to E' and so on.

Now $\quad A'D' = AD = BC = B'C'$

and $\quad A'B' = AB = CD = C'D'$

Therefore $A'B'C'D'$ is again a parallelogram.

Now let us prove that the triangles $OD'A'$ and $OC'E'$ are similar.

We know, $\quad \frac{OD'}{OC'} = \frac{OD}{OC} = \frac{DA}{CE} \quad \left[\because \frac{OD}{OC} = \frac{DA}{CE} \text{ from } (i)\right]$

$$= \frac{D'A'}{C'E'}$$

Also $\quad \angle OD'A' = \angle OC'E' \quad$ ($D'A' \parallel C'B'$ as $A'B'C'D'$ is a parallelogram)

Hence triangles $OD'A'$ and $OC'E'$ are similar.

$\therefore \quad \angle A'OD' = \angle E'OC'$

Hence O, A' and E' lie on a straight line.

Also from (*i*), we know that

$$\frac{OA}{OE} = \frac{OD}{OC}$$

$$= \frac{OD'}{OC'}$$

$$= \frac{OA'}{OE'} \quad (\because \text{Triangles } OD'A' \text{ and } OC'E' \text{ are similar})$$

The above result shows that the ratio of the distances of A and E from the fixed point remains same as the links are moved. This is true for all the positions of the links. Hence point E traces out the same path as described by point A.

Hence, if O is fixed to the frame of a machine by means of a turning pair and A is attached to a point in the machine which has rectilinear motion, then E will trace out exactly similar path but to different scale. Similarly if E is constrained to move in a straight line, then A will trace out similar path.

A pantograph is used to reproduce drawings to different scales (*i.e.* on enlarged or reduced scales). It is also used to guide cutting tools.

5.3. Straight Line Mechanism

The mechanisms used for producing straight line motion, are known as straight line mechanisms. These mechanisms may produce exactly straight line motion or approximate straight line motion. In these mechanisms either only turning pairs are used or one sliding pair is used.

5.4. Exact Straight Line Motion Mechanisms

The exact straight line motion can be obtained from the mechanism which is having

(*i*) Only turning pairs or

(*ii*) One sliding pair.

5.4.1. Exact Straight Line Motion Mechanism using only Turning Pairs

The following are the important mechanisms using only turning pairs and producing exact straight line motion :

1. Peaucellier mechanism
2. Hart's mechanism.

For exact straight line motion, the principle adopted is described as given below :

Let O is the centre of a circle of diameter AD. AB is any chord. The triangle inscribed on a semi-circle (*i.e.* ΔABD) will be a right angled triangle. The cord AB is produced upto point C. From C, draw a line CE perpendicular to the diameter AD produced.

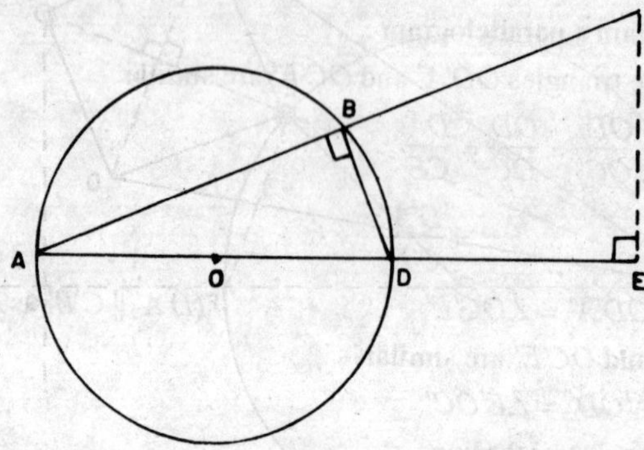

Fig. 5.2

The locus of the point C will be a *straight line*, perpendicular to the diameter AD (*i.e.* $\perp$ to AE) provided the product of $AB \times AC$ is a constant.

This can be proved as given below :

The Δs AEC and ABD are similar as $\angle DAB = \angle EAC$ (common angle) and $\angle ABD = \angle AEC = 90°$.

Hence
$$\frac{AD}{AC} = \frac{AB}{AE}$$

or
$$AB \times AC = AD \times AE$$

But AD is the diameter of the circle and hence it is constant. If AE is also constant then $AB \times AC$ will be constant. AE will be constant when the perpendicular from point C always coincides with point E. Hence the projection of C should be always at E.

Problems 5.1. *A circle with AD as diameter, has a point B on its circumference. On AB produced, there is a point C such that if B turns about A, the product AB × AC is constant. Prove that the point C moves in a straight line perpendicular to AD produced.*

Sol. Refer to Fig. 5.2.

From C, draw CE perpendicular to AD produced.

For any position of B on the circumference of the circle with diameter AD, the triangles ABD and ACE are similar as $\angle ABD = \angle AEC = 90°$ and $\angle DAB = \angle EAC$ (common angle)

$$\therefore \quad \frac{AB}{AD} = \frac{AE}{AC}$$

or
$$AD \times AE = AB \times AC$$

or
$$AE = \frac{AB \times AC}{AD}$$

But it is given that $AB \times AC$ is constant. Also AD is constant being the diameter of the circle. Hence AE is constant.

Hence AE will be constant for all positions of B. Therefore the location of E is fixed which means that C moves in a straight line perpendicular to AD produced.

1. Peaucellier Mechanism

It consists of eight links *i.e.* links AO, OE, AB, AD, EB, BC, CD and DE in which link AO is fixed and link OE is rotating about point O, as shown in Fig. 5.3. The links BC, CD, DE and EB are all of equal lengths and thus form a rhombus. All the links are connected to pin-joints.

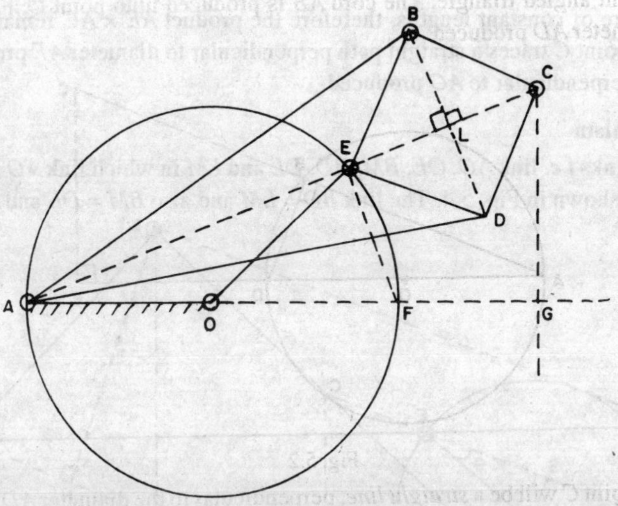

Fig. 5.3

The pin at E is constrained to move along the circumference of a circle of diameter AF by means of link OE. Also the length of link AD = length of link AB.

Hence in Fig. 5.3, we have $AD = AB$, $BC = CD = DE = EB$ and $OA = OE$.

It is required to prove that as the link OE moves around O, the point C moves in a straight line perpendicular to AO produced.

Since $BCDE$ is a rhombus, the diagonals EC and BD will bisect each other at right angles. Hence $\angle ELB = \angle CLB = 90°$. Also angle AEF will be a right angle for all position of E, at it is the angle subtended by a diameter of the circle on the circumference of the circle.

Let us first prove that AEC is a straight line. Consider triangles AEB and ADE. In these triangles, $AB = AD$, $BE = ED$ and AE is common.

$\therefore$ Triangles AEB and ADE are similar.

Hence $\angle EAB = \angle EAD$.

$\therefore$ Point E lies on the bisector of $\angle DAB$. ...(*i*)

Now consider triangles ACB and ACD. In these triangles, $AB = AD$, $BC = CD$ and AC is common.

$\therefore$ Triangles ACB and ACD are similar.

Hence $\angle CAB = \angle CAD$.

$\therefore$ Point C lies on the bisector of $\angle DAB$. ...(*ii*)

From (*i*) and (*ii*), it is clear that AEC is a straight line.

Now in right angled triangle ALB,

$$AB^2 = AL^2 + LB^2 \quad ...(iii)$$

And in right angled triangle, CLB,

$$BC^2 = CL^2 + LB^2 \quad ...(iv)$$

Subtracting equation (*iv*) from (*iii*), we get

$$AB^2 - BC^2 = AL^2 - CL^2$$
$$= (AL + CL)(AL - CL)$$
$$= AC \times (AL - EL) \qquad (\because \; CL = EL)$$
$$= AC \times AE$$

But AB and BC are of constant lengths, therefore the product $AE \times AC$ remains constant. Hence as proved in Art. 5.4.1, the point C traces a straight path perpendicular to diameter AF produced. Hence point C moves in a straight line perpendicular to AO produced.

2. Hart's Mechanism

It consists of six links *i.e.* link AO, OE, BM, BD, DL and LM in which link AO is fixed and link OE is rotating about point O as shown in Fig. 5.4. The link $BD = LM$ and also $BM = DL$ and $AO = OE$.

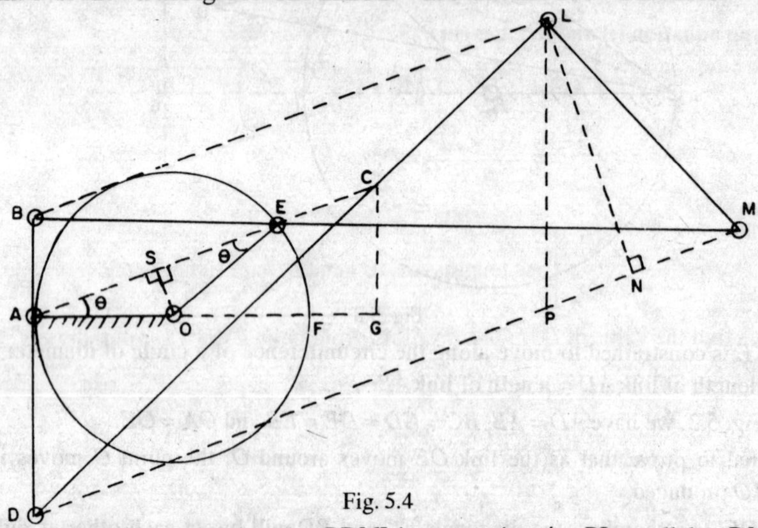

Fig. 5.4

The links are arranged in such a way that $BDML$ is a trapezium *i.e.* BL parallel to DM. The pins A and E on links BD and BM are located in such a way that

$$\frac{AB}{BD} = \frac{BE}{BM}$$

Also the point C on link LD divides this link in such a way that

$$\frac{AB}{BD} = \frac{BE}{BM} = \frac{CL}{LD}$$

It can be proved that as OE rotates about O, the point C moves in a line perpendcular to AO produced.

Let us first prove that AEC is a straight line. Consider triangle BDM. In this triangle

$$\frac{AB}{BD} = \frac{BE}{BM} \qquad \text{(Given)}$$

Therefore AE is parallel to DM. But $DM \parallel BL$. Hence AE is parallel to BL.

Now consider the triangle DBL. In this triangle,

$$\frac{AB}{BD} = \frac{CL}{LD} \qquad \text{(Given)}$$

Therefore AC is parallel to BL and thus parallel to DM. Now AE and AC are both parallel to BL and have a point E in common. Hence AEC is a straight line. This will be true for all configurations of the mechanism.

Now let us find the product of AE and AC.

The triangles ABE and DBM are similar as $AE \parallel DM$.

$$\therefore \quad \frac{AE}{DM} = \frac{AB}{BD}$$

$$\therefore \quad AE = \frac{AB}{BD} \times DM \quad \text{...}(i)$$

Also the triangles DAC and DBL are similar as $AC \parallel BL$

$$\therefore \quad \frac{AC}{BL} = \frac{AD}{BD}$$

or

$$AC = \frac{AD}{BD} \times BL \quad \text{...}(ii)$$

Multiplying equation (*i*) and (*ii*), we get

$$AE \times AC = \left(\frac{AB}{BD} \times DM\right) \times \left(\frac{AD}{BD} \times BL\right)$$

$$= \frac{AB \times AD}{BD^2} \times DM \times BL$$

$$= \text{Constant} \times DM \times BL \quad \text{...}(iii)$$

$$\left[\text{As lengths } AB, AD \text{ and } BD \text{ are fixed, hence } \left(\frac{AB \times AD}{BD^2}\right) \text{ is a constant}\right]$$

Now let us find the value of $DM \times BL$. For this, from point L draw LP parallel to LM. Also from point L, draw $LN \perp DM$.

Now

$$DM = DN + NM$$

$$BL = DP$$

$$= DN - PN$$

$$= DN - NM \quad (\because \; LP = BD = LM, \text{ hence } PN = NM)$$

$$\therefore \quad DM \times BL = (DN + NM)(DN - NM)$$

$$= DN^2 - NM^2$$

$$= (DL^2 - LN^2) - (LM^2 - LN^2)$$

[From right angled triangle DNL, $DN^2 = DL^2 - LN^2$ and from right angled triangle LNM, $NM^2 = LM^2 - LN^2$]

$$= DL^2 - LN^2 - LM^2 + LN^2$$

$$= DL^2 - LM^2$$

$$= \text{Constant} \quad \text{(as lengths } DL \text{ and } LM \text{ are fixed)}$$

Substituting this value in equation (*iii*), we get

$$AE \times AC = \text{constant} \quad \text{...}(iv)$$

But

$$AE = AS + SE$$

$$= AO \cos\theta + OE \cos\theta$$

$$= AO \cos\theta + AO \cos\theta \quad (\because \; AO = OE)$$

$$= 2AO \cos\theta$$

and

$$AC = AG/\cos\theta \quad \text{(consider triangle } AGC)$$

Substituting the values of AE and AC in equation (*iv*), we get

$$AE \times AC = 2AO \cos\theta \times \frac{AG}{\cos\theta}$$

$$= 2AO \times AG$$

But $AE \times AC$ is constant

$$\therefore \qquad 2AO \times AG = \text{constant}$$

But length AO is fixed and hence it is constant

$$\therefore \qquad AG = \text{constant.}$$

This means that the projection of point C is always the same point G at a distance AG from point A. Thus the trace of the point C is a straight line.

5.4.2. Exact Straight Line Motion Mechanism using One Sliding Pair

The exact straight line motion using one sliding pair is obtained by Scott-Russel Mechanism which is described below :

Scott-Russel Mechanism

Fig. 5.5 shows Scott-Russel mechanism, which consists of three links *i.e.* link OA, link AC (or extended link BC) and slider C. All the three links are movable links. The slider C is constrained to move in a horizontal straight line passing through O. The link OA is the crank and the link AC is the connecting rod. The link CA is extended upto B in such a way that

$$OA = AC = AB$$

It is required to prove that point B moves in a straight line perpendicular to line OC for all horizontal movements of slider C.

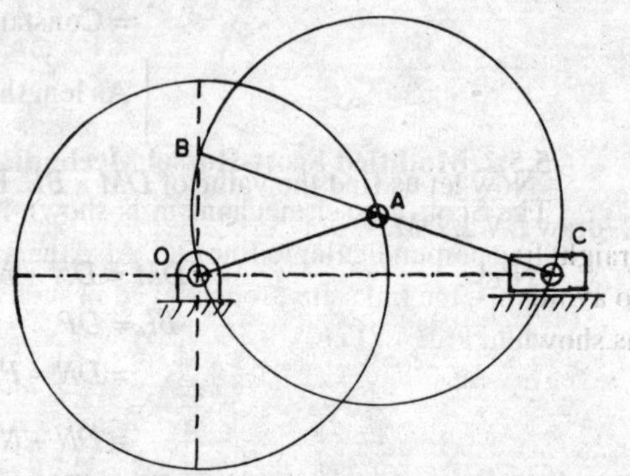

Fig. 5.5

Taking A as centre and radius equal to OA (or $AC = AB$) a circle can be drawn. This circle will pass through points C, O and B as $OA = AC = AB$. The diameter of this circle is BC. Diameter of a circle always subtends a right angle at any point on its circumference *i.e.* at point O. Hence COB will always be right angle for all positions of C. Now the line joining O to C is a horizontal straight line, for all positions of C, since C moves in a straight horizontal line. Therefore, a line joining O to B will always be a straight line perpendicular to OC. Thus locus of point B will be straight line perpendicular to OC.

5.5. Approximate Straight Line Motion Mechanism

The following are the important mechanisms which give approximate straight line motion :

(*i*) Watt's straight line mechanism
(*ii*) Modified Scott-Russel mechanism
(*iii*) Grass-Hopper mechanism
(*iv*) Tchebicheff's mechanism
(*v*) Roberts mechanism.

5.5.1. Watt's Straight Line Mechanism

It is a very simple mechanism originally used by James Watt for guiding the motion of the piston of his steam engine to have an approximate straight line motion. It has four links OA, AB, BC and CO. The link CO is fixed. The link AO can oscillate about the centre O whereas the link BC can oscillate about centre C as shown in Fig. 5.6.

In the mean position of the mechanism, the links AO and BC are parallel and the link AB is perpendicular to AO and BC. It is seen that if D is a point on the link AB such that $DA/DB = BC/OA$, then for small oscillations of OA and BC, the point D will trace an approximate straight line. Fig. 5.6 shows the three different positions in which the path of D is an approximate straight line.

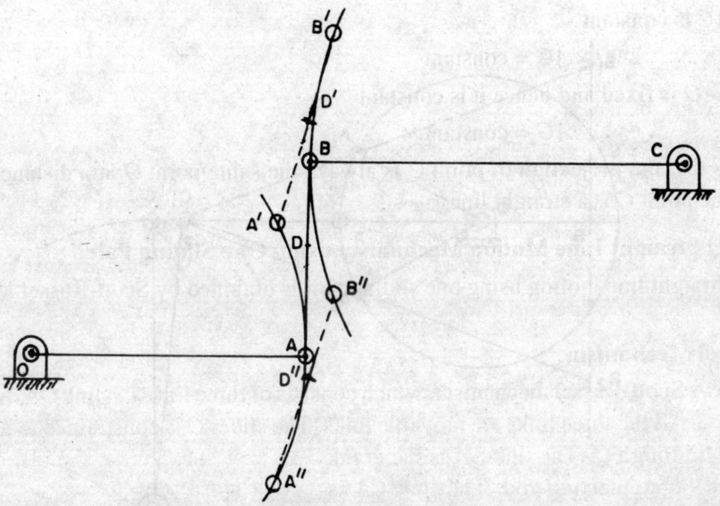

Fig. 5.6

5.5.2. Modified Scott-Russel Mechanism

The Scott-Russel mechanism is shown in Fig. 5.5. In that mechanism, the point B moves in a exact straight line perpendicular to line OC. Also the path of B is through the joint O, which is generally not desired. To avoid this, the links are proportioned in such a way that AC is the mean proportionate between OA and AB (as shown in Fig. 5.7) *i.e.*

$$\frac{OA}{AC} = \frac{AC}{AB}$$

or $$AC^2 = OA \times AB$$

or $$AC = (OA \times AB)^{1/2}$$

But in this case B will approximately traverse a straight line perpendicular to line OC. The mechanism shown in Fig. 5.7 is known as modified Scott-Russel mechanism. It has limitation and is true only for small movement of slider C. The mechanism is drawn in two different positions in which the path of point B is an approximate straight line.

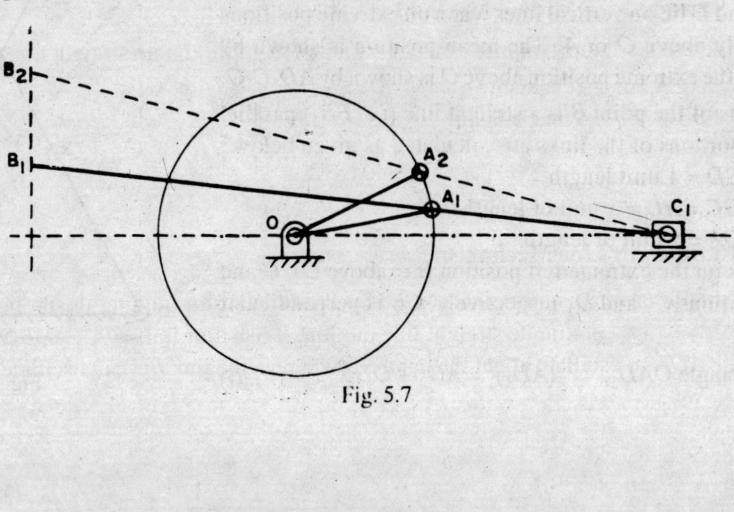

Fig. 5.7

5.5.3. Grass-Hopper Mechanism

Fig. 5.8 shows the Grass-Hopper mechanism. This mechanism is the modification of the modified Scott-Russel mechanism, in which the sliding pair at C is replaced by a turning pair DC as shown in Fig. 5.8. In the mean position the turning pair DC is perpendicular to OC. The turning pair DC is pin-joined at D which is fixed.

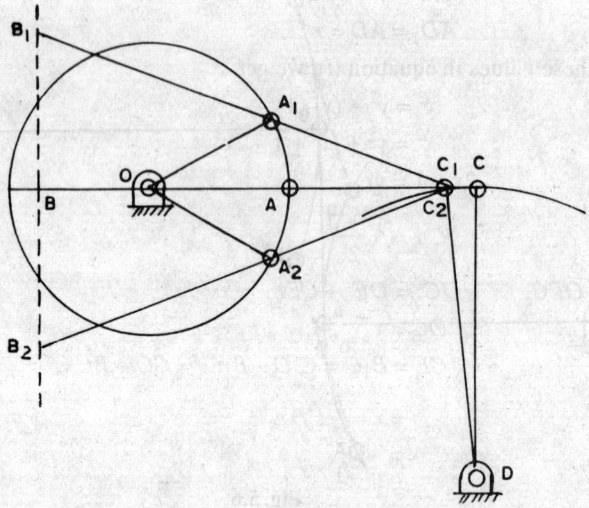

Fig. 5.8

For small angular movement of CD on each side from mean position, the point C moves approximately in a straight line perpendicular to CD if the length of CD is large enough. The point B will move again in an approximate straight line perpendicular to OC if AC is the mean proportionate between OA and AB *i.e.*

$$\frac{OA}{AC} = \frac{AC}{AB}$$

In Fig. 5.8, the mean position is shown by $DCAOB$ whereas the two other positions are shown by $DC_1A_1OB_1$ and $DC_2A_2OB_2$.

5.5.4. Tchebicheff's Mechanism

Fig. 5.9 shows Tchebicheff's mechanism which consists of four links namely OC, CD, AD and OA. The link OA is fixed. The links OC and AD are equal and crossed. B is the mid-point of CD. The point B is the tracing point. The proportions of the links are such that B, C and D lie on vertical lines when on extreme positions *i.e.* when directly above O or A. The mean position is shown by $ADCO$ whereas the extreme position above O is shown by AD_1C_1O.

The locus of the point B is a straight line (*i.e.* BB_1) parallel to OA. The proportions of the links are calculated as given below :

Let $CD = 1$ unit length

$OC = AD = x$ unit of length

$OA = y$ unit of length

when link CD is on the extreme left position (*i.e.* above O), C and D takes the positions C_1 and D_1 respectively. CE is perpendicular on OA.

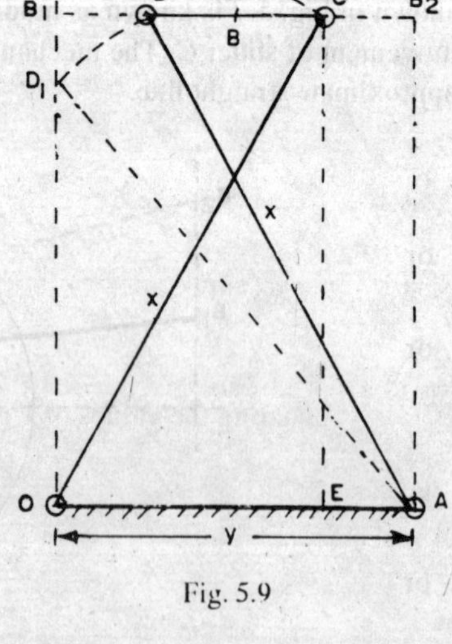

Fig. 5.9

From triangle OAD_1, $\quad (AD_1)^2 = AO^2 + OD_1^2 \quad$ ...(*i*)

But $OD_1 = OC_1 - C_1D_1$

$= OC - CD$ $\quad (\because\ OC_1 = OC = x \text{ and } C_1D_1 = CD = 1)$

$= (x - 1)$

$AO = y$

$AD_1 = AD = x$

Substituting these values in equation (*i*), we get

$$x^2 = y^2 + (x-1)^2$$

or $x^2 = y^2 + (x^2 + 1 - 2x)$

or $2x - 1 = y^2$

or $$x = \frac{y^2+1}{2} \quad \text{...(ii)}$$

Now triangle *OEC*, $OC^2 = OE^2 + CE^2$ ...(*iii*)

But $OC = x$

$CE = B_1O = C_1O - B_1C_1 = CO - BC$

$= x - \frac{1}{2}CD$

$= x - \frac{1}{2}$ $\quad (\because\ CD = 1 \text{ unit})$

$OE = OA - AE$

$= OA - CB_2$ $\quad (\because\ AE = CB_2)$

$= OA - (BB_2 - BC)$ $\quad (\because\ CB_2 = BB_2 - BC)$

$= y - \left(\frac{y}{2} - \frac{1}{2}\right)$ $\quad \left(\because\ BB_2 = \frac{1}{2}B_1B_2 = \frac{y}{2} \text{ and } BC = \frac{1}{2}\right)$

$= \frac{y}{2} + \frac{1}{2}$

Substituting the values of *OC*, *CE* and *OE* in equation (*iii*), we get

$$x^2 = \left(\frac{y}{2} + \frac{1}{2}\right)^2 + \left(x - \frac{1}{2}\right)^2$$

$$= \left(\frac{y^2}{4} + \frac{1}{4} + \frac{y}{2}\right) + \left(x^2 + \frac{1}{4} - x\right)$$

or $$0 = \frac{y^2}{4} + \frac{1}{4} + \frac{y}{2} + \frac{1}{4} - x$$

or $$x = \frac{y^2}{4} + \frac{y}{2} + \frac{1}{2} \quad \text{...(iv)}$$

Equating the values of *x* given by equations (*ii*) and (*iv*), we get

$$\frac{y^2+1}{2} = \frac{y^2}{4} + \frac{y}{2} + \frac{1}{2}$$

or $$\frac{y^2}{2} + \frac{1}{2} = \frac{y^2}{4} + \frac{y}{2} + \frac{1}{2}$$

or $$\frac{y^2}{2} - \frac{y^2}{4} = \frac{y}{2}$$

or $$\frac{y^2}{4} = \frac{y}{2}$$

or $$y = 2$$

Substituting this value of y in equation (*ii*), we get

$$x = \frac{2^2 + 1}{2} = 2.5$$

Hence $CD : OA : OC = 1 : 2 : 2.5$

If the links CD, OA and OC are made in the ratio of 1, 2 and 2.5 respectively, then point B will move in a horizontal straight line parallel to OA.

5.5.5. Robert's Mechanism

This mechanism is shown in Fig. 5.10. This is also four bar chain mechanism. In the mean position, it has the form of a trapezium. The links OC and AD are of equal length and link OA is fixed. A bar EB is rigidly attached to the link CD at the mid point B of link CD. The bar EB is at right angle to CD. The point E is the tracing point. When the mechanism is displaced as shown in Fig. 5.10 by the dotted lines, the point E will trace an approximate straight line.

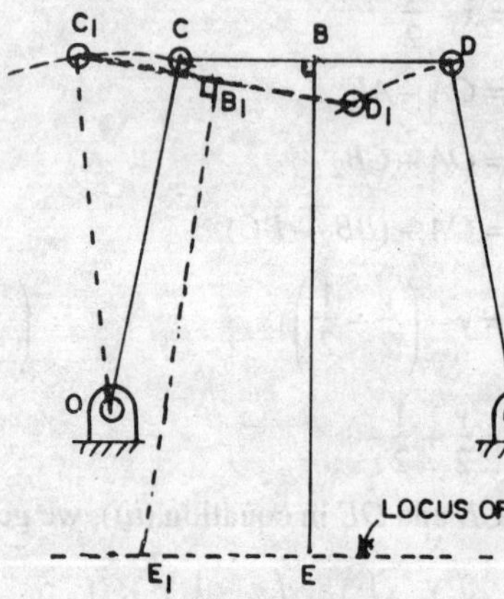

Fig. 5.10

Problem 5.2. *A torque of 100 N m is applied to the link OA at A of a Grass-Hopper mechanism shown in Fig. 5.11. The link OA makes an angle of 20° with the horizontal. Find the magnitude of the vertical force exerted at B to overcome the resisting torque of 100 N m. The lengths of links are given as OA = 40 mm, AC = 60 mm and CB = 150 mm.*

If the link OA makes an angle of 10° and 0° with the horizontal, what will be the vertical force at B to overcome the same torque of 100 N m.

Sol. Given :

$$OA = 40 \text{ mm}, AC = 60 \text{ mm}, CB = 150 \text{ mm}$$

$\therefore$ $$AB = BC - AC = 150 - 60 = 90 \text{ mm}$$

Resisting torque at A, $T_A = 100 \text{ N m} = 100 \times 1000 \text{ N mm} = 100{,}000 \text{ N mm}$

Angle, $\theta = 20°$

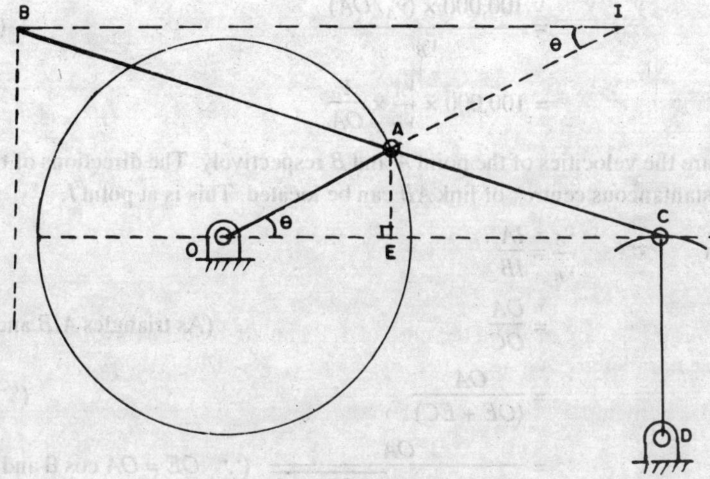

Fig. 5.11

The point B will have approximate straight line motion, if the condition for the dimensions of the Grass-Hopper mechanism is satisfied. The condition is

$$\frac{OA}{AC} = \frac{AC}{AB}$$

But $$\frac{OA}{AC} = \frac{40}{60} = \frac{2}{3}$$

And $$\frac{AC}{AB} = \frac{60}{90} = \frac{2}{3}$$

$$\therefore \quad \frac{OA}{AC} = \frac{AC}{AB}$$

As the condition is satisfied, the point B will have a straight line motion for small angles of OA with horizontal *i.e.* with small values of 'θ' as shown in Fig. 5.11.

Now we know that,

Power, $P = T \times \omega$

$\therefore$ Resisting power at A, $P = T_A \times \omega_A$ (ω_A = Angular velocity of link OA)

$= 100{,}000 \times \omega_A$ ...(*i*)

This power is to be overcomed by applying vertical force at B.

Let F_B = Vertical force at B

v_B = Velocity of point B in the vertical direction (*i.e.* in the direction of force F_B)

The power required at point B

= Force × Velocity at point B

$= F_B \times v_B$ ...(*ii*)

Equating the two values given by equations (*i*) and (*ii*), we get

$$F_B \times v_B = 100{,}000 \times \omega_A$$

or $$F_B = \frac{100{,}000 \times \omega_A}{v_B} \quad \text{...(iii)}$$

$$= \frac{100{,}000 \times (v_A/OA)}{v_B} \qquad (\because \; v_A = \omega_A \times OA)$$

$$= 100{,}000 \times \frac{v_A}{v_B} \times \frac{1}{OA} \qquad ...(iv)$$

As v_A and v_B are the velocities of the point A and B respectively. The directions of these velocities are known. Hence the instantaneous centre* of link AB can be located. This is at point I.

We know that
$$\frac{v_A}{v_B} = \frac{IA}{IB}$$

$$= \frac{OA}{OC} \qquad \text{(As triangles } AIB \text{ and } OAC \text{ are similar)}$$

$$= \frac{OA}{(OE + EC)} \qquad (\because \; OC = OE + EC)$$

$$= \frac{OA}{OA \cos\theta + \sqrt{AC^2 - AE^2}} \qquad (\because \; OE = OA\cos\theta \text{ and } EC^2 = AC^2 - AE^2)$$

$$= \frac{OA}{OA \cos\theta + \sqrt{[AC^2 - (OA \sin\theta)^2]}} \qquad (\because \; AE = OA \sin\theta)$$

$$= \frac{40}{40\cos\theta + \sqrt{[60^2 - (40\sin\theta)^2]}}$$

Substituting this value of $\frac{v_A}{v_B}$ in equation (*iv*), we get

$$F_B = \frac{100{,}000}{OA} \times \frac{40}{40\cos\theta + \sqrt{[60^2 - (40\sin\theta)^2]}}$$

$$= \frac{100{,}000}{40} \times \frac{40}{[40\cos\theta + \sqrt{60^2 - (40\sin\theta)^2}\,]} \qquad (\because \; OA = 40)$$

$$= \frac{100{,}000}{[40\cos\theta + \sqrt{60^2 - (40\sin\theta)^2}\,]} \qquad ...(v)$$

(*i*) When $\theta = 20°$,
$$F_B = \frac{100{,}000}{40\cos 20° + \sqrt{60^2 - (40\sin 20°)^2}}$$

$$= \frac{100{,}000}{37.58 + 58.42} = \mathbf{1041.6\ N.} \quad \textbf{Ans.}$$

(*ii*) When $\theta = 10°$,
$$F_B = \frac{100{,}000}{40\cos 10° + \sqrt{60^2 - (40\sin 10°)^2}}$$

$$= \frac{100{,}000}{39.39 + 59.94} = \mathbf{1006.7\ N.} \quad \textbf{Ans.}$$

(*iii*) When $\theta = 0°$,
$$F_B = \frac{100{,}000}{40\cos 0° + \sqrt{60^2 - (40\sin 0°)^2}}$$

$$= \frac{100{,}000}{40 + 60} = \mathbf{1000\ N.} \quad \textbf{Ans.}$$

---

*Velocity v_A is perpendicular to OA whereas the velocity v_B is vertical. The intersection of OA produced and perpendicular to the path of B, gives the location of instantaneous centre.

Problem 5.3. *In a modified Scott-Russel mechanism shown in Fig. 5.7, OA_1 = 3 cm, A_1C_1 = 5.1 cm. Find the length of the extended link A_1B_1 so that point B_1 is having approximate straight line motion.*

Sol. Given (Refer to Fig. 5.7)

$$OA_1 = 3 \text{ cm}, A_1C_1 = 5.1 \text{ cm}$$

For a modified Scott-Russel mechanism, the point B_1 will have an approximate straight line motion if

$$\frac{OA}{AC} = \frac{AC}{AB}$$

or

$$\frac{OA_1}{A_1C_1} = \frac{A_1C_1}{A_1B_1}$$

$\therefore$

$$A_1B_1 = \frac{(A_1C_1)^2}{OA_1}$$

$$= \frac{(5.1)^2}{3} = \mathbf{8.67 \text{ cm.}} \quad \textbf{Ans.}$$

5.6. Application of Straight Line Motion in Engine Indicators

The straight line motion is mostly used in Engine indicators, which are the instruments that keep the graphical records of pressure inside the engine cylinder during the piston stroke.

An indicator consists of a cylinder with a piston. The cylinder of indicator is known as indicator cylinder and piston is known as indicator piston. This indicator cylinder is connected to the engine cylinder. The indicator piston rises and falls in response to pressure variation within the engine cylinder. The displacement of the indicator piston is constrained by a spring to get a direct measure of the gas or the steam pressure acting upon it. The displacement is recorded by a pencil on paper, wrapped on a drum, to an enlarged scale with the help of a straight line mechanism.

The following are the important engine indicators which works on the straight line motion mechanism :

(*i*) Simplex indicator

(*ii*) Crosby indicator

(*iii*) Thomson indicator

(*iv*) Dobbie McInnes indicator.

5.6.1. Simplex Indicator

The principle of pantograph is used in this indicator. The Fig. 5.12 shows the simplex indicator in which *O* is the fixed pivot and *ABCD* is a parallelogram formed by four links.

OAB is one link. Link *BC* is extended to point *E* such that *ODE* fall in a straight line. The point *D* is attached to the piston rod of the indicator cylinder. Point *E* is the tracing point which traces the path of *D* or movement of piston.

The piston moves in a vertical straight line within the indicator cylinder. Its movement is controlled by the steam or gas pressure to be measured. The same displacement takes place with point *D*. The displacement of point *D* is reproduced on an enlarged scale, on the paper wrapped on the indicator drum, by the pencil fixed at point *E* which describe the path similar to that of *D*. The following relationships are used :

$$\frac{OE}{OD} = \frac{OB}{OA} = \frac{BE}{BC}$$

Each of the above ratio represents the magnification of this indicator.

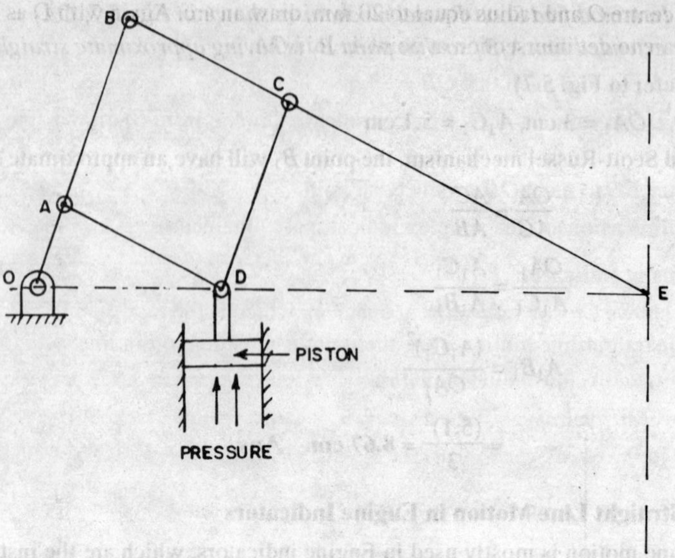

Fig. 5.12

Problem 5.4. *The simplex engine indicator is used to obtain the indicator diagram. The pencil point is located at a distance of 120 mm from the fixed point. The indicator diagram should be magnified by four times the displacement of the piston. Design the pantograph to be used in the indicator.*

Sol. Given :

Refer to Fig. 5.12.

Distance between fixed point and tracing point, $OE = 120$ mm

Magnification required $= 4$ times

or $$\frac{OE}{OD} = 4$$

$$\therefore \quad OD = \frac{OE}{4} = \frac{120}{4} = 30 \text{ mm}$$

The simplex indicator using pantograph has the following relationship :

$$\frac{OE}{OD} = \frac{OB}{OA} = \frac{BE}{BC} = 4 \qquad ...(i)$$

and $ABCD$ is a parallelogram.

Choose convenient dimensions of OA and AD.

Let $OA = 20$ mm and $AD = 35$ mm.

But from (*i*), $\frac{OB}{OA} = 4 \quad \therefore \quad OB = 4 \times OA = 4 \times 20 = 80$ mm

$\frac{BE}{BC} = 4 \quad \therefore \quad BE = 4 \times BC = 4 \times AD \qquad (\because \; BC = AD)$

$= 4 \times 35 = 140$ mm

Construct the diagram as given below :

(*i*) Draw a horizontal straight line $OE = 120$ mm.

(*ii*) Take $OD = 30$ mm. Hence point D is known.

(*iii*) With centre *O* and radius equal to 20 mm, draw an arc. Again with *D* as centre and radius equal to 35 mm, draw an arc to get intersection at point *A*. Join *OA*.

(*iv*) Produce *OA* to *B* such that *OB* = 80 mm.

(*v*) Join *AD* and from *D*, draw a line parallel to *OB* and from *B*, draw a line parallel to *DA*. These two lines meet at *C*.

(*vi*) Produce *BC* to meet *OD* produced at point *E*.

Thus the dimensions of the simplex indicator are obtained.

5.6.2. Crosby Indicator

Fig. 5.13 shows Crosby indicator mechanism, which is the modified form of the pantograph. In order to have a vertical straight line motion of *E*, the point *E* must remain in line with *O* and *D*, and also the links *OB* and *DC* must remain approximately parallel.

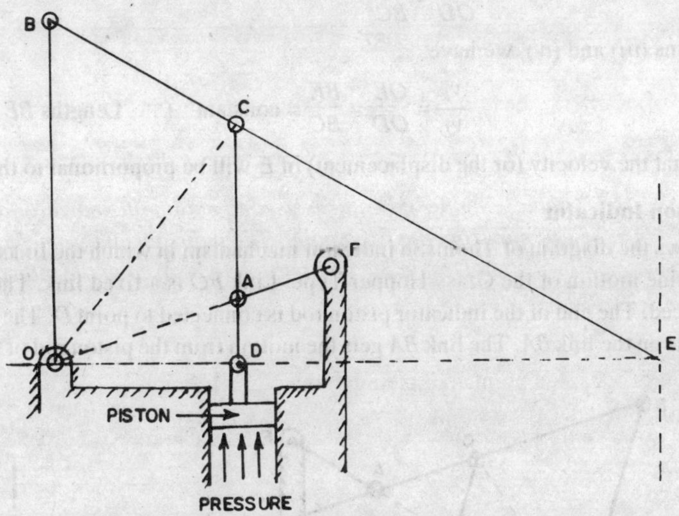

Fig. 5.13

Let us find the instantaneous centres for links *DC* and *BE* for the configuration shown in Fig. 5.13.

If the directions of velocities of any two points on a link, are known, then the instantaneous centre can be located easily which is the intersection of the perpendiculars to the direction of velocities at the two points.

On the link *DC*, there is a point *A* which is connected to the fixed point *F*. The directions of velocities at points *A* and *D* are known. The direction of velocity at *D* is vertical whereas the direction of velocity at *A* is perpendicular to *AF*. Hence draw a horizontal line through *D*, and produce line *FA* to meet the horizontal line at *O*. This point *O* is the instantaneous centre for link *DC*. The velocity of any point on this link is proportional to its distance from the instantaneous centre. The direction of velocity will be perpendicular to the line joining the point with the instantaneous centre.

For the link *BE*, the velocity of *B* is perpendicular to *OB* whereas the direction of velocity at *E* is vertical. Hence from *E* draw a horizontal line to meet the line *OB* at *O*. Then this point *O* will be the instantaneous centre for link *BE*. The velocities at *B* and *E* will be proportional to the distances of *B* and *E* from *O*.

Let v_D = Velocity of the point *D*,

v_C = Velocity of point *E*, and

v_B, v_E = Velocity of *B* and *E* respectively.

We know that $\frac{v_C}{v_D} = \frac{OC}{OD}$ ...(*i*)

and $\frac{v_E}{v_C} = \frac{OE}{OC}$ ...(*ii*)

Multiplying equations (*i*) and (*ii*), we get

$$\frac{v_C}{v_D} \times \frac{v_E}{v_C} = \frac{OC}{OD} \times \frac{OE}{OC}$$

or $\frac{v_E}{v_D} = \frac{OE}{OD}$ ...(*iii*)

Since *CD* is parallel to *OB*, hence triangles *EDC* and *EOB* are similar.

$\therefore$ $\frac{OE}{OD} = \frac{BE}{BC}$ ...(*iv*)

From equations (*iii*) and (*iv*), we have

$$\frac{v_E}{v_D} = \frac{OE}{OD} = \frac{BE}{BC} = \text{constant} \quad (\because \text{ Lengths } BE \text{ and } BC \text{ are constant})$$

This shows that the velocity (or the displacement) of *E* will be proportional to that of *D*.

5.6.3. Thomson Indicator

Fig. 5.14 shows the diagram of Thomson indicator mechanism in which the links *OB*, *BA*, *AF* and *FO* constitute a straight line motion of the Grass-Hopper Type. Link *FO* is a fixed link. The tracing point *E* lies on the link *BA* produced. The end of the indicator piston rod is connected to point *D*. The point *D* is connected to the point *C* which is on the link *BA*. The link *BA* gets the motion from the piston rod of the indicator through *D* and *C*.

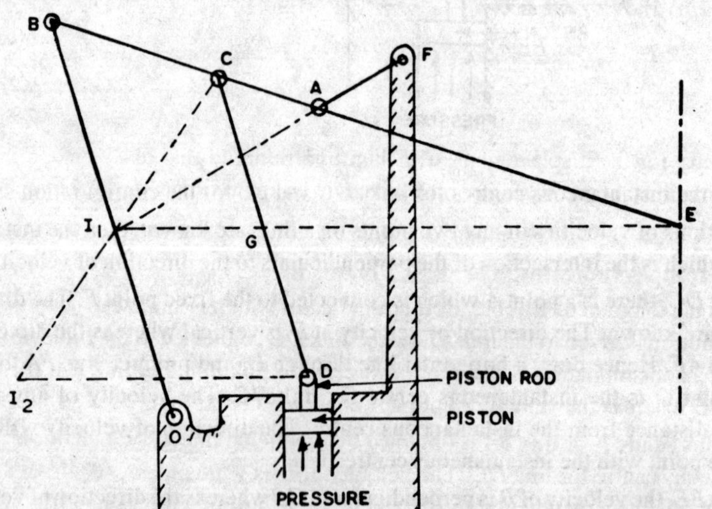

Fig. 5.14

The best position of the tracing point *E* is obtained by finding the instantaneous centres of link *BA* and *CD*.

For the link *BA*, the direction of velocities at *B* and *A* are known. The velocity at *A* is perpendicular to *AF* whereas the velocity at *B* is perpendicular to *BO*. Hence the instantaneous centre for link *BA* can be obtained by drawing lines *BO* and *FA* produced. It lies at I_1.

For the link CD, the direction of velocity at C is perpendicular to I_1C hence I_2 lies on I_1C produced. The direction of velocity at D is vertical, hence I_2 lies on a horizontal line through D. Hence I_2 is obtained.

Now the point E is to move in a vertical direction, it must lie on a horizontal line through I_1 which is the instantaneous centre for link BA (or BE). Hence the best position of the tracing point is obtained.

Let v_D = Velocity of point D,

v_C = Velocity of point C, and

v_E = Velocity of point E.

We know that, $$\frac{v_C}{v_D} = \frac{I_2C}{I_2D}$$

$$= \frac{I_1D}{I_1G} \quad \left(\because \ \text{As } CI_2D \text{ and } CI_1G \text{ are similar hence } \frac{I_2C}{I_2D} = \frac{I_1C}{I_1G}\right) \quad ...(i)$$

Also $$\frac{v_E}{v_C} = \frac{I_1E}{I_1C} \quad ...(ii)$$

Multiplying these above two equations, we get

$$\frac{v_C}{v_D} \times \frac{v_E}{v_C} = \frac{I_1C}{I_1G} \times \frac{I_1E}{I_1C}$$

or $$\frac{v_E}{v_D} = \frac{I_1E}{I_1G} \quad ...(iii)$$

Now if links CD and BO are parallel, then the triangles EGC and EI_1B will be similar.

$$\therefore \quad \frac{I_1E}{I_1G} = \frac{BE}{BC} \quad ...(iv)$$

From equations (*iii*) and (*iv*), we get

$$\frac{v_E}{v_D} = \frac{BE}{BC}$$

= constant (As lengths BE and BC are fixed)

Hence the velocity (or the displacement) of E is proportional to that of D. As CD and BO does not remain parallel for all positions, hence tracing point E moves in an approximate vertical line. But the variations are negligible.

5.6.4. Dobbie McInnes Indicator

Fig. 5.15 shows the diagram of Dobbie McInnes indicator which is similar to Thomson indicator with the difference that the motion is given to the link AF by the link DC connected to the indicator piston as shown in Fig. 5.15. [In case of Thomson indicator, the motion is given to link BA]. For link BA, the velocity of point B is perpendicular to OB whereas the velocity of point A is perpendicular to AF.

The instantaneous centre I_1 of link BA is obtained by producing line AF which intersects line BO at I_1. Now find the instantaneous centre for link DC. The velocity of D is vertical hence I_2 lies on a horizontal line through D. The velocity of C is perpendicular to I_1C and hence I_2 lies on CI_1 produced.

The tracing point E is moving in a vertical line and hence it must lie on a horizontal line through I_1 which is instantaneous centre for link BAE. Hence the best position of E is located.

The horizontal line I_1E cuts the link CD at G. From A, draw AH parallel to CD.

Let v_D = Velocity of point D,

v_C = Velocity of point C,

v_A = Velocity of point A, and

v_E = Velocity of point E.

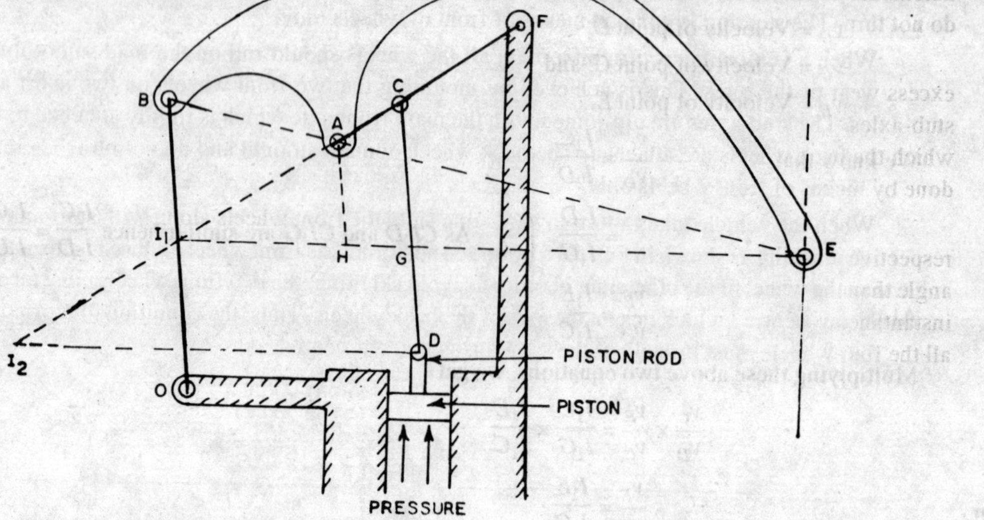

Fig. 5.15

We know that, $\frac{v_C}{v_D} = \frac{I_2C}{I_2D}$

$= \frac{I_1C}{I_1G}$ (As Δs CI_1G and CI_2D are similar)

$= \frac{I_1A}{I_1H}$ (As Δs CI_1G and AI_1H are similar) ...(*i*)

Since the link AF turns about F, the velocities at A and C are proportional to their distances from F.

$\therefore \quad \frac{v_A}{v_C} = \frac{AF}{CF}$ ...(*ii*)

Also we know that, $\frac{v_E}{v_A} = \frac{I_1E}{I_1A}$ ...(*iii*)

Multiplying equations (*i*), (*ii*) and (*iii*), we get

$$\frac{v_C}{v_D} \times \frac{v_A}{v_C} \times \frac{v_E}{v_A} = \frac{I_1A}{I_1H} \times \frac{AF}{CF} \times \frac{I_1E}{I_1A}$$

or $\frac{v_E}{v_D} = \frac{AF}{CF} \times \frac{I_1E}{I_1H}$

$= \frac{AF}{CF} \times \frac{BE}{BA}$ (As Δs EI_1B and EHA are similar)

= constant (As lengths of AF, CF, BE and BA are constant)

Hence velocity of E (or displacement of E) is proportional to that of D.

5.7. Steering Gears

The mechanism which is used for changing the direction of two or more of the wheel axles so as to move the automobile in any desired path, is known as steering gear. Generally the two back wheels of an automobile have a common axis, which is fixed in direction and hence the back wheels remain straight and do not turn. The steering is done by means of front of wheels only.

When a vehicle takes turn on a road, all the wheels should roll on the road smoothly preventing the excess wear of the tyres. This is achieved by mounting the two front wheels on two short axles, known as stub-axles. The stub-axles are pin-joined with the main front axle which is rigidly attached to the rear axle, to which the back wheels are attached. The back wheels remain straight and do not turn. Therefore, steering is done by means of front wheels only.

When the vehicle takes a turn towards one side, the front wheels along with stub-axles turn about the respective pin-joint as shown in Fig. 5.16 by dotted position. The front wheel of that side turns through a greater angle than the wheel of the otherside. In order to avoid skidding, the two front wheels must turn about the same instantaneous centre *I* which lies on the axis of the back wheels. Thus, the condition for correct steering is that all the four wheels must turn about the same instantaneous centre.

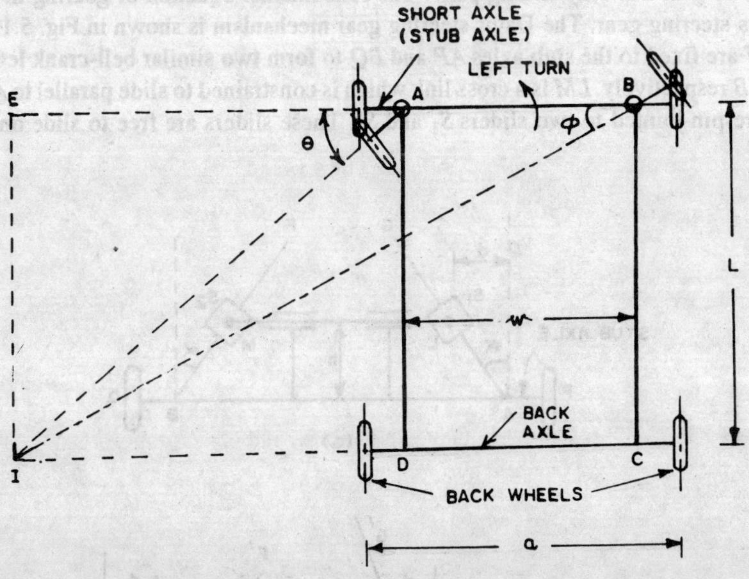

Fig. 5.16

Let a = Wheel track

L = Wheel base

w = Distance between the pivots of front axle

θ and ϕ = Angles turned by the stub axles.

Then $$\cot\theta = \frac{EA}{EI}$$

and $$\cot\phi = \frac{BE}{EI} = \frac{AE + AB}{EI} = \frac{AE + w}{EI}$$

$$\cot\phi - \cot\theta = \frac{AE + w}{EI} - \frac{EA}{EI}$$

$$= \frac{AE}{EI} + \frac{w}{EI} - \frac{AE}{EI}$$

$$= \frac{w}{EI} = \frac{w}{L}$$

or $$\cot\phi - \cot\theta = \frac{w}{L} \quad ...(4.1)$$

The above equation is known as *fundamental equation* of correct gearing. Steering gears are the mechanism which satisfies this fundamental equation.

Types of Steering Gears

Steering gears are of two main types, namely :

(*i*) Davis steering gear, and

(*ii*) Ackermann steering gear.

5.7.1. Davis Steering Gear

This type of gear has only sliding pairs. The fundamental equation of gearing in all the positions is satisfied in Davis steering gear. The Davis steering gear mechanism is shown in Fig. 5.17 (*a*) in which two arms *AG* and *BF* are fixed to the stub axles *AP* and *BQ* to form two similar bell-crank levers *PAG* and *QBF* pivoted at *A* and *B* respectively. *LM* is a cross link which is constrained to slide parallel to *AB*. The ends of the cross-link *LM* are pin-jointed to two sliders S_1 and S_2. These sliders are free to slide on links *AG* and *BM* respectively.

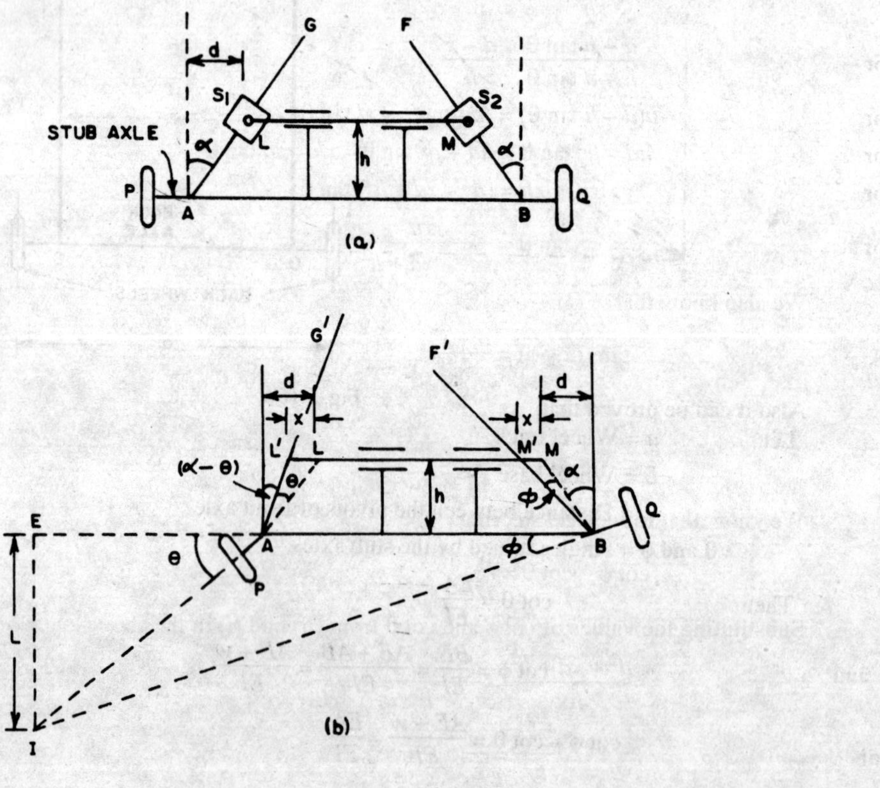

Fig. 5.17

The gear is in the mid-position, during the straight motion of the vehicle, with equal inclination of the arms *AG* and *FB* with *AB* as shown in Fig. 5.17 (*a*). The steering is affected by moving cross-link *LM* to the right or left of the mid-position. *L'M'* shows the position of *LM* for turning to the left as shown in Fig. 5.17 (*b*).

Let x = Distance moved by cross link from mid-position
= MM' or LL' [shown in Fig. 5.17 (*b*)]
h = Vertical distance between *AB* and *LM*
d = Horizontal distance of point *L* from *A* also
= Horizontal distance of point *M* from *B*
α = Angle of inclination of *AG* or *BF* with vertical when vehicle is in mid-position
θ and ϕ = Angle turned by stub axles.

From Fig. 5.17 (*b*), we get

$$\tan(\alpha - \theta) = \frac{d - x}{h}$$

or $$\frac{\tan\alpha - \tan\theta}{1 + \tan\alpha\tan\theta} = \frac{d-x}{h}. \text{ But } \tan\alpha = \frac{d}{h}$$

$$\therefore \quad \frac{\frac{d}{h} - \tan\theta}{1 + \frac{d}{h}\tan\theta} = \frac{d-x}{h}$$

or $$\frac{d - h\tan\theta}{h + d\tan\theta} = \frac{d-x}{h}$$

or $$h(d - h\tan\theta) = (d - x)(h + d\tan\theta)$$

or $$hd - h^2\tan\theta = hd + d^2\tan\theta - xh - xd\tan\theta$$

or $$xh = (d^2 - xd + h^2)\tan\theta$$

or $$\tan\theta = \frac{xh}{(d^2 - xd + h^2)} \qquad \text{...}(i)$$

We also know that

$$\tan(\alpha + \phi) = \frac{d + x}{h} \qquad \text{[Refer to Fig. 5.17 (b)]}$$

Also it can be proved that,

$$\tan\phi = \frac{hx}{(d^2 + xd + h^2)} \qquad \text{...}(ii)$$

We know that for correct steering,

$$\cot\phi - \cot\theta = \frac{w}{L} \qquad \text{[See equation (5.1)]}$$

Substituting the values of cot ϕ and cot θ from (*ii*) and (*i*) in the above equation, we get

$$\frac{d^2 + xd + h^2}{hx} - \frac{d^2 - xd + h^2}{hx} = \frac{w}{L}$$

or $$\frac{2xd}{hx} = \frac{w}{L}$$

or
$$\frac{d}{h} = \frac{w}{2L}$$

or
$$\tan \alpha = \frac{w}{2L} \qquad \left(\because \frac{d}{h} = \tan \alpha \right) \qquad ...(5.2)$$

As Davis steering gear has sliding pairs, there will be friction and more wear. Though this gear fulfills the fundamental equation of gearing in all the positions, yet it becomes inaccurate after some time due to easy wearing.

Problem 5.5. *A vehicle is moving in a straight path. The distance between the pivots of the front axle of a Davis steering gear is 1.25 m. The wheel base is 2.7 m. Find the inclination of the track arm to the longitudinal axis of the vehicle.*

Sol. Given :

Distance between pivots, $w = 1.25$ m

The wheel base, $L = 2.7$ m

Let α = Inclination of the track arm to the longitudinal axis. Using equation (5.2), we get

$$\tan \alpha = \frac{w}{2L} = \frac{1.25}{2 \times 2.7} = 0.2315$$

$$\therefore \qquad \alpha = \tan^{-1} 0.2315 = \mathbf{13.03°}. \quad \textbf{Ans.}$$

5.7.2. Ackermann Steering Gear

Fig. 5.18 shows the diagram of Ackermann steering gear, which has only turning pairs. The turning pairs are *AL*, *LM*, *MB* and *BA* as shown in Fig. 5.18 (*a*). The two arms *AL* and *BM* are of equal length and are connected by hinge joints with front wheel axle at *A* and *B* respectively. The links *AB* and *LM* are of unequal length.

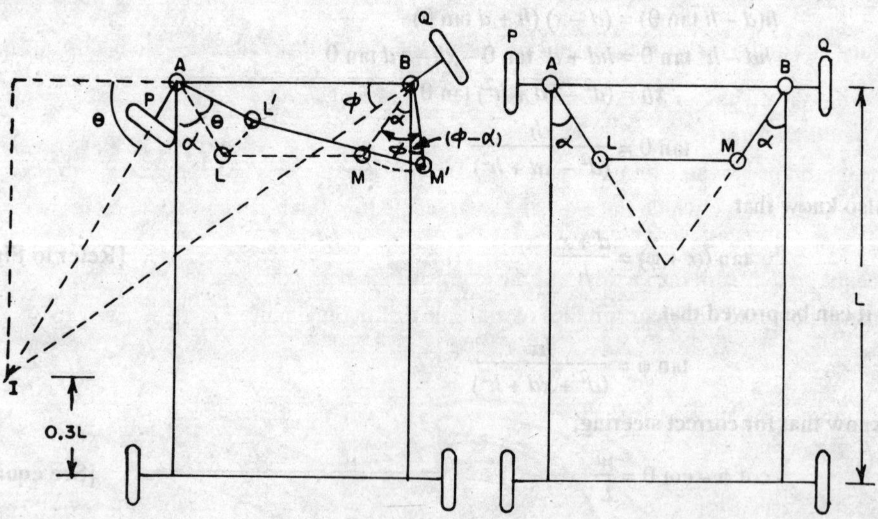

Fig. 5.18

The gear is in the mid-position with equal inclination of the arms *AL* and *BM* with *AB*, during the straight motion of the vehicle. In this position the link *LM* is parallel to the link *AB* as shown in Fig. 5.18 (*a*). The only three positions, in which Ackermann gear fulfills the fundamental equation of correct gearing, are :

(*i*) When the vehicle moves along a straight path

(*ii*) When the vehicle is steering to the left at a correct angle

(*iii*) When the vehicle is steering to the right at a correct angle.

In all other positions, the fundamental equation of correct gearing, is not fulfilled. The above two correct positions are obtained graphically by finding ($\cot\phi - \cot\theta$) at different positions. If the value of ($\cot\phi - \cot\theta$) or $\frac{w}{L} \simeq 0.45$, then the correct position is obtained. Hence the values of θ and ϕ are known for correct position.

The instantaneous centre I, in an Ackermann gear lies on a line parallel to rear axis approximately $0.3L$ above the rear axis. Also the whole mechanism of the Ackermann gear is on the back of the front wheels, whereas in Davis gear, it is in front of wheels.

The value of angle α, can now be obtained as the values of θ and ϕ are known for correct gearing.

Refer to Fig. 5.18 (*b*). To determine the angle α, it is assumed that the projection of LM and $L'M'$ on AB are equal. This means that the projection of LL' on AB will be equal to projection of MM' on AB.

or $$\text{Projection of } LL' \text{ on } AB = \text{Projection of } MM' \text{ on } AB$$

or $$AL' \cos[90 - (\alpha + \theta)] - AL \cos[90 - \alpha] = BM \cos[90 - \alpha] + BM' \sin[\phi - \alpha]$$

But $$AL' = AL = BM = BM'$$

$\therefore$ $$\sin(\alpha + \theta) - \sin\alpha = \sin\alpha + \sin(\phi - \alpha)$$

or $$(\sin\alpha\cos\theta + \cos\alpha\sin\theta) - \sin\alpha = \sin\alpha + (\sin\phi\cos\alpha - \cos\phi\sin\alpha)$$

or $$\sin\alpha\cos\theta - \sin\alpha - \sin\alpha + \sin\alpha\cos\phi = \cos\alpha\sin\phi - \cos\alpha\sin\theta$$

or $$\sin\alpha(\cos\theta + \cos\phi - 2) = \cos\alpha(\sin\phi - \sin\theta)$$

or $$\frac{\sin\alpha}{\cos\alpha} = \frac{\sin\phi - \sin\theta}{\cos\theta + \cos\phi - 2}$$

or $$\tan\alpha = \frac{\sin\phi - \sin\theta}{\cos\theta + \cos\phi - 2} \qquad \ldots(5.3)$$

As the values of θ and ϕ for correct gearing are known, hence value of α can be obtained from equation (5.3).

5.8. Universal or Hooke's Joint

An universal joint or Hooke's joint is used :

(*i*) to connect two shafts, which intersect at a small angle (or which are non-parallel and intersecting shafts),

(*ii*) to transmit power from the gear box of the engine of the rear axle, and

(*iii*) to transmit power to different spindles of multiple drilling machine.

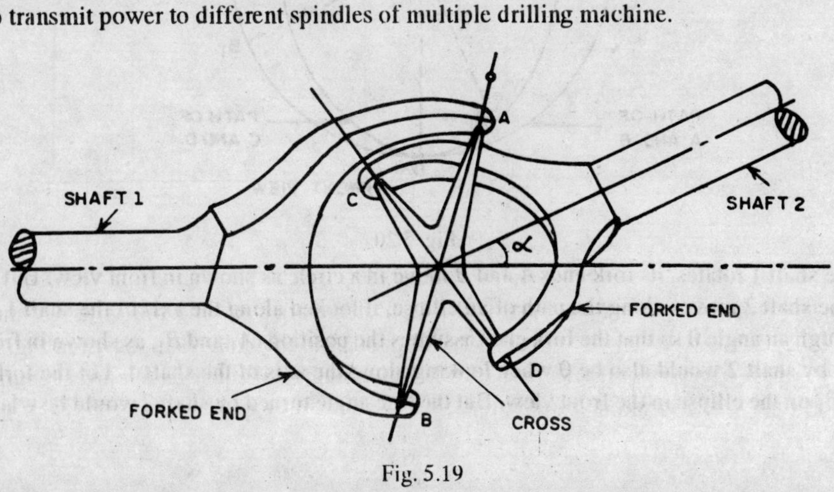

Fig. 5.19

The driving shaft rotates at a uniform angular speed whereas the driven shaft rotates at a continuously varying angular speed.

Fig. 5.19 shows the diagram of the Hooke's joint, which connects the two shafts, which are intersecting at a small angle 'α'. The end of each shaft is forked to U-type. The four ends of the two forks are connected by a centre piece, the arms of which rests in the bearings provided in the fork ends. The centre piece can be in the shape of a cross or sphere (having four arms). These four arms of the centre piece or cross are at right angles.

The motion is transmitted from the driving shaft to the driven shaft through a cross. The inclination of the two shafts may be constant, but usually there is a variation as the motion is being transmitted. Both the shafts rotate in the fixed bearings. A complete revolution of either shaft will cause the other to rotate through a complete revolution in the same time, but with varying angular speed.

5.8.1. Analysis of Hooke's Joint

Fig. 5.20 shows the front view and top view of the two shafts which are connected by a Hooke's joint. The inclination between the axes of the two shafts is angle α. If the joint is looked along the axis of shaft 1, the fork ends of this shaft will be A and B as shown in Fig. 5.20. And C and D are the positions assumed by the fork ends of shafts 2.

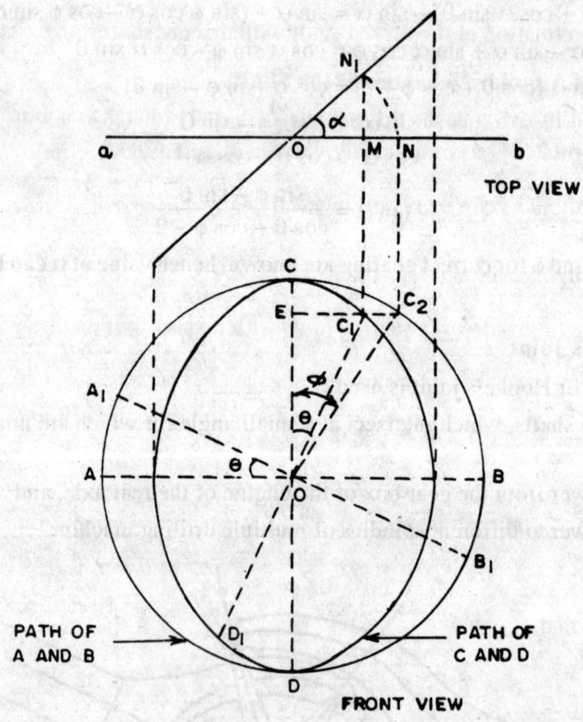

Fig. 5.20

As the shaft 1 rotates, its fork ends A and B move in a circle as shown in front view. But the fork ends C and D of the shaft 2 moves along the path of an ellipse, if looked along the axis of the shaft 1. Let the shaft 1 rotates through an angle θ so that the fork ends assumes the positions A_1 and B_1, as shown in front view. The angle moved by shaft 2 would also be θ when looking along the axis of the shaft 1. Let the fork end C takes the position C_1 on the ellipse in the front view. But the true angle turned by shaft 2 would be when it is looked along its own axis and in that case C and D will move in a circle. To find the true angle, project the point C_1 horizontally to intersect the circle at C_2. Therefore the angle COC_2 (or angle ϕ) is the true angle turned by shaft 2. Hence when shaft 1 turns through an angle θ, the shaft turns through an angle ϕ.

From Fig. 5.20, we have

$$\tan\theta = \frac{EC_1}{EO}$$

and

$$\tan\phi = \frac{EC_2}{EO}$$

$\therefore$

$$\frac{\tan\theta}{\tan\phi} = \frac{EC_1/EO}{EC_2/EO} = \frac{EC_1}{EC_2} \quad ...(i)$$

But

$$EC_1 = OM = ON_1 \cos\alpha$$

and

$$EC_2 = ON = ON_1$$

$\therefore$ Equation (*i*) becomes as

$$\frac{\tan\theta}{\tan\phi} = \frac{ON_1 \cos\alpha}{ON_1} = \cos\alpha$$

or

$$\tan\theta = \cos\alpha \tan\phi \quad ...(4.4)$$

where α = Angle of inclination of the driven shaft with driving shaft.

5.8.2. Ratio of Angular Velocities of the Shafts

It is mentioned in Art. 5.8, the driving shaft (shaft 1) rotates at a uniform angular speed whereas the driven shaft (shaft 2) rotates at a continuously varying angular speed.

Let ω_1 = angular velocity of driving shaft = $\frac{d\theta}{dt}$

ω_2 = angular velocity of driven shaft = $\frac{d\phi}{dt}$

Differentiating both sides of equation (5.4) with respect to time t,

$$\sec^2\theta \times \frac{d\theta}{dt} = \cos\alpha\left(\sec^2\phi \times \frac{d\phi}{dt}\right) \quad \text{(Angle } \alpha \text{ is constant)}$$

or

$$(\sec^2\theta) \times \omega_1 = \cos\alpha\,(\sec^2\phi) \times \omega_2$$

or

$$\frac{\omega_2}{\omega_1} = \frac{\sec^2\theta}{\cos\alpha \sec^2\phi}$$

$$= \frac{1}{\cos^2\theta \times \cos\alpha \times \sec^2\phi} \quad \left(\because \sec^2\theta = \frac{1}{\cos^2\theta}\right) \quad ...(i)$$

But we know that, $\sec^2\phi = 1 + \tan^2\phi$

$$= 1 + \frac{\tan^2\theta}{\cos^2\alpha} \quad \left[\because \text{From equation (5.4), } \tan\phi = \frac{\tan\theta}{\cos\alpha}\right]$$

$$= 1 + \frac{\sin^2\theta}{\cos^2\theta \times \cos^2\alpha}$$

$$= \frac{\cos^2\theta \times \cos^2\alpha + \sin^2\theta}{\cos^2\theta \times \cos^2\alpha}$$

$$= \frac{\cos^2\theta\,(1 - \sin^2\alpha) + \sin^2\theta}{\cos^2\theta \times \cos^2\alpha}$$

$$= \frac{\cos^2\theta - \cos^2\theta\sin^2\alpha + \sin^2\theta}{\cos^2\theta \times \cos^2\alpha}$$

$$= \frac{1 - \cos^2\theta\sin^2\alpha}{\cos^2\theta \times \cos^2\alpha} \qquad (\because \cos^2\theta + \sin^2\theta = 1)$$

Substituting the value of $\sec^2\phi$ in equation (*i*), we get

$$\frac{\omega_2}{\omega_1} = \frac{1}{\cos^2\theta \times \cos\alpha} \times \frac{\cos^2\theta \times \cos^2\alpha}{1 - \cos^2\theta\sin^2\alpha}$$

or
$$\frac{\omega_2}{\omega_1} = \frac{\cos\alpha}{1 - \cos^2\theta \times \sin^2\alpha} \qquad ...(5.5)$$

If N_1 and N_2 are the speeds of driving and driven shafts, then

$$\frac{N_2}{N_1} = \frac{\omega_2}{\omega_1}.$$

5.8.3. Condition for Equal Speeds of Driven and Driving Shafts

For equal speeds, $\omega_2 = \omega_1$

Hence equation (5.5) becomes as

$$1 = \frac{\cos\alpha}{1 - \cos^2\theta \times \sin^2\alpha}$$

or
$$\cos\alpha = 1 - \cos^2\theta \times \sin^2\alpha$$

or
$$\cos^2\theta \times \sin^2\alpha = 1 - \cos\alpha$$

or
$$\cos^2\theta = \frac{1 - \cos\alpha}{\sin^2\alpha}$$

$$= \frac{1 - \cos\alpha}{1 - \cos^2\alpha}$$

$$= \frac{(1 - \cos\alpha)}{(1 - \cos\alpha)(1 + \cos\alpha)}$$

$$= \frac{1}{1 + \cos\alpha}$$

$$= \frac{\sin^2\theta + \cos^2\theta}{1 + \cos\alpha} \qquad (\because 1 = \sin^2\theta + \cos^2\theta)$$

$$= \frac{\cos^2\theta\left(\frac{\sin^2\theta}{\cos^2\theta} + 1\right)}{1 + \cos\alpha}$$

or
$$(1 + \cos\alpha) = \frac{\sin^2\theta}{\cos^2\theta} + 1$$

or
$$\cos\alpha = \tan^2\theta$$

or
$$\tan^2\theta = \cos\alpha$$

$\therefore$
$$\tan\theta = \pm\sqrt{\cos\alpha} \qquad ...(5.6)$$

From equation (5.6) it is clear that there are two values of θ corresponding to positive sign and two values corresponding to negative sign. Hence there are four values of θ at which the speeds of driving and driven shafts will be equal.

5.8.4. Maximum and Minimum Speeds of Driven Shaft

The ratio of speeds of the driven and driving shafts is given by equation (5.5) as

$$\frac{\omega_2}{\omega_1} = \frac{\cos\alpha}{1 - \cos^2\theta \times \sin^2\alpha} \qquad ...(i)$$

where ω_2 = Angular speed of driven shaft, and

ω_1 = Angular speed of driving shaft.

(a) Maximum Speed

The speed of driven shaft (*i.e.* ω_2) will be maximum for a given inclination of the shafts (*i.e.* for a given angle α) when the denominator of the above equation (*i*) is minimum *i.e.* $(1 - \cos^2\theta \times \sin^2\alpha)$ is minimum. Now $(1 - \cos^2\theta \sin^2\alpha)$ for a given value of α, will be minimum if $\cos^2\theta = 1$ or $\theta = 0°$ or $180°$.

Substituting this value of $\cos^2\theta = 1$ in equation (*i*), we get

$$\frac{\omega_2}{\omega_1} = \frac{\cos\alpha}{(1 - \sin^2\alpha)}$$

$$\therefore \quad \omega_2 = \frac{\omega_1\cos\alpha}{(1 - \sin^2\alpha)} = \frac{\omega_1\cos\alpha}{\cos^2\alpha}$$

$$\therefore \quad \omega_2 = \frac{\omega_1}{\cos\alpha} \qquad ...(5.7)$$

Equation (5.7) gives the maximum angular speed of the driven shaft for a given value of 'α'.

(b) Minimum Speed

The speed of the driven shaft will be minimum if the denominator of equation (*i*), is maximum *i.e.* when $(1 - \cos^2\theta \times \sin^2\alpha)$ is maximum for a given value of 'α'.

This will happen when $\cos^2\theta$ is minimum,

or $$\theta = 90° \quad \text{or} \quad 270°$$

Substituting this value of θ is equation (*i*), we get

$$\frac{\omega_2}{\omega_1} = \frac{\cos\alpha}{1 - 0 \times \sin^2\alpha} \qquad (\text{as } \cos 90° = \cos 270° = 0)$$

$$= \cos\alpha$$

$$\therefore \quad \omega_2 = \omega_1 \times \cos\alpha \qquad ...(5.8)$$

The equation (5.8) gives the minimum value of angular speed of the driven shaft.

(c) Polar Velocity Diagram

The diagram, which shows the variation of the angular velocities of the driven shaft and driving shaft for one complete revolution, is known as *polar velocity diagram*. This is shown in Fig. 5.21.

From the Fig. 5.21, we see that

(*i*) The speed of the driven shaft is maximum when $\theta = 0°$ and $180°$ whereas the speed is minimum when $\theta = 90°$ and $270°$.

(*ii*) The speed of the driven shaft is equal to the speed of the driving speed at four points.

(*iii*) The speed of the driving shaft is constant and hence it is represented by a circle of radius equal to ω_1.

(*iv*) The speed of the driven shaft is not constant. The maximum value is '$\omega_1/\cos\alpha$' and minimum value is $\omega_1 \cos\alpha$. It is represented by an ellipse of semi-major axis $= \dfrac{\omega_1}{\cos\alpha}$ and semi-minor axis $= \omega_1 \times \cos\alpha$.

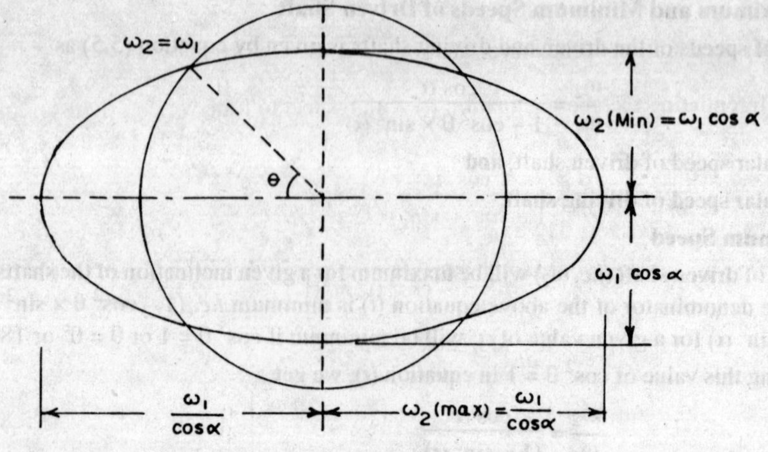

Fig. 5.21

(*d*) Maximum Fluctuation of Speed of Driven Shaft

The equation (5.7) gives the maximum angular speed of the driven shaft whereas equation (5.8) gives the minimum speed. These angular speeds are $\frac{\omega_1}{\cos\alpha}$ and $\omega_1 \times \cos\alpha$ respectively.

∴ Maximum fluctuation of speed of the driven shaft

$$= \text{Maximum angular speed} - \text{Minimum angular speed}$$

$$= \frac{\omega_1}{\cos\alpha} - \omega_1 \times \cos\alpha$$

$$= \omega_1\left(\frac{1}{\cos\alpha} - \cos\alpha\right)$$

$$= \omega_1\left(\frac{1-\cos^2\alpha}{\cos\alpha}\right) \qquad \text{...(5.8}A)$$

$$= \frac{\omega_1 \times \sin^2\alpha}{\cos\alpha} = \omega_1 \times \frac{\sin\alpha}{\cos\alpha} \times \sin\alpha$$

$$= \omega_1 \times \tan\alpha \times \sin\alpha \qquad \text{...(5.8}B)$$

If α is small, then $\sin\alpha \approx \tan\alpha \approx \alpha$

∴ Maximum fluctuation $= \omega_1 \times \alpha \times \alpha$

$$= \omega_1 \times \alpha^2 \qquad \text{...(5.9)}$$

From equation (5.9), it is clear that maximum fluctuation of the angular speed of the driven shaft is directly proportional to the square of the angle between the two shafts.

(*e*) Angular Acceleration of Driven Shaft

The ratio of the angular velocities of the shafts is given by equation (5.5) as

$$\frac{\omega_2}{\omega_1} = \frac{\cos\alpha}{1-\cos^2\theta\sin^2\alpha}$$

or the angular velocity of the driven shaft is given by,

$$\omega_2 = \frac{\omega_1\cos\alpha}{1-\cos^2\theta\sin^2\alpha} \qquad \text{...(}i)$$

The angular acceleration of the driven shaft = $\dfrac{d\omega_2}{dt}$

Hence differentiating the above equation (*i*) with respect to time we get, (here ω_1 and α are constant but θ is variable)

$$\frac{d\omega_2}{dt} = \omega_1 \frac{d}{dt}\left(\frac{\cos\alpha}{1-\cos^2\theta\sin^2\alpha}\right)$$

$$= \omega_1 \times \cos\alpha \frac{d}{dt}\left(\frac{1}{1-\cos^2\theta\sin^2\alpha}\right) \qquad (\because \cos\alpha \text{ is constant})$$

$$= \omega_1 \cos\alpha \frac{d}{dt}(1-\cos^2\theta\sin^2\alpha)^{-1}$$

$$= \omega_1 \cos\alpha \times (-1)(1-\cos^2\theta\sin^2\alpha)^{-2}\left[0-\sin^2\alpha\times 2\cos\theta\times(-\sin\theta)\times\frac{d\theta}{dt}\right]$$

$$= \frac{-\omega_1\cos\alpha}{(1-\cos^2\theta\sin^2\alpha)^2}\times\left[\sin^2\alpha\times 2\sin\theta\cos\theta\times\frac{d\theta}{dt}\right]$$

$$= \frac{-\omega_1\cos\alpha\times\sin^2\alpha\times\sin 2\theta\times\omega_1}{(1-\cos^2\theta\sin^2\alpha)^2}\quad\left(\because 2\sin\theta\cos\theta=\sin 2\theta \text{ and } \frac{d\theta}{dt}=\omega_1\right)$$

$$= \frac{-\omega_1^2\times\cos\alpha\times\sin^2\alpha\times\sin 2\theta}{(1-\cos^2\theta\sin^2\alpha)^2} \qquad ...(5.10)$$

The equation (5.10) gives the angular acceleration of the driven shaft. Negative sign does not mean that there is always retardation. The angular acceleration will be + ve if sin 2θ is – ve. And angular acceleration will be – ve if sin 2θ is + ve.

To find the condition when the angular acceleration will be maximum or minimum, differentiate equation (4.10) with respect to θ (as θ is the only variable in that equation) and equate the result to zero. The resulting expression is very cumbersome and the result can be approximated to

$$\cos 2\theta = \frac{2\sin^2\alpha}{2-\sin^2\alpha}. \qquad ...(5.11)$$

Problem 5.6. *A Hooke's joint is used to connect two shafts. The driving shaft is rotating uniformly with a speed of 400 r.p.m. The maximum speed of the driven shaft is 420 r.p.m. Determine the greatest permissible angle between the two shafts. Also find the minimum speed of the driven shaft.*

Sol. Given :

Speed of driving shaft, $N_1 = 400$ r.p.m.

∴ Angular speed, $\omega_1 = \dfrac{2\pi N_1}{60} = \dfrac{2\pi\times 400}{60}$

Max. speed of driven shaft, $N_2 = 420$ r.p.m.

Max. angular speed, $\omega_2 = \dfrac{2\pi N_2}{60} = \dfrac{2\pi\times 420}{60}$

Let α = greatest permissible angle between the two shafts

$(N_2)_{min}$ = Minimum speed of driven shaft

Maximum angular velocity of the driven shaft is given by equation (5.7).

$$\omega_2 = \frac{\omega_1}{\cos\alpha}$$

or $$\cos\alpha = \frac{\omega_1}{\omega_2}$$

$$= \frac{\left(\frac{2\pi \times 400}{60}\right)}{\left(\frac{2\pi \times 420}{60}\right)} = \frac{400}{420} = 0.9524$$

$\therefore$ $$\alpha = \cos^{-1} 0.9524 = \mathbf{17.75°}. \quad \textbf{Ans.}$$

Minimum angular speed of the driven shaft is given by equation (5.8).

$\therefore$ $$(\omega_2)_{min} = \omega_1 \times \cos\alpha$$

or $$\frac{2\pi \times (N_2)_{min}}{60} = \frac{2\pi N_1}{60} \times \cos 17.75° \qquad \left(\because \ \omega_1 = \frac{2\pi N_1}{60} \text{ and so on}\right)$$

or $$(N_2)_{min} = N_1 \times \cos 17.75°$$

$$= 400 \times 0.9524 = \mathbf{380.96\ r.p.m.} \quad \textbf{Ans.}$$

Problem 5.7. *Two shafts are connected by means of a Hooke's joint. The angle between the shafts is 20°. What will be the angle turned by the driving shaft when :*

(*i*) *Velocity ratio is maximum, minimum and unity.*

(*ii*) *Acceleration of the driven shaft is maximum and zero.*

Sol. Given :

Angle between the shafts, $\alpha = 20°$

(*i*) Let θ = Angle turned by the driving shaft.

(*a*) *When the velocity ratio is maximum*

Using equation (5.5) for velocity ratio,

$$\frac{\omega_2}{\omega_1} = \frac{\cos\alpha}{1 - \cos^2\theta \sin^2\alpha}$$

This velocity ratio for a given angle α will be maximum when $(1 - \cos^2\theta \sin^2\alpha)$ is minimum or $\cos^2\theta$ is maximum. But the maximum value of $\cos^2\theta$ is 1. Hence $\cos^2\theta$ will be 1 when $\theta = 0°$ or 180°.

Hence for maximum velocity ratio, θ = **0° or 180°**. **Ans.**

(*b*) *When the velocity ratio is minimum*

The velocity ratio given by equation (5.5) will be minimum for a given value of α, when $(1 - \cos^2\theta \sin^2\alpha)$ is maximum or $\cos^2\theta$ is minimum. But the minimum value of $\cos^2\theta$ is zero and it is when $\theta = 90°$ or 270°.

Hence for minimum velocity ratio, θ = **90° or 270°**. **Ans.**

(*c*) *When the velocity ratio is unity*

Velocity ratio is given by equation (5.5) as

$$\frac{\omega_2}{\omega_1} = \frac{\cos\alpha}{1 - \cos^2\theta \sin^2\alpha}$$

For velocity ratio to be unity*, we have $\frac{\omega_2}{\omega_1} = 1$

---

*For velocity ratio to be unity, equation (5.6) can also be used.

$\therefore$ $\tan\theta = \pm\sqrt{\cos\alpha} = \pm\sqrt{\cos 20°} = \pm\sqrt{0.9397} = \pm 0.9694$. When $\tan\theta = 0.9694$, then $\theta = \tan^{-1} 0.9694 = 44.11°$ and $180 + 44.11 = 224.11°$. When $\tan\theta = -0.9694$, then $\theta = \tan^{-1}(-0.9694) = 180 - 44.11 = 135.89°$ and $360 - 44.11 = 315.89°$. **Ans.**

Hence the above equation becomes as

$$1 = \frac{\cos\alpha}{1 - \cos^2\theta \sin^2\alpha}$$

or $$\cos\alpha = 1 - \cos^2\theta \sin^2\alpha$$

or $$\cos^2\theta = \frac{1-\cos\alpha}{\sin^2\alpha} = \frac{1-\cos\alpha}{1-\cos^2\alpha} \qquad (\because \sin^2\alpha = 1-\cos^2\alpha)$$

$$= \frac{1-\cos\alpha}{(1-\cos\alpha)(1+\cos\alpha)} = \frac{1}{(1+\cos\alpha)}$$

$$= \frac{1}{(1+\cos 20°)} \qquad (\because \alpha = 20°)$$

$$= \frac{1}{1+0.9397} = 0.5155$$

$\therefore$ $$\cos\theta = \pm\sqrt{0.5155} = \pm 0.718$$

There will be four values of θ. Two values will be for + ve sign and two values for – ve sign.

$\therefore$ **θ = 44.11°, 135.89°, 224.11°, 315.89°. Ans.**

(*ii*) Let θ = Angle turned by the driving shaft.

(a) When acceleration of the driven shaft is maximum

The condition for maximum or minimum acceleration of the driven shaft is given by equation (5.11) as

$$\cos 2\theta = \frac{2\sin^2\alpha}{2-\sin^2\alpha} = \frac{2\sin^2 20°}{2-\sin^2 20°}$$

$$= \frac{2\times(0.342)^2}{2-(0.342)^2} = 0.1242$$

(As cos 2θ is + ve, hence the value of 2θ lies in the 1st and 4th quadrant)

$\therefore$ $$2\theta = \cos^{-1} 0.1242 = 82.86°, (360° - 82.86°), (360° + 82.86°), (720° - 82.86°)$$

or $$\theta = 41.43°, 138.57°, 221.43°, 318.57°.$$

The four above values of θ make the acceleration as maximum or minimum. The acceleration of the driven shaft is given by equation (5.10) as

$$\text{Acceleration} = \frac{-\omega_1^2 \times \cos\alpha \times \sin^2\alpha \times \sin 2\theta}{(1-\cos^2\theta\sin^2\alpha)^2}$$

The above acceleration is + ve or maximum when sin 2θ is negative. sin 2θ will be negative when 2θ lies in 3rd and 4th quadrant *i.e.* when 2θ is equal to (360° – 82.86°) and (720° – 82.86°). This means θ will be equal to 138.57° and 318.57°.

$\therefore$ **θ = 138.57° and 318.57°. Ans.**

(b) When acceleration of the driven shaft is zero

The acceleration of the driven shaft is given by equation (5.10). This acceleration will be zero when

$$\sin 2\theta = 0$$

or $$2\theta = 0°, 180°, 360°, (360° + 180°)$$

or $$\theta = 0°, 90°, 180°, 270°$$

$\therefore$ Acceleration will be zero when

θ = 0°, 90°, 180°, 270°. Ans.

Problem 5.8. *Determine the greatest permissible angle between the axes of the two shafts which are connected by a Hooke's joint if the maximum variation in the speed of the driven shaft is ± 4% of the mean speed. The driving shaft is rotating at a uniform speed of 600 r.p.m. Also find the maximum and the minimum speeds of the driven shaft.*

Sol. Given :

Max. Variation in the speed of the driven shaft = ± 4% of mean speed.

i.e. 4% of speed of the driving shaft

∴ Total fluctuation of the speed of the driven shaft

$$= 4 + 4 = 8\% \text{ of mean speed}$$

$$= \frac{8}{100} \times \omega_1 = 0.08\omega_1$$

where ω_1 = speed of driving shaft

Speed of driving shaft, $N_1 = 600$ r.p.m.

∴ Angular speed, $\omega_1 = \dfrac{2\pi N_1}{60} = \dfrac{2\pi \times 600}{60} = 20\,\pi$ rad/s

Let α = Greatest permissible angle between the axes of the two shafts.

The maximum fluctuation (or total fluctuation) of the speed of the driven shaft is given by equation (5.8 *A*) as

$$\text{Total fluctuation} = \omega_1 \frac{(1 - \cos^2 \alpha)}{\cos \alpha}$$

or But total fluctuation $= 0.08\omega_1$

$$\therefore \quad 0.08\omega_1 = \omega_1 \frac{(1 - \cos^2 \alpha)}{\cos \alpha}$$

or $$0.08 \cos \alpha = 1 - \cos^2 \alpha$$

or $$\cos^2 \alpha + 0.08 \cos \alpha - 1 = 0$$

$$\therefore \quad \cos \alpha = \frac{-0.08 \pm \sqrt{(0.08)^2 + 4}}{2} = \frac{-0.08 \pm 2.0016}{2}$$

$$= 0.9608 \qquad \text{(Taking + ve sign as – ve sign is not possible)}$$

$$\therefore \quad \alpha = \cos^{-1} 0.9608 = \mathbf{16.09°}. \quad \textbf{Ans.}$$

(*i*) *Maximum speed* of driven shaft is given by equation (5.7) as

$$\omega_2 = \frac{\omega_1}{\cos \alpha}$$

or $$N_2 = \frac{N_1}{\cos \alpha} \quad \left(\because \omega_2 = \frac{2\pi N_2}{60} \text{ and } \omega_1 = \frac{2\pi N_1}{60}. \text{ Here } \frac{2\pi}{60} \text{ will be cancelled}\right)$$

$$= \frac{600}{\cos 16.09°}$$

$$= \frac{600}{0.9608} = \mathbf{624.48 \text{ r.p.m.}} \quad \textbf{Ans.}$$

(*ii*) *Minimum speed* of driven shaft is given by equation (5.8) as

$$\omega_2 = \omega_1 \times \cos \alpha$$

or $$N_2 = N_1 \times \cos \alpha$$

$$= 600 \times 0.9608 = \mathbf{576.48 \text{ r.p.m.}} \quad \textbf{Ans.}$$

Problem 5.9. *Two shafts are connected by a Hooke's joint. The driving shaft is rotating at a uniform speed of 1000 r.p.m. Find the greatest permissible angle between the axes of the two shafts so that the total fluctuation of speed does not exceed 90 r.p.m.*

Sol. Given :

Speed of driving shaft, $N_1 = 1000$ r.p.m.

Total fluctuation of speed $= 90$ r.p.m.

Let α = greatest permissible angle between the axes. Using fluctuation (5.8 *A*), we get

$$\text{Total fluctuation} = \frac{\omega_1(1 - \cos^2 \alpha)}{\cos \alpha}$$

The unit of the total fluctuation on L.H.S. will be rad/s. If ω_1 is replaced by N_1, then the unit of total fluctuation on L.H.S. will be in r.p.m.

$$\text{Hence total fluctuation in r.p.m.} = \frac{N_1(1 - \cos^2 \alpha)}{\cos \alpha}$$

But total fluctuation in r.p.m. $= 90$ r.p.m.

$$\therefore \quad 90 = \frac{N_1(1 - \cos^2 \alpha)}{\cos \alpha} = \frac{1000(1 - \cos^2 \alpha)}{\cos \alpha}$$

or $$\frac{90 \times \cos \alpha}{1000} = 1 - \cos^2 \alpha$$

or $$0.09 \cos \alpha = 1 - \cos^2 \alpha$$

or $$\cos^2 \alpha + 0.09 \cos \alpha - 1 = 0$$

$$\therefore \quad \cos \alpha = \frac{-0.09 \pm \sqrt{(0.09)^2 + 4}}{2} = \frac{-0.09 \pm 2.002}{2} = 0.956$$

$$\therefore \quad \alpha = \cos^{-1} 0.956 = \mathbf{17.06^\circ}. \quad \textbf{Ans.}$$

Problem 5.10. *A Hooke's joint connects two shafts which are having 165° as the included angle. The driving shaft rotates uniformly at 1000 r.p.m. Find the maximum angular acceleration of the driven shaft and the maximum torque required if the driven shaft carries a flywheel of mass of 10 kg and 100 mm radius of gyration.*

Sol. Given :

Included angle of the two shafts $= 165^\circ$

$\therefore \quad \alpha = 180^\circ - 165^\circ = 15^\circ$

Speed of driving shaft, $N_1 = 1000$ r.p.m.

$$\therefore \text{ Angular speed,} \quad \omega_1 = \frac{2\pi N_1}{60} = \frac{2\pi \times 1000}{60} = 104.72 \text{ rad/s}$$

Mass, $m = 10$ kg

Radius of gyration, $k = 100$ mm $= 0.1$ m

Mass moment of inertia of the driven shaft,

$$I = mk^2$$
$$= 10 \times (0.1)^2 = 0.1 \text{ kg-m}^2$$

The condition for maximum acceleration of the driven shaft is given by equation (5.11) as

$$\cos 2\theta = \frac{2 \sin^2 \alpha}{2 - \sin^2 \alpha}$$

$$= \frac{2\sin^2 15°}{2-\sin^2 15°} = \frac{2\times 0.2588^2}{2-0.2588^2}$$

$$= \frac{2\times 0.06697}{2-0.06697} = 0.06929$$

As cos 2θ is + ve, hence the value of 2θ lies in 1st and 4th quadrant

$$\therefore \qquad 2\theta = \cos^{-1} 0.06929 = 86.03° \text{ or } (360-86.03°)$$

$$= 86.03° \text{ and } 273.97°$$

or
$$\theta = 43.015° \text{ and } 136.98°$$

Hence substituting the value of 2θ or θ in equation (5.10), we get maximum acceleration of the driven shaft.

$$\therefore \quad \text{Maximum acceleration} = \frac{-\omega_1^2 \times \cos\alpha \times \sin^2\alpha \times \sin 2\theta}{(1-\cos^2\theta\sin^2\alpha)^2}$$

$$= \frac{-(104.72)^2 \times \cos 15° \times \sin^2 15° \times \sin(273.97°)}{(1-\cos^2 136.98° \times \sin^2 15°)^2}$$

(In the above expression, the value of 2θ is taken as 273.97° and not 86.03°. The reason being, the acceleration will be maximum when it is + ve. The acceleration will be + ve, when sin 2θ in equation (5.10) is – ve. sin 2θ will be – ve, if 2θ lies in 4th quadrant and not in first quadrant.)

$$\therefore \quad \text{Maximum acceleration} = \frac{-(104.72)^2 \times 0.9659 \times 0.2588^2 \times (-0.9976)}{[1-(-0.7311)^2 \times 0.2588^2]^2}$$

$$= \frac{+(104.72)^2 \times .9659 \times 0.06697 \times 0.9976}{(1-0.5345\times 0.06697)^2}$$

$$= 761.18 \text{ rad/s}^2$$

Maximum torque on the driven shaft will be when the acceleration is maximum. Hence maximum torque required will be equal to the product of maximum angular acceleration and mass moment of inertia.

$\therefore$ Maximum torque $= I \times \alpha$ where α = Maximum angular acceleration

$= 0.1 \times 761.18$ N m = **76.12 N m. Ans.**

5.9. Double Hooke's Joint

Double Hooke's joint means two Hooke's joint. In a single Hooke's joint, the driving shaft rotates at a uniform speed but the speed of the driven shaft is not uniform. It varies from maximum to minimum values. In order to have a constant velocity ratio of the driving and driven shafts, a double Hooke's joint is used. In a double Hooke's joint, two Hooke's joint and an intermediate shaft are used as shown in Fig. 5.22. The speed of the driving and driven shafts will be equal if :

(*i*) the driving shaft and driven shaft are equally inclined to the intermediate shaft and

(*ii*) the two forks on the intermediate shaft lie in the same plane.

Let for the same time interval,

θ = angle turned by driving shaft

ϕ = angle turned by driven shaft

γ = angle turned by intermediate shaft

α = angle of inclination of the driving shaft and driven shaft with intermediate shaft (equal inclination)

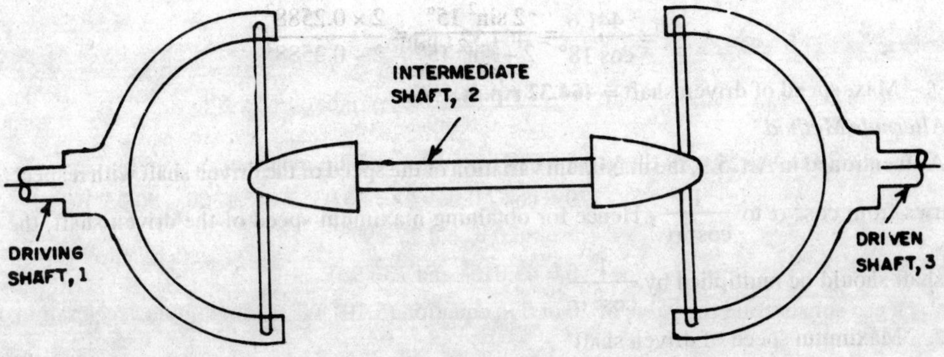

Fig. 5.22

Now from Art. 5.8.1 and equation (5.4), for driving and intermediate shaft, we have

$$\tan\theta = \cos\alpha \tan\gamma \quad \text{...}(i)$$

Similarly for driven and intermediate shaft, we have

$$\tan\phi = \cos\alpha \tan\gamma \quad \text{...}(ii)$$

From equations (*i*) and (*ii*), we have

$$\tan\theta = \tan\phi$$

or $$\theta = \phi.$$

As the angle turned by driving shaft and driven shaft in the same time are equal, hence their angular velocities will be equal.

If the forks of the intermediate shaft lie in planes perpendicular to each other, the variation of the speed of the driven shaft will be there. The maximum variation (or fluctuation) of the speed of the driven shaft will be from $\cos^2\alpha$ to $1/\cos^2\alpha$.

Problem 5.11. *In a double Hooke's joint, the angle of the driving and driven shaft with the intermediate shaft is 18°. The driving shaft is rotating at 420 r.p.m. If the forks of the intermediate shaft lie in plane perpendicular to each other, find the maximum and minimum speed of the driven shaft.*

Sol. Given :

Angle of inclination of driving and driven shaft with intermediate shaft,

$\therefore$ $$\alpha = 18°$$

Speed of driving shaft, $N_1 = 420$ r.p.m.

If the forks of the intermediate shaft are in the same plane, then the speed of the driven shaft will be same as that of driving shaft. But if the forks of the intermediate shaft lie in plane perpendicular to each other, then there will be variation in the speed of the driven shaft.

(a) Maximum speed of driven shaft

Consider first driving shaft (as shaft 1) and intermediate shaft (as shaft 2). The maximum speed of intermediate shaft is given by equation (5.7) as

$$N_2 = \frac{N_1}{\cos\alpha} = \frac{420}{\cos 18°} = 441.6 \text{ r.p.m.}$$

Now, consider intermediate shaft (shaft 2) and driven shaft (shaft 3). The maximum speed of driven shaft is given as

$$N_3 = \frac{N_2}{\cos\alpha}$$

$$= \frac{441.6}{\cos 18°} = 464.32 \text{ r.p.m.}$$

∴ Max. speed of driven shaft = **464.32 r.p.m. Ans.**

Alternate Method

As mentioned in Art. 5.9, the maximum variation of the speed of the driven shaft with respect to driving shaft varies from $\cos^2 \alpha$ to $\frac{1}{\cos^2 \alpha}$. Hence for obtaining maximum speed of the driven shaft, the speed of driving shaft should be multiplied by $\frac{1}{\cos^2 \alpha}$.

∴ Maximum speed of driven shaft

$$= N_1 \times \frac{1}{\cos^2 \alpha}$$

$$= \frac{420}{\cos^2 18°} = \mathbf{464.34 \text{ r.p.m.}} \quad \textbf{Ans.}$$

(b) Minimum speed of driven shaft

Consider first driving shaft (shaft 1) and intermediate shaft (shaft 2). The minimum speed of intermediate shaft is given by equation (5.8) as

$$N_2 = N_1 \times \cos \alpha$$

$$= 420 \times \cos 18° = 399.44 \text{ r.p.m.}$$

Now consider intermediate shaft (shaft 2) and driven shaft (shaft 3). The minimum speed of driven shaft is given by,

$$N_3 = N_2 \times \cos \alpha$$

$$= 399.44 \times \cos 18° = 379.89 \text{ r.p.m.}$$

∴ Minimum speed of driven shaft = **379.89 r.p.m. Ans.**

Alternate Method

The minimum speed of driven shaft is obtained by multiplying $\cos^2 \alpha$ to the speed of the driving shaft.

∴ Minimum speed of the driven shaft

$$= N_1 \times \cos^2 \alpha$$

$$= 420 \times \cos^2 18° = \mathbf{379.89 \text{ r.p.m.}} \quad \textbf{Ans.}$$

HIGHLIGHTS

1. Pantograph is a mechanism used to produce paths on an enlarged or reduced scale exactly similar to the one traced out by a point on the linkage.
2. The exact straight line motion can be obtained from the mechanism which is having only turning pairs or one sliding pair.
3. The Peaucellier and Hart's mechanisms produce exact straight line motion and use only turning pairs.
4. Scott-Russel mechanism produces exact straight line motion using one sliding pair.
5. The following are the mechanism which give the approximate straight line motion :
   (*i*) Watt's straight line mechanism
   (*ii*) Modified Scott-Russel mechanism
   (*iii*) Grass-Hopper mechanism
   (*iv*) Tchebicheff's mechanism
   (*v*) Roberts mechanism.
6. The straight line motion is mostly used in Engine indicators.

7. Engine indicators are the devices which keep the graphical records of pressure inside the engine cylinder during the piston stroke.
8. The following are the engine indicators, which works on the straight line motion mechanism :
   (*i*) Simplex indicator, (*ii*) Cross-by indicator,
   (*iii*) Thomson indicator, and (*iv*) Dobbie McInnes indicator.
9. Steering gear is used for changing the direction of two or more of the wheel axles so as to move the automobile in any desired path.
10. Steering is done by means of front wheel only.
11. The fundamental equation of correct gearing is

$$\cot \phi - \cot \theta = \frac{w}{L}$$

where θ and ϕ = Angles turned by stub-axles *i.e.* front short axles
w = Distance between the pivots of front axle
L = Wheel base.
12. Davis Steering Gear satisfies the fundamental equation of gearing in all the positions. The relation for this gear is

$$\tan \alpha = \frac{w}{2L}$$

where α = Inclination of the track arm with vertical when vehicles is in mid-position.
13. Ackermann gear fulfills the fundamental equation of correct gearing in three positions only.
14. Universal joint or Hooke's joint is used to connect two non-parallel and intersecting shafts.
15. If θ and ϕ are the angles turned by driving and driven shafts of a Hooke's joint at any time, then the relation between θ, ϕ and α is given by

$$\tan \theta = \cos \alpha \tan \phi$$

where α = Angle of inclination of the driven shaft with driving shaft.
16. The ratio of angular velocities (or speed) of the shafts is given by

$$\frac{\omega_2}{\omega_1}\left(\text{or } \frac{N_2}{N_1}\right) = \frac{\cos \alpha}{1 - \cos^2 \theta \times \sin^2 \alpha}$$

where θ = Angle turned by driving shaft.
17. In case of a Hooke's joint :
   (*i*) Speeds of driving and driven shafts are equal if $\tan \theta = \pm \sqrt{\cos \alpha}$
   (*ii*) Maximum speed of driven shaft = $\dfrac{\omega_1}{\cos \alpha}$
   (*iii*) Minimum speed of driven shaft = $\omega_1 \times \cos \alpha$
   (*iv*) Total fluctuation of speed of driven shaft = $\omega_1 \times \tan \alpha \times \sin \alpha$
   (*v*) Also total fluction is directly proportional to square of the angle between the two shafts.
   (*vi*) Angular acceleration = $\dfrac{-\omega_1^2 \times \cos \alpha \times \sin^2 \alpha \times \sin 2\theta}{(1 - \cos^2 \theta \sin^2 \alpha)^2}$
   (*vii*) Angular acceleration will be maximum or minimum if $\cos 2\theta = \dfrac{2 \sin^2 \alpha}{2 - \sin^2 \alpha}$.
18. Double Hooke's joint is used to have a constant velocity ratio of the driving and driven shafts.

EXERCISE 5

(A) Theoretical Questions

1. What do you mean by a pantograph and what are its uses ? Describe with a neat sketch the principle and working of the pantograph. *(AMIE, Winter 1979)*
2. What do you mean by straight line mechanism ? Name the different mechanism which are used for exact straight line motion.
3. Describe Hart's mechanism with a neat sketch and prove that the tracing point describes a straight line path. *(AMIE, Summer 1983)*
4. Name the different mechanisms which give the approximate straight line motion.
5. With a neat sketch, explain the working of Scott-Russel mechanism and modified Scott-Russel mechanism.
6. Describe Watt's straight line mechanism. What are the practical use of straight line mechanism ?
7. What do you mean by engine indicators ? Name the important engine indicators.
8. Sketch the Cross-by Indicator mechanism and show that the displacement of the point which traces the indicator diagram is proportional to the displacement of the indicator piston.
9. What do you mean by 'steering gear' ? Derive an expression for the fundamental equation of correct gearing.
10. Name the two different types of steering gear mechanism. What are their merits and demerits ?
11. Describe with a neat sketch the working of Davis Steering gear mechanism. Also prove that for Davis steering gear

$$\tan\alpha = \frac{w}{2L}$$

where w, α and L are having their usual meaning.

12. What is a Hooke's joint ? With a neat sketch, describe the working of a Hooke's joint. Also show that for a Hooke's joint

$$\tan\theta = \cos\alpha \tan\phi$$

where α = Angle of inclination of the driven shaft with driving shaft

ϕ and θ = Angles turned by driven and driving shafts at any instant.

13. Prove that the ratio of the angular velocities of the driven and driving shafts for a Hooke's joint is given by

$$\frac{\omega_2}{\omega_1} = \frac{\cos\alpha}{1 - \cos^2\theta \times \sin^2\alpha}.$$

14. For a Hooke's joint connecting the driving and driven shafts, prove that :
   (*i*) speeds of these two shafts will be equal if $\tan\theta = \pm\sqrt{\cos\alpha}$
   (*ii*) maximum speed of driven shaft = $\omega_1/\cos\alpha$
   (*iii*) minimum speed of driven shaft = $\omega_1 \times \cos\alpha$
   (*iv*) condition for maximum or minimum angular acceleration is given by

$$\cos 2\theta = \frac{2\sin^2\alpha}{2 - \sin^2\alpha}$$

where α = Angle of inclination of driven shaft with driving shaft

ω_1 = Angular velocity of driving shaft and

θ = Angle turned by driving shaft at any instant.

15. What is the use of Double Hooke's joint ? Explain why two Hooke's joints are used to transmit motion from the engine to the differential of an automobile.
16. Sketch the Dobbie-McInnes engine indicator mechanism and find the ratio of the displacement of the pencil to the displacement of the piston of the indicator mechanism. *(AMIE, Summer 1983)*

17. Draw a neat sketch of Peaucellier straight line mechanism. Explain with a proof how the tracing point describes a straight line path. *(AMIE, Winter 1983)*

18. What is the condition for correct steering ? Sketch and show the two main types of steering gears and discuss their relative advantages. *(AMIE, Summer 1984)*

B. Numerical Problems

1. A circle with *AD* as diameter, has a point *B* on its circumference on *AB* produced there is a point *C* such that if *B* turns about *A*, the product $AB \times AC$ is constant. Prove that the point *C* moves in a straight line perpendicular to *AD* produced. *(AMIE, Winter 1979)*

2. A torque of 50 N m is applied to the link *OA* at *A* of a Grass-Hopper mechanism shown in Fig. 5.11 at page 175. The link *OA* makes an angle of 20° with the horizontal. Find the magnitude of the vertical force exerted at *B* to overcome the resisting torque of 50 N m. The lengths of the links are given as OA = 8 cm, AC = 12 cm and CB = 30 cm. If the link *OA* makes an angle of 10° and 0° with the horizontal, what will be the vertical force at *B* to overcome the same torque of 50 N m. **[Ans. 260.4 N ; 252.5 N ; 250 N]**

3. In a modified Scott-Russel mechanism shown in Fig. 5.7 on page 171, OA_1 = 30 mm, A_1C_1 = 40 mm. Find the length of the extended link A_1B_1 so that point B_1 is having approximate straight line motion. **[Ans. 53.33 mm]**

4. The simplex engine indicator as shown in Fig. 5.12 is used to obtain the indicator diagram. The tracing point (or pencil point) is located at a distance of 160 mm from the fixed point *O*. The indicator diagram should be magnified by four times the displacement of the piston. Design the pantograph to be used in the indicator.
[Ans. If OA = 20 mm and AD = 40 mm, then OD = 40 mm ; OB = 80 mm and BE = 160 mm]

5. A vehicle is moving in a straight path. The distance between the pivots of the front axle of a Davis steering gear is 1.2 m and the wheel base is 2.7 m. Find the inclination of the track arm to the longitudinal axis of the vehicle. **[Ans. α = 12.5°]**

6. A Hooke's joint is used to connect two shafts. The driving shaft is rotating uniformly with a speed of 600 r.p.m. The maximum speed of the driven shaft is 630 r.p.m. Determine the greatest permissible angle between the two shafts. Also find the minimum speed of the driven shaft. **[Ans. α = 17.7°, 571.38 r.p.m.]**

7. Two shafts are connected by means of a Hooke's joint. The angle between the shafts is 18°. What will be the angle turned by the driving shaft when the velocity ratio is maximum and unity.
[Ans. For max. velocity ratio, θ = 0° or 180°, for unity velocity ratio, θ = 44.3°, (180 – 44.3), (180 + 44.3), (360 – 44.3) = 44.3°, 135.7°, 224.3°, 315.7°]

8. Determine the greatest permissible angle between the axes of the two shafts which are connected by a Hooke's joint if the maximum variation in the speed of the driven shaft is ± 6% of the mean speed. The driving shaft is rotating at a uniform speed of 500 r.p.m. Also find the maximum and the minimum speeds of the driven shaft. **[Ans. α = 19.64°, 530.89 r.p.m. ; 470.9 r.p.m.]**

9. A Hooke's joint connects two shafts which are having 160° as the included angle. The driving shaft rotates uniformly at 1500 r.p.m. Find the maximum angular acceleration of the driven shaft and the maximum torque required if the driven shaft carries a flywheel of mass 12 kg and 100 mm radius of gyration.
[Ans. 3090 rad/s^2 ; 371 N m]

10. In a double Hooke's joint, the angle of the driving and driven shaft with the intermediate shaft is 20°. The driving shaft is rotating at 500 r.p.m. If the forks of the intermediate shaft lie in plane perpendicular to each other, find the maximum and the minimum speed of the driven shaft. **[Ans. 566.25 r.p.m. ; 441.5 r.p.m.]**

6

Friction

6.1. INTRODUCTION

When a solid body slides over a stationary solid body, a force is exerted at the surface of contact by the stationary body on the moving body. This force is called the *force of friction* and is always acting in the direction opposite to the direction of motion. The property of the bodies by virtue of which a force is exerted by a stationary body on the moving body to resist the motion of the moving body is called *friction*. Friction acts parallel to the surface of contact and depends upon the nature of surface of contact.

In some cases, the force of friction should be reduced whereas in other cases it must be increased. For the first case, when the power transmitted is partially lost due to friction, the friction must be reduced by using lubricated surfaces. For the other cases where power has to be transmitted by friction as in case of belt and rope drive and other friction drives (friction clutches), the force of friction between the contact surfaces must be increased. In these cases the high friction materials are used.

The contact surfaces may be : (*i*) dry (*ii*) partially lubricated (greasy) and (*iii*) completely lubricated (film lubricated).

6.2. DEFINITIONS

For defining the terms like co-efficient of friction (μ) and angle of friction (ϕ), consider a solid body placed on a horizontal plane surface as shown in Fig. 6.1.

Let W = Weight of body acting through C.G. downward,

R = Normal reaction of body acting through C.G. upward,

P = Force acting on the body through C.G. and parallel to the horizontal surface.

If P is small, the body will not move as the force of friction acting on the body in the direction opposite to P will be more than P. But if the magnitude of P goes on increasing, a stage comes, when the solid body is on the point of motion. At this stage, the force of friction acting on the body is called *limiting force of friction*. The limiting force of friction is denoted by F.

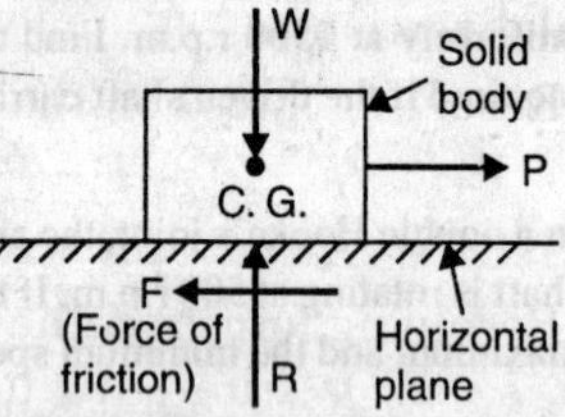

Fig. 6.1. Solid body on horizontal surface.

Resolving the forces on the body vertically and horizontally, we get

$$R = W$$

$$F = P.$$

If the magnitude of P is further increased the body will start moving. The force of friction, acting on the body when the body is moving, is called *kinetic friction*.

6.2.1. Co-efficient of Friction (μ). It is defined as the ratio of the limiting force of friction (F) to the normal reaction (R) between two bodies. It is denoted by the symbol μ. Thus

$$\mu = \frac{\text{Limiting force of friction}}{\text{Normal reaction}} = \frac{F}{R}$$

$$\therefore \quad F = \mu R \qquad \text{...(6.1)}$$

6.2.2. Angle of Friction (φ). It is defined as the angle made by the resultant of the normal reaction (R) and the limiting force of friction (F) with the normal reaction (R). It is denoted by φ. Fig. 6.2 shows a solid body resting on a rough horizontal plane.

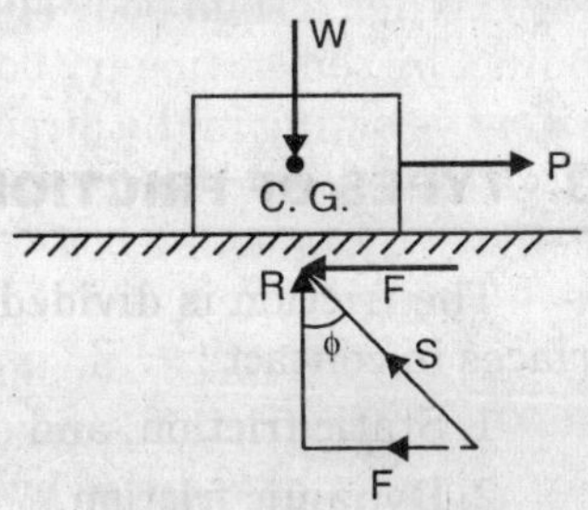

Fig. 6.2. Angle of friction.

Let S = Resultant of the normal reaction (R) and limiting force of friction (F)

Then angle of friction = ϕ

= Angle between S and R

From Fig. 6.2, we have

$$\tan\phi = \frac{F}{R} = \frac{\mu R}{R} \qquad [\because \; F = \mu R \text{ from (6.1)}]$$

$$= \mu = \text{Co-efficient of friction} \qquad \text{...(6.1A)}$$

Thus the tangent of the angle of friction is equal to the co-efficient of friction.

A block of weight W is placed on a rough horizontal plane surface as shown in Fig. 6.3 and a force P is applied at an angle θ with the horizontal such that the block just tends to move.

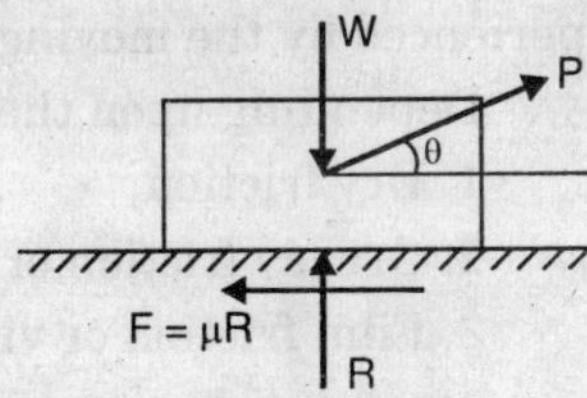

Fig. 6.3

Let R = Normal reaction

μ = Co-efficient of friction

F = Force of friction

= μR

In this case the normal reaction R will not be equal to weight of the body. The normal reaction is obtained by resolving the forces on the block horizontally and vertically. The force P is resolved in two components *i.e.*, $P\cos\theta$ in the horizontal direction and $P\sin\theta$ in the vertical direction.

Resolving forces on the block horizontally, we get

$$F = P\cos\theta$$

or

$$\mu R = P\cos\theta \qquad (\because \; F = \mu R) \qquad \text{...}(i)$$

Resolving forces on the block vertically, we get

$$R + P\sin\theta = W$$

$$\therefore \quad R = W - P\sin\theta \qquad \text{...}(ii)$$

From equation (*ii*), it is clear that normal reaction is not equal to the weight of the block.

If in equation (*ii*), the values of W, P and θ are known, then value of normal reaction (R) can be obtained. This value of R can be substituted in equation (*i*) to determine the value of co-efficient of friction μ.

Note. (*i*) The force of friction is always equal to μR (*i.e.*, $F = \mu R$).

(*ii*) The normal reaction (R) is not equal to the weight of the body always.

6.2.3. Cone of Friction. It is defined as the right circular cone with vertex at the point of contact of the two bodies (or surfaces), axis in the direction of normal reaction (R) and semi-vertical angle equal to angle of friction (ϕ). Fig. 6.4 shows the cone of friction in which,

O = Point of contact between two bodies

R = Normal reaction and also axis of the cone of friction

ϕ = Angle of friction.

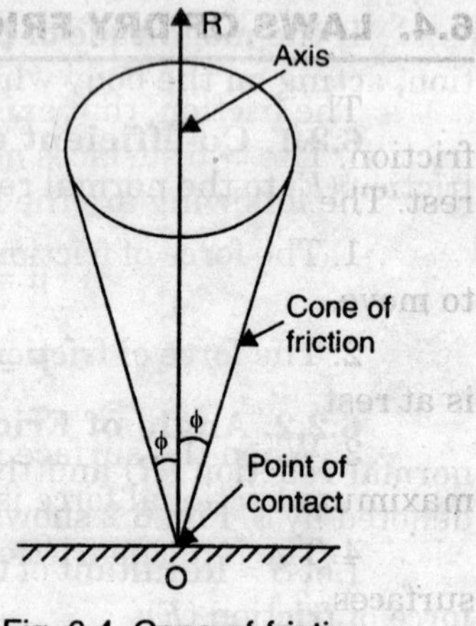

Fig. 6.4. Cone of friction.

6.3. TYPES OF FRICTION

The friction is divided into following two types depending upon the nature of the two surfaces in contact :

1. Static friction, and
2. Dynamic friction.

If the two surfaces, which are in contact, are at rest, the force experienced by one surface is called static friction. But if one surface starts moving and the other is at rest, the force experienced by the moving surface is called dynamic friction.

Depending upon the conditions of surfaces, the friction is also classified as

1. Dry friction,
2. Greasy friction or skin friction (or boundary friction),
3. Film friction or viscous friction.

6.3.1. Dry Friction. If between the two surfaces, no lubricant (oil or grease) is used, the friction that exists between the two surfaces, is called *Dry Friction*. Dry friction is further divided into two types namely

(*i*) Solid friction and (*ii*) Rolling friction.

Solid Friction. When the two surfaces have a sliding motion relative to each other, then the friction between the two surfaces is known as solid friction.

Rolling Friction. When one surface is rolling over another surface, then the friction between the two surfaces is known as rolling friction, for example the friction between ball and roller bearing is the case of rolling friction.

6.3.2. Greasy Friction or Skin Friction. When the two surfaces in contact have a very thin layer of lubricant between them, the friction between these surfaces is known as skin or greasy or boundary friction. The lubricant is absorbed on the surfaces and forms a very thin film between the surfaces.

6.3.3. Film Friction or Viscous Friction. When the two surfaces in contact have a thick layer of the lubricant between them, the friction between the surfaces is known as film friction or viscous friction. In this case, the actual surfaces do not come in contact and hence do not rub against each other. Each surface is having rubbing action with the thick layer of lubricant only.

6.4. LAWS OF DRY FRICTION

The friction, that exists between two surfaces which are not lubricated, is known as dry friction. The two surfaces may be at rest or one of the surface is moving and other surface is at rest. The following are the laws of solid friction :

1. The force of friction acts in the opposite direction in which surface is having tendency to move.

2. The force of friction is equal to the force applied to the surface, so long as the surface is at rest.

3. When the surface is on the point of motion, the force of friction is maximum and this maximum frictional force is called the limiting friction force.

4. The limiting frictional force bears a constant ratio to the normal reaction between two surfaces.

5. The limiting frictional force does not depend upon the shape and areas of the surfaces in contact.

6. The ratio between limiting friction and normal reaction is slightly less when the two surfaces are in motion.

7. The force of friction is independent of the velocity of sliding of one surface over the other.

The above laws of solid friction are also called laws of static and dynamic friction. The last law is valid only within the limits. It has been found that the friction force decreases slightly with the increase in velocity.

6.5. SCREW THREADS FRICTION

The screw threads are mainly of two types namely, V-threads and square threads. The V-threads are stronger and offer more frictional resistance to the motion than square threads. The square threads are used in screw jack whereas V-threads are used for the purpose of tightening pieces together. If the hypotenuse of a right angled triangle is wraped round the circumference of a cylinder, the screw thread is obtained as shown in Fig. 6.5. In this figure, the triangle *OLM* is the development of the helix of diameter d and lead l. The lead will be equal to the pitch p for a single-start thread. The angle *LOM* is known as helix angle and it is represented by 'α'. The relation between α, l and dia. d is given by

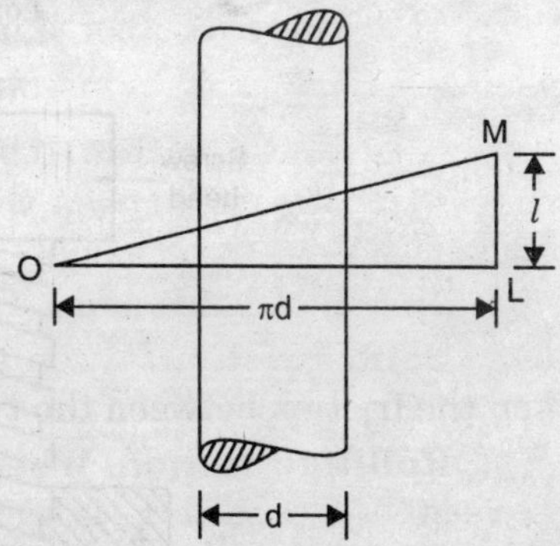

Fig. 6.5

$$\tan \alpha = \frac{1}{\pi d}$$

where l = lead and is the height of triangle

πd = circumference of the cylinder of screw threads and is the length of the base of triangle.

The following terms are mostly used for the study of screw threads :

1. **Helix.** The curve traced by a particle while moving along a screw thread is known as helix.

2. **Lead.** The distance which a screw thread advances axially in one turn is known as lead. It is represented by l.

3. **Pitch.** The distance between two adjacent threads parallel to the axis of the screw, is known as pitch. The relation between pitch (p) and lead (l) is given by

$$\text{Lead} = \text{Pitch} \times \text{Number of threads}$$

or $$l = p \times n$$

(i) For a single-start thread, $n = 1$ and hence $l = p$

or $$\text{lead} = \text{pitch}$$

(ii) For triple-start thread, $n = 3$ and hence $l = p \times 3$.

Therefore $$p = \frac{1}{3}.$$

4. **Helix angle.** The slope of the thread (or inclination of the thread) with horizontal is known as helix angle. It is represented by α, where

$$\tan \alpha = \frac{\text{Lead of screw}}{\text{Circumference of screw}}$$

$$= \frac{p}{\pi d} \quad \text{(For single-start thread)}$$

$$= \frac{n \times p}{\pi d} \quad \text{(For multi-start threads)}$$

6.6. SCREW-JACK

A screw-jack is a device used for lifting heavy weights or loads with the help of a small effort applied at its handle.

Fig. 6.6 shows the simple screw-jack, which consists of a nut, a screw with square threads and a handle fitted to the head of the screw. The nut also forms the body of the jack.

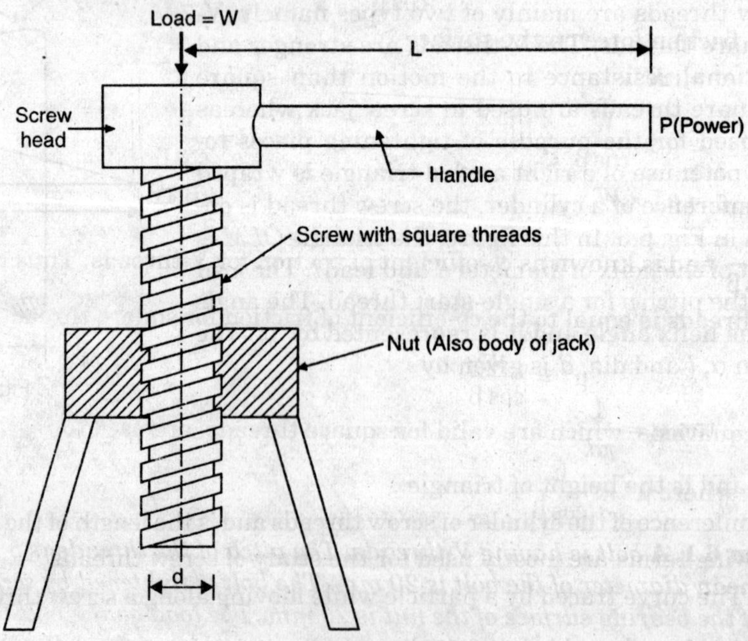

Fig. 6.6. Simple screw-jack.

The load to be lifted is placed on the head of the screw. At the end of the handle, fitted to the screw head, an effort P is applied in the horizontal direction to lift the load W. The screw-jack works on the same principle on which an inclined plane works.

Let W = Weight placed on the screw head,
P = Effort applied at the end of the handle,
L = Length of handle,
p = Pitch of the screw,
d = Mean diameter of the screw,
α = Angle of the screw or helix angle,
ϕ = Angle of friction,
μ = Co-efficient of friction between screw and nut = $\tan \phi$.

6.7. V-THREADS FRICTION

Fig. 6.7 shows a section of a V-thread in which 2β is the angle between the faces of the thread.

Let R = Normal reaction
W = Axial load acting
β = Semi-angle of the V-thread.

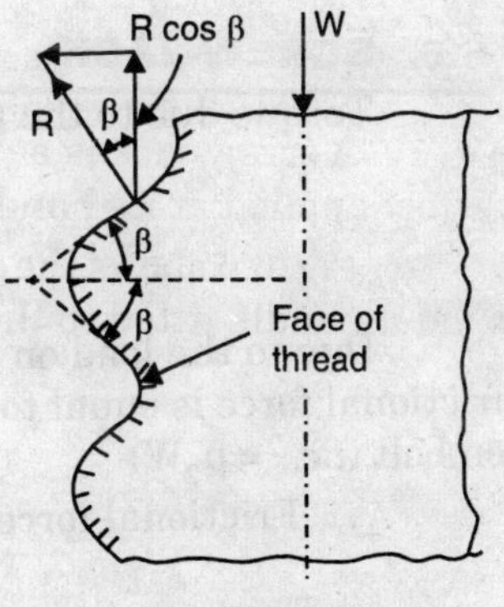

Fig. 6.7

The normal reaction has two components *i.e.*, axial and lateral. Axial component is equal to $R \cos \beta$ whereas the lateral component is $R \sin \beta$. For equilibrium, the axial component of R must be equal to W, *i.e.*,

$$R \cos \beta = W \quad \text{or} \quad R = \frac{W}{\cos \beta}$$

Now the friction force on the surface

$$= \mu R$$

$$= \mu \times \frac{W}{\cos \beta} \qquad \left(\because R = \frac{W}{\cos \beta}\right)$$

$$= \mu^* \times W \qquad \text{...(6.1B)}$$

where $\mu^* = \dfrac{\mu}{\cos \beta}$ and is known as co-efficient of friction for V-threads. Thus the co-efficient of friction for V-threads is equal to the co-efficient of friction for square threads divided by $\cos \beta$

$$\therefore \qquad \mu^* = \frac{\mu}{\cos \beta} \qquad \text{...(6.1C)}$$

All the equations, which are valid for square threads, are used for V-threads by substituting μ^* for μ, where $\mu^* = \dfrac{\mu}{\cos \beta}$.

Problem 6.1. *A bolt is having V-threads. The pitch of the threads is 5 mm and angle of V is 55°. The mean diameter of the bolt is 20 mm. The bolt is tightened by screwing a nut. The mean radius of the bearing surface of the nut is 25 mm. The load on the bolt is 5000 N. The co-efficient of friction for nut and bolt is 0.1 whereas for nut and bearing surface is 0.16. Determine the force required at the end of a spanner 0.6 m long.*

Sol. Given :

Pitch, $p = 5$ mm ; Angle of V-threads, $2\beta = 55°$ or $\beta = 27.5°$;

Mean diameter of bolt, $d = 25$ mm ; Mean radius of bearing surface, $R = 25$ mm ;

Load on bolt, $W = 5000$ N ;

Friction co-efficient for nut and bolt, $\mu = 0.1$

Friction co-efficient for nut and bearing surface, $\mu_2 = 0.16$

Length of spanner, $L = 0.6$ m $= 600$ mm

Co-efficient of friction for V-threads given by,

$$\mu^* = \frac{\mu}{\cos\beta} = \frac{0.1}{\cos 27.5°} = 0.113$$

$$\therefore \quad \tan\phi^* = \mu^* = 0.113$$

Also we know that $\tan\alpha = \dfrac{p}{\pi d} = \dfrac{5}{\pi \times 25} = 0.064$

Now the force on the screw is given by,

$$P = W\tan(\alpha + \phi^*) = W\frac{\tan\alpha + \tan\phi^*}{1 - \tan\alpha\tan\phi^*}$$

$$= 5000\left(\frac{0.064 + 0.113}{1 - 0.064 \times 0.113}\right) = 891.5 \text{ N}$$

Torque due to the force on the screw is given by,

$$T_1 = P \times \text{mean radius of the bolt}$$

$$= 891.5 \times \frac{d}{2} = 891.5 \times \frac{25}{2} \text{ N mm} \quad ...(i)$$

Due to the load on the bolt, there will be a frictional force on the bearing surface. This frictional force is equal to the product of co-efficient of friction for bearing surface and weight on bolt (*i.e.*, $= \mu_2 W$)

$\therefore$ Frictional force on bearing surface,

$$P_1 = \mu_2 \times W = 0.16 \times 5000 = 800 \text{ N}$$

$\therefore$ Torque due to frictional force on bearing surface,

$$T_2 = P_1 \times \text{Mean radius of bearing surface}$$

$$= 800 \times 25 = 20000 \text{ N mm} \quad ...(ii)$$

$$\therefore \text{ Total torque } = T_1 + T_2 = 891.5 \times \frac{25}{2} + 20000 \quad ...(iii)$$

Now torque required at the end of spanner

$$= \text{Force on spanner} \times \text{Length}$$

$$= P_2 \times L$$

$$= P_2 \times 600 \quad ...(iv)$$

Equating equations (*iii*) and (*iv*),

$$600P_2 = 891.5 \times \frac{25}{2} + 20000 = 31143.75$$

$$P_2 = \frac{31143.75}{600} = \mathbf{51.9\ N.}\quad \textbf{Ans.}$$

6.8. PIVOT AND COLLAR BEARING

The rotating shafts are frequently subjected to axial thrust. These shafts can be kept in correct axial position if bearing surfaces are provided. The bearing surfaces which are flat or conical carry the axial thrust. The bearing surfaces placed at the end of a shaft are known as **pivots**. The pivot may have a flat surface or a conical surface or truncated conical surface as shown in Fig. 6.8 (*a*), (*b*) and (*c*) respectively.

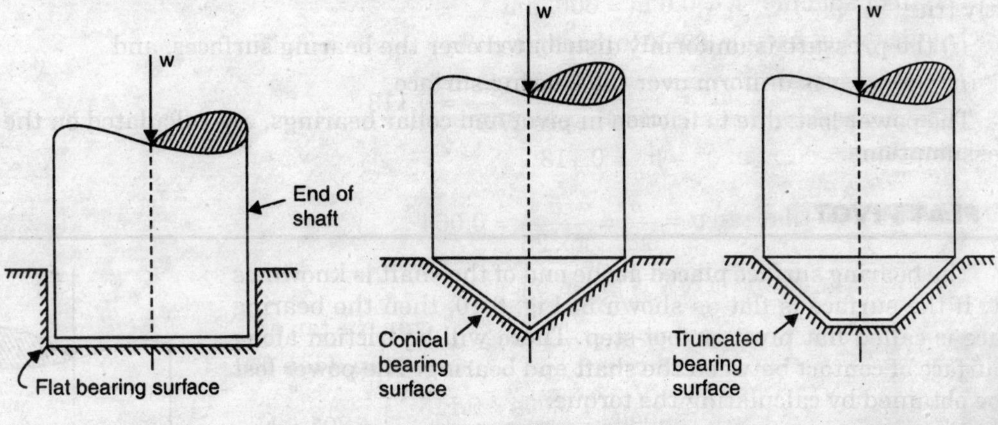

Fig. 6.8

The bearing surfaces provided at any position along the shaft (but not at the end of the shaft), to carry the axial thrust, is known as **collar**. The surface of the collar may be flat (normal to the axis of shaft) or of conical shape as shown in Fig. 6.9 (*a*) and (*b*) respectively. The collar bearings are also known as *thrust bearings*.

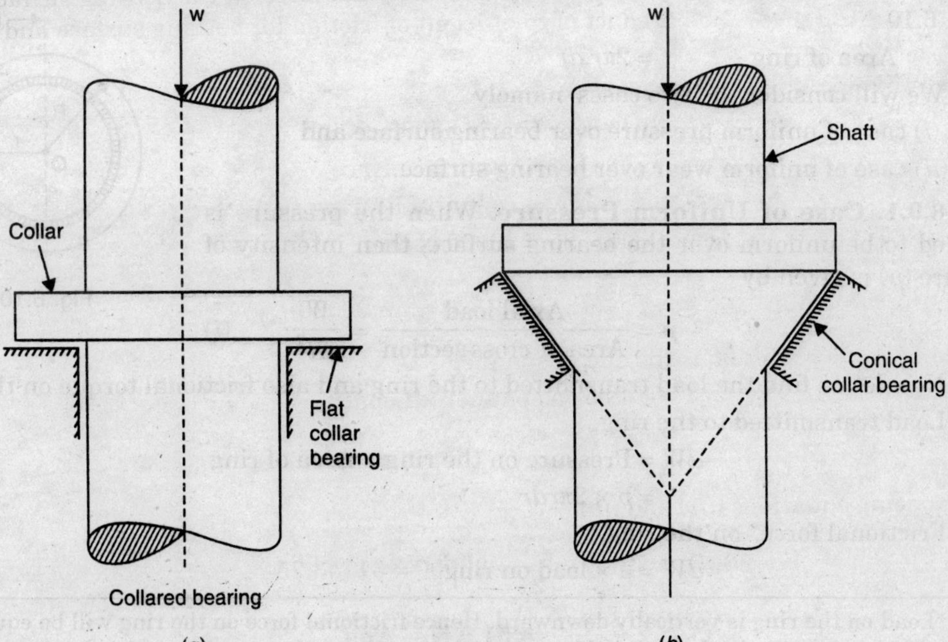

Fig. 6.9

For a new bearing, the contact between the shaft and bearing may be good over the whole surface. This means that the pressure over the rubbing surfaces may be assumed as uniformly distributed. But when the bearing becomes old, all parts of the rubbing surfaces will not move with the same velocity and hence the wear will be different at different radii. The pressure distribution will not be uniform. The rate of wear of surfaces depends upon the pressure and the rubbing velocities between the surfaces.

The design of bearings is based on the following assumptions though neither of them is strictly true :

(*i*) the pressure is uniformly distributed over the bearing surfaces, and

(*ii*) the wear is uniform over the bearing surface.

The power lost, due to friction in pivot and collar bearings, are calculated on the above two assumptions.

6.9. FLAT PIVOT

The bearing surface placed at the end of the shaft is known as pivot. If the surface is flat as shown in Fig. 6.10, then the bearing surface is called flat pivot or foot-step. There will be friction along the surface of contact between the shaft and bearing. The power lost can be obtained by calculating the torque.

Let W = Axial load, or load transmitted to the bearing surface,

R = Radius of pivot,

μ = Co-efficient of friction,

p = Intensity of pressure in N/m^2, and

T = Total frictional torque.

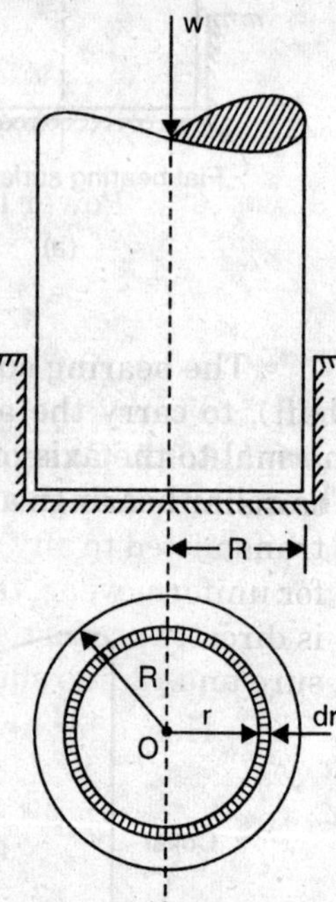

Fig. 6.10

Consider a circular ring of radius r and thickness dr as shown in Fig. 6.10.

$\therefore$ Area of ring $= 2\pi r.dr$

We will consider the two cases, namely

(*i*) case of uniform pressure over bearing surface and

(*ii*) case of uniform wear over bearing surface.

6.9.1. Case of Uniform Pressure. When the pressure is assumed to be uniform over the bearing surface, then intensity of pressure (p) is given by

$$p = \frac{\text{Axial load}}{\text{Area of cross-section}} = \frac{W}{\pi R^2} \qquad ...(i)$$

Now let us find the load transmitted to the ring and also frictional torque on the ring.

Load transmitted to the ring,

$$dW = \text{Pressure on the ring} \times \text{Area of ring}$$

$$= p \times 2\pi r dr$$

Frictional force* on the ring,

$$dW = \mu \times \text{load on ring}$$

*Load on the ring is vertically downward. Hence frictional force on the ring will be equal to $\mu \times$ normal reaction *i.e.*, $\mu \times$ load on ring. Here normal reaction on the ring is equal to load on the ring. Hence frictional force on ring, $dF = \mu \times dW$.

$= \mu \times p \times 2\pi r dr$

$\therefore$ Frictional torque on the ring

$= \text{Friction force} \times \text{Radius of ring} = dF \times r$

$\therefore$ Frictional torque, $dT = \mu \times p \times 2\pi r dr \times r$

$= 2\pi\mu p r^2 dr$

The total frictional torque (T) will be obtained by integrating the above equation from 0 to R.

$\therefore$ Total frictional torque, $T = \int_0^R 2\pi\mu p r^2 \, dr$

$$= 2\pi\mu p \int_0^R r^2 dr \qquad (\because \mu \text{ and } p \text{ are constant})$$

$$= 2\pi\mu p \left[\frac{r^3}{3}\right]_0^R = 2\pi\mu p \left[\frac{R^3}{3}\right] = \frac{2}{3}\pi\mu p R^3$$

$$= \frac{2}{3}\pi \times \mu \times \frac{W}{\pi R^2} \times R^3 \qquad \left(\because \text{From } (i),\ p \frac{W}{\pi R^2}\right)$$

$$= \frac{2}{3}\mu W R \qquad ...(6.2)$$

$\therefore$ Power lost in friction $= T \times \omega$

$$= T \times \frac{2\pi N}{60} \qquad \left(\because \omega = \frac{2\pi N}{60}\right)$$

$$= \frac{2\pi N T}{60} \qquad ...(6.3)$$

6.9.2. Case of Uniform Wear. For the uniform wear of the bearing surface, the load transmitted to the various circular rings should be same (or should be constant). But load transmitted to any circular ring is equal to the product of pressure and area of the ring. Hence for uniform wear, the product of pressure and area of ring should be constant. Area of the ring is directly proportional to the radius of the ring. Hence for uniform wear, the product of pressure and radius should be constant or $p \times r$ = constant.

Hence for uniform wear, we have

$$p \times r = \text{Constant} \quad (\text{say } C) \qquad ...(6.4)$$

$$\therefore \quad p = \frac{C}{r} \qquad ...(i)$$

Now we know that load transmitted to the ring

$= \text{Pressure} \times \text{Area or ring}$

$= p \times 2\pi r dr$

$$= \frac{C}{r} \times 2\pi r dr \qquad \left[\because \text{From } (i),\ p = \frac{C}{r}\right]$$

$$= 2\pi C \, dr \qquad ...(ii)$$

Total load transmitted to the bearing, is obtained by integrating the above equation from 0 to R.

$\therefore$ Total load transmitted to the bearing

$$= \int_0^R 2\pi C dr = 2\pi C \int_0^R dr = 2\pi C \left[r\right]_0^R$$

$$= 2\pi C R$$

But total load transmitted to the bearing = W

$\therefore \quad 2\pi CR = W$

or
$$C = \frac{W}{2\pi R} \qquad ...(iii)$$

Now frictional force on the ring,

$$dF = \mu \times \text{Load on ring} = \mu \times dW$$
$$= \mu \times 2\pi C dr \qquad [\because \text{ From } (ii), \text{ load on ring} = 2\pi C dr]$$

Hence frictional torque on the ring,

$$dT = \text{Frictional force on ring} \times \text{radius}$$
$$= \mu \times 2\pi C dr \times r$$

$\therefore$ Total frictional torque,
$$T = \int_0^R dT$$
$$= \int_0^R \mu \times 2\pi Cr\,dr$$
$$= 2\pi\mu C \int_0^R r\,dr \qquad [\mu \text{ and } C \text{ are constant}]$$
$$= 2\pi\mu C \left[\frac{r^2}{2}\right]_0^R = 2\pi\mu C \times \frac{R^2}{2}$$
$$= 2\pi\mu \times \frac{W}{2\pi R} \times \frac{R^2}{2} \qquad \left[\because \text{ From } (iii), C = \frac{W}{2\pi R}\right]$$

or
$$T = \frac{1}{2} \times \mu WR \qquad ...(6.5)$$

$\therefore$ Power lost in friction $= T \times \omega = \dfrac{2\pi NT}{60}$.

Problem 6.2. *Find the power lost in friction assuming (i) uniform pressure and (ii) uniform wear when a vertical shaft of 100 mm diameter rotating at 150 r.p.m. rests on a flat end foot step bearing. The co-efficient of friction is equal to 0.05 and shaft carries a vertical load of 15 kN.*

Sol. Given :

Diameter, $D = 100$ mm $= 0.1$ m $\quad \therefore \quad R = \dfrac{0.1}{2} = 0.05$ m

Speed, $N = 150$ r.p.m., Friction co-efficient, $\mu = 0.05$

Load, $W = 15$ kN $= 15 \times 10^3$ N

(i) Power lost in friction assuming uniform pressure

For uniform pressure, frictional torque is given by equation (6.2) as

$$T = \frac{2}{3} \mu WR$$
$$= \frac{2}{3} \times 0.05 \times 15 \times 10^3 \times 0.05 \text{ Nm} = 25 \text{ Nm}$$

Power lost in friction
$$= \frac{2\pi NT}{60}$$
$$= \frac{2\pi \times 150 \times 25}{60} \text{ W} = \textbf{392.7 W}. \quad \textbf{Ans.}$$

(ii) Power lost in friction assuming uniform wear

For uniform wear, the frictional torque is given by equation (6.5) as

$$T = \frac{1}{2} \mu WR$$

$$= \frac{1}{2} \times 0.05 \times 15 \times 10^3 \times 0.05 \text{ Nm} = 18.75 \text{ Nm}$$

$\therefore$ Power lost in friction $= \dfrac{2\pi NT}{60}$

$$= \frac{2\pi \times 150 \times 18.75}{60} \text{ W} = \mathbf{294.5\ W.} \quad \textbf{Ans.}$$

6.10. CONICAL PIVOT

The bearing surface placed at the end of a shaft and having a conical surface, is known as conical pivot as shown in Fig. 6.11.

Let W = Axial load, or load transmitted to the bearing surface

μ = Co-efficient of friction

R = Radius of shaft

α = Semi-angle of the cone

p = Pressure intensity normal to the cone surface.

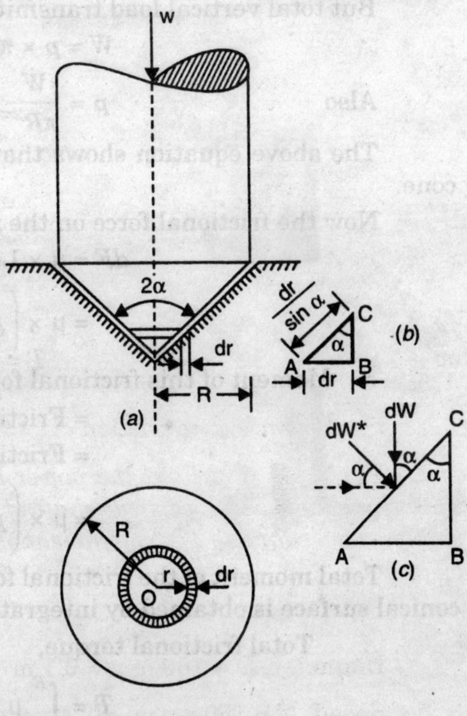

Fig. 6.11

Consider a circular ring of radius r and thickness dr. The actual thickness of the sloping ring will be $\dfrac{dr}{\sin \alpha}$ as shown in Fig. 6.11 (*b*) in which $AB = dr$ on enlarged scale, angle $ACB = \alpha$ and sloping length of ring $= AC = \dfrac{AB}{\sin \alpha} = \dfrac{dr}{\sin \alpha}$.

$\therefore$ Area of ring along conical surface

$= 2\pi r \times$ Actual thickness of sloping ring

$$= 2\pi r \times \frac{dr}{\sin \alpha}$$

Now we will consider two cases namely :

(*i*) Case of uniform pressure

(*ii*) Case of uniform wear.

6.10.1. Case of Uniform Pressure. Let us first find the load acting on the circular ring, normal to the conical surface.

$\therefore$ Load on the ring normal to conical surface,

dW^* = Pressure × Area of ring along conical surface

$$= p \times 2\pi r \times \frac{dr}{\sin \alpha}$$

Vertical component of the above load [Refer to Fig. 6.11 (*c*)]

$$dW = \left[p \times 2\pi r \times \frac{dr}{\sin \alpha} \right] \times \sin \alpha \qquad (\because \quad dW = dW^* \sin \alpha)$$

$$= p \times 2\pi r \times dr$$

∴ Total vertical load transmitted to the bearing

$$= \int_0^R p \times 2\pi r \times dr \qquad ...(6.6)$$

$$= p \times 2\pi \int_0^R r\, dr \quad (\because \text{ pressure is uniform and hence constant})$$

$$= p \times 2\pi \times \left[\frac{r^2}{2} \right]_0^R = p \times 2\pi \times \frac{R^2}{2} = p \times \pi R^2$$

But total vertical load transmitted is also = W

$$\therefore \qquad W = p \times \pi R^2 \qquad ...(i)$$

Also

$$p = \frac{W}{\pi R^2} \qquad ...(ii)$$

The above equation shows that pressure intensity is independent of the angle of the cone.

Now the frictional force on the ring along the conical surface,

$$dF = \mu \times \text{Loan on ring normal to conical surface} = \mu \times dW^*$$

$$= \mu \times \left(p \times 2\pi r \times \frac{dr}{\sin \alpha} \right)$$

∴ Moment of this frictional force about the shaft axis (dT)

= Frictional torque on the ring

= Frictional force × Radius = $dF \times r$

$$= \mu \times \left(p \times 2\pi r \times \frac{dr}{\sin \alpha} \right) \times r \qquad ...(6.7)$$

Total moment of the frictional force about the shaft axis or total frictional torque on the conical surface is obtained by integrating the above equation from 0 to R.

∴ Total frictional torque,

$$T = \int_0^R \mu \times p \times 2\pi r \times \frac{dr}{\sin \alpha} \times r$$

$$= \frac{2\pi \times \mu \times p}{\sin \alpha} \int_0^R r^2\, dr \qquad (\because \ \mu, p \text{ and } \alpha \text{ are constant})$$

$$= \frac{2\pi \times \mu \times p}{\sin \alpha} \left[\frac{r^3}{3} \right]_0^R = \frac{2\pi \times \mu \times p}{\sin \alpha} \times \frac{R^3}{3}$$

$$= \frac{2\pi \times \mu}{\sin \alpha} \times \frac{W}{\pi R^2} \times \frac{R^3}{3} \qquad \left[\because \text{ From } (ii),\ p = \frac{W}{\pi R^2} \right]$$

$$= \frac{2}{3} \times \frac{\mu W R}{\sin \alpha} \qquad ...(6.8)$$

$$\therefore \quad \text{Power lost in friction} = \frac{2\pi N T}{60}.$$

6.10.2. Case of Uniform Wear. From equation (6.4), for uniform wear, we know that

$$p \times r = \text{Constant} \quad (\text{say} = C)$$

$$\therefore \quad p \times r = C$$

or $$p = \frac{C}{r}$$

From equation (6.6), total vertical load transmitted to the bearing

$$= \int_0^R p \times 2\pi r \times dr$$

$$= \int_0^R \frac{C}{r} \times 2\pi r \times dr \qquad \left(\because \ p = \frac{C}{r}\right)$$

$$= 2\pi \times C \int_0^R dr = 2\pi \times C \Big[r \Big]_0^R$$

$$= 2\pi \times C \times R$$

But total vertical load transmitted to the bearing is also equal to W

$$\therefore \quad W = 2\pi \times C \times R$$

or $$C = \frac{W}{2\pi R}$$

Now the frictional torque on the ring is given by the equation (6.7) as

$$dT = \mu \times p \times 2\pi r \times \frac{dr}{\sin \alpha} \times r$$

$$= \mu \times \frac{C}{r} \times 2\pi r \times \frac{dr}{\sin \alpha} \times r \qquad \left(\because \ p = \frac{C}{r} \text{ for uniform wear}\right)$$

$$= 2\pi\mu C \times r \times \frac{dr}{\sin \alpha} \qquad \ldots(6.8A)$$

$$= 2\pi\mu \times \frac{W}{2\pi R} \times r \times \frac{dr}{\sin \alpha} \qquad \left(\because \ C = \frac{W}{2\pi R}\right)$$

$\therefore$ Total frictional torque,

$$T = \int_0^R dT = \int_0^R 2\pi\mu \times \frac{W}{2\pi R} \times r \times \frac{dr}{\sin \alpha}$$

$$= 2\pi\mu \times \frac{W}{2\pi R} \times \frac{1}{\sin \alpha} \int_0^R r\, dr = 2\pi\mu \times \frac{W}{2\pi R} \times \frac{1}{\sin \alpha} \times \frac{R^2}{2}$$

$$= \frac{1}{2} \times \frac{\mu W R}{\sin \alpha} \qquad \ldots(6.9)$$

$\therefore$ Power lost in friction $= \dfrac{2\pi N T}{60}$.

6.10.3. Truncated Conical Pivot. Fig. 6.12 shows the truncated conical pivot of external and internal radii as r_1 and r_2.

(*i*) Case of Uniform Pressure

Total vertical load transmitted to the bearing is obtained from equation (6.6) in which the limits of integration are from r_2 to r_1.

$\therefore$ Total vertical load transmitted to the bearing

$$= \int_{r_2}^{r_1} p \times 2\pi r \times dr$$

$$= p \times 2\pi \int_{r_2}^{r_1} r \times dr$$

(p is constant for uniform pressure)

$$= p \times 2\pi \left[\frac{r^2}{2}\right]_{r_2}^{r_1} = p \times 2\pi \left[\frac{r_1^2 - r_2^2}{2}\right]$$

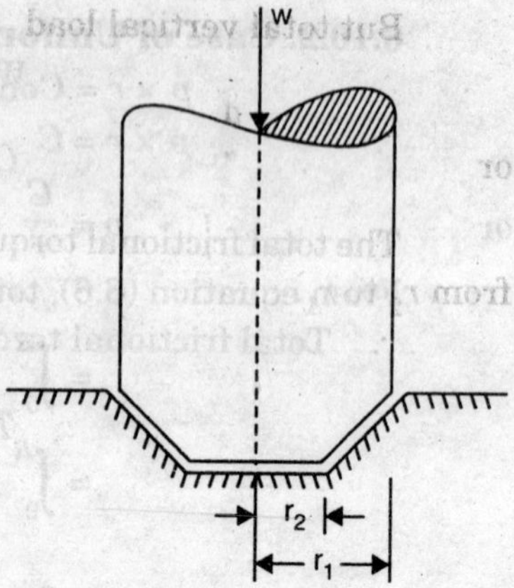

Fig. 6.12

But total vertical load $= W$

$$\therefore \quad W = p \times 2\pi \left[\frac{r_1^2 - r_2^2}{2}\right] = p \times \pi\,[r_1^2 - r_2^2]$$

or $$p = \frac{W}{\pi(r_1^2 - r_2^2)} \qquad \text{...(6.9 A)}$$

Total frictional torque on the truncated conical surface is obtained by integrating equation (6.7) from r_2 to r_1.

$$\therefore \quad T = \int_{r_2}^{r_1} \mu \times p \times 2\pi r \times \frac{dr}{\sin \alpha} \times r$$

$$= \frac{2\mu \times \pi \times p}{\sin \alpha} \int_{r_2}^{r_1} r^2\, dr \qquad (\mu, p \text{ and } \alpha \text{ are constant})$$

$$= \frac{2\pi \times \mu \times p}{\sin \alpha}\left[\frac{r^3}{3}\right]_{r_2}^{r_1} \frac{2\pi \times \mu \times p}{\sin \alpha}\left[\frac{r_1^3 - r_2^3}{3}\right]$$

$$= \frac{2\pi \times \mu}{\sin \alpha} \times \frac{W}{\pi(r_1^2 - r_2^2)} \times \left[\frac{r_1^3 - r_2^3}{3}\right] \qquad \left[\because \text{From (6.9A), } p = \frac{W}{\pi(r_1^2 - r_2^2)}\right]$$

$$= \frac{2}{3}\frac{\mu W}{\sin \alpha} \times \left(\frac{r_1^3 - r_2^3}{r_1^2 - r_2^2}\right) \qquad \text{...(6.10)}$$

(*ii*) **Case of Uniform Wear**

For uniform wear, $p \times r = C$

or $$p = \frac{C}{r}$$

The total vertical load transmitted to the bearing is obtained from equation (6.6) in which limits of integration are from r_2 to r_1.

$\therefore$ Total vertical load transmitted

$$= \int_{r_2}^{r_1} p \times 2\pi r \times dr$$

$$= \int_{r_2}^{r_1} \frac{C}{r} \times 2\pi r \times dr \qquad \left(\because p = \frac{C}{r}\right)$$

$$= 2\pi C \int_{r_2}^{r_1} dr = 2\pi C \left[r\right]_{r_2}^{r_1} = 2\pi C\,[r_1 - r_2]$$

But total vertical load $= W$

$$\therefore \quad W = 2\pi C\,[r_1 - r_2]$$

or
$$C = \frac{W}{2\pi[r_1 - r_2]}$$

The total frictional torque for uniform wear is obtained by integrating the equation (6.8A) from r_2 to r_1.

$\therefore$ Total frictional torque,

$$T = \int_{r_2}^{r_1} 2\pi\mu \times C \times r \times \frac{dr}{\sin\alpha}$$

$$= \int_{r_2}^{r_1} 2\pi\mu \times \frac{W}{2\pi(r_1 - r_2)} \times r \times \frac{dr}{\sin\alpha} \qquad \left(\because C = \frac{W}{2\pi(r_1 - r_2)}\right)$$

or
$$T = \frac{2\pi\mu \times W}{2\pi(r_1 - r_2)} \times \frac{1}{\sin\alpha}\int_{r_2}^{r_1} r\,dr$$

$$= \frac{\mu W}{(r_1 - r_2)} \times \frac{1}{\sin\alpha}\left[\frac{r^2}{2}\right]_{r_2}^{r_1} \frac{\mu W}{(r_1 - r_2)} \times \frac{1}{\sin\alpha}\left[\frac{r_1^2 - r_2^2}{2}\right]$$

$$= \frac{1}{2} \times \frac{\mu W}{\sin\alpha}(r_1 + r_2) \qquad \text{...(6.11)}$$

Problem 6.3. *A conical pivot with angle of cone as 120°, supports a vertical shaft of diameter 300 mm. It is subjected to a load of 20 kN. The co-efficient of friction is 0.05 and the speed of shaft is 210 r.p.m. Calculate the power lost in friction assuming (i) uniform pressure and (ii) uniform wear.*

Sol. Given :

$2\alpha = 120°$ $\quad\therefore\quad \alpha = 60°$;
$D = 300$ mm $= 0.3$ m $\quad\therefore\quad R = 0.15$ m ;
$W = 20$ kN $= 20 \times 10^3$ N ; $\quad \mu = 0.05$;
$N = 210$ r.p.m.

(i) Power lost in friction for uniform pressure

The frictional torque is given by equation (6.8) as

$$T = \frac{2}{3} \times \frac{\mu W R}{\sin\alpha}$$

$$= \frac{2}{3} \times \frac{0.05 \times 20 \times 10^3 \times 0.15}{\sin 60°} = 115.53 \text{ Nm}$$

$\therefore$ Power lost
$$= \frac{2\pi N T}{60}$$

$$= \frac{2\pi \times 210 \times 115.53}{60} = 2540.6 \text{ W} = \mathbf{2.54\ kW.} \quad \textbf{Ans.}$$

(ii) Power lost in friction for uniform wear

The friction torque is given by equation (6.9) as

$$T = \frac{1}{2} \times \frac{\mu W R}{\sin\alpha}$$

$$= \frac{1}{2} \times \frac{0.05 \times 20 \times 10^3 \times 0.15}{\sin 60^\circ} = 86.60 \text{ Nm}$$

$\therefore$ Power lost $= \dfrac{2\pi NT}{60}$

$$= \frac{2\pi \times 210 \times 86.6}{60} = 1904.4 \text{ W} = \mathbf{1.9044 \text{ kW}}. \quad \textbf{Ans.}$$

Problem 6.4. *A load of 25 kN is supported by a conical pivot with angle of cone as 120°. The intensity of pressure is not to exceed 350 kN/m². The external radius is 2 times the internal radius. The shaft is rotating at 180 r.p.m. and co-efficient of friction is 0.05. Find the power absorbed in friction assuming uniform pressure.*

Sol. Given :

Load, $W = 25$ kN $= 25 \times 10^3$ N ; Angle of cone, $2\alpha = 120°$ or $\alpha = 60°$

Pressure, $p = 350$ kN/m² $= 350 \times 10^3$ N/m² ; External radius $= 2 \times$ internal radius

Hence $r_1 = 2r_2$; Speed, $N = 180$ r.p.m. ; and $\mu = 0.05$

Using equation (6.9A) for uniform pressure, we get

$$p = \frac{W}{\pi(r_1^2 - r_2^2)}$$

or $$350 \times 10^3 = \frac{25 \times 10^3}{\pi[(2r_2)^2 - r_2^2]} \qquad (\because \ r_1 = 2r_2)$$

or $$[(2r_2)^2 - r_2^2] = \frac{25}{\pi \times 350} = 0.02273$$

or $$3r_2^2 = 0.02273$$

or $$r_2 = \sqrt{\frac{0.02273}{3}} = 0.087 \text{ m}$$

$\therefore$ $$r_1 = 2r_2 = 2 \times 0.087 = 0.174 \text{ m}$$

To find the power absorbed in friction, first calculate the total frictional torque when pressure is uniform.

Frictional torque when pressure is uniform is given by equation (6.10) as

$$T = \frac{2}{3} \times \frac{\mu W}{\sin \alpha} \times \left(\frac{r_1^3 - r_2^3}{r_1^2 - r_2^2}\right)$$

$$= \frac{2}{3} \times \frac{0.05 \times 25 \times 10^3}{\sin 60^\circ} \left(\frac{0.174^3 - 0.087^3}{0.174^2 - 0.087^2}\right)$$

$$= 962.278 \left(\frac{0.005268 - 0.0006585}{0.03027 - 0.007569}\right) = 962.278 \left(\frac{0.0046095}{0.0227}\right)$$

$$= 195.37 \text{ Nm}$$

$\therefore$ Power absorbed in friction,

$$P = \frac{2\pi NT}{60} = \frac{2\pi \times 180 \times 195.37}{60} = 3682.6 \text{ W} = \mathbf{3.6826 \text{ kW}}. \quad \textbf{Ans.}$$

6.10.4. Flat Collar. The bearing surface provided at any position along the shaft (but not at the end of the shaft), to carry axial thrust is known as collar which may be flat or conical. If the surface is flat, then bearing surface is known as flat collar as shown in Fig. 6.13. The collar bearings are also known as *thrust bearings*. The power lost in friction can be obtained by calculating the torque.

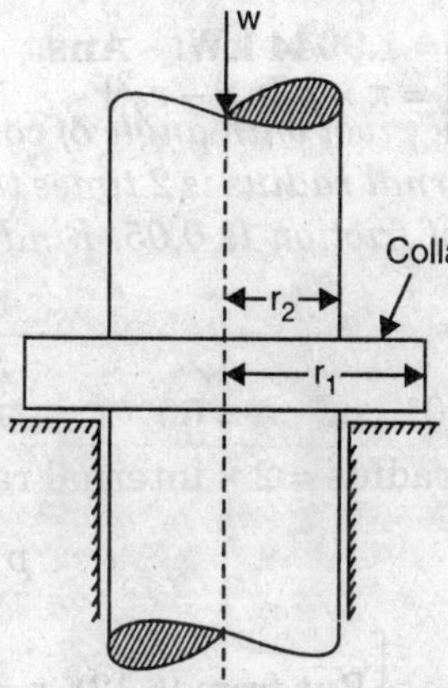

Fig. 6.13

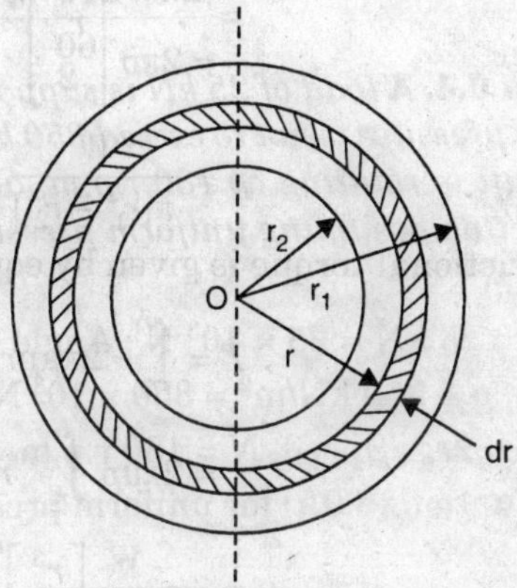

Fig. 6.13 (*a*)

Let r_1 = External radius of collar

r_2 = Internal radius of collar

p = intensity of pressure

W = Axial load or total load transmitted to the bearing surface

μ = Co-efficient of friction

T = Total frictional torque

Consider a circular ring of radius r and thickness dr as shown in Fig. 6.13 (*a*).

$\therefore$ Area of ring $= 2\pi r\, dr$.

Load on the ring $=$ Pressure $\times$ Area of ring

$= p \times 2\pi r\, dr$ ...(*i*)

Frictional force on the ring $= \mu \times$ Load on ring

$= \mu \times p \times 2\pi r\, dr$

Frictional torque on the ring, dT = Frictional force $\times$ Radius

$= (\mu \times p \times 2\pi r\, dr) \times r$

$= 2\pi\mu p r^2\, dr$

$\therefore$ Total frictional torque,

$$T = \int_{r_2}^{r_1} dT$$

$$= \int_{r_2}^{r_1} 2\pi\mu p r^2 dr \qquad ...(ii)$$

(*i*) **Uniform Pressure**

p = Constant

Total load transmitted to the bearing

$$= \int_{r_2}^{r_1} \text{Load on ring}$$

or $$W = \int_{r_2}^{r_1} p \times 2\pi r\, dr \qquad (\because \text{ Load on ring from } (i) = p \times 2\pi r\, dr)$$

$$= p \times 2\pi \int_{r_2}^{r_1} r\, dr \qquad (\because \ p \text{ is constant})$$

$$= 2\pi p \left[\frac{r^2}{2}\right]_{r_2}^{r_1} = 2\pi p \left[\frac{r_1^2 - r_2^2}{2}\right] = \pi \times p\ [r_1^{\ 2} - r_2^{\ 2}]$$

$$\therefore \quad p = \frac{W}{\pi[r_1^2 - r_2^2]} \qquad \text{...(6.12)}$$

Total frictional torque is given by equation (*ii*),

$$\therefore \quad T = \int_{r_2}^{r_1} 2\pi\mu p r^2\, dr$$

$$= 2\pi\mu p \int_{r_2}^{r_1} r^2\, dr \qquad (\because \ p \text{ is constant})$$

$$= 2\pi\mu p \left[\frac{r^3}{3}\right]_{r_2}^{r_1} \qquad \left[\text{But from (6.12), } p = \frac{W}{\pi(r_1^2 - r_2^2)}\right]$$

$$= 2\pi\mu p \left[\frac{r_1^3 - r_2^3}{3}\right] = 2\pi\mu \times \frac{W}{\pi(r_1^2 - r_2^2)} \times \left(\frac{r_1^3 - r_2^3}{3}\right)$$

$$= \frac{2}{3}\mu W \left[\frac{r_1^3 - r_2^3}{r_1^2 - r_2^2}\right] \qquad \text{...(6.13)}$$

$\therefore$ Power lost in friction,

$$P = \frac{2\pi NT}{60}$$

(*ii*) **Uniform Wear**

$$p \times r = \text{constant} \qquad \text{(say } C\text{)}$$

$$\therefore \quad p = \frac{C}{r}$$

Total load transmitted to the bearing

$$= \int_{r_2}^{r_1} \text{Load on ring} = \int_{r_2}^{r_1} p \times 2\pi r\, dr$$

$$\therefore \quad W = \int_{r_2}^{r_1} p \times 2\pi r\, dr$$

$$= \int_{r_2}^{r_1} \frac{C}{r} \times 2\pi r\, dr \qquad \left(\because \ p = \frac{C}{r}\right)$$

$$= 2\,\pi\, C \int_{r_2}^{r_1} dr \qquad (\because \ C \text{ is constant})$$

$$= 2\pi C \left[\ r\ \right]_{r_2}^{r_1} = 2\pi C\ [r_1 - r_2]$$

$$C = \frac{W}{2\pi(r_1 - r_2)} \qquad \text{...(}iii\text{)}$$

Total frictional torque is given by equation (*ii*),

$$T = \int_{r_2}^{r_1} 2\pi\mu p r^2\, dr$$

$$= 2\pi\mu \int_{r_2}^{r_1} p r^2\, dr \qquad \left(\text{Here } p \text{ is not constant it is } = \frac{C}{r}\right)$$

$$= 2\pi\mu \int_{r_2}^{r_1} \frac{C}{r} r^2\, dr$$

$$= 2\pi\mu \int_{r_2}^{r_1} Cr\, dr = 2\pi\mu C \int_{r_2}^{r_1} r\, dr \qquad (C \text{ is constant})$$

$$= 2\pi\mu C \left[\frac{r^2}{2}\right]_{r_2}^{r_1} = 2\pi\mu C \left[\frac{r_2^2 - r_1^2}{2}\right]$$

$$= 2\pi\mu \times \frac{W}{2\pi(r_1 - r_2)} \times \left[\frac{r_2^2 - r_1^2}{2}\right] \qquad \left[\because\ C = \frac{W}{2\pi(r_1 - r_2)} \text{ from } (iii)\right]$$

$$= \frac{\mu W}{2}(r_1 + r_2) \qquad \ldots(6.14)$$

∴ Power lost in friction

$$P = \frac{2\pi NT}{60}$$

If the axial load on the bearing is too great, then the bearing pressure on the collar will become more than the limiting bearing pressure which is approximately equal to 400 kN/m². Hence to reduce the intensity of pressure on collar, two or more collars are used (or multi-collars are used) as shown in Fig. 6.14.

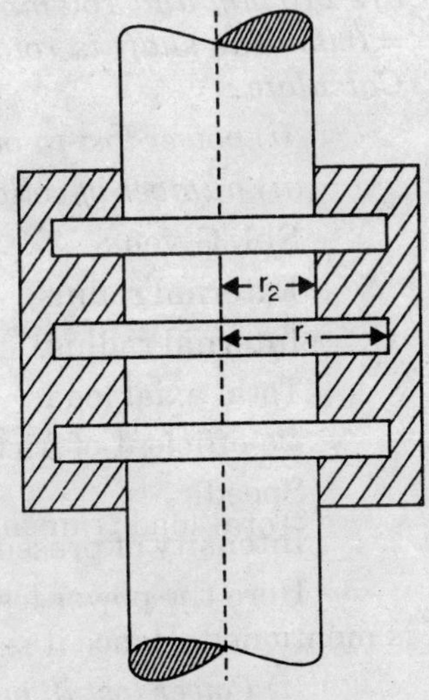

Fig. 6.14

If n = number of collars in multi-collar bearing, then

(*i*) $n = \dfrac{\text{Total load}}{\text{Load permissible on one collar}}$

(*ii*) p = Intensity of the uniform pressure

$$= \frac{\text{Load}}{\text{No. of collars} \times \text{Area of one-collar}}$$

$$= \frac{W}{n \times \pi\,[r_1^2 - r_2^2]}$$

(*iii*) Total torque transmitted remains constant *i.e.*,

$$T = \frac{2}{3} \times \mu W \left[\frac{r_1^3 - r_2^3}{r_1^2 - r_2^2}\right]$$

Note. The frictional torque for uniform pressure is greater than that of uniform wear. Hence for safe design of bearing surfaces when power lost in friction is to be determined and no assumption is mentioned, assume uniform pressure. But when power transmitted is to be determined and no assumption is stated, assume uniform wear.

Problem 6.5. *In a collar thrust bearing the external and internal radii are 250 mm and 150 mm respectively. The total axial load is 50 kN and shaft is rotating at 150 r.p.m. The co-efficient of friction is equal to 0.05. Find the power lost in friction assuming uniform pressure.*

Sol. Given :

External radius, $r_1 = 250 \text{ mm} = 0.25 \text{ m}$

Internal radius, $r_2 = 150 \text{ mm} = 0.15 \text{ m}$

Total axial load, $W = 50 \text{ kN} = 50 \times 10^3 \text{ N}$

Speed, $N = 150$ r.p.m.

Co-efficient of friction, $\mu = 0.05$

For uniform pressure, the total frictional torque is given by equation (6.13), as

$$T = \frac{2}{3} \mu W \left[\frac{r_1^3 - r_2^3}{r_1^2 - r_2^2} \right]$$

$$= \frac{2}{3} \times 0.05 \times 50 \times 10^3 \left[\frac{0.25^3 - 0.15^3}{0.25^2 - 0.15^2} \right]$$

$$= 1666.67 \times \left(\frac{0.015625 - 0.003375}{0.0625 - 0.0225} \right) = 1666.67 \times \frac{0.01225}{0.04}$$

$$= 510.42 \text{ Nm}$$

$\therefore$ Power lost in friction, $P = \frac{2\pi NT}{60}$

$$= \frac{2\pi \times 150 \times 510.42}{60} = 8017.6 \text{ W} = \mathbf{8.0176 \text{ kW}}. \quad \textbf{Ans.}$$

Problem 6.6. *In a thrust bearing the external and internal radii of the contact surfaces are 210 mm and 160 mm respectively. The total axial load is 60 kN and co-efficient of friction = 0.05. The shaft is rotating at 380 r.p.m. Intensity of pressure is not to exceed 350 kN/m². Calculate :*

(*i*) *power lost in overcoming the friction and*

(*ii*) *number of collars required for the thrust bearing.*

Sol. Given :

External radius, $r_1 = 210 \text{ mm} = 0.21 \text{ m}$

Internal radius, $r_2 = 160 \text{ mm} = 0.16 \text{ m}$

Total axial load, $W = 60 \text{ kN} = 60 \times 10^3 \text{ N}$

Co-efficient of friction, $\mu = 0.05$

Speed, $N = 380$ r.p.m.

Intensity of pressure, $p = 350 \text{ kN/m}^2 = 350 \times 10^3 \text{ N/m}^2$

Here the power lost in overcoming the friction is to be determined. Also no assumption is mentioned. Hence it is safe to assume uniform pressure.

(*i*) *Power lost in overcoming friction*

For uniform pressure, total frictional torque is given by equation (6.13) as

$$T = \frac{2}{3} \mu W \left[\frac{r_1^3 - r_2^3}{r_1^2 - r_2^2} \right]$$

$$= \frac{2}{3} \times 0.05 \times 60 \times 10^3 \left[\frac{0.21^3 - 0.16^3}{0.21^2 - 0.16^2} \right]$$

$$= 2000 \times \left[\frac{0.009261 - 0.004096}{0.0441 - 0.0256} \right]$$

$$= 2000 \times \frac{0.005165}{0.0185} = 558.378 \text{ Nm}$$

$\therefore$ Power lost in friction, $$P = \frac{2\pi NT}{60}$$

$\therefore$ $$P = \frac{2\pi \times 380 \times 558.378}{60} = 22219.8 \text{ W} = \mathbf{22.2198 \text{ kW}}. \quad \textbf{Ans.}$$

(*ii*) *Number of collars required.*

Number of collars, $$n = \frac{\text{Total load}}{\text{Load per collar}}$$

Now load per collar for uniform pressure is obtained from equation (6.12), as

$$p = \frac{W^*}{\pi (r_1^2 - r_2^2)}$$

where W^* is the load per collar.

$\therefore$ $$W^* = p \times \pi (r_1^2 - r_2^2)$$
$$= 350 \times 10^3 \times \pi (0.21^2 - 0.16^2)$$
$$= 350 \times 10^3 \times \pi (0.0441 - 0.0256) = 20341.8 \text{ N}$$

$\therefore$ Number of collars $$= \frac{\text{Total load}}{\text{Load per collar}}$$

$$= \frac{W}{W^*} = \frac{60 \times 10^3}{20341.8} = 2.95 = \mathbf{3 \text{ collars}}. \quad \textbf{Ans.}$$

6.11. FRICTION CLUTCHES

The device used to transmit the rotary motion of one shaft to another, the axes of which are coincident, is known a clutch or friction clutch. With the help of friction clutch, the power is transmitted from one shaft to another shaft which must be started and stopped frequently as in the case of automobile for automotive purposes. The engine shaft and gear box shaft is connected with the help of friction clutches.

The following types of friction clutches are mostly used :

(*i*) Disc clutch or single plate clutch, (*ii*) Multi-plate clutch, and

(*iii*) Cone clutch.

The principle of disc and cone clutches are same as that of the pivot and collar bearings. Though cone clutches and multiple-disc clutch are no longer in use for power transmission of power directly from the engine shaft by solid friction, multiple plate clutch is mostly used in automobiles. All modern cars have single plate clutch with a fabric facing on each side of the plate. The clutch plate is positioned between the flywheel and a solid plate, known as *pressure plate*.

6.11.1. Disc Clutch or Single Plate Clutch. Fig. 6.15 shows the diagram of a single plate clutch (or disc clutch) which consists of a single clutch plate with friction lining (*i.e.*, a lining of friction material) on both sides. This plate is attached to a hub (which is splined). The hub is free to move axially along the splines of the driven shaft. There is a pressure plate inside the clutch body. The pressure plate pushes the clutch plate towards the flywheel by a set of strong springs. The clutch body (or cover plate) is bolted to the flywheel. The pressure plate and the flywheel rotate with the driving shaft. The movement of the clutch pedal (not shown in Fig. 6.15) is transferred to the pressure plate through a thrust bearing.

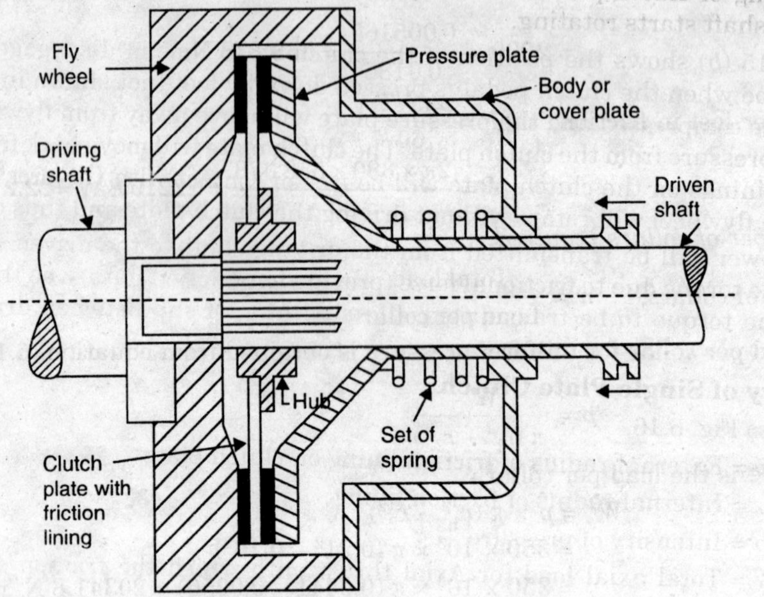

(*a*)

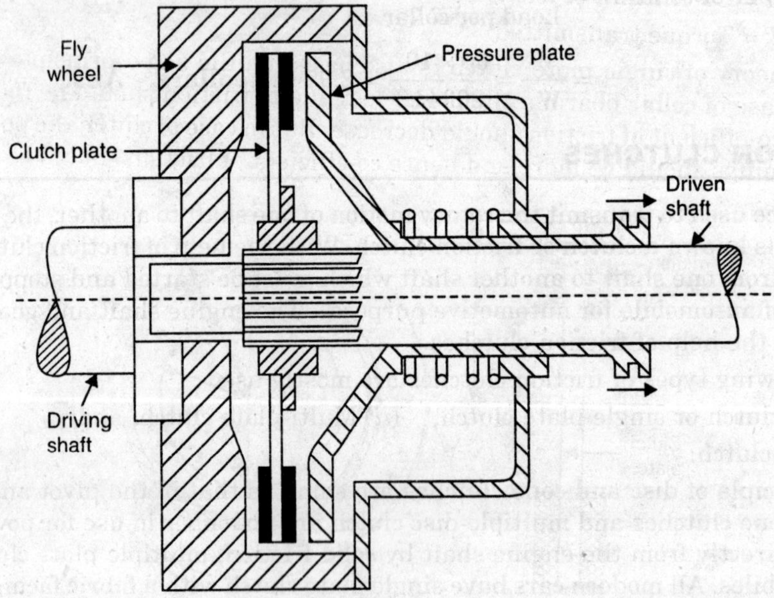

(*b*)

Fig. 6.15

Fig. 6.15 (*a*) shows the position of the clutch when it is in engaged position. This position will be when the foot is taken off from the clutch pedal. The set of strong springs will move the pressure plate to bring it in contact with the clutch plate which is attached to the hub. The hub moves axially along the splines of the driven shaft. Thus the clutch plate is tightly gripped between the pressure plate and the flywheel. There is a friction lining on both sides of the clutch plate. The friction lining on one side of plate is in contact with flywheel whereas the friction lining on other side of the clutch plate is in contact with pressure plate. Due to the tightly gripping of clutch plate between pressure plate and flywheel, the clutch plate and hence driven shaft starts rotating.

Fig. 6.15 (*b*) shows the position of the clutch when it is in disengaged position. This position will be when the clutch pedal is pressed down by foot (not shown in Fig.). The set of springs will be compressed and the pressure plate will move away from flywheel. This action removes the pressure from the clutch plate. The clutch plate will move back from the flywheel. The friction linings on the clutch plate will be free of contact with the pressure plate or the flywheel. The flywheel will rotate without driving the clutch plate and thus driven shaft.

The power will be transmitted from the driving shaft to the driven shaft in engaged position. If the torque due to frictional force (provided by friction linings on the clutch plate) is more than the torque to be transmitted, there will be no slip between driving and driven shafts.

Theory of Single Plate Clutch

Refer to Fig. 6.16.

Let r_1 = External radius of friction lining on clutch plate

r_2 = Internal radius of friction lining

p = Intensity of pressure

W = Total axial load (or Axial thrust with which the friction surfaces are held together)

μ = Co-efficient of friction

T = Torque transmitted.

The theory of single plate clutch is also based on the same principle as that of collar bearing. In case of collar bearing, the power lost due to friction should be reduced and hence the value of co-efficient of friction should decrease. But in case of clutch the power transmitted by friction linings should be more and hence co-efficient of friction should be increased.

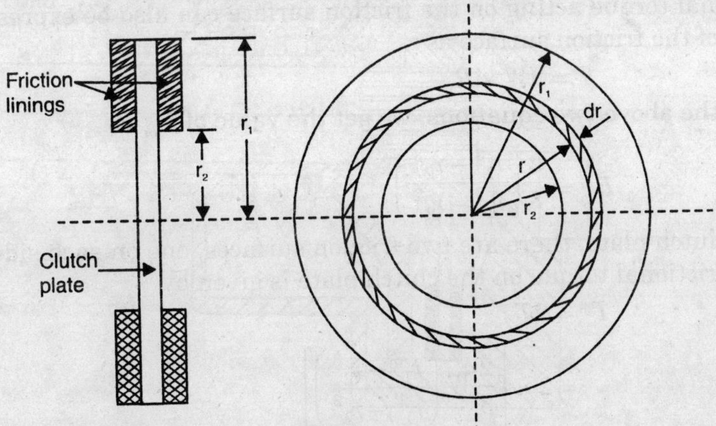

Fig. 6.16

Also in case of a new clutch, the intensity of pressure is approximately uniform over the entire surface whereas in an old clutch the uniform wear theory is more approximate.

Consider a circular ring of radius r and thickness dr as shown in Fig. 6.16.

Area of ring, $dA = 2\pi r\, dr$

Axial load on ring, dW = Pressure × Area of ring

$= p \times 2\pi r\, dr$

Frictional force on the ring, $dF = \mu \times$ Load on ring

$$= \mu \times (p \times 2\pi r\, dr)$$

Frictional torque on ring, dT = Frictional force × Radius

$$= dF \times r$$

$$= (\mu \times p \times 2\pi r\, dr) \times r$$

$$= \mu \times p \times 2\pi r^2\, dr \qquad \text{...(6.15)}$$

(i) For Uniform Pressure

$$p = \text{Constant}$$

$$\therefore \quad p = \frac{W}{\pi(r_1^2 - r_2^2)} \qquad \text{...(6.16)}$$

where W = Total axial thrust with which contact surfaces (or friction surfaces) are held together.

Total friction torque is obtained by integrating equation (6.14) from r_2 to r_1.

∴ Total friction torque acting on the friction surface,

$$T = \int_{r_2}^{r_1} dT = \int_{r_2}^{r_1} \mu \times p \times 2\pi r^2\, dr$$

$$= 2\pi\mu p \left[\frac{r^3}{3}\right]_{r_2}^{r_1} = 2\pi\mu p \left[\frac{r_1^3 - r_2^3}{3}\right]$$

$$= \frac{2}{3}\pi\mu \times \frac{W}{\pi(r_1^2 - r_2^2)} \times (r_1^3 - r_2^3) \qquad \left[\because P = \frac{W}{\pi(r_1^2 - r_2^2)}\right]$$

$$= \frac{2}{3}\mu W \left[\frac{r_1^3 - r_2^3}{r_1^2 - r_2^2}\right] \qquad \text{...(6.17)}$$

Total frictional torque acting on the friction surface can also be expressed in terms of mean radius (R_m) of the friction surface as

$$T = \mu W \times R_m \qquad \text{...(6.18)}$$

Comparing the above two equations, we get the value of R_m as

$$R_m = \frac{2}{3}\left(\frac{r_1^3 - r_2^3}{r_1^2 - r_2^2}\right) \qquad \text{...(6.19)}$$

In a single clutch plate, there are two friction surfaces, one on each side of the friction plate, hence total frictional torque on the clutch plate is given by

$$T^* = 2T$$

$$= 2 \times \left[\frac{2}{3} W \left(\frac{r_1^3 - r_2^3}{r_1^2 - r_2^2}\right)\right] \qquad \text{...(6.19A)}$$

where T^* = Total frictional torque on the clutch plate.

(ii) For Uniform Wear

$$p \times r = \text{constant} \qquad \text{(say } C\text{)}$$

$$\therefore \quad p = \frac{C}{r}$$

We know that axial load on ring [Refer to Fig. 6.16]

$$dW = p \times 2\pi r\, dr$$

$\therefore$ Total axial load is given by integrating the above equation

$$\therefore \qquad W = \int_{r_2}^{r_1} p \times 2\pi r\, dr$$

$$= \int_{r_2}^{r_1} \frac{C}{r} \times 2\pi r\, dr \qquad \left(\because\ p = \frac{C}{r}\right)$$

$$= 2\pi C\, [r]_{r_2}^{r_1} = 2\pi C\, [r_1 - r_2] \qquad \text{...(6.20)}$$

$$\therefore \qquad C = \frac{W}{2\pi (r_1 - r_2)}$$

The total frictional torque on the friction surface is obtained by integrating equation (6.15) from r_2 to r_1.

$$\therefore \qquad T = \int_{r_2}^{r_1} dT = \int_{r_2}^{r_1} \mu \times p \times 2\pi r^2\, dr$$

$$= \int_{r_2}^{r_1} \mu \times \frac{C}{r} \times 2\pi r^2\, dr = \mu \times C \times 2\pi \int_{r_2}^{r_1} r\, dr$$

$$= \mu \times C \times 2\pi \left[\frac{r^2}{2}\right]_{r_2}^{r_1} = \mu \times C \times 2\pi \times \left[\frac{r_1^2 - r_2^2}{2}\right]$$

$$= \mu \times C \times \pi\, [r_1^2 - r_2^2] = \mu \times \frac{W}{2\pi (r_1 - r_2)} \times \pi (r_1^2 - r_2^2)$$

$$= \frac{\mu \times W}{2} \times (r_1 + r_2) \qquad \text{...(6.21)}$$

$$= \mu \times W \times R_m \qquad \text{...(6.22)}$$

where R_m = Mean radius $= \dfrac{r_1 + r_2}{2}$ ...(6.23)

$\therefore$ Total torque on a single clutch plate, is given by

$$T^* = 2T$$

$$= 2 \times \left[\frac{\mu W}{2} (r_1 + r_2)\right] \qquad \text{...(6.24)}$$

N.B. (*i*) For power transmission by friction through a clutch, uniform wear theory gives safer result. Hence uniform wear should be assumed in case of friction clutch, unless it is specified otherwise.

Problem 6.7. *Calculate the power transmitted by a single plate clutch at a speed of 2000 r.p.m., if the outer and inner radii of friction surfaces are 150 mm and 100 mm respectively. The maximum intensity of pressure at any point of contact surface should not exceed 0.8 × 10⁵ N/m². Take both sides of the plate as effective and co-efficient of friction = 0.3. Assume uniform wear.*

Sol. Given :

Speed, N = 2000 r.p.m.

Outer radius of friction surface, r_1 = 150 mm = 0.15 m

Inner radius, r_2 = 100 mm = 0.1 m

Maximum pressure, $p_{max} = 0.8 \times 10^5$ N/m²

Co-efficient of friction, $\mu = 0.3$

No. of effective sides = 2

For uniform wear, we have

$$p \times r = \text{constant (say} = C)$$

or $$p_1 \times r_1 = p_2 r_2 = C$$

As for uniform wear, the product of pressure and radius is constant, hence pressure will be more where radius is less. Therefore at inner radius, the pressure will be more.

$\therefore$ $$p_{max} \times r_2 = C \qquad (\because \; r_2 \text{ is inner radius})$$

or $$(0.8 \times 10^5) \times 0.1 = C \qquad (\because \; p_{max} = 0.8 \times 10^5 \text{ and } r_2 = 0.1 \text{ m})$$

or $$C = 0.8 \times 10^4$$

Using equation (6.20) for uniform wear,

$$W = 2\pi C\,(r_1 - r_2)$$
$$= 2\pi \times 0.8 \times 10^4 \times (0.15 - 0.10) \text{ N} = 2513.27 \text{ N}$$

The torque due to both active surfaces is given by equation (6.24) as

$$T^* = 2 \times \left[\frac{\mu W}{2}(r_1 + r_2)\right]$$
$$= 2 \times \left[\frac{0.3 \times 2513.27}{2}(0.15 + 0.10)\right] \text{Nm} = 188.49 \text{ Nm}$$

$\therefore$ Power transmitted by the clutch is given by

$$P = \frac{2\pi NT}{60}$$
$$= \frac{2\pi \times 2000 \times 188.49}{60} = 39477.25 \text{ W} = \mathbf{39.477 \text{ kW}}. \quad \textbf{Ans.}$$

Problem 6.8. *Determine the external and internal radii of the friction plate of a single clutch if maximum torque transmitted is 90 Nm. The external radius of the friction plate is 1.5 times the internal radius and the maximum intensity of pressure at any point of contact surface should not exceed 0.8 × 10⁵ N/m². Take both sides of the plate as effective and co-efficient of friction = 0.3. Assume uniform wear. Also calculate the axial force exerted by the springs.*

Sol. Given :

Torque, $T = 90$ Nm ; external radius = 1.5 × internal radius *i.e.*, $r_1 = 1.5\, r_2$;

$p_{max} = 0.8 \times 10^5$ N/m² ; $\mu = 0.3$,

No. of effective sides = 2.

For uniform wear, $p \times r = \text{constant}$ (say = C)

or $$p_1 \times r_1 = p_2 r_2 = C$$

The pressure will be maximum at the inner radius.

$\therefore$ $$p_{max} \times r_2 = C$$

or $$0.8 \times 10^5 \times r_2 = C \qquad \text{...}(i)$$

Now using equation (6.20) for uniform wear,

$$W = 2\pi C\,(r_1 - r_2)$$
$$= 2\pi \times 0.8 \times 10^5 \times r_2\,(1.5\, r_2 - r_2)$$
$$[\because \text{ From } (i),\; C = 0.8 \times 10^5 \times r_2 \text{ and } r_1 = 1.5\, r_2]$$
$$= 2\pi \times 0.8 \times 10^5 \times r_2^2 \times (1.5 - 1) = 2\pi \times 0.8 \times 10^5 \times r_2^2 \times 0.5$$
$$= 251327.4\, r_2^2 \text{ N} \qquad \text{...}(ii)$$

The frictional torque due to both active surfaces is given by equation (6.24) as

$$T^* = 2 \times \left[\frac{\mu W}{2}(r_1 + r_2)\right]$$

$$= 2 \times \left[\frac{0.3 \times 251327.4\, r_2^2}{2}(1.5 r_2 + r_2)\right] \quad (\because \quad W = 251327.4\, r_2^2)$$

$$= 0.3 \times 251327.4\, r_2^3\,(1.5 + 1)$$

$$= 0.3 \times 251327.4 \times 2.5 \times r_2^3 \text{ Nm} \quad \text{...(iii)}$$

But maximum torque transmitted = 90 Nm ...(iv)

Hence equating the two values of the torque given by equations (*iii*) and (*iv*),

$$0.3 \times 251327.4 \times 2.5 \times r_2^3 = 90$$

or $$r_2 = \left(\frac{90}{0.3 \times 251327.4 \times 2.5}\right)^{1/3}$$

$$= (4.77465 \times 10^{-4})^{1/3} = 0.07818 \text{ m} \simeq \mathbf{78.2\ mm.}\quad \textbf{Ans.}$$

and $$r_1 = 1.5\, r_2 = 1.5 \times 78.2 = \mathbf{117.3\ mm.}\quad \textbf{Ans.}$$

Substituting the value of r_2 in equation (*ii*), we get

$$W = 251327.4 \times 0.07818^2 = \mathbf{1536.14\ N.}\quad \textbf{Ans.}$$

Problem 6.9. *The external radius of a friction plate of a single clutch having both sides as effective, is 150 mm. The power transmitted is 20 kW at a speed of 1000 r.p.m. The maximum intensity of pressure at any point of contact surface is 0.8×10^5 N/m². If the co-efficient of friction is 0.30, then find : (i) The internal radius of the friction plate, and (ii) Axial thrust with which the friction surfaces are held together.*

Sol. Given :

External radius, $r_1 = 150 \text{ mm} = 0.15 \text{ m}$

Power transmitted, $P = 20 \text{ kW} = 20 \times 10^3 \text{ W}$

Speed, $N = 1000$ r.p.m.

Max. pressure, $p_{max} = 0.8 \times 10^5 \text{ N/m}^2$

Friction co-efficient, $\mu = 0.3$

Find : (*i*) Internal radius *i.e.*, r_2

(*ii*) Axial thrust *i.e.*, W

(i) Internal radius (r_2)

In this question, nothing is mentioned about which theory (*i.e.*, uniform pressure or uniform wear) should be used. But as power is to be transmitted by friction through a clutch, it is safer to use uniform wear theory.

For uniform theory, $p \times r = C$ (constant)

Hence pressure will be maximum where radius is minimum (*i.e.*, at internal radius)

$\therefore$ $p_{max} \times \text{Internal radius} = C$

or $(0.8 \times 10^5) \times r_2 = C$ ...(*i*)

Now using the expression for power, we have

$$P = \frac{2\pi NT}{60}$$

or $$20 \times 10^3 = \frac{2\pi \times 1000 \times T}{60}$$

or $$T = \frac{20 \times 10^3 \times 60}{2\pi \times 1000} = 190.986 \text{ Nm} \quad \text{...}(ii)$$

Now using equation (6.20) for uniform wear,

$$W = 2\pi \times C \times (r_1 - r_2)$$
$$= 2\pi \times 0.8 \times 10^5 \times r_2 \times (0.15 - r_2)$$

[$\because$ From (*i*), $C = 0.8 \times 10^5\ r_2$ and $r_1 = 0.15$]

$$= 502654.8\ r_2\ (0.15 - r_2) \quad \text{...}(iii)$$

The frictional torque due to both sides active surfaces is given by equation (6.24) as,

$$\text{Friction torque} = 2 \times \left[\frac{\mu W}{2}(r_1 + r_2)\right]$$
$$= 2 \times \left[\frac{0.3 \times 502654.8\ r_2\ (0.15 - r_2)}{2}(0.15 + r_2)\right]$$

[$\because$ From (*iii*), $W = 502654.8\ r_2\ (0.15 - r_2)$ and $r_1 = 0.15$]

$$= 150796.44\ r_2\ (0.15^2 - {r_2}^2) \quad \text{...}(iv)$$

Equating the two values of the torque given by equations (*ii*) and (*iv*)

$$190.986 = 150796.44\ r_2\ (0.15^2 - {r_2}^2)$$

or $$\frac{190.986}{150796.44} = 0.0225\ r_2 - {r_2}^3$$

or $$0.0012665 = 0.0225\ r_2 - {r_2}^3$$

or $${r_2}^3 - 0.0225\ r_2 + 0.0012665 = 0 \quad \text{...}(v)$$

The above equation is a cubic equation. It will be solved by trial and error method. Consider equation (*v*) :

(*i*) Let $r_2 = 0.1$ m, then L.H.S. $= 0.1^3 - 0.0225 \times 0.1 + 0.0012665$
$= 0.001 - 0.00225 + .0012665 = 0.0000165$

(*ii*) Let $r_2 = 0.12$ m, then L.H.S. $= 0.12^3 - .0225 \times 0.12 + .0012665$
$= 0.001728 - 0.0027 + .0012665 = 0.0002945$

Hence if value of r_2 increases from 0.1 to 0.12, the value of L.H.S. equation (*v*) also increases. But for correct value of r_2, the L.H.S. should be zero. Hence the value of r_2 should be decreased from the value of 0.1.

(*iii*) Let $r_2 = 0.095$ m, then L.H.S. $= 0.095^3 - 0.0225 \times 0.095 + 0.0012665$
$= 0.0008573 - 0.0021375 + 0.0012665$
$= 0.0021238 - 0.0021375 = -0.0000137$

Now when $r_2 = 0.1$, L.H.S. is + ve. But when $r_2 = 0.095$, L.H.S. of equation (*v*) becomes – ve. Hence value of r_2 lies between 0.095 and 0.1.

(*iv*) Let $r_2 = 0.097$ m, then L.H.S. $= 0.097^3 - 0.0225 \times 0.097 + .0012665$
$= 0.00091267 - 0.0021825 + .0012665 = -0.00000333$

The value of r_2 will be slightly more than 0.097.

Let us take this value of $r_2 = 0.0975$ m

$\therefore$ Internal radius, $r_2 = 0.0975$ m = **97.5 mm. Ans.**

(ii) Axial Thrust (W)

Substituting this value of $r_2 = 0.0975$ m in equation (*iii*),

$$W = 502654.8 \times 0.0975\,(0.15 - 0.0975) = \mathbf{2572.96\ N.}\quad \textbf{Ans.}$$

Problem 6.10. *The external and internal radii of a friction plate of a single clutch are 120 mm and 60 mm respectively. The total axial thrust with which the friction surfaces are held together is 1500 N. For uniform wear, find the maximum, minimum and average pressure on the contact surfaces.*

Sol. Given :

External radius, $r_1 = 120$ mm $= 0.12$ m

Internal radius, $r_2 = 60$ mm $= 0.06$ m

Total axial thrust, $W = 1500$ N

For uniform wear, we have $p \times r = C$ (constant)

Pressure will be maximum where radius is minimum and pressure will be minimum where radius is maximum. Here pressure will be maximum at internal radius (*i.e.*, at r_2) and will be minimum at external radius (*i.e.*, at r_1).

Let p_{max} = Maximum pressure

p_{min} = Minimum pressure

(i) Maximum pressure

Pressure is maximum at internal radius.

$$\therefore \quad p_{max} \times r_2 = C$$

or $$p_{max} \times 0.06 = C$$

or $$C = 0.06\, p_{max} \qquad \ldots(i)$$

Now using equation (6.20) for total axial thrust (or load),

$$W = 2\pi \times C \times (r_1 - r_2)$$

or $$1500 = 2\pi \times 0.06\, p_{max} \times (0.12 - 0.06)$$

$$= 2\pi \times 0.06\, p_{max} \times 0.06 \qquad [\because\ C = 0.06\, p_{max}]$$

$$\therefore \quad p_{max} = \frac{1500}{2\pi \times 0.06 \times 0.06} = \mathbf{66314.56\ N/m^2.}\quad \textbf{Ans.}$$

$$\simeq \mathbf{0.663 \times 10^5\ N/m^2.}\quad \textbf{Ans.}$$

(ii) Minimum pressure

Pressure will be minimum at external radius.

$$\therefore \quad p_{min} \times r_1 = C$$

or $$p_{min} \times 0.12 = C$$

or $$C = 0.12 \times p_{min} \qquad \ldots(ii)$$

Using equation (6.20) for axial load, we get

$$W = 2\pi \times C \times (r_1 - r_2)$$

or $$1500 = 2\pi \times 0.12\, p_{min} \times (0.12 - 0.06) \qquad (\because\ C = 0.12\, p_{min})$$

$$\therefore \quad p_{min} = \frac{1500}{2\pi \times 0.12 \times 0.06}$$

$$= 33157.28\ \text{N/m}^2 \simeq \mathbf{0.3315 \times 10^5\ N/m^2.}\quad \textbf{Ans.}$$

(*iii*) *Average pressure* (p_{av})

Average pressure (p_{av}) is given by

$$p_{av} = \frac{\text{Total axial load or thrust}}{\text{Area of cross–section of contact surface}}$$

$$= \frac{W}{\pi(r_1^2 - r_2^2)} = \frac{1500}{\pi(0.12^2 - 0.06^2)} = \frac{1500}{\pi \times [0.0144 - 0.0036]}$$

$$= 44209.7 \text{ N/m}^2 \simeq \mathbf{0.442 \times 10^5 \text{ N/m}^2}. \quad \textbf{Ans.}$$

Problem 6.11. *The external and internal radii of a friction clutch of disc type are 90 mm and 50 mm respectively. Both sides of the friction clutch are effective and co-efficient of friction is equal to 0.25. The friction clutch is used to rotate a machine from a shaft which is rotating at a constant speed of 240 r.p.m. The moment of inertia of the rotating parts of the machine is 5.5 kg m². The intensity of pressure is not to exceed 0.8 × 10⁵ N/m². Assuming uniform wear, determine the time required for the machine to attain the full speed when the clutch is suddenly applied. Also determine the energy lost in slipping of the clutch.*

Sol. Given :

External radius, $r_1 = 90 \text{ mm} = 0.09 \text{ m}$

Internal radius, $r_2 = 50 \text{ mm} = 0.05 \text{ m}$

No. of effective sides = 2

Co-efficient of friction, $\mu = 0.25$

Constant speed of driving shaft = 240 r.p.m.

M.O.I. of machine parts, $I = 5.5 \text{ kg m}^2$

Max. pressure, $p_{max} = 0.8 \times 10^5 \text{ N/m}^2$

Theory assumed = Uniform wear

Find : (*i*) Time required for the machine to attain full speed of 240 r.p.m.

(*ii*) Energy lost during clutch slip.

(*i*) *Time required for the machine to attain full speed of 240 r.p.m.*

The driving shaft is rotating at a constant speed, whereas the machine is at rest. But when the clutch is engaged, the machine will attain its full speed not immediately but after some time. Let this time is '*t*' seconds.

Let us first find the axial thrust (or load) and total frictional torque for uniform wear.

For uniform wear, we have

$$p \times r = C \text{ (constant)}$$

Pressure will be maximum at the inner radius,

$$\therefore \quad p_{max} \times r_2 = C$$

or $$0.8 \times 10^5 \times 0.05 = C$$

or $$C = 4000 \qquad \text{...}(i)$$

Now using equation (6.20) for axial load, we have

$$W = 2\pi C\,(r_1 - r_2)$$

$$= 2\pi \times 4000 \times (0.09 - 0.05) \qquad (\because \; C = 4000)$$

$$= 1005.31 \text{ N} \qquad \text{...}(ii)$$

The frictional torque due to both active surfaces is given by equation (6.24) as

$$T^* = 2 \times \left[\frac{\mu W}{2}(r_1 + r_2)\right]$$

$$= 2 \times \left[\frac{0.25 \times 1005.31 \times (0.09 + 0.05)}{2}\right] \quad [\because \text{ From } (ii), W = 1005.31]$$

$$= 35.186 \text{ Nm.}$$

Now let us find the angular acceleration, when total torque is 35.186 Nm.

But Torque = M.O.I. × α where α is angular acceleration.

or $T = I \times \alpha$

$\therefore$ $35.186 = 5.5 \times \alpha$ $\quad (\because \; T = T^* = 35.186, \text{ and } I = 5.5 \text{ kg m}^2)$

$\therefore$ $\alpha = \dfrac{35.186}{5.5} = 6.397 \text{ rad/s}^2$

The machine starts from rest. After some time the final angular speed of the machine will be corresponding to speed of shaft (*i.e.*, corresponding to 240 r.p.m.).

$\therefore$ Final angular speed of machine, $\omega = \dfrac{2\pi N}{60} = \dfrac{2\pi \times 240}{60} = 8\pi$ rad/s

Initial angular speed of machine = $\omega_0 = 0$

Angular acceleration, $\alpha = 6.397 \text{ rad/s}^2$

Let t = time required for the machine to attain the full speed.

Using $\omega = \omega_0 + \alpha t$, we get $8\pi = 0 + 6.397 \times t$

$\therefore$ $t = \dfrac{8\pi}{6.397}$ = **3.928 s. Ans.**

Hence the full speed will be attained by the machine in 3.928 seconds.

(ii) Energy lost in slipping of the clutch

The driving shaft is rotating at a uniform speed of 240 r.p.m. *i.e.*, at a uniform angular speed of $\dfrac{2\pi \times 240}{60} = 8\pi$ rad/s, whereas the machine starts from rest and attains the full speed *i.e.*, 8π rad/s in time 3.928 seconds. Let us find the angles turned by driving shaft and the machine in time 3.928 seconds.

Angle turned by the driving shaft,

$\theta_1 = \omega \times t$ $\quad (\because$ shaft is rotating with a uniform angular velocity)

$= 8\pi \times 3.928 \text{ rad} = 98.72 \text{ rad.}$

The angle turned by the machine (or driven shaft) will be obtained by using

$$\theta = \omega_0 t + \frac{1}{2}\alpha t^2 \qquad \left(\text{corresponding to } s = ut + \frac{1}{2}at^2\right)$$

Let θ_2 = Angle turned by the driven shaft

$\therefore$ $\theta_2 = 0 \times 3.928 + \dfrac{1}{2} \times 6.397 \times (3.928)^2$

$(\because \; \omega_0 = 0, \alpha = 3.928, t = 3.928)$

$= 0 + 49.35 = 49.35$ rad.

Energy lost in friction due to clutch slip

= Frictional torque × Angle of slip

$= 35.186\,(\theta_1 - \theta_2) = 35.186\,(98.72 - 49.35) =$ **1737.13 Nm. Ans.**

Problem 6.12. *For the problem 6.11, determine the ratio of power transmitted with uniform wear to that with uniform pressure. Also find the intensity of pressure if the condition of uniform pressure is considered and the same axial load is developed as in case of uniform wear.*

Sol. Given :

From problem 6.11, we have $r_1 = 0.09$ m ; $r_2 = 0.05$ m ; $\mu = 0.25$, $N = 240$ r.p.m. ;

No. of effective surfaces = 2

Axial load for uniform wear = 1005.31 N (Calculated)

Also torque for uniform wear = 35.186 Nm (Calculated)

∴ Power transmitted with uniform wear,

$$P_1 = \frac{2\pi NT}{60}$$

$$= \frac{2\pi \times 240 \times 35.186}{60} = 884.32 \text{ W}$$

Now let us find the power transmitted assuming uniform pressure. For this, first find the axial load and then torque for uniform pressure. But axial load is same as in case of uniform wear.

∴ Axial load, $W = 1005.31$ N

For uniform pressure, W is also equal to $p \times \pi[r_1^2 \; r_2^2]$

$$\therefore \quad p = \frac{W}{\pi(r_1^2 - r_2^2)}$$

$$= \frac{1005.31}{\pi(0.09^2 - 0.05^2)} = \frac{1005.31}{\pi(0.0081 - 0.0025)} = \mathbf{57142.87 \ N/m^2}. \quad \textbf{Ans.}$$

Torque transmitted for uniform pressure is given by equation (6.19A) as

$$T^* = 2 \times \left[\frac{2}{3}\mu W\left(\frac{r_1^3 - r_2^3}{r_1^2 - r_2^2}\right)\right]$$

$$= 2 \times \left[\frac{2}{3} \times 0.25 \times 1005.31 \times \left(\frac{0.09^3 - 0.05^3}{0.09^2 - 0.05^2}\right)\right]$$

$$= 335.10 \times \left(\frac{0.000729 - 0.000125}{0.0081 - 0.0025}\right) = 335.10 \times \frac{0.000604}{0.0056}$$

$$= 36.143 \text{ Nm}$$

∴ Power transmitted with uniform pressure,

$$P_2 = \frac{2\pi NT^*}{60}$$

$$= \frac{2\pi \times 240 \times 36.143}{60} = 908.37 \text{ W}$$

$$\therefore \quad \frac{\text{Power transmitted by uniform wear}}{\text{Power transmitted by uniform pressure}} = \frac{P_1}{P_2} = \frac{884.32}{908.37} = \mathbf{0.9735}. \quad \textbf{Ans.}$$

Problem 6.13. *The capacity of a single plate clutch having both sides effective decreases by 16% during the initial wear period. Determine the minimum value of the ratio of internal radius to external radius* $\left(i.e., \text{ratio } \frac{r_2}{r_1}\right)$ *for the axial load.*

Sol. Given :

The new clutch has uniform pressure distribution. But after the initial wear, the clutch will have uniform wear.

Capacity of a clutch means the maximum torque (or maximum power) transmitted. During initial wear period, if capacity decreases by 16%, then it means that the torque transmitted (or power transmitted) will be 0.84 times the torque transmitted when the clutch is new. But when the clutch is new, torque is transmitted on the uniform pressure theory. After initial wear period, torque transmitted will be based on the uniform wear theory. Hence it is given that :

Torque based on uniform wear = 0.84 × Torque based on uniform pressure. ...(*i*)

Now torque based on uniform wear is given by,

$$T_{wear} = 2 \times \left[\frac{\mu W}{2}(r_1 + r_2)\right]$$

And torque based on uniform pressure is given by,

$$T_{pressure} = 2 \times \left[\frac{2}{3}\mu W\left(\frac{r_1^3 - r_2^3}{r_1^2 - r_2^2}\right)\right]$$

But, $$T_{wear} = 0.85 \times T_{pressure} \quad \text{(Given)}$$

$$\therefore \quad 2 \times \left[\frac{\mu W}{2}(r_1 + r_2)\right] = 0.84 \times 2 \times \left[\frac{2}{3}\mu W\left(\frac{r_1^3 - r_2^3}{r_1^2 - r_2^2}\right)\right]$$

or $$(r_1 + r_2) = 1.12\left(\frac{r_1^3 - r_2^3}{r_1^2 - r_2^2}\right)$$

or $$(r_1 + r_2)(r_1^2 - r_2^2) = 1.12(r_1^3 - r_2^3)$$

or $$(r_1 + r_2)(r_1 + r_2)(r_1 - r_2) = 1.12(r_1 - r_2)(r_1^2 + r_1 r_2 + r_2^2)$$

$$[\because \; r_1^3 - r_2^3 = (r_1 - r_2)(r_1^2 + r_1 r_2 + r_2^2)]$$

or $$(r_1 + r_2)(r_1 + r_2) = 1.12(r_1^2 + r_1 r_2 + r_2^2)$$

or $$r_1^2 + r_2^2 + 2r_1 r_2 = 1.12\, r_1^2 + 1.12\, r_1 r_2 + 1.12\, r_2^2$$

or $$0.12\, r_1^2 + 0.12\, r_2^2 - 0.88\, r_1 r_2 = 0$$

Dividing by r_1^2, we get

$$0.12 + 0.12\frac{r_2^2}{r_1^2} - 0.88\frac{r_2}{r_1} = 0$$

Dividing by 0.12, we get

$$1 + \frac{r_2^2}{r_1^2} - \frac{0.88}{0.12}\frac{r_2}{r_1} = 0$$

or $$\frac{r_2^2}{r_1^2} - 7.333\frac{r_2}{r_1} + 1 = 0$$

Let $\frac{r_2}{r_1} = x$, then above equation will be as

$$x^2 - 7.333x + 1 = 0$$

$$\therefore \quad x = \frac{7.333 \pm \sqrt{7.333^2 - 4}}{2} = \frac{7.333 \pm 7.055}{2}$$

$$= \frac{7.333 - 7.055}{2} \quad \left(\text{As +ve value is not possible as } \frac{r_2}{r_1} \text{ cannot be more than 1}\right)$$

$$= 0.139$$

$$\therefore \quad \frac{r_2}{r_1} = \mathbf{0.139.} \quad \textbf{Ans.}$$

6.11.2. Multi-plate Clutch. Fig. 6.17 shows the diagram of multi-plate clutch with friction plates having friction linings on both sides except the first plate which is adjacent to the flywheel. This plate is having friction lining on one side. The friction plates are connected on the top to the flywheel. Hence the friction plates rotate with the flywheel and hence with the driving shaft. The friction plates are also free to move axially.

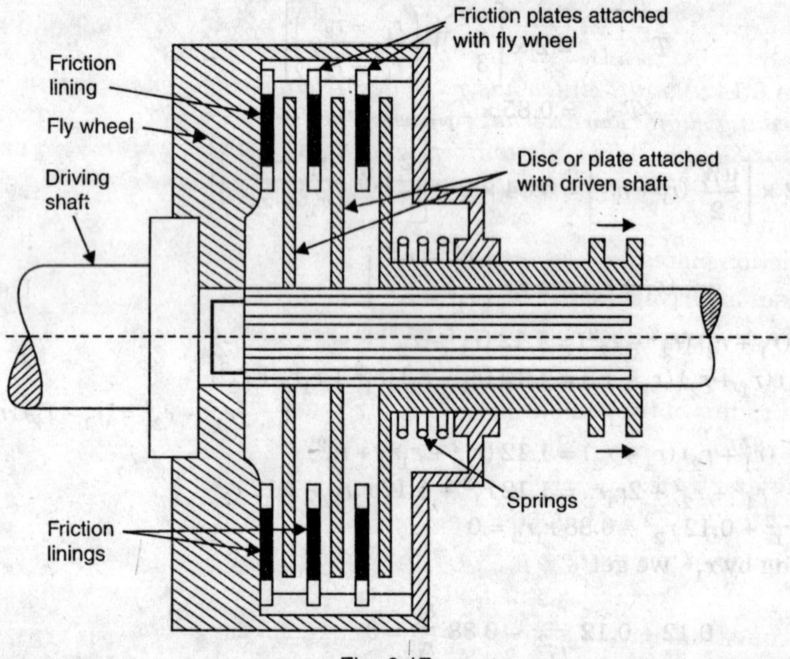

Fig. 6.17

The discs or plates are also supported on splines of the driven shaft. Hence these plates rotate with driven shaft. These plates are situated in between the friction plates and can also slide axially as shown in Fig. 6.17. Thus Fig. 6.17 shows the position of the friction plates and disc plates in disengaged position.

In the engaged position (which will be when the foot is taken off from the clutch pedal), the set of strong springs will press the discs into contact with the friction plates (or friction linings on the friction plates). Hence the power will be transmitted from the driving to the driven shaft.

Multi-plate clutch is used when a large torque is to be transmitted such as in case of motor cars and machine tools.

Theory of Multi-plate Clutch

Let r_1 = External radius of friction lining on friction plate,

r_2 = Internal radius of friction lining on friction plate,

W = Total axial load,

p = Intensity of pressure,

n_1 = Number of friction plates on driving shaft,

n_2 = Number of discs on the driven shaft

Then the number of active surfaces or friction surfaces (n) will be given as

$$n = n_1 + n_2 - 1 \quad ...(6.25)$$

Total torque transmitted is given by,

$$T = n \times \mu \times W \times R_m \quad ...(6.26)$$

where R_m = Mean radius of friction surfaces

$$= \left(\frac{r_1 + r_2}{2}\right) \quad \text{(For uniform wear)}$$

$$= \frac{2}{3}\left[\frac{r_1^3 - r_2^3}{r_1^2 - r_2^2}\right] \quad \text{(For uniform pressure)}$$

Problem 6.14. *A multi-clutch has six plates (friction rings) on the driving shaft and six plates on the driving shaft. The external radius of the friction surface is 115 mm whereas the internal radius is 80 mm. Assuming uniform wear and co-efficient of friction as 0.1, find the power transmitted at 2000 r.p.m. Axial intensity of pressure is not to exceed 0.16 N/mm².*

Sol. Given :

No. of friction plates, $n_1 = 6$

No. of discs on driven shaft, $n_2 = 6$

∴ No. of active surfaces (or friction surfaces) are given as,

$$n = n_1 + n_2 - 1 = 6 + 6 - 1 = 11$$

External radius of friction surface, r_1 = 115 mm = 0.115 m

Internal radius, r_2 = 80 mm = 0.8 m

Co-efficient of friction, μ = 0.1

Speed, N = 2000 r.p.m.

Max. intensity of pressure, p_{max} = 0.16 N/mm² = 0.16 × 10⁶ N/m²

Theory assumed = Uniform wear.

Total torque transmitted is given by equation (6.26) as

$$T = n \times \mu \times W \times R_m \quad ...(i)$$

where R_m = Mean radius of friction surface

$$= \frac{r_1 + r_2}{2} \quad \text{(For uniform wear)}$$

$$= \frac{0.115 + 0.08}{2} = 0.0975 \text{ m}$$

Let us now find the value of W.

For uniform wear, $p \times r$ = constant = C

For maximum pressure, radius is minimum. Hence pressure will be maximum at internal radius.

$$\therefore \quad p_{max} \times r_2 = C$$

or $$0.16 \times 10^6 \times 0.08 = C$$

or $$C = 128 \times 10^2$$

The expression for axial load (W) for uniform wear is given by equation (6.20) as

$$W = 2\pi C\,(r_1 - r_2)$$
$$= 2\pi \times 128 \times 10^2\,(0.115 - 0.08) = 2814.867 \text{ N}$$

Substituting the values of W, μ, n and R_m in equation (i),

$$T = 11 \times 0.1 \times 2814.867 \times 0.0975 = 301.894 \text{ Nm}$$

$\therefore$ Power transmitted is given by,

$$P = \frac{2\pi NT}{60}$$
$$= \frac{2\pi \times 2000 \times 301.894}{60} = 63228.5 \text{ W} = \mathbf{63.2285\ kW.}\quad \textbf{Ans.}$$

Problem 6.15. *A multi-plate clutch transmits 25 kW of power at 1600 r.p.m. It has three discs on the driving shaft and two on the driven shaft. Co-efficient of friction for the friction surfaces is 0.25. The external and internal radii of friction surfaces are 100 mm and 50 mm respectively. Find the maximum intensity of pressure between the discs. Assume uniform wear.*

Sol. Given :

$P = 25 \text{ kW} = 25 \times 10^3 \text{ W}$; $N = 1600$ r.p.m.

$\mu = 0.25$; $r_1 = 100 \text{ mm} = 0.1 \text{ m}$; $r_2 = 50 \text{ mm} = 0.05 \text{ m}$

No. of discs on driving shaft, $n_1 = 3$; No. of discs on driven shaft, $n_2 = 2$.

$\therefore$ No. of friction (or active) surfaces, $n = n_1 + n_2 - 1 = 3 + 2 - 1 = 4$

Theory assumed = Uniform wear.

Let p_{max} = Max. intensity of pressure.

We know that,
$$P = \frac{2\pi NT}{60}$$

or
$$25 \times 10^3 = \frac{2\pi \times 1600 \times T}{60}$$

or
$$T = \frac{25 \times 10^3 \times 60}{2\pi \times 1600} = 149.207 \text{ Nm}$$

Total torque transmitted is also given by equation (6.26) as

$$T = n \times \mu \times W \times R_m \qquad \ldots(i)$$

where R_m = Mean radius of friction surface

$$= \frac{r_1 + r_2}{2} \qquad \text{(For uniform wear)}$$
$$= \frac{0.1 + 0.05}{2} = 0.075 \text{ m}$$

$n = 4$ and $\mu = 0.25$

Substituting the values of T, n, μ and R_m in equation (i), we get

$$149.207 = 4 \times 0.25 \times W \times 0.075$$

$\therefore$
$$W = \frac{149.207}{4 \times .25 \times 0.075} = 1989.426 \text{ N}$$

The expression for axial load (W) for uniform wear is also given by equation (6.20) as

$$W = 2\pi C\,(r_1 - r_2)$$

Substituting the values of W, r_1 and r_2, we get

or $$C = \frac{1989.426}{2\pi \times 0.05} = 6332.54$$

For uniform wear, $p \times r = C$ (constant)

The pressure is maximum at internal radius

$\therefore$ $$p_{max} \times r_2 = C$$

or $$p_{max} \times 0.05 = 6332.54 \qquad [\because \ C = 6332.54 \text{ and } r_2 = 0.05]$$

or $$p_{max} = \frac{6332.54}{0.05} = 1266.50 \text{ N/m}^2 = \mathbf{0.12665 \text{ N/mm}^2}. \quad \textbf{Ans.}$$

Problem 6.16. *A power of 60 kW is transmitted by a multi-plate clutch at 1500 r.p.m. Axial intensity of pressure is not to exceed 0.15 N/mm². The co-efficient of friction for the friction surfaces is 0.15. The external radius of friction surface is 120 mm. Also the external radius is equal to 1.25 times the internal radius. Find the number of plates needed to transmit the required power. Assume uniform wear.*

Sol. Given :

$P = 60$ kW $= 60 \times 10^3$ W ; $N = 1500$ r.p.m. ; $p_{max} = 0.15$ N/mm² $= 0.15 \times 10^6$ N/m² ;

$\mu = 0.15$; $r_1 = 120$ mm $= 0.12$ m ; $r_1 = 1.25 \times r_2$

$\therefore$ $$r_2 = \frac{r_1}{1.25} = \frac{0.12}{1.25} = 0.096 \text{ m.}$$

Assume uniform wear. Find the number of plates required.

For uniform wear,

$$p \times r = \text{constant (say} = C)$$

$\therefore$ Pressure will be maximum, at the internal radius

$\therefore$ $$p_{max} \times r_2 = C$$

or $$(0.15 \times 10^6) \times 0.096 = C$$

$\therefore$ $$C = 0.15 \times 10^6 \times 0.096 = 14400$$

For uniform wear, the axial thrust or load (W) is given by equation (6.20) as

$$W = 2\pi C\,(r_1 - r_2)$$
$$= 2\pi \times 14400\,(0.12 - 0.096)$$
$$= 2171.47 \text{ N} \qquad ...(i)$$

Now let us find the total torque transmitted from the given power,

$$P = \frac{2\pi NT}{60}$$

or $$60 \times 10^3 = \frac{2\pi \times 1500 \times T}{60}$$

$\therefore$ $$T = \frac{60 \times 10^3 \times 60}{2\pi \times 1500} = 381.972 \text{ Nm} \qquad ...(ii)$$

But the total torque transmitted is also given by equation (6.26) as

$$T = n \times \mu \times W \times R_m \qquad ...(iii)$$

where n = no. of friction surfaces or active surfaces

R_m = Mean radius of friction surfaces

$$= \frac{r_1 + r_2}{2} \qquad \text{(For uniform wear)}$$

$$= \frac{0.12 + 0.096}{2} = 0.108 \text{ m}$$

and $W = 2171.47$ [From equation (*i*)]

Substituting the known values in equation (*iii*), we get

$$T = n \times 0.15 \times 2171.47 \times 0.108$$

$$= n \times 35.1778 \quad ...(iv)$$

Equating the two values of T given by equations (*ii*) and (*iv*),

$$381.972 = n \times 35.1778$$

$$\therefore \quad n = \frac{381.972}{35.1778} = 10.85 \text{ or } 11 \text{ surfaces}$$

∴ Number of friction surfaces required = **11 surfaces. Ans.**

But no. of friction surfaces,

$$n = n_1 + n_2 - 1 \quad \text{or} \quad 11 = n_1 + n_2 - 1$$

$$\therefore \quad n_1 + n_2 = 11 + 1 = 12$$

Hence there will be total 12 plates. The six plates (6) will be revolving with the driving shaft and other six with the driven shaft.

6.11.3. Cone Clutch. Fig. 6.18 shows the diagram of a cone clutch, in which the contact surfaces are in the form of cones. The driver cone is keyed to the driving shaft whereas the driven cone is keyed to the driven shaft. In the engaged position, the friction surfaces of the two cones are in complete contact due to spring pressure. In this position torque is transmitted from driving shaft to the driven shaft. For disengaging the clutch, the driven cone is pulled back through a lever system against the force of spring.

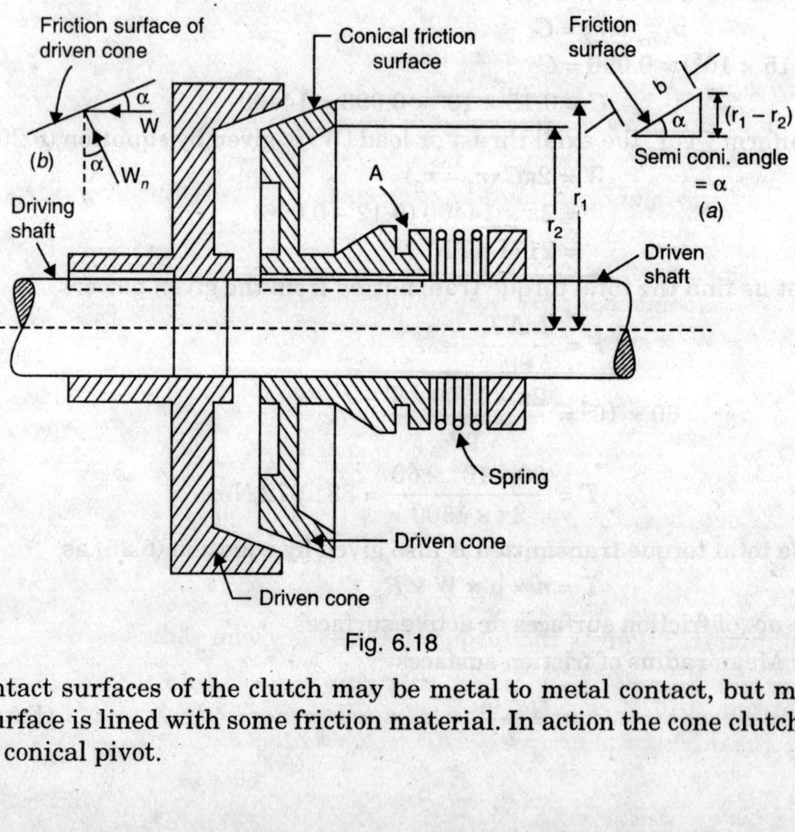

Fig. 6.18

The contact surfaces of the clutch may be metal to metal contact, but more often the driven cone surface is lined with some friction material. In action the cone clutch is similar to the truncated conical pivot.

Let r_1 = External radius of friction surface

r_2 = Internal radius of friction surface

α = Semi cone angle or the angle of the friction surface with the axis of the shaft

W = Total axial load required to engage the clutch supplied by spring

R_m = Mean radius of friction surface

μ = Co-efficient of friction

b = Width of contact surface or width of cone face

$$= \frac{(r_1 - r_2)}{\sin\alpha} \quad \text{[See Fig. 6.18 (a)]} \qquad \ldots(6.27)$$

(i) Case of Uniform Pressure

Similar to the truncated cone pivot, for uniform pressure we have the following equations :

$$p = \frac{W}{\pi(r_1^2 - r_2^2)}$$

or

$$W = p \times \pi (r_1^2 - r_2^2)$$

and

$$T = \frac{2}{3} \times \frac{\mu W}{\sin\alpha}\left(\frac{r_1^3 - r_2^3}{r_1^2 - r_2^2}\right) \qquad \ldots(6.28)$$

(ii) Case of Uniform Wear

$$p \times r = \text{constant (say } C)$$

$$W = 2\pi C (r_1 - r_2) \qquad \ldots(6.29)$$

and

$$p_{max} \times r_2 = C$$

Also

$$T = \frac{1}{2}\frac{\mu W}{\sin\alpha}(r_1 + r_2) \qquad \ldots(6.30)$$

(iii) Driving Torque based on Mean radius

Let p_m = Intensity of pressure at mean radius normal to friction surface

W_n = Total load normal to friction surface

= (pressure normal to friction surface) × Area of friction surface based on mean radius

$$= p_n \times (2\pi R_m \times b) \qquad \ldots(6.31)$$

W = Component of W_n in axial direction

$$= W_n \times \sin\alpha \quad \text{[See Fig. 6.18 (b)]} \qquad \ldots(6.32)$$

$$T = \frac{1}{2}\frac{\mu W}{\sin\alpha}(r_1 + r_2)$$

$$= \mu \times \frac{W}{\sin\alpha} \times \left(\frac{r_1 + r_2}{2}\right)$$

$$= \mu \times W_n \times R_m \qquad \ldots(6.33)$$

$$\left(\because \frac{W}{\sin\alpha} = W_n; \frac{r_1 + r_2}{2} = R_m\right)$$

Equation (6.33) gives the torque in terms of mean radius and load normal to friction surface.

Problem 6.17. *A cone clutch of cone angle 30° is used to transmit a power of 10 kW at 800 r.p.m. The intensity of pressure between the contact surfaces is not to exceed 85 kN/m². The* *width of the conical friction surface is half of the mean radius. If co-efficient of friction = 0.15, then find the dimensions of the contact surfaces. Assume uniform wear. Also find the axial load or force required to hold the clutch while transmitting the power. What is the width of the friction surface ?*

Sol. Given :

Cone angle = 30° ∴ Semi-cone angle, $\alpha = 15°$

Power, $P = 10$ kW $= 10 \times 10^3$ W ; $N = 800$ r.p.m.

Max. pressure, $p_{max} = 85$ kN/m² $= 85 \times 10^3$ N/m²

Width, $b = \frac{1}{2} \times$ Mean radius $= \frac{1}{2} \times R_m = \frac{1}{2}\left(\frac{r_1 + r_2}{2}\right) \left(\because \text{ For uniform wear}, R_m = \frac{r_1 + r_2}{2}\right)$

Co-efficient of friction, $\mu = 0.15$

Find : (*i*) dimensions of contact surfaces *i.e.*, r_1 and r_2.

(*ii*) Axial force of load required to keep the clutch engaging.

Assumed uniform wear.

We know that,
$$P = \frac{2\pi NT}{60}$$

or
$$10 \times 10^3 = \frac{2\pi \times 800 \times T}{60}$$

∴
$$T = \frac{60 \times 10 \times 10^3}{2\pi \times 800} = 119.366 \text{ Nm} \quad \text{...}(i)$$

Now width 'b' is given as
$$b = \frac{1}{2}\left(\frac{r_1 + r_2}{2}\right) = \frac{r_1 + r_2}{4} \quad \text{...}(ii)$$

Also the value of 'b' from equation (6.27) is given as
$$b = \frac{r_1 - r_2}{\sin \alpha} = \frac{r_1 - r_2}{\sin 15°} = \frac{r_1 - r_2}{0.2588} \quad \text{...}(iii)$$

Hence equating the two values of 'b' given by equations (*ii*) and (*iii*),
$$\frac{r_1 + r_2}{4} = \frac{r_1 - r_2}{0.2588}$$

or
$$r_1 + r_2 = \frac{4}{0.2588}(r_1 - r_2)$$
$$= 15.456\,(r_1 - r_2) = 15.456\, r_1 - 15.456\, r_2$$

or
$$16.456\, r_2 = 14.456\, r_1$$

or
$$r_1 = \frac{16.456}{14.456}\, r_2$$

or
$$r_1 = 1.138 r_2 \quad \text{...}(iv)$$

Now let us find the value of W (axial load) for uniform wear.

For uniform wear, $p \times r = C$ (constant)

The pressure will be maximum at internal radius

∴
$$p_{max} \times r_2 = C$$

or
$$85 \times 10^3 \times r_2 = C \quad \text{...}(v)$$

The value of W for uniform wear is given by equation (6.29) as

$$W = 2\pi C\,(r_1 - r_2)$$

$$= 2\pi \times 85 \times 10^3\, r_2\,(r_1 - r_2) \quad [\because \text{ From } (v),\ C = 85 \times 10^3 \times r_2]$$

$$= 534070\, r_2\,(r_1 - r_2) \qquad ...(vi)$$

The frictional torque for uniform wear is given by equation (6.30), as

$$T = \frac{1}{2}\frac{\mu W}{\sin\alpha}(r_1 + r_2)$$

$$= \frac{1}{2} \times \frac{0.15 \times 534070 r_2 (r_1 - r_2)}{\sin 15^\circ}(r_1 + r_2)$$

$$[\because \text{ From } (vi),\ W = 534070\, r_2\,(r_1 - r_2)]$$

$$= 154762 r_2\,(r_1 - r_2)\,(r_1 + r_2) \qquad ...(vii)$$

Substituting the value of T from equation (*i*) and value of r_1 from equation (*iv*) in the above equation, we get

$$119.366 = 154762 r_2\,(r_1^2 - r_2^2)$$

$$= 154762 r_2\,[(1.138 r_2)^2 - r_2^2] \qquad (\because\ r_1 = 1.138 r_2)$$

$$= 154762 r_2\,(1.295044 r^2 - r_2^2) = 45661 r_2^3$$

$$\therefore \quad r_2 = \left(\frac{119.366}{45661}\right)^{1/3} = 0.138 \text{ m} = \textbf{138 mm.} \quad \textbf{Ans.}$$

$$\therefore \quad r_1 = 1.138 r_2$$

$$= 1.138 \times 0.138 = 0.157 \text{ m} = \textbf{157 mm.} \quad \textbf{Ans.}$$

Substituting the values of r_1 and r_2 in equation (*vi*), we get

$$W = 534070 r_2\,(r_1 - r_2)$$

$$= 534070 \times 0.138\,(0.157 - 0.138) = \textbf{1400.3 N.} \quad \textbf{Ans.}$$

Width of the friction surface is given by equation (6.27) as

$$b = \frac{r_1 - r_2}{\sin\alpha} = \frac{157 - 138}{\sin 15^\circ} = \frac{19}{0.2588} = \textbf{73.4 mm.} \quad \textbf{Ans.}$$

Problem 6.18. *A cone clutch of semi-cone angle 15° is used to transmit a power of 30 kW at 800 r.p.m. The mean frictional surface radius is 150 mm. The normal intensity of pressure at the mean radius is not to exceed 0.15 N/mm². The co-efficient of friction is 0.2. Assuming uniform wear, determine :*

(*i*) *Width of the contact surface and*

(*ii*) *Axial load (or force) needed to engage the clutch.*

Sol. Given :

$\alpha = 15^\circ$; $P = 30$ kW $= 30 \times 10^3$ W ; $N = 800$ r.p.m. ;

$R_m = 150$ mm $= 0.15$ m ; $p_n = 0.15$ N/mm^2 $= 0.15 \times 10^6$ N/m^2 ; $\mu = 0.2$

Using,
$$P = \frac{2\pi N T}{60}$$

or
$$30 \times 10^3 = \frac{2\pi \times 800 \times T}{60}$$

$$\therefore \quad T = \frac{30 \times 10^3 \times 60}{2\pi \times 800} = 358.1 \text{ Nm}$$

Now using equation (6.33), we get

$$T = \mu \times W_n \times R_m$$

or $$358.1 = 0.2 \times W_n \times 0.15$$

$$\therefore \quad W_n = \frac{358.1}{0.2 \times 0.15} = 11936.67 \text{ N}$$

But W_n = Total load normal to friction surface of the cone

or $W_n = p_n \times (2\pi R_m \times b)$ [See equation (6.31)]

Substituting the known values in the above equation, we get

$$11936.67 = 0.15 \times 10^6 \times (2\pi \times 0.15 \times b)$$

$$\therefore \quad b = \frac{11936.67}{0.15 \times 10^6 \times 2\pi \times 0.15} = \mathbf{0.084 \text{ m or } 84 \text{ mm.}} \quad \textbf{Ans.}$$

Using equation (6.32) for axial load, we get

$$W = W_n \times \sin \alpha$$

$$= 11936.67 \times \sin 15° = \mathbf{3089.4 \text{ N.}} \quad \textbf{Ans.}$$

6.12. GREASY FRICTION

It is already mentioned in Art. 6.3.2, that when two surfaces in contact have a very thin layer of lubricant between them, the friction between these surfaces is known as greasy friction. The lubricant is absorbed on the surfaces and forms a very thin film between the surfaces. This thin film is of molecular thickness and is known as adsorbed film. The co-efficient of friction between these surfaces is considerably reduced. The laws for greasy friction are same as for dry or solid friction.

The following important cases of greasy friction will be studied :

(*i*) Greasy friction of a journal

(*ii*) Greasy friction of a slider-crank mechanism

(*iii*) Greasy friction of four-bar mechanism

6.12.1. Greasy Friction of a Journal. Fig. 6.19 (*a*) shows the diagram of a shaft inside a bearing. When the shaft is at rest in the bearing, the weight of the shaft *W*, passes through centre of gravity at *O*. A contact of the shaft and bearing is maintained at point *A* as shown in Fig. 6.19 (*a*). The contact point *A* is known as the seat of pressure for the bearing. The reaction of the bearing acts at *A* and is in line with *W* in the vertically upward direction.

When the shaft starts rotating (here rotation of the shaft is assumed in anticlockwise direction) because of the clearance the seat of pressure (*i.e.* contact point between the shaft and bearing) will roll or climb up the bearing in a direction opposite to that of rotation at point *B* as shown in Fig. 6.19 (*b*). Metal to metal contact exists at point *B* and greasy friction condition is applicable as oil film is having very thin layer of lubricant. The climbing or rolling up will stop when the following three forces are in equilibrium :

(*i*) Weight of shaft *W*, acting vertically downwards.

(*ii*) Normal reaction R_N at *B*, which is radial and passes through the point *O* as shown in Fig. 6.19 (*b*).

(*iii*) Frictional force, tangential to the shaft at *B* and acting in the opposite direction of motion of shaft. This force is equal to $\mu \times R_n$.

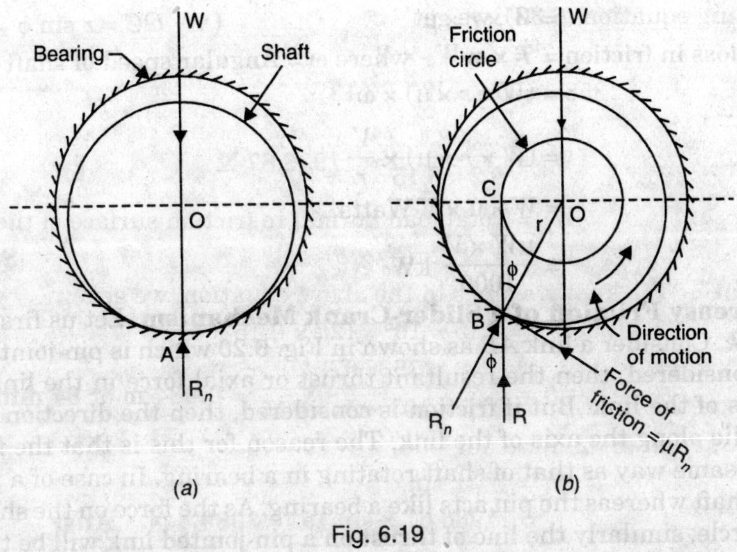

Fig. 6.19

The frictional force ($\mu \times R_N$) and normal reaction (R_N) can be combined into a single resultant force R which is inclined at an angle ϕ (*i.e.* angle of friction) with normal reaction R_N. Hence the shaft is in equilibrium now under the following two forces :

(*a*) Weight of shaft W, acting vertically downwards, and

(*b*) Single resultant reaction R.

For equilibrium, R must be equal to W and must act vertically upwards. Now these two forces (R and W) are equal and parallel. Hence they form a couple. This couple is known as friction couple.

Moment of friction couple

$$= W \times \perp \text{ distance between } R \text{ and } W$$
$$= W \times OC$$
$$= W \times r \sin\phi \quad (\because \text{ From triangle } OBC,\ OC = OB \sin\phi = r \sin\phi)$$

The angle ϕ is very small, hence $\sin\phi = \tan\phi$

$\therefore$ Moment of friction couple

$$= W \times r \times \tan\phi$$
$$= W \times r \times \mu \qquad (\because \mu = \tan\phi)$$

This friction couple acts in a direction opposite to the direction of rotation as is clear from Fig. 6.19 (*b*). Hence this friction couple oppose the driving torque on the shaft. And it will be equal to the driving torque for equilibrium.

Friction Circle. The circle of radius equal to $OC = r\tan\phi \simeq r \times \mu$ is known as friction circle. This radius of friction circle, which is equal to $r \times \mu$, will be constant as the values of r and μ are constant. Hence the radius of friction circle is independent of the load or weight of the shaft.

The effect of friction is that the reaction is displaced through a distance of $r \sin\phi$ or such that it is tangent to the friction circle.

Power loss in Friction

Frictional torque, $\quad T$ = Moment of friction couple

$$= W \times OC$$

$$= W \times r \times \mu \qquad (\because \; OC = r \sin \phi \simeq r \tan \phi \simeq r \times \mu)$$

$\therefore$ Power loss in friction $= T \times \omega$ where ω = Angular speed of shaft

$$= (W \times r \times \mu) \times \omega$$

$$= (W \times r \times \mu) \times \frac{V}{r} \qquad \left(\because \; \omega = \frac{V}{r}\right)$$

$$= W \times \mu \times V \text{ Watts}$$

$$= \frac{\mu W \times V}{1000} \text{ kW} \qquad \ldots(6.34)$$

6.12.2. Greasy Friction of a Slider-Crank Mechanism. Let us first define the *friction axis of a link*. Consider a link AB as shown in Fig. 6.20 which is pin-jointed at A and B. If friction is not considered, then the resultant thrust or axial force in the link acts along the longitudinal axis of the link. But if friction is considered, then the direction of the resultant thrust does not lie along the axis of the link. The reason for this is that the friction in a pin-joint acts in the same way as that of shaft rotating in a bearing. In case of a pin-jointed link, link acts like a shaft whereas the pin acts like a bearing. As the force on the shaft is tangential to the friction circle, similarly the line of thrust on a pin-jointed link will be tangential to the friction circles at the pin-joints. The net effect of this will be to shift the axis along which thrust acts. This new axis is known as the friction axis of the link. The radius of friction circles at A and B will be different.

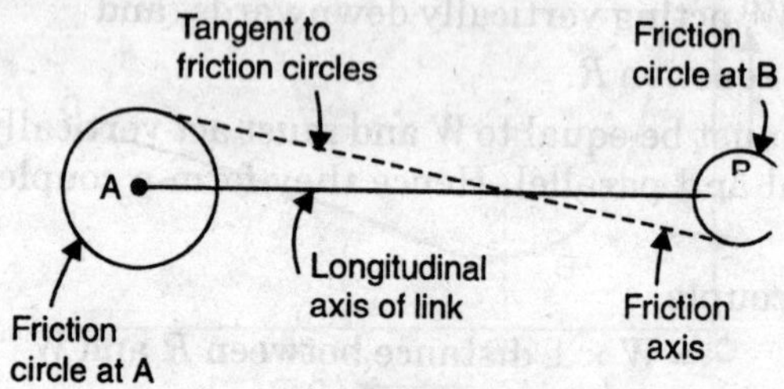

Fig. 6.20

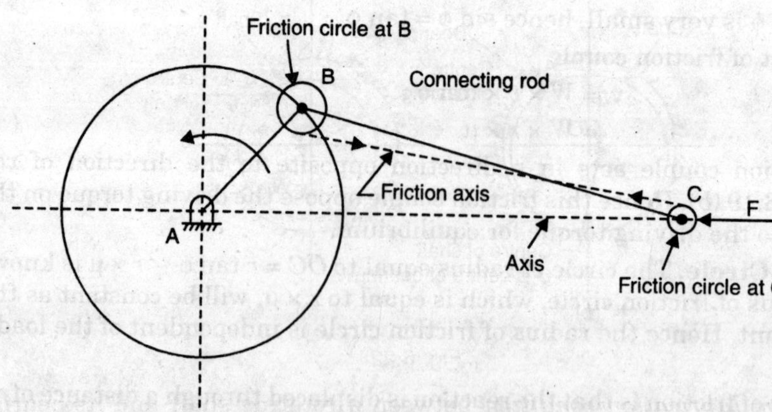

Fig. 6.21

6.13. FILM FRICTION OR VISCOUS FRICTION

It is already mentioned in Art. 6.3.3, that when two surfaces in contact have a thick layer of lubricant between them, the friction between the surfaces is known as film friction or viscous friction. In this case the actual metallic surfaces do not come in contact and hence do not rub against each other. Each surface is having rubbing action with the layers of lubricant only. Thus the friction resistance of two metallic surfaces are greatly reduced.

The layer of lubricant in contact with the stationary surface will be at rest whereas the layer of lubricant in contact with the moving surface, will move with the same speed as that of moving surface. The other layers between these two layers will be moving with the speed which is proportional to their distances from the surface at rest. The force required to slide one layer over the adjacent layer is the force of friction.

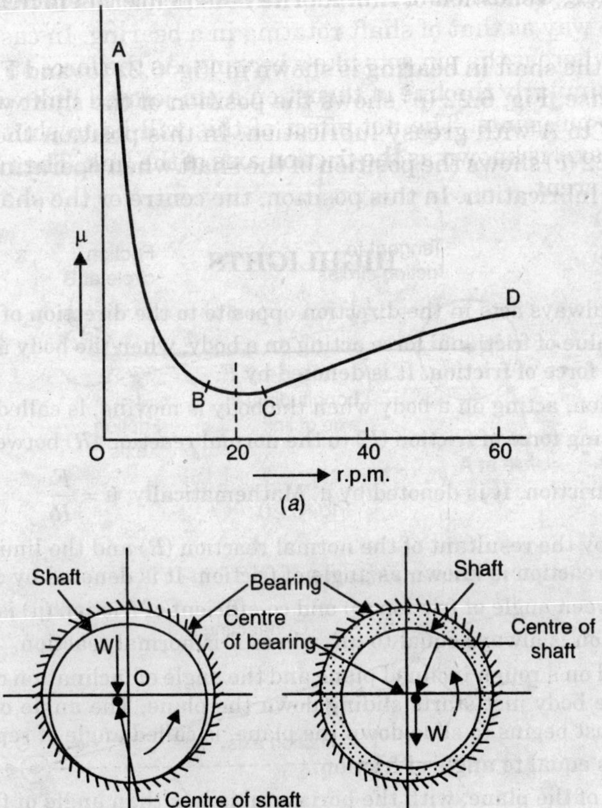

Fig. 6.22

For the *film lubrication condition* between a rotating shaft and its bearing, the frictional resistance is

(*i*) proportional to the speed of the shaft

(*ii*) proportional to the area

(*iii*) proportional to the viscosity

(*iv*) independent of the pressure

(*v*) independent of the materials of the shaft and the bearing.

Consider the case of a shaft running freely in a properly lubricated bearing. Due to the action of friction, the speed of shaft will decrease and ultimately, the shaft will cone to rest. If a graph is plotted between the speed and co-efficient of friction when the speed is varied from 60 r.p.m. to zero, a curve as shown in Fig. 6.22 (*a*) will be obtained. As the speed decreases from point *D* to point *C*, there is a gradual decrease in value of co-efficient of friction. After that if speed decreases, after a certain value of speed, the co-efficient of friction starts increasing rapidly. This is shown by curve from point *B* to *A*.

Actually at low speeds (*i.e.*, from point *B* to *A*), a metallic contact exists between shaft and bearing though there is a very thin layer of oil (*i.e.*, greasy friction condition) and hence co-efficient of friction increases rapidly. At high speeds (*i.e.*, from point *C* to *D*), the shaft is supported on oil film (*i.e.*, condition of film lubrication) hence co-efficient of friction increases slowly.

The position of the shaft in bearing is shown in Fig. 6.22 (*b*) and Fig. 6.22 (*c*) when shaft is rotating anticlockwise. Fig. 6.22 (*b*) shows the position of the shaft when operating on the curve position from *B* to *A* with greasy lubrication. In this position the centre of shaft is towards the left. Fig. 6.22 (*c*) shows the position of the shaft when operating on the curve portion from *C* to *D* with film lubrication. In this position, the centre or the shaft is towards right.

HIGHLIGHTS

1. Force of friction always acts in the direction opposite to the direction of motion.
2. The maximum value of frictional force acting on a body, when the body is on the point of motion, is called limiting force of friction. It is denoted by *F*.
3. The force of friction, acting on a body when the body is moving, is called dynamic friction.
4. The ratio of limiting force of friction (*F*) to the normal reaction (*R*) between two bodies is known as co-efficient of friction. It is denoted by μ. Mathematically, $\mu = \frac{F}{R}$.
5. The angle made by the resultant of the normal reaction (*R*) and the limiting force of friction (*F*) with the normal reaction is known as angle of friction. It is denoted by ϕ.
6. The relation between angle of friction (ϕ) and co-efficient of friction (μ) is expressed as $\tan \phi = \mu$.
7. The force of friction is always equal to μR, where *R* is normal reaction.
8. If a body is placed on a rough inclined plane and the angle of inclination of the plane is gradually increased, till the body just starts sliding down the plane. The angle of the inclined plane, at which the body just begins to slide down the plane, is called angle of repose.
9. Angle of repose is equal to angle of friction.
10. If the inclination of the plane, with the horizontal is less than angle of friction, the body placed on the inclined plane will be always in equilibrium without any external force.
11. A screw-jack is a device used for lifting heavy weights or loads with the help of a small effort applied at its handle.
12. The angle of screw or helix angle (α) in terms of pitch (*p*) of the screw and mean diameter (*d*) of the screw is given by $\tan \alpha = \frac{p}{\pi d}$.
13. The effort applied horizontally at the mean radius of the screw-jack of lift a load *W* is given by
$$P' = W \tan (\alpha + \phi)$$
and to lower the load *W* is given by $P' = W \tan (\phi - \alpha)$
where α = Angle of screw, ϕ = Angle of friction.

14. The effort applied at the end of the handle of a screw-jack is given by

$$P = \frac{Wd}{2L}\tan(\alpha + \phi) \quad \text{......to lift a load } W$$

$$= \frac{Wd}{2L}\tan(\phi - \alpha) \quad \text{......to lower a load } W$$

where d = Mean diameter of the screw, L = Length of the handle, α = Angle of screw, ϕ = Angle of friction.

15. Torque required to work the jack for lifting a load W is given by $T = \frac{Wd}{2}(\alpha + \phi)$.

16. Efficiency of a screw-jack for raising a load W is given by $\eta = \frac{\tan(\alpha)}{\tan(\alpha + \phi)}$.

The efficiency of the screw-jack is independent of the weight lifted or effort applied.

17. The efficiency of a screw-jack will be maximum if $\alpha = 45° - \frac{\phi}{2}$.

18. The maximum efficiency of a screw-jack is given by $\eta_{max} = \frac{1 - \sin\phi}{1 + \sin\phi}$.

19. The co-efficient of friction for V-threads is equal to co-efficient of friction for square threads divided by cos β where β is the semi-angle of V-threads or $\mu^* = \frac{\mu}{\cos\beta}$.

20. The bearing surfaces placed at the end of the shaft are known as pivots, whereas the bearing surfaces provided at any other positions are known as collar.

21. The friction torque is given by :

$$T = \frac{2}{3}\mu WR \quad \text{...... Flat pivot with uniform pressure}$$

$$= \frac{1}{2}\mu WR \quad \text{......Flat pivot with uniform wear}$$

$$= \frac{2}{3}\times\frac{\mu WR}{\sin\alpha} \quad \text{......Conical pivot with uniform pressure}$$

$$= \frac{1}{2}\times\frac{\mu WR}{\sin\alpha} \quad \text{......Conical pivot with uniform wear}$$

$$= \frac{2}{3}\times\frac{\mu WR}{\sin\alpha}\times\left(\frac{r_1^3 - r_2^3}{r_1^2 - r_2^2}\right) \quad \text{......Truncated pivot with uniform pressure}$$

$$= \frac{1}{2}\times\frac{\mu WR}{\sin\alpha}(r_1 + r_2) \quad \text{......Truncated pivot with uniform wear}$$

$$= \frac{2}{3}\times\mu W\times\left(\frac{r_1^3 - r_2^3}{r_1^2 - r_2^2}\right) \quad \text{......Flat collar with uniform pressure}$$

$$= \frac{\mu W}{2}(r_1 + r_2) \quad \text{......Flat collar with uniform wear.}$$

22. The device used to transmit the rotary motion of one shaft to another shaft is known as clutch or friction clutch. The types of friction clutches are :

(*i*) Disc or single plate clutch, (*ii*) Multi-plate clutch,

(*iii*) Cone clutch.

23. The expressions for frictional torque for single plate clutch are same as that of for collar bearing *i.e.*,

$$T = \frac{2}{3}\mu W\left[\frac{r_1^3 - r_2^3}{r_1^2 - r_2^2}\right] \quad \text{...... For uniform pressure}$$

$$= \frac{\mu W}{2}(r_1 + r_2) \quad \text{......For uniform wear.}$$

In a single plate clutch, there are two friction surfaces. Hence total torque is two times.

24. For new clutch, the intensity of pressure is approximately uniform over the entire surface whereas in old clutch the uniform wear theory is more approximate.

25. Multi-plate clutch is used when a large torque is to be transmitted such as in case of motor cars and machine tools. For multi-plate clutch :

$$n = n_1 + n_2 - 1, \qquad T = n \times \mu \times W \times R_m$$

where n = Number of friction surfaces or active surfaces,

n_1 = Number of friction plates on driving shaft,

n_2 = Number of friction plates on driven shaft.

R_m = Mean radius of friction surfaces

$$= \frac{r_1 + r_2}{2} \quad \text{...... For uniform wear}$$

$$= \frac{2}{3}\left(\frac{r_1^2 - r_2^3}{r_1^2 - r_2^2}\right) \quad \text{...... For uniform pressure.}$$

26. The cone-clutch, in action, is similar to the truncated conical pivot. Hence frictional torque,

$$T = \frac{2}{3} \times \frac{\mu W}{\sin\alpha} \times \left(\frac{r_1^3 - r_2^3}{r_1^2 - r_2^2}\right) \quad \text{...... For uniform pressure}$$

$$= \frac{1}{2} \times \frac{\mu W}{\sin\alpha}(r_1 + r_2) \quad \text{...... For uniform wear.}$$

27. When two surfaces in contact have a very thin layer of lubricant oil between them, the friction between the surfaces is known as greasy friction.

28. Friction circle is a circle of radius $= r \times \mu$ where r is the radius of the shaft (or journal) rotating in a bearing.

29. Film friction is the friction between two surfaces, when the surfaces in contact have a thick layer of lubricant between them.

EXERCISE 6

(A) Theoretical Questions

1. Define the following terms : Friction, limiting force of friction, co-efficient of friction and angle of friction.

2. Explain the difference between co-efficient of friction and angle of friction.

3. (*a*) State the laws of static and dynamic friction.

(*b*) State the laws of solid friction.

4. Prove that the angle of friction is equal to the angle of the inclined plane, when a solid body of weight W placed on the inclined plane, is about to slide down.

5. What do you mean by 'angle of repose' ? Prove that angle of repose is equal to the angle of friction.

6. A body of weight W is placed on an inclined rough plane. The inclination of the plane with the horizontal is less than the angle of friction. The body will
   (*a*) be in equilibrium (*b*) move downwards
   (*c*) move upwards (*d*) none of the above. **[Ans.** *a*]
7. If in the above question, the inclination of the plane with the horizontal is more than the angle of friction, the body will
   (*a*) be in equilibrium (*b*) move downwards
   (*c*) move upwards (*d*) none of the above. **[Ans.** *b*]
8. A body of weight W is placed on a rough inclined plane having inclination α to the horizontal. The force P is applied horizontally to drag the body. If the body is on the point of motion up the plane, prove that P is given by
   $$P = W \tan(\alpha + \phi) \qquad \text{where } \phi = \text{Angle of friction.}$$
9. In the above question, if the body is on the point of motion down the plane, prove that the force P is given by
   $$P = W \tan(\alpha - \phi).$$
10. Define statics and kinetic friction and state the laws of solid friction. (*AMIE, Summer 1983*)
11. Derive an expression for the least force required to drag a body on a rough horizontal plane.
12. What is a wedge ? Explain how a wedge is used to raise heavy loads. (*AMIE, Winter 1982*)
13. What is a screw-jack ? Show that for an ideal screw-jack, the efficiency is independent of the weight lifted. (*AMIE, Winter 1982*)
14. Show that the mechanical advantage on a screw nut arrangement is given by cot $(\alpha + \phi)$ where α is the angle of helix and ϕ is the friction angle. (*AMIE, Summer 1994*)
15. Differentiate between :
   (*i*) Pivot and collar bearing,
   (*ii*) Theory of uniform pressure and theory of uniform wear and
   (*iii*) Flat pivot and conical pivot.
16. From first principles, deduce an expression for the friction moment of a collar thrust bearing, stating clearly the assumptions made.
17. In a screw-jack, the helix angle of thread is 'α' and angle of friction is ϕ. Show that its efficiency is maximum when $2\alpha = (90° - \phi)$. (*AMIE, Summer 1987*)
18. Derive from first principles an expression for the effort required to raise a load with a screw-jack taking friction into consideration. (*AMIE, Summer 1984*)
19. Derive from first principles, the expression for the frictional moment (or torque due to friction) of a conical pivot assuming (*i*) uniform pressure and (*ii*) uniform wear.
20. Prove that the torque transmitted by a cone clutch when intensity of pressure is uniform, is given by
   $$T = \frac{2}{3}\frac{\mu W}{\sin\alpha}\left(\frac{r_1^3 - r_2^3}{r_1^2 - r_2^2}\right)$$
   where W = Axial load, α = Semi-cone angle, μ = Co-efficient of friction,
   r_1 = maximum radius, r_2 = minimum radius of contact surface. (*AMIE, Summer 1979*)
21. Explain clearly what are thick and thin film lubrications. (*AMIE, Summer 1986*)
22. What is meant by the expression friction circle ? Deduce an expression for the radius of friction circle in terms of radius of the journal and angle of friction.
23. Derive an expression for the friction moment for a flat collar bearing in terms of radius r_1 and r_2, axial thrust W and co-efficient of friction μ. Assume uniform intensity of pressure. (*AMIE, Winter 1983*)
24. Describe with a neat sketch the working of a single plate friction clutch.

25. Which of the two assumptions—uniform intensity of pressure or uniform rate of wear, would you make use of in designing friction clutch and why ? *(AMIE, Winter 84)*

[Ans. Uniform rate of wear]

26. What are the different types of friction clutches ? Describe with a neat sketch the working of a cone clutch.

27. What do you mean by greasy friction ? Describe in details the greasy friction of a journal and hence deduce the expression for friction circle.

(B) Numerical Problems

1. A screw-jack is used to lift a load of 4 kN. The screw of the screw-jack is square threaded with two threads to 1 cm. If the co-efficient of friction between the nut and screw is 0.095 and outer diameter of screw is 6 cm, find the force required at the end of the handle of length 50 cm to lift the load. **[Ans.** 28.29 N]

2. A screw-jack has a square thread, 6 cm mean diameter and 1 cm pitch. The load on the jack revolves with the screw. The co-efficient of friction at the screw threads is 0.045 :

   (*i*) Find the tangential force to be applied to the jack at 40 cm radius so as to lift a load of 700 N.

   (*ii*) State whether the jack is self-locking. If it is, find the torque necessary to lower the load. If not, find the torque, which must be applied to keep the load from descending.

   [Ans. (*i*) 5.16 N, (*ii*) No, .1675 Nm]

3. A screw-jack has a square thread of mean diameter 6 cm and pitch 9 mm. The co-efficient of friction at the screw-thread is 0.09. A load of 2 N is to be lifted through 10 cm. Determine the torque required and work done in lifting the load through 10 cm. Find the efficiency of the jack also. **[Ans.** .00829 kN ; 5792 kN ; 34.53%]

4. A vertical shaft 150 mm in diameter rotating at 100 r.p.m. rests on a flat foot-step bearing. The shaft carries a vertical load of 20 kN. Assuming uniform pressure and co-efficient of friction = 0.05, estimate the power lost in friction. Prove the formula used if any.

   [Ans. 523.6 W] *(AMIE, Winter 1977)*

5. A conical pivot with angle of cone as 120°, supports a vertical shaft of diameter 200 mm. It is subjected to a load of 15 kN. The co-efficient of friction is 0.05 and speed of the shaft is 140 r.p.m. Calculate the power lost in friction assuming (*i*) uniform pressure and (*ii*) uniform wear.

   [Ans. (*i*) 846 W and (*ii*) 634.8 W]

6. A truncated conical pivot of cone angle ϕ rotating at a speed N supports a load W. The smallest and the largest diameters of the pivot over the contact area are 'd' and 'D' respectively. Assuming uniform wear, derive the expression for the frictional torque, and hence determine the power lost in overcoming friction. If $W = 20$ kN, $d = 100$ mm, $D = 300$ mm, $\phi = 120°$, $N = 150$ r.p.m. and co-efficient of friction $\mu = 0.06$.

   Note. Here ϕ means $2\alpha = 120°$. **[Ans.** 2.176 kW]

7. A plain collar type thrust bearing having inner and outer diameters of 200 mm and 450 mm is subjected to an axial load of 40 kN. Assuming co-efficient of friction between the bearing surfaces as 0.25, find the power in kW absorbed in overcoming friction at a speed of 120 r.p.m. The rate of wear is considered to be proportional to the pressure and rubbing velocity.

   (AMIE, Summer 1985)

   Note. If rate of wear is proportional to pressure and rubbing velocity, then uniform rate of wear occurs *i.e.*, $p \times r$ = constant. **[Ans.** 20 kW]

8. A multiple collar thrust bearing has flat contact surfaces of 300 mm and 400 mm internal and external diameters respectively and support an axial thrust of 100 kN. Co-efficient of friction between bearing surfaces is 0.05. Assuming the condition of uniform pressure, determine : (*i*) number of collars required, and (*ii*) power lost in friction when the shaft rotates at 200 r.p.m. The maximum intensity of pressure is not to exceed 400 kN/m^2.

   [Ans. (*i*) 5, (*ii*) 18.326 kW] *(AMIE, Winter 1978)*

9. A shaft has a number of collars integral with it. External diameter of collar is 40 cm and shaft diameter is 25 cm. If the intensity of pressure is 350 kN/m^2 and is uniform, estimate : (*i*) power absorbed when shaft runs at 105 r.p.m. carrying a load of 150 kN, (*ii*) Number of collars required. Take co-efficient of friction as 0.05. **[Ans.** (*i*) 13.64 kW, (*ii*) 6] (*AMIE, Summer 1987*)

10. A multi-disc clutch has five plates having four pairs of active friction surfaces. If the intensity of pressure is not to exceed 127 kN/m^2, find the power in kW transmitted at 500 r.p.m. if the outer and inner radii of friction surfaces are 1.25 and 75 mm respectively. Assume uniform wear and take co-efficient of friction 0.3. **[Ans.** 18.8 kW] (*AMIE, Summer 1984*)

 Note. Here n = No. of active surfaces = 4.

11. Following data relate to a cone clutch :

Max. contact surface radius	= 150 mm
Min. contact surface radius	= 125 mm
Semi-cone angle	= 20°
Co-efficient of friction	= 0.25
Allowable normal pressure	= 140 kN/m^2

 For uniform intensity of pressure, estimate : (*i*) the axial load and (*ii*) power transmitted at 1000 r.p.m. **[Ans.** (*i*) 3.022 kN, (*ii*) 31.95 kW] (*AMIE, Summer 1979*)

7

Belts, Ropes and Chain Drives

7.1. Introduction

Belts are used for transmitting power from one shaft to another shaft. The belts may be flat belts, vee belts etc. Other methods of transmitting power are ropes and chains. The belts and ropes are running over the pulleys. The pulleys are mounted on the two shafts.

Belts, ropes and chains are used where the distance between the shafts is large. For small distances, gears are used. Belts and ropes are not having constant velocity ratios whereas chains and gears have constant velocity ratios.

The flat belt drives are of the following important types :

1. Open belt drive,
2. Cross belt drive and
3. Compound belt drive.

Selection of a belt-drive depends upon :

(*i*) centre distance between the shafts,

(*ii*) power to be transmitted,

(*iii*) speed of the driving and driven shafts and

(*iv*) speed reduction ratio.

Though there are many types of belts, yet the following types are more important :

(*i*) Flat belt—rectangular in section as shown in Fig. 7.1 (*a*)

(*ii*) V-belt—trapezoidal in section as shown in Fig. 7.1 (*b*)

(*iii*) Circular belt or rope—circular in section as shown in Fig. 7.1 (*c*).

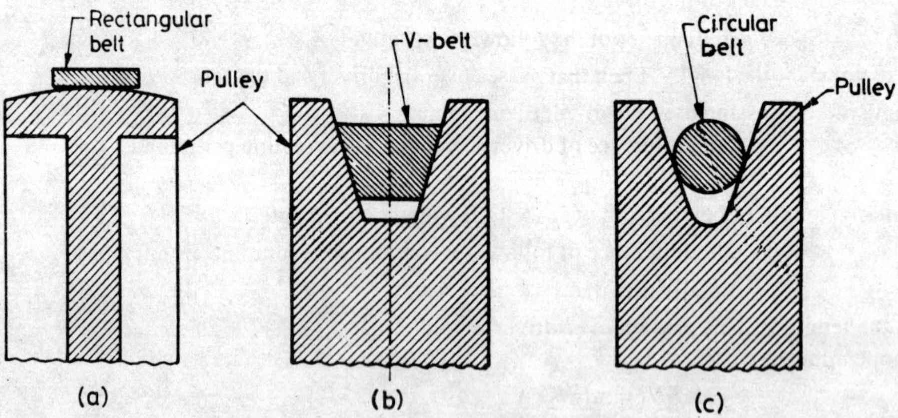

Fig. 7.1

7.2. Open Flat Belt Drive

An open belt drive consisting of pulleys *A* and *B* are shown in Fig. 7.2. The pulley which is keyed to the rotating shaft is known as *driver pulley* whereas the pulley which is keyed to the shaft which is to be rotated is known as *driven pulley* or follower pulley. Here pulley *A* is driver pulley and pulley *B* is driven pulley.

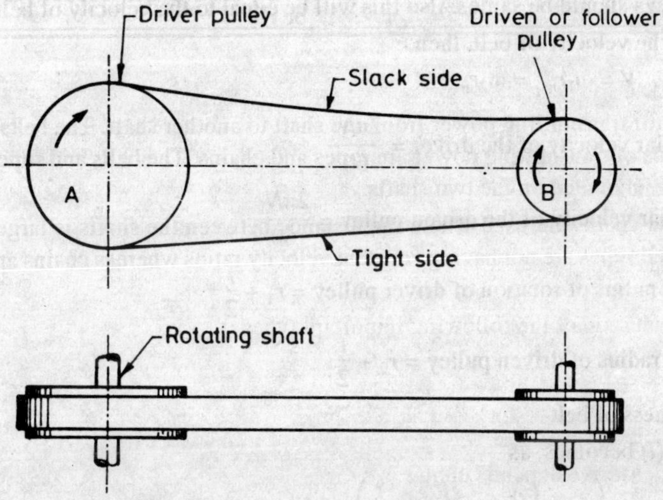

Fig. 7.2

When the shafts are arranged in parallel and rotating in the same direction, open belt drive as shown in Fig. 7.2 is used. The driver pulley *A* pulls the belt from lower side and delivers it to the upper side. Hence the tension in the lower side of the belt will be more than the tension in the upper side. The lower side belt is known as *tight side* whereas the upper side belt is known as slack side.

7.2.1. Velocity Ratio of Open Belt Drive. The ratio of the velocity of the follower (or driven) to the velocity of driver is known as velocity ratio. Mathematically, it is given as :

Let N_1 = Speed of the driver in r.p.m.

d_1 = Diameter of the driver

N_2, d_2 = Speed and diameter of the follower respectively

Let us consider the length of belt that passes over the driver and the follower in one minute.

Length of belt passing over the driver in one minute

= Circumference of driver × Number of revolution per minute

$= \pi d_1 \times N_1$

Similarly, length of belt passing over the following in one minute

= Circumference of follower × Number of revolution per minute

$= \pi d_2 \times N_2$

But the length of belt passing over driver in one minute is equal to the length of belt passing over follower in one minute.

$$\therefore \quad \pi d_1 N_1 = \pi d_2 N_2$$

or

$$\frac{N_2}{N_1} = \frac{d_1}{d_2}$$

$\therefore$ Velocity ratio,

$$\frac{N_2}{N_1} = \frac{d_1}{d_2}$$

Effect of Belt Thickness on Velocity Ratio

If belt thickness is considered, then for exact analysis the mean diameter of rotation is used for the calculation of the velocity ratio. The mean diameter of rotation will be equal to the diameter of the driver (or follower) plus the thickness of belt. Hence the mean radius of rotation will become as the radius of driver (or follower) plus half the belt thickness. If there is no slip between the belt and the pulleys, then the peripheral speed of the two pulleys should be same. Also this will be equal to the velocity of belt.

Hence if V is the velocity of belt, then

$$V = \omega_1 r_{m_1} = \omega_2 r_{m_2} \qquad \text{...}(i)$$

where ω_1 = Angular velocity of the driver = $\dfrac{2\pi N_1}{60}$

ω_2 = Angular velocity of the driven pulley = $\dfrac{2\pi N_2}{60}$

r_{m_1} = Mean radius of rotation of driver pulley = $r_1 + \dfrac{t}{2}$

r_{m_2} = Mean radius of driven pulley = $r_2 + \dfrac{t}{2}$

t = Thickness of belt

∴ Equation (i) becomes, as

$$\omega_1\left(r_1 + \frac{t}{2}\right) = \omega_2\left(r_2 + \frac{t}{2}\right)$$

$$\frac{\omega_2}{\omega_1} = \frac{r_1 + \frac{t}{2}}{r_2 + \frac{t}{2}} = \frac{2r_1 + t}{2r_2 + t} = \frac{d_1 + t}{d_2 + t}$$

or

$$\frac{N_2}{N_1} = \frac{d_1 + t}{d_2 + t} \qquad \text{...(7.2)}$$

Alternate method

Let N_1 = r.p.m. of driver pulley

N_2 = r.p.m. of follower (driven pulley)

d_1 = dia. of driver pulley = $2r_1$

d_2 = dia. of driven pulley = $2r_2$

ω_1 = Angular velocity of the driver = $\dfrac{2\pi N_1}{60}$

ω_2 = angular velocity of the driven pulley = $\dfrac{2\pi N_2}{60}$

If there is no slip between the belt and the pulleys and also the thickness of the belt is negligible, then peripheral speed of the two pulleys should be same. Also this will be equal to the velocity of belt. Hence if V is the velocity of belt, then

$$V = \omega_1 r_1 = \omega_2 r_2$$

or

$$\frac{2\pi N_1}{60} \times \frac{d_1}{2} = \frac{2\pi N_2}{60} \times \frac{d_2}{2} \qquad \left(\because\ \omega_1 = \frac{2\pi N_1}{60} \text{ and } r_1 = \frac{d_1}{2} \text{ etc.}\right)$$

or

$$N_1 d_1 = N_2 d_2$$

or $$\frac{N_2}{N_1} = \frac{d_1}{d_2} \quad \text{...(7.1)}$$

In the above equation $\frac{N_2}{N_1}$ is known as velocity ratio. Also from the above equation it is clear that speed of the pulley is inversely proportional to its diameter.

Problem 7.1. *Find the speed of a shaft which is driven with the help of a belt by an engine running at 200 r.p.m. The diameter of the engine pulley is 51 cm and that of shaft is 30 cm.*

Sol. Given :

Speed of engine (driver), $N_1 = 200$ r.p.m.

Dia. of engine (driver) pulley, $d_1 = 51$ cm

Dia. of driven shaft, $d_2 = 30$ cm

Let N_2 = speed of driven shaft.

Using equation (7.1), we get, $$\frac{N_2}{N_1} = \frac{d_1}{d_2}$$

$$\therefore \quad N_2 = \frac{d_1}{d_2} \times N_1 = \frac{51}{30} \times 200 = \mathbf{340\ r.p.m.} \quad \textbf{Ans.}$$

Problem 7.2. *If in the above equation, the thickness of the belt is 10 mm, then find the speed of the shaft.*

Sol. Given :

Thickness, $t = 10$ mm $= 1$ cm

$N_1 = 200$ r.p.m., $d_1 = 51$ cm, $d_2 = 30$ cm

Let N_2 = Speed of shaft.

Using equation (7.2), we get $$\frac{N_2}{N_1} = \frac{d_1 + t}{d_2 + t}$$

or $$N_2 = \left(\frac{d_1 + t}{d_2 + t}\right) \times N_1$$

$$= \left(\frac{51 + 1}{30 + 1}\right) \times 200 = \frac{52}{31} \times 200 = \mathbf{335.4\ r.p.m.} \quad \textbf{Ans.}$$

7.2.2. Slip of the Belt. When the driver pulley rotates, it carries the belt, due to a firm grip between its surface and the belt. The firm grip between the pulley and the belt is obtained by friction. This firm grip is known as frictional grip. But sometimes the frictional grip is not sufficient. This may cause some forward motion of the driver pulley without carrying the belt with it. This means that there is a relative motion between the driver pulley and the belt. The difference between the linear speeds of the pulley rim and the belt is a measure of slip. Generally, the slip is expressed as a percentage. In some cases, the belt moves faster in the forward direction, without carrying the *driven pulley* with it. Hence in case of driven pulley, the forward motion of the belt is more than that of driven pulley.

Let v = Velocity of belt, passing over the driver pulley/min.

N_1 = Speed in r.p.m. of driver

N_2 = Speed in r.p.m. of follower

s_1 = Slip between the driver and belt in percentage

s_2 = Percentage slip between follower and the belt

The peripheral velocity of the driver pulley

$$= \omega_1 r_1$$

$$= \frac{2\pi N_1}{60} \times \frac{d_1}{2} \qquad \left(\because \ \omega_1 = \frac{2\pi N_1}{60} \text{ and } r_1 = \frac{d_1}{2}\right)$$

$$= \frac{\pi N_1 d_1}{60} \text{ m/sec.}$$

Now due to slip between the driver pulley and the belt, the velocity of belt passing over the driver pulley will decrease.

$$\therefore \quad \text{Velocity of belt} = \frac{\pi N_1 d_1}{60} - \left(\frac{\pi N_1 d_1}{60}\right) \times \frac{s_1}{100}$$

$$= \frac{\pi N_1 d_1}{60}\left(1 - \frac{s_1}{100}\right)$$

Now this belt is passing over the follower pulley (*i.e.*, driven pulley). And as there is a slip between the belt and the driven pulley, the velocity (*i.e.*, peripheral velocity) of the follower pulley will decrease.

$\therefore$ Peripheral velocity of the follower

$$= \text{Velocity of belt} - \left(\text{Velocity of belt} \times \frac{s_2}{100}\right)$$

$$= \text{Velocity of belt}\left[1 - \frac{s_2}{100}\right]$$

$$= \frac{\pi N_1 d_1}{60}\left[1 - \frac{s_1}{100}\right]\left[1 - \frac{s_2}{100}\right] \qquad \ldots(i)$$

$$\left[\because \ \text{Velocity of belt} = \frac{\pi N_1 d_1}{60}\left(1 - \frac{s_1}{100}\right)\right]$$

But peripheral velocity of follower pulley

$$= \omega_2 \times r_2$$

$$= \frac{2\pi N_2}{60} \times \frac{d_2}{2} \qquad \left(\because \ \omega_2 = \frac{2\pi N_2}{60} \text{ and } r_2 = \frac{d_2}{2}\right)$$

$$= \frac{\pi N_2 d_2}{60} \qquad \ldots(ii)$$

Equating the two values given by equations (*i*) and (*ii*), we get

$$\frac{\pi N_2 d_2}{60} = \frac{\pi N_1 d_1}{60}\left(1 - \frac{s_1}{100}\right)\left(1 - \frac{s_2}{100}\right)$$

or $$N_2 d_2 = N_1 d_1\left[1 - \frac{s_2}{100} - \frac{s_1}{100} + \frac{s_1 s_2}{10000}\right]$$

$$= N_1 d_1\left[1 - \frac{s_1 + s_2}{100}\right] \qquad \left(\text{Neglecting } \frac{s_1 \times s_2}{10000}\right)$$

$$= N_1 d_1\left[1 - \frac{s}{100}\right]$$

(where $s = s_1 + s_2$ *i.e.*, total percentage of slip between driver and follower)

$$\therefore \qquad \frac{N_2}{N_1} = \frac{d_1}{d_2}\left[1 - \frac{s}{100}\right] \qquad ...(7.3)$$

If the belt thickness (t) is taken into account, then

$$\frac{N_2}{N_1} = \left(\frac{d_1 + t}{d_2 + t}\right)\left(1 - \frac{s}{100}\right). \qquad ...(7.4)$$

7.2.3. Creep of the Belt. In a belt drive, the belt is passing over the driver pulley and driven pulley (*i.e.*, follower). When the power is transmitted with the help of this belt, the belt is subjected to tensions. The part of the belt which leaves the follower and approaches the driver is known as the tight side of the belt and is subjected to tension T_1. But the part of the belt which leaves the driver and approaches the follower is known as slack side of the belt and is subjected to tension T_2. The two tensions are not equal in magnitude, *i.e.*, $T_1 > T_2$. Hence the stretch in the belt due to different tensions on the two sides of the pulley will be different. A certain portion of the belt when passes from slack side to the tight side, extends and the same portion contracts again when the belt passes from the tight side to the slack side. Due to these changes in length, there exists a relative motion between the belt and the pulley surfaces. This relative motion is known as **creep.** The creep reduces slightly the speed of the follower (*i.e.* driven pulley).

The velocity ratio when creep is considered is given by,

$$\frac{N_2}{N_1} = \frac{d_1}{d_2} \times \left[\frac{E + \sqrt{f_2}}{E + \sqrt{f_1}}\right] \qquad ...(7.5)$$

where E = Young's modulus for the material of the belt,

f_1 = Stress in the belt on the tight side, and

f_2 = Stress in the belt on the slack side.

Problem 7.3. *A shaft running at 200 r.p.m. is to drive a parallel shaft at 300 r.p.m. The pulley on the driving shaft is 60 cm diameter. Calculate the diameter of the pulley on the driven shaft :*

(i) Neglecting belt thickness.

(ii) Taking belt thickness into account, which is 5 mm thick.

(iii) Assuming in the latter case a total slip of 4%.

Sol. Given :

N_1 = 200 r.p.m., N_2 = 300 r.p.m.,

and d_1 = 60 cm, t = 5 mm = 0.5 cm

Total slip s = 4%.

(i) Neglecting belt thickness.

Let d_2 = Dia. of the pulley on driven shaft.

Using equation (7.1), we get $\dfrac{N_2}{N_1} = \dfrac{d_1}{d_2}$

or $d_2 = \dfrac{N_1}{N_2} \times d_1 = \dfrac{200}{300} \times 60 = 40$ **cm. Ans.**

(ii) Taking belt thickness into account only.

Using equation (7.2), we get $\dfrac{N_2}{N_1} = \dfrac{d_1 + t}{d_2 + t}$

or $(d_2 + t) = (d_1 + t) \times \dfrac{N_1}{N_2}$

$$= (60 + 0.5) \times \frac{200}{300} = 60.5 \times \frac{2}{3}$$

$$\therefore \quad d_2 = 60.5 \times \frac{2}{3} - 0.5 = \frac{121}{3} - 0.5$$

$$= 40.33 - 0.5 = \mathbf{39.83\ cm.} \quad \textbf{Ans.}$$

(iii) Considering the belt thickness and total slip.

Using equation (7.4), we get
$$\frac{N_2}{N_1} = \left(\frac{d_1 + t}{d_2 + t}\right)\left(1 - \frac{s}{100}\right)$$

$$\frac{300}{200} = \left(\frac{60 + 0.5}{d_2 + 0.5}\right)\left(1 - \frac{4}{100}\right)$$

$$\therefore \quad (d_2 + 0.5) = 60.5 \times 0.96 \times \frac{200}{300} = 38.72$$

$$\therefore \quad d_2 = 38.72 - 0.5 = \mathbf{38.22\ cm.} \quad \textbf{Ans.}$$

Problem 7.4. (S.I. Units). *The power is transmitted from a pulley 1 m diameter running at 200 r.p.m. to a pulley 2.5 m diameter by means of a belt. Find the speed lost by the driven pulley as a result of the creep, if the stress on the tight and slack side of the belt is 1.44 N/mm² and 0.49 N/mm² respectively. The Young's modulus for the material of the belt is 100 N/mm².*

Sol. Given :

Dia. of driver pulley, $d_1 = 1$ m

Speed of driver pulley, $N_1 = 200$ r.p.m.

Dia. of driven pulley, $d_2 = 2.5$ m

Stress on tight side, $f_1 = 1.44$ N/mm²

Stress on slack side, $f_2 = 0.49$ N/mm²

Young's modulus, $E = 100$ N/mm²

Let N_2 = speed of driven pulley.

Using equation (7.5), we get
$$\frac{N_2}{N_1} = \frac{d_1}{d_2} \times \left[\frac{E + \sqrt{f_2}}{E + \sqrt{f_1}}\right]$$

or
$$\frac{N_2}{200} = \frac{1}{2.5}\left[\frac{100 + \sqrt{0.49}}{100 + \sqrt{1.44}}\right] = \frac{1}{2.5}\left[\frac{100.7}{101.12}\right]$$

$$\therefore \quad N_2 = \frac{200}{2.5} \times \frac{100.7}{101.12} = 79.67 \text{ r.p.m.}$$

If the creep is neglected, then from equation (7.1), we get
$$\frac{N_2}{N_1} = \frac{d_1}{d_2}$$

$$\therefore \quad N_2 = \frac{d_1}{d_2} \times N_1 = \frac{1}{2.5} \times 200 = 80 \text{ r.p.m.}$$

$\therefore$ Speed lost by driven pulley due to creep

$$= 80 - 79.67 = \mathbf{0.33\ r.p.m.} \quad \textbf{Ans.}$$

Problem 7.5. (S.I. Units). *If in problem 7.4, the belt is 50 cm wide and 10 mm thick, find the speed in r.p.m. lost by the driven pulley as a result of the creep, if the tensions on the tight and slack sides of the belt* *are 10 N/mm width and 4 N/mm width respectively. Take the Young's modulus for the material of the belt = 100 N/mm².*

Sol. Given :

From problem 7.4, the data is $d_1 = 1$ m, $N_1 = 200$ r.p.m. and $d_2 = 2.5$ m.

Width of belt, $b = 50$ cm $= 500$ mm.

Thickness of belt, $t = 10$ mm.

∴ Area of cross-section, $A = b \times t = 500 \times 10 = 5000 \text{ mm}^2$

Tension on tight side = 10 N/mm width

= 10 × Width = 10 × 500 = 5000 N

Tension on slack side = 4 N/mm width = 4 × 500 = 2000 N

∴ Stress on tight side, $f_1 = \dfrac{\text{Tension on tight side}}{\text{Area of belt}} = \dfrac{5000}{5000} = 1 \text{ N/mm}^2$

Stress on slack side, $f_2 = \dfrac{\text{Tension on slack side}}{\text{Area of belt}} = \dfrac{2000}{5000} = 0.4 \text{ N/mm}^2$

Value of $E = 100 \text{ N/mm}^2$

Let N_2 = Speed of driven pulley

Now using equation (7.5), we get

$$\frac{N_2}{N_1} = \frac{d_1}{d_2} \times \left[\frac{E + \sqrt{f_2}}{E + \sqrt{f_1}}\right]$$

or

$$\frac{N_2}{200} = \frac{1}{2.5}\left[\frac{100 + \sqrt{0.4}}{100 + \sqrt{1.0}}\right]$$

∴

$$N_2 = \frac{200}{2.5}\left[\frac{100 + \sqrt{0.4}}{101}\right] = 79.99 \text{ r.p.m.}$$

If creep is neglected, then from equation (7.1), we get

$$\frac{N_2}{N_1} = \frac{d_1}{d_2}$$

or

$$N_2 = \frac{d_1}{d_2} \times N_1 = \frac{1}{2.5} \times 200 = 80 \text{ r.p.m.}$$

∴ Speed lost by driven pulley due to creep = 80 – 79.99 = **0.01 r.p.m.** **Ans.**

7.3. Cross Belt Drive

In an open belt drive both the pulleys rotate in the same direction, whereas in the cross-belt drive they rotate in the opposite direction. Hence the cross-beft drive is used when the shafts are arranged in parallel and rotate in the opposite direction as shown in Fig. 7.3. The driver pulley *A* pulls the belt from *CD* side and delivers

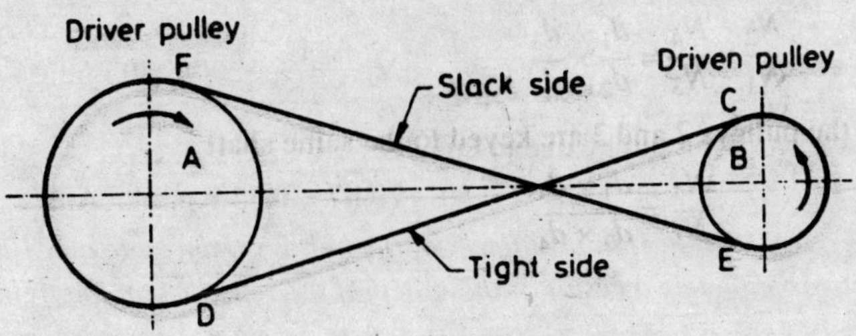

Fig. 7.3

it to *FE* side. Hence the tension in the belt *CD* will be more than that in the portion *FE*. And so the belt *CD* is known as tight side and portion *FE* is known as slack side.

7.4. Compound Belt Drive

When power is to be transmitted from one shaft to another through a number of pulleys, then a compound belt drive is used. Fig. 7.4 shows a compound belt-drive, in which the pulley 1 drives the pulley 2. But pulleys 2 and 3 are keyed (or fixed) to the same shaft, therefore the pulley 1 also drives pulley 3. But pulley 3 drives pulley 4 as shown in Fig. 7.4.

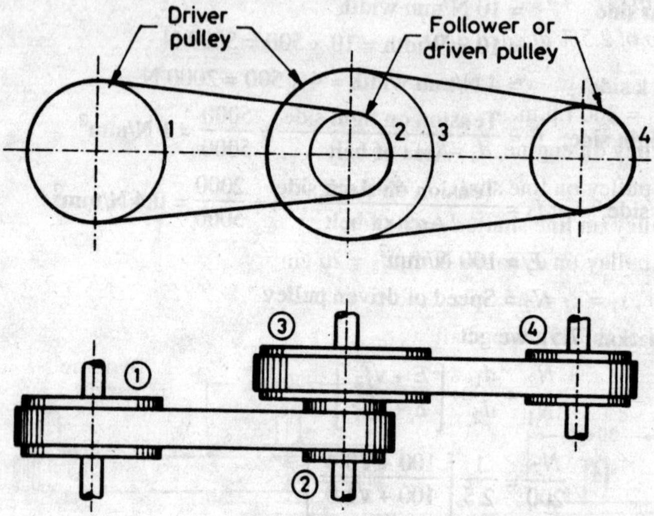

Fig. 7.4

Velocity Ratio of Compound Belt drive

Let N_1 = Speed of pulley 1 in r.p.m.

d_1 = Diameter of pulley 1

N_2, d_2 = Speed and dia. of pulley 2

N_3, d_3 and N_4, d_4 = Speed and dia. of pulleys 3 and 4 respectively.

From equation (7.1), we know that the velocity ratio of pulleys 1 and 2 is given by

$$\frac{N_2}{N_1} = \frac{d_1}{d_2} \qquad ...(i)$$

Similarly, the velocity ratio of pulleys 3 and 4 is given by

$$\frac{N_4}{N_3} = \frac{d_3}{d_4} \qquad ...(ii)$$

Multiplying equations (*i*) and (*ii*), we get

$$\frac{N_2}{N_1} \times \frac{N_4}{N_3} = \frac{d_1}{d_2} \times \frac{d_3}{d_4}$$

But $N_2 = N_3$ as the pulleys 2 and 3 are keyed to the same shaft

$$\therefore \quad \frac{N_4}{N_1} = \frac{d_1 \times d_3}{d_2 \times d_4} \qquad ...(7.6)$$

The above equation can be written as

$$\frac{\text{Speed of last follower}}{\text{Speed of first driver}} = \frac{\text{Product of dia. of drivers}}{\text{Product of dia. of followers}} \quad ...(7.7)$$

Problem 7.6. *With the help of a belt, an engine running at 200 r.p.m., drives a line shaft. The diameter of the pulley on the engine is 80 cm and the diameter of the pulley on the line shaft is 40 cm. A 100 cm diameter pulley on the line shaft drives a 20 cm diameter pulley keyed to a dynamo shaft. Find the speed of the dynamo shaft when :*

(*i*) *there is no slip*

(*ii*) *there is a slip of 2.5% at each drive.*

Sol. Given :

Engine speed, $N_1 = 200$ r.p.m.

Dia. of driver pulley on engine, $d_1 = 80$ cm

Dia. of follower pulley on line shaft, $d_2 = 40$ cm

Dia. of driver pulley on line shaft, $d_3 = 100$ cm

Dia. of follower pulley on dynamo shaft, $d_4 = 20$ cm

Slip at each drive, $s_1 = s_2 = 2.5$

Let N_4 = Speed of the dynamo shaft

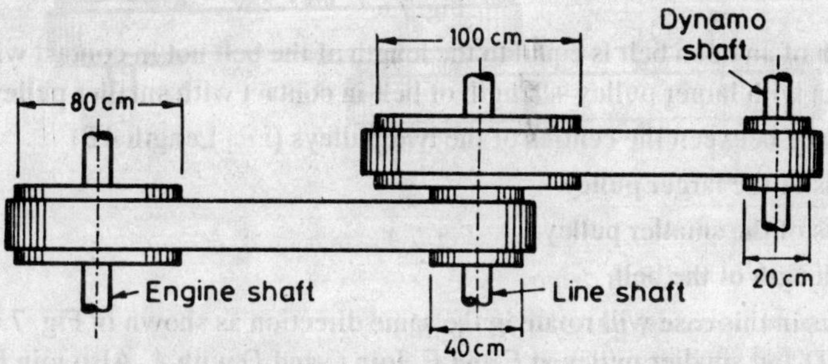

Fig. 7.5

(*i*) *When there is no slip.*

Using equation (7.6), we get

$$\frac{N_4}{N_1} = \frac{d_1 \times d_3}{d_2 \times d_4}$$

or

$$N_4 = \frac{d_1 \times d_3}{d_2 \times d_4} \times N_1 = \frac{80 \times 100}{40 \times 20} \times 200 = \mathbf{2000\ r.p.m.} \quad \textbf{Ans.}$$

(*ii*) *When there is a slip of 2.5% at each drive.*

In this case, we will have the equation

$$\frac{N_4}{N_1} = \frac{d_1 \times d_3}{d_2 \times d_4}\left(1 - \frac{s_1}{100}\right)\left(1 - \frac{s_2}{100}\right)$$

$$\therefore \quad N_4 = \frac{N_1 \times d_1 \times d_3}{d_2 \times d_4}\left(1 - \frac{s_1}{100}\right)\left(1 - \frac{s_2}{100}\right)$$

$$= \frac{200 \times 80 \times 100}{40 \times 20}\left(1 - \frac{2.5}{100}\right)\left(1 - \frac{2.5}{100}\right)$$

$$= 2000 \times \left(\frac{97.5}{100}\right)\left(\frac{97.5}{100}\right) = \mathbf{1901.25\ r.p.m.} \quad \textbf{Ans.}$$

7.5. Length of Belt

The length of a belt means the total length of the belt required to connect a driver and a follower. There are two cases to be considered. They are :

(*i*) Length of an open belt drive, and

(*ii*) Length of a cross-belt drive.

7.5.1. Length of an Open Belt Drive. If the distance between the centres of the two shafts and the diameter of pulleys are known, the required length of the belt may be easily calculated.

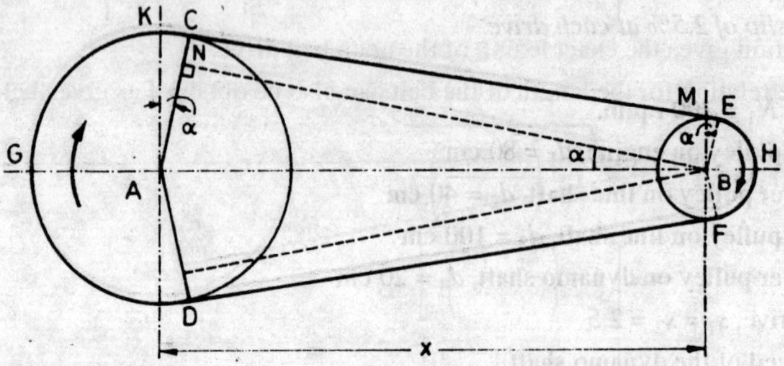

Fig. 7.6

The total length of an open belt is equal to the length of the belt not in contact with either pulley + the length of belt in contact with larger pulley + length of belt in contact with smaller pulley.

Let x = Distance between the centres of the two pulleys (*i.e.*, Length AB)

r_1 = Radius of the larger pulley

r_2 = Radius of the smaller pulley

L = Total length of the belt.

Both the pulleys in this case will rotate in the same direction as shown in Fig. 7.6. The belt leaves the larger pulley at C and D and smaller pulley at E and F. Join C and D with A. Also join E and F with B.

From B draw BN parallel to EC. But CE is tangent at C. Hence AC is perpendicular to CE, which means $\angle ACE = 90°$ or $\pi/2$, As BN is parallel to CE, hence $\angle ANB$ is also $90°$ or $\frac{\pi}{2}$ radians.

Let the angle $ABN = \alpha$ radians. Then angle $BAN = \left(\frac{\pi}{2} - \alpha\right)$. But angle $BAK = 90°$ or $\frac{\pi}{2}$. Hence angle $KAC = \alpha$. Similarly, it can be shown that angle $MBE = \alpha$.

Now length $\quad AN = AC - CN = r_1 - r_2$

$$BN = \sqrt{AB^2 - AN^2} = \sqrt{x^2 - (r_1 - r_2)^2}$$

or

$$BN = CE = \sqrt{x^2 - (r_1 - r_2)^2}$$

Also from triangle ABN,

$$\sin \alpha = \frac{AN}{AB} = \frac{r_1 - r_2}{x}$$

Since α is very small, hence $\sin \alpha = \alpha$

$$\therefore \quad \alpha = \sin \alpha = \frac{r_1 - r_2}{x} \quad ...(i)$$

Now total length (L) of the belt is given by,

$$L = \text{Arc } DGC + CE + \text{Arc } EHF + FD$$

$$= 2[\text{Arc } GC + CE + \text{Arc } EH] \qquad [\because \; CE = FD]$$

$$= 2\left[r_1\left(\frac{\pi}{2}+\alpha\right) + \sqrt{x^2-(r_1-r_2)^2} + r_2\left(\frac{\pi}{2}-\alpha\right)\right] \qquad ...(ii)$$

$$= 2\left[\frac{\pi}{2}(r_1+r_2) + \alpha\,(r_1-r_2) + \sqrt{x^2-(r_1-r_2)^2}\right] \qquad ...(7.7A)$$

The above equation gives the exact length of the open belt drive.

An approximate relation for the length of the belt can also be obtained as given below :

As $$\sqrt{x^2-(r_1-r_2)^2} = x\sqrt{1-\left(\frac{r_1-r_2}{x}\right)^2}$$

$$= x\left[1-\left(\frac{r_1-r_2}{x}\right)^2\right]^{1/2}$$

$$= x\left[1-\frac{1}{2}\left(\frac{r_1-r_2}{x}\right)^2 + ...\right] \qquad \text{(From Binomial expansion)}$$

$$= x\left[1-\frac{(r_1-r_2)^2}{2x^2}\right] \qquad \text{Neglecting smaller terms}$$

Substituting this value in equation (*ii*), we get

$$L = 2\left[r_1\left(\frac{\pi}{2}+\alpha\right) + x\left(1-\frac{(r_1-r_2)^2}{2x^2}\right) + r_2\left(\frac{\pi}{2}-\alpha\right)\right]$$

$$= 2\left[\frac{\pi}{2}(r_1+r_2) + \alpha(r_1-r_2) + \left(x-\frac{(r_1-r_2)^2}{2x}\right)\right]$$

$$= \pi(r_1+r_2) + 2\alpha(r_1-r_2) + 2x - \frac{(r_1-r_2)^2}{x}$$

But from equation (*i*), we have

$$\alpha = \frac{r_1-r_2}{x}$$

Substituting this value of α in the above equation, we get

$$L = \pi(r_1+r_2) + 2\times\left(\frac{r_1-r_2}{x}\right)(r_1-r_2) + 2x - \frac{(r_1-r_2)^2}{x}$$

$$= \pi(r_1+r_2) + \frac{2}{x}(r_1-r_2)^2 + 2x - \frac{(r_1-r_2)^2}{x}$$

$$= \pi(r_1+r_2) + \frac{(r_1-r_2)^2}{x} + 2x \qquad ...(7.8)$$

The above equation gives the approximate length of open belt drive.

From the above equation, we observe that the length of the open belt depends upon both the sum and difference of the radii.

7.5.2. Length of a Cross-Belt Drive. The cross-belt is shown in Fig. 7.7.

Let x = Distance between the centres of the two pulleys, length AB,

r_1 = Radius of the larger pulley,

r_2 = Radius of the smaller pulley, and

L = Total length of the cross-belt.

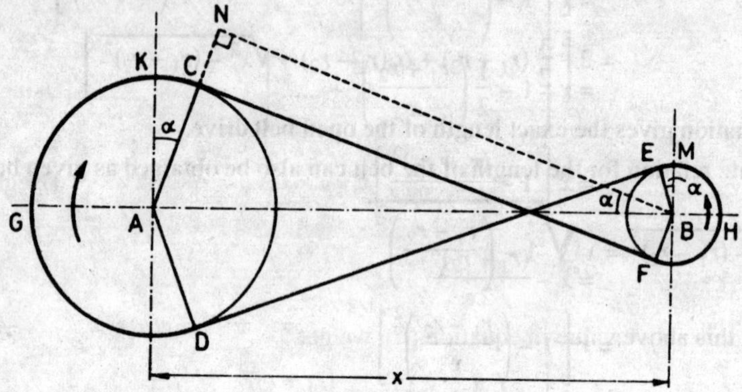

Fig. 7.7

Both the pulleys in this case will rotate in the opposite directions as shown in Fig. 7.7. The belt leaves the larger pulley at C and D and smaller pulley at E and F.

From B, draw BN parallel to FC. Now angle ACF = angle ANB = 90°. Let angle $ABN = \alpha$. Then it can be shown that angle $KAC = \alpha$ and also angle $EBM = \alpha$.

Length $AN = AC + CN = r_1 + FB$ $(\because\ CN = FB)$

$= r_1 + r_2$ $(\because\ FB = r_2)$

In triangle ABN, $BN = \sqrt{AB^2 - AN^2} = \sqrt{x^2 - (r_1 + r_2)^2}$

and $\sin\alpha = \dfrac{AN}{AB} = \dfrac{r_1 + r_2}{x}$

Since α is very small, hence

$$\sin\alpha = \alpha$$

$$\therefore \quad \alpha = \sin\alpha = \frac{r_1 + r_2}{x} \qquad ...(i)$$

Now total length (L) of the belt is given by,

$$L = \text{Arc } DGC + CF + \text{Arc } FHE + ED$$

$$= 2\,[\text{Arc } GC + CF + \text{Arc } HE] \qquad (\because\ ED = CF)$$

$$= 2\left[r_1\left(\frac{\pi}{2} + \alpha\right) + BN + r_2\left(\frac{\pi}{2} + \alpha\right)\right] \qquad (\because\ CF = BN)$$

$$= 2\left[r_1\left(\frac{\pi}{2} + \alpha\right) + \sqrt{x^2 - (r_1 + r_2)^2} + r_2\left(\frac{\pi}{2} + \alpha\right)\right] \qquad ...(ii)$$

$$= 2\left[\frac{\pi}{2}(r_1 + r_2) + \alpha\,(r_1 + r_2) + \sqrt{x^2 - (r_1 + r_2)^2}\right] \qquad ...(7.8A)$$

The above equation gives the exact length of the cross-belt drive.

An approximate relation for the length of the belt can be obtained as given below :

As $\sqrt{x^2-(r_1+r_2)^2} = x\sqrt{1-\left(\frac{r_1+r_2}{x}\right)^2}$

$$= x\left[1-\left(\frac{r_1+r_2}{x}\right)^2\right]^{1/2}$$

$$= x\left[1-\frac{1}{2}\left(\frac{r_1+r_2}{x}\right)^2+ \ldots\ldots\right] \quad \text{[Using Binomial expansion]}$$

$$= x\left[1-\frac{(r_1+r_2)^2}{2x^2}\right] \quad \text{[Neglecting smaller terms]}$$

$$= x-\frac{(r_1+r_2)^2}{2x}$$

Substituting this above value in equation (*ii*), we get

$$L = 2\left[r_1\left(\frac{\pi}{2}+\alpha\right)+x-\frac{(r_1+r_2)^2}{2x}+r_2\left(\frac{\pi}{2}+\alpha\right)\right]$$

$$= 2\left[\frac{\pi}{2}(r_1+r_2)+\alpha\,(r_1+r_2)+x-\frac{(r_1+r_2)^2}{2x}\right]$$

$$= \pi(r_1+r_2)+2\alpha(r_1+r_2)+2x-\frac{(r_1+r_2)^2}{x}$$

But from equation (*i*), $\alpha = \frac{r_1+r_2}{x}$

Substituting the value of α in the above equation, we get

$$L = \pi(r_1+r_2)+2\left(\frac{r_1+r_2}{x}\right)(r_1+r_2)+2x-\frac{(r_1+r_2)^2}{x}$$

$$= \pi(r_1+r_2)+\frac{2}{x}(r_1+r_2)^2+2x-\frac{(r_1+r_2)^2}{x}$$

$$= \pi(r_1+r_2)+\frac{(r_1+r_2)^2}{x}+2x \quad \ldots(7.9)$$

The above equation gives the approximate length for cross-belt drive.

From the above equation, we observe that the length of a crossed belt depends only upon the sum of the radii.

Hence for a belt connecting two stepped pulleys, the length of the crossed belt is constant if the sum of the radii of the corresponding steps is constant. If $(r_1 + r_2)$ is constant, then $(r_1 - r_2)$ is not a constant and hence this rule does not apply for an open belt.

Problem 7.7. *Two parallel shafts 6 m apart are to be connected by a belt running over pulleys of diameters 60 cm and 40 cm respectively. Determine the exact and approximate lengths of the belt required :*

(i) if the belt is open

(ii) if the belt is crossed.

Sol. Given :

Distance between the centre of shafts (or of pulleys), x = 6 m = 600 cm

Dia. of larger pulley $= 60$ cm

$\therefore$ Radius of larger pulley, $r_1 = \frac{60}{2} = 30$ cm

Dia. of smaller pulley $= 40$ cm

$\therefore$ Radius of smaller pulley, $r_2 = \frac{40}{2} = 20$ cm.

Let L = Required length of the belt.

(i) If belt is open

Exact length. Exact length for open belt drive is given by equation (7.7A), as

$$L = 2\left[\frac{\pi}{2}(r_1 + r_2) + \alpha\,(r_1 - r_2) + \sqrt{x^2 - (r_1 - r_2)^2}\right]$$

where α for open belt drive is given by

$$\sin\alpha = \frac{r_1 - r_2}{x}$$

$$\therefore \quad \alpha = \sin^{-1}\left(\frac{r_1 - r_2}{x}\right) = \sin^{-1}\left(\frac{30 - 20}{600}\right)$$

$$= \sin^{-1}\left(\frac{10}{600}\right) = \sin^{-1} 0.01666 = 0.955°$$

$$= 0.955 \times \frac{\pi}{180} \text{ radians} = 0.01666 \text{ radians.}$$

The value of α is 0.955° which is very small. Hence α in radians will be equal to $\sin\alpha$. Here also α in radian $= 0.01666$ and $\sin\alpha = \frac{1}{60} = 0.01666$.

Substituting, this value of α and also values of r_1, r_2 and x in the above equation, we get exact length as

$$L = 2\left[\frac{\pi}{2}(30 + 20) + 0.01666(30 - 20) + \sqrt{600^2 - (30 - 20)^2}\right]$$

$$= 2\,[\,25\pi + 0.1666 + 599.916\,] = 2\,[678.62] = \mathbf{1357.24 \text{ cm.}} \quad \textbf{Ans.}$$

Approximate length. Using equation (7.8), we get

$$L = \pi(r_1 + r_2) + \frac{(r_1 - r_2)^2}{x} + 2x$$

$$= \pi(30 + 20) + \frac{(30 - 20)^2}{600} + 2 \times 600 = \mathbf{1357.16 \text{ cm.}} \quad \textbf{Ans.}$$

(ii) If belt is crossed

Exact length. Exact length for cross-belt drive is given by equation (7.8A) as

$$L = 2\left[\frac{\pi}{2}(r_1 + r_2) + \alpha\,(r_1 + r_2) + \sqrt{x^2 - (r_1 + r_2)^2}\right]$$

where α is given by,

$$\alpha = \sin^{-1}\frac{r_1 + r_2}{x} = \sin^{-1}\left(\frac{30 + 20}{600}\right) = \sin^{-1} 0.08333 = 4.78°$$

$$= 4.78 \times \frac{\pi}{180} \text{ radians} = 0.08343 \text{ radians.}$$

Substituting this value of α in the above equation, we get exact length as

$$L = 2\left[\frac{\pi}{2}(30+20) + 0.08343(30+20) + \sqrt{600^2 - (30+20)^2}\right]$$

$$= 2\,[25\pi + 4.1715 + 597.913] = 2[680.6243] = \mathbf{1361.24\ cm.}\quad \textbf{Ans.}$$

Approximate Length. Using equation (7.9), we get

$$L = \pi(r_1 + r_2) + \frac{(r_1 + r_2)^2}{x} + 2x$$

$$= \pi(30+20) + \frac{(30+20)^2}{600} + 2 \times 600 = \mathbf{1361.31\ cm.}\quad \textbf{Ans.}$$

Problem 7.8. *A shaft which rotates at a constant speed of 160 r.p.m. is connected by belting to a parallel shaft 180 cm apart, which has to run at 60, 80 and 100 r.p.m. The smallest pulley on the driver shaft is 15 cm in radius. Determine the remaining radii of the two stepped pulleys for (a) a crossed belt, (b) an open belt. Neglect belt thickness and slip.* (A.M.I.E., Summer 1982)

Sol. Given :

Speed of driver shaft = 160 r.p.m.

$\therefore$ Speed of the pulleys 1, 3 and 5 keyed to the driver shaft = 160 r.p.m.

This means $N_1 = N_3 = N_5 = 160$ r.p.m.

Distance between the centre of shafts, $x = 180$ cm.

The speed of the driven shaft should be 60, 80 and 100 r.p.m. This means $N_2 = 60$ r.p.m. $N_4 = 80$ r.p.m. and $N_6 = 100$ r.p.m.

Radius of smallest pulley on the driver shaft, $r_1 = 15$ cm.

Let r_2, r_3, r_4, r_5 and r_6 be the radii of the pulleys 2, 3, 4, 5 and 6 respectively as shown in Fig. 7.8.

(*a*) *For a crossed belt.*

From equation (7.1), we get

$$\frac{N_2}{N_1} = \frac{d_1}{d_2} = \frac{2r_1}{2r_2} = \frac{r_1}{r_2}$$

$$\therefore \quad r_2 = r_1 \times \frac{N_1}{N_2} = 15 \times \frac{160}{60} = \mathbf{40\ cm.}\quad \textbf{Ans.}$$

The pulley 3 is on driver shaft and corresponding pulley 4 is on the driven shaft.

$$\therefore \quad \frac{N_4}{N_3} = \frac{d_3}{d_4} = \frac{2r_3}{2r_4} = \frac{r_3}{r_4}$$

$$\therefore \quad r_4 = r_3 \times \frac{N_3}{N_4} = r_3 \times \frac{160}{80} = 2r_3 \quad \ldots(i)$$

But in case of crossed-belt system, the length of the belt will be constant when the sum of radii of corresponding steps may be same, *i.e.*,

$$r_1 + r_2 = r_3 + r_4 = r_5 + r_6$$

But $\quad r_1 + r_2 = 15 + 40 = 55$ cm

$$\therefore \quad r_3 + r_4 = r_5 + r_6 = 55 \text{ cm} \quad \ldots(ii)$$

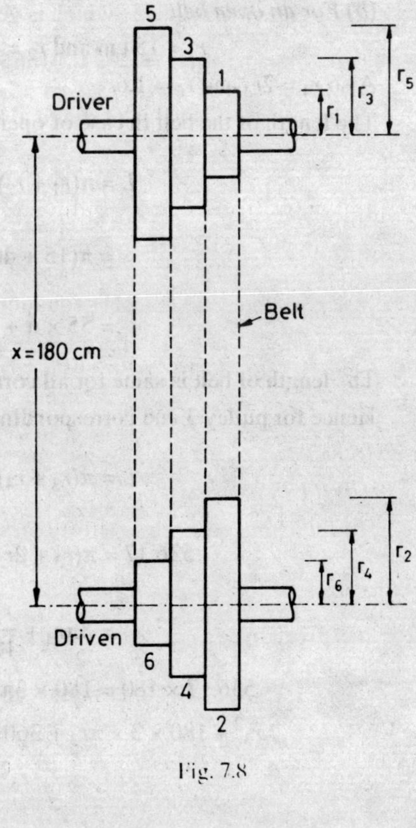

Fig. 7.8

or $$r_3 + r_4 = 55 \text{ cm}$$

or $$r_3 + 2r_3 = 55 \text{ cm} \qquad (\because \text{ From equation } (i), r_4 = 2r_3)$$

or $$3r_3 = 55 \text{ cm}$$

$$\therefore \quad r_3 = \frac{55}{3} = \mathbf{18.33 \text{ cm.}} \quad \textbf{Ans.}$$

Substituting this value in equation (*i*), we get

$$r_4 = 2 \times 18.33 = \mathbf{36.66 \text{ cm.}} \quad \textbf{Ans.}$$

Now again the pulley 5 is on the driver shaft and corresponding pulley 6 is on the driven shaft.

$$\therefore \quad \frac{N_6}{N_5} = \frac{r_5}{r_6}$$

$$\therefore \quad r_6 = r_5 \times \frac{N_5}{N_6} = r_5 \times \frac{160}{100} 1.6r_5 \qquad ...(iii)$$

But from equation (*ii*), we have

$$r_5 + r_6 = 55$$

or $$r_5 + 1.6r_5 = 55 \qquad (\because \quad r_6 = 1.6r_5)$$

or $$r_5 = \frac{55}{2.6} = \mathbf{21.15 \text{ cm.}} \quad \textbf{Ans.}$$

Substituting this value in equation (*iii*), we get

$$r_6 = 1.6 \times 21.15 = \mathbf{33.85 \text{ cm.}} \quad \textbf{Ans.}$$

(*b*) *For an open belt*

$$r_1 = 15 \text{ cm and } r_2 = 40 \text{ cm as before.}$$

Also $r_4 = 2r_3$ and $r_6 = 1.6r_5$ (See equations (*i*) and (*iii*) above)

The length of the belt in case of open-belt system is given by equation (7.8) as

$$L = \pi(r_1 + r_2) + \frac{(r_1 - r_2)^2}{x} + 2x$$

$$= \pi(15 + 40) + \frac{(15 - 40)^2}{180} + 2 \times 180$$

$$= 55 \times \pi + \frac{25}{180} + 2 \times 180 = 536.17 \text{ cm.}$$

The length of belt is same for all corresponding steps.

Hence for pulley 3 and corresponding pulley 4, we will have

$$L = \pi(r_3 + r_4) + \frac{(r_4 - r_3)^2}{x} + 2x$$

or $$536.17 = \pi(r_3 + 2r_3) + \frac{(2r_3 - r_3)^2}{180} + 2 \times 180 \qquad (\because \quad r_4 = 2r_3)$$

$$= 3\pi r_3 + \frac{r_3^{\,2}}{180} + 360$$

or $$536.17 \times 180 = 180 \times 3\pi r_3 + r_3^{\,2} + 360 \times 180$$

or $$r_3^{\,2} + 180 \times 3 \times \pi r_3 + 360 \times 180 - 536.17 \times 180 = 0$$

or $$r_3^{\,2} + 1696.5 r_3 - 31710.6 = 0$$

$$\therefore \quad r_3 = \frac{-1696.5 \pm \sqrt{1696.5^2 + 4 \times 31710.6}}{2}$$

$$= \frac{-1696.5 \pm 1733.5}{2} = \mathbf{18.5\ cm.} \quad \textbf{Ans.} \qquad \text{(Neglecting – ve root)}$$

$$\therefore \quad r_4 = 2 \times r_3 = 2 \times 18.5 = \mathbf{37.0\ cm.} \quad \textbf{Ans.}$$

Similarly, for the pulley 5 and corresponding pulley 6, we will have

$$L = \pi(r_6 + r_5) + \frac{(r_6 - r_5)^2}{x} + 2x$$

or
$$536.17 = \pi(1.6r_5 + r_5) + \frac{(1.6r_5 - r_5)^2}{180} + 2 \times 180 \qquad (\because \ r_6 = 1.6r_5)$$

$$= \pi \times 2.6r_5 + \frac{0.6^2 \times r_5^2}{180} + 360 = 8.168r_5 + 0.002r_5^2 + 360$$

or
$$0.002r_5^2 + 8.168r_5 + 360 - 536.17 = 0$$

or
$$r_5^2 + \frac{8.168}{0.002} r_5 + \frac{360 - 536.17}{0.002} = 0$$

or
$$r_5^2 + 4084r_5 - 88085 = 0$$

or
$$r_5 = \frac{-4084 \pm \sqrt{4084^2 + 4 \times 88085}}{2}$$

$$= \frac{-4084 \pm 4126.9}{2} = \mathbf{21.45\ cm.} \quad \textbf{Ans.} \qquad \text{(Neglecting – ve root)}$$

$$\therefore \quad r_6 = 1.6 \times r_5 = 1.6 \times 21.45 = \mathbf{34.32\ cm.} \quad \textbf{Ans.}$$

7.6. Ratio of Belt Tensions

Fig. 7.9 shows a driver pulley A and driven pulley B rotating in the clock-wise direction. Fig. 7.10 shows only the driven pulley B. Consider the driven pulley B.

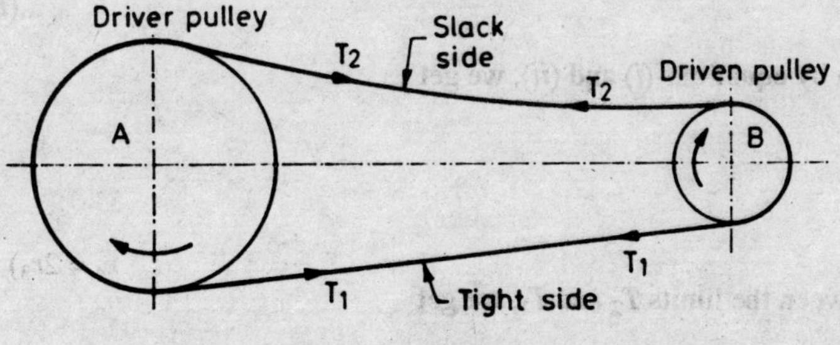

Fig. 7.9

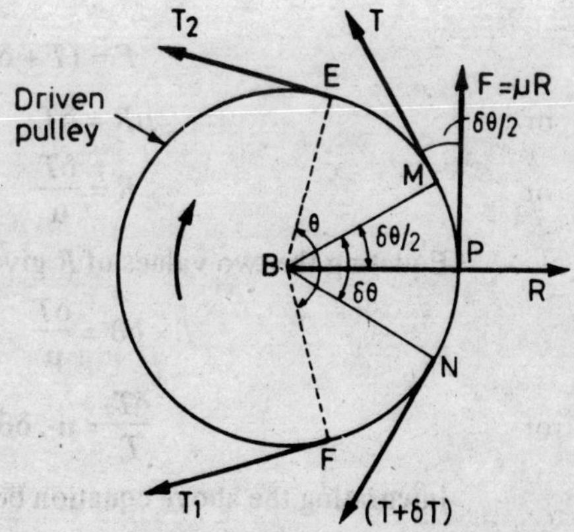

Fig. 7.10

Let T_1 = Tension in the belt on the tight side

T_2 = Tension in the belt on the slack side

θ = Angle of contact, *i.e.*, the angle subtended by the arc EF at the centre of the driven pulley.

μ = Co-efficient of friction between the belt and pulley.

The ratio of the two tensions may be found by considering an elemental piece of the belt *MN* subtending an angle $\delta\theta$ at the centre of the pulley *B* as shown in Fig. 7.10. The various forces which keep the elemental piece *MN* in equilibrium are :

(*i*) Tension *T* in the belt at *M* acting tangentially,

(*ii*) Tension $T + \delta T$ in the belt at *N* acting tangentially,

(*iii*) Normal reaction *R* acting radially outward at *P*, where *P* is the middle point of *MN*,

(*iv*) Frictional force $F = \mu R$ acting at right angles to *R* and in the opposite direction of the motion of pulley.

Now angle $PBM = \frac{\delta\theta}{2}$. Also angle $TPF = \frac{\delta\theta}{2}$.

Resolving all the forces acting on the belt *MN* in the horizontal direction, we get

$$R = T\sin\frac{\delta\theta}{2} + (T + \delta T)\sin\frac{\delta\theta}{2}$$

Since the angle $\delta\theta$ is very small, $\sin\frac{\delta\theta}{2}$ can be written as $\frac{\delta\theta}{2}$. Hence the above equation becomes as

$$R = T \times \frac{\delta\theta}{2} + (T + \delta T) \times \frac{\delta\theta}{2}$$

$$= T \times \frac{\delta\theta}{2} + T \times \frac{\delta\theta}{2} + \delta T \times \frac{\delta\theta}{2}$$

$$= T \times \delta\theta + \frac{\delta T \times \delta\theta}{2}$$

$$= T \times \delta\theta \qquad \left(\text{Neglecting the small quantity } \frac{\delta T \times \delta\theta}{2}\right) \qquad ...(i)$$

Now resolving all the forces vertically, we get

$$F = (T + \delta T)\cos\frac{\delta\theta}{2} - T\cos\frac{\delta\theta}{2}$$

Since $\delta\theta$ is very small, hence $\cos\frac{\delta\theta}{2}$ reduces to unity *i.e.* 1. Hence the above equation becomes as

$$F = (T + \delta T) - T = \delta T$$

or $$\mu R = \delta T \qquad (\because\ F = \mu R)$$

or $$R = \frac{\delta T}{\mu} \qquad ...(ii)$$

Equating the two values of *R* given by equations (*i*) and (*ii*), we get

$$T \times \delta\theta = \frac{\delta T}{\mu}$$

or $$\frac{\delta T}{T} = \mu \,.\, \delta\theta$$

Integrating the above equation between the limits T_2 and T_1, we get

$$\int_{T_2}^{T_1} \frac{dT}{T} = \int \mu \,.\, d\theta = \mu \int d\theta$$

or $$\log_e \frac{T_1}{T_2} = \mu \times \theta$$

or $$\frac{T_1}{T_2} = e^{\mu \times \theta} \quad ...(7.10)$$

In equation (7.10), θ should be taken in radians. Here θ is known as angle of contact. For an open belt or for a crossed belt the angle of contact is determined as given in Article 7.6.1.

7.6.1. Angle of contact for open bale drive. With an open belt drive, the belt will begin to slip on the smaller pulley, since the angle of lap is smaller on this pulley than on the large pulley. The angle θ should be taken as the minimum angle of contact. Hence in equation (7.10), the angle of contact of lap (θ) at the smaller pulley must be taken into consideration.

With reference to Fig. 7.6 for open belt in Art. 7.5.1 on page 288, the angle of lap on the smaller pulley is equal to angle *EBF* which is equal to (180° – 2α).

∴ Angle of contact, $\theta = (180 - 2\alpha)$ ...(7.11)

But the value of α is given by,

$$\sin \alpha = \frac{r_1 - r_2}{x} \quad ...(7.12)$$

where r_1 = Radius of larger pulley,

r_2 = Radius of smaller pulley, and

x = Distance between the centres of two pulleys.

7.6.2. Angle of contact for crossed belt drive. For a crossed belt drive, the angle of lap on both the pulleys is same. With reference to Fig. 7.7, for crossed belt in Art. 7.5.2 on page 290, the angle of lap on the smaller or larger pulley is equal to 180 + 2α.

∴ Angle of contact, $\theta = (180 + 2\alpha)$. ...(7.13)

The value of α is given by,

$$\sin \alpha = \frac{r_1 + r_2}{x} \quad ...(7.14)$$

where r_1 = Radius of larger pulley,

r_2 = Radius of smaller pulley, and

x = Distance between the centres of the two pulleys.

7.7. Power Transmitted by Belt

Let T_1 = Tension in the right side of the belt

T_2 = Tension in the slack side of the belt

v = Velocity of the belt in metre/sec.

The effective tension or force acting at the circumference of the driven pulley is the difference between the two tensions (*i.e.*, $T_1 - T_2$).

∴ Effective driving force $= (T_1 - T_2)$

∴ Work done per second = Force × Velocity

$= (T_1 - T_2) \times v$ Nm

∴ Power transmitted $= (T_1 - T_2) \times v$ Watts ...(7.15)

$$= \frac{(T_1 - T_2) \times v}{1000} \text{ kW} \quad ...(7.16)$$

Equation (7.15) gives the power in Watts, whereas equation (7.16) gives the power in kW. In case of equation (7.16), the tensions T_1 and T_4 are taken in Newtons.

Torque exerted on the driving pulley $= (T_1 - T_2) \times r_1$ ...(7.17)

and Torque exerted on the driven pulley $= (T_1 - T_2) \times r_2$ ...(7.18)

Problem 7.9. *A belt is running over a pulley of diameter 120 cm at 200 r.p.m. The angle of contact is 165° and co-efficient of friction between the belt and pulley is 0.3. If the maximum tension in the belt is 3000 N, find the power transmitted by the belt.*

Sol. Given :

Dia. of pulley, $d = 120 \text{ cm} = 1.2 \text{ m}$

Speed of pulley, $N = 200$ r.p.m.

Angle of contact, $\theta = 165° = 165 \times \dfrac{\pi}{180}$ radians. $\left(\because 1° = \dfrac{\pi}{180} \text{ rad.}\right)$

Co-efficient of friction, $\mu = 0.3$

Max. tansion, $T_1 = 3000$ N

Velocity of belt, $v = \dfrac{\pi d N}{60} = \dfrac{\pi \times 1.2 \times 200}{60} = 12.56$ m/s

Let T_2 = Tension on the slack side of the belt.

Using equation (7.10), we get

$$\frac{T_1}{T_2} = e^{\mu \times \theta}$$

$$= e^{0.3 \times 165 \times \pi/180} = e^{0.8635} = 2.3714$$

or $$\frac{3000}{T_2} = 2.3714$$

$\therefore$ $$T_2 = \frac{3000}{2.3714} = 1265 \text{ N}$$

Power transmitted is given by equation (7.16) as,

$$P = \frac{(T_1 - T_2) \times v}{1000}$$

$$= \frac{(3000 - 1265) \times 12.56}{1000} = \mathbf{21.79\ kW.} \quad \textbf{Ans.}$$

Problem 7.10. *An open-belt drive connects two pulleys 120 cm and 50 cm diameters, on parallel shafts 4 m apart. The maximum tension in the belt is 1855.3 N. The co-efficient of friction is 0.3. The driver pulley of diameter 120 cm runs at 200 r.p.m. Calculate :*

(*i*) *the power transmitted, and*

(*ii*) *torque on each of the two shafts.*

Sol. Given :

Dia. of larger pulley, $d_1 = 120 \text{ cm} = 1.20 \text{ m}$

$\therefore$ Radius of larger pulley, $r_1 = \dfrac{120}{2} = 60 \text{ cm} = 0.6 \text{ m}$

Dia. of smaller pulley, $d_2 = 50 \text{ cm} = 0.50 \text{ m}$

$\therefore$ Radius of smaller pulley, $r_2 = 25 \text{ cm} = 0.25 \text{ m}$

Distance between the shafts, $x = 4$ m

Max. tension, $T_1 = 1855.3$ N

Co-efficient of friction, $\mu = 0.3$

Speed of driver pulley, $N_1 = 200$ r.p.m.

We know that velocity of belt is given by,

$$v = \frac{\pi d_1 N_1}{60} = \frac{\pi \times 1.2 \times 200}{60} = 12.56 \text{ m/s}$$

Let us now calculate the angle of contact (θ). For an open belt drive angle of contact is given by equation (7.11) as

$$\theta = 180 - 2\alpha \qquad ...(i)$$

where α is given by equation (7.12) as

$$\sin \alpha = \frac{r_1 - r_2}{x} = \frac{0.6 - 0.25}{4} = 0.0875$$

$$\therefore \quad \alpha = \sin^{-1} 0.0875 = 5.02°$$

Substituting this value of α in equation (*i*), we get

$$\theta = 180 - 2 \times 5.02 = 169.96°$$

$$= 169.96 \times \frac{\pi}{160} \text{ radians} = 2.967 \text{ radians.}$$

Let T_2 = Tension on the slack side of the belt.

Now using equation (7.10),

$$\frac{T_1}{T_2} = e^{\mu \times \theta}$$

or

$$\frac{185.53}{T_2} = e^{0.3 \times 2.967} = e^{0.8901} = 2.435$$

$$\therefore \quad T_2 = \frac{1855.3}{2.435} = 761.8 \text{ N}$$

(*i*) *Power transmitted*. Using equation (7.16),

$$P = \frac{(T_1 - T_2) \times v}{1000}$$

$$= \frac{(1855.3 - 761.8) \times 12.56}{1000} = \mathbf{13.73\ kW.} \quad \textbf{Ans.}$$

(*ii*) *Torque on each of the two shafts*. Torque exerted on the driving shaft is given by equation (7.17).

$\therefore$ Torque $= (T_1 - T_2) \times r_1$

$= (1855.3 - 761.8) \times 0.6 =$ **656.1 Nm. Ans.**

Torque exerted on the driven shaft is given by equation (7.18)

$\therefore$ Torque $= (T_1 - T_2) \times r_2$

$= (1855.3 - 761.8) \times 0.25 =$ **273.4 Nm. Ans.**

7.8. Centrifugal Tension

The tension caused in the running belt by the centrifugal force is known as centrifugal tension. Whenever a particle of mass m is rotated in a circular path of radius r at a uniform velocity v, a centrifugal force is acting radially outward and its magnitude is equal to $\frac{mv^2}{r}$, where m is the mass of the particle.

The centrifugal tension in the belt can be calculated by considering the forces acting on an elemental length of the belt (*i.e.*, length MN) subtending an angle $\delta\theta$ at the centre as shown in Fig. 7.11.

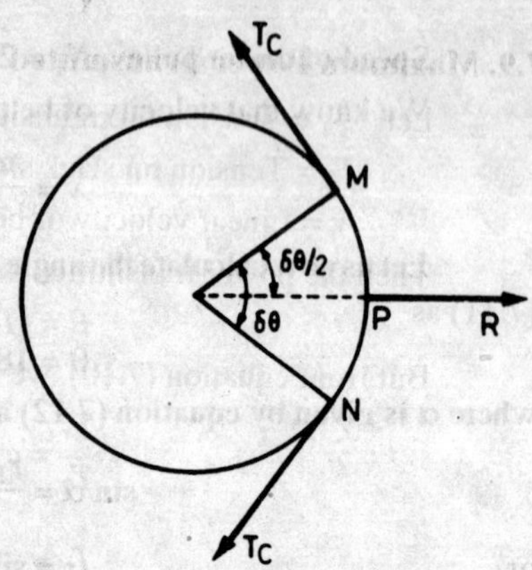

Fig. 7.11

Let v = Velocity of belt in metre/sec

r = Radius of pulley over which belt runs

M = Mass of elemental length of belt MN

m = Mass of belt per metre length

T_C = Centrifugal tension acting at M and N tangentially

R = Centrifugal force acting radially outward

The centrifugal force R acting radially outwards is balanced by the components of T_C acting radially inwards. Now elemental length of belt $MN = r \cdot \delta\theta$.

$\therefore$ Mass of belt MN

= (Mass per metre length) × length of MN

$$M = m \times r \times \delta\theta$$

$\therefore$ Centrifugal force,

$$R = M \times \frac{v^2}{r} = m \times r \times \delta\theta \times \frac{v^2}{r}$$

Now resolving the forces horizontally, we get

$$T_C \sin\frac{\delta\theta}{2} + T_C \sin\frac{\delta\theta}{2} = R$$

or

$$2T_C \sin\frac{\delta\theta}{2} = m \times r \times \delta\theta \times \frac{v^2}{r}$$

As the angle $\delta\theta$ is very small, hence

$$\sin\frac{\delta\theta}{2} = \frac{\delta\theta}{2}.$$

Then the above equation becomes as

$$2T_C \times \frac{\delta\theta}{2} = m \times r \times \delta\theta \times \frac{v^2}{r}$$

$$\therefore \quad T_C = m \times v^2 \qquad ...(7.19)$$

Note. (*i*) From the above equation, it is clear that the centrifugal tension is independent of T_1 and T_2. It depends upon the velocity of the belt. For lower belt speed (*i.e.*, belt speed less than 10 m/s) the centrifugal tension is very small and may be neglected.

(*ii*) When centrifugal tension is to be taken into consideration then total tensions on right side and slack side of the belt is given as :

For tight side $= T_1 + T_C$ ...(7.20)

For slack side $= T_2 + T_C$ ...(7.21)

(*iii*) Maximum tension (T_m) in the belt is equal to maximum safe stress in the belt multiplied by cross-sectional area of the belt.

$$\therefore \quad T_m = f \times (b \times t) \qquad ...(7.22)$$

where f = Maximum safe stress in the belt,

b = Width of belt, and

t = Thickness of belt.

Then $T_m = T_1 + T_C$ ...... if centrifugal tension is to be considered

$= T_1$ ...... if centrifugal tension is to be neglected.

7.9. Maximum Power Transmitted by a Belt

Let T_1 = Tension on right side,

T_2 = Tension on slack side,

v = Linear velocity of belt.

Then the power transmitted is given by equation (7.15) as

$$P = (T_1 - T_2) \times v \quad ...(i)$$

But from equation (7.10), we know that

$$\frac{T_1}{T_2} = e^{\mu \times \theta}$$

or
$$T_2 = \frac{T_1}{e^{\mu \times \theta}}$$

Substituting the value of T_2 in equation (*i*), we get

$$P = \left(T_1 - \frac{T_1}{e^{\mu \times \theta}}\right) \times v$$

$$= T_1\left(1 - \frac{1}{e^{\mu \times \theta}}\right) \times v \quad ...(ii)$$

Let $\left(1 - \frac{1}{e^{\mu \times \theta}}\right) = k$

The above equation becomes as

$$P = T_1 \times k \times v \quad \text{or} \quad kT_1 v \quad ...(iii)$$

Let T_{max} = Maximum tension, and

T_C = Centrifugal tension which is equal to $m \times v^2$.

Then
$$T_{max} = T_1 + T_C$$

$\therefore$
$$T_1 = T_{max} - T_C$$

Substituting this value of T_1 in equation (*iii*), we get

$$P = k(T_{max} - T_C) \times v$$

$$= k\,(T_{max} - m \times v^2) \times v \quad (\because T_C = m \times v^2)$$

$$= k\,[T_{max} \times v - m \times v^3] \quad ...(iv)$$

The power transmitted will be maximum, if

$$\frac{dP}{dv} = 0$$

Hence differentiating equation (*iv*) w.r.t. v and equating to zero for maximum horse power, we get

$$\frac{dP}{dv} = k\,(T_{max} - 3\,m \times v^2) = 0$$

or
$$T_{max} - 3m \times v^2 = 0$$

or
$$T_{max} = 3m \times v^2 \quad ...(v)$$

or
$$v = \sqrt{\frac{T_{max}}{3m}} \quad ...(7.23)$$

Equation (7.23) gives the velocity of the belt at which maximum power is transmitted.

From equation (v), $T_{max} = 3T_C$ $\quad (\because\ m \times v^2 = T_C)$

Hence when the power transmitted is maximum, centrifugal tension would be $\frac{1}{3}$ rd of the maximum tension.

We also know that $T_{max} = T_1 + T_C$

$$= T_1 + \frac{T_{max}}{3} \qquad \left(\because\ T_C = \frac{T_m}{3}\right)$$

$$\therefore \quad T_1 = T_{max} - \frac{T_{max}}{3}$$

$$= \frac{2}{3} T_{max} \qquad ...(vi)$$

Hence **conditions** for the transmission of maximum power are :

(i) $T_C = \frac{1}{3} T_{max}$ and (ii) $T_1 = \frac{2}{3} T_{max}$.

The maximum power transmitted is obtained by substituting the values of T_1 (from (vi)) and v (from equation 7.23) in equation (ii).

$$\therefore \quad \text{Max. power} = \frac{2}{3} T_{max} \left(1 - \frac{1}{e^{\mu \times \theta}}\right) \times \sqrt{\frac{T_{max}}{3m}} \qquad ...(7.24)$$

Problem 7.11. *A belt embraces the shorter pulley by an angle of 165° and runs at a speed of 1700 metres/min. Dimensions of the belt are : Width = 20 cm and thickness = 8 mm. Its density is 1 gm/cm³. Determine the maximum power that can be transmitted at the above speed, if the maximum permissible stress in the belt is not to exceed 250 N/cm² and μ = 0.25.* (A.M.I.E., Winter 1982)

Sol. Given :

Angle of contact, $\theta = 165° = 165 \times \frac{\pi}{180} = 2.88$ rad

Speed of belt, $v = 1700$ m/min $= \frac{1700}{60} = 28.33$ m/s

Width of belt, $b = 20$ cm

Thickness of belt, $t = 8$ mm $= 0.8$ cm

Density of belt, $\rho = 1$ gm/cm³ $= \frac{1}{1000}$ kg/cm³

Max. permissible stress, $f = 250$ N/cm²

Value of $\mu = 0.25$

Let us first find the maximum tension and centrifugal tensions in the belt.

Using equation (7.22), maximum tension (T_m) is given by,

$$T_m = f \times (\text{Area of belt}) = f \times b \times t$$

$$= 250 \times 20 \times 0.8 = 4000 \text{ N}.$$

Centrifugal tension (T_C) is given by equation (7.19),

$$T_C = m \times v^2 \qquad ...(i)$$

where m = Mass of belt per metre length

$$= \rho \times \text{Volume of belt of 1 m length}$$

$$= \rho \times (\text{Area of belt} \times \text{length})$$

$$= \rho \times (b \times t \times L)$$

$$= \frac{1}{1000} \times (20 \times 0.8 \times 100) \text{ kg} \qquad (\because \text{ Length of 1 m = 100 cm})$$

$$= 1.6 \text{ kg}$$

Substituting this value in the above equation (*i*), we get

$$T_C = 1.6 \times 28.33^2 = 1284 \text{ N}$$

But we know that, $T_m = T_1 + T_C$

$$\therefore \quad T_1 = T_m - T_C = 4000 - 1284 = 2716 \text{ N}$$

Let T_2 = Tension on the slack side of the belt.

Using equation (7.10), we get

$$\frac{T_1}{T_2} = e^{\mu \times \theta} = e^{0.25 \times 2.88} = 2.056$$

$$\therefore \quad T_2 = \frac{T_1}{2.056} = \frac{2716}{2.056} = 1321 \text{ N}$$

Now the maximum power transmitted is given by,

$$P = \frac{(T_1 - T_2) \times v}{1000} = \frac{(2716 \times 1321)}{1000} \times 28.83 = \mathbf{39.52\ kW.} \quad \textbf{Ans.}$$

Problem 7.12. *A belt of density 1 gm/cm³ has a maximum permissible stress of 250 N/cm². Determine the maximum power that can be transmitted by a belt of 20 cm × 1.2 cm if the ratio of the tight side to slack side tension is 2.* (A.M.I.E., Summer 1987)

Sol. Given :

Density of belt, $\rho = 1 \text{ gm/cm}^3 = \frac{1}{1000} \text{ kg/cm}^3$

Max. permissible stress, $f = 250 \text{ N/cm}^2$

Width of belt, $b = 20$ cm

Thickness of belt, $t = 1.2$ cm

Ratio of tensions, $\frac{T_1}{T_2} = 2.0.$

Let us first find the mass of 1 m length of the belt and also the maximum tension in the belt.

Let m = Mass of one metre length of belt

= (Density) × Volume of belt of one metre length

$$= \frac{1}{1000} \times (b \times t \times 100) \qquad (\because \text{ 1 m = 100 cm})$$

$$= \frac{1}{1000} \times 20 \times 1.2 \times 100 \text{ kg} = 2.40 \text{ kg}$$

Using equation (7.22),

T_m = Maximum tension

= (Max. stress) × Area of cross-section of belt

$$= 250 \times b \times t = 250 \times 20 \times 1.2 = 6000 \text{ N.}$$

Now for maximum power transmitted, the velocity of the belt is given by equation (7.23) as

$$v = \sqrt{\frac{T_m}{3m}} = \sqrt{\frac{6000}{3 \times 2.4}} \text{ m/s} = 28.86 \text{ m/s}$$

Now power transmitted is given by equation,

$$P = \frac{(T_1 - T_2) \times v}{1000} \quad ...(i)$$

Let us find the values of T_1 and T_2.

We know that $T_m = T_1 + T_C$ ...(ii)

where T_C = Centrifugal tension, and

T_1 = Tight side tension.

But for maximum power transmission,

$$T_C = \tfrac{1}{3} T_m.$$

Substituting this value in equation (ii), we get

$$T_m = T_1 + \tfrac{1}{3} T_m$$

or $$T_1 = T_m - \tfrac{1}{3} T_m = \tfrac{2}{3} T_m = \tfrac{2}{3} \times 6000 = 4000 \text{ N}$$

But $$\frac{T_1}{T_2} = 2 \quad \text{(given)}$$

$\therefore$ $$T_2 = \frac{T_1}{2} = \frac{4000}{2} = 2000 \text{ N.}$$

Substituting the values of T_1, T_2 and v in equation (i), we get

$$P = \frac{(4000 - 2000) \times 28.59}{1000} = \mathbf{57.18\ kW.} \quad \textbf{Ans.}$$

Note. In this question, speed of belt is not given. Hence for maximum power transmission, speed of the belt is given by equation (7.23). But in question 7.11, the speed of the belt is given and hence there is no need of calculating velocity of belt by using equation (7.23). In that question, maximum power transmitted at the given speed is to be calculated.

Problem 7.13. *A leather belt is required to transmit 9 kW from a pulley 120 cm in diameter running at 200 r.p.m. The angle embraced is 165° and the co-efficient of friction between leather belt and pulley is 0.3. If the safe working stress for the leather belt is 140 N/cm², the mass of leather is 1 gm/cm³ and the thickness of the belt is 10 mm, determine the width of the belt taking the centrifugal tension into account.*

(A.M.I.E., Summer 1976)

Sol. Given :

Power, $P = 9$ kW

Dia. of pulley, $d_1 = 120$ cm $= 1.20$ m

Speed of pulley, $N_1 = 200$ r.p.m.

Angle embraced, $\theta = 165° = 165 \times \frac{\pi}{180}$ radians

Co-efficient of friction, $\mu = 0.3$

Safe working stress, $f = 140$ N/cm^2

Density of leather, $\rho = 1$ gm/cm$^3 = \frac{1}{1000}$ kg/cm^3.

Thickness of belt, $t = 10$ mm $= 1.0$ cm.

Let b = Width of belt in cm.

Using equation (7.10),

$$\frac{T_1}{T_2} = e^{\mu\theta} = e^{0.3 \times 165 \times \frac{\pi}{180}}$$

$$= e^{0.8635} = 2.3714 \qquad ...(i)$$

Maximum tension in the belt is given by equation (7.22),

$$T_m = (\text{Safe stress}) \times \text{Area of belt} = 140 \times (b \times t)$$

$$= 140 \times (b \times 1) = 140\, b \text{ N}$$

Peripheral velocity (v) of the driver pulley (or belt) is given by,

$$v = \frac{\pi d_1 \times N_1}{60}$$

$$= \frac{\pi \times 1.2 \times 200}{60} = 12.56 \text{ m/s}$$

Centrifugal tension (T_C) is given by equation (7.19),

$$T_C = m \times v^2 \qquad ...(ii)$$

where m = Mass of belt per metre length

$$= \rho \times b \times t \times 100 \text{ N} \qquad (\because \ 1 \text{ m} = 100 \text{ cm})$$

$$= \frac{1}{1000} \times b \times 1 \times 100 = \frac{b}{10} \text{ N}$$

Substituting this value in equation (*ii*), we get

$$T_C = \frac{b}{10} \times v^2 = \frac{b}{10} \times (12.56)^2 = 16\, b \text{ N}$$

We know, $$P = \frac{(T_1 - T_2) \times v}{1000}$$

or $$9.0 = \frac{(T_1 - T_2) \times 12.56}{1000}$$

$\therefore$ $$T_1 - T_2 = \frac{9 \times 1000}{12.56} = 716.5 \qquad ...(iii)$$

But from equation (*i*), $T_1 = 2.3714\, T_2$.

Hence equation (*iii*) becomes as,

$$2.3714\, T_2 - T_2 = 716.5$$

or $$1.3714\, T_2 = 716.5$$

$\therefore$ $$T_2 = \frac{716.5}{1.3714} = 522.46 \text{ N}$$

and $$T_1 = 2.3714 \times 522.46 = 1238.96 \text{ N}$$

But $$T_m = T_1 + T_C$$

or $$140b = 1238.96 + 16b \qquad (\because \ T_m = 140b \text{ and } T_C = 16b)$$

or $$140b - 16b = 1238.96$$

or $$124b = 1238.96$$

$$b = \frac{1238.96}{124} = \mathbf{9.99 \text{ cm.}} \quad \textbf{Ans.}$$

Problem 7.14. (S.I. Units). *Determine the maximum power that can be transmitted using a belt of 100 mm × 10 mm with an angle of lap of 160°. The density of belt is 10^{-3} gm/mm^3 and co-efficient of friction may be taken as 0.25. The tension in the belt should not exceed 1.5 N/mm^2.*

(Converted to S.I. units, A.M.I.E., Summer 1984)

Sol. Given :

Width of belt, $b = 100$ mm

Thickness of belt, $t = 10$ mm

Angle of lap, $\theta = 160° = 160 \times \frac{\pi}{180} = 2.79$ radians

Density of belt, $\rho = 10^{-3}$ gm/mm^3 $= \frac{10^{-3}}{1000}$ kg/mm^3 $= 10^{-6}$ kg/mm^3

Co-efficient of friction, $\mu = 0.25$

Max. stress in the belt, $f = 1.5$ N/mm^2

Max. tension is given by equation (7.22),

$$T_m = f \times b \times t = 1.5 \times 100 \times 10 = 1500 \text{ N}$$

Now using equation (7.10), we get

$$\frac{T_1}{T_2} = e^{\mu \times \theta} = e^{0.25 \times 2.79} = e^{0.6975}$$

$$= 2.008 \simeq 2.0 \qquad \text{...(i)}$$

For maximum power transmission, we know that

$$T_C = \frac{1}{3} T_m = \frac{1500}{3} = 500 \text{ N}$$

But $T_m = T_1 + T_C$

or $1500 = T_1 + 500$

$\therefore$ $T_1 = 1500 - 50 = 1000$ N

From equation (*i*), $T_2 = \frac{T_1}{2} = \frac{1000}{2} = 500$ N

Let m = Mass of one metre length of the belt

$= \rho \times b \times t \times 1000$ $\qquad (\because 1 \text{ m} = 1000 \text{ mm})$

$= 10^{-6} \times 100 \times 10 \times 1000 = 1$ kg

For maximum power transmitted, the velocity of belt is given by,

$$v = \sqrt{\frac{T_m}{3m}} = \sqrt{\frac{1500}{3 \times 1}} = \sqrt{500} = 22.36 \text{ m/s.}$$

Now the power transmitted in S.I. unit is given by equation (7.15) as

$$P = (T_1 - T_2) \times v \text{ Watt}$$

$$= (1000 - 500) \times 22.36 = 11180 \text{ W}$$

$$= \mathbf{11.18 \text{ kW.}} \quad \textbf{Ans.}$$

7.10. Initial Tension in the Belt

The tension in the belt which is passing over the two pulleys (*i.e.*, driver and follower) when the pulleys are *stationary* is known as initial tension in the belt.

When power is transmitted from one shaft to another shaft with the help of the belt, passing over the two pulleys which are keyed to the driver and driven shafts, there should be firm grip between the pulleys and belt. When the pulleys are stationary, this firm grip is increased, by tightening the two ends of the belt. Hence the belt is subjected to some tension. This tension is known as *initial tension* in the belt.

Let T_0 = Initial tension in the belt,

T_1 = Tension in the tight side of the belt,

T_2 = Tension in the slack side of the belt,

α = Co-efficient of increase of belt length per unit force.

When power is supplied to one of the pulleys and transmitted to the other, the tension in the two free length of belts will be changed. The tight side of the belt stretches until the pull is increased from T_0 to T_1 and slack side shortens until the pull is decreased from T_0 to T_2.

$\therefore$ Increase of tension in tight side $= (T_1 - T_0)$

and decrease in tension in slack side $= T_0 - T_2$

Now increase in length of belt on tight side

$= \alpha \times$ increase in tension

$= \alpha \times (T_1 - T_0)$

Similarly, decrease in length of belt on slack side

$= \alpha$. Decrease in tension

$= \alpha \,.\, (T_0 - T_2)$

For a perfectly elastic material of the belt, the length of the belt remains constants when it is at rest or in motion.

Hence increase of length on the tight side should be equal to decrease of length on the slack side.

$$\therefore \quad \alpha \,.\, (T_1 - T_0) = \alpha \,.\, (T_0 - T_2)$$

or
$$T_1 - T_0 = T_0 - T_2$$

or
$$T_0 = \frac{T_1 + T_2}{2} \quad \text{...(7.25)}$$

If centrifugal tension is considered, then

$$T_0 = \frac{T_1 + T_2 + 2T_C}{2}. \quad \text{...(7.26)}$$

Problem 7.15. *An open belt running over two pulleys 24 cm and 60 cm diameters connects two parallel shafts 3 m apart and transmits 3.75 kW from the smaller pulley that rotates at 300 r.p.m. co- efficient of friction between the belt and the pulleys is 0.3 and the safe working tension in 100 N/cm width. Determine :*

(i) minimum width of the belt,

(ii) initial belt tension, and

(iii) length of the belt required (A.M.I.E., Winter 1978)

Sol. Given :

Dia. of larger pulley, $d_1 = 60$ cm $\therefore$ $r_1 = 30$ cm

Dia. of smaller pulley, $d_2 = 24$ cm $\therefore$ $r_2 = 12$ cm

Distance between two shafts, $x = 3$ m $= 300$ cm

Power transmitted = 3.75 kW

Speed of smaller pulley, $N_2 = 300$ r.p.m.

Co-efficients of friction, $\mu = 0.3$

Safe working tension = 100 N per cm width

Let b = Width of belt in cm

Then $T_{max} = 100 \times b = 100b$ N

Let us calculate the angle of contact (θ).

For an open belt, angle of contact is given by equation (7.11) as

$$\theta = 180 - 2\alpha \quad \text{...}(i)$$

where α is given by equation (7.12) as

$$\sin \alpha = \frac{r_1 - r_2}{x} = \frac{30 - 12}{300} = 0.06$$

$$\therefore \quad \alpha = \sin^{-1} 0.06 = 3.45°$$

Substituting this value of α in equation (*i*), we get

$$\theta = 180 - 2 \times 3.45° = 173.1°$$

$$= 173.1 \times \frac{\pi}{180} \text{ rad} = 3.02 \text{ radians.}$$

Now using equation (7.10),

$$\frac{T_1}{T_2} = e^{\mu \times \theta}$$

$$= e^{0.3 \times 3.02} = e^{0.906} = 2.474$$

$$\therefore \quad T_1 = 2.474\, T_2 \quad \text{...}(ii)$$

Power transmitted is given by,

$$P = \frac{(T_1 - T_2) \times v}{1000}$$

or

$$3.75 = \frac{(T_1 - T_2) \times v}{1000}$$

where v = velocity of belt

$$= \frac{\pi d_2 N_2}{60}$$

$$= \frac{\pi \times 24 \times 300}{60} = 376.9 \text{ cm/s} = 3.769 \text{ m/s} \simeq 3.77 \text{ m/s}$$

Substituting this value of v in the above equation, we get

$$3.75 = \frac{(T_1 - T_2) \times 3.77}{1000}$$

$$\therefore \quad (T_1 - T_2) = \frac{3.75 \times 1000}{3.77} = 994.7 \text{ N.}$$

Substituting the value of T_1 from equation (*ii*), we get

$$2.474\, T_2 - T_2 = 994.7$$

or

$$1.474\, T_2 = 994.7$$

$$\therefore \quad T_2 = \frac{994.7}{1.474} = 674.8 \text{ N}$$

and

$$T_1 = 2.474 \times T_2 = 2.474 \times 674.8 \text{ N} = 1669.5 \text{ N}$$

(*i*) *Minimum width of the belt.* Neglecting the effect of centrifugal tension, we have

$$T_{max} = T_1$$

or $$100 \times b = 1669.5$$

$\therefore$ $$b = \frac{1669.5}{100} = \textbf{16.7 cm.} \quad \textbf{Ans.}$$

(*ii*) *Initial belt tension (T_0).* Using equation (7.25), we get initial tension as

$$T_0 = \frac{T_1 + T_2}{2}$$

$$= \frac{1669.5 \times 674.8}{2} = \textbf{1172.15 N.} \quad \textbf{Ans.}$$

(*iii*) *Length of the belt required.* For an open belt system, the length of the belt (L) is given by equation (7.8),

$$L = \pi(r_1 + r_2) + \frac{(r_1 - r_2)^2}{x} + 2x$$

$$= \pi(30 + 12) + \frac{(30 - 12)^2}{300} + 2 \times 300 = 733 \text{ cm} = \textbf{7.33 m.} \quad \textbf{Ans.}$$

Problem 7.16. (S.I. Units). *An open belt running over two pulleys 1.5 m and 1.0 m diameters connects two parallel shafts 4.80 m apart. The initial tension in the belt when stationary is 3000 N. If the smaller pulley is rotating at 600 r.p.m. and co-efficient of friction between the belt and pulley is 0.3, determine the powe transmitted taking centrifugal tension into account. The mass of belt is given as 0.6703 kg/m length.*

Sol. Given :

Dia. of larger pulley, $d_1 = 1.5$ m $\therefore$ $r_1 = 0.75$ m

Dia. of smaller pulley, $d_2 = 1.0$ m $\therefore$ $r_2 = 0.5$ m

Distance between the two shafts, $x = 4.80$ m

Initial tension, $T_0 = 3000$ N

Speed of smaller pulley, $N_2 = 600$ r.p.m.

Co-efficient of friction, $\mu = 0.3$

Mass of belt/metre length, $m = 0.6703$ kg

The speed of belt is given by,

$$v = \frac{\pi d_2 N_2}{60} = \frac{\pi \times 1.0 \times 600}{60} = 10\,\pi \text{ m/s.}$$

The centrifugal tension is given by equation (7.19) as

$$T_C = m \times v^2$$

$$= 0.6703 \times (10\pi)^2 \text{ N} = 661.5 \text{ N}$$

Let T_1 = Tension on tight side

T_2 = Tension on slack side

Using equation (7.26), we get

$$T_0 = \frac{T_1 + T_2 + 2T_C}{2}$$

or $$3000 = \frac{T_1 + T_2 + 2 \times 661.5}{2}$$

$\therefore$ $$T_1 + T_2 = 2 \times 3000 - 2 \times 661.5 = 4677 \text{ N} \quad \ldots(i)$$

Let us calculate the angle of contact for open belt. Using equation (7.11) as

$$\theta = 180 - 2\alpha \qquad ...(ii)$$

where α is given by equation (7.12) as

$$\sin \alpha = \frac{r_1 - r_2}{x} = \frac{0.75 - 0.50}{4.80} = 0.0521$$

$\therefore$ $\alpha = \sin^{-1} 0.0521 = 3°$

$\therefore$ $\theta = 180 - 2\alpha = 180 - 2 \times 3° = 174°$

$$= 174 \times \frac{\pi}{180} = 3.04 \text{ rad.}$$

Now using equation (7.10),

$$\frac{T_1}{T_2} = e^{\mu \times \theta} = e^{0.3 \times 3.04} = 2.5$$

$\therefore$ $T_1 = 2.5T_2$ ...(iii)

Substituting this value in equation (*i*), we get

$$2.5T_2 + T_2 = 4677 \text{ N}$$

or $3.5T_2 = 4677$

$\therefore$ $T_2 = \frac{4677}{3.5} = 1336.3$ N

$\therefore$ $T_1 = 2.5 \times 1336.3 = 3340.7$ N

Now power transmitted is given by, equation (7.15) as

$$P = (T_1 - T_2) \times v \text{ W}$$

$$= (3340.7 - 1336.3) \times 10\pi \text{ W} = 62970 \; W = \mathbf{62.97 \; kW.} \quad \textbf{Ans.}$$

Problem 7.17. (S.I. Units). *An open belt connects two flat pulleys, the smaller pulley being of 400 mm in diameter. The angle of lap on the smaller pulley is 160° and co-efficient of friction between belt and pulley is 0.25.*

Which of the following alternatives would be more effective in increasing the power that could be transmitted :

(*i*) *increasing the initial tension by 10%.*

(*ii*) *increasing the co-efficient of friction by 10% by the application of suitable dressing to the belt ?*

(A.M.I.E., Winter 1981)

Sol. Given :

Dia. of smaller pulley, $d_2 = 400$ mm $= 0.4$ m

$\therefore$ Radius of smaller pulley, $r_2 = 0.2$ m

Angle of lap, $\theta = 160°$

Value of $\mu = 0.25$

We know the power transmitted is given by equation (7.15) as

$$P = (T_1 - T_2) \times v \text{ Watts} \qquad ...(i)$$

But $\frac{T_1}{T_2} = e^{\mu\theta}$

$\therefore$ $T_2 = \frac{T_1}{e^{\mu\theta}}$ ...(ii)

Substituting this value in (*i*), we get

$$P = \left(T_1 - \frac{T_1}{e^{\mu\theta}} \right) \times v = \frac{T_1 (e^{\mu\theta} - 1)}{e^{\mu\theta}} \times v \qquad ...(iii)$$

Let us substitute the value of T_1 in equation (*ii*) in terms of initial tension (T_0).

We know that $T_0 = \frac{T_1 + T_2}{2}$

or $T_1 = 2T_0 - T_2$

$$= 2T_0 - \frac{T_1}{e^{\mu\theta}} \qquad \left(\because \text{ From equation (ii), } T_2 = \frac{T_1}{e^{\mu\theta}} \right)$$

or $T_1 + \frac{T_1}{e^{\mu\theta}} = 2T_0$

or $T_1 (e^{\mu\theta} + 1) = 2T_0 \, e^{\mu\theta}$

$\therefore$ $T_1 = \frac{2T_0 \, e^{\mu\theta}}{(e^{\mu\theta} + 1)}$

Substituting this value in equation (*iii*), we get

$$P = \frac{2T_0 \, e^{\mu\theta}}{(e^{\mu\theta} + 1)} \frac{(e^{\mu\theta} - 1)}{e^{\mu\theta}} \times v$$

$$= 2T_0 \frac{(e^{\mu\theta} - 1)}{(e^{\mu\theta} + 1)} \times v \qquad ...(iv)$$

1st Case. If the initial tension (T_0) is increased by 10%, then for the given values of μ, θ and *v*, the power will increase by 10% as from equation (*iv*) it is clear that power varies directly with the initial tension.

2nd Case. If only μ is increased by 10% (keeping T_0, *v* and θ same), then power transmitted will change as given below :

Power transmitted when μ = 0.25 and θ = 160°, from equation (*iv*) is

$$P = \frac{2T_0 \left(e^{0.25 \times \frac{160 \times \pi}{180}} - 1 \right)}{\left(e^{0.25 \times 160 \times \frac{\pi}{180}} + 1 \right)} \times v$$

$$= \frac{2T_0 \, (2.01 - 1)}{(2.02 + 1)} \times v = \frac{2T_0 \times 1.01}{3.01} \times v$$

$$= 0.671 \; T_0 \times v \qquad ...(v)$$

Power transmitted when μ = 1.1 × 0.25 and θ = 160° is given by

$$P^* = \frac{2T_0 \left(e^{1.1 \times 0.25 \times \frac{160 \times \pi}{180}} - 1 \right)}{\left(e^{1.1 \times 0.25 \times \frac{160 \times \pi}{180}} + 1 \right)} \times 1$$

$$= \frac{2T_0 \, (2.154 - 1)}{(2.154 + 1)} \times v = \frac{2T_0 \times 1.154}{3.154} \times v$$

$$= 0.7317 \times T_0 \times v \qquad ...(vi)$$

% increase in power $= \dfrac{P* - P}{P} \times 100$

$= \dfrac{(0.7317 \times T_0 \times v - 0.671 \times T_0 \times v)}{0.671 \times T_0 \times v} \times 100$

$= \dfrac{(0.7317 - 0.671)}{0.671} \times 100 = 9.046\%.$

The power increase in first case is 10% whereas it is 9.046% in the second case. Thus power transmitted will be more in first case. Hence increasing the initial tension by 10% will be more effective for increasing the transmission of the power. **Ans.**

7.11. V-Belt Drive and Rope-Drive

The power from one shaft to another shaft is also transmitted with the help of V-belt drive and rope-drive.

1. V-Belt Drive. Fig. 7.12 shows a V-belt with a grooved pulley.

Let R_N = Normal reaction between belt and sides of the V-grooved pulley

2α = Angle of groove

μ = Co-efficient of friction between belt and pulley

R = Total reaction in the plane of groove

Resolving the forces vertically, we get

$R = R_N \sin \alpha + R_N \sin \alpha = 2R_N \sin \alpha$

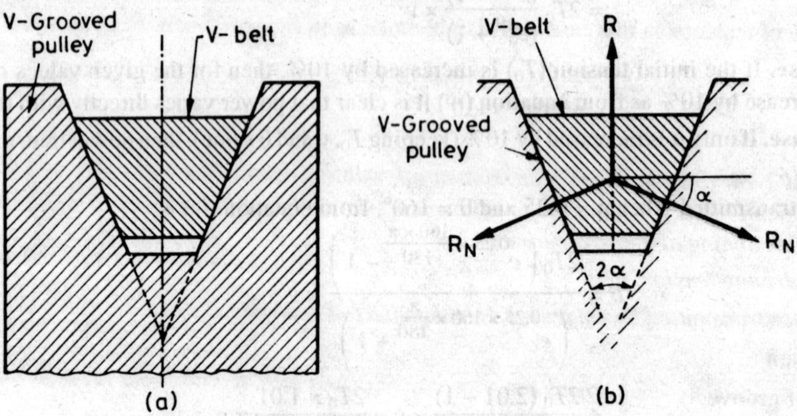

Fig. 7.12

$\therefore \quad R_N = \dfrac{R}{2 \sin \alpha} = \dfrac{R}{2} \operatorname{cosec} \alpha$

$\therefore$ Frictional resistance $= \mu R_N + \mu R_N = 2\mu R_N$

$= 2\mu \times \dfrac{R}{2} \operatorname{cosec} \alpha \qquad \left(\because R_N = \dfrac{R}{2} \operatorname{cosec} \alpha\right)$

$= \mu \times R \times \operatorname{cosec} \alpha = (\mu \operatorname{cosec} \alpha) \times R.$

In Art. 7.6, we have seen that frictional force (or resistance) for a flat belt is $\mu \times R$, whereas in case of V-belt it is equal to $(\mu \operatorname{cosec} \alpha) \times R$. Hence the ratio of tensions $\left(i.e., \dfrac{T_1}{T_2}\right)$ in case of V-belt can be obtained by replacing μ by '$\mu \operatorname{cosec} \alpha$' in equation (7.10).

$\therefore$ For V-belt, $$\frac{T_1}{T_2} = e^{(\mu \operatorname{cosec} \alpha) \times \theta}$$

or $$\frac{T_1}{T_2} = e^{\mu\theta \operatorname{cosec} \alpha} \qquad ...(7.27)$$

2. Rope Drive. Fig. 7.13 shows a rope-drive. The ropes are generally circular in section. Rope-drive is mostly used when the distance between the driving shaft and driven shaft is large. Frictional grip in rope-drive is more than that in V-belt drive.

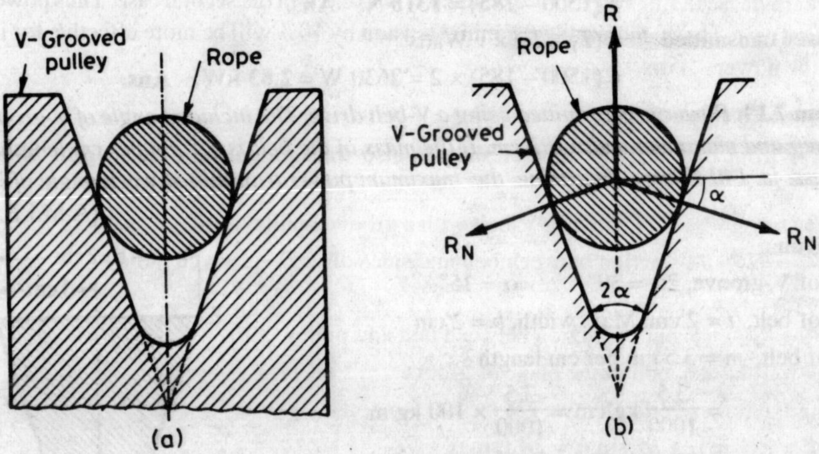

Fig. 7.13

The ratio of tensions in this case will also be same as in case of V-belt. Hence ratio of tension will be as

$$\frac{T_1}{T_2} = e^{\mu\theta \operatorname{cosec} \alpha} \qquad ...(7.28)$$

Problem 7.18. (S.I. Units). *The maximum allowable tension, in a V-belt of groove angle of 45°, is 1500 N. The angle of lap is 170° and the co-efficient of friction between the belt and material of the pulley is 0.27. If the belt is running at 2 m/s, determine :*

(i) Net driving tension, and

(ii) Power transmitted by the pulley. Neglect effect of centrifugal tension.

Sol. Given :

Angle of groove, $2\alpha = 45° \quad \therefore \quad \alpha = 22.5°$

Max. allowable tension, $T_{max} = 1500$ N

Angle of lap, $\theta = 170° = 170 \times \frac{\pi}{180}$ radians

Co-efficient of friction, $\mu = 0.27$

Belt speed, $v = 2$ m/s

We know $T_{max} = T_1 + T_C$

where T_C is the centrifugal tension, If centrifugal tension is neglected, then $T_{max} = T_1 = 1500$ N.

$\therefore \quad T_1 = 1500$ N.

Now for a V-belt drive, the ratio of tensions is given by equation (7.27) as

$$\frac{T_1}{T_2} = e^{\mu\theta \text{ cosec } \alpha} = e^{0.27 \times \frac{170 \times \pi}{180} \times \text{cosec } (22.5^\circ)} = 8.109$$

$$\therefore \quad T_2 = \frac{T_1}{8.109} = \frac{1500}{8.109} = 185 \text{ N}$$

(*i*) Net driving tension $= (T_1 - T_2)$

$= (1500 - 185) =$ **1315 N. Ans.**

(*ii*) Power transmitted $= (T_1 - T_2) \times v$ Watts

$= (1500 - 185) \times 2 = 2630 \text{ W} =$ **2.63 kW. Ans.**

Problem 7.19. *Power is transmitted using a V-belt drive. The included angle of V-grooves is 30°. The belt is 2 cm deep and maximum width is 2 cm. If the mass of the belt is 3.5 gm per cm length and maximum allowable stress is 140 N/cm², determine the maximum power transmitted when angle of lap is 140° and μ = 0.15.* (A.M.I.E., Winter 1984)

Sol. Given :

Angle of V-groove, $2\alpha = 30^\circ \quad \therefore \quad \alpha = 15^\circ$

Depth of belt, $t = 2$ cm, Max. width, $b = 2$ cm

Mass of belt, $m = 3.5$ gm per cm length

$$= \frac{3.5}{1000} \text{ kg/cm} = \frac{3.5}{1000} \times 100 \text{ kg/m}$$

$= 0.35$ kg/m length

Max. allowable stress, $f = 140 \text{ N/cm}^2$

Angle of lap, $\theta = 140^\circ = \frac{140 \times \pi}{180}$ radian $= 2.444$ rad

Co-efficient of friction, $\mu = 0.15$

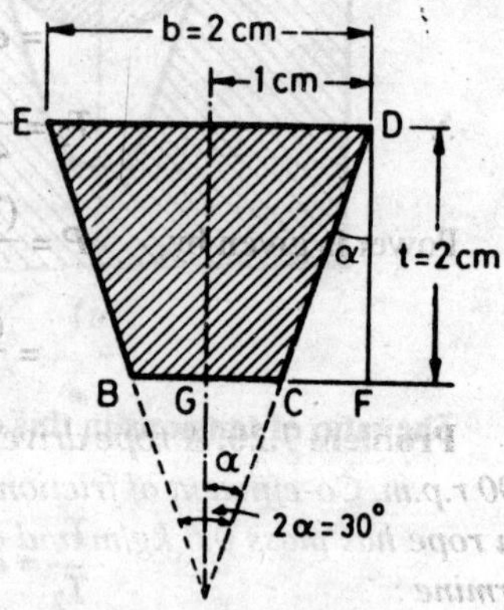

Fig. 7.14

Let us first calculate the cross-section area A of the V- belt, which is shown in Fig. 7.14. From triangle CFD,

$$\tan \alpha = \frac{CF}{DF} \quad \text{or} \quad \tan 15^\circ = \frac{CF}{2.0}$$

$\therefore \quad CF = 2 \times \tan 15^\circ$

Now $\quad GC = GF - CF = (1 - 2 \times \tan 15^\circ)$

and $\quad BC = 2 \times GC = 2(1 - 2 \tan 15^\circ)$

$\therefore$ Area of V-belt = Area of trapezium $BCDE$

$$= \frac{(ED + BC)}{2} \times DF$$

$$= \frac{[2 + 2(1 - 2 \tan 15^\circ)]}{2} \times 2$$

$$= 2 + 2(1 - 2 \tan 15^\circ) = 2 + 2(1 - 0.5358) = 2.9284 \text{ cm}^2$$

Now maximum permissible tension (T_{max}) is given by,

$$T_{max} = \text{Max. allowable stress} \times \text{Area of belt}$$

$$= f \times 2.9284 = 140 \times 2.9284 = 409.976 \text{ N}$$

Now for maximum power transmitted, centrifugal tension (T_C) is one-third of maximum tension.

$\therefore \quad T_C = \frac{1}{3} \times T_{max} = \frac{1}{3} \times 409.976 = \mathbf{136.66\ N.}$ **Ans.**

Using equation (7.19), we get

$$T_C = mv^2$$

or $\quad 136.66 = 0.35 \times v^2$

$$\therefore \quad v = \sqrt{\frac{136.66}{0.35}} = 19.76 \text{ m/s} \quad \ldots(i)$$

Also we know, $\quad T_{max} = T_1 + T_C$

$\therefore \quad T_1 = T_{max} - T_C$

$= 409.976 - 136.66 = 273.316$ N

For a V-belt drive, the ratio of tensions is given by equation (7.27) as

$$\frac{T_1}{T_2} = e^{\mu \times \theta \times \text{cosec}\,\alpha}$$

$$= e^{0.15 \times 2.444 \times \text{cosec}\,15^\circ} = e^{1.416} = 4.12$$

$$\therefore \quad T_2 = \frac{T_1}{4.12} = \frac{273.316}{4.12} = 66.34$$

Power is given by, $\quad P = \frac{(T_1 - T_2) \times v}{1000}$ kW

$$= \frac{(273.316 \times 66.34) \times 19.76}{1000} \text{ kW} = \mathbf{4.09\ kW.} \quad \textbf{Ans.}$$

Problem 7.20. *A rope drive transmits 75 kW through a 150 cm diameter 45° grooved pulley rotating at 200 r.p.m. Co-efficient of friction between the ropes and the pulley grooves is 0.3 and angle of lap is 160°. Each rope has mass 0.6 kg/m and can safely take a pull of 800 N. Taking centrifugal tension into account, determine :*

(i) the number of ropes required for the drive, and

(ii) initial rope tension.

Sol. Given :

Power = 75 kW

Dia. of driver pulley, d_1 = 150 cm = 1.50 m

Speed of driver pulley, N_1 = 200 r.p.m.

Angle of groove, $2\alpha = 45°$, $\therefore \quad \alpha = 22.5°$

Co-efficient of friction, $\mu = 0.3$

Angle of lap, $\theta = 160° = 160 \times \frac{\pi}{180}$ radian

Mass of one rope per m length, m = 0.6 kg

Max. safe pull or tension in one rope, T_{max} = 800 N

Now velocity of rope (v) is given by,

$$v = \frac{\pi d_1 N_1}{60} = \frac{\pi \times 1.5 \times 200}{60} = 15.7 \text{ m/s.}$$

(*i*) Let us find the power transmitted by one rope.

Centrifugal tension in one rope is given by equation (7.19) as

$$T_C = m \times v^2$$
$$= 0.6 \times (15.7)^2 = 147.894 \text{ N}$$

But $T_{max} = T_1 + T_C$

$\therefore$ $T_1 = T_{max} - T_C = 800 - 147.894 = 652.106$ N

Ratio of tensions in rope is given by equation (7.28) as,

$$\frac{T_1}{T_2} = e^{\mu\theta \text{ cosec } \alpha}$$
$$= e^{0.3 \times \frac{160 \times \pi}{180} \times \text{cosec}(22.5°)} = e^{2.1919} = 8.95$$

$\therefore$ $T_2 = \frac{T}{8.95} = \frac{652.106}{8.95}$ $(\because T_1 = 652.106)$

$$= 72.86 \text{ N}$$

$\therefore$ Power transmitted by one rope

$$= \frac{(T_1 - T_2) \times v}{1000} = \frac{(652.106 - 72.86) \times 15.7}{1000} = 9.094 \text{ kW}$$

But total power transmitted = 75 kW

$\therefore$ Number of ropes requied

$$= \frac{\text{Total power transmitted}}{\text{Power transmitted by one rope}}$$
$$= \frac{100}{12.07} = \textbf{8.24 say 9.} \quad \textbf{Ans.}$$

(*ii*) ***Initial rope tension*** **(T_0). If centrifugal tension is considered, then initial tension is given by equation (7.26) as**

$$T_0 = \frac{T_1 + T_2 + 2T_C}{2}$$
$$= \frac{652.106 + 72.86 + 2 \times 147.894}{2} = \textbf{510.377 N.} \quad \textbf{Ans.}$$

Problem 7.21. *Following data is given for a rope pulley transmitting 23.628 kW :*

Dia. of pulley	*= 40 cm ;*	*Speed*	*= 110 r.p.m.*
Angle of groove	*= 45° ;*	*Angle of lap*	*= 60°*
Co-efficient of friction	*= 0.28 ;*	*No. of ropes*	*= 10*

Mass in kg/m length of ropes = 0.0053 × C^2 and working tension is limited 12.2 C^2 N where C = grith of rope in cm.

Find : (i) initial tension, and (ii) diameter of each rope. (A.M.I.E., Summer 1979)

Sol. Given :

$$P = 23.628 \text{ kW}, d_1 = 40 \text{ cm} = 0.40 \text{ m}, N_1 = 110 \text{ r.p.m.}$$

$$2\alpha = 45°, \alpha = 22.5°, \theta = 160° = \frac{160 \times \pi}{180} \text{ radians}, \mu = 0.28, \text{No. of ropes} = 10$$

Mass (m) per metre length = 0.0053 × C^2

Mass, tension, T_{max} = 12.2 C^2 N

Velocity of rope is given by,

$$v = \frac{\pi d_1 N_1}{60} = \frac{\pi \times 0.4 \times 110}{60} = 2.303 \text{ m/s}$$

Using equation (7.19) for centrifugal tension,

$$T_C = m \times v^2$$

$$= 0.0053C^2 \times 2.303^2 = 0.0281C^2 \text{ N} \quad ...(i)$$

Now using equation (7.28) for the ratio of tension is rope,

$$\frac{T_1}{T_2} = e^{\mu\theta \text{ cosec } \alpha}$$

$$= e^{0.28 \times \frac{160 \times \pi}{180} \times \text{cosec } (22.5°)} = e^{2.0426} = 7.71$$

or $$T_1 = 7.71T_2 \quad ...(ii)$$

Now power transmitted by one rope

$$= \frac{\text{Total power}}{\text{No. of ropes}} = \frac{23.628}{10} = 2.3628 \text{ kW}$$

But power transmitted by one rope is also

$$= \frac{(T_1 - T_2) \times v}{1000}$$

$\therefore$ $$2.3628 = \frac{(T_1 - T_2) \times 2.303}{1000}$$

$\therefore$ $$(T_1 - T_2) = \frac{2.3628 \times 1000}{2.303} = 1026 \text{ N}$$

or $$7.71\,T_2 - T_2 = 1026 \quad (\because \text{ From equation } (ii),\ T_1 = 7.71\,T_2)$$

or $$6.71\,T_2 = 1026$$

$\therefore$ $$T_2 = \frac{1026}{6.71} = 153 \text{ N}$$

From equation (*ii*), $T_1 = 7.71 \times 153 = 1179$ N

Now $$T_{max} = T_1 + T_C$$

or $$12.2C^2 = 1179 + 0.0281C^2 \quad (\because \text{ From equation } (i),\ T_C = 0.0281C^2)$$

or $$(12.2 - 0.0281)C^2 = 1179$$

$\therefore$ $$C^2 = \frac{1179}{(12.2 - 0.0281)} = 96.86$$

$\therefore$ $$C = \sqrt{96.86} = 9.84 \text{ cm} \quad ...(iii)$$

Substituting the value of C in equation (*i*), we get

$$T_C = 0.0281 \times 9.84^2 = 2.72 \text{ N}$$

(*i*) *Initial tension* (T_0). Using equation (7.26), we get

$$T_0 = \frac{T_1 + T_2 + 2T_C}{2}$$

$$= \frac{1179 + 153 + 2 \times 2.72}{2} = \mathbf{668.72 \text{ N.}} \quad \textbf{Ans.}$$

(ii) Dia. of each rope (d). We know that

$$C = \text{grith of one rope} = \pi \times d$$

But from equation (*iii*),

$$C = 9.84$$

$$\therefore \quad 9.84 = \pi \times d$$

$$\therefore \quad d = \frac{9.84}{\pi} = \mathbf{3.132\ cm.} \quad \textbf{Ans.}$$

Problem 7.22. *A pulley, used to transmit power by means of ropes, has a diameter of 3.6 m and has 15 grooves of 45° angle. The angle of contact is 170° and the co-efficient of friction between the ropes and the groove side is 0.28. The maximum possible tension in the ropes is 960 N and the mass of the rope is 1.5 kg per m length. What is the speed of the pulley in r.p.m. and the power transmitted if the condition of maximum power prevail ?*

(A.M.I.E., Summer 1983)

Sol. Given :

Dia. of pulley, $D = 3.6$ m

No. of grooves or No. of ropes $= 15$

Angle of grooves, $2\alpha = 45°$ or $\alpha = 22.5°$

Angle of contact, $\theta = 170° = 170 \times \frac{\pi}{180}$ radians

Co-efficient of friction, $\mu = 0.28$

Max. tension, $T_{max} = 960$ N

Mass of rope, $m = 1.5$ kg per m length

Find : (*i*) Speed (N), and (*ii*) Max. power.

For maximum power :

(*i*) Centrifugal tension,

$$T_C = \frac{1}{3} \times T_m = \frac{960}{3} = 320 \text{ N}$$

(*ii*) The velocity of rope is given by equation (7.23) as

$$v = \sqrt{\frac{T_m}{3m}} = \sqrt{\frac{960}{3 \times 1.5}} = 14.60 \text{ m/s}$$

Velocity of rope is also given by,

$$v = \frac{\pi D N}{60}$$

or

$$14.6 = \frac{\pi \times 3.6 \times N}{60}$$

$$\therefore \quad N = \frac{60 \times 14.6}{\pi \times 3.6} = \mathbf{77.45\ r.p.m.} \quad \textbf{Ans.}$$

Now ratio of tensions in the rope is given by equation (7.28) as

$$\frac{T_1}{T_2} = e^{\mu\theta \operatorname{cosec} \alpha}$$

$$= e^{0.28 \times \frac{170 \times \pi}{180} \times \operatorname{cosec}(22.5°)} = e^{2.1698} = 8.756$$

$$\therefore \quad T_2 = \frac{T_1}{8.756} \quad \ldots(i)$$

But $\quad T_m = T_1 + T_C$

or $960 = T_1 + 320$ $(\because T_C = 320, T_{max} = 960)$

$\therefore$ $T_1 = 960 - 320 = 640$ N

Substituting this value of T_1 in equation (*i*),

$$T_2 = \frac{640}{8.756} = 73.08 \text{ N}$$

Maximum power transmitted by one rope

$$= \frac{(T_1 - T_2) \times v}{1000}$$

$$= \frac{(640 - 73.06) \times 14.6}{1000} = 8.277$$

$\therefore$ **Total maximum power transmitted**

= Power of one rope × No. of rope

= 8.277 × 15 = **124.16 kW. Ans.**

Problem 7.23. *A casting weighing 9000 N hangs freely from a rope which makes 2.5 turns round a drum of 30 cm diameter revolving at 20 r.p.m. The other end of the rope is pulled by a man. The co-efficient of friction is 0.25. Determine : (i) the force required by the man, and (ii) the power to raise the casting.*

(A.M.I.E., Winter 1986)

Sol. Given :

Weight of casting, $W = 9000$ N

No. of turns $= 2.5$

$\therefore$ Total angle covered, θ = No. of turns × angle covered in one turn = $2.5 \times 2\pi = 5\pi$ radians

Dia. of drum, $D = 30$ cm $= 0.3$ m

Speed, $N = 20$ r.p.m.

Co-efficient of friction, $\mu = 0.25$

(*i*) Let P = Force required by the man.

The two views of the given arrangement are shown in Fig. 7.15.

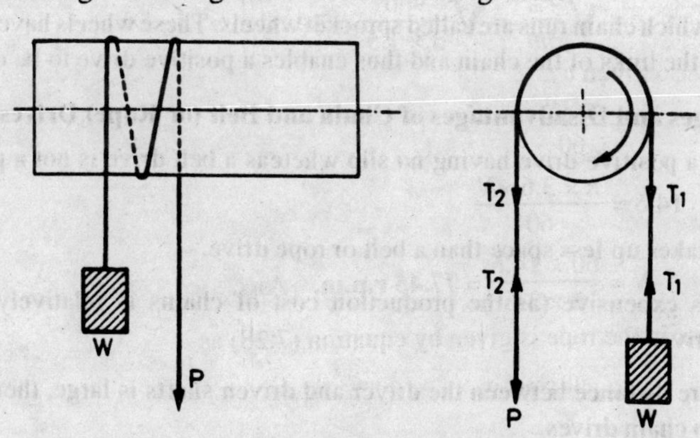

Fig. 7.15

Let T_1 = Tension in the tight side of the rope and is equal to W

T_2 = Tension in the slack side of the rope and is equal to P.

Using equation, we get

$$\frac{T_1}{T_2} = e^{\mu\theta}$$

$$= e^{0.25 \times 5\pi} = e^{3.925} = 50.65$$

or $$\frac{9000}{P} = 50.65 \qquad (\because \; T_1 = W = 9000 \text{ and } T_2 = P)$$

$$\therefore \qquad P = \frac{9000}{50.65} = 177.7 \text{ N. } \textbf{Ans.}$$

(*ii*) Let P^* = Power to raise the casting.

Using equation, $$P^* = \frac{(T_1 - T_2) \times v}{1000} \qquad ...(i)$$

But $$v = \frac{\pi \times D \times N}{60} \text{ m/s}$$

$$= \frac{\pi \times 0.3 \times 20}{60} = 0.1\,\pi \text{ m/s}$$

Substituting T_1, T_2 and v in equation (*i*), we get

$$P^* = \frac{(9000 - 177.7) \times 0.1 \times \pi}{1000} \qquad (\because \; T_1 = W = 9000, T_2 = P = 177.7)$$

$$= 2.77 \text{ kW. } \textbf{Ans.}$$

7.12. Chain Drive

When the power is transmitted over a shorter distance than in the case of belt or rope drives, chain-drive are used. Chains are used for transmission of power in bicycles, motor vehicles, road rollers and agricultural machinery etc. Chains are made up of rigid links and of various types.

In case of belt and rope-drives, slipping may occur. In order to avoid slipping steel chains are used. Hence chain drives are used where positive action is required.

The wheels over which chain runs are called sprocket wheels. These wheels have projecting teeth which fit into suitable recess in the links of the chain and thus enables a positive drive to be obtained.

7.13. Relative Advantages and Disadvantages of Chain and Belt (or Rope) Drives

1. Chain drive is a positive drive having no slip whereas a belt drive is not a positive drive in which slip occurs.

2. A chain drive takes up less space than a belt or rope drive.

3. Chain drive is expensive (as the production cost of chains is relatively high) but durable in comparison to the belt-drive.

4. When the centre distance between the driver and driven shafts is large, then the belt drive is most suitable in comparison to chain drives.

5. In case of chain drive no initial tension is required whereas in belt drive it is essential.

6. The chain drive gives a high transmission efficiency.

7. Wear and tear is much more in belt drives in comparison to that in chain drives. Correspondingly the maintenance in chain drive is less.

HIGHLIGHTS

1. Belt, ropes and chains are used for transmitting power from one shaft to another shaft.
2. The flat belts are of two types, *i.e.*, open belt and crossed belt.
3. Velocity ratio of a belt is given by

$$\frac{N_2}{N_1}=\frac{d_1}{d_2} \quad \text{...if thickness of belt is neglected}$$

$$=\frac{d_1+t}{d_2+t} \quad \text{...if thickness of belt is considered}$$

where N_1 = Speed of driver pulley,
d_1 = Dia. of driver pulley,
N_2 = Speed of driven pulley,
d_2 = Dia. of driven pulley and
t = Thickness of the belt.

4. Velocity ratio of a belt when creep is considered is given by

$$\frac{N_2}{N_1}=\frac{d_1}{d_2}\times\left[\frac{E+\sqrt{f_2}}{E+\sqrt{f_1}}\right]$$

where E = Young's modulus for belt material,
f_1 = Stress in the belt on tight side, and
f_2 = Stress in the belt on slack side.

5. The velocity ratio of a compound-belt drive is given by $\dfrac{\text{Speed of last follower}}{\text{Speed of first driver}}=\dfrac{\text{Product of dia. of drivers}}{\text{Product of dia. of followers}}$.

6. Length of a belt is given by

$$L=\pi(r_1+r_2)+\frac{(r_1-r_2)^2}{x}+2x \quad \text{...For an open belt}$$

$$=\pi(r_1+r_2)+\frac{(r_1+r_2)^2}{x}+2x \quad \text{...For a crossed belt}$$

where r_1 = Radius of larger pulley
r_2 = Radius of smaller pulley
x = Distance between the centres of two pulleys.

7. Ratio of the tensions on the two sides of a belt is given by

$$\frac{T_1}{T_2}=e^{\mu\times\theta}$$

where T_1 = Tension on the tight side,
T_2 = Tension on the slack side,
μ = Co-efficient of friction between belt and pulley,
and θ = Angle of contact in radians.

$$=(180-2\alpha) \quad \text{...For an open belt}$$

$$=(180+2\alpha) \quad \text{...For a crossed belt}$$

where $$\alpha=\sin^{-1}\left(\frac{r_1-r_2}{x}\right) \quad \text{...For an open belt}$$

$$=\sin^{-1}\left(\frac{r_1+r_2}{x}\right) \quad \text{...For a crossed belt.}$$

8. Power transmitted by a belt is given by $P = \frac{(T_1 - T_2) \times v}{1000}$ kW

$= (T_1 - T_2) \times v$ Watts

where T_1 and T_2 are in N.

9. Centrifugal tension in a belt is given by, $T_C = m \times v^2$ N ...in S.I. Units

where m = Mass of belt per metre length in kg.

10. Maximum tension (T_{max}) is given by,

$$T_{max} = f \times (\text{Area of belt}) = f \times (b \times t)$$

$= T_1 + T_C$ ...if centrifugal tension is considered

$= T_1$ ...if centrifugal tension is neglected.

11. For a maximum power transmission :

(*i*) Velocity of belt (v) is given by,

$$v = \sqrt{\frac{T_{max}}{3 \times m}}$$ ...in S.I. Units

where m = Mass in kg per metre length of belt.

(*ii*) $T_{max} = 3T_C$ or $T_C = \frac{1}{3} T_{max}$

(*iii*) $T_1 = \frac{2}{3} T_{max}$.

12. Initial tension (T_0) in a belt is given by,

$$T_0 = \frac{T_1 + T_2}{2}$$ ...if T_C is neglected

$$= \frac{T_1 + T_2 + 2T_C}{2}$$ ...if T_C is considered.

13. For a V-belt drive (or rope drive) the ratio of tension is given by,

$$\frac{T_1}{T_2} = e^{\mu\theta \text{ cosec } \alpha}$$

where α = Half of the angle of groove.

EXERCISE 7

(A) Theoretical Questions

1. Define and explain the terms : Belt-drive ; Rope-drive ; Chain-drive ; slip and creep of a belt.
2. Derive an expression for the velocity ratio of a belt-drive considering thickness of the belt.
3. Derive an expression for the ratio of tensions for a flat belt passing over a pulley, when it is just on the point of slipping. (*A.M.I.E., Winter 1981*)
4. Distinguish between initial tension and centrifugal tension in a belt. (*A.M.I.E., Winter 1987*)
5. Distinguish between slip and creep in a belt drive. Derive an expression for the ratio of tensions in the tight and slack sides in terms of μ and θ, when the belt is just on the point of slipping. (*A.M.I.E., Summer 1984*)
6. Derive the expression for optimum speed of flat belt for the transmission of maximum power considering the effect of centrifugal tension. (*A.M.I.E., Summer 1986*)
7. For a flat belt drive, prove that

$$\frac{T_1}{T_2} = e^{\mu\theta}$$

where T_1 = Tension on the tight side of the belt,

T_2 = Tension on the slack side of the belt,

μ = Co-efficient of friction between the belt and pulley surface,

θ = Angle of contact between the belt and the pulley. (*A.M.I.E., Summer 1976*)

8. Obtain the conditions for the maximum power transmitted by a belt from one pulley to another. (*A.M.I.E., Summer 1982*)

9. Derive an expression for the length of (*i*) an open belt, and (*ii*) a crossed belt.

10. Prove that the angle of contact (θ) is equal to $(180 - 2\alpha)$ for an open belt whereas it is equal to $(180 + 2\alpha)$ for a crossed-belt.

11. Derive an expression for the ratio of the driving tensions of a rope-drive assuming the angle of the groove of the pulley to be as 2α. (*A.M.I.E., Summer 1983*)

12. Show the construction of roller chain by a neat sketch and state its advantages over belt and gear drives. (*A.M.I.E., Winter 1986*)

13. What are the relative advantages and disadvantages of chain and belt drives ? (*A.M.I.E., Winter 1984*)

(B) Numerical Questions

1. Find the speed of a shaft which is driven by a belt an engine running at 300 r.p.m. The diameter of the engine pulley is 60 cm and that of shaft is 40 cm. [**Ans.** 450 r.p.m.]

2. If in the above question, thickness of the belt is 10 mm, then find the speed of shaft. $\left[\textbf{Ans.}\ \dfrac{61 \times 300}{41} = 446.34\right]$

3. A shaft running at 120 r.p.m. is to drive a parallel shaft at 180 r.p.m. The pulley on the driving shaft is 75 cm in diameter, calculate the diameter of the pulley on the driven shaft (*i*) neglecting belt thickness, (*ii*) taking belt thickness into account which is 15 mm, (*ii*) assuming in the latter case a total slip of 4%. [**Ans.** 50 cm, 49.7 cm, 47.8 cm]

4. The power is transmitted from a pulley 1 m diameter running at 200 r.p.m. to a pulley 2.25 m diameter by means of a belt. Find the speed lost by the driven pulley as a result of the creep, if the stress on the tight and slack side of the belt is 1.4 N/mm^2 and 0.5 N/mm respectively. The Young's modulus for the material of the belt is 100 N/mm^2. [**Ans.** 0.2 r.p.m.]

5. Two parallel shafts 120 cm apart are to have a velocity ratio of 2.5 and are to be connected by a flat belt. The pulley on the high speed shaft is 24 cm in diameter. Determine the length of the belt required ; (*i*) if the belt is open, and (*ii*) if the belt is crossed. [**Ans.** (*i*) 373.4 cm, (*ii*) 386 cm]

6. With the help of a belt, an engine running at 150 r.p.m., drives a line shaft. The diameter of the pulley on the engine is 70 cm and the diameter of the pulley on the line shaft is 35 cm. A 80 cm diameter pulley on the line shaft drives a 20 cm diameter pulley keyed to a dynamo shaft. Find the speed of the dynamo shaft when (*i*) there is no slip (*ii*) there is a slip of 3% at each drive. [**Ans.** (*i*) 1200 r.p.m. (*ii*) 1129 r.p.m.]

7. Two parallel shafts 12 metres apart are to be connected by a belt running over pulleys of diameters 480 cm and 80 cm respectively. Determine the length of the belt required ; (*i*) if the belt is open, and (*iii*) belt is crossed. [**Ans.** (*i*) 33.13 m, (*ii*) 33.45 m]

8. A shaft which rotates at a constant speed of 160 r.p.m. is connected by belling to a parallel shaft 72 cm apart which has to run at 60, 80 and 100 r.p.m. The smallest pulley on the driver shaft is 4 cm in radius. Determine the remaining radii of the two stepped pulleys for ; (*i*) a crossed belt, and (*ii*) an open belt. (*A.M.I.E., Winter 1976*)

[**Ans.** (*i*) r_2 = 10.67 cm, r_3 = 4.89 cm, r_4 = 9.78 cm, r_5 = 5.64 and r_6 = 9.02 cm ;
(*ii*) r_2 = 10.69, r_3 = 5, r_4 = 10, r_5 = 6 and r_6 = 9.6 cm]

9. A shaft rotating at 200 r.p.m. drives another shaft at 300 r.p.m., and transmits 6 kW through a belt. The belt is 10 cm wide and 1 cm thick. The distance between the shafts is 4 m. The smaller pulley is 50 cm in diameter. Calculate the stress in (*i*) open-belt, and (*ii*) crossed-belt. Take μ = 0.3. Neglect centrifugal tension. (*A.M.I.E., Summer 1985*)

[**Ans.** (*i*) 126.8 N/cm^2, (*ii*) 118.47 N/cm^2]

10. An open-belt drive connects two pulleys 120 cm and 50 cm diameter on parallel shafts 4 m apart. The belt weighs 0.9 kg/m length and maximum tension in it is not to exceed 2000 N. The co-efficient of friction is 0.3. The 120 cm pulley, which is the driver runs at 200 r.p.m. Due to belt slip on one of the pulleys, the velocity of the driven shaft is only 450 r.p.m. Calculate the torque on each of the two shafts, the power transmitted and power lost in friction. What is the efficiency of the drive. *(A.M.I.E., Winter 1979)*

[Ans. 656.1 Nm ; 273.4 N m ; 13.73 kW ; 1.0725 kW ; 93.7%]

11. The maximum allowable tension in a flat belt is 1500 N. The angle of lap is 170° and co-efficient of friction between the belt and material of the pulley is 0.27. Neglecting the effect of centrifugal tension, calculate the net driving tension and power transmitted if the belt speed is 2 m/s. **[Ans.** (*i*) 826.7 N ; (*ii*) 1.6534 kW]

12. Two pulleys on parallel shafts are connected by a crossed belt. The diameters of the pulleys are 450 mm and 200 mm. The shafts are 1.95 m apart. Find the length of the belt required and angle of contact between the belt and each pulley. **[Ans.** (*i*) L = 4.975 m, (*ii*) θ = 199.2°]

13. In question 12, what power can be transmitted by the above belt when the larger pulley rotates at 100 rev/min, if the maximum permissible tension in the belt is 1000 N and the co-efficient of friction between the belt and pulley is 0.25. **[Ans.** 1.37 kW]

14. The maximum permissible stress in a belt is 1.4 N/mm^2 and ratio of tensions is 2.0 Find the maximum power transmitted by a belt 150 mm × 10 mm if the density of leather is 1 Mg/m^3. **[Ans.** 14.75 kW]

15. An open flat belt drive connects two parallel shafts 1.2 m apart. The driving and driven shafts rotate at 350 r.p.m. and 140 r.p.m. respectively and the driven pulley is 40 cm in diameter. The belt is 5 mm thick and 8 cm wide. Co-efficient of friction between belt and pulley is 0.3 and maximum permissible tension in the belting is 140 N/cm^2. Determine : (*i*) diameter of driving pulley, (*ii*) maximum power that may be transmitted by the belting, and (*iii*) required initial tension in the belt. Neglect centrifugal tension. *(A.M.I.E., Winter 1977)*

[Ans. (*i*) 16 cm, (*ii*) 0.96 kW, (*iii*) 396 N]

16. A cross-belt drive is to transmit at 7.5 kW at 1000 r.p.m. of the smaller pulley. The diameter of the smallest pulley is 25 cm and velocity ratio is 2. The centre distance between the pulley is 125 cm. A flat belt of thickness 6 mm and of co-efficient friction is 0.3 is used over the pulleys. Determine the necessary width of the belt if the maximum allowable stress in the belt is 175 N/cm^2 and density of the belt is 1 gm/cm^3. **[Ans.** 9.0 cm]

17. A V-belt having a lap angle of 180° has a cross-sectional area of 6.2 cm^2 and runs in a groove of included angle of 40°. The density of belt is 0.0013 kg/cm^3 and maximum stress is limited to 400 N/cm^2 the co-efficient of friction being 0.2. Find the maximum power that can be transmitted, if the wheel has mean diameter of 30 cm and runs at 1000 r.p.m. **[Ans.** 30 kW]

18. Find the number of ropes required in a rope-drive which is used to transmit 230 kW from a pulley of 1 m diameter running at 450 r.p.m. The safe pull in each rope is 800 N and mass of the rope is 0.46 kg per metre length. The angle of lap is 160° whereas the angle of groove is 45°. The co-efficient of friction between the rope and the pulley is 0.3. **[Ans.** 21]

19. A V-belt of mass 0.9 kg/m length is used to transmit power between two shafts. The angle of lap is 170° and angle of groove is 45°. The co-efficient of friction between belt and pulley is 0.17 and maximum permissible tension in the belt is limited to 2.2 kN. Find : (*i*) velocity of belt for maximum power and (*ii*) power transmitted at this velocity.

[Ans. (*i*) 28.54 m/s, (*ii*) 30.67 kW]

20. Find (*i*) maximum power transmitted and (*ii*) initial tension in a flat belt 8 mm thick and 100 mm wide. The belt transmits power between two pulleys running at 31.67 m/sec. The angle of lap of smaller pulley is 165° and co-efficient of friction between the belt and pulley is 0.3. The mass of the belt is 0.9 kg/m length and maximum permissible stress in the belt is 2 MN/m^2. **[Ans.** (*i*) 14.82 kW, (*ii*) 1.322 kN]

8

Brakes and Dynamometers

8.1. INTRODUCTION

A *brake* is a device used either to bring to rest a body which is in motion or to hold a body in a state of rest or of uniform motion against the action of external forces or couples. Actually the brake offers the frictional resistance to the moving body and this frictional resistance retards the motion and the body comes to rest. In this process, the kinetic energy of the body is absorbed by brakes.

A *dynamometer* is a device used to measure the frictional resistance or frictional torque. This frictional resistance (or frictional torque) is obtained by applying a brake. Hence dynamometer is also a brake in addition it has a device to measure the frictional resistance (or frictional torque). This chapter deals with different types of brakes and dynamometers.

8.2. TYPES OF BRAKES

The brakes are classified as :

(*a*) Hydraulic brakes,

(*b*) Electric brakes,

(*c*) Mechanical brakes.

This chapter deals with mechanical brakes only. The following are the important types of mechanical brakes :

(*i*) Simple block or Shoe brake.

(*a*) Single block or Shoe brake

(*b*) Double block or Shoe brake.

(*ii*) Band brake

(*iii*) Band and block brake.

(*iv*) Internal expanding shoe brake.

8.2.1. Simple Block or Shoe Brake. A simple arrangement for applying a braking force is shown in Fig. 8.1. The face of a brake has a special friction material which has a high value of co-efficient of friction.

A single block or shoe brake consists of a block or shoe which is pressed against a rotating drum as shown in Fig. 8.1. The block is rigidly fixed to the lever. The force is applied at one end of the lever and the other end of the lever is pivoted on a fixed fulcrum *O*. As the force is applied to the lever, the block is pressed against the rotating drum. The friction between the block and the drum causes a tangential force to act on the drum, which tends to prevent its rotation.

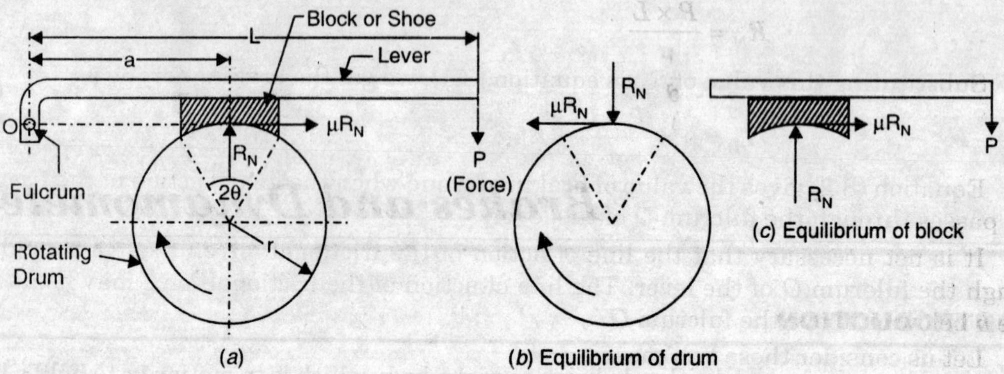

Fig. 8.1

The block is made of a softer material than that of the drum so that the block can be replaced easily on wearing. For light and slow vehicles, wood and rubber are used whereas for heavy and flat vehicles, cast steel is used.

Let P = Force applied at the lower end

r = Radius of the drum

μ = Co-efficient of friction

R_N = Normal reaction on the block

2θ = Angle made by contact surface of the block at the centre of the drum

F^* = Frictional force acting on block = μR_N

T_B = Braking torque.

When force P is applied at the lever end, the block is pressed against the rotating drum. The block exerts a radial force on the drum (*i.e.* this force passes through the centre of the drum). The drum will exert a normal reaction (R_N) on the block. Hence the radial force on the drum will be equal to the normal reaction (R_N) on the block.

Assuming that the normal reaction R_N and the frictional force F^* ($= \mu R_N$) act at the mid-point of the block, we have

Braking torque on the drum = Frictional force × radius

or
$$T_B = F^* \times r$$
$$= \mu R_N \times r \qquad (\because \quad F^* = \mu \times R_N) \quad ...(8.1)$$

The braking torque can be calculated if the value of R_N is known in equation (8.1). The value of R_N is obtained by considering the *equilibrium of the block*.

In Fig. 8.1, the drum is rotating clock-wise. Hence the frictional force on the drum will be acting in the opposite direction [*i.e.* in the anti-clockwise direction as shown in Fig. 8.1 (*b*)]. The frictional force on the block will be opposite to the direction of the frictional force on the drum. Hence the frictional force on the block will be in the clock-wise direction as shown in Fig. 8.1 (*c*) (*i.e.*, in the same direction in which drum is rotating). Let the line of action of this frictional force (μR_N) passes through the fulcrum O of the lever. The forces acting on the block are :

(*i*) R_N (Normal reaction), (*ii*) μR_N (Frictional force), (*iii*) P (Applied force). Taking moments of all forces about the pivot O, we have

$$R_N \times a = P \times L$$ (The frictional force μR_N passes through O, hence its moment is zero)

$$\therefore \qquad R_N = \frac{P \times L}{a}$$

Substituting this value of R_N is equation (8.1), we get the braking torque as,

$$T_B = \mu \times \frac{P \times L}{a} \times r \qquad \text{...(8.2)}$$

Equation (8.2) gives the value of braking torque when the line of action of the frictional force passes through the fulcrum O of the lever.

It is not necessary that the line of action of the frictional force ($\mu \times R_N$) should pass through the fulerum O of the lever. The line of action of the frictional force may be at a distance b below or above the fulcrum O.

Let us consider these two cases :

Case 1. *When the line of action of the frictional force (μR_N) is at a distance 'b' below the fulcrum O and the drum rotates clockwise as shown in Fig. 8.2.*

The forces acting on the block are : (i) R_N acting upwards, (ii) μR_N frictional force on block acting in the same direction in which drum is rotating and (iii) P (acting downwards.)

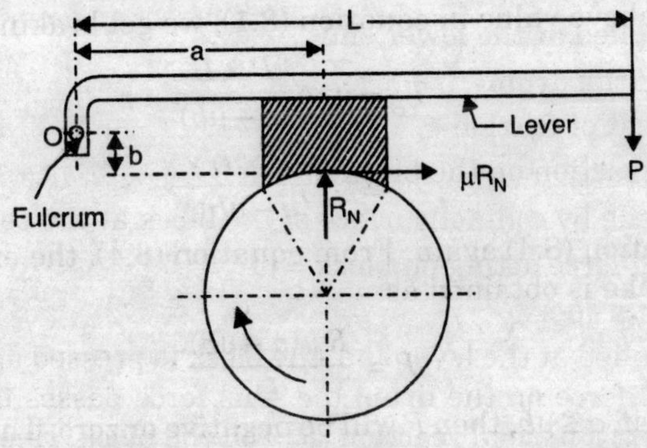

Fig. 8.2

Taking moments about the fulcrum O, we get

$$R_N \times a + \mu R_N \times b = P \times L$$

or

$$R_N (a + \mu \times b) = P \times L$$

$$\therefore \qquad R_N = \frac{P \times L}{(a + \mu b)}$$

Substituting this value of R_N is equation (8.1), we get braking torque (T_B) as,

$$T_B = \mu \times \frac{P \times L}{(a + \mu b)} \times r$$

$$= \frac{\mu \times P \times L \times r}{(a + \mu b)} \qquad \text{...(8.3)}$$

Now consider the above case when drum is rotating in anti-clockwise direction. If the drum is rotating in anti-clockwise direction as shown in Fig. 8.3 then the frictional force $\mu \times R_N$ will also be acting in anti-clockwise direction. The moment of all forces acting on the block (*i.e.* R_N, μR_N and P) about the fulcrum O will give,

$$R_N \times a = P \times L + \mu R_N \times b$$

or

$$R_N \times a - \mu R_N \times b = P \times L$$

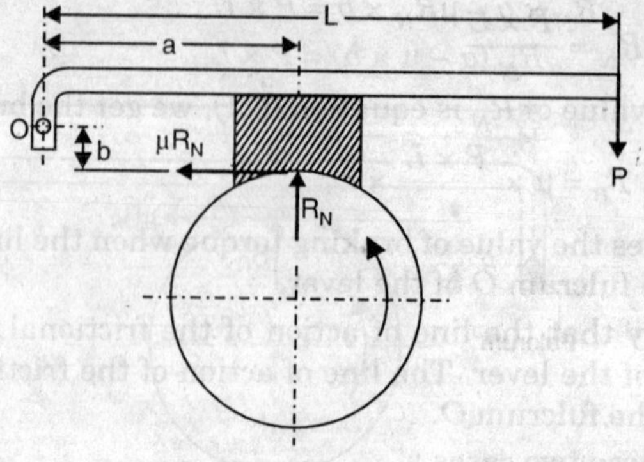

Fig. 8.3

or $$R_N(a - \mu b) = P \times L$$

$$R_N = \frac{P \times L}{(a - \mu b)} \quad ...(8.4)$$

Substituting the above value in equation (8.1), we get braking torque T_B, as

$$T_B = \mu \times \frac{P \times L}{(a - \mu b)} \times r$$

$$= \frac{\mu \times P \times L \times r}{(a - \mu b)} \quad ...(8.5)$$

Consider the equation (8.4) again. From equation (8.4), the expression for the force (P) required to apply the brake is obtained as

$$P = \frac{R_N (a - \mu b)}{L} \quad ...(8.5A)$$

In equation (8.5A) if, $a \le \mu b$, then P will be negative or zero. This means that no-external force is required to apply the brake and hence the brake is **self-locking.** Hence the condition for the brake to be self-locking is

$$a \le \mu b \quad ...(8.5B)$$

Again consider equation (8.4). From equation (8.4), we have the value of P as

$$P = \frac{R_N (a - \mu b)}{L}$$

$$= \frac{R_N \times a - \mu R_N \times b}{L} \quad ...(8.5C)$$

In the above equation '$R_N \times a$' is the moment of R_N about the fulcrum O whereas '$\mu R_N \times b$' is the moment of frictional force about the fulcrum. This moment is having negative sign.

Hence in this case the force P required to apply the brake decreases due to frictional force. Or in other words, the frictional force helps to apply the brake. Such types of brakes are known as **self-energised brakes.** In actual practice the brake should be self-energising and not self-locking. For the above case, the self-locking brake and self-energised brakes are possible. If $P = 0$ it is a self-locking brake. If $P > 0$ it is self-energising brake.

Case 2. *When the line of action of the frictional force* (μR_N) *is at a distance 'b' above the fulcrum O and the drum rotates clockwise as shown in Fig. 8.4*. The forces acting are : (*i*) R_N, (*ii*) μR_N and (*iii*) P. The frictional force ($\mu \times R_N$) on block is acting in the direction of rotation of drum. Taking the moments about the fulcrum, we get

$$R_N \times a = P \times L + \mu R_N \times b$$

or $$R_N \times a - \mu R_N \times b = P \times L$$
or $$R_N\,(a - \mu \times b) = P \times L$$

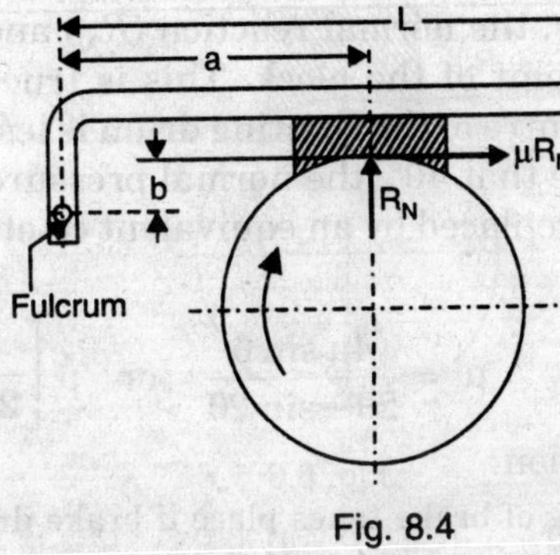

Fig. 8.4

$$\therefore \quad R_N = \frac{P \times L}{(a - \mu b)}$$

Substituting this value of R_N in equation (8.1), we get braking torque (T_B) as

$$T_B = \mu \times \frac{P \times L}{(a - \mu b)} \times r$$

$$= \frac{\mu \times P \times L \times r}{(a - \mu b)} \quad \text{...(8.6)}$$

In this case also, the brake may be self-locking or self-energised. If $P = 0$, the brake is self-locking and if $P > 0$ the brake is self-energised.

If the drum is rotating in anti-clockwise direction as shown in Fig. 8.5, then frictional force (μR_N) will also be acting in anti-clockwise direction.

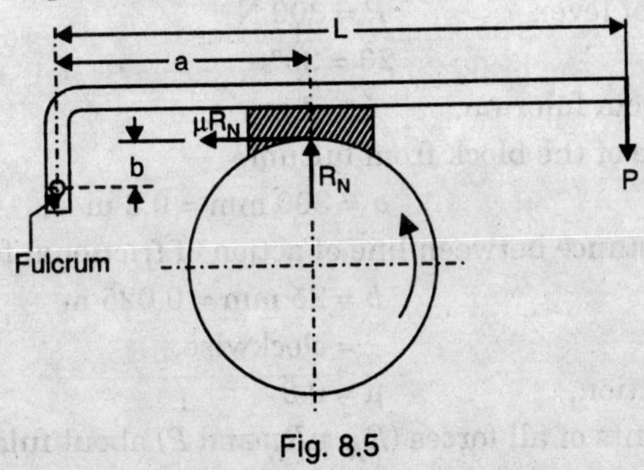

Fig. 8.5

Taking the moments of all forces about the fulcrum, we get

$$R_N \times a + \mu R_N \times b = P \times L$$
or $$R_N(a + \mu b) = P \times L$$
or $$R_N = \frac{P \times L}{(a + \mu b)}$$

Substituting this value in equation (8.1), we get braking torque as

$$T_B = \mu \times \frac{P \times L}{(a + \mu b)} \times r$$

$$= \frac{\mu \times P \times L \times r}{(a + \mu b)} \quad ...(8.7)$$

For all the above expressions, the normal reaction (R_N) and force of friction (μR_N) are assumed to be acting at the mid-point of the block. This is true only if the angle made by contact surface of the block at the centre of the rotating drum is less than or equal to 40° *i.e.* $2\theta \leq 40°$. But if angle of contact is more that 40°, the normal pressure is less at the ends than at the centre. In that case, μ has to be replaced by an equivalent co-efficient of friction μ′ as given by

$$\mu' = \frac{4\mu \sin \theta}{2\theta + \sin 2\theta} \quad \text{or} \quad \mu \left[\frac{4 \sin \theta}{2\theta + \sin 2\theta} \right] \quad ...(8.8)$$

where μ = Actual co-efficient of friction.

Note. (*i*) For Case 1, self-locking of brake takes place if brake drum is rotating anti-clockwise direction.

(*ii*) For Case 2, brake will be self locking if brake drum rotates clockwise.

Problem 8.1. *The brake drum of a single block brake is rotating at 500 r.p.m. in the clockwise direction. The diameter of the drum is 400 mm and the single block brake is of the type as shown in Fig. 8.2. The force required at the end of the lever to apply the brake is 300 N. If angle of contact is 30° and L = 1 m, a = 300 mm and b = 25 mm then determine the braking torque. The co-efficient of friction is equal to 0.3.*

Sol. Given (Refer to Fig. 8.2)

Speed, $N = 500$ r.p.m.

Dia. of drum = 400 mm = 0.4 m

∴ Radius of drum, $r = \frac{400}{2} = 200$ mm = 0.2 m

Force at the end of lever, $P = 300$ N

Angle of contact, $2\theta = 30°$

Length of lever from fulcrum, $L = 1$ m

Distance of centre of the block from fulcum,

$a = 300$ mm = 0.3 m

Perpendicular distance between line of action of frictional force and fulcrum,

$b = 25$ mm = 0.025 m

Rotation of drum = clockwise.

Co-efficient of friction, $\mu = 0.3$

Taking the moments of all forces (R_N, μR_N and P) about fulcrum, we get

$$R_N \times a + \mu R_N \times b = P \times L$$

or $$R_N \times 0.3 + 0.3 \times R_N \times 0.025 = 300 \times 1$$

or $$R_N(0.3 + 0.3 \times 0.025) = 300$$

or $$R_N = \frac{300}{0.3 + 0.3 \times 0.025}$$

$$= \frac{300}{0.3075} = 975.6 \text{ N}$$

Braking torque (T_B) is given by equation (8.1) as

$$T_B = \mu R_N \times r$$

$$= 0.3 \times 975.6 \times 0.2 = \mathbf{58.536 \text{ Nm.}}$$ **Ans.**

Alternately

When the drum is rotating clock-wise and line of action of the frictional force is at a distance 'b' below the fulcrum (Refer to Fig. 8.2), the braking torque is given by equation (8.3). Hence using equation (8.3), we get

$$T_B = \frac{\mu \times P \times L \times r}{(a + \mu b)}$$

$$= \frac{0.3 \times 300 \times 1 \times 0.2}{(0.3 + 0.3 \times 0.025)}$$

$$= \frac{18}{0.3075} = \mathbf{58.536\ Nm.} \quad \textbf{Ans.}$$

Problem 8.2. *If the brake drum in problem 8.1 rotates in the anti-clockwise direction as shown in Fig. 8.3 and all other data remain the same, then determine : (i) the braking torque and (ii) value of 'b' for self-locking of the brake.*

Sol. Refer to Fig. 8.3.

The data from Problem 8.1 :

$N = 500$ r.p.m., $r = 0.2$ m, $P = 300$ N, $L = 1$ m, $a = 0.3$ m, $b = 0.025$ m and $\mu = 0.3$.

(i) Braking Torque (T_B)

When the brake drum is rotating anti-clockwise and line of action of frictional force (μR_N) is at a distance 'b' below the fulcrum as shown in Fig. 8.3, the braking torque is given by equation (8.5). Hence using equation (8.5), we get

$$T_B = \frac{\mu \times P \times L \times r}{(a - \mu b)}$$

$$= \frac{0.3 \times 300 \times 1 \times 0.2}{0.3 - 0.3 \times 0.025}$$

$$= \frac{18}{0.2925} = \mathbf{61.538\ Nm.} \quad \textbf{Ans.}$$

(ii) Value of 'b' for self-locking of the brake

The condition for self-locking of the brake, is given by equation (8.5B) as

$$a \le \mu b$$

The values of 'a' and 'μ' are given. For self-locking of the brake, the value of 'b' is to be obtained. Substituting the values of a and μ in the equation, we get

$$0.3 \le 0.3 \times b$$

or $$\frac{0.3}{0.3} \le b$$

or $$1 \le b$$

or $$b \ge 1 \text{ m.} \quad \textbf{Ans.}$$

Problem 8.3. *The brake drum of a single block brake of diameter 300 mm is rotating at 400 r.p.m. as shown in Fig. 8.6. The force required at the end of the lever to apply the brake is 600 N. If angle of contact is 90° and co-efficient of friction between the drum and brake block is 0.3, find the braking torque.*

Sol. Given :

$P = 600$ N ; $d = 300$ mm or $r = 150$ mm $= 0.15$ m ; $2\theta = 90°$;

$\mu = 0.3$; $b = 40$ mm $= 0.04$ m ; $a = 250$ mm $= 0.25$ m ; $L = 550$ mm $= 0.55$ m.

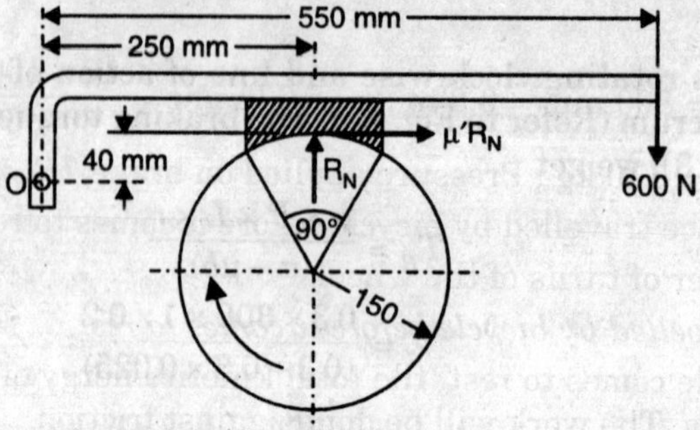

Fig. 8.6

As the angle of contact is more than 40°, hence equivalent co-efficient of friction (μ′) is given by equation (8.8) as

$$\mu' = \mu\left[\frac{4\sin\theta}{2\theta + \sin 2\theta}\right]$$

$$= 0.3\left[\frac{4 \times \sin 45^\circ}{\frac{\pi}{2} + \sin 90^\circ}\right] \qquad \left(\because\ 2\theta = 90^\circ = \frac{\pi}{2} \text{ radians and } \theta = 45^\circ\right)$$

$$= \frac{0.3 \times 4 \times 0.7071}{1.5708 + 1} = 0.33$$

The forces acting on the block are :

(*i*) Normal reaction, R_N

(*ii*) Frictional force $= \mu' \times R_N = 0.33 \times R_N$

(*iii*) Applied force, $P = 600$ N

Taking moments of all forces about the fulcrum O, we get

$$R_N \times 250 = \mu' \times R_N \times 40 + 600 \times 550$$

$$250R_N = 0.33 \times 40 \times R_N + 330000$$

$$= 13.20R_N + 330000$$

or $$250R_N - 13.2R_N = 330000$$

or $$236.8R_N = 330000$$

or $$R_N = \frac{330000}{236.8} = 1393.58 \text{ N}$$

Braking torque (T_B) is given by equation (8.1) as

$$T_B = \text{Frictional force} \times \text{radius of drum}$$

$$= (\mu' \times R_N) \times (r)$$

$$= (0.33 \times 1393.58) \times (0.15) = \mathbf{68.98\ Nm.}$$ **Ans.**

Problem 8.4. *The wheels of a bicycle are of diameter 800 mm. A rider on this bicycle is travelling at a speed of 16 km / hr on a level road. The total mass of rider and bicycle is 110 kg. A brake is applied to the rear wheel. The pressure applied. On the brake is 100 N and co-efficient of friction is 0.06. Before the cycle comes to rest, find :*

(*i*) *distance travelled by the bicycle and*

(*ii*) *number of turns of its wheel.*

Sol. Given :

$$d = 800 \text{ mm} = 0.8 \text{ m} ; v = 16 \text{ km/hr} = \frac{16 \times 1000}{60 \times 60} \text{ m/s} = 4.44 \text{ m/s} ;$$

mass (m) = 110 kg ; Pressure applied on brake, (R_N) = 100 N ; $\mu = 0.06$

Let s = distance travelled by bicycle before it comes to rest

n = number of turns of the wheel.

(i) Distance travelled by bicycle before it comes to rest

When the bicycle comes to rest, the total kinetic energy of the bicycle and rider will be absorbed by the brakes. The work will be done against friction.

$\therefore$ Work done against friction = K.E. of bicycle and rider

or $$\text{Frictional force} \times \text{distance} = \frac{1}{2} mv^2$$

or $$(\mu R_N) \times s = \frac{1}{2} mv^2 \qquad (\because \text{ Frictional force} = \mu R_N)$$

or $$0.06 \times 100 \times s = \frac{1}{2} \times 110 \times 4.44^2$$

$\therefore$ $$s = \frac{1 \times 110 \times 4.44^2}{2 \times 0.06 \times 100} = \mathbf{180.7 \text{ m.}} \quad \textbf{Ans.}$$

(ii) Number of turns of wheel

Total distance travelled by bicycle before it comes to rest

$$= 180.7 \text{ m}$$

Distance travelled in one turn = Circumference of wheel

$$= \pi \times d$$

$$= \pi \times 0.8$$

$\therefore$ $$\text{Number of turns of wheel} = \frac{\text{Total distance travelled}}{\text{Distance travelled in one turn}}$$

$$= \frac{180.7}{\pi \times 0.8} = \mathbf{71.89.} \quad \textbf{Ans.}$$

Problem 8.5. *The brake drum of a single block brake of diameter 300 mm is rotating at 400 r.p.m. in the clockwise direction as shown in Fig. 8.7. The lever OA is pivoted at O and brake shoe is rigidly fixed to the lever. The co-efficient of friction for the brake lining is 0.35. The force applied at the end of lever is 800 N. Find the braking torque.*

Sol. Given :

$d = 300 \text{ mm} = 0.3 \text{ m}$ or $r = 0.15 \text{ m}$; $N = 400$ r.p.m. ; $\mu = 0.35$; force at $A = 800$ N.

Let T_B = Braking torque.

R_N = Normal reaction acting upwards on the block

μR_N = Frictional force acting on the block in the direction of rotation of drum.

Taking the moments of all forces about the fulcrum O, we get

$$R_N \times 300 + \mu R_N \times 100 = 800 \times (300 + 400)$$

or $$300R_N + 0.35 \times R_N \times 100 = 800 \times 700$$

or $$300R_N + 350R_N = 800 \times 700$$

or $$750R_N = 800 \times 700$$

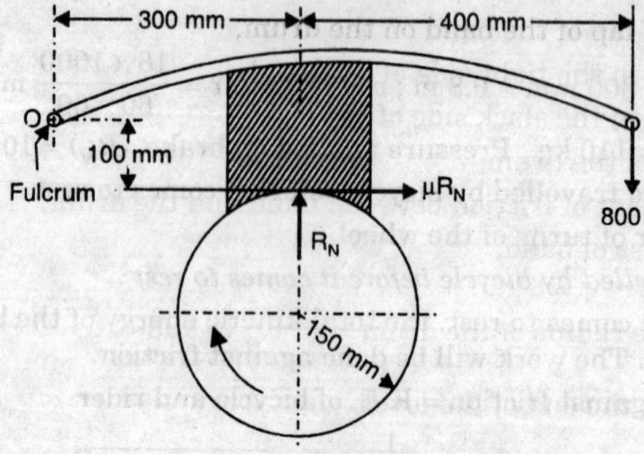

Fig. 8.7

$$\therefore \qquad R_N = \frac{800 \times 700}{750} = 746.67 \text{ N}$$

Braking torque = Frictional force × radius

$= (\mu R_N) \times r$

$= 0.35 \times 746.67 \times 0.15$ Nm = **39.2 Nm.** **Ans.**

Double Block or Shoe Brake

When a single block brake is pressed against a rotating drum, a side thrust on the bearing of the shaft supporting the drum will act due to normal reaction (R_N). This produces the bending of the shaft. This can be prevented by using two blocks on the two sides of the drum as shown in Fig. 8.8. The braking torque becomes two times. The braking torque is given by

$$T_B = \mu R_{N1} \times r + \mu R_{N2} \times r$$
$$= (\mu R_{N1} + \mu R_{N2}) \times r$$

The value of R_{N1} is obtained by taking moments of the forces R_{N1}, μR_{N1} and P about fulcrum O_1. Similarly the value of R_{N2} is obtained by taking moments of the forces R_{N2}, μR_{N2} and P about fulcrum O_2.

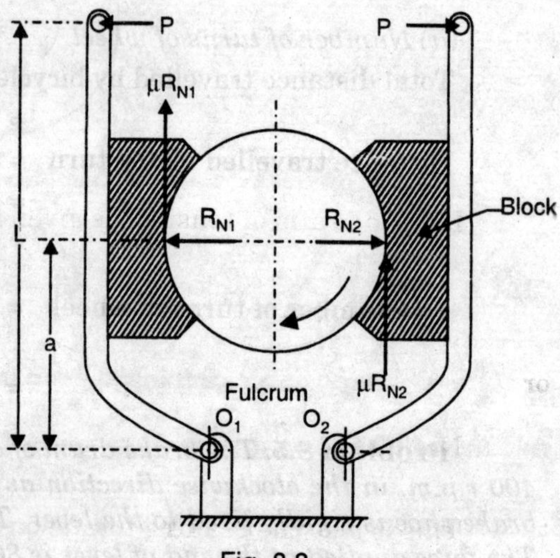

Fig. 8.8

8.2.2. Band Brake. If band is used for bringing a rotating body to rest, then it is known a band brake. A band brake may be a simple band brake or a differential brake.

(*a*) **Simple Band Brake.** It consists of one or more ropes, belt or flexible steel band lined with friction material, which embraces a part of the circumference of the rotating drum. Fig. 8.9 shows a simple band brake in which one end of the band is *attached with the fulcrum* (or fixed pin) of the lever while the other end is attached to the lever at a distance '*a*' from the fulcrum. In order to apply the brake, the band is tightened round the drum and the friction between the band and the drum provides the tangential braking torque.

The force *P* is applied at the free end of the lever which turns about the fulcrum *O*. This tightens the band on the drum and hence the brakes are applied. The braking force is provided by the friction between the band and the drum. The force *P* at the end of the lever for clockwise rotation and anti-clockwise rotation of drum is obtained as explained below :

Let θ = Angle of lap of the band on the drum,

T_1 = Tension in the tight side of the band,

T_2 = Tension in the slack side of the band,

r = Radius of the drum,

μ = Co-efficient of friction between band and the drum,

t = Thickness of band,

r_e = Effective radius of the drum = $\left(r + \frac{t}{2}\right)$, and

P = Force at the end of the lever.

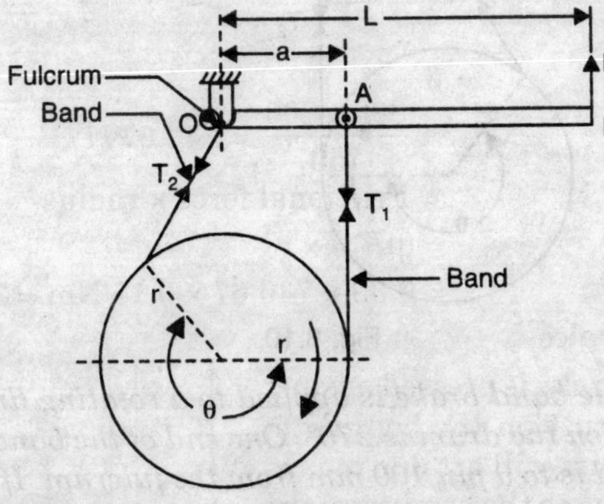

Fig. 8.9

Limiting ratio of tensions is given by,

$$\frac{T_1}{T_2} = e^{\mu\theta} \quad ...(8.9)$$

or

$$2.3 \log\left(\frac{T_1}{T_2}\right) = \mu\theta \quad ...(8.9A)$$

Net torque on drum $= T_1 \times r - T_2 \times r$

$= (T_1 - T_2) \times r$

This is also the braking torque on the drum

$\therefore$ Braking torque on the drum is given by

$T_B = (T_1 - T_2) \times r$ ...if thickness of belt is neglected

$= (T_1 - T_2) \times r_e$ ...if thickness of belt is considered

where r_e = Effective radius of band.

(*i*) **Value of P for Clock-wise rotation of drum.** For clock-wise rotation drum as shown in Fig. 8.9, the end of the band connected to the fulcrum O will be slack side with tension T_2 and the end of the band attached to A will be tight side with tension T_1.

Taking moments about the fulcrum O, we get

$$P \times L = T_1 \times a \quad (\because T_2 \text{ passes through } O) \quad ...(8.10)$$

where L = Distance OB and a = perpendicular distance from O to the line of action of T_1.

(*ii*) **Value of P for Anti-clockwise rotation of drum.** For anti-clockwise rotation of the drum as shown in Fig. 8.10, the end of the band connected to the fulcrum O will be tight side with tension T_1 and the end of the band attached to A will be slack side with tension T_2. Taking the moments about the fulcrum O, we get

$$P \times L = T_2 \times a \qquad (\because \; T_1 \text{ passes through } O) \qquad ...(8.11)$$

where L = Length of lever from fulcrum *i.e.* distance OB

a = Perpendicular distance from O to the line of action of T_2.

Note. For simple band brake, one end of band is always connected to the fulcrum.

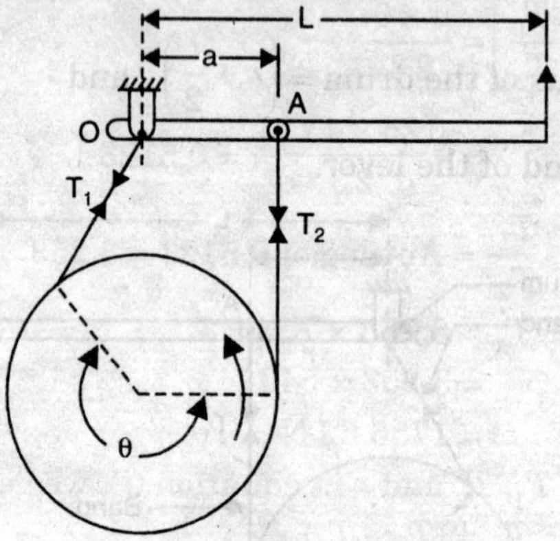

Fig. 8.10

Problem 8.6. *A simple band brake is applied to a rotating drum of diameter 500 mm. The angle of lap of the band on the drum is 270°. One end of the band is attached to a fulcrum pin of the lever and other end is to a pin 100 mm from the fulcrum. If the co-efficient of friction is 0.25 and a braking force of 90 N is applied at a distance of 600 mm from the fulcrum, find the braking torque when the drum rotates in the* (i) *anti-clockwise direction, and* (ii) *clockwise direction.*

Sol. Given :

Simple band brake. This means one end of brake is connected to fulcrum. Other data is :

$$d = 500 \text{ mm} = 0.5 \text{ m} \; ; \; r = 0.25 \text{ m} \; ; \; \theta = 270° = 270 \times \frac{\pi}{180} = 4.713 \text{ rad} \; ;$$

Distance $a = 100$ mm $= 0.1$ m ; $L = 600$ mm $= 0.6$ m ; $\mu = 0.25$; $P = 90$ N.

Let T_B = braking torque.

(i) Drum rotates in anti-clockwise direction

Refer to Fig. 8.10 in which the drum is rotating in anti-clockwise direction. The braking torque is the net torque on the drum and it is given by,

$$T_B = (T_1 - T_2) \times r \qquad ...(i)$$

In the above equation, the value of 'r' is known. But the values of T_1 and T_2 are unknown. Let us first find the values of T_1 and T_2.

Taking the moments of all forces (shown in Fig. 8.10) about O

$$T_2 \times a = P \times L$$

or

$$T_2 = \frac{P \times L}{a}$$

$$= \frac{90 \times 0.6}{0.1} = 540 \text{ N}$$

Now using equation (8.9) for limiting ratio of tensions,

$$\frac{T_1}{T_2} = e^{\mu \times \theta}$$

or $$2.3 \log \left(\frac{T_1}{T_2}\right) = \mu \times \theta$$

or $$\log \left(\frac{T_1}{T_2}\right) = \frac{\mu \times \theta}{2.3}$$

$$= \frac{0.25 \times 4.713}{2.3} = 0.5123$$

$\therefore$ $$\frac{T_1}{T_2} = \text{Antilog of } 0.5123 = 3.253$$

or $$T_1 = 3.253 \times T_2$$

$$= 3.253 \times 540 \qquad (\because \quad T_2 = 540 \text{ N})$$

$$= 1756.62 \text{ N}$$

Substituting the values of T_1, T_2 and r in equation (*i*), we get the braking torque as

$$T_B = (T_1 - T_2) \times r$$

$$= (1756.62 - 540) \times 0.25 \text{ Nm}$$

$$= \mathbf{304.155 \text{ Nm.}} \quad \textbf{Ans.}$$

(*ii*) *Drum rotates in clockwise direction*

Refer to Fig. 8.9 in which drum is rotating in clockwise direction. The braking torque is given by,

$$T_B = (T_1 - T_2) \times r$$

Let us first find the values of T_1 and T_2. Taking moments of all forces shown in Fig. 8.9 about O, we get

$$T_1 \times a = P \times L$$

or $$T_1 = \frac{P \times L}{a}$$

$$= \frac{90 \times 0.6}{0.1} = 540 \text{ N}$$

Also the know that

$$\frac{T_1}{T_2} = e^{\mu \times \theta}$$

or $$2.3 \log \frac{T_1}{T_2} = \mu \times \theta$$

or $$\log \left(\frac{T_1}{T_2}\right) = \frac{\mu \times \theta}{2.3}$$

$$= \frac{0.25 \times 4.713}{2.3} = 0.5123$$

$\therefore$ $$\frac{T_1}{T_2} = \text{Antilog of } 0.5123 = 3.253$$

$$\therefore \quad T_2 = \frac{T_1}{3.253} = \frac{540}{3.253} = 166 \text{ N}$$

$\therefore$ Braking torque is given by,

$$T_B = (T_1 - T_2) \times r$$
$$= (540 - 166) \times 0.25 = \mathbf{93.5\ Nm.} \quad \textbf{Ans.}$$

Problem 8.7. *Fig. 8.11 shows a simple band brake which is applied to a shaft carrying a flywheel (i.e. rotating drum) of mass 300 kg and of radius of gyration 350 mm. The flywheel rotates at 200 r.p.m. The brake drum diameter is 260 mm and co-efficient of friction is 0.20. The angle of lap of the band on the drum is 210°. If the braking torque is 39 Nm, find :*

(i) the force applied at the lever end,

(ii) the number of turns of the flywheel before it comes to rest,

(iii) the time taken by the flywheel to come to rest.

Sol. Given :

For a simple band brake, one end of the band should be connected to the fulcrum whereas the other end of the band may be connected to the lever either towards the same side is which force P is acting or towards the opposite side in which P is acting. Here the other end is in opposite direction.

The other given data is :

mass, m = 300 kg ; radius of gyration, k = 350 mm = 0.35 m ;

N = 200 r.p.m. ; d = 260 mm ; r = 130 mm = 0.13 m ;

$\mu = 0.20$; $\theta = 210°$ or $210 \times \frac{\pi}{180}$ rad = 3.666 rad.

braking torque, T_B = 39 Nm

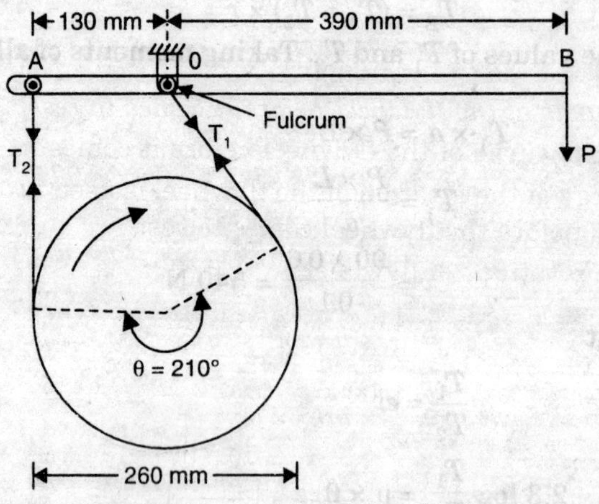

Fig. 8.11

(i) Force applied at the end of the lever

Let P = Force applied at the lever end.

The braking torque is given by,

$$T_B = (T_1 - T_2) \times r$$

or
$$39 = (T_1 - T_2) \times 0.13$$

$$(T_1 - T_2) = \frac{39}{0.13} = 300 \text{ N} \quad \text{...(i)}$$

Let us now find the tensions T_1 and T_2. We know that

$$\frac{T_1}{T_2} = e^{\mu \times \theta}$$

or $$2.3 \log \frac{T_1}{T_2} = \mu \times \theta = 0.2 \times 3.666 = 0.7322$$

$\therefore$ $$\log \frac{T_1}{T_2} = \frac{0.7322}{2.3} = 0.3188$$

or $$\frac{T_1}{T_2} = \text{Antilog of } 0.3188 = 2.08$$

or $$T_1 = 2.08 T_2 \quad \ldots(ii)$$

Substituting the value of T_1 in equation (*i*), we get

$$2.08T_2 - T_2 = 300$$

or $$1.08T_2 = 300$$

$\therefore$ $$T_2 = \frac{300}{1.08} = 277.77 \text{ N}$$

Substituting this value of T_2 in equation (*ii*), we get

$$T_1 = 2.08 \times 277.77 = 577.76 \text{ N}$$

To find the value of P, take the moments of all forces (*i.e.* T_1, T_2 and P) about the fulcrum O.

$\therefore$ $$P \times 390 = T_2 \times 130 \quad (\because T_1 \text{ passes through } O)$$

$\therefore$ $$P = \frac{T_2 \times 130}{390}$$

$$= \frac{277.77 \times 130}{390} = \mathbf{92.59\ N.}\ \ \textbf{Ans.}$$

(*ii*) *Number of turns of the flywheel before it comes to rest.*

Let n = Number of turns of the flywheel before it comes to rest.

The kinetic energy of the rotation of the flywheel is used to overcome the workdone due to braking torque (T_B), before the flywheel comes to rest.

Now K.E. of the rotation* of flywheel

$$= \frac{1}{2} \times I + \omega^2$$

$$= \frac{1}{2} \times mk^2 \times \omega^2 \quad (\because I = mk^2)$$

$$= \frac{1}{2} \times mk^2 \times \left(\frac{2\pi N}{60}\right)^2 \quad \left(\because \omega = \frac{2\pi N}{60}\right)$$

$$= \frac{1}{2} \times 300 \times 0.35^2 \times \left(\frac{2\pi \times 200}{60}\right)^2$$

$$= 8060.17 \text{ Nm} \quad \ldots(iii)$$

$\therefore$ Work done by the braking torque in 'n' number of turns of the flywheel

$$= T_B \times \text{Angular displacement in } n \text{ turns}$$

*K.E. due to linear velocity = $\frac{1}{2}$ mV^2 whereas the K.E. due to rotation = $\frac{1}{2}$ $I \times \omega^2$ where $I = mk^2$.

$= T_B \times 2\pi \times n$ ($\because$ Angular displacement for one turn $= 2\pi$)

$= 39 \times 2\pi \times n$ ...(*iv*)

But K.E. of flywheel = Work done by braking torque

$\therefore \quad 8060.17 = 39 \times 2\pi \times n$

$\therefore \quad n = \frac{8060.17}{39 \times 2\pi} = $ **32.89. Ans.**

(*iii*) *Time taken by flywheel to come to rest after applying the brake*

$N = 200$ r.p.m.

This means that 200 revolutions are made in one minute. The flywheel comes to rest after applying the brake in 32.89 revolution. Let us find the time for 32.89 revolution.

Time for 200 revolution = 1 minute

Time for 1 revolution $= \frac{1}{200}$ min

Time for 32.89 revolution $= \frac{1}{200} \times 32.89 = 0.16445$ min

$= 0.16445 \times 60$ seconds

= **9.867 seconds. Ans.**

Problem 8.8. *Fig. 8.12 shows a simple band brake which is applied on a drum of diameter 400 mm. The drum is rotating at 180 r.p.m. The angle of lap of the band on the drum is 270° and co-efficient of friction is 0.25. One end of the band is attached to a fixed pin (i.e. fulcrum) and other end to the lever arm at a distance of 100 mm from the fulcrum. The lever length is 600 mm. The lever arm is placed perpendicular to the diameter that bisects the angle of contact. Determine :*

(*i*) *the necessary force required at the end of the lever arm to stop the drum if a power of 30 kW is being absorbed. Also find the direction of this force.*

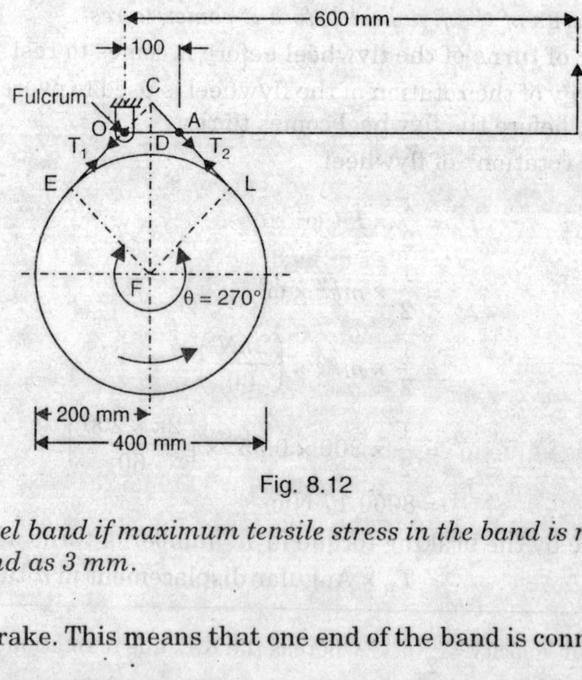

Fig. 8.12

(*ii*) *width of steel band if maximum tensile stress in the band is not to exceed 50 N/mm². Take thickness of band as 3 mm.*

Sol. Given :

Simple band brake. This means that one end of the band is connected to the fixed pin *O*

$d = 400$ mm or $r = 200$ mm $= 0.2$ m ;

$N = 180$ r.p.m. ; $\theta = 270° = 270 \times \frac{\pi}{180} = 4.713$ rad ; $\mu = 0.25$; Distance $OA = 100$ mm ;

Length $OB = 600$ mm ; Power $= 30$ kW $= 30 \times 10^3$ W.

(*i*) *Necessary force at the end of the lever arm*

Let P = Force applied at the end of the lever.

Fig. 8.12 shows the simple band brake in which one end is connected to a fixed point O. The other end of the band is attached at A. From Fig. 8.12, it is clear that force P must act upwards. When the drum is rotating anti-clockwise (as shown in Fig. 8.12), the end of the band attached to O will be tight with tension T_1. The other end of the band attached to A will be slack with tension T_2. The tensions T_1 and T_2 are obtained as given below :

We know that $\frac{T_1}{T_2} = e^{\mu \times \theta}$

or $2.3 \log \frac{T_1}{T_2} = \mu \times \theta = 0.25 \times 4.713 = 1.178$

or $\log \frac{T_1}{T_2} = \frac{1.178}{2.3} = 0.5122$

$\therefore$ $\frac{T_1}{T_2}$ = Antilog of $0.5122 = 3.25$

or $T_1 = 3.25 T_2$ ...(*i*)

Let T_B = Braking torque

The power absorbed is given by,

$$\text{Power} = T_B \times \omega$$

$$= T_B \times \frac{2\pi N}{60}$$

or $30 \times 10^3 = \frac{T_B \times 2\pi \times 180}{60} = 6\pi \times T_B$

or $T_B = \frac{30 \times 10^3}{6\pi} = 1591.55$ Nm ...(*ii*)

We know that braking torque T_B is given by,

$$T_B = (T_1 - T_2) \times r$$

or $1591.55 = (T_1 - T_2) \times 0.2$ $\quad (\because T_B = 1591.55 \text{ and } r = 0.2)$

$= (3.25T_2 - T_2) \times 0.2$ $\quad [\because T_1 = 3.25T_2 \text{ from } (i)]$

$= 2.25T_2 \times 0.2$

$\therefore$ $T_2 = \frac{1591.55}{2.25 \times 0.2} = 3536.77$ N

Substituting this value in equation (*i*), we get

$$T_1 = 3.25 \times 3536.77 = 11494.5 \text{ N}$$

To find the force P at the end of the lever, take the moments of all forces about O (*i.e.* fulcrum or fixed point).

$\therefore$ $P \times OB = T_2 \times OC$ ...(*iii*)

Let us find first the distance OC.

Distance $$OD = \frac{OA}{2} = \frac{100}{2} = 50 \text{ mm}$$

$$\angle EFC = \frac{1}{2}[360 - 270] = 45°$$

$\therefore$ $$\angle FCE = 45°$$

$[\because$ In ΔEFC, $\angle EFC = 45°$; $\angle FEC = 90°$ $\therefore$ $\angle FCE = 45°]$

In ΔOCD, $\angle DCO = 45°$, $\therefore$ $\cos 45° = \frac{OD}{OC}$

$\therefore$ $$OC = \frac{OD}{\cos 45°} = \frac{50}{\cos 45°} = 70.71 \text{ mm}$$

Substituting the values of OC, OB and T_2 in equation (*iii*),

$$P \times 600 = 3536.77 \times 70.71$$

$\therefore$ $$P = \frac{3536.77 \times 70.71}{600} = \mathbf{416.8\ N.}\quad \textbf{Ans.}$$

(*ii*) *Width of band*

Given :

Max. tensile stress = 50 N/mm^2. Thickness, $t = 3$ mm.

Let b = Width of band

The maximum tension in the band is T_1. The value of T_1 is 11494.5 N

$\therefore$ Maximum tension = 11494.5 N

But maximum tension = Max. tensile stress × Area of cross-section of band

or $$11494.5 = 50 \times [b \times t] = 50 \times [b \times 3]$$

$\therefore$ $$b = \frac{11494.5}{50 \times 3} = \mathbf{76.63\ mm.}\quad \textbf{Ans.}$$

(*b*) **Differential Band Brake.** In case of differential band brake no end of the band is connected to the fulcrum. Fig. 8.13 shows a differential band brake in which the ends of the band are connected at A and B which are on different sides of the fulcrum O.

When the drum rotates in the clock-wise direction, the end of the band attached at A will be tight with tension T_1 whereas the end of the band attached at B will be slack with tension T_2 as shown in Fig. 8.13 (*a*). The value of force P at the end of the lever, can be obtained as explained below :

Let a = perpendicular distance from fulcrum O on the line of action of tension T_2

b = perpendicular distance from O on line of action of tension T_1

L = length of lever from O.

Consider the equilibrium of lever BOC. The force acting on the lever BOC are (*i*) force, P, (*ii*) tension T_1 at A and (*iii*) tension T_2 at B as shown in Fig. 8.13 (*b*).

Taking moments of all forces about O, we get

$$P \times L + T_1 \times b = T_2 \times a$$

or $$P \times L = T_2 \times a - T_1 \times b$$

$\therefore$ $$P = \frac{T_2 \times a - T_1 \times b}{L} \quad ...(i)$$

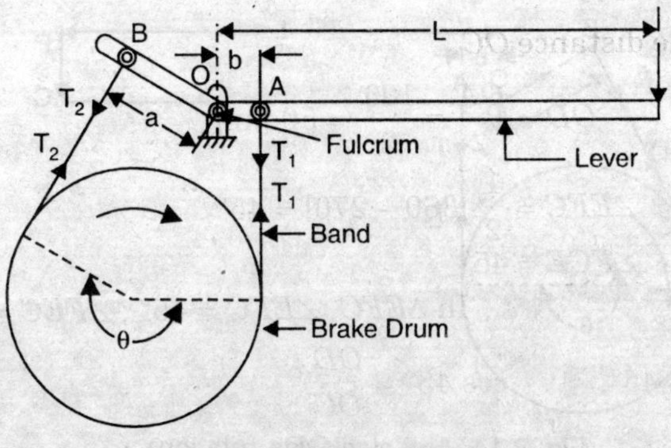

(*a*) Clockwise rotation.

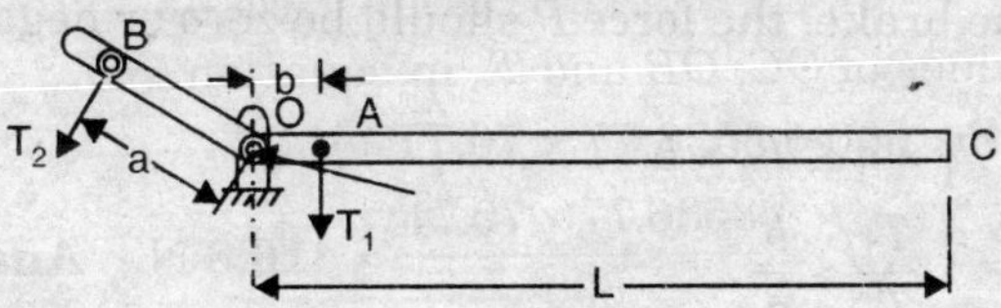

(*b*) Equilibrium of lever *BOC*

Fig. 8.13. Differential band brake.

But T_1 is always more than T_2. Hence in the above equation the force P will be positive if

$$T_2 \times a > T_1 \times b$$

or
$$T_1 \times b < T_2 \times a$$

or
$$\frac{T_1}{T_2} < \frac{a}{b} \qquad \text{...}(ii)$$

In equation (*i*), the force P will be zero or negative if

$$T_2 \times a \le T_1 \times b$$

or
$$\frac{T_2}{T_1} \le \frac{b}{a} \qquad \text{...}(iii)$$

If the force P is zero or negative, then the brake becomes as self locking. Hence for self-locking of the brake when drum rotates clock-wise the condition is

$$\frac{T_2}{T_1} \le \frac{b}{a} \qquad \text{...(8.12)}$$

Anti-clockwise rotation

Fig. 8.14 shows a differential band brake in which brake drum is rotating anti-clockwise. The end of the band connected to B will be tight with tension T_1 whereas the end of the band connecting to A will be slack with tension.

Taking the moments about the fulcrum O, we get

$$P \times L + T_2 \times b = T_1 \times a$$

or
$$P \times L = T_1 \times a - T_2 \times b$$

or
$$P = \frac{T_1 \times a - T_2 \times b}{L}$$

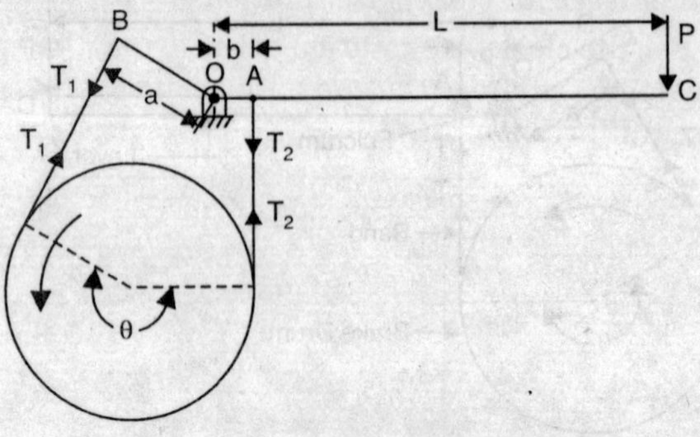

Fig. 8.14. Anti-clockwise rotation.

For self-locking of the brake, the force P should be zero or negative. But force P will be zero or negative if

$$T_1 \times a \le T_2 \times b$$

or $$\frac{T_1}{T_2} \le \frac{b}{a} \quad \text{...(8.13)}$$

Problem 8.9. *Fig. 8.15 shows a differential band brake of drum diameter 400 mm. The two ends of the band are fixed to the points on the opposite side of fulcrum of the lever at a distance of 50 mm and 160 mm from the fulcrum as shown in Fig. 8.15. The brake is to sustain a torque of 300 Nm. The co-efficient of friction between band and the brake is 0.2. The angle of contact is 210° and the length of lever from the fulcrum is 600 mm. Determine :*

(i) the force required at the end of the lever for the clockwise and anti-clockwise rotation of the drum.

(ii) value of OB for the brake to be self-locking for clockwise rotation.

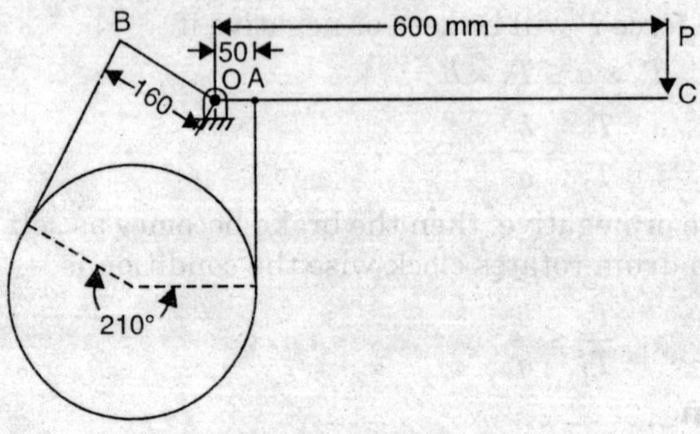

Fig. 8.15

Sol. Given :

Differential band brake, which means no ends of the band is connected to the fulcrum. The other data is :

$$d = 400 \text{ mm} \quad \text{or} \quad r = 200 \text{ mm} = 0.2 \text{ m} ;$$

Distance $OA = 50$ mm, distance $OB = 160$ mm, $T_B = 300$ Nm, $\mu = 0.2$, $L = 600$ mm,

$$\theta = 210° = 210 \times \frac{\pi}{180} = 3.665 \text{ rad.}$$

(i) Force at the end of lever for clock-wise rotation

Refer to Fig. 8.16. For the clockwise rotation of the drum, the end of the band connected to A will be tight with tension T_1 whereas the end of the band connected to B will be slack with tension T_2. Consider the equilibrium of BOC.

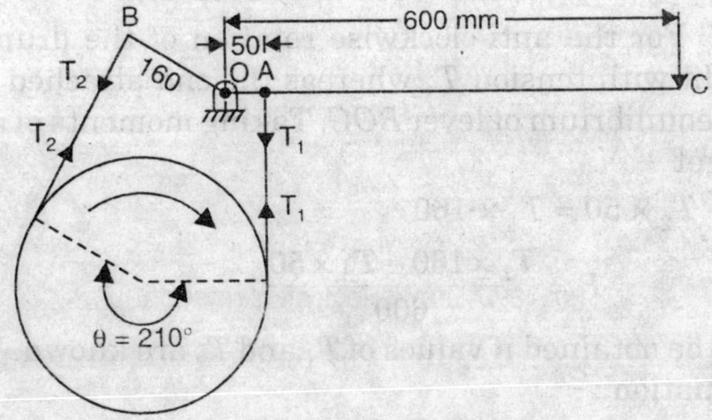

Fig. 8.16

Taking the moments of all forces acting on BOC,

(The focus acting on BOC are T_2, T_1 and P) about fulcrum O, we get

$$T_1 \times AO + P \times OC = T_2 \times OB \qquad ...(i)$$

or $$T_1 \times 50 + P \times 600 = T_2 \times 160 \qquad ...(ii)$$

Let us first find the values of T_1 and T_2, so that the value of P can be obtained.

Using equation,

$$\frac{T_1}{T_2} = e^{\mu \times \theta}$$

or $$2.3 \log \left(\frac{T_1}{T_2}\right) = \mu \times \theta = 0.2 \times 3.665 = 0.733$$

or $$\log \left(\frac{T_1}{T_2}\right) = \frac{0.733}{2.3} = 0.3187$$

$\therefore$ $$\frac{T_1}{T_2} = \text{..ti-log of } 0.3187 = 2.083$$

$\therefore$ $$T_1 = 2.083 T_2 \qquad ...(iii)$$

The braking torque is given by,

$$T_B = (T_1 - T_2) \times r$$

or $$300 = (T_1 - T_2) \times 0.2 \qquad (\because \ r = 0.2 \text{ m})$$

$$= (2.083 T_2 - T_2) \times 0.2 \qquad (\because \ T_1 = 2.083 T_2)$$

$$= 1.083 \times T_2 \times 0.2$$

$\therefore$ $$T_2 = \frac{300}{1.083 \times 0.2} = 1385 \text{ N}$$

Substituting the value of T_2 in equation (*iii*), we get

$\therefore$ $$T_1 = 2.083 \times 1385 = 2884.95 \text{ N}$$

Substituting the values of T_1 and T_2 in equation (*ii*), we get

$$2884.95 \times 50 + P \times 600 = 1385 \times 160$$

or $$144247.5 + 600P = 221600$$

or $$600P = 221600 - 144247.5 = 77352.5$$

or $$P = \frac{77352.5}{600} = \mathbf{128.92\ N.}\quad \textbf{Ans.}$$

Force at the end of the lever for anti-clockwise rotation

Refer to Fig. 8.17. For the anti-clockwise rotation of the drum, the end of the band attached to B will be tight with tension T_1 whereas the end attached to A will be slack with tension T_2. Consider the equilibrium of lever BOC. Taking moments of all forces (*i.e.* T_1, T_2 and P) about fulcrum O, we get

$$P \times 600 + T_2 \times 50 = T_1 \times 160$$

or $$P = \frac{T_1 \times 160 - T_2 \times 50}{600} \quad \text{...(A)}$$

The value of P can be obtained if values of T_1 and T_2 are known. The values of T_1 and T_2 are obtained by using equation :

$$\frac{T_1}{T_2} = e^{\mu\theta} \quad \text{or} \quad 2.3 \log \frac{T_1}{T_2} = \mu \times \theta$$

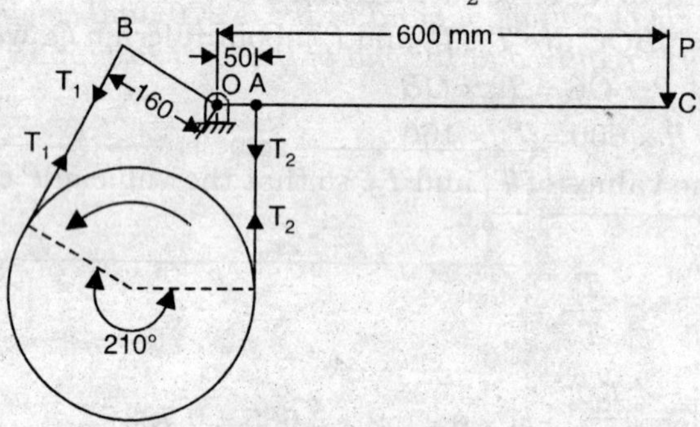

Fig. 8.17

or $$\log\left(\frac{T_1}{T_2}\right) = \frac{\mu \times \theta}{2.3} = \frac{0.2 \times 3.665}{2.3} = 0.3187$$

or $$\frac{T_1}{T_2} = \text{Antilog of } 0.3187 = 2.083$$

or $$T_1 = 2.083T_2$$

The braking torque is given by equation,

$$T_B = (T_1 - T_2) \times r$$

or $$300 = (2.083T_2 - T_2) \times 0.2 \qquad (\because\ T_1 = 2.083T_2 \text{ and } r = 0.2)$$

$$= 1.083T_2 - 0.2$$

$\therefore$ $$T_2 = \frac{300}{1.083 \times 0.2} = 1385\ \text{N}$$

and $$T_1 = 2.083 \times T_2 = 2.083 \times 1385 = 2884.95\ \text{N}$$

Substituting the values of T_1 and T_2 in equation (A), we get

$$P = \frac{2884.95 \times 160 - 1385 \times 50}{600}$$

$$= \frac{461592 - 69250}{600} = \mathbf{653.9\ N.}\quad \textbf{Ans.}$$

(ii) Value of OB for the brake to be self-locking for clockwise rotation

Refer to Fig. 8.16. For clockwise rotation of the drum, we have equation (*i*), as

$$T_1 \times AO + P \times OC = T_2 \times OB$$

The brake will be self-locking, if P is zero. Hence substituting $P = 0$ in the above equation, we get

$$T_1 \times OA = T_2 \times OB$$

The values of T_1, T_2 and OA are known, hence the value of OB can be obtained

$$\therefore \quad OB = \frac{T_1 \times OA}{T_2}$$

$$= \frac{2884.95 \times 50}{1385} = \mathbf{104.15\ mm.} \quad \textbf{Ans.}$$

Problem 8.10. *Fig. 8.18 shows a barrel and a differential band brake which are keyed to the same shaft. A rope is wound round a barrel and supports a load of 200 kN. The brake drum diameter is 600 mm and diameter of barrel is 300 mm. The two ends of the bands are attached to pins on opposite sides of the fulcrum of the brake lever at a distances of 20 mm and 80 mm as shown in Fig. 8.18. The angle of contact of band brake is 270° and co-efficient of friction 0.25. Determine the minimum force required at the end of the lever to support the load, if the length of the lever from the fulcrum is 2400 mm.*

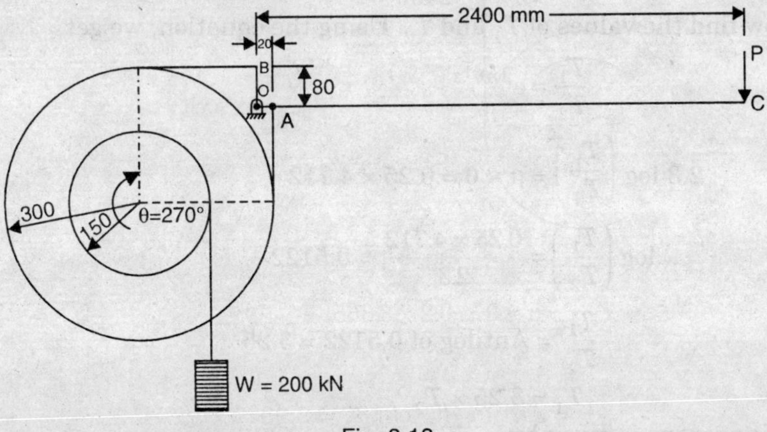

Fig. 8.18

Sol. Given :

Max. load, $W = 200$ kN $= 200 \times 10^3$ N ; $d = 600$ mm or $r = 300$ mm $= 0.3$ m ;

Dia. of barrel, $D = 300$ mm or radius of barrel, $R = 150$ mm $= 0.15$ m ;

Distance $OA = 20$ mm ; distance $OB = 80$ mm ; $\theta = 270° = 270 \times \dfrac{\pi}{180} = 4.712$ rad ;

$\mu = 0.25$, $L = 2400$ mm

Let P = Minimum force at the end of the lever to support the load

As OB is greater than OA, therefore the force P will act downward.

Let us find the two values of P when drum rotates clockwise and then anti-clockwise.

(i) Drum rotates clockwise

Refer to Fig. 8.19. For clockwise rotation, the end of the band attached to A will be tight with tension T_1 whereas the end attached to B will be slack with tension T_2.

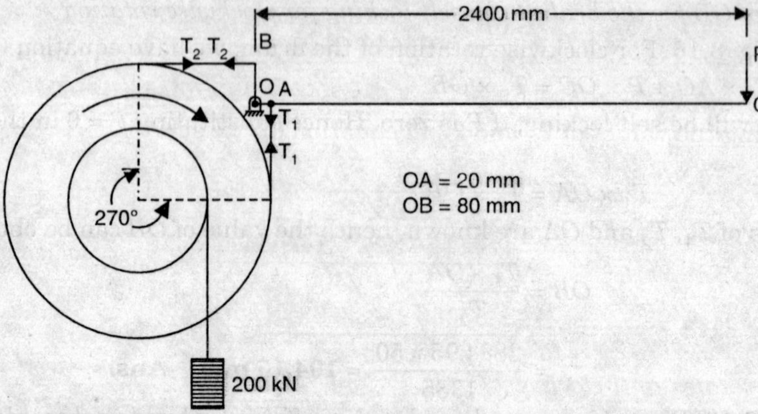

Fig. 8.19

Considering the forces acting on the lever and taking moments about the fulcrum O, we get

$$T_1 \times 20 + P \times 2400 = T_2 \times 80$$

or $$P = \frac{T_2 \times 80 - T_1 \times 20}{2400} \quad ...(i)$$

Let us now find the values of T_1 and T_2. Using the equation, we get

$$\frac{T_1}{T_2} = e^{\mu \times \theta}$$

or $$2.3 \log\left(\frac{T_1}{T_2}\right) = \mu \times \theta = 0.25 \times 4.712$$

or $$\log\left(\frac{T_1}{T_2}\right) = \frac{0.25 \times 4.712}{2.3} = 0.5122$$

$\therefore$ $$\frac{T_1}{T_2} = \text{Antilog of } 0.5122 = 3.25$$

or $$T_1 = 3.25 \times T_2 \quad ...(ii)$$

The braking torque is given by,

$$T_B = (T_1 - T_2) \times r = (T_1 - T_2) \times 0.3 \text{ Nm}$$

The torque due to load W = Load $\times R$

$$= W \times 0.15 = 200 \times 10^3 \times 0.15 \text{ Nm} \qquad (\because \quad W = 200 \times 10^3)$$

$$= 30 \times 10^3 \text{ Nm}$$

For equilibrium of the system, the braking torque should be equal to the torque due to load W, as the barrel (on which load is acting) and drum are keyed to the same shaft

$\therefore$ Braking torque = Torque due to load W

or $$(T_1 - T_2) \times 0.3 = 30 \times 10^3$$

or $$(T_1 - T_2) = \frac{30 \times 10^3}{0.3} = 10^5 \text{ N}$$

Substituting the value of $T_1 = 3.25T_2$ in the above equation, we get

$$(3.25T_2 - T_2) = 10^5$$

$$2.25T_2 = 10^5$$

or $$T_2 = \frac{10^5}{2.25} = 44444.44 \text{ N}$$

$\therefore$ $$T_1 = 3.25T_2 = 3.25 \times 44444.44 = 144444.44 \text{ N}$$

Substituting the values of T_1 and T_2 in equation (*i*), we get

$$P = \frac{T_2 \times 80 - T_1 \times 20}{2400}$$

$$= \frac{44444.44 \times 80 - 144444.44 \times 20}{2400}$$

$$= \frac{3555555.2 - 288888.8}{2400} = 277.77 \text{ N.}$$

(ii) Drum rotates anti-clockwise

Refer to Fig. 8.20. For anti-clockwise rotation, the end of the band attached to B will be tight with tension T_1. Whereas the end attached to A will be slack with tension T_2. Considering the forces acting on the lever and taking moments about the fulcrum O, we get

$$P \times 2400 + T_2 \times 20 = T_1 \times 80$$

or $$P = \frac{T_1 \times 80 - T_2 \times 20}{2400} \quad ...(iii)$$

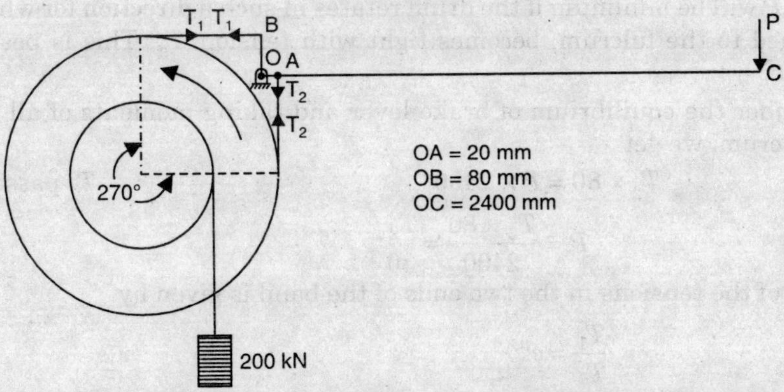

Fig. 8.20

The ratio of tensions T_1 and T_2 will be same as obtained in the 1st case *i.e.*

$$\frac{T_1}{T_2} = 3.25$$

or $$T_1 = 3.25T_2$$

Also braking torque = Torque due to load W

or $$(T_1 - T_2) \times r = W \times R$$

or $$(T_1 - T_2) \times 0.3 = (200 \times 10^3) \times 0.15 = 30 \times 10^3$$

$$= 30 \times 10^3 \text{ Nm}$$

or $$(T_1 - T_2) = \frac{30 \times 10^3}{0.3} = 10^5 \text{ N}$$

But $T_1 = 3.25T_2$. The values of T_1 and T_2 will be same as in 1st case *i.e.*

$$T_1 = 144444.44 \text{ and } T_2 = 44444.44$$

Substituting these values in equation (*iii*), we get

$$P = \frac{144444.44 \times 80 - 44444.44 \times 20}{2400}$$

$$= \frac{11555555.2 - 888888.8}{2400} = 4444.44 \text{ N}$$

From the above two cases, we find that minimum force required is 277.77 N. And the force is minimum when drum rotates clockwise.

$\therefore$ Minimum force = **277.77 N. Ans.**

Problem 8.11. *If in the above question, instead of a differential band brake, a simple band brake is used (which means one end of the brake is attached to the fulcrum and other end at B at a distance of 80 mm) and all other data remains same, then find the minimum force required at the end of the brake lever.*

Sol. Given :

The data from problem 8.10 is :

$W = 200 \times 10^3$ N ; $r = 0.3$ m ; $R = 0.15$ m ; $OB = 80$ mm ; $\theta = 270° = 4.712$ rad ;

$\mu = 0.25$ and $L = 2400$ mm

Let P = Minimum force.

The force P will be minimum if the drum rotates in such a direction for which the end of the band attached to the fulcrum, becomes tight with tension T_1. This is possible if drum rotates clockwise.

Now consider the equilibrium of brake lever and taking moments of all forces (*i.e.* T_2 and P) about fulcrum, we get

$$T_2 \times 80 = P \times 2400 \qquad (\because \ T_1 \text{ passes through } O)$$

$$\therefore \qquad P = \frac{T_2 \times 80}{2400} = \frac{T_2}{30} \qquad \text{...}(i)$$

The ratio of the tensions in the two ends of the band is given by

$$\frac{T_1}{T_2} = e^{\mu \times \theta}$$

or $$2.3 \log \frac{T_1}{T_2} = \mu \times \theta = 0.25 \times 4.712$$

or $$\log \frac{T_1}{T_2} = \frac{0.25 \times 4.712}{2.3} = 0.5122$$

or $$\frac{T_1}{T_2} = \text{Antilog of } 0.5122 = 3.25$$

or $$T_1 = 3.25 T_2 \qquad \text{...}(ii)$$

Also using the equation,

Braking torque = Torque due to load W

or $$(T_1 - T_2) \times r = W \times R$$

or $$(T_1 - T_2) = \frac{W \times R}{r}$$

$$= \frac{(200 \times 10^3) \times 0.15}{0.30} = 10^5 \text{ N}$$

$$\therefore \quad T_1 - T_2 = 10^5$$

or $$(3.25T_2 - T_2) = 10^5 \quad [\because \text{ From } (ii), T_1 = 3.25T_2]$$

or $$2.25T_2 = 10^5$$

$$\therefore \quad T_2 = \frac{10^5}{2.25} = 44444.44 \text{ N}$$

$$\therefore \quad T_1 = 3.25T_2 = 3.25 \times 44444.44$$

$$= 144444.44 \text{ N}$$

Substituting the value of T_2 in equation (i), we get

$$\therefore \quad P = \frac{44444.44}{30}$$

$$= \mathbf{1481.48\ N.} \quad \textbf{Ans.}$$

8.2.3. Band and Block Brake. If the band and also the blocks are used for applying brakes to a rotating body, then it is known as band and blockbrake. Fig. 8.21 shows the band and block brake which is the modification of the band brake and consists of a number of wooden blocks fixed inside a flexible steel band. The friction between the blocks and the drum provides braking action. When the brake is applied, the blocks are pressed against the drum. The wooden blocks have a higher co-efficient of friction. This increases the effectiveness of the brake. Also the wooden blocks can be easily replaced if worn out.

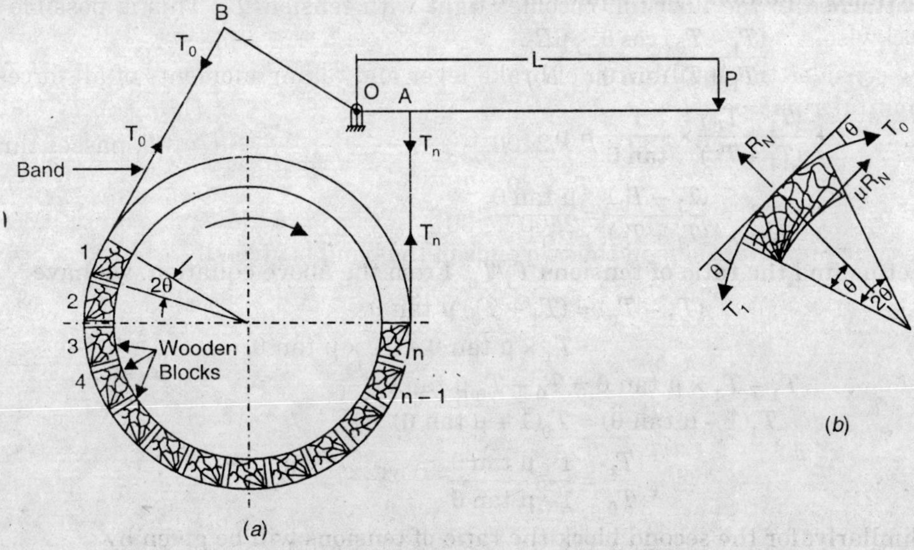

Fig. 8.21

Let there are 'n' number of blocks, each subtending on angle 2θ at the drum centre as shown in Fig. 8.21 (b). Let the drum rotates in clockwise direction. For clockwise rotation, the end of band attached to A will be tight with tension T_n whereas the end attached to B will be slack with tension, T_0.

Let n = Number of blocks

T_n = Tension on tight side after n block

T_0 = Tension on slack side

μ = Co-efficient of friction

T_1 = Tension in the band between first and second block

T_2 = Tension in the band between second and third block.

Consider one of the blocks (say first block) as shown in Fig. 8.21 (*b*). Each block embrass a short arc on the drum. The first block is in equilibrium under the action of following forces :

(*i*) Tension T_0 on the slack side

(*ii*) Tension T_1 on the tight side *i.e.* tension in the band between first and second block

(*iii*) Normal reaction R_N

(*iv*) Force of friction ($\mu \times R_N$) acting on the block in the direction of rotation of drum.

The frictional force on the drum will be in the opposite direction of the rotation of the drum (*i.e.* in anti-clockwise direction). And the frictional force on the *block* will be in the opposite direction of the frictional force on drum. Hence the friction force on the block will be in clockwise direction. This means that frictional force on the block will act in the direction of rotation of the drum.

Resolving the forces [acting on the block shown in Fig. 8.21 (*b*)] tangentially

$$T_1 \cos\theta - T_0 \cos\theta = \mu R_N \quad (\because \; T_1 \text{ is more than } T_0)$$

or $$(T_1 - T_0)\cos\theta = \mu R_N \quad \text{...}(i)$$

Resolving the force, radially

$$T_1 \sin\theta + T_0 \sin\theta = R_N$$

or $$(T_1 + T_0)\sin\theta = R_N \quad \text{...}(ii)$$

Dividing equation (*i*) by equation (*ii*),

$$\frac{(T_1 - T_0)\cos\theta}{(T_1 + T_0)\sin\theta} = \frac{\mu R_N}{R_N}$$

or $$\frac{(T_1 - T_0)}{(T_1 + T_0)} \times \frac{1}{\tan\theta} = \mu$$

or $$\frac{(T_1 - T_0)}{(T_1 + T_0)} = \frac{\mu \tan\theta}{1}$$

Let us find the ratio of tensions T_1/T_0. From the above equation, we have

$$(T_1 - T_0) = (T_1 + T_0)\,\mu \tan\theta$$

$$= T_1 \times \mu \tan\theta + T_0 \times \mu \tan\theta$$

or $$T_1 - T_1 \times \mu \tan\theta = T_0 + T_0\,\mu \tan\theta$$

or $$T_1(1 - \mu \tan\theta) = T_0(1 + \mu \tan\theta)$$

or $$\frac{T_1}{T_0} = \frac{1 + \mu \tan\theta}{1 - \mu \tan\theta}$$

Similarly, for the second block the ratio of tensions will be given by

$$\frac{T_2}{T_1} = \frac{1 + \mu \tan\theta}{1 - \mu \tan\theta} \text{ and so on.}$$

For the '*n*th' block, the ratio of tensions will be

$$\frac{T_n}{T_{n-1}} = \frac{1 + \mu \tan\theta}{1 - \mu \tan\theta}$$

Hence the ratio of tensions in the tight and slack sides of the complete band and block brake can be obtained as :

$$\frac{T_n}{T_0} = \frac{T_n}{T_{n-1}} \times \ldots\ldots \times \frac{T_3}{T_2} \times \frac{T_2}{T_1} \times \frac{T_1}{T_0}$$

$$= \left(\frac{1+\mu \tan \theta}{1-\mu \tan \theta}\right) \times \ldots\ldots \times \left(\frac{1+\mu \tan \theta}{1-\mu \tan \theta}\right) \times \left(\frac{1+\mu \tan \theta}{1-\mu \tan \theta}\right) \times \left(\frac{1+\mu \tan \theta}{1-\mu \tan \theta}\right)$$

$$= \left(\frac{1+\mu \tan \theta}{1-\mu \tan \theta}\right)^n \qquad \ldots(8.14)$$

where θ = Half of the angle subtended by each block on the centre of the drum.

Net torque on drum $= (T_n - T_0) \times r$

Now the braking torque on the drum will be

$$T_B = (T_n - T_0) \times r \qquad \ldots(8.15)$$

where r = Effective radius of band.

Problem 8.12. *The maximum braking torque acting on a band and block brake (shown in Fig. 8.22 (a)) is 2000 Nm. The band is lined with 15 blocks each of which subtends an angle of 12° at the centre of rotating drum. The co-efficient of friction between the band and block is 0.3. The diameter of the drum is 680 mm whereas the thickness of blocks is 60 mm. Find the least force required at the end of the lever which is 480 mm long.*

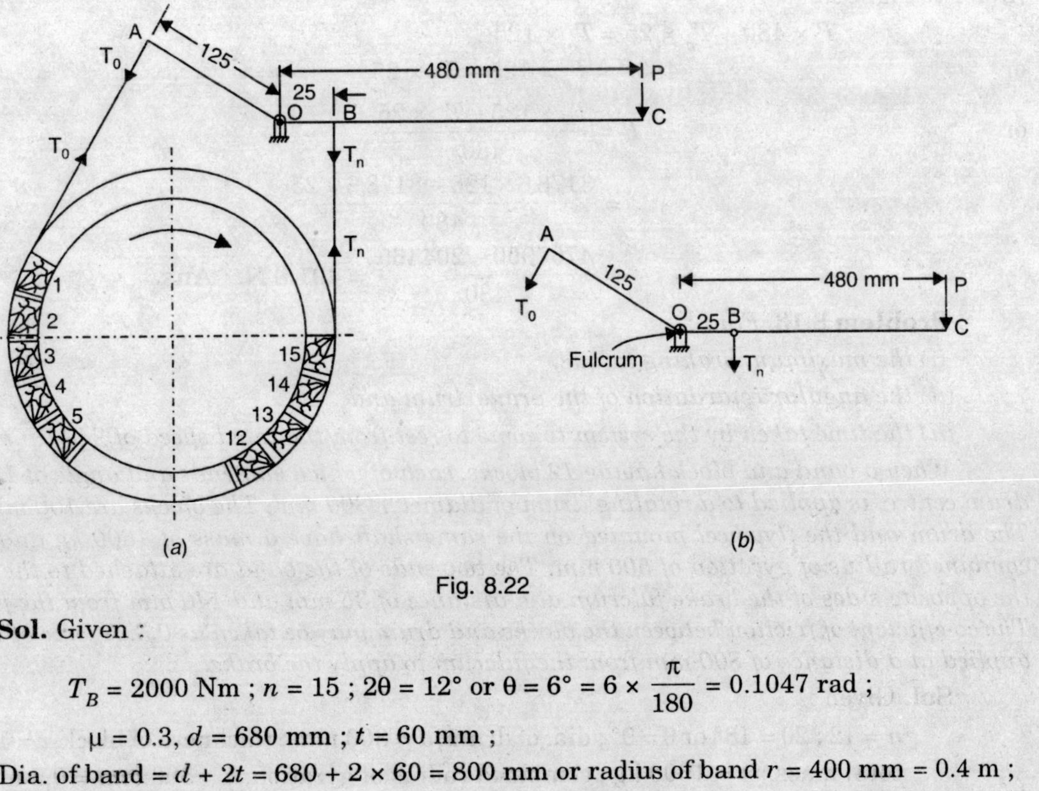

Fig. 8.22

Sol. Given :

$$T_B = 2000 \text{ Nm} \; ; \; n = 15 \; ; \; 2\theta = 12° \text{ or } \theta = 6° = 6 \times \frac{\pi}{180} = 0.1047 \text{ rad} \; ;$$

$$\mu = 0.3, \; d = 680 \text{ mm} \; ; \; t = 60 \text{ mm} \; ;$$

Dia. of band $= d + 2t = 680 + 2 \times 60 = 800$ mm or radius of band $r = 400$ mm $= 0.4$ m ;

$L = 480$ mm

As distance $OA >$ distance OB, hence force P must be applied at C downwards.

The force P will be least, if the end of the band attached to A is slack and the end attached to B is tight. This is only possible if drum rotates clockwise as shown in Fig. 8.22 (*a*).

For the band and block brake, using equation (8.14), we get

$$\frac{T_n}{T_0} = \left(\frac{1+\mu \tan \theta}{1-\mu \tan \theta}\right)^n$$

$$= \left(\frac{1+0.3\times\tan 6°}{1-0.3\times\tan 6°}\right)^{15} = \left(\frac{1+0.3\times 0.1051}{1-0.3\times 0.1051}\right)^{15}$$

$$= \left(\frac{1.0315}{0.9685}\right)^{15} = (1.065)^{15} = 2.573$$

$\therefore \quad T_n = 2.573T_o \quad ...(i)$

The braking torque is given by equation (8.15),

$$T_B = (T_n - T_0) \times r$$

or $\quad 2000 = (2.573T_o - T_o) \times 0.4 \quad (\because \; T_n = 2.573T_0 \text{ and } r = 0.4)$

$$= 1.573T_o \times 0.4$$

or $\quad T_o = \dfrac{2000}{1.573 \times 0.4} = 3178.6$

$\therefore \quad T_n = 2.573 \times T_o = 2.573 \times 3178.6 = 8178.5$ N

Now taking the moments of all force acting on the brake lever, about fulcrum, we get [Refer to Fig. 8.22 (*b*)]

$$P \times 480 + T_n \times 25 = T_o \times 125$$

or $\quad 480P = T_o \times 125 - T_n \times 25$

or $\quad P = \dfrac{T_0 \times 125 - T_n \times 25}{480}$

$$= \frac{3178.6 \times 125 - 8178.6 \times 25}{480}$$

$$= \frac{4767900 - 204465}{480} = \textbf{401.8 N.} \quad \textbf{Ans.}$$

Problem 8.13. *Find :*

(i) the maximum braking torque,

(ii) the angular retardation of the brake drum and

(iii) the time taken by the system to come to rest from the rated speed of 240 r.p.m.

When a band and block having 12 blocks, each of which subtends and angle of 18° at the drum centre, is applied to a rotating drum of diameter 800 mm. The blocks are 100 mm thick. The drum and the flywheel mounted on the same shaft have a mass of 1600 kg and have a combined radius of gyration of 500 mm. The two ends of the band are attached to the pins on the opposite sides of the brake fulcrum at a distance of 35 mm and 140 mm from the fulcrum. The co-efficient of friction between the blocks and drum may be taken as 0.3. A force of 150 N is applied at a distance of 800 mm from the fulcrum to apply the brake.

Sol. Given :

$n = 12$; $2\theta = 18°$ or $\theta = 9°$; dia. of drum, $d = 800$ mm ; thickness of block, $t = 100$ mm, total mass, $m = 1600$ kg ; combined radius of gyration, $k = 500$ mm $= 0.5$ m ; $OB = 35$ mm ; $OA = 140$ mm ; $\mu = 0.3$; $L = 800$ mm, $P = 150$ N

Dia. of band, $D = d + 2t = 800 + 2 \times 100 = 1000$ mm ; $r = 500$ mm $= 0.5$ m

(i) Maximum Braking Torque

As distance OA is greater than distance OB, the force P must act downwards at C. For maximum braking torque (or for least force to apply the brake), the brake should be arranged so that the tight side of the band is attached to the shorter distance *i.e.* tight side of band should be attached to B. This is possible if drum rotates clockwise, as shown in Fig. 8.23.

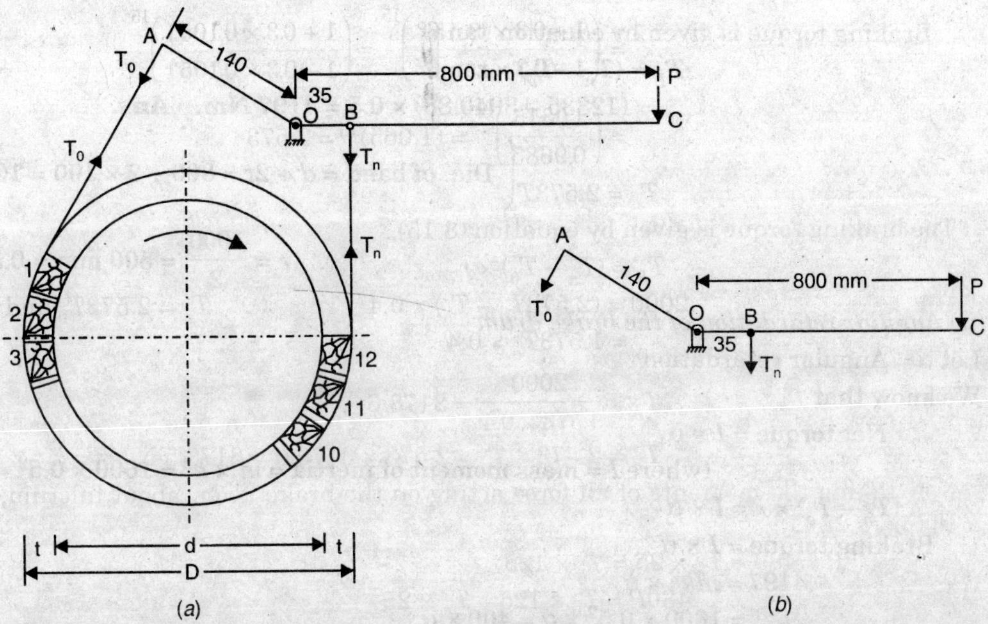

Fig. 8.23

Now using equation (8.14), we get

$$\frac{T_n}{T_o} = \left(\frac{1+\mu \tan\theta}{1-\mu\tan\theta}\right)^n$$

$$= \left(\frac{1+0.3\times\tan 9^\circ}{1-0.3\times\tan 9^\circ}\right)^{12} = \left(\frac{1+0.3\times 0.1584}{1-0.3\times 0.1584}\right)^{12}$$

$$= \left(\frac{1.0475}{0.9525}\right)^{12} = (1.0997)^{12} = 3.13$$

or $$T_n = 3.13T_o \qquad ...(i)$$

Now taking the moments of all forces acting on the brake lever as shown in Fig. 8.23 (*b*) about fulcrum, we get

$$P\times 800 + T_n \times 35 = T_o \times 140$$

or $$150\times 800 + 3.13T_o \times 35 = T_o\times 140 \qquad (\because\ P = 150\text{ N and } T_n = 3.13T_o)$$

or $$120000 + 109.55T_o = 140T_o$$

or $$120000 = (140 - 109.55)T_o = 30.45\times T_o$$

$$\therefore\quad T_o = \frac{120000}{30.45} = 3940.88\text{ N.}$$

Substituting the value of T_o in equation (*i*), we get

$$T_n = 3.13 \times 3940.88 = 12335\text{ N}$$

∴ Braking torque is given by equation (8.15) as

$$T_B = (T_n - T_o) \times r$$

$$= (12335 - 3940.88) \times 0.5 = \mathbf{4197\ Nm.}\quad \textbf{Ans.}$$

$$\left(\text{Dia. of band} = d + 2t = 800 + 2 \times 100 = 1000 \text{ m} \right.$$

$$\left. \therefore\quad r = \frac{1000}{2} = 500 \text{ mm} = 0.5 \text{ m} \right)$$

(ii) Angular retardation of the brake drum

Let α = Angular retardation

We know that

$$\text{Net torque} = I \times \alpha$$

(where I = mass moment of inertia = $m \times k^2 = 1600 \times 0.5^2 = 400$)

or $$(T_n - T_o) \times r = I \times \alpha$$

or $$\text{Braking torque} = I \times \alpha$$

∴ $$4197 = mk^2 \times \alpha$$

$$= 1600 \times 0.5^2 \times \alpha = 400 \times \alpha$$

∴ $$\alpha = \frac{4197}{400} = \mathbf{10.49\ rad/s^2.}\quad \textbf{Ans.}$$

(iii) Time taken by the system to come to rest from the rated speed of 240 r.p.m.

Let t = Time taken

Initial angular speed, $$\omega_0 = \frac{2\pi N_0}{60} = \frac{2\pi \times 240}{60} = 25.13 \text{ rad/s}$$

Final angular speed, $\omega = 0$

Using $\omega = \omega_0 - \alpha \times t$ (–ve sign is due to retardation)

∴ $$t = \frac{\omega_0 - \omega}{\alpha}$$

$$= \frac{25.13 - 0}{10.49} = \mathbf{2.39\ s.}\quad \textbf{Ans.}$$

8.2.4. Internal Expanding Shoe Brake. If the shoe is provided on the interior of the rotating drum and braking effect is produced due to the expansion of the shoe, then it is known as internal expanding shoe brake. Fig. 8.24 shows an internal expanding shoe brake which are commonly used in motor cars and light trucks. This consists of two shoes S_1 and S_2 whose outer surfaces are lined with some friction material (generally with Ferodo) to increase the co-efficient of friction. Each shoe is pivoted at one end about a fixed fulcrum O_1 and O_2 and the other end is having contact with a cam (or with each piston in a common hydraulic cylinder. There are two equal diameter pistons in a common hydraulic cylinder.) As the cam rotates, the shoes are pushed outwards and makes a contact with the drum. The friction between shoes and drum produces the braking torque and hence reduces the speed of the drum. The shoes are held in off position by a spring as shown in Fig. 8.24. Under normal running of the vehicle the drum rotates freely as the outer diameter of the shoes is a little less than the internal diameter of the drum.

The force required to operate such a brake can be calculated if we know the total forces acting on such a brake. Let us consider, the drum is rotating in anti-clockwise direction. For anti-clockwise rotation of the drum, the left hand shoe is known as *leading or primary shoe* whereas the right hand shoe is known as *trailing or secondary shoe*.

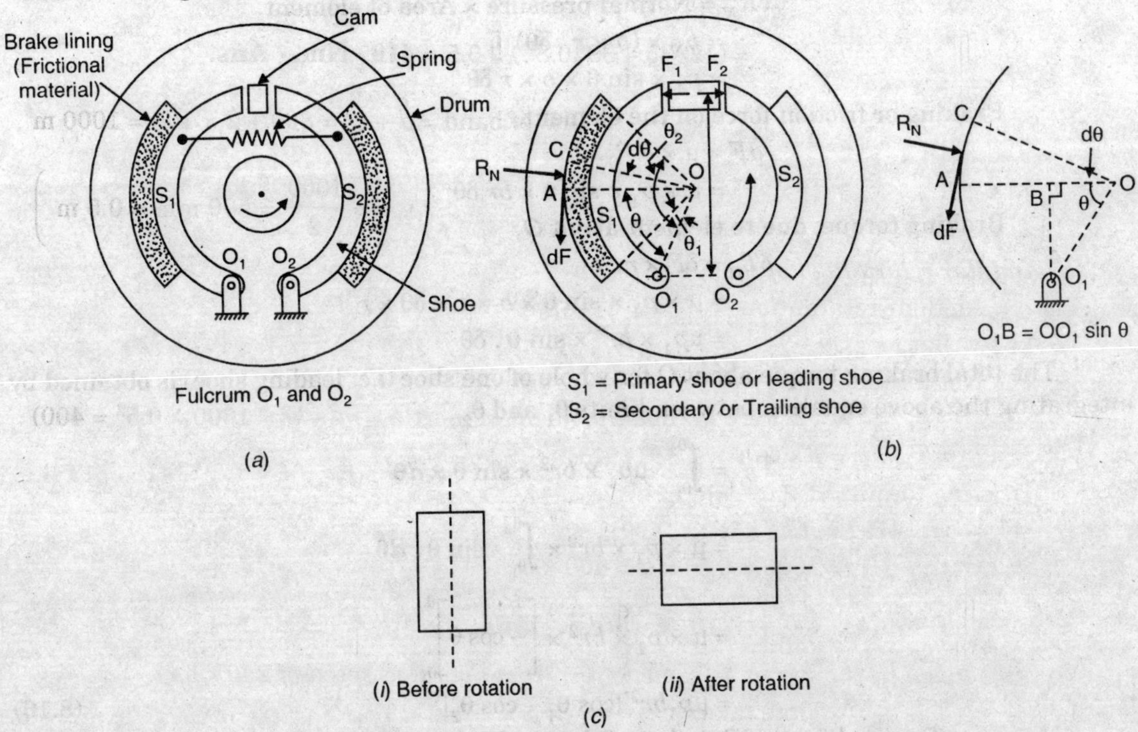

Fig. 8.24

Let F_1 = Force exerted by cam on the leading shoe

F_2 = Force exerted by cam on the trailing shoe

b = Width of brake lining

r = Internal radius of the wheel drum

p_n = Normal pressure

p_1 = Maximum intensity of normal pressure for the leading shoe

p_2 = Maximum intensity of normal pressure for trailing shoe

Consider a small length of brake lining say length AC which subtends an angle $\delta\theta$ at the centre as shown in Fig. 8.24 (*b*). Also let length OA makes an angle θ with OO_1. As the shoe turns about O_1, the rate of wear of the shoe lining at A will be directly proportional to the perpendicular distance from O_1 to OA *i.e.* distance O_1B as shown in Fig. 8.24 (*c*).

Now from the figure, we have

$$O_1B = OO_1 \sin\theta$$

Hence rate of wear at $A \propto O_1B$ or $OO_1 \sin\theta$ or $\sin\theta$

The normal pressure at A (*i.e.* p_N) is written as

$$p_N \propto \sin\theta$$

or

$$p_N = p_1 \sin\theta$$

where p_1 is constant of proportionality and is known as maximum intensity of pressure

∴ Normal force acting on the element,

$$\delta R_N = \text{Normal pressure} \times \text{Area of element}$$
$$= p_N \times (b \times r\,.\,\delta\theta)$$
$$= p_1 \times \sin\theta \times b \times r\,\delta\theta$$

∴ Braking or friction force on the element,

$$dF = \mu \times \delta R_N$$
$$= \mu \times p_1 \times \sin\theta \times br\,\delta\theta$$

∴ Braking torque, due to element about O,

$$\delta T_B = \delta F \times r$$
$$= \mu \times p_1 \times \sin\theta \times b \times r \times \delta\theta \times r$$
$$= \mu p_1 \times br^2 \times \sin\theta\,.\,\delta\theta$$

The total braking torque about O for whole of one shoe (*i.e.* leading shoe) is obtained by integrating the above equation between limits θ_1 and θ_2.

$$\therefore \quad T_{B_1} = \int_{\theta_1}^{\theta_2} \mu p_1 \times br^2 \times \sin\theta \times d\theta$$
$$= \mu \times p_1 \times br^2 \times \int_{\theta_1}^{\theta_2} \sin\theta\,.\,d\theta$$
$$= \mu \times p_1 \times br^2 \times \Big[-\cos\theta\Big]_{\theta_1}^{\theta_2}$$
$$= \mu p_1 br^2\,[\cos\theta_1 - \cos\theta_2] \qquad \text{...(8.16)}$$

Moment of normal force (δR_N) about fulcrum O_1,

$$\delta M_N = \delta R_N \times O_1B = \delta R_N \times (OO_1 \sin\theta)$$
$$= (p_1 \times \sin\theta \times b \times r \times \delta\theta) \times (OO_1 \sin\theta)$$
$$= p_1 \times \sin^2\theta \times (b \times r\,.\,\delta\theta) \times OO_1$$

Total moment of normal forces for the leading shoe about O_1,

$$M_{N_1} = \int_{\theta_1}^{\theta_2} p_1 \times \sin^2\theta \times (b \times r \times \delta\theta)\,OO_1$$
$$= p_1 \times b\,.\,r\,.\,OO_1 \int_{\theta_1}^{\theta_2} \sin^2\theta\,d\theta = p_1 \times b \times r$$
$$\times\, OO_1 \int_{\theta_1}^{\theta_2} \left[\frac{1-\cos 2\theta}{2}\right] d\theta \qquad \left[\because \sin^2\theta = \frac{1-\cos 2\theta}{2}\right]$$
$$= p_1 \times b \times r \times OO_1 \left[\frac{\theta - \dfrac{\sin 2\theta}{2}}{2}\right]_{\theta_1}^{\theta_2} = \frac{1}{2} \times p_1 \times b \times r \times OO_1 \left[\theta - \frac{\sin 2\theta}{2}\right]_{\theta_1}^{\theta_2}$$
$$= \frac{1}{2} \times p_1 \times b \times r \times OO_1 \left[\theta_2 - \frac{\sin 2\theta_2}{2} - \left(\theta_1 - \frac{\sin 2\theta_1}{2}\right)\right]$$
$$= \frac{1}{2} \times p_1 \times b \times r \times OO_1 \left[(\theta_2 - \theta_1) + \frac{1}{2}(\sin 2\theta_1 - \sin 2\theta_2)\right] \qquad \text{...(8.17)}$$

Moment of frictional force δF about the fulcrum O_1

$$\delta M_F = \delta F \times AB = \delta F\,[AO - OB] = \delta F\,[r - OO_1 \cos\theta]$$

$$= \mu p_1 \sin\theta \times br\,\delta\theta\,[r - OO_1 \cos\theta] \quad [\because \ \delta F = \mu p_1 \sin\theta \times b \times r \times \delta\theta]$$

$$= \mu p_1 \times br\,[r \sin\theta - OO_1 \cos\theta \sin\theta]\,\delta\theta$$

$$= \mu p_1 \times br\left[r\sin\theta - \frac{OO_1}{2}\sin 2\theta\right]\delta\theta \quad [\because \ 2\cos\theta\sin\theta = \sin 2\theta]$$

Total moment of frictional force about fulcrum O_1 for one shoe is obtained by integrating the above equation between limits θ_1 and θ_2.

$$\therefore \quad \text{Total moments,} \quad M_{F_1} = \int_{\theta_1}^{\theta_2} \mu p_1\, br\left(r\sin\theta - \frac{OO_1}{2}\sin 2\theta\right)\delta\theta$$

$$= \mu p_1 \times b \times r \int_{\theta_1}^{\theta_2}\left(r\sin\theta - \frac{OO_1}{2}\sin 2\theta\right)\delta\theta$$

$$= \mu p_1 \times b \times r\left[r(-\cos\theta) - \frac{OO_1}{2}\left(-\frac{\cos 2\theta}{2}\right)\right]_{\theta_1}^{\theta_2}$$

$$= \mu p_1 \times b \times r\left[-r\cos\theta + \frac{OO_1}{4}\cos 2\theta\right]_{\theta_1}^{\theta_2}$$

$$= \mu p_1 \times b \times r\left[-r\cos\theta_2 + \frac{OO_1}{4}\cos 2\theta_2 - \left(-r\cos\theta_1 + \frac{OO_1}{4}\cos 2\theta_1\right)\right]$$

$$= \mu p_1 \times b \times r\left[r(\cos\theta_1 - \cos\theta_2) + \frac{OO_1}{4}(\cos 2\theta_2 - \cos 2\theta_1)\right] \quad \text{...(8.18)}$$

Now the force required to operate the leading or primary shoe is obtained by taking moments of all forces acting on the leading shoe about fulcrum O_1. [See Fig. 8.24 (*c*)].

Taking moments about O_1, we get

$F_1 \times L$ + Total moment due to normal force + Total moment due to frictional force = 0

or $\quad F_1 \times L - M_{N_1} + M_{F_1} = 0 \quad$ (Moment due to R_N is clockwise, whereas due to F_1 and frictional force it is anti-clockwise)

or

$$F_1 \times L = M_{N_1} - M_{F_1} \quad \text{...(8.19)}$$

The force (F_2) required to operate the secondary shoe is obtained by taking moments of all forces acting on secondary shoe, about fulcrum O_2. [See Fig. 8.24 (*d*)].

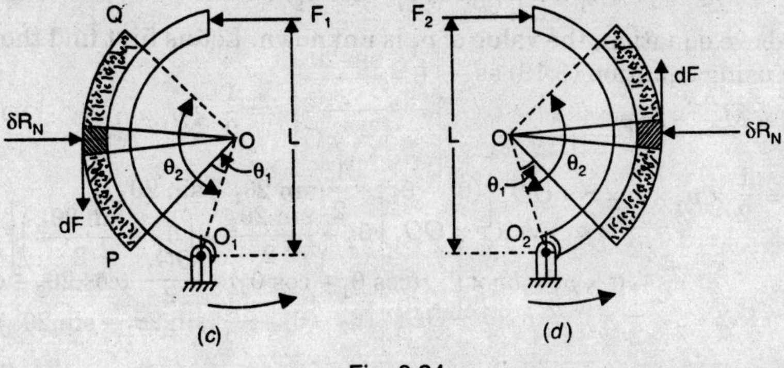

Fig. 8.24

$$\therefore \qquad F_2 \times L = M_{N_2} + M_{F_2} \qquad ...(8.20)$$

If $M_{F_2} > M_{N_2}$, the brake will be self-locking.

where $$M_{N_2} = \frac{1}{2} \times p_2 \times b \times r \times OO_2 \left[(\theta_2 - \theta_1) + \frac{1}{2}(\sin 2\theta_1 - \sin 2\theta_2) \right]$$

$$M_{F_2} = \mu \times p_2 \times b \times r \left[r(\cos\theta_1 - \cos\theta_2) + \frac{OO_2}{4}(\cos 2\theta_2 - \cos 2\theta_1) \right]$$

and $$T_{B_2} = \mu \times p_2 \times b \times r^2 [\cos\theta_1 - \cos\theta_2] \qquad ...(8.21)$$

and total braking torque on both shoes,

$$T_B = T_{B_1} + T_{B_2} \qquad ...(8.22)$$

$$= \mu \times p_1 \times b \times r^2 [\cos\theta_1 - \cos\theta_2] + \mu \times p_2 \times b \times r^2 [\cos\theta_1 - \cos\theta_2]$$

$$= \mu \times b \times r^2 \times [\cos\theta_1 - \cos\theta_2](p_1 + p_2) \qquad ...(8.23)$$

Problem 8.14. *Calculate the braking torque applied by an internal expanding shoe brake shown in Fig. 8.24 on the rotating drum of diameter 300 mm if the drum is rotating (i) anti-clockwise and (ii) clockwise. The other data given is :*

Force F on each shoe	*= 90 N*
Co-efficient of friction,	*$\mu = 0.3$*
Width of the brake lining,	*$b = 40$ mm*
Angles	*$\theta_1 = 30°$, $\theta_2 = 135°$*
Distance :	*$L = 200$ mm, $OO_1 = 120$ mm.*

Sol. Given : [Refer to Fig. 8.24]

Dia. of drum, $d = 300$ mm or radius $r = 150$ mm $= 0.15$ m ; $F_1 = F_2 = 90$ N ; $\mu = 0.3$; $b = 40$ mm $= 0.04$ m ; $\theta_1 = 30°$; $\theta_2 = 135°$; $L = 200$ mm $= 0.2$ m ; $OO_1 = 120$ mm $= 0.12$ m.

(i) Braking Torque for anti-clockwise rotation

Let T_B = Total braking torque

$= T_{B_1} + T_{B_2}$

where T_{B_1} = braking torque on leading shoe

T_{B_2} = braking torque on trailing shoe

The braking torque on leading shoe is given equation (8.16) as

$$T_{B_1} = \mu \times p_1 \times b \times r^2 (\cos\theta_1 - \cos\theta_2) \qquad ...(i)$$

In the above equation, the value of p_1 is unknown. Let us first find the value of p_1. This is obtained by using equation (8.19) as

$$F_1 \times L = M_{N_1} - M_{F_1}$$

$$= \frac{1}{2} \times p_1 \times b \times r \times OO_1 \left[(\theta_2 - \theta_1) + \frac{1}{2}(\sin 2\theta_1 - \sin 2\theta_2) \right]$$

$$- \mu \times p_1 \times br \times \left[r(\cos\theta_1 - \cos\theta_2) + \frac{OO_1}{4}(\cos 2\theta_2 - \cos 2\theta_1) \right]$$

or $$90 \times 0.2 = \frac{1}{2} \times p_1 \times 0.04 \times 0.15 \times 0.12 \left[\left(\frac{135 \times \pi}{180} - \frac{30 \times \pi}{180} \right) + \frac{1}{2} (\sin 2 \times 30° - \sin 2 \times 135°) \right]$$

$$- 0.3 \times p_1 \times 0.04 \times 0.15 \left[0.15 (\cos 30° - \cos 135°) + \frac{0.12}{4} (\cos 2 \times 135° - \cos 2 \times 30°) \right]$$

or $$18 = 0.00036 p_1 \left[(2.356 - 0.5236) + \frac{1}{2} (0.866 - (-1)) \right]$$

$$- 0.0018\, p_1 \left[.15\{0.866 - (-0.707)\} + 0.03\{(0) - 0.5\} \right]$$

$$= 0.00036 p_1 [2.7654] - 0.0018 p_1 [0.23595 - 0.015]$$

$$= 0.000995 p_1 - 0.0003977 p_1$$

$$= 0.0005973 p_1$$

$\therefore$ $$p_1 = \frac{18}{0.0005973} = 30135 \text{ N/m}^2$$

Substituting this value of p_1 in equation (*i*), we get

$$T_{B_1} = \mu \times 30135.4 \times b \times r^2 \times (\cos \theta_1 - \cos \theta_2)$$

$$= 0.3 \times 30135.4 \times 0.04 \times 0.15^2 \times (\cos 30° - \cos 135°)$$

$$= 0.3 \times 30135.4 \times 0.04 \times 0.0225 \times [0.866 - (-0.707)]$$

$$= 12.80 \text{ Nm} \qquad \ldots(ii)$$

The braking torque on trailing shoe is obtained by equation (8.21) as

$$T_{B_2} = \mu \times p_2 \times b \times r^2 \times (\cos \theta_1 - \theta_2) \qquad \ldots(iii)$$

In the above equation, the value of p_2 is obtained by using equation (8.20) as

$$F_2 \times L = M_{N_2} + M_{F_2}$$

Substituting the value of M_{N_2} and M_{F_2}, we get

$$F_2 \times L = \frac{1}{2} \times p_2 \times b \times r \times OO_2 \left[(\theta_2 - \theta_1) + \frac{1}{2} (\sin 2\theta_1 - \sin 2\theta_2) \right]$$

$$- \mu \times p_2 \times b \times r \times [r(\cos \theta_1 - \cos \theta_2) + \frac{OO_2}{4} (\cos 2\theta_2 - \cos 2\theta_1)]$$

or $$90 \times 0.2 = \frac{1}{2} \times p_2 \times 0.04 \times 0.15 \times 0.12 \left[\left(\frac{135 \times \pi}{180} - \frac{30 \times \pi}{180} \right) + \frac{1}{2} (\sin 60° - \sin 270°) \right]$$

$$- 0.3 \times p_2 \times 0.04 \times 0.15 \times \left[0.15 (\cos 30° - \cos 135°) + \frac{0.12}{4} (\cos 270° - \cos 60°) \right]$$

or $$18 = 0.000995 p_2 + 0.0003977 p_2 = 0.001393 p_2$$

$\therefore$ $$p_2 = \frac{18}{0.001393} = 12917.9 \text{ Nm}$$

Substituting the value of p_2 in equation (*iii*), we get

$$T_{B_2} = 0.3 \times 12917.9 \times 0.04 \times .15^2 \times (\cos 30° - \cos 135°)$$

$$= 5.485 \text{ Nm}$$

$\therefore$ For anti-clockwise direction, total torque is given by,

$$T_B = T_{B_1} + T_{B_2}$$
$$= 12.80 + 5.485 = \mathbf{18.285\ Nm.} \quad \textbf{Ans.}$$

Alternate Method

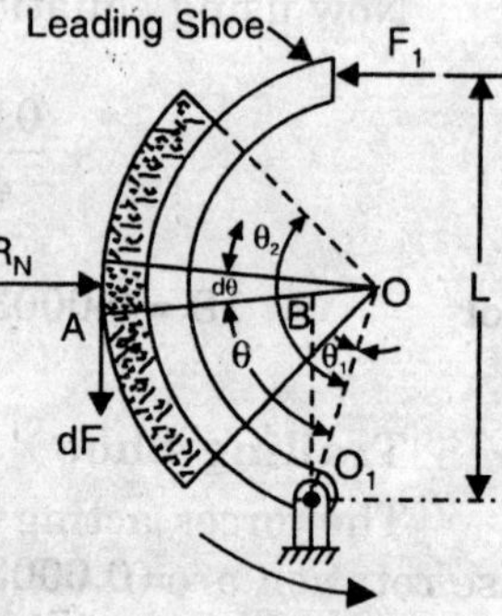

Fig. 8.25

For anti-clockwise rotation, the forces acting on the small element of brake lining of leading shoe are (*i*) Normal reaction δR_N and (*ii*) Frictional force δF.

where δR_N = Normal pressure × Area of element

$$= p_N \times b \times r \times \delta\theta$$
$$= (p_1 \sin\theta) \times b \times r \times \delta\theta \qquad (\because \; p_N = p_1 \sin\theta)$$

and $\delta F = \mu \times \delta R_N = \mu \times [p_1 \sin\theta \times b \times r \times \delta\theta]$

The operating force acting on leading shoe is F_1.

Taking moments of all forces (*i.e.*, total normal force, total frictional force and F_1) about fulcrum O_1, we get [Refer Fig. 8.25].

$$F_1 \times L - \int_{\theta_1}^{\theta_2} \delta R_N \times O_1B + \int_{\theta_1}^{\theta_2} \delta F \times AB = 0$$

or $F_1 \times L - \int_{\theta_1}^{\theta_2} (p_1 \sin\theta \times b \times r \times d\theta) \times OO_1 \sin\theta + \int_{\theta_1}^{\theta_2} \mu\, [p_1 \sin\theta \times b \times r \times \delta\theta) \times [r - OO_1 \cos\theta]$

$(\because \; O_1B = OO_1 \sin\theta \text{ and } AB = AO - BO = r - OO_1 \cos\theta)$

or $F_1 \times L - p_1 \times b \times r \times OO_1 \int_{\theta_1}^{\theta_2} \sin^2\theta \,.\, d\theta + \mu\, p_1 \times b \times r \times \int_{\theta_1}^{\theta_2} (r \sin\theta - OO_1 \sin\theta \cos\theta)\, d\theta = 0$

or $90 \times 0.2 - p_1 \times 0.04 \times 0.15 \times 0.12 \int_{30^\circ}^{135^\circ} \sin^2\theta \,.\, d\theta + 0.3 \times p_1 \times 0.04 \times 0.15$

$$\times \int_{30^\circ}^{135^\circ} (0.15 \sin\theta - 0.12 \times \sin\theta \cos\theta)\, d\theta = 0$$

or $18 - 0.00072 p_1 \int_{30^\circ}^{135^\circ} \sin^2\theta\, d\theta + 0.0018 p_1 \int_{30^\circ}^{135^\circ} (0.15 \sin\theta - 0.12 \sin\theta \cos\theta)\, d\theta = 0$

or $18 - 0.00072 p_1 \int_{30^\circ}^{135^\circ} \left(\dfrac{1 - \cos 2\theta}{2}\right) d\theta + 0.0018\, p_1 \int_{30^\circ}^{135^\circ} [0.15 \sin\theta - \dfrac{0.12}{2} \sin 2\theta]\, d\theta = 0$

or $18 - 0.00072 p_1 \left[\dfrac{\theta}{2} - \dfrac{\sin 2\theta}{4}\right]_{30^\circ}^{135^\circ} + 0.0018\, p_1 \left[0.15(-\cos\theta) - 0.06\left(\dfrac{-\cos 2\theta}{2}\right)\right]_{30^\circ}^{135^\circ} = 0$

or $18 - 0.00072 p_1 \left[\dfrac{1}{2}\left(\dfrac{135 \times \pi}{180} - \dfrac{30 \times \pi}{180}\right) - \dfrac{1}{4} (\sin 2 \times 135^\circ - \sin 2 \times 30^\circ\right]$

$$+ 0.0018\, p_1\, [-0.15\, \{\cos 135^\circ - \cos 30^\circ\} + 0.03\, \{\cos 270^\circ - \cos 60^\circ\}] = 0$$

or $18 - 0.00072 p_1 \left[\dfrac{1}{2} (2.356 - 0.523) - \dfrac{1}{4} (-1 - 0.866)\right]$

$$+ 0.0018 p_1\, [-0.15\, \{-0.707 - 0.866\} + 0.03\, \{0 - 0.5\}] = 0$$

or $18 - 0.00072 p_1 [0.9165 + 0.4665] + 0.0018 p_1\, [0.236 - .015] = 0$

or $18 - 0.000995 p_1 + 0.0003978 p_1 = 0$

or $18 - 0.0005972 p_1 = 0$

or $$18 = 0.0005972 p_1$$

or $$p_1 = \frac{18}{0.0005972} = 30140 \text{ N/m}^2$$

Now using equation (8.16) for braking torque on leading shoe, as

$$\begin{aligned} T_{B_1} &= \mu\, p_1 b r^2 (\cos\theta_1 - \cos\theta_2) \\ &= 0.3 \times 30140 \times 0.04 \times 0.15^2 \times (\cos 30° - \cos 135°) \\ &= 0.3 \times 30140 \times 0.04 \times 0.0225\, [0.866 - (-0.707)] \\ &= 8.1378(0.866 + 0.707) \\ &= 12.8 \text{ Nm} \end{aligned}$$

Trailing Shoe

The forces acting on the small element of brake lining of trailing shoe for anti-clockwise rotation are (*i*) δR_N and (*ii*) Frictional force δF [Refer Fig. 8.25 (*a*)].

where δR_N = Normal reaction
= Normal pressure × Area of element
$= p_N \times b \times r \times d\theta$

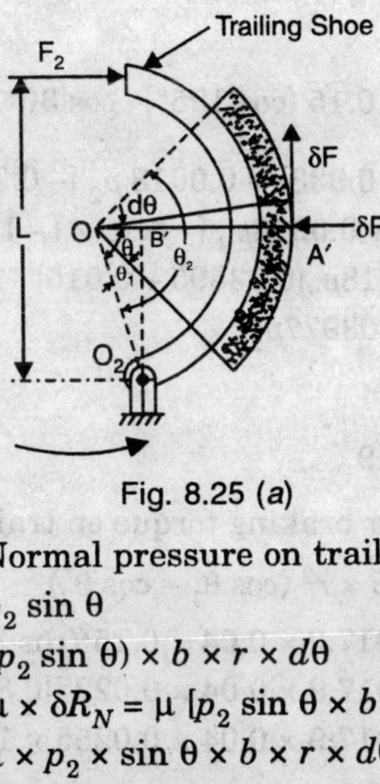

Fig. 8.25 (*a*)

But $$p_N = \text{Normal pressure on trailing shoe} = p_2 \sin\theta$$

∴ $$\delta R_N = (p_2 \sin\theta) \times b \times r \times d\theta$$

and $$\delta F = \mu \times \delta R_N = \mu\, [p_2 \sin\theta \times b \times r \times d\theta] = \mu \times p_2 \times \sin\theta \times b \times r \times d\theta$$

The operating force acting on trailing shoe is F_2. Taking moments of all forces (*i.e.*, total normal force, total frictional force and F_2) about fulcrum O_2, we get

$$F_2 \times L - \int_{\theta_1}^{\theta_2} \delta R_N \times O_2 B' - \int_{\theta_1}^{\theta_2} \delta F \times A'B' = 0$$

or $$F_2 \times L = \int_{\theta_1}^{\theta_2} \delta R_N \times O_2 B' + \int_{\theta_1}^{\theta_2} \delta F \times A'B'$$

$$90 \times 0.2 = \int_{30°}^{135°} (p_2 \sin\theta \times b \times r \times d\theta) \times OO_2 \sin\theta + \int_{\theta_1}^{\theta_2} (\mu p_2 \times \sin\theta \times b \times r \times d\theta)\,(A'O - OB')$$

$$18 = \int_{30°}^{135°} p_2 \times b \times r \times OO_2 \times \sin^2\theta\, d\theta + \int_{30°}^{135°} \mu p_2 \times b \times r \times \sin\theta\,(r - OO_2 \cos\theta)\, d\theta$$

$$= p_2 \times b \times r \times OO_2 \times \int_{30°}^{135°} \sin^2\theta\, d\theta + \mu \times p_2 \times b \times r \int_{30°}^{135°} (r \sin\theta - OO_2 \sin\theta \cos\theta)\, d\theta$$

$$= p_2 \times 0.04 \times 0.15 \times 0.12 \int_{30°}^{135°} \left(\frac{1 - \cos 2\theta}{2}\right) d\theta$$

$$+ 0.3 \times p_2 \times 0.04 \times 0.15 \int_{30°}^{135°} \left[r \sin\theta - \frac{OO_2}{2} \times 2 \sin\theta \cos\theta \right] d\theta$$

$$= 0.00072\, p_2 \left[\frac{\theta - \frac{\sin 2\theta}{2}}{2}\right]_{30°}^{135°} + 0.0018 p_2 \int_{30°}^{135°} \left(r \sin\theta - \frac{OO_2}{2} \times \sin 2\theta \right) d\theta$$

$$= 0.00036 p_2 \left[\left(\frac{135 \times \pi}{180} - \frac{30 \times \pi}{180}\right) - \frac{1}{2} (\sin 270° - \sin 60°\right]$$

$$+ 0.0018 p_2 \left[(-r \cos\theta) - \frac{OO_2}{2} \left(- \frac{\cos 2\theta}{2} \right) \right]_{30°}^{135°}$$

or $$18 = 0.00036\, p_2 \left[(2.356 - 0.523) - \frac{1}{2} (-1 - 0.866) \right]$$

$$+ 0.0018 p_2 \left[-0.15 (\cos 135° - \cos 30°) + \frac{0.12}{2} \times \frac{1}{2} (\cos 270° - \cos 60°) \right]$$

$$= 0.00036\, p_2 [1.833 + 0.933] + 0.0018\, p_2\, [-0.15(-0.707 - 0.866) + 0.03(0 - 0.5)]$$

$$= 0.00036 p_2 \times 2.766 + 0.0018 p_2\, [-0.15 \times (-1.573) - 0.015]$$

$$= 0.0009957 p_2 + 0.0018 p_2 [0.23595 - 0.015]$$

$$= 0.0009957 p_2 + 0.0003977 p_2$$

$$= 0.001393 p_2$$

$$\therefore \quad p_2 = \frac{18}{0.001393} = 12917.9$$

Now using equation (8.21) for braking torque on trailing shoe, we get

$$T_{B_2} = \mu \times p_2 \times b \times r^2 (\cos\theta_1 - \cos\theta_2)$$

$$= 0.3 \times 12917.9 \times 0.04 \times 0.15^2 (\cos 30° - \cos 135°)$$

$$= 0.3 \times 12917.9 \times 0.04 \times 0.0225 [0.866 - (-0.707)]$$

$$= 0.3 \times 12917.9 \times 0.04 \times 0.0255 \times 1.573$$

$$= 5.485 \text{ Nm}$$

$\therefore$ Total braking torque

$$= T_{B_1} + T_{B_1}$$

$= 12.8 + 5.485 =$ **18.285 Nm. Ans.**

(ii) Braking torque for clockwise rotation

When the rotation is reversed, the values of p_1 and p_2 are interchanged. Hence the value of braking torque will be same.

8.3. THE BRAKING OF A VEHICLE

When the brakes are applied to a moving vehicle, the speed decreases and retardation is produced. The retardation can be determined by reducing the system to an equivalent static system by applying d'Alembert's principle (*i.e.* by including the inertia force in the system of forces). Then the system of forces will be in equilibrium. After knowing retardation, the retarding force can be obtained. Retardation is the acceleration on the moving vehicle in the opposite direction of the motion of the vehicle.

The brakes in a car may be applied :

1. to the rear wheels only,
2. to the front wheels only, and
3. to all the four wheels.

Let us determine the retardation in all these cases. Fig. 8.26 shows a car *moving up* an inclined plane.

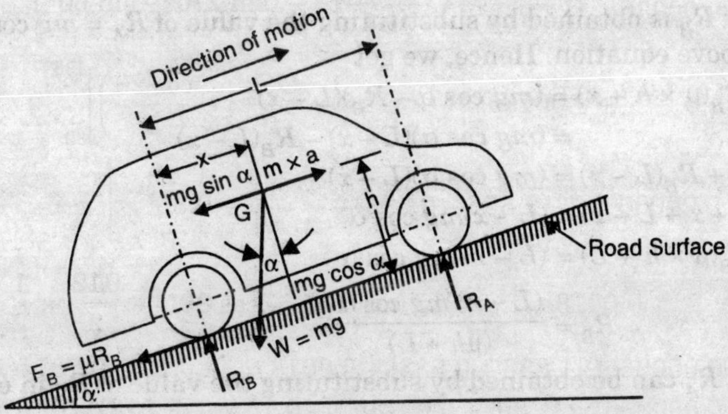

Fig. 8.26

Let α = Angle of inclination of the plane with the horizontal

W = Weight of the car = $m \times g$ where

m = Mass of the car

x = Perpendicular distance of C.G. of the car from the rear axle

h = Height of C.G. of car from road surface (*i.e.* from inclined surface)

L = Wheel-base of the car (*i.e.* distance between the centres of the rear and front wheels)

R_A = Reaction between front wheel and ground

R_B = Reaction between rear wheel and ground

μ = Co-efficient of friction between the tyres and road surface

a = Retardation of vehicle.

1. Value of retardation when the brakes are applied to rear wheels only.

Fig. 8.26 shows the various forces acting on the car in which F_B is the total braking force at the rims of the rear wheels due to the application of the brakes. The max. value or F_B is given by

$$F_B = \mu \times R_B$$

The car is in equilibrium* under the action of various forces. Resolving the forces parallel to the inclined plane,

$$F_B + mg \sin \alpha = m \times a$$

or $$\mu R_B + mg \sin \alpha = m \times a \qquad (\because F_B = \mu R_B) \qquad ...(i)$$

*For equilibrium, the equations *i.e.* $\Sigma F_x = 0$; $\Sigma F_y = 0$ and $\Sigma M = 0$ should be satisfied.

Resolving the forces perpendicular to the inclined plane,

$$R_A + R_B = mg \cos \alpha \qquad \text{...}(ii)$$

or

$$R_A = mg \cos \alpha - R_B \qquad \text{...}(A)$$

In equations (*i*) and (*ii*), there are three unknown *i.e.* R_A, R_B and retardation a. Hence we require a third equation. The third equation of equilibrium is obtained by taking moments of all forces about G, the centre of gravity of the car.

$\therefore$ Taking moments about G, we get

$$F_B \times h + R_B \times x = R_A(L - x) \qquad \text{...}(iii)$$

or

$$\mu \times R_B \times h + R_B \times x = R_A(L - x) \qquad (\because F_B = \mu R_B)$$

or

$$R_B(\mu \times h + x) = R_A(L - x)$$

The value of R_B is obtained by substituting the value of $R_A = mg \cos \alpha - R_B$ from equation (*A*) into the above equation. Hence, we get

$$R_B(\mu \times h + x) = (mg \cos \alpha - R_B)(L - x)$$

$$= (mg \cos \alpha)(L - x) - R_B(L - x)$$

or

$$R_B(\mu \times h + x) + R_B(L - x) = (mg \cos \alpha)(L - x)$$

or

$$R_B(\mu \times h + x + L - x) = (L - x) mg \cos \alpha$$

or

$$R_B(\mu \times h + L) = (L - x)\, mg \cos \alpha$$

or

$$R_B = \frac{(L - x)\, mg \cos \alpha}{(\mu h + L)}$$

The value of R_A can be obtained by substituting the value of R_B in equation (*A*).

$$\therefore \quad R_A = mg \cos \alpha - \frac{(L - x)\, mg \cos \alpha}{(\mu h + L)}$$

$$= \frac{mg \cos \alpha\, (\mu h + L) - (L - x)\, mg \cos \alpha}{(\mu h + L)}$$

$$= \frac{(mg \cos \alpha)\,(\mu h + L - L + x)}{(\mu h + L)}$$

$$= \frac{(mg \cos \alpha)\,(\mu h + x)}{(\mu h + L)}$$

The value of retardation '*a*' is obtained by substituting the value of R_B in equation (*i*).

$$\therefore \quad \frac{\mu \times (L - x)\, mg \cos \alpha}{(\mu h + L)} + mg \sin \alpha = m \times a$$

Dividing by '*m*' both sides, we get

$$\frac{\mu \times (L - x)\, g \cos \alpha}{(\mu h + L)} + g \sin \alpha = a$$

Hence the value of retardation '*a*' is given by

$$\therefore \quad a = \frac{(L - x)\, \mu g \cos \alpha}{(\mu h + L)} + g \sin \alpha \qquad \text{...}(8.24)$$

2. Value of retardation when brakes are applied to the front wheels only.

Fig. 8.27 shows the various forces acting on the car in which F_A is the total braking force at the rims of the front wheels due to the application of the brakes at the front wheels. Its maximum value is given by,

$$F_A = \mu R_A$$

The car is in equilibrium under the action of various forces shown in Fig. 8.27. Resolving the forces parallel to the inclined plane,

$$F_A + mg \sin \alpha = m \times a$$

or $$\mu R_A + mg \sin \alpha = m \times a \qquad (\because \quad F_A = \mu R_A) \qquad ...(i)$$

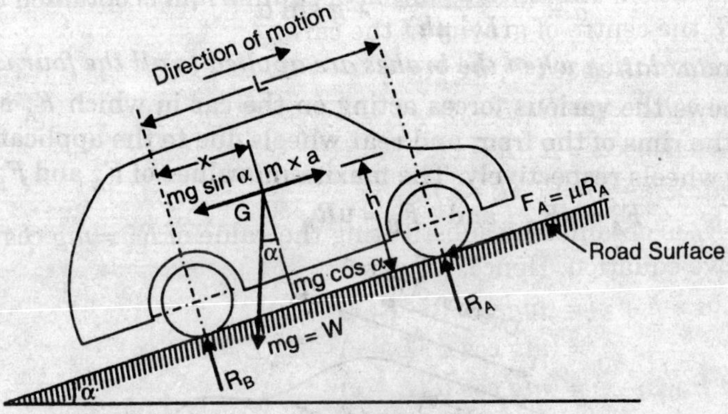

Fig. 8.27

Resolving the forces perpendicular to the inclined plane,

$$R_A + R_B = mg \cos \alpha \qquad ...(ii)$$

or $$R_B = mg \cos \alpha - R_A \qquad ...(A)$$

The third equation of equilibrium is obtained by taking moments of all forces about G. Hence we get

$$F_A \times h + R_B \times x = R_A \times (L - x) \qquad ...(iii)$$

or $$\mu R_A \times h + R_B \times x = R_A \times (L - x) \qquad (\because \quad F_A = \mu R_A)$$

The value of R_A is obtained by substituting the value of R_B from equation (A) into the above equation. Hence we get

$$\mu R_A \times h + (mg \cos \alpha - R_A) \times x = R_A (L - x)$$

or $$\mu \times h \times R_A + mg \cos \alpha \times x - R_A \times x = R_A (L - x)$$

or $$mg \cos \alpha \times x = R_A(L - x) + R_A x - \mu \times h \times R_A$$

$$= R_A(L - x + x - \mu h) = R_A (L - \mu h)$$

$$\therefore \qquad R_A = \frac{x \times mg \times \cos \alpha}{(L - \mu h)}$$

The value of R_B is obtained by substituting the value of R_A in equation (A). Hence we get

$$R_B = mg \cos \alpha - \frac{x \times mg \times \cos \alpha}{(L - \mu h)}$$

$$= (mg \cos \alpha) \left[1 - \frac{x}{(L - \mu h)} \right]$$

$$= (mg \cos \alpha) \left[\frac{L - \mu h - x}{L - \mu h} \right]$$

The value of retardation 'a' is obtained by substituting the value of R_A in equation (i). Hence we get

$$\frac{\mu \times x \times mg \times \cos \alpha}{(L - \mu h)} + mg \sin \alpha = m \times a$$

Dividing by 'm' to both sides, we get

$$\frac{\mu \times x \times g \cos \alpha}{(L - \mu h)} + g \sin \alpha = a$$

$$\therefore \quad a = \frac{\mu x \times g \cos \alpha}{(L - \mu h)} + g \sin \alpha \qquad \text{...(8.25)}$$

3. Value of retardation when the brakes are applied to all the four wheels.

Fig. 8.28 shows the various forces acting on the car in which F_A and F_B are the total braking forces at the rims of the front and rear wheels due to the application of the brakes at the front and rear wheels respectively. The maximum values of F_A and F_B are given by,

$$F_A = \mu R_A \quad \text{and} \quad F_B = \mu R_B$$

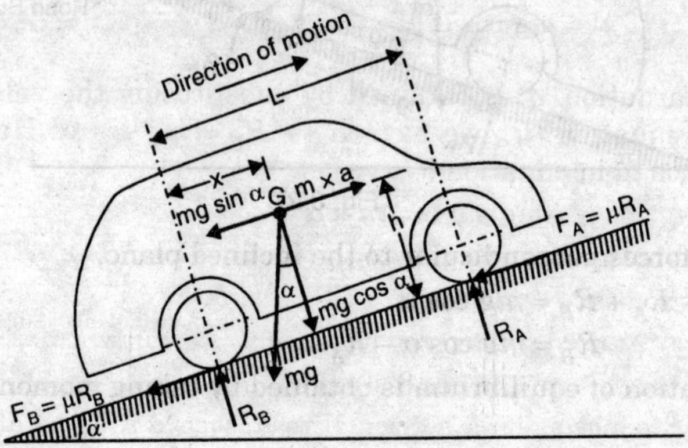

Fig. 8.28

The car is in equilibrium under the action of various forces shown in Fig. 8.28.

Resolving the forces parallel to the inclined plane,

$$F_A + F_B + mg \sin \alpha = m \times a$$

or $$\mu R_A + \mu R_B + mg \sin \alpha = m \times a \qquad (\because F_A = \mu R_A \text{ and } F_B = \mu R_B)$$

or $$\mu [R_A + R_B] + mg \sin \alpha = m \times a \qquad \text{...(i)}$$

Resolving the forces perpendicular to the inclined plane,

$$R_A + R_B = mg \cos \alpha \qquad \text{...(ii)}$$

or $$R_A = mg \cos \alpha - R_B \qquad \text{...(A)}$$

The third equation of equilibrium is obtained by taking moments of all forces about G, the centre of gravity. Hence we get

$$(F_A + F_B) \times h + R_B \times x = R_A(L - x) \qquad \text{...(iii)}$$

or $$(\mu R_A + \mu R_B) \times h + R_B \times x = R_A(L - x) \qquad (\because F_A = \mu R_A \text{ and } F_B = \mu R_B)$$

or $$\mu R_A \times h + \mu R_B \times h + R_B \times x = R_A(L - x)$$

or $$\mu R_B \times h + R_B \times x = R_A(L - x) - \mu R_A \times h$$

or $$R_B(\mu h + x) = R_A(L - x - \mu h)$$

Substituting the value of R_A from equation (A) into the above equation, we get

$$R_B(\mu h + x) = (mg \cos \alpha - R_B)(L - x - \mu h) = mg \cos \alpha \,(L - x - \mu h) - R_B(L - x - \mu h)$$

or $$R_B(\mu h + x) + R_B(L - x - \mu h) = mg \cos \alpha \,(L - x - \mu h)$$

or $$R_B(\mu h + x + L - x - \mu h) = mg \cos \alpha (L - x - \mu h)$$

or $$R_B \times L = mg \cos \alpha (L - x - \mu h)$$

$$\therefore \quad R_B = \frac{mg\cos\alpha\,(L-x-\mu h)}{L}$$

Substituting this value of R_B in equation (A), we get the value of R_A.

$$\therefore \quad R_A = mg\cos\alpha - \frac{mg\cos\alpha\,(L-x-\mu h)}{L}$$

$$= mg\cos\alpha\left[1-\frac{(L-x-\mu h)}{L}\right]$$

$$= mg\cos\alpha\left(\frac{L-L+x+\mu h}{L}\right)$$

$$= mg\cos\alpha\left(\frac{x+\mu h}{L}\right)$$

The value of retardation 'a' is obtained by substituting the values of R_A and R_B in equation (*i*). But from equation (*ii*), we have $R_A + R_B = mg\cos\alpha$. Hence substituting the value of $R_A + R_B = mg\cos\alpha$, in equation (*i*), we get

$$\mu \times mg\cos\alpha + mg\sin\alpha = m \times a$$

Dividing by 'm' to both sides, we get

$$\mu \times g \times \cos\alpha + g\sin\alpha = a$$

$$\therefore \quad \text{Retardation,} \quad a = \mu g\cos\alpha + g\sin\alpha$$

$$= g \times (\mu\cos\alpha + \sin\alpha) \qquad \text{...(8.26)}$$

Note. (*i*) If the car is moving on a horizontal road, then $\alpha = 0$ and all the equations will be modified.

(*ii*) If the car is moving down an inclined plane, then F_A and F_B will be acting upwards.

Problem 8.15. *A car moving along a level road is having the following data :*

Wheel base of car	= *2.85 m,*
Height of C.G. from road surface	= *600 mm,*
Perpendicular distance of C.G. from rear axle	= *1.2 m,*
Speed on level road	= *60 km/h,*
Co-efficient of friction between tyres and road	= *0.1.*

Determine the minimum distance in which the car may be stopped when brakes are applied :

(i) to the rear wheels, (ii) to the front wheels, and (iii) to all the four wheels.

Sol. Given :

Wheel base, $L = 2.85$ m ; $h = 600$ mm $= 0.6$ m ; $x = 1.2$ m ; $\alpha = 0$;

$\mu = 0.6$ and initial speed $u = 60$ km/h $= \frac{60 \times 1000}{60 \times 60}$ m/s $= 16.67$ m/s, Final speed, $v = 0$

Let S = Distance travelled by the car before coming to rest.

(*i*) *Brakes are applied to the rear wheels*

The initial speed and final speed of the car is given. The distance (S) can be calculated if the value of retardation 'a' is known. The value of 'a' (when brakes are applied to the rear wheels) is given by equation (8.24). Hence using equation (8.24), we get

$$\text{Retardation,} \quad a = \frac{(L-x)\mu \times g \times \cos\alpha}{(\mu h + L)} + g\sin\alpha$$

$$= \frac{(L-x)\times\mu\times g\times\cos 0}{(\mu h+L)} + g\sin 0 \qquad (\because \quad \alpha = 0 \text{ for level road})$$

$$= \frac{(L-x)\times\mu\times g}{(\mu h+L)} \qquad (\because \quad \cos 0 = 1 \text{ and } \sin 0 = 0)$$

$$= \frac{(2.85-1.2)\times 0.6\times 9.81}{(0.6\times 0.6+2.85)} = 3.025 \text{ m/s}^2$$

$\therefore$ Acceleration, $a = -3.025 \text{ m/s}^2$

Now using equation, $v^2 - u^2 = 2\times a\times S$, we get

$$0^2 - 16.67^2 = 2\times(-3.025)\times S \qquad (\because \quad v = 0, u = 16.67 \text{ and } a = -3.025 \text{ m/s}^2)$$

or $$16.67^2 = 2\times 3.025\times S$$

or $$S = \frac{16.67^2}{2\times 3.025} = \mathbf{45.93 \text{ m.}} \quad \textbf{Ans.}$$

(ii) Brakes are applied to the front wheels

The retardation '*a*' when brakes are applied to the front wheels, is given by equation (8.25) as

$$a = \frac{\mu\times x\times g\cos\alpha}{(L-\mu h)} + g\sin\alpha$$

$$= \frac{0.6\times 1.2\times 9.81\times\cos 0°}{(2.85-0.6\times 0.6)} + 9.81\sin 0° \qquad (\because \quad \alpha = 0)$$

$$= \frac{0.6\times 1.2\times 9.81}{(2.85-0.36)} = 2.836 \text{ m/s}^2$$

$\therefore$ Acceleration, $a = -2.836 \text{ m/s}^2$ (– ve sign is due to retardation)

Now using, $v^2 - u^2 = 2a\times S$, we get

$$0^2 - 16.67^2 = 2\times(-2.836)\times S$$

or $$S = \frac{-16.67^2}{2\times(-2.836)} = \mathbf{48.99 \text{ m.}} \quad \textbf{Ans.}$$

(iii) Brakes are applied to all the four wheels.

The retardation '*a*' when the brakes are applied to all the four wheels, is given by equation (8.26) as

$$a = g\times(\mu\cos\alpha + \sin\alpha)$$

$$= 9.81(0.6\cos 0° + \sin 0°) \qquad [\text{For a level road, } \alpha = 0]$$

$$= 9.81\times 0.6 = 5.886 \text{ m/s}^2$$

$\therefore$ Acceleration, $a = -5.886 \text{ m/s}^2$

Now using equation $v^2 - u^2 = 2aS$, we get

$$0^2 - 16.67^2 = 2\times(-5.886)\times S$$

$\therefore$ $$S = \frac{-16.67^2}{2(-5.886)} = \mathbf{23.6 \text{ m.}} \quad \textbf{Ans.}$$

Problem 8.16. *A car moving on a rough inclined plane is having the following data :*

Angle of inclination of plane	*= 15°*
Wheel base of the car	*= 2.0*
Height of C.G. of the car above the inclined plane	*= 1.0 m*

Perpendicular distance of C.G. from rear axle = 0.9 m
Speed of the car = 54 km / hour,
Co-efficient of friction between tyres and road = 0.45
Brakes are applied to all the four wheels.
Determine : (i) distance travelled by the car before coming to rest.
(ii) time taken in doing so if
(a) the car is moving up the plane, and
(b) the car is moving down the plane.

Sol. Given :

$\alpha = 15°$; $L = 2.0$ m ; $h = 1.0$ m ; $x = 0.9$ m ; $N = 54$ km/hour or $u = \dfrac{54 \times 1000}{60 \times 60}$ m/s = 15 m/s ;

$\mu = 0.45$; brakes are applied to all the four wheels. Final velocity of car, $v = 0$.

1st Case. *The car is moving up the plane.*

The retardation 'a' when the brakes are applied to all the four wheels is given by equation (8.26) as

$$a = g \times (\mu \cos \alpha + \sin \alpha)$$
$$= 9.81(0.6 \cos 15° + \sin 15°) \qquad (\because \ \alpha = 15° \text{ and } \mu = 0.6)$$
$$= 9.81(0.6 \times 0.966 + 0.2588) = 8.224 \text{ m/s}^2$$

$\therefore$ Acceleration, $a = -\,8.224 \text{ m/s}^2$ (– ve acceleration means retardation)

Let S = Distance travelled by the car before coming to rest
t = Time taken by car in coming to rest.

Now using equation, $v^2 - u^2 = 2aS$, we get

$$0^2 - 15^2 = 2 \times (-\,8.224) \times S$$

$$\therefore \quad S = \frac{-15^2}{2 \times (-\,8.224)} = \textbf{13.68 m.} \quad \textbf{Ans.}$$

The time 't' is obtained by using equation,

$$v = u + at$$

or $$0 = 15 + (-\,8.224) \times t$$

$$\therefore \quad t = \frac{15}{8.224} = \textbf{1.824 seconds.} \quad \textbf{Ans.}$$

2nd Case. *The car is moving down the plane.*

Fig. 8.29 shows all the forces acting on the car. The car is in equilibrium under these forces. The three equations of equilibrium are :

Along the plane, $$F_A + F_B = ma + mg \sin \alpha$$

or $$\mu R_A + \mu R_B = ma + mg \sin \alpha$$

or $$\mu (R_A + R_B) - mg \sin \alpha = ma \qquad ...(i)$$

Perpendicular to the plane,

$$R_A + R_B = mg \cos \alpha \qquad ...(ii)$$

Substituting the value of $(R_A + R_B)$ in equation (*i*), we get

$$\mu \times mg \cos \alpha - mg \sin \alpha = m \times a$$

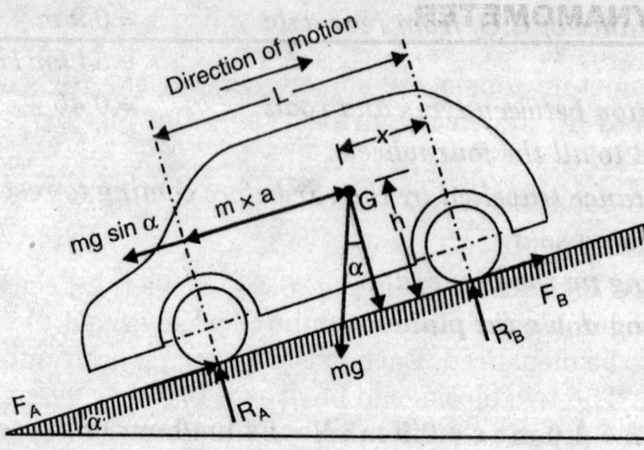

Fig. 8.29

Cancelling '*m*' to both sides, we get

$$\mu \times g \cos \alpha - g \sin \alpha = a$$

∴ Retardation, $a = \mu \times g \times \cos \alpha - g \sin \alpha$

$$= g(\mu \cos \alpha - \sin \alpha)$$

$$= 9.81(0.6 \times \cos 15° - \sin 15°)$$

$$= 9.81(0.6 \times 0.966 - 0.2588) = 3.147 \text{ m/s}^2$$

∴ Acceleration, $a = -3.147 \text{ m/s}^2$

Now using equation, $v^2 - u^2 = 2 \times a \times S$

or $0^2 - 15^2 = 2 \times (-3.147) \times S$

∴ $S = \dfrac{-15^2}{2 \times (-3.147)} = $ **35.75 m. Ans.**

The time '*t*' is obtained by using equation,

$$v = u + at$$

or $0 = 15 + (-3.147) \times t$

∴ $t = \dfrac{15}{3.147} = $ **4.76 seconds. Ans.**

8.4. DYNAMOMETER

A dynamometer is a device used to measure the frictional resistance or frictional torque. This frictional resistance or frictional torque is obtained by applying a brake. The dynamometer consists of a brake and also a device of measuring the braking force (or braking torque). Hence dynamometer is a brake with a device of measuring the frictional resistance or frictional torque. After knowing frictional torque, the power of the engine can be obtained.

Following are the two types of dynamometers :

(*i*) Absorption dynamometers and

(*ii*) Transmission dynamometer.

Absorption dynamometers absorb the available power in doing work against friction whereas transmission dynamometers transmit the available power to some other machines where the power is suitably measured.

8.5. ABSORPTION DYNAMOMETER

Absorption dynamometers consists of some form of brakes in which provision is made for measuring the frictional torque on the drum. The following are the important types of absorption dynamometer :

(*i*) Prony brake dynamometer

(*ii*) Rope brake dynamometer.

8.5.1. Prony Brake Dynamometer. Fig. 8.30 shows the Prony brake dynamometer which consists of two wooden blocks placed around a pulley fixed to the shaft of an engine whose power is required to be measured. Each of the wooden blocks embraces rather less than one half of the pulley rim. The two blocks can be drawn together by means of bolts, nuts and springs so as to increase the pressure on the pulley. The lower block carries an arm (lever) to the end of which a weight W can be applied. A second arm projects from the block in the opposite direction and carries a balance weight B, which balances the brake when unloaded. Two stops S, S are provided and the lever arm will float between these stops. The friction torque on the pulley may be increased by screwing up the bolts, until it balances the torque due to available power.

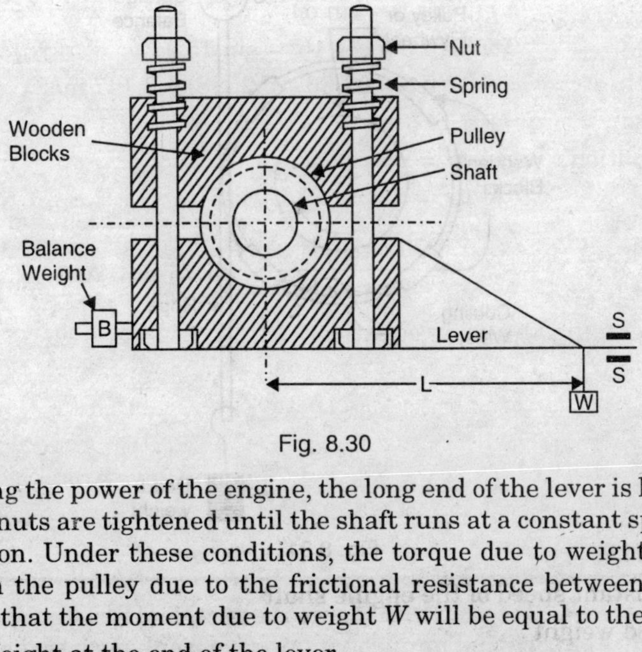

Fig. 8.30

For measuring the power of the engine, the long end of the lever is loaded with a known weight W. Now the nuts are tightened until the shaft runs at a constant speed and the lever is in horizontal position. Under these conditions, the torque due to weight W will balance the frictional torque on the pulley due to the frictional resistance between the blocks and the pulley. This means that the moment due to weight W will be equal to the frictional torque.

Let W = Weight at the end of the lever,

R = Radius of the pulley,

μ = Co-efficient of friction between pulley and blocks

L = Horizontal distance of weight W from the centre of the pulley,

N = Speed of the shaft in r.p.m.

Torque on the shaft, $T = W \times L$

$\therefore$ Power of the engine $= \text{Torque} \times \text{Angular speed}$

$= T \times \omega$

$$= T \times \frac{2\pi N}{60}$$

$$= W \times L \times \frac{2\pi N}{60} \text{ Watts.} \qquad (\because \quad T = W \times L)$$

From the above equation, it is clear that the power of the engine is independent of : (*i*) radius of the pulley, *R* (*ii*) co-efficient of friction between pulley and wooden blocks and (*iii*) pressure exerted by tightening the nuts.

8.5.2. Rope Brake Dynamometer. Fig. 8.31 shows the rope brake dynamometer which consists of one, two or more ropes wound round the rim of a pulley (or flywheel) fixed rigidly to the shaft of the engine whose power is required to be measured. The upper end of the ropes is attached to a spring balance (*S*) while the lower end carries the dead weight *W*. The ropes are spaced evenly across the width of the rim by means of three or four wooden blocks at different points round the rim (or around the circumference of the flywheel).

For measuring the power of an engine, the engine is made to run at a constant speed. Under this condition, the torque transmitted by the engine must be equal to the frictional torque due to the ropes.

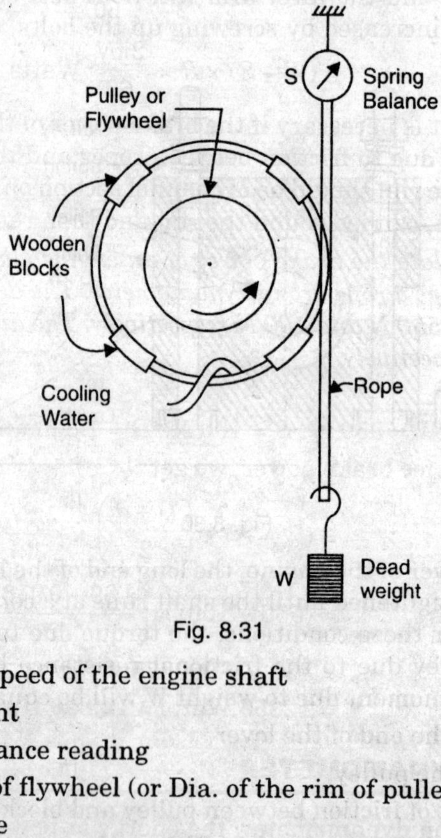

Fig. 8.31

Let N = Constant speed of the engine shaft

W = Deed weight

S = Spring balance reading

D = Diameter of flywheel (or Dia. of the rim of pulley)

d = Dia. of rope

Then net load on brake = $(W - S)$

$\therefore$ Frictional torque due to ropes

= (Net load on brake)

× Distance of load line from the centre of shaft

$$= (W - S) \times \left(\frac{D + d}{2}\right)$$

But torque transmitted by engine at constant speed

$$= \text{Frictional torque due to ropes}$$

$$= (W - S) \times \left(\frac{D + d}{2}\right)$$

$\therefore$ Brake power of engine = Torque transmitted by engine × Angular speed of engine

$$= (W - S) \times \left(\frac{D + d}{2}\right) \times \omega$$

$$= (W - S) \times \left(\frac{D + d}{2}\right) \times \frac{2\pi N}{60} \qquad \left(\because \ \omega = \frac{2\pi N}{60}\right) \qquad ...(8.27)$$

If dia. of rope (*i.e.* d) is neglected, then brake power of engine

$$= (W - S) \times \frac{D}{2} \times \frac{2\pi N}{60}$$

$$= (W - S) \times R \times \frac{2\pi N}{60} \text{ Watts} \qquad \left(\because \ R = \frac{D}{2}\right) \qquad ...(8.28)$$

A cooling arrangement is necessary if the brake power of the engine is very large, as in that case the heat produced due to friction between ropes and the flywheel will also be very large. For cooling the rim, the rim should be of channel section on the inside so that cold water may be supplied at one point, carried round the rim and then removed.

Problem 8.17. *Calculate the brake power of an engine which is running at a constant speed of 300 r.p.m. and carries a rope brake dynamometer. The dead weight on the engine and spring balance readings are 550 N and 100 N respectively. The diameters of flywheel and rope are 1.8 m and 18.75 mm respectively.*

Sol. Given :

$N = 300$ r.p.m. ; $W = 550$ N ; $S = 100$ N ; $D = 1.8$ m and $d = 18.75$ mm $= 0.01875$ m

Using equation (8.27) for brake power, we get

$$\text{Brake power} = (W - S) \times \left(\frac{D + d}{2}\right) \times \frac{2\pi N}{60}$$

$$= (550 - 100) \times \left(\frac{1.89 + 0.01875}{2}\right) \times \frac{2\pi \times 300}{60} \text{ Watts}$$

$$= 450 \times 0.909375 \times 10\pi$$

$= 12856$ Watts $=$ **12.856 kW. Ans.**

8.6. TRANSMISSION DYNAMOMETER

In case of transmission dynamometer, the energy or power is not absorbed. Hence the energy is not wasted in friction but is used for doing work. The energy or power produced by the engine is transmitted through the dynamometer to some other machines where the power developed is suitably measured. The following are the important types of transmission dynamometer :

(*i*) Epi-cyclic train dynamometer,

(*ii*) Belt transmission dynamometer, and

(*iii*) Torsion dynamometer.

8.6.1. Epi-cyclic Train Dynamometer. Fig. 8.32 show an epi-cyclic train dynamometer which consists of a simple epi-cyclic train of gears *i.e.* spur gear A, internal wheel D having internal teeth and an intermediate wheel C (*i.e.* pinion). The wheel A with spur gear is keyed to the *driving shaft* (*i.e.* engine shaft). Let it rotates in anti-clockwise direction. The wheel D having internal teeth is keyed to the *driven shaft* and it will rotate in clockwise direction. The gears of the wheel C (which is the intermediate wheel and known as pinion) meshes both the gears of wheel A and of wheel D. Thus the power is transmitted from wheel A to wheel D through the intermediate wheel C. The wheel C revolves freely on a pin fixed to the arm of a lever. The lever is pivoted about the common axis of the driving and driven shafts *i.e.* at point E. When the dynamometer is at rest, the weight B balances the lever arm.

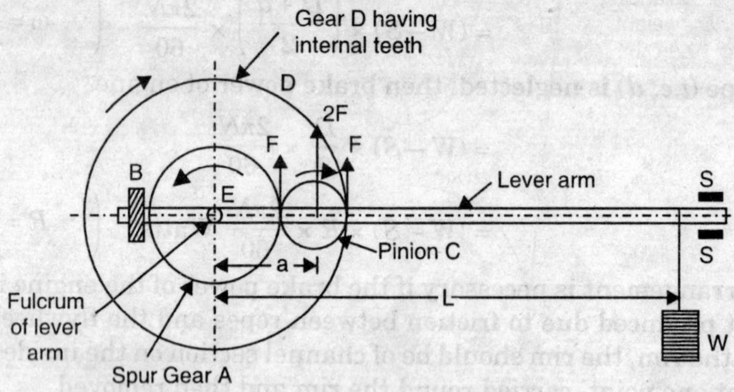

Fig. 8.32

When the dynamometer is in operation, the tangential force exerted by the wheel A on the wheel C and the tangential *reaction* of the wheel D on the wheel C will act in the upward direction. Also if the friction of the pin on which wheel C revolves is neglected, then the above two forces will be equal. Hence the total upward force on the lever arm acting through the axis of the wheel C is $2F$. This force tends to rotate the lever arm about its fulcrum E. The torque due to force $2F$ on the arm will be $2F \times a$. This torque will be balanced by the torque due to a dead weight W placed at the end of the lever as shown in Fig. 8.32. The lever arm floats between the stops S, S.

Hence for equilibrium of the lever arm,

Torque due to force $2F$ = Torque due to dead weight W

or $2F \times a = W \times L$

$$\therefore \quad F = \frac{W \times L}{2a}$$

Let R = Radius of wheel A

N = Speed of the rotation of driving shaft (*i.e.* speed of engine)

$\therefore$ Torque transmitted by engine

$$= F \times R = \frac{W \times L}{2a} \times R$$

$\therefore$ Power transmitted = Torque transmitted × ω

$$= \left(\frac{W \times L}{2a} \times R\right) \times \frac{2\pi N}{60} \text{ Watts.}$$

8.6.2. Belt Transmission Dynamometer. A belt transmission dynamometer measures the difference between the tensions on the tight and slack sides of a belt when it is running from one pulley to another pulley (*i.e.* when belt is transmitting power from one pulley to another pulley). This difference of tensions (*i.e.* $T_1 - T_2$) when multiplied by the speed of the belt, gives the power transmitted.

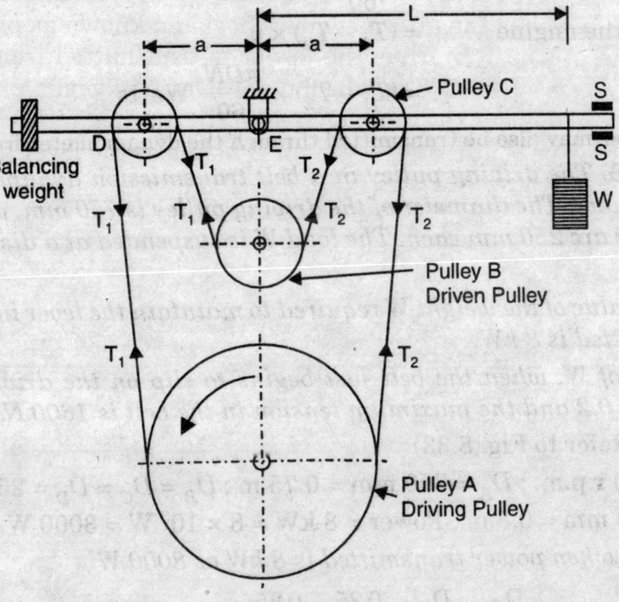

Fig. 8.33

Fig. 8.33 show a belt transmission dynamometer which is also called Tatham dynamometer. It consists of a driving pulley *A*, driven pulley *B* and intermediate pulleys *C* and *D*. The driving pulley *A* is rigidly fixed to the shaft of an engine whose power is to be measured. The driven pulley *B* is fixed to another shaft to which power is to be transmitted. The intermediate pulleys *C* and *D* rotates on pins fixed to the lever. The lever is pivoted at *E*, the mid point of the two intermediate pulley centres. A continuous belt runs over the driving and the driven pulleys through the two intermediate pulleys. The movement of lever is controlled between two stops *S* and *S* one on each side of the lever.

Let the driving pulley *A* rotates anti-clockwise. The tight and slack sides of the belt will be as shown in Fig. 8.33. The total downward force acting on pulley *D* is $2T_1$ whereas the total downward force on pulley *C* is $2T_2$. As $2T_1$ is greater than $2T_2$, therefore the lever starts rotating about *E* in anti-clockwise direction. In order to balance it, a weight *W* is suspended at a distance *L* from *E* on the lever as shown in Fig. 8.33.

When the lever is in horizontal position, the total moments of all the force about fulcrum *E* should be zero *i.e.*

Total anti-clockwise moment = Total clockwise moment

or $$2T_1 \times a = 2T_2 \times a + W \times L$$

or $$2T_1 \times a - 2T_2 \times a = W \times L$$

or $$2a(T_1 - T_2) = W \times L$$

or $$(T_1 - T_2) = \frac{W \times L}{2a} \qquad \text{...(8.29)}$$

Let v = Belt speed in m/s

D = Dia. of pulley A

N = Speed of engine shaft

Then $$v = \frac{\pi DN}{60}$$

$\therefore$ Power of the engine $= (T_1 - T_2) \times v$

$$= (T_1 - T_2) \times \frac{\pi DN}{60} \text{ Watts} \qquad ...(8.30)$$

Note. The power may also be transmitted through the dynamometer from pulley B to pulley A.

Problem 8.18. *The driving pulley in a belt transmission dynamometer shown in Fig. 8.33 rotates at 400 r.p.m. The diameter of the driving pulley is 750 mm, whereas the diameters of pulleys B, C and D are 250 mm each. The load W is suspended at a distance of 800 mm from the fulcrum E.*

Find : (i) the value of the weight W required to maintain the lever in a horizontal position when power transmitted is 8 kW.

(ii) the value of W, when the belt just begins to slip on the driving pulley A. The coefficient of friction is 0.2 and the maximum tension in the belt is 1600 N.

Sol. Given : (Refer to Fig. 8.33)

N_A = 400 r.p.m. ; D_A = 750 mm = 0.75 m ; $D_B = D_C = D_D$ = 250 mm = 0.25 m ;
L = 800 mm = 0.8 m ; Power = 8 kW = 8 × 10^3 W = 8000 W.

(i) Value of W when power transmitted is 8 kW or 8000 W

$$a = \frac{D_B}{2} + \frac{D_C}{2} = \frac{0.25}{2} + \frac{0.25}{2} = 0.25 \text{ m}$$

Let T_1 = Tension on tight side of the belt on pulley A

T_2 = Tension on slack side of the belt on pulley A.

For power transmitted, using equation (8.30), we get

$$\text{Power} = (T_1 - T_2) \times \frac{\pi DN}{60}$$

or $$8000 = (T_1 - T_2) \times \frac{\pi \times 0.75 \times 400}{60}$$

$\therefore$ $$(T_1 - T_2) = \frac{8000 \times 60}{\pi \times 0.75 \times 400} = 509.3$$

But from equation (8.0), we have

$$(T_1 - T_2) = \frac{W \times L}{2a}$$

or $$509.3 = \frac{W \times 0.8}{2 \times 0.25}$$

$\therefore$ $$W = \frac{509.3 \times 2 \times 0.25}{0.8} = \textbf{318.3 N.} \quad \textbf{Ans.}$$

(ii) Value of W, when the belt just begins to slip on driving pulley A.

μ = 0.2 and Max. tension *i.e.* T_1 = 1600 N. Also from Fig. 8.33, it is clear that lap angle θ = 180° = π

We know that $\frac{T_1}{T_2} = e^{\mu \times \theta}$

or $$2.3 \log\left(\frac{T_1}{T_2}\right) = \mu \times \theta = 0.2 \times \pi = 0.6284$$

or $$\log\left(\frac{T_1}{T_2}\right) = \frac{0.6284}{2.3} = 0.2732$$

or $$\frac{T_1}{T_2} = \text{Antilog of } 0.2732 = 1.876$$

or $$T_2 = \frac{T_1}{1.876} = \frac{1600}{1.876} = 852.87 \text{ N}$$

Using equation (8.29), we get

$$(T_1 - T_2) = \frac{W \times L}{2a}$$

or $$(1600 - 852.87) = \frac{W \times 0.8}{2 \times 0.25}$$

or $$747.13 = \frac{W \times 0.8}{0.5}$$

or $$W = \frac{747.13 \times 0.5}{0.8} = \mathbf{466.95 \text{ N}.} \quad \textbf{Ans.}$$

8.6.3. Torsion Dynamometer. The torsion dynamometer works on the principle of angle of twist in a shaft when power is transmitted along the shaft. Actually the torque transmitted is directly proportional to the angle of twist. Hence if angle of twist can be measured accurately, then the corresponding torque transmitted can be calculated.

The driving end of a shaft twists through a small angle relative to the driven end when power is transmitted along the shaft. The angle of twist is obtained from the torsion equation which is given below as

$$\frac{T}{J} = \frac{C \times \theta}{L}$$

where T = Torque transmitted,

J = Polar moment of inertia of shaft,

θ = Angle of twist in radians,

L = Length of shaft,

C = Modulus of rigidity of the shaft material.

From the equation, we have

$$T = \frac{C \times \theta}{L} \times J \text{ where } \quad J = \frac{\pi}{32} \times D^4$$

... For a solid shaft

$$= \frac{\pi}{32} (D_0^4 - D_i^4)$$

... For a hollow shaft

For a given shaft, the values of C, J and L are constant. Hence

$$T = k \times \theta \qquad \text{where } k = \frac{C \times J}{L} \text{ is a constant}$$

or $$\text{Torque transmitted} \propto \theta$$

Hence torque transmitted is directly proportional to the angle of twist. If angle of twist can be measured by some means then torque can be calculated. From the torque, the power transmitted can be obtained.

In actual practice, the angle of twist is measure for a small length of the shaft, therefore some magnifying device must be incorporated in the dynamometer for accurate measurement of the angle of twist. The Bevis-Gibson flash light torsion dynamometer uses this principle.

Bevis-Gibson Flash Light Torsion Dynamometer

Fig. 8.34 shows a Bevis-Gibson torsion dynamometer which consists of two discs *A* and *B* fixed on a shaft at points as far apart as possible. Each disc has a narrow radial slot and the two slots are in line when there is no torque transmitted along the shaft. A powerful electric lamp *L* is fixed to the bearing cap of the shaft behind disc *A*. The lamp is masked so as to throw a narrow pencil of light parallel to the axis of shaft. Also this lamp has a slot directly opposite to the slot of disc *A*. Behind the disc *B*, an eye-piece *E* is fitted to the shaft bearing. This eye-piece is capable of slight circumferential adjustment.

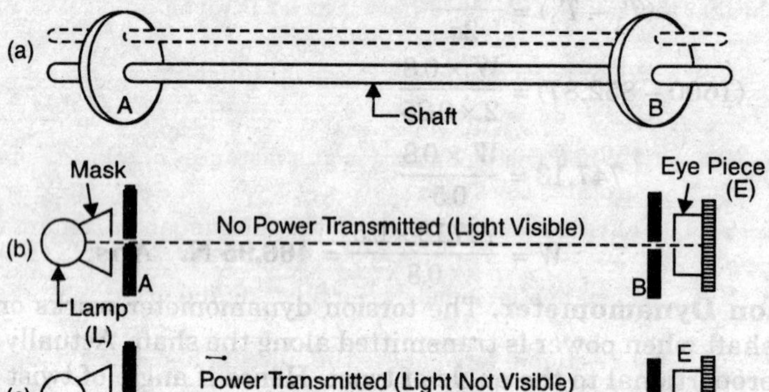

Fig. 8.34

The eye-piece is adjusted so as to receive the ray of light which passes from the lamp through the slots in the two discs, when the shaft is at rest. When the shaft rotates without transmitting any torque, a ray of light will be received in the eye-piece once per revolution as shown in Fig. 8.34 (*b*). But when shaft is rotating and torque is transmitted, the shaft twists and the slot in the disc *B* sights its position. Due to this, the ray of light does not reach to the eye-piece as shown in Fig. 8.34 (*c*). If now the eye-piece is displaced along the circular are by an amount equal to the lag of disc *B* by means of vernier, then the ray of light will be visible in eye- piece as shown in Fig. 8.34 (*d*). Hence the angular displacement of the eye-piece and therefore the angle of twist of the shaft may be measured upto $\frac{1}{100}$th of a degree.

HIGHLIGHTS

1. A brake is a device used to bring to rest a body or to hold a body in a state of uniform motion. A dynamometer is a device used to measure the frictional resistance or frictional torque.
2. The important-types of mechanical brakes are :
   (*i*) Simple block or Shoe brake
   (*ii*) Band brake
   (*iii*) Band and block brake
   (*iv*) Internal expanding shoe brake.
3. For a simple brake, the braking torque on the drum is given by,
$$T_B = F^* \times r = (\mu R_N) \times r$$
where R_N = Normal reaction on the block,
r = Radius of drum, and
μ = Co-efficient of friction.
4. For a simple brake, the value of R_N for different cases is given as
$$R_N = \frac{P \times L}{a}$$ ... When frictional forces passes through pivot.
$$= \frac{P \times L}{a + \mu b}$$ ...(*i*) When frictional force is at a distance '*b*' below the fulcrum and drum rotates clockwise
(*ii*) Frictional force is above the fulcrum and drum rotates anti-clockwise.
$$= \frac{P \times L}{a - \mu b}$$ ...(*i*) Friction force below the fulcrum and drum rotates anti-clockwise.
(*ii*) Frictional force above the fulcrum and drum rotates clockwise.
where a = Perpendicular distance between the line of action of R_N and fulcrum.
b = Perpendicular distance between the line of action of frictional force and fulcrum.
P = Force acting at the end of lever in vertically downward direction.
5. For self-locking brake, the force at the end of the lever should be zero or negative. But $P = \dfrac{R_N(a - \mu b)}{L}$. Hence for self locking brake, the condition is $(a - \mu b) \le 0$ or $a \le \mu b$.
6. The brake will be self-locking when $a \le \mu b$ for the cases :
   (*i*) Frictional force is below the fulcrum and drum rotates anti-clockwise.
   (*ii*) Frictional force is above the fulcrum and drum rotates clockwise.
7. For a self-energised brake, the frictional force helps in applying the brake. In equation $P = R_N (a - \mu b)$ if $P \le 0$, the brake is self-locking and if $P > 0$, the brake is self-energised.
8. If angle made by contact surface of the block at the centre of the rotating drum is more than 40° (*i.e.* $2\theta > 40°$) then equivalent co-efficient of friction (μ') is given by :
$$\mu' = \frac{4\mu \sin\theta}{2\theta + \sin 2\theta} \quad \text{or} \quad \mu\left[\frac{4 \sin\theta}{2\theta + \sin 2\theta}\right].$$
9. For band brake, braking torque is given by :
$$T_B = (T_1 - T_2) \times r \quad \text{... if thickness of belt is neglected}$$
$$= (T_1 - T_2) \times r_e \quad \text{... if thickness of belt is considered.}$$
10. In case of simple band brake, one end of the band is connected to the pivot.
11. In case of differential band brake, the two ends of the band are connected at *A* and *B* which are on the opposite side of the fulcrum.

12. A differential band brake will be self-locking :

(*i*) For clockwise rotation if $\frac{T_2}{T_1} \le \frac{b}{a}$

(*ii*) For anti-clockwise rotation if $\frac{T_1}{T_2} \le \frac{b}{a}$

where $b = \perp$ distance from fulcrum on the line of action of tension towards the force side P

$a = \perp$ distance from fulcrum on the line of action of tension opposite to the force side P.

13. For a band and block brake, the ratio of tensions in the tight and slack sides of the complete band and block brake is given by

$$\frac{T_n}{T_0} = \left(\frac{1+\mu \tan\theta}{-\mu \tan\theta}\right)^n$$

where θ = Half of the angle subtended by each block on the centre of drum,

T_n = Tension on the tight side after n blocks and

T_0 = Tension on the slack side.

14. Braking torque for band and block brake is given by

$$T_B = (T_n - T_0) \times r \qquad \text{where } r = \text{effective radius.}$$

15. For Internal expanding shoe brake,

$$p_N \propto \sin\theta \qquad \therefore \quad p_N = p_1 \sin\theta \ \text{ or } \ p_2 \sin\theta$$

$$\delta R_N = p_n \times (b \times r \times \delta\theta) \qquad \therefore \quad \delta R_{N_1} = p_1 \sin\theta \times (b \times r \times \delta\theta) \ \text{ or } \ \delta R_{N_2} = p_2 \sin\theta(b \times r \times \delta\theta)$$

$$\delta F = \mu\delta R_N \times (b \times r \times \delta\theta) \qquad \therefore \quad \delta F_1 = \mu p_1 \sin\theta \times (b \times r \times \delta\theta)$$

$$\delta T_B = \delta F \times r = \mu \times P_N \times (b \times r \times \delta\theta) \times r \quad \therefore \quad \delta T_{B_1} = \mu p_1 \times \sin\theta \times br^2 \times \delta\theta$$

$$\therefore \quad T_B = \int_{\theta_1}^{\theta_2} dT_B = \int_{\theta_1}^{\theta_2} \mu \times p_N \times br^2 \times \delta\theta \quad \therefore \quad T_{B_1} = \mu p_1 \times br^2 \times \int \sin\theta \, d\theta$$

$$= \mu p_1 \times br^2 (\cos\theta_1 - \cos\theta_2).$$

16. The brakes in a car may be applied to the :

(*i*) rear wheels only,

(*ii*) front wheels only, and

(*iii*) all the four wheels.

17. The value of retardation when brakes in a car (which is moving up an inclined plane) is applied :

$$a = \frac{\mu(L-x) \times g \cos\alpha}{(\mu h + L)} + g \sin\alpha \qquad \text{... to rear wheels.}$$

$$= \frac{\mu x \times g \cos\alpha}{(L - \mu h)} + g \sin\alpha \qquad \text{... to front wheels.}$$

$$= (\mu \cos\alpha + \sin\alpha) \times g \qquad \text{... to all the four wheels.}$$

If car is moving on a horizontal road, then $\alpha = 0$ and all the above equations will become as :

$$a = \frac{\mu[L-x] \times g}{(\mu h + L)} \qquad \text{... brakes applied to rear wheels}$$

$$= \frac{\mu \times x \times g}{(L - \mu h)} \qquad \text{... brakes applied to front wheels}$$

$$= \mu \times g \qquad \text{... brakes applied to all the four wheels.}$$

where μ = Co-efficient of friction between the tyres and road surface

L = wheel-base of the car (*i.e.*, distance between the centres of the rear and front wheels)

$x = \perp$ distance of C.G. of the car from the rear axle

h = Height of C.G. of car from road surface.

18. Absorption dynamometer absorb the available power in doing work against friction whereas transmission dynamometers transmit the available power to some other machines where the power is suitably measured.

19. Prony brake dynamometer and rope brake dynamometer are of absorption type dynamometers. The brake power of the engine measured by these dynamometers is given by :

Brake power $= T \times \omega$ where $T = W \times L$ and $\omega = \dfrac{2\pi N}{60}$

$$= (W \times L) \times \frac{2\pi N}{60} \quad \text{... For Prony brake dynamometer}$$

$$= (W - S) \times \frac{D}{2} \times \frac{2\pi N}{60} \quad \text{... For rope brake dynamometer.}$$

20. Epi-cyclic train dynamometer, belt transmission dynamometer and torsion dynamometer are of transmission type dynamometer.

21. For epi-cyclic train dynamometer, the power transmitted is given by :

Power transmitted = Torque × ω where Torque = Force × Radius = $F \times R$

$$= \left(\frac{W \times L}{2a}\right) \times R$$

$$= \left(\frac{W \times L}{2a}\right) \times \frac{2\pi N}{60}$$

where W = Dead weight suspended on lever

L = ⊥ distance between the line of action of W and centre of wheel A

a = Distance between pinion centre and centre of wheel A

R = Radius of driving wheel A

N = Speed of driving wheel.

22. A belt transmission dynamometer measures the difference between the tensions on the tight and slack sides of a belt, as given below :

$$(T_1 - T_2) = \frac{W \times L}{2a}$$

Power of engine $= (T_1 - T_2)\, v$

$$= (T_1 - T_2) \times \frac{\pi D N}{60}$$

where W = Dead weight at the end of the lever

a = Distance of each intermediate pulley from fulcrum

L = Distance of W from fulcrum

where v = speed of belt,

D = Dia. of pulley A, and

N = speed of engine shaft.

23. Torsion dynamometer measures the angle of twist in a shaft which is rotating transmitting torque. The torque and power can be obtained as :

$$\frac{T}{J} = \frac{C \times \theta}{L}$$ where J = Polar moment of inertia

Torque $= \dfrac{C \times \theta}{L} \times J$ $\quad = \dfrac{\pi}{32} D^4$ ... For solid shaft

$= \dfrac{C \times J}{L} \times \theta$ $\quad = \dfrac{\pi}{32} (D_0^5 - D_i^4)$... For hollow shaft

If θ is known, torque can be obtained

$$\therefore \quad \text{Power} = T \times \omega = T \times \frac{2\pi N}{60} \text{ Watts.}$$

24. Bevis-Gibson flash light torsion dynamometer is used to measure the angle of twist in a shaft which is rotating and transmitting torque.

EXERCISE 8

(A) Theoretical Questions

1. What is the difference between brakes and dynamometers ?
2. What are the different types of mechanical brakes ?
3. With a neat sketch, describe a single shoe brake. What is the advantage of double shoe brake over single shoe brake ?
4. Find an expression for the braking torque for a single shoe brake when brake drum is rotating (*a*) clockwise, (*b*) anti-clockwise and
   (*i*) Friction force passes through fulcrum,
   (*ii*) Friction force is at a distance '*b*' below the fulcrum, and
   (*iii*) Frictional force is at a distance '*b*' above the fulcrum.
5. What do you understand by 'self-locking brake' and 'self-energised brake'? Should we have self-locking brake or self-energised brake ?
6. What is the condition for : (*i*) self-locking of a brake and (*ii*) self-energised brake.
7. What is the difference between a simple band brake and a differential band brake ?
8. Find the condition for a differential band brake to be self-locking for clockwise rotation and anti-clockwise rotation of the brake drum.
9. Prove that the ratio of tensions in the tight and slack sides of a band and block brake is given by,

$$\frac{T_1}{T_0} = \left(\frac{1 + \mu \tan \theta}{1 - \mu \tan \theta}\right)^n$$

where θ = Half of the angle subtended by each block on the centre of the drum.

T_n = Tension on tight side after n blocks

T_0 = Tension on slack side.

10. What a neat sketch, describe the principles and working of an internal expanding shoe-brake. Derive the expression for the braking torque.
11. Differentiate between :
   (*i*) Brakes and dynamometer,
   (*ii*) Absorption dynamometer and transmission dynamometer
   (*iii*) Band brake and band-and-block brake
   (*iv*) Simple band brake and differential band brake.
12. Describe with the help of a neat sketch the construction and working of a prony brake or rope brake absorption dynamometer.
13. What is the principle on which a torsion dynamometer works ? Explain with details the calculations involved in finding the power transmitted.
14. What a neat sketch, describe the principle and working of a Bevis-Gibson flash light dynamometer.

(B) Numerical Problems

1. The brake drum of a single block brake is rotating at 400 r.p.m. in the clockwise direction. The diameter of drum is 800 mm and the single block brake is of the type as shown in Fig. 8.2. The force required at the end of the lever is 450 N. If angle contact is 25° and L = 1.5 m, a = 450 mm and b = 37.5 mm, then determine the braking torque. The co-efficient of friction is 0.3.

[**Ans.** 175.6 Nm]

2. If the brake drum in the above problem rotates in anti-clockwise direction as shown in Fig. 8.3 and all other data remains same, then determine :
   (*i*) the braking torque and (*ii*) value of *b* for self-locking of the brake.
   [Ans. (*i*) 184.6 Nm ; (*ii*) $b \geq 1.5$ m]

3. The brake drum of a single block brake of diameter 600 mm is rotating at 400 r.p.m. as shown in Fig. 8.6. The force required at the end of the lever to apply the brake is 300 N. If angle of contact is 90° and the co-efficient of friction between the drum and brake block is 0.3, find the braking torque. Take $b = 60$ mm ; $a = 375$ mm and $L = 825$ mm. **[Ans.** 69 Nm]

4. The brake drum of a single block brake of diameter 450 mm is rotating at 600 r.p.m. in the clockwise direction as shown in Fig. 8.7. The lever *OA* is pivoted at *O* and brake shoe is rigidly fixed to the lever. The co-efficient of friction for the brake lining is 0.35. The force applied at the end of the lever is 600 N. Find the braking torque.

5. A simple band brake is applied to a rotating drum of diameter 750 mm. The angle of lap of the band on the drum is 270. One end of the band is attached to the fulcrum and other and to a pin at a distance of 90 mm from the fulcrum. If the co-efficient of friction is 0.25 and a braking force of 54 N is applied at a distance of 900 mm from the fulcrum. Find the braking torque when the drum rotates in the (*i*) anti-clockwise direction and (*ii*) clockwise direction.
   [Ans. (*i*) 456.2 Nm ; (*ii*) 140.25 Nm]

6. Fig. 8.35 shows a simple band brake which is applied to a self carrying a flywheel of mass 200 kg and of radius of gyration 350 mm. The flywheel rotates at 300 r.p.m. The brake drum diameter is 280 mm and co-efficient of friction is 0.20. The angle of lap of band on the drum is 210°. If the braking torque is 63 Nm, find :
   (*i*) the force applied at the lever end,
   (*ii*) the number of turns of the flywheel before it comes to rest.
   (*iii*) the time taken by the flywheel to come to rest.
   [Ans. (*i*) 138.9 N ; (*ii*) 30.54 ; (*iii*) 6.1 seconds]

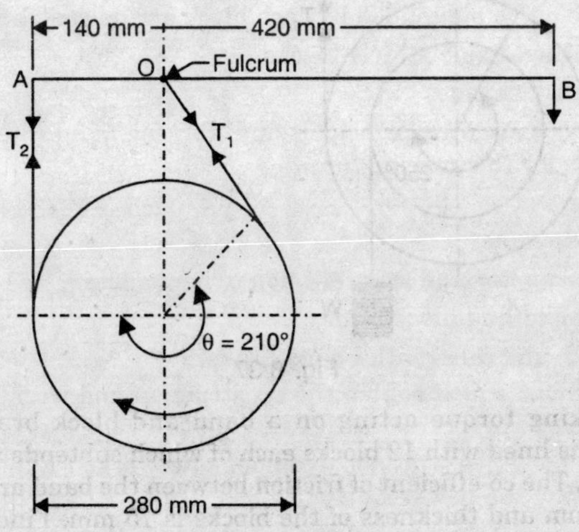

Fig. 8.35

7. Fig. 8.36 shows a differential band brake of drum diameter 600 mm. The two ends of the band are fixed to the pins on the opposite side of the fulcrum of the lever at a distance of 60 mm and 192 mm from the fulcrum as shown in Fig. 8.36. The brake is to sustain a torque of 450 Nm. If $\mu = 0.2$; angle of contact is 210° and length of lever from the fulcrum is 720 mm, then determine : (*i*) force required at the end of the lever for clockwise and anti-clockwise rotation of the drum (*ii*) value of *OB* for the brake to be self-locking for clockwise rotation.
   [Ans. (*i*) 128.9 N ; 653.9 N (*ii*) 125 mm]

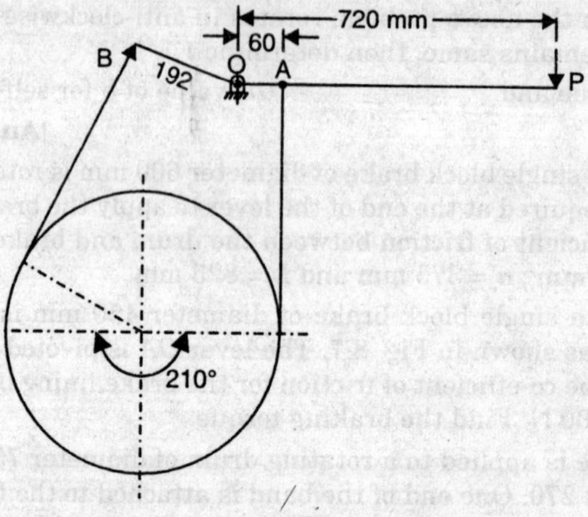

Fig. 8.36

8. **Fig. 8.37 shows a barrel and a differential band brake which are keyed to the same shaft. A rope is wound round a barrel which supports a load *W*. The brake drum diameter is 800 mm and diameter of barrel is 450 mm. The two ends of the bands are attached to pins on opposite sides of the fulcrum of the brake lever at a distance of 25 mm and 100 mm as shown in Fig. 8.37. The angle of contact of band brake is 250° and co-efficient of friction is 0.25. Determines the maximum load *W* which can be supported by the brake when a force of 750 N is applied to the lever at a distance of 3000 mm from the fulcrum.** [**Ans.** 309 kN]

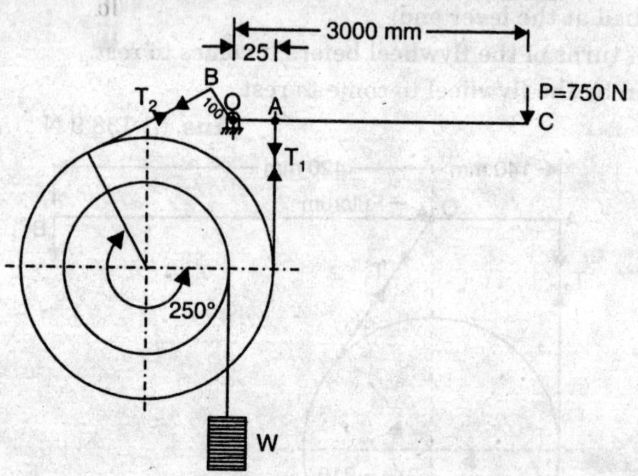

Fig. 8.37

9. **The maximum braking torque acting on a band and block brake is 8950 Nm (shown in Fig. 8.38). The band is lined with 12 blocks each of which subtends an angle of 15° at the centre of the rotating drum. The co-efficient of friction between the band and block is 0.4. The diameter of the drum is 850 mm and thickness of the blocks is 75 mm. Find the least force required at the end of the lever which is 500 mm long.** [**Ans.** 610.8 N]

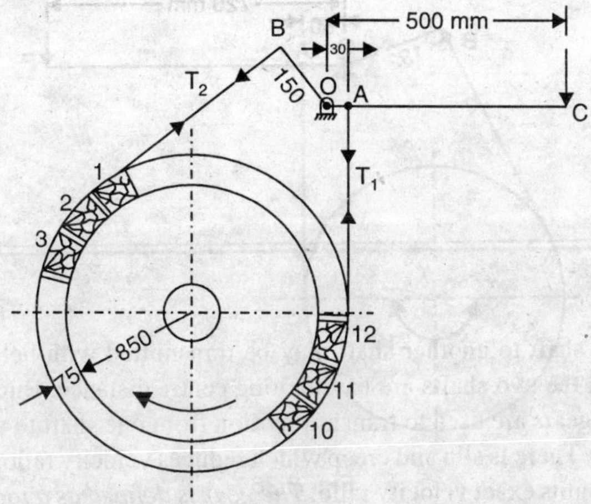

Fig. 8.38

10. Find : (*i*) the maximum braking torque, (*ii*) the angular retardation of the brake drum and (*iii*) the time taken by the system to come to rest from the rated speed of 300 r.p.m. when : a band and block brake having 12 blocks, each of which subtends an angle of 16° at the centre, is applied to a rotating drum of diameter 600 mm. The blocks are 75 mm thick. The drum and the flywheel momented on the same shaft have a mass of 1800 kg and have a combined radius of gyration on 600 mm. The two ends of the band are attached to the pins on the opposite sides of the brake fulcrum at distances of 40 mm and 150 mm from the fulcrum. A force of 250 N is applied at a distance of 900 mm from the fulcrum. Take co-efficient of friction between the blocks and drum as 0.3. [**Ans.** (*i*) 3703 Nm ; (*ii*) 5.71 rad/s^2 ; (*iii*) 5.5 s]

11. Calculate the braking torque applied by an internal expanding shoe brake on the rotating drum when it rotates (*i*) anti-clockwise and (*ii*) clockwise. The other data is given as :

Force on each shoe = 180 N ; Co-efficient of friction, $\mu = 0.3$; Internal radius of brake drum = 150 mm ; Width of brake lining, $b = 40$ mm ; Angles, $\theta_1 = 30°$ and $\theta_2 = 135°$; Distances : $L = 200$ mm and $OO_1 = 120$ mm. [**Ans.** (*i*) 36.54 Nm ; (*ii*) 36.54 Nm]

12. A car moving along a level road is having the following data :

Wheel base of the car = 2.8 m ; distance of C.G. from ground surface = 600 mm ; Perpendicular distance of C.G. from rear axle = 1.2 m ; Speed on level road = 50 km/h ; co-efficient of friction between the tyres and road = 0.6. Determine the minimum distance in which the car may be stopped when brakes are applied to the :

(*i*) rear wheels only, (*ii*) front wheels only and (*iii*) all the four wheels only.

[**Ans.** 32.4 m ; 33.26 m, 16.4 m]

13. A car moving on a rough inclined plane is having the following data :

Angle of inclination of the plane	= 10°
Wheel base of the car	= 1.8 m
Height of C.G. of the car above the inclined plane	= 0.9 m
Perpendicular distance of C.G. from rear axle	= 0.8 m
Speed of the car	= 36 km/h ; co-efficient of friction = 0.5 ;

Brakes are applied to all the four wheels. Determine :

(*i*) distance travelled by the car before coming to rest,

(*ii*) time taken in doing so if (*a*) car is moving up the plane and (*b*) car is moving down the plane.

[**Ans.** (*a*) (*i*) 7.657 m ; (*ii*) 1.53 s, (*b*) (*i*) 16 m ; (*ii*) 3.2 s]

9
Gears

9.1. Introduction

The motion from one shaft to another shaft may be transmitted with belts, ropes and chains. These method are mostly used when the two shafts are having long centre distance. But if the distance between the two shafts is very small, then gears are used to transmit motion from one shaft to another. In case of belts and ropes, the drive is not positive. There is slip and creep which reduces velocity ratio. But gear drive is a positive and smooth drive, which transmits exact velocity ratio. *The gear is defined as a toothed element which is used for transmitting rotary motion from one shaft to another.*

For small power transmission, the friction wheels as shown in Fig. 9.1 can be used. These wheels are mounted on the two shafts, having sufficient rough surfaces and pressing against each other.

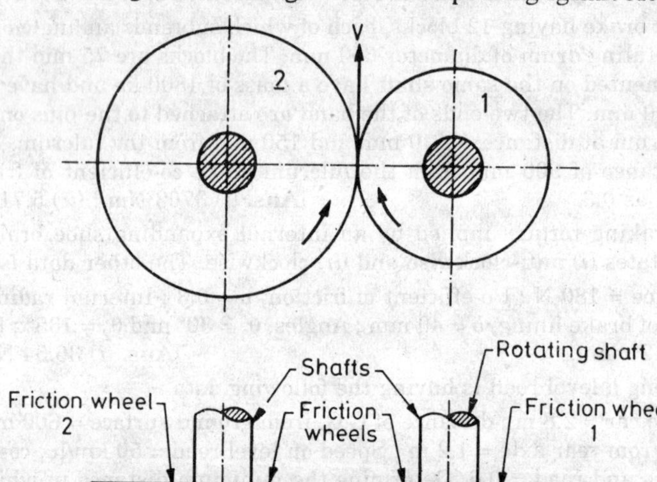

Fig. 9.1

The friction wheel 1 is keyed to the rotating shaft whereas the friction wheel 2 is keyed to the shaft which is to be rotated. When the friction wheel 1 rotates, it will rotate the friction wheel 2 in the opposite direction as shown in Fig. 9.1. There will be no slip between the two wheels for small power transmission. Hence the same motion can be transmitted easily as explained below :

For no slip of the two surfaces (*i.e.* no relative motion between the two surfaces of the friction wheels), their tangential velocities at the contact surfaces should be same.

i.e. $V_1 = V_2$

or $\omega_1 r_1 = \omega_2 r_2$ $(\because\ V_1 = \omega_1 r_1 \text{ and } V_2 = \omega_2 r_2)$

or $$\frac{2\pi N_1}{60} \times r_1 = \frac{2\pi N_2}{60} \times r_2 \qquad \left(\because\ \omega = \frac{2\pi N}{60}\right)$$

or $$\frac{N_1}{N_2} = \frac{r_2}{r_1}\left(\text{or} = \frac{d_2}{d_1}\right) \qquad ...(9.1)$$

where N_1 = Speed of friction wheel 1 in r.p.m.,

N_2 = Speed of friction wheel 2 in r.p.m.,

r_1 = Radius of wheel 1, and

r_2 = Radius of wheel 2.

From equation (9.1), it is clear that if the two wheels are rotating without slip then the speeds of the two wheels will be inversely proportional to their radii.

In order to prevent the slip between the two surfaces, a number of projections (known as teeth) are provided on the periphery of the wheel 1, which will fit into the corresponding recesses on the periphery of wheel 2 as shown in Fig. 9.2. Hence the teeth and corresponding recesses will mesh with each other and slip between them will be prevented.

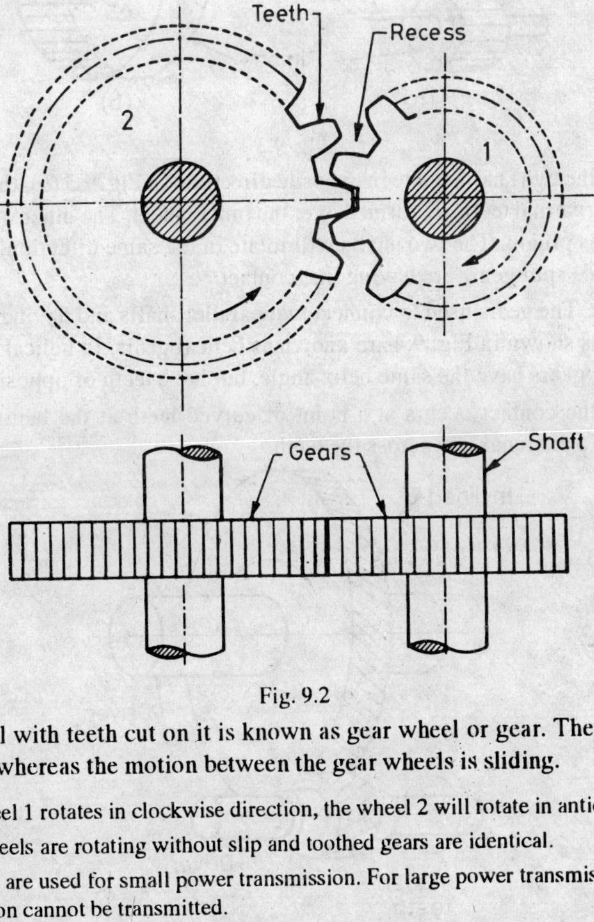

Fig. 9.2

The friction wheel with teeth cut on it is known as gear wheel or gear. The motion between the two friction wheels is rolling whereas the motion between the gear wheels is sliding.

Note : (*i*) If the wheel 1 rotates in clockwise direction, the wheel 2 will rotate in anticlock-wise direction.

(*ii*) The friction wheels are rotating without slip and toothed gears are identical.

(*iii*) Friction wheels are used for small power transmission. For large power transmission by friction wheels, slip occurs. Hence definite motion cannot be transmitted.

9.2. Classification of Gears

The following are the important classification of gears :

1. **Classification based on the position of axes of the shafts.** The axes of the two shafts between which motion is to be transmitted may be

(*a*) Parallel, (*b*) intersecting, and (*c*) non-parallel and non-intersecting.

(*a*) **Parallel Shafts**

The following are the main types of gears to join parallel shafts :

(*i*) Spur gears (*ii*) Helical gears

(*iii*) Double-helical gears.

(*i*) **Spur Gears.** The gears used to connect two parallel shafts and having straight teeth parallel to the axes of the wheel as shown in Fig. 9.3 are known as spur gears. In Fig. 9.3 (*a*) the gears have external teeth

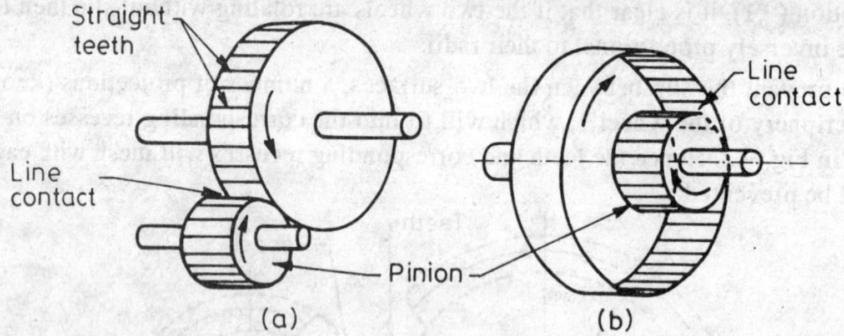

Fig. 9.3

on the outer surfaces and the two shafts rotate in opposite direction. In Fig. 9.3 (*b*), the internal teeth are formed over the outer wheel and external teeth are formed over the inner wheel. The inner wheel having external gears (smaller gear) is known as pinion. The two shafts will rotate in the same direction. In spur gears, the contact occurs across a line. Hence spur gears are having line contact.

(*ii*) *Helical Gears.* The gears used to connect two parallel shafts and having teeth inclined (or curved) to the axes of the shafts as shown in Fig. 9.4 are known as helical gears. In helical gears each teeth is helical in shape. The two mating gears have the same helix angle, but have teeth of opposite hands.

In helical gears, the contact occurs at a point of curved teeth at the beginning of engagement and afterwards extends along a diagonal line across the teeth.

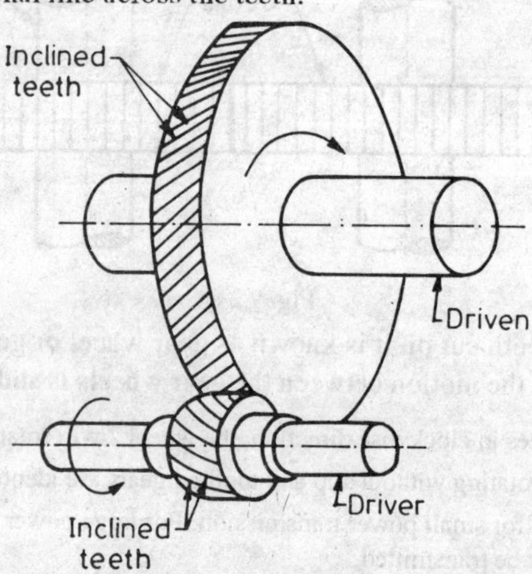

Fig. 9.4

(*iii*) **Double Helical Gears.** A pair of helical gears secured together, one having a right-hand helix and the other a left-hand helix is known as double helical gears as shown in Fig. 9.5. This pair is mounted on one shaft and the similar other pair (having teeth of opposite hand) is mounted on other parallel shaft which is to be connected.

The double helical gear is known as herringbone gear if the left and the right inclinations of the double-helical gears meet at a common apex and there is no groove in between as shown in Fig. 9.5.

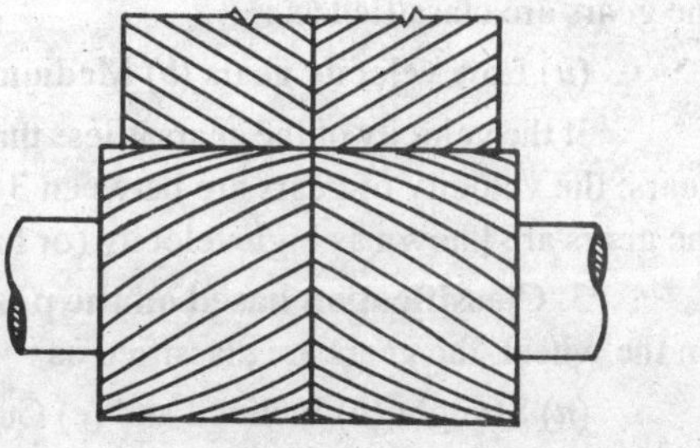

Fig. 9.5

(*b*) **Intersecting Shaft**

The gears used to connect two intersecting shafts are known as *bevel gears*. If the teeth on the gears are straight radial to the point of intersection of shaft axes then gears are known as straight bevel. But if teeth are inclined then gears are known as helical bevel (or spiral bevel) as shown in Fig. 9.6 and 9.7.

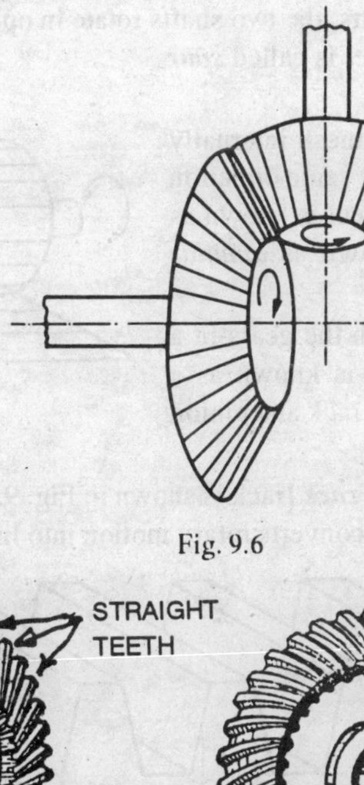

Fig. 9.6

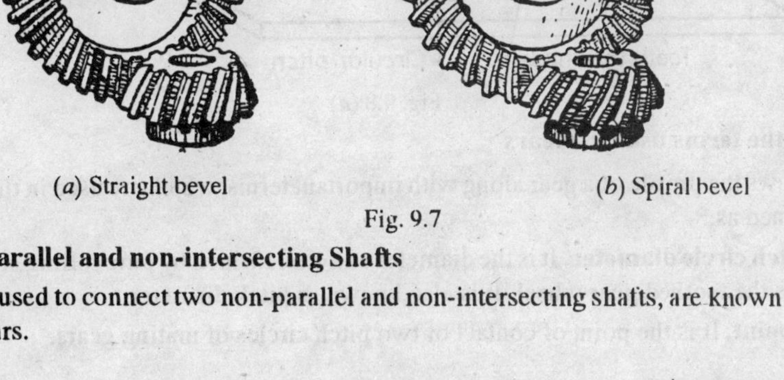

(*a*) Straight bevel (*b*) Spiral bevel

Fig. 9.7

(*c*) **Non-parallel and non-intersecting Shafts**

The gears used to connect two non-parallel and non-intersecting shafts, are known as skew bevel gears or skew spiral gears.

2. **Classification based on the peripheral velocity of the gears.** According to the peripheral velocity, the gears are classified as :

(*a*) Low velocity gears (*b*) Medium velocity gears and (*c*) High velocity gears.

If the velocity of the gears is less than 2 m/s, they are known as low velocity gears. For medium velocity gears, the velocity of gears are between 3 m/s and 15 m/s. But if the velocity of gears are more than 15 m/s, the gears are known as high velocity (or high speed) gears.

3. **Classification based on the position of teeth on the wheel.** According to the position of the teeth on the wheel, the gears are classified as :

(*a*) Straight (*b*) Inclined and (*c*) Curved.

The spur gears have straight teeth, helical gears have inclined teeth (which are inclined to the wheel rim surface) and spiral gears have curved teeth over the rim surface.

4. **Classification based on type of gearing.** According to this type, gears are classified as :

(*a*) External gear (*b*) Internal gear and (*c*) Rack and pinion.

If the gears of the two shafts mesh externally with each other, then the gears are known, external gears as shown in Fig. 9.3 (*a*). For external gears, the two shafts rotate in opposite direction. The smaller wheel is known as *pinion* whereas the larger wheel is called *spur wheel.*

For internal gears, the two gears mesh internally with each other and the shafts rotate in the same direction as shown in Fig. 9.3 (*b*). The smaller wheel is known as *pinion* whereas the larger wheel is known as *annular wheel.*

If the gear of a shaft meshes with the gears in a straight line (a wheel of infinite radius is known as a straight line), then the gears are known as rack and pinion as shown in Fig. 9.8.

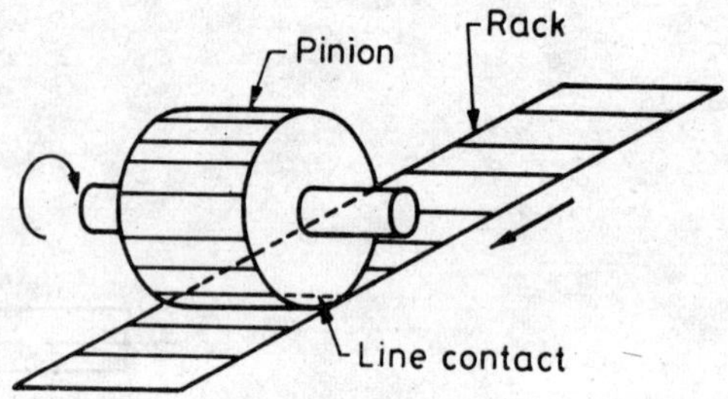

Fig. 9.8

The straight line gear is known as *rack* [rack is shown in Fig. 9.8 (*a*)] and the circular wheel is called pinion. The rack and pinion combination converts rotary motion into linear motion or vice-versa.

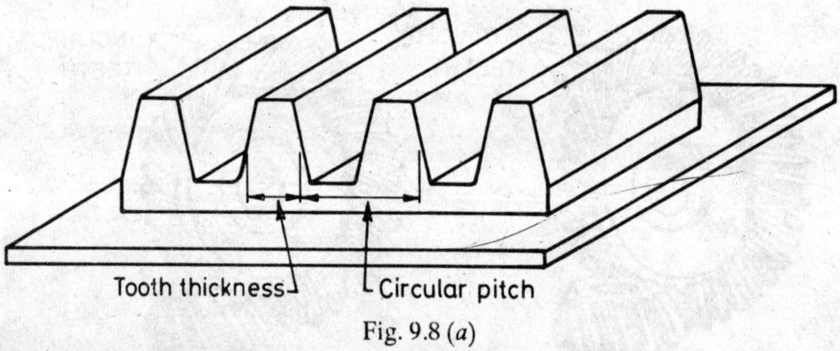

Fig. 9.8 (*a*)

9.3. Definition of the terms used in Gears

Fig. 9.9 shows the profile of a gear along with important terms, which are used in the study of the gears. The terms are defined as :

(*i*) **The pitch circle diameter.** It is the diameter of a circle which by pure rolling action would produce the same motion as the toothed gear wheel. It is also known as pitch diameter.

(*ii*) **Pitch point.** It is the point of contact of two pitch circles of mating gears.

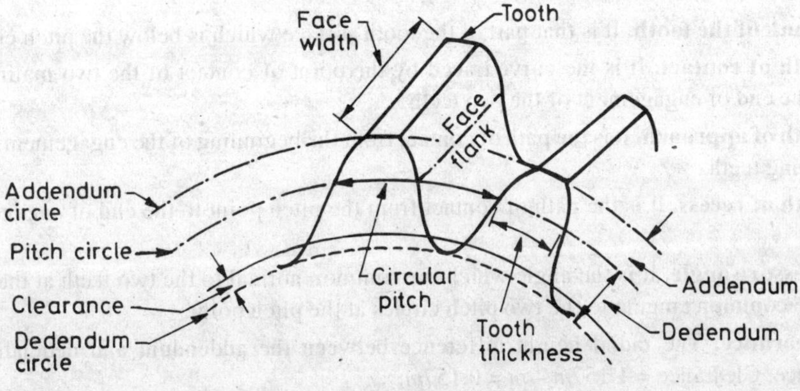

Fig. 9.9

(*iii*) **Circular pitch (p_c).** It is the distance measured along the circumference of the pitch circle from a point on one tooth to a corresponding point on the adjacent tooth. It will be equal to the pitch circle circumference divided by number of teeth on the wheel. It is denoted by p_c. Hence

$$\text{Circular pitch} = \frac{\pi D}{T} \quad \text{or} \quad p_c = \frac{\pi D}{T} \qquad ...(9.2)$$

where D = diameter of pitch circle

T = number of teeth

p_c = circular pitch.

(*iv*) **The diametral pitch (p_d).** It is equal to the number of teeth per unit length of pitch circle diameter. It is denoted by p_d.

$$\therefore \quad p_d = \frac{T}{D} \qquad ...(9.3)$$

If we multiply equations (9.2) and (9.3), we get

$$p_c \times p_d = \pi \qquad ...(9.4)$$

(*v*) **Module (m).** It is defined as the length of the pitch circle diameter per tooth. It is denoted by *m*. Hence

$$m = \frac{D}{T} \qquad ...(9.5)$$

m is generally expressed in millimeter. Module is the reciprocal of the diametral pitch.

Equation (9.2) can be written as :

$$p_c = \frac{\pi \times D}{T} = \pi \times m \qquad ...(9.5\ a)$$

(*vi*) **The addendum.** It is the radial distance of the tooth above pitch circle. Its value is generally one module. It is denoted by '*a*'.

(*vii*) **The dedendum.** It is the radial distance of the tooth below the pitch circle. Its value is generally 1.157 module or $\left(1 + \frac{\pi}{20}\right)$ module.

(*viii*) **Addendum circle.** It is the circle which passes through the top of the teeth. Diameter of addendum circle = *p.c.d.* + 2*m*.

(*ix*) **Dedendum circle.** It is the circle which passes through the bottom of the teeth. Dia. of dedendum circle = *p.c.d.* – 2 × 1.157*m*.

(*x*) **Face of the tooth.** It is that part of tooth surface which is above the pitch circle.

(*xi*) **Flank of the tooth.** It is that part of the tooth surface which is below the pitch circle.

(*xii*) **Path of contact.** It is the curve traced by the point of contact of the two mating teeth from the beginning to the end of engagement of the two teeth.

(*xiii*) **Path of approach.** It is the path of contact from the beginning of the engagement to the pitch point of the two mating teeth.

(*xiv*) **Path of recess.** It is the path of contact from the pitch point to the end of engagement of the two mating gears.

(*xv*) **Pressure angle.** It is the angle which the common normal to the two teeth at the point of contact makes with the common tangent to the two pitch circles at the pitch point.

(*xvi*) **Clearance.** The radial height difference between the addendum and dedendum is known as clearance. Hence, Clearance = $1.157m - m = 0.157m$.

(*xvii*) **Profile.** The curve forming face and flank is known as profile.

(*xviii*) **Pinion.** It is the smaller and usually the driving gear of a pair of mated gear.

(*xix*) **Rack.** A gear wheel of infinite diameter is known as rack as shown in Fig. 9.8 (*a*).

(*xx*) **Gear Ratio (G).** The ratio of number of teeth on the gear to that on the pinion is known as gear ratio. It is denoted by *G*. Hence

$$G = \frac{T}{t} \qquad \text{...(9.6)}$$

where T = number of teeth on the gear

t = number of teeth on the pinion and

G = gear ratio.

(*xxi*) **Velocity Ratio (V.R.).** The ratio of the angular velocity of the follower to the angular velocity of the driving gear, is known as velocity ratio. It is denoted by V.R. Hence velocity ratio,

$$V.R. = \frac{\text{Angular velocity of follower}}{\text{Angular velocity of driver}}$$

$$= \frac{\omega_2}{\omega_1}$$

$$= \frac{N_2}{N_1} \qquad (\omega_2 = 2\pi N_2 \text{ and } \omega_1 = 2\pi N_1)$$

$$= \frac{d_1}{d_2} \qquad \left(\because \ \pi d_1 N_1 = \pi d_2 N_2 \quad \therefore \ \frac{d_1}{d_2} = \frac{N_2}{N_1}\right)$$

$$= \frac{T_1}{T_2} \qquad \left(\because \ p_c = \frac{\pi d_1}{T_1} = \frac{\pi d_2}{T_2} \quad \therefore \ \frac{d_1}{d_2} = \frac{T_1}{T_2}\right) \qquad \text{...(9.7)}$$

(*xxii*) **Pressure line (or line of Action).** The common normal at the point of contact of the mating gears is known as pressure line (or line of action). This is also the line from the pitch point to the point of contact of the two gears. Actually the force is transmitted from the driving tooth to the driven tooth along this line.

Problem 9.1. *Determine the number of teeth and speed of the driver if the driven gear has 60 teeth of 8 mm module and rotates at 240 r.p.m. The two spur gears have a velocity ratio of* $\frac{1}{4}$. *Also calculate the pitch line velocities.*

Sol. Given :

No. of teeth on driven gear, $T_2 = 60$

Speed of driven gear, $N_2 = 240$ r.p.m.

Module, $m = 8$ mm

Velocity ratio, $V.R. = \frac{1}{4}$

Let T_1 = no. of teeth on driver gear

N_1 = speed of driver gear

Using the equation (9.7), we get

$$V.R. = \frac{N_2}{N_1} = \frac{T_1}{T_2}$$

$$\therefore \quad \frac{1}{4} = \frac{N_2}{N_1}$$

$$\therefore \quad N_1 = 4 \times N_2 = 4 \times 240 \quad (\because N_2 = 240 \text{ r.p.m.})$$

$$= \mathbf{960\ r.p.m.} \quad \textbf{Ans.}$$

Also $$V.R. = \frac{T_1}{T_2}$$

$$\therefore \quad T_1 = T_2 \times V.R. = 60 \times \frac{1}{4} = \mathbf{15.} \quad \textbf{Ans.}$$

Pitch line Velocity (V_p) is given by

$$V_p{}^* = \omega_1 r_1 \quad \text{or} \quad \omega_2 r_2$$

$$= \frac{2\pi N_1}{60} \times \frac{d_1}{2} \quad \left(\because \omega_1 = \frac{2\pi N_1}{60} \text{ and } r_1 = \frac{d_1}{2}\right)$$

$$= \frac{2\pi N_1}{60} \times \frac{mT_1}{2} \quad (\because d_1 = m \times T_1)$$

$$= \frac{2\pi \times 960}{60} \times \frac{8 \times 15}{2} \text{ mm/s}$$

$$= 6031.8 \text{ mm/s} = \mathbf{6.0318\ m/s.} \quad \textbf{Ans.}$$

Problem 9.2. *A spur gear has a module of 2 mm and its pitch line velocity is 0.6283 m/s. If the number of teeth of this spur gear is 30, find the speed of the gear. Also determine its circular pitch.*

Sol. Given :

Module, $m = 2$ mm

Pitch line velocity, $V_p = 0.6283$ m/s = 628.3 mm/s

No. of teeth, $T = 30$

Let N = speed of the gear

p_c = circular pitch

We know from equation (9.5 *a*) that,

$$p_c = \pi \times m = \pi \times 2 = \mathbf{6.28\ mm.} \quad \textbf{Ans.}$$

$^*V_p$ is also $= \omega_2 r_2 = \frac{2\pi N_2}{60} \times \frac{d_2}{2} = \frac{2\pi N_2}{60} \times \frac{mT_2}{2} = \frac{2\pi \times 240}{60} \times \frac{8 \times 60}{2} = 6031.8$ mm/s = 6.0318 m/s.

Pitch line velocity (V_p) is given by

$$V_p = \omega \times r$$

$$= \frac{2\pi N}{60} \times \frac{D}{2}$$

$$= \frac{2\pi N}{60} \times \frac{mT}{2} \qquad (\because \; D = m \times T)$$

or

$$628.3 = \frac{2\pi N}{60} \times \frac{2 \times 30}{2}$$

$\therefore$

$$N = \frac{628.3 \times 60 \times 2}{2\pi \times 2 \times 30}$$

$$= 199.99 \text{ r.p.m.} = \textbf{200 r.p.m.} \quad \textbf{Ans.}$$

9.4. Law of Gearing or Condition for Constant Velocity Ratio of Gear Wheels

Law of gearing states that the common normal to the two surfaces at the point of contact intersects the line joining the centres of rotation of the two surfaces at a fixed point, which divides the centre distance inversely as the ratio of angular velocities.

Let the two curved bodies 1 and 2 are rotating about the centres A and B as shown in Fig. 9.10. The two bodies are in contact at point C. The body 1 is rotating clockwise with angular velocity ω_1 and the body 2 is rotating anti-clockwise with angular velocity ω_2. Let t-t is the common tangent to the curves at point C and n-n is the common normal to the two surfaces at point C.

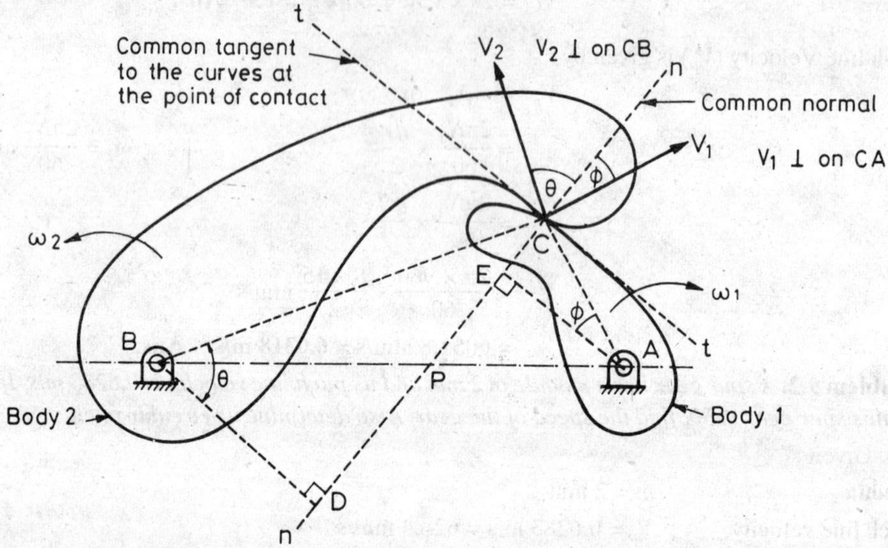

Fig. 9.10

Let V_1 = linear velocity of point C when point C is assumed to be on the surface of the body 1. This velocity will be $\perp$ to the line AC and will be equal to $\omega_1 \times AC$.

V_2 = linear velocity of point C when point C is assumed to be on the surface of the body 2. This velocity will be $\perp$ to the line BC and will be equal to $\omega_2 \times BC$.

Join the two centres A and B which cuts the common normal at point P.

Let θ = Angle made by V_2 with common normal n-n.

ϕ = Angle made by V_1 with normal n-n.

From B draw $BD \perp$ on n-n and from A draw $AE \perp$ on n-n.

As V_2 is perpendicular to BC hence $\angle BCV_2 = 90°$. Now

$$\angle DCB = 180 - \angle BCV_2 - \theta = 180 - 90 - \theta = 90 - \theta.$$

Similarly as V_1 is perpendicular to AC, hence $\angle ACV_1 = 90°$. Now

$$\angle ACE = 180 - \phi - \angle ACV_1 = 180 - \phi - 90 = 90 - \phi.$$

In ΔBDC, $\quad \angle DCB = 90 - \theta \quad \therefore \quad \angle DBC = \theta$

in ΔAEC, $\quad \angle ACE = 90 - \phi \quad \therefore \quad \angle CAE = \phi$

If the two surfaces are to remain in contact, one surface may slide relative to the other along the common tangent t-t. But the relative motion between the surfaces along the common normal n-n must be zero to avoid separation of the two surfaces or the penetration of the two surfaces into each other.

Component of V_1 along normal n-n = $V_1 \cos \phi$

Component of V_2 along normal n-n = $V_2 \cos \theta$

$\therefore$ Relative motion along normal n-n = $V_1 \cos \phi - V_2 \cos \theta$

For proper contact,

Relative motion along normal = zero

or $$V_1 \cos \phi - V_2 \cos \theta = 0$$

or $$V_1 \cos \phi = V_2 \cos \theta$$

or $$(\omega_1 \times AC) \times \cos \phi = (\omega_2 \times BC) \times \cos \theta \qquad (\because V_1 = \omega_1 \times AC \text{ and } V_2 = \omega_2 \times BC)$$

or $$(\omega_1 \times AC) \times \frac{AE}{AC} = (\omega_2 \times BC) \times \frac{BD}{BC}$$

$$\left(\because \text{ In } \Delta ACE, \cos \phi = \frac{AE}{AC} \text{ and in } \Delta BDC, \cos \theta = \frac{BD}{BC}\right)$$

or $$\omega_1 \times AE = \omega_2 \times BD$$

or $$\frac{\omega_1}{\omega_2} = \frac{BD}{AE}$$

$$= \frac{BP}{AP} \qquad \left(\because \Delta\text{s } BDP \text{ and } AEP \text{ are similar. Hence } \frac{BD}{AE} = \frac{BP}{AP}\right)$$

$\therefore$ $$\frac{\omega_1}{\omega_2} = \frac{BP}{AP} \qquad ...(9.8)$$

The above equation shows that the common normal to the two surfaces at the point of contact divides the line joining the centres of rotation in the inverse ratio of the angular velocities. But this ratio of angular velocities must be constant for all positions of the wheel. This will be, if the point P is a fixed point (*i.e.* pitch point).

Thus for constant angular velocity ratio of the two surfaces, the common normal at the point of contact must pass through the pitch point (fixed point) on the line joining the centres of rotation. *This statement is the law of gearing.*

The Δs BDP and AEP are similar

$\therefore$ $$\frac{BP}{AP} = \frac{DP}{PE}$$

Substituting the value of $\frac{BP}{AP}$ in equation (9.8), we get

$$\frac{\omega_1}{\omega_2} = \frac{DP}{PE}$$

or $$\omega_1 \times PE = \omega_2 \times DP \qquad ...(9.9)$$

9.5. Velocity of Sliding

If the two surfaces are to remain in contact, one surface may slide relative to the other along the common tangent t-t of Fig. 9.10.

The Velocity of sliding is the velocity of one surface relative to the other surface along the common tangent at the point of contact. Refer to Fig. 9.10.

Component of V_1 along t-t = $V_1 \sin \phi$

Component of V_2 along t-t = $-V_2 \sin \theta$ (–ve sign is due to opposite direction)

$\therefore$ Velocity of sliding = Relative velocity between two surface along t-t

$$= V_1 \sin \phi - (-V_2 \sin \theta)$$

$$= V_1 \sin \phi + V_2 \sin \theta$$

$$= (\omega_1 \times AC) \sin \phi + (\omega_2 \times BC) \times \sin \theta$$

$$(\because \; V_1 = \omega_1 \times AC \text{ and } V_2 = \omega_2 \times BC)$$

$$= (\omega_1 \times AC) \times \frac{EC}{AC} + (\omega_2 \times BC) \times \frac{DC}{BC}$$

$$\left(\because \; \text{In } \Delta ACE, \sin \phi = \frac{EC}{AC} \text{ and in } \Delta BDC, \sin \theta = \frac{DC}{BC}\right)$$

$$= \omega_1 \times EC + \omega_2 \times DC$$

$$= \omega_1 \times (PC - PE) + \omega_2 \times (DP + PC)$$

$$(\because \; EC = PC - PE \text{ and } DC = DP + PC)$$

$$= \omega_1 \times PC - \omega_1 \times PE + \omega_2 \times DP + \omega_2 \times PC$$

$$= \omega_1 \times PC + \omega_2 \times PC \qquad [\because \text{ From equation (9.9), } \omega_1 \times PE = \omega_2 \times DP]$$

$$= (\omega_1 + \omega_2) \times PC \qquad ...(9.10)$$

The above equation shows that the velocity of sliding is equal to the product of sum of angular velocities and the distance from the point of contact to the point of intersection of the common normal and the line joining the centres of rotation (*i.e.* pitch point).

9.6. Forms of Teeth

When two gears are in mesh, any arbitrary shape of the tooth can be choosen for the profile of teeth of one of the two gears, and the profile for the other may be obtained by applying the law of gearing. Such teeth are called conjugate teeth. Such conjugate gears with random profile, are difficult to manufacture, high in cost of production and difficult to replace on wearing. Therefore, conjugate teeth are not in normal use. In actual practice following are the two types of teeth commonly used which also satisfy the law of gearing :

1. Cycloidal profile teeth and
2. Involute profile teeth.

In the cycloidal system, the profile of a tooth consists of parts of an epicycloid and hypocycloid above and below the base circle. But in case of the involute system the part of the profile of a tooth above the base circle is an involute. The tooth profile for a rack meshing with a pinion is a part of cycloid.

9.7. Cycloidal Profile Teeth

Let us first define a cycloid, an epicycloid and a hypo-cycloid. A cycloid is the curve traced by a point on the circumference of a circle which rolls, without slipping, along a fixed straight line.

An epicycloid is the curve traced by a point on the circumference of a circle which rolls, without slipping, on the outside of another circle.

A hypo-cycloid is the curve traced by a point on the circumference of a circle which rolls, without slipping, on the inside of another circle.

9.7.1. Formation of a Cycloid Rack Tooth Profile

Fig. 9.11 (*a*) shows the fixed line or pitch line of a rack. There are two circles *M* and *N*, one above the pitch line and other below the pitch line. The curve generated by a point *P* on the circle *M*, when the circle rolls to the right without slipping above the pitch line, is the profile of the face of the cycloidal tooth. This

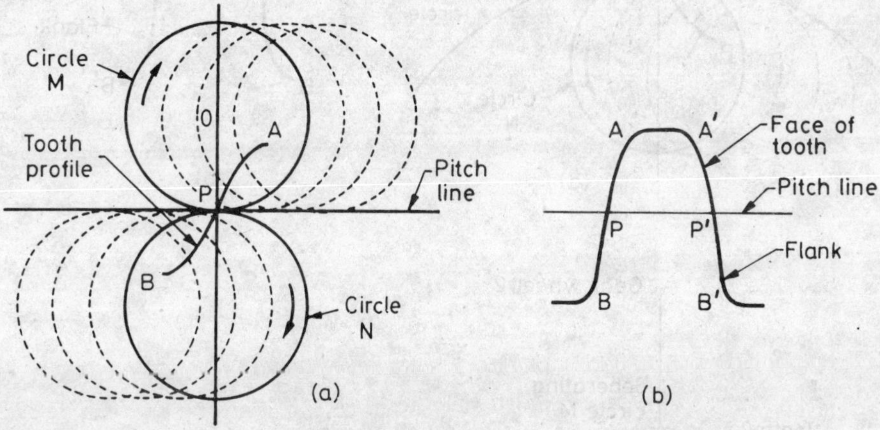

Fig. 9.11

profile is represented by *PA*. And the curve generated by the point *P* on the circle *N*, when the circle *N* rolls to the left without slipping below the pitch line, is the profile of the flank of the cycloidal tooth profile. This profile is represented by *PB*. The profile *BPA* is one side of the cycloidal rack tooth. Similarly, the two curves *P'A'* and *P'B'* forming the opposite side of the tooth profile can be traced by point *P'* when the circles *M* and *N* rolls in the opposite direction. The complete tooth profile is shown in Fig. 9.11 (*b*). A small portion of this curve near the pitch line is used for the face of the tooth. Similarly the points can be obtained on the flank.

9.7.2. Formation of a Cycloidal Teeth of a Gear

The construction of a tooth profile of a gear is shown in Fig. 9.12. The rolling circle *M*, rolls on the outside of the pitch circle to the right without slipping and the point *P* on the circumference of the circle traces an epicycloid. A small portion of this curve near the pitch circle say *PA* is the face portion of the tooth profile for cycloidal tooth. The rolling circle *N* rolls without slipping on the pitch circle to the left on the inside. The point *P* on the circumference of this circle traces a hypocycloid. A small portion of this curve near the pitch circle say *PB* is the flank portion of the tooth profile. The profile *BPA* is one side of the cycloid tooth. The other side *B'P'A'* is obtained as explained earlier. The complete tooth profile is shown in Fig. 9.12 (*b*).

The face of a tooth is obtained by a circle rolling outside the pitch circle whereas the flank of the tooth is obtained by a circle rolling inside the pitch circle.

The faces of the two meshing teeth should be generated by the rolling circle of the same diameter. But it is not necessary that the flank and face of the tooth are to be generated by the same rolling circle. The flanks and faces of the gear teeth in a set are generally generated by the same rolling circle for the interchangeability.

Fig. 9.12 (*c*) shows the construction of two mating cycloidal teeth. The circle *N* is rolling inside the pitch circle of wheel 1. The circle *N* is also rolling outside the pitch circle of wheel 2. Similarly the circle *M* is rolling inside the pitch circle of wheel 2 and outside the pitch circle of wheel 1. A point on the circle *N* will trace the flank of the tooth T_1 when circle *N* rolls without slipping on the inside of the pitch circle of wheel 1.

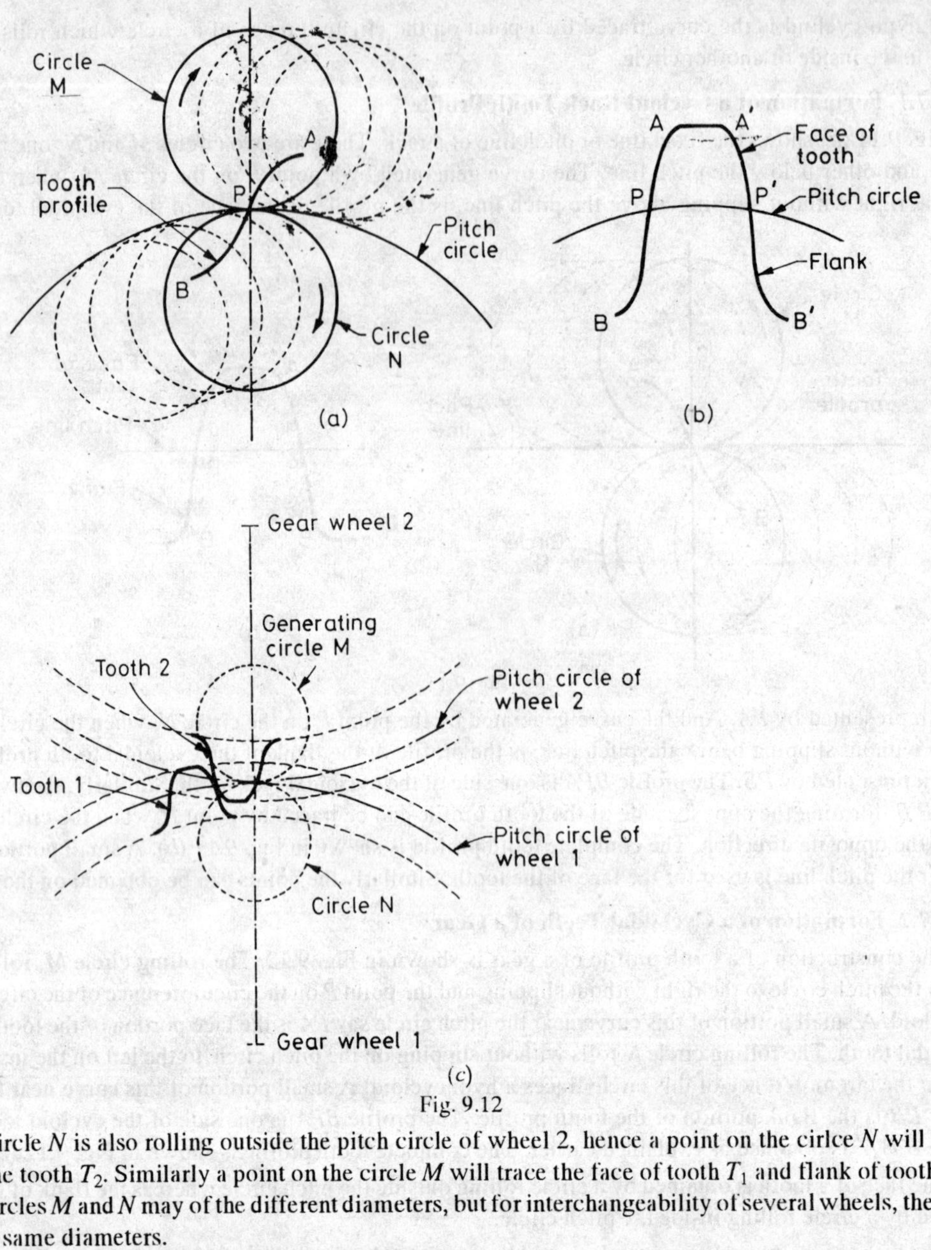

(c)

Fig. 9.12

As this circle N is also rolling outside the pitch circle of wheel 2, hence a point on the cirlce N will trace the face of the tooth T_2. Similarly a point on the circle M will trace the face of tooth T_1 and flank of tooth T_2. The rolling circles M and N may of the different diameters, but for interchangeability of several wheels, they should be of the same diameters.

9.7.3. Involute Profile Teeth

The curve traced by the end of a thread which is unwound from a stationary cylinder, is known as the involute of a circle. The thread should be kept taut. Fig. 9.13 shows the construction of an involute. Let the thread is wrapped on a stationary cylinder (or on a stationary circle as shown in Fig. 9.13. This circle is known as base circle for toothed wheels).

Let the starting point of the involute is A (*i.e.* the point on the circumference of the circle from which the end of the thread starts unwrapped). The circle is fixed. When the thread is at point 1 on the circumference of the base circle, the end of the taut thread will be at A_1. Similarly when the thread is at point 2 on the circumference of the circle, the end of the taut thread will be at A_2. Similarly the point A_3 can be obtained. By joining the points A, A_1, A_2, A_3 etc., the involute curve is obtained.

From Fig. 9.13, it is clear that when thread is taut, the length of thread $1A_1$ is tangent at point 1 on base circle. Similarly the length of thread $2A_2$ is tangent at point 2 on the base circle.

From Fig. 9.13, it is also clear that the line $A_1 - 1$ is normal at point A_1 which is on the involute curve. Similarly the line $A_2 - 2$ is normal to involute curve at point A_2. But the line $2A_2$ is tangent at point 2 on the base circle. Hence the normal at any point on the involute curve is the tangent on the base circle.

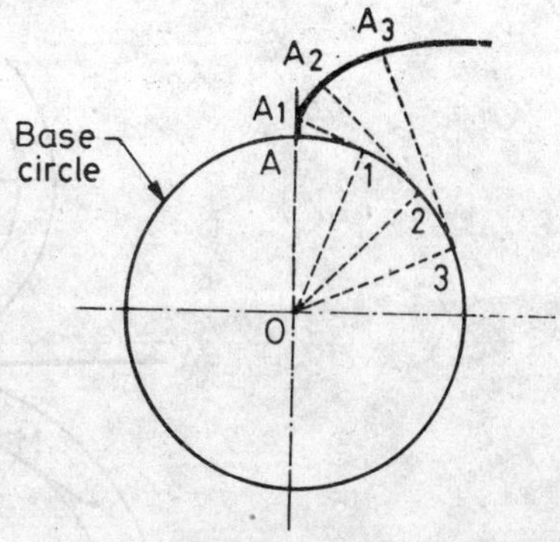

Fig. 9.13

The procedure of obtaining the involute curve (*i.e.* curve $A - A_1 - A_2 - A_3$ etc.) is given below :

The base circle is divided into equal numbers of parts such as $A1$, 12, 23 etc. The tangents are drawn at points 1, 2, 3 etc. To get point A_1, take length $1 - A_1$ equal to arc $A - 1$. Similarly to get point A_2, take length $2 - A_2$ equal to arc $A - 2$. Similarly point A_3 can be obtained. Then the involute curve will be obtained by joining points A, A_1, A_2, A_3 etc.

9.7.4. Involute Gear Teeth

Fig. 9.14 shows the two base circles of two wheels with O_1 and O_2 as fixed centres.

Let R_1 = Radius of base circle of wheel 1,

R_2 = Radius of base circle of wheel 2,

AB = Common tangent to the pitch circles

MN = Common tangent to the two base circles

ϕ = Angle made by line MN with line AB (*i.e.* angle made by common tangent of base circles with common tangent of pitch circles) *i.e.* pressure angle.

The angle ϕ is known as pressure angle or angle of obliquity.

In Article 9.7.3, it has been proved that tangent to the base circle is a normal to the involute. Hence a common tangent to the base circles is a common normal to the two involutes of the tooth profiles of the gears in mesh. Hence the line MN, which is the common tangent to the two base circles, is the common normal to the two involutes of the tooth profile of the two gears in mesh. But the common normal at the point of contact of two mating gears is known as pressure line. Hence the line MN is a pressure line, which is a straight line. Hence the path of contact is a straight line which means that the pressure angle for involute gears is always constant. The pressure angle varies from $14\frac{1}{2}°$ to $22\frac{1}{2}°$. The standard value of pressure angle is 20°.

Refer to Fig. 9.14, in triangle O_1MP,

$$\cos\phi = \frac{O_1M}{O_1P} = \frac{R_1}{O_1P} \qquad (\because O_1M = R_1)$$

$$\therefore \qquad O_1P = \frac{R_1}{\cos\phi}$$

But O_1P = Pitch circle radius of the wheel 1.

Similarly in triangle O_2NP,

$$\cos\phi = \frac{O_2N}{O_2P} = \frac{R_2}{O_2P} \qquad (\because O_2N = R_2)$$

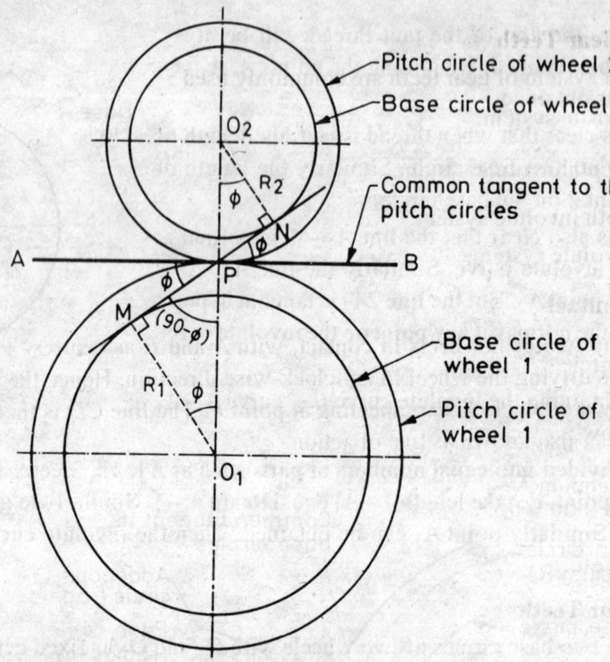

Fig. 9.14

$$\therefore \quad O_2P = \frac{R_2}{\cos\phi}$$

= Pitch circle radius of the wheel 2.

The centre distance between two gears

$$= O_1P + O_2P$$

$$= \frac{R_1}{\cos\phi} + \frac{R_2}{\cos\phi} = \frac{(R_1 + R_2)}{\cos\phi} \quad ...(9.10\ a)$$

The centre distance between two gears depends upon the base circle radii and the pressure angle.

9.7.5. Important Points for Involute Profile Toothed Gears in Mesh

The following are the important points for the involute profile toothed gears when they are in mesh :

1. The pressure angle, from the start of the engagement of the teeth to the end of the engagement, remains constant in involute gears.

2. When two involutes are in mesh, the angular velocity ratio is inversely proportional to the size of the base circles.

3. The common tangent to the two base circles of the two involutes, is the line of action and also the path of contact between the involutes.

4. The shape of the involute profile depends only on the dimensions of the base circle.

5. The base circle diameters and pitch diameters of two mating involutes are directly proportional.

6. The radii of the pitch circles of the meshing involutes are determined from the intersection of the common tangent to the two base circles.

7. No involute has a pitch circle or a pressure angle until it is brought in contact with other mating involute.

9.7.6. System of Gear Teeth

The following four system of gear teeth are commonly used :

1. The $14\frac{1}{2}°$ composite system.
2. The $14\frac{1}{2}°$ full depth involute system.
3. The 20° full depth involute system.
4. The 20° stub involute system.

9.8. Length of Path of Contact

Fig. 9.15 shows the two gear wheels in contact, with A and B as centres. The pinion is rotating in clock-wise direction and is driving the wheel in anticlock-wise direction. Hence the pinion is the driver and the wheel is driven. The two pitch circles are meeting at point P. The line CD is the common tangent to the two base circles. This line is also known as line of action.

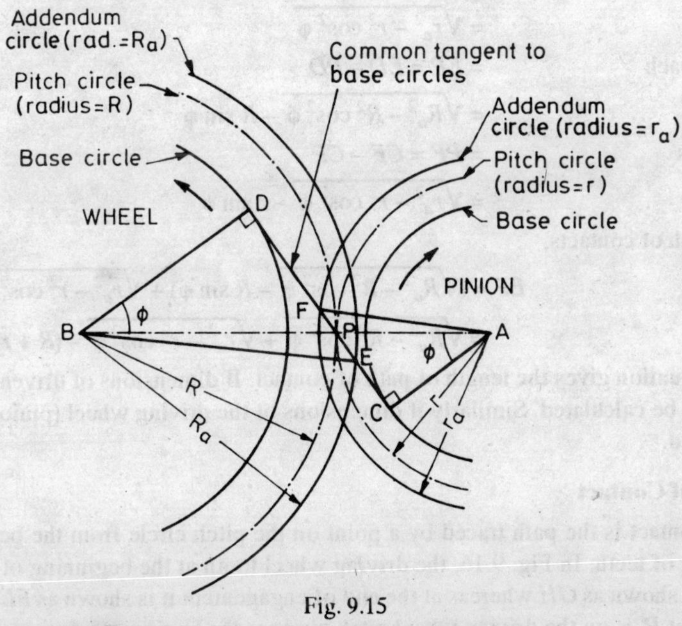

Fig. 9.15

The point E is the intersection of the common tangent and addendum circle of the wheel. The point F is the intersection of common tangent and addendum circle of pinion. The contact of two teeth begins where the addendum circle of the wheel meets the common tangent (*i.e.* point E) and ends where the addendum circle of the pinion meets the common tangent (*i.e.* point F). The line EF gives the length of path of contact. The length EP is known as *path of approach* whereas length PF is known as *path of recess*.

Let r = Radius of pitch circle of pinion *i.e.* length PA

R = Radius of pitch circle of wheel *i.e.* length BP

r_a = Addendum circle radius of pinion *i.e.* length AF

R_a = Addendum circle radius of wheel *i.e.* length BE

From Fig. 9.15, we know that

Length of path of contact = path of approach + path of recess

or
$$EF = EP + PF$$
$$= (ED - PD) + (CF - CP) \quad ...(i)$$

Let us now find the values of ED, PD, CF and CP.

In right angled triangle BPD,

$$PD = BP \sin \phi = R \sin \phi$$
$$BD = BP \cos \phi = R \cos \phi$$

In right angled triangle BED,

$$ED = \sqrt{(BE)^2 - (BD)^2} = \sqrt{R_a^2 - R^2 \cos^2 \phi} \qquad (\because \; BD = R \cos \phi)$$

In right angled triangle ACP,

$$CP = PA \sin \phi = r \sin \phi \qquad (\because \; PA = r)$$
$$CA = PA \cos \phi = r \cos \phi$$

In right angled triangle FCA,

$$CF = \sqrt{(AF)^2 - (CA)^2}$$
$$= \sqrt{r_a^2 - r^2 \cos^2 \phi}$$

Path of approach $= EP = ED - PD$

$$= \sqrt{R_a^2 - R^2 \cos^2 \phi} - R \sin \phi \qquad \text{...(9.10 } b)$$

Path of recess $= PF = CF - CP$

$$= \sqrt{r_a^2 - r^2 \cos^2 \phi} - r \sin \phi \qquad \text{...(9.10 } c)$$

Length of path of contacts,

$$EF = (\sqrt{R_a^2 - R^2 \cos^2 \phi} - R \sin \phi) + \sqrt{r_a^2 - r^2 \cos^2 \phi} - r \sin \phi$$
$$= \sqrt{R_a^2 - R^2 \cos^2 \phi} + \sqrt{r_a^2 - r^2 \cos^2 \phi} - (R + r) \sin \phi \qquad \text{...(9.11)}$$

The above equation gives the length of path of contact. If dimensions of driven wheel are given then path of approach can be calculated. Similarly if dimensions of the driving wheel (pinion) are known, path of recess can be obtained.

9.9. Length of Arc of Contact

The arc of contact is the path traced by a point on the pitch circle from the beginning to the end of engagement of a pair of teeth. In Fig. 9.16, the driving wheel tooth at the beginning of engagement with the driven wheel tooth is shown as GH whereas at the end of engagement it is shown as FL. The arc of contact is $P'PP''$ where the point P' is on the driving wheel pitch circle at the beginning of engagement and point P'' is on the same pitch circle at the end of engagement. This arc of contact is divided into two parts *i.e.* arc $P'P$ and arc PP''. The arc $P'P$ is known as arc of approach and arc PP'' is known as arc of recess.

From Fig. 9.16, we know that,

Arc of approach $= \text{Arc } P'P = \dfrac{EP}{\cos \phi}$

Arc of recess $= \text{Arc } PP'' = \dfrac{FP}{\cos \phi}$

Now length of arc of contact $= \text{Arc } P'P + \text{Arc } PP''$

$$= \frac{EP}{\cos \phi} + \frac{FP}{\cos \phi} = \frac{EP + FP}{\cos \phi}$$

$$= \frac{EF}{\cos \phi} \qquad \text{(9.11 } a)$$

$$= \frac{\text{Length of path of contact}}{\cos \phi} \qquad \text{...(9.12)}$$

($\because$ EF = length of path of contact)

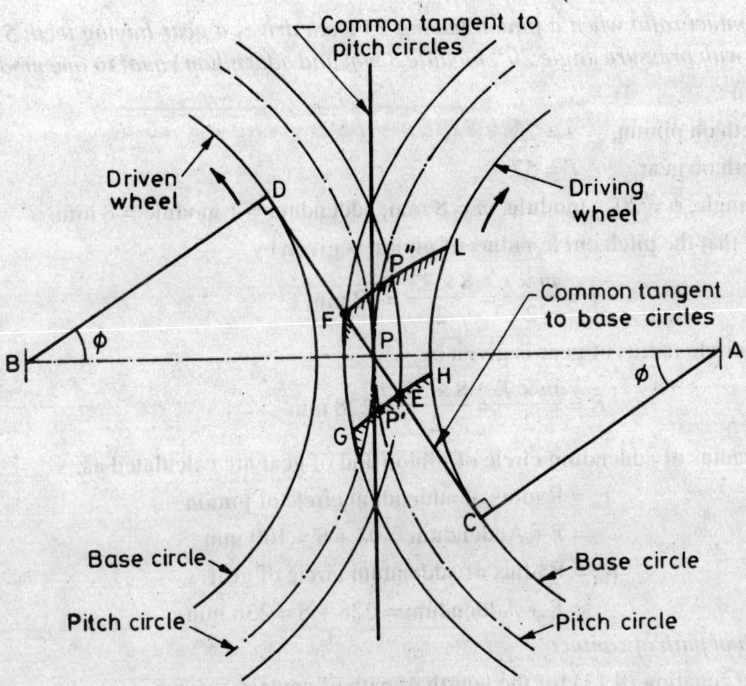

Fig. 9.16

9.10. Number of Pairs of Teeth in Contact (or Contact ratio)

The number of pairs of teeth in contact is defined as the ratio of length of arc of contact to the circular pitch. Hence,

Number of pairs of teeth in contact (or Contact ratio)

$$= \frac{\text{Length of arc of contact}}{\text{Circular pitch}} \qquad \text{...(9.12 a)}$$

But from equation (9.11 *a*) length of arc of contact is equal to $\frac{EF}{\cos \phi}$

and circular pitch from equation (9.5 *a*) is $p_c = \pi \times m$ where m = Module

$\therefore$ Number of pairs of teeth in contact

$$= \frac{EF}{\cos \phi} \times \frac{1}{\pi m} \qquad \text{...(9.13)}$$

For continuous transmission of motion, at least one tooth of one wheel must be in contact with another tooth of the second wheel. Hence number of pairs of teeth in contact (which is also known as contact ratio) must be greater than one. If contact ratio lies between 1 and 2, the number of teeth in contact at any time will be more than one but will never be more than two. If contact ratio is 1.6, this means that one pair of teeth is always in contact and second pair of teeth is in contact for 60% of the time. The gears will operate more quietly if contact ratio is large.

Problem 9.3. *Calculate :*

(i) length of path of contact,

(ii) arc of contact and

(iii) the contact ratio when a pinion having 23 teeth drives a gear having teeth 57. The profile of the gears is involute with pressure angle 20°, module 8 mm and addendum equal to one module.

Sol. Given :

No. of teeth on pinion, $t = 23$

No. of teeth on gear, $T = 57$

Pressure angle, $\phi = 20°$; module, $m = 8$ mm, addendum = 1 module = 8 mm.

We know that the pitch circle radius of pinion is given by,

$$r = \frac{m \times t}{2} = \frac{8 \times 23}{2} = 92 \text{ mm}$$

and pitch circle radius of gear is given by,

$$R = \frac{m \times T}{2} = \frac{8 \times 57}{2} = 228 \text{ mm}$$

Now the radius of addendum circle of pinion and of gear are calculated as,

r_a = Radius of addendum circle of pinion

$= r$ + Addendum = 92 + 8 = 100 mm

and R_a = Radius of addendum circle of gear

$= R$ + Addendum = 228 + 8 = 236 mm

(i) Length of path of contact

Now using equation (9.11) for the length of path of contact, we get

Length of path of contact

$$= \sqrt{R_a^2 - R^2 \cos^2 \phi} + \sqrt{r_a^2 - r^2 \cos^2 \phi} - (R + r) \sin \phi$$

$$= \sqrt{236^2 - 228^2 \cos^2 20°} + \sqrt{100^2 - 92^2 \cos^2 20°} - (228 + 92) \sin 20°$$

$$= \sqrt{55696 - 45903} + \sqrt{10000 - 7473.9} - (320) \times 0.342$$

$$= 98.96 + 50.26 - 109.44 = \mathbf{39.78 \text{ mm.}}$$ **Ans.**

(ii) Arc of Contact

Length of arc of contact is obtained from equation (9.12), as

$$\text{Length of arc of contact} = \frac{\text{Length of path of contact}}{\cos \phi} = \frac{39.78}{\cos 20°} = \mathbf{42.33 \text{ mm.}}$$ **Ans.**

(iii) The Contact Ratio

The contact ratio is given by equation (9.12 *a*). Hence using equation (9.12 *a*), we get

$$\text{Contact ratio} = \frac{\text{Length of arc of contact}}{\text{Circular pitch}} \quad ...(i)$$

where circular pitch, $p_c = \pi \times m = \pi \times 8 = 25.13$ mm

and length of arc of contact = 42.33 mm

$$\therefore \quad \text{Contact Ratio} = \frac{42.33}{25.13} = \mathbf{1.68 \text{ say } 2.}$$ **Ans.**

Problem 9.4. *The number of teeth on each of the two equal spur gears in mesh is 40. The teeth have 20° involute profile and the module is 6 mm. If the arc of contact is 1.75 times the circular pitch, find the addendum.* (AMIE, 1980 S)

Sol. Given :

No. of teeth on each gear = $T = t = 40$

The teeth have 20° involute profile, means the pressure angle = 20°

∴ The pressure angle, $\phi = 20°$,

Module, $m = 6$ mm

Length of arc of contact = 1.75 × Circular pitch

Find addendum.

From equation (9.5), we know that

$$m = \frac{D}{T} \quad \text{or} \quad D = m \times T \quad \text{or} \quad R = \frac{m \times T}{2} \quad \text{and} \quad r = m \times \frac{t}{2}.$$

∴ Pitch circle radii of each wheel is given by

$$R = r = \frac{T \times m}{2} = \frac{40 \times 6}{2} = 120 \text{ mm}$$

We know that the circular pitch (p_c) is given from equation (9.5 *a*) as,

$$p_c = \pi \times m = \pi \times 6 = 18.84 \text{ mm}$$

∴ Length of arc of contact $= 1.75 \times p_c = 1.75 \times 18.84 = 33$ mm

∴ Length of path of contact = Length of arc of contact × cos ϕ

$= 33 \times \cos 20° = 31$ mm

Let $R_a = r_a$ = Radius of addendum circle of each wheel.

The two gears are equal and have same no. of teeth. Hence their radius of addendum circle will be same.

But the length of path of contact from equation (9.11) is given as,

Length of path of contact $= \sqrt{R_a{}^2 - R^2 \cos^2 \phi} + \sqrt{r_a{}^2 - r^2 \cos^2 \phi} - (R + r) \sin \phi$

or
$$31 = \sqrt{R_a{}^2 - 120^2 \cos^2 20°} + \sqrt{R_a{}^2 - 120^2 \cos^2 \phi} - (120 + 120) \times \sin 20°$$

($\because$ $R = r = 120$, $\phi = 20°$ and $R_a = r_a$)

or
$$31 = 2\sqrt{R_a{}^2 - 120^2 \times (.9396)^2} - 240 \times 0.342$$

or
$$2\sqrt{R_a{}^2 - 12712.32} = 31 + 240 \times 0.342 = 113.08$$

∴
$$R_a{}^2 - 12712.32 = \left(\frac{113.08}{2}\right)^2 = 3196.77$$

∴
$$R_a{}^2 = 3196.77 + 12712.32 = 15909.092$$

∴
$$R_a = 126.13 \text{ mm}$$

We know that

Addendum on the wheels $= R_a - R = 126.13 - 120 =$ **6.13 mm. Ans.**

Problem 9.5. *Two mating spur gear with module pitch of 6.5 mm have 19 and 47 teeth of 20° pressure angle, and 6.5 mm addendum. Determine the number of pairs of teeth in contact and the angle turned through by the larger wheel for one pair of teeth in contact.*

Determine also the sliding velocity at the instant (i) engagement commences, (ii) the engagement terminates, and (iii) at the pitch point. When the pitch line velocity is 1.2 m/s. (A.M.I.E., 1976 W)

Sol. Given :

Module pitch, $m = 6.5$ mm

No. of teeth on smaller gear, $t = 19$

No. of teeth on larger gear, $T = 47$

Pressure angle, $\phi = 20°$

Pitch line velocity, $v = 1.2$ m/s

Addendum $= 6.5$ mm

Now pitch circle radius of smaller gear, $r = \dfrac{d}{2} = \dfrac{m \times t}{2} = \dfrac{6.5 \times 19}{2} = 61.75$ mm

Pitch circle radius of larger gear, $R = \dfrac{D}{2} = \dfrac{m \times T}{2} = \dfrac{6.5 \times 47}{2} = 152.75$ mm

Radius of the addendum circle of smaller gear (or of pinion),

$$r_a = r + \text{addendum} = 61.75 + 6.5 = 68.25 \text{ mm}$$

and radius of the addendum circle of larger gear

$$R_a = R + \text{addendum} = 152.75 + 6.5 = 159.25 \text{ mm}$$

We know that the length of path of approach (*i.e.* the path of contact when engagement occurs) is given by equation (9.10 *b*) as,

$$\text{Length of path of approach} = \sqrt{R_a^2 - R^2 \cos^2 \phi} - R \sin \phi$$

$$= \sqrt{159.25^2 - 152.75^2 \cos^2 20°} - 152.75 \times \sin 20°$$

$$= \sqrt{25360 - 23332 \times 0.883} - 52.24$$

$$= 68.97 - 52.24 = 16.73 \text{ mm}$$

The length of the path of recess (*i.e.* the path of contact when disengagement occurs) is given by equation (9.10 *c*) as

$$\text{Length of path of recess} = \sqrt{r_a^2 - r^2 \cos^2 \phi} - r \sin \phi$$

$$= \sqrt{68.25^2 - 61.75^2 \cos^2 20°} - 61.75 \times \sin 20°$$

$$= \sqrt{4658 - 38130 \times 0.883} - 21.12$$

$$= \sqrt{1291} - 21.12 = 35.93 - 21.12 = 14.81 \text{ mm}$$

$\therefore$ Length of path of contact = Length of path of approach + length of path of recess

$$= 16.73 + 14.81 = 31.54 \text{ mm}$$

Now the length of arc of contact is obtained from equation (9.12) as,

$$\text{Length of arc of contact} = \frac{\text{Length of path of contact}}{\cos \phi} = \frac{31.54}{\cos 20°} = 33.58 \text{ mm}$$

(i) No. of pairs of teeth in contact

We know from equation (9.12 *a*) that the no. of pairs of teeth in contact

$$= \frac{\text{Length of arc of contact}}{\text{Circular pitch}} = \frac{33.58}{\pi \times m}$$

$$= \frac{33.58}{\pi \times 6.5} = 1.645 \text{ or } \mathbf{2 \text{ pairs.}} \quad \textbf{Ans.}$$

(ii) Angle turned through by the gear for one pair of teeth in contact

We know that angle turned through by the gear

$$= \frac{\text{Length of arc of contact} \times 360°}{\text{Circumference of gear}}$$

Corresponding to circumference, angle turned = 360°

$$\text{Corresponding to one unit, angle turned} = \frac{360}{\text{Circumference}}$$

Corresponding to arc of contact, angle turned

$$= \frac{360}{\text{Circumference of gear}} \times \text{Length of arc of contact}$$

$$= \frac{360°}{2\pi R} \times 33.58 = \frac{360°}{2\pi \times 152.75} \times 33.58 = \mathbf{12.60°}. \quad \textbf{Ans.}$$

(iii) Sliding velocity

Let ω_1 = Angular velocity of the pinion

ω_2 = Angular velocity of the wheel

The pitch line velocity (or tangential velocity at the pitch point) is given by,

$$v = \omega_1 r = \omega_2 R$$

$$\therefore \quad \omega_1 = \frac{v}{r} = \frac{1.2 \times 1000}{61.75} \qquad (\because \; v = 1.2 \text{ m/s} = 1.2 \times 1000 \text{ mm/s and } r = 61.75 \text{ mm})$$

$$= 19.43 \text{ rad/s}$$

and

$$\omega_2 = \frac{v}{R} = \frac{1.2 \times 1000}{152.75} \qquad (\because \; R = 152.75 \text{ mm})$$

$$= 7.85 \text{ rad/s}$$

(a) The sliding velocity at the instant of engagement

$$V_{s_1} = (\omega_1 + \omega_2) \times \text{length of path of approach}$$

$$= (19.43 + 7.85) \times 16.73 \text{ mm/s} = \mathbf{456.39 \text{ mm/s}}. \quad \textbf{Ans.}$$

(b) The sliding velocity at the instant of engagement terminates,

$$V_{s_2} = (\omega_1 + \omega_2) \times \text{length of path of recess}$$

$$= (19.43 + 7.85) \times 14.81 = \mathbf{404 \text{ mm/s}}. \quad \textbf{Ans.}$$

(c) Since the velocity of sliding is proportional to the distance of contact point from the pitch point hence the sliding velocity at the pitch point is zero. **Ans.**

The velocity of sliding is maximum at the instant of engagement.

9.11. Interference in Involute Gears

Interference is the phenomenon between the two mating teeth when the tip of one tooth undercuts the root of other mating tooth.

Fig. 9.17 shows the two gears in mesh *i.e.* the pinion gear with centre *A* is in mesh with wheel gear with centre *B*. The line of action will be along *MN* which is the common tangent to the base circles of the two gears. The line *CD* is the path of contact between the two mating teeth, in which AC is the addendum radius of the pinion and *BD* is the addendum radius of the wheel. If pinion is driver and rotating clockwise then the teeth on the pinion and wheels are engaged at *D* and are disengaged at *C*.

Now if the radius of addendum circle of the pinion is increased, the point *C* shifts to *M*. The point *C* will coincide with *M* if the radius of addendum circle of the pinion is made equal to *AM*. If this radius *AM* is further increased, the point of contact will be on the inside of the base circle of the wheel. Since an involute can exist only outside the base circle, therefore any profile of teeth inside the base circle will of non-involute type. Hence the tip of tooth on the pinion will then under cut the tooth on the wheel at the root and remove part of involute profile of tooth on the wheel. Therefore, the interference occurs in the mating of two gears. *This phenomenon when the tip of the tooth undercuts the roots on its mating gears is known as interference.*

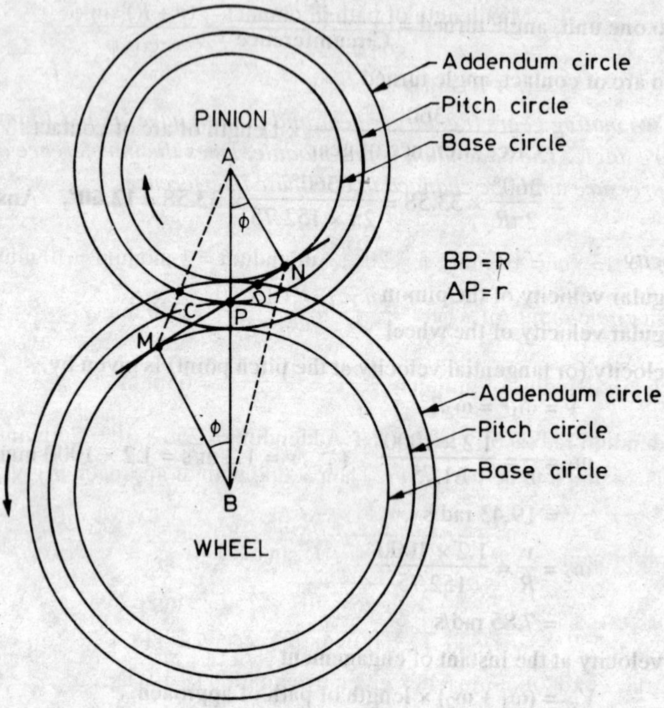

Fig. 9.17

Similarly if the radius of the addendum circle of the wheel is made greater than *BN*, then the tip of tooth on wheel will cause interference with the tooth on pinion. Hence interference of the teeth may be avoided if the addendum circles of the wheel and the pinion intersect the line of action between *M* and *N*. The points *M* and *N* are known as interference points. The limiting value of the radius of the addendum circle of the pinion is *AM* whereas of the wheel is *BN*. Hence in that case the maximum length of path of contact becomes as *MN* in which length *PN* is the maximum length of path of approach and length *PM* is the maximum length of path of recess.

From right angled triangle *APN*, we have

$$PN = AP \sin \phi$$

$$= r \sin \phi \qquad (\because \; AP = \text{Pitch circle radius of pinion} = r)$$

∴ Max. length of path of approach

$$= r \sin \phi \qquad \text{...(9.14)}$$

And from right angled triangle *BMP*, we have

$$PM = BP \sin \phi$$

$$= R \sin \phi \qquad (\because \; BP = \text{Pitch circle radius of wheel} = R)$$

$\therefore$ Max. length of path of recess $= R \sin \phi$ ...(9.15)

$\therefore$ Maximum length of path of contact,

$$MN = PN + PM = r \sin \phi + R \sin \phi$$

$$= (r + R) \sin \phi \quad ...(9.16)$$

Maximum length of arc of contact

$$= \frac{\text{Max length of path of contact}}{\cos \phi} = \frac{(r + R) \sin \phi}{\cos \phi}$$

$$= (r + R) \tan \phi \quad ...(9.17)$$

Problem 9.6. *Two mating gears (i.e. larger gear and pinion) have 50 and 13 involute teeth of module 10 mm and 20° pressure angle. The addendum is one module. Does the interference occur ? If it occurs, to what value should the pressure angle be changed to eliminate interference.*

Sol. Given :

$$T = 50 \; ; t = 13 \; ; m = 10 \text{ mm} \; ; \phi = 20° \; ; \text{Addendum} = 1 \text{ module} = 10 \text{ mm}.$$

Let R = Pitch circle radius of wheel $= \frac{m \times T}{2} = \frac{10 \times 50}{2} = 250$ mm

r = Pitch circle radius of pinion $= \frac{m \times t}{2} = \frac{10 \times 13}{2} = 65$ mm

R_a = Addendum radius of wheel $= R +$ Addendum $= 250 + 10 = 260$ mm

Let the pinion is assumed to be the driver. Then actual path of approach is given by equation (9.10 *b*) as

$$\text{Actual path of approach} = \sqrt{R_a^2 - R^2 \cos^2 \phi} - R \sin \phi$$

$$= \sqrt{260^2 - 250^2 \cos^2 20°} - 250 \times \sin 20°$$

$$= \sqrt{67600 - 62500 \times 0.883} - 85.5 = 111.4 - 85.50 = 25.91 \text{ mm}$$

Now max. length of path of approach is given by equation (9.14) as

$$= r \sin \phi$$

$$= 65 \times \sin 20° = 65 \times 0.342 = 22.23 \text{ mm}.$$

As actual length of path of approach is more than the max. permissible length of path of approach, hence interference will occur. **Ans.**

Value of pressure angle (ϕ) *to eliminate interference*

The interference will be eliminated if max. permissible length of path of approach is made equal to or more than actual length of path of approach.

$\therefore$ Max. permissible length = Actual length of path of approach.

But max. permissible length of path of approach $= r \sin \phi$

$\therefore \quad r \sin \phi = 25.91$

or $\quad 65 \sin \phi = 25.91$

or $\quad \sin \phi = \frac{25.91}{65} = 0.3986$

or $\quad \phi = \sin^{-1} 0.3986 = \mathbf{23.49°}$. **Ans.**

Alternate Method

Find the value of maximum permissible value of addendum radius of wheel *i.e.* value of $(R_a)_{max}$. This is given by BN in Fig. 9.17.

In right angled triangle *BMN*,

$$BN^2 = BM^2 + MN^2 = BM^2 + (MP + PN)^2 \qquad (\because \; MN = MP + PN)$$

$$\therefore \quad BN = (R_a)_{\max} = \sqrt{(R\cos\phi)^2 + (R\sin\phi + r\sin\phi)^2}$$

$$= \sqrt{(250\cos 20°)^2 + (250\sin 20° + 65\sin 20°)^2}$$

$$= \sqrt{250^2\cos^2 20° + (315\sin 20°)^2} = \sqrt{62500\cos^2 20° + 99225\sin^2 20°}$$

$$= \sqrt{62500 \times 0.883 + 99225 \times 0.117} = \sqrt{55187.5 + 11607}$$

$$= 258.44 \text{ mm.}$$

Actual addendum radius, R_a = 260 mm. As actual addendum radius is more than the maximum value of addendum radius of wheel, the interference will take place. **Ans.**

New Value of ϕ to eliminate interference

Interference can be avoided if $(R_a)_{max.}$ is increased. The value of $(R_a)_{max}$ can be increased by increasing the value of ϕ. The value of $(R_a)_{max.}$ must be atleast equal to R_a

$$\therefore \quad (R_a)_{max} = R_a = 260 \text{ mm}$$

But

$$(R_a)_{max} = \sqrt{R^2\cos^2\phi + (R\sin\phi + r\sin\phi)^2}$$

or

$$260 = \sqrt{250^2\cos^2\phi + (250\sin\phi + 65\sin\phi)^2}$$

$$= \sqrt{62500\cos^2\phi + 315^2\sin^2\phi}$$

$$= \sqrt{62500\cos^2\phi + 99225(1 - \cos^2\phi)}$$

$$= \sqrt{62500\cos^2\phi + 99225 - 99225\cos^2\phi}$$

Squaring both sides, we get

$$260^2 = 62500\cos^2\phi + 99225 - 99225\cos^2\phi$$

or

$$67600 = 99225 - 99225\cos^2\phi + 62500\cos^2\phi$$

or

$$99225\cos^2\phi - 62500\cos^2\phi = 99225 - 67600$$

or

$$36725\cos^2\phi = 31625$$

or

$$\cos^2\phi = \frac{31625}{36725} = 0.8611$$

or

$$\cos\phi = \sqrt{0.8611} = 0.9279$$

or

$$\phi = \cos^{-1} 0.9279 = \mathbf{21.88°}. \quad \textbf{Ans.}$$

Problem 9.7. *Two mating gears (i.e. larger gear and pinion) have 40 and 20 involute teeth of module 10 mm and 20° pressure angle. The addendum is one module. Does the interference occurs ?*

Sol. Given :

T = 40 ; t = 20 ; module, m = 10 mm ; ϕ = 20°, addendum = one module = 10 mm.

1st Method

To decide whether interference occurs or not, let us find the actual path of approach and max length of path of approach. If actual path of approach is less than the max. length of path of approach, no interference will occur.

Refer to Fig. 9.17. Actual path of approach (if pinion is assumed to be driver), is given by equation (9.10 *b*).

$$\therefore \quad \text{Actual path of approach,} \quad PD = \sqrt{R_a^{\,2} - R^2\cos^2\phi} - R\sin\phi \qquad ...(i)$$

where R = Pitch circle radius of wheel $= \frac{m \times T}{2}$

$$= \frac{10 \times 40}{2} = 200 \text{ mm} \quad (\because \ m = 10 \text{ mm} ; T = 40)$$

R_a = Addendum radius of wheel

$= R$ + addendum = 200 + 10 = 210 mm

Substituting the value of R and R_a in equation (i), we get

Actual path of approach, $PD = \sqrt{210^2 - 200^2 \cos^2 20°} - 200 \times \sin 20° = 25.3$ mm

Max. length of path of approach is given by equation (9.14) as

$$PN = r \sin \phi$$

where r = Pitch circle radius of pinion $= \frac{m \times t}{2}$

$$= \frac{10 \times 20}{2} = 100 \text{ mm}$$

$$\therefore \quad PN = 100 \sin 20° = 34.2 \text{ mm}$$

As actual path of approach is less than max. length of path of approach (*i.e.* 25.3 < 34.2), hence no interference will occur.

2nd Method

Find the actual addendum radius of wheel and maximum addendum radius of wheel. If actual addendum radius is less than the maximum addendum radius of wheel, there will be no interference.

Now actual addendum radius of wheel, R_a = 210 mm. (Already calculated above)

And max. addendum radius of wheel is BN. In Fig. 9.17, from right angled triangle BMN, we have

$$BN^2 = BM^2 + MN^2$$

$$\therefore \quad BN = \sqrt{(BP \cos \phi)^2 + (MP + PN)^2}$$

$$(\because \ BM = BP \cos \phi = R \cos \phi \ MN = MP + PN = R \sin \phi + r \sin \phi)$$

$$= \sqrt{(R \cos \phi)^2 + (R \sin \phi + r \sin \phi)^2}$$

$$= \sqrt{(200 \cos 20°)^2 + (200 \sin 20° + 100 \sin 20°)^2} = 214.1 \text{ mm}$$

As the actual addendum radius of wheel is lesser than the maximum permissible value of the addendum radius, no interference will occur. **Ans.**

Problem 9.8. *Two mating involute spur gears of 20° pressure angle have a gear ratio of 2. The number of teeth on the pinion is 20 and its speed is 250 r.p.m. The module pitch of the teeth is 12 mm.*

If the addendum on each wheel is such that the path of approach and the path of recess on each side are half the maximum possible length each, find :

(i) *the addendum for pinion and gear wheel ;*

(ii) *the length of arc of contact ;*

(iii) *the maximum velocity of sliding during approach and recess.*

Assume pinion to be driver. (AMIE, Winter, 1982)

Sol. Given :

$$\phi = 20° ; G = 2 ; t = 20 ; N_1 = 250 \text{ r.p.m.} ; m = 12 \text{ mm.}$$

Path of approach $= \frac{1}{2} \times$ Max. possible length of path of approach

Path of recess $= \frac{1}{2} \times$ Max. possible length of path of recess.

Pinion is to be assumed as driver.

We know that path of approach is given by equation (9.10 *b*) as,

Path of approach $= \sqrt{R_a^2 - R^2 \cos^2 \phi} - R \sin \phi$ ...(*i*)

Also the path of recess is given by equation (9.10 *c*) as,

Path of recess $= \sqrt{r_a^2 - r^2 \cos^2 \phi} - r \sin \phi$ ...(*ii*)

Also we know that max. possible length of path of approach is given by equation (9.14) as,

$= r \sin \phi$

and max. possible length of path of recess is given by equation (9.15) as

$= R \sin \phi$

But it is given that

Path of approach $= \frac{1}{2} \times$ Max. possible length of path of approach

$$\therefore \quad \sqrt{R_a^2 - R^2 \cos^2 \phi} - R \sin \phi = \frac{1}{2} \times r \sin \phi \quad ...(iii)$$

It is also given that

Path of recess $= \frac{1}{2} \times$ Max. possible length of path of recess

$$\therefore \quad \sqrt{r_a^2 - r^2 \cos^2 \phi} - r \sin \phi = \frac{1}{2} \times R \sin \phi \quad ...(iv)$$

(*i*) Addendum for pinion and gear wheel

Now let us find the values of *r* and *R* first

We know that, $r = \frac{m \times t}{2} = \frac{12 \times 20}{2} = 120$ mm

and $R = \frac{m \times T}{2} = \frac{12 \times (20 \times 2)}{2} = 240$ mm

$$\left(\because \frac{T}{t} = G = 2 \quad \therefore \quad T = 2 \times t = 2 \times 20 \right)$$

Substituting the values of *R*, *r* and ϕ in equation (*iii*), we get

$$\sqrt{R_a^2 - 240^2 \cos^2 20°} - 240 \sin 20° = \frac{1}{2} \times 120 \sin 20°$$

or

$$\sqrt{R_a^2 - 240^2 \cos^2 20°} = 240 \sin 20° + 60 \sin 20°$$

$$\sqrt{R_a^2 - 57600 \times (.09397)^2} = 300 \sin 20°$$

$$\sqrt{R_a^2 - 57600 \times 0.883} = 300 \times 0.342$$

$$\sqrt{R_a^2 - 50860.8} = 102.6$$

Squaring both sides, we get

$$R_a^2 - 50860.8 = 102.6^2 = 10526.7$$

or $$R_a^2 = 10526.7 + 50860.8 = 61387.5$$

$\therefore$ $$R_a = \sqrt{61387.5} = 247.7 \text{ mm}$$

$\therefore$ Addendum of wheel $= R_a - R = 247.7 - 240 =$ **7.7 mm. Ans.**

Now substituting values of r, R and ϕ is equation (*iv*), we get

$$\sqrt{r_a^2 - 120^2 \cos^2 20°} - 120 \sin 20° = \frac{1}{2} \times 240 \times \sin 20°$$

or $$\sqrt{r_a^2 - 14400 \times 0.883} = 120 \sin 20° + 120 \sin 20° = 240 \sin 20°$$

or $$\sqrt{r_a^2 - 12715.2} = 240 \times 0.342 = 82.08$$

Squaring both sides, we get

$$r_a^2 - 12715.2 = 82.08^2 = 6737.1$$

or $$r_a^2 = 6737.1 + 12715.2 = 19452.3$$

$\therefore$ $$r_a = \sqrt{19452.3} = 139.5 \text{ mm}$$

$\therefore$ Addendum of pinion $= r_a - r = 139.5 - 120 =$ **19.5 mm. Ans.**

(*ii*) The length of arc of contact

Path of contact = path of approach + path of recess

$$= \frac{1}{2} \times \text{Max. length of path of approach} + \frac{1}{2} \times \text{Max. length of path of recess}$$

$$= \frac{1}{2} \times r \sin \phi + \frac{1}{2} \times R \sin \phi = \frac{1}{2} [r + R] \sin \phi$$

$$= \frac{1}{2} [120 + 240] \sin 20° = 180 \times 0.342 = 61.56 \text{ mm}$$

We know from equation (9.12), that

Arc of contact $= \dfrac{\text{Path of contact}}{\cos \phi} = \dfrac{61.56}{\cos 20°} = \dfrac{61.56}{0.9397} =$ **65.51 mm. Ans.**

(*iii*) Max. Velocity of sliding during approach and recess

Angular velocity of pinion, $\omega_1 = \dfrac{2\pi N}{60} = \dfrac{2\pi \times 250}{60} = 26.18$ rad/s

Let ω_2 = Angular velocity of the wheel

We know that the angular velocities of two gears are inversely proportional to their no. of teeth

$\therefore$ $$\frac{\omega_1}{\omega_2} = \frac{T}{t}$$

$\therefore$ $$\omega_2 = \omega_1 \times \frac{t}{T} = 26.18 \times \frac{20}{40} = 13.09 \text{ rad/s}$$

$\therefore$ Max. Velocity of sliding during approach

$$= (\omega_1 + \omega_2) \times \text{Length of path of approach}$$

$$= (26.18 + 13.09) \times \frac{r \sin \phi}{2}$$

$$\left(\because \text{Length of path of approach} = \frac{1}{2} \times \text{Max. length of approach} = \frac{1}{2} \times r \sin \phi \right)$$

$$= \frac{39.27 \times 120 \times \sin 20°}{2} = \textbf{805.8 mm/s.} \quad \textbf{Ans.}$$

and maximum Velocity of sliding during recess

$= (\omega_1 + \omega_2) \times$ Length of path of recess

$= (26.18 + 13.09) \times \frac{1}{2} \times R \sin \phi \quad \left(\because \text{ length of path of recess} = \frac{R \sin \phi}{2} \right)$

$= 39.27 \times \frac{1}{2} \times 240 \times \sin 20° =$ **1611.6 mm/s. Ans.**

9.12. The Minimum Number of Teeth required on the Pinion in Order to Avoid Interference

In the previous article, it has already been pointed out that in order to avoid interference, the addendum circles for the two mating gears must cut the common tangent to the base circles between the points of tangency. The limiting condition will be reached when the addendum circles of pinion and wheel passes through points *M* and *N* respectively in Fig. 9.18.

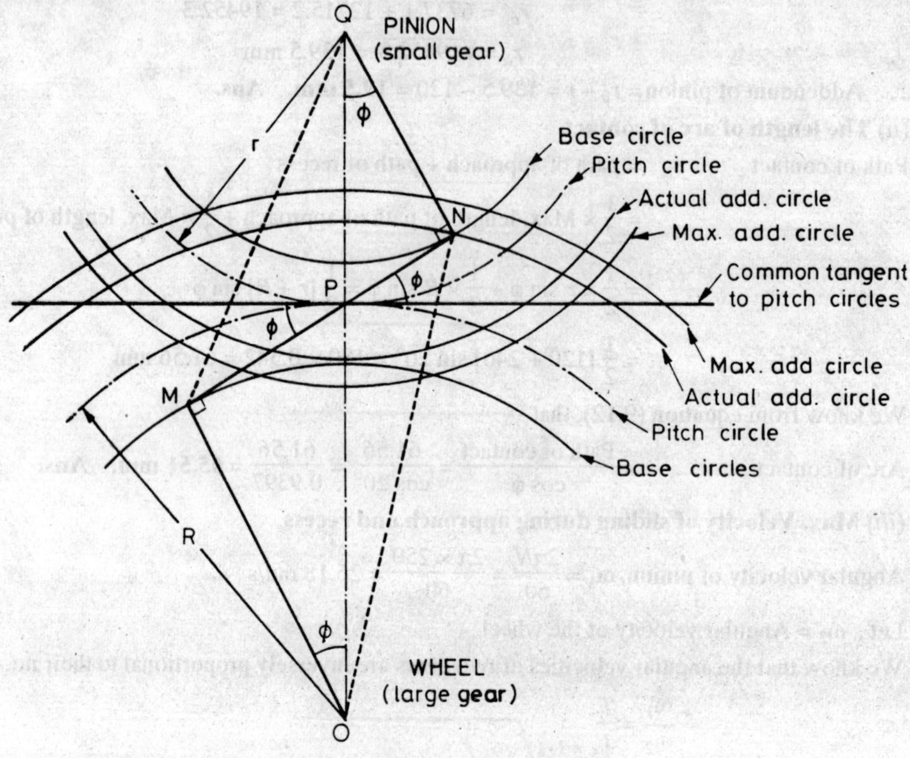

Fig. 9.18

Let T = Number of teeth on the wheel (or on larger gear)

t = Number of teeth on the pinion (or on smaller gear)

m = Module of teeth

d = Pitch circle diameter of pinion = $m \times t$

r = Pitch circle radius of pinion = $\frac{d}{2} = \frac{m \times t}{2}$

D = Pitch circle diameter of the wheel = $m \times T$

R = Pitch circle radius of the wheel $= \dfrac{D}{2} = \dfrac{m \times T}{2}$

G = Gear ratio $= \dfrac{T}{t}$ also $= \dfrac{R}{r}$

ϕ = Pressure angle or angle of obliquity.

The point Q is the centre of pinion whereas point O is the centre of wheel.

From triangle MPQ, which is towards pinion side, we have

$$QM^2 = QP^2 + PM^2 - 2QP \times PM \times \cos QPM$$

$$= r^2 + R^2 \sin^2 \phi - 2r \times R \sin \phi \times \cos (90 + \phi)$$

$(\because\ QP = r, PM = R \sin \phi \text{ and } \angle QPM = 90 + \phi)$

$$= r^2 + R^2 \sin^2 \phi - 2\,rR \sin \phi \times (-\sin \phi) \qquad [\because\ \cos(90 + \phi) = -\sin \phi]$$

$$= r^2 + R^2 \sin^2 \phi + 2rR \sin^2 \phi$$

$$= r^2 \left[1 + \frac{R^2 \sin^2 \phi}{r^2} + \frac{2rR \sin^2 \phi}{r^2}\right] \qquad \text{(Taking } r^2 \text{ common)}$$

$$= r^2 \left[1 + \frac{R^2 \sin^2 \phi}{r^2} + \frac{2R \sin^2 \phi}{r}\right] = r^2 \left[1 + \frac{R}{r}\left(\frac{R}{r} + 2\right) \sin^2 \phi\right]$$

$$\therefore \quad QM = r\sqrt{1 + \frac{R}{r}\left(\frac{R}{r} + 2\right) \sin^2 \phi}$$

$$= \frac{m \times t}{2}\sqrt{1 + \frac{R}{r}\left(\frac{R}{r} + 2\right) \sin^2 \phi} \qquad \left(\because\ r = \frac{m \times t}{2}\right)$$

$$= \frac{m \times t}{2}\sqrt{1 + \frac{T}{t}\left(\frac{T}{t} + 2\right) \sin^2 \phi} \qquad \left(\because\ \frac{R}{r} = \frac{T}{t} = G\right)$$

Now we know that the addendum of the pinion

$$= QM - QP$$

$$= QM - r \qquad \left(\because\ QP = r = \frac{m \times t}{2}\right)$$

$$= QM - \frac{m \times t}{2}$$

$$= \frac{m \times t}{2}\sqrt{1 + \frac{T}{t}\left(\frac{T}{t} + 2\right) \sin^2 \phi} - \frac{m \times t}{2}$$

$$\left(\because\ QM = \frac{m \times t}{2}\sqrt{1 + \frac{T}{t}\left(\frac{T}{t} + 2\right) \sin^2 \phi}\right)$$

$$= \frac{m \times t}{2}\left[\sqrt{1 + \frac{T}{t}\left(\frac{T}{t} + 2\right) \sin^2 \phi} - 1\right] \qquad \text{...(9.18)}$$

Let $A_p \times m$ = Addendum of the pinion to avoid interference

where A_p = Fraction by which the standard addendum of one module for the pinion should be multiplied in order to avoid interference

$\therefore$ Addendum of pinion $= A_p \times m$

Equating the two values of addendum of pinion given by equations (9.18), and (*i*), we get

$$A_p \times m = \frac{m \times t}{2}\left[\sqrt{1 + \frac{T}{t}\left(\frac{T}{t} + 2\right)\sin^2\phi} - 1\right]$$

or

$$A_p = \frac{t}{2}\left[\sqrt{1 + \frac{T}{t}\left(\frac{T}{t} + 2\right)\sin^2\phi} - 1\right]$$

or

$$t = \frac{2A_p}{\left[\sqrt{1 + \frac{T}{t}\left(\frac{T}{t} + 2\right)\sin^2\phi} - 1\right]}$$

$$= \frac{2A_p}{[\sqrt{1 + G(G+2)\sin^2\phi} - 1]} \quad \text{...(9.19)}$$

The above equation is used for finding the minimum number of teeth required on the pinion in order to avoid interference.

If $G = 1$ which means number of teeth on the wheel and pinion are equal, then equation (9.19) becomes as

$$t = \frac{2A_p}{[\sqrt{1 + 1(1+2)\sin^2\phi} - 1]} = \frac{2A_p}{[\sqrt{1 + 3\sin^2\phi} - 1]} \quad \text{...(9.20)}$$

The minimum number of teeth on the pinion which will mesh with any gear (also rack) without interference are given as :

S. No.	*System of gear teeth*	*Minimum No. of teeth on pinion*
1.	$14\frac{1}{2}^\circ$ Composite	12
2.	$14\frac{1}{2}^\circ$ Full depth involute	32
3.	20° Full depth involute	18
4.	20° Stub involute	14

9.13. The Minimum Number of Teeth required on the Wheel in order to avoid Interference

Refer to Fig. 9.18 and using the same notation as in Art. 9.12, we have from triangle *OPN*, which is towards wheel side,

$$ON^2 = OP^2 + PN^2 - 2OP \times PN \times \cos OPN$$

$$= R^2 + r^2\sin^2\phi - 2 \times R \times r\sin\phi\cos(90 + \phi)$$

$$(\because \; OP = R,\; PN = QP\sin\phi = r\sin\phi \text{ and angle } OPN = 90 + \phi)$$

$$= R^2 + r^2\sin^2\phi - 2R \times r\sin\phi \times (-\sin\phi)$$

$$= R^2 + r^2\sin^2\phi + 2R \times r\sin^2\phi$$

$$= R^2\left[1 + \frac{r^2\sin^2\phi}{R^2} + \frac{2R \times r \times \sin^2\phi}{R^2}\right] \quad \text{(Taking } R^2 \text{ common)}$$

$$= R^2\left[1 + \frac{r^2\sin^2\phi}{R^2} + \frac{2r\sin^2\phi}{R}\right]$$

$$= R^2 \left[1 + \frac{r}{R} \left(\frac{r}{R} + 2 \right) \sin^2 \phi \right]$$

$$\therefore \quad ON = R \sqrt{\left[1 + \frac{r}{R} \left(\frac{r}{R} + 2 \right) \sin^2 \phi \right]}$$

We know that the addendum of the wheel

$$= ON - OP$$

$$= ON - R \qquad \left(\because \; OP = R = \frac{m \times T}{2} \right)$$

$$= ON - \frac{m \times T}{2}$$

$$= R \sqrt{1 + \frac{r}{R} \left(\frac{r}{R} + 2 \right) \sin^2 \phi} - \frac{m \times T}{2}$$

$$= \frac{m \times T}{2} \sqrt{1 + \frac{r}{R} \left(\frac{r}{R} + 2 \right) \sin^2 \phi} - \frac{m \times T}{2} \qquad \left(\because \; R = \frac{m \times T}{2} \right)$$

$$\therefore \quad \text{Addendum of wheel} = \frac{m \times T}{2} \left[\sqrt{1 + \frac{r}{R} \left(\frac{r}{R} + 2 \right) \sin^2 \phi} - 1 \right]$$

$$= \frac{m \times T}{2} \left[\sqrt{1 + \frac{1}{G} \left(\frac{1}{G} + 2 \right) \sin^2 \phi} - 1 \right] \qquad \left(\because \; \frac{r}{R} = \frac{1}{G} \right) \qquad ...(9.21)$$

Let $A_w \times m$ = Addendum of the wheel to avoid interference.

where A_w = Fraction by which the standard addendum of one module for the wheel should be multiplied in order to avoid interference.

$\therefore$ Addendum of the wheel = $A_w \times m$

Equating the two values of addendum of the wheel, we get

$$A_w \times m = \frac{m \times T}{2} \sqrt{1 + \frac{r}{R} \left(\frac{r}{R} + 2 \right) \sin^2 \phi} - \frac{m \times T}{2}$$

$$= \frac{m \times T}{2} \left[\sqrt{1 + \frac{r}{R} \left(\frac{r}{R} + 2 \right) \sin^2 \phi} - 1 \right]$$

or

$$A_w = \frac{T}{2} \left[\sqrt{1 + \frac{r}{R} \left(\frac{r}{R} + 2 \right) \sin^2 \phi} - 1 \right]$$

or

$$T = \frac{2 \times A_w}{\sqrt{1 + \frac{r}{R} \left(\frac{r}{R} + 2 \right) \sin^2 \phi} - 1}$$

or
$$T = \frac{2 \times A_w}{\sqrt{1 + \frac{1}{G}\left(\frac{1}{G} + 2\right)\sin^2 \phi} - 1} \qquad \left(\because \frac{r}{R} = \frac{1}{G}\right) \qquad ...(9.22)$$

The above equation gives the minimum number of the teeth on the wheel if interference is to be avoided.

If $G = 1$, which means no. of teeth on the wheel and on pinion are equal, then equation (9.22) becomes as

$$T = \frac{2 \times A_w}{\sqrt{1 + 1(1 + 2)\sin^2 \phi} - 1} = \frac{2 \times A_w}{\sqrt{1 + 3\sin^2 \phi} - 1} \qquad ...(9.23)$$

From equation (9.22), the minimum no. of teeth on pinion can also be obtained by multiplying both sides by $\frac{t}{T}$.

Hence
$$T \times \frac{t}{T} = \frac{2 \times A_w \times \frac{t}{T}}{\left[\sqrt{1 + \frac{1}{G}\left(\frac{1}{G} + 2\right)\sin^2 \phi} - 1\right]}$$

or
$$t = \frac{2 \times A_w}{\frac{T}{t}\left[\sqrt{1 + \frac{1}{G}\left(\frac{1}{G} + 2\right)\sin^2 \phi} - 1\right]}$$

$$= \frac{2 \times A_w}{G\left[\sqrt{1 + \frac{1}{G}\left(\frac{1}{G} + 2\right)\sin^2 \phi} - 1\right]} \qquad ...(9.24)$$

Problem 9.9. *A pair of spur gear with involute teeth is to give a gear ratio of 4 : 1. The arc of approach is not to be less than the circular pitch and the smaller wheel is the driver. The angle of pressure is $14\frac{1}{2}$ degrees :*

(i) What is the least number of teeth that can be used on each wheel ?

(ii) What is the addendum of the wheel in terms of the circular pitch ? (AMIE, Summer, 1987)

Sol. Given :

Gear ratio, $G = 4 : 1$. But $G = \frac{T}{t} = \frac{R}{r} = 4$.

Pressure angle, $\phi = 14\frac{1}{2}°$.

Arc of approach is not less than circular pitch.

$\therefore$ Length of arc of approach = Circular pitch (p_c).

(i) Least number of teeth on each wheel

Let us find the maximum length of the arc of approach and the value of circular pitch.

$$\text{Max. length of arc of approach} = \frac{\text{Max. length of the path of approach}}{\cos \phi}$$

But max. length of path of approach is given by equation (9.14).

$\therefore$ Max. length of path of approach = $r \sin \phi$

$\therefore$ Max. length of Arc of approach $= \dfrac{r \sin \phi}{\cos \phi} = r \tan \phi$

Now circular pitch is given by equation (9.5 *a*) as, $p_c = \pi \times m$

$$\therefore \quad p_c = \pi \times \frac{2r}{t} \qquad \left(\because \; r = \frac{m \times t}{2} \text{ or } m = \frac{2r}{t}\right)$$

It is given that arc of approach is not to be less than the circular pitch or in the limiting case

Max. length of Arc of approach = Circular pitch

or $$r \tan \phi = \pi \times \frac{2r}{t}$$

or $$t = \frac{\pi \times 2r}{r \tan \phi} = \frac{2\pi}{\tan \phi} = \frac{2\pi}{\tan 14.5^\circ} = 24.3 \text{ say } \mathbf{25}.$$

and $$T = G \times t$$

$$= 4 \times 25 = 100 \qquad (\because \; G = 4)$$

$\therefore$ Least no. of teeth on smaller wheel, t = **25.**

and least no. of teeth on larger wheel, T = **100. Ans.**

(ii) Addendum of the wheel in terms of Circular pitch

The addendum of wheel is given by equation (9.21) as,

$$\text{Addendum of wheel} = \frac{m \times T}{2}\left[\sqrt{1 + \frac{1}{G}\left(\frac{1}{G} + 2\right)\sin^2 \phi} - 1\right]$$

$$= \frac{m \times 100}{2}\left[\sqrt{1 + \frac{1}{4}\left(\frac{1}{4} + 2\right)\sin^2 14.5^\circ} - 1\right]$$

$$= 50m\left[\sqrt{1 + 0.5625 \times 0.25038^2} - 1\right]$$

$$= 50m\,[1.01747 - 1] = 0.8735 \times m$$

$$= 0.8735 \times \frac{p_c}{\pi} \qquad \left(\because \; p_c = \pi \times m \quad \therefore \quad m = \frac{p_c}{\pi}\right)$$

$= \mathbf{0.278p_c}$. **Ans.**

Problem 9.10. *A pair 20° full depth involute spur gears having 30 and 50 teeth respectively of module 4 mm are in mesh, the smaller gear rotates at 1000 r.p.m. Determine : (a) sliding velocities at engagement and at disengagement of a pair of teeth and (b) contact ratio.* (A.M.I.E., 1977 W)

Sol. Given :

Pressure angle, $\phi = 20^\circ$

No. of teeth on smaller gear, $t = 30$

No. of teeth on large gear, $T = 50$

$\therefore$ Gear ratio, $G = \dfrac{T}{t} = \dfrac{50}{30}$

Module, $m = 4$ mm

Speed of smaller gear, $N_1 = 1000$ r.p.m.

$\therefore$ Angular speed of smaller gear, $\omega_1 = \dfrac{2\pi N_1}{60} = \dfrac{2\pi \times 1000}{60} = 104.67$ rad/s.

(*a*) *Sliding velocities at engagement and at disengagement*

First calculate the length of path of approach and length of path of recess. For these we require the values of R_a, R, r_a and r. Also we require the values of addendum of the wheel and addendum of the pinion.

Now addendum of wheel (or of larger gear) is given by equation (9.21) as,

$$\text{Addendum of wheel} = \frac{m \times T}{2}\left[\sqrt{\left\{1+\frac{1}{G}\left(\frac{1}{G}+2\right)\sin^2\phi\right\}}-1\right]$$

$$= \frac{4 \times 50}{2}\left[\sqrt{\left\{1+\frac{30}{50}\left(\frac{30}{50}+2\right)\sin^2 20°\right\}}-1\right] \text{ mm}$$

$$\left(\because\ G = \frac{T}{t} = \frac{50}{30}\ \therefore\ \frac{1}{G} = \frac{30}{50}\right)$$

$$= 100\left[\sqrt{\{1 + 0.6 \times 2.6 \times 0.342^2\}} - 1\right]$$

$$= 100\left[\sqrt{1.1824} - 1\right] = 9 \text{ mm}$$

Also the addendum of the pinion (or of smaller gear) is given by equation (9.18) as,

$$\text{Addendum of pinion} = \frac{m \times t}{2}\left[\sqrt{\left\{1+\frac{T}{t}\left(\frac{T}{t}+2\right)\sin^2\phi\right\}}-1\right]$$

$$= \frac{4 \times 30}{2}\left[\sqrt{\left\{1+\frac{50}{30}\left(\frac{50}{30}+2\right)\sin^2 20°\right\}}-1\right]$$

$$= 60\left[\sqrt{\{1 + 1.667(3.667) \times 0.342^2\}} - 1\right]$$

$$= 60\left[\sqrt{1 + 0.7154} - 1\right] = 60(1.309 - 1) \text{ mm} = 18.54 \text{ mm}$$

Now pitch circle radius of wheel,

$$R = \frac{m \times T}{2} = \frac{4 \times 50}{2} = 100 \text{ mm}$$

$\therefore$ Radius of addendum circle of wheel,

$$R_a = R + \text{addendum of wheel} = 100 + 9 = 109 \text{ mm}$$

Now, pitch radius of pinion,

$$r = \frac{m \times t}{2} = \frac{4 \times 30}{2} = 60 \text{ mm}$$

$\therefore$ Radius of addendum circle of pinion

$$r_a = r + \text{addendum of pinion} = 60 + 18.54 = 78.54 \text{ mm}$$

Now length of the path of approach (*i.e.* the path of contact when engagement occurs) from equation (9.10 *b*)

$$= \sqrt{R_a^2 - R^2\cos^2\phi} - R\sin\phi$$

$$= \sqrt{109^2 - 100^2\cos^2 20°} - 100 \times \sin 20°$$

$$= \sqrt{11881 - 8830} - 34.2 = 55.235 - 34.2 = 21.03 \text{ mm}$$

Length of the path of recess (*i.e.* the path of contact when disengagement occurs) [Refer to equation (9.10 *c*)]

$$= \sqrt{r_a^2 - r^2\cos^2\phi} - r\sin\phi$$

$$= \sqrt{78.54^2 - 60^2 \cos^2 20°} - 60 \sin 20°$$

$$= \sqrt{6168.53 - 3174.19} - 20.52 = 54.72 - 20.52 = 34.2 \text{ mm}$$

Now let us find the sliding velocities at engagement and disengagement of a pair of teeth

Let ω_2 = Angular speed of the wheel (or large gear)

We know from equation (9.7) that

$$\frac{\omega_2}{\omega_1} = \frac{t}{T}$$

$$\therefore \quad \omega_2 = \omega_1 \times \frac{t}{T} = 104.67 \times \frac{30}{50} \text{ rad/s} = 62.8 \text{ rad/s}$$

$\therefore$ Sliding velocity at engagement of a pair of teeth

$= (\omega_1 + \omega_2) \times$ length of path of approach

$= (104.67 + 62.8) \times 21.03$ mm/s

= 3521.9 mm/s = **3.522 m/s. Ans.**

And sliding velocity at disengagement position

$= (\omega_1 + \omega_2) \times$ length of path of recess

$= (104.67 + 62.8) \times 34.2$ mm/s

= 5727.5 mm/s = **5.727 m/s. Ans.**

(b) Contact ratio

$$\text{Contact ratio} = \frac{\text{Length of arc of contact}}{\text{Circular pitch}} \quad ...(i)$$

But length of arc of contact

$$= \frac{\text{Length of path of contact}}{\cos \phi}$$

$$= \frac{\text{Length of path of approach + Length of path of recess}}{\cos \phi}$$

$$= \frac{21.03 + 34.2}{\cos 20°} = 58.8 \text{ mm}$$

Also we know that

Circular pitch $= \pi \times m = \pi \times 4 = 12.568$ mm

$\therefore$ Substituting the above values in equation (*i*), we get

$$\text{Contact ratio} = \frac{58.8}{12.568} = 4.67 \text{ say } 5. \quad \textbf{Ans.}$$

Problem 9.11. *Two gear wheels mesh externally and are to give a velocity ratio of 3 to 1. The teeth are of involute form ; module = 6 mm, addendum = one module, pressure angle = 20°. The pinion rotates at 90 r.p.m. Find :*

(a) Number of teeth on pinion to avoid interference on it and the corresponding number on the wheel.

(b) The length of path and arc of contact.

(c) The number of pairs of teeth in contact.

(d) The maximum velocity of sliding. (A.M.I.E., 1980 W)

Sol. Given :

Velocity ratio, $V.R. = \frac{3}{1}$.

Module, $m = 6$ mm ; addendum = 1 module = $1 \times m = 6$ mm ; pressure angle $\phi = 20°$; pinion speed = 90 r.p.m.

$\therefore$ Addendum of pinion or of wheel = $A_p = A_w = 1$; module = 6 mm.

(*a*) *Number of teeth on pinion to avoid interference*

Let T = Number of teeth on wheel (or of larger gear)

t = Number of teeth on pinion (or of smaller gear)

Now $V.R. = \dfrac{T}{t}$ or $G = \dfrac{T}{t} = \dfrac{3}{1}$ where G = Gear ratio

or $\dfrac{3}{1} = \dfrac{T}{t}$

$\therefore$ $T = 3t$ ...(*i*)

To avoid interference, the no. of teeth on pinion is given by equation (9.19) as

$$t = \frac{2 \times A_p}{[\sqrt{\{1 + G(G+2)\sin^2\phi\}} - 1]}$$

where A_p = Addendum on pinion = 6 mm

$$= \frac{2 \times 6}{[\sqrt{\{1 + 3(3+2)\sin^2 20°\}} - 1]} = \frac{12}{[\sqrt{1 + 15 \times .342^2} - 1]}$$

$$= \frac{12}{[\sqrt{2.754} - 1]} = \frac{12}{(1.66 - 1)} = \frac{12}{0.66} = 18.2 \text{ say } \mathbf{19}. \quad \textbf{Ans.}$$

and corresponding number of teeth on wheel,

$$T = 3 \times t = 3 \times 19 = \mathbf{57}. \quad \textbf{Ans.}$$

(*b*) *Length of path and arc of contact*

We know that pitch circle radius of pinion,

$$r = \frac{m \times t}{2} = \frac{6 \times 19}{2} = 57 \text{ mm}$$

$\therefore$ Radius of addendum circle of pinion,

$$r_a = r + \text{addendum on pinion } (A_p) = 57 + 6 = 63 \text{ mm}$$

Now pitch circle radius of wheel,

$$R = \frac{m \times T}{2} = \frac{6 \times 57}{2} = 171 \text{ mm}$$

$\therefore$ Radius of addendum circle of wheel,

$$R_a = R + \text{addendum of wheel } (A_w) = 171 + 6 = 177 \text{ mm}$$

We know that the path of approach (*i.e.* path of contact when engagement occurs) [Refer to equation (9.10 *b*)]

$$= \sqrt{R_a^2 - R^2\cos^2\phi} - R\sin\phi = \sqrt{177^2 - 171^2\cos^2 20°} - 171\sin 20°$$

$$= \sqrt{31329 - 25819} - 58.482 = 74.229 - 58.482 = 15.747 \text{ mm}$$

and path of recess (*i.e.* path of contact when disengagement occurs). [Refer to equation (9.10 *c*)]

$$= \sqrt{r_a^2 - r^2\cos^2\phi} - r\sin\phi = \sqrt{63^2 - 57^2\cos^2 20°} - 57\sin 20°$$

$$= \sqrt{3969 - 2862} - 19.494 = \sqrt{1107} - 19.494$$

$$= 33.271 - 19.494 = 13.777 \text{ mm}$$

$\therefore$ **Length of path of contact**

$$= \text{Path of approach} + \text{path of recess}$$

$$= 15.747 + 13.777 = \mathbf{29.524\ mm.}\quad \textbf{Ans.}$$

$$\text{Length of arc of contact} = \frac{\text{Length of path of contact}}{\cos\phi}$$

$$= \frac{29.524}{\cos 20°} = \mathbf{31.42\ mm.}\quad \textbf{Ans.}$$

(c) Number of pairs of teeth in contact

Number of pairs of teeth in contact

$$= \frac{\text{Length of arc of contact}}{\text{Circular pitch } (p_c)}$$

$$= \frac{31.42}{\pi \times m} \qquad (\because\ p_c = \pi \times m)$$

$$= \frac{31.42}{\pi \times 6} \qquad (\because\ m = 6)$$

$$= \mathbf{1.667\ or\ 2.}\quad \textbf{Ans.}$$

(d) Maximum velocity of sliding

Let ω_2 = Angular speed of wheel in rad/s

$$\omega_1 = \text{Angular speed of pinion} = \frac{2\pi N}{60} = \frac{2\pi \times 90}{60} = 9.42 \text{ rad/s}$$

We know that $\dfrac{\omega_1}{\omega_2} = \dfrac{T}{t} = 3.0$

$$\therefore \quad \omega_2 = \frac{\omega_1}{3} = \frac{9.42}{3} = 3.14 \text{ rad/s}$$

Maximum velocity of sliding (*i.e.* V_s) is given by,

$$V_s = (\omega_1 + \omega_2) \times \text{path of approach} \qquad (\because\ \text{Path of approach} > \text{path of recess})$$

$$= (9.42 + 3.14) \times 15.747 = \mathbf{197.94\ mm/s.}\quad \textbf{Ans.}$$

9.14. Interference between Rack and Pinion and Minimum Number of Teeth on a Pinion for Involute Rack in order to avoid Interference

Fig. 9.19 shows a rack and pinion in mesh. The pinion is rotating in clockwise direction and driving the rack.

Let r = Pitch circle radius of the pinion

$$= \frac{m \times t}{2}$$

where t = No. of teeth on pinion and m = Module

ϕ = Pressure angle

$A_R \times m$ = Addendum for rack where A_R is the addendum co-efficient by which the standard addendum of one module for the rack is to be multiplied.

Rack is a part of toothed wheel of infinite diameter. Hence the base circle diameter and profile of involute teeth of Rack will be straight lines. In Fig. 9.19, P is the pitch point and PE is the line of action. The rack tooth starts engaging with the pinion tooth at point E. The point E gives the maximum value of the addendum of the rack. Hence addendum of rack must be less than GE for no interference. Consider triangle PEG.

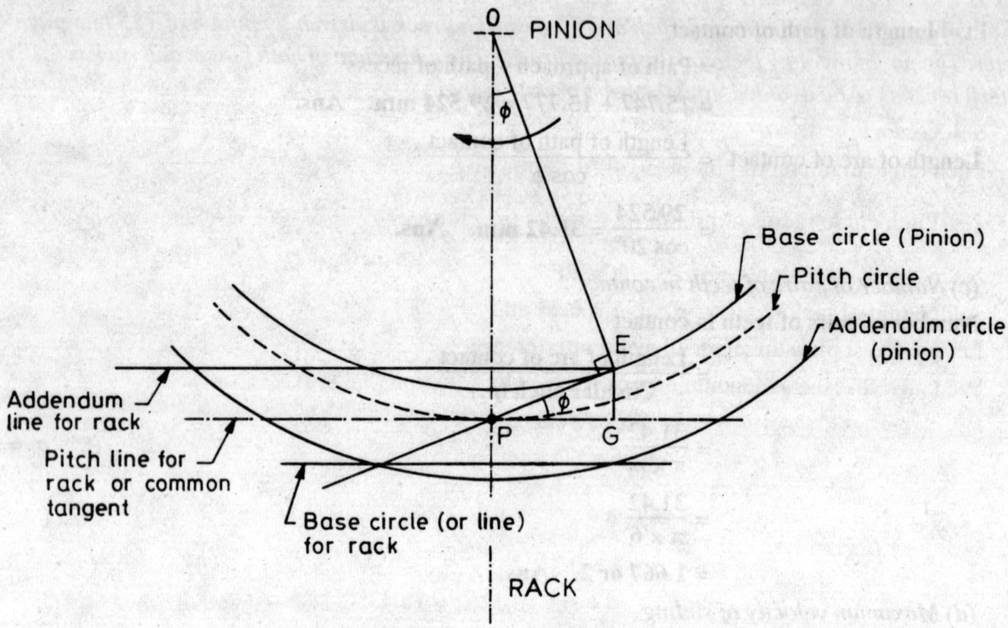

Fig. 9.19

Now $GE = PE \sin \phi$

$= (r \sin \phi) \sin \phi$ $\quad$ ($\because$ *PE* from right angled triangle $PAE = r \sin \phi$)

$= r \sin^2 \phi$ $\quad$ ...(9.25)

$= \dfrac{m \times t}{2} \sin^2 \phi$ $\quad$ $\left(\because \; r = \dfrac{m \times t}{2} \right)$

There will be no interference if

$$GE \geq A_R \times m$$

or $$\frac{m \times t}{2} \sin^2 \phi \geq A_R \times m$$

or $$t \geq \frac{2A_R \times m}{m \times \sin^2 \phi}$$

$$\geq \frac{2 \times A_R}{\sin^2 \phi}$$

$\therefore$ **Minimum no. of teeth on pinion for no interference is given by**

$$t_{min} = \frac{2 \times A_R}{\sin^2 \phi}$$

If $A_R = 1$ which is true for standard addendum,

$$t_{min} = \frac{2}{\sin^2 \phi}$$

If pressure angle $\phi = 20°$, then

$$t_{min} = \frac{2}{\sin^2 20°} = 17.1 \quad \text{or} \quad 18$$

$\therefore$ **The minimum no. of teeth required on the pinion to avoid interference is 18.**

Problem 9.12. *A pinion and rack are in mesh. The rack is driven by pinion of 125 mm pitch circle diameter. The no. of involute teeth on the pinion are 20. The addendum of both pinion and rack is 6.25 mm. If the interference is to be avoided, find the least pressure angle.*

Sol. Given :

Pitch circle diameter of pinion, $d = 125$ mm

$\therefore$ Pitch circle radius of pinion, $r = \frac{125}{2} = 62.5$ mm

No. of teeth on pinion, $t = 20$

Addendum of rack, $GE = 6.25$ mm

Let ϕ = Least pressure angle to avoid interference

We know that the addendum of rack is given by equation (9.25) as

$$GE = r \sin^2 \phi$$

or $$6.25 = 62.5 \sin^2 \phi$$

or $$\sin^2 \phi = \frac{6.25}{62.5} = 0.1$$

or $$\sin \phi = 0.3162$$

or $$\phi = \sin^{-1} 0.3162 = 18.435°.$$ **Ans.**

9.15. Helical Gears

In Art. 9.2, it is already mentioned that the gears used to connect two parallel shafts and having teeth inclined to the axes of shafts are known as helical gears. Helical gears are already shown in Fig. 9.4.

In helical gears, the contact occurs at a point of curved teeth at the beginning of the engagement and afterwards extends along a line across the teeth.

Fig. 9.20 shows two helical gears. The gear 1 in Fig. 9.20 (*a*) is a left-handed helical gear, as the helix slopes towards the left of the viewer when this gear is viewed parallel to the axis of the shaft. Similarly gear 2 in Fig. 9.20 (*b*) is a right-handed helical gear.

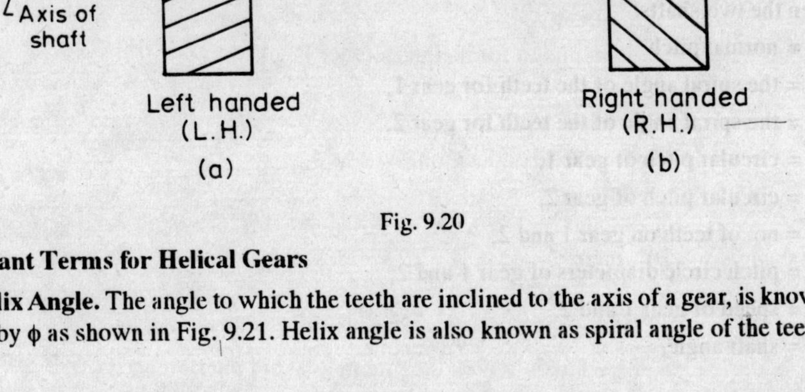

Fig. 9.20

9.16. Important Terms for Helical Gears

1. **Helix Angle.** The angle to which the teeth are inclined to the axis of a gear, is known as helix angle. It is denoted by ϕ as shown in Fig. 9.21. Helix angle is also known as spiral angle of the teeth.

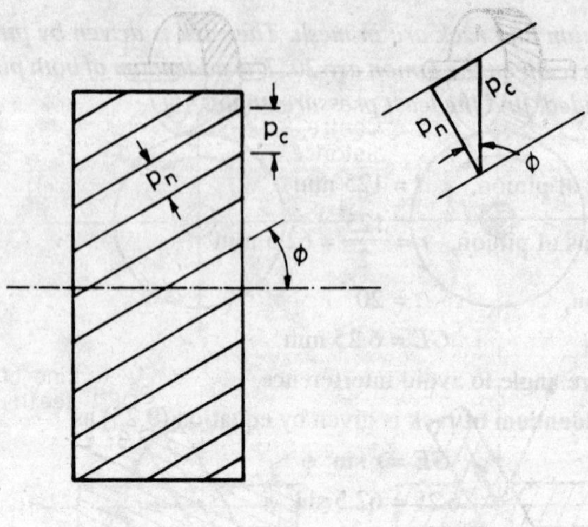

Fig. 9.21

2. **Normal pitch (p_n).** The shortest distance between similar faces of the adjacent teeth is known as normal pitch. It is denoted by p_n. The normal pitch of two mating gears must be same.

3. **Circular pitch (p_c).** The distance measured parallel to the axis between similar faces of adjacent teeth is known as circular pitch. It is denoted by p or p_c. It is also known as axial pitch or transverse pitch.

From Fig. 9.19, we have $$p_n = p_c \cos \phi \qquad \text{...(9.25 } a)$$

9.17. Spiral Gears

The gears used to connect two non-parallel and non intersecting shafts are known as spiral gears. They are used where small power is to be transmitted. The spiral gears may be of the same hand or of opposite hand. There is a point of contact between two mating gears of spiral gears. The pitch surfaces are cylindrical. The shortest distance between the two shafts gives the centre distance for a pair of spiral gears. Two mating spiral gears have a point of contact. This is proved by having two cylinders. When the axes of the two cylinders are parallel, they have a line contact. But if one cylinder is rotated through same angle, so that their axes are no longer parallel, then they have a point contact only.

Fig. 9.22 shows two pairs of spiral gears (*i.e.* gear 1 and gear 2). Let us first define the shaft angle. The shaft angle is the angle through which the axis of one shaft must be rotated in order to bring it parallel to the axis of other shaft and the two shafts revolving in opposite directions. The normal pitch of the teeth must be same for both wheels, but the circular pitch will be different.

Expression for Centre Distance. The shortest distance between the two shafts is known as centre distance between the two shafts.

Let p_n = normal pitch

α = the spiral angle of the teeth for gear 1,

β = the spiral angle of the teeth for gear 2,

p_1 = circular pitch of gear 1,

p_2 = circular pitch of gear 2,

T_1 and T_2 = no. of teeth on gear 1 and 2,

D_1 and D_2 = pitch circle diameters of gear 1 and 2,

N_1 and N_2 = speed of gear 1 and 2,

θ = shaft angle,

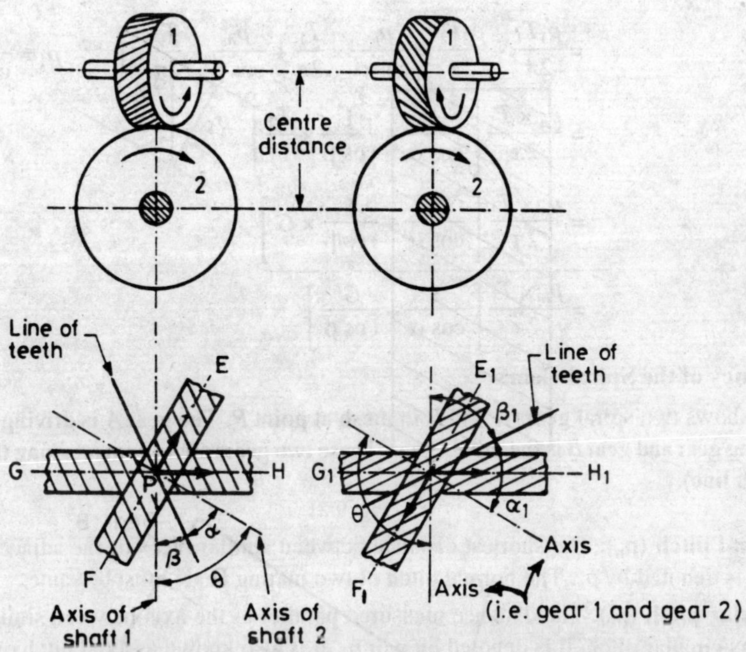

Fig. 9.22

C = centre distance

$$G = \text{Gear ratio} = \frac{T_2}{T_1} = \frac{N_1}{N_2}$$

We know that, normal pitch in terms of helix angle (or spiral angle) and circular pitch is given by,

$$p_n = p_c \cos \alpha$$

$\therefore$ For gear 1, $\quad p_n = p_1 \cos \alpha \quad \therefore \quad p_1 = \dfrac{p_n}{\cos \alpha}$

and for gear 2, $\quad p_n = p_2 \cos \beta \quad \therefore \quad p_2 = \dfrac{p_n}{\cos \beta}$

We also know that circular pitch in terms of pitch circle diameter and number of teeth on the wheels, is given by

$$p_c = \frac{\pi D}{T} \qquad \text{...(9.26)}$$

$\therefore$ For gear 1, we have, $p_1 = \dfrac{\pi D_1}{T_1}$ or $D_1 = \dfrac{p_1 \times T_1}{\pi}$

and for gear 2, we have, $p_2 = \dfrac{\pi D_2}{T_2}$ or $D_2 = \dfrac{p_2 \times T_2}{\pi}$

The centre distance (c) is given by

$$C = \frac{D_1 + D_2}{2} = \frac{\dfrac{p_1 \times T_1}{\pi} + \dfrac{p_2 \times T_2}{\pi}}{2}$$

$$= \frac{p_1 T_1}{2\pi} + \frac{p_2 T_2}{2\pi} = \frac{p_n}{\cos\alpha} \times \frac{T_1}{2\pi} + \frac{p_n}{\cos\beta} \times \frac{T_2}{2\pi} \quad \left(\because \; p_1 = \frac{p_n}{\cos\alpha}; p_2 = \frac{p_n}{\cos\beta} \right)$$

$$= \frac{p_n \times T_1}{2\pi} \left[\frac{1}{\cos\alpha} + \frac{1}{\cos\beta} \times \frac{T_2}{T_1} \right]$$

$$= \frac{p_n \times T_1}{2\pi} \left[\frac{1}{\cos\alpha} + \frac{1}{\cos\beta} \times G \right] \quad \left(\because \; \frac{T_2}{T_1} = G \right)$$

$$= \frac{p_n \times T_1}{2\pi} \left[\frac{1}{\cos\alpha} + \frac{G}{\cos\beta} \right] \quad \text{...(9.27)}$$

9.18. The Efficiency of the Spiral Gears

Fig. 9.23 shows two spiral gears *A* and *B* in mesh at point *P*. The gear *A* is driving the gear *B*. Hence gear *A* is the driving gear and gear *B* is the driven gear. These two gears are in contact along the inclined surface *CC* (or along teeth line).

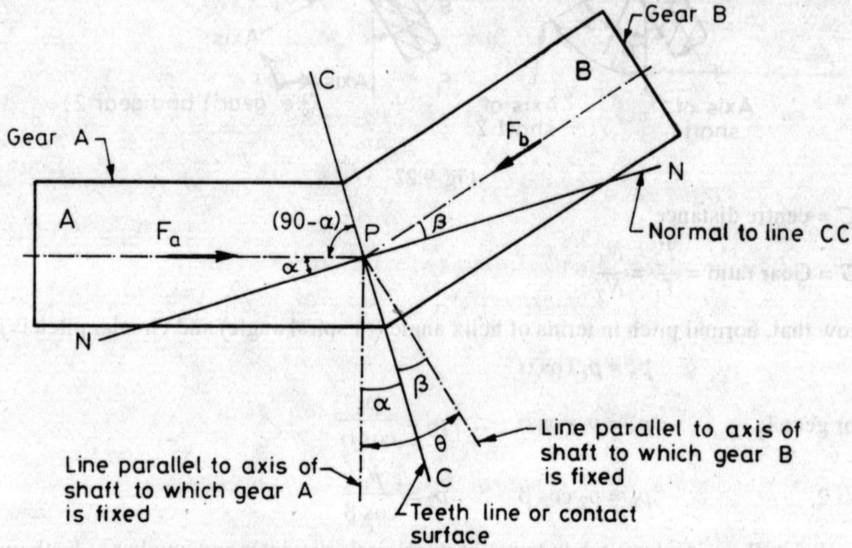

Fig. 9.23

Let F_a = Axial force on gear *A*

F_b = Axial force on gear *B*

R_N = Normal reaction at the point of contact

ϕ = Angle of friction *i.e.* angle made by resultant reaction with the normal.

R = Resultant reaction at the point of contact, and

θ = Shaft angle

$= \alpha + \beta$

where α = Angle between axis of shaft (to which gear *A* is fixed) and inclined surface *CC*.

β = Angle between axis of shaft (to which gear *B* is fixed) and inclined surface *CC*

N-N = Line normal to *CC*.

(A) Idle Case i.e. No Friction

Let us first consider that there is no friction between the mating gears. Then force of friction (*i.e.* $\mu \times R_N$) will be zero. At the point of contact, the following forces will be acting on gear *A* as shown in Fig. 9.23 (*a*) :

(*i*) The normal reaction R_N, which is normal to surface *CC* and towards the surface *CC* at the point of contact.

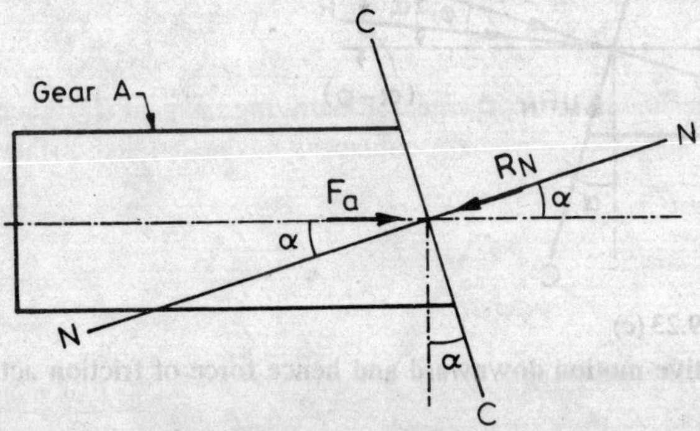

Fig. 9.23 (*a*)

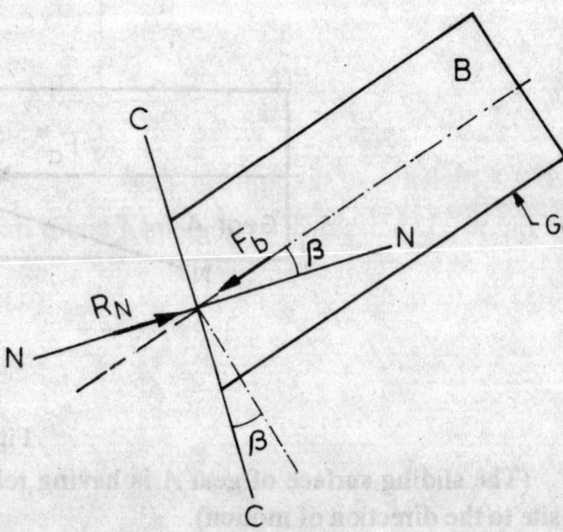

Fig. 9.23 (*b*)

(*ii*) The axial force F_a.

For the equilibrium of gear *A* in the direction of axial force, we have

$$F_b = R_N \cos \alpha$$

or

$$R_N = \frac{F_a}{\cos \alpha} \qquad ...(i)$$

Now consider the equilibrium of gear *B*. The following forces will be acting on gear *B* at the point of contact as shown in Fig. 9.23 (*b*) :

(*i*) Normal reaction R_N, which is normal to surface *CC* and towards the surface *CC* at the point of contact.

(*ii*) The axial force F_b.

For the equilibrium of gear *B* in the direction of axial force (F_b), we have

$$F_b = R_N \cos \beta$$

or

$$R_N = \frac{F_b}{\cos \beta} \qquad ...(ii)$$

From equations (*i*) and (*ii*), we have

$$\frac{F_a}{\cos \alpha} = \frac{F_b}{\cos \beta}$$

(**Note.** The reaction of *B* on *A* is equal to reaction of *A* on *B*.)

$$\therefore \qquad F_a = \frac{F_b}{\cos \beta} \times \cos \alpha \qquad ...(9.28)$$

(B) Considering Friction

In actual case, there is a friction between the contact surfaces of the mating gears. Hence the force of friction ($\mu \times R_N$) will be acting on the contact surfaces in the direction opposite to that of relative sliding.

The resultant reaction (R) will not be normal to the surface but will inclined at an angle of friction (ϕ) with the normal to the surfaces. Hence the following forces will be acting on gear A at the point of contact as shown in Fig. 9.23 (c) :

(i) The normal reaction, R_N

(ii) Force of friction, $\mu \times R_N$

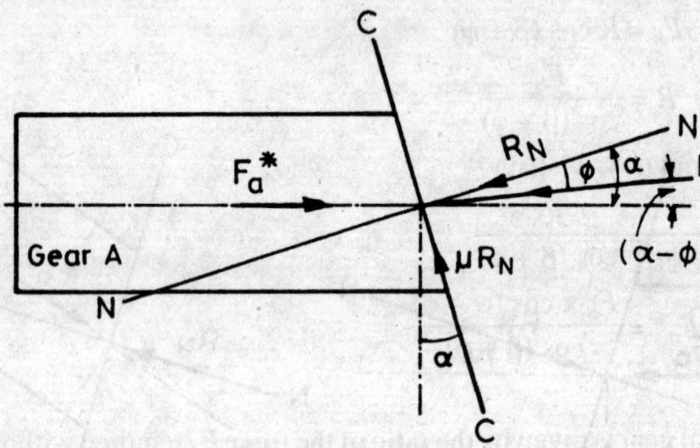

Fig. 9.23 (c)

(The sliding surface of gear A is having relative motion downward and hence force of friction acts opposite to the direction of motion).

(iii) The axial force F_a*.

(Axial force F_a* is the force required with friction when F_b is constant.)

The normal reaction R_N and force of friction $\mu \times R_N$ combine into a single resultant force R, which is inclined at an angle ϕ with the normal.

For the equilibrium of gear A in the direction of axial force (F_a*), we have

$$F_a^* = R \cos (\alpha - \phi)$$

or
$$R = \frac{F_a^*}{\cos (\alpha - \phi)} \qquad \ldots(iii)$$

Now consider the equilibrium of gear B. The following forces will be acting on gear B at the point of contact as shown in Fig. 9.23 (d) :

(i) The normal reaction, R_N.

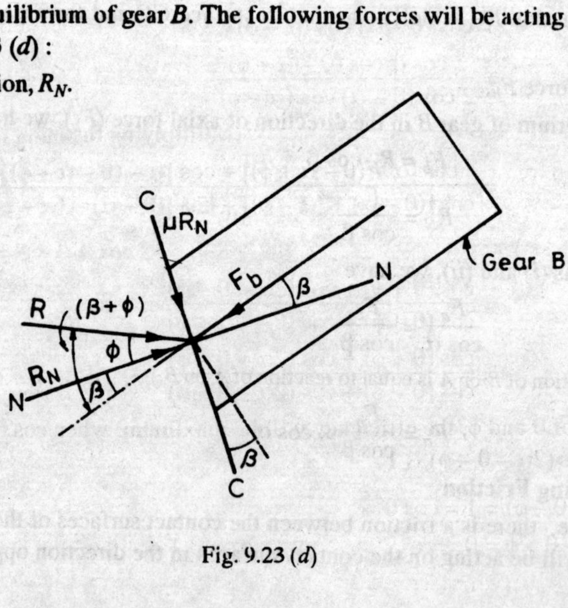

Fig. 9.23 (d)

(*ii*) Force of friction, μR_N.

(*iii*) The axial force, F_b.

The normal reaction R_N and force of friction μR_N combine into a single resultant force R, which is inclined at an angle ϕ with the normal.

For the equilibrium of gear B in the direction of axial force (F_b), we have

$$F_b = R\cos(\beta + \phi)$$

or
$$R = \frac{F_b}{\cos(\beta + \phi)} \quad \text{...(iv)}$$

From equations (*iii*) and (*iv*), we have

$$\frac{F_a{}^*}{\cos(\alpha - \phi)} = \frac{F_b}{\cos(\beta + \phi)}$$

or
$$F_a{}^* = \frac{F_b \times \cos(\alpha - \phi)}{\cos(\beta + \phi)} \quad \text{...(9.29)}$$

Efficiency of Spiral Gears

The efficiency of spiral gear is given by the ratio of the force F_a required without friction to that required with friction when F_b is constant. Hence from equations (9.28) and (9.29), we have

$$\eta = \frac{F_a}{F_a{}^*} = \frac{\text{Force without friction}}{\text{Force with friction}}$$

$$= \frac{\left(\dfrac{F_b \cos\alpha}{\cos\beta}\right)}{F_b \times \dfrac{\cos(\alpha - \phi)}{\cos(\beta + \phi)}}$$

$$= \frac{\cos\alpha\cos(\beta + \phi)}{\cos\beta\cos(\alpha - \phi)} \quad \text{...(9.30)}$$

But $\beta = \theta - \alpha$, hence above equation becomes as

$$\eta = \frac{\cos\alpha\cos(\theta - \alpha + \phi)}{\cos(\theta - \alpha)\cos(\alpha - \phi)}$$

$$= \frac{2\cos\alpha\cos(\theta - \alpha + \phi)}{2\cos(\theta - \alpha)\cos(\alpha - \phi)}$$

(Multiplying the numerator and denominator by 2)

$$= \frac{\cos[\alpha + (\theta - \alpha + \phi)] + \cos[\alpha - (\theta - \alpha + \phi)]}{\cos[(\theta - \alpha) + (\alpha - \phi)] + \cos[(\theta - \alpha) - (\alpha - \phi)]}$$

$[\because\ 2\cos A\cos B = \cos(A + B) + \cos(A - B)]$

$$= \frac{\cos(\theta + \phi) + \cos(2\alpha - \theta - \phi)}{\cos(\theta - \phi) + \cos(\theta - 2\alpha + \phi)}$$

$$= \frac{\cos(\theta + \phi) + \cos(2\alpha - \theta - \phi)}{\cos(\theta - \phi) + \cos(2\alpha - \theta - \phi)} \quad [\because\ \cos(-A) = \cos A] \quad \text{...(9.31)}$$

For a given value of θ and ϕ, the efficiency will be maximum, when $\cos(2\alpha - \theta - \phi)$ is a maximum. But maximum value of $\cos(2\alpha - \theta - \phi)$ is 1

$\therefore \quad \cos(2\alpha - \theta - \phi) = 1 = \cos 0° \qquad (\because\ \cos 0° = 1)$

$\therefore \quad 2\alpha - \theta - \phi = 0 \qquad \text{or} \qquad 2\alpha = \theta + \phi$

or $$\alpha = \frac{\theta + \phi}{2} \quad \text{...(9.32)}$$

Substituting the value of α in equation (9.31), we get the maximum efficiency.

$$\therefore \quad \eta_{max} = \frac{\cos(\theta + \phi) + \cos(\theta + \phi - \theta - \phi)}{\cos(\theta - \phi) + \cos(\theta + \phi - \theta - \phi)}$$

$$= \frac{\cos(\theta + \phi) + \cos 0}{\cos(\theta - \phi) + \cos 0}$$

$$= \frac{\cos(\theta + \phi) + 1}{\cos(\theta - \phi) + 1} \quad (\because \cos 0 = 1) \quad \text{...(9.33)}$$

Problem 9.13. *The approximate centre distance between two meshing spiral gears is 375 mm and gear ratio is 2. The angle between the shafts is 50° and the normal circular pitch is 19 mm. If the driving and driven wheels are having same spiral angles and friction angle = 6°, then determine :*

(i) Number of teeth on each wheel,

(ii) The exact centre distance and

(iii) Efficiency of the drive.

Sol. Given :

Approximate centre distance, $C = 375$ mm ; Gear ratio, $G = 2$; Shaft angle, $\theta = 50°$;

Normal pitch, $p_n = 19$ mm ; friction angle, $\phi = 6°$.

Spiral angle of driving wheel = Spiral angle of driven wheel.

Let α = Spiral angle of driving wheel

β = Spiral angle of driven wheel

$$\therefore \quad \alpha = \beta \quad \text{(given)}$$

But shaft angle $$\theta = \alpha + \beta$$

or $$50° = \alpha + \beta$$

$$\therefore \quad \alpha = \beta = \frac{50}{2} = 25°$$

(i) Number of teeth on each wheel

Let T_1 = Number of teeth on driving wheel

T_2 = Number of teeth on driven wheel

The centre distance (C) between two shafts is given by equation (9.27) as,

$$C = \frac{p_n \times T_1}{2\pi}\left[\frac{1}{\cos\alpha} + \frac{G}{\cos\beta}\right]$$

or $$375 = \frac{19 \times T_1}{2\pi}\left[\frac{1}{\cos 25°} + \frac{2}{\cos 25°}\right] \quad (\because C = 375 ; \alpha = \beta = 25° ; p_n = 19)$$

$$= \frac{19 \times T_1}{2\pi}\left[\frac{1+2}{\cos 25°}\right] = \frac{19 \times T_1}{2\pi} \times \frac{3}{\cos 25°}$$

$$\therefore \quad T_1 = \frac{375 \times 2\pi \times \cos 25°}{19 \times 3} = 37.46 \text{ say } \mathbf{38.} \quad \textbf{Ans.}$$

and $$T_2 = G \times T_1 = 2 \times 38 = \mathbf{76.} \quad \textbf{Ans.} \quad \left(\because G = \frac{T_2}{T_2}\right)$$

(ii) Exact Centre distance

Using equation (9.27), we get

$$C = \frac{p_n \times T_1}{2\pi}\left[\frac{1}{\cos\alpha} + \frac{G}{\cos\beta}\right]$$

$$= \frac{19 \times 38}{2\pi}\left[\frac{1}{\cos 25^\circ} + \frac{2}{\cos 25^\circ}\right]$$

$$= \frac{19 \times 38}{2\pi} \times \frac{3}{\cos 25^\circ} = \textbf{380.4 mm.} \quad \textbf{Ans.}$$

(iii) Efficiency of the drive

The efficiency of the spiral gears is given by equation (9.30).

$$\therefore \qquad \eta = \frac{\cos\alpha \cos(\beta + \phi)}{\cos\beta \cos(\alpha - \phi)}$$

$$= \frac{\cos 25^\circ \cos(25 + 6)}{\cos 25^\circ \cos(25 - 6)} \qquad (\because \ \alpha = \beta = 25^\circ \text{ and } \phi = 6^\circ)$$

$$= \frac{\cos 31^\circ}{\cos 19^\circ} = \frac{0.8571}{0.9455} = 0.9065 \text{ or } \textbf{90.65\%}. \quad \textbf{Ans.}$$

Problem 9.14. *The angle between two meshing spiral gears is 90° and centre distance is 150 mm approximately. The normal circular pitch of the gears is 10 mm and the gear ratio is 2.5. The frictional angle is 6° and efficiency of the drive is maximum. Determine :*

(i) the spiral angles of the teeth,

(ii) the circular pitches of the gears,

(iii) the number of teeth on each wheel,

(iv) the exact centre distance and pitch circle diameters of the two wheels and

(v) the efficiency of the drive.

Sol. Given :

$$\theta = 90^\circ \ ; C = 150 \text{ mm approx.} \ ; p_n = 10 \text{ mm} \ ; G = 2.5 \ ; \phi = 6^\circ \ ; \eta \text{ is max.}$$

(i) Spiral angles of the teeth

Let α = spiral angle of driving wheel

β = spiral angle of driven wheel

θ = shaft angle = $\alpha + \beta$

or $\qquad 90^\circ = \alpha + \beta \qquad$ ...(*i*)

For maximum efficiency, the value of angle 'α' from equation (9.32) is given by

$$\alpha = \frac{\theta + \phi}{2} = \frac{90 + 6}{2} = \frac{96}{2} = \textbf{48°}. \quad \textbf{Ans.}$$

$$\therefore \qquad \beta = 90 - \alpha = 90 - 48 = \textbf{42°}. \quad \textbf{Ans.}$$

(ii) The circular pitches of the gears

Let p_{c_1} = Circular pitch of the gear on driving wheel

p_{c_2} = Circular pitch of the gear on driven wheel

Now using equation (9.25 *a*), we have

$$p_n = p_c \text{ cosine of spiral angle}$$

$$\therefore \qquad p_n = p_{c_1} \cos\alpha = p_{c_2} \cos\beta$$

$\therefore \quad p_{c_1} = \dfrac{p_n}{\cos \alpha} = \dfrac{10}{\cos 48°} = \mathbf{14.94\ mm.}$ **Ans.**

and $\quad p_{c_2} = \dfrac{p_n}{\cos \beta} = \dfrac{10}{\cos 42°} = \mathbf{13.45\ mm.}$ **Ans.**

(*iii*) *The number of teeth on each wheel*

Let T_1 = No. of teeth on driving wheel

T_2 = No. of teeth on driven wheel.

Using equation (9.27), we have

$$C = \frac{p_n \times T_1}{2\pi}\left[\frac{1}{\cos \alpha} + \frac{G}{\cos \beta}\right]$$

or
$$150 = \frac{10 \times T_1}{2\,\pi}\left[\frac{1}{\cos 48°} + \frac{2.5}{\cos 42°}\right]$$

($\because$ $C = 150$ mm, $p_n = 10$, $\alpha = 48°$, $\beta = 42°$, $G = 2.5$)

$$= \frac{5T_1}{\pi}[1.4944 + 3.364] = \frac{5T_1}{\pi} \times 4.8584$$

$\therefore \quad T_1 = \dfrac{150 \times \pi}{5 \times 4.8584} = 19.4$ **say 20. Ans.**

and $\quad T_2 = G \times T_1 = 2.5 \times 20 = \mathbf{50.}$ **Ans.** $\quad \left(\because\ G = \dfrac{T_2}{T_1}\right)$

(*iv*) *Exact Centre distance and pitch circle diameter of each wheel*

For centre distance, using equation (9.27), we have

$$C = \frac{p_n \times T_1}{2\pi}\left[\frac{1}{\cos \alpha} + \frac{G}{\cos \beta}\right]$$

$$= \frac{10 \times 20}{2\pi}\left[\frac{1}{\cos 48°} + \frac{2.5}{\cos 42°}\right] = \frac{100}{\pi}[1.4944 + 3.364]$$

$$= \frac{100}{\pi} \times 4.8584 = \mathbf{154.64\ mm.}\ \ \textbf{Ans.}$$

From equation (9.26), we know the relation between circular pitch, pitch circle diameter and number of teeth on the wheels, as

$$p_c = \frac{\pi \times D}{T}$$

$\therefore \quad p_{c_1} = \dfrac{\pi D_1}{T_1}$ and $p_{c_2} = \dfrac{\pi D_2}{T_2}$

where D_1 = Pitch circle diameter of driving wheel and

D_2 = Pitch circle diameter of driven wheel.

From the above relations, we have

$D_1 = \dfrac{p_{c_1} \times T_1}{\pi} = \dfrac{14.94 \times 20}{\pi}$ $\quad$ ($\because$ $p_{c_1} = 14.94$ mm, $T_1 = 20$)

$= \mathbf{95.11\ mm.}$ **Ans.**

and $$D_2 = \frac{p_{c_2} \times T_2}{\pi} = \frac{13.45 \times 50}{\pi} \qquad (\because \; p_{c_2} = 13.45 \text{ mm}, T_2 = 50)$$

$$= \mathbf{214.06 \text{ mm}}. \quad \textbf{Ans.}$$

(v) The efficiency of the drive

The efficiency of the drive is given by equation (9.30) as

$$\eta = \frac{\cos \alpha \cos (\beta + \phi)}{\cos \beta \cos (\alpha - \phi)}$$

$$= \frac{\cos 48° \cos (42 + 6)}{\cos 42° \cos (48 - 6)} = \frac{\cos 48° \cos 48°}{\cos 42° \cos 42°}$$

$$= \frac{0.6691 \times 0.6691}{0.7431 \times 0.7431} = 0.81 \text{ or } \mathbf{81\%}. \quad \textbf{Ans.}$$

Problem 9.15. *The centre distance between the two shafts, which are connected by two left-handed helical gears is 37 cm. The shaft angle is 60° and normal module is 6 mm. If the gear ratio is 2 and larger gear is having 70 teeth, then find the helix angles of the two gears.*

Sol. Given :

$C = 37$ cm $= 370$ mm ; $\theta = 60°$, m_n = Normal module = 6 mm, $G = 2$ and $T_2 = 70$.

Let α = Helix angle of the driving gear

β = Helix angle of the driven gear

We know that, $\theta = \alpha + \beta$

or $60 = \alpha + \beta$

$\therefore \quad \beta = (60 - \alpha) \qquad ...(i)$

We also know that,

Normal circular pitch $= \pi \times$ normal module

$\therefore \quad p_n = \pi \times m_n$

$= \pi \times 6 \qquad (\because \; m_n = 6 \text{ mm}) \qquad ...(ii)$

We also know that, $G = \dfrac{T_2}{T_1}$

$\therefore \quad T_1 = \dfrac{T_2}{G} = \dfrac{70}{2} = 35$

Now using equation (9.27) for centre distance, we have

$$C = \frac{p_n T_1}{2\pi}\left[\frac{1}{\cos \alpha} + \frac{G}{\cos \beta}\right]$$

$$= \frac{\pi \times 6 \times T_1}{2\pi}\left[\frac{1}{\cos \alpha} + \frac{2}{\cos (60 - \alpha)}\right] \qquad (\because \; p_n = 6\pi, G = 2, \beta = 60 - \alpha)$$

$$= 3T_1\left[\frac{1}{\cos \alpha} + \frac{2}{\cos (60 - \alpha)}\right]$$

or $$370 = 3 \times 35\left[\frac{1}{\cos \alpha} + \frac{2}{\cos (60 - \alpha)}\right]$$

or $$\frac{370}{3 \times 35} = \left[\frac{1}{\cos \alpha} + \frac{2}{\cos (60 - \alpha)}\right]$$

or
$$3.524 = \frac{1}{\cos\alpha} + \frac{2}{\cos(60-\alpha)}$$

The above equation is solved by trial and error method.

(*i*) Let $\alpha = 30°$, then

$$\text{R.H.S.} = \frac{1}{\cos 30°} + \frac{2}{\cos(60-30)} = \frac{1}{\cos 30°} + \frac{2}{\cos 30°}$$

$$= \frac{3}{\cos 30°} = \frac{3}{0.866} = 3.464$$

But L.H.S. = 3.524

(*ii*) Let $\alpha = 28°$, then

$$\text{R.H.S.} = \frac{1}{\cos 28°} + \frac{2}{\cos(60-28)} = \frac{1}{\cos 28°} + \frac{2}{\cos 32°}$$

$$= 1.1325 + 2.3583 = 3.4908$$

(*iii*) Let $\alpha = 26°$, then

$$\text{R.H.S.} = \frac{1}{\cos 26°} + \frac{2}{\cos(60-26)} = \frac{1}{\cos 26°} + \frac{2}{\cos 34°}$$

$$= 1.1126 + 2.4124 = 3.525$$

The value of R.H.S. is very very close to L.H.S.

$\therefore$ $\alpha =$ **26°. Ans.**

and $\beta = 60 - 26 =$ **34°. Ans.**

HIGHLIGHTS

1. Gear is a toothed element, used for transmitting rotary motion from one shaft to another.
2. For no slip between two gears, the tangential velocity at the contact surfaces should be same

i.e. $V_1 = V_2$ $\left(\because\ V = \frac{\pi d N}{60}\right)$

or $\frac{N_1}{N_2} = \frac{d_2}{d_1}$ $\left(\text{or} = \frac{r_2}{r_1}\right)$

3. Gears are classified according to the

(*a*) **Position of shafts :**

(*i*) **Spur gears**—used for connecting two parallel shafts and having straight teeth parallel to the axis of the shaft.

(*ii*) **Helical gear**—used for connecting two parallel shafts and having teeth inclined or curved to the axis of the shaft.

(*iii*) **Bevel gear**—used for connecting two intersecting shafts straight bevel gears have straight and radial gears whereas helical or spiral bevel gears have inclined or curved gears.

(*b*) **Peripheral Velocities of the gears :**

(*i*) **Low Velocity gears**—if velocity of gears is less than 3 m/s.

(*ii*) **Medium Velocity gears**—if velocity of gears is between 3 m/s to 15 m/s.

(*iii*) **High Velocity gears**—if velocity of gears is more than 15 m/s.

(*c*) **Position of teeth on the Wheel :**

(*i*) **Straight gears**—spur gears have straight teeth.

(*ii*) **Inclined gears**—helical gears have inclined teeth.

(*iii*) **Curved gears**—spiral gears have curved teeth.

(*d*) **Type of gearing :**

(*i*) **External gear**—gears rotate in opposite directions.

(*ii*) **Internal gear**—gears rotate in same direction.

(*iii*) **Rack and pinion**—converts rotary motion into linear motion.

4. Circular pitch (p_c) is equal to the pitch circle circumference divided by number of teeth on the wheel *i.e.*

$$p_c = \frac{\pi D}{T}$$

where D = Diameter of pitch circle and

T = Number of teeth.

5. Diametral pitch (p_d) is given by,

$$p_d = \frac{T}{D}.$$

6. Module (m) is given by (module is generally given in mm)

$$m = \frac{D}{T}.$$

7. Gear ratio (G) is the ratio of number of teeth on the gear to that on the pinion. Hence

$$G = \frac{T}{t}$$

where T = Number of teeth on the gear,

t = Number of teeth on the pinion.

Gear ratio is always more than one.

8. Pinion is the smaller and usually the driving gear of a pair of mated gear.

9. The centre distance between two gears in terms of base circle radii is given by,

$$C = \frac{R_1 + R_2}{\cos \phi}$$

where R_1 and R_2 = Base circle radii of two meshing gears and

ϕ = pressure angle.

10. Path of approach and path of recess are given as,

Path of approach $= \sqrt{R_a^2 - R^2 \cos^2 \phi} - R \sin \phi$

and path of recess $= \sqrt{r_a^2 - r^2 \cos^2 \phi} - r \sin \phi$

where R = Pitch circle radius of wheel,

r = Pitch circle radius of pinion, (*i.e.* smaller gear)

R_a = Addendum circle radius of wheel,

r_a = Addendum circle radius of pinion and

ϕ = Pressure angle.

11. The path of contact is given by,

Path of contact $= \sqrt{R_a^2 - R^2 \cos^2 \phi} + \sqrt{r_a^2 - r^2 \cos^2 \phi} - (R + r) \sin \phi.$

12. Length of arc of contact (arc of contact is the path traced by a point on the pitch circle from the beginning to the end of engagement of a pair of teeth) is given by,

Length of arc of contact $= \dfrac{\text{Length of path of contact}}{\cos \phi}.$

13. Interference is the phenomenon between two mating teeth when the tip of one tooth under cuts the root of other mating tooth.

14. Maximum length of arc of contact is

$= (r + R) \tan \phi$

where r = pitch circle radius of pinion

R = pitch circle radius of wheel (of larger radius).

15. The minimum number of teeth (t) required on the pinion in order to avoid interference is given by

$$t = \frac{2A_p}{[\sqrt{1 + G(G+2)\sin^2\phi} - 1]}$$

where A_p = Fraction by which the standard addendum of one module for the pinion should be multiplied in order to avoid interference,

G = Gear ratio, and

ϕ = pressure angle.

For $G = 1 \quad \left(i.e. \frac{T}{t} = 1\right), \quad t = \frac{2A_p}{[\sqrt{1 + 3\sin^2\phi} - 1]}$.

16. Minimum number of teeth on the wheel if interference is to be avoided is,

$$T = \frac{2 \times A_w}{\left[\sqrt{1 + \frac{1}{G}\left(\frac{1}{G} + 2\right)\sin^2\phi} - 1\right]} \quad \text{and} \quad \text{for } G = 1, T = \frac{2 \times A_w}{[\sqrt{1 + 3\sin^2\phi} -]}.$$

17. There will be no interference between pinion and rack if the addendum of rack is less than $r \sin^2 \phi$

where r = Pitch circle radius of the pinion and

ϕ = Pressure angle

$\therefore \quad r \sin^2 \phi \geq$ Addendum of rack (or $A_R \times m$).

18. Minimum no. of teeth on pinion for no interference is given by

$$t_{min} = \frac{2 \times A_R}{\sin^2 \phi}$$

If $A_R = 1$ and $\phi = 20°$, then $t_{min} = 18$.

19. Helical gears are used to connect two parallel shafts and they are having their teeth inclined to the axes of the shaft.

20. Helix angle (or spiral angle of the teeth) is the angle to which the teeth are inclined to the axis of a gear.

21. Normal pitch (p_n) is the shortest distance between two adjacent teeth. Normal pitch of two mating gears should be same.

22. Circular pitch (p_c) is the distance between two adjacent teeth measured parallel to the axis of the gear.

23. The relation between normal pitch (p_n), circular pitch (p_c) and helix angle (ϕ) is given by $p_n = p_c \cos \phi$.

24. Spiral gears are used to connect two non-parallel and non-intersecting shafts.

25. Shaft angle (θ) is the angle through which the axis of one shaft must be rotated in order to bring it parallel to the axis of other shaft and the two shafts revolving in opposite direction.

26. The two spiral gears in mesh must have the same normal pitch. The circular pitch may be different.

27. The shortest distance between the two shafts is known as the centre distance between the two shafts. It is given by

$$C = \frac{D_1 + D_2}{2} \quad \text{where } D_1 = \frac{p_1 \times T_1}{\pi} \text{ and } D_2 = \frac{p_2 \times T_2}{\pi}$$

$$= \frac{p_n \times T_1}{2\pi}\left[\frac{1}{\cos \alpha} + \frac{G}{\cos \beta}\right]$$

where G = gear ratio = $\frac{T_2}{T_1} = \frac{N_1}{N_2}$

α = Helix angle of first gear (driving)

β = Helix angle of second gear (driven)

C = Centre distance

p_n = Normal pitch.

28. Efficiency of spiral gears is given by,

$$\eta = \frac{\cos \alpha \cos (\beta + \phi)}{\cos \beta \cos (\alpha - \phi)} \qquad \text{where } \phi = \text{friction angle.}$$

29. The efficiency of spiral gears will be maximum if

$$\alpha = \frac{\theta + \phi}{2} \qquad \text{where } \theta = \text{shaft angle} = \alpha + \beta$$

And the maximum efficiency is given by $\eta_{max} = \frac{\cos (\theta + \phi) + 1}{\cos (\theta - \phi) + 1}$.

EXERCISE 9

A. THEORETICAL QUESTIONS

1. Define and explain the terms : Gears ; friction wheels ; spur gears ; bevel gears and helical gears.
2. What are important types or classfications of the gears ?
3. What do you mean by pitch point ; circular pitch ; module, addendum and dedendum of a Gear ? What is the relation between circular pitch, diameter of pitch circle and number of teeth.
4. Define and explain the terms : path of approach, path of recess and path of contact between two mating gears.
5. Explain the terms :
   (*i*) module, (*ii*) pressure angle and
   (*iii*) addendum. (*A.M.I.E., Winter, 1976*)
6. State and prove the law of gear tooth action for constant velocity ratio and show how the involute teeth profile satisfies the condition. Derive an expression for the velocity of sliding between a pair of involute teeth. State the advantages of involute profile as a gear tooth profile. (*A.M.I.E., Winter, 1978*)
7. Explain what is interference and how it is prevented. (*A.M.I.E., 1986 S*)
8. What are the different forms of a tooth ?
9. Prove that the centre distance between two involute gear teeth which are in mesh is given by,

$$C = \frac{(R_1 + R_2)}{\cos \phi}$$

   where R_1 = Radius of the base circle of wheel 1,

   R_2 = Radius of the base circle of wheel 2, and

   ϕ = Pressure angle.
10. Name the four important system of gear teeth that are commonly used.
11. Find an expression for the length of the path of contact between two mating spur gears.
12. Define the term ; 'length of arc of contact' and prove that it is equal to length of path of contact divided by cosine of the pressure angle.
13. What do you mean by 'interference' between two mating gears ?
14. If interference between two involute gears is to be avoided then prove that the maximum length of arc of contact will be equal to $(r + R) \tan \phi$,

   where r = pitch circle radius of pinion,

   R = pitch circle radius of wheel, and

   ϕ = pressure angle.

15. Prove that the minimum number of teeth (t) required on the pinion in order to avoid interference is given by

$$t = \frac{2\,A_p}{[\sqrt{1 + G(G+2)\sin^2\phi} - 1]}$$

where A_p = Fraction by which the standard addendum of one module for the pinion should be multiplied in order to avoid interference,

G = Gear ratio, and

ϕ = Pressure angle.

16. Find an expression for minimum number of teeth (T) on the wheel if interference is to be avoided between two mating gears. Also find the above expression if gear ratio = 1.

17. Find an expression for the minimum number of teeth on a pinion which is in mesh with a rack in order to avoid interference.

18. Prove that the minimum no. of teeth required on a pinion to avoid interference with a rack is 18 if pressure angle $\phi = 20°$ and addendum co-efficient (A_R) is 1.

19. Define and explain the terms : Helical gears, helix angle, normal pitch, circular pitch and spiral gears.

20. What is the relation between normal pitch, circular pitch and helix angle.

21. Prove that the centre distance (C) between the two shafts connected by spiral gears is given by,

$$C = \frac{p_n \times T_1}{2\pi}\left[\frac{1}{\cos\alpha} + \frac{G}{\cos\beta}\right]$$

where p_n = Normal pitch of the spiral gears,

T_1 = No. of teeth on driving gear,

α = Helix angle or spiral angle of driving gear,

β = Spiral angle of driven gear, and

G = Gear ratio.

22. Find an expression for the efficiency of the spiral gears which connects two shafts.

23. Find the condition for the maximum efficiency of the spiral gears which connects two shafts. Also prove that the maximum efficiency is given by

$$\eta_{max} = \frac{\cos(\theta + \phi) + 1}{\cos(\theta - \phi) + 1}$$

where θ = shaft angle = $\alpha + \beta$ and

ϕ = friction angle.

B. NUMERICAL PROBLEMS

1. Determine the number of teeth and speed of the driver if the driven gear has 80 teeth of 8 mm module and rotates at 360 r.p.m. The two spur gears have a velocity ratio of $\frac{1}{4}$. Also calculate the pitch line velocities.

[Ans. 20 ; 1440 r.p.m. ; 12.06 m/s]

2. A spur gear has a module of 3 mm and its pitch line velocity is 942.45 mm/s. If the number of teeth of this spur gear is 20, find the speed of the gear. Also determine its circular pitch. **[Ans.** 300 r.p.m. ; 9.42 mm]

3. Calculate : (*i*) length of path of contact, (*ii*) arc of contact and (*iii*) contact ratio when a pinion having 17 teeth drives a gear having 49 teeth. The profile of the gear is involute with pressure angle 20°, module = 6 mm and addenda on pinion and gear wheel = 1 module. **[Ans.** 28.91 mm ; 30.8 mm and 1.6 or 2]

4. The number of teeth on each of the two equal spur gears in mesh is 30. The teeth have 20° involute profile and the module is 8 mm. If arc of contact is 1.3125 times the circular pitch, find the addendum. **[Ans.** 6.13 mm]

5. Two mating involute spur gears with module pitch of 5 mm have 20 and 40 teeth of 20° pressure angle and 5 mm addendum. Determine the max. velocity sliding and the angle turned through by pinion, when one pair of teeth is in mesh and pitch line speed is 1.2 m/s. **[Ans.** 455.4 mm/s and 29.45°]

6. Two mating involute spur gears with module pitch of 8 mm have 23 and 57 teeth of 20° pressure angle. The addenda on pinion and gear wheel are equal to one module. Find : (*i*) number of pairs of teeth in contact, (*ii*) angle turned through by pinion and gear wheels, and (*iii*) ratio of sliding velocity to rolling velocity at the beginning of the contact, at the pitch point and at the end of contact. [**Ans.** (*i*) 1.68, (*ii*) 26.3° and 10.63°, (*iii*) 0.32 ; 0 ; 0.287]

7. Two mating involute spur gears with module pitch of 10 mm have 20 teeth (on pinion) and 40 teeth (on gear wheel) of 20° pressure angle. The addendum is equal to one module. Does the interference occur. [**Ans.** No]

8. Two 20° involute spur gears mesh externally and have a gear ratio equal to 2. The number of teeth on smaller gear (*i.e.* pinion) is 20. The module is 10 mm. The addendum on each wheel is to be made of such a length that the line of contact on each side of the pitch point has half the maximum possible length. Determine : (*i*) the addendum height for each gear wheel, (*ii*) length of path of contact (*iii*) arc of contact and (*iv*) contact ratio.

[**Ans.** (*i*) 6.2 mm and 6.5 mm (*ii*) 51.3 mm, (*iii*) 54.6 mm, (*iv*) 1.75]

9. Find the minimum number of teeth required in order to avoid interference on a pinion which is to gear with : (*i*) an equal pinion (*ii*) a gear wheel to give a gear ratio of 3 and (*iii*) a rack. [**Ans.** (*i*) 13, (*ii*) 16 and (*iii*) 18]

10. Two 20° involute spur gears have 25 teeth each. The arc of contact is equal to 1.6 times the circular pitch. Find the addendum required in terms of circular pitch. [**Ans.** 0.317 times the circular pitch]

11. Two mating involute spur gears have 28 and 45 teeth and a standard addendum of one module. Find the length of path of contact and length of arc of contact in terms of module when pressure angle is 20°.

[**Ans.** 6.113 m and 6.314 m]

12. The angle between two meshing spiral gears is 90° and centre distance is 160 mm approximately. The normal circular pitch of the gears is 8 mm and the gear ratio is 3. The frictional angle is 5° and efficiency of the drive is maximum. Determine :

(*i*) the spiral angles of the teeth,

(*ii*) the number of teeth on each wheel,

(*iii*) the exact centre distance, and pitch circle diameters of the two wheels and

(*iv*) the efficiency of the drive. [**Ans.** (*i*) $\alpha = 47.5°$, $\beta = 42.5°$, (*ii*) $T_1 = 23$, $T_2 = 69$, (*iii*) $C = 162.5$ mm, $D_1 = 86.7$ mm and $D_2 = 238.3$ mm, (*iv*) $\eta = 84\%$]

13. The centre distance between two meshing spiral gears is 400 mm approximately and the angle between the two shafts is 50°. The normal circular pitch of the gears is 18 mm and gear ratio = 3. The frictional angle is 6° and spiral angles for the driving and driven wheels are equal. Find : (*i*) number of teeth on each wheel, (*ii*) Exact centre distance and (*iii*) efficiency of the drive. [**Ans.** (*i*) $T_1 = 32$, $T_2 = 96$, (*ii*) $C = 404.5$ mm, (*iii*) 90.7%]

10
Gear Trains

10.1. Introduction

A combination of two or more gears, which are arranged in such a way that power is transmitted from a driving shaft to a driven shaft, is known a *gear train.* This term is generally applied to mean more than two gears in mesh between the driving shaft and the driven shaft. The gear train may consist of spur, bevel or spiral gears.

10.2. Types of Gear Trains

The following are the different types of gear trains :

1. *Ordinary gear train* which is further classified as

(*i*) Simple gear train and

(*ii*) Compound gear train. The compound gear train may be reverted gear train or non-reverted gear train.

2. *Epicyclic gear train* which may have one fixed gear or no fixed gear (which is also known as differential).

10.2.1. Simple Gear Train

If the axes of all gears (or the axes of the shafts on which gears are mounted) remain fixed relative to each other, the gear train is known as simple gear train or *ordinary gear train.*

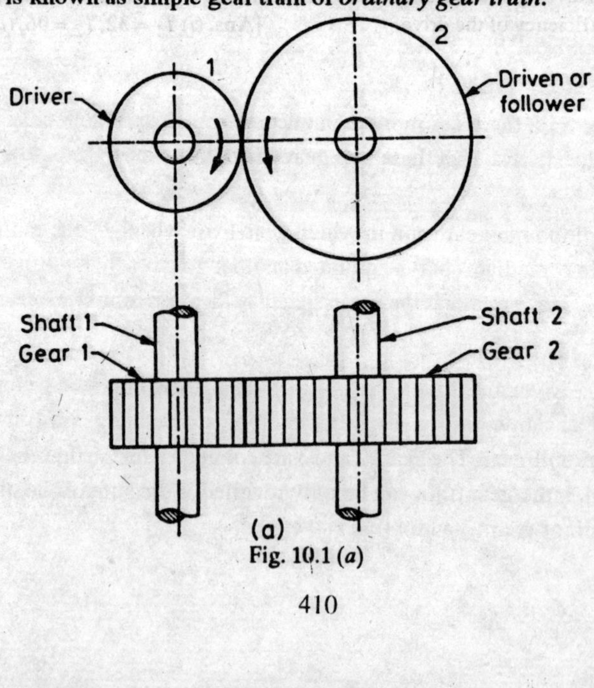

Fig. 10.1 (*a*)

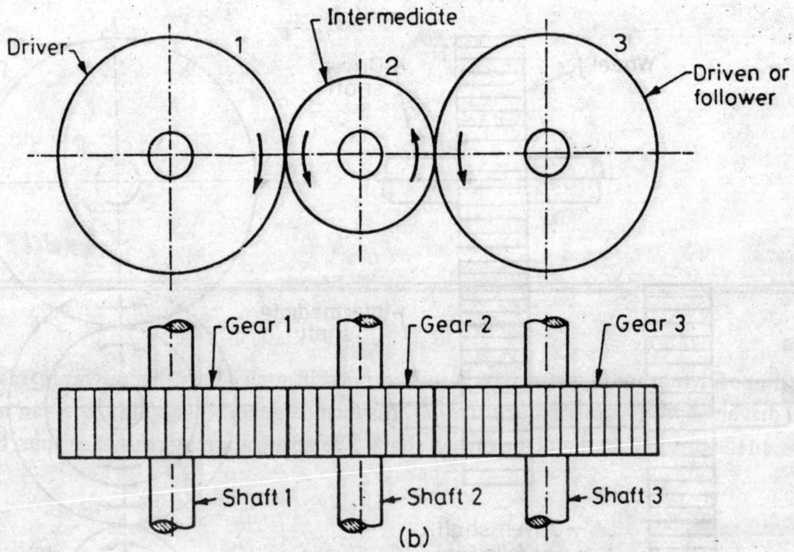

Fig. 10.1 (*b*). Simple Gear Train.

In case of a simple gear train, each gear is on a separate shaft as shown in Fig. 10.1 (*a*). In Fig. 10.1 (*a*), there are only two gears. Each gear is mounted on the separate shaft. The combination of these two gears is known as simple gear train. If power is transmitted from gear 1 to gear 2, then gear 1 is driver whereas gear 2 is driven or follower. These two gears rotate in opposite direction.

In Fig. 10.1 (*b*), there are three gears. The shaft 1 carries only one gear, shaft 2 also carries only one gear and shaft 3 also has only one gear. The combination of these gears, is known as simple gear train, when they are arranged in such a way that power is transmitted from a driving shaft to a driven shaft. The shaft 1 is a driver whereas the shaft 3 is known as driven shaft or follower. The shaft 2 is called intermediate shaft. It may be noted that when the number of intermediate shafts are *odd*, the motion of both the shafts or gears (*i.e.* driver and follower) is like. But if the number of intermediate shafts are *even*, the motion of follower will be in the opposite direction of the driver.

10.2.2. Compound Gear Train

In a compound gear train there are more than one gear on a shaft, generally intermediate shafts have two gears rigidly fixed to the shaft so that these two gears have the same speed as they, are mounted on same shaft.

Fig 10.2 shows a compound gear train in which gear 1 (or wheel 1) drives the gear 2 mounted on the intermediate shaft. On the intermediate shaft is mounted another gear 3 which meshes with the gear 4 mounted on the driven shaft. Gears 2 and 3 rotate at the same speed as they are mounted on the same shaft.

10.2.3. Reverted Gear Train

When the axes of the driver and driven shafts are co-axial, the compound gear train is known as reverted gear train or co-axial train as shown in Fig. 10.3. It should be noted that gears 1 and 4 are on the driver and driven shafts whose axes are collinear. The gears 2 and 3 are connected on the intermediate shaft. In the reverted gear train shown in Fig. 10.3, the gear train can be only reverted if the sum of the pitch radii of gears 1 and 2 equals the sum of pitch radii of gears 3 and 4 (*i.e.* $r_1 + r_2 = r_3 + r_4$).

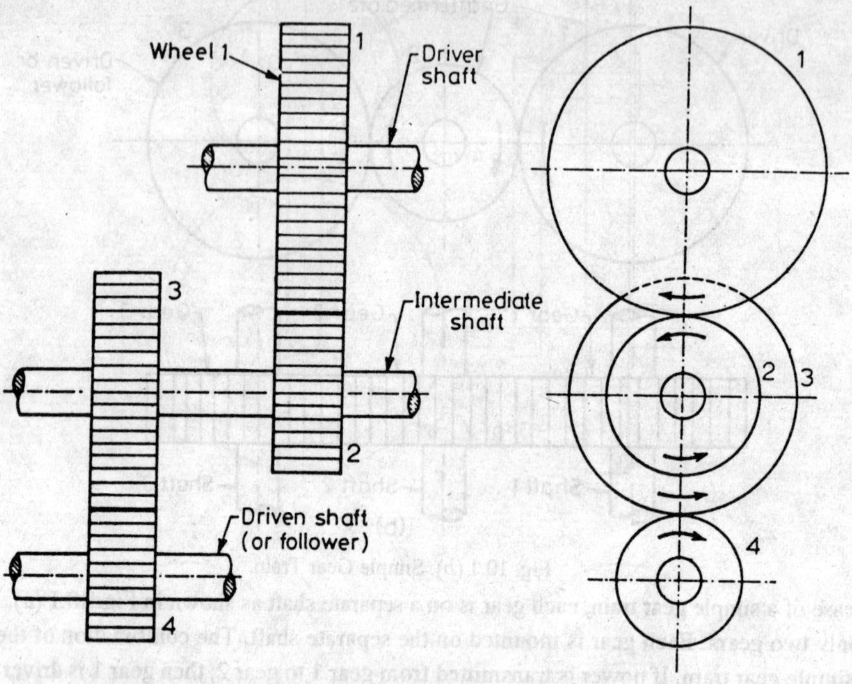

Fig. 10.2. Compound Gear Train.

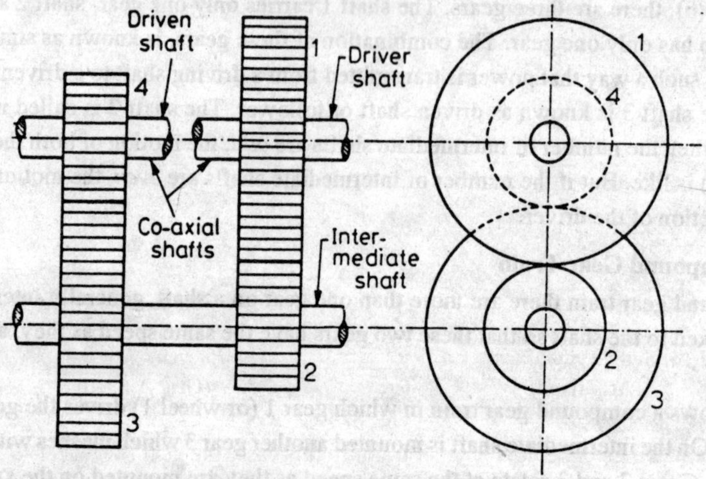

Fig. 10.3. Reverted Gear Train.

10.2.4. Epicyclic Gear Train

If the axes of the shafts, over which the gears are mounted are moving relative to a fixed axis, the gear train is known as an epicyclic gear train. Hence in an epicyclic train, at least one of the gear axis is in motion relative to the frame. The Fig. 10.4 shows two gears *A* and *B* which are meshing with each other along with an arm. The arm and gear *A* are having a common axis at O_1 about which they can rotate. The gear *B* is having its axis at O_2 on the arm. If the arm is fixed, then the axis of the two shaft on which wheels *A* and *B* are mounted will be fixed and we will get a simple gear train as shown in Fig. 10.4 (*a*). Here gear *A* can drive gear *B* or vice versa. But if the wheel *A* is fixed and arm is rotated about O_1 as shown in Fig. 10.4 (*b*) then gear *B* will be

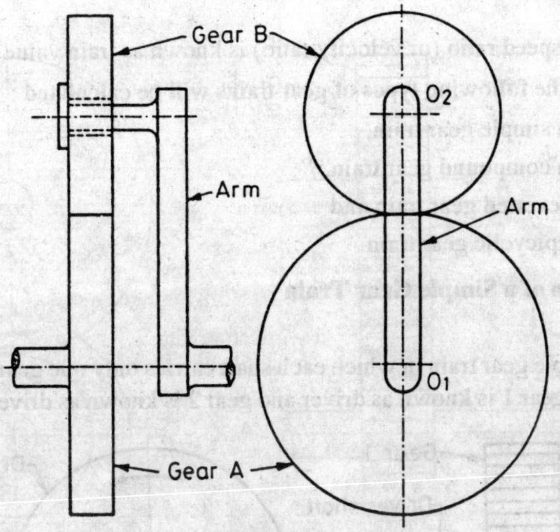

Fig. 10.4

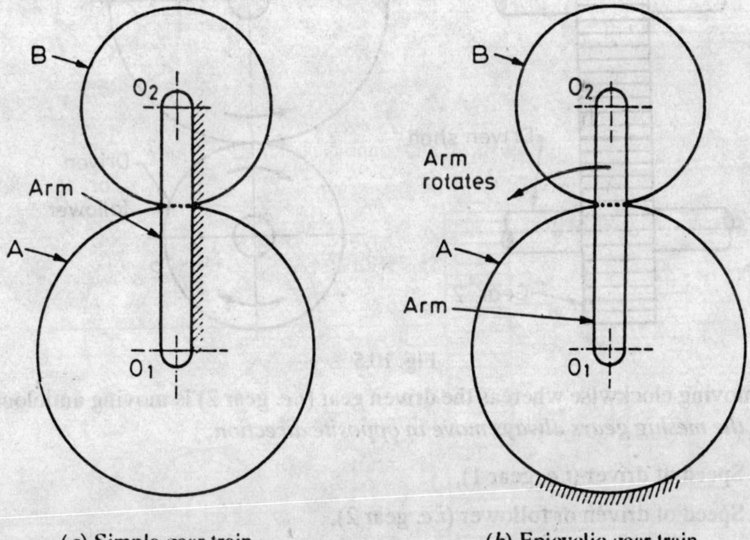

(*a*) Simple gear train (Arm fixed)

(*b*) Epicyclic gear train (Gear A fixed)

Fig. 10.4

rotating about gear *A* and we get an epicyclic gear train. In this case, the gear (or wheel) *B* rolls about gear *A*. Hence the axis of gear *B* moves relative to the axis of gear *A*. The epicyclic gear train may be simple or compound.

10.3. Velocity Ratio of Gear Trains

The ratio of the speed of the driver to the speed of the follower is known as velocity ratio or speed ratio of gear train. Mathematically, it is written as

$$\text{Velocity Ratio (V.R.)} = \frac{\text{Speed of driver}}{\text{Speed of follower}}.$$

The reciprocal of the speed ratio (or velocity ratio) is known as train value of gear train.

The velocity ratio of the following types of gear trains will be calculated :

(*i*) Velocity ratio of a simple gear train,

(*ii*) Velocity ratio of a compound gear train,

(*iii*) Velocity ratio of reverted gear train and

(*iv*) Velocity ratio of epicyclic gear train.

10.3.1. Velocity Ratio of a Simple Gear Train

First Case

Fig. 10.5 shows a simple gear train in which each shaft carries only one gear. The power is transmitted from gear 1 to gear 2. Hence gear 1 is known as driver and gear 2 is known as driven or follower. The driving

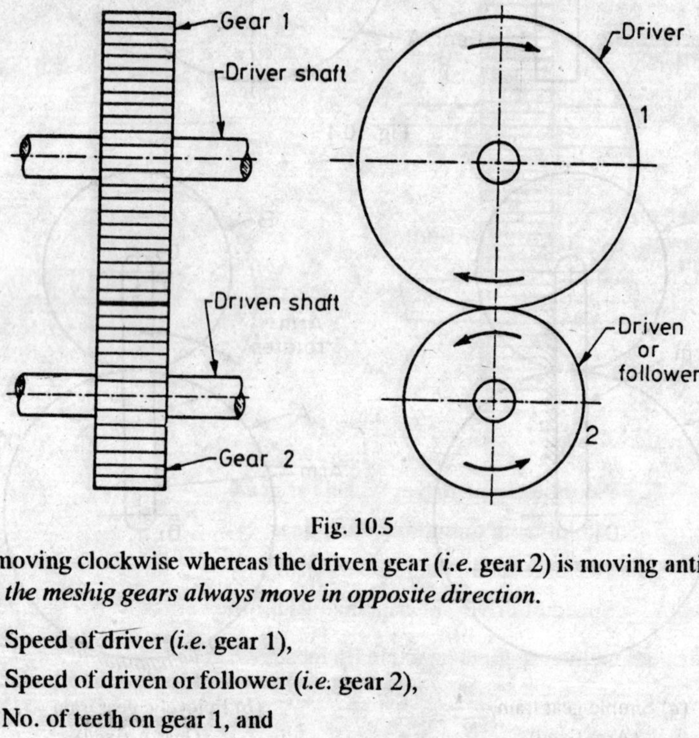

Fig. 10.5

gear (*i.e.* gear 1) is moving clockwise whereas the driven gear (*i.e.* gear 2) is moving anticlockwise. *Hence in a simple gear train, the meshig gears always move in opposite direction.*

Let N_1 = Speed of driver (*i.e.* gear 1),

N_2 = Speed of driven or follower (*i.e.* gear 2),

T_1 = No. of teeth on gear 1, and

T_2 = No. of teeth on gear 2.

Speed ratio (or velocity ratio) $= \dfrac{\text{Speed of driver}}{\text{Speed of follower}} = \dfrac{N_1}{N_2}$

But the ratio of speeds of any pair of gears in mesh is the inverse of their number of teeth.

$$\therefore \quad \frac{N_1}{N_2} = \frac{T_2}{T_1}$$

$$\therefore \quad \text{Speed ratio} = \frac{N_1}{N_2} = \frac{T_2}{T_1}$$

The reverse of the speed ratio is known as the train value of the gear train.

$\therefore$ Train value $= \dfrac{N_2}{N_1} = \dfrac{T_1}{T_2}$.

Second Case

Fig 10.6 shows a simple gear train with one intermediate shaft

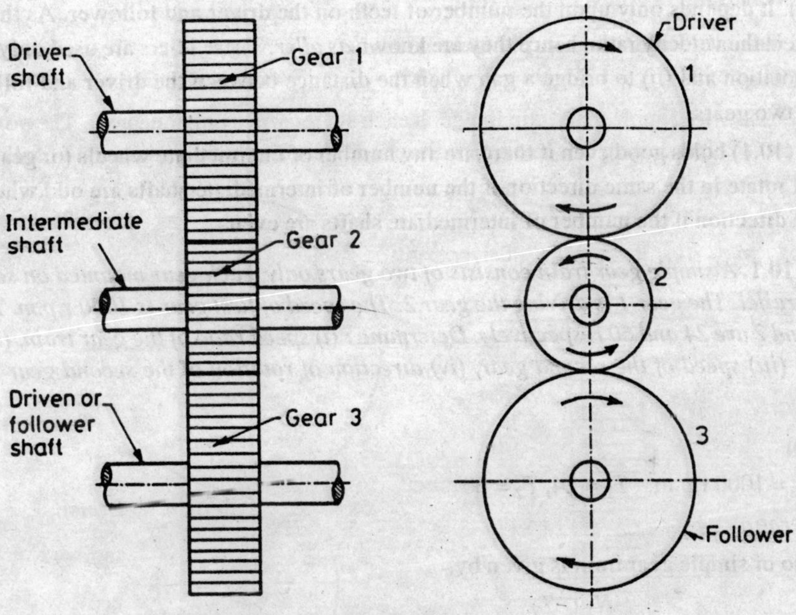

Fig. 10.6

Let T_1 = No. of teeth on driver wheel (or gear)

T_2 = No. of teeth on intermediate gear

T_3 = No. of teeth on the follower gear

N_1, N_2 and N_3 = Speed of driver, intermediate and follower respectively in r.p.m.

When the driver and intermediate gears are in mesh, we get

$$\frac{N_2}{N_1} = \frac{T_1}{T_2} \quad ...(i)$$

Similarly, when the intermediate gear and follower are in mesh, we get

$$\frac{N_3}{N_2} = \frac{T_2}{T_3} \quad ...(ii)$$

Multiplying equation (*i*) and (*ii*), we get

$$\frac{N_2}{N_1} \times \frac{N_3}{N_2} = \frac{T_1}{T_2} \times \frac{T_2}{T_3}$$

or

$$\frac{N_3}{N_1} = \frac{T_1}{T_3}$$

$$\therefore \quad \frac{\text{Speed of follower}}{\text{Speed of driver}} = \frac{\text{No. of teeth on driver}}{\text{No. of teeth on follower}}$$

or

$$\text{Speed ratio} = \frac{\text{Speed of driver}}{\text{Speed of follower}} = \frac{\text{No. of teeth on follower}}{\text{No. of teeth on driver}} \quad ...(10.1)$$

From equation (10.1), we see that the velocity ratio is independent of the no. of teeth on intermediate wheels (or gears). It depends only upon the number of teeth on the driver and follower. As the intermediate wheels do not effect the velocity ratio, hence they are known *as idler*. These idlers are used only : (*i*) to change the direction of rotation and (*ii*) to bridge a gap when the distance between the driver and follower shafts is too great for just two gears.

Equation (10.1) holds good even if there are any number of intermediate wheels (or gears). The driver and follower will rotate in the same direction if the number of intermediate shafts are odd whereas they will rotate in opposite direction if the number of intermediate shafts are even.

Problem 10.1. *A simple gear train consists of two gears only, each gear mounted on separate shafts. The shafts are parallel. The gear 1 is driving the gear 2. The speed of first gear is 1000 r.p.m. The number of teeth of gears 1 and 2 are 24 and 60 respectively. Determine : (i) speed ratio of the gear train, (ii) Train value of the gear train, (iii) speed of the second gear, (iv) direction of rotation of the second gear if first gear is rotating clockwise.*

Sol. Given :

$N_1 = 1000$ r.p.m. ; $T_1 = 24$, $T_2 = 60$

(*i*) *Speed ratio*

Speed ratio of simple gear train is given by,

$$\text{Speed ratio} = \frac{N_1}{N_2} = \frac{T_2}{T_1}$$

Here the values of T_2 and T_1 are given.

$$\therefore \quad \text{Speed ratio} = \frac{T_2}{T_1} = \frac{60}{24} = \mathbf{2.5.} \quad \textbf{Ans.}$$

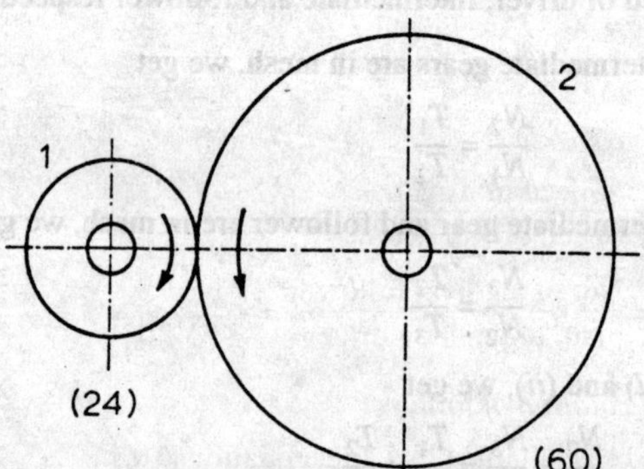

Fig. 10.7

(*ii*) *Train value*

$$\text{Train value} = \frac{1}{\text{Speed ratio}} = \frac{1}{2.5} = \mathbf{0.4.}$$

(*iii*) *Speed of second gear*

We know that

$$\text{Speed ratio} = \frac{N_1}{N_2}$$

or $$2.5 = \frac{1000}{N_2}$$

$$\therefore \quad N_2 = \frac{1000}{2.5} = \mathbf{400\ r.p.m.}\quad \textbf{Ans.}$$

(*iv*) In a simple gear train, the two meshing gears always move in opposite directions. As the first gear is rotating clockwise hence second gear will rotate in **anticlockwise direction. Ans.**

Problem 10.2. *A simple gear train consists of three gears, each mounted on separate shaft as shown in Fig. 10.8. The shafts are parallel. The gear 1 is the driver and gear 3 is follower. The gear 1 is rotating*

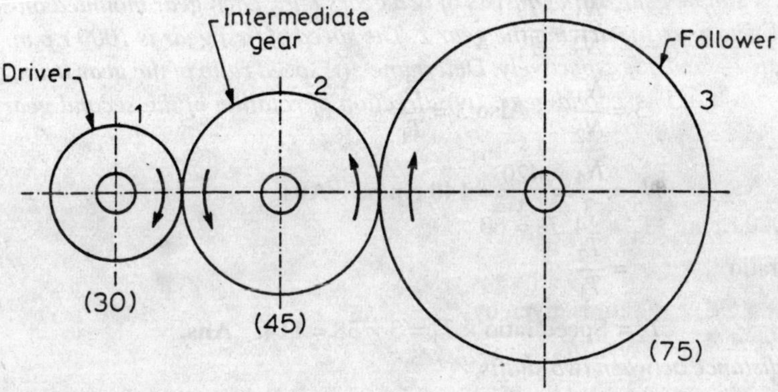

Fig. 10.8

clockwise at a speed of 750 r.p.m. The no. of teeth on gears 1, 2 and 3 are 30, 45 and 75 respectively. Find : (i) speed ratio of the gear train, (ii) direction of rotation and speed of rotation of follower.

Sol. Given :

$N_1 = 750$ r.p.m. ; $T_1 = 30$; $T_2 = 45$ and $T_3 = 75$

(*i*) *Speed ratio*

Using equation (10.1), we get

$$\text{Speed ratio} = \frac{\text{Speed of driver}}{\text{Speed of follower}}$$

$$= \frac{\text{No. of teeth on follower}}{\text{No. of teeth on driver}} = \frac{T_3}{T_1} = \frac{75}{30} = \mathbf{2.5.}\quad \textbf{Ans.}$$

(*ii*) *Direction and speed of rotation of follower*

Speed of rotation of follower is obtained by using equation (10.1).

$$\text{Speed ratio} = \frac{\text{Speed of driver}}{\text{Speed of follower}}$$

$$\therefore \quad \text{Speed of follower} = \frac{\text{Speed of driver}}{\text{Speed ratio}} = \frac{N_1}{2.5} = \frac{750}{2.5} = \mathbf{300\ r.p.m.}\quad \textbf{Ans.}$$

Direction of rotation

In a simple gear train with intermediate gears, the driver and follower will rotate in the same direction if the number of intermediate shafts (or gears) are odd. Here no. of intermediate gear is only one (*i.e.* no. of gear is odd), hence follower will rotate **clockwise. Ans.**

Problem 10.3. *Two parallel shafts are connected with the help of two gears one gear on each shaft. The number of teeth on one gear is 38 and speed of the shaft is 420 r.p.m. If the speed ratio is equal to 3 and circular pitch of the gears is 25 mm, then, find : (i) number of teeth and speed of other shaft, (ii) centre distance between two shafts.*

Sol. Given :

$T_1 = 38$; $N_1 = 420$ r.p.m. ; speed ratio = 3 ; circular pitch*, $p_c = 25$ mm.

Find : (*i*) T_2 and N_2, (*ii*) centre distance between two shafts.

(*i*) *Value of T_2 and N_2*

We know that speed ratio is given by,

$$\text{Speed ratio} = \frac{N_1}{N_2} = \frac{T_2}{T_1}$$

or

$$3 = \frac{N_1}{N_2}. \quad \text{Also } 3 = \frac{T_2}{T_1}$$

$$\therefore \quad N_2 = \frac{N_1}{3} = \frac{420}{3} = \mathbf{140\ r.p.m.}\quad \textbf{Ans.}$$

$$\text{Also speed ratio} = \frac{T_2}{T_1}$$

$$\therefore \quad T_2 = \text{Speed ratio} \times T_1 = 3 \times 38 = \mathbf{114.}\quad \textbf{Ans.}$$

(*ii*) *Centre distance between two shafts*

Centre distance between two parallel shafts

$$= r_1 + r_2$$

$$= \frac{d_1}{2} + \frac{d_2}{2} = \frac{d_1 + d_2}{2} \quad \ldots(i)$$

Also we know that

$$T_1 = \frac{\pi d_1}{p_c} \qquad \therefore \quad d_1 = \frac{T_1 \times p_c}{\pi} = \frac{38 \times 25}{\pi} = 302.39 \text{ mm}$$

Similarly,

$$T_2 = \frac{\pi d_2}{p_c}$$

$$\therefore \quad d_2 = \frac{T_2 \times p_c}{\pi} = \frac{114 \times 25}{\pi} = 907.18 \text{ mm}$$

Substituting the value of d_1 and d_2 is equation (*i*), we get,

$$\text{Centre distance} = \frac{302.39 + 907.18}{2} = \mathbf{604.78\ mm.}\quad \textbf{Ans.}$$

Problem 10.4. *Determine the number of teeth and pitch for two toothed wheels to transmit a velocity ratio of 4 to 1 between two parallel shafts, the centre of which are at a distance of 676 mm approximately. The drive must satisfy the following conditions :*

(*i*) *the standard pitch in module (mm) must be chosen from the following :*

24, 22, 18, 16, 15, 14, 13, 12 and 11 ;

*Refer equation (9.2) for circular pitch, $p_c = \dfrac{\pi D}{T}$.

(ii) the actual distance between shaft centres must not vary by more than 1% from that given above ;

(iii) the number of teeth must be as small as possible. (AMIE, S 1994)

Sol. Given :

$$\text{V.R.} = 4:1 \qquad \therefore \quad \frac{N_1}{N_2} = \frac{T_2}{T_1} = \frac{4}{1}; \quad \text{Centre distance} = 676 \text{ mm (approx.)};$$

We know that centre distance between a gear pair $= r_1 + r_2$

or centre distance $$= \frac{d_1}{2} + \frac{d_2}{2} = \frac{d_1 + d_2}{2} \qquad ...(i)$$

We also know that, module*, $m = \dfrac{\text{Pitch circle diameter}}{\text{No. of teeth}} = \dfrac{d}{T}$

$\therefore$ $$d = m \times T$$

or $$d_1 = m \times T_1 \text{ and } d_2 = m \times T_2$$

Substituting the values of d_1 and d_2 in equation (*i*), we get

Centre distance $$= \frac{mT_1 + mT_2}{2} \qquad \left(\because \text{V.R.} = \frac{T_2}{T_1} = 4 \therefore T_2 = 4T_1 \right) \qquad ...(ii)$$

$\therefore$ Centre distance $$= \frac{m \times T_1 + m \times 4T_1}{2} = \frac{5mT_1}{2}$$

or $$676 = \frac{5m \times T_1}{2} \qquad (\because \text{Centre distance} = 676)$$

or $$m \times T_1 = \frac{676 \times 2}{5} = 270.4 \qquad ...(iii)$$

Let us now select the suitable value of module (m) from the standard given values. But we must remember that :

(*i*) The value of 'm' should be such that no. of teeth is an integer

(*ii*) Smaller of the two gears must have enough teeth to avoid interference.

Let us select $m = 15$ from standard given values.

Substituting this value of $m = 15$ in equation (*iii*), we get

$$15 \times T_1 = 270.4$$

$\therefore$ $$T_1 = \frac{270.4}{15} = 18.02 \simeq 18$$

Then, $$T_2 = 4T_1 = 4 \times 18 = 72.$$

And the exact centre distance from equation (*ii*), will be

$$= \frac{mT_1 + mT_2}{2} = \frac{m}{2}(T_1 + T_2) = \frac{15}{2}(18 + 72) = \mathbf{675 \text{ mm}}. \quad \textbf{Ans.}$$

Percentage variation of centre distance

$$= \left(\frac{\text{Final} - \text{Initial}}{\text{Initial}} \right) \times 100 = \left(\frac{675 - 676}{676} \right) \times 100 = -0.145\%$$

The variation is within limits.

10.3.2. Velocity Ratio of a Compound Gear Train

In a compound gear train as shown in Fig. 10.2, the intermediate shafts have two wheels (or gears) rigidly fixed to it so that these two gears have the same speed. One of these two gears meshes with the driver and the other with the follower.

*For module m see equation (9.5).

In Fig. 10.2, the gear 1 (or wheel 1) drives the gear 2 mounted on the intermediate shaft. On the intermediate shaft is mounted another gear 3 which meshes with the gear 4 mounted on the driven (or follower) shaft. Gears 2 and 3 rotate at the same speed as they are mounted on the same intermediate shaft.

Let T_1 = No. of teeth on driver 1.

T_2, T_3 and T_4 = No. of teeth on gears 2, 3 and 4 respectively.

N_1 = speed of driver 1.

N_2, N_3 and N_4 = speed of gears 2, 3 and 4 respectively.

The gear 1 drives the gear 2, therefore,

$$\frac{N_2}{N_1} = \frac{T_1}{T_2} \qquad \text{...}(i)$$

Gear 3 drives the gear 4, therefore

$$\frac{N_4}{N_3} = \frac{T_3}{T_4} \qquad \text{...}(ii)$$

Multiplying equations (*i*) and (*ii*), we get

$$\frac{N_2}{N_1} \times \frac{N_4}{N_3} = \frac{T_1}{T_2} \times \frac{T_3}{T_4}.$$

But $N_2 = N_3$ as gears 2 and 3 are mounted on the same shaft.

Hence the above equation becomes are

$$\frac{N_4}{N_1} = \frac{T_1 \times T_3}{T_2 \times T_4}.$$

As gear 1 drives gear 2 whereas gear 3 drives gear 4 hence gears 1 and 3 becomes as drivers whereas gears 2 and 4 becomes as followers. Hence the above equation can be stated as

$$\frac{\text{Speed of follower}}{\text{Speed of driver}} = \frac{\text{Product of teeth on the drivers}}{\text{Product of teeth on the followers}} \qquad \text{...(10.2)}$$

Problem 10.5. *In a compound gear train shown in Fig. 10.9, the power is transmitted from a motor shaft to output shaft. The motor shaft is connected to gear 1 whereas the output shaft is connected to gear 4.*

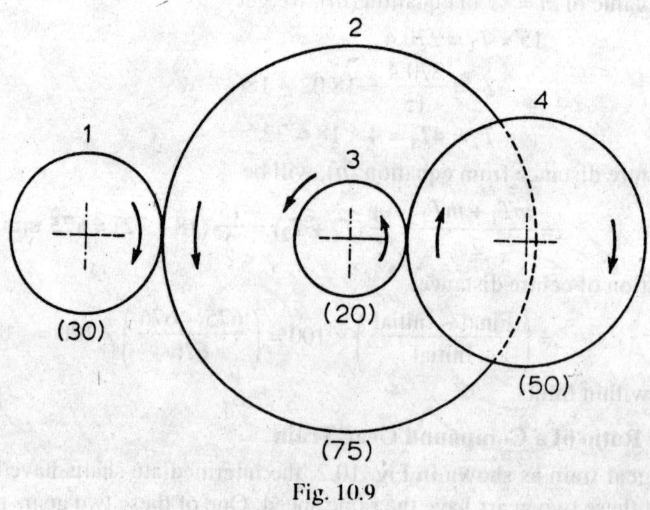

Fig. 10.9

The gears 2 and 3 are mounted on the same shaft. The motor shaft is rotating at 1250 r.p.m. in the clockwise direction. Find the direction and speed of the output shaft. The no. of teeth on each gear are given as :

Gear	*1*	*2*	*3*	*4*
No. of teeth	*30*	*75*	*20*	*50*

Sol. Given :

$$N_1 = 1250 \text{ r.p.m.} ; T_1 = 30 ; T_2 = 75 ; T_3 = 20 \text{ and } T_4 = 50$$

From Fig. 10.9 it is clear that gears 1 and 3 are driving gears (*i.e.* drivers) whereas gears 2 and 4 are driven gears or followers. The gears 2 and 3 are mounted on the same shaft and hence they will rotate in the same direction.

The gear 1 is rotating in clockwise direction. The gear 2 is in mesh with gear 1. Hence gear 2 will rotate in opposite direction *i.e.* in anticlockwise direction. Gear 3 will rotate in the direction of gear 2 *i.e.* in anticlockwise direction. The gear 4 is in mesh with gear 3. Hence gear 4 will rotate in opposite direction *i.e.* clockwise direction. Gear 4 is connected to output shaft.

∴ Direction of rotation of output shaft = **Clockwise. Ans.**

Speed of output shaft

Let N_4 = speed of output shaft (or speed of gear 4)

Using the relation,

$$\frac{\text{Speed of first driver}}{\text{Speed of last driven}} = \frac{\text{Product of no. of teeth on drivens}}{\text{Product of no. of teeth on drivers}}$$

or

$$\frac{N_1}{N_4} = \frac{T_2 \times T_4}{T_1 \times T_3} = \frac{75 \times 50}{30 \times 20} = 6.25$$

$$\therefore \quad N_4 = \frac{N_1}{6.25} = \frac{1250}{6.25} = \mathbf{200 \text{ r.p.m.}} \quad \textbf{Ans.}$$

Problem 10.6. *In a compound gear train, shown in Fig. 10.10, the power is transmitted from a motor shaft to output shaft. The motor shaft is connected to gear 1 whereas the output shaft is connected to gear 6. The motor shaft is rotating at 1125 r.p.m. in the clockwise direction, find the direction and speed of output shaft. The number of teeth on each gear are given as :*

Gear	*1*	*2*	*3*	*4*	*5*	*6*
No. of teeth	*30*	*75*	*40*	*120*	*28*	*70*

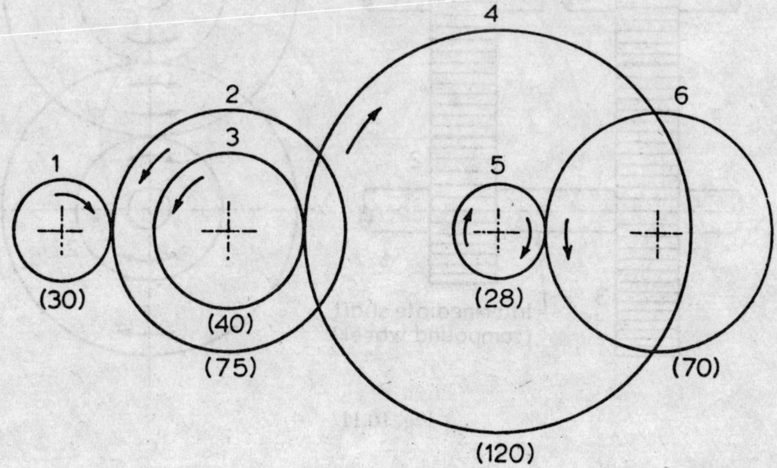

Fig. 10.10

Sol. Given :

$N_1 = 1125$ r.p.m. ; $T_1 = 30$; $T_2 = 75$; $T_3 = 40$; $T_4 = 120$; $T_5 = 28$ and $T_6 = 70$.

(*i*) *Direction of rotation of output shaft*

Refer to Fig. 10.10. The gears 1, 3 and 5 are driving gears (*i.e.* driver) whereas the gears 2, 4 and 6 are driven gears or followers. The gears 2 and 3 are mounted on the same shaft and hence they rotate in the same direction. Also gears 4 and 5 are mounted on the same shaft and hence they will also rotate in the same direction. The gear 1 is rotating in clockwise direction. Then gear 2 which is in mesh with gear 1 will rotate in anticlockwise direction. Gear 3 will rotate in the direction of gear 2 (*i.e.* anticlockwise). The gear 4 is in mesh with gear 3 and hence gear 4 will rotate in clockwise direction. The gear 5 will also rotate in clockwise direction. The gears 5 and 6 are in mesh and hence gear 6 will rotate in anticlockwise direction.

∴ Direction of rotation of output shaft = **Anticlockwise. Ans.**

(*ii*) *Speed of output shaft*

Let N_6 = Speed of gear 6 or speed of output shaft or speed of last driven gear.

Using the relation,

$$\frac{\text{Speed of first driver}}{\text{Speed of last driven}} = \frac{\text{Product of no. of teeth on drivens}}{\text{Product of no. of teeth on drivers}}$$

or
$$\frac{N_1}{N_6} = \frac{T_2 \times T_4 \times T_6}{T_1 \times T_3 \times T_5} = \frac{75 \times 120 \times 70}{30 \times 40 \times 28} = 18.75$$

∴
$$N_6 = \frac{N_1}{18.75} = \frac{1125}{18.75} = \mathbf{60\ r.p.m.}\quad \textbf{Ans.}$$

10.3.3. Velocity Ratio of Reverted Gear Train

Fig. 10.11 shows a reverted gear train in which driver and driven shafts are having their axes collinear. The gear 1 is mounted on the driver shaft whereas gear 4 is mounted on driven shaft. The gears 2 and 3 are on the intermediate shaft and this is a compound wheel. In case of reverted gear train shown in Fig. 10.11, the

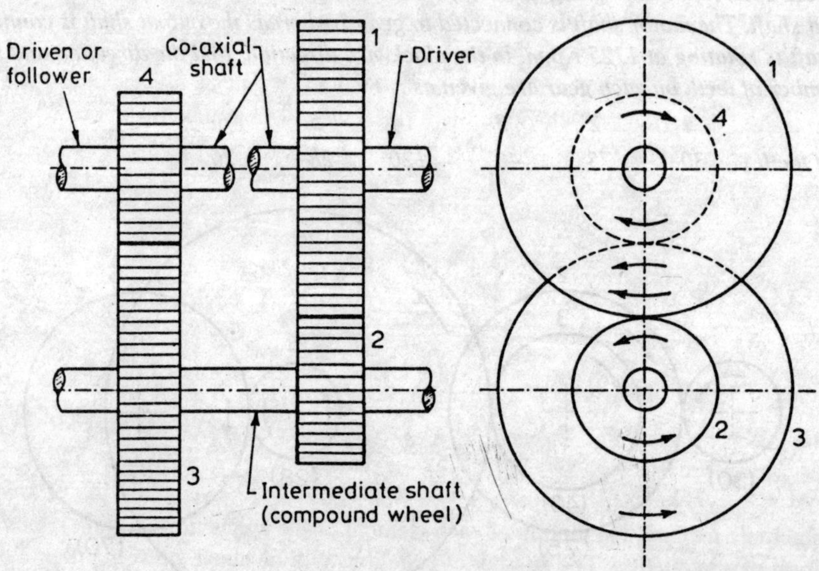

Fig. 10.11

sum of the pitch radii of gears 1 and 2 should be equal to the sum of the pitch radii of gears 3 and 4 (or the distance between the centres of shafts of gears 1 and 2 as well as gears 3 and 4 is same).

$$\therefore \quad r_1 + r_2 = r_3 + r_4 \quad \text{...(10.3)}$$

where r_1 = Radius of gear 1.

r_2, r_3 and r_4 = Radii of gears 2, 3 and 4 respectively.

If the circular pitch of all the gears is assumed to be same, then number of teeth on each wheel will be proportional to its circumference or radius. Hence the equation (10.3) can be written as

$$T_1 + T_2 = T_3 + T_4 \quad \text{...(10.4)}$$

where T_1 = No. of teeth on wheel 1.

T_2, T_3 and T_4 = No. of teeth on wheel 2, 3 and 4 respectively.

The speed ratio is given by,

$$\text{Speed ratio} = \frac{\text{Product of number of teeth on drivens}}{\text{Product of number of teeth on drivers}}$$

or

$$\frac{N_1}{N_4} = \frac{T_2 \times T_4}{T_1 \times T_3} \quad \text{...(10.5)}$$

All the above three equations *i.e.* (10.3), (10.4) and (10.5) are used for finding the number of teeth on each gear.

10.3.4. Velocity Ratio of Epicyclic Gear Train

The velocity ratio of epicyclic gear train may be determined by various methods. We shall consider only the two following methods :

1. Algebraic method or relative velocity method, and
2. Tabular method.

The basic principle of these method is that the relative motion between a pair of mating gears is always the same whether the axes of rotation are fixed or moving. And the ratio of the relative motion is equal to the inverse ratio of number of teeth on the gears.

1. Algebraic Method

Fig. 10.12 shows two gears *A* and *B* and an arm *C*. Gear *A* is fixed and the system becomes an example of epicyclic gear train. Let us find the velocity ratio of this epicyclic gear train by algebraic method.

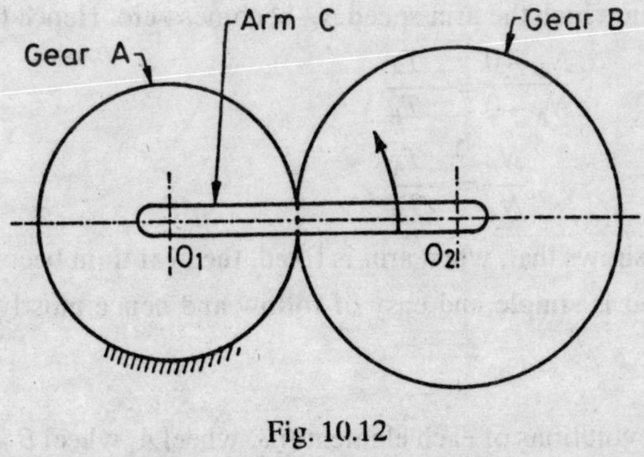

Fig. 10.12

In the algebraic method, the motion of each element of the epicyclic gear train relative to the arm is written in the form of an equation. The number of equations will be equal to the number of elements in the gear train. The fixed element will have zero speed. Other element has specified or given speed. These two conditions are used for solving all the equations.

Let N_A = Speed of gear A in r.p.m.,

N_B = Speed of gear B, and

N_C = Speed of arm C in r.p.m.

T_A = No. of teeth on gear A

T_B = No. of teeth on gear B

The speed of gear A relative to arm C $= N_A - N_C$

and, the speed of gear B relative to arm C $= N_B - N_C$

Since the gears A and B are meshing directly, therefore they will revolve in opposite direction.

But the relative motion between a part of mating gears is always same whether the axes of rotation are fixed or moving. And this ratio of relative motion is equal to the inverse ratio of the number of teeth on the gears. Therefore for mating gears A and B, we have,

$$\frac{\text{Speed of } B \text{ rel. to arm}}{\text{Speed of } A \text{ rel. to arm}} = -\frac{T_A}{T_B} \qquad \text{(– ve sign is due to opposite rotation)}$$

or
$$\frac{N_B - N_C}{N_A - N_C} = -\frac{T_A}{T_B} \qquad ...(i)$$

Since wheel A is fixed, hence $N_A = 0$. Therefore the above equation becomes as

$$\frac{N_B - N_C}{0 - N_C} = -\frac{T_A}{T_B}$$

or
$$-\frac{N_B}{N_C} + 1 = -\frac{T_A}{T_B}$$

or
$$\frac{N_B}{N_C} = 1 + \frac{T_A}{T_B}$$

or
$$N_B = N_C \left[1 + \frac{T_A}{T_B} \right]$$

Hence, if the speed of arm C is known then speed of wheel B can be calculated. The no. of teeth on gears A and B are already given.

If instead of gear A, the arm C is fixed, then the ratio of the speed of gear B to the speed of gear A is obtained from equation (i), in which the arm speed N_C becomes zero. Hence from equation (i), we have

$$\frac{N_B - 0}{N_A - 0} = -\frac{T_A}{T_B}$$

or
$$\frac{N_B}{N_A} = -\frac{T_A}{T_B}$$

The above equation shows that, when arm is fixed, the gear train becomes a simple gear train.

The algebraic method is simple and easy of follow and hence mostly used for solving problems on epicyclic gear train.

2. Tabular Method

In this method, the revolutions of each element (*i.e.* wheel A, wheel B and arm C) along with the steps of motion are written in a tabular form as shown in Table 10.1.

Let us apply this method to the epicyclic gear train shown in Fig. 10.12. If the arm is supposed to be fixed, then the axes of both the wheels are also fixed relative to each other. Let the wheel A is rotated by one revolution anti-clockwise, then the wheel B will make $\frac{T_A}{T_B}$ revolution clockwise $\left(\text{This is because } \frac{N_B}{N_A} = \frac{T_A}{T_B} \text{ and if } N_A = 1 \text{ revolution, then } N_B = \frac{T_A}{T_B} \right)$. If anticlockwise rotation is assumed positive and clockwise rotation as negative then wheel A makes + 1 revolution whereas wheel B makes $-\frac{T_A}{T_B}$ revolution. This relative motion is shown in the first row of the table. It also means that if the wheel A makes $+x$ revolution, then the wheel B will make $-x\frac{T_A}{T_B}$ revolutions. This statement is shown in the second row of the table, [*i.e.* second row is obtained by multiplying x to the first row of the element].

Now give each element of the epicyclic train + y revolution. This means that whole system is bodily revolved by y revolution. This statement is shown in third row. The resultant motion of each element of the gear is obtained by adding up the second and third row. This statement is shown in fourth row.

Table 10.1

S. No.	*Steps of Motion*	*Revolution of elements*		
		Arm C	*Wheel A*	*Wheel B*
1.	Arm is fixed. Rotate A through + 1 revolution *i.e.* 1 revolution Anticlockwise.	0	+ 1	$-\frac{T_A}{T_B}$
2.	Arm is fixed. Rotate A through + x revolution.	0	+ x	$-x \cdot \frac{T_A}{T_B}$
3.	Add + y revolution to all elements.	+ y	+ y	+ y
4.	Resultant motion (*i.e.* add second and third row)	+ y	$x + y$	$-x\frac{T_A}{T_B} + y$
		First column	Second column	Third column

In the third column of revolution of elements of the fourth row, there are four unknowns *i.e.* x, y, T_A and T_B. If two conditions of rotation of any two elements are known, then the speed of third element may be obtained by substituting the given data in the third column of the fourth row.

Note. (*i*) If anticlockwise rotation is + ve, then clockwise rotation will be negative.

(*ii*) The term '+ x' revolution means the element is rotated anticlockwise by x revolution.

(*iii*) Add '+ y' revolutions to all elements, it means that all elements are rotated anti-clockwise by 'y' revolution. This is equivalent to locking the whole system and bodily revolving the whole system by 'y' revolution.

(*iv*) If the no. of teeth on a gear in a gear train is not given, then the no. of teeth on that gear is obtained by assuming all gears of the same pitch.

Let us apply the above method to a specific problem.

Suppose it is given that arm C is rotating through 5 r.p.m. anticlockwise and wheel A is fixed, then find the speed of wheel B.

Now according to 1st condition as wheel A is fixed, hence $N_A = 0$. But the resultant motion of wheel A from table 10.1 is $(x + y)$. Hence,

$$x + y = 0 \qquad \text{...(i)}$$

The second condition is that arm C is rotating at 5 r.p.m. anticlockwise.

Hence $N_C = 5$. (Anticlockwise rotation is assumed positive). But the resultant motion of arm C from table 10.1 is y. Hence

$$y = 5 \qquad \text{...(ii)}$$

Substituting this value of y in equation (i), we get

$$x + 5 = 0 \quad \text{or} \quad x = -5$$

Now the resultant motion of wheel B from table 10.1 is given as $-x \dfrac{T_A}{T_B} + y$. Hence the speed of B will be

$$= -x \frac{T_A}{T_B} + y$$

$$= -(-5) \frac{T_A}{T_B} + 5 \qquad (\because \; x = -5 \text{ and } y = 5)$$

$$= 5 \frac{T_A}{T_B} + 5$$

From the above equation the speed of B will be obtained if the number of teeth on wheel A and B are known.

Note. (i) In tabular method the arm is fixed and the fixed member of the train (here wheel A) is given one revolution in the anticlockwise direction. The motion of the remaining members of the epicyclic gear train is calculated like ordinary gearing.

(ii) The resultant motion of any element is obtained from the last row (*i.e.* 4th row here) of the table.

Problem 10.7. *The arm of an epicyclic gear train rotates at 100 r.p.m. in the anti-clockwise direction. The arm carries two wheels A and B having 36 and 45 teeth respectively. The wheel A is fixed and the arm rotates about the centre of wheel A. Find the speed of wheel B. What will be the speed of B, if the wheel A instead of being fixed, makes 200 r.p.m. clockwise ? Solve by both the methods i.e. algebraic method and tabular method.*

Sol. Given :

Arm speed, $N_C = 100$ r.p.m. (Anticlockwise)

Anticlockwise rotation is taken + ve whereas clockwise rotation as – ve.

No. of teeth on A, $T_A = 36$

No. of teeth on B, $T_B = 45$

(i) Wheel A is fixed. Hence $N_A = 0$, find N_B.

(ii) Wheel A makes 200 r.p.m. clockwise, find N_B.

(a) Solution by Algebraic Method

The speed of wheel A relative to arm C $= N_A - N_C$

The speed of wheel B relative to arm C $= N_B - N_C$.

Since the wheels A and B are meshing directly, therefore they will revolve in opposite direction.

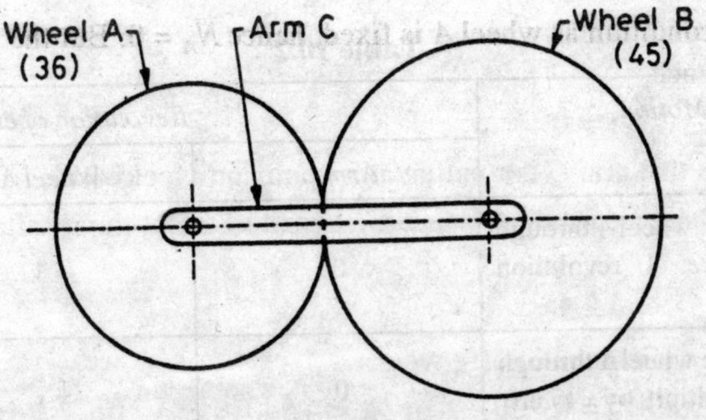

Fig. 10.13

But the relative motion between a pair of mating gears is always same whether the axes of rotation are fixed or moving. And this ratio of relative motion is equal to the inverse ratio of no. of teeth on the gears. Hence for mating gears A and B, we have

$$\frac{\text{Speed of } B \text{ relative to arm } C}{\text{Speed of } A \text{ relative to arm } C} = -\frac{T_A}{T_B}$$ (– ve sign in due to rotation in opposite direction)

or

$$\frac{N_B - N_C}{N_A - N_C} = -\frac{T_A}{T_B} \quad ...(i)$$

(i) Speed of wheel B when A is fixed

Hence $N_A = 0$. But speed of arm C, $N_C = +100$ r.p.m. $T_A = 36$, $T_B = 45$.

Substituting these values in equation (*i*), we get

$$\frac{N_B - 100}{0 - 100} = -\frac{36}{45} = -\frac{4}{5}$$

or

$$N_B - 100 = -100\left(-\frac{4}{5}\right) = 80$$

$\therefore$ $N_B = 100 + 80 = 180$ **r.p.m. anticlockwise. Ans.**

($\because$ Positive value means the direction is anticlockwise).

(ii) Speed of wheel B when A makes 200 r.p.m. clockwise

Here $N_A = -200$ r.p.m. (– ve sign due to clockwise rotation)

$N_C = +100$

Substituting these values in equation (*i*), we get

$$\frac{N_B - 100}{-200 - 100} = -\frac{36}{45} = -\frac{4}{5}$$

or

$$N_B - 100 = -300 \times \left(-\frac{4}{5}\right) = 240$$

$\therefore$ $N_B = 240 + 100 =$ **340 r.p.m. anticlockwise. Ans.**

(*b*) **Solution by Tabular method**

The table of motion is prepared as given in Table 10.2 below :

Table 10.2

S. No.	*Steps of Motion*	*Revolution of elements*		
		Arm C	*Wheel A*	*Wheel B*
1.	Arm is fixed. Rotate wheel *A* through + 1 revolution (*i.e.* 1 revolution anti-clockwise.	0	– 1	$-\frac{T_A}{T_B}$
2.	Arm is fixed. Rotate wheel *A* through + *x* revolution. (Multiply by *x* to all)	0	+ *x*	$-x\frac{T_A}{T_B}$
3.	Add + *y* revolutions to all elements.	+ *y*	+ *y*	+ *y*
4.	Resultant motion (*i.e.* add second and third row)	+ *y*	*x* + *y*	$-x\frac{T_A}{T_B}+y$

(*i*) *Speed of wheel B when A is fixed*

Speed of arm is given as 100 r.p.m. anticlockwise. But from the table 10.2, the resultant motion of the arm is *y*. Hence

$$y = +\ 100 \text{ r.p.m.}$$

As wheel *A* is fixed, hence $N_A = 0$. But from the table 10.2 the resultant motion of wheel *A* is $(x + y)$.

$\therefore$ $x + y = 0$ or $x = -y = -100$ r.p.m.

From the table 10.2, the resultant motion of wheel *B* is $-x\frac{T_A}{T_B}+y$.

$$\therefore \text{ Speed of wheel } B = -x\frac{T_A}{T_B}+y$$

$$= -(-100)\frac{36}{45} + 100 \qquad (\because\ x = -100 \text{ and } y = 100)$$

$$= 80 + 100 = 180 \text{ r.p.m. (Anticlockwise).} \quad \textbf{Ans.}$$

(*ii*) *Speed of wheel B when A makes 200 r.p.m. clockwise*

Here $N_A = -200$ r.p.m. (– ve sign is due to clockwise motion)

But from the table 10.2 the resultant motion of wheel *A* is $(x + y)$. Hence

$$x + y = -200 \qquad ...(i)$$

Arm speed is 100 r.p.m. anticlockwise. But arm speed from the table is *y*

$$\therefore \quad y = +\ 100$$

Substituting the value of *y* in equation (*i*), we get

$$x = -200 - y = -200 - 100 = -300$$

Now from the table 10.2 the resultant motion of wheel *B* is $-x\frac{T_A}{T_B}+y$.

$$\therefore \text{ Speed of wheel } B = -x\frac{T_A}{T_B}+y$$

$$= -(-300)\times\frac{36}{45} + 100 \qquad (\because\ x = -300)$$

$$= 240 + 100 = 340 \text{ r.p.m. anticlockwise.} \quad \textbf{Ans.}$$

($\because$ positive value means, that the direction of rotation is anticlockwise).

Problem 10.8. *In a reverted epicyclic train, the arm F carries two wheels A and D and a compound wheel B-C. The wheel A meshes with wheel B and the wheel D meshes with wheel C. The numbers of teeth on wheel A, D and C are 80, 48 and 72 respectively. Find the speed and direction of wheel D when wheel A is fixed and arm F, makes 200 r.p.m. clockwise.*

Sol. Given :

No. of teeth on wheel A, $T_A = 80$

No. of teeth on wheel D, $T_D = 48$

No. of teeth on wheel C, $T_C = 72$

Wheel A is fixed and speed of arm F,

$$N_F = 200 \text{ r.p.m. clockwise}$$
$$= -200 \text{ r.p.m. anticlockwise} \quad \text{(– ve sign is due to clockwise rotation).}$$

Find the speed and direction of wheel D. Let us first find the no. of teeth on wheel B.

From the Fig. 10.14, it is clear that $r_A + r_B = r_D + r_C$ or $\rightarrow d_A + d_B = d_D + d_C$.

The number of teeth are proportional to their pitch circle diameters, hence for the **reverted** epicyclic train shown in Fig. 10.14, we get

$$T_A + T_B = T_D + T_C \qquad \text{where } T_B = \text{No. of teeth on wheel } B$$

$$\therefore \quad T_B = T_D + T_C - T_A = 48 + 72 - 80 = 40$$

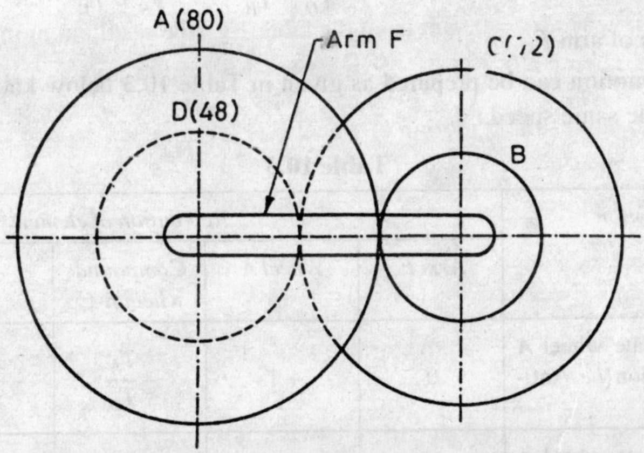

Fig. 10.14

Before making the table of motion, let us first find the revolution of each element when arm is fixed and wheel A makes one revolution anticlockwise.

As wheels A and B are having external teeth in mesh, hence they will rotate in opposite direction. The ratio of their speeds will be inversely proportional to their number of teeth.

$$\therefore \quad \frac{N_B}{N_A} = -\frac{T_A}{T_B} \qquad \text{(– ve sign is due to opposite direction)}$$

But $\quad N_A = +1,$

$$\therefore \quad N_B = -\frac{T_A}{T_B} \qquad ...(i)$$

Wheels B and C is a compound wheel (or wheels B and C are mounted on the same shaft). Hence their speeds will be same.

$$N_C = N_B = -\frac{T_A}{T_B} \qquad ...(ii)$$

The wheels C and D are having external teeth and hence they will rotate in opposite direction. The ratio of their speeds is inversely proportional to their number of teeth,

$$\therefore \quad \frac{N_D}{N_C} = -\frac{T_C}{T_D}$$

or
$$N_D = \left(-\frac{T_C}{T_D}\right) \times N_C$$

$$= \left(-\frac{T_C}{T_D}\right) \times \left(-\frac{T_A}{T_B}\right) \qquad \left[\because N_C = \left(-\frac{T_A}{T_B}\right)\right]$$

$$= \frac{T_C}{T_D} \times \frac{T_A}{T_B}$$

$\therefore$ For one revolution of wheel A when arm is fixed, we have

Revolution of wheel A $= +1$

Revolution of wheel B or C $= \left(-\frac{T_A}{T_B}\right)$

Revolution of wheel D $= \frac{T_C}{T_D} \times \frac{T_A}{T_B}$ or $\frac{T_A}{T_B} \times \frac{T_C}{T_D}$

Revolution of arm F $= 0$

Now the table of motion can be prepared as given in Table 10.3 below knowing that the compound wheel B-C will rotate at the same speed.

Table 10.3

S. No.	Steps of Motion	Revolution of elements			
		Arm F	Wheel A	Compound wheel B-C	Wheel D
1.	Arm is fixed. Rotate wheel A through + 1 revolution (*i.e.* Anti-clockwise)	0	+ 1	$-\frac{T_A}{T_B}$	$-\frac{T_A}{T_B} \times \left(-\frac{T_C}{T_D}\right) = \frac{T_A}{T_B} \times \frac{T_C}{T_D}$
2.	Arm is fixed. Rotate wheel A through $+x$ revolution. (Multiply by x)	0	$+x$	$-x \times \frac{T_A}{T_B}$	$x \times \frac{T_A}{T_B} \times \frac{T_C}{T_D}$
3.	Add $+y$ revolutions to all	$+y$	$+y$	$+y$	$+y$
4.	Resultant motion (*i.e.* Add second and third row)	$+y$	$x+y$	$-x\frac{T_A}{T_B} + y$	$x\frac{T_A}{T_B} \times \frac{T_C}{T_D} + y$

But it is given that speed of arm is 200 r.p.m. clockwise or – 200 r.p.m. anticlockwise. But from the table 10.3 the resultant motion of arm is y r.p.m. anticlockwise.

$$\therefore \quad y = -200 \text{ r.p.m.}$$

The wheel A is fixed. Hence speed of wheel A is zero. But from the table 10.3 the resultant motion of wheel A is $(x + y)$.

$$\therefore \quad x + y = 0$$

or
$$x = -y = -(-200) = 200 \qquad (\because y = -200)$$

Now resultant motion of wheel D from the table 10.3 is $x\dfrac{T_A}{T_B}\times\dfrac{T_C}{T_D}+y$.

$\therefore$ Speed of wheel D $=x\dfrac{T_A}{T_B}\times\dfrac{T_C}{T_D}+y$

$=200\times\dfrac{80}{40}\times\dfrac{72}{48}-200$ ($\because$ $x=200$, $y=-200$)

$=600-200=400$ r.p.m. **Ans.**

The positive sign shows that wheel D rotates in the anti-clockwise direction.

Problem 10.9. *An epicyclic train as shown in Fig. 10.15 is composed of a fixed annular wheel A having 150 teeth. Meshing with A is a wheel B which drives wheel D through an idle wheel C, D being concentric*

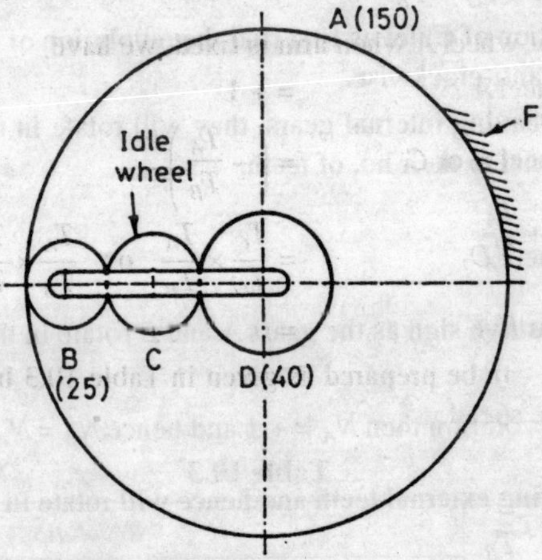

Fig. 10.15

with A. Wheels B and C are carried on an arm which revolves clockwise at 100 r.p.m. about the axis of A or D. If the wheels B and D are having 25 teeth and 40 teeth respectively, find the number of teeth of C and the speed and sense of rotation of C. (A.M.I.E., Winter 1984)

Sol. Given :

No. of teeth on A, $T_A = 150$

Speed of arm = 100 r.p.m. clockwise = – 100 r.p.m. anticlockwise

No. of teeth on B, $T_B = 25$

No. of teeth on D, $T_D = 40$

Wheel A is fixed. Hence speed of A, $N_A = 0$

Find : (*i*) No. of teeth on C, (*ii*) Speed of C and (*iii*) Its sense *i.e.* direction of rotation.

(*i*) *No. of teeth on C*

Let T_C = No. of teeth of C and N_C = speed of C.

Assuming the diametral pitch to be the same for all the wheels.

From Fig. 10.15, we see that radius of A = Dia. of B + Dia. of C + Radius of D

or $$\frac{d_A}{2}=d_B+d_C+\frac{d_D}{2}$$

Since number of teeth on the wheels are proportional to their pitch circle diameters, hence the above equation becomes as

$$\frac{T_A}{2} = T_B + T_C + \frac{T_D}{2}$$

$$\left(\because \pi d = p \times T \text{ or } d = \frac{p \times T}{\pi} \text{ or } d \propto T \text{ if } p \text{ is same}\right)$$

or

$$T_C = \frac{T_A}{2} - T_B - \frac{T_D}{2}$$

$$= \frac{150}{2} - 25 - \frac{40}{2} = 75 - 25 - 20 = \mathbf{30.} \quad \textbf{Ans.}$$

(ii) Speed and direction of C

To find the speed and direction of C, let us first find the revolution of each element when arm is fixed and wheel A makes one revolution anti-clockwise.

As the wheels A and B are having internal gears, they will rotate in the same direction. The ratio of their speeds are inversely proportional to their no. of teeth.

$$\therefore \quad \frac{N_B}{N_A} = + \frac{T_A}{T_B}$$

Before $\frac{T_A}{T_B}$, we will have positive sign as the gears A and B rotate in the same direction.

When wheel A makes + 1 revolution then $N_A = +1$ and hence $N_B = N_A \times \frac{T_A}{T_B} = 1 \times \frac{T_A}{T_B} = \frac{T_A}{T_B}$.

The wheels B and C are having external teeth and hence will rotate in opposite direction.

$$\therefore \quad \frac{N_C}{N_B} = -\frac{T_B}{T_C} \quad \text{(– ve sign is due to rotation in opposite direction)}$$

$$\therefore \quad N_C = -\frac{T_B}{T_C} \times N_B$$

$$= -\frac{T_B}{T_C} \times \frac{T_A}{T_B} \quad \left(\because N_B = \frac{T_A}{T_B}\right)$$

$$= -\frac{T_A}{T_C}$$

The wheels C and D are having external teeth and hence will rotate in opposite direction.

$$\therefore \quad \frac{N_D}{N_C} = -\frac{T_C}{T_D} \quad \text{(– ve sign is due to rotation in opposite direction)}$$

$$\therefore \quad N_D = \left(-\frac{T_C}{T_D}\right) \times N_C$$

$$= \left(-\frac{T_C}{T_D}\right) \times \left(-\frac{T_A}{T_C}\right) \quad \left(\because N_C = -\frac{T_A}{T_C}\right)$$

$$= \frac{T_A}{T_D}$$

Now the first row of the table 10.4 can be filled.

Prepare the table of motion as given below :

Table 10.4

S. No.	Operation	Revolution of				
		Arm	Wheel A	Wheel B	Wheel C	Wheel D
1.	Arm fixed and give one revolution to wheel A in anti-clockwise direction.	0	+ 1	$+\frac{T_A}{T_B}$	$-\frac{T_A}{T_B} \times \frac{T_B}{T_C} = -\frac{T_A}{T_C}$	$\left(-\frac{T_A}{T_C}\right) \times \left(-\frac{T_C}{T_D}\right) = \frac{T_A}{T_D}$
2.	Multiply by x to all	0	$+x$	$-x \cdot \frac{T_A}{T_B}$	$-x \cdot \frac{T_A}{T_C}$	$+x \cdot \frac{T_A}{T_D}$
3.	Add y revolution to all	$+y$	$x+y$	$+x \cdot \frac{T_A}{T_B} + y$	$-x \cdot \frac{T_A}{T_C} + y$	$+x \cdot \frac{T_A}{T_B} + y$

In the problem, it is given that the arm rotates at 100 r.p.m. clockwise or – 100 r.p.m. anticlockwise. But from the above table 10.4 we see that arm rotates through y revolution anti-clockwise.

$$\therefore \quad y = -100$$

Also it is given that the wheel A is fixed. This means the revolution of A is zero. But the revolution of A from the above table 10.4 is $x + y$.

$$\therefore \quad x + y = 0$$

$$\therefore \quad x = -y = -(-100) = 100$$

Speed of C

From the above table 10.4, the speed of C is $-x\frac{T_A}{T_C} + y$.

$$\therefore \quad \text{Speed of } C = -x\frac{T_A}{T_C} + y$$

$$= -100 \times \frac{150}{30} + (-100) \qquad [\because T_A = 150, T_C = 30 \text{ and } y = -100]$$

$$= -500 - 100 = -600 \text{ r.p.m.}$$

– ve sign means the speed is in clockwise direction, as anti-clockwise rotation was taken positive.

Hence wheel C will rotate at 600 r.p.m. clockwise. **Ans.**

Problem 10.10. *An epicyclic train is shown in Fig. 10.16. Internal gear A is keyed to the driving shaft and has 30 teeth. Compound wheel C and D of 20 and 22 teeth respectively are free to rotate on the pin fixed to the arm P which is rigidly connected to the driven shaft. Internal gear B which has 32 teeth is fixed. If the driving shaft runs at 60 r.p.m. clockwise, determine the speed of the driven shaft. What is the direction of rotation of driven shaft with reference to driving shaft ? (Figure on next page)* (A.M.I.E., Summer, 1986)

Sol. Given :

No. of teeth on gear A, $T_A = 30$

No. of teeth on C, $T_C = 20$

No. of teeth on D, $T_D = 22$

No. of teeth on B, $T_B = 32$

Speed of driving shaft = 60 r.p.m. clockwise.

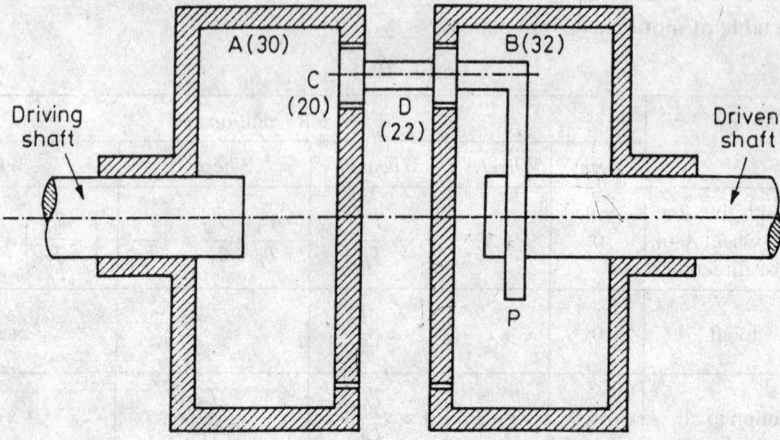

Fig. 10.16

As gear A is keyed to the driving shaft,

$\therefore \quad N_A = 60$ r.p.m. clockwise

Find : (*i*) Speed of driven shaft, (*ii*) Direction of rotation of driven shaft.

To prepare the table of motion, let us find the no. of revolution of each element for one revolution of wheel A, keeping arm fixed.

The gears C and A are in mesh internally, hence they will rotate in the same direction. Also their speeds are inversely proportional to their no. of teeth.

$$\therefore \quad \frac{N_C}{N_A} = \frac{T_A}{T_C} \quad \text{or} \quad N_C = \frac{T_A}{T_C} \times N_A = \frac{T_A}{T_C} \times 1 = \frac{T_A}{T_C} \qquad (\because \; N_A = 1)$$

Wheel C-D is a compound wheel. Hence speed of C is equal to speed of D

$$\therefore \quad N_D = N_C = \frac{T_A}{T_C}$$

The gears D and B are in mesh internally and hence they will rotate in the same direction.

$$\therefore \quad \frac{N_B}{N_D} = \frac{T_D}{T_B} \quad \text{or} \quad N_B = \frac{T_D}{T_B} \times N_D \quad \text{or} \quad N_B = \frac{T_D}{T_B} \times \frac{T_A}{T_C}.$$

Now prepare the table of motion as given below :

Table 10.5

S. No.	*Operation*	*Revolution of elements*			
		Arm	*Wheel A*	*Wheel compound C-D*	*Wheel B*
1.	Arm fixed and give + 1 revolution to wheel A in anticlockwise direction.	0	+ 1	$+\frac{T_A}{T_C}$	$+\frac{T_A}{T_C} \times \frac{T_D}{T_B}$
2.	Multiply by x to all elements.	0	$+x$	$+x \cdot \frac{T_A}{T_C}$	$+x \cdot \frac{T_A}{T_C} \cdot \frac{T_D}{T_B}$
3.	Add y revolution to all elements	y	$x + y$	$x \cdot \frac{T_A}{T_C} + y$	$+x \cdot \frac{T_A}{T_C} \cdot \frac{T_D}{T_B} + y$

In the problem, it is given that internal gear B is fixed. Hence revolution of wheel B is zero. But from the table 10.5, the revolution of wheel B is $x \cdot \frac{T_A}{T_C} \cdot \frac{T_D}{T_B} + y$.

$$\therefore \qquad x \cdot \frac{T_A}{T_C} \cdot \frac{T_D}{T_B} + y = 0$$

or $$x \times \frac{30}{20} \times \frac{22}{32} + y = 0$$

or $$\frac{33x}{32} + y = 0 \qquad ...(i)$$

Also it is given in the problem that driving shaft is rotating at 60 r.p.m. clockwise. This means the speed of wheel A is also 60 r.p.m. clockwise. This speed can also be stated as – 60 r.p.m. anticlockwise. Hence speed of wheel A (*i.e.* N_A) is – 60 r.p.m. anticlockwise. But from table 10.5 the revolution (or speed) of wheel A is $(x + y)$ anticlockwise.

$$\therefore \qquad x + y = -60 \qquad ...(ii)$$

or $$y = -60 - x \qquad ...(iii)$$

Substituting the value of y in equation (*i*), we get

$$\frac{33x}{32} - 60 - x = 0$$

or $$33x - 32 \times 60 - 32x = 0$$

or $$x = 32 \times 60 = 1920 \text{ r.p.m.}$$

From equation (*iii*), $y = -60 - 1920 = -1980$ r.p.m.

(*i*) *Speed of driven shaft*

The arm P is rigidly connected to the driven shaft. Hence the speed of driven shaft will be same as the speed of the arm. But from table 10.5, the revolution (or speed) of arm is $+ y$.

$\therefore$ Speed of driven shaft $= + y$

$= -1980$ r.p.m. $\qquad (\because \; y = -1980)$

$= 1980$ r.p.m. clockwise. **Ans.**

(*ii*) *Direction of rotation of driven shaft with reference to driving shaft*

The driving shaft is rotating clockwise. Also driven shaft is rotating clockwise. Hence the direction of rotation of driven shaft is the same as that of driving shaft. **Ans.**

Problem 10.11. *Fig. 10.17 shows below an epicyclic gear train known as Ferguson's paradox. Gear A is fixed to the frame and is therefore stationary. The arm B and gear C and D are free to rotate on the shaft. Gears A, C and D have 100, 101 and 99 teeth respectively. Planet gear P has 20 teeth. Pitch circle diameters of all are the same so that the planet gear P meshes with all of them. Determine the revolutions of gears C and D for one revolution of the arm B. (Figure on next page)* (A.M.I.E., Winter 1986, 79)

Sol. Given :

No. of teeth on A, $\quad T_A = 100$

No. of teeth on C, $\quad T_C = 101$

No. of teeth on D, $\quad T_D = 99$

No. of teeth on P, $\quad T_P = 20$

Gear A is fixed and stationary.

$\therefore$ Speed of A, $\quad N_A = 0$

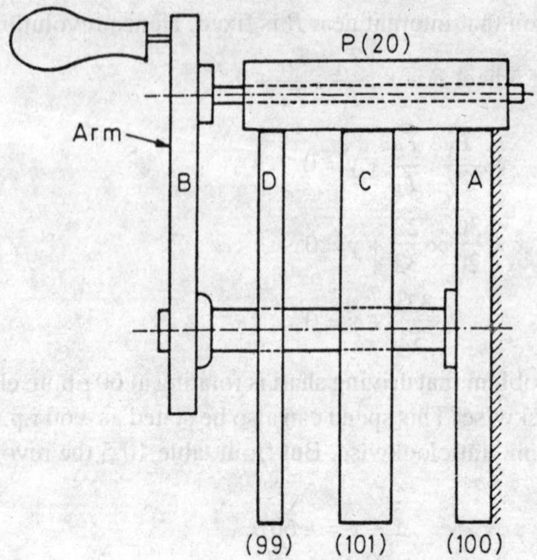

Fig. 10.17

Find revolutions of C and D for one revolution of arm B *i.e.* for $N_B = 1$.

To prepare the table of motion, let us find the no. of revolution of each element for one revolution of gear A, keeping arm fixed.

Gears A and C are in mesh through planet gear P, which will act as an idler gear. The direction of rotation of gear C will be same as that of gear A. The ratio of speeds of gears C and A will be as

$$\frac{N_C}{N_A} = \frac{T_A}{T_C} \qquad \text{or} \quad N_C = N_A \times \frac{T_A}{T_C} = 1 \times \frac{T_A}{T_C} = \frac{T_A}{T_C}.$$

The gears C and D are in mesh also through planet gear P.

The direction of rotation of gear D will be same as that of gear C.

$$\therefore \quad \frac{N_D}{N_C} = \frac{T_C}{T_D} \qquad \text{or} \quad N_D = N_C \times \frac{T_C}{T_D} = \frac{T_A}{T_C} \times \frac{T_C}{T_D} = \frac{T_A}{T_D}.$$

Now prepare the table of motion as given below :

Table 10.6

S. No.	Operations	Revolutions of elements			
		Arm B	Gear A	Gear C	Gear D
1.	Arm B is fixed. Rotate gear A through + 1 revolution	0	+ 1	$\frac{T_A}{T_C}$	$\frac{T_A}{T_C} \times \frac{T_C}{T_D} = \frac{T_A}{T_D}$
2.	Multiply by x to all	0	$+x$	$x \cdot \frac{T_A}{T_C}$	$x \cdot \frac{T_A}{T_D}$
3.	Add y revolution to all	$+y$	$x+y$	$x \cdot \frac{T_A}{T_C} + y$	$x \cdot \frac{T_A}{T_D} + y$

In the problem, it is given that arm B makes one revolution. But from the last row of the table 10.6, we find that arm B makes y revolution.

$$y = 1$$

In the problem it is also given that gear A is fixed *i.e.* revolution of gear A is zero. But from the table 10.6, we find that gear A has $(x + y)$ revolution.

$$\therefore \quad x + y = 0$$

or $$x = -y = -1 \qquad (\because \; y = 1)$$

Now from the table 10.6, the revolution of C

$$= x \cdot \frac{T_A}{T_C} + y$$

$$= (-1) \times \frac{100}{101} + 1 = \frac{-100 + 101}{101} = +\frac{1}{101}. \quad \textbf{Ans.}$$

and revolution of D $$= x \frac{T_A}{T_D} + y$$

$$= (-1) \times \frac{100}{99} + 1 = \frac{-100 + 99}{99} = -\frac{1}{99}. \quad \textbf{Ans.}$$

From above we find that for one revolution of the arm B, the gear C rotates through $\frac{1}{101}$ revolution in the same direction as the arm B, and gear D rotates through $\frac{1}{99}$ revolution in the opposite direction to the arm B. **Ans.**

Problem 10.12. *In an epicyclic gear train a gear C is keyed to the driving shaft A which rotates at 900 r.p.m. Gears D and E are fixed together and rotate freely on a pin carried by the arm M which is keyed to the driven shaft B. Gear D is in mesh with gear C while the gear E is in mesh with a fixed annular wheel F. The annular wheel is concentric with the driven shaft B. If the shafts A and B are collinear and number of teeth on gears C, D, E and F are respectively 21, 28, 14 and 84. Determine the speed and sense of rotation of the driven shaft B.* (A.M.I.E., Winter 1985)

Sol. Given :

No. of teeth on gear C, $T_C = 21$

No. of teeth on gear D, $T_D = 28$

No. of teeth on gear E, $T_E = 14$

No. of teeth on gear F, $T_F = 84$

Speed of driving shaft = 900 r.p.m.

As gear C is keyed to the driving shaft. The speed of gear C will be equal to the speed of driving shaft.

$\therefore \quad N_C = 900$ r.p.m.

The annular wheel F is fixed. Hence speed of F is zero or $N_F = 0$

Find speed of driven shaft and its sense of rotation.

Now prepare a table of motion as given in Table 10.7 below.

To prepare the table of motion, let us find the no. of revolution of each element for one revolution of gear C, keeping arm M fixed.

Gears C and D are in mesh externally, hence the direction of D will be opposite to that of gear C. Hence

$$\frac{N_D}{N_C} = -\frac{T_C}{T_D} \quad \text{or} \quad N_D = -\frac{T_C}{T_D} \times N_C = -\frac{T_C}{T_D} \times 1 = -\frac{T_C}{T_D}$$

The speed of gear E will be same as that of gear D.

$$\therefore \quad N_E = N_D = -\frac{T_C}{T_D}$$

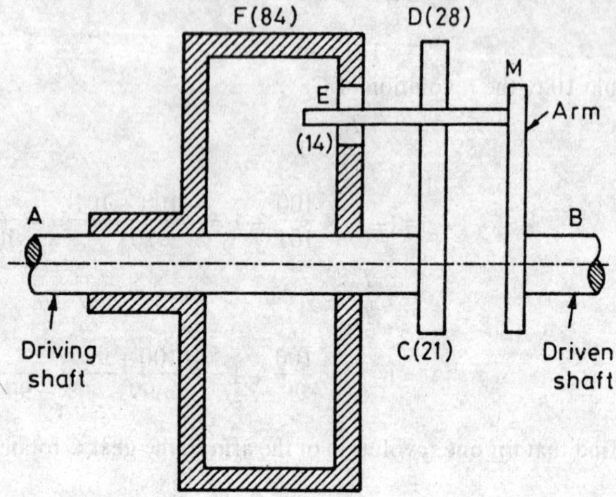

Fig. 10.18

The gears E and F are in mesh internally. Hence the direction of rotation of F will be same as that of E. Hence

$$\frac{N_F}{N_E} = \frac{T_E}{T_F}$$

or

$$N_F = \frac{T_E}{T_F} \times N_E = \frac{T_E}{T_F} \times \left(-\frac{T_C}{T_D}\right) = -\frac{T_C}{T_D} \times \frac{T_E}{T_F}$$

Table 10.7

S. No.	*Steps of Motion*	*Revolution of elements*			
		Arm M	*Wheel C*	*Wheel D-E*	*Wheel F*
1.	Arm is fixed and rotate wheel C through + 1 revolution (*i.e.* in the anticlockwise direction)	0	+1	$-\frac{T_C}{T_D}$	$-\frac{T_C}{T_D} \times \frac{T_E}{T_F}$
2.	Multiply by x to all	0	$+x$	$-x \cdot \frac{T_C}{T_D}$	$-x \cdot \frac{T_C}{T_D} \times \frac{T_E}{T_F}$
3.	Add y revolution to all	$+y$	$x+y$	$-x \cdot \frac{T_C}{T_D} + y$	$-x \cdot \frac{T_C}{T_D} \times \frac{T_E}{T_F} + y$

In the problem it is given that speed (or revolution) of F is zero. But from the table 10.7 the revolution of wheel F is $-x \cdot \frac{T_C}{T_D} \times \frac{T_E}{T_F} + y$.

$$\therefore \quad -x \cdot \frac{T_C}{T_D} \times \frac{T_E}{T_F} + y = 0$$

or

$$-x \times \frac{21}{28} \times \frac{14}{84} + y = 0$$

or $$-\frac{x}{8}+y=0$$

$\therefore$ $$x = 8y \quad ...(i)$$

The speed of gear C is given as 900 r.p.m. But from the table 10.7 the speed of C is $x + y$.

$\therefore$ $$x + y = 900$$

or $$8y + y = 900 \quad [\because \text{ From } (i),\ x = 8y]$$

or $$9y = 900$$

$\therefore$ $$y = \frac{900}{9} = 100$$

(i) Speed of driven shaft

The arm M is keyed to the driven shaft. Hence speed of driven shaft will be equal to the speed of the arm M. But speed (or revolution) of the arm M from the table 10.7 is $+ y$.

$\therefore$ Speed of driven shaft = Speed of arm $M = + y$

= + 100 r.p.m. ($\because$ y = 100 r.p.m.). **Ans.**

Its sense of rotation is anti-clockwise or same as the sense of rotation of the driving shaft. **Ans.**

10.4. Sun and Planet Gear

Sun gear is a gear at the centre of a gear train, whereas the planet gears are the gears whose axes move around the sun gear. If an annular gear D is added to the epicyclic gear train of Fig. 10.4, we get the sun and planet gear as shown in Fig. 10.19.

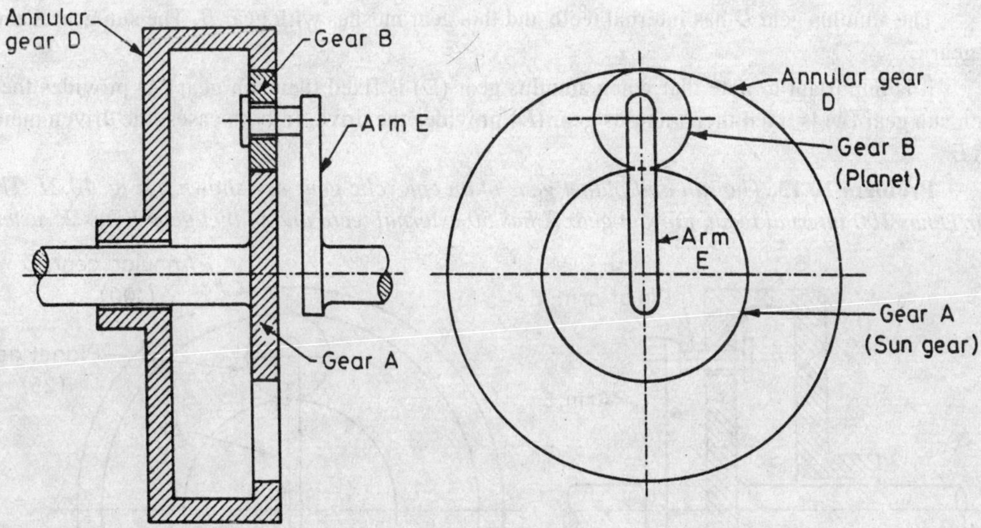

Fig. 10.19

The gear A at the centre is known as sun gear. The gear B which revolves round the gear A, is known as planet gear. The annular gear D which has internal teeth, meshes with gear B. The gear B can rotate freely on the arm E. Also gear A and annular gear D can rotate freely. It is also possible to fix either gear D or gear A. If gear D is fixed then gear A is the driving member. And if gear A is fixed then gear D is the driving member. In both cases, the arm E is the driven member.

Fig. 10.20 shows a compound epicyclic gear train which consists of an annulus gear *D*, the compound gear *B-C*, the sun gear *A* and the arm *E*. The annulus gear *D* is fixed. The compound gear *B-C* is also known as planet gear as this gear moves around the sun gear *A*. The sun gear *A* is co-axial with annulus gear *D* and the arm. The compound gear *B-C* is carried by the arm and revolves freely on a pin of the arm *E*.

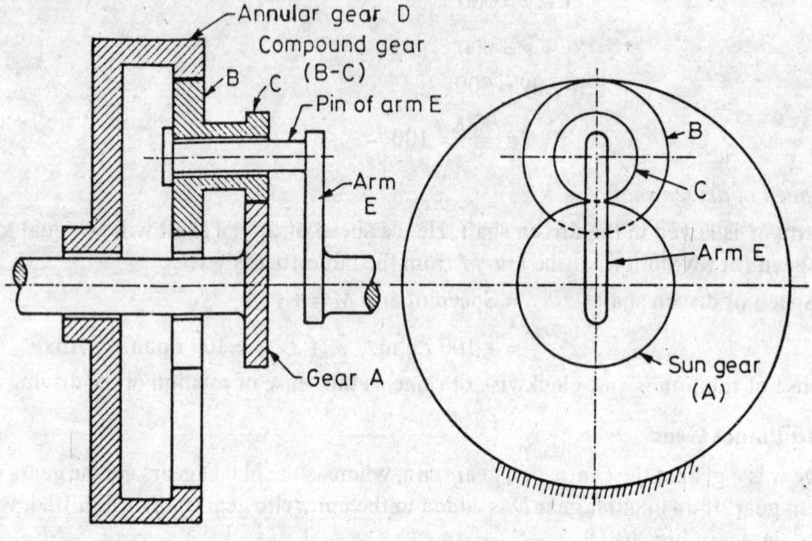

Fig. 10.20

The annulus gear *D* has internal teeth and this gear meshes with gear *B*. The sun gear *A* meshes with the gear *C*.

It is important to note that when annulus gear (*D*) is fixed then sun gear (*A*) provides the drive and when sun gear (*A*) is fixed then annulus gear (*D*) provides the drive. In both cases, the driven member is the arm *E*.

Problem 10.13. *The sun and planet gear of an epicyclic gear are shown in Fig. 10.21. The annular gear D has 100 internal teeth, the sun gear A has 50 external teeth and planet gear B has 25 external teeth.*

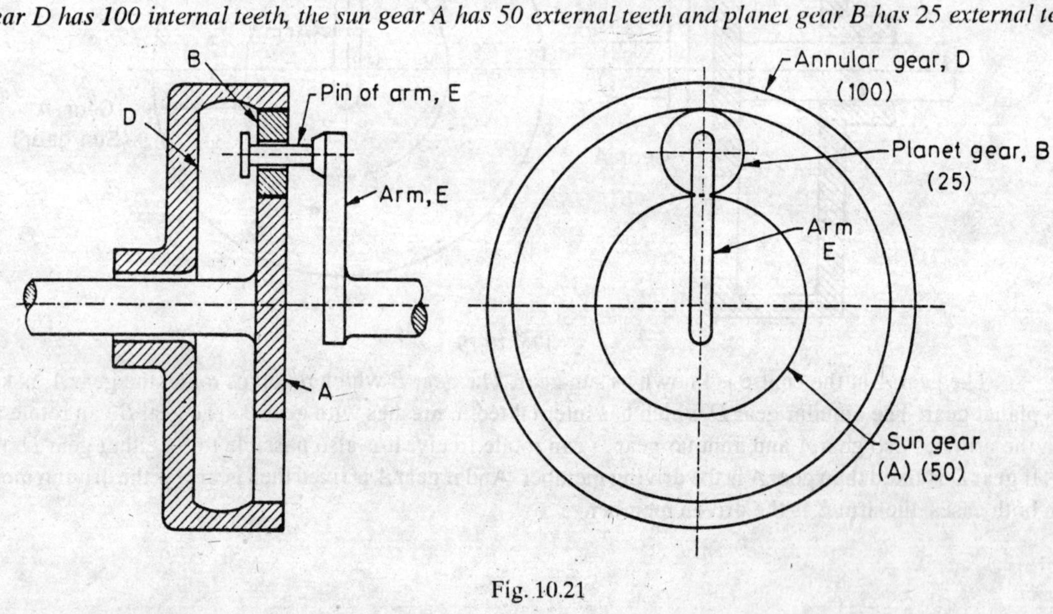

Fig. 10.21

The gear B meshes with gear D and gear A. The gear B is carried on the arm E, which rotates about the centre of annular gear D. If the gear D is fixed and arm rotates at 20 r.p.m., then find the speeds of gears A and B.

Sol. Given :

$T_D = 100$; $T_A = 50$ and $T_B = 25$; Speed of arm E, $N_E = 20$ r.p.m.

Gear D is fixed hence $N_D = 0$. Find N_A and N_B.

The table of motion can be prepared by keeping the arm, E fixed. It is also clear that if arm is fixed and sun gear A is rotated anticlockwise, then the planet gear B will rotate in clockwise direction. The gear B meshes with gear D internally, hence gear D will rotate in the same direction as that of gear B. Hence gear D will rotate in clockwise direction.

Now the table of motion can be prepared as shown in Table 10.8.

Table 10.8

S. No.	*Operations*	*Revolution of elements*			
		Arm E	*Gear A*	*Gear B*	*Gear D*
1.	Arm fixed. Sun gear A rotates through + 1 revolution (*i.e.* anti-clockwise)	0	+ 1	$-\frac{T_A}{T_B}$	$-\frac{T_A}{T_B}\times\frac{T_B}{T_D}=-\frac{T_A}{T_D}$
2.	Multiply by x to all	0	$+x$	$-x\times\frac{T_A}{T_B}$	$-x\times\frac{T_A}{T_D}$
3.	Add + y revolution to all	$+y$	$+y$	$+y$	$+y$
4.	Resultant Motion	$+y$	$x+y$	$y-x\frac{T_A}{T_B}$	$y-x\frac{T_A}{T_D}$

The given conditions are : (*i*) Speed of arm, $N_E = 20$ r.p.m. and (*ii*) gear D is fixed or $N_D = 0$.

From table 10.8, the resultant motion of arm is $+ y$ and resultant motion of gear D is $y - x \times \frac{T_A}{T_D}$.

Equating the motion of arm, we get

$\therefore \quad + y = 20 \quad$ or $\quad y = 20$

Equating the motion of gear D, we get

$$y - x \times \frac{T_A}{T_D} = 0$$

or $$y - x \times \frac{50}{100} = 0 \qquad (\because T_A = 50 \text{ and } T_D = 100)$$

or $$20 - x \times \frac{1}{2} = 0 \qquad (\because y = 20)$$

or $$40 - x = 0$$

$\therefore$ $$x = 40$$

(*i*) *Speed of Gear B*

The resultant motion of gear B from table is $\left(y - x \times \frac{T_A}{T_B}\right)$.

$\therefore$ Speed of gear B $= y - x \times \frac{T_A}{T_B}$

$$= 20 - 40 \times \frac{50}{25} \qquad (\because \; T_A = 50 \text{ and } T_B = 25)$$

$$= 20 - 80$$

$$= \mathbf{-60 \text{ r.p.m.}} \quad \textbf{Ans.} \qquad (\text{– ve sign means, the gear } B \text{ rotates clockwise})$$

(ii) Speed of Gear A

The resultant motion of gear A from Table 10.8 is $(x + y)$

$\therefore$ Speed of gear A $= x + y = 40 + 20$

$= \mathbf{60 \text{ r.p.m.}}$ **Ans.** (+ ve sign means, the gear A rotates clockwise)

Problem 10.14. *If in the above problem, two planet gears B and C having 30 teeth each are attached to the arm E as shown in Fig. 10.22, and gear A is having 40 teeth instead of 50, then find the number of revolutions made by the arm when :*

(i) gear A makes one revolution clockwise and D makes half a revolution anticlockwise and

(ii) gear A makes one revolution clockwise and D is stationary.

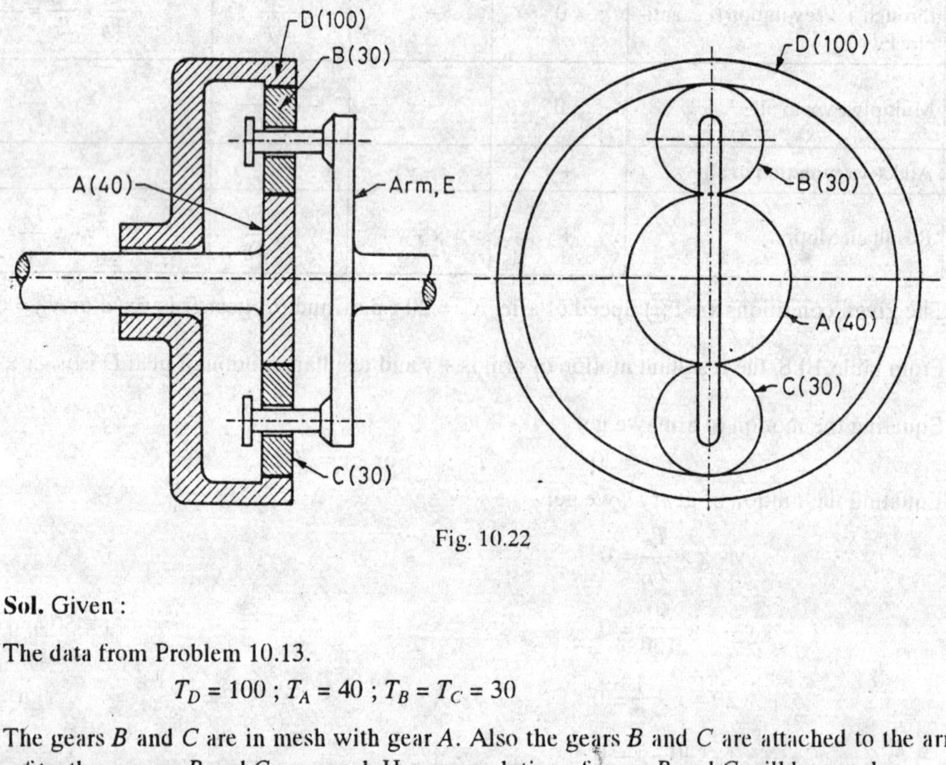

Fig. 10.22

Sol. Given :

The data from Problem 10.13.

$$T_D = 100 \; ; \; T_A = 40 \; ; \; T_B = T_C = 30$$

The gears B and C are in mesh with gear A. Also the gears B and C are attached to the arm D. The number of teeth on gears B and C are equal. Hence revolution of gears B and C will be equal.

The table of motion can be prepared by keeping arm (E) fixed. It is also clear that if arm (E) is fixed and gear A is rotated anticlockwise, then gears B and C will rotate in clockwise direction as the gears B and C are meshing with gear A externally. The gears B and C are meshing with gear D internally. Hence gear D will rotate in clockwise direction.

Now table of motion can be prepared as shown in Table 10.9.

Table 10.9

S. No.	Operations	Revolution of			
		Arm E	Gear A	Gear B or Gear C	Gear D
1.	Arm fixed. Gear A rotates through + 1 revolution.	0	+ 1	$-\frac{T_A}{T_B}$	$-\frac{T_A}{T_B} \times \frac{T_B}{T_D} = -\frac{T_A}{T_D}$
2.	Arm fixed. Gear A rotates through + x revolution (Multiply by x to all)	0	+ x	$-x \times \frac{T_A}{T_B}$	$-x \times \frac{T_A}{T_D}$
3.	Add + y to all	+ y	+ y	+ y	+ y
4.	Resultant Motion	y	x + y	$y - x \times \frac{T_A}{T_B}$	$y - x \frac{T_A}{T_D}$

Take + ve sign for anti-clockwise and
– ve sign for clockwise.

1st Case. *Gear A makes one revolution clockwise and D makes half a revolution anticlockwise*

Given conditions are : $N_A = -1$ (– ve sign due to clockwise)

$N_D = +\frac{1}{2}$ (+ ve sign due to anticlockwise)

But from Table 10.9, the resultant motion of gear *A* is $(x + y)$ or $N_A = (x + y)$

Hence equating the motion of gear *A* (*i.e.* two values of N_A), we get

$$x + y = -1 \qquad ...(i)$$

Again from Table 10.9 the resultant motion of gear *D* is $\left(y - x \times \frac{T_A}{T_D}\right)$

or
$$N_D = \left(y - x \times \frac{T_A}{T_D}\right)$$

Equating the two values of N_D, we get

$$y - x \times \frac{T_A}{T_D} = \frac{1}{2}$$

or
$$y - x \times \frac{40}{100} = \frac{1}{2} \qquad (\because\ T_A = 40 \text{ and } T_D = 100)$$

or
$$y - 0.4x = \frac{1}{2} \qquad ...(ii)$$

We have to find the revolution of arm. But the revolution of the arm from Table 10.9 is given by '*y*'. Hence let us find the value of '*y*' from equations (*i*) and (*ii*)

From equation (*i*), $x = -1 - y$

Substituting this value of *x* in equation (*ii*), we get

$$y - 0.4(-1 - y) = \frac{1}{2} \text{ or } 0.5$$

or
$$y + 0.4 + 0.4y = 0.5$$

or
$$1.4y = 0.5 - 0.4 = 0.1$$

or $$y = \frac{0.1}{1.4} = \frac{1}{14} = 0.071 \text{ revolution}$$

(+ ve sign means the revolution will be anticlockwise)

∴ Speed of arm = y = **0.071 revolution anti-clockwise. Ans.**

2nd Case. *Gear A makes one revolution clockwise and gear D is stationary*

∴ Given conditions are : $N_A = -1$

$N_D = 0$

But from Table 10.9 resultant motions of gears A and D are,

$$N_A = (x + y) \text{ and } N_D = \left(y - x\frac{T_A}{T_D}\right)$$

∴ $$N_A = (x + y) = -1$$

and $$N_D = \left(y - x\frac{T_A}{T_D}\right) = 0$$

or $$x + y = -1 \quad ...(iii)$$

and $$y - x\frac{T_A}{T_D} = 0$$

or $$y - x\frac{40}{100} = 0 \quad (\because T_A = 40, T_D = 100)$$

or $$y - 0.4x = 0 \quad ...(iv)$$

Find the value of 'y' (*i.e.* speed of arm) from equations (*iii*) and (*iv*).

From equation (*iii*), $x = -1 - y$

Substituting this value of x in equation (*iv*), we get

$$y - 0.4(-1 - y) = 0$$

or $$y + 0.4 + 0.4y = 0$$

or $$1.4y = -0.4$$

or $$y = -\frac{0.4}{1.4} = -0.285 \text{ revolutions}$$

(– ve sign means the revolution will be clockwise)

Speed of arm = y = **0.285 revolution clockwise. Ans.**

Problem 10.15. *Fig. 10.20 given on page 474 shows a compound epicyclic gear which consists of an annular gear D having internal teeth, a compound gear B-C, the sun gear A and the arm E. The compound gear B-C is also known as planet gear as this gear moves around the sun gear A. The sun gear A and annular gear D are co-axial. One shaft is connected to sun gear A and another shaft is connected to the arm. These two shafts are also co-axial. The sun gear A meshes with compound gear C and annular gear D meshes with compound gear B. The gears A, B and C are having 60, 40 and 25 external teeth respectively. If all the gears are having the same module, find the number of teeth on the fixed annular gear D. Also find the speed of the shaft connected to the arm, if the speed of the shaft connected to sun gear A is 100 r.p.m.*

Sol. Given : Refer to Fig. 10.20 given on page 474.

$T_A = 60$; $T_B = 40$; $T_C = 25$. Speed of shaft connected to sun gear A = 100 r.p.m. ∴ N_A = 100 r.p.m.

All the gears have same module. Hence no. of teeth on the gears are proportional to their pitch circle diameters.

Find : (*i*) No. of teeth on gear D *i.e.* T_D.

(*ii*) Speed of shaft connected to arm *i.e.* N_{arm} or N_E

(*i*) *No. of teeth on annular gear D (i.e. value of T_D)*

Let d_A, d_B, d_C and d_D are the pitch circle diameters of gears *A*, *B*, *C* and *D* respectively. Then from Fig. 10.20, we have

$$r_D = r_A + r_C + r_B$$

where r_A, r_B, r_C and r_D are the pitch circle radii of the gears *A*, *B*, *C* and *D*.

The above equation can be written as

$$2r_D = 2r_A + 2r_C + 2r_B \quad \text{(Multiplying by 2)}$$

or

$$d_D = d_A + d_C + d_B \quad (\because \; 2r_D = d_D, 2r_A = d_A \text{ and so on})$$

The no. of teeth on the gears are proportional to their pitch circle diameters. Hence the above equation in terms of no. of teeth on gears, becomes as

$$T_D = T_A + T_C + T_B = 60 + 25 + 40 = \mathbf{125}. \quad \textbf{Ans.}$$

(*ii*) *Speed of shaft connected to arm (i.e. value of N_E)*

The speed of the arm (*i.e.* value of N_E) can be obtained from the table of motion. The table of motion can be prepared by keeping arm (*E*) fixed. It is also clear that if arm is fixed and sun gear *A* is rotated anticlockwise, the compound gear *C* which meshes with gear *A* with external teeth, with rotate clockwise. The gear *B*-*C* is a compound gear and hence gear *B* will also rotate with the same speed as that of gear *C*, in the clockwise direction. The gear *B* meshes with the gear *D* with internal teeth. Hence gear *D* will rotate in the same direction as that of gear *B* (*i.e.* in the clockwise direction).

Now the table of motion can be prepared as shown in Table 10.10.

Table 10.10

S. No.	*Operations*	*Revolution of*			
		Arm E	*Gear A*	*Compound gear B-C*	*Gear D*
1.	Arm fixed. Gear *A* rotates through + 1 revolution	0	+ 1	$-\frac{T_A}{T_C}$	$-\frac{T_A}{T_C} \times \frac{T_B}{T_D}$
2.	Arm fixed. Gear *A* rotates through + *x* revolution (Multiply by *x* to all)	0	+ *x*	$-x \times \frac{T_A}{T_C}$	$-x \times \frac{T_A}{T_C} \times \frac{T_B}{T_D}$
3.	Add + *y* to all	+ *y*	+ *y*	+ *y*	+ *y*
4.	Resultant Motion (Add S. No. 2 and S. No. 3)	+ *y*	*x* + *y*	$y - x\frac{T_A}{T_C}$	$y - x\frac{T_A}{T_C} \times \frac{T_B}{T_D}$

Take + ve sign for anticlockwise and

– ve sign for clockwise

Given conditions are : (*i*) Gear *D* is fixed. Hence $N_D = 0$.

(*ii*) Speed of shaft connected to gear *A* = 100 r.p.m. or N_A = 100 r.p.m.

From Table 10.10, the resultant speed of gear *D* is $\left(y - x\frac{T_A}{T_C} \times \frac{T_B}{T_D}\right)$

∴ Equating the two values of N_D, we get

$$N_D = \left(y - x \frac{T_A}{T_C} \times \frac{T_B}{T_D} \right) = 0$$

or $$y - x \times \frac{60}{25} \times \frac{40}{125} = 0 \qquad (\because \; T_A = 60, T_B = 40, T_C = 25 \text{ and } T_D = 125)$$

or $$y - 0.768x = 0 \qquad ...(i)$$

From Table 10.10, the resultant speed of gear A is $(x + y)$ or $N_A = (x + y)$

∴ Equating the two values of N_A, we get

$$x + y = 100 \qquad ...(ii)$$

The resultant motion of arm from table 10.10 is 'y'. Hence let us find the value of 'y' from equations (*i*) and (*ii*)

From equation (*ii*), $x = 100 - y$

Substituting this value of x in (*i*), we get

$$y - 0.768\,(100 - y) = 0$$

or $$y - 76.8 + 0.768y = 0$$

or $$1.768y = 76.8$$

or $$y = \frac{76.8}{1.768} = 43.43 \text{ r.p.m. anticlockwise}$$

∴ Speed of shaft connected to arm = y = 43.43 r.p.m. anticlockwise. **Ans.**

10.5. Torques and Tooth Loads in Epicyclic Gear Train

When a geared system transmits power, torques are transmitted from one element to another. If all the elements of an epicyclic gear train are rotating at uniform speeds then the angular acceleration will be zero. And the algebraic sum of all the external torques (or moments) acting on the gear train will be zero, or

$$\Sigma M = 0 \qquad (\because \; \Sigma M = \Sigma T = I \times \alpha = I \times 0 = 0)$$

Where ΣM is the summation of all external torques.

For most of the cases, there are only three external torques namely :

(*i*) Input torque on the driving element, M_i ;

(*ii*) Resisting torque or load torque on the driven element, M_r

(*iii*) Holding or braking torque on the fixed element, M_b

For no acceleration acting on the gear train,

$$M_i + M_r + M_b = 0 \qquad ...(10.6)$$

But, torque = force × radius or $M = F \times r$, then above equation becomes as

$$F_i \times r_i + F_r \times r_r + F_b \times r_b = 0 \qquad ...(10.7)$$

where F_i = External force applied at the driving element,

F_r = Resisting force,

F_b = Braking or holding force, and

r_i, r_r and r_b = Corresponding radii.

Assuming no losses in power transmission *i.e.* there are no internal friction losses at the bearings and at the contact surfaces of the gear teeth, the net energy dissipated by the gear train must be zero or

$$\Sigma\,(M \times \omega) = 0 \qquad (\because \; T \times \omega \text{ or } M \times \omega = \text{Power or Energy})$$

or $$M_i \times \omega_i + M_r \times \omega_r + M_b \times \omega_b = 0 \qquad ...(10.8)$$

where ω_i = Angular velocity of driving element,

ω_r = Angular velocity of resisting element, and

ω_b = Angular velocity of braking or holding element.

For the fixed or holding element, $\omega_b = 0$

∴ equation (10.8) becomes as

$$M_i \times \omega_i + M_r \times \omega_r = 0$$

∴ Resisting torque or load torque becomes as,

$$M_r = (-M_i \times \omega_i)/\omega_r \qquad ...(10.9)$$

Substituting this value in equation (10.6), we get

$$M_i + \left(-\frac{M_i \times \omega_i}{\omega_r}\right) + M_b = 0$$

or

$$M_b = -M_i + \frac{M_i \times \omega_i}{\omega_r}$$

$$= M_i\left(\frac{\omega_i}{\omega_r} - 1\right) \qquad ...(10.10)$$

If the value of M_i *i.e.* the torque at the input element is known, then the value of braking torque or holding torque can be obtained.

Hence equations (10.9) and (10.10) are used to find the values of resisting torque (M_r) and holding torque or braking torque (M_b), if the input torque (M_i) applied to the driving member is known. From the resisting torque and braking torque the resisting force (F_r) and braking load (F_b) may be obtained.

Note. (*i*) For torque the symbol M is used and not T.

(*ii*) The symbol T has been used for no. of teeth on a gear.

(*iii*) The input and output torques will be in opposite directions if the input and output shafts rotate in the same direction. But if input and output shafts rotate in the opposite direction then input and output torques will be in the same direction.

Problem 10.16. *Fig. 10.23 (figure on next page) shows an epicyclic speed reduction gear. The driving shaft is attached to an arm E. The arm carries a pin on which the compound gear B-C is free to revolve. The gear A is keyed to the driven shaft and gear D is a fixed gear. The gear B meshes with gear A whereas gear C meshes with the fixed gear D. The no. of teeth on the gears A, B, C and D are 24, 27, 30 and 21 respectively. Determine :*

(*i*) *speed of driven shaft if driving shaft is rotating at 800 r.p.m. anticlockwise and*

(*ii*) *load torque (or resisting torque) on driven shaft and holding torque on fixed gear D, if the input torque to driving shaft is 10 Nm. Assume that there is no friction losses and that the members are revolving at uniform speeds.*

Sol. Given :

$$T_A = 24 \; ; T_B = 27 \; ; T_C = 30 \text{ and } T_D = 21.$$

Speed of driving shaft (or speed of arm E), N_E = 800 r.p.m. anticlockwise.

Gear D is fixed. ∴ $N_D = 0$

Find : (*i*) Speed of driven shaft (or speed of gear A) or N_A

(*ii*) Resisting torque (M_r) and holding torque (M_b).

(*i*) *Speed of driven shaft or speed of gear A (i.e. N_A)*

The speed of gear A can be obtained by *Tabular Method* for which the table of motion can be prepared by keeping the arm (E) fixed. It is also clear that when gear A is rotated anticlockwise, the gear B will rotate

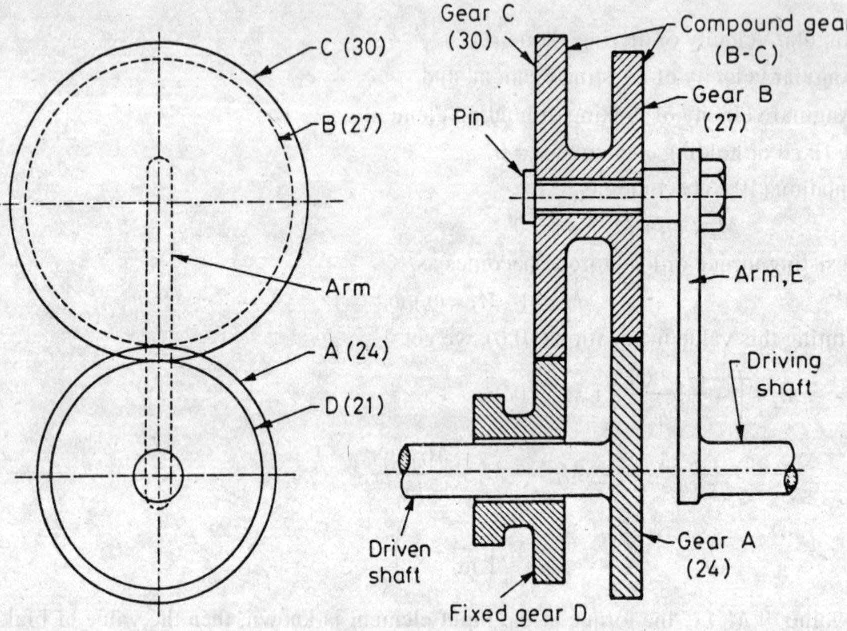

Fig. 10.23

in clockwise direction as gear A meshes with gear B with external teeth. The gear B-C is a compound gear and hence gear C will rotate with the same speed as that of gear B and in the same direction (*i.e.* clockwise). The gear C meshes with gear D with external teeth. Hence gear D will rotate in anticlockwise direction. Now the table of motion can be prepared as shown in Table 10.11.

[When arm is fixed and gear is rotated through + 1 revolution, then compound gear will make $-\frac{T_A}{T_B}$ revolution and gear D will make $\left(-\frac{T_A}{T_B}\right)\times\left(-\frac{T_C}{T_D}\right)$ revolution.]

Table 10.11

S. No.	*Operations*	*Revolution of*			
		Arm E	*Gear A*	*Compound gear B-C*	*Gear D*
1.	Arm fixed. Rotate gear A through + 1 revolution	0	+ 1	$-\frac{T_A}{T_B}$	$\left(-\frac{T_A}{T_B}\right)\times\left(-\frac{T_C}{T_D}\right)$
2.	Arm fixed. Rotate gear A through $+x$ revolution (Multiply by x to all)	0	$+x$	$-x\times\frac{T_A}{T_B}$	$x\times\left(-\frac{T_A}{T_B}\right)\left(-\frac{T_C}{T_D}\right)$
3.	Add $+y$ revolution to all	$+y$	$+y$	$+y$	$+y$
4.	Resultant Motion	$+y$	$x+y$	$y-x\times\frac{T_A}{T_B}$	$y+x\times\frac{T_A}{T_B}\times\frac{T_C}{T_D}$

Take + ve sign for anti-clockwise and

– ve sign for clockwise.

The given conditions are : (*i*) speed of driving shaft (or speed of arm *E*), N_E = 800 r.p.m. anticlockwise.

(*ii*) speed of gear *D* is zero or $N_D = 0$

But from table 10.11, resultant motion of arm $E = +y$ or $N_E = y$

and resultant motion of gear *D* is $\left(y + x \times \frac{T_A}{T_B} \times \frac{T_C}{T_D}\right)$.

$$\therefore \quad N_D = \left(y + x \times \frac{T_A}{T_B} \times \frac{T_C}{T_D}\right)$$

Equating the two values of N_E, we get

$$N_E = y = 800 \quad \therefore \quad y = 800 \text{ r.p.m. anticlockwise} \qquad ...(i)$$

Similarly equating the two values of N_D, we get

$$N_D = 0 = \left(y + x \frac{T_A}{T_B} \times \frac{T_C}{T_D}\right)$$

or $$0 = y + x \times \frac{24}{27} \times \frac{30}{21} \qquad (\because \; T_A = 24, T_B = 27, T_C = 30, T_D = 21)$$

$$0 = y + 1.27x \qquad ...(ii)$$

$$0 = 800 + 1.27x \qquad [\because \; y = 800 \text{ from equation } (i)]$$

or $$1.27x = -800$$

or $$x = -\frac{800}{1.27} = -629.92 \qquad ...(iii)$$

From table 10.11, the resultant speed of gear *A* (or speed of driven shaft) is $(x + y)$

$\therefore$ Speed of gear *A* $= x + y$

$= -629.92 + 800 \qquad (\because \; x = -629.92 \text{ and } y = 800)$

$= 170.08$ r.p.m.

$\therefore$ Speed of driven shaft (or speed of gear *A*) = 170.08 r.p.m. anticlockwise.

(*ii*) *Resisting torque and holding torque (i.e. values of M_r and M_b)*

Input torque (or torque applied to dividing shaft) = 10 Nm

$$\therefore \quad M_i = 10 \text{ Nm}$$

As the members as revolving at uniform speed and hence from equation (10.6), we get

$$M_i + M_r + M_b = 0$$

where M_i = Input torque (or torque on driving shaft) = 10 Nm (given)

M_r = Resisting torque (or torque on driven shaft)

M_b = Braking torque or holding torque (or torque on fixed member)

For no friction losses, the net energy dissipated by gear train should be zero.

Hence $\Sigma T \times \omega$ should be zero, as $T \times \omega$ is Energy. Hence from equation (10.8), we get

$$M_i \times \omega_i + M_r \times \omega_r + M_b \times \omega_b = 0$$

For the fixed member, $\omega_b = 0$

$$\therefore \quad M_i \times \omega_i + M_r \times \omega_r = 0$$

$$\therefore \qquad M_r = -M_i \times \frac{\omega_i}{\omega_r}$$

where ω_i = Angular velocity of driving shaft $= \dfrac{2\pi N_i}{60}$

$= \dfrac{2\pi \times 800}{60}$ $\qquad$ [$\because$ N_i = speed of driving shaft = 800 r.p.m. (given)]

ω_r = Angular velocity of driven shaft $= \dfrac{2\pi N_r}{60}$

$= \dfrac{2\pi \times 170.08}{60}$ $\qquad$ [$\because$ N_r = speed of driven shaft = 170.08 r.p.m. (calculated)]

Substituting these values in the above equation, we get

$$M_r = -10 \times \frac{\left(\dfrac{2\pi \times 800}{60}\right)}{\left(\dfrac{2\pi \times 170.08}{60}\right)} = -10 \times \frac{800}{170.08} = \mathbf{-47.03\ Nm.} \quad \textbf{Ans.}$$

$\therefore$ Resisting torque $= \mathbf{-47.03\ Nm.}$ **Ans.**

We also know from equation (10.6) that

$$M_i + M_r + M_b = 0$$

$$\therefore \qquad M_b = -M_i - M_r = -10 - (-47.03) = 37.03 \text{ Nm}$$

$\therefore$ Braking torque or holding torque = **37.03 Nm. Ans.**

Note. In the above problem, the driving shaft (input shaft) and driven shaft (output shaft) are rotating in the same direction. Hence the input torque (torque on driving shaft) and output torque (torque on driven shaft) are in opposite direction.

Problem 10.17. *Fig. 10.24 shows an epicyclic gear train. Pinion A has 15 teeth and is rigidly fixed to the motor shaft. The wheel B has 20 teeth and gears with A and also with the annular fixed wheel D. Pinion C*

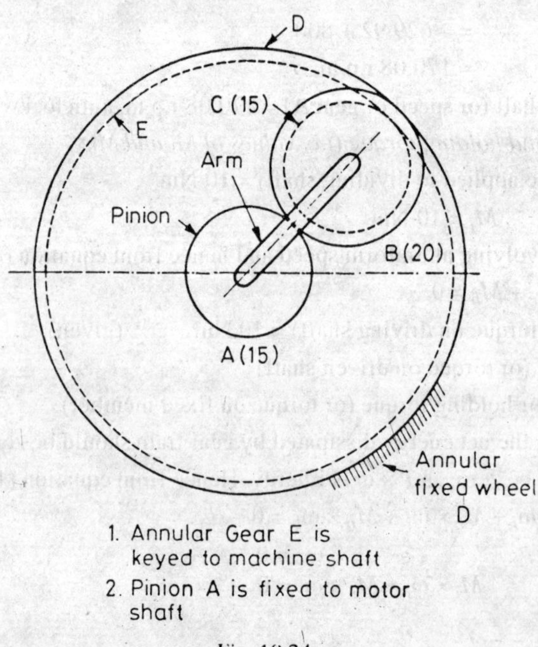

Fig. 10.24

has 15 teeth and is integral with B (B, C being a compound gear wheel). Gear C meshes with annular wheel E, which is keyed to the machine shaft. The arm rotates about the same shaft on which A is fixed and carries the compound wheel B, C. If the motor runs at 1000 r.p.m., find the speed of the machine shaft. Find the torque exerted on the machine shaft if the motor develops a torque of 100 Nm.

(Converted to S.I. Units, A.M.I.E., Summer 1976)

Sol. Given :

No. of teeth on A, $T_A = 15$

No. of teeth on B, $T_B = 20$

No. of teeth on C, $T_C = 15$

B, C is a compound gear wheel.

Speed of motor or of pinion A, $N_A = 1000$ r.p.m.

Torque developed by motor (or pinion A) = 100 N m.

Find (*i*) Speed of machine shaft (or of E).

(*ii*) Torque exerted on machine shaft (or on E).

Let us first calculate the number of teeth on wheel D and E.

Let T_D = No. of teeth on wheel D,

T_E = No. of teeth on wheel E.

Now prepare the table of motion as given below :

Let d_A = Pitch circle dia. of wheel A,

d_B = Pitch circle dia. of wheel B and

d_D = Pitch circle dia. of wheel D.

Now from Fig. 10.24,

$$\frac{d_D}{2} = d_B + \frac{d_A}{2}$$

or

$$d_D = 2d_B + d_A$$

Since the number of teeth are proportional to their pitch circle diameters, hence the above equation can be written in terms of respective no. of teeth as :

$$T_D = 2T_B + T_A$$

$$= 2 \times 20 + 15 \qquad (\because\ T_B = 20, T_A = 15)$$

$$= 55$$

Also from Fig. 10.24, the radius of wheel E is given by,

$$\frac{d_E}{2} = \frac{d_A}{2} + \frac{d_B}{2} + \frac{d_C}{2}$$

or

$$d_E = d_A + d_B + d_C$$

The above equation in terms of respective no. of teeth becomes as

$$\therefore \qquad T_E = T_A + T_B + T_C = 15 + 20 + 15 = 50$$

Now the table of motion can be prepared. Let us find the no. of revolution of each element for one revolution of wheel A, keeping arm fixed.

Revolution of compound wheel B-C

The wheels A and B are having external teeth. Hence they will rotate in opposite direction. Also their speeds are inversely proportional to their no. of teeth.

$$\therefore \qquad \frac{N_B}{N_A} = -\frac{T_A}{T_B} \qquad \text{(– ve sign due to opposite direction)}$$

or $$N_B = \left(-\frac{T_A}{T_B}\right) \times N_A$$

$$= -\left(\frac{T_A}{T_B}\right) \times 1 \qquad (\because \; N_A = +1)$$

$$= -\frac{T_A}{T_B}$$

The wheels B and C is a compound wheel or they are mounted on the same shaft. Hence the speed of wheel B will be equal to the speed of wheel C

$$\therefore \qquad N_C = N_B = -\frac{T_A}{T_B}.$$

Revolution of Wheel E

The wheels C and E are in mesh with internal gears. Hence they will rotate in the same direction. Also their speeds are inversely proportional to their no. of teeth.

$$\therefore \qquad \frac{N_E}{N_C} = \frac{T_C}{T_E} \qquad \text{(+ ve sign due to same direction)}$$

$$\therefore \qquad N_E = N_C \times \frac{T_C}{T_E}$$

$$= -\frac{T_A}{T_B} \times \frac{T_C}{T_E} \qquad \left(\because \; N_C = -\frac{T_A}{T_B}\right)$$

Revolution of Wheel D

The wheels B and D are in mesh with internal gears. Hence they will rotate in the same direction. Also their speeds are inversely proportional to their no. of teeth

$$\frac{N_D}{N_B} = \frac{T_B}{T_D}$$

or $$N_D = N_B \times \frac{T_B}{T_D}$$

$$= -\frac{T_A}{T_B} \times \frac{T_B}{T_D} \qquad \left(\because \; N_B = N_C = -\frac{T_A}{T_B}\right)$$

$$= -\frac{T_A}{T_D}$$

$\therefore$ For one revolution of wheel A, the revolutions of other elements are as :

Wheel $B = -\frac{T_A}{T_B}$; Wheel $E = -\frac{T_A}{T_B} \times \frac{T_C}{T_E}$ and Wheel $D = -\frac{T_A}{T_D}$.

Table 10.12

S. No.	*Operation*	*Revolution of elements*				
		Arm	*Pinion A*	*Compound wheel B-C*	*Wheel E*	*Wheel D*
1.	Arm fixed. Pinion A rotates through one revolution.	0	+ 1	$-\frac{T_A}{T_B}$	$-\frac{T_A}{T_B} \times \frac{T_C}{T_E}$	$-\frac{T_A}{T_B} \times \frac{T_B}{T_D} = -\frac{T_A}{T_D}$

2.	Multiply by x to all	0	$+x$	$-x \cdot \frac{T_A}{T_B}$	$-x \cdot \frac{T_A}{T_B} \times \frac{T_C}{T_E}$	$-x \cdot \frac{T_A}{T_E}$
3.	Add y revolution to all elements	y	$x+y$	$y-x \cdot \frac{T_A}{T_B}$	$y-x \cdot \frac{T_A}{T_B} \times \frac{T_C}{T_E}$	$y-x \frac{T_A}{T_D}$

(i) Speed of the Machine shaft or of wheel E

In the problem, it is given that the speed of the motor or the speed of the pinion A is 1000 r.p.m. But from table 10.12, the revolution of the pinion A is $x + y$. Therefore

$$x + y = 1000 \qquad ...(i)$$

It is also given that annular wheel D is fixed. Hence $N_D = 0$. But from table the revolution of D is $\left(y - x\frac{T_A}{T_D}\right)$. Therefore

$$\therefore \quad y - x \cdot \frac{T_A}{T_D} = 0$$

or
$$y = x \cdot \frac{T_A}{T_D} = x \cdot \frac{15}{55} \qquad (\because \; T_A = 15, T_D = 55)$$

$$= \frac{3x}{11} \qquad ...(ii)$$

Substituting the value of y in equation (i), we get

$$x + \frac{3x}{11} = 1000$$

or
$$11x + 3x = 11 \times 1000$$

or
$$14x = 11 \times 1000$$

$$\therefore \quad x = \frac{11 \times 1000}{14} = \frac{11000}{14} = 785.71 \text{ say } 786 \text{ r.p.m.}$$

From equation (ii),
$$y = \frac{3 \times 786}{11} = 214.36 \text{ say } 214 \text{ r.p.m.}$$

$\therefore$ Speed of machine shaft (or speed of wheel E) (from table 10.12) is $\left(y - x\frac{T_A}{T_B} \times \frac{T_C}{T_E}\right)$

$$\therefore \quad N_E = y - x \cdot \frac{T_A}{T_B} \times \frac{T_C}{T_E}$$

$$= 214 - 786 \times \frac{15}{20} \times \frac{15}{50} = 214 - 176.85$$

$$= \mathbf{37.15 \text{ r.p.m. anticlockwise.}} \quad \textbf{Ans.}$$

(ii) Torque exerted on the machine shaft

Let M_i = Input torque or torque applied to driving shaft *i.e.* motor shaft connected to pinion A.

M_r = Resisting torque or torque on driven shaft or machine shaft connected to wheel E.

Input torque (or torque applied to driving shaft *i.e.* motor shaft) is 100 Nm $\quad \therefore \quad M_i = 100$ Nm

The resisting torque *i.e.* torque on driven shaft or machine shaft will be obtained by using equation (10.9),

$$M_r = -\frac{M_i \times \omega_i}{\omega_r}$$

where ω_i = Angular velocity of driving shaft (*i.e.* pinion A)

$$= \frac{2\pi N_i}{60} \qquad [N_i = \text{speed of driving shaft} = 1000 \text{ r.p.m. (given)}]$$

$$= \frac{2\pi \times 1000}{60}$$

ω_r = Angular velocity of driven shaft (*i.e.* of gear E)

$$= \frac{2\pi N_E}{60} \qquad [N_E = \text{speed of gear } E = 37.15 \text{ r.p.m.}]$$

$$= \frac{2\pi \times 37.15}{60}$$

Substituting these values in the above equation, we get

$$M_r = -\frac{100 \times \frac{2\pi \times 1000}{60}}{\frac{2\pi \times 37.15}{60}} = -100 \times \frac{1000}{37.15} = \mathbf{-2691.79\ N\ m.} \quad \textbf{Ans.}$$

Problem 10.18. *An epicyclic gear train consists of a sunwheel S, a stationary internal gear E and three identical planet wheels P carried on a star-shaped planet carrier C. The size of different toothed wheels are such that the planet C rotates at $\frac{1}{5}$ of the speed of the sunwheel S. The minimum number of teeth on any wheel is 16. The driving torque on the sun wheel is 100 Nm. Determine : (a) number of teeth on different wheels of the train and (b) torque necessary to keep the internal gear stationary.* (A.M.I.E., Summer 1978)

Sol. Given :

Speed of planet carrier, $C = \frac{1}{5}$ th speed of sun wheel S

or $$N_C = \frac{1}{5} \times N_S$$

Minimum number of teeth on any wheel = 16.

Torque on wheel S = 100 Nm (or Input torque)

Let T_S = Number of teeth on S and

T_E = Number of teeth on internal gear E

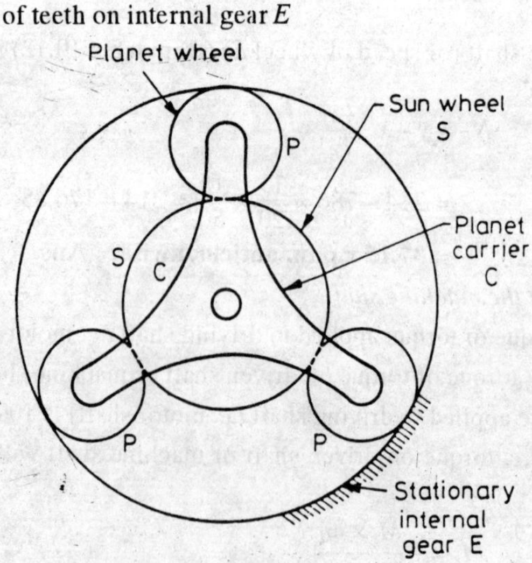

Fig. 10.25

Now prepare the table to motion as given below :

Table 10.13

S. No.	Operation	Revolution of elements			
		Planet carrier C	Sun wheel S	Planet wheel P	Internal gear E
1.	Planet carrier C fixed. Sun wheel S rotates through + 1 revolution. (*i.e.* 1 rev. anticlockwise)	0	+ 1	$-\frac{T_S}{T_P}$	$-\frac{T_S}{T_P} \times \frac{T_P}{T_E} = -\frac{T_S}{T_E}$
2.	Multiply by x to all	0	$+x$	$-x \times \frac{T_S}{T_P}$	$-x \times \frac{T_S}{T_E}$
3.	Add y revolution to all elements	$+y$	$x+y$	$-x\frac{T_S}{T_P}+y$	$-x\frac{T_S}{T_E}+y$

In the problem, it is given that when the sun wheel S makes 5 revolution, the planet carrier makes one revolution. Therefore from the last row of the table 10.13, we see that the revolutions of planet carrier C and sun wheel S, are : $N_C = y$ and $N_S = x + y$.

But $$N_C = \frac{1}{5} N_S \quad \text{(given)}$$

$\therefore$ $$y = \frac{1}{5}(x+y)$$

or $$5y = x + y$$

or $$4y = x$$

It is also given that gear E is stationary (*i.e.* revolution is zero).

But from column six (*i.e.* last column) of Table 10.13, the revolution of gear E is $\left(-x\frac{T_S}{T_E}+y\right)$.

$\therefore$ $$-x\frac{T_S}{T_E}+y=0$$

or $$-4y \times \frac{T_S}{T_E}+y=0 \quad (\because\ x = 4y)$$

or $$-4\frac{T_S}{T_E}+1=0 \quad \text{(Cancelling 'y' to both sides)}$$

or $$\frac{T_S}{T_E}=\frac{1}{4}$$

Assuming the number of teeth on sunwheel as 16 (*i.e.* minimum number of teeth on any wheel), we get

$$T_S = 16$$

$\therefore$ $$T_E = 4 \times T_S = 4 \times 16 = 64.$$ **Ans.**

Let us now find the number of teeth on planet wheel, P.

Let D_S = Pitch circle diameter of wheel S,

D_P = Pitch circle diameter of wheel P, and

D_E = Pitch circle diameter of wheel E.

Assuming the pitch of all the gears to be same, then from Fig. 10.25, we find

$$\frac{D_S}{2} + D_P = \frac{D_E}{2}$$

or $$D_S + 2D_P = D_E \qquad ...(i)$$

We know that the number of teeth are proportional to their pitch circle diameters and hence equation (*i*) can also be written as

$$T_S + 2T_P = T_E$$

or $$16 + 2T_P = 64$$

or $$T_P = \frac{64 - 16}{2} = 24. \quad \textbf{Ans.}$$

∴ No. of teeth on different wheels are $T_S = 16$; $T_E = 64$ and $T_P = 24$. **Ans.**

Torque necessary to keep the internal gear stationary.

Input torque (or driving torque on sunwheel), M_i = 100 N m

Let M_r = Resisting torque on driven wheel (*i.e.* planet carrier *C*)

M_b = Holding torque to keep the internal gear (*E*) stationary.

The value of resisting torque (M_r) is obtained by using equation (10.9) as

$$M_r = -M_i \times \frac{\omega_i}{\omega_r}$$

where ω_i = Angular velocity of driving wheel (*i.e.* sunwheel S)

$$= \frac{2\pi \times N_s}{60}$$

ω_r = Angular velocity of driven wheel (*i.e.* planet carrier *C*)

$$= \frac{2\pi N_C}{60}$$

Substituting these values in the above equation, we get

$$M_r = -M_i \times \left(\frac{\frac{2\pi N_s}{60}}{\frac{2\pi N_C}{60}} \right) = -M_i \times \frac{N_s}{N_C}$$

$$= -100 \times 5 \qquad \left[\because \frac{N_s}{N_C} = 5 \text{ (given)} \right]$$

$$= -500 \text{ N m}$$

To find holding torque (M_b), the equation (10.6) is used

∴ $$M_i + M_r + M_b = 0$$

or $$M_b = -M_i - M_r = -100 - (-500) = -100 + 500 = 400 \text{ Nm}$$

∴ Torque necessary to keep the internal gear stationary = **400 Nm. Ans.**

10.6. Compound Epicyclic Gear Train

When two or more epicyclic gear train are connected in series, the combination is known a compound epicyclic gear train. The compound epicyclic gear train is analysed by considering each epicyclic gear train separately. That epicyclic gear train is first analysed where two conditions (or the speeds of two elements) are known. After that second epicyclic gear train is solved and so on.

Problem 10.19. *Fig. 10.26 shows a compound epicyclic gear train in which the two gears S_1 and S_2 are integral with the input shaft, B. The arm, A_2 is integral with the output shaft, C. The planet gear P_2 revolves*

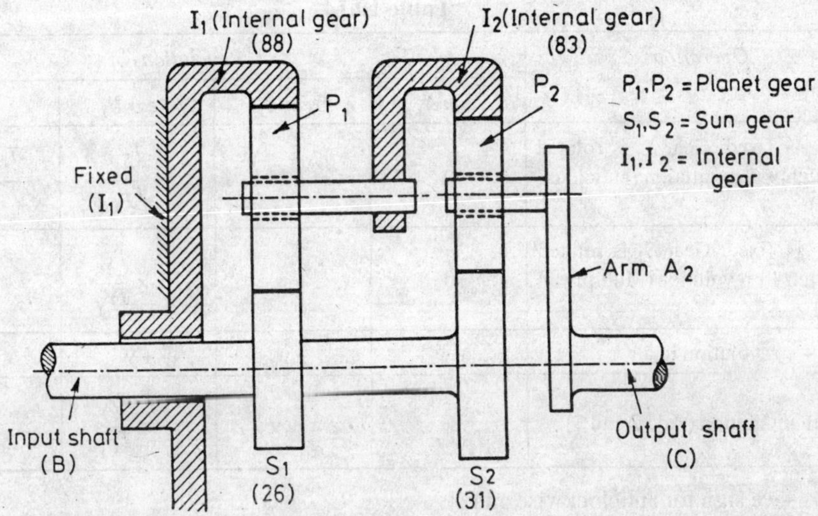

Fig. 10.26

on a pin attached to the arm A_2. The gear P_2 meshes with the sun gear S_2 and internal gear I_2. The internal gear I_2 is co-axial with the input shaft. The planet gear P_1 meshes with the fixed internal gear I_1 and sun gear S_1. The planet gear P_1 revolves on a pin fixed to internal gear I_2. The number of teeth on the gears S_1, S_2, I_1 and I_2 are 26, 31, 88 and 83 respectively. The input power on the shaft (B) is 10.47 kW at 1000 r.p.m. Find :

(*i*) *the speed and torque at output shaft*

(*ii*) *the torque required to hold the internal gear I_1 stationary. The input shaft B rotates anticlockwise.*

Sol. Given :

The gears S_1 and S_2 are integral with input shaft.

∴ Speed of S_1 = speed of S_2 = speed of input shaft = 1000 r.p.m. or $N_{S_1} = N_{S_2} = 1000$ r.p.m.

The arm, A_2 is integral with the output shaft, C

∴ Speed of arm, A_2 = speed of output, C or $N_{A_2} = N_C$

The internal gear I_1 is fixed. Hence speed of I_1 is zero.

∴ $$N_{I_1} = 0$$

The gear P_1 revolves on a pin fixed to internal gear I_2. Hence speed of arm A_1 = speed of gear I_2.

The no. of teeths are $T_{S_1} = 26$, $T_{S_2} = 31$, $T_{I_1} = 88$ and $T_{I_2} = 83$.

Input power, $P_i = 10.47$ kW = (10.47 × 1000) W.

The no. of teeths on planet gears P_1 and P_2 are not given. These values can be obtained from the geometry of the Fig. 10.26. But here these values are not required.

This problem consists of two epicyclic gear trains. They are gears S_1, P_1, I_1 and arm A_1 and gears S_2, P_2, I_2 and arm A_2. The arm A_1 is integral with internal gear I_2. Let us first analyse that epicyclic gear train where two conditions (or speeds of two elements) are known. In case of first epicyclic gear train, the internal gear I_1 is fixed and sun gear S_1 is rotating at 1000 r.p.m. Hence for this epicyclic gear train the two conditions are known. Hence this epicyclic gear train is analysed first.

The no. of revolutions of various elements (*i.e.* sun gear S_1, planet gear P_1, internal gear I_1 and arm A_1) are shown in the table of motion given in Table 10.14 as :

Table 10.14

S. No.	*Operations*	*Revolutions of*			
		Arm A_1	*Gear* S_1	*Gear* P_1	*Gear* I_1
1.	Arm A_1 fixed. Gear S_1 is rotated through + 1 revolution (anticlockwise)	0	+ 1	$-\frac{T_{S_1}}{T_{P_1}}$	$-\frac{T_{S_1}}{T_{P_1}} \times \frac{T_{P_1}}{T_{I_1}} = -\frac{T_{S_1}}{T_{I_1}}$
2.	Arm A_1 fixed. Gear S_1 is rotated through $+x$ revolution (Multiply all by x)	0	$+x$	$-x \times \frac{T_{S_1}}{T_{P_1}}$	$-x \times \frac{T_{S_1}}{T_{I_1}}$
3.	Add $+y$ revolution to all	$+y$	$+y$	$+y$	$+y$
4.	Resultant motion (Add 2 and 3)	y	$x+y$	$y - x\frac{T_{S_1}}{T_{P_1}}$	$y - x\frac{T_{S_1}}{T_{I_1}}$

Take + ve sign for anticlockwise and
– ve sign for clockwise.

Let us find the speed of arm A_1 which will be the speed of I_2 also. The arm A_1 is integral with I_2

The two conditions given are : (*i*) Gear I_1 is fixed. Hence $N_{I_1} = 0$

(*ii*) Gear S_1 rotates at 1000 r.p.m. $\therefore$ $N_{S_1} = 1000$ r.p.m.

But from Table 10.14, the resultant motion of gear I_1 is $\left(y - x\frac{T_{S_1}}{T_{I_1}}\right)$

$$\therefore \quad y - x\frac{T_{S_1}}{T_{I_1}} = 0$$

or
$$y - x \times \frac{26}{88} = 0 \qquad [\because \; T_{S_1} = 26 \text{ and } T_{I_1} = 88 \text{ (given)}] \qquad ...(i)$$

Also from table 10.14, the resultant motion of gear S_1 is $(x + y)$

$$\therefore \quad x + y = 1000 \qquad ...(ii)$$

or
$$x = 1000 - y$$

Substituting the value of x in equation (*i*), we get

$$y - (1000 - y) \times \frac{26}{88} = 0$$

or
$$y - 1000 \times \frac{26}{88} + \frac{26}{88}y = 0$$

or $$y + \frac{26}{88} y = 1000 \times \frac{26}{88}$$

or $$y\left(1 + \frac{26}{88}\right) = 1000 \times \frac{26}{88}$$

or $$y\left(\frac{88 + 26}{88}\right) = 1000 \times \frac{26}{88}$$

or $$\frac{114y}{88} = 1000 \times \frac{26}{88}$$

or $$y = \frac{88 + 1000 + 26}{114 \times 88} = 228.07 \text{ r.p.m.} \simeq 228 \text{ r.p.m.}$$

Now from the Table 10.14, the resultant motion of arm A_1 is (y).

$\therefore$ Speed of arm $A_1 = y = 228$ r.p.m.

As arm A_1 is integral with gear I_2, hence speed of gear $I_2 = 228$ r.p.m.

Now the second epicyclic gear train (*i.e.* train consisting of gears S_2, P_2, I_2 and arm A_2) can be analysed as the two conditions of motions are known. These conditions of motions are : (*i*) speed of gear $S_2 = 1000$ r.p.m. and (*ii*) speed of gear $I_2 = 228$ r.p.m.

(*i*) *Speed and torque at output shaft*

The no. of revolutions of various elements (*i.e.* gears S_2, P_2, I_2 and arm A_2) are shown in the table of motion given in Table 10.15, as

Table 10.15

S. No.	Operations	Revolutions of			
		Arm A_2	*Gear* S_2	*Gear* P_2	*Gear* I_2
1.	Arm A_2 fixed. Rotate gear S_2 through + 1 revolution (anticlockwise)	0	+ 1	$-\frac{T_{S_2}}{T_{P_2}}$	$-\frac{T_{S_2}}{T_{P_2}} \times \frac{T_{P_2}}{T_{I_2}} = -\frac{T_{S_2}}{T_{I_2}}$
2.	Arm A_2 fixed. Rotate gear S_2 through $+x$ revolution (Multiply by x to all)	0	$+x$	$-x\frac{T_{S_2}}{T_{P_2}}$	$-x \times \frac{T_{S_2}}{T_{I_2}}$
3.	Add + y revolution to all	$+y$	$+y$	$+y$	$+y$
4.	Resultant motion (Add 2 and 3)	y	$x + y$	$y - x\frac{T_{S_2}}{T_{P_2}}$	$y - x\frac{T_{S_2}}{T_{I_2}}$

The two conditions of motions are : (*i*) speed of gear $S_2 = 1000$ r.p.m. and

(*ii*) speed of gear $I_2 = 228$ r.p.m.

But from Table 10.15, the resultant motion of S_2 is $(x + y)$ and resultant motion of gear I_2 is $\left(y - x\frac{T_{S_2}}{T_{I_2}}\right)$.

$\therefore$ $$x + y = 1000 \quad \text{...(iii)}$$

and $$y - x \times \frac{T_{S_2}}{T_{I_2}} = 228 \quad \text{...(iv)}$$

or $$y - x \times \frac{31}{83} = 228 \qquad (\because\ T_{S_2} = 31 \text{ and } T_{I_2} = 83)$$

or $$y - (1000 - y) \times \frac{31}{83} = 228 \qquad [\because \text{ From equation } (iii),\ x = 1000 - y]$$

or $$y - 1000 \times \frac{31}{83} + y \times \frac{31}{83} = 228$$

or $$y + y \times \frac{31}{83} = 228 + 1000 \times \frac{31}{83}$$

or $$y\left(y + \frac{31}{83}\right) = 228 + 373.49 = 601.49$$

or $$y\left(\frac{83+31}{83}\right) = 601.49$$

or $$y = \frac{601.49 \times 83}{(83+31)} = 437.92 \text{ r.p.m. (Anticlockwise)}$$

$$\simeq 438 \text{ r.p.m. (Anticlockwise)}$$

But from the table 10.15, the resultant motion of arm A_2 is 'y'

$\therefore$ Speed of Arm, $A_2 = y = 438$ r.p.m.

But arm A_2 is integral with output shaft.

$\therefore$ Speed of output shaft = **438 r.p.m. (Anticlockwise)**. **Ans.**

Torque at output shaft

Let us first find the torque at input shaft from input power.

As power, $P = \dfrac{2\pi NT}{60}$ $\qquad \therefore \quad T = \dfrac{P \times 60}{2\pi N}$

or torque at input shaft, T_i (or M_i) $= \dfrac{P_i \times 60}{2\pi N_i}$

where P_i = Power at input shaft = 10.47×1000 W

N_i = Speed of input shaft = 1000 r.p.m.

$$= \frac{(10.47 \times 1000) \times 60}{2\pi \times 1000} = 99.98 \simeq 100 \text{ Nm}$$

Let M_r = Torque at output shaft (or resisting torque at driven shaft)

M_i = Torque at input shaft = 100 Nm

ω_r = Angular velocity of output shaft

$= \dfrac{2\pi N_r}{60}$ where N_r = speed of output shaft = 438 (calculated above)

$= \dfrac{2\pi \times 438}{60}$ = rad/s

ω_i = Angular velocity of input shaft = $\dfrac{2\pi N_i}{60}$

$= \dfrac{2\pi \times 1000}{60}$ $\qquad [\because$ speed of input shaft, $N_i = 1000$ r.p.m. (given)]

Now using equation (10.9), we get

$$M_r = -M_i \times \frac{\omega_i}{\omega_r} = -100 \times \frac{\left(\frac{2\pi \times 1000}{60}\right)}{\left(\frac{2\pi \times 438}{60}\right)} = -\frac{100 \times 1000}{438} = -228.31 \text{ Nm}$$

∴ Torque at output shaft = **– 228.31 N m. Ans.**

(*ii*) *Torque required to hold the gear I_1 stationary*

Let M_b = torque required to hold the gear stationary.

Using equation (10.6), we get, $M_i + M_b + M_r = 0$

or $M_b = -M_i - M_r = -100 - (-228.31) =$ **128.31 N m. Ans.**

10.7. Epicyclic Gear Train with Bevel Gear

Bevel gears are used for high speed reduction with few gears. In previous articles, we have discussed the method which are used for solution of epicyclic trains of spur gears. These methods are also applicable to the epicyclic trains consisting of bevel gears. But for the wheels (or gears) whose axes are inclined to the main axis, the term clockwise and anticlockwise are not applicable. Hence in case of tabular method, plus or minus sign is omitted for these wheels. Also the addition of y is not convenient to make as the rotation is about a different axes.

The epicyclic gear train with bevel gears is mostly used in Humpage's speed reduction gear and differential gear of an automobile.

Problem 10.20. *Fig. 10.27 shows an epicyclic bevel gear train (Humpage's reduction gear) in which gear B is connected to the input shaft and gear F is connected to the output shaft. The arm A, carrying the compound wheels D and E, turns freely on the output shaft. If the input speed is 1000 r.p.m. anticlockwise when seen from the right, determine the speed of the output shaft when : (i) gear C is fixed, and (ii) gear C is rotated at 10 r.p.m. anticlockwise.*

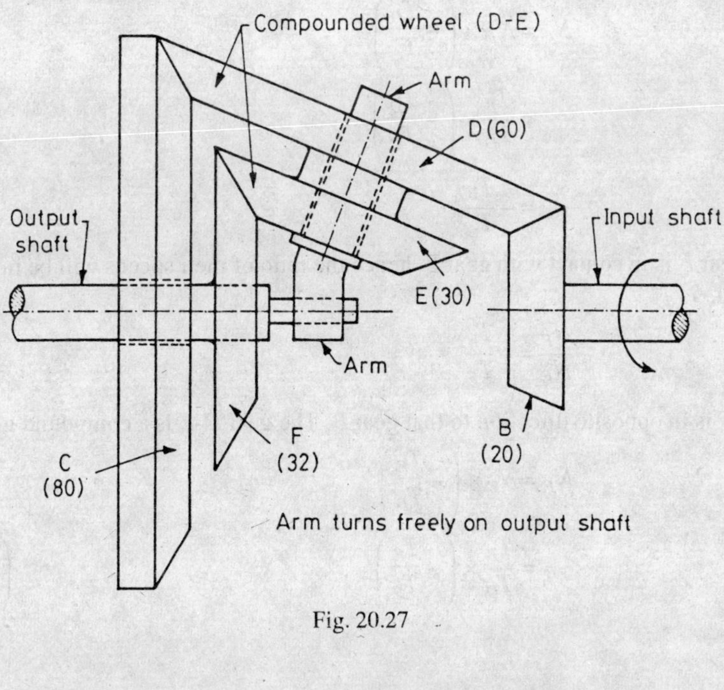

Fig. 20.27

Sol. Given :

Gear B with input shaft, Gear F with output shaft ; Arm A carries compound wheel D-E ; Arm A turns freely on output shaft ; $N_B = 1000$ r.p.m. anticlockwise.

(i) Speed of output shaft when gear C is fixed

The axis of the compound wheel D-E is included to the main axis (*i.e.* the axis of wheels B, F and C). Therefore, no sign is fixed to the direction of rotation of compound wheel D-E, as mentioned in Art. 10.6. But these wheels (D-E) surely influence the direction of rotation of other wheels.

Let us fix the arm A and rotate the gear B (*i.e.* input shaft) through + 1 revolution (*i.e.* 1 rotation anticlockwise). The no. of revolution of each elements will be obtained as :

(*i*) The gear B is in contact with bevel gear D. Hence the ratio of their speed will be inversely proportional to their no. of teeth.

$$\therefore \quad \frac{N_D}{N_B} = \frac{T_B}{T_D} \qquad \left(\text{Due to bevel gear, no sign is given in front of } \frac{T_B}{T_D}\right)$$

$$\therefore \quad N_D = N_B \times \frac{T_B}{T_D} = \frac{T_B}{T_D} \text{ (as } N_B = +1)$$

The speed of gear E will be same as that of gear D as D-E is a compound wheel.

$$\therefore \quad N_E = N_D = \frac{T_B}{T_D}$$

(*ii*) Gear C is in contact with bevel gear D. Hence ratio of their speeds will be inversely proportional to their no. of teeth.

$$\therefore \quad \frac{N_C}{N_D} = \frac{T_D}{T_C}$$

(– ve sign is due to the fact that speed of C will be in opposite direction to that of B. The gear B is given anticlockwise rotation. Hence gear C will rotate clockwise)

$$\therefore \quad N_C = N_D \times \left(-\frac{T_D}{T_C}\right)$$

$$= \frac{T_B}{T_D} \times \left(-\frac{T_D}{T_C}\right) \qquad \left(\because N_D = \frac{T_B}{T_D}\right)$$

$$= -\frac{T_B}{T_C}$$

(*iii*) The gear F is in contact with gear E, hence the ratio of their speeds will be inversely proportional to their no. of teeth.

$$\therefore \quad \frac{N_F}{N_E} = -\frac{T_E}{T_F}$$

(speed of F is in opposite direction to that gear B. The gear D-E is a compound gear)

$$\therefore \quad N_F = N_E \times \left(-\frac{T_E}{T_F}\right)$$

$$= \frac{T_B}{T_D} \times \left(-\frac{T_E}{T_F}\right) \qquad \left(\because N_E = N_D = \frac{T_B}{T_D}\right)$$

$$= -\frac{T_B}{T_D} \times \frac{T_E}{T_F}$$

Hence no. of revolution of each element when gear B is rotated through + 1 revolution and arm A is kept fixed, are $N_D = \frac{T_B}{T_D}$; $N_C = -\frac{T_B}{T_C}$ and $N_F = -\frac{T_B}{T_D} \times \frac{T_E}{T_F}$. This gives the first row of the table 10.16.

Now the table of motion can be prepared as given in table 10.16.

Table 10.16

S. No.	Slips of Motion	Revolution of elements				
		Arm A	*Gear B Input*	*Wheel D-E compound*	*Gear C*	*Gear F output*
1.	Arm fixed, Gear B is rotated through + 1 revolution	0	+ 1	$* \frac{T_B}{T_D}$	$-\frac{T_B}{T_C}$	$-\frac{T_B}{T_D} \times \frac{T_E}{T_F}$
2.	Arm fixed, Gear B rotated through + x revolution (Multiply the 1st row by x)	0	$+x$	$x\frac{T_B}{T_D}$	$-x\frac{T_B}{T_C}$	$-x\frac{T_B}{T_D} \times \frac{T_E}{T_F}$
3.	Add + y revolution to all elements and get the resultant motion.	...	$x + y$	...	$-x\frac{T_B}{T_C} + y$	$-x\frac{T_B}{T_D} \times \frac{T_E}{T_F} + y$

*No sign is given in front of $\frac{T_B}{T_D}$ due to bevel gears.

N.B. The compute system can be rotated about **one** axis only. Thus addition of y to compound wheel D-E is not done.

When gear C is fixed, then speed of gear C *i.e.* $N_C = 0$.

But the resultant motion of the gear C from the table 10.16 is $\left(-x\frac{T_B}{T_C} + y\right)$.

$$\therefore \qquad -x\frac{T_B}{T_C} + y = 0$$

or $$-x\frac{20}{80} + y = 0$$

or $$-\frac{x}{4} + y = 0$$

or $$-x + 4y = 0 \qquad \text{...(i)}$$

Also the speed of input shaft (*i.e.* speed of gear B) is given as 1000 r.p.m. anticlockwise. But the resultant motion from the table 10.16 of gear B is $(x + y)$.

$$\therefore \qquad x + y = 1000 \qquad \text{...(ii)}$$

Adding equations (*i*) and (*ii*), we get

$$5y = 1000 \qquad \text{or} \qquad y = 200$$

Substituting this value of y in equation (*ii*), we get

$$x + 200 = 1000 \qquad \text{or} \quad x = 800$$

Speed of output shaft = Speed of Gear F

$$= -x\frac{T_B}{T_D} \times \frac{T_E}{T_F} + y \qquad \text{(From table 10.16)}$$

$$= -800 \times \frac{20}{60} \times \frac{30}{32} + 200 = -250 + 200$$

$$= \mathbf{-50 \text{ r.p.m. (clockwise)}}. \quad \textbf{Ans.}$$

(*ii*) *Speed of output shaft when gear C is rotated at 10 r.p.m. anticlockwise*

∴ Speed of $C = +10$ r.p.m. (+ ve sign due to anticlockwise rotation)

But the resultant motion of gear C from table 10.16 is $\left(-x\frac{T_B}{T_C} + y\right)$.

$$\therefore \qquad -x\frac{T_B}{T_C} + y = +10$$

or $$-x\frac{20}{80} + y = 10$$

or $$-\frac{x}{4} + y = 10$$

or $$-x + 4y = 40 \qquad \text{...(iii)}$$

But the speed input E shaft (*i.e.* speed of gear B) is given as 1000 r.p.m. anticlock-wise. But from the table 10.16, the resultant motion of gear B is $(x + y)$

$$\therefore \qquad x + y = 1000 \qquad \text{...(iv)}$$

Adding equation (*iii*) and (*iv*), we get

$$5y = 1000 + 40 = 1040$$

$$\therefore \qquad y = \frac{1040}{5} = 208 \text{ r.p.m.}$$

From (*iv*), $x = 1000 - y = 1000 - 208 = 792$

Now from table 10.16 the speed of output shaft (*i.e.* speed of gear F)

$$= -x\frac{T_B}{T_D} \times \frac{T_E}{T_F} + y = -792 \times \frac{20}{60} \times \frac{30}{32} + 208$$

$$= -247.5 + 208 = \mathbf{-39.5 \text{ r.p.m. (clockwise)}}. \quad \textbf{Ans.}$$

Problem 10.21. *For the Humpage's speed reduction gear shown in fig. 10.27, prove that (i) if the ratio of the tooth numbers* $\frac{T_D}{T_E}$ *is greater than* $\frac{T_C}{T_F}$, *the wheel F will rotate in the same direction as wheel B and (ii) if the ratio* $\frac{T_D}{T_E}$ *is less than* $\frac{T_C}{T_F}$, *the direction of F is reversed.*

If no. of teeth on wheels B, D, E, F and C are 17, 60, 19, 25 and 75 respectively and the input power at the input shaft is 7.5 kW when input shaft is rotating at 500 r.p.m., find the output torque at the output shaft and the forces tangential to the pitch cones at the contact points between wheels E and F and between wheels D and C if the module of all the wheels is 6.5 mm.

Sol. Given : Refer to Fig. 10.27.

The gear B is connected to input shaft whereas gear F is connected to output shaft. The gear C is fixed.

Now refer to table of motion given in table 10.16.

Let the input shaft rotates by one revolution anticlockwise. Hence the speed of gear B is also one revolution anticlockwise.

But from table 10.16, the resultant motion of gear B is $(x + y)$

$$\therefore \quad x + y = 1 \quad \text{...}(i)$$

The gear C is fixed. Hence speed of gear C is zero. But the resultant motion of gear C from table 10.16 is $\left(y - x \times \frac{T_B}{T_C}\right)$.

$$\therefore \quad y - x \times \frac{T_B}{T_C} = 0 \quad \text{...}(ii)$$

From equation (i), $\quad x = (1 - y) \quad \text{...}(iii)$

Substituting this value in equation (ii), we get

$$y - (1 - y) \times \frac{T_B}{T_C} = 0$$

or $$y - \frac{T_B}{T_C} + y\frac{T_B}{T_C} = 0 \quad \text{or} \quad y + y\frac{T_B}{T_C} = \frac{T_B}{T_C}$$

or $$y\left(1 + \frac{T_B}{T_C}\right) = \frac{T_B}{T_C} \quad \text{or} \quad y\left(\frac{T_B + T_C}{T_C}\right) = \frac{T_B}{T_C} \quad \text{or} \quad y = \frac{T_B}{(T_B + T_C)}$$

Substituting the value of y in equation (iii), we get

$$x = 1 - \frac{T_B}{T_B + T_C} = \frac{T_B + T_C - T_B}{(T_B + T_C)} = \frac{T_C}{(T_B + T_C)}$$

The speed of the output shaft = speed of gear F.

But speed of gear F from table 10.16, is $\left(-x \times \frac{T_B}{T_D} \times \frac{T_E}{T_F} + y\right)$.

$\therefore$ Speed of output shaft

$$= \text{Speed of } F$$

$$= -x\frac{T_B}{T_D} \times \frac{T_E}{T_F} + y$$

$$= -\frac{(T_C)}{(T_B + T_C)} \times \frac{T_B}{T_D} \times \frac{T_E}{T_F} + \frac{T_B}{(T_B + T_C)} \quad \left[\because \frac{T_C}{T_B + T_C} \text{ and } y = \frac{T_B}{T_B + T_C}\right]$$

$$= \left(\frac{T_B}{T_B + T_C}\right)\left[1 - \frac{T_C}{T_D} \times \frac{T_E}{T_F}\right]$$

$$= \left(\frac{T_B}{T_B + T_C}\right)\left[1 - \frac{(T_C/T_F)}{(T_D/T_E)}\right]$$

$$= \left(\frac{T_B}{T_B + T_C}\right)\left[1 - \frac{(T_C/T_F)}{(T_D/T_E)}\right] \quad \text{...}(iv)$$

(*i*) If $(T_D/T_E) > (T_C/T_F)$, then equation (*iv*) will be positive and hence output shaft will rotate in the same direction as input shaft.

(*ii*) If $(T_D/T_E) < (T_C/T_F)$, then equation (*iv*) will be negative and hence output shaft will rotate in the opposite direction as input shaft. **Ans.**

Numerical Problem

Given :

Number of teeth on various gears are :

$$T_B = 17 \,;\, T_D = 60 \,;\, T_E = 19 \,;\, T_F = 25 \text{ and } T_C = 75$$

Power at input shaft, $P_i = 7.5 \text{ kW} = 7.5 \times 1000 \text{ W}$

Speed of input shaft, $N_i = 500$ r.p.m.

The above is the speed of gear B also. Hence $N_B = 500$ r.p.m.

From table 10.16, the speed of gear B is $(x + y)$

$$\therefore \quad x + y = 500 \qquad \ldots(i)$$

The gear C is fixed. Hence speed of gear C is zero.

But from table 10.16, the speed of C is $\left(y - x \times \dfrac{T_B}{T_C}\right)$

$$\therefore \quad y - x \times \frac{T_B}{T_C} = 0$$

or $$y - x \times \frac{17}{75} = 0 \qquad (\because \; T_B = 17 \text{ and } T_C = 75) \qquad \ldots(ii)$$

But from equation (*i*), $x = (500 - y)$ ...(*iii*)

Substituting the value of x in equation (*ii*), we get

$$y - (500 - y) \times \frac{17}{75} = 0$$

or $$75y - (8500 - 17y) = 0$$

or $$75y + 17y = 8500$$

or $$92y = 8500$$

or $$y = \frac{8500}{92} = 92.39 \simeq 92.4 \text{ r.p.m.}$$

Substituting this value in equation (*iii*), we get

$$x = 500 - 92.4 = 407.6 \text{ r.p.m.}$$

Now the speed of output shaft is equal to speed of gear F. But from table 10.16, the speed of gear F is $\left(y - x \times \dfrac{T_B}{T_D} \times \dfrac{T_E}{T_F}\right)$.

$\therefore$ Speed of output shaft = Speed of gear F

$$= y - x \times \frac{T_B}{T_D} \times \frac{T_E}{T_F}$$

$$= y - x \times \frac{17}{60} \times \frac{19}{25} \qquad (\because \; T_B = 17, T_D = 60, T_E = 19 \text{ and } T_F = 25)$$

$$= 92.4 - 407.6 \times \frac{17}{60} \times \frac{19}{25} \qquad (\because \; y = 92.4 \text{ and } x = 407.6)$$

$$= 92.4 - 87.77 = 4.63 \text{ r.p.m.}$$

(*i*) *Torque at output shaft*

The efficiency of the gear train, $\eta = \dfrac{\text{Output power}}{\text{Input power}}$

For 100 percent efficiency,

$$\text{Output power} = \text{Input power} = 7.5 \text{ kW} = 7.5 \times 1000 \text{ W} \qquad \text{(given)}$$

But speed of output shaft = 4.63 r.p.m. (Calculated above)

$$\text{But, power} = \frac{2\pi NT}{60}$$

$$\therefore \quad \text{Torque} = \frac{60 \times \text{power}}{2\pi \times N}$$

$$\therefore \quad \text{Output torque} = \frac{60 \times \text{power at output shaft}}{2\pi \times \text{speed of output shaft}} = \frac{60 \times (7.5 \times 1000)}{2\pi \times 4.63} = \textbf{15468.6 Nm.} \quad \textbf{Ans.}$$

(*ii*) *Tangential force at contact point between Gears E-F*

Module, $m = 6.5$ mm (given)

We know that torque transmitted at the contact point between two gear is equal to the product of tangential force and pitch circle radius of that gear.

$\because$ Torque on gear F = (Tangential force at contact point between F-E) × pitch circle radius of gear F

or Output torque = (Tangential force) × Pitch circle radius of F

($\because$ Torque on F = Output torque)

$$\therefore \quad 15468.6 = (\text{Tangential force}) \times \text{P.C. radius of gear } F \qquad ...(i)$$

$$\text{We also known that module} = \frac{\text{Pitch circle diameter}}{\text{Number of teeth}}$$

$$\therefore \quad 6.5 = \frac{\text{P.C.D. of gear } F}{\text{Number of teeth of } F \text{ (i.e. } T_F)}$$

$$= \frac{\text{P.C.D. of gear } F}{25} \qquad (\because \; T_F = 25)$$

$$\therefore \quad \text{P.C.D. of gear } F = 6.5 \times 25 \text{ mm} = 162.5 \text{ mm} = 0.1625 \text{ m}$$

$$\therefore \quad \text{P.C. radius of gear } F = \frac{0.1625}{2} = 0.08125 \text{ m}$$

Substituting this value in equation (*i*), we get

$$15468.6 = \text{Tangential force} \times 0.08125$$

$$\therefore \quad \text{Tangential force} = \frac{15468.6}{0.08125} = 190382 \text{ N} = \textbf{190.382 KN.} \quad \textbf{Ans.}$$

(*iii*) *Tangential force at contact point between gears D-C*

First of all, we must know that torque at the contact point of gear D-C. As gear C is fixed, hence the torque at C will be fixing torque. Let us find the fixing (or holding torque).

Using equation (10.6), we have $M_i + M_r + M_b = 0$ ...(*ii*)

where M_i = Input torque, which can be obtained from input power and speed at input shaft

$$\left(i.e.\ P_i = \frac{2\pi N_i \times T}{60} \text{ or } T = \frac{60 \times P_i}{2\pi \times N_i} = \frac{60 \times (7.5 \times 1000)}{2\pi \times 500} = 143.24 \text{ N m} \right)$$

$$= 143.24 \text{ Nm}$$

M_r = Resisting torque or torque at output shaft

= 15468.6 Nm (Calculated above)

M_b = Holding torque or fixing torque.

We also know that if input shaft and output shaft rotate in the same direction, then input torque and output torque are in opposite direction. In this problem the input shaft is having + 500 r.p.m. and output shaft is having + 4.63 r.p.m. Hence input and output shafts rotate in the same direction. Hence input and output torques are in opposite direction.

Hence if $M_r = -15468.6$ Nm, then $M_i = +143.24$ Nm

From equation (*ii*), we have

$$M_b = -(M_i + M_r) = -M_i - M_r$$

$$= -143.24 - (-15468.6) = -143.24 + 15468.6 = 15325.36 \text{ Nm}$$

∴ Tangential force between gears *C-D*

$$= \frac{\text{Fixing Torque at } C}{\text{P.C. radius of gear } C} = \frac{15325.36}{\text{P.C. radius of gear } C} \quad ...(iii)$$

But module, $m = \frac{\text{P.C.D. of gear } C}{\text{No. of teeth on C}}$

or $6.5 = \frac{\text{P.C.D. of gear } C}{75}$ $(\because \; T_C = 75)$

∴ P.C.D. of gear *C* $= 6.5 \times 75$ mm $= 487.5$ mm $= 0.4875$ m

∴ P.C. radius of gear $C = \frac{0.4875}{2} = 0.24375$ m

Substituting this value in equation (*iii*), we get

Tangential force between gears $C\text{-}D = \frac{15325.36}{0.24375} = 62873$ N = **62.873 kN. Ans.**

Problem 10.22. *Fig. 10.28 shows an epicyclic gear train in which the gear B is a fixed one and meshes with pinion C. The arm A is keyed to the driving shaft. The gear E is keyed to the driven shaft and meshes with pinion D. The pinions C and D are keyed to a shaft, which revolves in bearing on the arm A. Find the speed of the driven shaft when : (i) the driving shaft makes 400 r.p.m. ; (ii) the wheel B turns in the same sense as the driving shaft at 80 r.p.m., the driving shaft still makes 400 r.p.m.*

Take the number of teeth on the gears as : $T_B = 70$, $T_C = 18$, $T_D = 17$ *and* $T_E = 67$.

Sol. Given :

Gear *B* is fixed. Hence speed of gear *B* is zero or $N_B = 0$

Arm *A* is keyed to driving shaft. Hence speed of arm and driving shaft is same.

Gear *E* is keyed to driven shaft. Hence speed of gear *E* and driven shaft is same.

Pinions *C* and *D* are keyed to a shaft. Hence speed of *C* and *D* will be same. No. of teeth on gears are :

$$T_B = 70, T_C = 18, T_D = 17 \text{ and } T_E = 67.$$

The axis of the pinions *C* and *D* is inclined to the main axis (*i.e.* the axis of the gear *B*, Arm *A* and gear *E*). Therefore no sign is fixed to the direction of rotation of pinions *C* and *D*.

Let us fix the arm *A* and rotate the gear *B* through + 1 revolution (*i.e.* one revolution anti-clockwise). Then the number of revolution of each element will be obtained as :

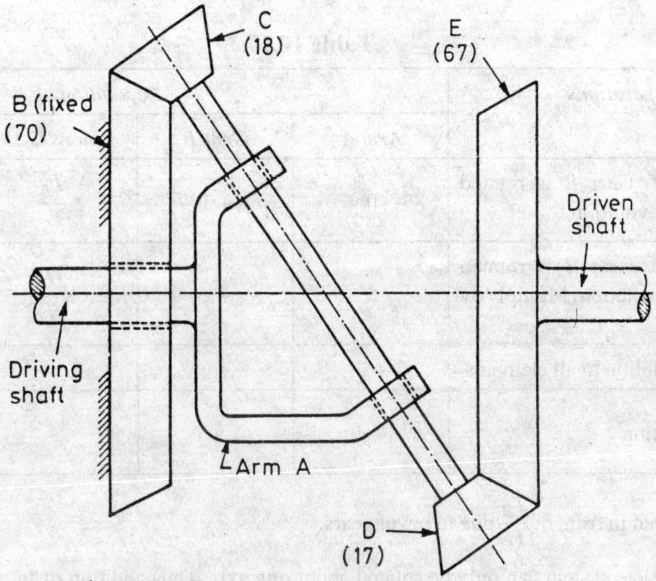

Fig. 10.28

(*i*) The gear B is in contact with bevel gear C. Hence the ratio of their speed will be inversely proportional to their number of teeth

$$\therefore \quad \frac{N_C}{N_B} = \frac{T_B}{T_C} \qquad \text{or} \quad N_C = N_B \times \frac{T_B}{T_C} = 1 \times \frac{T_B}{T_C} = \frac{T_B}{T_C}$$

$$\left(\text{No sign in given is front of } \frac{T_B}{T_C} \text{ due to bevel gear}\right)$$

The speed of D will be same as that of C as C and D is keyed to the same shaft.

$$\therefore \quad N_D = N_C = \frac{T_B}{T_C}.$$

(*ii*) The bevel gear D is in contact with gear E. Ratio of their speeds will be inversely proportional to their no. of teeth.

$$\therefore \quad \frac{N_E}{N_D} = \frac{T_D}{T_E} \qquad \text{or} \quad N_E = N_D \times \frac{T_D}{T_E} = \frac{T_B}{T_C} \times \frac{T_D}{T_E} \qquad \left(\because \; N_D = \frac{T_B}{T_C}\right)$$

Hence for one revolution of gear B, keeping arm fixed, we get

$$N_C = \frac{T_B}{T_C} \qquad \text{and} \quad N_E = \frac{T_B}{T_C} \times \frac{T_D}{T_E}$$

Now the table of motion can be prepared as given as table 10.17.

Table 10.17

S. No.	Operations	Revolution of			
		Arm A	*Gear B*	*Pinons C-D*	*Gear E*
1.	Arm A fixed. Gear B is rotated through + 1 revolution	0	+ 1	$* \frac{T_B}{T_C}$	$\frac{T_D}{T_E} \times \frac{T_B}{T_C}$
2.	Arm A fixed. Gear B is rotated through $+x$ revolution (Multiply by x)	0	$+x$	$x \times \frac{T_B}{T_C}$	$x \times \frac{T_D}{T_E} \times \frac{T_B}{T_C}$
3.	Add + y revolution to all elements	$+y$	$+y$	...	$+y$
4.	Resultant motion	$+y$	$x+y$	...	$y + x\frac{T_C}{T_E} \times \frac{T_B}{T_C}$

*No sign is given in front of $\frac{T_B}{T_C}$ due to bevel gears.

N.B. The complete system can only be rotated about **one** axis. Thus addition of 'y' to bevel gears (C-D) is not convenient as it is having different axis.

(*i*) *Speed of driven shaft when the driving shaft makes 400 r.p.m. and gear B is fixed.*

As driving shaft is keyed to the arm, hence speed of the driving shaft is equal to the speed of arm.

∴ Speed of arm, N_A = Speed of driving shaft = 400 r.p.m.

But resultant motion of arm from table 10.17 is 'y'

∴ $y = 400$ ...(*i*)

As gear B is fixed. Hence $N_B = 0$. But the resultant motion of the gear B from table 10.17 is $(x + y)$

∴ $x + y = 0$ ...(*ii*)

or $x + 400 = 0$ ($\because$ $y = 400$)

or $x = -400$ ...(*iii*)

Speed of driven shaft = Speed of gear E

$$= y + x\frac{T_D}{T_E} \times \frac{T_B}{T_C}$$ (Resultant motion of E from Table)

($\because$ $y = 400$ and $x = -400$)

$$= 400 - 400 \times \frac{17}{67} \times \frac{70}{18} = 400\left(1 - \frac{17}{67} \times \frac{70}{18}\right)$$

$= 400(1 - 0.9867) =$ **5.3 r.p.m. Ans.**

(*ii*) *Speed of driven shaft when driving shaft makes 400 r.p.m. and the gear B 80 r.p.m.*

Hence, $N_A = 400$ and $N_B = 80$.

But from table 10.17, the resultant motion of arm A is '$+ y$' and resultant motion of gear B is $(x + y)$.

∴ $y = 400$ ...(*iv*)

and $x + y = 80$ ...(*v*)

or $x = 80 - y = 80 - 400$ ($\because$ $y = 400$)

$= -320$

∴ Speed of driven shaft or speed of gear E

$$= y + x\frac{T_D}{T_E} \times \frac{T_B}{T_C} \quad \text{[Resultant motion of gear } E \text{ from Table 10.17]}$$

$$= 400 - 320 \times \frac{17}{67} \times \frac{70}{18}$$

($\because$ $y = 400, x = -320, T_D = 17, T_E = 67, T_B = 70$ and $T_C = 18$)

$= 400 - 315.75 =$ **84.25 r.p.m. Ans.**

Humpage's Gear. Fig. 10.29 shows a Humpage's Gear in which the driving shaft and driven shaft are co-axial. The gear B is keyed to the driving shaft whereas the gear F is keyed to the driven shaft. The gears D-E is a compound bevel gear. The gear B on the driving shaft meshes with bevel gear D. The bevel gear D

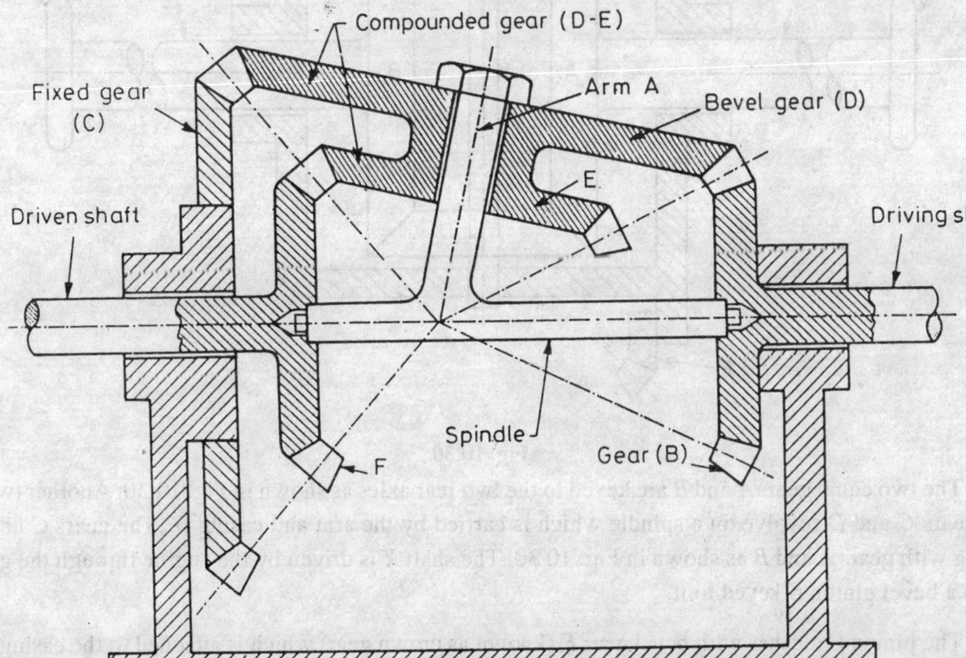

Fig. 10.29

also meshes with the fixed gear C. The gear F on the driven shaft meshes with bevel gear E. The compound gear D-E revolves freely on the arm (A) which is attached to the spindle. The spindle revolves freely about the same axis as the driving and driven shaft.

Originally the Humpage's speed reduction gear was designed as a substitute for back gearing of a lathe, but its use is now considerably extended to all kinds of workshop machines.

Differential Gear. Fig. 10.30 shows a differential gear which is used in the back axle of an automobile. The purpose of the differential gear is (*i*) to enable the rear wheels to revolve at different speeds when the automobile is taking a turn round a corner and (*ii*) to enable the rear wheels to revolve at the same speed when the automobile is moving on straight path. The speed of the outer wheel should be more than the inner wheel during a turn as the distance covered by outer wheel is more than the distance covered by inner wheel. This is achieved by using differential gear. But when the car is moving on a straight path then the speeds of both rear wheels should be same.

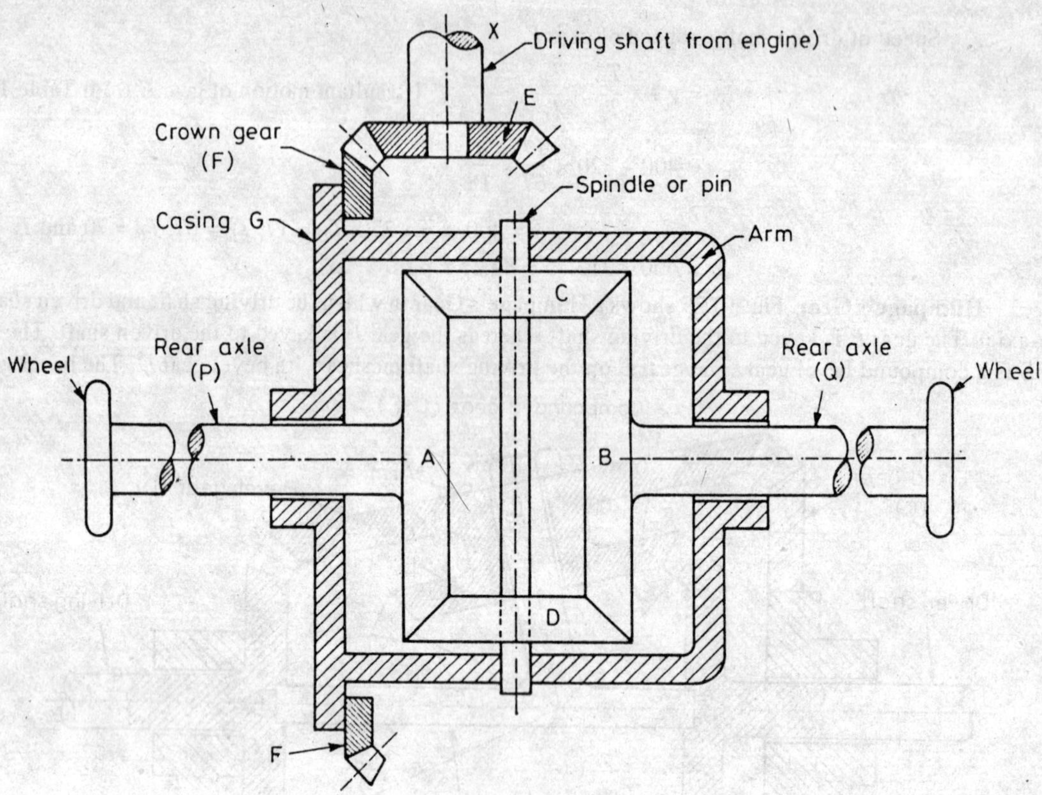

Fig. 10.30

The two equal gears *A* and *B* are keyed to the two rear axles as shown in Fig. 10.30. Another two equal bevel gears *C* and *D* revolve on a spindle which is carried by the arm and casing *G*. The gears *C* and *D* are meshing with gears *A* and *B* as shown in Fig. 10.30. The shaft *X* is driven by the engine through the gear box and has a bevel pinion *E* keyed to it.

The pinion *E* meshes with bevel gear *F* (known as crown gear) which is attached to the casing *G*. The casing *G* revolves freely on the rear axles.

If the car is moving along a straight path the gears *A* and *B* and also the casing (*G*) will revolve at the same speed and the bevel gears *C* and *D* will remain stationary relative to the casing. Hence gears *C* and *D* donot revolve about their own axes. But if the car is moving round a curved path, the gears *A* and *B* will revolve at different speeds and hence the gears *C* and *D* start rotating about the spindle axis and at the same time revolve about the axle axis. Casing will revolve at a speed which is the *arithmetic mean* of the speeds of *A* and *B*. At the same time the bevel gears *C* and *D* will revolve on their pins.

During the turn, *C* and *D* rotate about their own axes and the system works as an epicyclic gear giving two outputs at *A* and *B* with one input at *F* (or *G*). The table of motion can be drawn as shown in Table 10.18.

Table 10.18

S. No.	Operations	Revolutions of			
		Gear F (or G)	Gear A	Gear C	Gear B
1.	Gear F (or G) is fixed. Gear A is rotated through + 1 revolution	0	+ 1	$+\frac{T_A}{T_C}$	$-\frac{T_A}{T_C}\times\frac{T_C}{T_B}=-\frac{T_A}{T_B}=-1$ (∵ $T_A = T_B$ = equal gears)
2.	Gear F (or G) is fixed. Gear A is rotated through + x revolution (multiply by x to all)	0	+ x	$x\times\frac{T_A}{T_C}$	− x
3.	Add y to all	+ y	+ y	+ y	+ y
4.	Resultant motion	+ y	x + y	$y + x\times\frac{T_A}{T_C}$	y − x

From the above table 10.18, it is clear that the speed of *F* (or *G*) *i.e.* Casing is the arithmetic mean of the speeds of gears *A* and *B*. Speed of casing = *y*, speed of gear *A* = (*x* + *y*) and speed of gear *B* (*y* − *x*).

$$\therefore \quad \text{Speed of casing,} \quad y = \frac{\text{Speed of gear } A + \text{Speed of gear } B}{2}$$

$$= \frac{(x+y)+(y-x)}{2} = y$$

Hence it is clear that during a turn, if the speed of *A* decreases than that of casing, there will be a corresponding increase in the speed of *B*. Hence during a turn, the two gears *A* and *B* rotate at different speeds.

Problem 10.23. *Fig. 10.30 shows a differential gear used in motor car. P and Q are the shafts (or rear axles) to which the rear wheels are attached. The gear A is keyed to rear axle P whereas the gear B is keyed to rear axle Q. The pinion E is fixed to the driving shaft (which is connected to the engine through universal coupling not shown in the diagram). The pinion E is having 12 teeth. The pinion E meshes with the crown gear F, which is having 60 teeth. The driving shaft X is rotating at 1000 r.p.m. While taking a turn, the rear axle Q is having a speed of 205 r.p.m. Find the speed of the rear axle P.*

Sol. Given :

$T_E = 12$; $T_F = 60$; $N_Q = 205$ r.p.m., $N_E = 1000$ r.p.m.

The pinion *E* meshes with gear *F*. The gearing between *E* and *F* is external. Hence the speed of gear *F* is obtained by using.

$$\frac{N_F}{N_E} = \frac{T_E}{T_F} \quad \text{(Ratio of speeds are inversely proportional to ratio of their teeth)}$$

$$\therefore \quad N_F = N_E \times \frac{T_E}{T_F}$$

$$= 1000 \times \frac{12}{60} = 200 \text{ r.p.m.}$$

The table of motion is prepared as given in table 10.19.

Table 10.19

S. No.	Operations	Revolution of			
		Gear F (or G)	Gear A	Gear C	Gear B
1.	Gear F (or G) is fixed. Gear A is rotated through + 1 revolution	0	+ 1	$-\frac{T_A}{T_C}$	$-\frac{T_A}{T_C} \times \frac{T_C}{T_B} = -\frac{T_A}{T_B}$
2.	Gear F (or G) is fixed. Gear A is rotated through + x revolution (Multiply all by x)	0	$+x$	$-x \times \frac{T_A}{T_C}$	$-x \times \frac{T_A}{T_B}$
3.	Add + y revolution to all	y	y	y	$+y$
4.	Resultant motion	y	$x+y$	$y - x\frac{T_A}{T_C}$	$y - x\frac{T_A}{T_B}$

Now the two conditions of motion are : (*i*) speed of gear F = 200 r.p.m. and (*ii*) speed of rear axle Q (or speed of gear B as gear B is keyed to rear axle Q) = 315 r.p.m.

But from table 10.19, the resultant speed of gear F is 'y'.

$$\therefore \quad y = 200 \text{ r.p.m.} \quad \quad ...(i)$$

Also from table 10.19, the resultant motion of gear B is $\left(y - x\frac{T_A}{T_B}\right)$

$$\therefore \quad y - x\frac{T_A}{T_B} = 205$$

or $y - x = 205$ ($\because$ $T_A = T_B$ as the two gears A and B are equal and have same number of teeth) ...(*ii*)

or $y - 205 = x$

But from (*i*), $y = 200$. Hence above equation becomes as

$$200 - 205 = x$$

or $-5 = x$

or $x = -5$

$\therefore$ Speed of the rear axle P (or of gear A)

$= x + y$ (From table resultant motion of Gear A)

$= -5 + 200$ ($\because$ $x = -115$ and $y = 200$)

$= $ **195 r.p.m. Ans.**

Problem 10.24. *Fig. 10.31 shows a gear train consisting of bevel gears, spur gears and a worm and worm gear. The bevel pinion is mounted on a shaft which is driven by a V-belt on a pulley. If pulley 2 rotates at 1200 r.p.m. in the direction shown, find the speed and direction of rotation of gear 9.*

(AMIE, W 1994)

Sol. Given :

Dia. of pulley 2, d_2 = 120 mm

Dia. of pulley 3, d_3 = 200 mm

The no. of teeth on gears are $T_4 = 18$, $T_5 = 38$, $T_6 = 20$, $T_7 = 48$, $T_8 = 3$ and $T_9 = 36$.

Speed of pulley 2 or N_2 = 1200 r.p.m. clockwise.

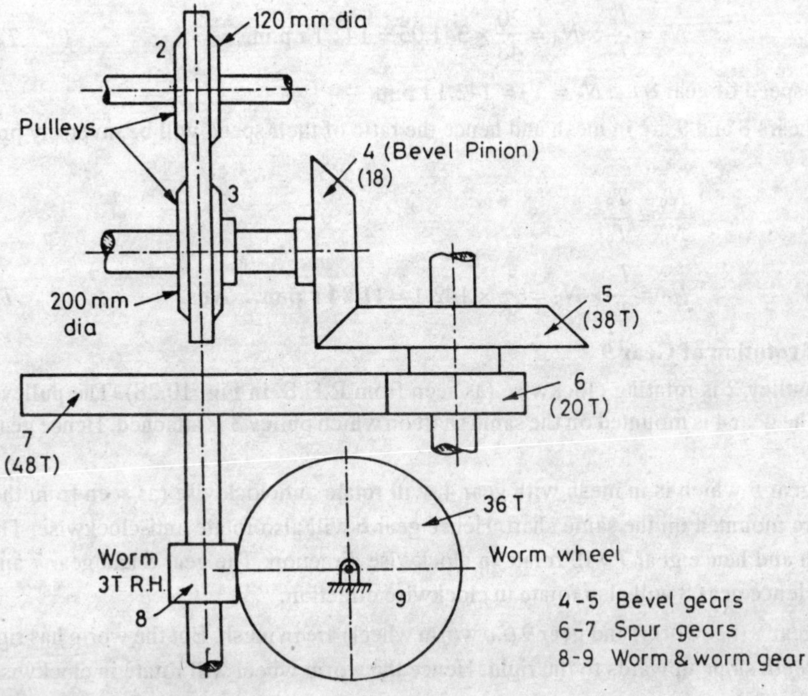

Fig. 10.31

Assuming zero slip in the belt drive, the speed of pulley 3 is obtained by

$$\frac{N_2}{N_3} = \frac{d_3}{d_2}$$

($\because$ The ratio of speeds are inversely proportional to dia. of pulleys)

$$\therefore \quad N_3 = \frac{N_2 \times d_2}{d_3} = 1200 \times \frac{120}{200} = 720 \text{ r.p.m.}$$

The speed of pulley 3 and bevel pinion (*i.e.* gear 4) will be same as they are mounted on the same shaft.

$$\therefore \quad N_3 = N_4 = 720 \text{ r.p.m.}$$

The speeds of gear 5 and 6 will be same as they are mounted on the same shaft. But speed of gear 5 will be obtained as

$$\frac{N_5}{N_4} = \frac{T_4}{T_5}$$

($\because$ The ratio of speeds of gears are inversely proportional to their no. of teeth)

$$\therefore \quad N_5 = \frac{T_4}{T_5} \times N_4 = \frac{18}{38} \times 720 = 341.05 \text{ r.p.m.} \qquad (\because \; T_4 = 18, T_5 = 38)$$

$\therefore$ Speed of gear 6 *i.e.* $N_6 = N_5 = 341.05$ r.p.m.

The speed of gears 7 and 8 will be same as they are mounted on the same shaft. But gear 7 is in mesh with gear 6, hence speed of gear 7 is obtained from equation,

$$\frac{N_7}{N_6} = \frac{T_6}{T_7}$$

($\because$ The ratio of speeds of gears are inversely proportional to their no. of teeth)

$$\therefore \quad N_7 = \frac{T_6}{T_7} \times N_6 = \frac{20}{48} \times 341.05 = 142.1 \text{ r.p.m.} \qquad (\because \; T_6 = 20 \text{ and } T_7 = 48)$$

Also speed of gear 8 *i.e.* $N_8 = N_7 = 142.1$ r.p.m.

The gears 8 and 9 are in mesh and hence the ratio of their speed will be inversely proportional to their no. of teeth.

$$\frac{N_9}{N_8} = \frac{T_8}{T_9}$$

$$\therefore \quad N_9 = \frac{T_8}{T_9} \times N_8 = \frac{3}{36} \times 142.1 = \mathbf{11.84 \text{ r.p.m.}} \quad \textbf{Ans.} \qquad (\because \; T_8 = 3 \text{ and } T_9 = 36)$$

Direction of rotation of Gear 9

The pulley 2 is rotating clockwise (as seen from R.H.S. in Fig. 10.28). The pulley 3 will also rotate clockwise. The gear 4 is mounted on the same shaft on which pulley 3 is attached. Hence gear 4 will also rotate clockwise.

The gear 5 which is in mesh with gear 4, will rotate anticlockwise (as seen from the top). The gear 6 and gear 5 are mounted on the same shaft. Hence gear 6 will also rotate anti-clockwise. The gear 7 and gear 6 are in mesh and hence gear 7 will rotate in clockwise direction. The gear 8 and gear 7 and mounted on the same shaft. Hence gear 8 will also rotate in clockwise direction.

The gear 8 (*i.e.* worm) and gear 9 (*i.e* worm wheel) are in mesh. But the worm has right hand teeth (*i.e.* 3*T RH*), the teeth slope upwards to the right. Hence the worm wheel will rotate in clockwise direction.

$\therefore$ Direction of rotation of gear 9 = **clockwise. Ans.**

Problem 10.25. *Fig. 10.32 shows a planetary gear train that has two inputs ; sun gear 2 rotates at 500 r.p.m. and arm 6 rotates at 750 r.p.m. both clockwise, as viewed from left. Determine the speed and direction of rotation of gear 5.* (AMIE, W 1994)

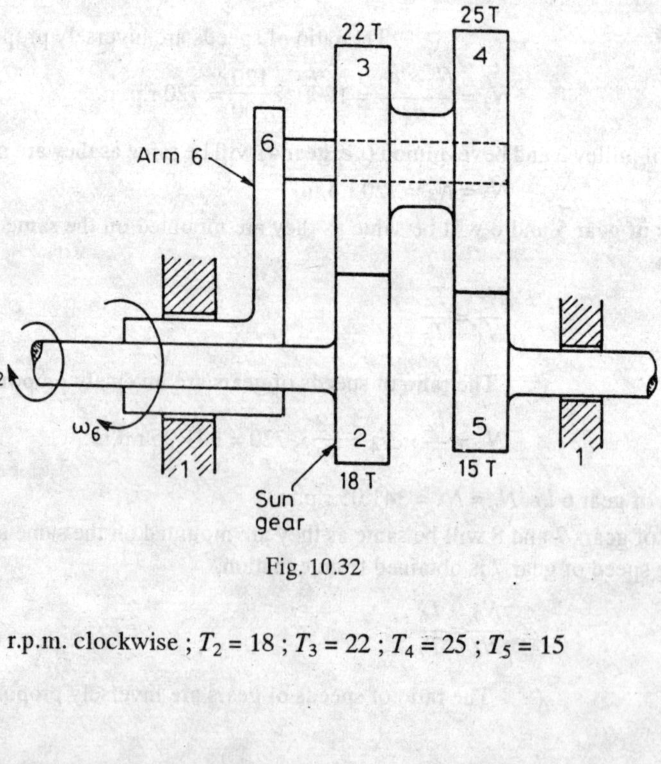

Fig. 10.32

Sol. Given :

$N_2 = 500$ r.p.m. clockwise ; $T_2 = 18$; $T_3 = 22$; $T_4 = 25$; $T_5 = 15$

Speed of arm 6, $N_6 = 750$ r.p.m. clockwise.

Find N_5 in magnitude and direction.

We know that, $\omega = \dfrac{2\pi N}{60}$

$$\therefore \quad \omega_2 = \frac{2\pi N_2}{60} = \frac{2\pi \times 500}{60} = 52.36 \text{ rad/s}$$

and

$$\omega_6 = \frac{2\pi N_6}{60} = \frac{2\pi \times 750}{60} = 78.54 \text{ rad/s}$$

Let us find first the angular velocities of gear 5 and gear 2 with respect to arm.

Angular velocity of gear 5 w.r.t. to arm $= \omega_5 - \omega_{arm}$

$= \omega_5 - \omega_6$ $(\because \ \omega_{arm} = \omega_6)$

$= \omega_5 - 78.54$ ...(*i*)

Angular velocity of gear 2 w.r.t. to arm $= \omega_2 - \omega_{arm}$

$= \omega_2 - \omega_6$

$= 52.36 - 78.54$ ...(*ii*)

Dividing equation (*i*) by equation (*ii*), we get

$$\therefore \quad \frac{\text{Angular velocity of Gear 5 w.r.t. to arm}}{\text{Angular velocity of Gear 2 w.r.t. to arm}} = \frac{\omega_5 - 78.54}{52.36 - 78.54}$$

$$= \frac{\omega_5 - 78.54}{-26.18} \quad \text{...(iii)}$$

If arm is fixed, then

$$\frac{\text{Angular velocity of Gear 5 w.r.t. to arm}}{\text{Angular velocity of Gear 2 w.r.t. to arm}} = \frac{\omega_5 - \omega_{arm}}{\omega_2 - \omega_{arm}}$$

$$= \frac{\omega_5}{\omega_2} \quad (\because \ \omega_{arm} = 0 \text{ being fixed}) \quad \text{...(iv)}$$

Let us find the value of (ω_5/ω_2) from the no. of teeth given on various gears.

As gear 2 and gear 3 are in mesh and hence their angular velocities are inversely proportional to their no. of teeth.

$$\therefore \quad \frac{\omega_3}{\omega_2} = \frac{T_2}{T_3} = \frac{18}{22}$$

$$\therefore \quad \omega_2 = \frac{18}{22} \times \omega_2$$

The gears 3 and 4 are compound gear and are mounted on the same shaft. Hence angular velocity of gear 4 will be same as that of gear 3.

$$\therefore \quad \omega_4 = \omega_3 = \frac{18}{22} \times \omega_2 \quad \text{...(v)}$$

Again the gears 4 and 5 are in mesh and hence their angular velocities will be inversely proportional to their no. of teeth.

$$\frac{\omega_5}{\omega_4} = \frac{T_4}{T_5} = \frac{25}{15}$$

$$\therefore \quad \omega_5 = \frac{25}{15} \times \omega_4$$

$$= \frac{25}{15} \times \left(\frac{18}{22} \times \omega_2\right) \qquad \left[\because \quad \omega_4 = \frac{18}{22} \times \omega_2 \text{ from equation } (v)\right]$$

$$\therefore \quad \frac{\omega_5}{\omega_2} = \frac{25}{15} \times \frac{18}{22} = \frac{15}{11}$$

Substituting the above value of (ω_5/ω_2) in equation (*iv*), we get

$$\frac{\text{Angular velocity of Gear 5 w.r.t. Arm}}{\text{Angular velocity of Gear 2 w.r.t. Arm}} = \frac{15}{11} \qquad ...(vi)$$

Equating the two values of $\left(\frac{\text{Angular velocity of 5 w.r.t. Arm}}{\text{Angular velocity of 2 w.r.t. Arm}}\right)$ given by equations (*iii*) and (*vi*), we get

$$\frac{\omega_5 - 78.54}{-26.18} = \frac{15}{11}$$

or
$$\omega_5 - 78.54 = (-26.18) \times \frac{15}{11}$$

or
$$\omega_5 = 78.54 - 26.18 \times \frac{15}{11} = 42.84 \text{ rad/s}$$

But
$$\omega_5 = \frac{2\pi N_5}{60}$$

$$\therefore \quad N_5 = \frac{\omega_5 \times 60}{2\pi} = \frac{42.84 \times 60}{2\pi} = \textbf{409.09 r.p.m.} \quad \textbf{Ans.}$$

Direction of rotation

The gear 2 is rotating clockwise. But gear 2 and gear 3 are in mesh. Hence gear 3 will rotate anticlockwise. Gear 4 and gear 3 are mounted on the same shaft. Hence gear 4 will also rotate anticlockwise. Now gear 4 and gear 5 are in mesh. Hence gear 5 will rotate clockwise.

∴ Rotation of gear 5 = **clockwise. Ans.**

HIGHLIGHTS

1. A combination of two or more gears is known as gear train. The gear train may consists of spur gears, bevel gears or spiral gears.
2. Types of gear trains are : (*i*) ordinary gear train and (*ii*) epicyclic gear train. Ordinary gear train may be simple or compound.
3. In simple gear train, the axes of shafts on which gears are mounted remain fixed relative to each other.
4. In compound gear train, there is more than one gear on a shaft. Generally intermediate shafts have two gears rigidily fixed to the shaft.
5. The compound gear train may be reverted gear train or non-reverted gear train.
6. In reverted gear train, the axes of the driver and driven shafts are co-axial.
7. In case of epicyclic gear train, atleast one of the gear axis is in motion relative to frame.
8. Velocity ratio (V.R.) of gear train is the ratio of the speed of the driver to the speed of the follower

$$\text{Velocity ratio (V.R.)} = \frac{N_1}{N_2} \qquad \text{...For a simple gear}$$

$$= \frac{\text{Product of teeth on followers}}{\text{Product of teeth on drivers}} \qquad \text{...For a compound gear}$$

9. Velocity ratio of an epicyclic gear train are obtained by
   (*i*) Algebraic method or relative velocity method, and
   (*ii*) Tabular method.
   The basic principle of these method is that the relative motion between a pair of mating gears is always the same whether the axes of rotation are fixed or moving. And the ratio of relative motion is equal to the inverse ratio of the number of teeth on the gears.
10. The gear at the centre of a gear train is known as sun gear whereas the gear whose axis moves round the sun gear, is known as planet gear.
11. The sun and planet gear is obtained by adding annulus gear to the epicyclic gear train.
12. In case of sun and planet gear, either sun gear or annular gear may be fixed. If sun gear is fixed then annular gear provides the drive whereas if annular gear is fixed then sun gear provides the drive. But in both cases, the driven member is the arm.
13. For torque for epicyclic gear train the following equations are used :
   (*i*) $\Sigma M = 0$ or $M_i + M_r = M_b = 0$
   (*ii*) $\Sigma F \times r = 0$ or $F_i \times r_i + F_r \times r_r + F_b \times r_b = 0$
   (*iii*) $\Sigma M \times \omega = 0$ or $M_i \times \omega_i + M_r \times \omega_r + M_b \times \omega_b = 0$
   (*iv*) $M_r = -\dfrac{M_i \times \omega_i}{\omega_r}$ ($\because$ For fixed member, $\omega_b = 0$)
   (*v*) $M_b = M_i\left(\dfrac{\omega_i}{\omega_r} - 1\right)$
   where M_i = Input torque or torque at driving shaft
   M_r = Resisting torque or torque at the driven shaft
   M_b = Holding torque or fixing torque or braking torque,
   F_i, F_r and F_b = corresponding forces at driving, driven and fixing elements,
   ω_i, ω_r and ω_b = corresponding angular velocities.
14. If input and output shafts rotate in the same direction, the input and output torques will be in opposite directions.
15. For compound epicyclic gear train, each epicyclic gear train is analysed separately. That epicyclic gear train is first analysed where two conditions (or the speeds of two elements) are known.
16. The most important types of epicyclic gear train with bevel gears are (*i*) Humpage's speed reduction gear and (*ii*) Differential gear of an automobile.
17. The bevel gear whose axis is inclined to the main axis, the term clockwise and anticlockwise is not applicable in tabular method. Hence for the speeds of these gears, the plus or minus sign is omitted. But these gears surely influence the direction of rotation of other gears.
18. The differential gear is used in the back axle of an automobile :
   (*i*) to enable the rear wheels to revolve at different speeds when a car is taking a turn round a corner and
   (*ii*) to enable the rear wheels to move at the same speed when the car is moving on a straight path.

EXERCISE 10

A. Theoretical Questions

1. What do you mean by gear train ? Mention the different types of the gear train.
2. Differentiate between : (*i*) Simple gear train and compound gear train (*ii*) Reverted gear train and epicyclic gear train.
3. Explain the term train value. How is it related to velocity ratio ? (*AMIE, S 1994*)
4. Explain the procedure of obtaining the velocity ratio of epicyclic gear train by tabular method.
5. Name the different methods of obtaining the velocity ratio of an epicyclic gear train.
6. What are the advantages of epicyclic gearing ? (*A.M.I.E., Winter, 1984*)

7. With the help of a neat sketch, explain the working of a reverted gear train. Give at least two applications of the same.

8. What do you understand by epicyclic gear train ? What are the merits and demerits of an epicyclic gear train as compared to reverted gear train and compound gear train.

9. Prove that the resisting torque or torque on a driven shaft in case of an epicyclic gear train is given by,

$$M_r = -M_i \times \left(\frac{\omega_i}{\omega_r}\right)$$

where M_i = Input torque

ω_i = Input angular velocity

ω_r = Angular velocity at resisting shift.

10. Describe the procedure of calculating the fixing torque in case of fixed wheel in case of an epicyclic gear train.

11. Explain the working principle of any two of the following, with good diagrammatic sketches :

(*i*) Differential gears ;

(*ii*) Reverted gears ;

(*iii*) Epicyclic gear train ;

(*iv*) Humpage gear. *(AMIE, W 1987)*

12. Explain the function of differential gear in an automobile. *(AMIE, S 1994)*

13. Explain the term, Sun and Planet gears. With a neat sketch explain the working of an epicyclic gear train with a sun and planet gear.

14. What are the different equations which are used for calculating the torques and tooth loads in an epicyclic gear train.

15. What is a compound epicyclic gear train ?

16. Explain in details the procedure of analysing the compound epicyclic gear train.

17. With a neat sketch, explain the working of a Humpage's speed reduction gear.

18. What are the function of a differential gear in an automobiles ? Explain the working of a differential gear with a neat sketch.

19. Fig. 10.27 shows a Humpage reduction gear. Prove that in Humpage reduction gear the driving and driven shafts (*i.e.* input shaft and output shafts) rotate in the same direction if T_D/T_E is more than T_C/T_F and in the opposite direction if the same is less than T_C/T_F.

B. Numerical Problems

1. A simple gear train consists of two gears only. Each gear is mounted on separate shaft which are parallel. The first gear is driving the second gear. The speed of first gear is 600 r.p.m. The number of teeth on gears first and second are 20 and 60 respectively. Determine : (*i*) speed ratio of gear train, (*ii*) Train value of the gear train, (*iii*) speed of second gear and (*iv*) direction of rotation of the second gear if first gear is rotating clockwise.

[**Ans.** (*i*) 3, (*ii*) 0.33, (*iii*) 200 r.p.m., (*iv*) Anticlockwise]

2. A simple gear train consists of three gears, each mounted on a separate shaft. The shafts are parallel. The gear 1 meshes with gear 2 whereas gear 2 meshes with gear 3. The gear 1 is the driver rotating clockwise at 1000 r.p.m. and gear 3 is follower. The number of teeth on gears 1, 2 and 3 are 20, 30 and 50 respectively. Find (*i*) speed ratio of the gear train (*ii*) speed of follower and (*iii*) direction of rotation of follower.

[**Ans.** (*i*) 2.5, (*ii*) 400 r.p.m., (*iii*) clockwise]

3. Two parallel shafts are connected with the help of two gears one gear on each shaft. The number of teeth on one gear is 40 and speed of the shaft is 500 r.p.m. If the speed ratio is 2.5 and circular pitch of the gears is 24 mm, then find : (*i*) number of teeth and speed of other shaft and (*ii*) centre distance between the two shafts.

[**Ans.** (*i*) 100 and 200 r.p.m. (*ii*) 534.75 mm]

4. Two parallel shafts are to be connected by spur gears. The shafts are 600 mm apart approximately. Speed of one shaft is 360 r.p.m. and of the other is 120 r.p.m. The circular pitch is 25 mm, find : (*i*) No. of teeth on each gear, (*ii*) pitch circle diameters of the gears and (*iii*) Exact centre distance between two shafts.

[**Ans.** (*i*) 38 and 114, (*ii*) 302.36 and 907.1 mm, (*iii*) 604.73 mm]

5. The Fig. 10.9 shows a compound gear train. The power is transmitted from a motor shaft to output shaft. The motor shaft is connected to gear 1 whereas the output shaft is connected to gear 4. The gears 2 and 3 are mounted on the same shaft. The gear 1 meshes with gear 2 whereas gear 3 meshes with gear 4. The motor shaft is rotating at 1200 r.p.m. in the clockwise direction. Find the direction and speed of output shaft if no. of teeth on gears 1, 2, 3 and 4 are 25, 50, 20, 40. [**Ans.** 300 r.p.m. clockwise]

6. Fig. 10.10 shows a compound gear train. The power is transmitted from a motor shaft to output shaft. The motor shaft is connected to gear 1 whereas the output shaft is connected to gear 6. The motor shaft is rotating at 1200 r.p.m. in the clockwise direction, find the speed and direction of output shaft. The number of teeth on gears 1, 2, 3, 4, 5 and 6 are 20, 40, 30, 60, 25 and 50 respectively. The gears 2 and 3 are mounted on the same shaft. The gears 4 and 5 are also mounted on the same shaft. Gear 1 meshes with gear 2 ; gear 3 meshes with gear 4 and gear 5 meshes with gear 6. [**Ans.** 150 r.p.m. Anticlockwise.]

7. Fig. 10.13 shows an epicyclic gear train in which the arm C is rotating at 80 r.p.m. anticlockwise. The arm carries two gears A and B having 32 and 40 teeth respectively. The gear A is fixed and the arm rotates about the centre of gear A. Find the speed of gear B. What will be the speed f B, if the gear A instead of being fixed, makes 160 r.p.m. clockwise. [**Ans.** 144 r.p.m. anticlockwise ; 272 r.p.m. anticlockwise]

8. Fig. 10.14 shows a reverted epicyclic gear train in which arm F carries two gears A and D and a compound gear B-C. The gear A meshes with gear B and the gear D meshes with gear C. The number of teeth on gears A, D and C are 60, 36, 54 respectively. Find the speed and direction of gear D when gear A is fixed and arm F makes 150 r.p.m. clockwise. [**Ans.** 300 r.p.m. anticlockwise]

9. Fig. 10.15 shows an epicyclic train consisting of a fixed annular wheel A having 180 teeth. Meshing with wheel A is a gear B which drives gear D through an idle gear C. The gear D is concentric with wheel A. Gears B and C are carried on an arm which revolves clockwise at 80 r.p.m., about the axis of A or D. If the gears B and D are having 30 teeth and 48 teeth respectively, find the number of teeth of C and speed and sense of rotation of C.

[**Ans.** 36, 480 r.p.m., clockwise]

10. The Fig. 10.21 shows an epicyclic gear train with a sun and planet gear. The annular gear D has 80 internal teeth whereas the sun and planet gears have 40 and 20 external teeth respectively. The gear B meshes with annular gear D and also with sun gear A. The arm E rotates, about the centre of annular gear D. The gear B is carried on the arm E. If the gear D is fixed and arm rotates at 30 r.p.m., find the speeds of gears A and B.

[**Ans.** N_A = 90 r.p.m. ; N_B = – 90 r.p.m.]

11. Fig. 10.20 shows a compound epicyclic gear which consists of an annular gear D having internal teeth, a compound gear B-C, the sun gear A and the arm E. The compound gear B-C is also known as planet gear as this gear moves around the sun gear A. The sun gear A and annular gear D are co-axial. One shaft is connected to sun gear A and another shaft is connected to the arm. These two shafts are co-axial. The sun gear A meshes with compound gear C whereas annular gear D meshes with compound gear B. The gears A, B and C are having 50, 35 and 20 external teeth respectively. If all the gears are having the same module, find the number of teeth on the fixed annular gear D. Also find the speed of the shaft connected to the arm, if the speed of the shaft connected to sun gear A is 110 r.p.m. [**Ans.** (*i*) 105 ; (*ii*) 50 r.p.m. anticlockwise]

12. Fig. 10.23 shows an epicyclic speed reduction gear. The driving shaft is attached to an arm E. The arm carries a pin on which the compound gear B-C is free to revolve. The gear A is keyed to the driven shaft and gear D is a fixed gear. The number of teeth on the gears A, B, C and D are 21, 24, 27 and 18 respectively. Determine :

(*i*) Speed of driven shaft if driving shaft is rotating at 1000 r.p.m. anticlockwise, and

(*ii*) Resisting torque (or load torque) on driven shaft and holiday torque on fixed gear D if the input torque to driving shaft is 12 Nm. Assume that there is no friction losses and members are revolving at uniform speeds.

[**Ans.** (*i*) 238.1 r.p.m. anticlockwise (*ii*) – 50.4 Nm and 38.4 Nm]

13. Fig. 10.24 shows an epicyclic gear train. Pinion *A* has 18 teeth and is rigidly fixed to the motor shaft. The wheel *B* has 24 teeth and gears with *A* and also with the annular fixed wheel *D*. Pinion *C* has 18 teeth and is integral with *B* (*B*-*C* being a compound gear wheel). Gear *C* meshes with annular wheel *E*, which is keyed to the machine shaft. The arm rotates about the same shaft on which *A* is fixed and carries the compound wheel *B*- *C*. If the motor runs at 1200 r.p.m., find the speed of the machine shaft. Find the torque exerted on the machine shaft if the motor develops a torque of 120 Nm. **[Ans.** (*i*) 44.825 r.p.m. anticlockwise (*ii*) – 3212.5 Nm]

14. Fig. 10.25 shows an epicyclic gear train which consists of a sunwheel *S*, a stationary internal gear *E* and three identical planet wheels *P* carried on a star-shaped planet carrier *C*. The size of the different toothed wheels are such that the planet carrier *C* rotates at $\frac{1}{4}$ of the speed of the sunwheel *S*. The minimum number of teeth on any wheel is 18. The driving torque on the sun wheel is 80 Nm. Determine : (*i*) number of teeth on different wheels of the train and (*ii*) torque necessary to keep the internal gear stationary.

[Ans. (*i*) $T_S = 18, T_E = 54, T_P = 18$; (*ii*) 240 Nm]

15. Fig. 10.26 shows a compound epicyclic gear train in which the two gears S_1 and S_2 are integral with the input shaft *B*. The arm A_2 is integral with the output shaft *C*. The planet gear P_2 revolves on a pin attached to the arm A_2. The gear P_2 meshes with the sun gear S_2 and internal gear I_2. The internal gear I_2 is co-axial with the input shaft. The planet gear P_1 meshes with the fixed internal gear I_1 and sun gear S_1. The planet gear P_1 revolves on a pin fixed to internal gear I_2. The number of teeth on the gears S_1, S_2, I_1 and I_2 are 24, 29, 86 and 81 respectively. The input power on the shaft B is 8 kW at 1200 r.p.m. Find : (*i*) the speed and torque at output shaft (*ii*) the torque required to hold the internal gear I_1 stationary. The input shaft *B* rotates anticlockwise.

[Ans. (*i*) 509 r.p.m. anticlockwise ; (*ii*) 86.34 Nm]

11

Inertia Forces in Reciprocating Parts

11.1. Introduction

Let us first define 'Inertia'. Inertia is a property of matter by virtue of which a body resists any change in velocity. Inertia force is an imaginary force, which when acts upon a rigid body, brings the body in an equilibrium position. The *inertia force* is numerically equal to the accelerating force in magnitude but opposite in direction. Hence,

Inertia force, $F_i = -$ Accelerating force $= -m \times a$ ...(11.1)

where m = Mass of the body, and

a = Linear acceleration of the centre of the gravity of the body.

Similarly the *inertia torque* resists any change in the angular velocity of the body. Hence inertia torque is an imaginary torque, which when applied to a body, brings it in equilibrium position. It is equal to accelerating couple in magnitude but opposite in direction. Hence,

Inertia torque, $T_i = -I \times \alpha$ ...(11.2)

where I = mass moment of inertia of the body about an axis passing through the C.G. of the body and ⊥ to the plane of rotation of the body

$= m \times k^2$ where m = mass of the body and k = radius of gyration

α = Angular acceleration.

11.2. D-Alembert's Principle

D-Alembert's principle states that the resultant force acting on a body together with the inertia force are in equilibrium. Hence according to D-Alembert's principle :

$$F_r + F_i = 0 \qquad ...(11.3)$$

where F_r = Resultant external force acting on the body, and

F_i = Inertia force.

The equation (11.3) can also be derived by considering a rigid body, which is acted upon by a number of external forces. These external forces can be reduced to a single resultant force. But according to Newton's second law of motion, the resultant force is equal to the product of mass of the body and acceleration of the centre of the mass of the body.

Hence resultant force = Mass × acceleration

or $F_r = m \times a$

The above equation can also be written as

$$F_r - m \times a = 0$$

But from equation (11.1), we know that $(-m \times a)$ is equal to inertia force (F_i). Hence above equation becomes as

$$F_r + F_i = 0 \qquad (\because \; -m \times a = F_i)$$

D-Alembert's principle also states that the resultant torque acting on a body together with the inertia torque are in equilibrium. Hence according to this principle :

$$T_r + T_i = 0 \quad ...(11.4)$$

where T_r = Resultant external torque acting on the body ; and

T_i = Inertia torque.

The equation (11.4) can also be derived by considering a rigid body, which is acted upon by a number of external torques. These external torques can be reduced to a single resultant torque. But according to Newton's second law of motion, the resultant torque is equal to the product of mass moment of inertia and angular acceleration.

∴ Resultant torque = Mass moment of inertia × Angular Acceleration.

or $T_r = I \times \alpha$

The above equation can also be written as

$$T_r - I \times \alpha = 0$$

But from equation (11.2), we know that $(-I \times \alpha)$ is equal to inertia torque (T_i). Hence above equation becomes as

$$T_r + T_i = 0.$$

Note. (*i*) The above principle is used to reduce a dynamic problem into an equivalent static problem.

11.3. Effect of a Number of Forces Acting on a Rigid Body

Fig. 11.1 shows a rigid body on which a number of forces are acting (These forces are not shown in the Fig. 11.1). These forces may be reduced to a single resultant force. Let F is the resultant force, whose line of action may or may not pass through the centre of gravity. Let us consider it is at a distance of h from the centre of gravity G, of the body. Now apply two equal and opposite forces (of magnitude F) through point G. These two forces are parallel to the resultant force. As the two forces are equal and opposite, the resultant force on the body will be same as before (*i.e.* the resultant force will be F).

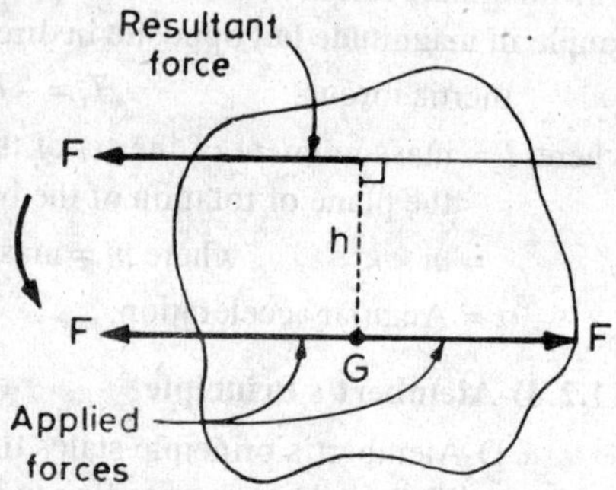

Fig. 11.1

The rigid body is now considered to be acted upon :

(*i*) by a couple of magnitude equal to $F \times h$ having anticlock rotation and

(*ii*) a force of magnitude F passing through G.

The couple $(F \times h)$ causes angular acceleration of the rigid body about an axis passing through G and perpendicular to the plane of couple. The force F causes the linear acceleration of the c.g. of the body. Hence corresponding to couple and force, we will have two equations as given below :

(*i*) Corresponding to couple, we have

Couple = $I \times \alpha$

or $F \times h = mk^2 \times \alpha$

where I = Mass moment of inertia of the body about an axis passing through c.g. of the body and perpendicular to the plane of couple

$= mk^2$

α = Angular acceleration of the rigid body due to couple.

where k = radius of gyration about an axis through G.

(*ii*) Corresponding to force, we have

Force = mass × acceleration

or $F = m \times a$

where a = linear acceleration of rigid body through G.

From the above two equation (*i.e.* $F \times h = mk^2 \times \alpha$ and $F = m \times a$), we can calculate the magnitude of linear acceleration 'a' and angular acceleration 'α' for given values of F, h, m and k.

In some cases, the magnitude and direction of the linear acceleration (a) and angular acceleration (α) are given or obtained as discussed in velocity and acceleration chapter. Then it is required to find the magnitude, direction and line of action of the resultant force F.

The magnitude of resultant force is known from equation

$$F = m \times a \qquad (\because \; m \text{ is known and '}a\text{' is also known})$$

The line of action of F is calculated from the expression

$$h = \frac{I \times \alpha}{F} = \frac{mk^2 \times \alpha}{F} \qquad (m, k, \alpha \text{ and } F \text{ are known})$$

The value of h is the distance of the resultant force from the centre of gravity of the body.

The direction of resultant force is obtained from the direction of linear acceleration and angular acceleration.

Problem 11.1. *Fig. 11.2 shows a connecting rod of mass 12 kg and of length 400 mm between the two centres. The C.G. (centre of gravity) of the connecting rod is at a distance of 150 mm from the centre of big end. The radius of gyration of the connecting rod is 120 mm about an axis passing through C.G. Find the magnitude, direction and line of action of the resultant force acting on the connecting rod if the linear acceleration of the C.G. of the connecting rod is 60 m/s² in the direction shown in the Fig. 11.2 and angular acceleration of the rod is 100 rad/s² clockwise.*

Sol. Given : Mass, m = 12 kg ; length, l = 400 mm ;

Distance of C.G. from big end = 150 mm = 0.15 m

Radius of gyration, k = 120 mm = 0.12 m

Linear acceleration, $a = 60 \text{ m/s}^2$

Angular acceleration, $\alpha = 100 \text{ rad/s}^2$

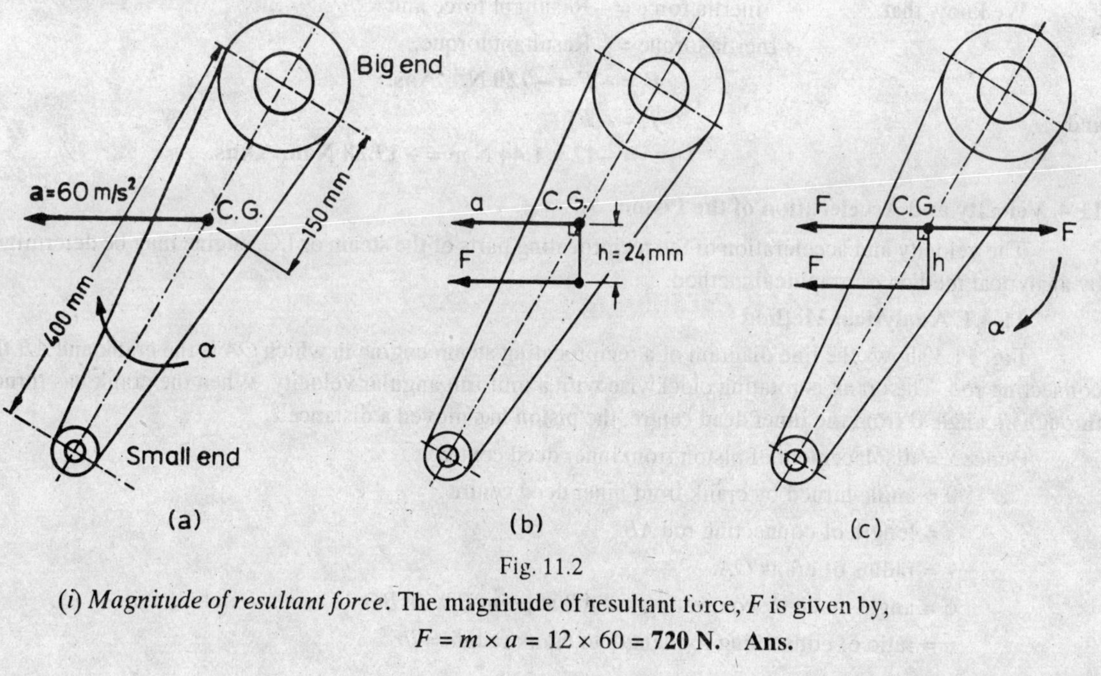

Fig. 11.2

(*i*) *Magnitude of resultant force*. The magnitude of resultant force, F is given by,

$$F = m \times a = 12 \times 60 = \textbf{720 N.} \quad \textbf{Ans.}$$

(*ii*) *Direction and line of action of the resultant force.* The resultant force will act in the direction of linear acceleration. Direction of linear acceleration is shown in Fig. 11.2. Hence direction of resultant force is known.

To find the line of action of the resultant force (or the distance of the resultant force from the C.G. *i.e.* the distance 'h'), we must first calculate the couple required to produce angular acceleration. This couple is given by

$$T = I \times \alpha$$

$$= (mk^2) \times \alpha \qquad (\because \ I = mk^2)$$

$$= (12 \times 0.12^2) \times 100 \text{ kg m}^2/\text{s}^2 = 12 \times 1.44 \text{ kg m}^2/\text{s}^2$$

$$= 12 \times 1.44 \frac{\text{kg m}}{\text{s}^2} \times \text{m}$$

$$= 12 \times 1.44 \text{ N m} \qquad (\because \text{ kg m/s}^2 = N = \text{Mass} \times \text{Acceleration})$$

But couple due to resultant force is also

$$= F \times h = 720 \times h \text{ N m}$$

$$\therefore \quad F \times h = I \times \alpha$$

or $$720 \times h = 12 \times 1.44$$

or $$h = \frac{12 \times 1.44}{720} = 0.024 \text{ m} = \mathbf{24 \text{ mm}}. \quad \textbf{Ans.}$$

Hence the resultant force must act in the downward direction at a distance of 24 mm from line of acceleration as shown in Fig. 11.2 (*c*). Then the connecting rod will have angular acceleration clockwise.

Problem 11.2. *For the above problem, find the values of inertia force and inertia torque which must be applied on the given connecting rod to bring it in equilibrium.*

Sol. Let F_i = Inertia force, and T_i = Inertia torque.

We know that inertia force = – Resultant force and

Inertia torque = – Resultant torque.

$$\therefore \quad F_i = -F = \mathbf{-720 \text{ N}}. \quad \textbf{Ans.}$$

and $$T_i = -T$$

$$= -12 \times 1.44 \text{ N m} = \mathbf{-17.28 \text{ N m}}. \quad \textbf{Ans.}$$

11.4. Velocity and Acceleration of the Piston

The velocity and acceleration of the reciprocating parts of the steam or I.C. engine may be determined by analytical method or graphical method.

11.4.1. Analytical Method

Fig. 11.3 shows the line diagram of a reciprocating steam engine in which OA is the crank and AB the connecting rod. The crank is rotating clockwise with a uniform angular velocity. When the crank has turned through an angle θ from the inner dead centre, the piston has moved a distance x.

Hence x = displacement of piston from inner deed centre

θ = angle turned by crank from inner deed centre

l = length of connecting rod AB

r = radius of crank OA.

ϕ = angle made by connecting rod AB at point B with BO

n = ratio of connecting rod length to crank radius = l/r

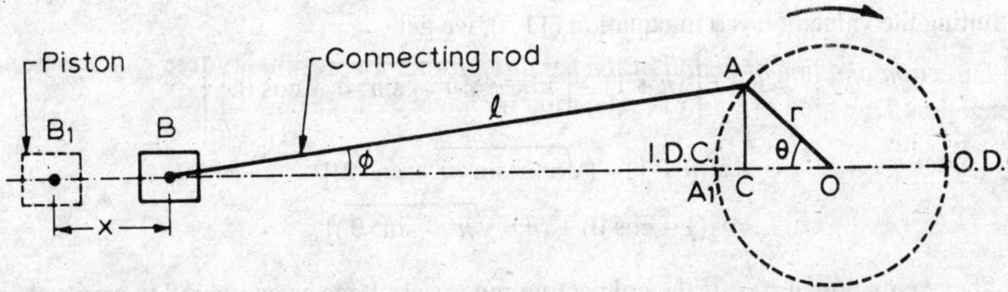

Fig. 11.3

When crank angle $\theta = 0$, the point A coincides with A_1 and B coincides with B_1. Then A_1B_1 is the length of connecting rod and A_1O is the length of the crank.

When crank has turned through an angle θ, the distance moved by piston is 'x'. And AB is the position of connecting rod and AO is the position of crank. The value of 'x' is given by

$$x = B_1B = B_1O - BO$$

$$= (B_1A_1 + A_1O) - (BC + CO) \quad (\because \; B_1O = B_1A_1 + A_1O \text{ and } BO = BC + CO)$$

$$= (l + r) - (BC + CO) \quad (\because \; B_1A_1 = l \,;\, A_1O = r)$$

$$= (l + r) - (BA \cos\phi + AO \cos\theta) \quad (\because \; BC = AB \cos\phi \,;\, CO = AO \cos\theta)$$

$$= (l + r) - (l \cos\phi + r \cos\theta) \quad (\because \; BA = l \,;\, AO = r)$$

$$= \frac{r}{r}[(l + r) - (l \cos\phi + r \cos\theta)] \quad \text{[Multiplying and dividing by } r\text{]}$$

$$= r\left[\left(\frac{l}{r} + 1\right) - \left(\frac{l}{r}\cos\phi + \cos\theta\right)\right]$$

$$= r[(n + 1) - (n \cos\phi + \cos\theta)] \quad \left(\text{Taking } \frac{l}{r} = n\right) \quad \text{...(11.5)}$$

In the above equation, the angle ϕ can also be converted into angle θ as :

$$\cos\phi = \sqrt{1 - \sin^2\phi}$$

In triangle ABC, $\quad AC = BA \sin\phi = l \sin\phi$

Also in triangle ACO, $AC = r \sin\theta$

$\therefore \quad l \sin\phi = r \sin\theta$

or $$\sin\phi = \frac{r}{l}\sin\theta \quad \text{...(11.5A)}$$

Substituting the value of $\sin\phi$ in above equation, we get

$$\cos\phi = \sqrt{1 - \left(\frac{r}{l}\sin\theta\right)^2} = \sqrt{1 - \frac{r^2}{l^2}\sin^2\theta}$$

$$= \sqrt{1 - \frac{\sin^2\theta}{n^2}} \quad \left(\because \; \frac{r^2}{l^2} = \frac{1}{n^2}\right)$$

$$= \sqrt{\frac{n^2 - \sin^2\theta}{n^2}}$$

$$= \frac{1}{n}\sqrt{n^2 - \sin^2\theta} \quad \text{...(11.5B)}$$

Substituting the value of $\cos\phi$ in equation (11.5), we get

$$x = r\left[(n+1) - \left(n \times \frac{1}{n}\sqrt{n^2 - \sin^2\theta} + \cos\theta\right)\right]$$

$$= r[(n+1) - (\sqrt{n^2 - \sin^2\theta} + \cos\theta)]$$

$$= r[(1 - \cos\theta) + (n - \sqrt{n^2 - \sin^2\theta})] \qquad ...(11.6)$$

In the above equation, $n = \frac{l}{r}$. If the connecting rod is very large as compared to crank, then n^2 will be very large. The term $\sqrt{n^2 - \sin^2\theta}$ will be approximately equal to $\sqrt{n^2}$ or n as the maximum value of $\sin^2\theta$ will be unity (*i.e.* $\sqrt{n^2 - 1} \simeq \sqrt{n^2} \simeq n$ if n is very large). Then

$$x = r[(1 - \cos\theta) + (n - n)] = r[1 - \cos\theta] \qquad ...(11.7)$$

The above equation is the expression of a simple harmonic motion. Hence piston is having simple harmonic motion, if connecting rod is very large.

Alternate approximate expression for piston displacement

The equation (11.6) gives the exact displacement of the piston whereas equation (11.7) gives the displacement when n is very very large. But if n (*i.e.* ratio of length of connecting rod to the length of crank) is not very very large, then the displacement of piston is obtained approximately as given below :

From equation (11.5), we have

$$x = r[(n+1) - (n\cos\phi + \cos\theta)]$$

$$= r[n + 1 - n\cos\phi - \cos\theta] \qquad ...(i)$$

Now

$$\cos\phi = [1 - \sin^2\phi]^{1/2}$$

$$= \left[1 - \frac{\sin^2\theta}{n^2}\right]^{1/2} \qquad \left[\because \text{ From equation (11.5}A\text{), } \sin\phi = \frac{r}{l}\sin\theta = \frac{\sin\theta}{n}\right]$$

$$= 1 - \frac{1}{2} \times \frac{\sin^2\theta}{n^2} + \ldots\ldots$$

[Expanding by binomial theorem and neglecting higher terms]

Substituting the value of $\cos\phi$ in equation (*i*), we get

$$x = r\left[(n+1) - n\left(1 - \frac{\sin^2\theta}{2n^2}\right) - \cos\theta\right]$$

$$= r\left[n + 1 - n + \frac{\sin^2\theta}{2n} - \cos\theta\right]$$

$$= r\left[(1 - \cos\theta) + \frac{\sin^2\theta}{2n}\right] \qquad ...(11.7\,A)$$

Velocity of Piston

The velocity of piston (v) is obtained from the displacement as given below [*i.e.* by differentiating equation (11.6) w.r.t. time] :

$$v = \frac{dx}{dt} = \frac{dx}{d\theta} \cdot \frac{d\theta}{dt}$$

$$= \frac{dx}{d\theta} \cdot \omega \qquad \left(\because \frac{d\theta}{dt} = \text{rate of change of angular displacement} = \omega\right)$$

$$= \frac{d}{d\theta}[r\{(1-\cos\theta) + (n - \sqrt{n^2 - \sin^2\theta})\}] \,.\, \omega$$

[Substituting the value of x from equation (11.6)]

$$= r\left[(0 + \sin\theta) + \left\{0 - \frac{1}{2}(n^2 - \sin^2\theta)^{-1/2} \,.\, (0 - 2\sin\theta\cos\theta)\right\}\right] . \, \omega$$

$$= r\left[\sin\theta + \frac{\sin 2\theta}{2\sqrt{n^2 - \sin^2\theta}}\right] . \, \omega \qquad ...(11.8)$$

If n^2 is large compared to $\sin^2\theta$, then $\sqrt{n^2 - \sin^2\theta}$ approaches to $\sqrt{n^2}$ or n. Hence equation (11.8) becomes as

$$v = r\left[\sin\theta + \frac{\sin 2\theta}{2n}\right] . \, \omega$$

$$= r\omega\left[\sin\theta + \frac{\sin 2\theta}{2n}\right] \qquad ...(11.9)$$

When n is very very large, then $\frac{\sin 2\theta}{2n}$ becomes negligible. Hence neglecting $\frac{\sin 2\theta}{n}$, we get

$$v = r\omega \sin\theta \qquad ...(11.10)$$

Acceleration of Piston

The acceleration (a) of the piston is obtained by differentiating the velocity w.r.t. time [*i.e.* equation (11.9) with respect to time]

$\therefore$ Acceleration, $a = \frac{dv}{dt}$

$$= \frac{dv}{d\theta} \cdot \frac{d\theta}{dt} = \frac{d}{d\theta}\left[r\omega\left(\sin\theta + \frac{\sin 2\theta}{2n}\right)\right] \cdot \frac{d\theta}{dt}$$

$$= r\omega\left(\cos\theta + \frac{2\cos 2\theta}{2n}\right) . \, \omega \qquad \left(\because \frac{d\theta}{dt} = \omega\right)$$

$$= r\omega^2\left(\cos\theta + \frac{\cos 2\theta}{n}\right) \qquad ...(11.11)$$

When $\theta = 0°$ *i.e.* when crank is at I.D.C., $a = r\omega^2\left(1 + \frac{1}{n}\right)$

When $\theta = 180°$ *i.e.* when crank is at O.D.C., $a = rw^2\left(-1 + \frac{1}{n}\right)$

From O.D.C., as crank is moving towards I.D.C., the motion is reversed and sign will be changed in the above expression.

$$\therefore \qquad a = -(r\omega^2)\left(-1 + \frac{1}{n}\right)$$

$$= r\omega^2\left[\frac{1}{n} - 1\right]$$

The acceleration given by equation (11.11) is not exact but approximate as it has been obtained by differentiating the approximate expression for velocity. To differentiate the exact expression for velocity, is usually very cumbersome.

11.4.2. Angular Velocity and Angular Acceleration of Connecting Rod

As the angular velocity of crank is $\frac{d\theta}{dt}$, similarly the angular velocity of the connecting rod is $\frac{d\phi}{dt}$.

Let ω = Angular velocity of crank = $\frac{d\theta}{dt}$

ω_c = Angular velocity of crank = $\frac{d\phi}{dt}$

From equation (11.5 *A*), we have

$$\sin\phi = \frac{r}{l}\sin\theta$$

Differentiating the above equation w.r.t., time, we get

$$\cos\phi \, . \, \frac{d\phi}{dt} = \frac{r}{l}\cos\theta \, . \, \frac{d\theta}{dt}$$

or $$\cos\phi \, . \, \omega_c = \frac{r}{l}\cos\theta \, . \, \omega \qquad \left(\because \frac{d\phi}{dt} = \omega_c \text{ and } \frac{d\theta}{dt} = \omega\right)$$

or $$\omega_c = \frac{r}{l} \, . \, \omega \, . \, \frac{\cos\theta}{\cos\phi}$$

But from equation (11.5 *B*), $\cos\phi$ is given by

$$\cos\phi = \frac{1}{n}\sqrt{n^2 - \sin^2\theta}$$

Substituting this value of $\cos\phi$ in the above equation, we get

$$\omega_c = \frac{r \, . \, \omega}{l} \, . \, \frac{\cos\theta}{\frac{1}{n}\sqrt{n^2 - \sin^2\theta}}$$

$$= \frac{\omega}{n} \, . \, \frac{\cos\theta}{\frac{1}{n}\sqrt{n^2 - \sin^2\theta}} \qquad \left(\because \frac{r}{l} = \frac{1}{n}\right)$$

$$= \frac{\omega\cos\theta}{\sqrt{n^2 - \sin^2\theta}} \qquad \text{...(11.12)}$$

Equation (11.12) gives the angular acceleration of the connecting rod.

Let α_c = Angular acceleration of the connecting rod

$$= \frac{d\omega_c}{dt} = \frac{d\omega_c}{d\theta} \cdot \frac{d\theta}{dt}$$

$$= \frac{d}{d\theta}\left(\frac{\omega\cos\theta}{\sqrt{n^2 - \sin^2\theta}}\right) \cdot \frac{d\theta}{dt} \qquad \left(\because \omega_c = \frac{\omega\cos\theta}{\sqrt{n^2 - \sin^2\theta}}\right)$$

$$= \omega \frac{d}{d\theta}[\cos\theta \, . \, (n^2 - \sin^2\theta)^{-1/2}] \, . \, \omega \qquad \left(\because \frac{d\theta}{dt} = \omega\right)$$

$$= \omega^2\left[\cos\theta \, . \left(-\frac{1}{2}\right)(n^2 - \sin^2\theta)^{-3/2}(-2\sin\theta\cos\theta) + (n^2 - \sin^2\theta)^{-1/2}(-\sin\theta)\right]$$

$$= \omega^2[\cos^2\theta \, . \, \sin\theta\,(n^2 - \sin^2\theta)^{-3/2} - \sin\theta\,(n^2 - \sin^2\theta)^{-1/2}]$$

$$= \omega^2\sin\theta\,[\cos^2\theta\,(n^2 - \sin^2\theta)^{-3/2} - (n^2 - \sin^2\theta)^{-1/2}]$$

$$= \omega^2 \sin\theta \left[\frac{\cos^2\theta}{(n^2 - \sin^2\theta)^{+3/2}} - \frac{1}{(n^2 - \sin^2\theta)^{1/2}} \right]$$

$$= \omega^2 \sin\theta \left[\frac{\cos^2\theta - (n^2 - \sin^2\theta)}{(n^2 - \sin^2\theta)^{3/2}} \right] = \omega^2 \sin\theta \left[\frac{\cos^2\theta - n^2 + \sin^2\theta}{(n^2 - \sin^2\theta)^{3/2}} \right]$$

$$= \omega^2 \sin\theta \left[\frac{1 - n^2}{(n^2 - \sin^2\theta)^{3/2}} \right] \qquad (\because \cos^2\theta + \sin^2\theta = 1)$$

$$= -\omega^2 \sin\theta \left[\frac{n^2 - 1}{(n^2 - \sin^2\theta)^{3/2}} \right] \qquad ...(11.13)$$

The negative sign shows that the sense of the acceleration of the connecting rod is such that it tends to reduce the angle ϕ. Actually $\alpha_c = \frac{d}{dt}(\omega_c) = \frac{d}{dt}\left(\frac{d\phi}{dt}\right) = \frac{d^2\phi}{dt^2}$. Hence if α_c is positive then angle ϕ is increasing and if α_c is negative then ϕ is reducing. From Fig. 11.3, it is clear that if ϕ is increasing, the connecting rod is rotating anticlockwise and if ϕ is decreasing, the connecting rod is rotating clockwise.

Note. (*i*) In comparison to n^2, the value of $\sin^2\theta$ is very small (max. value of $\sin^2\theta$ is 1 only). Hence $\sin^2\theta$ may be neglected. Then equation (11.12) and (11.13) will become as

$$\omega_c = \frac{\omega \cos\theta}{n} \qquad ...(11.14)$$

and

$$\alpha_c = \frac{-\omega^2 \sin\theta\,(n^2 - 1)}{(n^2)^{3/2}} = \frac{-\omega^2 \sin\theta\,(n^2 - 1)}{n^3} \qquad ...(11.15)$$

Also in comparison to n^2, the value of unity is small and can be neglected. Hence the above expression of α_c becomes as

$$\alpha_c = \frac{-\omega^2 \sin\theta \times n^2}{n^3} = \frac{-\omega^2 \sin\theta}{n} \qquad ...(11.16)$$

Problem 11.3. *The lengths of crank and connecting rod of a vertical reciprocating engine are 300 mm and 1.5 m respectively. The crank is rotating at 200 r.p.m. clockwise as shown in Fig. 11.4. Find analytically, (i) acceleration of piston, (ii) velocity of piston and (iii) angular acceleration of the connecting rod when the crank has turned through 40° from the top dead centre and piston is moving downwards.*

Sol. Given : $r = 300$ mm $= 0.3$ m ; $l = 1.5$ m, N = 200 r.p.m.

$$\therefore \quad \omega = \frac{2\pi N}{60} = \frac{2\pi \times 200}{60} = 20.944 \text{ rad/s}, \ \theta = 40°$$

$$\text{Now, } n = \frac{l}{r} = \frac{1.5}{0.3} = 5$$

(i) Acceleration of piston (a)

Acceleration of piston is given by equation (11.11) as

$$a = r\omega^2 \left(\cos\theta + \frac{\cos 2\theta}{n} \right)$$

$$= 0.3 \times \left(\frac{2\pi \times 200}{60} \right)^2 \left(\cos 40° + \frac{\cos(2 \times 40)}{5} \right)$$

$$= 0.3 \times 20.944^2 \times \left(0.766 + \frac{0.1736}{5} \right)$$

$$= 0.3 \times 438.65 \times 0.8 = \mathbf{105.27 \ m/s^2}. \quad \textbf{Ans.}$$

(ii) Velocity of piston (v)

Using equation (11.8), we get the velocity of piston as,

$$v = r\omega\left(\sin\theta + \frac{\sin 2\theta}{2\sqrt{n^2 - \sin^2\theta}}\right)$$

$$= 0.3 \times \left(\frac{2\pi \times 200}{60}\right)\left(\sin 40° + \frac{\sin(2 \times 40)}{2 \times \sqrt{5^2 - \sin^2 40}}\right)$$

$$= 0.3 \times 20.944 \times \left(0.6428 + \frac{0.9848}{2\sqrt{25 - 0.413}}\right)$$

$$= 0.3 \times 20.944 \times 0.742 = \mathbf{4.66\ m/s.} \quad \textbf{Ans.}$$

(iii) Angular acceleration of the connecting rod (α_c)

Angular acceleration of the connecting rod is given by equatio (11.13) as

$$\alpha_c = -\omega^2 \sin\theta\left[\frac{n^2 - 1}{(n^2 - \sin^2\theta)^{3/2}}\right]$$

$$= -20.944^2 \times \sin 40°\left[\frac{5^2 - 1}{(5^2 - \sin^2 40)^{3/2}}\right]$$

$$= -438.65 \times 0.6428\left[\frac{24}{(25 - 0.4132)^{3/2}}\right]$$

$$= -438.65 \times .6428 \times \frac{24}{121.91} = \mathbf{-55.5\ rad/s^2.} \quad \textbf{Ans.}$$

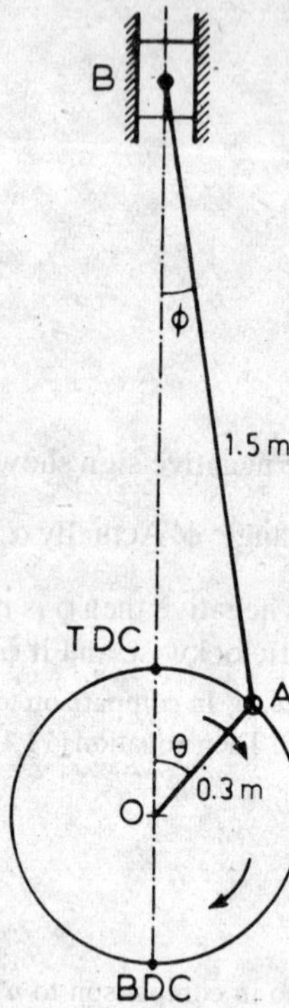

Fig. 11.4

(– ve sign shows that the sense of the acceleration of the connecting rod is such that it tends to reduce the angle ϕ. Hence connecting rod is rotating clockwise)

Problem 11.4. *For the problem 11.3, find (i) the crank angle at which the velocity of piston is maximum and (ii) value of maximum velocity of piston.*

Sol. Given : Data from Problem 11.3,

$$r = 0.3\text{ m} ;\ l = 1.5\text{ m} ;\ \omega = 20.94\text{ rad/s} ;\ n = 5$$

(i) Crank angle (θ) at which velocity of piston is maximum

The velocity of piston is given by equation (11.8) as,

$$v = r\omega\left(\sin\theta + \frac{\sin 2\theta}{2\sqrt{n^2 - \sin^2\theta}}\right)$$

The value of $\sin^2\theta$ is small* in comparison to n^2 and hence it may be neglected. Then the above expression becomes as

$$v = r\omega\left(\sin\theta + \frac{\sin 2\theta}{2n}\right) \qquad \text{[See equation (11.9)]} \qquad ...(i)$$

For maximum velocity of the piston, $\dfrac{dv}{d\theta}$ should be zero.

But $$\frac{dv}{d\theta} = r\omega\left(\cos\theta + \frac{2\cos 2\theta}{2n}\right)$$

---

*If $\theta = 90°$ then $\sin\theta = 1$ and $\sin^2\theta = 1$. If $n = 5$, then $n^2 = 25$. Hence max. value of $\sin^2\theta = 1$ whereas $n^2 = 25$.

$\therefore$ For maximum velocity, $\dfrac{dv}{d\theta} = 0$

or $$r\omega\left(\cos\theta + \frac{2\cos 2\theta}{2n}\right) = 0$$

or $$\cos\theta + \frac{\cos 2\theta}{n} = 0$$

or $$n\cos\theta + \cos 2\theta = 0$$

or $$n\cos\theta + (2\cos^2\theta - 1) = 0 \qquad (\because \cos 2\theta = 2\cos^2\theta - 1)$$

or $$2\cos^2\theta + n\cos\theta - 1 = 0$$

The above equation is a quadratic equation in $\cos\theta$.

$$\therefore \quad \cos\theta = \frac{-n \pm \sqrt{n^2 + 4 \times 2 \times 1}}{2 \times 2}$$

$$= \frac{-5 \pm \sqrt{5^2 + 8}}{4} \qquad (\because n = 5)$$

$$= \frac{-5 \pm 5.744}{4} = \frac{0.744}{4} \qquad \text{(Taking + ve value)}$$

$$= 0.186$$

$$\therefore \quad \theta = \cos^{-1} 0.186 = \mathbf{79.28^\circ}. \quad \textbf{Ans.}$$

(ii) Maximum velocity of piston

Substituting the value of $\theta = 79.28^\circ$ in equation (*i*) above, we get the value of max. velocity.

$$\therefore \quad v_{max} = r\omega\left(\sin 79.28^\circ + \frac{\sin(2 \times 79.28^\circ)}{2 \times n}\right)$$

$$= 0.3 \times 20.94\left(0.9825 + \frac{0.3655}{2 \times 5}\right) \qquad (\because n = 5, r = 0.3, \omega = 20.94)$$

$$= \mathbf{6.4\ m/s.} \quad \textbf{Ans.}$$

Problem 11.5. *The lengths of crank and connecting rod of a horizontal reciprocating engine are 100 mm and 500 mm respectively. The crank is rotating at 400 r.p.m. When the crank has turned 30° from the inner deed centre, find analytically (i) acceleration of the piston, (ii) velocity of the piston, (iii) angular velocity of the connecting rod and (iv) angular acceleration of the connecting rod.*

Sol. Given : r = 100 mm = 0.1 m ; l = 500 mm = 0.5 m ; N = 400 r.p.m.

or $$\omega = \frac{2\pi N}{60} = \frac{2 \times \pi \times 400}{60} = 41.88 \text{ rad/s}, \theta = 30^\circ. \text{ The value of } n = \frac{l}{r} = \frac{0.5}{0.1} = 5.$$

(i) Acceleration of piston (a)

Using equation (11.11), we get acceleration of piston (a) as

$$a = r\omega^2\left(\cos\theta + \frac{\cos 2\theta}{n}\right)$$

$$= 0.1 \times 41.88^2\left(\cos 30^\circ + \frac{\cos 60^\circ}{5}\right) = 0.1 \times 1753.93\left(0.866 + \frac{0.5}{5}\right)$$

$$= 0.1 \times 1753.93 \times 0.966 = \mathbf{169.43\ m/s^2}. \quad \textbf{Ans.}$$

(ii) Velocity of piston (v)

Using equation (11.8), we get velocity of piston (v) as

$$v = r\omega\left(\sin\theta + \frac{\sin 2\theta}{2\sqrt{n^2 - \sin^2\theta}}\right)$$

$$= 0.1 \times 41.88 \times \left(\sin 30° + \frac{\sin 60°}{2\sqrt{5^2 - \sin^2 30°}}\right)$$

$$= 0.1 \times 41.88 \times \left(0.5 + \frac{0.866}{2\sqrt{25 - 0.25}}\right)$$

$$= 0.1 \times 41.88\,(0.5 + 0.087) = \mathbf{2.458\ m/s.} \quad \textbf{Ans.}$$

(iii) Angular velocity of connecting rod (ω_c)

Using equation (11.12), we get angular velocity of connecting rod (ω_c) as

$$\omega_c = \omega\frac{\cos\theta}{\sqrt{n^2 - \sin^2\theta}}$$

$$= 41.88\frac{\cos 30°}{\sqrt{5^2 - \sin^2 30°}} = 41.88 \times \frac{0.866}{\sqrt{25 - 0.25}} = \mathbf{7.29\ rad/s.} \quad \textbf{Ans.}$$

(iv) Angular acceleration of the connecting rod (α_c)

Using equation (11.13), we get the angular acceleration (α_c) of connecting rod as,

$$\alpha_c = -\omega^2\sin\theta\left[\frac{n^2 - 1}{(n^2 - \sin^2\theta)^{3/2}}\right]$$

$$= -41.88^2 \times \sin 30°\left[\frac{5^2 - 1}{(5^2 - \sin^2 30°)^{3/2}}\right] = -1753.93 \times 0.5\left[\frac{25 - 1}{(25 - 0.25)^{3/2}}\right]$$

$$= -1753.95 \times 0.5 \times \left[\frac{24}{123.13}\right] = \mathbf{-170.93\ rad/s^2.} \quad \textbf{Ans.}$$

Problem 11.6. *For the problem 11.5, find the position of the crank from the inner deed centre for zero acceleration of the piston.*

Sol. Given : Data from Problem 11.5.

$$r = 0.1\text{ m},\ l = 0.5\text{ m},\ \omega = 41.88\text{ rad/s},\ n = 5$$

Acceleration of the piston is given by equation (11.11) as

$$a = r\omega^2\left(\cos\theta + \frac{\cos 2\theta}{n}\right)$$

For zero acceleration, the value of a should be zero in the above equation. Hence we get

$$0 = r\omega^2\left(\cos\theta + \frac{\cos 2\theta}{n}\right)$$

or $$\cos\theta + \frac{\cos 2\theta}{n} = 0$$

or $$n\cos\theta + \cos 2\theta = 0$$

or $$n\cos\theta + (2\cos^2\theta - 1) = 0 \qquad (\because\ \cos 2\theta = 2\cos^2\theta - 1)$$

or $$n\cos\theta + 2\cos^2\theta - 1 = 0$$

or $2\cos^2\theta + n\cos\theta - 1 = 0$

or $2\cos^2\theta + 5\cos\theta - 1 = 0$ $(\because\ n = 5)$

The solution of the above equation is

$$\cos\theta = \frac{-5 \pm \sqrt{5^2 + 4 \times 2 \times 1}}{2 \times 2} = \frac{-5 \pm \sqrt{25 + 8}}{4} = \frac{-5 \pm 5.744}{4}$$

$$= \frac{0.744}{4} = 0.186 \quad \text{(Taking + ve value)}$$

$\therefore$ $\theta = \cos^{-1} 0.186 =$ **79.28°. Ans.**

Problem 11.7. *For the problem 11.5, if crank angle is not given but displacement of piston is given as 20% of the outstroke. Then find the angle turned by crank from inner deed centre.*

Sol. Given : Data from Problem 11.5 (assuming θ is not given)

$$r = 0.1 \text{ m}, l = 0.5 \text{ m}, \omega = 41.88 \text{ rad/s}, n = 5$$

Displacement of piston = 20% of outstroke

or $x = \frac{20}{100} \times \text{stroke}$

$= 0.2 \times (2r)$ $(\because\ \text{stroke} = 2r)$

$= 0.4r$ ...(*i*)

Displacement of piston is also given by equation (11.6) as

$$x = r[(1 - \cos\theta) + (n - \sqrt{n^2 - \sin^2\theta})] \quad ...(ii)$$

Equating the two values of 'x', we get

$$0.4r = r[(1 - \cos\theta) + (n - \sqrt{n^2 - \sin^2\theta})]$$

or $0.4 = 1 - \cos\theta + n - \sqrt{n^2 - \sin^2\theta}$

or $0.4 = 1 - \cos\theta + 5 - \sqrt{5^2 - \sin^2\theta}$ $(\because\ n = 5)$

or $\sqrt{5^2 - \sin^2\theta} = 1 - \cos\theta + 5 - 0.4 = 5.6 - \cos\theta$

Squaring we get $5^2 - \sin^2\theta = (5.6 - \cos\theta)^2$

or $25 - \sin^2\theta = 5.6^2 + \cos^2\theta - 2 \times 5.6 \times \cos\theta$

or $25 - 5.6^2 + 2 \times 5.6 \times \cos\theta = \cos^2\theta + \sin^2\theta$

or $25 - 31.36 + 11.2\cos\theta = 1$ $(\because\ \cos^2\theta + \sin^2\theta = 1)$

or $11.2\cos\theta = 1 - 25 + 31.36 = 7.36$

or $\cos\theta = \frac{7.36}{11.2} = 0.657$

or $\theta = \cos^{-1} 0.657 =$ **48.93°. Ans.**

11.4.3. Graphical Method for velocity and acceleration of the reciprocating parts of steam engine or I.C. engine (i.e. internal combustion engine)

The followings are the important graphical methods (or construction) of determining the velocity and acceleration of the reciprocating parts of a steam engine or internal combustion engine :

(*i*) Klein construction ;

(*ii*) Ritterhau's construction ;

(*iii*) Bennett's construction.

(*i*) **Klein Construction.** This construction was given by professor Klein. Fig. 11.5 (*a*) shows the line diagram of a reciprocating steam engine in which OA is the crank and AB is the connecting rod. The crank is rotating clockwise with a uniform angular velocity. Let the crank makes an angle θ with the line of stroke BO. The following is the method of drawing velocity and acceleration by Klein's construction.

(*a*) *Velocity diagram by Klein's Construction.* The procedure is as given below :

From point O, draw OC perpendicular to OB. Produce the line BA to intersect the line OC at C. Then OAC is known as Klein's velocity diagram.

In Chapter 3 (*i.e.* velocity of Mechanism chapter) the velocity diagram for the reciprocating steam engine has been drawn. The velocity diagram is a triangle oab as shown in Fig. 11.5 (*b*). In triangle oab, oa represents v_{AO} (*i.e.* velocity of A with respect to O) and is perpendicular to AO, ob represents the velocity v_{BO} (*i.e.* the velocity of B with respect to O) and is parallel to BO and ab represents velocity v_{BA} (*i.e.* the velocity of B with respect to A) and is perpendicular to BA.

Let us now rotate the velocity triangle oab by 90°, we will get triangle $oa'b'$.

In this triangle oa_1 represents the velocity of A with respect to O (or velocity v_{AO}) ; ob_1 represents the velocity of B with respect to O (or velocity v_{BO}) and a_1b_1 represents the velocity of B with respect to A (or velocity v_{BA}).

In velocity diagram shown in Fig. 11.5 (*b*), oa (*i.e.* v_{AO}) is perpendicular to AO. But oa_1 is perpendicular to oa (as velocity triangle oab is rotated through 90°). Hence oa_1 is parallel to AO. Similarly we can say that ob_1 is perpendicular to BO and a_1b_1 is parallel to BA.

In Klein's velocity diagram OAC, AC is parallel to BA, CO is perpendicular to BO and AO may be considered parallel to crank AO. Hence the triangles AOC and oa_1b_1 are similar.

$$\therefore \quad \frac{oa_1}{OA} = \frac{ob_1}{OC} = \frac{a_1b_1}{AC} \qquad ...(i)$$

But
$$\frac{oa_1}{OA} = \frac{v_{AO}}{OA}\left(=\frac{v_A}{OA}\right) = \omega \qquad \left(\because \frac{V}{r} = \omega\right)$$

($\because$ Crank OA is rotating with a constant angular velocity, ω)

Hence equation (*i*) can be written as

$$\frac{oa_1}{OA} = \frac{ob_1}{OC} = \frac{a_1b_1}{AC} = \omega$$

or
$$\frac{v_{AO}}{OA} = \frac{v_{BO}}{OC} = \frac{v_{BA}}{AC} = \omega$$

$$\therefore \quad v_{AO} = \omega \times OA \qquad ...(11.17A)$$

$$v_{BO} = \omega \times OC \qquad ...(11.17B)$$

and
$$v_{BA} = \omega \times AC \qquad ...(11.17C)$$

The value of ω is given and lengths OA, AC and CO are known from Klein's velocity diagram. Hence the velocities v_{AO}, v_{BO} and v_{BA} can be obtained. In equation (11.17), ω is the value of constant angular velocity of crank.

(*b*) *Acceleration diagram by Klein's Construction*

The procedure of drawing acceleration diagram by Klein construction is shown in Fig. 11.5 (*d*) and is as given below :

(*i*) Draw a circle with A as centre and AC as radius.

(*ii*) Draw another circle with AB as diameter. Let this circle intersects the previous circle at points D and E.

(*iii*) Join DE, which intersects the line BO at F. The line DE also intersects line BA at H. Then the quadrilateral $OAHF$ is known *as Klein's acceleration diagram.*

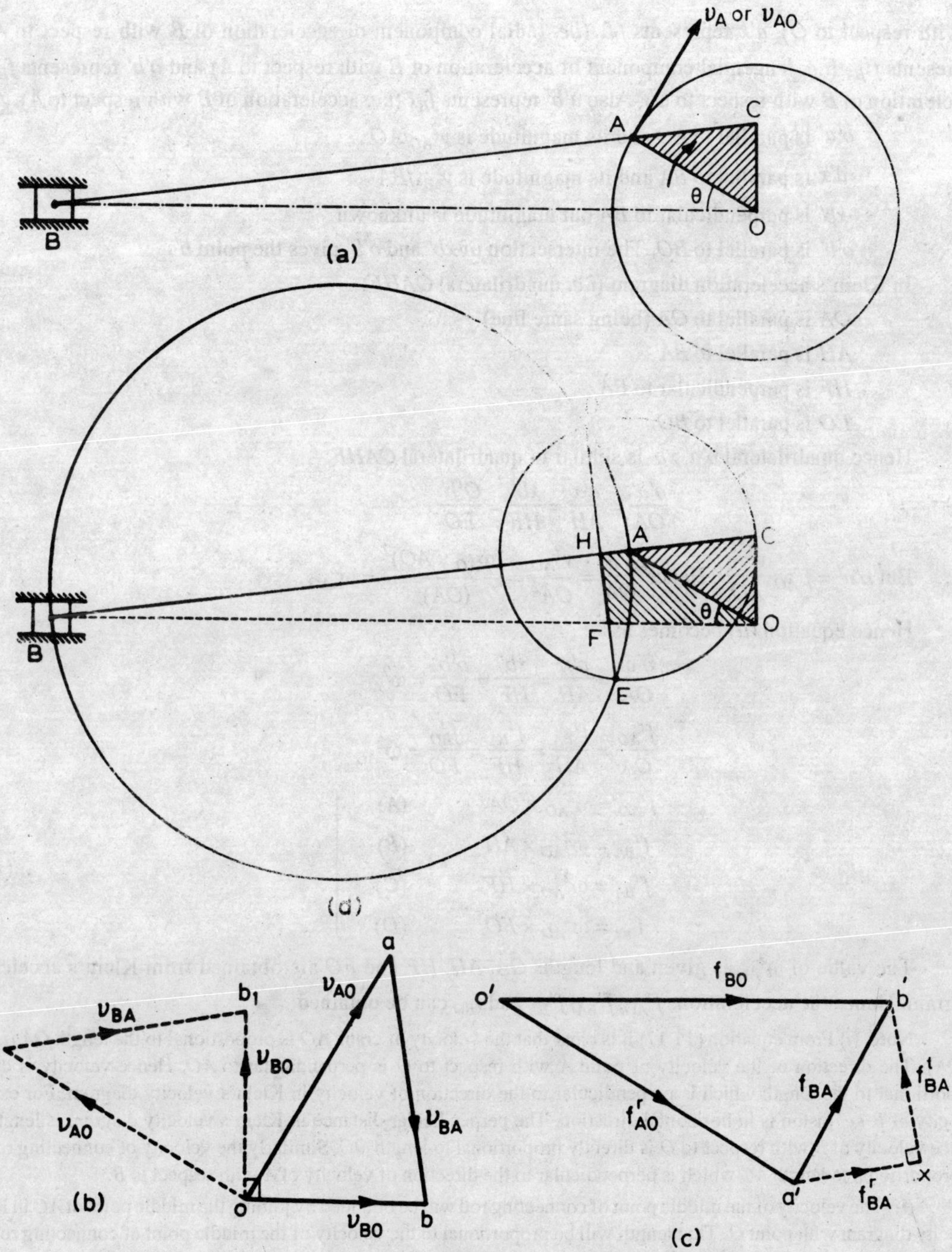

Fig. 11.5

In chapter 4 (*i.e.* Acceleration of Mechanism Chapter) the acceleration diagram for the reciprocating steam engine (*i.e.* Slider Crank Mechanism) has been drawn. The acceleration diagram is a quadrilateral *o'a' xb'* as shown in Fig. 11.5 (*c*). In this quadrilateral, *o'a'* represents f_{AO}^r (*i.e.* radial component of acceleration of

A with respect to O), $a'x$ represents f^r_{BA} (*i.e.* radial component of acceleration of B with respect to A), xb' represents f^t_{BA} (*i.e.* tangential component of acceleration of B with respect to A) and $o'b'$ represents f_{BO} (*i.e.* acceleration of B with respect to O). Also $a'b'$ represents f_{BA} (*i.e.* acceleration of B with respect to A). Also

$o'a'$ is parallel to AO and its magnitude is v^2_{AO}/AO

$a'x$ is parallel to BA and its magnitude is v^2_{BA}/BA

xb' is perpendicular to BA but magnitude is unknown

$o'b'$ is parallel to BO. The intersection of xb' and $o'b'$ gives the point b'.

In Klein's acceleration diagram (*i.e.* quadrilateral $OAHF$)

OA is parallel to OA (being same line)

AH is parallel to BA

HF is perpendicular to BA

FO is parallel to BO

Hence quadrilateral $o'a'\,xb'$ is similar to quadrilateral $OAHF$.

$$\therefore \quad \frac{o'a'}{OA} = \frac{a'x}{AH} = \frac{xb'}{HF} = \frac{O'b'}{FO} \qquad \text{...(ii)}$$

But $o'a' = f^r_{AO}\,\dfrac{v^2_{AO}}{AO}$. Hence $\dfrac{o'a'}{OA} = \dfrac{v^2_{AO}}{OA^2} = \dfrac{(\omega_{AO} \times AO)^2}{(OA)^2} = \omega^2_{AO}$

Hence equation (*ii*) becomes as

$$\frac{o'a'}{OA} = \frac{a'x}{AH} = \frac{xb'}{HF} = \frac{o'b'}{FO} = \omega^2_{AO}$$

or

$$\frac{f^r_{AO}}{OA} = \frac{f^r_{BA}}{AH} = \frac{f^t_{BA}}{HF} = \frac{f_{BO}}{FO} = \omega^2_{AO}$$

$$\therefore \quad \left.\begin{aligned} f^r_{AO} &= \omega^2_{AO} \times OA \qquad (A)\\ f^r_{BA} &= \omega^2_{AO} \times AH \qquad (B)\\ f^t_{BA} &= \omega^2_{AO} \times HF \qquad (C)\\ f_{BO} &= \omega^2_{AO} \times FO \qquad (D)\end{aligned}\right\} \qquad \text{...(11.18)}$$

The value of ω^2_{AO} is given and lengths OA, AH, HF and FO are obtained from Klein's acceleration diagram. Hence the accelerations $f^r_{AO}, f^r_{BA}, f^t_{BA}$ and f_{BO} can be obtained.

Note. (*i*) From equation (11.17) it is clear that the velocity of crank AO is proportional to the length OA (*i.e.* $v_{AO} \propto OA$). The direction of the velocity of point A with respect to O is perpendicular to AO. Hence velocity is directly proportional to the length which is perpendicular to the direction of velocity in Klein's velocity diagram. For example velocity of B *i.e.* piston is in horizontal direction. The perpendicular distance in Klein's velocity diagram is length OC. Hence velocity of B with respect to O is directly proportional to length OC. Similarly the velocity of connecting rod will be proportional to length AC which is perpendicular to the direction of velocity of A with respect to B.

(*ii*) The velocity of the middle point of connecting rod will be obtained by joining the middle point of AC in Klein's velocity diagram with point O. That length will be proportional to the velocity of the middle point of connecting rod with respect to O. If G is the middle point of connecting rod and L is the middle point of length AC, then

$$v_{GO} = \omega_{OA} \times \text{Length } OL$$

where length OL will be known from Klein's velocity diagram.

(*iii*) From equation (11.18), it is clear that the radial component of acceleration is directly proportional to the length (of Klein's acceleration diagram) which is in the direction of radial component. This means

$$f^r_{AO} \propto AO \qquad (\because \; AO \text{ of Klein's acceleration diagram is in the direction of } AO)$$

$$f^r_{BA} \propto HA \qquad (\because \; HA \text{ of Klein's acceleration diagram is in the direction of } BA)$$

(*iv*) In acceleration diagram ; $HA \propto f^r_{BA}$ and $HF \propto f^t_{BA}$. Also lengths AH and HF are perpendicular. Hence total acceleration of connecting rod (*i.e.* f_{BA}) is proportional to length AF which is the resultant of AH and HF.

(*v*) The total acceleration of the middle point (say G) of connecting rod wll be obtained by drawing from that point a line parallel to the line of stroke and cutting the line AF at some point (say J). Join J to O in Klein's acceleration diagram. Then length OJ will be proportional to the acceleration of middle point of connecting rod.

$$\therefore \qquad f_{GO} = \omega^2_{OA} \times OJ$$

where f_{GO} = Acceleration of middle of point of connecting rod

ω_{OA} = Angular velocity of crank

OJ = Length from Klein's acceleration diagram.

Problem 11.8. ***The lengths of the crank and connecting rod of a horizontal reciprocating engine are 100 mm and 500 mm respectively. The crank is rotating at 400 r.p.m. Using Klein's construction, find : (i) velocity of piston, (ii) acceleration of piston, (iii) angular velocity of the connecting rod and (iv) angular acceleration of the connecting rod when the crank has turned 30° from the inner dead centre.***

Sol. Given : (Refer to Fig. 11.6)

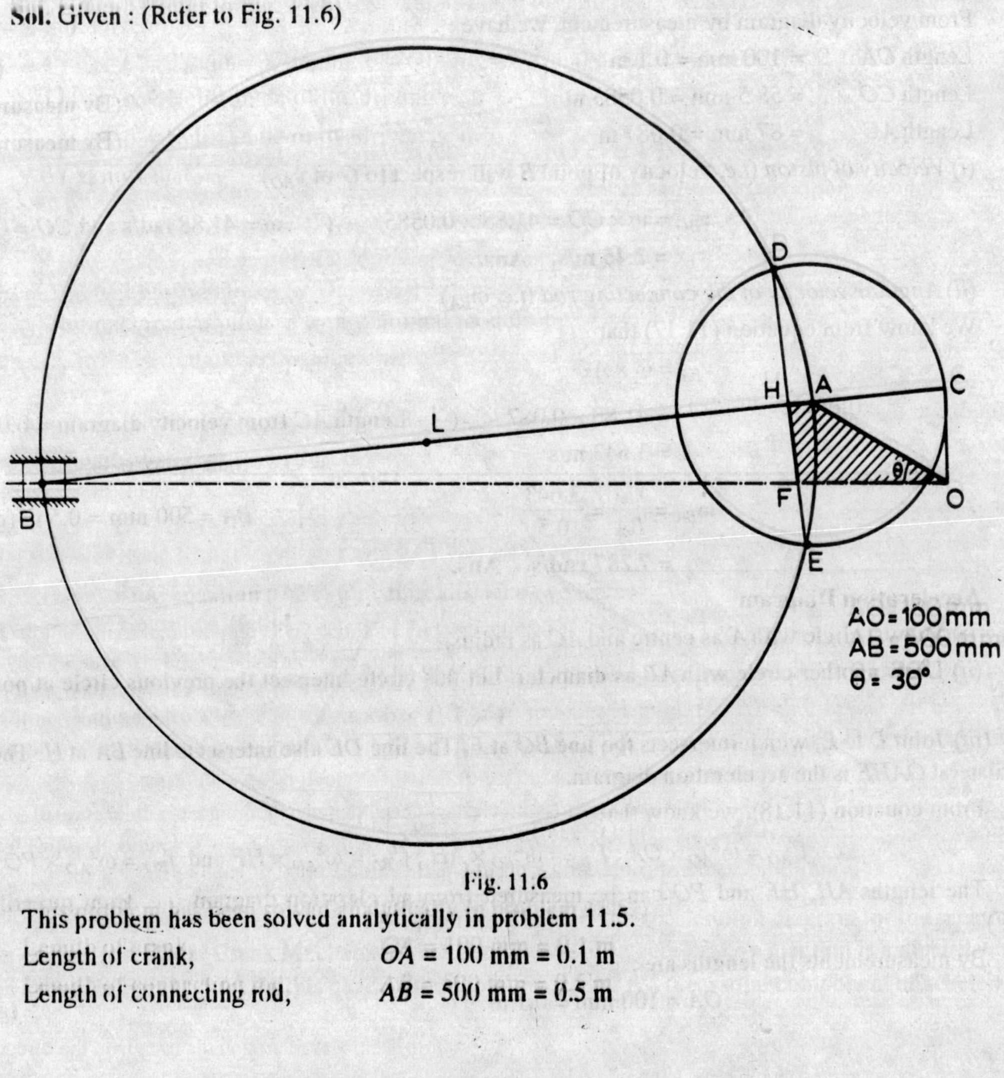

Fig. 11.6

This problem has been solved analytically in problem 11.5.

Length of crank, OA = 100 mm = 0.1 m

Length of connecting rod, AB = 500 mm = 0.5 m

Speed of crank, $N_{AO} = 400$ r.p.m.

$\therefore$ Angular velocity of crank, $\omega_{AO} = \frac{2\pi N_{AO}}{60} = \frac{2\pi \times 400}{60} = 41.88$ rad/s

Angle turned by crank from I.D.C., $\theta = 30°$.

With the given dimensions, draw the line diagram (*i.e. OAB*) of the horizontal reciprocating engine in which $OA = 100$ mm, $\angle BOA = 30°$ and length $AB = 500$ mm. In the diagram AO represents crank, AB represents connecting rod and point B represents the position of piston.

Velocity Diagram

From point O, draw OC perpendicular to OB. Produce the line BA to intersect the line OC at C. Then OAC is the velocity diagram. From equation (11.17) we know that

v_{AO} = Velocity of A with respect to $O = \omega \times AO$

v_{BO} = Velocity of B with respect to $O = \omega \times CO$

v_{BA} = Velocity of B with respect to $A = \omega \times AC$

where ω = constant angular velocity of crank = $\omega_{AO} = 41.88$ rad/s

From velocity diagram by measurement, we have :

Length OA = 100 mm = 0.1 m (given)

Length CO = 58.5 mm = 0.0585 m (By measurement)

Length AC = 87 mm = 0.087 m (By measurement)

(*i*) *Velocity of piston* (*i.e.* velocity of point B will respect to O or v_{BO})

$$v_{BO} = \omega \times CO = 41.88 \times 0.0585 \quad (\because \omega = 41.88 \text{ rad/s and } CO = 0.0585)$$

$$= \mathbf{2.45 \text{ m/s.}} \quad \textbf{Ans.}$$

(*ii*) *Angular velocity of the connecting rod* (*i.e.* ω_{BA})

We know from equation (11.17) that

$$v_{BA} = \omega \times AC$$

$$= 41.88 \times 0.087 \quad (\because \text{ Length } AC \text{ from velocity diagram} = 0.087 \text{ m})$$

$$= 3.643 \text{ m/s}$$

$$\therefore \quad \omega_{BA} = \frac{v_{BA}}{BA} = \frac{3.643}{0.5} \quad [\because BA = 500 \text{ mm} = 0.5 \text{ m (given)}]$$

$$= \mathbf{7.287 \text{ rad/s.}} \quad \textbf{Ans.}$$

Acceleration Diagram

(*i*) Draw a circle with A as centre and AC as radius.

(*ii*) Draw another circle with AB as diameter. Let this circle intersect the previous circle at points D and E.

(*iii*) Joint D to E, which intersects the line BO at F. The line DE also intersects line BA at H. Then the quadrilateral $OAHF$ is the acceleration diagram.

From equation (11.18), we know that

$$f^r_{AO} = \omega^2_{AO} \times OA \; ; f^r_{BA} = \omega^2_{AO} \times AH \; ; f^t_{BA} = \omega^2_{AO} \times HF \text{ and } f_{BO} = \omega^2_{AO} \times FO$$

The lengths AH, HF and FO can be measured from acceleration diagram (*i.e.* from quadrilateral $OAHF$).

By measurements the lengths are :

OA = 100 mm = 0.1 m (given)

$AH = 17.5$ mm $= 0.0175$ m (By measurement)

$HF = 48.5$ mm $= 0.0485$ m (By measurement)

$FO = 97$ mm $= 0.097$ m (By measurement)

(iii) Acceleration of the piston (i.e. f_{BO} or acceleration of B with respect to O)

We know from equation (11.18) that

$$f_{BO} = \omega^2_{AO} \times \text{Length } FO$$

$$= 41.88^2 \times 0.097 \qquad (\because \omega_{AO} = 41.88 \text{ and } FO = 0.097 \text{ m})$$

$$= \mathbf{170.13 \text{ m/s}^2}. \quad \textbf{Ans.}$$

(iv) Angular acceleration of connecting rod (i.e. α_{BA} or angular acceleration of B with respect to A)

We know that $$\alpha_{BA} = \frac{f^t_{BA}}{BA} \qquad \left(\text{or } \frac{\text{Tangential acceleration of } B \text{ with respect to } A}{\text{Length of connecting rod}}\right)$$

But $$f^t_{BA} = \omega^2_{AO} \times HF \qquad \text{[See equation (11.18)]}$$

$$= 41.88^2 \times 0.0485 \text{ m/s}^2 = 85.066 \text{ m/s}^2$$

$\therefore$ $$\alpha_{BA} = \frac{85.066}{0.5} \qquad (\because BA = 0.5 \text{ m})$$

$$= \mathbf{170.132 \text{ rad/s}^2}. \quad \textbf{Ans.}$$

Problem 11.9. *The lengths of the crank and connecting rod of a reciprocating engine are 200 mm and 800 mm respectively. The crank is rotating at a uniform speed of 480 r.p.m. Using Klein's construction find (i) the acceleration of the piston (ii) the acceleration of the middle point of the connecting rod and (iii) angular acceleration of the connecting rod when the crank has turned through 45° from the inner dead centre.*

Sol. Given : (Refer to Fig. 11.7)

Length of crank, $OA = 200$ mm $= 0.2$ m

Length of connecting rod, $AB = 800$ mm $= 0.8$ m

Speed of crank, $N_{AO} = 480$ r.p.m.

$\therefore$ $$\omega_{AO} = \frac{2\pi N_{AO}}{60} = \frac{2\pi \times 480}{60} = 50.265 \text{ rad/s}$$

Angle turned by crank from I.D.C., $\theta = 45°$.

First draw the line diagram of the reciprocating engine with the given dimensions. Draw any horizontal line BO. Make $\angle BOA = 45°$ and take length $OA = 200$ mm $= 0.2$ m. From point A, cut line $AB = 800$ mm $= 0.8$ m. In this diagram, OA represents crank, AB represents connecting rod and point B represents the piston. Now draw the acceleration diagram.

Acceleration diagram

(*i*) From point O, draw OC perpendicular to BO. Produce line BA to intersect the line OC at C.

(*ii*) Draw a circle with A as centre and AC as radius.

(*iii*) Draw another circle with AB as diameter. Let this circle intersects the previous circle at points D and E.

(*iv*) Joint D to E. The line DE intersects the line BO at F. Actually in this particular case, the points F and E coincides together. The line DE also intersects the line AB at H. Then the quadrilateral $OAHF$ is Klein's acceleration diagram.

From equation (11.18), we know that

$$f^r_{AO} = \omega^2_{AO} \times AO \, ; f^r_{BA} = \omega^2_{AO} \times AH \, ; f^t_{BA} = \omega^2_{AO} \times HF \text{ and } f_{BO} = \omega^2_{AO} \times FO$$

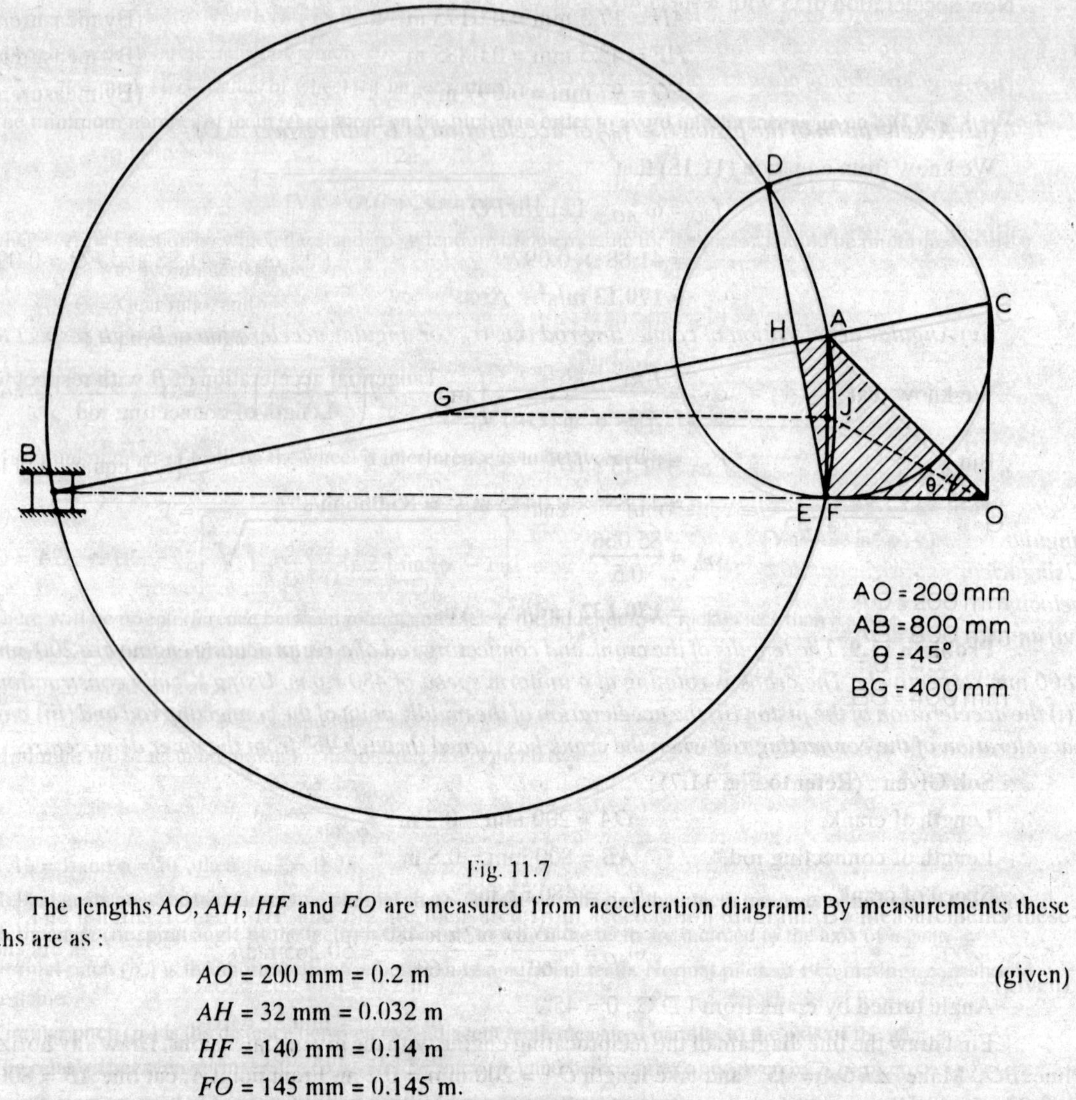

Fig. 11.7

The lengths AO, AH, HF and FO are measured from acceleration diagram. By measurements these lengths are as :

$$AO = 200 \text{ mm} = 0.2 \text{ m} \qquad \text{(given)}$$
$$AH = 32 \text{ mm} = 0.032 \text{ m}$$
$$HF = 140 \text{ mm} = 0.14 \text{ m}$$
$$FO = 145 \text{ mm} = 0.145 \text{ m}.$$

(i) Acceleration of piston (i.e. acceleration of B or value of f_{BO})

From equation (11.18), we know that

$$f_{BO} = \text{Acceleration of } B \text{ with respect to } O$$
$$= \omega_{AO}^2 \times FO$$
$$= (50.265)^2 \times 0.145 \qquad (\because \omega_{AO} = 50.265 \text{ and } FO = 0.145 \text{ m})$$
$$= \mathbf{366.35 \ m/s^2}. \quad \textbf{Ans.}$$

(ii) Acceleration of middle point of connecting rod

The mid-point of connecting rod is point G. The total acceleration of G will be f_{GO} *i.e.* acceleration G with respect to O. From point G, draw line GJ parallel to the line of stroke to cut the line AF at J. Join J to O. Measure the length JO.

By measurement the length $JO = 164 \text{ mm} = 0.164 \text{ m}$

Now acceleration of G with respect to $O = \omega^2_{AO} \times JO$

or $f_{GO} = (50.265)^2 \times 0.164 = \mathbf{414.35\ m/s^2}$. **Ans.**

(iii) Angular acceleration of connecting rod (i.e. value of α_{BA})

We know that

$$\alpha_{BA} = \frac{f^t_{BA}}{BA} \qquad \left(i.e. = \frac{\text{Tangential acceleration of connecting rod}}{\text{Length of connecting rod}}\right)$$

But from equation (11.18),

$$f^t_{BA} = \omega^2_{OA} \times HF$$

$$= (50.265)^2 \times 0.14 \qquad (\because \ HF = 0.14$$

$$= 353.72 \text{ m/s}^2$$

$$\therefore \qquad \alpha_{BA} = \frac{353.72}{0.8} \qquad (\because \ BA = 0.8 \text{ m})$$

$$= \mathbf{442.15\ rad/s^2}. \quad \textbf{Ans.}$$

Problem 11.10. *The crank of a reciprocating engine is rotating in clockwise direction with a constant angular velocity of 60 rad/s. The lengths of crank and connecting rod are 100 mm and 350 mm respectively. Using Klein's construction find : (i) velocity of piston (ii) velocity of mid point of connecting rod (iii) angular velocity of connecting rod (iv) acceleration of piston (v) acceleration of mid-point of connecting rod (vi) angular acceleration of connecting rod when crank has turned through 30° from inner dead centre (I.D.C.).*

Sol. Given : (Refer to Fig. 11.8)

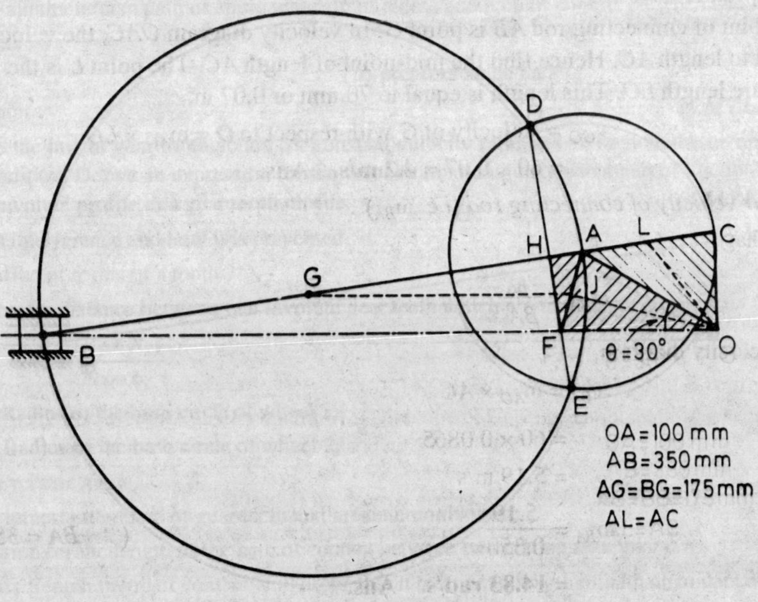

Fig. 11.8

Angular velocity of crank,	$\omega_{AO} = 60$ rad/s
Crank length,	$OA = 100$ mm $= 0.1$ m,
Connecting rod length,	$AB = 350$ mm $= 0.35$ m
Angle turned by crank,	$\theta = 30°$.

First draw the line diagram of the reciprocating engine with the given dimensions. Draw any horizontal line BO. At point O, make $\angle BOA = 30°$. Take $OA = 100$ mm. From point A, cut line $AB = 350$ mm. In this diagram, OA represents crank, AB represents connecting rod and point B represents the position of piston. Now draw the Klein's velocity diagram.

Klein's Velocity diagram

From point O, draw OC perpendicular to BO. Produce the line BA to intersect the line OC at C. Then OAC is the Klein's velocity diagram. From equation (11.17) we know that

$$v_{AO} = \text{Velocity of } A \text{ with respect to } O = \omega_{AO} \times AO$$

$$v_{BO} = \text{Velocity of } B \text{ respect to } O = \omega_{AO} \times CO$$

$$v_{BA} = \text{Velocity of } B \text{ with respect to } A = \omega_{AO} \times AC.$$

The values of AO, CO and AC are obtained from velocity diagram.

By measurements, we have

$AO = 100$ mm $= 0.1$ m (Given)

$CO = 63$ mm $= 0.063$ mm (By measurement)

$AC = 86.5$ mm $= 0.0865$ mm (By measurement)

(i) Velocity of piston (i.e. v_{BO} or velocity of piston B with respect to O)

$$v_{BO} = \omega_{AO} \times CO$$

$$= 60 \times 0.063 \quad (\because CO = 0.063 \text{ m})$$

$$= \mathbf{3.78 \text{ m/s.}} \quad \textbf{Ans.}$$

(ii) Velocity of mid-point of connecting rod (or v_{GO})

The mid-point of connecting rod AB is point G. In velocity diagram OAC, the velocity of connecting rod is proportional to length AC. Hence find the mid-point of length AC. The point L is the mid-point of AC. Join L to O. Measure length LO. This length is equal to 70 mm or 0.07 m.

$$\therefore \quad v_{GO} = \text{Velocity of } G \text{ with respect to } O = \omega_{AO} \times LO$$

$$= 60 \times 0.07 = \mathbf{4.2 \text{ m/s.}} \quad \textbf{Ans.}$$

(iii) Angular velocity of connecting rod (i.e. ω_{BA})

We know that

$$\omega_{BA} = \frac{v_{BA}}{BA}$$

But from velocity diagram,

$$v_{BA} = \omega_{AO} \times AC$$

$$= 60 \times 0.0865 \quad (\because AC = 0.0865 \text{ m})$$

$$= 5.19 \text{ m/s}$$

$$\therefore \quad \omega_{BA} = \frac{5.19}{0.350} \quad (\because BA = 350 \text{ mm} = 0.35 \text{ m})$$

$$= \mathbf{14.83 \text{ rad/s.}} \quad \textbf{Ans.}$$

To find the accelerations of different points, draw the Klein's acceleration diagram.

Klein's Acceleration Diagram

(*i*) Draw a circle with A as centre and AC as radius.

(*ii*) Draw another circle with AB as diameter. Let this circle intersects the previous circle at point D and E.

(iii) Join point *D* to *E*. The line *DE* intersects the line *BO* at *F*. The line *DE* also intersects the line *AB* at *H*. Then the quadrilateral *OAHF* is Klein's acceleration diagram.

From equation (11.18), we know that

$$f^r_{AO} = \omega^2_{AO} \times OA \,;\, f^r_{BA} = \omega^2_{AO} \times AH \,;\, f^t_{BA} = \omega^2_{AO} \times HF \text{ and } f_{BO} = \omega^2_{AO} \times FO$$

The lengths *OA*, *AH*, *HF* and *FO* are measured from acceleration diagram. By measurements, we have

$OA = 100 \text{ mm} = 0.1 \text{ m}$ (given)

$AH = 23 \text{ mm} = 0.023 \text{ m}$ (measured)

$HF = 46.5 \text{ mm} = 0.0465 \text{ m}$ (measured)

$FO = 100 \text{ mm} = 0.1 \text{ m}$ (measured)

(iv) Acceleration of piston (i.e. f_{BO})

The acceleration of piston or acceleration of point *B* with respect to *O* is given by,

$$f_{BO} = \omega^2_{AO} \times FO$$
$$= 60^2 \times 0.1 = \mathbf{360 \text{ m/s}^2}. \quad \textbf{Ans.}$$

(v) Acceleration of mid-point (G) of the connecting rod (i.e. value of f_{GO})

From point *G*, draw the line *GJ* parallel to the line of stroke and cutting the line *AF* at *J*. Joint point *J* to *O*. Measure length *JO*. By measurement,

length $JO = 96.5 \text{ mm} = 0.0965 \text{ m}$

Acceleration of *G*, $f_{GO} = \omega^2_{AO} \times JO = 60^2 \times 0.0965 = \mathbf{347.4 \text{ m/s}^2}.$ **Ans.**

(vi) Angular acceleration of connecting rod (i.e. value of α_{BA})

We know that, $$\alpha_{BA} = \frac{f^t_{BA}}{BA} \quad \left(i.e. = \frac{\text{Tangential acceleration of connecting rod}}{\text{Length of connecting rod}}\right)$$

But the value of f^t_{BA} from acceleration diagram is given by,

$$f^t_{BA} = \omega^2_{AO} \times \text{Length } HF$$
$$= 60^2 \times 0.0465 \qquad (\because \; HF = 46.5 \text{ mm} = 0.0465 \text{ m})$$
$$= 167.4 \text{ m/s}^2$$

$\therefore$ Angular acceleration of connecting rod *BA*,

$$\alpha_{BA} = \frac{f^t_{BA}}{BA} = \frac{167.4}{0.350} = \mathbf{478.285 \text{ rad/s}^2}. \quad \textbf{Ans.}$$

***(ii)* Ritterhaus's Construction**

Fig. 11.9 shows the line diagram of a reciprocating engine in which *OA* is the crank and *AB* is the connecting rod. The crank is rotating clockwise with a uniform angular velocity. Let the crank makes an angle θ

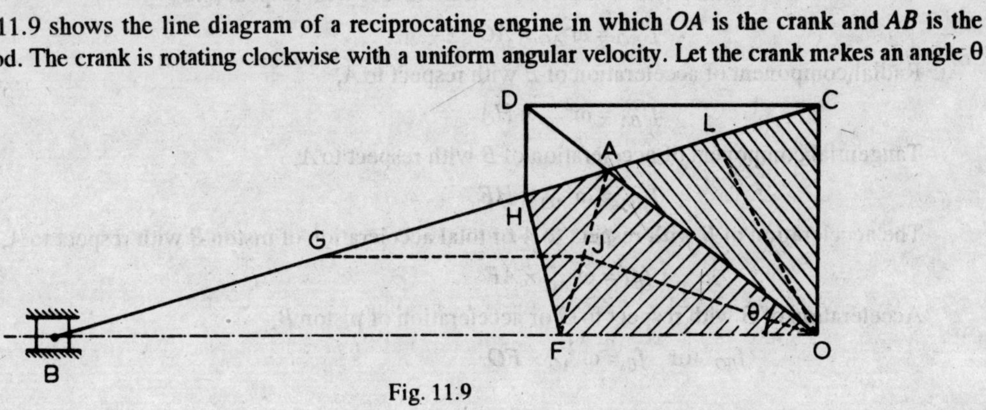

Fig. 11.9

with the line of stroke BO. The following is the method of drawing velocity diagram and acceleration diagram by Ritterhaus's construction :

Velocity Diagram

Velocity diagram by this method is same as given by Klein's construction. From point O, draw OC perpendicular to OB. Produce the line BA to intersect the line OC at C. Then OAC is known as Ritterhaus's velocity diagram.

$\therefore$ Velocity of A with respect to O or velocity of crank pin A,

$$v_{AO} = v_A = \omega_{AO} \times OA$$

where ω_{AO} = Angular velocity of crank

OA = length of crank

Velocity of piston B or velocity of B with respect to O,

$$v_B = v_{BO} = \omega_{AO} \times OC$$

Velocity of connecting rod BA *or* velocity of B with respect to A,

$$v_{BA} = \omega_{AO} \times AC$$

The velocity of any point G on the connecting rod BA, is obtained by dividing AC at L in the same ratio as G divides BA. In other words,

$$\frac{AL}{AC} = \frac{AG}{AB}$$

After obtaining point L, joint L to O.

Then velocity of G with respect to O,

$$v_{GO} = \omega_{AO} \times OL.$$

Acceleration Diagram

The procedure of drawing the acceleration diagram is given below :

(*i*) From point C, draw the line CD parallel to the line of stroke BO. Produce line OA to intersect the line CD at D.

(*ii*) From point D, draw the line DH parallel to line CO. The line DH cuts the connecting rod at point H.

(*iii*) From point H, draw the line HF perpendicular to the connecting rod (or line BA). The line HF cuts the line of stroke at point F. The quadrilateral $OAHF$ is known as *Ritterhous's acceleration diagram.*

From this diagram we have

Acceleration of crank pin A or radial acceleration of A with respect to O,

$$f^r_{AO} = \omega^2_{AO} \times AO \quad \text{...(11.18A)}$$

Radial component of acceleration of B with respect to A,

$$f^r_{BA} = \omega^2_{AO} \times HA \quad \text{...(11.18B)}$$

Tangential component of acceleration of B with respect to A,

$$f^t_{BA} = \omega^2_{AO} \times HF \quad \text{...(11.18C)}$$

The acceleration of B with respect to A or total acceleration of piston B with respect to A,

$$f_{BA} = \omega^2_{AO} \times AF \quad \text{...(11.18D)}$$

Acceleration of B with respect to O or acceleration of piston B,

$$f_{BO} \text{ or } f_B = \omega^2_{AO} \times FO \quad \text{...(11.18E)}$$

The acceleration of any point G on the connecting rod BA, is obtained by drawing the line GJ parallel to the line of stroke BO. The line GJ meets the line AF at point J. Join J to O.

Then acceleration of G with respect to O,

$$f_{GO} = \omega^2_{AO} \times JO \quad \text{...(11.18F)}$$

(iii) **Bennett's Construction**

Fig. 11.10 shows the line diagram of a reciprocating engine, in which OA is the crank and AB is the connecting rod. The crank is rotating clockwise with a uniform angular velocity, ω rad/s. Let the crank makes an angle θ with the line of stroke BO. The following is the method of drawing velocity diagram and acceleration diagram by *Bennett's Construction.*

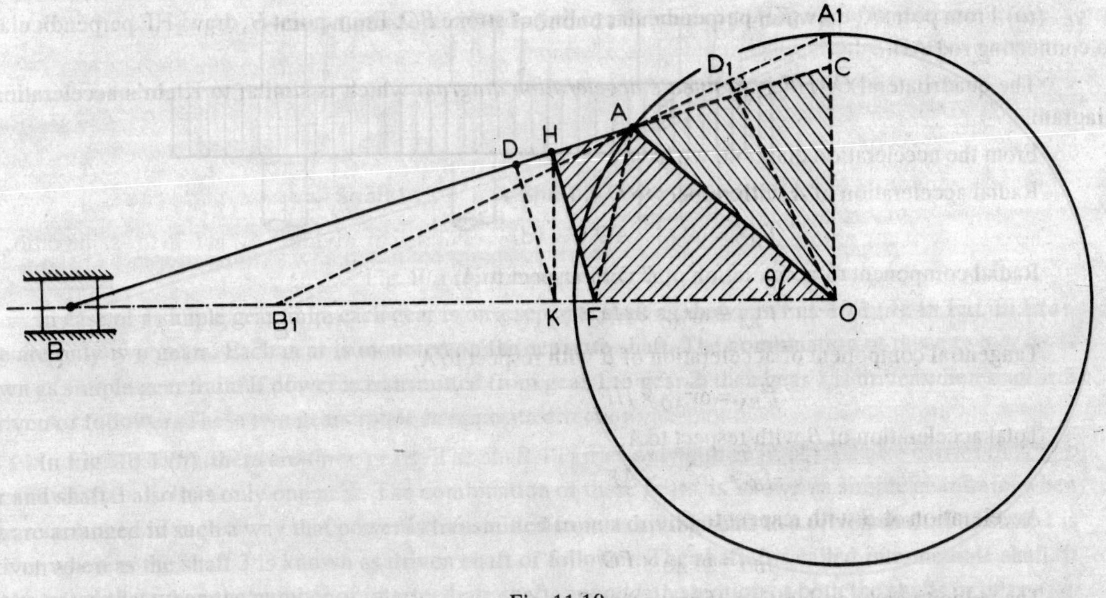

Fig. 11.10

Velocity Diagram

(i) OA is the given position of crank and AB is the corresponding position of connecting rod.

(ii) Let the crank is at right angle to the line of stroke. Then OA_1 is the position of crank and A_1B_1 is the corresponding position of the connecting rod.

(iii) Produce the line BA which intersects the line OA_1 at C. Then OAC is the velocity diagram by Bennett. It is similar to Klein's velocity diagram.

From the velocity diagram, we have

Velocity of A with respect to O or velocity of crank pin A,

$$v_{AO} = v_A = \omega_{AO} \times AO \text{ where } \omega_{AO} = \text{Angular velocity of crank.}$$

Velocity of piston B or velocity of B with respect to O,

$$v_B = v_{BO} = \omega_{AO} \times OC$$

Velocity of connecting rod BA or velocity of B with respect to A,

$$v_{BA} = \omega_{AO} \times AC.$$

Acceleration Diagram

The procedure of drawing the acceleration diagram is given below :

(*i*) From O, draw OD_1 perpendicular to connecting rod B_1A_1 (*i.e.* the position of connecting rod when crank is at right angle). Hence D_1 is a point on the connecting rod when crank is at right angle to line of stroke. Locate this point on the connecting rod when crank is at a given position.

Hence take AD on the connecting rod AB equal to A_1D_1. Hence take $AD = A_1D_1$.

(*ii*) From point D, draw DK perpendicular to connecting rod AB. The line DK intersects the line of stroke BO at point K.

(*iii*) From point K, draw KH perpendicular to line of stroke BO. From point H, draw HF perpendicular to connecting rod AB.

The quadrilateral $OAHF$ is *Bennett's acceleration diagram* which is similar to Klein's acceleration diagram.

From the acceleration diagram, we have

Radial acceleration of A with respect to O,

$$f^r_{AO} = \omega^2_{AO} \times AO$$

Radial component of acceleration of B with respect to A,

$$f^r_{BA} = \omega^2_{AO} \times HA$$

Tangential component of acceleration of B with respect to A,

$$f^t_{BA} = \omega^2_{AO} \times HF$$

Total acceleration of B with respect to A,

$$f_{BA} = \omega^2_{AO} \times AF$$

Acceleration of B with respect to O,

$$f_{BO} = \omega^2_{AO} \times FO$$

The velocity and acceleration of any point G on the connecting rod is obtained in the similar way as discussed before.

11.5. Torque Exerted on the Crank-shaft when Friction and Inertia of Moving Parts are Neglected

Fig. 11.11 shows the diagram of a single cylinder horizontal steam engine in which OA is the crank and AB is the connecting rod. The end B is connected to cross-head to which one end of piston rod is also connected. The other end of piston rod is connected to piston. The piston is moving inside the cylinder.

The force will be exerted by the steam on the piston and piston will start moving towards the right. Due to this the crank will start rotating clockwise. Let the angle turned by crank from inner dead centre (I.D.C.) is θ for the given position. And ω is the angular velocity of the crank at the given instant. The energy is supplied by the steam to the piston and energy is delivered by the crank to the crank-shaft.

Neglecting friction and inertia of the moving parts, we have

Energy supplied by steam on piston = Energy delivered by crank ...(*i*)

But energy supplied by steam on piston at the given instant

= Force exerted by steam on piston × Velocity of piston at that instant

($\because$ Energy $= F \times v$)

$= F_P \times v_P$

where F_P = Force exerted by steam on piston

$= p \times A$ ...For single acting

$= p_1A_1 - p_2A_2$ ...For double acting

(p_1 = pressure of steam on cover end

p_2 = pressure of steam on crank end

A_1 = Area of piston on cover end

A_2 = Area of piston on crank end)

v_P = Velocity of piston at the given instant

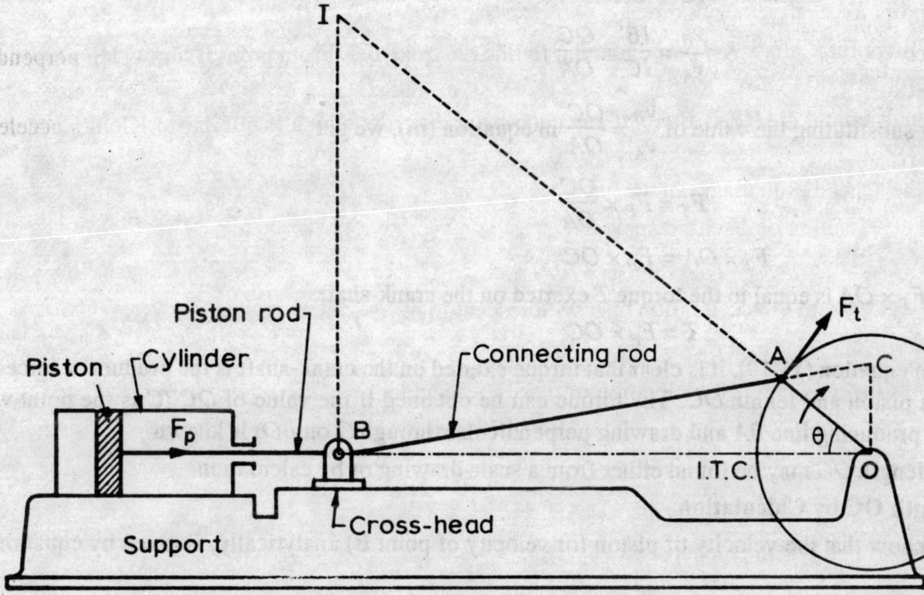

Fig. 11.11

Now the energy delivered by the crank on crank-shaft

$$= F_T \times v_A$$

where F_T = Tangential force acting on the crank at A which is perpendicular to crank OA. The value of F_T is not known. The product of $F_T \times OA$ gives the torque on crank-shaft.

v_A = Velocity of crank OA. The velocity v_A is also perpendicular to OA.

$= \omega_A \times OA$ (ω_A = Angular velocity of crank)

Substituting the values of energy supplied and energy delivered in equation (*i*), we get

$$F_P \times v_P = F_T \times v_A$$

or

$$F_T = F_P \times \frac{v_P}{v_A} \quad \text{...(ii)}$$

The piston and point B are connected with the help of piston rod, which is moving horizontally in a straight line. Hence velocity of piston = Velocity of point B

$$\therefore \quad v_P = v_B$$

Substituting the value of v_P is equation (*ii*), we get

$$F_T = F_P \times \frac{v_B}{v_A} \quad \text{...(iii)}$$

Now in Fig. 11.11, I is the instantaneous centre of the connecting rod AB (I is obtained by producing line OA and drawing perpendicular through B on BO), and hence

$$\frac{v_B}{v_A} = \frac{IB}{IC}$$

But from the similar triangles IBA and OAC, we have

$$\frac{IB}{IC} = \frac{OC}{OA}$$

Hence
$$\frac{v_B}{v_A} = \frac{IB}{IC} = \frac{OC}{OA}. \quad ...(iv)$$

Now substituting the value of $\frac{v_B}{v_A} = \frac{OC}{OA}$ in equation (*iii*), we get

$$F_T = F_P \times \frac{OC}{OA}$$

or
$$F_T \times OA = F_P \times OC$$

But $F_T \times OA$ is equal to the torque T exerted on the crank-shaft.

$$\therefore \quad T = F_P \times OC \quad ...(11.19)$$

From equation (11.19), it is clear that torque exerted on the crank-shaft is the product of force exerted by steam on piston and length OC. The torque can be obtained if the value of OC (C is the point which is obtained by producing line BA and drawing perpendicular through O on BO) is known.

The length OC may be found either from a scale drawing or by calculation.

Length OC by Calculation

We know that the velocity of piston (or velocity of point B) analytically, is given by equation (11.9) as

$$v_B = \omega r\left[\sin\theta + \frac{\sin 2\theta}{2n}\right] \quad \left(\text{If } n \text{ is very large where } n = \frac{AB}{AO} = \frac{l}{r}\right)$$

and
$$v_A = \omega \times r$$

$$\therefore \quad \frac{v_B}{v_A} = \frac{\omega \times r\left[\sin\theta + \frac{\sin 2\theta}{2n}\right]}{\omega \times r} = \sin\theta + \frac{\sin 2\theta}{2n}$$

But from equation (*iv*), $\frac{v_B}{v_A}$ is also equal to $\frac{OC}{OA}$

$$\therefore \quad \frac{v_B}{v_A} = \frac{\sin\theta + \sin 2\theta}{2n} = \frac{OC}{OA}$$

or
$$\frac{OC}{OA} = \frac{\sin\theta + \sin 2\theta}{2n}$$

or
$$OC = OA\left(\sin\theta + \frac{\sin 2\theta}{2n}\right)$$

$$= r\left(\sin\theta + \frac{\sin 2\theta}{2n}\right) \quad (\because \ OA = r) \quad ...(11.20)$$

Substituting this value in equation (11.19), we get

$$T = F_P \times r\left(\sin\theta + \frac{\sin 2\theta}{2n}\right) \quad ...(11.21)$$

Problem 11.11. *A horizontal steam engine has a piston of diameter 225 mm and a stroke of 600 mm. When the crank has turned through 40° from the I.D.C., the steam pressure on the cover and crank sides are 1.1 N/mm² ane 0.220 N/mm² abs. respectively. The length of the connecting rod is 1.5 m and diameter of piston rod is 50 mm. The r.p.m. of the crank is 240. Neglecting the friction and inertia of the moving parts, find the torque exerted on the crank-shaft.*

Sol. Given :

Dia. of piston D = 225 mm ; stroke L = 600 mm or crank radius $r = \frac{L}{2}$ = 300 mm ; θ = 40° ; steam pressure on coverside p_1 = 1.1 N/mm², and on crank-side p_2 = 0.220 N/mm² ; length of connecting rod l = 1.5 m = 1500 mm ; dia. of piston d = 50 mm ; r.p.m. of crank N = 240 r.p.m. or angular velocity of crank $\omega = \frac{2\pi N}{60}$ $= \frac{2\pi \times 240}{60} = 8\pi$ rad/s ; Friction and inertia of moving parts is to be neglected.

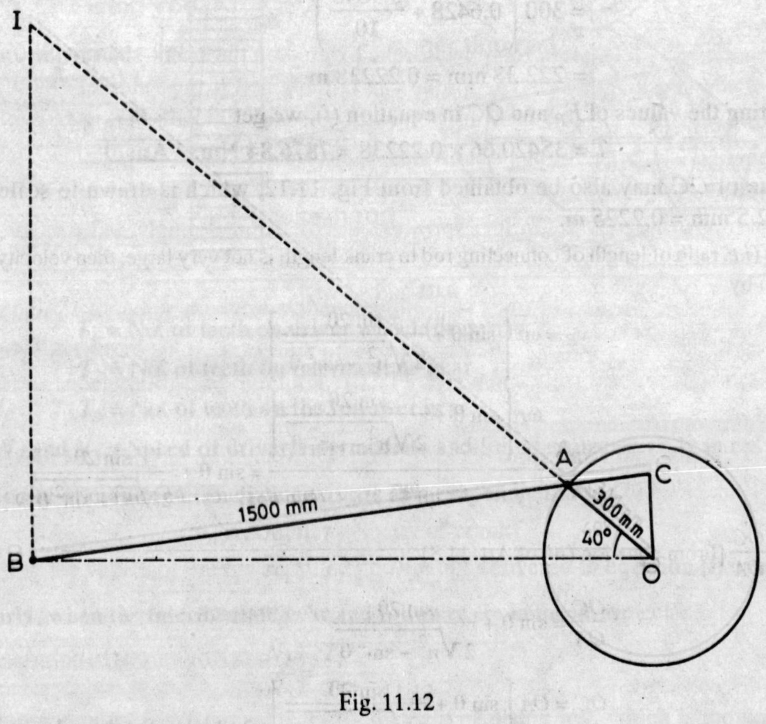

Fig. 11.12

Torque exerted on the crank-shaft

Torque exerted on the crank-shaft is given by equation (11.19) as

$$T = F_P \times OC \quad ...(i)$$

where F_P = Force exerted by steam on piston

$= p_1A_1 - p_2A_2$

Now p_1 = 1.1 N/mm² ; p_2 = 0.22 N/mm²

$$A_1 = \frac{\pi}{4} D^2 = \frac{\pi}{4} \times 225^2 = 39760 \text{ mm}^2$$

$$A_2 = \frac{\pi}{4} (D^2 - d^2) = \frac{\pi}{4} (225^2 - 50^2)$$

$$= \frac{\pi}{4} (50625 - 2500) = 37797 \text{ mm}^2$$

$$\therefore \quad F_P = p_1 A_1 - p_2 A_2$$

$$= (1.1 \times 39760 - 0.22 \times 37797) \text{ N}$$

$$= (43736 - 8315.34) \text{ N} = 35420.66 \text{ N} \qquad ...(ii)$$

The value of OC by calculation is obtained from equation (11.20) as

$$OC = r \left(\sin \theta + \frac{\sin 2\theta}{2n} \right)$$

$$= 300 \left(\sin 40° + \frac{\sin 80°}{2 \times 5} \right) \qquad \left(\because \ \theta = 40° \text{ and } n = \frac{l}{r} = \frac{1500}{300} = 5 \right)$$

$$= 300 \left(0.6428 + \frac{0.9848}{10} \right)$$

$$= 222.38 \text{ mm} = 0.22238 \text{ m} \qquad ...(iii)$$

Substituting the values of F_P and OC in equation (*i*), we get

$$T = 35420.66 \times 0.22238 = \mathbf{7876.84\ Nm.} \quad \textbf{Ans.}$$

The value of OC may also be obtained from Fig. 11.12, which is drawn to scale. By measurement length OC = 222.5 mm = 0.2225 m.

Note. If n (*i.e.* ratio of length of connecting rod to crank length is not very large, then velocity of piston or velocity of point B is given by

$$v_B = \omega r \left[\sin \theta + \frac{\sin 2\theta}{2\sqrt{n^2 - \sin^2 \theta}} \right]$$

$$\therefore \quad \frac{v_B}{v_A} = \frac{\omega r \left[\sin \theta + \dfrac{\sin 2\theta}{2\sqrt{n^2 - \sin^2 \theta}} \right]}{\omega r} = \sin \theta + \frac{\sin 2\theta}{2\sqrt{n^2 - \sin^2 \theta}}$$

But $\dfrac{v_B}{v_A} = \dfrac{OC}{OA}$ [From equation (*iv*) of Art. 11.5]

$$\therefore \quad \frac{OC}{OA} = \sin \theta + \frac{\sin 2\theta}{2\sqrt{n^2 - \sin^2 \theta}}$$

or

$$OC = OA \left[\sin \theta + \frac{\sin 2\theta}{2\sqrt{n^2 - \sin^2 \theta}} \right]$$

$$= r \left[\sin \theta + \frac{\sin 2\theta}{2\sqrt{n^2 - \sin^2 \theta}} \right] \qquad ...(11.22)$$

11.6. Forces on the Reciprocating Parts of an Engine Considering Friction and Inertia of Moving Parts but Neglecting Weight of the Connecting Rod

Fig. 11.13 shows the various forces acting on a reciprocating parts of a steam or I.C. engine for which the weight of connecting rod is neglected.

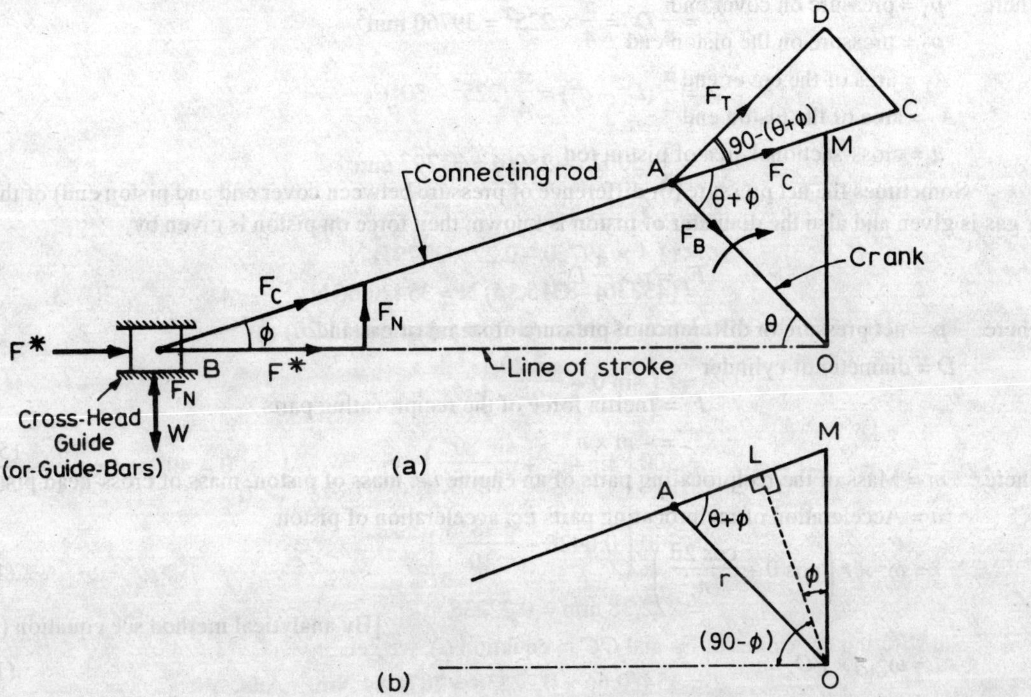

Fig. 11.13

These forces are :

(*i*) Piston effort (F^*)

(*ii*) Force acting along the connecting rod (F_c)

(*iii*) Thrust on the sides of the cylinder walls or normal reaction (F_N)

(*iv*) Crank effort (F_T) and

(*v*) Thrust on crank shaft bearings (F_B).

11.6.1. Piston Effort (F*)

The net force acting on the piston or cross-head pin *along the line of stroke* is known as *piston effort*. The piston effort is also known as effective driving force. The piston is reciprocating with simple harmonic motion. During the first half of the stroke, the reciprocating masses are accelerating. The inertia force due to acceleration of the masses, opposes the force on the piston (the force on the piston is due to steam or gas) and hence the net force on the piston decreases. But during the later half of the stroke, the reciprocating masses are retarding and inertia force opposes this retardation or inertia force acts in the direction of the applied gas pressure and thus increases the effective force on the piston. Mathematically piston effort for a *horizontal engine* is given by

Piston effort = Force on piston + Inertia force (– ve sign is taken during first half of stroke and + ve sign during later half of stroke)

or $$F^* = F_P + F_i \qquad ...(11.23)$$

where F^* = Piston effort

F_P = Force on piston due to the pressure of steam or gas

$= p_1A_1 - p_2A_2 = p_1A_1 - p_2(A_1 - a)$

where p_1 = pressure on cover end

p_2 = pressure on the piston end

A_1 = area of the cover end

A_2 = area of the piston end

a = cross-sectional area of piston rod

Sometimes the net pressure (or difference of pressure between cover end and piston end) of the steam or gas is given and also the diameter of piston is known, then force on piston is given by,

$$F_P = p \times \frac{\pi}{4} D^2$$

where p = net pressure or difference of pressure of steam or gas and

D = diameter of cylinder

$$F_i = \text{Inertia force of the reciprocating parts}$$

$$= -m \times a$$

where m = Mass of the reciprocating parts of an engine *i.e.* mass of piston, mass of cross-head pin etc.

a = Acceleration of reciprocating parts *i.e.* acceleration of piston

$$= \omega^2 \times r \left(\cos\theta + \frac{\cos 2\theta}{n} \right) \quad \text{...(11.23A)}$$

[By analytical method see equation (11.11)]

$$= \omega^2_{AO} \times FO \quad \text{...(11.23B)}$$

(By Klein's construction)

where ω or ω_{AO} = Uniform angular velocity of the crank,

θ = Angle turned by crank from I.D.C.,

n = Ratio of length of connecting rod to length of crank,

FO = Length obtained from Klein's acceleration diagram.

If frictional resistance is also taken into account, then piston effort (F^*) is given by,

$$F^* = F_P + F_i - F_R \quad \text{...(11.24)}$$

where F_R = Frictional resistance.

For a *vertical engine,* when piston is moving downwards (*i.e.* from the top dead centre to bottom dead centre), the weight (W) of the reciprocating parts assists the piston effort. But when the piston is moving upwards, the weight of reciprocating parts opposes the piston effort.

$\therefore$ Piston effort, $F^* = F_P + F_i \pm W - F_R$ ...(11.25)

where W = weight of reciprocating parts = $m \times g$.

11.6.2. Force acting along the connecting rod (F_c)

Let F_c = Force acting along the connecting rod

ϕ = Angle made by connecting rod with line of stroke.

From Fig. 11.13, we have

$$F_c \times \cos\phi = F^*$$

or

$$F_c = \frac{F^*}{\cos\phi} \quad \text{...(11.26)}$$

But we know from equation (11.5*B*) that

$$\cos\phi = \frac{1}{n}\sqrt{n^2 - \sin^2\theta}$$

$$= \sqrt{\frac{n^2 - \sin^2\theta}{n^2}} = \sqrt{1 - \frac{\sin^2\theta}{n^2}}$$

$$\therefore \quad F_c = \frac{F^*}{\sqrt{1 - \frac{\sin^2\theta}{n^2}}} \quad \text{...(11.26A)}$$

11.6.3. Thrust on the sides of the cylinder walls or normal reaction (F_N)

From Fig. 11.13, we have

$$F_N = F_c \times \sin\phi$$

$$= \frac{F^*}{\cos\phi} \times \sin\phi \qquad \left(\because F_c = \frac{F^*}{\cos\phi} \text{ or } F_c\cos\phi = F^*\right)$$

$$= F^* \times \tan\phi \quad \text{...(11.27)}$$

11.6.4. Crank Effort (F_T)

The net effort (or net force) applied at the crank-pin perpendicular to the crank is known as crank effort. It is represented by F_T. In Fig. 11.13, *AO* is the crank and line *AD* is perpendicular to crank *AO*. The force acting along the connecting rod (F_c) can be resolved into two components. One component is perpendicular to crank *AO* and the other component is along the crank.

Crank effort, F_T = Component of F_c perpendicular to crank

$$= F_c \cos[90 - (\theta + \phi)]$$

$$= F_c \sin(\theta + \phi) \quad \text{...(11.28)}$$

$$= \frac{F^*}{\cos\phi}\sin(\theta + \phi) \qquad \left(\because F_c = \frac{F^*}{\cos\phi}\right)$$

$$= F^* \frac{\sin(\theta + \phi)}{\cos\phi} \quad \text{...(11.28A)}$$

11.6.5. Thrust on Crank-shaft Bearing (F_B)

The component of F_c along the crank, is known as thrust on crank-shaft bearing. It is represented by F_B.

$\therefore$ F_B = Component of F_c along the crank

$$= F_c \times \cos(\theta + \phi)$$

$$= \frac{F^*}{\cos\phi} \times \cos(\theta + \phi) \qquad \left(\because F_c = \frac{F^*}{\cos\phi}\right)$$

$$= F^* \times \frac{\cos(\theta + \phi)}{\cos\phi} \quad \text{...(11.29)}$$

11.7. Torque on the Crank-shaft or Turning Moment on Crank-Shaft

Torque on the crank shaft is equal to the product of crank-effort (F_T) and radius of crank (r). Torque on the crank-shaft is also known as turning moment on crank-shaft. It is denoted by T.

$\therefore$ Torque on crank-shaft, T = crank effort × crank radius

$$= F_T \times r$$

$$= F^* \frac{\sin(\theta + \phi)}{\cos\phi} \times r \qquad \left(\because \; F_T = F^* \frac{\sin(\theta+\phi)}{\cos\phi}\right) \quad ...(i)$$

$$= F^* \frac{(\sin\theta\cos\phi + \cos\theta\sin\phi)}{\cos\phi} \times r$$

$$= F^* \left(\sin\theta + \frac{\cos\theta\sin\phi}{\cos\phi}\right) \times r$$

$$= F^* (\sin\theta + \cos\theta \times \tan\phi) \times r \quad ...(ii)$$

Let us find the value of tan ϕ in terms of angle θ.

From equation (11.5*A*) and (11.5*B*), we have

$$\sin\phi = \frac{r}{l}\sin\theta = \frac{\sin\theta}{\left(\frac{l}{r}\right)} = \frac{\sin\theta}{n} \qquad \left(\because \; \frac{l}{r} = n\right)$$

and

$$\cos\phi = \frac{1}{n}\sqrt{n^2 - \sin^2\theta}$$

$$\therefore \quad \tan\phi = \frac{\sin\phi}{\cos\phi} = \frac{\left(\frac{\sin\theta}{n}\right)}{\frac{1}{n}\sqrt{n^2 - \sin^2\theta}} = \frac{\sin\theta}{\sqrt{n^2 - \sin^2\theta}}$$

Substituting the value of tan ϕ in equation (*ii*), we get

$$T = F^* \left(\sin\theta + \cos\theta \times \frac{\sin\theta}{\sqrt{n^2 - \sin^2\theta}}\right) \times r$$

$$= F^* \left(\sin\theta + \frac{2\cos\theta\sin\theta}{2\sqrt{n^2 - \sin^2\theta}}\right) \times r = F^* \left(\sin\theta + \frac{\sin 2\theta}{2\sqrt{n^2 - \sin^2\theta}}\right) \times r$$

$$= F^* \times r \left(\sin\theta + \frac{\sin 2\theta}{2\sqrt{n^2 - \sin^2\theta}}\right) \quad ...(11.30)$$

From Fig. 11.13 (*b*), in right angled triangle *AOL* we have

$$OL = OA \sin(\theta + \phi)$$

$$= r\sin(\theta + \phi) \qquad (\because \; OA = r) \quad ...(iii)$$

Also in right angled triangle *OLM*, we have

$$OL = OM \cos\phi \quad ...(iv)$$

From equations (*iii*) and (*iv*), we have

$$r\sin(\theta + \phi) = OM\cos\phi$$

Substituting the value of $r\sin(\theta + \phi)$ in equation (*i*), we get

$$T = F^* \frac{OM \times \cos\phi}{\cos\phi} \qquad [\because \; r\sin(\theta+\phi) = OM\cos\phi]$$

$$= F^* \times OM \quad ...(11.31)$$

where *OM* is the length obtained by drawing perpendicular from *O* on *BO* and producing connecting rod *BA* upto *M*.

Problem 11.12. *The lengths of crank and connecting rod of a horizontal reciprocating engine are 200 mm and 1.0 m respectively. The crank is rotating at 400 r.p.m. When the crank has turned 30° from the inner dead centre, the difference of pressure between the cover end and piston end is 0.4 N/mm². If the mass of the reciprocating parts is 100 kg and cylinder bore is 0.4 m, then calculate : (i) Inertia force, (ii) Force on piston, (iii) Piston effort, (iv) Thrust on the sides of cylinder walls, (v) Thrust in the connecting rod, (vi) Crank-effort and (vii) Turning moment on the crank shaft. Neglect the effect of piston rod diameter and frictional resistance.*

Sol. Given :

$r = 200$ mm $= 0.2$ m ; $l = 1.0$ m ; $N = 400$ r.p.m. or $\omega = \frac{2\pi N}{60} = \frac{2\pi \times 400}{60} = 41.88$ rad/s ;

$\theta = 30°$, $(p_1 - p_2) = 0.4$ N/mm² ; $m = 100$ kg ; $D = 0.4$ m

(i) Inertia Force (F_i)

Inertia force in given by,

$$F_i = -m \times a$$

(– ve sign is due to the fact that inertia force opposes the accelerating force)

where a = Acceleration of the piston which is analytically given by equation (11.11) as

$$= r\omega^2\left(\cos\theta + \frac{\cos 2\theta}{n}\right) \text{ where } \theta = 30° \text{ and } n = \frac{l}{r} = \frac{1.0}{0.20} = 5$$

$$= 0.2 \times 41.88^2\left(\cos 30° + \frac{\cos(2 \times 30)}{5}\right) = 0.2 \times 1753.93\left(0.866 + \frac{0.5}{5}\right)$$

$$= 0.2 \times 1753.93 \times 0.966 = 338.86 \text{ m/s}^2$$

As the value of a is + ve, hence piston is accelerating

$$\therefore \quad F_i = -m \times a$$

$$= -100 \times 338.86 = \mathbf{-33886\ N.} \quad \textbf{Ans.}$$

(– ve sign means that inertia force is acting opposite to accelerating force)

The acceleration of piston can also be calculated graphically by Klein's construction (*i.e.* Klein's acceleration diagram). Refer to Fig. 11.6.

The acceleration of piston = $\omega^2 \times$ Length FO.

The length FO can be obtained from Klein's acceleration diagram.

(ii) Force on piston (F_P)

The force on piston by steam or gas is given by,

$$\therefore \quad \text{Force on piston, } F_P = p \times \frac{\pi}{4} D^2$$

where p = difference of pressure = $p_1 - p_2 = 0.4$ N/mm²

and D = Dia. of piston or bore of cylinder = 0.4 m = 400 mm

$$\therefore \quad F_P = 0.4 \times \frac{\pi}{4}(400)^2 \text{ N} = \mathbf{50265\ N.} \quad \textbf{Ans.}$$

(iii) Piston effort (F^*)

For a horizontal reciprocating engine, the piston effort neglecting frictional resistance is given by equation (11.23) as

$$F^* = F_P + F_i$$

$$= 50265 - 33886 \qquad (\because \ F_i = -33886)$$

$$= \mathbf{16379\ N.} \quad \textbf{Ans.}$$

(vi) Thrust on the sides of cylinder walls or normal reaction (F_N)

The thrust on the sides of cylinder walls (or normal reaction), F_N is given by equation (11.27) as

$F_N = F^* \tan \phi$ where F^* = Piston effort = 16379 N

ϕ = Angle made by connecting rod with line of stroke.

The value of ϕ in terms of θ is given by equation (11.5 *A*) as

$$\sin \phi = \frac{r}{l} \sin \theta$$

$$= \frac{0.2}{1.0} \times \sin 30° \qquad (\because \; r = 0.2, l = 1.0 \text{ and } \theta = 30°)$$

$$= \frac{0.2}{1.0} \times 0.5 = 0.1$$

$$\therefore \qquad \phi = \sin^{-1} 0.1 = 5.739°$$

$$\therefore \qquad F_N = F^* \tan \phi = 16379 \times \tan 5.739°$$

$$= 16379 \times 0.1005 = \mathbf{1646.1 \text{ N.}} \quad \textbf{Ans.}$$

(v) Thrust in the connecting rod (F_c)

The thrust in the connecting rod is given by,

$$F_c = \frac{F^*}{\cos \phi}$$

$$= \frac{16379}{\cos 5.739°} \qquad (\because \phi = 5.739°)$$

$$= \mathbf{16461.5 \text{ N.}} \quad \textbf{Ans.}$$

(vi) Crank-effort (F_T) or Tangential force on crank-pin

The crank-effort (F_T) is given by equation (11.28) as

$$F_T = F_c \sin (\theta + \phi) = 16461.5 \sin (30 + 5.739)$$

$$= 16461.5 \times \sin 35.739 = \mathbf{9615 \text{ N.}} \quad \textbf{Ans.}$$

(vii) Turning moment (or Torque) on the crank shaft

Turning moment on the crank-shaft is given by,

$$T = F_T \times r = 9615 \times 0.2 = \mathbf{1923 \text{ Nm.}} \quad \textbf{Ans.}$$

Problem 11.13. *A vertical petrol engine 150 mm diameter and 200 mm stroke has a connecting rod 350 mm long. The mass of the piston is 1.6 kg and engine speed is 1800 r.p.m. On the expansion stroke with crank angle 30° from the top dead centre, the gas pressure is 750 kN/m². Determine the net thrust on the engine.* (A.M.I.E., Summer 1987)

Sol. Given :

Engine diameter D = 150 mm = 0.15 m ; stroke L = 200 mm = 0.2 m or radius of crank $r = \frac{L}{2} = \frac{0.2}{2} = 0.1$ m ; Connecting rod length l = 350 mm = 0.35 m ; mass of piston m = 1.6 kg ; engine speed N = 1800 r.p.m. or angular velocity $\omega = \frac{2\pi N}{60} = \frac{2\pi \times 1800}{60} = 188.49$ rad/s, crank angle $\theta = 30°$, gas pressure $p = 750 \text{ kN/m}^2 = 75 \times 10^3 \text{ N/m}^2$

Engine is vertical.

The net thrust for vertical engine is given by equation (11.25) as

$$F^* = F_P + F_i \pm W \qquad ...(i)$$

Now F_P = Force exerted on the piston by the pressure of gas

$= p \times \text{Area of piston}$

$= p \times \frac{\pi}{4} D^2 = (750 \times 10^3) \times \left(\frac{\pi}{4} \times 0.15^2 \right) \text{ N}$

$= 750000 \times 0.01767 = 13253.59 \text{ N}$

$F_i = \text{Inertia force}$

$= -(\text{mass of piston}) \times \text{Acceleration of piston}$

The acceleration of piston 'a' is given by equation (11.11) as

$$a = \omega^2 r \left(\cos\theta + \frac{\cos 2\theta}{n} \right)$$

$$= (188.49)^2 \times 0.1 \left(\cos 30° + \frac{\cos 60°}{3.5} \right)$$

$$\left[\because \; r = \frac{L}{2} = 0.1 \text{ m and } n = \frac{l}{r} = \frac{0.35}{0.1} = 3.5 \right]$$

$$= 3552.848 \left(0.866 + \frac{0.5}{3.5} \right) = 3584.31 \text{ m/s}^2$$

$\therefore \quad F_i = -m \times a = -1.6 \times 3584.31 = -5734.89 \text{ N}$

$W = \text{Weight of piston or weight of reciprocating parts}$

$= m \times g$ where m = mass of piston or mass of reciprocating parts

$= 1.6 \times 9.81 = 15.696 \text{ N}$

As $\theta = 30°$, hence piston position is during first half of stroke. Hence during this position, piston is accelerating and inertia force opposes the accelerating force. Hence inertia force acts opposite to the acceleration. The piston is moving downward when $\theta = 30°$. Hence weight is acting downward *i.e.* in the direction of acceleration

$\therefore \quad F^* = F_P + F_i + W$

Substituting the values of F_P, F_i and W, we get

$F^* = 13253.59 + (-5734.89) + 15.696 = 13253.59 - 5734.89 + 15.696$

$= 7534.396 \text{ N}$

$\therefore$ Net thrust on piston $= F^* =$ **7534.396 N. Ans.**

Problem 11.14. *The length of the connecting rod of a vertical double acting steam engine is 1.5 m. The diameter of the cylinder is 400 mm and stroke of the engine is 600 mm. The crank is rotating at 200 r.p.m. in the clockwise direction. The crank has turned through 40° from the top dead centre and piston is moving downwards. The steam pressure above the piston is 0.6 N/mm² and below the piston is 0.05 N/mm². The mass of the reciprocating parts is 200 kg. The diameter of piston rod is given as 50 mm. Find the thrust on guide bars and crank-shaft bearing and also turning moment on crank-shaft.*

Sol. Given : Length of connecting rod, $l = 1.5$ m ; Dia. of cylinder, $D = 400$ mm ; 0.4 m ; stroke of engine, $L = 600$ mm. Hence radius of crank, $r = \frac{L}{2} = \frac{600}{2} = 300$ mm $= 0.3$ m ; $N = 200$ r.p.m. or $\omega = \frac{2\pi N}{60}$

$= \frac{2\pi \times 200}{60} = 20.944$ rad/s ; $\theta = 40°$; $p_1 = 0.6 \text{ N/mm}^2 = 0.6 \times 10^6 \text{ N/m}^2$; $p_2 = 0.05 \text{ N/mm}^2 = 0.05 \times 10^6 \text{ N/m}^2$;

$m = 200$ kg, dia. of piston rod, $d = 50$ mm $= 0.05$ m

Find : (*i*) Thrust on guide bars *i.e.* F_N

(*ii*) Thrust on crank-shaft bearing *i.e.* F_B and

(*iii*) Turning moment on crank-shaft.

(i) Thrust on guide bars i.e. F_N (or normal reaction)

Let us first find the piston effort (F^*)

For a vertical engine, the piston effort *i.e.* net force acting on piston (or F^*) is given by equation (11.25) as

$$F^* = F_P + F_i \pm W - F_R$$

$$= F_P + F_i \pm W \quad (F_R \text{ or frictional resistance is neglected})$$

$$= F_P + F_i + W \quad \text{...}(i)$$

[When $\theta = 40°$, the piston is accelerating during first half of stroke and hence inertia force (F_i) opposes the acceleration, and net force on the piston decreases. Also when $\theta = 40°$, the piston is moving downward, the weight of reciprocating parts helps the driving force.]

Now let us find the values of F_P, F_i and W.

F_P = Force on piston due to steam pressure

$$= p_1A_1 - p_2A_2$$

where $p_1 = 0.6 \times 10^6 \text{ N/m}^2$; $p_2 = 0.05 \times 10^6 \text{ N/m}^2$; $A_1 = \frac{\pi}{4}D^2 = \frac{\pi}{4}(0.4^2) = 0.12566 \text{ m}^2$;

$$A_2 = \frac{\pi}{4}D^2 - \frac{\pi}{4}d^2 = \frac{\pi}{4}[D^2 - d^2] = \frac{\pi}{4}[0.4^2 - 0.05^2] = 0.1237 \text{ m}^2$$

$\therefore$
$$F_P = 0.6 \times 10^6 \times 0.12566 - 0.05 \times 10^6 \times 0.1237$$

$$= 10^6\,(0.6 \times .12566 - 0.05 \times 0.1237)$$

$$= 10^6\,(0.075396 - 0.006185) = 69211 \text{ N}$$

Inertia force, $F_i = -m \times a$

where a = Acceleration of piston and it is given by equation (11.11) as

$$= r\omega^2\left(\cos\theta + \frac{\cos 2\theta}{n}\right)$$

$$= 0.3 \times 20.944^2\left[\cos 40° + \frac{\cos(2 \times 40)}{5}\right] \quad \left(\because\ n = \frac{l}{r} = \frac{1.5}{0.3} = 5 ;\ \theta = 40°\right)$$

$$= 0.3 \times 438.65\left[0.766 + \frac{0.1736}{5}\right]$$

$$= 0.3 \times 438.65 \times 0.8 = 105.27 \text{ m/s}^2$$

As value of a is + ve hence piston is accelerating. This is also true from the given value of θ. As $\theta = 40°$, hence during first half of the stroke, piston is accelerating.

$$m = 200 \text{ kg}$$

$\therefore$
$$F_i = -m \times a = -200 \times 105.27 = -21054 \text{ N}$$

Also W = Weight of reciprocating parts

$$= m \times g = 200 \times 9.81 \text{ N} = 1962 \text{ N}$$

Substituting the values of F_P, F_i and W in equation (*i*), we get

$$F^* = F_P + F_i + W$$

$$= 69211 + (-21054) + 1962 = 50119 \text{ N}$$

(*i*) *Thrust on guide bars* (*i.e.* normal reaction F_N)

Using equation (11.27), we have

$$F_N = F^* \tan\phi \qquad ...(ii)$$

Let us find the value of ϕ in terms of given value of θ.

From equation (11.5 *A*), we have,

$$\sin\phi = \frac{r}{l}\sin\theta$$

$$= \frac{0.3}{1.5}\sin 40° = \frac{0.3}{1.5} \times 0.64278 = 0.12855$$

$$\therefore \quad \phi = \sin^{-1} 0.12855 = 7.3858°$$

Substituting the value of ϕ in equation (*ii*), we get

$$F_N = F^* \tan 7.3858 = 50119 \times 0.1296 = 6496.7 \text{ N}$$

$\therefore$ Thrust on guide bars = F_N = **6496.7 N. Ans.**

(*ii*) *Thrust on crank-shaft bearing,* F_B

The thrust on crank-shaft bearing (*i.e.* F_B) is given by equation (11.29)

$$\therefore \quad F_B = F^* \times \frac{\cos(\theta + \phi)}{\cos\phi}$$

$$= 50119 \times \frac{\cos(40 + 7.3858)}{\cos(7.3858)} = \frac{50119 \times 0.677}{0.9917} = \mathbf{34214.4 \text{ N}}. \quad \textbf{Ans.}$$

(*iii*) *Turning moment on crank-shaft*

The turning moment (or torque *T*) on the crank-shaft is given by,

$$T = F_T \times r \qquad ...(iii)$$

where F_T, the net tangential force at the crank-pin is given by equation (11.28*A*) as

$$F_T = F^* \frac{\sin(\theta + \phi)}{\cos\phi}$$

$$= 50119 \times \frac{\sin(40 + 7.3858)}{\cos(7.3858)} = \frac{50119 \times \sin(47.3858)}{\cos(7.3858)}$$

$$= \frac{50119 \times 0.7359}{0.9917} = 37192.7 \text{ N}$$

Substituting the value of F_T in equation (*iii*), we get

$$T = 37192.7 \times 0.3 \qquad (\because r = 0.3 \text{ m})$$

$$= \mathbf{11157.8 \text{ N m}}. \quad \textbf{Ans.}$$

Problem 11.15. *If in Problem 11.14, the crank instead of turning at angle of 40° from the top dead centre, has turned through 140° from the top dead centre and piston is moving downwards, then find the piston effort (or net force acting on the piston). All other data is same. Neglect frictional resistance.*

Sol. From Problem 11.14, the data which is common is :

$p_1 = 0.6$ N/mm^2 ; $p_2 = 0.05$ N/mm^2 ; $\omega = 20.944$ rad/s ; $r = 0.3$ m, $l = 1.5$ m, $m = 200$ kg ;

$A_1 = 0.12566$ m^2 ; $A_2 = 0.1237$ m^2, $n = \frac{l}{r} = \frac{1.5}{0.3} = 5.$

The angle turned by crank from top-dead centre, $\theta = 140°$.

The piston effort (F^*) for a vertical engine, neglecting frictional resistance is given by equation (11.25) as

$$F^* = F_P + F_i \pm W$$

As piston is having simple harmonic motion hence during first half of the stroke the piston is accelerating and during later half of the stroke, the piston is retarding.

When $\theta = 140°$, the piston is retarding and hence the inertia force opposes this retardation. This means the inertia force helps the driving force. Also when $\theta = 140°$, the piston is moving downwards and hence the weight (W) which is acting downward also helps the driving force.

$\therefore$ $$F^* = F_P + F_i + W \quad ...(i)$$

Now F_P = Force on piston due to steam pressure

$$= p_1A_1 - p_2A_2$$

$$= 0.6 \times 10^6 \times 0.12566 - 0.05 \times 10^6 \times 0.1237 = 69211 \text{ N}$$

$$F_i = -m \times a \quad ...(ii)$$

where $$a = r\omega^2\left(\cos\theta + \frac{\cos 2\theta}{n}\right)$$ [Using equation (11.11)]

$$= 0.3 \times 20.944^2 \left[\cos 140° + \frac{\cos(2 \times 140)}{n}\right] \quad (\because \theta = 140°)$$

$$= 0.3 \times 438.65 \left[-0.766 + \frac{0.17365}{5}\right] \quad (\because n = 5)$$

$$= 0.3 \times 438.65\,[-0.73127] = -96.23 \text{ m/s}^2$$

As the value of a is $-$ve, hence piston is retarding. This is also true from the given value of θ. As $\theta = 140°$, hence during the later half of the stroke, the piston is retarding.

Substituting the value of a in equation (*ii*), we get

$$F_i = -m \times (-96.23)$$

$$= -200(-96.23) \quad (\because m = 200 \text{ kg})$$

$$= 19246 \text{ N}$$

and W = Weight of reciprocating parts

$$= m \times g = 200 \times 9.81 = 1962 \text{ N}$$

Substituting the values of F_P, F_i and W is equation (*i*), we get

$$F^* = F_P + F_i + W$$

$$= 69211 + 19246 + 1962 = \mathbf{90419\ N.} \quad \textbf{Ans.}$$

Problem 11.16. *The lengths of crank and connecting rod of a horizontal steam engine are 300 mm and 1.2 m respectively. When the crank has moved 30° from the inner dead centre, the acceleration of piston is 35 m/s². The average frictional resistance to the motion of piston is equivalent to a force of 550 N and net effective steam pressure on piston is 500 kN/m². The diameter of piston is 0.3 m and mass of reciprocating parts is 160 kg. Determine :*

(i) Reaction on the cross-head guides ;

(ii) Thrust on the crank-shaft bearings ; and

(iii) Torque on the crank shaft.

Sol. Given : r = 300 mm = 0.3 m ; l = 1.2 m ; θ = 30° ; f_B = Acceleration of piston = 35 m/s² ; average frictional resistance, F_R = 550 N ; net effective steam pressure, p = 500 kN/m² = 500 × 10³ N/m² ; dia of piston D = 0.3 m ; mass of reciprocating parts m_R = 160 kg.

(i) Reaction on the cross-head guides (F_N)

The piston effort (or net force acting on piston) for a horizontal engine is given by

$$F^* = F_P + F_i - F_R \quad \text{...}(i)$$

where F_P = Force exerted by steam on piston

$= p \times A$ where p = net effective steam pressure $= 500 \times 10^3$ N/m^2

A = Area of piston $= \frac{\pi}{4} D^2 = \frac{\pi}{4} \times 0.3^2$

$$= (500 \times 10^3) \times \left(\frac{\pi}{4} \times 0.3^2 \right) = 35342.9 \text{ N}$$

F_i = Inertia force due to reciprocating mass

= – (Mass of reciprocating parts) × (Acceleration of piston)

= – (160) × (35) N = – 5600 N

F_R = Frictional force = 550 N

∴ Equation (*i*) becomes as,

$$F^* = 35342.9 - 5600 - 550 = 29192.9 \text{ N}$$

The reaction on the cross-head (F_N) is given by equation (11.27) as

$$F_N = F^* \tan \phi \quad \text{...}(ii)$$

where ϕ = Angle made by connecting rod to the line of stroke.

The value of angle ϕ is obtained from equation (11.5 *A*) as

$$l \sin \phi = r \sin \theta$$

or

$$\sin \phi = \frac{r}{l} \sin \theta = \frac{\sin \theta}{n} \qquad \left(\because \; n = \frac{l}{r} = \frac{1.2}{0.3} = 4 \right)$$

$$= \frac{\sin 30°}{4} = 0.125$$

$$\therefore \quad \phi = \sin^{-1} 0.125 = 7.18°$$

Substituting the values of ϕ and F^* in equation (*ii*), we get

$$F_N = 29192.9 \tan 7.18° = \mathbf{3677.56 \text{ N.}} \quad \textbf{Ans.}$$

(ii) Thrust on the crank-shaft bearing (F_B)

The thrust on crank-shaft bearing is given by equation (11.29) as

$$F_B = \frac{F^* \cos (\theta + \phi)}{\cos \phi}$$

$$= \frac{29192.9 \cos (30 + 7.18)}{\cos (7.18)} = \frac{29192.9 \times 0.7967}{0.9921} = \mathbf{23441.8 \text{ N.}} \quad \textbf{Ans.}$$

(iii) Torque on the crank-shaft (T)

Using equation for torque we get,

$$T = F_T \times r$$

$$= \frac{F^* \sin (\theta + \phi)}{\cos \phi} \times r \qquad \left[\because \; \text{From equation (11.28 } A\text{)}, \; F_T = F^* \frac{\sin (\theta + \phi)}{\cos \phi} \right]$$

$$= 29192.9 \frac{\sin (30 + 7.18)}{\cos (7.18)} \times 0.3$$

$$= 29192.9 \times \frac{0.6043}{0.9921} \times 0.3 = \mathbf{5334.52\ N\ m.}\quad \textbf{Ans.}$$

Problem 11.17. *A single cylinder vertical engine has a bore of 300 mm and a stroke of 400 mm. The connecting rod is 1 m long. The mass of the reciprocating parts is 140 kg. On the expansion stroke with the crank at 30° from the top dead centre the gas pressure is 0.7 MPa. If the engine runs at 250 r.p.m., determine :*

(i) net force acting on the piston,

(ii) resultant load on the gudgeon pin,

(iii) thrust on the cylinder walls, and

(iv) the speed above which, other thing remaining same, the gudgeon pin load would be reversed in direction.

(AMIE, Winter 1994)

Sol. Given :

$D = 300$ mm $= 0.3$ m ; stroke $= 400$ mm $= 0.4$ m or crank radius $r = \frac{\text{stroke}}{2} = \frac{0.4}{2} = 0.2$ m ; $l = 1$ m ; mass of reciprocating parts $m_R = 140$ kg ; $\theta = 30°$, gas pressure $p = 0.7$ MPa* $= 0.7 \times 10^6$ N/m², $N = 250$ r.p.m.

or $\omega = \frac{2\pi N}{60} = \frac{2\pi \times 250}{60} = 26.18$ rad/s.

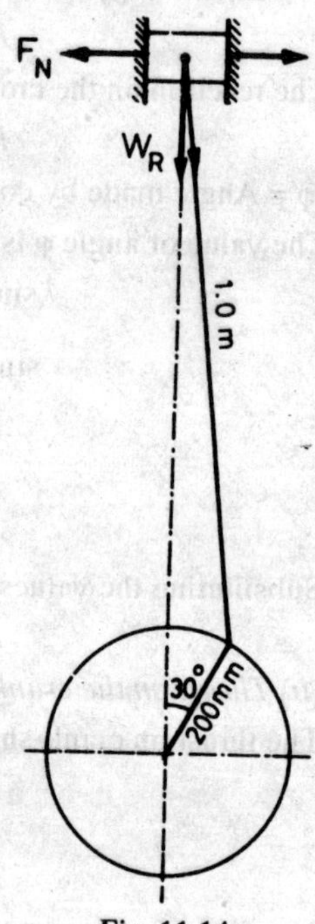

Fig. 11.14

(i) Net force acting on the piston (F^*)

During expansion stroke when $\theta = 30°$, the pressure force due to gas acts downwards, weight of reciprocating parts also acts downwards and inertia force acts upwards.

For a vertical engine, the net force (F^*) acting on piston is given by,

$$F^* = F_P + F_i + W \qquad ...(i)$$

where F_P = Force exerted by the pressure of the gas

$= p \times A$ where p = pressure of gas $= 0.7 \times 10^6$ N/m²

A = Area of piston $= \frac{\pi}{4} D^2 = \frac{\pi}{4}(0.3^2)$

$$= (0.7 \times 10^6) \times \left(\frac{\pi}{4} \times 0.3^2\right) = 49480 \text{ N}$$

F_i = Inertia force due to reciprocating mass

= – (mass of reciprocating parts) × (Acceleration of piston)

= – (140) × (Acceleration of piston)

But acceleration of piston

$$= r \times \omega^2 \times \left(\cos\theta + \frac{\cos 2\theta}{n}\right) \quad \text{[See equation (11.11)]}$$

$$= 0.2 \times 26.18^2 \times \left(\cos 30° + \frac{\cos 60°}{5}\right) \qquad \left(\because\ n = \frac{l}{r} = \frac{1.0}{0.2} = 5\right)$$

$$= 137.078\left(0.866 + \frac{0.5}{5}\right) = 132.41 \text{ m/s}^2$$

$\therefore$ $F_i = -(140) \times 132.41 = -18537.4$ N

Now $W = m_R \times g = 140 \times 9.81 = 1373.4$ N

*M stands for Mega $= 10^6$ and P_a stands for Pascal = N/m².

Substituting the values of F_P, F_i and W in equation (*i*), we have

$$F^* = 49480 - 18537.4 + 1373.4 = \mathbf{32316\ N.}\quad \textbf{Ans.}$$

(ii) Resultant load on gudgeon pin

On the gudgeon pin, the following two forces are acting :

(*i*) The net force (F^*) acting along the line of stroke and

(*ii*) The normal reaction (F_N)

The resultant of F^* and F_N is the resultant load on gudgeon pin.

The resultant of F^* and F_N is the force acting along the connecting rod (*i.e.* force F_c). But F_c is also given by,

$$F_c = \frac{F^*}{\cos\phi} \qquad (\because\ F^* = F_c \cos\phi)$$

where ϕ = Inclination of connecting rod with line of stroke. Let us now find the value of ϕ. The value of ϕ is obtained from relation (11.5*A*) as

$$l \sin\phi = r \sin\theta$$

or

$$\sin\phi = \frac{r}{l}\sin\theta = \frac{\sin\theta}{\left(\frac{l}{r}\right)}$$

$$= \frac{\sin\theta}{n} \qquad \left(\because\ n = \frac{l}{r} = \frac{1.0}{0.2} = 5\right)$$

$$= \frac{\sin 30°}{5} = \frac{0.5}{5} = 0.1$$

$\therefore$
$$\phi = \sin^{-1} 0.1 = 5.739°$$

$\therefore$
$$F_c = \frac{F^*}{\cos\phi}$$

$$= \frac{32316}{\cos(5.739)} \qquad (\because\ F^* = 32316)$$

$$= 32478.8\ \text{N}$$

$\therefore$ Resultant load on gudgeon pin = F_c = **32478.8 N. Ans.**

(iii) Thrust on the cylinder walls (F_N)

We known that thrust on cylinder walls,

$$F_N = F^* \tan\phi \quad \text{or} \quad F_c \sin\phi$$

$$= 32316 \tan 5.739 \quad \text{or} \quad 32478.8 \sin 5.739$$

$$= 3247.78\ \text{N} \quad \text{or} \quad 3247.78\ \text{N}$$

$\therefore$
$$F_N = \mathbf{3247.78\ N.}\quad \textbf{Ans.}$$

(iv) Speed above which other thing remaining same, the gudgeon pin load would be reversed in direction.

The gudgeon pin load is the force in connecting rod (*i.e.* F_c).

$\therefore$ Gudgeon pin load = F_c.

Now the gudgeon pin load would be reversed in direction if F_c becomes negative. But F_c will be negative if F^* is negative as $F_c = F^*/\cos\phi$. Now F^* will become negative if in equation (*i*), F_i is more than $(F_P + W)$,

or
$$F_i > F_P + W \qquad ...(ii)$$

Now let N_* is the speed for which F_i is more than $F_P + W$.

Then F_i = mass × Acceleration

$$= m_R \times r \times \omega^2 \times \cos\left(\theta + \frac{\cos 2\theta}{n}\right)$$

$$= 140 \times 0.2 \times \left(\frac{2\pi N_*}{60}\right)^2 \times \cos\left(30° + \frac{\cos 60°}{5}\right) \qquad \left(\because \text{ Here } \omega = \frac{2\pi N_*}{60}\right)$$

$$= \frac{28 \times 4 \times \pi^2}{3600} \times N_*^2 \times \left(0.866 + \frac{0.5}{5}\right)$$

$$= \frac{28 \times 4 \times \pi^2}{3600} \times 0.966 \times N_*^2 = 0.2966 N_*^2$$

Substituting the values of F_i, F_P and W in equation (*ii*), we get

$$0.2966 N_*^2 > 49480 + 1373.4$$

$$> 50853.4$$

or

$$N_* > \sqrt{\frac{50853.4}{0.2966}}$$

$$> \textbf{414.07 r.p.m.} \quad \textbf{Ans.}$$

Problem 11.18. *The diameter of cylinder of a vertical single cylinder single acting diesel engine is 300 mm. The lengths of crank and connecting rods are 250 mm and 1.125 m respectively. Reciprocating parts are having a mass of 140 kg and engine is running at 270 r.p.m. The ratio of compression is 14 and pressure remains constants during injection of oil for $\frac{1}{10}$th of the stroke. The index of law for compression and expansion is 1.35 (i.e. $pV^{1.35}$ = constant for compression and expansion). Find the torque on the crank-shaft when the crank makes an angle of 45° with the inner dead centre during expansion stroke. Suction may be assumed at a pressure of 100 kN/m^2.*

Sol. Given : D = 300 mm = 0.3 m ; r = 250 mm = 0.25 m ; stroke $L = 2r = 2 \times 0.25 = 0.5$ m ;

l = 1.125 m ; m_R = 140 kg ; N = 270 r.p.m. or $\omega = \frac{2\pi N}{60} = \frac{2\pi \times 270}{60} = 28.27$ rad/s ;

compression ratio $= \frac{V_2}{V_3} = 14$; $\theta = 45°$; index of compression and expansion = 1.35 or $pV^{1.35}$ = constant ;

suction pressure $p_1 = p_2$ = 100 kN/m^2 = 100 × 10^3 N/m^2.

Fig. 11.15 shows the pressure volume (*i.e. p-V*) diagram for a diesel engine in which the compression and expansion are according to the law $pV^{1.35}$ = constant. The injection of fuel is at constant pressure. In this Fig. 11.15,

1-2 represents suction,

2-3 represents compression,

3-4 represents injection of oil,

4-5 represents expansion,

5-2-1 represents exhaust.

We know that swept volume (V_s) is given by,

$$V_s = \frac{\pi}{4} D^2 \times L = \frac{\pi}{4}(.3^2) \times 0.5 = 0.0353 \text{ m}^3$$

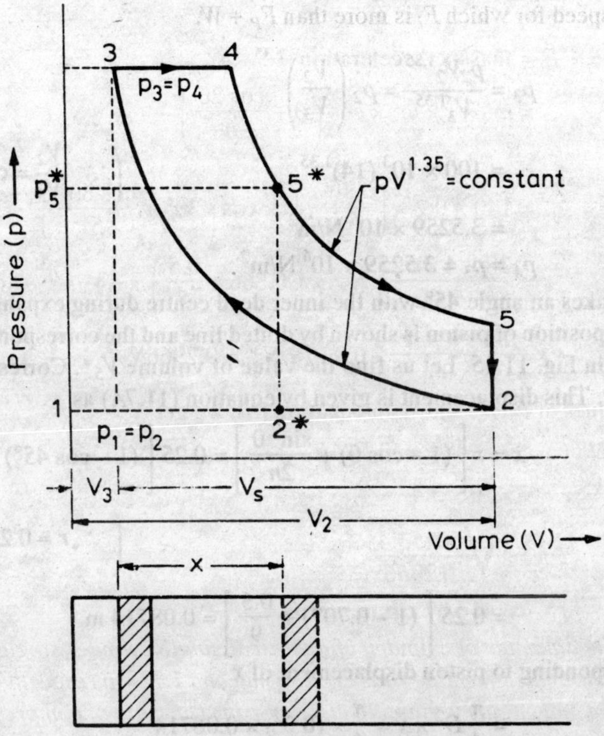

Fig. 11.15

Also compression ratio $= \dfrac{V_2}{V_3} = \dfrac{V_3 + V_s}{V_3} = 1 + \dfrac{V_s}{V_3}$

or $\quad 14 = 1 + \dfrac{V_s}{V_3} \qquad \left(\because \dfrac{V_2}{V_3} = 14\right)$

or $\quad \dfrac{V_s}{V_3} = 14 - 1 = 13$

$\therefore \quad V_3 = \dfrac{V_s}{13} = \dfrac{0.0353}{13} = 0.0027 \text{ m}^3$

Let us now find the volume at the end of the injection of oil [*i.e.* value of V_4].

Volume, $\quad V_4 = V_3 + \text{Volume between 3-4} \qquad ...(i)$

$= 0.0027 + \text{Volume between 3-4} \qquad (\because V_3 = 0.0027)$

But volume between 3-4 $= \dfrac{1}{10}$ of the swept volume

$= \dfrac{1}{10} \times V_s = \dfrac{0.0353}{10} = 0.00353 \text{ m}^3$

$\therefore \quad V_4 = 0.0027 + 0.00353 = 0.00623 \text{ m}^3$

Now let us find the value of p_4. The injection of oil is at constant pressure. Hence $p_3 = p_4$. The value of p_4 is known if value of p_3 is known. The value of p_3 is obtained from the law $pV^{1.35}$ = constant. The process 2-3 represents the compression,

$$p_2 V_2^{1.35} = p_3 V_3^{1.35}$$

or $$p_3 = \frac{p_2 V_2^{1.35}}{V_3^{1.35}} = p_2 \left(\frac{V_2}{V_3} \right)^{1.35}$$

$$= 100 \times 10^3 (14)^{1.35} \qquad \left(\because \frac{V_2}{V_3} = \text{compression ratio} = 14 \right)$$

$$= 3.5259 \times 10^6 \text{ N/m}^2$$

$$\therefore \quad p_4 = p_3 = 3.5259 \times 10^6 \text{ N/m}^2 .$$

When the crank makes an angle 45° with the inner dead centre during expansion stroke [process 4-5 represents expansion], the position of piston is shown by dotted line and the corresponding point 5* is marked on p-V diagram as shown in Fig. 11.15. Let us find the value of volume $V_5{}^*$. Corresponding to $\theta = 45°$, the displacement of piston is x. This displacement is given by equation (11.7A) as

$$x = r \left[(1 - \cos \theta) + \frac{\sin^2 \theta}{2n} \right] = 0.25 \left[(1 - \cos 45°) + \frac{\sin^2 45°}{2 \times 4.5} \right]$$

$$\left(\because r = 0.25, n = \frac{l}{r} = \frac{1.125}{0.2} = 4.5 \right)$$

$$= 0.25 \left[(1 - 0.707) + \frac{0.5}{9} \right] = 0.08714 \text{ m.}$$

$\therefore$ Volume corresponding to piston displacement of x

$$= \frac{\pi}{4} D^2 \times x = \frac{\pi}{4} \times (0.3^2) \times 0.08714 \qquad (\because x = 0.08714 \text{ m})$$

$$= 0.00616 \text{ m}^3.$$

$\therefore$ Volume $V_5{}^* = V_3$ + Volume corresponding to x displacement

$$= 0.0027 + 0.00616 = 0.00886 \text{ m}^3$$

Now the expansion between 4 and 5* follows the law $pV^{1.35}$ = constant,

$$\therefore \quad p_4 V_4^{1.35} = p_5{}^* V_5{}^{*1.35}$$

$$\therefore \quad p_5{}^* = p_4 \times \left(\frac{V_4}{V_5{}^*} \right)^{1.35}$$

$$= 3.5259 \times 10^6 \times \left(\frac{0.00623}{0.00886} \right)^{1.35}$$

$$[\because p_4 = 3.5259 \times 10^6 \text{ N/m}^2 ; V_4 = 0.00623 \text{ m}^3 \text{ and } V_5{}^* = 0.00886 \text{ m}^3]$$

$$= 2.192 \times 10^6 \text{ N/m}^2$$

Hence when the crank position is at an angle of 45° from I.D.C. during expansion stroke, the pressure on one side of the piston is $p_5{}^*$ and on the other side is $p_2{}^*$. But $p_2{}^*$ equal to suction pressure $p_2 = 100 \times 10^3$ N/m^2 or 0.1×10^6 N/m^2

$\therefore$ Difference of pressure on two sides of the piston,

$$p = p_5{}^* - p_2{}^* \text{ (or } p_2)$$

$$= (2.192 \times 10^6 - 0.1 \times 10^6) \text{ N/m}^2 = 2.092 \times 10^6 \text{ N/m}^2$$

$\therefore$ Force due to gas pressure on piston,

$$F_P = p \times A = p \times \frac{\pi}{4} D^2 = (2.092 \times 10^6) \times \left(\frac{\pi}{4} \times 0.3^2 \right) = 147874 \text{ N}$$

The net force on the piston for vertical engine is given as

$$F^* = F_P + F_i + W$$

(When $\theta = 45°$, piston is acceleration and moving downwards, hence inertia force F_i acts upwards but weight W is acting downwards)

where F_i = Inertia force due to mass of reciprocating parts

= – (Mass of reciprocating parts) × (Acceleration of piston)

$$= -(m_R) \times r \times \omega^2 \left[\cos\theta + \frac{\cos 2\theta}{n}\right] \qquad \left[\because \text{ Acceleration of piston} = r\omega^2\left(\cos\theta + \frac{\cos 2\theta}{n}\right)\right]$$

$$= -140 \times 0.25 \times (28.27)^2 \left[\cos 45° + \frac{\cos 90°}{4.5}\right] \qquad (\because\ m_R = 140,\ \omega = 28.27)$$

$$= -19776 \text{ N}$$

and W = Weight of reciprocating parts

$$= m_R \times g = 140 \times 9.81 = 1373.4 \text{ N}$$

Substituting the values of F_P, F_i and W, we get net force on piston as

$$F^* = F_P + F_i + W$$

$$= 147874 - 19776 + 1373.4 = 129471.4 \text{ N}$$

Torque on the crank-shaft

We know that torque (T) on crank-shaft is given by,

$$T = F_t \times r \qquad \text{...(ii)}$$

where F_t = Force normal to the crank at crank-pin

$$= F^* \frac{\sin(\theta + \phi)}{\cos\phi} \qquad \text{[Refer to equation 11.28A]}$$

The value of ϕ (*i.e.* angle of inclination of the connecting rod with the line of stroke) is obtained from relation,

$$l \sin\phi = r \sin\theta$$

or

$$\sin\phi = \frac{r}{l}\sin\theta = \frac{\sin\theta}{n} \qquad \left(\because\ n = \frac{l}{r}\right)$$

$$= \frac{\sin 45°}{4.5} = 0.1571 \qquad (\because\ n = 4.5)$$

$$\therefore \quad \phi = \sin^{-1} 0.1571 = 9.04°$$

$$\therefore \quad F_t = F^* \frac{\sin(\theta + \phi)}{\cos\phi} = 129471.4 \frac{\sin(45 + 9.04)}{\cos(9.04°)}$$

$$= 129471.4 \times \frac{0.8094}{0.9876} = 106112.2 \text{ N}$$

$\therefore$ Torque, $T = F_t \times r = 106112.2 \times 0.25 =$ **26528.05 Nm. Ans.**

Problem 11.19. *A horizontal double acting steam engine has cylinder diameter 220 mm, length of stroke 360 mm and runs at 240 r.p.m. The length of connecting rod is 720 mm. The mass of reciprocating parts is 100 kg. Steam is admitted at a pressure of 6 bar gauge and cut off takes place at $\frac{1}{4}$th of the stroke. After this, expansion takes place according to the law pV = constant. The exhaust is at a pressure of – 0.80 bar gauge. Determine the torque exerted on the crank-shaft when crank has turned through 75° from the inner dead centre. Neglect the effect of clearance and assume atmospheric pressure as 1 bar.*

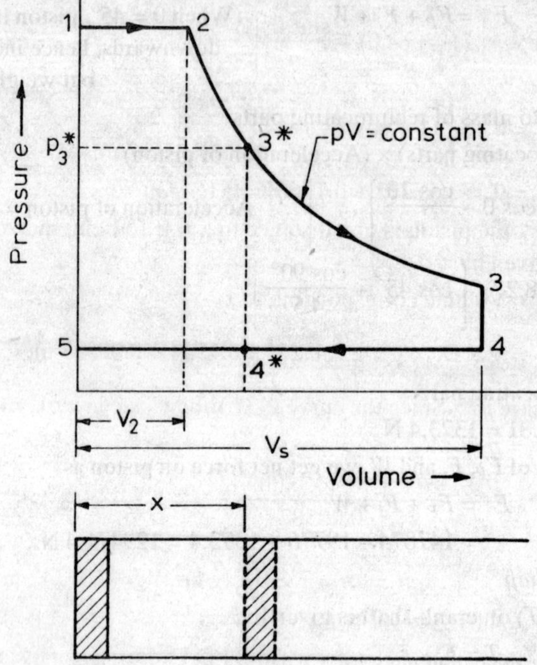

Fig. 11.16

Sol. Given :

D = 220 mm = 0.22 m ; stroke length L = 360 mm = 0.36 m or $r = \frac{L}{2} = \frac{0.36}{2}$ = 0.18 m ; N = 240 r.p.m. or $\omega = \frac{2\pi N}{60} = \frac{2\pi \times 240}{60} = 8\pi$ rad/s ; connecting rod length l = 720 mm = 0.72 m ; m_R = 100 kg ; steam admission pressure p_1 = 6 + 1 = 7 bar* = 7×10^5 = N/m² (abs.) ; exhaust pressure p_5 = – 0.80 + 1.0 = 0.2 bar = 0.2×10^5 N/m² (abs.) ; $\theta = 75°$; steam cut off = at $\frac{1}{4}$ th stroke ; law of expansion pV = constant ; $p_1 = p_2 = 7 \times 10^5$ N/m².

Fig. 11.16 shows the pressure volume (*i.e. p-V*) diagram for a steam engine neglecting clearance in which,

1-2 represents the admission of steam,
2-3 represents the expansion, and
4-5 represents the exhaust of steam.

Point 2 represents the cut off point *i.e.* the steam is cut off at 2. Let us first find the stroke volume and volume of steam at cut off.

Stroke volume, $V_s = \frac{\pi}{4} D^2 \times L$

$$= \frac{\pi}{4} (.22^2) \times 0.36 = 0.01368 \text{ m}^3$$

Since the steam is cut off at $\frac{1}{4}$ th of the stroke, hence volume of steam at cut off,

$$V_2 = \frac{1}{4} \text{ th of stroke volume} = \frac{1}{4} \times 0.01368 = 0.00342 \text{ m}^3$$

*One bar = 1×10^5 N/m².

Corresponding to $\theta = 75°$ from I.D.C. the piston displacement (x) is given by equation (11.7A)

$$x = r\left[(1 - \cos\theta) + \frac{\sin^2\theta}{2n}\right]$$

$$= 0.18\left[(1 - \cos 75°) + \frac{\sin^2 75°}{2 \times 4}\right] \qquad \left(\because n = \frac{l}{r} = \frac{0.72}{0.18} = 4\right)$$

$$= 0.18\,[0.7412 + 0.1166] = 0.1544 \text{ m}$$

Corresponding to $\theta = 75°$, the piston is at a distance of $x = 0.1544$ m, the volume of point 3* which lies on expansion curve 2-3, is given by

$$V_3^* = \text{Volume corresponding to } x$$

$$= \frac{\pi}{4} \times D^2 \times x = \frac{\pi}{4} \times 0.22^2 \times 0.1544 = 0.00587 \text{ m}^3$$

Now let us find the pressure p_3^*. Since the curve 2-3* follows the law pV = constant, hence

$$p_2V_2 = p_3^* \times V_3^*$$

$$\therefore \quad p_3^* = \frac{p_2V_2}{V_3^*} = \frac{(7 \times 10^5) \times 0.00342}{0.00587}$$

$$[\because p_2 = p_1 = 7 \times 10^5 \text{ N/m}^2,\ V_2 = 0.00342 \text{ m}^3 \text{ and } V_3^* = 0.00587 \text{ m}^3]$$

$$= 4.0783 \times 10^5 \text{ N/m}^2$$

Hence when crank position is at an angle of 75° from I.D.C. during expansion stroke, the pressure on one side of the piston is p_3^* and on the other side is p_4^* which is exhaust pressure (*i.e.* 0.2×10^5 N/m^2).

$\therefore$ Difference of pressure on two sides of the piston,

$$p = p_3^* - p_4^*$$

$$= (4.0783 \times 10^5 - 0.2 \times 10^5) \text{ N/m}^2 = 3.8783 \times 10^5 \text{ N/m}^2$$

$\therefore$ Force due to steam on piston,

$$F_P = p \times A = p \times \frac{\pi}{4} D^2$$

$$= 3.8783 \times 10^5 \times \frac{\pi}{4}(.22^2) \text{ N} = 14742 \text{ N}$$

Acceleration of the reciprocating parts

$$= \omega^2 r\left(\cos\theta + \frac{\cos 2\theta}{n}\right)$$

$$= (8\pi)^2 \times 0.18\left(\cos 75° + \frac{\cos 150°}{4}\right) \qquad \left(\because n = \frac{l}{r} = \frac{0.72}{0.18} = 4\right)$$

$$= 113.7\,[0.2588 + (-0.2165)] = 4.808 \text{ m/s}^2$$

$\therefore$ Inertia force on the reciprocating parts,

$$F_i = -m_R \times \text{Acceleration of reciprocating parts}$$

$$= -100 \times 4.808 = -480.8 \text{ N}$$

The net force on the piston for horizontal engine is given by,

$$F^* = F_P + F_i = 14742 - 480.8 = 14261.2 \text{ N}$$

Torque exerted on the crank-shaft is given by,

$$T = F_t \times r \qquad ...(i)$$

where F_t = Force normal to the crank at the crank-pin,

$$= F^* \frac{\sin(\theta + \phi)}{\cos \phi} \qquad \text{where } \sin\phi = \frac{r}{l}\sin\theta = \frac{\sin\theta}{n} \qquad \left(\because\ n = \frac{l}{r} = 4\right)$$

$$= \frac{\sin 75°}{4} = 0.2415$$

$$= 14261.2 \frac{\sin(75 + 13.974)}{\cos(13.974)} \qquad \therefore\ \phi = \sin^{-1} 0.2415 = 13.974°$$

$$= 14261.2 \times \frac{0.9998}{0.9704} = 14693.18 \text{ N}$$

$\therefore$ $T = 14693.18 \times 0.18 =$ **2644.77 Nm. Ans.**

11.8. Dynamically Equivalent System

In previous articles we have calculated the piston effort, crank-effort and torque on the crank-shaft without considering the weight of the connecting rod. But the weight of connecting rod and the inertia of the connecting rod is to be taken into account in actual practice. As the motion of the connecting rod is not linear, hence it is somewhat difficult to find the inertia of the connecting rod.

To find the inertia of the connecting rod (or any rigid body) it is convenient to replace the rigid body by two masses assumed to be concentrated at points and connected rigidly together. The two-mass system will be dynamically equivalent to the rigid body if :

(*i*) The total mass of the two masses is equal to the mass of the rigid body ;

(*ii*) The C.G. of the two masses coincides with that of the rigid body ; and

(*iii*) The total moment of inertia of the two masses about an axis passing through the C.G. is equal to that of the rigid body.

Fig. 11.17 shows a rigid body having its centre of gravity at G.

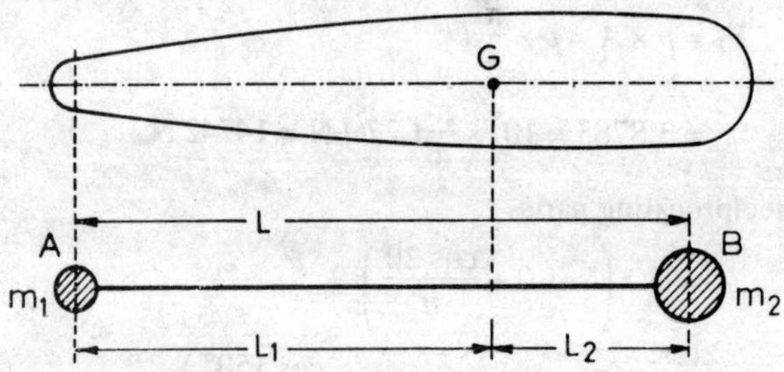

Fig. 11.17

Let m = Mass of the rigid body

k = Radius of gyration of the rigid body about an axis through G

m_1 and m_2 = Two masses which form an equivalent dynamical system and connected rigidly together.

L = Distance between two masses, m_1 and m_2,

L_1 = Distance of mass m_1 from G,

L_2 = Distance of mass m_2 from G.

For the two masses to be dynamically equivalent, we will get the three equations which are given below :

$$\left.\begin{aligned} &m_1 + m_2 = m && \text{...}(i)\\ &m_1 \times L_1 = m_2 \times L_2 && \text{...}(ii)\\ \text{and}\quad &m_1(L_1)^2 + m_2(L_2)^2 = m \times k^2 && \text{...}(iii)\end{aligned}\right] \quad \text{...(11.32)}$$

From equations (*i*) and (*ii*), find the values of m_1 and m_2. Then substitute the values of m_1 and m_2 in equation (*iii*).

Now from equation (*i*), $m_1 = m - m_2$

And from equation (*ii*), $$m_1 = \frac{m_2 \times L_2}{L_1} \quad \text{...}(iv)$$

Equating the two values of m_1, we get

$$m - m_2 = \frac{m_2 \times L_2}{L_1}$$

or $$m = \frac{m_2 \times L_2}{L_1} + m_2 = m_2\left[\frac{L_2}{L_1} + 1\right] = m_2\left[\frac{L_2 + L_1}{L_1}\right]$$

$\therefore$ $$m_2 = \frac{mL_1}{L_1 + L_2} \quad \text{...(11.32}A\text{)}$$

Substituting the value of m_2 in equation (*iv*), we get

$$m_1 = \frac{m_2 \times L_2}{L_1}$$

$$= \frac{mL_1}{L_1 + L_2} \times \frac{L_2}{L_1} = \frac{mL_2}{(L_1 + L_2)} \quad \text{...(11.32}B\text{)}$$

Now substituting the values of m_1 and m_2 from equations (11.32 *B*) and (11.32 *A*) into equation (*iii*), we get

$$\left[\frac{mL_2}{(L_1 + L_2)}\right] \times L_1^2 + \left[\frac{mL_1}{(L_1 + L_2)}\right] \times L_2^2 = mk^2$$

or $$\frac{mL_2 \times L_1^2 + mL_1L_2^2}{(L_1 + L_2)} = mk^2$$

or $$\frac{mL_1L_2\,(L_1 + L_2)}{(L_1 + L_2)} = mk^2$$

or $$mL_1L_2 = mk^2$$

or $$k^2 = L_1 \times L_2 \quad \text{...(11.33)}$$

From equation (11.33) it is clear that either L_1 or L_2 may be chosen arbitrarily. If L_1 is chosen then L_2 will be obtained from the above equation. The value of '*k*' already is known or given.

Note. (*i*) Either L_1 or L_2 may be chosen arbitrarily but not both.

(*ii*) Sometimes the radius of gyration *k* is unknown. Then if L_1 is chosen ; L_2 can not be obtained. To find the value L_2 the body is suspended vertically, so that it is free to swing about the axis passing through *A*. The length *L* of the simple pendulum which has the same period of oscillation as the body will then be equal to $(L_1 + L_2)$. Hence length L_2 [which is equal to $(L - L_1)$] is known and the second mass can be placed at a distance L_2 from the C.G. of the body.

Problem 11.20. *A connecting rod of an engine is 225 mm long between its centres. The mass of the connecting rod is 1.2 kg. The distance of the centre of gravity from the small end is 150 mm and radius of* *gyration about an axis through the centre of gravity is 95 mm. Find the equivalent dynamical system if one of the masses coincides in position with the small end centre.*

Sol. Given : $L = 225$ mm ; $m = 1.2$ kg ; $L_1 = 150$ mm ; $k = 95$ mm.

Find equivalent dynamical system. It means to find the values of m_1, m_2 and L_2.

Fig. 11.18 (*a*) shows the given connecting rod whereas Fig. 11.18 (*b*) shows the equivalent dynamical system.

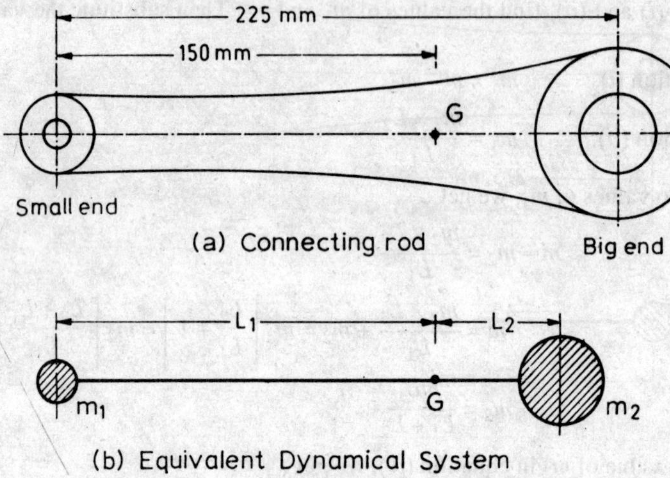

Fig. 11.18

We know that for equivalent dynamical system,

$$k^2 = L_1 \times L_2$$

or $$95^2 = 150 \times L_2$$

$\therefore$ $$L_2 = \frac{95^2}{150}$$

$$= \frac{95 \times 95}{150} = \mathbf{60.16\ mm.}\quad \textbf{Ans.}$$

The value of mass m_1 is given by equation (11.32 *B*) as

$$m_1 = \frac{m \times L_2}{(L_1 + L_2)}$$

$$= \frac{1.2 \times 60.16}{(150 + 60.16)} = \mathbf{0.3435\ kg.}\quad \textbf{Ans.}$$

The value of mass m_2 is obtained from

$$m_1 + m_2 = m$$

or $$0.3435 + m_2 = 1.2$$

$\therefore$ $$m_2 = 1.2 - .3435 = \mathbf{0.8565\ kg.}\quad \textbf{Ans.}$$

Hence the mass m_1 (= 0.3435 kg) must be placed at the centre of small end at a distance of 150 mm from the centre of gravity. The mass m_2 (= 0.8565) must be placed at a distance of 60.16 mm from the centre of gravity.

Problem 11.21. *The length of a connecting rod of an engine is 500 mm measured between the centres and its mass is 18 kg. The centre of gravity is 125 mm from the crank-pin centre and crank radius is 100 mm.*

Determine the dynamically equivalent system keeping one mass at the small end. The frequency of oscillations of the rod when suspended from the centre of the small end is 43 vibrations per minute.

(A.M.I.E., Summer 1980)

Sol. Given : L = 500 mm ; m = 18 kg ; Distance of C.G. from big end = 125 mm

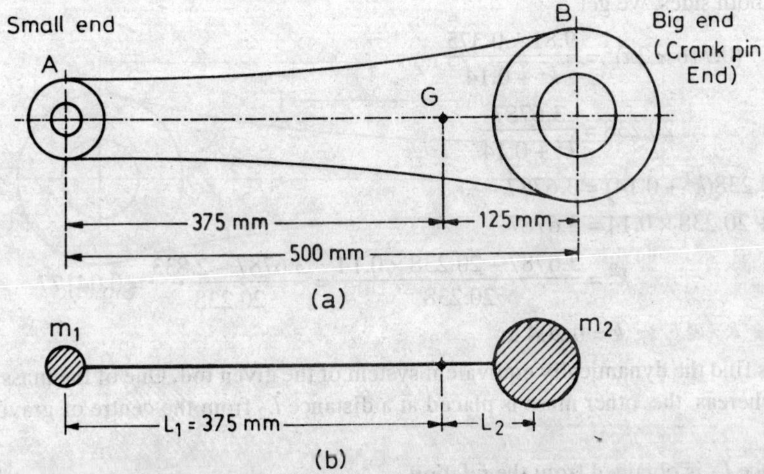

Fig. 11.19

$\therefore$ Distance of C.G. from small end = 500 − 125 = 375 mm

$\therefore$ L_1 = 375 mm

Crank radius*, r = 100 mm

No. of vibrations of the rod when suspended from end A = 43 vibrations/min

$$= \frac{43}{60} \text{ vibration/second} = 0.716 \text{ vibrations/second} = 0.716 \text{ cycles/s.}$$

Let us first find the value of radius of gyration, k.

Let k = radius of gyration of connecting rod (considering it a compound pendulum) about an axis passing through G.

We know that, the frequency of oscillation,

$$n = \frac{1}{2\pi}\sqrt{\frac{g \times h}{k^2 + h^2}}$$

where h = Distance between C.G. and point of suspension

$= L_1$ = 375 mm = 0.375 m

$$= \frac{1}{2\pi}\sqrt{\frac{g \times L_1}{k^2 + {L_1}^2}},$$

*In this problem, the data of crank radius is superflux.

$$\therefore \quad 0.716 = \frac{1}{2\pi}\sqrt{\frac{9.81 \times 0.375}{k^2 + 0.375^2}}$$

or
$$0.716 \times 2\pi = \sqrt{\frac{9.81 \times 0.375}{k^2 + 0.14}}$$

Squaring both sides, we get

$$(0.716 \times 2\pi)^2 = \frac{9.81 \times 0.375}{k^2 + 0.14}$$

or
$$20.238 = \frac{3.6787}{k^2 + 0.14}$$

or
$$20.238(k^2 + 0.14) = 3.6787$$

or
$$20.238k^2 + 20.238 \times 0.14 = 3.6787$$

or
$$k^2 = \frac{3.6787 - 20.238 \times 0.14}{20.238} = \frac{3.6787 - 2.833}{20.238} = 0.04177$$

$$\therefore \quad k = 0.204 \text{ m}$$

Now let us find the dynamically equivalent system of the given rod. One of the mass is to be placed at the small end whereas the other mass is placed at a distance L_2 from the centre of gravity G as shown in Fig. 11.19 (*b*).

The distance L_2 is obtained from the relation,

$$L_1 \times L_2 = k^2$$

$$\therefore \quad L_2 = \frac{k^2}{L_1} = \frac{0.04177}{0.375} \qquad (\because \; k^2 = 0.04177 \text{ and } L_1 = 0.375 \text{ m})$$

$$= 0.111 \text{ m}$$

Now let m_1 = mass placed at the small end A

m_2 = mass placed at a distance L_2 from G.

We know that the value of m_1 from equation (11.32*B*) is given by,

$$m_1 = \frac{m \times L_2}{L_1 + L_2}$$

$$= \frac{18 \times 0.111}{0.375 + 0.111} \qquad (\because \; m = 18 \text{ kg})$$

$$= \mathbf{4.11 \text{ kg.}} \quad \textbf{Ans.}$$

and
$$m_2 = m - m_1 = 18 - 4.11 = \mathbf{13.89 \text{ kg.}} \quad \textbf{Ans.}$$

Problem 11.22. *The following data relate to a connecting rod of a reciprocating engine :*

Mass = 55 kg ; distance between bearing centres = 850 mm ; diameter of small end bearing = 75 mm ; diameter of big end bearing = 100 mm ; time of oscillations when the connecting rod is suspended from the top of small end = 1.83 seconds ; time of oscillation when the connecting rod is suspended from the top end of big end = 1.68 second, determine :

(i) the radius of gyration of the rod about an axis passing through the centre of gravity perpendicular to the plane of oscillation ;

(ii) the moment of inertia of the rod about the same axis ;

(iii) the dynamically equivalent system for the connecting rod, constituted of two masses one of which is situated at the small end centre. (AMIE, Summer 1976)

Sol. Given : $m = 55$ kg ; $l = 850$ mm $= 0.85$ m ; $d_1 = 75$ mm $= 0.075$ m ;

$d_2 = 100$ mm $= 0.1$ m ; $t_1 = 1.83$ s ; $t_2 = 1.68$ s.

Refer to Fig. 11.20.

(i) The radius of gyration of the rod about on axis passing through G and perpendicular to the plane of oscillation

Let k = Radius of gyration of the rod.

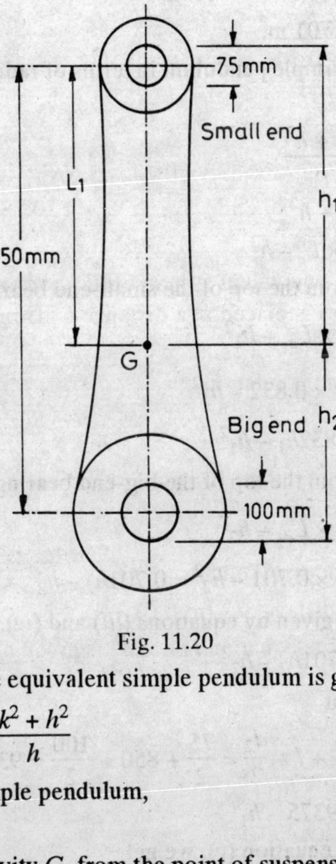

Fig. 11.20

We know that the length of the equivalent simple pendulum is given by,

$$L_e = \frac{k^2 + h^2}{h} \quad ...(i)$$

where L_e = Length of equivalent simple pendulum,

k = Radius of gyration and

h = Distance of centre of gravity G, from the point of suspension.

Let h_1 = Distance of G from the top of small-end bearing (*i.e.* point of suspension of the rod),

h_2 = Distance of G from the top of big-end bearing (*i.e.* point of suspension of the rod),

L_{e_1} = Length of equivalent simple pendulum when suspended from the top of small-end bearing

L_{e_2} = Length of equivalent simple pendulum when suspended from the top of big-end bearing.

The values of L_{e_1} and L_{e_2} are obtained from the time of oscillation. Time of oscillation of equivalent simple pendulum when suspended from small end is given by

$$t_1 = 2\pi\sqrt{\frac{L_{e_1}}{g}} \text{ or squaring both sides } t_1^2 = 4\pi^2 \times \frac{L_{e_1}}{g}$$

$\therefore \qquad L_{e_1} = \dfrac{t_1^2 \times g}{4\pi^2} = \dfrac{(1.83)^2 \times 9.81}{4 \times \pi^2} \qquad (\because \; t_1 = 1.83 \text{ s})$

$= 0.832 \text{ m}$

Similarly $\qquad t_2 = 2\pi \sqrt{\dfrac{L_{e_2}}{g}}$ or $t_2^2 = 4\pi^2 \times \dfrac{L_{e_2}}{g}$

$\therefore \qquad L_{e_2} = \dfrac{t_2^2 \times g}{4 \times \pi^2} = \dfrac{1.68^2 \times 9.81}{4 \times \pi^2} \qquad (\because \; t_2 = 1.68 \text{ s})$

$= 0.701 \text{ m.}$

The length of the equivalent simple pendulum in terms of radius of gyration is also given by equation (*i*) as

$$L_e = \frac{k^2 + h^2}{h}$$

or $\qquad h \times L_e = k^2 + h^2$

or $\qquad k^2 = h \times L_e - h^2 \qquad \ldots(ii)$

When the rod is suspended from the top of the small-end bearing, the above equation becomes as

$k^2 = h_1 \times L_{e_1} - h_1^2 \qquad (\because \; h = h_1 \text{ and } L_e = L_{e_1})$

$= h_1 \times 0.832 - h_1^2 \qquad (\because \; L_{e_1} = 0.832)$

$= 0.832 h_1 - h_1^2 \qquad \ldots(iii)$

When the rod is suspended from the top of the big-end bearing, the equation (*ii*) becomes as

$k^2 = h_2 \times L_{e_2} - h_2^2 \qquad (\because \; h = h_2 \text{ and } L_e = L_{e_2})$

$= h_2 \times 0.701 - h_2^2 = 0.701 h_2 - h_2^2 \qquad \ldots(iv)$

Equating the two values of k^2 given by equations (*iii*) and (*iv*), we have

$0.832 h_1 - h_1^2 = 0.701 h_2 - h_2^2 \qquad \ldots(v)$

From Fig. 11.20, it is clear that

$$h_1 + h_2 = \frac{d_1}{2} + l + \frac{d_2}{2} = \frac{75}{2} + 850 + \frac{100}{2} = 937.5 \text{ mm} = 0.9375 \text{ m}$$

$\therefore \qquad h_2 = 0.9375 - h_1 \qquad \ldots(vi)$

Substituting the value of h_2 in equation (*v*), we get

$0.832 h_1 - h_1^2 = 0.701(0.9375 - h_1) - (0.9375 - h_1)^2$

$= 0.701 \times 0.9375 - 0.701 h_1 - (0.9375^2 + h_1^2 - 2 \times 0.9375 \times h_1)$

$= 0.6571 - 0.701 h_1 - 0.879 - h_1^2 + 1.875 h_1$

or $\qquad 0.832 h_1 - h_1^2 + 0.701 h_1 + h_1^2 - 1.875 h_1 = 0.6571 - 0.879$

or $\qquad -0.342 h_1 = -0.2219$

or $\qquad h_1 = \dfrac{0.2219}{0.342} = 0.6488 \simeq 0.65 \text{ m}$

Substituting the value of h_1 in equation (*iii*), we get

$k^2 = 0.832 h_1 - h_1^2$

$$= 0.832 \times 0.65 - 0.65^2 = 0.5408 - 0.4225 = 0.1183$$

$$\therefore \quad k = \sqrt{0.1183} = 0.3439 \simeq \mathbf{0.344\ m.} \quad \textbf{Ans.}$$

(ii) Moment of inertia of the rod

Let I = Moment of inertia of the rod

We know that moment of inertia is given by,

$$I = mk^2$$

$$= 55 \times 0.1183 \qquad (\because\ k^2 = 0.1183)$$

$$= \mathbf{6.5065\ kg\ m^2.} \quad \textbf{Ans.}$$

(iii) Dynamically equivalent system for the connecting rod

Let m_1 = one of the mass situated at the centre of small end bearing

m_2 = magnitude of second mass

L_1 = Distance of mass m_1 from the centre of gravity, G

$$= h_1 - \frac{d_1}{2} = 0.65 - \frac{0.075}{2} = 0.6125 \text{ m}$$

L_2 = Distance of mass m_2 from the G

For a dynamically equivalent system, we have

$$L_1 \times L_2 = k^2$$

$$\therefore \quad L_2 = \frac{k^2}{L_1} = \frac{0.1183}{0.6125} \qquad (\because\ k^2 = 0.1183 \text{ and } L_1 = 0.6125)$$

$$= 0.193 \text{ m}$$

The mass (m_1) and mass (m_2) are obtained from equations (11.32*A*) and (11.32*B*)

$$m_1 = \frac{L_2}{(L_1 + L_2)} \times m = \frac{0.193}{(0.6125 + 0.193)} \times 55 \qquad (\because\ m = 55)$$

$$= \mathbf{13.18\ kg.} \quad \textbf{Ans.}$$

and $$m_2 = \frac{L_1}{(L_1 + L_2)} \times m = \frac{0.6125}{(0.6125 + 0.193)} \times 55 = \mathbf{41.82\ kg.} \quad \textbf{Ans.}$$

11.9. Magnitude of Correction Couple Applied to Two Mass System which is Dynamically Equivalent to Rigid Body

The two-mass system will be dynamically equivalent to a rigid body if the three conditions given by equations (*i*), (*ii*) and (*iii*) are satisfied by two-mass system. Also the two mass system and rigid body should have same acceleration. In this case the position of one mass is fixed and the position of second mass is obtained by the three equations (*i*), (*ii*) and (*iii*) of equation (11.32).

But in many cases, the position of the second mass is fixed at some convenient point (position of first mass is already fixed), then all the three conditions given by (*i*), (*ii*), (*iii*) of equation (11.32) are not satisfied. Only the first two conditions [*i.e.* sum of two masses is equal to the mass of rigid body and the C.G. of the two masses coincides with the C.G. of the rigid body] are satisfied. Hence the two-mass system is not dynamically equivalent to rigid body. For making this system dynamically equivalent, the third condition (*i.e.* the moment of inertia of the two masses about an axis passing through G should be same as the moment of inertia of the rigid body about the axis passing through G) should be satisfied. The third condition will be satisfied, if a couple is applied to the two-mass system. This couple is known as *correction couple*. It is represented by T_c. The magnitude of correction couple is obtained as given below :

Fig. 11.21 shows a rigid body and two masses m_1 and m_3 which are placed at a distance L_1 and L_3 from the centre of gravity (G). The distances L_1 and L_3 are *quite arbitrarily*. Hence the two mass system (*i.e.* mass m_1 and mass m_3 connected rigidly to-gether) will not be dynamically equivalent to the given rigid body. Let the two mass system (m_1 and m_2) is dynamically equivalent to the rigid body.

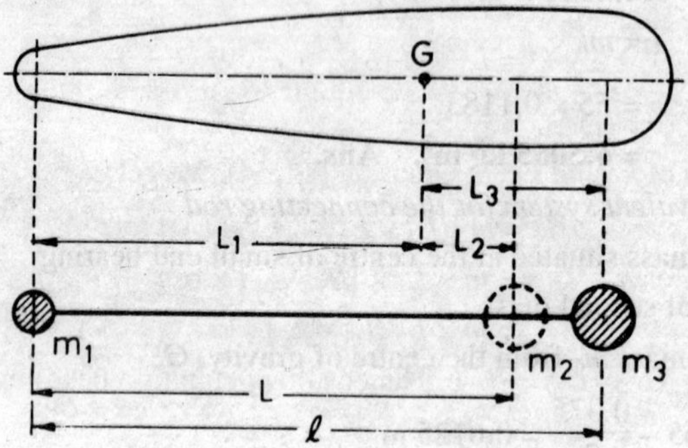

Fig. 11.21

Let I_1 = New mass moment of inertia of masses (m_1 and m_3 which are not dynamically equivalent to rigid body)

k_1 = New radius of gyration

α = Angular acceleration of the rigid body

I = Mass moment of inertia of masses (m_1 and m_2 which are dynamically equivalent to rigid body)

k = Radius of gyration of dynamically equivalent system

m = Mass of rigid body

Now we have $\quad I = mk^2$

and $\quad I_1 = mk_1^2$

where k = radius of gyration of dynamically equivalent system and

k_1 = radius of gyration of two mass system placed arbitrarily.

Also the values of k and k_1 are different.

We know that the torque required to accelerate the rigid body,

$$T = I \times \alpha = mk^2 \times \alpha \quad ...(i)$$

Similarly the torque required to accelerate the two mass system placed arbitrarily,

$$T_1 = I_1 \times \alpha = mk_1^2 \times \alpha \quad ...(ii)$$

$\therefore$ Difference between the torque required to accelerate the two mass system placed arbitrarily and that required to accelerate the rigid body is given by,

$$T_c = T_1 - T = mk_1^2 \times \alpha - mk^2 \times \alpha$$
$$= m(k_1^2 - k^2) \times \alpha \quad ...(11.34)$$

The difference of the torques (*i.e.* $T_1 - T$) is known as *correction couple*. It is represented by T_c. This couple must be applied to the two mass system in which the two masses are placed arbitrarily. Then the two-mass system will be dynamically equivalent to the rigid body.

Let L = Distance between the two masses m_1 and m_2 which form a true dynamically equivalent system (or = $L_1 + L_2$)

l = Distance between the two masses m_1 and m_3 which are placed arbitrarily (or = $L_1 + L_3$)

then
$$(l - L) = (L_1 + L_3) - (L_1 + L_2)$$
$$= L_3 - L_2 \quad ...(iii)$$

Also we know that,
$$k^2 = L_1 \times L_2$$
and
$$k_1^2 = L_1 \times L_3$$

Substituting the values of k^2 and k_1^2 in equation (11.34), we get
$$T_c = m\,[(L_1 \times L_3) - (L_1 \times L_2)] \times \alpha$$
$$= m \times L_1 \times [L_3 - L_2] \times \alpha$$
$$= m \times L_1\,(L_3 - L_2) \times \alpha$$

But from equation (*iii*), $(L_3 - L_2) = l - L$. Hence above equation becomes as
$$T_c = m \times L_1\,(l - L) \times \alpha \quad ...(11.35)$$

where L = Distance between two masses which are dynamically equivalent to the rigid body. It is also the length of the **equivalent** simple pendulum when the body is suspended from an axis which passes through the position of mass m_1.
$$= \frac{k^2 + L_1^2}{L_1}.$$

Problem 11.23. *A small connecting rod 220 mm long between centres, has a mass of 2 kg and moment of inertia of 2 × 10^4 kg mm^2 about its centre of gravity. Centre of gravity is located at a distance of 150 mm from the small end centre. Determine the dynamically two mass system when one mass is located at the small end centre.*

If the connecting rod is replaced by two masses located at the two centres, find the correction couple that must be applied for complete dynamical equivalence of the system, when the angular acceleration of the connecting rod is 20000 rad/s^2 clockwise. (A.M.I.E., Winter 1985)

Sol. Given : Distance between centres, l = 220 mm

Mass, m = 2 kg ; Moment of inertia, $I = 2 \times 10^4$ kg mm^2,

Distance of C.G. from the small end centre L_1 = 150 mm

∴ Distance of C.G. of rod from big end centre = 220 – 150 = 70 mm

Let m_1 = First mass which is placed at small end centre

m_2 = Second mass which is to be placed in such a way that the system becomes dynamically equivalent.

But distance of G from small end centre = 150 mm

∴ L_1 = 150 mm.

Let L_2 = Distance of mass m_2 from the centre of gravity of the rod.

Fig. 11.22 (*a*) shows the given connecting rod whereas Fig. 11.22 (*b*) shows the equivalent dynamical system of two masses m_1 and m_2.

We know that for equivalent dynamical system,
$$k^2 = L_1 \times L_2 \quad ...(i)$$

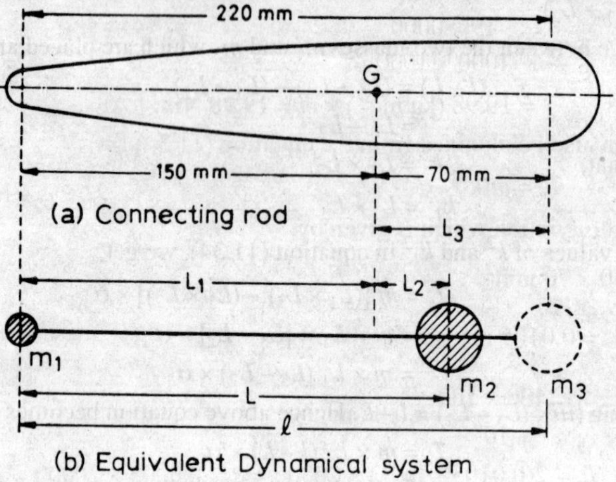

Fig. 11.22

Let us find the value of 'k' from mass moment of inertia of the connecting rod.

$$I = mk^2$$

or $$2 \times 10^4 = 2 \times k^2 \qquad (\because\ I = 2 \times 10^4 \text{ kg mm}^2 \text{ and } m = 2 \text{ kg})$$

$$\therefore \quad k = 10^2 = 100 \text{ mm}$$

Substituting the value of k in equation (i), we get

$$100^2 = L_1 \times L_2$$

$$= 150 \times L_2 \qquad (\because\ L_1 = 150 \text{ mm})$$

$$\therefore \quad L_2 = \frac{100^2}{150} = \mathbf{66.67\ mm.}\ \ \textbf{Ans.}$$

The value of m_1 is given by equation (11.32B) as

$$m_1 = \frac{m \times L_2}{(L_1 + L_2)} = \frac{2 \times 66.67}{(150 + 66.67)} = \mathbf{0.615\ kg.}\ \ \textbf{Ans.}$$

The value of m_2 is obtained from,

$$m_1 + m_2 = m$$

or $$m_2 = m - m_1 = 2 - 0.615 = \mathbf{1.385\ kg.}\ \ \textbf{Ans.}$$

Correction Couple (T_c)

Angular acceleration of connecting rod = 20000 rad/s^2 clockwise

$\therefore \quad \alpha = 20000$ rad/s^2

When the two masses are located at the two centres of the connecting rod (*i.e.* mass m_1 at the small end centre and mass m_3 at the big end centre) then for complete dynamical equivalence of the system, the correction couple (T_c) that must be applied is given by equation (11.35) as

$$T_c = m \times L_1 \times (l - L) \times \alpha$$

where L_1 = 150 mm

l = Distance between two centres = 220 mm

L = Distance between two masses which are dynamically equivalent to the given rigid body

$= L_1 + L_2 = 150 + 66.67 = 216.67$ mm

$\therefore \quad T_c = 2 \times 150 \times (220 - 216.67) \times 20000 = 19980000 \text{ kg mm}^2/\text{s}^2$

$$= \frac{19980000}{1000 \times 1000} \text{ kg m}^2/\text{s}^2 \qquad \left(\because \ \text{kg} \times \frac{m}{s^2} = N\right)$$

$$= 19.98 \text{ (kg m/s}^2) \times m = \mathbf{19.98 \text{ Nm.}} \quad \textbf{Ans.}$$

The value of T_c can also be obtained from the equation (11.34) as

$$T_c = m(k_1^2 - k^2) \times \alpha$$

where k_1 = Now radius of gyration which is given by,

$$k_1^2 = L_1 \times L_3 = 150 \times 70 \text{ mm}^2 \qquad (\because \ L_3 = 220 - 150 = 70)$$

$$= \frac{150 \times 70}{10^6} \text{ m}^2 = 0.0105 \text{ m}^2$$

and $k^2 = 100 \times 100 \text{ mm}^2 = \dfrac{100 \times 100}{10^6} = 0.01 \text{ m}^2$

$\therefore \quad T_c = 2(0.0105 - 0.01) \times 20000 = 2 \times 0.0005 \times 20000 = \mathbf{20 \text{ Nm.}}$ **Ans.**

This correction couple has the same sense as the angular acceleration and therefore acts clockwise.

11.10. Torque Exerted on the Crank-shaft, Considering the Weight of the Connecting Rod

Fig. 11.23 (a) shows the line diagram of a reciprocating engine in which OA is the crank and AB is the connecting rod. The crank is rotating clockwise with an uniform angular velocity ω rad/s. Let the crank makes an angle θ with the line of stroke BO. The C.G. of the connecting rod lies at point G. As the weight of the connecting rod is not negligible, hence there will be the inertia force of the connecting rod in addition to the inertia force of the reciprocating masses. The effect of the inertia of the connecting rod on the crank-shaft torque may be obtained by :

(i) Graphical Method and

(ii) Analytical Method.

(i) **Graphical Method.**The inertia forces may be obtained graphically as given below :

(i) The first step is to draw the acceleration diagram for the mechanism. The Klein's construction may be used and quadrilateral $OAHF$ may be obtained. This quadrilateral is the acceleration diagram. The same quadrilateral is shown in Fig. 11.23 (b).

The acceleration of piston B with respect to O is

$$f_{BO} = \omega^2 \times FO$$

and acting in the direction from F to O. The inertia force F_i of the reciprocating parts therefore acts in the opposite direction as shown in Fig.11.23 (b).

(ii) The second step is to replace the connecting rod by dynamically equivalent system of two masses as explained in Article 11.8. Let one of the masses is placed arbitrarily at B. To find the position of the second mass, draw GL perpendicular to BA and take $GL = k$ where k is the radius of gyration of the connecting rod. Join BL. From point L, draw LJ perpendicular to BL. The line LJ meets the line BA at point J. Then point J gives the position* of the second mass.

(iii) The third step is to locate the points G and J on FA, which is the acceleration image of the connecting rod. This is done by drawing parallel lines from G and J parallel to line of stroke BO. The parallel line from J intersects the line FA at j whereas the parallel line from G intersects the line FA at g. Now join j to O and point g to O. The jO and gO are proportional to the acceleration of J and G respectively.

*The position of second mass is also given by equation, $GB \times GJ = k^2$.

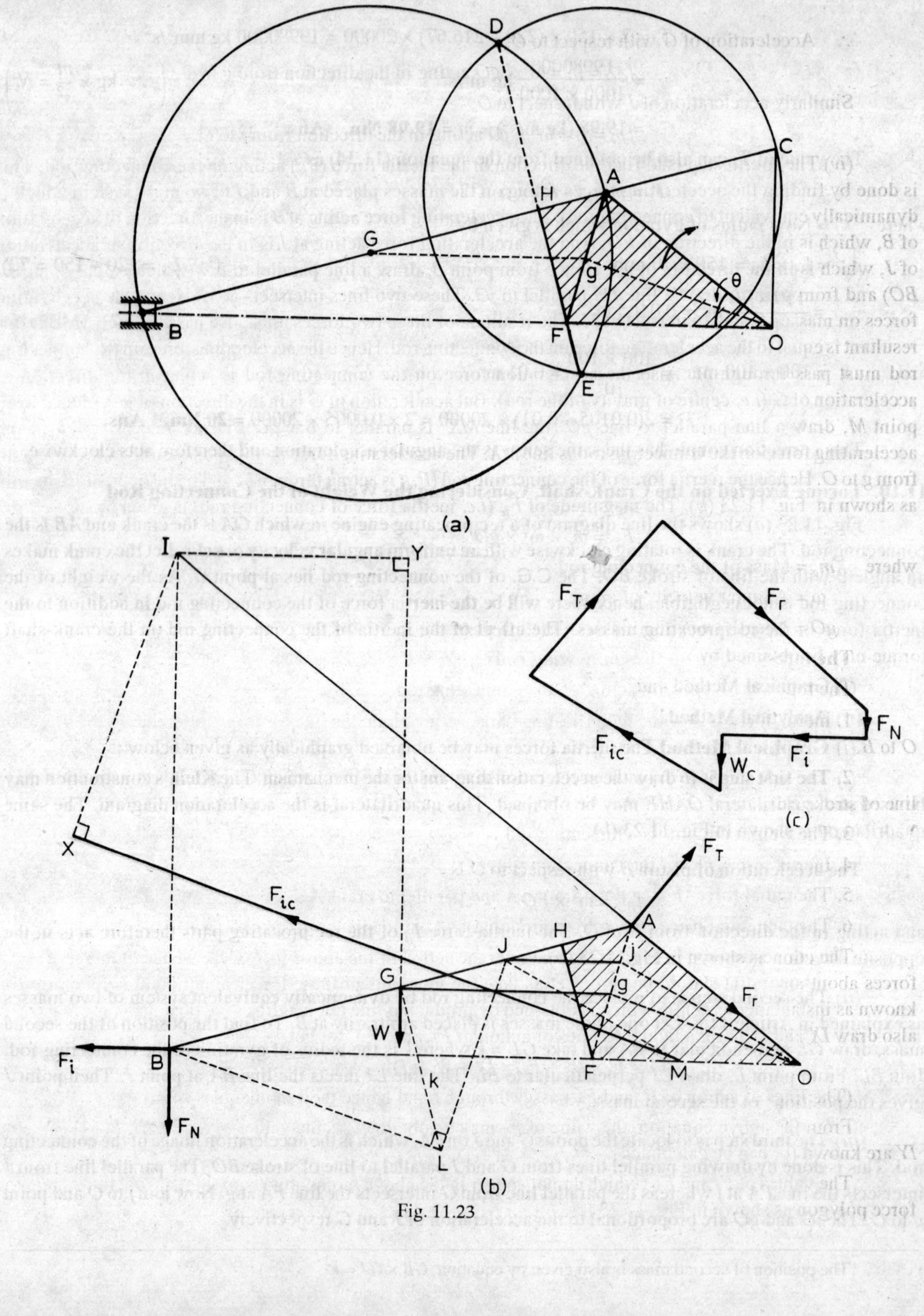

Fig. 11.23

∴ Acceleration of G with respect to O,

$$f_{GO} = f_G = \omega^2 \times gO \text{ acting in the direction from } g \text{ to } O.$$

Similarly acceleration of J with respect to O,

$$f_{JO} = f_J = \omega^2 \times jO \text{ acting in the direction from } j \text{ to } O.$$

(*iv*) The fourth step is to find the direction of the inertia force (F_{ic}) acting on the connecting rod. This is done by finding the accelerating forces acting on the masses placed at B and J (Two mass system which is dynamically equivalent to connecting rod). The accelerating force acting at B is in the direction of acceleration of B, which is in the direction of FO. And the accelerating force acting at J is in the direction of acceleration of J, which is in the direction of jO. Hence from point B, draw a line parallel to FO (which is line of stroke BO) and from point J, draw a line JM parallel to jO. These two lines intersects at M. Hence the accelerating forces on masses B and J intersect at M. The resultant of these two forces must also pass through M. But this resultant is equal to the accelerating force on the connecting rod. Hence the accelerating force on the connecting rod must pass through M. Also the acceleration force on the connecting rod is acting in the direction of acceleration of G (*i.e.* centre of gravity of the rod). But acceleration of G is in the direction of gO. Hence from point M, draw a line parallel to line gO (the line MX is parallel to line gO). Hence line of action of the accelerating force on the connecting rod is line MX. The accelerating force on the rod is acting in the direction from g to O. Hence the inertia force of the connecting rod (F_{ic}) is acting through M and in the opposite direction as shown in Fig. 11.23 (*b*). The magnitude of F_{ic} (*i.e.* inertia force of connecting rod) is given by,

$$F_{ic} = m_c \times \omega^2 \times gO$$

where m_c = Mass of the connecting rod,

ω = Angular velocity of crank,

gO = Length from Klein's Acceleration diagram.

The forces acting on the connecting rod

The following forces are acting on the connecting rod :

1. Inertia force (F_i) of the reciprocating parts, acting along the line of stroke through B in the direction O to B.

2. The side thrust (F_N) between the cross-head and guide bars, acting through B perpendicular to the line of stroke BO.

3. The weight (W_c) of the connecting rod.

4. Inertia force (F_{ic}) of the connecting rod.

5. The radial force (F_r), acting through A and parallel to crank AO.

6. The tangential force (F_t), acting through A and perpendicular to crank AO.

The connecting rod is in equilibrium under the action of the above forces. Hence the moment of these forces about any point should be zero. Let us take the moments of these forces about point I (the point I is known as instantaneous centre which is obtained by producing line OA and line of action of F_N. From point I, also draw IX and IY perpendicular to lines of action of F_{ic} and W_c), we get

$$F_T \times IA = F_i \times IB + F_{ic} \times IX + W_c \times IY \qquad ...(11.36)$$

(The lines of action of F_r and F_N passes through I and hence their moments is zero.)

From the above equation, the value of F_T may be obtained, as the values of IA, F_i, IB, F_{ic}, IX, W_c and IY are known (or can be calculated).

The values of F_r and F_N (which do not appear in the above momentum equation) are obtained from the force polygon as shown in Fig. 11.23 (*c*).

The torque which must be exerted on the crank-shaft in order to overcome the inertia of the moving parts is then given by,

$$T = F_T \times OA \qquad \text{...(11.36A)}$$

(ii) Analytical Method

Fig. 11.24 shows the line diagram of a reciprocating engine in which OA is the crank and AB is the connecting rod. The crank is rotating clockwise with an uniform angular velocity ω rad/s. Let the crank makes an angle θ with the line of stroke BO. The point B shows the position of cross-head pin and point A shows the position of crank-pin. The effect of the inertia of the connecting rod on the crank-shaft torque may be obtained analytically also.

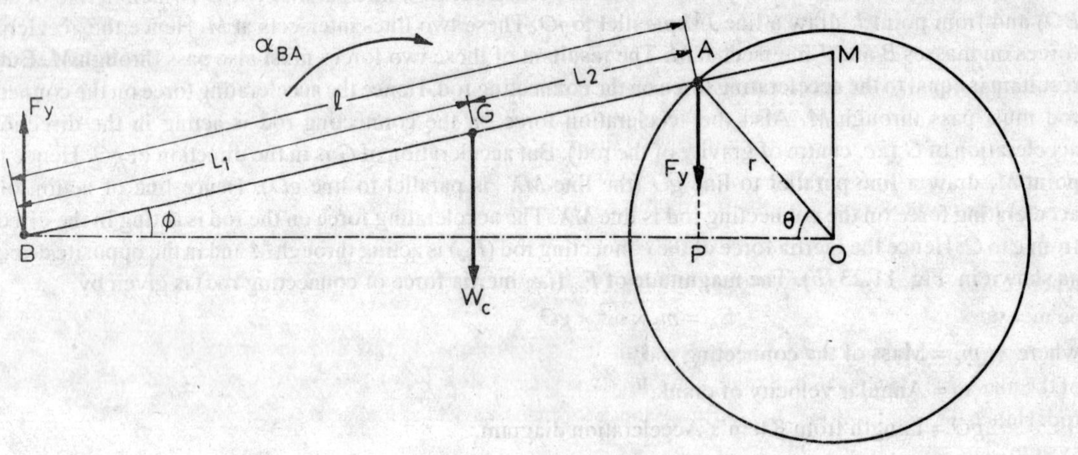

Fig. 11.24

For the analytical method, following are the steps :

(i) The total mass of the connecting rod (m_c) is divided into two masses. One of the masses (say m_1) is placed at the cross-head pin B and the other at the crank pin A. In this way, the C.G. of these two masses coincides with the C.G. of the rod *i.e.* at point G. The mass at A has no effect on the crank-shaft torque as the inertia force due to mass at A acts radially* outwards along the crank OA. The mass at B may be included with the reciprocating masses.

(ii) The total inertia force of total mass at B is obtained as given below :

Let W_c = weight of connecting rod

$= m_c \times g$ where m_c = total mass of connecting rod or $m_c = \dfrac{W_c}{g}$

l = length of the connecting rod,

L_1 = Distance of C.G. of the rod from point B *i.e.* distance BG.

m = Mass of the reciprocating parts which is also acting at B.

The mass (m_1) of the connecting rod which is placed at B is given by

$$m_1 = \frac{(l - L_1)}{l} \times m_c$$

$$\left[\text{Taking moments of } m_1 \text{ and } m_c \text{ about } A,\ m_1 \times l = m_c \times (l - L_1) \text{ hence } m_1 = m_c \times \frac{(l - L_1)}{l}\right]$$

---

*The accelerating force at A is in the direction of acceleration of A with respect to O. The acceleration of A is in the direction of AO *i.e.* from A to O. The inertia force due to mass at A will be in the opposite direction *i.e.* from O to A.

∴ Total equivalent reciprocating mass at B

$$= m + m_1 = m + \left(\frac{l - L_1}{l}\right) \times m_c$$

The acceleration of B with respect to $O = f_{BO}$ or f_B

∴ Total accelerating force at B = (Total equivalent reciprocating mass) × acceleration

$$= \left[m + \left(\frac{l - L_1}{l}\right) \times m_c\right] \times f_B$$

∴ Total inertia force (F_i) acting at B

$$= -\left[m + \left(\frac{l - L_1}{l}\right) \times m_c\right] \times f_B \qquad ...(11.37)$$

The theoretical value of acceleration of piston B is given by

$$f_B = \omega^2 r\left(\cos\theta + \frac{\cos 2\theta}{n}\right) \qquad \text{[Refer to equation (11.11)]}$$

The torque exerted by the inertia force (F_i) on the crank-shaft

$$T_B = (F_i) \times OM \qquad ...(11.38)$$

where OM is obtained by producing line BA and drawing a perpendicular at O on BO. The distance OM can be measured.

(*iii*) The equation gives the approximate inertia torque. In this derivation, we have assumed that one of the masses is placed at B and other at A. This two mass system is not dynamically equivalent to the connecting rod. Hence the assumption is not correct. Hence it is necessary to apply a correction couple (T_c) to the two-mass system. The correction couple is given by equation (11.35) as

$$T_c = m \times l_1 \times (l - L) \times \alpha$$

$$= m_c \times l_1 (l - L) \times \alpha_{BA} \qquad \text{(For connecting rod, } m = m_c \text{ and } \alpha = \alpha_{BA}\text{)}$$

where L = length of the equivalent simple pendulum when swung about an axis through B

$$= \frac{k^2 + l_1^2}{l_1}$$

and α_{BA} = Angular acceleration of the connecting rod and it is given by equation (11.16) as

$$\alpha_{BA} = \frac{-\omega^2 \sin\theta}{n} \text{ if } n^2 \text{ is very large in comparison to } \sin^2\theta \text{ and } 1.$$

The magnitude of the correction couple can be obtained. This correction couple may be applied to the system by two equal and opposite vertical forces F_y acting through B and A as shown in Fig. 11.24. Then

$$F_y \times BP = T_c$$

$$\therefore \quad F_y = \frac{T_c}{BP}$$

and the corresponding torque $(T_c{}^*)$ on the crank-shaft,

$$T_c{}^* = F_y \times PO$$

$$= \frac{T_c}{BP} \times PO \qquad \left(\because \; F_y = \frac{T_c}{BP}\right)$$

But $PO = r\cos\theta$ and $BP = l\cos\phi$

$$\therefore \quad T_c{}^* = \frac{T_c \times r\cos\theta}{l\cos\phi}$$

$$= \frac{T_c}{n} \times \frac{\cos\theta}{\cos\phi} \qquad \left(\because \frac{r}{l} = \frac{1}{n}\right)$$

$$= \frac{T_c}{n} \times \frac{\cos\theta}{\sqrt{\frac{n^2 - \sin^2\theta}{n^2}}} \qquad \left(\because \cos\phi = \sqrt{\frac{n^2 - \sin^2\theta}{n^2}}\right)$$

$$= \frac{T_c \times \cos\theta}{\sqrt{n^2 - \sin^2\theta}}$$

As $\sin^2\theta$ is very small in comparison to n^2. Hence neglecting $\sin^2\theta$,

$$T_c^* = \frac{T_c \times \cos\theta}{n} \qquad \text{...(11.38A)}$$

$$= [m_c \times l_1 (l - L) \times \alpha_{BA}] \times \frac{\cos\theta}{n} \qquad [\because T_c = m_c\, l_1 (l - L)\, \alpha_{BA}]$$

$$= \left[m_c \times l_1 (l - L) \times \frac{(-\omega^2 \sin\theta)}{n}\right] \times \frac{\cos\theta}{n} \qquad \left(\because \alpha_{BA} = \frac{-\omega^2 \sin\theta}{n}\right)$$

$$= m_c \times l_1 (l - L)(-\omega^2) \times \frac{\sin\theta\cos\theta}{n^2}$$

$$= -m_c \times l_1 (l - L)\, \omega^2 \times \frac{2\sin\theta\cos\theta}{2n^2}$$

$$= -m_c \times l_1 (l - L)\, \omega^2 \frac{\sin 2\theta}{2n^2} \qquad \text{...(11.39)}$$

(iv) *Torque on crank-shaft due to weight of connecting rod (T_w).* The connecting rod has been replaced by two mass system. The mass (m_1) is placed on cross-head pin B and mass (m_2) is placed on crank-pin A. The weight (W_2) due to m_2 will be acting vertically downward at A. Due to this weight, there will be a torque on the crank-shaft.

∴ Torque due to weight (W_2) on crank-shaft is given by

$$T_W = -(W_2) \times PO \qquad \text{(– ve sign is due to anticlock direction)}$$

$$= -(m_2 \times g) \times r\cos\theta \qquad (\because W_2 = m_2 \times g \text{ and } PO = r\cos\theta)$$

The value of m_2 will be given by,

$$m_2 = \frac{l - L_2}{l} \times m_c = \frac{L_1}{l} \times m_c \qquad (\because l - L_2 = L_1)$$

$$\therefore \quad T_W = -\frac{L_1}{l} \times m_c \times g \times r \times \cos\theta \qquad \text{...(11.40)}$$

∴ Total torque exerted on crank-shaft

$$= T_B + T_c^* + T_W.$$

Problem 11.24. *The crank of a horizontal reciprocating engine is rotating at a speed of 300 r.p.m. in the clockwise direction. The lengths of crank and connecting rod are 100 mm and 400 mm respectively. Find : (i) acceleration of piston and (ii) the magnitude, position and direction of inertia force if the mass of connecting rod is 40 kg and the centre of gravity of the connecting rod is at a distance of 180 mm from the crank pin centre. The radius of gyration of connecting rod about the centre of gravity is 120 mm. The crank position is 45° from the inner dead centre.*

Sol. Given : $N = 300$ r.p.m. or $\omega = \frac{2\pi N}{60} = \frac{2\pi \times 300}{60} = 10\pi$ rad/s ; $r = OA = 100$ mm $= 0.1$ m ;

$l = AB = 400$ mm $= 0.4$ m ; mass $m_c = 40$ kg ; distance of G from $A = 180$ mm $= 0.18$ m ;

$k = 120$ mm $= 0.12$ m ; $\theta = 45°$.

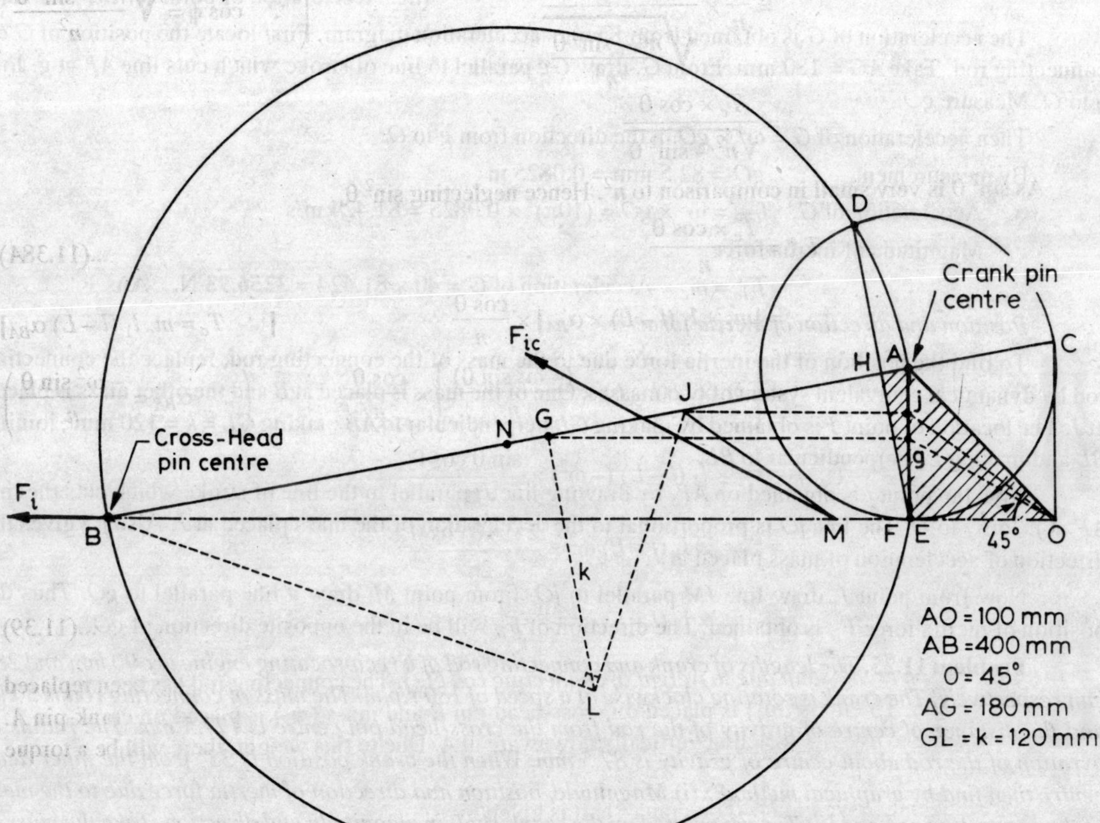

Fig. 11.25

(*i*) *Acceleration of piston*

Acceleration of piston is the acceleration of point B. The acceleration of point B can be obtained by drawing Klein's acceleration diagram as shown in Fig. 11.25. First draw the configuration diagram OAB to some suitable scale.

Take $OA = 100$ mm ; $AB = 400$ mm ; and $\theta = 45°$.

The Klein's acceleration diagram $OAHF$ is drawn in the same manner as already given in Art. 11.4.3. Then

Acceleration of piston, $\quad f_{BO} = \omega^2 \times FO$.

The length FO is obtained by measurement.

∴ By measurement, $\quad FO = 72.5$ mm $= 0.0725$ m

∴ Acceleration of piston $\quad = \omega^2 \times FO = (10\pi)^2 \times 0.0725 = 71.55$ m/s. **Ans.**

(ii) Magnitude, position and direction of inertia force due to mass of the connecting rod.

Magnitude of inertia force due to mass of connecting rod

= Mass of connecting rod × Acceleration of centre of gravity of connecting rod

(*i.e.* Acceleration of point G) ...(*i*)

The acceleration of G is obtained from Klein's acceleration diagram. First locate the position of G on connecting rod. Take AG = 180 mm. From G, draw Gg parallel to line of stroke which cuts line AF at g. Join g to O. Measure gO.

Then acceleration of $G = \omega^2 \times gO$ in the direction from g to O

By measurement, gO = 82.5 mm = 0.0825 m

$\therefore$ Acceleration of G, $f_{GO} = \omega^2 \times gO = (10\pi)^2 \times 0.0825 = 81.424$ m/s

$\therefore$ Magnitude of inertia force,

$$F_{ic} = m_c \times \text{Acceleration of } G = 40 \times 81.424 = \mathbf{3256.98\ N.} \quad \textbf{Ans.}$$

Position and direction of inertia force

To find the position of the inertia force due to the mass of the connecting rod, replace the connecting rod by dynamical equivalent system of two masses. One of the mass is placed at B and the other mass is placed at J. The location of point J is obtained by making GL perpendicular to AB ; taking $GL = k$ = 120 mm ; joining BL and making LJ perpendicular to BL.

Now the point j is obtained on AF, by drawing line Jj parallel to the line of stroke which cuts the line AF at j. Join j to O. The line jO is proportional to the acceleration of the mass placed at J. Also jO gives the direction of acceleration of mass placed at J.

Now from point J, draw line JM parallel to jO. From point M, draw a line parallel to gO. Thus the position of inertia force F_{ic} is obtained. The direction of F_{ic} will be in the opposite direction of gO.

Problem 11.25. *The lengths of crank and connecting rod of a reciprocating engine are 90 mm and 280 mm respectively. The crank is rotating clockwise at a speed of 180 r.p.m. The mass of connecting rod is 50 kg and the distance of centre of gravity of the rod from the cross-head pin centre is 177.5 mm. The radius of gyration of the rod about centre of gravity is 87.5 mm. When the crank position is 33° from the inner dead centre then find by graphical method : (i) Magnitude, position and direction of inertia force due to the mass of the connecting rod and (ii) Torque exerted on the crank-shaft in magnitude and direction. Take the mass of reciprocating parts = 60 kg.*

Sol. Given : $r = OA$ = 90 mm ; $l = AB$ = 280 mm ; N = 180 r.p.m. or $\omega = \dfrac{2\pi N}{60} = \dfrac{2\pi \times 180}{60} = 6\pi$ rad/s ;

m_c = 50 kg ; distance BG = 177.5 mm ; θ = 33° ; k = 87.5 mm and mass of reciprocating parts = 60 kg.

(i) Magnitude, position and direction of inertia force due to mass of connecting rod.

With the given dimensions, first draw the configuration diagram OAB in which OA = 90 mm ; AB = 280 mm and angle BOA = 33° to some suitable scale.

Now draw the Klein's acceleration diagram $OAHF$ as shown in Fig. 11.26 in the same manner as given in Art. 11.4.3. The acceleration of G is obtained by drawing line Gg parallel to the line of stroke which cuts line AF to g. Then gO gives the acceleration of G. Hence measure gO.

By measurement,

Length gO = 84 mm = 0.084 mm

$\therefore$ Acceleration of G, $f_{GO} = \omega^2 \times gO = (6\pi)^2 \times 0.084 = 29.84$ m/s^2

$\therefore$ Magnitude of inertia force due to mass of connecting rod,

$$F_{ic} = \text{Mass of connecting rod} \times \text{Acceleration of } G$$

$$= m_c \times f_{GO} = 50 \times 29.84 = \mathbf{1492.3\ N.} \quad \textbf{Ans.}$$

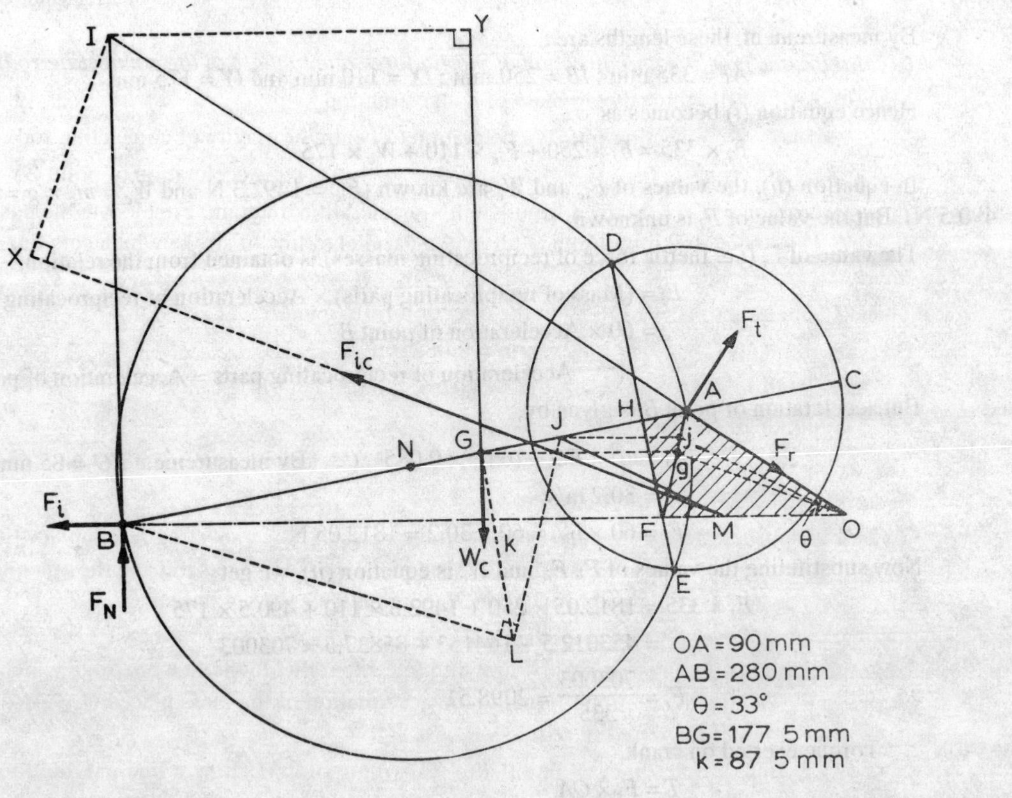

Fig. 11.26

To find the position of the inertia force, replace the connecting rod by dynamical equivalent system of two masses. One of the mass is placed at B and other mass is placed at J. The location of point J on the connective rod is obtained by making GL perpendicular to AB ; taking $GL = k = 87.5$ mm ; joining BL and making LJ perpendicular to BL.

Now locate the position of j on AF. This is done by drawing line Jj parallel to line of stroke BO which cuts the line AF at j. Then the length jO will be proportional to the acceleration of the mass placed at J. Also jO gives the direction of acceleration of mass placed at J.

To find the direction of inertia force (F_{ic}), from point J, draw line JM parallel to jO which cuts the line BO at M. From point M, draw a line MX parallel to gO. The line MX gives the position of F_{ic}. The direction of F_{ic} will be in the opposite direction of gO.

(ii) Torque exerted on crank-shaft in magnitude and direction

In Fig. 11.26, the point I is the instantaneous centre of the connecting rod for the given crank-position. The line IX is perpendicular on the line of action of F_{ic} whereas IY is perpendicular on the line of action W_c (*i.e.* weight of connecting rod).

Let F_T = Force acting perpendicular to crank OA.

Taking moments of all forces [*i.e.* F_i, F_{ic}, W_c, F_N, F_T and F_r] about point I,

$$F_t \times AI = F_i \times IB + F_{ic} \times IX + W_c \times IY \qquad ...(i)$$

(Forces F_N and F_r passes through point I)

The values of AI, IB, IX and IY are obtained from Fig. 11.26.

By measurement, these lengths are :

$$AI = 335 \text{ mm} ; IB = 250 \text{ mm} ; IX = 110 \text{ mm and } IY = 175 \text{ mm}$$

Hence equation (*i*) becomes as

$$F_t \times 335 = F_i \times 250 + F_{ic} \times 110 + W_c \times 175 \quad ...(ii)$$

In equation (*ii*), the values of F_{ic} and W_c are known (F_{ic} = 1492.3 N and $W_c = m_c \times g = 50 \times 9.81 =$ 490.5 N). But the value of F_i is unknown.

The value of F_i (*i.e.* inertia force of reciprocating masses) is obtained from the relation,

$$F_i = \text{(Mass of reciprocating parts)} \times \text{Acceleration of reciprocating parts}$$
$$= 60 \times \text{Acceleration of point } B$$

($\because$ Acceleration of reciprocating parts = Acceleration of point $B = f_{BO}$)

But acceleration of point B is given by,

$$f_{BO} = \omega^2 \times FO = (6\pi)^2 \times 0.085 \quad (\because \text{ By measurement } FO = 85 \text{ mm} = 0.085 \text{ m})$$
$$= 30.2 \text{ m/s}^2$$

$$\therefore \quad F_i = 60 \times f_{BO} = 60 \times 30.2 = 1812.05 \text{ N}$$

Now substituting the values of F_i, F_{ic} and W_c is equation (*ii*), we get

$$F_t \times 335 = 1812.05 \times 250 + 1492.3 \times 110 + 490.5 \times 175$$
$$= 453012.5 + 164153 + 85837.5 = 703003$$

$$\therefore \quad F_t = \frac{703003}{335} = 2098.51$$

$\therefore$ Torque exerted on crank,

$$T = F_t \times OA$$
$$= 2098.51 \times 0.09 \quad (\because \; OA = 90 \text{ mm} = 0.09 \text{ m})$$
$$= \mathbf{188.86 \text{ Nm.}} \quad \textbf{Ans.}$$

Problem 11.26. *Find the torque on the crank-shaft in magnitude and direction analytically for the Problem 11.25.*

Sol. The data from Problem 11.25 is :

$r = OA$ = 90 mm ; $l = AB$ = 280 mm ; N = 180 r.p.m. or $\omega = 6\pi$ rad/s ; m_c = 50 kg ; BG = 177.5 mm ; $\theta = 33°$; k = 87.5 mm and mass of reciprocating parts m_R = 60 kg.

Divide the mass of the connecting rod (*i.e.* m_c = 50 kg) into two parts. One of the mass is placed at B and other at A.

The mass at B (or at gudgeon pin) is given by,

$$m_b = m_c \times \frac{(l - l_1)}{l} \quad \text{where } l_1 = BG = 177.5 \text{ mm}$$

$$= 50 \times \frac{(280 - 177.5)}{280} = 18.3 \text{ kg}$$

The mass at A (or at crank pin),

$$m_a = 50 - 18.3 = 31.7 \text{ kg}$$

$\therefore$ Total equivalent reciprocating mass

$$= m_b + m_R = 18.3 + 60 = 78.3 \text{ kg}$$

The acceleration of the reciprocating parts is given by,

$$f_B = \omega^2 \times r \times \left(\cos\theta + \frac{\cos 2\theta}{n} \right)$$

$$= (6\pi)^2 \times 0.09 \left(\cos 33° + \frac{\cos 66°}{3.11} \right)$$

$$\left[\because \quad \omega = 6\pi,\ r = 90 \text{ mm} = 0.09 \text{ m},\ \theta = 33° \text{ and } n = \frac{l}{n} = \frac{280}{90} = 3.11 \right]$$

$$= 31.97 \left(0.8386 + \frac{0.4067}{3.11} \right) = 30.99 \text{ m/s}^2 \simeq 31 \text{ m/s}^2.$$

$\therefore$ Total equivalent inertia force,

$$F_i = -\text{Total equivalent reciprocating mass} \times f_B$$

$$= -78.3 \times 31 = -2427.3 \text{ N}$$

(The inertia force acts in the opposite direction of acceleration. Hence – ve sign is taken.)

The torque exerted on crank-shaft is given by,

$$T_B = F_i \times OC$$

(where from Fig. 11.26 by measurement OC = 62.75 mm = 0.06275 m)

$$= -2427.3 \times 0.06275 = -152.31 \text{ Nm} \qquad ...(i)$$

As the mass of connecting rod is replaced by two masses, one mass is placed at A and other is at B. This two-mass system is not dynamically equivalent. Hence a correction couple (T_c) is to be applied. The value of correction couple (T_c) is given by equation (11.35). The corresponding torque (T_c*) on the crank-shaft is given by equation (11.39) as

$$T_c^* = -m_c \times L_1 (l - L) \frac{\omega^2 \sin^2\theta}{2n^2}$$

where L_1 = Distance of G from B = distance BG = 177.5 mm = 0.1775 m

l = length of connecting rod = 280 mm (given) = 0.28 m

L = Length of equivalent simple pendulum when swung about an axis through B

$$= \frac{k^2 + L_1^2}{L_1} = \frac{87.5^2 + 177.5^2}{177.5} = \frac{7656.25 + 31506.25}{177.5}$$

$$= 220.63 \text{ mm} = 0.22063 \text{ m} \simeq 0.220 \text{ m}$$

$$\theta = 33°,\ n = \frac{l}{r} = 3.11,\ m_c = 50$$

Hence the value of corresponding torque (T_c*) on crank shaft becomes as

$$T_c^* = -50 \times 0.1775(0.280 - 0.220)\,(6\pi)^2 \times \frac{\sin 66°}{2 \times 3.11^2}$$

$$= -8.935 \text{ Nm} \qquad ...(ii)$$

The mass placed at A (*i.e.* mass m_a = 31.7 kg) has no effect on the crank-shaft torque since its inertia force acts radially outwards along the crank, [the acceleration of A (*i.e.* f_{AO}) is radially inwards along the crank and so the inertia force will be opposite to the direction of acceleration]. But due to the pull of gravity at mass

*The value of OC by analytical method is given by

$$OC = r\left(\sin\theta + \frac{\sin 2\theta}{2\sqrt{n^2 - \sin^2\theta}} \right) = 0.09\left(\sin 33° + \frac{\sin 66°}{2\sqrt{3.11^2 + \sin^2 33°}} \right) = 0.0624 \text{ m.}$$

placed at A, there will be a torque on crank-shaft. This means the weight of mass placed at A will exert a torque on crank-shaft. This torque (T_w) is given by,

$$T_W = (m_a \times g) \times r \times \cos\theta$$

$$= (31.7 \times 9.81) \times 0.09 \times \cos 33° \quad (\because\ r = 0.09 \text{ m}, \theta = 33°)$$

$$= 24.47 \text{ Nm (anticlockwise)}$$

$$= -24.47 \text{ Nm (– ve sign due to anticlockwise)} \quad ...(iii)$$

The total torque (T) exerted on the crank-shaft by the inertia of the moving parts is the algebraic sum of T_B, T_C and T_W given by equations (*i*), (*ii*) and (*iii*) respectively.

$$\therefore \quad T = T_B + T_C{}^* + T_W$$

$$= -152.31 + (-8.935) + (-24.47) = \mathbf{-185.718\ Nm.}\quad \textbf{Ans.}$$

Problem 11.27. *The lengths of crank and connecting rod of a horizontal reciprocating engine are 300 mm and 1.5 m respectively. The crank is rotating at 120 r.p.m. clockwise. The mass of the reciprocating parts of the engine is 290 kg whereas the mass of the connecting rod is 250 kg. The C.G. of the connecting rod is 475 mm from the crank-pin centre and radius of gyration of the connecting rod about an axis passing through the C.G. is 625 mm. Find the inertia torque on the crank-shaft by (i) analytical method and (ii) graphical method, when* $\theta = 40°$.

Sol. Given : $r = 300$ mm $= 0.3$ m ; $l = 1.5$ m $= 1500$ mm ; $N = 120$ r.p.m.

or $\omega = \dfrac{2\pi N}{60} = \dfrac{2\pi \times 120}{60} = 4$p rad/s ; mass of reciprocating parts, $m = 290$ kg, mass of connecting rod, $m_c = 250$ kg ; Distance of G from crank-pin $A = 475$ mm ; $\therefore$ Distance of G from cross-head pin B, $l_1 = 1500 - 475 = 1025$ mm, radius of gyration, $k = 625$ mm, $\theta = 40°$.

(*i*) *Analytical Method*

Let m_1 = A part of mass of connecting rod placed at cross-head pin, B

$$= \left(\frac{l - L_1}{l}\right) \times m_c$$

$$= \frac{(1500 - 1025)}{1500} \times 250 \quad (\because\ l = 1500 \text{ mm and } L_1 = 925 \text{ mm})$$

$$= \frac{475}{1500} \times 250 = 79.16 \text{ kg}$$

$\therefore$ Total equivalent reciprocating mass

$$= \text{mass of reciprocating parts} + m_1$$

$$= 290 + 79.16 = 369.16 \text{ kg}$$

The acceleration of point B or of reciprocating parts is given by equation (11.11) as [$\because$ The mass of reciprocating parts and m_1 are concentrated at B]

$$f_B \text{ (or } a) = r\omega^2\left(\cos\theta + \frac{\cos 2\theta}{n}\right)$$

$$= 0.3 \times (4\pi)^2\left[\cos 40° + \frac{\cos(2 \times 40)}{5}\right] \quad \left(\because\ \theta = 40°, n = \frac{l}{r} = \frac{1500}{300} = 5\right)$$

$$= 0.3 \times 16 \times \pi^2\left[0.766 + \frac{0.1736}{5}\right]$$

$$= 0.3 \times 16 \times \pi^2 \times 0.8 = 37.89 \text{ m/s}^2$$

∴ Total equivalent inertia force,

F_i = – (Total equivalent reciprocating mass) × acceleration of reciprocating parts

$$= -(369.16) \times 37.89 \text{ N} = -13987.47 \text{ N}$$

The approximate inertia torque exerted on crank-shaft is given by equation as,

$$T_B = F_i \times OC$$

$$= -13987.47 \times OC \quad \text{...(i)}$$

The value of OC is obtained from Fig. 11.27. By measurement, we have

$$OC = 222.4 \text{ mm} = 0.2224 \text{ m}$$

Substituting the value of OC in equation (*i*), we get

$$T_B = -13987.47 \times 0.2224 \text{ Nm} = -3110.8 \text{ Nm}$$

Exact Inertia Torque

Let us first find the value of correction couple which is given by equation as

$$T_c = m_c \times l_1 (l - L) \times \alpha_{BA} \quad \text{...(ii)}$$

The value of L, which is the equivalent length of a simple pendulum when swung about an axis through B is given by,

$$L = \frac{k^2 + (l_1)^2}{l_1} = \frac{(0.625)^2 + 1.025^2}{1.025} = \frac{0.390625 + 1.050625}{1.025} = 1.4061$$

and value of α_{BA} *i.e.* angular acceleration of connecting rod from equation (11.16) is given by,

$$\alpha_{BA} = -\frac{\omega^2 \times \sin\theta}{n}$$

where $n = \frac{l}{r} = \frac{1.5}{0.3} = 5 = -\frac{(4\pi)^2 \times \sin\theta}{n} = -\frac{16\pi^2 \times \sin 40°}{5} = -20.3 \text{ rad/s}^2$

Substituting the values of L and α_{BA} in equation (*ii*), we get

$$T_c = 250 \times 1.025\,(1.5 - 1.4061)(-20.3) = -488.45 \text{ Nm}$$

The above is the value of correction couple. The torque on the crank-shaft due to the correction couple is given by equation (11.38*A*) as

$$T_c{}^* = \frac{T_c \times \cos\theta}{n} = \frac{(-488.45) \times \cos 40°}{5} = -74.83 \text{ Nm}$$

Now let us calculate the torque (T_w) due to the weight of the connecting rod on the crank-shaft.

The connecting rod has been replaced by two mass system, one mass (m_1) is placed on cross-head pin B and other mass (m_2) is placed on crank-pin A.

∴ The mass (m_2) of the connecting rod which is placed at A is given by equation as

$$m_2 = \left(\frac{l - l_2}{l}\right) \times m_c$$

$$= \frac{L_1}{l} \times m_c \qquad (\because\ l - L_2 = L_1)$$

$$= \frac{1.025}{1.5} \times 250 \text{ kg} = 170.83 \text{ kg}$$

∴ The weight (W_2) of the connecting rod which is placed at $A = m_2 \times g$

or $$W_2 = m_2 \times g = 170.83 \times 9.81 = 1675.84 \text{ N}$$

This weight is acting vertically downward at A.

∴ Torque exerted by weight (W_2) on the crank-shaft,

$T_W = -W_2 \times PO$ (– ve sign is due to anti-clockwise direction)

$= -1675.84 \times r \cos\theta$ ($\because$ $PO = r\cos\theta = 0.3 \cos 40°$)

$= -1675.84 \times 0.3 \times \cos 40° = -340$ Nm

The total torque exerted on the crank-shaft due to the inertia of moving parts

$= T_B + T_c^* + T_W$

$= -3110.8 - 74.83 - 385.13 =$ **– 3552.8 Nm. Ans.**

(– ve sign shows that the torque exerted on the crank-shaft is anticlockwise)

(*ii*) **Graphical Method.** The followings are the steps for graphical method :

(*i*) With the given dimensions, first of all draw the configuration diagram *OAB* to some suitable scale as shown in Fig. 11.27 in which $OA = 300$ mm ; $AB = 1.5$ m and $\angle BOC = 40°$.

(*ii*) Now draw the acceleration diagram *OAHF* by Klein's construction by the method as given in Article 11.4.3. Join point *A* to *F*. Then *AF* is the acceleration image of the connecting rod.

(*iii*) In Fig. 11.27, *N* is the middle point of connecting rod. The location of centre of gravity (*i.e.* point *G*) is obtained by taking $AG = 475$ mm to the chosen scale.

(*iv*) From *G*, draw $GL \perp$ to *BA*. Take $GL = k = 625$ mm. Join *B* and *L*. Draw *LJ* perpendicular to *BL*. This gives the the position of the point *J*. From *J* draw a line *Jj* parallel to the line of stroke *BO*. The line *Jj* cuts the line *AF* at *j*. Similarly locate the position *g* on *AF*. Joins *j* and *g* to *O*. Measure *jO*, *gO* and *FO*. By measurements : $jO = 295$ mm ; $gO = 265$ mm and $FO = 240$ mm.

The acceleration of piston (or point *B*) and acceleration of the C.G. of the connecting rod are obtained as :

Acceleration of piston (or point *B*), f_{BO} or f_B

$= \omega^2 \times FO$

$= (4\pi)^2 \times (0.240)$ ($\because$ $FO = 240$ mm $= 0.24$ m)

$= 37.9$ m/s^2

Acceleration of the *C.G.* of the connecting rod, f_{GO} or f_G

$= \omega^2 \times gO$ ($\because$ $gO = 267.5$ mm $= 0.2675$ m)

$= (4\pi)^2 \times 0.2675 = 42.24$ m/s^2

∴ Inertia force due to reciprocating parts,

$F_i = m \times$ acceleration of piston

$= 290 \times f_{BO} = 290 \times 37.9 = 10991$ N

Inertia force due to connecting rod,

F_{ic} = mass of connecting rod × acceleration of *C.G.* of the connecting rod

$= m_c \times f_{GO} = 250 \times 42.24 = 10560$ N

Now in Fig. 11.27, draw from point *J* a line *JM* parallel to *jO*, which cuts the line *BO* at *M*. Now from point *M*, draw a line *MX* parallel to *gO*. From point *B*, draw a line *BI* perpendicular to *BO* and produce the *OA* which cuts the line *BI* at *I*. Now from point *I*, draw *IX* perpendicular to *MX*.

From point *G*, draw a vertical line *GY*. From point *I*, draw *IY* perpendicular to *GY*. Now measure the lengths *IB*, *IX*, *IY* and *IA* on measurements these lengths are :

$IB = 1440$ mm $= 1.44$ m

$IX = 465$ mm $= 0.465$ m

$IY = 1012$ mm $= 1.012$ m

$IA = 1940$ mm $= 1.94$ m

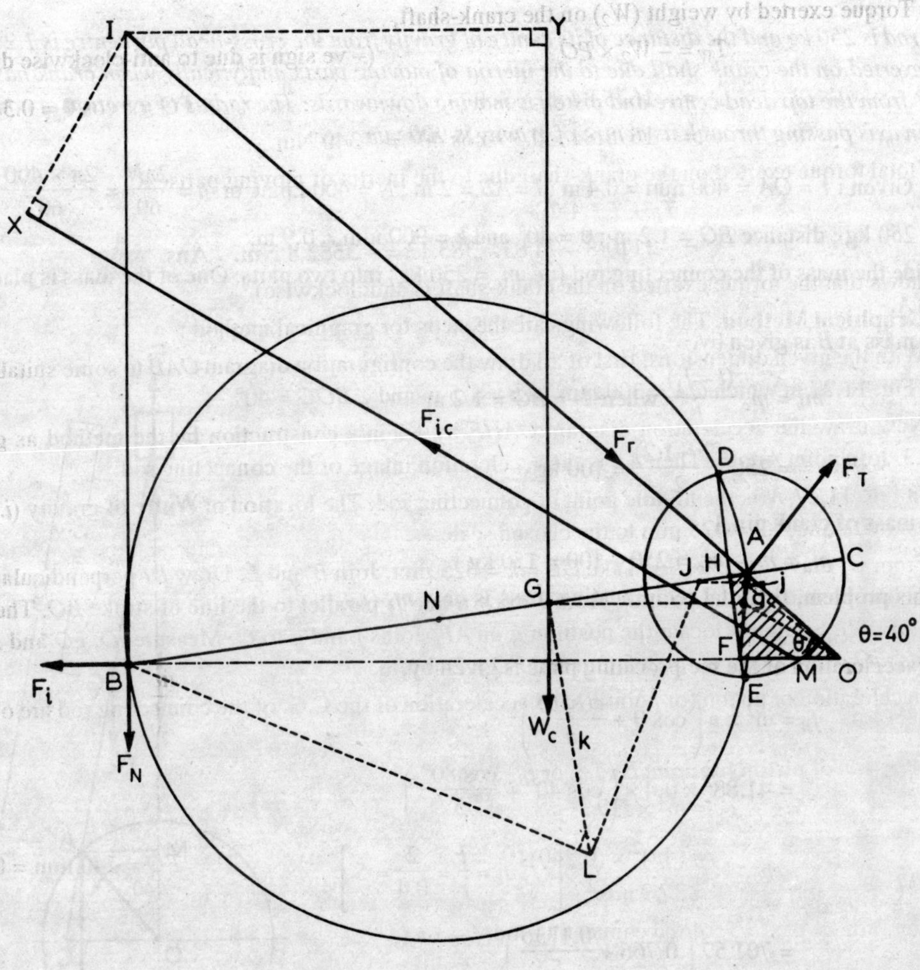

Fig. 11.27

Now using equation (11.36), we get

$$F_T \times IA = F_i \times IB + F_{ic} \times IX + W_c \times IY$$

$$F_T \times 1.94 = 10991 \times 1.44 + 10560 \times 0.465 + 250 \times 9.81 \times 1.012$$

$$(\because \quad W_c = m_c \times g = 250 \times 9.81)$$

$$= 15827 + 4910.4 + 2481.9 = 23219.3$$

$$\therefore \qquad F_T = \frac{23219.3}{1.94} = 11968.7 \text{ N}$$

The torque which must be exerted on the crank-shaft in order to overcome the inertia of the moving parts is given by equation (11.36A) as

$$T = F_T \times OA$$

$$= 11968.7 \times 0.3 \qquad (\because \quad OA = 300 \text{ mm} = 0.3 \text{ m})$$

$$= \mathbf{3590.6 \text{ Nm.}} \quad \textbf{Ans.}$$

Problem 11.28. *The lengths of crank and connecting rod of a vertical reciprocating engine are 400 mm and 2 m respectively. The crank is rotating clockwise at a speed of 400 r.p.m. The mass of the* *connecting rod is 250 kg and the distance of its centre of gravity from the cross-head pin centre is 1.2 m. Find the torque exerted on the crank-shaft due to the inertia of moving parts analytically when crank has turned through 40° from the top dead centre and piston is moving downwards. The radius of gyration of connecting rod about an axis passing through its centre of gravity is 900 mm.*

Sol. Given : $r = OA = 400$ mm $= 0.4$ m ; $l = AB = 2$ m ; $N = 400$ r.p.m. or $\omega = \frac{2\pi N}{60} = \frac{2\pi \times 400}{60} = 41.88$ rad/s ; $m_c = 250$ kg ; distance $BG = 1.2$ m ; $\theta = 40°$ and $k = 900$ mm $= 0.9$ m.

Divide the mass of the connecting rod (*i.e.* $m_c = 250$ kg) into two parts. One of the mass is placed at B and other at A.

The mass at B is given by,

$$m_b = m_c \frac{(l - l_1)}{l} \text{ where } l_1 = BG = 1.2 \text{ m}$$

$$= 250 \frac{(2 - 1.2)}{2} = 100 \text{ kg}$$

The mass of crank pin A,

$$m_a = m_c - m_b = 250 - 100 = 150 \text{ kg}$$

In this problem, the total reciprocating mass is only m_b (*i.e.* 100 kg).

The acceleration of the reciprocating parts is given by,

$$f_B = \omega^2 \times r \left(\cos\theta + \frac{\cos 2\theta}{n} \right)$$

$$= 41.88^2 \times 0.4 \times \left(\cos 40° + \frac{\cos 80°}{5} \right)$$

$$\left[\because n = \frac{l}{r} = \frac{2}{0.4} = 5 \right]$$

$$= 701.57 \left(0.766 + \frac{0.1736}{5} \right)$$

$$= 561.76 \text{ m/s}^2$$

$\therefore$ Inertia force, $F_i = -$ Total reciprocating mass $\times f_B$

$$= -100 \times 561.76 = -56176 \text{ N}$$

(The inertia force acts in the opposite direction of acceleration. Hence – ve sign is taken)

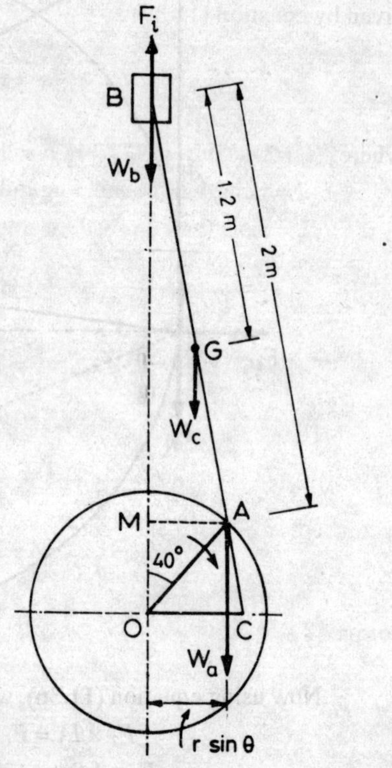

Fig. 11.28

The torque exerted on the crank-shaft is given by

$$T_B = F_i \times OC \qquad ...(i)$$

The value of OC may be obtained graphically or analytically. The value of OC by analytical method is given by,

$$OC = r \left(\sin\theta + \frac{\sin 2\theta}{2\sqrt{n^2 - \sin^2\theta}} \right)$$

$$= 0.4 \left(\sin 40° + \frac{\sin 80°}{2\sqrt{25 - \sin^2 40°}} \right) \qquad (\because n = 5 \quad \therefore \quad n^2 = 25)$$

$$= 0.4\left(0.6428 + \frac{0.9848}{2 \times 4.9585}\right) = 0.2968 \text{ m} \quad ...(ii)$$

Hence equation (*i*) becomes as

$$T_B = F_i \times OC$$
$$= -56176 \times 0.2968 = -16673 \text{ Nm} \quad ...(iii)$$

As the mass of the connecting rod is replaced by two masses, one mass is placed at A and other is at B. The two-mass system is not dynamically equivalent. Hence a correction couple (T_c) is to be applied. The value of correction couple is given by equation (11.35). The corresponding torque (T_c*) on the crank-shaft is given by equation (11.39) as

$$T_c^* = -m_c \times l_1 \times (l - L)\frac{\omega^2 \sin 2\theta}{2n^2}$$

where l_1 = Distance of G from B = distance BG = 1.2 m ;

l = Length of connecting rod = AB = 2 m (given) ;

L = Length of equivalent simple pendulum when swung about an axis through B

$$= \frac{k^2 + l_1^2}{l_1} = \frac{0.9^2 + 1.2^2}{1.2} = 1.875 \text{ m} ;$$

$$\theta = 40° ; m_c = 250 ; n = \frac{l}{r} = \frac{2}{0.4} = 5$$

Hence the value of correction couple becomes as,

$$T_c = -250 \times 1.2 \times (2 - 1.875) \times \frac{41.88^2 \sin 80°}{2 \times 5^2}$$

$$= -1295.46 \text{ Nm} \quad ...(iv)$$

The weight of mass placed at A (*i.e.* mass m_a = 150 kg) will exert a torque on the crank-shaft. This torque (T_{Wa}) is given by

$$T_{Wa} = W_a \times r \sin\theta \text{ (clockwise)}$$
$$= (m_a \times g) \times r \sin\theta = 150 \times 9.81 \times 0.4 \times \sin 40°$$
$$= 378.34 \text{ N m (clockwise)} \quad ...(v)$$

The torque due to weight of mass placed at B is given by,

$$T_{Wb} = W_b \times OC$$

$$\left[\text{where the value of } OC \text{ from equation } (ii) \text{ is } 0.2978 \text{ m } i.e.\ OC = r\left(\sin\theta + \frac{\sin 2\theta}{2\sqrt{n^2 - \sin^2\theta}}\right) = 0.2968 \text{ m}\right]$$

$$= (m_b \times g) \times OC = (100 \times 9.81) \times 0.2968$$
$$= 291.16 \text{ Nm (clockwise)} \quad ...(vi)$$

∴ Total inertia torque exerted on crank-shaft

$$= T_B + T_c^* + T_{Wa} + T_{Wb}$$
$$= -16673 - 1295.46 + 378.34 + 291.16 \text{ Nm}$$

$$= \mathbf{-17298.96 \text{ Nm (Anti-clockwise).}} \quad \textbf{Ans.}$$

HIGHLIGHTS

1. Inertia force $(F_i) = -m \times a$ and inertia torque $(T_i) = -I \times \alpha$
   where a = acceleration, α = Angular acceleration, and
   I = mass moment of inertia = $m \times k^2$; k = radius of gyration.
2. Inertia force acts in the opposite direction of acceleration whereas inertia torque acts in opposite direction of angular acceleration.
3. D-Alemberts principle states that : the resultant force (or resultant torque) acting on a body together with the inertia force (or inertia torque) are in equilibrium. Hence according to this principle :
   $$F_r + F_i = 0 \quad \text{or} \quad T_r + T_i = 0$$
   where F_r = resultant force acting on a body, T_r = resultant torque acting on the body.
4. The displacement (x), velocity (v) and acceleration (a) of the piston of the reciprocating engine are given by,

   (*i*) $x = r[(1 - \cos\theta) + (n - \sqrt{n^2 - \sin^2\theta})]$ ...... Exact dislacement

   $= r\left[(1 - \cos\theta) + \dfrac{\sin 2\theta}{2n}\right]$ ...... Approximate displacement

   $= r[1 - \cos\theta]$ ...... When 'n' is very very large

   (*ii*) $v = \omega r\left[\sin\theta + \dfrac{\sin 2\theta}{2\sqrt{n^2 - \sin^2\theta}}\right]$ ...... Exact velocity

   $= \omega r\left[\sin\theta + \dfrac{\sin 2\theta}{2n}\right]$ ...... Approximate velocity

   $= \omega r \sin\theta$ ...... When 'n' is very very large.

   (*iii*) $a = r\omega^2\left(\cos\theta + \dfrac{\cos 2\theta}{n}\right)$ ...... Approximate acceleration (Exact expression for acceleration is very difficult

   where r = crank radius, θ = Inclination of crank with I.D.C. (inner dead centre)

   $n = \dfrac{l}{r}$ = ratio of length of connecting rod to the length of crank radius

   ω = Angular velocity of crank
5. The angular velocity (ω_c) and angular acceleration (α_c) of the connecting rod are given by,

   (*i*) $\omega_c = \dfrac{\omega\cos\theta}{\sqrt{n^2 - \sin^2\theta}}$ .... Exact expression

   $= \dfrac{\omega\cos\theta}{n}$ ...... Approximate expression

   (*ii*) $\alpha_c = -\omega^2\sin\theta\left[\dfrac{n^2 - 1}{(n^2 - \sin^2\theta)^{3/2}}\right]$ ...... Exact expression

   $= \dfrac{-\omega^2\sin\theta\,(n^2 - 1)}{n^3}$ ...... Approximate expression

   $= \dfrac{-\omega^2\sin\theta}{n}$ ...... When 'n' is very very large

   where ω = Angular velocity of the crank.
6. The relation between angle (θ) and angle (ϕ) is given by
   $$\sin\phi = \frac{\sin\theta}{n} \text{ where } n = \frac{l}{r}.$$
7. If the value of α_c is positive, then angle ϕ is increasing. This means connecting rod is rotating anti-clockwise.

8. The velocity and acceleration of the reciprocating parts of the steam engine or I.C. engine can be determined graphically by :
(*i*) Klein's construction
(*ii*) Ritterhau's construction and
(*iii*) Bennett's construction.

9. According to Klein's construction (if OA = crank length ; BA = connecting rod length and ω = angular velocity of crank)

$$v_A = v_{AO} = \omega \times OA \; ; \; v_B = v_{BO} = \omega \times OC \text{ and } v_{BA} \times AC \text{ and}$$

$$f_A = f_{AO} = f^r_{AO} = \omega^2 \times OA, f_B = f_{BO} = f^r_{BO} = \omega^2 \times FO \; ; f^r_{BA} = \omega^2 \times AH \text{ and } f^t_{BA} = \omega^2 \times HF$$

The values of OC and AC are obtained from Klein's velocity diagram whereas the values of FO, AH and HF are obtained from Klein's acceleration diagram.

10. Torque (T) exerted on the crank-shaft is given by,
(*i*) $T = F_t \times r$ where F_t = Tangential force on crank
$= F_p \times OC$ ...... if friction and inertia of moving parts are neglected

where F_P = Force exerted by steam on piston $= p \times \frac{\pi}{4} D^2$ or $(p_1 A_1 - p_2 A_2)$

$$OC = r\left(\sin\theta + \frac{\sin 2\theta}{2n}\right)$$

The value of OC can also be obtained from scale drawing.

11. The torque (T) on the crank-shaft when friction and inertia of moving parts are considered but weight of connecting rod is neglected, is given by

$$\left.\begin{aligned} T &= F_t \times r \\ &= F^* \frac{\sin(\theta+\phi)}{\cos\phi} \times r \\ &= F^* \times r \times \left(\sin\theta + \frac{\sin 2\theta}{2\sqrt{n^2 - \sin^2\theta}}\right) \\ &= F^* \times OM \end{aligned}\right]$$ Considering friction and inertia of moving parts, but neglecting the weight of the connecting rod.

where F^* = Piston effort or net force acting on the piston or cross-head pin along the line of stroke
$= F_P + F_i - F_R$ .... For horizontal engine
$= F_P + F_i - F_R \pm W$ ...... For vertical engine

where F_i = Inertia force and F_R = Frictional resistance and W = weight of reciprocating parts.
OM = Length obtained from scale drawing.

12. The torque (T) on the crank-shaft when weight of the connecting rod is considered, is given by,
(*a*) $T = F_t \times r$
where F_t is obtained from equation

$$F_t \times IA = F_i \times IB + F_{ic} \times IX + W_c \times IY$$

The values of IA, IB, IX and IY are obtained by graphical method and
F_i = Inertia force of reciprocating parts
F_{ic} = Inertia force of connecting rod and
W_c = Weight of connecting rod.

13. The torque (T) on the crank-shaft when weight of the connecting rod is considering can also be obtained by analytical method as given below :

$$T = T_B + T_c^* + T_W$$

where $T_B = F_i \times OC$ where F_i = total inertia force of equivalent reciprocating mass and OC = length obtained from scale drawing.

T_c^* = corresponding torque on the crank-shaft due to correction couple T_c

$= \frac{T_c \times \cos\theta}{n}$ where T_c = correction couple applied

$= m_c \times L_1 (l - L) \times \alpha_{BA}$

T_W = Torque due to one part of weight of connecting rod placed at A

$= W_2 \times r \times \cos\theta$

where $W_2 = m_2 \times g$ in which m_2 is the one part of mass of connecting rod placed at A.

14. The expression for various forces acting on the reciprocating parts of an engine, neglecting the weight of the connecting rod, are given by :

$$F_c = \text{Force along the connecting rod} = \frac{F^*}{\cos\phi}$$

$$F_N = \text{Thrust on the sides of the cylinder walls} = F^* \tan\phi$$

$$F_t = \text{Tangential force on crank (or crank effort)} = F^* \frac{\sin(\theta + \phi)}{\cos\phi}$$

$$F_B = \text{Thrust on crank-shaft bearing} = F^* \frac{\cos(\theta + \phi)}{\cos\phi}$$

where F^* = Piston effort $= F_P + F_i - F_R$.

15. The two mass system having masses m_1 and m_2 will be dynamically equivalent to the rigid body of mass m if

$$m_1 + m_2 = m \text{ ; } m_1 \times L_1 = m_2 \times L_2 \text{ ; and } m_1 \times L_1^2 + m_2 L_2^2 = mk^2$$

Then $L_1 \times L_2 = k^2$; $m_1 = \frac{mL_2}{L_1 + L_2}$ and $m_2 = \frac{mL_1}{L_1 + L_2}$.

16. If two masses (m_1 and m_2) are placed at arbitarily distances from the C.G. of the solid body, then the two masses will not be dynamically equivalent to the solid body. Hence a correction couple (T_c) will be applied to make the two masses dynamically equivalent. The value of correction couple (T_c) is given by

$$T_c = m \times L_1 \times (l - L) \times \alpha$$

where m = mass of the solid body (or connecting rod)

L_1 = Distance of C.G. from the point where mass m_1 is placed

l = Distance between two masses which are placed arbitrarily

L = Distance between two masses m_1 and m_2 which form a true dynamically equivalent system

α = Angular acceleration of solid body.

EXERCISE 11

A. THEORETICAL QUESTIONS

1. Define and explain the terms : Inertia force and inertia torque.
2. State D-Alembert's principle.
3. For a reciprocating engine, prove that the displacement, velocity and acceleration of the piston are given respectively by the expression :

Displacement, $x_P = r\,[(1 - \cos\theta) + (n - \sqrt{n^2 - \sin^2\theta})]$

Velocity, $v_P = \omega r \left(\sin\theta + \frac{\sin 2\theta}{2n} \right)$

Acceleration, $a_P = \omega^2 r \left(\cos\theta + \frac{\cos 2\theta}{n} \right)$

where r = length of crank, l = length of connecting rod, $n = \frac{l}{r}$ and

θ = Inclination of the crank to the inner dead centre position. *(AMIE, Summer 1976)*

4. Name the different graphical methods for obtaining velocity and acceleration of the reciprocating parts of a steam engine or internal combustion engine.

5. Describe in details the Klein's construction for obtaining velocity and acceleration of the reciprocating parts of a steam engine.

6. Find an expression for the torque exerted on the crank-shaft when friction and inertia of moving parts are neglected.

7. Define and explain the terms : piston effort, crank effort, thrust on the sides of cylinder walls and thrust on the crank- shaft bearing.

8. Prove that the torque exerted on the crank-shaft, when friction and inertia of the moving parts of a reciprocating engine is considered but weight of the connecting rod is neglected, is given by

$$T = F^* \times r \times \left(\sin\theta + \frac{\sin 2\theta}{2\sqrt{n^2 - \sin^2\theta}} \right)$$

where F^* = Piston effort, r = length of crank, θ = Inclination of crank from I.D.C., and

$n = \frac{l}{r}$ = ratio of length of connecting rod to crank length.

9. What is meant by a dynamically equivalent system. *(AMIE, Winter 1985)*

10. The two mass system (m_1, m_2) is dynamically equivalent to a rigid body of mass (m). Then prove that

$$m_1 = \frac{m \times L_2}{L_1 + L_2},\ m_2 = \frac{m \times L_1}{L_1 + L_2} \text{ and } L_1 \times L_2 = k^2$$

where L_1 = Distance of mass m_1 from the C.G. of the body,

L_2 = Distance of mass m_2 from the C.G. of the body and k = Radius of gyration.

11. Prove that the correction couple (T_c) is given by,

$$T_c = m \times L_1 \times (l - L) \times \alpha$$

where m = mass of rigid body,

L_1 = Distance of mass m_1 from the C.G. of the body

L = Distance between two masses m_1 and m_2 which form a true dynamically equivalent system

l = Distance between two masses which are placed arbitarily.

α = Angular acceleration of the rigid body.

12. Describe is details the analytical and graphical method of obtaining the torque exerted on the crank-shaft when weight of the connecting rod is considered.

B. NUMERICAL PROBLEMS

1. Fig. 11.2 (*a*) shows a connecting rod. If the connecting rod length is 500 mm between the two centres, mass = 15 kg and distance of the C.G. of the rod from the big end is 200 mm. The radius of gyration of the connecting rod is 140 mm about an axis passing through C.G. Find the magnitude, direction and line of action of the resultant force acting on the connecting rod if the linear acceleration of the C.G. of the connecting rad is 50 m/s^2 in the direction shown in Fig. 11.2 and angular acceleration of the rod is 150 rad/s^2 clockwise.

   [**Ans.** (*i*) 750 N (*ii*) the resultant force will act in the direction of linear acceleration (*iii*) The line of action of the resultant force will be at a distance 58.8 mm downward]

2. For the problem 1, find the inertia force and inertia torque which must be applied on the given connecting rod to bring it in equilibrium. [**Ans.** $F_i = -750$ N and $T_i = -44.1$ Nm]

3. The lengths crank and connecting rod of a vertical reciprocating engine are 150 mm and 750 mm respectively. The crank is rotating at 400 r.p.m. clockwise. Find analytically : (*i*) acceleration of piston, (*ii*) velocity of piston

and (*iii*) angular acceleration of the connecting rod when the crank has turned through 40° from the top dead centre and piston is moving downwards. **[Ans.** (*i*) 210.54 m/s^2 ; (*ii*) 4.66 m/s ; (*iii*) – 222 rad/s^2]

4. For the problem 3, find : (*i*) the crank angle at which the velocity of the piston is maximum and (*ii*) the value of the maximum velocity of the piston. **[Ans.** (*i*) 79.28° and (*ii*) 6.4 m/s]

5. The lengths of crank and connecting rod of a horizontal reciprocating engine are 200 mm and 1.0 m respectively. The crank is rotating at 200 r.p.m. When the crank has turned 30° from the inner dead centre, find analytically : (*i*) acceleration of the piston (*ii*) velocity of the piston (*iii*) angular velocity of the connecting rod and (*iv*) angular acceleration of the connecting rod. **[Ans.** (*i*) 84.72 m/s^2, (*ii*) 2.458 m/s, (*iii*) 3.645 rad/s, (*iv*) – 42.73 rad/s^2]

6. For the problem 5, find the position of the crank from the inner dead centre for zero acceleration of the piston.

[Ans. θ = 79.28°]

7. Using Klein's construction for the problem 5, find : (*i*) velocity of the piston (*ii*) acceleration of piston (*iii*) angular velocity of the connecting rod and (*iv*) angular acceleration of the connecting rod when the crank has turned 30° from the inner dead centre. **[Ans.** (*i*) 84.7 m/s^2 (*ii*) 2.46 m/s (*iii*) 3.65 rad/s (*iv*) – 42.7 rad/s^2]

8. The lengths of crank and connecting rod of a reciprocating engine are 100 mm and 400 mm respectively. The crank is rotating at a uniform speed of 240 r.p.m. Using Klein's construction find : (*i*) the acceleration of the piston (*ii*) the acceleration of the middle point of the connecting rod and (*iii*) angular acceleration of the connecting rod when the crank has turned through 45° from the inner dead centre.

[Ans. (*i*) 45.92 m/s^2 (*ii*) 51.92 m/s^2 (*iii*) 110.51 rad/s^2]

9. The crank of a reciprocating engine is rotating in clockwise direction with a constant angular velocity of 30 rad/s. The lengths of crank and connecting rod are 200 mm and 750 mm respectively. Using Klein's construction, find : (*i*) velocity of piston (*ii*) velocity of mid-point of connecting rod (*iii*) angular velocity of connecting rod (*iv*) acceleration of piston (*v*) acceleration of mid-point of connecting rod (*vi*) angular acceleration of connecting rod when crank has turned through 30° from inner dead centre.

[Ans. (*i*) 3.8 m/s (*ii*) 4.2 m/s (*iii*) 7.42 rad/s (*iv*) 180 m/s^2 (*v*) 173.7 m/s^2 (*vi*) 119.57 rad/s^2]

12

Turning Moment Diagrams

12.1. Introduction

Turning moment diagram is a diagram which shows the variation of turning moment (or torque) on a crank for various position of the crank. Hence turning moment diagram is a graph between T and θ. The turning moment (or torque T) is taken along y-axis and crank angle (θ) along x-axis. Generally the effect of inertia of connecting rod is neglected while drawing the turning moment diagram. Turning moment diagram is also written as T-M diagram.

The variation of torque (T) during one revolution of the crank-shaft of a steam engine or an I.C. engine is given by

$$T = F_t \times r \quad \text{...}(i)$$

$$= F^* \times r \times \left[\sin\theta + \frac{\sin 2\theta}{2\sqrt{n^2 - \sin^2\theta}} \right] \quad \text{...}(ii)$$

where F_t = Tangential force (or force normal to crank) on crank,

r = Crank radius,

F^* = Net force acting on the piston = $F_p \pm W$ when effect of inertia of connecting rod is neglected,

θ = Angle turned by crank from I.D.C. (or from top dead centre), and

n = Ratio of the length of connecting rod to crank radius.

From equation (*ii*), it is clear that torque (T) on crank shaft can be obtained if F^* (*i.e.*, the net force on the piston) and θ are known. After knowing the values of T and θ for one revolution of crank, T.M. diagram can be drawn.

12.2. Turning Moment Diagrams for Different Types of Engines

The turning moment diagram for the following types of engines will be discussed :

(*i*) Turning moment diagram for a single-cylinder double acting steam engine,

(*ii*) Turning moment diagram for a four stroke cycle internal combustion engine and

(*iii*) Turning moment diagram for a multi-cylinder engine.

12.2.1. Turning Moment Diagram for a Single-Cylinder Double Acting Steam Engine

Fig. 12.1 shows the line diagram of a single cylinder, double-acting horizontal steam engine. Along with this diagram, the indicator diagram is also drawn for the cover end only. [There will be an indicator diagram for crank end also.] For the given position of the piston when crank is rotating-clockwise, the pressure of steam on the cover side of the piston is given by ordinate p_1 whereas the pressure on the crank side of the piston is given by ordinate p_2.

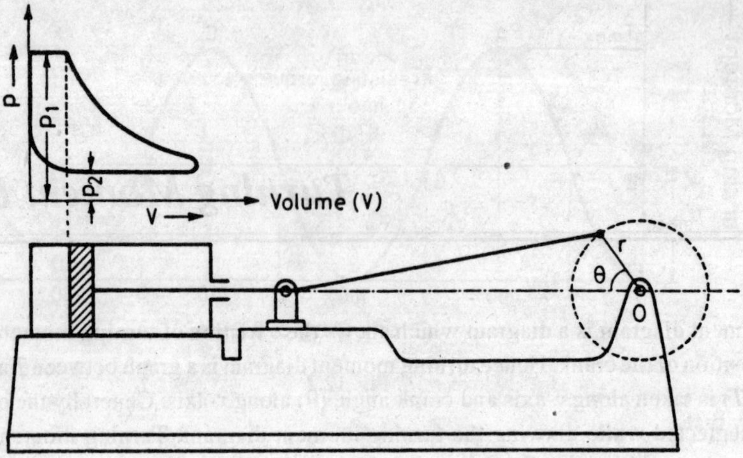

Fig. 12.1

Now the net force exerted by steam on piston is given by

$$F_P = p_1 A_1 - p_2 A_2$$

where A_1 = Area of piston on cover side and

A_2 = Area of piston on crank side.

Thus net force exerted by steam on piston, tends to move the piston towards the right. If the effect of inertia of the moving parts is neglected, the force F_P will produce a torque (or turning moment) on the crank-shaft. The magnitude of the torque (T) or turning moment is given by,

$$T = F^* \times OC$$

$$= F_P \times OC \qquad (\because\ F^* = F_P \text{ for horizontal engine, if inertia force of moving parts is neglected})$$

$$= F_P \times r \times \left(\sin\theta + \frac{\sin 2\theta}{2\sqrt{n^2 - \sin^2\theta}} \right)$$

[The value of OC may be obtained from scale drawing or from analytical method]

$$\left[\because\ OC = r\left(\sin\theta + \frac{\sin 2\theta}{2\sqrt{n^2 - \sin^2\theta}} \right) \right]$$

$$= F_P \times r \times \left(\sin\theta + \frac{\sin 2\theta}{2n} \right) \qquad [\text{If } n^2 \text{ is very large as compared to } \sin^2\theta]$$

Now if for a complete revolution of the crank-shaft, the variation of turning moment (or torque T) for various values of θ is plotted, we get the turning moment diagram (*i.e.*, *T-M* diagram). For the above case the variation of turning moment against θ for one revolution of crank shaft is shown in Fig. 12.2. [The turning moment is represented along y-axis and crank angle θ along the x-axis].

It is clear from the above expression that if θ = 0, the turning moment (T) is zero. When θ = 90°, the turning moment is maximum. When θ = 180°, the turning moment is again zero. For the outstroke of the piston, θ varies from 0 to 180°. The turning moment diagram for the outstroke of the piston (*i.e.*, when θ varies from 0 to 180°) is shown by curve *OAB* in Fig. 12.2. For the instroke of the piston, θ varies from 180° to 360°. The turning moment diagram for the instroke of the piston is shown by curve *BCD*.

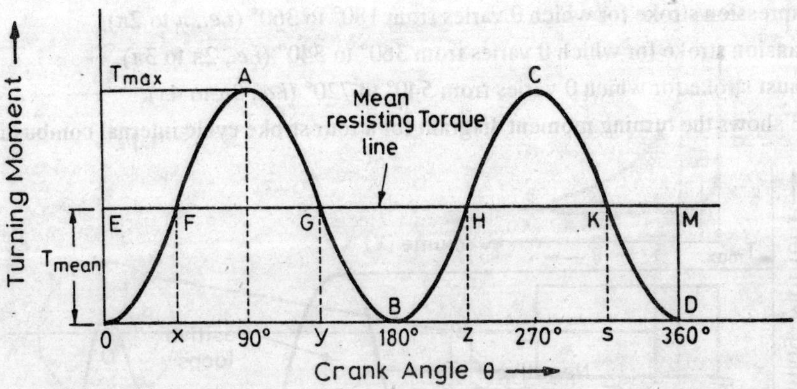

Fig. 12.2

We know that,

$$\text{Work done} = T \times \theta \qquad ...(12.1)$$

where θ = angle turned.

$\therefore$ Work done per revolution $= T \times 2\pi$ ($\because$ For one revolution of crank, the angle turned is 2π)

But $T \times \theta$ = Area of turning moment diagram

or $T \times 2\pi$ = Area of turning moment diagram for one revolution of the crank.

$\therefore$ Work done per revolution = Area of turning moment diagram for one revolution.

= Area $OABCD$

In actual practice, the engine is assumed to work against the uniform mean resisting torque. The mean torque (T_{mean}) against which the engine works is given by,

$$T_{mean} = \frac{\text{Area } OABCD}{2\pi}$$

where T_{mean} is the mean torque and is the mean height of the turning moment diagram which is equal to ordinate OE.

Now the area of rectangle $OEMD = OE \times OD$

$= T_{mean} \times 2\pi$ ($\because$ $OE = T_{mean}$ and $OD = 2\pi$)

= Area of turning moment diagram $OABCD$

= Work done per revolution

$\therefore$ Area of rectangle $OEMD$ represents the work done against the mean resisting torque.

Let T = Torque at any instant on the crank-shaft

T_{mean} = Mean resisting torque

Then $T - T_{mean}$ = Accelerating torque on the rotating parts of the engine (*i.e.*, flywheel).

Hence if T is more than T_{mean}, the flywheel accelerates whereas if T is less than T_{mean}, the flywheel retards.

12.2.2. Turning Moment Diagram for a four stroke cycle internal Combustion Engine

In case of steam engine, the turning moment diagram repeats itself after every one revolution whereas in case of four stroke internal combustion engine, the turning moment diagram repeats itself after every two revolution. The four strokes are :

(*i*) Suction stroke for which angle θ varies from 0 to 180° (*i.e.*, 0 to π)

(*ii*) Compression stroke for which θ varies from 180° to 360° (*i.e.*, π to 2π)

(*iii*) Expansion stroke for which θ varies from 360° to 540° (*i.e.*, 2π to 3π)

(*iv*) Exhaust stroke for which θ varies from 540° to 720° (*i.e.*, 3π to 4π).

Fig. 12.3 shows the turning moment diagram for a four stroke cycle internal combustion engine.

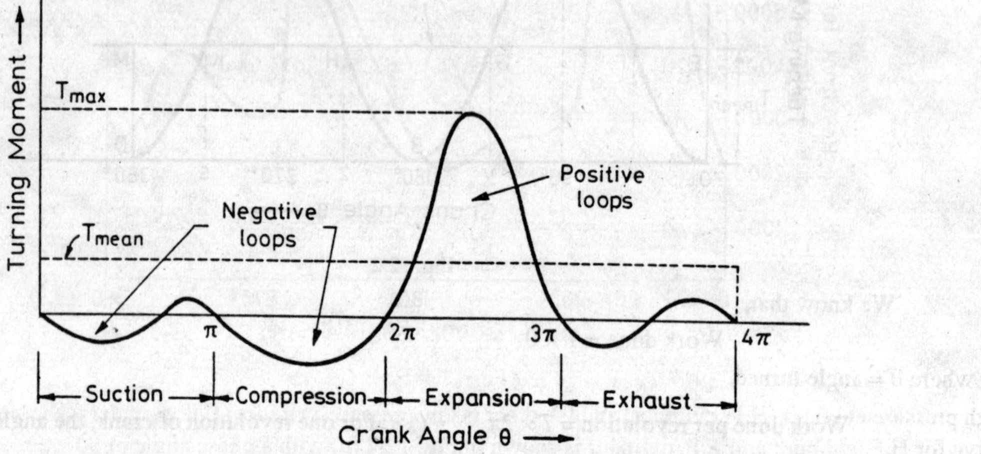

Fig. 12.3

During suction stroke, the pressure inside the cylinder is less than the atmospheric pressure and hence the turning moment on the crank is negative for most of the suction stroke. Hence a negative loop is formed as shown in Fig. 12.3. During compression stroke, the work is done by piston on the gases and hence a large negative loop is obtained. During expansion stroke the work is done by the gases on the piston and due to this the turning moment on the crank is positive. Hence during this stroke a large positive loop is obtained. During exhaust stroke, the work is done by the piston on gases and hence turning moment on crank is negative. Therefore, a negative loop is formed during this stroke also as shown in Fig. 12.3.

12.2.3. Turning Moment Diagram for a Multi-cylinder Engine

For multi-cylinder engine, the turning moment diagram is obtained by combining the turning moment diagram for each cylinder. Let us consider the case of two cylinders of a steam engine. One of the cylinder is

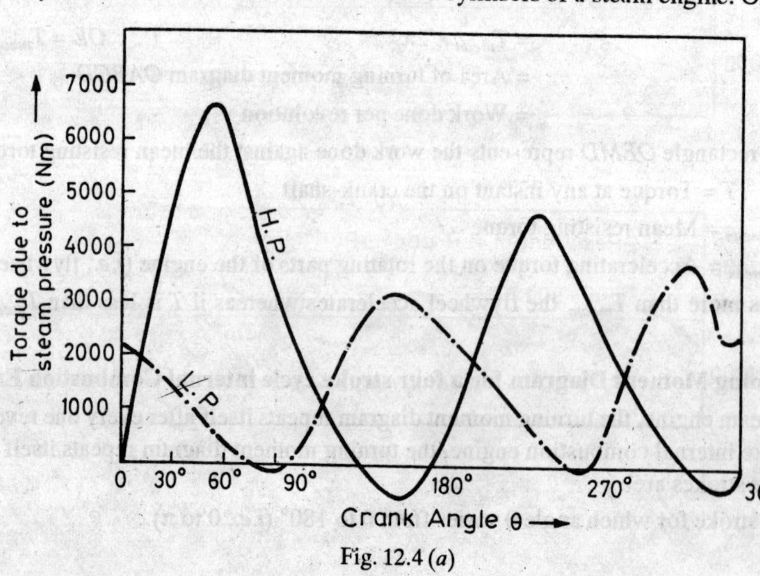

Fig. 12.4 (*a*)

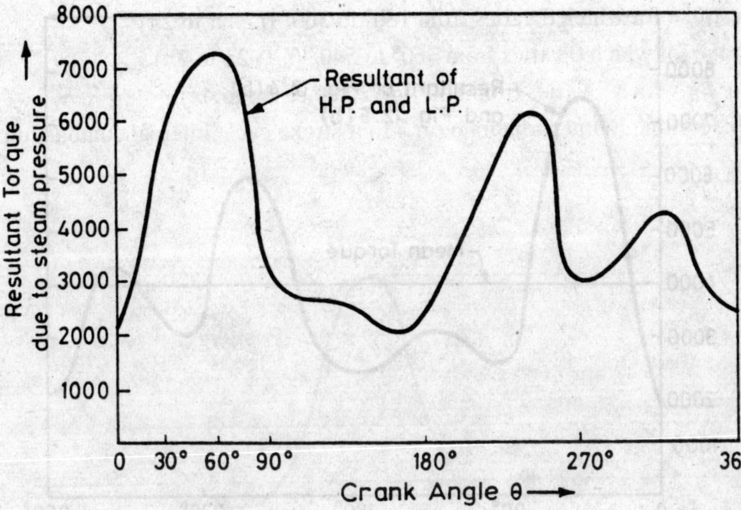

Fig. 12.4 (*b*)

high pressure cylinder (H.P. cylinder) and the other is low pressure-cylinder (L.P. cylinder). The steam torque curve for H.P. cylinder and L.P. cylinder is shown in Fig. 12.4 (*a*) with a phase angle of 90°.

The resultant turning moment (or torque) diagram due to steam pressure is obtained by combining these two turning moment diagrams. This resultant turning moment diagram due to steam pressure is shown in Fig. 12.4 (*b*).

Fig. 12.5 (*a*) shows the variation of torque due to inertia for H.P. cylinder and L.P. cylinder whereas Fig. 12.5 (*b*) shows the resultant turning moment diagram due to inertia for H.P. cylinder and L.P. cylinder.

Now the turning moment diagrams of Fig. 12.4 (*b*) and of Fig. 12.5 (*b*) are combined and the resultant turning moment diagram due to steam pressure and inertia is obtained as shown in Fig. 12.5 (*c*).

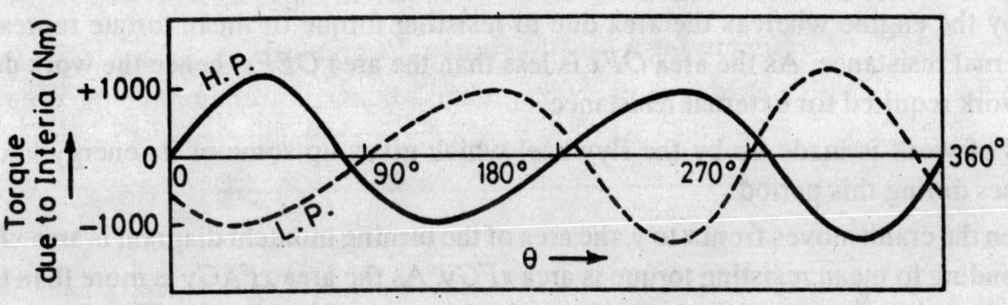

(*a*)

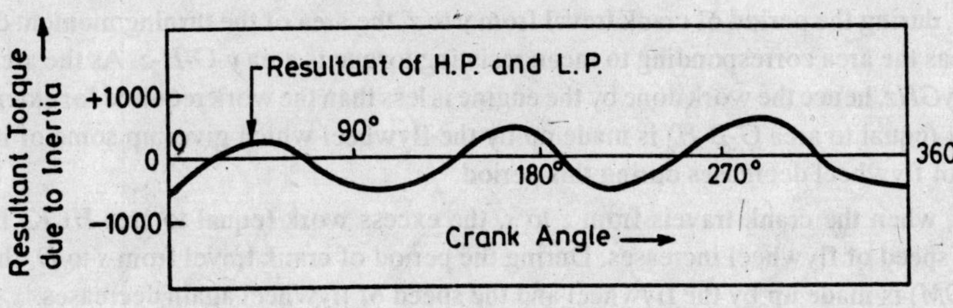

(*b*)

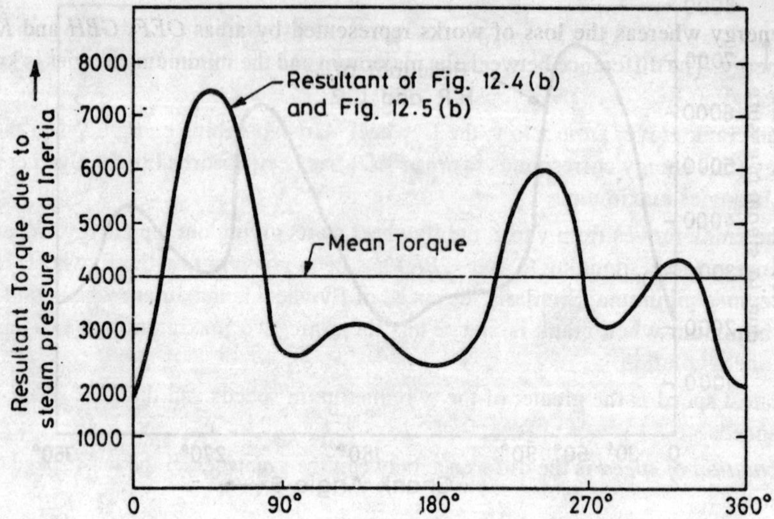

(c)

Fig. 12.5

12.3. Fluctuation of Energy and Fluctuation of Speed of Crank-shaft

The torque which resists the rotation of the crank-shaft is approximately uniform. But the torque exerted on the crank-shaft by steam (or gas) fluctuate considerably. Hence there is unbalanced torque which tends either to increase or to decrease the speed of rotation of the crank-shaft.

The turning moment diagram for a single cylinder double acting steam engine is shown in Fig. 12.2. The mean resisting torque line *EM* cuts the turning moment diagram at points *F*, *G*, *H* and *K*. Let us consider the movement of crank from *O* to *x*. The area of the turning moment diagram is equal to area *OFx*. But the area due to resisting torque (or mean torque) is area *OEFx*. The area of turning moment diagram represents the work done by the engine whereas the area due to resisting torque or mean torque represents the work required for external resistance. As the area *OFx* is less than the area *OEFx*, hence the work done by engine is less than the work required for external resistance.

This loss of work is made up by the flywheel which gives up some of its energy and the speed of flywheel decreases during this period.

Now when the crank moves from *x* to *y*, the area of the turning moment diagram is area *xFAGy* whereas the area corresponding to mean resisting torque is area *xFGy*. As the area *xFAGy* is more than the area *xFGy*, hence work done by engine is more than the work required for external resistance. The excess work (equal to area *FAG*) is stored in the flywheel and hence the speed of flywheel increases during this period.

Similarly, during the period of crank travel from *y* to *z*, the area of the turning moment diagram is area *y-G-B-H-z* whereas the area corresponding to mean resisting torque is area *y-G-H-z*. As the area *y-G-B-H-z* is less than the area *yGHz*, hence the work done by the engine is less than the work required for external resistance. This loss of work (equal to area *G-B-H*) is made up by the flywheel which gives up some of its energy. Due to this the speed of flywheel decreases during this period.

Similarly, when the crank travels from *z* to *s*, the excess work (equal to area *HCK*) is stored in the flywheel and the speed of flywheel increases. During the period of crank travel from *s* to *D*, the loss of work (equal to area *KDM*) is made up by the flywheel and the speed of flywheel again decreases.

The excess works and loss of works represented by the areas *FAG, HCK, OEF, GBH* and *KDM* represent the *fluctuation of energy*. The excess works represented by the areas *FAG* and *HCK* are positive fluctuation of energy whereas the loss of works represented by areas *OEF*, *GBH* and *KDM* are negative fluctuation of energy. The difference between the maximum and the minimum energies is known as *maximum fluctuation of energy*.

When the crank moves from *x* to *y*, the flywheel starts absorbing energy. When the crank is at *y*, the maximum energy (this energy corresponds to area *FAG*) has been absorbed in the flywheel. Hence the speed of the flywheel becomes maximum.

When the crank moves from *y* to *z*, the flywheel starts giving out the energy. When the crank is at *z*, the maximum energy (corresponding to area *GBH*) has been given out by the flywheel. Hence the speed of the flywheel becomes minimum. Similarly the speed of flywheel is maximum when crank is at *s*. The speed of flywheel is minimum when crank is at *x*. Thus there are two maximum and two minimum speeds of crank-shaft for one revolution.

The greatest speed is the greater of the two maximum speeds and the least speed is the lesser of the two minimum speeds.

The fluctuation of speed is the difference between the greatest and the least speeds of the crank-shaft for one revolution.

12.4. Co-efficient of Fluctuation of Energy (K_e)

The ratio of maximum fluctuation of energy to the work done per cycle is known as co-efficient of fluctuation of energy. It is represented by K_e. Mathematically, co-efficient of fluctuation of energy is given by,

$$K_e = \frac{\text{Maximum fluctuation of energy}}{\text{Work done per cycle}} \quad \text{...(12.2)}$$

Now let us find the values of maximum fluctuation of energy and work-done per cycle.

12.4.1. Maximum Fluctuation of Energy (ΔE)

The difference between the maximum and minimum energies is known as maximum fluctuation of energy. It is represented by ΔE. Hence mathematically,

Maximum fluctuation of Energy,

$$\Delta E = \text{Maximum energy} - \text{Minimum energy}. \quad \text{...(12.3)}$$

Fig. 12.6 shows the turning moment diagram for a multi-cylinder engine. The mean torque line is represented by the horizontal line *AG*.

Let a_1, a_3 and a_5 = Areas above the mean torque line in work units *i.e.* Nm

a_2, a_4 and a_6 = Areas below the mean torque line in Nm

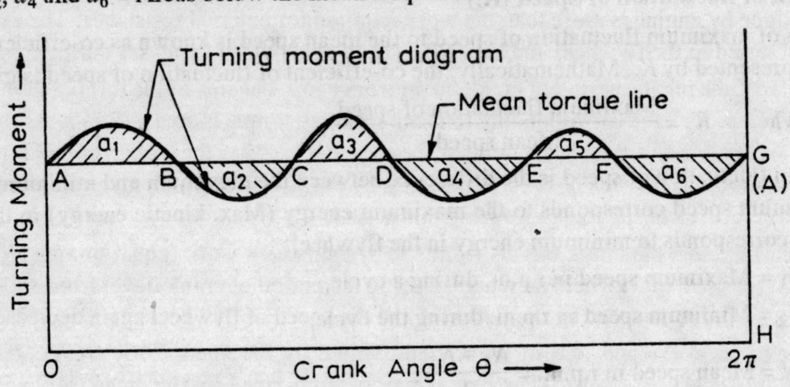

Fig. 12.6

The areas a_1, a_3 and a_5 represents the quantities of energies added to the flywheel whereas the areas a_2, a_4 and a_6 represents the quantities of energies taken from the flywheel.

Let E = Energy in the flywheel corresponding to point A.

The energies of the flywheel corresponding to points B, C, D, E, F and G are obtained from Fig. 12.5 as,

Energy at B = Energy at A + Area $a_1 = E + a_1$

Energy at C = Energy at B – Area $a_2 = E + a_1 - a_2$

Energy at D = Energy at $C + a_3 = E + a_1 - a_2 + a_3$

Energy at E = Energy at $D - a_4 = E + a_1 - a_2 + a_3 - a_4$

Energy at F = Energy at $E + a_5 = E + a_1 - a_2 + a_3 - a_4 + a_5$

Energy at G = Energy at $F - a_6 = E + a_1 - a_2 + a_3 - a_4 + a_5 - a_6$

The cycle repeats after G. Hence energy at G is equal to energy at A.

The greatest of these energies is the maximum kinetic energy of the flywheel. And the least of these energies is the least kinetic energy of the flywheel. The difference between the greatest and least energies is known as maximum fluctuation of energy.

∴ Maximum fluctuation of energy = Greatest energy – Least energy.

Work-done per cycle

Work-done per cycle is obtained from,

Work-done per cycle = Area corresponding to mean torque line

= Area of $OAGH = OA \times OH$

$= T_{mean} \times \theta$ ...(12.4)

where θ = Angle turned (in radians), in one revolution

$= 2\pi$ ... For steam engine and two stroke I.C. Engines

$= 4\pi$ ... For four stroke I.C. Engines.

T_{mean} = Mean torque and is obtained from $P = T_{mean} \times \omega$ ...(12.5)

$= \dfrac{P}{\omega}$ where P = Shaft-power and $\omega = \dfrac{2\pi N}{60}$

After knowing the values of maximum fluctuation of energy and work-done per cycle, the co-efficient of fluctuation of energy can be obtained.

12.5. Co-efficient of fluctuation of Speed (K_s)

The ratio of maximum fluctuation of speed to the mean speed is known as co-efficient of fluctuation of speed. It is represented by K_s. Mathematically, the co-efficient of fluctuation of speed is given by,

$$K_s = \frac{\text{Maximum fluctuation of speed}}{\text{Mean speed}} \qquad \text{...(12.6)}$$

Maximum fluctuation of speed is the difference between the maximum and minimum speed during a cycle. The maximum speed corresponds to the maximum energy (Max. kinetic energy) in the flywheel and minimum speed corresponds to minimum energy in the flywheel.

Let N_1 = Maximum speed in r.p.m. during a cycle,

N_2 = Minimum speed in r.p.m. during the cycle,

N = Mean speed in r.p.m. $= \dfrac{N_1 + N_2}{2}$.

ω_1, ω_2 and ω = Corresponding maximum angular speed, minimum and mean angular speeds during the cycle. Then co-efficient of fluctuation of speed,

$$K_s = \frac{\text{Max. speed} - \text{Min. speed}}{\text{Mean speed}} \quad \left(\text{or } \frac{\omega_1 - \omega_2}{\omega}\right) \quad ...(12.7)$$

$$= \frac{N_1 - N_2}{N} = \frac{N_1 - N_2}{\left(\frac{N_1 + N_2}{2}\right)} = \frac{2(N_1 - N_2)}{(N_1 + N_2)}$$

$$= \frac{2(\omega_1 - \omega_2)}{\omega_1 + \omega_2} \quad \left(\because \omega_1 = \frac{2\pi N_1}{60} \text{ and } \omega_2 = \frac{2\pi N_2}{60}\right) \quad ...(12.8)$$

12.6. Flywheels

A flywheel is used to control the speed variations caused by the fluctuation of energy during each cycle of operation. It acts as a reservoir of energy which stores energy during the period when the supply of energy is more than the requirement and releases the energy during the period when the supply of energy is less than the requirement. When the flywheel absorbs the energy, the speed of flywheel increases whereas when the flywheel releases the energy, the speed of flywheel decreases.

Let N_1 = Maximum speed of flywheel in r.p.m. during the cycle,

N_2 = Minimum speed during the cycle,

ω_1 and ω_2 = Corresponding maximum and minimum angular speeds,

I = Moment of inertia of flywheel,

m = Mass of flywheel,

k = Radius of gyration of the flywheel,

ΔE = Maximum fluctuation of energy,

K_e = Co-efficient of fluctuation of energy,

K_s = Co-efficient of fluctuation of speed,

N = Mean speed of flywheel during the cycle = $\frac{N_1 + N_2}{2}$

ω = Mean angular speed of flywheel during the cycle = $\frac{\omega_1 + \omega_2}{2}$

E = Kinetic energy of flywheel at mean speed.

We know that the kinetic energy of the flywheel corresponding to mean angular velocity is given by,

$$E = \frac{1}{2} \times I \times \omega^2$$

$$= \frac{1}{2} \times mk^2 \times \omega^2 \quad (\because I = mk^2)$$

The speed of flywheel changes from ω_1 to ω_2, hence maximum fluctuation of energy is given by,

$$\Delta E = \text{Maximum K.E.} - \text{Minimum K.E.}$$

$$= \frac{1}{2} \times I \times \omega_1^2 - \frac{1}{2} \times I \times \omega_2^2$$

$$= \frac{1}{2} I (\omega_1^2 - \omega_2^2) = \frac{1}{2} I (\omega_1 + \omega_2)(\omega_1 - \omega_2)$$

$$= I \times \omega \times (\omega_1 - \omega_2) \quad ...\left(\because \frac{\omega_1 + \omega_2}{2} = \omega\right)$$

$$= I \times \omega^2 \left(\frac{\omega_1 - \omega_2}{\omega}\right) \quad \text{...[Multiplying and dividing by } \omega]$$

But $\dfrac{\text{Max. speed} - \text{Min. speed}}{\text{Mean speed}} = \dfrac{\omega_1 - \omega_2}{\omega} = K_s$

$\therefore$ The above equation becomes as,

$$\Delta E = I \times \omega^2 \times K_s \quad ...(12.9)$$

$$= \frac{1}{2} I \times \omega^2 \times 2 \times K_s \quad \text{...[Multiplying and dividing by 2]}$$

$$= E \times 2 \times K_s \quad ...\left(\because \frac{1}{2} I\omega^2 = E\right)$$

$$= 2E \times K_s \quad ...(12.10)$$

The equation (12.9) can be written as

$$\Delta E = I \times \omega^2 \times K_s = mk^2 \times \omega^2 \times K_s \quad (\because I = mk^2) \quad ...(12.11)$$

If the thickness of the rim of the flywheel is very small as compared to the diameter of flywheel, then radius of gyration may be taken equal to mean radius of the flywheel *i.e.* $k = R$. Hence substituting the value of $k = R$ in equation (12.11), we get

$$\Delta E = m \times R^2 \times \omega^2 \times K_s \quad ...(12.12)$$

$$= m \times v^2 \times K_s \quad (\because v = \omega \times R) \quad ...(12.13)$$

Problem 12.1. *The maximum and minimum speed of a flywheel are 242 r.p.m. and 238 r.p.m. respectively. The mass of flywheel is 2600 kg and radius of gyration is 1.8 m. Find (i) mean speed of flywheel, (ii) maximum fluctuation of energy and (iii) co-efficient of fluctuation of speed.*

Sol. Given :

$$N_1 = 242 \text{ r.p.m. or } \omega_1 = \frac{2\pi N_1}{60} = \frac{2\pi \times 242}{60} = 25.34 \text{ rad/s ;}$$

$$N_2 = 238 \text{ r.p.m. or } \omega_2 = \frac{2\pi N_2}{60} = \frac{2\pi \times 238}{60} = 24.92 \text{ rad/s ; } m = 2600 \text{ kg ; } k = 1.8 \text{ m}$$

Let N = mean speed of the flywheel ; and

ΔE = Max. fluctuation of energy

I = Mass moment of inertia of flywheel = $mk^2 = 2600 \times 1.8^2 = 8424$ kg m^2

(i) Mean speed (N)

Mean speed, $N = \dfrac{N_1 + N_2}{2} = \dfrac{242 + 238}{2} =$ **240 r.p.m. Ans.**

(ii) Max. fluctuation of energy (ΔE)

Max. fluctuation of energy is given by,

$$\Delta E = \text{Max. K.E. of flywheel} - \text{Min. K.E. of flywheel}$$

$$= \frac{1}{2} I \times \omega_1^2 - \frac{1}{2} \times I \times \omega_2^2$$

$$= \tfrac{1}{2} \times 8424 \times 25.34^2 - \tfrac{1}{2} \times 8424 \times 24.92^2$$

$$(\because \; I = 8424 \text{ kg m}^2 \; ; \; \omega_1 = 25.34 \text{ rad/s} \; ; \; \omega_2 = 24.92 \text{ rad/s})$$

$$= 2704590 - 2615679 = \mathbf{88911 \text{ Nm}}. \quad \textbf{Ans.}$$

(iii) Co-efficient of fluctuation of speed (K_s)

Using equation (12.9), we get

$$\Delta E = 2 \times E \times K_s$$

or $$K_s = \frac{\Delta E}{2E} \qquad ...(i)$$

where E = Kinetic energy of flywheel corresponding to mean speed

$$= \frac{1}{2} \times I \times \omega^2 \quad ... \left(\text{where } \omega = \frac{2\pi N}{60} = \frac{2\pi \times 240}{60} = 25.13 \text{ rad/s} \right.$$

$$\left. \text{or } \omega = \frac{\omega_1 + \omega_2}{2} = \frac{25.34 + 24.92}{2} = 25.13 \text{ rad/s} \right)$$

$$= \frac{1}{2} \times 8424 \times 25.13^2 \quad ...(\because \; I = 8424 \text{ kg m}^2)$$

$$= 2659949 \text{ Nm}$$

and $\Delta E = 88911$ Nm

Substituting the values of ΔE and E in equation (*i*), we get

$$K_s = \frac{88911}{2 \times 2659949} = \mathbf{0.0167}. \quad \textbf{Ans.}$$

Alternate Method

Using equation (12.7), we get

$$K_s = \frac{\omega_1 - \omega_2}{\omega} = \frac{25.34 - 24.92}{25.13} \qquad (\because \; \omega = 25.13)$$

$$= \mathbf{0.0167}. \quad \textbf{Ans.}$$

Problem 12.2. *Find the maximum and minimum speeds of a flywheel of mass 5200 kg and radius of gyration 1.8 m when the fluctuation of energy is 100800 Nm. The mean speed of the engine is 180 r.p.m.*

Sol. Given :

$m = 5200$ mg ; $k = 1.8$ m ; $\Delta E = 100800$ Nm ; N = mean speed = 180 r.p.m.

or $$\text{mean angular speed } \omega = 2\pi N/60 = \frac{2\pi \times 180}{60} = 18.85 \text{ rad/s}$$

$$\therefore \qquad \omega = \frac{\omega_1 + \omega_2}{2} = 18.85 \text{ rad/s}$$

Let N_1 = Maximum speed and

N_2 = Minimum speed.

Fluctuation of energy (ΔE) is given by,

$$\Delta E = \text{Difference of max. and min. kinetic energy of flywheel}$$

$$= \text{Max. K.E. of flywheel} - \text{Min. K.E. of flywheel}$$

$$= \tfrac{1}{2} \times I \times \omega_1^2 - \tfrac{1}{2} \times I \times \omega_2^2$$

$$= \tfrac{1}{2} \times I\,(\omega_1^2 - \omega_2^2) = \tfrac{1}{2} I\,(\omega_1 + \omega_2)\,(\omega_1 - \omega_2)$$

$$= I \times \left(\frac{\omega_1 + \omega_2}{2}\right)(\omega_1 - \omega_2)$$

$$= I \times 18.85\,(\omega_1 - \omega_2) \qquad \left(\because \frac{\omega_1 + \omega_2}{2} = \omega = 18.85\right)$$

$$= mk^2 \times 18.85\,(\omega_1 - \omega_2) \qquad (\because I = mk^2 = 5200 \times 1.8^2)$$

or $$100800 = 5200 \times 1.8^2 \times 18.85\,(\omega_1 - \omega_2) \qquad (\because \Delta E = 100800 \text{ Nm})$$

$\therefore$ $$\omega_1 - \omega_2 = \frac{100800}{5200 \times 1.8^2 \times 18.85} = 0.3174 \qquad ...(i)$$

We know that mean angular speed (ω),

$$\omega = \frac{\omega_1 + \omega_2}{2}$$

or $$18.85 = \frac{\omega_1 + \omega_2}{2} \qquad (\because \omega = 18.85 \text{ rad/s})$$

or $$\omega_1 + \omega_2 = 18.85 \times 2 = 37.70 \qquad ...(ii)$$

Adding equation (*i*) and (*ii*), we get

$$2\omega_1 = 37.70 + 0.3174 = 38.0174$$

or $$\omega_1 = \frac{38.0174}{2} = 19.0087 \text{ rad/s}$$

From equations (*ii*), $\omega_2 = 37.70 - \omega_1 = 37.70 - 19.0087 = 18.6913$ rad/s

Max. speed $$N_1 = \frac{60 \times \omega_1}{2\pi} \qquad \left(\because \omega_1 = \frac{2\pi N_1}{60} \text{ hence } N_1 = \frac{60 \times \omega_1}{2\pi}\right)$$

$$= \frac{60 \times 19.0087}{2\pi} = \mathbf{181.519 \simeq 181.5 \text{ r.p.m.}} \quad \textbf{Ans.}$$

Minimum speed $$N_2 = \frac{60 \times \omega_2}{2\pi} = \frac{60 \times 18.6913}{2\pi} = 178.48 \simeq \mathbf{178.5 \text{ r.p.m.}} \quad \textbf{Ans.}$$

Problem 12.3. *The turning moment diagram for a multi-cylinder engine has been drawn to a scale of 1 mm = 325 Nm vertically and 1 mm = 3° horizontally. The areas above and below the mean torque line are – 26, + 378, – 256, + 306, – 302, + 244, – 380, + 261 and – 225 mm². The engine is running at a mean speed of 600 r.p.m. The total fluctuation of speed is not to exceed ± 1.8% of the mean speed. If the radius of flywheel is 0.7 m, find the mass of the flywheel.*

Sol. Given :

Turning moment scale : 1 mm = 325 Nm

Crank angle scale, 1 mm = 3° $= 3 \times \frac{\pi}{180} = 0.05236$ rad

$\therefore$ 1 mm² on turning moment diagram = 325 × 0.05236 Nm = 17.017 Nm

Mean speed, $N = 600$ r.p.m.

$\therefore$ Mean angular speed, $\omega = \frac{2\pi N}{60} = \frac{2\pi \times 600}{60} = 62.84$ rad/s ; Radius, $R = 0.7$ m

As the total fluctuation of speed is not to exceed ± 1.8% of the mean speed, hence $(\omega_1 - \omega_2) = (1.8 \times 2)\%$ of ω. This means $\frac{\omega_1 - \omega_2}{\omega} = \frac{3.6}{100} = 0.036$. $\therefore$ Co-efficient of fluctuation of speed, $K_s = 0.036$.

Let 1 mm^2 on turning moment diagram = x Nm

$\therefore$ $x = 17.017$ Nm $\quad (\because \; 1\text{ mm}^2 = 17.017\text{ Nm})$ ...(*i*)

Then the areas above and below the mean torque line in work units (*i.e.* Nm) will be $-26x$; $+378x$, $-256x$, $+306x$, $-302x$, $+244x$, $-380x$, $+261x$ and $-225x$ respectively.

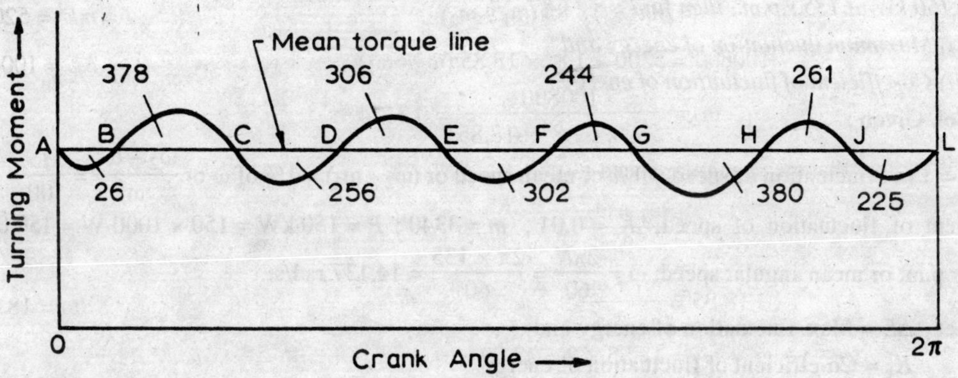

Fig. 12.7

Let E = total energy at A (Refer to Fig. 12.7)

Then energy at B $= E - 26x$

Energy at C $= E - 26x + 378x = E + 352x$

Energy at D $= E + 352x - 256x = E + 96x$

Energy at E $= E + 96x + 306x = E + 402x$ ...(Max. Energy)

Energy at F $= E + 402x - 302x = E + 100x$

Energy at G $= E + 100x + 244x = E + 344x$

Energy at H $= E + 344x - 380x = E - 36x$ ...(Max. Energy)

Energy at J $= E - 36x + 261x = E + 225x$

Energy at L $= E + 225x - 225x = E$ = Energy at A

The cycle repeats after L and hence energy at L is equal to energy at A.

The maximum energy is at E and minimum energy is at H.

$\therefore$ Max. energy $= E + 402x$

Min. energy $= E - 36x$

$\therefore$ Max. fluctuation of energy,

$$\Delta E = \text{Max. energy} - \text{Min. energy}$$

$$= (E + 402x) - (E - 36x) = 438x$$

$$= 438 \times 17.017 \text{ Nm} \quad [\because \text{ From equation } (i),\ x = 17.017 \text{ Nm}]$$

$$= 7453.446 \text{ Nm}$$

Let m = Mass of flywheel.

Now using equation (12.12), we get

$$\Delta E = mR^2 \times \omega^2 \times K_s$$ [In this equation, $k = R$ which is true if thickness of rim of flywheel is very small as compared to dia. of wheel]

But $\Delta E = 7453.446$ Nm, $R = 0.7$ m, $\omega = 62.84$ rad/s and $K_s = 0.036$. Hence above equation becomes as

$$7453.446 = m \times 0.7^2 \times 62.84^2 \times 0.036$$

or $$m = \frac{7453.446}{0.7^2 \times 62.84^2 \times 0.036} = \mathbf{107.00.} \quad \textbf{Ans.}$$

Problem 12.4. *The radius of gyration of a flywheel is 1 meter and the fluctuation of speed is not to exceed 1% of the mean speed of the flywheel. If the mass of the flywheel is 3340 kg and the steam engine develops 150 kW at 135 r.p.m., then find :*

(*i*) *Maximum fluctuation of energy and*

(*ii*) *Co-efficient of fluctuation of energy.*

Sol. Given :

$k = 1$ m ; fluctuation of speed = 1% of mean speed or $(\omega_2 - \omega_1) = 1\%$ of ω or $\frac{\omega_2 - \omega_1}{\omega} = \frac{1}{100} = 0.01$ or

Co-efficient of fluctuation of speed, $K_s = 0.01$; $m = 3340$; $P = 150$ kW = 150 × 1000 W = 150,000 W ;

$N = 135$ r.p.m. or mean angular speed, $\omega = \frac{2\pi N}{60} = \frac{2\pi \times 135}{60} = 14.137$ rad/s.

Let ΔE = Max. fluctuation of energy and

K_e = Co-efficient of fluctuation of energy.

The maximum fluctuation of energy is given by equation (12.11) as,

$$\Delta E = mk^2 \times \omega^2 \times K_s$$

$$= 3340 \times 1^2 \times 14.137^2 \times 0.01$$

($\because$ $m = 3340$ kg ; $k = 1$ m ; $\omega = 14.137$ rad/s and $K_s = 0.01$)

$$= \mathbf{6675.13\ Nm.} \quad \textbf{Ans.}$$

The co-efficient of fluctuation of energy (K_e) is given by equation (12.2) as,

$$K_e = \frac{\text{Maximum fluctuation of energy}}{\text{Work done per cycle}} \quad ...(i)$$

Let us now find the work done per cycle in case of steam engine.

Using equation (12.4),

Work done per cycle $= T_{mean} \times \theta$

$= T_{mean} \times 2\pi$ ($\because$ $\theta = 2\pi$ radians for steam engine)

The value of T_{mean} is obtained from,

$$P = T_{mean} \times \omega$$

$\therefore$ $$T_{mean} = \frac{P}{\omega} = \frac{150000}{14.137}$$ ($\because$ $P = 150000$ W and $\omega = 14.137$ rad/s)

$$= 10610.45 \text{ Nm}$$

$\therefore$ Work done per cycle $= T_{mean} \times 2\pi = 10610.45 \times 2\pi = 66667.42$ Nm/cycle

Substituting the values in equation (*i*), we get

$$K_e = \frac{\Delta E}{66667.42} = \frac{6675.13}{66667.42} = 0.1001 \simeq \mathbf{0.1.} \quad \textbf{Ans.}$$

Problem 12.5. *The areas above and below the mean torque line for an I.C. engine are – 25, + 200, – 100, + 150, – 300, + 150 and – 75 mm² taken in order. The scale for the turning moment diagram is 1 mm vertical scale = 10 Nm and 1 mm horizontal scale = 1.5°. The mass of the rotating parts are 45 kg with a radius of gyration of 150 mm. If the engine speed is 1500 r.p.m., find the co-efficient of fluctuation of speed.*

Sol. Given :

Turning moment scale : 1 mm = 10 Nm and crank angle scale 1 mm = 1.5° = $1.5 \times \frac{\pi}{180}$ rad = 0.02618 rad ; m = 45 kg ; k = 150 mm = 0.15 m ; N = 1500 r.p.m. or $\omega = \frac{2\pi N}{60} = \frac{2\pi \times 1500}{60}$ = 50 π rad/s. Now 1 mm^2 on turning moment diagram = 10 × 0.02618 = 0.2618 Nm.

Let 1 mm^2 on turning moment diagram = x Nm

Then x = 0.2618 Nm ...(*i*)

The areas above and below the mean torque line in works units (*i.e.* in Nm) will be – 25x, + 200x, – 100x, + 150x, – 300x, + 150x and – 75x.

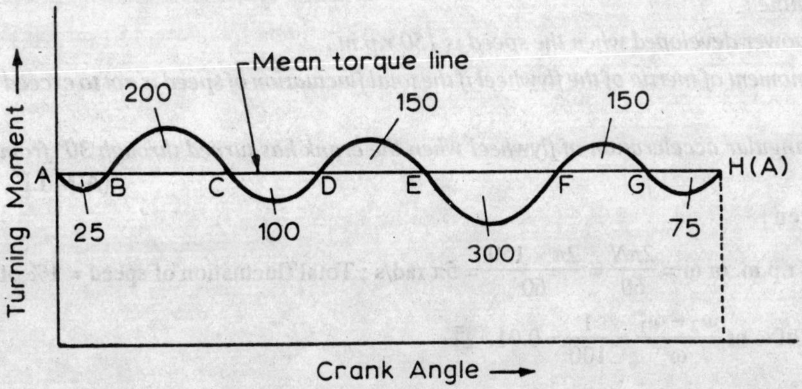

Fig. 12.8

Let E = Total energy at A (Refer to Fig. 12.8). Then

Energy at B = $E - 25x$

Energy at C = $E - 25x + 200x = E + 175x$

Energy at D = $E + 175x - 100x = E + 75x$

Energy at E = $E + 75x + 150x = E + 225x$ ...(Max. Energy)

Energy at F = $E + 225x - 300x = E - 75x$ ...(Min. Energy)

Energy at G = $E - 75x + 150 = E + 75x$

Energy at H = $E - 75x - 75x = E$ = Energy at A

The cycle repeats after H and hence energy at H is equal to energy at A.

The maximum energy is at E whereas the minimum energy is at F.

∴ Max. energy = $E + 225x$ and min. energy = $E - 75x$

∴ Max. fluctuation of energy,

$$\Delta E = \text{Max. Energy} - \text{Min. Energy}$$

$$= (E + 225x) - (E - 75x)$$

$$= 300x = 300 \times 0.2618 \text{ Nm}$$ [∵ From equation (*i*), x = 0.2618 Nm]

$$= 78.54 \text{ Nm}$$

Let K_s = Co-efficient of fluctuation of speed.

Using equation (12.11), we get

$$\Delta E = mk^2 \times \omega^2 \times K_s$$

$$\therefore \quad K_s = \frac{\Delta E}{mk^2 \times \omega^2}$$

$$= \frac{78.54}{45 \times 0.15^2 \times (50\pi)^2}$$

[$\because$ $\Delta E = 78.54$ Nm ; $m = 45$ kg ; $k = 15$ m and $\omega = 50\,\pi$ rad/s]

$= \mathbf{0.0031.}$ **Ans.**

Problem 12.6. *The torque exerted on the crank-shaft of a two stroke engine is given by the equation :*

$$T\ (N\text{-}m) = 14500 + 2300 \sin 2\theta - 1900 \cos 2\theta$$

where θ is the crank angle displacement from the inner dead centre. Assuming the resisting torque to be constant, determine :

(i) the power developed when the speed is 150 r.p.m.,

(ii) the moment of inertia of the flywheel if the total fluctuation of speed is not to exceed 1% of the mean speed ;

(iii) the angular acceleration of flywheel when the crank has turned through 30° from the inner dead centre.

(A.M.I.E., Summer 1982)

Sol. Given :

$N = 150$ r.p.m. or $\omega = \dfrac{2\pi N}{60} = \dfrac{2\pi \times 150}{60} = 5\pi$ rad/s ; Total fluctuation of speed = 1% of mean speed or

$(\omega_2 - \omega_1) = 1\%$ of ω or $\dfrac{\omega_2 - \omega_1}{\omega} = \dfrac{1}{100} = 0.01.$

$$\therefore \quad K_s = \frac{\omega_2 - \omega_1}{\omega} = 0.01$$

The equation of torque-exerted on the crank-shaft,

$$T = 14500 + 2300 \sin 2\theta - 1900 \cos 2\theta$$

The T-θ diagram is a function of 2θ, hence the cycle will be repeated after every crank rotation of 180°. The turning moment diagram is shown in Fig. 12.9.

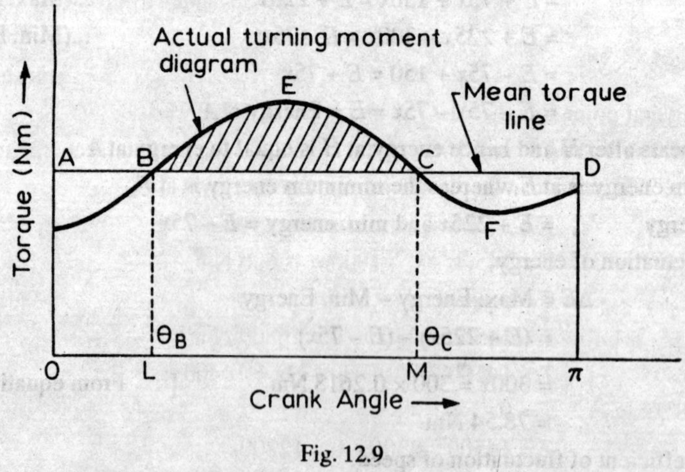

Fig. 12.9

(*i*) Work done per cycle = Area of turning moment diagram for one cycle

$$= \int_0^{\pi} T \times d\theta \qquad \text{where } T \text{ is torque at any instant}$$

$$= \int_0^{\pi} (14500 + 2300 \sin 2\theta - 1900 \cos 2\theta)\, d\theta$$

$$= \left[14500\,\theta + 2300 \frac{(-\cos 2\theta)}{2} - 1900 \times \frac{\sin 2\theta}{2} \right]_0^{\pi}$$

$$= 14500\pi - 1150[\cos 2\pi - \cos 0] - 950\,[\sin 2\pi - \sin 0]$$

$$= 14500\pi - 1150[1 - 1] - 950[0 - 0] = 14500\pi \text{ Nm}$$

But area of turning moment diagram corresponding to mean torque

$$= T_{mean} \times \pi$$

$$\therefore \quad T_{mean} \times \pi = 14500\pi$$

$$\therefore \quad T_{mean} = 14500 \text{ Nm}$$

$\therefore$ Power developed $= T_{mean} \times \omega$

$$= 14500 \times 5\pi \qquad \left[\because \ \omega = \frac{2\pi N}{60} = \frac{2\pi \times 150}{60} = 5\pi \text{ rad/s} \right]$$

$= 227765$ W $=$ **227.765 kW. Ans.**

(*ii*) *Moment of Inertia of flywheel (I)*

Let I = moment of inertia of flywheel

The Fig. 12.9 show the turning moment diagram for one cycle. At the points B and C, the torque exerted on the crank-shaft is equal to the mean resisting torque. Hence for these points

$$T = T_{mean}$$

or $\quad 14500 + 2300 \sin 2\theta - 1900 \cos 2\theta = 14500$

or $\quad 2300 \sin 2\theta = 1900 \cos 2\theta$

or $\quad \dfrac{\sin 2\theta}{\cos 2\theta} = \dfrac{1900}{2300}$

or $\quad \tan 2\theta = 0.826$

or $\quad 2\theta = \tan^{-1} 0.826 = 39.56°$ and $219.56°$

or $\quad \theta = 19.78°$ and $109.78°$

i.e. $\quad \theta_B = 19.78°$ and $\theta_C = 109.78°$

$\therefore$ Crank angle at point B is $\theta_B = 19.78°$ and at point C is $\theta_C = 109.78°$.

Hence at crank angles 19.78° and 109.78°, the actual *T-M* curve intersects the mean torque line.

Now the actual *T-M* curve can be plotted as explained below :

$$T = 14500 + 2300 \sin 2\theta - 1900 \cos 2\theta$$

At $\theta = 0$, $\quad T = 14500 + 0 - 1900 = 12600$ Nm

At $\theta = 19.78°$, $\quad T = 0$

At $\theta = 45°$, $\quad T = 14500 + 2300 - 0 = 16800$ Nm

At $\theta = 90°$, $\quad T = 14500 + 0 - 1900\,(-1) = 16400$

At $\theta = 109.78°$, $\quad T = 0$

At $\theta = 135°$, $\quad T = 14500 - 2300 - 0 = 12200$

At $\theta = 180°$, $\quad T = 14500 + 0 - 1900 = 12600$

Now plot *T-M* diagram as shown in Fig. 12.9.

Now the fluctuation of energy

$= \text{Area } BEC = \text{Area } LBECM - \text{Area } LBCM$

$$= \int_{\theta_B}^{\theta_C} T \, . \, d\theta - \int_{\theta_B}^{\theta_C} T_{mean} \times d\theta = \int_{\theta_B}^{\theta_C} (T - T_{mean}) \, d\theta$$

$$= \int_{19.78^\circ}^{109.78^\circ} [(14500 + 2300 \sin 2\theta - 1900 \cos 2\theta) - 14500] \, d\theta$$

$$= \int_{19.78^\circ}^{109.78^\circ} (2300 \sin 2\theta - 1900 \cos 2\theta) \, d\theta$$

$$= \left[2300 \left(\frac{-\cos 2\theta}{2} \right) - 1900 \left(\frac{\sin 2\theta}{2} \right) \right]_{19.78^\circ}^{109.78^\circ}$$

$= [-1150 \{\cos (2 \times 109.78) - \cos (2 \times 19.78)\} - 950 \{\sin (2 \times 109.78^\circ) - \sin (2 \times 19.78)\}]$

$= [-1150 (\cos 219.56^\circ - \cos 39.56^\circ) - 950(\sin 219.56^\circ - \sin 39.56^\circ)]$

$= [-1150(-0.771 - 0.771) - 950(-0.637 - 0.637)]$

$= -1150 (-1.542) - 950 (-1.274)$

$= 1773.3 + 1210.3 = 2983.6$ Nm

$\therefore$ Maximum fluctuation of energy, $\Delta E = 2983.6$ Nm

But the maximum fluctuation of energy is also given by equation (12.11) as,

$$\Delta E = mk^2 \times \omega^2 \times K_s \text{ or } I \times \omega^2 \times K_s$$

Equating the two values of ΔE, we get

$\therefore \quad 2983.6 = I \times \omega^2 \times K_s$

or $\quad I = \dfrac{2983.6}{\omega^2 \times K_s}$

$= \dfrac{2983.6}{(5\pi)^2 \times 0.01}$ $\quad (\because \ \omega = 5\pi$ rad/s and $K_s = 0.01)$

$=$ **1209.2 kg m²**. **Ans.**

(iii) Angular acceleration flywheel when $\theta = 30^\circ$

Let α = Angular acceleration

The excess torque over the mean torque, produces the acceleration in the flywheel. The excess torque at any instant is given by,

$$T_{Ex} = T - T_{mean}$$

$= (14500 + 2300 \sin 2\theta - 1900 \cos 2\theta) - 14500 \quad (\because \ T_{mean} = 14500 \text{ Nm})$

$= 2300 \sin 2\theta - 1900 \cos 2\theta$

$\therefore$ T excess when $\theta = 30$ is given by

$T_{Ex} = 2300 \sin (2 \times 30^\circ) - 1900 \cos (2 \times 30^\circ)$

$= 2300 \times 0.866 - 1900 \times 0.5 = 1991.8 - 950 = 1041.8$ Nm

But $\quad T_{Ex} = I \times \alpha \quad$ where I = M.O.I. of flywheel = 1209.2 kg m²

$\therefore \quad 1041.8 = 1209.2 \times \alpha$

$\therefore \quad \alpha = \dfrac{1041.8}{1209.2} =$ **0.8615 rad/s²**. **Ans.**

Problem 12.7. *The torque exerted on the crank-shaft of a two-stroke engine is given by the equation :*

$$T\ (Nm) = 7000 + 1000 \sin 2\theta - 2000 \cos 2\theta$$

where θ is the crank displacement from the inner dead centre. Assuming the resisting torque to be constant, determine :

(*i*) *the power developed when the engine speed is 300 r.p.m.*

(*ii*) *the total fluctuations in speed in percentage and*

(*iii*) *the maximum retardation of the flywheel.*

The mass of flywheel is 500 kg and its radius of gyration is 750 mm.

Sol. Given :

$N = 300$ r.p.m. or $\omega = \frac{2\pi N}{60} = \frac{2\pi \times 300}{60} = 10\pi$ rad/s ; $m = 500$ kg and $k = 750$ mm $= 0.75$ m

The equation of torque exerted on the crank shaft,

$$T = 7000 + 1000 \sin 2\theta - 2000 \cos 2\theta$$

The *T*-θ diagram is a function of 2θ, hence the cycle will be repeated after every crank rotation of 180°. The turning moment diagram is shown in Fig. 12.10.

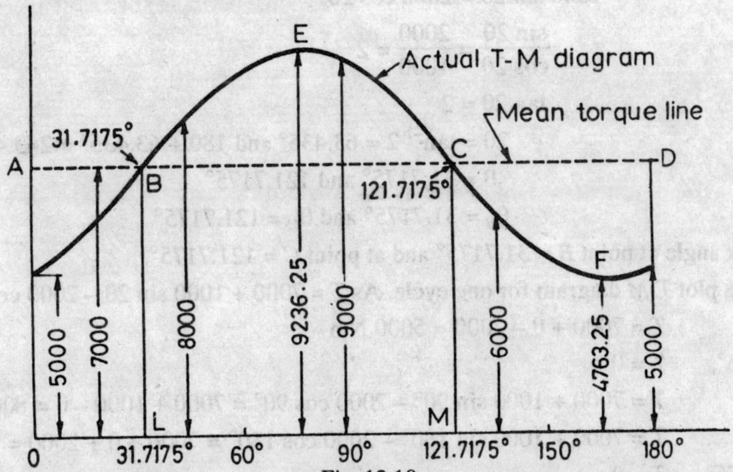

Fig. 12.10

(*i*) Work done per cycle = Area of turning moment diagram for one cycle

$$= \int_0^{\pi} T \times d\theta \quad \text{where } T \text{ is torque at any instant}$$

$$= \int_0^{\pi} (7000 + 1000 \sin 2\theta - 2000 \cos 2\theta)\, d\theta$$

$$= \left[7000 \times \theta + 1000 \left(\frac{-\cos 2\theta}{2} \right) - 2000 \left(\frac{\sin 2\theta}{2} \right) \right]_0^{\pi}$$

$$= \left[7000\,\theta - 500 \cos 2\theta - 1000 \sin 2\theta \right]_0^{\pi}$$

$$= [7000\,(\pi - 0) - 500\,(\cos 2\pi - \cos 0) - 1000\,(\sin 2\pi - \sin 0)]$$

$$= [7000\pi - 500\,(1 - 1) - 1000\,(0 - 0)] = 7000\pi \text{ Nm}$$

But area of *T-M* diagram corresponding to mean torque

$$= T_{mean} \times \pi \quad \text{(Area of rectangle } OADH = OA \times OH\text{)}$$

$\therefore$ $$T_{mean} \times \pi = 7000\pi$$

or $$T_{mean} = 7000 \text{ Nm}$$

$\therefore$ Power developed $= T_{mean} \times \omega$

$$= 7000 \times 10\pi \quad (\because \ \omega = 10\pi \text{ rad/s})$$

$= 219911$ W = **219.911 kW. Ans.**

(*ii*) *Total fluctuation in speed in percentage*

Let us first find the maximum fluctuation in energy. For this, we require the points where the mean torque line intersects the actual turning moment curve. Let *B* and *C* are the points of intersection as shown in Fig. 12.10. Let us find the crank angles corresponding to points *B* and *C* where the actual *T-M* diagram intersects the mean torque line. Hence at points *B* and *C*

$$T = T_{mean}$$

or $$7000 + 1000 \sin 2\theta - 2000 \cos 2\theta = 7000$$

or $$1000 \sin 2\theta = 2000 \cos 2\theta$$

or $$\frac{\sin 2\theta}{\cos 2\theta} = \frac{2000}{1000} = 2$$

or $$\tan 2\theta = 2$$

$\therefore$ $$2\theta = \tan^{-1} 2 = 63.435° \text{ and } 180 + 63.435° = 243.435°$$

or $$\theta = 31.7175° \text{ and } 121.7175°$$

$\therefore$ $$\theta_B = 31.7175° \text{ and } \theta_C = 121.7175°$$

$\therefore$ Crank angle at point $B = 31.7175°$ and at point $C = 121.7175°$.

Now let us plot *T-M* diagram for one cycle. As $T = 7000 + 1000 \sin 2\theta - 2000 \cos 2\theta$ hence

when $\theta = 0$, $T = 7000 + 0 - 2000 = 5000$ Nm

when $\theta = 31.7175°$, $T = 0$

when $\theta = 45°$, $T = 7000 + 1000 \sin 90° - 2000 \cos 90° = 7000 + 1000 - 0 = 8000$ Nm

when $\theta = 90°$, $T = 7000 + 1000 \sin 180° - 2000 \cos 180° = 7000 + 0 + 2000 = 9000$

when $\theta = 121.7175°$, $T = 0$

when $\theta = 135°$, $T = 7000 + 1000 \sin 270° - 2000 \cos 270° = 7000 - 1000 - 0 = 6000$

when $\theta = 180°$, $T = 7000 + 1000 \sin 360° - 2000 \cos 360° = 7000 + 0 - 2000 = 5000$ Nm

Now *T-M* diagram can be plotted as shown in Fig. 12.10.

Now fluctuation of energy

$$= \text{Area } BEC = \text{Area } LBECM - \text{Area } LBCM$$

$$= \int_{\theta_B}^{\theta_C} T \, . \, d\theta - \int_{\theta_B}^{\theta_C} T_{mean} \times d\theta$$

$$= \int_{\theta_B}^{\theta_C} (T - T_{mean}) d\theta = \int_{\theta_B}^{\theta_C} [(7000 + 1000 \sin 2\theta - 2000 \cos 2\theta) - 7000] \, d\theta$$

$$= \int_{31.7175°}^{121.7175°} (1000 \sin 2\theta - 2000 \cos 2\theta) \, d\theta$$

$$= \left[1000 \left(\frac{-\cos 2\theta}{2} \right) - 2000 \left(\frac{\sin 2\theta}{2} \right) \right]_{31.7175°}^{121.7175°}$$

$$= \left[-500 (\cos 2\theta) - 1000 (\sin 2\theta) \right]_{31.7175°}^{121.7175°}$$

$$= [-500 \{\cos (2 \times 121.7175) - \cos (2 \times 31.7175)\} - 1000 \{\sin (2 \times 121.7175) - \sin (2 \times 31.7175)\}]$$

$$= [-500 (\cos 243.435° - \cos 63.435) - 1000 (\sin 243.435° - \sin 63.435°)]$$

$$= [-500 (-0.4472 - 0.4472) - 1000 (-0.8944 - 0.8944)]$$

$$= [-500 (-0.8944) - 1000 (-1.7888)]$$

$$= 500 \times 0.8944 + 1000 \times 1.7888 = 447.2 + 1788.8 = 2236 \text{ Nm}$$

$\therefore \quad \Delta E = 2236$ Nm.

But max. fluctuation of energy is also given by equation (12.11) as

$$\Delta E = I \times \omega^2 \times K_s \quad \text{or} \quad mk^2 \times \omega^2 \times K_s$$

Equating the two values of ΔE, we get

$$2236 = mk^2 \times \omega^2 \times K_s$$

$$= 500 \times 0.75^2 \times (10\pi)^2 \times K_s \qquad (\because \; m = 500 \text{ kg}, k = 0.75 \text{ m}, \omega = 10\pi \text{ rad/s})$$

$$\therefore \quad K_s = \frac{2236}{500 \times 0.75^2 \times (10\pi)^2}$$

$$= 0.00805 \simeq 0.008 \quad \text{or} \quad \mathbf{0.8\%.} \quad \textbf{Ans.}$$

(iii) The maximum retardation of flywheel

Let α = Acceleration of flywheel.

The excess torque over mean torque produces the acceleration in the flywheel. The excess torque over the mean torque at any instant is given by,

$$T^* = T - T_{mean}$$

$$= (7000 + 1000 \sin 2\theta - 2000 \cos 2\theta) - 7000$$

$$= 1000 \sin 2\theta - 2000 \cos 2\theta \qquad ...(i)$$

The acceleration will be maximum, when excess torque is maximum. The excess torque will be maximum, when $\frac{d}{d\theta}(T^*) = 0$

or $\quad \frac{d}{d\theta}(1000 \sin 2\theta - 2000 \cos 2\theta) = 0 \qquad (\because \; T^* = 1000 \sin 2\theta - 2000 \cos 2\theta)$

or $\quad 1000 \times 2 \times \cos 2\theta - 2000 \times 2 (-\sin 2\theta) = 0$

or $\quad \cos 2\theta + 2 \sin 2\theta = 0 \qquad$ [Cancelling 1000×2 to both sides]

or $\quad 2 \sin 2\theta = -\cos 2\theta$

or $\quad \frac{\sin 2\theta}{\cos 2\theta} = -\frac{1}{2} = -0.5$

or $\quad \tan 2\theta = -0.5$

$\therefore \quad 2\theta = \tan^{-1}(-0.5) = 153.435° \quad$ and $\quad 153.435 + 180 = 333.435°$

or $\quad \theta = 76.718 \quad$ and $\quad 166.718°$

Hence excess torque will be maximum, when $\theta = 76.718°$ or $166.718°$.

When $\theta = 76.718°$, the excess torque is obtained by substituting the value of $\theta = 76.718°$ in equation (*i*) as

$$T^* = 1000 \sin 2\theta - 2000 \cos 2\theta$$
$$= 1000 \sin (2 \times 76.718°) - 2000 \cos (2 \times 76.718°)$$
$$= 1000 \sin 153.435° - 2000 \cos 153.435°$$
$$= 1000 \times 0.4472 - 2000 \times (-0.8944)$$
$$= 447.2 + 1788.85 = 2236.05 \text{ Nm}$$

As excess torque is having positive sign and hence it will produce acceleration.

When $\theta = 166.718°$, the excess torque is given by,

$$T^* = 1000 \sin (2 \times 166.718°) - 2000 \cos (2 \times 166.718°)$$
$$= 1000 \sin (333.435°) - 2000 \cos (333.435°)$$
$$= 1000 \times (-0.4472) - 2000 \times (0.8944)$$
$$= -447.2 - 1788.85 = -2236.05 \text{ Nm}$$

As the excess torque is negative, it will produce retardation.

Now using the equation,

$$\text{Torque} = I \times \alpha$$

or
$$T^* = mk^2 \times \alpha$$

$\therefore$
$$\alpha = \frac{T^*}{mk^2} = \frac{-2236.05}{500 \times 0.75^2} \qquad (\because m = 500 \text{ kg and } k = 0.75 \text{ m})$$
$$= -7.95 \text{ rad/s}^2$$

$\therefore$ Max. retardation $= \mathbf{7.95\ rad/s^2}$. **Ans.**

Problem 12.8. *The turning moment diagram (T-M diagram) of an engine rotating at a mean speed of 200 r.p.m. is given by the relation,*

$$T \text{ (in Nm)} = 15000 + 8000 \sin 2\phi - 2000 \cos 2\phi$$

where ϕ is the crank angle. External resistance is constant. A flywheel of mass 2000 kg is fitted on the engine shaft so that the total fluctuation of speed does not exceed by 1%. Determine the least-value of the moment of inertia and radius of gyration of the flywheel. (AMIE, Summer 1978)

Sol. Given :

$N_{mean} = 200$ r.p.m. Hence $\omega_{mean} = \dfrac{2\pi N_{mean}}{60} = \dfrac{2\pi \times 200}{60} = 20.94$ rad/s ;

Mass of flywheel, $m = 2000$ kg ; Total fluctuation of speed $K_s = 1\% = 0.01$.

$$T = 15000 + 8000 \sin 2\phi - 2000 \cos 2\phi$$

From the above relation, it is clear that turning moment T is a function of 2ϕ. Hence the cycle will be repeated after every crank rotation of 180° or π radians.

Work done per cycle = Area of turning moment diagram

$$= \int_0^{\pi} T \, d\phi$$

But
$$T = 15000 + 8000 \sin 2\phi - 2000 \cos 2\phi$$

$$\therefore \text{ Work done per cycle} = \int_0^{\pi} (15000 + 8000 \sin 2\phi - 2000 \cos 2\phi)\, d\phi$$

$$= \left[15000\phi + \frac{8000(-\cos 2\phi)}{2} - \frac{2000 \times \sin 2\phi}{2}\right]_0^{\pi}$$

$$= [15000 \times \pi - 400 (\cos 2\pi - \cos 0) - 1000 (\sin 2\pi - \sin 0)]$$

$$= [15000 \times \pi - 400 (1 - 1) - 1000 (0 - 0)]$$

$$= [15000 \times \pi - 4000 \times 0 - 1000 \times 0] = 15000 \times \pi \text{ Nm}$$

But area of *T-M* diagram corresponding to mean torque for one cycle = $T_{mean} \times \pi$

$\therefore \quad T_{mean} \times \pi = 15000 \times \pi$

or $\quad T_{mean} = 15000$ Nm

Let us now find the maximum fluctuation of energy. For this we require the points where the mean torque line intersects the actual turning moment curve. At the intersection points, actual torque is equal to mean torque

or $\quad T = T_{mean}$

or $\quad 15000 + 8000 \sin 2\phi - 2000 \cos 2\phi = 15000$

or $\quad 8000 \sin 2\phi - 2000 \cos 2\phi = 15000 - 15000 = 0$

or $\quad 8000 \sin 2\phi = 2000 \cos 2\phi$

or $\quad \dfrac{\sin 2\phi}{\cos 2\phi} = \dfrac{2000}{8000}$

or $\quad \tan 2\phi = 0.25$

$\therefore \quad 2\phi = \tan^{-1} 0.25 = 14.03°$ and $14.03 + 180 = 194.03°$

or $\quad \phi = 7.015°$ and $97.015°$

Hence at crank angles 7.015° and 97.015°, the actual turning moment curve intersects the mean torque line. Now let us plot the *T-M* diagram for one cycle. As

$$T = 15000 + 8000 \sin 2\phi - 2000 \cos 2\phi$$

When $\theta = 0$, $\quad T = 15000 + 0 - 2000 = 13000$

When $\theta = 7.015°$, $\quad T = T_{mean} = 15000$

When $\theta = 45°$, $\quad T = 15000 + 8000 \sin (90°) - 2000 \cos (90°) = 15000 + 8000 - 0 = 23000$

When $\theta = 90°$, $\quad T = 15000 + 8000 \sin 180° - 2000 \cos 180° = 15000 + 0 + 2000 = 17000$

When $\theta = 97.015°$, $\quad T = T_{mean} = 15000$

When $\theta = 135°$, $\quad T = 15000 + 8000 \sin 270° - 2000 \cos 270° = 15000 - 8000 - 0 = 7000$

When $\theta = 180°$, $\quad T = 15000 + 8000 \sin 360° - 2000 \cos 360° = 15000 - 2000 = 13000$

Fig. 12.11 shows the *T-M* diagram for one cycle. To find the maximum fluctuation of energy, we must consider the area above and below mean torque line.

Now fluctuation of energy = Area *BEC* = Area *LBECM* – Area *LBCM*

$$= \int_{\phi_B}^{\phi_C} T\, d\phi - \int_{\phi_B}^{\phi_C} T_{mean}\, d\phi = \int_{\phi_B}^{\phi_C} (T - T_{mean})\, d\phi$$

$$= \int_{7.015°}^{97.015°} [(15000 + 8000 \sin 2\phi - 2000 \cos 2\phi) - 15000]\, d\phi$$

$$= \int_{7.015°}^{97.015°} (8000 \sin 2\phi - 2000 \cos 2\phi)\, d\phi$$

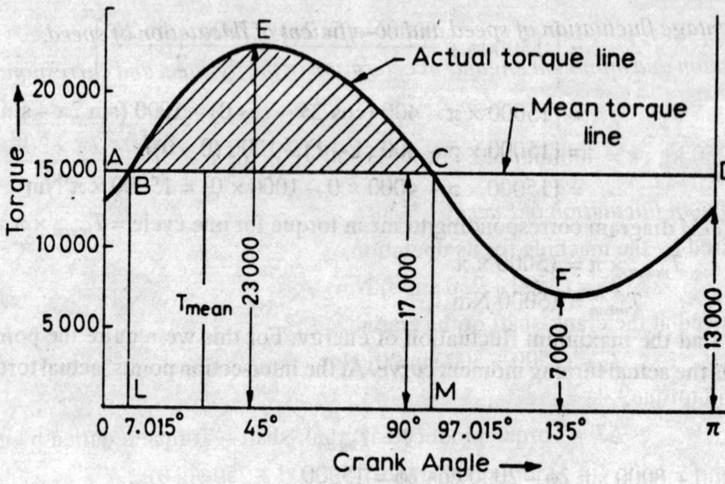

Fig. 12.11

$$= \left[8000 \times \frac{(-\cos 2\phi)}{2} - 2000 \times \frac{(\sin 2\phi)}{2} \right]_{7.015°}^{97.015°}$$

$$= \left[-4000 \cos 2\phi - 1000 \sin 2\phi \right]_{7.015°}^{97.015°}$$

$$= -4000 (\cos 2 \times 97.015° - \cos 2 \times 7.015°) - 1000 (\sin 2 \times 97.015 - \sin 2 \times 7.015)$$

$$= -4000 (\cos 194.03° - \cos 14.03°) - 1000 (\sin 194.03° - \sin 14.03°)$$

$$= -4000 (-0.97 - 0.97) - 1000 (-0.2424 - 0.2424)$$

$$= 7760 + 484.8 = 8244.8 \text{ Nm}$$

$\therefore \quad \Delta E = 8244.8$ Nm

But fluctuation of energy is also given by equation (12.11) as

$$\Delta E = I \times \omega^2 \times K_s$$

Equating the two values of ΔE,

$$8244.8 = I \times 20.94^2 \times 0.01 \qquad (\because \ \omega = \omega_{mean} = 20.94 \text{ rad/s} ; K_s = 0.01)$$

$$\therefore \quad I = \frac{8244.8}{20.94^2 \times 0.01} = \mathbf{1880.3 \text{ kg m}^2}. \quad \textbf{Ans.}$$

Let k = Radius of gyration of flywheel

Now $\quad I = mk^2$

or $\quad k = \sqrt{\frac{I}{m}} = \sqrt{\frac{1880.3}{2000}} = \mathbf{0.9696 \text{ m}}.$ **Ans.**

Problem 12.9. *The torque at the crank-shaft of an engine is given by equation,*

$$T = (2500 + 300 \sin 2\theta) \text{ Nm}.$$

The mass and radius of gyration of flywheel and other rotating parts attached to the engine are 250 kg and 400 mm respectively. The mean speed is 300 r.p.m. The engine is directly coupled to a machine which requires a torque of (2500 + 250 sin θ) Nm for its operation, where θ is the angle of rotation of the shaft measured from a certain datum. Determine :

(i) the maximum fluctuation of energy,

(*ii*) *the percentage fluctuation of speed and co-efficient of fluctuation of speed,*

(*iii*) *the maximum and minimum angular acceleration of the flywheel and corresponding shaft position.*

Sol. Given :

$$m = 250 \text{ kg} ; k = 400 \text{ mm} = 0.4 \text{ m} ; N = 300 \text{ r.p.m. or } \omega = \frac{2\pi N}{60} = \frac{2\pi \times 300}{60} = 10\pi \text{ rad/s} ;$$

(*i*) *The maximum fluctuation of Energy*

Torque required by the machine for its operation,

$$T_1 = (2500 + 250 \sin \theta) \text{ Nm}$$

Torque produced at the crank-shaft of the engine,

$$T_2 = (2500 + 300 \sin 2\theta) \text{ Nm}$$

∴ Change in torque,

$$\Delta T = \text{Torque produced at crank-shaft} - \text{Torque required by machine}$$
$$= (2500 + 300 \sin 2\theta) - (2500 + 250 \sin \theta)$$
$$= 300 \sin 2\theta - 250 \sin \theta \quad \text{...(i)}$$

Let us find now the points where the *T-M* diagram for the engine intersects the *T-M* diagram for the machine. At the intersection points, the change in torque should be zero.

or $$\Delta T = 0$$

∴ $$300 \sin 2\theta - 250 \sin \theta = 0$$

or $$300 \times (2 \sin \theta \cos \theta) - 250 \sin \theta = 0 \qquad (\because \sin 2\theta = 2 \sin \theta \cos \theta)$$

or $$\sin \theta (600 \cos \theta - 250) = 0$$

∴ Either $\sin \theta = 0$ or $600 \cos \theta - 250 = 0$

When $\sin \theta = 0$, then $\theta = 0°, 180°$ and $360°$

When $600 \cos \theta - 250 = 0$, then

$$\cos \theta = \frac{250}{600} = 0.4167 \quad \text{or} \quad \theta = \cos^{-1} 0.4167 = 65.37° \text{ and } 360 - 65.37 = 294.63°$$

Hence the points of intersection of the *T-M* diagram for the engine and *T-M* diagram for the machine are given by,

$$\theta = 0°, 65.37°, 180°, 294.63° \text{ and } 360°.$$

From equation (*i*), $\Delta T = 300 \sin 2\theta - 250 \sin \theta$

When $\theta = 0$, $\Delta T = 0$

When $\theta = 45°$, $\Delta T = 300 \sin 90° - 250 \sin 45° = 300 - 176.77 = 123.23$ Nm

When $\theta = 65.37°$, $\Delta T = 0$

When $\theta = 90°$, $\Delta T = 300 \sin 180° - 250 \sin 90° = 0 - 250 = -250$ Nm

When $\theta = 135°$, $\Delta T = 300 \sin 270° - 250 \sin 135° = -300 - 176.77 = -476.77$

When $\theta = 180°$, $\Delta T = 300 \sin 360° - 250 \sin 180° = 0$

When $\theta = 225°$, $\Delta T = 300 \sin 450 - 250 \sin 225 = 300 - 250 (-0.7071) = 476.77$ Nm

When $\theta = 270°$, $\Delta T = 300 \sin 540 - 250 \sin 270° = 0 - 250 (-1) = 250$ Nm

When $\theta = 294.63°$, $\Delta T = 0$

When $\theta = 315°$, $\Delta T = 300 \sin 630 - 250 \sin 315 = -300 - 250 (-0.7071)$

$$= -300 + 176.77 = -123.23 \text{ Nm}$$

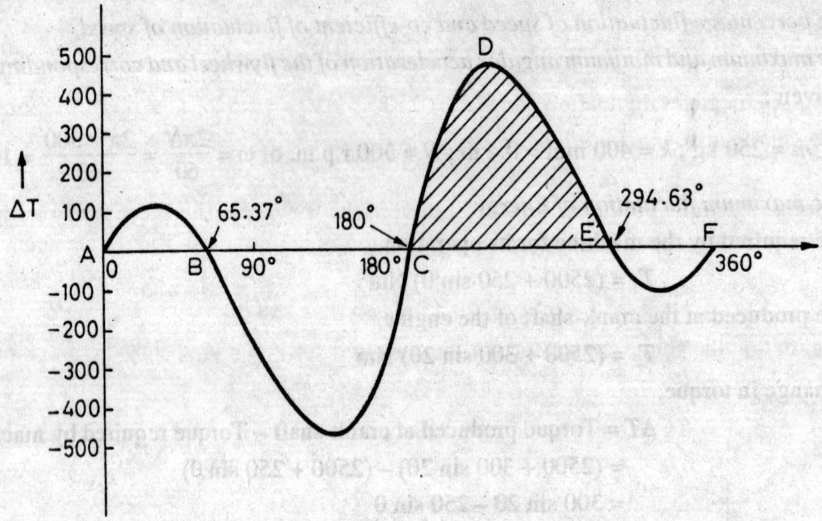

Fig. 12.12

Now the variation of ΔT against θ can be plotted as shown in Fig. 12.12. The maximum fluctuation of energy is represented by the largest positive area. The largest positive area lies between 180° and 294.63°.

$\therefore$ **Maximum fluctuation of energy,**

$$\Delta E = \text{Shaded area } CDE$$

$$= \int_{180^\circ}^{294.63^\circ} \Delta T \times d\theta$$

$$= \int_{180^\circ}^{294.63^\circ} (300 \sin 2\theta - 250 \sin \theta)\, d\theta \qquad (\because \ \Delta T = 300 \sin 2\theta - 250 \sin \theta)$$

$$= \left[300 \frac{(-\cos 2\theta)}{2} - 250 (-\cos \theta) \right]_{180^\circ}^{294.63^\circ}$$

$$= \left[-150 \cos 2\theta + 250 \cos \theta \right]_{180^\circ}^{294.63^\circ}$$

$$= -150 (\cos 2 \times 294.63^\circ - \cos 2 \times 180^\circ) + 250 (\cos 294.63^\circ - \cos 180^\circ)$$

$$= -150 (\cos 589.26^\circ - \cos 360^\circ) + 250 [0.4167 - (-1)]$$

$$= -150 [(-0.6526 - 1)] + 250 [(1.4167)]$$

$$= -150 (-1.6526) + 354.175 = 247.89 + 354.175 = \mathbf{602.065 \ Nm.} \quad \textbf{Ans.}$$

(ii) The percentage fluctuation of speed

Let K_s = Co-efficient of fluctuation of speed or total percentage fluctuation of speed.

Now using equation (12.11), we get

$$\Delta E = mk^2 \times \omega^2 \times K_s$$

or
$$602.065 = 250 \times 0.4^2 \times (10\pi)^2 \times K_s$$

$$(\because \ \Delta E = 602.065 \text{ Nm}, m = 250 \text{ kg}, k = 0.4 \text{ m}, \omega = 10\pi \text{ rad/s})$$

$\therefore \quad K_s = \dfrac{602.065}{250 \times 0.4^2 \times (10\pi)^2} = 0.0152$

$\therefore$ Total percentage fluctuation of speed = 0.0152 × 100 = **1.52%. Ans.**

(iii) The maximum and minimum angular acceleration of the flywheel and corresponding shaft position :

The change in torque produces the acceleration in the flywheel. The acceleration in the flywheel will be maximum or minimum, when change in torque is maximum or minimum. But the change in torque is given by,

$$\Delta T = 300 \sin 2\theta - 250 \sin \theta$$

Differentiating the above expression with respect to θ and equating to zero for maximum or minimum values, we get

$$\frac{d}{d\theta}(\Delta T) = 0$$

or $\dfrac{d}{d\theta}(300 \sin 2\theta - 250 \sin \theta) = 0$

or $300 \times 2 \times \cos 2\theta - 250 \times \cos \theta = 0$

or $600 \cos 2\theta - 250 \cos \theta = 0$

or $12 \cos 2\theta - 5 \cos \theta = 0$ (Dividing by 50 to both sides)

or $12 (2 \cos^2 \theta - 1) - 5 \cos \theta = 0$ ($\because \cos 2\theta = 2 \cos^2 \theta - 1$)

or $24 \cos^2 \theta - 12 - 5 \cos \theta = 0$

or $24 \cos^2 - 5 \cos \theta - 12 = 0$

The above equation is a quadratic equation in cos θ. Hence the solution is

$$\cos \theta = \frac{5 \pm \sqrt{5^2 - 4 \times 24 \times (-12)}}{2 \times 24} = \frac{5 \pm \sqrt{25 + 1152}}{48}$$

$$= \frac{5 \pm 34.3}{48} = \frac{39.3}{48} \text{ and } \frac{-29.3}{48} = 0.81875 \text{ and } -0.6104$$

$\therefore \quad \theta = \cos^{-1} 0.81875$ and $\cos^{-1}(-0.6104)$ = **35.04° and 127.62°. Ans.**

Hence the shaft position for maximum and minimum acceleration is given by θ = 35.04° and 127.62°.

For maximum change in torque, the value of θ = 35.04° is substituted in equation (*i*).

$$\therefore \quad (\Delta T)_{max.} = 300 \sin (2 \times 35.04) - 250 \sin (35.04)$$
$$= 300 \sin (70.08) - 250 \sin (35.04)$$
$$= 300 \times 0.94 - 250 \times 0.5741$$
$$= 282 - 143.537 = 138.463 \text{ Nm}$$

For minimum change in torque, the value of θ = 127.62° is substituted in equation (*i*),

$$\therefore \quad (\Delta T)_{min.} = 300 \sin (2 \times 127.62°) - 250 \sin (127.62°)$$
$$= 300 \sin (255.24°) - 250 \times 0.792$$
$$= 300 \times (-0.967) - 198 = -290.1 - 198 = -488.1 \text{ Nm}$$

We know that maximum acceleration is given by

$$\alpha_{max.} = \frac{(\Delta T)_{max.}}{I} \qquad \left(\because \text{ Torque} = I \times \alpha \quad \therefore \quad \alpha = \frac{\text{Torque}}{I}\right)$$

$$= \frac{138.463}{m \cdot k^2} \qquad (\because \ I = mk^2 = 250 \times 0.4^2)$$

$$= \frac{138.463}{250 \times 0.4^2} = \mathbf{3.46\ rad/s^2}. \quad \textbf{Ans.}$$

Minimum acceleration (or maximum retardation) is given by,

$$\alpha_{min.} = \frac{(\Delta T)_{min.}}{I} = \frac{-488.1}{m \cdot k^2} = \frac{-488.1}{250 \times 0.4^2} = \mathbf{-12.2\ rad/s^2}. \quad \textbf{Ans.}$$

Problem 12.10. *The variation of crank-shaft torque of a four cylinder petrol engine may be approximately represented by taking the torque as zero for crank angles 0° and 180° and as 260 Nm for crank angles 20° and 45°, the intermediate portions of the torque graph being straight lines. The cycle is being repeated in every half revolution. The average speed is 600 r.p.m. Supposing that the engine drives a machine requiring a constant torque, determine the mass of the flywheel of radius of gyration 250 mm, which must be provided, so that the total variation of speed shall be one percent.* **(A.M.I.E., Winter 1980)**

Sol. Given :

Torque, $T = 0$ at crank angles $\theta = 0°, 180°$

$T = 260$ Nm at $\theta = 20°$ and 45°. Intermediate portions are straight lines.

Average speed, $N = 600$ r.p.m. or $\omega = \frac{2\pi N}{60} = \frac{2\pi \times 600}{60} = 20\pi$ rad/s ; $k = 250$ mm $= 0.25$ m ; total variation of speed $= 1\% = \frac{1}{100} = 0.01$ or co-efficient of fluctuation of speed, $K_s = 0.01$.

The turning moment (*T-M*) diagram for half revolution (*i.e.* for 180°) is shown in Fig. 12.13.

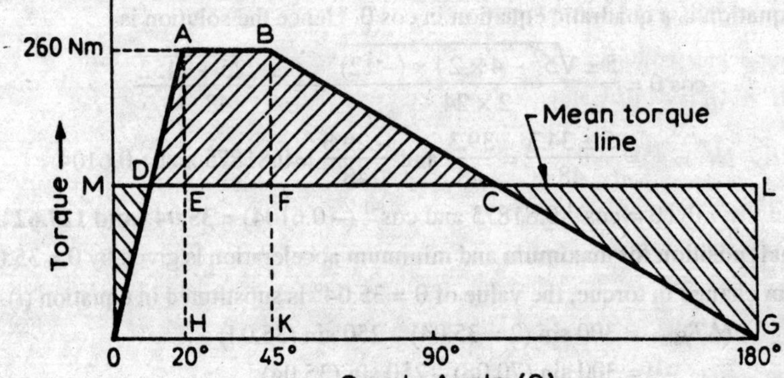

Fig. 12.13

The points *H* and *K* represents the crank positions at an angle of 20° and 45° respectively. The ordinates $AH = 260$ Nm and $BK = 260$ Nm. The point A is joined to *O* *i.e.* *AO* is a straight line. The point B is joined to *G*.

Work-done for half revolution = Area of *T-M* diagram

= Area of *OABG*

= Area of triangle *OAH* + Area of rectangle *HABK* + Area of triangle *KBG*

$$= \frac{1}{2} \times OH \times AH + AH \times HK + \frac{1}{2} \times BK + KG$$

$$= \frac{1}{2} \times \left(\frac{20 \times \pi}{180}\right) \times 260 + 260 \times \frac{(45-20)\pi}{180°} + \frac{1}{2} \times 260 \times \frac{(180-45)\pi}{180°}$$

$$\left(\because \ AH = BK = 260 \text{ Nm} ;\ OH = 20° = \frac{20 \times \pi}{180} \text{ rad} ;\ HK = (45-20)° = \frac{(45-20) \times \pi}{180} \text{ rad.} ;\right.$$

$$\left. \text{and } \ KG = (180-45)° = (180-45) \times \frac{\pi}{180} \text{ rad}\right)$$

$$= 260 \times \frac{\pi}{180}\left(\frac{20}{2} + 25 + \frac{135}{2}\right) = 260 \times \frac{\pi}{180} \times 102.5 = 465.13 \text{ Nm}$$

Let T_{mean} = Mean torque in Nm = Ordinate OM

Then work done corresponding of mean torque for half revolution

= Area of rectangle $OMLG$ where OM = Mean torque = T_{mean}

$= OM \times OG = T_{mean} \times \pi$ $\quad (\because \ OG = 180°$ or π radian and OM = torque)

$\therefore \quad T_{mean} \times \pi$ = Workdone for half revolution = 465.13 Nm

$$\therefore \quad T_{mean} = \frac{465.13}{\pi} = 148.05 \text{ Nm}$$

Since the area above the mean torque line represents the maximum fluctuation of energy, therefore the maximum fluctuation of energy,

ΔE = Area $DABC$

= Area DAE + Area $EABF$ + Area BFC

$$= \left(\frac{1}{2} \times DE \times AE\right) + (AB \times AE) + \left(\frac{1}{2} \times BF \times FC\right) \quad ...(i)$$

But $\quad AE = AH - EH = 260 - 148.05 \quad (\because \ EH = OM = T_{mean} = 148.05)$

= 111.95 Nm

Also $\quad BF = AE = 111.95$ Nm

Now $\quad AB = HK = 45° - 20° = 25° = 25 \times \frac{\pi}{180}$ radians = 0.4363 radians

Now let us find the values of DE and FC

The triangles AOH and ADE are similar.

$$\therefore \quad \frac{DE}{OH} = \frac{AE}{AH}$$

or

$$DE = OH \times \frac{AE}{AH} = \left(20 \times \frac{\pi}{180}\right) \times \frac{111.95}{260}$$

$$\left(\because \ OH = 20° = 20 \times \frac{\pi}{180} \text{ rad}, AE = 111.95 \text{ and } AH = 260\right)$$

= 0.15 radians.

The triangles BGK and BFC are similar.

$$\therefore \quad \frac{FC}{KG} = \frac{BF}{BK}$$

or

$$FC = KG \times \frac{BF}{BK} = \left[(180-45)\frac{\pi}{180}\right] \times \frac{111.95}{260}$$

$$\left(\because \ KG = 180° - 45° = (180-45) \times \frac{\pi}{180} \text{ rad.}, BF = AE = 111.95 \text{ and } BK = AH = 260\right)$$

$$= 135 \times \frac{\pi}{180} \times \frac{111.95}{260} = 1.014 \text{ rad.}$$

Now substituting all values in equation (*i*), we get

Maximum fluctuation of energy,

$$\Delta E = \tfrac{1}{2} \times 0.15 \times 111.95 + 0.4363 \times 111.95 + \tfrac{1}{2} \times 111.95 \times 1.014$$

$$(\because \; DE = 0.15 \; ; AE = 111.95 \; ; AB = 0.4363 \; ; BF = 111.95 \text{ and } FC = 1.014)$$

$$= 111.95\,(0.075 + 0.4363 + 0.507) = 111.95 \times 1.0183 = 114 \text{ Nm}$$

Let m = Mass of flywheel

Now using equation (12.11), we get

$$\Delta E = I \times \omega^2 \times K_s \text{ or } mk^2 \times \omega^2 \times K_s$$

or $\quad 114 = m \times 0.25^2 \times (20\pi)^2 \times 0.01 \quad (\because \; k = 0.25 \text{ m} \; ; \omega = 20\pi \text{ rad/s} \; ; K_s = 0.01)$

or $\quad m = \dfrac{114}{0.25^2\,(20\pi)^2 \times 0.01} = \mathbf{46.2 \text{ kg.}}$ **Ans.**

Problem 12.11. *A flywheel of mass 250 kg and radius of gyration of 600 mm is attached to a shaft. The shaft rotates at a speed of 200 r.p.m. and drives a machine. The torque of the machine varies in a cyclic manner over a period of 3 revolutions. The torque rises from 250 Nm to 1000 Nm uniformly during first* $\frac{1}{2}$ *revolution and remains constant for next one revolution. It then falls uniformly to 250 Nm during the next* $\frac{1}{2}$ *revolution and remains constant for one revolution, the cycle being repeated thereafter. Determine* :

(*i*) *the power required to drive the machine and*

(*ii*) *percentage fluctuation in speed if the driving torque applied to the shaft is constant.*

Sol. Given :

$$m = 250 \text{ kg} \; ; k = 600 \text{ mm} = 0.6 \text{ m} \; ; N = 200 \text{ r.p.m. or } \omega = \frac{2\pi N}{60} = \frac{2\pi \times 200}{60} = 20.94 \text{ rad/s} \; ;$$

Torque of machine

(*i*) rises uniformly from 250 Nm to 1000 Nm during $\frac{1}{2}$ revolution

(*ii*) remains constant (*i.e.* 1000 Nm) for one revolution

(*iii*) falls uniformly from 1000 Nm to 250 Nm during next $\frac{1}{2}$ revolution

(*iv*) remains constant (*i.e.* 250 Nm) for next one revolution.

Fig. 12.14 shows the turning moment diagram for the complete cycle. One cycle consists of three revolutions (or 6π radians) as one revolution is equal to 2π radians.

Work-done per cycle = Area of Turning Moment diagram

= Area of *OABCDEF*

= Area of *OAEF* + Area *ABG* + Area *BCHG* + Area *CDH*

$$= OA \times OF + \tfrac{1}{2} \times AG \times BG + GH \times BG + \tfrac{1}{2} \times CH \times HD$$

$$= 250 \times 6\pi + \tfrac{1}{2} \times \pi \times (1000 - 250) + 2\pi \times (1000 - 250) + \tfrac{1}{2} \times (1000 - 250) \times \pi$$

$$= 1500\pi + 375\pi + 1500\pi + 375\pi$$

$$= 3750\pi \text{ Nm} \quad \ldots(i)$$

Let T_{mean} = Mean torque in Nm = Ordinate *OM*

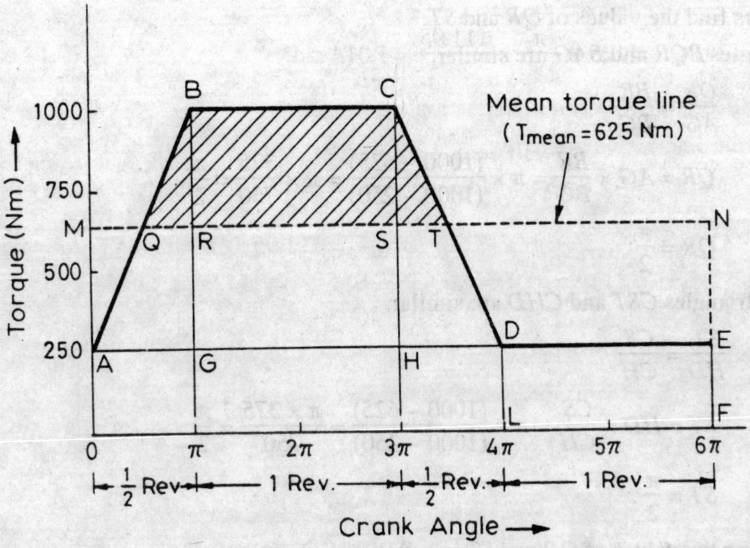

Fig. 12.14

∴ Work done corresponding to mean torque for one cycle

$$= \text{Area of rectangle } OMNF$$

$$= OM \times OF$$

$$= T_{mean} \times 6\pi \qquad (\because \; OM = T_{mean} \text{ and } OF = 6\pi \text{ radian}) \qquad ...(ii)$$

Equating the two work done per cycle given by equations (*i*) and (*ii*), we get

$$3750 \times \pi = T_{mean} \times 6\pi$$

or $$T_{mean} = \frac{3750 \times \pi}{6\pi} = 625 \text{ Nm}.$$

(*i*) *Power required to drive the machine*

The power (P) required to drive the machine is given by,

$$P = T_{mean} \times \omega = 625 \times 13.09 \text{ W} = \mathbf{8181.25 \text{ W}}. \quad \textbf{Ans.}$$

(*ii*) *Percentage fluctuation in speed*

Let K_s = Co-efficient of fluctuation of speed.

Let us first find the maximum fluctuation of energy. We know that maximum fluctuation of energy is given by the area above the mean torque line.

∴ Maximum fluctuation of energy,

$$\Delta E = \text{Shaded area } QBCT$$

$$= \text{Area } QBR + \text{Area } RBCS + \text{Area } CST$$

$$= \frac{1}{2} QR \times BR + BC \times BR + \frac{1}{2} CS \times ST$$

$$= \frac{1}{2} \times QR \times (1000 - 625) + 2\pi \times (1000 - 625) + \frac{1}{2} \times (1000 - 625)\, ST$$

$$= \frac{1}{2} \times QR \times 375 + 2\pi \times 375 + \frac{1}{2} \times 375 \times ST$$

$$= 375 \left[\frac{QR}{2} + 2\pi + \frac{ST}{2} \right] \qquad ...(iii)$$

Now let us find the values of QR and ST.

The triangles BQR and BAG are similar.

$$\therefore \quad \frac{QR}{AG} = \frac{BR}{BG}$$

or $$QR = AG \times \frac{BR}{BG} = \pi \times \frac{(1000 - 625)}{(1000 - 250)} = \pi \times \frac{375}{750} = \frac{\pi}{2}$$

$$\therefore \quad QR = \frac{\pi}{2}$$

Now the triangles CST and CHD are similar.

$$\therefore \quad \frac{ST}{HD} = \frac{CS}{CH}$$

or $$ST = HD \times \frac{CS}{CH} = \pi \times \frac{(1000 - 625)}{(1000 - 250)} = \frac{\pi \times 375}{750} = \frac{\pi}{2}$$

$$\therefore \quad ST = \frac{\pi}{2}.$$

Substituting the values of QR and ST in equation (*iii*), we get

$$\Delta E = 375\left[\frac{\pi}{2 \times 2} + 2\pi + \frac{\pi}{2 \times 2}\right] = 375 \times 2.5\pi \text{ Nm} = 2945.24 \text{ Nm}$$

Now using equation (12.11), we have

$$\Delta E = I \times \omega^2 \times K_s = mk^2 \times \omega^2 \times K_s$$

$$\therefore \quad K_s = \frac{\Delta E}{mk^2 \times \omega^2} = \frac{2945.24}{250 \times 0.6^2 \times 20.94} = \mathbf{0.0746.} \quad \textbf{Ans.}$$

Problem 12.12. *The equation of turning moment curve of a three crank engine is 2500 + 750 sin 3θ Nm, where θ is the crank angle in radians. The mean speed of engine is 300 r.p.m. The flywheel and other rotating parts attached to the engine have a mass of 500 kg at a radius of gyration of 1 m. Calculate :*

(*i*) *the power of the engine ;*

(*ii*) *the total fluctuation of the speed of the flywheel in percentage when*

(*a*) *the resisting torque is constant,*

(*b*) *the resisting torque is 2500 + 300 sin θ Nm.*

Sol. Given :

$$N = 300 \text{ r.p.m. or } \omega = \frac{2\pi N}{60} = \frac{2\pi \times 300}{60} = 10\pi \text{ rad/s} \; ; \; m = 500 \text{ kg} \; ; \; k = 1 \text{ m} \; ; \; T = 2500 + 750 \sin 3\theta$$

The expression for torque being a function of 3θ, the cycle is repeated after every 120° $\left(\frac{360}{3} = 120°\right)$ or $\frac{2\pi}{3}$ radians of the crank rotation.

Work done per cycle = Area of T-M diagram for one cycle

$$= \int_0^{2\pi/3} T \times d\theta$$

$$\left(\because \text{ Cycle is repeated after every crank rotation of } \frac{2\pi}{3} \text{ radians}\right)$$

$$= \int_0^{2\pi/3} (2500 + 750 \sin 3\theta)\, d\theta$$

$$= \left[2500 \times \theta + 750 \frac{(-\cos 3\theta)}{3} \right]_0^{2\pi/3} = \left[2500 \times \theta - 250 \cos 3\theta \right]_0^{2\pi/3}$$

$$= 2500\left(\frac{2\pi}{3} - 0\right) - 250\left[\cos 3\left(\frac{2\pi}{3}\right) - \cos(3 \times 0)\right]$$

$$= \frac{5000\pi}{3} - 250\,[\cos 2\pi - \cos 0]$$

$$= \frac{5000\pi}{3} - 250[1 - 1] \qquad (\because \cos 2\pi = \cos 360° = 1 \text{ and } \cos 0° = 1)$$

$$= \frac{5000 \times \pi}{3} \text{ Nm}$$

Let T_{mean} = Mean torque

Then area of *T-M* diagram corresponding to mean torque for one cycle

$= T_{mean} \times$ Length of one cycle in radians

$= T_{mean} \times \frac{2\pi}{3}$ $\left[\because \text{Cycle is repeated after every crank rotation of } \frac{2\pi}{3} \text{ radians. Hence length of one cycle in radian} = \frac{2\pi}{3}\right]$

$$T_{mean} \times \frac{2\pi}{3} = \frac{5000 \times \pi}{3}$$

or $$T_{mean} = \frac{5000 \times \pi}{3} \times \frac{3}{2\pi} = 2500 \text{ Nm.}$$

(i) Power of engine

Let P = power of the engine

Then $P = T_{mean} \times \omega = 2500 \times 10\pi$ Nm/s = 78539.8 W $\qquad (\because 1 \text{ W} = 1 \text{ Nm/s})$

= **78.5398 kW. Ans.**

(ii) Total fluctuation of the speed of the flywheel in percentage when the resisting torque is constant.

Let K_s = Total fluctuation of speed of flywheel in percentage.

Since the resisting torque is constant, therefore the torque exerted on the shaft is equal to mean resisting torque on the flywheel.

$\therefore$ $T = T_{mean}$

or $2500 + 750 \sin 3\theta = 2500$

or $750 \sin 3\theta = 2500 - 2500 = 0$

or $\sin 3\theta = 0$

or $3\theta = 0$ or $180°$

or $\theta = 0$ or $60°$

$\therefore$ Maximum fluctuation of energy,

$$\Delta E = \int_0^{60°} (T - T_{mean})\, d\theta = \int_0^{60°} [(2500 + 750 \sin 3\theta) - 2500]\, d\theta$$

$$= \int_0^{60°} 750 \sin (3\theta)\, d\theta = \left[750 \left(\frac{-\cos 3\theta}{3} \right) \right]_0^{60°}$$

$$= \left[-250 \cos 3\theta \right]_0^{60°} = -250 [\cos (3 \times 60°) - \cos (3 \times 0)]$$

$$= -250 [\cos 180° - \cos 0°] = -250 [-1-1] \quad (\because \cos 180° = -1)$$

$$= -250 (-2) = 500 \text{ Nm}$$

The maximum fluctuation of energy (ΔE) is also given by equation (12.11) as

$$\Delta E = I \times \omega^2 \times K_s = m \times k^2 \times \omega^2 \times K_s$$

$$\therefore \quad 500 = m \times k^2 \times \omega^2 \times K_s \quad (\because \Delta E = 500 \text{ Wm})$$

or $$K_s = \frac{500}{m \times k^2 \times \omega^2} = \frac{500}{500 \times 1^2 \times (10\pi)^2} = \frac{1}{100 \times \pi^2}$$

$$= 0.00101 \text{ or } \mathbf{0.101\%}. \quad \textbf{Ans.}$$

(iii) Value of K_s when the resisting torque is (2500 + 300 sin θ)

Let us first find the points at which the curve representing the torque exerted on the shaft, intersects the resisting torque line. At the points of intersection, the two torques should be equal *i.e.*

Torque exerted on the shaft = Resisting torque

or $$2500 + 750 \sin 3\theta = 2500 + 300 \sin \theta$$

or $$750 \sin 3\theta = 300 \sin \theta$$

or $$\frac{750}{300} \sin 3\theta = \sin \theta$$

or $$2.5 \sin^* 3\theta = \sin \theta$$

or $$2.5 (3 \sin \theta - 4 \sin^3 \theta) = \sin \theta \quad [\because \sin 3\theta = 3 \sin \theta - 4 \sin^3 \theta]$$

$$[\because \sin 3\theta = \sin (2\theta + \theta) = \sin 2\theta \cos \theta + \cos 2\theta \sin \theta$$
$$= (2 \sin \theta \cos \theta) \cos \theta + [\cos^2 \theta - \sin^2 \theta] \sin \theta$$
$$= 2 \sin \theta \cos^2 \theta + \cos^2 \theta \sin \theta - \sin^3 \theta = 3 \sin \theta \cos^2 \theta - \sin^3 \theta]$$

or $$\frac{3 \sin \theta - 4 \sin^3 \theta}{\sin \theta} = \frac{1}{2.5} = 0.4$$

or $$\frac{\sin \theta (3 - 4 \sin^2 \theta)}{\sin \theta} = 0.4$$

or $$3 - 4 \sin^2 \theta = 0.4$$

or $$\frac{3 - 0.4}{4} = \sin^2 \theta$$

or $$\sin \theta = \pm \sqrt{\frac{3 - 0.4}{4}} = \pm 0.8062$$

If $\sin \theta = 0.8062$, then $\theta = \sin^{-1} 0.8062 = 53.72°$ and $(180 - 53.72) = 126.28°$.

If $\sin \theta = -0.8062$, then $\theta = 233.72°$ and $306.28°$.

$^*\sin 3\theta = \sin (2\theta + \theta) = \sin 2\theta \cos \theta + \cos 2\theta \sin \theta = (2 \sin \theta \cos \theta) \cos \theta + (\cos^2 \theta - \sin^2 \theta) \sin \theta$
$= 2 \sin \theta \cos^2 \theta + \cos^2 \theta \sin \theta - \sin^3 \theta = 3 \sin \theta \cos^2 \theta - \sin^3 \theta = 3 \sin \theta (1 - \sin^2 \theta) - \sin^3 \theta$
$= 3 \sin \theta - 3 \sin^3 \theta - \sin^3 \theta = 3 \sin \theta - 4 \sin^3 \theta.$

But the cycle is repeated after every cranks angle of 120°. Hence the possible values of θ are 53.72° and 126.28°.

∴ Maximum fluctuation of energy,

$$\Delta E = \int_{53.72°}^{126.28°} (\text{Torque on shaft} - \text{Resisting torque})\, d\theta$$

$$= \int_{53.72°}^{126.28°} [(2500 + 750 \sin 3\theta) - (2500 + 300 \sin \theta)]\, d\theta$$

$$= \int_{53.72°}^{126.28°} (750 \sin 3\theta - 300 \sin \theta)\, d\theta = \left[750 \left(\frac{-\cos 3\theta}{3} \right) + 300 \cos \theta \right]_{53.72°}^{126.28°}$$

$$= -250\,[\cos (3 \times 126.28°) - \cos (3 \times 53.72°)] + 300\,(\cos 126.28° - \cos 53.72°)$$

$$= -250\,[\cos 378.84° - \cos 161.16°] + 300\,(\cos 126.28° - \cos 53.72°)$$

$$= -250\,[0.9464 - (-0.9464)] + 300\,(-0.5917 - 0.5917)$$

$$= -250\,[0.9464 + 0.9464] - 177.52 - 177.52$$

$$= -473.2 - 355.04 = -828.24 \text{ Nm}$$

The maximum fluctuation of energy (ΔE) is also given by equation (12.11) as

$$\Delta E = I \times \omega^2 \times K_s = mk^2 \times \omega^2 \times K_s$$

or
$$-828.24 = mk^2 \times \omega^2 \times K_s$$

∴
$$K_s = \frac{-828.28}{mk^2 \times \omega^2} = \frac{-828.28}{500 \times 1^2 \times (10\pi)^2} = -0.00168 \quad \text{or} \quad \mathbf{0.168\%.} \quad \textbf{Ans.}$$

12.7. Flywheel's Rim Dimensions

Fig. 12.15 show a rim of a flywheel.

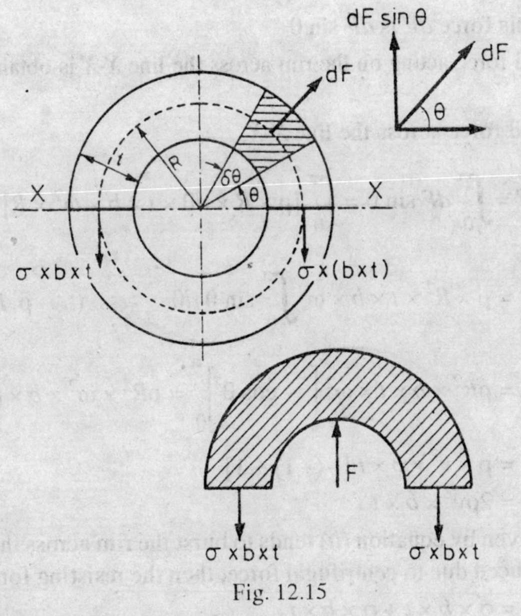

Fig. 12.15

Let N = speed of flywheel in r.p.m. ;

ω = Angular velocity of flywheel in rad/s = $\frac{2\pi N}{60}$

v = Linear velocity at the mean radius in m/s = $\omega \times R = \frac{2\pi N}{60} \times R = \frac{\pi N D}{60}$

R = Mean radius of rim

D = Mean diameter of rim = $2R$

ρ = Density of rim material

σ = Tensile stress or hoop stress in N/m^2 due to centrifugal force.

Consider a small element of the rim as shown shaded in Fig. 12.15. Let the small element subtends an angle $\delta\theta$ at the centre of the flywheel. The centrifugal force acting radially outwards on the element is dF. This force dF acting on the centre of element is given by

$$dF = \frac{mv^2}{R}$$ where m = Mass of element

v = linear velocity at mean radius $R = \omega \times R$

$$= m \times \frac{\omega^2 R^2}{R}$$

$$= m \times \omega^2 \times R \qquad ...(i)$$

Now mass of element, m = Density × Volume of element

$$= \rho \times [(R \times \delta\theta) \times t \times b]$$

[$\because$ Volume of element = Area of element × Width = $[(R \times \delta\theta) \times t] \times b$]

where b = Width of element perpendicular to plane of paper.

Substituting the value of 'm' in equation (i), we get

$$dF = [\rho \times R \times \delta\theta \times t \times b] \times \omega^2 \times R$$

Vertical component of this force dF is $dF \sin\theta$

$\therefore$ Total vertical upward force acting on the rim across the line X-X is obtained by integrating $dF \sin\theta$ from the angle 0 to π.

$\therefore$ Total vertical upward force across the line X-X,

$$F = \int_0^{\pi} dF \sin\theta = \int_0^{\pi} [\rho \times R \times d\theta \times t \times b \times \omega^2 \times R] \sin\theta$$

$$= \rho \times R^2 \times t \times b \times \omega^2 \int_0^{\pi} \sin\theta \, d\theta \qquad (\because \rho, R, t, b \text{ and } \omega \text{ are constants})$$

$$= \rho R^2 \times b \times t \times \omega^2 \Big[-\cos\theta \Big]_0^{\pi} = \rho R^2 \times \omega^2 \times b \times t\,[-\cos\pi - (-\cos 0°)]$$

$$= \rho \times v^2 \times b \times t\,[-(-1) + 1] \qquad (\because v^2 = \omega^2 R^2)$$

$$= 2\rho v^2 \times b \times t \qquad ...(ii)$$

The total vertical force given by equation (ii) tends to burst the rim across the diameter X-X. If σ is the tensile stress or hoop stress produced due to centrifugal force, then the resisting force will be

$$= \sigma \times b \times t + \sigma \times b \times t$$

$$= 2\sigma \times b \times t \qquad ...(iii)$$

Equating the two force (*i.e.* bursting force and resisting force) given by equations (*ii*) and (*iii*), we get

$$2\sigma \times b \times t = 2\rho v^2 \times b \times t$$

or $$\sigma = \rho v^2 \quad ...(12.14)$$

or $$v = \sqrt{\frac{\sigma}{\rho}} \quad ...(12.15)$$

Problem 12.13. *The turning moment diagram in Fig. 12.16, for a multi-cylinder engine has been drawn to a scale of 1 mm to 500 N-m torque and 1 mm to 6° of crank displacement. The intercepted area between output torque curve and mean resistance line taken in order from one end, in sq. mm are :*

– 30, + 410, – 280, + 320, – 330, + 250, – 360, + 280, – 260 sq. mm when the engine is running at 800 r.p.m.

The engine has a stroke of 300 mm and fluctuation of speed is not to exceed 2% of the mean speed.

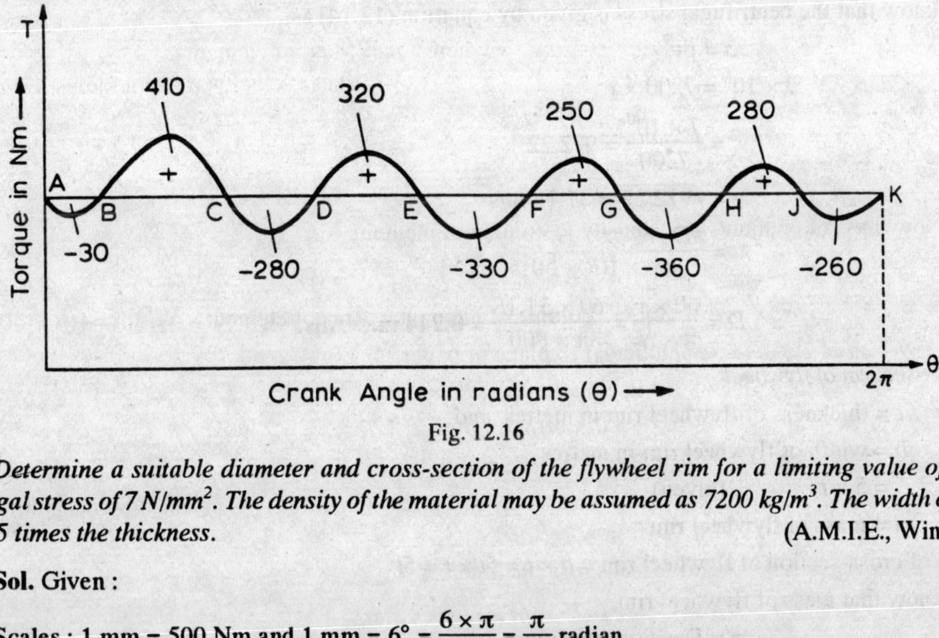

Fig. 12.16

Determine a suitable diameter and cross-section of the flywheel rim for a limiting value of the safe centrifugal stress of 7 N/mm². The density of the material may be assumed as 7200 kg/m³. The width of the rim is to be 5 times the thickness. (A.M.I.E., Winter 1976)

Sol. Given :

Scales : 1 mm = 500 Nm and 1 mm = 6° $= \frac{6 \times \pi}{180} = \frac{\pi}{30}$ radian.

Engine stroke, L = 300 mm = 0.3 m

$\therefore$ Crank radius, $r = \frac{L}{2} = \frac{0.3}{2} = 0.15$ m

Mean speed of engine, N_{mean} = 800 r.p.m.

Max. fluctuation of speed = 2% of $N_{mean} = 0.02 \times N_{mean}$

Centrifugal stress, σ = 7 N/mm² = 7×10^6 N/m²

Density of material of flywheel, ρ = 7200 kg/m³

Now mean angular speed of engine,

$$\omega_{mean} = \frac{2\pi N_{mean}}{60} = \frac{2\pi \times 800}{60} = 83.77 \text{ rad/s}$$

Now turning moment scale is 1 mm = 500 Nm and crank angle scale is 1 mm = $6° = \frac{6 \times \pi}{180} = \frac{\pi}{30}$ radians

$\therefore$ 1 mm^2 on the turning moment diagram

= 1 mm for turning moment × 1 mm for crank angle

$$= (500 \text{ Nm}) \times \left(\frac{\pi}{30}\right) = 500 \times \frac{\pi}{30} \text{ Nm} = 52.36 \text{ Nm}$$

Let 1 mm^2 on *T-M* diagram = x Nm

Then $x = 52.36$ Nm ($\because$ 1 mm^2 = 52.36 Nm)

Diameter of the flywheel rim

Let D = Diameter of flywheel rim in meter and

v = Peripheral velocity of flywheel rim in m/s

We know that the centrifugal stress is given by equation (12.14) as,

$$\sigma = \rho v^2$$

$$\therefore \quad 7 \times 10^6 = 7200 \times v^2$$

$$\therefore \quad v^2 = \frac{7 \times 10^6}{7200} = 972.22$$

or $v = \sqrt{972.22} = 31.18$ m/s

But $v = \frac{\pi D \,.\, N}{60}$

or $D = \frac{60 \times v}{\pi \times N} = \frac{60 \times 31.18}{\pi \times 800} =$ **0.744 m. Ans.**

Cross-section of flywheel

Let t = thickness of flywheel rim in metres, and

b = width of flywheel rim in metres

$= 5 \times t$ (given)

m = mass of flywheel rim

Area of cross-section of flywheel rim = $b \times t = 5t \times t = 5t^2$

We know that mass of flywheel rim,

m = Density × Volume

= $\rho \times [\pi D \times$ Area of cross-section]

= $7200 \times [\pi \times 0.744 \times 5t^2]$ ($\because$ $D = 0.744$ m, Area = $5t^2$)

= $84144.4t_2$

$\therefore$ Mass, $m = 84144.4t^2$ ...(*i*)

Hence if the value of mass is known, then thickness t can be obtained. After that the value of b (which is equal to $5t$) can be known.

The value of mass will be obtained from the relation of maximum fluctuation of energy (ΔE) which is given by equation (12.13) as,

$$\Delta E = m \times v^2 \times K_s \quad \text{...(ii)}$$

where K_s = Max. fluctuation of speed = 0.02 (given)

In the above relation ΔE and m are unknown. The value of ΔE (*i.e.* maximum fluctuation of energy) is obtained from the turning moment diagram, which is shown in Fig. 12.16.

Let 1 mm^2 on *T-M* diagram = x Nm

Then $x = 52.36$ Nm ($\because$ 1 mm^2 = 52.36 Nm)

Then the areas below and above the mean torque line in works units (*i.e.* Nm units) will be $-30x$, $+410x$, $-280x$, $+320x$, $-330x$, $+250x$, $-360x$, $+280x$ and $-260x$ respectively.

Let the energy at $A = E$.

Now refer to Fig. 12.16. The energy at B will be (E + Area in work units between A to B). Area in work units between A and B is $-30x$, hence energy at $B = E - 30x$

Energy at B $= E - 30x$ ...(Minimum Energy)

Energy at C $= (E - 30x) + 410x = E + 380x$

Energy at D $= (E + 38x) - 280x = E + 100x$

Energy at E $= (E + 100x) + 320x = E + 420x$ ...(Maximum Energy)

Energy at F $= (E + 420x) - 330x = E + 90x$

Energy at G $= (E + 90x) + 250x = E + 340x$

Energy at H $= (E + 340x) - 360x = E - 20x$

Energy at K $= (E - 20x) + 280x = E + 260x$

Energy at L $= (E + 260x) - 260x = E$ = Energy at A

The maximum fluctuation of energy (ΔE) is given by

$$\Delta E = \text{Maximum Energy} - \text{Minimum Energy}$$

$$= (E + 420x) - (E - 30x) = 450x$$

$$= 450 \times 52.36 \quad (\because \; x = 52.36 \text{ Nm})$$

$$= 23562 \text{ Nm}$$

Substituting the value of ΔE in equation (*ii*), we get

$$\Delta E = m \times v^2 \times K_s$$

or $$23562 = m\,(31.18)^2 \times 0.02 \quad (\because \; v = 31.18 \text{ m/s and } K_s = 0.02)$$

$\therefore$ $$m = \frac{23562}{(31.18)^2 \times 0.02} = 1211.79 \text{ kg}$$

Now substituting the value of m in equation (*i*), we get

$$1211.79 = 84144.4t^2$$

or $$t = \sqrt{\frac{1211.79}{84144.4}} = 0.12 \text{ m} = \mathbf{120\ mm.}\quad \textbf{Ans.}$$

and $$b = 5 \times t = 5 \times 120 = \mathbf{600\ mm.}\quad \textbf{Ans.}$$

Problem 12.14. *A steam engine runs at 150 r.p.m. Its turning moment diagram gave the following area moments taken in order above and below mean torque line :*

500, – 250, 270, – 390, 190, – 340, 270, – 250 (all in square mm)

With the scale : Turning moment 1 mm = 500 Nm

Crank displacement 1 mm = 5°

If the total fluctuation of speed is 1.5% of the mean speed, determine the cross-section of the rim of the flywheel assumed rectangular with axial dimension equal to $1\frac{1}{2}$ times the radial dimension. The hoop stress is limited to 3 N/mm^2 and density of the material of the flywheel is 7500 kg/m^3. (A.M.I.E., Summer 1985)

Sol. Given :

$N = 150$ r.p.m. or $\omega = \frac{2\pi N}{60} = \frac{2\pi \times 150}{60} = 5\pi$ rad/s ; Scale : *T-M* scale is 1 mm = 500 Nm and crank angle scale is 1 mm = $5° = 5 \times \frac{\pi}{180} = \frac{\pi}{36}$ radians

$\therefore$ 1 mm^2 on *T-M* diagram = 1 mm for turning moment × 1 mm for crank angle

$= 500 \times \frac{\pi}{36}$ Nm = 43.63 Nm

Total fluctuation of speed = 1.5% of mean speed = 0.015 of N_{mean}

$\therefore$ K_s = Co-efficient of fluctuation of speed = $\frac{\text{Total fluctuation of speed}}{\text{Mean speed}} = 0.015$

$b = 1.5t$; σ = hoop stress = 3 N/mm^2 = 3×10^6 N/m^2 ; $\rho = 7500$ kg/m^3

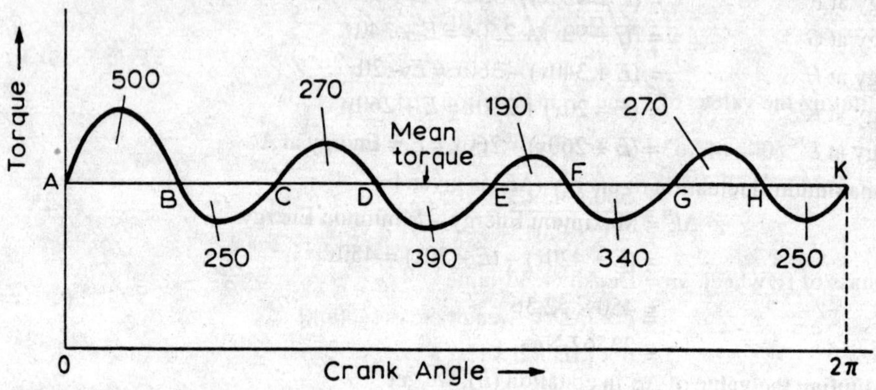

Fig. 12.17

Let us first find the maximum fluctuation of energy (ΔE) from the turning moment diagram which is shown in Fig. 12.17.

Let 1 mm^2 on *T-M* diagram = x Nm

Then $x = 43.63$ Nm ($\because$ 1 mm^2 on *T-M* diagram = 43.63 Nm)

Now the areas above and below the mean torque line in works units (*i.e.* in Nm) will be $500x$, $-250x$, $270x$, $-390x$, $190x$, $-340x$, $270x$ and $-250x$ respectively.

Let the energy at $A = E$ Nm

Then energy at B will be (E + area in works units between A and B). But area is works units between A and B is $500x$ Nm. Hence energy at B will be $E + 500x$

$\therefore$ Energy at B $= E + 500x$ Nm

Energy at C $= (E + 500x) - 250x = E + 250x$

Energy at D $= (E + 250x) + 270x = E + 520x$ ...(Maximum Energy)

Energy at E $= (E + 520x) - 390x = E + 130x$

Energy at F $= (E + 130x) + 190x = E + 320x$

Energy at G $= (E + 320x) - 340x = E - 20x$ ...(Minimum Energy)

Energy at H $= (E - 20x) + 270x = E + 250x$

Energy at K $= (E + 250x) - 250x = E$ = Energy at A.

The maximum fluctuation of energy (ΔE) is given by,

$$\Delta E = \text{Maximum Energy} - \text{Minimum Energy}$$
$$= (E + 520x) - (E - 20x) = 500x$$
$$= 500 \times 43.63 \text{ Nm} \quad (\because \; x = 43.63 \text{ Nm}) \quad ...(i)$$

But maximum fluctuation of energy (ΔE) is also given by equation (12.13) as

$$\Delta E = m \times v^2 \times K_s \quad ...(ii)$$

Equating the two values of ΔE given by therefore equations (*i*) and (*ii*), we get

$$500 \times 43.63 = m \times v^2 \times K_s \quad ...(iii)$$

The value of V is obtained from the hoop stress which is given by equation (12.14) as,

$$\sigma = \rho \times v^2$$

or
$$v = \sqrt{\frac{\sigma}{\rho}} = \sqrt{\frac{3 \times 10^6}{7500}} = 20 \text{ m/s} \quad ...(iv)$$

Substituting the values of V and K_s in equation (*iii*), we get

$$500 \times 43.63 = m \times 20^2 \times 0.015 \quad (\because \; K_s = 0.015)$$

$$\therefore \quad m = \frac{500 \times 43.63}{20^2 \times 0.015} = 3635.8 \text{ kg}$$

But mass of flywheel, m = Density × Volume

$$= \rho \times [\pi D \times \text{Area of cross-section}]$$
$$= \rho \times [\pi D \times b \times t]$$
$$= \rho \times [\pi D \times 1.5t \times t] \quad (\because \; b = 1.5t)$$

or
$$3635.8 = \rho \times [\pi D \times 1.5t^2] \quad (\because \; m = 3635.8 \text{ kg}) \quad ...(v)$$

The value of D will be obtained from $v = \frac{\pi D N}{60}$

or
$$20 = \frac{\pi \times D \times 150}{60} \quad (\because \; v = 20 \text{ m/s and } N = 150 \text{ r.p.m.})$$

$$\therefore \quad D = \frac{20 \times 60}{\pi \times 150} = 2.546 \text{ m}$$

Substituting the value of D in equation (v), we get

$$3635.8 = \rho\, [\pi \times 2.546 \times 1.5 \times t^2]$$
$$= 7500\, [\pi \times 2.546 \times 1.5 \times t^2] \quad (\because \; \rho = 7500 \text{ kg/m}^3)$$

$$\therefore \quad t = \sqrt{\frac{3635.8}{7500 \times \pi \times 2.546 \times 1.5}} = 0.201 \text{ m} = \mathbf{201 \text{ mm}}. \quad \textbf{Ans.}$$

and
$$b = 1.5t = 1.5 \times 201 = \mathbf{301.5 \text{ mm}}. \quad \textbf{Ans.}$$

12.8. Operation of a Flywheel in a Punching Press

When a flywheel is attached to the crank-shaft of an engine, the load on crank-shaft is constant but the input torque varies during a cycle. But when flywheel is attached to a punching press or a rivetting machine, the input torque is constant but load varies during the cycle.

Fig. 12.18 shows the diagram of a punching press. A motor which supplies constant torque drives the crank. The crank is connected to the connecting rod at the end of which a punching tool is fixed. The punch is at the position of the slider of the slider-crank mechanism. When the crank rotates from $\theta = \theta_1$ to $\theta = \theta_2$, the actual punching of the plate takes place. For the rest of the cycle, the load (or force exerted) on the punch is zero. This mean that when $\theta = \theta_2$ to $\theta = 2\pi$ (or $\theta = 0$) and again from $\theta = 0$ to $\theta = \theta_1$, there will be no load. But the input energy from the motor is supplied continuously. If a flywheel is not used, then the speed of crank will increase too much during the rotation of the crank from $\theta = 0$ to $\theta = \theta_1$ and again from $\theta = \theta_2$ to $\theta = 2\pi$. But during the rotation of the crank from $\theta = \theta_1$ to $\theta = \theta_2$ (*i.e.* during actual punching of the plate), the drop in speed of the crank will be very large due to much more load than the energy supplied. But if a flywheel is used, it will absorb excess energy during $\theta = 0$ to $\theta = \theta_1$ and again $\theta = \theta_2$ to $\theta = 2\pi$ and supply the energy during punching. Hence by using flywheel, the fluctuations of speed will be kept within permissible limits.

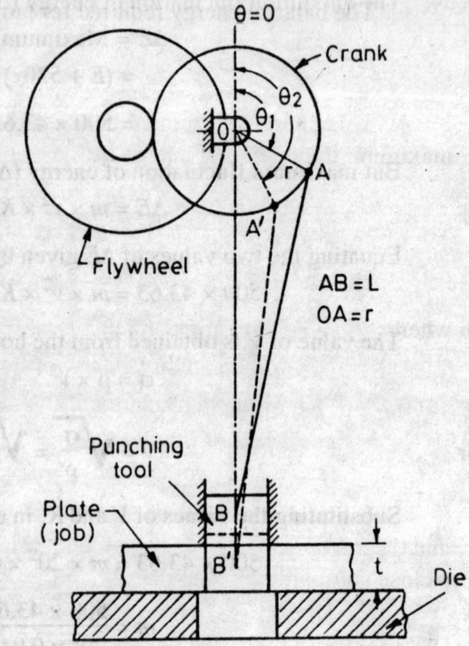

Fig. 12.18

Let E_1 = Energy required for one punch (or for one punching operation)

d = Diameter of the hole punched

t = Thickness of the plate in which hole is to be punched

τ = Shear stress for the plate material

F_s = Max. shear force required for punching

= Shear stress × Area sheared

$= \tau \times \pi d \times t$

E_2 = Energy supplied by motor for actual punching.

If it is assumed that one punching operation takes place per revolution or for crank rotation of 360° or 2π radians, then energy supplied by the motor to crank-shaft will also be equal to E_1 per revolution. But when crank rotates from $\theta = \theta_1$ to $\theta = \theta_2$, the actual punching of plates takes place.

∴ During actual punching of the plate, the motor will supply energy equal to $\dfrac{E_1 \times (\theta_2 - \theta_1)}{2\pi}$. This is because :

For one revolution (or crank rotation of 2π radians), energy supplied by motor = E_1

For crank rotation of $(\theta_2 - \theta_1)$ radians, the energy supplied by motor will be

$$= \frac{E_1}{2\pi} \times (\theta_2 - \theta_1) \quad \text{or} \quad \frac{E_1 (\theta_2 - \theta_1)}{2\pi}$$

∴ Energy supplied by motor during actual punching,

$$E_2 = \frac{E_1 (\theta_2 - \theta_1)}{2\pi}$$

The balance energy required for punching is equal to

$$= (E_1 - E_2) = E_1 - \frac{E_1(\theta_2 - \theta_1)}{2\pi} = E_1\left(1 - \frac{\theta_2 - \theta_1}{2\pi}\right)$$

This energy is supplied by the flywheel. Hence the kinetic energy of the flywheel decreases. Hence maximum fluctuation of energy of flywheel

$$\Delta E = E_1 - E_2 = E_1\left[1 - \frac{(\theta_2 - \theta_1)}{2\pi}\right] \quad \text{...(12.16)}$$

But $$\Delta E = \frac{1}{2} I(\omega_1^2 - \omega_2^2) \quad \text{...(12.17)}$$

where ω_1 = Max. angular velocity of flywheel,

ω_2 = Min. angular velocity of flywheel, and

I = Moment of inertia of flywheel = $m \times k^2$,

m = Mass of flywheel and

k = Radius of gyration of flywheel.

In equation (12.16), the values of θ_1 and θ_2 are determined if the lengths of connecting rod, crank radius and the relative position of the job with respect to the crank-shaft axis are known. In the absence of these data, a close approximation may be

$$\frac{(\theta_2 - \theta_1)}{2\pi} \simeq \frac{t}{2s} = \frac{t}{4r} \quad \text{...(12.18)}$$

where t = Thickness of plate

s = Stroke of the punch = $2 \times r$ where r = Crank radius.

Problem 12.15. *A punching press is required to punch 30 mm diameter holes in a plate of 20 mm thickness at the rate of 20 holes per minute. It requires 6 Nm of energy per mm² of sheared area. If punching takes place in $\frac{1}{10}$ of a second and the r.p.m. of the flywheel varies from 160 to 140, determine the mass of the flywheel having radius of gyration of 1 meter.* (A.M.I.E., Summer 1987)

Sol. Given :

Dia. of holes, d = 30 mm ; plate thickness, t = 20 mm ; No. of holes = 20 holes per minute ; Energy required, E = 6 Nm/mm² of sheared area ; punching time = $\frac{1}{10}$ second = 0.1 s ; Variation of r.p.m. of flywheel = 160 to 140 or N_1 = 160 and N_2 = 140 ; radius of gyration of flywheel, k = 1 m.

Let m = Mass of flywheel

We know that sheared area per hole = $\pi d \times t = \pi \times 30 \times 20 = 600\pi$ mm²

$\therefore$ Energy required to punch a hole,

E_1 = (Energy required per mm² sheared area) × sheared area

$= 6 \times 600\pi$ Nm ($\because$ Energy/mm² of sheared area = 6 Nm)

$= 11309.73$ Nm

No. of holes punched per minute = 20

or 20 holes are punched is 60 seconds

$\therefore$ Time required to punch a hole = $\frac{60}{20}$ = 3 seconds.

$\therefore$ Energy required for punching work per second

$$= \frac{\text{Energy required to punch a hole}}{\text{Time required to punch a hole}} = \frac{11309.73}{3} = 3769.91 \text{ Nm/s}$$

Since the punching takes place $\frac{1}{10}$ of a second, therefore energy supplied by motor in $\frac{1}{10}$ second,

$$E_2 = (\text{Energy required for punching per second}) \times \text{Time}$$

$$= 3769.91 \times \frac{1}{10} = 376.991 \text{ Nm}$$

$\therefore$ Energy to be supplied by the flywheel during punching a hole (or maximum fluctuation of energy of the flywheel),

$$\Delta E = E_1 - E_2$$

$$= 11309.73 - 376.991 = 10932.739 \text{ Nm} \quad ...(i)$$

But the maximum fluctuation of energy (ΔE) is also given by,

$$\Delta E = \frac{1}{2} \times I \times \omega_1^2 - \frac{1}{2} I \times \omega_2$$

$$= \frac{1}{2} I (\omega_1^2 - \omega_2^2) = \frac{1}{2} I \left[\left(\frac{2\pi \times N_1}{60} \right)^2 - \left(\frac{2\pi N_2}{60} \right)^2 \right]$$

$$= \frac{1}{2} I \left[\left(\frac{2\pi \times 160}{60} \right)^2 - \left(\frac{2\pi \times 140}{60} \right)^2 \right] = \frac{1}{2} I \,[280.735 - 214.938]$$

$$= 32.898 \times I$$

$$= 32.898 \times m \times k^2 \quad (\because I = mk^2)$$

$$= 32.898 \times m \times 1^2 \quad (\because k = 1 \text{ m})$$

$$= 32.898 \text{ m} \quad ...(ii)$$

Equating the two values of ΔE given by equations (*i*) and (*ii*), we get

$$10932.739 = 32.898 \text{ m}$$

or $$m = \frac{10932.739}{32.898} = \mathbf{332.32 \text{ kg.}}$$ **Ans.**

Problem 12.16. *A punching machine operates at the rate of 600 holes/hr. It does 45 Nm of work per sq. mm of sheared area in cutting 25 mm diameter hole in a 3 mm thick plate. The machine is operated by a constant torque motor. The speed of the machine fluctuates between 250 and 230 r.p.m. The frictional losses are 20% of the work done during punching and actual punching time per hole is 2 seconds. Find :*

(i) the power required to drive punching machine

(ii) the maximum fluctuation of energy

(iii) the mass of the flywheel required to keep the speed fluctuation in the given range. Radius of gyration of flywheel is 500 mm. (AMIE, Winter 1983)

Sol. Given :

No. of holes/hr. = 600 or 600 holes are punched in 60 × 60 seconds.

Hence one hole will be punched in $\frac{60 \times 60}{600} = 6$ seconds.

But actual punching time per hole = 2 seconds. Hence machine will be idle for (6 – 2) = 4 seconds. This means that a cycle is repeated in 6 seconds.

Work done = 45 Nm per mm^2 of sheared area ; dia. of hole d = 25 mm, thickness of plate, t = 3 mm. Fluctuation of speeds is between 250 r.p.m. and 230 r.p.m. or N_1 = 250 r.p.m. and N_2 = 230 r.p.m.

$$\therefore \quad \omega_1 = \frac{2\pi N_1}{60} = \frac{2\pi \times 250}{60} = 26.18 \text{ rad/s} \quad \text{and} \quad \omega_2 = \frac{2\pi N_2}{60} = \frac{2\pi \times 230}{60} = 24.08 \text{ rad/s}$$

Frictional losses = 20% of work done during punching.

Input torque to the machine is constant.

Now sheared area per hole = $\pi d \times t = \pi \times 25 \times 3 = 75\pi$ mm^2

Work done (or energy required) during punching 1 hole

$$E_1 = (\text{Work done per mm}^2 \text{ of sheared area}) \times \text{Sheared area}$$
$$= 45 \times 75\pi \text{ Nm} = 10602.8 \text{ Nm}$$

Work done per second = Work done per hole × No. of holes/second

$$= 10602.8 \times \frac{600}{3600} \quad \left(\because \text{ No. of holes/hr.} = 600 \ \therefore \text{ No. of holes/s} = \frac{600}{3600}\right)$$
$$= 1767.13 \text{ Nm/s}$$

Work dissipated per second as frictional losses

$$= 20\% \text{ Work done during punching/s}$$
$$= \frac{20}{100} \times 1767.13 = 353.426 \text{ Nm/s}$$

$\therefore$ Work required for punching holes/s

$$= 1767.13 + 353.426 = 2120.556 \text{ Nm/s}$$

(i) Power required to drive the punching machine

Let P = Power required

$\therefore$ Power required to drive punching machine,

$$P = \text{Work required for punching/s}$$
$$= 2120.556 \text{ Nm/s} = 2120.556 \text{ W} \qquad (\because \text{ Watt} = \text{Nm/s})$$
$$= 2.120556 \simeq \mathbf{2.12 \ kW.} \quad \textbf{Ans.}$$

(ii) The maximum fluctuation of energy

It is given that 600 holes are punched in one hour or in 60 × 60 seconds. Hence one hole will be punched in $\frac{60 \times 60}{600}$ = 6 seconds. But actual punching time per hole is 2 seconds. Hence energy supplied by motor in two seconds.

Let E_1 = Energy required during actual punching time for a cycle and

E_2 = Energy consumed during corresponding idle time of 4 seconds.

E_2 = (Energy (or work) required for punching per second) × actual punching time

$= 2120.556 \times 2 = 4241.112$ Nm

and E_2 = (Energy consumed (or work dissipated as frictional losses) per second) × Idle time for a cycle

$= 353.426 \times 4 = 1413.704$ Nm

$\therefore$ Maximum fluctuation of energy per cycle,

$$\Delta E = E_1 - E_2 = 4241.112 - 1413.704 = \mathbf{2827.408 \ Nm.} \quad \textbf{Ans.}$$

(iii) Mass of flywheel

Let m = Mass of flywheel and

k = Radius of gyration of wheel = 500 mm = 0.5 m (given)

We know that, $\Delta E = \frac{1}{2} \times I \times (\omega_1^2 - \omega_2^2)$

$= \frac{1}{2} \times mk^2 \times (\omega_1^2 - \omega_2^2)$ $(\because \; I = mk^2)$

or $2827.408 = \frac{1}{2} \times m \times 0.5^2 \times (26.18^2 - 24.08^2)$

$(\because \; \Delta E = 2827.408, \omega_1 = 26.18 \text{ and } \omega_2 = 24.08 \text{ rad/s})$

$= \frac{1}{2} \times m \times 0.25 \times (685.392 - 579.846) = 13.193 \times m$

$\therefore$ $m = \dfrac{2827.408}{13.193} = \mathbf{214.3 \text{ kg.}}$ **Ans.**

Problem 12.17. *A flywheel, which is rotating at a maximum speed of 250 r.p.m. and is having radius of gyration as 0.5 m, is attached to a punching press. The press is driven by a constant torque electric motor and punches 750 holes per hour. Each punching operation requires 14000 Nm of energy and takes 1.8 seconds. If the speed of the flywheel is not to fall below 225 r.p.m. find : (i) power of the motor and (ii) mass of flywheel.*

Sol. Given :

$N_1 = 250$ r.p.m. or $\omega_1 = \dfrac{2\pi N_1}{60} = \dfrac{2\pi \times 250}{60} = 26.18$ rad/s ; $k = 0.5$ m ; no. of holes/ hr. = 750 ; energy for one punching operation, $E_1 = 14000$ Nm ; punching time = 1.8 s ; $N_2 = 225$ r.p.m. or $\omega_2 = \dfrac{2\pi N_2}{60} = \dfrac{2\pi \times 225}{60} = 23.56$ rad/s.

(i) Power of motor

No. of holes/hr. = 750.

This means 750 holes are punched in one hour or in 60 × 60 seconds

$\therefore$ One hole will be punched in $\dfrac{60 \times 60}{750} = 4.8$ seconds

Let us now find the energy required for punching per second.

$\therefore$ Energy required/s $= \dfrac{\text{Energy required/hole}}{\text{Time required/hole}} = \dfrac{14000}{4.8}$ Nm/s = 2916.67 Nm/s

$\therefore$ Power of the motor = Energy required/s = 2916.67 Nm/s

= 2916.67 W $(\because \; 1 \text{ Nm/s} = 1 \text{ W})$

= **2.91667 kW. Ans.**

(ii) Mass of flywheel

Let m = Mass of flywheel

Since each punching operation takes 1.8 seconds, hence energy supplied by the motor in 1.8 seconds,

E_2 = (Energy/s) × Time = 2916.67 × 1.8 = 5250.006 ≃ 5250 Nm

$\therefore$ Energy to be supplied by the flywheel during punching or maximum fluctuation of energy,

$\Delta E = E_1 - E_2 = 14000 - 5250 = 8750$ Nm ...(i)

But maximum fluctuation of energy is also given by,

$$\Delta E = \frac{1}{2} \times I\,(\omega_1^2 - \omega_2^2) \quad ...(ii)$$

Equating the two values of ΔE, we get

$$8750 = \frac{1}{2} \times I \times (\omega_1^2 - \omega_2^2) \quad \text{...But } I = mk^2$$

$$= \frac{1}{2} \times mk^2 \times (26.18^2 - 23.56^2) \quad (\because \omega_1 = 26.18 \text{ rad/s}, \omega_2 = 23.56 \text{ rad/s})$$

$$= \frac{1}{2} \times m \times 0.5^2 \times (685.4 - 555) \quad (\because k = 0.5 \text{ m})$$

$$= 16.3 \times m$$

$$\therefore \quad m = \frac{8750}{16.3} = \mathbf{536.8 \text{ kg.}} \quad \textbf{Ans.}$$

Problem 12.18. *A constant torque 4 kW motor drives a rivetting machine. A flywheel of mass 140 kg and radius of gyration of 0.5 m is fitted to the rivetting machine. Each rivetting operation takes 1 seconds and requires 9000 Nm of energy. If the speed of flywheel is 420 r.p.m. before rivetting, then find :*

(*i*) *the fall in speed of the flywheel after the rivetting,*

(*ii*) *the number of rivets closed per hour.*

Sol. Given :

$P = 4$ kW $= 4000$ W ; $m = 140$ kg ; $k = 0.5$ m, energy required during rivetting, $E_1 = 9000$ Nm ;

$N_1 = 420$ r.p.m. or $\omega_1 = \frac{2\pi N_1}{60} = \frac{2\pi \times 420}{60} = 43.98 \simeq 44$ rad/s, time for one rivetting = 1 s.

(*i*) *Fall of speed of the flywheel after the rivetting*

Let ω_2 = Angular speed of flywheel after rivetting

ω_1 = Angular speed of flywheel before rivetting = 44 rad/s (given)

Energy required during rivetting operation which takes one second,

$$E_1 = 9000 \text{ Nm per second}$$

Energy supplied by motor in one second,

$$E_2 = 4000 \text{ W} = 4000 \text{ Nm/s} \quad (\because 1 \text{ W} = 1 \text{ Nm/s})$$

$\therefore$ Energy to be supplied by the flywheel for each rivetting operation per second or the maximum fluctuation of energy,

$$\Delta E = E_1 - E_2 = 9000 - 4000 = 5000 \text{ Nm}$$

But maximum fluctuation of energy is also given by,

$$\Delta E = \frac{1}{2} \times I\,(\omega_1^2 - \omega_2^2)$$

Equating the two values of ΔE, we get

$$5000 = \frac{1}{2} \times I \times (\omega_1^2 - \omega_2^2) \quad \text{where } I = mk^2 = 140 \times 0.5^2 = 35 \text{ kg m}^2$$

$$= \frac{1}{2} \times 35 \times (44^2 - \omega_2^2)$$

or $$\frac{5000 \times 2}{35} = 44^2 - \omega_2^2$$

or $$\omega_2^2 = 44^2 - \frac{5000 \times 2}{35} = 1936 - 285.7 = 1650.3$$

or $$\omega_2 = \sqrt{1650.3} = 40.624 \text{ rad/s}$$

$\therefore$ Corresponding speed in r.p.m.,

$$N_2 = \frac{60 \times \omega_2}{2\pi} \qquad \left(\because \ \omega_2 = \frac{2\pi N_2}{60} \text{ or } N_2 = \frac{60 \times \omega_2}{2\pi} \right)$$

$$= \frac{60 \times 40.624}{2\pi} = 387.9 \text{ r.p.m.}$$

$\therefore$ Fall in speed after rivetting

$$= N_1 - N_2 = 420 - 387.9 = \mathbf{32.1 \text{ r.p.m.}} \quad \textbf{Ans.}$$

(*ii*) *No. of rivets closed per hour*

Energy required for one rivetting operation which takes one second is

$$E_1 = 9000 \text{ Nm}$$

Energy supplied by motor = 4000 W = 4000 Nm/s

Energy supplied by motor in one hour = 4000 × (60 × 60) Nm/hour

$\therefore$ No. of rivets closed/hr.

$$= \frac{\text{Energy supplied by motor in one hour}}{\text{Energy required for one rivetting}} = \frac{4000 \times 60 \times 60}{9000} = \mathbf{1600 \text{ rivets/hr.}} \quad \textbf{Ans.}$$

Problem 12.19. *A punching press is required to punch 40 mm diameter holes in a plate of 30 mm thickness at the rate of 4 holes per minute. It requires 6 Nm of energy per mm² of sheared area. The punch has a stroke of 100 mm. The r.p.m. of the flywheel varies from 320 to 280. If the radius of gyration of flywheel is 1 m, find :*

(*i*) *mass of the flywheel and*

(*ii*) *power of the motor for punching machine.*

Sol. Given :

$d = 40$ mm ; $t = 30$ mm ; no. of holes/minute = 4 ; energy required = 6 Nm/mm² of sheared area ; Stroke, $s = 100$ mm ; $N_1 = 320$ r.p.m. or $\omega_1 = \frac{2\pi N_1}{60} = \frac{2\pi \times 320}{60} = 33.51$ rad/s ; $N_2 = 280$ r.p.m. or $\omega_2 = \frac{2\pi N_2}{60} = \frac{2\pi \times 280}{60} = 29.32$ rad/s ; $k = 1$ m.

(*i*) *Mass of the flywheel*

Let m = Mass of flywheel

Now sheared area per hole = $\pi \times d \times t = \pi \times 40 \times 30 = 1200\pi$ mm²

$\therefore$ Energy required to punch a hole,

$$E_1 = (\text{Energy required per mm}^2 \text{ of sheared area}) \times \text{Sheared area in mm}^2$$
$$= 6 \times 1200\pi = 22619.5 \text{ Nm}$$

Now from equation (12.16), we know that maximum fluctuation of energy,

$$\Delta E = E_1 \left[1 - \frac{(\theta_2 - \theta_1)}{2\pi} \right] \qquad ...(i)$$

Also the value of $\left(\frac{\theta_2 - \theta_1}{2\pi} \right)$ from equation (12.18) is

$$\frac{(\theta_2 - \theta_1)}{2\pi} = \frac{t}{2s}$$ where t = thickness of plate = 30 mm and s = Stroke length = 100 mm

$$= \frac{30}{2 \times 100} = \frac{3}{20}$$

Substituting the above value in equation (*i*), we get

$$\Delta E = E_1\left[1 - \frac{3}{20}\right] = E_1 \times \frac{17}{20} = 22619.5 \times \frac{17}{20} \qquad (\because\ E_1 = 22619.5 \text{ Nm})$$

$$= 19226.6 \text{ Nm} \qquad ...(ii)$$

But maximum fluctuation of energy is also given by,

$$\Delta E = \frac{1}{2} \times I \times (\omega_1^2 - \omega_2^2)$$

$$= \frac{1}{2} \times mk^2 \times (\omega_1^2 - \omega_2^2) \qquad (\because\ I = mk^2) \qquad ...(iii)$$

Equating the two values of ΔE given by equations (*ii*) and (*iii*), we get

$$19226.6 = \frac{1}{2} \times mk^2 (\omega_1^2 - \omega_2^2)$$

$$= \frac{1}{2} \times m \times 1^2 \times (33.51^2 - 29.32^2)$$

$$(\because\ k = 1 \text{ m}, \omega_1 = 33.51 \text{ rad/s} ; \omega_2 = 29.32 \text{ rad/s})$$

$$= \frac{m}{2} \times (1222.92 - 859.66)$$

$$= 181.63m$$

$$\therefore \quad m = \frac{19226.6}{181.63} = \mathbf{105.85\ kg.} \quad \textbf{Ans.}$$

(*ii*) *Power of the motor*

Energy required to punch a hole, $E_1 = 22619.5$ Nm

$\therefore$ Energy required to punch 4 holes = 4 × 22619.5 Nm

But 4 holes are punched in one minute.

$\therefore$ Energy required for punching in one minute = 4 × 22619.5 Nm

$\therefore$ Energy required/second for punching $= \frac{4 \times 22619.5}{60} = 1507.96$ Nm/s

$\therefore$ Power of motor = 1507.96 Nm/s

= 1507.96 W $\qquad (\because\ 1 \text{ W} = 1 \text{ Nm/s})$

= 1.50796 kW $\simeq$ **1.508 kW. Ans.**

HIGHLIGHTS

1. The turning moment diagram (*i.e. T-M* Diagram) is a graph between torque (*T*) and crank angle θ. The torque exerted on the crank is taken along *y*-axis where as the crank angle θ along *x*-axis.
2. The mean torque (T_{mean}) against which the engine works, is given by,

$$T_{mean} = \frac{\text{Area of } T\text{-}M \text{ diagram for one revolution (or for one cycle)}}{\text{Angle turned in radians for one revolution (or for one cycle)}}$$

$$= \frac{\int_0^{2\pi} T \times d\theta}{2\pi} \qquad \text{if angle turned for one revolution} = 2\pi \text{ radians}$$

where T = torque at any instant on the crank-shaft

The angle turned for one revolution

= 2π ... For steam engine and two storkes I.C. Engines

= 4π ... For four strokes I.C. Engines.

3. If T is more than T_{mean} (*i.e.* $T > T_{mean}$) the flywheel accelerates whereas if T is less than T_{mean} *i.e.* ($T < T_{mean}$) the flywheel retards.
4. If during a cycle for a particular period the work done by the engine is less that the work required for external resistance than this loss of work is made up the flywheel which gives up some of its energy and speed of flywheel decreases during that period.
5. If during a cycle for a particular period the work done by the engine is more than the work required for external resistance than the excess work is stored in the flywheel and hence the speed of flywheel increases during that period.
6. The excess work and loss of work are known as positive fluctuation of energy and negative fluctuation of energy. The difference between the maximum and minimum fluctuation of energy is known as maximum fluctuation of energy. It is represented by ΔE.
7. The difference between the greatest and least speeds of the crank-shaft for one revolution (or for one cycle) is known as fluctuation of speed.
8. The co-efficient of fluctuation of energy (K_e) is defined as the ratio of maximum fluctuation of energy to the work done per cycle. Hence mathematically,

$$K_e = \frac{\text{Maximum fluctuation of energy}}{\text{Work done per cycle (or per revolution)}}$$

where maximum fluctuation of energy (ΔE)

$$= \text{Maximum energy} - \text{Minimum energy}$$

and work done per cycle $= T_{mean} \times \theta$

The value of θ *i.e.* angle turned in radians in one revolution is given by,

$$\theta = 2\pi \quad \text{... For steam engine and two stroke I.C. Engines}$$

$$= 4\pi \quad \text{... For four stroke I.C. Engines.}$$

9. The co-efficient of fluctuation of speed is defined as the ratio of maximum fluctuation of speed to the mean speed. It is represented by K_s. Mathematically,

$$K_s = \frac{\text{Maximum fluctuation of speed}}{\text{Mean speed}}$$

$$= \frac{\text{Max. speed} - \text{Min. speed}}{\text{Mean speed}}$$

$$= \frac{N_1 - N_2}{N} \qquad \text{where } N = \frac{N_1 + N_2}{2}$$

$$= \frac{\omega_1 - \omega_2}{\omega}.$$

10. Flywheel is used to control the speed variation caused by the fluctuation of energy during a cycle of operation

$$\Delta E = \text{Max. K.E.} - \text{Min. K.E.}$$

$$= \frac{1}{2} I \omega_1^2 - \frac{1}{2} I \omega_2^2.$$

11. The maximum fluctuation of energy (ΔE) is given by,

$$\Delta E = I \times \omega^2 \times K_s = mk^2 \times \omega^2 \times K_s$$

$$= 2E \times K_s \qquad \text{where } E = \frac{1}{2} I \omega^2$$

If $k = R$ which is true for a rim of very small thickness, then

$$\Delta E = mR^2 \times \omega^2 \times K_s$$

$$= m \times v^2 \times K_s \qquad (\because \ v = \omega \times R)$$

12. The relation between hoop stress (σ), linear velocity of the rim at the mean radius (v) and density of rim material (ρ) is given by

$$v = \sqrt{\frac{\sigma}{\rho}} \quad \text{or} \quad \sigma = \rho v^2.$$

13. The maximum fluctuation of energy of flywheel of a punching press is given by,

$$\Delta E = E_1 \left[1 - \frac{(\theta_2 - \theta_1)}{2\pi} \right]$$

where E_1 = Energy required for one punch operation,

$(\theta_2 - \theta_1)$ = The crank angle through which actual punching takes place

Also $\frac{(\theta_2 - \theta_1)}{2\pi} \simeq \frac{t}{2\pi} = \frac{t}{4r}$

where t = thickness of plate and

r = Crank radius and

s = Stroke of the punch = $2r$.

EXERCISE 12

(A) Theoretical Questions

1. What do you mean by the term, 'Turning Moment Diagram' ?
2. Draw the turning moment diagrams for the following different types of engines neglecting the effect of inertia of the connecting rod :
   (*i*) For a single cylinder double acting steam engine
   (*ii*) For a four stroke cycle I.C. Engine and
   (*iii*) For a multi-cylinder engine.
3. What do you mean by mean resisting torque ? What are the conditions for a flywheel of a steam engine to accelerate and to retard ?
4. What do you mean by fluctuation of energy and fluctuation of speed of crank-shaft. Also explain the terms maximum fluctuation of energy and maximum fluctuation of speed.
5. Define the terms 'co-efficient of fluctuation of energy' and 'co-efficient of fluctuation of speed'. *(AMIE, Winter 1976)*
6. Prove that co-efficient of fluctuation of speed (K_s) is given by,

$$K_s = \frac{2(\omega_1 - \omega_2)}{\omega_1 + \omega_2}$$

   where ω_1 = Max. angular velocity of flywheel and

   ω_2 = Min. angular velocity of flywheel.
7. What is the function of a flywheel in a prime mover ? *(AMIE, Summer 1994)*
8. Explain clearly how the functions of flywheel and governor differ from each other in a steam engine. *(AMIE, Summer 1986)*
9. Derive the following relations for the maximum fluctuation of energy (ΔE),
   (*i*) $\Delta E = 2 \times E \times K_s$
   (*ii*) $\Delta E = m \times v^2 \times K_s$

   where E = K.E. of flywheel corresponding to mean speed = $\frac{1}{2} \times I \times \omega^2$,

   K_s = Co-efficient of fluctuation of speed,

   v = Linear velocity of the rim of flywheel at the mean radius and

   m = Mass of flywheel.

10. Derive an expression for the hoop stress (σ) produced in the rim of the flywheel due to centrifugal forces in terms of linear velocity and density of the material of the rim.

11. What is a punching press ? Describe with a neat sketch the operation of the flywheel in a punching press.

B. Numerical Problems

1. The maximum and minimum speed of a flywheel are 121 r.p.m. and 119 r.p.m. respectively. The mass of the flywheel is 2600 kg and radius of gyration is 1.8 m. Find : (*i*) mean speed of flywheel, (*ii*) maximum fluctuation of energy and (*iii*) co-efficient of fluctuation of speed. [**Ans.** (*i*) 120 r.p.m., (*ii*) 22227.8 Nm and (*iii*) 0.0167]

2. Find the maximum and minimum speeds of a flywheel of mass 3250 kg and radius of gyration 1.8 m when the fluctuation of energy is 112 kN m. The mean speed of engine is 240 r.p.m. [**Ans.** 242 r.p.m. and 238 r.p.m.]

3. The turning moment diagram for a multi-cylinder engine has been drawn to a scale of 1 mm = 650 Nm vertically and 1 mm = 6° horizontally. The areas above and below the mean torque line are – 30, + 382, – 260, + 310, – 306, + 248, – 380, + 265 and – 229 mm^2 when taken in order from one end. The engine is running at a mean speed of 900 r.p.m. The total fluctuation of speed is not to excead ± 1.8% of the mean speed. If the radius of flywheel is 1.05 m, find the mass of the flywheel. [**Ans.** 84.54 kg]

4. The radius of gyration of a flywheel is 1 metre and the fluctuation of speed is not to exceed 1% of the mean speed of the flywheel. If the mass of the flywheel is 1670 kg and the steam engine develops 600 kW at 270 r.p.m., then find : (*i*) maximum fluctuation of energy and (*ii*) co-efficient of fluctuation of energy.

[**Ans.** 13350.26 Nm ; 0.1]

5. The areas above and below the mean torque line for an I.C. Engine are : – 20, + 195, – 95, + 145, – 390, + 145 and – 70 mm^2 taken in order. The scale for turning moment diagram is 1 mm vertical scale = 5 Nm and 1 mm horizontal scale = 3°. The mass of the rotating parts are 36 kg with a radius of gyration of 150 mm. If the engine speed is 1000 r.p.m., find the co-efficient of fluctuation of speed. [**Ans.** 0.0087]

6. The torque exerted on the crank-shaft of a two stroke engine is given by the equation :

$$T(\text{N-m}) = 20000 + 9500 \sin 2\theta - 5700 \cos 2\theta$$

where θ is the crank angle displacement from the inner dead centre. Assuming the resisting torque to be constant, determine :

(*i*) the power developed by the engine when the speed of engine is 270 r.p.m,

(*ii*) the moment of inertia of flywheel if the total fluctuation of the speed is not to exceed 1% of the mean speed which is 270 r.p.m., and

(*iii*) the angular acceleration of the flywheel when the crank has turned through 45° from inner dead centre.

[**Ans.** (*i*) 565.5 kW ; (*ii*) 3121 kg-m^2 and (*iii*) 3.044 rad/s^2]

7. The torque exerted on the crank-shaft of a two-stroke engine is given by the equation :

$$T(\text{Nm}) = 3500 + 500 \sin 2\theta - 1000 \cos 2\theta$$

where θ is the crank displacement from inner dead centre. Assuming the resisting torque to be constant, determine :

(*i*) the power developed when the engine speed is 450 r.p.m.

(*ii*) the total fluctuations in speed in percentage and

(*iii*) the maximum retardation of the flywheel. The mass of the flywheel is 250 kg and its radios of gyration is 750 mm. [**Ans.** (*i*) 164.835 kW, (*ii*) 0.355%, (*iii*) 7.95 rad/s^2]

8. The turning moment diagram (*T-M* diagram) of an engine rotating at a mean speed of 300 r.p.m. is given by the relation,

$$T \text{ (in Nm)} = 7500 + 4000 \sin 2\phi - 1000 \cos 2\phi$$

where ϕ is the crank angle. External resistance is constant. A flywheel of mass 1000 kg is fitted on the engine shaft so that the total fluctuations of speed does not exceed by 1%. Determine the least value of the moment of inertia and radius of gyration of the flywheel. [**Ans.** 835.7 kg m^2 ; 0.914 m]

9. The torque at the crank-shaft of an engine is given by the equation,

$$T = (5000 + 600 \sin 2\theta) \text{ Nm}$$

The mass and radius of gyration of flywheel and other rotating parts attached to the engine are 500 kg and 400 mm respectively. The mean speed is 150 r.p.m. The engine is directly coupled to a machine which requires a torque of (5000 + 500 sin θ) Nm for its operation, where θ is the angle of rotation of the shaft measured from a certain datum. Determine :

(*i*) the maximum fluctuation of energy,

(*ii*) the percentage fluctuation of speed and co-efficient of fluctuation of speed, and

(*iii*) the maximum and minimum angular acceleration of the flywheel and corresponding shaft position.

[Ans. (*i*) 1204 Nm, (*ii*) 6.1% and 0.061, (*iii*) 3.46 rad/s^2 and 12.2 rad/s^2 ; θ = 35° or 127.6°**]**

10. The variation of crank-shaft torque of a four cylinder petrol engine may be approximately represented by taking the torque as zero for crank angles 0° and 180° and as 520 Nm for crank angles 20° and 45°, the intermediate portions of the torque graph being straight lines. The cycle is being repeated in every half revolution. The average speed is 1200 r.p.m. Supposing that the engine drives a machine requiring a constant torque, determine the mass of the flywheel of radius of gyration 250 mm, which must be provided, so that the total variation of speed should be one percent. **[Ans.** 23.1 kg**]**

11. A flywheel of mass 900 kg and radius of gyration of 500 mm is fitted to a shaft which rotates at a speed of 500 r.p.m. The shaft drives a machine for which the torque varies in a cyclic manner over a period of 4 revolutions. The torque rises uniformly from 400 Nm to 1500 Nm during the first half-revolution. During the next one revolution, the torque remains constant. During next one revolution, the torque decreases uniformly from 1500 Nm to 400 Nm. During the last $1\frac{1}{2}$ revolution, the torque remains constant. The cycle being repeated thereafter.

Determine :

(*i*) the power required to drive the machine, and

(*ii*) percentage fluctuation in speed. **[Ans.** 46.142 kW ; 1.79%**]**

12. A two-stroke engine produces a torque of (400 + 90 sin 3θ) Nm, where θ is the crank angle. The mean engine speed is 200 r.p.m. The flywheel and other rotating parts attached to the engine have a mass of 525 kg at a radius of gyration of 220 mm. Calculate : (*i*) the power of the engine and (*ii*) the total fluctuation of speed of the flywheel when (*a*) the resisting torque is constant, (*b*) the resisting torque is (400 + 40 sin θ) Nm.

[Ans. (*i*) 8.378 kW, (*ii*) (*a*) 0.00538 or 0.538%, (*b*) 0.00937 or 0.937%**]**

13. The turning moment diagram for a multi-cylinder engine has been drawn to a vertical scale of 1 mm = 325 N-m and a horizontal scale of 1 mm = 4.5. The areas above and below the mean torque line are – 30, + 382, – 262, + 312, – 302, + 244, – 380, + 267 and – 231 mm^2.

The fluctuation of speed is limited to 1.8% of the mean speed which is 300 r.p.m. Determine a suitable diameter and cross-section of the flywheel rim for a limiting value of the safe centrifugal stress of 6 N/mm^2. The density of the rim material may be assumed as 7000 kg/m^3. The width of the rim is 4.5 times its thickness. Neglect the effect of the boss and arms. **[Ans.** 1.864 m ; t = 62.67 mm ; b = 282 mm**]**

14. A punching machine is required to punch 40 mm diameter hole in a plate of 35 mm thickness at the rate of 6 holes per minute. It requires 8 Nm of energy/mm^2 of the sheared area. If the punch has a stroke of 9.5 cm and the mean speed of the flywheel is 20 m/s, find the power of the motor required. Determine the mass of the flywheel also if the total fluctuation of the speed is not to exceed 3% of the mean speed. **[Ans.** 3.519 kW ; 2392 kg**]**

15. An electric motor drives a punching press, which is capable of punching 40 holes in 3 minutes. The radius of gyration of the flywheel fitted to the press is 500 mm. The flywheel is running at 250 r.p.m. Each punching operation requires 12,000 Nm of work and takes 1.5 seconds. Determine : (*i*) rating of machine in kW and (*ii*) mass of the flywheel if the speed of the flywheel does not drop below 230 r.p.m. **[Ans.** 2.67 kW ; 608 kg**]**

13

Balancing of Rotating Masses

13.1. Introduction

Balancing is defined as the process of designing (or modifying) a machine in which unbalance force is minimum. The rotating and reciprocating parts of a high speed engine (or of other machines) if are not properly balanced, the dynamic forces will be set up. These dynamic forces will :

(*a*) increase the loads on the bearings

(*b*) increase the stresses in the members of the machine and

(*c*) produce the unpleasant and even dangerous vibrations.

Hence all the rotating and reciprocating parts of the high speed engine (or of other machines) should be as completely balanced as possible. This chapter deals with the balancing of unbalanced forces caused by revolving masses.

13.2. Balancing of Rotating Masses

We know that when a mass is rotating in a circular path, a centrifugal force (a force acting radially outwards) is acting on the mass. If the mass is attached to a rotating shaft, the mass will experience a centrifugal force whose effect will be to bend the shaft and produce vibrations. But if another mass is attached to the opposite side of the shaft at such a distance that the centrifugal force of both the masses are equal and opposite, then there will be no bending and also no vibrations in the shaft . This process of providing the second mass in order to counteract the effect of centrifugal force of the first mass, is known as balancing of rotating masses.

Let us consider the following cases of balancing :

(*i*) Balancing of a single rotating mass,

(*ii*) Balancing of several masses rotating in the same plane,

(*iii*) Balancing of several masses rotating in different planes.

13.2.1. Balancing of a Single Rotating Mass

The single rotating mass can be balanced by two different ways namely :

(*a*) the balance mass may be rotating in the same plane and

(*b*) the balance mass is not rotating in the same plane.

(*a*) The balance mass is rotating in the same plane

Fig. 13.1 shows a disturbing mass (m_1) attached to a shaft which is rotating with angular velocity ω rad/s. In Fig. 13.1 (*b*), the shaft is not shown. The axis of shaft is perpendicular to the plane of paper and passing through *O*.

Let r_1 = the distance of centre of gravity of mass m_1 from the axis of rotation.

m_2 = Balancing mass in the same plane of rotation.

r_2 = the distance of centre of gravity of mass m_2 from the axis of rotation.

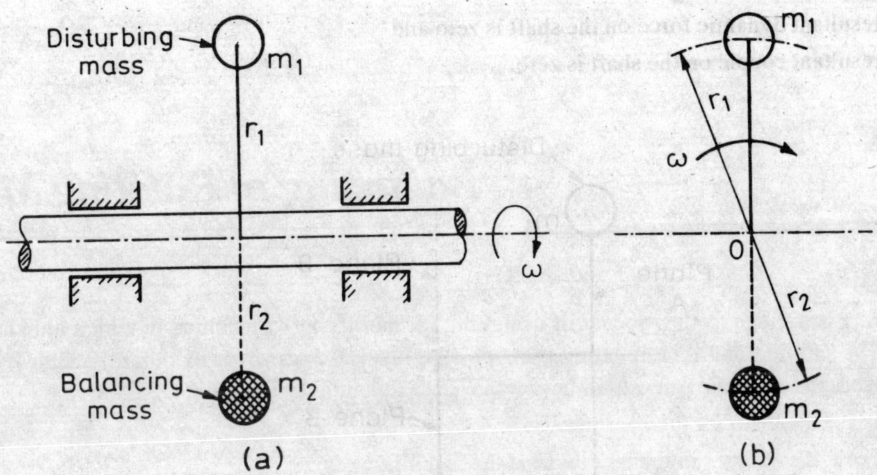

Fig. 13.1

The centrifugal force acting radially outwards on the disturbing mass

$$= \frac{mV^2}{r} = \frac{m(\omega r)^2}{r} \qquad (\because \ V = \omega \times r)$$

$$= m\omega^2 r$$

$$= m_1\omega^2 r_1 \qquad (\because \ \text{Mass} = m_1 \text{ and radius} = r_1) \qquad ...(i)$$

This centrifugal force, acting radially outwards will produce a bending moment on the shaft. To counteract the effect of this force, a balancing mass (m_2) may be attached to the shaft in the same plane of rotation as that of the disturbing mass (m_1), such that the centrifugal forces of the two masses are equal and opposite.

Centrifugal force due to balancing mass (m_2)

$$= m_2 \times \omega^2 \times r_2 \qquad ...(ii)$$

($\because$ Mass m_2 is also rotating with angular velocity ω rad/s)

Equating the two centrifugal forces given by equations (*i*) and (*ii*), we get

$$m_1 \times \omega^2 \times r_1 = m_2 \times \omega^2 r_2$$

or
$$m_1 \times r_1 = m_2 \times r_2 \qquad ...(13.1)$$

The equation (13.1) shows that the product of mass and radius is constant.

Hence the balancing mass (m_2) can be reduced by increasing the radius (r_2).

Note. The disturbing mass (m_1) and balancing mass (m_2) are in the same plane *i.e.* in the plane of paper.

(*b*) The Balance Mass is not rotating in the same plane

If the balance mass is not in the same plane of rotation as the disturbing mass, then it is not sufficient to use a single balancing mass.

Fig. 13.2 shows the disturbing mass (m_1) and balancing mass (m_2) rotating in two different planes *i.e.* mass m_1 is rotating in plane *A* and mass m_2 is rotating in plane *B*. The centrifugal forces of the two masses may be made equal in magnitude. But the directions of these two forces are opposite. Thus two forces also have different line of action. Hence they will give rise to a couple which tends to rock the shaft in its bearings. As a couple is balanced by a couple, hence a second balancing mass will be required to produce the couple of opposite sense so that the system is in perfect balance. Hence for perfect balancing, two balancing masses must be used. Now the three masses (one disturbing mass and two balancing masses) should be so arranged that :

(*i*) the resultant dynamic force on the shaft is zero and

(*ii*) the resultant couple on the shaft is zero.

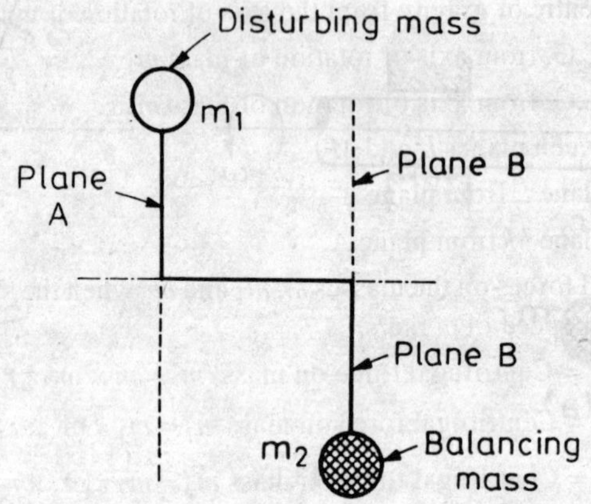

Fig. 13.2

This is only possible if the lines of action of three centrifugal forces are parallel and the algebraic sum of their moments about any point in the same plane is zero.

The two balancing masses which are in different planes may be arranged in two possible ways as given below :

(*i*) The planes of the balancing masses may lie on the same side of the plane of the disturbing mass.

(*ii*) The planes of the balancing masses may be on either side of the plane of the disturbing mass.

Both the above cases are discussed below :

(*i*) The planes of the balancing masses may lie on the same side of the plane of the disturbing mass

Fig. 13.3 shows the disturbing mass along with two balancing masses in such a way that the planes of balancing masses are on the same side of the plane of the disturbing mass.

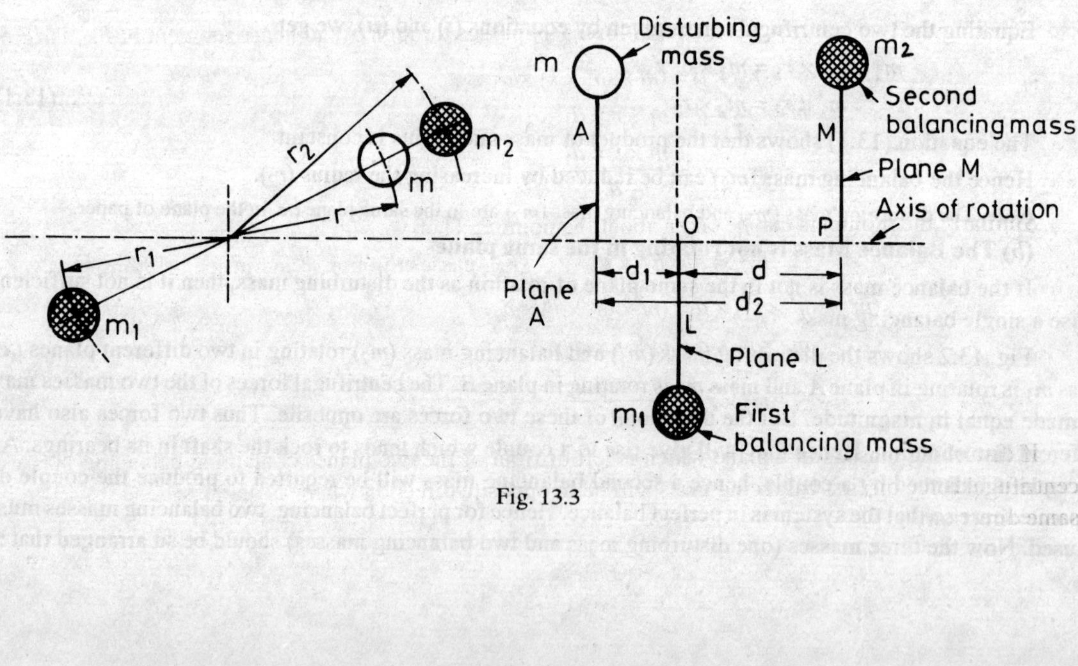

Fig. 13.3

Let m = disturbing mass in the plane A

m_1 = first balancing mass in the plane L

m_2 = second balancing mass in the plane M

r = distance of centre of gravity from the axis of rotation of mass m

r_1 = distance of C.G. from axis of rotation of mass m_1

r_2 = distance of C.G. from axis of rotation of mass m_2

d = distance between planes L and M

d_1 = distance of plane L from plane A

d_2 = distance of plane M from plane A.

Let us find the centrifugal forces on the masses m, m_1 and m_2 when the shaft to which these masses are attached, is rotating at an angular speed of ω rad/s.

Let F_C = Centrifugal force on mass 'm' $= m \times \omega^2 \times r$

F_{C_1} = Centrifugal force on mass $m_1 = m_1 \times \omega^2 \times r_1$

F_{C_2} = Centrifugal force on mass $m_2 = m_2 \times \omega^2 \times r_2$

Now for perfect balancing, the resultant dynamic force on the shaft should be zero. Hence from Fig. 13.3, it is clear that the sum of the centrifugal forces on masses m and m_2 must be equal to the centrifugal force on mass m_1.

$$\therefore \quad F_C + F_{C_2} = F_{C_1}$$

or
$$m \times \omega^2 \times r + m_2 \times \omega^2 \times r_2 = m_1 \times \omega^2 \times r_1$$

or
$$m \times r + m_2 \times r_2 = m_1 \times r_1 \qquad \text{...(13.2)}$$

Also for perfect balancing, the algebraic sum of the moments of the centrifugal forces about any point on the shaft should be zero.

Let O is the point of intersection of plane L and axis of rotation and P is the point of intersection of plane M and axis of rotation.

Now taking moments about point O, we get

$$F_{C_2} \times d = F_C \times d_1$$

(Centrifugal force F_{C_1} passes through O and hence moment of F_{C_1} is zero)

or
$$(m_2 \times \omega^2 \times r_2) \times d = (m \times \omega^2 \times r) \times d_1$$

or
$$m_2 \times r_2 \times d = m \times r \times d_1 \qquad \text{...(13.3)}$$

or
$$m_2 \times r_2 = \frac{m \times r \times d_1}{d}$$

Similarly the moments can be taken about the point P. Then we will get

$$F_{C_1} \times d = F_C \times d_2$$

(F_{C_2} passes through P and hence its moment is zero)

or
$$(m_1 \times \omega^2 \times r_1) \times d = (m \times \omega^2 \times r) \times d_2$$

or
$$m_1 \times r_1 \times d = m \times r \times d_2 \qquad \text{...(13.4)}$$

or
$$m_1 \times r_1 = \frac{m \times r \times d_2}{d}$$

If disturbing mass is in a plane which is left or right of the two planes containing balancing masses, the centrifugal force on first balancing mass will be in opposite direction where as on the second it will be in the same direction.

***(ii)* The planes of the balancing masses may be on either side of the plane of the disturbing mass**

Fig. 13.4 shows the disturbing mass m and two balancing masses m_1 and m_2 which are on either side of the plane of disturbing mass m.

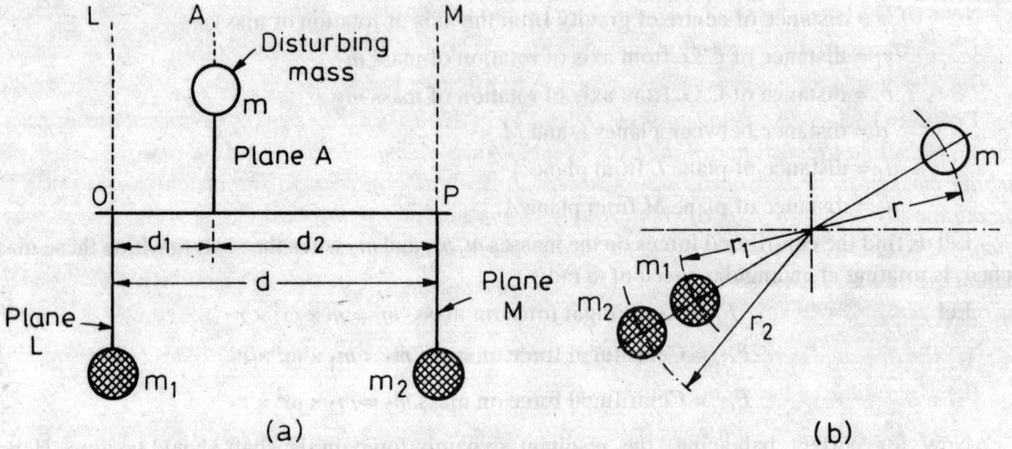

Fig. 13.4

The disturbing mass m is lying in plane A whereas the balancing masses m_1 and m_2 are in planes L and M respectively. The radii of rotation of masses m, m_1 and m_2 are r, r_1 and r_2 respectively as shown in Fig. 13.4 (*b*).

Let d = distance between planes L and M

d_1 = distance of plane L from plane A

d_2 = distance of plane M from plane A

F_C = centrifugal force on mass $m = m \times \omega^2 \times r$

F_{C_1} = centrifugal force on mass $m_1 = m_1 \times \omega^2 \times r_1$

F_{C_2} = centrifugal force on mass $m_2 = m_2 \times \omega^2 \times r_2$

For perfect balancing, the net force (*i.e.* net centrifugal force) on the shaft should be zero and also the moment of the forces about any point on the shaft should be zero.

Hence from Fig. 13.4, the net force (*i.e.* net centrifugal force) will be zero on the shaft, if

$$F_{C_1} + F_{C_2} = F_C$$

or $$m_1 \times \omega^2 \times r_1 + m_2 \times \omega^2 \times r_2 = m \times \omega^2 \times r$$

or $$m_1 \times r_1 + m_2 \times r_2 = m \times r \quad \text{...(13.5)}$$

Taking moments of the forces (*i.e.* centrifugal forces) about point O, we get

$$F_C \times d_1 = F_{C_2} \times d \quad (F_{C_1} \text{ passes through } O)$$

or $$(m \times \omega^2 \times r) \times d_1 = (m_2 \times \omega^2 \times r_2) \times d$$

or $$m \times r \times d_1 = m_2 \times r_2 \times d \quad \text{...(13.6)}$$

or $$m_2 r_2 = \frac{m \times r \times d_1}{d}$$

Similarly taking moments of the forces about point P, we get

$$F_C \times d_2 = F_{C_1} \times d$$

or $$(m \times \omega^2 \times r) \times d_2 = (m_1 \times \omega^2 \times r_1) \times d$$

or $$m \times r \times d_2 = m_1 \times r_1 \times d \quad \text{...(13.7)}$$

If disturbing mass is in a plane in between the planes of the balancing masses, then the centrifugal force on balancing masses will be opposite in direction to the direction of disturbing mass.

Note. (*i*) The line of action of three centrifugal forces are parallel.

(*ii*) The three parallel forces are in equilibrium.

(*iii*) The middle force is in opposite direction to the two outer forces.

Problem 13.1. *A disturbing mass 600 kg is attached to a shaft. The shaft is rotating at a uniform angular velocity ω rad/s and the distance of the C.G. of the disturbing mass from the axis of rotation is 270 mm. The disturbing mass is to be balanced by two masses in two different planes. The distances of the C.G. of the balancing masses from the axis of rotation is 450 mm each. The distances between the two planes of the balancing masses is 1.5 m and the distance between the plane of the disturbing mass and one of the planes of the balancing masses is 300 mm.*

Determine :

(*i*) *the distance between the plane of disturbing mass and the plane of the other balancing mass.*

(*ii*) *Magnitude of balancing masses when :*

(*a*) *the planes of balancing masses are on the same side of the plane of the disturbing mass*

(*b*) *the planes of the balancing masses are on either side of the plane of the disturbing mass.*

Sol. Given :

Disturbing mass, $m = 600$ kg

Distance of C.G. of disturbing mass from axis of rotation, $r = 270$ mm $= 0.27$ m

Distance of C.G. of balancing masses from axis of rotation, $r_1 = r_2 = 450 = 0.45$ m

Distance between the two planes of balancing masses, $d = 1.5$ m

Distance between the plane of disturbing mass and the plane of first balancing mass,

$$d_1 = 300 \text{ mm} = 0.3 \text{ m}.$$

1st Case. *When the planes of the balancing masses are on the same side of the plane of disturbing mass.*

Refer to Fig. 13.3.

Let d_2 = distance between the plane of disturbing mass and the plane of second balancing mass.

m_1 and m_2 = Magnitude of balancing masses.

For perfect balancing, the resultant centrifugal force should be zero.

Hence using, equation (13.2), for resultant force to be zero, we get

$$m \times r + m_2 r_2 = m_1 r_1$$

or $$600 \times 0.27 + m_2 \times 0.45 = m_1 \times 0.45 \quad \text{...(}i\text{)}$$

The resultant moment about any point should be zero. The resultant moment about point O is given by equation (13.3) as

$$m_2 \times r_2 \times d = m \times r \times d_1$$

or $$m_2 \times r_2 = \frac{m \times r \times d_1}{d}$$

or $$m_2 \times 0.45 = \frac{600 \times 0.27 \times 0.3}{1.5}$$

or $$m_2 = \frac{600 \times 0.27 \times 0.3}{1.5 \times 0.45} = \textbf{72 kg.} \quad \textbf{Ans.}$$

Substituting the value of m_2 is equation (*i*), we get

$$600 \times 0.27 + 72 \times 0.45 = m_1 \times 0.45$$

or $$\frac{600 \times 0.27}{0.45} + \frac{72 \times 0.45}{0.45} = m_1$$

or $$360 + 72 = m_1$$

or $$m_1 = \mathbf{432\ kg.} \quad \textbf{Ans.}$$

To find the value of 'd_2' *i.e.* the distance between the plane of the disturbing mass and the plane of second balancing mass the equation (13.4) is used. Hence by using equation (13.4), we get

$$m_1 \times r_1 \times d = m \times r \times d_2$$

or $$d_2 = \frac{m_1 \times r_1 \times d}{m \times r}$$

$$= \frac{432 \times 0.45 \times 1.5}{600 \times 0.27} = \mathbf{1.8\ m.} \quad \textbf{Ans.}$$

2nd Case. *When the planes of the balancing masses are on either side of the plane of the disturbing mass.*

Refer to Fig. 13.4.

For resultant force to be zero, the equation (13.5) is used. Hence from equation (13.5), we get

$$m_1 \times r_1 + m_2 r_2 = m \times r$$

or $$m_1 \times 0.45 + m_2 \times 0.45 = 600 \times 0.27 \quad \text{...}(ii)$$

For resultant moment to be zero about point O, the equation (13.6) is used. Hence using equation (13.6), we get

$$m \times r \times d_1 = m_2 \times r_2 \times d$$

or $$600 \times 0.27 \times 0.3 = m_2 \times 0.45 \times 1.5$$

or $$m_2 = \frac{600 \times 0.27 \times 0.3}{0.45 \times 1.5} = \mathbf{72\ kg.} \quad \textbf{Ans.}$$

Substituting the value of m_2 is equation (*ii*), we get

$$m_1 \times 0.45 + 72 \times 0.45 = 600 \times 0.27$$

or $$m_1 \times 0.45 = 600 \times 0.27 - 72 \times 0.45$$

or $$m_1 = \frac{600 \times 0.27}{0.45} - \frac{72 \times 0.45}{0.45}$$

$$= 360 - 72 = \mathbf{288\ kg.} \quad \textbf{Ans.}$$

To find the value of 'd_2' the equation (13.7) is used. Hence using equation (13.7), we get

$$m \times r \times d_2 = m_1 \times r_1 \times d$$

or $$d_2 = \frac{m_1 \times r_1 \times d}{m \times r}$$

$$= \frac{288 \times 0.45 \times 1.5}{600 \times 0.27} = \mathbf{1.2\ m.} \quad \textbf{Ans.}$$

13.2.2. Balancing of Several Masses Rotating in the Same Plane

Fig. 13.5 (*a*) shows a number of masses m_1, m_2, m_3 and m_4 which are rigidly attached to a shaft which is revolving at a uniform angular speed of ω rad/s about an axis passing through O and perpendicular to the plane of the paper. The radii of rotation of the masses are r_1, r_2, r_3 and r_4. The angles made by these masses with the horizontal line OX are θ_1, θ_2, θ_3 and θ_4 respectively.

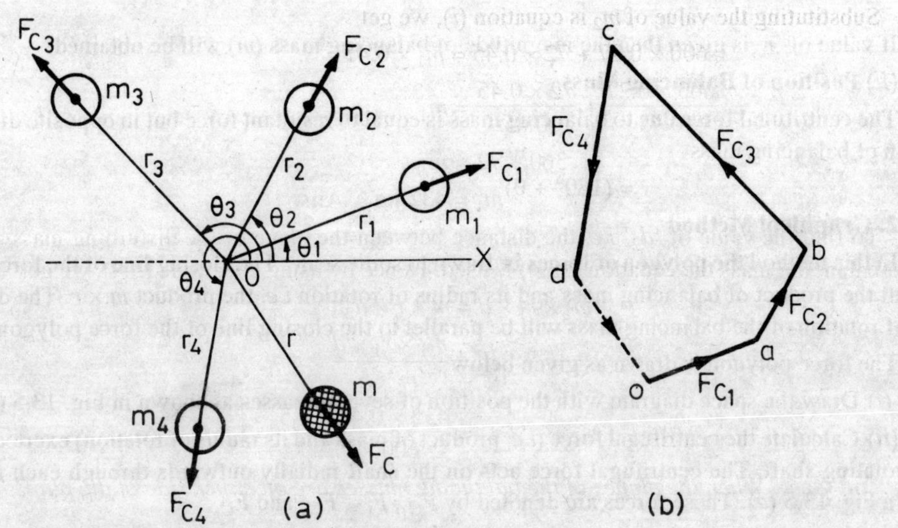

Fig. 13.5

When the shaft revolves, a centrifugal force acts on the shaft radially outwards through each mass centre. These centrifugal forces are denoted by F_{C_1}, F_{C_2}, F_{C_3} and F_{C_4} in Fig. 13.5 (*a*). These forces are proportional to the corresponding product of mass and radius *i.e.* $F_{C_1} \propto m_1 r_1$; $F_{C_2} \propto m_2 r_2$; $F_{C_3} \propto m_3 r_3$ and $F_{C_4} \propto m_4 r_4$. The resultant of these forces is obtained either by analytical method or by graphical method as given below. After knowing the resultant force, the magnitude and position of balancing mass is obtained.

1. Analytical Method

Resolve each centrifugal* force (*i.e.* product of mass and radius) horizontally and vertically and find the total horizontal force and total vertical force as :

Total horizontal force $\quad \Sigma H = F_{C_1} \cos \theta_1 + F_{C_2} \cos \theta_2 + F_{C_3} \cos \theta_3 + F_{C_4} \cos \theta_4$

or $\quad \Sigma H = m_1 r_1 \cos \theta_1 + m_2 r_2 \cos \theta_2 + m_3 r_3 \cos \theta_3 + m_4 r_4 \cos \theta_4 \quad$ ...(13.8)

Similarly total vertical force,

$$\Sigma V = m_1 r_1 \sin \theta_1 + m_2 r_2 \sin \theta_2 + m_3 r_3 \sin \theta_3 + m_4 r_4 \sin \theta_4 \quad \text{...(13.9)}$$

$\therefore$ Resultant centrifugal force,

$$F_C = \sqrt{(\Sigma H)^2 + (\Sigma V)^2}$$

And the angle made by the resultant centrifugal force with horizontal is given by,

$$\tan \theta = \frac{\Sigma V}{\Sigma H} \quad \text{...(13.10)}$$

(*i*) Magnitude of Balancing Mass

The magnitude of balancing mass (m) is obtained from resultant force as given below :

Resultant force, $\quad F_C = m \times r$

where $\quad m$ = Balancing mass and

r = Radius of rotation of balancing mass

*Actual centrifugal force is $m \times \omega^2 \times r$. Since ω^2 is same for each mass, hence magnitude of centrifugal force for each mass is proportional to $m \times r$.

If value of 'r' is given then the magnitude of balancing mass (m) will be obtained.

(*ii*) **Position of Balancing Mass**

The centrifugal force due to balancing mass is equal to resultant force but in opposite direction. Hence direction of balancing mass

$$= (180° + \theta) \qquad ...(13.11)$$

2. Graphical Method

In this method the polygon of forces is drawn to some scale. The closing line of the force polygon will represent the product of balancing mass and its radius of rotation *i.e.* the product $m \times r$. The direction of the radius of rotation of the balancing mass will be parallel to the closing line of the force polygon.

The force polygon is drawn as given below :

(*i*) Draw the space diagram with the position of several masses as shown in Fig. 13.5 (*a*).

(*ii*) Calculate the centrifugal force (*i.e.* product of mass and its radius of rotation) exerted by each mass on the rotating shaft. The centrifugal force acts on the shaft radially outwards through each mass centre as shown in Fig. 13.5 (*a*). These forces are denoted by F_{C_1}, F_{C_2}, F_{C_3} and F_{C_4}.

(*iii*) Now the vector diagram is drawn as shown in Fig. 13.5 (*b*). Take any point *o*. From *o*, draw *oa* parallel to F_{C_1} and take $oa = m_1 r_1$ to some suitable scale. From *a*, draw *ab* parallel to F_{C_2} and take $ab = m_2 r_2$. From *b*, draw *bc* parallel to F_{C_3} and take $bc = m_3 r_3$. From *c*, drawn *cd* parallel to F_{C_4} and take $cd = m_4 r_4$.

(*iv*) Now join point *d* to *o*. The closing line *do* will represent the centrifugal force to scale in magnitude and direction due to balancing mass. Hence

$$m \times r = \text{line } do$$

where m = Balancing mass and

r = its radius of rotation.

The direction of radius of rotation of the balance mass will be parallel to line *do* of vector diagram of Fig. 13.5 (*b*).

Problem 13.2. *A shaft is rotating at a uniform angular speed. Four masses m_1, m_2, m_3 and m_4 of magnitudes 300 kg, 450 kg, 360 kg and 390 kg respectively are attached rigidly to the shaft. The masses are rotating in the same plane. The corresponding radii of rotation are 200 mm, 150 mm, 250 mm and 300 mm respectively. The angles made by these masses with horizontal are 0°, 45°, 120° and 255° respectively. Find :*

(*i*) *the magnitude of the balancing mass and*

(*ii*) *the position of the balancing mass if its radius of rotation is 200 mm.*

Sol. Given :

$m_1 = 300$ kg ; $m_2 = 450$ kg ; $m_3 = 360$ kg ; $m_4 = 390$ kg ;

$r_1 = 200$ mm $= 0.2$ m ; $r_2 = 150$ mm $= 0.15$ m ; $r_3 = 250$ mm $= 0.25$ m ; $r_4 = 300$ mm $= 0.3$ m ;

$\theta_1 = 0°$; $\theta_2 = 45°$; $\theta_3 = 120°$; $\theta_4 = 255°$; and $r = 200$ mm $= 0.2$ m

Let us first find the product of mass and radius of rotation for each mass.

Hence, $m_1 r_1 = 300 \times 0.2 = 60$ kg m

$m_2 r_2 = 450 \times 0.15 = 67.5$ kg m

$m_3 r_3 = 360 \times 0.25 = 90$ kg m

$m_4 r_4 = 390 \times 0.3 = 117$ kg m

The problem may be solved by analytical or graphical method.

1. Analytical Method

Calculate the values of ΣH and ΣV by using equations (13.8) and (13.9).

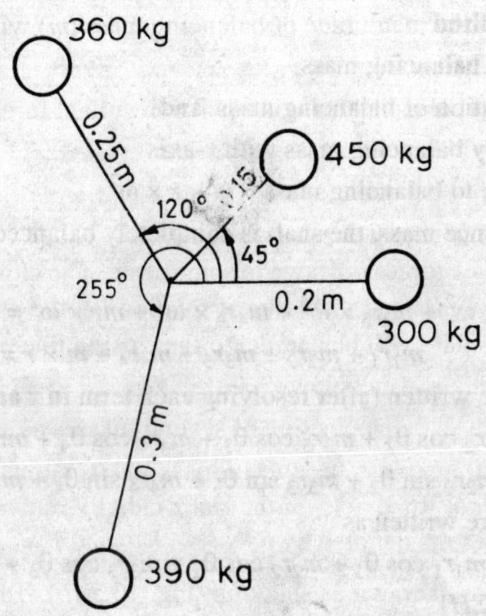

Fig. 13.6

$$\therefore \quad \Sigma H = m_1r_1 \cos \theta_1 + m_2r_2 \cos \theta_2 + m_3r_3 \cos \theta_3 + m_4r_4 \cos \theta_4$$
$$= 60 \cos 0° + 67.5 \cos 45° + 90 \cos 120° + 117 \cos 255°$$
$$= 60 + 47.73 - 45 - 30.28 = 32.45 \text{ kg m}$$

and
$$\Sigma V = m_1r_1 \sin \theta_1 + m_2r_2 \sin \theta_2 + m_3r_3 \sin \theta_3 + m_4r_4 \sin \theta_4$$
$$= 60 \sin 0° + 67.5 \sin 45° + 90 \sin 120° + 117 \sin 255°$$
$$= 0 + 47.73 + 77.94 - 113.01 = 12.65 \text{ kg-m}$$

$\therefore$ Resultant force $= \sqrt{(\Sigma H)^2 + (\Sigma V)^2} = \sqrt{32.45^2 + 12.65^2} = 34.82$ kg m.

(a) Magnitude of Balancing Mass

Let m = Magnitude of balancing mass, and

r = Radius of rotation of balancing mass = 0.2 m (given)

We know that

Resultant force $= m \times r$

$$\therefore \quad 34.82 = m \times 0.2$$

$$\therefore \quad m = \frac{34.82}{0.2} = \mathbf{174.1 \text{ kg.}} \quad \textbf{Ans.}$$

(b) Position of Balancing Mass

The angle made by resultant centrifugal force with horizontal is given be equation (13.10) as

$$\tan \theta = \frac{\Sigma V}{\Sigma H} = \frac{12.65}{32.45} = 0.3898$$

$$\therefore \quad \theta = \tan^{-1} 0.3898 = \mathbf{21.30°}$$

The centrifugal force due to balancing mass is equal to resultant force, but in *opposite direction*. Hence direction of balancing mass is given by equation (13.11) as :

Fig. 13.6 (*a*)

Direction of balancing mass $= 180° + \theta$

$= 180° + 21.30° = \mathbf{201.30°}$. **Ans.**

Alternate Analytical Method

Let m = Magnitude of balancing mass,

r = Radius of rotation of balancing mass, and

θ = Angle made by balancing mass with x-axis

Then centrifugal force due to balancing mass = $m \times r \times \omega^2$

After introducing the balance mass, the shaft is completely balanced. Hence the resultant force acting on the shaft should be zero

i.e. $$m_1r_1 \times \omega^2 + m_2r_2 \times \omega^2 + m_3r_3 \times \omega^3 + m_4r_4 \times \omega^2 + mr \times \omega^2 = 0$$

or $$m_1r_1 + m_2r_2 + m_3r_3 + m_4r_4 + m \times r = 0$$

The above equation can be written (after resolving each term in x and y directions) as

$$m_1r_1 \times \cos\theta_1 + m_2r_2 \cos\theta_2 + m_3r_3 \cos\theta_3 + m_4r_4 \cos\theta_4 + mr\cos\theta = 0$$

and $$m_1r_1 \sin\theta_1 + m_2r_2 \sin\theta_2 + m_3r_3 \sin\theta_3 + m_4r_4 \sin\theta_4 + mr\sin\theta = 0$$

or the above two equations are written as

$$m \times r \times \cos\theta = -[m_1r_1 \cos\theta_1 + m_2r_2 \cos\theta_2 + m_3r_3 \cos\theta_3 + m_4r_4 \cos\theta_4]$$
$$= -[\Sigma H] \quad ...(i)$$

and $$m \times r \times \sin\theta = -[m_1r_1 \sin\theta_1 + m_2r_2 \sin\theta_2 + m_3r_3 \sin\theta_3 + m_4r_4 \sin\theta_4]$$
$$= -[\Sigma V] \quad ...(ii)$$

where $\Sigma H = m_1r_1 \cos\theta_1 + m_2r_2 \cos\theta_2 + m_3r_3 \cos\theta_3 + m_4r_4 \cos\theta_4$

and $\Sigma V = m_1r_1 \sin\theta_1 + m_2r_2 \sin\theta_2 + m_3r_3 \sin\theta_3 + m_4r_4 \sin\theta_4$

Squaring and adding (*i*) and (*ii*), we get

$$m \times r = \sqrt{[(\Sigma H)^2 + (\Sigma V)^2]} \quad ...(13.12)$$

Dividing (*ii*) by (*i*), we get,

$$\frac{\sin\theta}{\cos\theta} = \tan\theta = \frac{-\Sigma V}{-\Sigma H} \quad ...(13.13)$$

The quadrant in which θ lies depends upon the signs of numerator and denominator of the above equation.

Now $\Sigma H = 32.45$ kg-m (already obtained)

$\Sigma V = 12.65$ kg-m (already obtained)

Using equation (13.12), we get

$$m \times r = \sqrt{(\Sigma H)^2 + (\Sigma V)^2} = \sqrt{32.45^2 + 12.65^2} = 34.82 \text{ kg-m}$$

Using equation (13.13),

$$\tan\theta = \frac{-\Sigma V}{-\Sigma H} = \frac{-12.65}{-32.45} = 0.3898$$

(As the sign of numerator is – ve and also the sign of denominor is – ve hence θ will lie in third quadrant. In equation 13.13, numerator corresponds to $\sin\theta$ and denominator corresponds to $\cos\theta$. In third quadrant, the $\sin\theta$ and $\cos\theta$ are both – ve.)

$\therefore \quad \theta = \tan^{-1} 0.3898 = 21.3° + 180° =$ **201.30°. Ans.**

2. Graphical Method

The followings are the steps of finding the magnitude and position of the balancing mass by graphical method :

(*i*) Draw the space diagram with the position of several masses as shown in Fig. 13.7 (*a*).

(*ii*) **Calculate the centrifugal force (*i.e.* product of mass and its radius of rotation) exerted by each mass on the rotating shaft as given below :**

$$m_1 r_1 = 300 \times 0.2 = 60 \text{ kg-m}$$
$$m_2 r_2 = 450 \times 0.15 = 67.5 \text{ kg-m}$$
$$m_3 r_3 = 360 \times 0.25 = 90 \text{ kg-m}$$
$$m_4 r_4 = 390 \times 0.3 = 117 \text{ kg-m.}$$

(*iii*) The centrifugal forces act on the shaft radially outwards through each mass centre as shown in Fig. 13.7 (*a*). These forces are denoted by F_{C_1}, F_{C_2}, F_{C_3} and F_{C_4}.

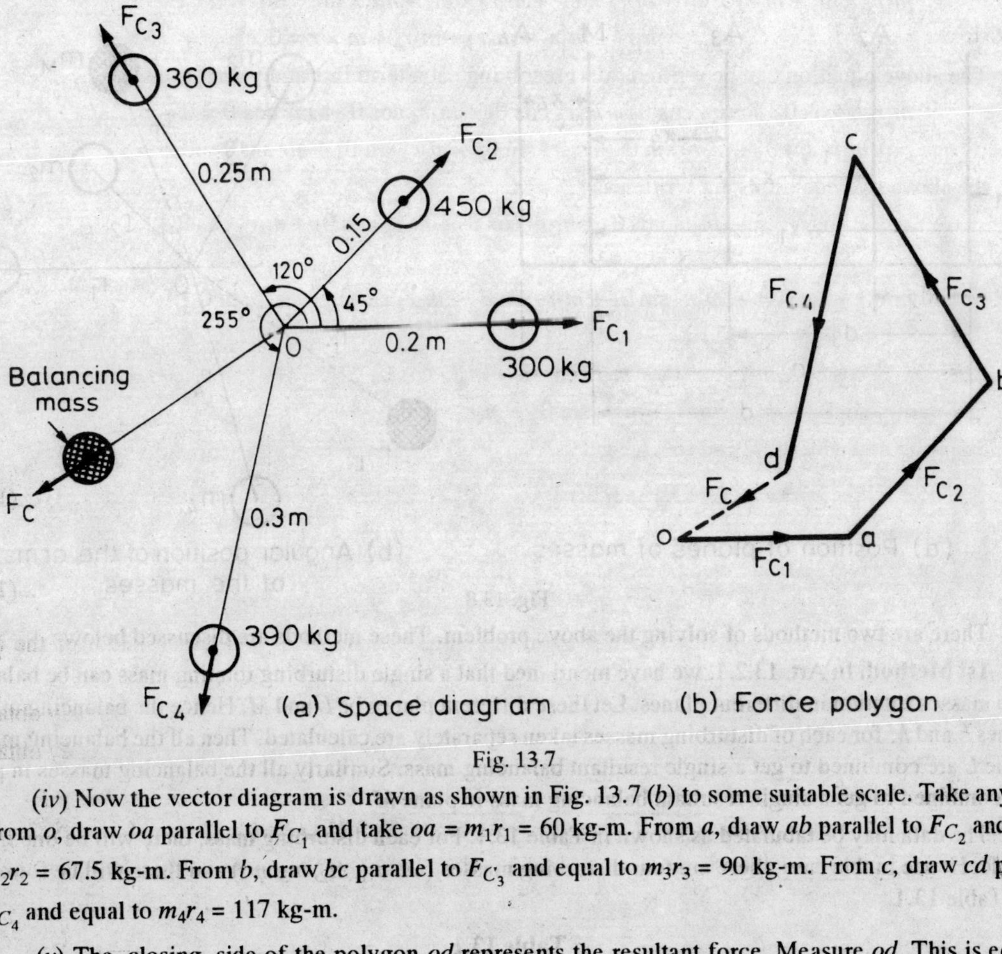

(a) Space diagram (b) Force polygon

Fig. 13.7

(*iv*) Now the vector diagram is drawn as shown in Fig. 13.7 (*b*) to some suitable scale. Take any point *o*. From *o*, draw *oa* parallel to F_{C_1} and take $oa = m_1 r_1 = 60$ kg-m. From *a*, draw *ab* parallel to F_{C_2} and equal to $m_2 r_2 = 67.5$ kg-m. From *b*, draw *bc* parallel to F_{C_3} and equal to $m_3 r_3 = 90$ kg-m. From *c*, draw *cd* parallel to F_{C_4} and equal to $m_4 r_4 = 117$ kg-m.

(*v*) The closing side of the polygon *od* represents the resultant force. Measure *od*. This is equal to 34.8 kg-m on the chosen scale. Hence *od* = 34.8 kg-m.

(*vi*) The centrifugal force due to balancing mass is equal to the resultant force but *opposite in direction*. But balancing force is proportional to $m \times r$, therefore

$$m \times r = 34.8$$

or $$m \times 0.2 = 34.8$$

or $$m = \frac{34.8}{0.2} = \mathbf{174 \text{ kg.}} \quad \textbf{Ans.}$$

(*vii*) Draw the force F_C parallel to line *do* in space diagram. This gives the position of the balancing mass. Measure the angle made by F_C with *x*-axis. This is equal to **201°**. **Ans.**

∴ Position of balancing mass = 201° with *x*-axis. **Ans.**

13.2.3. Balancing of Several Masses Rotating in Different Planes

Fig. 13.8 (*a*) shows several masses (*i.e.* m_1, m_2, m_3 and m_4) rotating in different planes (*i.e.* in planes A_1, A_2, A_3 and A_4 respectively) at radii of r_1, r_2, r_3 and r_4 respectively. The relative angular positions of the arms of these masses are shown in the end view *i.e.* Fig. 13.8 (*b*). It is required to find the masses and angular positions of the two balancing masses in planes *L* and *M* that will balance completely the given system.

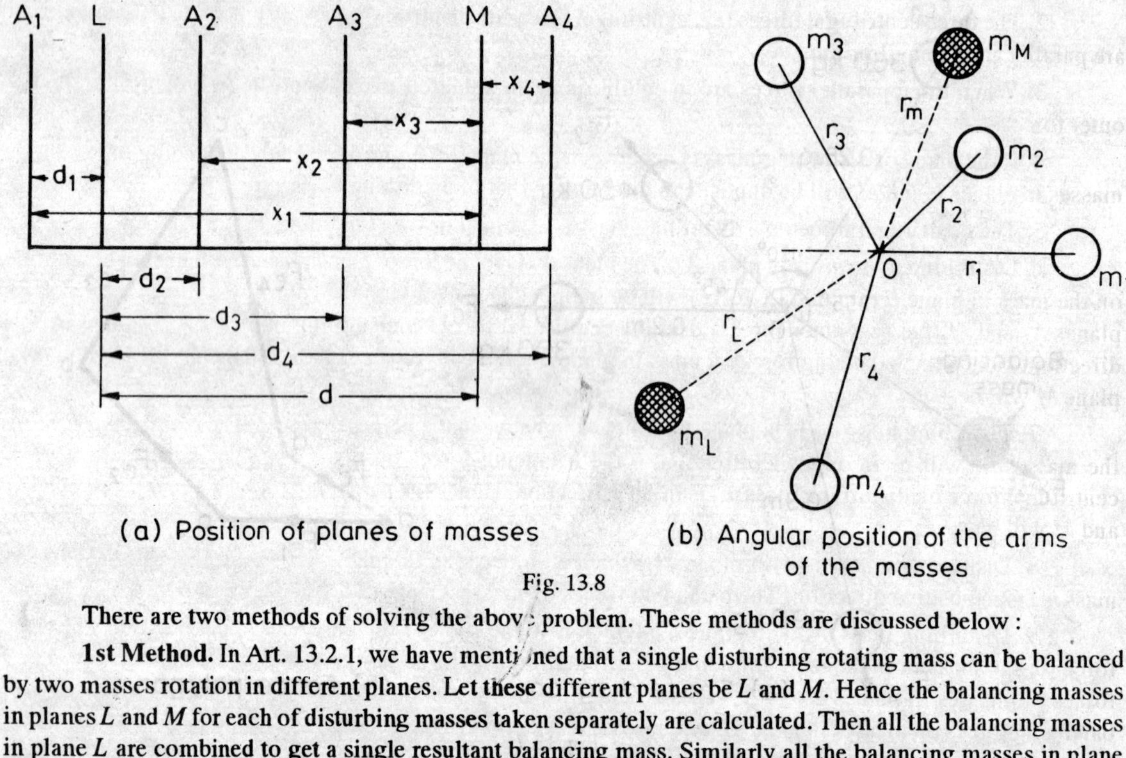

(a) Position of planes of masses (b) Angular position of the arms of the masses

Fig. 13.8

There are two methods of solving the above problem. These methods are discussed below :

1st Method. In Art. 13.2.1, we have mentioned that a single disturbing rotating mass can be balanced by two masses rotation in different planes. Let these different planes be *L* and *M*. Hence the balancing masses in planes *L* and *M* for each of disturbing masses taken separately are calculated. Then all the balancing masses in plane *L* are combined to get a single resultant balancing mass. Similarly all the balancing masses in plane *M* are combined to get a single resultant balancing mass in plane *M*.

The data may be tabulated as shown in Table 13.1. For each disturbing mass, there will be one line in the table. Hence in this case, there are four disturbing masses (*i.e.* m_1, m_2, m_3, m_4) and there will be four lines in the Table 13.1.

Table 13.1

Plane	Mass (m) (in kg)	Radius (r) (in meter)	Centrifugal force ÷ ω^2 $(m \times r)$	Distance from		Balancing Force ÷ ω^2	
				Plane L (d*)	Plane M (x)	Plane L $\frac{m \times r \times x}{d}$	Plane M $\frac{m \times r \times d^*}{d}$
A_1	m_1	r_1	$m_1 r_1$	d_1	x_1	$-\frac{m_1 r_1 x_1}{d}$	$+\frac{m_1 r_1 d_1}{d}$

A_2	m_2	r_2	m_2r_2	d_2	x_2	$-\frac{m_2r_2x_2}{d}$	$-\frac{m_2r_2d_2}{d}$
A_3	m_3	r_3	m_3r_3	d_3	x_3	$-\frac{m_3r_3x_3}{d}$	$-\frac{m_3r_3d_3}{d}$
A_4	m_4	r_4	m_4r_4	d_4	x_4	$+\frac{m_4r_4x_4}{d}$	$-\frac{m_4r_4d_4}{d}$

The sign convention in the last two columns of the table are decided as given below :

1. A single disturbing mass is balanced by two masses in planes L and M.

2. The three centrifugal forces (*i.e.* centrifugal forces on disturbing mass and on two balancing masses) are parallel and in equilibrium.

3. When three parallel forces are in equilibrium, the middle force is opposite in direction to the two outer forces.

4. If the plane of disturbing mass is in between the planes of L and M, then the balancing forces on the masses in planes L and M will be opposite in direction to that of disturbing mass.

5. The centrifugal force on a disturbing mass is assumed in + ve direction.

6. Disturbing mass m_1 is in plane A_1. The plane A_1 is to the left of planes L and M. The balancing force on the mass in plane L (*i.e.* middle force) will be in opposite direction to the centrifugal forces on masses in planes A_1 and M (*i.e.* two outer forces). But the centrifugal force on disturbing mass m_1 is assumed in + ve direction. Hence the balancing force on mass in plane L will be – ve whereas the balancing force on mass in plane M will be + ve.

7. Disturbing mass m_2 is in plane A_2 which is in between the planes L and M. The centrifugal force on the mass 'm_2' will be in opposite direction to the balancing forces on masses in planes L and M. But the centrifugal force on disturbing mass m_2 is in + ve direction. Hence the balancing forces on masses in planes L and M will be – ve.

8. Disturbing mass m_3 is in plane A_3, which is in between the planes L and M. Centrifugal force on mass m_3 is in positive direction. The balancing forces on masses in planes L and M will be – ve.

9. Disturbing mass m_4 is in plane A_4, which is to the right of the planes of L and M. The centrifugal force on the mass in plane M (*i.e.* middle centrifugal force) will be in opposite direction to the centrifugal forces on masses in planes L and A_4 (*i.e.* two outer forces). Force on disturbing mass m_4 is + ve and hence balancing forces on masses in planes L will also be + ve, whereas the balancing force on mass in plane M will be – ve.

Problem 13.3. *Four masses 200 kg, 300 kg, 240 kg and 260 kg are attached to a shaft. These masses are revolving at radii 270 mm, 210 mm, 300 mm and 360 mm respectively in planes measured from A_1 at 270 mm, 420 mm and 720 mm respectively. The angles measured anticlockwise are m_1 to m_2 45°, m_2 and m_3 75°, m_3 to m_4 135° and the distance between the planes L and M in which the balance masses are to be placed is 500 mm. The distance between planes A_1 and L is 120 mm and M and A_4 is 100 mm. If the balancing masses revolve at a radius of 72 mm, find their magnitude and angular positions.*

Sol. Given :

$m_1 = 200$ kg ; $m_2 = 300$ kg ; $m_3 = 240$ kg ; $m_4 = 260$ kg ;

$r_1 = 270$ mm $= 0.27$ m ; $r_2 = 210$ mm $= 0.21$ m ; $r_3 = 300$ mm $= 0.3$ m ; $r_4 = 360$ mm $= 0.36$ m.

Distance between planes A_1 and $A_2 = 270$ mm ; between A_1 and $A_3 = 420$ mm and between A_1 and $A_4 = 720$ mm ;

Now distance ; $L_1 = 120$ mm ; $L_2 = 270 - 120 = 150$ mm ; $L_3 = 420 - 120 = 300$ mm ; $L_4 = 500 + 100 = 600$ mm ; $d = 500$ mm $= 0.5$ m.

Also distance from plane M, $x_1 = 620$ mm, $x_2 = 620 - 270 = 350$, $x_3 = 500 - 300 = 200$ mm, $x_4 = 100$ mm.

Radius of Balancing masses *i.e.* $r_L = r_m = 72$ cm $= 0.72$ m.

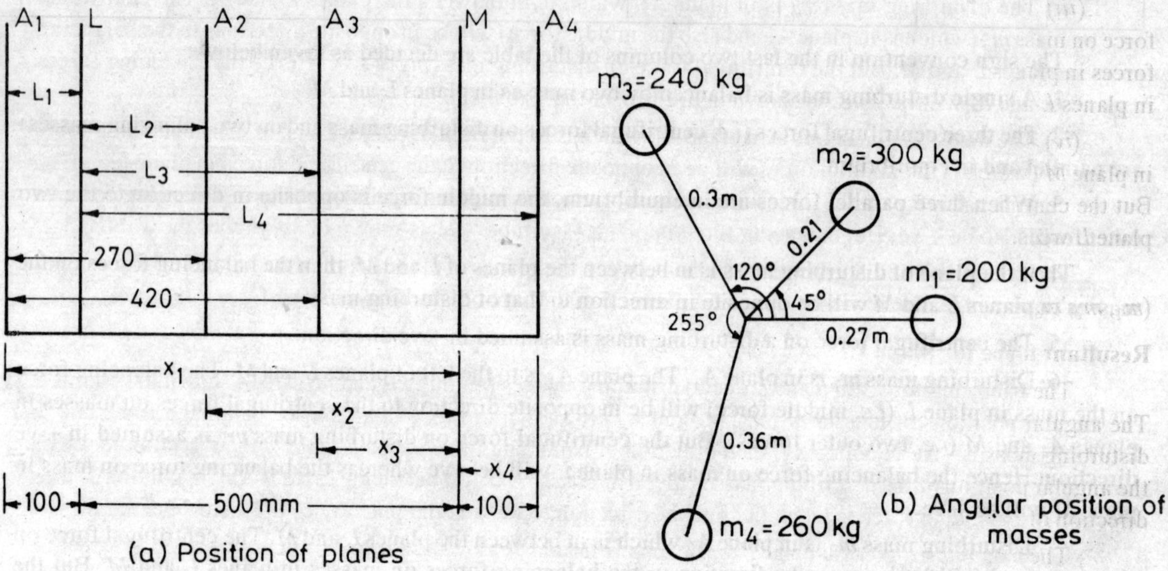

Fig. 13.9. (*a*) and (*b*)

Assuming the mass m_1 along *x*-axis *i.e.* horizontal. Then the positions of planes and angular position of masses are shown in Fig. 13.9 (*a*) and (*b*) respectively. The position of balancing planes *L* and *M* are also shown in Fig. 13.9 (*a*). Let us construct the Table 13.2 as given below :

Table 13.2

Planes A	*Mass (m) in kg*	*Radius (r) in meter*	*Centrifugal Force ÷ ω²* $(m \times r)$	*Distance from Plane L (L)*	*Distance from Plane M (x)*	*Balancing force ÷ ω² Plane L* $\frac{m \times r \times x}{d}$	*Balancing force ÷ ω² Plane M* $\frac{m \times r \times L}{d}$
A_1	$m_1 = 200$ kg	$r_1 = 0.27$ m	$m_1r_1 = 54$ kg-m	$L_1 = 0.12$	$x_1 = 0.62$	$-\frac{m_1r_1 \times x_1}{d} = -66.96$	$+\frac{m_1r_1 \times L}{d} = 12.96$
A_2	300	0.21	$300 \times 0.21 = 63$	$L_2 = 0.15$	$x_2 = 0.35$	– 44.1	– 18.9
A_3	240	0.30	$240 \times 0.3 = 72$	$L_3 = 0.30$	$x_3 = 0.20$	– 28.8	– 43.2
A_4	260	0.36	$260 \times 0.36 = 93.6$	$L_4 = 0.60$	$x_4 = 0.10$	+ 18.72	– 112.32

The steps for the above tables are self-explanatory. The sign convention for the balancing forces in planes *L* and *M* due to disturbing masses m_1, m_2, m_3 and m_4 are as given below :

(*i*) The disturbing mass m_1 is in plane A_1, which is to the left of planes *L* and *M*. The balancing force in plane *L* (which is middle force) will be in opposite direction to the forces in planes *M* and A_1. But the

centrifugal force on the disturbing mass m_1 (which is in plane A_1) is positive. Hence the balancing force in plane M is also + ve. But balancing force in plane L will be negative as shown in first row of the Table 13.2

(*ii*) The disturbing mass m_2 is in plane B_2, which is in between the plane L and M. The centrifugal force on mass m_2 will be in opposite direction to the balancing forces in planes L and M. But centrifugal force on disturbing mass m_2 is + ve. Hence the balancing forces in planes L and M will be – ve, as shown in second row of the Table 13.2.

(*iii*) The disturbing mass m_3 is in plane A_3, which is in between the planes L and M. The centrifugal force on mass m_3 (which is in plane A_3 and also the middle force) will be in opposite directions to the balancing forces in planes L and M. But the centrifugal force on disturbing mass m_3 is + ve. Hence the balancing forces in planes L and M will be – ve, as shown in 3rd row of the Table 13.2.

(*iv*) The disturbing mass m_4 is in plane A_4, which is to the right of planes L and M. The balancing force in plane M (which is the middle force) will be in opposite direction to the centrifugal forces in plane L and A_4. But the centrifugal force on the disturbing mass m_4 (which is in plane A_4) is + ve. Hence balancing force in plane L will also be + ve. But balancing force in plane M will be – ve as shown in 4th row of the Table 13.2.

The columns 7 and 8 gives the values of balancing forces in plane L and M due to disturbing masses (m_1, m_2, m_3 and m_4). The resultant forces for planes L and M are obtained by drawing force polygons.

Resultant force for plane L

The balancing forces in plane L due to disturbing masses (m_1, m_2, m_3 and m_4) are as shown in column 7. The angular positions of masses m_1, m_2, m_3 and m_4 are shown in Fig. 13.9 (*b*). If the balancing force due to disturbing mass is + ve in column 7, then direction of balancing force is taken in the same direction in which the angular position of the disturbing mass is given in Fig. 13.9 (*b*). If balancing force is – ve in column 7, then direction of balancing force is taken in opposite direction to the angular position of the disturbing mass.

The balancing forces for plane L due to disturbing masses are shown to scale in magnitude and direction in Fig. 13.9 (*c*) in which oa = – 66.96 (opposite to direction m_1) ; ob = – 44.1 (opposite to direction m_2) ; oc = – 28.8 (opposite to direction m_3) ; od = 18.72 (same direction as of m_4).

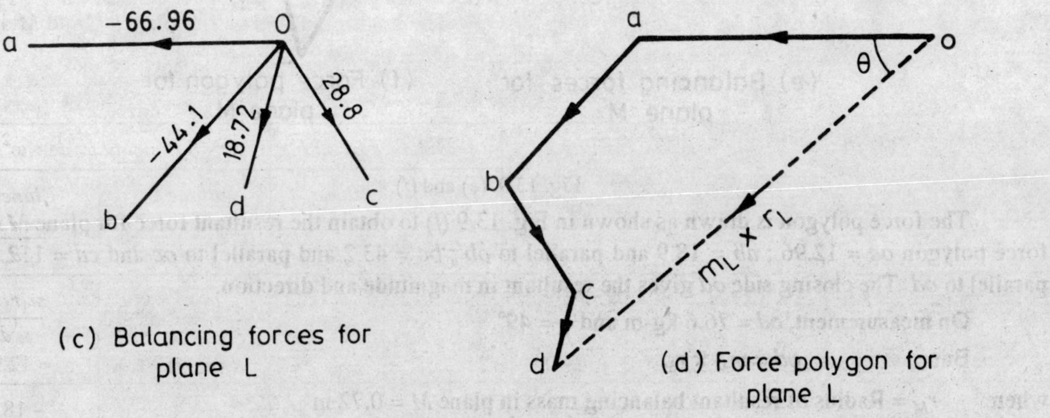

Fig. 13.9. (*c*) and (*d*)

The force polygon is drawn as shown in Fig. 13.9 (*d*) to obtain the result force for plane L. In this force polygon, oa = 66.96, ab is parallel to ob and equal to 44.1 ; bc is parallel to oc and = 28.8 and cd is parallel to od and = 18.72. The closing side od gives the resultant force or $od = m_L \times r_L$

where m_L = Balancing mass for plane L and

r_L = Radius of balancing mass in plane L = 0.72 m (given)

Measure od. This is on measurement is equal to 115.56.

$\therefore$ $od = 115.56$ kg-m

$\therefore$ $m_L \times r_L = od = 115.56$

$$\therefore \quad m_L = \frac{115.56}{r_L} = \frac{115.56}{0.72} = \mathbf{160.5\ kg.} \quad \textbf{Ans.}$$

Measure the angle θ also. This is the angle which the resultant makes with horizontal. On measurement,

$$\theta = 40°. \quad \textbf{Ans.}$$

Resultant force for Plane M

Consider column 8th of the table 13.2. The balancing forces for plane *M* due to disturbing masses are drawn in magnitude and direction as shown in Fig. 13.9 (*e*) in which *oa* = 12.96 (same direction as of m_1) ; *ob* = – 18.9 (opposite to the direction of m_2) ; *oc* = – 43.2 (opposite to direction of m_3) and *ad* = – 112.32 (opposite to direction of m_4).

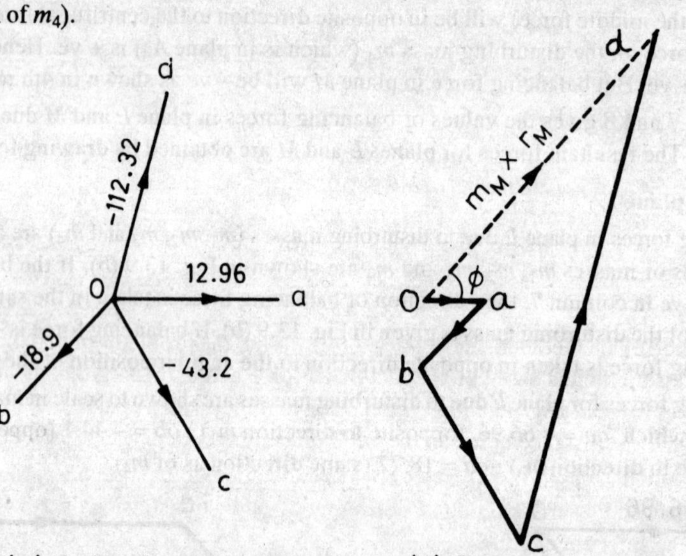

(e) Balancing forces for plane M

(f) Force polygon for plane M

Fig. 13.9. (*e*) and (*f*)

The force polygon is drawn as shown in Fig. 13.9 (*f*) to obtain the resultant force for plane *M*. In this force polygon *oa* = 12.96 ; *ab* = 18.9 and parallel to *ob* ; *bc* = 43.2 and parallel to *oc* and *cd* = 112.32 and parallel to *od*. The closing side *od* gives the resultant in magnitude and direction.

On measurement, *od* = 76.6 kg-m and φ = 49°.

But $od = m_M \times r_M$

where r_M = Radius of resultant balancing mass in plane *M* = 0.72 m

m_M = Resultant balancing mass

or $76.6 = m_M \times 0.72$

$$\therefore \quad m_M = \frac{76.6}{0.72} = \mathbf{106.38\ kg.} \quad \textbf{Ans.}$$

The actual angular positions of disturbing masses (m_1, m_2, m_3 and m_4) and balancing masses in planes *L* and *M* are shown in Fig. 13.9 (*g*).

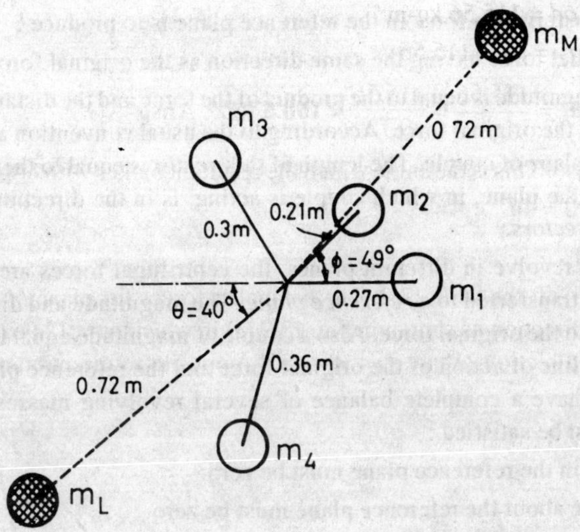

(g) Actual angular positions of disturbing masses (m_1, m_2, m_3, m_4) and balancing masses (m_M, m_L)

Fig. 13.9. (*g*)

Second Method. This method is mostly used for solving the problems on balancing as it involves less arithmetical work. This method is based on the fact that a force acting on a rigid body at a point, is equivalent to an equal and parallel force acting at another fixed point together with a couple, the magnitude of the couple is equal to the force and the distance between the two points.

We have already discussed that it is not always convenient to balance a rotating mass in the plane of rotation. It can be balanced by a mass acting in a different plane by transferring the rotating mass in this plane. This plane is known as **reference plane.**

Let a force F is acting at a point A on a shaft which is at a distance of L from the reference plane as shown in Fig. 13.10. The equilibrium of the system does not change if two equal and opposite force 'F' are added in the reference plane at point B. The net effect would be a single force F (shown by dotted line) in the reference plane having the same direction as the original force at A, along with a couple of magnitude $F \times L$. The couple is acting in a plane containing the lines of action of forces (at A and B) and the shaft.

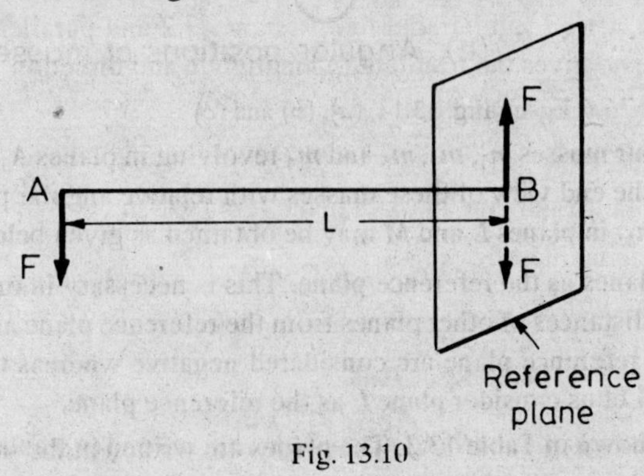

Fig. 13.10

Thus the effect of transferring a force in the reference plane is to produce :

(*i*) an equal and parallel force having the same direction as the original force and

(*ii*) a couple whose magnitude is equal to the product of the force and the distance between the reference plane and the line of action of the original force. According to the usual convention a couple is represented by a vector at right angles to the plane of couple. The length of this vector is equal to the magnitude of the couple. The vector at right angles to the plane, in which couple is acting, is in the direction of force. *Hence couple vectors are parallel to force vectors.*

When several masses revolve in different planes, the centrifugal forces are acting on these masses. Each centrifugal force can be transferred to *a reference plane*. The magnitude and direction of the force in the reference plane will be equal to the original force. Also a couple of magnitude equal to the product of the force and the distance between the line of action of the original force and the reference plane, will be acting in the reference plane. In order to have a complete balance of several revolving masses in different planes, the following two conditions must be satisfied :

1. The resultant force in the reference plane must be zero.

2. The resultant couple about the reference plane must be zero.

The resultant force will be zero if force polygon is closed. Also the resultant couple will be zero if couple polygon is closed.

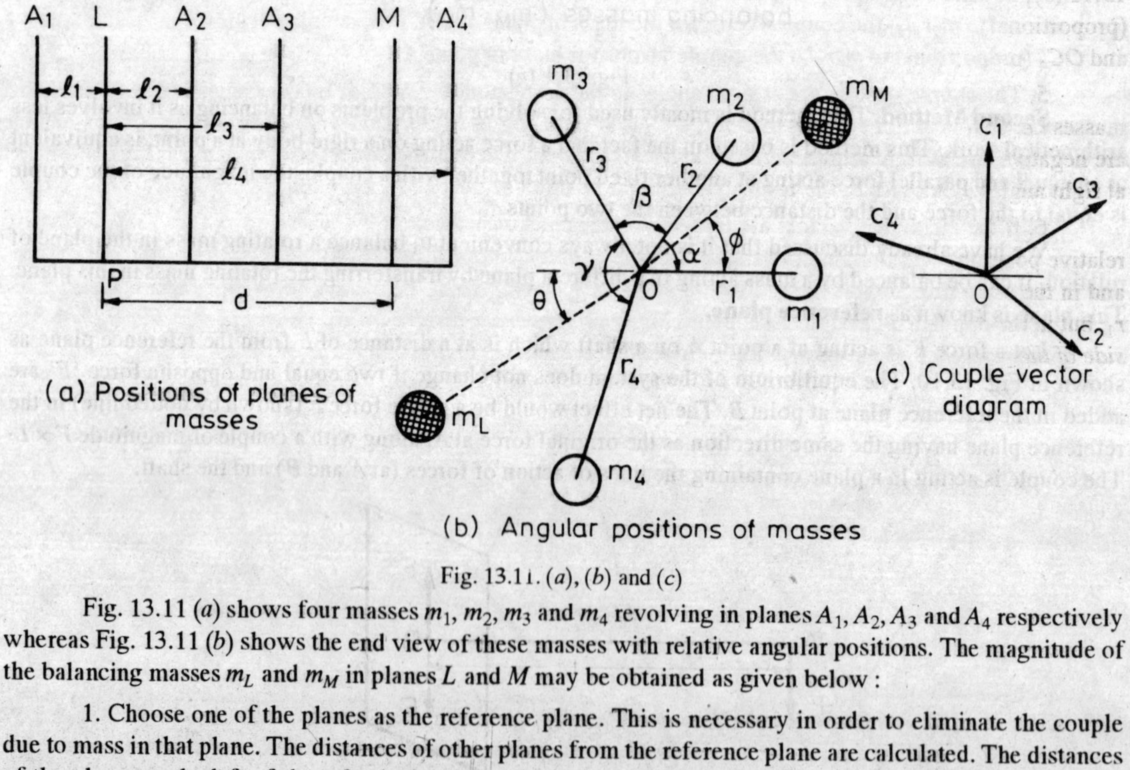

Fig. 13.11. (*a*), (*b*) and (*c*)

Fig. 13.11 (*a*) shows four masses m_1, m_2, m_3 and m_4 revolving in planes A_1, A_2, A_3 and A_4 respectively whereas Fig. 13.11 (*b*) shows the end view of these masses with relative angular positions. The magnitude of the balancing masses m_L and m_M in planes L and M may be obtained as given below :

1. Choose one of the planes as the reference plane. This is necessary in order to eliminate the couple due to mass in that plane. The distances of other planes from the reference plane are calculated. The distances of the planes to the left of the reference plane are considered negative whereas to the right of the reference plane are considered positive. Let us consider plane L as the reference plane.

2. Prepare the data as shown in Table 13.3. The planes are written in the same order in which they are given, starting from left to right.

3. Either plane L or plane M should be taken as reference plane so that the couple is zero due to the mass in this plane.

Table 13.3

Planes	*Mass (m) in kg*	*Radius (r) in meter*	*Centrifugal Force* $\div \omega^2$ *(m × r) kg-m*	*Distance from Reference Plane l*	*Couple* $\div \omega^2$ *(m × r × l) kg-m²*
(i)	(ii)	(iii)	(iv)	(v)	(vi)
A_1	m_1	r_1	$m_1 r_1$	$-l_1$	$-m_1 r_1 l_1$
L	m_L	r_L	$M_L \times r_L$	0	0
A_2	m_2	r_2	$m_2 r_2$	l_2	$m_2 r_2 l_2$
A_3	m_3	r_3	$m_3 r_3$	l_3	$m_3 r_3 l_3$
M	m_M	r_M	$m_M \times r_M$	l_M	$m_M \times r_M \times l_M$
A_4	m_4	r_4	$m_4 \times r_4$	l_4	$m_4 r_4 l_4$

4. The couples given in column 6th are represented by the vectors shown in the end view [*i.e.* Fig. 13.11 (*c*)] in which OC_1 (proportional to $m_1 r_1 l_1$) is the couple vector for the revolving mass in plane A_1, OC_2 (proportional to $m_2 r_2 l_2$) the couple vector for the mass in plane A_2, OC_3 the couple vector for mass in plane A_3 and OC_4 (proportional to $m_4 r_4 l_4$) the couple vector for mass in plane A_4.

5. The couple vectors are at right angles to the corresponding radii or relative angular positions of the masses *i.e.* OC_1 is at right angles to Om_1 ; OC_2 is at right angles to Om_2 and so on. The couple vectors, which are negative, are at right angles anti-clockwise to the radii whereas the couple vectors which are positive arc at right angles clockwise to the respective angular position of masses as shown in Fig. 13.11 (*c*).

6. If the all couple vectors are turned counter-clockwise through 90° as shown in Fig. 13.11 (*d*), their relative positions will remain unchanged. But the couple vectors OC_2, OC_3 and OC_4 will now be parallel to, and in the same direction as the radii r_2, r_3 and r_4 respectively. But the couple vector OC_1 will be parallel to r_1, but in the opposite direction. Hence the *couple vectors are drawn radially outwards for the masses on one side of the reference plane and radially inward for the masses on the other side of the reference plane.*

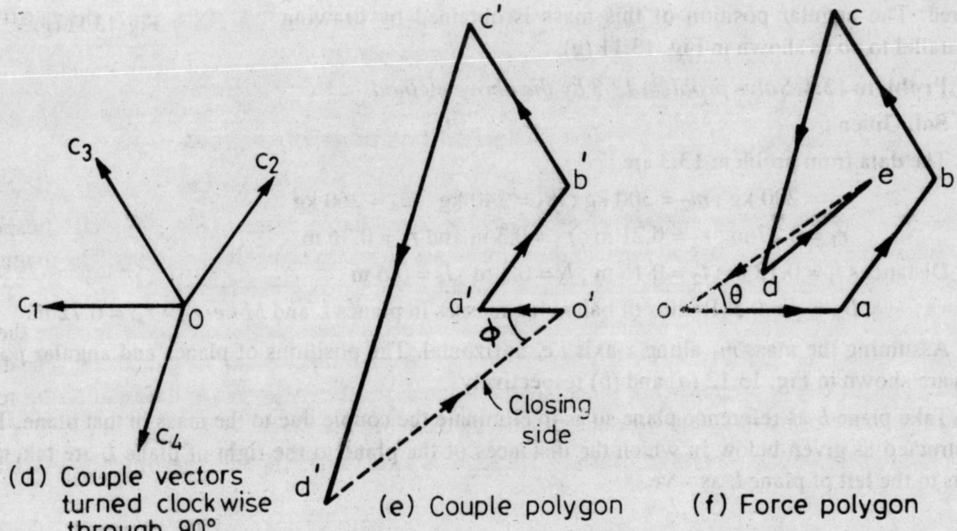

Fig. 13.11. (*d*), (*e*) and (*f*)

7. Now the couple polygon is drawn as shown in Fig. 13.11 (*e*). Take any point o'. Draw vector $o'a'$ parallel and = OC_1 (or $m_1r_1l_1$). Vector $a'b'$ is parallel and equal to OC_2 (or $m_2r_2l_2$) ; Vector $b'c'$ is parallel and equal to OC_3 ; Vector $c'd'$ is parallel and equal to OC_4. The closing side of the polygon (*i.e.* vector $d'o'$) will represent the couple due to unknown balancing mass in plane *M*. This couple is proportional to $m_M \times r_M \times l_M$.

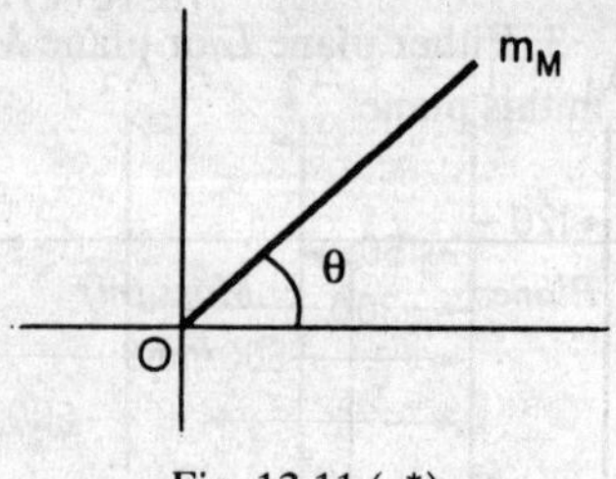

Fig. 13.11 (*e**)

∴ Vector $d'o' = m_M \times r_M \times l_M$

∴ Unknown balancing mass in plane *M* in the above equation is given as,

$$m_M = \frac{\text{Vector } d'o'}{r_M \times l_M}$$

Hence the magnitude and direction (*i.e.* angle ϕ) of the unknown balancing mass in plane *M* is determined. The angular position of this mass is shown in Fig. 13.11 (*b*) and is obtained by drawing Om_M parallel to $d'o'$.

8. Now from the masses m_1, m_2, m_3, m_4 and m_M which are known in magnitude and direction, the force polygon is drawn as shown in Fig. 13.11 (*f*). The closing side of the force polygon will give the magnitude and direction of the force due to balancing mass required in plane *L*.

Force-polygon. Take any point *o*. Draw vector *oa* equal to m_1r_1 and parallel to Om_1. The vector $ab = m_2r_2$ and parallel to Om_2 ; vector $bc = m_3r_3$ and parallel to Om_3 ; vector $cd = m_4r_4$ and parallel to Om_4 and vector $de = m_M \times r_M$ and parallel to Om_M. The closing vector *eo* gives the value of $m_L \times r_L$

∴ Vector $eo = m_L \times r_L$

∴ Balancing mass in plane *L* is obtained as

$$m_L = \frac{\text{Vector } eo}{r_L}$$

The direction of this mass is given by angle θ, which can be measured. The angular position of this mass is obtained by drawing Om_L parallel to *eo* as shown in Fig. 13.11 (*g*).

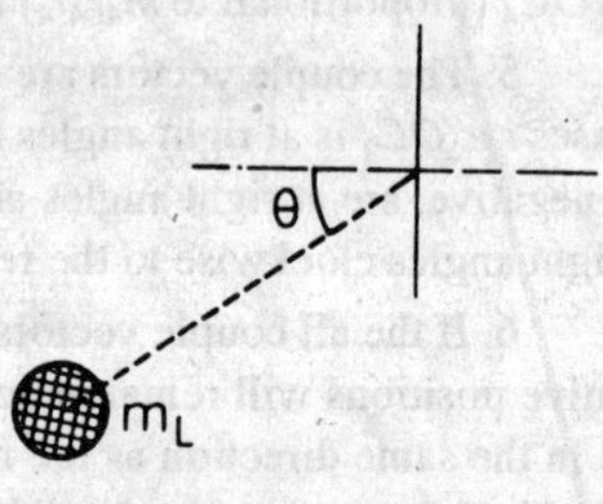

(g) Angular position of balancing mass m_L

Fig. 13.11 (*g*)

Problem 13.4. *Solve problem 13.3 by the above method.*

Sol. Given :

The data from problem 13.3 are :

$m_1 = 200$ kg ; $m_2 = 300$ kg ; $m_3 = 240$ kg ; $m_4 = 260$ kg

$r_1 = 0.27$ m ; $r_2 = 0.21$ m ; $r_3 = 0.3$ m and $r_4 = 0.36$ m

Distances $l_1 = 0.12$ m ; $l_2 = 0.15$ m ; $l_3 = 0.3$ m ; $l_4 = 0.6$ m

and $L_m = d = 0.5$. Radius of balancing masses in planes *L* and *M* *i.e.* $r_L = r_M = 0.72$ m.

Assuming the mass m_1 along *x*-axis *i.e.* horizontal. The positions of planes and angular position of masses are shown in Fig. 13.12 (*a*) and (*b*) respectively.

Take plane *L* as reference plane so as to eliminate the couple due to the mass in that plane. The table is constructed as given below in which the distances of the plane to the right of plane *L* are taken as + ve whereas to the left of plane *L* as – ve.

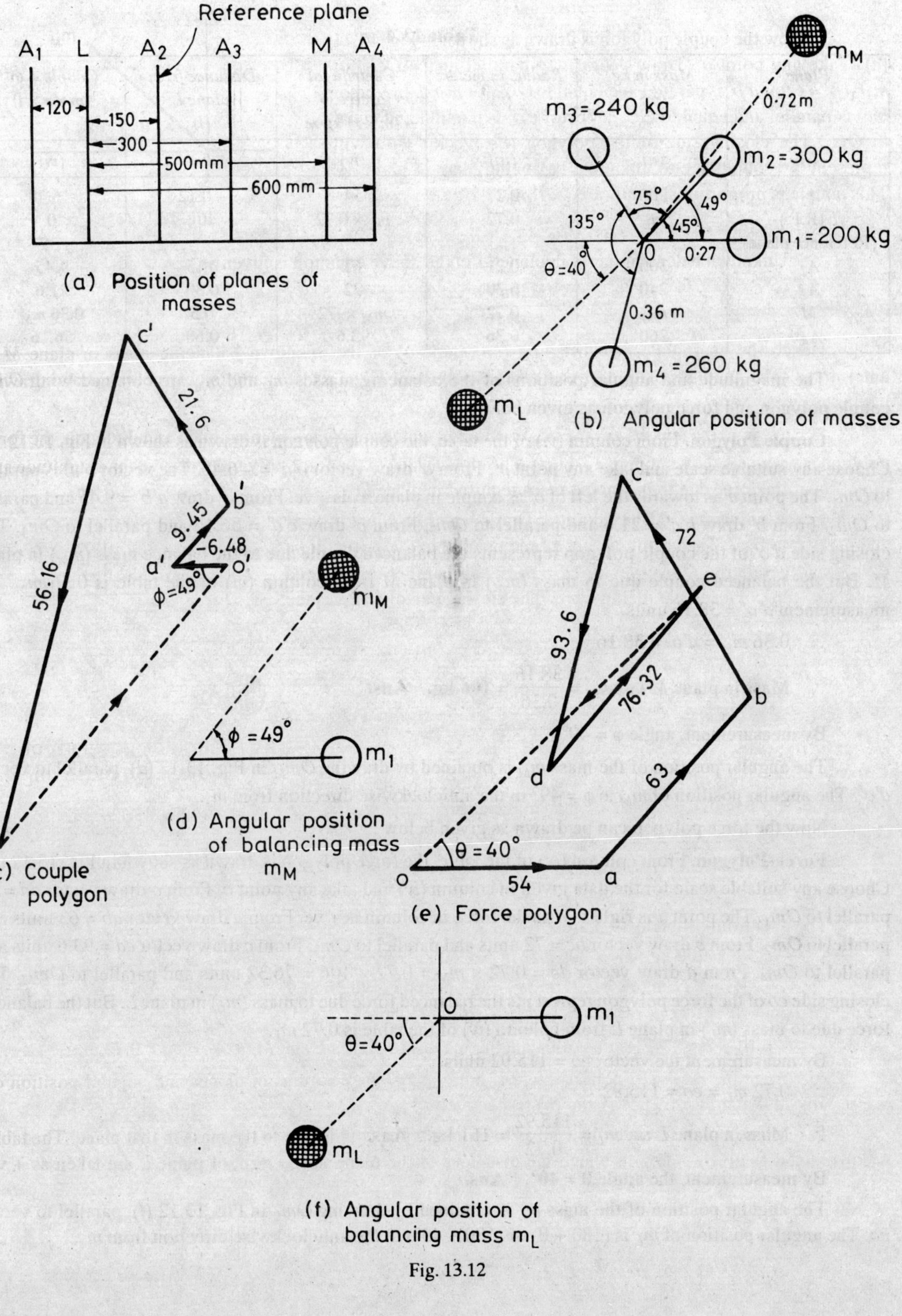
Reference plane
A1 L A2 A3 M A4
120
150
300
500mm
600 mm
(a) Position of planes of masses
mM
0.72m
m3 = 240 kg
m2 = 300 kg
75°
49°
135°
45°
m1 = 200 kg
0
0.27
θ = 40°
0.36 m
m4 = 260 kg
mL
(b) Angular position of masses
c'
21.6
b'
9.45
-6.48
a'
o'
φ = 49°
56.16
d'
(c) Couple polygon
mM
φ = 49°
m1
(d) Angular position of balancing mass mM
c
72
e
93.6
76.32
b
d
63
θ = 40°
o
54
a
(e) Force polygon
O
m1
θ = 40°
mL
(f) Angular position of balancing mass mL

Fig. 13.12

Table 13.4

Plane	*Mass in kg (m)*	*Radius in meter (r)*	*Centrifugal force* $\div \omega^2$ $(m \times r)$ *kg-m*	*Distance from plane L (l)*	*Couple* $\div \omega^2$ $(m \times r \times l)$
(i)	*(ii)*	*(iii)*	*(iv)*	*(v)*	*(vi)*
A_1	200	0.27	54	– 0.12	– 6.48
L (R.P.) (Reference plane)	m_L	0.72	$m_L \times 0.72$	0	0
A_2	300	0.21	63	0.15	9.45
A_3	240	0.30	72	0.30	21.6
M	m_M	0.72	$m_M \times .72$	0.5	$0.36\ m_M$
A_4	260	0.36	93.6	0.60	56.16

The magnitude and angular positions of the balancing masses m_L and m_M are obtained by drawing couple polygon and force polygon as given below :

Couple Polygon. From column (*vi*) of the table, the couple polygon is drawn as shown in Fig. 13.12 (*c*). Choose any suitable scale and take any point o'. From o' draw vector $o'a' = -6.48$. The vector $o'a'$ is parallel to Om_1. The point a' is towards the left of o' as couple in plane A_1 is – ve. From a' draw $a'b' = 9.45$ and parallel to Om_2. From b' draw $b'c' = 21.6$ and parallel to Om_3. From c' draw $c'd' = 56.16$ and parallel to Om_4. The closing side $d'o'$ of the couple polygon represents the balanced couple due to the balance mass (m_M) in plane M. But the balanced couple due to mass (m_M) in plane M from column (*vi*) of the table is $0.36\ m_M$. By measurement $d'o' = 38.16$ units.

$\therefore \quad 0.36\ m_M = d'o' = 38.16$

$$\therefore \quad \text{Mass in plane } M \text{ i.e. } m_M = \frac{38.16}{0.36} = \mathbf{106\ kg.} \quad \textbf{Ans.}$$

By measurement, angle $\phi = 49°$

The angular position of the mass m_M is obtained by drawing Om_M in Fig. 13.12 (*d*), parallel to vector $d'o'$. The angular position of m_M is $\phi = 49°$ in the anticlockwise direction from m_1.

Now the force polygon can be drawn as given below :

Force-Polygon. From column (*iv*) of the table, the force polygon is drawn as shown in Fig. 13.12 (*e*). Choose any suitable scale for the data given in column (*iv*) and take any point o. From o draw vector $oa = 54$ parallel to Om_1. The point a is right to o as the force in column is + ve. From a draw vector $ab = 63$ units and parallel to Om_2. From b draw vector $bc = 72$ units and parallel to Om_3. From c draw vector $cd = 93.6$ units and parallel to Om_4. From d draw vector $de = 0.72 \times m_M = 0.72 \times 106 = 76.32$ units and parallel to Om_M. The closing side eo of the force polygon represents the balanced force due to mass (m_L) in plane L. But the balanced force due to mass (m_L) in plane L from column (*iv*) of the table is $0.72\ m_L$.

By measurement the vector $eo = 115.92$ units.

$\therefore \quad 0.72\ m_L = eo = 115.92$

$$\therefore \quad \text{Mass in plane } L \text{ i.e. } m_L = \frac{115.92}{0.72} = \mathbf{161\ kg.} \quad \textbf{Ans.}$$

By measurement, the angle $\theta = \mathbf{40°}$. **Ans.**

The angular position of the mass m_L is obtained by drawing Om_L in Fig. 13.12 (*f*), parallel to vector eo. The angular position of m_L is $(180 + \theta)$ or $(180 + 40°)$ in the anticlockwise direction from m_1.

Analytical Method

Given : $m_1 = 200$ kg ; $m_2 = 300$ kg ; $m_3 = 240$ kg ; $m_4 = 260$ kg ;

$r_1 = 0.27$ m ; $r_2 = 0.21$ m ; $r_3 = 0.3$ m ; $r_4 = 0.36$ m ;

$l_1 = 0.12$ m ; $l_2 = 0.15$ m ; $l_3 = 0.3$ m ; $l_4 = 0.6$ m ; $l_m = d = 0.5$ m ; $r_L = r_m = 0.72$ m.

Assuming the mass m_1 along x-axis. The position of planes and angular position of masses are shown in Fig. 13.12 (*a*) and 13.12 (*b*) respectively. Also the plane L is assumed as the reference plane, so that the couple due to unknown mass m_L is zero in that plane. To find :

(*i*) Magnitude of balancing masses *i.e.* m_L and m_M.

(*ii*) Angular positions of balancing masses *i.e.* θ_L and θ_M. For complete balancing, the resultant force and resultant couple both should be zero.

(*a*) *For resultant force to be zero,* we have

$$m_1r_1\omega^2 + m_2r_2\omega^2 + m_3r_3\omega^2 + m_4r_4\omega^2 + m_L \times r_L \times \omega^2 + m_M \times r_M \times \omega^2 = 0$$

or
$$m_1r_1 + m_2r_2 + m_3r_3 + m_4r_4 + m_L \times r_L + m_M \times r_M = 0$$

The above equation can be written (after resolving each term in x and y direction) as

$$m_1r_1 \cos\theta_1 + m_2r_2 \cos\theta_2 + m_3r_3 \cos\theta_3 + m_4r_4 \cos\theta_4 + m_L \times r_L \times \cos\theta_L + m_M \times r_M \times \cos\theta_M = 0 \quad ...(13.14)$$

and
$$m_1r_1 \sin\theta_1 + m_2r_2 \sin\theta_2 + m_3r_3 \sin\theta_3 + m_4r_4 \sin\theta_4 + m_L \times r_L \times \sin\theta_L + m_M \times r_M \times \sin\theta_M = 0 \quad ...(13.15)$$

(*b*) *For resultant couple to be zero,* (about reference plane L), we have

$$(m_1r_1\omega^2) \times l_1 + (m_2r_2\omega^2) \times l_2 + (m_3 \times r_3 \times \omega^3) \times l_3 + (m_4r_4 \times \omega^2) \times l_4 + (m_M \times r_M \times \omega^2)\, l_M = 0$$

or
$$m_1r_1l_1 + m_2r_2l_2 + m_3r_3l_3 + m_4r_4l_4 + m_Mr_M \times l_M = 0$$

The above equation can be written (after resolving each term in x and y direction) as

$$m_1r_1l_1 \cos\theta_1 + m_2r_2l_2 \cos\theta_2 + m_3r_3l_3 \cos\theta_3 + m_4r_4l_4 \cos\theta_4 + m_Mr_Ml_M \times \cos\theta_M = 0 \quad ...(A)$$

and
$$m_1r_1l_1 \sin\theta_1 + m_2r_2l_2 \sin\theta_2 + m_3r_3l_3 \sin\theta_3 + m_4r_4l_4 \sin\theta_4 + m_M \times r_M \times l_M \times \sin\theta_M = 0 \quad ...(B)$$

The above equations can be written as

$$m_M \times r_M \times l_M \times \cos\theta_M = -(m_1r_1l_1 \cos\theta_1 + m_2r_2l_2 \cos\theta_2 + m_3r_3l_3 \cos\theta_3 + m_4r_4l_4 \cos\theta_4)$$
$$= -(\Sigma\, mrl \cos\theta) \quad ...(i)$$

and
$$m_M \times r_M \times l_M \times \sin\theta_M = -(m_1r_1l_1 \sin\theta_1 + m_2r_2l_2 \sin\theta_2 + m_3r_3l_3 \sin\theta_3 + m_4r_4l_4 \sin\theta_4)$$
$$= -(\Sigma\, mrl \sin\theta) \quad ...(ii)$$

Squaring and adding equations (*i*) and (*ii*), we get

$$m_M \times r_M \times l_M = \sqrt{(\Sigma mrl \cos\theta)^2 + (\Sigma mrl \sin\theta)^2} \quad ...(13.16)$$

where $\Sigma\, mrl \cos\theta = m_1r_1l_1 \cos\theta_1 + \ldots\ldots + m_4r_4l_4 \cos\theta_4$

$\Sigma\, mrl \sin\theta = m_1r_1l_1 \sin\theta_1 + \ldots\ldots + m_4r_4l_4 \sin\theta_4$

In equation (13.16) the values of r_M and l_M are known and hence value of m_M can be calculated.

Dividing equation (*ii*) by (*i*), we get

$$\frac{\sin\theta_M}{\cos\theta_M} = \tan\theta_M = \frac{-(\Sigma\, mrl \sin\theta)}{-(\Sigma\, mrl \cos\theta)} \quad ...(13.17)$$

The quadrant in which θ_M lies depends upon the signs of numerator and denominator of the above equation. Hence the value of θ_M is obtained from the above equation.

Now the values of m_M and θ_M are known.

The equations (13.14) and (13.15) are written as

$$m_L \times r_L \times \cos\theta_L = -[m_1 r_1 \cos\theta_1 + m_2 r_2 \cos\theta_2 + m_3 r_3 \cos\theta_3 + m_4 r_4 \cos\theta_4 + m_M \times r_M \times \cos\theta_M]$$

$$= -[\Sigma mr \cos\theta + m_M \times r_M \times \cos\theta_M] \quad ...(iii)$$

and $$m_L \times r_L \times \sin\theta_L = -[m_1 r_1 \sin\theta_1 + m_2 r_2 \sin\theta_2 + m_3 r_3 \sin\theta_3 + m_4 r_4 \sin\theta_4 + m_M \times r_M \times \sin\theta_M]$$

$$= -[\Sigma\, mr \sin\theta + m_M \times r_M \times \sin\theta_M] \quad ...(iv)$$

Squaring and adding equations (*iii*) and (*iv*), we get

$$m_L \times r_L = \sqrt{[\Sigma mr \cos\theta + m_M \times r_M \times \cos\theta_M]^2 + [\Sigma mr \sin\theta + m_M \times r_M \times \sin\theta_M]^2} \quad ...(13.18)$$

The R.H.S. of the above equation is known completely, also the value of r_L is given, hence value of m_L is calculated.

To find the value of θ_L, divide equation (*iv*) by equation (*iii*), we get

$$\frac{\sin\theta_L}{\cos\theta_L} = \tan\theta_L = \frac{-[\Sigma mr \sin\theta + m_M \times r_M \times \sin\theta_M]}{-[\Sigma mr \cos\theta + m_M \times r_M \times \cos\theta_M]} \quad ...(13.19)$$

The quadrant in which θ_L lies depends upon the signs of numerator and denominator of the above equation.

Values of m_M and θ_M

From Fig. 13.12 (*b*), we have $\theta_1 = 0°$; $\theta_2 = 45°$; $\theta_3 = 45 + 75 = 120°$; $\theta_4 = 45 + 75 + 135 = 255°$

Using equation (13.16),

$$m_M \times r_M \times l_M = \sqrt{(\Sigma mrl \cos\theta)^2 + (\Sigma mrl \sin\theta)^2} \quad ...(i)$$

where $\Sigma mrl \cos\theta = m_1 r_1 l_1 \cos\theta_1 + m_2 r_2 l_2 \cos\theta_2 + m_3 r_3 l_3 \cos\theta_3 + m_4 r_4 l_4 \times \cos\theta_4$

$= 200 \times 0.27 \times -0.12 \cos 0° + 300 \times 0.21 \times 0.15 \times \cos 45° + 240 \times 0.3 \times 0.3 \times \cos 120° + 260 \times 0.36 \times 0.6 \times \cos 255°$

(As plane A_1 is to the left of reference plane L and hence distance l_1 is – ve)

$= -6.48 + 6.682 + (-10.8) + (-14.53) = -25.128$

and $\Sigma mrl \sin\theta = m_1 r_1 l_1 \sin\theta_1 + m_2 r_2 l_2 \sin\theta_2 + m_3 r_3 l_3 \sin\theta_3 + m_4 r_4 l_4 \sin\theta_4$

$= 200 \times 0.27 \times (-0.12) \sin 0° + 300 \times 0.21 \times 0.15 \sin 45° + 240 \times 0.3 \times 0.3 \times \sin 120° + 260 \times .36 \times .6 \times \sin 255°$

$= 0 + 6.682 + 18.7 - 54.24 = -28.86$

Now from equation (*i*), we get

$$m_m \times r_m \times l_m = \sqrt{(-25.128)^2 + (-28.86)^2} = \sqrt{631.4 + 832.8} = 38.26$$

But $r_m = 0.72m$ and $l_m = d = 0.5m$

$\therefore \quad m_m \times 0.72 \times 0.5 = 38.26$

or $$m_m = \frac{38.26}{0.72 \times 0.5} = \mathbf{106.2\ kg.} \quad \textbf{Ans.}$$

The angle θ_m is obtained by using equation (13.17) as

$$\tan\theta_m = \frac{-(\Sigma\, mrl \sin\theta)}{-(\Sigma\, mrl \cos\theta)} = \frac{-(-28.86)}{-(-25.128)} = \frac{28.86}{25.128}$$

As the numerator and denominator of $\tan\theta_m$ (*i.e.* of above equation), are + ve, hence angle θ_m lies in first quadrant. [$\because$ numerator corresponds to $\sin\theta_m$ and denominator corresponds to $\cos\theta_m$]

$$\therefore \qquad \theta_m = \tan^{-1}\frac{28.128}{25.128} = \tan^{-1} 1.1485 = \mathbf{48°.95 \simeq 49°}. \quad \textbf{Ans.}$$

(iii) **Value of m_L and θ_L**

Using equation (13.18),

$$m_L \times r_L = \sqrt{[\Sigma\, m \times r \times \cos\theta + m_M \times r_M \times \cos\theta_M]^2 + [\Sigma\, m \times r \times \sin\theta + m_M \times r_M \times \sin\theta_M]^2} \quad ...(ii)$$

where $\Sigma\, m \times r \times \cos\theta = m_1 \times r_1 \times \cos\theta_1 + m_2 r_2 \times \cos\theta_2 + m_3 r_3 \cos\theta_3 + m_4 r_4 \times \cos\theta_4$

$= 200 \times 0.27 \times \cos 0° + 300 \times 0.21 \times \cos 45° + 240 \times 0.3 \times \cos 120° + 260 \times 0.36 \times \cos 255°$

$= 54 + 44.54 + (-36.00) + (-24.22) = 38.32$

$\Sigma\, m \times r \times \sin\theta = m_1 r_1 \sin\theta_1 + m_2 r_2 \sin\theta_2 + m_3 r_3 \sin\theta_3 + m_4 r_4 \sin\theta_4$

$= 200 \times 0.27 \times \sin 0° + 300 \times 0.21 \times \sin 45° + 240 \times 0.3 \times \sin 120° + 260 \times 0.36 \times \sin 255°$

$= 0 + 44.54 + 62.35 - 90.4 = 16.94$

$m_M \times r_M \times \cos\theta_M = 106.2 \times 0.72 \times \cos 49° = 50.16$ and

$m_M \times r_M \times \sin\theta_M = 106.2 \times 0.72 \times \sin 49° = 57.7$

Now the equation (*ii*) becomes as

$$m_L \times r_L = \sqrt{(38.32+50.16)^2 + (16.49 + 57.7)^2}$$

$$= \sqrt{7828.7 + 5504.1} = 115.46$$

$$\therefore \qquad m_L = \frac{115.46}{0.72} \qquad (\because\ r_L = 0.72)$$

$$= \mathbf{160.37\ kg}. \quad \textbf{Ans.}$$

Now using equation (13.19), we get

$$\tan\theta_L = \frac{-[\Sigma\, mr \sin\theta + m_M \times r_M \times \sin\theta_M]}{-[\Sigma\, mr \cos\theta + m_M \times r_M \times \cos\theta_M]}$$

$$= \frac{-[16.49 + 57.7]}{-[38.32 + 50.16]} = \frac{-74.19}{-88.48}$$

In the above equation the numerator (*i.e.* $\sin\theta_L$) is –ve and also the denominator (*i.e.* $\cos\theta_L$) is negative, hence θ_L lies in third quadrant as $\sin\theta_L$ and $\cos\theta_L$ are – ve in third quadrant.

$$\therefore \quad \theta_L = \tan^{-1}\left(\frac{-74.19}{-88.48}\right) = \tan^{-1} 0.8385 = 39.97 + 180 = 219.97 \simeq \mathbf{220°}. \quad \textbf{Ans.}$$

Problem 13.5. *A shaft carries four rotating masses A, B, C and D in this order along its axis. The mass A may be assumed to be concentrated at radius of 18 cm, B of 24 cm, C of 12 cm and D of 15 cm. The masses of B, C and D are 30 kg, 50 kg and 40 kg respectively. The planes containing B and C are 30 cm apart. The angular spacing of the planes containing C and D are 90° and 210° respectively relative to B measured in the same plane. If the shaft and masses are to be in complete dynamic balance, find :*

(i) the mass and the angular position of mass A

(ii) the position of planes A and D. (AMIE, Summer 1976)

Sol. Given :

Masses, $m_B = 30$ kg ; $m_C = 50$ kg ; $m_D = 40$ kg, $m_A = ?$

Radii, $r_A = 18$ cm ; $r_B = 24$ cm ; $r_C = 12$ cm and $r_D = 15$ cm

Angular positions of planes, $\theta_C = 90°$; $\theta_D = 210°$ relative to *B*.

Distance between planes *B* and *C* = 30 cm.

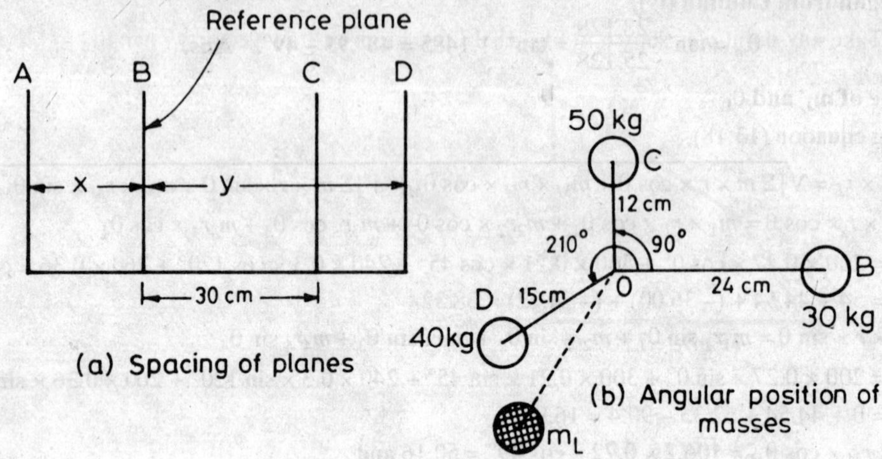

Fig. 13.13. (*a*) and (*b*)

The shaft and masses are in complete balance. Find :

(*i*) m_A and angular position of mass *A*

(*ii*) Position of planes *A* and *D*.

Assuming plane *B* as reference plane and mass *B* (m_B) along the horizontal line.

The angular position of masses *B*, *C* and *D* are shown in Fig 13.13 (*b*). The spacing of planes are shown in Fig. 13.13 (*a*). The angular spacing of mass *C* and *D* with respect to *B* are known. Let us find the angular position of mass *A* also w.r.t. to the angular position of mass *B*. Hence take plane *B* as the reference plane. Let mass *B* is along *x*-axis as shown in Fig. 13.13 (*b*).

Now prepare the table as given below :

Table 13.5

Planes	*Mass (m) kg*	*Radius (r) cm*	*Force ÷ ω^2 (m × r) kg-cm*	*Distance of planes from (R.P.) B (l) cm*	*Balance couple ÷ ω^2 (m × r × l) kg-cm^2*
(*i*)	(*ii*)	(*iii*)	(*iv*)	(*v*)	(*vi*)
A	m_A	18	$18m_A$	$-x$	$-18x \times m_A$
B (R.P.)	30	24	720	0	0
C	50	12	600	30	18,000
D	40	15	600	y	$600y$

Column (*iv*) of the above table represents centrifugal forces whereas the column (*vi*) represents the couples. In column (*iv*) only m_A is unknown whereas in column (*vi*) *x*, *y* and m_A are unknown.

From the values written in column (*iv*) in the above table, the force polygon can be drawn first. Hence it is drawn to a convenient scale as shown in Fig. 13.13 (*c*). As the system is in complete balance, hence force polygon must close. The procedure of drawing force-polygon is given below :

Force-Polygon from Column (iv)

(*i*) Take any point *o* as shown in Fig. 13.13 (*c*). From *o*, draw vector *ob* parallel to *OB* and equal to 720 [*i.e.* $(F_C)_B = 720$].

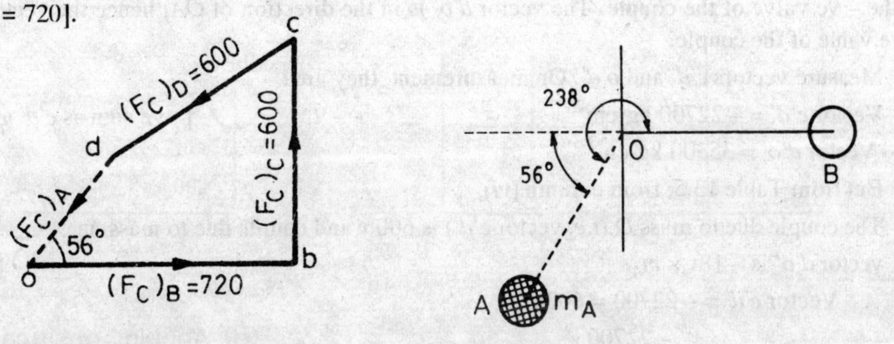

(c) Force polygon (d) Angular position of mass A

Fig. 13.13. (*c*) and (*d*)

(*ii*) From *b*, draw vector *bc* parallel to *OC* and equal to 600 to represent the centrifugal force at mass *C* *i.e.* $(F_C)_C = 600$.

(*iii*) From point *c*, draw vector *cd* parallel to *OD* and equal to 600 to represent the centrifugal force at mass *D* *i.e.* $(F_C)_D = 600$.

(*iv*) Join *d* to *O*. Then the vector *dO* represents the centrifugal force at mass *A*. Measure *dO* which is = 360.

But from column (*iv*) of the table, the centrifugal force at mass *A* is = 18 m_A.

$$\therefore \qquad 18\, m_A = 360$$

or $$m_A = \frac{360}{18} = \mathbf{20\ kg.} \quad \textbf{Ans.}$$

In order to obtain the angular position of mass *A*, a vector line parallel to '*dO*' from Fig. 13.13 (*c*) is drawn in Fig. 13.13 (*d*). The mass *A* makes 26° with mass *D* in the anti-clockwise direction.

or $$\angle BOA = 210 + 26 = 236°.$$

(*ii*) **Position of planes A and D**

Let x = Distance of plane *A* from plane *B* and

y = Distance of plane *D* from plane *B*.

Then $l_A = -x$ and $l_D = y$.

Now draw the **couple polygon** as explained below :

The values of the couple-polygon are taken from column (*vi*). Take any point *o'*. From *o'* draw a vector *o'c'* parallel to mass *C* or parallel to *OC* [Refer to Fig. 13.13 (*b*)] and equal to 18000 kg-cm² to a convenient scale. From the two ends of this line, draw two lines *c'd'* and *o'd'* parallel to masses *A* and *D* both meeting and closing the couple polygon as shown in Fig. 13.13 (*e*).

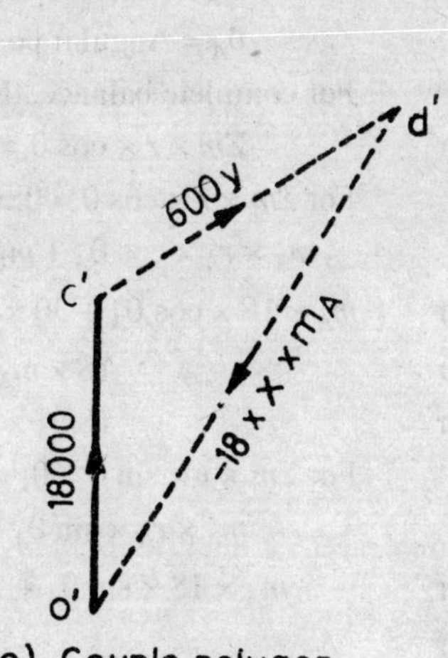

(e) Couple polygon

Fig. 13.13 (*e*)

The direction of vector *o'c'* is vertically upwards. The shaft and masses are in complete balance. Hence the couple polygon must be a closed figure. For closed polygon the direction of vector *c'd'* should be from *c'* to *d'*. And the direction of vector *d'o'* should be from *d'* to *o'*.

The vector $c'd'$ is parallel to mass D (*i.e.* line OD), whereas the vector $d'o'$ is parallel to the direction of mass A *i.e.* line OA. Also the vector $c'd'$ is opposite to the line OD (*i.e.* line of mass D). Hence vector $c'd'$ gives the – ve valve of the couple. The vector $d'o'$ is in the direction of OA, hence the vector $d'o'$ gives the positive value of the couple.

Measure vectors $c'd'$ and $o'd'$. On measurement, they are :

Vector $c'd' = -22700$ kg-cm^2 [– ve sign as $c'd'$ gives – ve value]

and Vector $d'o' = 35300$ kg-cm^2

But from Table 13.5, from column (*vi*),

The couple due to mass D (*i.e.* vector $c'd'$) is $600\,y$ and couple due to mass A

i.e. vector $d'o'$ is $-18x \times m_A$.

$\therefore$ Vector $c'd' = -22700 = 600y$

or $$y = \frac{-22700}{600} = -37.83 \text{ cm} \simeq -37.8 \text{ cm}$$

Also vector $d'o' = 35300 = -18x \times m_A$

or $$x = \frac{35300}{-18 \times m_A} = \frac{35300}{-18 \times 20} = -98.05 \text{ cm} \simeq -98 \text{ cm}$$

Now the distances of planes A and D from plane B are

$$l_A = -x = -(-98.05) = 98.05 \text{ cm} \simeq 98 \text{ cm}$$

and $$l_D = y = -37.83 \text{ cm} \simeq -37.8 \text{ cm}$$

The distance of plane A (*i.e.* $l_A = 98$) is + ve, hence plane A lies towards right of plane B at a distance of 98 cm.

The distance of plane D (*i.e.* $l_D = -37.8$) is – ve, hence plane D lies towards left of plane B at a distance of 37.8 cm from plane B. The correct positions of planes are shown in Fig. 13.13 (*f*) which is given on page 682.

Analytical Method

Fig. 13.13 (*a*) shows the position of planes A, B, C and D. The Fig. 13.13 (*b*) shows the angular position of masses m_B, m_C and m_D in which mass m_B is assumed along x-axis.

Let m_A = mass of A and

θ_A = Angular position of mass A with respect to angular position of mass B.

For complete balance, the resultant force should be zero.

or $$\Sigma m \times r \times \cos\theta = 0 \text{ and } \Sigma m \times r \times \sin\theta = 0$$

For $\Sigma m \times r \times \cos\theta = 0$, we have

$$m_A \times r_A \times \cos\theta_A + m_B \times r_B \times \cos\theta_B + m_C \times r_C \times \cos\theta_C + m_D \times r_D \times \cos\theta_D = 0$$

or $$m_A \times 18 \times \cos\theta_A + 30 \times 24 \times \cos 0° + 50 \times 12 \times \cos 90° + 40 \times 15 \times \cos 210° = 0$$

or $$18 \times m_A \times \cos\theta_A + 720 + 0 + (-519.6) = 0$$

or $$18 m_A \times \cos\theta_A = -720 + 519.6 = -200.4 \quad \text{...(i)}$$

For $\Sigma m \times r \times \sin\theta = 0$, we have

$$m_A \times r_A \times \sin\theta_A + m_B \times r_B \times \sin\theta_B + m_C \times r_C \times \sin\theta_C + m_D \times r_D \times \sin\theta_D = 0$$

or $$m_A \times 18 \times \sin\theta_A + 30 \times 24 \times \sin 0° + 50 \times 12 \times \sin 90° + 40 \times 15 \times \sin 210° = 0$$

or $$18 m_A \times \sin\theta_A + 0 + 600 + (-300) = 0$$

or $$18 m_A \times \sin\theta_A = -600 + 300 = -300 \quad \text{...(ii)}$$

Squaring and adding (*i*) and (*ii*), we get

$$324 \times m_A^2 = (-200.4)^2 + (-300)^2 = 40160 + 90000 = 130160.16$$

or $$m_A^2 = \frac{130160.16}{324} = 401.728$$

or $$m_A = \sqrt{401.728} = \mathbf{20.04\ kg.}\quad \textbf{Ans.}$$

Dividing (*ii*) by (*i*), we get

$$\frac{18\, m_A \times \sin\theta_A}{18\, m_A \times \cos\theta_A} = \frac{-300}{-200.4}$$

or $$\frac{\sin\theta_A}{\cos\theta_A} = \tan\theta_A = \frac{-300}{-200.4}$$

As the numerator (*i.e.* $\sin\theta_A$) and denominator (*i.e.* $\cos\theta_A$) of the above equation are – ve, hence θ_A lies in 3rd quadrant. [∵ In third quadrant the sine and cosine of an angle are – ve]

∴ $$\tan\theta_A = 1.497$$

or $$\theta_A = 180 + 56.25 = \mathbf{236.25°}.\quad \textbf{Ans.}$$

For complete balance, the resultant couple should be zero

or $$\Sigma m \times r \times l \times \cos\theta = 0 \text{ and } \Sigma m \times r \times l \times \sin\theta = 0$$

Take B as the reference plane

Then $l_A = -x$, $l_B = 0$, $l_C = 30$ and $l_D = y$.

For $\Sigma m \times r \times l \times \cos\theta = 0$ about plane B, we have

$$m_A \times r_A \times l_A \times \cos\theta_A + m_B \times r_B \times l_B \cos\theta_B + m_C \times r_C \times l_C \times \cos\theta_C + m_D \times r_D \times l_D \times \cos\theta_D = 0$$

or $$20.04 \times 18 \times -x \times \cos 236.25° + 30 \times 24 \times 0 \times \cos 0° + 50 \times 12 \times 30$$
$$\times \cos 90° + 40 \times 15 \times y \times \cos 210° = 0$$

or $$200x + 0 + 0 + (-519.6y) = 0$$

or $$200x - 519.6y = 0$$

or $$200x = 519.6y$$

or $$x = \frac{519.6y}{200} = 2.593y \qquad ...(iii)$$

For $\Sigma m \times r \times l \times \sin\theta = 0$, we have

$$m_A \times r_A \times l_A \times \sin\theta_A + m_B \times r_B \times l_B \times \sin\theta_B + m_C \times r_C \times l_C \times \sin\theta_C + m_D \times r_D \times l_D \times \sin\theta_D = 0$$

or $$20.04 \times 18 \times (-x) \times \sin 236.25° + 30 \times 24 \times 0 \times \sin 0° + 50 \times 12$$
$$\times 30 \times \sin 90° + 40 \times 15 \times y \times \sin 210° = 0$$

or $$299.33x + 0 + 18000 - 300y = 0$$

or $$299.33x + 18000 - 300y = 0$$

or $$299.33x = -18000 + 300y \qquad ...(iv)$$

or $$299.33 \times 2.593y = -18000 + 300y \qquad [\because \text{ From } (iii),\ x = 2.593y]$$

or $$776.16y = -18000 + 300y$$

or $$776.16y - 300y = -18000$$

or $$476.16y = -18000$$

or $$y = \frac{-18000}{476.16} = -37.8 \text{ cm}$$

Substituting the value of y in equation (*iii*), we get

$$x = 2.593 \times (-37.8 \text{ cm}) = -98.02 \simeq -98 \text{ cm}$$

The distances of planes A, C and D from plane B are : $l_A = -x = -(-98) = 98$ cm, $l_C = 30$ and $l_D = y = -37.8$. As the distance of plane A is + ve, hence plane A lies towards right of plane B at a distance of 98 cm. Also as the distance of plane D is – ve, hence plane D lies towards left of plane B at a distance of 37.8 cm from plane B. The correct position of all planes are shown in Fig. 13.13 (*f*).

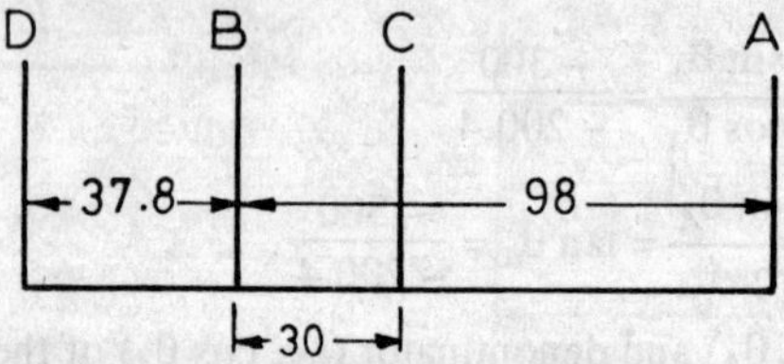

(f) Actual spacing of planes

Fig. 13.13 (*f*)

Problem 13.6. *Four masses are attached to a shaft at planes A, B, C and D at equal radii. The distance of the planes B, C and D from A are 40 cm, 50 cm and 120 cm respectively. The masses at A, B and C are 60 kg, 45 kg and 70 kg respectively. If the system is in complete balance, determine the mass at D and the position of masses B, C and D with respect to A.* (AMIE, Winter 1986)

Sol. Given :

Masses, $m_A = 60$ kg ; $m_B = 45$ kg ; $m_C = 70$ kg

Distances of planes from A *i.e.* distance between planes A and $B = 40$ cm ; between planes A and $C = 50$ cm and between planes A and $D = 120$ cm,

Radii of masses are equal.

∴ Radius of masses, $r_A = r_B = r_C = r_D = 1$ cm (assumed)

The system is in complete balance. Find :

(*i*) mass at D *i.e.* m_D,

(*ii*) position of masses B, C and D with respect to A.

First draw Fig. 13.14 (*a*), showing planes A, B, C and D at the distances as given in the problems. The radius has been assumed as 1 cm.

For angular position of masses assume mass m_A along x-axis as shown in Fig. 13.14 (*c*).

Let α = Angle made by mass m_B with mass m_A

β = Angle made by mass m_C with mass m_A.

Take the plane D as reference plane (R.P.) so as to eliminate the couple due to mass in that plane. Prepare the following table on the basis of the given data :

Table 13.6

Planes	*Mass (m)* *kg*	*Radius (r)* *cm*	*Force* ÷ ω^2 = $(m \times r)$ *kg cm*	*Distance from R.P. (D)* *(l cm)*	*Balancing couple* ÷ ω^2 = $(m \times r \times l)$ *kg cm*^{2}
A	60	1	60	– 120	– 7200
B	45	1	45	– 80	– 3600
C	70	1	70	– 70	– 4900
D (R.P.)	m_D	1	m_D	0	0

As the system is in complete balance, hence couple polygon must close. Hence draw the couple polygon with the data given in column 6th of the table as shown in Fig. 13.14 (*b*). Assume m_A = 60 kg at *A* along *x*-axis.

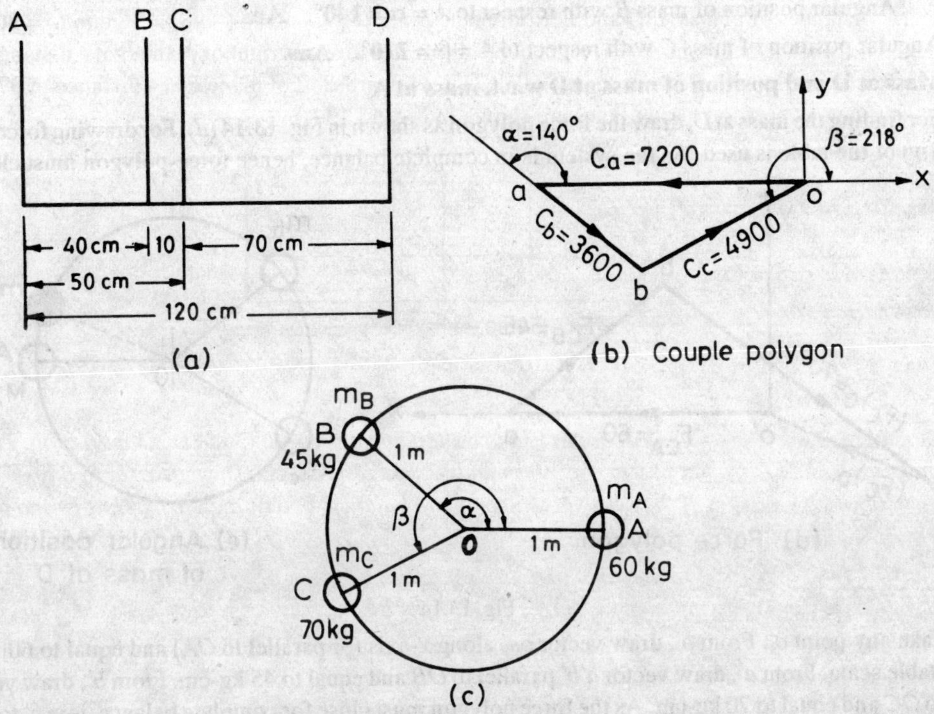

Fig. 13.14

Couple Polygon

The magnitude of couples are given in column (*vi*) of the table 13.6. Only three couples are acting hence couple polygon will be a triangle in this case. Take any point *o*. From *o*, draw a vector *oa* = – 7200 kg-cm^2. As couple due to m_A is – ve, hence point *a* should be in the opposite direction of *OA* (*i.e.* point *a* is towards the left of point *o*). Now with *a* as centre and radius equal to 3600 kg-cm^2, draw an arc downward. With *o* as centre and radius equal to 4900 kg-cm^2, draw another arc downward which cuts the previous arc at '*b*'. Join *a* to *b* and *o* to *b*. Then triangle *oab* is a couple polygon.

For complete balance, the couple polygon (here triangle) should be a closed figure. In the couple polygon, the direction of vector *oa* is from *o* to *a*. Hence the direction of vector *ab* should be from *a* to *b*. Also the direction of vector *bo* should be from *b* to *o*.

Also the vector *ab* shows the couple due to mass m_B so it should be parallel to line showing the angular position of mass m_B whereas the vector *bo* shows the couple to mass m_C so it should be parallel to the line showing the angular position of mass m_C.

From column (*vi*), the couples due to mass m_B and mass m_C are – ve, hence vector *ab* is in opposite direction to the line showing angular position of m_B. Similarly the vector *bo* is in opposite direction to the line showing angular position of m_C.

Hence the angular position of mass m_B is along vector *ba* and angular position of mass m_C is along vector *ob*.

Positions of masses at B and C with respect to mass at A

Refer to Fig. 13.14 (*c*). Draw a circle with *O* as centre and radius equal to 1 cm. Position of mass at *A* is shown along *x*-axis at a radius 1 cm. Circle has been drawn as radii are equal. Draw *OB* parallel to vector *ba*. Also draw *OC* parallel to vector *ob*. Thus the position of masses at *B* and *C* are fixed as shown in Fig. 13.14 (*c*). Measure angle *AOB* and *AOC* which are 140° and 218° respectively.

∴ Angular position of mass *B* with respect to $A = \alpha =$ **140°**. **Ans.**

Angular position of mass *C* with respect to $A = \beta =$ **210°**. **Ans.**

Mass at D and position of mass at D w.r.t. mass at A

For finding the mass at *D*, draw the force polygon as shown in Fig. 13.14 (*d*). For drawing force-polygon column (*iv*) of the table is used. As the system is in complete balance, hence force-polygon must close.

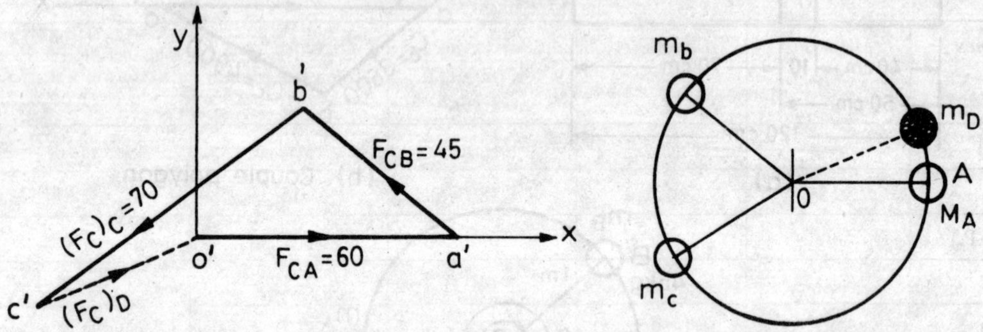

(d) Force polygon

(e) Angular position of mass at D

Fig. 13.14

Take any point *o'*. From *o'*, draw vector *oa'* along *x*-axis (or parallel to *OA*) and equal to 60 kg-cm to some suitable scale. From *a'*, draw vector *a'b'* parallel to *OB* and equal to 45 kg-cm. From *b'*, draw vector *b'c'* parallel to *OC* and equal to 70 kg-cm. As the force polygon must close for complete balance, join *c'* to *o'*. Then vector *c'o'* represents the product of mass and radius of mass *D*. Measure *c'o'*, which is equal to 36 kg-cm.

∴ $m_D \times r_D = 36$ kg-cm

But $r_D = 1$ cm

∴ $m_D = 36$ kg.

∴ Mass at $D =$ **36 kg. Ans.**

The angular position of mass at *D* is obtained by drawing a line from *O*, parallel to *c'o'* as shown in Fig. 13.14 (*c*). Measure the angle *AOD* which is equal to 7°.

∴ Angular position of mass *D* with respect to mass at $A = 7°$. **Ans.**

Problem 13.7. *A shaft carries four rotating masses A, B, C and D in this order as shown in Fig. 13.15 along its axis. The mass A may be assumed concentrated at a radius of 12 cm, B at 15 cm, C at 14 cm and D at 18 cm. The masses A, C and D are 15 kg, 10 kg and 8 kg respectively. The planes of rotation of A and B are 15 cm apart and of B and C are 18 cm apart. The angle between the radii of A and C is 90°. If the shaft is in complete dynamic balance, determine :*

(*i*) *the angles between the radii of A, B and D*

(*ii*) *the distance between the planes of revolution of C and D, and*

(*iii*) *the mass B.* (AMIE, Winter 1994)

Sol. Given :

Radius of masses *i.e.* $r_A = 12$ cm ; $r_B = 15$ cm ; $r_C = 14$ cm and $r_D = 18$ cm

Mass *i.e.* $m_A = 15$ kg ; $m_C = 10$ kg and $m_D = 8$ kg

Planes of rotation of *A* and $B = 15$ cm apart ; of *B* and $C = 18$ cm apart.

Angles between radii of A and C *i.e.* between r_A and $r_C = 90°$

Shaft is in complete dynamic balance and hence the force and couple polygon must close.

Let m_B = Mass of B and l_d = Distance of plane D from reference plane B.

Let us consider the plane B as the reference plane (*i.e.* R.P.) so as to eliminate the couple due to unknown mass in that plane.

Now the table will be constructed as given below :

Table 13.7

Planes	*Mass (m) kg*	*Radius (r) cm*	*Force ÷ ω² = (m × r) kg cm*	*Distance from reference plane (l)*	*Couple ÷ ω² = (m × r × L)*
(*i*)	(*ii*)	(*iii*)	(*iv*)	(*v*)	(*vi*)
A	15	12	180	15	2700
B (R.P.)	m_B	15	$15\ m_B$	0	0
C	10	14	140	– 18	– 2520
D	8	18	144	$-\ l_d$	$-\ 144\ L_d$

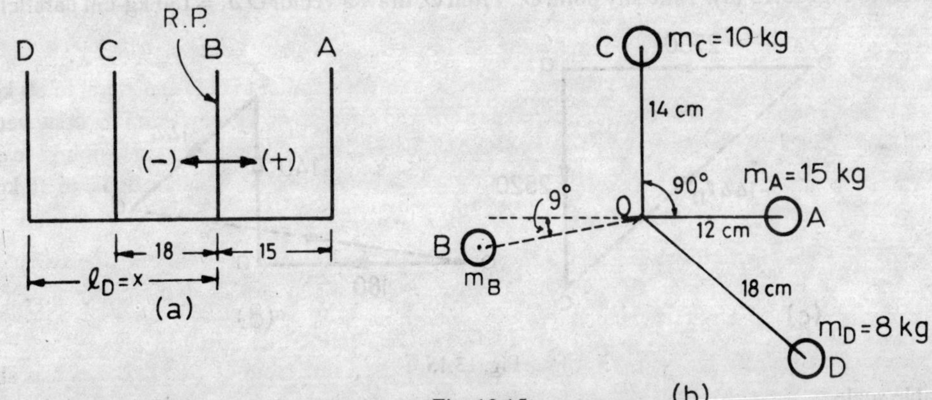

Fig. 13.15

As the system is in complete balance, hence couple polygon must close. Hence draw a couple polygon with the date given in column (*vi*) of the table 13.7. Assume the mass m_A along x-axis as shown in Fig. 13.15 (*b*). The angles between radii of A and C *i.e.* between r_A and r_C is given as 90°. Hence position of mass m_C will be at 90° with respect to m_A as shown in Fig. 13.15 (*b*).

Couple Polygon

From column (*vi*) of the table 13.7, we see that only three couples are acting and system is in complete balance. Hence couple polygon will be a triangle in this case. The couple due to mass m_A is + ve whereas the couple due to mass m_C is – ve. Hence couple vector due to mass m_A will be in the direction of OA whereas the couple vector due to mass m_C will be in the opposite direction of OC.

To draw the couple polygon, take any point o as shown in Fig. 13.15 (*c*). From o, draw vector oa parallel to OA and equal to 2700 kg-cm² to a suitable scale. From a, draw vector ac parallel to OC and in the opposite direction of OC (*i.e.* in the downward direction) and equal to 2520 kg-cm² to the same scale. Now join point c to o. The closing side co represents the couple due to mass m_D in magnitude and direction.

Measure *co*.

As the couple vector due to mass m_D is – ve, hence angular position of mass m_D will be in the direction of *oc* and not in the direction of *co*.

On measurement, the vector

$$oc = -3700 \text{ kg-cm}^2$$

But from column (*vi*), the couple due to mass m_D is $-144\ l_d$. As the couple due to mass m_D is – ve, hence the angular position of mass m_D will be in opposite direction of couple vector *CO*. Hence the position of mass m_D is in the direction of vector *OC*. Hence from *O* draw *OD* || to *OC*.

$$\therefore \quad -144\ l_d = -3700$$

or $$l_d = \frac{-3700}{-144} = +\mathbf{25.69\ cm.} \quad \textbf{Ans.}$$

The distance of plane *D* from plane $B = -l_D = -(+25.69) = -25.69$

As the distance of plane *D* from plane *B* is – ve, hence plane *D* lies towards left of plane *B*.

The angular position of mass m_D is obtained by drawing a line *OD* parallel to vector *OC* [of Fig. 13.15 (*c*)] in Fig. 13.15 (*b*).

To find the mass m_B, the force polygon is drawn using column (*iv*) of the table 13.7.

Force-Polygon

[Refer to Fig. 13.15 (*d*). Take any point *O′*. From *O*, draw a vector *O′a′* = 180 kg-cm parallel to *OA* to

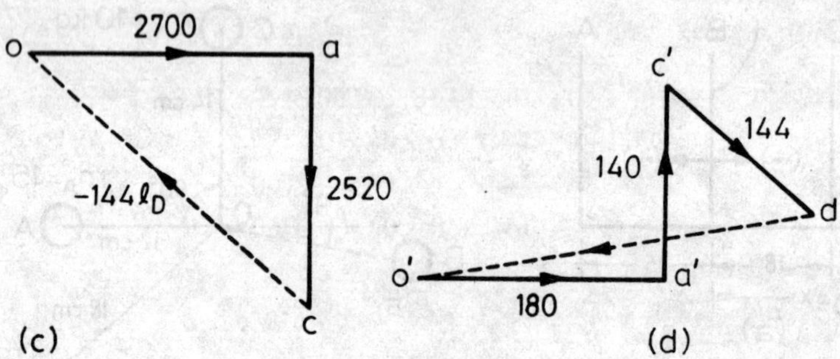

Fig. 13.15

some suitable scale.

From point *a′*, draw vector *a′C′* = 140 kg-cm parallel to *OC*. From *C′*, draw vector *C′d′* = 144 and parallel to *OD*. Join point *d′* to *O′*. The closing side *d′O′* represents the force in magnitude and direction due to mass m_B. Measure *d′O′*. On measurement, vector *d′O′* = 290 kg-cm.

But from column (*iv*), corresponding to mass m_B, the force is 15 m_B

$$\therefore \quad 15\ m_B = 270$$

$$\therefore \quad m_B = \frac{270}{15} = 19.33 \text{ kg}$$

$\therefore$ Mass m_B = **19.33 kg. Ans.**

The angular position of mass m_B is obtained by drawing a line *OB* in Fig. 13.15 (*b*) parallel to *d′O′* from Fig. 13.15 (*d*).

From Fig. 13.15 (*b*),

(*i*) Angle between radii *A* and *B* = 180 + 9° = 189° Anticlock w.r.t. to *A*.

Angle between radii *A* and *D* = 43° clockwise w.r.t. to *A*.

From Fig. 13.15 (*a*).

(*ii*) The distance between planes *C* and *D* = 25.69 – 18 = 7.69 cm.

The mass m_B = **19.33 kg. Ans.**

Analytical Method

Consider plane *B* as the reference plane so as to make the couple zero due to unknown mass m_B in that plane. Then, l_A = 15 cm ; l_C = – 18 cm and l_B = 0.

Let m_B = Mass of *B* and

x = Distance of plane *D* from plane *B*

Then $l_D = -x$.

For locating angular positions, assume mass m_A along *x*-axis. Then $\theta_A = 0°$ and $\theta_C = 90°$ anticlockwise to m_A.

Let θ_B = Angle made by m_B with *x*-axis anticlockwise

θ_D = Angle made by mass m_D with mass m_A anticlockwise

For complete balance, the resultant couple about plane *B* should be zero.

or $\Sigma mrl \cos\theta = 0$ and $\Sigma mrl \sin\theta = 0$ about plane *B*

For $\Sigma mrl \cos\theta = 0$ about plane *B*, we have

$$m_A \times r_A \times l_A \times \cos\theta_A + m_C \times r_C \times l_C \times \cos\theta_C + m_D \times r_D \times l_D \times \cos\theta_D = 0$$

$$(\because\ l_B = 0,\ \therefore\ m_B \times r_B \times l_B \times \cos\theta_B \approx 0)$$

or $$15 \times 12 \times 15 \times \cos 0° + 10 \times 14 \times (-18) \cos 90° + 8 \times 18 \times (-x) \times \cos\theta_D = 0 \quad (\because\ l_D = -x)$$

or $$2700 + 0 - 144x \times \cos\theta_D = 0$$

or $$-144x \times \cos\theta_D = -2700$$

or $$144x \times \cos\theta_D = +2700 \quad \ldots(i)$$

For $\Sigma mrl \sin\theta = 0$ about plane *B*, we have

$$m_A \times r_A \times l_A \times \sin\theta_A + m_C \times r_C \times l_C \times \sin\theta_C + m_D \times r_D \times l_D \times \sin\theta_D = 0$$

or $$15 \times 12 \times 15 \times \sin 0° + 10 \times 14 \times (-18) \sin 90° + 8 \times 18 \times (-x) \times \sin\theta_D = 0 \quad (\because\ l_D = -x)$$

or $$-2520 - 144x \sin\theta_D = 0$$

or $$-144x \sin\theta_D = 2520$$

or $$144x \sin\theta_D = -2520 \quad \ldots(ii)$$

Dividing equation (*ii*) by (*i*), we get

$$\frac{144x \sin\theta_D}{144x \cos\theta_D} = \frac{\sin\theta_D}{\cos\theta_D} = \tan\theta_D = \frac{-2520}{+2700}$$

In the above equation, numerator is – ve (*i.e.* $\sin\theta_D$ is – ve) whereas the denominator is + ve (*i.e.* $\cos\theta_D$ is + ve). Hence θ lies in 4th quadrant.

$$\therefore\quad \tan\theta_D = \frac{-2520}{+2700} = -0.9333$$

$$\therefore\quad \theta_D = \tan^{-1}(-0.9333) = -43° = 360 - 43 = \mathbf{317°}.\ \textbf{Ans.}$$

Substituting the value of θ_D in equation (*ii*), we get

$$144x \times \sin 317° = -2520$$

or $$x = \frac{-2520}{144 \times \sin 317°} = 25.66 \text{ cm}$$

Distance of plane D from plane $B = -x = -25.66$.

As l_D is – ve, hence plane D lies towards left of reference plane B at a distance of 25.66 cm from plane B as shown in Fig. 13.15 (*a*).

∴ Distance between plane C and $D = 25.66 - 18 =$ **7.66. Ans.**

For complete balance, the resultant force should be zero

or $\Sigma m \times r \cos\theta = 0$ and $\Sigma m \times r \times \sin\theta = 0$

For $\Sigma m \times r \times \cos\theta = 0$, we have

$$m_A \times r_A \times \cos\theta_A + m_B \times r_B \times \cos\theta_B + m_C \times r_C \times \cos\theta_C + m_D \times r_D \times \cos\theta_D = 0$$

or $$15 \times 12 \times \cos 0° + m_B \times 15 \times \cos\theta_B + 10 \times 14 \times \cos 90° + 8 \times 18 \times \cos 317° = 0$$

or $$180 + 15\, m_B \times \cos\theta_B + 0 + (+105.3) = 0$$

or $$15\, m_B \times \cos\theta_B = -105.3 - 180 = -285.3 \quad ...(iii)$$

For $\Sigma m \times r \times \sin\theta = 0$, we have

$$m_A \times r_A \times \sin\theta_A + m_B \times r_B \times \sin\theta_B + m_C \times r_C \times \sin\theta_C + m_D \times r_D \times \sin\theta_D = 0$$

$$15 \times 12 \times \sin 0° + m_B \times 15 \times \sin\theta_B + 10 \times 14 \times \sin 90° + 8 \times 18 \times \sin 317° = 0$$

or $$0 + 15\, m_B \times \sin\theta_B + 140 - 98.2 = 0$$

or $$15\, m_B \times \sin\theta_B = -140 + 98.2 = -41.79 \quad ...(iv)$$

Squaring and adding equations (*iii*) and (*iv*), we have

$$(15\, m_B)^2 (\cos^2\theta_B + \sin^2\theta_B) = (-285.3)^2 + (-41.79)^2$$

or $$225 \times m_B^2 = 81396 + 1746.6$$

∴ $$m_B = \sqrt{\frac{83142.69}{225}} = \mathbf{19.2\ kg.}\quad \textbf{Ans.}$$

Dividing equation (*iv*) by (*iii*), we get

$$\frac{\sin\theta_B}{\cos\theta_B} = \frac{-41.79}{-285.3}$$

As in the above equation, the numerator (*i.e. sin* θ_B) is – ve and the denominator (*i.e.* cos θ_B) is also – ve, hence θ_B lies in 3rd quadrant.

∴ $$\tan\theta_B = \frac{-41.79}{-285.3} = 0.1464$$

or $$\theta_B = \tan^{-1} 0.1464 = 8.33° + 180 = \mathbf{188.33°.}\quad \textbf{Ans.}$$

Problem 13.8. *A 270 cm long shaft carries three pulleys, two at its ends and the third at mid-point. The two end pulleys have masses 48 kg and 20 kg and their centre of gravities are 1.5 cm and 1.25 cm respectively from the axis of the shaft, the middle pulley has mass 56 kg and its C.G. is 1.5 cm from the shaft axis. The pulleys are so keyed to the shaft that the assembly is in static balance. The shaft rotates at 300 r.p.m. in two bearings 180 cm apart with equal overlungs on either sides. Determine :*

(*a*) *relative angular position of the pulleys.* (AMIE, Summer 1978)

Sol. Given :

Length of shaft = 270 cm ; Mass of end pulleys = 48 kg and 20 kg.

Their C.G. = 1.5 cm and 1.25 cm ; Mass of middle pulley, B = 56 kg, C.G. = 1.5 cm ; N = 300 r.p.m., Distance between bearings = 180 cm.

The weight of pulleys are acting at their C.G. Hence the distances of the C.G. of the pulleys from the axis of rotation represents the radius of the pulleys.

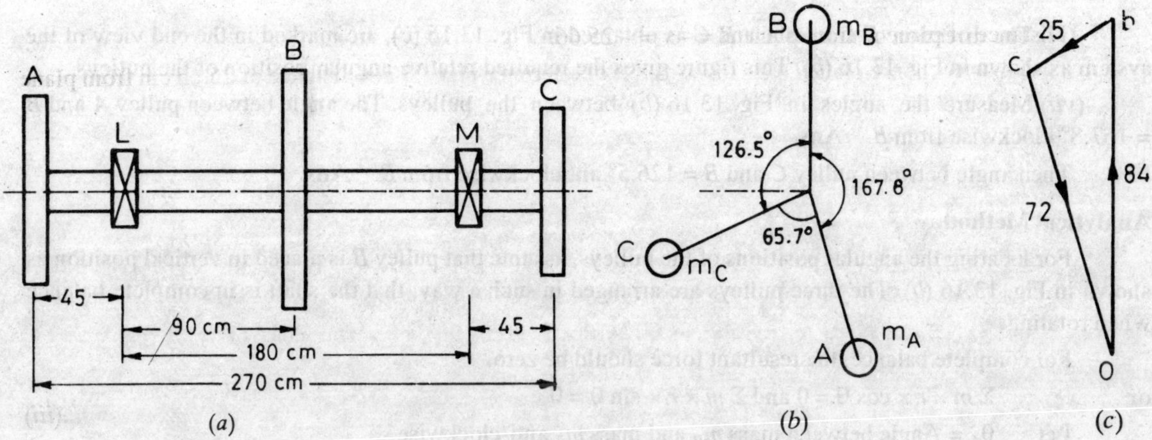

Fig. 13.16

Let A and C = Pulleys at the end of the shaft

B = Middle pulley

L and M = Two bearings.

Distance of two over hungs = 270 – 180 = 90.

∴ Distance of end pulleys from bearing = $\frac{90}{2}$ = 45 cm each.

The given data are :

$m_A = 48$ kg ; $m_C = 20$ kg ; $m_B = 56$ kg ; $r_A = 1.5$ cm ; $r_C = 1.25$ cm ; $r_B = 1.5$ cm.

Let us prepare the table from the above data.

Table 13.8

Plane	*Mass (m) kg*	*Radius (r) cm*	*Centrifugal force ÷ ω^2 = $(m \times r)$ kg-cm*
A	48	1.5	72
B	56	1.5	84
C	20	1.25	25

(*a*) *Relative angular position of the pulleys*

As the assembly is in static balance, the centre of gravity of the system must lie on the axis of the rotation. Therefore the force polygon must be a closed figure.

Now let us first draw the centrifugal force polygon using column (*iv*) of the above table. The following are the steps :

(*i*) Choose a suitable centrifugal force scale.

(*ii*) Let us assume that the pulley B is placed in vertical position as shown in Fig. 13.16 (*b*).

(*iii*) The centrifugal force acting on mass B is 84 kg-cm and is acting vertically up. Hence take a vertical line *ob* vertically up and cut off its length equal to 84 kg-cm to the chosen scale as shown in Fig. 13.16 (*c*).

(*iv*) With centre *b* and radius equal to 25 kg-cm, draw an arc of circle. Next with centre *o* and radius equal to 72 kg-cm, draw another arc of circle to intersect the former arc at *c*.

(*v*) Join *o* to *c* and *c* to *b*. Hence *oac* is the force triangle which gives the magnitude and direction of the masses A, B and C as shown in Fig. 13.16 (*c*).

(*vi*) The direction of masses *A* and *C* as obtained in Fig. 13.16 (*c*), are marked in the end view of the system as shown in Fig. 13.16 (*b*). This figure gives the required relative angular position of the pulleys.

(*vii*) Measure the angles in Fig. 13.16 (*b*) between the pulleys. The angle between pulley *A* and *B* = 167.8° clockwise from *B*. **Ans.**

Then angle between pulley *C* and *B* = 126.5° anticlockwise from *B*. **Ans.**

Analytical Method

For locating the angular positions of the pulleys, assume that pulley *B* is placed in vertical position as shown in Fig. 13.16 (*b*). The three pulleys are arranged in such a way, that the shaft is in complete balance when rotating.

For complete balance, the resultant force should be zero.

or $\Sigma\, m \times r \times \cos\theta = 0$ and $\Sigma\, m \times r \times \sin\theta = 0$

Let θ_A = Angle between mass m_B and mass m_A anti-clockwise

θ_C = Angle between mass m_B and mass m_C anti-clockwise.

Both the angles should be taken in the same sense.

The angle $\theta_B = 0$ as all angles are measured from the angular position of pulley *B*.

For $\Sigma\, m \times r \times \cos\theta = 0$, we have

$$m_B \times r_B \times \cos\theta_B + m_A \times r_A \times \cos\theta_A + m_C \times r_C \times \cos\theta_C = 0$$

or $$56 \times 1.5 \times \cos 0° + 48 \times 1.5 \times \cos\theta_A + 20 \times 1.25 \times \cos\theta_C = 0$$

or $$84 + 72 \times \cos\theta_A + 25 \times \cos\theta_C = 0$$

or $$72\cos\theta_A = -84 - 25\cos\theta_C \qquad ...(i)$$

For $\Sigma\, m \times r \times \sin\theta = 0$, we have

$$m_B \times r_B \times \sin\theta_B + m_A \times r_A \times \sin\theta_A + m_C \times r_C \times \sin\theta_C = 0$$

or $$56 \times 1.5 \times \sin 0° + 48 \times 1.5 \times \sin\theta_A + 20 \times 1.25 \times \sin\theta_C = 0$$

or $$0 + 72\sin\theta_A + 25\sin\theta_C = 0$$

or $$72\sin\theta_A = -25\sin\theta_C \qquad ...(ii)$$

Squaring and adding equations (*i*) and (*ii*),

$$72^2(\cos^2\theta_A + \sin^2\theta_A) = (-84 - 25\cos\theta_C)^2 + (-25\sin\theta_C)^2$$

or $$72^2 = (-84)^2 + (-25\cos\theta_C)^2 + 2(-84)(-25\cos\theta_C) + (-25\sin\theta_C)^2$$

or $$5184 = 7056 + 625\cos^2\theta_C + 4200\cos\theta_C + 625\sin^2\theta_C$$

$$= 7056 + 625(\cos^2\theta_C + \sin^2\theta_C) + 4200\cos\theta_C$$

$$= 7056 + 625 + 4200\cos\theta_C$$

or $$5184 - 7056 - 625 = 4200\cos\theta_C$$

or $$-2497 = 4200\cos\theta_C$$

or $$\cos\theta_C = \frac{-2497}{4200} = -0.5945$$

$\therefore$ $$\theta_C = \cos^{-1}(-0.5945) = 126.47° \text{ Anti-clockwise to mass } m_B$$

Dividing equation (*ii*) by equation (*i*), we get

$$\frac{72\sin\theta_A}{72\cos\theta_A} = \frac{-25\sin\theta_C}{-84 - 25\cos\theta_C} = \frac{-25 \times \sin 126.47°}{-84 - 25\cos 126.47°} = \frac{-20.1}{-69.14}$$

or $$\tan\theta_A = \frac{\sin\theta_A}{\cos\theta_A} = \frac{-20.1}{-69.14}$$

In the above equation the numerator (*i.e.* sin θ_A) is – ve and also the denominator (*i.e.* cos θ_A) is also – ve, have θ_A lies in third quadrant with respect to mass m_B.

$$\therefore \quad \theta_A = \tan^{-1}\left(\frac{-20.1}{-69.14}\right) = \tan^{-1} 0.29 = 16.2° + 180 = 196.2° \text{ anti-clockwise to } m_B$$

$$= (360 - 196.2) = 167.8° \text{ clockwise to } m_B$$

Hence the relative angular positions of pulleys *A*, *B* and *C* are shown in Fig. 13.16 (*b*).

Problem 13.9. *A shaft carries four rotating masses A, B, C and D which are completely balanced. The masses B, C and D are 50 kg, 80 kg and 70 kg respectively. The masses C and D make angles of 90° and 195° respectively with mass B in the same sense. The masses A, B, C and D are concentrated at radius 75 mm, 100 mm, 50 mm and 90 mm respectively. The plane of rotation of masses B and C are 250 mm apart. Determine :*

(*i*) *the mass A and its angular position*

(*ii*) *the position of planes of A and D.*

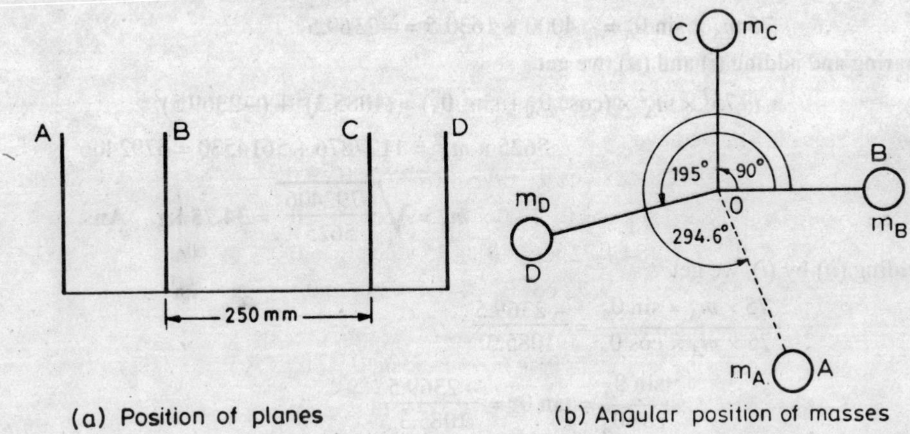

Fig. 13.17

Sol. Given :

Masses, $m_B = 50$ kg ; $m_C = 80$ kg ; $m_D = 70$ kg.

Radius of masses, $r_A = 75$ mm ; $r_B = 100$ mm ; $r_C = 50$ mm ; $r_D = 90$ mm.

Angles between masses, $\angle BOC = 90°$ and $\angle BOD = 195°$ or $\theta_C = 90°$ and $\theta_D = 195°$.

Distance between planes *B* and *C* = 250 mm.

Find : (*i*) m_A and θ_A.

(*ii*) Positions of planes *A* and *D*.

Analytical Method

Fig. 13.17 (*a*) shows the position of planes *A*, *B*, *C* and *D* whereas Fig. 13.17 (*b*) shows the angular position of masses *B*, *C* and *D* in which the angular position of mass m_B is assumed in horizontal direction. Hence $\theta_B = 0°$.

(*i*) *The mass A and its angular position.*

Let $\quad m_A$ = Mass *A*

θ_A = Angular position of mass *A* with respect to mass *B*.

The four masses are completely balanced, hence resultant force should be zero.

or $$\Sigma\, m \times r \times \cos\theta = 0 \text{ and } \Sigma\, m \times r \times \sin\theta = 0$$

Let us first find the product of known mass and corresponding radius.

$$\therefore \quad m_B \times r_B = 50 \times 100 = 5000$$

$$m_C \times r_C = 80 \times 50 = 4000$$

$$m_D \times r_D = 70 \times 90 = 6300$$

Also we know that $\theta_B = 0$, $\theta_C = 90°$ and $\theta_D = 195°$

Now for $\Sigma\, m \times r \times \cos\theta = 0$, we have

$$m_A \times r_A \times \cos\theta_A + m_B \times r_B \times \cos\theta_B + m_C \times r_C \times \cos\theta_C + m_D \times r_D \times \cos\theta_D = 0$$

or $$m_A \times 75 \times \cos\theta_A + 5000 \times \cos 0° + 4000 \times \cos 90° + 6300 \times \cos 195° = 0$$

or $$75\, m_A \times \cos\theta_A + 5000 + 0 + (-6085.3) = 0$$

or $$75\, m_A \times \cos\theta_A = 6085.3 - 5000 = 1085.3 \quad ...(i)$$

For $\Sigma\, m \times r \times \sin\theta = 0$, we have

$$m_A \times r_A \times \sin\theta_A + m_B \times r_B \times \sin\theta_B + m_C \times r_C \times \sin\theta_C + m_D \times r_D \times \sin\theta_D = 0$$

or $$m_A \times 75 \times \sin\theta_A + 5000 \times \sin\ 0° + 4000 \sin 90° + 6300 \sin 195° = 0$$

or $$75\, m_A \times \sin\theta_A + 0 + 4000 + (-1630.5) = 0$$

or $$75\, m_A \times \sin\theta_A = -4000 + 1630.5 = -2369.5 \quad ...(ii)$$

Squaring and adding (*i*) and (*ii*), we get

$$75^2 \times m_A^{\,2} \times (\cos^2\theta_A + \sin^2\theta_A) = (1085.3)^2 + (-2369.5)^2$$

or $$5625 \times m_A^{\,2} = 1177876 + 5614530 = 6792406$$

$\therefore$ $$m_A = \sqrt{\frac{6792406}{5625}} = \mathbf{34.75\ kg.}\quad \textbf{Ans.}$$

Dividing (*ii*) by (*i*), we get

$$\frac{75 \times m_A \times \sin\theta_A}{75 \times m_A \times \cos\theta_A} = \frac{-2369.5}{1085.3}$$

or $$\frac{\sin\theta_A}{\cos\theta_A} = \tan\theta_A = \frac{-2369.5}{1085.3}$$

In the above equation the numerator (*i.e.* $\sin\theta_A$) is – ve whereas the denominator (*i.e.* $\cos\theta_A$) is + ve. Hence θ_A lies in fourth quadrant as sine of an angle is – ve and cosine of the angle is + ve in fourth quadrant.

$$\therefore\quad \theta_A = \tan^{-1}\frac{-2369.5}{1085.3} = \tan^{-1}(-2.183) = -65.4° = 360 - 65.4 = \mathbf{294.6°}.\quad \textbf{Ans.}$$

(ii) Position of planes of A and D

Take plane *A* as the reference plane.

Then $l_A = 0$

l_B = distance of plane *B* from plane *A*

l_C = distance of plane *C* from plane *A* = $l_B + 250$

l_D = distance of plane *D* from plane *A*

As the four masses are completely balanced, hence the resultant couple about the reference plane should be zero.

or $$\Sigma\, m \times r \times l \cos\theta = 0 \text{ and } \Sigma m \times r \times l \times \sin\theta = 0$$

For $\Sigma m \times r \times l \times \cos\theta = 0$ about plane *A*, we have

$$m_B \times r_B \times l_B \times \cos\theta_B + m_C \times r_C \times l_C \times \cos\theta_C + m_D \times r_D \times l_D \times \cos\theta_D = 0$$

or $$5000 \times l_B \times \cos 0° + 4000 \times (l_B + 250) \cos 90° + 6300 \times l_D \times \cos 195° = 0$$

$$(\because\ m_B \times r_B = 5000\ ;\ m_C \times r_C = 4000\ ;\ m_D \times r_D = 6300\ ;\ l_C = l_B + 250)$$

or $$5000\, l_B + 0 + 6300 \times l_D \times (-0.966) = 0$$

or $$5000\, l_B - 6085.3\, l_D = 0$$

or $$l_B = \frac{6085.3 l_D}{5000} = 1.217 l_D \quad \text{...(iii)}$$

$\Sigma m \times r \times l \times \sin\theta = 0$ about plane A, we have

$$m_B \times r_B \times l_B \times \sin\theta_B + m_C \times r_C \times l_C \times \sin\theta_C + m_D \times r_D \times l_D \times \sin\theta_D = 0$$

or $$5000 \times l_B \times \sin 0° + 4000 \times (l_B + 250) \sin 90° + 6300 \times l_D \times \sin 195° = 0 \quad (\because \; l_C = l_B + 250)$$

or $$0 + 4000\,(l_B + 250) + 6300 \times l_D \times (-0.2588) = 0$$

or $$4000 l_B + 1000000 - 1630.5 l_D = 0$$

or $$4000\,(1.217 l_D) + 1000000 - 1630.5 l_D = 0 \quad [\because \; l_B = 1.217\, l_D \text{ from equation (iii)}]$$

or $$4868 l_D + 1000000 - 1630.5 l_D = 0$$

or $$3237.5 l_D = -1000000$$

$$\therefore \quad l_D = \frac{-1000000}{3237.5} = -308.8 \text{ mm}$$

From equation (*iii*), $l_B = 1.217 l_D = 1.217\,(-308.8) = -375.8$ mm

But $l_C = l_B + 250 = -375.8 + 250 = -125.8$ mm

The correct position of planes A, B, C and D are shown in Fig. 13.17 (*c*) whereas the angular position of mass m_A is shown in Fig. 13.17 (*b*).

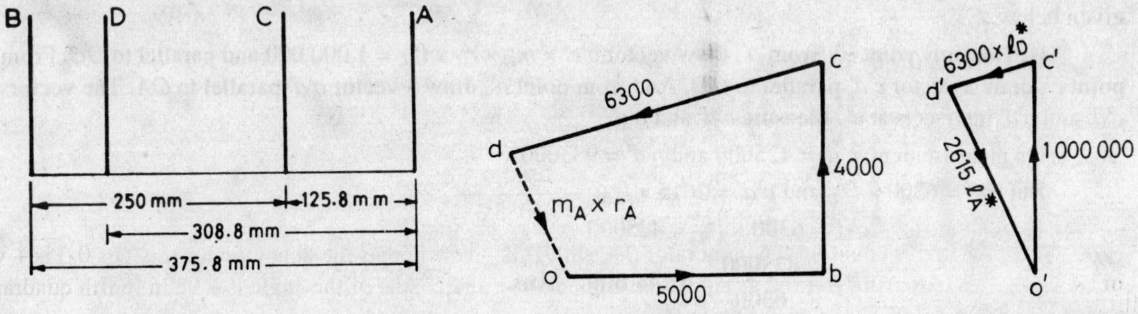

(c) Correct position of planes (d) Force polygon (e) Couple polygon

Fig. 13.17

Graphical Method

We know that

$$m_B \times r_B = 5000 \; ; \; m_C \times r_C = 4000 \text{ and } m_D \times r_D = 6300.$$

(*i*) *The mass m_A and its angular position*

The four masses are completely balanced, hence the force polygon should be a closed figure. The force polygon is drawn as shown in Fig. 13.17 (*d*) by the method as given below :

(*i*) Take any point *o*. From *o*, draw vector *ob* = 5000 and parallel to *OB* of Fig. 13.17 (*b*) to some suitable scale.

(*ii*) From point *b*, draw vector *bc* = 4000 and parallel to *OC*.

(*iii*) From point *c*, draw vector *cd* = 6300 and parallel to *OD*. Join point *d* to *o*.

The closing side *do* represents the product of mass and radius of balancing mass (*i.e.* $m_A \times r_A$). Measure *do*. On measurement, *do* = 2615.

$$\therefore \quad m_A \times r_A = 2615$$

$$\therefore \quad m_A = \frac{2615}{r_A} = \frac{2615}{75} = \mathbf{34.86\ kg.}\quad \textbf{Ans.}$$

To find the value of θ_A (i.e. angle made by mass m_A with mass m_B), draw the line *OA* in Fig. 13.17 (*b*) parallel to line *do* from Fig. 13.17 (*d*). On measurement angle $\theta_A = 294.6°$

$$\therefore \quad \theta_A = \mathbf{294.6°}.\quad \textbf{Ans.}$$

(*ii*) *The position of planes of A and D*

Take plane *B* of Fig. 13.17 (*c*) as the reference plane.

Then l^*_D = Distance of plane *D* from plane *B*

l^*_C = Distance of plane *C* from plane *B* = 250 mm (given)

l^*_A = Distance of plane *A* from plane *A*

We know that,

$m_C \times r_C \times l^*_C = 4000 \times 250 = 1000000$; $m_D \times r_D \times l^*_D = 6300 \times l^*_D$ and $m_A \times r_A \times l^*_A = 2615 \times l^*_A$

($\because$ $m_C \times r_C = 4000$, $l^*_C = 250$, $m_D \times r_D = 6300$ and $m_A \times r_A$ from force polygon = 2615)

The product $m_B \times r_B \times l^*_B = 0$ about reference plane *B*.

The four masses are completely balanced, hence the couple polygon should be a closed figure. About reference plane *B*, there are only three couples. Hence the couple polygon will be a triangle. Also the couples are acting in the direction of forces. The couple polygon is drawn as shown in Fig. 13.17 (*e*) by the method as given below :

(*i*) Take any point *o'*. From *o'*, draw vector $o'c' = m_C \times r_C \times l^*_C = 1{,}000{,}000$ and parallel to *OC*. From point *c'*, draw a vector *c'd'* parallel to *OD*. And from point *o'*, draw a vector *o'd'* parallel to *OA*. The vectors *c'd'* and *o'd'* intersects at *d'*. Measure *c'd'* and *o'd'*.

On measurement, $c'd' = 425000$ and $o'd' = 983000$

But $c'd' = 6300 \times l^*_D$ and $o'd' = 2615 \times l^*_A$

$$\therefore \quad c'd' = 6300 \times l^*_D = 425000$$

or
$$l^*_D = \frac{425000}{6300} = \mathbf{67.46\ mm.}\quad \textbf{Ans.}$$

and
$$o'd' = 2615 \times l^*_A = 983000$$

or
$$l^*_A = \frac{983000}{2615} = \mathbf{375.9\ mm.}\quad \textbf{Ans.}$$

The values of l^*_A and l^*_D by analytical method are :

$l^*_A = 375.8$ mm and $l^*_D = 375.8 - 308.8 = 67$ mm.

The values given by analytical method and graphical method are closely equal.

HIGHLIGHTS

1. A mass attached to a shaft which is rotating with angular velocity ω rad/s, can be balanced by another single mass which is attached to the opposite side of the shaft in the same plane if

$$m_1 \times r_1 = m_2 \times r_2$$

where m_1 = Mass attached to the shaft at a radius r_1

m_2 = Balancing mass placed at the radius r_2.

2. A single mass can also be balanced by two masses placed in different planes. The planes of the balancing masses may lie on the same side of the plane of the disturbing mass or the planes of the balancing masses may be on either side of the plane of the disturbing mass.

(i) In the first case, the centrifugal force on the first balancing mass will be in the opposite direction whereas on the second balancing mass will be in the direction of disturbing mass.

(ii) In the second case, the centrifugal forces on the balancing masses will be in the opposite direction of the direction of disturbing mass.

3. The balancing of several masses rotating in the same plane is obtaining by finding the resultant force in magnitude and direction by analytical method or graphical method as given below :

(i) Resultant centrifugal force,

$$F_C = \sqrt{(\Sigma H)^2 + (\Sigma V)^2}$$

where $\Sigma H = m_1 r_1 \cos \theta_1 + m_2 r_2 \cos \theta_2 + m_3 r_3 \cos \theta_3 + ...$

and $\Sigma V = m_1 r_1 \sin \theta_1 + m_2 r_2 \sin \theta_2 + m_3 r_3 \sin \theta_3 + ...$

(ii) The angle made by the resultant centrifugal force with the horizontal is given by

$$\tan \theta = \frac{\Sigma V}{\Sigma H}.$$

(iii) The resultant centrifugal force, $F_C = m \times r$

where m = Balancing mass and

r = Radius of rotation of balancing mass

If r is known than balancing mass 'm' can be obtained.

(iv) Position of balancing mass = (180 + θ).

(v) By graphical method, the force polygon is drawn, the closing side gives the resultant in magnitude and direction. The direction of balancing mass will be in the opposite direction of closing side.

4. The balancing of several masses rotating in different planes can be done by transferring the centrifugal force corresponding to each rotating mass in a reference plane. Corresponding to the centrifugal force of each rotating mass, there will be an equal and parallel force in the reference plane along with a couple whose magnitude is equal to the product of force and the distance between the reference plane and line of action of original force. For complete balance,

(i) the resultant force in the reference plane must be zero and

(ii) the resultant couple about the reference plane must be zero.

EXERCISE 13

(A) Theoretical Questions

1. Define and explain the term 'Balancing of Rotating Masses'. What will be the harm if the rotating parts of a high speed engine are not properly balanced ?

2. A mass is attached to a shaft which is rotating at an angular speed of ω rad/s. Describe the procedure of balancing this mass by

   (i) a single mass only

   (ii) two masses in different planes.

3. A number of masses (say four masses) are attached to a shaft which is rotating at an angular speed of ω rad/s. If all the masses are in the same plane, then describe the analytical and graphical method of balancing these four masses by a single mass only.

4. What do you mean by static balancing and dynamic balancing ? What are the necessary conditions to achieve them ? *(A.M.I.E., Summer 1994)*

5. A number of masses are attached to a shaft which is rotating at an angular speed of ω rad/s. If all the masses are in different planes, then describe the analytical method and graphical method of balancing these masses.

6. What do you mean by reference plane ?

7. The several revolving masses in different planes will have a complete balance if
(a) the net couple due to dynamic forces acting on the shaft is equal to zero
(b) the net dynamic forces acting on the shaft is equal to zero
(c) both (a) and (b)
(d) none of the above. **[Ans.** (c)**]**

(B) Numerical Problems

1. A shaft is rotating at a speed of 240 r.p.m. Four masses m_1, m_2, m_3 and m_4 of magnitudes 100 kg, 150 kg, 120 kg and 130 kg respectively are rigidly attached to the shaft. The masses are rotating in the same plane. The corresponding radii of rotation are 400 mm, 300 mm, 500 mm and 600 mm respectively. The angles made by these masses with the horizontal are 0°, 45°, 120° and 255° respectively. Find analytically and graphically :
(i) the magnitude of the balancing mass and
(ii) the position of the balancing mass if its radius of rotation is 400 mm. **[Ans.** (i) 58 kg (ii) 201.30°**]**
2. Four masses (m_1, m_2, m_3 and m_4) 300 kg, 450 kg, 360 kg and 390 kg are attached to a shaft. These masses are revolving at radii 135 mm, 105 mm, 150 mm and 180 mm respectively in planes measured from *A* at 135 mm, 210 mm and 360 mm respectively. The angles measured anti-clockwise are m_1 to m_2 45°, m_2 to m_3 75°, m_3 to m_4 135° and the distance between the planes *L* and *M* in which the balance masses are to be placed is 250 mm. The distance between planes *A* and *L* is 60 mm and *M* and *D* is 50 mm. If the balancing masses revolve at a radius of 36 mm, find their magnitudes and angular positions.
[Ans. (i) $m_L = 240.75$ kg ; $\theta_L = 220°$ with m_1 (ii) $m_M = 159.57$ kg ; $\theta_M = 49°$ with m_1**]**
3. A shaft carries four masses *A*, *B*, *C* and *D* of magnitude 300 kg, 450 kg, 600 kg and 300 kg respectively and revolving at radii 120 mm, 105 mm, 90 mm and 120 mm in planes measured from *A* at 350 mm, 600 mm and 1050 mm. The angles between the cranks measured anti-clockwise are *A* to *B* 45°, *B* to *C* 70° and *C* to *D* 120°. The balancing masses are to be placed in planes *X* and *Y*.
The distance between the planes *A* and *X* is 150 mm, between *X* and *Y* is 600 mm and between *Y* and *D* is 300 mm. If the balancing masses revolve at a radius of 150 mm, find their magnitudes and angular positions.
[Ans. $m_X = 427.5$ kg ; $\theta_X = 283°$ anti-clockwise from m_A ; $m_Y = 273.75$ kg ; $\theta_Y = 348°$ anti-clockwise from m_A**]**
4. A shaft is rotating at a speed of 360 r.p.m. Three masses m_1, m_2 and m_3 of magnitudes 8 kg, 6 kg and 5 kg respectively are attached rigidly to the shaft. The masses are rotating in the same plane. The corresponding radii of rotation are 150 mm, 170 mm and 100 mm respectively. The angles made by these masses with the horizontal are 45°, 135° and 240° respectively. Find :
(i) magnitude of the balancing mass and
(ii) the position of the balancing mass if its radius of rotation is 150 mm. **[Ans.** (i) 7.62 kg (ii) $\theta = 276.2°$**]**
5. A shaft carries four masses *A*, *B*, *C* and *D* of magnitudes 18 kg, 15 kg, 27 kg, and 22.5 kg respectively and revolving at radii 20 mm, 25 mm, 30 mm and 15 mm respectively. The masses are rotating in the same plane. The angular position of masses *B*, *C* and *D* are 60°, 135° and 270° from mass *A*. Find the magnitude and position of the balancing mass at a radius of 50 mm. **[Ans.** 11.34 kg ; 273° anti-clockwise from *A***]**
6. Four masses *A*, *B*, *C* and *D* are rigidly attached to a shaft. The radii of rotation are 60 mm, 76 mm, 80 mm and 70 mm respectively from the axis of rotation. The magnitude of masses *A*, *C* and *D* are 15 kg, 10 kg and 8 kg respectively. The axial distance between the planes of rotation of *A* and *B* is 400 mm whereas between *B* and *C* is 500 mm. The masses *A* and *C* are at right angles to each other. If the masses are in complete balance, then find :
(i) the angles of masses *B* and *D* from mass *A*.
(ii) the axial distance between the planes of rotation of *C* and *D*.
(iii) the magnitude of mass *B*. **[Ans.** (i) 162.5° and 47.5°, (ii) 511 mm, (iii) 18.48 kg**]**
7. Four masses *A*, *B*, *C* and *D* are rigidly attached to a shaft. The masses *B*, *C* and *D* are 37.5 kg, 60 kg and 52.5 kg respectively. The masses *C* and *D* make an angle of 90° and 195° respectively with mass *B* anti-clockwise. The

masses A, B, C and D are concentrated at radii 150 mm, 200 mm, 100 mm and 180 mm respectively. The planes of rotation of masses B and C are 250 mm apart. If the shaft is in complete balance, then determine :

(*i*) the mass A and its angular position, and

(*ii*) the position of planes of A and D. **[Ans.** (*i*) 26.1 kg, 294.6° (*ii*) l_A = 376 mm ; l_D = 67 mm**]**

8. Four masses A, B, C and D are attached to a rotating shaft with radii 50 mm, 62.5 mm, 100 mm and 75 mm respectively. The distances between planes A and B ; between planes B and C and between planes C and D are 600 mm each. The masses B, C and D are 20 kg, 10 kg and 8 kg respectively. If the shaft is in complete balance, then find :

(*i*) Magnitude of mass A and

(*ii*) Angular positions of the four masses. **[Ans.** (*i*) 14 kg, (*ii*) θ_A = 155° ; θ_B = 0 ; θ_C = 240° and θ_D = 100°**]**

14

Balancing of Reciprocating Masses

14.1. Introduction

In chapter 13, the balancing of rotating masses have been discussed. The balancing of reciprocating masses is equally important. Any lack of balance of a engine and other machine is responsible for dynamic forces which increases the bearing loads and stresses in the members of the engine or machine. These dynamic forces may even set up the unpleasent and even dangerous vibrations in the members of the machine. Hence the reciprocating parts should also be as completely balanced as possible. This chapter deals with the balancing of reciprocating masses.

14.2. Balancing of Reciprocating Engine

Fig. 14.1 shows the line diagram of a horizontal reciprocating engine, in which *OA* represents the crank and *AB* represents the connecting rod. Let the crank is rotating with a uniform velocity ω rad/s in the clockwise direction.

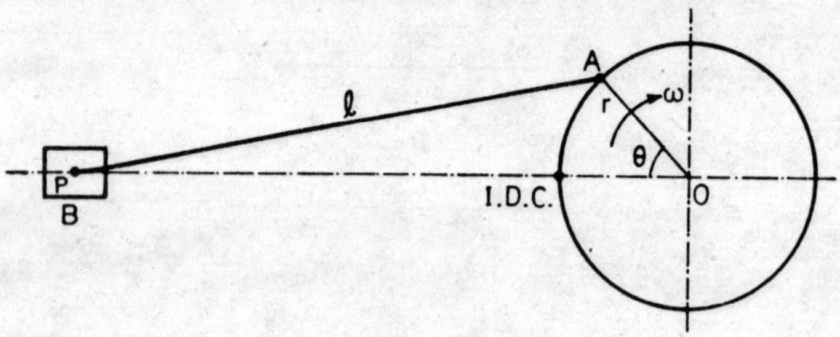

Fig. 14.1

Let r = radius of crank *OA*

l = length of connecting rod *AB*

n = ratio of length of connecting rod to crank radius = $\dfrac{l}{r}$

θ = angle turned by crank from I.D.C.

m_R = mass of the reciprocating parts of the engine

f_P = acceleration of piston (or of reciprocating masses)

F = force required to accelerate the reciprocating masses.

The acceleration of piston (or of reciprocating masses) may be expressed by equation (11.11) with sufficient accuracy as

$$f_P = \omega^2 r\left(\cos\theta + \frac{\cos 2\theta}{n}\right)$$

Now the force required to accelerate the reciprocating masses is given by,

$$F = \text{Mass of reciprocating parts} \times \text{Acceleration}$$
$$= m_R \times f_P$$
$$= m_R \times \omega^2 r \left(\cos\theta + \frac{\cos 2\theta}{n} \right) \quad \text{...(14.1)}$$

The inertia force (F_i) is equal and opposite to the force required to accelerate the reciprocating masses

$$\therefore \quad F_i = -F = -m_R \times \omega^2 r \left(\cos\theta + \frac{\cos 2\theta}{n} \right)$$

The inertia force is divided into two parts corresponding to the two terms in the brackets of equation (14.1).

$$\therefore \quad F_i = -\left[m_R \times \omega^2 r \cos\theta + m_R \times \omega^2 r \frac{\cos 2\theta}{n} \right]$$
$$= -[F_P + F_S]$$

where $F_P = m_R \times \omega^2 r \cos\theta$ = primary disturbing force of the reciprocating mass ...(14.1A)

$F_s = m_R \times \omega^2 r \dfrac{\cos 2\theta}{n}$ = secondary disturbing force of the reciprocating mass ...(14.1B)

As the inertia force (F_i) acts along the line of stroke of reciprocating engine hence the primary and secondary disturbing forces will be acting along the line of stroke.

When $\theta = 0$, $\quad F_P = m_R \times \omega^2 \times r \cos 0° = m_R \times \omega^2 \times r$ ... Max. value

$F_s = m_R \times \omega^2 \times r \dfrac{\cos 0°}{n} = \dfrac{m_R \times \omega^2 \times r}{n}$ ... Max. value

When $\theta = 90°$, $\quad F_P = m_R \times \omega^2 \times r \cos 90° = 0$

$F_s = m_R \times \omega^2 \times r \dfrac{\cos 180°}{n} = -\dfrac{m_R \times \omega^2 \times r}{n}$ ... Magnitude maximum

When $\theta = 180°$, $F_P = m_R \times \omega^2 \times r \cos 180° = -m_R \times \omega^2 \times r$ ... Magnitude maximum

$F_s = m_R \times \omega^2 \times r \dfrac{\cos 360°}{n} = \dfrac{m_R \times \omega^2 \times r}{n}$ ... Max. value

When $\theta = 270°$, $F_P = m_R \times \omega^2 \times r \cos 270° = 0$

$F_s = m_R \times \omega^2 \times r \dfrac{\cos (540°)}{n} = -\dfrac{m_R \times \omega^2 \times r}{n}$ ... Magnitude maximum

From the above, we see that

(*i*) Max. value of secondary force, $F_s = \dfrac{m_R \times \omega^2 \times r}{n}$ whereas

max. value of primary force, $F_P = m_R \times \omega^2 \times r$

$\therefore$ Max. value of secondary force, $F_s = \dfrac{1}{n}$ [Max. value of primary force]

As n is much greater than unity, the secondary force is small compared with the primary force.

(*ii*) For one revolution of the crank, the maximum value of primary force occurs two times whereas the max. value of secondary force occurs four times.

(*iii*) The unbalanced force due to reciprocating mass (*i.e.* primary and secondary disturbing forces) varies in magnitude but is constant in direction (*i.e.* acting along the line of stroke). Whereas the unbalanced force due to rotating mass is constant in magnitude but varies in direction.

Hence a single rotating mass cannot balance completely a reciprocating mass. Also a reciprocating mass cannot be used to balance a rotating mass. But in some cases the partial balance of a reciprocating mass can be obtained by a rotating balance mass.

(*iv*) In case of moderate speeds, the secondary unbalanced force is so small that it may be neglected as compared to primary unbalanced force.

(*v*) The inertia force acts along the line of stroke from O to P, hence the primary and secondary unbalanced forces will also act along the line of stroke from O to P.

14.3. Partial Balancing of Primary Force

In Art. 14.2, we have seen that the primary disturbing force is given by equation (14.1A) as,

$$F_P = m_R \times \omega^2 \times r \times \cos\theta$$

where m_R = mass of the reciprocating parts of the engine,

ω = uniform angular velocity of the crank,

r = radius of crank, and

θ = inclination of crank with I.D.C.

The primary disturbing force is acting along the line of stroke from O to P.

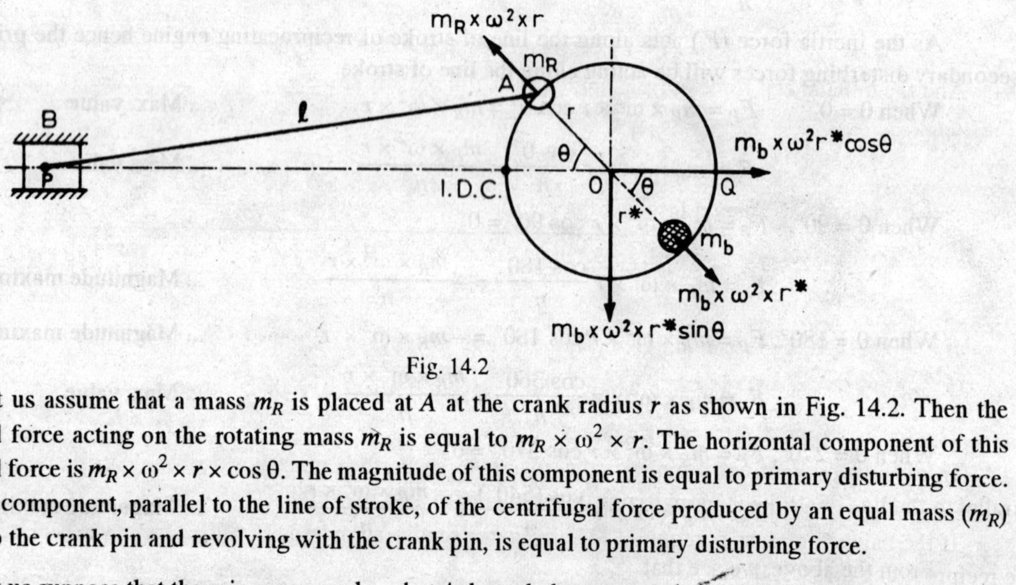

Fig. 14.2

Let us assume that a mass m_R is placed at A at the crank radius r as shown in Fig. 14.2. Then the centrifugal force acting on the rotating mass m_R is equal to $m_R \times \omega^2 \times r$. The horizontal component of this centrifugal force is $m_R \times \omega^2 \times r \times \cos\theta$. The magnitude of this component is equal to primary disturbing force. Hence the component, parallel to the line of stroke, of the centrifugal force produced by an equal mass (m_R) attached to the crank pin and revolving with the crank pin, is equal to primary disturbing force.

Let us suppose that there is no mass placed at A, but a balance mass (m_b) is fixed at radius r^* directly opposite to the crank. The centrifugal force of this mass will be equal to $(m_b) \times \omega^2 \times r^*$.

The component of this force parallel to the line of stroke $= m_b \times \omega^2 \times r^* \times \cos\theta$.

This component acts along the line of stroke from O to Q. Hence this component is opposite in direction to the primary disturbing force (*i.e.* primary disturbing force acts from O to P).

$\therefore$ Resultant disturbing force along the line of stroke from O to P

$$= m_R \times \omega^2 \times r \times \cos\theta - m_b \times \omega^2 \times r^* \cos\theta \quad \text{...(14.2)}$$

$$= (m_R \times r - m_b \times r^*)\, \omega^2 \times \cos\theta \quad \text{...(14.2A)}$$

The resultant disturbing force along the line of stroke will be zero, if

$$m_R \times r = m_b \times r^* \qquad ...(14.3)$$

This means that if $m_R \times r = m_b \times r^*$, then the primary force is completely balanced. But the centrifugal force produced by the rotating mass (m_b) has a vertical component perpendicular to the line of stroke, the magnitude of which is $m_b \times \omega^2 \times r^* \sin\theta$. This component remains unbalanced. The maximum value of this force is $m_b \times \omega^2 \times r^*$ when crank is at right angles to the line of stroke (*i.e.* when $\theta = 90°$ and $270°$). The maximum value of primary disturbing force is equal to $m_R \times \omega^2 \times r$ or $m_b \times \omega^2 \times r^*$ when $m_R \times r = m_b \times r^*$ and $\theta = 0°$ and $180°$.

Hence by introducing the rotating balance mass (m_b) has only served to change the direction of the disturbing force. The disturbing force which was previously along the line of stroke, now exists perpendicular to the line of stroke. A compromise is made and only a fraction of the reciprocating masses is balanced. This is done by making

$$m_b \times r^* = c \times m_R \times r \text{ where } c < 1$$

(The value of c varies from 0.5 to 0.75 for most of the cases)

Now the resultant unbalanced force along the line of stroke given by equation (14.2) will become

$$= m_R \times \omega^2 \times r \cos\theta - c m_R \times r \times \omega^2 \times \cos\theta \qquad (\because\ m_b \times r^* = c \times m_R \times r)$$

$$= m_R \times \omega^2 \times r \times \cos\theta \times (1 - c)$$

$$= (1 - c) \times m_R \times \omega^2 \times r \times \cos\theta \qquad ...(14.4)$$

And the unbalanced force at right angles to the line of stroke

$$= m_b \times r^* \times \omega^2 \times \sin\theta \qquad ...(14.4A)$$

$$= c \times m_R \times r \times \omega^2 \times \sin\theta \qquad (\because\ m_b \times r^* = c \times m_R \times r) \qquad ...(14.4B)$$

∴ Resultant unbalanced force on the engine frame at any instant

$$= \sqrt{[(1-c)\, m_R \times \omega^2 \times r \times \cos\theta]^2 + [c \times m_R \times r \times \omega^2 \times \sin\theta]^2}$$

$$= m_R \times \omega^2 \times r \sqrt{(1-c)^2 \cos^2\theta + c^2 \sin^2\theta} \qquad ...(14.5)$$

If $c = 0.5$, then the resultant unbalanced force on the engine frame is least and is equal to

$$= m_R \times \omega^2 \times r \sqrt{0.5^2 (\cos^2\theta + \sin^2\theta)}$$

$$= 0.5 \times m_R \times \omega^2 \times r \qquad ...(14.5B)$$

In case of locomotives a higher value of c may be used as in case of locomotives disturbing force parallel to the line of stroke is more harmful than one perpendicular to it.

If the balancing mass is required to balance the revolving masses as well as give a partial balance of the reciprocating masses, then :

$$m_b \times r^* = M \times r + c \times m_R \times r$$

$$= (M + c \times m_R) \times r \qquad ...(14.6)$$

where M = Magnitude of revolving masses

m_R = Mass of reciprocating parts

c = Fraction of reciprocating parts which is to be balanced

r = radius of crank.

Hence $(M + C \times m_R)$ is the total equivalent revolving mass at crank radius r which has to be balanced.

Note. (*i*) Due to primary unbalanced force, the mechanism slides to and fro on its mounting.

(*ii*) Due to unbalanced force perpendicular to line of stroke, the mechanism jumps up and down.

Problem 14.1. *A single cylinder reciprocating engine has the following data :*

Speed of engine = 120 r.p.m. ; stroke = 320 mm ; mass of reciprocating parts = 45 kg and mass of revolving parts = 35 kg at crank radius. If 60% of the reciprocating parts and all the revolving parts are to be balanced, then find :

(i) the balance mass required at a radius of 300 mm, and

(ii) the unbalanced force when the crank has rotated 60° from top dead centre.

Sol. Given :

$N = 120$ r.p.m. or $\omega = \frac{2\pi N}{60} = \frac{2\pi \times 120}{60} = 4\pi$ rad/s ;

stroke = 320 mm = 0.32 m or crank radius $r = \frac{0.32}{2} = 0.16$ m ; mass of reciprocating parts,

$m_R = 45$ kg ; mass of revolving parts, $M = 35$ kg ; radius of revolving parts, $r = 160$ mm $= 0.16$ m ;

fraction of reciprocating parts to be balanced = 60% $= \frac{60}{100} = 0.6$

or $c = 0.6$; all revolving parts are to be balanced.

(i) Balance mass required at a radius of 300 mm

Let m_b = Balance mass required, and

r^* = Radius of rotation of the balance mass = 300 mm = 0.3 m

Using equation (14.6), we have

$$m_b \times r^* = (M + c \times m_R) \times r$$

or $$m_b \times 0.3 = (35 + 0.6 \times 45) \times 0.16 = (35 + 27) \times 0.16$$

or $$m_b = \frac{(35 + 27) \times 0.16}{0.3} = \mathbf{33.06\ kg.}\quad \textbf{Ans.}$$

(ii) Unbalanced force when crank has rotated 60° from top-dead centre

Here $\theta = 60°$ (given)

The unbalanced force at any instant is given by equation (14.5) as

$$= m_R \times \omega^2 \times r\sqrt{(1 - c)^2 \cos^2\theta + c^2 \sin^2\theta}$$

$$= 4.5 \times (4\pi)^2 \times 0.16\sqrt{(1 - 0.6)^2 \times \cos^2 60° + 0.6 \sin^2 60°}$$

$$= 1136.98\sqrt{.16 \times 0.25 + 0.6 \times 0.75} = 1136.98\sqrt{0.04 + 0.45}$$

$$= 1136.98 \times 0.7 = \mathbf{795.9.}\quad \textbf{Ans.}$$

14.4. Partial Balancing of Locomotives

There are two cylinders of the same dimensions placed symmetrically either between the frames or outside the frames in many locomotives. The two cranks are placed at right angles to each other, so that one crank, at least, is away from the dead centre and it is always possible to start the locomotive. The followings are the classification of the two cylinder locomotives.

(*i*) Inside cylinder locomotives and

(*ii*) Outside cylinder locomotives.

If the two cylinders are placed symmetrically between the frames (or wheels) the two cylinder locomotive is known as inside cylinder locomotive whereas if the two cylinders are placed symmetrically outside the frames or wheels it is called outside cylinder locomotive.

The ratio of the length of the connecting rod to the crank length is generally large, so that secondary forces are small. Moreover, the cranks are at right angles, the secondary forces for one set of the reciprocating parts are equal and opposite to those for the other set.

A locomotive may also be known as

(*a*) Uncoupled locomotive ; and (*b*) Coupled locomotive.

In case of uncoupled locomotive, the effort is transmitted to one pair of wheels whereas in case of coupled locomotives, the driving wheels are connected to leading and trailing wheel by an outside coupling rod which connects the crank pins of the wheels. The coupling rod revolves with the crank-pin and hence the proportionate mass of coupling rod can be considered as a revolving mass.

14.5. Effect of Partial Balancing of Locomotives

The reciprocating parts of a reciprocating engine are only partially balanced. This means that there is an unbalanced primary force along the line of stroke and also an unbalanced primary force perpendicular to the line of stroke. The unbalanced primary force along the line of stroke produces :

(*i*) Variation of tractive force along the line of stroke ; and

(*ii*) Swaying couple.

The unbalanced primary force perpendicular to the line of stroke produces :

(*iii*) Hammer blow (*i.e.* the maximum magnitude of the unbalanced force perpendicular to the line of stroke).

14.5.1. Variation of Tractive Force (or Effort)

Variation of tractive force (effort) of an engine is caused by the unbalanced part of the primary disturbing force along the line of stroke. Variation of tractive force is the resultant unbalanced force along the line of stroke. For a locomotive engine of two cylinders, let us find the resultant unbalanced force due to two cylinders along the line of stroke. Let θ is the angle of inclination of the crank of the first cylinder with the line of stroke. Since the crank for the second cylinder is at right angles to the first crank, hence (90 + θ) will be angle of inclination of the second crank with line of stroke.

Let c = Fraction of the reciprocating mass to be balanced

m_R = Mass of the reciprocating parts per cylinder

ω = Angular velocity of crank (both cranks rotates at the same angular velocity)

F_T = Tractive force

r = Radius of each crank.

The unbalanced primary force along the line of stroke for first cylinder

$$= (1 - c) \times m_R \times \omega^2 \times r \times \cos\theta$$

Similarly unbalanced primary force along the line of stroke for second cylinder

$$= (1 - c) \times m_R \times \omega^2 \times r \times \cos(90 + \theta) = -(1 - c)\, m_R \times \omega^2 \times r \sin\theta$$

∴ Resultant unbalanced force along the line of stroke

$$= (1 - c)\, m_R \times \omega^2 \times r \cos\theta - (1 - c)\, m_R \times \omega^2 \times r \sin\theta$$

$$= (1 - c)\, m_R \times \omega^2 \times r\,(\cos\theta - \sin\theta)$$

But the resultant unbalanced force along the line of stroke is equal to variation of tractive force.

∴ Variation of tractive force $= (1 - c)\, m_R \times \omega^2 \times r\,(\cos\theta - \sin\theta)$ ...(14.7)

This is maximum when $(\cos\theta - \sin\theta)$ is maximum

or when $\frac{d}{d\theta}(\cos\theta - \sin\theta) = 0$

or $$-\sin\theta - \cos\theta = 0$$

or $$-\sin\theta = \cos\theta \quad \text{or} \quad \frac{\sin\theta}{\cos\theta} = -1 \quad \text{or} \quad \tan\theta = -1$$

or $$\theta = 135° \quad \text{or} \quad 315°$$

When $\theta = 135°$, maximum variation in tractive force [substitute $\theta = 135°$ is equation (14.7)]

$$= (1-c)\, m_R \times \omega^2 \times r\,(\cos 135° - \sin 135°)$$

$$= (1-c)\, m_R \times \omega^2 \times r\left(-\frac{1}{\sqrt{2}} - \frac{1}{\sqrt{2}}\right) = (1-c)\, m_R \times \omega^2 \times r\left(-\frac{2}{\sqrt{2}}\right)$$

$$= -\sqrt{2}\,(1-c)\, m_R \times \omega^2 \times r$$

When $\theta = 315°$, the maximum variation in tractive force

$$= (1-c)\, m_R \times \omega^2 \times r\,(\cos 315° - \sin 315°)$$

$$= (1-c)\, m_R \times \omega^2 \times r\left[\frac{1}{\sqrt{2}} - \left(-\frac{1}{\sqrt{2}}\right)\right]$$

$$= \sqrt{2}\,(1-c)\, m_R \times \omega^2 \times r$$

$\therefore$ Maximum variation of tractive force

$$= \pm\sqrt{2}\,(1-c)\, m_R \times \omega^2 \times r \qquad \text{...(14.8)}$$

14.5.2. Swaying Couple

The unbalanced parts of the primary disturbing forces along the line of stroke for the two cylinders constitute a horizontal couple about the engine centre line. This horizontal couple is known as swaying couple. This tends to make the leading wheels sway from side to side.

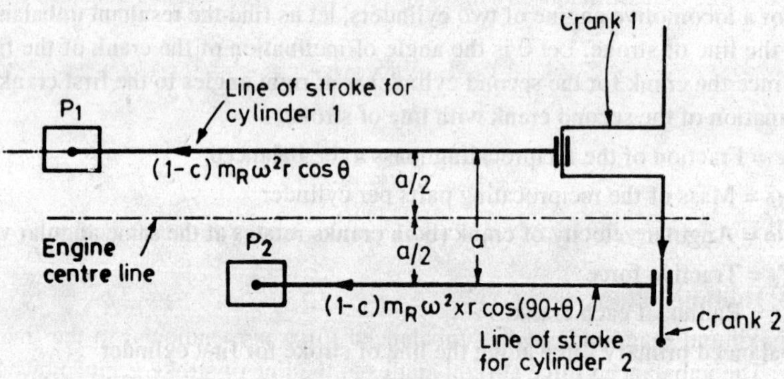

Fig. 14.3

Let a = Distance between the centre lines of the two cylinders

Unbalanced force along the line of stroke for cylinder 1

$$= (1-c)\, m_R\, \omega^2 r \cos\theta$$

Unbalanced force along the line of stroke for cylinder 2

$$= (1-c)\, m_R\, \omega^2 r \cos(90 + \theta)$$

Taking moments about the engine centre line, the resultant unbalanced couple or swaying couple

$$= [(1-c)\, m_R\, \omega^2 r \cos\theta] \times \frac{a}{2} - [(1-c)\, m_R\, \omega^2 r \cos(90+\theta)] \times \frac{a}{2}$$

[Negative sign is used because the two forces lie on opposite sides of the engine centre line about which moments are taken.]

$$= [(1-c)\, m_R\, \omega^2 r \cos\theta] \times \frac{a}{2} - [(1-c)\, m_R\, \omega^2 r\, (-\sin\theta)] \times \frac{a}{2}$$

$$= [(1-c)\, m_R\, \omega^2 r \cos\theta] \times \frac{a}{2} + [(1-c)\, m_R\, \omega^2 r \sin\theta] \times \frac{a}{2}$$

$$= (1-c)\, m_R\, \omega^2 r \times \frac{a}{2} (\cos\theta + \sin\theta)$$

$\therefore$ Swaying couple $= (1-c)\, m_R\, \omega^2 r \times \dfrac{a}{2} (\cos\theta + \sin\theta)$

The swaying couple will be maximum when $(\cos\theta + \sin\theta)$ is maximum.

Now $(\cos\theta + \sin\theta)$ will be maximum if

$$\frac{d}{d\theta}(\cos\theta + \sin\theta) = 0$$

or $\quad -\sin\theta + \cos\theta = 0 \quad$ or $\quad -\sin\theta = -\cos\theta \quad$ or $\quad \tan\theta = 1$

or $\quad \theta = 45^\circ \quad$ or $\quad 225^\circ$

When $\theta = 45^\circ$, maximum swaying couple $= (1-c)\, m_R\, \omega^2 r \times \dfrac{a}{2} (\cos 45^\circ + \sin 45^\circ)$

$$= (1-c)\, m_R\, \omega^2 r \times \frac{a}{2}\left[\frac{1}{\sqrt{2}} + \frac{1}{\sqrt{2}}\right] = (1-c)\, m_R\, \omega^2 r \times \frac{a}{2} \times \frac{2}{\sqrt{2}}$$

$$= \frac{a}{\sqrt{2}} (1-c)\, m_R \times \omega^2 r$$

When $\theta = 225^\circ$, maximum swaying couple $= (1-c)\, m_R \times \omega^2 r \times \dfrac{a}{2} (\cos 225^\circ + \sin 225^\circ)$

$$= (1-c)\, m_R\, \omega^2 r \times \frac{a}{2}\left(-\frac{1}{\sqrt{2}} - \frac{1}{\sqrt{2}}\right) = (1-c)\, m_R \times \omega^2 r \times \frac{a}{2}\left(-\frac{2}{\sqrt{2}}\right)$$

$$= -\frac{a}{\sqrt{2}} (1-c)\, m_R \times \omega^2 \times r$$

$\therefore$ Maximum swaying couple $= \pm \dfrac{a}{\sqrt{2}} (1-c)\, m_R \times \omega^2 \times r.$ ...(14.9)

14.5.3. Hammer Blow

The maximum magnitude of the unbalanced force perpendicular to the line of stroke is known as *hammer blow*. The unbalanced force at right angles to the line of stroke is given by equation (14.4*A*) as

$$= m_b \times r^* \times \omega^2 \times \sin\theta$$

This force will be maximum when $\sin\theta$ is maximum or when $\theta = 90^\circ$ or 270°.

$\therefore$ Maximum unbalanced force perpendicular to the line of stroke (or hammer blow)

$$= m_b \times r^* \times \omega^2 \times \sin 90^\circ \quad \text{or} \quad m_b \times r^* \times \omega^2 \sin 270^\circ$$

$$= m_b \times r^* \times \omega^2 \quad \text{or} \quad -m_b \times r^* \times \omega^2$$

$$= \pm\, m_b \times r^* \times \omega^2 \qquad \text{...(14.10)}$$

Due to hammer blow, there will be variation in pressure between the wheel and the rail. This variation is shown in Fig. 14.4 for one revolution of the wheel, where P = static wheel load (acting downwards).

$\therefore$ Net pressure between the wheel and the rail

$$= P \pm m_b \times r^* \times \omega^2 \qquad \text{...(14.10A)}$$

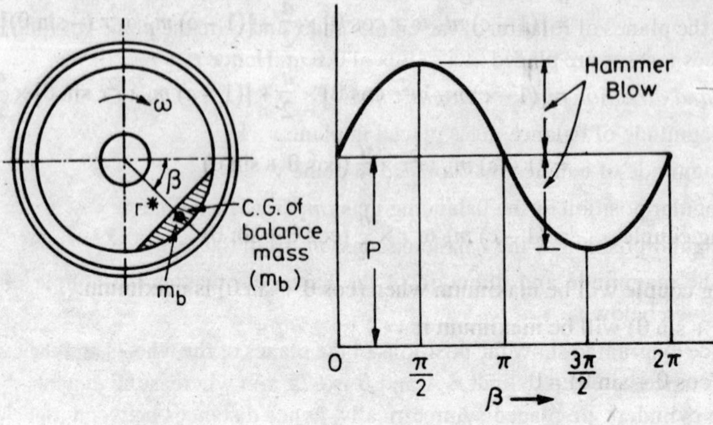

Fig. 14.4

The wheel will be lifted from the rails if $(P - m_b \times r^* \times \omega^2)$ is negative. The limiting condition when the wheel does not lift from the rails is

$$P - m_b \times r^* \times \omega^2 = 0$$

or $$P = m_b \times r^* \times \omega^2$$

or $$\omega = \sqrt{\frac{P}{m_b \times r^*}} \quad ...(14.11)$$

The above equation gives the permissible value of the angular speed.

Note. At high speeds, the hammer blow which varies as the square of the speed, may be greater than the static load P and the wheel may actually lift from the rails.

Problem 14.2. *The cranks of a two-cylinder, uncoupled inside cylinder locomotive, are at right angles and are 325 mm long. The cylinders are 675 mm apart. The rotating mass per cylinder is 200 kg at the crank pin and the mass of the reciprocating parts per cylinder is 240 kg. The wheel centre lines are 1.5 m apart. The whole of the rotating and two-third of the reciprocating masses are to be balanced and the balance masses are to be placed in the planes of the rotation of the driving wheels at a radius of 800 mm. Find :*

(i) the magnitude and direction of the balancing masses ;

(ii) the magnitude of hammer blow ;

(iii) variation in tractive force ; and

(iv) maximum swaying couple at a crank speed of 240 r.p.m.

Sol. Given :

Inside cylinder locomotive means the two cylinders are placed symmetrically between the wheels. Let the two cylinder are A and B.

Crank radius, $r_A = r_B = 325$ mm $= 0.325$ m ; distance between the cylinders, $a = 675$ mm $= 0.675$ m ; rotating mass, $M = 200$ kg ; reciprocating mass, $m_R = 240$ kg ; distance between the centre lines of wheel $= 1.5$ m ; radius of balance mass, $r^* = 800$ mm $= 0.8$ m ; $N = 240$ r.p.m. or $\omega = \frac{2\pi N}{60} = \frac{2\pi \times 240}{60} = 8\pi$ rad/s.

Fraction of reciprocating mass to be balanced $= \frac{2}{3}$ or $c = \frac{2}{3}$.

Now the total equivalent revolving mass per cylinder at crank radius, which has to be balanced

$$= M + c \times m_R$$

$$= 200 + \frac{2}{3} \times 240 = 200 + 160 = 360 \text{ kg}$$

$\therefore \qquad m_A = m_B = 360$ kg.

Let A and B are the planes of rotation of the cranks and x and y are the planes of rotation of wheels and balance masses. Balances masses are placed at a radius of 0.8 m. Hence $r_x = r_y = 0.8$ m.

(i) Magnitude and direction of the balancing masses

Let $\quad m_x$ = Magnitude of balance mass placed in plane x

m_y = Magnitude of balance mass placed in plane y.

θ_x = Angular position of the balancing mass m_x from the crank A

θ_y = Angular position of the balancing mass m_y from the crank A.

To determine the magnitude and direction of the balancing masses the couple polygon and force polygon are drawn as given below :

1. Draw the space diagram to show the positions of the planes of the wheels and the cylinders as shown in Fig. 14.5 (*a*). The distance between cylinders A and B is 675 mm whereas the distance between wheels x and y is 1500 mm. The cylinders are placed symmetrically, hence distances between planes x and A is equal to the distance between planes B and y. Hence $xA = By = \dfrac{1500 - 675}{2} = \dfrac{825}{2} = 412.5$ mm.

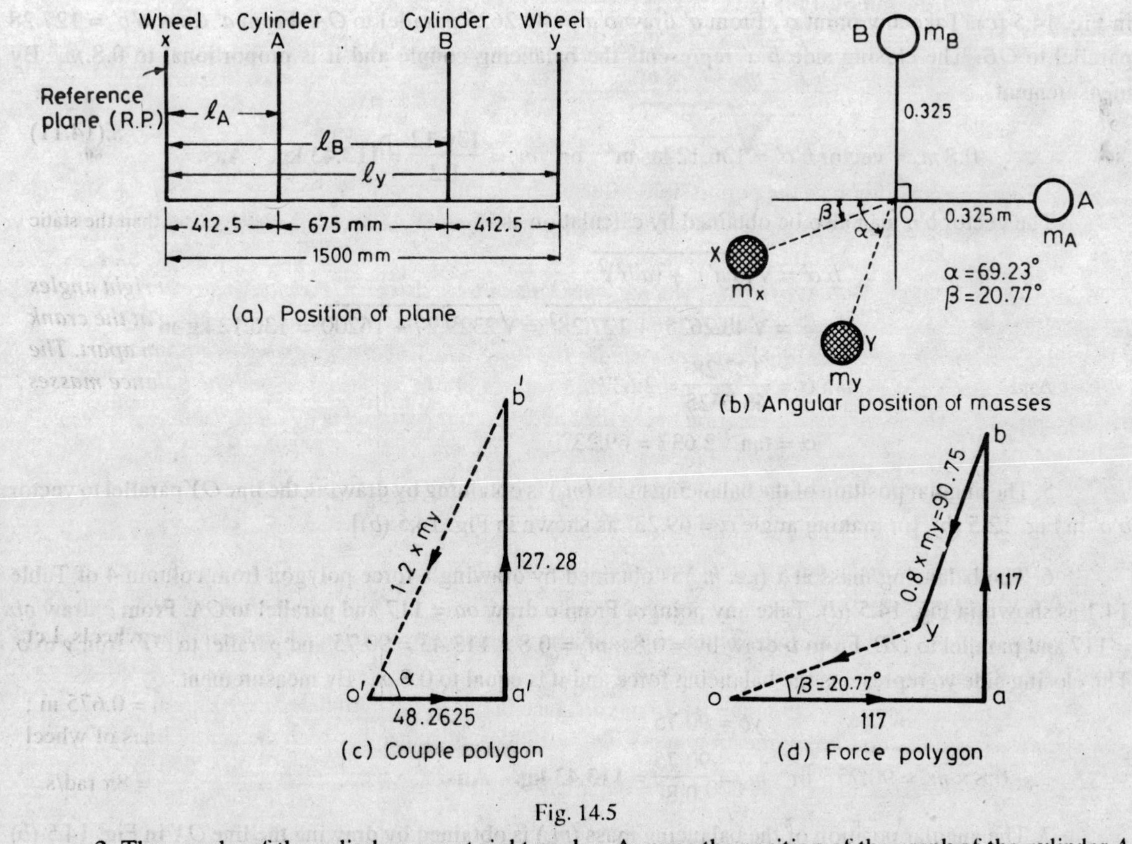

Fig. 14.5

2. The cranks of the cylinders are at right angles. Assume the position of the crank of the cylinder A in the horizontal direction. Then the position of crank of the cylinder B will be at right angles as shown in Fig. 14.5 (*b*).

3. Assuming the plane x as the reference plane*, so as to eliminate the couple due to unknown mass in that plane, the data can be tabulated as shown in Table 14.1 below :

Table 14.1

Plane	*Mass in kg (m)*	*Radius in m (r)*	*Cent. force* $\div \omega^2$ *in kg- m* $(m \times r)$	*Distance from Reference Plane x in m (l)*	*Couple* $\div \omega^2$ *in kg-m*^{2} $(m \times r \times l)$
(1)	*(2)*	*(3)*	*(4)*	*(5)*	*(6)*
x (R.P.)	m_x	$r_x = 0.8$ m	$0.8\ m_x$	0	0
A	360	$r_A = 0.325$ m	117	0.4125 m	48.2625
B	360	$r_B = 0.325$ m	117	412.5 + 675 = 1087.5 mm = 1.0875 m	127.28
y	m_y	$r_y = 0.8$ m	$0.8\ m_y$	1.5 m	$1.2 \times m_y$

4. From column (6) of the Table 14.1, the couple polygon can be drawn to some suitable scale as shown in Fig. 14.5 (*c*). Take any point o'. From o' draw $o'a'$ = 48.2625 parallel to OA. From a' draw $a'b'$ = 127.28 parallel to OB. The closing side $b'o'$ represents the balancing couple and it is proportional to $0.8\ m_y$. By measurement.

$$0.8\, m_y = \text{vector } b'o' = 136.12 \text{ kg-m}^2 \quad \text{or} \quad m_y = \frac{136.12}{1.2} = \mathbf{113.43\ kg.} \quad \textbf{Ans.}$$

The vector $b'o'$ can also be obtained by calculation as

$$b'o' = \sqrt{(o'a')^2 + (a'b')^2}$$

$$= \sqrt{48.2625^2 + 127.28^2} = \sqrt{2329.27 + 16200} = 136.12 \text{ kg-m}^2$$

And $$\tan \alpha = \frac{127.28}{48.2625} = 2.637$$

$$\therefore \quad \alpha = \tan^{-1} 2.637 = 69.23°.$$

5. The angular position of the balancing mass (m_y) is obtaining by drawing the line OY parallel to vector $b'o'$ in Fig. 12.5 (*b*). [or making angle α = 69.23° as shown in Fig. 14.5 (*b*)].

6. The balancing mass at x (*i.e.* m_x) is obtained by drawing a force polygon from column 4 of Table 14.1 as shown in Fig. 14.5 (*d*). Take any point o. From o draw oa = 117 and parallel to OA. From a draw ab = 117 and parallel to OB. From b draw by = $0.8 \times m_y = 0.8 \times 113.43 = 90.75$ and parallel to OY. Join y to o. The closing side yo represents the balancing force and it is equal to $0.8\ m_x$. By measurement,

$$yo = 90.75$$

$$\therefore \quad 0.8 \times m_x = 90.75 \quad \text{or} \quad m_x = \frac{90.75}{0.8} = \mathbf{113.43\ kg.} \quad \textbf{Ans.}$$

7. The angular position of the balancing mass (m_x) is obtained by drawing the line OX in Fig. 14.5 (*b*) parallel to vector yo. The angle β made by OX with x-axis is 20.77°.

*The magnitude of balance masses placed in planes x and y are unknown. Hence either plane x or plane y may be taken as reference plane.

(ii) *Magnitude of hammer blow*

Hammer blow is the maximum unbalanced force perpendicular to the line of stroke and it is given by equation (14.10) as

$$\text{Hammer blow} = \pm m_b \times r^* \times \omega^2$$

where m_b = Part of balance mass which is required for balancing reciprocating parts only.

The unbalanced force perpendicular to the line of stroke is due to the part of the reciprocating mass which has been balanced.

Now total balancing mass for each cylinder (*i.e.* m_x or m_y) = 113.43 kg.

Part of this total balancing mass is used for balancing rotating parts and the remaining part is used for balancing reciprocating parts.

Total equivalent mass which has to be balanced per cylinder is given as

$$= (M + c \times m_R) = 200 + \frac{2}{3} \times 240 = 200 + 160 = 360 \text{ kg.}$$

Out of 360 kg mass only $\frac{2}{3}$ of 240 = 160 kg mass is used for balancing reciprocating parts.

Hence out of total balancing mass of 113.43 kg the mass required for balancing reciprocating parts will be $\frac{160}{360} \times 113.43 = 50.41$ kg.

The vertical unbalanced force will be due to 50.41 kg.

$$\therefore \quad m_b = 50.41 \text{ kg.}$$

$$\therefore \quad \text{Hammer blow} = \pm m_b \times r^* \times \omega^2 \qquad (\because \; r^* = 0.8 \text{ m and } \omega = 8\pi \text{ rad/s})$$

$$= \pm 50.41 \times 0.8 \times (8\pi)^2 = \pm \mathbf{25473.37 \text{ N.}} \quad \textbf{Ans.}$$

(iii) *Variation in tractive force*

Maximum variation of tractive force is given by equation (14.8) as

$$= \pm \sqrt{2}(1 - c)\, m_R \times \omega^2 \times r$$

$$= \pm \sqrt{2}\left(1 - \frac{2}{3}\right) 240 \times (8\pi)^2 \times 0.325 \text{ N} \qquad (\because \; m_R = 240 \text{ kg} \; ; r = 0.325 \text{ m})$$

$$= \pm \mathbf{23225.6 \text{ N.}} \quad \textbf{Ans.}$$

(iv) *Swaying Couple*

Using equation (14.9),

$$\text{Max. swaying couple} = \pm \frac{a}{\sqrt{2}}(1 - c)\, m_R \times \omega^2 \times r$$

$$= \pm \frac{0.675}{\sqrt{2}}\left(1 - \frac{2}{3}\right) \times 240 \times (8\pi)^2 \times 0.325 \qquad (\because \; a = 0.675 \text{ m})$$

$$= \pm \mathbf{7838.6 \text{ N m.}} \quad \textbf{Ans.}$$

Analytical Method

The magnitude and direction of the balancing masses (*i.e.* m_x, m_y, θ_x and θ_y) can also be determined by analytical method as given below :

(i) For complete balancing, the resultant force and resultant couple both should be zero

i.e. $\quad F_R = 0$ and $C_R = 0$.

where $\quad F_R$ = resultant force = $\Sigma m \times r \times \omega^2$ $\qquad (\because \; m.r.\,\omega^2$ = Centrifugal force)

C_R = resultant couple = $\Sigma m \times r \times l \times \omega^2$ $\qquad (\because \; m.r.\,\omega^2 \times l$ = Force × distance = couple)

As the resultant force is zero, the components of the resultant force (and of resultant couple) in x and y direction should also be zero.

$\therefore$ $(F_R)_x = 0$ or $\Sigma m \times r \times \omega^2 \times \cos\theta = 0$ or $\Sigma m \times r \times \cos\theta = 0$ ...(*i*)

$(F_R)_y = 0$ or $\Sigma m \times r \times \omega^2 \times \sin\theta = 0$ or $\Sigma m \times r \times \sin\theta = 0$ ...(*ii*)

$(C_R)_x = 0$ or $\Sigma m \times r \times l \times \omega^2 \times \cos\theta = 0$ or $\Sigma m \times r \times l \times \cos\theta = 0$ ...(*iii*)

and $(C_R)_y = 0$ or $\Sigma m \times r \times l \times \omega^2 \times \sin\theta = 0$ or $\Sigma m \times r \times l \times \sin\theta = 0$ ...(*iv*)

Assume the angular position of the crank of the cylinder A in the horizontal direction. Then the position of the crank of the cylinder B will be at right angles as shown in Fig. 14.5 (*b*).

The magnitude and direction of balance mass m_y (i.e. in the plane y)

Take the plane x of Fig. 14.5 (*a*) as the reference plane so as to eliminate the couple due to the unknown mass in that plane. Now from equation (*iii*), we get

$\Sigma m \times r \times l \times \cos\theta = 0$ or $m_A \times r_A \times l_A \times \cos\theta_A + m_B \times r_B \times l_B \times \cos\theta_B + m_y \times r_y \times l_y \times \cos\theta_y = 0$

But $m_A = m_B = 360$ kg ; $r_A = r_B = 0.325$ m ; $r_y = 0.8$ m ; $l_A = 0.4125$ m ; $l_B = 1.0875$ m ; $l_y = 1.5$ m

$\theta_A = 0$ and $\theta_B = 90°$.

$\therefore$ $360 \times 0.325 \times 0.4125 \times \cos 0° + 360 \times 0.325 \times 1.0875 \times \cos 90° + m_y \times 0.8 \times 1.5 \times \cos\theta_y = 0$

or $$48.2625 \times 1 + 127.28 \times 0 + 1.2\, m_y \times \cos\theta_y = 0$$

or $$48.2625 + 0 + 1.2\, m_y \cos\theta_y = 0$$

or $$1.2\, m_y \times \cos\theta_y = -48.2625 \quad \text{or} \quad m_y \times \cos\theta_y = \frac{-48.2625}{1.2} = -40.21875 \qquad ...(v)$$

Similarly from equation (*iv*), we have

$\Sigma m \times r \times l \times \sin\theta = 0$ or $m_A \times r_A \times l_A \times \sin\theta_A + m_B \times r_B \times l_B \times \sin\theta_B + m_y \times r_y \times l_y \times \sin\theta_y = 0$

or $360 \times 0.325 \times 0.4125 \times \sin 0° + 360 \times 0.325 \times 1.0875 \times \sin 90° + m_y \times 0.8 \times 1.5 \times \sin\theta_y = 0$

or $$0 + 127.28 + 1.2 \times m_y \times \sin\theta_y = 0 \quad \text{or} \quad m_y \times \sin\theta_y = \frac{-127.28}{1.2} = -106.066 \qquad ...(vi)$$

Dividing equation (*vi*) by equation (*v*), we get

$$\frac{m_y \times \sin\theta_y}{m_y \times \cos\theta_y} = \frac{-106.066}{-40.21875} \quad \text{or} \quad \frac{\sin\theta_y}{\cos\theta_y} = \frac{-106.066}{-40.21875} \qquad ...(vii)$$

In equation (*vii*), the sign of numerator and denominator will give the quadrant in which θ_y will lie. The numerator corresponds to sine and denominator corresponds to cosine. As sine is – ve and also cosine is – ve, hence θ_y will lie in third quadrant.

From equation (*vii*),

$$\tan\theta_y = 2.6372 \quad \text{or} \quad \theta_y = \tan^{-1} 2.6372 = 69.23°$$

As θ_y will lie in third quadrant, hence angular position of mass m_y will be at an angle of 180 + 69.23 = 249.23° with respect to the angular position of crank A.

Squaring and adding equations (*v*) and (*vi*), we get,

$$m_y^2 \cos^2\theta_y + m_y^2 \sin^2\theta_y = (-40.21875)^2 + (-106.066)^2 = 1617.5 + 11245 = 12862.5$$

or $$m_y^2 (\cos^2\theta_y + \sin^2\theta_y) = 12862.5$$

or $$m_y^2 = 12862.5 \quad \text{or} \quad m_y = \sqrt{12862.5} = \mathbf{113.41\ kg.}$$ **Ans.**

Magnitude and direction of balance mass in the plane x

Taking the plane y of Fig. 14.5 (*a*) as the reference plane and writing the couple equations (*iii*) and (*iv*) as

$$m_A \times r_A \times l_A^* \times \cos\theta_A + m_B \times r_B \times l_B^* \cos\theta_B + m_x \times r_x \times l_x^* \times \cos\theta_x = 0$$

where $l_B^* = 0.4125$ m ; $l_A^* = 0.4175 + 0.675 = 1.0875$ m ; $l_x^* = 1.5$ m

$$\therefore \quad 360 \times 0.325 \times 1.0875 \times \cos 0° + 360 \times 0.325 \times 0.4125 \times \cos 90° + m_x \times 0.8 \times 1.5 \times \cos\theta_x = 0$$

or $$127.28 + 0 + 1.2\, m_x \times \cos\theta_x = 0$$

or $$m_x \times \cos\theta_x = \frac{-127.28}{1.2} = -106.066 \qquad ...(viii)$$

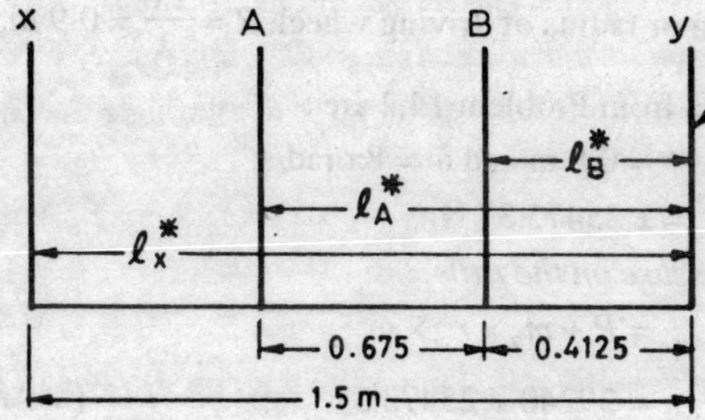

Fig. 14.5 (*e*)

From couple equation (*iv*), we have

$$m_A \times r_A \times l_A^* \times \sin\theta_A + m_B \times r_B \times l_B^* \sin\theta_B + m_x \times r_x \times l_x^* \times \sin\theta_x = 0$$

or $$360 \times 0.325 \times 1.0875 \times \sin 0° + 360 \times 0.325 \times 0.4125 \times \sin 90° + m_x \times 0.8 \times 1.5 \times \sin\theta_x = 0$$

or $$0 + 48.2625 + m_x \times 1.2 \times \sin\theta_x = 0$$

or $$m_x \times \sin\theta_x = \frac{-48.2625}{1.2} = -40.21875 \qquad ...(ix)$$

Dividing equation (*ix*) by equation (*viii*), we get

$$\frac{m_x \times \sin\theta_x}{m_x \times \cos\theta_x} = \frac{-40.21875}{-106.066}$$

or $$\frac{\sin\theta_x}{\cos\theta_x} = \frac{-40.21875}{-106.066}.$$

As the numerator is – ve and also denominator is – ve in the above equation, hence θ_x lies in third quadrant.

Also $\tan\theta_x = \dfrac{\sin\theta_x}{\cos\theta_x} = \dfrac{-40.21875}{-106.066} = 0.3791$ or $\theta_x = \tan^{-1} 0.3791 = 20.76°$.

As θ_x lies in third quadrant, hence angular position of mass m_x will be at an angle of 180 + 20.76 = 200.76° with respect to angular position of crank *A*.

Squaring and adding equations (*viii*) and (*ix*), we get

$$m_x^2 [\cos^2\theta_x + \sin^2\theta_x] = (-106.066)^2 + (-40.21875)^2$$
$$= 11245 + 1617.5 = 12862.5$$

or $$m_x = \sqrt{12862.5}$$
$$= \mathbf{113.4\ kg.} \quad \textbf{Ans.}$$

The correct angular positions of balancing masses (m_x and m_y) are shown in Fig. 14.5 (*f*).

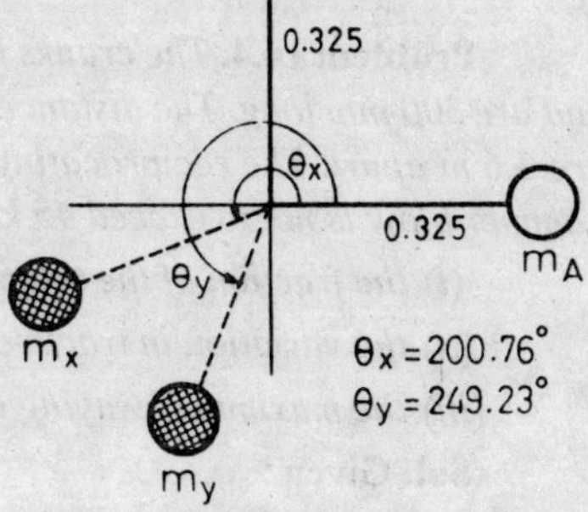

Fig. 14.5 (*f*)

Problem 14.3. *If in problem 14.2, dead load (or static load) on each wheel is 4 tonne then find :*

(*i*) *the maximum and minimum pressure on the rails,*

(*ii*) *the maximum speed of the locomotive (in km/hr) without lifting the wheels from the rails. Take the diameter of the driving wheels as 1.8 m.*

Sol. Given :

Dead load (or static load) on each wheel, $P = 4$ tonne $= 4 \times 1000$ kgf $= 4 \times 1000 \times 9.81$ N $= 39240$ N ;

dia. of driving wheel, $D = 1.8$ m or radius of driving wheel, $R = \frac{1.8}{2} = 0.9$ m.

Data and other calculated values from Problem 14.2 are :

$$m_b = 50.41 \text{ kg}, r^* = 0.8 \text{ m and } \omega = 8\pi \text{ rad/s}$$

Hammer blow $= \pm m_b \times r^* \times \omega^2 = \pm 25473.37$ N

(*i*) *Maximum and minimum pressure on the rails*

Maximum pressure on rails $= P + m_b \times r^* \times \omega^2$ [See equation (14.10A)]

$= 39240 + 25473.37$ $(\because \; m_b \times r^* \times \omega^2 = 25473.37)$

$=$ **64713.37 N. Ans.**

Minimum pressure on rails $= P - m_b \times r^* \times \omega^2$

$= 39240 - 25473.37 =$ **13766.63 N. Ans.**

(*ii*) *Maximum speed of the locomotive without lifting the wheels from the rails*

Maximum speed of the locomotive without lifting the wheel from the rails will be when the dead (or static) load becomes equal to hammer blow,

i.e. Hammer blow $= P$(static load)

or $m_b \times r^* \times \omega^2 = 39240$

or $50.41 \times 0.8 \times \omega^2 = 39240$ $(\because \; m_b = 50.41 \text{ kg}, r^* = 0.8 \text{ m})$

or $\omega = \sqrt{\dfrac{39240}{50.41 \times 0.8}}$

$\left[\omega \text{ can also be obtained from equation (14.11) as } \omega = \sqrt{\dfrac{P}{m_b \times r^*}}\right]$

$= 31.19$ rad/s

$\therefore$ Velocity of wheel $= \omega \times R$ m/s $= 31.19 \times 0.9 = 28.071$ m/s

$= \frac{28.071}{1000} \times 60 \times 60$ km/hr $=$ **101.05 km/hr. Ans.**

Problem 14.4. *The cranks of a two-cylinder uncoupled inside cylinder locomotive are at right angles and are 300 mm long. The distance between the centre lines of the cylinder is 650 mm. The wheel centre lines are 1.6 m apart. The reciprocating mass per cylinder is 300 kg. The driving wheel diameter is 1.8 m. If the hammer blow is not to exceed 45 kN at 100 km/hr, determine :*

(*i*) *the fraction of the reciprocating masses to be balanced,*

(*ii*) *the variation in tractive effort*

(*iii*) *the maximum swaying couple.*

Sol. Given :

Inside cylinder locomotive.

This means the two cylinders are placed symmetrically between the driving wheels.

Let the two cylinders are *A* and *B*.

Now crank, radius of cylinders, $r_A = r_B = 300$ mm = 0.3 m ; distance between the centre lines of the cylinder, a = 650 mm = 0.65 m, distance between driving wheels centre = 1.6 m ; reciprocating mass per cylinder, m_R = 300 kg ; driving wheel diameter, D = 1.8 m or radius of driving wheel, $R = \frac{1.8}{2} = 0.9$ m ;

Maximum hammer blow = 45 kN = 45 × 1000 = 45000 N

$$\text{Speed of wheel} \quad = 100 \text{ km/hr} = \frac{100 \times 1000}{60 \times 60} \text{ m/s} = 27.77 \text{ m/s}$$

$$\therefore \quad \text{Angular velocity of driving wheel, } \omega = \frac{V}{R} = \frac{27.77}{0.9} = 30.86 \text{ rad/s.}$$

(i) The fraction of the reciprocating masses to be balanced

Let $\quad c$ = The fraction of the reciprocating masses to be balanced

m_b = Balancing mass placed at each of the driving wheel at a radius r*.

If c is the fraction of the reciprocating masses to be balanced, then the total mass of the reciprocating parts to be balanced = $c \times m_R = c \times 300 = 300c$ kg.

Fig. 14.6 (*a*) shows the position of the driving wheels and cylinders. The cylinders are placed

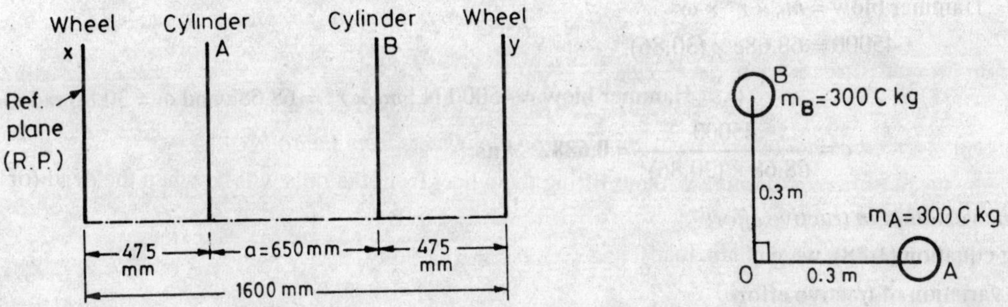

(*a*) Position of planes (*b*) Angular position of masses

Fig. 14.6

symmetrically between the driving wheels. Hence the distance between planes x and A is equal to the distance between planes B and y. Hence $xA = By = \frac{1600 - 650}{2} = \frac{950}{2} = 475$ mm. The cranks are at right angles. Assume the position of the crank of cylinder A in the horizontal direction. Then the position of crank of cylinder B will be at right angles as shown in Fig. 14.6 (*b*).

Assume the plane x as the reference plane*. Now the data can be tabulated as shown in Table 14.2 below.

Table 14.2

Plane	*Mass in kg* (m)	*Radius in m* (r)	*Cent. force* ÷ ω^2 *in kg-m* $(m \times r)$	*Distance from reference plane x in m* (l)	*Couple* ÷ ω^2 *in kg-m*^{2} $(m \times r \times l)$
(1)	*(2)*	*(3)*	*(4)*	*(5)*	*(6)*
x (R.P.)	$m_x = m_b$	r*	$m_b \times r$*	0	0
A	300c	0.3	90c	0.475 m	42.75c
B	300c	0.3	90c	475 + 650 = 1125 mm = 1.125 m	101.25c
y	$m_y = m_b$	r*	$m_b \times r$*	1600 mm = 1.6 m	1.6 × $m_b \times r$*

*The magnitude of balance masses placed in planes x and y are unknown. Hence either plane x or plane y may be taken as reference plane.

From the data given in Table 14.2 (column 6), the couple polygon can be drawn to some suitable scale as shown in Fig. 14.6 (*c*). Take any point o'. From o' draw $o'a' = 42.75c$ parallel to OA. From a', draw $a'b' = 101.25c$ parallel to OB [or $a'b'$ is perpendicular to $o'a'$]. The closing side $b'o'$ represents the balancing couple and it is proportional to $1.6 \times m_b \times r^*$. The vector $b'o'$ can be measured or it can be obtained by calculation as

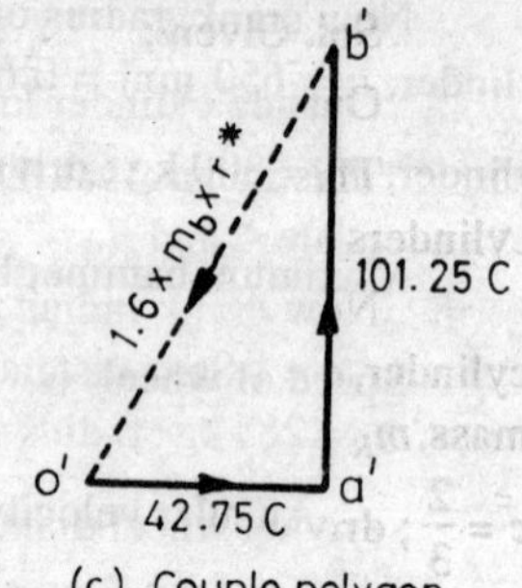

Fig. 14.6 (*c*)

$$b'o' = \sqrt{(o'a')^2 + (a'b')^2}$$

$$= \sqrt{(42.75c)^2 + (101.25c)^2} = c \times \sqrt{42.75^2 + 101.25^2}$$

$$= c \times \sqrt{1827.56 + 10251.56} = 109.9c$$

But $\qquad b'O' = 1.6 \times m_b \times r^*$

$\therefore \qquad 1.6 \times m_b \times r^* = 109.9c$

or $\qquad m_b \times r^* = \dfrac{109.9c}{1.6} = 68.68c$

Now using equation (14.10) for hammer blow, we get

$$\text{Hammer blow} = m_b \times r^* \times \omega^2$$

or $\qquad 45000 = 68.68c \times (30.86)^2$

($\because$ Hammer blow = 45000 N ; $m_b \times r^* = 68.68c$ and $\omega = 30.86$ rad/s)

or $\qquad c = \dfrac{45000}{68.68 \times (30.86)^2} = \mathbf{0.688.}$ **Ans.**

(*ii*) *The variation in tractive effort*

Using equation (14.8), we get

Max. variation of tractive effort

$$= \pm \sqrt{2}\,(1 - c)\, m_R \times \omega^2 \times r$$

$$= \pm \sqrt{2}\,(1 - 0.688) \times 300 \times (30.86)^2 \times 0.3 = \mathbf{37818.4\ N.}$$ **Ans.**

(*iii*) *The maximum swaying couple*

Using equation (14.9), we get

Max. swaying couple $\qquad = \pm \dfrac{a}{\sqrt{2}}(1 - c) \times m_R \times \omega^2 \times r$

$$= \pm \frac{0.65}{\sqrt{2}}(1 - 0.688) \times 300 \times (30.86)^2 \times 0.3 = \pm\, \mathbf{12291\ N.}$$ **Ans.**

Problem 14.5. *The cranks of a two cylinder uncoupled outside cylinder locomotive are at right angles and are 300 mm long. The distance between the centre lines of the cylinder is 1.8 m. The wheel centre lines are 1.4 m apart. The rotating mass per cylinder is 350 kg and the mass of the reciprocating parts per cylinder is 285 kg. The whole of the rotating and two third of the reciprocating masses are to be balanced in the plane of the driving wheels at a radius of 800 mm, then find :*

(*i*) *The magnitude and angular positions of balance masses,*

(*ii*) *The maximum speed of the locomotive in km/hr without lifting the wheels from the rails if the dead load (or static load) on each driving wheel is 28000 N and dia. of the driving wheel is 1.8 m and*

(*iii*) *Swaying couple at the maximum speed.*

Sol. Given :

Outside cylinder locomotive.

This means that the two cylinders are placed symmetrically outside the driving wheels. Let the two cylinders are A and B.

Now crank radius of cylinder, $r_A = r_B = 300$ mm $= 0.3$ m ; distance between the centre lines of the cylinder, $a = 1.8$ m ; distance between driving wheels centre = 1.4 m ; rotating mass, $M = 350$ kg ; reciprocating mass, $m_R = 285$ kg ; radius of balance mass, $r^* = 800$ mm $= 0.8$ m, fraction of reciprocating mass to be balanced, $c = \frac{2}{3}$; driving wheel diameter, $D = 1.8$ m or radius $R = \frac{1.8}{2} = 0.9$ m ; dead load, $P = 28000$ N.

The total equivalent revolving mass per cylinder at crank radius, which has to be balanced

$$= M + c \times m_R = 350 + \frac{2}{3} \times 285 = 350 + 190 = 540 \text{ kg.}$$

(i) The magnitude and angular positions of balance masses

The two cylinders are A and B. The mass per cylinder to be balanced is 540 kg. Hence

$$m_A = m_B = 540 \text{ kg.}$$

Let A and B are the planes of rotation of the cranks and x and y are the planes of rotation of wheels (or of balancing masses). The balancing masses are to be placed at a radius of 0.80 m. Hence $r_x = r_y = 0.80$ m.

Let m_x and m_y = Magnitude of balancing masses in planes x and y respectively.

θ_x and θ_y = Angular position of balancing masses (m_x and m_y) from the crank A respectively.

To determine the magnitude and direction of the balancing masses the couple polygon and force polygon are drawn as given below :

1. Draw the space diagram to show the positions of the planes of the driving wheels and the cylinders as shown in Fig. 14.7 (*a*). The distance between the centre line of cylinders A and B is 1.8 m whereas the distance between the centre lines of the wheels x and y is 1.4 m. The cylinders are placed symmetrically outside the wheels and distance between planes A and x is equal to the distance between planes y and B. Hence

$$Ax = yB = \frac{1.8 - 1.4}{2} = \frac{0.4}{2} = 0.2 \text{ m.}$$

2. The cranks of the cylinders are at right angles. If the angular position of the crank of the cylinder A is assumed horizontal, then the angular position of the crank of cylinder B will be at right angle to OA as shown in Fig. 14.7 (*b*).

3. Assume the plane x as the reference plane, then the data can be tabulated as shown in Table 14.3 below :

Table 14.3

Plane	*Mass in kg (m)*	*Radius in m (r)*	*Cent. force ÷ ω^2 in kg-m ($m \times r$)*	*Distance from reference plane in m (l)*	*Couple ÷ ω^2 in kg m^2 ($m \times r \times l$)*
(1)	*(2)*	*(3)*	*(4)*	*(5)*	*(6)*
A	540	0.3	162	– 0.2 m	– 32.4
x (R.P.)	m_x	0.8	$0.8\, m_x$	0	0
y	m_y	0.8	$0.8\, m$	+ 1.4 m	$1.12\, m_y$
B	540	0.3	162	+ 1.6 m	259.2

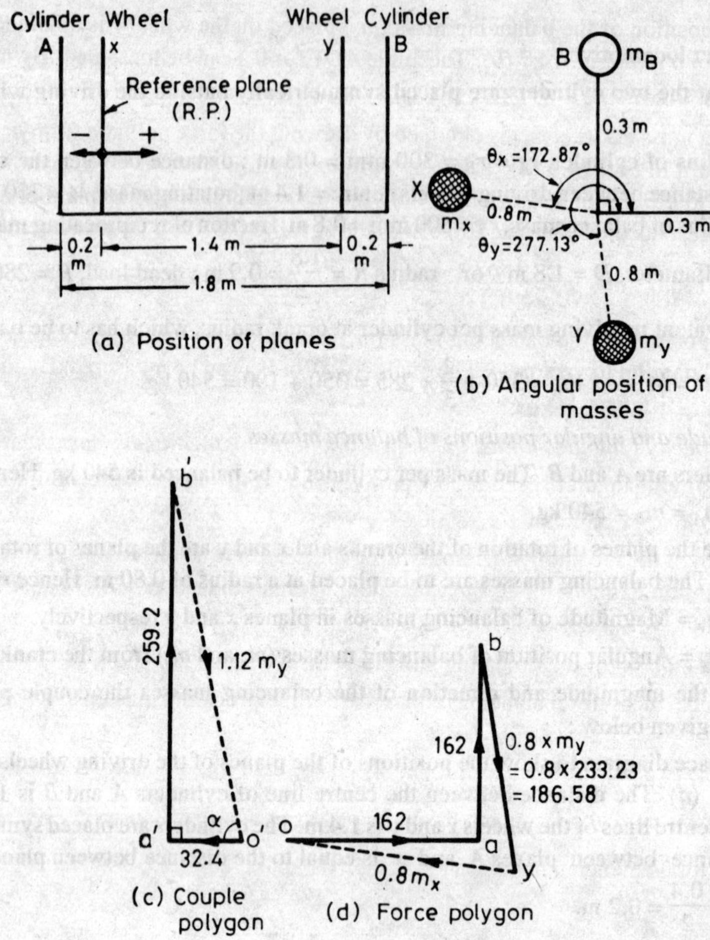

Fig. 14.7

4. From the data given in Table 14.3 (column 6), the couple polygon can be drawn as shown in Fig. 14.7 (*c*). Take any point *o'*. From *o'* draw *o'a'* = – 32.4 to some suitable scale parallel to *OA* but in opposite direction as couple for cylinder *A* is negative. From *a'* draw *a'b'* = 259.2 parallel to *OB*. The vector *a'b'* is at right angle to vector *o'a'*. The closing vector *b'o'* represents the balancing couple and it is proportional to 1.12 m_y. The vector *b'o'* can be measured or can be obtained by calculation as

$$b'o' = \sqrt{(o'a')^2 + (a'b')^2} = \sqrt{(-32.4)^2 + (259.2)^2}$$

$$= \sqrt{1049.76 + 67184.64} = \sqrt{68234.4} = 261.22$$

But *b'o'* is also = 1.12 m_y

$$\therefore \quad 1.12\, m_y = 261.22 \quad \text{or} \quad m_y = \frac{261.22}{1.12} = 233.23 \text{ kg}$$

∴ Balancing mass is plane *y*, m_y = **233.23 kg. Ans.**

From Fig. 14.7 (*c*), $\tan \alpha = \dfrac{259.2}{32.4} = 8$

$$\therefore \quad \alpha = \tan^{-1} 8 = 82.87°.$$

5. The angular position of the balancing mass (m_y) placed on the wheel *y* is obtained by drawing the line *OY* parallel to vector *b'o'* in Fig. 14.7 (*b*). The angle *AOY* (or θ_y) can be measured. By measurement

$$\theta_y = \mathbf{277.13°}. \quad \textbf{Ans.}$$

6. The balancing mass at *x* (*i.e.* m_x) is obtained by drawing the force polygon from data given in Table 14.3 (column 4) as shown in Fig. 14.7 (*d*). Take any point *o*. From *o* draw *oa* = 162 parallel to *OA*. From *a* draw *ab* = 162 parallel to *OB*. From *b* drawn $by = 0.8\, m_y = 0.8 \times 233.23 = 186.58$ parallel to *OY*. Join *y* to *o*. The closing side *yo* represents the balancing force and it is equal to $0.8 \times m_x$. By measurement, *yo* = 186.6.

$$\therefore \quad 0.8 \times m_x = 186.6 \quad \text{or} \quad m_x = \frac{186.6}{0.8} = \mathbf{233.25\ kg.} \quad \textbf{Ans.}$$

7. The angular position of the balancing mass (m_x) placed on the wheel *x* is obtained by drawing the line *OX* in Fig. 14.7 (*b*) parallel to vector *yo*. The angle *AOX* (or θ_x) can be measured. By measurement.

Angle $AOX = \theta_x = 172.87°$. **Ans.**

(ii) Maximum speed of the locomotive in km/hr without lifting the wheels from the rails

Let ω = Maximum angular speed of locomotive without lifting the wheels from the rails.

P = Dead load or static load = 28000 N (given)

Maximum speed of the locomotive without lifting the wheels from the rails will be when the dead load becomes equal to maximum hammer blow. But maximum hammer blow is given by equation (14.10) as,

$$\text{Max. hammer blow} = m_b \times r^* \times \omega^2 \qquad ...(1)$$

where m_b = the balance mass for reciprocating parts only *or*

= the balance mass for balancing either part of reciprocating masses or whole of reciprocating masses.

In this problem, two third of the reciprocating masses are to be balanced. We know that the total equivalent revolving mass per cylinder which has to be balanced $= M + c \times m_R = 350 + \frac{2}{3} \times 285 = 350 + 190$ = 540 kg. Hence if the total balancing mass is 540 kg, then the mass required for balancing reciprocating parts is $\frac{2}{3} \times 285 = 190$ kg.

But total balancing mass, m_x or m_y = 233.23 kg.

Now let us find out of 233.23 kg how much mass will be required for balancing reciprocating parts only. This value will be equal to $\frac{190}{540} \times 233.23 = 82.06$ kg.

$$\therefore \quad m_b = 82.06 \text{ kg.}$$

Substituting this value in equation (*i*), we get,

$$\text{Max. hammer blow} = 82.06 \times 0.8 \times \omega^2 \qquad (\because \ r^* = 0.8)$$

The wheels will not be lifted from the rails if

Max. hammer blow = Dead load

or $82.06 \times 0.8 \times \omega^2 = 28000$

$$\therefore \quad \omega = \sqrt{\frac{28000}{82.06 \times 0.8}} = \sqrt{426.51} = 20.65 \text{ rad/s} \qquad (\because \ R = 0.9)$$

Now velocity, $V = \omega \times R = 20.65 \times 0.9 = 18.585$ m/s

$$\therefore \quad \text{Speed of locomotive in km/hr} = \frac{18.585}{1000} \times 3600 \text{ km/hr} = \mathbf{66.906\ km/hr.} \quad \textbf{Ans.}$$

(iii) Swaying Couple at the maximum speed or when $\omega = 20.65$ *rad/s*

From equation (14.9), we know that

$$\text{Max. swaying couple} = \pm \frac{a}{\sqrt{2}}(1-c) \times m_R \times \omega^2 \times r$$

$$= \pm \frac{1.8}{\sqrt{2}}\left(1 - \frac{2}{3}\right) \times 285 \times (20.65)^2 \times 0.3 \qquad (\because \; r = \text{crank radius} = 0.3 \text{ m})$$

$$= \pm \mathbf{15468.3 \ N.} \quad \textbf{Ans.}$$

14.6. Coupled Locomotives

The uncoupled locomotives are obsolete these days. Hence the coupled locomotives are used instead of uncoupled locomotives. In case of coupled locomotives two or three pairs of wheels are coupled together to increase the adhesive weight. The coupling rods cranks are placed at 180° to the adjacent driving cranks or wheels. In such locomotives, each coupled axle is separately considered for determining the position and magnitude of the balance masses required. For the driving axle, there will be two sets of coupling rod masses to consider, as well as two sets of the cylinder masses and the balancing masses in the planes of the wheels. The driving wheel is also known as leading wheel and coupled wheel is known as trailing wheel.

Hammer blow can also be reduced in case of coupled locomotives, by equally distributing the balance masses required for the reciprocating parts between the coupled wheels instead of being concentrated in the driving wheels.

Problem 14.6. *The cranks of a two cylinder locomotive with two coupled wheels on each side, are at right angles and are 300 mm long. The distance between the centre lines of the cylinder is 600 mm. The wheel centre lines are 1.5 m apart whereas the distance between coupling rods is 1.8 m. The rotating mass per cylinder is 190 kg and the mass of the reciprocating parts per cylinder is 232 kg. Mass of each coupling rod is 244 kg and the radius of centre of coupling rod pin is 240 mm.*

The coupling rod pins are at 180° to their respective main cranks. The whole of the rotating and three fourth of the reciprocating masses are to be balanced in the plane of the wheels at a radius of 700 mm. The balance mass of the reciprocating parts is to be divided equally between the driving wheels and the coupled wheels. Determine :

(i) The magnitudes and angular positions of the masses required for driving (or leading wheels) and trailing wheels, and

(ii) The hammer blow at 100 km/hr. Take the diameter of the wheels as 1.8 m.

Sol. Given :

Let A and B are two cylinders ; x and y are two driving wheels and L and M are two coupling rods.

Crank radius of cylinders, $r_A = r_B = 300$ mm $= 0.3$ m ; distance between cylinders *i.e.* $AB = 600$ mm $= 0.6$ m ; distance between wheels *i.e.* $xy = 1.5$ m $= 1500$ mm ; distance between coupling rods *i.e.* $LM = 1.8$ m $= 1800$ mm ; rotating mass, $M = 190$ kg ; reciprocating mass, $m_R = 232$ kg ; mass of each coupling rod $= 244$ kg ; radius of coupling rod pins *i.e.* $r_L = r_M = 240$ mm $= 0.24$ m ; cranks of cylinders A and B are at right angles (this means that angular position of m_A and m_B will be at 90° angle) ; coupling rod pins of L and M are at 180° to their respective main cranks (this means the angular position of coupling rod L will be at 180° to the crank of cylinder A and angular position of coupling rod M will be at 180° to the crank position of cylinder B) ; fraction of reciprocating masses to be balanced $= \frac{3}{4}$ or $c = \frac{3}{4}$; radius of balancing masses in the planes of wheels, $r^* = 700$ mm or $r_x = r_y = 700$ mm $= 0.7$ m ; balance mass of reciprocating parts are to be divided equally between driving wheels and coupled wheels.

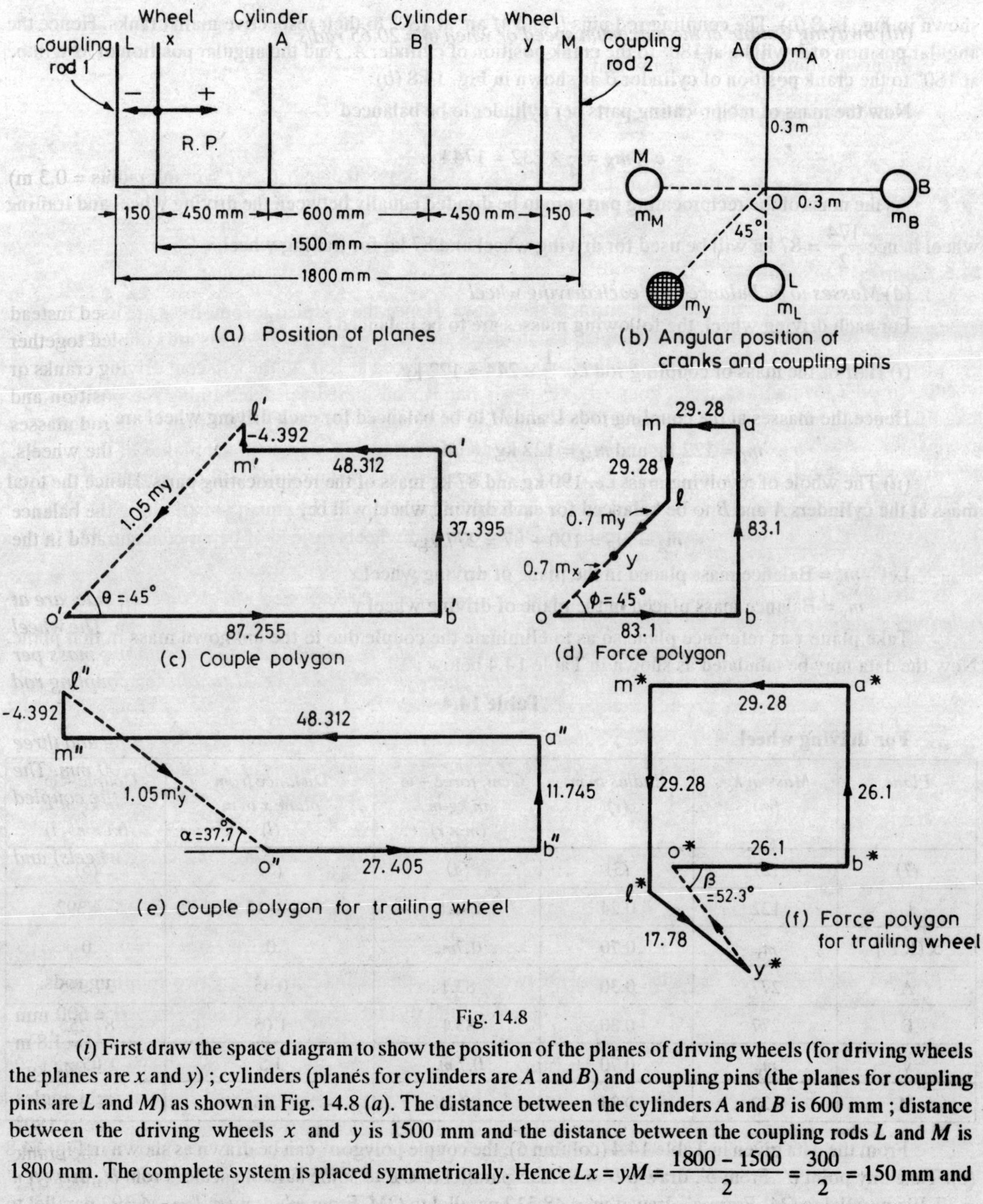

Fig. 14.8

(*i*) First draw the space diagram to show the position of the planes of driving wheels (for driving wheels the planes are x and y) ; cylinders (planes for cylinders are A and B) and coupling pins (the planes for coupling pins are L and M) as shown in Fig. 14.8 (*a*). The distance between the cylinders A and B is 600 mm ; distance between the driving wheels x and y is 1500 mm and the distance between the coupling rods L and M is 1800 mm. The complete system is placed symmetrically. Hence $Lx = yM = \frac{1800-1500}{2} = \frac{300}{2} = 150$ mm and also $xA = By = \frac{1500-600}{2} = \frac{900}{2} = 450$ mm.

(*ii*) The cranks of the cylinders A and B are at right angles. If the angular position of the crank of cylinder B is assumed horizontal, then the angular position of the crank of cylinder A will be at right angles to OB as shown in Fig. 14.8 (*b*). The coupling rod pins L and M are at 180° to their respective main cranks. Hence the angular position of L will be at 180° to the crank position of cylinder A. And the angular position of M will be at 180° to the crank position of cylinder B as shown in Fig. 14.8 (*b*).

Now the mass of reciprocating parts per cylinder to be balanced

$$= c \times m_R = \frac{3}{4} \times 232 = 174 \text{ kg.}$$

As the mass of the reciprocating parts are to be divided equally between the driving wheel and trailing wheel hence $\frac{174}{2} = 87$ kg will be used for driving wheel and 87 kg for trailing wheel.

(*a*) *Masses to be balanced for each driving wheel*

For each driving wheel, the following masses are to be balanced :

(*i*) Half of the mass of coupling rod *i.e.* $\frac{1}{2} \times 244 = 122$ kg.

Hence the masses at the coupling rods L and M to be balanced for each driving wheel are :

$$m_L = 122 \text{ kg and } m_M = 122 \text{ kg}$$

(*ii*) The whole of revolving mass *i.e.* 190 kg and 87 kg mass of the reciprocating parts. Hence the total mass at the cylinders A and B to be balanced for each driving wheel will be,

$$m_A = m_B = 190 + 87 = 277 \text{ kg.}$$

Let m_x = Balance mass placed in the plane of driving wheel x

m_y = Balance mass placed in the plane of driving wheel y.

Take plane x as reference plane so as to eliminate the couple due to the unknown mass in that plane. Now the data may be tabulated as shown in Table 14.4 below :

Table 14.4

For driving wheel

Plane	*Mass in kg* *(m)*	*Radius in m* *(r)*	*Cent. force* ÷ ω^2 *in kg-m* $(m \times r)$	*Distance from plane x in m* *(l)*	*Couple* ÷ ω^2 *in kg* m^2 $(m \times r \times l)$
(1)	(2)	(3)	(4)	(5)	(6)
L	122	0.24	29.28	– 0.15	– 4.392
x (R.P.)	m_x	0.70	$0.7 m_x$	0	0
A	277	0.30	83.1	0.45	37.395
B	277	0.30	83.1	1.05	87.255
y	m_y	0.70	$0.7 m_y$	1.5	$1.05 m_y$
M	122	0.24	29.28	1.65	48.312

From the data given in Table 14.4 (column 6), the couple polygon* can be drawn as shown in Fig. 14.8 (*c*). Take any point o'. From o', draw $o'b'$ = 87.255 parallel to OB to some suitable scale. From b', draw $b'a'$ = 37.395 parallel to OA. From a', draw $a'm'$ = 48.312 parallel to OM. From m', draw $m'l'$ = – 4.392 parallel to OL. As the couple for plane L is negative, the vector $m'l'$ should be drawn vertically upward and not downward. The closing side $l'o'$ represents the balancing couple and it is proportional to $1.05 m_y$. Measure $l'o'$.

By measurement, $l'o' = 57$ kg-m^2

$\therefore$ $l'o' = 1.05\, m_y = 57$

$$\therefore \quad m_y = \frac{57}{1.05} = \mathbf{54.3\ kg.}\ \ \textbf{Ans.}$$

and by measurement, $\theta = 45°$. **Ans.**

Now from the data given in Table 14.4 (column 4), the force polygon can be drawn as shown in Fig. 14.8 (*d*). Take any point *o*. From *o*, draw $ob = 83.1$ parallel to OB to some suitable scale. From *b*, draw $ba = 83.1$ parallel to OA. From *a*, draw $am = 29.28$ parallel to OM. From *m*, draw $ml = 29.28$ parallel to OL. From *l*, draw *ly* parallel to $l'o'$ [from Fig. 14.8 (c)]. Take $ly = 0.7 \times m_y = 0.7 \times 54.3 = 38.01$. The closing side *yo* represents the balancing force and it is proportional to $0.7 \times m_x$. Measure *yo*. By measurement, $yo = 38.01$.

$$\therefore \quad 0.7\, m_x = 38.01 \quad \text{or} \quad m_x = \frac{38.01}{0.7} = \mathbf{54.3\ kg.}\ \ \textbf{Ans.}$$

By measurement, the angle $\phi = 45°$. **Ans.**

(b) Masses to be balanced for each trailing wheel

For each trailing wheel, the following masses are to be balanced :

(*i*) Half of the mass of the coupling rod *i.e.* $\frac{1}{2} \times 244 = 122$ kg.

Hence the masses at the coupling rods *L* and *M* to be balanced for each trailing wheel are :

$$m_L = 122 \text{ kg and } m_M = 122 \text{ kg.}$$

(*ii*) The mass of 87 kg of the reciprocating parts. Hence the mass at the cylinders *A* and *B* to be balanced for each trailing wheel are :

$$m_A = m_B = 87 \text{ kg.}$$

Let m_x' = Balance mass placed in the plane of trailing wheel *x*

m_y' = Balance mass placed in the plane trailing wheel *y*.

Take plane *x* as the reference plane so as to eliminate the couple due to the unknown mass in that plane. Now the data may be tabulated as shown in Table 14.5 below :

Table 14.5

For Trailing Wheels

Plane	*Mass in kg (m)*	*Radius in m (r)*	*Cent. force ÷ ω^2 in kg-m (m × r)*	*Distance from plane x in m (l)*	*Couple ÷ ω^2 in kg m^2 (m × r × l)*
(1)	(2)	(3)	(4)	(5)	(6)
L	122	0.24	29.28	– 0.15	– 4.392
x (R.P.)	m_x'	0.70	$0.7\, m_x'$	0	0
A	87	0.30	26.1	0.45	11.745
B	87	0.30	26.1	1.05	27.405
y	m_y'	0.70	$0.7\, m_y'$	1.5	$1.05\, m_y'$
M	122	0.24	29.28	1.65	48.312

*For starting couple polygon, the angular position of known masses from Fig. 14.8 (*b*) are taken into account. The angular position of mass *B* is in horizontal direction and hence from *o'*, the vector *o'b'* should be drawn first in horizontal direction. Then the vectors for masses *A*, *M* and *L* are drawn respectively.

The balance mass m_y' placed in the plane of trailing wheel *y*, can be obtained by drawing a *couple polygon* from the data given in Table 14.5 (column 6) to some suitable scale as shown in Fig. 14.8 (*e*). Take any point o''. From o'', draw vector $o''b''$ = 27.405 parallel to *OB*. From point b'', draw vector $b''a''$ = 11.745 parallel to *OA*. From point a'', draw vector $a''m''$ = 48.312 parallel to *OM*. From point m'', draw vector $m''l''$ = – 4.392 parallel to *OL* but in opposite direction of *OL*. The closing vector $l''o''$ represents the balancing couple and it is proportional to 1.05 m_y'. Measure vector $l''o''$. By measurement,

Vector $l''o'' = 26.67 = 1.05\, m_y'$

$$\therefore \quad m_y' = \frac{26.67}{1.05} = \mathbf{25.4\ kg.} \quad \textbf{Ans.}$$

and by measurement, $\alpha = 37.7°$.

The balance mass m_x' placed in the plane of trailing wheel *x*, can now be obtained by drawing a *force polygon* from the data given in Table 14.5 (column 4) to some suitable scale as shown in Fig. 14.8 (*f*). Take any point o^*. From o^*, draw vector o^*b^* = 26.1 parallel to *OB*. From b^* draw vector b^*a^* = 26.1 parallel to *OA*. From a^*, draw vector a^*m^* = 29.28 parallel to *OM*. From point m^*, draw vector m^*l^* = 29.28 parallel to *OL*. From point l^*, draw vector $l^*y^* = 0.7 \times m_y' = 0.7 \times 25.4 = 17.78$ and parallel to vector $l''o''$ from Fig. 14.8 (*e*). Join y^* to o^*. The closing side y^*o^* represents the balancing force and it is proportional to $0.7 \times m_x'$. By measurement, $y^*o^* = 17.78$.

$$\therefore \quad 0.7 \times m_x' = 17.78 \quad \text{or} \quad m_x' = \frac{17.78}{0.7} = \mathbf{25.4\ kg.} \quad \textbf{Ans.}$$

And by measurement, $\beta = \mathbf{52.3°}$. **Ans.**

Fig. 14.9 shows the position of balanced masses in driving wheels and trailing wheels.

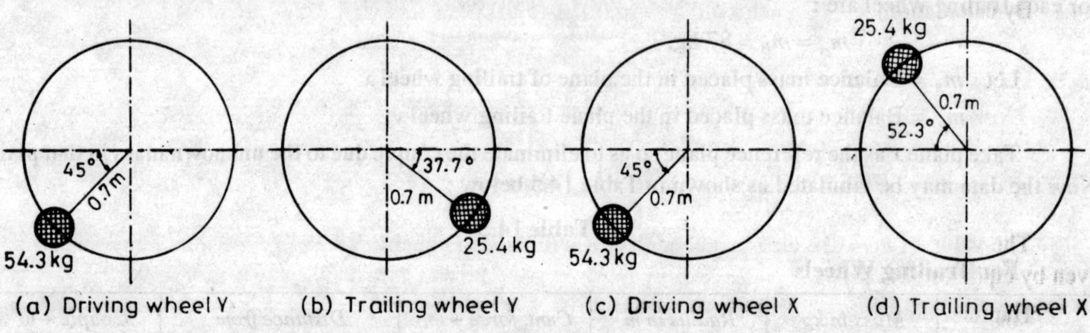

Fig. 14.9

(ii) Hammer blow at 100 km/hr

$$\text{Speed} = 100 \text{ km/hr} = \frac{100 \times 1000}{3600} \text{ m/s} = 27.77 \text{ m/s}$$

$$\text{Dia. of wheel, } D = 1.8 \text{ m} \quad \text{or} \quad \text{radius, } R = \frac{1.8}{2} = 0.9 \text{ m.}$$

$$\therefore \quad \text{Angular speed, } \omega = \frac{27.77}{R} = \frac{27.77}{0.9} = 30.86 \text{ rad/s.}$$

The hammer blow can be obtained if the balanced masses for reciprocating parts only are known.

Let m_Y'' = Balanced mass for reciprocating parts only in plane *Y* and

m_X'' = Balanced mass for reciprocating parts only in the plane *X*.

The mass at the cylinders *A* and *B* to be balanced for reciprocating parts only is 87 kg each. Hence $m_A = m_B = 87$ kg.

Take plane X as the reference plane so as to eliminate the couple due to the unknown mass in that plane. Now the data may be tabulated as shown in Table 14.6 below :

Table 14.6

For hammer blow

Plane	*Mass in kg (m)*	*Radius in m (r)*	*Cent. force ÷ ω^2 in kg-m $(m \times r)$*	*Distance from plane x in m (l)*	*Couple ÷ ω^2 in kg m^2 $(m \times r \times l)$*
(1)	*(2)*	*(3)*	*(4)*	*(5)*	*(6)*
X (R.P.)	m_x''	0.7	$0.7\,m_x''$	0	0
A	87	0.3	26.1	0.45	11.745
B	87	0.3	26.1	1.05	27.405
Y	m_Y''	0.7	$0.7\,m_y''$	1.5	$1.05\,m_Y''$

From the data given in Table 14.6 (column 6), the couple polygon can be drawn as shown in Fig. 14.10 to some suitable scale. Take any point o'. From point o', draw vector $o'b'$ = 27.405 parallel to OB. From point b', draw vector $b'a'$ = 11.745 parallel to OA. The closing side $a'o'$ represents the balancing couple and it is proportional to $1.05\,m_y''$. The vector $a'o'$ can be obtained either by measurement or by calculation.

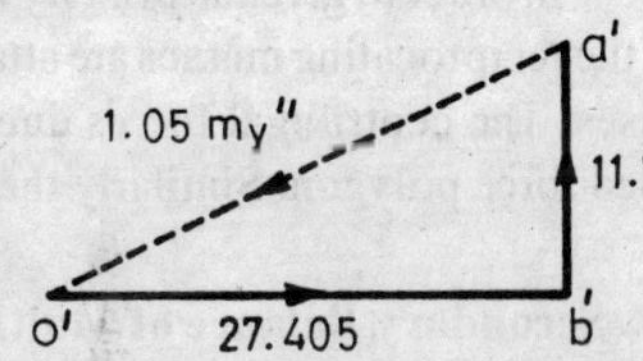

Fig. 14.10

By calculation, $\quad$ Vector $a'o' = \sqrt{(o'b')^2 + (b'a')^2}$

$$= \sqrt{27.405^2 + 11.745^2}$$

$$= \sqrt{751 + 137.94} = 29.81$$

$$\therefore \quad a'o' = 29.81 = 1.05\,m_y''$$

$$\therefore \quad m_y'' = \frac{29.81}{1.05} = 28.39 \text{ kg.}$$

The value of m_X'' can also be calculated, it will be also equal to 28.39 kg. Now the hammer blow is given by equation (14.10) as,

Hammer blow $= \pm\, m_b \times r^* \times \omega^2$

where m_b = balance mass for reciprocating parts only = m_y'' = 28.39 kg

r^* = Radius where the balance mass is placed = 700 mm = 0.7 m (given)

$\therefore$ Hammer blow $= \pm\, 28.39 \times 0.7 \times (30.86)^2 \qquad (\because\ \omega = 30.86 \text{ rad/s})$

$= \pm$ **18925.8 N. Ans.**

14.7. Primary Balance of Multi-cylinder In-line Engines

If the cylinder centre lines of the multi-cylinder engines are in the same plane and also on the same side of the centre line of the crank shaft, then the multi-cylinder engines are known as *In-line engines.* In order to give the primary balance of the reciprocating parts of a multi-cylinder engine the following two conditions must be satisfied :

(*i*) The algebraic sum of the forces (primary forces) must be zero

or $\quad \Sigma m_R \times \omega^2 \times r \times \cos\theta = 0$

or $\quad$ primary force polygon must close.

(*ii*) The algebraic sum of the couples about any point in the plane of the forces (primary forces) must be zero.

or $\Sigma m_R \times \omega^2 \times r \times a \times \cos\theta = 0$

or primary couple polygon must close.

where a = distance of the plane of rotation of the crank from a parallel reference plane.

The first condition will be satisfied if the *primary force polygon* is a closed figure. If the polygon is not closed, then the closing side of the polygon will give the maximum unbalanced primary force. Similarly the second condition will be satisfied if the *primary couple polygon* is a closed figure. If the polygon is not a closed figures, then the closing side of the polygon will give the maximum unbalanced couple.

It has already been shown that the primary unbalanced force due to a reciprocating mass is equal to the component parallel to the line of stroke of the centrifugal force produced by an equal mass attached to, and revolving with, the crank pin. [Refer to equation (14.1A), in which the primary force $F_P = m_R \times \omega^2 \times r \times \cos\theta$. The centrifugal force due to mass m_R placed at the crank pin and revolving with crank-pin at an angular velocity ω is $m_R \times \omega^2 \times r$. Component of this centrifugal force parallel to the line of stroke is $m_R \times \omega^2 \times r \times \cos\theta$].

In order to give the primary balance of the reciprocating masses of a multi-cylinder engine, it is assumed that the reciprocating masses are attached to their respective crankpins and the problem is assumed as of rotating masses. The centrifugal forces due to the rotating masses attached to the respective crankpins, must form a closed force polygon. Similarly the couple polygon for the corresponding centrifugal forces must be a closed one.

14.8. Secondary Balance of Multi-cylinder In-line Engines

It has already been shown that the secondary force (F_S) due to the reciprocating mass m_R is given by equation (14.1B) as

$$F_S = m_R \times \omega^2 \times r \times \frac{\cos 2\theta}{n} \qquad \text{...}(i)$$

where m_R = Mass of the reciprocating parts,

n = ratio of length of connecting rod to the length of crank = $\frac{l}{r}$.

We have already discussed that it is convenient to consider the primary force as equal to the component of the centrifugal force along the line of strokes. Similarly for convenience, the secondary force can also be considered as equal to the component of the centrifugal force along the line of stroke. For this the equation (*i*) can be written as

$$F_S = m_R \times (2\omega)^2 \times \left(\frac{r}{4n}\right) \times \cos 2\theta \qquad \text{...}(ii)$$

The equation (*ii*) is identical with the component parallel to the line of stroke of the centrifugal force of mass m_R, attached to an imaginary crank of length $\left(\frac{r}{4n}\right)$, revolving at twice the speed of the actual crank (*i.e.* 2ω). The two cranks coincides in position at the I.D.C., so that the imaginary secondary crank always makes an angle 2θ with I.D.C. Fig. 14.10 (*a*) shows the two cranks *i.e.* actual crank and imaginary crank.

In order to give the secondary balance of the reciprocating parts of a multi-cylinder engine, the following two conditions must be satisfied :

(*i*) The algebraic sum of the secondary forces must be equal to zero,

or $\Sigma m_R \times (2\omega)^2 \times \left(\frac{r}{4n}\right) \times \cos 2\theta = 0$

or the secondary force polygon must close.

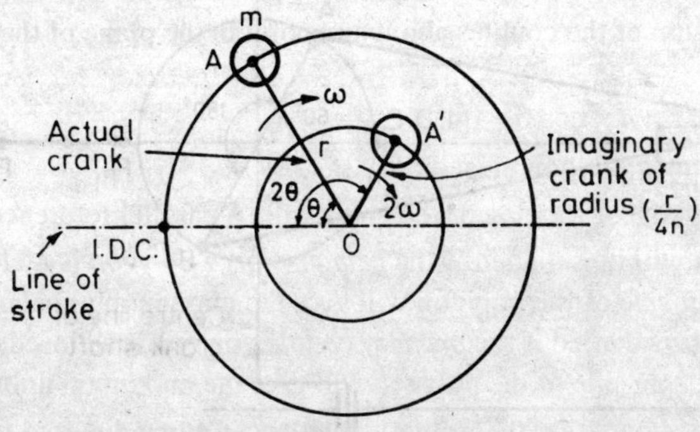

Fig. 14.10 (*a*)

(*ii*) The algebraic sum of the couples about any point in the plane of secondary forces must be zero,

or $$\Sigma m_R \times (2\omega)^2 \times \left(\frac{r}{4n}\right) \times a \times \cos 2\theta = 0$$

or the secondary couple polygon must close

where a = the distance of the plane of rotation of the secondary crank from a parallel reference plane.

Note. (*i*) If actual crank makes an angle θ with I.D.C., then the imaginary secondary crank will make an angle 2θ with I.D.C.

(*ii*) The actual crank and imaginary secondary crank coincides with 1 *D.C.*

(*iii*) Imaginary secondary crank revolves at twice the speed of the actual crank.

(*iv*) To each imaginary secondary crank, a revolving mass is attached, equal to the corresponding reciprocating mass.

(*v*) The radius of the imaginary secondary crank is less than the radius of actual crank $\left(i.e.\ r_s = \frac{r_{\text{actual}}}{4 \times n}\right)$ where r_s is the radius of imaginary secondary crank.

Problem 14.7. *A two-cylinder engine with the cranks at 180° and the cylinders on the same side of the crank-shaft centre line, is shown in Fig. 14.10 (b). If the crank of the first cylinder, makes an angle of 60° with I.D.C. and the reciprocating masses, crank lengths and connecting rod lengths are identical for each cylinder, then to what extent the engine is balanced for*

(*i*) *primary and secondary forces and*

(*ii*) *primary and secondary couples.*

Sol. Given :

Let *OA* is the crank of first cylinder which makes an angle of 60° with I.D.C.

OB is the crank of second cylinder which makes an angle of 60 + 180 = 240° with I.D.C.

m_{R_1} = Mass of reciprocating parts of 1st cylinder

m_{R_2} = Mass of reciprocating parts of 2nd cylinder

r_1 = Crank length of 1st cylinder

r_2 = Crank length of 2nd cylinder

l_1 = Connecting rod length of 1st cylinder

l_2 = Connecting rod length of 2nd cylinder

θ_1 = Angle made by the crank of 1st cylinder with I.D.C. = 60° (given)

θ_2 = Angle made by the crank of 2nd cylinder with I.D.C. = 60 + 180 = 240°

ω = Angular speed of the each crank.

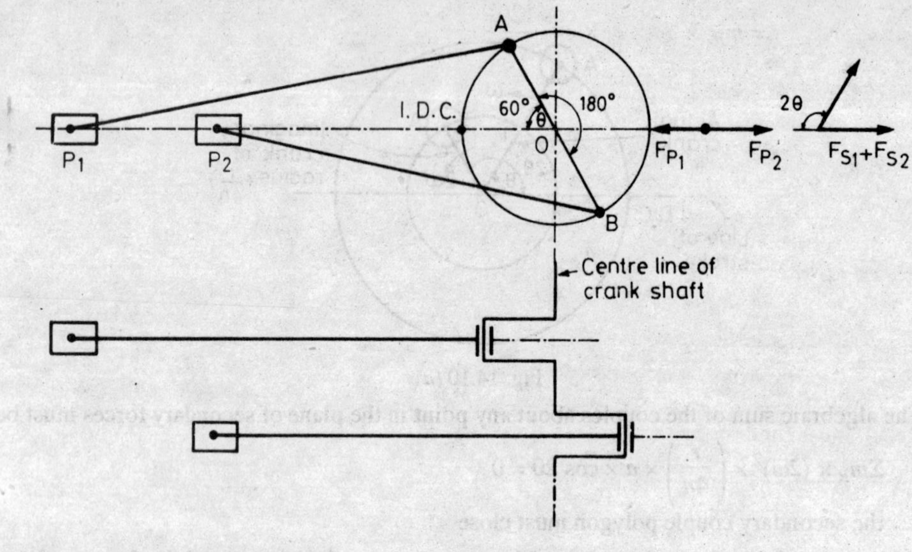

Fig. 14.10 (*b*)

(*i*) *Balancing of the engine for primary and secondary forces*

(*a*) *For primary force*

The primary force (F_P) is given by equation (14.1A) as,

$$F_P = m_R \times \omega^2 \times r \times \cos\theta$$

Hence primary force for the first cylinder,

$$F_{P_1} = m_{R_1} \times \omega^2 \times r_1 \times \cos\theta_1 = m_{R_1} \times \omega^2 \times r_1 \times \cos 60° \qquad (\because \quad \theta_1 = 60°)$$

Primary force for the second cylinder,

$$F_{P_2} = m_{R_2} \times \omega^2 \times r_2 \times \cos\theta_2 = m_{R_2} \times \omega^2 \times r_2 \times \cos(60 + 180) \qquad (\because \quad \theta_2 = 60 + 180)$$

$$= m_{R_1} \times \omega^2 \times r_1 \times \cos(60 + 180) \qquad (\because \quad m_{R_1} = m_{R_2}, r_1 = r_2)$$

$$= m_{R_1} \times \omega^2 \times r_1 \times (-\cos 60°)$$

$$= -m_{R_1} \times \omega^2 \times r_1 \times \cos 60°$$

$\therefore$ Resultant primary force

$$= F_{P_1} + F_{P_2}$$

$$= m_{R_1} \times \omega^2 \times r_1 \times \cos 60° - m_{R_1} \times \omega^2 \times r_1 \cos 60° = 0$$

The resultant primary force is zero and it is true for all positions of crank-shaft relative to I.D.C. Hence primary forces are balanced.

(*b*) *For secondary force*

The secondary force (F_S) is given by equation (14.1B) as $F_S = m_R \times \omega^2 \times r \times \dfrac{\cos 2\theta}{n}$

Hence secondary force for first cylinder,

$$F_{S_1} = m_{R_1} \times \omega^2 \times r_1 \times \frac{\cos(2\theta_1)}{n_1} = m_{R_1} \times \omega^2 \times r_1 \times \frac{\cos(2 \times 60)}{n_1} \qquad (\because \quad \theta_1 = 60°)$$

$$= m_{R_1} \times \omega^2 \times r_1 \times \frac{\cos 120°}{n_1}$$

Secondary force for second cylinder,

$$F_{S_2} = m_{R_2} \times \omega^2 \times r_2 \times \frac{\cos (2\theta_2)}{n_2} = m_{R_2} \times \omega^2 \times r_2 \times \frac{\cos (2 \times 240)}{n_2} \qquad (\because \ \theta_2 = 240°)$$

$$= m_{R_1} \times \omega^2 \times r_1 \times \frac{\cos 480°}{n_2} \qquad (\because \ m_{R_2} = m_{R_1} \,;\, r_2 = r_1)$$

$$= m_{R_1} \times \omega^2 \times r_1 \times \frac{\cos (360 + 120)}{n_1} \left(\because \ 480 = 360 + 120 \,;\, n_2 = n_1 = \frac{l}{r} = \frac{l_1}{r_1} = \frac{l_2}{r_2} \right)$$

$$= m_{R_1} \times \omega^2 \times r_1 \times \frac{\cos 120}{n_1}$$

∴ Resultant secondary force

$$= F_{S_1} + F_{S_2}$$

$$= m_{R_1} \times \omega^2 \times r_1 \times \frac{\cos 120°}{n_1} + m_{R_1} \times \omega^2 \times r_1 \times \frac{\cos 120°}{n_1}$$

$$= 2m_{R_1} \times \omega^2 \times r_1 \times \frac{\cos 120°}{n_1}$$

As the resultant secondary force is not zero. Hence secondary forces are not balanced.

(ii) Balancing of the engine for primary and secondary couples

(a) For primary couple

The primary forces are equal and opposite. But the two cranks of the two cylinders are revolving in different planes. Hence the primary forces will form a couple, which is unbalanced.

(b) For secondary couple

The resultant secondary force is not zero. The resultant secondary force will have a moment about any point (except a point on its own line of action). Hence secondary couple is not balanced.

Problem 14.8. *A four-cylinder vertical engine has cranks 300 mm long. The planes of rotation of the first, third and fourth cranks are 750 mm, 1050 mm and 1650 mm respectively from that of the second crank and their reciprocating masses are 150 kg, 400 kg and 250 kg respectively. Find the mass of the reciprocating parts for the second cylinder and the relative angular positions of the cranks in order that the engine may be in complete primary balance.*

Sol. Given :

Crank lengths of all cylinder = 300 mm = 0.3 m or $r_1 = r_2 = r_3 = r_4 = 0.3$ m ; mass of reciprocating parts $m_1 = 150$ kg, $m_3 = 400$ kg and $m_4 = 250$ kg. For complete primary balance find : (*i*) mass of the reciprocating parts for the second cylinder *i.e.* m_2 and (*ii*) relative angular positions of the cranks.

We know that in order to give the primary balance of the reciprocating parts of a multi-cylinder engine, the problem may be treated as that of revolving masses with the reciprocating masses transferred to their respective crank pins.

First draw the position of the planes of the four cylinders as shown in Fig. 14.11 (*a*). These planes are represented by *A*, *B*, *C* and *D* of four cylinders 1, 2, 3 and 4. The distances between planes *B* and *A* is 750 mm ; between *B* and *C* is 1050 mm and between *B* and *D* is 1650 mm as shown in Fig. 14.11 (*a*).

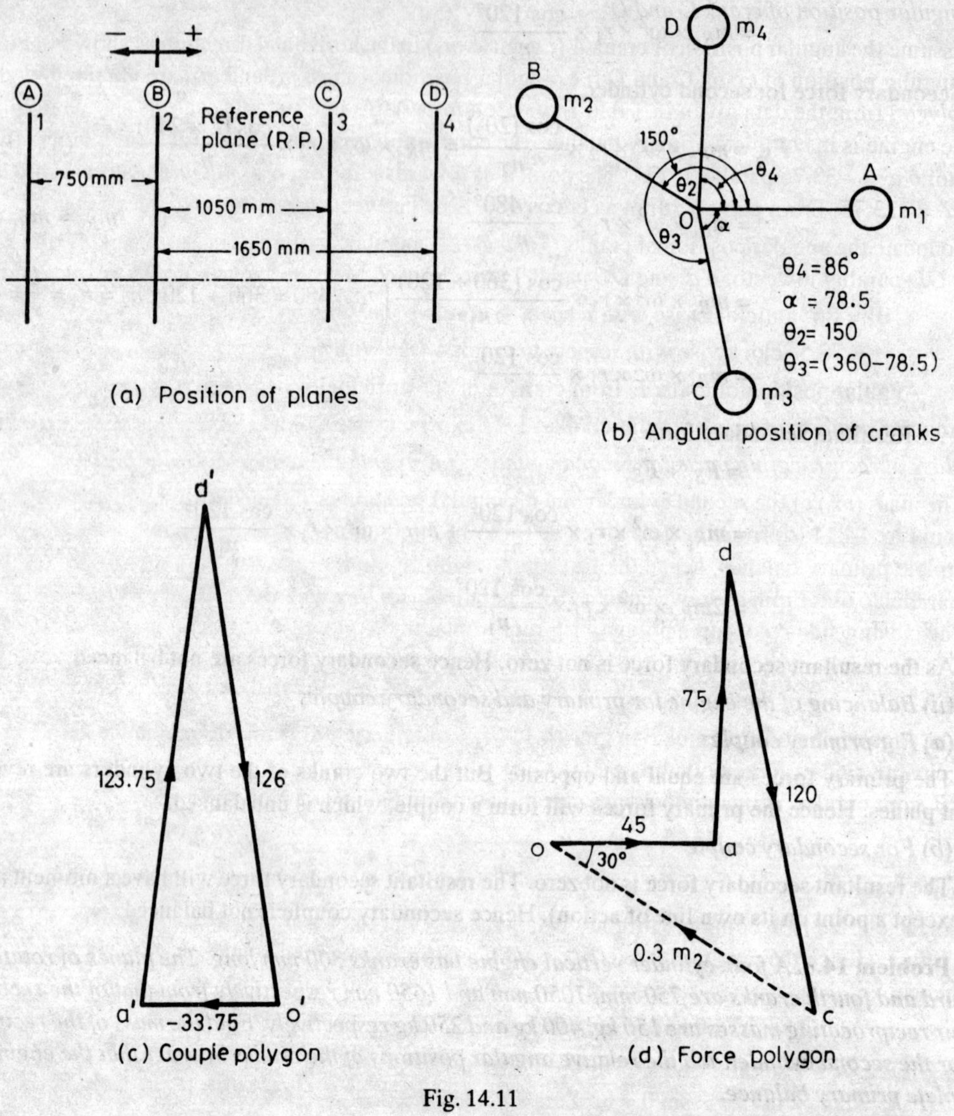

Fig. 14.11

Take plane *B* (*i.e.* plane of second cylinder) as reference plane (R.P.) so as to eliminate the couple due to the unknown mass in that plane. The data may be now tabulated as given in Table 14.7 below :

Table 14.7

Plane	*Mass in kg (m)*	*Radius in m (r)*	*Cent. force ÷ ω^2 in kg-m (m × r)*	*Distance from plane 2 in m (l)*	*Couple ÷ ω^2 in kg m^2 (m × r × l)*
(i)	*(ii)*	*(iii)*	*(iv)*	*(v)*	*(vi)*
A	150	0.3	45	– 0.75	– 33.75
B (R.P.)	m_2	0.3	0.3 m_2	0	0
C	400	0.3	120	1.05	126
D	250	0.3	75	1.65	123.75

Angular position of crank C and D

Assume the angular position of crank A (*i.e.* mass m_1) in the horizontal direction as shown in Fig. 14.12 (*b*). The angular position of crank C and D (*i.e.* angular position of mass m_3 and m_4) are obtained by drawing *couple polygon* from the data given in Table 14.7. (Column *vi*) to some suitable scale as shown in Fig. 14.11 (*c*). As the engine is in complete primary balance, the couple polygon must close. Take any point o'. From o', draw vector $o'a' = -33.75$ parallel to OA. The point a' is towards left from o' as $o'a'$ is negative. From a', draw vector $a'd' = 123.75$. From point o', draw vector $o'd' = 126$. The vectors $a'd'$ and $o'd'$ intersect at d'.

To locate the angular position of cranks C and D (*i.e.* angular position of mass m_3 and m_4) in Fig. 14.12 (*b*), draw OD parallel to vector $a'd'$ and OC parallel to vector $d'o'$. Measure angle θ_4 and α. By measurement ;

$$\theta_4 = 86° \text{ anticlockwise with respect to crank } A \text{ (i.e. with } m_1)$$

and $$\alpha = 78.5° \text{ clockwise with respect to crank } A \text{ (i.e. with } m_1)$$

∴ Angular position of crank D from crank A is **86° anticlockwise. Ans.**

Angular position of crank C from crank A is **78.5° clockwise. Ans.**

Mass of reciprocating parts for second cylinder (i.e. cylinder B) and its angular position.

The mass (m_2) of the second cylinder and its angular position is obtained by drawing the force polygon as shown in Fig. 14.11 (*d*) from the data given in Table 14.7 (column *iv*) to some suitable scale. As the engine is in complete primary balance, hence the force polygon must close. Take any point o. From o, draw vector $oa = 45$ parallel to OA. From a, draw vector $ad = 75$ parallel to OD. From point d, draw vector $dc = 120$ parallel to OC. The closing side co of force polygon is proportional to $0.3\ m_2$. By measurement, vector $co = 85.5$.

$$\therefore \quad 0.3m_2 = 85.5 \quad \text{or} \quad m_2 = \frac{85.5}{0.3} = \mathbf{285\ kg.}\ \ \textbf{Ans.}$$

The angular position of crank B in Fig. 14.11 (*b*) is obtained by drawing OB parallel to vector co [vector co is the closing side of force polygon of Fig. 14.11 (*d*)]. By measurement, we find that the angular position of crank B from crank A is in anticlockwise direction with angle $\theta_2 = 150°$.

$$\therefore \quad \theta_2 = \mathbf{150°.}\ \ \textbf{Ans.}$$

Analytical Method

As the engine is in complete primary balance hence the force polygon and couple polygon should be a closed figure. Hence the resultant force (F_R) as well as its components will always be zero. As the resultant is zero, hence

$$(F_R)_x = 0 \text{ and } (F_R)_y = 0$$

where $$(F_R)_x = \Sigma F_P = \Sigma m_R \times \omega^2 \times r \times \cos\theta = \omega^2\, \Sigma m_R \times r \times \cos\theta = 0 \quad ...(i)$$

$$(F_R)_y = \Sigma m_R \times \omega^2 \times r \times \sin\theta = \omega^2\, \Sigma m_R \times r \times \sin\theta = 0 \quad ...(ii)$$

Similarly for the closed couple polygon, the resultant couple (C_R) should be zero or

$$(C_R)_x = 0 \text{ and } (C_R)_y = 0$$

where $$(C_R)_x = \Sigma m_R \times \omega^2 \times r \times l \times \cos\theta = \omega^2\, \Sigma m_R \times r \times l \times \cos\theta = 0 \quad ...(iii)$$

$$(C_R)_y = \Sigma m_R \times \omega^2 \times r \times l \sin\theta = \omega^2\, \Sigma m_R \times r \times l \times \sin\theta = 0 \quad ...(iv)$$

Now refer to Fig. 14.11. The Fig. 14.11 (*a*) shows the position of planes of four cylinder whereas Fig. 14.11 (*b*) shows the angular positions of cranks A, B, C and D. The angular position of crank A is assumed in the horizontal direction (*i.e.* $\theta_1 = 0°$).

Let θ_2 = Angle made by crank B with the position of crank A

θ_3 = Angle made by crank C with the position of crank A

θ_4 = Angle made by crank D with the position of crank A.

Take plane *B* (*i.e.* plane of second cylinder) as reference plane so as to eliminate the couple due to the unknown mass in that plane.

Now : $m_1 = 150$ m ; $m_3 = 400$ kg ; $m_4 = 250$ kg ; $r_1 = r_2 = r_3 = r_4 = 0.3$ m ; $l_1 = -0.75$ m ; $l_2 = 0$; $l_3 = 1.05$ m and $l_4 = 1.65$ m.

$$\Sigma m_R \times r \times l \times \cos\theta = m_1 \times r_1 \times l_1 \times \cos\theta_1 + m_2 \times r_2 \times l_2 \times \cos\theta_2 + m_3 \times r_3 \times l_3 \times \cos\theta_3 + m_4 \times r_4 \times l_4 \times \cos\theta_4$$

$$= 150 \times 0.3 \times (-0.75) \times \cos 0 + 0 + 400 \times 0.3 \times 1.05 \times \cos\theta_3 + 250 \times 0.3 \times 1.65 \times \cos\theta_4$$

$$= -150 \times 0.3 \times 0.75 \times 1 + 0 + 126 \cos\theta_3 + 123.75 \cos\theta_4$$

$$= -33.75 + 126 \cos\theta_3 + 123.75 \cos\theta_4.$$

Similarly $\Sigma m_R \times r \times l \times \sin\theta = m_1 \times r_1 \times l_1 \times \sin\theta_1 + m_2 r_2 l_2 \sin\theta_2 + m_3 r_3 l_3 \times \sin\theta_3 + m_4 r_4 l_4 \sin\theta_4$

$$= 150 \times 0.3 \times (-0.75) \times \sin 0 + m_2 \times 0.3 \times 0 \times \sin\theta_2 + 400 \times 0.3 \times 1.05 \sin\theta_3 + 250 \times 0.3 \times 1.65 \times \sin\theta_4$$

$$= 0 + 0 + 126 \sin\theta_3 + 123.75 \sin\theta_4$$

For complete primary balance, from equations (*iii*) and (*iv*), we have

$$\omega^2 \Sigma m_R \times r \times l \times \cos\theta = 0 \text{ and } \omega^2 \Sigma m_R \times r \times l \times \sin\theta = 0$$

or $\Sigma m_R \times r \times l \times \cos\theta = 0$ and $\Sigma m_R \times r \times l \times \sin\theta = 0$

or $-33.75 + 126 \cos\theta_3 + 123.75 \cos\theta_4 = 0$ and $126 \sin\theta_3 + 123.75 \sin\theta_4 = 0$

or $126 \cos\theta_3 = +33.75 - 123.75 \cos\theta_4$ ...(A)

and $126 \sin\theta_3 = -123.75 \sin\theta_4$ ...(B)

Squaring and adding these two equations (A) and (B), we get

$$126^2 (\cos^2\theta_3 + \sin^2\theta_3) = (+33.75 - 123.75 \cos\theta_4)^2 + (-123.75 \sin\theta_4)^2$$

or $126^2 = (+33.75)^2 + (-123.75 \cos\theta_4)^2 + 2(+33.75)(-123.75 \cos\theta_4) + (-123.75 \sin\theta_4)^2$

$$15876 = 1139.06 + (-123.75)^2 (\cos^2\theta_4 + \sin^2\theta_4) - 8353.125 \cos\theta_4$$

$$= 1139.06 + 15314 \times 1 - 8353.125 \cos\theta_4 \qquad (\because \cos^2\theta_4 + \sin^2\theta_4 = 0)$$

or $8353.125 \cos\theta_4 = 1139.06 + 15314 - 15876$

or $\cos\theta_4 = \dfrac{577.06}{8353.125} = 0.069$

$\therefore$ $\theta_4 = \cos^{-1}(0.069) = \mathbf{86.03}$. **Ans.**

Substituting the value of θ_4 in equation (B), we get

$$126 \sin\theta_3 = -123.75 \sin 86.03°$$

or $126 \sin\theta_3 = -123.75 \times 0.9976$

or $126 \sin\theta_3 = -123.453$

or $\sin\theta_3 = \dfrac{-123.453}{126} = -0.9797$

$\therefore$ $\theta_3 = \sin^{-1}(-0.9797) = -78.46° = 360 - 78.46 = \mathbf{281.54°}$. **Ans.**

$$\Sigma m_R \times r \times \cos\theta = m_{R_1} \times r_1 \times \cos\theta_1 + m_{R_2} \times r_2 \times \cos\theta_2 + m_{R_3} \times r_3 \times \cos\theta_3 + m_{R_4} \times r_4 \times \cos\theta_4$$

$$= m_1 \times r_1 \times \cos 0° + m_2 \times r_2 \times \cos\theta_2 + m_3 \times r_3 \times \cos\theta_3 + m_4 \times r_4 \times \cos\theta_4$$

$$= 150 \times 0.3 \times 1 + m_2 \times 0.3 \times \cos\theta_2 + 400 \times 0.3 \times \cos\theta_3 + 250 \times 0.3 \times \cos\theta_4$$

$$= 45 + 0.3 m_2 \cos\theta_2 + 120 \cos\theta_3 + 75 \cos\theta_4$$

$$= 45 + 0.3 m_2 \times \cos\theta_2 + 120 \cos 281.54° + 75 \cos 86.03°$$

$$= 45 + 0.3m_2 \times \cos\theta_2 + 24 + 5.19 = 74.19 + 0.3m_2 \times \cos\theta_2$$

$$(\because \theta_3 = 281.54° ; \theta_4 = 86.03°)$$

Similarly $\Sigma m_R \times r \times \sin\theta = m_1r_1 \sin\theta_1 + m_2r_{21} \sin\theta_2 + m_3r_3 \sin\theta_3 + m_4r_4 \sin\theta_4$

$$= 150 \times 0.3 \times \sin 0° + m_2 \times 0.3 \times \sin\theta_2 + 400 \times 0.3 \times \sin 281.54° + 250 \times 0.3 \times \sin 86.04°$$

$$= 0 + 0.3m_2 \sin\theta_2 + (-117.57) + 74.82 = -42.75 + 0.3m_2 \sin\theta_2$$

For complete primary balance from equations (*i*) and (*ii*), we have

$$\omega^2 \Sigma m_R \times r \times \cos\theta = 0 \text{ and } \omega^2 \Sigma m_R \times r \times \sin\theta = 0$$

or $\Sigma m_R \times r \times \cos\theta = 0$ and $\Sigma m_R \times r \times \sin\theta = 0$

or $74.19 + 0.3m_2 \cos\theta_2 = 0$ and $-42.75 + 0.3m_2 \sin\theta_2 = 0$

or $0.3m_2 \cos\theta_2 = -74.19$ ...(C)

and $0.3m_2 \sin\theta_2 = 42.75$ ...(D)

Squaring and adding these equations

$$(0.3m_2)^2 \times (\cos^2\theta_2 + \sin^2\theta_2) = (-74.19)^2 + (42.75)^2 = 5504.15 + 1827.56 = 7331.71$$

or $0.09\, m_2^2 = 7331.71$ or $m_2 = \sqrt{\dfrac{7331.71}{0.09}} = \mathbf{285.4}$

From equation (C), $0.3 \times 285.4 \sin\theta_2 = 42.75$.

Substituting the value of m_2 in equation (C), we get

$$0.3 \times 285.4 \times \cos\theta_2 = -74.19$$

or $\cos\theta_2 = \dfrac{-74.19}{0.3 \times 285.4} = -0.866 \quad \therefore \quad \theta_2 = \cos^{-1}(-0.866) = \mathbf{150°}$. **Ans.**

Problem 14.9. *If in Problem 14.8, each connecting rod of all four cylinders is 1.35 m long and the speed is 300 r.p.m., find :*

(*i*) *maximum unbalanced secondary force and couple and*

(*ii*) *crank positions at which maximum unbalanced secondary force and couple occur.*

Sol. Data from Problem 14.8 :

$r_1 = r_2 = r_3 = r_4 = 0.3$ m ; $m_1 = 150$ kg ; $m_2 = 285$ kg (calculated in Problem 14.8) ; $m_3 = 400$ kg ; $m_4 = 250$ kg.

Data given in Problem 14.9 :

$$l_1 = l_2 = l_3 = l_4 = 1.35 \text{ m}, N = 300 \text{ r.p.m.} \quad \text{or} \quad \omega = \frac{2\pi N}{60} = \frac{2\pi \times 300}{60} = 10\pi \text{ rad/s}$$

$$\text{Now } n_1 = n_2 = n_3 = n_4 = \frac{l_1}{r_1} = \frac{l_2}{r_2} = \frac{l_3}{r_3} = \frac{l_4}{r_4} = \frac{1.35}{0.3} = 4.5$$

(*i*) *Maximum unbalanced secondary force and couple*

First find the relative positions of the secondary cranks. The secondary crank positions are found from the fact that each secondary crank must be inclined at an angle equal to two times the angle for the corresponding actual crank. Taking the position of actual crank *A* in the horizontal direction as shown in Fig. 14.11 (*b*). The actual angular positions of other cranks from crank *A* are given by [Refer to Fig. 14.11 (*b*)] :

Angular position of crank *B*, $\theta_2 = 150°$

Angular position of crank *C*, $\theta_3 = (360 - 78.5) = 281.5°$

Angular position of crank *D*, $\theta_4 = 86°$

The positions of secondary crank from crank A are obtained as :

Angular position of secondary crank $B = 2 \times \theta_2 = 2 \times 150 = 300°$

Angular position of secondary crank $C = 2 \times \theta_3 = 2 \times 281.5 = 563° = 563 - 360 = 203°$

Angular position of secondary crank $D = 2 \times \theta_4 = 2 \times 86 = 172°$.

Now the angular position of secondary cranks can be obtained as shown in Fig. 14.12 (*a*). Now draw the force polygon for secondary cranks as shown in Fig. 14.12 (*b*) from the data given in Table 14.7 (column *iv*) to some suitable scale. Take any point *o*. From *o*, draw *oa* = 45 parallel to *OA*. From *a*, draw *ab* = 85.5

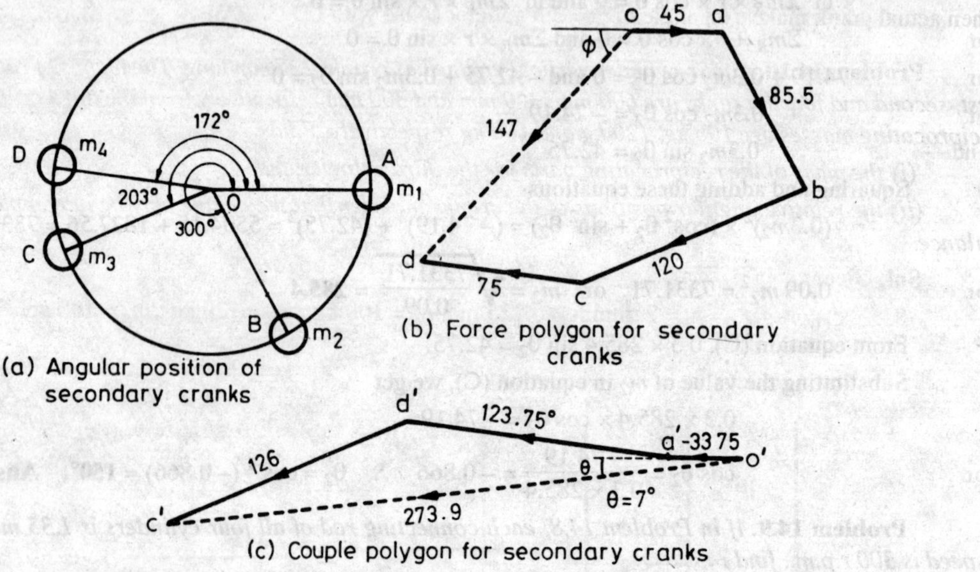

(a) Angular position of secondary cranks

(b) Force polygon for secondary cranks

(c) Couple polygon for secondary cranks

Fig. 14.12

parallel to *OB*. From *b*, draw *bc* = 120 parallel to *OC*. From *c*, draw *cd* = 75 parallel to *OD*. As force polygon for secondary cranks is not a closed figure. Hence there will be an unbalanced secondary force. The closing side *od* of the force polygon shown dotted in the figure will be proportional to some unbalanced force. The angle made by *od* with line of stroke is ϕ. By measurement *od* = 147 kg m and $\phi = 49°$.

$$\therefore \quad \text{Maximum unbalanced force} = 147 \times \frac{\omega^2}{n} \text{ where } \omega = 10\pi \text{ rad/s and } n = 4.5$$

$$= 147 \times \frac{(10\pi)^2}{4.5} \text{ N} = \mathbf{32240\ N = 32.24\ kN.} \quad \textbf{Ans.}$$

Maximum unbalanced secondary couple

The maximum unbalanced secondary couple can be obtained by drawing couple polygon for secondary cranks shown in Fig. 14.12 (*a*). The couple polygon for secondary cranks is shown in Fig. 14.12 (*c*), which is drawn from the data given in Table 14.7 (column *vi*). Take any point *o'*. From *o'*, draw *o'a'* = – 33.75 parallel to *OA* to some suitable scale. The point *a'* is towards left from *o'* as *o'a'* is negative. From *a'*, draw *a'd'* = 123.75 parallel to *OD*. From *d'*, draw *d'c'* = 126 parallel to *OC*. As the secondary couple polygon is not a closed figure. Hence there will be an unbalanced secondary couple. The closing side *o'c'* of the couple polygon shown dotted in Fig. 14.12 (*c*) will be proportional to maximum unbalanced secondary couple. By measurement *o'c'* = 273.9 kg m^2.

$$\therefore \quad \text{Max. unbalanced couple} = 273.9 \times \frac{\omega^2}{n}$$

$$= 273.9 \times \frac{(10\pi)^2}{4.5} \text{ N m} = 60073 \text{ N m} = \mathbf{60.073 \text{ kN m.}} \quad \textbf{Ans.}$$

(ii) Crank positions at which maximum unbalanced secondary force and couple occur

From Fig. 14.12 (*b*) which is force polygon for secondary cranks the angle $\phi = 49°$. The angle of actual crank will be $\frac{\phi}{2} = \frac{49}{2} = 24.5°$. Hence max. unbalanced force occurs when actual crank makes an angle of 24.5° anti-clockwise with I.D.C. Also from Fig. 14.12 (*c*), the angle $\theta = 7°$. Hence max. unbalanced couple occurs when actual crank makes, an angle of $\frac{7}{2} = 3.5$ anticlockwise with I.D.C. **Ans.**

Problem 14.10. *A four-cylinder vertical engine has cranks 225 mm long. The planes of rotation of the first, second and fourth cranks are 600 mm, 300 mm and 300 mm respectively from the third crank and their reciprocating masses are 100 kg, 120 kg and 100 kg respectively. Find :*

(i) the mass of the reciprocating parts for the third cylinder and

(ii) the relative angular positions of the cranks in order that the engine may be in complete primary balance.

Sol. Given :

$$r_1 = r_2 = r_3 = r_4 = 225 \text{ mm} = 0.225 \text{ m} \; ; m_1 = 100 \text{ kg} \; ; m_2 = 120 \text{ kg} \; ; m_4 = 100 \text{ kg}.$$

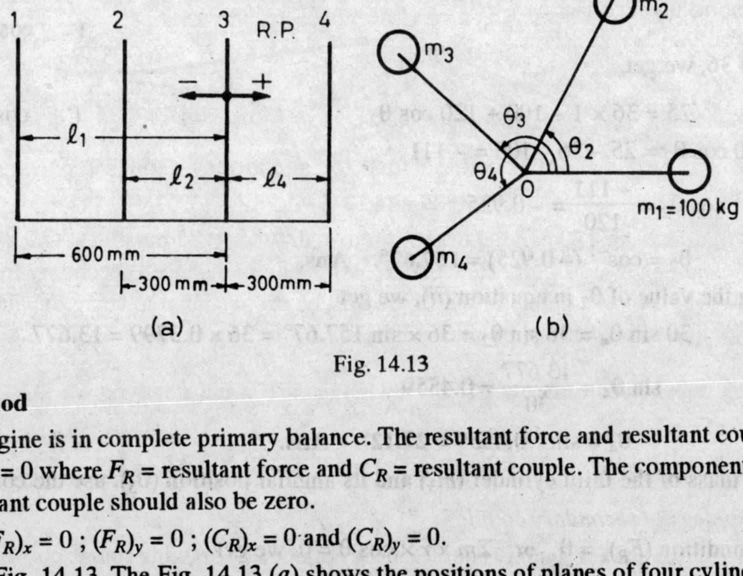

Fig. 14.13

Analytical Method

As the engine is in complete primary balance. The resultant force and resultant couple should be zero or $F_R = 0$ and $C_R = 0$ where F_R = resultant force and C_R = resultant couple. The components of resultant force and that of resultant couple should also be zero.

or $\qquad (F_R)_x = 0 \; ; (F_R)_y = 0 \; ; (C_R)_x = 0$ and $(C_R)_y = 0$.

Refer to Fig. 14.13. The Fig. 14.13 (*a*) shows the positions of planes of four cylinders. Let us assume the angular position of crank 1 of the first cylinder in horizontal position as shown in Fig. 14.13 (*b*).

Let $\quad \theta_2$ = Angle made by the crank of second cylinder with crank 1

θ_3 = Angle made by the crank of third cylinder with crank 1

θ_4 = Angle made by the crank of fourth cylinder with crank 1

m_3 = mass of reciprocating parts of third cylinder.

Take plane of the third cylinder (*i.e.* plane 3) as the reference plane so as to eliminate the couple due to unknown mass in that plane (*i.e.* $m_3 \times r_3 \times l_3 \times \cos\theta_3 = 0$) as shown in Fig. 14.13 (*a*).

Now using the condition, $(C_R)_x = 0$ or $\Sigma m \times r \times l \times \cos\theta = 0$, we get

$$m_1 r_1 l_1 \cos\theta_1 + m_2 r_2 l_2 \cos\theta_2 + m_4 r_4 \times l_4 \times \cos\theta_4 = 0$$

where $l_1 = -600$ mm $= -0.6$ m ; $l_2 = -300$ mm $= -0.3$ m and $l_4 = 300$ mm $= 0.3$ m

or $$100 \times 0.225 \times (-0.6) \times \cos 0° + 120 \times 0.225 \times (-0.3) \times \cos\theta_2 + 100 \times 0.225 \times 0.3 \times \cos\theta_4 = 0$$

Dividing by 0.225, we get

$$100 \times (-0.6) \times 1 + 120 \times (-0.3) \cos\theta_2 + 100 \times 0.3 \cos\theta_4 = 0$$

or $$-60 - 36\cos\theta_2 + 30\cos\theta_4 = 0 \quad \text{or} \quad 30\cos\theta_4 = 36\cos\theta_2 + 60 \qquad ...(i)$$

Now using the condition, $(C_R)_y = 0$ or $\Sigma m \times r \times l \times \sin\theta = 0$, we get

$$m_1 \times r_1 \times l_1 \sin\theta_1 + m_2 r_2 l_2 \sin\theta_2 + m_4 r_4 l_4 \sin\theta_4 = 0$$

or $$100 \times 0.225 \times (-0.6) \sin 0° + 120 \times 0.225 \times (-0.3) \times \sin\theta_2 + 100 \times 0.225 \times 0.3 \times \sin\theta_4 = 0$$

Dividing by 0.225, we get

$$-100 \times 0.6 \times \sin 0° + 120 \times (-0.3) \sin\theta_2 + 100 \times 0.3 \sin\theta_4 = 0$$

or $$0 - 36\sin\theta_2 + 30\sin\theta_4 = 0 \quad \text{or} \quad 30\sin\theta_4 = 36\sin\theta_2 \qquad ...(ii)$$

Squaring and adding equations (*i*) and (*ii*), we get

$$900\cos^2\theta_4 + 900\sin^2\theta_4 = (36\cos\theta_2 + 60)^2 + (36\sin\theta_2)^2$$

or $$900(\cos^2\theta_4 + \sin^2\theta_4) = 36^2 \times \cos^2\theta_2 + 3600 + 2 \times 36 \times 60 \times \cos\theta_2 + 36^2 \sin^2\theta_2$$

or $$900 = 36^2(\cos^2\theta_2 + \sin^2\theta_2) + 3600 + 2 \times 36 \times 60\cos\theta_2$$

$$(\because \cos^2\theta_4 + \sin^2\theta_4 = 1)$$

Dividing by 36, we get

$$25 = 36 \times 1 + 100 + 120\cos\theta_2 \qquad (\because \cos^2\theta_2 + \sin^2\theta_2 = 1)$$

or $$120\cos\theta_2 = 25 - 36 - 100 = -111$$

or $$\cos\theta_2 = \frac{-111}{120} = -0.925$$

$$\therefore \quad \theta_2 = \cos^{-1}(-0.925) = \mathbf{157.67°}. \quad \textbf{Ans.}$$

Substituting the value of θ_2 in equation (*ii*), we get

$$30\sin\theta_4 = 36\sin\theta_2 = 36 \times \sin 157.67° = 36 \times 0.3799 = 13.677$$

$$\therefore \quad \sin\theta_4 = \frac{13.677}{30} = 0.4559$$

or $$\theta_4 = \sin^{-1} 0.4559 = \mathbf{27.12°}. \quad \textbf{Ans.}$$

To find the mass of the third cylinder (m_3) and its angular position (θ_3), use the conditions $(F_R)_x = 0$ and $(F_R)_y = 0$.

Using the condition $(F_R)_x = 0$ or $\Sigma m \times r \times \cos\theta = 0$, we get

$$m_1 r_1 \cos\theta_1 + m_2 r_2 \cos\theta_2 + m_3 r_3 \cos\theta_3 + m_4 r_4 \cos\theta_4 = 0$$

or $$100 \times 0.225 \times \cos 0° + 120 \times 0.225 \times \cos 157.67° + m_3 \times 0.225 \times \cos\theta_3 + 100 \times 0.225 \times \cos 27.12° = 0$$

Dividing by 0.225, we get

$$100 \times 1 + 120 \times (-0.925) + m_3 \times \cos\theta_3 + 100 \times 0.89 = 0$$

or $$100 - 111 + m_3\cos\theta_3 + 89 = 0 \quad \text{or} \quad m_3\cos\theta_3 = -100 + 111 - 89 = -78 \qquad ...(iii)$$

Now using the condition $(F_R)_y = 0$ or $\Sigma m \times r \times \sin\theta = 0$, we get

$$m_1 r_1 \sin\theta_1 + m_2 r_2 \sin\theta_2 + m_3 r_3 \sin\theta_3 + m_4 r_4 \sin\theta_4 = 0$$

$$100 \times 0.225 \times \sin 0 + 120 \times 0.225 \times \sin 157.67 + m_3 \times 0.225 \sin\theta_3 + 100 \times 0.225 \times \sin 27.12 = 0$$

Dividing by 0.225, we get

$$100 \times 0 + 120 \times 0.3799 + m_3 \sin \theta_3 + 100 \times 0.4559 = 0$$

or $$0 + 45.59 + m_3 \sin \theta_3 + 45.59 = 0 \quad \text{or} \quad m_3 \sin \theta_3 = -45.59 - 45.59 = -91.18 \qquad ...(iv)$$

Squaring and adding equations (*iii*) and (*iv*), we get

$$m_3^2 (\cos^2 \theta_3 + \sin^2 \theta_3) = (-78)^2 + (-91.18)^2 = 6084 + 8313.8 = 14397.8$$

or $$m_3^2 \times 1 = 14397.8 \qquad (\because \cos^2 \theta_3 + \sin^2 \theta_3 = 1)$$

or $$m_3 = \sqrt{14397.8} = \mathbf{120\ kg.} \quad \textbf{Ans.}$$

Now dividing equation (*iv*) by equation (*iii*), we get

$$\frac{m_3 \sin \theta_3}{m_3 \cos \theta_3} = \frac{-91.18}{-78} \quad \text{or} \quad \frac{\sin \theta_3}{\cos \theta_3} = \frac{-91.18}{-78}$$

(As the numerator is – ve and denominator is also – ve which means $\sin \theta_3$ is – ve and also $\cos \theta_3$ is – ve. This is possible only when θ_3 lies in third quadrant)

$$\therefore \quad \tan \theta_3 = \frac{-91.18}{-78} = 1.1689$$

$$\therefore \quad \theta_3 = \tan^{-1} 1.1689 = 49.45°$$

As θ_3 is in third quadrant, hence

$$\theta_3 = 180 + 49.45 = \mathbf{229.45°}. \quad \textbf{Ans.}$$

The correct angular positions of all cylinders are shown in Fig. 14.13 (*c*).

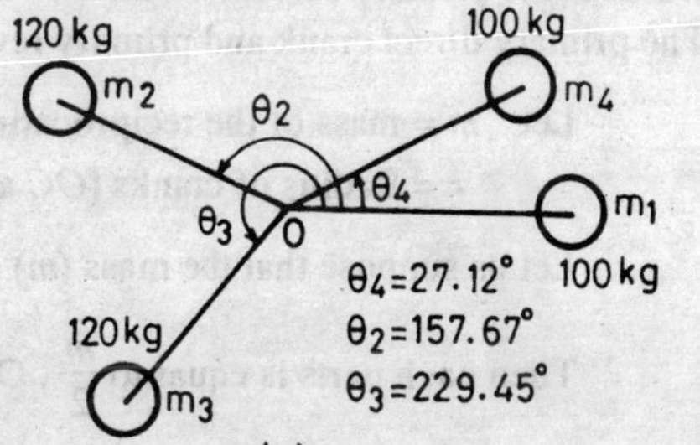

(c)

Fig. 14.13 (*c*)

Graphical Method

The above problem can be done graphically Fig. 14.13 (*c*) by drawing force and couple polygons as discussed earlier.

14.9. Method of Direct and Reverse Cranks

The method of direct and reverse cranks is used in balancing the multi-cylinder radial engine in which all the connecting rods are connected to a *common crank*. This method is very useful for determining any primary and secondary forces in radial or *V*-engines.

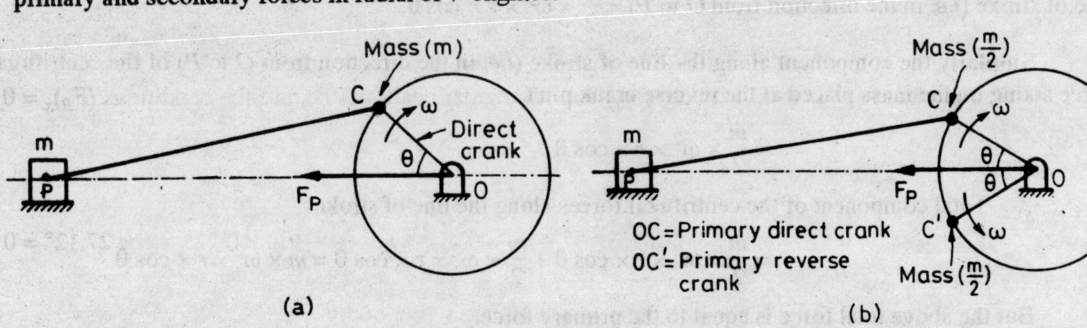

Fig. 14.14

Fig. 14.14 (*a*) shows a reciprocating engine in which the crank *OC* rotates uniformly at ω radians per second in clockwise direction. Let θ is the angle made by the crank *OC* at any instant with *OP*.

Primary force

We know that the primary force, $F_P = m\omega^2 \times r \times \cos\theta$

where m = Mass of reciprocating parts, placed at P as shown in Fig. 14.14.

This force is equal to the horizontal component of the centrifugal force produced by mass m placed at crank pin C.

Fig. 14.14 (*b*) shows a reciprocating engine in which OC is the actual crank and OC' is imaginary crank which makes an angle $(-\theta)$ with OP and is rotating uniformly at ω rad/s in anticlockwise direction.

The actual crank OC is known as primary direct crank whereas the imaginary crank OC' is known as primary reverse crank. The primary direct crank OC is rotating clockwise uniformly at a speed of ω rad/s whereas the primary reverse crank OC' is rotating *anti-clockwise* uniformly at the same speed *i.e.* at ω rad/s. The primary direct crank and primary reverse crank coincides at I.D.C.

Let m = mass of the reciprocating parts placed at P

r = Radius of cranks (OC and OC')

Let us suppose that the mass (m) of the reciprocating parts is divided equally into two parts.

Then each parts is equal to $\frac{m}{2}$. One of the part $\left(i.e. \text{ one mass } \frac{m}{2}\right)$ is placed at the direct crank pin C and the other part $\left(i.e. \text{ second mass } \frac{m}{2}\right)$ is placed at the reverse crank pin C' as shown in Fig. 14.14 (*b*).

The centrifugal force acting on the mass placed at the direct crank pin C

$$= \frac{m}{2} \times \omega^2 \times r$$

Similarly the centrifugal force acting on the mass placed at the reverse crank pin C'

$$= \frac{m}{2} \times \omega^2 \times r$$

Now the component of the centrifugal force acting on the mass placed at direct crank pin C along the line of stroke (*i.e.* in the direction from O to P) $= \frac{m}{2} \times \omega^2 \times r \times \cos\theta$.

Similarly the component along the line of stroke (*i.e.* in the direction from O to P) of the centrifugal force acting on the mass placed at the reverse crank pin C'

$$= \frac{m}{2} \times \omega^2 \times r \times \cos\theta$$

$\therefore$ Total component of the centrifugal forces along the line of stroke

$$= \frac{m}{2} \times \omega^2 \times r \times \cos\theta + \frac{m}{2} \times \omega^2 \times r \times \cos\theta = m \times \omega^2 \times r \times \cos\theta$$

But the above total force is equal to the primary force.

Hence for determining the primary force. The mass m of the reciprocating parts at P can be replaced by two masses $\left(\frac{m}{2}\right)$ each at C and C'.

The components of the centrifugal forces of these two masses normal to the line of stroke will be equal and opposite.

Secondary force

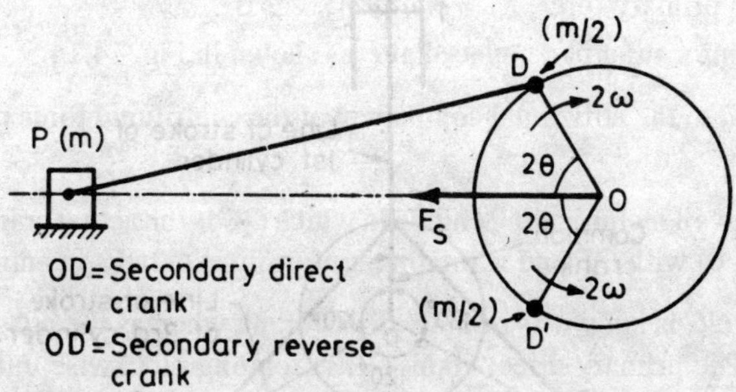

Fig. 14.15

We know that the secondary force (F_S) is given by,

$$F_S = m \times \omega^2 \times r \times \frac{\cos 2\theta}{n}$$

$$= m \times (2\omega)^2 \times \frac{r}{4n} \times \cos 2\theta$$

For secondary force, the mass m of the reciprocating parts is replaced by two masses and equal to $\frac{m}{2}$ placed at D and D' such that OD is the direct secondary crank making an angle $(+ 2\theta)$ and OD' is the reverse secondary crank making an angle $(- 2\theta)$ with the line of stroke as shown in Fig. 14.15. The crank OD is rotating clockwise uniformly at 2ω rad/s whereas the crank OD' is rotating anti-clockwise uniformly at 2ω rad/s. The length of the crank OD and OD' is equal to $\frac{r}{4n} = \frac{r}{4 \times \frac{l}{r}} = \frac{r^2}{4l}$.

$$\therefore \quad OD = OD' = \frac{r}{4n} = \frac{r^2}{4l}.$$

Problem 14.11. *A radial engine has three cylinders whose axes are spaced at angular interval of 120°. The three connecting rods are coupled directly to a single crank. The stroke is 120 mm, and the length of each connecting rod is 180 mm. The mass of the reciprocating parts per cylinder is 2 kg. Find the resultant primary and secondary forces acting on the frame of the engine when running at 2100 r.p.m.*

Sol. Given :

Stroke, L = 120 mm or $r = \frac{L}{2} = \frac{120}{2}$ = 60 mm = 0.06 m ; length of connecting rods, l = 180 mm = 0.18 m ; mass of reciprocating parts per cylinder, m = 2 kg ; N = 2100 r.p.m. or $\omega = \frac{2\pi \times N}{60} = \frac{2\pi \times 2100}{60}$ = 70π rad/s.

Fig. 14.16 shows the positions of three cylinders in which OX, OY and OZ are the lines of strokes (or centre lines) of cylinder 1, 2 and 3 respectively. Let the common crank OC is along the inner dead centre of cylinder 1. The common crank OC is rotating clockwise and hence θ will be positive when measured clockwise. Also θ is the angle made by the common crank with the line of stroke of the respective cylinder.

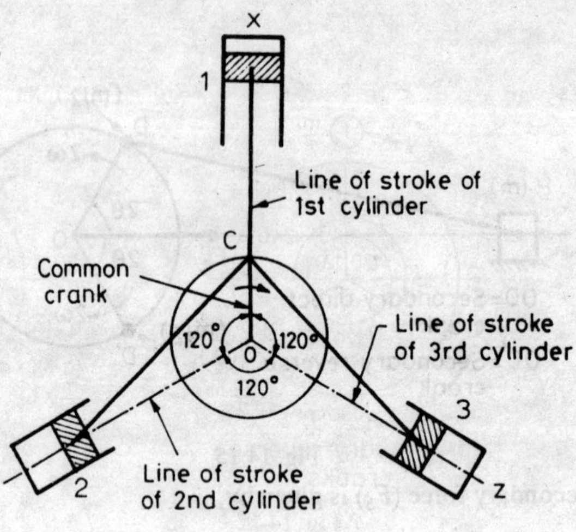

Fig. 14.16

Resultant primary force acting on the frame of the engine

To find the resultant primary force, the positions of the primary direct cranks and reverse cranks are to be obtained first by the method as given below :

(*i*) For cylinders 1, $\theta = 0$, therefore both the primary direct and reverse cranks will coincide and they will be along the common crank *OC*.

(*ii*) For cylinder 2, the line of stroke is along OY. The clockwise angle made by the common crank *OC* with the line of stroke of 2nd cylinder (*i.e.* with axis *OY*) is 120°. The crank *OC* with respect to axis *OY* is making an angle 120° clockwise. Hence θ for second cylinder is + 120°. Hence the primary direct crank for cylinder 2 will be at an angle of + 120° (*i.e.* clockwise from axis *OY*) from axis *OY*.

Hence the primary direct crank will be along *OC*. The primary reverse crank for cylinder 2 will be at an angle of – 120° from the axis *OY* (*i.e.* anticlockwise from axis *OY*).

(*iii*) For cylinder 3, the line of stroke is along *OZ*. The clockwise angle made by the common crank *OC* with the line of stroke of 3rd cylinder (*i.e.* with axis *OZ*) is 120 + 120 = 240°. The crank *OC* with respect to axis *OZ* is making an angle of 240° clockwise [or 120° anticlockwise]. Hence θ for third cylinder is + 240°. Hence the primary direct crank for cylinder 3 will be at an angle of + 240° clockwise from axis *OZ*. Hence the primary direct crank will be along *OC*. The primary reverse crank for cylinder 3 will be at an angle of – 240° from the axis *OZ*.

The primary direct cranks for all the three cylinders will be along the common crank *OC* as shown in Fig. 14.17 (*a*). All the three primary direct cranks will be rotating clockwise at a speed of ω rad/s.

The total primary force for direct cranks [Refer to Fig. 14.17 (*a*)]

= Sum of primary forces for direct cranks of cylinders 1, 2 and 3

$$= \frac{m}{2} \times \omega^2 \times r + \frac{m}{2} \omega^2 \times r + \frac{m}{2} \omega^2 \times r$$

$$= 3 \times \frac{m}{2} \times \omega^2 \times r$$

$$= 3 \times \frac{2}{2} \times (70\pi)^2 \times 0.06 \text{ N} \qquad (\because \quad m = 2 \text{ kg} ; \omega = 70\pi \text{ rad/s})$$

$$= 8705 \text{ N}$$

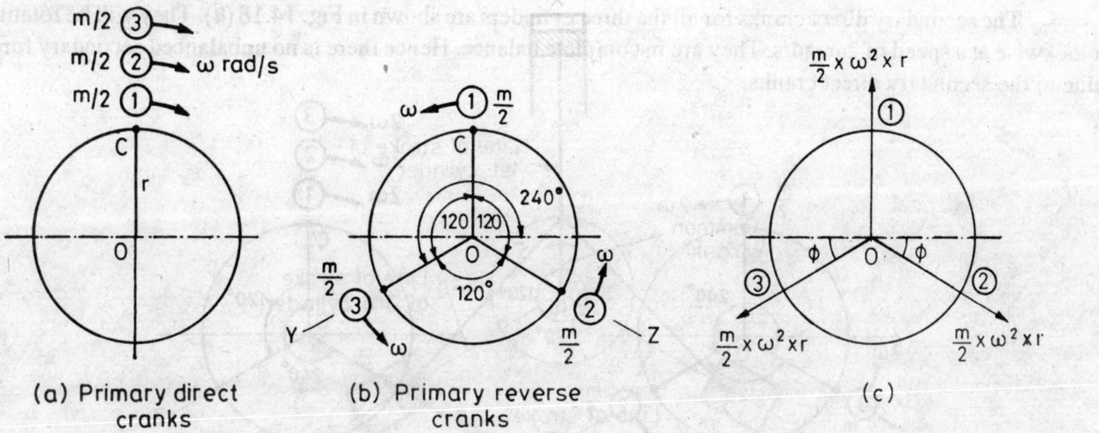

Fig. 14.17

The primary reverse cranks for all the three cylinders are shown in Fig. 14.17 (*b*). All the three primary reverse cranks will be rotating anti-clockwise at a speed of ω rad/s. Fig. 14.17 (*c*) shows the directions and magnitudes of centrifugal forces acting on the three primary reverse cranks.

Total force upward $= \frac{m}{2} \times \omega^2 \times r - \frac{m}{2} \times \omega^2 \times r \times \sin\phi - \frac{m}{2} \times \omega^2 \times r \times \sin\phi$

[Here $\phi = 120° - 90° = 30°$]

$$= \frac{m}{2} \times \omega^2 \times r - 2 \times \frac{m}{2} \times \omega^2 \times r \times \sin 30°$$

$$= \frac{m}{2} \times \omega^2 \times r - 2 \times \frac{m}{2} \times \omega^2 \times r \times \frac{1}{2} = 0$$

Total force horizontally $= \frac{m}{2} \times \omega^2 \times r \cos\phi - \frac{m}{2} \times \omega^2 \times r \times \cos\phi = 0$

∴ Resultant primary force for reverse cranks is zero. Hence the reverse cranks for primary forces are completely balanced.

∴ Total primary unbalanced force is due to direct cranks only and it is equal to **8705 N.**

∴ Resultant primary force acting on the frame of the engine = **8705 N. Ans.**

Resultant secondary force acting on the frame of the engine

To find the resultant secondary force, the positions of the secondary direct cranks and reverse cranks are to be obtained first, by the method as given below :

(*a*) *Positions of secondary direct cranks*

(*i*) For cylinder 1, θ = 0. Hence 2θ will also be zero. This means the secondary direct crank for cylinder 1, will coincide with common crank *OC*.

(*ii*) For cylinder 2, θ = 120°. Hence 2θ will be 2 × 120 = 240°. Hence the secondary direct crank for cylinder 2, will make an angle of 240° clockwise with the axis *OY* (*i.e.* centre line of cylinder 2) as shown in Fig. 14.18 (*a*).

(*iii*) For cylinder 3, θ = 240°. Hence 2θ will be 2 × 240° = 480° or (480° – 360°) = 120°. Hence the secondary direct crank for cylinder 3, will make an angle of 480° or 120° clockwise with the axis *OZ* (*i.e.* centre line of cylinder 3 or line of stroke of cylinder 3).

The secondary direct cranks for all the three cylinders are shown in Fig. 14.18 (*a*). They will be rotating clockwise at a speed of 2ω rad/s. They are in complete balance. Hence there is no unbalanced secondary force due to the secondary direct cranks.

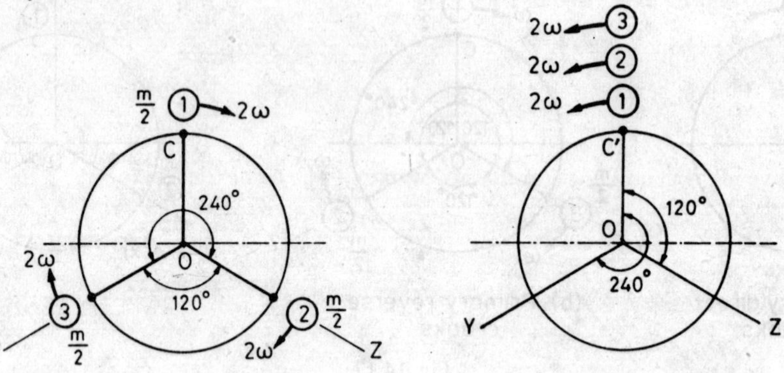

(a) Positions of secondary direct cranks (b) Position of secondary reverse cranks

Fig. 14.18

(b) Positions of secondary reverse cranks

(*i*) For cylinder 1, $\theta = 0$. Hence secondary reverse crank for cylinder 1, will coincide with common crank *OC*. This secondary crank will be rotating anti-clockwise at a speed of 2ω rad/s.

(*ii*) For cylinder 2, $\theta = 120°$. Hence 2θ will be $2 \times 120° = 240°$. Hence the secondary reverse crank for cylinder 2, will make an angle of $-240°$ with the axis *OY*.
(*i.e.* an angle of 240° anticlockwise with the axis *OY*) as shown in Fig. 14.18 (*b*).

(*iii*) For cylinder 3, $\theta = 240°$. Hence 2θ will be $2 \times 240 = 480°$ or $(480° - 360°)$ 120°. Hence the secondary reverse crank for cylinder 3, will make an angle of $-480°$ or $-120°$ with the axis *OZ* (*i.e.* an angle of 480° or 120° anticlockwise with the axis *OZ*).

The secondary reverse cranks for all the three cylinders are shown in Fig. 14.18 (*b*). The reverse cranks for all the three cylinders coincide at *C* and rotating anticlockwise at a speed of 2ω rad/s. The length of the secondary reverse cranks is equal to $\dfrac{r}{4n}$ or $\dfrac{r}{4\dfrac{l}{r}} = \dfrac{r^2}{4l}$.

The total secondary force for reverse cranks

$$= \text{Sum of secondary force for reverse cranks of cylinders 1, 2 and 3}$$

$$= \frac{m}{2} \times (2\omega)^2 \times \frac{r}{4n} + \frac{m}{2} \times (2\omega)^2 \times \frac{r}{4n} + \frac{m}{2} \times (2\omega)^2 \times \frac{r}{4n}$$

$$= 3 \times \frac{m}{2} (2\omega)^2 \times \frac{r}{4n}$$

$$= 3 \times \frac{2}{2} (2 \times 70\pi)^2 \times \frac{0.06}{4 \times 3} \text{ N}$$

$$\left(\because \quad m = 2 \text{ kg} ; r = 0.06 ; n = \frac{l}{r} = \frac{0.18}{0.06} = 3 ; \omega = 70\pi \text{ rad/s}\right)$$

$$= 2901.66 \text{ N.}$$

Hence the total secondary unbalanced force is due to reverse cranks only. Hence resultant secondary force acting on the frame of the engine is **2901.66 N. Ans.**

Problem 14.12. *Fig. 14.19 shows a 60° V-engine having two cylinders placed symmetrically. The two connecting rods are coupled directly to a single crank. The stroke is 100 mm and length of each connecting rod is 165 mm. The mass of the reciprocating parts per cylinder is 1.5 kg. Determine the values of primary force when crank is rotating at a speed of 2000 r.p.m. clockwise.*

Sol. Given :

Stroke $L = 100$ mm hence $r = \frac{L}{2} = \frac{100}{2} = 50$ mm $= 0.05$ m ; $l = 165$ mm $= 0.165$ m ; $m = 1.5$ kg ; $N = 2000$ r.p.m.

Angle between the centre lines (or lines of stroke) of the two cylinders = 60°.

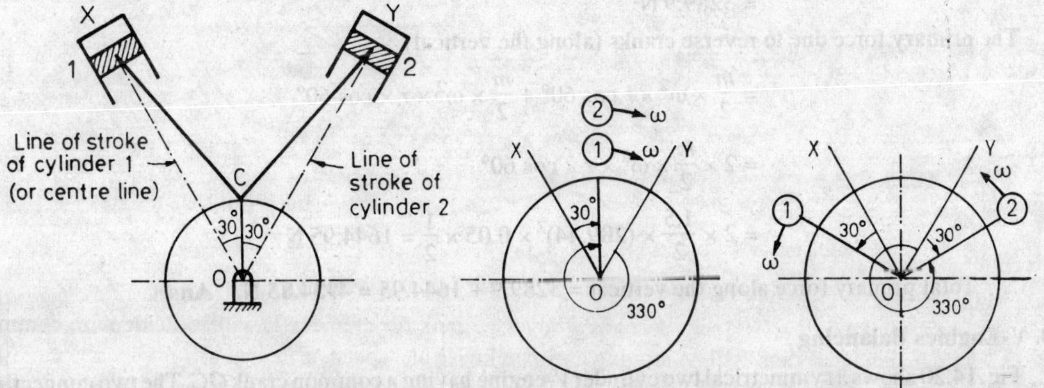

(a) Positions of cylinder (b) Positions of primary direct cranks $2 \times \frac{m}{2} \times \omega^2 \times r$ (c) Positions of primary reverse cranks

Fig. 14.19

The cylinders are placed symmetrically. Hence common crank will make an angle $\frac{60}{2} = 30°$ with the centre lines of the cylinders as shown in Fig. 14.19 (*a*). In this figure *OX* and *OY* are the lines of stroke (or centre lines) of cylinders 1 and 2 respectively.

1. *Value of primary force*

To find the value of primary force, first find the positions of primary direct cranks and primary reverse cranks.

(*i*) The common crank *OC* makes an angle of 30° in clockwise direction to the centre line of cylinder 1 (*i.e.* to axis *OX*). Hence $\theta = 30°$ for cylinder 1. The primary direct crank of cylinder 1, will make an angle of 30° clockwise with axis *OX*. Hence the primary, direct crank of cylinder 1, will coincide with common crank *OC* as shown in Fig. 14.19 (*b*).

The primary reverse crank of cylinder 1, will make an angle of (– 30°) with axis *OX* (*i.e.* an angle of 30°) anticlockwise with axis *OX*) as shown in Fig. 14.19 (*c*).

(*ii*) The common crank *OC* makes an angle of 30° anticlockwise to the centre line of cylinder 2 (*i.e.* to the axis *OY*). This means that common crank *OC* makes an angle of (360 – 30) = 330° clockwise to the axis *OY*. Hence $\theta = 330°$ for cylinder 2. The primary direct crank of cylinder 2, will make an angle of 330° clockwise with axis *OY*. This means that the primary direct crank of cylinder 2, will coincide with common crank *OC* as shown in Fig. 14.19 (*b*).

The primary reverse crank of cylinder 2, will make an angle of – 330° with axis *OY* (*i.e.* an angle of 330° anticlockwise with axis *OY*) as shown in Fig. 14.19 (*c*).

The primary direct cranks are rotating clockwise at ω rad/s where as the primary reverse cranks are rotating anticlockwise at ω rad/s.

The primary force due to direct cranks (along the vertical)

$$= \frac{m}{2} \times \omega^2 \times r + \frac{m}{2} \times \omega^2 \times r$$

$$= 2 \times \frac{m}{2} \times \omega^2 \times r \qquad \left(\because \quad \omega = \frac{2\pi N}{60} = \frac{2\pi \times 2000}{60} = 209.44 \text{ rad/s} \right)$$

$$= 2 \times \frac{1.5}{2} \times (209.44)^2 \times 0.05 \qquad (\because \quad m = 1.5 \text{ kg and } r = 0.05 \text{ m})$$

$$= 3289.9 \text{ N}$$

The primary force due to reverse cranks (along the vertical)

$$= \frac{m}{2} \times \omega^2 \times r \cos 60° + \frac{m}{2} \times \omega^2 \times r \times \cos 60°$$

$$= 2 \times \frac{m}{2} \times \omega^2 \times r \times \cos 60°$$

$$= 2 \times \frac{1.5}{2} \times (209.44)^2 \times 0.05 \times \frac{1}{2} = 1644.95 \text{ N}$$

∴ Total primary force along the vertical = 3289.9 + 1644.95 = **4934.85 N. Ans.**

14.10. V-Engines Balancing

Fig. 14.20 shows a symmetrical two cylinder *V*-engine having a common crank *OC*. The two connecting rods *CP* and *CQ* are connected to the common crank *OC* at crank-pin *C*. The line of stroke of cylinder 1 is represented by *OP* whereas the line of stroke of cylinder 2 is represented by line *OQ*. These lines of strokes are equally inclined to the vertical *OY* at an angle α. The position of the crank at any instant is shown by *OC* which makes an angle θ with the vertical *OY*. The connecting rod lengths are equal.

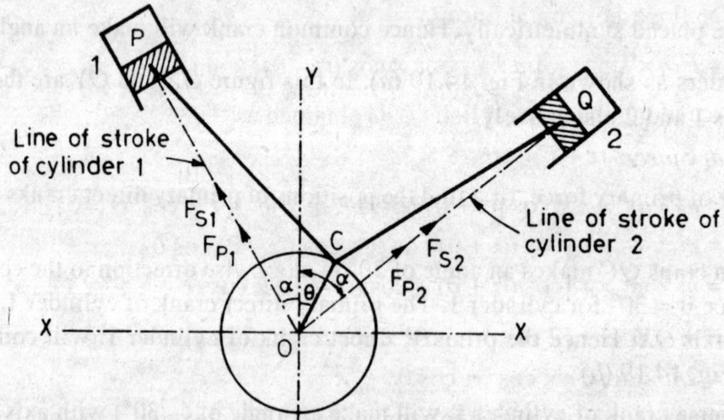

Fig. 14.20

Let r = radius of crank OC

l = length of connecting rod

m = mass of reciprocating parts per cylinder

ω = angular velocity of crank

n = ratio of connecting rod length to crank radius = $\frac{l}{r}$

Then $(\alpha + \theta)$ = Angle made by crank OC with line of stroke of cylinder 1, and
$(\alpha - \theta)$ = Angle made by crank OC with line of stroke of cylinder 2.

Balancing for primary forces

The primary force acting along the line of stroke of cylinder 1 is given by,

$$F_{P_1} = m \times \omega^2 \times r \times \cos(\alpha + \theta)$$

[∵ The angle made by crank OC with line of stroke of cylinder 1 is $(\alpha + \theta)$]

This force F_{P_1} is resolved in two components, one along vertical axis OY and second along horizontal axis OX.

Component of F_{P_1} along vertical line OY

$$= F_{P_1} \times \cos\alpha$$

$$= [m \times \omega^2 \times r \times \cos(\alpha + \theta)] \times \cos\alpha$$

Component of F_{P_1} along horizontal line OX

$$= F_{P_1} \times \sin\alpha$$

$$= [m \times \omega^2 \times r \times \cos(\alpha + \theta)] \times \sin\alpha$$

Now the primary force acting along the line of stroke of cylinder 2, is given by,

$$F_{P_2} = m \times \omega^2 \times r \times \cos(\alpha - \theta)$$

(∵ The crank OC makes an angle $(\alpha - \theta)$ with the line of stroke of cylinder 2)

Component of F_{P_2} along vertical line OY

$$= F_{P_2} \times \cos\alpha$$

$$= [m \times \omega^2 \times r \times \cos(\alpha - \theta)] \times \cos\alpha$$

Component of F_{P_2} along horizontal line OX

$$= F_{P_2} \times \sin\alpha = [m \times \omega^2 \times r \times \cos(\alpha - \theta)] \times \sin\alpha$$

Total primary force along the vertical line OY is obtained as,

$$(F_P)_y = F_{P_1} \cos\alpha + F_{P_2} \cos\alpha$$

$$= m \times \omega^2 \times r \times \cos(\alpha + \theta)\cos\alpha + m \times \omega^2 \times r \times \cos(\alpha - \theta)\cos\alpha$$

$$= m \times \omega^2 \times r\,[\cos(\alpha + \theta)\cos\alpha + \cos(\alpha - \theta)\cos\alpha]$$

$$= m \times \omega^2 \times r\,[\cos(\alpha + \theta) + \cos(\alpha - \theta)] \times \cos\alpha$$

$$= m \times \omega^2 \times r\,[2\cos\alpha\cos\theta] \times \cos\alpha \quad [\because \cos(\alpha + \theta) + \cos(\alpha - \theta) = 2\cos\alpha\cos\theta]$$

$$= 2m \times \omega^2 \times r \times \cos^2\alpha\cos\theta$$

Similarly total primary force along the horizontal line OX is obtained as,

$$(F_P)_x = F_{P_2}\sin\alpha - F_{P_1}\sin\alpha$$

$$= m \times \omega^2 \times r\cos(\alpha - \theta) \times \sin\alpha - m \times \omega^2 \times r \times \cos(\alpha + \theta)\sin\alpha$$

$$= m \times \omega^2 \times r \times \sin\alpha \times [\cos(\alpha - \theta) - \cos(\alpha + \theta)]$$

$$= m \times \omega^2 \times r \times \sin\alpha \times [2\sin\alpha\sin\theta] \quad [\because \cos(\alpha - \theta) - \cos(\alpha + \theta) = 2\sin\alpha\sin\theta]$$

$$= 2m \times \omega^2 \times r \times \sin^2\alpha \times \sin\theta$$

$\therefore$ Resultant primary force,

$$(F_P)_R = \sqrt{(F_P)_x{}^2 + (F_P)_y{}^2}$$

$$= 2m \times \omega^2 \times r\sqrt{(\sin^2\alpha \times \sin\theta)^2 + (\cos^2\alpha \cos\theta)^2} \qquad \text{...(A)}$$

The angle (ϕ) made by the resultant primary force with vertical line OY is given by

$$\tan\phi = \frac{(F_P)_x}{(F_P)_y} = \frac{\sin^2\alpha \times \sin\theta}{\cos^2\alpha \times \cos\theta} \qquad \text{...(B)}$$

The resultant primary force for different V-engines are obtained from equation (A) by substituting the value of α in equation (A) as given below :

(*i*) For 60°—V engine, $2\alpha = 60°$ or $\alpha = 30°$

$$\therefore \quad (F_P)_R = 2m \times \omega^2 \times r \times \sqrt{(\sin^2 30° \sin\theta)^2 + (\cos^2 30° \times \cos\theta)^2}$$

$$= 2m \times \omega^2 \times r \times \sqrt{\left(\frac{\sin\theta}{4}\right)^2 + \left(\frac{3}{4} \times \cos\theta\right)^2}$$

$$= \frac{2m \times \omega^2 \times r}{4} \times \sqrt{\sin^2\theta + 9\cos^2\theta}$$

$$= \frac{m}{2} \times \omega^2 \times r \times \sqrt{\sin^2\theta + 9\cos^2\theta} \qquad \text{...(14.12)}$$

(*ii*) For 90°—V engine, $2\alpha = 90°$ or $\alpha = 45°$

$$\therefore \quad (F_P)_R = 2m \times \omega^2 \times r \times \sqrt{(\sin^2 45° \times \sin\theta)^2 + (\cos^2 45° \times \cos\theta)^2}$$

$$= 2m \times \omega^2 \times r \times \sqrt{\left(\frac{\sin\theta}{2}\right)^2 + \left(\frac{\cos\theta}{2}\right)^2}$$

$$= \frac{2m \times \omega^2 \times r}{2}\sqrt{\sin^2\theta + \cos^2\theta} = m \times \omega^2 \times r \qquad \text{...(14.13)}$$

From equation (B), $\tan\phi = \dfrac{\sin^2 45 \times \sin\theta}{\cos^2 45 \times \cos\theta} = \tan\theta$ or $\phi = \theta$. This means the resultant primary force acts along the crank.

(*iii*) For 120°—V engine, $2\alpha = 120°$ or $\alpha = 60°$

$$\therefore \quad (F_P)_R = 2m \times \omega^2 \times r \times \sqrt{(\sin^2 60° \sin\theta)^2 + (\cos^2 60° \times \cos\theta)^2}$$

$$= 2m \times \omega^2 \times r \times \sqrt{\left(\frac{3\sin\theta}{4}\right)^2 + \left(\frac{\cos\theta}{4}\right)^2}$$

$$= \frac{2m \times \omega^2 \times r}{4}\sqrt{9\sin^2\theta + \cos^2\theta}$$

$$= \frac{m}{2} \times \omega^2 \times r\sqrt{9\sin^2\theta + \cos^2\theta} \qquad \text{...(14.14)}$$

Balancing for secondary force

The secondary force along the line of stroke of cylinder 1 is given by,

$$F_{S_1} = m \times \omega^2 \times r \times \frac{\cos[2(\alpha + \theta)]}{n}$$

[$\because$ $(\alpha + \theta)$ is the angle made by crank OC with the line of stroke of cylinder 1]

This force can be resolved in two components.

Components of F_{S_1} along the vertical line OY

$$= F_{S_1} \times \cos\alpha = \left[m \times \omega^2 \times r \times \frac{\cos 2(\alpha + \theta)}{n} \right] \times \cos\alpha$$

Component of F_{S_1} along the horizontal line OX'

$$= F_{S_1} \times \sin\alpha = \left[m \times \omega^2 \times r \times \frac{\cos 2(\alpha + \theta)}{n} \right] \times \sin\alpha$$

The secondary force along the line of stroke of cylinder 2, is given by,

$$F_{S_2} = m \times \omega^2 \times r \times \frac{\cos [2(\alpha - \theta)]}{n}$$

[$\because$ $(\alpha - \theta)$ is the angle made by the crank OC with the line of stroke of cylinder 2]

This force is also resolved in two components.

Component of F_{S_2} along the vertical line OY

$$= F_{S_2} \times \cos\alpha = \left[m \times \omega^2 \times r \times \frac{\cos 2(\alpha - \theta)}{n} \right] \times \cos\alpha$$

Component of F_{S_2} along the horizontal line OX

$$= F_{S_2} \times \sin\alpha = \left[m \times \omega^2 \times r \times \frac{\cos 2(\alpha - \theta)}{n} \right] \times \sin\alpha$$

Now total secondary force along the vertical line OY is obtained as,

$$(F_S)_y = F_{S_1} \times \cos\alpha + F_{S_2} \times \cos\alpha$$

$$= m \times \omega^2 \times r \times \frac{\cos 2(\alpha + \theta)}{n} \times \cos\alpha + m \times \omega^2 \times r \times \frac{\cos 2(\alpha - \theta)}{n} \times \cos\alpha$$

$$= \frac{m \times \omega^2 \times r}{n} \times \cos\alpha \, [\cos 2(\alpha + \theta) + \cos 2(\alpha - \theta)]$$

$$= \frac{m \times \omega^2 \times r \times \cos\alpha}{n} \times 2 \cos 2\alpha \cos 2\theta$$

[$\because$ $\cos 2(\alpha + \theta) + \cos 2(\alpha - \theta) = 2 \cos 2\alpha \cos 2\theta$]

$$= \frac{2m \times \omega^2 \times r}{n} \times \cos\alpha \times \cos 2\alpha \times \cos 2\theta.$$

Similarly the total secondary force along the horizontal line OX is obtained as,

$$(F_S)_X = F_{S_2} \times \sin\alpha - F_{S_1} \times \sin\alpha$$

$$= \frac{m \times \omega^2 \times r}{n} \times \cos 2(\alpha - \theta) \times \sin\alpha - \frac{m \times \omega^2 \times r}{n} \cos 2(\alpha + \theta) \times \sin\alpha$$

$$= \frac{m \times \omega^2 \times r}{n} \times \sin\alpha \, [\cos 2(\alpha - \theta) - \cos 2(\alpha + \theta)]$$

$$= \frac{m \times \omega^2 \times r}{n} \times \sin\alpha \, [(\cos 2\alpha \times \cos 2\theta + \sin 2\alpha \sin 2\theta) - (\cos 2\alpha \cos 2\theta - \sin 2\alpha \sin 2\theta)]$$

$$= \frac{m \times \omega^2 \times r}{n} \times \sin\alpha \times 2 \sin 2\alpha \sin 2\theta$$

$$= \frac{2m \times \omega^2 \times r}{n} \times \sin\alpha \times \sin 2\alpha \times \sin 2\theta.$$

∴ Resultant secondary force,

$$(F_S)_R = \sqrt{(F_S)_x^{\ 2} + (F_S)_y^{\ 2}}$$

$$= \frac{2m \times \omega^2 \times r}{n} \sqrt{(\sin\alpha \times \sin 2\alpha \times \sin 2\theta)^2 + (\cos\alpha \times \cos 2\alpha \times \cos 2\theta)^2} \quad \text{...(C)}$$

The angle (ϕ^*) made by the resultant secondary force with vertical line OY is given by,

$$\tan\phi^* = \frac{(F_S)_x}{(F_S)_y} = \frac{\sin\alpha \times \sin 2\alpha \times \sin 2\theta}{\cos\alpha \times \cos 2\alpha \times \cos 2\theta} \quad \text{...(D)}$$

The resultant secondary force for different V-engines are obtained from equation (C) by substituting the value of α in equation (C), as given below :

(*i*) For 60°—V engine, $2\alpha = 60°$ or $\alpha = 30°$

$$\therefore \quad (F_S)_R = \frac{2m \times \omega^2 \times r}{n} \sqrt{(\sin 30° \times \sin 60° \times \sin 2\theta)^2 + (\cos 30° \times \cos 60° \times \cos 2\theta)^2}$$

$$= \frac{2m \times \omega^2 \times r}{n} \sqrt{\left(\frac{1}{2} \times \frac{\sqrt{3}}{2} \times \sin 2\theta\right)^2 + \left(\frac{\sqrt{3}}{2} \times \frac{1}{2} \times \cos 2\theta\right)^2}$$

$$= \frac{2m \times \omega^2 \times r}{n} \times \frac{\sqrt{3}}{4} \sqrt{\sin^2 2\theta + \cos^2 2\theta} = \frac{\sqrt{3}}{2} \times \frac{m \times \omega^2 \times r}{n}$$

(*ii*) For 90°—V engine, $2\alpha = 90°$ or $\alpha = 45°$

$$\therefore \quad (F_S)_R = \frac{2m \times \omega^2 \times r}{n} \sqrt{(\sin 45° \times \sin 90° \sin 2\theta)^2 + (\cos 45° \times \cos 90° \times \cos 2\theta)^2}$$

$$= \frac{2m \times \omega^2 \times r}{n} \sqrt{\left(\frac{1}{\sqrt{2}} \times 1 \times \sin 2\theta\right)^2 + \left(\frac{1}{\sqrt{2}} \times 0 \times \cos 2\theta\right)^2}$$

$$= \frac{2m \times \omega^2 \times r}{n} \times \frac{1}{\sqrt{2}} \times \sin 2\theta$$

$$= \frac{\sqrt{2} \times m \times \omega^2 \times r}{n} \times \sin 2\theta$$

For this case, the angle ϕ^* is obtained from equation (D) as

$$\tan\phi^* = \frac{\sin 45° \times \sin 90° \times \sin 2\theta}{\cos 45° \times \cos 90° \times \cos 2\theta} = \frac{0.707 \times 1 \times \sin 2\theta}{0.707 \times 0 \times \cos 2\theta} = \infty$$

or $$\phi^* = 90°.$$

This means that resultant secondary force acts along horizontal line OX.

(*iii*) For 120°—V engine, $2\alpha = 120°$ or $\alpha = 60°$

$$\therefore \quad (F_S)_R = \frac{2m \times \omega^2 \times r}{n} \sqrt{(\sin 60° \sin 120° \sin 2\theta)^2 + (\cos 60° \cos 120° \cos 2\theta)^2}$$

$$= \frac{2m \times \omega^2 \times r}{n} \sqrt{\left(\frac{\sqrt{3}}{2} \times \frac{\sqrt{3}}{2} \times \sin 2\theta\right)^2 + \left(\frac{1}{2} \times \left(-\frac{1}{2}\right) \times \cos 2\theta\right)^2}$$

$$= \frac{2m \times \omega^2 \times r}{n} \sqrt{\frac{9}{16} \times \sin 2\theta + \frac{1}{16} \times \cos^2 2\theta}$$

$$= \frac{2m \times \omega^2 \times r}{n} \times \frac{1}{4}\sqrt{9 \sin^2 2\theta + \cos^2 2\theta}$$

$$= \frac{m \times \omega^2 \times r}{2n} \times \sqrt{9 \sin^2 2\theta + \cos^2 2\theta}\,.$$

Problem 14.13. *A 90°—V engine has two cylinders which are placed symmetrically. The two connecting rods operate a common crank. The length of connecting rods are 320 mm each and crank radius is 80 mm. The reciprocating mass per cylinder is 12 kg. If the engine speed is 600 r.p.m., then find the resultant primary and resultant secondary forces. Also find the maximum resultant secondary force.*

Sol. Given :

Angle of V-engine = 90° or $2\alpha = 90°$ or $\alpha = 45°$

length of each connecting rod, $l = 320$ mm = 0.32 m ; crank radius, $r = 80$ mm = 0.08 m

$$m = 12 \text{ kg} ; N = 600 \text{ r.p.m.} \quad \text{or} \quad \omega = \frac{2\pi N}{60} = \frac{2\pi \times 600}{60} = 20\pi \text{ rad/s} ; n = \frac{l}{r} = \frac{320}{80} = 4.$$

(*i*) The resultant primary force $[(F_P)_R]$ for any V-engine is given by equation (A) as

$$(F_P)_R = 2m \times \omega^2 \times r\sqrt{(\sin^2 \alpha \times \sin \theta)^2 + (\cos^2 \alpha \cos \theta)^2}$$

$$= 2m \times \omega^2 \times r\sqrt{(\sin^2 45° \times \sin \theta)^2 + (\cos^2 45° \cos \theta)^2} \qquad (\because \alpha = 45°)$$

$$= 2m \times \omega^2 \times r\sqrt{\left(\frac{\sin \theta}{2}\right)^2 + \left(\frac{\cos \theta}{2}\right)^2} = \frac{2m \times \omega^2 \times r}{2}\sqrt{\sin^2 \theta + \cos^2 \theta}$$

$$= m \times \omega^2 \times r$$

$$= 12 \times (20\pi)^2 \times 0.08 \text{ N} = \mathbf{3789.9\ N.} \quad \textbf{Ans.}$$

(*ii*) The resultant secondary force $[(F_S)_R]$ for any engine is given by equation (C) as

$$(F_S)_R = \frac{2m \times \omega^2 \times r}{n}\sqrt{(\sin \alpha \sin 2\alpha \times \sin 2\theta)^2 + (\cos \alpha \times \cos 2\alpha \times \cos 2\theta)^2}$$

$$= \frac{2m \times \omega^2 \times r}{n}\sqrt{(\sin 45° \sin 90° \times \sin 2\theta)^2 + (\cos 45° \cos 90° \times \cos 2\theta)^2}$$

$$= \frac{2m \times \omega^2 \times r}{n}\sqrt{\left(\frac{1}{\sqrt{2}} \times 1 \times \sin 2\theta\right)^2 + \left(\frac{1}{\sqrt{2}} \times 0 \times \cos 2\theta\right)^2}$$

$$= \frac{2m \times \omega^2 \times r}{n} \times \sqrt{\left(\frac{1}{\sqrt{2}} \times 1 \times \sin 2\theta\right)^2 + 0} = \frac{2m \times \omega^2 \times r}{n} \times \frac{\sin 2\theta}{\sqrt{2}}$$

$$= \frac{\sqrt{2} \times m}{n} \times \omega^2 \times r \times \sin 2\theta = \frac{\sqrt{2} \times 12}{4} \times (2\pi)^2 \times 0.08 \times \sin 2\theta$$

$$= \mathbf{1339.94 \sin 2\theta\ N.} \quad \textbf{Ans.}$$

Maximum resultant secondary force

The secondary force 1339.94 sin 2θ will be maximum when sin 2θ is maximum *i.e.* when $\sin 2\theta = \pm 1$ or $\theta = 45°$ and 135°.

$$\therefore \qquad [(F_S)_R]_{max} = 1339.94\,[\pm 1] = \pm\,\mathbf{1339.94\ N.} \quad \textbf{Ans.}$$

HIGHLIGHTS

1. The force required to accelerate the reciprocating masses is given by,

$$F = \text{mass of reciprocating parts} \times \text{Acceleration}$$

$$= m_R \times \omega^2 . r \left(\cos\theta + \frac{\cos 2\theta}{n} \right).$$

2. Inertia force is equal and opposite to force F.

$$\therefore \quad F_i = -F = -m_R \times \omega^2 . r \left(\cos\theta + \frac{\cos 2\theta}{n} \right)$$

$$= -\left[m_R \times \omega^2 . r \cos\theta + m_R \times \omega^2 . r \frac{\cos 2\theta}{n} \right]$$

$$= -[F_P + F_s]$$

where F_P = Primary disturbing force $= m_R \times \omega^2 . r \cos\theta$ and

F_s = Secondary disturbing force $= m_R \times \omega^2 . r \dfrac{\cos 2\theta}{n}$.

3. As inertia force acts along the line of stroke, hence primary and secondary disturbing forces also act along the line of stroke.

4. Maximum value of secondary force $= \dfrac{1}{n} \times$ Max. value of primary force.

5. In case of moderate speeds, the secondary unbalanced force is so small that it may be neglected as compared to primary unbalanced force.

6. For complete balance of primary disturbing force, we must have

$$m_R \times r = m_b \times r^*$$

where m_R = Mass of reciprocating parts,

r = Radius of crank,

m_b = Balancing mass placed at a radius r^*.

7. If the primary force is completely balanced, then a disturbing force perpendicular to the line of stroke will be acting. Hence only a fraction of reciprocating mass is balanced. Hence for partial balancing of reciprocating mass, we have

$$C \times m_R \times r = m_b \times r^*$$

where C is the fraction of the reciprocating parts which is less than 1.

8. If the balancing mass is required to balance the revolving mass (M) as well as give a partial balance of the reciprocating masses, then we have

$$M \times r + C \times m_R \times r = m_b \times r^*.$$

9. If the two cylinders are placed symmetrically between the wheels, then the two cylinder locomotive is known as inside cylinder locomotive. But if the two cylinders are placed symmetrically outside the frame, it is called outside cylinder locomotive.

10. Hammer blow is the maximum magnitude of the unbalanced force perpendicular to the line of stroke. Mathematically,

$$\text{Hammer blow} = \pm\, m_b \times r^* \times \omega^2.$$

11. The net pressure between the wheel and the rail

$$= \text{Static wheel load} + \text{Hammer blow}$$

$$= P \pm m_b \times r^* \times \omega^2.$$

12. The limiting condition when the wheel does not lift from the rails is given by

$$\omega = \sqrt{\frac{P}{m_b \times r^*}}.$$

13. Maximum variation of tractive force = $\pm \sqrt{2}\,(1 - C) \times m_R \times \omega^2 \times r$.

14. Maximum swaying couple = $\pm \frac{a}{\sqrt{2}}(1 - C) \times m_R \times \omega^2 \times r$

where a = Distance between the centre lines of the two cylinders.

15. In case of coupled locomotives two or three pairs of wheels are coupled together to increase the adhesive weight.

16. For primary balance of multi-cylinder in-line engines :

(*i*) the algebraic sum of primary forces must be zero or $\Sigma m_R \times \omega^2 \times r \times \cos\theta = 0$

(*ii*) the algebraic sum of the couples about any point due to primary forces must be zero or

$$\Sigma m_R \times \omega^2 \times r \times a \times \cos\theta = 0.$$

17. For secondary balance of multi-cylinder in-line engines :

(*i*) the algebraic sum of the secondary forces must be zero or $\Sigma m_R \times (2\omega)^2 \times \left(\frac{r}{4n}\right) \times \cos 2\theta = 0$

(*ii*) the algebraic sum of the couples about any point due to secondary forces must be zero or

$$\Sigma m_R \times (2\omega)^{2*}\left(\frac{r}{4n}\right) \times a \times \cos 2\theta = 0.$$

18. Imaginary secondary crank is having a radius of $\frac{r}{4n}$. As the value of n is more than 1, hence the radius of imaginary secondary crank is less than the radius of actual crank.

19. For balancing the multi-cylinder radial engines, method of direct and reverse cranks is used.

EXERCISE 14

A. Theoretical Questions

1. What do you mean by 'balancing of reciprocating engine' ?
2. Explain the terms : Primary disturbing force and secondary disturbing force.
3. For a reciprocating engine, prove that

   (*i*) maximum value of secondary disturbing force = $\frac{1}{n}$ × Primary disturbing force

   where n = ratio of length of connecting rod to crank length.

   (*ii*) for one revolution of the crank, the maximum value of primary force occurs two times whereas the maximum value of secondary force occurs four times.
4. Explain the term 'Partial balancing of primary force' ? Why is it necessary ?
5. What do you understand by inside cylinder locomotives and outside cylinder locomotives ? What is the effect of partial balancing of locomotive ?
6. Explain the terms : variation of tractive force ; swaying couple and hammer blow.
7. Prove that the maximum variation of tractive force is obtained when θ = 135° or 315°. Also prove that maximum variation of tractive is equal to $\pm \sqrt{2}\,(1 - C)\, m_R \times \omega^2 \times r$ where C is the fraction of the mass of reciprocating parts which is to be balanced.
8. Prove that maximum swaying couple is equal to $\pm \frac{a}{\sqrt{2}}(1 - C)\, m_R \times \omega^2 \times r$ where a = distance between the centre lines of the two cylinders and C = the fraction of the mass of the reciprocating parts which is to be balanced.
9. Prove that the limiting condition for the angular speed when the wheel does not lift from the rails is given by,

   $$\omega = \sqrt{\frac{P}{m_b \times r^*}}$$

   where P = static wheel load,

   m_b = Balancing mass placed at a radius of r^*.

10. Prove the relation that
Hammer blow = $\pm m_b \times r^* \times \omega^2$.

11. What do you mean by coupled locomotives and uncoupled locomotives ? What is the advantage of coupled locomotive ?

12. Explain the procedure of :
(*i*) primary balancing of multi-cylinder In-line engines, and
(*ii*) secondary balancing of multi-cylinder In-line engines.

13. Explain in details the procedure of balancing multi-cylinder radial engines by direct and reverse cranks method.

14. Explain the procedures of balancing *V*-engine.

B. NUMERICAL PROBLEMS

1. A single cylinder reciprocating engine has the following data :
speed of engine = 240 r.p.m. ; stroke = 320 mm ; mass of reciprocating parts = 67.5 kg and mass of revolving parts = 52.5 kg at crank radius. If 60% of the reciprocating parts and all the revolving parts are to be balanced, then find :
(*i*) the balancing mass required at a radius of 300 mm, and
(*ii*) the unbalanced force when the crank has rotated 60° from top dead centre. [**Ans.** (*i*) 49.59 kg (*ii*) 4775.4 N]

2. The cranks of a two-cylinder, uncoupled inside cylinders locomotive, are at right angles and are 300 mm long. The centre lines of the cylinders are 700 mm apart. The rotating mass per cylinder is 150 kg at the crank-pin and the mass of the reciprocating parts per cylinder is 180 kg. The wheel centre lines are 1.5 m apart. The whole of the rotating and two-third of the reciprocating masses are to be balanced and the balance masses are to be placed in the planes of rotation of the driving wheels at a radius of 600 mm. Find :
(*i*) the magnitude and direction of the balancing masses ;
(*ii*) the magnitude of hammer blow ;
(*iii*) variation in tractive force ; and
(*iv*) maximum swaying couple of the crank speed is 150 r.p.m.
[**Ans.** (*i*) 105 kg each ; 200° and 250° (*ii*) ± 6900 N ; (*iii*) ± 6281.7 N ; (*iv*) 2199.25 N]

3. If in problem 2, dead weight on each wheel is 10 kN, then find :
(*i*) maximum and minimum pressure on then rails :
(*ii*) the maximum speed of the locomotive (in km/hr) without lifting the wheels from the rails. Take the diameter of the driving wheel as 1.8 m. [**Ans.** (*i*) 16900 N ; 3100 N ; (*ii*) 61.27 km/hr]

4. The cranks of a two cylinder uncoupled outside cylinder locomotives are at right angles and are 300 mm long. The distance between the centre lines of the cylinder is 1.75 m. The wheel centre lines are 1.45 m apart. The rotating mass per cylinder is 360 kg and the mass of the reciprocating parts are cylinder is 300 kg. The whole of the rotating and two-thirds of the reciprocating masses are to be balanced in the plane of the driving wheels at a radius of 750 mm, then find :
(*i*) the magnitude and angular positions of the balance masses,
(*ii*) the maximum speed of the locomotive in km/hr without lifting the wheels from the rails if the dead weight on each driving wheel is 30 kN and dia. of driving wheel is 1.8 m and
(*iii*) swaying couple at the maximum speed.
[**Ans.** (*i*) 249 kg each, 174.5° and 275° (*ii*) 68.7 km/hour (*iii*) 16687 Nm]

5. The cranks of a two cylinder locomotive with two coupled wheels on each side, are at right angles and are 300 mm long. The distance between the centre lines of the cylinder is 600 mm. The wheel centre lines are 1.6 m apart whereas the distance between coupling rods is 2*m*. The rotating mass per cylinder is 260 kg and the mass of the reciprocating parts are cylinder is 315 kg. Mass of each coupling rod is 130 kg and the radius of centre of coupling rod pin is 240 mm.

The coupling rod pins are at 180° to their respective main cranks. The whole of the rotating and two-third of the reciprocating masses are to be balanced in the plane of the wheels at a radius of 750 mm. The balance mass of the reciprocating parts is to be divided equally between the driving wheels and coupled wheels. Determine :

(*i*) the magnitude and angular positions of the masses required for driving and trailing wheels, and

(*ii*) the hammer blow at 80 km/hr. Take the diameter of the wheels as 1.9 m.

[**Ans.** (*i*) For driving wheels, mass = 73.85 kg each at 223.5° and 226.5° ; for trailing wheels, mass = 25.63 kg each at 135.4° and 314.4° ; (*ii*) 13.015 kN]

6. A two-cylinder engine with cranks at 180° and the cylinders on the same side of the crank-shaft centre-line is having identical reciprocating masses, crank lengths and connecting rod lengths for each cylinder. If the crank of the first cylinder makes an angle of 30° with I.D.C. then to what extent, the engine is balanced for : (*i*) primary and secondary forces and (*ii*) primary and secondary couples.

[**Ans.** (*i*) Primary forces are balanced but secondary forces are not balanced (*ii*) Primary and secondary couples are not balanced]

7. A four-cylinder vertical engine is in complete primary balance. The length of the cranks are 150 mm. The planes of rotation of the first, second and fourth cranks are 400 mm, 200 mm and 200 mm respectively from the third crank and their reciprocating masses are 50 kg, 60 kg and 50 kg respectively. Find : (*i*) the mass of the reciprocating parts for the third cylinder and (*ii*) the relative angular positions of the cranks.

[**Ans.** (*i*) 60 kg (*ii*) θ_2 = 160°, θ_3 = 227° and θ_4 = 26° and θ_1 = 0°]

8. A radial engine has three cylinders whose axes are spaced at an angular interval of 120°. The three connecting rods of the cylinders are coupled directly to a single crank. The stroke is 120 mm and the length of each connecting rod is 180 mm. The mass of the reciprocating parts per cylinder is 1 kg. Find the resultant primary and secondary forces acting on the frame of the engine when the engine is running at 3150 r.p.m. [**Ans.** 9793.1 N ; 3264.36 N]

9. A 60°-V engine has two cylinders which are placed symmetrically. The connecting rod of each the cylinder is connected to a single crank. The stroke is 120 mm and length of each connecting rod is 2 m. The mass of the reciprocating parts per cylinder is 1.25 kg. Determine the value of primary force when crank is rotating at a speed of 2000 r.p.m. [**Ans.** 4934.85 N]

10. A 90°-V engine has two cylinders which are placed symmetrically. The two connecting rods operate a common crank. The length of connecting rods are 400 mm each and crank radius is 100 mm. The reciprocating mass per cylinder is 4.8 kg. If the engine speed is 1200 r.p.m., then find the resultant primary and secondary forces. Also find the maximum resultant secondary force. [**Ans.** (*i*) 7579.8 N (*ii*) 2679.88 × sin 2θ (*iii*) ± 2679.88 N]

15

Governors

15.1. Introduction

Governor is a device used for maintaining a constant mean speed of rotation of the crank-shaft over long periods during which the load on the engine may vary. When the load on the engine increases, the speed of the engine will decrease. The governor will act in such a way that it will increase the supply of working fluid. Similarly when the load on the engine decreases, the speed of the engine increases. Then the governor will act in such a way that the supply of working fluid decreases. Thus the mean speed of rotation of the engine will be maintained constant as closely as possible over a long period.

The function of a flywheel is to limit the fluctuation of speed during each cycle which arises from the fluctuations of turning moment on the crank-shaft. The flywheel does not control the speed variations caused by a varying load.

The function of governor is to control the mean speed of rotation over a long period due to the variation of load. The governor has no influence over cyclic speed fluctuation.

15.2. Types of Governors

The following are the important types of governors :

(*i*) Centrifugal governors and

(*ii*) Inertia governors.

In case of centrifugal governors, the two or more masses known as governor balls are caused to revolve about the axis of the shaft, which is driven by the engine crank-shaft through bevel gears. When governor balls are revolving at a uniform speed, the centrifugal force on the balls is equal to the inward controlling force. The inward controlling force is provided by a dead weight, a spring or a combination of the two.

In case of inertia governors, the governor balls are so arranged that the inertia forces caused by an angular acceleration or retardation of the governor shaft, tend to alter their positions.

15.2.1. Centrifugal Governors

Fig. 15.1 shows the line diagram of a centrifugal governor, which consists of two balls of equal masses (the balls are also known as governor balls) attached to the two arms. The upper ends of the arms are pivoted to a spindle, which is driven by the engine through bevel gears. The lower arms (or links) are connected to a sleeve which is keyed to the spindle. The sleeve revolves with the spindle, but can slide up and down. The two stops S_1 and S_2 on the spindle prevents the upwards and downward motion of the sleeve. The sleeve is connected by a bell crank lever to a throttle valve, which controls the supply of the working fluid. When sleeve rises, the supply of the working fluid decreases and when sleeve falls, the supply of the working fluid increases.

When the load on the engine decreases, the speed of the engine increases. As the spindle of the governor is driven by the engine, hence the speed of the spindle also increases. This will increase the centrifugal force on the governor balls and the balls will move outwards. Due to the movement of balls outwards the sleeve will rise upwards. The upward movement of the sleeve will operate a throttle valve at the other end of the bell crank lever to reduce the supply of the working fluid by reducing the throttle valve opening.

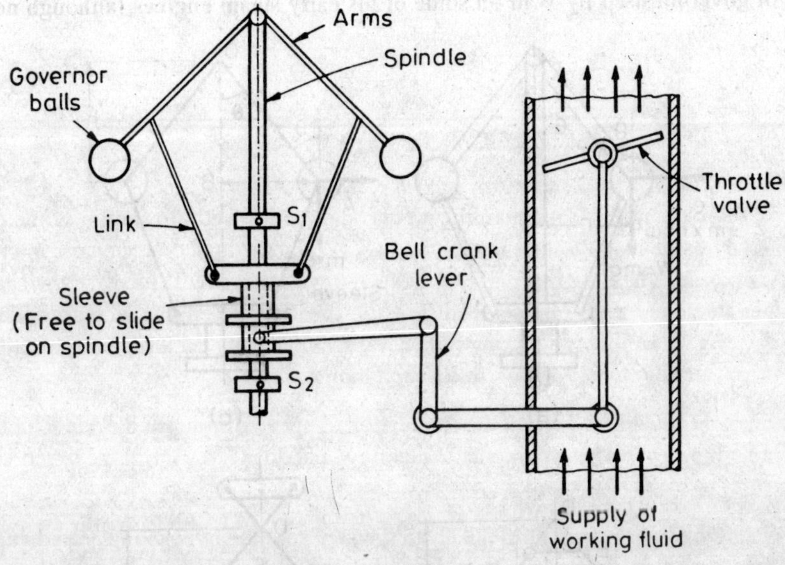

Fig. 15.1

Similarly when the load on the engine increases, the speed of the engine decreases. Also the speed of the spindle of the governor decreases. Hence the centrifugal force on the governor balls will also decrease. The balls of the governor will move inwards and hence the sleeve will move downwards. The downward movement of the sleeve will increase the supply of the working fluid by increasing the opening of the throttle valve and thus the engine speed is increased.

15.3. Types of Centrifugal Governors

The followings are the classifications of the centrifugal governors :

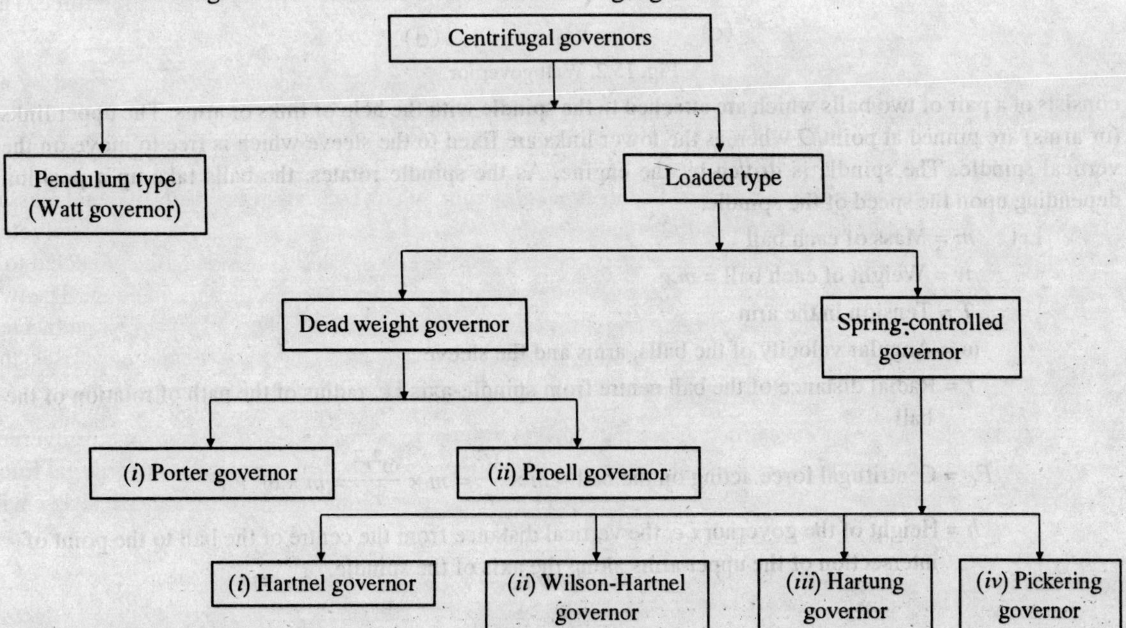

15.4. Watt Governor

Fig. 15.2 shows the simplest form of a centrifugal governor, which is known as 'Watt Governor'. It is the original form of governor used by Watt on some of his early steam engines (although now obsolete). It

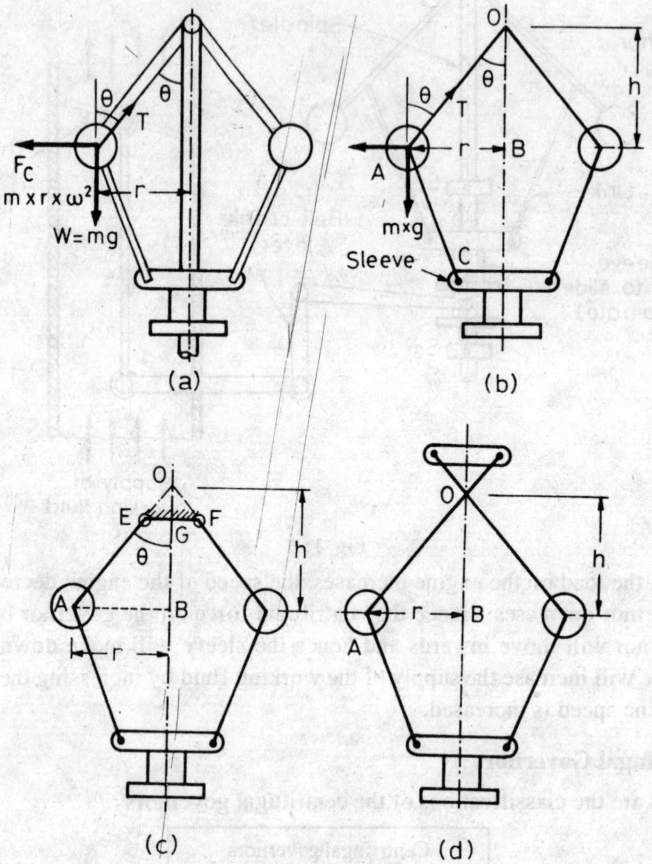

Fig. 15.2. Watt governor.

consists of a pair of two balls which are attached to the spindle with the help of links or arms. The upper links (or arms) are pinned at point O whereas the lower links are fixed to the sleeve which is free to move on the vertical spindle. The spindle is driven by the engine. As the spindle rotates, the balls take up a position depending upon the speed of the spindle.

Let m = Mass of each ball

w = Weight of each ball = $m.g$

T = Tension in the arm

ω = Angular velocity of the balls, arms and the sleeve

r = Radial distance of the ball centre from spindle-axis *i.e.* radius of the path of rotation of the ball

F_C = Centrifugal force acting on the ball $= m \times \dfrac{V^2}{r} = m \times \dfrac{\omega^2 r^2}{r} = m \times \omega^2 \times r$

h = Height of the governor *i.e.* the vertical distance from the centre of the ball to the point of intersection of the upper arms along the axis of the spindle.

With the increase of the speed, the height of governor (*i.e.* h) decreases, whereas with the decrease of the speed, the height h increases.

Assuming the weight of the arms, links and the sleeve to be negligible as compared to the weight of balls, each ball will be in equilibrium under the action of following forces :

(*i*) the centrifugal force, F_C acting on the ball where $F_C = m \times \omega^2 \times r$

(*ii*) the weight of ball, w where $w = m \times g$

(*iii*) the tension T in the upper arm.

There will be no tension in the lower link if sleeve is assumed to be massless and also friction is neglected.

Resolving the forces acting on the ball in horizontal direction,

$$T \sin \theta = F_C = m \times \omega^2 \times r \quad \text{...}(i)$$

Resolving forces is vertical direction,

$$T \cos \theta = w = m \times g \quad \text{...}(ii)$$

Dividing equation (*i*) and (*ii*), we get

$$\frac{T \sin \theta}{T \cos \theta} = \frac{m \times \omega^2 \times r}{m \times g} = \frac{\omega^2 \times r}{g}$$

or

$$\tan \theta = \frac{\omega^2 \times r}{g} \quad \text{...}(iii)$$

But from Fig. 15.2, we have

$$\tan \theta = \frac{r}{h} \quad \text{...}(iv)$$

Equating the two values of tan θ, given by equations (*iii*) and (*iv*),

$$\frac{\omega^2 \times r}{g} = \frac{r}{h}$$

or

$$h = \frac{r \times g}{r \times \omega^2} = \frac{g}{\omega^2} \quad \text{...(15.1)}$$

If g is taken in m/s^2 and ω in rad/s, then h will be in meters. If N is the speed in r.p.m., then $\omega = \frac{2\pi N}{60}$

$$\therefore \quad h = \frac{9.81}{\left(\frac{2\pi N}{60}\right)^2} = 9.81 \times \left(\frac{60}{2\pi}\right)^2 \times \frac{1}{N^2} \quad (\because \; g \text{ in m/s}^2 = 9.81)$$

or

$$h = \frac{895}{N^2} \quad \text{...(15.2)}$$

From equation (15.2) it is clear that height of a Watt governor is inversely proportional to the square of the speed. Therefore at high speeds, the value of h is very small. For example if $N = 50$ r.p.m. then $h = \frac{895}{50^2} = 0.358$ m = 35.8 cm. But if $N = 300$ then $h = \left(\frac{895}{300^2}\right) = 0.0099$ m = 0.99 cm. Hence this governor works satisfactorily at low speeds *i.e.* from 50 to 85 r.p.m.

Problem 15.1. *Calculate the change in vertical height of a Watt governor when its speed*

(*i*) *increases from 50 r.p.m. to 51 r.p.m., and*

(*ii*) *increases from 200 r.p.m. to 201 r.p.m.*

Sol. Given :

(*i*) Initial speed = 50 r.p.m. and final speed 51 r.p.m.

The initial and final height of governor is given by equation (15.2).

$\therefore$ Initial height $= \frac{895}{N^2} = \frac{895}{50^2} = \frac{895}{2500} = 0.358 \text{ m} = 358 \text{ mm}$

Final height $= \frac{895}{(\text{Final speed})^2} = \frac{895}{51 \times 51} = 0.344 \text{ m} = 344 \text{ mm}$

$\therefore$ Change in vertical height $= 358 - 344 =$ **14 mm. Ans.**

(*ii*) Initial speed = 200 r.p.m. and final speed = 201 r.p.m.

$\therefore$ Initial vertical height $= \frac{895}{200^2} = 0.0224 \text{ m} = 22.4 \text{ mm}$

Final vertical height $= \frac{895}{201^2} = 0.0221 \text{ m} = 22.1 \text{ mm}$

$\therefore$ Change in vertical height $= 22.4 - 22.1 =$ **0.3 mm. Ans.**

Problem 15.2. *Determine : (i) maximum speed, (ii) minimum speed and (iii) range of speed of a Watt governor of open-arm type shown in Fig. 15.2 (c), in which length of arm AE = 200 mm and length EF = 30 mm when angle θ changes from 40° to 30°.*

Sol. Given : (Refer to Fig. 15.2 (*c*))

Length AE = 200 mm ; length EF = 30 mm or length $EG = \frac{EF}{2} = \frac{30}{2} = 15$ mm, $\theta = 40°$ and $30°$ which means $\theta_1 = 40°$ and $\theta_2 = 30°$.

Let N_1 = Maximum speed corresponding to $\theta_1 = 40°$

N_2 = Minimum speed corresponding to $\theta_2 = 30°$.

(*i*) *Maximum speed* (N_1)

The relation between height and speed of Watt governor is given by equation (15.2) as

$$h = \frac{895}{N^2} \quad ...(i)$$

Refer to Fig. 15.3.

Height of governor, $h = BO = BG + OG$

But $BG = EL = AE \cos\theta = 200 \cos\theta$

and OG from triangle OEG is given by, $\tan\theta = \frac{EG}{OG}$

or $OG = \frac{EG}{\tan\theta} = \frac{15}{\tan\theta}$ ($\because$ EG = 15 mm)

$$\therefore \quad h = BG + OG = 200\cos\theta + \frac{15}{\tan\theta} \quad ...(ii)$$

When speed is N_1, the height is h_1. Hence above equation (*i*) becomes as

$$h_1 = \frac{895}{N_1^2}$$

or

$$N_1^2 = \frac{895}{h_1} \quad ...(iii)$$

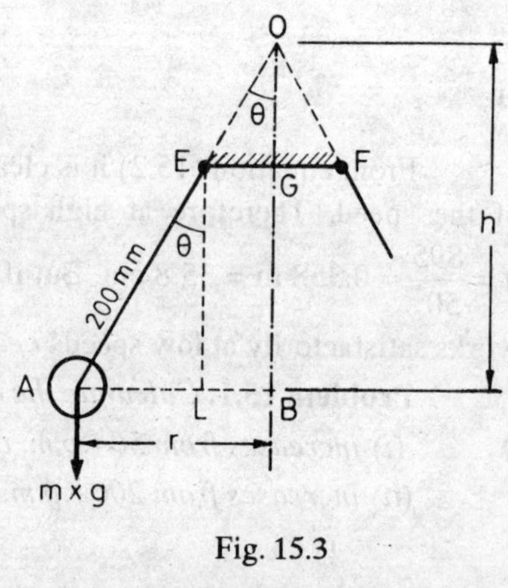

Fig. 15.3

Let us find the value of h_1 when $\theta = 40°$.

From equation (*ii*), when $\theta = 40°$, the value of h_1 will be

$$h_1 = 200 \cos 40° + \frac{15}{\tan 40°}$$

$$= 153.2 + 17.87 = 171.07 \text{ mm} = 0.17107 \text{ m}$$

Substituting the value of h_1 in equation (*iii*), we get

$$N_1^2 = \frac{895}{0.17107} = 5231.776$$

or $$N_1 = \sqrt{5231.776} = \mathbf{72.33 \text{ r.p.m.}} \quad \textbf{Ans.}$$

(*ii*) *Minimum speed* (N_2)

When $\theta = 30°$, the speed is minimum. The corresponding height (h_2) of governor is obtained by substituting $\theta = 30°$ in equation (*ii*).

$$\therefore \quad h_2 = 200 \times \cos 30° + \frac{15}{\tan 30°}$$

$$= 173.2 + 25.98 = 199.18 \text{ mm} = 0.19918 \text{ m}$$

Now the speed (N_2) is obtained as

$$N_2^2 = \frac{895}{h_2} = \frac{895}{0.19918} = 4493.42$$

$$\therefore \quad N_2 = \sqrt{4493.42} = \mathbf{67.03 \text{ r.p.m.}} \quad \textbf{Ans.}$$

(*iii*) *Range of speed*

Range of speed = Max. speed – Min. speed

$= N_1 - N_2$

$= 72.33 - 67.03 = \mathbf{5.3 \text{ r.p.m.}}$ **Ans.**

15.5. Porter Governor

Fig. 15.4 (*a*) shows the diagram of a Porter governor. In case of Porter governor, a central heavy load is attached to the sleeve. The central load and sleeve moves up and down the spindle.

Let M = Mass of central load

W = Weight of central load = $M \times g$

w = Weight of each ball = $m \times g$

m = Mass of each ball

h = Height of governor

r = Radius of rotation

F_c = Centrifugal force on the ball = $m \times \omega^2 \times r$

ω = Angular speed of ball = $\frac{2\pi N}{60}$ rad/s

N = Speed of ball in r.p.m.

T_1 = Tension in upper arm

T_2 = Tension in lower link

α = Angle of inclination of the upper arm to the vertical

β = Angle of inclination of the lower link to the vertical

F_f = Force of friction between sleeve and spindle.

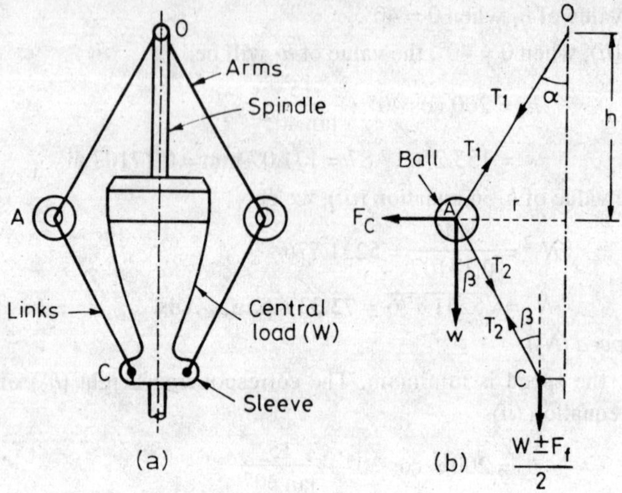

Fig. 15.4. Porter Governor.

The force of friction always acts in a direction opposite to that of the motion. When sleeve moves up, the force of friction acts in the downward direction. Then the total force acting on the sleeve in the downward direction will be $(W + F_f)$. Similarly when the sleeve moves down, the total force on the sleeve will be $(W - F_f)$. In general, the total force acting on the sleeve will be $(W \pm F_f)$ depending upon whether the sleeve moves upwards or downwards.

The relation between the height of governor and angular speed of ball

Fig,. 15.4 (*b*) shows the forces acting on left-hand half of the governor *i.e.* on the sleeve and on each ball.

Firstly, considering the equilibrium of left-half of sleeve, we have

$$T_2 \cos \beta = \frac{W \pm F_f}{2} \quad \text{or} \quad T_2 = \frac{W \pm F_f}{2 \cos \beta} \qquad \ldots(i)$$

Now, considering the equilibrium of the left ball, which is acted upon by (*i*) centrifugal force (F_C), (*ii*) weight ($w = m \times g$), (*iii*) tension in upper arm (T_1), and (*iv*) tension (T_2) in lower link, we have

$$T_1 \cos \alpha = w + T_2 \cos \beta \quad \text{(Forces in vertical direction)} \qquad \ldots(ii)$$

$$T_1 \sin \alpha + T_2 \sin \beta = F_C \quad \text{(Forces in horizontal direction)} \qquad \ldots(iii)$$

or
$$T_1 \sin \alpha + \left(\frac{W \pm F_f}{2 \cos \beta}\right) \times \sin \beta = F_C \qquad \left(\because \text{ From equation } (i),\ T_2 = \frac{W \pm F_f}{2 \cos \beta}\right)$$

or
$$T_1 \sin \alpha = F_C - \left(\frac{W \pm F_f}{2 \cos \beta}\right) \times \sin \beta$$

$$= F_C - \left(\frac{W \pm F_f}{2}\right) \times \tan \beta \qquad \ldots(iv)$$

The tension T_1 can be eliminated from equations (*ii*) and (*iv*)

Dividing equation (*iv*) by equation (*ii*),

$$\frac{T_1 \sin \alpha}{T_1 \cos \alpha} = \frac{F_C - \left(\frac{W \pm F_f}{2}\right) \times \tan \beta}{w + T_2 \cos \beta}$$

$$= \frac{F_C - \left(\frac{W \pm F_f}{2}\right) \times \tan\beta}{w + \left(\frac{W \pm F_f}{2}\right)} \qquad \left(\because \quad T_2 \cos\beta = \frac{W \pm F_f}{2} \text{ from } (i)\right)$$

or
$$\tan\alpha = \frac{F_C - \left(\frac{W \pm F_f}{2}\right) \tan\beta}{w + \left(\frac{W \pm F_f}{2}\right)}$$

or
$$\left(w + \frac{W \pm F_f}{2}\right) \tan\alpha = F_C - \left(\frac{W \pm F_f}{2}\right) \tan\beta$$

or
$$\left(w + \frac{W \pm F_f}{2}\right) = \frac{F_C}{\tan\alpha} - \left(\frac{W \pm F_f}{2}\right) \frac{\tan\beta}{\tan\alpha}$$

Let $\frac{\tan\beta}{\tan\alpha} = k$, then the above equation becomes as

$$\left(w + \frac{W \pm F_f}{2}\right) = \frac{F_C}{\tan\alpha} - \left(\frac{W \pm F_f}{2}\right) \times k$$

But from Fig. 15.4 (*b*), we have, $\tan\alpha = \frac{r}{h}$

Now the above equation becomes as

$$\left(w + \frac{W \pm F_f}{2}\right) = \frac{F_C \times h}{r} - \left(\frac{W \pm F_f}{2}\right) \times k \qquad \left(\because \quad \tan\alpha = \frac{r}{h}\right)$$

$$= \frac{m \times \omega^2 \times r \times h}{r} - \left(\frac{W \pm F_f}{2}\right) \times k \qquad (\because \quad F_C = m \times \omega^2 \times r)$$

$$= m \times \omega^2 \times h - \left(\frac{W \pm F_f}{2}\right) \times k$$

or
$$\left(m \times g + \frac{M \times g \pm F_f}{2}\right) = m \times \omega^2 \times h - \left(\frac{M \times g \pm F_f}{2}\right) \times k \qquad (\because \quad w = m \times g \text{ and } W = M \times g)$$

or
$$m \times g + \frac{M \times g \pm F_f}{2} + \left(\frac{M \times g \pm F_f}{2}\right) \times k = m \times \omega^2 \times h$$

or
$$m \times g + \frac{M \times g \pm F_f}{2}(1 + k) = m \times \omega^2 \times h$$

or
$$\omega^2 = \frac{m \times g + \left(\frac{M \times g \pm F_f}{2}\right)(1 + k)}{mh} \qquad \text{...(15.3)}$$

$$= \frac{m \times g + \left(\frac{M \times g \pm F_f}{2}\right)(1 + k)}{m \times g} \times \frac{g}{h} \qquad \text{(Multiplying and dividing by } g\text{)}$$

$$= \frac{w + \left(\frac{W \pm F_f}{2}\right)(1 + k)}{w} \times \frac{g}{h} \quad \text{...(15.4)}$$

($\because$ $m \times g = w$ and $M \times g = W$)

From equation (15.3), the height of governor (h) is obtained as

$$h = \frac{m \times g + \left(\frac{M \times g \pm F_f}{2}\right)(1 + k)}{m \times \omega^2} \quad \text{...(15.5)}$$

If there is no friction between sleeve and spindle, then F_f (*i.e.* force of friction becomes zero and hence equations (15.3) and (15.4) reduces to

$$\omega^2 = \frac{m \times g + \frac{M \times g}{2}(1 + k)}{m \times h} \quad \text{...(15.6)}$$

$$= \frac{w + \frac{W}{2}(1 + k)}{w} \times \frac{g}{h}. \quad \text{...(15.7)}$$

Note. (*i*) In equations (15.3), (15.4) and (15.5), the force of friction (F_f) is taken with + ve sign when the sleeve moves upwards (or the speed of governor increases). If sleeve moves downwards (or the speed of governor decreases) then – ve sign is taken with force of friction F_f.

(*ii*) From equation (15.5), it is clear that for the same value of h, there will be two values of ω.

(*iii*) In these equations the value of 'h' should be taken in meter and value of 'g' as equal to 9.81 m/s^2.

Instantaneous Centre Method [Refer to Fig. 15.4 (*c*)]

In this method, the equilibrium of the lower arm AC is considered. The forces acting on the lower arm AC are :

(*i*) Centrifugal force F_C through A,

(*ii*) The weight of ball ($m \times g$) through A, and

(*iii*) Half of the weight of the sleeve *i.e.* $\frac{W}{2}$.

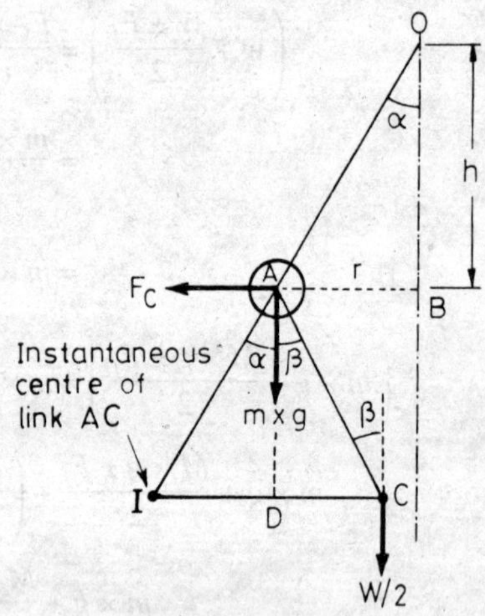

Fig. 15.4 (*c*)

First, the instantaneous centre of the lower arm AC is obtained. As the point A moves along a circular arc which has O as centre and AO is radius and point C moves parallel to the axis of the governor, the instantaneous* centre I lies at the point of intersection of OA produced and a line drawn through C perpendicular to the governor axis.

Taking moments of forces $\left(i.e.\ F_C, mg \text{ and } \frac{W}{2}\right)$ acting on lower arm AC, about the point I,

$$F_C \times AD = (m \times g) \times ID + \frac{W}{2} \times IC$$

or

$$F_C = (m \times g) \times \frac{ID}{AD} + \frac{W}{2} \times \frac{IC}{AD}$$

*To locate instantaneous centre of a link, the lines are drawn perpendicular to the direction of velocity at the two ends. The point of intersection of these lines gives instantaneous centre. In Fig. 15.4 (*c*), CI is perpendicular to the velocity of point C, whereas OA produced is the line perpendicular to the velocity of point A.

$$= m \times g \times \tan\alpha + \frac{W}{2} \times \left(\frac{ID + CD}{AD}\right) \qquad (\because \; IC = ID + DC)$$

$$= m \times g \times \tan\alpha + \frac{W}{2} \times \left(\frac{ID}{AD} + \frac{CD}{AD}\right)$$

$$= m \times g \times \tan\alpha + \frac{M \times g}{2} (\tan\alpha + \tan\beta)$$

$$\left(\because \; \frac{ID}{AD} = \tan\alpha, \frac{CD}{AD} = \tan\beta \text{ and } W = M \times g\right)$$

$$= m \times g \times \tan\alpha + \frac{M \times g}{2} \times \tan\alpha \left(1 + \frac{\tan\beta}{\tan\alpha}\right)$$

$$= m \times g \times \tan\alpha + \frac{M \times g}{2} \tan\alpha\,(1 + k) \qquad \left(\text{where } k = \frac{\tan\beta}{\tan\alpha}\right)$$

$$= \tan\alpha \left[m \times g + \frac{M \times g}{2}(1 + k)\right]$$

But from triangle OAB, $\tan\alpha = \frac{r}{h}$ and $F_C = m \times \omega^2 \times r$.

Substituting these values in the above equation, we get

$$m \times \omega^2 \times r = \frac{r}{h}\left[m \times g + \frac{M \times g}{2}(1 + k)\right]$$

or

$$\omega^2 = \frac{m \times g + \frac{M \times g}{2}(1 + k)}{m \times h} \qquad ...(15.7A)$$

If $k = 1$ which is true when $\tan\alpha = \tan\beta$, then above equation becomes as

$$\omega^2 = \frac{m \times g + M \times g}{m \times h} = \frac{(m + M) \times g}{m \times h}$$

But

$$\omega = \frac{2\pi N}{60}$$

$$\therefore \quad \left(\frac{2\pi N}{60}\right)^2 = \frac{(m + M) \times g}{m \times h}$$

or

$$h = \frac{(m + M) \times g}{m \times \left(\frac{2\pi N}{60}\right)^2} = \frac{(m + M) \times g \times 3600}{m \times 4\pi^2 \times N^2} = \frac{m + M}{m} \times \frac{9.81 \times 3600}{4\pi^2 \times N^2}$$

$$\therefore \quad h = \frac{m + M}{m} \times \frac{894.56}{N^2} \qquad ...(15.7B)$$

In the above equation, h is in meter.

If frictional force at the sleeve is taken into consideration, then total force in general acting on C when sleeve moves upwards or downwards is equal to $\frac{1}{2}(M \times g \pm F)$. Then equation (15.7A) becomes as

$$\omega^2 = \frac{m \times g + \frac{(M \times g \pm F)}{2}(1 + k)}{m \times h}$$

The above equation is same as equation (15.3).

Problem 15.3. *Calculate the minimum speed, maximum speed and range of the speed of a Porter governor, which has equal arms each 200 mm long and pivoted on the axis of rotation. The mass of each ball is 4 kg and the central mass on the sleeve is 20 kg. The radius of rotation of the ball is 100 mm when the governor begins to lift and 130 mm when the governor is at maximum speed.*

Sol. Given :

Length of all arms = 200 mm ; hence $AO = AC = 200$ mm ; arm are pivoted on the axis ; mass of each ball, $m = 4$ kg ; central mass on sleeve, $M = 20$ kg ; minimum radius, $r_1 = 100$ mm and maximum radius, $r_2 = 130$ mm.

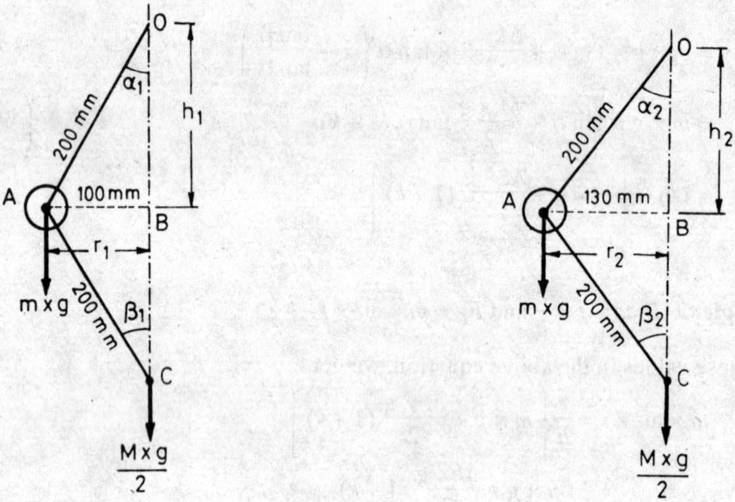

(*a*) Minimum radius (*b*) Maximum radius

Fig. 15.5

Let N_1 = minimum speed when $r_1 = 100$ mm

N_2 = maximum speed when $r_2 = 130$ mm.

(*i*) *Minimum speed* (N_1) (Refer to Fig. 15.5 (*a*))

When there is no friction between sleeve and spindle of a Porter governor, the relation between angular speed and height of governor is given by equation (15.6) as

$$\omega^2 = \frac{m \times g + \frac{M \times g}{2}(1 + k)}{m \times h} \quad \text{where } k = \frac{\tan \beta}{\tan \alpha}.$$

When the arms are of equal length and points O and C lie on the same vertical line, then

$$\tan \alpha = \tan \beta \quad \text{or} \quad k = \frac{\tan \alpha}{\tan \beta} = 1$$

Hence the above equation becomes, as

$$\omega^2 = \frac{m \times g + \frac{M \times g}{2}(1 + 1)}{m \times h} \qquad (\because \; k = 1 \text{ in this problem})$$

$$= \frac{m \times g + M \times g}{m \times h} = \frac{(m + M) \times g}{m \times h} \qquad ...(i)$$

When minimum speed is N_1, then angular speed is ω_1 and height is h_1.

Hence in terms of minimum angular speed, the above equation becomes as

$$\omega_1^2 = \frac{(m + M) \times g}{m \times h_1} \qquad ...(ii)$$

The value of h_1 from Fig. 15.5 (*a*) is given by,

$$h_1 = \sqrt{AO^2 - AB^2}$$

$$= \sqrt{200^2 - 100^2} = \sqrt{40000 - 10000} = 173.2 \text{ mm} \quad \text{or} \quad 0.1732 \text{ mm.}$$

Substituting the values of h_1, m and M in equation (*ii*), we get

$$\omega_1^2 = \frac{(4 + 20) \times 9.81}{4 \times 0.1732} = 339.838$$

or $$\omega_1 = \sqrt{339.838} = 18.43 \text{ rad/s.}$$

But $\omega_1 = \frac{2\pi N_1}{60}$ hence $N_1 = \frac{60 \times \omega_1}{2\pi}$

$$\therefore \quad N_1 = \frac{60 \times 18.43}{2\pi} = 175.99 \simeq 176 \text{ r.p.m.}$$

(*ii*) *Maximum speed* (N_2) (Refer to Fig. 15.5 (*b*))

$$h_2 = \sqrt{AO^2 - AB^2}$$

$$= \sqrt{200^2 - 130^2} = \sqrt{40000 - 16900} = 151.98 \text{ mm} \quad \text{or} \quad 0.15198 \text{ m}$$

In terms of maximum angular speed (ω_2), the equation (*i*) is written as

$$\omega_2^2 = \frac{(m \times M) \times g}{m \times h_2}$$

$$= \frac{(4 + 20) \times 9.81}{4 \times 0.15198} \qquad (\because \ h_2 = 0.15198 \text{ m})$$

$$= 387.287$$

$$\therefore \quad \omega_2 = \sqrt{387.287} = 19.68 \text{ rad/s}$$

$$\therefore \quad N_2 = \frac{60 \times \omega_2}{2\pi} = \frac{60 \times 19.68}{2\pi} = 187.9 \text{ r.p.m.}$$

(*iii*) *Range of speed*

Range of speed $= N_2 - N_1 = 187.9 - 176 =$ **11.9 r.p.m. Ans.**

Problem 15.4. *If in Problem 15.3, the friction at the sleeve is equivalent to 15 N, then determine the range of speed.*

Sol. Data from Problem 15.3 are :

$m = 4$ kg ; $M = 20$ kg (Given)

$h_1 = 0.1732$ m ; $h_2 = 0.15198$ m and $k = 1$ (Calculated)

Friction at sleeve, $F = 15$ N

Let N_1 = Minimum speed when $r_1 = 100$ mm and friction at sleeve is considered

N_2 = Max. speed when $r_2 = 130$ mm and friction is considered.

(*i*) *Minimum speed* (N_1)

Speed will be minimum, when radius of rotation is minimum. The radius of rotation will be minimum when sleeve moves downwards. When sleeve moves downwards, the frictional force acts upwards. The relation between angular speed and height of governor (h) when frictional force is considered is given by equation (15.3) as

$$\omega^2 = \frac{m \times g + \frac{(M \times g \pm F)}{2}(1 + k)}{m \times h}$$

In this problem, $k = 1$ hence the above equation becomes as

$$\omega^2 = \frac{m \times g \frac{(M \times g \pm F)}{2}(1 + 1)}{m \times h} = \frac{m \times g + (M \times g \pm F)}{m \times h} \quad \text{...(A)}$$

When radius is minimum, angular speed is minimum (*i.e.* ω_1) and sleeve moves downwards. Frictional force acts upwards and hence (– ve) sign is taken with frictional force. The above equation (A) becomes as

$$\omega_1^2 = \frac{m \times g + (M - F)}{m \times h_1}$$

$$= \frac{4 \times 9.81 + (20 \times 9.81 - 15)}{4 \times 0.1732} \quad (\because\ F = 15 \text{ N and } h_1 = 0.1732 \text{ m})$$

$$= \frac{39.24 + 181.2}{4 \times 0.1732} = 318.187$$

$$\therefore \quad \omega_1 = \sqrt{318.187} = 17.837 \text{ rad/s} \quad \left(\text{But } \omega_1 = \frac{2\pi N_1}{60} \quad \therefore \quad N_1 = \frac{60 \times \omega_1}{2\pi}\right)$$

$$\therefore \quad N_1 = \frac{60 \times \omega_1}{2\pi} = \frac{60 \times 17.837}{2\pi} = 170.33 \text{ r.p.m.}$$

(ii) Maximum speed (N_2)

Speed will be maximum, when radius of rotation is maximum. The radius of rotation will be maximum, when sleeve moves upwards. When sleeve moves upwards, the frictional force acts downwards and hence (+ ve) sign will be taken with frictional force in equation (A). The values of w and h in equation (A) will be ω_2 and h_2.

Hence equation (A) becomes as

$$\omega_2^2 = \frac{m \times g + (M \times g + F)}{m \times h_2}$$

$$= \frac{4 \times 9.81 + (20 \times 9.81 + 15)}{4 \times 0.15198} \quad (\because\ h_2 = 0.15198)$$

$$= \frac{39.24 + 211.2}{4 \times 0.15198} = 411.962$$

$$\therefore \quad \omega_2 = \sqrt{411.962} = 20.297 \text{ rad/s}$$

$$\therefore \quad N_2 = \frac{60 + \omega_2}{2\pi} = \frac{60 \times 20.297}{2\pi} = 193.82 \text{ r.p.m.}$$

$\therefore$ Range of speed = $N_2 - N_1$ = 193.83 – 170.33 = **23.49 r.p.m. Ans.**

Problem 15.5. *A Porter governor has all four arms 300 mm long. The upper arms are pivoted on the axis of rotation and lower arms are attached to the sleeve at a distance of 3.5 mm from the axis. The mass of each ball is 7 kg and the mass on the sleeve is 54 kg. If the extreme radii of rotation of the balls are 200 mm and 250 mm, find the range of speed of the governor.* (AMIE, S 1976)

Sol. Given :

Length of all arms = 300 mm

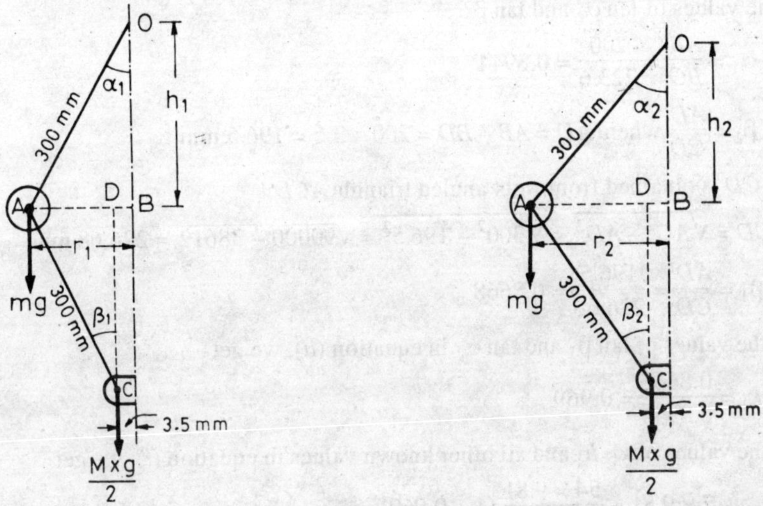

(*a*) Minimum radius (*b*) Maximum radius

Fig. 15.6

$\therefore \quad AO = AC = 300$ mm

Mass of each ball, $m = 7$ kg.

Mass of sleeve, $M = 54$ kg.

The lower arms are attached to the sleeve at a distance of 3.5 mm from the axis.

$\therefore$ Distance $BD = 3.5$ mm.

Extreme radii of rotation = 200 mm and 250 mm or minimum radius, $r_1 = 200$ mm and maximum radius = 250 mm.

Let N_1 = Minimum speed when $r_1 = 200$ mm

N_2 = Maximum speed when $r_2 = 250$ mm.

(i) Value of Minimum speed (N_1)

Refer to Fig. 15.6 (*a*). There is no friction between sleeve and spindle and hence using equation (15.6), we get

$$\omega^2 = \frac{m \times g + \frac{M \times g}{2}(1 + k)}{m \times h} \text{ where } k = \frac{\tan \beta}{\tan \alpha}.$$

Corresponding to minimum speed N_1, the minimum angular speed is ω_1.

$$\therefore \quad \omega_1^2 = \frac{m \times g + \frac{M \times g}{2}(1 + k_1)}{m \times h_1} \quad ...(i)$$

Let us now find the value of k_1 and h_1.

In Fig. 15.6 (*a*) *ABO* is a right angled triangle in which $AO = 300$ mm and $AB = r_1 = 200$ mm.

$$\therefore \quad h_1 = BO = \sqrt{AO^2 - AB^2}$$

$$= \sqrt{300^2 - 200^2} = \sqrt{90000 - 40000} = \sqrt{50000} = 223.6 \text{ mm} \quad \text{or} \quad = 0.2236 \text{ m}$$

Now $$k_1 = \frac{\tan \beta_1}{\tan \alpha_1} \quad ...(ii)$$

Let us find the values of $\tan \alpha_1$ and $\tan \beta_1$.

$$\tan \alpha_1 = \frac{AB}{BO} = \frac{200}{223.6} = 0.8944$$

and $$\tan \beta_1 = \frac{AD}{CD} \text{ where } AD = AB - BD = 200 - 3.5 = 196.5 \text{ mm}$$

The value of CD is obtained from right angled triangle ACD.

$$\therefore \quad CD = \sqrt{AC^2 - AD^2} = \sqrt{300^2 - 196.5^2} = \sqrt{90000 - 38612} = 226.68 \text{ mm}$$

$$\therefore \quad \tan \beta_1 = \frac{AD}{CD} = \frac{196.5}{226.68} = 0.8668$$

Substituting the values of $\tan \beta_1$ and $\tan \alpha_1$ in equation (*ii*), we get

$$k_1 = \frac{0.8668}{0.8944} = 0.969$$

Substituting the values of k_1, h_1 and all other known values in equation (*i*), we get

$$\omega_1^2 = \frac{7 \times 9.81 + \frac{54 \times 9.81}{2}(1 + 0.969)}{7 \times 0.2236}$$

($\because$ The value of h_1 here should be taken in meter, hence $h_1 = 0.2236$ m)

$$= \frac{68.67 + 521.53}{7 \times 223.6} = 377.076$$

$$\therefore \quad \omega_1 = \sqrt{377.076} = 19.418 \text{ rad/s}$$

But $\omega_1 = \frac{2\pi N_1}{60}$ hence $N_1 = \frac{60 \times \omega_1}{2\pi}$

$$\therefore \quad N_1 = \frac{60 \times 19.418}{2\pi} = 185.4 \text{ r.p.m.}$$

(ii) Value of Maximum Speed (N_2)

Refer to Fig. 15.6 (*b*).

Corresponding to maximum speed (N_2), maximum angular speed (ω_2) is given by

$$\omega_2^2 = \frac{m \times g + \frac{M \times g}{2}(1 + k_2)}{m \times h_2} \quad \text{...(iii)}$$

Let us now find the values of h_2 and k_2.

In Fig. 15.6 (*b*), ABO is a right angled triangle in which $AO = 300$ mm and $AB = r_2 = 250$ mm

$$\therefore \quad h_2 = BO = \sqrt{AO^2 - AB^2}$$

$$= \sqrt{300^2 - 250^2} = \sqrt{90000 - 62500} = 165.83 \text{ mm} \quad \text{or} \quad 0.16583 \text{ m}$$

Now $$\tan \alpha_2 = \frac{AB}{BO} = \frac{250}{165.83} = 1.5075$$

and $$\tan \beta_2 = \frac{AD}{CD} \text{ where } AD = AB - BD = 250 - 3.5 = 246.5 \text{ mm}$$

$$= \frac{246.5}{CD} \text{ where } CD = \sqrt{AC^2 - AD^2} = \sqrt{300^2 - 246.5^2} = \sqrt{90000 - 60762} = 170.99$$

$$= \frac{246.5}{170.99} = 1.4416$$

$$\therefore \quad k_2 = \frac{\tan \beta_2}{\tan \alpha_2} = \frac{1.4416}{1.5075} = 0.956$$

Substituting the values of k_2, h_2 (in meter) and other known values in equation (*iii*), we get

$$\omega_2^2 = \frac{7 \times 9.81 + \frac{54 \times 9.81}{2}(1 + 0.956)}{7 \times 0.16583} = \frac{68.67 + 518.085}{7 \times 0.16583} = 505.47$$

$$\therefore \quad \omega_2 = \sqrt{505.47} = 22.4826 \text{ rad/s}$$

$$\therefore \quad N_2 = \frac{60 \times \omega_2}{2\pi} = \frac{60 \times 22.4826}{2\pi} = 214.7 \text{ r.p.m.}$$

Now range of speed $= N_2 - N_1 = 214.7 - 185.4 =$ **29.3 r.p.m. Ans.**

Problem 15.6. *The lengths of the upper and lower arms of a Porter governor are 200 mm and 250 mm respectively. Both the arms are pivoted on the axis of the rotation. The central load is 150 N, the weight of each ball is 20 N and the friction of the sleeve together with the resistance of the operating gear is equivalent to a force of 30 N at the sleeve. If the limiting inclinations of the upper arms to the vertical are 30° and 40°, determine the range of the speed of the governor.* (AMIE, Winter 1994)

Sol. Given :

Weight of each ball, $w = 20$ N ; central load, $W = 150$ N ; frictional force, $F_f = 30$ N ; limiting inclinations of upper arm with verticals = 30° and 40° *i.e.* $\alpha_1 = 30°$ and $\alpha_2 = 40°$; length of upper arm = 200 mm and length of lower arm = 250 mm. Hence $AO = 200$ mm and $AC = 250$ mm.

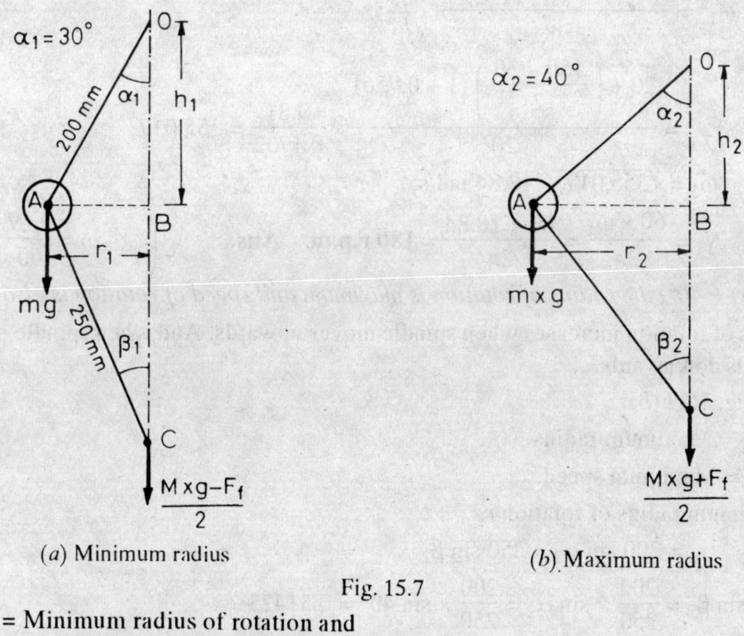

(*a*) Minimum radius (*b*) Maximum radius

Fig. 15.7

Let r_1 = Minimum radius of rotation and

r_2 = Maximum radius of rotation.

(*i*) *When* $\alpha_1 = 30$, *the radius of rotation is minimum and speed of rotation is also minimum*

The radius of rotation decreases when spindle moves downwards. And when sleeve moves downwards the frictional force acts upwards. Refer to Fig. 15.7 (*a*).

Let r_1 = Minimum radius

N_1 = Minimum speed

Min. radius of rotation, $r_1 = 200 \sin \alpha_1$ or $250 \sin \beta_1$

$\therefore \quad r_1 = 200 \sin \alpha_1 = 250 \sin \beta_1$

or $$\sin \beta_1 = \frac{200 \sin \alpha_1}{250} = 0.8 \times \sin 30° = 0.8 \times 0.5 = 0.40$$

$\therefore \quad \beta_1 = \sin^{-1} 0.40 = 23.58°$

Now $$k_1 = \frac{\tan \beta_1}{\tan \alpha_1} = \frac{\tan 23.58°}{\tan 30°} = 0.756$$

When $\alpha_1 = 30°$, the values of r_1 and h_1 are given by

$r_1 = 200 \sin \alpha_1 = 200 \times \sin 30° = 100$ mm, and

$h_1 = 200 \cos \alpha = 200 \times \cos 30° = 173.2$ mm or 0.1732 m.

Now using equation (15.4), we have

$$\omega^2 = \frac{w + \left(\frac{W \pm F}{2}\right)(1 + k)}{w} \times \frac{g}{h}$$

For minimum speed, the radius of rotation is minimum and spindle moves downwards and frictional force acts upwards. Hence – ve sign is taken with frictional force

$$\therefore \quad \omega_1^2 = \frac{w + \left(\frac{W - F}{2}\right)(1 + k_1)}{w} \times \frac{g}{h_1}$$

$$= \frac{20 + \left(\frac{150 - 30}{2}\right)(1 + 0.756)}{20} \times \frac{9.81}{0.1732} = 355.017$$

$\omega_1 = \sqrt{355.017} = 18.84$ rad/s

and $$N_1 = \frac{60 \times \omega_1}{2\pi} = \frac{60 \times 18.84}{2\pi} = \textbf{180 r.p.m.} \quad \textbf{Ans.} \qquad \left(\because \; \omega = \frac{2\pi N}{60} \text{ or } N = \frac{60 \times \omega}{2\pi}\right)$$

(ii) When $\alpha = 40°$, the radius of rotation is maximum and speed of rotation is also maximum.

The radius of rotation increases when spindle moves upwards. And when spindle moves upwards, the frictional force acts downwards.

Refer to Fig. 15.7 (*b*)

Let r_2 = Maximum radius

N_2 = Maximum speed.

Now maximum radius of rotation r_2

$r_2 = 200 \sin \alpha_2 = 250 \sin \beta_2$

or $$\sin \beta_2 = \frac{200}{250} \times \sin \alpha_2 = \frac{200}{250} \times \sin 40° = 0.51423$$

$\therefore \quad \beta_2 = \sin^{-1} 0.51423 = 30.95°$

and $$k_2 = \frac{\tan \beta_2}{\tan \alpha_2} = \frac{\tan 30.95°}{\tan 40°} = 0.715$$

The values of r_2 and h_2 corresponding to $\alpha_2 = 40°$ are given by

$r_2 = 200 \sin 40° = 128.56$ mm and

$h_2 = 200 \cos 40° = 153.2$ mm or 0.1532 m.

Now using equation (15.4), we have

$$\omega^2 = \frac{w + \left(\frac{W \pm F}{2}\right)(1 + k)}{w} \times \frac{g}{h}$$

For maximum speed, the radius of rotation is maximum and sleeve moves upwards. The frictional force acts downwards. Hence + ve sign is taken with frictional force. Hence for maximum speed, the equation becomes as

$$\omega_2^2 = \frac{w + \left(\frac{W + F}{2}\right)(1 + k_2)}{w} \times \frac{g}{h_2}$$

$$= \frac{20 + \left(\frac{150 + 30}{2}\right)(1 + 0.715)}{20} \times \frac{9.81}{0.1532} = 558.22$$

$\therefore \quad \omega_2 = \sqrt{558.22} = 23.626$ rad/s

Corresponding speed, $N_2 = \frac{60 \times \omega_2}{2\pi} = \frac{60 \times 23.626}{2\pi} = 225.6$ r.p.m.

$\therefore$ Range of speed = 225.6 – 180 = **45.6 r.p.m. Ans.**

Problem 15.7. *All the arms of a Porter governor are 250 mm long. The upper arms are hinged at a distance of 40 mm from the axis of rotation whereas the lower arms are pivoted at a distance of 50 mm from the axis of rotation. The mass of each ball is 2.5 kg and mass of sleeve is 25 kg. The force of friction on the sleeve is 20 N. The masses revolve at a radius of 125 mm at minimum speed and at a radius of 150 mm at maximum speed. Determine the range of speed.*

Sol. Given : (Refer to Fig. 15.8 (*a*))

Length of all arms = 250 mm

$\therefore \quad AE = AC = 250$ mm ; Distance $EF = 40$ mm ; Distance $CD = 50$ mm.

Mass of each ball, $m = 2.5$ kg

Mass of sleeve, $M = 25$ kg

Force of friction, $F = 20$ N

Minimum radius, $r_1 = 125$ mm

Maximum radius, $r_2 = 150$ mm.

Let N_1 = Minimum speed when $r_1 = 125$ mm, and

N_2 = Max. speed when $r_2 = 150$ mm

(*i*) *Minimum speed* (N_1) (Refer to Fig. 15.8 (*a*))

Let us first find the values of h_1 and k_1 where $k_1 = \frac{\tan \beta_1}{\tan \alpha_1}$. The value k_1 can be obtained if $\tan \beta_1$ and $\tan \alpha_1$ are known.

For minimum speed, the radius is $r_1 = 125$ mm.

$\therefore \quad r_1 = AB = 125$ mm

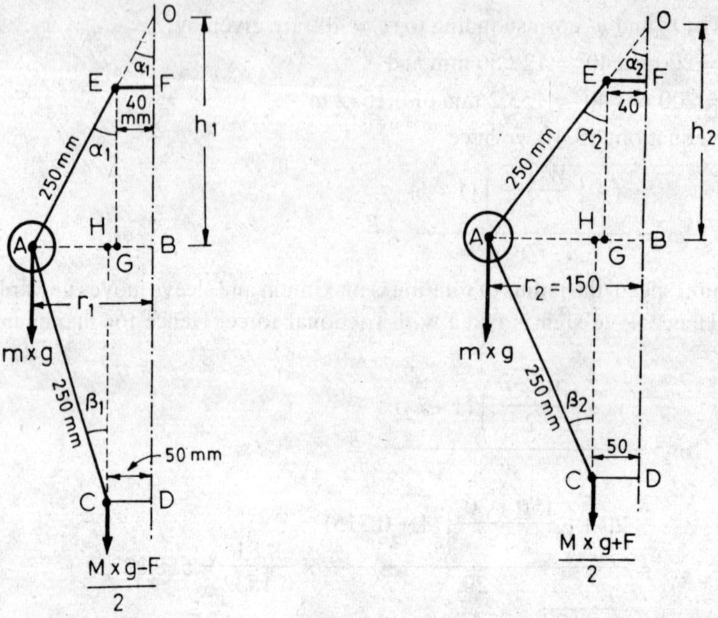

(*a*) Minimum radius (*b*) Maximum radius

Fig. 15.8

$\therefore$ Distance $AG = AB - BG = AB - EF = 125 - 40 = 85$ mm

Distance $AE = 250$ mm (Given)

$\therefore$ From triangle AEG, $\sin \alpha_1 = \dfrac{AG}{AE} = \dfrac{85}{250} = 0.34$ $\quad (\because \ \alpha_1 = \sin^{-1} 0.34 = 19.877°)$

From triangle ACH, $\sin \beta_1 = \dfrac{AH}{AC} = \dfrac{125 - 50}{250} = \dfrac{75}{250} = 0.3$

$(\because \ AH = AB - BH = AB - CD = 125 - 50 = 75 \text{ mm})$

$\therefore$ $\beta_1 = \sin^{-1} 0.3 = 17.457°$

Now the value of k_1 can be obtained as,

$$k_1 = \frac{\tan \beta_1}{\tan \alpha_1} = \frac{\tan 17.457°}{\tan 19.877°} = \frac{0.3145}{0.3615} = 0.8698 \simeq 0.87$$

The value of h_1 is obtained from triangle ABO in which $\tan \alpha_1 = \dfrac{AB}{BO}$

or $\tan 19.877° = \dfrac{125}{h_1}$ $\quad (\because \ \alpha_1 = 19.877°, AB = r_1 = 125 \text{ mm}, BO = h_1)$

$\therefore$ $h_1 = \dfrac{125}{\tan 19.877°} = \dfrac{125}{0.3615} = 345.78$ mm or 0.34578 m

Now using equation (15.3), we get

$$\omega^2 = \frac{m \times g + \left(\dfrac{M \times g \pm F}{2}\right)(1 + k)}{m \times h}$$

For minimum speed, the radius of rotation is minimum and sleeve moves downwards. The frictional force acts upwards whereas the weight of sleeve acts downwards. Hence frictional force is in the opposite direction of weight. Hence – ve sign is taken with frictional force. Also when speed is minimum, $\omega = \omega_1$, $k = k_1$ and $h = h_1$. Now the above equation can be written as

$$\omega_1^2 = \frac{m \times g + \left(\dfrac{M \times g - F}{2}\right)(1 + k_1)}{m \times h_1}$$

$$= \frac{2.5 \times 9.81 + \left(\dfrac{25 \times 9.81 - 20}{2}\right)(1 + 0.87)}{2.5 \times 0.34578}$$

($\because$ $m = 2.5$ kg ; $M = 25$ kg ; $k_1 = 0.87$ and $h_1 = 0.34578$ m)

$$= \frac{24.525 + 210.6}{2.5 \times 0.34578} = 272$$

$\therefore$ $\omega_1 = \sqrt{272} = 16.492$ rad/s. But $\omega_1 = \dfrac{2\pi N_1}{60}$ $\therefore$ $N_1 = \dfrac{60 \times \omega_1}{2\pi}$

or $N_1 = \dfrac{60 \times 16.492}{2\pi} = 157.5$ r.p.m.

(ii) Maximum speed (N_2) (Refer to Fig. 15.8 (*b*))

Let us find the values of h_2 and k_2 where $k_2 = \dfrac{\tan \beta_2}{\tan \alpha_2}$.

Here radius $r_2 = 150$ mm or $AB = 150$ mm

Now distance $AG = AB - BG = AB - EF = 150 - 40 = 110$ mm

Distance $AH = AB - BH = AB - CD = 150 - 50 = 100$ mm

Distance AE = Distance $AC = 250$ mm.

In triangle AEG, $\sin \alpha_2 = \dfrac{AG}{AE} = \dfrac{110}{250} = 0.44$ $\therefore$ $\alpha_2 = \sin^{-1} 0.44 = 26.1°$

In triangle ACH, $\sin \beta_2 = \dfrac{AH}{AC} = \dfrac{100}{250} = 0.4$ $\therefore$ $\beta_2 = \sin^{-1} 0.4 = 23.58°$

$\therefore$ Value of $k_2 = \dfrac{\tan \beta_2}{\tan \alpha_2} = \dfrac{\tan 23.58°}{\tan 26.1°} = \dfrac{0.4365}{0.4899} = 0.89$

The value of h_2 is obtained from triangle ABO in which $\tan \alpha_2 = \dfrac{AB}{BO}$

$\therefore$ $\tan 26.1° = \dfrac{150}{h_2}$ ($\because$ $\alpha_2 = 26.1°, AB = r_2 = 150$ mm and $BO = h_2$)

$\therefore$ $h_2 = \dfrac{150}{\tan 26.1°} = \dfrac{150}{0.4899} = 306.185$ mm or 0.306185 m.

For maximum speed, the radius of rotation is maximum and sleeve moves upwards. The frictional forces acts downwards. The weight of sleeve is also acting downwards. Hence frictional force is in the direction of weight of sleeve. Hence + ve sign is taken with frictional force. Also when speed is maximum, $\omega = \omega_2$, $k = k_2$ and $h = h_2$. Hence equation (15.3) becomes as

$$\omega_2^2 = \frac{m \times g + \left(\dfrac{M \times g + F}{2}\right)(1 + k_2)}{m \times h_2}$$

$$= \frac{2.5 \times 9.81 + \left(\dfrac{25 \times 9.81 + 20}{2}\right)(1 + 0.89)}{2.5 \times 0.306185} \quad (\because\ k_2 = 0.89 \text{ and } h_2 = 0.306185 \text{ m})$$

$$= \frac{24.525 + 250.661}{2.5 \times 0.306185} = 359.5$$

$$\therefore \quad \omega_2 = \sqrt{359.5} = 18.96 \text{ rad/s.}$$

Now $$N_2 = \frac{60 \times \omega_2}{2\pi} = \frac{60 \times 18.96}{2\pi} = 181 \text{ r.p.m.}$$

$\therefore$ Range of speed $= N_2 - N_1 = 181 - 157.5 =$ **23.5 r.p.m. Ans.**

15.6. Proell Governor

Fig. 15.9 (*a*) shows the line diagram of a Proell governor, which is similar to the Porter governor having a heavy central load at the sleeve. But it differs from Porter governor in the arrangement of balls. The balls are carried on the extension of the lower arms instead of at the junction of the upper and lower arms. The action of this governor is similar to the Watt's governor. An increase in the speed of rotation increases the radius of rotation and raise the sleeve, thus reducing the amount of energy supplied to the engine. Conversely a decrease in speed results in decrease in radius of rotation, thus lowering the sleeve and increasing the amount of energy supplied to the engine.

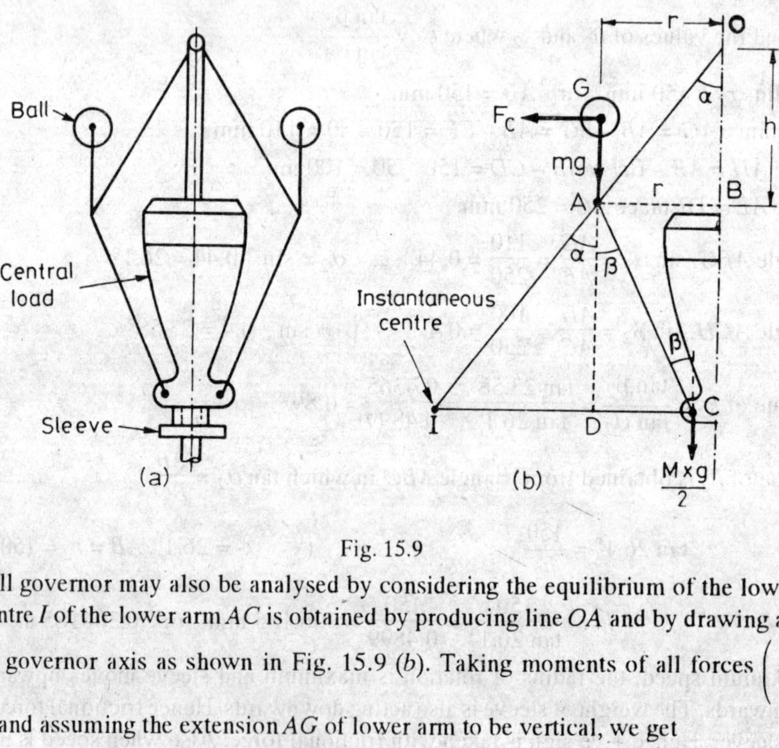

Fig. 15.9

The Proell governor may also be analysed by considering the equilibrium of the lower arm AC. The instantaneous centre I of the lower arm AC is obtained by producing line OA and by drawing a line through C perpendicular to governor axis as shown in Fig. 15.9 (*b*). Taking moments of all forces $\left(\text{i.e. } F_C,\ mg \text{ and } \dfrac{M \times g}{2}\right)$ about I and assuming the extension AG of lower arm to be vertical, we get

$$F_C \times GD = m \times g \times ID + \frac{M \times g}{2} \times IC$$

Dividing both sides by AD, we get

$$F_C \times \frac{GD}{AD} = m \times g \times \frac{ID}{AD} + \frac{M \times g}{2} \times \frac{IC}{AD}$$

$$= m \times g \times \tan\alpha + \frac{M \times g}{2} \times \left(\frac{ID + DC}{AD}\right) \qquad \left(\because\ IC = ID + DC \text{ and } \frac{ID}{AD} = \tan\alpha\right)$$

$$= m \times g \times \tan\alpha + \frac{M \times g}{2}\left(\frac{ID}{AD} + \frac{DC}{AD}\right)$$

$$= m \times g \times \tan\alpha + \frac{M \times g}{2}(\tan\alpha + \tan\beta) \qquad \left(\because\ \frac{DC}{AD} = \tan\beta\right)$$

$$= m \times g \times \tan\alpha + \frac{M \times g}{2} \times \tan\alpha\left(1 + \frac{\tan\beta}{\tan\alpha}\right)$$

$$= \tan\alpha\left[m \times g + \frac{M \times g}{2}(1 + k)\right] \qquad \left(\text{where } k = \frac{\tan\beta}{\tan\alpha}\right)$$

or

$$F_C = \frac{AD}{GD} \times \tan\alpha \times \left[m \times g + \frac{M \times g}{2}(1 + k)\right]$$

But from triangle OAB, $\tan\alpha = \frac{r}{h}$ in which r = radius of governor and $F_C = m \times \omega^2 \times r$, here r = radius of rotation of ball. Hence substituting these values in the above equation, we get

$$m \times \omega^2 \times r = \frac{AD}{GD} \times \frac{r}{h}\left[m \times g + \frac{M \times g}{2}(1 + k)\right]$$

or

$$\omega^2 = \frac{AD}{GD} \times \frac{r}{h}\left[m \times g + \frac{M \times g}{2}(1 + k)\right] \times \frac{1}{m \times r}$$

$$= \frac{AD}{GD}\left[\frac{m \times g + \frac{M \times g}{2}(1 + k)}{m \times h}\right] \qquad \text{...(15.8)}$$

If $k = 1$ which is true when $\tan\alpha = \tan\beta$, then the above equation becomes as

$$\omega^2 = \frac{AD}{GD}\left(\frac{m \times g + M \times g}{m \times h}\right)$$

$$= \frac{AD}{GD}\,\frac{(m + M) \times g}{m \times h} \qquad \text{...(15.9)}$$

Comparing equation (15.8) with equation (15.6) of the Porter governor, we find that the angular speed of Proell governor reduces for a given values of m, M and h. In equation (15.8), GD is more than AD. We can also say that for the same angular speed for the given values of M and h, the balls of smaller masses can be used in the Proell governor. In equations (15.8) and (15.9), the radius of governor is equal to the radius of rotation of balls.

Problem 15.8. *Calculate the minimum speed of a Proell governor, which has equal arms each 200 mm and pivoted on the axis of rotation. The mass of each ball is 4 kg and the central mass on the sleeve is 20 kg. The extension arms of the lower links are each 60 mm long and parallel to the axis when the minimum radius of the ball is 100 mm.*

Sol. Given :

Lengths $AO = AC = 200$ mm ; $m = 4$ kg ; $M = 20$ kg ; length of extension arm $AG = 60$ mm ; also AG is parallel to axis ; minimum radius, $r_1 = 100$ mm.

Let N_1 = Minimum speed when r_1 = 100 mm.

(*i*) *Minimum speed* (N_1) (Refer to Fig. 15.10)

AG is parallel to the axis. Hence radius of rotation of ball is equal to the radius of governor.

The value of h_1 is given by,

$$h_1 = \sqrt{AO^2 - AB^2}$$

$$= \sqrt{200^2 - 100^2} = \sqrt{40000 - 10000}$$

$$= \sqrt{30000} = 173.2 \text{ mm} \quad \text{or} \quad 0.1732 \text{ m}$$

Length $AD = BC$

$$= \sqrt{AC^2 - AB^2}$$

$$= \sqrt{200^2 - 100^2} = 173.2 \text{ mm} \quad \text{or} \quad 0.1732 \text{ m}$$

Length $GD = AD + AG$

$$= 173.2 + 60 = 233.2 \text{ mm} = 0.2332 \text{ m}.$$

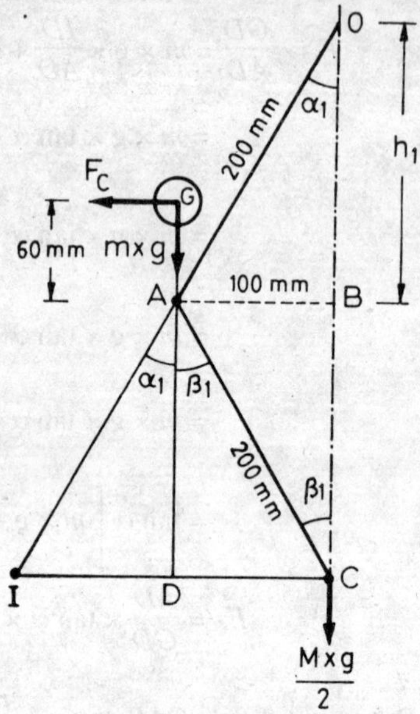

Fig. 15.10. Minimum radius.

As the lengths of arms *AO* and *CO* are equal and arms are pivoted on the axis. Hence $\alpha_1 = \beta_1$ or $\tan \alpha_1 = \tan \beta_1$ or $k_1 = \frac{\tan \beta_1}{\tan \alpha_1} = 1$. As the length *AG* is parallel to the axis, hence radius of governor is equal to the radius of rotation of the ball. Hence equation (15.9) can be used.

The angular speed is given by equation (15.9) as

$$\omega^2 = \frac{AD}{GD} \frac{(m + M) \times g}{m \times h} \qquad \text{...(A)}$$

For minimum speed, $\omega = \omega_1$ and $h = h_1$. Hence the above equation becomes as

$$\omega_1^2 = \frac{AD}{GD} \frac{(m + M) \times g}{m \times h_1}$$

$$= \frac{0.1732}{0.2332} \left(\frac{4 + 20}{4 \times 0.1732} \right) \times 9.81 = 252.4$$

$$\therefore \qquad \omega_1 = \sqrt{252.4} = 15.887 \text{ rad/s} \qquad \left(\text{But } \omega_1 = \frac{2\pi N_1}{60} \text{ hence } N_1 = \frac{60 \times \omega_1}{2\pi} \right)$$

and $$N_1 = \frac{\omega_1 \times 60}{2\pi} = \frac{15.887 \times 60}{2\pi} = \mathbf{151.7 \text{ r.p.m.}} \quad \textbf{Ans.}$$

Problem 15.9. *A Proell governor has all four arms of length 300 mm. The upper arms are pivoted on the axis of rotation and lower arms are attached to a sleeve at a distance of 40 mm from the axis of rotation. The mass of each ball of the governor is 3.75 kg and the mass on the sleeve is 40 kg. The extension of the lower arms to which the balls are attached are 100 mm long and are parallel to the governor axis at the minimum radius. The minimum and maximum radii of governor are 180 mm and 240 mm. Find the corresponding equilibrium speeds.*

Sol. Given :

Length of all four arms = 300 mm or $OA = AC$ = 300 mm ; m = 3.75 kg ; M = 40 kg ; AG = 100 mm ; radii of governors *i.e.* $AB = r_1^*$ = 180 mm and $A_1B_1 = r_2^*$ = 240 mm.

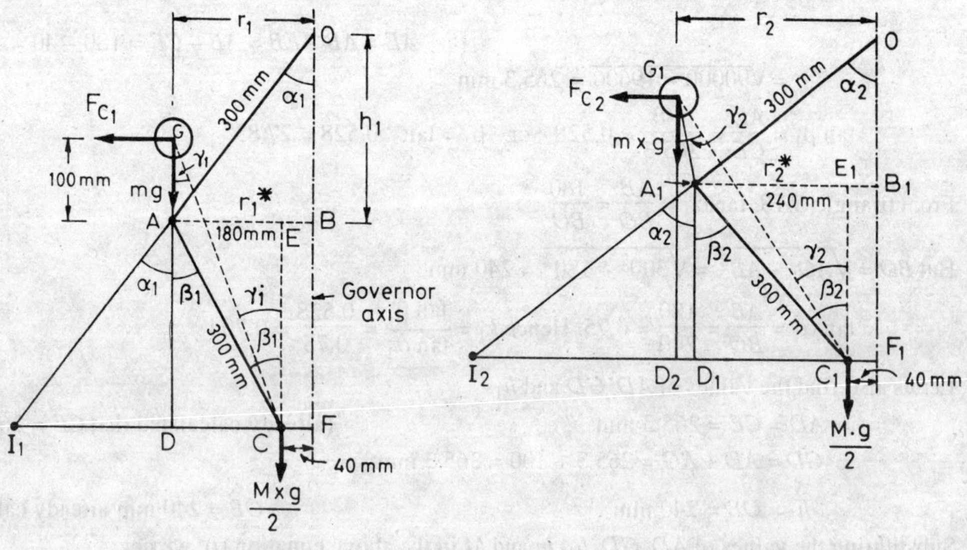

(*a*) Minimum radius (*b*) Maximum radius

Fig. 15.11

Let r_1 = Minimum radius of rotation of balls,

r_2 = Maximum radius of rotation of ball,

N_1 = Minimum equilibrium speed,

N_2 = Maximum equilibrium speed.

(*i*) *Equilibrium speed at the minimum radius of governor*

The radius of the governor is the distance of the point of intersection of the upper and lower arms from the governor axis. In Fig. 15.11 (*a*) this is the distance $AB = r_1^*$. When the extensions of the lower arms are parallel to the governor axis (*i.e.* length *AG* is parallel to governor axis), then the radius of the governor is equal to the radius of rotation of ball. Hence $r_1 = r_1^*$.

Fig. 15.11 (*a*) shows the governor configuration at the minimum radius of the governor. As in this case the radius of governor is equal to radius of rotation of the ball, hence equation (15.8) can be used.

∴ Using equation (15.8), we get

$$\omega^2 = \frac{AD}{GD}\left[\frac{m \cdot g + \frac{M \cdot g}{2}(1+k)}{m \cdot h}\right]$$

For the minimum radius of the governor, $\omega = \omega_1$; $k = k_1$ and $h = h_1$. Hence the above equation becomes as

$$\omega_1^2 = \frac{AD}{GD}\left[\frac{m \times g + \frac{M \times g}{2}(1+k_1)}{m \times h_1}\right] \quad ...(i)$$

where $k_1 = \frac{\tan \beta_1}{\tan \alpha_1}$. In triangle *ACE*, $\tan \beta_1 = \frac{AE}{CE} = \frac{AB - EB}{CE} = \frac{180 - 40}{CE} = \frac{140}{CE}$

Now $CE = \sqrt{AC^2 - AE^2} = \sqrt{300^2 - 140^2}$

($\because$ $AE = AB - EB = AB - CF = 180 - 40 = 140$ mm)

$= \sqrt{90000 - 19600} = 265.3$ mm

$\therefore \quad \tan \beta_1 = \dfrac{AE}{CE} = \dfrac{140}{265.3} = 0.528$ or $\beta_1 = \tan^{-1} 0.528 = 27.83°$

From triangle OAB, $\tan \alpha_1 = \dfrac{AB}{BO} = \dfrac{180}{BO}$.

But $BO = \sqrt{AO^2 - AB^2} = \sqrt{300^2 - 180^2} = 240$ mm

$\therefore \quad \tan \alpha_1 = \dfrac{AB}{BO} = \dfrac{180}{240} = 0.75$. Hence $k_1 = \dfrac{\tan \beta_1}{\tan \alpha_1} = \dfrac{0.528}{0.75} = 0.704$

Let us also find the values of AD, GD and h_1

$AD = CE = 265.3$ mm (already calculated that $CE = 265.3$ mm)

$GD = AD + AG = 265.3 + 100 = 365.3$ mm

$h_1 = OB = 240$ mm ($\because$ $OB = 240$ mm already calculated)

Substituting the values of AD, GD, h_1, m and M in the above equation (*i*), we get

$$\omega_1^2 = \frac{265.3}{365.3}\left[\frac{3.75 \times 9.81 + \dfrac{40 \times 9.81}{2}(1 + 0.704)}{3.75 \times 0.240}\right]$$

($\because$ $h_1 = 240$ mm $= 0.24$ m. As $g = 9.81$ m/s^2 hence h_1 should be taken in meter)

$$= \frac{265.3}{365.3}\left[\frac{36.75 + 334.32}{3.75 \times 0.24}\right] = 299.44$$

$\therefore \quad \omega_1 = \sqrt{299.44} = 17.304$ rad/s.

But $\quad \omega = \dfrac{2\pi N}{60}$ or $\omega_1 = \dfrac{2\pi N_1}{60}$ or $N_1 = \dfrac{60 \times \omega_1}{2\pi}$

$\therefore \quad N_1 = \dfrac{60 \times 17.304}{2\pi} =$ **165.24 r.p.m. Ans.**

The extension AG is rigidly connected to the lower arm AC. Hence CAG is a continuous link. Hence the length CG will remain same in every position of the governor. Also the angle GCA will remain unchanged in every position of the governor. The length CG is obtained from triangle GDC in Fig. 15.11 (*a*) as

$CG = \sqrt{GD^2 + DC^2}$

$= \sqrt{365.3^2 + 140^2}$ ($\because$ $GD = 365.3$ mm and $DC = AE = 140$ mm)

$= \sqrt{133444 + 19600} = 391.2$ mm

Also in triangle GDC, $\angle DGC = \angle GCE = \gamma_1$

$\therefore \quad \tan \gamma_1 = \dfrac{CD}{GD} = \dfrac{140}{365.3} = 0.3832$ or $\gamma_1 = \tan^{-1} 0.3832 = 20.97°$

$\therefore$ Angle $GCA = \beta_1 - \gamma_1 = 27.83 - 20.97 = 6.86°$.

(ii) Equilibrium speed at the maximum radius of governor [Refer to Fig. 15.11 (*b*)]

Fig. 15.11 (*b*) shows the governor configuration at the maximum radius of governor (*i.e.* $r_2^* = 240$ mm). In this position, the radius of governor is different than the radius of rotation of the ball (*i.e.* $r_2 \neq r_2^*$). Let us first find r_2.

From Fig. 15.11 (*b*), we know

$$r_2 = D_2F_1 = D_2C_1 + C_1F_1 = D_2C_1 + 40 \qquad (\because\ C_1F_1 = 40 \text{ mm}) \qquad ...(ii)$$

The value of D_2C_1 will be obtained from triangle $G_1D_2C_1$. We know that $C_1A_1G_1$ is a continuous link which is same as link CAG. Length $C_1G_1 = CG = 391.2$ mm (already calculated).

Also angle $G_1C_1A_1$ = Angle $GCA = 6.86°$ (already calculated)

But angle $G_1C_1A_1 = \beta_2 - \gamma_2$

$$\therefore \quad 6.86 = \beta_2 - \gamma_2 \qquad ...(iii)$$

From triangle $A_1D_1C_1$,

$$\sin\beta_2 = \frac{C_1D_1}{A_1C_1} = \frac{240-40}{300} \qquad (\because\ C_1D_1 = A_1B_1 - B_1E_1 = 240 - 40 = 200)$$

$$= 0.667$$

$$\therefore \quad \beta_2 = \sin^{-1} 0.667 = 41.83°$$

From equation (*iii*), we get

$$6.86 = 41.83° - \gamma_2 \quad \text{or} \quad \gamma_2 = 41.83 - 6.86 = 34.97°$$

Now from triangle $C_1G_1D_2$,

$$\sin\gamma_2 = \frac{D_2C_1}{C_1G_1} = \frac{D_2C_1}{391.2} \qquad (\because\ C_1G_1 = CG = 391.2 \text{ mm})$$

$$\therefore \quad D_2C_1 = 391.2 \times \sin\gamma_2 = 391.2 \times \sin 34.97° = 224.2 \text{ mm.}$$

Substituting the above value in equation (*ii*), we get

$$r_2 = D_2C_1 + 40 = 224.2 + 40 = 264.2 \text{ mm} = 0.2642 \text{ m}$$

Let us also find the angle α_2. In triangle OA_1B_1, $\sin\alpha_2 = \frac{A_1B_1}{A_1O} = \frac{240}{300} = 0.8 \quad \therefore \quad \alpha_2 = 53.13°$.

Taking moments about I_2, we get

$$F_{C_2} \times G_1D_2 = m.g \times I_2D_2 + \frac{M.g}{2} \times I_2C_1 \qquad ...(iv)$$

Now G_1D_2 is obtained from triangle $C_1G_1D_2$. Hence

$$G_1D_2 = C_1G_1 \times \cos\gamma_2 = 391.2 \times \cos 34.97° = 320.56 \text{ mm}$$

$$I_2D_2 = I_2C_1 - D_2C_1 = I_2C_1 - 224.2 \qquad (\because\ D_2C_1 = 224.2 \text{ mm})$$

But $I_2C_1 = I_2D_1 + D_1C_1$ where in triangle $A_1D_1I_2$, $\tan\alpha_2 = \frac{I_2D_1}{A_1D_1}$. Hence $I_2D_1 = A_1D_1 \times \tan\alpha_2$

But $\quad A_1D_1 = A_1C_1 \cos\beta_2 = 300 \times \cos 41.83° = 223.53$ mm

$\therefore \quad I_2D_1 = A_1D_1 \times \tan\alpha_2 = 223.54 \times \tan 53.13° = 298.05$ mm

and $\quad D_1C_1 = A_1E_1 = 240 - 40 = 200$ mm

$\therefore \quad I_2C_1 = I_2D_1 + D_1C_1 = 298.05 + 200 = 498.05$ mm

And $\quad I_2D_2 = I_2C_1 - D_2C_1 = 498.05 - 224.2 = 273.85$ mm.

Substituting the values of G_1D_2, I_2D_2 and I_2C_1 in equation, we get

$$F_{C_2} \times 320.56 = 3.75 \times 9.81 \times 273.85 + \frac{40 \times 9.81}{2} \times 498.05$$

$$= 10074.3 + 97717.4 = 107791.7$$

$$\therefore \qquad F_{C_2} = \frac{107791.7}{320.56} = 336.26$$

But $\qquad F_{C_2} = m \times \omega_2^2 \times r_2 = 3.75 \times \omega_2^2 \times 0.2642 \qquad (\because \; r_2 = 264.2 \text{ mm} = 0.2642 \text{ m})$

$$\therefore \qquad 336.26 = 3.75 \times \omega_2^2 \times 0.2642$$

or $$\omega_2 = \sqrt{\frac{336.26}{3.75 \times 0.2642}} = 18.42 \text{ rad/s}$$

and $$N_2 = \frac{60 \times \omega_2}{2\pi} = \frac{60 \times 18.42}{2\pi} = \mathbf{175.92 \text{ r.p.m.}} \quad \textbf{Ans.}$$

15.7. Hartnell Governor

Fig. 15.12 shows a Hartnell governor, which is a spring loaded governor. Two bell-crank levers, each carrying a ball at one end and a roller at the other, are pivoted at points O and O' to the frame. The rollers fit into a groove in the sleeve. The frame is attached to the governor spindle and hence rotates with it. A helical spring in compression provides equal downward forces on the two rollers through a collar on the sleeve.

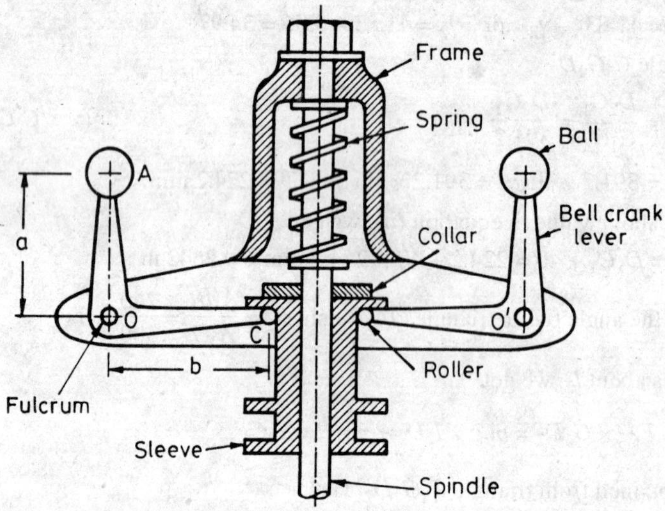

Fig. 15.12

When the speed increases the radius of rotation of balls increases and the balls move away from the spindle axis. The bells are connected to the bell-crank levers which are pivoted at points O and O'. As the balls move away from the spindle axis, the rollers (connected at the other end of the bell-crank lever) lift the sleeve against the spring force. If the speed decreases, the sleeve moves downwards. The movement of the sleeve is transferred to the throttle of the engine to control the amount of energy supplied to the engine.

Let r_1 = Minimum radius of rotation of ball centre from spindle axis,

r_2 = Maximum radius of rotation of ball centre from spindle axis,

S_1 = Spring force exerted on sleeve at minimum radius,

S_2 = Spring force exerted on sleeve at maximum radius,

m = Mass of each ball,

M = Mass of sleeve,

N_1 = Minimum speed of governor at minimum radius,

N_2 = Maximum speed of governor at maximum radius,

ω_1 and ω_2 = Corresponding minimum and maximum angular velocities

$(F_C)_1$ = Centrifugal force corresponding to minimum speed = $m \times \omega_1^2 \times r_1$

$(F_C)_2$ = Centrifugal force corresponding to maximum speed = $m \times \omega_2^2 \times r_2$

s = Stiffness of spring or the force required to compress the spring by one mm,

r = Distance of fulcrum O from the governor axis or radius of rotation when the governor is in mid-position,

a = Length of ball arm of bell-crank lever *i.e.* distance OA

b = Length of sleeve arm of bell-crank lever *i.e.* distance OC.

Figs. 15.13 (*a*) and 15.13 (*b*) shows the forces acting on the bell-crank lever in two positions *i.e.* at minimum radius position and at maximum radius position.

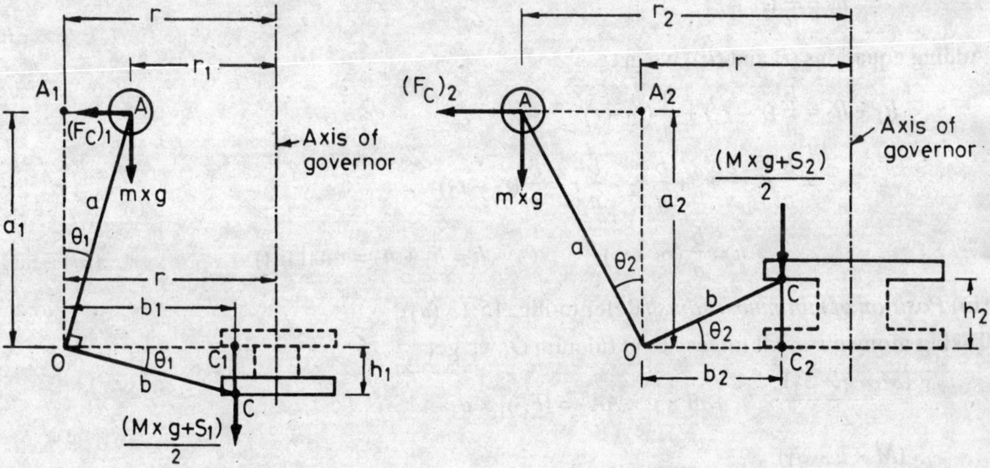

(*a*) Position of minimum radius (*b*) Position of maximum radius

Fig. 15.13

Let h = compression of the spring when radius of rotation changes from r_1 to r_2. This is also known as lift of the sleeve.

(*i*) *Position of minimum radius* (Refer to Fig. 15.13 (*a*)).

The position of bell-crank lever at the minimum radius is shown by AOC whereas the position of bell-crank lever when governor is in mid-position is shown by dotted line A_1OC_1.

Let h_1 = lift of sleeve *i.e.* vertical distance CC_1.

The angle turned by bell-crank lever between mid-position and minimum radius position is θ_1. This means the angle between OA and OA_1 is same as between OC and OC_1

$$\therefore \quad \theta_1 = \frac{CC_1}{OC} = \frac{AA_1}{OA} \qquad \left(\because \ \theta_1 = \frac{\text{Arc}}{\text{Radius}}. \text{ For } OCC_1, \text{ radius is } OC \text{ and arc is } CC_1 \right)$$

or
$$\frac{CC_1}{OC} = \frac{AA_1}{OA}$$

or
$$\frac{h_1}{b} = \frac{(r - r_1)}{a} \qquad (\because \ AA_1 = r - r_1 \,;\, OA = a \text{ and } OC = b)$$

or
$$h_1 = \frac{b}{a}(r - r_1) \qquad \text{...}(i)$$

(ii) Position of maximum radius (Refer to Fig. 15.13 *(b)*)

The position of the bell-crank lever at the maximum radius is shown by AOC whereas the position of bell-crank when governor is in mid-position is shown by dotted line A_2OC_2.

Let h_2 = lift of sleeve from mid-position *i.e.* vertical distance C_2C.

The angle turned by bell-crank lever between mid-position and maximum radius position is θ_2, *i.e.* $\angle C_2OC = \angle A_2OA = \theta_2$.

$$\therefore \quad \theta_2 = \frac{h_2}{OC} = \frac{AA_2}{OA}$$

or $$\frac{h_2}{b} = \frac{(r_2 - r)}{a} \qquad (\because \; AA_2 = r_2 - r_1)$$

$$\therefore \quad h_2 = \frac{b}{a}(r_2 - r) \qquad \text{...}(ii)$$

Adding equations *(i)* and *(ii)*, we get

$$h_1 + h_2 = \frac{b}{a}(r - r_1) + \frac{b}{a}(r_2 - r)$$

$$= \frac{b}{a}r - \frac{b}{a}r_1 + \frac{b}{a}r_2 - \frac{b}{a}r = \frac{b}{a}(r_2 - r_1)$$

or $$h = \frac{b}{a}(r_2 - r_1) \qquad (\because \; h = h_1 + h_2 = \text{total lift}) \qquad \text{...}(iii) \quad \text{...}(15.10)$$

(iii) Position of minimum radius (Refer to Fig. 15.13 *(a)*)

Taking moments of all forces about fulcrum O, we get

$$\frac{(M \times g + S_1)}{2} \times b_1 + m \times g \times AA_1 = (F_C)_1 \times a_1 \qquad \text{...(A)}$$

or $$\frac{(M \times g + S_1)}{2} \times b_1 = (F_C)_1 \times a_1 - m \times g \times AA_1$$

or $$(M \times g + S_1) = \frac{2}{b_1}[(F_C)_1 \times a_1 - m \times g \times AA_1]$$

$$= \frac{2}{b_1}[(F_C)_1 \times a_1 - mg(r - r_1)] \qquad (\because \; AA_1 = r - r_1) \qquad \text{...}(iv)$$

(iv) Position of maximum radius (Refer to Fig. 15.13 *(b)*)

Taking moments of all forces about the fulcrum, we get

$$\frac{(M \times g + S_2)}{2} \times b_2 = (F_C)_2 \times a_2 + mg \times AA_2 \qquad \text{...(B)}$$

$$= (F_C)_2 \times a_2 + mg(r_2 - r) \qquad (\because \; AA_2 = r_2 - r)$$

or $$(M \times g + S_2) = \frac{2}{b_2}[(F_C)_2 \times a_2 + mg\,(r_2 - r)] \qquad \text{...}(v)$$

Substracting equation *(iv)* from equation *(v)*, we get

$$S_2 - S_1 = \frac{2}{b_2}[(F_C)_2 \times a_2 + mg\,(r_2 - r)] - \frac{2}{b_1}[(F_C)_1 \times a_1 - mg(r - r_1)]$$

But spring stiffness(s) is given by

$$s = \frac{S_2 - S_1}{\text{Total lift}} = \frac{S_2 - S_1}{h}$$

But from equation (15.10), $h = \frac{b}{a}(r_2 - r_1)$

$$\therefore \quad s = \frac{S_2 - S_1}{\frac{b}{a}(r_2 - r_1)} = \frac{a}{b}\left[\frac{S_2 - S_1}{r_2 - r_1}\right] \quad ...(15.11)$$

Value of spring stiffness if obliquity of the arms of bell-crank lever is neglected and also the moment due to the weight of the balls is neglected.

(*i*) If obliquity of the arms is neglected, then $b_1 = b_2 = b$ and $a_1 = a_2 = a$.

(*ii*) If the moment due to weight of the ball is neglected, then $mg \times AA_1 = 0$ and $m \times g \times AA_2 = 0$.

Substituting these values (*i.e.* $b_1 = b_2 = b$; $a_1 = a_2 = a$ and $m \times g \times AA_1 = m \times g \times AA_2 = 0$) in above equations (A) and (B), we get

$$\left(\frac{M \times g + S_1}{2}\right) \times b + 0 = (F_C)_1 \times a$$

and

$$\frac{M \times g + S_2}{2} \times b + 0 = (F_C)_2 \times a$$

or

$$M \times g + S_1 = \frac{2a}{b} \times (F_C)_1 \quad ...(vi)$$

and

$$M \times g + S_2 = \frac{2a}{b}(F_C)_2 \quad ...(vii)$$

Substracting equation (*vi*) from equation (*vii*), we get

$$S_2 - S_1 = \frac{2a}{b}[(F_C)_2 - (F_C)_1] \quad ...(viii)$$

But spring stiffness s is given by,

$$s = \frac{S_2 - S_1}{h} \text{ where } h = \frac{b}{a}(r_2 - r_1)$$

$$= \frac{\left(\frac{2a}{b}\right)[(F_C)_2 - (F_C)_1]}{\frac{b}{a}(r_2 - r_1)} \quad \left(\because \; S_2 - S_1 = \frac{2a}{b}[(F_C)_2 - (F_C)_1]\right)$$

$$= 2\left(\frac{a}{b}\right)^2 \left[\frac{(F_C)_2 - (F_C)_1}{(r_2 - r_1)}\right] \quad ...(C)$$

The stiffness of the given spring is constant for all positions. Hence stiffness of the spring for minimum and intermediate positions can be obtained from equation (C) by substituting $(F_C)_2 = F_C$ and $r_2 = r$ as for intermediate position, the centrifugal force is F_C and radius is r.

$$\therefore \quad s = 2\left(\frac{a}{b}\right)^2 \left[\frac{F_C - (F_C)_1}{r - r_1}\right] \quad ...(D)$$

Similarly the spring stiffness for intermediate and maximum position is obtained from equation (C) by substituting $(F_C)_1 = F_C$ and $r_1 = r$.

$$\therefore \quad s = 2\left(\frac{a}{b}\right)^2 \left[\frac{(F_C)_2 - F_C}{r_2 - r}\right] \quad ...(E)$$

The three values of s given by equation (C), (D) and (E) can be equated as

$$2\left(\frac{a}{b}\right)^2\left[\frac{(F_C)_2-(F_C)_1}{r_2-r_1}\right]=2\left(\frac{a}{b}\right)^2\left[\frac{F_C-(F_C)_1}{r-r_1}\right]=2\left(\frac{a}{b}\right)^2\left[\frac{(F_C)_2-F_C}{r_2-r}\right]$$

or

$$\left[\frac{(F_C)_2-(F_C)_1}{r_2-r_1}\right]=\frac{F_C-(F_C)_1}{r-r_1}=\frac{(F_C)_2-F_C}{r_2-r} \quad ...(F)$$

From first two equations, we have

$$F_C-(F_C)_1=[(F_C)_2-(F_C)_1]\left[\frac{r-r_1}{r_2-r_1}\right]$$

or

$$F_C=(F_C)_1+[(F_C)_2-(F_C)_1]\left[\frac{r-r_1}{r_2-r_1}\right] \quad ...(15.12)$$

Similarly from 1st and last part of equation (F), we have

$$F_C=(F_C)_2-[(F_C)_2-(F_C)\left[\frac{r_2-r}{r_2-r_1}\right].$$

Note. 1. The weight of sleeve $M \times g$ is replaced by $(M \times g \pm F)$ when friction is taken into account.

2. The obliquity effect of the arms and moment due to the weight of the balls is neglected, unless otherwise stated.

Problem 15.10. *A Hartnell governor having a central sleeve spring and two right-angled bell crank levers operates between 290 r.p.m. and 310 r.p.m. for a sleeve lift of 15 mm. The sleeve arms and the ball arms are 80 mm and 120 mm respectively. The levers are pivoted at 120 mm from the governor axis and mass of each ball is 2.5 kg. The ball arms are parallel to the governor axis at the lowest equilibrium speed. Determine : (i) loads on the spring at the lowest and the highest equilibrium speeds and (ii) stiffness of the spring.*

(AMIE, S 1978)

Sol. Given :

$N_1 = 290$ r.p.m. ; $N_2 = 310$ r.p.m. $\therefore\ \omega_1 = \frac{2\pi N_1}{60} = \frac{2\pi \times 290}{60} = 30.4$ rad/s and $\omega_2 = \frac{2\pi \times 310}{60} = 32.5$ rad/s, $h = 15$ mm ; sleeve arm of bell crank lever, $b = 80$ mm ; ball arm, $a = 120$ mm ; distance of pivot of lever from governor axis, $r = 120$ mm, $m = 2.5$ kg.

Find : (*i*) Load on springs *i.e.* S_1 and S_2 and (*ii*) Stiffness of spring(s).

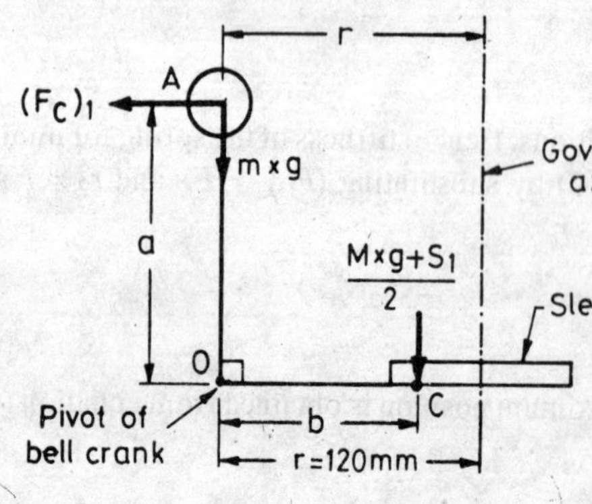

(*a*) Minimum radius

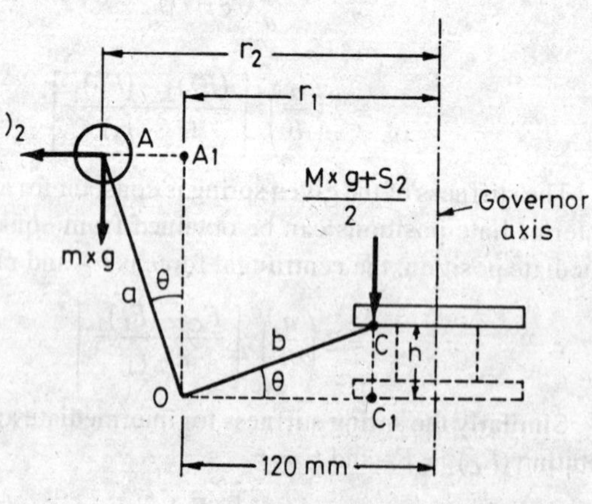

(*b*) Maximum radius

Fig. 15.14

The ball arms are parallel to the governor axis at the lowest equilibrium speed of 290 r.p.m. as shown in Fig. 15.14 (*a*), hence $r_1 = r = 120$ mm = 0.12 m.

(*i*) *Loads on the spring at the lowest and highest equilibrium speeds*

Let S_1 = Spring load at the lowest equilibrium speed, and

S_2 = Spring load at the highest equilibrium speed.

The centrifugal force at the lowest equilibrium speed,

$$(F_C)_1 = m \times \omega_1^2 \times r_1 = 2.5 \times (30.4)^2 \times 0.12 = 277 \text{ N}$$

The centrifugal force at the highest equilibrium speed,

$$(F_C)_2 = m \times \omega_2^2 \times r_2 = 2.5 \times 32.5^2 \times r_2 \quad ...(i)$$

In the above equation, the value of r_2 is unknown. This value is obtained by considering the position of ball arm and sleeve arm at the highest equilibrium speed as shown in Fig. 15.14 (*b*).

In Fig. 15.14 (*b*), the triangles OCC_1 and OAA_1 are similar (the angle turned by bell-crank lever *i.e.* θ is same)

$$\therefore \quad \frac{CC_1}{OC} = \frac{AA_1}{OA} \quad \text{or} \quad \frac{h}{b} = \frac{(r_2 - r_1)}{a}$$

or $$h = (r_2 - r_1) \times \frac{b}{a}$$

or $$r_2 = \frac{a}{b} \times h + r_1 = \frac{120}{80} \times 15 + 120 = 142.5 \text{ mm} \quad \text{or} \quad 0.1425 \text{ m.}$$

Substituting the value of r_2 in equation (*i*), we get

$$(F_C)_2 = 2.5 \times 32.5^2 \times 0.1425 = 376 \text{ N.}$$

Case I. Taking moments about O for the lowest equilibrium speed, (Refer to Fig. 15.14 (*a*)), we have

$$(F_C)_1 \times a + mg \times 0 = \left(\frac{M \times g + S_1}{2}\right) \times b$$

or $$(F_C)_1 \times a = \frac{S_1}{2} \times b \quad (\because \; M = 0)$$

or $$S_1 = 2 \times (F_C)_1 \times \frac{a}{b}$$

$$= 2 \times 277 \times \frac{120}{80} = \textbf{831 N. Ans.} \quad (\because \; (F_C)_1 = 277)$$

Case II. Taking moments about O for the highest equilibrium speed, (Refer to Fig. 15.14 (*b*)), we have

$$(F_C)_2 \times a + mg \times AA_1 = \frac{(M \times g + S_2)}{2} \times b$$

$$376 \times 0.12 + (2.5 \times 9.81) \times 0.0225 = \frac{S_2}{2} \times 0.08$$

$(\because \; AA_1 = r_2 - r_1 = 142.5 - 120 = 22.5 \text{ mm} = 0.0225 \text{ m};$
$a = 120 \text{ mm} = 0.12 \text{ m and } b = 80 \text{ mm} = 0.08 \text{ m})$

or $$45.12 + 0.552^* = S_2 \times 0.04$$

or $$45.672 = 0.04 \times S_2$$

or $$S_2 = \frac{45.672}{0.04} = \textbf{1141.8 N. Ans.}$$

*The moment due to the weight of ball is $m \times g \times AA_1 = 0.552$ N. This is very small in comparison to the moment due to $(F_C)_1$ which is 45.12 N.

(ii) Stiffness of the spring(s)

The stiffness of the spring is given by,

$$s = \frac{S_2 - S_1}{\text{Sleeve lift}} = \frac{S_2 - S_1}{h} = \frac{1141.8 - 831}{15} = \mathbf{20.72\ N/mm.}\quad \textbf{Ans.}$$

Problem 15.11. *In a spring loaded Hartnell type of governor, the mass of each ball is 4 kg and the lift of the sleeve is 50 mm. The governor begins to float at 240 r.p.m., when radius of the ball path is 110 mm. The mean working speed of the governor is 20 times the range of the speed when friction is neglected. The lengths of the ball and roller arms of the bell-crank lever are 120 mm and 100 mm respectively. The pivot centre and the axis of the governor are 140 mm apart. Determine the initial compression of the spring, taking into account the obliquity of arms.*

If the friction is equivalent to a force of 20 N at the sleeve, find the total alteration in speed before the sleeve begins to move from mid-position.

Sol. Given :

$$m = 4\text{ kg}\ ;\ h = 50\text{ mm} = 0.05\text{ m}\ ;\ N_1 = 240\text{ r.p.m.}\quad\text{or}\quad \omega_1 = \frac{2\pi N_1}{60} = \frac{2\pi \times 240}{60} = 8\pi\text{ rad/s}\ ;$$

$r_1 = 110$ mm $= 0.11$ m ; mean speed $= 20 \times$ range of speed ; $a = 120$ mm $= 0.12$ m ; $b = 100$ mm $= 0.1$ m ; $r = 140$ mm $= 0.14$ m ; $F = 20$ N.

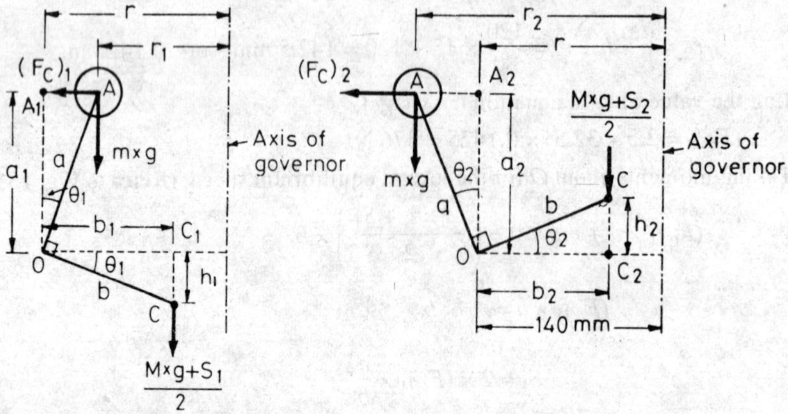

(*a*) Position of minimum radius. (*b*) Position of maximum radius.

Fig. 15.15

Let N_1 = Minimum speed at minimum radius, r_1

N_2 = Maximum speed at maximum radius, r_2.

We know that mean speed, $N = \dfrac{N_1 + N_2}{2}$ and range of speed $= N_2 - N_1$

But mean speed $= 20 \times$ range of speed (given)

$$\therefore\quad N = 20 \times (N_2 - N_1)$$

or
$$\frac{N_1 + N_2}{2} = 20 \times (N_2 - N_1)$$

or
$$N_1 + N_2 = 40(N_2 - N_1) = 40N_2 - 40N_1$$

or
$$41N_1 = 39N_2$$

or
$$N_2 = \frac{41 \times N_1}{39} = \frac{41 \times 240}{39} = 252.3\text{ r.p.m.}$$

$$\therefore \qquad \omega_2 = \frac{2\pi N_2}{60} = \frac{2\pi \times 252.3}{60} = 26.42 \text{ rad/s}$$

Using equation (15.10) for total lift, we get

$$h = \frac{b}{a}(r_2 - r_1)$$

or $$50 = \frac{100}{120}(r_2 - 110) \qquad (\because\ h = 50 \text{ mm} ; b = 100 \text{ mm} ; a = 120 \text{ mm and } r_1 = 110 \text{ mm})$$

or $$\frac{120}{100} \times 50 = (r_2 - 110)$$

or $$r_2 = \frac{120}{100} \times 50 + 110 = 60 + 110 = 170 \text{ mm} = 0.17 \text{ m}.$$

From Fig. 15.15 (*a*) which shows the position of bell-crank lever at minimum radius, we have the angle turned by bell-crank lever between mid-position and minimum radius position $= \theta_1 = \frac{h_1}{b} = \frac{AA_1}{a}$

or $$h_1 = \frac{b}{a} \times AA_1 \qquad (\text{But } AA_1 = r - r_1 = 140 - 110 = 30 \text{ mm})$$

$$= \frac{100}{120} \times 30 = 25 \text{ mm}$$

Now from triangle OCC_1, we have $b_1 = \sqrt{OC^2 - h_1^2} = \sqrt{b^2 - h_1^2} = \sqrt{100^2 - 25^2}$

$$= 96.8 \text{ mm} = 0.0968 \text{ m}$$

From triangle AOA_1, $a_1 = \sqrt{OA^2 - (AA_1)^2} = \sqrt{a^2 - 30^2}$ $\quad (\because\ AA_1 = r - r_1 = 140 - 110 = 30 \text{ mm})$

$$= \sqrt{120^2 - 30^2} = 116.2 \text{ mm} = 0.1162 \text{ m}$$

We know that $h = h_1 + h_2$

or $$h_2 = h - h_1 = 50 - 25 = 25 \text{ mm}.$$

From Fig. 15.15 (*b*), which shows the position of bell-crank lever at maximum radius, in triangle AOA_2,

$$OA_2 = \sqrt{OA^2 - (AA_2)^2} = \sqrt{a^2 - (r_2 - r)^2} = \sqrt{120^2 - (170 - 140)^2} = 116.2 \text{ mm}$$

$$(\because\ AA_2 = r_2 - r = 170 - 140 = 30 \text{ mm} = 0.03 \text{ m})$$

or $$a_2 = 116.2 \text{ mm} = 0.116 \text{ m}$$

Also from triangle COC_2,

$$OC_2 = \sqrt{OC^2 - (CC_2)^2} = \sqrt{b^2 - h_2^2} = \sqrt{100^2 - 25^2} = 96.8 \text{ mm}$$

or $$b_2 = 96.8 \text{ mm} = 0.0968 \text{ m} \qquad (\because\ OC_2 = b_2)$$

The centrifugal forces at the minimum and maximum radii are given by,

$$(F_C)_1 = m \times \omega_1^2 \times r_1 = 4 \times (8\pi)^2 \times 0.11 = 277.93 \text{ N}$$

and $$(F_C)_2 = m \times \omega_2^2 \times r_2 = 4 \times 26.42^2 \times 0.17 = 474.65 \text{ N}.$$

Now take the moments about fulcrum O for minimum and maximum radii positions. Taking moments about point O for the minimum radius as shown in Fig. 15.15 (*a*),

$$(F_C)_1 \times a_1 = \left(\frac{M \times g \times S_1}{2}\right) \times b_1 + mg \times AA_1$$

or $$277.93 \times 0.1162 = \left(\frac{O + S_1}{2}\right) \times 0.0968 + 4 \times 9.81 \times 0.03$$

$$(\because \ M = 0 \ ; a_1 = 0.1162 \text{ m} \ ; b_1 = 0.0968 \text{ m} \ ; AA_1 = 30 \text{ mm} = 0.03 \text{ m})$$

or $$32.295 = 0.0484 S_1 + 1.177$$

or $$0.0484 S_1 = 32.295 - 1.177$$

$\therefore$ $$S_1 = \frac{32.295 - 1.177}{0.0484} = \mathbf{642.93\ N.} \quad \textbf{Ans.}$$

Now taking moments about point O for the maximum radius as shown in Fig. 15.15 (*b*), we have

$$(F_C)_2 \times a_2 = \left(\frac{M \times g + S_2}{2}\right) \times b_2 - m \times g \times AA_2$$

or $$474.65 \times 0.1162 = \left(\frac{0 + S_2}{2}\right) \times 0.0968 - 4 \times 9.81 \times 0.03$$

or $$55.148 = 0.0484 \times S_2 - 1.177$$

or $$S_2 = \frac{55.148 + 1.177}{0.0484} = \mathbf{1163.75\ N.} \quad \textbf{Ans.}$$

The stiffness of the spring is given by relation,

$$s = \frac{S_2 - S_1}{h} = \frac{1163.75 - 642.93}{50} = 10.4 \text{ N/mm}$$

$\therefore$ Initial compression of the spring $= \dfrac{S_1}{s} = \dfrac{642.93}{10.4} = \mathbf{61.82\ mm.}$ **Ans.**

Alternation in speed before the sleeve begins to move from mid-position when friction is considered

From the mid-position, the sleeve can move upwards or can move downwards. When sleeve starts moving up, the frictional force acts downwards *i.e.* in the direction of weight of sleeve whereas when sleeve starts moving downwards, the frictional force acts upwards *i.e.* in the opposite direction of weight of sleeve.

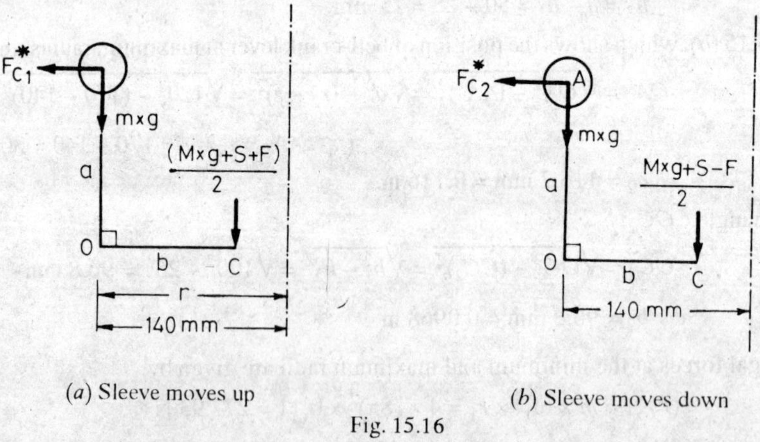

(*a*) Sleeve moves up (*b*) Sleeve moves down

Fig. 15.16

Let us first find the spring force at mid-position.

Let S = spring force at mid-position

The spring force (or force of compression in the spring) increases as the sleeve moves upwards from minimum radius position.

$\therefore$ $$S = S_1 + h_1 \times s = 642.93 + 25 \times 10.4 \qquad (\because \ h_1 = 25 \text{ mm}, s = 10.4 \text{ N/mm})$$

$$= 902.93 \text{ N}$$

When sleeve starts moving up from mid-position the total force at C is $\frac{M \times g + S + F}{2}$ whereas when sleeve starts moving down, the force at C is $\left(\frac{M \times g + S - F}{2}\right)$.

Let ω_1^* = Angular speed when sleeve starts moving upwards from mid-position

ω_2^* = Angular speed when sleeve starts moving downwards from mid-position.

N_1^* and N_2^* = Corresponding r.p.m. when sleeve starts moving up and down from mid-position respectively.

The values of centrifugal forces when sleeve starts moving up and down are given by,

$$F_{C_1}^* = m \times (\omega_1^*)^2 \times r = 4 \times (\omega_1^*)^2 \times 0.14 N = 0.56 \times (\omega_1^*)^2 \ N$$

$(\because \ m = 4 \text{ kg and } r = 140 \text{ mm} = 0.14 \text{ m})$

and
$$F_{C_2}^* = m \times (\omega_2^*)^2 \times r = 4 \times (\omega_2^*)^2 \times 0.14 N = 0.56 \times (\omega^*_2)^2 \ N$$

Taking moments about O when sleeve starts up as shown in Fig. 15.16 (a), we have

$$F_{C_1}^* \times a = m \times g \times O + \left(\frac{M \times g + S + F}{2}\right) \times b$$

or
$$0.56 \times (\omega_1^*)^2 \times 0.12 = 0 + \left(\frac{0 + 902.93 + 20}{2}\right) \times 0.1$$

$(\because \ a = 0.12 \text{ m} ; b = 0.1 \text{ m} ; M = 0, S = 902.93 \text{ N and } F = 20 \text{ N})$

$$= \frac{922.93}{2} \times 0.1 = \frac{92.293}{2}$$

$$\therefore \quad \omega_1^* = \sqrt{\frac{92.293}{2} \times \frac{1}{0.56 \times 0.12}} = 26.205 \text{ rad/s}$$

$$\therefore \quad N_1^* = \frac{60 \times \omega_1^*}{2\pi} = 250.24 \text{ r.p.m.}$$

Taking moments about O, when sleeve starts moving down as shown in Fig. 15.16 (b),

$$F_{C_2}^* \, a = m \times g \times O + \left(\frac{M \times g + S - F}{2}\right) \times b$$

or
$$0.56 \times (\omega_2^*)^2 \times 0.12 = 0 + \left(\frac{0 + 902.93 - 20}{2}\right) \times 0.1 = \frac{882.93}{2} \times 0.1 = \frac{88.293}{2}$$

$$\therefore \quad \omega_2^* = \sqrt{\frac{88.293}{2} \times \frac{1}{0.56 \times 0.12}} = 25.63 \text{ rad/s}$$

$$\therefore \quad N_2^* = \frac{60 \times \omega^*_2}{2\pi} = \frac{60 \times 25.63}{2\pi} = 244.75 \text{ r.p.m.}$$

$\therefore$ Alteration in speed = $N_1^* - N_2^*$ = 250.24 – 244.75 = **5.49 r.p.m. Ans.**

Problem 15.12. *The mass of each ball of a Hartnell type governor is 1.4 kg. The length of ball arm of the bell-crank lever is 100 mm whereas the length of arm towards sleeve is 50 mm. The distance of the fulcrum of bell-crank lever from the axis of rotation is 80 mm. The extreme radii of rotation of the balls are 75 mm and 112.5 mm. The maximum equilibrium speed is 6% greater than the minimum equilibrium speed which is 300 r.p.m. Determine : (i) stiffness of the spring and (ii) equilibrium speed when radius of rotation of the ball is 90 mm. Neglect the obliquity of the arms.*

Sol. Given :

$m = 1.4$ kg ; $a = 100$ mm $= 0.1$ m ; $b = 50$ mm $= 0.05$ m ; $r = 80$ mm $= 0.08$ m ; $r_1 = 75$ mm $= 0.075$ m ;

$r_2 = 112.5$ mm $= 0.1125$ m ; minimum speed, $N_1 = 300$ r.p.m. or $\omega_1 = \frac{2\pi N_1}{60} = \frac{2\pi \times 300}{60} = (10\pi)$ rad/s ;

maximum speed = 6% greater than minimum speed $= 1.06 \times N_1 = 1.06 \times 300 = 318$ r.p.m.

or $\omega_2 = \frac{2\pi N_2}{60} = \frac{2\pi \times 318}{60} = (10.6\pi)$ rad/s.

Let we first calculate the centrifugal forces at minimum and maximum speeds *i.e.*

$$(F_C)_1 = m \times \omega_1^2 \times r_1 = 1.4 \times (10\pi)^2 \times 0.075 = 103.63 \text{ N}$$

and

$$(F_C)_2 = m \times \omega_2^2 \times r_2 = 1.4 \times (10.6\pi)^2 \times 0.1125 = 174.66 \text{ N}.$$

(i) Stiffness of spring(s)

The stiffness of spring (when effect of obliquity of arms is to be neglected) is given by equation (15.11) as

$$s = 2\left(\frac{a}{b}\right)^2 \left[\frac{(F_C)_2 - (F_C)_1}{r_2 - r_1}\right] = 2\left(\frac{100}{50}\right)^2 \left[\frac{174.66 - 103.63}{112.5 - 75}\right] = \mathbf{15.15 \text{ N/mm.}} \quad \textbf{Ans.}$$

(ii) Equilibrium speed when $r = 90$ mm $= 0.09$ m

Let ω = Equilibrium angular velocity when $r = 90$ mm $= 0.09$ m. When the obliquity of the arms is neglected, then the centrifugal force at any instant is given by equation (15.12) as

$$F_C = (F_C)_1 + [(F_C)_2 - (F_C)_1]\left[\frac{r - r_1}{r_2 - r_1}\right]$$

$$= 103.63 + [174.66 - 103.63]\left[\frac{90 - 75}{112.5 - 75}\right]$$

$$= 103.63 + 28.41 = 132.04 \text{ N}$$

But the centrifugal force F_C is also given by,

$$F_C = m \times \omega^2 \times r = 1.4 \times \omega^2 \times 0.09$$

$\therefore$ $$1.4 \times \omega^2 \times 0.09 = 132.04$$

or $$\omega = \sqrt{\frac{132.04}{1.4 \times 0.09}} = 32.37 \text{ rad/s}$$

$\therefore$ $$N = \frac{60 \times \omega}{2\pi} = \frac{60 \times 32.37}{2\pi} = 309.13 \text{ r.p.m.} \simeq \mathbf{309 \text{ r.p.m.}} \quad \textbf{Ans.}$$

Problem 15.13. *A spring loaded governor of a Hartnell type has arms of equal length. When the sleeve is in the mid-position, the masses rotate in a circle of diameter 150 mm (the ball arms are vertical in the mid-position). Neglecting friction, the equilibrium speed for this position is 360 r.p.m. the maximum sleeve movement is 30 mm and maximum variation of speed taking account of friction is ± 6% of the mid-position speed. The mass of the sleeve is 5 kg and friction may be considered equivalent to 35 N at the sleeve.*

Assume that the power of the governor is sufficient to overcome the friction by 1% change of speed either way at mid-position. Neglecting obliquity effect of arms, determine : (i) the value of each rotating mass, (ii) the spring stiffness in N/mm and (iii) the initial compression of the spring.

Sol. Given :

$a = b$; $d = 150$ mm or $r = \frac{d}{2} = \frac{150}{2} = 75$ mm $= 0.075$ m ; $N = 360$ r.p.m.

or $\omega = \frac{2\pi N}{60} = \frac{2\pi \times 360}{60} = 12\pi$ rad/s ; $h = 30$ mm = 0.03 m ; Mass of sleeve $M = 5$ kg ; $F = 35$ N ;

Neglect obliquity effect of arms.

(*i*) *Value of each rotating mass*

Let m = Value of rotating mass

S = Spring force on the sleeve at mid-position

Change of speed at mid-position is 1% either way (*i.e.* ± 1%) to overcome friction.

∴ Minimum speed at mid-position,

$$\omega_1^* = \omega - \frac{1}{100} \times \omega = \frac{99}{100} \times \omega = \frac{99}{100} \times 12\pi \qquad (\because \; \omega = 12\pi \text{ rad/s})$$

$$= 11.88\pi \text{ rad/s}$$

and maximum speed at mis-position,

$$\omega_2^* = \omega + \frac{1}{100} \times \omega = \frac{101}{100} \times \omega = \frac{101}{100} \times 12\pi = 12.12\pi \text{ rad/s.}$$

The centrifugal forces corresponding to minimum and maximum speeds at the mid-positions are

$$F_{C_1}^* = m \times (\omega_1^*)^2 \times r = m \times (11.88\pi)^2 \times 0.075 = m \times 104.47 \text{ N}$$

and

$$F_{C_2}^* = m \times (\omega_2^*)^2 \times r = m \times (12.12\pi)^2 \times 0.075 = m \times 108.73 \text{ N}$$

From the mid-position, as sleeve starts moving up, the speed is maximum (*i.e.* ω_2^*) and frictional force acts downward *i.e.* in the direction of sleeve weight. In this position, the centrifugal force is $F_{C_2}^*$.

Hence taking moments about fulcrum, we have

$$F_{C_2}^* \times a = \left(\frac{M \times g + S + F}{2}\right) \times b \qquad \text{[In mid-position, ball arm is vertical]}$$

or

$$F_{C_2}^* = \left(\frac{M \times g + S + F}{2}\right) \qquad [\because \; a = b = \text{(given)}]$$

or

$$m \times 108.73 = \frac{5 \times 9.81 + S + 35}{2} \qquad (\because \; F_{C_2}^* = m \times 108.73 \text{ N}) \qquad ...(i)$$

From the mid-position, as sleeve starts moving downwards, the speed is minimum (*i.e.* ω_1^*) and frictional force acts upwards *i.e.* opposite to the direction of sleeve weight. In this position centrifugal force is $F_{C_1}^*$.

Hence taking moments about fulcrum, we have

$$F_{C_1}^* \times a = \left(\frac{M \times g + S - F}{2}\right) \times b$$

or

$$F_{C_1}^* = \frac{M \times g + S - F}{2} \qquad [\because \; a = b]$$

$$m \times 104.47 = \frac{5 \times 9.81 + S - 35}{2} \qquad (\because \; F_{C_1}^* = m \times 104.47) \qquad ...(ii)$$

Subtracting equation (*ii*) from (*i*), we get

$$m \times 108.73 - m \times 104.47 = \frac{5 \times 9.81 + S + 35}{2} - \frac{5 \times 9.81 + S - 35}{2}$$

or

$$4.26 \times m = 35$$

or

$$m = \frac{35}{4.26} = \mathbf{8.216 \text{ kg.}} \quad \textbf{Ans.}$$

(ii) The spring stiffness in N/mm (i.e. s)

Let us first find the spring forces at the minimum radius (*i.e.* at r_1) and at the maximum radius (*i.e.* at r_2).

Let S_1 = Spring force at minimum radius at the minimum speed

S_2 = Spring force at maximum radius at the maximum speed

N_1 = Minimum speed at radius r_1

N_2 = Maximum speed at radius r_2

$(F_C)_1$ = Centrifugal force at minimum speed at radius r_1

$(F_C)_2$ = Centrifugal force of maximum speed at radius r_2.

Since maximum variation of speed taking friction into account is ± 6% of mid-position speed.

$$\therefore \quad \text{Maximum speed,} \quad N_2 = N + \frac{6}{100} \times N = \frac{106}{100} \times N = \frac{106}{100} \times 360 = 381.6 \text{ r.p.m.}$$

and minimum speed,
$$N_1 = N - \frac{6}{100} \times N = \frac{94}{100} \times N = \frac{94}{100} \times 360 = 338.4 \text{ r.p.m.}$$

$$\therefore \quad \text{Maximum angular speed,} \quad \omega_2 = \frac{2\pi N_2}{60} = \frac{2\pi \times 381.6}{60} = 39.96 \text{ rad/s}$$

and minimum angular speed,
$$\omega_1 = \frac{2\pi N_1}{60} = \frac{2\pi \times 338.4}{60} = 35.43 \text{ rad/s.}$$

Let us now find the values of maximum radius and minimum radius from the mid-position, the lift of the sleeve upwards and downwards are equal

$$\therefore \quad h_1 = h_2 = \frac{h}{2} = \frac{30}{2} = 15 \text{ mm}$$

But
$$h_1 = \frac{b}{a}(r - r_1) \quad \text{or} \quad 15 = (r - r_1) \qquad (\because \ a = b)$$

or
$$15 = 75 - r_1 \qquad \therefore \quad r_1 = 75 - 15 = 60 \text{ mm} = 0.06 \text{ m}$$

Similarly,
$$h_2 = \frac{b}{a}(r_2 - r_1) = (r_2 - r) \qquad (\because \ b = a)$$

or
$$15 = r_2 - 75 \quad \therefore \quad r_2 = 15 + 75 = 90 \text{ mm} = 0.09 \text{ m}$$

Now the values of centrifugal forces at minimum speed and at maximum speed are given as

$$(F_C)_1 = m \times \omega_1^2 \times r_1 = 8.21 \times (35.43)^2 \times 0.06 = 618.35 \text{ N}$$

and
$$(F_C)_2 = m \times \omega_2^2 \times r_2 = 8.21 \times (39.96)^2 \times 0.09 = 1179.87$$

Neglecting the affect of obliquity and considering friction and taking moments about fulcrum (for minimum radius position),

$$(F_C)_1 \times a = \left(\frac{M \times g + S_1 - F}{2}\right) \times b$$

[For minimum radius, sleeve moves downwards and frictional force acts upwards]

or
$$(F_C)_1 = \frac{M \times g + S_1 - F}{2} \qquad (\because \ a = b)$$

or
$$618.35 = \frac{5 \times 9.81 + S_1 - 35}{2} \qquad [\because \ (F_C)_1 = 618.35]$$

or
$$S_1 = 618.35 \times 2 - 5 \times 9.81 + 35 = 1236.70 - 49.05 + 35 = 1222.65 \text{ N}$$

Now taking the moments about fulcrum for maximum radius position,

$$(F_C)_2 \times a = \left(\frac{M \times g + S_2 + F}{2} \right) \times b$$

[For maximum radius, sleeve moves upwards and frictional force acts downwards]

or
$$(F_C)_2 = \frac{M \times g + S_2 + F}{2} \qquad (\because \; a = b)$$

or
$$1179.87 = \frac{5 \times 9.81 + S_2 + 35}{2} \qquad [\because \; (F_C)_2 = 1179.87]$$

or
$$S_2 = 2 \times 1179.87 - 5 \times 9.81 - 35 = 2359.74 - 49.05 - 35 = 2275.69 \text{ N}$$

$\therefore$ Stiffness, $s = \frac{S_2 - S_1}{h} = \frac{2275.69 - 1222.65}{30}$ = **35.1 N/mm. Ans.**

(iii) Initial compression of the spring

Initial compression of the spring is $= \frac{S_1}{s} = \frac{1222.65}{35.1}$ = **34.83 mm. Ans.**

15.8. Wilson-Hartnell Governor

Fig. 15.17 shows a Wilson-Hartnell governor which is a spring loaded type of governor. In this governor, the balls are connected by two springs which are known as main springs. The main springs are arranged symmetrically on either side of the sleeve. The balls are attached to the vertical arms of the two bell-crank levers. The horizontal arms of the bell-crank levers carry two rollers at their ends. The rollers at the horizontal arm press against the sleeve. *The bell-cranks rotate with the spindle.* When speed increases, the ball-radius increases, the springs exert an inward pull P on the balls and the rollers press against the sleeve which is raised.

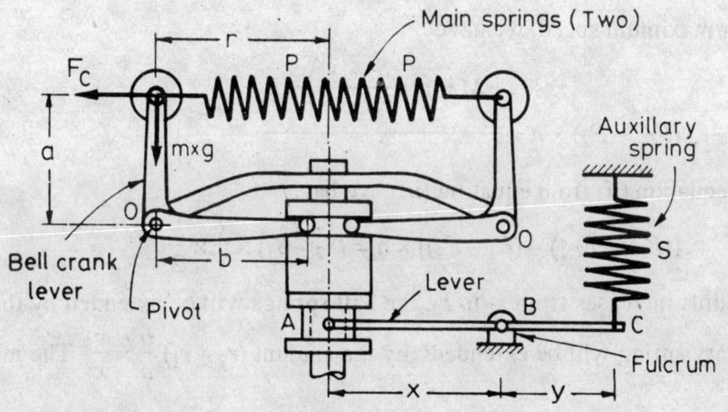

Fig. 15.17

An adjustable auxiliary spring (S) is attached to the sleeve through a lever. The lever is pivoted at a fulcrum B. One end of the lever is connected to the auxiliary spring whereas the other end of the lever fits into the groove in the sleeve. The auxiliary spring tends to keep the sleeve down.

Let m = Mass of each ball

M = Mass of sleeve

W = Weight of sleeve = $M \times g$

P = Tension (or Pull) in the main spring

S = Tension in the auxiliary spring

F_C = Centrifugal force of each ball

r = Radius of rotation of balls

s = Stiffness of each ball spring

s^* = Stiffness of auxiliary spring.

The total downward force on the sleeve

= Weight of sleeve + force at lever end A due to tension S in auxiliary spring (*i.e.* at point C)

$$= W + \frac{S \times y}{x} = \left(M \times g + \frac{S \times y}{x} \right)$$

Taking the moments about the pivot O of the bell-crank lever and neglecting the effect of the pull of gravity on the balls, we have

$$(F_C - P) \times a = \left(\frac{W + \frac{S \times y}{x}}{2} \right) \times b \qquad \text{...}(i)$$

Let corresponding to minimum speed, F_{C_1} = Centrifugal force = $m \times \omega_1^2 \times r_1$

P_1 = Tension in main spring, S_1 = Tension in auxiliary spring

and F_{C_2}, P_2 and S_2 = corresponding values of centrifugal force, tension in main spring and tension in auxiliary spring corresponding to maximum speed.

Substituting these values in equation (*i*), we have for minimum speed

$$(F_{C_1} - P_1) \times a = \frac{\left(M \times g + \frac{S_1 \times y}{x} \right) \times b}{2} \qquad (\because \quad W = m \times g) \qquad \text{...}(ii)$$

Similarly for maximum speed, we have

$$(F_{C_2} - P_2) \times a = \frac{\left(M \times g + \frac{S_2 \times y}{x} \right) \times b}{2} \qquad \text{...}(iii)$$

Subtracting equation (*ii*) from equation (*iii*), we have

$$[(F_{C_2} - F_{C_1}) - (P_1 - P_2)] \times a = (S_2 - S_1) \times \frac{y}{x} \times \frac{b}{2} \qquad \text{...}(iv)$$

When the radius increases from r_1 to r_2, the ball springs will be extended by the amount $(d_2 - d_1)$ or $2(r_2 - r_1)$ and auxiliary spring will be extended* by the amount $(r_2 - r_1)\,\frac{b}{a} \times \frac{y}{x}$. The main spring consists of two springs.

$\therefore \quad P_2 - P_1$ = Net pull (or tension) in two main spring when radius increases from r_1 to r_2

= 2 × Force exterted by each main spring

= 2 × [stiffness of main spring × extension of ball springs]

$= 2 \times [s \times 2(r_2 - r_1)]$

$= 4 \times s \times (r_2 - r_2) = 4\,.\,s\,.\,(r_2 - r_1)$

*Let h = total lift of sleeve when radius increases from r_1 to r_2. The angle turned by ball arm and sleeve arm are same. Hence $\frac{h}{b} = \frac{(r_2 - r_1)}{a}$ or $h = \frac{b}{a}(r_2 - r_1)$. But h is the lift of lever at point A. The point C will move down. Let h^* = downward movement of C [See Fig. 15.17 (*b*)]. Then $\frac{h}{x} = \frac{h^*}{y}$ or $h^* = h \times \frac{y}{x} = \frac{b}{a}(r_2 - r_1) \times \frac{y}{x}$.

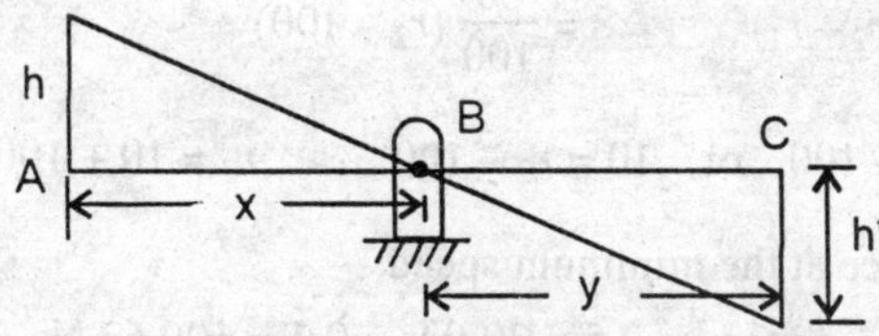

Fig. 15.17 (*b*)

The net force in auxiliary spring is given by

$$S_2 - S_1 = \text{stiffness of auxiliary spring} \times \text{extension of auxiliary spring}$$

$$= s^* . (r_2 - r_1) . \frac{b}{a} . \frac{y}{x}$$

Substituting the values of $(P_2 - P_1)$ and $(S_2 - S_1)$ in equation (*iv*), we get

$$[(F_{C_2} - F_{C_1}) - 4 . s . (r_2 - r_1)] . a = \left[s^* . (r_2 - r_1) . \frac{b}{a} . \frac{y}{x} \right] . \frac{y}{x} . \frac{b}{2}$$

or $$(F_{C_2} - F_{C_1}) - 4 . s . (r_2 - r_1) = s^* . (r_2 - r_1) . \frac{b}{a} . \frac{y}{x} . \frac{y}{x} . \frac{b}{2} . \frac{1}{a}$$

$$= \frac{s^*}{2} . (r_2 - r_1) . \left(\frac{b}{a} . \frac{y}{x} \right)^2$$

Dividing by $(r_2 - r_1)$ to both sides, we get

$$\frac{F_{C_2} - F_{C_1}}{r_2 - r_1} - 4 . s = \frac{s^*}{2} . \left(\frac{b}{a} . \frac{y}{x} \right)^2$$

or $$\frac{F_{C_2} - F_{C_1}}{r_2 - r_1} = 4 . s + \frac{s^*}{2} \left(\frac{b}{a} . \frac{y}{x} \right)^2 \quad ...(15.13)$$

If auxiliary spring is not used, then $s^* = 0$, then

$$\frac{F_{C_2} - F_{C_1}}{r_2 - r_1} = 4 . s \quad ...(15.14)$$

Problem 15.14. *The mass of each ball in a Wilson-Hartnell type of governor is 2.5 kg. The length of ball arm of each bell-crank lever is 100 mm whereas the length of the sleeve arm of bell-crank lever is 80 mm. The minimum equilibrium speed is 200 r.p.m. when radius of rotation is 100 mm. When the sleeve is lifted by 8 mm, the equilibrium speed is 212 r.p.m. The stiffness of each of the springs connected to the balls is 200 N/m. The lever for the auxiliary spring is pivoted at the mid-point. Find the stiffness of the auxiliary spring.*

Sol. Given :

$m = 2.5$ kg ; $a = 100$ mm $= 0.1$ m ; $b = 80$ mm $= 0.08$ m ; $N_1 = 200$ r.p.m.

or $\omega_1 = \frac{2\pi N_1}{60} = \frac{2\pi \times 200}{60} = 20.94$ rad/s ; $r_1 = 100$ mm $= 0.1$ m ; $h = 8$ mm ; $N_2 = 212$ r.p.m

or $\omega_2 = \frac{2\pi N_2}{60} = \frac{2\pi \times 212}{60} = 22.2$ rad/s ; $s = 200$ N/m ; lever of auxiliary spring is pivoted at mid-point hence $x = y$; $s^* = ?$

Let us first find the values of centrifugal forces. The radius of rotation r_1 is known, but r_2 is unknown. The radius r_2 will be obtained by using,

$$\text{lift} = \frac{b}{a} (r_2 - r_1) \quad \text{[See equation (15.10)]}$$

or $$h = \frac{b}{a}(r_2 - r_1) \quad \therefore \quad 8 = \frac{80}{100}(r_2 - 100)$$

or $$\frac{800}{80} = r_2 - 100 \quad \text{or} \quad 10 = r_2 - 100 \quad \therefore \quad r_2 = 10 + 100 = 110 \text{ mm} = 0.11 \text{ m}$$

Hence the centrifugal force at the minimum speed,

$$F_{C_1} = m \times \omega_1^2 \times r_1 = 2.5 \times 20.94^2 \times 0.1 = 109.62 \text{ N}$$

and centrifugal force at the maximum speed,

$$F_{C_2} = m \times \omega_2^2 \times r_2 = 2.5 \times 22.2^2 \times 0.11 = 135.53 \text{ N}$$

Now using equation (15.13), we get

$$\frac{F_{C_2} - F_{C_1}}{r_2 - r_1} = 4 \,.\, s + \frac{s*}{2}\left(\frac{b}{a} \cdot \frac{y}{x}\right)^2$$

or $$\frac{135.53 - 109.62}{0.11 - 0.10} = 4 \times 200 + \frac{s*}{2}\left(\frac{0.08}{0.10} \cdot \frac{x}{2}\right)^2 \qquad (\because \; x = y, b = 0.08 \text{ m and } a = 0.10 \text{ m})$$

$$2591 = 800 + s* \times 0.32$$

or $$s* = \frac{2591 - 800}{0.32} = 5596.875 \text{ N/m} \quad \text{or} \quad \mathbf{5.597 \text{ N/mm.}} \quad \textbf{Ans.}$$

15.8.1. Spring-controlled Governor of the Type shown in Fig. 15.18

Fig. 15.18 (*a*) shows the line-diagram of a spring-controlled governor which consists of two bell-crank levers having fulcrum *O* which is attached to the moving sleeve. The spring is compressed between the sleeve and the cap which is fixed to the end of governor spindle. This type of governor may be analysed most simply by taking moments about the instantaneous centre of all the forces which act on one of the bell-crank levers. The instantaneous centre can be located easily as the path of the fulcrum *O* is parallel to governor axis whereas the path of the centre of roller *B* is at right angles to governor axis.

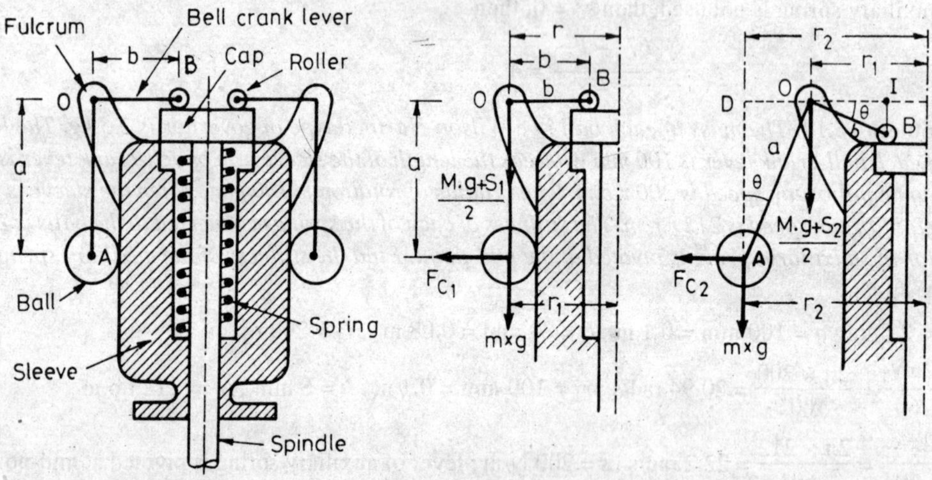

(*a*) Line diagram (*b*) Minimum radius (*c*) Maximum radius

Fig. 15.18

For the minimum radius position shown in Fig. 15.18 (*b*), the instantaneous centre lies at *B* whereas for the maximum radius position shown in Fig. 15.18 (*c*), the instantaneous centre lies at *I*, above point *B*.

Two forces are acting through the ball centre A. They are

(*i*) Weight of ball $(m \times g)$ and

(*ii*) Centrifugal force $(F_C = m \times \omega^2 \times r)$.

In addition to these forces, there is one half of the total sleeve load, which act through the fulcrum O. The total sleeve load will consists of the weight of sleeve $(M \times g)$ together with the load due to the compression spring which surrounds the sprindle. The load due to the compression spring is denoted by S.

$$\therefore \quad \text{Total load through fulcrum } O = \frac{M \times g + S}{2}$$

For minimum radius position [shown in Fig. 15.18 (*b*)], taking moments about B (point B is the instantaneous centre for this position), we get

$$\left[mg + \left(\frac{M \times g + S_1}{2}\right)\right] \times OB = F_{C_1} \times a \qquad \text{...(i)}$$

where S_1 = spring force at minimum radius position.

For maximum radius position [shown in Fig. 15.18 (*c*)], taking moments about instantaneous centre I, we get

$$F_{C_2} \times AD = m \cdot g \times ID + \left(\frac{M \cdot g + S_2}{2}\right) \times OI \qquad \text{...(ii)}$$

where S_2 = spring force at maximum radius position,

a = Length of ball-arm of bell-crank lever and

b = Length of roller-arm of bell-crank lever.

Problem 15.15. *The following particulars refer to a governor of the type shown in Fig. 15.18 on page 794 :*

The mass of each ball = 0.75 kg ; mass of sleeve = 4 kg ; the lengths of the arms a and b of the bell-crank lever are 110 mm and 44 mm respectively ; the distance of the fulcrum O of each bell-crank lever from the axis of rotation = 55 mm ; minimum radius of rotation of ball = 55 mm ; equilibrium speed at minimum radius = 240 r.p.m. ; sleeve lift is 10 mm and corresponding equilibrium speed is 252 r.p.m. Find the stiffness of the spring.

Sol. Given :

$m = 0.75$ kg ; $M = 4$ kg ; $a = 110$ mm $= 0.11$ m ; $b = 44$ mm $= 0.044$ m ; $r = 55$ mm $= 0.055$ m ;

$r_1 = 55$ mm $= 0.055$ m ; $N_1 = 240$ r.p.m. or $\omega_1 = \frac{2\pi N_1}{60} = \frac{2\pi \times 240}{60} = 25.13$ rad/s ; $h = 10$ mm $= 0.01$ m ;

$N_2 = 252$ r.p.m. or $\omega_2 = \frac{2\pi N_2}{60} = \frac{2\pi \times 252}{60} = 26.4$ rad/s. Find stiffness of spring.

Let S_1 = Spring force at minimum equilibrium speed and

S_2 = Spring force at maximum equilibrium speed

s = Stiffness of spring

r_2 = Radius of rotation of ball corresponding to speed N_2.

But stiffness of spring is given by

$$s = \frac{S_2 - S_1}{h} \text{ where } h \text{ is lift of the sleeve} \qquad \text{...(i)}$$

The centrifugal force acting on the ball at minimum equilibrium speed,

$$F_{C_1} = m \times \omega_1^2 \times r_1 = 0.75 \times (25.13)^2 \times 0.055 = 26.05 \text{ N}$$

The minimum radius position is shown in Fig. 15.18 (*b*). For this position the instantaneous centre for the ball-crank lever coincides with the roller centre *B*. Taking moments of all forces about *B*, we get

$$F_{C_1} \times a = \left(mg + \frac{M \, . \, g + S_1}{2} \right) \times b$$

or
$$26.05 \times 0.11 = \left(0.75 \times 9.81 + \frac{4 \times 9.81 + S_1}{2} \right) \times 0.044$$

$$= 0.75 \times 9.81 \times 0.044 + \left(\frac{4 \times 9.81 + S_1}{2} \right) \times 0.045$$

or
$$2.865 = 0.323 + 0.8632 + 0.0225 S_1$$

or
$$S_1 = \frac{2.865 - 0.323 - 0.8632}{0.0225} = 74.6 \text{ N}$$

The maximum radius position of the governor is shown in Fig. 15.18 (*c*). For this position the instantaneous centre lies at *I* above the roller centre *B*. Let us first find the value of r_2. From equation (15.10), we know that

$$h = \frac{b}{a}(r_2 - r_1)$$

[From Fig. 15.18 (*c*) it is clear that the arms of bell-crank lever turns by the same angle θ.

Hence $\frac{h}{b} = \left(\frac{r_2 - r_1}{a} \right)$ or $h = \frac{b}{a}(r_2 - r_1)$]

or
$$0.01 = \frac{0.044}{0.11}(r_2 - 0.055)$$

or
$$\frac{0.01 \times 0.11}{0.044} = r_2 - 0.055 \quad \text{or} \quad r_2 = \frac{0.01 \times 0.11}{0.044} + 0.055 = 0.08 \text{ m.}$$

Now the centrifugal force acting on the ball at maximum equilibrium speed,

$$F_{C_2} = m \times \omega_2{}^2 \times r_2 = 0.75 \times 26.4^2 \times 0.08 = 41.8 \text{ N}$$

Taking moments about *I* [Refer to Fig. 15.18 (*c*)], we get

$$F_{C_2} \times AD = m \, . \, g \times ID + \left(\frac{M \, . \, g + S_2}{2} \right) \times IO \quad \text{...(ii)}$$

where $AD = \sqrt{AO^2 - OD^2} = \sqrt{a^2 - (r_2 - r_1)^2} = \sqrt{0.11^2 - (0.08 - 0.055)^2} = \sqrt{0.0121 - 0.000625} = 0.107$

$IO = \sqrt{BO^2 - BI^2} = \sqrt{b^2 - h^2} = \sqrt{0.044^2 - 0.01^2} = 0.0428$ m

$ID = IO + OD = 0.0428 + (r_2 - r_1) = 0.0428 + (0.08 - 0.055) = 0.0678$ m

Substituting these values and other known values in equation (*ii*), we get

$$41.8 \times 0.107 = 0.75 \times 9.81 \times 0.0678 + \left(\frac{4 \times 9.81 + S_2}{2} \right) \times 0.0428$$

or
$$4.4726 = 0.4988 + 0.8397 + 0.0214 S_2$$

or
$$S_2 = \frac{4.4726 - 0.4988 - 0.8397}{0.0214} = 146.45 \text{ N.}$$

From equation (*i*), we have stiffness of spring

$$s = \frac{S_2 - S_1}{h} = \frac{146.45 - 74.6}{0.01}$$

$$= 7185 \text{ N/m} \quad \text{or} \quad \frac{7185}{1000} \text{ N/mm} = \mathbf{7.185 \text{ N/mm.}}$$ **Ans.**

15.9. Some Important Definitions

Let us define the followings most important terms which are used in connection with governors :

(*i*) Sensitiveness (*ii*) Stability

(*iii*) Isochronism and (*iv*) Hunting.

15.9.1. Sensitiveness

A governor is said to be sensitive if with a given fractional change of speed, the displacement of the sleeve is bigger. Hence the movement of the sleeve for a fractional change of speed is the measure of sensitivity of a governor. A governor is also said to be sensitive if for a given displacement of the sleeve, the fractional change of speed is smaller. This definition of sensitiveness is quite satisfactory when the governor is considered as an independent mechanism. But when the governor is fitted to an engine the practical requirement is simply that the change of equilibrium speed from the full load to zero load position of the sleeve, should be as small a fraction as possible of the mean equilibrium speed. The actual displacement of the sleeve is immaterial, provided that it is sufficient to change the energy supplied to the engine by the required amount. For this reason sensitiveness is also *defined as the ratio of difference between the maximum and minimum equilibrium speeds to the mean equilibrium speed.*

Let N_1 = Minimum equilibrium speed corresponding to full-load condition,

N_2 = Maximum equilibrium speed corresponding to zero load condition

N = Mean equilibrium speed = $\frac{N_1 + N_2}{2}$

Then sensitiveness of the governor

$$= \frac{\text{Difference of maximum and minimum equilibrium speeds}}{\text{Mean equilibrium speed}}$$

$$= \frac{N_2 - N_1}{N} = \frac{N_2 - N_1}{\frac{N_1 + N_2}{2}} = \frac{2(N_2 - N_1)}{N_1 + N_2}.$$

15.9.2. Stability

A governor is said to be stable when for each speed there is only one radius of rotation of the governor balls at which the governor is in equilibrium. The speed should be within the working range of the governor.

15.9.3. Isochronism

A governor is said to be *isochronous* if the equilibrium speed is constant for all radii of rotation of the balls within the working range. This means that when radius of rotation changes from minimum radius to maximum radius, the equilibrium speed remains constant.

Let r_1 = Minimum radius of rotation

r_2 = Maximum radius of rotation

N_1 and N_2 = corresponding speeds.

Then for isochronism, $N_1 = N_2$.

15.9.4. Hunting

If the speed of the engine controlled by the governor fluctuates continuously above and below the mean speed, the governor is said to be hunting. This is caused by a too sensitive governor which changes the fuel supply by a large amount when a small change in speed of rotation takes place. For example, if a slight load increases on the engine, the speed of the engine will decrease. If the governor is very sensitive, the governor sleeve immediately falls to its lowest position. This will open the control valve wide and the supply of fuel to the engine will now be in excess of its requirements, so that the speed will rapidly increase again and the sleeve will rise to its highest position. Due to this movement of the sleeve, the control valve will cut off the fuel supply to the engine and thus the engine speed will begin to fall once again. This cycle is repeated indefinitely. Such a governor would admit either the maximum or minimum amount of fuel and could not possibly admit an amount of fuel between these two extremes. The effect of this will be to cause wide fluctuations in the engine speed or in other words, the engine will hunt.

15.10. Governor Effort and Power

Governor Effort. The effort of the governor is the force exerted by the governor at the sleeve and the sleeve tends to move. When the speed is constant, the force exerted to the sleeve is zero as the sleeve does not tend to move and hence at constant speed, the effort of the governor is zero. But when the speed changes, the sleeve tends to move to its new equilibrium position and hence a force is exerted on the sleeve. This force gradually diminishes to zero as the sleeve moves to the equilibrium position corresponding to new speed. *The mean force exerted on the sleeve during a given change of speed, is known as the effort of the governor.* The given change of speed is taken generally as 1%. Hence the effort is defined as the force exerted on the sleeve for 1% change of speed.

Governor Power. The power of a governor is defined as the work done at the sleeve for a given percentage change of speed. Hence the power of a governor is the product of the governor effort and the displacement of the sleeve. Mathematically,

Power of a governor = Governor effort × displacement of sleeve.

15.10.1. Method of Determining the Effort and Power of a Governor

The effort and power of a governor may be determined by the following method. Let us apply this method on Porter governor. The same principle will be used for any other type of governor.

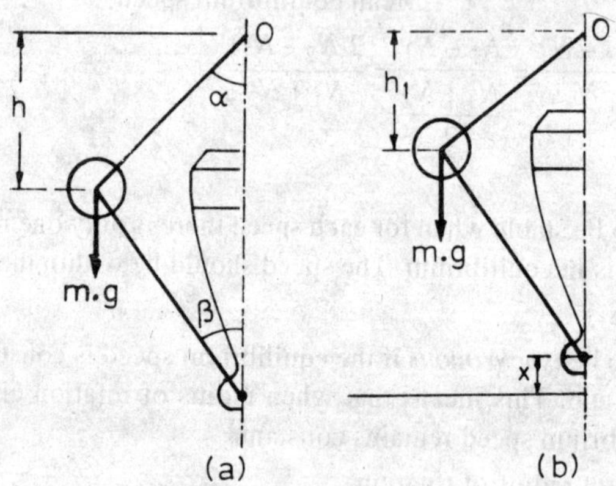

Fig. 15.19

Fig. 15.19 shows the two positions of a Porter governor.

Let N = Equilibrium speed corresponding to configuration shown in Fig. 15.19 (*a*)

W = Weight of sleeve = $M \times g$ where M is the mass of the sleeve

h = Height of governor corresponding to speed N

$c.N$ = Increase of speed

c = A factor which when multiplied to equilibrium speed, gives the increase in speed.

$\therefore$ Increased speed $= \text{Equilibrium speed} + \text{increase of speed}$

$= N + c.N = (1 + c)N$

h_1 = Height of governor corresponding to increased speed $(1 + c)N$.

The equilibrium position of the governor for the increased speed is shown in Fig. 15.19 (*b*). In order to prevent the sleeve from rising when the increase of speed takes place, a downward force will have to be exerted on the sleeve.

Let W_1 = Weight of sleeve (which is more than W) so that the rising of sleeve is prevented when the speed is $(1 + c)N$. This means that W_1 is the weight of sleeve when height of governor is h. Also $W_1 = M_1 \times g$

$\therefore$ Downward force to be applied when the rising of sleeve is to be prevented when speed increases from N to $(1 + c)N = W_1 - W$.

The values of W_1 and W can be obtained by calculating corresponding masses M_1 and M by equation (15.7B).

When speed is N r.p.m. and the angles α and β are equal so that $k = 1$, the height h is given by equation (15.7B) as

$$h = \frac{m + M}{m} \times \frac{894.56}{N^2} \qquad ...(i)$$

If the speed increases to $(1 + c)N$ and height remains the same and mass of sleeve is M_1, then we have

$$h = \left(\frac{m + M_1}{m}\right) \times \frac{894.56}{(1 + c)^2 N^2} \qquad ...(ii)$$

where M_1 is the mass of sleeve corresponding to weight W_1.

Equating the two values of h given by equations (*i*) and (*ii*), we get

$$\left(\frac{m + M}{m}\right) \times \frac{894.56}{N^2} = \left(\frac{m + M_1}{m}\right) \times \frac{894.56}{(1 + c)^2 N^2}$$

or
$$m + M = (m + M_1) \times \frac{1}{(1 + c)^2} \qquad \left(\text{Cancelling } \frac{894.56}{mN^2} \text{ to both sides}\right)$$

or
$$(m + M)(1 + c)^2 = m + M_1$$

$\therefore$
$$M_1 = (m + M)(1 + c)^2 - m$$

Subtracting M to both sides, we get

$$M_1 - M = (m + M)(1 + c)^2 - m - M$$
$$= (m + M)(1 + c)^2 - (m + M)$$
$$= (m + M)[(1 + c)^2 - 1] \qquad ...(15.14)$$

Now
$$W_1 - W = M_1 \times g - M \times g = (M_1 - M) \times g$$
$$= (m + M)[(1 + c)^2 - 1] \times g \qquad (\because\ M_1 - M = (m + M)[(1 + c)^2 - 1)])$$

But $W_1 - W$ is the downward force which must be applied in order to prevent the sleeve from rising when the increase of speed takes place. This is also the force exerted by the governor on the sleeve when the speed changes from N to $(1 + c)N$ and before the sleeve begins to move. As the sleeve rises to the new equilibrium position as shown in Fig. 15.18 (*b*), this force gradually diminishes to zero. The mean force P exerted on the sleeve during the change of speed from N to $(1 + c)N$ is therefore given by

$$P = \frac{W_1 - W}{2} = \frac{(m + M)[(1 + c)^2 - 1] \times g}{2}$$

$$= \frac{(m + M) \times g \times [1 + c^2 + 2c - 1]}{2}$$

$$= (m + M) \times g \left[\frac{c^2 + 2c}{2} \right]$$

$$= (m + M) \times g \left(\frac{2c}{2} \right) \quad \text{(Neglecting } c^2 \text{ which is very small)}$$

$$= (m + M) \times g \times c = c(m + M) \times g \quad ...(15.15)$$

$\therefore$ Governor effort, $P = c(m + M) \times g$ ...(15.16)

Power of Governor

Power of a governor is defined as the product of governor effort and displacement of the sleeve.

Let x = Displacement of sleeve when speed increases from N to $(1 + c)N$,

h = Height of governor when speed is N,

h_1 = Height of governor when speed is $(1 + c)N$. The mass of sleeve is not changed *i.e.* mass of sleeve is M.

Then $x = 2(h - h_1)$ ...(A)

Now $$h = \frac{m + M}{m} \times \frac{894.56}{N^2} \quad (\because \text{ When speed is } N, \text{ height is } h)$$

and $$h_1 = \left(\frac{m + M}{m} \right) \times \frac{894.56}{(1 + c)^2 N^2} \quad (\because \text{ When speed is } (1 + c)N, \text{ height is } h_1)$$

$$\therefore \quad \frac{h_1}{h} = \frac{\left(\frac{m + M}{m} \right) \times \frac{894.56}{(1 + c)^2 N^2}}{\left(\frac{m + M}{m} \right) \times \frac{894.56}{N^2}} = \frac{1}{(1 + c)^2}$$

or $$h_1 = \frac{h}{(1 + c)^2}$$

Substituting the value of h_1 is equation (A), we get

$$x = 2\left[h - \frac{h}{(1 + c)^2} \right] = 2h \left[\frac{(1 + c)^2 - 1}{(1 + c)^2} \right] = 2\text{h} \left[\frac{1 + c^2 + 2c - 1}{(1 + c)^2} \right] = \frac{2h\,[c^2 + 2c]}{1 + c^2 + 2c}$$

$$= \frac{2h \times 2c}{(1 + 2c)} \quad \text{[Neglecting } C^2 \text{ which is very small]}$$

$$= 2h \,.\, \frac{2c}{(1 + 2c)}$$

Now governor power = Governor effort × displacement of sleeve

$= P \times x$ ($\because$ Governor effort = P and displacement of sleeve = x)

$$= [c(m + M) \times g] \times 2h \,.\, \frac{2c}{1 + 2c}$$

$$\left[\because \text{ From equation (15.16), } P = c(m + M) \times g \text{ and } x = 2h \,.\, \frac{2c}{1 + 2c} \right]$$

$\therefore$ Governor power $$= \frac{4c^2}{(1 + 2c)} (m + M)g \,.\, h \quad ...(15.17)$$

If angles α and β are not equal, then equation (15.17) for governor effort will become as,

$$\text{Governor effort*, } P \simeq c\left(\frac{2m}{1+k}+M\right)\times g, \text{ where } k=\frac{\tan\beta}{\tan\alpha} \quad \text{...(15.18)}$$

Similarly, the equation for the lift of the sleeve will no longer apply as x is not equal to $2(h-h_1)$. The value of x will approximately be equal to $(1+k)(h-h_1)$

$$\therefore \quad x \simeq (1+k)(h-h_1)$$

$$\simeq (1+k)\left(h-\frac{h}{(1+c)^2}\right) \qquad \left[\because \; h_1=\frac{h}{(1+c)^2}\right]$$

$$=(1+k)\,.\,h\left(1-\frac{1}{(1+c)^2}\right)=(1+k)\,.\,h\,.\left(\frac{1+c^2+2c-1}{1+c^2+2c}\right)$$

$$=(1+k)\,.\,h\,.\left(\frac{c^2+2c}{1+c^2+2c}\right)$$

$$\simeq (1+k)\,.\,h\,.\,\frac{2c}{1+2c} \qquad \text{[Neglecting } C^2 \text{ which is very small]}$$

$$\therefore \quad \text{Governor Power} = P\times x \simeq \left[c\left(\frac{2m}{1+k}+M\right)\times g\right]\times\left[(1+k)\,.\,h\,.\,\frac{2c}{1+2c}\right]$$

$$=\frac{2c^2}{(1+2c)}[2m+M(1+k)]\,g\,.\,h$$

$$=\frac{4c^2}{(1+2c)}\left[m+\frac{M}{2}(1+k)\right].\,g\,.\,h \quad \text{...(15.19)}$$

Problem 15.16. *A Porter governor has equal arms each 200 mm in length and pivoted on the axis of rotation. The mass of each ball is 4 kg and mass of sleeve is 20 kg. The radius of rotation of the ball is 100 mm when the governor begins to lift. If the fractional increase of speed is 1%, then determine the governor effort and power of the governor.*

Sol. Given :

$m = 4$ kg ; $M = 20$ kg ; $r = 100$ mm $= 0.1$ m ; fractional increase of speed = 1% or $\frac{1}{100} = 0.01$ hence $c = 0.01$.

The value of h from Fig. 15.19 (*c*) is given by,

$$h=\sqrt{AO^2-AB^2}=\sqrt{200^2-100^2}=173.2 \text{ mm} = 0.1732 \text{ m}$$

As the arms are of equal length and are pivoted on the axis of rotation, hence $\alpha = \beta$.

---

*If α and β are not equal, then $\frac{\tan\beta}{\tan\alpha}=k$. When speed is N, height is 'h'

$$\therefore \quad h=\frac{m+\frac{M}{2}(1+k)}{m}\times\frac{894.56}{N^2}$$

When speed increases to $(1+c)N$, the height of governor will be same if the mass of the sleeve is increased to M_1

$$h=\frac{m+\frac{M_1}{2}(1+k)}{m}\times\frac{894.56}{(1+c)^2\,.\,N^2}.$$

(*i*) Governor effort (P) is given by equation (15.16) as

$$P = c(m + M) \cdot g$$
$$= 0.01\,(4 + 20) \times 9.81$$
$$= \mathbf{2.354\ N.}\quad \textbf{Ans.}$$

(*ii*) Power of the governor is given by equation (15.17) as

Governor power

$$= \frac{4c^2}{(1 + 2c)}(m + M) \cdot g \cdot h$$
$$= \frac{4 \times 0.01^2}{1 + 2 \times 0.01}(4 + 20) \times 9.81 \times 0.1732 \text{ N m}$$
$$= \mathbf{0.016\ N\ m}$$

or $\quad 0.016 \times 1000$ N mm = **16 N mm. Ans.**

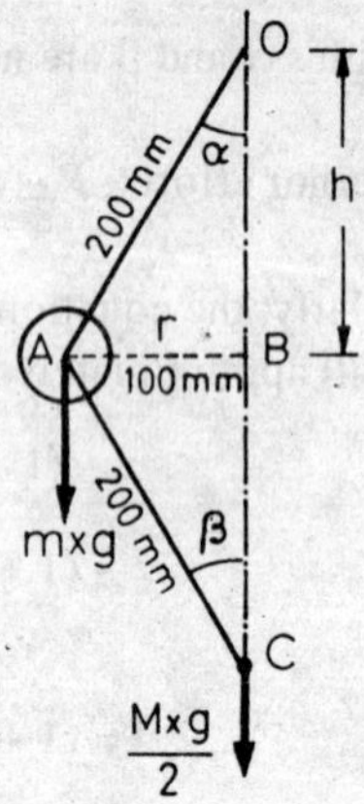

Fig. 15.19 (*c*)

15.11. Controlling Force

When the speed of rotation is uniform, the balls of a governor are subjected to outward centrifugal force, which will tend to move them outwards. But the outward movement of the balls is resisted by an equal and opposite force acting radially inwards. This inward force is known as controlling force. Hence controlling force is equal to centrifugal force but acting in the opposite direction. It is denoted by F_C.

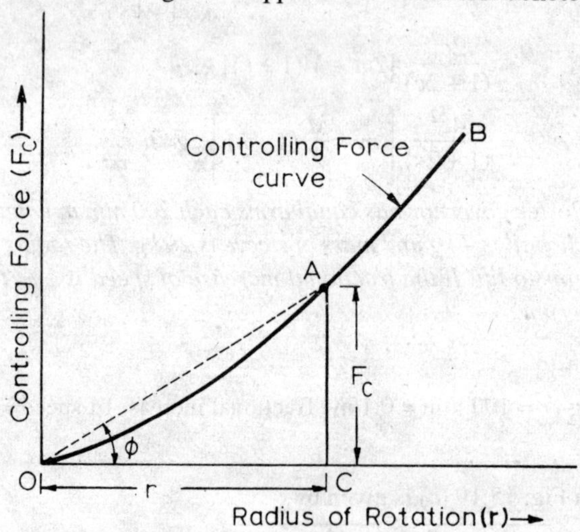

Fig. 15.20

∴ Controlling force, $F_C = m \times \omega^2 \times r$

The controlling force is provided by :

(*i*) Weight of rotating mass in a Watt governor,

(*ii*) Weight of rotating mass and weight of sleeve in a Porter governor, and

(*iii*) Compressed spring in the case of a Hartnell governor.

Controlling force Curve. Controlling force curve is a curve which shows the variation of the controlling force with the radius of rotation. The radius of rotation (r) is taken along x-axis whereas the controlling force (F_C) is taken along y-axis. This curve is useful in finding the stability and sensitiveness of a governor. This curve also shows the effect of friction clearly. The controlling force curve is shown in Fig. 15.20. In Fig. 15.20, the curve *OAB* represents the controlling force curve.

We know that controlling force,

$$F_C = m \times \omega^2 \times r = m \times \left(\frac{2\pi N}{60}\right)^2 \times r \qquad \left(\because \omega = \frac{2\pi N}{60}\right)$$

$$= m \times \left(\frac{2\pi}{60}\right)^2 \times N^2 \times r$$

or
$$N^2 = \frac{F_C}{m \times \left(\frac{2\pi}{60}\right)^2 \times r} = \frac{1}{m} \times \left(\frac{60}{2\pi}\right)^2 \times \frac{F_C}{r}$$

or
$$N = \left(\frac{60}{2\pi}\right) \times \sqrt{\frac{F_C}{m \times r}} = \left(\frac{60}{2\pi}\right)\sqrt{\frac{1}{m} \times \frac{F_C}{r}}$$

or
$$N = \left(\frac{60}{2\pi}\right)\sqrt{\frac{1}{m} \times \tan\phi} \qquad ...(15.20)$$

$$\left(\because \frac{F_C}{r} = \tan\phi \text{ from Fig. 15.20, where in triangle } AOC,\ \tan\phi = \frac{AC}{OC} = \frac{F_C}{r}\right)$$

where ϕ is the inclination to the r axis of the line joining a given point on the controlling force curve to the origin.

Let us find the shape of the controlling force curve for : (*i*) stable governor, (*ii*) sensitive governor and (*iii*) isochronous governor.

(*i*) *For stable governor*, the shape of the controlling force curve must be such that the angle ϕ increases continuously as r increases. This is due to the fact that for a stable governor, the equilibrium speed must increase as the radius of rotation of the governor balls increases. From equation (15.20) it is clear that the speed N increases as angle ϕ is increasing.

(*ii*) For sensitive governor, the change in the value of ϕ over the change of radius of rotation should be as small as possible.

(*iii*) For the governor to be isochronous, the controlling force curve should be a straight line passing through the origin. In that case, the angle ϕ will be constant for all values of the radius.

Problem 15.17. *The following data is related to the Porter governor :*

The mass of each ball = 2.5 kg ; mass of sleeve = 15 kg ; length of each arm = 300 mm and all arms are pivoted on the axis of rotation ; the extreme radii of rotation are 150 mm and 225 mm ;

Draw the controlling force curve and set off a speed scale along the ordinate (y-axis) corresponding to a radius of 250 mm.

Sol. Given :

$m = 2.5$ kg ; $M = 15$ kg ; $l = 300$ mm $= 0.3$ m ; $r_1 = 150$ mm $= 0.15$ m ; $r_2 = 225$ mm $= 0.225$ m.

Taking moments about the instantaneous centre I (Refer to Fig. 15.21)

$$F_C \times AD = m \times g \times ID + \frac{M \times g}{2} \times IC$$

$$= m \times g \times ID + \frac{M \cdot g}{2} \cdot (ID + CD)$$

or
$$F_C = m \cdot g \frac{ID}{AD} + \frac{M \cdot g}{2}\left(\frac{ID}{AD} + \frac{CD}{AD}\right)$$

$$= m \cdot g \cdot \tan\alpha + \frac{M \cdot g}{2}(\tan\alpha + \tan\beta)$$

$$= m \cdot g \cdot \tan\alpha + \frac{M \cdot g}{2} \cdot \tan\alpha\left(1 + \frac{\tan\beta}{\tan\alpha}\right)$$

$$= \tan\alpha\left[m \cdot g + \frac{M \cdot g}{2}\left(1 + \frac{\tan\beta}{\tan\alpha}\right)\right]$$

$$= \tan\alpha\left[m \cdot g + \frac{M \cdot g}{2}(1 + k)\right] \text{ where } k = \frac{\tan\beta}{\tan\alpha}$$

$$= \left[m \cdot g + \frac{M \cdot g}{2}(1 + k)\right] \cdot \tan\alpha$$

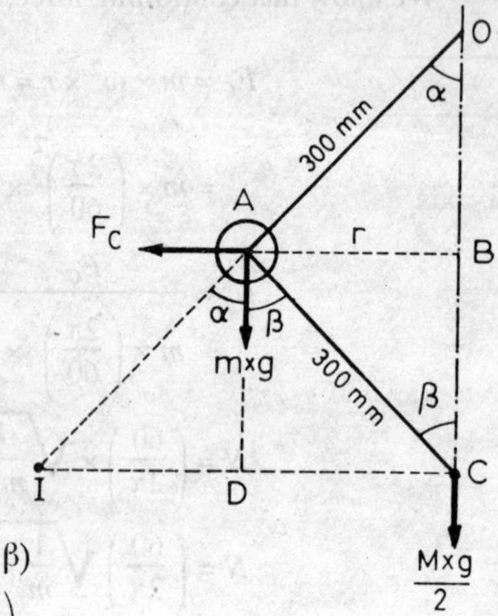

Fig. 15.21

Here $\tan\alpha = \tan\beta$

($\because$ In triangle OAC, $OA = AC$ hence $\alpha = \beta$)

and $\tan\alpha = \dfrac{r}{h}$ $\left(\because \text{ In triangle } OAB, \tan\alpha = \dfrac{r}{h}\right)$

Hence $k = \dfrac{\tan\beta}{\tan\alpha} = 1$

$$\therefore \quad F_C = \left[m \cdot g + \frac{M \cdot g}{2}(1 + 1)\right]\frac{r}{h} = (m \cdot g + M \cdot g)\frac{r}{h} = (m + M) \times g \times \frac{r}{h} \quad ...(i)$$

From triangle OAB, $h = \sqrt{OA^2 - AB^2} = \sqrt{0.3^2 - r^2}$ ($\because$ $OA = 300$ mm $= 0.3$ m)

$$= \sqrt{0.09 - r^2}$$

Substituting the value of h is equation (*i*), we have

$$F_C = (m + M) \cdot g \cdot \frac{r}{\sqrt{0.09 - r^2}}$$

$$= \frac{(2.5 + 15) \times 9.81 \times r}{\sqrt{0.09 - r^2}} = \frac{171.675 \times r}{\sqrt{0.09 - r^2}} \quad ...(ii)$$

Let us vary the radius 'r' upto 250 mm or 0.25 m, with an increment of 50 mm, *i.e.* $r = 50$ mm, 100 mm, 150 mm, 200 mm, 250 mm. Substituting the values of r in equation (*ii*), the values of F_C can be obtained.

At $r = 0.05$ m, $F_C = \dfrac{171.675 \times 0.05}{\sqrt{0.09 - 0.05^2}} = 29$ N

At $r = 0.1$ m, $F_C = \dfrac{171.675 \times 0.1}{\sqrt{0.09 - 0.1^2}} = 60.7$ N

At $r = 0.15$ m, $F_C = \dfrac{171.675 \times 0.15}{\sqrt{0.09 - 0.15^2}} = 99.1$ N

At $r = 0.2$ m, $F_C = \dfrac{171.675 \times 0.2}{\sqrt{0.09 - 0.04}} = 153.6$ N

At $r = 0.25$ m, $F_C = 258.8$ N.

The values of F_C for different values of r are shown in Table 15.1.

Table 15.1

r (m)	0.05	0.1	0.15	0.2	0.25
F_C (N)	29	60.7	99.1	153.6	258.8

The controlling force curve is drawn by taking the values of F_C along y-axis and values of r along x-axis, as shown in Fig. 15.22. The speed scale along the y-axis corresponding to radius r = 250 mm = 0.25 m

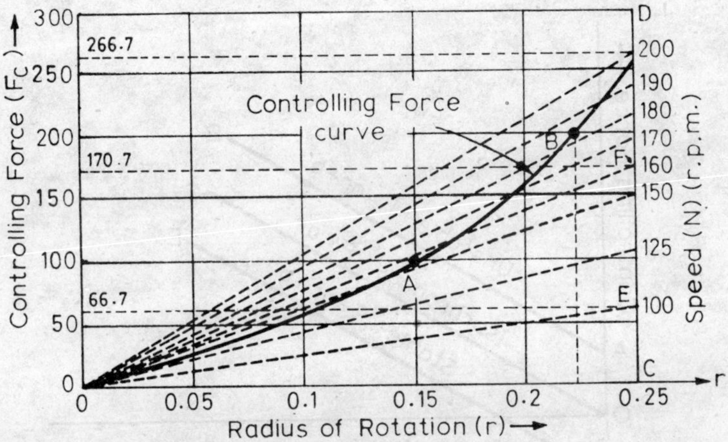

Fig. 15.22. Controlling Force Curve.

is set-off by obtaining a relation between centrifugal force (F_C) and speed (N) corresponding to known radius r = 0.25 m. The relation between F_C and N for r = 0.25 m is given by

$$F_C = m \cdot \omega^2 \cdot r = 2.5 \times \left(\frac{2\pi N}{60}\right)^2 \times 0.25 \qquad \left(\because \omega = \frac{2\pi N}{60}\right)$$

or

$$F_C = 0.00667\ N^2 \qquad ...(iii)$$

Substituting the different values of N in equation (iii), the values of F_C can be obtained. The values of F_C for different values of N are shown in Table 15.2.

Table 15.2

N (r.p.m.)	100	125	150	160	170	180	190	200
F_C (N)	66.7	104.16	150	170.7	192.7	216	240.7	266.7

The speed scale is obtained as explained below and shown in Fig. 15.22 :

(i) Corresponding to radius r = 0.25 m, draw the ordinate CD.

(ii) Corresponding to F_C = 66.7 N, draw a horizontal line to intersect the ordinate CD at E. The point E (on ordinate CD) represents the speed of 100 r.p.m. Join O to E.

(iii) Similarly corresponding to values of F_C = 104.16, 150, 170.7,......, draw horizontal lines to intersect the ordinate CD. The corresponding points of intersections will represent a speed of 125, 150, 160,....., respectively. Join these points with origin O.

Note. The speed of governor corresponding to any radius of rotation can be obtained from controlling force diagram having speed scale. Let the radius of rotation is 150 mm = 0.15 m. Corresponding to r = 0.15, get the point A on controlling force curve ($i.e.$ Fig. 15.22). Join point A to the origin O. Produce line OA to intersect the speed scale line ($i.e.$ ordinate CD) at some point (say F). From point F, the speed can be obtained. This speed is approximately 160 r.p.m.

15.11.1. Stability of Spring-Controlled Governors

Fig. 15.23 shows the controlling force curves for stable, isochronous and unstable spring-controlled governors. The controlling force curve is a straight line or an approximately straight line. As controlling force curve represents the variation of controlling force (F_C) with radius of rotation (r). Hence straight line equation can be,

$$F_C = a \cdot r \pm b$$

where a and b are constants. In the above equation b may be + ve, or – ve or zero.

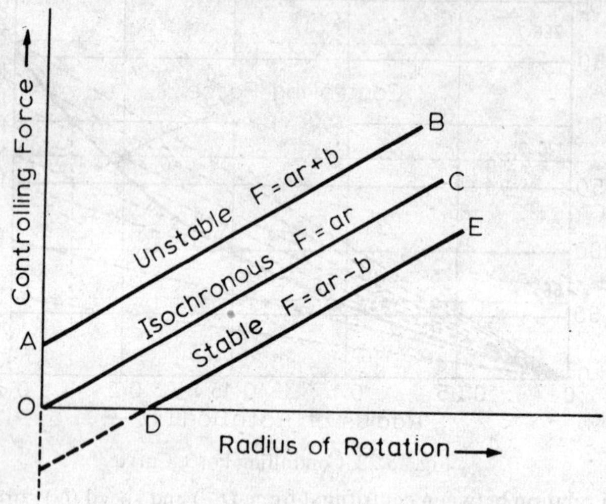

Fig. 15.23

The controlling force is given by,

$$F_C = m \times \omega^2 \times r$$

or

$$\frac{F_C}{r} = m \times \omega^2$$

We know that for a stable governor, the ratio $\frac{F_C}{r}$ must increase as r increases. Hence the controlling force curve DE for a stable governor must intersect the controlling force axis (*i.e.* y-axis) below the origin, when produced. Then the equation of the curve will be of the form

$$F_C = a \cdot r - b \qquad \text{...(i)}$$

where a and b are constants.

By increasing the initial tension of the spring, the curve may be raised parallel to itself so as to make b either zero or positive.

(*i*) If b in equation (*i*) is zero then the controlling force curve OC will pass through the origin. The ratio $\frac{F_C}{r}$ will be constant for all radius of rotation and hence the governor will become isochronous. Hence for isochronous, the equation will be

$$F_C = ar \qquad \text{...(ii)}$$

(*ii*) If b is positive, then controlling force curve AB will intersect the controlling force axis (*i.e.* y-axis) above the origin. The equation of the curve will be

$$F_C = ar + b \qquad \text{...(iii)}$$

The ratio $\frac{F_C}{r}$ will decrease as r increases $\left(\because \frac{F_C}{r} = a + \frac{b}{r}\right)$. This means that with the increase of r, the equilibrium speed decreases, which is impracticable. Hence the governor becomes as unstable. Hence the equation for unstable governor is

$$F_C = a \cdot r + b \qquad \text{...(iii)}$$

Problem 15.18. *Fig. 15.24 shows a spring controlled governor. The following are the particulars of this governor :*

The mass of each ball is 2.25 kg and the mass of the sleeve is 11.25 kg. The arms a and b (i.e. ball arm and sleeve arm) of the bell-crank lever are 112.5 mm and 50 mm long respectively and are at right angles to each other. The extreme radii of rotation are 62.3 mm and 112.5 mm. The ball arm is vertically below the pivot of the bell-crank lever at the minimum radius and the spring load is 240 N. The stiffness of the spring is 15.75 N/mm. Draw the controlling force curve and erect a speed scale along the ordinate through r = 125 mm.

Sol. Given :

$m = 2.25$ kg ; $M = 11.25$ kg ; $a = 112.5$ mm $= 0.1125$ m ; $b = 50$ mm $= 0.05$ m ; $r_1 = 62.5$ mm $= 0.0625$ m ; $r_2 = 112.5$ mm $= 0.1125$ m ; $S_1 = 240$ N ; $s = 15.75$ N/mm $= 12 \times 1000$ N/m $= 1200$ N/m.

Let S_1 = Spring load at the minimum radius of rotation ; and

S_2 = Spring load at the maximum radius of rotation.

Fig. 15.23 (*a*) and (*b*) shows the position of the balls at the minimum and maximum radius positions. For minimum radius position, instantaneous centre of bell-crank lever lies at B whereas for maximum radius

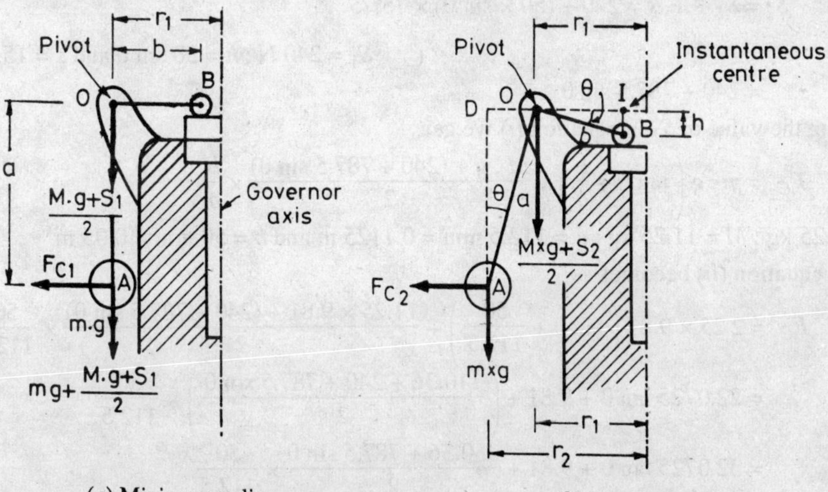

(*a*) Minimum radius (*b*) Maximum radius

Fig. 15.24

position the instantaneous centre I lies above point B as shown in Fig. 15.24 (*b*).

For maximum radius position [shown in Fig. 15.24 (*b*)], taking moment about I,

$$F_{C_2} \times AD = m \cdot g \times ID + \left(\frac{M \cdot g + S_2}{2}\right) \times IO$$

(The pivot O is moving with sleeve. The weight of sleeve and spring load act through O)

Let us find the values of AD, ID and IO in terms of θ and bell-crank lever lengths.

Now $AD = a \cdot \cos\theta$; $ID = DO + OI = a \sin\theta + b \cos\theta$ and $IO = b \cos\theta$

Substituting these values, we get

$$F_{C_2} \times a \cos\theta = m \cdot g \times (a \sin\theta + b \cos\theta) + \left(\frac{M \cdot g + S_2}{2}\right) \times b \cos\theta$$

$$\therefore \quad F_{C_2} = \frac{m \cdot g \times (a \sin\theta + b \cos\theta) + \left(\frac{M \cdot g + S_2}{2}\right) \times b \cos\theta}{a \cdot \cos\theta}$$

$$= m \cdot g \left(\frac{a \sin\theta}{a \cos\theta} + \frac{b \cos\theta}{a \cdot \cos\theta}\right) + \left(\frac{M \cdot g + S_2}{2}\right) \times \frac{b \cos\theta}{a \cdot \cos\theta}$$

$$= m \cdot g \left(\tan\theta + \frac{b}{a}\right) + \left(\frac{M \cdot g + S_2}{2}\right) \times \frac{b}{a} \quad \text{...}(i)$$

Let us find the value of S_2 in terms of θ and given value of spring load from Fig. 15.24 (*b*), the value of h (from triangle *BOI*) is given by,

$$h = b \sin\theta = (50 \sin\theta) \text{ mm} \quad (\because \; b = 50 \text{ mm})$$

But we know that, spring stiffness, $s = \frac{S_2 - S_1}{h}$

or $h \cdot s = S_2 - S_1$

or $S_2 = S_1 + h \cdot s = 240 + (50 \times \sin\theta) \times 15.75$

$(\because \; S_1 = 240 \text{ N} ; h = 50 \sin\theta \text{ and } s = 15.75 \text{ N/mm})$

$= 240 + 787.5 \sin\theta$

Substituting the value of S_2 in equation (*i*) we get

$$F_{C_2} = m \cdot g \left(\tan\theta + \frac{b}{a}\right) + \frac{M \cdot g + (240 + 787.5 \sin\theta)}{2} \times \frac{b}{a} \quad \text{...}(ii)$$

But $m = 2.25$ kg ; $M = 11.25$ kg ; $a = 112.5$ mm $= 0.1125$ m and $b = 50$ mm $= 0.05$ m.

Hence the equation (*ii*) becomes as

$$F_{C_2} = 2.25 \times 9.81 \left(\tan\theta + \frac{50}{112.5}\right) + \frac{11.25 \times 9.81 + (240 + 787.5 \sin\theta)}{2} \times \frac{50}{112.5}$$

$$= 22.0725 \tan\theta + 9.81 + \left(\frac{110.36 + 240 + 787.5 \sin\theta}{2}\right) \times \frac{50}{112.5}$$

$$= 22.0725 \tan\theta + 9.81 + \frac{350.36 + 787.5 \sin\theta}{2} \times \frac{50}{112.5}$$

$$= 22.0725 \tan\theta + 9.81 + \frac{350.36}{2} \times \frac{50}{112.5} + \frac{787.5 \sin\theta}{2} \times \frac{50}{112.5}$$

$$= 22.0725 \tan\theta + 9.81 + 77.85 + 175 \sin\theta$$

or $$F_{C_2} = 22.0725 \tan\theta + 175 \sin\theta + 87.66 \quad \text{...}(iii)$$

Now from Fig. 15.24 (*b*), the relation between r_2 and r_1 in terms of θ is given by

$$r_2 = r_1 + OD = r_1 + a \sin\theta$$

or $r_2 = (0.0625 + 0.1125 \sin\theta)$ m $\quad (\because \; r_1 = 0.0625 \text{ m and } a = 0.1125 \text{ m}) \quad$ ...(*iv*)

For drawing the controlling force curve, the different values of F_C (here F_{C_2}) and different values of radius of rotation r (here r_2) must be known. The values of F_C (or F_{C_2}) are obtained from equation (*iii*) and values of r (or r_2) are obtained from equation (*iv*) for different values of θ.

θ°	0	5	10	15	20	25	30
F (or F_2)	87.66	104.83	121.93	138.86	155.54	171.90	187.9
r (or r_2)	0.0625	0.0723	0.082	0.092	0.101	0.11	0.1187

From equation (*iii*), $F_{C_2} = 22.0725 \tan\theta + 175 \sin\theta + 87.66$

When $\theta = 0$, $F_{C_2} = 22.0725 \tan 0° + 175 \sin 0° + 87.66 = 87.66$

When $\theta = 5°$, $F_{C_2} = 22.0725 \tan 5° + 175 \sin 5° + 87.66 = 1.93 + 15.2425 + 87.66 = 104.83$

When $\theta = 10°$, $F_{C_2} = 22.0725 \tan 10° + 175 \sin 10° + 87.66 = 3.89 + 30.38 + 87.66 = 121.93$

When $\theta = 15°$, $F_{C_2} = 22.0725 \tan 15° + 175 \sin 15° + 87.66 = 5.91 + 45.29 + 87.66 = 138.86$

When $\theta = 20°$, $F_{C_2} = 22.0725 \tan 70° + 175 \sin 20° + 87.66 = 8.03 + 59.85 + 87.66 = 155.54$

When $\theta = 25°$, $F_{C_{23}} = 22.0725 \tan 25° + 175 \sin 25° + 87.66 = 10.29 + 73.95 + 87.66 = 171.90$

When $\theta = 30°$, $F_{C_2} = 22.0725 \tan 30° + 175 \sin 30° + 87.66 = 12.74 + 87.5 + 87.66 = 187.9.$

From equation (*iv*), $r_2 = 0.0625 + 0.1125 \sin\theta$

When $\theta = 0$, $r_2 = 0.0625 + 0.1125 \sin 0° = 0.0625$

When $\theta = 5°$, $r_2 = 0.0625 + 0.1125 \sin 5° = 0.0723$

$\theta = 10°$, $r_2 = 0.0625 + 0.1125 \sin 10° = 0.082$

$\theta = 15°$, $r_2 = 0.0625 + 0.1125 \sin 15° = 0.092$

$\theta = 20°$, $r_2 = 0.0625 + 0.1125 \sin 20° = 0.101$

$\theta = 25°$, $r_2 = 0.0625 + 0.1125 \sin 25° = 0.11$

$\theta = 30°$, $r_2 = 0.0625 + 0.1125 \sin 30° = 0.1187.$

Fig. 15.25 shows the graph between controlling force (F_C) and radius of rotation (r). The controlling force curve is approximately a straight line.

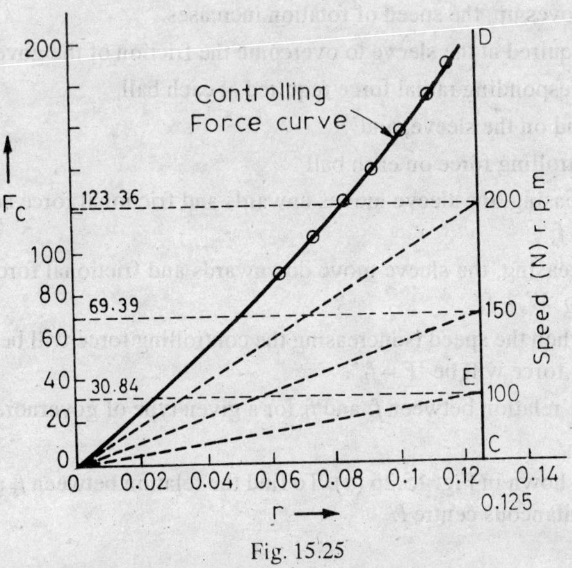

Fig. 15.25

The speed scale along the y-axis corresponding to the radius r = 125 mm = 0.125 m is set-off by obtaining a relation between F_C and N corresponding to known radius r = 0.125 m.

The relation between F_C and N for r = 0.125 m is given by,

$$F_C = m \times \omega^2 \times r = 2.25 \times \left(\frac{2\pi N}{60}\right)^2 \times 0.125$$

$$\left(\because \quad m = 2.25 \text{ kg}, r = 0.125 \text{ m, and } \omega = \frac{2\pi N}{60}\right)$$

$$= 2.25 \times \frac{4\pi^2}{3600} \times N^2 \times 0.125$$

$$= 0.003084N^2 \qquad \ldots(v)$$

Substituting the different values of N in equation (v), the values of F_C can be obtained. The values of F_C for different values of N are given in Table 15.3.

Table 15.3

N (r.p.m.)	100	150	200	210	220	230	240	250	260
F_C (N)	30.84	69.39	123.36	136	149.2	163.1	177.6	192.75	208.5

The speed scale is obtained as explained below and shown in Fig. 15.25 :

(i) Corresponding to radius r = 0.125 m, draw the ordinate CD.

(ii) Corresponding to F_C = 30.84 N, draw a horizontal line to intersect the ordinate CD at E. The point E (on ordinate CD) represents the speed of 100 r.p.m. Join O to E.

(iii) Similarly corresponding to the values of F_C = 69.39, 123.36, 136, 149.2 ......, 208.5 draw horizontal lines to intersect the ordinate CD. The corresponding points of intersection on CD will represent a speed of 150 r.p.m., 200 r.p.m., 210 r.p.m., ..., 260 r.p.m. respectively. Join these points with origin O.

15.12. Friction and Insensitiveness

The governor has been assumed to be frictionless. In actual practice, there is always friction in the joints and operating mechanism of the governor. Friction always opposes motion and therefore, it acts in downward direction when sleeve moves up. Similarly the friction acts in upward direction when sleeve moves downward. When sleeve moves downward, the radius of rotation of ball decreases and also the speed of rotation decreases. When sleeve moves up, the speed of rotation increases.

Let f_s = Force required at the sleeve to overcome the friction of the governor,

f_b = The corresponding radial force required at each ball,

W_s = Total load on the sleeve, and

F = The controlling force on each ball

If the speed is increasing, the sleeve moves upwards and frictional force acts downwards. The sleeve load will be equal to '$W_s + f_s$'.

If the speed is decreasing, the sleeve move downwards and frictional force acts upwards. Hence the sleeve load will be equal to '$W_s - f_s$'.

In the same way, when the speed is increasing the controlling force will be '$F + f_b$' and when speed is decreasing, the controlling force will be '$F - f_b$'.

Let us now find the relation between f_s and f_b for a given type of governor.

Porter Governor

Porter governor is shown in Fig. 15.26 (a). To find the relation between f_s and f_b, take the moments of these force about the instantaneous centre I.

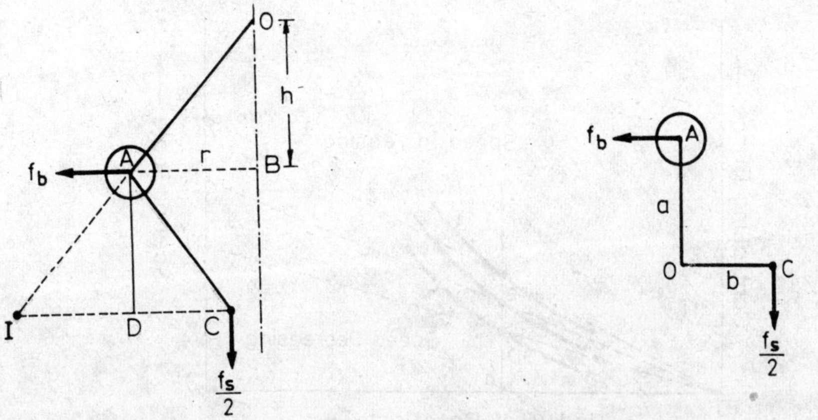

(*a*) Porter Governor (*b*) Spring loaded governor

Fig. 15.26

$$\therefore \quad f_b \times AD = \frac{f_s}{2} \times IC$$

$$= \frac{f_s}{2}(ID + CD)$$

or
$$f_b = \frac{f_s}{2}\left(\frac{ID + CD}{2}\right) = \frac{f_s}{2}\left(\frac{ID}{AD} + \frac{CD}{AD}\right)$$

$$= \frac{f_s}{2}(\tan\alpha + \tan\beta) = \frac{f_s}{2} \times \tan\alpha \times \left(1 + \frac{\tan\beta}{\tan\alpha}\right)$$

$$= \frac{f_s}{2} \times \tan\alpha + (1 + k) \text{ where } k = \frac{\tan\beta}{\tan\alpha}$$

$$= \frac{f_s}{2} \times \frac{r}{h} \times (1 + k) \qquad \left(\because \text{ In triangle } ABO, \tan\alpha = \frac{r}{h}\right) \qquad \text{...(15.20)}$$

Spring loaded governor

Spring loaded governor is shown in Fig. 15.26 (*b*). To find the relation between f_s and f_b, take the moments about the fulcrum *O* of the bell-crank lever.

$$\therefore \quad f_b \times a = \frac{f_s}{2} \times b$$

or
$$f_b = \frac{f_s}{2} \times \frac{b}{a} \qquad \text{...(15.21)}$$

Due to friction, the controlling force diagram (or graph) is modified as shown in Fig. 15.27.

There are three curves in the Fig. 15.27. The upper curves is for increasing speeds and the ordinates of this curve are proportional to the controlling force '$F + f_b$'. The middle curve is obtained when friction is neglected and the ordinates of this curve are proportional to controlling force *F*. The lower curve is for decreasing speeds and ordinates of this curve are proportional to the controlling force '$F - f_b$'. From Fig. 15.27, it is clear that for one radius of rotation, there will be three values of controlling forces and three values of corresponding speeds on the speed scale. Thus for radius *OA*, the three values of controlling force are :

(*i*) Ordinate *AB* which is the controlling force when friction is neglected.

(*ii*) Ordinate *AC* which is the controlling force when speed is increasing. When speed is increasing, the effect of friction is to increase the controlling force.

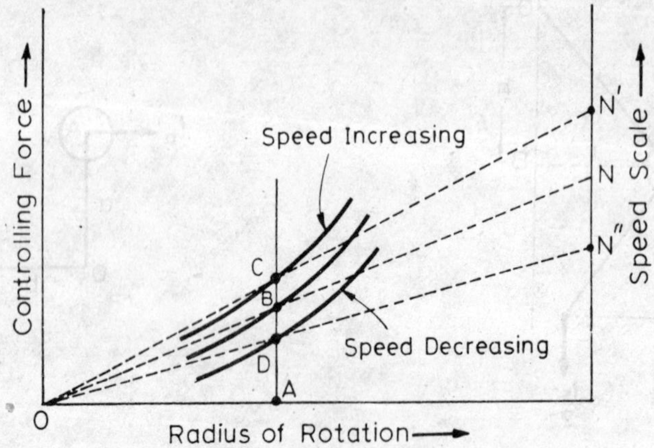

Fig. 15.27

(*iii*) Ordinate AD which is the controlling force when speed is decreasing. When speed is decreasing, the effect of friction is to decrease the controlling force.

The corresponding three values of speeds for the same radius OA are :

1. The speed N when friction is neglected,
2. The speed N' when speed is increasing or sleeve is on the verge of moving up, and
3. The speed N'' when speed is decreasing or sleeve is on the verge of moving down.

This means that, when radius is OA, the speed of rotation may vary between the limits N'' and N', without causing any displacement of the governor sleeve. The governor is said to be insensitive over this range of speed. *The ratio* $\left(\frac{N' - N''}{N}\right)$ *is known as co-efficient of insensitiveness of the governor.*

$$\therefore \quad \text{Co-efficient of insensitiveness} = \left(\frac{N' - N''}{N}\right) \qquad \text{...(15.22)}$$

where N = speed when friction is neglected for a given radius,

N' = maximum speed for the same radius or the speed when sleeve is on the verge of moving upwards,

N'' = minimum speed for the same radius or the speed when sleeve is on the verge of moving downwards.

Co-efficient of insensitiveness in terms of force 'f_b' and controlling force F

We know that the controlling force is proportional to the square of the speed for a given radius. Hence we have

$$F \propto N^2 \qquad \text{where } F \text{ is controlling force and } N \text{ is the speed}$$

$$\therefore \quad F = k \cdot N^2 \qquad \text{...(15.23)}$$

Similarly when controlling force is $(F + f_b)$ the speed is N' and when controlling force is $(F - f_b)$ then speed is N''. Therefore, we have

$$F + f_b = k(N')^2 \qquad \text{...(15.24)}$$

and

$$F - f_b = k(N'')^2 \qquad \text{...(15.25)}$$

Subtracting equation (15.25) from (15.24), we get

$$2f_b = k(N')^2 - k(N'')^2$$

Dividing the above equation by equation (15.23),

$$\frac{2f_b}{F} = \frac{k(N')^2 - k(N'')^2}{k \cdot N^2} = \frac{(N')^2 - (N'')^2}{N^2} = \frac{(N' + N'')(N' - N'')}{N \cdot N}$$

$$= \frac{(N' - N'')}{N} \cdot \frac{(N' - N'')}{N}$$

But $\frac{N' + N''}{2}$ is approximately equal to N. Hence above equation becomes as

$$\frac{2f_b}{F} = 2 \times \frac{N' - N''}{N} \qquad \left(\because \frac{N' + N''}{2} \simeq N \text{ or } \frac{N' + N''}{N} = 2\right)$$

or

$$\frac{f_b}{F} = \frac{N' - N''}{N}$$

$\therefore$ Co-efficient of insensitiveness $= \frac{N' - N''}{N} \simeq \frac{f_b}{F}$ ...(15.26)

Problem 15.19. *The following data is related to a Porter governor :*

Mass of each ball = 2.5 kg ; mass of sleeve = 15 kg ; length of each arm = 300 mm and all arms are pivoted on the axis of rotation ; the extreme radii of rotation are 150 mm and 225 mm.

If the friction at the sleeve is equivalent to 5 N, find the co-efficient of insenstiveness at the extreme radii of rotation.

Sol. Given :

$m = 2.5$ kg ; $M = 15$ kg ; $l = 300$ mm $= 0.3$ m ; all arms are pivoted on the axis of rotation ;

$r_1 = 150$ mm $= 0.15$ m ; $r_2 = 225$ mm $= 0.225$ m and $f_s = 5\,N$ *i.e.* force required at the sleeve to overcome the friction.

Let f_b = Radial force required at each ball corresponding to the force required at the sleeve to overcome the friction.

For Porter governor, the relation between f_s and f_b is given by equation (15.20) as

$$f_b = \frac{f_s}{2} \times \frac{r}{h} \times (1 + k) \text{ where } k = \frac{\tan \beta}{\tan \alpha}$$

But in the present problem, all the arms are of equal length and are pivoted on the axis of rotation.

Hence $\tan \alpha = \tan \beta$ and $k = \frac{\tan \beta}{\tan \alpha} = 1$. Hence above relation reduces to

$$f_b = \frac{f_s}{2} \times \frac{r}{h} (1 + 1) = f_s \times \frac{r}{h} \quad ...(i)$$

The data given in the above problem is same as in Problem 15.17. There we have shown in equation (*i*) that

$$F_C = (m + M) \times g \times \frac{r}{h}$$

Here F_C is the centrifugal force. The controlling force (F) is equal to centrifugal force in magnitude.

$$\therefore \quad F = F_C = (m + M) \times g \times \frac{r}{h} \quad ...(ii)$$

Dividing equation (*i*) by equation (*ii*), we get

$$\frac{f_b}{F} = \frac{f_s \times \frac{r}{h}}{(m + M) \times g \times \frac{r}{h}} = \frac{f_s}{(m + M) \times g}$$

Co-efficient of insensitiveness is given by equation (15.26) as

Co-efficient of insensitiveness $= \frac{f_b}{F}$

$$= \frac{f_s}{(m+M)\times g} = \frac{5}{(2.5+15)\times 9.81} = \mathbf{0.0291} \quad \textbf{or} \quad \mathbf{2.91\%.} \quad \textbf{Ans.}$$

Note. In the above problem, the co-efficient of insensitiveness is given by $\left(\frac{f_s}{(m+M)\times g}\right)$. This shows that the co-efficient of insensitiveness in case of Porter governor is independent of radius of rotation.

Problem 15.20. *In problem 15.18, a spring controlled governor is shown in Fig. 15.24, for which the following is the data :*

Mass of each ball = 2.25 kg ; Mass of sleeve = 11.25 kg ; the arms of bell-crank levers are a = 112.5 mm and b = 50 mm and are at right angles to each other ; extreme radii of rotations are 62.5 mm and 112.5 mm ; at the minimum radius the ball arm is vertically below the pivot of the bell-crank lever and the spring load is 240 N. The stiffness of the spring is 15.75 N/mm. If the friction in the governor and mechanism is equivalent to 11 N at the sleeve, find the co-efficient of insensitiveness at the extreme radii of rotation.

Sol. Given :

$m = 2.25$ kg ; $M = 11.25$ kg ; $a = 112.5$ mm ; $b = 50$ mm ; $r_1 = 62.5$ mm ; $r_2 = 112.5$ mm ; $S_1 = 240$ N, $s = 15.75$ N/mm ; frictional force = 11 N.

The friction in the governor and mechanism is equivalent to 11 N at the sleeve. This means that the force required at the sleeve to overcome friction of the governor is 11 N.

$\therefore \qquad f_s = 11$ N

Let f_b = Radial force required at each ball corresponding to the force required at the sleeve to overcome the friction.

For a spring loaded governor of the type given in the above problem, the relation between f_s and f_b is given by equation (15.21) as

$$f_b = \frac{f_s}{2}\times\frac{b}{a} = \frac{f_s}{2}\times\frac{50}{112.5} = \frac{11}{2}\times\frac{50}{112.5} \qquad (\because\ f_s = 11N,\ b = 50,\ a = 112.5)$$
$$= 2.444 \text{ N}$$

In case of spring controlled governor, the ratio of $\left(\frac{f_b}{F}\right)$ is not independent of radius of rotation. Hence we will have to calculate the value of F (controlling force or centrifugal force) at the minimum and maximum radius of rotation. The controlling force curve for the above problem is drawn in Fig. 15.25. From the Fig. 15.25, the values of F at radius $r_1 = 62.5$ mm = 0.0625 m and at radius $r_2 = 112.5$ mm = 0.1125 m can be read. These values are :

F_1 (corresponding to $r = 0.0625$ m) = 87.66 N

F_2 (corresponding to $r = 0.1125$ m) = 177 N

Now using equation (15.26) for co-efficient of insensitiveness, we get

Co-efficient of insensitiveness $= \frac{f_b}{F}$

$\therefore$ At the minimum radius, co-efficient of insensitiveness $= \frac{f_b}{F} = \frac{2.44}{87.66} = \mathbf{0.0278}$ **or 2.78%. Ans.**

At the maximum radius, co-efficient of insensitiveness $= \frac{f_b}{F} = \frac{2.44}{177} = \mathbf{0.0138}$ **or 1.38%. Ans.**

Alternate Method

In this governor, the pivot of the bell-crank lever is connected to the sleeve. The sleeve is moving vertically up and down, hence pivot O is also moving vertically. All the loads (*i.e.* load of sleeve, spring load and frictional load) are transferred to both the pivots of the bell-crank lever *i.e.* $\frac{M.g+S\pm F}{2}$ at each pivot.

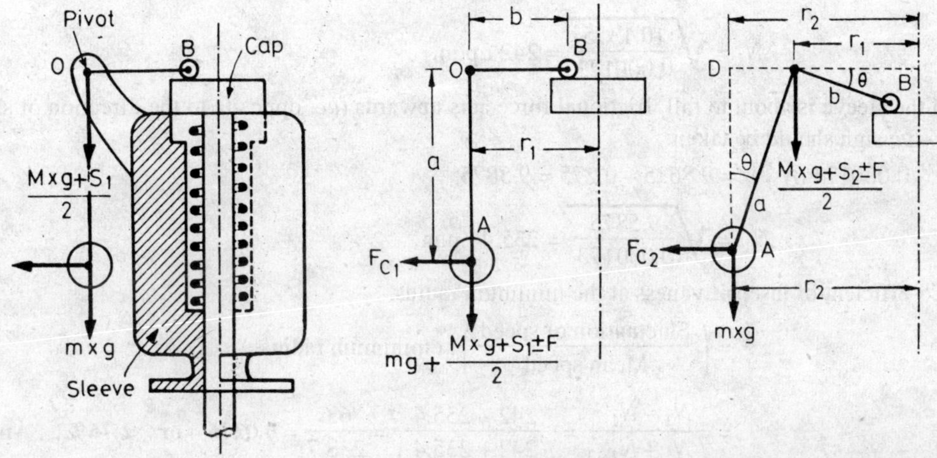

(*a*) (*b*) Minimum radius (*c*) Maximum radius

Fig. 15.28

(*i*) *Co-efficient of insensitiveness at minimum radius*

The position of the bell-crank lever at the minimum radius is shown in Fig. 15.28 (*b*). Since the pivot O moves vertically and the roller B moves horizontally, hence point B is the instantaneous centre of the bell-crank lever in this position.

Taking the moments about the point B (*i.e.* instantaneous centre),

$$F_{C_1} \times a = \left(mg + \frac{M \times g + S_1 \pm F}{2} \right) \times b \qquad ...(i)$$

where $F_{C_1} = m \times \omega_1^2 \times r_1 = 2.25 \times \left(\frac{2\pi N_1}{60} \right)^2 \times 0.0625 \quad \left(\because \; m = 2.25 \text{ kg} ; \omega_1 = \frac{2\pi N_1}{60} \text{ and } r_1 = 0.0625 \text{ m} \right)$

$$= \frac{2.25 \times 4 \times \pi^2 \times N_1^2}{3600} \times 0.0625 = 0.00154\, N_1^2$$

Substituting the value of F_{C_1} in equation (*i*), we get

$$(0.00154\, N_1^2) \times a = \left(m.g + \frac{M.g + S_1 \pm F}{2} \right) \times b$$

or $$(0.00154\, N_1^2) \times 0.1125 = \left(2.25 \times 9.81 + \frac{11.25 \times 9.81 + 240 \pm 11}{2} \right) \times 0.05$$

($\because \; a = 0.1125$ m, $b = 0.05$ m, $S_1 = 240$ N, $F = 11$ N and $M = 11.25$ kg)

$$0.000173\, N_1^2 = (22.0725 + 175.18 \pm 5.5)\; 0.05$$

$$= (197.25 \pm 5.5) \times 0.05 = 197.25 \times 0.05 \pm 5.5 \times 0.05$$

or $$0.000173\, N_1^2 = 9.8625 \pm 0.275 \qquad ...(ii)$$

At the minimum radius, the sleeve may either rise or fall.

When the sleeve is about to rise, frictional force acts downward (*i.e.* in the direction of sleeve load). Hence + ve sign should be taken.

$$\therefore \quad 0.000173\,(N_1)^2 = 9.8625 + 0.275 = 10.1375$$

or $$N_1 = \sqrt{\frac{10.1375}{0.000173}} = 242 \text{ r.p.m.}$$

When the sleeve is about to fall, frictional force acts upwards (*i.e.* opposite to the direction of sleeve load). Hence – ve sign should be taken.

$$\therefore \quad 0.000173\,(N_1')^2 = 9.8625 - 0.275 = 9.5875$$

$$\therefore \quad N_1' = \sqrt{\frac{9.5875}{0.000173}} = 235.4 \text{ r.p.m.}$$

$\therefore$ Co-efficient of insensitiveness at the minimum radius.

$$= \left(\frac{\text{Fluctuation of speed}}{\text{Mean speed}}\right) \text{at minimum radius}$$

$$= \frac{N_1 - N_1'}{\left(\frac{N_1 + N_1'}{2}\right)} = \frac{242 - 235.4}{\left(\frac{242 + 235.4}{2}\right)} = \frac{6.6}{238.7} = \mathbf{0.0276 \quad or \quad 2.76\%.\quad Ans.}$$

(iii) Co-efficient of insensitiveness at maximum radius

The position of the bell-crank lever at the maximum radius is shown in Fig. 15.28 (*c*). Since the pivot *O* moves vertically and roller *B* moves horizontally, hence the instantaneous centre lies at *I* in this position (*i.e.* point of intersection of a horizontal line through *O* and a vertical line through *B*).

Taking moments about *I*,

$$F_{C_2} \times AD = m \,.\, g \times DI + \left(\frac{M \,.\, g + S_2 \pm F}{2}\right) \times OI \qquad ...(ii)$$

where $F_{C_2} = m \times \omega_2^2 \times r_2 = 2.25 \times \left(\frac{2\pi N_2}{60}\right)^2 \times 0.1125 \qquad \left(\because \; m = 2.25 \text{ kg} \,;\, \omega_2 = \frac{2\pi N_2}{60} \,;\, r_2 = 0.1125 \text{ m}\right)$

$$= 2.25 \times \frac{4\pi^2 \times N_2^2}{3600} \times 0.1125 = 0.002776\, N_2^2$$

$$AD = \sqrt{AO^2 - OD^2} = \sqrt{a^2 - (r_2 - r_1)^2} \qquad (\because \; OD = r_2 - r_1)$$

$$= \sqrt{0.1125^2 - (0.1125 - 0.0625)^2} = \sqrt{0.012656 - 0.0025} = 0.1 \text{ m}$$

$$DI = OD + OI \qquad (\because \; OD = r_2 - r_1 = 0.1125 - 0.0625 = 0.05 \text{ m},$$
$$OI = OB \times \cos\theta = b\cos\theta = 0.0625 \times \cos\theta)$$

But from triangle *OAD*, $\cos\theta = \frac{AD}{AO} = \frac{AD}{AO} = \frac{0.1}{0.1125}$

$$\therefore \quad OI = 0.0625 \times \frac{0.1}{0.1125} = 0.0555 \text{ m} = 0.05 + 0.0555 = 0.1055 \text{ m}$$

$$OI = 0.0625 \times \cos\theta = 0.0555 \text{ m.}$$

To find the value of S_2, first find the value of *BI* which is equal to lift.

Now $$BI = \sqrt{OB^2 - OI^2} = \sqrt{b^2 - OI^2} = \sqrt{0.0625^2 - 0.0555^2} = 0.0277 \text{ m} = 27.7 \text{ mm}$$

We know that $S_2 - S_1 = BI \times s$ (where BI = lift = h)

$\therefore \quad S_2 = S_1 + BI \times s$

$= 240 + 28.7 \times 15.75$ (Here BI should be taken in mm as the value of s is given in N/mm)

$= 692$ N

Substituting the above values in equation (*ii*), we get

$$0.002776\, N_2^2 \times 0.1 = 2.25 \times 9.81 \times 0.1055 + \left(\frac{11.25 \times 9.81 + 692 \pm 11}{2}\right) \times 0.0555$$

$$= 0.0002776\, N_2^2 = 2.328 + (401.18 \pm 5.5) \times 0.0555$$

$$= 2.328 + 22.265 \pm 0.305$$

or $$0.0002776\, N_2^2 = 24.593 \pm 0.305$$

If the sleeve is about to rise, + ve sign is taken. Let the speed is N_2

$\therefore$ $$0.0002776\, N_2^2 = 24.593 + 0.305 = 24.898$$

or $$N_2 = \sqrt{\frac{24.898}{0.0002776}} = 299.5 \text{ r.p.m.}$$

When the sleeve is about to fall, – ve, sign is taken. Let the speed in N_2'

$\therefore$ $$0.0002776\, (N_2')^2 = 24.593 - 0.305 = 24.288$$

or $$N_2' = \sqrt{\frac{24.288}{0.0002776}} = 295.8 \text{ r.p.m.}$$

$\therefore$ Co-efficient of insensitiveness at the maximum radius

$$= \left(\frac{\text{Fluctuation of speed}}{\text{Mean speed}}\right) \text{at maximum radius}$$

$$= \frac{N_2 - N_2'}{\left(\frac{N_2 + N_2'}{2}\right)} = \frac{299.5 - 295.8}{\left(\frac{299.5 + 295.8}{2}\right)} = \frac{3.7}{297.65} = \mathbf{0.0124 \text{ or } 1.24\%.} \quad \textbf{Ans.}$$

Problem 15.21. *The controlling force in a spring-controlling governor is 1500 N when radius of rotation is 200 mm and 887.5 N when radius of rotation is 130 mm. The mass of each ball is 8 kg. If the controlling force curve is a straight line, then find :*

(*i*) *controlling force when radius of rotation is 150 mm,*

(*ii*) *the speed of the governor when radius of rotation is 150 mm,*

(*iii*) *increase in initial tension so that governor is isochronous, and*

(*iv*) *isochronous speed.*

Sol. Given :

$F = 1500$ when $r = 200$ mm $= 0.2$ m ; $F = 887.5\ N$ when $r = 130$ mm $= 0.13$ m ; $m = 8$ kg.

The controlling force curve is a straight line curve. As controlling force curve gives the variation of controlling force (F) with radius of rotation (r), hence the controlling force can be expressed as

$$F = ar + b \quad ...(i)$$

where a and b are constant.

$F = 1500$ N when $r = 0.2$ m, hence equation (*i*) becomes as

$$1500 = a \times 0.2 + b \quad ...(ii)$$

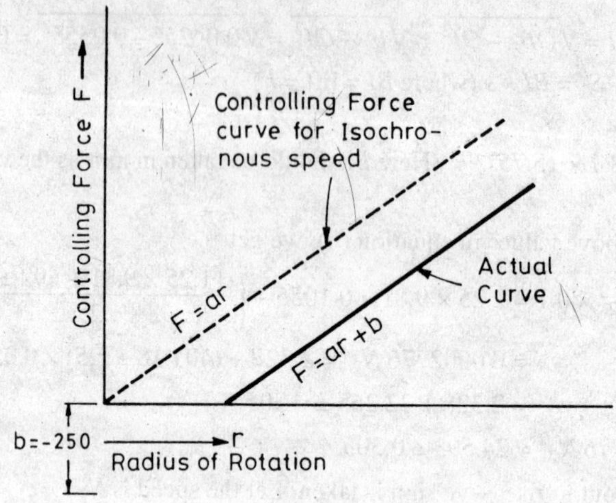

Fig. 15.29

$F = 887.5\ N$ when $r = 0.13$ m, hence equation (*i*) becomes as

$$887.5 = a \times 0.13 + b \qquad \text{...(iii)}$$

Subtracting equation (*iii*) from equation (*ii*), we get

$$1500 - 887.5 = 0.2a - 0.13a$$

or $$612.5 = 0.07a$$

$\therefore$ $$a = \frac{612.5}{0.07} = 8750$$

Substituting the value of a in equation (*ii*), we get

$$1500 = 8750 \times 0.2 + b$$

or $$b = 1500 - 8750 \times 0.2 = 1500 - 1750 = -250$$

Substituting the values of a and b in equation (*i*), we get

$$F = 8750 \times r - 250. \qquad \text{...(iv)}$$

(*i*) *Controlling force when r = 150 mm = 0.15 m*

Substituting $r = 0.15$ m in equation (*iv*), we get controlling force as

$$F = 8750 \times 0.15 - 250 = \mathbf{1062.5\ N.}$$ **Ans.**

(*ii*) *Speed of governor when r = 150 mm = 0.15 m*

Let N = speed of governor

F = controlling force = 1062.5 N

We know that $F = F_C = m \times \omega^2 \times r$

or $$1062.5 = 8 \times \omega^2 \times 0.15$$

$\therefore$ $$\omega^2 = \frac{1062.5}{8 \times 0.15} = 885.4$$

or $$\omega = \sqrt{885.4} = 29.75 \text{ rad/s}$$

But $\omega = \frac{2\pi N}{60}$ or $N = \frac{60 \times \omega}{2\pi} = \frac{60 \times 29.75}{2\pi} = 284.1$ r.p.m.

$\therefore$ N = **284.1 r.p.m. Ans.**

(iii) *Increase in initial tension so that governor is isochronous*

The governor will be isochronous if the controlling force curve passes through the origin. The controlling force curve (or line) will pass through the origin if b is zero. The value of b will be zero if initial tension is increased to **250 N. Ans.**

(iv) *Isochronous speed*

Let N = isochronous speed.

For isochronous speed, b is zero.

$\therefore$ Controlling force, $F = ar + b = ar + 0 = a \times r = 8750 \times r\ N$

But controlling force $F = m \times \omega^2 \times r = 8 \times \omega^2 \times r$

Equating the two values of F, we get

$$8 \times \omega^2 \times r = 8750 \times r$$

or $$\omega^2 = \frac{8750}{8} = 1093.75$$

$$\therefore \quad \omega = \sqrt{1093.75} = 33.07 \text{ rad/s}$$

$\therefore$ Isochronous speed, $N = \dfrac{60 \times \omega}{2\pi} =$ **315.8 r.p.m. Ans.**

HIGHLIGHTS

1. The function of the governor is to maintain a constant mean speed of rotation over a long period due to the variation of load on the engine.
2. The function of flywheel is to limit the fluctuation of speed during each cycle.
3. In centrifugal governors, the centrifugal force on the balls which are revolving about the axis of the shaft, is equal to the inward controlling force which is provided by a dead weight or a spring or a combination of the two.
4. The height (h) of a Watt governor is given by,

$$h = \frac{g}{\omega^2} \quad \text{... in terms of angular speed } \omega$$

$$= \frac{895}{N^2} \quad \text{... in terms of r.p.m. } (N)\text{. Here '}h\text{' is in meter.}$$

5. The height of a Watt governor is inversely proportional to the square of speeds. At high speeds, the value of 'h' is very small. Hence Watt governor works satisfactorily at low speeds.
6. For a Porter governor, the equations are :

$$h = \frac{m \cdot g + \left(\dfrac{M \cdot g \pm F_f}{2}\right)(1 + k)}{m \times \omega^2} \quad \text{... in terms of friction between sleeve and spindle}$$

$$= \frac{m \cdot g + \dfrac{M \cdot g}{2}(1 + k)}{m \cdot \omega^2} \quad \text{... if there is no friction between sleeve and spindle}$$

where m = Mass of the ball, M = Mass of central load, F_f = Force of friction between sleeve and spindle

ω = Angular speed of governor, $k = \dfrac{\tan \beta}{\tan \alpha}$.

If $k = 1$ (which is true when the arms are of equal length and both pivots lie on the same vertical line) and there is no friction, then

$$h = \left(\frac{m + M}{m}\right) \times \frac{894.56}{N^2}.$$

7. Proell governor is similar to Porter governor except the arrangement of the balls. The balls in Proell governor are carried on the extension of lower arms whereas in Porter governor they are carried at the function of upper and lower arms.

8. Hartnell governor is a spring loaded governor. The followings are the equations for this governor :

$$h_1 = \frac{b}{a}(r - r_1) \quad \text{... For minimum radius } r_1$$

$$h_2 = \frac{b}{a}(r_2 - r) \quad \text{... For maximum radius } r_2$$

$$h = \frac{b}{a}(r_2 - r_1) \quad \text{... Total lift } (h_1 + h_2)$$

$$s = \frac{a}{b}\left(\frac{S_2 - S_1}{r_2 - r_1}\right) \quad \text{... In terms of spring forces } S_1 \text{ and } S_2$$

where a = Length of ball-arm of bell-crank lever, b = Length of sleeve arm of bell-crank lever
s = Stiffness of the spring.

9. Wilson-Hartnell governor is also a spring loaded governor in which the balls are connected by two springs.

10. A governor is said to be sensitive if with a given fractional change of speed, the displacement of sleeve is very large. Mathematially sensitiveness

$$= \frac{N_2 - N_1}{N} \text{ where } N_2 = \text{Maximum speed, } N_1 = \text{Minimum speed and } N = \text{mean speed.}$$

11. A governor is said to be stable when for each speed there is only one radius of rotation of the governor balls.

12. A governor is said to be isochronous if the equilibrium speed is constant for all radii of rotation of the balls.

13. A governor is said to be hunting, if the speed of the engine fluctuates continuously above and below the mean speed.

14. The force exerted by the governor at the sleeve, is known as the effort of the governor whereas the work done at the sleeve is known as the power of the governor.

Governor Power = Governor effort × displacement of sleeve

where governor effort $= c(m + M) \cdot g$ ... if $k = 1 \left(i.e. \frac{\tan \beta}{\tan \alpha} = 1 \right)$ or angles α and β are equal

$$= c\left(\frac{2m}{1+k} + M\right) . g \text{... if angles } \alpha \text{ and } \beta \text{ are not equal.}$$

and governor power $= \frac{4c^2}{(1+2c)}\left[m + \frac{M}{2}(1+k)\right] . g . h$ ... if α and β are not equal

$$= \frac{4c^2}{(1+2c)}[m + M] . g . h \quad \text{... if } \alpha = \beta$$

15. The outwards movement of the balls is resisted by an equal and opposite force acting radially inwards, which is known as controlling force (F_C).

∴ Controlling force $= F_C = m . \omega^2 . r$.

EXERCISE 15

A. THEORETICAL QUESTIONS

1. What is the difference between a governor and a flywheel ?
2. What are the important types of the governors ? Differentiate between centrifugal governors and inertia governors.
3. Derive an expression for the height of Watt governor and prove that the height of the governor is inversely proportional to the square of the speed of governor.

4. Prove that the height of Porter governor is given by

$$h = \frac{m \times g + \left(\frac{M \times g \pm F_f}{2}\right)(1 + k)}{m \times \omega^2}$$

where m = Mass of each ball, M = Mass of central load, F_f = Force of friction,

ω = Angular velocity of rotation, and $k = \frac{\tan \beta}{\tan \alpha}$.

If the arms of the governor are of equal lengths and both are pivoted on the same vertical line, then the above equation will reduce to (when friction is neglected)

$$h = \left(\frac{m + M}{m}\right) \times \frac{894.56}{N^2}.$$

5. What is the difference between Porter governor and Proell governor ? Prove that in case of Proell governor, the angular speed reduces for a given value of m, M and h.

6. With a neat sketch, explain the working of Hartnell governor. Prove that for a Hartnell governor :

(*i*) the total lift, $h = \frac{b}{a}(r_2 - r_1)$ and

(*ii*) stiffness of spring, $s = \frac{a}{b}\left[\frac{S_2 - S_1}{r_2 - r_1}\right]$

where a and b = Lengths of bell-crank lever of ball-arm and sleeve arm respectively S_2 and S_1 = Spring forces at maximum and minimum radii.

7. With a neat sketch, explain the working of Wilson-Hartnell governor.

8. Define and explain the terms : governor effort and governor power. Prove that these terms are given by,

(*i*) governor effort $= c(m + M) \times g$ ... when $\alpha = \beta$ or $k = 1$

$= c\left(\frac{2m}{1+k} + M\right) \times g$ ... when $\alpha \neq \beta$ or $k = \frac{\tan \beta}{\tan \alpha} \neq 1$

(*ii*) governor power $= \frac{4c^2}{(1 + 2c)}(m + M) \times g \times h$ ... when $\alpha = \beta$ or $k = 1$

$= \frac{4c^2}{(1 + 2c)}\left[m + \frac{M}{2}(1 + k)\right] \times g \times h$ ... when $\alpha \neq \beta$.

9. What is meant by the term 'stability of a governor' ? Derive the necessary condition of stability for a centrifugal governor. (*AMIE, Winter 1994*)

10. What are the criteria that are taken into account to judge the quality of governing ? Show that the sensitiveness for a Watt and a Porter governor having all arms equal in length and intersecting on axis, is the same. (*AMIE, Summer 1994*)

11. Explain the terms : Sensitiveness ; stability ; isochronism and hunting in connection with governors. (*AMIE, S 1976*)

12. What do you mean by controlling force and controlling force curve.

B. NUMERICAL PROBLEMS

1. Calculate the change in vertical height of a Watt governor when its speed increases from

(*i*) 60 r.p.m. to 61 r.p.m. and

(*ii*) 150 r.p.m. to 151 r.p.m. **[Ans.** (*i*) 8.1 mm (*ii*) 0.52 mm]

2. Determine : (*i*) maximum speed, (*ii*) minimum speed and (*iii*) range of speed of a Watt governor of open arm type shown in Fig. 15.2 (*b*) in which the length of arm AE = 400 mm and length EF = 50 mm when angle θ changes from 35° to 30°. **[Ans.** (*i*) 49.6 r.p.m., (*ii*) 47.9 r.p.m. ; (*iii*) 1.7 r.p.m.]

3. Calculate the minimum speed, maximum speed and range of speed of a Porter governor, which has equal arms each 250 mm long and pivoted on the axis of rotation. The mass of each ball is 6 kg and the central mass on the sleeve is 18 kg. The radius of rotation of ball is 150 mm when the governor begins to lift and 200 mm when the governor is at the maximum speed. **[Ans.** 133.8 r.p.m. ; 154.5 r.p.m. and 20.7 r.p.m.]

4. If in the above question the central mass of the sleeve is 36 kg instead of 18 kg and all other data remains the same, then find minimum speed, maximum speed and range of speed. **[Ans.** 177 r.p.m. ; 204.4 r.p.m. and 27.4 r.p.m.]

5. If in question 4, the friction at the sleeve is equivalent to 24 *N*, then determine the range of the speed. **[Ans.** 38 r.p.m.]

6. Determine the range of speed of a Porter governor which has all four arms 250 mm long. The upper arms are attached to the axis of rotation and lower arms are attached to the sleeve at a distance of 30 mm from the axis. The mass of each ball is 6 kg and central mass on the sleeve is 60 kg. The extreme radii of rotation are 150 mm and 200 mm. **[Ans.** 30 r.p.m.]

7. Determine the minimum speed, maximum speed and range and speed of a Porter governor which has all four arms 250 mm long. The upper and lower arms are pivoted to the links at a distance of 40 mm and 50 mm respectively from the axis of rotation. The mass of each ball is 6 kg and central mass on the sleeve is 60 kg. The force of friction on the sleeve is 48 N. The radius of rotation of the ball is 125 mm when governor begins to lift and 150 mm when the governor is at the maximum speed. **[Ans.** 157.6 r.p.m. ; 181.1 r.p.m. and 23.5 r.p.m.]

8. The lengths of the upper and lower arms of a Porter governor are 200 mm and 250 mm respectively. Both the arms are pivoted on the axis of rotation. The weight of each ball is 30 N, the central load on sleeve is 225 N and the friction of the sleeve together with the resistance of the operating gear is equivalent to a force of 45 N at the sleeve. If the limiting inclinations of the upper arms to the vertical are 30° and 40°, determine the range of speed of the governor. **[Ans.** N_1 = 180 r.p.m. ; N_2 = 225.6 r.p.m. ; and range of speed = 45.6 r.p.m.]

9. Calculate the minimum speed of a Proell governor, which has equal arms each 200 mm and pivoted on the axis of the rotation. The mass of each ball is 5 kg and the central mass on the sleeve is 25 kg. The extension arms of lower links are each 60 mm long and parallel to the axis when the minimum radius of the ball is 100 mm. **[Ans.** 151.7 r.p.m.]

10. The mass of each ball of a Proell governor is 9.6 kg and the mass on the sleeve is 90 kg. Each of the arm is 305 mm long. The upper arms are pivoted on the axis of rotation whereas the lower arms are attached to a sleeve at a distance of 38 mm from the axis. The mass of the ball is attached to the extension of the lower arms which are 102 mm long. The minimum and maximum radii of governor are 165 mm and 216 mm. Assuming that the extensions of the lower arms are parallel to the governor axis at the minimum radius, find the corresponding equilibrium speeds. **[Ans.** 165 r.p.m. and 170 r.p.m.]

16

Vibrations (Longitudinal and Transverse)

16.1. Introduction

When an elastic body (like shaft, springs etc.) which is fixed at one end and is displaced at other end from its equilibrium position by the application of an external force, the body starts to move to and fro (or up and down). Then the body is said to be in vibrations. The vibrations are due to internal elastic forces within the body.

Whenever the body is displaced from its equilibrium position by the external force, work is done by the external force in producing the initial displacement against the internal elastic forces which resist deformation. This work is stored as strain energy. If the external force is removed and body is released, *the internal elastic forces tend to restore the body to its equilibrium position.* If the resistance to the motion of the vibrating body is neglected, then the whole of the strain energy is converted into kinetic energy when the body reaches to its original equilibrium position. Due to the maximum kinetic energy at the equilibrium position, the body continues to move in the opposite direction until the whole of the kinetic energy is absorbed in doing work against the internal elastic forces. This work against the internal elastic forces is again stored as strain energy. Due to the strain energy again the body begins to return to the equilibrium position and vibrations are repeated indefinitely if the friction is neglected.

16.2. Types of Vibration

The followings are the important types of vibrations :

(*i*) Free vibrations or natural vibrations

(*ii*) Forced vibrations, and

(*iii*) Damped vibrations.

16.2.1. Free Vibrations. A vibration in which no external force acts on the body, after giving an initial displacement and the motion is maintained by internal elastic forces (when friction and any other resistance offered to the motion is neglected), is known as free or natural vibration.

16.2.2. Forced Vibrations. A vibration in which external force (periodic disturbing force) is applied to the body and the vibrations have the same frequency as the applied force, is known as forced vibration.

16.2.3. Damped Vibrations. A vibration in which there is a reduction in amplitude over every cycle of vibration, is known as damped vibration. In any vibrating body, when external force is removed, the energy possessed by the body is gradually dissipated in over coming internal and external resistances to the motion, and the amplitude goes on reducing over every cycle of vibration and the body finally comes to rest in its original equilibrium position.

16.3. Important Definitions for Vibrating Motion

The followings are the important terms commonly used for the vibratory motions :

1. Period of vibration or time period,

2. Cycle,

3. Frequency, and

4. Resonance.

1. **Period of Vibration or time period.** It is the time taken by a motion to repeat itself. It is measured in seconds.

2. **Cycle.** It is the motion completed during one time period.

3. **Frequency.** It is the number of cycles completed in one second. It is expressed in hertz (which is written as Hz) and is equal to one cycle per second.

4. **Resonance.** When the frequency of the external force is the same as that of the natural frequency of the system, resonance takes place. Resonance results in large amplitudes of vibrations and this may be dangerous.

16.4. Free Vibrations

In case of free-vibration, no external force (or external torque) acts on the body, after giving an initial displacement. The motion is maintained by internal elastic forces. The following three are the important types of free-vibrations :

(*i*) Longitudinal vibrations,

(*ii*) Transverse vibrations, and

(*iii*) Torsional vibrations.

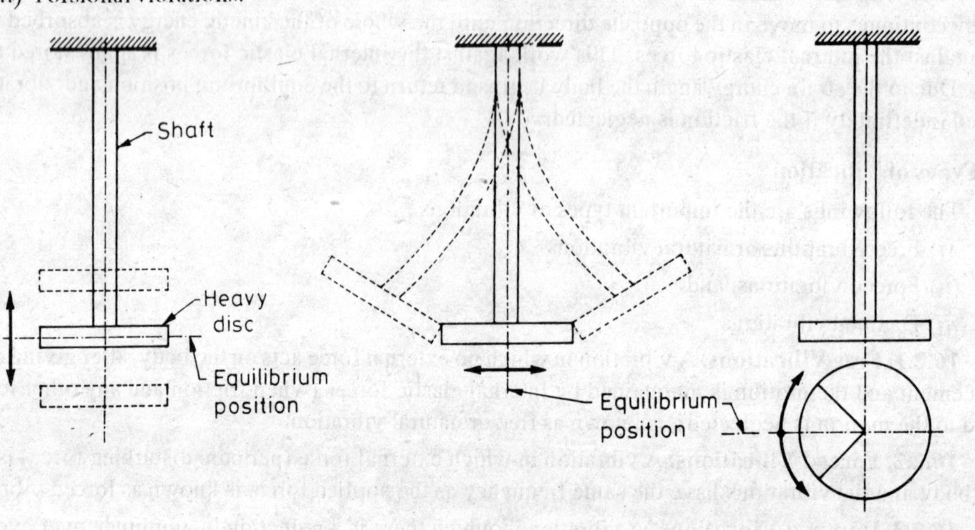

(*a*) Longitudinal vibrations (*b*) Transverse vibrations (*c*) Torsional vibrations

Fig. 16.1

Consider a weightless vibrating body *i.e.* a shaft or spring or a rod. Fig. 16.1 shows a shaft whose one end is fixed and other end carries a heavy disc. The shaft is assumed to be weightless. This system may be made to vibrate in one of three above mentioned types of vibrations :

1. *Longitudinal Vibrations.* The disc may be given initial displacement from the equilibrium position by external force in the downward direction and then released. The disc will be moving up and down as shown in Fig. 16.1 (*a*). All the particles of the disc will vibrate along straight paths parallel to the axis of the shaft. This type of vibration is known as longitudinal vibration. In longitudinal vibration, the shaft is elongated and shortened alternately resulting in tensile and compressive stresses in the shaft.

2. *Transverse Vibrations.* The disc may be given initial displacement from equilibrium position by external force perpendicular to the axis of the shaft (*i.e.* towards left or right) as shown in Fig. 16.1 (*b*) and then released. All the particles of the disc will vibrate along straight* paths perpendicular to the axis of the shaft. This type of vibration is known as transverse vibration. In transverse vibration, the shaft is alternately bent and straight resulting in bending stresses in the shaft.

3. *Torsional Vibrations.* The disc may be given initial angular displacement from the equilibrium position by an external torque (*i.e.* the disc may be twisted in a circular arc) as shown in Fig. 16.1 (*c*) and then released. All the particles of the disc will vibrate along circular arcs whose centres lie on the axis of the shaft. This type of vibration is known as torsional vibration. In torsional vibration, the shaft is alternately twisted and untwisted resulting in torsional shear stresses in the shaft.

In each of the above case if the limit of proportionality (*i.e.* stress proportional to strain) is not exceeded, then the *restoring force* in longitudinal and transverse vibrations or the *restoring couple* in torsional vibrations which is exerted on the disc by the shaft, is directly proportional to the displacement of the disc from its equilibrium (or mean) position. The force is the product of mass and acceleration. Hence it follows that acceleration towards equilibrium position is directly proportional to the displacement from that position and the *vibration is therefore simple harmonic*.

16.5. Methods of Finding the Natural Frequency of Free Longitudinal Vibrations

The following three methods are used for finding the natural frequency of free longitudinal vibrations :

1. Equilibrium method,
2. Energy method, and
3. Rayleigh's method.

16.5.1. Equilibrium method. Consider a spring of negligible mass suspended vertically from a rigid support as shown in Fig. 16.2 (*a*). The other end of the spring is free. If a weight W is suspended from the free

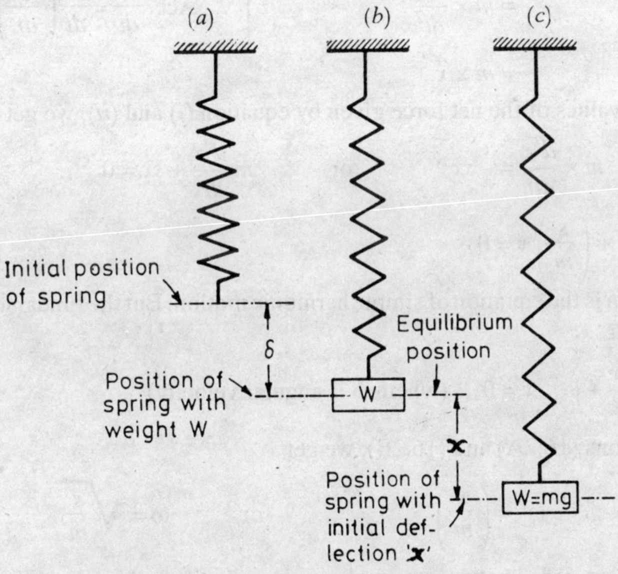

Fig. 16.2. Equilibrium method for natural frequency.

*As the amplitude of the vibrations is generally small in comparison with the length of the shaft hence the particles of the disc will approximately vibrate perpendicular to the axis of the shaft.

end, the spring is stretched by a distance δ as shown in Fig. 16.2 (*b*). The distance δ is known as static deflection of the spring under the weight W.

Let s = stiffness of the spring *i.e.* the force required to produce unit deflection in the spring

m = mass of the weight suspended

$\therefore \quad W = m \times g$

For the equilibrium position shown in Fig. 16.2 (*b*),

Weight suspended = upward force of spring

(The force of spring acts in the direction opposite to deflection)

or $$m \times g = s \,.\, \delta \qquad ...(16.1)$$

Now the weight W is given an initial displacement 'x' in the downward direction and then released as shown in Fig. 16.2 (*c*). The weight W will start vibrating above and below the equilibrium position. These vibrations are known as longitudinal vibrations. If the frictional resistance is neglected then these vibrations will continue forever. Then the vibrations will be known as free longitudinal vibrations.

When the weight W is moving downward and is at a distance of x from equilibrium position, the net force acting on the weight

= Downward force – upward force of spring

$= W - s \,.\, (\delta + x)$

$= mg - s \,.\, \delta - s \,.\, x$

$= s \,.\, \delta - s \,.\, \delta - s \,.\, x$ $\qquad [\because$ From equation (16.1), $m \,.\, g = s \,.\, \delta]$

$= -x \,.\, s$ $\qquad ...(i)$

But the net force is equal to the product of mass and acceleration.

$\therefore$ Net force = mass × acceleration

$$= m \times \frac{d^2x}{dt^2} \qquad \left[\because \text{ Acc. } \frac{dv}{dt} = \frac{d}{dt}\left(\frac{dx}{dt}\right) = \frac{d^2x}{dt^2} = \ddot{x}\right] \qquad ...(ii)$$

$$= m \times \ddot{x}$$

Equating the two values of the net force given by equation (*i*) and (*ii*), we get

$$m \times \frac{d^2x}{dt^2} = -sx \quad \text{or} \quad m\frac{d^2x}{dt^2} + sx = 0 \qquad ...(16.2)$$

or $$\frac{d^2x}{dt^2} + \left(\frac{s}{m}\right)x = 0 \qquad ...(16.2\text{A})$$

The above equation is the equation of simple harmonic motion. But the fundamental equation of simple harmonic equation is

$$\frac{d^2x}{dt^2} + \omega^2 \,.\, x = 0 \quad \text{(where } \omega \text{ is angular velocity)} \qquad ...(16.2\text{B})$$

Comparing equations (16.2A) and (16.2B), we get

$$\omega^2 = \left(\frac{s}{m}\right) \quad \text{or} \quad \omega = \sqrt{\frac{s}{m}}$$

In the above equation, ω is known as natural angular velocity and is generally denoted by ω_n.

$$\therefore \quad \omega_n = \sqrt{\frac{s}{m}} \qquad ...(16.3)$$

Now the time period, $$t = \frac{2\pi}{\omega_n} = \frac{2\pi}{\sqrt{\frac{s}{m}}} = 2\pi\sqrt{\frac{m}{s}} \qquad ...(16.4)$$

∴ Natural frequency (f_n) is given by,

$$f_n = \frac{1}{t} = \frac{1}{2\pi\sqrt{\frac{m}{s}}} = \frac{1}{2\pi}\sqrt{\frac{s}{m}} \qquad ...(16.5)$$

The natural frequency can also be expressed in terms of static deflection 'δ' by substituting the value of s in equation (16.5).

From equation (16.1), $m \times g = s \,.\, \delta$ or $s = \frac{m \,.\, g}{\delta}$

∴ Equation (16.5) becomes as

$$f_n = \frac{1}{2\pi}\sqrt{\left(\frac{m \,.\, g}{\delta}\right) \times \frac{1}{m}} = \frac{1}{2\pi}\sqrt{\frac{g}{\delta}} \qquad ...(16.6)$$

In the above equation, if δ is taken in meter than value of g will be 9.81 m/s^2.

Hence the above equation becomes as

$$f_n = \frac{1}{2\pi}\sqrt{\frac{9.81}{\delta}} = \frac{0.4985}{\sqrt{\delta}} \text{ cycles/s} = \frac{0.4985}{\sqrt{\delta}} \text{ Hz} \qquad ...(16.7)$$

In the above equation, δ is the extension of the spring (or shaft) produced by the load W. The value of δ can be obtained from the given conditions of the problem.

We know, $$E = \frac{\text{Stress}}{\text{Strain}} = \frac{\left(\frac{W}{A}\right)}{\left(\frac{\delta}{L}\right)} = \frac{W}{A} \times \frac{L}{\delta}$$

$$\therefore \quad \delta = \frac{W \times L}{A \times E} \qquad ...(16.8)$$

Hence if W = weight suspended at free end ; L = length of spring or shaft ; A = cross-sectional area and E = Young's modulus of the material of the spring or shaft are known, then value of δ can be obtained.

16.5.2. Energy method. In free vibrations, the total energy of the system (*i.e.* sum of kinetic energy and potential energy) remains constant. Potential energy is taken with respect to a certain datum position. Datum position is taken as equilibrium position in case of vibration.

$$\therefore \quad \text{K.E.} + \text{P.E.} = \text{Constant} \quad \text{or} \quad \frac{d}{dt}[\text{K.E.} + \text{P.E.}] = 0 \qquad ...(A)$$

But $$\text{K.E.} = \frac{1}{2}mv^2 = \frac{1}{2}m\left(\frac{dx}{dt}\right)^2$$

and P.E. = Mean force × displacement from equilibrium position

$$= \left(\frac{0 + s \,.\, x}{2}\right) \times x = \frac{sx^2}{2}$$

Now equation (A) becomes as $$\frac{d}{dt}\left[\frac{1}{2}m\left(\frac{dx}{dt}\right)^2 + \frac{sx^2}{2}\right] = 0$$

or $$\frac{1}{2} m \times 2 \left(\frac{dx}{dt} \right) \times \frac{d^2x}{dt^2} + \frac{1}{2} s \times 2x \times \frac{dx}{dt} = 0$$

or $$m \times \frac{d^2x}{dt^2} + sx = 0 \qquad \left(\text{Cancelling } \frac{dx}{dt} \text{ to both sides} \right)$$

This equation is same as equation (16.2). Hence the time period and natural frequency will be same as given in equilibrium method.

16.5.3. Rayleigh's method. In this method, the maximum kinetic energy at the mean position (where potential energy is zero) is equal to the maximum potential energy at the extreme position (where kinetic energy is zero). As the motion of the vibration is simple harmonic, hence

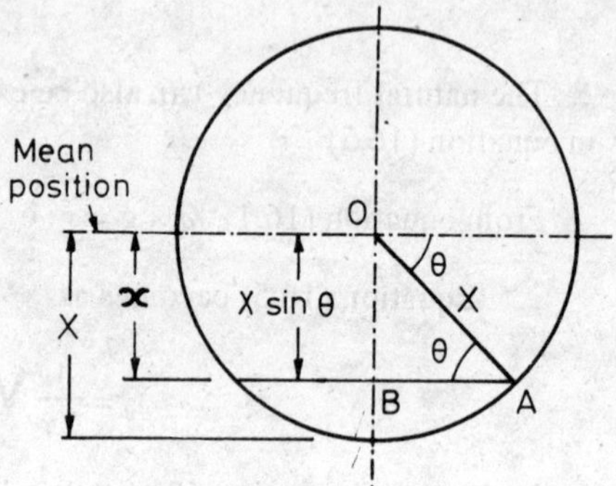

Fig. 16.3

$$OB = OA \sin\theta \text{ (see Fig. 16.3)}$$

or $$x = X \sin(\omega t) \quad \text{or} \quad X \sin\theta \qquad ...(i)$$

where x = displacement (of a body having S.H.M.) from mean position

X = Maximum displacement from mean position to extreme position

Differentiating equation (*i*), we have

$$\frac{dx}{dt} = X \times \omega \cos(\omega \,.\, t)$$

At the mean position, $t = 0$, hence velocity at mean position

$$v = \left(\frac{dx}{dt} \right)_{t=0} = X \,.\, \omega \quad \text{or} \quad \omega \,.\, X$$

$\therefore$ Maximum K.E. at mean position $= \frac{1}{2} \times m \times v^2 = \frac{1}{2} \,.\, m \,.\, (\omega \,.\, X)^2$

$$= \frac{1}{2} \,.\, m \,.\, \omega^2 \,.\, X^2 \qquad ...(ii)$$

Maximum potential energy at the external position

$$= \left(\frac{0 \times sX}{2} \right) \times X = \frac{1}{2} \,.\, s \,.\, X^2 \qquad ...(iii)$$

($\because$ P.E. = Mean force × Displacement)

Equating equations (*ii*) and (*iii*), we get

$$\frac{1}{2} \,.\, m \,.\, \omega^2 \,.\, X^2 = \frac{1}{2} \,.\, s \,.\, X^2$$

or $$m \,.\, \omega^2 = s \qquad \left(\text{Cancelling } \frac{1}{2} \,.\, X^2 \text{ to both sides}\right)$$

or $$\omega = \sqrt{\frac{s}{m}} \quad \text{or} \quad \omega_n = \sqrt{\frac{s}{m}}$$

$\therefore$ Time period, $$t = \frac{2\pi}{\omega} = 2\pi \sqrt{\frac{m}{s}}$$

and natural frequency, $$f_n = \frac{1}{t} = \frac{1}{2\pi \sqrt{\frac{m}{s}}} = \frac{1}{2\pi} \sqrt{\frac{s}{m}} \qquad \text{...(same as before)}$$

16.6. Method for Natural Frequency of Free Transverse Vibrations

Fig. 16.4 (*a*) shows a shaft of negligible mass whose one end is fixed and other end is free. If a body of weight W is attached at the free end, then the shaft will be deflected by a distance δ as shown in Fig. 16.4 (*b*).

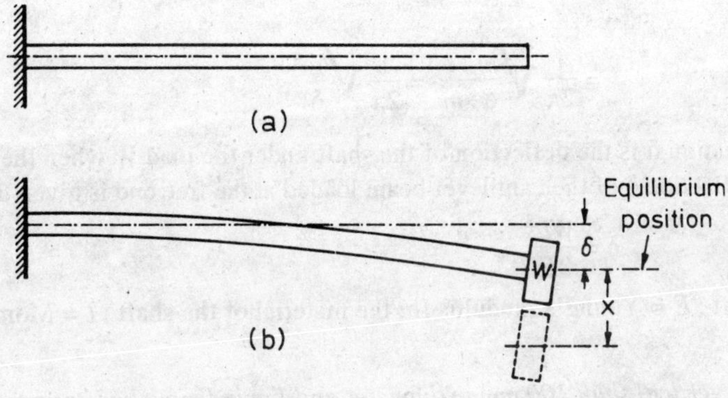

Fig. 16.4

The distance δ is known as static deflection of the shaft under the weight W.

Let s = Stiffness of the shaft *i.e.* force required to produce unit deflection in the shaft.

Then $s \,.\, \delta$ = Total force required to produce δ deflection

But δ is the deflection produced by weight W.

$$\therefore \qquad W = s \,.\, \delta \quad \text{or} \quad m \,.\, g = s \,.\, \delta \qquad (\because \; W = m \,.\, g) \qquad ...(i)$$

Now the weight W is given an initial deflection 'x' in the downward direction and then released as shown in Fig. 16.4 (*b*). The weight W will start vibrating above and below the equilibrium position. As all the particles of the body are vibrating along straight paths perpendicular to the axis of the shaft, hence the vibrations are known as transverse vibrations. If frictional resistance is neglected, these vibrations will continue for ever. Then the vibrations will be known as free transverse vibrations.

The net force acting on the body, when body is at a distance of 'x' below the equilibrium position

= Weight of body – upward force in the shaft due to stiffness

$$= W - s \,.\, (x + \delta)$$

$$= m \,.\, g - sx - s \,.\, \delta$$

$$= -sx \qquad [\because \; m \,.\, g = s \,.\, \delta \text{ from equation } (i)] \qquad ...(ii)$$

But net force = mass × acceleration

$$= m \times \frac{d^2x}{dt^2} \qquad ...(iii)$$

Equating the two values of net force given by equations (*ii*) and (*iii*), we get

$$m \,.\, \frac{d^2x}{dt^2} = -sx$$

or

$$m \,.\, \frac{d^2x}{dt^2} + s \,.\, x = 0 \quad \text{or} \quad \frac{d^2x}{dt^2} + \left(\frac{s}{m}\right) . \, x = 0$$

This equation is same as equation (16.2A).

Hence the time period and natural frequency of the transverse vibrations will be same as that of longitudinal vibrations which are given by equations (16.4) and (16.5) respectively.

$\therefore$ Time period, $t = 2\pi\sqrt{\frac{m}{s}}$ and

Natural frequency, $f_n = \frac{1}{t} = \frac{1}{2\pi}\sqrt{\frac{s}{m}}$ $\left(\text{But } m \cdot g = s \cdot \delta \text{ or } s = \frac{m \cdot g}{\delta}\right)$

$$= \frac{1}{2\pi}\sqrt{\frac{m \cdot g}{\delta \cdot m}} = \frac{1}{2\pi}\sqrt{\frac{g}{\delta}} \quad \text{...(16.9)}$$

In the above equation δ is the deflection of the shaft under the load W when the shaft is treated as a cantilever. The static deflection δ of the cantilever beam loaded at the free end is given by

$$\delta = \frac{WL^3}{3EI} \quad \text{...(16.10)}$$

where L = length of shaft ; E = Young's modulus for the material of the shaft ; I = Moment of inertia of the shaft.

Problem 16.1. *A vertical shaft 100 mm in diameter and 1 m in length has its upper end fixed at the top as shown in Fig. 16.5. At the other end it carries a disc of weight 5000 N. The modulus of elasticity (or Young's modulus) of the material of the shaft is 2×10^5 N/mm². Neglecting the weight of the shaft, determine the frequency of longitudinal vibrations and transverse vibrations.*

Sol. Given :

d = 100 mm = 0.1 m ; L = 1 m ; W = 5000 N ; $E = 2 \times 10^5$ N/mm² $= 2 \times 10^5 \times 10^6$ N/m² $= 2 \times 10^{11}$ N/m² ;

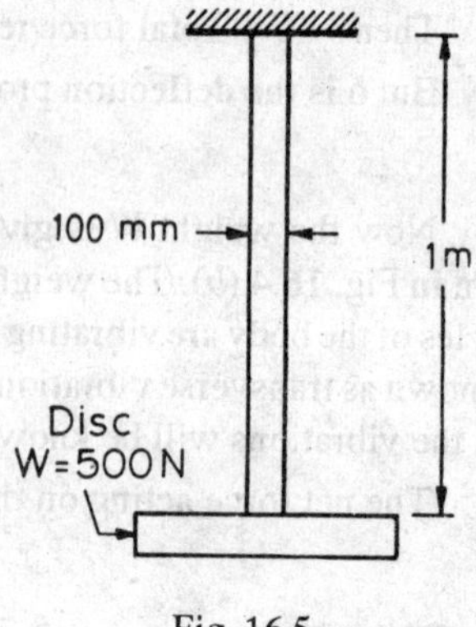

Fig. 16.5

(*i*) *Frequency of longitudinal vibrations*

It is given by equation (16.6) as

$$f_n = \frac{1}{2\pi}\sqrt{\frac{g}{\delta}} \quad \text{...}(i)$$

where δ is the static deflection produced by load W. The value of δ for longitudinal vibration is given by equation (16.8) as

$$\delta = \frac{W \times L}{A \times E}$$

$$= \frac{5000 \times 1}{\frac{\pi}{4}(.1)^2 \times 2 \times 10^{11}}$$

$$\left(\because A = \frac{\pi}{4}d^2 = \frac{\pi}{4}(.1)^2 \text{ m}^2 ; E = 2 \times 10^{11} \text{ N/m}^2 ; L = 1 \text{ m}\right)$$

$$= \frac{5000 \times 4}{\pi \times 0.01 \times 2 \times 10^{11}} = 3.183 \times 10^{-6} \text{ m}$$

Substituting the value of δ in equation (*i*), we get

$$f_n = \frac{1}{2\pi}\sqrt{\frac{9.81}{3.183 \times 10^{-6}}}$$

(As δ is in m, hence g should be taken in m/s². Hence g = 9.81 m/s²)

= 279.5 cycles/s or 279.5 Hz. Ans.

(ii) Frequency of transverse vibrations

It is given by equation (16.9) or (16.6) as

$$f_n = \frac{1}{2\pi}\sqrt{\frac{g}{\delta}} \qquad ...(ii)$$

where δ is the static deflection of the shaft under the load W when the shaft is placed horizontally as a cantilever. The value of δ is given by equation (16.10) as

$$\delta = \frac{W \times L^3}{3EI} \quad \text{where } I = \text{Moment of inertia} = \frac{\pi}{64} \times d^4 = \frac{\pi}{64}(0.1)^4 \text{ m}^4$$

$$= \frac{5000 \times 1^3}{3 \times 2 \times 10^{11} \times \frac{\pi}{64} \times (0.1)^4} = \frac{5000 \times 64}{6 \times \pi \times 10^7} = 1.7 \times 10^{-3} \text{ m}$$

Substituting the value of δ in equation (*ii*), we get

$$f_n = \frac{1}{2\pi} \times \sqrt{\frac{9.81}{1.7 \times 10^{-3}}} = 12 \text{ cycles/s} = \mathbf{12\ Hz.} \quad \textbf{Ans.}$$

16.7. Effect of the Inertia of Shaft on Longitudinal and Transverse Vibrations

In the previous articles, we have neglected the weight of the shaft while deriving the expression for longitudinal and transverse vibrations. Let us now consider the weight of the shaft thus the effect of inertia also.

(i) Longitudinal Vibration

Fig. 16.6 shows a shaft whose upper end is fixed and lower end carries a disc of weight W.

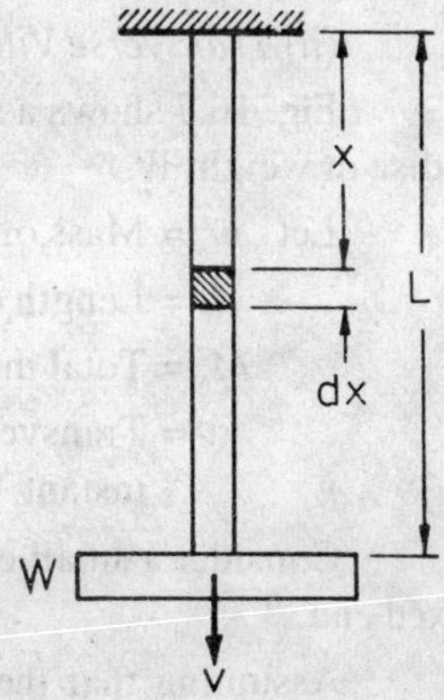

Fig. 16.6

Let m_s = Mass of shaft per unit length

L = Length of shaft

M_s = Total mass of shaft

$= m_s \times L$

v = Longitudinal velocity of disc (or free end of the shaft) at a given instant.

Consider a small element of shaft of length 'dx' at a distance x from the fixed end. Then the velocity of small element at the given instant is approximately

$$= \frac{x}{L} \cdot v$$

Mass of the element = mass per unit length × length of element = $m_s \cdot dx$

∴ K.E. of the small element at the given instant

$$= \frac{1}{2} \times (\text{mass of element}) \times (\text{velocity of element})^2$$

$$= \frac{1}{2} \cdot (m_s\, dx) \times \left(\frac{x}{L} \cdot v\right)^2 = \frac{1}{2} \cdot m_s \cdot dx \cdot \frac{x^2}{L^2} \cdot v^2$$

$$= \frac{m_s v^2 \cdot x^2 \cdot dx}{2L^2}$$

$$\therefore \quad \text{Total K.E. of the shaft} = \int_0^L \frac{m_s v^2 . x^2 . dx}{2L^2} = \frac{m_s v^2}{2L^2} \int_0^L x^2 . dx$$

$$= \frac{m_s . v^2}{2L^2}\left[\frac{x^3}{3}\right]_0^L = \frac{m_s . v^2}{2L^2} . \frac{L^3}{3} = \frac{m_s . L}{6} . v^2$$

$$= \frac{1}{2} . \frac{m_s L}{3} . v^2 = \frac{1}{2} . \left(\frac{M_s}{3}\right) v^2 \qquad (\because \; M_s = m_s . L)$$

But $\frac{1}{2} . \left(\frac{M_s}{3}\right) . v^2$ is the K.E. of one third of the mass of the shaft concentrated at the free-end where longitudinal velocity is v. Hence for taking the inertia of the shaft into account, one third of the mass of the shaft should be added to the mass of the disc attached to the shaft.

The natural frequency of longitudinal vibration in terms of mass attached neglecting mass of the shaft (or spring) is given by equation (16.5) as

$$f_n = \frac{1}{2\pi}\sqrt{\frac{s}{m}} \qquad \text{where } m = \text{mass attached}$$

The natural frequency of longitudinal vibration, considering mass of shaft will be given by,

$$f_n = \frac{1}{2\pi}\sqrt{\frac{s}{(m + M_s/3)}} \qquad \text{...(16.11)}$$

(ii) Transverse Vibrations

Fig. 16.7 shows a shaft whose upper end is fixed and lower end carries a disc of weight W.

Let m_s = Mass of shaft per unit length

L = Length of shaft

M_s = Total mass of shaft = $m_s \times L$

v = Transverse velocity at free end which carries disc at a given instant.

Consider a small element of shaft of length dx at a distance x from the fixed end.

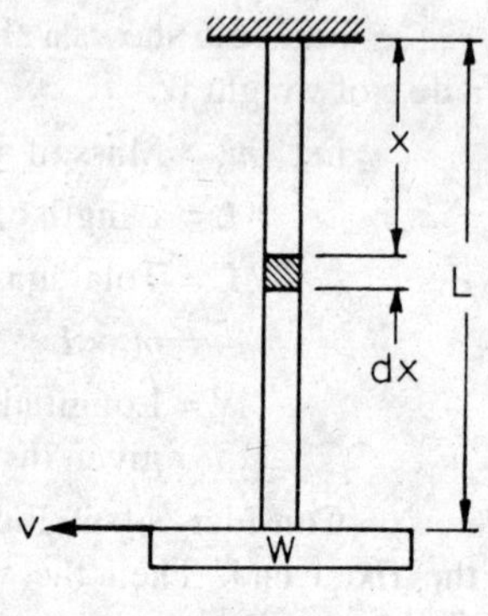

Fig. 16.7

Assuming that the shape of the curve into which the vibrating shaft deflects is identical with the static deflection curve of a cantilever loaded at the free end, then the velocity at the distance x from the fixed end at the same instant will be

$$= \frac{3Lx^2 - x^3}{2L^3} \times v$$

$$\therefore \quad \text{Velocity of element} = \left(\frac{3Lx^2 - x^3}{2L^3}\right) \times v$$

$$\text{Mass of element} = m_s . dx$$

$$\therefore \quad \text{K.E. of element} = \frac{1}{2} . (\text{Mass of element}) . (\text{Velocity of element})^2$$

$$= \frac{1}{2} . (m_s . dx) \times \left(\frac{3Lx^2 - x^3}{2L^3} \times v\right)^2$$

$$\therefore \quad \text{K.E. of the whole shaft} = \int_0^L \frac{1}{2} \cdot (m_s \cdot dx) \times \left(\frac{3Lx^2 - x^3}{2L^3} \times v \right)^2$$

$$= \frac{1}{2} \cdot m_s \cdot \frac{v^2}{4L^6} \int_0^L (9L^2x^4 + x^6 - 6Lx^5) \cdot dx = \frac{m_s \cdot v^2}{8L^6} \left[\frac{9L^2x^5}{5} + \frac{x^7}{7} - \frac{6Lx^6}{6} \right]_0^L$$

$$= \frac{m_s \cdot v^2}{8L^6} \left[\frac{9L^7}{5} + \frac{L^7}{7} - \frac{L^7}{1} \right] = \frac{m_s \cdot v^2}{8L^6} \left[\frac{63L^7 + 5L^7 - 35L^7}{35} \right]$$

$$= \frac{m_s \cdot v^2}{8L^6} \times \frac{33L^7}{35} = \frac{m_s \cdot v^2 \times 33L}{280} = \frac{1}{2} \times \left(\frac{33m_s \cdot L}{140} \right) \cdot v^2$$

$$= \frac{1}{2} \times \left(\frac{33M_s}{140} \right) \times v^2 \qquad (\because \; m_s \cdot L = M_s)$$

But $\frac{1}{2} \cdot \left(\frac{33M_s}{140} \right) \cdot v^2$ is the K.E. of the fraction $\frac{33}{140}$ of the mass of the shaft concentrated at the free end of the shaft where transverse velocity is v. Hence for taking the inertia of the shaft into account, $\frac{33}{140}$ of the mass of the shaft should be added to the mass of the disc attached to the shaft.

Hence the natural frequency of transverse vibrations, considering mass of the shaft will be given by,

$$f_n = \frac{1}{2\pi} \sqrt{\left(\frac{s}{m + 33M_s/140} \right)} \qquad \text{...(16.12)}$$

16.8. Natural Frequency of Transverse Vibrations of Shafts or Beams under Different Types of Loads and End Conditions

The following types of loads may be acting on a shaft or on a beam :

(*i*) Single concentrated load,

(*ii*) Uniformly distributed load.

The natural frequency of transverse vibrations may be obtained as explained below :

16.8.1. Natural frequency of a shaft carrying a single concentrated load. The single concentrated load may be acting on a cantilever, or on a simply supported beam or on a fixed beam as shown in Figs. 16.8 (*a*), 16.8 (*b*) and 16.8 (*c*) respectively. In case of shafts and beams of negligible mass carrying a concentrated load the force is proportional to the deflection of the weight from the equilibrium position and the equation for natural frequency of transverse vibration *i.e.* $f_n = \frac{1}{2\pi}\sqrt{\frac{g}{\delta}}$ is applicable. The value of δ will be different for each case.

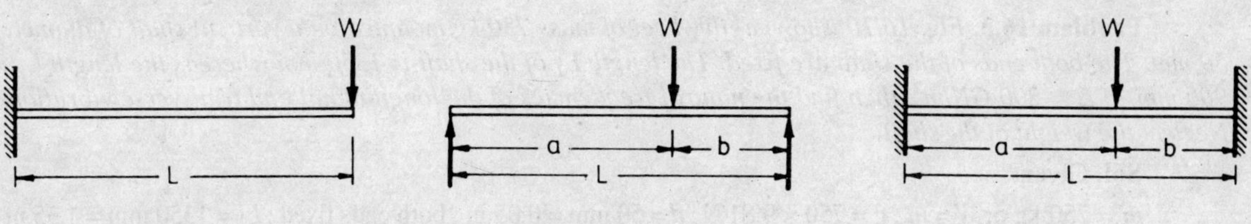

(*a*) Cantilever (*b*) Simply supported beam (*c*) Fixed beam

Fig. 16.8

$\therefore$ Natural frequency of transverse vibrations for a single concentrated load is given by

$$f_n = \frac{1}{2\pi}\sqrt{\frac{g}{\delta}} \quad ...(16.13)$$

where $\delta = \frac{WL^3}{3EI}$ ... for cantilever as shown in Fig. 16.8 (*a*) ...(16.14)

$= \frac{Wa^2b^2}{3EIL}$ ... for simply supported beam as shown in Fig. 16.8 (*b*) ...(16.15)

$= \frac{Wa^3b^3}{3EIL^3}$ ... for beams fixed at both ends as shown in Fig. 16.8 (*c*) ...(16.16)

Problem 16.2. *A simply supported shaft of length 800 mm carries a mass of 60 kg placed 250 mm from one end. If E = 200 GN/m² and diameter of shaft is 50 mm, then find the natural frequency of transverse vibrations.*

Sol. Given :

$L = 800$ mm $= 0.8$ m ; $m = 60$ kg ; $a = AC = 250$ mm $= 0.25$ m ; $b = BC = 800 - 250 = 550$ mm ; $E = 200$ GN/m² $= 200 \times 10^9$ N/m² ; $d = 50$ mm $= 0.05$ m.

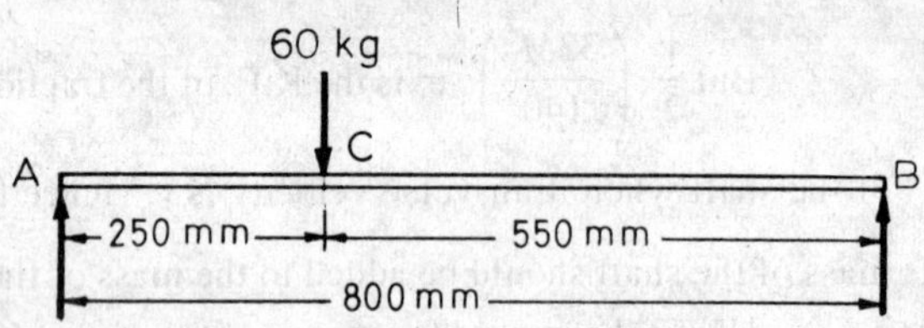

Fig. 16.9

The natural frequency of transverse vibrations of a simply supported shaft is given by, equation (16.13) as

$$f_n = \frac{1}{2\pi}\sqrt{\frac{g}{\delta}} \quad ...(i)$$

where the value of δ for a simply supported beam from equation (16.15) is

$$\delta = \frac{Wa^2b^2}{3EIL} \quad ...(ii)$$

Now $W = m \cdot g = 60 \times 9.81$ N ; $a = 0.25$ m ; $b = 0.55$ m ; $E = 200 \times 10^9$ N/m² ; $L = 0.8$ m and I = Moment of inertia of shaft

$$= \frac{\pi}{64}d^4 = \frac{\pi}{64} \times (0.05)^4 = 3.07 \times 10^{-7} \text{ m}^4$$

Substituting the above values in equation (*ii*), we get

$$\delta = \frac{(60 \times 9.81) \times (0.25)^2 \times (0.55)^2}{3 \times 200 \times 10^9 \times 3.07 \times 10^{-7} \times 0.80} = 0.755 \times 10^{-4} \text{ m}$$

Substituting the value of 'δ' in equation (*i*), we get

$$f_n = \frac{1}{2\pi}\sqrt{\frac{9.81}{0.755 \times 10^{-4}}} = \mathbf{57.37 \text{ Hz.}} \quad \textbf{Ans.}$$

Problem 16.3. *Fig. 16.10 show a flywheel of mass 750 kg mounted on a vertical shaft of diameter 50 mm. The both ends of the shaft are fixed. The length L_1 of the shaft is 1350 mm whereas the length L_2 is 900 mm. If E = 200 GN/m², then find the natural frequencies of the longitudinal and transverse vibrations. Neglect the weight of the shaft.*

Sol. Given :

$m = 750$ kg or $W = m \cdot g = 750 \times 9.81$ N ; $d = 50$ mm $= 0.05$ m ; both ends fixed ; $L_1 = 1350$ mm $= 1.35$ m ; $L_2 = 900$ mm $= 0.9$ m ; $E = 200$ GN/m² $= 200 \times 10^9$ N/m².

(i) Natural frequency of longitudinal vibrations

For longitudinal vibrations, the flywheel is pulled downward (*i.e.* along the axis of the shaft) and then released. The upper part of the shaft will be under extension and lower part will be under compression and *vice-versa*. The shaft is fixed at both ends, hence the total length of the shaft will remain same.

∴ Extension of upper part = Compression of lower part.

The total weight of flywheel is 750 × 9.81 N. Some of the load will be carried by length L_1 and the rest of load will be carried by length L_2.

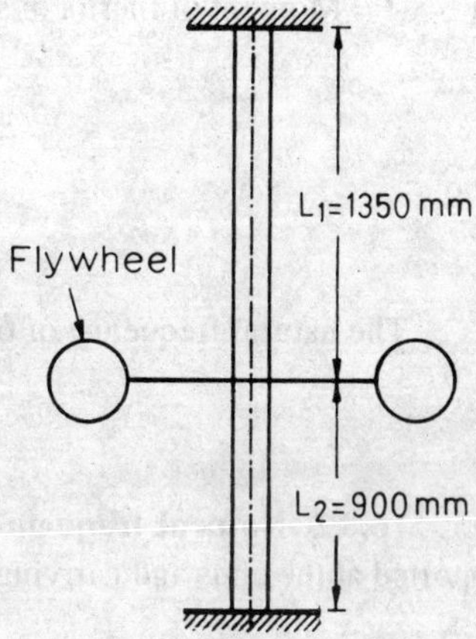

Fig. 16.10

Let W_1 = Load carried by length L_1, and

$(W - W_1)$ = Load carried by length L_2.

Now extension of length L_1 due to load W_1

$$= \frac{W_1 L_1}{A \cdot E} \quad \text{where } A = \text{Cross-sectional area of the shaft}$$

Compression of length L_2 due to load $(W - W_1)$

$$= \frac{(W - W_1) \cdot L_2}{A \cdot E}$$

But extension of length L_1 is equal to compression of length L_2

$$\therefore \quad \frac{W_1 L_1}{A \cdot E} = \frac{(W - W_1) \cdot L_2}{A \cdot E}$$

or $$W_1 L_1 = (W - W_1) \cdot L_2$$

or $$W_1 \times 1.35 = (750 \times 9.81 - W_1) \times 0.9$$

$(\because \; W = 750 \times 9.81 \text{ N}, L_1 = 1.35 \text{ m and } L_2 = 0.9 \text{ m})$

or $$\frac{W_1 \times 1.35}{0.9} = 750 \times 9.81 - W_1$$

or $$1.5 W_1 = 750 \times 9.81 - W_1 \quad \text{or} \quad 1.5 W_1 + W_1 = 750 \times 9.81$$

or $$2.5 W_1 = 750 \times 9.81 \quad \text{or} \quad W_1 = \frac{750 \times 9.81}{2.5} = 300 \times 9.81 \text{ N}$$

∴ Extension of length L_1,

$$\delta = \frac{W_1 \cdot L_1}{A \cdot E} \quad \text{where } A = \frac{\pi}{4} d^2 = \frac{\pi}{4} (0.05)^2 = 0.00196 \text{ m}^2$$

$$= \frac{(300 \times 9.81) \times 1.35}{0.00196 \times 200 \times 10^9} = 1.0135 \times 10^{-5} \text{ m}$$

The natural frequency of longitudinal vibrations is given by,

$$f_n = \frac{1}{2\pi}\sqrt{\frac{g}{\delta}} = \frac{1}{2\pi}\sqrt{\frac{9.81}{1.0135 \times 10^{-5}}} = \mathbf{156.58 \text{ Hz.}} \quad \textbf{Ans.}$$

(ii) Natural frequency of transverse vibrations

We know that the static deflection under the point load, for a horizontal shaft fixed at both ends is given by equation (16.16),

$$\delta = \frac{W \cdot a^3 \cdot b^3}{3EIL^3}$$

where $a = L_1 = 1.35$ m ; $b = L_2 = 0.9$ m, $W = 750 \times 9.81$ N

and I = Moment of inertia of shaft $= \frac{\pi}{64} d^4 = \frac{\pi}{64} \times (0.05)^4 = 0.307 \times 10^{-6}$ m^4

$$\therefore \quad \delta = \frac{750 \times 9.81 \times 1.35^3 \times 0.9^3}{3 \times 200 \times 10^9 \times 0.307 \times 10^{-6} \times 2.25^3}$$

$(\because \; L = L_1 + L_2 = 1.35 + 0.9 = 2.25 \text{ m})$

$$= 6.2895 \times 10^{-3} \text{ m}$$

The natural frequency of transverse vibrations is given by,

$$f_n = \frac{1}{2\pi}\sqrt{\frac{g}{\delta}} = \frac{1}{2\pi}\sqrt{\frac{9.81}{6.2895 \times 10^{-3}}} = \mathbf{6.28\ Hz.} \quad \textbf{Ans.}$$

16.8.2. Natural frequency of a shaft carrying a uniformly distributed load. A shaft (or beam) supported at the ends and carrying a uniformly load is shown in Fig. 16.11.

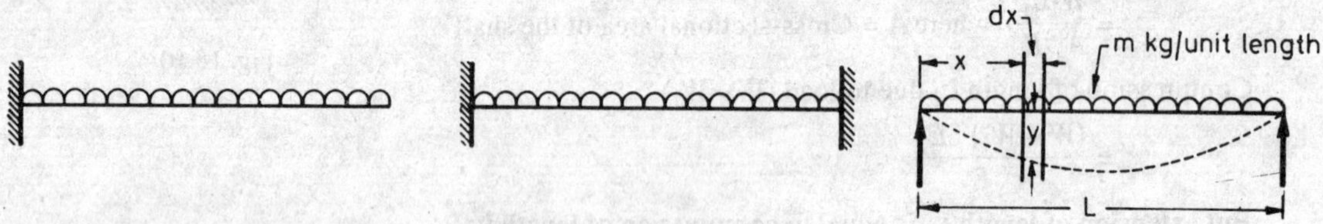

(c) Cantilever (b) Fixed beam (or shaft) (a) Simply supported beam (or shaft)

Fig. 16.11

Let L = Length of the shaft, and

m = Mass in kg per unit length distributed on the shaft.

If the shaft is pulled downwards and then released, it will start making transverse vibrations. Consider a small length 'dx' of the shaft at a distance x from one end. Let y be the displacement of length dx from the equilibrium position. Then the dynamic load on length dx will be equal to centrifugal force due to transverse vibrations [*i.e.* $(m \times dx) \times \omega^2 \times y$ where ω is the angular velocity of transverse vibrations in rad/s]. The dynamic load per unit length will be equal to centrifugal force per unit length *i.e.* $(m \times 1) \times \omega^2 \times y$.

But from the theory of bending of shaft, we know

$$EI \frac{d^4y}{dx^4} = \text{dynamic load per unit length}$$

$$= \text{centrifugal force per unit length}$$

$$= m \times \omega^2 \times y$$

or $$EI \frac{d^4y}{dx^4} - m\,\omega^2 y = 0 \qquad \text{or} \qquad \frac{d^4y}{dx^4} - \frac{m}{EI}\omega^2 . y = 0$$

or $$\frac{d^4y}{dx^4} - (m^*)^4\, y = 0 \qquad \text{...(16.17)}$$

where $$(m^*)^4 = \frac{m\,.\,\omega^2}{EI} \qquad \text{...(16.18)}$$

The solution of the differential equation (16.17) is of the form,

$$y = A \cos m^*x + B \sin m^*x + C \cosh m^*x + D \sinh m^*x \qquad \text{...(16.19)}$$

where A, B, C and D are constants of integration. Their values are obtained from the end conditions.

(*i*) **Case of Simply Supported Shaft** (as shown in Fig. 16.11 (*a*))

The end conditions are :

(*a*) $y = 0$ when $x = 0$ and $x = L$

(*b*) $\frac{d^2y}{dx^2} = 0$ when $x = 0$ and $x = L$ bending moment is zero at ends

When $x = 0$, $y = 0$. Then from equation (16.19), we get

$$0 = A \cos 0 + B \sin 0 + C \cosh 0 + D \cosh 0 \qquad (y = 0 \text{ at } x = 0)$$

or
$$0 = A + C \qquad ...(i)$$

When $x = L$, $y = 0$. Then from equation (16.19), we get

$$0 = A \cos m^*L + B \sin m^*L + C \cosh m^*L + D \sinh m^*L \qquad ...(ii)$$

Now differentiating equation (16.19) with respect to x twice, we get

$$\frac{dy}{dx} = A \,.\, m^* \,.\, (-\sin m^*x) + B \,.\, m^* \,.\, \cos m^*x + C \,.\, m^* \,.\, \sinh m^*x + D \,.\, m^* \,.\, \cosh m^*x$$

$$= m^* \left[-A \sin m^*x + B \cos m^*x + C \sinh m^*x + D \cosh m^*x\right]$$

and
$$\frac{d^2y}{dx^2} = m^* \left[-A \,.\, m^* \cos m^*x + B \,.\, m^* (-\sin m^*x) + C \,.\, m^* \cosh m^*x + D \,.\, m^* \sinh m^*x\right]$$

$$= m^{*2} \left[-A \cos m^*x - B \sin m^*x + C \cosh m^*x + D \sinh m^*x\right] \qquad ...(A)$$

When $x = 0$, $\frac{d^2y}{dx^2} = 0$, hence above equation becomes

$$0 = m^{*2} [-A - 0 + C + 0] = m^{*2} [-A + C]$$

or
$$0 = -A + C \qquad ...(iii)$$

Also when $x = L$, $\frac{d^2y}{dx^2} = 0$, hence equation (A) becomes as

$$0 = m^{*2} \left[-A \cos m^*L - B \sin m^*L + C \cosh m^*L + D \sinh m^*L\right] \qquad ...(iv)$$

Adding equations (*i*) and (*iii*), we get

$$2C = 0 \text{ or } C = 0. \text{ From } (i), \text{ we have } A + C = 0 \text{ or } A + 0 = 0 \text{ or } A = 0$$

$$\therefore \quad A = C = 0$$

Substituting $A = C = 0$ in equations (*ii*) and (*iv*), we get

$$0 = B \sin m^*L + D \sinh m^*L \quad \text{[From equation } (ii)] \qquad ...(v)$$

and
$$0 = m^{*2} [-B \sin m^*L + D \sinh m^*L] \quad \text{[From equation } (iv)]$$

or
$$0 = -B \sin m^*L + D \sinh m^*L \qquad ...(vi)$$

Adding equati ons (*v*) and (*vi*), we get, $2D \sinh m^*L = 0$

Either $D = 0$ or $\sinh m^*L = 0$. But $\sinh m^*L$ cannot be zero. Because if $\sinh m^*L = 0$, then $m^*L = 0$ or $m^* = 0$ or from equation (16.18) $\frac{m \times \omega^2}{EI} = 0$ or $\frac{m}{EI} (2\pi f_n)^2 = 0$ or $f_n = 0$ which means the system does not vibrate. But system is vibrating.

$$\therefore \quad D = 0$$

Substituting the value of D in equation (*v*), we get

$$0 = B \sin m^*L$$

From above, either $B = 0$ or $\sin m^*L = 0$. But B cannot be zero, because if B is zero (A, C and D are already zero), then from equation (16.19) $y = 0$.

This means, shaft is not vibrating.

$$\therefore \qquad \sin m^*L = 0 \quad \text{or} \quad m^* . L = 0, \pi, 2\pi, 3\pi \text{ etc.}$$

But m^*L cannot be zero. Because if $m^*L = 0$, then $m^* = 0$ and system will not vibrate.

$$\therefore \qquad m^*L = \pi, 2\pi, 3\pi, \text{etc.}$$

or
$$m^* = \frac{\pi}{L}, \frac{2\pi}{L}, \frac{3\pi}{L}, \text{etc.} \qquad ...(16.20)$$

Substituting the smallest value of m^* in equation (16.18), we get

$$\left(\frac{\pi}{L}\right)^4 = \frac{m \times \omega^2}{EI}$$

or
$$\omega = \sqrt{\frac{EI}{m}} \times \left(\frac{\pi}{L}\right)^2 = \frac{\pi^2}{L^2} \times \sqrt{\frac{EI}{m}}$$

Hence the natural frequency of a simply supported shaft carrying a uniformly distributed load is given by,

$$f_n = \frac{\omega}{2\pi} = \frac{\pi^2}{L^2} \times \sqrt{\frac{EI}{m}} \times \frac{1}{2\pi}$$

$$= \frac{\pi}{2L^2}\sqrt{\frac{EI}{m}} = \frac{\pi}{2}\sqrt{\frac{EI}{mL^4}} \qquad ...(16.21)$$

A simply supported shaft carrying a uniformly distributed load has maximum deflection at the centre, which is given as,

$$\delta = \frac{5}{384}\frac{wL^4}{EI} = \frac{5}{384}\frac{m . g . L^4}{EI} \qquad (\because \quad w = m . g)$$

or
$$\frac{EI}{m . L^4} = \frac{5}{384} \times \frac{g}{\delta} \quad \text{or} \quad \sqrt{\frac{EI}{m . L^4}} = \sqrt{\frac{5}{384} \times \frac{g}{\delta}}$$

Substituting the value of $\sqrt{\frac{EI}{m . L^4}}$ in equation (16.21), we get

$$f_n = \frac{\pi}{2}\sqrt{\frac{5}{384} \times \frac{g}{\delta}} \qquad ...(16.22)$$

where δ = Static deflection of the shaft.

From equation (16.20), we have used the smallest value of m^* in equation (16.18), and obtained the *lowest or fundamental frequency of vibration* in equation (16.21). By using the second, third ...... values of m^* from equation (16.20) in equation (16.18), the other frequencies of vibrations can be obtained. These frequencies will be 4, 9, 16 ...... etc. times the fundamental frequency respectively. Thus a simply supported shaft will have an infinite number of frequencies. Similarly, the shafts fixed at both ends and cantilevers can be analysed.

(*ii*) *Case of a shaft fixed at both ends* (as shown in Fig. 16.11 (*b*))

The end conditions are :

(*a*) $y = 0$ when $x = 0$ and $x = L$

(*b*) $\frac{dy}{dx} = 0$ when $x = 0$ and $x = L$ (slope is zero at fixed ends)

These end conditions are substituted in equation (16.19) and the solution is obtained.

The static deflection at the centre is given by, $\delta = \frac{m \cdot g \cdot L^4}{384 \cdot E \cdot I}$.

(*iii*) *Case of a Cantilever* (as shown in Fig. 16.11 (*c*))

The end conditions are :

(*a*) $y = 0$ when $x = 0$ (deflection is zero at fixed end)

(*b*) $\frac{dy}{dx} = 0$ at $x = 0$ (slope is zero at fixed end)

(*c*) $\frac{d^2y}{dx^2} = 0$ at $x = L$ (bending moment is zero at free end)

(*d*) $\frac{d^3y}{dx^3} = 0$ at $x = L$ (shear force is zero at free end)

These conditions are substituted in equation (16.19) and the solution is obtained.

The static deflection at the free end is given by, $\delta = \frac{m \cdot g \cdot L^4}{8EI}$.

16.9. Natural Frequency of Transverse Vibrations of a System of Several Loads attached to the Same Shaft

When several loads are attached to the same shaft, the natural frequency of transverse vibrations of the system is obtained by the following two methods :

(*i*) Energy method (or Rayleigh's method). This method gives accurate results.

(*ii*) Dunkerley's method. This method was suggested by Professor Dunkerley. This method is semi-empirical which gives approximate results. This method is used when the diameter of the shaft is uniform.

16.9.1. Energy (or Rayleigh's) method. Fig. 16.12 shows a shaft with negligible mass carrying point loads W_1, W_2, W_3 and W_4 due to masses m_1, m_2, m_3 and m_4 respectively. Let y_1, y_2, y_3 and y_4 be the total deflections under these loads. Let the system is vibrating with angular velocity ω rad/s.

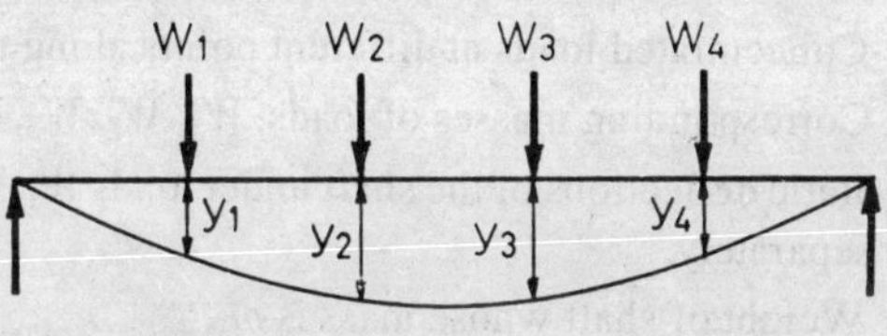

Fig. 16.12

The system possesses the maximum potential energy in the extreme position with no kinetic energy. This means that total energy of the system, when shaft is in extreme position is only potential energy. Similarly the system possesses the maximum kinetic energy when the shaft is in mean position and no potential energy. Hence in mean position, the total energy of the system is only kinetic energy.

For extreme position of the shaft,

$$\begin{aligned}\text{Maximum potential energy} &= \Sigma \text{ Mean load} \times \text{deflection of load}\\ &= \tfrac{1}{2} W_1 \cdot y_1 + \tfrac{1}{2} W_2 \cdot y_2 + \tfrac{1}{2} W_3 \cdot y_3 + \tfrac{1}{2} W_4 \cdot y_4\\ &= \tfrac{1}{2} (m_1 \cdot g) \cdot y_1 + \tfrac{1}{2} (m_2 \cdot g) \cdot y_2 + \tfrac{1}{2} (m_3 \cdot g) \cdot y_3 + \tfrac{1}{2} (m_4 \cdot g) \cdot y_4\\ &= \frac{g}{2} [m_1 y_1 + m_2 y_2 + m_3 y_3 + m_4 y_4]\end{aligned}$$

$$= \frac{g}{2}\,\Sigma my \qquad ...(i)$$

For the mean position of the shaft,

Maximum kinetic energy $= \frac{1}{2} m_1 \times v_1^2 + \frac{1}{2} m_2 \times v_2^2 + \frac{1}{2} m_3 \times v_3^2 + \frac{1}{2} m_4 \times v_4^2$

$$= \frac{1}{2} m_1 \,.\, (\omega \,.\, y_1)^2 + \frac{1}{2} m_2 \,.\, (\omega \,.\, y_2)^2 + \frac{1}{2} m_3 \,.\, (\omega \,.\, y_3)^2 + \frac{1}{2} m_4 \,.\, (\omega \,.\, y_4)^2$$

$$= \frac{1}{2} m_1 \,.\, \omega^2 \,.\, y_1^2 + \frac{1}{2} m_2 \,.\, \omega^2 \,.\, y_2^2 + \frac{1}{2} m_3 \,.\, \omega^2 \,.\, y_3^2 + \frac{1}{2} m_4 \,.\, \omega^2 \,.\, y_4^2$$

$$= \frac{1}{2} \,.\, \omega^2 \,.\, [m_1 y_1^2 + m_2 y_2^2 + m_3 y_3^2 + m_4 y_4^2]$$

$$= \frac{1}{2} \,.\, \omega^2 \,.\, \Sigma my^2 \qquad ...(ii)$$

Equating the maximum potential energy to maximum kinetic energy [*i.e.* equations (*i*) and (*ii*)],

$$\frac{g}{2}\,\Sigma my = \frac{1}{2} \,.\, \omega^2\, \Sigma my^2$$

$$\therefore \quad \omega^2 = \frac{g\,\Sigma\, m \,.\, y}{\Sigma\, m \,.\, y^2}$$

$$\therefore \quad \omega = \sqrt{\frac{g\,\Sigma\, m \,.\, y}{\Sigma\, m \,.\, y^2}}$$

$\therefore$ Natural frequency of transverse vibrations,

$$f_n = \frac{\omega}{2\pi} = \frac{1}{2\pi} \times \sqrt{\frac{g\,\Sigma\, m \,.\, y}{\Sigma\, m \,.\, y^2}} \qquad ...(16.23)$$

16.9.2. Dunkerley's Method. This is a semi-empirical method, which can only be applied to a shaft of uniform diameter.

Let W_1, W_2, W_3, etc. = Concentrated loads at different points along the shaft.

m_1, m_2, m_3, etc. = Corresponding masses of loads, W_1, W_2, W_3, etc.

$\delta_1, \delta_2, \delta_3$, etc. = Static deflections of the shaft under loads W_1, W_2, etc. when each load is acting separately.

W_s = Weight of shaft whose mass is m_s.

δ_s = Maximum deflection of the shaft under its own weight.

$f_{n_1}, f_{n_2}, f_{n_3}$, etc. = Natural frequency of transverse vibrations with each load acting separately

f_{ns} = Natural frequency of transverse vibrations of the shaft under its own weight

f_n = Natural frequency of transverse vibrations of the system as a whole.

Then according to Dunkerley's empirical formula,

$$\frac{1}{f_n^2} = \frac{1}{f_{n_1}^2} + \frac{1}{f_{n_2}^2} + \frac{1}{f_{n_3}^2} + \ldots\ldots + \frac{1}{f_{n_s}^2} \qquad ...(16.24)$$

where $f_{n_1} = \frac{1}{2\pi}\sqrt{\frac{g}{\delta_1}} = \frac{1}{2\pi} \times \sqrt{\frac{9.81}{\delta_1}} = \frac{0.4985}{\sqrt{\delta_1}}$

Similarly $f_{n_2} = \frac{0.4985}{\sqrt{\delta_2}}, f_{n_3} = \frac{0.4985}{\sqrt{\delta_3}}$ and $f_{ns} = \frac{\pi}{2}\sqrt{\frac{5 \times g}{384 \times \delta_s}} = \frac{\pi}{2}\sqrt{\frac{5 \times 9.81}{384 \times \delta_s}} = \frac{0.5615}{\sqrt{\delta_s}}$

Substituting the above values in equation (16.24), we get

$$\frac{1}{f_n^2} = \frac{1}{\left(\frac{0.4985}{\sqrt{\delta_1}}\right)^2} + \frac{1}{\left(\frac{0.4985}{\sqrt{\delta_2}}\right)^2} + \frac{1}{\left(\frac{0.4985}{\sqrt{\delta_3}}\right)^2} + \ldots\ldots + \frac{1}{\left(\frac{0.5615}{\sqrt{\delta_s}}\right)^2}$$

$$= \frac{\delta_1}{(0.4985)^2} + \frac{\delta_2}{(0.4985)^2} + \frac{\delta_3}{(0.4985)^2} + \ldots\ldots + \frac{\delta_s}{(0.5615)^2}$$

$$= \frac{1}{(0.4985)^2}\left[\delta_1 + \delta_2 + \delta_3 + \ldots\ldots \delta_s \times \frac{(0.4985)^2}{(0.5615)^2}\right]$$

$$= \frac{1}{(0.4985)^2}\left[\delta_1 + \delta_2 + \delta_3 + \ldots\ldots \frac{\delta_s}{\left(\frac{0.5615}{0.4985}\right)^2}\right]$$

$$= \frac{1}{(0.4985)^2}\left[\delta_1 + \delta_2 + \delta_3 + \ldots\ldots + \frac{\delta_s}{1.27}\right]$$

$$\therefore \quad f_n = \frac{0.4985}{\sqrt{\delta_1 + \delta_2 + \delta_3 + \ldots\ldots + \frac{\delta_s}{1.27}}} \quad \ldots(16.25)$$

Note. 1. The deflection of the shaft under the load when each load acts separately is given by,

$$\delta = \frac{W \cdot a^2 \cdot b^2}{3EIL}$$

where a and b = Distances of the load W from the two supports

L = Length of the shaft,

E = Young's modulus, and

I = Moment of inertia of the shaft (second moment of area of the shaft).

2. If the mass of the shaft is negligible (or there is no uniformly distributed load on the shaft), then $\delta_s = 0$.

Problem 16.4. *Find the frequency of the transverse vibrations of a shaft which is simply supported at the ends and is of 40 mm in diameter and 2.5 m in length. The shaft carries three point loads of masses 30 kg, 70 kg and 45 kg at 0.5 m, 1 m and 1.7 m respectively from the left support. The Young's modulus for the material of the shaft is 200 GN/m². Neglect the weight of the shaft.*

Sol. Given :

d = 40 mm = 0.04 m ; L = 2.5 m ; m_1 = 30 kg or $W_1 = m_1 \cdot g = 30 \times 9.81$ N ; m_2 = 70 kg or $W_2 = 70 \times 9.81$ N ; m_3 = 45 kg or $W_3 = 45 \times 9.81$ N ; E = 200 GN/m² = 200×10^9 N/m². Weight of shaft is negligible.

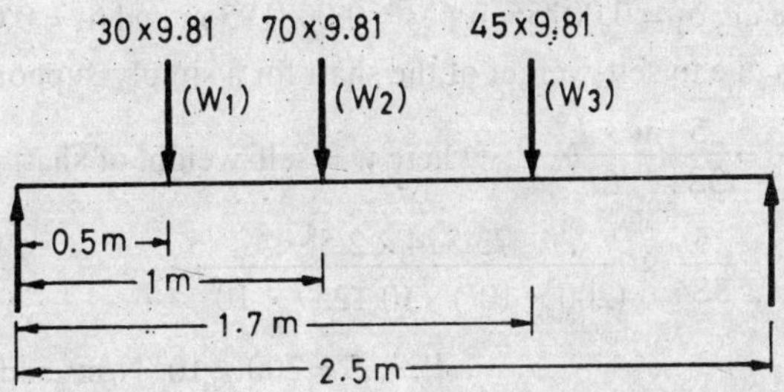

Fig. 16.13

The moment of inertia of the shaft,

$$I = \frac{\pi}{64} \times d^4 = \frac{\pi}{64} \times (0.04)^4 = 0.1257 \times 10^{-6} \text{ m}^4$$

Let us now find the static deflections under the loads when each load acts separately. In each case

a = Distance of the load from left support and b = distance of the same load from right support.

∴ Static deflection under load W_1 (*i.e.* 30 × 9.81 N),

$$\delta_1 = \frac{W_1 \cdot a^2 \cdot b^2}{3EIL} \qquad \text{(Here } a = 0.5 \text{ m and } b = 2 \text{ m)}$$

$$= \frac{(30 \times 9.81) \times 0.5^2 \times 2^2}{3 \times (200 \times 10^9) \times (0.1257 \times 10^{-6}) \times 2.5} = 0.00156 \text{ m}$$

Similarly static deflection under load W_2 (*i.e.* 70 × 9.81 N),

$$\delta_2 = \frac{W_2 \cdot a^2 \cdot b^2}{3EIL} \qquad \text{(Here } a = 1 \text{ m and } b = 1.5 \text{ m)}$$

$$= \frac{70 \times 9.81 \times 1^2 \times 1.5^2}{3 \times (200 \times 10^9) \times (0.1257 \times 10^{-6}) \times 2.5} = 0.008195 \text{ m}$$

and static deflection under load W_3 (*i.e.* 45 × 9.81 N),

$$\delta_3 = \frac{W_3 \cdot a^2 \cdot b^2}{3EIL} \qquad \text{(Here } a = 1.7 \text{ m and } b = 2.5 - 1.7 = 0.8 \text{ m)}$$

$$= \frac{(45 \times 9.81) \times 1.7^2 \times 0.8^2}{3 \times (200 \times 10^9) \times (0.1257 \times 10^{-6}) \times 2.5} = 0.00433 \text{ m}$$

As the weight of the shaft is negligible, hence static deflection due to self-weight of the shaft is zero *i.e.* $\delta_s = 0$.

The frequency of transverse vibrations of the shaft as a whole is given by equation (16.25) as

$$f_n = \frac{0.4985}{\sqrt{\delta_1 + \delta_2 + \delta_3 + \ldots\ldots \dfrac{\delta_s}{1.27}}}$$

$$= \frac{0.4985}{\sqrt{0.00156 + 0.008195 + 0.00433}} = \mathbf{4.2\ Hz.}\quad \textbf{Ans.} \qquad (\because \text{ Here } \delta_s = 0)$$

Problem 16.5. *If in Problem 16.4, the weight of the shaft is not negligible but it is given as 73.575 N per metre length. All other data remains the same, then find the frequency of the transverse vibrations of the shaft.*

Sol. Given : Weight of shaft per metre length, w = 73.575 N.

As the data in this problem is same as given in problem 16.4. Hence the static deflections due to loads W_1, W_2 and W_3 will be same *i.e.* δ_1 = 0.00156 m ; δ_2 = 0.008195 m and δ_3 = 0.00433 m.

The static deflection due to self-weight of the shaft for a simply-supported shaft is given by,

$$\delta_s = \frac{5}{384} \frac{w \times L^4}{EI} \qquad \text{where } w = \text{self-weight of shaft per metre length} = 73.575 \text{ N}$$

$$= \frac{5}{384} \times \frac{73.574 \times 2.5^4}{(200 \times 10^9) \times (0.1257 \times 10^6)}$$

$$[\because E = 200 \times 10^9 \text{ N/m}^2 \text{ and } I = 0.1257 \times 10^{-6} \text{ m}^4 \text{ (given)}]$$

$$= 0.00149 \text{ m}$$

Now using equation (16.25), we have

$$f_n = \frac{0.4985}{\sqrt{\delta_1 + \delta_2 + \delta_3 + \dfrac{\delta_s}{1.27}}} = \frac{0.4985}{\sqrt{0.00156 + 0.008195 + 0.00433 + \dfrac{0.00149}{1.27}}}$$

$$= \mathbf{4.03 \simeq 4\ Hz.}\quad \textbf{Ans.}$$

16.10. Whirling Speeds or Critical Speeds

The centre of gravity of a loaded shaft will always be displaced from axis of rotation, although the amount of displacement may be very small.

Fig. 16.14 (*a*) shows a horizontal shaft whose C.G. coincides with axis of rotation whereas the Fig. 16.14 (*b*) shows a loaded shaft *i.e.* a shaft fixed with a rotor. The centre of gravity of the loaded shaft *i.e.* the C.G. of the shaft and rotor is displaced from the axis of rotation. This displacement is shown by '*h*'. When the shaft begins to rotate, the centre of gravity of the loaded shaft is subjected to radially outwards centrifugal force which is proportional to *h*. This radially outward force bends the shaft in the direction of initial displacement which increases the displacement of the centre of gravity further from the axis of rotation as shown in Fig. 16.14 (*c*). This new centre of gravity [*i.e.* point *G* in Fig. 16.14 (*c*) which is at a distance of '$y + h$'] will be subjected to increased outward centrifugal force* (proportional to $y + h$), which will again bend the shaft and will again increase the displacement of the centre of gravity further from the axis of rotation. In this way the effect is therefore cumulative and ultimately the shaft may even fail. The bending of shaft depends upon : (*i*) initial displacement of the C.G. of the loaded shaft from the axis of rotation and (*ii*) speed at which the shaft rotates.

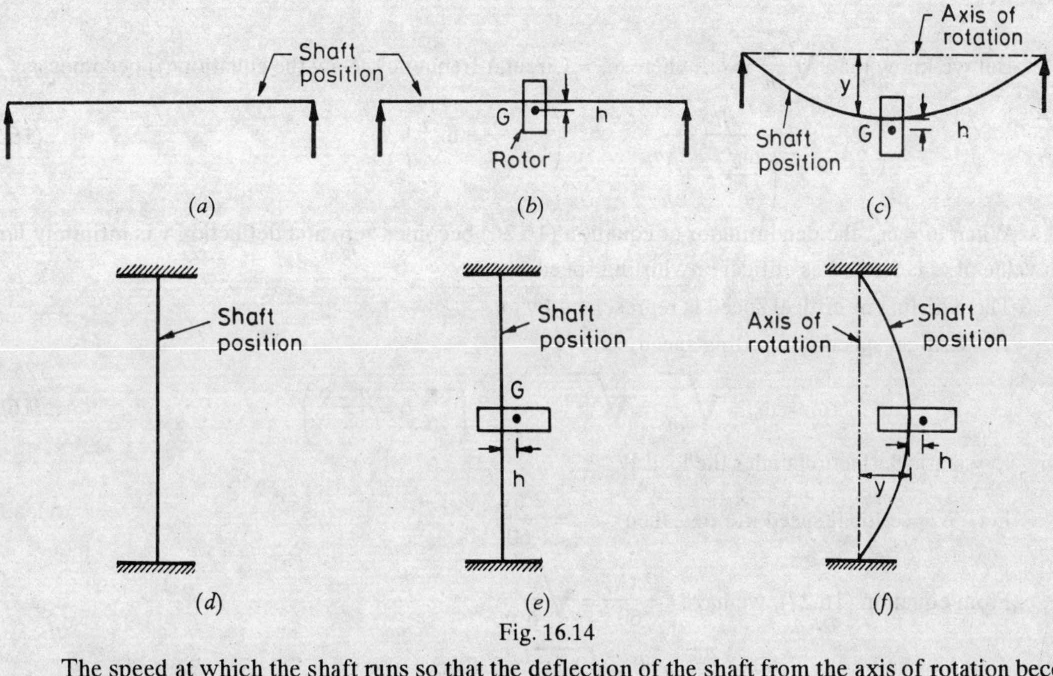

Fig. 16.14

The speed at which the shaft runs so that the deflection of the shaft from the axis of rotation becomes infinite, is known as *whirling speed or critical speed or whipping speed.* At the critical speed, the shaft tends to vibrate violently in the transverse direction.

*When shaft begins to rotate, initial, centrifugal force is proportional to *h*. After that for the new centre of gravity, the centrifugal force will be proportional to '$y + h$' and so on.

Let W = Weight of the rotor attached to a shaft of negligible mass,

m = Mass of the rotor = $\frac{W}{g}$

h = Initial displacement of the C.G. of the shaft and rotor from the axis of rotation,

ω = Angular velocity of the shaft,

s = Stiffness of the shaft *i.e.* force required to produce unit deflection,

y = Additional displacement of the C.G. from the axis of rotation due to centrifugal force, when the shaft rotates at a uniform speed of ω rad/s.

Now the centrifugal force through G, acting radially outwards

$$= m \cdot \omega^2 \cdot (y + h)$$

This is balanced by the inward elastic pull exerted by the shaft which behaves like a spring. Hence the force resisting the deflection y

$$= s \cdot y$$

For equilibrium, $m \cdot \omega^2 \cdot (y + h) = s \cdot y$

or $$m \cdot \omega^2 \cdot y + m \cdot \omega^2 \cdot h = s \cdot y$$

or $$m \cdot \omega^2 \cdot h = s \cdot y - m \cdot \omega^2 \cdot y = y(s - m \cdot \omega^2)$$

or $$y = \frac{m \cdot \omega^2 \cdot h}{(s - m \cdot \omega^2)} = \frac{h}{\left(\frac{s}{m \cdot \omega^2} - \frac{m \cdot \omega^2}{m \cdot \omega^2}\right)} = \frac{h}{\left(\frac{s}{m \cdot \omega^2} - 1\right)} \quad ...(i)$$

But we know that, $\sqrt{\frac{s}{m}} = \omega_n$ where ω_n = Circular frequency. Now the equation (*i*) becomes as

$$y = \frac{h}{\left(\frac{\omega_n^2}{\omega^2} - 1\right)} \qquad \left(\because \frac{s}{m} = \omega_n^2\right) \quad ...(16.26)$$

When $\omega = \omega_n$, the denominator of equation (16.26) becomes zero and deflection y is infinitely large. This value of ω is termed as critical or whirling speed.

The whirling or critical speed is represented by ω_c.

∴ Critical speed or whirling speed,

$$\omega_c = \omega_n = \sqrt{\frac{s}{m}} = \sqrt{\frac{g}{\delta}} \qquad \left(\because \delta = \frac{m \cdot g}{s}\right) \quad ...(16.27)$$

where δ = static deflection under the load W

Let N_c = critical speed in r.p.s., then $\omega_c = \frac{2\pi N_c}{60}$

From equation (16.27), we have $\frac{2\pi N_c}{60} = \sqrt{\frac{g}{\delta}}$

or $$N_c = \frac{60}{2\pi}\sqrt{\frac{g}{\delta}} = \frac{60}{2\pi}\sqrt{\frac{9.81}{\delta}} = \frac{0.4985}{\sqrt{\delta}} \quad ...(16.28)$$

Note. (*i*) A shaft supported in short bearings is assumed to be simply supported whereas a shaft in long bearings is assumed to be fixed at both ends.

(*ii*) The critical speed in r.p.s. of a shaft which carries point loads or uniformly distributed load or combination of both, is equal to the frequency of transverse vibration.

Problem 16.6. *A shaft is simply supported at the ends and is of 20 mm in diameter and 600 mm in length. The shaft carries a load of 19.62 N at its centre. The weight of shaft per metre length is 248.2 N. Find the critical speed of the shaft. Take Young's modulus = 200 GN/m².*

Sol. Given :

d = 20 mm = 0.02 m ; L = 600 mm = 0.6 m ; load at centre, W = 19.62 N, the weight of shaft per metre length, w = 248.2 N ; E = 200 GN/m² = 200 × 10^9 N/m²

We know that the critical speed in r.p.s. of a shaft which carries point loads or uniformly distributed load or combination of both, is equal to frequency of transverse vibration in Hz. The frequency of transverse vibration of the shaft may be obtained by using Dunkerley's method. Refer to Fig. 16.15.

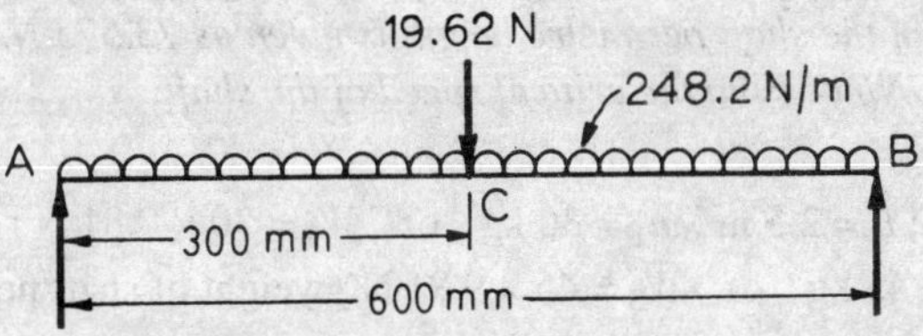

Fig. 16.15

The frequency of transverse vibrations is given by equation (16.25), as

$$f_n = \frac{0.4985}{\sqrt{\delta_1 + \delta_2 + \ldots\ldots + \frac{\delta_s}{1.27}}} \quad \ldots(i)$$

In this problem, there is only one point load. The weight of the shaft per metre length is also given. Hence the above equation can be written as

$$f_n = \frac{0.4985}{\sqrt{\delta_1 + \frac{\delta_s}{1.27}}} \quad \ldots(ii)$$

Now let us find the values of δ_1 and δ_s.

δ_1 = static deflection due to point load at the centre

$$= \frac{WL^3}{48EI} \quad \text{(For a simply supported shaft)}$$

$$= \frac{19.62 \times 0.6^3}{48 \times (200 \times 10^9) \times 7.855 \times 10^{-9}}$$

$\left(\because \; W = 19.62 \text{ N} ; L = 0.6 \text{ m} ; E = 200 \times 10^9 \text{ N/m}^2 \text{ and} \right.$

$\left. I = \text{Moment of inertia of shaft} = \frac{\pi}{64} \times d^4 = \frac{\pi}{64} \times (0.02)^4 = 7.855 \times 10^{-9} \text{ m}^4 \right)$

$= 0.000056$ m

and δ_s = static deflection at the centre due to weight of the shaft

$$= \frac{5}{384} \times \frac{w \times L^4}{EI} \quad \text{(Simply supported shaft with uniformly distributed load)}$$

$$= \frac{5}{384} \times \frac{248.2 \times 0.6^4}{(200 \times 10^9) \times (7.855 \times 10^{-9})} \quad (\because \; w = 248.2 \text{ N/m} ; I = 7.855 \times 10^{-9} \text{ m}^4)$$

$= 0.000266$ m

Substituting the values of δ_1 and δ_s is equation (*ii*), we get

$$f_n = \frac{0.4985}{\sqrt{0.000056 + \dfrac{0.000266}{1.27}}} = 30.6 \text{ Hz}$$

The critical speed in r.p.s. of a shaft is equal to frequency of transverse vibrations in Hz.

∴ Critical speed of shaft, $N_c = f_n = 30.6$ r.p.s. $= 30.6 \times 60$ r.p.m. $= 1836$ r.p.m.

∴ N_c **= 1836 r.p.m. Ans.**

Problem 16.7. *A shaft is simply supported at its ends and is of 40 mm in diameter and 2.5 m in length. The shaft carries three point loads of masses 30 kg, 70 kg and 45 kg at 0.5 m, 1 m and 1.7 m respectively from the left support. The weight of the shaft per metre length is given as 73.575 N. The Young's modulus for the material of the shaft is 200 GN/m². Find the critical speed of the shaft.*

Sol. Given :

$d = 40$ mm $= 0.04$ m ; $L = 2.5$ m ; $m_1 = 30$ kg or $W_1 = 30 \times 9.81$ N ; $m_2 = 70$ kg

or $W_2 = 70 \times 9.81$ N ; $m_3 = 45$ kg or $W_3 = 45 \times 9.81$ N ; weight of shaft per metre length $w = 73.575$ N ; $E = 200$ GN/m² $= 200 \times 10^9$ N/m².

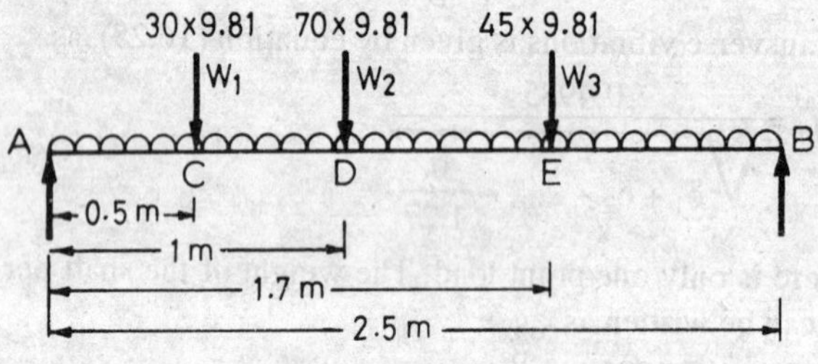

Fig. 16.16

This problem is the combination of problem 16.4 and 16.5.

The critical speed in r.p.s. of the given shaft will be equal to the frequency of transverse vibrations of the shaft in Hz.

The transverse vibrations of the shaft is given by equation (16.25) as

$$f_n = \frac{0.4985}{\sqrt{\delta_1 + \delta_2 + \delta_3 + \dfrac{\delta_s}{1.27}}} \quad ...(i)$$

where δ_1 = static deflection due to point load W_1

$$= \frac{W_1 \cdot a^2 \cdot b^2}{3EIL} \quad \left(\text{Here } a = 0.5 \text{ m} ; b = 2.5 - 0.5 = 2 \text{ m and } I = \frac{\pi}{64} d^4 = \frac{\pi}{64} \times (0.04)^4 = 0.1257 \times 10^{-6} \text{ m}^4\right)$$

$$= \frac{(30 \times 9.81) \times 0.5^2 \times 2^2}{3 \times (200 \times 10^9) \times (0.1257 \times 10^{-6}) \times 2.5} = 0.00156 \text{ m}$$

The values of δ_2 and δ_3 are calculated in Problem 16.4. These values are $\delta_2 = 0.008195$ m and $\delta_3 = 0.00433$ m.

The value of δ_s is calculated in Problem 16.5. This value is $\delta_s = 0.00149$ m.

Substituting the values of δ_1, δ_2, δ_3 and δ_s in equation (*i*), we get

$$f_n = \frac{0.4985}{\sqrt{0.00156 + 0.008195 + 0.00433 + \dfrac{0.00149}{1.27}}} = 4.03 \text{ Hz}$$

$\therefore$ The critical speed of the shaft in r.p.s. $= f_n = 4.03$ r.p.s. $= 4.03 \times 60$ r.p.m. $= 241.8$ r.p.m.

$\therefore$ Critical speed, $N_c =$ **241.8 r.p.m. Ans.**

Problem 16.8. *A vertical shaft is held in long bearings and a disc is attached to the shaft at its mid-point. The centre of gravity of the disc does not coincide with the axis of the shaft. Determine :*

(i) the critical speed of the shaft and

(ii) the range of the speed over which it is unsafe to run the shaft.

The diameter of the shaft is 15 mm and the span of the shaft between the bearings is 1 metre. The mass of the disc is 10 kg and the centre of gravity of the disc is 0.30 mm from the axis of the shaft. Take E = 200 GN/m² and permissible stress in the shaft material is 70 × 10⁶ N/m².

Sol. Given :

The shaft is held in long bearings hence it is assumed to be fixed at both ends.

$d = 15$ mm $= 0.015$ m ; $L = 1$ m ; $m = 10$ kg or $W = 10 \times 9.81$ N ; initial displacement of the C.G. of disc from axis of shaft, $h = 0.30$ mm $= 0.0003$ m ; $E = 200$ GN/m² $= 200 \times 10^9$ N/m² ; $\sigma = 70 \times 10^6$ N/m².

(i) Critical speed of the sheft (N_c)

The critical speed of shaft in r.p.s. is equal to the natural frequency of transverse vibrations in Hz. The natural frequency of transverse vibrations of the shaft which carries a point load is given by,

$$f_n = \frac{0.4985}{\sqrt{\delta}} \qquad ...(i)$$

where δ = static deflection due to static load (W)

$$= \frac{WL^3}{192EI}$$ (the shaft is assumed to be fixed at both ends. The load W is acting at the centre. Hence δ is the deflection at the centre)

$$= \frac{(10 \times 9.81) \times 1^3}{192 \times (200 \times 10^9) \times (2.5 \times 10^{-9})}$$ $\left(\because W = 10 \times 9.81 \text{ N}, L = 1 \text{ m and } I = \text{Moment of inertia of shaft} = \frac{\pi}{64} d^4 = \frac{\pi}{64} \times (0.015)^4 = 2.5 \times 10^{-9} \text{ m}^4\right)$

$= 0.00102$ m

Substituting this value in equation (*i*), we get

$$f_n = \frac{0.4985}{\sqrt{0.00102}} = 15.6 \text{ Hz}$$

$\therefore$ Critical speed of shaft, $N_c = f_n = 15.6$ r.p.s. $= 15.6 \times 60$ r.p.m. $=$ **936 r.p.m. Ans.**

(ii) The range of the speed

Let N_1 = Minimum speed and

N_2 = Maximum speed.

Then range of speed is from N_1 to N_2.

When the shaft starts rotating, the shaft is subjected to the additional, dynamic load. This dynamic load is obtained from the bending equation

$$\frac{M}{I} = \frac{\sigma}{y^*} \qquad ...(ii)$$

where M = Bending moment due to additional dynamic load

$= \frac{W_1 \times L}{8}$ where W_1 = Additional dynamic load at the centre and

$\frac{W_1 \times L}{8}$ = Max. bending moment for a shaft fixed at both ends and carrying a point load W_1 at the centre.

I = M.O.I. of shaft

$$= \frac{\pi}{64} \times d^4 = \frac{\pi}{64}(0.015)^4 = 2.5 \times 10^{-9} \text{ m}^4$$

σ = Maximum permissible stress in the shaft material

$$= 70 \times 10^6 \text{ N/m}^2$$

$$y^* = \frac{d}{2} = \frac{0.015}{2} = 0.0075 \text{ m}$$

Substituting all the values in equation (*ii*), we get

$$\frac{\left(\frac{W_1 \times L}{8}\right)}{2.5 \times 10^{-9}} = \frac{70 \times 10^6}{0.0075}$$

$$\therefore \quad W_1 = \frac{70 \times 10^6 \times 2.5 \times 10^{-9} \times 8}{L \times 0.0075} = \frac{70 \times 2.5 \times 8 \times 10^{-3}}{1 \times 0.0075} \qquad (\because \; L = 1 \text{ m})$$

$$= 186.66 \text{ N}$$

$\therefore$ Additional deflection due to load W_1

$$= \frac{\delta}{W} \times W_1$$

$$\left(\because \; \text{When load is } W, \text{ deflection} = \delta, \text{ When load is 1, deflection} = \frac{\delta}{W}\right.$$

$$\left.\text{When load is } W_1, \text{ deflection} = \frac{\delta}{W} \times W_1\right)$$

$$= \frac{0.00102}{10 \times 9.81} \times 186.66 = 1.94 \times 10^{-3} \text{ m}$$

Additional deflection is represented by 'y'.

$\therefore$ Additional deflection, $y = 1.94 \times 10^{-3}$ m

But additional deflection is given by equation (16.26) as

$$y = \frac{\pm h}{\left(\frac{\omega_n^2}{\omega^2} - 1\right)}$$

or

$$1.94 \times 10^{-3} = \frac{\pm 0.0003}{\left(\frac{\omega_n^2}{\omega^2} - 1\right)} \qquad (\because \; h = 0.0003 \text{ m})$$

$$= \frac{\pm 0.0003}{\left(\frac{\omega_c^2}{\omega^2} - 1\right)} \qquad (\because \; \omega_n = \omega_c)$$

$$= \frac{\pm 0.0003}{\left(\frac{N_c^2}{N^2} - 1\right)} \qquad \left(\because \; \frac{\omega_c^2}{\omega^2} = \frac{(2\pi N_c)^2}{(2\pi N)^2} = \frac{N_c^2}{N^2}\right)$$

or $$\left(\frac{N_c^2}{N^2}-1\right)=\frac{\pm 0.0003}{1.94\times 10^{-3}}=\pm 0.1546$$

or $$\frac{N_c^2}{N^2}=1\pm 0.1546=1.1546 \quad \text{or} \quad 0.8454$$

If $\frac{N_c^2}{N^2}=1.1546$, then $N^2=\frac{N_c^2}{1.1546}$ or $N=\frac{N_c}{\sqrt{1.1546}}$

If $\frac{N_c^2}{N^2}=0.8454$, then $N=\frac{N_c}{\sqrt{0.8454}}$

$\therefore$ Maximum speed, $$N_2=\frac{N_c}{\sqrt{0.8454}}=\frac{936}{\sqrt{0.8454}}=1018 \text{ r.p.m.} \qquad (\because N_c=936 \text{ r.p.m.})$$

and minimum speed, $$N_1=\frac{N_c}{\sqrt{1.1546}}=\frac{936}{\sqrt{1.1546}}=871 \text{ r.p.m.}$$

$\therefore$ Range of unsafe speed is from **871 r.p.m. to 1018 r.p.m. Ans.**

16.11. Damped Vibrations

The vibrations in which the amplitude of vibrations of a vibrating body gradually diminishes, are known as damped vibrations. This is due to the fact that a certain amount of energy is dissipated in overcoming frictional resistance of the medium in which the body vibrates. Hence the vibrations die out after sometime. These vibrations are said to be damped.

The word damping means the diminishing of vibrations with time. The resistance to the motion of the body is provided partly by the medium in which the body vibrates, partly by the internal friction and partly in some cases by a dashpot or other external damping device.

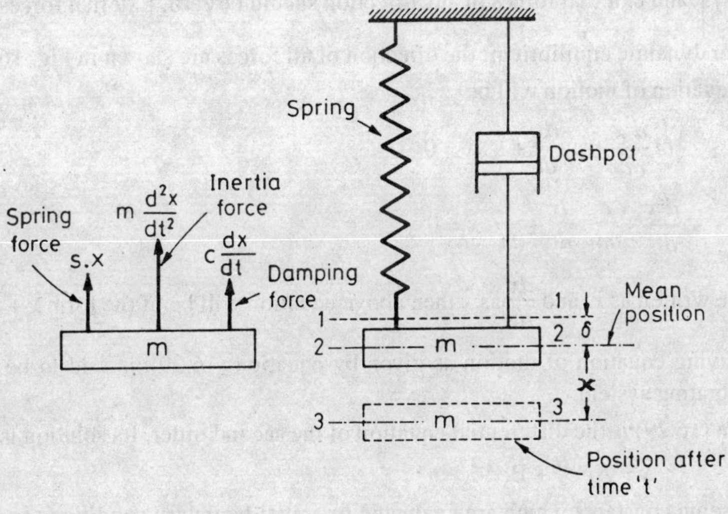

Fig. 16.17

Fig. 16.17 shows a spiral spring, whose one end is fixed and other end is free. A body of mass m is suspended from the free end of the spring. The horizontal line 1-1 shows the level of free end of the spring before mass m is suspended whereas the horizontal line 2-2 shows the level of static equilibrium under the weight of the body. A dashpot is provided between the mass and rigid support. Hence mass is attached to the dashpot as shown in Fig. 16.17.

Let s = stiffness of the spring *i.e.* force required to produce unit extension of the spring.

δ = static deflection of spring under weight of the body

c = damping co-efficient or the damping force per unit velocity (it is assumed that damping force is proportional to the velocity of vibration. Damping force is the resistance to the motion of the body)

x = displacement of the mass from mean position at time 't'

t = time taken by the body when it is displaced by distance x from mean position.

Let the body is displaced by a distance x downward from mean position in time t. Then the forces acting on the body are :

(*i*) Accelerating force in the direction of motion (*i.e.* downwards)

$$= \text{Mass} \times \text{Acceleration} = m \times \frac{d^2x}{dt^2}$$

(*ii*) Damping force (or frictional force) in the opposite direction of motion (*i.e.* upwards)

= (damping force per unit velocity) × velocity of vibration

$$= c \times \frac{dx}{dt} \qquad \left(\because \text{ Velocity} = \frac{dx}{dt}\right)$$

(*iii*) Spring force on the mass acting in the opposite direction of motion (*i.e.* upwards)

= stiffness of spring × displacement from mean position

$= s \,.\, x$

For the dynamic equilibrium of the body, the sum of inertia* force $\left[-\text{accelerating force} = -m\frac{d^2x}{dt^2} = m \times \frac{d^2x}{dt^2} \text{(upwards)}\right]$ and external forces in any direction should be zero. External forces are damping force and spring force. For dynamic equilibrium, the direction of all forces are shown in Fig. 16.17 (*b*).

Hence the equation of motion will be

$$\therefore \qquad m\frac{d^2x}{dt^2} + c\,.\frac{dx}{dt} + s\,.\,x = 0$$

or

$$\frac{d^2x}{dt^2} + \frac{c}{m}\,.\,\frac{dx}{dt} + \frac{s}{m}\,.\,x = 0 \qquad \text{...(16.29)}$$

$\left(\frac{d^2x}{dt^2} \text{ may be written as } \ddot{x}, \text{ and } \frac{dx}{dt} \text{ as } \dot{x} \text{ then above equation will be of the form } \ddot{x} + \frac{c}{m}\dot{x} + \frac{s}{m}x = 0\right)$

A system having equation of motion as given by equation (16.29), is said to be a single degree of freedom damped vibrating system.

The equation (16.29) is the differential equation of the second order. Its solution is of the form

$$x = \text{A}e^{\alpha_1 t} + \text{B}e^{\alpha_2 t} \qquad \text{...(16.30)}$$

where A and B are some constants which are evaluated by initial boundary conditions and α_1 and α_2 are the roots of the auxiliary equation

$$\alpha^2 + \frac{c}{m}\alpha + \frac{s}{m} = 0 \qquad \text{...(16.31)}$$

*The magnitude of inertia force is same as of accelerating force. But inertia force acts in the opposite direction. Hence here inertia force will act in upward direction.

The values of roots are

$$\alpha = \frac{-\frac{c}{m} \pm \sqrt{\left(\frac{c}{m}\right)^2 - 4 \,.\, 1 \,.\left(\frac{s}{m}\right)}}{2 \times 1} = -\frac{c}{2m} \pm \sqrt{\left(\frac{c}{2m}\right)^2 - \left(\frac{s}{m}\right)}$$

or $$\alpha_1 = -\frac{c}{2m} + \sqrt{\left(\frac{c}{2m}\right)^2 - \left(\frac{s}{m}\right)} \text{ and } \alpha_2 = -\frac{c}{2m} - \sqrt{\left(\frac{c}{2m}\right)^2 - \left(\frac{s}{m}\right)} \quad ...(i)$$

But from equation (16.3), we know that $\sqrt{\frac{s}{m}} = \omega_n$ hence $\left(\frac{s}{m}\right) = \omega_n^2$. ...(*ii*)

Also the ratio of $\left(\frac{c}{2m}\right)^2$ to $\left(\frac{s}{m}\right)$ represents the degree of dampness provided in the system and square root of this ratio is known as damping factor. The damping factor is represented by the symbol ζ.

$$\therefore \text{ Damping factor, } \quad \zeta = \sqrt{\frac{\left(\frac{c}{2m}\right)^2}{\left(\frac{s}{m}\right)}} = \sqrt{\frac{\left(\frac{c}{2m}\right)^2}{\omega_n^2}} \quad \left(\because \frac{s}{m} = \omega_n^2\right)$$

$$= \frac{\left(\frac{c}{2m}\right)}{\omega_n} = \frac{c}{2m \,.\, \omega_n} \quad ...(16.31\text{A})$$

or $$\frac{c}{2m} = \omega_n \,.\, \zeta \quad ...(iii)$$

Substituting the values of $\left(\frac{s}{m}\right)$ and $\left(\frac{c}{2m}\right)$ from equations (*ii*) and (*iii*) respectively in equation (*i*), we get the values of roots as

$$\alpha_1 = -\omega_n \,.\, \zeta + \sqrt{\omega_n^2 \,.\, \zeta^2 - \omega_n^2} \text{ and } \alpha_2 = -\omega_n \,.\, \zeta - \sqrt{\omega_n^2 \,.\, \zeta^2 - \omega_n^2}$$

$$= \omega_n \left(-\zeta + \sqrt{\zeta^2 - 1}\right) \text{ and } \alpha_2 = \omega_n \left(-\zeta - \sqrt{\zeta^2 - 1}\right)$$

or $$\alpha_{1,\,2} = \omega_n \left(-\zeta \pm \sqrt{\zeta^2 - 1}\right) \quad ...(16.32)$$

Depending upon whether ζ is greater than, equal to, or less than one, these roots may be real, or imaginary.

If (*i*) $\zeta > 1$, the roots are real and both the roots are negative

(*ii*) $\zeta < 1$, the roots are imaginary and both roots are complex conjugates

(*iii*) $\zeta = 1$, the roots are real and equal.

Also if (*i*) damping factor $\zeta = 1$, the damping is known as critical.

(*ii*) $\zeta > 1$, the system is over-damped, and

(*iii*) $\zeta < 1$, the system is under-damped.

From equation (*iii*), the damping co-efficient c is given by

$$c = 2m \,.\, \omega_n \,.\, \zeta \quad ...(16.33)$$

When $\zeta = 1$, the damping is known as critical. The corresponding value of damping co-efficient is known as critical damping co-efficient. This value is represented by c_c. Hence the value of c_c is obtained by substituting $\zeta = 1$ in equation (16.33).

$\therefore \qquad c_c = 2m \cdot \omega_n$ ...(16.34)

Dividing equation (16.33) by equation (16.34), we get

$$\frac{c}{c_c} = \zeta$$

$\therefore$ Damping factor, $\zeta = \dfrac{c}{c_c} = \dfrac{\text{Actual damping co-efficient}}{\text{Critical damping co-efficient}}$ ...(16.35)

Hence the damping factor is also defined as the ratio of actual damping co-efficient to critical damping co-efficient.

16.11.1. Expression for displacement for over-damped, under-damped and critical-damped system

The displacement is given by equation (16.30) as

$$x = Ae^{\alpha_1 t} + Be^{\alpha_2 t} \qquad ...(i)$$

where A and B are some constants which are evaluated from initial given conditions and α_1 and α_2 are roots. The values of α_1 and α_2 are given by equation (16.32),

$$\alpha_{1,\,2} = \omega_n \left(-\zeta \pm \sqrt{\zeta^2 - 1}\right) \qquad ...(ii)$$

(*a*) *Over-damped system i.e.* $\zeta > 1$

For over-damped system, the damping factor $\zeta > 1$. The roots α_1 and α_2 will be real as is clear from equation (*ii*). Hence the values of α_1 and α_2 are :

$$\alpha_1 = \omega_n \left(-\zeta + \sqrt{\zeta^2 - 1}\right) \text{ and } \alpha_2 = \omega_n \left(-\zeta - \sqrt{\zeta^2 - 1}\right)$$

Substituting the values of α_1 and α_2 in equation (*i*), we get the displacement as

$$x = Ae^{\omega_n \left(-\zeta + \sqrt{\zeta^2 - 1}\right) \cdot t} + Be^{\omega_n \left(-\zeta - \sqrt{\zeta^2 - 1}\right) \cdot t} \qquad ...(16.36)$$

The above equation is an equation of a periodic motion *i.e.* the system will not vibrate due to over-damping. The resultant displacement approaches to zero with time as shown in Fig. 16.18.

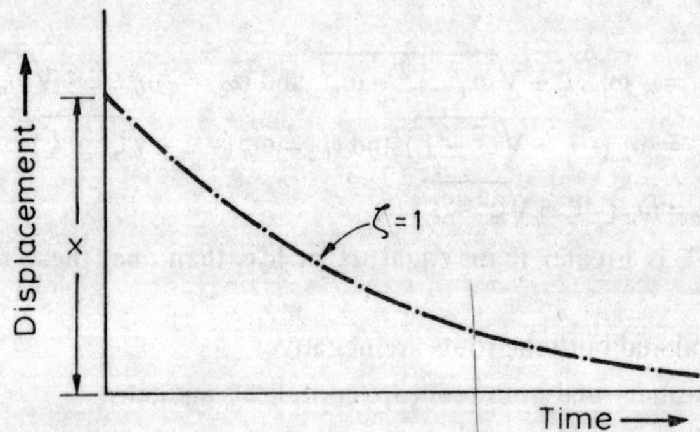

Fig. 16.18

(*b*) *Under-damped system i.e.* $\zeta < 1$

For under-damped system, the damping factor $\zeta < 1$. When this condition is substituted in equation (*ii*) [or equation (16.32)], the roots α_1 and α_2 will be imaginary as the value of $\zeta^2 - 1$ will be negative. The values of α_1 and α_2 are

$$\alpha_1 = \omega_n\left(-\zeta + \sqrt{\zeta^2 - 1}\right) \quad \text{and} \quad \alpha_2 = \omega_n\left(-\zeta - \sqrt{\zeta^2 - 1}\right)$$

$$= \omega_n\left(-\zeta + \sqrt{(-1)(1 - \zeta^2)}\right) \quad \text{and} \quad \alpha_2 = \omega_n\left(-\zeta - \sqrt{(-1)(1 - \zeta^2)}\right)$$

$$= \omega_n(-\zeta + i\sqrt{1-\zeta^2}) \text{ and } \alpha_2 = \omega_n(-\zeta - i\sqrt{1-\zeta^2}) \qquad (\because \sqrt{(-1)} = \text{iota } i)$$

Substituting the values of α_1 and α_2 in equation (i) above [or equation (16.30)], we get the displacement

$$x = Ae^{\omega_n(-\zeta+i\sqrt{1-\zeta^2}).t} + Be^{\omega_n(-\zeta-i\sqrt{1-\zeta^2}).t}$$

$$= Ae^{-\omega_n.\zeta.t+\omega_n.i\sqrt{1-\zeta^2}.t} + Be^{-\omega_n.\zeta.t-\omega_n-i\sqrt{1-\zeta^2}.t}$$

$$= Ae^{-\omega_n.\zeta.t}.e^{\omega_n.i\sqrt{1-\zeta^2}.t} + Be^{-\omega_n.\zeta.t}.e^{-\omega_n.i\sqrt{1-\zeta^2}.t}$$

$$= e^{-\omega_n.\zeta.t}[Ae^{\omega_n i\sqrt{1-\zeta^2}.t} + Be^{-\omega_n.i\sqrt{1-\zeta^2}.t}]$$

Let $\omega_n . \sqrt{1-\zeta^2} = \omega_d$ (Frequency of under-damped vibration) ...(16.36A)

Then the above equation becomes as,

$$x = e^{-\omega_n.\zeta.t}[Ae^{i.\omega_d.t} + Be^{-i.\omega_d.t}] \qquad ...(A)$$

But from the Euler's formula, we know that

$$e^{i.\theta} = \cos\theta + i\sin\theta \text{ and } e^{-i\theta} = \cos\theta - i\sin\theta$$

Hence $e^{i(\omega_d.t)} = \cos(\omega_d.t) + i\sin(\omega_d.t)$ and $e^{-i(\omega_d t)} = \cos(\omega_d.t) - i\sin(\omega_d.t)$

Substituting these values in equation (A), we get

$$x = e^{-\omega_n.\zeta.t}[A(\cos\omega_d.t + i\sin\omega_d.t) + B(\cos\omega_d.t - i\sin\omega_d.t)]$$

$$= e^{-\zeta.\omega_n.t}[(A+B)\cos(\omega_d.t) + i(A-B)\sin(\omega_d.t)] \qquad ...(16.37)$$

$$= e^{-\zeta.\omega_n.t}[C\cos(\omega_d.t) + D\sin(\omega_d.t)] \qquad ...(16.38)$$

where $C = A + B$ and $D = i(A - B)$ ...(16.38A)

Hence the values of C and D can be obtained if values of A and B are known. But the values of A and B are known from initial conditions.

From equation (16.37), we know that

$$x = e^{-\zeta.\omega_n.t}[(A+B)\cos(\omega_d.t) + i(A-B)\sin(\omega_d.t)]$$

Let us put $A + B = X\sin\phi$ and $i(A - B) = X\cos\phi$, then the above equation becomes as

$$x = e^{-\zeta.\omega_n.t}[X\sin\phi\cos(\omega_d.t) + X\cos\phi\sin(\omega_d.t)]$$

$$= Xe^{-\zeta.\omega_n.t}\sin(\omega_d.t + \phi) \qquad ...(16.39)$$

$$[\because \sin\phi\cos(\omega_d.t) + \cos\phi\sin(\omega_d.t)] = \sin(\omega_d.t + \phi)]$$

We have put, $A + B = X\sin\phi$ and $i(A - B) = X\cos\phi$

$$\therefore \quad (A+B)^2 = X^2\sin^2\phi \text{ and } (i)^2(A-B)^2 = X^2\cos^2\phi$$

or $$(A+B)^2 + (i)^2(A-B)^2 = X^2\sin^2\phi + X^2\cos^2\phi = X^2(\sin^2\phi + \cos^2\phi) = X^2$$

$$\therefore \quad X = \sqrt{(A+B)^2 - (A-B)^2} \qquad [\because (i)^2 = (\sqrt{-1})^2 = -1] \qquad ...(16.39A)$$

Also $$\frac{X\sin\phi}{X\cos\phi} = \frac{(A+B)}{i(A-B)}$$

or $$\tan\phi = \frac{(A+B)}{i(A-B)} \quad \text{or} \quad \phi = \tan^{-1}\frac{(A+B)}{i(A-B)} \qquad ...(16.39B)$$

The equation (16.39) is an equation of harmonic motion in which the amplitude is $Xe^{-\zeta.\omega_n.t}$ and angular frequency is ω_d. The factor $e^{-\zeta.\omega_n.t}$ indicates that amplitude goes on decreasing exponentially with time. Fig. 16.18 (a) explains this equation. Hence in case of under-damped system, the system is vibrating

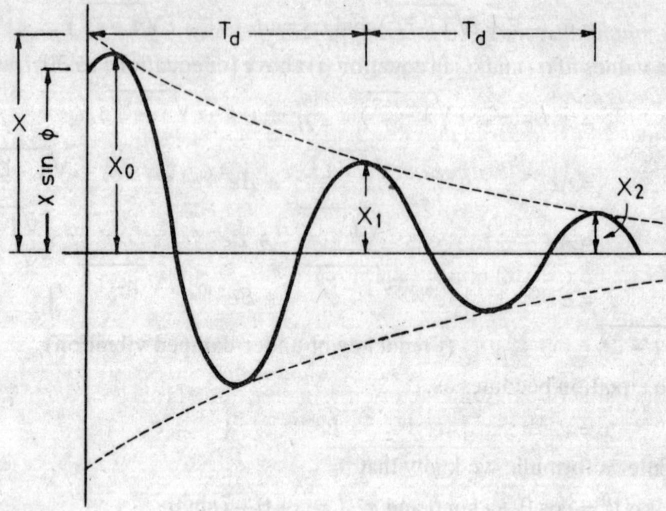

Fig. 16.18 (*a*)

The time period of vibration is given by,

$$T_d = \frac{2\pi}{\omega_d} \qquad \text{where } \omega_d = \text{Angular frequency of under-damped vibrations}$$

Linear frequency is given by, $f_d = \dfrac{1}{T_d} = \dfrac{\omega_d}{2\pi}$

(c) *Critically-damped system i.e.* $\zeta = 1$

Substituting $\zeta = 1$ in equation (16.32), the two roots are $\alpha_{1,\,2} = \omega_n(-1 \pm \sqrt{1^2 - 1}) = \omega_n(-1) = -\omega_n$.

Both the roots are equal. Hence $\alpha_1 = \alpha_2 = -\omega_n$.

Substituting the values of α_1 and α_2 in equation (16.30), we get the displacement, as

$$x = Ae^{-\omega_n \cdot t} + Be^{-\omega_n \cdot t} \quad \text{or} \quad x = (A + B)\, e^{-\omega_n \cdot t} \qquad \text{...(16.40)}$$

The above equation shows that motion is aperiodic. Since $e^{-\omega_n \cdot t}$ approaches zero as $t \to \infty$, the motion will eventually diminish to zero.

Fig. 16.19 shows the variation of displacement with time for over-damped system (*i.e.* $\zeta > 1$), under-damped system (*i.e.* $\zeta < 1$) and critically-damped system (*i.e.* $\zeta = 1$) and undamped system ($\zeta = 0$).

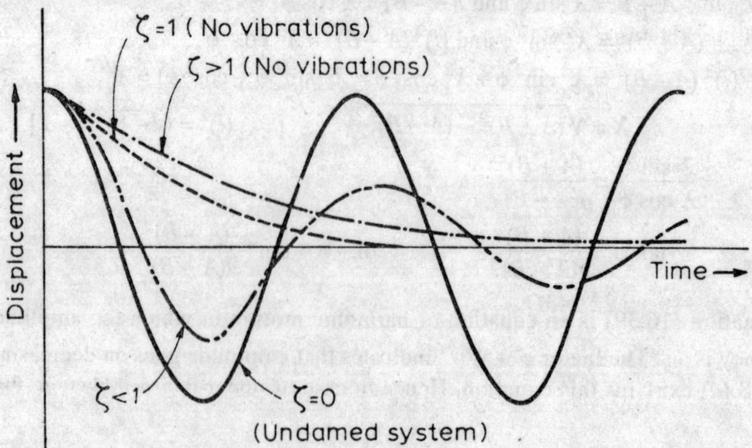

Fig. 16.19

Problem 16.9. *A machine of mass 10 kg is supported on springs and dashpots. The total stiffness of springs is 5 N/mm and total damping is 0.075 N/mm/s. The system is initially at rest and a velocity of 100 mm/s is imparted to the mass. Determine :*

(*i*) *the displacement and velocity of mass as a function of time and*

(*ii*) *the displacement and velocity after one second.*

Sol. Given :

$m = 10$ kg ; $s = 5$ N/mm $= 5 \times 100$ N/m $= 5000$ N/m ; $c = 0.075$ N/mm/s $= 0.075 \times 1000$ N/m/s $=$ 75 N/m/s ; when $x = 0$, $t = 0$ and $v = 100$ mm/s $= 0.1$ m/s

We know that, $\omega_n = \sqrt{\dfrac{s}{m}} = \sqrt{\dfrac{5000}{10}} = 22.36$ rad/s

Let us now find the value of damping factor ζ. Using equation (16.31A), we have

$$\zeta = \frac{c}{2m \cdot \omega_n} = \frac{75}{2 \times 10 \times 22.36} = 0.1677$$

As damping factor is less than one, hence the system is under-damped. The frequency of under-damped vibrations is given by equation (16.36A) as

$$\omega_d = \omega_n \sqrt{1 - \zeta^2} = 22.36 \sqrt{1 - 0.1677^2} = 22.043 \text{ rad/s}$$

(*i*) *Displacement and velocity of mass as a function of time*

The displacement is given by equation (16.38),

$$x = e^{-\zeta \cdot \omega_n \cdot t} [C \cos (\omega_d \cdot t) + D \sin (\omega_d \cdot t)]$$

$$= e^{-0.1677 \times 22.36 \times t} [C \cos (22.043t) + D \sin (22.043t)]$$

$$(\because \ \zeta = 0.1677, \omega_n = 22.36 \text{ and } \omega_d = 22.043)$$

$$= e^{-3.75t} [C \cos (22.043t) + D \sin (22.043t)] \quad ...(i)$$

When $x = 0$, $t = 0$. Hence above equation becomes as

$$0 = e^{-3.75 \times 0} [C \cos 0 + D \sin 0]$$

$$= e^0 [C \times 1 + D \times 0] = 1 \times C = C$$

$$\therefore \quad C = 0$$

Hence substituting the value of C in equation (*i*), we get

$$x = e^{-3.75t} [0 + D \sin (22.043t)]$$

$$= e^{-3.75t} \cdot D \cdot \sin (22.043t) \quad ...(ii)$$

Differentiating equation (*ii*) w.r.t. to t, we get the velocity (v),

$$v = \frac{dx}{dt} = De^{-3.75t} \times 22.043 \cos (22.043t) + D \sin (22.043t) \times (-3.75) \cdot e^{-3.75t}$$

$$= 22.043\, De^{-3.75t} \cdot \cos (22.043t) - 3.75\, D \cdot e^{-3.75t} \sin (22.043t) \quad ...(iii)$$

When $t = 0$, $v = 0.1$ m/s. Hence equation (*iii*) becomes as

$$0.1 = 22.043\, De^0 \cdot \cos (22.043 \times 0) - 3.75\, D \cdot e^0 \sin 0$$

$$= 22.043D \times 1 \times 1 - 3.75D \times 1 \times 0 \quad (\because \ \cos 0 = 1, \sin 0 = 0 \text{ and } e^0 = 1)$$

$$= 22.043D$$

or $$D = \frac{0.1}{22.043} = 4.536$$

Substituting the value of D in equation (*ii*) and (*iii*), we get displacement and velocity as

$$x = e^{-3.75t} \times 4.536 \,.\, \sin 22.043t = \mathbf{4.536\, e^{-3.75t} \,.\, \sin 22.043t}. \quad \textbf{Ans.} \qquad ...(iv)$$

$$v = 22.043 \times 4.536 \,.\, e^{-3.75t} \times \cos 22.043t - 3.75 \times 4.536 \times e^{-3.75t} \sin 22.043t$$

$$= 100 \times e^{-3.75t} \times \cos 22.043t - 17 \times e^{-3.75t} \sin 22.043t$$

$$= \mathbf{e^{-3.75t} (100 \cos 22.043t - 17 \sin 22.043t)}. \quad \textbf{Ans.} \qquad ...(v)$$

(*ii*) *Displacement and velocity after one second*

The displacement as a function of time is given by equation (*iv*) as

$$x = 4.536\, e^{-3.75t} \,.\, \sin 22.043t$$

After one second, $t = 1$, hence the above equation becomes as

$$x = 4.536\, e^{-3.75 \times 1} \,.\, \sin (22.043 \times 1) \qquad (\because\ t = 1)$$

$$= 4.536 \times 0.0235177^* \times 0.3753 = \mathbf{0.04\ mm}. \quad \textbf{Ans.}$$

The velocity as a function of time is given by equation (v) as

$$v = e^{-3.75t} (100 \cos 22.043t - 17 \sin 22.043t)$$

After one second, $t = 1$, hence the above equation becomes as

$$v = e^{-3.75 \times t} [100 \cos (22.043 \times 1) - 17 \sin (22.043 \times 1)]$$

$$= 0.0235177\, [100 \times 0.9269 - 17 \times 0.3753]$$

$$= 0.0235177\, [92.69 - 6.3801] = \mathbf{2.03\ mm/s}. \quad \textbf{Ans.}$$

Problem 16.10. *Find the damping factor of a vibrating system which consists of a mass of 3.5 kg, a spring of stiffness 2.5 N/mm and a damper of damping co-efficient 0.018 N/mm/s. What is the natural frequency of damped vibration ?*

Sol. Given :

$m = 3.5$ kg ; $s = 2.5$ N/mm $= 2.5 \times 1000$ N/m $= 2500$ N/m ;

$c = 0.018$ N/mm/s $= 0.018 \times 1000$ N/m/s $= 18$ N/m/s

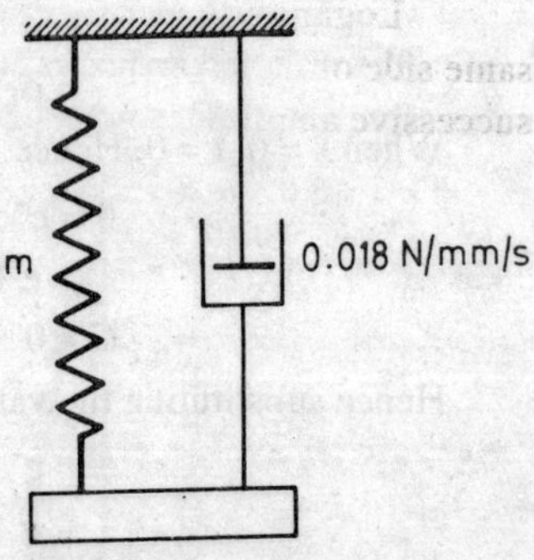

Fig. 16.19 (*a*)

We know that $\quad \omega_n = \sqrt{\dfrac{s}{m}} = \sqrt{\dfrac{2500}{3.5}} = 26.72$ rad/s

The damping factor (ζ) can be obtained by using equation (16.31A).

$$\therefore \qquad \zeta = \frac{c}{2 \,.\, m \,.\, \omega_n} = \frac{18}{2 \times 3.5 \times 26.72}$$

$$= \mathbf{0.0962}. \quad \textbf{Ans.}$$

Let ω_d = Natural frequency of damped vibration.

Using equation (16.36A), we get

$$\omega_d = \omega_n \,.\, \sqrt{1 - \zeta^2} = 26.72 \sqrt{1 - 0.0962^2} = \mathbf{26.596\ rad/s}. \quad \textbf{Ans.}$$

Problem 16.11. *For the problem 16.10, determine :*

(*i*) *the ratio of frequencies of damped and undamped vibrations, and*

(*ii*) *the periodic time of damped vibrations.*

---

*Value of $e = 2.71828$. Hence $e^{-3.75} = 2.71828^{-3.75} = 0.0235177$.

The value of $e^{-3.75}$ can also be obtained by taking – 3.75 on calculator, pressing '1 NV' key and then pressing Ln e^x key.

Sol. Given :

Data from problem 16.10,

$m = 3.5$ kg ; $s = 2500$ N/m ; $c = 0.018$ N/mm/s $= 0.018 \times 1000$ N/m/s $= 18$ N/m/s

$$\omega_n = \sqrt{\frac{s}{m}} = \sqrt{\frac{2500}{3.5}} = 26.72 \text{ rad/s}$$

$$\zeta = \frac{c}{2 \,.\, m \,.\, \omega_n} = \frac{18}{2 \times 3.5 \times 26.72} = 0.0962$$

$$\omega_d = \omega_n \sqrt{1 - \zeta^2} = 26.72 \sqrt{1 - 0.0962^2} = 26.596.$$

(*i*) *Ratio of frequencies of damped and undamped vibrations*

Let ω_d = Frequency of damped vibrations, and

ω_n = Frequency of undamped vibrations

$$\therefore \quad \frac{\text{Damped vibrations frequency}}{\text{Undamped vibrations frequency}} = \frac{\omega_d}{\omega_n} = \frac{\omega_n \sqrt{1 - \zeta^2}}{\omega_n} = \sqrt{1 - \zeta^2}$$

$$= \sqrt{1 - 0.0962^2} = \mathbf{0.99}. \quad \textbf{Ans.}$$

(*ii*) *The periodic time of damped vibrations*

Let T_d = Periodic time of damped vibrations

$$\text{Then} \quad T_d = \frac{2\pi}{\omega_d} = \frac{2\pi}{26.596} = \mathbf{0.236 \text{ s.}} \quad \textbf{Ans.}$$

16.12. Logarithmic Decrement

Logarithmic decrement is defined as the natural logarithm of any two successive amplitudes on the same side of the mean position in an under-damped system. It is denoted by δ. Actually, the ratio of any two successive amplitudes in an under-damped system is always constant.

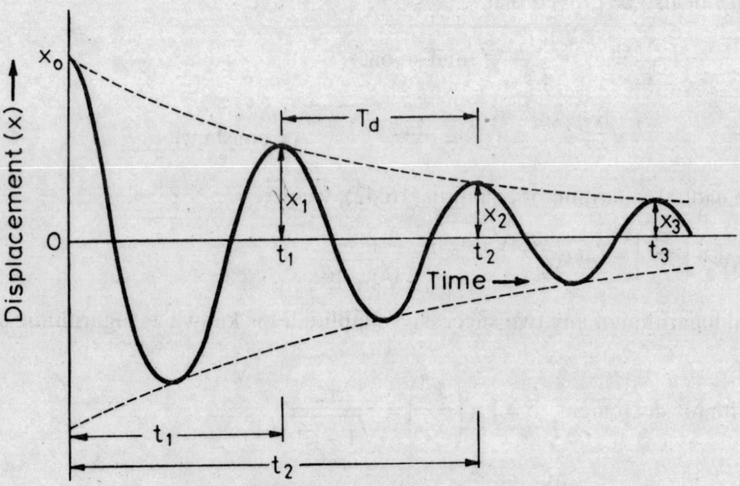

Fig. 16.20

The displacement in an under-damped system is given by equation (16.39) as

$$x = Xe^{-\zeta \,.\, \omega_n \,.\, t} \sin(\omega_d \,.\, t + \phi) \qquad \text{...}(i)$$

This equation is an equation of harmonic motion in which $Xe^{-\zeta \cdot \omega_n \cdot t}$ is the amplitude and ω_d is the angular frequency. Maximum amplitude in a cycle will occur when $\sin(\omega_d \cdot t + \phi)$ is equal to unity. Also the factor $e^{-\zeta \cdot \omega_n \cdot t}$ indicates that amplitude will go on decreasing exponentially with time. Fig. 16.20 shows the shape of decaying curve for this type of motion.

Maximum amplitude, $x = X \cdot e^{-\zeta \cdot \omega_n \cdot t}$ ...(*ii*)

Let x_1 = Max. amplitude when time is t_1 and

x_2 = Max. amplitude when time is t_2

Substituting these values in equation (*ii*), we get

$$x_1 = X \cdot e^{-\zeta \cdot \omega_n \cdot t_1} \text{ and } x_2 = X \cdot e^{-\zeta \cdot \omega_n \cdot t_2}$$

Hence the ratio of two successive amplitudes,

$$\frac{x_1}{x_2} = \frac{X \cdot e^{-\zeta \cdot \omega_n \cdot t_1}}{X \cdot e^{-\zeta \cdot \omega_n \cdot t_2}} = \frac{e^{-\zeta \cdot \omega_n \cdot t_1}}{e^{-\zeta \cdot \omega_n \cdot t_2}} = e^{-\zeta \cdot \omega_n \cdot t_1 - (-\zeta \cdot \omega_n \cdot t_2)}$$

$$= e^{-\zeta \cdot \omega_n \cdot t_1 + \zeta \cdot \omega_n \cdot t_2} = e^{\zeta \cdot \omega_n \cdot (t_2 - t_1)} \qquad \text{...(iii)}$$

where $(t_2 - t_1)$ is the period of oscillation (or time period) or time difference between two consecutive amplitudes

$$\therefore \qquad t_2 - t_1 = T_d = \frac{2\pi}{\omega_d}$$

Hence equation (*iii*) becomes as, $\dfrac{x_1}{x_2} = e^{\zeta \cdot \omega_n \cdot 2\pi/\omega_d}$ ...(16.41)

$$= e^{\zeta \cdot \omega_n \cdot 2\pi/\omega_n \cdot \sqrt{1-\zeta^2}} \qquad (\because \ \omega_d = \omega_n \cdot \sqrt{1-\zeta^2})$$

$$= e^{\zeta \cdot 2\pi/\sqrt{1-\zeta^2}} = e^{2\pi\zeta/\sqrt{1-\zeta^2}} \qquad \text{...(16.42)}$$

Similarly it can also be proved that

$$\frac{x_2}{x_3} = e^{2\pi\zeta/\sqrt{1-\zeta^2}} \text{ and so on.}$$

$$\therefore \qquad \frac{x_1}{x_2} = \frac{x_2}{x_3} = \frac{x_3}{x_4} = \text{......} = e^{2\pi\zeta/\sqrt{1-\zeta^2}} \text{ (= constant)} \qquad \text{...(16.43)}$$

Taking the natural logarithm of equation (16.42), we get

$$\text{Ln}\left(\frac{x_1}{x_2}\right) = \text{Ln } e^{(2\pi\zeta/\sqrt{1-\zeta^2})} = \frac{2\pi\zeta}{\sqrt{1-\zeta^2}}$$

The natural logarithm of any two successive amplitudes is known as logarithmic decrement, which is denoted by δ.

$$\therefore \quad \text{Logarithmic decrement, } \delta = \text{Ln}\left(\frac{x_1}{x_2}\right) = \frac{2\pi\zeta}{\sqrt{1-\zeta^2}} \qquad \text{...(16.43A)}$$

$$\therefore \qquad \delta = \frac{2\pi\zeta}{\sqrt{1-\zeta^2}} \qquad \text{...(16.44)}$$

$$\simeq 2\pi\zeta \qquad \text{if } \zeta \text{ is very very less than 1.} \qquad \text{...(16.45)}$$

Problem 16.12. *Find the logarithmic decrement and the ratio of any two consecutive amplitude of a vibrating system, which consists of a mass of 3.5 kg, a spring of stiffness 2.5 N/mm and a damper of damping co-efficient 0.018 N/mm/s.*

Sol. Given :

$m = 3.5$ kg ; $s = 2.5$ N/mm $= 2.5 \times 1000 = 2500$ N/m ; $c = 0.018$ N/mm/s $= 0.018 \times 1000 = 18$ N/m/s.

(*i*) *Logarithmic decrement* (δ)

Using equation (16.44), we get logarithmic decrement as,

$$\delta = \frac{2\pi\zeta}{\sqrt{1-\zeta^2}} \qquad ...(i)$$

Let us find the value of damping factor ζ. Using equation (16.31A),

$$\zeta = \frac{c}{2 \cdot m \cdot \omega_n}$$

where $\omega_n = \sqrt{\frac{s}{n}} = \sqrt{\frac{2500}{3.5}} = 26.72$ rad/s

$$\therefore \quad \zeta = \frac{18}{2 \times 3.5 \times 26.72} = 0.0962$$

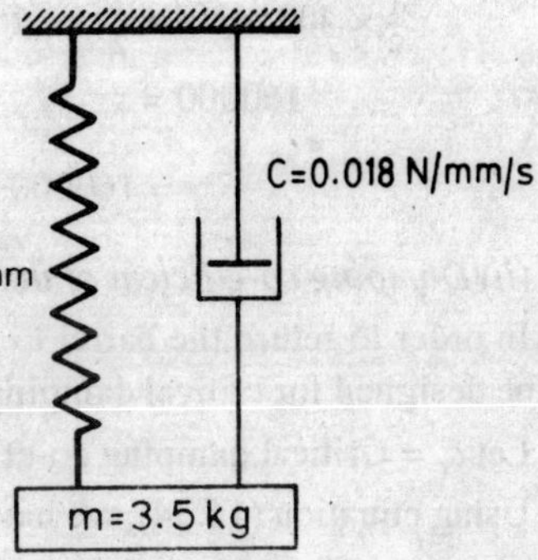

Fig. 16.21

Substituting this value in equation (*i*), we get

$$\delta = \frac{2\pi \times 0.0962}{\sqrt{1-0.0962^2}} = \frac{2\pi \times 0.0962}{0.99} = \mathbf{0.6105}. \quad \textbf{Ans.}$$

(*ii*) *Ratio of any two consecutive amplitudes*

Let $\frac{x_1}{x_2}$ = ratio of any two consecutive amplitudes

Now using equation (16.43A), we have

$$\delta = \text{Ln}\left(\frac{x_1}{x_2}\right)$$

$$\therefore \quad \frac{x_1}{x_2} = e^{\delta} = e^{0.6105} \qquad (\because \quad \delta = 0.6105)$$

$= \mathbf{1.84}.$ **Ans.** [or $e^{0.6105} = 2.71828^{0.6105} = 1.84$ ($\because$ $e = 2.71828$)]

Problem 16.13. *The barrel of a large gun recoils against a spring on firing. At the end of the recoil, a dashpot is engaged that allows the barrel to return to its initial position in the minimum time without oscillation. A gun barrel has a mass of 400 kg and the initial recoil velocity of the gun barrel at the instant of firing is 20 m/s. The barrel recoils 1 m on firing. Find :*

(*i*) *proper spring stiffness in N/mm, and*

(*ii*) *damping co-efficient of the dashpot.*

Sol. Given :

$m = 400$ kg ; $v = 20$ m/s ; $x = 1$ m

(*i*) *Proper stiffness of spring*

Let s = stiffness of the spring

Since the dashpot is not operating during recoil, the initial kinetic energy of the barrel is equal to the work done on the spring.

$\therefore$ K.E. of the barrel = work done on the spring

or $\frac{1}{2} . m . v^2 =$ (Average force on spring) × distance moved

$$= (\tfrac{1}{2} \times sx) \times x \qquad [\because \; sx = \text{Force on spring and } \tfrac{1}{2} sx = \text{Average force}]$$

$$= \frac{1}{2} . s . x^2$$

or $$\frac{1}{2} \times 400 \times 20^2 = \frac{1}{2} . s . 1^2 \qquad (\because \; x = 1 \text{ m and } v = 20 \text{ m/s})$$

or $$160000 = s$$

or $$s = 160000 \text{ N/m} = \frac{160000}{1000} \text{ N/mm} = \mathbf{160 \text{ N/mm}.} \quad \textbf{Ans.}$$

(ii) Damping co-efficient of the dashpot

In order to return the barrel to the original position in the least time without oscillation, the dashpot should be designed for critical damping. Hence the damping co-efficient of the dashpot should be critical.

Let c_c = Critical damping co-efficient of the dashpot.

Using equation (16.34), we have

$$c_c = 2 . m . \omega_n$$

$$= 2 . m . \sqrt{\frac{s}{m}} \qquad \left(\because \; \omega_n = \sqrt{\frac{s}{m}}\right)$$

$$= 2 \times 400 \sqrt{\frac{160,000}{400}} \qquad (\because \; s = 160000 \text{ N/m})$$

$$= 2 \times 400 \times 20 = 16000 \text{ N/m/s} = \frac{16000}{1000} = \mathbf{16 \text{ N/mm/s}.} \quad \textbf{Ans.}$$

Problem 16.14. *Determine : (i) the critical damping co-efficient, (ii) the damping factor, (iii) the natural frequency of damped vibrations, (iv) the logarithmic decrement and (v) the ratio of two consecutive amplitudes of a vibrating system which consists of a mass of 25 kg, a spring of stiffness 15 kN/m and a damper. The damping provided is only 15% of the critical value.*

Sol. Given :

$m = 25$ kg ; $s = 15$ kN/m $= 15 \times 10^3$ N/m $= 15000$ N/m ; damping provided = 15% of critical value of damping. This means damping co-efficient $= \frac{15}{100} \times$ critical damping co-efficient or $c = 0.15 \times c_c$.

(i) Critical damping co-efficient (i.e. c_c)

Critical damping co-efficient is given by equation (16.34) as

$$c_c = 2 . m . \omega_n$$

$$= 2 . m \times \sqrt{\frac{s}{m}} \qquad \left(\because \; \omega_n = \sqrt{\frac{s}{m}}\right)$$

$$= 2 \times 25 \times \sqrt{\frac{15000}{25}} = 1224.7 \text{ N/m/s.}$$

(ii) Damping factor (i.e. ζ)

Using equation (16.35), we get

$$\zeta = \frac{c}{c_c} = \frac{0.15 c_c}{c_c} = \mathbf{0.15.} \quad \textbf{Ans.} \qquad (\because \; c = 0.15\, c_c)$$

(iii) Natural frequency of damped vibrations (i.e. ω_d)

Using equation (16.36A), we get

$$\omega_d = \omega_n \sqrt{1 - \zeta^2} \qquad \text{where } \omega_n = \sqrt{\frac{s}{m}} = \sqrt{\frac{15000}{25}} = 24.5 \text{ rad/s}$$

$$= 24.5\sqrt{1 - 0.15^2} = 24.5 \times 0.9886 = \mathbf{24.22 \text{ rad/s.}} \quad \textbf{Ans.}$$

(iv) The logarithmic decrement (i.e. δ)

Using equation (16.44), we get

$$\delta = \frac{2\pi\zeta}{\sqrt{1-\zeta^2}} = \frac{2\pi \times 0.15}{0.9886} = \mathbf{0.9533.} \quad \textbf{Ans.}$$

(v) The ratio of two consecutive amplitudes $\left(i.e. \ \frac{x_1}{x_2}\right)$

Using equation (16.43A), we have

$$\delta = L_n\left(\frac{x_1}{x_2}\right)$$

$$\therefore \quad \frac{x_1}{x_2} = e^{\delta} = e^{0.9533} = \mathbf{2.594.} \quad \textbf{Ans.}$$

Problem 16.15. *For a vibrating system with viscous damping, the mass of the vibrating body is 3.75 kg and the stiffness of the spring is 4.5 N/mm. If the amplitude decreases to 0.33 of the initial value after four consecutive cycles, then find the value of the damping co-efficient of the damper in the system.*

Sol. Given :

$m = 3.75$ kg ; $s = 4.5$ N/mm $= 4.5 \times 1000$ N/m $= 4500$ N/m ; $x_5 = 0.33x_1$

Let x_1 = Initial amplitude

x_5 = Final amplitude after four consecutive cycles $= 0.33x_1$

c = Damping co-efficient

From equation (16.43), we know that

$$\frac{x_1}{x_2} = \frac{x_2}{x_3} = \frac{x_3}{x_4} = \frac{x_4}{x_5}$$

$$\therefore \quad \frac{x_1}{x_5} = \frac{x_1}{x_2} \times \frac{x_2}{x_3} \times \frac{x_3}{x_4} \times \frac{x_4}{x_5} = \frac{x_1}{x_2} \times \frac{x_1}{x_2} \times \frac{x_1}{x_2} \times \frac{x_1}{x_2} = \left(\frac{x_1}{x_2}\right)^4$$

But $$\frac{x_1}{x_5} = \frac{x_1}{0.33x_1} = 3.03 \qquad (\because \ x_5 = 0.33x_1)$$

$$\therefore \quad 3.03 = \left(\frac{x_1}{x_2}\right)^4$$

$$\therefore \quad \left(\frac{x_1}{x_2}\right) = (3.03)^{1/4} = (3.03)^{0.25} = 1.3193$$

From equation (16.43A), we know that

$$\delta = \text{Ln}\left(\frac{x_1}{x_2}\right) = \text{Ln } 1.3193 = 0.277$$

Now using equation (16.44), we get

$$\delta = \frac{2\pi\zeta}{\sqrt{1-\zeta^2}} \qquad \text{or} \quad 0.277 = \frac{2\pi\zeta}{\sqrt{1-\zeta^2}}$$

Squaring to both sides, we get

$$0.277^2 = \frac{4\pi^2\zeta^2}{(1-\zeta^2)} \qquad \text{or} \quad (1-\zeta^2) = \frac{4\pi^2}{0.277^2} \times \zeta^2 = 514.517\zeta^2$$

or $$1 = 515.517\zeta^2 \qquad \text{or} \quad \zeta = \sqrt{\frac{1}{515.517}} = 0.044$$

The damping co-efficient (c) is given by equation (16.33) as

$$c = 2m \,.\, \omega_n \,.\, \zeta \qquad \text{where } \omega_n = \sqrt{\frac{s}{m}} = \sqrt{\frac{4500}{3.75}} = 34.64 \text{ rad/s}$$

$$= 2 \times 3.75 \times 34.64 \times 0.044 = \mathbf{11.4312 \text{ N/m/s.}} \quad \textbf{Ans.}$$

Problem 16.16. *In a single-degree damped vibrating system, a suspended mass of 3.75 kg makes 12 oscillation in 7 seconds when disturbed from its equilibrium position. The amplitude of vibration reduces to 0.33 of its initial value after four oscillations. Determine : (i) stiffness of the spring (ii) logarithmic decrement (iii) damping factor and (iv) the damping co-efficient.*

Sol. Given :

$m = 3.75$ kg ; no. of oscillation = 12 in time = 7 seconds ; hence no. of oscillations in one second $= \frac{12}{7} = 1.71$ cycles/s or $f_n = 1.71$ Hz.

If x_1 = Initial amplitude and x_5 = Final amplitude after four oscillations = $0.33x_1$ (given)

then $$\frac{x_1}{x_5} = \frac{x_1}{0.33x_1} = \frac{1}{0.33} = 3.03 \qquad ...(i)$$

(i) Stiffness of spring

Let s = Stiffness of the spring.

We know that $$\omega_n = \sqrt{\frac{s}{m}} \qquad ...(ii)$$

But $$\omega_n = 2\pi \times f_n = 2\pi \times 1.71 \qquad (\because \; f_n = 1.71)$$

$$= 10.744 \text{ rad/s} \qquad ...(iii)$$

Equating the two values of ω_n given by equations *(ii)* and *(iii)*, we get

$$\sqrt{\frac{s}{m}} = 10.744 \qquad \text{or} \quad \frac{s}{m} = 10.744^2$$

or $$s = 10.744^2 \times m = 10.744^2 \times 3.75 = \mathbf{432.875 \text{ N/m.}} \quad \textbf{Ans.}$$

(ii) Logarithmic decrement (δ)

From equation (16.43), we know that

$$\frac{x_1}{x_2} = \frac{x_2}{x_3} = \frac{x_3}{x_4} = \frac{x_4}{x_5}$$

$$\frac{x_1}{x_5} = \frac{x_1}{x_2} \times \frac{x_2}{x_3} \times \frac{x_3}{x_4} \times \frac{x_4}{x_5} = \frac{x_1}{x_2} \times \frac{x_1}{x_2} \times \frac{x_1}{x_2} \times \frac{x_1}{x_2} = \left(\frac{x_1}{x_2}\right)^4$$

But $\frac{x_1}{x_5} = 3.03$ **[It is from equation (*i*)]**

$$\therefore \quad 3.03 = \left(\frac{x_1}{x_2}\right)^4$$

or $$\frac{x_1}{x_2} = (3.03)^{1/4} = (3.03)^{0.25} = 1.3193$$

But from equation (16.43A), the logarithmic decrement (δ) is given by

$$\delta = \text{Ln}\left(\frac{x_1}{x_2}\right) = \text{Ln } 1.3193 = \mathbf{0.277.} \quad \textbf{Ans.}$$

(*iii*) *Damping factor* (ζ)

From equation (16.44), we know that

$$\delta = \frac{2\pi\zeta}{\sqrt{1-\zeta^2}} \quad \text{or} \quad 0.277 = \frac{2\pi\zeta}{\sqrt{1-\zeta^2}}$$

or $$0.277^2 = \frac{4\pi^2\zeta^2}{1-\zeta^2} \quad \text{(Squaring both sides)}$$

or $$1-\zeta^2 = \frac{4\pi^2}{0.277^2}\times\zeta^2 = 514.517\zeta^2 \quad \text{or} \quad 1 = 515.517\zeta^2$$

or $$\zeta = \sqrt{\frac{1}{515.517}} = \mathbf{0.044.} \quad \textbf{Ans.}$$

(*iv*) *Damping co-efficient* (*c*)

From equation (16.33), we know that

$$c = 2\,.\,m\,.\,\omega_n\,.\,\zeta$$
$$= 2 \times 3.75 \times 10.744 \times 0.044 = \mathbf{3.545\ N/m/s.} \quad \textbf{Ans.}$$

16.13. Forced-Damped Vibrations

Fig 16.22 shows a spiral spring, whose one end is fixed and other end is free to which a body of mass *m* is suspended. The mass is also attached to a dashpot which is fixed to the rigid support. The horizontal line

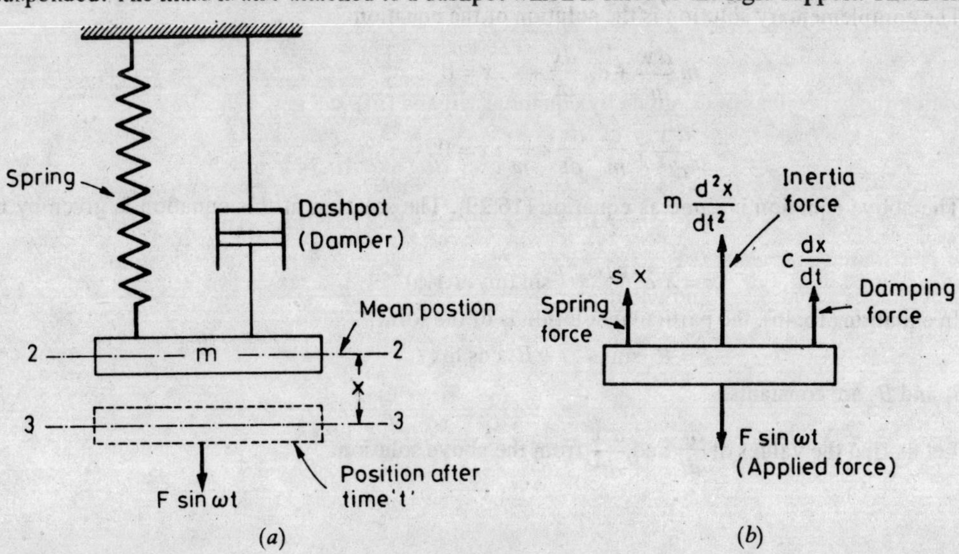

Fig. 16.22

2-2 shows the level of static equilibrium under the weight of the body. Let the system is acted upon by an external harmonic force = $F \sin \omega . t$.

Let the body is displaced by a distance x downward from the mean position in time 't'. Then the forces acting on the body are :

(*i*) Accelerating force in the direction of motion (*i.e.* downwards)

$$= m \times \frac{d^2x}{dt^2}$$

(*ii*) Damping force in the opposite direction of motion (*i.e.* upwards)

= damping force per unit velocity × velocity of vibration

$$= c \times \frac{dx}{dt}$$

(*iii*) Spring force in the opposite direction of motion (*i.e.* upwards)

= stiffness of spring × extension of spring

$$= s . x$$

(*iv*) External applied force in the direction of motion (*i.e.* downwards)

$$= F \sin \omega . t$$

For dynamic equilibrium of the body, the sum of inertia force $\left(i.e. - \text{accelerating force or } m \frac{d^2x}{dt^2} \text{ acting}\right.$ upwards) and all external forces (*i.e.* damping force, spring force and applied force) in any direction should be zero. For dynamic equilibrium, the direction of all forces are shown in Fig. 16.22 (*b*).

Hence the equation of motion will be

$$m . \frac{d^2x}{dt^2} + c\frac{dx}{dt} + s . x - F \sin \omega . t \quad = 0$$

or
$$m\frac{d^2x}{dt^2} + c . \frac{dx}{dt} + s . x \quad = F \sin \omega . t \qquad ...(16.46)$$

This equation of motion is a second order linear differential equation with right hand side in some function in t. The general solution of such type of differential equation consists of two parts. One part is the complementary solution and the second part is a particular solution.

The complementary solution is the solution of the equation

$$m\frac{d^2x}{dt^2} + c . \frac{dx}{dt} + s . x = 0$$

or
$$\frac{d^2x}{dt^2} + \frac{c}{m} . \frac{dx}{dt} + \frac{s}{m} . x = 0$$

The above equation is same as equation (16.29). The solution of this equation is given by equation (16.39) as

$$x = X e^{-\zeta . \omega_n . t} \sin (\omega_d . t + \phi) \qquad ...(i)$$

In equation (16.46), the particular solution is of the form

$$x = B_1 \sin \omega . t + B_2 \cos \omega . t \qquad ...(ii)$$

where B_1 and B_2 are constants.

Let us find the values of $\frac{dx}{dt}$ and $\frac{d^2x}{dt^2}$ from the above solution.

Differentiating equation (*ii*), w.r.t. to t, we get

$$\frac{dx}{dt} = B_1 . \omega . \cos \omega . t + B_2 . \omega . (-\sin \omega . t)$$

$$= B_1 . \omega . \cos \omega . t - B_2 . \omega . \sin \omega . t$$

and
$$\frac{d^2x}{dt^2} = B_1 . \omega . \omega . (-\sin \omega t) - B_2 . \omega . \omega \cos \omega t$$

$$= -B_1 . \omega^2 . \sin \omega . t - B_2 . \omega^2 . \cos \omega . t$$

Substituting the values of x, $\frac{dx}{dt}$ and $\frac{d^2x}{dt^2}$ in equation (16.46), we get

$$m . (-B_1 . \omega^2 . \sin \omega . t - B_2 . \omega^2 . \cos \omega . t) + c(B_1 . \omega . \cos \omega . t - B_2 \omega . \sin \omega t)$$
$$+ s(B_1 \sin \omega . t + B_2 \cos \omega . t) = F \sin \omega t$$

Collecting the co-efficients of sine and cosine terms, we get

$$(-m . B_1 . \omega^2 - c . B_2 . \omega + s . B_1) \sin \omega . t + (-m . B_2 . \omega^2$$
$$+ c . B_1 . \omega + s . B_2) \cos \omega . t = F \sin \omega . t$$

Equating the co-efficients of $\sin \omega . t$ and $\cos \omega . t$ on both sides, we get

$$-m . B_1 . \omega^2 - c . B_2 . \omega + s . B_1 = F \quad ...(iii)$$

and
$$-m . B_2 . \omega^2 + c . B_1 . \omega + s . B_2 = 0 \quad ...(iv)$$

From equations (*iii*) and (*iv*), let us find the values of B_1 and B_2.

From equation (*iv*), we have

$$B_2(-m . \omega^2 + s) = -cB_1 . \omega \quad \text{or} \quad B_2(s - m\omega^2) = -c . B_1 . \omega$$

or
$$B_1 = \frac{B_2(s - m . \omega^2)}{-c . \omega} \quad ...(v)$$

Substituting the value of B_1 in equation (*iii*),

$$-m . \frac{B_2(s - m . \omega^2)}{-c . \omega} . \omega^2 - c . B_2 . \omega + \frac{s . B_2(s - m . \omega^2)}{-c . \omega} = F$$

or
$$B_2\left[\frac{-m(s - m . \omega^2)}{-c . \omega} . \omega^2 - c . \omega + \frac{s(s - m\omega^2)}{-c\omega}\right] = F$$

or
$$B_2\left[\frac{-m(s - m . \omega^2) . \omega^2 + c^2\omega^2 + s(s - m\omega^2)}{-c . \omega}\right] = F$$

or
$$B_2\left[\frac{(s - m . \omega^2)(s - m . \omega^2) + c^2\omega^2}{-c . \omega}\right] = F$$

or
$$B_2\left[\frac{(s - m . \omega^2)^2 + c^2 . \omega^2}{-c . \omega}\right] = F$$

or
$$B_2 = \frac{-c . \omega . F}{[(s - m . \omega^2)^2 + c^2 . \omega^2]}$$

Substituting this value in equation (*v*), we get

$$B_1 = \frac{-c . \omega . F}{[(s - m . \omega^2)^2 + c^2\omega^2]} \times \frac{(s - m\omega^2)}{-c . \omega} = \frac{(s - m\omega^2) . F}{[(s - m . \omega^2)^2 + c^2 . \omega^2]}$$

Substituting the values of B_1 and B_2 in equation (*ii*), we get the particular solution as

$$x = \frac{(s - m\,.\,\omega^2)\,.\,F}{[(s - m\omega^2)^2 + c^2\,.\,\omega^2]} \times \sin \omega\,.\,t + \frac{-c\,.\,\omega\,.\,F}{[(s - m\omega^2)^2 + c^2\,.\,\omega^2]} \cos \omega t$$

$$= \frac{F}{[(s - m\,.\,\omega^2)^2 + c^2\,.\,\omega^2]}\,[(s - m\,.\,\omega^2) \sin \omega\,.\,t - c\,.\,\omega \cos \omega t]$$

$$= \frac{F}{[(s - m\omega^2)^2 + c^2\,.\,\omega^2]}\,[X \cos \phi\,.\,\sin \omega\,.\,t - X \sin \phi \cos \omega\,.\,t]$$

(where let $(s - m\omega^2) = X \cos \phi$ and $c\,.\,\omega = X \sin \phi$)

$$= \frac{F\,.\,X}{[(s - m\omega^2)^2 + c^2\,.\,\omega^2]}\,[\sin (\omega\,.\,t - \phi)] \qquad ...(vi)$$

It has been assumed that $X \sin \phi = c\,.\,\omega$ and $X \cos \phi = (s - m\,.\,\omega^2)$

$$\therefore \quad X^2 \sin^2 \phi + X^2 \cos^2 \phi = c^2\,.\,\omega^2 + (s - m\,.\,\omega^2)^2$$

or $$X^2(\sin^2 \phi + \cos^2 \phi) = c^2.\,\omega^2 + (s - m\,.\,\omega^2)^2$$

or $$X = \sqrt{c^2\,.\,\omega^2 + (s - m\,.\,\omega^2)^2} \qquad (\because \sin^2 \phi + \cos^2 \phi = 1)$$

and $$\frac{X \sin \phi}{X \cos \phi} = \frac{c\,.\,\omega}{(s - m\,.\,\omega^2)}$$

or $$\tan \phi = \frac{c\,.\,\omega}{(s - m\,.\,\omega^2)} \quad \text{or} \quad \phi = \tan^{-1} \frac{c\,.\,\omega}{(s - m\,.\,\omega^2)} \qquad ...(16.47)$$

Substituting the value of X in equation (*vi*), we get

$$x = \frac{F \times \sqrt{c^2\,.\,\omega^2 + (s - m\,.\,\omega^2)^2}}{[(s - m\,.\,\omega^2)^2 + c^2\,\omega^2]} \sin (\omega\,.\,t - \phi)$$

$$= \frac{F \sin (\omega\,.\,t - \phi)}{\sqrt{(s - m\,.\,\omega^2)^2 + c^2\,.\,\omega^2}} \qquad ...(vii)$$

$\therefore$ Complete solution of the differential equation (16.46) is the sum of complimentary solution given by equation (*i*) and particular solution given by equation (*vii*)

$$x = \text{Complimentary solution} + \text{Particular solution}$$

$$= Xe^{-\zeta\,.\,\omega_n\,.\,t} \sin (\omega_d\,.\,t + \phi) + \frac{F \sin (\omega\,.\,t - \phi)}{\sqrt{(s - m\,.\,\omega^2)^2 + c^2\,.\,\omega^2}} \qquad ...(16.48)$$

The damped-free vibrations of complimentary solution represented by the first term dies out rapidly owing to the damping effect of friction as the amplitude of the vibrations is given by $Xe^{-\zeta\,.\,\omega_n\,.\,t}$. With the passage of time, $e^{-\infty} = 0$. The second term represents the forced vibration which is maintained by the periodic force. Hence when a steady-state of vibration has been reached, the displacement x is given by the second term (or by particular solution) only.

$$\therefore \quad x = \frac{F \sin (\omega\,.\,t - \phi)}{\sqrt{(s - m\,.\,\omega^2)^2 + c^2\,.\,\omega^2}} \qquad ...(16.49)$$

The above equation shows that the motion is simple harmonic whose circular frequency is ω and the amplitude is $\frac{F}{\sqrt{(s - m\,.\,\omega^2)^2 + c^2\,.\,\omega^2}}$. And ϕ is the phase lag for the displacement with respect to applied force $F \sin \omega\,.\,t$.

Maximum displacement or amplitude of forced vibration,

$$x_{max} = \frac{F}{\sqrt{(s - m\,.\,\omega^2)^2 + c^2\,.\,\omega^2}} \qquad ...(16.50)$$

$$= \frac{(F/s)}{\sqrt{\frac{(s - m\omega^2)^2}{s^2} + \frac{c^2\,.\,\omega^2}{s^2}}}$$ (Dividing the numerator and denominator by s)

$$= \frac{\delta}{\sqrt{\left(1 - \frac{m\,.\,\omega^2}{s}\right)^2 + \left(\frac{c}{s}\,.\,\omega\right)^2}}$$

$$\left(\because \frac{F}{s} = \delta, \text{ the deflection produced by static load } F\right)$$

$$= \frac{\delta}{\sqrt{\left(1 - \frac{\omega^2}{(s/m)}\right)^2 + \left(\frac{c}{s}\,.\,\omega\right)^2}}$$

$$= \frac{\delta}{\sqrt{\left(1 - \frac{\omega^2}{\omega_n^2}\right)^2 + \left[\frac{(2m\,.\,\omega_n\,.\,\zeta)\,.\,\omega}{s}\right]^2}}$$

$$\left[\because \frac{s}{m} = \omega_n^2 \text{ and } c = 2m\,.\,\omega_n\,.\,\zeta \text{ from equation (16.33)}\right]$$

$$= \frac{\delta}{\sqrt{\left[1 - \left(\frac{\omega}{\omega_n}\right)^2\right]^2 + \left[\frac{2\,.\,\omega_n\,.\,\zeta}{(s/m)}\,.\,\omega\right]^2}}$$

$$= \frac{\delta}{\sqrt{\left[1 - \left(\frac{\omega}{\omega_n}\right)^2\right]^2 + \left[\frac{2\,.\,\omega_n\,.\,\zeta}{\omega_n^2}\,.\,\omega\right]^2}} \qquad \left(\because \frac{s}{m} = \omega_n^2\right)$$

$$= \frac{\delta}{\sqrt{\left[1 - \left(\frac{\omega}{\omega_n}\right)^2\right]^2 + \left(2\zeta\,.\,\frac{\omega}{\omega_n}\right)^2}} \qquad ...(16.51)$$

Note. (*i*) Frequency of steady-state forced vibrations is same as that of impressed vibrations (*i.e.* frequency of external applied force).

(*ii*) The equation (16.49), (16.50) and (16.51) holds good when steady state of forced vibrations have reached.

Graphical Method

The equation of motion for a forced-damped vibrations can also be solved by graphical method as given below :

Let us assume that the displacement of the vibrating mass shown in Fig. 16.23 under the action of the applied simple harmonic force is also simple harmonic and lags by an amount ϕ, so that it can be represented by the equation

$$x = A \sin(\omega t - \phi) \qquad \text{where } A = \text{Amplitude of vibrations} \qquad ...(16.51A)$$

Differentiating the above equation,

$$\frac{dx}{dt} = A \, . \, [\cos(\omega \, . \, t - \phi)] \, . \, \omega = A \, . \, \omega \cos(\omega t - \phi) = A \, . \, \omega \, . \, \sin[90° + (\omega \, . \, t - \phi)] \quad ...(16.51B)$$

and

$$\frac{d^2x}{dt^2} = A \, . \, \omega \, . \, [-\sin(\omega \, . \, t - \phi)] \, . \, \omega = -A \, . \, \omega^2 \sin(\omega t - \phi) = A \, . \, \omega^2 \, . \, \sin[180° + (\omega \, . \, t - \phi)]$$

As we have assumed that $x = A \sin(\omega t - \phi)$ is the solution of the equation (16.46), hence the above values can be substituted in equation (16.46). Hence substituting the above values in equation (16.46), we get

$$m \, . \, A \, . \, \omega^2 \, . \, \sin[180° + (\omega t - \phi)] + c \, . \, A \, . \, \omega \, . \, \sin[90° + (\omega t - \phi)] + s \, . \, A \sin(\omega t - \phi) = F \sin(\omega \, . \, t)$$

i.e. Inertia force + Damping force + Spring force = Applied force ...(16.51C)

This means the algebraic sum of these three forces (*i.e.* inertia force, damping force and spring force) at any given instant must be equal to applied force $F \sin(\omega t)$. The inertia force is leading by 180° to spring force whereas the damping force is leading by 90° to spring force. Also the spring force is laging by ϕ with respect to applied force. These forces are represented graphically in Fig. 16.23 (*a*). The vector OP represents

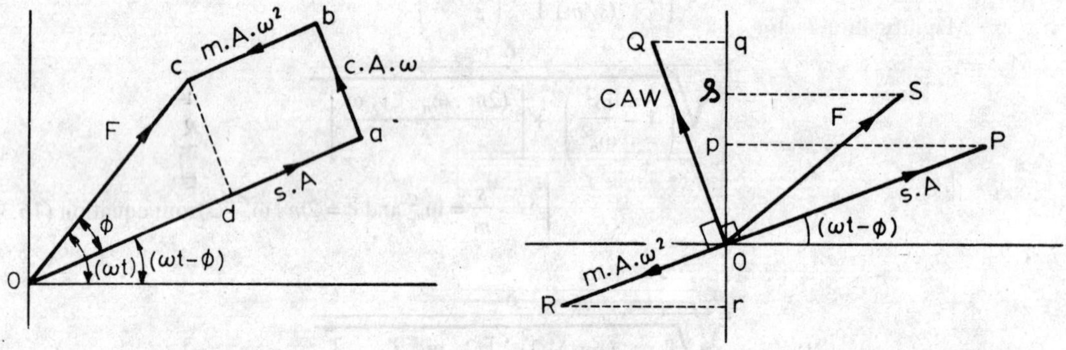

(*b*) (*a*)

Fig. 16.23

the spring force (of maximum values $s \, . \, A$) to some suitable scale, at an inclination $(\omega \, . \, t - \phi)$ to the horizontal. The damping force and inertia force are represented by vectors OQ (of maximum value $c \, . \, A \, . \, \omega$) and by vector OR (of maximum value $m \, . \, A \, . \, \omega^2$) respectively to the same scale. The vector OQ is at 90° to vector OP whereas the vector OR is at 180° to vector OP (or at 90° to vector OQ).

The projected lengths of these vectors on vertical axis represents the instantaneous values of these forces at any time t. Thus, the projected lengths Op, Oq and Or are values of spring force, damping force and inertia force respectively at time t. The algebraic sum of these length is Os. Hence Os must represent the value of applied force *i.e.* $F \sin \omega t$ at that instant. Hence the force vector OS is the vector sum of OP, OQ and OR. Thus the force F is the vector sum of $s \, . \, A$, $c \, . \, A \, . \, \omega$, and $m \, . \, A \, . \, \omega^2$ as shown in Fig. 16.23 (*b*). In Fig. 16.23 (*b*), Oa is parallel to vector OP and is equal to $s \, . \, A$, $ab = c \, . \, A \, . \, \omega$ and is parallel to vector OQ and $bc = m \, . \, A \, . \, \omega^2$ and is parallel to vector OR. Then Oc is the resultant of these forces and will be equal to F.

From Fig. 16.23 (*b*), $F = Oc = \sqrt{od^2 + dc^2} = \sqrt{(oa - da)^2 + dc^2} = \sqrt{(oa - bc)^2 + dc^2}$

$$= \sqrt{(s \, . \, A - m \, . \, A \, . \, \omega^2)^2 + (c \, . \, A \, . \, \omega)^2} \quad (\because \; dc = ab = c \, . \, A \, . \, \omega)$$

$$= A\sqrt{(s - m \, . \, \omega^2)^2 + (c \, . \, \omega)^2}$$

or

$$A = \frac{F}{\sqrt{(s - m \, . \, \omega^2)^2 + (c \, . \, \omega)^2}} \quad ...(16.51D)$$

But A is the amplitude of assumed vibrations. This amplitude is same as given by equation (16.50). Also from Fig. 16.23 (*b*),

$$\tan\phi = \frac{cd}{od} = \frac{cd}{(oa - da)} = \frac{c \,.\, A \,.\, \omega}{(s \,.\, A - m \,.\, A \,.\, \omega^2)} = \frac{c \,.\, \omega}{(s - m\omega^2)}$$

This is also same as given by equation (16.47).

16.14. Magnification Factor

The ratio the amplitude (or the maximum displacement) of the forced vibration to the deflection due to static load F, is known as magnification factor. It is represented by MF.

The amplitude (or maximum displacement) of the forced vibration is given by equation (16.50) as

$$x_{max} = \frac{F}{\sqrt{(s - m \,.\, \omega^2)^2 + (c \,.\, \omega)^2}}$$

And deflection due to static load F is given by,

$$\delta = \frac{F}{s} \qquad \text{where } s = \text{stiffness of spring.}$$

∴ Magnification factor,

$$MF = \frac{x_{max}}{\delta} \qquad \text{...(16.52)}$$

$$= \frac{\left(\dfrac{F}{\sqrt{(s - m\omega^2)^2 + (c \,.\, \omega)^2}}\right)}{(F/s)} = \frac{s}{\sqrt{(s - m\omega^2)^2 + (c\omega)^2}}$$

$$= \frac{1}{\sqrt{\left(1 - \dfrac{m}{s}\omega^2\right)^2 + \left(\dfrac{c}{s}\omega\right)^2}}$$

[Dividing the numerator and denominator by s]

$$= \frac{1}{\sqrt{\left(1 - \dfrac{\omega^2}{\omega_n^2}\right)^2 + \left(2\zeta\dfrac{\omega}{\omega_n}\right)^2}} \qquad \text{...(16.53)}$$

$$\left(\because \sqrt{\frac{s}{m}} = \omega_n \text{ or } \frac{s}{m} = \omega_n^2 \text{ and } c = 2m \,.\, \omega_n \,.\, \zeta\right)$$

From equation (16.53) it is clear that the magnification factor depends upon the ratio of frequencies ω/ω_n and the damping factor ζ.

Refer to equation (16.52).

The equation (16.52) can be written as $x_{max} = \delta \times$ Magnification Factor ...(16.54)

Hence the magnification factor is the factor which is when multiplied to the static deflection (produced by static load F) gives the maximum displacement (or amplitude) of forced vibrations.

Important conclusions :

(*i*) If there is no damping, then damping factor $\zeta = 0$. Then equation (16.53) reduces to

$$\text{Magnification factor, } MF = \frac{1}{\sqrt{\left(1-\frac{\omega^2}{\omega_n^2}\right)^2}} = \frac{1}{1-\frac{\omega^2}{\omega_n^2}} = \frac{\omega_n^2}{\omega_n^2-\omega^2} \quad \text{...(16.54A)}$$

(*ii*) If there is resonance, then $\omega = \omega_n$ and equation (16.53) becomes as

$$\text{Magnificator factor, } MF = \frac{1}{\sqrt{\left(1-\frac{\omega_n^2}{\omega_n^2}\right)^2+\left(2\zeta\frac{\omega_n}{\omega_n}\right)^2}} = \frac{1}{2\zeta} \quad \text{...(16.54B)}$$

Fig. 16.24 shows the relationship between magnification factor (*MF*) and ratio of frequencies $\left(\frac{\omega}{\omega_n}\right)$ for different values of damping factor ζ. From the curve it is clear that as damping increases (*i.e.* ζ increases), the value of magnification factor decreases. And when ζ decreases, the value of *MF* increases. At $\zeta = 0$ (*i.e.* when there is no damping) the value of magnification factor reaches to infinity at $\frac{\omega}{\omega_n} = 1$. At $\frac{\omega}{\omega_n} = 1$, the frequency of free vibration is equal to the frequency of forced vibrations. This condition is known as resonance.

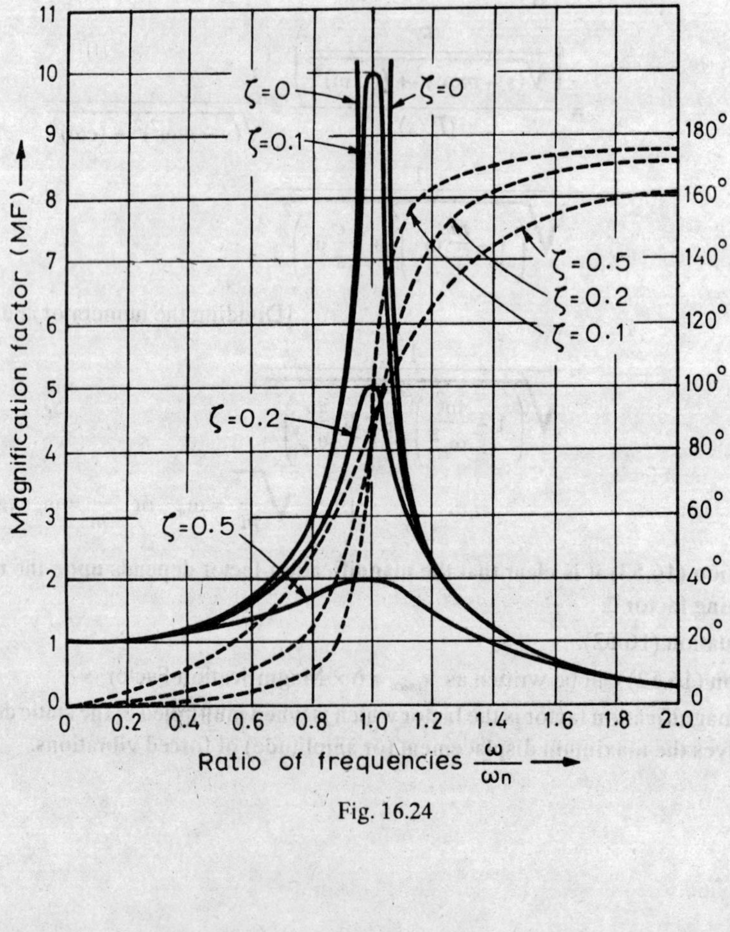

Fig. 16.24

Fig 16.24 also shows the relationship between phase angle (ϕ) and ratio of frequencies $\left(\frac{\omega}{\omega_n}\right)$ for different values of damping factor (ζ).

Problem 16.17. *A spring of stiffness 0.3 N/mm is attached to a mass which has viscous damping device. When the mass was displaced and released, the period of vibration was 1.8 second and the ratio of consecutive amplitudes was 4.2 : 1. Determine : (i) the amplitude and (ii) the phase angle when a force F = 2 sin (3t) N acts on the system.*

Sol. Given :

$s = 0.3$ N/mm $= 0.3$ N $\times$ 1000 N/m $= 300$ N/m ; $T_d = 1.8$ s ; $\frac{x_1}{x_2} = \frac{4.2}{1}$, force applied $= 2 \sin (3t)$

The frequency of damped free-vibration in rad/s is given by,

$$\omega_d = \frac{2\pi}{T_d} = \frac{2\pi}{1.8} = 3.5 \text{ rad/s}$$

We know from equation (16.36A) that

$$\omega_d = \omega_n \sqrt{1 - \zeta^2} \quad \text{...(i)}$$

Let us find the value of damping factor ζ from the given data.

From equation (16.44), we know that, logarithmic decrement

$$\delta = \frac{2\pi\zeta}{\sqrt{1 - \zeta^2}} \quad \text{...(ii)}$$

The value of δ is also given by equation (16.43A), in terms of the ratio of consecutive amplitudes as

$$\delta = \text{Ln}\left(\frac{x_1}{x_2}\right) = \text{Ln } 4.2 \qquad \left(\because \frac{x_1}{x_2} = 4.2\right)$$

$$= 1.43508 = 1.4351$$

Substituting the above value in equation (*ii*), we get

$$1.4351 = \frac{2\pi\zeta}{\sqrt{1 - \zeta^2}}$$

or $$\sqrt{1 - \zeta^2} = \frac{2\pi\zeta}{1.4351} = 4.3782\zeta$$

Squaring both sides, we get

$$(1 - \zeta^2) = (4.3782\zeta)^2 = 19.1866\zeta^2$$

or $$1 = 19.1866\zeta^2 + \zeta^2 = 20.1866\zeta^2$$

or $$\zeta = \sqrt{\frac{1}{20.1866}} = 0.2226$$

Substituting this value of ζ in equation (*i*),

$$\omega_d = \omega_n \sqrt{1 - 0.2226^2}$$

or $$3.5 = \omega_n \sqrt{1 - 0.2226^2} \qquad (\because \omega_d = 3.5)$$

$$= \omega_n \times 0.9749$$

$\therefore$ $$\omega_n = \frac{3.5}{0.9749} = 3.59 \text{ rad/s}$$

Now the applied force is given as $= 2 \sin (3t)$ N.2 sin (90 + 3t)

If we compare this applied force with $F \sin \omega t$, we get $F = 2$ N and $\omega = 3$ rad/s

(i) Amplitude of forced-damped vibrations

Amplitude of forced-damped vibrations (or maximum displacement) is given by equation (16.51) as

$$x_{max} = \frac{\delta}{\sqrt{\left[1-\left(\frac{\omega}{\omega_n}\right)^2\right]^2+\left(2\zeta \cdot \frac{\omega}{\omega_n}\right)^2}} \quad \text{where } \delta = \frac{F}{s} = \frac{2}{300} = 0.006667 \text{ m}$$

$$= \frac{0.006667}{\sqrt{\left[1-\left(\frac{3}{3.59}\right)^2\right]^2+\left(2 \times 0.2226 \times \frac{3}{3.59}\right)^2}}$$

$$= \frac{0.00667}{\sqrt{0.091+0.1384}} = 0.01392 \text{ m} = \mathbf{13.92 \text{ mm}}. \quad \textbf{Ans.}$$

(ii) The phase angle (ϕ)

Using equation (16.47), we get

$$\tan \phi = \frac{c \cdot \omega}{(s - m \cdot \omega^2)} \quad \text{or} \quad \phi = \tan^{-1} \frac{c \cdot \omega}{(s - m \cdot \omega^2)}$$

$$= \frac{\left(\frac{c \cdot \omega}{s}\right)}{\left(1 - \frac{m}{s} \cdot \omega^2\right)} \quad \text{(Dividing the numerator and denominator by } s\text{)}$$

$$= \frac{\frac{(2m \cdot \omega_n \cdot \zeta) \times \omega}{s}}{\left(1 - \frac{m}{s}\omega^2\right)} \quad (\because \; c = 2m \cdot \omega_n \cdot \zeta)$$

$$= \frac{\frac{2 \cdot \omega_n \cdot \zeta}{\omega_n^2} \cdot \omega}{\left(1 - \frac{\omega^2}{\omega_n^2}\right)} \quad \left(\because \; \sqrt{\frac{s}{m}} = \omega_n \text{ or } \frac{s}{m} = \omega_n^2 \text{ or } \frac{m}{s} = \frac{1}{\omega_n^2}\right)$$

$$= \frac{2 \cdot \zeta \cdot \frac{\omega}{\omega_n}}{\left(1 - \frac{\omega^2}{\omega_n^2}\right)} = \frac{2 \times 0.2226 \times \frac{3}{3.59}}{1 - \left(\frac{3}{3.59}\right)^2} \quad (\because \; \omega = 3 \text{ rad/s} ; \omega_n = 3.59 \text{ rad/s})$$

$$= 1.2332$$

$$\therefore \quad \phi = \tan^{-1} 1.2332 = \mathbf{50.96°}. \quad \textbf{Ans.}$$

Problem 16.18. *A single cylinder vertical petrol engine has a mass of 200 kg and is mounted upon a steel chasis frame. The vertical static deflection of the frame is 2.4 mm due to the weight of the engine. The reciprocating parts of the engine has a mass of 9 kg and move through a vertical stroke of 160 mm with simple harmonic motion. A dashpot with a damping co-efficient of 1 N/mm/s is also used to dampen the vibrations. Considering that the steady-state of vibration is reached determine :*

(i) amplitude of the forced vibration if the driving shaft rotates at 500 r.p.m. and

(ii) the speed of the driving shaft at which resonance will occur.

Sol. Given :

$m = 200$ kg ; $\delta = 2.4$ mm $= 2.4 \times 10^{-3}$ m ; mass of reciprocating parts, $m_1 = 9$ kg ; $L = 160$ mm $= 0.16$ m ;

$c = 1$ N/mm/s $= 1 \times 1000$ N/m/s $= 1000$ N/m/s ; N = 500 r.p.m. or $\omega = \frac{2\pi N}{60} = \frac{2\pi \times 500}{60} = 52.36$ rad/s

(*i*) *Amplitude of the forced vibration*

Let us first find the stiffness of the frame and the applied force.

The stiffness of the frame is given by,

$$s = \frac{W}{\delta} \quad \text{where } W = \text{weight of engine which is mounted on the frame}$$

$$= \frac{m \times g}{\delta} = \frac{200 \times 9.81}{2.4 \times 10^{-3}} = 8.175 \times 10^5 \text{ N/m}$$

The reciprocating parts of the engine move with a simple harmonic motion. Hence the centrifugal force due to reciprocating parts will be having simple harmonic motion. This centrifugal force is the applied force.

Since the length of the stroke is $L = 160$ mm $= 0.16$ m, hence radius of crank,

$$r = \frac{L}{2} = \frac{0.16}{2} = 0.08 \text{ m.}$$

$\therefore$ Centrifugal force due to reciprocating parts (or applied force)

$$F = m_1 \cdot \omega^2 \cdot r = 9 \times (52.36)^2 \times 0.08 = 1973.93 \text{ N}$$

The amplitude of the forced vibration (or maximum displacement) is given by equation (16.50) as

$$x_{max} = \frac{F}{\sqrt{(s - m \cdot \omega^2)^2 + c^2 \cdot \omega^2}}$$

$$= \frac{1973.93}{\sqrt{(8.175 \times 10^5 - 200 \times 52.36^2)^2 + 1000^2 \times 52.36^2}}$$

$= \mathbf{0.0072}$ **m** or **7.2 mm. Ans.**

(*ii*) *Speed of the driving shaft at which resonance will occur*

The angular speed at which resonance occurs is given by

$$\omega = \omega_n = \sqrt{\frac{s}{m}} = \sqrt{\frac{8.175 \times 10^5}{200}} = 63.93 \text{ rad/s.}$$

But $\omega = \frac{2\pi N}{60}$. Hence $N = \frac{60 \times \omega}{2\pi} = \frac{60 \times 63.93}{2\pi} = \mathbf{610.5}$ **r.p.m. Ans.**

Problem 16.19. *A harmonic exciting force of 25 N is acting on a machine part, which is having a mass of 2 kg and is vibrating in a viscous medium. This exciting force causes a resonant amplitude of 12.5 mm with a period of 0.20 seconds. Determine the damping co-efficient.*

If the system is excited by a harmonic force of frequency 4 Hz, find the increase in amplitude of forced vibration when damper is removed.

Sol. Given :

Harmonic exciting force, $F = 25$ N ; $m = 2$ kg ; resonant amplitude = 12.5 mm ; period of vibration, $T = 0.2$ s.

(*i*) *Damping co-efficient (c)*

Let c = damping co-efficient.

The frequency of the exciting force is given by, $\omega = \frac{2\pi}{T} = \frac{2\pi}{0.2} = 10\pi$ rad/s

At resonance, $\omega = \omega_n$ and amplitude = 12.5 mm

$$\therefore \quad \omega_n = \omega = 10\pi \text{ rad/s}$$

But $\omega_n = \sqrt{\dfrac{s}{m}}$. Hence $10\pi = \sqrt{\dfrac{s}{2}}$ where s = stiffness of spring

or $\quad 100 \times \pi^2 = \dfrac{s}{x} \quad$ or $\quad s = 2 \times 100 \times \pi^2 = 1973.92$ N/m

The amplitude of forced vibration is given by equation (16.51) as

$$x_{max} = \frac{\delta}{\sqrt{\left[1 - \left(\frac{\omega}{\omega_n}\right)^2\right]^2 + \left(2\zeta \, . \, \frac{\omega}{\omega_n}\right)^2}} \quad \text{where } \delta = \frac{F}{s} = \frac{25}{1973.92} = 0.012665$$

At resonance, x_{max} = 12.5 mm = 0.0125 m and $\omega = \omega_n$ or $\dfrac{\omega}{\omega_n} = 1$. The above equation becomes as

$$0.0125 = \frac{0.012665}{\sqrt{(1-1)^2 + (2\zeta \, . \, 1)^2}} \quad (\because \ \delta = 0.012665)$$

$$= \frac{0.012665}{2\zeta}$$

or $$\zeta = \frac{0.012665}{2 \times 0.0125} = 0.5066$$

But from equation (16.31A), we know that

$$\zeta = \frac{c}{2 \, . \, m \, . \, \omega_n}$$

$$\therefore \quad 0.5066 = \frac{c}{2 \times 2 \times 10\pi} \quad (\because \ \omega_n = 10\pi \text{ rad/s})$$

$$\therefore \quad c = 0.5066 \times 2 \times 2 \times 10\pi = 63.66 \text{ N/m/s} = \frac{63.66}{1000} \text{ N/mm/s}$$

$$= \mathbf{0.06366 \ N/mm/s.} \quad \textbf{Ans.}$$

(ii) Increase in amplitude when frequency of exciting force is 4 Hz and damper is removed

Frequency of exciting force = 4 Hz

But frequency $= \dfrac{\omega}{2\pi}$

$$\therefore \quad \omega = 2\pi \times \text{frequency} = 2\pi \times 4 = 8\pi \text{ rad/s}$$

But $\omega_n = 10\pi$ rad/s

The amplitude is given by equation (16.51) (when damper is in the system) as

$$x_{max} = \frac{\delta}{\sqrt{\left[1 - \left(\frac{\omega}{\omega_n}\right)^2\right]^2 + \left(2\zeta \, . \, \frac{\omega}{\omega_n}\right)^2}}$$

$$= \frac{0.012665}{\sqrt{\left[1 - \left(\frac{8\pi}{10\pi}\right)^2\right]^2 + \left(2 \times 0.5066 \times \frac{8\pi}{10\pi}\right)^2}}$$

$$(\because \ \delta = 0.012665 \text{ and } \zeta = 0.5066)$$

$$= \frac{0.012665}{\sqrt{0.1296 + 0.657}} = 0.01428 \text{ m} = 14.28 \text{ mm}$$

When damper is removed ($\zeta = 0$), the amplitude is given as

$$x_{max} = \frac{\delta}{\sqrt{\left[1 - \left(\frac{\omega}{\omega_n}\right)^2\right]^2}} \qquad (\because \ \delta = 0)$$

$$= \frac{0.012665}{1 - \left(\frac{\omega}{\omega_n}\right)^2} = \frac{0.012665}{1 - \left(\frac{8\pi}{10\pi}\right)^2} = \frac{0.012665}{0.36} = 0.035175 \text{ m} = 35.2 \text{ mm}$$

$\therefore$ **Increase in amplitude = 35.2 – 14.28 = 20.92 mm. Ans.**

Problem 16.20. *Determine the frequency of the free vibrations, when a body of mass 20 kg is suspended from a spring which deflects 15 mm under the weight of the body. Also find the viscous damping force required to make the motion aperiodic at a speed of 1 m/s.*

If when damped to this extent, a distinguish force having a maximum value of 187.5 N and vibrating at 8 Hz is made to act on the body, find the amplitude of the ultimate motion.

Sol. Given :

m = 20 kg ; δ = 15 mm = 0.015 m ; speed = 1 m/s ; F = 187.5 N and frequency, f = 8 Hz.

Frequency of free vibration is given by,

$$f_n = \frac{1}{2\pi}\sqrt{\frac{g}{\delta}} = \frac{1}{2\pi}\sqrt{\frac{9.81}{0.015}} = \mathbf{4.07 \text{ Hz.}} \quad \textbf{Ans.}$$

The motion will become aperiodic when the damped frequency is zero or when the motion is critically damped. For motion to be critically damped, the damping factor ζ should be one. *i.e.* $\zeta = 1$.

Now we know that, $\omega_n = \sqrt{\frac{s}{m}}$ where $s = \frac{m \times g}{\delta} = \frac{20 \times 9.81}{0.015} = 13080$ N/m

$$= \sqrt{\frac{13080}{20}} = 25.57 \text{ rad/s}$$

Now using equation (16.31A), we have

$$\zeta = \frac{c}{2 \cdot m \cdot \omega_n}$$

The motion will be aperiodic if $\zeta = 1$.

$$\therefore \qquad 1 = \frac{c}{2 \cdot m \cdot \omega_n}$$

or $$c = 2 \cdot m \cdot \omega_n = 2 \times 20 \times 25.57 = 1022.8 \text{ N/m/s}$$

But c is the damping co-efficient *i.e.* damping force per unit velocity. Hence the damping force required to make the motion aperiodic at a speed of 1 m/s is **1022.8 N. Ans.**

Amplitude of ultimate motion

The amplitude of ultimate motion or the maximum displacement of the forced vibration is given by equation (16.50) as

$$x_{max} = \frac{F}{\sqrt{(s - m\omega^2)^2 + c^2 \cdot \omega^2}} \qquad ...(i)$$

where F = Maximum value of disturbing force = 187.5 N

s = Stiffness of spring = 13080 N/m (calculated above)

m = Mass of body suspended = 20 kg,

c = Damping co-efficient = 1022.8 N/m/s (calculated above)

ω = Angular speed of disturbing force = 2π × frequency of disturbing force

= $2\pi \times 8$ = 50.3 rad/s

Substituting the above values in equation (*i*), we get

$$x_{max} = \frac{187.5}{\sqrt{(13080 - 20 \times 50.3^2)^2 + 1022.8^2 \times 50.3^2}}$$

$$= \frac{187.5}{1407885475 + 2646777345} = \frac{187.5}{63676.2}$$

$$= 2.94 \times 10^{-3} \text{ m} = \mathbf{2.94 \text{ mm.}} \quad \textbf{Ans.}$$

16.15. Vibration Isolation and Transmissibility

If the machines are having some unbalanced masses, the vibrations will be produced in the machines. These vibrations will be transmitted to the foundation upon which the machines are installed. But if between the machines and foundation the springs and dampers are mounted then the vibrating forces will reach the foundation through these springs and dampers. Instead of springs and dampers some other vibration isolating material can also be used between machines and foundation.

The ratio of the force transmitted to the foundation to the force applied to the machine is known as transmissibility. It is represented by ε.

If springs and damper are used between the machine and foundation as shown in Fig. 16.25, then the resultant force transmitted to the foundation will be the vector some of spring force and damping force. The machine is supposed to have one degree of freedom *i.e.* it can move up and down only. In this arrangement the springs are in compression instead of in tension.

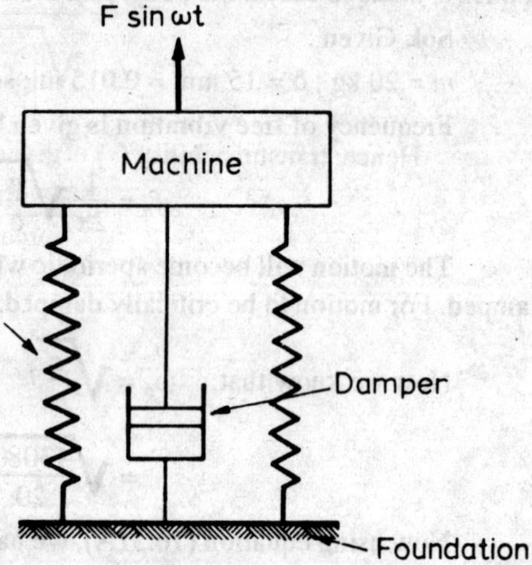

Fig. 16.25

The spring force (maximum) = stiffness of spring × Max. displacement

$= s \cdot A$ ($\because$ From equation (16.51A) x = displacement = $A \sin(\omega t - \phi)$

$\therefore$ Max. displacement = A)

The damping force (maximum) $= c \cdot \frac{dx}{dt}$ $\left(\because \frac{dx}{dt} = A \cdot \omega \cdot \cos(\omega t - \phi) = A \cdot \omega \cdot \sin[90° + (\omega t - \phi)]\right.$

$\left.\text{Max. velocity} = A \cdot \omega\right)$

$= c \cdot A \cdot \omega$ [See equation (16.51B)]

The spring force and damping force are perpendicular to each other. Hence the resultant of these two forces $= \sqrt{(s \cdot A)^2 + (c \cdot A \cdot \omega)^2}$.

Let $\qquad F_t$ = Resultant force transmitted to the foundation

$$= \sqrt{(s\,.\,A)^2 + (c\,.\,A\,.\,\omega)^2} = \sqrt{A^2\,(s^2 + c^2\omega^2)} = A\sqrt{s^2 + c^2\omega^2}$$

But from equation (16.51D), we have

$$A = \frac{F}{\sqrt{s - m\,.\,\omega^2)^2 + (c\omega)^2}}$$

$$\therefore \qquad F_t = \frac{F}{\sqrt{(s - m\omega^2)^2 + (c\omega)^2}} \times \sqrt{s^2 + c^2\omega^2} = \frac{F\sqrt{s^2 + c^2\omega^2}}{\sqrt{(s - m\omega^2)^2 + (c\omega)^2}}$$

$$= \frac{F\sqrt{1 + \left(\frac{c}{s}\,.\,\omega\right)^2}}{\sqrt{\left(1 - \frac{m}{s}\omega^2\right)^2 + \left(\frac{c}{s}\omega\right)^2}} \quad \text{[Dividing numerator and denominator by } s\text{]}$$

$$= \frac{F\sqrt{1 + \left(\frac{2\zeta\omega}{\omega_n}\right)^2}}{\sqrt{\left[1 - \left(\frac{\omega}{\omega_n}\right)^2\right]^2 + \left(2\zeta\,.\,\frac{\omega}{\omega_n}\right)^2}} \qquad \left(\because\ c = 2m\,.\,\omega_n\,.\,\zeta\,;\ s = m\,.\,\omega_n^2 \quad \therefore\ \frac{c}{s} = \frac{2m\,.\,\omega_n\,.\,\zeta}{m\,.\,\omega_n^2} = \frac{2\zeta}{\omega_n}\right)$$

Hence, transmissibility (ε) is given by

$$\varepsilon = \frac{\text{Resultant force transmitted}}{\text{Force applied (Maximum)}}$$

$$= \frac{F_t}{F} \qquad (\because\ \text{Force applied} = F \sin \omega t \text{ and force applied (maximum)} = F)$$

$$= \frac{F\sqrt{1 + \left(2\zeta\frac{\omega}{\omega_n}\right)^2}}{\sqrt{\left[1 - \left(\frac{\omega}{\omega_n}\right)^2\right]^2 + \left(2\zeta\,.\,\frac{\omega}{\omega_n}\right)^2} \times F} = \frac{\sqrt{1 + \left(2\zeta\frac{\omega}{\omega_n}\right)^2}}{\sqrt{\left[1 - \left(\frac{\omega}{\omega_n}\right)^2\right]^2 + \left(2\zeta\,.\,\frac{\omega}{\omega_n}\right)^2}} \qquad ...(16.55)$$

If no damper is used, then damping factor $\zeta = 0$ and above equation becomes as

$$\varepsilon = \frac{1 + 0}{\sqrt{\left[1 - \left(\frac{\omega}{\omega_n}\right)^2\right]^2 + 0}} = \frac{1}{\pm\left[1 - \left(\frac{\omega}{\omega_n}\right)^2\right]} \qquad ...(16.56)$$

If damper is used, but resonance takes place then

$$\frac{\omega}{\omega_n} = 1$$

and $\qquad$ equation (16.55) becomes as

$$\varepsilon = \frac{\sqrt{1 + (2\zeta \times 1)^2}}{\sqrt{[1 - 1]^2 + (2\zeta \times 1)^2}} = \frac{\sqrt{1 + (2\zeta)^2}}{2\zeta} \qquad ...(16.57)$$

Fig. 16.26 shows the variation of transmissibility (ε) with $\left(\frac{\omega}{\omega_n}\right)$ for different values of ζ. From the Fig. 16.26 it is clear that

1. The transmissibility (ε) is more than 1 if $\left(\frac{\omega}{\omega_n}\right) < \sqrt{2}$ for all values of damping factor (ζ).

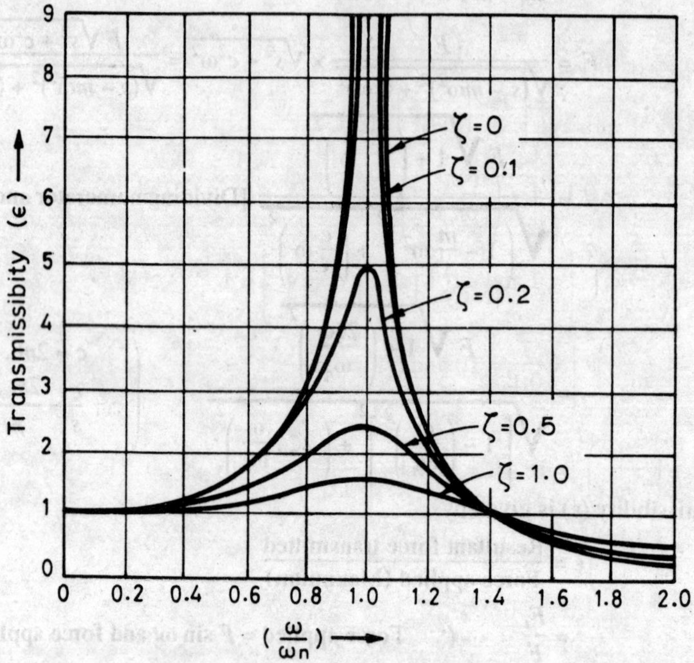

Fig. 16.26

2. The transmissibility (ε) is less than 1, if $\left(\frac{\omega}{\omega_n}\right) > \sqrt{2}$ for all values of ζ.

3. The transmissibility (ε) is equal to 1, if $\left(\frac{\omega}{\omega_n}\right) = \sqrt{2}$ for all values of ζ.

4. The transmissibility (ε) is infinity when $\left(\frac{\omega}{\omega_n}\right) = 1$, and no damper is used (*i.e.* when ζ = 0). But if damper is used then the value of (ε) can be reduced.

5. The transmissibility (ε) increases when $\left(\frac{\omega}{\omega_n}\right) > \sqrt{2}$ and the damping factor (ζ) increases.

Problem 16.21. *Find the stiffness of each spring when a refrigerator unit having a mass of 30 kg is to be supported on three springs. The force transmitted to the supporting structure is only 10% of the impressed force. The refrigerator unit operates at 420 r.p.m.*

Sol. Given :

$m = 30$ kg ; no. of springs = 3 ; force transmitted, F_t = 10% of impressed force $= \frac{10}{100} \times F = 0.1$ F ;

$N = 420$ r.p.m. or $\omega = \frac{2\pi N}{60} = \frac{2\pi \times 420}{60} = 14\pi$ rad/s.

Transmissibility, $\varepsilon = \frac{F_t}{F} = \frac{0.1F}{F} = 0.1.$

In this problem, no damper is used. Hence using equation (16.56), we get

$$\varepsilon = \frac{1}{\pm\left[1 - \left(\frac{\omega}{\omega_n}\right)^2\right]}$$

or $$0.1 = \frac{1}{\pm\left[1 - \left(\frac{14\pi}{\omega_n}\right)^2\right]} \qquad (\because\ \omega = 14\pi \text{ rad/s})$$

or $$\pm 0.1\left[1 - \left(\frac{14\pi}{\omega_n}\right)^2\right] = 1$$

or $$0.1\left[1 - \left(\frac{14\pi}{\omega_n}\right)^2\right] = 1 \text{ and } -0.1\left[1 - \left(\frac{14\pi}{\omega_n}\right)^2\right] = 1$$

or $$\left[1 - \left(\frac{14\pi}{\omega_n}\right)^2\right] = \frac{1}{0.1} = 10 \text{ and } \left[1 - \left(\frac{14\pi}{\omega_n}\right)^2\right] = \frac{1}{-0.1} = -10.$$

or $$-\left(\frac{14\pi}{\omega_n}\right)^2 = 10 - 1 = 9 \text{ and } -\left(\frac{14\pi}{\omega_n}\right)^2 = -10 - 1 = -11$$

or $$\left(\frac{14\pi}{\omega_n}\right)^2 = -9 \text{ and } \left(\frac{14\pi}{\omega_n}\right)^2 = 11$$

or $$\left(\frac{14\pi}{\omega_n}\right) = \sqrt{-9} \text{ and } \frac{14\pi}{\omega_n} = \sqrt{11}$$

But $\frac{14\pi}{\omega_n} = \sqrt{-9}$ is not possible.

Hence $$\frac{14\pi}{\omega_n} = \sqrt{11} \quad \text{or} \quad \omega_n = \frac{14\pi}{\sqrt{11}} = 13.26 \text{ rad/s}$$

We know that $$\omega_n = \sqrt{\frac{s}{m}}$$ where s = stiffness of all springs

$\therefore$ $$13.26 = \sqrt{\frac{s}{30}}$$

or $$(13.26)^2 = \frac{s}{30}$$ or $s = 30 \times (13.26)^2 = 5274.828$ N/m

$\therefore$ Stiffness of each spring $= \frac{5274.828}{3} = 1758.276$ N/m = **1.758 N/mm. Ans.**

Problem 16.22. *A machine has a total mass of 90 kg and unbalanced reciprocating parts of mass 2 kg which moves through a vertical stroke of 100 mm with simple harmonic motion. The machine is mounted on four springs. The machine is having only one degree of freedom and can undergo vertical displacement only.*

Calculate the combined stiffness of the springs if the force transmitted to the foundation is one-thirtieth of the applied force. Neglect damping and take the speed of rotation of the machine crank-shaft as 1000 r.p.m.

When the machine is actually supported on the springs, it is found that the damping reduces the amplitude of the successive free vibrations by 30%. Find :

(*i*) *the force transmitted to the foundation at 900 r.p.m.,*

(*ii*) *the force transmitted to the foundation at resonance, and*

(*iii*) *the amplitude of the forced vibrations at resonance.*

Sol. Given :

$m_1 = 90$ kg ; $m_2 = 2$ kg ; $L = 100$ mm $= 0.1$ m or $r = \frac{L}{2} = \frac{0.1}{2} = 0.05$ m ; no. of springs = 4 ; force transmitted, $F_t = \frac{1}{30}$ of force applied $= \frac{1}{30} \times F$; $N = 900$ r.p.m. or $\omega = \frac{2\pi N}{60} = \frac{2\pi \times 900}{60} = 30\pi$ rad/s, damping is neglected *i.e.* $\zeta = 0$; $\varepsilon = \frac{F_t}{F} = \frac{\frac{1}{30}F}{F} = \frac{1}{30}$

Combined stiffness of springs

Let s = Combined stiffness of springs

When damping is neglected (*i.e.* $\zeta = 0$), the transmissibility (ε) is given by equation (16.56), as

$$\varepsilon = \frac{1}{\pm\left[1 - \left(\frac{\omega}{\omega_n}\right)^2\right]}$$

or
$$\frac{1}{30} = \frac{1}{\pm\left[1 - \left(\frac{30\pi}{\omega_n}\right)^2\right]} \qquad \left(\because \ \varepsilon = \frac{1}{30} \text{ and } \omega = 30\pi \text{ rad/s}\right)$$

or
$$\pm\left[1 - \left(\frac{30\pi}{\omega_n}\right)^2\right] = 30$$

With + ve sign, $1 - \left(\frac{30\pi}{\omega_n}\right)^2 = 30$ or $-\left(\frac{30\pi}{\omega_n}\right)^2 = 29$ or $\left(\frac{30\pi}{\omega_n}\right) = \sqrt{-29}$ which is not possible.

With – ve sign, $-\left[1 - \left(\frac{30\pi}{\omega_n}\right)^2\right] = 30$ or $-\left(\frac{30\pi}{\omega_n}\right)^2 = -31$ or $\left(\frac{30\pi}{\omega_n}\right) = \sqrt{31}$ which is possible.

$$\therefore \quad \omega_n = \frac{30\pi}{\sqrt{31}} = 16.93 \text{ rad/s}$$

But $\omega_n = \sqrt{\frac{s}{m}}$ or $16.93 = \sqrt{\frac{s}{90}}$ (Here $m = m_1$ = Mass of machine)

or
$$(16.93)^2 = \frac{s}{90}$$

or
$$s = 90 \times (16.93)^2 = 25796.24 \text{ N/m} = \mathbf{25.796 \text{ N/mm}}. \quad \textbf{Ans.}$$

(*i*) *The force transmitted to the foundation at 900 r.p.m.*

Let F_t = Force transmitted, and

x_1 = Initial amplitude of vibration.

Then x_2 = Next successive amplitude $= x_1 - 30\%$ of $x_1 = x_1 - 0.3x_1 = 0.7x_1$

Then
$$\frac{x_1}{x_2} = \frac{x_1}{0.7x_1} = \frac{1}{0.7} = 1.4285 \qquad (\because \ x_2 = 0.7x_1)$$

But from equation (16.42), we have

$$\frac{x_1}{x_2} = e^{2\pi\zeta/\sqrt{1-\zeta^2}} \quad \text{or} \quad \text{Ln}\left(\frac{x_1}{x_2}\right) = \frac{2\pi\zeta}{\sqrt{1-\zeta^2}}$$

or $$\text{Ln } 1.4285 = \frac{2\pi\zeta}{\sqrt{1-\zeta^2}} \qquad \left(\because \frac{x_1}{x_2} = 1.4285\right)$$

or $$0.3567 = \frac{2\pi\zeta}{\sqrt{1-\zeta^2}} \quad \text{or} \quad \frac{\zeta}{\sqrt{1-\zeta^2}} = \frac{0.3567}{2\pi} = 0.05677$$

Squaring both sides, we get

$$\frac{\zeta^2}{1-\zeta^2} = (0.05677)^2 = 0.00322$$

or $$\zeta^2 = (1-\zeta^2) \times 0.00322 = 0.00322 - 0.00322\zeta^2$$

or $$\zeta^2 + 0.00322\zeta^2 = 0.00322$$

or $$1.00322\zeta^2 = 0.00322$$

or $$\zeta = \sqrt{\frac{0.00322}{1.00322}} = 0.0566$$

Now using equation (16.55), we have

$$\varepsilon = \frac{\sqrt{1 + \left(2\zeta\frac{\omega}{\omega_n}\right)^2}}{\sqrt{\left[1 - \left(\frac{\omega}{\omega_n}\right)^2\right]^2 + \left(2\zeta\frac{\omega}{\omega_n}\right)^2}}$$

$$= \frac{\sqrt{1 + \left(2 \times 0.0566 \times \frac{30\pi}{16.93}\right)^2}}{\sqrt{\left[1 - \left(\frac{30\pi}{16.93}\right)^2\right]^2 + \left(2 \times 0.00566 \times \frac{30\pi}{16.93}\right)^2}} = \frac{1.182}{29.99} = 0.0394$$

We know that the maximum unbalanced force on the machine due to reciprocating parts is given by,

$$F = m_2 \, . \, \omega^2 \, . \, r = 2 \times (30\pi)^2 \times 0.05 = 888.26 \text{ N}$$

But transmissibility, $\varepsilon = \frac{F_t}{F}$

$\therefore$ Force transmitted to the foundation is given by,

$$F_t = \varepsilon \times F$$

$$= 0.0394 \times 888.26 = 34.997 \simeq \mathbf{35 \text{ N.}} \quad \textbf{Ans.}$$

(ii) The force transmitted to the foundation at resonance

At resonance, $\omega = \omega_n$ or $\frac{\omega}{\omega_n} = 1.$

The transmissibility at resonance is given by equation (16.57) as

$$\varepsilon = \frac{\sqrt{1 + (2\zeta)^2}}{2\zeta} = \frac{\sqrt{1 + (2 \times 0.0566)^2}}{2 \times 0.0566} = \frac{1.0064}{0.1132} = 8.89$$

And maximum unbalanced force on the machine due to reciprocating parts at resonance speed (ω_n) is given by,

$$F = m_2 \times (\omega_n)^2 \times r = 2 \times (16.93)^2 \times 0.05 = 28.66 \text{ N} \qquad (\because \ \omega_n = 16.93 \text{ rad/s})$$

$\therefore$ Force transmitted to the foundation at resonance,

$$F_t = \varepsilon \times F = 8.89 \times 28.66 = \mathbf{254.78 \ N.} \quad \textbf{Ans.}$$

(iii) Amplitude of the forced vibrations at resonance

The amplitude of the forced vibration at resonance is given by,

$$\text{Amplitude} = \frac{\text{Force transmitted at resonance}}{\text{Combined stiffness}}$$

$$= \frac{254.78}{25796.24} = 9.876 \times 10^{-3} \text{ m} = \mathbf{9.876 \ mm.} \quad \textbf{Ans.}$$

HIGHLIGHTS

1. The vibrations are of three types *i.e.* Free vibrations (or natural vibrations), forced vibrations and damped vibrations.
2. The vibrations, without any external force, are known as free vibrations. The vibrations with external force, are known as forced vibrations and vibrations in which there is a reduction in amplitude over every cycle, are known as damped vibrations.
3. No. of cycles completed per second is known as frequency. It is expressed in hertz (which is written as Hz).
4. When the frequency of the external force is the same as that of the natural frequency of the system, resonance takes place.
5. Free vibrations are of three types :
   (*i*) Longitudinal vibrations,
   (*ii*) Transverse vibrations and
   (*iii*) Torsional vibrations.
6. The natural frequency of free longitudinal vibrations can be obtained by :
   (*i*) Equilibrium method,
   (*ii*) Energy method, and
   (*iii*) Rayleigh's method.
7. Natural frequency (f_n) of free longitudinal vibration is given by

   (*i*) $f_n = \frac{1}{2\pi}\sqrt{\frac{s}{m}}$ ...... in terms of stiffness and mass

   (*ii*) $f_n = \frac{1}{2\pi}\sqrt{\frac{g}{\delta}}$ ...... in terms of static deflection

   $= \frac{0.4985}{\sqrt{\delta}}$ Hz where δ = static deflection in metre

   The value of $\delta = \frac{W \times L}{A \times E}$ ...... for longitudinal free vibration

   $= \frac{WL^3}{3EI}$ ...... for transverse vibrations.
8. If the mass of the shaft is to be taken into consideration, then :

   (*i*) Natural frequency of longitudinal vibrations is given by, $f_n = \frac{1}{2\pi}\sqrt{\frac{s}{(m + M_s/3)}}$

   where s = Stiffness, m = Mass attached to the shaft, and M_s = Mass of the shaft.

(*ii*) Natural frequency of transverse vibrations is given by, $f_n = \frac{1}{2\pi}\sqrt{\frac{s}{(m + 33M_s)/140)}}$

9. Natural frequency of vibrations when a shaft carries a single concentrated load is given by,

$$f_n = \frac{1}{2\pi} \times \sqrt{\frac{g}{\delta}}$$

where $\delta = \frac{WL^3}{3EI}$ ...... For a cantilever when the point load W is at free end

$= \frac{W.a^2.b^2}{3EIL}$ ...... For a simply supported beam and point load is at a distance of a and b from supports.

$= \frac{W.a^3.b^3}{3EIL^3}$ ...... For a beam fixed at both ends and point load is at a distance of a and b from the supports.

10. Natural frequency of vibrations when a shaft carries an uniformly distributed load is given by,

$f_n = \frac{\pi}{2}\sqrt{\frac{EI}{mL^4}}$ ...... For a simply supported beam in terms of mass distributed per unit length

$= \frac{\pi}{2}\sqrt{\frac{5}{384} \times \frac{g}{\delta}}$ ...... For a simply supported beam in terms of static deflection due to distributed load.

11. Natural frequency of transverse vibrations of a system of several loads attached to the same shaft is given by (*a*) energy method as :

$$f_n = \frac{1}{2\pi}\sqrt{\frac{g\Sigma m.y}{\Sigma m.y^2}}$$

where m = Mass attached to the shaft, and

y = Corresponding deflection of the shaft under the weight attached to the shaft

(*b*) By Dunkerley's method as :

$$f_n = \frac{0.4985}{\sqrt{\delta_1 + \delta_2 + \delta_3 + \ldots\ldots \frac{\delta_s}{1.27}}}$$

where $\delta_1, \delta_2, \ldots\ldots$ = Deflection of shaft under loads when each load acts separately

$= \frac{W.a^2.b^2}{3EIL}$

δ_s = Deflection due to the mass of the shaft

$= \frac{5}{384}\frac{w \times L^4}{EI}$ where w = self-weight of shaft per meter length.

12. The speed at which the shaft runs so that the deflection of the shaft from the axis of rotation becomes infinite, is known as whirling speed or critical speed. It is represented by N_c and is given by,

$N_c = \frac{0.4985}{\sqrt{\delta}}$ where δ = static deflection under the load.

13. Critical speed in r.p.s. is equal to the frequency of transverse vibration of a shaft when the shaft carries a point load, or a uniformly distributed load or combination of both.

14. In case of damped vibrations, the amplitude of vibrations gradually diminishes (*i.e.* amplitude goes on decreasing and finally becomes zero).

15. Damping factor is defined as the square root of the ratio of $\left(\frac{c}{2m}\right)^2$ to $\frac{s}{m}$. It is represented by ζ.

$$\therefore \quad \zeta = \sqrt{\frac{(c/2m)^2}{(s/m)}} = \sqrt{\frac{(c/2m)^2}{\omega_n^2}} = \frac{c}{2 \cdot m \cdot \omega_n}$$

where c = Co-efficient of damping or damping force per unit velocity

m = Mass of the vibrating body, and

ω_n = Natural circular frequency or Natural angular velocity of the vibrating body.

16. The damping may be over-damped, under-damped or critical if $\zeta > 1, \zeta < 1$ or $\zeta = 1$.

17. The co-efficient of damping corresponding to damping factor equal to one (*i.e.* $\zeta = 1$), is known as critical co-efficient of damping. It is represented by C_c. It is given by $C_c = 2 \cdot m \cdot \omega_n$.

18. Damping factor is also defined as the ratio of actual damping co-efficient to critical damping co-efficient.

19. For critically-damped system and over-damped system, the motion is aperiodic and eventually diminishes to zero.

20. For under-damped system, the motion is simple harmonic with amplitude $Xe^{-\zeta \cdot \omega_n \cdot t}$ and angular frequency of ω_d. The amplitude goes on decreasing exponentially with time. The displacement is given by,

$$x = Xe^{-\zeta \cdot \omega_n \cdot t} \sin(\omega_d \cdot t + \phi).$$

21. The natural logarithm of any two successive amplitudes on the same side of the mean position in an under-damped system is known as logarithmic decrement. It is denoted by δ.

$$\therefore \quad \delta = \text{Ln}\left(\frac{x_1}{x_2}\right) = \text{Ln}\left(\frac{x_2}{x_3}\right) = \text{Ln}\left(\frac{x_3}{x_4}\right) = \ldots\ldots = \frac{2\pi\zeta}{\sqrt{1-\zeta^2}}.$$

22. The displacement of forced-damped vibration is given by,

$$x = \frac{F \sin(\omega \cdot t - \phi)}{\sqrt{(s - m \cdot \omega)^2 + c^2 \cdot \omega^2}}$$

where applied force = $F \sin \omega \cdot t$

ϕ = Phase lag with respect to applied force $F \sin \omega \cdot t$

s = Stiffness of spring

c = Damping co-efficient

ω = Circular frequency

m = Mass of vibrating body.

23. The ratio of maximum displacement of forced vibration to the deflection due to static load F is known as magnification factor. It is represented by M.F.

$$\therefore \quad \text{M.F.} = \frac{1}{\sqrt{\left(1 - \frac{\omega^2}{\omega_n^2}\right)^2 + \left(2\zeta \frac{\omega}{\omega_n}\right)^2}}.$$

24. The ratio of resultant force transmitted to the foundation to the maximum force applied is known as transmissibility. It is represented by ε. It is given by,

$$\varepsilon = \frac{\sqrt{1 + \left(2\zeta \frac{\omega}{\omega_n}\right)^2}}{\sqrt{\left[1 - \left(\frac{\omega}{\omega_n}\right)^2\right]^2 + \left(2\zeta \frac{\omega}{\omega_n}\right)^2}} \quad \text{...with damper}$$

$$= \frac{1}{\pm\left[1-\left(\frac{\omega}{\omega_n}\right)^2\right]} \quad \text{...with no damper } (i.e.\ \zeta = 0)$$

$$= \frac{\sqrt{1+(2\zeta)^2}}{2\zeta} \quad \text{...with damper and also with resonance } \left(i.e.\ \omega = \omega_n \text{ or } \frac{\omega}{\omega_n} = 1\right)$$

EXERCISE 16

A. THEORETICAL QUESTIONS

1. Define and explain the terms : vibrations, free vibration, forced vibration and damped vibration.
2. What do you mean by period of vibration, cycle, frequency and resonance as applied to vibratory motions ?
3. What are the important types of free vibrations ? What type of stresses are produced in case of free-vibrations of different types ?
4. What are the different methods of finding the natural frequency of free longitudinal vibrations ? Explain *f* ' in details any two methods.
5. Prove that the natural angular velocity (or circular frequency) of a body having free longitudinal vibration is given by

$$\omega_n = \sqrt{\frac{s}{m}}$$

where ω_n = Circular frequency or natural angular velocity,
s = Stiffness of the spring to which the body of mass m is attached and
m = Mass of vibrating body.

6. Prove that the natural frequency of free-longitudinal vibrations, is given by,

$$f_n = \frac{0.4985}{\sqrt{\delta}} \text{ Hz} \quad \text{where } \delta = \text{static deflection in meter.}$$

7. Prove that the natural frequency of longitudinal vibrations, when the mass of the shaft (or of the spring) is taken into consideration, is given by

$$f_n = \frac{1}{2\pi}\sqrt{\frac{s}{(m + M_s/3)}}$$

where m = Mass attached to the shaft (or to the spring),
M_s = Total mass of the shaft, and
s = Stiffness of the shaft (or of the spring).

8. A simply supported beam of span L carries an uniformly distributed mass m kg per meter length. Prove that the natural frequency of transverse vibration is given by,

$$f_n = \frac{\pi}{2}\sqrt{\frac{EI}{mL^4}} \quad \text{...in terms of uniformly distributed mass}$$

$$= \frac{\pi}{2}\sqrt{\frac{5}{384} \times \frac{g}{\delta}} \quad \text{...in terms of static deflection } (\delta).$$

9. What are the different methods of finding the natural frequency of transverse vibrations of a system of several loads attached to the same shaft ? Explain in details anyone of the methods.
10. Prove that the natural frequency of transverse vibrations of a system of several loads attached to the same shaft, is given by

$$f_n = \frac{1}{2\pi}\sqrt{\frac{g \times \Sigma m\,.\,y}{\Sigma m\,.\,y^2}}$$

where m = Mass of load attached to the shaft, and
y = deflection of the shaft under the corresponding load.

11. Prove the relation,

$$f_n = \frac{0.4985}{\sqrt{\delta_1 + \delta_2 + \delta_3 + \ldots\ldots \frac{\delta_s}{1.27}}}$$

where $\delta_1, \delta_2, \ldots\ldots$ = Deflection of the shaft under the load when each load acts separately

δ_s = Deflection of the shaft due to self-weight.

12. Define the term "Whirling speed of a shaft". Prove that the whirling speed of a shaft in r.p.s. (resolution per second) is given by the relation,

$$N_c = \frac{0.4985}{\sqrt{\delta}}$$

13. Explain the term, 'Damped vibrations'. Prove that the equation of motion for a damped vibration is given by

$$\ddot{x} + \frac{c}{m}\dot{x} + \frac{s}{m}x = 0$$

where c = Damping co-efficient or damping force per unit velocity,

m = Mass of the vibrating body,

s = Stiffness of spring, and

x = Displacement from mean position.

14. Define the terms : damping factor, co-efficient of damping and critical damping co-efficient. Find an expression for damping co-efficient in terms of circular efficiency mass of vibrating body and damping factor.

15. What do you mean by over-damped system, under-damped system and critical damped system ? Plot the variation of displacement with time on a single graph for over-damped system, under-damaged, critical damped and un-damped system.

16. What do you mean by Logarithmic decrement ? Find an expression for logarithmic decrement in terms of damping factor.

17. Prove that the ratio of two successive amplitudes in case of under-damped system is constant.

18. Define the term, "Magnification factor".

19. Define the terms : vibration isolation and transmissibility. Prove that the transmissibility is given by,

$$\varepsilon = \frac{1}{\pm\left[1 - \left(\frac{\omega}{\omega_n}\right)^2\right]} \quad \text{...when no damper is used in the system}$$

$$= \frac{\sqrt{1 + (2\zeta)^2}}{2\zeta} \quad \text{...when damper is used but resonance takes place.}$$

B. NUMERICAL PROBLEMS

1. A vertical shaft 100 mm in diameter and 1 m in length has its upper end fixed at the top. At the other end it carries a disc of weight 20 kN. The Young's modulus of the material of the shaft is 2×10^5 N/mm^2. Neglecting the weight of the shaft, determine the frequency of longitudinal vibrations and transverse vibrations.

[**Ans.** (*i*) 139.75 Hz, (*ii*) 6 Hz]

2. A simply supported shaft of length 1.6 m carries a mass of 120 kg placed 500 mm from one end. If E = 200 GN/m^2 and diameter of shaft is 50 mm, then find the natural frequency of transverse vibrations. [**Ans.** 14.34 Hz]

3. Find the frequency of transverse vibrations of a shaft which is simply supported at the ends and is of 40 mm in diameter. The length of the shaft is 5 m. The shaft carries three point loads of masses 15 kg, 35 kg and 22.5 kg at 1 m, 2 m and 3.4 m respectively from the left support. The Young's modulus for the material of the shaft is 200 GN/m^2. Neglect the weight of the shaft. [**Ans.** 2.1 Hz]

4. If in question 3, the weight of shaft is not negligible but it is given as 18.394 N per meter length. All other data remains the same, then find the frequency of the transverse vibrations of the shaft. [**Ans.** 2 Hz]

5. A shaft is simply supported at the ends and is of 20 mm is diameter and 600 mm in length. The shaft carries a load of 9.81 N at its centre. The weight of shaft per metre length is 124.1 N. Find the critical speed of the shaft. Take Young's modulus = 200 GN/m^2. **[Ans.** 2596.5 r.p.m.]

6. A shaft is simply supported at its ends and is of 40 mm is diameter and of length 5 m. The shaft carries three point loads of masses 15 kg, 35 kg and 22.5 kg at 1 m, 2 m, and 3.4 m respectively from the left support. The weight of the shaft per metre length is given as 18.394 N. The Young's modulus for the material of the shaft is 200 GN/m^2. Find the critical speed of the shaft. **[Ans.** 120 r.p.m.]

7. A vertical shaft is held in long bearings and a disc is attached to the shaft at its mid-point. The centre of gravity of the disc does not coincide with the axis of the shaft. Determine : (*i*) the critical speed of the shaft and (*ii*) the range of the speed over which it is unsafe to 'run the shaft'.

   The diameter of the shaft is 15 mm and the span of the shaft between the bearing is 1 metre. The mass of the disc is 20 kg and the C.G. of disc from the axis of shaft is 0.30 mm. Take E = 200 GN/m^2 and permissible stress in the shaft material is 70 MN/m^2. **[Ans.** (*i*) 661.85 r.p.m. ; (*ii*) 615.9 r.p.m. to 719.8 r.p.m.]

8. A machine of mass 20 kg is supported on springs and dashpot. The total stiffness of the spring is 10 N/mm and total damping is 0.15 N/mm/s. If the system is initially at rest and a velocity of 100 mm/s is imported to the mass, determine (*i*) the displacement and velocity of the mass as a function of time and (*ii*) the displacement and velocity after one second. **[Ans.** (*i*) $x = 4.536\, e^{-3.75t} \sin(22.043t)$; $v = e^{-3.75t}$ (100 cos 22.043t – 17 sin 22.043t) ; (*ii*) x = 0.04 mm and v = 2.03 mm/s]

9. Find the damping factor of a vibrating system which consists of a mass of 7 kg, a spring of stiffness 5 N/mm and a damper of damping co-efficient of 0.036 N/mm/s. Find also the logarithmic decrement and the ratio of any two consecutive amplitude. **[Ans.** 0.0962 ; 0.604 and 1.829]

10. The guns are designed so that on firing, the barrel recoils against a spring. A dashpot is engaged that allows the barrel to return to its position at the end of each recoil. A gun barrel has a mass of 500 kg and a recoil spring constant of 300 N/mm. The barrel recoils 1 m on firing. Determine : (*i*) the initial recoil velocity of the gun barrel, and (*ii*) the critical damping co-efficient of the dashpot engaged at the end of the recoil stroke. **[Ans.** (*i*) 24.5 m/s and (*ii*) 24.5 N/mm/s]

17

Torsional Vibrations

17.1. Introduction

A shaft, assumed to be weightless and fixed at one end and carrying a heavy disc or flywheel at the free end, can be made to vibrate in a circle about the axis of the shaft by applying an external torque on the disc (the external torque is removed after initial displacement). The shaft is twisted and untwisted alternately and shear stresses are induced, these vibrations are known as torsional vibrations. This chapter deals with the determination of the frequency of torsional vibrations of the various systems such as :

(*i*) Natural frequency of free torsional vibrations,

(*ii*) Effect of inertia on torsional vibrations,

(*iii*) Free torsional vibrations of a single rotor system,

(*iv*) Free torsional vibration of a two rotor system,

(*v*) Free torsional vibration of a three rotor system, and

(*vi*) Free torsional vibration of a geared system.

17.2. Natural Frequency of Free Torsional Vibrations

Fig. 17.1 shows a shaft of negligible mass whose upper end is fixed and lower end carries a heavy disc. If the disc is given a twist about its vertical axis and then released, it will start oscillating about the axis, which are known as torsional vibrations.

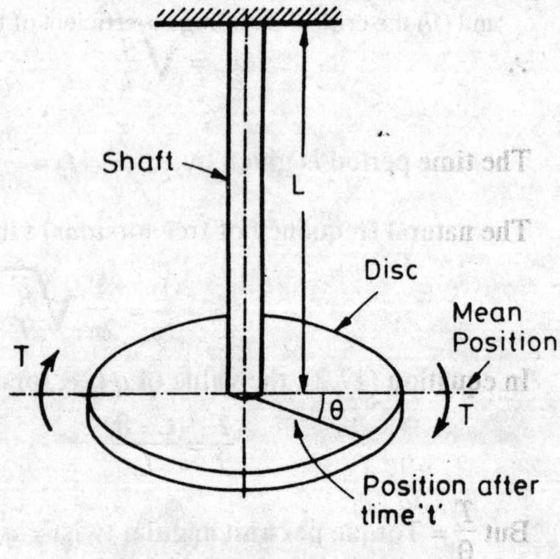

Fig. 17.1

Let I = Mass moment of inertia of disc = $m \cdot k^2$,

m = Mass of the disc,

k = Radius of gyration in meter,

θ = Angular displacement of the disc from its mean position after time t, and

q = Torsional stiffness of the shaft or torque required to twist the shaft through one radian.

At any instant, the torques acting on the disc are :

(*i*) inertia torque and

(*ii*) restoring torque (or spring torque)

The inertia torque is equal to accelerating torque but opposite in direction. The accelerating torque

$$= \text{Moment of inertia} \times \text{Angular acceleration}$$

$$= I \times \alpha$$

$$= I \times \frac{d^2\theta}{dt^2} \qquad \left(\because \ \alpha = \frac{d^2\theta}{dt^2}\right)$$

$\therefore$ Inertia torque $= -$ Acceleration torque $= -I \times \dfrac{d^2\theta}{dt^2}$

and restoring torque $= -$ Torsional stiffness $\times$ Angular displacement in radians

(– ve sign is used as restoring torque acts opposite to the angular displacement)

$= -q \times \theta$

For equilibrium, the sum of all torques acting on the disc must be zero.

$\therefore$ Inertia torque + restoring torque = 0

or $$-I \times \frac{d^2\theta}{dt^2} + (-q \times \theta) = 0 \quad \text{or} \quad I \times \frac{d^2\theta}{dt^2} + q \times \theta = 0$$

or $$\frac{d^2\theta}{dt^2} + \frac{q}{I} \times \theta = 0 \qquad ...(i)$$

The fundamental equation of the simple harmonic motion is

$$\frac{d^2\theta}{dt^2} + \omega^2 \,.\, \theta = 0 \qquad ...(ii)$$

Comparing equations (*i*) and (*ii*), we get

$$\omega^2 = \frac{q}{I} \quad \text{or} \quad \omega = \sqrt{\frac{q}{I}}$$

But ω is the angular velocity of the simple harmonic motion. This is also known as natural angular velocity of torsional vibrations and is represented by ω_n.

$$\therefore \quad \omega_n = \sqrt{\frac{q}{I}} \qquad ...(17.1)$$

The time period is given by, $$t_p = \frac{2\pi}{\omega_n} = 2\pi\sqrt{\frac{I}{q}}$$

The natural frequency of free torsional vibrations is given by,

$$f_n = \frac{1}{t_p} = \frac{1}{2\pi}\sqrt{\frac{q}{I}} \qquad ...(17.2)$$

In equation (17.2), the value of q (*i.e.* torsional stiffness) is obtained from the torsional equation,

$$\frac{T}{J} = \frac{C\,.\,\theta}{L} \quad \text{or} \quad \frac{T}{\theta} = \frac{C\,.\,J}{L}$$

But $\dfrac{T}{\theta}$ = Torque per unit angular twist $= q$

$$\therefore \quad q = \frac{C\,.\,J}{L} \qquad ...(17.3)$$

where C = Modulus of rigidity for shaft material,

J = Polar moment of inertia of shaft cross-section $= \dfrac{\pi}{32}\,.\,d^4$,

d = Diameter of shaft, and

L = Length of shaft.

Problem 17.1. *A vertical shaft is fixed at the top and carries a flywheel of weight 5000 N welded at its bottom. The diameter of the shaft is 100 mm and its length is 1 meter. The modulus of rigidity for the material*

of the shaft is 8.16 × 10⁴ N/mm². If the radius of gyration of flywheel is 250 mm, then determine the frequency of torsional vibrations.

Sol. Given :

$W = 5000$ N or mass, $m = \frac{W}{g} = \frac{5000}{9.81}$ kg ; $d = 100$ mm = 0.1 m ; $L = 1$ m ;

$C = 8.16 \times 10^4$ N/mm² = $8.16 \times 10^4 \times 10^6$ N/m² = 8.16×10^{10} N/m² ; $k = 250$ mm = 0.25 m.

The frequency of torsional vibrations is given by equation (17.2) as

$$f_n = \frac{1}{2\pi}\sqrt{\frac{q}{I}} \qquad \text{...(i)}$$

where q = Torsional stiffness = $\frac{C.J}{L}$ [Refer to equation (17.3)]

$$= \frac{(8.16 \times 10^{10}) \times \left(\frac{\pi}{32} \times d^4\right)}{1} \qquad \left(\because\ C = 8.16 \times 10^{11} \text{ N/m}^2 \text{ and } J = \frac{\pi}{32}d^4, L = 1 \text{ m}\right)$$

$$= (8.16 \times 10^{10}) \times \left(\frac{\pi}{32} \times 0.1^4\right) \qquad (\because\ d = 0.1 \text{ m})$$

$$= 8.011 \times 10^5 \text{ N m}$$

and I = Mass moment of inertia of shaft

$$= m \times k^2$$

$$= \left(\frac{5000}{9.81}\right) \times (0.25^2) \text{ kg m}^2 = 31.855 \text{ kg m}^2$$

Substituting the values of q and I in equation (*i*), we get

$$f_n = \frac{1}{2\pi}\sqrt{\frac{8.011 \times 10^5}{31.855}} = 25.25 \text{ cycles/s}$$ or **25.25 Hz. Ans.**

17.3. Effect of Inertia of Shaft on Torsional Vibrations

Fig. 17.2 shows a shaft whose upper end is fixed and lower end is free.

Let L = Length of the shaft

ω = Angular velocity of free end of the shaft

I_1 = Mass moment of inertia of the shaft.

Consider an element of length dx at a distance x from the fixed end.

Now the angular velocity* of the element

$$= \frac{\omega}{L} \times x$$

Mass moment of inertia** of element $= \frac{I_1}{L} \times dx$

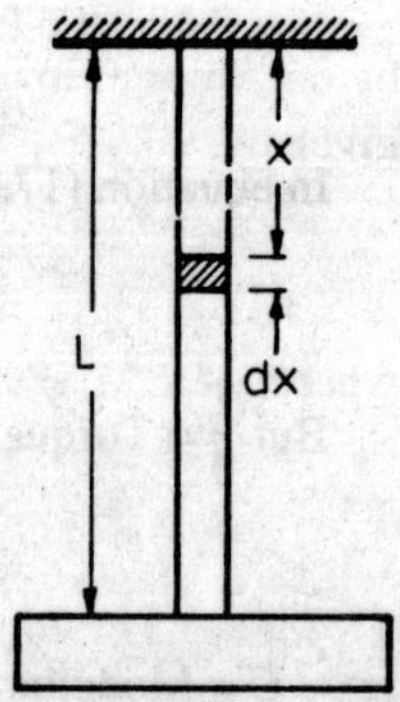

Fig. 17.2

*At a distance L from fixed end, the angular velocity is ω. Hence at a distance x from fixed end, the angular velocity will be $\frac{\omega}{L} \times x$.

**The mass moment of inertia of shaft of length L is I_1. Hence mass moment of inertia of element of length dx will be $\frac{I_1}{L} \times dx$.

The kinetic energy possessed by the element

$$= \frac{1}{2} \times (\text{Moment of inertia of element}) \times (\text{Angular Vel. of element})^2$$

$$\left(\because \quad \text{K.E.} = \frac{1}{2} \times I \times \omega^2 \right)$$

$$= \frac{1}{2} \times \left(\frac{I_1}{L} \times dx \right) \times \left(\frac{\omega}{L} \times x \right)^2$$

$$= \frac{1}{2} \times \frac{I_1}{L} \times \frac{\omega^2}{L^2} \times x^2 \times dx = \frac{1}{2} \times \frac{I_1}{L^3} \times \omega^2 \times x^2 \times dx$$

Total K.E. of the shaft is obtained by integrating the above equation from 0 to L.

$$\therefore \quad \text{Total kinetic energy of the shaft} = \int_0^L \frac{1}{2} \times \frac{I_1}{L^3} \times \omega^2 \times x^2 \times dx$$

$$= \frac{1}{2} \times \frac{I_1}{L^3} \times \omega^2 \times \int_0^L x^2 \, dx$$

$$= \frac{1}{2} \times \frac{I_1}{L^3} \times \omega^2 \left[\frac{x^3}{3} \right]_0^L$$

$$= \frac{1}{2} \times \frac{I_1}{L^3} \times \omega^2 \times \frac{L^3}{3} = \frac{1}{2} \times \left[\frac{I_1 \times \omega^2}{3} \right] = \frac{1}{2} \left[\left(\frac{I_1}{3} \right) \times \omega^2 \right]$$

In the above expression, ω is the angular velocity of the free end of the shaft.

If a mass whose mass moment of inertia is equal to $\left(\frac{I_1}{3} \right)$ is placed at the free end and shaft is assumed to be of negligible mass, then the K.E. of the mass placed at the free-end will be $\frac{1}{2} \times \frac{I_1}{3} \times \omega^2$.

Hence to consider the effect of inertia of the shaft, the mass moment of inertia of the disc is increased by one-third of mass moment of inertia of the shaft. The natural frequency of the torsional vibrations then is given by,

$$f_n = \frac{1}{2\pi} \sqrt{\frac{q}{I + I_1/3}} \quad \text{...(17.4)}$$

where I = Mass moment of inertia of disc,

I_1 = Mass moment of inertia of shaft, and

q = Torsional rigidity.

17.4. Free Torsional Vibrations of a Single Rotor System

Fig. 17.3 shows a shaft whose one end is fixed and other end carries a rotor. This is known as single rotor system. When torsional vibrations are produced in this system, the natural frequency of torsional vibrations of this system is given by,

$$f_n = \frac{1}{2\pi} \sqrt{\frac{q}{I}} .$$

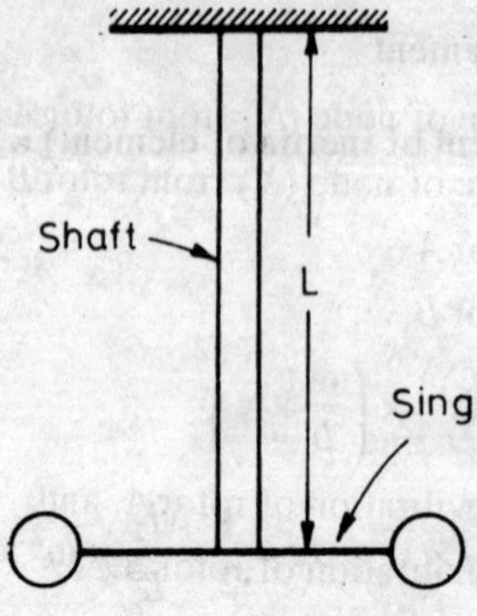

Fig. 17.3

The amplitude of vibrations is maximum at the free end where single rotor is attached whereas the amplitude of vibrations is zero at the fixed end of the shaft. The section of shaft where the amplitude is zero, is known as *node*. Hence at node, the shaft remains undisturbed by the vibrations.

17.5. Free Torsional Vibrations of a Two Rotor System

Fig. 17.4 (*a*) shows a shaft to which two rotors are attached at its free ends. This system is known as two rotor system. This system will vibrate torsionally if the two rotors twist* in opposite* directions. Thus some length of the shaft is twisted in one direction while the rest is twisted in the other.

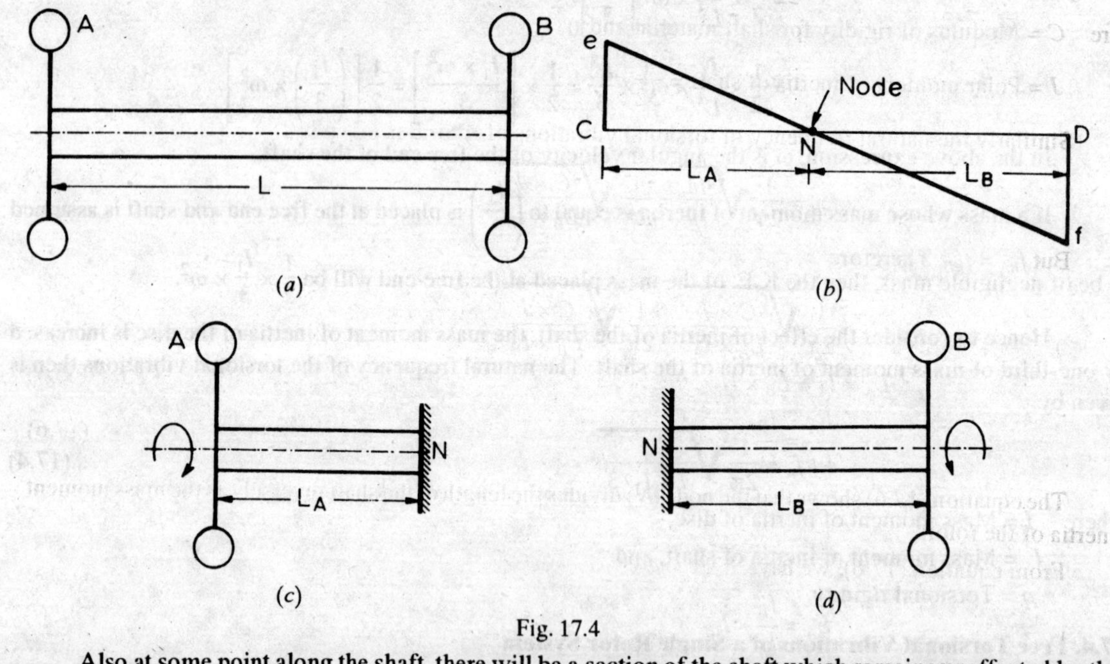

Fig. 17.4

Also at some point along the shaft, there will be a section of the shaft which remains unaffected by the vibrations. This section is known as node. In Fig. 17.4 (*b*), the node is shown by point *N*. The shaft behaves as if clamed at the node and the two sections vibrate as two separate shafts with the **same** frequency but opposite in phase or opposite in direction. Hence the sections *NC* and *ND* can be considered as two separate shafts fixed at *N* as shown in Fig. 17.4 (*c*).

---

*If the two rotors twist in the same direction, the system will rotate only. The system will not vibrate torsionally then.

Let L = Length of the shaft,

L_A = Length of shaft NC *i.e.* distance of node (N) from rotor A

L_B = Length of shaft ND *i.e.* distance of node (N) from rotor B

I_A = Mass moment of inertia of rotor A,

I_B = Mass moment of inertia of rotor B,

q_A = Torsional stiffness of length NC,

q_B = Torsional stiffness of length ND, and

f_{nA} = Natural frequency of torsional vibration of rotor A, and

f_{nB} = Natural frequency of torsional vibration of rotor B.

The natural frequency of torsional vibrations of rotor A is given by,

$$f_{nA} = \frac{1}{2\pi}\sqrt{\frac{q_A}{I_A}}$$

$$= \frac{1}{2\pi}\sqrt{\frac{(C \times J/L_A)}{I_A}} \qquad \left(\because \; q_A = \frac{C \times J}{L_A}\right)$$

$$= \frac{1}{2\pi}\sqrt{\frac{C \times J}{L_A \times I_A}} \qquad \ldots(17.5)$$

where C = Modulus of rigidity for shaft material and

J = Polar moment of inertia of shaft $= \frac{\pi}{32} d^4$.

Similarly the natural frequency of torsional vibrations of rotor B is given by,

$$f_{nB} = \frac{1}{2\pi}\sqrt{\frac{q_B}{I_B}} = \frac{1}{2\pi}\sqrt{\frac{C \times J}{L_B \times I_B}}$$

But $f_{nA} = f_{nB}$. Therefore

$$\frac{1}{2\pi}\sqrt{\frac{C \times J}{L_A \times I_A}} = \frac{1}{2\pi}\sqrt{\frac{C \times J}{L_B \times I_B}}$$

or
$$L_A \times I_A = L_B \times I_B \qquad \ldots(17.5A)$$

or
$$\frac{L_A}{L_B} = \frac{I_B}{I_A} \qquad \ldots(17.6)$$

The equation (17.6) shows that the node (N) divides the length of the shaft inversely as the mass moment of inertia of the rotors.

From equation (17.6), we have

$$L_A = \frac{I_B}{I_A} \times L_B \qquad \ldots(i)$$

We also know that, $L = L_A + L_B$ or $L_B = (L - L_A)$

Substituting the value of L_B in equation (i), we get

$$L_A = \frac{I_B}{I_A} \times (L - L_A) = \frac{I_B}{I_A} \times L - \frac{I_B}{I_A} \times L_A$$

or
$$L_A + \frac{I_B}{I_A} . L_A = \frac{I_B}{I_A} . L \qquad \text{or} \qquad L_A\left(1 + \frac{I_B}{I_A}\right) = \frac{I_B}{I_A} . L$$

or $$L_A\left(\frac{I_A+I_B}{I_A}\right)=\frac{I_B}{I_A}\cdot L \quad \text{or} \quad L_A=\frac{I_B}{(I_A+I_B)}\cdot L \qquad ...(17.7)$$

Hence the value of L_A (or position of node N) can be obtained from equation (17.7). This value is substituted in equation (17.5) and the frequency of torsional vibration is obtained from equation (17.5).

In Fig. 17.4 (*b*), *Ce* and *Df* are the amplitudes of oscillations of rotor *A* and *B* respectively. The straight line *ef* will pass through the node *N*. The line *ef* is known as *elastic line* of the shaft.

17.6. Free Torsional Vibrations of a Three Rotor System

Fig. 17.5 shows a shaft to which three rotors *A*, *B* and *C* are attached. The rotors *A* and *C* are attached to the ends of the shaft, whereas the rotor *B* is attached in between *A* and *C*. There are two possible natural modes of vibrations as given below :

(*i*) In one mode, there are two nodes, one between rotor *A* and rotor *B* and other node between rotor *B* and rotor *C*. In this mode, the outside rotors (*i.e.* rotor *A* and rotor *C*) rotate in the same direction whereas the rotor *B* rotate in opposite direction as shown in Fig. 17.5 (*a*).

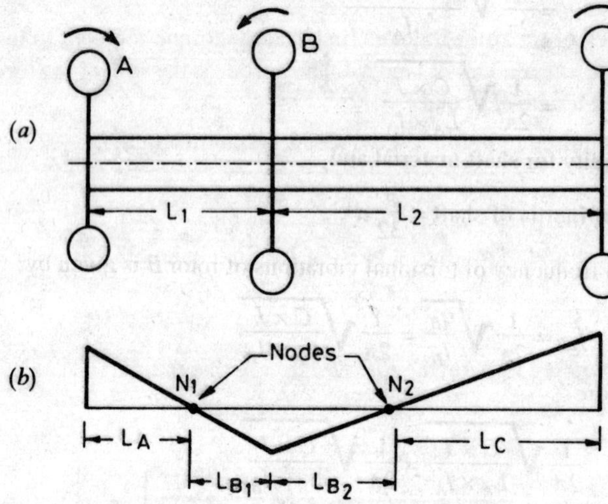

Fig. 17.5

(*ii*) In second mode, there is a single node between rotors *A* and *B* or between rotors *B* and *C*. In this mode, the outside rotors (*i.e.* rotors *A* and *C*) rotate in opposite direction.

Let us consider the case when the rotors *A* and *C* (*i.e.* outside rotors) rotate in the same direction whereas the rotor *B* rotate in opposite direction as shown in Fig. 17.5 (*a*). There will be two nodes in this case. One node will be between *A* and *B* and the second node will be between *B* and *C*. Let these two nodes are N_1 and N_2 as shown in Fig. 17.5 (*b*). At the nodes, the shaft may be assumed as fixed.

Let I_A, I_B and I_C = Mass moment of inertia of rotors *A*, *B* and *C* respectively,

L_1 = Distance between rotor *A* and *B*

L_2 = Distance between rotor *B* and *C*

L_A = Distance of node N_1 from rotor *A*,

L_C = Distance of node N_2 from rotor *C*,

L_{B1} = Distance of node N_1 from rotor $B = (L_1 - L_A)$

L_{B2} = Distance of node N_2 from rotor $B = (L_2 - L_C)$

d = Diameter of shaft

C = Modulus of rigidity for shaft material

J = Polar moment of inertia of shaft = $\frac{\pi}{32} d^4$

As the shaft is assumed as fixed at the nodes N_1 and N_2, the frequency is given by equation (17.2) as,

$$f_n = \frac{1}{2\pi}\sqrt{\frac{q}{I}} \qquad \text{where } q = \frac{C \times J}{L} \text{ from equation (17.3)}$$

∴ The frequency of the rotor A is given as

$$f_{nA} = \frac{1}{2\pi}\sqrt{\frac{q_A}{I_A}} = \frac{1}{2\pi}\sqrt{\frac{C \times J}{L_A \times I_A}} \qquad \left(\because \; q_A = \frac{C \times J}{L_A}\right) \qquad ...(i)$$

Similarly the frequency of the rotor C is given by,

$$f_{nC} = \frac{1}{2\pi}\sqrt{\frac{q_c}{I_c}} = \frac{1}{2\pi}\sqrt{\frac{C \times J}{L_c \times I_c}} \qquad \left(\because \; q_C = \frac{C \times J}{L_C}\right) \qquad ...(ii)$$

To find the frequency of the rotor B, let us first find the torque required to twist the rotor B through 1 radian when the shaft is fixed at the nodes N_1 and N_2. This torque is the sum of the torques required to produce a twist of 1 radian in each of lengths L_{B1} and L_{B2}.

Let q_{B1} = Torque required to produce a twist of 1 radian in length L_{B1}

$$= \frac{C \times J}{L_{B1}}$$

q_{B2} = Torque required to produce a twist of 1 radian in length L_{B2}

$$= \frac{C \times J}{L_{B2}}$$

q_B = Torque required to twist the rotor B through one radian.

∴ Torque required to twist the rotor B through one radian,

$$q_B = q_{B1} + q_{B2}$$

$$= \frac{C \times J}{L_{B1}} + \frac{C \times J}{L_{B2}} = C \times J\left[\frac{1}{L_{B1}} + \frac{1}{L_{B2}}\right]$$

$$= C \times J\left[\frac{1}{(L_1 - L_A)} + \frac{1}{(L_2 - L_C)}\right]$$

Hence the frequency of rotor B is given by

$$f_{nB} = \frac{1}{2\pi}\sqrt{\frac{q_B}{I_B}}$$

$$= \frac{1}{2\pi}\sqrt{\frac{C \times J\left[\frac{1}{(L_1 - L_A)} + \frac{1}{(L_2 - L_C)}\right]}{I_B}}$$

$$= \frac{1}{2\pi}\sqrt{\frac{C \times J}{I_B}\left[\frac{1}{(L_1 - L_A)} + \frac{1}{(L_2 - L_C)}\right]} \qquad ...(iii)$$

Since the frequencies f_{nA}, f_{nB} and f_{nC} must all be equal or

$$f_{nA} = f_{nB} = f_{nC}$$

Now equating equations (*i*) and (*ii*), we get

$$\frac{1}{2\pi}\sqrt{\frac{C \times J}{L_A \times I_A}} = \frac{1}{2\pi}\sqrt{\frac{C \times J}{L_C \times I_C}} \qquad \left(\because \text{ Cancelling } \frac{1}{2\pi}\sqrt{C \times J} \text{ to both sides}\right)$$

or
$$L_A \times I_A = L_C \times I_C \qquad ...(17.8A)$$

or
$$L_A = \frac{L_C \times I_C}{I_A} \qquad ...(iv)$$

Now equating equations (*ii*) and (*iii*), we get

$$\frac{1}{2\pi}\sqrt{\frac{C \times J}{L_C \times I_C}} = \frac{1}{2\pi}\sqrt{\frac{C \times J}{I_B}\left[\frac{1}{(L_1 - L_A)} + \frac{1}{(L_2 - L_C)}\right]}$$

or
$$\frac{1}{L_C \times I_C} = \frac{1}{I_B}\left[\frac{1}{(L_1 - L_A)} + \frac{1}{(L_2 - L_C)}\right] \qquad ...(17.8B)$$

If the value of L_A from equation (*iv*) is substituted in equation (17.8B), a quadratic equation in L_C is obtained. There will be two values of L_C and correspondingly two values of L_A. One value of L_A and corresponding value of L_C gives the position of two nodes. The value of frequency is obtained by substituting the value of L_A or L_C is equation (*i*) or (*ii*). This frequency is known *as two node frequency*.

In the other set of values (*i.e.* the second value of L_A and corresponding value of L_C), one gives the position of a real single node and the other is beyond the physical limits of the equation. The frequency corresponding to real single node is known as *single node frequency* or *fundamental frequency*. Also it is important to note that :

(*i*) If rotors A and B rotate in the same direction and C rotates in opposite direction as shown in Fig. 17.6 (*a*) by thick arrows, a single node is obtained between B and C as shown in Fig. 17.6 (*b*). In this case $L_A > L_1$. Also L_A does not give the actual node point.

(*ii*) Similarly if rotors B and C rotate in the same direction and A rotates in opposite direction (as shown in Fig. 17.6 (*a*) by dotted arrows downwards), a single node is obtained between A and B as shown in Fig. 17.6 (*c*). In this case $L_C > L_2$. Also L_C does not give the actual node point.

Let a_1 = Amplitude of vibration of rotor A,

a_2, a_3 = Amplitude of vibrations of rotor B and C respectively,

Then we have
$$\frac{a_1}{L_A} = \frac{a_2}{(L_A - L_1)} \quad \text{[From Fig. 17.6 (b)]}$$

and
$$\frac{a_3}{L_C} = \frac{a_2}{(L_C - L_2)} \quad \text{[From Fig. 17.6 (c)].}$$

Note. (*i*) A two-rotor system has one natural frequency of vibration.

(*ii*) A three-rotor system has two natural frequency.

(*iii*) Hence the number of different natural frequencies of a given system is one less than the number of rotors in the system.

(*iv*) In a three rotor system, for single node if $L_A > L_1$, then node lies between B and C. On the other hand if $L_C > L_2$, then node lies between A and B.

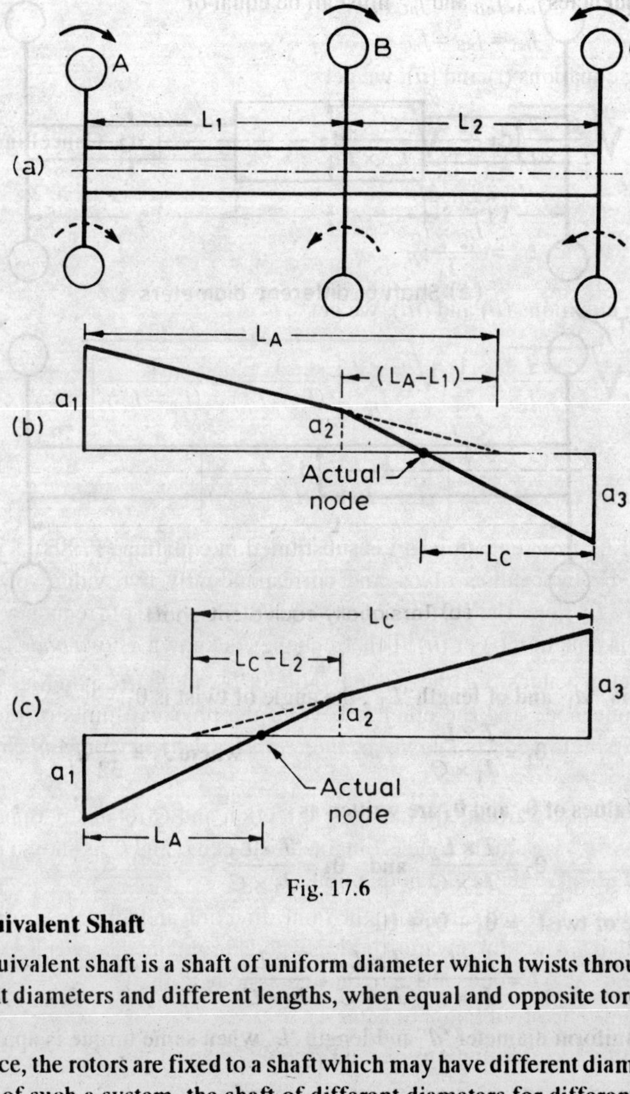

Fig. 17.6

17.7. Torsionally Equivalent Shaft

Torsionally equivalent shaft is a shaft of uniform diameter which twists through the same angle as the actual shaft of different diameters and different lengths, when equal and opposite torques of given amount are applied.

In actual practice, the rotors are fixed to a shaft which may have different diameter for different lengths. To find the frequency of such a system, the shaft of different diameters for different lengths is replaced by a torsionally equivalent shaft of uniform diameter.

Fig. 17.7 (*a*) shows a shaft of different diameters and different lengths whereas the Fig. 17.7 (*b*) shows a shaft of uniform diameter '*d*' and of length *L*. The shaft shown in Fig. 17.7 (*b*) will become torsionally equivalent shaft if angle of twist is same in both the shaft when equal and opposite torque *T* is applied at their opposite ends.

Let d_1, d_2 and d_3 = Diameters of lengths L_1, L_2 and L_3 respectively.

θ_1, θ_2 and θ_3 = Angle of twist for lengths L_1, L_2 and L_3 respectively.

θ = Angle of twist for a shaft of uniform dia. '*d*' and length '*L*'.

We know the torsional equation as

$$\frac{T}{J} = \frac{C \times \theta}{L} \quad \text{or} \quad \theta = \frac{T \times L}{J \times C}$$

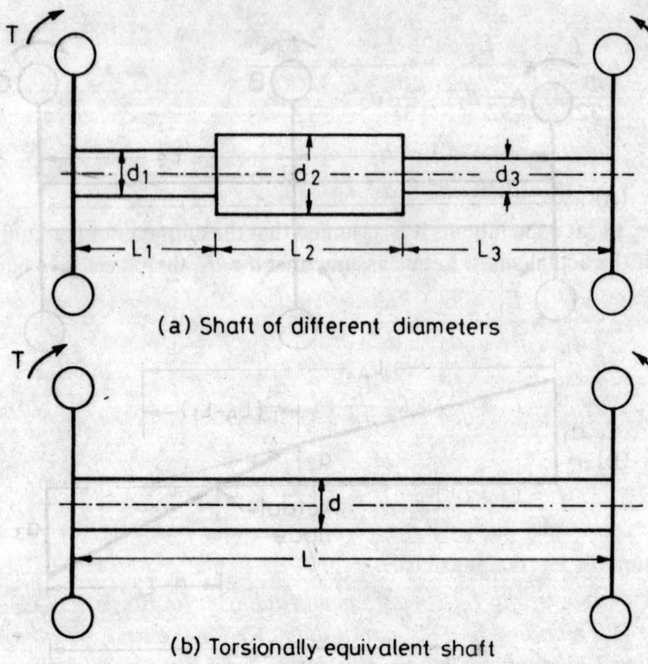

Fig. 17.7

For a shaft of dia. 'd_1' and of length 'L_1', the angle of twist is θ_1.

$$\therefore \quad \theta_1 = \frac{T \times L_1}{J_1 \times C} \qquad \text{where } J_1 = \frac{\pi}{32} d_1^4$$

Similarly, the values of θ_2 and θ_3 are written as

$$\theta_2 = \frac{T \times L_2}{J_2 \times C} \quad \text{and} \quad \theta_3 = \frac{T \times L_3}{J_3 \times C}$$

$$\therefore \quad \text{Total angle of twist} = \theta_1 + \theta_2 + \theta_3$$

$$= \frac{T \times L_1}{J_1 \times c} + \frac{T \times L_2}{J_2 \times C} + \frac{T \times L_3}{J_3 \times C}$$

For a shaft of uniform diameter 'd' and length 'L' when same torque is applied, the angle of twist is given by

$$\theta = \frac{T \times L}{J \times C} \qquad \text{where } J = \frac{\pi}{32} d^4$$

For torsionally equivalent shaft,

Angle of twist in a shaft of uniform diameter

= Total angle of twist in a shaft of different diameters of different lengths

or $$\theta = \theta_1 + \theta_2 + \theta_3$$

or $$\frac{T \times L}{J \times C} = \frac{T \times L_1}{J_1 \times C} + \frac{T \times L_2}{J_2 \times C} + \frac{T \times L_3}{J_3 \times C}$$

or $$\frac{L}{J} = \frac{L_1}{J_1} + \frac{L_2}{J_2} + \frac{L_3}{J_3} \qquad \left(\text{Cancelling } \frac{T}{C} \text{ to both sides} \right)$$

or $$\frac{L}{\frac{\pi}{32}d^4}=\frac{L_1}{\frac{\pi}{32}d_1^4}+\frac{L_2}{\frac{\pi}{32}d_2^4}+\frac{L_3}{\frac{\pi}{32}d_3^4}$$

or $$\frac{L}{d^4}=\frac{L_1}{d_1^4}+\frac{L_2}{d_2^4}+\frac{L_3}{d_3^4}$$

In actual practice, to save the labour, it is assumed that the diameter of the equivalent shaft is equal to that of one of sections of the actual shaft. Let us assume that $d = d_1$, then the above equation becomes as

$$\frac{L}{d_1^4}=\frac{L_1}{d_1^4}+\frac{L_2}{d_2^4}+\frac{L_3}{d_3^4}$$

or $$L=L_1\times\frac{d_1^4}{d_1^4}+L_2\frac{d_1^4}{d_2^4}+L_3\frac{d_1^4}{d_3^4}$$

or $$L=L_1+L_2\left(\frac{d_1}{d_2}\right)^4+L_3\left(\frac{d_1}{d_3}\right)^4+\ldots\ldots \quad \ldots(17.9)$$

where L = length of torsionally equivalent shaft.

Problem 17.2. *A shaft of length 1.25 m is 75 mm in diameter for the first 275 mm of its length, 125 mm in diameter for the next 500 mm length, 87.5 mm in diameter for the next 375 mm length and 175 mm in diameter for the remaining 100 mm of its length. The shaft carries two rotors at two ends. The mass moment of inertia of the first rotor is 75 kg m² whereas of the second rotor is 50 kg m². Find the frequency of natural torsional vibrations of the system. The modulus of the rigidity of shaft material may be taken as 80 GN/m².*

Sol. Given :

Length of shaft = 1.25 m ; d_1 = 75 mm = 0.075 m ; L_1 = 275 mm = 0.275 m ; d_2 = 125 mm = 0.125 m ; L_2 = 500 mm = 0.5 m ; d_3 = 87.5 mm = 0.0875 m ; L_3 = 375 mm = 0.375 m ; d_4 = 175 mm = 0.175 mm ; L_4 = 100 mm ; I_A = 75 kg m^2 ; I_B = 50 kg m^2 and C = 80 GN/m^2 = 80 × 10^9 N/m^2.

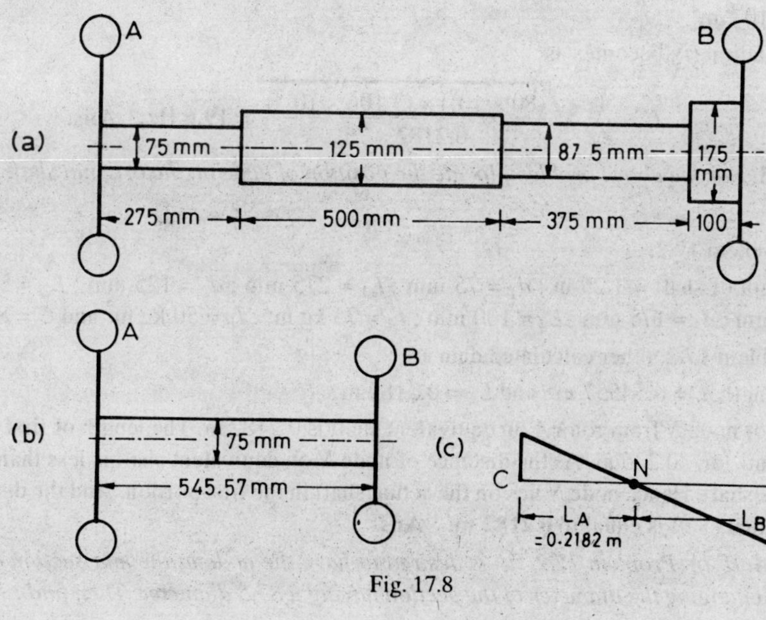

Fig. 17.8

Actual shaft is shown in Fig. 17.8 (*a*). First of all, let us find the length of the equivalent shaft assuming its diameter as d_1 = 75 mm.

Using equation (17.9), we get

$$L = L_1 + L_2\left(\frac{d_1}{d_2}\right)^4 + L_3\left(\frac{d_1}{d_3}\right)^4 + L_4\left(\frac{d_1}{d_4}\right)^4$$

$$= 275 + 500\left(\frac{75}{125}\right)^4 + 375\left(\frac{75}{87.5}\right)^4 + 100\left(\frac{75}{175}\right)^4$$

$$= 275 + 64.8 + 202.4 + 3.37 = 545.57 \text{ mm} = 0.54557 \text{ m}.$$

The equivalent shaft is shown in Fig. 17.8 (*b*). The diameter of equivalent shaft is 75 mm whereas the length is 545.57 mm. Let the node of equivalent shaft is at N as shown in Fig. 17.8 (*c*).

The position of node N is obtained by using equation (17.7).

$$\therefore \quad L_A = \frac{I_B}{(I_A + I_B)} . L \quad \text{where } L \text{ is the length of equivalent shaft} = 0.54557 \text{ m}$$

$$= \frac{50}{(75+50)} \times 0.54557 = \frac{50}{125} \times 0.54557 = 0.2182 \text{ m}$$

The natural frequency of torsional vibrations is given by equation (17.5) as

$$f_n = f_{nA} \quad \text{or} \quad f_{nB}$$

$$= \frac{1}{2\pi}\sqrt{\frac{C \times J}{L_A \times I_A}} \qquad \text{...}(i)$$

where J = Polar moment of inertia of the equivalent shaft

$$= \frac{\pi}{32} d_1^4 = \frac{\pi}{32} \times (0.075)^4 \text{ m}^4 \qquad (\because \; d_1 = 75 \text{ mm} = 0.075 \text{ m})$$

$$= 3.106 \times 10^{-6} \text{ m}^4$$

Hence the equation (*i*), becomes as

$$f_n = \frac{1}{2\pi}\sqrt{\frac{(80 \times 10^9) \times (3.106 \times 10^{-6})}{0.2182 \times 75}} = \mathbf{19.6 \text{ Hz.}} \quad \textbf{Ans.}$$

Problem 17.3. *For the above problem, locate the position of node on the original shaft from rotor A.*

Sol. Given :

Data from problem 17.2.

Original length of shaft = 1.25 m ; d_1 = 75 mm ; L_1 = 275 mm ; d_2 = 125 mm ; L_2 = 500 mm ; d_3 = 87.5 mm ; L_3 = 375 mm ; d_4 = 175 mm ; L_4 = 100 mm ; I_A = 75 kg m^2 ; I_B = 50 kg m^2 and $C = 80 \times 10^9$ N/m^2

From the problem 17.2, other calculated data are :

Equivalent length, L = 0.54557 m ; and L_A = 0.2182 m.

The distance of node N from rotor A on equivalent shaft is 0.2182 m. The length of first portion of the actual shaft is 275 mm or 0.275 m. As the distance of node N on equivalent shaft is less than the length of the first portion of the shaft. Hence node N lies on the actual shaft in the first portion. And the distance of node N from rotor A on actual shaft is equal to **0.2182 m. Ans.**

Problem 17.4. *If in Problem 17.2, it is desired to have the node at the mid-section of the shaft of 125 mm diameter by changing the diameter of the section having a 87.5 diameter. Then find :*

(*i*) *the new diameter of the section and*

(*ii*) *natural frequency of the torsional vibrations.*

Sol. Given :

$d_1 = 75$ mm ; $L_1 = 275$ mm ; $d_2 = 125$ mm ; $L_2 = 500$ mm ; $d_3 = d$; $L_3 = 375$ mm ; $d_4 = 175$ mm ; $L_4 = 100$ mm ; $I_A = 75$ kg m^2 ; $I_B = 50$ kg m^2 and $C = 80 \times 10^9$ N/m^2.

(*i*) *The new diameter of the section of length 375 mm*

If the node point '*N*' is to be at the mid-section of the shaft with diameter 125 mm, then the distance of node *N*, from rotor *A* on equivalent shaft will be,

$$L_A = L_1 + \frac{L_2}{2}\left(\frac{d_1}{d_2}\right)^4 = 275 + \frac{500}{2}\left(\frac{75}{125}\right)^4$$

$$= 275 + 32.4 = 307.4 \text{ mm} = 0.3074 \text{ m}$$

Now using the relation given by equation (17.5A),

$$L_A \times I_A = L_B \times I_B$$

$$0.3074 \times 75 = L_B \times 50$$

$$\therefore \quad L_B = \frac{0.3074 \times 75}{50} = 0.4611 \text{ m} = 461.1 \text{ mm} \quad \text{...}(i)$$

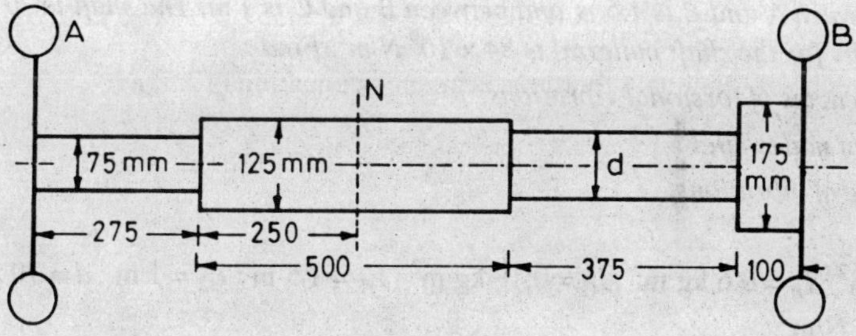

Fig. 17.9

Let '*d*' be the new diameter of the section as shown in Fig. 17.9. Then the distance of node *N*, from rotor *B* on equivalent shaft will be given as

$$L_B = \frac{L_2}{2}\left(\frac{d_1}{d_2}\right)^4 + L_3\left(\frac{d_1}{d_3}\right)^4 + L_4\left(\frac{d_1}{d_4}\right)^4$$

$$= \frac{500}{2}\left(\frac{75}{125}\right)^4 + 375\left(\frac{75}{d}\right)^4 + 100\left(\frac{75}{175}\right)^4 \qquad (\because \; d_3 = d)$$

$$= 32.4 + 375\left(\frac{75}{d}\right)^4 + 3.37 \quad \text{...}(ii)$$

But from equation (*i*), $L_B = 461.1$

$\therefore$ Equation (*ii*) becomes as

$$461.1 = 32.4 + 375\left(\frac{75}{d}\right)^4 + 3.37$$

or $$461.1 - 32.4 - 3.37 = 375\left(\frac{75}{d}\right)^4$$ or $$425.33 = 375\left(\frac{75}{d}\right)^4$$

or $$\left(\frac{425.33}{375}\right)^{1/4} = \frac{75}{d}$$ or $$1.032 = \frac{75}{d}$$

or $$d = \frac{75}{1.032} = \mathbf{72.67\ mm.}\quad \textbf{Ans.}$$

(ii) Natural frequency of the torsional vibrations

The natural frequency of torsional vibrations is given by equation (17.5) as

$$f_n = f_{nA} \text{ or } f_{nB}$$

$$= \frac{1}{2\pi}\sqrt{\frac{C \times J}{L_A \times I_A}}$$

where J = Polar moment of inertia of equivalent shaft of diameter 75 mm

$$= \frac{\pi}{32} d_1^4 = \frac{\pi}{32} \times (0.075)^4 = 3.106 \times 10^{-6}\ \text{m}^4 \qquad (\because\ d_1 = 75\ \text{mm} = 0.075\ \text{m})$$

$L_A = 0.3074$ m and $I_A = 75$ kg m^2

$$\therefore \qquad f_n = \frac{1}{2\pi}\sqrt{\frac{(80 \times 10^9) \times (3.106 \times 10^{-6})}{0.3074 \times 75}} = \mathbf{16.5\ Hz.}\quad \textbf{Ans.}$$

Problem 17.5. *The moments of inertia of three rotors A, B and C are respectively 0.3, 0.6 and 0.18 kg m². The distance between A and B is 1.5 m and between B and C is 1 m. The shaft is 70 mm in diameter and the modulus of rigidity for the shaft material is 84 × 10⁹ N/m². Find :*

(i) the frequencies of torsional vibrations,

(ii) position of nodes, and

(iii) amplitude of vibrations.

Sol. Given :

$I_A = 0.3$ kg m^2 ; $I_B = 0.6$ kg m^2 ; $I_C = 0.18$ kg m^2 ; $L_1 = 1.5$ m ; $L_2 = 1$ m ; $d = 70$ mm $= 0.07$ m ;
$C = 84 \times 10^9$ N/m^2.

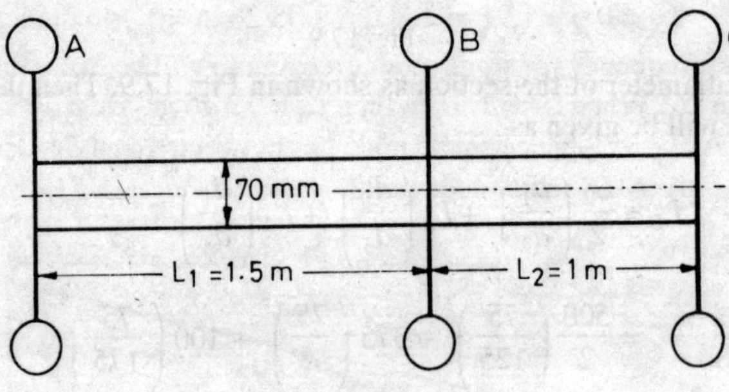

Fig. 17.10 (*a*)

(i) The frequencies of torsional vibrations

Let L_A = Distance of node N_1 from rotor A and

L_C = Distance of node N_2 from rotor C

From equation (17.8A), we know that

$$L_A \times I_A = L_C \times I_C$$

or $$L_A = \frac{L_C \times I_C}{I_A} = \frac{L_C \times 0.18}{0.30} = 0.6\, L_C$$

Also from equation (17.8B), we know that

$$\frac{1}{L_C \times I_C} = \frac{1}{I_B}\left[\frac{1}{(L_1 - L_A)} + \frac{1}{(L_2 - L_C)}\right]$$

or $$\frac{1}{L_C \times 0.18} = \frac{1}{0.6}\left[\frac{1}{(1.5 - L_A)} + \frac{1}{(1 - L_C)}\right]$$

or $$\frac{0.6}{L_C \times 0.18} = \left[\frac{1}{(1.5 - L_A)} + \frac{1}{(1 - L_C)}\right]$$

There are two unknown (*i.e.* L_A and L_C) in the above equation. Substituting the value of $L_A = 0.6\,L_C$ from equation (*i*), we get one unknown only. Hence the above equation becomes as

$$\frac{0.6}{L_C \times 0.18} = \frac{1}{(1.5 - 0.6L_C)} + \frac{1}{1 - L_C} \qquad [\because\ L_A = 0.6\,L_C \text{ from } (i)]$$

$$= \frac{1 - L_C + 1.5 - 0.6L_C}{(1.5 - 0.6L_C)(1 - L_C)} = \frac{2.5 - 1.6L_C}{1.5 - 1.5L_C - 0.6L_C + 0.6L_C^2} = \frac{2.5 - 1.6L_C}{1.5 - 2.1L_C + 0.6L_C^2}$$

or $$\frac{10}{3L_C} = \frac{2.5 - 1.6L_C}{1.5 - 2.1L_C + 0.6L_C^2} \qquad \left(\because\ \frac{0.6}{0.18} = \frac{0.60}{0.18} = \frac{10}{3}\right)$$

or $$15 - 21L_C + 6L_C^2 = 7.5L_C - 4.8L_C^2$$

or $$15 - 21L_C + 6L_C^2 - 7.5L_C + 4.8L_C^2 = 0$$

or $$10.8L_C^2 - 28.5L_C + 15 = 0$$

The above equation is a quadratic equation, whose roots are given as

$$L_C = \frac{28.5 \pm \sqrt{28.5^2 - 4 \times 10.8 \times 15}}{2 \times 10.8}$$

$$= \frac{28.5 \pm 12.81}{21.6} = \frac{28.5 + 12.81}{21.6} \text{ and } \frac{28.5 - 12.81}{21.6} = 1.91 \text{ m and } 0.726 \text{ m}$$

Substituting these values in equation (*i*), we get

$$L_A = 0.6 \times 1.91 \text{ m and } 0.6 \times 0.726 \text{ m} = 1.146 \text{ m and } 0.4356 \text{ m}$$

(*a*) The fundamental frequency (or single node frequency) will be that which corresponds to the larger of these two values of L_C or L_A. When $L_C = 1.91$ m then $L_A = 1.146$ m. This gives the position of single node as shown in Fig. 17.10 (*b*). As $L_C > L_2$ hence single node lies between *A* and *B*. Also L_C will not give the actual node point. The distance L_A gives the actual node point.

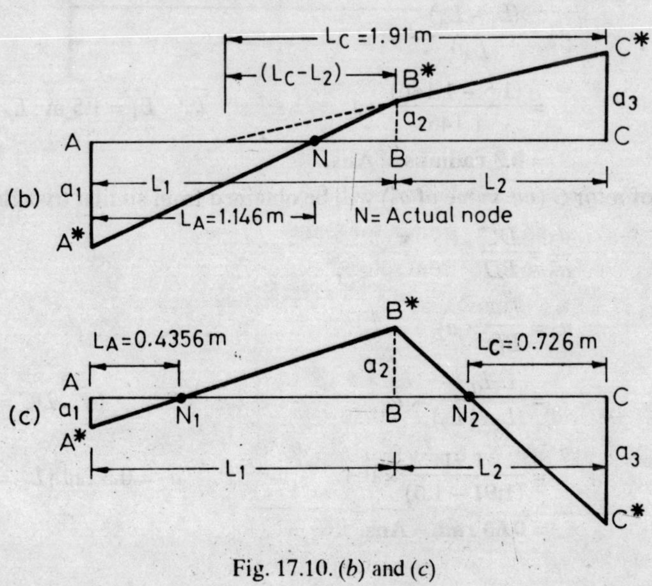

Fig. 17.10. (*b*) and (*c*)

The natural frequency of torsional vibrations for single node is given by,

$$f_{nA} = \frac{1}{2\pi}\sqrt{\frac{C \times J}{L_A \times I_A}}$$

where J = Polar moment of inertia of shaft

$$= \frac{\pi}{32} d^4 = \frac{\pi}{32} (0.07)^4 = 2.36 \times 10^{-6} \text{ m}^4$$

$$\therefore \quad f_{nA} = \frac{1}{2\pi}\sqrt{\frac{84 \times 10^9 \times 2.36 \times 10^{-6}}{1.146 \times 0.3}} = \textbf{120.9 Hz. Ans.}$$

(*b*) The two-node frequency will be obtained when $L_C = 0.726$ m and $L_A = 0.4356$ m as shown in Fig. 17.10 (*c*). Hence the natural frequency of torsional vibrations for two node is given by,

$$f_{nA} = \frac{1}{2\pi}\sqrt{\frac{C \times J}{L_A \times I_A}} = \frac{1}{2\pi}\sqrt{\frac{(84 \times 10^9) \times (2.36 \times 10^{-6})}{0.4356 \times 0.3}}$$

$$= \textbf{195.8 Hz. Ans.}$$

(*ii*) *Position of nodes*

For single node, the position of node lies at a distance of $L_A = 1.146$ m from rotor *A* as shown in Fig. 17.10 (*b*).

For two-nodes, the positions of nodes N_1 and N_2 lies at a distance of 0.4356 from rotor *A* and at a distance of 0.726 m from rotor *C*.

(*iii*) *Amplitude of vibrations*

(*a*) For a single node refer to Fig. 17.10 (*b*). Also assume the amplitude of rotor *A* equal to 1 radian.

Let a_1, a_2 and a_3 = Amplitudes of rotors *A*, *B* and *C* respectively, and

Now a_1 = Amplitude of rotor *A*

= 1 radian (assumed)

From similar triangles AA^*N and BB^*N, we have

$$\frac{a_1}{a_2} = \frac{L_A}{(L_1 - L_A)}$$

or
$$a_2 = \frac{(L_1 - L_A)}{L_A} \times a_1$$

$$= \frac{(1.5 - 1.146)}{1.146} \times 1 \qquad (\because \; L_1 = 1.5 \text{ m} ; L_A = 1.146 \text{ for one node})$$

$$= \textbf{0.3 radians. Ans.}$$

The amplitude of rotor *C* (*i.e.* value of a_3) will be obtained from similar triangles DCC^* and DBB^*.

$$\therefore \quad \frac{a_3}{a_2} = \frac{DC}{BD}$$

or
$$a_3 = \frac{DC}{BD} \times a_2$$

$$= \frac{L_C}{(L_C - L_2)} \times a_2 \qquad (\because \; DC = L_C \text{ and } BD = L_C - L_2)$$

$$= \frac{1.91}{(1.91 - 1.0)} \times 0.3 \qquad (\because \; a_2 = 0.3 \text{ rad} ; L_C = 1.91 \text{ m and } L_2 = 1 \text{ m})$$

$$= \textbf{0.63 rad. Ans.}$$

(*b*) For two-node refer to Fig. 17.10 (*c*). Let N_1 and N_2 are the two-nodes from similar triangles AN_1A^* and N_1BB^*, we have

$$\frac{a_2}{a_1} = \frac{BN_1}{AN_1} \quad \text{or} \quad a_2 = \frac{BN_1}{AN_1} \times a_1 \quad ...(i)$$

But $BN_1 = AB - AN_1 = L_1 - L_A = 1.5 - 0.4356, \quad AN_1 = L_A = 0.4356$

and $a_1 = 1$ rad. (assumed)

Hence the above equation (*i*) becomes as

$$a_2 = \frac{(1.5 - 0.4356)}{0.4356} \times 1 = \mathbf{2.44\ rad.} \quad \textbf{Ans.}$$

To find the amplitude of rotor *C* (*i.e.* the value of a_3), consider the similar triangles BB^*N_2 and N_2CC^*. We have,

$$\frac{a_3}{a_2} = \frac{CN_2}{BN_2} \quad \text{or} \quad a_3 = \frac{CN_2}{BN_2} \times a_2 \quad ...(ii)$$

But $CN_2 = L_C = 0.726$ m, $BN_2 = L_2 - L_C = 1 - 0.726 = 0.274$ m, and $a_2 = 2.44$ rad (calculated above)

Hence the above equation (*ii*) becomes as

$$a_3 = \frac{0.726}{0.274} \times 2.44 = \mathbf{6.46\ rad.} \quad \textbf{Ans.}$$

Problem 17.6. *The moment of inertia of three rotors A, B and C are respectively 400 kg m², 160 kg m² and 10 kg m². The distance between rotor A and B is 2 m and they are connected by a shaft of diameter 50 mm. The distance between rotor B and C is also 2 m and they are connected by a shaft of diameter 25 mm. Determine :*

(i) Natural frequencies of torsional vibrations and

(ii) position of nodes.

Take modulus of rigidity as 80 kN/mm² and neglect the inertia of the shaft.

Sol. Given :

$I_A = 400$ kg m² ; $I_B = 160$ kg m² ; $I_C = 10$ kg m² ; $L_1 = 2$ m ; $L_2 = 2$ m ; $d_1 = 50$ mm $= 0.05$ m ; $d_2 = 25$ mm $= 0.025$ m ; $C = 80$ kN/mm² $= 80 \times 10^3$ N/mm² $= 80 \times 10^3 \times 10^6$ N/m² $= 80 \times 10^9$ N/m².

(i) Natural frequencies of torsional vibrations

To find the natural frequency of torsional vibrations of the system the original system should be replaced by an equivalent system. Let us assume that the diameter of equivalent shaft is $d_1 = 50$ mm.

The length of equivalent system is given by equation (17.9) as

$$L = L_1 + L_2\left(\frac{d_1}{d_2}\right)^4 = 2 + 2\left(\frac{50}{25}\right)^4 = 34 \text{ m}$$

The equivalent shaft is shown in Fig. 17.11 (*b*) in which

$$L_1 + L_2^* = 34$$

$$\therefore \quad L_2^* = 34 - L_1 = 34 - 2 = 32 \text{ m.}$$

The position of nodes for the equivalent system is obtained as given below :

Let L_A = Distance of node N_1 from rotor A, and

L_C = Distance of node N_2 from rotor C.

Now using equation (17.8A), we get

$$L_A \times I_A = L_C \times I_C$$

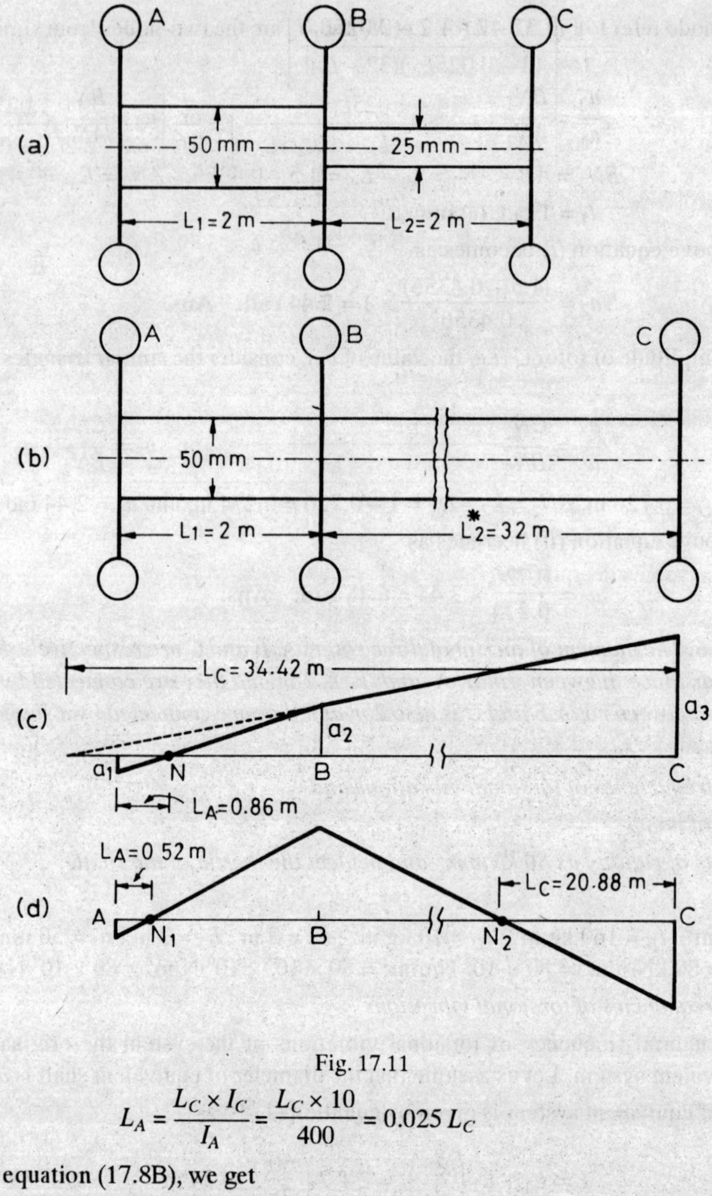

Fig. 17.11

or
$$L_A = \frac{L_C \times I_C}{I_A} = \frac{L_C \times 10}{400} = 0.025\, L_C \qquad ...(i)$$

Now using equation (17.8B), we get

$$\frac{1}{L_C \times I_C} = \frac{1}{I_B}\left[\frac{1}{(L_1 - L_A)} + \frac{1}{(L_2{}^* - L_C)}\right]$$

(In equation (17.8B), L_2 is the length of equivalent second shaft which is $L_2{}^*$ here)

or
$$\frac{1}{L_C \times 10} = \frac{1}{160}\left[\frac{1}{(2 - L_A)} + \frac{1}{32 - L_C)}\right] \qquad (\because\ L_2{}^* = 32 \text{ m})$$

or
$$\frac{1}{L_C} = \frac{1}{16}\left[\frac{1}{(2 - 0.025 L_C)} + \frac{1}{32 - L_C}\right] \qquad [\because \text{ From equation } (i),\ L_A = 0.025\, L_C]$$

$$= \frac{1}{16}\left[\frac{32 - L_C + 2 - 0.025L_C}{(2 - 0.025L_C)(32 - L_C)}\right]$$

$$= \frac{1}{16}\left[\frac{34 - 1.025L_C}{64 - 0.8L_C - 2L_C + 0.025L_C^2}\right] = \frac{1}{16}\left[\frac{34 - 1.025L_C}{64 - 2.8L_C + 0.025L_C^2}\right]$$

or $$16[64 - 2.8L_C + 0.025\,L_C^2] = 34L_C - 1.025\,L_C^2$$

or $$1024 - 44.8L_C + 0.4L_C^2 = 34L_C - 1.025\,L_C^2$$

or $$1024 - 44.8L_C + 0.4\,L_C^2 - 34L_C + 1.025\,L_C^2 = 0$$

or $$1024 - 78.8L_C + 1.425\,L_C^2 = 0$$

or $$1.425\,L_C^2 - 78.8L_C + 1024 = 0$$

The roots of the above quadratic equation are :

$$L_C = \frac{78.8 \pm \sqrt{78.8^2 - 4 \times 1.425 \times 1024}}{2 \times 1.425} = \frac{78.8 \pm 19.3}{2.85}$$

$$= 34.42 \text{ m} \quad \text{and} \quad 20.88 \text{ m}.$$

Substituting these values in equation (*i*), we get

$$L_A = 0.025 \times L_C = 0.025 \times 34.42 = 0.86 \text{ m and } 0.025 \times 20.88 = 0.52 \text{ m}.$$

(*a*) The fundamental frequency (or single node frequency) will be that which corresponds to the larger of these two values of L_C or L_A. Of the two values of L_C or L_A, L_C is having larger value. Hence when $L_C = 34.42$ m then $L_A = 0.86$ m. This set gives the position of the single node as shown in Fig. 17.11 (*c*). As $L_C > L_2$*, hence single node lies between *A* and *B*. Also L_C will not give the actual node point. The distance L_A will give the actual node point.

The natural frequency of torsional vibrations for single node is given by,

$$f_{nA} = \frac{1}{2\pi}\sqrt{\frac{C \times J}{L_A \times I_A}}$$

where J = Polar moment of inertia of equivalent shaft

$$= \frac{\pi}{32}(d_1)^4 = \frac{\pi}{32} \times (0.05)^4 = 0.614 \times 10^{-6} \text{ m}^4$$

$C = 80 \times 10^9$ N/m^2, $L_A = 0.86$ m and $I_A = 400$ kg m^2

$$\therefore \quad f_{nA} = \frac{1}{2\pi}\sqrt{\frac{(80 \times 10^9) \times (0.614 \times 10^{-6})}{0.86 \times 400}} \text{ Hz} = \mathbf{1.9\ Hz.} \quad \textbf{Ans.}$$

(*b*) The two-node frequency will be obtained when $L_C = 20.88$ m and corresponding value of $L_A = 0.52$ m as shown in Fig. 17.11 (*d*). Hence natural frequency of torsional vibrations for two-node system is given by,

$$f_{nA} = \frac{1}{2\pi}\sqrt{\frac{C \times J}{L_A \times I_A}} = \frac{1}{2\pi}\sqrt{\frac{(80 \times 10^9) \times (0.614 \times 10^{-6})}{0.52 \times 400}} = \mathbf{2.44\ Hz.} \quad \textbf{Ans.}$$

(*ii*) *Position of nodes*

(*a*) The position of single node is given by the distance L_A which is at a distance of 0.86 m from rotor *A* as shown in Fig. 17.11 (*c*).

(*b*) The position of the two nodes is given by the distance L_A and L_C. The distance $L_A = 0.52$ from rotor *A*. This gives the position of first node N_1 as shown in Fig. 17.11 (*d*). The distance $L_C = 20.88$ m from rotor *C*.

This gives the position of second node N_2 on the equivalent shaft as shown in Fig. 17.11 (*d*). The corresponding value of the second node N_2 in the original system is obtained as given below :

$$\text{Distance of } N_2 \text{ from rotor } C \text{ in original system} = 20.88\left(\frac{d_2}{d_1}\right)^4 = 20.88\left(\frac{25}{50}\right)^4 = 1.3 \text{ m. Hence } N_2 \text{ lies}$$

at a distance of 1.3 m from rotor C in original system. **Ans.**

17.8. Free Torsional Vibrations of a Geared System

Fig. 17.12 (*a*) shows a gear system, which is having two shafts (*i.e.* shaft 1 and shaft 2) of diameters d_1 and d_2 and of lengths L_1 and L_2 respectively. The shaft of diameter d_1 carries a rotor A at one end and a

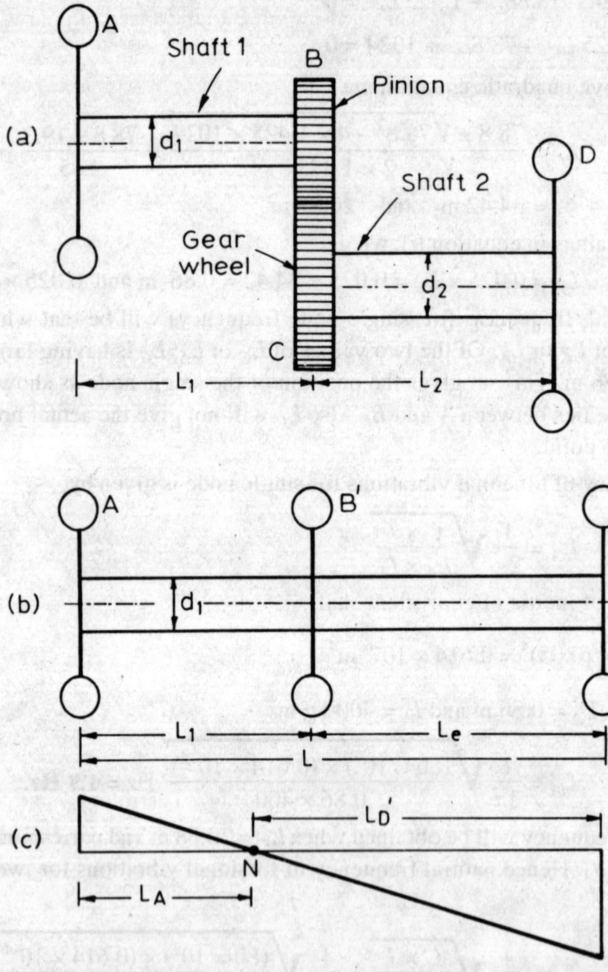

Fig. 17.12

pinion b at other end. The shaft of diameter d_2 carries a rotor D at one end and a gear wheel C at the other end. The gear wheel C meshes with the pinion B. This system may be reduced to an equivalent single shaft system as shown in Fig. 17.12 (*b*) if it is assumed that :

(*i*) the gear teeth are rigid and do not distort under the tooth loads ;

(*ii*) there is no backlash in the gearing, and

(*iii*) the inertia of the shafts and gears is negligible.

With these assumptions, the system reduces to a two-rotor system.

The equivalent single shaft system shown in Fig. 17.12 (*b*) carries a rotor *A* at one end and a rotor *D*′ at the other end.

If the shafts are not strained beyond the limit of proportionality, each rotor in the geared system will oscillate with simple harmonic motion and there will be a node either in length L_1 or in the length L_2.

A single shaft system will be equivalent to the geared system if for a given amplitude of oscillation of the rotor *A* :

(*i*) the maximum strain energies stored in the two systems are equal, and

(*ii*) the maximum kinetic energies of the rotors in the two systems are equal.

Let G = Gear ratio *i.e.* ratio of the speed of pinion *B* to the speed of gear wheel *C*

$$= \frac{\text{Speed of pinion B}}{\text{Speed of gear wheel C}} = \frac{\omega_B}{\omega_C}$$

$$= \frac{\omega_A}{\omega_D} \qquad (\because \text{ Speed of rotor } A = \text{speed of pinion } B \text{ and speed of rotor } D = \text{speed of gear wheel } C)$$

where ω_A = Angular speed of rotor *A* and

ω_D = Angular speed of rotor *D*.

(*a*) The maximum kinetic energy of the rotors in the two system will be equal if

Kinetic energy of original rotors = Kinetic energy of equivalent rotors

K.E. of rotor *A* + K.E. of rotor *D* = K.E. of rotor *A* + K.E. of rotor *D*′

or K.E. of rotor *D* = K.E. of rotor *D*′

or
$$\frac{1}{2} \times I_D \times \omega_D^2 = \frac{1}{2} \times I_D' \times \omega_D'^2 \qquad \left(\because \text{ K.E.} = \frac{1}{2} \times I \times \omega^2\right) \qquad ...(i)$$

where I_D = Mass moment of inertia of rotor *D*,

I_D' = Mass moment of inertia of rotor *D*′,

ω_D' = Angular speed of rotor *D*′.

From equation (*i*), we have

$$I_D' = \frac{I_D \times \omega_D^2}{\omega_D'^2}$$

$$= I_D \times \frac{\omega_D^2}{\omega_A^2} \qquad [\because \omega_D' = \omega_A. \text{ The rotors } A \text{ and } D' \text{ are connected to a single equivalent shaft. Hence speed of shaft is equal to the speed of rotor } A \text{ or speed of rotor } D'.]$$

$$= \frac{I_D}{\left(\dfrac{\omega_A}{\omega_D}\right)^2}$$

$$= \frac{I_D}{G^2} \qquad \left(\because \frac{\omega_A}{\omega_D} = G\right) \qquad ...(17.10)$$

(*b*) The maximum strain energy stored in the two systems will be equal if

Max. strain energy stored in original system

= Max. strain energy stored in equivalent system.

or Strain energy stored in rotor A + strain energy stored in rotor D

= strain energy stored in rotor A + strain energy stored in rotor D'

or Strain energy stored in rotor D

= str n energy stored in rotor D

$$\therefore \quad \frac{1}{2} T_D \times \theta_D = \frac{1}{2} T_D' \times \theta_D' \qquad \left(\because \text{ Strain energy stored} = \frac{1}{2} T \times \theta \right) \qquad ...(ii)$$

where T_D = Torque on rotor D,

T_D' = Torque on rotor D',

θ_D = Angle of twist of rotor D,

θ_D' = Angle of twist of rotor D,

d_D = Diameter of shaft 2 to which rotor D is attached = d_2 (given)

d_D' = Diameter of equivalent shaft to which rotor D' is attached = d_1 (assumed)

Now the equation (*ii*) becomes as

$$T_D \times \theta_D = T_D' \times \theta_D' \qquad ...(iii)$$

From torsional equation, we know that

$$\frac{T}{J} = \frac{C \times \theta}{L}$$

or
$$T = \frac{C \times \theta \times J}{L} = \frac{C \times \theta \times \frac{\pi}{32} d^4}{L} \qquad \left(\because J = \frac{\pi}{32} d^4 \right)$$

$\therefore$ Values of T_D and T_D' becomes as

$$T_D = \frac{C \times \theta_D \times \frac{\pi}{32} d_D^4}{L_2} \qquad (\because \text{ For rotor } D, L = L_2)$$

and
$$T_D' = \frac{C \times \theta'_D \times \frac{\pi}{32} \times d_D'^4}{L_e} \qquad (\because \text{ For rotor } D', L = L_e)$$

Substituting the values of T_D and T_D' in equation (*ii*), we get

$$\frac{C \times \theta_D \times \frac{\pi}{32} \times d_D^4}{L_2} \times \theta_D = \frac{C \times \theta_D' \times \frac{\pi}{32} \times d_D'^4}{L_e} \times \theta_D'$$

or
$$\frac{\theta_D^2 \times d_D^4}{L_2} = \frac{\theta_D'^2 \times d_D'^4}{L_e}$$

or
$$L_e = \frac{\theta_D'^2 \times d_D'^4}{\theta_D^2 \times d_D^4} \times L_2$$

$$= \left(\frac{\theta_D'}{\theta_D} \right)^2 \times \left(\frac{d_D'}{d_D} \right)^4 \times L_2 \qquad ...(iv)$$

But angle of twist, $\theta = \omega \times t$

$\therefore$ $\theta_D = \omega_D \times t$

and $\theta_D' = \omega_D' \times t$

$= \omega_A \times t$ (∵ $\omega_D' = \omega_A$. As rotor A and rotor D' are connected to single equivalent shaft)

$$\therefore \quad \frac{\theta_D'}{\theta_D} = \frac{\omega_A \times t}{\omega_D \times t} = \frac{\omega_A}{\omega_D} = G$$

Substituting this value of $\frac{\theta'_D}{\theta_D}$ in equation (*iv*), we get

$$L_e = G^2 \times \left(\frac{d_D'}{d_D}\right)^4 \times L_2 \quad \text{...(17.11)}$$

Assuming the diameter of the equivalent shaft equal to that of shaft 1 (*i.e.* $d_D' = d_1$), the above equation becomes as

$$\therefore \quad L_e = G^2 \times \left(\frac{d_1}{d_2}\right)^4 \times L_2 \qquad (\because \; d_D = d_2) \quad \text{...(17.12)}$$

The single shaft is equivalent to the original geared system, if mass moment of inertia of the rotor D' satisfies the equation (17.10) and additional length of the equivalent shaft L_e satisfies the equation (17.12).

Total length of the equivalent shaft becomes as

$$\therefore \quad L = L_1 + L_e = L_1 + G^2 \times \left(\frac{d_1}{d_2}\right)^4 \times L_2 \quad \text{...(17.13)}$$

Natural frequency of the torsional vibration

The original geared system has been reduced to two-rotor system. The natural frequency of the torsional vibration is obtained as given below :

Let N shows the position of node of the equivalent system as shown in Fig. 17.12 (*c*). In this figure, L_A is the distance of node N from rotor A and L_D' is the distance of node N from rotor D'. These distances are unknown. Their values can be obtained as given below :

The natural frequency of torsional vibration of rotor A, is given by

$$f_{nA} = \frac{1}{2\pi}\sqrt{\frac{C \times J}{L_A \times I_A}} \quad \text{...(}v\text{)}$$

Similarly the natural frequency of torsional vibration of rotor D' is given by,

$$f_{nD}' = \frac{1}{2\pi}\sqrt{\frac{C \times J}{L_D' \times I_D'}} \quad \text{...(}vi\text{)}$$

But, we know that $\quad f_{nA} = f_{nD}'$

$$\therefore \quad \frac{1}{2\pi}\sqrt{\frac{C \times J}{L_A \times I_A}} = \frac{1}{2\pi}\sqrt{\frac{C \times J}{L_D' \times I_D'}}$$

or $$L_A \times I_A = L_D' \times I_D' \quad \text{...(}vii\text{)} \quad \text{...(17.14)}$$

Also we know that $\quad L = L_A \times L_D' \quad$ ...(*viii*)

From equations (*vii*) and (*viii*), the values of L_A and L_D' can be obtained. Then the natural frequency of torsional vibration can be calculated.

Where the inertia of the gearing is not negligible, there must be an additional rotor B' on the equivalent single shaft system to allow for the inertia of the pinion and gear wheel. This rotor is shown dotted in Fig. 17.12 (*b*) at a distance L_1 from the rotor A. The mass moment of inertia of this rotor is given by,

I_C

where I_B = Mass moment of inertia of pinion and

I_C = Mass moment of inertia of the gear wheel.

The system then becomes a three rotor system and the frequency of such a system may be obtained as given in Article 17.6.

Problem 17.7. *A shaft carries a motor at one end and a pinion on the other end. The length and diameter of this shaft are 500 mm and 60 mm respectively. There is another shaft of length 900 mm and of diameter 100 mm. This shaft carries a gear wheel at one end and a centrifugal pump on another end. The gear wheel and pinion are meshing together so that the centrifugal pump is driven by the motor. The mass moment of inertia of motor and centrifugal pump are 200 kg m² and 750 kg m² respectively. If the inertia of gears and shaft is neglected and pump speed is one third of the motor then find the frequency of torsional vibrations of the system. Take the value of modulus of rigidity as 80 kN/mm².*

Sol. Given :

L_1 = 500 mm = 0.5 m ; d_1 = 60 mm = 0.06 m ; L_2 = 900 mm = 0.9 m ; d_2 = 100 mm = 0.1 m ;

Mass moment of inertia of motor, I_A = 200 kg m^2 ; mass moment of inertia of pump, I_D = 750 kg m^2 ;

pump speed = $\frac{1}{3}$ of motor speed. Hence speed of motor = 3 × pump speed ; C = 80 kN/mm^2 = 80 × 10^3 N/mm^2 = 80 × 10^3 × 10^6 N/m^2 = 80 × 10^9 N/m^2.

The inertia of gears and shaft is neglected. Then this becomes the two-rotor system. The original and equivalent system is shown in Fig. 17.13 (*a*) and (*b*) respectively.

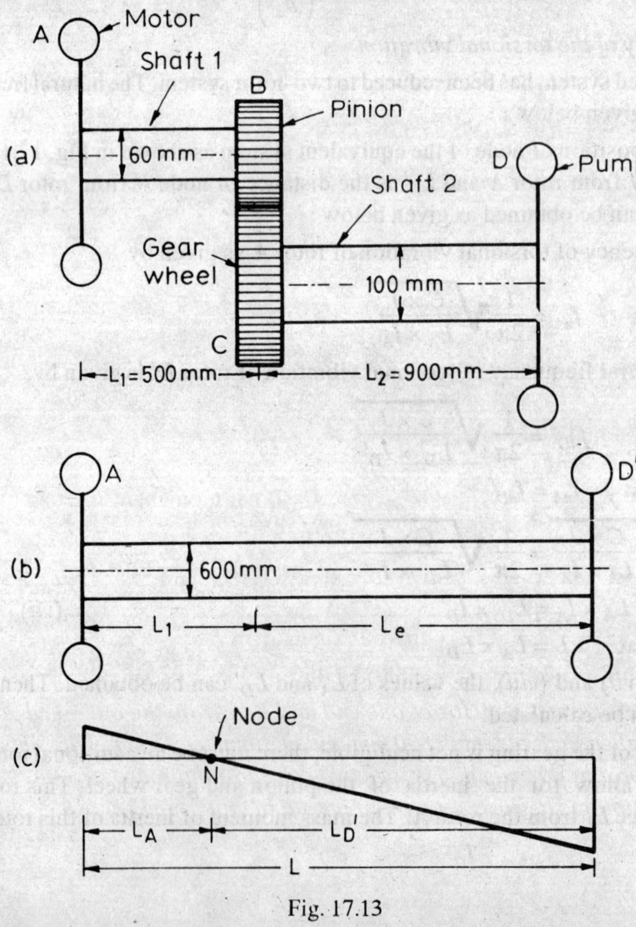

Fig. 17.13

Let us first find the value of gear ratio, G.

The gear ratio, $$G = \frac{\text{Speed of pinion } B}{\text{Speed of gear wheel } C} = \frac{\text{Speed of motor}}{\text{Speed of pump}} = 3 \quad \text{(given)}$$

Now let us find the mass moment of inertia of the equivalent rotor D' and the additional length of the equivalent shaft.

The mass moment of inertia of the equivalent rotor D' is given by equation (17.10) as

$$I_D' = \frac{I_D}{G^2} = \frac{750}{3^2} = \frac{750}{9} = 83.33 \text{ kg m}^2$$

Additional length of the equivalent shaft, assuming its diameter as $d_1 = 60$ mm, is given by equation (17.12) as

$$L_e = G^2 \times \left(\frac{d_1}{d_2}\right)^4 \times L_2 = 3^2 \times \left(\frac{60}{100}\right)^4 \times 900 = 1049.76 \text{ mm} \simeq 1050 \text{ mm}$$

∴ Total length of equivalent shaft,

$$L = L_1 + L_e = 500 + 1050 = 1550 \text{ mm} = 1.55 \text{ m}$$

Fig. 17.13 (*c*) shows the position of node N on the equivalent system.

Let L_A = Distance of node N from rotor A,

L_D' = Distance of node N from rotor D'.

From equation (17.14), we know that

$$L_A \times I_A = L_D' \times I_D'$$

or $$L_A \times 200 = L_D' \times 83.33 \quad (\because I_D' = 83.33 \text{ calculated above})$$

$$= (L - L_A) \times 83.33 \quad (\because L_D' = L - L_A)$$

$$= (1.55 - L_A) \times 83.33 \quad (\because L = 1.55 \text{ m calculated above})$$

or $$\frac{L_A \times 200}{83.33} = 1.55 - L_A$$

or $$2.4 L_A = 1.55 - L_A$$

or $$3.4 L_A = 1.55$$

or $$L_A = \frac{1.55}{3.4} = 0.4558 \simeq 0.456 \text{ m} = 456 \text{ mm}$$

The frequency of torsional vibration is given by,

$$f_{nA} = \frac{1}{2\pi}\sqrt{\frac{C \times J}{I_A \times L_A}}$$ where J = Polar moment of inertia of the equivalent shaft

$$= \frac{\pi}{34} d_1^4 = \frac{\pi}{32} \times (0.06)^4 = 1.27 \times 10^{-6} \text{ m}^4$$

$$= \frac{1}{2\pi}\sqrt{\frac{(80 \times 10^9) \times (1.27 \times 10^{-6})}{200 \times 0.456}} \text{ Hz} = \mathbf{5.31 \text{ Hz.}} \quad \textbf{Ans.}$$

Problem 17.8. *A shaft carries a motor at one end and a gear on the other end. The diameter of this shaft is 54 mm. There is another shaft of length 800 mm and of diameter 45 mm. This shaft carries a pinion at one end and a centrifuge on the other end. The gear wheel and pinion are meshing together so that the centrifuge is driven by the motor. The mass moment of inertia of motor and centrifuge are 0.1875 kg m² and*

0.299 kg m² respectively. If the inertia of gears and shaft is neglected and the speed of the centrifuge is four times the speed of the motor, then find :

(i) the length of the motor shaft if the gears are at the node for torsional vibration of the system and

(ii) frequency of torsional vibrations.

Take the value of modulus of rigidity for shaft material as 84 kN/mm².

Sol. Given :

d_1 = 54 mm = 0.054 m ; L_1 = ? ; L_2 = 800 mm ; d_2 = 45 mm = 0.045 m ; mass moment of inertia of motor, I_A = 0.1875 kg m^2 ; mass moment of inertia of centrifuge, I_D = 0.299 kg m^2 ; speed of centrifuge = 4 × speed of motor ; node (N) lies at gears or distance of node (N) from rotor A = L_1 ; C = 84 kN/mm^2 = 84 × 10^3 N/mm^2 = 84 × 10^3 × 10^6 N/m^2 = 84 × 10^9 N/m^2.

The inertia of gears and shaft is neglected. This is a case of two-rotor system. The original and equivalent system is shown in Fig. 17.14 (*a*) and (*b*) respectively.

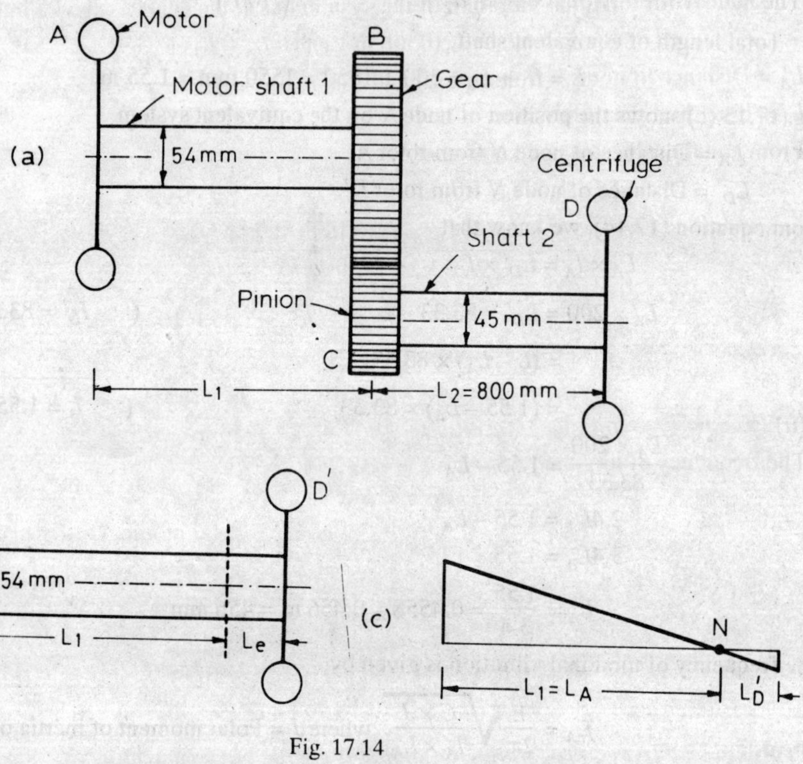

Fig. 17.14

Let us first find the value of gear ratio, G.

The gear ratio, $$G = \frac{\text{Speed of motor}}{\text{Speed of centrifuge}} = \frac{\text{Speed of motor}}{4 \times \text{speed of motor}}$$

($\because$ Speed of centrifuge = 4 × speed of motor)

$$= \frac{1}{4}$$

Now let us find the mass moment of inertia of the equivalent rotor D' and the additional length of the equivalent shaft.

The mass moment of inertia of the equivalent rotor D' is given by equation (17.10) as

$$I_D' = \frac{I_D}{G^2} = \frac{0.299}{\left(\frac{1}{4}\right)^4} = 0.299 \times 16 = 4.784 \text{ kg m}^2$$

Additional length of the equivalent shaft, assuming its diameter as $d_1 = 54$ mm, is given by equation (17.12) as

$$L_e = G^2 \times \left(\frac{d_1}{d_2}\right)^4 \times L_2 = \left(\frac{1}{4}\right)^2 \times \left(\frac{54}{45}\right)^4 \times 800 = 103.68 \text{ mm} = 0.10368 \text{ m}.$$

(i) The length of the motor shaft if the node lies at the gears for torsional vibrations of the system

Let L_1 = Length of the motor shaft

The node N for torsional vibration of the system lies at the gears as shown in Fig. 17.14 (c) hence,

$$L_D' = L_e = 0.10368 \text{ m} \quad \text{and} \quad L_A = L_1$$

where L_A = Distance of node N from rotor A and

L_1 = Length of the motor shaft

From equation (17.14), we know that

$$L_A \times I_A = L_D' \times I_D'$$

$$\therefore \quad L_A = \frac{L'_D \times I'_D}{I_A} = \frac{0.10368 \times 4.784}{0.1875}$$

$[\because L_D' = L_e = 0.10368 \text{ m}, I_D' = 4.784 \text{ kg m}^2 \text{ (calculated)}]$

$$= 2.64 \text{ m}$$

$$\therefore \quad L_1 = L_A = \mathbf{2.64 \text{ m.}} \quad \textbf{Ans.}$$

(ii) Frequency of torsional vibrations

The frequency of torsional vibrations is given by,

$$f_{nA} = \frac{1}{2\pi}\sqrt{\frac{C \times J}{I_A \times L_A}}$$ where J = Polar moment of inertia of the equivalent shaft

$$= \frac{\pi}{32} d_1^4 = \frac{\pi}{32}(0.054)^4 = 0.835 \times 10^{-6} \text{ m}^4$$

$$= \frac{1}{2\pi}\sqrt{\frac{(84 \times 10^9) \times (0.835 \times 10^{-6})}{0.1875 \times 2.64}} = \mathbf{59.9 \text{ Hz.}} \quad \textbf{Ans.}$$

Problem 17.9. *A reciprocating I.C. engine is coupled to a centrifugal pump through a pair of gears. The shaft from the flywheel of the engine to the gear wheel has a 45 mm diameter and is 1 m long. The shaft from the pinion to the pump has a 30 mm diameter and is 810 mm long. Pump speed is four times the engine speed. The moment of inertia of flywheel, gear wheel, pinion and pump are 400 kg m², 8 kg m², 2 kg m² and 10 kg m² respectively. If the modulus of rigidity of shaft material is 80 GN/m², then find the natural frequency of torsional vibrations of the system.*

Sol. Given :

$d_1 = 45$ mm $= 0.045$ m ; $L_1 = 1.0$ m ; $d_2 = 30$ mm $= 0.03$ m ; $L_2 = 810$ mm $= 0.81$ m ; pump speed $= 4 \times$ engine speed ; $C = 80 \text{ GN/m}^2 = 80 \times 10^9 \text{ N/m}^2$.

Let the engine flywheel and centrifugal pump are represented by rotors *A* and *D* respectively. The gear wheel and pinion are represented by symbol *B* and *C* respectively. Then moment of inertia of engine flywheel, gear wheel, pinion and pump will be represented as :

$$I_A = 400 \text{ kg m}^2 \text{ ; } I_B = 8 \text{ kg m}^2 \text{ ; } I_C = 2 \text{ kg m}^2 \text{ and } I_D = 10 \text{ kg m}^2.$$

The original and the equivalent system is shown in Fig. 17.15 (*a*) and (*b*) respectively.

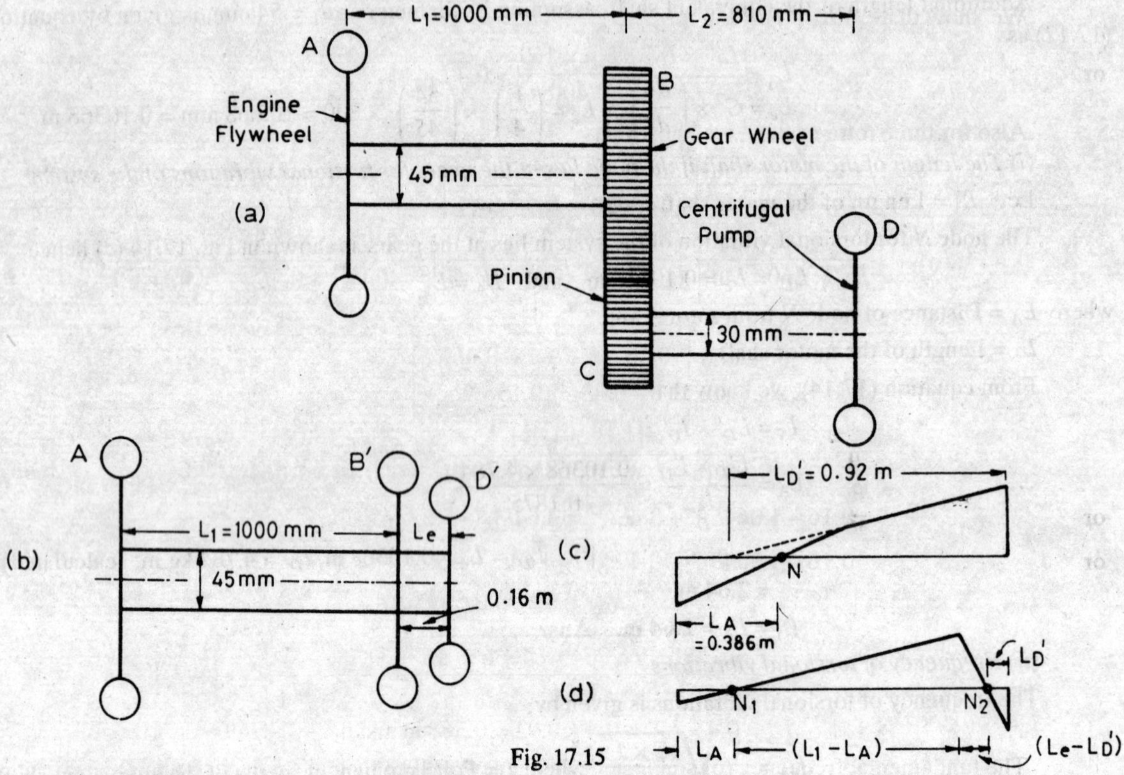

Fig. 17.15

The pump speed is given as four times the engine speed, hence pinion is supported on the pump shaft and the gear on the engine shaft. Let us now find the value of gear ratio, *G*.

$$\text{The gear ratio, } G = \frac{\text{Speed of engine}}{\text{Speed of pump}} = \frac{\text{Speed of engine}}{4 \times \text{speed of engine}} = \frac{1}{4} = 0.25$$

Now let us find the mass moment of inertia of equivalent rotor D', equivalent gearing B' and the additional length of the equivalent shaft.

The mass moment of inertia of the equivalent rotor D' is given by equation (17.10) as

$$I_D' = \frac{I_D}{G^2} = \frac{10}{0.25^2} = 160 \text{ kg m}^2$$

The mass moment of inertia of the equivalent gearing B' is given by equation (17.15) as

$$I_B' = I_B + \frac{I_C}{G^2} = 8 + \frac{2}{0.25^2} = 40 \text{ kg m}^2$$

Additional length of the equivalent shaft, assuming its diameter as d_1 = 45 mm is given by equation (17.12) as

$$L_e = G^2 \times \left(\frac{d_1}{d_2}\right)^4 \times L_2 = 0.25^2 \times \left(\frac{45}{30}\right)^4 \times 0.81 = 0.16 \text{ m}$$

As the inertia of the gearing is not negligible, the original system reduces to three rotor system as shown in Fig. 17.15 (*b*). The position of nodes for the equivalent system is obtained as given below :

Let N_1 and N_2 are the two nodes on the equivalent system as shown in Fig. 17.15 (*d*) where

L_A = Distance of node N_1 from rotor A, and

L_D' = Distance of node N_2 from rotor D'.

We know that $L_A \times I_A = L_D' \times I_D'$

or $$L_A = \frac{L_D' \times I_D'}{I_A} = \frac{L'_D \times 160}{400} = 0.4\,L_D' \quad ...(i)$$

Also for three rotor system, we know that

$$\frac{1}{L_D' \times I_D'} = \frac{1}{I_B'}\left[\frac{1}{(L_1 - L_A)} + \frac{1}{(L_e - L_D')}\right]$$

$$\frac{1}{L_D' \times 160} = \frac{1}{40}\left[\frac{1}{(1 - L_A)} + \frac{1}{(0.16 - L_D')}\right] = \frac{1}{40}\left[\frac{1}{(1 - 0.4L_D')} + \frac{1}{(0.16 - L_D')}\right]$$

$(L_A = 0.4L_D')$

$$\frac{1}{L_D'} = \frac{160}{40}\left[\frac{0.16 - L_D' + 1 - 0.4L_D'}{(1 - 0.4L_D')(0.16 - L_D')}\right] \quad [\because L_1 = 1.0 \text{ m}, L_e = 0.16 \text{ m}]$$

$$= \frac{4[1.16 - 1.4L_D']}{0.16 - L_D' - 0.064L_D' + 0.4L_D'^2} = \frac{(4.64 - 5.6L_D')}{0.16 - 1.064L_D' - 4L_D'^2}$$

or $$0.16 - 1.064L_D' + 0.4L_D'^2 = 4.64L_D' - 5.6\,L_D'^2$$

or $$0.16 - 1.064L_D' + 0.4\,L_D'^2 - 4.64L_D' + 5.6L_D'^2 = 0$$

$$6L'^2_D - 5.704L_D' + 0.16 = 0$$

$$L_D' = \frac{5.704 \pm \sqrt{5.704^2 - 4 \times 6 \times .16}}{2 \times 6} = \frac{5.704 \pm 5.357}{12} = 0.92 \text{ m and } 0.0284 \text{ m}$$

$\therefore$ $$L_A = 0.4 \times L_D' = 0.4 \times 0.92 = 0.368 \text{ m and } 0.4 \times 0.0289 = 0.01156 \text{ m.}$$

The fundamental frequency (or single node frequency) will be that which corresponds to the larger of these two values of L_D' or L_A. Of the two values of L_D' or L_A, L_D' is having larger value which is 0.92 m. Hence when L_D' = 0.92 m then L_A = 0.368 m. This set gives the position of single node. Also L_D' will not give the actual node point. The distance L_A will give the actual node point as shown in Fig. 17.15 (*c*). The natural frequency of torsional vibrations for single node is given by,

$$f_{n1} = \frac{1}{2\pi}\sqrt{\frac{C \times J}{L_A \times I_A}}$$ where J = Polar moment of inertia of equivalent shaft

$$= \frac{\pi}{32}(d_1^4) = \frac{\pi}{32}(0.045)^4 = 4.0258 \times 10^{-7} \text{ m}^4$$

$$= \frac{1}{2\pi}\sqrt{\frac{(80 \times 10^9) \times 4.0258 \times 10^{-7}}{0.368 \times 400}} = \textbf{2.35 Hz.} \quad \textbf{Ans.}$$

The other set *i.e.* L_D' = 0.0289 m and L_A = 0.01156 m will give the two-node frequency. The positions of two nodes N_1 and N_2 is shown in Fig. 17.15 (*d*).

$$f_{n2} = \frac{1}{2\pi}\sqrt{\frac{C \times J}{L_A \times I_A}}$$

$$= \frac{1}{2\pi}\sqrt{\frac{(80 \times 10^9) \times (4.0258 \times 10^{-7})}{0.01156 \times 400}} \qquad \text{(Here } L_A = 0.01156 \text{ m)}$$

$= \mathbf{13.28\ Hz.}$ **Ans.**

HIGHLIGHTS

1. The natural frequency of free torsional vibrations is given by,

$$f_n = \frac{1}{2\pi}\sqrt{\frac{q}{I}} \quad \text{or} \quad \frac{1}{2\pi}\sqrt{\frac{C \times J}{L \times I}}$$

where q = Torsional stiffness of the shaft or torque required to twist the shaft through one radian

$$= \frac{T}{\theta} = \frac{C \times J}{L} \qquad \left(\because \frac{T}{J} = \frac{C \times \theta}{L}\right)$$

I = Mass moment of inertia of disc attached to the shaft

and L = Length of shaft.

2. If the inertia of shaft is to be considered, then the frequency of torsional vibrations is given by

$$f_n = \frac{1}{2\pi}\sqrt{\frac{q}{(I + I_1/3)}}$$

where I = Mass moment of inertia of the disc,

I_1 = Mass moment of inertia of the shaft, and

q = Torsional rigidity of shaft $= \frac{C \times J}{L}$.

3. For a free torsional vibrations of a single rotor system, the frequency of vibration is given by,

$$f_n = \frac{1}{2\pi}\sqrt{\frac{q}{I}}$$

4. For a two-rotor system there is a only one node and we have the equations as

(*i*) $L_A \times I_A = L_B \times I_B$

(*ii*) $L_A = \frac{I_B}{(I_A + I_B)} \times L$

(*iii*) $f_{nA} = f_{nB} = \frac{1}{2\pi}\sqrt{\frac{C \times J}{I_A \times L_A}} = \frac{1}{2\pi}\sqrt{\frac{C \times J}{I_B \times L_B}}$

where L_A = Distance of node from rotor A,

L_B = Distance of node from rotor B,

L = Total length between rotor A and rotor B,

I_A, I_B = Mass moment of inertia of rotor A and rotor B respectively,

J = Polar moment of inertia of equivalent shaft,

C = Modulus of rigidity of shaft.

5. For a three-rotor system, there is maximum two nodes (one less than the number of rotors). Also there will be a single node, which will give the fundamental frequency of the torsional vibrations. Let the three rotors are A, B and C where rotors A and C are at the ends and rotor B is in between.

For a three rotor system, we have the following equations :

(*i*) $L_A \times I_A = L_C \times I_C$ ...... where rotors A and C are at the ends.

$$(ii)\ \frac{1}{L_C \times I_C} = \frac{1}{I_B}\left[\frac{1}{(L_1 - L_A)} + \frac{1}{(L_2 - L_C)}\right] \quad \text{..... where rotor } B \text{ is in between}$$

where L_1 = Distance between rotors A and B,

L_2 = Distance between rotors B and C,

L_A = Distance of node N_1 from rotor A,

L_C = Distance of node N_2 from rotor C.

6. A shaft of uniform diameter, which twists through the same angle as the actual shaft of different diameters and different lengths, when equal and opposite torques of the given amounts are applied, is known as torsionally equivalent shaft.

7. The length of torsionally equivalent shaft of uniform diameter d_1 is given by,

$$L = L_1 + L_2\left(\frac{d_1}{d_2}\right)^4 + L_3\left(\frac{d_1}{d_3}\right)^4 + \ldots\ldots$$

where L = Length of torsionally equivalent shaft having uniform diameter d_1.

$L_1, L_2, L_3, \ldots\ldots$ = Lengths of different shafts.

$d_1, d_2, d_3, \ldots\ldots$ = Diameters of different shafts.

8. A gear system may be reduced to an equivalent single shaft system if :

(*i*) the maximum strain energy stored in the two systems are equal, and

(*ii*) maximum kinetic energies of the rotors in the two systems are equal.

9. The single shaft is equivalent to the original geared system if the mass moment of inertia of rotor D' is given by

(*i*) $I_D' = \dfrac{I_D}{G^2}$ and

additional length of the equivalent shaft is given by

(*ii*) $L_e = G^2 \times \left(\dfrac{d_1}{d_2}\right)^4 \times L_2$ where G = Gear ratio.

10. If the inertia of the gearing is negligible, then the gear system reduces to two rotor system. But if the inertia of the gearing is not negligible, then the system becomes a three rotor system.

EXERCISE 17

(A) THEORETICAL QUESTIONS

1. Define and explain the term 'torsional vibration'.
2. Find an expression for natural frequency of free torsional vibrations when
   (*i*) effect of inertia of shaft is not considered, and
   (*ii*) effect of inertia of shaft is considered.
3. Prove that the natural frequency of free torsional vibrations of a single rotor system is given by,

$$f_n = \frac{1}{2\pi}\sqrt{\frac{q}{I}}$$

where q = Torsional stiffness of the shaft (*i.e.* torque required to twist the shaft through one radian) and

I = Mass moment of inertia of single rotor.

4. Define the term 'the elastic line of the shaft'. In case of a two rotor system, there is one node which divides the length of the shaft inversely as the mass moment of inertia of the rotors. Prove the above statement.
5. Describe in details the method of finding the frequency of torsional vibration of a two rotor system.
6. Explain the procedure of finding : (*i*) the positions of two nodes in a three rotor system and (*ii*) the position of a real single node in a three rotor system.
7. What do you understand by 'Torsionally equivalent shaft' ?

8. Prove that the length of torsionally equivalent shaft of uniform diameter 'd_1' is given by

$$L = L_1 + L_2\left(\frac{d_1}{d_2}\right)^4 + L_3\left(\frac{d_1}{d_3}\right)^4 + ...$$

where $d_1, d_2, d_3, ...$ = Diameters of different shafts and

$L_1, L_2, L_3, ...$ = Corresponding lengths of different shafts.

9. Describe in details the procedure of reducing a gear system to an equivalent single shaft when : (*i*) the effect of inertia of the gearing is neglected and (*ii*) the effect of inertia of the gearing is considered.

10. For a single shaft which is equivalent to the original geared system, prove that the mass moment of inertia of the rotor D' is given by

$$I_D' = \frac{I_D}{G^2}$$

and the additional length of the equivalent shaft is given by

$$L_e = G^2 \times \left(\frac{d_1}{d_2}\right)^4 \times L_2$$

where G_2 = gear ratio.

(B) NUMERICAL PROBLEMS

1. A vertical shaft 100 mm in diameter and 1 m in length has its upper end fixed to the ceiling. At the other end it carries a disc of weight 5000 N having a radius of gyration of 450 mm. The modulus of rigidity for the material of the shaft is 0.8×10^5 N/mm^2. Determine the frequency of torsional vibrations. **[Ans.** 13.88 Hz]

2. A shaft of length 1.5 m is 0.095 m in diameter for the first 0.6 m of its length, 0.06 m in diameter for the next 0.5 m of the length and 0.05 m in diameter for the remaining 0.4 m of its length. The shaft carries two rotors at two ends. The mass moment of inertia of the first rotor is 650 kg m^2 whereas of the second rotor is 212 kg m^2. Determine the natural frequency of free torsional vibration of the system. The modulus of rigidity of shaft material may be taken as 80 GN/m^2. **[Ans.** 3.38 Hz]

3. For the above problem, locate the position of node on the original shaft from first rotor. **[Ans.** 0.855 m from rotor *A*]

4. A shaft of length 985 mm is 100 mm in diameter for the first 300 mm of its length, 150 mm in diameter for the next 160 mm of the length, 120 mm in diameter for the next 125 mm of the length and 90 mm in diameter for the remaining 400 mm of its length. The shaft carries two rotors at two ends. The mass moment of inertia of the first rotor is 168.75 kg m^2 whereas of the second rotor is 405 kg m^2. Determine the frequency of the torsional vibrations. It is desired to have the node at the mid-section of the shaft of 120 mm diameter by changing the diameter of the section having a 90 mm diameter. What will be the new diameter ? **[Ans.** (*i*) 13.2 Hz ; (*ii*) 135 mm]

5. The moment of inertia of three rotors *A*, *B* and *C* are respectively 7.5 kg m^2, 22.5 kg m^2 and 12.5 kg m^2. The distance between *A* and *B* is 1.25 m and between *B* and *C* is 4 m. The shaft is 40 mm in diameter and modulus of rigidity for the shaft material is 85 GN/m^2. Find :

(*i*) the frequency of torsional vibrations,

(*ii*) position of nodes, and

(*iii*) amplitude of vibrations.

[Ans. (*i*) 8.81 Hz for two nodes and 3.84 Hz for single-node vibration.

(*ii*) For two nodes, the position of nodes N_1 and N_2 are at a distance of 0.91 m and 0.55 m from rotor *A* and rotor *C* respectively. For single node, the position of node lies at a distance of 2.9 m from rotor *C*.

(*iii*) For two nodes ; a_1 = 1 radian, a_2 = 0.366 radian and a_3 = 0.06 radian.

For single node : a_1 = 1 radian, a_2 = 0.74 radian and a_3 = 1.94 radian]

6. The moment of inertia of three rotors A, B and C are respectively 200 kg m^2, 80 kg m^2 and 5 kg m^2. The distance between rotor A and B is 2 m and they are connected by a shaft of diameter 50 mm. The distance between rotors B and C is also 2 m and they are connected by a shaft of diameter 25 mm. Determine : (*i*) Natural frequencies of torsional vibrations and (*ii*) position of nodes.

[**Ans.** (*i*) For a single node frequency is 1.69 Hz whereas for two nodes the frequency is 3.46 Hz.
(*ii*) For single node, the position of node is at a distance of 0.86 m from rotor A. For two nodes, the position of nodes N_1 and N_2 are at a distance of 0.52 m from rotor A and 1.3 m from rotor C.]

7. A shaft carries a motor at one end and a pinion on the other end. The length and diameter of this shaft is 300 mm and 60 mm respectively. There is another shaft of length 600 mm and of diameter 100 mm. This shaft carries a gear wheel at one end and a centrifugal pump on another end. The gear wheel and pinion are meshing together so that the centrifugal pump is driven by the motor. The mass moment of inertia of the motor and centrifugal pump are 800 kg m^2 and 3000 kg m^2 respectively. If the inertia of gears and shaft is neglected and pump speed is one third of the motor then find the frequency of torsional vibrations of the system. Take the value of modulus of rigidity of shaft material as 80 GN/m^2. [**Ans.** 6.6 Hz]

8. A shaft carries a motor at one end and a gear wheel on the other end. The diameter of this shaft is 54 mm and length is L meter. There is another shaft of length 400 mm and of diameter 45 mm. This shaft carries a pinion at one end and a centrifuge on the other end. The gear wheel and pinion are meshing together so that the centrifuge is driven by the motor. The mass moment of inertia of motor and of centrifuge are 0.375 kg m^2 and 0.59 kg m^2 respectively. Neglecting the inertia effect of the gears, find the value of L if the gears are to be at the node for the torsional vibrations of the system. The speed of the centrifuge is four times the speed of the motor. Find also the frequency of torsional vibrations. Take the value of modulus of rigidity for shaft material as equal to 84 kN/mm^2.

[**Ans.** 1.3 m ; 60.4 Hz]

18

Cams

18.1. Introduction

A rotating machine element, which gives reciprocating or oscillating motion to a second element is known as a cam. The second element is called a follower. The cam rotates usually at constant speed and drives the follower whose motion depends upon the shape of the cam. Almost always the cam is the driver and the follower is the driven. The cams are commonly used in internal combustion engines (for operating the inlet and exhaust valves), in printing machinery, in machine tools, in automatic machines etc.

In a cam mechanism, the three essential members are : (*i*) *the cam* which has a curved surface or straight surface, (*ii*) the *follower* and (*iii*) *the frame* which supports and guides, the follower and the cam. Fig. 18.1 shows all these three members.

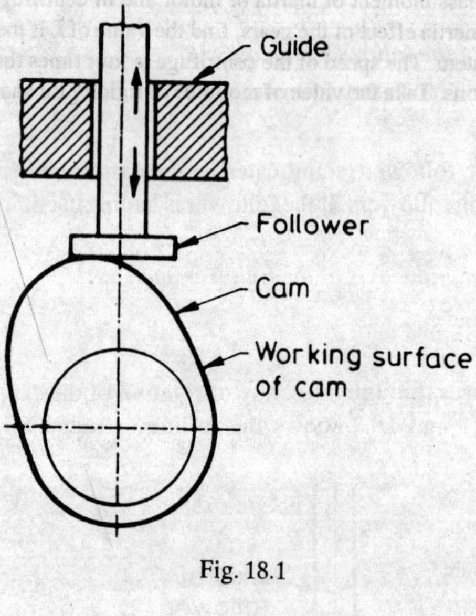

Fig. 18.1

18.2. Types of Followers

The followers which are having either reciprocating or oscillating motion, are generally classified as :

(*i*) *According to their shape of that part which is in contact with the cam such as*

(*a*) knife-edge follower,

(*b*) roller follower, and

(*c*) flat faced follower or mushroom follower.

Fig. 18.2 (*a*) shows a knife-edge follower, Fig. 18.2 (*b*) shows a roller follower and Fig. 18.2 (*c*) a flat-faced follower. The knife-edge follower shown in Fig. 18.2 (*a*) is not a practical one owing to the rapid rate at which the knife-edge wears. The roller followers are mostly used.

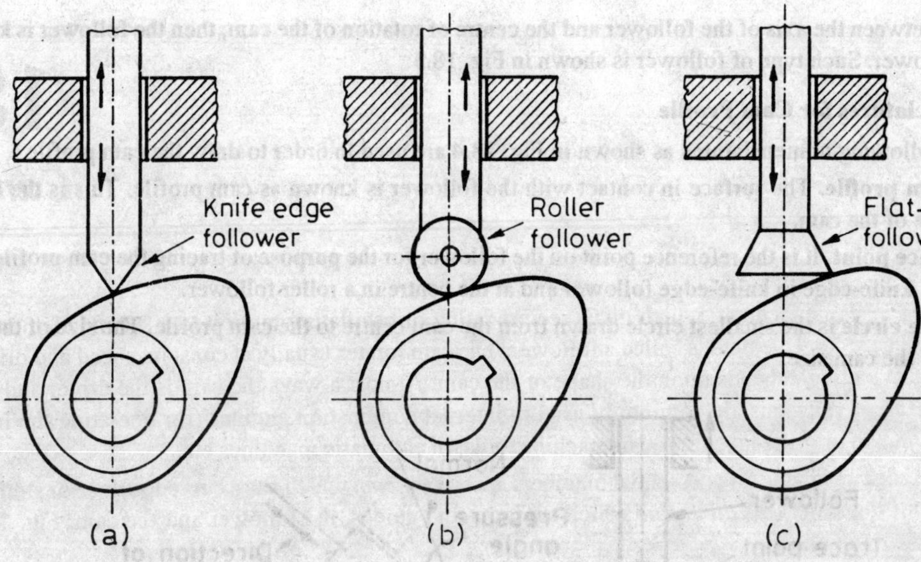

Fig. 18.2

(*ii*) *According to the motion of the follower such as*

(*a*) reciprocating follower (or Translatory follower)

(*b*) oscillating follower.

In reciprocating follower, the follower reciprocates as the cam rotates uniformly. In Figs. 18.1 and 18.2, all the followers are reciprocating followers. If the follower is having oscillating motion, then the follower is known as oscillating follower.

(*iii*) *According to the location of the axis of the follower such as*

(*a*) In-line or radial followers,

(*b*) Off-set follower.

If the axis of the follower passes through the centre of rotation of the cam, the followers are known as in line or radial follower. Figs. 18.1 and 18.2 shows the in-line or radial follower. But if there is same

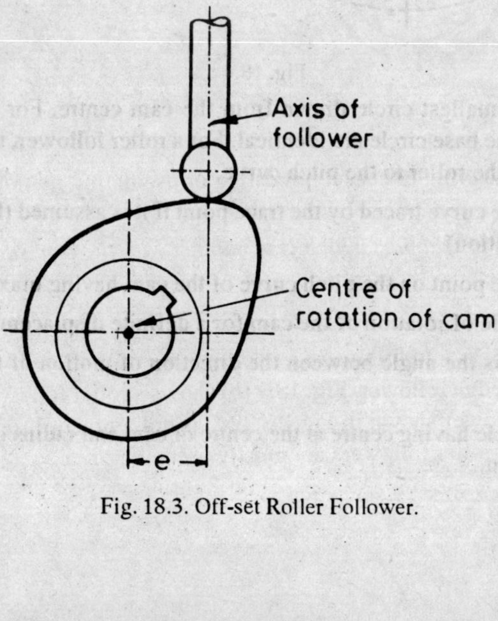

Fig. 18.3. Off-set Roller Follower.

eccentricity between the axis of the follower and the centre of rotation of the cam, then the follower is known as off-set follower. Such type of follower is shown in Fig. 18.3.

18.3. Nomenclatures for Cam Profile

The following nomenclatures as shown in Fig. 18.4 are used in order to draw the cam profile :

1. **Cam profile.** The surface in contact with the follower is known as cam profile. This is the actual working curve of the cam.

2. **Trace point.** It is the reference point on the follower for the purpose of tracing the cam profile. It is situated at the knife-edge in knife-edge follower and at the centre in a roller follower.

3. **Base circle** is the smallest circle drawn from the cam centre to the cam profile. The size of the cam circle decides the cam size.

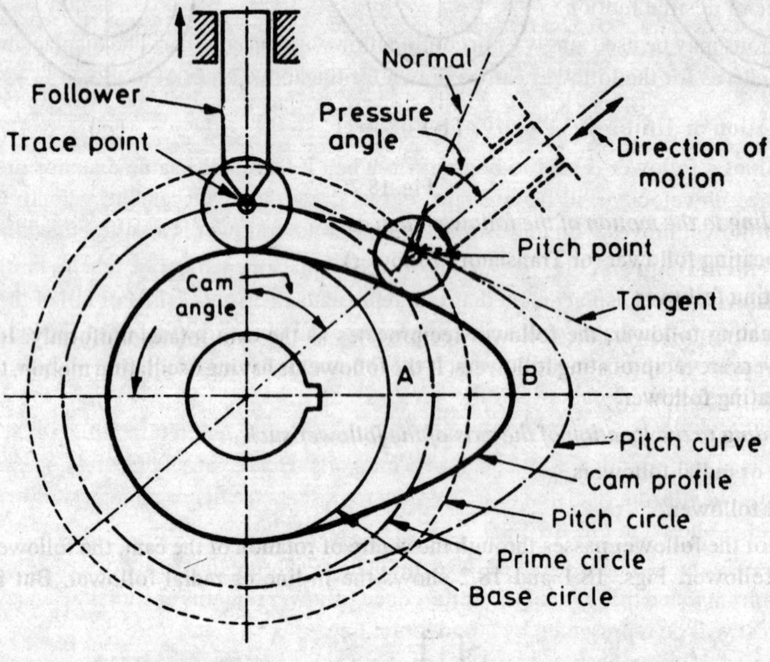

Fig. 18.4

4. **Prime circle** is the smallest circle drawn from the cam centre. For a knife-edge and a flat face follower, the prime circle and the base circle are identical. For a roller follower, the prime circle is larger than the base circle by the radius of the roller to the pitch curve.

5. **The pitch curve** is the curve traced by the trace point if it is assumed that follower is rotating round the cam (instead of the cam rotation).

6. **The pitch point** is the point on the pitch curve of the cam having maximum pressure angle.

7. **Cam angle** is the angle of rotation of the cam for a definite displacement of the follower.

8. **The pressure angle** is the angle between the direction of motion of follower and a normal to the pitch curve.

9. **Pitch circle** is the circle having centre at the centre of cam and radius equal to the distance between the centre of cam and pitch point.

10. Life or stroke is the maximum travel of the follower from its lowest position to the top most position. It is equal to the distance AB as shown in Fig. 18.4.

11. Period of dwell is the period during which the follower remains stationary during some finite rotation of the cam.

18.4. Motions of the Follower

A follower may have the following types of motions :

1. Uniform motion or uniform velocity,
2. Simple harmonic motion,
3. Uniform acceleration and uniform retardation,
4. Cycloidal motion, and
5. Any type of desired motion.

These motions may be used singly or in combination with one another. The displacement, velocity and acceleration time curves for the follower can be drawn for the above types of motions.

18.5. Uniform Motion or Uniform Velocity of a Follower

The motion of a follower is said to be uniform when it travels the same distance in each succeeding time interval. Hence its velocity will be constant. Let us draw the displacement, velocity and acceleration diagrams when a follower moves with uniform velocity. Let the follower is a knife-edged follower.

1. Displacement Diagram. Fig. 18.5 (*a*) shows the displacement diagram when a knife-edged follower moves with uniform velocity. Displacement diagram represents the rise (*i.e.* lift) or fall of the follower during one revolution of the cam. The base (or abscissa) represents the time (*i.e.* number of seconds required for the cam to complete one revolution) or it may represent the angular displacement of the cam in degrees. The ordinate represents the displacement of the follower.

The follower moves with the uniform velocity during its rise and return stroke, that means the equal distances are traversed by the follower in equal time intervals. Hence the slopes of AE (which represents the displacement of follower during rise) and of FG (which represents the displacement of follower during return stroke) are constant as shown in Fig. 18.5 (*a*).

The follower is at rest during a part of the cam rotation. During the period of rest, the distance between the centre of the cam shaft and the point of the knife-edged follower remains unaltered. Hence the displacement during period of rest will be represented by a horizontal line.

This period of rest of the follower is known as dwell period. The line CD represents the dwell period after rise and line GB represents the dwell period after return stroke. The line EF (which is horizontal) represents the displacement of the follower after rise stroke.

Let S = Stroke of the follower or maximum lift of the follower (or Maximum displacement of follower)

x = Lift of the follower (displacement of follower) when angle turned by cam is θ at any time 't'

θ_0 = Angle turned by cam during outstroke

θ = Angle turned by cam in time 't'

ω = Angular velocity of cam.

We know that $$\theta = \omega \times t \qquad ...(i)$$

From Fig. 18.5 (*a*), the displacement of follower (or life of follower) during outstroke in time 't' is given by,

$$x = \frac{S}{\theta_0} \times \theta \qquad \left(\because \ \frac{x}{\theta} = \frac{S}{\theta_0} \quad \therefore \ x = \frac{S}{\theta_0} \times \theta\right)$$

$$= \frac{S}{\theta_0} \times \omega \times t \qquad (\because \ \theta = \omega \times t)$$

The velocity of follower (v_0) during outstroke is given by

$$v_0 = \frac{dx}{dt} = \frac{d}{dt}\left(\frac{S}{\theta_0} \times \omega \times t\right)$$

$$= \frac{S}{\theta_0} \times \omega = \text{constant} \qquad ...(18.1)$$

($\because$ S, θ_0 and ω are constant)

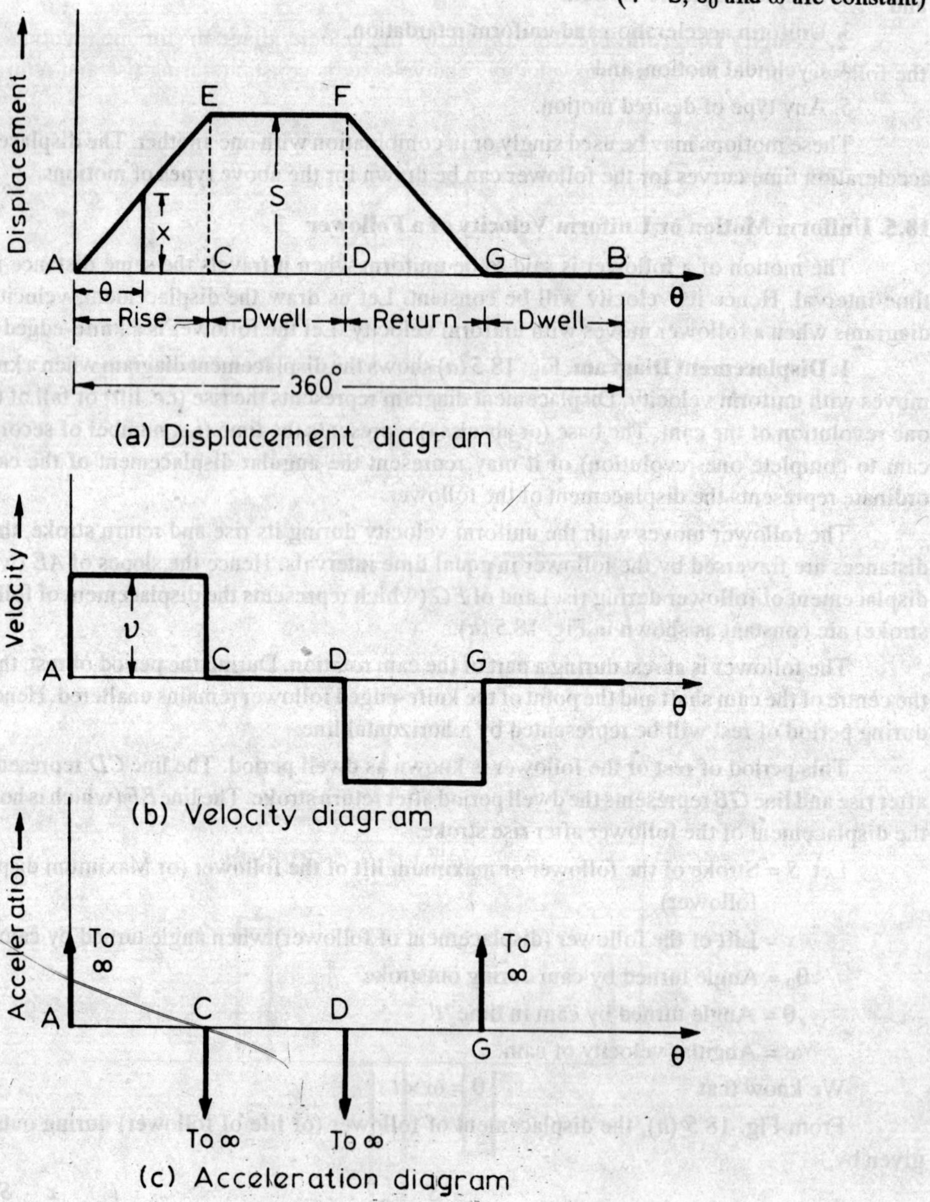

Fig. 18.5

The acceleration of the follower (f_0) during outstroke is given by,

$$f_0 = \frac{dv}{dt} = 0 \qquad \left(\because \ v \text{ is constant. Hence } \frac{dv}{dt} = 0 \right)$$

$$\therefore \qquad f_0 = 0 \qquad \text{...(18.1 A)}$$

Similarly, the velocity and acceleration during return stroke are obtained changing θ_0 to θ_r.

$$v_r = \frac{S}{\theta_r} \times \omega \qquad \text{...(18.1 B)}$$

and $$f_r = 0 \qquad \text{...(18.1 C)}$$

2. Velocity Diagram. Fig. 18.5 (*b*) shows the velocity diagram (for one revolution of the cam) when the follower moves with uniform velocity. The velocity is constant during rise and return stroke. Hence the

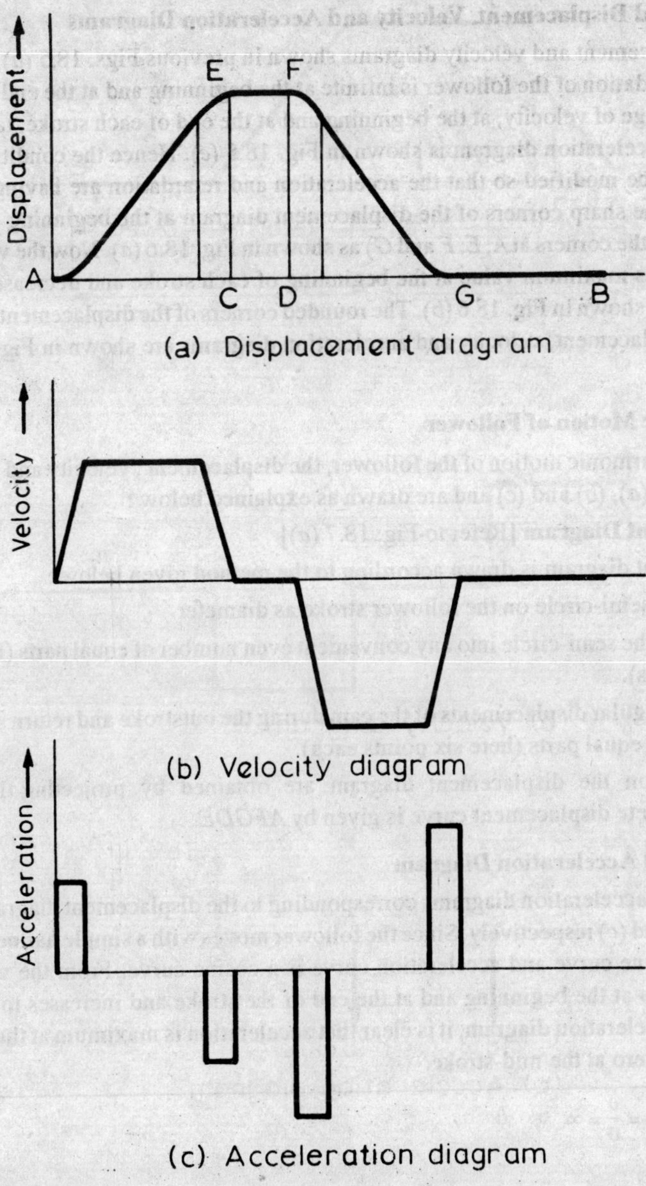

Fig. 18.6

velocity will be represented by horizontal lines during rise and return stroke. The displacement of the follower increases during rise stroke and hence velocity will be positive whereas the displacement decreases during return stroke and hence velocity will be negative as shown in Fig. 18.5 (*b*). During the periods of dwell, the velocity is zero as rate of change of displacement during this period is zero.

3. Acceleration Diagram. Fig. 18.5 (*c*) shows the acceleration diagram when the follower moves with uniform velocity. At the beginning and at the end of the rise stroke, the velocity changes suddenly (*i.e.* change in velocity takes place in zero time). This means the acceleration* (or retardation) is infinite in magnitude at the beginning (or at the end) of the rise stroke, which is physically impossible. There will be acceleration at the beginning of the rise stroke and there will be retardation at the end of the rise stroke. The change in velocity (except the beginning and ending of rise stroke) will be zero and hence acceleration will be zero. Same is the case with return stroke.

18.5.1. Modified Displacement, Velocity and Acceleration Diagrams

From the displacement and velocity diagrams shown in previous Figs. 18.5 (*a*) and (*b*), it is clear that the acceleration or retardation of the follower is infinite at the beginning and at the end of each stroke. This is due to the fact that change of velocity, at the beginning and at the end of each stroke, takes place in zero time (*i.e.* in no time). The acceleration diagram is shown in Fig. 18.5 (*c*). Hence the conditions which govern the follower motion, must be modified so that the acceleration and retardation are having finite values. This is done by rounding off the sharp corners of the displacement diagram at the beginning and at the end of each stroke (*i.e.* rounding off the corners at *A*, *E*, *F* and *G*) as shown in Fig. 18.6 (*a*). Now the velocity of the follower increases gradually to its maximum value at the beginning of each stroke and decreases gradually to zero at the end of each stroke as shown in Fig. 18.6 (*b*). The rounded corners of the displacement diagram are parabolic arcs. The modified displacement, velocity and acceleration diagrams are shown in Figs. 18.6 (*a*), (*b*) and (*c*) respectively.

18.6. Simple Harmonic Motion of Follower

For the simple harmonic motion of the follower, the displacement, velocity and acceleration diagrams are shown in Figs. 18.7 (*a*), (*b*) and (*c*) and are drawn as explained below :

(A) Displacement Diagram [Refer to Fig. 18.7 (*a*)]

The displacement diagram is drawn according to the method given below :

(*i*) First draw a semi-circle on the follower stroke as diameter.

(*ii*) Then divide the semi-circle into any convenient even number of equal parts (in the present example it is divided into six parts).

(*iii*) Divide the angular displacements of the cam during the outstroke and return strokes of the follower into the same number of equal parts (here six points each).

(*iv*) The points on the displacement diagram are obtained by projecting the points as shown in Fig. 18.7 (*a*). The complete displacement curve is given by *AFGDE*.

(B) Velocity and Acceleration Diagram

The velocity and acceleration diagrams corresponding to the displacement diagram of Fig. 18.7 (*a*) are shown at Figs. 18.7 (*b*) and (*c*) respectively. Since the follower moves with a simple harmonic motion, therefore the velocity curve is a sine curve and acceleration curve is a cosine curve. From the velocity diagram, it is clear that velocity is zero at the beginning and at the end of the stroke and increases to a maximum value at the mid-stroke. From acceleration diagram, it is clear that acceleration is maximum at the beginning and at the end of the stroke and is zero at the mid-stroke.

*Acceleration $= \frac{dv}{dt} = \frac{v}{0} = \infty$.

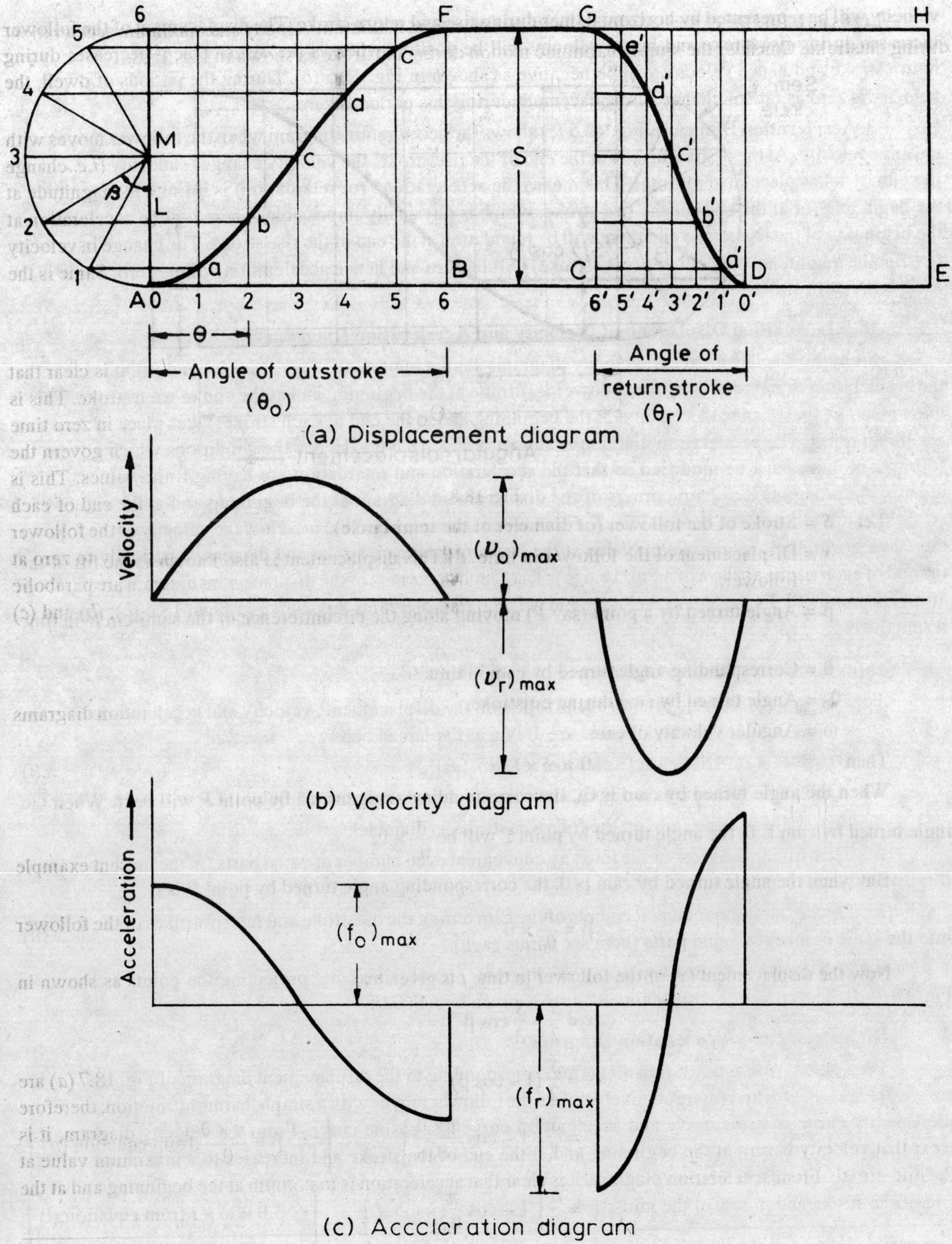
6
5
4
3
2
1
M
L
β
A 0 1 2 3 4 5 6
a
b
c
d
e
F
G
H
S
B
C
D
E
e'
d'
c'
b'
a'
6' 5' 4' 3' 2' 1' 0'
θ
Angle of outstroke
(θ_o)
Angle of returnstroke
(θ_r)
(a) Displacement diagram
Velocity
(v_o)max
(v_r)max
(b) Velocity diagram
Acceleration
(f_o)max
(f_r)max
(c) Acceleration diagram

Fig. 18.7

Let us now find the expression for velocity, acceleration, maximum velocity and maximum acceleration during outstroke. Consider the simple harmonic motion during outstroke as shown in Fig. 18.8.

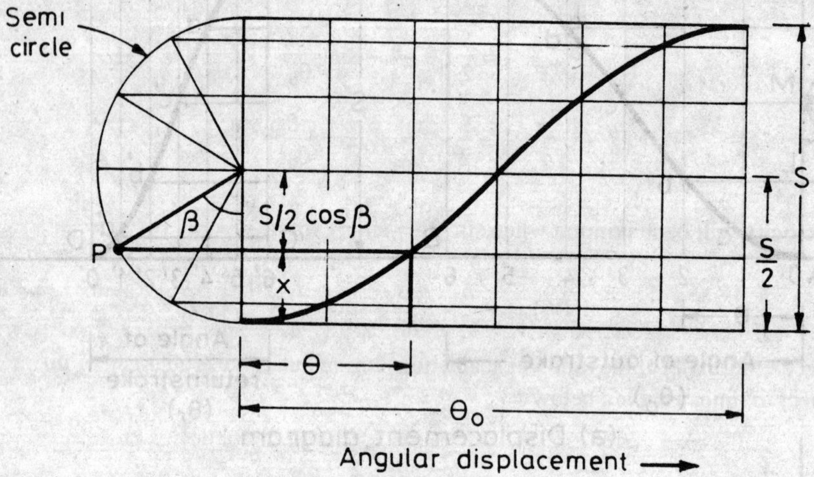

Fig. 18.8

Let S = Stroke of the follower (or diameter of the semi-circle).

x = Displacement of the follower in time 't'. This displacement is also known as the lift of the follower.

β = Angle turned by a point (say P) moving along the circumference of the semi-circle in time 't'.

θ = Corresponding angle turned by cam in time 't'.

θ_0 = Angle turned by cam during outstroke.

ω = Angular velocity of cam.

Then
$$\theta = \omega \times t \qquad ...(i)$$

When the angle turned by cam is θ_0, the corresponding angle turned by point P will be π. When the angle turned by cam is θ, the angle turned by point P will be $\frac{\pi}{\theta_0} \times \theta$.

But when the angle turned by cam is θ, the corresponding angle turned by point P is β

$$\therefore \quad \beta = \frac{\pi}{\theta_0} \times \theta \qquad ...(ii)$$

Now the displacement (x) of the follower in time t is given by,

$$x = \frac{S}{2} - \frac{S}{2} \cos \beta$$

$$= \frac{S}{2}(1 - \cos \beta)$$

$$= \frac{S}{2}\left(1 - \cos \frac{\pi \times \theta}{\theta_0}\right) \qquad \left[\because \beta = \frac{\pi \times \theta}{\theta_0} \text{ from equation } (ii)\right]$$

$$= \frac{S}{2}\left(1 - \cos \frac{\pi}{\theta_0} \times \omega \times t\right) \qquad \left[\because \theta = \omega \times t \text{ from equation } (i)\right]$$

The velocity (v_0) of the follower at any time during outstroke is obtained as,

$$v_0 = \frac{dx}{dt} = \frac{d}{dt}\left[\frac{S}{2}\left(1 - \cos\frac{\pi \times \omega \times t}{\theta_0}\right)\right]$$

$$= \frac{S}{2}\left[0 - \left(-\sin\frac{\pi\omega t}{\theta_0}\right) \times \frac{\pi\omega}{\theta_0}\right] = \frac{S \times \pi \times \omega}{2\theta_0} \times \frac{\sin \pi\omega t}{\theta_0} \qquad ...(iii)$$

$$= \frac{S}{2} \times \frac{\pi\omega}{\theta_0} \times \sin\left(\frac{\pi\theta}{\theta_0}\right) \qquad [\because \ \omega \times t = \theta]$$

The velocity will be maximum when $\sin\frac{\pi\theta}{\theta_0} = \sin\frac{\pi}{2}$ or $\frac{\pi\theta}{\theta_0} = \frac{\pi}{2}$ or $\theta = \frac{\theta_0}{2}$.

$$(v_0)_{max} = \frac{S}{2} \times \frac{\pi\omega}{\theta_0} \times \sin\left(\frac{\pi}{\theta_0} \times \frac{\theta_0}{2}\right) = \frac{S}{2} \times \frac{\pi \times \omega}{\theta_0} \qquad ...(18.2)$$

The acceleration (f_0) of the follower at any time during outstroke is obtained by differentiating equation (*iii*) with respect to time, as given below

$$f_0 = \frac{dv_0}{dt} = \frac{d}{dt}\left(\frac{S}{2} \times \frac{\pi\omega}{\theta_0} \times \sin\frac{\pi\omega t}{\theta_0}\right)$$

$$= \frac{S}{2} \times \frac{\pi \times \omega}{\theta_0} \times \frac{\pi \times \omega}{\theta_0} \times \cos\frac{\pi\omega t}{\theta_0}$$

$$= \frac{S}{2} \times \left(\frac{\pi \times \omega}{\theta_0}\right)^2 \times \cos\frac{\pi\theta}{\theta_0} \qquad (\because \ \omega \times t = \theta)$$

The acceleration will be maximum when $\theta = 0$. Hence maximum acceleration will be

$$(f_0)_{max} = \frac{S}{2} \times \left(\frac{\pi \times \omega}{\theta_0}\right)^2 \qquad ...(18.2\,A)$$

Similarly the maximum velocity and maximum acceleration of follower during return stroke are obtained as

$$(v_r)_{max} = \frac{S}{2} \times \frac{\pi \times \omega}{\theta_r} \qquad ...(18.3)$$

and

$$(f_r)_{max} = \frac{S}{2}\left(\frac{\pi \times \omega}{\theta_r}\right)^2 \qquad ...(18.4)$$

where θ_r = Angle turned by cam during return stroke.

Note. In Fig. 18.7 (*a*), the angle of return (θ_r) is assumed less than angle of outstroke (θ_0). Hence $(v_r)_{max}$ and $(f_r)_{max}$ will be more than $(v_0)_{max}$ and $(f_0)_{max}$ as shown in Figs. 18.7 (*b*) and (*c*) respectively.

18.7. Uniform Acceleration and Uniform Retardation

For uniform acceleration and uniform retardation of the follower, the displacement, velocity and acceleration diagrams are shown in Figs. 18.9 (*a*), (*b*), (*c*) and are drawn as explained below :

(A) Displacement Diagram. The displacement diagram is drawn according to the method given below :

(*i*) The angular displacement of the cam during outstroke is divided into any even number of equal parts. Here the angular displacement (θ_0) during outstroke is divided into six equal parts. These parts are marked as 1, 2, 3, 4, 5 and 6. Draw vertical lines through these points. During the first three intervals the follower is accelerated and during the second three intervals it is retarded.

(*ii*) **Divide the middle vertical line (*i.e.* vertical line 3-N) into the same number of equal parts *i.e.* six equal parts. These points are shown as *I, J, K, L, M* and *N*.**

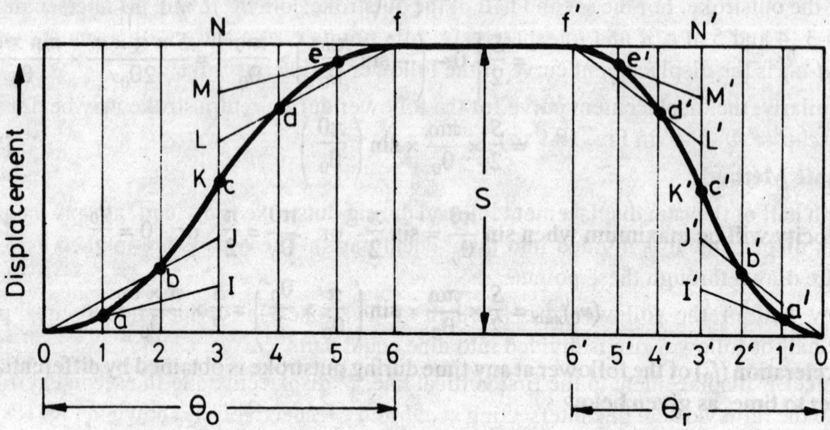

(a) Displacement diagram

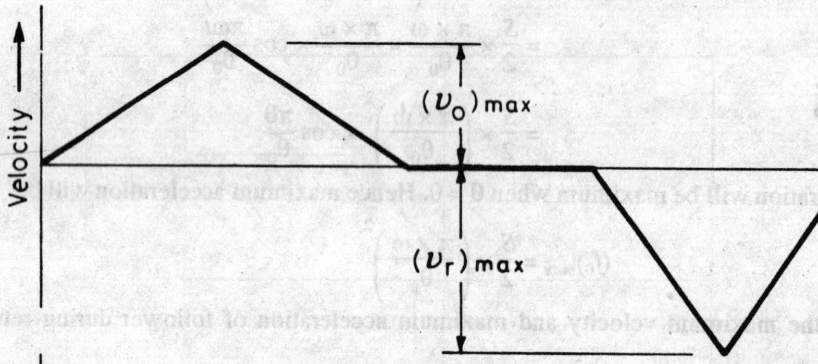

(b) Velocity diagram

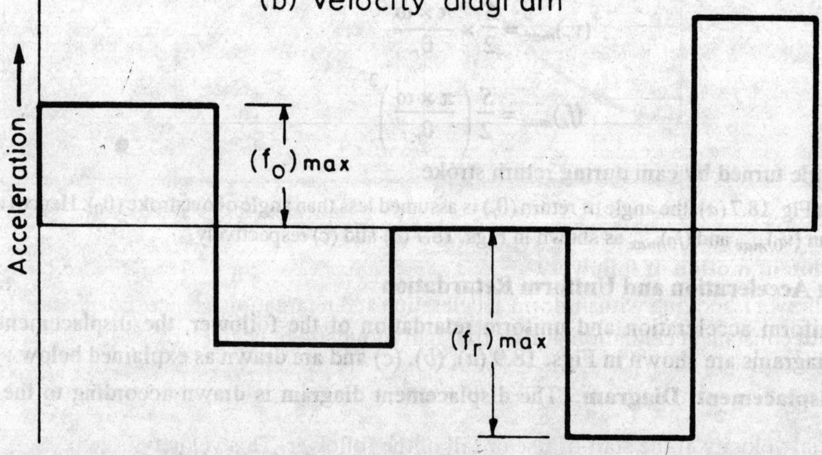

(c) Acceleration diagram

Fig. 18.9

(*iii*) Join *OI, OJ* and *OK* intersecting the vertical lines drawn through 1, 2 and 3 at *a*, *b* and *c* respectively. Join points 0, *a*, *b* and *c* with a smooth curve. The curve 0-*a*-*b*-*c* is a displacement curve of the follower for the first half of the outstroke. For the second half of the outstroke, join *fK*, *fL* and *fM* intersecting vertical lines drawn through 3, 4 and 5 at *c*, *d* and *e* respectively. Join points *c*, *d*, *e* and *f* with a smooth curve. Then the curve 0-*a*-*b*-*c*-*d*-*e*-*f* is the displacement curve of the follower during outstroke.

(*iv*) Similarly, the displacement curve for the follower during return stroke may be drawn. This curve is shown by *f*′-*e*′-*d*′-*c*′-*b*′-*a*′-0′ in Fig. 18.9 (*a*).

Alternate Method

(*i*) Each half of the cam displacement interval during outstroke is divided into any *n* equal parts (here half of the cam displacement is divided into three equal parts). These parts are marked as 1, 2 and 3. The vertical lines are drawn through these points.

(*ii*) Now half of the follower rise (*i.e.* half of the stroke *S*) is divided into n^2 equal parts (here it is $3^2 = 9$). Hence half of follower rise is divided into nine equal parts.

(*iii*) Project 1^2 displacement to the first vertical line, 2^2 displacement to the second vertical line and 3^2 displacement to the third vertical line intersecting at *a*, *b* and *c* respectively as shown in Fig. 18.10. Join points 0, *a*, *b* and *c* with smooth curve. The curve 0-*a*-*b*-*c* is a displacement curve of the follower for the first half of the outstroke. Similarly the displacement curve for the second half is drawn.

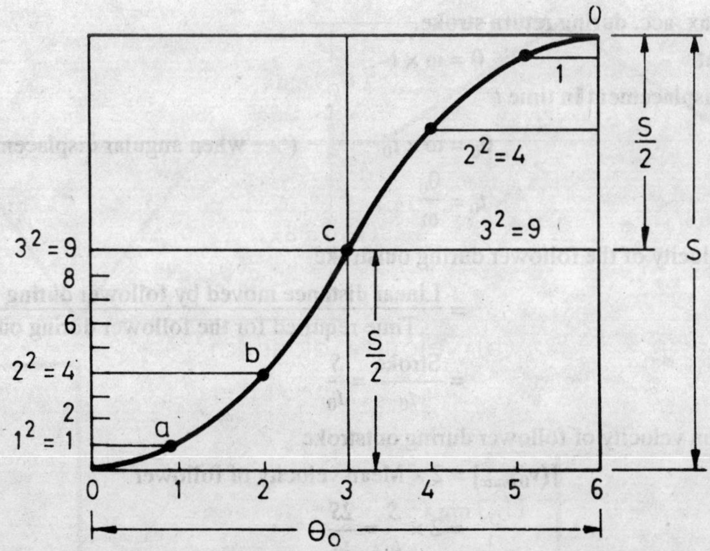

Fig. 18.10

Equation of motion of follower

The follower is moving with uniform acceleration and retardation. Hence the equation for linear motion of follower with constant acceleration during first half of outstroke is given by

$$S = ut + \frac{1}{2}ft^2$$

where *u* = Initial velocity at the start of rise or fall of the follower. This velocity is zero.

Hence $S = \frac{1}{2}ft^2$ where *f* = acceleration and is constant.

The above equation is a equation of parabola.

The velocity is given by, $v = \frac{ds}{dt} = \frac{d}{dt}\left(\frac{1}{2}ft^2\right)$ $\left(\because \ s = \frac{1}{2}ft^2\right)$

$= \frac{1}{2}f \times 2t$ ($\because$ f is constant)

$= f \times t$ ...(*i*)

In the above equation, f is constant.

Hence velocity varies directly with time. The velocity diagram is shown in Fig. 18.9 (*b*).

Let S = Stroke of the follower

ω = Angular velocity of cam

θ_0 = Angular displacement of cam during outstroke

θ_r = Angular displacement of cam during return stroke

t_0 = Time required for the follower during outstroke

$(v_0)_{max}$ = Max. velocity during outstroke

$(v_r)_{max}$ = Max. velocity during return stroke

$(f_0)_{max}$ = Max. acc. during outstroke, and

$(f_r)_{max}$ = Max. acc. during return stroke.

We know that $\theta = \omega \times t$

where θ is angular displacement in time t

$\therefore$ $\theta_0 = \omega \times t_0$ ($\because$ when angular displacement is θ_0, time is t_0)

or $t_0 = \frac{\theta_0}{\omega}$ ...(*ii*)

The mean velocity of the follower during outstroke

$$= \frac{\text{Linear distance moved by follower during outstroke}}{\text{Time required for the follower during outstroke}}$$

$$= \frac{\text{Stroke}}{t_0} = \frac{S}{t_0}$$

The maximum velocity of follower during outstroke

$$[(v_0)_{max}] = 2 \times \text{Mean velocity of follower}$$

$$= 2 \times \frac{S}{t_0} = \frac{2S}{t_0}$$

$$= \frac{2 \times S}{\left(\frac{\theta_0}{\omega}\right)} \qquad \left[\because \text{ From equation (ii), } t_0 = \frac{\theta_0}{\omega}\right]$$

$$= \frac{2S \times \omega}{\theta_0}$$

$\therefore$ $$(v_0)_{max} = \frac{2S \times \omega}{\theta_0} \qquad ...(18.5)$$

Similarly the maximum velocity of follower during return stroke is obtained by substituting θ_r for θ_0 in the above equation.

$$\therefore \qquad (v_r)_{max} = \frac{2S \times \omega}{\theta_r} \qquad ...(18.6)$$

From the velocity diagram, it is clear that the maximum velocity of the follower during outstroke is reached in the mid-stroke *i.e.* after time $\left(\frac{t_0}{2}\right)$.

Substituting $v = (v_0)_{max}$ and $t = \frac{t_0}{2}$ in equation (*i*), we get the maximum acceleration [*i.e.* $(f_0)_{max}$] during outstroke.

$$\therefore \qquad (v_0)_{max} = (f_0)_{max} \times \frac{t_0}{2}$$

or

$$(f_0)_{max} = \frac{2 \times (v_0)_{max}}{t_0}$$

$$= \frac{2 \times \left(\frac{2S \times \omega}{\theta_0}\right)}{\left(\frac{\theta_0}{\omega}\right)} \qquad \left[\because \text{ From equation } (v_0)_{max} = \frac{2S \times \omega}{\theta_0} \text{ and from equation } (ii),\ t_0 = \frac{\theta_0}{\omega}\right]$$

$$= \frac{4 \times S \times \omega^2}{{\theta_0}^2} \qquad ...(18.7)$$

Similarly the maximum acceleration of the follower during return stroke is obtained by substituting θ_r for θ_0 in the above equation

$$\therefore \qquad (f_r)_{max} = \frac{4 \times S \times \omega^2}{{\theta_r}^2} \qquad ...(18.8)$$

The acceleration diagram is shown in Fig. 18.9 (*c*).

18.8. Cycloidal Motion

A cycloid is a curve traced by a point on a circle when the circle rolls without slipping on a straight line.

The follower is having linear motion. Hence in case of cams the straight line is the stroke of the follower and the circumference of the rolling circle is equal to the stroke (S) of the follower. Hence the radius of the rolling circle will be ($S/2\pi$).

For the cycloidal motion of the follower the displacement, the velocity and acceleration diagrams are shown in Figs. 18.11 (*a*), (*b*), (*c*) and are drawn as explained below :

(A) Displacement Diagram

The displacement diagram is shown in Fig. 18.11 (*a*) and is drawn according to the method given below :

(*i*) Draw a circle of radius ($S/2\pi$) with 0 as centre.

(*ii*) Divide the circle into any number of equal even parts (say eight) and number them as shown in the Fig. 18.11 (*a*). These points are shown as 1′, 2′, 3′, ... 8′. Project these points horizontally on the vertical centre line of the circle. These points are 2′, *L*, *O*, *M* and 6′.

(*iii*) The angular displacement of the cam during outstroke (*i.e.* θ_0) is now divided into the same number of equal even parts as the circle is divided. These points are shown as 1, 2, 3, ... , 8. Draw vertical lines through these points.

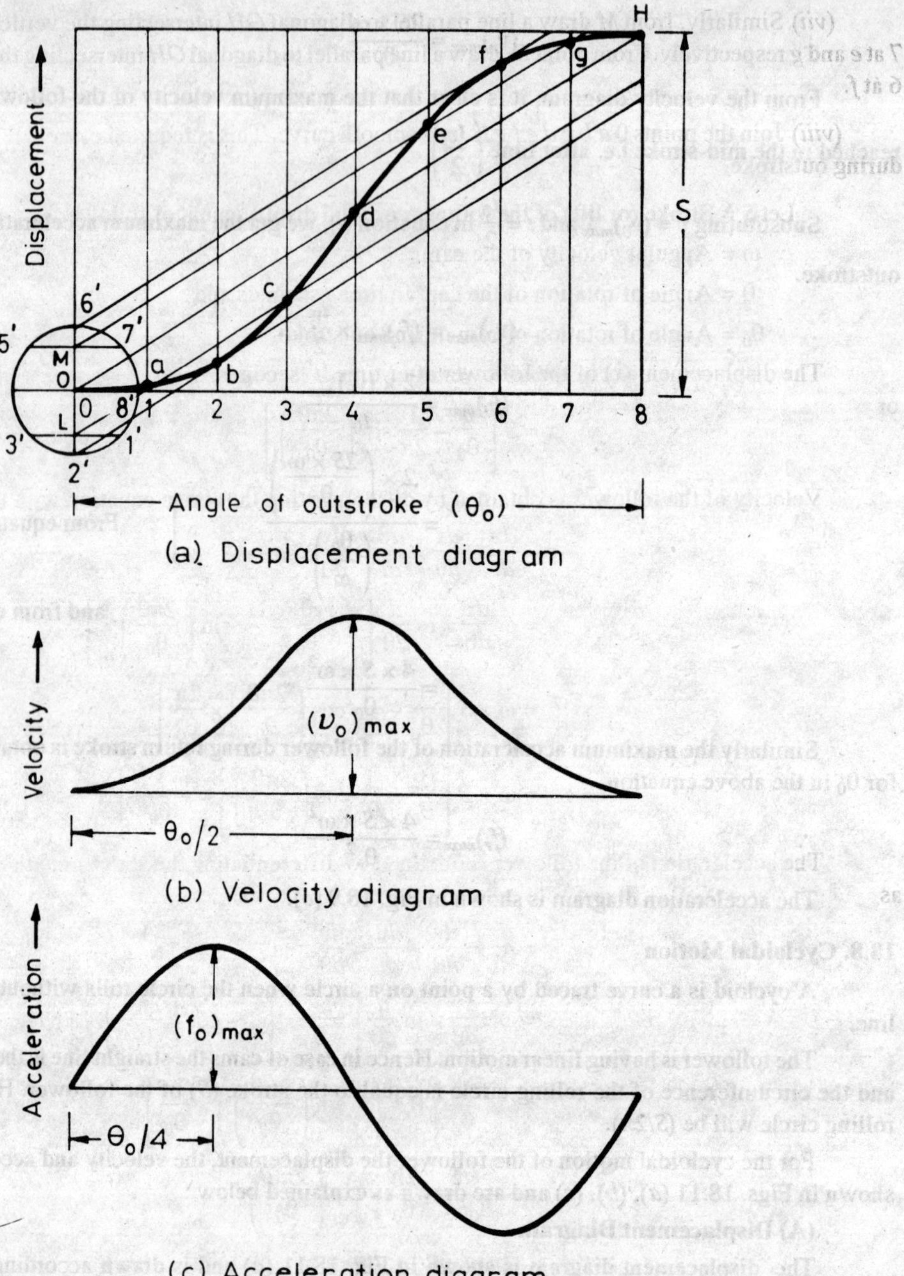

(a) Displacement diagram
(b) Velocity diagram
(c) Acceleration diagram

Fig. 18.11

(*iv*) Draw the diagonal of the displacement diagram passing through O (*i.e.* join line OH). This diagonal intersects the vertical line through 4 at d.

(*v*) From 2′, draw a line parallel to OH intersecting the vertical line through 2 at b.

(*vi*) Now from point L, draw a line parallel to diagonal OH intersecting the vertical lines through 1 and 3 at a and c respectively.

(*vii*) Similarly, from *M* draw a line parallel to diagonal *OH* intersecting the vertical lines through 5 and 7 at *e* and *g* respectively. From point 6′, draw a line parallel to diagonal *OH* intersecting the vertical line through 6 at *f*.

(*viii*) Join the points 0 *a b c d e f g H* by a smooth curve. This is required cycloidal curve for the follower during outstroke.

Let S = Stroke (or lift) of the follower or total displacement of follower,

ω = Angular velocity of the cam,

θ = Angle of rotation of the cam in time t seconds and

θ_0 = Angle of rotation of cam during outstroke.

The displacement (x) of the follower after time 't' seconds,

$$x = S\left[\frac{\theta}{\theta_0} - \frac{1}{2\pi}\sin\left(\frac{2\pi\theta}{\theta_0}\right)\right] \qquad ...(i) \quad ...(18.8\,A)$$

Velocity of the follower is obtained by differentiating the above equation with respect to time, as

$$v = \frac{dx}{dt} = \frac{dx}{d\theta} \times \frac{d\theta}{dt} = \frac{dx}{d\theta} \times \omega \qquad \left(\because \ \frac{d\theta}{dt} = \omega\right)$$

$$= \omega\frac{dx}{d\theta} = \omega\frac{d}{d\theta}\left[S\left\{\frac{\theta}{\theta_0} - \frac{1}{2\pi}\sin\left(\frac{2\pi\theta}{\theta_0}\right)\right\}\right]$$

$$= \omega \times S\left[\frac{1}{\theta_0} - \frac{1}{2\pi}\cos\left(\frac{2\pi\theta}{\theta_0}\right) \times \frac{2\pi}{\theta_0}\right]$$

$$= \omega \times S\left[\frac{1}{\theta_0} - \frac{1}{\theta_0}\cos\left(\frac{2\pi\theta}{\theta_0}\right)\right] = \frac{\omega \times S}{\theta_0}\left[1 - \cos\left(\frac{2\pi\theta}{\theta_0}\right)\right] \qquad ...(18.8\,B)$$

The acceleration of the follower is obtained by differentiating the above equation with respect to time, as

$$f = \frac{dv}{dt} = \frac{dv}{d\theta} \times \frac{d\theta}{dt} = \frac{dv}{d\theta} \times \omega$$

$$= \omega \,.\, \frac{dv}{d\theta} = \omega\frac{d}{d\theta}\left[\frac{\omega \times S}{\theta_0}\left\{1 - \cos\left(\frac{2\pi\theta}{\theta_0}\right)\right\}\right]$$

$$= \omega \,.\, \frac{\omega S}{\theta_0}\left[0 - \left(-\sin\left(\frac{2\pi\theta}{\theta_0}\right)\right) \times \frac{2\pi}{\theta_0}\right]$$

$$= \frac{\omega^2 \times S}{\theta_0} \times \frac{2\pi}{\theta_0} \times \sin\left(\frac{2\pi\theta}{\theta_0}\right)$$

$$= \frac{2\pi\omega^2 \,.\, S}{\theta_0^2}\sin\left(\frac{2\pi\theta}{\theta_0}\right) \qquad ...(18.8\,C)$$

From equation (18.8 *B*), it is clear that velocity will be maximum when

$$\cos\left(\frac{2\pi\theta}{\theta_0}\right) = -1 \quad \text{or} \quad \frac{2\pi\theta}{\theta_0} = \pi \quad \text{or} \quad \theta = \frac{\theta_0}{2}$$

Substituting $\theta = \frac{\theta_0}{2}$ in equation (18.8 *B*), we get the maximum velocity of the follower during outstroke as

$$v_{max} \text{ or } (v_0)_{max} = \frac{\omega \times S}{\theta_0}\left[1 - \cos\left(\frac{2\pi \times \frac{\theta_0}{2}}{\theta_0}\right)\right] = \frac{\omega \times S}{\theta_0}[1 - \cos \pi]$$

$$= \frac{\omega \times S}{\theta_0}[1 - (-1)] = \frac{2\omega \times S}{\theta_0} \quad \text{...(18.8 } D)$$

Similarly the maximum velocity of the follower during return stroke is given by,

$$(v_r)_{max} = \frac{2 \times \omega \times S}{\theta_r} \quad \text{...(18.8 } E)$$

From equation (18.8 *C*), it is clear that acceleration will be maximum when

$$\sin\frac{2\pi\theta}{\theta_0} = 1 \quad \text{or} \quad \frac{2\pi\theta}{\theta_0} = \frac{\pi}{2} \quad \text{or} \quad \theta = \frac{\theta_0}{4}$$

Substituting $\theta = \frac{\theta_0}{4}$ in equation (18.8 *C*), we get the maximum acceleration of the follower during outstroke as

$$f_{max} \text{ or } (f_0)_{max} = \frac{2\pi \cdot \omega^2 \cdot S}{{\theta_0}^2}\sin\left(\frac{2\pi}{\theta_0} \times \frac{\theta_0}{4}\right) = \frac{2\pi\omega^2 \times S}{{\theta_0}^2} \times \sin\frac{\pi}{2} = \frac{2\pi\omega^2 \times S}{{\theta_0}^2} \quad \text{...(18.8 } F)$$

Similarly the maximum acceleration of the follower during return stroke is obtained as

$$(f_r)_{max} = \frac{2\pi \times \omega^2 \times S}{{\theta_r}^2} \quad \text{...(18.8 } G)$$

The velocity and acceleration diagrams are shown in Figs. 18.11 (*b*) and 18.11 (*c*) respectively.

18.9. Cam Profile Construction

The shape of cam profile is obtained from the displacement diagram of the follower. Hence the first step in constructing the cam profile is to draw the displacement diagram according to the desired motion of follower. Now the actual conditions are reversed. The cam is supposed to remain fixed while the line of stroke of the follower is revolved round the cam in the opposite direction to that in which the cam actually turns on its axis. The profile of the contact surfaces of the follower with the cam for the successive positions of the follower gives the actual profile of the cam.

Problem 18.1. *Draw the profile of a cam operating a knife-edge follower when the axis of the follower passes through the axis of cam shaft from the following data :*

(a) Follower to move outwards through 40 mm during 60° of cam rotation,

(b) Follower to dwell for the next 45°,

(c) Follower to return to its original position during next 90°,

(d) Follower to dwell for the rest of the cam rotation.

The displacement of the follower is to take place with simple harmonic motion during both the outward and the return strokes. The least radius of cam is 50 mm. If the cam rotates at 300 r.p.m., determine the maximum velocity and acceleration of the follower during the outward stroke and return stroke.

(A.M.I.E., Summer 1976)

Sol. Given :

Lift of follower, $S = 40$ mm

Outstroke angle, $\theta_0 = 60°$

Angle of dwell after outstroke, $\theta_{DO} = 45°$

Angle of return, $\theta_r = 90°$

Angle of dwell after return stroke, $\theta_{DR} = 360 - (60 + 45 + 90) = 165°$

Least radius of cam $= 50$ mm

Speed of cam, $N = 300$ r.p.m.

$\therefore$ Angular velocity of cam, $\omega = \frac{2\pi N}{60} = \frac{2\pi \times 300}{60} = 31.42$ rad/s.

First draw a displacement diagram as shown in Fig. 18.12 (*a*) as per method given below :

1. Choose a suitable scale for displacement diagram. Let horizontal scale is 10 mm = 30° and vertical scale is 10 mm = 10 mm. Now draw horizontal line $AB = 360°$. On this line, take $AC = 60°$ to represent outstroke, $CD = 45°$ to represent dwell after outstroke, $DE = 90°$ to represent return stroke and $EF = 165°$ to represent dwell after return stroke.

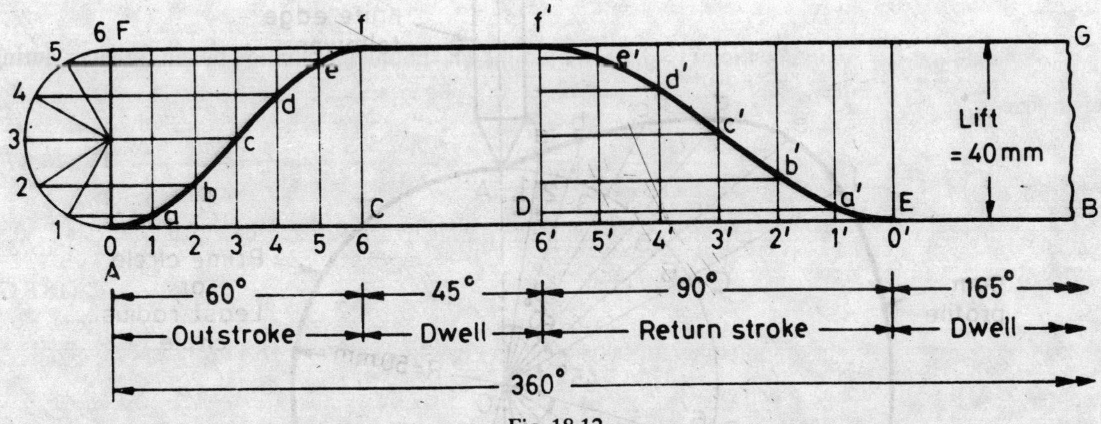

Fig. 18.12

2. From A, draw vertical line $AF = 40$ mm to represent the lift (or stroke) of the follower. Complete the rectangle $AFGB$ as shown in Fig. 18.12 (*a*).

3. Now draw a semi-circle with AF as diameter. Divide the semi-circle into any equal number of even parts (say in this case six equal parts marked as 1, 2, 3, 4, 5, 6 on semi-circle). Join them with the centre of semi-circle.

4. Through each division points 1, 2, 3, 4, 5, 6 draw horizontal lines as shown.

5. Divide the distances AC and DE into six equal parts marked as 1, 2, 3, ... , 6 for outstroke and 6', 5', 4', 3', 2', 1' for return stroke. Draw vertical lines through these points as shown in Fig. 18.12 (*a*). These vertical lines intersect the horizontal lines at a, b, c, d, e and f for outstroke and at a', b', c', d', e', f' for return stroke.

6. Join the points $A, a, b, c, d, e, f, f', e'$, ... $a', 0', B$ by a smooth curve as shown in Fig. 18.12 (*a*). This is the required displacement diagram.

To draw the cam profile

The cam profile is drawn as shown in Fig. 18.13 as per method given below : (Assuming the axis of the follower passing through the axis of cam shaft and also assuming that the cam rotates in the clockwise direction).

1. With O as centre and radius equal to 50 mm *i.e.* minimum radius of the cam, draw a prime circle.

2. Mark angles of 60°, 45° and 90° on the prime circle to represent the outstroke, dwell and return stroke respectively such that angle AOC = 60°, angle COD = 45° and angle DOE = 90°.

3. Divide the angles AOC and DOE into the same number of even parts (*i.e.* six parts) as in the displacement diagram.

4. From O, draw radial lines 0-1, 0-2, 0-3, ... 0-6 for outstroke and 0-6′, 0-5′, ... 0-1′ for return stroke. Produce these radial lines as shown in Fig. 18.13.

5. Now mark distances 1-*a*, 2-*b*, 3-*c*, ... , 6-*f* and 6′-*f*′, 5′-*e*′, ... 2′-*b*′, 1′-*a*′ equal to the corresponding displacement as measured from the displacement diagram shown in Fig. 18.12.

6. A smooth curve passing through *A*, *a*, *b*, *c*, *d*, *e*, *f* in Fig. 18.13 shows the profile of the cam during the outstroke whereas the curve *f*′, *e*′, *d*′, *c*′, *b*′, *a*′, *E* shows the profile of the cam during return stroke. For the dwell period, with *O* as centre and radius equal to *of*, draw an arc *ff*′ as shown in Fig. 18.13. This arc of circle will be parallel to the prime circle.

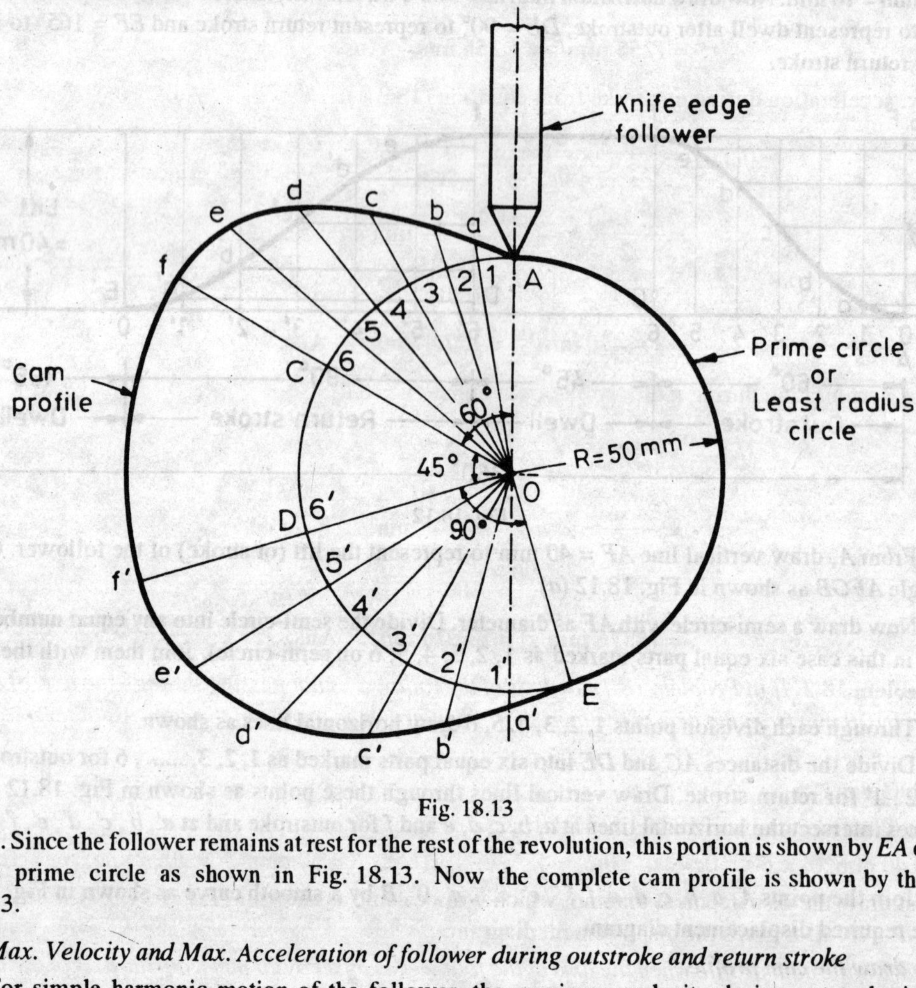

Fig. 18.13

7. Since the follower remains at rest for the rest of the revolution, this portion is shown by *EA* coinciding with the prime circle as shown in Fig. 18.13. Now the complete cam profile is shown by thick line in Fig. 18.13.

Max. Velocity and Max. Acceleration of follower during outstroke and return stroke

For simple harmonic motion of the follower, the maximum velocity during outstroke is given by equation (18.2) and during return stroke is given by equation (18.3).

Max. velocity during outstroke,

$$(v_0)_{max} = \frac{\pi S}{2} \cdot \frac{\omega}{\theta_0}$$

$$= \frac{\pi \times 40}{2} \times \frac{31.42}{\left(60 \times \frac{\pi}{180}\right)} \qquad \left(\because \theta_0 = 60° = \frac{60 \times \pi}{180} \text{ radian}\right)$$

$$= \mathbf{1885\ mm/s = 1.885\ m/s.} \quad \textbf{Ans.}$$

Maximum velocity during return stroke,

$$(v_r)_{max} = \frac{\pi S}{2} \times \frac{\omega}{\theta_R}$$

$$= \frac{\pi \times 40}{2} \times \frac{31.42}{\left(90 \times \frac{\pi}{180}\right)} \qquad \left(\because \theta_R = 90° = 90 \times \frac{\pi}{180}\right)$$

$$= \mathbf{1256\ mm/s = 1.256\ m/s.} \quad \textbf{Ans.}$$

Max. acceleration during outstroke from equation (18.2*A*),

$$(f_0)_{max} = \frac{\pi^2 \times S}{2} \times \frac{\omega^2}{\theta_0^2}$$

$$= \frac{\pi^2 \times 40}{2} \times \frac{31.42^2}{\left(60 \times \frac{\pi}{180}\right)^2} \text{ mm/s}^2$$

$$= \mathbf{177470\ mm/s^2 = 177.47\ m/s^2.} \quad \textbf{Ans.}$$

Max. acceleration during return stroke from equation (18.4),

$$(f_r)_{max} = \frac{\pi^2 \times S}{2} \times \frac{\omega^2}{\theta_R^2}$$

$$= \frac{\pi^2 \times 40}{2} \times \frac{31.42^2}{\left(90 \times \frac{\pi}{180}\right)^2} \text{ mm/s}^2$$

$$= \mathbf{78876\ mm/s^2 = 78.876\ m/s^2.} \quad \textbf{Ans.}$$

Problem 18.2. *If in Problem 18.1, the axis of the follower is not passing through the axis of cam shaft but is offset by 20 mm from the axis of the cam shaft, then draw the profile of the cam. All other data remains the same.*

Sol. Given :

The data from problem 18.1 is :

$S = 40$ mm, $\theta_0 = 60°$; $\theta_r = 90°$; $\theta_{DO} = 45°$; $\theta_{DR} = 165°$; $R = 50$ mm ; $N = 300$ r.p.m. ; $\omega = 31.42$ rad/s.

First draw the displacement diagram which will be same as shown in Fig. 18.12. The curve *A-a-b-c-d-e-f-f '-e'-d'-c'-b'-a'-E* gives the displacement diagram.

To draw the cam profile when the axis of the follower is offset by 20 mm from the axis of the cam shaft.

The cam profile is drawn as shown in Fig. 18.14 as per method given below :

(*i*) With *O* as centre and radius equal to 50 mm, draw a prime circle.

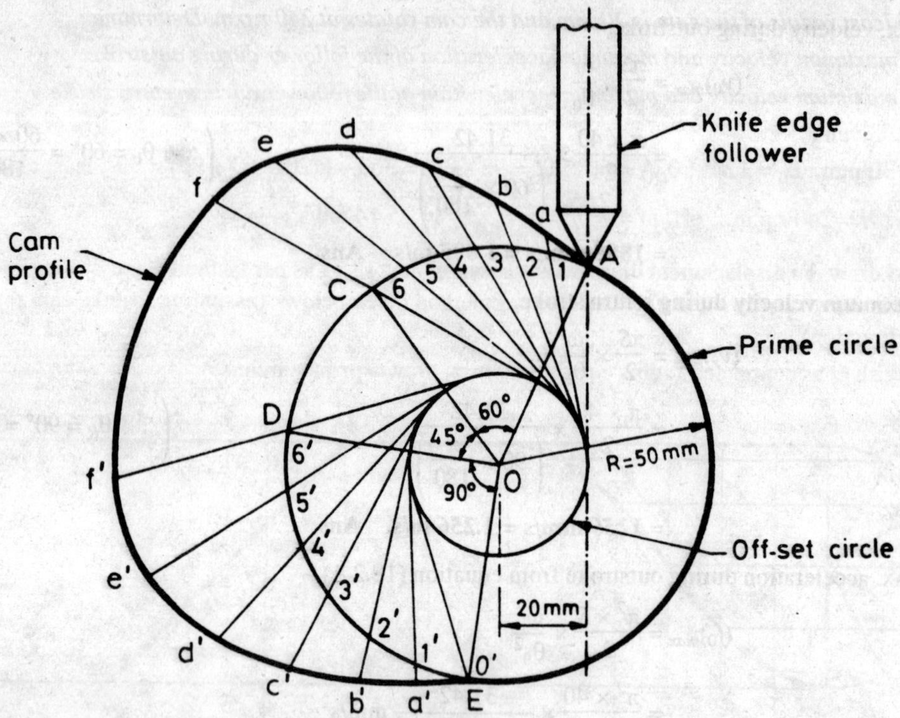

Fig. 18.14

(*ii*) Draw the axis of the follower at a distance of 20 mm from the axis of the cam, which intersects the prime circle at *A*.

(*iii*) With *O* as centre, draw an offset circle of radius 20 mm.

(*iv*) Join *A* to *O*. Make angle *AOC* = 60° to represent outstroke, angle *COD* = 45° to represent dwell and angle *DOE* = 90° to represent return stroke.

(*v*) Divide the angle *AOC* and *DOE* into the same number of even parts (*i.e.* six parts) as in the displacement diagram.

(*vi*) From points, 1, 2, 3, ..., 6 for outstroke and 6′, 5′, 4′, ..., 1′ for return stroke on the prime circle draw tangents to the offset circle and produce these tangents beyond the prime circle as shown in Fig. 18.14.

(*vii*) Now mark distances 1-*a*, 2-*b*, 3-*c*, ..., 6-*f* and 6′-*f*, 5′-*e*′, ... 2′-*b*′, 1′-*a*′ equal to the corresponding displacement as measured from the displacement diagram shown in Fig. 18.12.

(*viii*) A smooth curve passing through *A*, *a*, *b*, *c*, *d*, *e*, *f* in Fig. 18.14 shows the profile of the cam during the outstroke whereas the curve *f* ′, *e*′, *d*′, *c*′, *b*′, *a*′, *F*. shows the profile of the cam during return stroke. For the dwell period, with *O* as centre and radius equal to 0-*f* ′, draw an arc *ff* ′ as shown in Fig. 18.14.

(*ix*) Since the follower remains at rest for the rest of the revolution, this portion is shown by *EA* coinciding with the prime circle as shown in Fig. 18.14. Now the complete cam profile is shown by curve *A-a-b-c-d-e- f-f* ′*- e*′*-d*′*- c*′*-b*′*-a*′*-E-A*.

Problem 18.3. *Draw the profile of a cam operating a knife-edge follower (when the axis of the follower passes through the axis of cam shaft) from the following data :*

(*i*) *follower to move outward through 30 mm with Simple Harmonic motion during 120° of cam rotation*

(*ii*) *follower to dwell for the next 60°*

(*iii*) *follower to return to its original position with uniform velocity during 90° of cam rotation*

(*iv*) *follower to dwell for the rest of the cam rotation.*

The least radius of the cam is 20 mm and the cam rotates at 240 r.p.m. Determine :

(i) maximum velocity and maximum acceleration of the follower during outstroke.

(ii) maximum velocity and maximum acceleration of the follower during return stroke.

Sol. Given :

$S = 30$ mm ; $\theta_0 = 120°$; $\theta_{DO} = 60°$; $\theta_r = 90°$, $\theta_{Dr} = 360 - (120 + 60 + 90) = 90°$, least radius of cam, $R = 20$ mm ; $N = 240$ r.p.m. or $\omega = \frac{2\pi N}{60} = \frac{2\pi \times 240}{60} = 25.14$ rad/s.

First draw the displacement diagram as shown in Fig. 18.15 as per method discussed earlier. Then the cam profile is drawn as shown in Fig. 18.16 as per method given below : (assuming that the cam rotates in the clockwise direction).

1. With O as centre and radius equal to 20 mm, draw a prime circle.

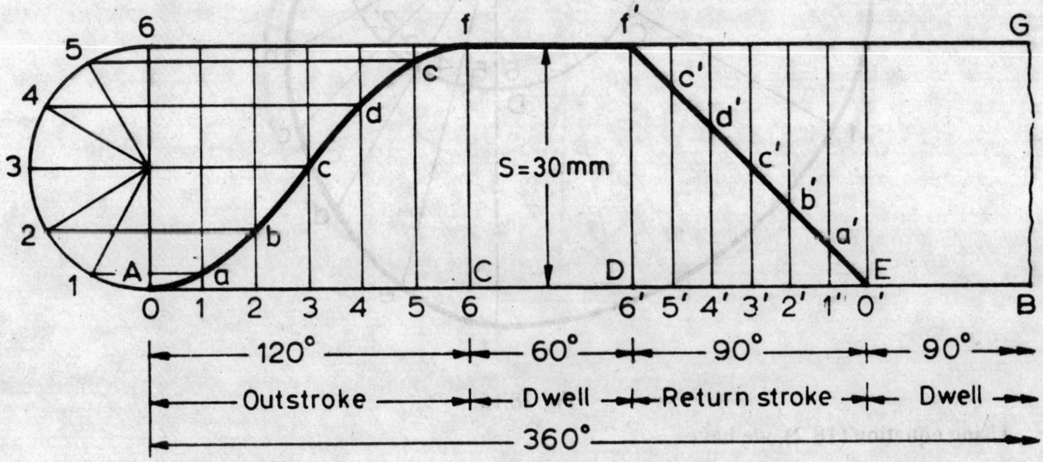

Fig. 18.15

2. Mark angles of 120°, 60° and 90° on the prime circle to represent the outstroke, dwell and return stroke respectively such that angle $AOC = 120°$, angle $COD = 60°$ and angle $DOE = 90°$.

3. Divide the angles AOC and DOE into the same number of even parts (*i.e.* six parts) as in the displacement diagram.

4. From O, draw radial lines 0-1, 0-2, 0-3, ..., 0-6 for outstroke and 0-6′, 0-5′, ..., 0-1′ for return stroke. Produce these radial lines as shown in Fig. 18.16.

5. Now mark distances 1-*a*, 2-*b*, 3-*c*, ..., 6-*f* and 6′-*f* ′, 5′-*e*′, ..., 2′-*b*′, 1′-*a*′ equal to corresponding displacement as measured from the displacement diagram shown in Fig. 18.15.

6. A smooth curve passing through *A*, *a*, *b*, *c*, *d*, *e*, *f* in Fig. 18.16 shows the profile of the cam during outstroke whereas the curve *f* ′, *e*′, *d*′, *c*′, *b*′, *a*′, *E* shows the profile of the cam during return stroke. For the dwell period, with *O* as centre and radius equal to 0-*f* (or equal to 0-*f* ′) draw an arc *ff* ′ as shown in Fig. 18.16. This arc of circle will be parallel to the prime circle.

7. Since the follower remains at rest for the rest of the revolution, this portion is shown by *EA* coinciding with the prime circle. Now the complete cam profile is shown by thick line in Fig. 18.16.

(i) Maximum velocity and maximum acceleration of the follower during outstroke

During outstroke, the follower moves with simple harmonic motion. The maximum velocity is given by equation (18.2) whereas maximum acceleration is given by equation (18.2*A*).

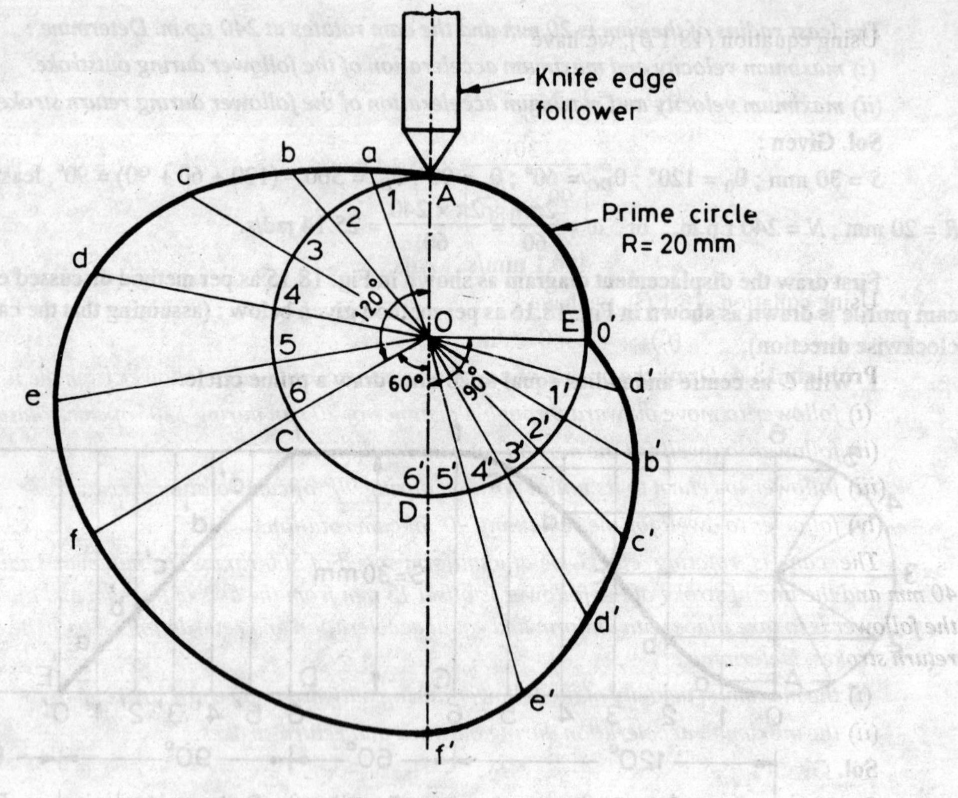

Fig. 18.16

Using equation (18.2), we have

$$(v_0)_{max} = \frac{S}{2} \times \frac{\pi \times \omega}{\theta_0}$$

$$= \frac{30}{2} \times \frac{\pi \times 25.14}{\left(120 \times \frac{\pi}{180}\right)}$$

$$\left(\because \quad \theta_0 = 120° = \frac{120 \times \pi}{180} \text{ radians} ; \omega = 25.14 \text{ rad/s} ; S = 30 \text{ mm} \right)$$

$= \mathbf{565.65}$ **mm/s. Ans.**

Using equation (18.2 *A*), we have

$$(f_0)_{max} = \frac{S}{2} \times \left(\frac{\pi \times \omega}{\theta_0}\right)^2$$

$$= \frac{30}{2} \times \left[\frac{\pi \times 25.14}{\left(120 \times \frac{\pi}{180}\right)} \right]^2 = 21330 \text{ mm/s}^2 = \mathbf{21.33 \text{ m/s}^2}. \quad \textbf{Ans.}$$

(ii) Maximum velocity and maximum acceleration of the follower during return stroke

During return stroke, the follower moves with uniform velocity. The velocity is constant during the return stroke. The maximum velocity (which is equal to constant velocity) is given by equation (18.1 *B*), whereas the acceleration (or maximum acceleration) is given by equation (18.1 *C*).

Using equation (18.1 *B*), we have

$$(v_r)_{max} = v_r = \frac{S}{\theta_r} \times \omega$$

$$= \frac{30}{\left(90 \times \frac{\pi}{180}\right)} \times 25.14 \text{ mm/s} \qquad \left(\because \theta_r = 90° = 90 \times \frac{\pi}{180} \text{ radians}\right)$$

$$= \mathbf{480.1 \text{ mm/s.}} \quad \textbf{Ans.}$$

Using equation (18.1 *C*), we have

$$(f_r)_{max} = f_r = 0. \quad \textbf{Ans.}$$

Problem 18.4. *Draw the profile of a cam operating a knife-edge follower from the following data :*

(i) follower to move outward through a distance of 20 mm during 120° of cam rotation

(ii) follower to dwell for the next 60° of cam rotation

(iii) follower to return to its initial position during 90° of cam rotation

(iv) follower to dwell for the remaining 90° of cam rotation.

The cam is rotating clockwise at a uniform speed of 500 r.p.m. The minimum radius of the cam is 40 mm and the line of stroke of the follower is offset 15 mm from the axis of the cam and the displacement of the follower is to take place with uniform and equal acceleration and retardation on both the outward and the return strokes. Determine :

(i) the maximum velocity of the follower during outward and return strokes,

(ii) the maximum acceleration during outward and return strokes.

Sol. Given :

S = 20 mm ; $\theta_0 = 120°$; $\theta_{DO} = 60°$; $\theta_r = 90°$; $\theta_{Dr} = 360 - (120 + 60 + 90) = 90°$; least radius of cam R = 40 mm ; N = 500 r.p.m. or $\omega = \frac{2\pi N}{60} = \frac{2\pi \times 500}{60} = 52.36$ rad/s ; offset of line of stroke of the follower = 15 mm.

First draw the displacement diagram as shown in Fig. 18.17 as per method discussed in Article 18.7. As the displacement of the follower is to take place with uniform acceleration and retardation on both the outstroke and return strokes, hence the displacement curves will be parabolic for both the strokes. (The curve 0-*a*-*b*-*c*-*d*-*e*-*f* is the displacement curve for outstroke whereas curve *f* '-*e*'-*d*'-*c*'-*b*'-*a*'-0' is the displacement curve for return stroke).

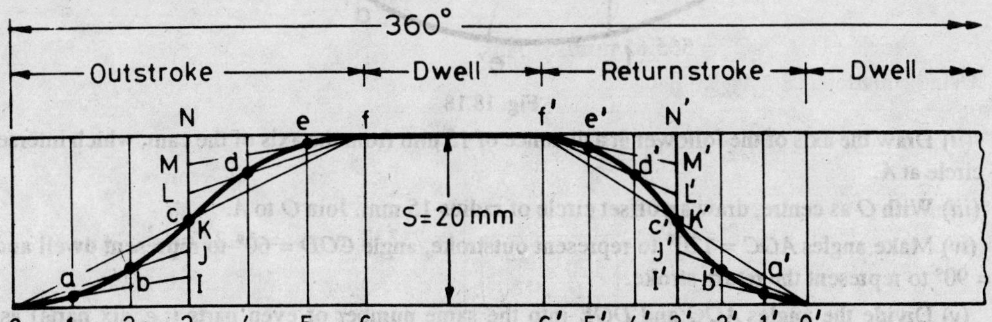

Fig. 18.17

Cam Profile

The cam profile is drawn as shown in Fig. 18.18 as per method given below :

(*i*) With *O* as centre and radius equal to 40 mm, draw a prime circle.

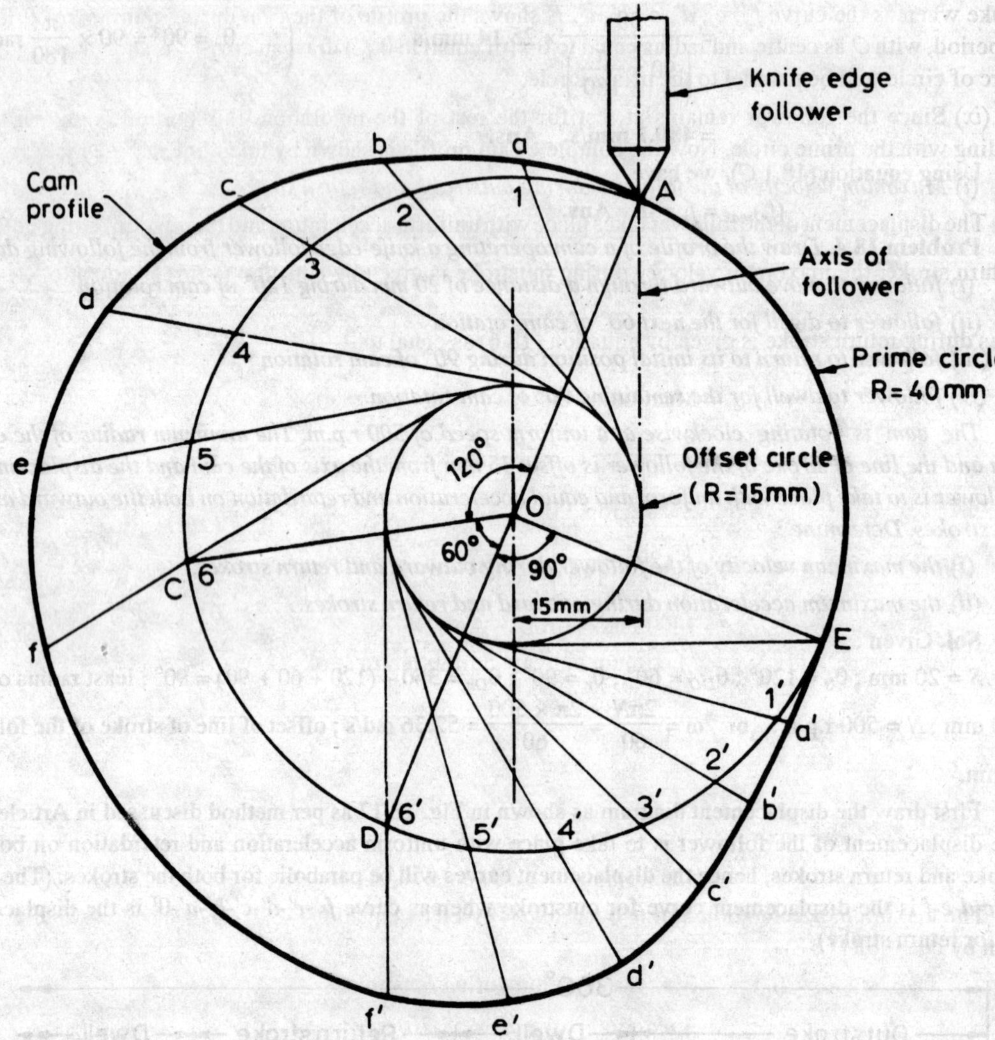

Fig. 18.18

(*ii*) Draw the axis of the follower at a distance of 15 mm from the axis of the cam, which intersects the prime circle at *A*.

(*iii*) With *O* as centre, draw an offset circle of radius 15 mm. Join *O* to *A*.

(*iv*) Make angles *AOC* = 120° to represent outstroke, angle *COD* = 60° to represent dwell and angle *DOE* = 90° to represent the return stroke.

(*v*) Divide the angles *AOC* and *DOE* into the same number of even parts (*i.e.* six parts) as in the displacement diagram.

(*vi*) From points 1, 2, ..., 5, 6 for outstroke and 6′, 5′, ..., 2′, 1′ for return stroke on the prime circle, draw tangents to the offset circle and produce these tangents beyond the prime circle as shown in Fig. 18.18.

(*vii*) Now mark distances 1-*a*, 2-*b*, 3-*c*, ..., 6-*f* and 6′-*f*′, 5′-*e*′, ... 2′-*b*′, 1′-*a*′ equal to the corresponding displacement as measured from the displacement diagram shown in Fig. 18.17.

(*viii*) A smooth curve passing through *A*, *a*, *b*, *c*, *d*, *e*, *f* in Fig. 18.18 shows the profile of the cam during outstroke whereas the curve *f*′, *e*′, *d*′, *c*′, *b*′, *a*′, *E* shows the profile of the cam during return stroke. For the dwell period, with *O* as centre and radius equal to 0-*f* (or equal to 0-*f*′) draw an arc *f*-*f*′ as shown in Fig. 18.18. This arc of circle will be parallel to the prime circle.

(*ix*) Since the follower remains at rest for the rest of the revolution, this portion is shown by *EA* coinciding with the prime circle. Now the complete cam profile is shown by thick line in Fig. 18.18.

(*i*) *Maximum velocity of the follower during outstroke and return stroke*

The displacement of the follower takes place with uniform acceleration and retardation during outstroke and return stroke, the maximum velocity during outstroke is given by equation (18.5) as equal to $\frac{2 \times \omega \times S}{\theta_0}$ whereas during return stroke is given by equation (18.6) as equal to $\frac{2 \times \omega \times S}{\theta_r}$.

$$\therefore \quad (v_0)_{max} = \text{Maximum velocity during outstroke}$$

$$= \frac{2 \times \omega \times S}{\theta_0}$$

$$= \frac{2 \times 52.36 \times 20}{\left(120 \times \frac{\pi}{180}\right)} \qquad \left(\because \ \theta_0 = 120^\circ = 120 \times \frac{\pi}{180} \text{ radians}\right)$$

$$= 1000 \text{ m/s} = \mathbf{1 \ m/s.} \quad \textbf{Ans.}$$

and
$$(v_r)_{max} = \text{Maximum velocity during return stroke}$$

$$= \frac{2S \times \omega}{\theta_r}$$

$$= \frac{2 \times 20 \times 52.36}{\left(90 \times \frac{\pi}{180}\right)} \qquad \left(\because \ \theta_r = 90^\circ = 90 \times \frac{\pi}{180} \text{ radians}\right)$$

$$= 1333.3 \text{ mm/s} = \mathbf{1.333 \ m/s.} \quad \textbf{Ans.}$$

The maximum acceleration during outstroke is given by equation (18.7) whereas during return stroke is given by equation (18.8).

$$\therefore \quad (f_0)_{max} = \text{Max acceleration during outstroke}$$

$$= \frac{4 \times S \times \omega^2}{\theta_0^2} = \frac{4 \times 20 \times 52.36^2}{\left(120 \times \frac{\pi}{180}\right)^2} = \frac{80 \times 52.36^2}{\left(\frac{2\pi}{3}\right)^2}$$

$$= \frac{80 \times 52.36^2 \times 9}{4\pi^2} = 50000 \text{ mm/s}^2 = \mathbf{50 \ m/s^2.} \quad \textbf{Ans.}$$

and
$$(f_r)_{max} = \text{Max. acceleration during return stroke}$$

$$= \frac{4 \times S \times \omega^2}{\theta_r^2} = \frac{4 \times 20 \times 52.36^2}{\left(90 \times \frac{\pi}{180}\right)^2}$$

$$= \frac{4 \times 20 \times 52.36^2}{\pi^2} \times 4 = 88889.3 \text{ mm/s}^2 = \mathbf{88.8893 \ m/s^2.} \quad \textbf{Ans.}$$

18.10. Cam Profile for Roller Followers

Sometimes, a roller is fitted to a follower in order to reduce friction and wear. The displacement of the follower is determined by the displacement of the roller centre. The path of the roller centre is determined by the same method as for the knife-edge follower. Here the distance *OA* is equal to the sum of the radius of the roller and minimum radius of the cam. From the path of the roller centre which has been already determined the circles of radius equal to the radius of roller are drawn. Then a smooth curve passing through the bottoms of these circles, is drawn. This smooth curve will represent the cam profile.

Problem 18.5. *A cam with 30 mm as minimum diameter is rotating clock-wise at a uniform speed of 1200 r.p.m. and has to give the following motion to a roller-follower 10 mm in diameter :*

(a) Follower to complete outward stroke of 25 mm during 120° of cam rotation with equal uniform acceleration and retardation.

(b) Follower to dwell for 60° of cam rotation.

(c) Follower to return to its initial position during 90° of cam rotation with equal uniform acceleration and retardation.

(d) Follower to dwell for the remaining 90° of cam rotation.

Draw the cam profile if the axis of the roller-follower passes through the axis of the cam. Determine the maximum velocity of the follower during the outstroke and the return stroke and also the uniform acceleration of the follower on the outstroke and return stroke. (A.M.I.E., Summer 1982)

Sol. Given :

Min. dia. of cam, $D_{min} = 30$ mm or minimum radius of cam, $R = \frac{30}{2} = 15$ mm ;

Speed of cam, $N = 1200$ r.p.m.

$\therefore$ Angular velocity of cam, $\omega = \frac{2\pi N}{60} = \frac{2\pi \times 1200}{60} = 125.6$ rad/s

Dia. of roller-follower = 10 mm or radius of roller = $\frac{10}{2} = 5$ mm

Lift, $s = 25$ mm

Outstroke angle, $\theta_0 = 120°$

Dwell angle after outstroke, $\theta_{DO} = 60°$

Return stroke angle, $\theta_r = 90°$

Angle of dwell after return stroke, $\theta_{Dr} = 90°$

The outstroke and return stroke has uniform acceleration and retardation.

First draw the displacement diagram as shown in Fig. 18.19 as per method given in Article 18.7. As the displacement of the follower is to take place with uniform acceleration and retardation on both the strokes (outstroke and return stroke) hence the displacement curves will be parabolic for both the strokes.

Cam Profile. The cam profile is drawn as shown in Fig. 18.20 as per method given below :

(*i*) With *O* as centre, draw a base circle with radius equal to the minimum radius of the cam (*i.e.* radius = 15 mm).

(*ii*) With *O* as centre, draw a prime circle with radius equal to the sum of minimum radius of the cam and radius of roller (*i.e.* radius = 15 + 5 = 20 mm)

$\therefore$ Radius *OA* = 20 mm.

(*iii*) Mark angles of 120°, 60° and 90° on the prime circle to represent outstroke, dwell and return stroke respectively such that angle *AOC* = 120°, angle *COD* = 60° and angle *DOE* = 90°.

(*iv*) Divide the angles *AOC* and *DOE* into the same number of even parts (*i.e.* six parts) as in the displacement diagram.

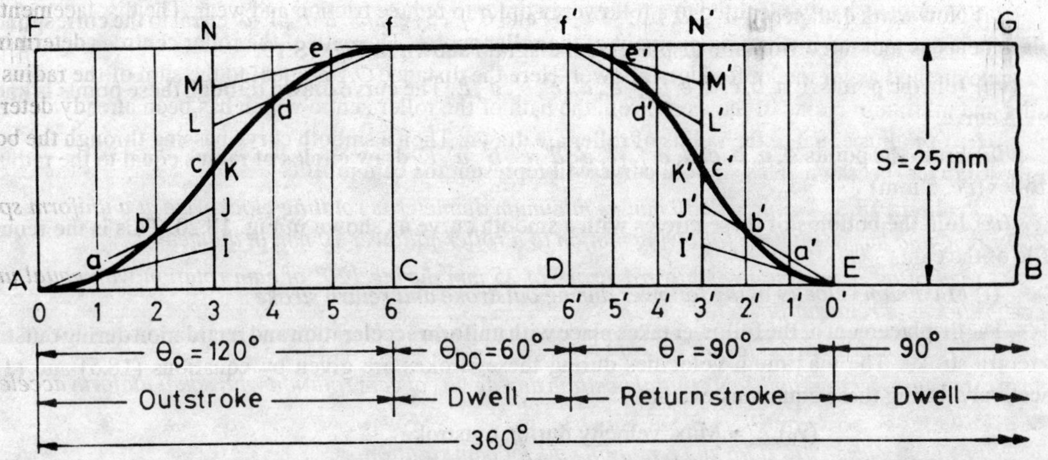

Fig. 18.19

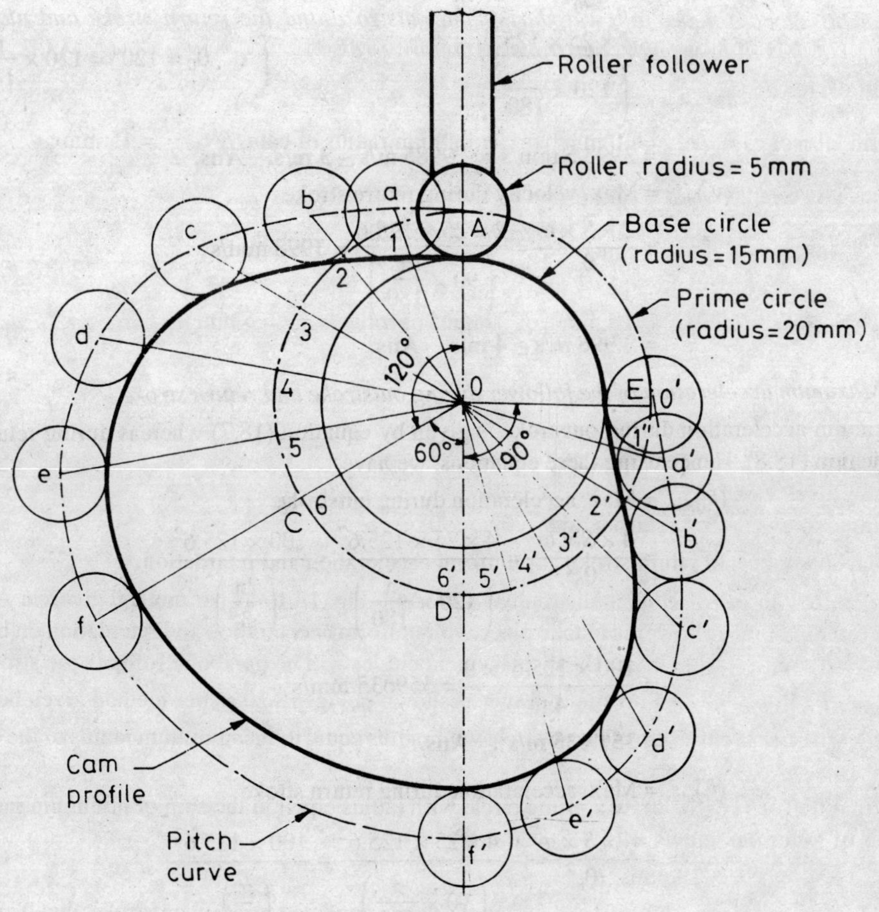

Fig. 18.20

(*v*) From O, draw radial lines 0-1, 0-2, 0-3, ..., 0-6 for outstroke and 0-6′, 0-5′, ..., 0-1′ for return stroke. Produce these lines beyond prime circle as shown in Fig. 18.20.

(*vi*) Now mark distances 1-*a*, 2-*b*, ..., 5-*e*, 6-*f* and 6′-*f*′, 5′-*e*′, ... 2′-*b*′, 1′-*a*′ equal to the corresponding displacements as measured from the displacement diagram shown in Fig. 18.19.

(*vii*) Join the points $A, a, b, c, d, e, f, f', e', d', c', b', a', E$. The curve drawn through these points is known as pitch curve.

(*viii*) From the points $A, a, b, c, d, e, f, f', e', d', c', b', a', E$, draw circles of radius equal to the radius of the roller (*i.e.* 5 mm).

(*ix*) Join the bottoms of these circles with a smooth curve as shown in Fig. 18.20. This is the required profile of the cam.

(*i*) *Maximum velocity of the follower during outstroke and return stroke*

The displacement of the follower takes place with uniform acceleration and retardation during outstroke and return stroke. The maximum velocities during these strokes are given by equations (18.5) and (18.6) respectively. Using these equations, we get

$$(v_0)_{max} = \text{Max. velocity during outstroke}$$

$$= \frac{2 \times S \times \omega}{\theta_0}$$

$$= \frac{2 \times 25 \times 125.6}{\left(120 \times \dfrac{\pi}{180}\right)} \qquad \left(\because \ \theta_0 = 120° = 120 \times \frac{\pi}{180} \text{ radians}\right)$$

$$= 2998.5 \text{ mm/s} = 2.9985 \text{ m/s} \simeq \mathbf{3 \ m/s.} \quad \textbf{Ans.}$$

and

$$(v_r)_{max} = \text{Max. velocity during return stroke}$$

$$= \frac{2 \times S \times \omega}{\theta_r} = \frac{2 \times 25 \times 125.6}{\left(90 \times \dfrac{\pi}{180}\right)} = 3998 \text{ mm/s}$$

$$= 3.998 \text{ m/s} \simeq \mathbf{4 \ m/s.} \quad \textbf{Ans.}$$

(*ii*) *Maximum acceleration of the follower during outstroke and return stroke*

Maximum acceleration during outstroke is given by equation (18.7) whereas during return stroke is given by equation (18.8). Hence using these equations, we have

$$(f_0)_{max} = \text{Max. acceleration during outstroke}$$

$$= \frac{4 \times S \times \omega^2}{\theta_0^{\,2}} = \frac{4 \times 25 \times 125.6^2}{\left(120 \times \dfrac{\pi}{180}\right)^2} = \frac{100 \times 125.6^2}{\left(\dfrac{2\pi}{3}\right)^2}$$

$$= \frac{100 \times 125.6^2 \times 9}{4\pi^2} = 359635 \text{ mm/s}^2$$

$$= \mathbf{359.635 \ m/s^2.} \quad \textbf{Ans.}$$

and

$$(f_r)_{max} = \text{Max. acceleration during return stroke}$$

$$= \frac{4 \times S \times \omega^2}{\theta_r^{\,2}} = \frac{4 \times 25 \times 125.6^2}{\left(90 \times \dfrac{\pi}{180}\right)^2} = \frac{100 \times 125.6^2}{\left(\dfrac{\pi}{2}\right)^2}$$

$$= \frac{100 \times 125.6^2 \times 4}{\pi^2} = 639351 \text{ mm/s}^2$$

$$= \mathbf{639.351 \text{ m/s}^2}. \quad \textbf{Ans.}$$

Problem 18.6. *From the following data draw the profile of a cam in which the follower moves with simple harmonic motion (S.H.M.) during ascent while it moves with uniformly accelerated and decelerated motion during descent :*

Least radius of cam = 50 mm, Angle of ascent, θ_0 = 48°

Angle of dwell between ascent and descent, θ_{DO} = 42°

Angle of descent, θ_r = 60°, the lift of follower = 40 mm

Diameter of roller = 30 mm

Distance between line of action of the follower and axis of cam = 20 mm

If the cam rotates at 360 r.p.m. anti-clockwise, find the maximum velocity and acceleration of the follower during descent. (A.M.I.E., Winter 1987)

Sol. Given :

Least radius of cam,	R_{min} = 50 mm
Angle of ascent (outstroke),	θ_0 = 48°
Angle of dwell between ascent and descent,	θ_{DO} = 42°
Angle of descent (or return),	θ_r = 60°
∴ Angle of dwell after return stroke	= 360 – (48 + 42 + 60) = 210°
Lift of follower,	s = 40 mm
Dia. of roller	= 30 mm
Distance of off-set	= 20 mm

Follower moves with S.H.M. during ascent while it moves with uniform acceleration and deceleration during descent.

First draw the displacement diagram as shown in Fig. 18.21 as per method given below :

1. Choose a suitable scale for displacement diagram. Let horizontal scale is 1 mm = 1° and vertical scale is 1 mm = 1 mm. Now draw horizontal line AB = 360°. On this line take AC = 48° to represent ascent (or outstroke), CD = 42° to represent dwell after outstroke, DE = 60° to represent return stroke and EB = 210° to represent dwell after return stroke.

2. From A, draw vertical line AF = 40 mm to represent the lift of the follower. Complete the rectangle $AFGB$ as shown in Fig. 18.21.

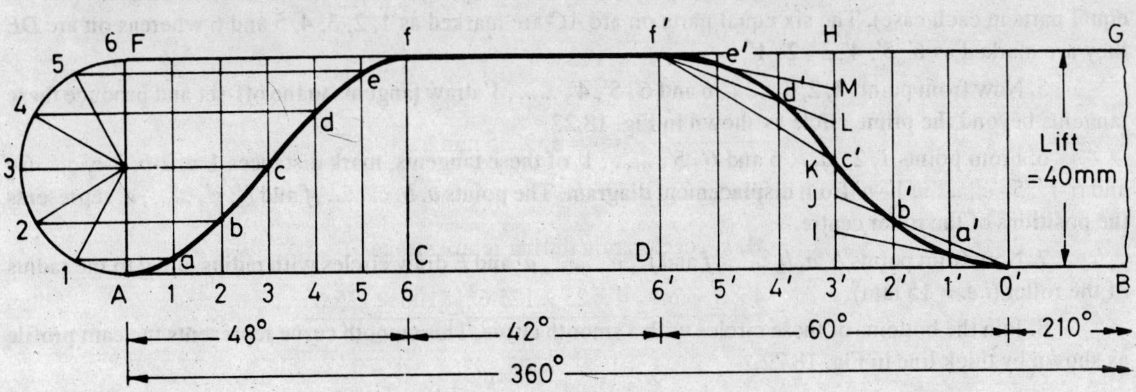

Fig. 18.21

3. Now draw a semi-circle with AF as diameter to represent the S.H.M. of the follower during ascent. Divide the semi-circle into any equal number of even parts (say six). Name these parts as 1, 2, 3, ..., 6. Join them with the centre of semi-circle by radial lines.

4. Through each points 1, 2, 3, ..., 6 draw horizontal lines as shown.

5. Now divide the distance AC into same equal number of parts as the number of parts on the semi-circle (*i.e.* six). Mark these parts as 1, 2, ..., 6. Draw vertical lines through these points. These vertical lines intersect the already drawn horizontal lines at $a, b, c,$ ...... , f.

6. Join points A, a, b, c, d, e and f. The curve A-a-b- ...... -f represents the displacement diagram for outstroke.

7. To draw the displacement diagram for return stroke in which case the follower moves with uniform acceleration and deceleration, divide the distance DE into six equal parts. Mark these points as 6′, 5′, 4′, ... , 1′. Draw the vertical lines through these points.

8. Divide the vertical line 3′-H into six equal parts as shown by points I, J, K, L, M and H. The displacement diagram for return stroke will consist of double parabola. Join EI, EJ and EK intersecting the vertical lines at a', b' and c' respectively. Join points E, a', b' and c' with a smooth curve.

9. Join $f'M$, $f'L$ and $f'K$ intersecting vertical lines drawn through 5′, 4′ and 3′ at e', d' and c' respectively. Join points f', e', d' and c' with a smooth curve. The curve A-a-b-c-d-e-f-f'-e' ...... a'-E is the required displacement diagram.

Profile of the Cam

First calculate the minimum, pitch circle radius, which is equal to the sum of radii of the minimum radius of the cam and radius of the roller *i.e.* $50 + \frac{30}{2} = 65$ mm

$\therefore$ Minimum pitch circle radius = 65 mm.

Now the profile of the cam is drawn as shown in Fig. 18.22 as per method given below :

1. Choose any point O. With O as centre and radius OA equal to minimum pitch circle radius (*i.e.* 65 mm) draw a prime circle (or min. pitch circle). With O as centre, also draw another circle of radius 20 mm (*i.e.* equal to eccentricity). This circle is known as off-set circle.

2. Draw the axis of the follower at a distance of 20 mm from point O (*i.e.* axis of the cam). The axis of the follower meets the prime circle at A. Join AO.

3. From OA, draw angle $AOC = 48°$ to represent the ascent (or outstroke), angle $COD = 42°$ to represent the dwell after outstroke and angle $DOE = 60°$ to represent return stroke.

4. Divide the arc AC and arc DE into same number of equal parts as in displacement diagram (*i.e.* six equal parts in each case). The six equal parts on arc AC are marked as 1, 2, 3, 4, 5 and 6 whereas on arc DE they are marked as 6′, 5′, 4′, 3′, 2′, 1′.

5. Now from points 1, 2, 3, ... , 6 and 6′, 5′, 4′, ... , 1′ draw tangents to the off-set and produce these tangents beyond the prime circle as shown in Fig. 18.22.

6. From points 1, 2, ... , 6 and 6′, 5′, ... , 1′ of these tangents, mark distances 1-a, 2-b, 3-c, ... 6-f and 6′-f', 5′-e', ... , 1′-a' from displacement diagram. The points $a, b, c,$ ...... , f and $f', e',$ ...... , a' represents the positions of the roller centre.

7. Now from points $A, a, b,$ ...... , f and $f', e',$ ...... , a' and E draw circles with radius equal to the radius of the roller (*i.e.* = 15 mm).

8. Join the bottoms of these circles with a smooth curve. This smooth curve represents the cam profile as shown by thick line in Fig. 18.22.

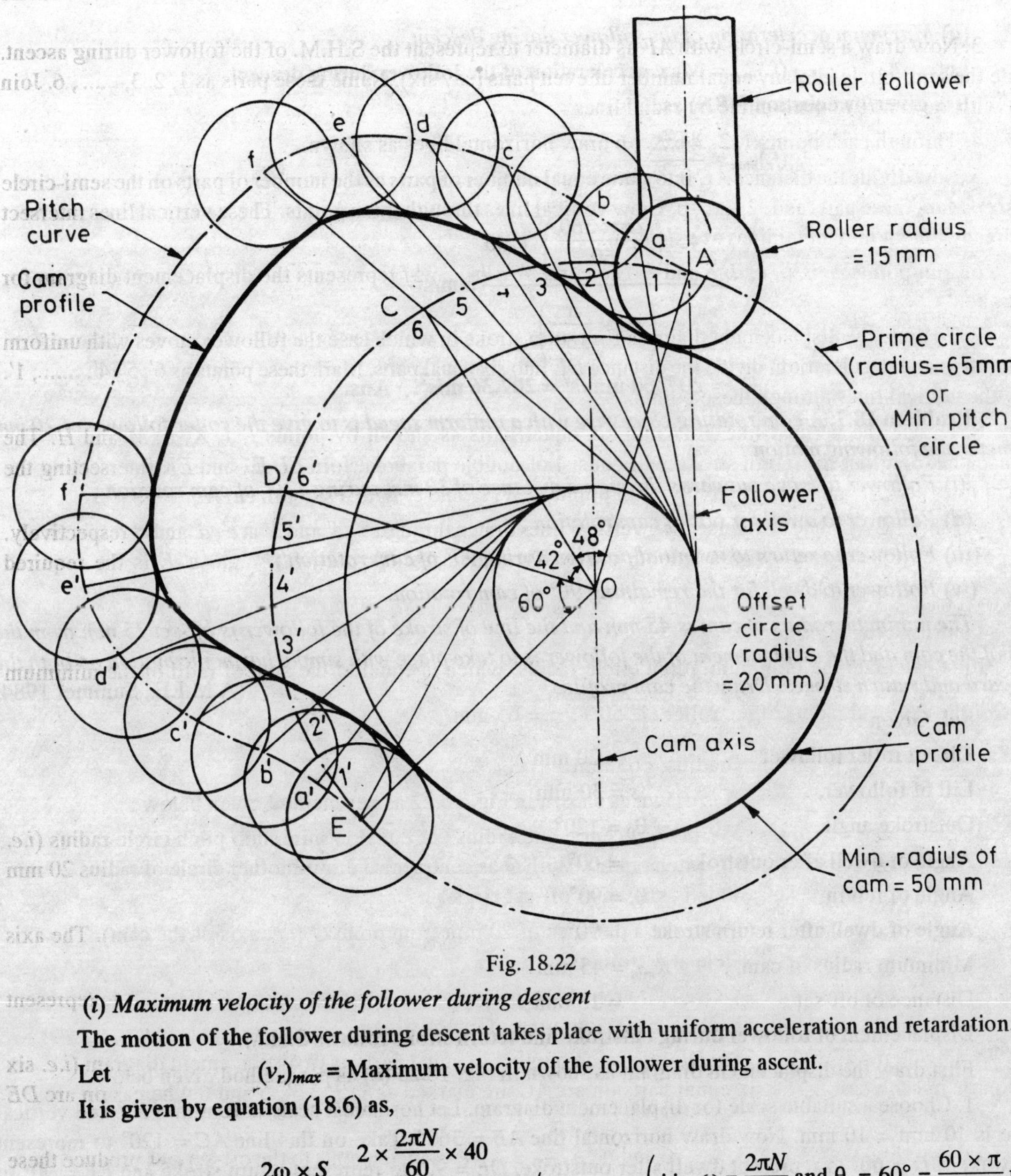

Fig. 18.22

(i) Maximum velocity of the follower during descent

The motion of the follower during descent takes place with uniform acceleration and retardation.

Let $(v_r)_{max}$ **= Maximum velocity of the follower during ascent.**

It is given by equation (18.6) as,

$$(v_r)_{max} = \frac{2\omega \times S}{\theta_r} = \frac{2 \times \frac{2\pi N}{60} \times 40}{\left(60 \times \frac{\pi}{180}\right)} \text{ mm/s} \qquad \left(\because \quad \omega = \frac{2\pi N}{60} \text{ and } \theta_r = 60° = \frac{60 \times \pi}{180} \text{ rad}\right)$$

$$= \frac{2 \times \frac{2\pi \times 360}{60} \times 40}{\left(60 \times \frac{\pi}{180}\right)} = 2880 \text{ mm/s}$$

$$= \mathbf{2.88 \text{ m/s.}} \quad \textbf{Ans.}$$

(*ii*) *Maximum acceleration of the follower during descent*

Let $(f_r)_{max}$ = Max. acceleration of the follower during descent.

It is given by equation (18.8) as,

$$(f_r)_{max} = \frac{4\omega^2 S}{\theta_r^2}$$

$$= \frac{4 \times \left(\frac{2\pi \times 360}{60}\right)^2 \times 40}{\left(60 \times \frac{\pi}{180}\right)^2} \text{ mm/s}^2$$

$= 207360 \text{ mm/s}^2 = \mathbf{207.36 \text{ m/s}^2}$. **Ans.**

Problem 18.7. *A cam rotating clockwise with a uniform speed is to give the roller follower of 20 mm diameter the following motion :*

(*i*) *Follower to move outwards through a distance of 30 mm during 120° of cam rotation ;*

(*ii*) *Follower to dwell for 60° of cam rotation ;*

(*iii*) *Follower to return to its initial position during 90° of cam rotation ;*

(*iv*) *Follower to dwell for the remaining 90° of cam rotation.*

The minimum radius of cam is 45 mm and the line of stroke of the follower is off-set 15 mm from the axis of the cam and the displacement of the follower is to take place with simple harmonic motion on both the outward and return strokes. Draw the cam profile. (A.M.I.E., Summer 1984)

Sol. Given :

Dia. of roller follower = 20 mm

Lift of follower, s = 30 mm

Outstroke angle, θ_0 = 120°

Angle of dwell after outstroke = 60°

Angle of return, θ_r = 90°

Angle of dwell after return stroke = 90°

Minimum radius of cam, R_{min} = 45 mm

Distance of off-set = 15 mm

Displacement of follower during outstroke and return stroke follows S.H.M.

First draw the displacement diagram as shown in Fig. 18.23 (*a*) as per method given below :

1. Choose a suitable scale for displacement diagram. Let horizontal scale is 10 mm = 30° and vertical scale is 10 mm = 10 mm. Now draw horizontal line *AB* = 360°. Take on this line *AC* = 120° to represent outstroke, *CD* = 60° to represent dwell after outstroke, *DE* = 90° to represent return stroke and *EF* = 90° to represent dwell after return stroke.

2. From *A*, draw vertical line *AF* = 30 mm to represent the lift or stroke of the follower. Complete the rectangle *AFGB* as shown in Fig. 18.23 (*a*).

3. Now draw a semi-circle with *AF* as diameter. Divide the semi-circle into any equal number of even parts (say six). Name these parts as 1, 2, 3, ... , 6. Join them with the centre of semi-circle.

4. Through each points 1, 2, 3, ... , 6, draw horizontal lines as shown.

5. Now divide the distances *AC* and *DE* into the same equal number of parts as the number of parts on the semi-circle (*i.e.* six equal parts). The six equal parts on distance *AC* are marked as 1, 2, 3, ... , 6 whereas

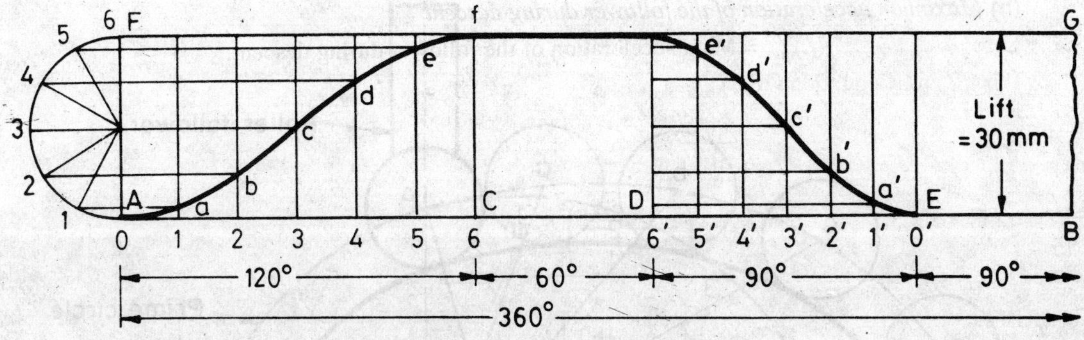

(a) Displacement diagram

Fig. 18.23

the six equal parts on distance *DE* are marked as 6′, 5′, ... , 1′. Now draw the vertical lines through these points. These vertical line intersect the horizontal lines at *a*, *b*, *c*, ... , *f* for outstroke and at *a*′, *b*′, *c*′, ... , *f*′ for return stroke.

6. Join points 0, *a*, *b*, *c*, ... , *f*, *f*′, *e*′, ... , *a*′, 0′ by a smooth curve as shown in Fig. 18.23 (*a*). This is the required displacement diagram.

To draw Cam Profile

First calculate the minimum pitch circle radius. This is equal to sum of radii of the minimum radius of the cam and radius of the roller *i.e.* $45 + \frac{20}{2} = 55$ mm

∴ Minimum Pitch circle radius = 55 mm.

Now the profile of the cam is drawn as shown in Fig. 18.23 (*b*) as per method given below :

1. Choose any point *O*. With *O* as centre and radius *OA* equal to minimum pitch circle radius (*i.e.* 55 mm) draw a prime circle. Also draw another circle of radius equal to eccentricity of 15 mm with *O* as centre. This circle is known as off-set circle.

2. Draw the axis of the follower at a distance of 15 mm from point *O* (*i.e.* axis of the cam). The axis of follower meets the prime circle at *A*. Join *AO*.

3. From *OA*, draw angle *AOC* = 120° to represent outstroke, angle *COD* = 60° to represent dwell after outstroke and angle *DOE* = 90° to represent return stroke.

4. Divide angle *AOC* (or Arc *AC*) and angle *DOE* (or Arc *DE*) into same number of equal parts as in displacement diagram (*i.e.* six equal parts in each case). The six equal parts on arc *AC* are marked as 1, 2, 3, ... , 6 whereas on arc *DE* they are marked as 6′, 5′, ... , 1′.

5. Now from points 1, 2, 3, ... , 6 and 6′, 5′, ... , 1′ draw tangents to the off-set circle and produce these tangents beyond the prime circle as shown in Fig. 18.24.

6. From points 1, 2, ... , 6 and 6′, 5′, ... , 1′ of these tangents, mark distances 1-*a*, 2-*b*, 3-*c*, ... , 6-*f* and 6′-*f*′, 5′-*e*′, ... , 1′-*a*′ from displacement diagram. The points *a*, *b*, *c*, ... , *f* and *f*′, *e*′, ... , *a*′ represents the positions of the roller centre.

7. Now from points *A*, *a*, *b*, ... , *f* and *f*′, *e*′, ... *a*′ and *E*, draw circles with radius equal to the radius of the roller.

8. Join the bottoms of these circles with a smooth curve. This smooth curve represents the cam profile as shown by a thick line in Fig. 18.24.

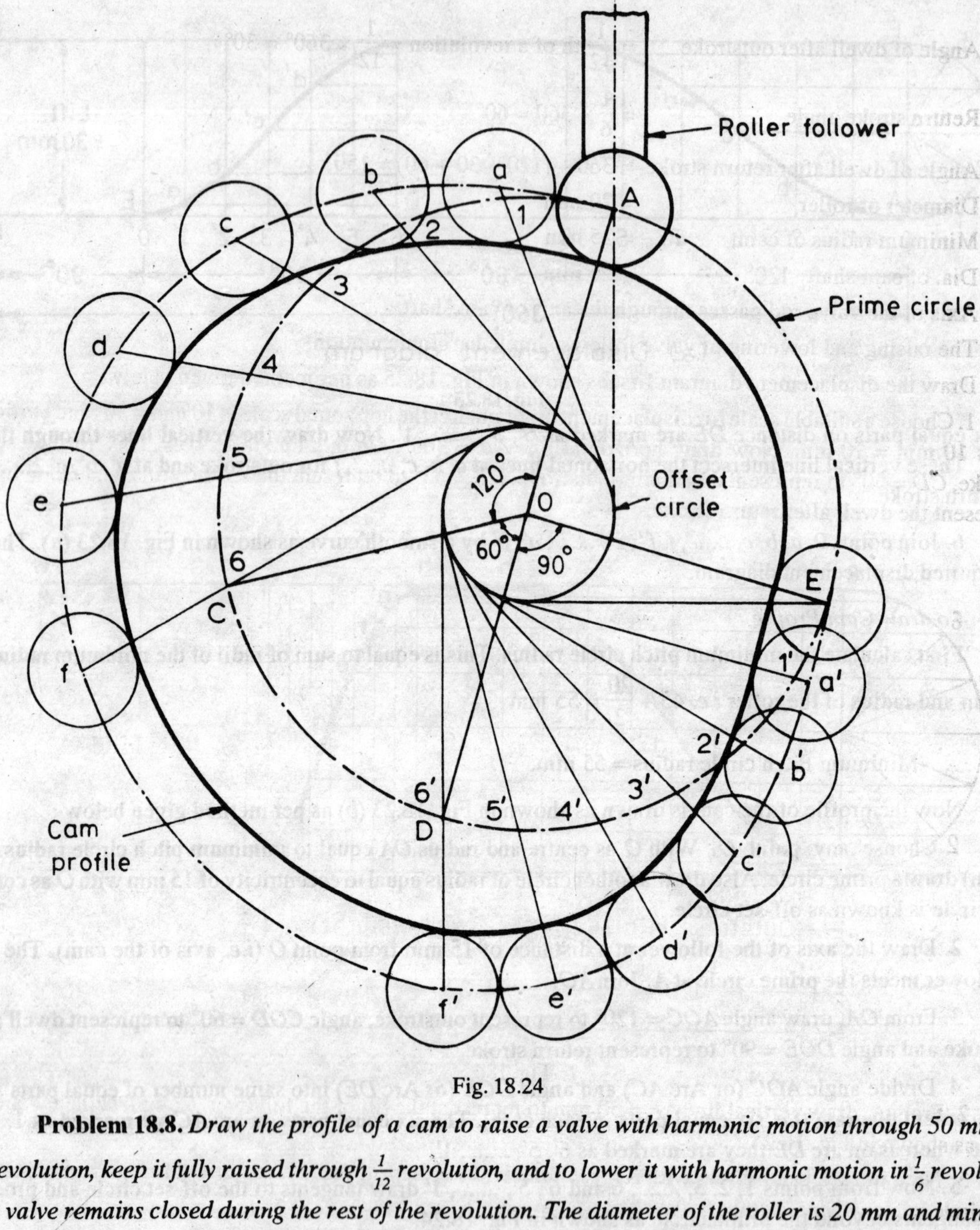

Fig. 18.24

Problem 18.8. *Draw the profile of a cam to raise a valve with harmonic motion through 50 mm in $\frac{1}{3}$ of revolution, keep it fully raised through $\frac{1}{12}$ revolution, and to lower it with harmonic motion in $\frac{1}{6}$ revolution. The valve remains closed during the rest of the revolution. The diameter of the roller is 20 mm and minimum radius of the cam is to be 25 mm. The diameter of the cam shaft is 25 mm. The axis of the valve rod passes through the axis of the cam shaft. Assume the cam shaft to rotate with a uniform velocity.*

(A.M.I.E., Summer 1980)

Sol. Given :

Lift, s = 50 mm

Outstroke angle, $\theta_0 = \frac{1}{3}$ rd of revolution $= \frac{1}{3} \times 360° = 120°$

Angle of dwell after outstroke $= \frac{1}{12}$ th of a revolution $= \frac{1}{12} \times 360° = 30°$

Return stroke angle $= \frac{1}{6} \times 360 = 60°$

Angle of dwell after return stroke $= 360 - (120 + 30 + 60) = 150°$

Diameter of roller $= 20$ mm

Minimum radius of cam, $R_{min} = 25$ mm

Dia. of cam shaft $= 25$ mm

Axis of the valve rod passes through the axis of cam shaft.

The raising and lowering of valve follows simple harmonic motion.

Draw the displacement diagram first as shown in Fig. 18.25 as per method given below :

1. Choose a suitable scale for displacement diagram. Let the horizontal scale is 10 mm = 30° and vertical scale is 10 mm = 10 mm. Now draw horizontal line $AB = 360°$. On this line take $AC = 120°$ to represent outstroke, $CD = 30°$ to represent dwell after outstroke, $DE = 60°$ to represent the return stroke and $EB = 150°$ to represent the dwell after return stroke.

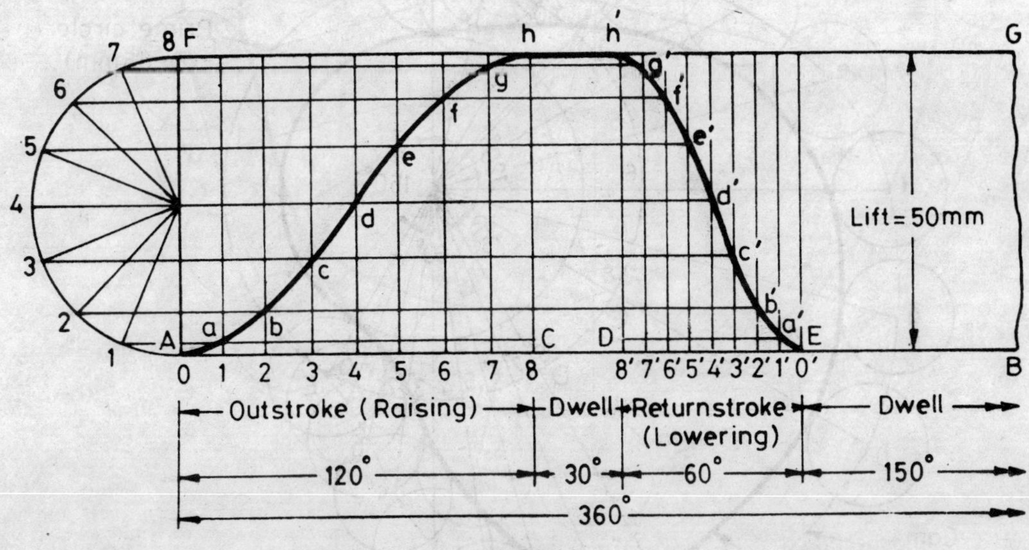

Fig. 18.25

2. From A, draw vertical line $AF = 50$ mm to represent the lift of the follower. Complete the rectangle $AFGB$ as shown in Fig. 18.25.

3. Now draw a semi-circle with AF as diameter to represent the S.H.M. of the follower. Divide the semi-circle into any equal number of even parts (say eight). Name these parts as 1, 2, 3, ... , 8. Join them with the centre of the semi-circle by radial lines.

4. Through each points 1, 2, 3, ... , 8 draw horizontal lines as shown in Fig. 18.25.

5. Now divide the distance AC and DE into the same equal number of parts as the number of parts on the semi-circle (*i.e.* eight). Mark these parts as 1, 2, 3, ... , 8 for outstroke and 8′, 7′, ... , 1′, 0′ for return stroke. Draw vertical lines through these points. These vertical lines intersect the already drawn horizontal lines at *a*, *b*, *c*, ... , *g*, *h* for outstroke and at *h*′, *g*′, ... *b*′, *a*′ for return stroke.

6. Join parts *A*, *a*, *b*, ... , *g*, *h*, *h*′, *g*′, ... *b*′, *a*′, 0′, *B* with a smooth curve. This curve represents the displacement diagram for the follower.

Profile of the cam

First calculate the minimum pitch circle radius which is equal to the sum of radii of the minimum radius of the cam and radius of the roller *i.e.* $25 + \frac{20}{2} = 35$ mm.

$\therefore$ Minimum pitch circle radius = 35 mm.

Now the profile of the cam is drawn as shown in Fig. 18.26 as per method given below :

1. Choose any point *O*. With *O* as centre and radius equal to *OA* (*i.e.* minimum pitch circle radius = 35 mm) draw a prime circle.

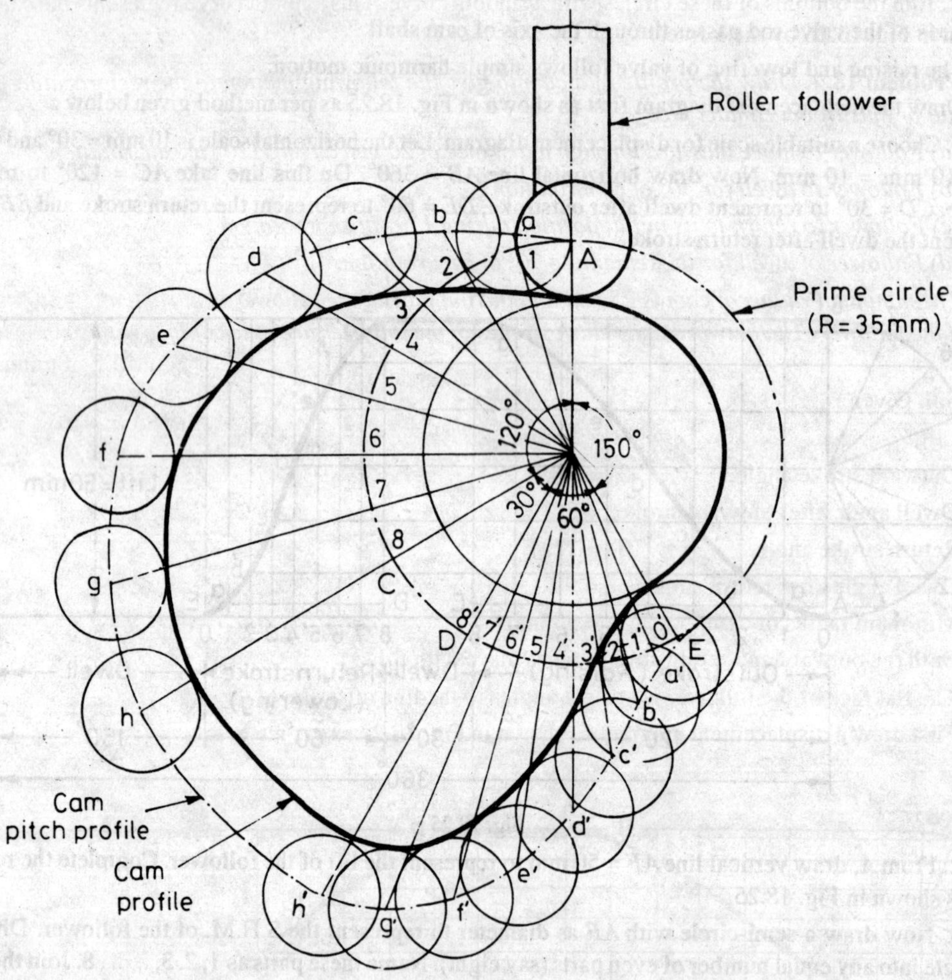

Fig. 18.26

2. From *OA*, draw angle *AOC* = 120° to represent the outstroke, angle *COD* = 30° to represent the dwell after outstroke and angle *DOE* = 60° to represent the return stroke. The points *A*, *C*, *D* and *E* lie on the prime circle.

3. Divide the arc *AC* and arc *DE* into the same number of equal parts as in the displacement diagram (*i.e.* eight equal parts in each case). The eight equal parts on the arc *AC* are marked as 1, 2, 3, 4, 5, 6, 7, 8 whereas on the arc *DE* they are marked as 8′, 7′, ... , 2′, 1′.

4. Join the above points with O and produce these radial lines beyond the prime circle as shown in Fig. 18.26.

5. On these radial lines, mark distances 1-a, 2-b, ... , 7-g, 8-h and 8′- h', 7′- g', ... , 2′- b', 1′- a' equal to the corresponding displacements as measured from the displacement diagram shown in Fig. 18.25. The points a, b, c, ... , h and h', g', ... , b', a' represents the position of the roller centre.

6. Now from points A, a, b, ... , h and h', g', ... , a' and E draw circles with radius equal to the radius of the roller (*i.e.* = 10 mm).

7. Join the bottoms of these circles with a smooth curve. This smooth curve represents the cam profile as shown in Fig. 10.26 by thick line.

Problem 18.9. *Draw the profile of a cam to give the following motion to the reciprocating follower with a flat, or mushroom contact face :*

(*a*) *Follower to move outward through a distance of 20 mm during 120° of cam rotation* ;

(*b*) *Follower to dwell for 30° of cam rotation* ;

(*c*) *Follower to return to its initial position during 120° of cam rotation* ;

(*d*) *Follower of dwell for the remaining 90° of cam rotation.*

The minimum radius of cam is 25 mm and the flat face of the follower is at right angles to the line of stroke of the follower. The outward and return strokes of the follower are to take place with simple harmonic motion. (A.M.I.E., Summer 1983)

Sol. Given :

Lift, $s = 20$ mm

Outward stroke angle, $\theta_0 = 120°$

Dwell angle after outward stroke = 30°

Return stroke angle, $\theta_R = 120°$

Dwell angle after return stroke = 90°

Minimum radius of cam, $R_{min} = 25$ mm

Both the outward and return strokes takes place with S.H.M.

The flat face of the follower is at right angles to the line of stroke of the follower.

First draw a displacement diagram as shown in Fig. 18.27 as per method given below :

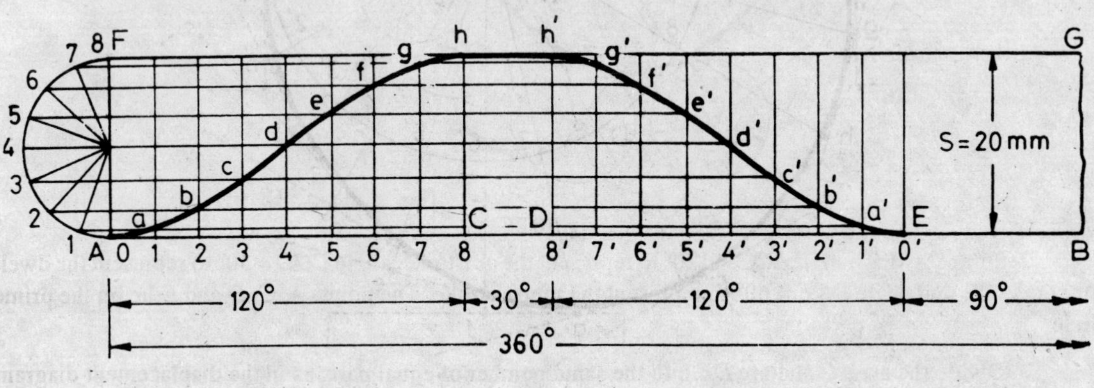

Fig. 18.27

Draw a semi-circle of 20 mm (lift) and divide it into any number of equal parts, say in this case eight equal parts.

Though these parts, draw horizontal lines as shown. On the abscissa take cam displacements *AC*, *CD*, *DE*, and *EB* equal to 120° for ascent, 30° for dwell, 120° for descent and 90° for dwell respectively.

Next divide the distance *AC* and *DE* into eight equal parts and obtain the corresponding points *a*, *b*, *c*, *d*, *e*, *f*, *g*, *h* for ascent and *h'*, *g'*, *f'*, *e' d'*, *c'*, *b'*, *a'* for descent.

Join these points by a smooth curve and complete the displacement diagram as shown in Fig. 18.27.

The profile of the cam driving a flat reciprocating follower is drawn as shown in Fig. 18.28 and as per method as given below :

(*i*) Draw a base circle with centre *O* and radius *OA* equal to the minimum radius of the cam (*i.e.* 25 mm).

(*ii*) Draw angles *AOC*, *COD*, *DOE* and *EOA* equal to 120°, 30°, 120° and 90° to represent the outward stroke, dwell, inward stroke and dwell respectively.

(*iii*) Divide angles *AOE* and *DOE* into the same number of equal parts as in the displacement diagram (*i.e.* eight equal parts each). Name these parts 1, 2, ... 8, 8', 7', ... 1'.

(*iv*) Join the points 1, 2, 3, ... 8, 8', 7', ... 2', 1' with centre *O* and produce beyond the base circle.

(*v*) From points 1, 2, 3, ... 8, 8', 7', ... 2', 1' set off 1*a*, 2*b*, 3*c*, ... 7*g*, 8*h*, 8' *h*', 7'*g*', ... 2'*b*' and 1' *a*' equal to the distances measured from the displacement diagram.

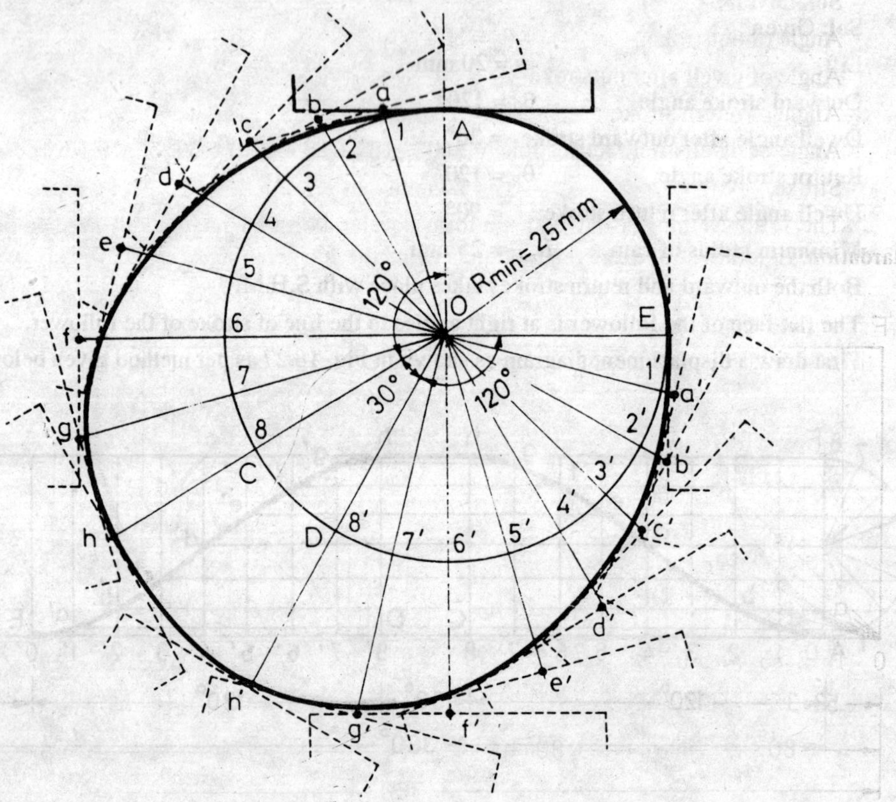

Fig. 18.28

(*vi*) Now at points *a*, *b*, *c*, ... *h*, *h'*, *g'*, ... *b'*, *a'* draw position of the flat-faced follower. The axis of the follower at all these points passes through the cam centre.

(*vii*) The cam profile is obtained by drawing a curve which touches the successive positions of the contact face of the follower.

The flat face of the follower occupies the position shown dotted relative to the cam and the profile of the cam must be so shaped so as to touch the dotted profile of the follower. It should be noted that, except at the beginning and at the end of each stroke, the point of contact of the cam profile and the follower face does not lie on the line of stroke of the follower.

Problem 18.10. *A flat faced reciprocating follower has the following motion :*

(i) The follower moves out for 80° of cam rotation with uniform acceleration and retardation, acceleration being twice the retardation.

(ii) The follower dwells for the next 80° of cam rotation.

(iii) The follower moves in for the next 120° of cam rotation with uniform acceleration and retardation, the retardation being twice the acceleration.

(iv) The follower dwells for the remaining period.

The base circle diameter of the cam is 60 mm and the stroke of the follower is 20 mm. The line of movement of the follower passes through the cam centre. Draw displacement diagram.

(A.M.I.E., Summer 1986)

Sol. Given :

Angle of outstroke, $\theta_0 = 80°$

Angle of dwell after outstroke $= 80°$

Angle of return stroke $= 120°$

Angle of dwell after return stroke $= 360° - (80 + 80 + 120) = 80°$

Stroke, $s = 20$ mm.

The follower moves out with uniform acceleration and retardation, acceleration being twice the retardation. In return stroke, the retardation is twice the acceleration.

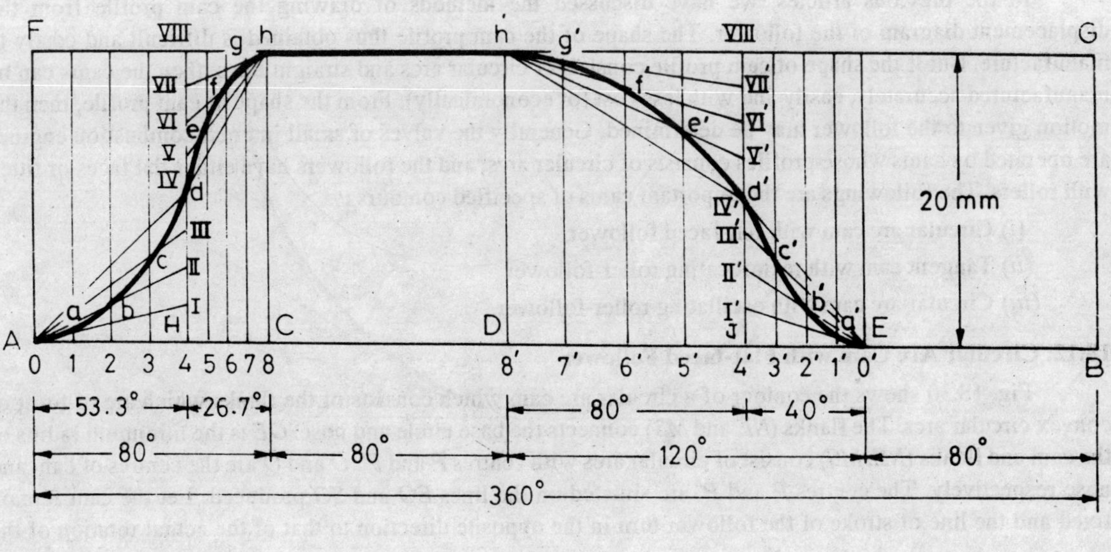

Fig. 18.29

First draw the displacement diagram as shown in Fig. 18.29 as per method given below :

1. Choose a suitable scale for displacement diagram. Let horizontal scale is 10 mm = 20° and vertical scale is 1 mm = 1mm. Now draw horizontal line *AB* of length 180 mm representing 360°. On this line, mark *AC* = 80° to represent outstroke, *CD* = 80° to represent dwell after outstroke, *DE* = 120° to represent return stroke and *EB* = 80° to represent dwell after return-stroke.

2. From *A*, draw vertical line *AF* = 20 mm to represent the stroke of the follower. Complete the rectangle *AFGB* as shown in Fig. 18.29.

3. In the outstroke, the follower moves with the uniform acceleration and retardation, acceleration being twice the retardation. Hence mark $AH = \frac{2}{3} \times 80° = 53.3°$ and $HC = \frac{1}{3} \times 80 = 26.7°$. Same is the case with return stroke. Hence mark $DJ = \frac{2}{3} \times 120 = 80°$ and $JE = \frac{1}{3} \times 120 = 40°$.

4. Divide distance *AH* in four equal parts, distance *HC* also in four equal parts, distance *DJ* in four equal parts and distance *JE* in four equal parts. These parts are marked as 1, 2, 3, 4, 5, ... , 8 for outstroke and 8′, 7′, 6′, ... , 1′ for return stroke. Draw vertical lines through these points.

5. Divide the vertical lines through 4 and 4′ into eight equal parts. The eight equal parts on vertical line through 4 are marked as I, II, III, ... , VIII whereas the eight equal parts on vertical line through 4′ are marked as I′, II′, III′, ... , VIII′.

6. For the first half of outstroke, join *A*-I, *A*-II, *A*-III and *A*-IV. The intersection of *A*-I and vertical line drawn through 1, gives point *a*. The intersection of *A*-II and vertical line draw through 2, gives point *b*. Similarly obtain points *c* and *d*.

7. For the second half of the stroke, join *h*-VII, *h*-VI, *h*-V and *h*-IV. The intersection of *h*-V and vertical line drawn through 5 gives point *e*. Similarly obtain points *f* and *g*. Join points *A*-*a*-*b*-*c*-*d*-*e*-*f*-*g*-*h* with a smooth curve. This curve represents the displacement of the follower during outstroke.

8. Similarly obtain points *h*′, *g*′, *f* ′, *e*′, *d*′, *c*′, *b*′ and *a*′ for return stroke. Join these points also with smooth curve. Then curve *A*, *a*, *b*, ... , *h*, *h*′, *g*′, ... , *a*′ *E* and *B* represent the required displacement diagram.

18.11. Cams with Specified Contours

In the previous articles, we have discussed the methods of drawing the cam profile from the displacement diagram of the follower. The shape of the cam profile thus obtained is difficult and costly to manufacture. But if the shape of cam profile consists of circular arcs and straight lines, then the cams can be manufactured accurately, easily and with less cost (or economically). From the shape of cam profile, then the motion given to the follower may be determined. Generally the valves of small internal combustion engines are operated by cams whose profiles consists of circular arcs, and the followers have either flat faces or fitted with rollers. The followings are the important cams of specified contours :

(*i*) Circular arc cam with flat-faced follower

(*ii*) Tangent cam with reciprocating roller-follower

(*iii*) Circular arc cam with oscillating roller-follower.

18.12. Circular Arc Cam with Flat-faced Follower

Fig. 18.30 shows the contour of a circular arc cam which consists of the flanks which are of form of convex circular arcs. The flanks (*NE* and *MS*) connects the base circle and nose. *OE* is the minimum radius of the cam and flanks (*NE*, *MS*) consist of circular arcs with centres *P* and *P*′. *O* and *Q* are the centres of cam and nose respectively. The centres *P* and *P*′ are situated on the lines *EO* and *SO* produced. Let the cam remain fixed and the line of stroke of the follower turn in the opposite direction to that of the actual rotation of the cam. Hence if cam rotates in clockwise direction, then line of stroke of the follower turns in anti-clockwise direction, as shown in the Fig. 18.30.

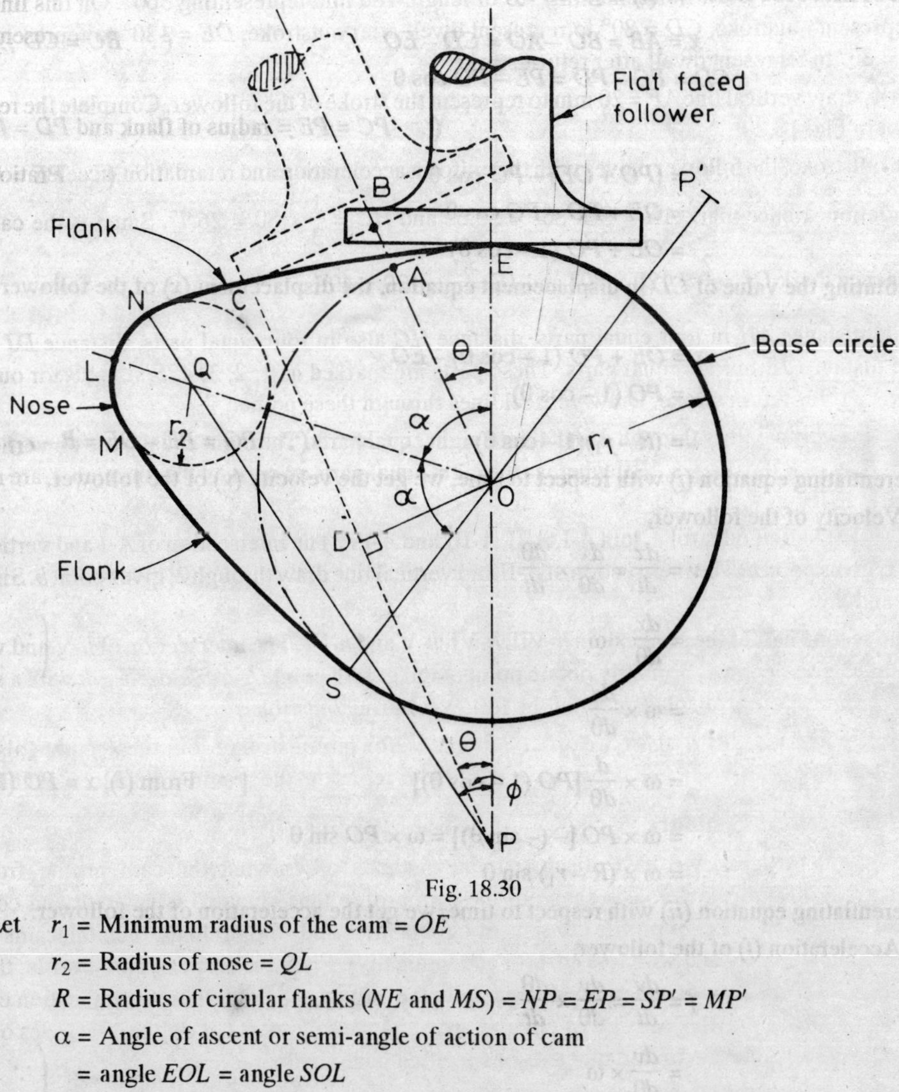

Fig. 18.30

Let r_1 = Minimum radius of the cam = OE

r_2 = Radius of nose = QL

R = Radius of circular flanks (NE and MS) = $NP = EP = SP' = MP'$

α = Angle of ascent or semi-angle of action of cam

= angle EOL = angle SOL

2α = Total angle of action of cam = angle EOS

ϕ = Angle of action of cam on the circular flank = angle OPQ or angle OPN

Let us consider the following two cases :

(*i*) Follower is in contact with any point on the circular flank (NE)

(*ii*) Follower is in contact with any point on the nose.

(*i*) *Follower is in contact with any point on the circular flank (NE).* (Refer to Fig. 18.30)

Let us consider first that the flat face of the follower has contact at E (*i.e.* the common point of the base circle and circular flank). When the line of stroke of the follower turns through the angle θ relative to cam; then the flat face of the follower will be in contact with the cam profile at the point C, where PC is perpendicular to the face of the follower. In this position, the displacement of the follower is given by AB. Since OB is perpendicular to BC, therefore OB is parallel to PC. From O, draw OD perpendicular to PC. Then the displacement or lift of the follower (x) is given by,

$$x = AB = BO - AO = CD - EO \qquad (\because\ BO = CD\ ;AO = EO)$$

But $$CD = PC - PD = PE - PO\cos\theta$$

$$(\because\ PC = PE = \text{radius of flank and } PD = PO\cos\theta)$$

$$= (PO + OE) - PO\cos\theta \qquad (\because\ PE = PO + OE)$$

$$= OE + PO - PO\cos\theta$$

$$= OE + PO\,(1-\cos\theta)$$

Substituting the value of CD in displacement equation, the displacement (x) of the follower becomes as

$$x = OE + PO\,(1-\cos\theta) - EO$$

$$= PO\,(1-\cos\theta) \qquad \text{...}(i)$$

$$= (R - r_1)(1-\cos\theta) \qquad (\because\ PO = PE - OE = R - r_1) \quad \text{...(18.9)}$$

Differentiating equation (i) with respect to time, we get the velocity (v) of the follower.

∴ Velocity of the follower,

$$v = \frac{dx}{dt} = \frac{dx}{d\theta}\times\frac{d\theta}{dt}$$

$$= \frac{dx}{d\theta}\times\omega \qquad \left(\because\ \frac{d\theta}{dt} = \omega\right)$$

$$= \omega\times\frac{dx}{d\theta}$$

$$= \omega\times\frac{d}{d\theta}[PO\,(1-\cos\theta)] \qquad [\because\ \text{From } (i),\ x = PO\,(1-\cos\theta)]$$

$$= \omega\times PO\,[-(-\sin\theta)] = \omega\times PO\sin\theta \qquad \text{...}(ii)$$

$$= \omega\times(R - r_1)\sin\theta \qquad \text{...(18.10)}$$

Differentiating equation (ii) with respect to time, we get the acceleration of the follower.

∴ Acceleration (f) of the follower,

$$f = \frac{dv}{dt} = \frac{dv}{d\theta}\times\frac{d\theta}{dt}$$

$$= \frac{dv}{d\theta}\times\omega \qquad \left(\because\ \frac{d\theta}{dt} = \omega\right)$$

$$= \omega\times\frac{dv}{d\theta}$$

$$= \omega\times\frac{d}{d\theta}(\omega\times PO\sin\theta)$$

$$= \omega^2\times PO\,.\cos\theta$$

$$= \omega^2\times(R - r_1)\cos\theta \qquad \text{...(18.11)}$$

The above three equations (18.9), (18.10) and (18.11) are applicable when the follower is in contact with any point on the circular flank. This means that for the above three equations, the values of θ varies from 0 to φ where φ is the angle OPQ or OPN. From equation (18.10) it is also clear that velocity of follower is zero when $\theta = 0$. As the value of θ increases, velocity also increases. The velocity will be maximum when $\theta = \phi$. Hence maximum velocity is given by

$$v_{max} = \omega \times OP \times \sin\phi = \omega \times (R - r_1) \times \sin\phi \quad ...(18.12)$$

[Substitute ϕ for θ in equation (18.10)]

From equation (18.11) it is clear that acceleration of the follower will be maximum when $\theta = 0$. This is due to the fact that when $\theta = 0$, then $\cos\theta = \cos 0 = 1$. The maximum value of acceleration is given by,

$$f_{max} = \omega^2 \times PO = \omega^2 \times (R - r_1) \quad ...(18.13)$$

$(\because \; PO = PE - OE = R - r_1)$

(ii) Follower is in contact with any point on the nose

Fig. 18.31 shows the position of the follower when the flat face is in contact at any point *C* which lies on the nose. The centre of the nose is at point *Q*. The circular arc *NE* shows the position of flank. The

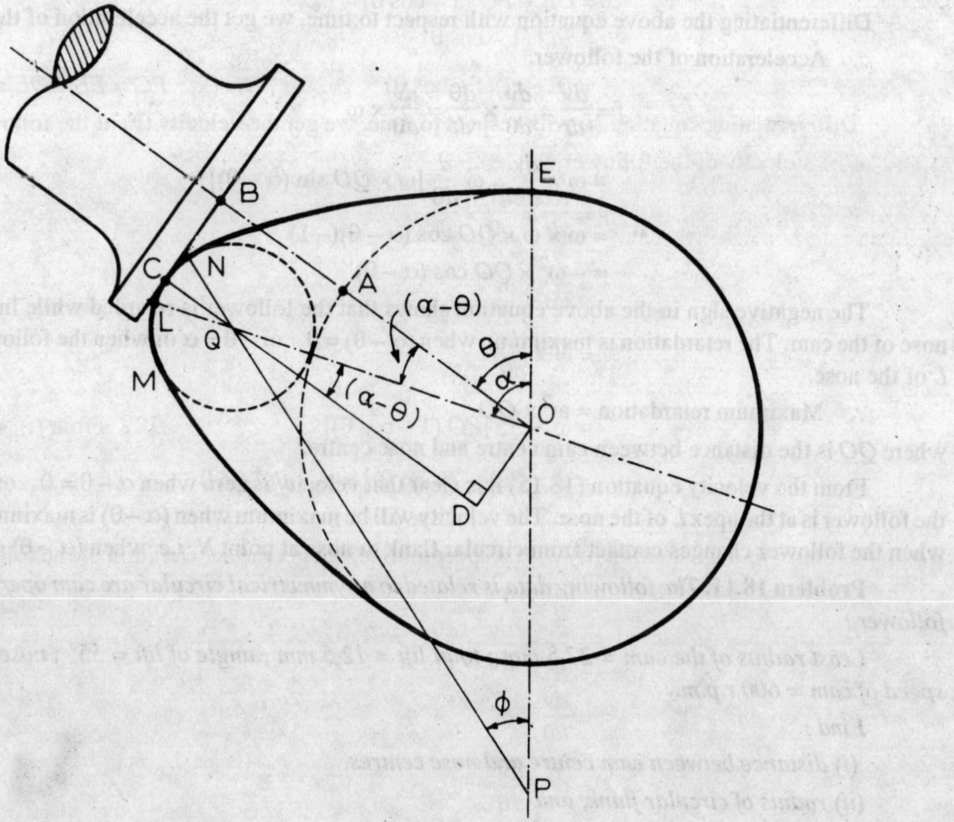

Fig. 18.31

displacement or lift of the follower (x) when follower has turned through an angle θ (greater than ϕ) is given by,

$$x = AB = BO - AO = CD - AO \quad (\because \; BO = CD) \quad ...(i)$$

But $\quad CD = CQ + QD = CQ + QO\cos(\alpha - \theta) \quad [\because \; QD = QO\cos(\alpha - \theta)]$

Substituting the value of *CD* in equation (*i*), we get

$$x = CQ + QO\cos(\alpha - \theta) - AO \quad ...(18.14)$$

In the above equation, $CQ(=r_2), QO, AO(=r_1)$ and α are constant for a given cam. Hence differentiating the above equation with respect to time we get the velocity of the follower.

$\therefore$ Velocity of the follower

$$v = \frac{dx}{dt} = \frac{dx}{d\theta} \times \frac{d\theta}{dt} = \frac{dx}{d\theta} \times \omega \qquad \left(\because \frac{d\theta}{dt} = \omega\right)$$

$$= \omega \times \frac{dx}{d\theta}$$

$$= \omega \times \frac{d}{d\theta}[CQ + QO\cos(\alpha - \theta) - AO]$$

$$= \omega \times [0 + QO(-)\sin(\alpha - \theta)(-1) - 0]$$

$$= \omega \times QO \sin(\alpha - \theta) \qquad \text{...(18.15)}$$

where QO is the distance between cam centre and nose centre.

Differentiating the above equation with respect to time, we get the acceleration of the follower.

$\therefore$ Acceleration of the follower,

$$f = \frac{dv}{dt} = \frac{dv}{d\theta} \times \frac{d\theta}{dt} = \frac{dv}{d\theta} \times \omega$$

$$= \omega \times \frac{dv}{d\theta} = \omega \frac{d}{d\theta}[\omega \times QO \sin(\alpha - \theta)]$$

$$= \omega \times \omega \times QO \cos(\alpha - \theta)(-1)$$

$$= -\omega^2 \times QO \cos(\alpha - \theta) \qquad \text{...(18.16)}$$

The negative sign in the above equation shows that the follower is retarded while in contact with the nose of the cam. The retardation is maximum, when $(\alpha - \theta) = 0$ or $\theta = \alpha$ or when the follower is at the apex L of the nose.

$\therefore$ Maximum retardation $= \omega^2 \times QO$ ...(18.17)

where QO is the distance between cam centre and nose centre.

From the velocity equation (18.15) it is clear that velocity is zero when $\alpha - \theta = 0$ or $\theta = \alpha$ or when the follower is at the apex L of the nose. The velocity will be maximum when $(\alpha - \theta)$ is maximum. This happens when the follower changes contact from circular flank to nose at point N, *i.e.* when $(\alpha - \theta) = \phi$.

Problem 18.11. *The following data is related to a symmetrical circular arc cam operating a flat-faced follower :*

Least radius of the cam = 27.5 mm ; total lift = 12.5 mm ; angle of lift = 55° ; nose radius = 3 mm ; speed of cam = 600 r.p.m.

Find :

(i) distance between cam centre and nose centres,

(ii) radius of circular flank, and

(iii) angle of contact on the circular flank.

Sol. Given :

$r_1 = OE = 30$ mm ; Lift $= AL = 12.5$ mm ; $\alpha = 55°$; $r_2 = 3$ mm ; $N = 600$ r.p.m.

Refer to Fig. 18.32. In this figure,

PE = Radius of circular flank = R

OQ = Distance between cam centre and nose centre

ϕ = Angle OPQ or angle OPN
= Angle of contact on the circular flank
AL = total lift of the follower.

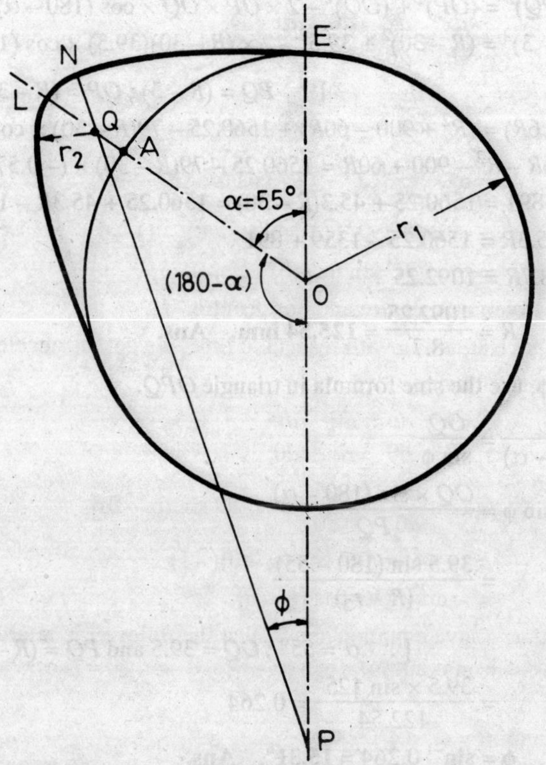

Fig. 18.32

From Fig. 18.32, we know that

$$AL = OL - OA$$

$$= OL - r_1 \qquad (\because\ OA = OE = r_1)$$

$$= (OQ + LQ) - r_1 \qquad (\because\ OL = OQ + QL)$$

$$= OQ + r_2 - r_1 \qquad (\because\ QL = r_2)$$

or $$12.5 = OQ + r_2 - r_1 \qquad (\because\ AL = \text{total lift} = 12.5 \text{ mm})$$

$\therefore$ $OQ = 12.5 - r_2 + r_1 = 12.5 - 3 + 30 =$ **39.5 mm. Ans.**

We know that

$PN = PE = R$ = Radius of circular flank.

Let us now find the values of PQ and OP in terms of R

$$PN = PQ + QN = PQ + r_2 \qquad (\because\ QN = r_2)$$

$\therefore$ $$PQ = PN - r_2 = R - r_2 \qquad (\because\ PN = R)$$

or $$PQ = R - 3 \qquad (\because\ r_2 = 3)$$

Also $$PE = PO + OE$$

or $$R = PO + r_1$$

$\therefore$ $$PO = R - r_1 = R - 30 \qquad (\because\ r_1 = 30)$$

Now from triangle OPQ,

$$(PQ)^2 = (OP)^2 + (OQ)^2 - 2 \times OP \times OQ \times \cos(180 - \alpha)$$

$$(R-3)^2 = (R-30)^2 + 39.5^2 - 2 \times (R-30)(39.5) \times \cos(180-55)$$

$$[\because\ PQ = (R-3)\ ;\ OP = (R-30)\ ;\ OQ = 39.5 \text{ and } \alpha = 55]$$

$\therefore$ $$(R^2 + 9 - 6R) = (R^2 + 900 - 60R) + 1560.25 - 79(R-30) \times \cos 125°$$

or $$R^2 + 9 - 6R - R^2 - 900 + 60R = 1560.25 - 79(R-30) \times (-0.5735)$$

or $$54R - 891 = 1560.25 + 45.3(R-30) = 1560.25 + 45.3R - 1359$$

or $$54R - 45.3R = 1560.25 - 1359 + 891$$

or $$8.7R = 1092.25$$

or $$R = \frac{1092.25}{8.7} = \mathbf{125.54\ mm.} \quad \textbf{Ans.}$$

To find the value of ϕ, use the sine formula in triangle OPQ.

$\therefore$ $$\frac{PQ}{\sin(180-\alpha)} = \frac{OQ}{\sin\phi}$$

or $$\sin\phi = \frac{OQ \times \sin(180-\alpha)}{PQ}$$

$$= \frac{39.5 \sin(180-55)}{(R-r_2)}$$

$$[\because\ \alpha = 55°\ ;\ OQ = 39.5 \text{ and } PQ = (R - r_2) = 125.54 - 3 = 122.54]$$

$$= \frac{39.5 \times \sin 125°}{122.54} = 0.264$$

$\therefore$ $$\phi = \sin^{-1} 0.264 = \mathbf{15.31°}. \quad \textbf{Ans.}$$

Problem 18.12. *Following is the data for a circular arc cam working with a flat faced reciprocating follower. Minimum radius of the cam = 30 mm. Total angle of cam action, $2\alpha = 120°$. Radius of circular arc = 80 mm. Nose radius = 10 mm :*

(a) Find the distance of the centre of nose circle from the centre of the cam.

(b) Draw the profile of the cam to full scale.

(c) Find the angle through which the cam turns when the point of contact moves from the junction of minimum radius arc and circular arc to the junction of nose radius arc and circular arc.

(d) Find the velocity in m/s and acceleration in m/s² of the follower when the cam has turned through an angle of $\theta = 20°$. Angle θ is measured from the point where the follower just starts moving away from the cam. The angular velocity of the cam = 10 rad/s. (A.M.I.E., Winter 1983)

Sol. Given :

Minimum radius of the cam, $r_1 = 30$ mm

Total angle of cam action, $2\alpha = 120°$

$\therefore$ Semi-angle of cam action, $\alpha = \frac{120}{2} = 60°$

Radius of circular arc (Flank radius), $R = 80$ mm

Nose radius, $r_2 = 10$ mm and $\omega = 10$ rad/s.

(*a*) *Distance of the centre of the nose circle from cam axis* (Refer to Fig. 18.33)

Let O = Centre of cam axis

Q = Centre of nose circle

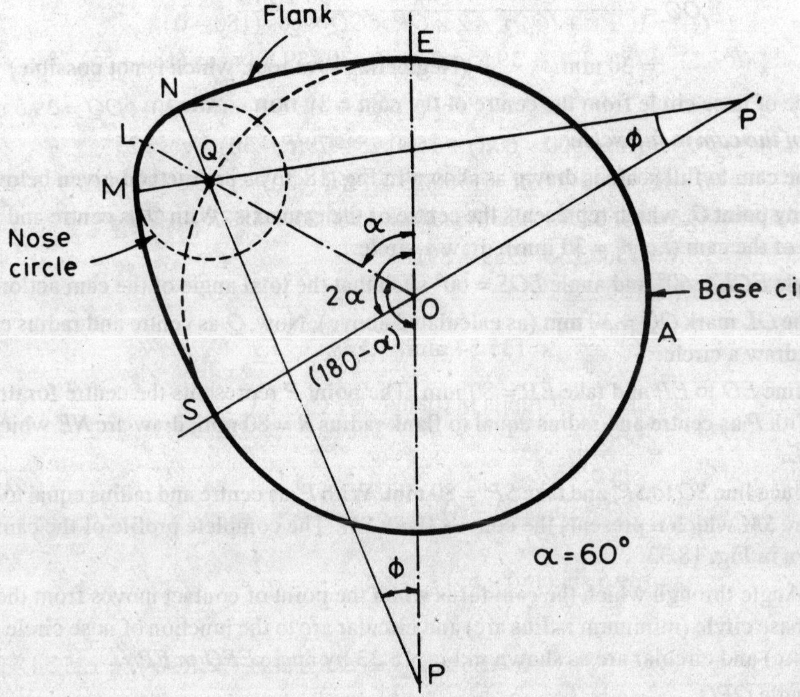

Fig. 18.33

Then OQ = Distance of the nose circle from cam centre.

The point Q can lie anywhere on the axis OL. It is not necessary that the point Q is on the circumference of the cam circle of minimum radius (*i.e.* r_1).

Also we know

$$OP = PE - OE$$
$$= \text{Flank radius} - \text{least radius}$$
$$= R - r_1$$
$$= 80 - 30 = 50 \text{ mm}$$

Also

$$PQ = PN - QN$$
$$= \text{Flank radius} - \text{nose radius}$$
$$= R - r_2$$
$$= 80 - 10 = 70 \text{ mm}$$

Now from triangle QOP, we get

$$PQ^2 = OQ^2 + PO^2 - 2 \times OQ \times PO \times \cos(180 - \alpha)$$
$$70^2 = OQ^2 + 50^2 - 2 \times OQ \times 50 \times \cos(180 - 60)$$
$$4900 = OQ^2 + 2500 - 100 \times OQ \times \cos 120°$$
$$= OQ^2 + 2500 - 100 \times OQ \times (-0.5)$$
$$= OQ^2 + 2500 + 50\, OQ$$

or $$OQ^2 + 50OQ + 2500 - 4900 = 0$$

or $$OQ^2 + 50OQ - 2400 = 0$$

$$\therefore \quad OQ = \frac{-50 \pm \sqrt{50^2 + 4 \times 2400}}{2} = \frac{-50 \pm 110}{2}$$

$$= 30 \text{ mm.} \qquad \text{(Neglecting – ve root, which is not possible)}$$

∴ Distance of nose circle from the centre of the cam = **30 mm. Ans.**

(b) Profile of the cam to full scale

Profile of the cam to full scale is drawn as shown in Fig. 18.33 as per method given below :

1. Choose any point O, which represents the centre of the cam axis. With O as centre and radius equal to minimum radius of the cam (*i.e.* $r_1 = 30$ mm), draw a circle.

2. Draw angle $EOL = 60°$ and angle $LOS = 60°$ such that the total angle of the cam action is 120°.

3. On the line OL mark $OQ = 30$ mm (as calculated above). Now, Q as centre and radius equal to nose radius $r_2 = 10$ mm, draw a circle.

4. Produce line EO to EP and take $EP = 80$ mm. The point P represents the centre for drawing flank (NE) of the cam. With P as centre and radius equal to flank radius $R = 80$ mm, draw arc NE which represents the convex flank NE.

5. Now produce line SO to SP' and take $SP' = 80$ mm. With P' as centre and radius equal to flank radius $R = 80$ mm, draw arc SM which represents the convex flank SM. The complete profile of the cam is given by $ENLMSAE$ as shown in Fig. 18.33.

(c) Let ϕ = Angle through which the cam turns when the point of contact moves from the junction of base circle (minimum radius arc) and circular arc to the junction of nose circle (nose radius arc) and circular arc as shown in Fig. 18.33 by angle EPQ or EPN.

Now in triangle OPQ,

Angle $OPQ = \phi$

Using sine formula, we get

$$\frac{OQ}{\sin \phi} = \frac{QP}{\sin (180 - \alpha)}$$

$$\therefore \quad \frac{30}{\sin \phi} = \frac{70}{\sin \alpha} \qquad (\because \; QP = 70, OQ = 30 \text{ mm})$$

$$\therefore \quad \sin \phi = \frac{30}{70} \sin \alpha = \frac{3}{7} \sin 60° \qquad (\because \; \alpha = 60°)$$

$$= \frac{3}{7} \times 0.866 = 0.3711$$

$$\therefore \quad \phi = \sin^{-1} 0.3711 = \mathbf{21.78°} \text{ or } \mathbf{21° \, 47'}. \text{ **Ans.**}$$

(d) Velocity and acceleration of the follower

Angle turned by cam, $\theta = 20°$

Angular velocity of cam, $\omega = 10$ rad/s

As angle θ is less than ϕ, hence the follower is in contact with the circular flank. The velocity at any point on circular flank (NE) is given by equation (18.10) as

$$v = \omega \times PO \times \sin \theta$$

$$= \omega \times (R - r_1) \times \sin \theta$$

$$= 10 \times (80 - 30) \times \sin 20° = 10 \times 50 \times 0.342 = \mathbf{171 \text{ mm/s}}. \text{ **Ans.**}$$

(*ii*) The acceleration (*f*) is given by equation (18.11) as

$$f = \omega^2 \times (R - r_1) \times \cos\theta$$
$$= 10^2 \times (80 - 30) \times \cos 20°$$
$$= 100 \times 50 \times 0.9397 = \mathbf{4698.46\ mm/s^2}. \quad \textbf{Ans.}$$

18.13. Tangent Cam with Roller-follower

Fig. 18.34 shows a cam which has straight flanks *EN* and *MS*. A cam in which the flanks are straight and tangential to the base circle and nose circle, is known as a tangent cam. A tangent cam is symmetrical about the centre line of the cam-shaft as shown in Fig. 18.34. The path of the roller centre is shown by dotted

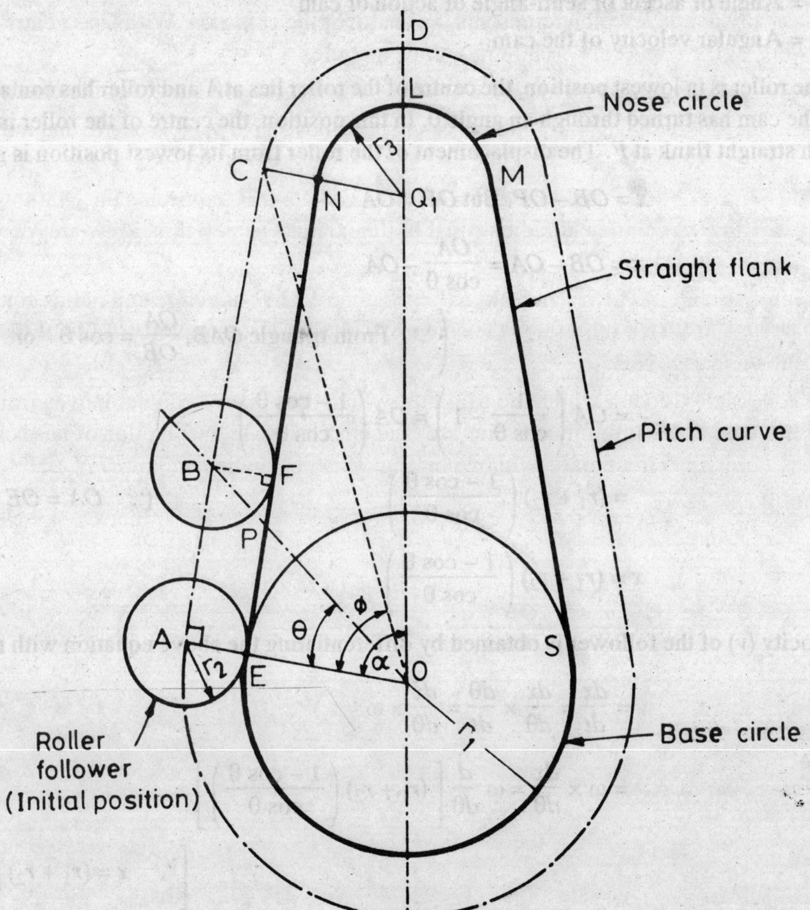

Fig. 18.34

line. Since the whole of the cam profile is formed of circular arcs and straight lines, the working surface of the cam can be machined accurately. For operating the inlet and exhaust valves of internal combustion engines, the tangent cams are mostly used. Let us find the expressions for displacement, velocity and acceleration of the follower for the following two cases :

(*i*) The roller is in contact with the straight flanks ; and

(*ii*) The roller is in contact with the nose.

(i) The roller is in contact with the straight flank

Fig. 18.34 shows a tangent cam in which the roller has contact with straight flank *EN*. The centre of cam is shown by point *O* whereas the centre of nose is shown by point *Q*.

Let r_1 = Minimum radius of the cam or radius of the base circle

r_2 = Radius of the roller

r_3 = Radius of nose

θ = Angle turned by the cam from the beginning of the roller displacement

ϕ = Angle turned by the cam for contact of roller with the straight flank

α = Angle of ascent or semi-angle of action of cam

ω = Angular velocity of the cam.

When the roller is in lowest position, the centre of the roller lies at *A* and roller has contact with straight flank at *E*. Let the cam has turned through an angle θ. In this position, the centre of the roller is at *B* and roller has contact with straight flank at *F*. The displacement of the roller from its lowest position is given by

$$\therefore \quad x = OB - OP. \text{ But } OP = OA$$

$$x = OB - OA = \frac{OA}{\cos\theta} - OA$$

$$\left(\because \text{ From triangle } OAB, \frac{OA}{OB} = \cos\theta \quad \text{or} \quad OB = \frac{OA}{\cos\theta}\right)$$

$$= OA\left(\frac{1}{\cos\theta} - 1\right) = OA\left(\frac{1-\cos\theta}{\cos\theta}\right)$$

$$= (r_1 + r_2)\left(\frac{1-\cos\theta}{\cos\theta}\right) \qquad (\because\ OA = OE + EA = r_1 + r_2)$$

$$\therefore \quad x = (r_1 + r_2)\left(\frac{1-\cos\theta}{\cos\theta}\right) \qquad \text{...(18.18)}$$

The velocity (v) of the follower is obtained by differentiating the above equation with respect to time.

$$\therefore \quad v = \frac{dx}{dt} = \frac{dx}{d\theta} \times \frac{d\theta}{dt} = \frac{dx}{d\theta} \times \omega \qquad \left(\because\ \frac{d\theta}{dt} = \omega\right)$$

$$= \omega \times \frac{dx}{d\theta} = \omega\frac{d}{d\theta}\left[(r_1 + r_2)\left(\frac{1-\cos\theta}{\cos\theta}\right)\right]$$

$$\left[\because\ x = (r_1 + r_2)\left(\frac{1-\cos\theta}{\cos\theta}\right)\right]$$

$$= \omega(r_1 + r_2)\frac{d}{d\theta}\left(\frac{1-\cos\theta}{\cos\theta}\right)$$

$$= \omega(r_1 + r_2)\frac{\cos\theta \times [0 - (-\sin\theta)] - (1 - \cos\theta)(-\sin\theta)}{\cos^2\theta}$$

$$= \omega(r_1 + r_2)\frac{\cos\theta \times \sin\theta + (\sin\theta - \cos\theta\sin\theta)}{\cos^2\theta}$$

$$= \omega(r_1 + r_2)\frac{\sin\theta}{\cos^2\theta} \qquad \text{...(18.19)}$$

The acceleration (f) of the follower is obtained by differentiating the above equation with respect to time.

$$\therefore \quad f = \frac{dv}{dt} = \frac{dv}{d\theta} \times \frac{d\theta}{dt} = \frac{dv}{d\theta} \times \omega \qquad \left(\because \frac{d\theta}{dt} = \omega\right)$$

$$= \omega \frac{dv}{d\theta} = \omega \frac{d}{d\theta}\left[\omega(r_1 + r_2)\frac{\sin\theta}{\cos^2\theta}\right] \qquad \left[\because \quad v = \omega(r_1 + r_2)\frac{\sin\theta}{\cos^2\theta}\right]$$

$$= \omega^2(r_1 + r_2)\frac{d}{d\theta}\left(\frac{\sin\theta}{\cos^2\theta}\right)$$

$$= \omega^2(r_1 + r_2)\frac{\cos^2\theta\,(\cos\theta) - \sin\theta\,(2\cos\theta)(-\sin\theta)}{\cos^4\theta}$$

$$= \omega^2(r_1 + r_2)\frac{(\cos\theta)(\cos^2\theta + 2\sin^2\theta)}{\cos^4\theta}$$

$$= \omega^2(r_1 + r_2)\frac{[\cos^2\theta + 2(1 - \cos^2\theta)]}{\cos^3\theta} \qquad (\because \quad \sin^2\theta = 1 - \cos^2\theta)$$

$$= \omega^2(r_1 + r_2)\left(\frac{2 - \cos^2\theta}{\cos^3\theta}\right) \qquad \text{...(18.20)}$$

The above three equations are applicable only for the part of the follower motion during which the roller is in contact with the straight flank EN, *i.e.* for which θ varies from 0 to ϕ.

From equation (18.19) it is clear that velocity will be maximum, when $\frac{\sin\theta}{\cos^2\theta}$ is maximum. But when θ increases then $\sin\theta$ also increases but $\cos\theta$ decreases. This means that with the increase of θ, the term $\frac{\sin\theta}{\cos^2\theta}$ increases. But for the straight flank EN, the maximum value of θ is ϕ. Hence when $\theta = \phi$ (*i.e.* when the roller just leaves contact with the straight flank at N), the velocity is maximum. Hence substituting $\theta = \phi$ in equation (18.19), we get the maximum velocity as

$$v_{max} = \omega(r_1 + r_2)\frac{\sin\phi}{\cos^2\phi} \qquad \text{...(18.21)}$$

From equation (18.20) it is clear that acceleration will be maximum when $\frac{2 - \cos^2\theta}{\cos^3\theta}$ is maximum. But when θ increases, $\cos\theta$ decreases. This means that the term $\frac{2 - \cos^2\theta}{\cos^3\theta}$ increases. The maximum value of θ for straight flank is ϕ. Hence when $\theta = \phi$ (*i.e.* when the roller just leaves contact with the straight flank at N), the acceleration is maximum. Hence for maximum acceleration, substitute θ equal to ϕ in equation (18.20).

$\therefore$ Maximum acceleration of the follower,

$$f_{max} = \omega^2(r_1 + r_2)\left(\frac{2 - \cos^2\phi}{\cos^3\phi}\right) \qquad \text{...(18.22)}$$

Acceleration will be minimum, when θ is minimum. But minimum value of θ is zero. Hence for minimum acceleration substitute $\theta = 0$ in equation (18.20).

$\therefore$ Minimum acceleration of the follower,

$$f_{min} = \omega^2(r_1 + r_2)\frac{(2 - 1)}{1} \qquad (\because \quad \cos\phi = \cos 0 = 1)$$

$$= \omega^2(r_1 + r_2) \qquad \text{...(18.23)}$$

(ii) The roller is in contact with the nose

Fig 18.35 shows a tangent cam in which the roller has contact with the nose. When the roller is in the lowest position, the centre of roller lies at A and roller has contact with the cam at point E. Let the cam has

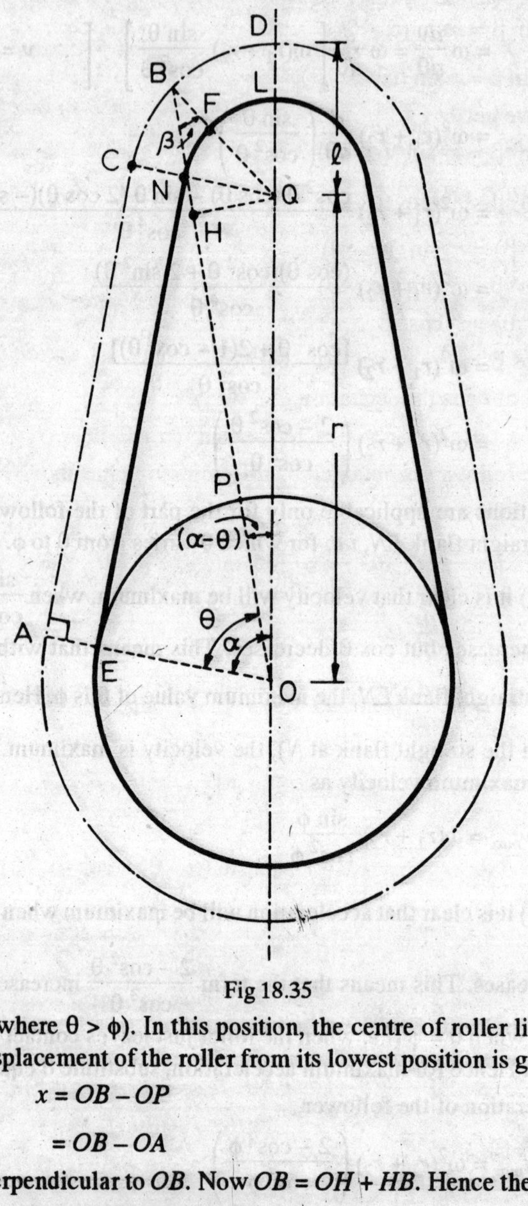

Fig 18.35

turned through an angle θ (where $\theta > \phi$). In this position, the centre of roller lies at B and roller has contact with cam at point F. The displacement of the roller from its lowest position is given by

$$x = OB - OP$$

$$= OB - OA \qquad (\because \; OP = OA)$$

From Q draw QH perpendicular to OB. Now $OB = OH + HB$. Hence the above equation becomes as

$$x = (OH + HB) - OA$$

But $OA = OE + EA = (r_1 + r_2)$; $OH = OQ \cos(\alpha - \theta) = r \cos(\alpha - \theta)$ $(\because \; r = OQ)$

and $HB = BQ \cos\beta = QD \cos\beta = l \cos\beta$ $(\because \; QD = l)$

Hence the displacement (x) becomes as

$$x = [r \cos(\alpha - \theta) + l \cos\beta] - (r_1 + r_2) \qquad \text{...(A)}$$

Now let us find the value of β in terms of angles α and θ.

From right angled triangles *QBH* and *OQH*, we have

$$QH = BQ \sin\beta = OQ \sin(\alpha-\theta)$$

or $$BQ \sin\beta = r \sin(\alpha-\theta)$$

or $$DQ \sin\beta = r \sin(\alpha-\theta) \qquad (\because\ BQ = DQ)$$

Squaring both sides, we get

$$(DQ)^2 \sin^2\beta = r^2 \sin^2(\alpha-\theta)$$

or $$l^2 \sin^2\beta = r^2 \sin^2(\alpha-\theta) \qquad (\because\ DQ = l)$$

or $$l^2 (1-\cos^2\beta) = r^2 \sin^2(\alpha-\theta) \qquad (\because\ \sin^2\beta = 1-\cos^2\beta)$$

or $$l^2 - l^2\cos^2\beta = r^2 \sin^2(\alpha-\theta)$$

or $$l^2 - r^2\sin^2(\alpha-\theta) = l^2\cos^2\beta$$

or $$l\cos\beta = [l^2 - r^2\sin^2(\alpha-\theta)]^{1/2}$$

Substituting the value of $l \cos\beta$ in equation (*A*), we get

$$x = r\cos(\alpha-\theta) + [l^2 - r^2\sin^2(\alpha-\theta)]^{1/2} - (r_1+r_2) \qquad \text{...(18.24)}$$

The velocity (*v*) of the follower is obtained by differentiating the above equation with respect to time.

Hence velocity of the follower,

$$v = \frac{dx}{dt} = \frac{dx}{d\theta} \times \frac{d\theta}{dt} = \frac{dx}{d\theta} \times \omega \qquad \left(\because\ \frac{d\theta}{dt} = \omega\right)$$

$$= \omega \frac{dx}{d\theta}$$

$$= \omega \frac{d}{d\theta}[r\cos(\alpha-\theta) + \{l^2 - r^2\sin^2(\alpha-\theta)\}^{1/2} - (r_1+r_2)]$$

$$= \omega \left[r\{-\sin(\alpha-\theta)\}(-1) + \frac{1}{2}[l^2 - r^2\sin^2(\alpha-\theta)]^{-1/2} \times [0 - r^2 \times 2\sin(\alpha-\theta) \times \cos(\alpha-\theta)(-1)\} - 0] \right]$$

$$\left[\because\ r_1 + r_2 \text{ is constant} \quad \therefore\ \frac{d}{d\theta}(r_1+r_2) = 0\right]$$

$$= \omega \left[r\sin(\alpha-\theta) + \frac{r^2 \times 2\sin(\alpha-\theta)\times\cos(\alpha-\theta)}{2\{l^2 - r^2\sin^2(\alpha-\theta)\}^{1/2}} \right]$$

$$= \omega \left[r\sin(\alpha-\theta) + \frac{r^2 \times \sin 2(\alpha-\theta)}{2\{l^2 - r^2\sin^2(\alpha-\theta)\}^{1/2}} \right]$$

$$= \omega \times r \left[\sin(\alpha-\theta) + \frac{r\sin 2(\alpha-\theta)}{2\{l^2 - r^2\sin^2(\alpha-\theta)\}^{3/2}} \right] \qquad \text{...(18.25)}$$

The acceleration (*f*) of the follower is obtained by differentiating the above equation with respect to time.

Hence acceleration of the follower,

$$f = \frac{dv}{dt} = \frac{dv}{d\theta} \times \frac{d\theta}{dt} = \frac{dv}{d\theta} \times \omega \qquad \left(\because\ \frac{d\theta}{dt} = \omega\right)$$

$$= \omega \times \frac{dv}{d\theta} = \omega \frac{d}{d\theta}\left[\omega \times r \sin(\alpha - \theta) + \frac{r \sin 2(\alpha - \theta)}{2\{l^2 - r^2 \sin^2(\alpha - \theta)\}^{1/2}}\right]$$

$$= \sin 2(\alpha - \theta) \times \frac{1}{2}\{l^2 - r^2 \sin^2(\alpha - \theta)\}^{-1/2} \times \{0 - r^2 \times 2 \sin(\alpha - \theta) \times \cos(\alpha - \theta)(-1)\}$$

$$= \omega \times \omega \times r\left[\cos(\alpha - \theta)(-1) + \frac{\begin{array}{c}\{l^2 - r^2 \sin^2(\alpha - \theta)\}^{1/2}\{r \times \cos 2(\alpha - \theta) \times 2(-1) + r \sin 2(\alpha - \theta)\} \\ \times \frac{1}{2}\{l^2 - r^2 \sin^2(\alpha - \theta)\}^{-1/2}\{-r^2 \times 2 \sin(\alpha - \theta) \times \cos(\alpha - \theta)(-1)\}\end{array}}{2\{l^2 - r^2 \sin^2(\alpha - \theta)\}}\right]$$

$$= \omega^2 \times r\left[-\cos(\alpha - \theta) + \frac{\{l^2 - r^2 \sin^2(\alpha - \theta)\}^{1/2}\{-2r \cos 2(\alpha - \theta)\} - \dfrac{r \sin 2(\alpha - \theta)\{r^2 \sin 2(\alpha - \theta)\}}{2\{l^2 - r^2 \sin^2(\alpha - \theta)\}^{1/2}}}{2\{l^2 - r^2 \sin^2(\alpha - \theta)\}}\right]$$

$$= \omega^2 \times r\left[-\cos(\alpha - \theta) + \frac{\{l^2 - r^2 \sin^2(\alpha - \theta)\}\{-2r \cos 2(\alpha - \theta)\} - \dfrac{r^3}{2} \sin 2(\alpha - \theta) \times \sin 2(\alpha - \theta)}{2\{l^2 - r^2 \sin^2(\alpha - \theta)\}^{3/2}}\right]$$

$$= \omega^2 \times r\left[-\cos(\alpha - \theta) + \frac{-l^2 \times 2r \cos 2(\alpha - \theta) + 2r^3 \sin^2(\alpha - \theta) \cos 2(\alpha - \theta) - \dfrac{r^3}{2}\{2 \sin(\alpha - \theta) \cos(\alpha - \theta)\}^2}{2\{l^2 - r^2 \sin^2(\alpha - \theta)\}^{3/2}}\right]$$

$$\{\because \sin 2(\alpha - \theta) = 2 \sin(\alpha - \theta) \times \cos(\alpha - \theta)\}$$

$$= +\omega^2 r\left[-\cos(\alpha - \theta) + \frac{-l^2 \times 2r \times \cos 2(\alpha - \theta) + 2r^3 \sin^2(\alpha - \theta)\{1 - 2\sin^2(\alpha - \theta)\} - 2r^3 \sin^2(\alpha - \theta) \cos^2(\alpha - \theta)}{2\{l^2 - r^2 \sin^2(\alpha - \theta)\}^{3/2}}\right]$$

$$\{\because \cos 2(\alpha - \theta) = 1 - 2\sin^2(\alpha - \theta)\}$$

$$= +\omega^2 r\left[-\cos(\alpha - \theta) + \frac{-2l^2 r \cos 2(\alpha - \theta) + 2r^3 \sin^2(\alpha - \theta) - 4r^3 \sin^4(\alpha - \theta) - 2r^3 \sin^2(\alpha - \theta)\{1 - \sin^2(\alpha - \theta)\}}{2\{l^2 - r^2 \sin^2(\alpha - \theta)\}^{3/2}}\right]$$

$$\{\because \cos^2(\alpha - \theta) = 1 - \sin^2(\alpha - \theta)\}$$

$$= +\omega^2 r\left[-\cos(\alpha - \theta) + \frac{\begin{array}{c}-2l^2 r \cos 2(\alpha - \theta) + 2r^3 \sin^2(\alpha - \theta) - 4r^3 \sin^4(\alpha - \theta) \\ - 2r^3 \sin^2(\alpha - \theta) + 2r^3 \times \sin^4(\alpha - \theta)\end{array}}{2\{l^2 - r^2 \sin^2(\alpha - \theta)\}^{3/2}}\right]$$

$$= +\omega^2 r\left[-\cos(\alpha - \theta) + \frac{-2l^2 \times r \cos 2(\alpha - \theta) - 2r^3 \sin^4(\alpha - \theta)}{2\{l^2 - r^2 \sin^2(\alpha - \theta)\}^{3/2}}\right]$$

$$= +\omega^2 \times r\left[-\cos(\alpha - \theta) - \frac{l^2 \times r \cos 2(\alpha - \theta) + r^3 \sin^4(\alpha - \theta)}{l\{l^2 - r^2 \sin^2(\alpha - \theta)\}^{3/2}}\right]$$

$$= -\omega^2 \times r\left[\cos(\alpha - \theta) + \frac{l^2 \times r \cos 2(\alpha - \theta) + r^3 \sin^4(\alpha - \theta)}{2\{l^2 - r^2 \sin^2(\alpha - \theta)\}^{3/2}}\right] \quad \text{...(18.26)}$$

In the equations of displacement, velocity and acceleration, the value of θ varies from φ to α. When the value of θ is equal to φ, then the roller has contact where the straight flank merges into a nose *i.e.* at point *N*. When the value of θ = α, then the roller has contact with the apex of the nose *i.e.* at point *L*.

When $\theta = \alpha$ (*i.e.* when the roller has control with the apex of the nose), then velocity and acceleration of the follower become equal to

$$v = \omega \times r \left[\sin(\alpha - \alpha) + \frac{r \sin 2(\alpha - \alpha)}{2\{l^2 - r^2 \sin^2(\alpha - \alpha)\}^{3/2}} \right]$$

$$= 0$$

and
$$f = -\omega^2 \times r \left[\cos(\alpha - \alpha) + \frac{l^2 \times r \cos 2(\alpha - \alpha) + r^3 \sin^4(\alpha - \alpha)}{\{l^2 - r^2 \sin^2(\alpha - \alpha)\}^{3/2}} \right]$$

$$= -\omega^2 \times r \left[1 + \frac{l^2 \times r + r^3 \times 0}{\{l^2 - r^2 \times 0\}^{3/2}} \right]$$

$$= -\omega^2 r \left[1 + \frac{l^2 \times r}{l^3} \right] = -\omega^2 r \left[1 + \frac{r}{l} \right] \quad ...(18.27)$$

Problem 18.13. *A symmetrical tangent cam with a least radius of 25 mm operates a roller follower of radius 10 mm. The angle of ascent is 60° and total lift is 15 mm. If the speed of the cam is 400 r.p.m., then calculate :*

(i) the principal dimensions of the cam i.e. the distance between the cam centre and nose centre ; nose radius and angle of control of cam with straight flank and

(ii) the acceleration of the follower at the beginning of the lift, where the roller just touches the nose and at the apex of the circular nose.

Assume that there is no dwell between ascent and decent.

Sol. Given :

$r_1 = 25$ mm ; $r_2 = 10$ mm ; $\alpha = 60°$; total lift = 15 mm ; $N = 400$ r.p.m.

or
$$\omega = \frac{2\pi N}{60} = \frac{2\pi \times 400}{60} = 41.88 \text{ rad/s.}$$

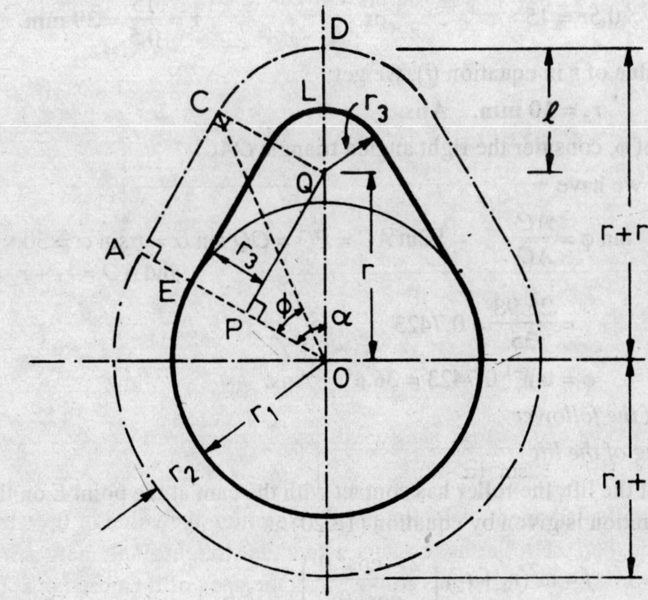

Fig. 18.36

(i) Principle dimensions of the cam

Let r = Distance between cam centre and nose centre

r_3 = Nose radius

ϕ = Angle of contact of cam with straight flanks.

The lift or displacement (x) of the cam is given by equation (18.24) as

$$x = r\cos(\alpha - \theta) + \{l^2 - r^2\sin^2(\alpha - \theta)\}^{1/2} - (r_1 + r_2)$$

For total lift (or for maximum displacement) $\theta = \alpha$. Hence above equation becomes as

Total lift $(x) = r\cos(\alpha - \alpha) + \{l^2 - r^2\sin^2(\alpha - \alpha)\}^{1/2} - (r_1 + r_2)$

or Total lift $= r + l - (r_1 + r_2)$

$= r + (r_2 + r_3) - (r_1 + r_2)$ $(\because\ l = r_2 + r_3)$

$= r + r_3 - r_1$

or $15 = r + r_3 - 25$ $(\because\ r_1 = 25)$

or $r + r_3 = 15 + 25 = 40$ ...(*i*)

In the above equation both r and r_3 are unknown. Let us make another equation in terms of r and r_3. Refer to Fig. 18.36. From Q, draw QP perpendicular on OE.

Now $OE = OP + PE$

$= (r\cos\alpha) + r_3$

($\because$ In triangle OPQ, $OP = OQ\cos\alpha = r\cos 60°$ and $PE = r_3$)

or $r_1 = r\cos 60° + r_3$ $(\because\ OE = r_1 = 25 \text{ mm})$

or $25 = r \times \frac{1}{2} + r_3$

or $25 = 0.5r + r_3$ ...(*ii*)

Subtracting equation (*ii*) from equation (*i*), we get

$0.5r = 15$ or $r = \frac{15}{0.5} = \mathbf{30\ mm}$. **Ans.**

Substituting the value of r in equation (*i*), we get

$r_3 = \mathbf{10\ mm}$. **Ans.**

To find the value of ϕ, consider the right angled triangle OAC.

From this triangle, we have

$\tan\phi = \frac{AC}{AO}$ [But $AC = PQ = OQ\sin\alpha = r\sin\alpha = 30 \times \sin 60° = 25.98$ mm and $AO = r_1 + r_2 = 25 + 10 = 35$ mm]

$= \frac{25.98}{35} = 0.7423$

$\therefore$ $\phi = \tan^{-1} 0.7423 = \mathbf{36.6°}$. **Ans.**

(ii) Acceleration of the follower

(a) At the beginning of the lift

At the beginning of the lift, the roller has contact with the cam at the point E on the straight flank and value of $\theta = 0$. The acceleration is given by equation (18.20) as

$$f = \omega^2(r_1 + r_2)\left(\frac{2 - \cos^2\theta}{\cos^3\theta}\right)$$

$$= \omega(r_1 + r_2)\left(\frac{2 - \cos^2 0°}{\cos^3 0°}\right) \qquad (\because \ \theta = 0)$$

$$= \omega^2(r_1 + r_2)\left(\frac{2 - 1}{1}\right) = \omega^2 (r_1 + r_2)$$

$$= 41.88^2 (25 + 10) = 61387.7 \text{ mm/s}^2 = \mathbf{61.3877 \ m/s^2}. \quad \textbf{Ans.}$$

(b) Acceleration of the follower when the roller just touches the nose i.e. when $\theta = \phi = 36.6°$.

In this position the roller has contact with the cam at the point N. The point N can be considered on the straight flank EN or on the circular nose NL. When the point N lies on the straight flank EN, the acceleration is given by equation (18.20) as

$$f = \omega^2 (r_1 + r_2)\left(\frac{2 - \cos^2 \theta}{\cos^3 \theta}\right)$$

$$= \omega^2 (r_1 + r_2)\left(\frac{2 - \cos^2 \phi}{\cos^3 \phi}\right) \qquad (\because \ \theta = \phi = 36.6°)$$

$$= 41.88^2 (25 + 10)\left(\frac{2 - \cos^2 36.6°}{\cos^3 36.6°}\right) = 41.88^2 \times 35 \times \left(\frac{2 - 0.6445}{0.5174}\right)$$

$$= 160825 \text{ mm/s}^2 = \mathbf{160.825 \ m/s^2}. \quad \textbf{Ans.}$$

When the point N lies on the circular nose, the acceleration is given by equation (18.26) as

$$f = -\omega^2 r\left[\cos(\alpha - \theta) + \frac{l^2 \times r \cos 2(\alpha - \theta) + r^3 \sin^4 (\alpha - \theta)}{\{l^2 - r^2 \sin^2 (\alpha - \theta)\}^{3/2}}\right]$$

$$= -41.88^2 \times 30\left[\cos(60 - 36.6) + \frac{20^2 \times 30 \times \cos 2(60 - 36.6) + 30^3 \times \sin^4 (60 - 36.6)}{\{20^2 - 30^2 \sin^2 (60 - 36.6)\}^{3/2}}\right]$$

($\because$ When N lies on the circular nose, $\theta = \phi = 36.6°$. The other values are $r = 30$ mm ; $l = r_2 + r_3 = 10 + 10 = 20$ mm)

$$\therefore \quad f = -41.88^2 \times 30\left[\cos 23.4° + \frac{12000 \times \cos 46.8° + 27000 \times \sin^4 (23.4°)}{\{400 - 900 \times \sin^2 (23.4°)\}^{3/2}}\right]$$

$$= -41.88^2 \times 30\left[0.9177 + \frac{12000 \times 0.6845 + 27000 \times 0.3971^4}{\{400 - 900 \times 0.3971^2\}^{3/2}}\right]$$

$$= -52618\left[0.9177 + \frac{8214 + 671.37}{\{400 - 141.919\}^{3/2}}\right]$$

$$= -161053.3 \text{ mm/s}^2 = \mathbf{-161.053 \ m/s^2}. \quad \textbf{Ans.}$$

(c) Acceleration of the follower at the apex of the nose

When the follower is at the apex of the nose, *i.e.* when the follower is having contact with the cam at point L, $\alpha = 60°$. (In this position, the centre of the roller of the follower is at point D and contact of roller with cam is at point L). The acceleration is given by equation (18.27) as

$$f = -\omega^2 r\left[1 + \frac{r}{l}\right]$$

$$= -41.88^2 \times 30 \left[1 + \frac{30}{20} \right]$$

[$\because$ $r = 30$ mm and $l = r_2 + r_3 = 10 + 10 = 20$ mm]

$$= -41.88^2 \times 30 \times 2.5$$

$$= 131545 \text{ mm/s}^2 = \mathbf{131.545 \text{ m/s}^2}. \quad \textbf{Ans.}$$

HIGHLIGHTS

1. The cam rotates usually at constant speed and drives the follower.

2. The motion of the follower depends upon the shape of the cam.

3. The follower are of the following types :

(*i*) Knife-edge follower,
(*ii*) Roller follower,
(*iii*) Flat-faced follower,
} This classification is based on the shape of that part of the follower which is in contact with the cam.

(*iv*) Reciprocating follower,
(*v*) Oscillating follower,
} This classification is based on the motion of the follower.

(*vi*) In-line follower and
(*vii*) Off-set follower.
} This classification is based on the location of the axis of the follower.

4. A follower may have the following types of motions :

(*i*) Uniform motion or uniform velocity
(*ii*) Simple harmonic motion
(*iii*) Uniform acceleration and uniform retardation
(*iv*) Cycloidal motion, or
(*v*) Any type of desired motion.

5. For the uniform velocity of the follower, the displacement, velocity and acceleration (during outstroke and return stroke for any angle turned by cam) are given by

$$x = \frac{S}{\theta_0} \times \theta \qquad \textbf{or} \qquad x = \frac{S}{\theta_r} \times \theta$$

$$v = \frac{S}{\theta_0} \times \omega \qquad \qquad v = \frac{S}{\theta_r} \times \omega$$

and $$f = 0 \qquad \qquad f = 0$$

where θ_0 = Outstroke angle in radians, S = Lift or total displacement of follower,
ω = Angular velocity of cam, θ_r = Return-stroke angle,
θ = Any angle turned by cam during any time t.

6. For simple harmonic motion of the follower the displacement, velocity and acceleration of the follower during outstroke and return stroke (for any angle θ turned by cam) are given by,

$$x = \frac{S}{2}\left(1 - \cos\frac{\pi \times \theta}{\theta_0}\right) \qquad \textbf{or} \qquad x = \frac{S}{2}\left(1 - \cos\frac{\pi\theta}{\theta_r}\right)$$

$$v = \frac{S}{2} \times \frac{\pi \times \omega}{\theta_0} \times \sin\left(\frac{\pi\theta}{\theta_0}\right) \qquad v = \frac{S}{2} \times \frac{\pi \times \omega}{\theta_r} \times \sin\left(\frac{\pi\theta}{\theta_r}\right)$$

and $$f = \frac{S}{2} \times \left(\frac{\pi \times \omega}{\theta_0}\right)^2 \times \cos\left(\frac{\pi\theta}{\theta_0}\right) \qquad f = \frac{S}{2} \times \left(\frac{\pi \times \omega}{\theta_r}\right)^2 \times \cos\left(\frac{\pi\theta}{\theta_r}\right)$$

The velocity is maximum when $\theta = \frac{\theta_0}{2}$ or $\frac{\theta_r}{2}$. And acceleration is maximum when $\theta = 0$.

7. For uniform acceleration and uniform retardation of the follower, the displacement, velocity and acceleration during any stroke at any angle θ is given by

$$x = \frac{1}{2} f t^2 \text{ where } t = \frac{\theta}{\omega} \qquad (\because \ \theta = \omega \times t)$$

$$= \frac{1}{2} \times f \times \frac{\theta^2}{\omega^2}$$

$$(v)_{max} = \frac{2S \times \omega}{\theta_0} \quad \text{or} \quad \frac{2S \times \omega}{\theta_r}$$

$$(f)_{max} = \frac{4 \times S \times \omega^2}{\theta_0^2} \quad \text{or} \quad \frac{4S \times \omega^2}{\theta_r^2}.$$

8. For cycloidal motion of the follower, the displacement, velocity and acceleration at any time is given by

$$x = S\left[\frac{\theta}{\theta_0} - \frac{1}{2\pi}\sin\left(\frac{2\pi\theta}{\theta_0}\right)\right]$$

$$v = \frac{\omega \times S}{\theta_0}\left[1 - \cos\left(\frac{2\pi\theta}{\theta_0}\right)\right]$$

$$f = \frac{2\pi \times \omega^2 \times S}{\theta_0^2}\sin\left(\frac{2\pi\theta}{\theta_0}\right)$$

For return stroke, replace θ_0 by θ_r.

9. For a circular arc cam with flat-faced follower, the displacement velocity and acceleration are given by :

(i) When point of contact is on circular flank	*(ii) When point of contact is on the circular nose*
$x = (R - r_1)(1 - \cos\theta)$	$x = r_2 + QO\cos(\alpha - \theta) - r_1$
$v = \omega \times (R - r_1)\sin\theta$	$v = \omega \times QO\sin(\alpha - \theta)$
$f = \omega^2 \times (R - r_1)\cos\theta$	$f = -\omega^2 \times QO\cos(\alpha - \theta)$
$v_{max} = \omega \times (R - r_1)\sin\phi$	$v_{max} = \omega \times QO\sin\phi$
$f_{max} = \omega^2 \times (R - r_1)$	$f_{max} = -\omega^2 \times QO$ at the apex of nose.

where R = Radius of the circular flank

r_1 = Minimum radius of the cam

QO = Distance between cam centre and nose centre.

10. For a tangent cam with a roller-follower, the displacement, velocity and acceleration are given by :

(i) When the roller is in contact with the straight flank	*(ii) When the roller is in contact with the nose*
$x = (r_1 + r_2)\left(\frac{1 - \cos\theta}{\cos\theta}\right)$	$x = r\cos(\alpha - \theta) + [l^2 - r^2\sin^2(\alpha - \theta)]^{1/2} - (r_1 + r_2)$
$v = \omega(r_1 + r_2)\frac{\sin\theta}{\cos^2\theta}$	$v = \omega \times r\left[\sin(\alpha - \theta) + \frac{r\sin 2(\alpha - \theta)}{2\{l^2 - r^2\sin^2(\alpha - \theta)\}^{3/2}}\right]$
$f = \omega^2(r_1 + r_2)\left(\frac{2 - \cos^2\theta}{\cos^3\theta}\right)$	$f = -\omega^2 \times r\left[\cos(\alpha - \theta) + \frac{l^2 \times r\cos 2(\alpha - \theta) + r^3\sin^4(\alpha - \theta)}{\{l^2 - r^2\sin^2(\alpha - \theta)\}^{3/2}}\right]$
$v_{max} = \omega(r_1 + r_2)\frac{\sin\phi}{\cos^2\phi}$	At the apex of the nose, $v = 0$

$f_{max} = \omega^2 (r_1 + r_2) \left(\frac{2 - \cos^2 \phi}{\cos^3 \phi} \right)$	$f = -\omega^2 \times r \left[1 + \frac{r}{l} \right]$

where r_1 = Minimum radius of the cam, r_2 = Radius of the roller, r_3 = Radius of the nose

θ = Angle turned by the cam from the beginning of the roller displacement

ϕ = Angle turned by the cam for contact of the roller with the straight flank

α = Angle of ascent or semi-angle of action of the cam

ω = Angular velocity of the cam

$l = (r_2 + r_3)$

r = Distance between cam centre and nose centre.

EXERCISE 18

(A) THEORETICAL QUESTIONS

1. What do you understand by the terms cam and follower ? Name the essential members of a cam mechanism.
2. What are the different types of followers ? Classify them according to the motion of follower, location of axis of the follower and according to the shape of that part which is in contact with the cam.
3. Define and explain the terms : cam profile, base circle, prime circle, pitch curve, lift and period of dwell.
4. Differentiate between :
   (*i*) Base circle and prime circle
   (*ii*) Cam angle and pressure angle
   (*iii*) Pitch point and trace point
   (*iv*) Period of ascent and period of decent.
5. Name the different motions that a follower can have.
6. Draw the displacement, velocity and acceleration diagrams for a follower when it moves with uniform velocity. Also draw the modified displacement, velocity and acceleration diagrams. Why the modification in these diagrams are done ?
7. For a simple harmonic motion of the follower, prove that the maximum velocity and maximum acceleration during outstroke is given by

$$v_{max} = \frac{S}{2} \times \frac{\pi \times \omega}{\theta_0} \quad \text{and} \quad f_{max} = \frac{S}{2} \times \left(\frac{\pi \times \omega}{\theta_0} \right)^2.$$

where S = Stroke of the follower,

θ_0 = Angle turned by cam during outstroke, and

ω = Angular velocity of the cam.

8. Draw the displacement, velocity and acceleration diagrams for a follower when it moves with S.H.M. Derive the expression for velocity and acceleration during outstroke and return stroke of the follower.
(*A.M.I.E., Summer 1987*)
9. Draw the displacement, velocity and acceleration diagrams for a follower when it moves with uniform acceleration and uniform retardation. Derive expression for velocity and acceleration during outstroke and return stroke of the follower.
10. For a tangent cam with circular nose and roller follower prove that the velocity of the follower is given by :

$$v = \omega (r + r_1) \frac{\sin \theta}{\cos^2 \theta}$$

where r = Minimum radius of the cam, r_1 = Roller radius,

θ = any angle by which cam has rotated, ω = the uniform angular velocity of the cam.

(*A.M.I.E., Winter 1986*)

11. Draw the following :

(*i*) Tangent cam (*ii*) Convex circular arc cam. (*A.M.I.E., Summer 1986*)

12. What is the method of constructing the profile of a cam ?

(B) NUMERICAL PROBLEMS

1. Draw the profile of a cam operating a knife-edge follower when the axis of the follower passes through the axis of the cam shaft from the following data :

(*i*) Follower to move outwards through 30 mm during 90° of cam rotation,

(*ii*) Follower to dwell for the next 45°,

(*iii*) Follower to return to its original position during next 60°,

(*iv*) Follower to dwell for the rest of the cam rotation.

The displacement of the follower is to take place with simple harmonic motion during both the outward and the return strokes. The least radius of the cam is 50 mm. If the cam rotates at 600 r.p.m., determine the maximum velocity and acceleration of the follower during outward stroke and return stroke.

[**Ans.** 1.884 m/s ; 2.8275 m/s ; 236.6 m/s^2 ; 532.41 m/s^2]

2. If in Problem 1, the axis of the follower is not passing through the axis of the cam shaft but is off-set by 20 mm from the axis of the cam-shaft, then draw the profile of the cam. All other data remains the same.

3. Draw the profile of a cam operating a knife-edge follower (when the axis of the follower passes through the axis of the cam shaft) from the following data :

(*i*) follower to move outward through 22.5 mm with simple harmonic motion during 90° of cam rotation

(*ii*) follower to dwell for the next 60°,

(*iii*) follower to return to its original position with uniform velocity during 120° of cam rotation,

(*iv*) follower to dwell for the rest of the cam rotation.

The least radius of cam is 30 mm and cam rotates at 360 r.p.m. Determine :

(*i*) maximum velocity and maximum acceleration of the follower during outstroke, and

(*ii*) maximum velocity and maximum acceleration of the follower during return stroke.

[**Ans.** (*i*) 848.47 mm/s ; 64 m/s^2, (*ii*) 405.08 mm/s ; 0]

4. Draw the profile of a cam operating a knife-edge follower from the following data :

(*i*) follower to move outward through a distance of 30 mm during 90° of the cam rotation,

(*ii*) follower to dwell for the next 60° of cam rotation,

(*iii*) follower to return to its original position during 120° of cam rotation and

(*iv*) follower to dwell for the remaining 90° of cam rotation.

The cam is rotating clockwise at a uniform speed of 375 r.p.m. The minimum radius of the cam is 40 mm and the line of stroke of the follower is offset 15 mm from the axis of the cam and the displacement of the follower is to take place with uniform and equal acceleration and retardation on both the outward and the return strokes. Determine :

(*i*) maximum velocity of the follower during outstroke and return strokes

(*ii*) maximum acceleration during outstroke and return stroke.

[**Ans.** (*i*) 1.5 m/s ; 1.125 m/s, (*ii*) 75 m/s^2 ; 42.187 m/s^2]

5. A cam with 40 mm as minimum diameter is rotating clockwise at a uniform speed of 900 r.p.m. and has to give the following motion to a roller-follower 10 mm in diameter :

(*i*) follower to complete outward stroke of 30 mm during 90° of cam rotation with equal uniform acceleration and retardation,

(*ii*) follower to dwell for 60° of cam rotation,

(*iii*) follower to return to its initial position during 120° of cam rotation with equal uniform acceleration and retardation, (*iv*) follower to dwell for the remaining 90° of the cam rotation.

Draw the cam profile if the axis of the roller-follower passes through the axis of the cam. Determine the maximum velocity of the follower during the outstroke and return stroke and also the uniform acceleration of the follower on the outstroke and return stroke. **[Ans.** 3.595 m/s ; 4.795 m/s ; 431.75 m/s^2 ; 136.55 m/s^2]

6. From the following data, draw the profile of a cam in which the follower moves with simple harmonic motion (S.H.M.) during ascent while it moves with uniformly accelerated and decelerated motion during descent :

Least radius of cam = 60 mm ; angle of ascent = 72° ; angle of dwell between ascent and descent = 42° ; angle of descent = 90° ; the lift of the follower = 60 mm ; diametef of roller = 40 mm ; distance between line of action of follower and axis of cam = 30 mm.

If the cam rotates at 540 r.p.m. anti-clockwise find the maximum velocity and acceleration of the follower during ascent. **[Ans.** 4.32 m/s ; 311 m/s^2]

7. A cam rotating clockwise with a uniform speed is to give the roller-follower of 20 mm diameter the following motion :

(*i*) Follower to move outward through a distance of 45 mm during 90° of cam rotation

(*ii*) Follower to dwell for 60° of cam rotation

(*iii*) Follower to return to its original position during 120° of cam rotation

(*iv*) Follower to dwell for the remaining 90° of cam rotation.

The minimum radius of the cam is 50 mm and the line of stroke of the follower is offset by 20 mm from the axis of the cam and the displacement of the follower is to take place with simple harmonic motion on both the outward and return stroke. Draw the cam profile.

8. Draw the profile of a cam to raise a value with harmonic motion through 40 mm in $\frac{1}{4}$ th of revolution, keep it fully raised through $\frac{1}{10}$ th revolution and to lower it with harmonic motion in $\frac{1}{6}$ th revolution. The valve remains closed during the rest of the revolution. The diameter of the roller is 20 mm and minimum radius of the cam to be 30 mm. The diameter of the cam shaft is 30 mm. The axis of the valve rod passes through the axis of the cam shaft. Assume the cam shaft to rotate with a uniform velocity.

9. Draw the profile of a cam to give the following motion to the reciprocating follower with a flat (or mushroom) contact face :

(*i*) follower to move outwards through a distance of 30 mm during 90° of cam rotation,

(*ii*) follower to dwell for 60° of cam rotation,

(*iii*) follower to return to its initial position during 90° of cam rotation,

(*iv*) follower to dwell for the remaining 120° of cam rotation.

The minimum radius of the cam is 30 mm and the flat face of the follower is at right angles to the line of stroke of the follower. The outward and return strokes of the follower are to take place with S.H.M.

10. A flat faced reciprocating follower has the following motion :

(*i*) the follower moves outward for 90° of cam rotation with uniform acceleration and retardation, acceleration being twice the retardation,

(*ii*) the follower dwell for the next 60° of cam rotation,

(*iii*) the follower moves to the initial position for the next 120° of cam rotation with uniform acceleration and retardation, acceleration being twice the retardation,

(*iv*) the follower dwells for the remaining period.

The base circle diameter of the cam is 80 mm and stroke of the follower is 30 mm. The line of movement of the follower passes through the cam centre. Draw the displacement diagram.

11. The following data is related to a symmetrical circular arc cam operating a flat-faced follower :
Least radius of the cam = 30 mm ; total lift = 20 mm ; angle of lift = 75°, nose radius = 5 mm ;
speed of cam = 300 r.p.m. Find :

(*i*) Distance between cam centre and nose centre ;

(*ii*) Radius of circular flank ; and

(*iii*) Angle of contact on the circular flank. [**Ans.** (*i*) 45 mm ; (*ii*) 82.4 mm and (*iii*) 34.2°]

12. For the problem 11, find the acceleration of the follower at the beginning of the lift, at the end of contact with the circular flank, at the beginning of contact with nose and at the apex of the nose.

[**Ans.** 51.73 m/s^2 ; 42.78 m/s^2 ; – 33.5 m/s^2 and – 44.4 m/s^2]

13. For a symmetrical tangent cam operating a roller follower, find :

(*i*) the distance between cam centre and nose centre

(*ii*) nose radius

(*ii*) angle of contact of cam with straight flank

(*iii*) the acceleration of the follower at the beginning of the lift, where the roller just touches the nose and at the apex of the circular nose if

the least radius of cam = 30 mm ; roller radius = 17.5 mm ; angle of ascent = 75° and total lift = 17.5 mm. The speed of the cam shaft is 300 r.p.m. Assume that there is no dwell between ascent and decent.

[**Ans.** (*i*) 23.6 mm, (*ii*) 23.88 mm, (*iii*) 25.6°, (*iv*) 46.9 m/s^2, 75.95 m/s^2, 36.63 m/s^2]

19

Computer-Aided Analysis of Mechanisms

19.1. Introduction

The displacement, velocity and acceleration of a point in a mechanism or link can be determined by graphical method for very few positions say one or two of the crank. Moreover the accuracy of these methods is limited. Also if these quantities are required for more positions or for their several configuration diagrams are required, it will be very convenient to use desk calculator or a digital computer. In this chapter, expressions for displacement, velocity and acceleration shall be derived in terms of general parameters of the mechanisms. It facilitates forming a computer program and thereby desired quantities can be determined by using computer.

19.2. Coupler Curve

The path traced by a point on the coupler link for different positions of a mechanism linkages, is known as the coupler curve. Or a coupler curve is the locus of a point on the coupler link of a mechanism. This point is called as *coupler point*. The coupler curve for a mechanism can be drawn if the co-ordinates of the coupler point are known in terms of the lengths of the links of the mechanism and angle of input link. The coupler curve for the following mechanisms will be discussed :

(*i*) Coupler curve for slider-crank mechanism, and

(*ii*) Coupler curve for a four-bar mechanism.

19.3. Coupler Curve for Slider Crank Chain

Coupler curve is the locus of a point on the coupler. Fig. 19.1 shows a slider crank mechanism *OAB* in which $OA = q$ is a crank or input link and $AB = s$ is a coupler link (or connecting rod link). It represents a very general case where the slider motion has eccentricity with respect to centre of rotation (*i.e.*, point *O*) of crank *OA*. The point *D* on the coupler has also been considered away from coupler link *AB*. The coupler is represented by *ABD*.

Drop *AE* perpendicular to the *X*-axis. The line of stroke of the slider meets *AE* at *F*. Let $OA = q$, $AB = s$, $AD = f$ and eccentricity $= t$. Assuming the point D as (X_D, Y_D) whose locus is to be obtained. In Fig. 19.1, *AC* is parallel to *X*-axis and *DC* is perpendicular to *AC*.

Let ϕ = Angle made by *AD* with *AB*

ψ = Angle made by *AB* with *AC*

Then $(\phi - \psi)$ = Angle made by *AD* with *AC*.

From Fig. 19.1, we have

$$X_D = OE + AC = q \cos \theta + f \cos (\phi - \psi) \quad \text{...(19.1)}$$

and

$$Y_D = AE + CD = q \sin \theta + f \sin (\phi - \psi) \quad \text{...(19.2)}$$

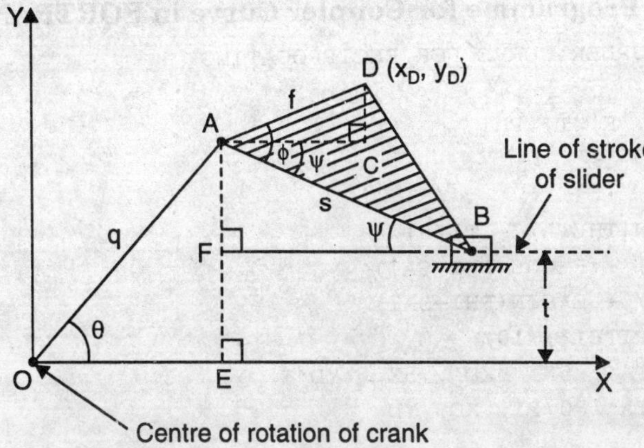

Fig. 19.1

In the figure ψ is negative and it can be determined as follows :

$$\sin\psi = \frac{AF}{AB} = \frac{AE - EF}{AB} = \frac{q\sin\theta - t}{s}$$

or $$\psi = \sin^{-1}\left\{\frac{q\sin\theta - t}{s}\right\} \qquad ...(19.3)$$

In equations (19.1) and (19.2), the values of q, f, θ and ϕ are given. From equation (19.3), the value of ψ can be determined as the values of q, θ, s and t are known.

The above expressions can be programmed and calculations can be done by the computer.

Input data : The lengths of the links and angle φ.

Here : The lengths of the links in the programme are represented by :

$Q = q = OA$ = crank

$S = s = AB$ = connecting rod

$F = f = AD$

$T = t$ = Eccentricity = EF

$PHI = \phi$

Also we know $\tan 45° = \tan\left(\frac{\pi}{4}\right) = 1.0$

$$\therefore \qquad \tan^{-1}(1.0) = \frac{\pi}{4}$$

or $$\pi = 4\tan^{-1}(1.0)$$

If $\pi = PI$ and $\tan^{-1} = ATAN$

Then $PI = 4 * ATAN(1.0)$

Output data : The co-ordinates of the point D with the variation of angle θ. The co-ordinates of point D in the program are represented by XD and YD. Hence output data is: θ, XD and YD.

The above expressions can be programmed and calculations can be done by the computer.

19.3.1. Computer Programme for Coupler Curve in FORTRAN

```
C   PROGRAMME FOR COUPLER CURVE FOR RECIPROCATING
C   ENGINE MECHANISM
    READ (*, *) Q, S, F, T, PHI
    PI = 4*ATAN(1.0)
    THETA = 0.0
8   SAI = ASIN((Q*SIN(THETA) - T)/S)
    XD = Q*COS(THETA) + F*COS(PHI-SAI)
    YD = Q*SIN(THETA) + F*SIN(PHI-SAI)
    IF(THETA.EQ.0) WRITE(*, 10)
10  FORMAT (4X, 'THETA', 5X, 'XD', 5X, 'YD')
    WRITE(*, 20) THETA*180/PI, XD, YD
20  FORMAT(3F 10.3)
    THETA = THETA + PI/6
    THP = 2*PI-PI/6
    IF(THETA.GT.THP) GO TO 30
    GO TO 8
30  STOP
    END
```

19.3.2. Computer Programme for Coupler Curve for Reciprocating Engine Mechanism (or Slider Crank Mechanism) in C++

Programme for coupler curve for reciprocating engine mechanism

```
#include<iostream.h>
#include<conio.h>
#include<math.h>
void main()
{
cout<<"ALL DISTANCE MEASUREMENTS ARE IN CENTIMETRES"<<endl,
float Q=48, S=160, F=40, T=10, PHI=(3.14/180)*30, THETA=0.0;
double long SAI,XD,YD;
float PI=4*atan(1.0);
for(int i=0;i<=12;i++)
{
SAI=asin((Q*sin(THETA)-T)/S);
XD=Q*cos(THETA)+F*cos(PHI-SAI);
YD=Q*sin(THETA)+F*sin(PHI-SAI);
cout<<"THETA=\t"<<THETA<<endl<<"XD=\t"<<XD<<endl<<"YD=\t"<<YD<<en
dl<<endl;
THETA=(THETA+PI/6);
}
getch();
}
```

Input variables are :

Q = Length of link OA (*i.e.*, length of crank)

S = Length of link AB (*i.e.* length of connecting rod)

T = Magnitude of eccentricity

F = Length of AD

PHI = Angle BAD

Output variables are :

THETA = Angle made by link AO with X-axis

XD = X-coordinate of the point D

YD = Y-coordinate of the point D

Problem 19.1. Draw a coupler curve of coupler point D of the slider-crank mechanism OAB shown in Fig. 19.1 in which lengths of link OA = 48 cm, link AB = 160 cm and eccentricity t = 10 cm. The location of coupler point D is given by AD = 40 cm and $\angle BAD = 30°$.

Sol. Given : (Refer to Fig. 19.1)

Length of link OA = 48 cm ;

Length of link AB = 160 cm ;

Eccentricity t = 10 cm ;

Length AD = 40 cm and

$\angle BAD = 30°$

The programme shown in Art. 19.3.2 was used for the given data. The input and the output is :

Input Data

$Q = q = 48$ cm ; $S = s = 160$ cm ; $F = f = 40$ cm ;

$$T = t = 10 \text{ cm}, PHI = \phi = 30° = \frac{30 \times \pi}{180} = \frac{30 \times 3.14}{180} \text{ radians.}$$

Output Data

THETA (θ)	*XD in cm*	*YD in cm*
0°	81.3292	22.1171
30°	77.8319	40.8826
60°	61.9096	54.3311
90°	38.4028	59.1903
120°	13.9096	54.3311
150°	– 5.3066	40.8826
180°	– 14.6708	22.1171
210°	– 11.9622	2.8966
240°	2.3542	– 11.4784
270°	25.0431	– 16.8096
300°	50.3542	– 11.4785
330°	71.1672	2.8966

After knowing the co-ordinates of the coupler point D for various values of theta (θ) (*i.e.*, after knowing the values of X_D and Y_D) these points are plotted. The plotted points are joined by a smooth curve. This curve is known as coupler curve as shown in Fig. 19.1 (*a*).

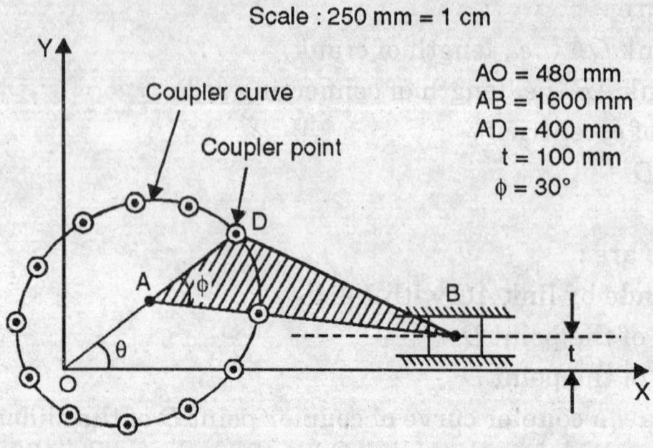

Fig. 19.1 (*a*)

19.4. Coupler Curve for Four Bar Kinematic Chain

As stated in the earlier section, coupler curve is the locus of a point on the coupler. Fig. 19.2 shows a four bar kinematic chain. $O_1 ABO_2$. Let $O_1 A$ act as a crank. The coupler is

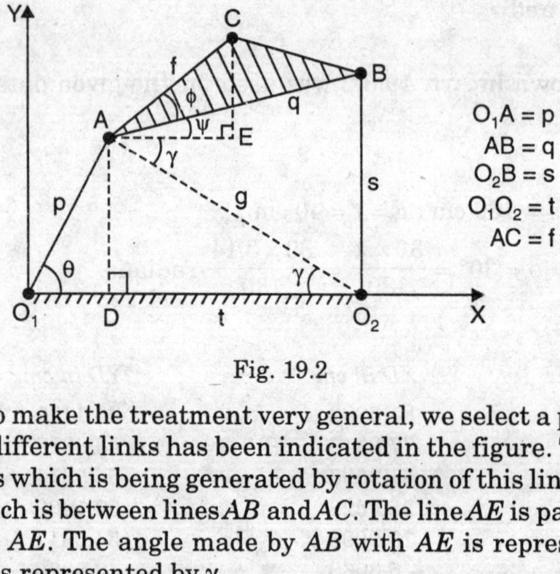

Fig. 19.2

represented by *ABC*. To make the treatment very general, we select a point *C* which is outside line *AB*. The length of different links has been indicated in the figure. The angle θ is the angle of the crank with X-axis which is being generated by rotation of this link. Angle φ is the built in angle of the coupler which is between lines *AB* and *AC*. The line *AE* is parallel to *X*-axis whereas *CE* is perpendicular to *AE*. The angle made by *AB* with *AE* is represented by ψ. The angle made by AO_2 with *AE* is represented by γ.

Let X_C and Y_C be the coordinates of the point *C* at any instant. Then

∴ φ = Angle between *AB* and *AC*

ψ = Angle between *AB* and *AE*

γ = Angle between AO_2 and *AE*

$p = O_1A$; $f = AC$; $q = AB$;

$g = O_2A$, $t = O_1O_2$

$$X_C = O_1D + AE = O_1A \cos\theta + AC \cos(\phi + \psi) = p \cos\theta + f \cos(\phi + \psi) \quad \text{...(19.4)}$$

$$Y_C = AD + CE = O_1A \sin\theta + AC \sin(\phi + \psi) = p \sin\theta + f \sin(\phi + \psi) \quad \text{...(19.5)}$$

In equations (19.4) and (19.5) p, θ, f, ϕ are known. Angle ψ is to be determined for using these equations. Applying cosine law in ΔAO_2B

$$s^2 = q^2 + g^2 - 2qg \cos(\psi + \gamma) \qquad ...(19.6)$$

or
$$\cos(\psi + \gamma) = \frac{(q^2 + g^2 - s^2)}{2qg}$$

or
$$(\psi + \gamma) = \cos^{-1}\left\{\frac{q^2 + g^2 - s^2}{2qg}\right\} \qquad ...(19.7)$$

From ΔAO_2D,
$$\tan\gamma = \frac{AD}{O_2D} = \frac{p\sin\theta}{t - p\cos\theta} \quad \text{or} \quad \gamma = \tan^{-1}\left\{\frac{p\sin\theta}{t - p\cos\theta}\right\} \qquad ...(19.8)$$

Applying cosine law in ΔO_1AO_2

$$g^2 = p^2 + t^2 - 2pt\cos\theta \qquad ...(19.9)$$

The value of g can be determined by equation (19.9) and angle γ can be determined by equation (19.8). Therefore

$$\psi = \cos^{-1}\left\{\frac{q^2 + g^2 - s^2}{2qg}\right\} - \gamma \qquad ...(19.10)$$

Angle ψ can be determined from equation (19.10). For different values of angle θ, values of X_C, Y_C can be determined and they can be plotted to get coupler curve for point C.

The above equations (19.4), (19.5), (19.8), (19.9) and (19.10) can be programmed and calculations can be made by the computer.

Input Data

The lengths of the links and angle ϕ. The lengths of the links in the programme are represented by *P, Q, S, F, T* and angle ϕ by *PHI*. Hence $P = p$, $Q = q$, $S = s$, $F = f$, $T = t$ and $PHI = \phi$ in the programme.

Output Data

The co-ordinates of the point C with the variation of angle θ. The co-ordinates in the programme are represented by *XC* and *YC*.

19.4.1. Computer Programme for Coupler Curve in FORTRAN

```
C    PROGRAMME FOR COUPLER CURVE FOR FOUR
C    BAR KINEMATIC CHAIN
     READ (*, *) P, Q, S, F, T, PHI
     PI = 4*ATAN (1.0)
     THETA = 0.0
8    GAMA = ATANP*SIN(THETA)/(T-P*COS(THETA))
     SQG = P**2 + T**2 - 2.0*P*T*COS(THETA)
     G = SQRT(SQG)
     SAI = ACOS(Q**2 + G**2 - S**2)/(2.0*Q*G) - GAMA
     XC = P*COS(THETA) + F*COS(PHI + SAI)
     YC = P*SIN(THETA) + F*SIN(PHI + SAI)
     IF(THETA . EQ . 0) WRITE(*, 10)
10   FORMAT (5X, THETA, 5X, 'XC', 5X, 'YC')
     WRITE (*, 20) THETA*180/PI, XC, YC
```

```
20 FORMAT(3F10.3)
   THETA = THETA*PI/6
   TH = 2*PI-PI/6
   IF(THETA.GT.TH) GO TO 30
   GO TO 8
30 STOP
   END
```

19.4.2. Computer Programme for Coupler Curve for Four Bar Kinematic Chain in C++

Programme for coupler curve for four bar kinematics chain

```
#include<iostream.h>
#include<math.h>
#include<iomanip.h>
void main()
{
cout<<"ALL DISTANCE MEASUREMENTS ARE IN CENTIMETRES"<<endl;
float P=40,Q=55, S=81, F=22, T=90, PHI=(3.14/180)*40,THETA=0.0;
double long SAI,XC,YC;
float PI=4*atan(1.0);
flaot gamma,SQG,G;
for(int i=0;i<=12;i++)
{
gamma=atan(P*sin(THETA)/(T-P*cos(THETA)));
SQG=(P*P)+(T*T)-(2.0*P*T*cos(THETA));
G=sqrt(SQG);
SAI=(a cos((Q*Q+G*G-S*S)/(2.0*Q*C))-gamma);
XC=P*cos(THETA)+F*cos(PHI+SAI);
YC=P*sin(THETA)+F*sin(PHI+SAI);
cout<<"theta=\t"<<(THETA*180/PI)<<endl<<"XC=\t"<<XC<<endl<<"YC=\t"<<YC
<<endl<<endl;
THETA=(THETA+PI/6);
}
}
```

Problem 19.2. Draw a coupler curve of coupler point C of the four bar chain O_1ABO_2 shown in Fig. 19.2 in which lengths of link O_1A = 40 cm, AB = 55 cm, BO_2 = 81 cm and O_1O_2 = 90 cm. The location of coupler point C is given by AC = 22 cm and $\angle BAC$ = 40°.

Sol. Given. (Refer to Fig. 19.2)

Length of links O_1A = 40 cm ; AB = 55 cm ;

BO_2 = 81 cm ; O_1O_2 = 90 cm ; AC = 22 cm and ϕ = 40°.

The computer programme shown in Art. 19.4.2 was used for the given data. The input and output data is :

Input Data

P = Link O_1A = 40 cms

Q = Link AB = 55 cms

$S = \text{Link } BO_2 = 81$ cms

$T = \text{Link } O_1O_2 = 90$ cms

$F = AC = 22$ cms

$$PHI = \phi = 40° = \frac{40 \times 3.14}{180} \text{ radians}$$

Output Data

THETA (θ)	X_C	Y_C
0	22.9422	13.8936
30	26.8476	40.5733
60	21.4250	56.5948
90	7.0392	60.8435
120	– 9.7308	54.0972
150	– 22.8585	38.5788
180	– 29.2754	19.2090
210	– 29.2628	1.3325
240	– 22.7325	– 12.8114
270	– 10.7545	– 20.8078
300	3.4102	– 20.1918
330	15.4979	– 9.1598

After calculating the co-ordinates of the coupler point C for various values of theta (θ) (*i.e.*, after calculating the values of X_C and Y_C) these points are plotted. These plotted points are joined by a smooth curve as shown in Fig. 19.2 (*a*).

This curve is known as coupler curve.

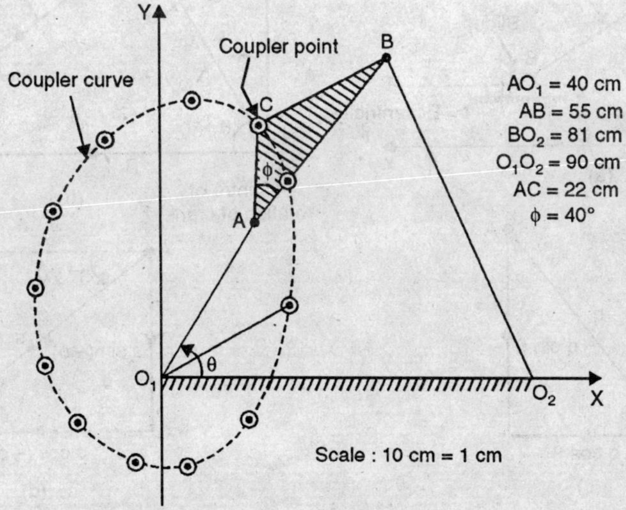

Fig. 19.2 (*a*)

19.5. Computer-Aided Analysis of Mechanism

Analytical methods have high degree of accuracy. They can be programmed and calculations can be done by using a computer. In this section, we shall discuss a method which can be easily used for simple mechanisms of undergraduate level :

A link can be represented by a vector. These vectors are resolved along X- and Y-axes. The displacement equations can be obtained by writing equations about X- and Y-axes. The vectors representing links form closed loops in a mechanism. It is similar to the case of triangle of forces and a closed polygon of forces. Since they do not have any resultant, therefore, closed loop of vectors also represent case of no resultant. Therefore the net components along X- and Y-axes shall be zero for a closed loop of vectors.

The computer-aided analysis of the following mechanisms are considered :

(*i*) Slider-Crank Mechanism, and

(*ii*) Four-bar Mechanism.

19.6. Computer-Aided Analysis of Slider Crank Mechanism

A slider crank mechanism is shown in Fig. 19.3 (*a*). There are two links in this mechanics which are crank OA and connecting rod AB. The slider is offset with eccentricity 't'. Fig. 19.3 (*b*) shows these links as vectors. Also in this Fig. 19. 3 (*b*) BG and OG are shown as vectors. Hence BG and OG can be considered as links. The length of link OG is not constant. With the movement of slider, this length varies. All other lengths are constant. Let $OG = p$. Hence p is the horizontal distance of slider from Y-axis. This represents a general case when axis of motion of slider is offset with respect to the centre of rotation of the crank. The angle of connecting rod ϕ is in the negative* sense as shown in the figure. The expressions for displacement, velocity and acceleration are derived as given in the next articles.

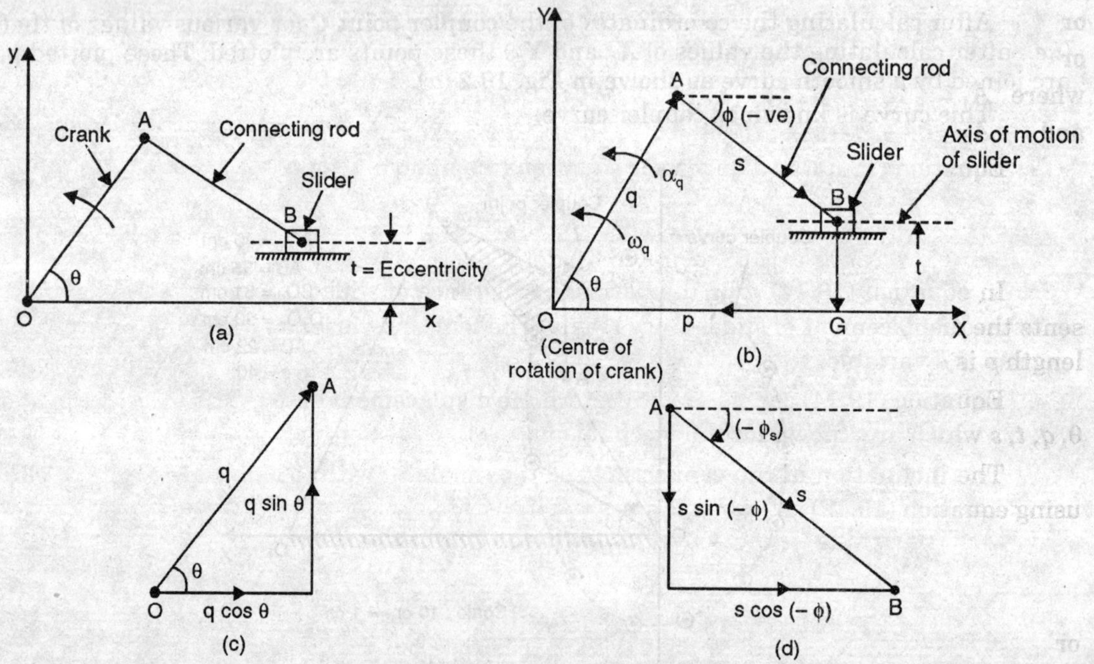

Fig. 19.3. Slider crank mechanism.

---

*If the angle made by the link with x-axis is anticlockwise, it is considered positive. But if the angle is clockwise with X-axis, it is considered negative. Hence angle θ is taken +ve whereas angle ϕ is taken (–ve).

Let $AO = q$

$BA = s$

$GB = t$

$GO = p$

The angle made by GO with x-axis is zero whereas angle made by GB is 90°.

19.6.1. Expression for displacement

For a closed loop in a mechanism, the sum of components along X-axis and Y-axis must be equal to zero. In Fig. 19.3 (b), $OABGO$ is a closed loop. Taking the sum of components along X-axis, we get

$$q \cos \theta + s \cos (-\phi) - p = 0 \quad \text{[Refer to Figs. (b), (c) and (d)]}$$

or $$q \cos \theta + s \cos \phi - p = 0$$

or $$s \cos \phi = p - q \cos \theta \quad \text{...(19.11)}$$

Taking the sum of components along Y-axis, we have

$$q \sin \theta - t - s \sin (-\phi) = 0$$

or $$q \sin \theta - t + s \sin \phi = 0$$

or $$s \sin \phi = t - q \sin \theta \quad \text{...(19.12)}$$

Squaring both sides of equations (19.11) and (19.12) and then adding them provides.

$$s^2 (\cos^2 \phi + \sin^2 \phi) = p^2 + q^2 (\cos^2 \theta + \sin^2 \theta) - 2pq \cos \theta + t^2 - 2tq \sin \theta$$

or $$s^2 = p^2 + q^2 + t^2 - 2pq \cos \theta - 2tq \sin \theta$$

or $$p^2 - (2q \cos \theta)p + q^2 + t^2 - s^2 - 2tq \sin \theta = 0$$

or $$p^2 + A_1 p + A_2 = 0 \quad \text{...(19.13)}$$

where $A_1 = -2q \cos \theta$

and $A_2 = q^2 + t^2 - s^2 - 2tq \sin \theta$

Equation (19.13) is a quadratic equation in p and value of p is given by

$$p = \frac{(-A_1 \pm \sqrt{A_1^2 - 4A_2})}{2} \quad \text{...(19.14)}$$

In equation (19.14), p is the horizontal distance of slider from Y-axis. Hence p represents the displacement of slider from Y-axis. The length p varies as the crank rotates. Hence length p is a variable.

Equation (19.14) can be used to determine displacement of the slider for given values of θ, q, t, s which are known for the mechanism.

The inclination of the connecting rod (*i.e.* angle ϕ) with X-axis can be determined by using equation (19.12) as

$$\sin \phi = \frac{(t - q \sin \theta)}{s}$$

or $$\phi = \sin^{-1} \left(\frac{t - q \sin \theta}{s} \right) \quad \text{...(19.15)}$$

Note. If the slider lies on x-axis, then eccentricity $t = 0$. Then the values of A_2 and ϕ becomes as $A_2 = q^2 - s^2$ and $\phi = \sin^{-1} \left(\frac{-q \sin \theta}{s} \right)$.

19.6.2. Expression for Velocity

Let v_p = Velocity of slider = $\dfrac{dp}{dt}$

ω_q = Angular velocity of crank $OA = \dfrac{d\theta}{dt}$

ω_s = Angular velocity of connecting rod $AB = \dfrac{d\phi}{dt}$

For determination of velocity, differentiating equation (19.11), we get

$$s \times (-\sin\phi) \times \frac{d\phi}{dt} = \frac{dp}{dt} - q(-\sin\theta)\frac{d\theta}{dt}$$

($\because$ p is a variable but s and q are constant)

or $$-s\sin\phi\omega_s = v_p + q\sin\theta\omega_q \qquad \left(\because \frac{d\phi}{dt} = \omega_s, \frac{dp}{dt} = v_p \text{ and } \frac{d\theta}{dt} = \omega_q\right)$$

Therefore $$v_p = -q\sin\theta\,\omega_q - s\sin\phi\,\omega_s \qquad \text{...(19.16)}$$

Now differentiating equation (19.12), we get

$$s\cos\phi \times \frac{d\phi}{dt} = 0 - q\cos\theta\frac{d\theta}{dt}$$

But $$\frac{d\phi}{dt} = \omega_s \text{ and } \frac{d\theta}{dt} = \omega_q$$

$\therefore$ $$s\cos\phi\omega_s = -q\cos\theta\,\omega_q$$

or $$s\cos\phi\omega_s + q\cos\theta\,\omega_q = 0 \qquad \text{...(19.17)}$$

Multiplying equation (19.16) by $\cos\phi$ and equation (19.17) by $\sin\phi$ and then by adding, we get

$$q\omega_q(-\sin\theta\cos\phi + \cos\theta\sin\phi) - v_p\cos\phi = 0$$

or $$v_p = \frac{q\omega_q\sin(\phi-\theta)}{\cos\phi} \qquad \text{...(19.18)}$$

From equation (19.17)

$$\omega_s = \left(-\frac{q\cos\theta}{s\cos\phi}\right)\omega_q \qquad \text{...[19.18 (a)]}$$

From equations (19.18) and (19.18a), the velocity of slider and angular velocity of connecting rod can be determined.

19.6.3. Expression for Acceleration

Let a_p = acceleration of slider = $\dfrac{d^2p}{dt^2} = \dfrac{dv_p}{dt}$

α_q = Angular acceleration of crank = $\dfrac{d\omega_q}{dt}$

α_s = Angular acceleration of connecting rod = $\dfrac{d\omega_s}{dt}$

Differentiating equations (19.16) and (19.17) once again, we get accelerations. Differentiation of equation (19.16) gives

$$\frac{dv_p}{dt} = -q\left[\sin\theta \times \frac{d\omega_q}{dt} + \omega_q \times \cos\theta \times \frac{d\theta}{dt}\right] - s\left[\sin\phi \times \frac{d\omega_s}{dt} + \omega_s \times \cos\phi \times \frac{d\phi}{dt}\right]$$

$$= -q[\sin\theta \times \alpha_q + \omega_q \times \cos\theta \times \omega_q] - s[\sin\phi \times \alpha_s + \omega_s \times \cos\phi \times \omega_s]$$

$$\left[\because \frac{d\omega_q}{dt} = \alpha_q, \frac{d\omega_s}{dt} = \alpha_s\right]$$

$$= -q\sin\theta\alpha_q - q\cos\theta\omega_q^{\;2} - s\sin\phi\alpha_s - s\cos\phi\omega_s^{\;2}$$

or $$a_p = -q\cos\theta\,(\omega_q)^2 - q\sin\theta\,\alpha_q - s\cos\phi\,(\omega_s)^2 - s\sin\phi\alpha_s \qquad \text{...[19.18 }(b)]$$

where $$\alpha_q = \left(\frac{d\omega_q}{dt}\right), \alpha_s = \frac{d\omega_s}{dt}, \alpha_p = \frac{dv_p}{dt}$$

Now the differentiation of equation (19.17) gives

$$s\left[\cos\phi\frac{d\omega_s}{dt} + \omega_s(-\sin\phi)\frac{d\phi}{dt}\right] + q\left[\cos\theta\frac{d\omega_q}{dt} + \omega_q(-\cos\theta)\frac{d\theta}{dt}\right] = 0$$

or $$s[\cos\phi\alpha_s - \omega_s\sin\phi\omega_s] + q[\cos\theta\alpha_q - \omega_q\cos\theta\omega_q] = 0$$

or $$s\cos\phi\,\alpha_s - s\sin\phi\,\omega_s^{\;2} + q\cos\theta\,\alpha_q - q\cos\theta\,\omega_q^{\;2} = 0$$

or $$-s\sin\phi(\omega_s)^2 + s\cos\phi\,\alpha_s + q\cos\theta\,\alpha_q - q\sin\theta\,\omega_q = 0 \qquad \text{...(19.19)}$$

Multiplying equation (19.18*b*) by cos ϕ and equation (19.19) by sin ϕ and then adding, we get

$$-q\cos\theta\cos\phi\,\omega_q^{\;2} - q\sin\theta\cos\phi\,\alpha_q - s\cos^2\phi\,\omega_s^{\;2} - s\,\alpha_s\sin\phi\cos\phi - a_p\cos\phi$$

$$-s\sin^2\phi\,\omega_s^{\;2} + s\cos\phi\sin\phi\,\alpha_s + q\cos\theta\sin\phi\,\alpha_q - q\sin\theta\sin\phi\,(\omega_q) = 0$$

$$q\alpha_q(\cos\theta\sin\phi - \sin\theta\cos\phi) - q\omega_q^{\;2}(\cos\theta\cos\phi + \sin\theta\sin\phi) - s\,\omega_s^{\;2} - a_p\cos\phi = 0$$

or $$q\alpha_q\sin(\phi-\theta) - q\omega_q^{\;2}\cos(\phi-\theta) - s\omega_s^{\;2} = a_p\cos\phi$$

Therefore, acceleration of piston is given by

$$a_p = \frac{q\alpha_q\sin(\phi-\theta) - q\omega_q^{\;2}\cos(\phi-\theta) - s\omega_s^{\;2}}{\cos\phi} \qquad \text{...(19.20)}$$

The equation (19.20) gives the acceleration of slider. The angular acceleration of connecting rod is obtained from equation (19.19).

From equation (19.19), the angular acceleration of connecting rod is given by,

$$\alpha_s = \frac{+s\sin\phi\,\omega_s^{\;2} - q\cos\theta\,\alpha_q + q\sin\theta\,\omega_q^{\;2}}{s\cos\phi} \qquad \text{...(19.21)}$$

Therefore, to determine displacement of the slider we need to solve quadratic equation (19.14) by using known qualities q, t, s and θ. Value of ϕ can be calculated from equation (19.15) by using known values of t, q, s and θ. To determine velocity of slider equation (19.18) can be used by substituting q, ω_q, ϕ and θ.

Angular speed of connecting rod can be calculated by equation (19.18*a*) by using values of q, θ, ω_q, s and ϕ. To determine acceleration of slider equation (19.20) can be used by substituting values of q, α_q, ϕ, θ, ω_s, s. To determine angular acceleration of connecting rod, equation (19.21) can be solved by substituting values of q, α_q, ϕ, θ, q, ω_s, s, ω_q.

Input Data

The lengths of the links, angular velocity of the crank, angular acceleration of crank and angle of rotation of crank θ. The lengths of the links and other input data in the programme are represented by

Q = Crank length
S = Connecting rod length
T = Eccentricity
VELQ = Angular velocity of crank = ω_q
ACCQ = Angular acc. of the crank = α_q.

Output Data

The angle ϕ, displacement of the slider, linear velocity of slider, linear acc. of the slider, angular velocity of connecting rod, angular acceleration of connecting rod with the variation of crank angle θ. These parameters are represented by :

PHI = Angle ϕ
DISP = Displacement of slider
VELP = Velocity of slider
ACCP = Acceleration of slider
VELS = Angular velocity of connecting rod and
ACCS = Angular acceleration of connecting rod.

19.7. Computer Programme for a Slider Crank Mechanism in FORTRAN

The following is a computer programme for a slider crank mechanism.

```
C    PROGRAMME FOR DISPLACEMENT, VELOCITY
C    AND ACCELERATION FOR SLIDER CRANK
C    MECHANISM
     READ (*, *) Q, S, T, VELQ, ACCQ, THETA
     PI = 4*ATAN(1.0)
     THETAX = 0.0
     IE = 180/THETA
     DTHETA = PI/IE
     DO 40 I = 1, 2*IE
     THETAX = (I-1)*DTHETA
     PHI = ASIN ((T-Q*SIN(THETAX))/S)
     AI = - 2*Q*COS(THETAX)
     A2 = Q**2 + T**2-S² - 2.0*T*Q × SIN (THETAX)
     A12 = A1**2 - 4.0*A2
     A21 = SQRT (A12)
     DISP = (- A1 + A21)/2.0
     VELP = (Q*VELQ*SIN (PHI-THETAX))/COS(PHI)
     VELS = (- VELQ*Q*COS(THETAX))/(S*COS(PHI))
     ACCP = (Q*ACCQ*SIN (PHI-THETAX) - Q*(VELQ**2)*
            COS(PHI-THETAX) - S*VELS**2)/COS(PHI)
     ACCS = (S*SIN(PHI)*(VELS**2) - Q*COS(THETAX)*ACCQ
            + Q*SIN (THETAX)*(VELQ**2))/(S*COS(PHI))
     IF(I.EQ.1) WRITE(*, 10)
10   FORMAT (4X, 'THETA', 5X, 'PHI', 4X, 'DISP', 4X, 'VELP', 4X, 'VELS', 4X,
            'ACCP', 4X, 'ACCS')
```

```
20 WRITE (*, 30) THETAX*180/PI, PHI*180/PI, DISP, VELP, VELS, ACCP, ACCS
30 FORMAT(6F8.2)
40 CONTINUE
   STOP
   END
```

19.7.1. Computer Programme for a Slider Crank Mechanism in C++

Computer programme to find velocity and acceleration for slider crank mechanism

```
#include<iostream.h>
#include<math.h>
void main()
{
float q=.36, s=1.2,t=.075, velq=20, accq=10, theta=0.0,pi;
pi=4*atan(1.0);
for (int i=0;i<=12;i++)
{
float phi=asin((t-q*sin(theta))/s);
float a1=-2*q*cos(theta)
float a2=q*q+t*t-s*s-2*t*q*sin(theta),
float a12=a1*a1-4*a2,
float a21=sqrt(a12);
float disp=(-a1+a21)/2;
float velp=(q*velq*sin(phi-theta))/cos(phi);
float vels=-(velq*q*cos(theta))/(s*cos(phi));
float accp=(q*accq*sin(phi-theta)-q*velq*velq*cos(phi-theta)-s*vels*vels)/
cos(phi);
float accs=(s*sin(phi)*vels*vels-
q*cos(theta)*accq+q*sin(theta)*velq*velq)/(s*cos(phi));
cout<<"theta="<<(theta*180/pi)<<"phi="<<(phi*180/pi)<<",disp="<<disp<<"
velp="<<velp<<",vels="<<vels<<",accp="<<accp<<",accs="<<accs<<endl<<endl;
theta=theta+pi/6;
}
}
```

Problem 19.3. In a slider crank mechanism, the length of crank is 360 mm whereas the length of connecting rod is 1.2 m. If the eccentricity of the slider is 75 mm and the crank is rotating with an angular speed of 20 rad/s and angular acceleration of 10 rad/s^2 at all positions of the crank, then calculate at an interval of 30° of the crank :

(*i*) the displacement, the velocity and acceleration of slider.

(*ii*) the angular velocity and angular accelerate of connecting rod.

Sol. Given. (Refer to Fig. 19.3)

$$q = 360 \text{ mm}, s = 1.2 \text{ m}, t = 75 \text{ mm}, \omega_q = 20 \text{ rad/s}, \alpha_q = 10 \text{ rad/s}^2.$$

Input Data

Q = Length of crank OA = 360 mm = 0.36 m

S = Length of connecting rod AB = 1200 mm = 1.2 m

T = Eccentricity = 75 mm = 0.075 m

$VELQ$ = Angular velocity of crank = 20 rad/s

$ACCQ$ = Angular Acc. of crank = 10 rad/s^2

The computer programme shown in Art. 19.7.1 was used for the given data. The output from the programme is :

Output Data

THETA(θ)	PHI	DISP	VELP	VELS	ACCP	ACCS
0	3.58332	1.55765	.450882	–6.01175	– 187.229	– .7426
30	– 5.0198	1.50717	– 4.1477	– 5.21616	– 153.233	55.233
60	– 11.3796	1.3564	– 6.9599	– 3.0602	– 61.8436	102.592
90	– 13.7390	1.1657	– 7.2000	– 0.0	+ 31.6074	123.535
120	– 11.3796	0.9964	– 5.51083	3.0601	82.8810	105.065
150	– 5.0198	.883628	– 3.0523	– 5.21616	96.7301	60.4492
180	3.58332	.837654	– .450882	6.01175	100.3	5.26914
210	12.2689	.860824	2.24401	5.3176	106.762	– 52.5943
240	18.8025	.955962	5.00966	3.16912	104.234	– 104.77
270	21.2538	1.11838	7.2	2.99×10^{-6} = (0)	59.6096	– 128.75
300	18.8025	1.31596	7.4611	– 3.16912	– 38.5406	– 107.94
330	12.2689	1.48436	4.95599	– 5.3176	– 141.297	– 57.912

19.8. Computer Aided Analysis of Four Bar Kinematic Chain

A four bar kinematic chain is shown Fig. 19.4. The length of the respective link has been indicated in the figure. Hence length of link $AO_1 = q$, length of link $BA = s$, length of link $BO_2 = t$ and length of link $O_1O_2 = p$.

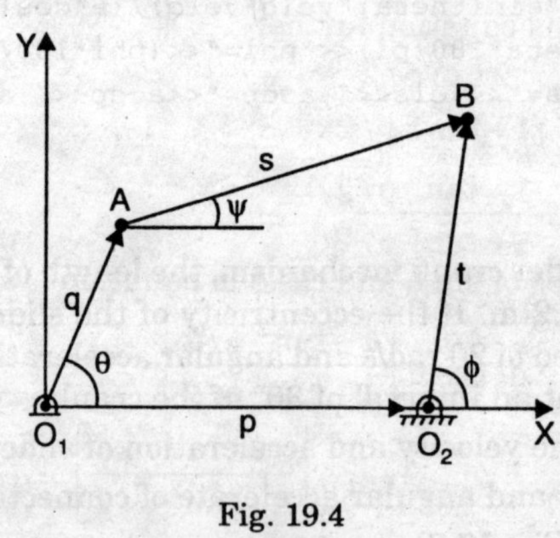

Fig. 19.4

The angle made by the links with x-axis are given as :

Angle made by link $AO_1 = \theta$

Angle made by link $BA = \psi$

Angle made by $BO_2 = \phi$ (+ve)

Angle made by $O_2O_1 = 0$

19.8.1. Expression for Displacement

Refer to Fig. 19.4. The links have been represented by vectors. The vectors can be resolved along X-axis and Y-axis.

The equation for displacement along X-axis is obtained by resolving vectors along X-axis. Sum of components of vectors q and s along X-axis should be equal to the sum of components of vectors p and t along X-axis.

$\therefore$ $$q \cos \theta + s \cos \psi = p \cos 0° + t \cos \phi$$

or $$q \cos \theta + s \cos \psi = p + t \cos \phi \qquad (\because \cos 0° = 1)$$

or $$s \cos \psi = p + t \cos \phi - q \cos \theta \qquad ...(19.22)$$

Now resolving the vectors along Y-axis. The sum of components of vectors q and s along Y-axis should be equal to sum of components of vectors p and t alone Y-axis.

$\therefore$ $$q \sin \theta + s \sin \psi = p \sin 0° + t \sin \phi$$

or $$q \sin \theta + s \sin \psi = t \sin \phi$$

or $$q \sin \theta + s \sin \psi - t \sin \phi = 0 \qquad ...(19.23)$$

In equations (19.22) and (19.23), there are two unknown *i.e.,* ϕ and ψ. Let us eleminate ψ from these equations.

Equation (19.22) and (19.23) can be rearranged such that

$$s \cos \psi = t \cos \phi - q \cos \theta + p \qquad ...(19.24)$$

$$s \sin \psi = t \sin \phi - q \sin \theta \qquad ...(19.25)$$

Squaring equations (19.24) and (19.25) and adding, the following is obtained

$$s^2 = q^2 + t^2 + p^2 - 2qt \cos \theta \cos \phi - 2qp \cos \theta + 2tp \cos \phi - 2qt \sin \theta \sin \phi$$

$$- 2qt (\cos \theta \cos \phi + \sin \theta \sin \phi) + 2tp \cos \phi - 2qp \cos \theta + p^2 + t^2 + q^2 - s^2 = 0$$

Let $$2K = p^2 + t^2 + q^2 - s^2$$

Therefore, the above equation becomes

$$- 2qt (\cos \theta \cos \phi + \sin \theta \sin \phi) + 2tp \cos \phi - 2qp \cos \theta + 2K = 0$$

or $$- qt (\cos \theta \cos \phi + \sin \theta \sin \phi) + tp \cos \phi - qp \cos \theta + K = 0 \qquad ...(19.26)$$

In equation (19.26), only ϕ is unknown. But it is very difficult to find the value of ϕ from this equation. Let us simply this equation further.

Since, $$\sin \phi = \frac{2 \tan \phi/2}{1 + \tan^2 \phi/2}$$

and $$\cos \phi = \frac{1 - \tan^2 \phi/2}{1 + \tan^2 \phi/2}$$

Substituting these values of $\sin \phi$ and $\cos \phi$ in equation (19.26)

$$tp \left[\frac{1 - \tan^2 \phi/2}{1 + \tan^2 \phi/2}\right] - qt \cos \theta \left[\frac{1 - \tan^2 \phi/2}{1 + \tan^2 \phi/2}\right]$$

$$- qt \sin \theta \left[\frac{2 \tan \phi/2}{1 + \tan^2 \phi/2}\right] - qp \cos \theta + K = 0 \qquad ...(19.27)$$

Multiplying equation (19.27) by $\left(1 + \tan^2 \frac{\phi}{2}\right)$, we get

$$t_p \left(1 - \tan^2 \frac{\phi}{2}\right) - qt \cos \theta \left(1 - \tan^2 \frac{\phi}{2}\right) - qt \sin \theta \times 2 \tan \frac{\phi}{2}$$

$$- qp \cos \theta \left(1 + \tan^2 \frac{\phi}{2}\right) + K \left(1 + \tan^2 \frac{\phi}{2}\right) = 0$$

or $$\left(\tan^2\frac{\phi}{2}\right)(-tp + qt\cos\theta - qp\cos\theta + K) + \left[\left(\tan\frac{\phi}{2}\right)\right](-2qt\sin\theta) + tp - qt\cos\theta - qp\cos\theta + K = 0$$

or $$\left[\tan^2\left(\frac{\phi}{2}\right)\right](qt\cos\theta - qp\cos\theta - tp + K) + \left[\tan\left(\frac{\phi}{2}\right)\right](-2qt\sin\theta) + tp - qt\cos\theta + K - qp\cos\theta = 0$$

or $$(qt\cos\theta - qp\cos\theta - tp + K)\tan^2\frac{\phi}{2} - 2qt\sin\theta\tan\frac{\phi}{2} + (tp - qt\cos\theta + K - qp\cos\theta) = 0 \quad ...(19.28)$$

Let $A = q(t-p)\cos\theta - tp + K$

$B = -2qt\sin\theta$

and $C = tp + K - q(t+p)\cos\theta.$

Substituting above parameters in the preceeding equation, it becomes

$$A\tan^2\frac{\phi}{2} + B\tan\frac{\phi}{2} + C = 0 \quad ...(19.29)$$

It is a quadratic equation in $\tan\frac{\phi}{2}$ and value of ϕ is given as following :

$$\tan\left(\frac{\phi}{2}\right) = \frac{-B \pm \sqrt{B^2 - 4AC}}{2A}$$

or $$\left(\frac{\phi}{2}\right) = \tan^{-1}\frac{-B \pm \sqrt{B^2 - 4AC}}{2A}$$

or $$\phi = 2\tan^{-1}\left[\frac{-B \pm \sqrt{B^2 - 4AC}}{2A}\right] \quad ...(19.30)$$

This angle gives position of the output link. After knowing the value of ϕ, the value of angle ψ can be obtained from equation (19.25) as

$$\sin\psi = \frac{t\sin\phi - q\sin\theta}{s}$$

$\therefore$ $$\psi = \sin^{-1}\left[\frac{t\sin\phi - q\sin\theta}{s}\right]$$

The value of ψ can also be obtained by eliminating ϕ from equations (19.22) and (19.23)

Equations (19.22) and (19.23) can be written as follows :

$$t\cos\phi = -p + q\cos\theta + s\cos\psi \quad ...(19.31)$$

and $$t\sin\phi = q\sin\theta + s\sin\psi \quad ...(19.32)$$

Squaring and adding equations (19.30) and (19.31), we get

$$t^2 = q^2 + s^2 + p^2 + 2qs\cos\theta\cos\psi - 2sp\cos\psi - 2qp\cos\theta + 2qs\sin\theta\sin\psi \quad ...(19.33)$$

Let $q^2 + s^2 + p^2 - t^2 = 2k$

$\therefore$ $2qs(\cos\theta\cos\psi + \sin\theta\sin\psi) - 2qp\cos\theta - 2s\cos\psi + 2k = 0$

$$qs(\cos\theta\cos\psi + \sin\theta\sin\psi - qp\cos\theta - sp\cos\psi + k) = 0 \quad ...(19.34)$$

By substituting $\cos\psi = \dfrac{1-\tan^2\frac{\psi}{2}}{1+\tan^2\frac{\psi}{2}}$ and $\sin\psi = \dfrac{2\tan\frac{\psi}{2}}{1+\tan^2\frac{\psi}{2}}$

In equation (19.33), it can be arranged similar to equation (19.29) as follows :

$$D\tan^2\frac{\psi}{2} + E\tan\frac{\psi}{2} + F = 0 \qquad ...(19.35)$$

where $D = k - q(p+s)\cos\theta + sp$

$E = 2qs\sin\theta$

$F = k - q(p-s)\cos\theta - sp$

Equation (19.35) is a quadratic equation and it can be used to determine ψ by the following equation :

$$\psi = 2\tan^{-1}\left[\frac{-E \pm \sqrt{E^2 - 4DF}}{2D}\right] \qquad ...(19.36)$$

Equations (19.30) and (19.36) provide output angle ϕ and angle of inclination of coupler ψ for given values of θ and link lengths p, q, s and t.

19.8.2. Expression for Velocity

For velocity analysis,

Let ω_q = Angular velocity of crank $O_1A = \dfrac{d\theta}{dt}$

ω_s = Angular velocity of link $AB = \dfrac{d\psi}{dt}$

ω_t = Angular velocity of output link $BO_2 = \dfrac{d\phi}{dt}$

Equations (19.22) and (19.23) can be written as follows :

$$s\cos\psi = p + t\cos\phi - q\cos\theta \qquad ...(19.37)$$

and

$$q\sin\theta + s\sin\psi = t\sin\phi \qquad ...(19.38)$$

Differentiating equation (19.37) with respect to time, we get

$$s(-\sin\psi)\times\frac{d\psi}{dt} = 0 + t(-\sin\phi)\times\frac{d\phi}{dt} - q(-\sin\theta)\times\frac{d\theta}{dt}$$

$$\left(p \text{ is constant. Hence } \frac{dp}{dt} = 0\right)$$

or

$$-s\sin\psi\times\omega_s = -t\sin\phi\times\omega_t + q\sin\theta\times\omega_q$$

$$\left(\because \frac{d\psi}{dt} = \omega_s, \frac{d\phi}{dt} = \omega_t, \frac{d\theta}{dt} = \omega_q\right)$$

or

$$-s\omega_s\sin\psi = -t\omega_t\sin\phi + q\omega_q\sin\theta \qquad ...(i)$$

or

$$-s\omega_s\sin\psi + t\omega_t\sin\phi - q\omega_q\sin\theta = 0 \qquad ...(19.39)$$

Differentiating equation (19.38) with respect to time,

$$q\cos\theta\times\frac{d\theta}{dt} + s\cos\psi\times\frac{d\psi}{dt} = t\cos\phi\times\frac{d\phi}{dt}$$

or

$$q\cos\theta\times\omega_q + s\cos\psi\times\omega_s = t\cos\phi\times\omega_t$$

$$\left(\because \frac{d\theta}{dt} = \omega_q, \frac{d\psi}{dt} = \omega_s \text{ and } \frac{d\phi}{dt} = \omega_t\right)$$

or $$q \times \omega_q \cos\theta + s\omega_s \cos\psi = t\omega_t \cos\phi$$

or $$q \times \omega_q \cos\theta + s\omega_s \cos\psi - t\omega_t \cos\phi = 0 \qquad ...(19.40)$$

In equations (19.39) and (19.40), the two velocities *i.e.*, ω_s and ω_t are unknown. These two velocities can be calculated from these two equations.

(A) Expression for ω_t

First eliminate 'ω_s' from these two equations to get the value of ω_t. This can be done by multiplying equation (19.39) by cos ψ and equation (19.40) by sin ψ, we get

$$-s\omega_s \sin\psi \cos\psi + t\omega_t \sin\phi \cos\psi - q\omega_q \sin\theta \cos\psi = 0 \qquad ...(i)$$

and $$q\omega_q \cos\theta \sin\psi + s\omega_s \cos\psi \sin\psi - t\omega_t \cos\phi \sin\psi = 0 \qquad ...(ii)$$

Adding equations (*i*) and (*ii*), we get

$$t\omega_t \sin\phi \cos\psi - q\omega_q \sin\theta \cos\psi + q\omega_q \cos\theta \sin\psi - t\omega_t \cos\phi \sin\psi = 0$$

or $$t\omega_t [\sin\phi \cos\psi - \cos\phi \sin\psi] + q\omega_q [\cos\theta \sin\psi - \sin\theta \cos\psi] = 0$$

or $$t\omega_t \sin(\phi - \psi) + q\omega_q \sin(\psi - \theta) = 0$$

or $$\omega_t = \frac{-q \times \omega_q \times \sin(\psi - \theta)}{t \sin(\phi - \psi)} \qquad ...(19.41)$$

(B) Expression for ω_s

To find the velocity ω_s, eliminate ω_t in equations (19.39) and (19.40). Hence multiply equation (19.39) by cos φ and equation (19.40) by sin φ, we get

$$-s\omega_s \sin\psi \cos\phi + t\omega_t \sin\phi \cos\phi - q\omega_q \sin\theta \cos\phi = 0 \qquad ...(iii)$$

and $$q\omega_q \cos\theta \sin\phi + s\omega_s \cos\psi \sin\phi - t\omega_t \cos\phi \sin\phi = 0 \qquad ...(iv)$$

Adding equations (*iii*) and (*iv*), we get

$$-s\omega_s \sin\psi \cos\phi - q\omega_q \sin\theta \cos\phi + q\omega_q \cos\theta \sin\phi + s\omega_s \cos\psi \sin\phi = 0$$

or $$s\omega_s (\sin\phi \cos\psi - \cos\phi \sin\psi) + q\omega_q (\sin\phi \cos\theta - \cos\phi \sin\theta) = 0$$

or $$s\omega_s \sin(\phi - \psi) + q\omega_q \sin(\phi - \theta) = 0$$

or $$\omega_s = \frac{-q\omega_q \sin(\phi - \theta)}{s \sin(\phi - \psi)} \qquad ...(19.42)$$

From equations (19.41) and (19.42) ω_t and ω_s can be determined.

19.8.3. Expression for Acceleration

For acceleration determination,

Let α_q = Angular acceleration of crank $O_1A = \dfrac{d(\omega_q)}{dt} = \dfrac{d^2\theta}{dt^2}$

α_s = Angular acceleration of link $AB = \dfrac{d\omega_s}{dt} = \dfrac{d^2\psi}{dt^2}$

α_t = Angular acceleration of $BO_2 = \dfrac{d\omega_t}{dt} = \dfrac{d^2\phi}{dt^2}$

Differentiating equation (19.39), with respect to time, we get

$$-s\left[\omega_s \cos\psi \times \frac{d\psi}{dt} + \sin\psi \frac{d\omega_s}{dt}\right] + t\left[\omega_t \cos\phi \times \frac{d\phi}{dt} + \sin\phi \frac{d\omega_t}{dt}\right]$$
$$- q\left[\omega_q \cos\theta \times \frac{d\theta}{dt} + \sin\theta \frac{d\omega_q}{dt}\right] = 0$$

or $$-s[\omega_s \cos\psi \times \omega_s + \sin\psi \times \alpha_s] + t[\omega_t \cos\phi \times \omega_t + \sin\phi \times \alpha_t] - q[\omega_q \cos\theta \times \omega_q + \sin\theta \times \alpha_q] = 0$$

$$\left[\because \frac{d\psi}{dt} = \omega_s, \frac{d\phi}{dt} = \omega_t, \frac{d\theta}{dt} = \omega_q, \frac{d\omega_s}{dt} = \alpha_s, \frac{d\omega_t}{dt} = \alpha_t \text{ and } \frac{d\omega_q}{dt} = \alpha_q\right]$$

or $$-s[\omega_s^2 \cos\psi + \alpha_s \sin\psi] + t[\omega_t^2 \cos\phi + \alpha_t \sin\phi] - q[\omega_q^2 \cos\theta + \alpha_q \sin\theta] = 0$$

or $$-s\omega_s^2 \cos\psi - s\alpha_s \sin\psi + t\omega_t^2 \cos\phi + t\alpha_t \sin\phi - q\omega_q^2 \cos\theta - q\alpha_q \sin\theta = 0 \quad ...(19.43)$$

Now differentiating equation (19.40) with respect to time, we get

$$q\left[\omega_q(-\sin\theta) \times \frac{d\theta}{dt} + \cos\theta \frac{d\omega_q}{dt}\right] + s\left[\omega_s(-\sin\psi) \times \frac{d\psi}{dt} + \cos\psi \frac{d\omega_s}{dt}\right] - t\left[\omega_t(-\sin\phi) \times \frac{d\phi}{dt} + \cos\phi \frac{d\omega_t}{dt}\right] = 0$$

or $$q[\omega_q(-\sin\theta) \times \omega_q + \cos\theta \times \alpha_q] + s[\omega_s(-\sin\psi)\omega_s + \cos\psi \times \alpha_s] - t[\omega_t(-\sin\phi) \times \omega_t + \cos\phi \times \alpha_t] = 0$$

or $$-q\omega_q^2 \sin\theta + q \times \alpha_q \times \cos\theta - s\omega_s^2 \sin\psi + s\alpha_s \cos\psi + t\omega_t^2 \sin\phi - t\alpha_t \cos\phi = 0 \quad ...(19.44)$$

Equations (19.43) and (19.44) can be rearranged as follows :

$$-q\alpha_q \sin\theta - q\omega_q^2 \cos\theta - s\alpha_s \sin\psi - s\omega_s^2 \cos\psi - t\alpha_t \sin\phi + t\omega_t^2 \cos\phi = 0 \quad ...(19.45)$$

and $$q\alpha_q \cos\theta - q\omega_q^2 \sin\theta + s\alpha_s \cos\psi - s\omega_s^2 \sin\psi - t\alpha_t \cos\phi + t\omega_t^2 \sin\phi = 0 \quad ...(19.46)$$

In equations (19.45) and (19.46), there are two unknowns *i.e.*, α_s and α_t. To find the value of α_s from these equations, eliminate α_t from these equations. This can be done by multiplying equation (19.45) by cos ϕ and equation (19.46) by sin ϕ and then adding both the resultant equations.

Multiplying equation (19.45) by cos ϕ and equation (19.46) by sin ϕ and adding both, following is obtained.

$$q\alpha_q(\cos\theta \sin\phi - \sin\theta \cos\phi) - q\omega_q^2(\cos\theta \cos\phi + \sin\theta \sin\phi) + s\alpha_s(\cos\psi \sin\phi - \sin\psi \cos\phi) - s\omega_s^2(\cos\psi \cos\phi + \sin\phi \sin\psi) + t\omega_t^2(\cos^2\phi + \sin^2\phi) = 0$$

or $$q\alpha_q \sin(\phi - \theta) - q\omega_q^2 \cos(\phi - \theta) + s\alpha_s \sin(\phi - \psi) - s\omega_s^2 \cos(\phi - \psi) + t\omega_t^2 = 0$$

$$(\because \cos^2\phi + \sin^2\phi = 1)$$

or $$\alpha_s = \frac{q\omega_q^2 \cos(\phi - \theta) + s\omega_s^2 \cos(\phi - \psi) - q\alpha_q \sin(\phi - \theta) - t\omega_t^2}{s \sin(\phi - \psi)} \quad ...(19.47)$$

To find the value of α_t from equations (19.45) and (19.46), eliminate α_s from these equations. This can be done by multiplying equation (19.45) by cos ψ and equation (19.46) by sin ψ and then adding both the resultant equations.

Now multiplying equation (19.45) by cos ψ and equation (19.46) by sin ψ and by adding both, we obtain

$$q\alpha_q(\sin\psi \cos\theta - \cos\psi \sin\theta) - q\omega_q^2(\sin\psi \sin\theta + \cos\psi \cos\theta) - s\omega_s^2(\cos^2\psi + \sin^2\psi) + t\alpha_t(\sin\phi \cos\psi - \cos\phi \sin\psi) + t\omega_t^2(\sin\psi \sin\phi + \cos\psi \cos\phi) = 0$$

or $$q\alpha_q \sin(\psi-\theta) - q\omega_q^2 \cos(\psi-\theta) - S\omega_s^2 + t\alpha_t \sin(\phi-\psi) + t\omega_t^2 \cos(\phi-\psi) = 0$$
$(\because \cos^2\psi + \sin^2\psi = 1)$

or $$\alpha_t = \frac{q\omega_q^2 \cos(\psi-\theta) - q\alpha_q \sin(\psi-\theta) + s\omega_s^2 - t\omega_t^2 \cos(\phi-\psi)}{t \sin(\phi-\psi)} \quad ...(19.48)$$

Equations (19.47) and (19.48) can be used to determine angular accelerations of the coupler and output links respectively.

19.9. Computer Programme for a Four Bar Kinematic Chain

Input Data

The lengths of links, angular velocity of crank, angular acceleration of crank and angle of rotation of crank. the lengths of the links and other input data in the programme are represented by :

Refer to Fig. 19.4

$P = p$ = Fixed length of link O_1O_2

$Q = q$ = Crank length of link O_1A

$S = s$ = Length of link AB

$T = t$ = length of link BO_2

$VELQ = \omega_q$ = Angular velocity of crank

$ACCQ = \alpha_q$ = Angular acceleration of crank

Output Data

The angle ϕ, the angle ψ, angular velocity of link AB, angular velocity of link BO_2, angular acceleration of link AB, angular acceleration of link BO_2 with the variation of crank angle θ. Corresponding to a single value of θ, there will be two values of ϕ, two values of ψ, two values of angular velocities and angular accelerations of links AB and BO_2. These parameters are represented by,

PHI = angle ϕ is radians

PP = angle ϕ in degrees

SHI = angle ψ in radians

SH = angle ψ in degrees

VELT = angular velocity of output link $BO_2 = \omega_t$

VELS = angular velocity of link $AB = \omega_s$

ACCT = angular acceleration of output link $BO_2 = \alpha_t$

ACCS = angular acceleration of link $AB = \alpha_s$

The following is a computer programme for a four bar kinematic chain :

```
C     PROGRAMME FOR FINDING VELOCITY AND
C     ACCELERATION IN A FOUR BAR KINEMATIC
C     CHAIN
      DIMENSION PH(2), PHI(2), PP(2), PPX(2), SHI(2), SH(2)
      1, VELT(2), VELS(2), ACCT(2), ACCS(2), T1(2)
      2, T2(2), T3(2), T4(2), S1(2), S2(2), S3(2), S4(2)
      READ (*,*) P, Q, S, T, VELQ, ACCQ, THETA
      PI = 4.0*ATAN(1.0)
      THETAX = 0.0
```

```
      IT = 180/THETA
      DHE = PI/IT
      DO 20 J=1, 2*IT
      THETAX = (J-1)*DHE
      AK1 = (0.5*(P**2.0 + T**2.0 + Q**2.0 - S**2.0))
      THX = THETAX**180/PI
      AA1 = Q*(T-P)* COS(THETAX) - T*P + AK1
      BB1 = - 2.0*Q*T* sin θ
      CC1 = T*P + AK1 - Q*(T+P)* COS(THETAX)
      AB1 = BB1**2.0 - 4*AA1*CC1
      IF(AB1.LT.0.0) GO TO 20
      PHH = SQRT(AB1)
      PH(1) = - BB1 + PHH
      PH(2) = - BB1 - PHH
      AK2 = (Q**2.0 + S**2.0 + P**2.0 - T**2.0)/2
      AA2 = AK2 - Q*(P + S)* COS(THETAX) + S*P
      BB2 = 2.0*Q*S*SIN(THETAX)
      CC2 = AK2-Q*(P - S)* COS(THETAX) - S*P
      AB2 = BB2**2.0 - 4.0*AA2*CC2
      IF(AB2 . LT . 0.0) GO TO 20
      PHP = SQRT(AB2)
      PPX(1) = - BB2 + PHP
      PPX(2) = - BB2 - PHP
      DO 10 I = 1,2
      PHI(I) = 2*ATAN ((PH(I)*0.5)/AA1)
      PP(I) = PHI(I)*180/PI
      SHI(I) = 2*ATAN((PPX(I)*0.5)/AA2)
      SH(I) = PPX(I)*180/PI
      VELT(I) = - (Q*VELQ* SIN(SHI(I) - THETAX)/(T*SIN (PHI(I) - SHI(I)))
      VELS(I) = - (Q*VELQ*SIN(PHI(I) - THETAX)/(S*SIN(PHI(I) - SHI(I)))
      S1(I) = Q*VELQ**2.0*(COS(PHI(I) - THETAX))
      S2(I) = S*VELS**2.0*(COS(PHI(I) - SHI(I)))
      S3(I) = Q*ACCQ*SIN(PHI(I) - THETAX) + T*VELT(I) **2.0
      S4(I) = S*SIN(PHI(I) - SHI(I))
      ACCS(I) = (S1(I) + S2(I) - S3(I))/S4(I)
      T1(I) = Q*VELQ**2.0 * COS(SHI(I) - THETAX)
      T2(I) = Q*ACCQ*SIN(SHI(I) - THETAX)
      T3(I) = S*VELS**2.0 - T*VELT(I)**2.0*COS (PHI(I) - SHI(I))
      T4(I) = T*SIN(PHI(I) - SHI(I))
   10 ACCT(I) = (T1(I) - T2(I) + T3(I))/T4(I)
      IF(J.NE.1) GO TO 15
      WRITE (*,18)
   18 FORMAT (4X, 'THETA', 4X, 'PHI', 4X, 'SHAI', 4X, 'VELT', 4X,
               'VELS', 4X, 'ACCT', 4X, 'ACCS')
   15 WRITE(*,19) THX, PP(1), SH(1), VELT(1), VELS(1), ACCT(1), ACCS(1)
```

```
19 FORMAT (7F8.2)
   WRITE(*,21) PP(2), SH(2), VELT(2), VELS(2), ACCT(2), ACCS(2)
21 FORMAT(8X, 6F8.2)
20 CONTINUE
   STOP
   END
```

Problem 19.4. The length of all the four links in a four bar kinematic chain (Refer to Fig. 19.4) are p = 80 mm, q = 30 mm, s = 70 mm and t = 60 mm. The link $O_1O_2 = p$ is a fixed link. The crank O_1A is rotating with an angular velocity of 10.5 rad/s and with an angular retardation of 26 rad/s^2 both anticlockwise at all positions of the crank. Determine :

(*i*) angle ϕ of output link BO_2 and angle ψ of link AB.

(*ii*) angular velocity of output link and angular velocity of link AB.

(*iii*) angular acceleration of output link BO_2 and angular acceleration of link AB.

Sol. Given. (Refer to Fig. 19.4)

p = 80 mm = 0.08 m, q = 30 mm = 0.03 mm,

s = 70 mm = 0.07 m, t = 60 mm = 0.06 m,

Angular velocity of crank = ω_q = 10.5 rad/s

Angular retardation of crank = 26 rad/s^2

$\therefore$ Angular acceleration of crank = – 26 rad/s^2

(–ve sign is due to retardation)

The computer programme shown in Art. 19.9 was used. The following is the input and output from this programme.

Input Data

$P = p$ = length of fixed link O_1O_2 = 0.08 m,

$Q = q$ = length of crank O_1A = 0.03 m,

$S = s$ = length of link AB = 0.07 m

$T = t$ = length of output link BO_2 = 0.06 m

$VELQ = \omega_q$ = Angular velocity of crank = 10.5 rad/s

$ACCQ = \alpha_q$ = Angular acceleration of crank = – 26 rad/s^2

Output Data

THETA	*PHI*	*SHAI*	*VELT*	*VELS*	*ACCT*	*ACCS*
.00	– 101.54	– 57.12	– 6.30	– 6.30	– 52.81	37.20
	101.54	57.12	– 6.30	– 6.30	84.01	– 6.00
30.00	– 121.42	– 71.04	– 6.69	2.79	55.49	104.12
	90.38	40.00	– 1.18	– 5.08	111.32	62.69
60.00	– 137.16	– 72.54	– 4.28	1.47	57.75	62.69
	93.59	28.96	3.00	– 2.76	50.41	45.47
90.00	– 146.37	– 64.58	– 2.27	3.78	39.79	19.48
	105.26	23.47	4.86	– 1.19	8.41	28.73

120.00	– 150.59	– 52.37	– 0.70	4.54	32.51	– 8.34
	120.00	21.78	5.25	9.00	– 16.98	23.87
150.00	– 150.31	– 39.70	0.94	4.15	33.90	– 28.46
	134.20	23.59	4.51	1.30	– 36.33	26.03
180.00	– 144.90	– 29.52	2.86	2.86	31.52	– 38.21
	144.90	29.52	2.86	2.86	– 45.70	24.02
210.00	– 134.20	– 23.59	4.51	1.31	13.98	32.51
	150.31	39.70	0.94	4.15	– 38.58	7.90
240.00	– 120.00	– 21.38	– 5.25	0.00	– 9.02	– 23.87
	150.59	52.38	– 0.70	4.54	– 29.02	– 14.18
270.00	– 105.26	– 23.47	4.86	– 1.19	– 32.50	– 22.80
	146.37	64.58	– 2.27	3.78	– 28.52	– 38.23
300.00	– 93.590	– 28.96	3.00	2.75	– 65.24	– 31.83
	137.16	72.54	– 4.28	1.47	– 36.55	– 69.97
330.00	– 90.38	– 40.00	– 1.18	– 5.07	– 105.45	– 37.53
	121.42	71.04	6.68	– 2.79	– 22.36	– 90.28

EXERCISE 19

1. **Draw coupler curve of a middle point of coupler of slider crank mechanism O_1AB in which eccentricity of piston is zero. The lengths of links O_1A and AB are 10 cm and 35 cm respectively.**

[Hint. Use computer programme of Art 19.3.2. Take $Q = q = 10$ cm, $S = s = 35$ cm, $F = f = 17.5$ cm, $T = t = 0$ and $PHI = \phi = 0.0$. The output is :

THETA (θ)	*XD*	*YD*
0	27.5	0.0
30	25.98	2.5
60	21.956	4.33
90	16.77	5.0
120	11.956	4.33
150	8.66	2.50
180	7.50	0.00
210	8.66	– 2.5
240	11.956	– 4.33
270	16.77	– 5.00
300	21.956	– 4.33
330	25.98	– 2.50

2. **Draw coupler curve of a point C on the line AB of four bar chain O_1ABO_2. This point divides AB in the proportion of 1 : 2. The link length for O_1A, AB, BO_2 and O_1O_2 are 10 cm, 25 cm, 15 cm and 30 cm respectively.**

[Hint. Use computer programme of Art. 19.4.2. Take $P = p = 30$ cm, $Q = q = 10$ cm, $S = s = 25$ cm, $T = t = 15$ cm, $F = f = 8.33$ and $PHI = \phi = 0.0$. The output is :

THETA (θ)	*XC*	*YC*
0	21.67	0.0
30	17.70	16.06
60	22.975	23.576
90	6.66	25.002

The lengths of the links are such that only rotation is possible upto 90°.]

3. **The length of all the four links in a four bar kinematic chain are p = 21 cm, q = 12 cm, S = 20 cm and t = 18 cm. Determine angular acceleration of the output link for rpm of input link equal to 320 and θ = 50°.**

[Hint. Use computer programme of Art. 19.9. Take $P = p$ = 21 cm, $Q = q$ = 12 cm, $S = s$ = 20 cm, $T = t$ = 18 cm, VELQ = $\frac{2\pi \times 320}{60}$ = 50.265 rad/s, ACCQ = α_q = 0.0 and theta (θ) = $\frac{50 \times \pi}{180}$ rad. The output is

THETA	*PHI*	*SHAI*	*VELT*	*VELS*	*ACCT*	*ACCS*
50.00	– 143.22	– 86.75	– 27.56	8.26	1883.81	2545.56
	73.86	23.89	19.24	– 15.95	2034.15	1588.94

4. **In a slider crank mechanism, length of different links are :**
q = 15 cm, S = 45 cm and c = 5 cm.
Determine acceleration of slider and angular acceleration of connecting rod plot them against angle of crank.

20

Synthesis of Mechanism

20.1. INTRODUCTION

Kinematics is concerned with study of motion of mechanism. It is termed as kinematic analysis. The design of a mechanism to satisfy prescribed motion is termed as kinematic synthesis. In the chapter of cams, the cam profile was designed to produce the prescribed motion of the follower. In this chapter, we shall limit ourself to the introduction of kinematic synthesis of planar linkages namely four bar kinematic chain and slider crank chain. There are so many types of mechanisms and their kinematic synthesis shall form separate field in itself. The kinematic synthesis involves several stages. In the first stage, one has to select the type of mechanism which comes under the class of type synthesis. The second stage is the number synthesis which refers to the study of the movability of the mechanism. The third stage is the dimensional synthesis.

20.1.1. Number Synthesis

It refers to the study of movability of linkages in a mechanism. The movability is related to the degrees of freedom. The degrees of freedom in a mechanism is equal to the number of independent coordinates which have to be assigned to a kinematic chain in order to define its motion.

20.2. DEGREES OF FREEDOM OF A PLANAR KINEMATIC CHAIN (Grübler Criterion)

Let the kinematic chain consists of

n = number of links

j = number of simple hinges

h = number of higher pairs

A link when executes plane motion has three degrees of freedom because two independent coordinates are required to define its motion. Therefore, n links shall have $3n$ degrees of freedom if they freely execute plane motion. If one of the links is fixed, $(n-1)$ links shall have $3(n-1)$ degrees of freedom. If the link is hinged, it can only rotate in the plane. That means a hinged link shall have only one degree of freedom. Therefore, one simple hinge reduces two degrees of freedom of link and, thereby, 'j' simple hinges shall reduce the degrees of freedom by '$2j$'. Therefore, a kinematic chain having 'n' links and 'j' simple hinges shall have degrees of freedom given by

$$F = 3(n-1) - 2j \quad ...(20.1)$$

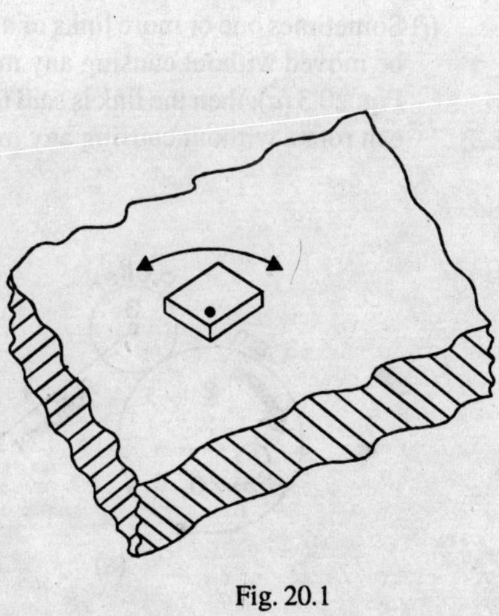

Fig. 20.1

Like simple hinge, a higher pair also reduces one degree of freedom. Therefore, the degrees of freedom of a kinematic chain having 'n' links, 'j' simple hinges, 'h' higher pairs is given by

$$F = 3(n-1) - 2j - h \quad \text{...(20.1 A)}$$

For a kinematic chain, having 'n' links may have different type of links. Let it has

n_2 = number of binary links

(a)

n_3 = number of ternary links

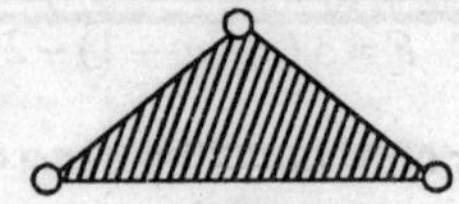

(b)

n_4 = number of quarternary links and

(c)

Fig. 20.2

so on

Therefore, $n = n_2 + n_3 + n_4 + \ldots\ldots + n_i$ ...(20.1B)

Similarly, number of simple hinges is given by

$$2j = 2n_2 + 3n_3 + 4n_4 + \ldots\ldots + n\, n_n. \quad \text{...(20.1C)}$$

If a mechanism includes higher order hinges also, the corresponding number of simple hinges can be determined by the following equation

$$j = j_1 + 2j_2 + 3j_3 + \ldots + i\, j_i \quad \text{(20.1D)}$$

where j_i is the number of hinges connecting $(i + 1)$ links.

In order to determine, the number of degrees of freedom of a mechanism, the following also may be considered :

(*i*) Sometimes one or more links of a mechanism may have redundant degrees of freedom. If a link can be moved without causing any movement in the rest of the mechanism as shown in the following Fig. 20.3 (*a*), then the link is said to have redundant degree of freedom (F_r). In this mechanism, roller can rotate without causing any movement in the rest of the mechanism.

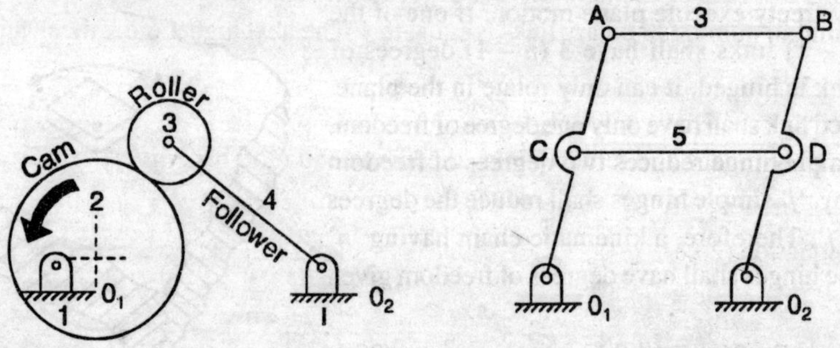

Fig. 20.3

(*ii*) A mechanism may posses one or more links which do not introduce any extra constraint. Such links are redundant. A mechanism shown in Fig. 20.3 (*b*) has links *AB* and *CD* which are identical and therefore, each of the links leads to same constraint. Thus, when determining the degrees of freedom the redundant links (n_r) should not be taken into account. Similarly redundant joints (*jr*) should also not be taken into account.

Thus, finally, if a mechanism has redundant links n_r, redundant kinematic pairs j_r, and redundant degrees of freedom F_r, the degrees of freedom can be determined by the following expression :

$$F = 3\,(n - n_r - 1) - 2\,(j - j_r) - h - F_r \qquad ...(20.2)$$

20.2.1. Minimum Number of Binary Links in a Constrained Mechanism with Simple Hinges

If a mechanism has simple hinges, the number of joints are also given by the following equation :

$$2\,j = 2n_2 + 3\,n_3 + 4\,n_4 + ... + i\,n_i \qquad ...(20.3)$$

It is because a ternary link joins three links and similarly for other links.

Substituting for *n* from Eq. (20.1 B) and for 2*j*-from Eq. (20.3) in the expression for degrees of freedom , [*i.e.*, equation (20.1)], the following equation is obtained :

$$F = 3\,[(n_2 + n_3 + ... + n_i) - 1] - [2\,n_2 + 3\,n_3 + 4\,n_4 + ... + i\,n_i]$$

Since for a fully constrained linkage has degrees of freedom is equal to one. Therefore,

$$1 = n_2 - n_4 + ... + (3 - i)\,n_i - 3 \qquad ...(20.4)$$

or $$n_2 = 4 + n_4 + ... + (i - 3)\,n_i \qquad ...(20.5)$$

From Eq. (20.5), it is quite evident, the minimum number of binary links is four.

Therefore, the four bar kinematic chain is the simplest mechanism.

20.2.2. Fully Constrained Motion of a Kinematic Chain

Let there be *n* links in a kinematic chain and they are connected through *j* simple hinges. The degree of freedom of a fully constrained kinematic chain is one. Therefore, in equation (20.1), $F = 1$.

$\therefore$ $$1 = 3\,(n - 1) - 2j \quad \text{or} \quad 1 = 3n - 3 - 2j$$

or $$3\,n = 2j + 4 \qquad ...(20.6)$$

From Eq. (20.6), it is evident that whatever be the value of *j*, the value of 3*n* has to be even according to Eq. (20.6). This means that a mechanism with simple hinges or pairs should have even number of links. Therefore, the value of *n* cannot be 5, 7, 9 etc. for a mechanism having simple lower pairs. It has already been observed that minimum number of binary links is equal to 4. The next higher order mechanism starts with $n = 6$. From Eq. (20.1 B)

$$6 = n_2 + n_3 + n_4 + ...$$

Assuming only binary and ternary links, the above equation becomes

$$n_2 + n_3 = 6 \qquad ...(20.7)$$

But from equation (20.6),

or $$3n = 2j + 4$$

$$2j = 3\,n - 4$$

or $$2j = 3 \times 6 - 4 = 14 \qquad (\because \; n = 6)$$

or $$j = 7$$

In Eq. (20.3), substituting for j, we get

$$2 \times 7 = 2\,n_2 + 3\,n_3 \quad \text{[For binary and ternary links, only } n_2 \text{ and } n_3 \text{ exist]} \quad ...(20.8)$$

In Eqs. (20.7) and (20.8) n_2 and n_3 are unknowns and they can be solved. It gives $n_2 = 4$ and $n_3 = 2$.

Therefore, a six-link mechanism will have four binary, two ternary links and seven simple hinges. A possible mechanism is shown in Fig. 20.4 below in which links 2, 3, 4 and 6 are binary links and links 1 and 5 are ternary links :

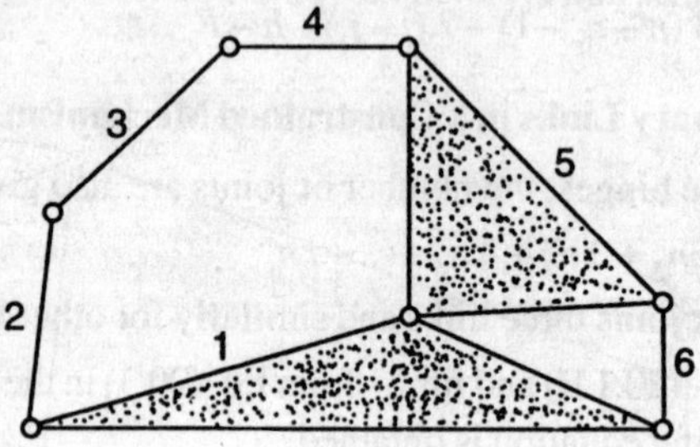

Fig. 20.4

Let us now consider an eight-link chain and considering it has binary, ternary and quarternary links, from Eq. (20.1 B) the following expression is obtained :

$$8 = n_2 + n_3 + n_4 \quad (\because \ n = 8) \quad ...(20.9)$$

But from equation (20.6),

$$3n = 2j + 4 \quad \text{or} \quad 2j = 3n - 4$$

or

$$2j = 3 \times 8 - 4 = 20 \quad \text{or} \quad j = 10 \quad (\because \ n = 8)$$

From Eq. (20.3),

$$2j = 2\,n_2 + 3\,n_3 + 4\,n_4$$

$$2\,n_2 + 3\,n_3 + 4\,n_4 = 2 \times 10 = 20 \quad (\because \ j = 10) \quad ...(20.10)$$

Eqs. (20.9) and (20.10) shall provide several solutions as equations are two but unknowns are three (*i.e.*, n_2, n_3 and n_4).

If $n_2 = 4$ they give $n_3 = 4$ and $n_4 = 0$

If $n_2 = 5$ they give $n_3 = 2$ and $n_4 = 1$

If $n_2 = 6$ they give $n_3 = 0$ and $n_4 = 2$

Similarly, for ten-link chain or twelve-link chain (*i.e.*, $n = 10, 12$ etc.) several solutions shall be available and it is not possible to get unique solution.

20.3. DIMENSIONAL SYNTHESIS

It deals with the determination of actual dimensions of the mechanism to satisfy the specified motion characteristics. There are many types of mechanisms which are being used in our daily life. All of them cannot be covered at this stage. We shall limit our discussion only to four bar kinematic chain and slider crank mechanism. In majority of cases, you may find that either they are used as it is or with some modifications. In some cases, a mechanism, which is combination of these two are being used. The procedure discussed in this section can be extended to these combined mechanisms.

20.3.1 Graphical Method

For a certain class of problems in synthesis of mechanisms, the graphical methods are more convenient and also quicker. Sometimes, the approximate solution obtained by using a graphical method can serve as the first choice while carrying out analytical synthesis iteratively.

Pole and Relative Pole of a Four Bar Linkage

The Fig. 20.5 shows the two positions of a four bar linkage. These positions are $O_2A_1B_1O_4$ and $O_2A_2B_2O_4$.

Let us consider a finite movement of the coupler AB of the four bar linkage from position A_1 B_1 to A_2 B_2. The perpendicular bisectors can be drawn for segments A_1 B_1 and A_2 B_2. These two perpendiculars meet at point P_{12}. (The $\perp$ bisectors for segments A_1B_1 and A_2B_2 are not shown in the Fig.) This point P_{12} is the centre for finite rotation from A_1 B_1 to A_2 B_2 and is called the 'pole'.

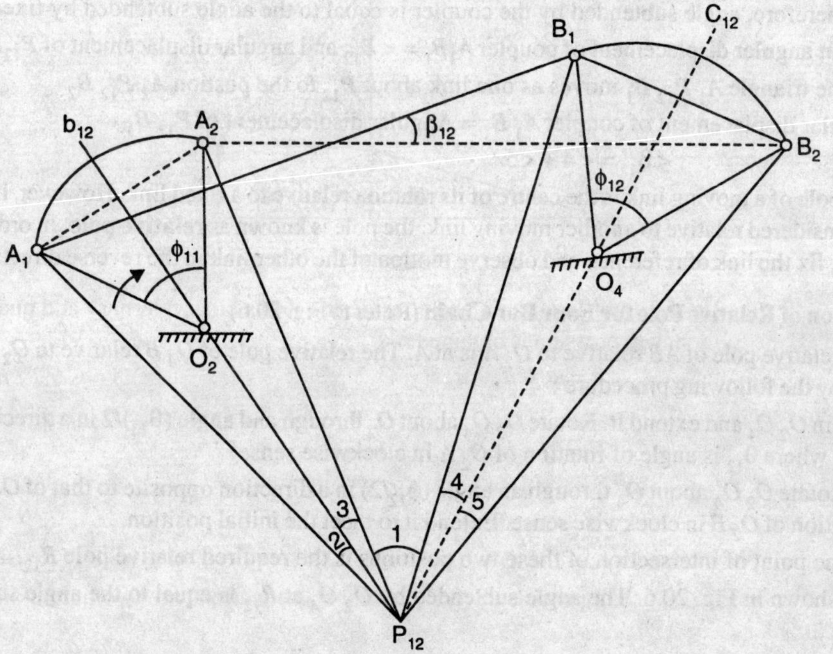

Fig. 20.5

Properties of Pole

(*i*) Since $AB = A_1B_1 = A_2B_2$, the perpendicular bisectors to A_1 A_2 and B_1 B_2 pass through fixed centres O_2 and O_4.

(*ii*) Since coupler AB rotates about P_{12} from position A_1 B_1 to A_2 B_2 and, therefore $\Delta A_1P_{12}B_1 \equiv \Delta A_2P_{12}B_2$

$\therefore$ $<2+<3+<1=<1+4+<5$

Therefore, Angle subtended by A_1 B_1 at P_{12} = Angle subtended by A_2 B_2 at P_{12}.

(*iii*) Since $<2+<3+<1=<1+<4+<5$

or, $<2+<3=<4+<5$

This means A_1 A_2 and B_1 B_2 subtend equal angles at P_{12}.

(*iv*) Since P_{12} lies on perpendicular bisectors of A_1 A_2 and B_1 B_2

$\therefore$ $<2=<3$ and $<4=<5$ are equal

(*v*) Since $<2+<3=<4+<5$

But $<2=<3$ and $<4=<5$

$\therefore$ $\quad < 2 = < 4$

and $\quad < 3 = < 5$

Therefore, input and output links subtend equal angles at pole P_{12} as they move from one position to another.

(*vi*) Since $< 2 + < 3 + < 1 = < 1 + < 4 + < 5 = < 1 + < 4 + < 3$ $\quad (\because < 5 = < 3)$

But $< 2 + < 3 + < 1$ is the angle subtended by coupler at P_{12}. And $< 1 + < 4 + < 3$ or $(< 3 + < 1 + < 4)$ is the angle subtended by fixed link O_2O_4 at P_{12}.

Therefore, angle subtended by the coupler is equal to the angle subtended by fixed link.

But angular displacement of coupler $A_1B_1 = < B_{12}$ and angular displacement of $P_{12}B_1 = <4 + < 5$.

(*vii*) The triangle $A_1\ P_{12}\ B_1$ moves as one link about P_{12} to the postion $A_2\ P_{12}\ B_2$

Angular displacement of coupler $A_1\ B_1$ = Angular displacement of $P_{12}\ B_1$

$\therefore$ $\quad < \beta_{12} = < 4 + < 5$

The pole of a moving link is the centre of its rotation relative to a fixed link. However, if the rotation of the link is considered relative to another moving link, the pole is known as **relative pole.** In order to determine relative pole, fix the link of reference and observe motion of the other link in the reverse direction.

Determination of Relative Pole for Four Bar Chain (Refer to Fig. 20.6)

The relative pole of *AB* relative to O_2A is at *A*. The relative pole of O_4B relative to O_2A is at R_{12}. It is determined by the following procedure :

(*i*) Join $O_2\ O_4$ and extend it. Rotate $O_2\ O_4$ about O_2 through and angle $(\theta_{12})/2$ in a direction opposite to that of O_2A, where θ_{12} is angle of rotation of O_2A in clockwise sense.

(*ii*) Rotate $O_2\ O_4$ about O_4 through an angle $(\phi_{12}/2)$ in a direction opposite to that of O_2B where ϕ_{12} is angle of rotation of O_2B in clockwise sense. Extend it to meet the initial position.

(*iii*) The point of intersection of these two positions is the required relative pole R_{12}.

It is shown in Fig. 20.6. The angle subtended by $O_2\ O_4$ at R_{12} is equal to the angle subtended by the coupler at R_{12}.

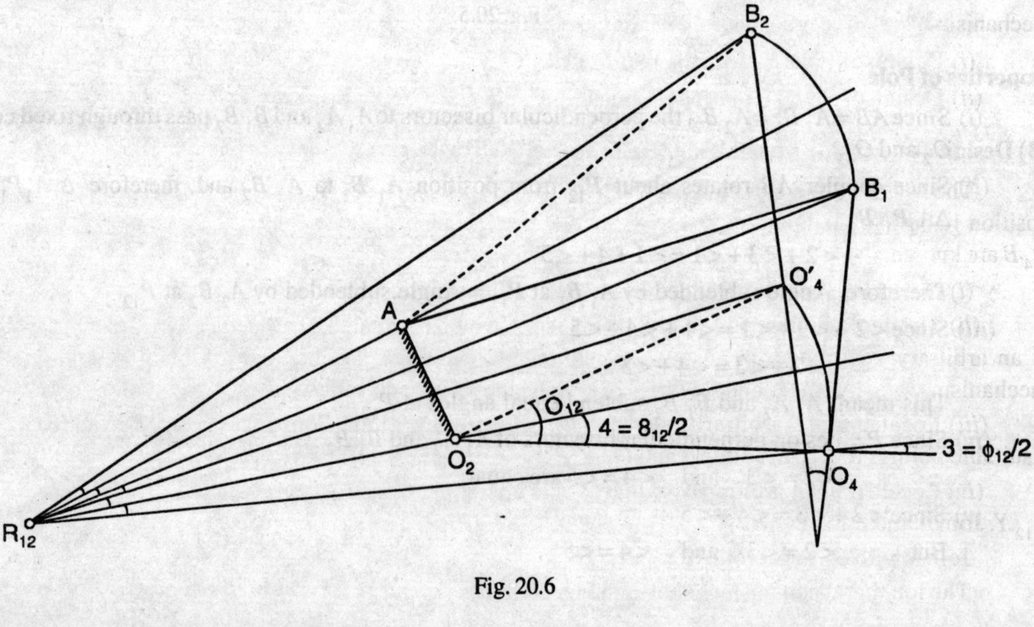

Fig. 20.6

Determination of Relative Pole for Slider Crank Mechanism

The point B reciprocates with slider in Fig. 20.7 through a horizontal distance say 's'. The procedure for determining the relative pole R_{12} is as follows :

(*i*) Draw two parallel lines l_1 and l_2 at a distance equal to the eccentricity 'e'.

(*ii*) Plot line segment $O_2 E$ of length $(s/2)$ on line l_2 such that this segment is measured in a direction opposite to the motion of the slider.

(*iii*) Draw perpendicular lines p_1 and p_2 at O_2 and E respectively to the line l_1.

(*iv*) Rotate p_1 about O_2 by angle $(\theta_{12})/2$ at point O_2 in a sense opposite to the rotation of $O_2 A$. It meets perpendicular p_2 at R_{12} as shown in Fig. 20.7.

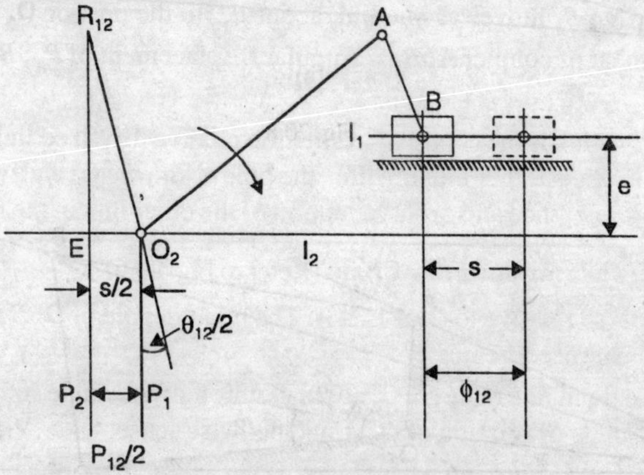

Fig. 20.7

(A) Design of Mechanism By Relative Pole Method

Use of Relative Pole for Design of Mechanism

The relative pole method can be used to design a mechanism graphically. The design of following mechanism by relative pole method will be considered :

(*i*) Design of Four Bar Mechanism, and

(*ii*) Design of Slider Crank Mechanism.

(B) Design of Four Bar Mechanism [Refer to Fig. 20.8 (a)]

The relative pole can be used if length of the fixed link $O_2 O_4$, angular displacement θ_{12} (between position 1 and 2) of in input link $O_2 A$ and angular displacement ϕ_2 (between position 1 and 2) of follower link $O_4 B$ are known.

(*i*) Determine position R_{12} as explained in earlier section.

(*ii*) The angle subtended by $O_2 O_4$ at R_{12} is ψ_{12} which is equal to $(\theta_{12} - \phi_{12}/2)$. Construct the same angle at an arbitrary position. Join any two points on the two lines so obtained to obtain coupler AB of the mechanism. Join A with O_2 and B with O_4 to get the input link and follower link respectively.

(*iii*) Locate point B arbitrarily so that $B O_4$ is the follower link. Construct $< B R_{12} Z$ equal to ψ_{12}. Select a suitable point A to R_{12} Z. Join AB and $O_2 A$.

(*iv*) Locate point A arbitrarily so that $O_2 A$ is input link. Construct $< A R_{12} Y = \psi_{12}$. Select any point B on R_{12} Y. Join A and B to get AB.

Join B with O_4 to get follower link $O_4 B$.

The four bar mechanism is shown in Fig. 20.8 (*a*).

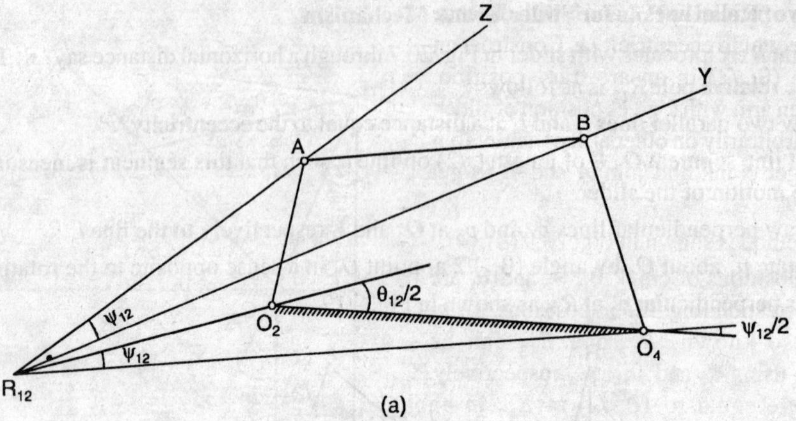

Fig. 20.8

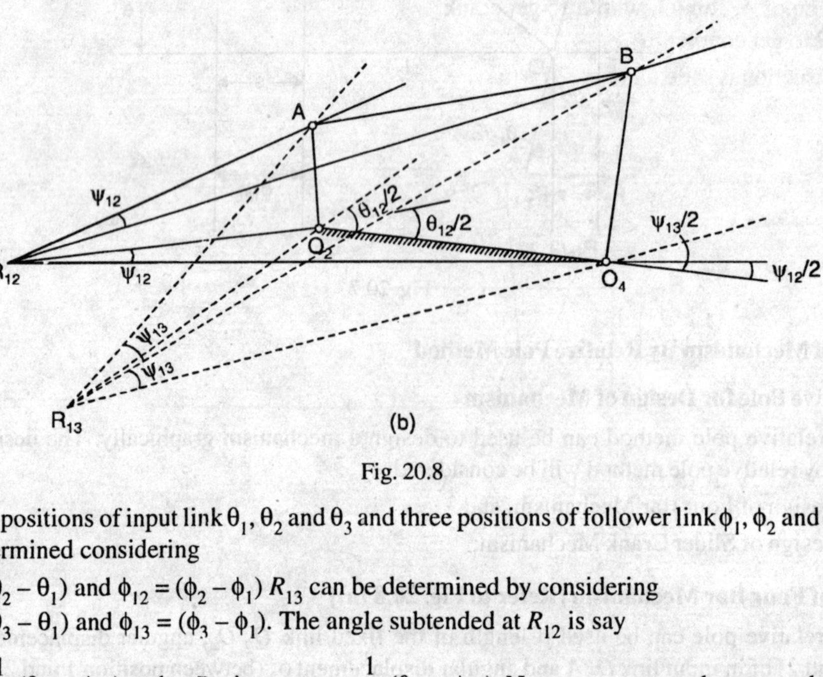

Fig. 20.8

If three positions of input link θ_1, θ_2 and θ_3 and three positions of follower link ϕ_1, ϕ_2 and ϕ_3 are known, R_{12} can be determined considering

$\theta_{12} = (\theta_2 - \theta_1)$ and $\phi_{12} = (\phi_2 - \phi_1)$ R_{13} can be determined by considering
$\theta_{13} = (\theta_3 - \theta_1)$ and $\phi_{13} = (\phi_3 - \phi_1)$. The angle subtended at R_{12} is say

$\psi_{12} = \dfrac{1}{2}(\theta_{12} - \phi_{12})$ and at R_{13} is say $\psi_{13} = \dfrac{1}{2}(\theta_{13} - \phi_{13})$. Now construct angles ψ_{12} and ψ_{13} at R_{12} and R_{13} respectively in arbitrary positions such that the arms of the angles intersect at A and B. Join A and B to get coupler AB. Join A with O_2 and B with O_4 to get input and follower links. This methods is demonstrated in Fig. 20.8 (*b*).

(C) Design of Slider Crank Mechanism

Let θ_{12} be the angular displacement of crank between positions 1 and 2 and s_{12} be the corresponding displacement of the slider. Let θ_1 be eccentricity.

(*i*) Locate relative pole R_{12} as explained in earlier section.

Draw two parallel lines l_1 and l_2 with distance between them equal to eccentricity e. Construct an angle equal to $(\theta_{12}/2)$ in an arbitrary position. Intersection of an arm with l_2 gives position of slider. Select point A arbitrarily on other arm of angle. Join A with O_2 to get crank and join A and B to get coupler.

This method is shown in Fig. 20.9 (*a*).

Three positions of crank θ_1, θ_2 and θ_3 are known and corresponding slider positions s_1, s_2 and s_3 are also known, R_{12} and R_{13} can be determined by using s_1 and θ_{13}, s_{13} respectively. Construct angle equal to $(\theta_{12}/2)$ at R_{12} in an arbitrary position to locate point B which is position of slider B. Intersection of the two remaining arms locates position of A. Join O_2 with A to get crank. Join A with B to get coupler AB.

Construction is shown in Fig. 20.9 (*b*).

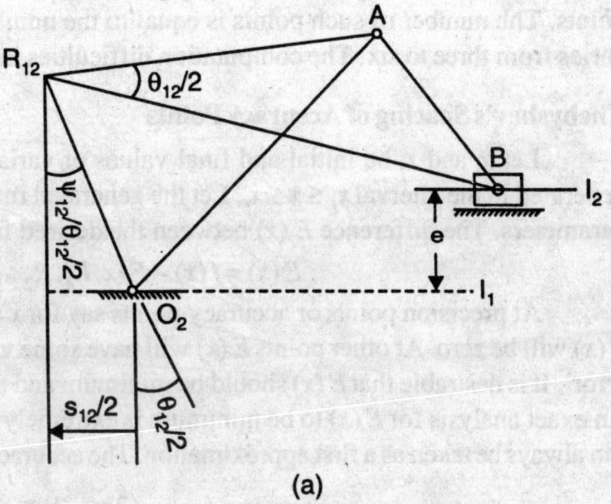

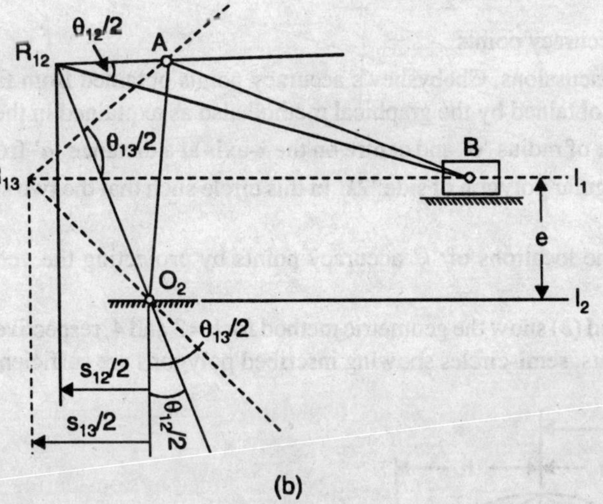

Fig. 20.9

20.3.2. Analytical Method

The movement of the driver or input link can be represented by the input variable '*x*' and that of the driven link or follower by the dependent variable '*y*'. Both of them may have some prescribed functional relationship, say $y = f(x)$.

There are two approaches : approximates synthesis and exact synthesis. By approximate synthesis, we mean that the function generated by the mechanism fits the defined function only at a finite number of points in the interval and intersects the desired path at a finite number of points. By exact synthesis, we mean that the generated function or path fits the desired function or path at all points in the interval. Exact synthesis is limited to a few arbitrary functions, on the other hand, approximate synthesis pertains to almost all functions.

The points at which the generated and desired functions meet are called precision points or accuracy points. The number of such points is equal to the number of design parameters at our disposal. This number varies from three to six. The computation difficulties increase with increase in the number of accuracy points.

Chebyshev's Spacing of Accuracy Points

Let x_i and x_f be initial and final values of variable x respectively. The function $f(x)$ is desired to be generated in the interval $x_i \le x \le x_f$. Let the generated function be $F(x, R_1, R_2 \ldots R_k)$ where $R_1, R_2 \ldots R_k$ be design parameters. The difference $E(x)$ between the desired function and generated function can be represented by

$$E(x) = f(x) - F(x, R_1, R_2 \ldots R_k) \qquad \text{...(20.11)}$$

At precision points or accuracy points say for $x = x_1, x_2 \ldots x_k$ desired and generated functions agree and $E(x)$ will be zero. At other points $E(x)$ will have some value and that will be the error. $E(x)$ is called 'structural error'. It is desirable that $E(x)$ should be minimum and therefore, spacing of accuracy points in very important. An exact analysis for $E(x)$ to be minimum is extremely difficult. The Chebyshev's spacing of accuracy points can always be taken as a first approximation. The accuracy points according to Chebyshev's spacing are given by

$$x_i = a + b\cos\frac{(2m-1)\pi}{2k} \quad \text{where } m = 1, 2, 3 \ldots \qquad \text{...(20.12)}$$

$$\text{where } a = \frac{(x_i + x_f)}{2}, \qquad b = \frac{(x_f - x_i)}{2} \quad \text{and} \qquad \text{...(20.12 A)}$$

k = number of accuracy points

In the further discussions, Chebyshev's accuracy points obtained from Eq. (20.12) will be used. The accuracy points can be obtained by the graphical method also as explained in the following steps :

(*i*) Draw a circle of radius 'b' and centre on the x-axis at a distance 'a' from O.

(*ii*) Inscribe a regular polygon of side '$2k$' in this circle such that the two sides are perpendicular to the x-axis.

(*iii*) Determine the locations of 'k' accuracy points by projecting the vertices on x-axis as shown in Fig. 20.10.

Fig. 20.10 (*a*) and (*b*) show the geometric method for $k = 3$ and 4, respectively. Since the aim is to get the values of accuracy points, semi-circles showing inscribed polygons are sufficient.

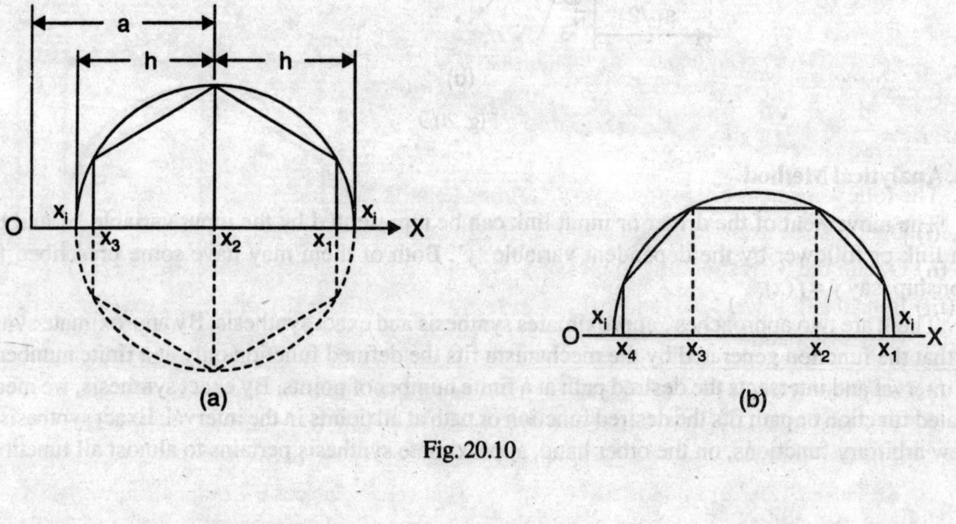

Fig. 20.10

Function Generation by Mechanisms

Let θ and ϕ be the angles of rotation of input or driver link and driven or follower link respectively. Let $y = f(x)$ be the function to be generated. The angle of rotation of the input or driver link $O_2 A$ represents independent variable x. The angle of rotation ϕ of the follower link $O_4 B$ represents dependent variable y as indicated in Fig. 20.11. The relation between x and θ and that between y and ϕ is usually assumed to be linear. Let θ_i and ϕ_i be the initial values of θ and ϕ representing x_i and y_i respectively. On the basis of an assumed linear relationship, we have

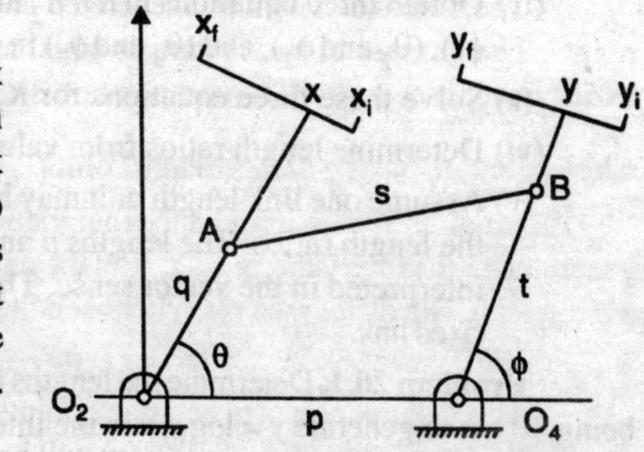

Fig. 20.11

$$\frac{\theta - \theta_i}{x - x_i} = \text{constant} = r_x \text{ (say)} = \frac{\theta_f - \theta_i}{x_f - x_i} \qquad \text{...(20.12 B)}$$

Similarly,
$$\frac{\phi - \phi_i}{y - y_i} = \text{constant} = r_y \text{ (say)} = \frac{\phi_f - \phi_i}{y_f - y_i} \qquad \text{...(20.12 C)}$$

where r_x and r_y are scale factors. The subscripts i and f denote the initial and final values.

Freudenstein's Equation for Four Bar Chain for Three Accuracy Points

There are seven design parameters. There are θ, ϕ, r_x, r_y and three ratios of link lengths. In the present section we shall consider three accuracy points. The parameters θ, ϕ, r_x, and r_y are either given or assumed and the three length ratios p/q, p/s and p/t are to be determined as design parameters.

In reference to Fig. 20.11, the coordinates of the point A are $(q \cos \theta, q \sin \theta)$ and for point B are

$$[(p + t \cos \phi), t \sin \phi]$$

$$AB^2 = s^2 = [\text{diff. in } x \text{ coordinates}]^2 + [\text{diff. in } y \text{ coordinates}]^2$$

or
$$s^2 = [q \cos \theta - (p + t \cos \phi)]^2 + [q \sin \theta - t \sin \phi]^2 *$$

$$= p^2 + q^2 + t^2 + 2\, p\, t \cos \phi - 2\, p\, q \cos\, \theta - 2\, q\, t\, (\cos \theta \cos \phi + \sin \theta \sin \phi)$$

Dividing this expression by $2\,qt$ and after rearranging, the following equation is obtained :

$$\cos(\theta - \phi) = \frac{p}{q} \cos \phi - \frac{p}{t} \cos \theta + \frac{p^2 + q^2 - s^2 + t^2}{2qt}$$

or
$$\cos(\theta - \phi) = K_1 \cos \phi - K_2 \cos \theta + K_3 \qquad \text{...(20.13)}$$

where
$$K_1 = \frac{p}{q},\ K_2 = \frac{p}{t},\ K_3 = \frac{p^2 + q^2 - s^2 + t^2}{2qt} \qquad \text{...(20.13 A)}$$

The following steps are followed to determine length ratios :

(*i*) Obtain three accuracy points by using Chebyshev's spacing *i.e.*, x_1, x_2 and x_3 from Eq. (20. 12).

(*ii*) Obtain the corresponding values of y_1, y_2 and y_3 by using given function $y = f(x)$.

(*iii*) By using Eqs. (20.12 B) and (20.12 C), determine (θ_1, θ_2 and θ_3) and (ϕ_1, ϕ_2 and ϕ_3) corresponding to values of x and y.

$$*s^2 = [q^2 \cos^2 \theta + (p + t \cos \phi)^2 - 2q \cos \theta\, (p + t \cos \phi)] + [q^2 \sin^2 \theta + t^2 \sin^2 \phi - 2qt \sin \theta \sin \phi]$$
$$= [q^2 \cos^2 \theta + (p + t \cos \phi)^2 - 2q \cos \theta\, (p + t \cos \phi)] + [q^2 \sin^2 \theta + t^2 \sin^2 \phi - 2qt \sin \theta \sin \phi]$$
$$= [q^2 \cos^2 \theta + (p^2 + t^2 \cos^2 \phi + 2pt \cos \phi) - 2qp \cos \theta - 2qt \cos \theta \cos \phi] + [q^2 \sin^2 \theta + t^2 \sin^2 \phi - 2qt \sin \theta \sin \phi]$$
$$= q^2 (\cos^2 \theta + \sin^2 \theta) + p^2 + t^2 (\cos^2 \phi + \sin^2 \phi) + 2pt \cos \phi - 2pq \cos \theta - 2qt (\cos \theta \cos \phi + \sin \theta \sin \phi)$$
$$= q^2 + p^2 + t^2 + 2pt \cos \phi - 2pq \cos \theta - 2qt (\cos \theta \cos \phi + \sin \theta \sin \phi)$$

(*iv*) Obtain three equations in K_1, K_2 and K_3 by substituting the three related pairs of the angles (θ_1 and ϕ_1), (θ_2 and ϕ_2), and (θ_3 and ϕ_3) in equation (20.13).

(*v*) Solve these three equations for K_1, K_2 and K_3.

(*vi*) Determine length ratios from values of K_1, K_2 and K_3.

(*v*) Assume one link length or it may be given. Determine other lengths which can be determined from the length ratios. The lengths *p* and *s* will always be positive. The negative values of *q* or *t* can be interpreted in the vector sense. The negative sense will mean that their layouts are opposite to the fixed link.

Problem 20.1. Determine the lengths of all the four links in a four bar chain for the length of the smallest being 10 cm to generate $y = \log_{10} x$ in the interval $1 \le x \le 10$ for three accuracy points. The range of angles of input link and output link are

$$45° \le \theta \le 105° \text{ and } 135° \le \phi \le 225°.$$

Solution. The initial and final values of *x* are :

$x_i = 1.0$ and $x_f = 10.0$; The values of '*a*' and '*b*' for equation (20.12) are given by equation (20.12 A)

as $$a = \frac{x_i + x_f}{2} = \frac{1+10}{2} = 5.5$$

and $$b = \frac{x_f - x_i}{2} = \frac{10-1}{2} = 4.5$$

Number of accuracy points, $k = 3$

From Eq. (20.12), $$x_1 = a + b\cos\frac{(2-1)\pi}{2\times 3} = 5.5 + 4.5\cos\frac{\pi}{6} = 9.4$$

$$\left[\because\ x_i = a + b\cos\frac{(2m-1)\pi}{2k} \text{ where } m = 1, 2, 3 \ldots\right]$$

$$x_2 = 5.5 + 4.5\cos\frac{(2\times 2-1)\pi}{2\times 3} = 5.5 \qquad (\because\ \text{Here } m = 2, k = 3)$$

$$x_3 = 5.5 + 4.5\cos\frac{(2\times 3-1)\pi}{2\times 3} = 1.6 \qquad (\because\ \text{Here } m = 3)$$

The corresponding values of y_1, y_2 and y_3 can be determined by using given function (*i.e.*, $y = \log_{10} x$)

$$\therefore \quad y_1 = \log_{10}(x_1) = \log_{10}(9.4) = 0.974$$
$$y_2 = \log_{10}(x_2) = \log_{10}(5.5) = 0.741$$
$$y_3 = \log_{10}(x_3) = \log_{10}(1.6) = 0.204$$
$$y_i = \log_{10}(x_i) = \log_{10}(1.0) = 0$$
$$y_f = \log_{10}(x_f) = \log_{10}(10) = 1.0$$

The values of r_x and r_y are obtained from equations (20.12 B) and (20.12 C) as

$$r_x = \frac{\theta_f - \theta_i}{x_f - x_i} = \frac{105-45}{10-1} = \frac{60}{9}$$

and $$r_y = \frac{\phi_f - \phi_i}{y_f - y_i} = \frac{225-135}{1.0-0} = \frac{90}{1} = 90$$

The values of (θ_1, ϕ_1), (θ_2, ϕ_2) and (θ_3, ϕ_3) are obtained from equations (20. 12 B) and (20.12 C) as

$$r_x = \frac{\theta - \theta_i}{x - x_i} = \frac{\theta_1 - \theta_i}{x_1 - x_i} = \frac{\theta_2 - \theta_i}{x_2 - x_i} = \frac{\theta_3 - \theta_i}{x_3 - x_i} = \text{constant}$$

and

$$r_y = \frac{\phi - \phi_i}{y - y_i} = \frac{\phi_1 - \phi_i}{y_1 - y_i} = \frac{\phi_2 - \phi_i}{y_2 - y_i} = \frac{\phi_3 - \phi_i}{y_3 - y_i} = \text{constant}$$

$$\therefore \quad \theta_1 = \theta_i + r_x (x_1 - x_i) = 45° + \frac{60}{9}(9.4 - 1) = 45 + \frac{60}{9} \times 8.4 = 101°$$

$$\phi_1 = \phi_i + r_y (y_1 - y_i) = 135° + 90 \times 0.974 = 222.7°$$

$$\theta_2 = 45° + \frac{60}{9}(x_2 - x_i) = 45° + \frac{60}{9}(5.5 - 1.0) = 45° + 30° = 75°$$

$$\phi_2 = 135° + 90(y_2 - y_i) = 135° + 90(0.741 - 0) = 201.7$$

$$\theta_3 = 45° + \frac{60}{9}(x_3 - x_i) = 45° + \frac{60}{9}(1.6 - 1.0) = 45° + \frac{60}{9} \times 0.6 = 49.0$$

$$\phi_3 = 135° + 90(y_3 - y_i) = 135° + 90(0.204 - 0) = 153.40$$

For one pair of rotation (101°, 222.7°), the Freudenstein Eq. (20.13) becomes

$$\cos(101 - 222.7) = K_1 \cos 222.7° - K_2 \cos 101° + K_3 \qquad [\because \ \theta = 101° \text{ and } \phi = 222.7°]$$

or

$$-0.5255 = -0.7349 K_1 + 0.1908 K_2 + K_3 \qquad ...(20.14)$$

Substituting second pair of rotation (75°, 201.7°) in Eq. (20.13), it becomes

$$\cos(75 - 201.7) = K_1 \cos 201.7° - K_2 \cos 75° + K_3$$

or

$$-0.5976 = -0.9291 K_1 - 0.2588 K_2 + K_3 \qquad ...(20.15)$$

Substituting third pair of rotation (49°, 153.4°) in Eq. (20.13), it becomes

$$\cos(49 - 153.4) = K_1 \cos 153.4° - K_2 \cos 49° + K_3$$

or

$$-0.2487 = -0.8942 K_1 - 0.6561 K_2 + K_3 \qquad ...(20.16)$$

In equations (20.14), (20.15) and (20.16), there are three unknown which can be solved as :

Subtracting Eq. (20.15) from Eq. (20.14) and Eq. (20.16), the following equations are obtained.

$$0.0721 = 0.1942 K_1 + 0.4496 K_2 \qquad ...(20.17)$$

$$0.3489 = 0.0349 K_1 - 0.3973 K_2 \qquad ...(20.18)$$

Multiplying Eq. (20.17) by 3.49 and Eq. (20.18) by 19.42 the following equations are obtained.

$$0.2516 = 0.6778 K_1 + 1.6691 K_2 \qquad ...(20.19)$$

$$6.7756 = 0.6778 K_1 - 7.7156 K_2 \qquad ...(20.20)$$

Subtracting Eq. (20.20) from Eq. (20.19), the following equation is obtained.

$$-6.524 = 9.2847 K_2 \quad \text{or} \quad K_2 = -0.703$$

From Eq. (20.19) after substituting for K_2

$$0.2516 = 0.6778 K_1 + 1.5691 \times (-0.703)$$

or

$$K_1 = +1.0772$$

But from equation (20.13 A), we have

$$K_1 = \frac{p}{q}, K_2 = \frac{p}{t} \text{ and } K_3 = \frac{p^2 + q^2 - s^2 + t^2}{2qt}$$

since

$$K_1 = 1.9986 \quad \therefore \quad \frac{p}{q} = 1.9986$$

since $K_2 = -0.703 \quad \therefore \quad \dfrac{p}{t} = -0.703$

$K_3 = 1.0772 \quad \therefore \quad \dfrac{p^2+q^2-s^2+t^2}{2qt} = 1.0772$

To find the value of $\left(\dfrac{p}{s}\right)$, substitute the value of q and t in terms of p in the above equation.

$$\therefore \quad \frac{p^2+\left(\dfrac{p}{1.9986}\right)^2 - s^2 + \left(\dfrac{p}{-0.0703}\right)^2}{2\times\left(\dfrac{p}{1.9986}\right)\times\left(\dfrac{p}{-0.703}\right)} = 1.0772 \qquad \left(\because\ q = \frac{p}{1.9986} \text{ and } t = \frac{p}{-0.703}\right)$$

or $$p^2 + \frac{p^2}{3.9944} - s^2 + \frac{p^2}{(-0.703)^2} = 1.0772\times\frac{2p^2}{(1.9986)(-0.703)}$$

or $$p^2 + \frac{p^2}{3.9944} + \frac{p^2}{(-0.703)^2} + \frac{1.0772\times 2p^2}{(1.9986)(0.703)} = s^2$$

or $$p^2 + 0.25p^2 + 2.023p^2 + 1.533p^2 = s^2$$

or $$4.806p^2 = s^2$$

or $$\frac{p}{s} = \left(\frac{1}{4.806}\right)^{\frac{1}{2}} = 0.456$$

Already we have calculated the values of $\dfrac{p}{q}$ and $\dfrac{p}{t}$.

These values are : $\dfrac{p}{q} = 1.9986$ and $\dfrac{p}{t} = -0.703$ (Negative value only shows negative sense)

Considering the magnitudes only, we have

$$\frac{p}{q} > \frac{p}{t} > \frac{p}{s} \qquad \left[\text{As } \frac{p}{q} = 1.9986, \frac{p}{t} = 0.703 \text{ and } \frac{p}{s} = 0.456\right]$$

or $$q < p < t < s$$

This means that the length of link q is smallest.

But length of smallest link = 10 cm (given)

$\therefore \qquad q = 10 \text{ cm}$

As $\dfrac{p}{q} = 1.9986 \quad \therefore \quad p = 1.9986 \times q = 1.9986 \times 10 = 19.986$ cm

As $\dfrac{p}{t} = -0.703 \quad \therefore \quad t = \dfrac{p}{(-0.703)} = \dfrac{-19.986}{0.703} = -28.43$ cm (Negative)

As $\dfrac{p}{s} = 0.456 \quad \therefore \quad s = \dfrac{p}{0.456} = \dfrac{19.986}{0.456} = 43.83$ cm

$\therefore$ Length of all four links are :

$p = 19.986$ cm, $q = 10$ cm, $s = 43.83$ cm and $t = -28.43$ cm. **Ans.**

Freudenstein's Equation for Slider Crank Chain for Three Accuracy Points

In slider crank mechanism, the displacement of the slider '*B*' has to be coordinated with the rotation of crank O_2A. The mechanism is shown in Figure 20.12. As in case of four bar chain, if the displacement of the slider is assumed proportional to the crank rotation, it can be expressed as follows :

$$p - p_i = C(\theta - \theta_i) \qquad ...(i)$$

For an interval $\theta_i \leq \theta \leq \theta_f$

$$p_f - p_i = C(\theta_f - \theta_i)$$

[In equation (*i*), when $\theta = \theta_f$, then $p = p_f$]

Where C = constant of proportionality,

p_i = distance of slider from origin, and

θ = angle of rotation of crank from the line of stroke.

Subscript '*i*' refers to the initial value and '*f*' to final value.

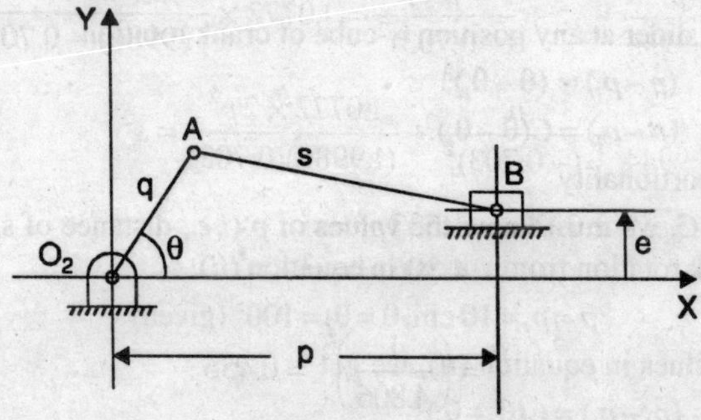

Fig. 20.12

There are five design parameters which are θ, p, q, s and e. For three accuracy points the three parameters q, s and e are determined. The coordinates of points A and B are $(q\cos\theta, q\sin\theta)$ and (p, e) respectively.

The distance

$$AB^2 = s^2 = (x_A - x_B)^2 + (y_A - y_B)^2$$

$$AB^2 = (q\cos\theta - p)^2 + (q\sin\theta - e)^2$$

$$= q^2\cos^2\theta + p^2 - 2pq\cos\theta + q^2\sin^2 + e^2 - 2qe\sin\theta$$

$$= q^2(\cos^2\theta + \sin^2\theta) + p^2 - 2pq\cos\theta + e^2 - 2qe\sin\theta$$

$$s^2 = q^2 + p^2 - 2pq\cos\theta - 2qe\sin\theta + e^2$$

$$\therefore \quad p^2 = K_1 p\cos\theta + K_2\sin\theta - K_3 \qquad ...(20.21)$$

where $K_1 = 2q,\ K_2 = 2qe$

and $K_3 = q^2 - s^2 + e^2$ $\qquad ...(20.21\text{ A})$

Eq. (20.21) is the Freudenstein's equation for slider crank chain.

The following steps are followed to determine lengths :

(*i*) Obtain three accuracy points by using Chebyshev's spacing *i.e.*, θ_1, θ_2 and θ_3.

(*ii*) Obtain corresponding values of p_1, p_2 and p_3.

(*iii*) Obtain three equations in K_1, K_2 and K_3.

(*iv*) Obtain q, s and e from values of K_1, K_2 and K_3.

Problem 20.2. *Design a slider crank mechanism so that displacement of the slider is proportional to cube of the crank rotation in the interval of 30° ≤ θ ≤ 100°. Assume initial distance of the slider equal to 15 cm and final distance to be 10 cm.*

Solution. Given : (Refer to Fig. 20.12)

Displacement of slider ∝ cube of crank rotation ...(*i*)

Interval of crank rotation is $30° \le \theta \le 100°$

This means, initial angle of crank with *x*-axis, $\theta_i = 30°$

Final angle of crank with *x*-axis, $\theta_f = 100°$

Initial distance of slider from *y*-axis, $p_i = 15$ cm

Final distance of slider from *y*-axis, $p_f = 10$ cm

Let p = distance of slider from *y*-axis, when θ is the angle turned by crank with respect to *x*-axis.

∴ Displacement of slider from initial position $= (p - p_i)$

and crank rotation from initial position $= (\theta - \theta_i)$

But displacement of slider at any position ∝ cube of crank rotation

$$\therefore \quad (p - p_i) \propto (\theta - \theta_i)^3$$

or
$$(p - p_i) = C(\theta - \theta_i)^3 \quad \text{...(ii)}$$

where C is constant of proportionality

To find the value of C, we must know the values of p (*i.e.*, distance of slider from *y*-axis) and θ(*i.e.*, corresponding angle of crank rotation from *x*-axis) in equation (*ii*).

When $p = p_f = 10$ cm, $\theta = \theta_f = 100°$ (given)

Substituting these values in equation (*ii*), we get

$$(p_f - p_i) = C(\theta_f - \theta_i)^3$$

or
$$C = \frac{(p_f - p_i)}{(\theta_f - \theta_i)^3}$$

$$= \frac{10 - 15}{(100 - 30)^3} = \frac{-5}{70^3} = -\frac{5}{343000}$$

Substituting the value of C in equation (*ii*), we get

$$(p - p_i) = \left(\frac{-5}{343000}\right)(\theta - \theta_i)^3 \quad \text{...(iii)}$$

The accuracy points according to Chebyshev's spacing are given by equation (20.12) as

$$\theta_i = a + b \cos \frac{(2m - 1)\pi}{2k}$$

where $a = \frac{\theta_i + \theta_f}{2}$, $b = \frac{\theta_f - \theta_i}{2}$, $m = 1, 2, 3, \ldots$, and k = No. of accuracy points = 3 in present case.

$$a = \frac{\theta_i + \theta_f}{2} = \frac{30 + 100}{2} = 65° \; ; \; b = \frac{\theta_f - \theta_i}{2} = \frac{100 - 30}{2} = 35°$$

$$\therefore \quad \theta_1 = a + b \cos \frac{(2m - 1)\pi}{2k} = 65 + 35 \cos \frac{(2 \times 1 - 1)\pi}{6} \quad (\because \; m = 1, k = 3)$$

$$= 65 + 35 \cos \frac{5\pi}{6} = 95.3$$

$$\theta_2 = 65 + 35 \cos \frac{\pi}{2} = 65° \qquad (m = 2, k = 3)$$

$$\theta_3 = 65 + 35 \cos \frac{5\pi}{6} = 34.69° \qquad (m = 3, k = 3)$$

The distance of slider from y-axis for three points (*i.e.*, distance p_1, p_2 and p_3) is obtained by using equation (*iii*) given above as

$$p - p_i = \frac{-5}{34300} (\theta - \theta_i)^3$$

or

$$p = p_i + \left(\frac{-5}{34300}\right)(\theta - \theta_i)^3$$

When $\theta = \theta_1$ then $p = p_1$

$$\therefore \quad p_1 = p_i + C(\theta_1 - \theta_i)^3 = 15 + \left(\frac{-5}{343000}\right)(95.3 - 30)^3 = 10.94 \text{ cm}$$

When $\theta = \theta_2$ then $p = p_2$

$$\therefore \quad p_2 = 15 + \left(\frac{-5}{343000}\right)(65 - 30)^3 = 14.375 \text{ cm}$$

When $\theta = \theta_3$ then $p = p_3$

$$\therefore \quad p_3 = 15 + \left(\frac{-5}{343000}\right)(34.69 - 30)^3 = 14.998 \text{ cm}$$

Substituting (95.3, 10.94) in Eq. (20.21) in which $p = p_1 = 10.94$ and $\theta = \theta_1 = 95.3°$

$$10.94^2 = K_1 \times 10.94 \cos 95.3 + K_2 \sin 95.3 - K_3$$

or

$$109.68 = -K_1 + K_2 - K_3 \qquad ...(1)$$

Similarly substituting (65°, 14.375) in Eq. (20.21) in which $p = p_2 = 14.375°$ and $\theta = \theta_2 = 65°$

$$14.375^2 = K_1 \times 14.375 \cos 65° + K_2 \sin 65° - K_3$$

$$206.64 = 6.08 K_1 + 0.9 K_2 - K_3 \qquad ...(2)$$

For (34.69°, 14.998) equation (20.21) becomes as [Here $p = p_3 = 14.998$ and $\theta = \theta_3 = 34.69°$]

$$14.998^2 = K_1 \times 14.998 \cos 34.69 + K_2 \sin 34.69 - K_3$$

$$224.94 = 12.3 K_1 + 0.57 K_2 - K_3 \qquad ...(3)$$

There are three unknowns in equations 1, 2 and 3 and they can be solved as :

Subtracting Eqs. (1) and (2) from Eq. (3)

$$115.26 = 13.3 K_1 - 0.43 K_2 \qquad ...(4)$$

$$18.30 = 6.22 K_1 - 0.33 K_2 \qquad ...(5)$$

From Eq. (4), $\quad 268 = 30.93 K_1 - K_2 \qquad ...(6)$

From Eq. (5), $\quad 55.45 = 18.8 K_1 - K_2 \qquad ...(7)$

Subtracting Eq. (7) from Eq. (6)

$$212.55 = 12.13K_1$$

or $$K_1 = 17.52$$

From Eq. (7), $$K_2 = 18.8 \times 17.52 - 55.45 = 273.9$$

From Eq. (1), $$K_3 = -17.52 + 273.9 - 109.68 = 146.72$$

But from equation (20.21 A),

$$K_1 = 2q = 17.52 \quad \therefore \quad q = 8.76 \quad \textbf{Ans.}$$

$$K_2 = 2qe = 273.9 \quad \therefore \quad e = \frac{273.9}{2 \times 8.76} = 15.6 \text{ cm} \quad \textbf{Ans.}$$

$$K_3 = q^2 - s^2 + e^2 = 146.7$$

or $$s = 13.2 \text{ cm} \quad \textbf{Ans.}$$

20.4. MECHANISMS FOR POSITION GUIDANCE

Mechanism may be required to provide guidance to a body in a prescribed manner. In a problem where a rigid body has to be so guided that it attains three positions during its movement as shown in Fig. 20.13.

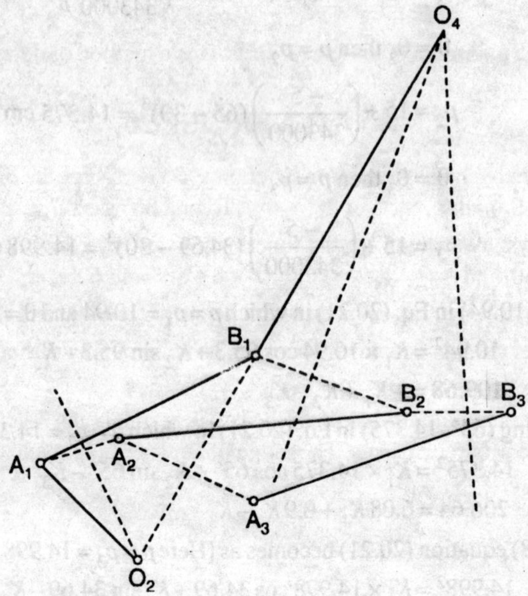

Fig. 20.13

This can be achieved by a four bar linkage for which this body acts as the coupler. The fixed hinges O_2 and O_4 are the centres of the circles passing through the points A_1, A_2, A_3 and B_1, B_2, B_3 respectively. Join A_1 with A_2 and A_2 with A_3. Draw perpendicular bisectors of $A_1 A_2$ and $A_2 A_3$. These perpendicular bisectors will meet at O_2. Join B_1 with B_2 and B_2 with B_3. Draw perpendicular bisectors of $B_1 B_2$ and $B_2 B_3$. These bisectors will meet at O_4. Thus the four bar mechanism $O_2 A_1 B_1 O_4$ so constituted ensures that the guided body takes up the given positions in succession.

20.5. SUMMARY

The mechanism requires several links and their proportional lengths to perform a desired function. The degree of freedom of a mechanism should be equal to one. The number of links and joints are related to degree of freedom. With simple hinges, the number of links can be 4, 6, 8 etc. but for number of links more than six, unique solution cannot be obtained. For a mechanism which has higher order hinges, equivalent number of simple hinges can be determined by means of a relation given in this chapter.

To determine lengths of the links in a mechanism graphical or analytical method of synthesis can be used. The approximate synthesis methods have zero error at accuracy points which can be selected to minimise the error at other points.

The mechanism can also be determined to provide guided motion. If three position of a body are given, a four bar mechanism can be designed to provide such a motion by a graphical method explained in this chapter.

HIGHLIGHTS

(*i*) **Pole.** It is the centre of finite rotation of the coupler.

(*ii*) **Relative Pole.** It is the pole of coupler relative to other moving link.

(*iii*) **Accuracy Points.** The points at which desired function and generated functions meet.

(*iv*) **Structural Error.** The differences between desired and generated functions is called structural error.

(*v*) **Chebyshev's Spacing of Accuracy Points.** It is a method of determining accuracy points.

EXERCISE 20

1. Determine the Chebyshev spacing for function $y = x + 3x^2$ for the domain $0 \le x \le 3$. For these points determine $\theta_1, \theta_2, \theta_3$ and ϕ_1, ϕ_2, ϕ_3 if $20° \le \theta \le 80°$ and $30° \le \phi \le 130°$.
2. Determine Chebyshev spacing for the function $y = 3x^3 - 2x$ for $0 \le x \le 1$ and specify three precision points. Also find θ_j and ϕ_j $(j = 1, 2, 3)$ for $20° < \theta \le 80°$ and $60° \le \phi \le 190°$.
3. Synthesise a four bar mechanism such that $\theta_{12} = 50°$, $\theta_{23} = 40°$ and $\phi_{12} = 80°$, $\phi_{23} = 50°$.
4. Synthesise a slider crank mechanism for which position of slider is proportional to crank rotation for the domain $20° \le \theta \le 80°$ and $12 \text{ cm} \le p \le 7 \text{ cm}$.
5. Synthesise a four bar chain for length of the smallest link being 15 cm to generate $y = e^x$ in the interval $0 \le x \le 1$ for three accuracy points. The range of angles of input and output link are $30° \le \theta \le 180°$ and $40° \le \phi \le 120°$.

21

Gyroscopic Effects and Gyroscope

21.1. INTRODUCTION

Newton carried out studies on the motion of earth which rotates about its axis and at the same time it revolves around the sun. These studies were further carried out by Euler and he laid down the mathematical foundation of the principles of gyroscopic motion.

To understand gyroscopic effect, a body can be considered which rotates about its axis of symmetry. It will be found to offer a resistance to any change in the direction of the axis of rotation. This is gyroscopic effect. A gyroscope is an instrument which consists a heavy spinning rotor. It resists any change in the axis of the rotor. One of the popular example of gyroscope is the spinning top.

The use of gyroscope was earlier limited to stabilisation of seaborne ships and for gyroscopic compasses. After World War II, it found application in bomb-sights, aeroplanes, guided missiles etc. The gyroscopic effects generates forces and couples which act on the vehicles and other means of transport like ships, aeroplanes etc. These effects must be taken into account while designing them especially in selection of bearings etc.

21.2. SPINNING AND PRECESSION

The angular motion of a body about its axis of symmetry is spinning motion. The angular velocity (ω) is a vector quantity and to specify it the following is required : magnitude, direction of axis of spin and sense whether clockwise or anticlockwise. The spin vector is normal to the plane of spinning. To determine sense of spin vector right hand screw rule can be used : keeping fingers curving on the inside pointing towards the sense of rotation, stretch out thumb of right hand outwards perpendicular to fingers as shown in the Fig. 21.1. The thumb indicates direction of spin vector. The reverse sense of rotation will give opposite sense of spin vector.

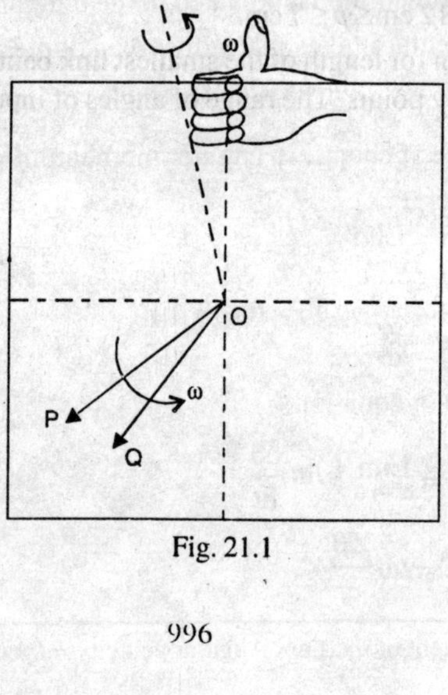

Fig. 21.1

Precessional motion is also angular motion but in a plane other than spin plane. To differentiate between spinning and precession consider a disc of massmoment of inertia '*I*' which is spinning with angular velocity 'ω' as shown in Fig. 21.2. The plane of the disc is normal to the X-axis. Therefore, X-axis is the spinning axis. If X-axis rotates about Y-axis with angular velocity 'ω_p' as shown in Fig. 21.2, then the rotation of X-axis about Y-axis is precessional motion and precessional angular velocity is 'ω_p'.

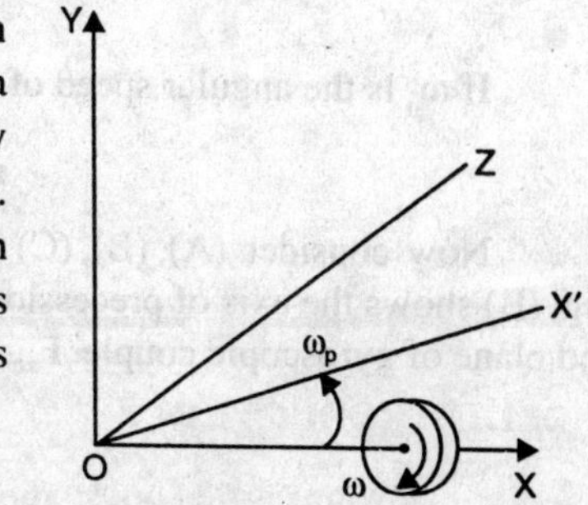

Fig. 21.2

21.3. GYROSCOPIC COUPLE

As discussed in the preceding section, consider a disc of mass moment of inertia '*I*' which is spinning with angular speed 'ω' along the axis of spin *OX* in clockwise direction as seen from *O* towards *X* as shown in Fig. 21.3. The plane in which the disc lies is normal to X-axis and this plane is known as spinning plane *i.e., YOZ* is the spinning plane. The plane *XOZ* is a horizontal plane. Let the spinning axis *OX* rotate in *XOZ* plane. Therefore, *OY* axis is the axis of precession and *XOZ* is the plane of precession. The angular momentum* of the spinning disc is '*I*ω', the direction of it is given by right hand screw rule. Looking from *O* towards *X* and keeping fingers of right hand curving on the inside pointing towards the sense of rotation, stretch out thumb outwards. The thumb will indicate the direction of angular momentum. In the present case, thumb is in the direction of *OX* (*i.e.* vector in the direction from *O* to *X*). By assuming suitable scale it can be plotted as shown in Fig. 21.3 (*a*). It is represented by vector *oa*. Let after time 'δ*t*' axis of precession occupies position *OX′* making angle δθ with initial position. The angular momentum vector is now represented by '*ob*' and the change in angular momentum is indicated in the diagram by vector '*ab*'.

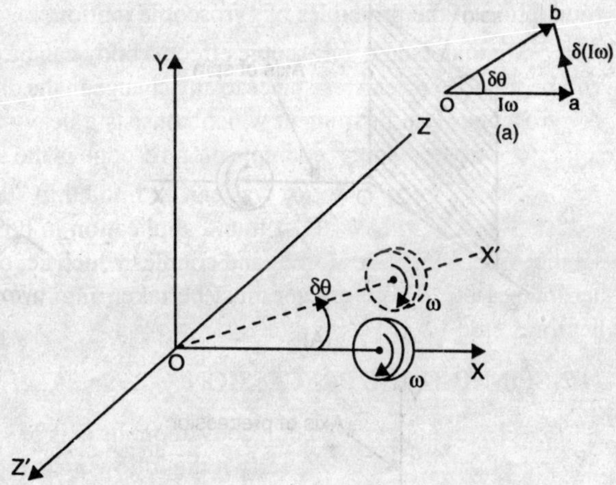

Fig. 21.3

The change in angular momentum takes place due to a couple and the couple is equal to the rate of change of angular momentum. Let the couple is represented by *C*.

Therefore, couple (*C*) = Rate of change in angular momentum = $\frac{\text{Change of angular momentum}}{\text{Time}}$

or
$$C = \lim_{\delta t \to 0} \frac{\delta (I\omega)}{\delta t} \quad \text{...}(i)$$

From vector diagram $\delta(I\omega) \rightarrow ab \simeq oa\,\delta\theta = I\omega\,\delta\theta$ $\quad (\because\ oa = I\omega)$

or
$$\delta(I\omega) \simeq I\omega\,\delta\theta$$

Substituting the above value in equation (*i*), we get

$$C = \lim_{\delta t \to 0} \left(I\omega \frac{\delta\theta}{\delta t} \right)$$

or
$$C = I\omega \frac{d\theta}{dt}$$

*Angular momentum = Mass moment of inertia × Angular velocity = $I \times \omega$.

If ω_p is the angular speed of precession, then, $\frac{d\theta}{dt} = \omega_p$

$$\therefore \qquad C = I\omega\,\omega_p \qquad \text{...(21.1)}$$

Now consider (A), (B), (C) and (D) of Fig. 21.3 (*b*). Fig. (A) shows the axis of spin and plane of spin. Fig. (B) shows the axis of precession and plane of precession. Fig. (C) shows the axis of gyroscopic couple and plane of gyroscopic couple. Fig. (D) shows all the three axis and three planes.

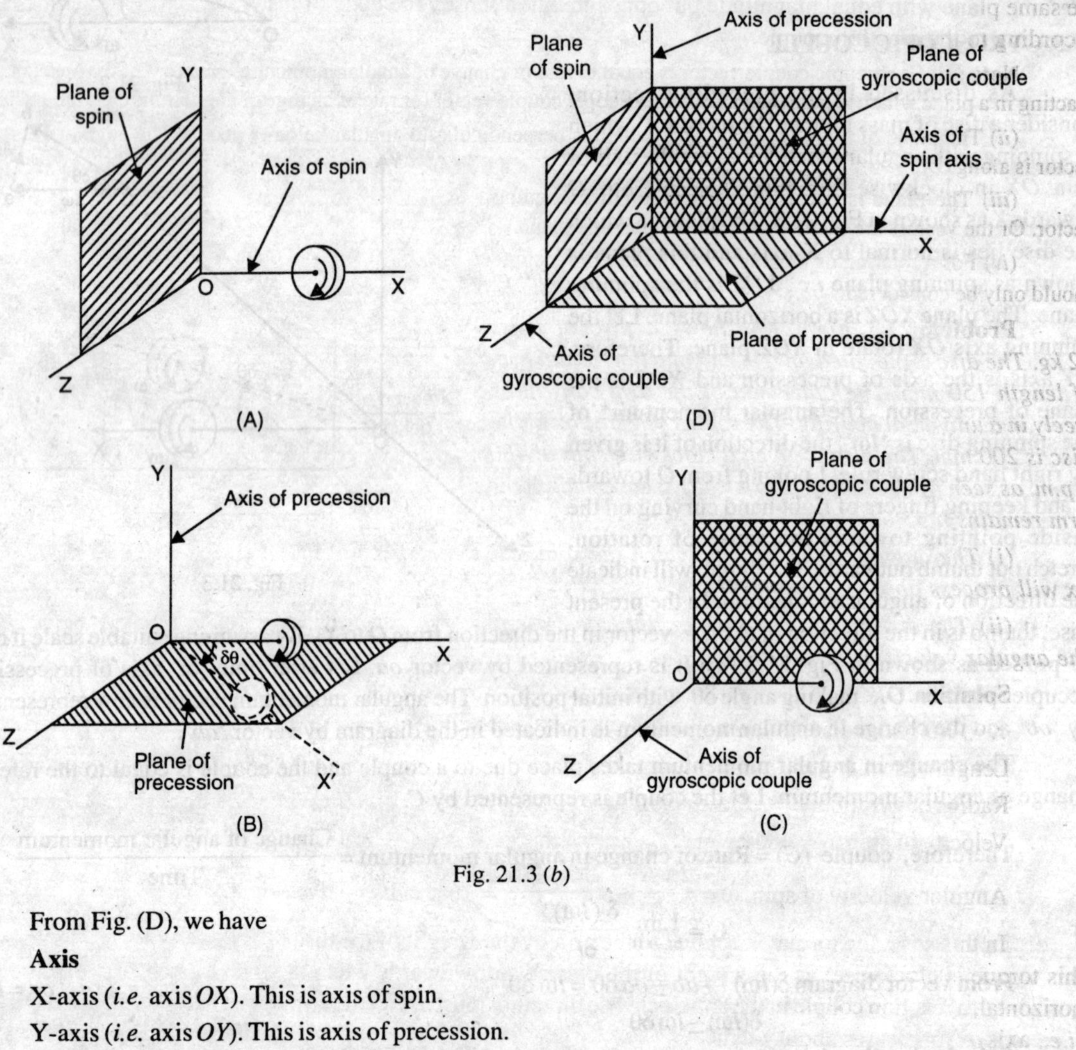

Fig. 21.3 (*b*)

From Fig. (D), we have

Axis

X-axis (*i.e.* axis *OX*). This is axis of spin.

Y-axis (*i.e.* axis *OY*). This is axis of precession.

Z-axis (*i.e.* axis *OZ*). This is axis of gyroscopic couple.

Planes

XOZ is a horizontal plane. It is also plane of precession.

YOZ is a vertical plane. It is also known as plane of spinning.

XOY is a vertical plane. It is also known as plane of gyroscopic couple.

Every plane is normal to the other two planes.

The couple represented by 'C' in equation (21.1) is the gyroscopic couple. Since the angular momentum change vector (*ab*) in the present case lies in the *XOZ* plane (or the horizontal plane), therefore, gyroscopic couple vector will also lie in this plane. The plane of gyroscopic couple vector and the plane of gyroscopic couple will be at right angles*. In case of a very small displacement, the vector '*ab*' will be perpendicular to the vertical plane *XOY*. Hence the gyroscopic couple will be in the vertical plane *XOY*. Therefore, the couple plane will be *XOY* plane, which is perpendicular to spinning plane and precessional plane. The sense of the couple can be determined with the help of right hand screw rule. Align thumb parallel to '*ab*' and fingers will indicate the sense of the couple. This couple is accompanied by the **reaction couple** which acts in the same plane with equal magnitude but opposite in the sense. The body or disc in this article shall behave according to the reaction couple.

Note : (*i*) Gyroscopic couple vector is equal to rate of change of angular momentum vector. Gyroscopic couple is acting in a plane which is perpendicular to gyroscopic couple vector (or rate of change of angular momentum vector).

(*ii*) The plane of rotating disc is *YOZ* which is perpendicular to angular velocity vector. The angular velocity vector is along *OX*.

(*iii*) The plane is perpendicular to the corresponding vector. Or the vectors are normal to the corresponding plane.

(*iv*) For a rotating or moving body the reaction couple should only be considered.

Problem 21.1. *Fig. 21.4 shows a disc of mass 12 kg. The disc is mounted on one end of a horizontal arm of length 150 mm. The other end of the arm can rotate freely in a universal bearing. The radius of gyration of the disc is 200 mm. The disc is given a clockwise spin of 480 r. p.m. as seen from O about its axis which is horizontal. If the arm remains horizontal, then determine :*

(i) The angular velocity with which the spin axis ox will process about y-axis and

(ii) The direction (clockwise or anticlockwise of the angular velocity of precession.

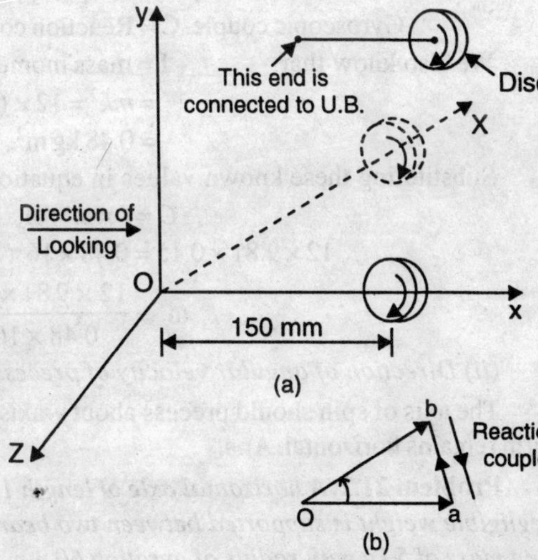

Fig. 21.4

Solution. Given : Mass of disc, $m = 12$ kg.

$\therefore$ Weight of disc, $W = m \times g = 12 \times 9.81$ N

Length of arm, $L = 150$ mm $= 0.15$ m

Radius of gyration, $k = 200$ mm $= 0.2$ m

Velocity of spin, $N = 480$ r.p.m.

Angular velocity of spin, $\omega = \dfrac{2\pi N}{60} = \dfrac{2\pi \times 480}{60} = 16\pi$ rad/s.

In this case, the torque is applied to the arm by the weight of the disc (Torque = W × L). The sense of this torque is clockwise, as the weight of the disc acts downwards when the disc rotates. To keep the arm horizontal, a reaction couple in the anticlockwise direction must act on the arm, when the arm or the axis of spin (*i.e.*, axis *ox*) precesses about *y*-axis.

Let us assume that the axis of spin precesses about *y*-axis in the anticlockwise direction looking from top as shown in Fig. 21.4 (*a*). The vector *oa* will rotate to the position *ob*. Then vector *ab* is the applied couple and vector *ba* will be reaction couple and perpendicular to *oa* in the limit. This couple acts in the vertical plane. The sense of the vector *ab* can be determined by stretching the thumb of the right hand outwards from *a* to *b*, keeping fingers curving on the inside and looking from the front. Then the curve of the fingers points the sense of rotation. In this case the sense of vector *ab* is clockwise. Hence the sense of the reaction couple is

*Also see Article 2.7 on page 35.

opposite to vector *ab*. Hence the sense of the reaction couple (*i.e.*, vector *ba*) is anticlock. We have already seen that if the reaction couple is acting on the arm in the anti-clockwise direction, the arm will remain horizontal. Hence our assumption that axis of spin should process about *y*-axis in the anticlockwise direction is true.

(*i*) *Angular velocity with which the spin axis ox precesses about y-axis.*

Let ω_p = Angular velocity of precession

Using equation (21.1), we get

$$C = I \times \omega \times \omega_p \qquad \ldots(i)$$

where C = gyroscopic couple or reaction couple.

Now torque (or couple) applied to the arm

= Weight of disc × Distance of weight from *y*-axis

$= W \times L = m \times g \times L$

$= 12 \times 9.81 \times 0.15$ Nm. $\quad (\because \; m = 12\text{ kg}, L = 0.15\text{ m})$

∴ Reaction couple = Torque applied = $12 \times 9.81 \times 0.15$ Nm

or Gyroscopic couple, C = Reaction couple = $12 \times 9.81 \times 0.15$ Nm

We also know that I = mass moment of inertia of disc

$= mk^2 = 12 \times (0.2)^2 \quad (\because \; k = 0.2\text{ m})$

$= 0.48\text{ kg m}^2.$

Substituting these known values in equation (*i*) above, we get

$$C = I \times \omega \times \omega_p$$

or $$12 \times 9.81 \times 0.15 = 0.48 \times 16\pi \times \omega_p \qquad (\because \; \omega = 16\pi \text{ rad/s})$$

∴ $$\omega_p = \frac{12 \times 9.81 \times 0.15}{0.48 \times 16\pi} = \mathbf{0.731\ rad/s.}$$ **Ans.**

(*ii*) *Direction of angular velocity of precession.*

The axis of spin should precess about *y*-axis in the anticlockwise direction (looking from top) so that the arm remains horizontal. **Ans.**

Problem 21.2. *A horizontal axle of length 100 mm and of negligible weight is supported between two bearing. A disc having mass of 5 kg with radius of gyration 60 mm is mounted centrally on the axle. The centre line of the axle, disc and the bearings lie in a horizontal plane. The disc is rotating at 720 r.p.m. in the clockwise direction looking from the left hand side bearing. The axle precesses about a vertical axis at 60 r.p.m. in the clockwise direction looking from the top. Determine the resultant reaction at each bearing due to the weight of the disc and gyroscopic effect.*

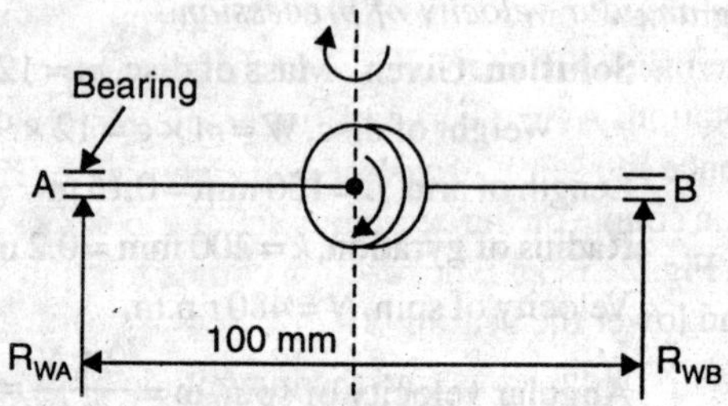

Solution. Given : Length of axle, L = 100 mm = 0.1 m

Mass of disc, m = 5 kg.

∴ Weight of disc, $W = m \times g = 5 \times 9.81$ N

Radius of gyration, k = 60 mm = 0.06 m

∴ Mass moment of inertia of disc, $I = mk^2 = 5 \times (0.6)^2\text{ kg m}^2 = 0.018\text{ kg m}^2$

Speed of disc, N = 720 r.p.m.

∴ Angular velocity of disc, $\omega = \frac{2\pi N}{60} = \frac{2\pi \times 720}{60} = 24\,\pi$ rad/s.

Speed of precession, N_p = 60 r.p.m.

$\therefore$ Angular velocity of precession, $\omega_p = \dfrac{2\pi N_p}{60} = \dfrac{2\pi \times 60}{60} = 2\pi$ rad/s.

(i) Reaction at each bearing due to weight

Refer to Fig. 21.5.

Let $(R_w)_A$ = Reaction due to weight of disc at bearing A

$(R_w)_B$ = Reaction due to weight at bearing B.

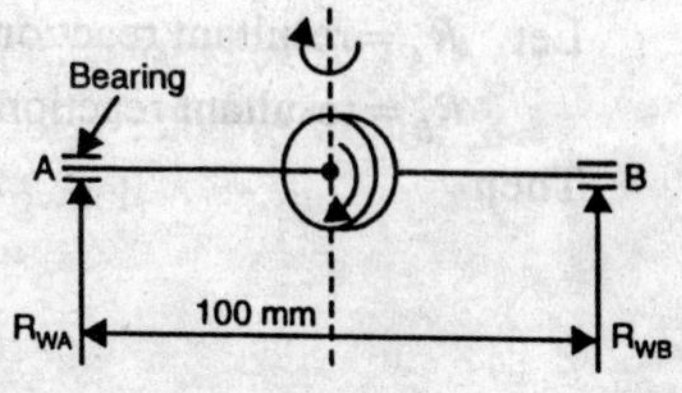

Fig. 21.5

As the disc is mounted on the centre of the axle, hence the reaction at each bearing due to weight will be equal to half of the weight of the disc and these reactions will be acting upwards.

$$\therefore \quad (R_w)_A = (R_w)_B = \frac{W}{2} = \frac{5 \times 9.81}{2} = 24.525 \text{ N (upwards)} (\uparrow)$$

(ii) Reaction at each bearing due to gyroscopic couple

Refer Fig. 21.6.

Let $(R_C)_A$ = Reaction due to gyroscopic couple at bearing A

$(R_C)_B$ = Reaction due to gyroscopic couple bearing B.

Initially the disc is rotating (or spinning) at 720 r.p.m. [with an angular velocity $\omega = 24\pi$ rad/s] in the clockwise direction looking from the left hand side bearing. This angular momentum ($I \times \omega$) is represented by vector *oa* according to right hand screw rule. When the axle precesses about a vertical axis in the clockwise direction looking from the top, the vector *oa* would rotate to the position *ob* in a short period. Then vector *ab* is the applied couple and the vector *ba* will be reaction couple and perpendicular to *oa* in the limit. This couple acts in the vertical plane. The sense of the vector *ba* can be determined by stretching the thumb of the right hand outwards from *b* to *a*, keeping fingers curving on the inside. Then the curve of the fingers points the sense of rotation. As the curve of fingers in this case, is clockwise, hence the sense of vector *ba* is clockwise. Hence the reaction couple on the bearing is acting clockwise. This is shown in Fig. 21.6 (*a*). The reaction couple tends to raise bearing *A* and lower the bearing *B*.

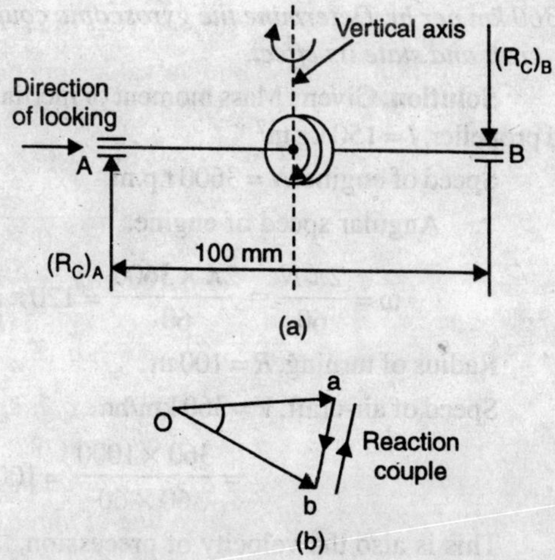

Fig. 21.6

Now let us find the magnitude of gyroscopic couple.

Gyroscopic couple is given by equation (21.1) as

$$C = I \times \omega \times \omega_p$$
$$= 0.018 \times 24\pi \times 2\pi \text{ Nm} \qquad (\because \; I = 0.018, \omega = 24\pi \text{ and } \omega_p = 2\pi)$$
$$= 8.527 \text{ Nm.}$$

But gyroscopic couple is equal to reaction couple. Hence reaction couple = 8.527 Nm.

But reaction couple (Refer to Fig. 21.6 (*a*)) = Force × Distance between forces

$$= [(R_c)_A \text{ or } (R_e)_B] \times L$$

$$\therefore \quad 8.527 = (R_c)_A \times 0.1 \qquad [\because \; L = 0.1 \text{ m}]$$

$$\therefore \quad (R_c)_A = \frac{8.527}{0.1} = 85.27 \text{ N}$$

Now $(R_C)_A = (R_C)_B$. But $(R_C)_A$ is acting upwards whereas $(R_C)_B$ is acting downwards as shown in Fig. 21.6 (*a*).

$\therefore$ $(R_C)_A = 85.27\text{ N}\uparrow$ and $(R_C)_B = 85.27\text{ N}\downarrow$.

Let R_A = resultant reaction due to weight and gyroscopic effect at bearing A.

R_B = resultant reaction due to weight and gyroscopic effect at bearing B.

Then
$$R_A = (R_W)_A + (R_C)_A$$
$$= 24.525\uparrow + 85.27\ (\uparrow) \text{ (Both are acting upward)}$$
$$= 109.795\text{ N (upwards) }(\uparrow). \textbf{ Ans.}$$

and
$$R_B = (R_W)_B + (R_C)_B$$
$$= 24.525 + (-85.27)\text{ N} \quad [\because \ (R_C)_B \text{ is acting downwards}]$$
$$= -60.745\text{ N} = 60.745\text{ N (downwards) }(\downarrow). \textbf{ Ans.}$$

Problem 21.3. *An aircraft consists of a propeller. It also consists of engine and propeller of mass moment of inertia 150 kg m². The engine rotates at 3600 r.p.m. in a sense clockwise looking from rear. The aircraft completes half circle of radius 100 m towards left when flying at 360 km per hr. Determine the gyroscopic couple on the air-craft and state its effect.*

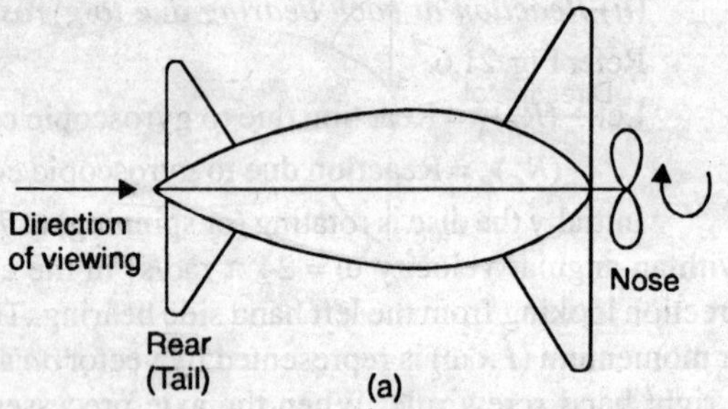

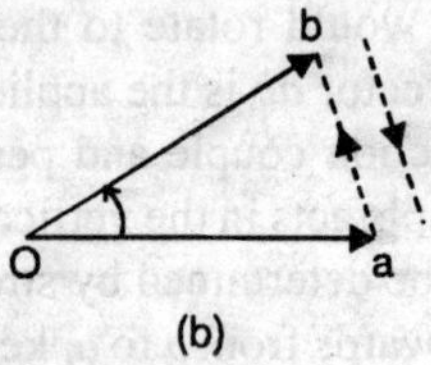

Fig. 21.7

Solution. Given : Mass moment of inertia of engine and propeller, $I = 150\text{ kg m}^2$

Speed of engine, $N = 3600$ r.p.m.

$\therefore$ Angular speed of engine,

$$\omega = \frac{2\pi N}{60} = \frac{2\pi \times 3600}{60} = 120\pi \text{ rad/s}$$

Radius of turning, $R = 100$ m

Speed of air-craft, $V = 360$ km/hr.

$$= \frac{360 \times 1000}{60 \times 60} = 100 \text{ m/s}$$

This is also the velocity of precession.

$\therefore$ $V_p = 100$ m/s

Now angular velocity of precession, $\omega_p = \dfrac{V_p}{R} = \dfrac{100}{100} = 1$ rad/s.

(*i*) *Gyroscopic couple on the air-craft*

Gyroscopic couple is given by equation (21.1) as

$$C = I \times \omega \times \omega_p = 150 \times 120\pi \times 1 \text{ Nm} \quad (\because \ \omega = 120\pi \text{ rad/s}, \omega_p = 1 \text{ rad/s})$$
$$= \mathbf{56548.67\ Nm.} \quad \textbf{Ans.}$$

(*ii*) *Effect of gyroscopic couple on the air-craft.*

Refer to Fig. 21.7 (*a*) and (*b*).

Before turning to the left, the engine in rotating (or spinning) at 3600 r.p.m. (with an angular velocity $\omega = 120\pi$ rad/s) in the clockwise direction looking from rear. The angular momentum of the engine and propeller (*i.e.*, $I \times \omega$) before turning is represented by vector *oa* according to right hand screw rule as shown in Fig. 21.7 (*b*). When the air-craft turns towards left (this means the

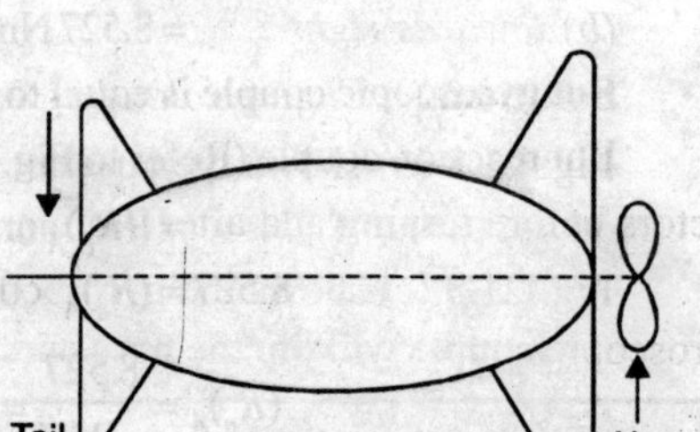

Fig. 21.7 (*c*)

axis of the engine precesses towards left), the vector *oa* would rotate to the position *ob* in a short period. Then vector *ab* is the applied couple to the air-craft. And vector *ba* will be reaction couple (or gyroscopic couple) and is perpendicular to *oa* in the limit. This couple acts in the vertical plane. The sense of vector *ba* can be determined by stretching the thumb of the right-hand outwards from *b* to *a*, keeping fingers curving on the inside and looking from front. Then curve of the fingers points the sense of rotation. As the curve of the fingers in this case, is anti-clockwise, hence the sense of vector *ba* is anti-clockwise*. Hence the gyroscopic couple will be acting on the air-craft in the vertical plane in anti-clockwise direction as shown in Fig. 21.7 (*c*). The effect of this couple will be to lift the nose and dip the nail.

Problem 21.4. *For the problem 21.3, what will be the effect of gyroscopic couple on the air-craft if :*

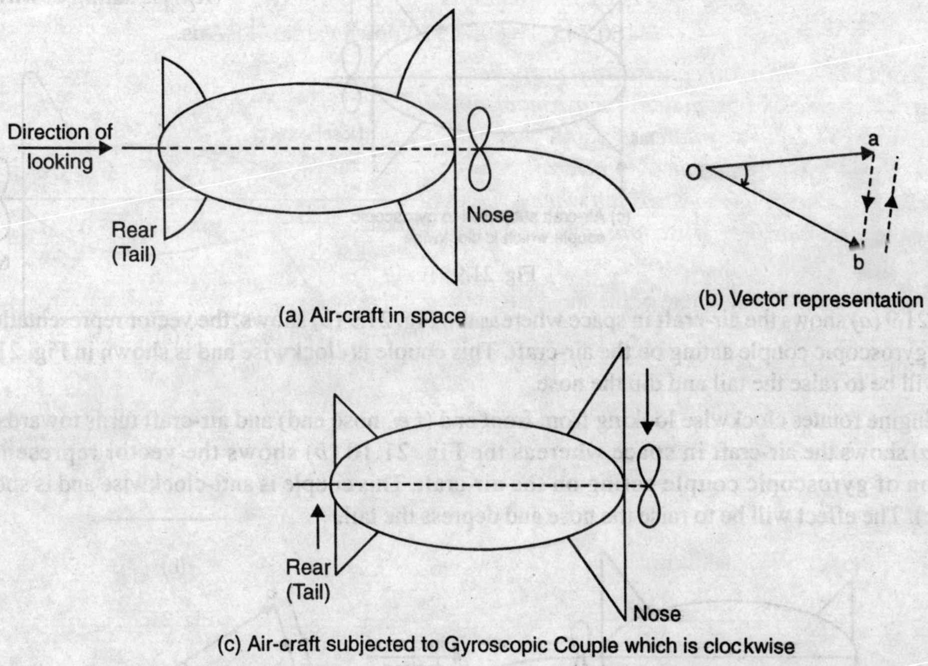

Fig. 21.8

(i) the air-craft while flying at 360 km/hr turns towards right while the engine rotates at 3600 r.p.m. in a sense clockwise looking from rear.

(ii) the engine rotates at 3600 r.p.m. in a sense clockwise looking from the front (nose end) and air-craft turns :

(a) towards left and

(b) towards right ?

Solution. (*i*) *Engine rotates clockwise looking from rear and air-craft turns to right.*

Fig. 21.8 (*a*) shows the air-craft in space. Fig. 21.8 (*b*) shows the positions of angular momentum vectors before turning and after turning.

Fig. 21.8 (*c*) shows the gyroscopic couple acting on the air-craft. From Fig. 21.8 (*c*) it is clear that gyroscopic couple will dip the nose and raise the tail.

*The sense of vector *ab* is clockwise according to right hand screw rule and hence the sense of vector *ba* will be anti-clockwise.

(ii) Engine rotates clockwise looking from the front end (i.e., nose end) and air-craft turns towards left.

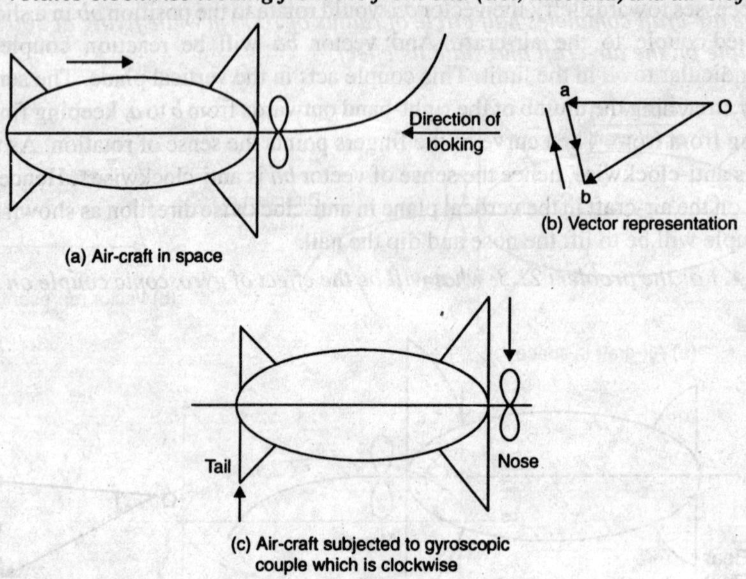

(a) Air-craft in space

(b) Vector representation

(c) Air-craft subjected to gyroscopic couple which is clockwise

Fig. 21.9

Fig. 21.9 (*a*) shows the air-craft in space whereas the Fig. 21.9 (*b*) shows, the vector representation and direction of gyroscopic couple acting on the air-craft. This couple is clockwise and is shown in Fig. 21.9 (*c*). The effect will be to raise the tail and dip the nose.

(*b*) Engine rotates clockwise looking from front end (*i.e.,* nose end) and air-craft turns towards right. Fig. 21.10 (*a*) shows the air-craft in space whereas the Fig. 21.10 (*b*) shows the vector representation and direction of gyroscopic couple acting on the air-craft. This couple is anti-clockwise and is shown in Fig. 21.10 (*c*). The effect will be to raise the nose and depress the tail.

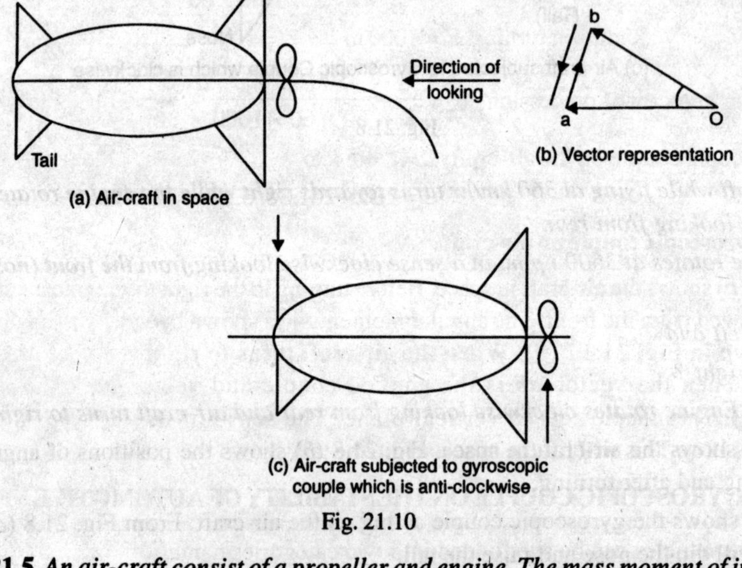

(a) Air-craft in space

(b) Vector representation

(c) Air-craft subjected to gyroscopic couple which is anti-clockwise

Fig. 21.10

Problem 21.5. *An air-craft consist of a propeller and engine. The mass moment of inertia of propeller and engine is 100 kg m*^{2}*. The engine rotates at 2500 r.p.m in the clockwise direction if viewed from the front* *of the air-craft. The air-craft completes half circle of radius of 1000 m while flying at 500 km/hr. Determine the gyroscopic couple on the air-craft and state its effect.*

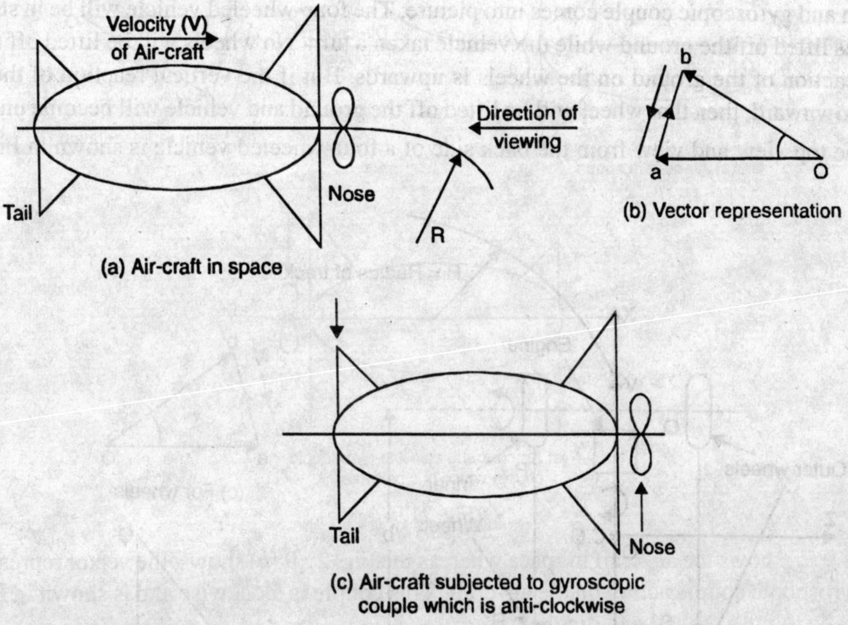

Fig. 21.11

Solution. Given : $I = 100\text{ kg m}^2$; $N = 2500$ r.p.m. hence $\omega = \dfrac{2\pi N}{60} = \dfrac{2\pi \times 2500}{60} = 261.8$ rad/s

$$\text{Velocity of air-craft, } V = 500 \text{ km/hr} = \frac{500 \times 1000}{60 \times 60} \text{ m/s}$$

Radius of turning, R = 1000 m

$$\therefore \quad \text{Angular velocity of precession, } \omega_p = \frac{V}{R} = \frac{500 \times 1000}{60 \times 60 \times 1000} = \frac{5}{36} = 0.139 \text{ rad/s}$$

$\therefore$ Gyroscopic couple on the air-craft, $C = I \times \omega \times \omega_p$

or $C = 100 \times 261.8 \times 0.139 =$ **3639 Nm. Ans.**

Effect of gyroscopic couple on air-craft

Fig. 21.11 (a) shows the air-craft in space. Before turning to the right the engine is rotating in clockwise direction when viewed from the front. The angular momentum is shown by vector oa according to right hand screw rule as shown in Fig. 21.11 (b). When the air-craft turns to right the vector oa would rotate to the position ob. Then the vector ab is the applied couple and vector ba is the reaction couple (or gyroscopic couple). This couple acts in vertical plane. The sense of this couple is anti-clockwise as shown in Fig. 21.11 (c). This will lift the nose and dip the tail.

21.4. EFFECT OF GYROSCOPIC COUPLE ON THE STABILITY OF AUTOMOTIVE VEHICLES

Automotive vehicles can be subdivided into two categories namely

1. Cars, buses, trucks etc. *i.e.*, four wheelers.
2. Motor cycles, scooters etc. *i.e.*, two wheelers.

21.4.1. Stability of Four Wheelers

When a four wheeler negotiates a curve of say radius '*R*', the axis of rotating parts undergoes precession and gyroscopic couple comes into picture. The four-wheeled vehicle will be in stable condition if no wheel is lifted off the ground while the vehicle taken a turn. No wheels will be lifted off the ground if the vertical reaction of the ground on the wheels is upwards. But if the vertical reaction of the ground on any wheel is downward, then that wheel will be lifted off the ground and vehicle will become unstable.

The top view and view from the back side of a four-wheeled vehicle is shown in Fig. 21.12 (*a*) and 21.12 (*b*).

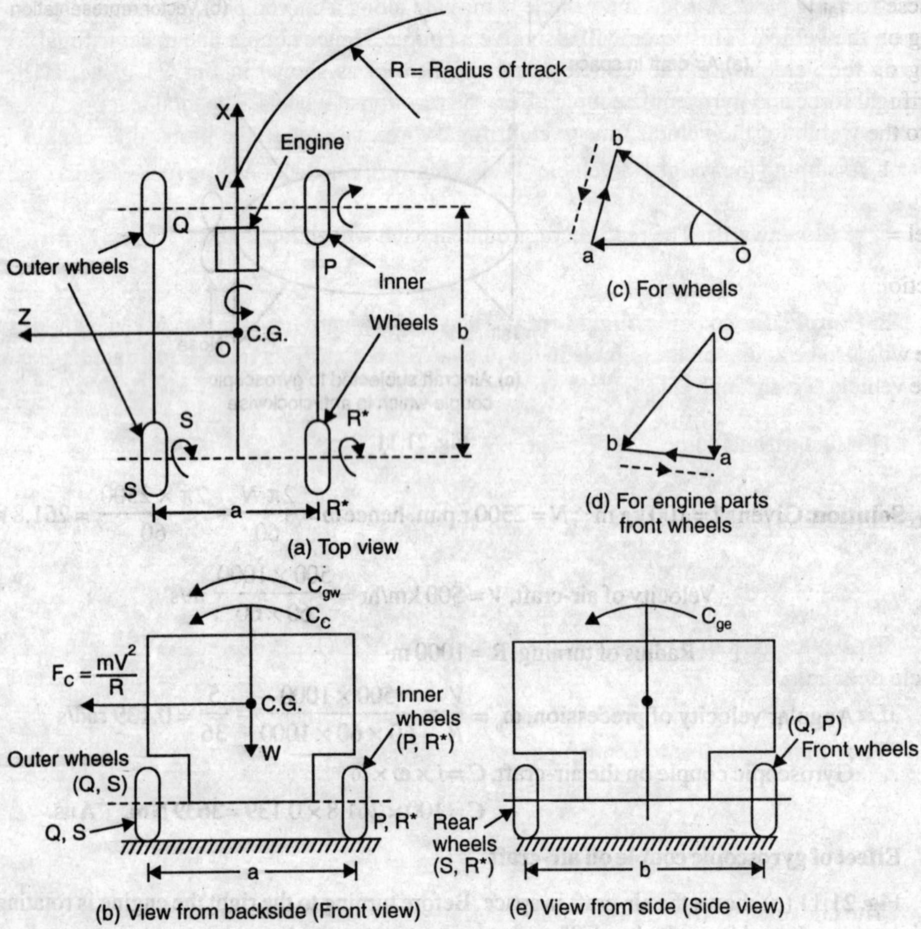

Fig. 21.12

Let h = the height of C.G. of four wheeler above the gound

W = the weight of the four wheeler

I_w = the mass moments of inertia of each wheel

I_E = the mass moment of inertia of parts rotating at engine speed

G = the gear ratio between engine and wheel

(*i.e.*, G = Speed of Engine/Speed of wheel)

r = the radius of the wheel

a = distance between two wheels on an axle.

Distance a is also known as **track width**

b = Perpendicular distance between front and rear wheel

= This distance is also known as **wheel base.**

For simplifying the analysis, let the centre of gravity be considered to lie on a vertical line through the geometric centre. As a convection, X-axis is always considered along the forward motion. The origin of the axis system is at the C.G. of the four wheeler.

When the four wheeler negotiates a curve of radius R, the axis of the rotating parts [Here rotating parts are (*i*) four wheels and (*ii*) engine rotating parts] undergoes precession and gyroscopic couple will be acting on these rotating parts. Also as the vehicle is moving along a curved path, the centrifugal force will also be acting on the vehicle. This force will also give a couple. Hence couple due to centrifugal force will also be acting on the vehicle. Let the vehicle takes a right turn as shown in Fig. 21.12 (*a*). The couple due to centrifugal force and gyroscopic couple affect the reaction at wheels. The total reaction at the wheels will be due to the weight of the vehicle, due to centrifugal force and due to the gyroscopic couple.

1. Assuming the weight of vehicle (W) is equally divided among the four wheels, then weight on each wheel = $\frac{W}{4}$ (downward). The reaction of ground at each wheel due to the weight 'W' = $\frac{W}{4}$ (in the upward direction.)

2. **Couple due to centrifugal force.** The vehicle is moving along a curved path, hence a centrifugal force will also be acting on the vehicle in the outward direction. This force acts through the centre of the mass of the vehicle (*i.e.* through C.G.).

Now centrifugal force, $F_c = m\frac{V^2}{R}$

where m = Mass of vehicle = $\frac{W}{g}$

V = Velocity of vehicle and

R = Radius of curvature (or radius of curved path)

This force is acting radially outward through the C.G. of the vehicle and will tend to overturn the vehicle outwards.

Let C_c = Couple due to centrifugal force.

Then the couple due to centrifugal force is given by

$$C_c = \text{Centrifugal force} \times \text{Distance of C.G. of vehicle from ground}$$

$$= F_c \times h = \left(\frac{mV^2}{R}\right) \times h \quad \ldots(21.3)$$

The above is the overturning couple acting on the vehicle. The resisting couple will be acting on the wheels which is equal to the couple due to equal and opposite forces acting on the outer and inner wheels. The inner wheels are towards the centre of curved path whereas the outer wheels are away from the centre. Here inner wheels are P and R* whereas outer wheels are Q and S.

But the perpendicular distance between outer and inner wheels = a

∴ Resisting couple on wheels = Force on outer wheels (or force on inner wheels) × perpendicular distance between outer and inner wheels

= [Force on outer wheels (or force on inner wheels)] × a

For no overturning of vehicle, overturning couple = Resisting couple

or C_c = [Force on outer wheels (or force on inner wheels)] × a

or $\left(\frac{mV^2}{R}\right) \times h =$ Force on outer wheels (or force on inner wheels)] $\times a$

$\therefore$ Force on outer wheels (or force on inner wheels) $= \left(\frac{mV}{R}\right)^2 \times \frac{h}{a} = \frac{C_c}{a}$ $\left(\because \frac{mV^2}{R} \times h = C_c\right)$

There are two outer wheels and two inner wheels.

$\therefore$ Force on each outer wheel (or force on each inner wheel) $= \frac{1}{2}\left(\frac{C_c}{a}\right) = \frac{C_c}{2a}$

Hence the force due to centrifugal couple will be equal to $\frac{C_c}{2a}$ on each wheel. But the force on each outer wheel will be acting downwards* whereas on each inner wheel it will be acting upward. The reaction of the ground on each wheel will be opposite to the corresponding force on each wheel. The reaction of the ground on each outer wheels

(*i.e.*, on wheel Q and S) $= \frac{C_c}{2a}$ (upwards)

Similarly the reaction of the ground on each inner wheels

(*i.e.*, on wheel P and R*) $= \frac{C_c}{2a}$ (downwards)

3. **Effect of Gyroscopic couple.** The gyroscopic couple will be acting on the vehicle when the vehicle takes a right turn as shown in Fig. 21.12 (*a*). The gyroscopic couple will be due to four wheels and due to engine rotating parts.

Let C_{gw} = gyroscopic couple due to four wheels

C_{ge} = gyroscopic couple due to engine rotating parts.

(*a*) Gyroscopic couple due to four wheels. The gyroscopic couple due to four wheels is given by, $C_{gw} = 4 \times I_w \times \omega_w \times \omega_p$

where ω_w = Angular velocity of wheels $= \frac{V}{r} = \frac{\text{Velocity of vehicle}}{\text{Radius of wheel}}$

ω_p = Angular velocity of precession $= \frac{V}{R}$ (R = Radius of curvature)

$$\therefore \quad C_{gw} = 4 \times I_w \times \frac{V}{r} \times \frac{V}{R} = \frac{4I_w \times V^2}{rR} \quad \text{...(21.4)}$$

Fig. 21.12 (*c*) shows the vector representation of the gyroscopic couple on each wheel. The wheel is rotating in clockwise direction (vehicle moving forwards) looking from the inner wheel side. The angular momentum ($I \times \omega$) of the wheel is represented by vector *oa* according to right hand screw rule before taking a right turn.

When vehicle takes right turn, the vector *oa* of Fig. 21.12 (*c*) will rotate to position *ob*. Then vector *ab* is the applied couple and vector *ba* (shown by dotted line) is the gyroscopic couple. The sense of this couple is anti-clockwise. This couple is also shown in Fig. 21.12 (*b*). Due to this couple, the overturning of vehicle will take place. This couple is equal to the couple due to equal and opposite forces acting on the outer and inner wheels. Outer wheels are Q and S whereas inner wheels are P and R*.

*The centrifugal force is acting radially outwards through the C.G. of the vehicle. As this force is radially out wards, it will try to press the outer wheels to the ground and lift the inner wheels from ground.

$\therefore$ C_{gw} = [Force on outer wheels (or force on inner wheels)] × perpendicular distance between outer and inner wheels

= [Force on outer wheels (or force on inner wheels)] × a

$\therefore$ Force on outer wheels [or force on inner wheels] $= \dfrac{C_{gw}}{a}$

There are two outer wheels and two inner wheels.

$\therefore$ Force on each outer wheel (or force on each inner wheel)

$$= \frac{1}{2} \times \frac{C_{gw}}{a} = \frac{C_{gw}}{2a}$$

Hence the force due to gyroscopic couple on the wheels will be equal to $\dfrac{C_{gw}}{2a}$ on each wheel. But the force on each outer wheel will be acting downwards whereas on each inner wheel it will be acting upwards. [$\because$ The couple is anti-clockwise in Fig. 21.12 (*b*), hence corresponding force will be pulling the inner wheels from the ground and pressing the outer wheels towards ground]. The reaction of the ground on each wheel will be opposite to the corresponding force on each wheel.

$\therefore$ Reaction of ground on each outer wheels

(*i.e.*, on wheel Q and S) $= \dfrac{C_{gw}}{2a}$ (upwards)

Reaction of ground on each inner wheels

(*i.e.*, on wheel P and R^*) $= \dfrac{C_{gw}}{2a}$ (downwards)

(*b*) Gyroscopic couple due to engine rotating parts

The gyroscopic couple due to engine rotating parts is given by,

$$C_{ge} = I_e \times \omega_e \times \omega_p$$

where C_{ge} = Gyroscopic couple due to engine rotating parts

I_e = Mass moment of inertia of engine rotating parts

ω_e = Angular velocity of engine

$$= G \times \omega_w \qquad \left[\because\ G = \frac{\omega_e}{\omega_w}\ \therefore\ \omega_e = G \times \omega_w\right]$$

$$= G \times \frac{V}{r} \qquad \left(\because\ \omega_w = \frac{V}{r},\ r = \text{Radius of wheel}\right)$$

ω_p = Angular velocity of precession $= \dfrac{V}{R}$

$$\therefore \quad C_{ge} = I_e \times \left(G \times \frac{V}{r}\right) \times \frac{V}{R} = \frac{I_e \times G \times V^2}{rR} \qquad \text{...(21.5)}$$

Fig. 21.12 (*d*) shows the vector representation of the gyroscopic couple acting on the engine rotating parts. Looking from back side of the vehicle, the engine is rotating anti-clockwise. The spin vector of the engine rotating parts is shown by vector *oa* according to right hand screw rule (–ve side of X-axis) before taking a right turn. When vehicle takes a right turn, the vector *oa* will change to vector *ob*. Then vector *ab* is the applied couple and vector *ba* (shown by dotted line) is the gyroscopic couple. The sense of this couple looking from the back of the vehicle is anticlockwise. This couple is also shown in Fig. 21.12(*e*). Due to this couple, the lifting of the vehicle from the front wheels will take place. This couple is equal to the couple due to equal and opposite forces acting on the front and rear wheels. Here the front wheels are P and Q whereas the rear wheels are R^* and S.

$$C_{ge} = [\text{Force on front wheels (or force on rear wheels)}] \times \text{perpendicular distance between front and rear wheels}$$
$$= \text{Force on front wheels (or force on rear wheels]} \times b$$

$\therefore$ Force on front wheels (or force on rear wheels) $= \dfrac{C_{ge}}{b}$

There are two front wheels and two rear wheels.

$\therefore$ Force on each front wheel (or force on each rear wheel) $= \dfrac{1}{2}\left[\dfrac{C_{ge}}{b}\right] = \dfrac{C_{ge}}{2b}$

Hence the force on each wheel due to gyroscopic couple on rotating engine parts, will be equal to $\dfrac{C_{ge}}{2b}$. The force on front wheels will be acting upwards whereas the force on rear wheels will be downwards. The reaction of the ground on each wheel will be opposite to the corresponding force on each wheel.

$\therefore$ Reaction of ground on each front wheels

(*i.e.*, wheel P and Q) $= \dfrac{C_{ge}}{2b}$ (downward)

Reaction of ground on each rear wheels

(*i.e.*, wheel S and R*) $= \dfrac{C_{ge}}{2b}$ (upwards)

4. The spin axis of the engine rotating parts is perpendicular to that of wheels. The spin vector is towards negative side of X-axis. The change in angular momentum vector is shown in Fig. 21.12 (*d*).

Table 21.1 shows the reaction of ground on each wheel due to weight of vehicle (W), due to centrifugal couple (C_c), due to gyroscopic couple acting on wheels [(C_{gw})] and due to gyroscopic couple on engine rotating parts (C_{ge}).

Table 21.1

S. No.	*Action*	*Contribution to reaction at each wheel with sense*			
		P	*Q*	*R**	*S*
1.	Weight of Vehicle	$\frac{W}{4}\uparrow$	$\frac{W}{4}\uparrow$	$\frac{W}{4}\uparrow$	$\frac{W}{4}\uparrow$
2.	Couple due to centrifugal force	$\frac{C_c}{2a}\downarrow$	$\frac{C_c}{2a}\uparrow$	$\frac{C_c}{2a}\downarrow$	$\frac{C_c}{2a}\uparrow$
3.	Gyroscopic couple due to wheels	$\frac{C_{gw}}{2a}\downarrow$	$\frac{C_{gw}}{2a}\uparrow$	$\frac{C_{gw}}{2a}\downarrow$	$\frac{C_{gw}}{2a}\uparrow$
4.	Gyroscopic couple due to engine parts	$\frac{C_{ge}}{2b}\downarrow$	$\frac{C_{ge}}{2b}\downarrow$	$\frac{C_{ge}}{2b}\uparrow$	$\frac{C_{ge}}{2b}\uparrow$

The four-wheeled vehicle will be in stable condition if the total vertical reaction of the ground on any of the wheels is positive (or upwards). If the total vertical reaction of the ground is downwards on any wheel, than that wheel will be lifted off the ground and vehicle will become unstable.

According to the table 21.1, the minimum reaction is at wheel 'P' which is

$$R_P = \frac{W}{4} - \frac{C_c}{2a} - \frac{C_{gw}}{2a} - \frac{C_{ge}}{2b}$$

$$= \frac{W}{4} - \frac{1}{2a}\left\{m\frac{V^2}{R'}\times h + 4I_w\frac{V^2}{rR}\right\} - \frac{I_e G V^2}{2rRb}$$

Reaction at other wheels are :

$$R_Q = \frac{W}{4} + \frac{C_c}{2a} + \frac{C_{gw}}{2a} - \frac{C_{ge}}{2b}$$

$$R_R{}^* = \frac{W}{4} - \frac{C_c}{2a} - \frac{C_{gw}}{2a} + \frac{C_{ge}}{2b}$$

and $$R_S = \frac{W}{4} + \frac{C_c}{2a} + \frac{C_{gw}}{2a} + \frac{C_{ge}}{2b}$$

It is obvious from the above expressions that the reaction will be maximum at wheel '*S*'. The values of C_c, C_{gw} and C_{ge} are given by equations (21.3), (21.4) and (21.5) respectively. Substituting these values in the above equation we get the reaction at

$$R_p = \frac{W}{4} - \left(\frac{mV^2}{R}\right)\times\frac{h}{2a} - \frac{4\,I_w\times V^2}{rR}\times\frac{1}{2a} - I_e.\frac{GV^2}{rR}\times\frac{1}{2b}$$

$$= \frac{W}{4} - \frac{V^2}{2aR}\left[m\times h + \frac{4I_w}{r}\right] - \frac{I_e\,GV^2}{2r\,Rb}$$

In the above equation the only variable is the speed of the vehicle. Hence this reaction depends upon the speed of the vehicle. At a particular speed, the reaction can becomes zero. Let us find that speed at which R_p is zero. For this speed

$$R_p = 0$$

or $$\frac{W}{4} - \frac{V^2}{2aR}\left[m\times h + \frac{4I_w}{r}\right] - \frac{I_e\times GV^2}{2r\,Rb} = 0$$

or $$\frac{V^2}{R}\left\{\frac{1}{2a}\left(m\times h + \frac{4I_w}{r}\right) + \frac{I_eG}{2rb}\right\} = \frac{W}{4}$$

or $$V = \sqrt{\frac{WR}{4\left\{\frac{1}{2a}\left(m\times h + \frac{4I_w}{r}\right) + \frac{I_eG}{2rb}\right\}}} \quad \text{...(21.6)}$$

The speed given by equation (21.6) is also known as **critical speed.** If vehicle moves at a speed more than this, wheel *P* will leave ground and it gives rise to the instability.

Problem 21.6. *A car is of total mass 3000 kg. It has wheel base equal to 2.5 m and track width equal to 1.5 m. The effective diameter of each wheel is 80 cm. and moment of inertia of each wheel is 1.0 kg m². The rear axle ratio is 4. The mass moment of inertia of engine rotating parts is 3 kg m² and spin axis of engine parts is perpendicular to the spin axis of wheels. Determine the reaction at each wheel if car takes right turn of 100 m radius at 108 km/hr speed. Also determine critical speed. The height of C.G. is 0.5 m from ground and it is placed on the vertical line through geometric centre of wheels.*

Solution. Given : Speed of car = 108 km/hr.

$\therefore$ Vehicle speed 'V' = 108 km/hr.

$$= \frac{108 \times 1000}{60 \times 60} = 30 \text{ m/sec.}$$

Wheel base, $b = 2.5$ m

Track width, $a = 1.5$ m

Total mass of car, $m = 3000$ kg

$\therefore$ Weight of car, $W = m \times g = 3000 \times 9.81$ N

Mass moment of inertia of each wheel 'I_w' = 1.0 kg m^2

Mass moment of inertia of engine parts 'I_e' = 3 kg m^2

Diameter of wheel, $d = 80$ cm = 0.8 m

Wheel radius 'r' $= \frac{0.8}{2} = 0.4$ m

Track radius 'R' = 100 m

Height of C.G. 'h' = 0.5 m

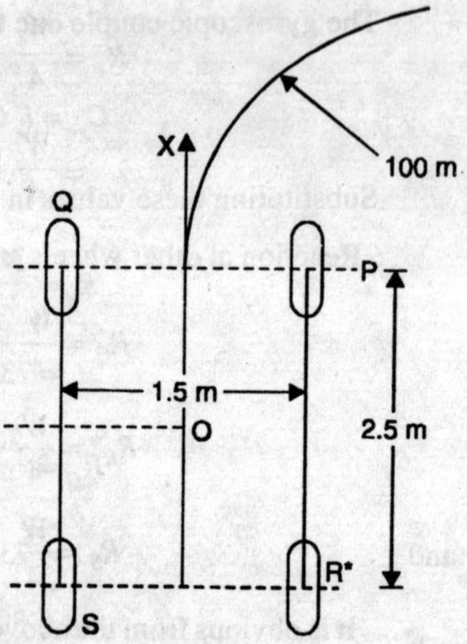

Fig. 21.13

At this speed car is moving in top gear and therefore, overall gear ratio between engine and wheel, 'G' = 4.0.

As the C.G. of the car is at a height from ground on the geometric centre of wheels, hence the weight of the car will be equally divided among the four wheels.

With reference to the table 21.1 of preceding article, the expression of wheel reaction is given as follows :

The reaction at wheel P, $R_P = \frac{W}{4} - \frac{C_c}{2a} - \frac{C_{gw}}{2a} - \frac{C_{ge}}{2b}$

The reaction at wheel Q, $R_Q = \frac{W}{4} + \frac{C_c}{2a} + \frac{C_{gw}}{2a} - \frac{C_{ge}}{2b}$

The reaction at wheel R^*, $R_{R*} = \frac{W}{4} - \frac{C_c}{2a} - \frac{C_{gw}}{2a} + \frac{C_{ge}}{2b}$

The reaction at wheel S, $R_S = \frac{W}{4} + \frac{C_c}{2a} + \frac{C_{gw}}{2a} + \frac{C_{ge}}{2b}$

Centrifugal couple is given by equation (21.3), as $C_c = m\frac{V^2}{R} \times h$

Since, $m = 3000$ kg $\therefore$ $C_c = 3000 \times \frac{30^2}{100} \times 0.5$

or $$C_c = 13500 \text{ Nm}$$

The gyroscopic couple due to wheels is given by equation (21.4) as $C_{gw} = 4I_w \frac{V^2}{r \times R}$

or $$C_{gw} = 4 \times 1 \times \frac{30^2}{0.4 \times 100} = 90 \text{ Nm}$$

The gyroscopic couple due to engine is given by equation (21.5)

$$C_{ge} = I_e G \frac{V^2}{rR} = 3.0 \times 4 \times \frac{30^2}{0.4 \times 100} = \frac{12 \times 9}{0.4} = 270 \text{ Nm}$$

Substituting these values in reactions at wheels *P, Q, R** and *S*, we get

$$\therefore \quad R_p = \frac{3000 \times 9.81}{4} - \frac{13500}{2 \times 1.5} - \frac{90}{2 \times 1.5} - \frac{270}{2 \times 2.5} \qquad (\because \quad a = 1.5 \text{ and } b = 2.5)$$

$$= 7357.5 - 4500 - 30 - 54 = \mathbf{2773.5\ N.}\quad \textbf{Ans.}$$

$$R_Q = \frac{3000 \times 9.81}{4} + \frac{13500}{2 \times 1.5} + \frac{90}{2 \times 1.5} - \frac{270}{2 \times 2.5}$$

$$= 7357.5 + 4500 + 30 - 54 = \mathbf{11833.5\ N.}\quad \textbf{Ans.}$$

$$R_{R*} = \frac{3000 \times 9.81}{4} - \frac{13500}{2 \times 1.5} - \frac{90}{2 \times 1.5} + \frac{270}{2 \times 2.5} = \mathbf{2881.5\ N.}\quad \textbf{Ans.}$$

$$R_s = \frac{3000 \times 9.81}{4} + \frac{13500}{2 \times 1.5} + \frac{90}{2 \times 1.5} + \frac{270}{2 \times 2.5} = \mathbf{11887.5\ N.}\quad \textbf{Ans.}$$

Critical speed is given by equation (21.6) as

$$\text{Critical speed} \qquad \text{`}V\text{'} = \sqrt{\frac{WR}{4\left\{\frac{1}{2a}\left(m \times h + \frac{4 I_w}{r}\right) + \frac{I_e G}{2rb}\right\}}}$$

$$V = \sqrt{\frac{3000 \times 9.81 \times 100}{4\left\{\frac{1}{2 \times 1.5}\left(3000 \times 0.5 + \frac{4 \times 1}{0.4}\right) + \frac{3 \times 4}{2 \times 0.4 \times 2.5}\right\}}} = \sqrt{\frac{3000 \times 9.81 \times 100}{4\left\{\frac{1}{3}(1500 + 10) + 6\right\}}}$$

$$= \sqrt{\frac{300000 \times 9.81}{2037.33}} = \mathbf{37.82\ m/s.} = \frac{\mathbf{37.82}}{\mathbf{1000}} \times \mathbf{60 \times 60\ km/hr} \quad \text{or} \quad \mathbf{136.16\ km/hr.} \quad \textbf{Ans.}$$

Important Points (*i*) Some times the C.G. of the car may not be at the geometric centre of the wheels. Then the weight of the car will not be equally divided among the four wheels. But the weight on each rear wheels will be equal. Also the weight on front wheels will be equal. But the weight on rear wheels will be different from the weight on front wheels.

(*ii*) Sometimes, the spin axis of the engine (and its rotating parts) may be parallel to the spin axis of the four wheels. Then the gyroscopic couple on the rotating parts of the engine will be in the same direction as that of the gyroscopic couple on four wheels, provided engine is rotating in the same direction as wheels.

If engine is rotating in opposite direction, then gyroscopic couple on engine rotating parts will be in opposite direction.

Problem 21.7. *A car is of total mass 200 kg. It has wheel base equal to 2.5 m and track width = 1.5 m. The centre of gravity lies at 500 mm above ground level and 1.5 m from the rear axle. The effective diameter of each wheel is 800 mm and moment of inertia of each wheel is 1.0 kgm². The rear axle ratio is 4. The equivalent mass of engine rotating parts (i.e., drive shaft, engine flywheel and transmission etc.) is*

140 kg with radius of gyration of 150 mm. The spin axis of the rotating engine parts is perpendicular to the spin axis of wheels. The engine parts are rotating in clockwise direction when viewed from the front. Determine the reaction at each wheel if car takes a right turn of 100 m radius at 90 km/hr speed.

Also find the reaction at each wheel if car takes a left turn.

Solution. Given : $m = 2000$ kg

$\therefore$ $W = m \times g = 2000 \times 9.81$ N

Wheel base, $b = 2.5$ m

Track width, $a = 1.5$ m

Height of C.G., $h = 500$ mm $= 0.5$ m

Distance of C.G. from rear axle $= 1.5$ m

Effective dia. of each wheel $= 800$ mm $= 0.8$ m

$\therefore$ Radius of each wheel, $r = 0.4$ m

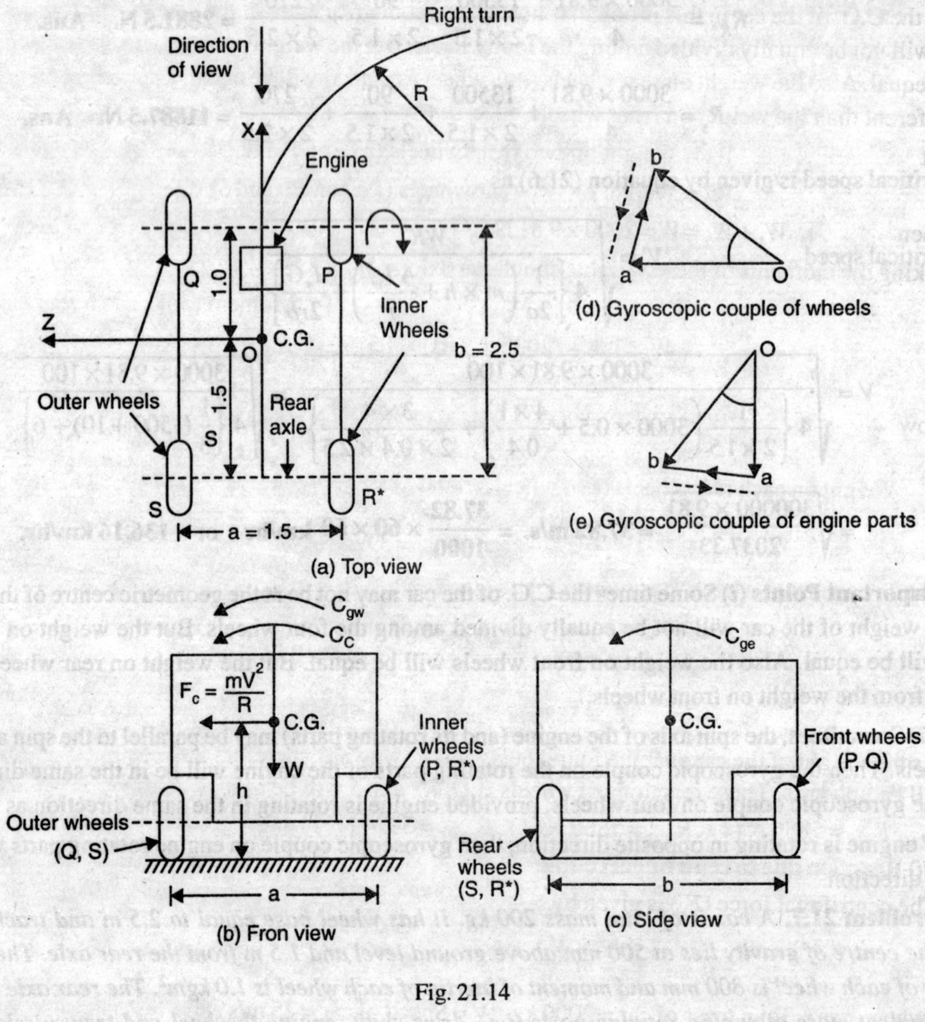

Fig. 21.14

Moment of inertia of each wheel, $I_w = 1.0 \text{ kg m}^2$

Rear axle ratio or gear ratio, $G = 4$

Mass of engine rotating parts, $m_e = 140$ kg

Radius of gyration of rotating parts, $k_e = 150 \text{ mm} = 0.15$ m

$\therefore$ Mass moment of inertia of engine rotating parts, $I_e = m_e \times k_c^2 = 140 \times 0.15^2 = 3.15 \text{ kg m}^2$.

Spin axis of engine rotating parts is perpendicular to the spin axis of the wheels. Engine parts are rotating clockwise when viewed from the front.

Radius of track, $R = 100$ m

$$\text{Speed of car} = 90 \frac{\text{km}}{\text{hr}} = \frac{90 \times 1000}{60 \times 60} = 25 \text{ m/s}$$

$\therefore \quad V = 25$ m/s.

(*a*) Car takes a right turn

(*i*) Reaction due to weight of car

As the C.G. of the car in this problem is not at the geometric centre of the wheels, the total weight (*W*) of the car will not be equally divided among the four wheels. But the weight on front wheels (*i.e.* wheel *P* and *Q*) will be equal. Also the weight on rear wheels (*i.e.*, wheel *R** and *S*) will be equal. The weight on rear wheels will be different than the weight on front wheels.

Let W_1 = Weight on two front wheels (*i.e.*, on *P* and *Q*)

W_2 = Weight on two rear wheels (*i.e.*, on *R** and *S*)

Then $W_1 + W_2 = W = 2000 \times 9.81$ N

Taking the moments of these weights about rear axle, $W_1 \times 2.5 = W \times 1.5$

($\because$ Moment of W_2 about rear axle is zero)

$$\therefore \quad W_1 = \frac{W \times 1.5}{2.5} = \frac{(2000 \times 9.81) \times 1.5}{2.5} = 1200 \times 9.81 \text{ N} \qquad (\because W = 2000 \times 9.81)$$

Now $W_2 = W - W_1 = 2000 \times 9.81 - 1200 \times 9.8 \text{ N} = 800 \times 9.81$ N

$$\therefore \quad \text{Weight on each rear wheels (i.e., wheel } R^* \text{ and } S) = \frac{W_2}{2} = \frac{800 \times 9.81}{2} = 440 \times 9.81 \text{ N} = 3924 \text{ N}$$

$\therefore \quad W_s = 3924$ N and $W_{R*} = 3924$ N

and weight on each front wheels (*i.e.*, wheel *P* and *Q*)

$$= \frac{W_1}{2} = \frac{1200 \times 9.81}{2} = 600 \times 9.81 \text{ N} = 5886 \text{ N}$$

$\therefore \quad W_P = 5886$ N and $W_Q = 5886$ N

As the weight of the car over the four wheels acts downward, therefore the reaction of ground on each wheel will be upwards. These reactions of wheels *P*, *Q*, *R** and *S* will be as :

$\therefore \quad R_{WP} = 5886 \text{ N} (\uparrow)$; $R_{WQ} = 5886 \text{ N} (\uparrow)$; $R_{WR*} = 3924 \text{ N} (\uparrow)$ and $R_{WS} = 3924 \text{ N} (\uparrow)$

(*ii*) Reaction due to centrifugal couple

The centrifugal force (F_c) is given by

$$F_c = m \frac{V^2}{R}$$

(where m = Total mass of car = 2000 kg ; V = Speed of car = 25 m/s and R = Radius of track = 100 m)

$$= 2000 \times \frac{25^2}{100} = 12500\text{ N}$$

This force is acting radially outward through the centre of gravity of the car. The C.G. of car is at a height 'h'.

∴ Centrifugal couple tending to overturn the car or overturning couple.

$$C_c = F_c \times h \text{ where } h = 0.5\text{ m}$$
$$= 12500 \times 0.5 = 6250\text{ Nm}$$

This overturning couple is acting anti-clockwise as shown in Fig. 21.14 (*b*) and will press the outer wheels (*i.e.*, wheels Q and S) and lift the inner wheels (*i.e.*, wheels P and R*). Hence the reaction of the ground on the outer wheels will be upwards whereas on the inner wheels will be downwards (*i.e.*, in the opposite direction due to overturning couple). Thus the overturning couple will be balanced by vertical reactions which are vertically upwards on the outer wheels and vertically downwards on the inner wheels.

∴ Overturning couple = Resisting couple

or C_c = [Force reaction of ground on outer wheels (or force (or reaction) on inner wheels)] × perpendicular distance between outer wheels and inner wheels

$$= F \times a$$

∴ $$F = \frac{C_c}{a}$$

∴ Reaction of ground on outer wheels (or inner wheels) $= \frac{C_c}{a}$

There are two outer wheels and two inner wheels

∴ Reaction on each outer wheel (or on each inner wheel) $= \frac{C_c}{2a} = \frac{6250}{2 \times 1.5} = 2083.33\text{ N}$

(∵ $C_c = 6250$ Nm and $a = 1.5$ m)

But the reaction of ground on each outer wheel is vertically upwards whereas on each inner wheel is vertically downwards. The wheels Q and S are outer wheels and P and R* are inner wheels.

∴ The reaction of ground on wheels P, Q, R* and S are as :

$R_{CP} = 2083.33$ N (↓), $R_{CQ} = 2083.33$ N (↑), $R_{CR}{}^* = 2083.33$ N (↓) and $R_{CS} = 2083.33$ N (↑)

***(iii)* Reaction due to gyroscopic couple**

The gyroscopic couple will be acting on the four wheels and also on the rotating parts of the engine, when engine takes a turn.

1. The gyroscopic couple due to four wheels is given by, $C_{gw} = 4I_w \times \omega_w \times \omega_p$

where ω_w = angular velocity of wheels $= \frac{V}{r} = \frac{25}{0.4} = 62.5$ rad/s

ω_p = Angular velocity of precession $= \frac{V}{R} = \frac{25}{100} = 0.25$ rad/s and

I_w = Mass moment of inertia of each wheel = 1 kg m^2

∴ Gyroscopic couple due to four wheels becomes as

$$C_{gw} = 4 \times 1 \times 62.5 \times 0.25 = 62.5\text{ Nm.}$$

This gyroscopic couple is shown in Fig. 21.14 (*b*). This will tend to lift the inner wheels and press the outer wheels. This means the reaction of ground will be vertically downwards on the inner wheels and vertically upwards on the outer wheels in the same way as that of centrifugal couple.

∴ Reaction of ground on outer wheels (or on inner wheels)

$$= \frac{\text{Gyroscopic couple}}{\text{Perpendicular Distance between outer and inner wheels}} = \frac{C_{gw}}{a}$$

There are two outer wheels and two inner wheels.

∴ Reaction on each outer wheel (or on each inner wheel) $= \frac{1}{2}\left[\frac{C_{gw}}{a}\right]$

$$= \frac{C_{gw}}{2a} = \frac{62.5}{2\times 1.5} = 20.83\text{ N}$$

But reaction of ground on outer-wheels (Q and S) is vertically upwards.

∴ $R_{gQ} = R_{gS} = 20.83\text{ N}\,(\uparrow)$

And reaction of ground on inner wheels (*i.e.*, wheel P and R*) is vertically downwards

∴ $R_{gP} = R_{gR}{*} = 20.83\text{ N}\,(\downarrow)$

2. Gyroscopic couple due to engine rotating parts.

It is given by, $C_{ge} = I_e \times \omega_e \times \omega_p$

where I_e = Mass moment of inertia of engine rotating parts = 3.15 kg m^2 (calculated)

ω_e = Angular velocity of engine rotating parts

$$= G \times \omega_w \qquad \left(\because\ G = \frac{\omega_e}{\omega_w}\ \therefore\ \omega_e = G.\omega_w\right)$$

$$= 4\times 62.5 \qquad (\because\ G = 4 \text{ and } \omega_w = 62.5\text{ rad/s})$$

$$= 250\text{ rad/s}$$

and ω_p = Angular velocity of precession $= \frac{V}{R} = \frac{25}{100} = 0.25$ rad/s

∴ Gyroscopic couple due to engine rotating parts becomes as

$$C_{ge} = 3.5 \times 250 \times 0.25 = 196.875\text{ Nm}.$$

The spin axis of engine rotating parts is at right angles to the spin axis of rotating wheels. Fig. 21.14 (*e*) shows the vector representation looking from the front of the car. This gyroscopic couple is acting in the clockwise direction when seen from the front. This is also shown in Fig. 21.14 (*c*). This gyroscopic couple will tend to lift the front wheels and press the rear wheels. Hence the reaction will be vertically downwards on the front wheels and vertically upwards on the rear wheels.

∴ C_{ge} = Vertical reactions on front wheels (or on rear wheels) × b

∴ Vertical reactions on front wheels (or on rear wheels) $= \frac{C_{ge}}{b}$

∴ Vertical reaction on each front wheel (or on each rear wheel)

$$= \frac{1}{2}\left(\frac{C_{ge}}{b}\right) = \frac{C_{ge}}{2b} = \frac{196.875}{2\times 2.5} = 39.375\text{ N} \qquad (\because\ b = 2.5\text{ m})$$

But reaction on front wheel is downwards and on rear wheels upwards.

$\therefore \quad R_{eP} = R_{eQ} = 39.375 \text{ N}(\uparrow). \quad$ **And** $\quad R_{eS} = R_{eR}* = 39.375 \text{ N}(\uparrow).$

Hence reaction at each wheel when car takes a right turn is the sum of reactions due to weight, due to centrifugal couple and due to gyroscopic couple (*i.e.,* due to wheels and due to rotating engine parts).

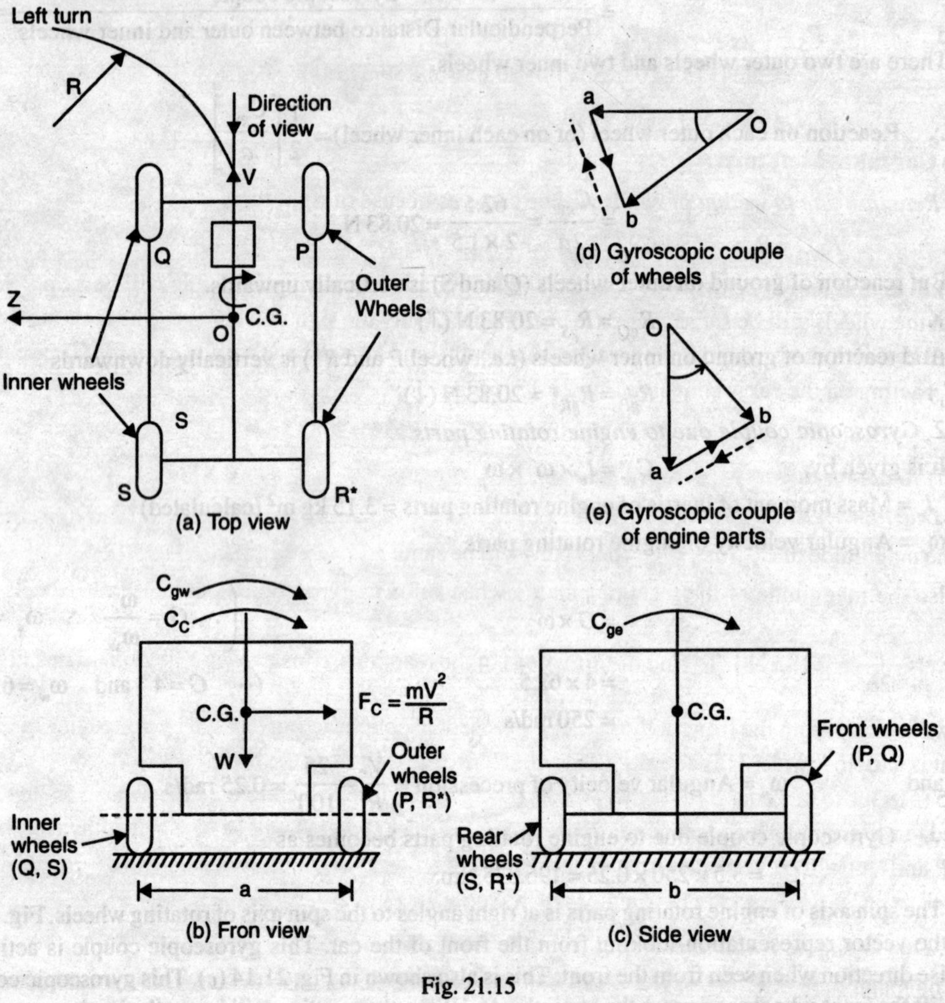

Fig. 21.15

$$\therefore \quad R_Q = R_{WQ} + R_{cQ} + R_{gQ} + R_{eQ}$$

$$= 5886\,(\uparrow) + 2083.33 \text{ N}\,(\uparrow) + 20.83\,(\uparrow) + 39.375\,(\downarrow)$$

$$= 5886 + 2083.33 + 20.83 - 39.375$$

(Taking upward reaction + ve and downward reaction – ve)

$$= 7950.785 \text{ N}\,(\uparrow). \quad \textbf{Ans.}$$

$$R_P = R_{WP} + R_{cP} + R_{gP} + R_{eP}$$

$$= 5886\,(\uparrow) + 2083.33 \text{ N}\,(\downarrow) + 20.83\,(\downarrow) + 39.375 \text{ N}\,(\downarrow)$$

$$= 5886 - 2083.33 - 20.83 - 39.375$$

$$= 3742.465 \text{ N}\,(\uparrow). \quad \textbf{Ans.}$$

$$R_R{}^* = R_{WR}{}^* + R_{cR}{}^* + R_{gR}{}^* + R_{eR}{}^*$$
$$= 3924\,(\uparrow) + 2083.33\,(\downarrow) + 20.83\,(\downarrow) + 39.375\,(\uparrow)$$
$$= 3924 - 2083.33 - 20.83 + 39.375$$
$$= 1859.215\text{ N}\,(\uparrow).\quad \textbf{Ans.}$$

and
$$R_S = R_{WS} + R_{cS} + R_{gS} + R_{eS}$$
$$= 3924\,(\uparrow) + 2083.33\,(\uparrow) + 20.83\,(\uparrow) + 39.375\,(\uparrow)$$
$$= 6067.535\text{ N}(\uparrow).\quad \textbf{Ans.}$$

(*b*) Car takes a left turn

(*i*) *Reaction due to weight of car.* It will be same as in case of right turn.

$$\therefore\qquad R_{WQ} = R_{WP} = 5886\text{ N}\,(\uparrow)\quad\text{and}\quad R_{WR}{}^* = R_{WS} = 3924\text{ N}\,(\uparrow)$$

(*ii*) *Reaction due to centrifugal couple.* The magnitude to centrifugal couple will be same. Also the reaction on the wheels will be same as in case of right turn. Only the sign will change. The reaction on wheel *Q* and *S* will be downwards whereas the reaction on wheels *P* and *R** will be vertically upwards. The centrifugal couple (C_c) acting on the car is shown in Fig. 21.15 (*b*).

$$\therefore\qquad R_{CQ} = R_{CS} = 2083.33\text{ N}(\downarrow)\ \text{and}\quad R_{CP} = R_{CR}{}^* = 2083.33\text{ N}\,(\uparrow)$$

(*iii*) *Reaction due to gyroscopic couple.*

1. *Due to wheels.* (C_{gw})

The magnitude of C_{gw} (*i.e.*, $C_{gw} = 62.5$ Nm) will be same.

Also the magnitude of the reactions on the wheels due to gyroscopic couple of four wheels will be same $\left(i.e., = \frac{C_{gw}}{2b} = 20.83\text{ N}\right)$. To find the directions of the reactions, consider Fig. 21.15 (*d*). Thus Figure has been drawn looking from the front of the car. The gyroscopic couple acting (shown by dotted line *ba*) looking from front is anticlockwise. This couple will be clockwise looking from the back of the car, which is shown in Fig. 21.15 (*b*). Due to this couple, the inner wheels (*Q*, *S*) will be lifted whereas the outer wheels (*P* and *R**) will be pressed to the ground. Hence reactions of ground on wheels (*Q* and *S*) will be vertically downwards and on wheels (*P* and *R**) will be vertically upwards.

$$\therefore\quad R_{gQ} = R_{gS} = 20.83\text{ N}\,(\downarrow)\ \text{and}\ R_{gP} = R_{gR}{}^* = 20.83\text{ N}\,(\uparrow)$$

2. *Due to engine rotating parts* (C_{ge})

The magnitude of C_{ge} (*i.e.*, C_{ge} = 196.875 Nm) will be same. Also the magnitude of reactions on the wheels due to gyroscopic couple of engine rotating parts will be same $\left(i.e., = \frac{C_{ge}}{2b} = \frac{196.875}{2 \times 2.5} = 39.375\right)$. To find the directions of reactions, consider Fig. 21.15 (*e*). This figure has been drawn looking from the front of the car. The gyroscopic couple acting (shown by dotted line by vector *ba* in Fig. 21.15 (*e*)) looking from the front of the car is anticlockwise. Hence this couple will be clockwise looking from back of the car which is shown in Fig. 21.10 (*c*). Due to this couple the rear wheels (*S* and *R**) will be lifted whereas the front wheels (*P* and *Q*) will be pressed to the ground. Hence reaction of ground on wheels (*S* and *R**) will be vertically downwards and on wheels (*P* and *Q*) will be upwards.

$$\therefore\qquad R_{eS} = R_{eR}{}^* = 39.375\text{ N}\,(\downarrow)\ \text{and}\ R_{eP} = R_{eQ} = 39.375\text{ N}\,(\uparrow)$$

Hence for left turn, the total reaction on different wheels will be as

$$R_Q = R_{wQ} + R_{cQ} + R_{gQ} + R_{eQ}$$
$$= 5886\text{ N}(\uparrow) + 2083.33\text{ N}(\downarrow) + 20.83\text{ N}(\downarrow) + 39.375\text{ N}(\uparrow)$$
$$= 5886 - 2083.33 - 20.83 + 39.375$$
$$= 3821.215\text{ N}(\uparrow).\quad \textbf{Ans.}$$

$$R_p = R_{wP} + R_{cP} + R_{gP} + R_{eP}$$
$$= 5886\text{ N}(\uparrow) + 2083.33\text{ N}(\uparrow) + 20.83\text{ N}(\uparrow) + 39.375\text{ N}(\uparrow)$$
$$= 5886 + 2083.33 + 20.83 + 39.375$$
$$= 8029.535\text{ N}(\uparrow).\quad \textbf{Ans.}$$

$$R_R{}^* = R_{WR}{}^* + R_{CR}{}^* + R_{gR}{}^* + R_{eR}{}^*$$
$$= 3924\text{ N}(\uparrow) + 2083.33(\uparrow) + 20.83\text{ N}(\uparrow) + 39.375\text{ N}(\downarrow)$$
$$= 3924 + 2083.33 + 20.83 - 39.375$$
$$= 5988.785\text{ N}(\uparrow).\quad \textbf{Ans.}$$

and
$$R_S = R_{WS} + R_{CS} + R_{gs} + R_{es}$$
$$= 3924\text{ N}(\uparrow) + 2083.33\text{ N}(\downarrow) + 20.83\text{ N}(\downarrow) + 39.375\text{ N}(\downarrow)$$
$$= 3924 - 2083.33 - 20.83 - 39.375$$
$$= 1780.465\text{ N}(\uparrow).\quad \textbf{Ans.}$$

Problem 21.8. *A car is of total mass 2000 kg. It has wheel base equal to 2.4 m and track width = 1.4 m. The centre of gravity lies at 500 mm above ground level on the vertical line through geometric centre of wheels. The effective diameter of each wheel is 800 mm and mass moment of inertia of each wheel is 1 kg m². The rear axle ratio is 4. The mass moment of inertia of engine rotating parts is 3 kg m². The spin axis of the rotating engine parts is parallel to the spin axis of wheels. The engine is rotating in the same sense as the wheels. Determine the critical speed of the car when it takes a right turn of 100 m radius.*

Solution. Given : $m = 2000$ kg $\therefore$ $W = m \times g = 2000 \times 9.81$ N ; $b = 2.4$ m ; $a = 1.4$ m, $h = 500$ mm $= 0.5$ m. The C.G. is on the vertical line through geometric centre of wheels. Hence weight on each wheel will be equal to $\frac{W}{4}$.

$d = 800$ mm $= 0.8$ m $\therefore$ $r = 0.4$ m, $I_w = 1$ kg m² ; $G = 4$; $I_e = 3$ kg m². The spin axis of the engine rotating parts is parallel to the spin axis of the wheels. Also the rotation of wheels and of engine parts are in the same sense, hence the gyroscopic couple of wheels and of engine rotating parts will of the same sense.

Car takes a right turn with $R = 100$ m.

Let V = critical speed of the car.

If the reaction on the wheels of the car is positive (*i.e.*, in the upward direction), the vehicle will be stable. But if the reaction on any wheel becomes negative (*i.e.*, in the downward direction), then that wheel will be lifted from the ground and car will become unstable. For critical speed, the minimum reaction on any wheel should be zero.

Now let us find the reaction on each wheel.

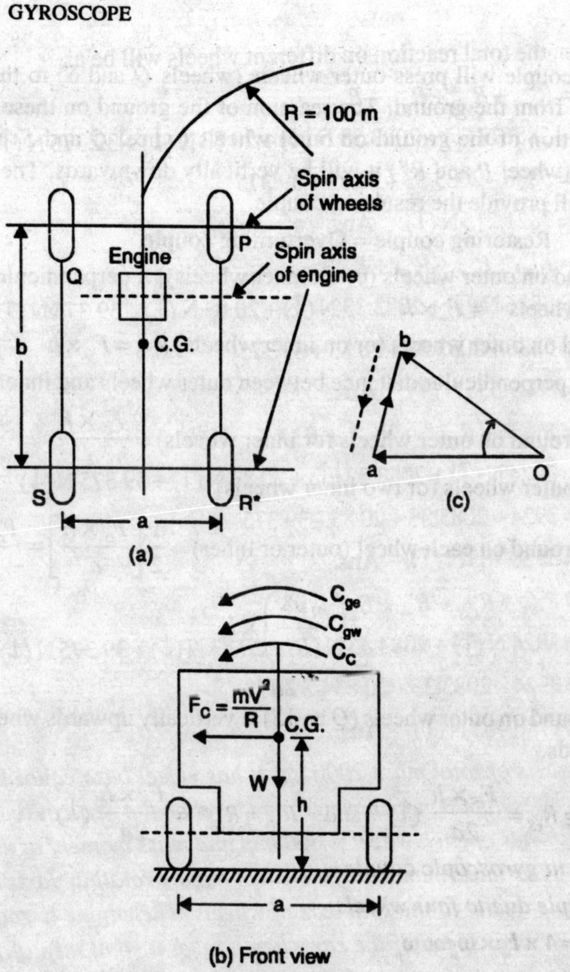

Fig. 21.16

(*i*) ***Reaction due to weight of car.*** The weight on each wheel is equal to $\frac{W}{4}$ and this weight is acting vertically downwards. Hence the reaction of ground on each wheel due to weight will be equal to $\frac{W}{4}$ upwards.

$$\therefore \qquad R_{wP} = R_{wQ} = R_{WS} = R_{WR}{*} = \frac{W}{4}\ (\uparrow) \qquad \ldots(i)$$

(*ii*) ***Reaction due to centrifugal couple.*** Centrifugal force, $F_c = \frac{mV^2}{R}$

When car takes a right turn, this force will be acting radially outward through the C.G. of the car as shown in Fig. 21.16 (*b*). The C.G. of the car is at a height '*h*' from the ground. This force will produce a couple of magnitude $F_c \times h$. This couple is known as centrifugal couple and is represented by C_c. The centrifugal couple tending to overturn the car or overturning couple

$$C_c = F_c \times h = \frac{mV^2}{R} \times h$$

This overturning couple will press outer wheels (wheels Q and S) to the ground and lift the inner wheels (wheel P and R^*) from the ground. The reaction of the ground on these wheels will be in opposite direction. Hence the reaction of the ground on outer wheels (wheel Q and S) will be vertically upwards, whereas on inner wheels (wheel P and R^*) it will be vertically downwards. The reactions of the ground on outer and inner wheels will provide the restoring couple.

∴ Restoring couple = Overturning couple

or [Reaction of ground on outer wheels (or on inner wheels)] × perpendicular distance between outer wheels and inner wheels $= F_c \times h$

or Reaction of ground on outer wheels (or on inner wheels) $\times a = F_c \times h$

(∵ perpendicular distance between outer wheels and inner wheels = Track width = a)

$$\therefore \text{ Reaction of ground on outer wheels (or inner wheels)} = \frac{F_c \times h}{a}$$

But there are two outer wheels (or two inner wheels)

$$\therefore \text{ Reaction of ground on each wheel (outer or inner)} = \frac{1}{2}\left[\frac{F_c \times h}{a}\right] = \frac{F_c \times h}{2a}$$

$$= \frac{\left(\frac{mV^2}{R}\right) \times h}{2a} \qquad \left(\because F_c = \frac{mV^2}{R}\right)$$

But reaction of ground on outer wheels (Q and S) is vertically upwards whereas on inner wheels (P and R^*) is vertically downwards.

$$\therefore \quad R_{cQ} = R_{cS} = \frac{F_c \times h}{2a} \; (\uparrow) \quad \text{and} \quad R_{cP} = R_{cR^*} = \frac{F_c \times h}{2a} \; (\downarrow) \qquad \ldots(ii)$$

(iii) Reaction due to gyroscopic couple.

1. *Gyroscopic couple due to four wheels*

$$C_{gw} = 4 \times I_w \times \omega_w \times \omega_p$$

$$= 4 \times 1 \times \frac{V}{r} \times \frac{V}{R} = \frac{4V^2}{rR}$$

$$\left(\because \omega_w = \text{Angular velocity of wheel} = \frac{V}{r}, \; \omega_p = \text{Angular velocity of precession} = \frac{V}{R}\right)$$

The vector diagram for gyroscopic couple is drawn Fig. 21.16 (*c*) looking from the back of the vehicle. The direction of this gyroscopic couple is given by vector *ba* shown dotted in Fig. 21.16 (*c*). The sense of this couple is anti-clockwise [according to right hand screw rule and looking from the back of the vehicle]. This couple (*i.e.,* C_{gw}) which is acting anti-clockwise on the vehicle is also shown in Fig. 21.16 (*b*). This couple will try to lift the inner wheels (P and R^*) from the ground and press the outer wheels (Q and S) to the ground. The reaction of the ground on these wheels will be in opposite direction. This means the reaction on inner wheels (P and R^*) will be vertically downwards whereas on outer wheels (Q and S) will be vertically upwards.

Also, we know,

C_{gw} = Reaction of ground on inner wheels (or on outer wheels) × perpendicular distance between inner wheels and outer wheels

= Reaction of ground on inner wheels (or on outer wheels) × a

$$\therefore \text{ Reaction of ground on inner wheels (or on outer wheels)} = \frac{C_{gw}}{a}$$

or Reaction of ground on each inner wheel (or on each outer wheel) $= \frac{1}{2}\left[\frac{C_{gw}}{a}\right] = \frac{C_{gw}}{2a}$

But reaction of ground on inner wheels [*P* and *R**] is vertically downward and on outer wheels [*Q* and *S*] is vertically upwards.

$$\therefore \quad R_{gP} = R_{gR}{}^* = \frac{C_{gw}}{2a}\ (\downarrow) \quad \text{and} \quad R_{gQ} = R_{gS} = \frac{C_{gw}}{2a}\ (\uparrow) \qquad \ldots(iii)$$

2. *Gyroscopic couple due to engine rotating parts*

$$C_{ge} = I_e \times \omega_e \times \omega_p$$

$$= 3 \times (G \times \omega_w) \times \frac{V}{R} \qquad \left(\because \; I_e = 3 \text{ kg m}^2;\ G = \frac{\omega_e}{\omega_w} \text{ or } \omega_e = G\omega_w \text{ and } \omega_p = \frac{V}{R}\right)$$

$$= 3 \times \left(4 \times \frac{V}{r}\right) \times \frac{V}{R} \qquad \left(\because \; G = 4 \text{ and } \omega_w = \frac{V}{R}\right)$$

$$= \frac{12V^2}{rR}$$

As the spin axis of engine rotating parts is parallel to the spin axis of the rotating wheels and also the sense of rotation and engine parts and wheel are given same, hence the vector diagram for this gyroscopic couple will also be same as shown in Fig. 21.16 (*c*). The direction of this gyroscopic couple will also be given by vector *ba*, which is anti-clock. This couple (*i.e.*, C_{ge}) which is acting anti-clockwise on the vehicle is shown in Fig. 21.16 (*b*). This couple will have the same effect as C_{gw}. The reaction on each wheel will be $\frac{C_{ge}}{2a}$. The reaction will be vertically downwards on inner wheels [*P* and *R**] and vertically upwards on outer wheels (*Q* and *S*).

$$\therefore \quad R_{eP} = R_{eR}{}^* = \frac{C_{ge}}{2a}\ (\downarrow) \quad \text{and} \quad R_{eQ} = R_{eS} = \frac{C_{ge}}{2a}\ (\uparrow) \qquad \ldots(iv)$$

For stable equilibrium of the vehicle, the reaction on any wheel should not be negative (*i.e.*, in the downward direction). From equations (*i*), (*ii*), (*iii*) and (*iv*) we find that the minimum reactions is on inner wheels. The total reaction on any inner wheel (*P* or *R**)

$$= \frac{W}{4} - \frac{F_c \times h}{2a} - \frac{C_{gw}}{2a} - \frac{C_{ge}}{2a}$$

[Taking upward reaction + ve and downward reaction – ve]

For critical speed, the total minimum reaction on any wheel should be zero or more than zero. Hence for critical speed, we have

$$\therefore \quad \frac{W}{4} - \frac{F_c \times h}{2a} - \frac{C_{gw}}{2a} - \frac{C_{ge}}{2a} \geq 0 \quad \text{or} \quad \frac{W}{4} \geq \frac{F_c \times h}{2a} + \frac{C_{gw}}{2a} + \frac{C_{ge}}{2a}$$

or
$$\frac{2000 \times 9.81}{4} \geq \frac{\left(\frac{mV^2}{R}\right) \times h}{2a} + \frac{4V^2}{rR} \times \frac{1}{2a} + \frac{12V^2}{rR} \times \frac{1}{2a}$$

$$\left[\because \; F_c = \frac{mV^2}{R};\ C_{gw} = \frac{4V^2}{rR},\ C_{ge} = \frac{12V^2}{rR}\right]$$

or $$500\times 9.81 \geq \frac{(2000\times V^2)\times 0.5}{100\times 2\times 1.4}+\frac{4V^2}{0.4\times 100}\times\frac{1}{2\times 1.4}+\frac{12V^2}{0.4\times 100}\times\frac{1}{2\times 1.4}$$

$[\because\ R = 100\text{ m},\ \ h = 0.5\text{ m},\ \ a = 1.4\text{ and } r = 0.4\text{ m}]$

or $$4905 \geq 3.571\,V^2 + 0.0357\,V^2 + 0.1071\,V^2$$

or $$4905 \geq 3.7138\,V^2$$

or $$V \leq \sqrt{\frac{4905}{3.7138}} = 36.34\text{ m/s}$$

$$\leq \frac{36.34}{1000}\times 60\times 60 = 130.83\text{ km/hr.}$$

Hence for stable equilibrium of the vehicle, the speed should be less than 130.83 km/hr.

$\therefore$ Critical speed = **130.83 km/hr.**

Problem 21.9. *A four wheeled trolley car has total mass 2000 kg. The car runs on rails of 1.4 m gauge and rounds a curve of 30 m radius at 36 km/hr on a track of embankment slope of 10°. The external diameter of wheels is 0.6 m and each pair of axle has a mass of 200 kg with radius of gyration of 250 mm. The height of the centre of gravity of the car above the wheel base in 1 m. Determine the pressure on each rail allowing for centrifugal force any gyroscopic couple action.*

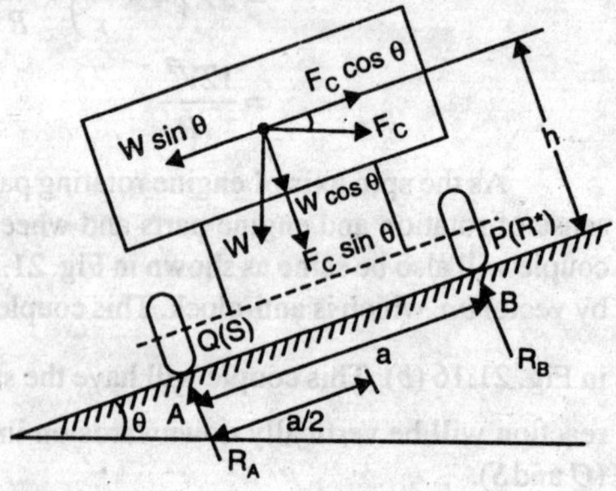

Fig. 21.17

Solution. Given : Total mass, $m = 2000$ kg

$\therefore$ $W = m\times g = 2000\times 9.81$ N

Distance between the rails, $a = 1.4$ m

Radius of curve, $R = 30$ m, speed of car,

$V = 36$ km/hr

$$= \frac{36\times 1000}{60\times 60}\text{ m/s} = 10\text{ m/s}$$

Angle at which track in banked, $\theta = 10°$

Diameter of wheel = 0.6 m hence $r = 0.3$ m

Mass of each pair of axle, $m_1 = 200$ kg and radius of gyration, $k = 0.25$ m

$\therefore$ Moment of inertia of each pair, $I = m_1k^2 = 200\times(0.25)^2 = 12.5\text{ kg m}^2$

The height of C.G. of car above wheel base, $h = 1$ m.

Refer to Fig. 21.17. Let us first consider the effect of weight (W) and centrifugal force acting on the car. The reactions R_A and R_B at the wheels Q and P can be determined as given below :

(*i*) The weight W is acting vertically downwards at the C.G. of the car. The components of this weight parallel to the base of the wheel (*i.e.*, parallel to the track) and perpendicular to the track are $W\sin\theta$ and $W\cos\theta$

(*ii*) The centrifugal force F_c is acting in the horizontal direction outward through the C.G. of the car. The components of F_c parallel to the track and perpendicular to the track are $F_c\cos\theta$ and $F_c\sin\theta$ respectively.

Now resolving these forces perpendicular to the track, we get

$$R_A + R_B = W\cos\theta + F_c \sin\theta$$

$$= (2000 \times 9.81)\cos 10° + \left(\frac{mV^2}{R}\right)\sin 10° \qquad \left(\because \ \theta = 10° \text{ and } F_c = \frac{mV^2}{R}\right)$$

$$= 19321.9 + 2000 \times \frac{10^2}{30} \times 0.1736 \qquad (\because \ m = 2000\,\text{kg}, V = 10\,\text{m/s}, R = 30\,\text{m})$$

$$= 19321.9 + 1157.3 = 20479.2\,\text{N} \qquad \ldots(i)$$

Now taking moments of all forces about B, we get

$$R_A \times a = (W\cos\theta) \times \frac{a}{2} + (F_c \sin\theta) \times \frac{a}{2} + (W\sin\theta) \times h - (F_c \cos\theta) \times h$$

Now $\qquad F_c = \dfrac{mV^2}{R} = 2000 \times \dfrac{10^2}{30} = 6666.67\,\text{N}.$

Substituting the value of F_c in the above equation, we get

$$R_A \times 1.4 = (2000 \times 9.81 \times \cos 10°) \times \frac{1.4}{2} + (6666.67 \times \sin 10°)$$

$$\times \frac{1.4}{2} + (2000 \times 9.81 \times \sin 10°) \times 1 - (6666.67 \times \cos 10°) \times 1$$

$$[\because \ a = 1.4\,\text{m}, h = 1\,\text{m and } \theta = 10°]$$

or $\qquad 1.4 R_A = 13525.33 + 810.13 + 3406 - 6565.3 = 11176.16\,\text{N}$

$\therefore \qquad R_A = \dfrac{11176.16}{1.4} = 7982.9\,\text{N} \qquad \ldots(ii)$

Substituting this value of R_A in equation (*i*), we get

$$R_B = 20479.2 - 7982.9 = 12496.3\,\text{N} \qquad \ldots(iii)$$

Not let us find the reactions due to gyroscopic couple.

The angular momentum about axle $= I \times \omega_w$

$\therefore$ Angular momentum about horizontal $= I \times \omega_w \times \cos\theta$

$\therefore$ Gyroscopic couple = (Angular momentum about horizontal) $\times \omega_p$

$$= (I \times \omega_w \times \cos\theta) \times \omega_p$$

where I = Mass moment of inertia of each pair of axle $(= m_1 k^2 = 200 \times 0.25^2 = 12.5\,\text{kg m}^2) = 12.5\,\text{kg m}^2$.

ω_w = Angular velocity of wheels $= \dfrac{V}{r} = \dfrac{10}{0.4} = 24$ rad/s

and $\quad \omega_p$ = Angular velocity of precession $= \dfrac{V}{R} = \dfrac{10}{30} = 0.333$ rad/s

$\therefore$ Gyroscopic couple, $\quad C_{gw} = (12.5 \times 25 \times \cos 10°) \times 0.333 = 12.5 \times 25 \times 0.9848 \times 0.333 = 102.48$ Nm.

Due to this couple, the car will tend to overturn about the outer wheels. This means that the inner Wheel will be lifted from the ground and outer wheel will be pressed to the ground. Let F is the force at each pair of wheels or each rail.

Then $\qquad F \times a = C_{gw}$

or $$F=\frac{C_{gw}}{a}=\frac{102.48}{1.4}=73.2\text{ N}$$

This force will be vertically downwards on the inner wheel whereas this force will be vertically upwards on outer wheels.

$\therefore$ Total reaction on the inner rail $=R_A-73.2=7982.9-73.2=7909.7$ N (↑). **Ans.**

and total reaction on the outer rail $=R_B+73.2=12496.3+73.2=12569.5$ N (↑). **Ans.**

The pressure is opposite to the reaction.

$\therefore$ Pressure on inner rail = 7909.7 N (↓) **Ans.**

and Pressure on outer rail = 12569.5 N (↓) **Ans.**

21.4.2. Stability of Two Wheelers

When a two wheeler moves straight on road, vehicle remains stable under the action of various forces like tractive effort, resistance forces etc. As the speed of two wheeler increases it becomes more stable. Rider can, sometimes, even leave the handle of motor cycle because gyroscopic couple maintains front wheel straight on road.

When two wheeler is steered to the left or right, the axis of the wheels and other rotating parts undergoes precession along with spinning. Because of this, gyroscopic couple also acts along with other forces.

Let I_w = mass moment of inertia of each wheel in kg m^2 (There are two wheels)

I_E = mass moment of inertia of engine parts in kg m^2.

r = radius of wheel

R = the radius of track

G = the gear ratio between engine and the wheel $=\dfrac{\text{Engine speed}}{\text{Wheel speed}}=\dfrac{\omega_e}{\omega}$

ω_e = angular velocity of engine rotating parts in rad/sec.

ω = angular velocity of wheels in rad/sec $=\dfrac{V}{r}$

h = the height of C.G. of two wheel along with the rider n 'm'

V = the linear velocity of two wheeler in m/s $=\omega\times r$

W = total weight of two wheeler with the rider.

θ = the inclination of the vehicle to the vertical during equilibrium.

The two couples *i.e.,* gyroscopic couple and couple due to centrifugal force will be acting on the moving vehicle when it takes a right or left turn. When the two wheeler takes a turn, then due to precession a gyroscopic couple will act on the vehicle. To find the gyroscopic couple on the vehicle, let us first find the velocity of precession and total angular momentum of the rotating parts.

Velocity of precession, $\omega_p=\dfrac{V}{R}$ and angular velocity of wheels, $\omega=\dfrac{V}{r}$

Angular momentum of two wheels $=2I_w\times\omega=2I_w\times\dfrac{V}{r}$ $\quad\left(\because\ \omega=\dfrac{V}{r}\right)$

Angular momentum of engine rotating parts $=I_E\times\omega_E=I_E\times G\times\omega=I_E\times G\times\dfrac{V}{r}$

$$\left(\because\ G=\frac{\omega_e}{\omega}\quad\therefore\ \omega_e=G\times\omega\right)$$

Let H = Total angular momentum of rotating parts.

Total angular momentum of the rotating parts

$$H = 2I_w \frac{V}{r} + I_E\left(\frac{GV}{r}\right) = \frac{V}{r}\,(2I_w + G\,I_E)$$

As two wheeler takes a turn, it does not remain in the vertical plane but its plane becomes inclined to the vertical plane. The angular momentum vector 'H' is perpendicular to the plane of the two wheeler. It can be resolved into 'ob' and 'ba' as shown in Fig. 21.18 (*c*). The component '$H\cos\theta$' undergoes precession in the horizontal plane as two wheeler moves along the curved path. Therefore, gyroscopic couple 'C_g' is given by

C_g = (Horizontal component of total angular momentum) × ω_p

or $$C_g = H\cos\theta \times \frac{V}{R} \qquad (\because\ \omega_p = V/R)$$

or $$C_g = \frac{V}{r}\,(2I_w + GI_E)\cos\theta \times \frac{V}{R} \qquad \left[\because\ H = \frac{V}{r}\,(2I_w + GI_E)\right]$$

$$= \frac{V^2}{Rr}\,(2I_w + GI_E)\cos\theta$$

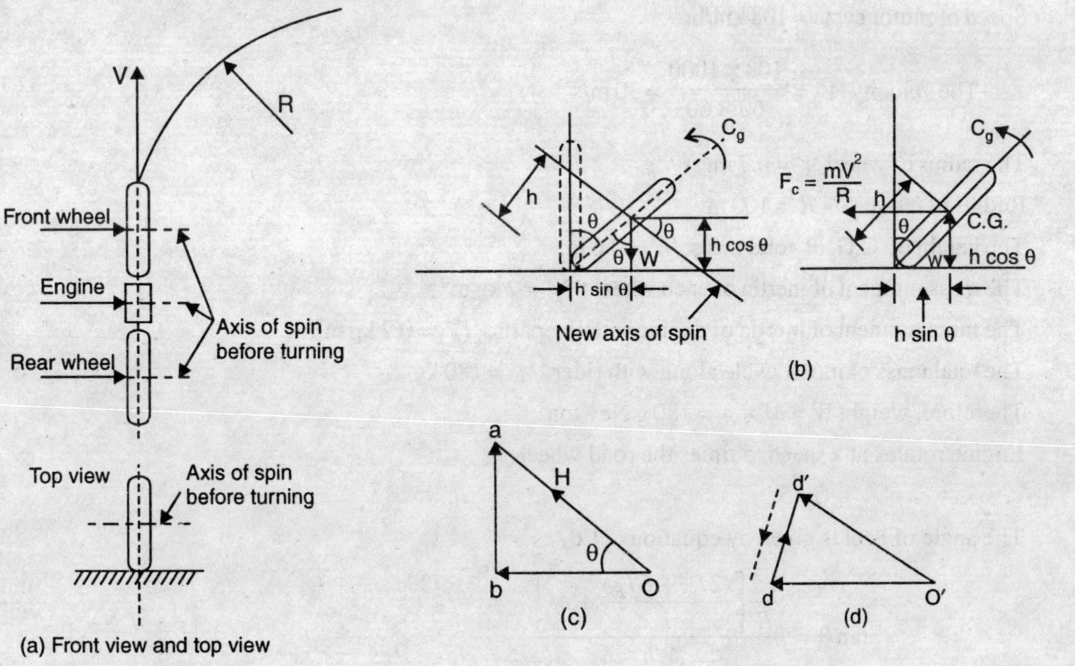

Fig. 21.18

The reaction couple vector is shown by dotted line in Fig. 21.18 (*d*). Its action is to bring two wheeler in the vertical plane back as shown in Fig. 21.18 (*b*). The centrifugal force, $F_c = \dfrac{mV^2}{R}$ is acting outward through the C.G. of the vehicle as shown in Fig. 21.18 (*b*). The weight W is acting downward through the C.G. of the vehicle. Taking moment about the point of contact equilibrium equation can be written as follows :

$$C_g + M\frac{V^2}{R} \times h\cos\theta = Wh\sin\theta.$$

or
$$\frac{V^2}{Rr}(2I_w + GI_E)\cos\theta + M\frac{V^2}{R}h\cos\theta = Wh\sin\theta$$

$$\therefore \quad \tan\theta = \frac{\frac{V^2}{R}\left[\frac{(2I_w + GI_E)}{r} + M\right]}{Wh} \qquad \text{... (21.6)}$$

The angle 'θ' in the above equation is also known as angle of heel.

Problem 21.10. *The mass of the motor cycle along with the rider is 180 kg. The height of the centre of gravity of total mass is 60 cm above the ground when it moves straight. Each wheel has diameter equal to 70 cm and polar mass moment of inertia of each wheel is 2 kgm². The engine rotates at a speed 5 times the road wheel and engine rotating parts have polar mass moment of inertia equal to 0.2 kgm². Determine the angle of heel required if motor cycle negotiates a curve of radius 100 m at a speed of 108 km/hr.*

Solution. Let angle of heel be θ. It is the angle of inclination of the plane of motor cycle with vertical plane.

Speed of motor cycle = 108 km/hr

$$\therefore \quad \text{The velocity '}V\text{'} = \frac{108 \times 1000}{60 \times 60} = 30 \text{ m/s}$$

The radius of wheel 'r' = 0.7 m.

Radius of curve, R = 100 m.

The height of C.G. of total mass, 'h' = 0.6 m.

The mass moment of inertia of each wheel, 'I_w' = 2 kg m²

The mass moment of inertia of engine rotating parts, 'I_E' = 0.2 kg m²

The total mass of motor cycle along with rider 'M' = 180 kg

Therefore, weight $W = M \times g$ = 180 g Newton

Engine rotates at a speed 5 times the road wheel

$\therefore \quad G = 5$

The angle of heel is given by equation (21.6) as

$$\tan\theta = \frac{\frac{V^2}{R}\left[\frac{(2I_w + GI_E)}{r} + M\right]}{Wh}$$

$$= \frac{\frac{30^2}{100}\left[\frac{(2 \times 2 + 5 \times 0.2)}{0.7} + 180 \times 0.6\right]}{108 \times g \times 0.6} = \frac{9}{180 \times 9.81}\left\{\frac{5}{0.7} + 180 \times 0.6\right\}$$

or $\tan\theta = 0.978$

or $\theta = $ **44.36°. Ans.**

21.5. GYROSCOPIC EFFECTS ON SHIPS AND AEROPLANES

Ships and aeroplanes undergo rotations about all the three certesian axes. These rotations are identified by special names. As a convention X-axis is always aligned along the forward motion. Then the layout of Z-axis shall be along the side in the horizontal plane and Y-axis shall be in the vertical plane.

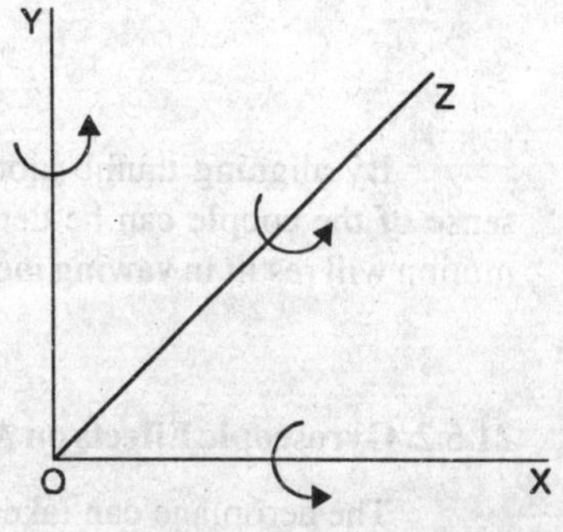

Fig. 21.19

The rotation about X-axis is called rolling motion. The rotation about Z-axis is called pitching motion and rotation about Y-axis is called yawing motion.

21.5.1. Gyroscopic Effects on Ships

Two views of ship are shown in Fig. 21.20. Let the turbine rotor rotate in clockwise sense when viewing from rear. Let V be the velocity of ship in m/s.

Let I be the mass moment of inertia of turbine rotor and ω be the angular speed. Let the ship steers to the left as shown in Fig. 21.20.

Angular momentum $= I\omega$.

According to right hand screw rule, it is directed along X-axis. When ship moves along curved path, angular momentum vector also turns to left. The reaction couple vector is shown by dotted line. The precessional angular velocity

$$\omega_p = \frac{V}{R}.$$

$\therefore$ Gyroscopic couple 'C_g' $= I\omega\, \omega_p$

$$= I\omega \frac{V}{R} \text{ Nm}$$

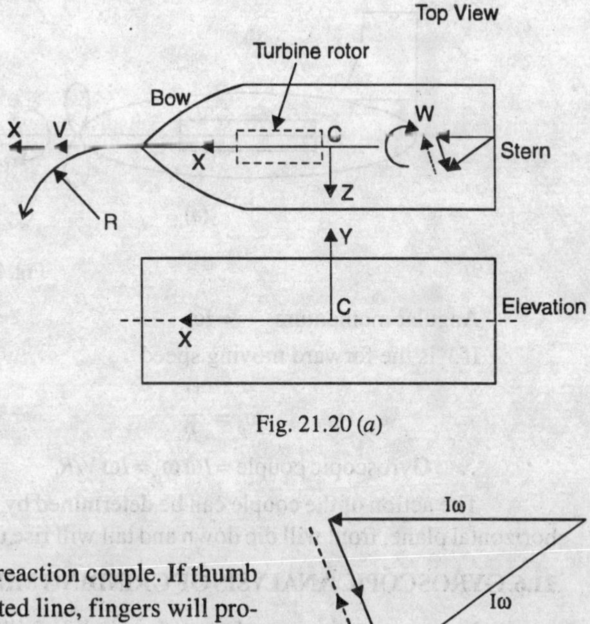

Fig. 21.20 (*a*)

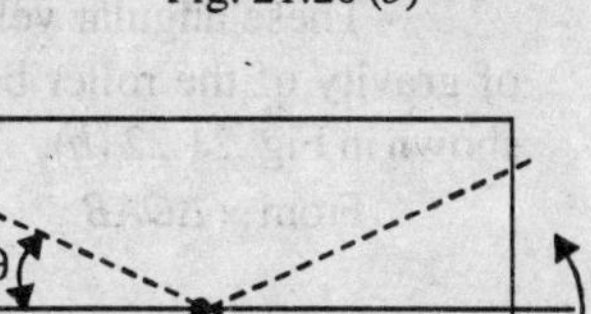

Fig. 21.20 (*b*)

It's action on the ship shall be according to reaction couple. If thumb is aligned along the reaction couple vector *i.e.*, dotted line, fingers will provide sense of the couple. According to this, bow will lift and stern shall dip down.

When ship rolls, precession does not take place and therefore, no gyroscopic couple will act.

When ship undergoes pitching, let θ degrees be the amplitude of pitching. Let the pitching takes place with simple harmonic motion. Let T be the time period of S.H.M.

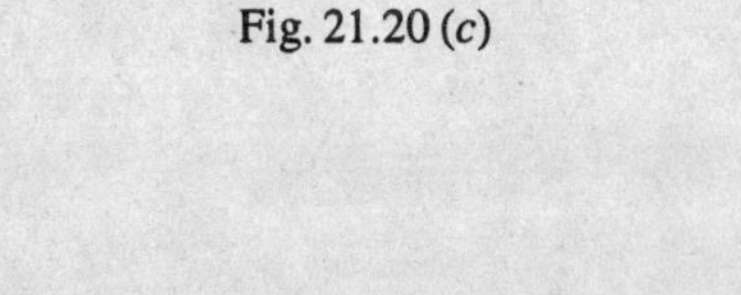

Fig. 21.20 (*c*)

$$\text{Angular frequency of motion} = \frac{2\pi}{T} \text{ rad/s}$$

$$\text{Max. angular precessional velocity } \omega_p = \frac{2\pi}{T} \times \theta \times \frac{\pi}{180}$$

Therefore, max. gyroscopic couple 'C_g' $= I\omega\, \omega_p$

$$= I\omega \times \frac{2\pi}{T} \times \theta \times \frac{\pi}{180}$$

By aligning thumb along dotted line shown in Fig. 21.20 (*d*) sense of the couple can be determined. It is obvious that pitching motion will result in yawing motion of ship.

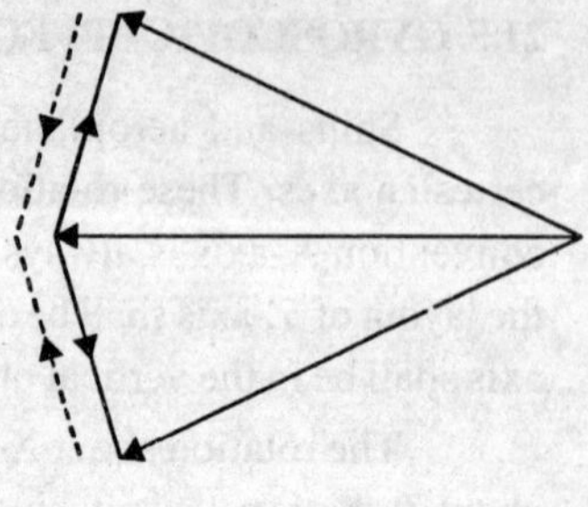

Fig. 21.20 (*d*)

21.5.2. Gyroscopic Effects on Aeroplanes

The aeroplane can take turn in the horizontal plane. It can also move along curved path in vertical plane. It can undergo pitching motion like ships. If I is the mass moment of inertia of the propeller and ω is angular speed.

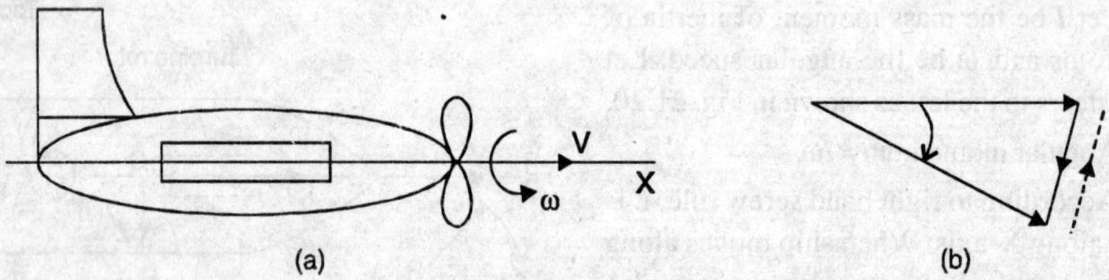

Fig. 21.21

Angular momentum $= I\omega$

If V is the forward moving speed

$$\omega_p = \frac{V}{R}$$

$\therefore$ Gyroscopic couple $= I\omega\, \omega_p = I\omega\, V/R$.

The action of the couple can be determined by the right hand screw rule. If it steers to right side in the horizontal plane, front will dip down and tail will rise up.

21.6. GYROSCOPIC ANALYSIS OF GRINDING MILL

Let us assume a general case of a grinding mill which has a conical roller. The gyroscopic effect in the grinding mill increases crushing force. Let ω be angular speed of the roller and ω' be angular speed of the driving shaft.

These angular velocities can be plotted and the resultant meets the floor of roller at C. Let the centre of gravity of the roller be located on the radius passing through C. The free body diagram of the roller is shown in Fig. 21.22 (*b*).

From ΔOAB,

$$\frac{AO}{\sin(\theta - \phi)} = \frac{AB}{\sin\phi}$$

$$\frac{AO}{AB} = \frac{\sin(\theta - \phi)}{\sin\phi} = \frac{\omega}{\omega'}$$

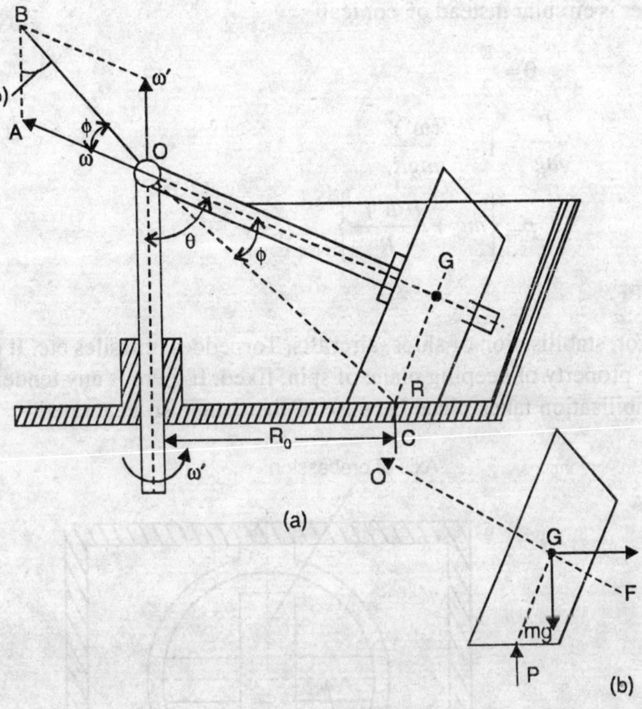

Fig. 21.22

Also,
$$R\omega = R_o\omega'$$

or
$$\frac{\omega}{\omega'} = \frac{R_o}{R}.$$

The gyroscopic couple is given by

$$C_g = I\,\omega\,\omega' \sin\theta + (I - I')\,(\omega')^2 \sin\theta\cos\theta$$

$$= I\,(\omega')^2 \sin\theta\,\frac{R_0}{R} + (I - I')\,(\omega')^2 \sin\theta\cos\theta$$

$$= \left\{ I\,\frac{R_0}{R} + (I - I')\cos\theta \right\} (\omega')^2 \sin\theta.$$

By taking moment of all forces

$$C_g = F\,(R_0 + R\cos\theta)\cot\theta - mg\,(R_0 + R\cos\theta) + PR_0 \qquad \text{...(21.7)}$$

$$\therefore \quad F\,(R_0 + R\cos\theta)\cot\theta - mg\,(R_0 + R\cos\theta) + PR_0$$

$$= \left[I\,\frac{R_0}{R} + (I - I')\cos\theta \right] (\omega')^2 \sin\theta.$$

$$\therefore \quad \frac{P}{mg} = \frac{(\omega')^2 \sin\theta}{mg\,R_0}\left[I\,\frac{R_0}{R} + (I - I')\cos\theta \right] - \frac{F\cot\theta}{mg}\left(1 + \frac{R}{R_0}\cos\theta\right)$$

If the roller is circular instead of conical

$$\theta = \frac{\pi}{2}$$

$$\frac{P}{mg} = 1 + \frac{I(\omega')^2}{mgR}$$

or $$P = \left\{ mg + \frac{I(\omega')^2}{R} \right\} \qquad \ldots (21.8)$$

21.7. GYROSCOPE

It is used for stabilisation of ships, aircrafts, Torpedoes, missiles etc. It consists off heavy rotor. The gyroscope has the property of keeping plane of spin, fixed. If there is any tendency to change plane, precession starts and stabilisation takes place because of the couple so generated.

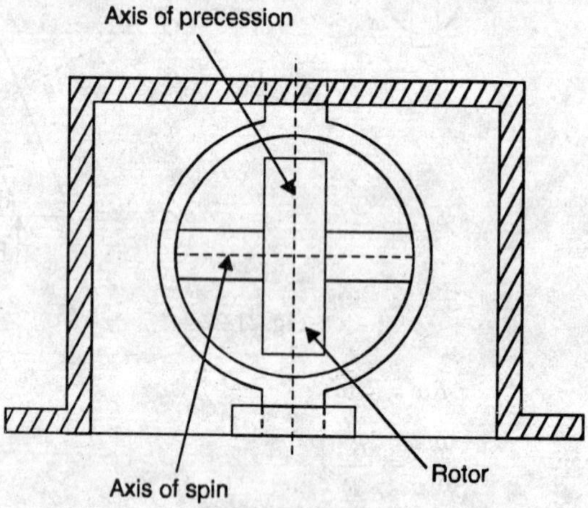

Fig. 21.23

Problem 21.11. *A railway car has two axles. The mass of car is 5000 kg. It has two axles. One of these axles is driven by a motor directly. The axis of motor is parallel to the axle. The distance between the rails is 1.5 m. The mass moment of the inertia of axles along with wheels is 35 kg m². The mass moment of inertia of motor armature is 25 kg m². The radius of each wheel is 50 cm. Determine limiting speed for the railway car when it moves around a curve of 200 m that no wheel leaves the rail. The centre of gravity of car is at 1.2 m height.*

Solution. Given : Mass of car, $M = 500$ kg ; Weight $W = 5000 \times 9.81$ N

Mass moment of inertia of the axles along with wheels = 35 kg m²

Mass moment of inertia of motor armature = 25 kg m²

Since motor armature spin axis and wheel spin axis are parallel and both are rotating in same sense, there moment of inertia can be clubbed.

Total Mass moment of inertia = 35 + 25 = 60 kg m²

Distance between rails, $a = 1.5$ m

Let V = Velocity of wheel, r = radius of wheel = 50 cm = 0.5 m.

Angular speed for spin (or ω) $= \frac{V}{r} = \frac{V}{0.5}$ rad/s $= 2V$ rad/s

Radius of curve, $R = 200\,\text{m}$

Angular speed for precession (or ω_p) $= \dfrac{V}{R} = \dfrac{V}{200}$ rad/s

Gyroscopic couple is given by equation (21.1),

Therefore, gyroscopic couple 'C_g' $= I\omega\,\omega_p$

$$= 60 \times 2V \times \frac{V}{200}$$

$$= 0.6V^2 \qquad \left(\because \;\; \omega = 2V \text{ and } \omega_p = \frac{V}{200}\right)$$

Mass of railway car, $M = 5000$ kg

Height of C.G. of railway car, $h = 1.2$ m

Couple due to centrifugal force is given by equation (21.3) as

$$C_c = \frac{W}{g}\frac{V^2}{R} \times h = M \;\; \frac{V^2}{R} \times h \qquad \left(\because \;\; M = \frac{W}{g}\right)$$

$\therefore$ Couple due to centrifugal force

$$C_c = M\frac{V^2}{R} \times h = 5000\,\frac{V^2}{200} \times 1.2 \qquad (\because \;\; M = 5000\,\text{kg}, R = 200\,\text{m}, h = 1.2\,\text{m})$$

$$= 30.0\,V^2$$

This is a overturning couple and due to this couple, the inner wheels will be lifted from the ground and outer wheels will be pressed to the ground. The vertical reactions due to this couple at inner wheel will be downwards and at outer wheels upwards.

But the magnitude will be equal to $C_g/2a$ on all wheels.

This is also overturning couple and magnitude of reactions on each wheel $= C_c/2a$.

Total couple $= C_g + C_c = 0.6\,V^2 + 30.0\,V^2$

$= 30.6\,V^2$

This total couple is overturning couple and magnitude of vertical reactions on each wheel

$$= \frac{C_g + C_c}{2a} = \frac{30.6V^2}{2 \times 1.5} \qquad (\because \;\; a = 1.5\,\text{m and } C_g + C_c = 30.6\,V^2)$$

Due to this overturning couple the vertical reactions on inner wheels will be downwards and on outer wheels upwards.

$\therefore$ Total vertical reaction on any outer wheels $= \dfrac{W}{4}(\uparrow) + \dfrac{C_g + C_c}{2a}(\uparrow) = \left[\dfrac{W}{4} + \dfrac{30.6V^2}{3}\right](\uparrow)$

and total vertical reaction on any inner wheels $= \dfrac{W}{4}(\uparrow) + \dfrac{C_g + C_c}{2a}(\downarrow) = \left[\dfrac{W}{4} - \dfrac{30.6V^2}{3}\right](\uparrow)$

From the above, it can be observed that the reaction on the outer wheels is always positive (upwards) but there are chances that the reaction of the ground on the inner wheels may not be upwards and thus the inner wheels are lifted from the ground.

For limiting speed, the condition will be $\frac{W}{4} - \frac{C_g + C_c}{2a} \geq 0$

or $\frac{5000 \times 9.81}{4} - \frac{30.6\, V^2}{3} \geq 0 \quad \left[\because\ W = M \times g = 5000 \times 9.81 \text{ and } \frac{C_g + C_c}{2a} = \frac{30.6\, V^2}{3}\right]$

or $\frac{5000 \times 9.81}{4} \geq \frac{30.6\, V^2}{3}$ or $\frac{30.6\, V^2}{3} \leq \frac{5000 \times 9.81}{4}$

or $V^2 \leq \frac{5000 \times 9.81 \times 3}{4 \times 30.6}$

or $V \leq \sqrt{\frac{5000 \times 9.81 \times 3}{4 \times 30.6}}$ or $V \leq 34.67$ m/s

$\leq \frac{34.67}{1000} \times 60 \times 60$ km/hr

≤ 124.82 km/hr. **Ans.**

∴ Limiting speed is 124.82 km/hr.

Problem 21.12. *A ship has a propeller of mass moment of inertia 2000 kg m². The propeller rotates at a speed of 360 rpm in clockwise sense looking from stern. Determine,*

(a) *gyroscopic couple and its effect when ship moves at 30 km/hr. and steers to the left at a radius of 200 m.*

(b) *max. gyroscopic couple and its effect when ship pitches and moving up having amplitude 10° and time period 20 seconds. The motion occurs with SHM.*

Solution. Given :

$I = 2000$ kg m²

$N = 360$ rpm

∴ $\omega = \frac{2\pi N}{60}$ rad/s

or $\omega = \frac{2\pi\, 360}{60} = 12\pi$ rad/s

∴ $I \times \omega = 2000 \times 12\pi$

$= 75398.224$

(a) Speed of ship = 30 km/hr.

∴ $V = 30$ km/hr

or $\frac{30 \times 1000}{60 \times 60} = 8.33$ m/s

∴ $R = 200$ m

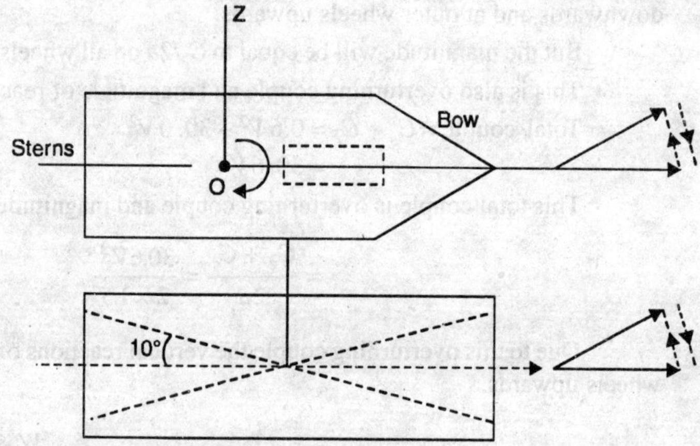

Fig. 21.24

∴ $\omega_p = \frac{V}{R} = \frac{8.33}{200} = 0.0417$ rad/s

Gyroscopic couple is given by equation (21.1) as,

$C_g = I\omega\,\omega_p = 75398.224 \times 0.41$

$= 3141.6$ Nm. $\qquad (\because\ I \times \omega = 75398.22)$

It will lift bow and dip stern as determined by right hand screw rule.

(*b*) Time period = 20 seconds

$\therefore$ Angular frequency $= \dfrac{2\pi}{\text{Time period}}$

$$= \frac{2\pi}{20} = \frac{\pi}{10} \text{ rad/s}$$

Max. gyroscopic couple will be at maximum angular velocity of precession.

$\therefore$ Max. ω_p = angular frequency × amplitude

$$= \frac{\pi}{10} \times \frac{10 \times \pi}{180} = 0.0548 \text{ rad/s.}$$

Max. Gyroscopic couple, $C_g = I\omega\,\omega_p$

$$= 75398.224 \times 0.0548 \qquad (\because \quad I \times \omega = 75398.22)$$

$$= \mathbf{4134.17\ Nm.}$$

It will result in yawing in clockwise sense looking from top.

Determine the gyroscopic couple on the air-craft and state its effect.

It tends to lift nose and dip tail.

Problem 21.13. *A trolley car weighing 3 tonnes run on rails 1 m apart at speed of 40 kmph. The track is curved with a radius of 80 m towards right side of the driver. The car has four wheels each of diameter 70 cm and mass moment of inertia 15 kgm². The car is driven by a motor of mass moment of inertia 10 kgm². The motor is placed centrally in the car and rotates nine times the speed of the axle in the same sense. The height of C.G. of the car is 1 m above the rail level. Determine the height by which the outer rail should be lifted up as compared to the inner rail to overcome the effect of gyroscopic couple and couple due to centrifugal force.* (D.U. 2001)

Solution. Given : Distance between rails = 1 m

Speed of trolley car, $V = 40 \text{ km ph} = \dfrac{40 \times 1000}{60 \times 60} = 11.11 \text{ m/s}$

Track radius, $R = 80 \text{ m}$

Diameter of each wheel, $d_w = 70$ cm $\therefore$ Radius of each wheel, $r_w = \dfrac{70}{2} = 35 \text{ cm} = 0.35 \text{ m}$

Mass moment of Inertia of each wheel, $I_w = 15 \text{ kg m}^2$

Mass moment of inertia motor, $I_m = 10 \text{ kg m}^2$

Speed of motor = 9 × speed of axle

$\therefore$ Speed ratio, $G = \dfrac{\text{Speed of motor}}{\text{Speed of axle}} = 9$

Height of C.G. = 1 m

Let C_{gw} = gyroscopic couple due to wheels

ω = Angular speed of wheel $= \dfrac{V}{r_w} = \dfrac{11.11}{0.35} = 31.746$ rad/s

C_{gm} = gyroscopic couple due to motor

C_g = Total gyroscopic couple = $C_{gw} + C_{gm}$

C_c = Couple due to centrifugal force

C = Total couple = $C_g \pm C_c$

However nothing has been mentioned about placement of C.G. with respect to two axles and side wheel, we can assume that C.G. is equidistant with respect to the wheels.

The gyroscopic couple due to wheels

$$(C_{gw}) = 4 \times I_w \times \omega \times \omega_p$$

The mass moment of inertia of each wheel

$$(I_w) = 15 \text{ kg m}^2$$

The angular speed of the wheel $(\omega) = \dfrac{V}{r_w}$

$$\text{speed } 'V' = 40 \text{ kmph} = \frac{40}{3.6} \text{ m/s} = 11.11 \text{ m/s}$$

$$\text{wheel radius } 'r_w' = \frac{70}{2} = 35 \text{ cm} = 0.35 \text{ m}$$

$$\therefore \quad \omega = \frac{11.11}{0.35} = 31.746 \text{ rad/s}$$

The speed of precession $(\omega_p) = \dfrac{V}{R}$

The track radius $'R' = 80 \text{ m}$

$$\therefore \quad \omega_p = \frac{11.11}{80} = 0.139 \text{ rad/s}$$

$$\therefore \quad C_{gw} = 4 \times 15 \times 31.746 \times 0.139 = 264.52 \text{ Nm}$$

The gyroscopic couple due to motor $(C_{gm}) = I_m \times G \times \omega \times \omega_p$

The ratio between motor and axle, $(G) = 9$

The mass moment of inertia of motor parts $(I_m) = 10 \text{ kgm}^2$

$$\therefore \quad C_{gm} = 10 \times 9 \times 31.746 \times 0.139 = 397.14 \text{ Nm}$$

Since sense of rotation of motor and wheels is same and their axes are parallel, therefore these gyroscopic couples can be added. The total gyroscopic couple $(C_g) = C_{gw} + C_{gm}$.

$$\therefore \quad C_g = 264.52 + 397.14 = 661.66 \text{ Nm.}$$

Centrifugal force, $(F_c) = \dfrac{M.V^2}{R}$

Trolley mass, $'M' = 3 \text{ tonnes} = 3000 \text{ kg}$

$$\therefore \quad F_c = 3000 \times \left(\frac{11.11}{80}\right)^2 = 4628.7 \text{ N}$$

The couple due to centrifugal force

$$C_c = F_c \times \text{height of C.G.}$$
$$= 4628.7 \times 1 = 4628.7 \text{ Nm.}$$

Both these couples tend to increase reaction on outer wheels and reduce reaction at inner wheels. Since they act in same plane and with same sense, therefore, they can be added.

The total couple $(C) = C_g + C_c$

or $C = 661.66 + 4628.7 = 5290.36 \text{ Nm}$

The corresponding force due to this couple at C.G. $(F) = \dfrac{C}{\text{height of C.G.}} = \dfrac{5290.36}{1} = 5290.36$ N.

As shown in the diagram

$$F\cos\theta = W\sin\theta$$

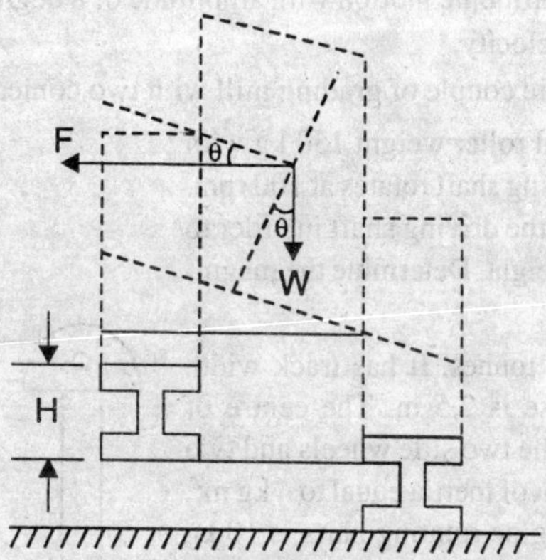

Fig. 21.25

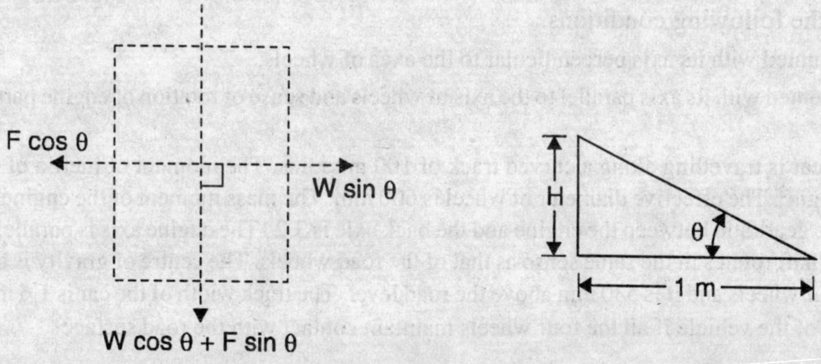

Fig. 21.26

$$\therefore \qquad \tan\theta = \frac{F}{W} = \frac{5290.36}{3000\times 9.81}$$

$$\tan\theta = 0.1798$$

$$\theta = 10.19°$$

$$\tan\theta = \frac{\text{H}}{\text{Distance between rails}} = \frac{H}{1.0} \quad \text{or} \quad H = 0.1798 \text{ m}$$

or $\quad H$ = **17.98 cm.**

EXERCISE 21

1. Explain the terms spin and precession. How do they differ from each other ?
2. Derive expression of gyroscopic couple.

3. Explain the gyroscopic effect on four wheeled vehicles.
4. Explain the gyroscopic effect on two wheeled vehicles.
5. A ship has a turbine rotor of mass 5 tonnes and radius of gyration 60 cm. It rotates at 1500 rpm in clockwise sense when looking from stern. Determine the gyroscopic couple and its effect under the following conditions :
   (*a*) The ship steers to the left at a radius of 100 m when moving at 2 km/hr.
   (*b*) The ship pitches with simple harmonic motion with amplitude of 8 degrees and time period 12 sec. The ship is depending with maximum velocity.
6. Derive the expression of gyroscopic couple of grading mill with two conical rollers and state the effect.
7. A grinding mill has one cylindrical roller weight 150 kg with diameter equal to 100 cm. The driving shaft rotates at 100 rpm. Determine the sense of rotation of the driving shaft in order to have crushing force more that its weight. Determine the magnitude of crushing force.

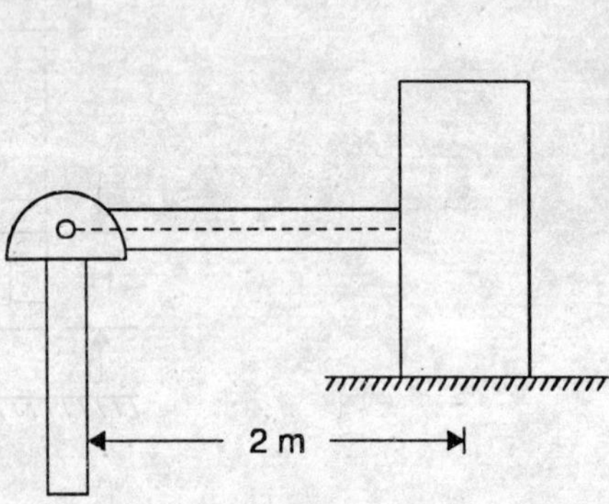

Fig. 21.27

8. A racing car has mass equal to 2 tonnes. It has track width equal to 110 cm. The wheel base is 2.5 m. The centre of gravity lies at the centre between the two side wheels and two axles. Each wheel has mass moment of inertia equal to 4 kg m^2. The mass moment of inertia of engine rotating parts is 10 kg m^2. The speed of the car is 40 km/hr. Determine critical speed and reaction at each wheel if car is steered to left at a radius of 60 m and the ratio of engine speed is 5 times the wheel speed under the following conditions.
   (*a*) engine mounted with its axis perpendicular to the axes of wheels.
   (*b*) engine mounted with its axis parallel to the axis of wheels and sense of rotation of engine parts is same as that of wheels.
9. A rear engine car is travelling along a curved track of 100 m radius. The moment of inertia of each of the four wheels is 2.2 kgm^2. The effective diameter of wheel is 600 mm. The mass moment of the engine rotating parts is 1.25 kgm^2. The gear ratio between the engine and the back axle is 3.2 . The engine axis is parallel to the near axle and the crank shaft rotates in the same sense as that of the road wheels. The centre of gravity is at equal distance from all the four wheels and it is 530 mm above the road level. The track width of the car is 1.6 m. Determine the limiting speed of the vehicle if all the four wheels maintain contact with the road surface.

OBJECTIVE TYPE QUESTIONS

Objective Type Questions

Tick mark the most appropriate answer :

1. Theory of machines is the branch of science, which deals with the study of
   (*a*) the relative motion between the parts of a machine and the study of forces which acts on those parts
   (*b*) the relative motion between the parts of a machine
   (*c*) the forces acting on the parts of the machine
   (*d*) none of the above.
2. Link or element is a
   (*a*) part of a machine
   (*b*) stationary part of a machine
   (*c*) part of a machine which has motion relative to some other part
   (*d*) none of the above.
3. When the relative motion between two elements is completely or successfully constrained, then these two elements from a
   (*a*) mechanism (*b*) machine (*c*) kinematic pair (*d*) kinematic chain.
4. Kinematic pairs are classified as
   (*a*) sliding pair (*b*) rolling pair (*c*) turning pair (*d*) spherical pair
   (*e*) all of the above (*f*) none of the above.
5. Piston and cylinder of a reciprocating steam engine forms a
   (*a*) turning pair (*b*) rolling pair (*c*) sliding pair (*d*) spherical pair
   (*e*) none of the above.
6. A ball and a socket joint form a
   (*a*) turning pair (*b*) rolling pair (*c*) sliding pair (*d*) spherical pair.
7. Shaft with collars at both ends fitted into a circular hole forms a
   (*a*) turning pair (*b*) rolling pair (*c*) sliding pair (*d*) spherical pair.
8. Ball-bearing and roller bearing form a
   (*a*) turning pair (*b*) rolling pair (*c*) sliding pair (*d*) spherical pair.
9. A bolt and nut form a
   (*a*) turning pair (*b*) rolling pair (*c*) screw pair (*d*) spherical pair.
10. When the two elements of a pair have a surface contact when in motion and the surface of one element slides over the surface of the other element, the pair formed is called a
    (*a*) higher pair (*b*) lower pair (*c*) force-closed pair (*d*) none of the above.
11. When the two elements of a pair have line or point contact when in motion, the pair formed is called a
    (*a*) higher pair (*b*) lower pair (*c*) unclosed pair (*d*) none of the above.

12. The lower pair is

(*a*) closed pair (*b*) unclosed pair (*c*) point contact pair

(*d*) surface contact pair (*e*) both (*a*) and (*b*) (*f*) both (*a*) and (*c*).

13. The higher pair is a

(*a*) closed pair (*b*) unclosed pair (*c*) point contact pair

(*d*) surface contact pair (*e*) both (*a*) and (*d*) (*f*) both (*a*) and (*c*).

14. Which one of the following is a lower pair ?

(*a*) ball and roller bearing (*b*) automobile steering gear

(*c*) cam and follower (*d*) belt and chain drives.

15. Which of the following is a higher pair

(*a*) ball and roller bearing (*b*) automobile steering gear

(*c*) cam and follower (*d*) belt and chain drives

(*e*) all of the above except (*b*) (*f*) none of the above.

16. Choose the correct statement

(*a*) A sliding pair has incompletely constrained motion.

(*b*) A pair of friction discs constitutes a lower pair.

(*c*) Rectilinear motion of a piston is converted into rotary motion by slider crank.

(*d*) Automobile steering gear is an example of higher pair.

17. Choose the wrong statement

(*a*) An element needs not be a rigid body, but it must be a resistant body.

(*b*) Cam and follower is an example of lower pair.

(*c*) A sliding pair has a completely constrained motion.

(*d*) The motion of a shaft in a circular hole is an example of incompletely constrained motion.

18. The function of an element is to

(*a* transmit motion (*b*) to serve as a support

(*c*) to guide other elements (*d*) all of the above

(*e*) none of the above.

19. The motion of a shaft between a foot-step bearing is an example of

(*a*) incompletely constrained motion (*b*) completely constrained motion

(*c*) successfully constrained motion (*d*) none of the above.

20. The motion of a shaft with collars at each end in a circular hole is an example of

(*a*) incompletely constrained motion (*b*) completely constrained motion

(*c*) successfully constrained motion (*d*) none of the above.

21. The motion of a shaft in a circular hole is an example of

(*a*) incompletely constrained motion (*b*) completely constrained motion

(*c*) successfully constrained motion (*d*) none of the above.

22. A kinetic chain is

(*a*) a combination of kinetic pair in which one link is fixed

(*b*) a combination of kinetic pair in which each link forms part of two pairs and the relative motion between the links is completely constrained (*c*) the same as mechanism

(*d*) also called inversion.

23. In a kinetic chain, if the specification of one co-ordinate, or dimension or position of a single link is sufficient to define the position of all other links, then the chain is called a kinematic chain of

(*a*) two degree of freedom (*b*) one degree of freedom

(*c*) three degrees of freedom (*d*) none of the above.

24. The differential mechanism of an automobile is having

(*a*) one degree of freedom (*b*) two degrees of freedom

(*c*) three degrees of freedom (*d*) zero degree of freedom.

25. If J = Number of binary joints in a chain, H = Number of unclosed pairs, and L = Number of links, then to determine whether a chain is a locked, constrained or constrained, the relation used is given by

(*a*) $J + H = \frac{3}{2}[L - 2]$ (*b*) $J + H = 3L - 2$

(*c*) $J + \frac{H}{2} = \frac{3}{2}L - 2$ (*d*) $\frac{J}{2} + H = \frac{3}{2}L - 2$.

26. If the right-hand side of the equation (which is used to determine whether a chain is locked, constrained or unconstrained) is greater than the left-hand side, then the chain is

(*a*) locked (*b*) constrained (*c*) unconstrained (*d*) none of the above.

27. If R.H.S. = L.H.S. of the equation (which is used to determine whether a chain is locked, constrained or unconstrained), then the chain is

(*a*) locked (*b*) constrained (*c*) unconstrained (*d*) none of the above.

28. If R.H.S. is less than L.H.S. of the equation (used for determining whether a chain is locked, constrained or unconstrained), then the chain is

(*a*) locked (*b*) constrained (*c*) unconstrained (*d*) none of the above.

29. Ternary joint is a joint at which three links are joined at the same connection. It is equivalent to

(*a*) one binary joint (*b*) two binary joints

(*c*) three binary joints (*d*) combination of (*a*) and (*b*).

30. Quaternary joint is a joint at which four links are joined at the same connection and it is equivalent to

(*a*) one binary joint (*b*) two binary joints

(*c*) three binary joints (*d*) combination of (*a*) and (*b*).

31. If a kinematic chain has 'l' links, then the number of mechanism obtained are

(*a*) $l - 1$ (*b*) $l - 2$ (*c*) $l + 1$ (*d*) l.

32. Choose the wrong statement.

(*a*) A chain consisting of three links and three joints is known as locked chain.

(*b*) A chain consisting of four links and four joints is known as kinematic chain.

(*c*) Quorternary joint is equivalent to three binary joints.

(*d*) Rectangular bar in a rectangular hole is the type of partially constrained motion.

33. Choose the correct statement.

(*a*) If a kinematic chain has '*l*' links, then '*l*' different mechanism are obtained by fixing each of the links in turn.

(*b*) A machine serves to modify and transmit energy (or force and motion) while the structure modifies and transmits force only.

(*c*) When one of the links of a kinematic chain is fixed, the chain is known as a mechanism.

(*d*) All of the above

(*e*) None of the above.

34. In a kinematic chain with four lower pairs, if all the four lower pairs are turning pairs, the mechanism is classified into the chain known as

(*a*) crossed slider crank chain (*b*) four bar chain

(*c*) slider crank chain (*d*) double slider crank chain.

35. In a kinematic chain with four lower pairs, if one is sliding pair and three turning pairs, the mechanism is classified into the chain known as

(*a*) crossed slider crank chain (*b*) four bar chain

(*c*) slider crank chain (*d*) double slider crank chain.

36. In a kinematic chain with four lower pairs, if there are two sliding pairs and two turning pairs and two similar pairs are adjacent then the mechanism is classified into the chain known as

(*a*) crossed slider crank chain (*b*) four bar chain

(*c*) slider crank chain (*d*) double slider crank chain.

37. In a kinematic chain with four lower pairs, if there are two sliding pairs and two turning pairs and two similar pairs are not adjacent, then the mechanism is classified into the chain known as

(*a*) crossed slider crank chain (*b*) four bar chain

(*c*) slider crank chain (*d*) double slider crank chain.

38. The Ackermann steering gear is the inversion of a

(*a*) slider crank chain (*b*) four bar chain

(*c*) double slider crank chain (*d*) crossed slider crank chain.

39. Whitworth quick return mechanism is an inversion of

(*a*) double slider crank chain (*b*) single slider crank chain

(*c*) four bar chain (*d*) crossed slider crank chain.

40. Oldham's coupling and elliptic trammels are the inversion of

(*a*) double slider crank chain (*b*) single slider crank chain

(*c*) four bar chain (*d*) crossed slider crank chain.

41. Rapson's slide is an inversion of

(*a*) double slider crank chain (*b*) single slider crank chain

(*c*) four bar chain (*d*) crossed slider crank chain.

42. Choose the correct statement.

(*a*) The ordinary reciprocating steam engine is an inversion of double slider crank chain.

(*b*) Scotch Yoke mechanism is an inversion of slider crank chain.

(*c*) Quick return mechanism is an inversion of slider crank chain.

(*d*) The number of inversions in a mechanism having number of links equal to l, will be $(l + 1)$.

43. A rigid body can be replaced by two masses, connected rigidly together. The system of two masses will be dynamically equivalent to the rigid body if

(*a*) the mass of the two systems are the same

(*b*) the centre of gravity of the two mass system coincides with the centre of gravity of the rigid body

(*c*) the total mass moment of inertia of the two mass system about an axis through the centre of gravity is equal to that of rigid body about the same axis

(*d*) all of the above

(*e*) none of the above.

44. In simple harmonic motion, acceleration is proportional to

(*a*) displacement (*b*) velocity

(*c*) square of displacement (*d*) square of velocity

(*e*) none of the above.

45. A body moving with simple harmonic motion has an amplitude of 30.0 cm. The maximum velocity of the body is 3 m/sec. The period of vibration will be equal to

(*a*) 0.314 sec (*b*) 0.628 sec (*c*) 1.59 sec (*d*) 3.18 sec.

46. In question 54, the frequency of vibration will be equal to

(*a*) 2.0 vibrations/sec (*b*) 1.8 vibrations/sec

(*c*) 1.59 vibrations/sec (*d*) 1.30 vibrations/sec.

47. The time period of a compound pendulum is equal to

(*a*) $2\pi\sqrt{\dfrac{h}{g}}$ (*b*) $2\pi\sqrt{\dfrac{k^2+h^2}{h\times g}}$

(*c*) $2\pi\sqrt{\dfrac{k\times h}{g}}$ (*d*) $2\pi\sqrt{\dfrac{k^2+h^2}{g}}$

where k = Radius of gyration of the pendulum about an axis through its C.G.

h = Distance of its C.G. from the axis of suspension.

48. Equivalent simple pendulum means a simple pendulum which

(*a*) has time period equal to one second (*b*) has time period equal to two seconds

(*c*) has length equal to 99.56 cm

(*d*) has the same periodic time of oscillation as a given compound pendulum.

49. The length of the equivalent simple pendulum is given by

(*a*) $\dfrac{k^2}{h}+h$ (*b*) $\dfrac{k^2}{h^2}+h$ (*c*) $\dfrac{k}{h^2}+h$ (*d*) $\dfrac{k^2}{h^2}+h^2$

where k = Radius of gyration of compound pendulum about its C.G.

h = Distance of the C.G. of the compound pendulum from the axis of suspension.

50. When the distance of suspension of a compound pendulum from its centre of gravity is equal to the radius of gyration about its centre of gravity, then the time period of compound pendulum is

(*a*) 1 sec (*b*) maximum (*c*) minimum (*d*) zero.

51. The minimum period of a compound pendulum is equal to

(*a*) $2\pi\sqrt{\frac{h}{g}}$ (*b*) $2\pi\sqrt{\frac{k}{g}}$ (*c*) $2\pi\sqrt{\frac{2k}{g}}$ (*d*) $2\pi\sqrt{\frac{k}{2g}}$

where k = Radius of gyration of the compound pendulum about its C.G.

52. Choose the wrong statement.

(*a*) The distance between the centre of suspension and centre of percussion is equal to the equivalent length of a simple pendulum.

(*b*) The centre of suspension and centre of percussion are interchangeable.

(*c*) The periodic time of a compound pendulum when suspended from the point of suspension is the same as when suspended from the centre of percussion.

(*d*) The equivalent length of a simple pendulum is equal to the distance of its C.G. from the point of suspension.

53. Bifilar suspension is an experimental method of determining

(*a*) periodic time of a body (*b*) moment of inertia of the body

(*c*) centre of gravity of a body (*d*) centre of percussion of the body.

54. Three springs having stiffness k_1, k_2 and k_3 are connected in series. The stiffness of a single spring equivalent to these three springs, would be equal to

(*a*) $k_1 + k_2 + k_3$ (*b*) $\frac{k_1 + k_2 + k_3}{3}$

(*c*) $\frac{k_1 \times k_2 \times k_3}{k_1k_2 + k_2k_3 + k_3k_1}$ (*d*) $\frac{k_1k_2 + k_2k_3 + k_3k_1}{k_1 \times k_2 \times k_3}$.

55. In question 54, if the three springs are connected in parallel, then the stiffness of a single spring which is equivalent to these three springs will be equal to

(*a*) $k_1 + k_2 + k_3$ (*b*) $\frac{k_1 + k_2 + k_3}{3}$

(*c*) $\frac{k_1k_2k_3}{k_1k_2 + k_2k_3 + k_3k_1}$ (*d*) $\frac{k_1k_2 + k_2k_3 + k_3k_1}{k_1k_2k_3}$.

56. Choose the wrong statement.

(*a*) The instantaneous centre is the point, about which one body rotates relative to another body for the configuration being considered.

(*b*) At the instantaneous centre the two bodies do not have any linear velocity relative to each other.

(*c*) At the instantaneous centre the two bodies have the same linear velocity to the third body.

(*d*) All of the above.

57. A mechanism is having n links, then the number of instantaneous centres would be equal to

(*a*) n (*b*) $\frac{n(n-1)}{2}$ (*c*) $\frac{n}{2}$ (*d*) $(n - 1)$.

58. The total number of instantaneous centre of a mechanism having 8 links, is equal to

(*a*) 8 (*b*) 28 (*c*) 4 (*d*) 7.

59. Kinnedy's theorem states that any three bodies moving relatively to each other
(*a*) having three instantaneous centres and these three centres lie on a curved line
(*b*) have three instantaneous centres and these three centres lie on a straight line
(*c*) have only one instantaneous centre
(*d*) have two instantaneous centre and these two centres lie on a circle.

60. Choose the wrong statement.
(*a*) When two links are connected by a pin joint, their instantaneous centre will lie on the centre of the pin joint.
(*b*) When a link slides over another fixed link which is having flat surface, then the instantaneous centre of the two links lies at infinity.
(*c*) When a link slides over another fixed link which is having curved surface, then the instantaneous centre of the two links lies at the centre of curvature of the curved surface.
(*d*) All of the above
(*e*) None of the above.

61. Choose the correct statement.
(*a*) Bifilar suspension-method is used to determine the moment of inertia of the body.
(*b*) The instantaneous centres, which remain in the same place for all configuration of the mechanism, are called fixed instantaneous centres.
(*c*) The instantaneous centres, which move when the mechanism moves but the joints are of permanent nature, are called permanent instantaneous centres.
(*d*) All of the above
(*e*) None of the above.

62. For a rigid link, the velocity of one end of the link relative to other end will be
(*a*) at 45° to the link (*b*) at right angles to the link
(*c*) parallel to the link (*d*) none of the above.

63. In the design of many machines, a knowledge of methods of determining acceleration is essential. The acceleration of any point in a mechanism is determined by
(*a*) instantaneous centre method (*b*) analytical method
(*c*) acceleration diagram method (*d*) only (*b*) and (*c*)
(*e*) none of the above.

64. In a rigid link *AB*, the point *B* is moving with respect to *A*. Then the acceleration of *B* will be equal to
(*a*) acceleration of *A* × distance *AB*
(*b*) (acceleration of *A*) ÷ distance *AB*
(*c*) vector sum of acceleration of *A* and acceleration of *B*, relative to *A*
(*d*) acceleration of *A* × square of distance *AB*.

65. In question 64, if ω is the angular velocity of the link *AB* about *A*, then normal component of acceleration of *B* relative to *A* will be equal to
(*a*) V_{BA}^2/AB (*b*) $V_{BA} \times AB$
(*c*) $V_{BA}^2 \times AB$ (*d*) V_{BA}/AB
where V_{BA} = Linear velocity of *B* relative to *A*.

66. In question 64, the angular acceleration of the link AB will be equal to

(*a*) (tangential acceleration) ÷ AB (*b*) (normal acceleration) ÷ AB

(*c*) (total acceleration) ÷ AB (*d*) (tangential acceleration) × AB.

67. The velocity of the piston of reciprocating engine is equal to

(*a*) $\omega^2 r\left(\cos\theta + \dfrac{\cos 2\theta}{n}\right)$ (*b*) $\omega^2 r\left(\sin\theta + \dfrac{\sin 2\theta}{n}\right)$

(*c*) $\omega r\left(\sin\theta + \dfrac{\sin 2\theta}{n}\right)$ (*d*) $\omega r\left(\cos\theta + \dfrac{\cos 2\theta}{n}\right)$

where ω = Angular velocity of crank, θ = Angle turned by crank from inner dead centre

r = Radius of crank, $n = \dfrac{l}{r}$ = Ratio of length of connecting rod to crank radius.

68. The displacement of the piston of a reciprocating engine is equal to

(*a*) $r(1 + n - \cos\theta - \sqrt{n^2 - \sin^2\theta})$ (*b*) $r(1 - \cos\theta - n)$

(*c*) $r(1 - \cos\theta - \sqrt{n^2 - \sin^2\theta})$ (*d*) $r(1 + \sin\theta - n)$.

69. The angular velocity of the connecting rod of a reciprocating engine is equal to

(*a*) $\dfrac{\omega\sin\theta}{\sqrt{n^2 + \sin^2\theta}}$ (*b*) $\dfrac{\omega\cos\theta}{\sqrt{n^2 - \sin^2\theta}}$ (*c*) $\dfrac{\omega\cos\theta}{\sqrt{n^2 - \cos^2\theta}}$ (*d*) $\dfrac{\omega\sin\theta}{\sqrt{n^2 - \sin^2\theta}}$.

70. The angular acceleration of the connecting rod of a reciprocating engine is equal to

(*a*) $\dfrac{(n^2 - \sin^2\theta)}{\omega^2(1 - n^2)}$ (*b*) $\dfrac{-\omega^2(n^2 - 1)\sin\theta}{(n^2 - \sin^2\theta)^{3/2}}$

(*c*) $\dfrac{n^2 + \sin^2\theta}{\omega^2(1 + n^2)}$ (*d*) none of the above.

71. Klein's construction is a graphical method of determining

(*a*) acceleration of various parts (*b*) velocity of various parts

(*c*) displacement of various parts (*d*) all of the above

(*e*) none of the above.

72. Klein's construction is used when

(*a*) crank has uniform acceleration (*b*) crank has uniform angular velocity

(*c*) crank has non-uniform acceleration (*d*) crank has non-uniform angular velocity.

73. When a point is moving in a curved path, the tangential acceleration is equal to

(*a*) rate of change in speed (*b*) rate of change of square of speed

(*c*) square of speed/radius of curvature (*d*) speed/radius of curvature.

74. When a point is moving in a curved path, the normal acceleration is equal to

(*a*) rate of change in speed (*b*) rate of change of square of speed

(*c*) square of speed/radius of curvature (*d*) speed/radius of curvature.

75. Which one of the following statement is wrong ?

(*a*) The normal component of acceleration of a particle moving in a straight path is zero.

(*b*) The tangential component of acceleration of a particle moving with constant speed is zero.

(*c*) The normal component of acceleration of a particle moving in a circular path with constant velocity (v) is v^2/r.

(*d*) The tangential component of acceleration of a particle moving in a circular path with constant velocity (v) is v^2/r.

76. If a point moves along a straight line which is rotating, then the normal component of acceleration is equal to

(*a*) $\frac{v^2}{r}$ (*b*) $\frac{dv}{dt} - r\omega^2$ (*c*) $\frac{dv}{dt}$ (*d*) $2v\omega + r\alpha$

where v = Velocity of the point along the straight line

r = Radius of the point, ω = Angular velocity of the line

α = Angular acceleration of the line.

77. In question 76, the tangential component of acceleration is equal to

(*a*) $\frac{v^2}{r}$ (*b*) $\frac{dv}{dt} - r\omega^2$ (*c*) $\frac{dv}{dt}$ (*d*) $2\omega v + r\alpha$.

78. In question 76, the Coriolis component of acceleration is equal to

(*a*) $r\omega^2$ (*b*) $\frac{dv}{dt}$ (*c*) $2v\omega$ (*d*) $r\alpha$

(*e*) none of the above.

79. The Coriolis component of acceleration exists only whenever a point

(*a*) moves along a circular path

(*b*) moves in a straight line

(*c*) moves along a straight line which has rotational motion

(*d*) none of the above.

80. The Coriolis law states that if a point moves along a path that has rotation, the absolute acceleration of the point is

(*a*) equal to the absolute acceleration of the point relative to coincident point in the path

(*b*) equal to the absolute acceleration of the coincident point in the path

(*c*) equal to the vector sum of (*a*), (*b*) and a third component known as Coriolis' component acceleration

(*d*) zero.

81. Which of the following statement concerning the Coriolis component is correct ?

(*a*) It is equal to twice the product of the instantaneous velocity of sliding of the point on the link and instantaneous angular velocity of the link on which the point slides.

(*b*) It acts perpendicular to the sliding surface.

(*c*) The direction of the Coriolis component of acceleration is the direction of the relative velocity vector when it has been rotated 90° in the direction of the angular velocity of the link.

(*d*) All of the above

(*e*) None of the above.

82. In which of the following mechanism, the Coriolis component of acceleration exists ?

(*a*) Shaper mechanism (*b*) Whitworth Quick Return mechanism

(*c*) Tangent cam mechanism (*d*) All of the above

(*e*) None of the above.

83. A disc is rotating about an axis with an angular velocity ω (also known as angular velocity of spin) and the axis of spin is changing its position, then the tangential component of acceleration is equal to

(*a*) $\omega \dfrac{d\theta}{dt}$ (*b*) $\dfrac{d\omega}{dt}$ (*c*) $r\omega^2$ (*d*) $r \dfrac{d\omega}{dt}$

where $\dfrac{d\theta}{dt} = \Omega$ *i.e.*, the rate at which the axis of rotation changes its position

r = Radius.

84. In question 83, the normal component of acceleration is equal to

(*a*) $\omega \dfrac{d\theta}{dt}$ (*b*) $\dfrac{d\omega}{dt}$ (*c*) $r\omega^2$ (*d*) $r \dfrac{d\omega}{dt}$.

85. A disc is rotating about an axis with an angular velocity and the axis of rotation is changing its position. The change in direction of the plane of rotation of the disc is called

(*a*) rotational motion (*b*) precessional motion
(*c*) rectilinear motion (*d*) none of the above.

86. Gyroscopic acceleration is the

(*a*) tangential component of the acceleration of a disc rotating with an angular velocity about an axis and the axis of rotation is changing its position.
(*b*) normal component of the acceleration of a disc rotating with an angular velocity about an axis and axis of rotation is changing its position.
(*c*) tangential component of the acceleration of a point moving in a straight line which has rotational motion.
(*d*) normal component of the acceleration of a point moving in a straight line which has rotational motion.

87. The rate at which the axis of rotation (about which a disc is rotating) changes its position is known as

(*a*) angular velocity (*b*) angular velocity of precession
(*c*) linear velocity (*d*) sliding velocity.

88. The gyroscopic acceleration depends upon

(*a*) instantaneous value of ω
(*b*) the rate at which the axis of spin changes its speed
(*c*) both (*a*) and (*b*)
(*d*) none of the above.

89. The slotted lever of a quick return mechanism at a certain instant has an angular velocity of 5 rad/sec clockwise and angular acceleration of 2 rad/sec^2 clockwise. The slider on the lever at this instant is 0.5 m from the axis of oscillation of the lever and it moves away from the axis of rotation at a velocity of 1.5 m/sec and its acceleration is 0.5 m/sec^2 outwards. Then the Coriolis component of acceleration of the slider will be

(*a*) zero (*b*) 7.5 m/sec^2 (*c*) 10 m/sec^2 (*d*) 15 m/sec^2.

90. In question 89, the tangential component of acceleration of slider will be

(*a*) 16 m/sec^2 (*b*) 11.0 m/sec^2 (*c*) 8.5 m/sec^2 (*d*) zero.

91. In question 89, the normal component of acceleration of slider will be

(*a*) 12.5 m/sec^2 (*b*) 13 m/sec^2 (*c*) – 12.0 m/sec^2 (*d*) – 10 m/sec^2.

92. In question 89, the total acceleration of the slider will be
(*a*) 10 m/sec^2 (*b*) 20 m/sec^2 (*c*) 25 m/sec^2 (*d*) 30 m/sec^2.

93. The flywheel of a motor car rotates about a horizontal axis at 2100 r.p.m., while the car moves in a horizontal circle of radius 40 m at a speed of 72 km/hour. Then the angular velocity of precession will be
(*a*) 1 rad/sec (*b*) 0.5 rad/sec (*c*) 1.5 rad/sec (*d*) 2.0 rad/sec.

94. In the equation 93, the gyroscopic acceleration will be
(*a*) 100 rad/sec^2 (*b*) 105 rad/sec^2 (*c*) 110 rad/sec^2 (*d*) 120 rad/sec^2.

95. The total kinetic energy of a body whose centre of gravity is moving with a linear velocity 'v' and body is also rotating about its centre of gravity with angular velocity 'ω', is equal to

(*a*) $\frac{v^2}{2g} + \frac{1}{2} I\omega^2$ (*b*) $\frac{v^2}{2g} + I\omega^2$ (*c*) $\frac{1}{2}\frac{W}{g} v^2 + \frac{1}{2} I\omega^2$ (*d*) $\frac{1}{2}\frac{W}{g} v^2 + I\omega^2$

where W = Weight of body, and
I = Moment of inertia of the body about the axis of rotation.

96. The product of moment of inertia and angular velocity is known as
(*a*) kinetic energy (*b*) angular momentum
(*c*) angular torque (*d*) none of the above.

97. Two co-axial rotors having moments of inertia, I_1, I_2 and angular speeds ω_1 and ω_2 respectively are engaged together. The common speed of the two rotors after engagement will be

(*a*) $\frac{\omega_1 + \omega_2}{2}$ (*b*) $\frac{I_1\omega_1 + I_2\omega_2}{I_1 + I_2}$ (*c*) $\frac{I_1\omega_1 - I_2\omega_2}{I_1}$ (*d*) $\frac{\omega_1 + \omega_2}{I_1 + I_2}$.

98. In question 97, the loss of energy during engagement is equal to

(*a*) $\frac{I_1 I_2(\omega_1 - \omega_2)^2}{2(I_1 + I_2)}$ (*b*) $\frac{I_1 I_2({\omega_1}^2 - {\omega_2}^2)}{2(I_1 + I_2)}$ (*c*) $\frac{2I_1 I_2(\omega_1 - \omega_2)^2}{(I_1 + I_2)}$ (*d*) $\frac{I_1{\omega_1}^2 + I_2{\omega_2}^2}{(I_1 + I_2)}$.

99. A machine is said to be self-locking if efficiency of the machine is
(*a*) more than 50% (*b*) equal to 50% (*c*) less than 50% (*d*) equal to 100%.

100. Choose the correct statement.
(*a*) If the angle of inclined plane is less than the friction angle, the arrangement is known as self-locking.
(*b*) The efficiency of a square threaded screw is maximum when the helix angle of the screw is equal to 45° minus half of friction angle.
(*c*) Self-locking is a desired property of a machine.
(*d*) All the above.
(*e*) None of the above.

101. The equivalent co-efficient of friction (μ_1) for angular threads is equal to

(*a*) $\mu \cos \beta$ (*b*) $\mu \sin \beta$ (*c*) $\frac{\mu}{\cos \beta}$ (*d*) $\frac{\mu}{\sin \beta}$

where μ = Actual co-efficient of friction, and β = Semi-angle of the threads.

102. The equivalent co-efficient of friction for *V*-threads is
(*a*) equal to actual co-efficient of friction (*b*) less than actual co-efficient of friction
(*c*) greater than the actual co-efficient of friction (*d*) none of the above.

103. The total angle of metric threads is 60°. If the actual co-efficient of friction between the thread and the nut is 0.15, then the value of equivalent co-efficient of friction will be equal to
(*a*) 0.12 (*b*) 0.1730 (*c*) 0.225 (*d*) 0.245.

104. In question 103, if the screw is self-locking, then the value of helix angle will be equal to approximately
(*a*) 8° (*b*) 12° (*c*) 10° (*d*) 6°.

105. Which one is a correct statement ?
(*a*) The rubbing surface in case of pivot and collar bearing is always flat.
(*b*) The rubbing surface in case of pivot and collar bearing is always conical.
(*c*) The rubbing surface in case of pivot bearing is flat and in case of collar bearing it is conical.
(*d*) The rubbing surface in case of pivot bearing is either flat or conical but in case of collar bearing it is generally flat.

106. In case of flat pivot bearing, the rubbing velocity is
(*a*) maximum at the centre of the contact area
(*b*) zero at the centre and maximum at the outer radius
(*c*) uniform throughout the contact area
(*d*) zero at the outer radius.

107. The rate of wearing in case of flat pivot bearing is
(*a*) inversely proportional to the rubbing velocity
(*b*) inversely proportional to the pressure intensity
(*c*) directly proportional to the rubbing velocity
(*d*) directly proportional to the pressure intensity
(*e*) both (*c*) and (*d*).

108. For calculating the friction power for bearing, the assumptions made in practice are
(*a*) the rate of wearing at the surface is uniform (*b*) the rubbing velocity is constant
(*c*) the pressure intensity is uniform (*d*) (*a*) and (*b*) only
(*e*) (*a*) and (*c*) only.

109. In case of flat pivot bearing, the frictional torque transmitted for uniform pressure as compared to the frictional torque transmitted for uniform wear
(*a*) is more (*b*) is less
(*c*) is constant (*d*) may be more or less.

110. The frictional torque, transmitted in case of flat pivot bearing for uniform pressure, is equal to
(*a*) μWR (*b*) $\frac{2}{3}\mu WR$ (*c*) $\frac{1}{3}\mu WR$ (*d*) $\frac{1}{2}\mu WR$
where W = Total axial load carried by pivot, μ = Co-efficient of friction, and
R = Radius of bearing surface.

111. The frictional torque, transmitted in case of flat pivot bearing for uniform wear, is equal to

(*a*) μWR (*b*) $\frac{2}{3}\mu WR$ (*c*) $\frac{1}{3}\mu WR$ (*d*) $\frac{1}{2}\mu WR$.

112. The ratio of frictional torque transmitted for uniform pressure to the frictional torque for uniform wear in case of flat pivot, is equal to

(*a*) 1/3 (*b*) 2/3 (*c*) 1.0 (*d*) 4/3.

113. The frictional torque transmitted in case of a flat collar bearing for uniform pressure, is equal to

(*a*) $\mu W\left(\frac{R_1^3 - R_2^3}{R_1^2 - R_2^2}\right)$ (*b*) $\frac{2}{3}\mu W\left(\frac{R_1^3 - R_2^3}{R_1^2 - R_2^2}\right)$

(*c*) $\frac{\mu W}{2}(R_1 + R_2)$ (*d*) $\frac{\mu W}{2}(R_1 - R_2)$

where R_1 = External radius of the collar, R_2 = Internal radius of the collar, and W = Total axial load carried by the shaft.

114. The frictional torque transmitted for uniform wear, in case of flat collar bearing, is equal to

(*a*) $\mu W\left(\frac{R_1^3 - R_2^3}{R_1^2 - R_2^2}\right)$ (*b*) $\frac{2}{3}\mu W\left(\frac{R_1^3 - R_2^3}{R_1^2 - R_2^2}\right)$

(*c*) $\frac{\mu W}{2}(R_1 + R_2)$ (*d*) $\frac{\mu W}{2}(R_1 - R_2)$.

115. When the axial load carried by the shaft in case of flat collar bearing, is large ; a number of collars are provided on the shaft. This is done

(*a*) to reduce the frictional torque (*b*) to increase the frictional torque

(*c*) to reduce the intensity of pressure (*d*) to increase the intensity of pressure.

116. The value of limiting bearing pressure on a collar is about

(*a*) 1 kgf/cm² (*b*) 5 kgf/cm² (*c*) 10 kgf/cm² (*d*) 15 kgf/cm².

117. Choose the correct statement regarding the safe design of a bearing

(*a*) when horse power lost in friction is to be determined, uniform rate of wear is assumed

(*b*) when horse power transmitted is to be determined, uniform pressure is assumed

(*c*) when horse power lost in friction is to be determined, uniform pressure is assumed and when horse power transmitted is to be determined, uniform wear is assumed

(*d*) none of the above.

118. A flat foot step bearing 20 cm in diameter supports a load of 2500 kgf. The shaft is running at 420 r.p.m. and the co-efficient of friction is 0.09. The frictional torque, for uniform pressure, will be equal to

(*a*) 11.25 kgf-m (*b*) 15 kgf-m (*c*) 18.50 kgf-m (*d*) 20.0 kgf-m.

119. In question 118, the frictional torque, for uniform wear, will be equal to

(*a*) 11.25 kgf-m (*b*) 15 kgf-m (*c*) 18.50 kgf-m (*d*) 20.0 kgf-m.

120. In question 118, the horse power lost in overcoming friction at the bearing, will be equal to

(*a*) 6.6 h.p. (*b*) 8.8 h.p. (*c*) 4.4 h.p. (*d*) 2.2 h.p.

121. In question 118, the horse power transmitted by the friction between the surface, will be equal to

(*a*) 6.6 h.p. (*b*) 8.8 h.p. (*c*) 4.4 h.p. (*d*) 2.2 h.p.

122. The frictional torque transmitted for uniform pressure, in case of a conical pivot bearing having semi-angle of the cone as θ, is equal to

(*a*) $\frac{\mu WR}{\sin\theta}$ (*b*) $\frac{\mu WR}{2\sin\theta}$ (*c*) $\frac{2\mu WR}{\sin\theta}$ (*d*) $\frac{2\mu WR}{3\sin\theta}$

where W = Total axial load, and R = Radius of sbaft.

123. The frictional torque transmitted for uniform wear, in case of a conical pivot bearing having semi-angle of the cone as θ, is equal to

(*a*) $\frac{\mu WR}{\sin\theta}$ (*b*) $\frac{\mu WR}{2\sin\theta}$ (*c*) $\frac{2\mu WR}{\sin\theta}$ (*d*) $\frac{2\mu WR}{3\sin\theta}$

124. For uniform pressure the ratio of the frictional torque for a flat collar bearing to the frictional torque for a conical collar bearing, is equal to

(*a*) 1.33 (*b*) 1.0 (*c*) sin θ (*d*) cos θ

where θ = Semi-angle of the cone.

125. For uniform wear, the ratio of the frictional torque for a flat collar bearing to the frictional torque for a conical collar bearing, is equal to

(*a*) 1.33 (*b*) 1.0 (*c*) sin θ (*d*) cos θ.

126. The frictional torque transmitted by a conical collar bearing is

(*a*) less than that of flat collar bearing (*b*) equal to that of flat collar bearing

(*c*) more than that of flat collar bearing (*d*) none of the above.

127. Choose the wrong statement

(*a*) The frictional torque transmitted by a shaft fitted with a large number of collars is the same as transmitted by a shaft with one collar.

(*b*) The frictional force for uniform pressure, in case of a flat pivot bearing, can be assumed to be acting at a distance of two-thirds of the radius of the shaft from the axis of shaft.

(*c*) When horse power lost in friction is to be determined, uniform pressure is assumed for safe design of bearing.

(*d*) The frictional force for uniform pressure, in case of a flat collar pivot bearing, can be assumed to be acting at a distance of half of the sum of internal and external radii of the shaft.

128. A thrust bearing, having collars of external and internal diameters as 40 cm and 20 cm respectively, carries an axial load of 3000 kgf. The co-efficient of friction is 0.09 and shaft runs at 420 r.p.m. The horse power lost in friction is to be determined. The corresponding value of friction torque would be

(*a*) 40.5 kgf-m (*b*) 42 kgf-m (*c*) 45.0 kgf-m (*d*) 48.5 kgf-m.

129. In question 128, the horse power lost in friction would be equal to

(*a*) 22.8 (*b*) 24.8 (*c*) 26.2 (*d*) 28.1.

130. Which of the following is an example of friction clutch ?

(*a*) disc clutch (*b*) cone clutch (*c*) centrifugal clutch (*d*) plate clutch

(*e*) all of the above (*f*) none of the above.

131. When a journal is rotating in its bearing, the resultant of the normal force (R_N) and frictional force (μR_N)
(*a*) passes through the centre of the journal
(*b*) is tangent to the circle of the journal at the point of contact
(*c*) is tangent to a small circle of radius $r \times \sin \phi$, where $\phi = \tan^{-1} \mu$
(*d*) none of the above.

132. Friction circle in case of journal bearing is
(*a*) the circle of the journal
(*b*) the circle to which resultant force *F* is tangent
(*c*) the circle which is having radius equal to half of the radius of the journal
(*d*) the circle which is having radius equal to $\mu \times R$, where *R* is the radius of the journal
(*e*) both (*b*) and (*d*)
(*f*) none of the above.

133. The radius of friction circle increases
(*a*) with the increase of load (*b*) with the increase of radius of journal
(*c*) with the increase of speed
(*d*) with the increase of co-efficient of sliding friction
(*e*) all of the above (*f*) (*b*) and (*d*) only.

134. The screw jack is having maximum efficiency, when the helix angle (α) is equal to
(*a*) $\frac{\pi}{4} + \phi$ (*b*) $\frac{\pi}{4} + \frac{\phi}{2}$ (*c*) $\frac{\pi}{4} - \phi$ (*d*) $\frac{\pi}{4} - \frac{\phi}{2}$
where ϕ = Angle of friction.

135. The maximum efficiency of a screw jack is a function of
(*a*) helix angle (*b*) angle of friction (*c*) load lifted (*d*) effort.

136. The maximum efficiency of a screw jack is given by
(*a*) $\frac{1+\cos\phi}{1-\cos\phi}$ (*b*) $\frac{1-\sin\phi}{1+\sin\phi}$ (*c*) $\frac{1-\cos\phi}{1+\cos\phi}$ (*d*) $\frac{1-\tan\phi}{1+\tan\phi}$
where ϕ = Angle of friction.

137. Choose the wrong statement.
(*a*) Angle of friction is the angle made by the normal reaction with the resultant of normal reaction and force of friction.
(*b*) Static frictions is more than dynamic friction.
(*c*) Limiting friction is the maximum frictional force, which comes into play when a body just begins to slide over the surface of other body.
(*d*) The radius of friction circle for a shaft rotating inside a bearing is $r \sin \phi$.
(*e*) None of the above.

138. Choose the correct statement.
(*a*) When the helix angle is equal to $\left(45° - \frac{\phi}{2}\right)$, the efficiency of a screw jack is maximum.
(*b*) Angle of repose is the angle of inclination of the plane, at which the body begins to move down the plane.

(c) The ratio of the limiting friction to the normal reaction is called co-efficient friction.

(d) If the angle of friction is less than the angle of inclination of the plane, the body will begin to move down the inclined plane.

(e) All of the above.

139. The length of an open belt, connecting two pulleys of radii r_1 and r_2 and at a centre distance D apart, is equal to

(a) $\pi(r_1 - r_2) + \dfrac{(r_1 + r_2)^2}{D} + 2D$ (b) $\pi(r_1 - r_2) + \dfrac{(r_1 + r_2)^2}{D} + 2D$

(c) $\pi(r_1 + r_2) + \dfrac{(r_1 - r_2)^2}{D} + 2D$ (d) $\pi(r_1 + r_2) + \dfrac{(r_1 + r_2)^2}{D} + 2D.$

140. The length of a crossed belt, connecting two pulleys of radii r_1 and r_2 and at a centre distance D apart, is equal to

(a) $\pi(r_1 + r_2) + \dfrac{(r_1 + r_2)^2}{D} + 2D$ (b) $\pi(r_1 - r_2) + \dfrac{(r_1 - r_2)^2}{D} + 2D$

(c) $\pi(r_1 + r_2) + \dfrac{(r_1 - r_2)^2}{D} + 2D$ (d) $\pi(r_1 + r_2)^2 + \dfrac{(r_1 + r_2)^2}{D} + 2D.$

141. The angular velocities of the two pulleys, connected either by an open belt or a crossed belt, are

(a) directly proportional to their diameters

(b) inversely proportional to their diameters

(c) proportional to the square of their diameters

(d) proportional to the square root of their diameters.

142. Choose the wrong statement.

(a) With an open belt the pulleys rotate in the same direction.

(b) The angle of contact in crossed belt drive is more than that in an open belt drive.

(c) With a crossed belt the pulleys rotate in opposite direction.

(d) The length of crossed belt depends on the sum and difference of the radii of the pulleys.

143. When two pulleys, of unequal sizes, the connected by a belt drive the angle θ is taken as

(a) angle of contact on the bigger pulley

(b) angle of contact on the smaller pulley

(c) average angle of the contact on the two pulleys

(d) none of the above.

144. If T_1 and T_2 are the tensions on the tight and slack side of a belt and θ is the angle of contact, then ratio of tension is given by

(a) $\dfrac{T_1}{T_2} = \mu\theta$ (b) $\dfrac{T_1}{T_2} = e^{\mu\theta}$ (c) $\dfrac{T_1}{T_2} = e^{1/\mu\theta}$ (d) $\dfrac{T_1}{T_2} = \mu e^{\theta}$

where μ = Co-efficient of friction between the belt and pulley.

145. The effect of slip on velocity ratio of the belt is to

(*a*) increase the velocity ratio (*b*) decrease the velocity ratio

(*c*) keep the velocity ratio constant (*d*) none of the above.

146. The horse power transmitted by a belt is equal to

(*a*) $\frac{T_1 \times v}{75}$ (*b*) $\frac{T_2 \times v}{75}$ (*c*) $\frac{(T_1 + T_2)}{2} \times \frac{v}{75}$ (*d*) $\frac{(T_1 - T_2) \times v}{75}$

where T_1 = Tension to tight side, T_2 = Tension on slack side, and

v = Linear velocity of belt.

147. The tension caused by centrifugal force on the belt is equal to

(*a*) $\frac{w}{g} \frac{V^2}{r}$ (*b*) $\frac{w}{g} V^2$ (*c*) $w \times \frac{V^2}{r}$ (*d*) $\frac{1}{g} \times \frac{V^2}{r}$

where w = Weight of belt per unit length, v = Velocity of belt, and

r = Radius of bigger pulley.

148. The tension caused by centrifugal force on the belts

(*a*) increases power transmission (*b*) decreases power transmission

(*c*) does not affect power transmission (*d*) none of the above.

149. If the effect of centrifugal tension in thc belt is considered, the horse power transmitted is zero when the velocity of the belt is zero. The horse power will be again zero when the velocity of belt is equal to

(*a*) $\sqrt{\frac{w}{T_1 \times g}}$ (*b*) $\sqrt{\frac{T_1}{w \times g}}$ (*c*) $\sqrt{\frac{T_1 \times g}{w}}$ (*d*) $\sqrt{\frac{w \times g}{T}}$

where T_1 = Total tension on the tight side of the belt, and

w = Weight of belt per unit length.

150. If the effect of centrifugal tension in the belt is considered then the velocity of the belt, at which the maximum horse power is transmitted, is equal to

(*a*) $\sqrt{\frac{T_1 \times g}{3w}}$ (*b*) $\sqrt{\frac{3w}{T_1 \times g}}$ (*c*) $\sqrt{\frac{w}{3T_1 \times g}}$ (*d*) $\sqrt{\frac{T_1}{3w \times g}}$

where T_1 = Total tension on the tight side of the belt, and

w = Weight of belt per unit length.

151. When maximum horse power is transmitted by a belt, the centrifugal tension in the belt is

(*a*) half the maximum tension allowed in the belt

(*b*) one-third the maximum tension allowed in the belt

(*c*) two-thirds the maximum tension allowed in the belt

(*d*) one-fourth the maximum tension allowed in the belt.

152. When maximum horse power is transmitted, the effective tension in the tight side of the belt is equal to

(*a*) twice the centrifugal tension (*b*) three times the centrifugal tension

(*c*) half the centrifugal tension (*d*) one-third the centrifugal tension.

153. The belts are installed with initial tension, which is generally equal to

(*a*) sum of the tensions on the tight side and slack side of the belt

(*b*) average tension of the tight side and slack side of the belt

(c) half of tension on the tight side of the belt
(d) half of tension on the slack side of the belt.

154. Creep in belt drive is due to
(a) weak material of the belt (b) weak material of the pulley
(c) uneven extensions and contractions of the belt when it passes from tight side to slack side
(d) none of the above.

155. Crowning of pulley is done
(a) to avoid the slipping of the belt (b) to make them more study
(c) to enable pulley rigidly fixed to the shaft
(d) to make pulley look more pleasant in appearance.

156. A shaft running at 100 r.p.m. is to drive a parallel shaft at 200 r.p.m. If the diameter of the pulley on the driving shaft is 80 cm, then the diameter of the pulley on the driven shaft will be
(a) 160 cm (b) 80 cm (c) 40 cm (d) 80 cm.

157. If the question 156, the thickness of belt is 4 mm, then the diameter of the pulley on the driven shaft will be
(a) 40.40 cm (b) 80.40 cm (c) 39.60 cm (d) 29.80 cm.

158. Choose the wrong statement
(a) When two pulleys are connected by means of a cross-belt drive, then both the pulleys will rotate in opposite directions.
(b) The centrifugal tension in belts reduces power transmission.
(c) Abbreviation P.I.V. drive stands for positive, infinitely variable drive.
(d) The power transmitted by a belt is maximum when the maximum tension in the belt is one-third of centrifugal tension.

159. Choose the correct statement
(a) In a belt drive, the speeds are directly proportional to the diameters of the pulley.
(b) Considering the effect of centrifugal tension in the belt, the horse power transmitted increase directly with the increase of speed.
(c) The length of open belt depends upon the sum and difference of the radii of the pulleys.
(d) Crowning of pulley is done to make them more sturdy.

160. Two parallel shafts, having centre distance 5 metres, are connected by an open belt. The diameters of the pulley fixed to the shafts are 40 cm and 30 cm respectively. The length of the belt will be
(a) 1000 cm **(b) 1110.05 cm** (c) 1112.45 cm (d) 1120 cm.

161. If in the question 160, the shafts are connected by a cross belt, then the length of the belt would be
(a) 1000 cm (b) 110.05 cm (c) 1112.45 cm (d) 1120 cm.

162. The effective tensions on the tight side and slack in a belt drive are 60 kgf and 30 kgf respectively. If the linear velocity of the belt is 300 metre/sec, then the horse power transmitted by the belt would be
(a) 120 h.p. (b) 0.5 h.p. (c) 4 h.p. (d) 2 h.p.

163. If in question 162, the tension on the tight side only is increased to 1.5 times, then the horse power transmitted will be
(*a*) 120 h.p. (*b*) 0.5 h.p. (*c*) 4 h.p. (*d*) 2 h.p.

164. If in question 162, tension on the slack side only is increased to 1.5 times then the horse power transmitted will be
(*a*) 120 h.p. (*b*) 0.5 h.p. (*c*) 4 h.p. (*d*) 2 h.p.

165. If in question 162, the tensions on the tight and slack side are increased to 1.5 times then the horse power transmitted will be
(*a*) 180 h.p. (*b*) 0.75 h.p. (*c*) 6 h.p. (*d*) 3 h.p.

166. The centrifugal tension in the belt depends upon
(*a*) tight side tension of the belt (*b*) slack side tension of the belt
(*c*) velocity of the belt (*d*) weight per unit length of the belt
(*e*) (*a*) and (*b*) (*f*) (*c*) and (*d*).

167. When centrifugal tension (T_c) is considered, the belt is designed for the tension
(*a*) T_1 (*b*) $T_1 - T_c$ (*c*) $T_1 + T_c$ (*d*) none of the above.

168. The horse power transmitted by a belt, when the effect of centrifugal tension is considered and belt is running with a velocity equal to $\left(\dfrac{T_1 \times g}{3w}\right)^{1/2}$, will be
(*a*) zero (*b*) maximum (*c*) minimum (*d*) none of the above.

169. Choose the wrong statement
(*a*) When the shafts are at long centre distances, the power is transmitted with the help of belts, ropes and chains.
(*b*) When the shafts are at short centre distances, the power is transmitted by gears.
(*c*) Gear drive is a positive and smooth drive.
(*d*) The belts and rope drives are not positive drive.
(*e*) none of the above.

170. Two intersecting and co-planar shafts are connected by gears. This type of gear is called
(*a*) helical gear (*b*) spur gear (*c*) bevel gear (*d*) spiral gear.

171. Two parallel and coplanar shafts are connected by gears. This type of gear is called
(*a*) helical gear (*b*) spur gear (*c*) bevel gear (*d*) spiral gear.

172. If the peripheral velocity of the gears are more than 15 m/sec, the gears are known as
(*a*) low velocity type (*b*) medium velocity type
(*c*) high velocity type (*d*) none of the above.

173. Choose the wrong statement
(*a*) The gears are termed as medium velocity gears if their peripheral velocity lies in the range of 3 to 14 m/sec.
(*b*) When two intersecting and coplanar shafts are connected by gears, the arrangement is called bevel gearing.
(*c*) The radial distance of a tooth from the pitch circle to the bottom of tooth is called addendum.
(*d*) The face of the tooth is the surface of the tooth above the pitch surface.

174. Choose the correct statement

(*a*) The gears are termed as high velocity gears if their peripheral velocity lies in the range of 3 to 10 m/sec.

(*b*) The flank of the tooth is the surface of the tooth above the pitch surface.

(*c*) The distance measured along the circumference of the pitch circle from a point on one tooth to a corresponding point on the adjacent tooth is known as circular pitch.

(*d*) The dedendum is the radial distance of the tooth from the pitch circle to the top of the tooth.

175. Diametral pitch is defined as the ratio of

(*a*) number of teeth to pitch circle diameter

(*b*) pitch circle diameter to number of teeth

(*c*) circumference of pitch circle to number of teeth

(*d*) none of the above.

176. For a spur gear, the circular pitch is equal to the ratio of

(*a*) number of teeth to pitch circle diameter

(*b*) pitch circle diameter to number of teeth

(*c*) circumference of pitch circle to number of teeth

(*d*) none of the above.

177. The module (m) is defined as equal to the ratio of

(*a*) number of teeth to pitch circle diameter

(*b*) pitch circle diameter to number of teeth

(*c*) circumference of the pitch circle to number of teeth

(*d*) none of the above.

178. The product of the circular pitch and diametral pitch is equal to

(*a*) 2π (*b*) π (*c*) $\frac{\pi}{2}$ (*d*) 1.0.

179. The product of the module and diametral pitch is equal to

(*a*) 2π (*b*) π (*c*) $\frac{\pi}{2}$ (*d*) 1.0.

180. The pressure angles varies from

(*a*) 40° to 50° (*b*) 30° to 40° (*c*) 20° to 30° (*d*) 14° to 20°.

181. Law of gearing states that

(*a*) the normal at the point of contact between a pair of teeth divides the line joining the centre of rotation directly in the ratio of angular velocities

(*b*) the common normal at the point of contact between a pair of teeth does not pass, always through the pitch point

(*c*) the common normal at the point of contact between a pair of teeth always passes through the pitch point

(*d*) none of the above.

182. The velocity of sliding of the mating teeth of the two gears is equal to

(*a*) product of angular velocity of first gear and the distance of the point of contact from the pitch point (*b*) product of angular velocity of second gear and the distance of the point of contact from the pitch point

(*c*) product of the sum of the angular velocities of the two gears and the distance of the point of contact from the pitch point

(*d*) none of the above.

183. The teeth, which are normally used and satisfy the law of gearing are

(*a*) conjugate teeth (*b*) cycloidal teeth (*c*) involute teeth

(*d*) only (*b*) and (*c*) (*e*) none of the above.

184. The involute function [*INV* (ϕ)] in terms of pressure angle (ϕ) is expressed as

(*a*) $INV(\phi) = \phi - \tan\phi$ (*b*) $INV(\phi) = 1 - \tan\phi$

(*c*) $INV(\phi) = \tan\phi - \phi$ (*d*) $INV(\phi) = 1 + \tan\phi$.

185. If the pressure angle is 20, the involute function will be equal to

(*a*) 0.702 (*b*) 0.636 (*c*) 0.0149 (*d*) 1.3639.

186. In involute teeth, the tangent to the base circle is a

(*a*) tangent to the involute (*b*) normal to the involute

(*c*) normal to the pitch circle (*d*) none of the above.

187. The path of contact in involute gears is a

(*a*) circle (*b*) straight line (*c*) curved line (*d*) none of the above.

188. Choose the wrong statement

(*a*) The product of the circular pitch and diametral pitch is equal to π.

(*b*) The pressure angle for involute gears is always constant.

(*c*) The angular velocity ratio when two involutes are in mesh, is inversely proportional to the radius of base circles.

(*d*) The involute function in terms of pressure angle (ϕ) is equal to ($\phi - \tan\phi$).

189. Choose the correct statement

(*a*) The range of pressure angle is usually from 20° to 30°.

(*b*) The involute of circle is the locus of a point on the circumference of a circle which rolls without slipping on a fixed straight line.

(*c*) The module is the reciprocal of diametral pitch.

(*d*) The pressure angle for involute gears always depends upon the size of teeth.

190. The centre distance between two meshing involute gears is

(*a*) directly proportional to the sum of base circle radii of the meshing gears

(*b*) inversely proportional to the cosine of pressure angle

(*c*) directly proportional to the cosine of pressure angle

(*d*) both (*a*) and (*b*)

(*e*) none of the above.

191. The locus of a point on the circumference of a circle, which rolls without slipping on a fixed straight line, is known as

(*a*) involute (*b*) cycloid (*c*) hypo-cycloid (*d*) epicycloid.

192. The locus of a point on the circumference of a circle, which rolls without slipping on the outside of another circle, is known as

(*a*) involute (*b*) cycloid (*c*) hypo-cycloid (*d*) epicycloid.

193. The locus of a point on the circumference of a circle, which rolls without slipping on the inside of another circle, is known as

(*a*) involute (*b*) cycloid (*c*) hypo-cycloid (*d*) epicycloid.

194. Which is correct statement about cycloidal gears ?

(*a*) The pressure angle is constant from the commencement of the engagement to the end of engagement.

(*b*) The gears are manufactured with simplicity.

(*c*) There is no interference in these gears.

(*d*) Variation in centre distance within limits does not affect the velocity ratio of the two mating gears.

195. The work wasted in friction, for a given total arc of action between gear teeth, will be minimum when arc of approach is

(*a*) half of arc of recess (*b*) one-fourth of arc of recess

(*c*) equal to arc of recess (*d*) twice the arc of recess.

196. For helical gears, the normal circular pitch is equal to

(*a*) $\dfrac{p_c}{\cos\alpha}$ (*b*) $p_c \times \cos\alpha$ (*c*) $\dfrac{\cos\alpha}{p_c}$ (*d*) $\dfrac{1}{p_c \times \cos\alpha}$

where p_c = Circular pitch, and α = Helix angle.

197. For helical gears, the axial pitch is equal to

(*a*) $p_c \times \tan\alpha$ (*b*) $\dfrac{\tan\alpha}{p_c}$ (*c*) $\dfrac{p_c}{\tan\alpha}$ (*d*) $\dfrac{1}{p_c \times \tan\alpha}$

where p_c = Circular pitch and α = Helix angle.

198. The efficiency of spiral gears will be maximum, if

(*a*) $\alpha = \dfrac{\theta - \phi}{2}$ (*b*) $\alpha = \dfrac{\theta + \phi}{2}$ (*c*) $\alpha = \theta - 2\phi$ (*d*) $x = 2\theta - \phi$

where α = Spiral angle, and θ = Shaft angle and ϕ = Friction angle.

199. The maximum efficiency of spiral gears is equal to

(*a*) $\dfrac{1-\cos(\theta+\phi)}{1+\cos(\theta-\phi)}$ (*b*) $\dfrac{1+\cos(\theta+\phi)}{1+\cos(\theta-\phi)}$ (*c*) $\dfrac{1+\cos(\theta-\phi)}{1+\cos(\theta+\phi)}$ (*b*) $\dfrac{1-\cos(\theta-\phi)}{1+\cos(\theta+\phi)}$

where θ = Shaft angle and ϕ = Friction angle.

200. In case of worm and worm gear, the maximum efficiency is equal to

(*a*) $\dfrac{1+\sin\phi}{1-\sin\phi}$ (*b*) $\dfrac{1-\sin\phi}{1+\sin\phi}$ (*c*) $\dfrac{1+\sin\phi}{\sin\phi}$ (*d*) $\dfrac{1-\sin\phi}{\sin\phi}$

where ϕ = Friction angle.

201. Two gears having involute teeth are in mesh. If the path of approach and path of recess are half of their maximum possible values, then the length of path of contact is equal to

(*a*) $\dfrac{(R-r)\sin\phi}{2}$ (*b*) $\dfrac{(R+r)\sin\phi}{2}$ (*c*) $\dfrac{(R-r)\cos\phi}{2}$ (*d*) $\dfrac{(R+r)\cos\phi}{2}$

where R = Pitch circle radius of wheel, r = Pitch circle radius of pinion, and ϕ = Pressure angle.

202. In question 201, the number of involute teeth on the pinion and wheels are 20 and 40 respectively. If the pressure angle is equal to 20° and module pitch 10 mm, then pitch circle radius of wheel would be equal to

(*a*) 100 mm (*b*) 150 mm (*c*) 175 mm (*d*) 200 mm.

203. In equation 202, the pitch circle radius of the pinion would be equal to

(*a*) 100 mm (*b*) 150 mm (*c*) 175 mm (*d*) 200 mm.

204. In equation 202, the required path of approach would be equal to

(*a*) 88.2 mm (*b*) 44.1 mm (*c*) 34.2 mm (*d*) 17.1 mm.

205. In question 202, the required path of recess would be equal to

(*a*) 88.2 mm (*b*) 44.1 mm (*c*) 34.2 mm (*d*) 17.1 mm.

206. In question 202, the path of contact would be

(*a*) 88.2 mm (*b*) 51.3 mm (*c*) 34.2 mm (*d*) 17.1 mm.

207. In question 202, the arc of contact would be equal to

(*a*) 88.2 mm (*b*) 51.3 mm (*c*) 34.2 mm (*d*) 58.73 mm.

208. The addendum as compared to dedendum of a toothed gear wheel is

(*a*) more (*b*) less (*c*) same (*d*) can be more or less.

209. The centre distance between two involute teeth gears is equal to

(*a*) $(R + r) \cos \phi$ (*b*) $(R + r) \sin \phi$ (*c*) $\dfrac{(R+r)}{\cos \phi}$ (*d*) $\dfrac{(R+r)}{\sin \phi}$

where R and r = Radii of the base circle of two gears and ϕ = Pressure angle.

210. In a simple trains of three wheels, the third wheel will rotate

(*a*) in the opposite direction to the first wheel

(*b*) in the same direction as the first wheel

(*c*) in any direction

(*d*) none of the above.

211. In question 210, the second wheel is known as ideal wheel (or the idler). This wheel

(*a*) has no influence on the velocity ratio. It only affects the direction of rotation

(*b*) changes the velocity ratio and also direction of rotation

(*c*) may change the velocity ratio but may not affect the direction of rotation

(*d*) none of the above.

212. In question 210, the ratio of the speed of the third wheel to that of first wheel is equal to the ratio of

(*a*) number of teeth on the first wheel to the number of teeth on the third wheel

(*b*) number of teeth on the third wheel to the number of teeth on the first wheel

(*c*) product of number of teeth on the first and third wheels to the number of wheel

(*d*) none of the above.

213. When the axes of the first and last wheels of a compound gear train are co-axial, then the train is known as

(*a*) non-reverted gear train (*b*) reverted gear train

(*c*) epicyclic gear train (*d*) none of the above.

214. If in a gear trains, the first wheel is a driver and the last wheel is a driven wheel, then train value is defined as the

(*a*) product of the speeds of the driver and driven wheels

(*b*) ratio of the speed of the driver wheel to the speed of the driven wheel

(*c*) ratio of the speed of the driven wheel to the speed of the driver wheel

(*d*) products of number of teeth on driver and driven wheels.

215. In question 214, the speed ratio of the gear train is defined as

(*a*) product of the speeds of the driver and driven wheels

(*b*) ratio of the speed of the driver wheel to the speed of the driven wheel

(*c*) ratio of the speed of the driven wheel to the speed of the driver wheel

(*d*) products of number of teeth on driver and driven wheels.

216. Train value of a gear train is

(*a*) equal to the speed ratio (*b*) half of the speed ratio

(*c*) reciprocal of the speed ratio (*d*) double the speed ratio.

217. The train values and speed ratios are specified as their plus or minus. The negative sign of speed ratio indicates that

(*a*) speed of driver wheel is less than the speed of driven wheel

(*b*) speed of driven wheel is less than the speed of driver wheel

(*c*) driver and driven wheels are rotating in opposite direction

(*d*) driver and driven wheels are rotating in the same direction.

218. A gear train, in which at least one of the gear axes is in motion relative to the frame, is known as

(*a*) reverted gear train (*b*) non-reverted gear train

(*c*) epicyclic gear (*d*) none of the above.

219. A dynamometer is a device used for measuring

(*a*) speed of a machine (*b*) torque exerted by or on a machine

(*c*) kinetic energy of the machine (*d*) all of the above

(*e*) none of the above.

220. Which of the followings are the correct statement for clutch and brake ?

(*a*) clutch connects a moving member of stationary frame

(*b*) clutch connects a moving member to another moving member

(*c*) brake connects a moving member to another moving member

(*d*) brake connects a moving members to stationary frame.

221. Which one of the followings is an absorption dynamometer ?

(*a*) gear dynamometer (*b*) torsion dynamometer

(*c*) belt dynamometer (*d*) prony brake dynamometer.

222. Which one of the followings is an transmission dynamometer ?

(*a*) prony brake dynamometer (*b*) torsion dynamometer

(*c*) fluid friction dynamometer (*d*) band brake dynamometer.

223. The abbreviation FMR in brakes and dynamometer stands for

(*a*) friction motion ratio (*b*) friction multiplication ratio

(*c*) force multiplication ratio (*d*) force mean rate.

224. The smallest circle drawn to the cam profile from the cam centre is known as
(*a*) prime circle (*b*) base circle (*c*) pitch circle (*d*) pitch curve.

225. The smallest circle drawn to the pitch curve from the centre of rotation of the cam is known as
(*a*) prime circle (*b*) base circle (*c*) pitch circle (*d*) none of the above.

226. The circle, with centre as the centre of the cam axis and radius such that it passes through the pitch point, is known as
(*a*) prime circle (*b*) base circle (*c*) pitch circle (*d*) pitch curve.

227. The path, followed by the trace point if the follower moved about the cam, is known as
(*a*) prime circle (*b*) base circle (*c*) pitch circle (*d*) pitch curve.

228. For the knife edge follower, the cam profile and pitch curve
(*a*) are different (*b*) are same
(*c*) may be same or different (*d*) none of the above.

229. On which circle or curve the size of the cam depends ?
(*a*) prime circle (*b*) base circle (*c*) pitch circle (*d*) pitch curve.

230. The reference point, on the follower for tracing the cam profile, is called
(*a*) the pitch point (*b*) the trace point (*c*) the centre of cam (*d*) none of the above.

231. The point, on the cam pitch curve having the maximum pressure angle, is called
(*a*) the pitch point (*b*) the trace point
(*c*) the centre of cam (*d*) none of the above.

232. The pressure angle of a cam is the angle at any point on the pitch curve.
(*a*) made by line of motion of the follower with horizontal axis
(*b*) made by line of motion of the follower with vertical axis
(*c*) between the normal to that point on the curve and line of motion of follower at that instant
(*d*) between the tangent to that point on the curve and line of motion of follower at that instant.

233. The angle, which represents the cam profile and is most important in cam design, is
(*a*) cam angle (*b*) pressure angle (*c*) angle of dwell (*d*) angle of descent.

234. The stroke of the follower is equal to
(*a*) half of the maximum travel of the follower from the lowest position to the topmost position
(*b*) the maximum travel of the follower from the lowest position of the topmost position
(*c*) half of the diameter of base circle
(*d*) half of the diameter of the prime circle.

235. Choose the wrong statement
(*a*) The reference point for a knife edge follower is the edge of the knife.
(*b*) The reference point for a roller follower is the centre of the roller.
(*c*) The pitch curve and cam profile, for a knife edge follower, are the same
(*d*) The pressure angle is the angle between the direction of the follower motion and a tangent to the pitch curve.

236. A follower of a cam has

(*a*) uniform velocity (*b*) simple harmonic motion

(*c*) cycloidal motion (*d*) uniform acceleration and retardation

(*e*) any one of the above (*f*) none of the above.

237. Which factor affects the size of a cam ?

(*a*) base circle (*b*) pressure angle (*c*) hub size

(*d*) undercutting (*e*) all of the above (*f*) none of the above.

238. The pressure angle of a cam

(*a*) decreases if radius of base circle decreases

(*b*) decreases if radius of base circle increase

(*c*) is independent of the radius of base circle

(*d*) first decreases then increases if the radius of base circle decreases.

239. The lift of a roller follower, when it is an contact with the straight flank of a tangent cam, is equal to

(*a*) $(r + r_1)(1 - \cos\theta)$ (*b*) $(r + r_1)\left(\frac{1-\cos\theta}{\cos\theta}\right)$

(*c*) $(r - r_1)(1 + \cos\theta)$ (*d*) $(r - r_1)\left(\frac{1-\cos\theta}{\cos\theta}\right)$

where θ = Any angle turned by the cam from the beginning of the lift of the follower

r = Radius of the base circle of the cam, r_1 = Radius of the roller follower.

240. For the question 239, the velocity of the follower is equal to

(*a*) $\omega (r + r_1) \sin\theta \sec^2\theta$ (*b*) $\omega (r + r_1) \sin\theta \cos\theta$

(*c*) $\omega (r - r_1) \sin\theta \sec^2\theta$ (*d*) $\omega (r - r_1) \cos\theta \sin^2\theta$.

241. For the question 239, the acceleration of the follower is equal to

(*a*) $\omega^2 (r - r_1)\left(\frac{2-\cos^2\theta}{\cos^3\theta}\right)$ (*b*) $\omega^2 (r + r_1)\left(\frac{2-\cos^2\theta}{\cos^3\theta}\right)$

(*c*) $\omega^2 (r - r_1)(1 + \cos^2 r)$ (*d*) $\omega^2 (r - r_1)(1 + \sin^2\theta)$.

242. A roller follower, the diameter 30 mm, is in contact with the straight flask of a tangent cam of base circle diameter 70 mm. The cam has rotated through 30° when the roller is just to leave contact of the flank on its ascent. If the angular speed of the cam is 10 rad/sec, then left the roller, when it is just to leave contact of the flank will be

(*a*) 10 mm (*b*) 7.73 mm (*c*) 5 mm (*d*) 4 mm.

243. In question 241, the velocity of the roller, when the roller is just to leave contact of the flank, will be

(*a*) 10 cm/sec (*b*) 77.3 cm/sec (*c*) 33.33 cm/sec (*d*) 50 cm/sec.

244. In question 242, the acceleration of the roller, when the roller is just to leave contact of the flank, will be

(*a*) 50.8 m/sec^2 (*b*) 25.3 m/sec^2 (*c*) 15.4 m/sec^2 (*d*) 9.615 m/sec^2.

245. The term $\omega^2 \frac{d^3y}{d\theta^3}$, in a cam follower motion, represents

(*a*) acceleration of follower (*b*) velocity of follower

(*c*) jerk (*d*) displacement.

246. The left of a flat-faced follower, when it is in contact with the flank of a circular arc cam, is equal to

(a) $(R + r)(1 - \cos \theta)$ (b) $(R - r)(1 - \cos \theta)$

(c) $(R - r)(1 + \cos \theta)$ (d) $(R + r)(1 + \cos \theta)$

where R = Flank radius, r = Minimum radius of cam

θ = Angle turned by cam from beginning of the left of follower.

247. The acceleration of a flat-faced follower, when it is in contact with the flank of a circular arc cam, is equal to

(a) $\omega^2 (R - r) \sin \theta$ (b) $\omega^2 (R + r) \sin \theta$ (c) $\omega^2 (R - r) \cos \theta$ (d) $\omega^2 (R + r) \cos \theta$.

248. The velocity of a flat-faced follower, when it is in contact with the flank of a circular arc cam, is equal to

(a) $\omega (R - r) \sin \theta$ (b) $\omega (R - r) \sin \theta$ (c) $\omega (R + r) \cos \theta$ (d) $\omega (R + r) \cos \theta$.

249. Period of dwell (in case of cam and follower) is the period

(a) during which cam rotates

(b) during which follower moves from its lower position to highest position

(c) during which follower remains stationary during some finite rotation of the cam

(d) none of the above.

250. Which one is a correct statement regarding governor ?

(a) The governor reduces the fluctuation of speed during a cycle for constant output load.

(b) The governor controls the mean speed over a period for output load variations.

(c) The governor has no influence on mean speed for prime mover.

(d) The governor has no influence over the varying load demand on prime mover.

251. Which one is a correct statement regarding flywheel ?

(a) The flywheel reduces the fluctuation of speed during a cycle for constant output load.

(b) The flywheel controls the mean speed of the prime mover.

(c) The flywheel adjusts supply when output is varying.

(d) The flywheel is no influence over cyclic speed fluctuation.

252. The inertia and flywheel governors as compared to centrifugal governors are

(a) more sensitive and more simple (b) less sensitive and more simple

(c) more sensitive and less simple (d) less sensitive and less simple.

253. Which one of the following is a spring loaded type governor ?

(a) Watt governor (b) Hartnell governor

(c) Porter governor (d) Proell governor.

254. Which one of the following is a gravity controlled type governor ?

(a) Hartung governor (b) Hartnell governor

(c) Pickering governor (d) Watt governor.

255. The height (h) of a Watt's governor is given by

(a) $h = g\omega^2$ (b) $h = \dfrac{\omega^2}{g}$ (c) $h = \dfrac{g}{\omega^2}$ (d) $h = \dfrac{1}{g\omega^2}$.

256. The height (h) of a Porter governor (when the upper and lower arms are equal and pivoted either on the governor axis or at equal distance from it) is given by

(*a*) $h = \left(\dfrac{w+W}{w}\right) \times \dfrac{g}{\omega^2}$ (*b*) $h = \left(\dfrac{w+W}{w}\right) \times \dfrac{\omega^2}{g}$

(*c*) $h = \left(\dfrac{w}{w+W}\right) \times \dfrac{g}{\omega^2}$ (*d*) $h = \left(\dfrac{w}{w+W}\right) \times g\omega^2$

where w = Weight of the balls of governor, and W = Weight of the sleeve.

257. The height of a Watt's governor is

(*a*) inversely proportional to speed (*b*) inversely proportional to square of speed

(*c*) directly proportional to speed (*d*) directly proportional to square of speed.

258. The angular speed of a Watt's governor, when its height is 20 cm, will be equal to

(*a*) 20 rad/sec (*b*) 10 rad/sec (*c*) 6 rad/sec (*d*) 7 rad/sec.

259. If, in question 258, the height of governor is 5 cm, then the angular speed of the governor will be

(*a*) 40 rad/sec (*b*) 5 rad/sec (*c*) 14 rad/sec (*d*) 20 rad/sec.

260. If the upper and lower arms of a Porter governor are equal and are pivoted either on the governor axis or at equal distance from it, then the ratio of height of Porter governor to that of Watt's governor is equal to

(*a*) $\dfrac{W}{W-w}$ (*b*) $\dfrac{W-w}{W}$ (*c*) $\dfrac{w}{w+W}$ (*d*) $\dfrac{W+w}{w}$

where W = Weight of sleeve, and w = Weight of the balls of governor.

261. When the speed of governor increases

(*a*) height of governor and radius of rotation increases

(*b*) height of governor and radius of rotation decreases

(*c*) height of governor decreases but radius of rotation increases

(*d*) height of governor increases but radius of rotation decreases.

262. When the speed of governor decreases

(*a*) height of governor and radius of rotation increases

(*b*) height of governor and radius of rotation decreases

(*c*) height of governor decreases but radius of rotation increases

(*d*) height of governor increases but radius of rotation decreases.

263. For a Porter governor (which the upper and lower arms are equal and pivoted either on the governor axis or at equal distance from it) the height of the governor, when speed is increasing and effect of friction is considered, is given by

(*a*) $\left(\dfrac{W+R+w}{w}\right) \times \omega^2 \times g$ (*b*) $\left(\dfrac{W+R+w}{w}\right) \times \dfrac{g}{\omega^2}$

(*c*) $\left(\dfrac{W-R+w}{w}\right) \times \dfrac{g}{\omega^2}$ (*d*) $\left(\dfrac{W-R+w}{w}\right) \times \omega^2 \times g$

where W = Weight of sleeve, w = Weight of balls of governor, and

R = Frictional resistance of sleeve and gears etc.

264. For question 263, when speed is decreasing and effect of friction is considered, then height of governor is given by

(a) $\left(\dfrac{W+R+w}{w}\right)\times\omega^2\times g$ (b) $\left(\dfrac{W+R+w}{w}\right)\times\dfrac{g}{\omega^2}$

(c) $\left(\dfrac{W-R+w}{w}\right)\times\dfrac{g}{\omega^2}$ (d) $\left(\dfrac{W-R+w}{w}\right)\times\omega^2\times g.$

265. The friction, at the sleeve of a governor

(a) increases the sensitiveness of the governor

(b) decreases the sensitiveness of the governor

(c) has no effect on sensitiveness of the governor

(d) none of the above.

266. Proell governor as compared to Porter governor

(a) is more sensitive (b) is less sensitive

(c) requires weights of smaller size (d) requires weights of bigger size

(e) both (a) and (c).

267. The spring loaded governors as compared to gravity controlled governors

(a) may be operated at very high speeds

(b) are much smaller in overall size

(c) may revolve about a vertical or inclined axis

(d) all of the above

(e) none of the above.

268. If F = centrifugal force on the ball of a Hartnell governor, S = spring force exerted on the sleeve, a and b are the lengths of vertical and horizontal arms of the bell crank levers, then spring force exerted on the sleeve (neglecting the effect of the pull of gravity on governor balls and arms) is given by

(a) $S=\dfrac{F\times a}{2b}$ (b) $S=\dfrac{F\times b}{2a}$ (c) $S=2F\times\dfrac{a}{b}$ (d) $S=\dfrac{b}{2F\times a}$.

269. The lift of the sleeve in a Hartnell governor of question 268 is equal to

(a) $\dfrac{a}{b}(r_1-r_2)$ (b) $\dfrac{b}{a}(r_1-r_2)$ (c) $\dfrac{a}{b}(r_1+r_2)$ (d) $\dfrac{b}{a}(r_1+r_2)$

where r_1 = Maximum radius of rotation, and r_2 = Minimum radius of rotation.

270. The stiffness of the spring in Hartnell governor of question 268 is equal to

(a) $2\left(\dfrac{a}{b}\right)^2\left(\dfrac{F_1+F_2}{r_1-r_2}\right)$ (b) $2\left(\dfrac{a}{b}\right)^2\left(\dfrac{F_1+F_2}{r_1-r_2}\right)$

(c) $2\left(\dfrac{a}{b}\right)^2\left(\dfrac{F_1-F_2}{r_1+r_2}\right)$ (d) $2\left(\dfrac{a}{b}\right)^2\left(\dfrac{F_1+F_2}{r_1+r_2}\right)$

where F_1 = Centrifugal force at maximum radius of rotation, and

F_2 = Centrifugal force at minimum radius of rotation.

271. The stiffness of the spring in a Hartnell governor is also equal to

(*a*) $\dfrac{S_1 - S_2}{h}$ (*b*) $\dfrac{2(S_1 - S_2)}{h}$ (*c*) $\dfrac{S_1 - S_2}{2h}$ (*d*) $\dfrac{S_1 + S_2}{h}$

where S_2 = Spring force exerted on the sleeve at maximum radius of rotation,

S_2 = Spring force exerted on the sleeve at minimum radius of rotation, and

h = Lift of sleeve.

272. In a Hartnell governor, the weight of each ball is 2.45 kgf. The maximum and minimum speed of rotations are 10 rad/sec and 8 rad/sec. The maximum and minimum radii of rotations are 20 cm and 12 cm respectively. The lengths of vertical and horizontal arms of bell crank levers are 20 cm and 10 cm respectively. If the effect of pull of gravity is neglected then the lift of the sleeve will be equal to

(*a*) 10 cm (*b*) 8 cm (*c*) 6 cm (*d*) 4 cm.

273. For the question 272, the spring force exerted on the sleeve at maximum radius of rotation will be

(*a*) 20 kgf (*b*) 10 kgf (*c*) 7.68 kgf (*d*) 3.08 kgf.

274. For the question 272, the spring force exerted on the sleeve at minimum radius of rotation will be

(*a*) 20 kgf (*b*) 10 kgf (*c*) 7.68 kgf (*d*) 3.08 kgf.

275. For the question 272, the stiffness of the spring will be

(*a*) 20 kgf (*b*) 10 kgf (*c*) 7.68 kgf (*d*) 3.08 kgf.

276. The range of speed a governor is equal to

(*a*) sum of maximum and minimum equilibrium speeds

(*b*) difference of maximum and minimum equilibrium speeds

(*c*) average of sum of maximum and minimum equilibrium speeds

(*d*) average of difference of maximum and minimum equilibrium speeds.

277. Sensitiveness of a governor is defined as the ratio of

(*a*) mean speed to range of speed

(*b*) range of speed to mean speed

(*c*) maximum equilibrium speed to mean speed

(*d*) minimum equilibrium speed to mean speed.

278. A governor is said to be sensitive when

(*a*) for a small change of speed, the displacement of sleeve is maximum

(*b*) the governor readily responds to a small alternation of speed

(*c*) the ratio of range of speed to mean speed is maximum

(*d*) any one of the above

(*e*) none of the above.

279. Which one of the following is a correct statement regarding the effect of friction on governors ?

(*a*) Due to friction, there is a range of speed for any sleeve position over which the governor is sensitive.

(*b*) Due to friction, the governor will have same configuration whether it approaches equilibrium from the higher or lower speed side.

(c) Due to friction, the governor will have different configuration whether it approaches equilibrium from the higher or lower speed side.
(d) Due to friction, the load on sleeve decreases when sleeve is rising.

280. A governor is said to isochronous when
(a) the equilibrium speed is constant for all radii of rotation of the balls within working range
(b) the range of speed is zero for all radii of rotation of the balls within working range
(c) any one of the above
(d) none of the above.

281. The mean force exerted by the governor on the sleeve for a given change of speed is known as
(a) sensitiveness of governor (b) effort of governor
(c) stability of governor (d) frictional force of governor.

282. If for each speed within working range, there is only one radius of rotation of the governor's balls, then the governor is said to be
(a) hunting (b) isochronous (c) stable (d) sensitive.

283. When a governor is over sensitive, the sleeve will oscillate between two extreme positions when there is a slightest change of speed. Then the governor is said to be
(a) hunting (b) isochronous (c) stable (d) none of the above.

284. The power of governor is
(a) the work done at the sleeve for a given percentage change of speed
(b) equal to the product of governor effort and maximum equilibrium speed
(c) equal to product of governor effort and the displacement of sleeve
(d) both (a) and (c)
(e) none of the above.

285. The effort of a Porter governor (when the arms of the governor are equal and have equal inclination with the axis of the governor spindle) is equal to
(a) $C(w + W)$ (b) $\frac{(w \times W)}{C}$ (c) $\frac{(w + W \times C)}{C}$ (d) $\frac{(C \times w + W)}{C}$
where w = Weight of ball of governor, W = Weight on sleeve, and
C = Percentage increase in speed.

286. The effort of a Hartnell governor (where the moments due to weight of the arms and ball are neglected) is equal to
(a) $\frac{C}{S}$ (b) CS (c) $\frac{1}{CS}$ (d) $\frac{S}{C}$
where C = Percentage increase in speed, and S = Spring force exerted on the sleeve.

287. The power of a Porter governor (when the arms of the governor are equal and have equal inclination with the axis of the governor spindle) is equal to
(a) $\left(\frac{1+2C}{4C}\right)(w \times W) \times h$ (b) $\left(\frac{1-2C}{4C^2}\right)(w + W) \times h$
(c) $\left(\frac{4C^2}{1+2C}\right)(w \times W) \times h$ (d) $\left(\frac{4C^2}{1-2C}\right)(w + W) \times h$
where h = Height of governor, and C = Percentage increase in speed.

288. The controlling force in case of Porter governor is provided by
(*a*) weight of balls (*b*) springs (*c*) both (*a*) and (*b*) (*d*) none of the above.

289. The controlling force, in case of Hartnell governor, is provided by
(*a*) weight of balls (*b*) springs (*c*) both (*a*) and (*b*) (*d*) none of the above.

290. The controlling force curve is a graph between controlling force and
(*a*) speed of rotation (*b*) radius of rotation (*c*) range of speed (*d*) lift of governor.

291. The controlling force curve is used to find out
(*a*) stability of a governor (*b*) sensitiveness of a governor
(*c*) hunting of a governor (*d*) (*a*) and (*b*) only.

292. The condition for stability of a governor is stated as
(*a*) the rate of increase of controlling force is equal to the rate of increase of radius
(*b*) the rate of increase of controlling force must be greater than the rate of increase of radius
(*c*) the rate of increase of controlling force must be less than the rate of increase of radius
(*d*) none of the above.

293. The equation for controlling curve for a stable Hartnell governor is given by
(*a*) $F_c = mr + C$ (*b*) $F_c = mr$ (*c*) $F_c = mr - C$ (*d*) none of the above
where F_c = Controlling force, and m and C = are constants.

294. Choose the wrong statement
(*a*) Static balance between revolving masses exists if the masses are in equilibrium when not running.
(*b*) Dynamic balance exists when the inertia forces and couples exerted by the moving masses are in equilibrium among themselves.
(*c*) All revolving and reciprocating masses should be balanced as completely as possible.
(*d*) Balancing of rotating and reciprocating parts of an engine is necessary when it runs at slow speed.

295. A weight attached to a rotating shaft is to be balanced by a weight B attached to the shaft and revolving in the same plane as the weight W. If 'r' is the distance of centre of gravity of the revolving weight W from the axis of shaft and 'b' is the distance of centre of gravity of balancing weight B from the axis of shaft, then the necessary condition for balancing if
(*a*) $W \times b = B \times r$ (*b*) $W \times B = b \times r$ (*c*) $W \times r = B \times b$ (*d*) $W = B \times b \times r$.

296. Which of the following is a correct statement for balancing a revolving disturbing weight by a weight revolving in the same plane ?
(*a*) The condition for balancing is independent of speed.
(*b*) The shaft is in static and dynamic balance.
(*c*) The centre of gravity is on the axis of the shaft and remains in a fixed position relative to the frame whether the shaft is rotating or standing still.
(*d*) All of the above.
(*e*) none of the above.

297. A weight W attached to a rotating shaft is to be balanced. But it is not possible to introduce the single balance weight in the plane of rotation of the disturbing weight. Then the single balance weight is introduced in a plane parallel to the plane of rotation of the disturbing weight. The system will be in

(*a*) static balance (*b*) dynamic balance (*c*) complete balance (*d*) none of the above.

298. When the balance weights are introduced in a plane parallel to the plane of rotation of the disturbing weight, then the minimum number of balance weights, for balancing a single revolving disturbing weight, will be

(*a*) one (*b*) two (*c*) three (*d*) four.

299. Reference plane is a plane which is

(*a*) passing through the plane of rotation of the rotating weight

(*b*) passing through the plane of rotation of the balancing weight

(*c*) at an angle of 45° to the rotating weight

(*d*) at an angle of 45° to the balancing weight.

300. The effect of transferring the rotating weight in the reference plane is to produce

(*a*) the same magnitude of the centrifugal force as produced by the rotating weight

(*b*) the same magnitude of the centrifugal force as produced by the balancing weight

(*c*) a couple equal to the product of centrifugal force and distance of the reference plane of rotation from disturbing weight

(*d*) both (*a*) and (*c*)

(*e*) none of the above.

301. For complete balance of the several revolving weights in different planes, the conditions is

(*a*) the vector sum of the forces must be zero

(*b*) the vector sum of all the couples must be zero

(*c*) both (*a*) and (*b*)

(*d*) none of the above.

302. The acceleration of the piston of a reciprocating engine is equal to

(*a*) $\omega^2 r\left(\cos\theta + \dfrac{\sin 2\theta}{n}\right)$ (*b*) $\omega^2 r(\cos\theta + \cos 2\theta)$

(*c*) $\omega^2 r\left(\cos\theta + \dfrac{\cos 2\theta}{n}\right)$ (*d*) $\omega^2 r\left(\sin\theta + \dfrac{\sin 2\theta}{n}\right)$

where ω = Angular speed of crank, r = Crank radius,

θ = Inclination of crank with inner dead centre, and

n = Ratio of the length of connecting rod to crank radius $= \dfrac{l}{r}$.

303. The inertia force due to weight of reciprocating parts in a reciprocating engine is equal to

(*a*) $\dfrac{W}{g}\,\omega^2 r\left(\cos\theta + \dfrac{\sin 2\theta}{n}\right)$ (*b*) $\dfrac{W}{g}\,\omega^2 r(\cos\theta + \cos 2\theta)$

(*c*) $\dfrac{W}{g}\,\omega^2 r\left(\cos\theta + \dfrac{\cos 2\theta}{n}\right)$ (*d*) $\dfrac{W}{g}\,\omega^2 r\left(\sin\theta + \dfrac{\sin 2\theta}{n}\right)$

where W = Weight of reciprocating parts.

304. The term $\left(\frac{W}{g}\omega^2 r \cos 2\theta\right)$ in reciprocating engine is called

(*a*) primary couple (*b*) secondary couple
(*c*) primary disturbing force (*d*) secondary disturbing force.

305. The term $\left(\frac{W}{g}\omega^2 r \frac{\cos 2\theta}{n}\right)$ in reciprocating engine is called

(*a*) primary couple (*b*) secondary couple
(*c*) primary disturbing force (*d*) secondary disturbing force.

306. The frequency of the secondary force as compared to primary force is

(*a*) one-half (*b*) double (*c*) one-fourth (*d*) one-third.

307. The condition of balancing the reciprocating masses of a reciprocating engine is to balance

(*a*) primary forces couple (*b*) secondary forces only (*c*) primary couples only
(*d*) secondary couples only (*e*) all of the above (*f*) none of the above.

308. When the angle of inclination of crank with inner dead centre of a reciprocating engine is 0°, then

(*a*) primary force is maximum (*b*) secondary force is maximum
(*c*) primary force is minimum (*d*) secondary force is minimum.
(*e*) both (*a*) and (*b*) (*f*) both (*c*) and (*d*).

309. When the angle of inclination of crank with inner dead centre of a reciprocating engine is 90°, then

(*a*) primary force is maximum (*b*) secondary force is maximum
(*c*) primary force is zero (*d*) secondary force is zero
(*e*) both (*b*) and (*c*) (*f*) both (*a*) and (*d*).

310. When the angle of inclination of crank with inner dead centre of a reciprocating engine is 45°, then

(*a*) primary force is maximum (*b*) secondary force is maximum
(*c*) primary force is zero (*d*) secondary force is zero
(*e*) none of the above.

311. Choose the correct statement.

(*a*) The secondary disturbing force is more than primary disturbine force.
(*b*) For one revolution of crank, the secondary force becomes maximum two times.
(*c*) For one revolution of crank, the primary force becomes maximum four times.
(*d*) The primary disturbing force due to inertia of reciprocating parts in an engine is equal to $\frac{W}{g}\omega^2 r \cos\theta$.

312. The secondary disturbing force due to inertia of reciprocating parts of an engine is equal to the centrifugal produced by the equivalent crank

(*a*) having an equivalent radius of $\frac{r^2}{4l}$ (*b*) running at twice the speed
(*c*) having a weight attached to its end (*d*) all of the above
(*e*) none of the above.

313. If a weight equal to the weights of reciprocating parts is attached diametrically opposite to crank at crank radius, then

(*a*) primary disturbing force is completely balanced
(*b*) primary disturbing force is not balanced
(*c*) primary disturbing force is completely balanced but an unbalanced vertical force is introduced
(*d*) none of the above.

314. The primary forces are balanced partially. The balance weight placed diametrically opposite to the crank is cW, where W is the weight of reciprocating parts. The usual value of c for locomotive is

(*a*) one-fourth (*b*) one-half (*c*) one-third (*d*) two-thirds.

315. If the balance weight placed diametrically opposite to the crank at crank radius (r) is cW, then unbalanced force along the line of stroke of the reciprocating engine, is equal to

(*a*) $\dfrac{cW}{g}\,\omega^2 r\cos\theta$ (*b*) $(1-c)\,\dfrac{W}{g}\,\omega^2 r\cos\theta$
(*c*) $\dfrac{cW}{g}\,\omega^2 r\sin\theta$ (*d*) $(1-c)\,\dfrac{W}{g}\,\omega^2 r\sin\theta$.

316. In question 315, the unbalanced force normal to the line of stroke of the reciprocating engine, is equal to

(*a*) $\dfrac{cW}{g}\,\omega^2 r\cos\theta$ (*b*) $(1-c)\,\dfrac{W}{g}\,\omega^2 r\cos\theta$
(*c*) $\dfrac{cW}{g}\,\omega^2 r\sin\theta$ (*d*) $(1-c)\,\dfrac{W}{g}\,\omega^2 r\sin\theta$.

317. In question 315, the resultant unbalanced force at any instant, is equal to

(*a*) $\dfrac{cW}{g}\,\omega^2 r$ (*b*) $(1-c)\,\dfrac{W}{g}\,\omega^2 r$
(*c*) $\dfrac{W}{g}\,\omega^2 r\sqrt{(1-c)^2\cos^2\theta + c^2\sin^2\theta}$ (*d*) $\dfrac{W}{g}\,\omega^2 r\sqrt{(1-c)^2\cos^2\theta + \sin^2\theta}$.

318. If the ratio of the length of connecting rod to the crank radius is very large, then

(*a*) primary forces are negligible (*b*) primary forces are maximum
(*c*) secondary forces are negligible (*d*) secondary forces are maximum.

319. In locomotives, with two cylinders, the cranks of two cylinders are placed at right angles to each other in order to

(*a*) start the locomotive in any position (*b*) reduce the effect of secondary forces
(*c*) reduced the effect of primary forces (*d*) all of the above
(*e*) (*a*) and (*b*) only.

320. The swaying couple and variation in the tractive effort is due to

(*a*) primary disturbing force (*b*) secondary disturbing force
(*c*) both (*a*) and (*c*) (*d*) none of the above.

321. The unbalanced force acting vertically upward or downward due to balancing of the reciprocating parts in a locomotive varies

(*a*) directly with the speed (*b*) directly with the square of the speed
(*c*) inversely with the speed (*d*) inversely with the square of the speed.

322. The dead weight of the wheel and locomotive is

(*a*) a function of speed (*b*) a function of square of the speed

(*c*) independent of the speed (*d*) none of the above.

323. The lifting of the wheels from the rails and striking back on the rails take place at

(*a*) low speed (*b*) normal speed (*c*) high speed (*d*) none of the above.

324. When the centre of gravity of the balance weights is either directly below the wheel centre or directly above the wheel centre, then

(*a*) unbalanced horizontal force is maximum

(*b*) unbalanced vertical force is maximum

(*c*) unbalanced vertical force is minimum (*d*) none of the above.

325. Hammer blow is

(*a*) maximum value of unbalanced force along the line of stroke

(*b*) maximum value of unbalanced force perpendicular to the line of stroke

(*c*) resultant value of the unbalanced force

(*d*) resultant value of unbalanced force perpendicular to the line of stroke.

326. If B is the balance weight for reciprocating parts placed at a radius (r) in a locomotive for which W is the dead load acting on each wheel, then the force trying to lift to the wheels from the rails will be

(*a*) $W + \frac{B}{g}\,\omega^3 r \sin\theta$ (*b*) $W - \frac{B}{g}\,\omega^1 r^2 \sin\theta$

(*c*) $W - \frac{B}{g}\,\omega^2 r$ (*d*) $W - \frac{B}{g}\,\omega^2 r \cos\theta$.

327. The lifting of wheel can be prevented if the angular speed is less than

(*a*) $\sqrt{\frac{W}{g}\frac{r}{B}}$ (*b*) $\sqrt{\frac{Wg}{B \times r}}$ (*c*) $\sqrt{\frac{g \times B}{W \times r}}$ (*d*) $\sqrt{\frac{W}{g \times B \times r}}$

where W = Dead weight acting on a wheel,

r = Radius at which balance weight is placed, and

B = Balance weight for reciprocating masses.

328. For two cylinder locomotives having their cranks at right angles to each other, the tractive force is equal to

(*a*) $(1 - c)\frac{W}{g}\,\omega^2 r \cos\theta$ (*b*) $(1 - c)\frac{W}{g}\,\omega^2 r \sin\theta$

(*c*) $(1 - c)\frac{W}{g}\,\omega^2 r (\cos\theta - \sin\theta)$ (*d*) $(1 - c)\frac{W}{g}\,\omega^2 r (\cos\theta - 1)$

where W = Weight of reciprocating parts, C = Fraction of the reciprocating parts

r = Radius of crank.

329. The tractive force will be maximum or minimum when angle of inclination of crank with the line of stroke is equal to

(*a*) 45° and 90° (*b*) 90° and 180° (*c*) 135° and 315° (*d*) 0° and 180°.

330. If $C = 2/3$, the maximum value of tractive force is equal to

(a) $\sqrt{2}\ \frac{W}{g}\ \omega^2 r$ (b) $\frac{\sqrt{2}}{3}\ \frac{W}{g}\ \omega^2 r$ (c) $\frac{3}{\sqrt{2}}\ \frac{W}{g}\ \omega^2 r$ (d) $3\sqrt{2}\ \frac{W}{g}\ \omega^2 r$.

331. Choose the wrong statement.

(a) Swaying couple is maximum or minimum when angle of inclination of crank to the line of stroke is 45° or 225.

(b) Tractive force is maximum or minimum when angle of inclination of crank to the line of stroke is 135° and 315°.

(c) For starting the locomotive in any position, the cranks of the two cylinders of a locomotive are placed at right angle to each other.

(d) The sway couple is due to secondary unbalanced force.

332. Choose the correct statement.

(a) The effect of hammer blow can be reduced by decreasing the speed and using two or three pairs of wheels coupled together.

(b) Partial balancing in locomotive results in variation in tractive force, hammer blow and swaying couple.

(c) In locomotives, the effect of secondary forces is minimised by making the ratio of the connecting rod length to the crank radius very large.

(d) all of the above.

(e) none of the above.

333. The number of cycles of motion completed in a unit interval of time is known as

(a) period (b) frequency (c) resonance (d) damping.

334. The time taken, by a vibrating body to complete one cycle, is known as

(a) period (b) frequency (c) resonance (d) damping.

335. Natural frequency of a system is due to

(a) free vibration (b) forced vibration (c) resonance (d) damping.

336. Resonance is a phenomenon in which the frequency of the external exciting force

(a) is half the natural frequency of the system

(b) is double the natural frequency of the system

(c) coincides with the natural frequency of the system

(d) does not coincide with the natural frequency of the system.

337. A vibrating body is said to be number the action of damping when

(a) the amplitude of the body gradually increases with time

(b) the amplitude of the body is constant with respect to time

(c) the amplitude of the body gradually decreases with time

(d) none of the above.

338. If the particles of a body vibrate parallel to the axis of the body, then the body is said to have

(a) transverse vibration (b) longitudinal vibration

(c) torsional vibration (d) none of the above.

339. If the particles of a body vibrate perpendicular to the axis of the body, then the body is said to have

(*a*) transverse vibration (*b*) longitudinal vibration

(*c*) torsional vibration (*d*) none of the above.

340. If the particles of a body vibrates along a circular arc, whose centre lies on the axis of the shaft, then the body is said to have

(*a*) transverse vibration (*b*) longitudinal vibration

(*c*) torsional vibration (*d*) none of the above.

341. Choose the wrong statement.

(*a*) Bending stresses are induced when a body is having transverse vibrations.

(*b*) Shear stresses are induced when a body is having torsional vibrations.

(*c*) Tensile or compressive stresses are induced when a body is having longitudinal vibrations.

(*d*) all of the above.

(*e*) none of the above.

342. The equation of motion for the free vibration of the spring mass system without viscous damping is given by

(*a*) $\frac{d^2x}{dt^2} + \frac{c}{m} + \frac{dx}{dt} + \frac{k}{m}x = 0$ (*b*) $\frac{d^2x}{dt^2} + \frac{k}{m}x = 0$

(*c*) $\frac{d^2x}{dt^2} + \frac{c}{m} \cdot \frac{dx}{dt} + \frac{k}{m}x = F \sin \omega t$ (*d*) $\frac{d^2x}{dt^2} + \frac{k}{m}x = F \sin \omega t$

where c = Co-efficient of viscous damping, m = Mass of the vibrating body

k = Stiffness of the spring, $F \sin \omega t$ = Harmonic force acting on the body.

343. The equation of motion for the free vibration of the spring mass system with viscous damping is given by

(*a*) $\frac{d^2x}{dt^2} + \frac{c}{m}\frac{dx}{dt} + \frac{k}{m}x = 0$ (*b*) $\frac{d^2x}{dt^2} + \frac{k}{m}x = 0$

(*c*) $\frac{d^2x}{dt^2} + \frac{c}{m}\frac{dx}{dt} + \frac{k}{m}x = F \sin \omega t$ (*d*) $\frac{d^2x}{dt^2} + \frac{k}{m}x = F \sin \omega t$.

344. The equation of motion for the forced vibration of spring mass system with viscous damping is given by

(*a*) $\frac{d^2x}{dt^2} + \frac{c}{m} + \frac{dx}{dt} + \frac{k}{m}x = 0$ (*b*) $\frac{d^2x}{dt^2} + \frac{k}{m}x = 0$

(*c*) $\frac{d^2x}{dt^2} + \frac{c}{m}\frac{dx}{dt} + \frac{k}{m}x = F \sin \omega t$ (*d*) $\frac{d^2x}{dt^2} + \frac{k}{m}x = F \sin \omega t$.

345. The natural frequency of the free longitudinal vibrations is given by

(*a*) $2\pi\sqrt{\frac{m}{k}}$ (*b*) $\frac{1}{2\pi}\sqrt{\frac{k}{m}}$ (*c*) $2\pi\sqrt{\frac{g}{\delta}}$ (*d*) $2\pi\sqrt{\frac{\delta}{g}}$

(*e*) both (*b*) and (*c*) (*f*) both (*a*) and (*d*)

where m = Mass of the body, k = Stiffness of spring, and

δ = Static deflection of the spring.

346. The natural period of the free longitudinal vibrations is given by

(a) $2\pi\sqrt{\frac{m}{k}}$ (b) $\frac{1}{2\pi}\sqrt{\frac{k}{m}}$ (c) $\frac{1}{2\pi}\sqrt{\frac{g}{\delta}}$ (d) $2\pi\sqrt{\frac{\delta}{g}}$

(e) both (b) and (c) (f) both (a) and (d).

347. The natural frequency of the free longitudinal vibrations is the

(a) same as the natural period of vibrations
(b) reciprocal of the natural period of vibrations
(c) twice the natural period of vibrations
(d) half the natural period of vibrations.

348. Two springs of stiffness k_1 and k_2 are connected in series. The resultant stiffness will be equal to

(a) $k_1 + k_2$ (b) $\frac{k_1 + k_2}{2}$ (c) $\frac{k_1 + k_2}{k_1 k_2}$ (d) $\frac{k_1 k_2}{k_1 + k_2}$.

349. The natural frequency of a system is a function of

(a) stiffness of the system (b) mass of the system
(c) amplitude of oscillation (d) both (a) and (b)
(e) none of the above.

350. A mass m is attached to one end of a spring, whose upper end is fixed. The mass and stiffness of the spring are m_1 and k respectively. If the mass oscillates longitudinally, then natural frequency of the mass will be equal to

(a) $\frac{1}{2\pi}\sqrt{\frac{k}{(m+m_1)}}$ (b) $\frac{1}{2\pi}\sqrt{\frac{2k}{(m+m_1)}}$ (c) $\frac{1}{2\pi}\sqrt{\frac{k}{\left(m+\frac{m_1}{2}\right)}}$ (d) $\frac{1}{2\pi}\sqrt{\frac{2k}{\left(m+\frac{m_1}{3}\right)}}$.

351. If in question 350, the mass of spring is negligible then the natural frequency of the mass (attached to the spring) as compared to that of question 350, will be

(a) less (b) more
(c) same (d) may be more or less.

352. At certain speed, the rotating shaft ends to vibrate violently in transverse direction. Then speed is called

(a) whirling speed (b) critical speed (c) whipping speed
(d) all of the above (e) none of the above.

353. The critical speed of a shaft carrying a weight (W) at the centre is given by

(a) $\sqrt{\frac{k \times W}{g}}$ (b) $\sqrt{\frac{k \times g}{W}}$ (c) $\sqrt{\frac{g}{kW}}$ (d) $\sqrt{\frac{W}{k \times g}}$

where k = Stiffness of the shaft.

354. The natural frequency of shaft carrying a weight (W) at the centre and vibrating laterally is given by

(a) $\frac{5.623}{\sqrt{\delta}}$ (b) $\frac{4.987}{\sqrt{\delta}}$ (c) $\frac{4.5}{\sqrt{\delta}}$ (d) $\frac{3.25}{\sqrt{\delta}}$

where δ = Static deflection in cm due to load W.

355. A shaft carries a weight (W) at the centre. Let the C.G. of the weight is displaced by an amount 'e' from the axis of the rotation. When the shaft is rotating, the C.G. will be subjected to centrifugal force and it will be displaced further. If 'y' is the additional displacement of C.G. from the axis of rotation due to centrifugal force, then the ratio of 'y' to 'e' is given by

(*a*) $\dfrac{1}{\left(\dfrac{\omega_c}{\omega}\right)^2 + 1}$ (*b*) $\dfrac{1}{\left(\dfrac{\omega_c}{\omega}\right)^2 - 1}$ (*c*) $\left(\dfrac{\omega_c}{\omega}\right)^2 + 1$ (*d*) $\dfrac{\omega}{\left(\dfrac{\omega_c}{\omega}\right)^2 - 1}$.

where = ω_c = Critical speed of the shaft and ω = Angular speed of the shaft.

356. In question 355, the additional displacement of C.G., from the axis of rotation due to centrifugal force, will be positive when the shaft is rotating with a speed which is

(*a*) equal to critical speed (*b*) less than critical speed
(*c*) more than critical speed (*d*) none of the above.

357. In question 355, the additional displacement of C.G. from the axis of rotation due to centrifugal force will be negative where the shaft is rotation with a speed which is

(*a*) equal to critical speed (*b*) less than critical speed
(*c*) more than critical speed (*d*) none of the above.

358. When a shaft is rotating at a speed which is less than critical speed the phase difference between displacement and centrifugal force is

(*a*) 180° (*b*) 90° (*c*) 45° (*d*) 0°.

359. When a shaft is rotating at a speed which is more than critical speed, the phase difference between the displacement and centrifugal force is

(*a*) 180° (*b*) 90° (*c*) 45° (*d*) 0°.

360. In rotating horizontal shafts, there are large vibrations when the shafts are rotating at a speed which is

(*a*) one-fourth of critical speed (*b*) one-third of critical speed
(*c*) one-half of critical speed (*d*) two-third of critical speed.

361. The natural frequency of free transverse vibrations of a uniformly distributed loaded shaft, having ends simply supported, is given by

(*a*) $\dfrac{5.623}{\sqrt{\delta}}$ (*b*) $\dfrac{4.987}{\sqrt{\delta}}$ (*c*) $\dfrac{4.5}{\sqrt{\delta}}$ (*d*) $\dfrac{3.25}{\sqrt{\delta}}$

where δ = Maximum static deflection of the shaft due to uniformly distributed load in cm.

362. Two heavy rotating masses are connected by shafts of lengths l_1, l_2, l_3 and corresponding diameters d_1, d_2 and d_3. This system is reduced to a torsionally equivalent system having uniform diameter 'd_1' of the shaft. The equivalent length of the shaft is equal to

(*a*) $l_1 + l_2 + l_3$ (*b*) $\dfrac{l_1 + l_2 + l_3}{3}$

(*c*) $l_1 + l_2\left(\dfrac{d_1}{d_2}\right)^3 + l_3\left(\dfrac{d_1}{d_3}\right)^3$ (*d*) $l_1 + l_2\left(\dfrac{d_1}{d_2}\right)^4 + l_3\left(\dfrac{d_1}{d_3}\right)^4$.

363. In question 362, the system will become torsionally equivalent when equal and opposite torque are applied to the two rotors and the system twists through an angle which is

(*a*) half of the angle of twist of actual shafts

(*b*) one-third of the angle twist of actual shafts

(*c*) equal to the angle of twist of actual shafts

(*d*) three times of the angle of twist of actual shafts.

364. The angle of twist of a shaft is given by

(*a*) $\frac{T}{J} \times \frac{C}{l}$ (*b*) $\frac{T}{J} \times \frac{l}{C}$ (*c*) $\frac{T \times J}{C \times l}$ (*d*) $\frac{T}{J \times l \times C}$

where T = Torque, l = Length of shaft, C = Modulus of rigidity, and J = Polar moment of inertia of the shaft.

365. In question 362, the twist of one rotating mass with respect to the second rotating mass is equal to

(*a*) $\frac{32}{\pi C}\left(\frac{d_1^{\,4}}{l_1} + \frac{d_2^{\,4}}{l_2} + \frac{d_3^{\,4}}{l_3}\right)$ (*b*) $\frac{32}{\pi C}\left(\frac{d_1^{\,3}}{l_1} + \frac{d_2^{\,3}}{l_2} + \frac{d_3^{\,3}}{l_3}\right)$

(*c*) $\frac{32}{\pi C}\left(\frac{l_1}{d_1^{\,4}} + \frac{l_2}{d_2^{\,4}} + \frac{l_3}{d_3^{\,4}}\right)$ (*d*) none of the above.

366. A shaft with two rotors at its ends will have

(*a*) three nodes (*b*) two nodes (*c*) one node (*d*) zero node.

367. A shaft with three rotors will have

(*a*) three nodes (*b*) two nodes (*c*) one node (*d*) zero node.

368. A node means a section where the amplitude of vibration is

(*a*) maximum (*b*) half of maximum

(*c*) zero (*d*) one-fourth of the maximum.

369. The number of node points is equal to

(*a*) number of rotors (*b*) one more than the number of rotors

(*c*) one less than the number of rotors (*d*) half the number of rotors.

370. Transmissibility is defined as

(*a*) the ratio of the force transmitted through rigid support to the force transmitted through elastic support

(*b*) the ratio of the force transmitted through elastic support to the force transmitted through rigid support

(*c*) the product of the force transmitted through the elastic support to the force transmitted through rigid support

(*d*) none of the above.

371. When damping is negligible and frequency ratio $\frac{\omega}{\omega_n} > \sqrt{2}$, the transmissibility is given by

(*a*) $\frac{1}{1-\left(\frac{\omega}{\omega_n}\right)^2}$ (*b*) $\frac{1}{\left(\frac{\omega}{\omega_n}\right)^2 - 1}$ (*c*) $1-\left(\frac{\omega}{\omega_n}\right)^2$ (*d*) $\left(\frac{\omega}{\omega_n}\right)^2 - 1$.

372. In vibration isolation systems, the transmissibility (*i.e.*, the ratio of the force transmitted through elastic support to the force transmitted through rigid support) will be less than unity, for all values of damping factors, if the frequency ratio $\left(\frac{\omega}{\omega_n}\right)$ is

(*a*) more than $\sqrt{2}$ (*b*) less than $\sqrt{2}$ (*c*) equal to 1.0 (*d*) less than 1.0.

373. The transmitted force will be greater than that transmitted through rigid supports for all values of damping factors, if the frequency ratio $\left(\frac{\omega}{\omega_n}\right)$ is

(*a*) more than $\sqrt{2}$ (*b*) less than $\sqrt{2}$ (*c*) equal to 1.0 (*d*) less than 1.0.

374. Choose the wrong statement

(*a*) For vibration isolation, dampers need not be provided.

(*b*) When frequency ratio $\left(\frac{\omega}{\omega_n}\right)$ is equal to $\sqrt{2}$, transmissibility for all damping factors is equal to unity.

(*c*) When frequency ratio $\left(\frac{\omega}{\omega_n}\right)$ is less than $\sqrt{2}$, the transmissibility increases as damping factor decreases.

(*d*) When frequency ratio $\left(\frac{\omega}{\omega_n}\right)$ is more than $\sqrt{2}$, the transmissibility decreases as damping factor increases.

(*e*) all of the above (*f*) none of the above.

375. A system is said to be critically damped if the damping factor for a vibrating system is

(*a*) more than one (*b*) equal to one (*c*) less than one (*d*) equal to zero.

376. The amount of damping, necessary for a system to be critically damped, is known as

(*a*) damping factor (*b*) magnification factor

(*c*) critical damping co-efficient (*d*) logarithmic decrement.

377. The ratio of the actual damping co-efficient to the critical damping co-efficient is known as

(*a*) damping factor (*b*) magnification factor

(*c*) critical damping co-efficient (*d*) logarithmic decrement.

378. In a light or under damped vibrating system, the natural logarithm of the ratio of any two successive amplitudes is known as

(*a*) damping factor (*b*) magnification factor

(*c*) critical damping co-efficient (*d*) logarithmic decrement.

379. A system is said to be overdamped if the damping factor for the system is

(*a*) more than one (*b*) equal to one (*c*) less than one (*d*) equal to zero.

380. An overdamped system is disturbed from equilibrium position with some initial velocity. The system will

(*a*) not cross the equilibrium position

(*b*) vibrate about equilibrium position

(*c*) immediately return to the equilibrium position

(*d*) none of the above.

381. Choose the wrong statement.

(*a*) Critical damping is a function of mass and stiffness.

(*b*) The amplitude in steady state forced vibration at resonance is directly proportional to the damping co-efficient.

(*c*) For a vibrating system if the damping factor is equal to one, the system is known as critically damped.

(*d*) For an overdamped system, motion is a periodic.

382. The shafts (*i.e.*, driving and driven shafts) are connected by a Hooke's joint. The relation between the angular displacements of the driver (θ) and driven (ϕ) shafts and the inclination (α) between the axes of the driver and driven shafts is given by

(*a*) $\tan\theta = \tan\alpha \times \tan\phi$ (*b*) $\tan\theta = \tan\alpha \times \cos\phi$

(*c*) $\tan\theta = \cos\alpha \times \tan\phi$ (*d*) $\tan\theta = \sin\alpha \times \cos\phi$.

383. If the inclination (α) between the axes of the driver and driven shafts of a Hooke's joint is constant, then the velocity ratio of the driven shaft to the driving shaft is given by

(*a*) $\dfrac{\cos\alpha}{1+\cos^2\theta\sin^2\alpha}$ (*b*) $\dfrac{\cos\alpha}{1-\cos^2\theta\sin^2\alpha}$ (*c*) $\dfrac{\cos\alpha}{1-\cos^2\theta\sin^2\alpha}$ (*d*) $\dfrac{\sin\alpha}{1-\cos^2\theta\sin^2\alpha}$

where θ = Angular displacement of the driving shaft.

384. The velocity of the driven shaft of a Hooke's joint is minimum when θ is equal to

(*a*) 0° (*b*) 90° (*c*) 180°

(*d*) 270° (*e*) (*b*) and (*d*) (*f*) (*a*) and (*c*).

385. The velocity of the driven shaft of a Hooke's joint is maximum when the value of angular displacement (θ) of the driving shaft is equal to

(*a*) 0° (*b*) 90° (*c*) 180°

(*d*) 270° (*e*) (*b*) and (*d*) (*f*) (*a*) and (*c*).

386. The value of angular displacement (θ) of the driving shaft, for which is velocity of the driven shaft is equal to the driver shaft, is given by the relation

(*a*) $\tan\theta = \cos\alpha$ (*b*) $\tan\theta = \pm\sqrt{\cos\alpha}$

(*c*) $\tan\theta = \sqrt{\sin\alpha}$ (*d*) $\tan\theta = \pm\sqrt{\cos\alpha \times \sin\alpha}$

where α = Inclination between the two shafts.

387. The co-efficient of fluctuation of speed of the driven shaft of a Hooke's joint, when the angle between the two shafts is small and is measured in radian

(*a*) is proportional to the angle between the two shafts

(*b*) is proportional to the square of the angle between the two shafts

(*c*) is inversely proportional to the angle between the two shafts

(*d*) is inversely proportional to the square of the angle between the two shafts.

388. The acceleration on the driven shaft of a Hooke's joint is maximum for a value of θ. This approximate value of θ is given by

(*a*) $\cos 2\alpha = \dfrac{2\sin^2\alpha}{2+\sin^2\alpha}$ (*b*) $\cos 2\theta = \dfrac{\sin^2\alpha}{1-2\sin^2\alpha}$

(*c*) $\cos 2\theta = \dfrac{2\sin^2\alpha}{1-\sin^2\alpha}$ (*d*) $\cos 2\theta = \dfrac{2\sin^2\alpha}{2-\sin^2\alpha}$.

Fill in the Blanks :

389. The two pulleys will rotate in the ...... direction, if they are connected by an open belt drive.
(*a*) opposite (*b*) same.

390. When the maximum tension in the belt is ...... of centrifugal tension, the power transmitted by the belt will be maximum.
(*a*) two times (*b*) three times.

391. The ratio of the driving tensions for *V*-belt is ...... times that of flat belts.
(*a*) sin α (*b*) cosec α.

392. The initial tension in the belt is equal to ...... tension of the tight side and slack side of the belt.
(*a*) product of (*b*) average.

393. The product of diametrical pitch and ...... is equal to one.
(*a*) circular pitch (*b*) module.

394. The pitch circle in case of gears is always ...... than the base circle.
(*a*) less (*b*) greater.

395. The contact ratio is ...... for gears.
(*a*) less than one (*b*) greater than one.

396. The expression *L* = ... gives the relation between number of pairs (*p*) forming a kinematic chain and number of links (*L*).
(*a*) $2p - 3$ (*b*) $2p - 4$.

397. A chain consists of four links and ...... joints is called a kinematic chain.
(*a*) three (*b*) four.

398. A simple mechanism consists of ...... links.
(*a*) three (*b*) four.

399. If ...... the links is fixed, then a kinematic chain is known as mechanism.
(*a*) two of (*b*) one of.

400. A body is moving with simple harmonic motion, then its velocity will be ...... at the mean position.
(*a*) zero (*b*) maximum.

401. A body is moving with simple harmonic motion. The product of its periodic time and ...... will be equal to one.
(*a*) amplitude (*b*) frequency.

402. The total number of instantaneous centres are ...... for a mechanism consisting of *n* links.
(*a*) $n(n + 1)$ (*b*) $n(n - 1)/2$.

403. The coriolis component of acceleration acts ...... to the sliding surface.
(*a*) parallel to (*b*) perpendicular to.

404. In a reciprocating engine, usually ...... of the reciprocating masses are balanced.
(*a*) two-third (*b*) one-half.

ANSWERS

1. (*a*)	**2.** (*c*)	**3.** (*c*)	**4.** (*e*)	**5.** (*c*)	**6.** (*d*)
7. (*a*)	**8.** (*b*)	**9.** (*c*)	**10.** (*b*)	**11.** (*a*)	**12.** (*e*)
13. (*f*)	**14.** (*b*)	**15.** (*e*)	**16.** (*c*)	**17.** (*b*)	**18.** (*d*)
19. (*c*)	**20.** (*b*)	**21.** (*a*)	**22.** (*b*)	**23.** (*b*)	**24.** (*b*)
25. (*c*)	**26.** (*c*)	**27.** (*b*)	**28.** (*a*)	**29.** (*b*)	**30.** (*c*)
31. (*d*)	**32.** (*d*)	**33.** (*d*)	**34.** (*b*)	**35.** (*c*)	**36.** (*d*)
37. (*a*)	**38.** (*b*)	**39.** (*b*)	**40.** (*a*)	**41.** (*d*)	**42.** (*c*)
43. (*d*)	**44.** (*a*)	**45.** (*b*)	**46.** (*c*)	**47.** (*b*)	**48.** (*d*)
49. (*a*)	**50.** (*c*)	**51.** (*c*)	**52.** (*d*)	**53.** (*b*)	**54.** (*c*)
55. (*a*)	**56.** (*d*)	**57.** (*b*)	**58.** (*b*)	**59.** (*b*)	**60.** (*e*)
61. (*d*)	**62.** (*b*)	**63.** (*d*)	**64.** (*c*)	**65.** (*a*)	**66.** (*c*)
67. (*c*)	**68.** (*a*)	**69.** (*b*)	**70.** (*b*)	**71.** (*a*)	**72.** (*b*)
73. (*a*)	**74.** (*c*)	**75.** (*d*)	**76.** (*b*)	**77.** (*d*)	**78.** (*c*)
79. (*c*)	**80.** (*c*)	**81.** (*d*)	**82.** (*d*)	**83.** (*b*)	**84.** (*a*)
85. (*b*)	**86.** (*b*)	**87.** (*b*)	**88.** (*c*)	**89.** (*d*)	**90.** (*a*)
91. (*c*)	**92.** (*b*)	**93.** (*b*)	**94.** (*c*)	**95.** (*c*)	**96.** (*b*)
97. (*b*)	**98.** (*a*)	**99.** (*c*)	**100.** (*d*)	**101.** (*c*)	**102.** (*c*)
103. (*b*)	**104.** (*c*)	**105.** (*d*)	**106.** (*b*)	**107.** (*e*)	**108.** (*e*)
109. (*a*)	**110.** (*b*)	**111.** (*d*)	**112.** (*d*)	**113.** (*b*)	**114.** (*c*)
115. (*c*)	**116.** (*b*)	**117.** (*c*)	**118.** (*b*)	**119.** (*a*)	**120.** (*b*)
121. (*a*)	**122.** (*d*)	**123.** (*b*)	**124.** (*c*)	**125.** (*c*)	**126.** (*c*)
127. (*d*)	**128.** (*b*)	**129.** (*b*)	**130.** (*e*)	**131.** (*c*)	**132.** (*e*)
133. (*f*)	**134.** (*d*)	**135.** (*b*)	**136.** (*b*)	**137.** (*e*)	**138.** (*e*)
139. (*c*)	**140.** (*d*)	**141.** (*b*)	**142.** (*d*)	**143.** (*b*)	**144.** (*b*)
145. (*b*)	**146.** (*d*)	**147.** (*b*)	**148.** (*b*)	**149.** (*c*)	**150.** (*a*)
151. (*b*)	**152.** (*a*)	**153.** (*b*)	**154.** (*c*)	**155.** (*a*)	**156.** (*c*)
157. (*d*)	**158.** (*d*)	**159.** (*c*)	**160.** (*b*)	**161.** (*c*)	**162.** (*d*)
163. (*c*)	**164.** (*b*)	**165.** (*d*)	**166.** (*f*)	**167.** (*c*)	**168.** (*b*)
169. (*e*)	**170.** (*c*)	**171.** (*b*)	**172.** (*c*)	**173.** (*c*)	**174.** (*c*)
175. (*a*)	**176.** (*c*)	**177.** (*b*)	**178.** (*b*)	**179.** (*d*)	**180.** (*d*)
181. (*c*)	**182.** (*c*)	**183.** (*d*)	**184.** (*c*)	**185.** (*c*)	**186.** (*b*)
187. (*b*)	**188.** (*d*)	**189.** (*c*)	**190.** (*d*)	**191.** (*b*)	**192.** (*d*)
193. (*c*)	**194.** (*c*)	**195.** (*c*)	**196.** (*b*)	**197.** (*c*)	**198.** (*b*)

199. (*b*)	**200.** (*b*)	**201.** (*b*)	**202.** (*d*)	**203.** (*a*)	**204.** (*d*)
205. (*c*)	**206.** (*b*)	**207.** (*d*)	**208.** (*b*)	**209.** (*c*)	**210.** (*b*)
211. (*a*)	**212.** (*a*)	**213.** (*b*)	**214.** (*c*)	**215.** (*b*)	**216.** (*c*)
217. (*c*)	**218.** (*c*)	**219.** (*b*)	**220.** (*b*), (*d*)	**221.** (*b*)	**222.** (*b*)
223. (*c*)	**224.** (*b*)	**225.** (*a*)	**226.** (*c*)	**227.** (*d*)	**228.** (*b*)
229. (*b*)	**230.** (*b*)	**231.** (*a*)	**232.** (*c*)	**233.** (*b*)	**234.** (*b*)
235. (*d*)	**236.** (*e*)	**237.** (*e*)	**238.** (*b*)	**239.** (*b*)	**240.** (*a*)
241. (*b*)	**242.** (*b*)	**243.** (*c*)	**244.** (*d*)	**245.** (*c*)	**246.** (*b*)
247. (*c*)	**248.** (*a*)	**249.** (*c*)	**250.** (*b*)	**251.** (*a*)	**252.** (*c*)
253. (*b*)	**254.** (*d*)	**255.** (*c*)	**256.** (*a*)	**257.** (*b*)	**258.** (*d*)
259. (*c*)	**260.** (*d*)	**261.** (*c*)	**262.** (*d*)	**263.** (*b*)	**264.** (*c*)
265. (*b*)	**266.** (*e*)	**267.** (*d*)	**268.** (*c*)	**269.** (*b*)	**270.** (*a*)
271. (*a*)	**272.** (*d*)	**273.** (*a*)	**274.** (*c*)	**275.** (*d*)	**276.** (*b*)
277. (*b*)	**278.** (*d*)	**279.** (*c*)	**280.** (*c*)	**281.** (*b*)	**282.** (*c*)
283. (*a*)	**284.** (*d*)	**285.** (*a*)	**286.** (*b*)	**287.** (*c*)	**288.** (*a*)
289. (*b*)	**290.** (*b*)	**291.** (*d*)	**292.** (*b*)	**293.** (*c*)	**294.** (*d*)
295. (*c*)	**296.** (*d*)	**297.** (*a*)	**298.** (*b*)	**299.** (*b*)	**300.** (*d*)
301. (*c*)	**302.** (*c*)	**303.** (*c*)	**304.** (*c*)	**305.** (*d*)	**306.** (*d*)
307. (*e*)	**308.** (*e*)	**309.** (*e*)	**310.** (*d*)	**311.** (*d*)	**312.** (*d*)
313. (*c*)	**314.** (*d*)	**315.** (*b*)	**316.** (*c*)	**317.** (*c*)	**318.** (*c*)
319. (*e*)	**320.** (*c*)	**321.** (*b*)	**322.** (*c*)	**323.** (*c*)	**324.** (*b*)
325. (*b*)	**326.** (*c*)	**327.** (*b*)	**328.** (*c*)	**329.** (*c*)	**330.** (*b*)
331. (*d*)	**332.** (*d*)	**333.** (*b*)	**334.** (*a*)	**335.** (*a*)	**336.** (*c*)
337. (*c*)	**338.** (*b*)	**339.** (*a*)	**340.** (*c*)	**341.** (*e*)	**342.** (*b*)
343. (*a*)	**344.** (*a*)	**345.** (*e*)	**346.** (*f*)	**347.** (*b*)	**348.** (*d*)
349. (*d*)	**350.** (*d*)	**351.** (*b*)	**352.** (*d*)	**353.** (*b*)	**354.** (*b*)
355. (*b*)	**356.** (*b*)	**357.** (*c*)	**358.** (*d*)	**359.** (*a*)	**360.** (*c*)
361. (*a*)	**362.** (*d*)	**363.** (*c*)	**364.** (*d*)	**365.** (*c*)	**366.** (*c*)
367. (*b*)	**368.** (*c*)	**369.** (*c*)	**370.** (*b*)	**371.** (*b*)	**372.** (*a*)
373. (*b*)	**374.** (*f*)	**375.** (*b*)	**376.** (*c*)	**377.** (*a*)	**378.** (*d*)
379. (*a*)	**380.** (*a*)	**381.** (*b*)	**382.** (*c*)	**383.** (*c*)	**384.** (*e*)
385. (*e*)	**386.** (*b*)	**387.** (*b*)	**388.** (*d*)	**389.** (*b*)	**390.** (*b*)
391. (*b*)	**392.** (*b*)	**393.** (*b*)	**394.** (*b*)	**395.** (*b*)	**396.** (*b*)
397. (*b*)	**398.** (*b*)	**399.** (*b*)	**400.** (*b*)	**401.** (*b*)	**402.** (*b*)
403. (*b*)	**404.** (*a*)				

Subject Index

U

V

W